W0245973

STRUCTURE REPORTS

Volume 47B

Structure Reports is prepared under the guidance of a Commission of the International Union of Crystallography.

STRUCTURE REPORTS

Volume 47B (Part 1)

ORGANIC INDEXES

SUBJECT/FORMULA INDEXES (1913-1980)
AUTHOR INDEX (1971-1980)

General editor

G. Ferguson

with the assistance of

S.J. Rettig
J. Trotter

Published for the

INTERNATIONAL UNION OF CRYSTALLOGRAPHY

Springer Science+Business Media, B.V.

First published in 1993

ISSN 0166–6983

Originally published by Kluwer Academic Publishers in 1993
Softcover reprint of the hardcover 1st edition 1993

Printed on acid-free paper

ISBN 978-94-017-3165-2
DOI 10.1007/978-94-017-3163-8
ISBN 978-94-017-3163-8 (eBook)

TABLE OF CONTENTS

Introduction VI

Arrangement VII

Classified Index 1

INTRODUCTION

This Cumulative Index of Structure Reports contains all the structures for the years 1913-1980 (Volumes 1 to 46B) and the authors for the latest 10-year period, 1971-1980 (Volumes 37B to 46B). The arrangement in the classified structure index is as outlined on pages VII-VIII.

University of Guelph,
Guelph, Ontario, Canada

G. FERGUSON

ARRANGEMENT

Classified Organic Index (Page 1)

Each entry gives the formula, name, and Strukturbericht and Structure Reports references. The entries are arranged in accordance with the classification scheme based on chemical structure of the Cambridge Crystallographic Data Centre; within each class the order is by formula. The formula index will help to locate a compound for which the classification is doubtful. The classification is:

(0) Aliphatic hydrocarbons and derivatives
1 Aliphatic carboxylic acid derivatives
2 Aliphatic carboxylic acid salts (ammonium, IA, IIA metals)
3 Aliphatic amines
4 Aliphatic (N and S) compounds
7 Nitriles (aliphatic and aromatic)
8 Urea compounds (aliphatic and aromatic)
9 Nitrogen-nitrogen compounds (aliphatic and aromatic)
10 Nitrogen-oxygen compounds (aliphatic and aromatic)
11 Sulphur and selenium compounds
12 Carbonium ions, carbanions, radicals
13 Benzoic acid derivatives
14 Benzoic acid salts (ammonium, IA, IIA metals)
15 Benzene nitro compounds
16 Anilines
17 Phenols and ethers
18 Benzoquinones
19 Benzene miscellaneous
20 Monocyclic hydrocarbons (3,4,5-membered rings)
21 Monocyclic hydrocarbons (6-membered rings)
22 Monocyclic hydrocarbons (7,8-membered rings)
23 Monocyclic hydrocarbons (9- and higher-membered rings)
24 Naphthalene compounds
25 Naphthoquinones
26 Anthracene compounds
27 Hydrocarbons (2 fused rings)
28 Hydrocarbons (3 fused rings)
29 Hydrocarbons (4 fused rings)
30 Hydrocarbons (5 or more fused rings)
31 Bridged-ring hydrocarbons
32 Hetero-nitrogen (3,4,5-membered monocyclic)
33 Hetero-nitrogen (6-membered monocyclic)
34 Hetero-nitrogen (7- and higher-membered monocyclic)
35 Hetero-nitrogen (2 fused rings)
36 Hetero-nitrogen (more than 2 fused rings)
37 Hetero-nitrogen (bridged-ring systems)
38 Hetero-oxygen
39 Hetero-sulphur and hetero-selenium
40 Hetero-(nitrogen and oxygen)
41 Hetero-(nitrogen and sulphur)
42 Hetero-mixed miscellaneous
43 Barbiturates
44 Pyrimidines and purines
45 Carbohydrates
46 Phosphates
47 Nucleosides and Nucleotides
48 Amino-acids and peptides
49 Porphyrins and corrins
50 Antiobiotics
51 Steroids

52 Monoterpenes
53 Sesquiterpenes
54 Diterpenes
55 Sesterterpenes
56 Triterpenes
57 Tetraterpenes
58 Alkaloids
59 Miscellaneous natural products
60 Molecular complexes
61 Clathrates

62 Boron compounds
63 Silicon compounds
64 Phosphorus compounds
65 Arsenic compounds
66 Antimony and bismuth compounds
67 Groups IA and IIA compounds
68 Group III compounds
69 Germanium, tin, lead compounds
70 Tellurium compounds

Transition metal complexes:
71 Transition metal-carbon compounds
72 Metal π-complexes (open-chain)
73 Metal π-complexes (cyclopentadiene)
74 Metal π-complexes (arene)
75 Metal π-complexes (miscellaneous ring systems)
76 Metal complexes (ethylenediamine)
77 Metal complexes (acetylacetone)
78 Metal complexes (salicylic derivatives)
79 Metal complexes (thiourea)
80 Metal complexes (thiocarbamate or xanthate)
81 Metal complexes (carboxylic acid)
82 Metal complexes (amino-acid)
83 Metal complexes (nitrogen ligand)
84 Metal complexes (oxygen ligand)
(84') Metal complexes (nitrogen and oxygen ligands)
85 Metal complexes (sulphur or selenium ligand)
86 Metal complexes (phosphine or arsine ligand)
(89) Metal complexes (other ligands)

(87) Complex inorganic anions
(88) Polymers

ORGANIC CLASSIFIED INDEX

The entries are arranged in accordance with the classification scheme described in the Introduction; within each class the order is by formula.

0. ALIPHATIC HYDROCARBONS AND DERIVATIVES

$CBrF_3$, Bromotrifluoromethane (gas-ed), 18, 631; 22, 554
$CBrF_3$, Bromotrifluoromethane (gas-mw), 13, 444; 16, 417
CBr_3F, Tribromofluoromethane (gas-ed), 6, 82
CBr_4, Carbon tetrabromide (gas-ed), 5, 62, 155; 8, 245, 246; 9, 311; 11, 514
CBr_4, Carbon tetrabromide (phase I), 43B, 64
CBr_4, Carbon tetrabromide (phase II), 43B, 64
CBr_4, Tetrabromomethane, 1, 615, 641
$CClF_3$, Trifluorochloromethane (gas-ed), 18, 631; 19, 504; 22, 554
$CClF_3$, Trifluorochloromethane (gas-mw), 13, 444
CCl_2F_2, Dichlorodifluoromethane (gas-ed), 20, 467; 22, 554
CCl_2F_2, Dichlorodifluoromethane (gas-mw), 43B, 1485
CCl_3F, Trichlorofluoromethane (gas-ed), 6, 82
CCl_4, Carbon tetrachloride (gas-ed), 4, 125; 11, 514; 12, 298; 19, 504; 22, 553; 30B, 400
CCl_4, Carbon tetrachloride (phase II), 45B, 31
CCl_4, Carbon tetrachloride (phase III), 39B, 26
CD_3NO_2, Nitromethane, 46B, 39
CD_4, Deuteromethane (cubic), 38B, 51
CD_4, Deuteromethane (tetragonal), 46B, 39
CD_4, Deuteromethane (gas-ed), 24, 513; 26, 543
CD_4, Deuteromethane (gas-irr), 19, 499
CF_3ClO_2, Trifluoromethyl chloroperoxide (gas-ed), 43B, 1484
CF_3I, Iodotrifluoromethane (gas-ed), 18, 631; 38B, 1064
CF_3I, Iodotrifluoromethane (gas-mw), 13, 445; 16, 417
CF_4, Carbon tetrafluoride (gas-ed), 3, 327; 4, 125; 17, 594, 609; 18, 631
CF_4, α-Tetrafluoromethane, 38B, 52
CF_4O, Trifluoromethyl hypofluorite (gas-ed), 37B, 692
CF_4O_2, Trifluoromethyl fluoroperoxide (gas-ed), 43B, 1484
CI_4, Carbon tetraiodide (gas-ed), 5, 62, 155; 8, 245; 10, 184
CI_4, Tetraiodomethane, 1, 614, 641
$CHBr_3$, Bromoform (gas-mw), 16, 417
$CHBr_3$, Bromoform, 38B, 52
$CHClF_2$, Chlorodifluoromethane (gas-mw), 27, 729
$CHCl_2F$, Dichlorofluoromethane (gas-mw), 29, 449
$CHCl_3$, Chloroform (gas-ed), 3, 706; 4, 276
$CHCl_3$, Chloroform (gas-mw), 13, 444; 16, 416
$CHCl_3$, Chloroform, 31B, 33
CHF_3, Fluoroform (gas-mw), 13, 444; 16, 416
CHF_3O_2, Trifluoromethyl hydroperoxide (gas-ed), 43B, 1484
CHI_3, Iodoform, 2, 805, 820; 15, 374; 17, 608
$CH_xCl_zF_y$, Fluorochloromethanes (gas-ed), 5, 153
CH_2Br_2, Dibromomethane (gas-ed), 10, 184; 40B, 1137
CH_2Br_2, Dibromomethane, 39B, 27
CH_2ClF, Chlorofluoromethane (gas-mw), 17, 609

CH_2Cl_2, Dichloromethane (gas-ed), 3, 706; 4, 276; 10, 184
CH_2Cl_2, Dichloromethane (gas-mw), 16, 415
CH_2Cl_2, Methylene dichloride, 39B, 28
CH_2F_2, Difluoromethane (gas-mw-ed), 5, 154; 10, 184; 16, 416
CH_2I_2, Methylene diiodide, 39B, 27
CH_2O, Formaldehyde (gas-ed), 7, 249; 40B, 1141
CH_2O, Formaldehyde (gas-mw), 28, 664
CH_3Br, Bromomethane (gas-mw), 13, 443; 16, 414; 22, 602
CH_3Br, Methyl bromide, 39B, 28
CH_3Cl, Chloromethane (gas-ed), 3, 705; 4, 276; 19, 503
CH_3Cl, Chloromethane (gas-mw), 13, 443; 15, 374; 16, 414; 22, 602
CH_3Cl, Chloromethane, 16, 414; 17, 607
CH_3F, Fluoromethane (gas-ed), 5, 154
CH_3F, Fluoromethane (gas-mw), 13, 444; 22, 602; 42B, 1035
CH_3I, Iodomethane (gas-mw), 13, 443; 16, 414; 22, 602
CH_3I, Methyl iodide, 39B, 28
CH_4, Methane (gas-ed), 24, 513; 26, 543
CH_4, Methane, 1, 613, 640; 2, 819; 38B, 51
CH_4O, Methanol (gas-ed), 13, 446; 20, 469; 23, 523
CH_4O, Methanol (gas-irr), 19, 516
CH_4O, Methanol (gas-mw), 15, 378
CH_4O, Methanol, 16, 421
C_2BrCl, Bromochloroacetylene (gas-ed), 42B, 1025
C_2BrCl, Bromochloroacetylene (gas-mw), 35B, 903
C_2BrI, Bromoiodoacetylene (gas-ed), 42B, 1025
C_2Br_4, Tetrabromoethylene, 33B, 237
C_2Br_6, Hexabromoethane, 9, 316; 38B, 53
C_2ClF_3, Chlorotrifluoroethene (gas-ed), 23, 522; 26, 587
C_2Cl_2, Dichloroacetylene (gas-ed), 8, 264
$C_2Cl_2F_4$, 1,1,2,2-Tetrafluoro-1,2-dichloroethane (gas-ed), 13, 521; 22, 555
$C_2Cl_3F_3$, 1,1,1-Trichloro-2,2,2-trifluoroethane (gas-mw), 33B, 539
$C_2Cl_3F_3$, 1,1,2-Trifluoro-1,2,2-trichloroethane (gas-ed), 28, 632
C_2Cl_4, Tetrachloroethylene (gas-ed), 3, 705; 16, 419
$C_2Cl_4F_2$, 1,2-Difluoro-1,1,2,2-tetrachloroethane (gas-mw), 21, 500
C_2Cl_6, Hexachloroethane (cubic), 17, 614
C_2Cl_6, Hexachloroethane (gas-ed), 9, 317; 10, 184; 12, 313; 17, 614; 18, 631; 22, 556; 29, 449
C_2Cl_6, Hexachloroethane (orthorhombic), 17, 614; 23, 520; 45B, 31
C_2F_4, Tetrafluoroethylene (gas-ed), 13, 447; 16, 424
C_2F_5I, Perfluoroethyl iodide (gas-ed), 38B, 1064
C_2F_6, Hexafluoroethane (gas-ed), 18, 632; 19, 504
C_2F_6, Hexafluoroethane (gas-mw), 40B, 1146
C_2I_2, Diiodoacetylene, 38B, 54
C_2I_4, Tetraiodoethylene, 17, 613; 43B, 65; 44B, 36
C_2Na_2, Sodium acetylide, 40A, 120
C_2HBr, Bromoacetylene (gas-ed), 6, 231
C_2HCl, Chloroacetylene (gas-ed), 6, 231
C_2HCl_3, Trichloroethylene (gas-ed), 3, 705
C_2HF, Fluoroacetylene (gas-mw), 24, 515
C_2HF_3, Trifluoroethylene (gas-mw), 39B, 895
C_2HF_3O, Trifluoroacetaldehyde (gas-ed), 22, 553
C_2H_2, Acetylene (cubic), 16, 411; 45B, 31
C_2H_2, Acetylene (gas-irr), 21, 499
C_2H_2, Acetylene (orthorhombic), 41B, 48
$C_2H_2Br_2$, trans-1,2-Dibromoethene (gas-ed), 6, 231
$C_2H_2Br_4$, 1,1,2,2-Tetrabromoethane (gas-ed), 9, 317; 10, 184

$C_2H_2Cl_2$, cis-1,2-Dichloroethylene (gas-ed), 3, 705; 22, 557; 30B, 402
$C_2H_2Cl_2$, cis-1,2-Dichloroethylene (gas-mw), 24, 515
$C_2H_2Cl_2$, trans-1,2-Dichloroethylene (gas-ed), 3, 705; 23, 522
$C_2H_2Cl_2$, 1,1-Dichloroethylene (gas-ed), 3, 705; 22, 557; 23, 522
$C_2H_2Cl_2$, 1,1-Dichloroethylene (gas-mw), 21, 500
$C_2H_2Cl_4$, 1,1,2,2-Tetrachloroethane (gas-ed), 8, 264; 9, 317; 10, 184
$C_2H_2F_2$, cis-1,2-Difluoroethylene (gas-ed), 40B, 1138; 42B, 1027
$C_2H_2F_2$, cis-1,2-Difluoroethylene (gas-mw), 26, 544; 28, 644
$C_2H_2F_2$, trans-1,2-Difluoroethylene (gas-ed), 40B, 1138; 42B, 1027
$C_2H_2F_2$, 1,1-Difluoroethylene (gas-ed), 13, 447
$C_2H_2F_2$, 1,1-Difluoroethylene (gas-mw), 21, 500; 26, 545; 28, 664
$C_2H_2I_2$, cis-1,2-Diiodoethene (gas-ed), 6, 231
$C_2H_2I_2$, trans-1,2-Diodoethylene, 3, 666, 711; 6, 231
C_2H_2O, Ketene (gas-mw-ed), 6, 234; 16, 424; 17, 616
$C_2H_2O_2$, Glyoxal (gas-ed), 7, 250; 33B,530
$C_2H_2O_2$, cis-Glyoxal (gas-mw), 38B, 1070; 42B, 1034
C_2H_3Br, Vinyl bromide (gas-ed), 6, 231
C_2H_3Br, Vinyl bromide (gas-mw), 28, 664
$C_2H_3Br_3$, 1,1,2-Tribromoethane (gas-ed), 9, 317; 10, 184
C_2H_3Cl, Vinyl chloride (gas-ed), 3, 705; 23, 522; 24, 516; 38B, 1064
$C_2H_3Cl_3$, 1,1,1-Trichloroethane (gas-ed), 7, 242; 17, 622
$C_2H_3Cl_3$, 1,1,1-Trichloroethane (gas-mw), 16, 416; 20, 469; 33B, 539; 39B, 895
$C_2H_3Cl_3$, 1,1,1-Trichloroethane, 38B, 55
$C_2H_3Cl_3$, 1,1,2-Trichloroethane (gas-ed), 7, 243; 9, 317; 10, 184; 40B, 1138
$C_2H_3Cl_3O_2$, Chloral hydrate, 13, 448; 28, 400; 39B, 29
C_2H_3F, Vinyl fluoride (gas-mw), 23, 522; 26, 544
$C_2H_3F_3$, 1,1,1-Trifluoroethane (gas-ed), 20, 468; 22, 553
$C_2H_3F_3$, 1,1,1-Trifluoroethane (gas-mw), 21, 499
$C_2H_3F_3O$, 2,2,2-Trifluoroethanol (gas-ed), 20, 470
C_2H_3I, Vinyl iodide (gas-ed), 6, 231
C_2H_4, Ethylene (gas-ed), 21, 498; 23, 517; 30B, 401, 402
C_2H_4, Ethylene (gas-irr), 19, 499; 22, 561
C_2H_4, Ethylene (monoclinic), 3, 665, 708; 6, 230; 9, 315; 45B, 32
C_2H_4, Ethylene (orthorhombic), 43B, 65
C_2H_4BrCl, 1-Bromo-2-chloroethane (gas-ed), 7, 241
$C_2H_4Br_2$, 1,1-Dibromoethane (gas-ed), 13, 448
$C_2H_4Br_2$, 1,2-Dibromoethane (gas-ed), 7, 241; 9, 317; 10, 184
$C_2H_4Cl_2$, 1,1-Dichloroethane (gas-ed), 13, 448; 23, 520
$C_2H_4Cl_2$, 1,2-Dichloroethane (gas-ed), 7, 241; 9, 317; 10, 184; 16, 418; 40B, 1135
$C_2H_4Cl_2$, 1,2-Dichloroethane, 15, 375; 17, 612; 26, 545
$C_2H_4Cl_2O$, Bis(chloromethyl) ether (gas-ed), 42B, 1024
$C_2H_4F_2$, 1,2-Difluoroethane (gas-ed), 39B, 890
$C_2H_4F_2$, 1,2-Difluoroethane (gas-mw), 37B, 696
$C_2H_4I_2$, 1,2-Diiodoethane, 3, 665, 709
C_2H_4O, Acetaldehyde (gas-ed), 4, 280; 7, 245; 22, 553; 40B, 1141
$C_2H_4O_2$, Glycolaldehyde (gas-mw), 37B, 697; 39B, 895
C_2H_5Br, Ethyl bromide (gas-ed), 7, 242
C_2H_5BrO, Monobromodimethyl ether (gas-ed), 24, 517
C_2H_5Cl, Ethyl chloride (gas-ed), 7, 242
C_2H_5Cl, Ethyl chloride (gas-mw), 27, 732
C_2H_5ClO, Chlorodimethyl ether (gas-ed), 24, 517; 30B, 402
C_2H_5ClO, 2-Chloroethanol (gas-ed), 20, 471; 37B, 686; 43B, 1484
C_2H_5F, Ethyl fluoride (gas-mw), 19, 504; 24, 515

C_2H_5FO, 2-Fluoroethanol (gas-ed), 20, 471; 39B, 889
C_2H_6, Ethane (cubic), 44B, 36
C_2H_6, Ethane (monoclinic), 44B, 36
C_2H_6, Ethane (gas-ed-irr), 15, 338; 19, 499; 27, 732; 30B, 401; 33B, 529; 39B, 888
C_2H_6, Ethane (gas-mw), 43B, 1485
C_2H_6, Ethane (hexagonal), 1, 658
$C_2H_6BaO_2$, Barium methoxide, 44B, 37
$C_2H_6CaO_2$, Calcium methoxide, 44B, 37
C_2H_6O, Dimethyl ether (gas-ed), 3, 705; 23, 523
C_2H_6O, Dimethyl ether (gas-mw), 28, 664
C_2H_6O, Ethanol (gas-ed), 12, 313; 13, 446
C_2H_6O, Ethanol (gas-mw), 33B, 539
C_2H_6O, Ethanol, 42B, 29
$C_2H_6O_2$, Ethane-1,2-diol (gas-ed), 39B, 887
$C_2H_6O_2Sr$, Strontium methoxide, 44B, 37
C_3Cl_6O, Hexachloroacetone (gas-ed), 40B, 1135
$C_3Cl_6O_3$, Bistrichloromethylcarbonate (triphosgene), 37B, 23
C_3Cl_8, Octachloropropane (gas-ed), 40B, 1136
C_3F_3Cl, 1-Chloro-3,3,3-trifluoropropyne (gas-mw), 38B, 1071
C_3F_6, Hexafluoropropene (gas-ed), 13, 450
C_3F_6O, Hexafluoroacetone (gas-ed), 35B, 891
C_3F_7I, Perfluoroisopropyl iodide (gas-ed), 38B, 1064
C_3HF_3, Trifluoromethylacetylene (gas-mw), 15, 377
$C_3H_2F_2$, 1,1-Difluoroallene (gas-mw), 40B, 1144
C_3H_2O, Propynal (gas-ed), 38B, 1066
C_3H_2O, Propynal (gas-mw), 23, 525
C_3H_3Br, 1-Bromo-1-propyne (gas-mw), 16, 420
C_3H_3Br, 3-Bromopropyne (gas-ed), 9, 320
C_3H_3Br, 3-Bromopropyne (gas-mw), 26, 549
$C_3H_3BrO_2$, Bromomalonaldehyde, 41B, 49
$C_3H_3BrO_2$, 2-Bromopropane-1,3-dione, 19, 518; 29, 454
C_3H_3Cl, 1-Chloroprop-1-yne (gas-mw), 19, 505
C_3H_3Cl, 3-Chloropropyne (gas-ed), 9, 320
C_3H_3Cl, 3-Chloropropyne (gas-mw), 22, 567; 26, 549
C_3H_3ClO, *a*-Chloroacrolein (gas-ed), 26, 550
$C_3H_3Cl_5$, 1,1,2,3,3-Pentachloropropane (gas-ed), 43B, 1484
$C_3H_3F_3$, 3,3,3-Trifluoropropene (gas-ed), 39B, 891
$C_3H_3F_3$, 3,3,3-Trifluoropropene (gas-mw), 40B, 1143
$C_3H_3F_3O$, Trifluoroacetone (gas-ed), 38B, 1066
C_3H_3I, 1-Iodo-1-propyne (gas-mw), 16, 420
C_3H_3I, 3-Iodopropyne (gas-ed), 9, 320
C_3H_4, Allene (gas-ed), 23, 517
C_3H_4, Allene (gas-irr), 19, 500; 22, 567
C_3H_4, Allene (gas-mw), 39B, 894
C_3H_4, Methylacetylene (propyne) (gas-mw-ed), 7, 240; 13, 450; 19, 502; 22, 602
C_3H_4ClF, cis-1-Chloro-2-fluoropropene (gas-mw), 33B, 540
$C_3H_4Cl_2O$, 1,3-Dichloroacetone, 40B, 50
C_3H_4O, Acrolein (gas-ed), 15, 380; 33B, 530; 35B, 892; 40B, 1142
C_3H_4O, Acrolein (gas-mw), 21, 506
C_3H_4O, Prop-2-yne-1-ol (gas-ed), 29, 453
$C_3H_4O_2$, Methylglyoxal (gas-ed), 26, 550
$C_3H_4O_3$, 2-Propen-2,3-diol-1-one, 40B, 50
C_3H_5Br, Allyl bromide (gas-ed), 19, 505
C_3H_5Br, 3-Bromopropene (gas-mw), 42B, 1035
$C_3H_5Br_3$, 1,2,3-Tribromopropane (gas-ed), 40B, 1135

C_3H_5Cl, Allyl chloride (gas-ed), 19, 505

$C_3H_5Cl_3$, 1,2,3-Trichloropropane (gas-ed), 40B, 1136

C_3H_5I, Allyl iodide (gas-ed), 19, 505

C_3H_6, Propene (gas-ed), 39B, 891

C_3H_6, Propene (gas-mw), 26, 549

$C_3H_6Br_2$, 1,2-Dibromopropane (gas-ed), 8, 265

$C_3H_6Br_2$, 1,3-Dibromopropane (gas-ed), 40B, 1135

$C_3H_6Cl_2$, 1,2-Dichloropropane (gas-ed), 16, 420

$C_3H_6Cl_2$, 1,3-Dichloropropane (gas-ed), 42B, 1025

$C_3H_6Cl_2$, 2,2-Dichloropropane (gas-ed), 17, 622

$C_3H_6Cl_2O$, 1,3-Dichloro-2-hydroxypropane (gas-ed), 20, 471

C_3H_6O, Acetone (gas-mw-ed), 11, 534; 15, 380; 16, 423; 20, 472; 21, 506; 38B, 1063; 40B, 1141

C_3H_6O, Prop-2-ene-1-ol (gas-ed), 29, 453

$C_3H_6O_3$, Glycol monoformate (gas-ed), 39B, 891

C_3H_7Br, 2-Bromopropane (gas-ed), 7, 242

C_3H_7Cl, Propyl chloride (gas-ed), 22, 570

C_3H_7Cl, 2-Chloropropane (gas-ed-mw), 7, 42; 43B, 1484

C_3H_7ClO, 3-Chloro-1-propanol (gas-ed), 38B, 1067

C_3H_8, Propane (gas-ed), 38B, 1063

C_3H_8, Propane (gas-mw), 24, 517

C_3H_8O, 1-Propanol (gas-ed), 26, 550; 29, 453

C_3H_8O, 2-Propanol (gas-ed), 29, 453

C_3H_8O, 2-Propanol (gas-mw), 40B, 1146

$C_3H_8O_2$, Dimethoxymethane (gas-ed), 39B, 887

C_4F_6, Hexafluoro-2-butyne (gas-ed), 16, 420; 37B, 686, 694

C_4F_6, Perfluorobuta-1,3-diene (gas-ed), 37B, 694

C_4HBr, Monobromodiacetylene (gas-ed), 35B, 893

C_4HF_9, Tris(trifluoromethyl)methane (gas-ed), 37B, 687

C_4H_2, Diacetylene (gas-ed), 3, 711; 7, 240; 37B, 687

C_4H_2, Diacetylene (gas-irr), 21, 499

$C_4H_2Cl_4$, 1,2,3,4-Tetrachlorobutadiene, 38B, 56

$C_4H_2F_6$, Hexafluoroisobutene (gas-ed), 35B, 891

C_4H_3Cl, trans-1-Chlorobuten-3-yne (gas-mw), 39B, 893

$C_4H_3F_3$, 1,1,1-Trifluoro-2-butyne (gas-mw), 21, 508; 24, 518

C_4H_4, Butatriene (gas-ed), 26, 553

C_4H_4, Vinylacetylene (gas-ed), 40B, 1141

$C_4H_4Br_2$, 1,4-Dibromobut-2-yne (gas-ed), 21, 509

$C_4H_4Cl_2$, 1,4-Dichlorobut-2-yne (gas-ed), 21, 508

$C_4H_4O_2$, Fumaraldehyde (gas-ed), 40B, 1136

C_4H_5Cl, 2-Chlorobuta-1,3-diene (gas-ed), 22, 574

C_4H_5Cl, 4-Chloro-1,2-butadiene (gas-mw), 43B, 1485

C_4H_6, Butadiene (gas-ed), 7, 270; 22, 577; 33B, 530

C_4H_6, Dimethylacetylene (gas-ed), 40B, 1141

C_4H_6, Dimethylacetylene (gas-irr), 21, 499

C_4H_6, Dimethylacetylene, 19, 500; 23, 518

C_4H_6O, Crotonaldehyde (gas-ed), 15, 380

C_4H_6O, Divinyl ether (gas-ed), 12, 314

C_4H_6O, Ethoxyethyne (gas-mw), 40B, 1145

C_4H_6O, Methyl allenyl ether (gas-ed), 38B, 1066

$C_4H_6O_2$, Dimethylglyoxal (gas-ed), 7, 250

$C_4H_6O_2$, 2,3-Butanedione (gas-ed), 39B, 889

C_4H_7Cl, 1-Chloro-2-methylpropene (gas-ed), 7, 242

C_4H_8, cis-2-Butene (gas-ed), 4, 277; 35B, 894

C_4H_8, trans-2-Butene (gas-ed), 4, 277; 35B, 894

C_4H_8, 2-Methylpropene (gas-ed), 24, 518; 26, 553; 40B, 1139

C_4H_8, 2-Methylpropene (gas-mw), 28, 666

$C_4H_8Br_2$, 2,3-Dibromobutane (gas-ed), 7, 243
C_4H_8O, Isopropyl carboxaldehyde (gas-ed), 30B, 402
C_4H_8O, 2-Butanone (gas-ed), 20, 472; 40B, 1142
C_4H_9Br, n-Butyl bromide (gas-ed), 28, 666
C_4H_9Br, t-Butyl bromide (gas-ed), 6, 236; 19, 505
C_4H_9Br, t-Butyl bromide (gas-mw), 13, 451
C_4H_9Cl, n-Butyl chloride (gas-irr), 42B, 1037
C_4H_9Cl, t-Butyl chloride (gas-mw-ed), 6, 236; 13, 451; 17, 622, 629; 19, 505; 28, 666; 37B, 691; 38B, 1070
C_4H_9Cl, t-Butyl chloride, 33B, 28
C_4H_9Cl, 1-Chlorobutane (gas-ed), 27, 733
C_4H_9Cl, 2-Chlorobutane (gas-ed), 26, 554; 27, 733; 29, 450
C_4H_9F, t-Butyl fluoride (gas-mw-ed), 17, 629
C_4H_9I, t-Butyl iodide (gas-ed), 19, 505
C_4H_9I, t-Butyl iodide (gas-mw), 13, 451
C_4H_{10}, Isobutane (gas-ed), 6, 236; 8, 264; 39B, 890
C_4H_{10}, Isobutane (gas-mw), 24, 517
C_4H_{10}, n-Butane (gas-ed), 23, 519
$C_4H_{10}O$, Diethyl ether (gas-mw), 40B, 1144
$C_4H_{10}O$, Diethyl ether, 38B, 144
$C_4H_{10}O$, sec-Butyl alcohol (gas-ed), 26, 554
$C_4H_{10}O_4$, Erythritol, 46B, 39
$C_4H_{16}Br_2CaO_4$, Calcium bromide methanol solvate, 44B, 38
C_5F_{12}, Perfluoro-n-pentane (gas-ed), 16, 421
C_5F_{12}, Perfluoroisopentane (gas-ed), 16, 421
$C_5H_2F_6O_2$, Hexafluoroacetylacetone (gas-ed), 37B, 690
C_5H_4, Methyldiacetylene (gas-mw), 17, 633; 19, 500
$C_5H_5F_3O_2$, Trifluoroacetylacetone (gas-ed), 38B, 1065
C_5H_8, Isoprene (gas-ed), 26, 558
C_5H_8, 1-Pentyne (gas-mw), 38B, 1069
$C_5H_8Br_4$, Tetrakis(bromomethyl)methane (gas-ed), 42B, 1024
$C_5H_8Cl_4$, Tetrakis(chloromethyl)methane (gas-ed), 40B, 1135
C_5H_8O, trans-1-Methoxy-1,3-butadiene (gas-mw), 43B, 1485
$C_5H_8O_2$, Acetylacetone (gas-ed), 37B, 691; 38B, 1065
$C_5H_9Cl_3$, 1,1,1-Tris(chloromethyl)ethane (gas-ed), 40B, 1135
C_5H_{10}, 2-Methyl-1-butene (gas-ed), 33B, 531
$C_5H_{10}Cl_2$, 1,3-Dichloro-2,2-dimethylpropane (gas-ed), 40B, 1135
$C_5H_{10}O$, 3-Pentanone (gas-ed), 20, 472
$C_5H_{11}Cl$, Neopentyl chloride (gas-ed), 13, 456
$C_5H_{11}ClO_2$, DL-3-Chloro-2,4-pentanediol, 41B, 49
C_5H_{12}, Neopentane (gas-ed), 24, 519
C_5H_{12}, Neopentane (gas-ed), 40B, 1142
C_5H_{12}, n-Pentane (gas-ed), 23, 519
C_5H_{12}, n-Pentane, 29, 449; 32B, 40
$C_5H_{12}O_4$, Tetramethoxymethane (gas-ed), 40B, 1137
C_6H_6, Dimethyldiacetylene (2,4-hexadiyne) (gas-ed), 7, 240; 20, 467
C_6H_6, Dimethyldiacetylene (2,4-hexadiyne), 22, 587
C_6H_6O, 2,4-Hexadiyn-1-ol, 44B, 38
C_6H_8, 1,3,5-cis-Hexatriene (gas-ed), 33B, 533
C_6H_8, 1,3,5-trans-Hexatriene (gas-ed), 33B, 533
C_6H_{10}, t-Butylacetylene (gas-ed), 40B, 1143
C_6H_{10}, t-Butylacetylene (gas-mw), 27, 735
C_6H_{10}, 2,3-Dimethylbutadiene (gas-ed), 33B, 534
C_6H_{12}, (E)-3-Methyl-2-pentene (gas-ed), 39B, 890
C_6H_{12}, (Z)-3-Methyl-2-pentene (gas-ed), 39B, 890
C_6H_{12}, 2,3-Dimethyl-2-butene (gas-ed), 40B, 1137
$C_6H_{12}Br_2$, 2,3-Dibromo-2,3-dimethylbutane, 35B, 41

$C_6H_{12}Br_2O_4$, Dibromodulcitol, 37B, 238
$C_6H_{12}Cl_2O_4$, Dichlorodulcitol, 37B, 238
C_6H_{14}, n-Hexane (gas-ed), 23, 519
C_6H_{14}, n-Hexane, 26, 558
C_6H_{14}, 1,1,2,2-Tetramethylethane (gas-ed), 42B, 1028
$C_6H_{14}O$, Dipropyl ether (gas-ed), 43B, 1485
$C_6H_{14}O_2$, 2,3-Dimethyl-2,3-butanediol, 44B, 39
$C_6H_{14}O_6$, D-Glucitol, 37B, 241
$C_6H_{24}Br_2CaO_6$, Calcium bromide methanol solvate, 44B, 39
$C_6H_{24}Br_2MgO_6$, Magnesium bromide methanol solvate, 44B, 39
$C_6H_{26}O_8$, Pinacol hexahydrate, 35B, 29
C_7F_{16}, Perfluoro-n-heptane (gas-ed), 16, 421
C_7H_{10}, cis-2-Methyl-1,3,5-hexatriene (gas-ed), 40B, 1134
C_7H_{10}, trans-2-Methyl-1,3,5-hexatriene (gas-ed), 40B, 1136
$C_7H_{10}Cl_6$, 1,1,3,5,7,7-Hexachloroheptane, 46B, 40
C_7H_{16}, n-Heptane, 44B, 40
C_7H_{16}, n-Heptane (gas-ed), 23, 519
$C_7H_{16}O_3$, meso-(2S,4S,6R)-2,4,6-Heptanetriol, 40B, 51
C_8H_6, Octa-2,4,6-triyne, 16, 413
$C_8H_{10}Cl_8$, meso-(RS)-1,1,1,3,6,8,8,8-Octachlorooctane, 45B, 33
$C_8H_{10}O_2$, 3,5-Octadiyne-1,8-diol, 44B, 40
C_8H_{14}, 3,4-Dimethyl-2,4-hexadienes (gas-ed), 35B, 900
$C_8H_{16}O_2 \cdot 0.5\ H_2O$, trans-2,5-Dimethyl-3-hexene-2,5-diol hemihydrate, 40B, 52; 45B, 34
$C_8H_{16}O_2$, cis-2,5-Dimethyl-3-hexene-2,5-diol, 40B, 52
$C_8H_{16}O_2$, trans-2,5-Dimethyl-3-hexene-2,5-diol, 40B, 52; 45B, 34
C_8H_{18}, Hexamethylethane (gas-ed), 42B, 1028
C_8H_{18}, n-Octane, 26, 560; 32B, 40
C_8H_{18}, 2,2,3,3-Tetramethylbutane (gas-ed), 9, 321
$C_8H_{18}O_2$, 2,5-Dimethylhexane-2,5-diol, 42B, 30
$C_8H_{24}Br_2CaO_4$, Calcium bromide ethanol solvate, 44B, 38
$C_8H_{26}O_6$, 2,5-Dimethyl-2,5-hexanediol tetrahydrate, 38B, 56
$C_{10}H_{12}O_4$, Tetraacetylethylene, 44B, 85
$C_{10}H_{14}Br_3Cl_3$, 1,1,7-Tribromo-2,6,8-trichloro-3,7-dimethyloct-3-ene, 45B, 35
$C_{10}H_{14}O_4$, Tetraacetylethane, 31B, 33; 41B, 50
$C_{10}H_{18}$, 2,3,4,5-Tetramethyl-2,4-hexadiene (gas-ed), 43B, 1485
$C_{10}H_{18}O_2$, 2,3-Diisopropylidene-1,4-butanediol, 40B, 53
$C_{10}H_{30}O_6$, 2,7-Dimethyl-2,7-octanediol tetrahydrate, 44B, 41
$C_{11}H_{23}BrO$, 11-Bromoundecanol, 38B, 57
$C_{13}H_{28}$, Tri-t-butylmethane (gas-ed), 38B, 1064
$C_{14}H_8Br_2F_2$, p,p-Dibromo-α,α'-difluorostilbene, 32B, 41
$C_{14}H_9Cl_5$, 1-(o-Chlorophenyl)-1-(p-chlorophenyl)-2,2,2-trichloroethane, 38B, 58
$C_{14}H_9Cl_5$, 1,1-Bis-(p-chlorophenyl)-2,2,2-trichloroethane, 38B, 58
$C_{15}H_{13}Br$, 2-Bromo-1,1-diphenyl-prop-1-ene, 31B, 34; 32B, 41
$C_{16}H_{15}Br$, 2-Bromo-1,1-di-p-tolylethylene, 32B, 42
$C_{16}H_{25}NO$, N-t-Butyl-2-hydroxy-3,3-dimethyl-2-phenylbutanimine, 42B, 30
$C_{16}H_{34}$, n-Hexadecane (gas-ed), 42B, 1026
$C_{16}H_{34}O$, n-Hexadecanol, 24, 524
$C_{18}H_{18}Cl_2$, 4,4'-Dichloro-α,β-diethylstilbene, 32B, 43
$C_{18}H_{38}$, n-Octadecane, 27, 743; 38B, 59
$C_{19}H_{38}O$, Tetra-t-butylacetone, 39B, 30
$C_{20}H_{18}Cl_2$, 1,6-Di-o-chlorophenyl-3,4-dimethylhexatriene, 31B, 35
$C_{20}H_{18}Cl_2$, 1,6-Di-p-chlorophenyl-3,4-dimethylhexatriene, 31B, 35
$C_{20}H_{38}Cl_4$, 2,6,11,15-Tetrachloro-2,6,11,15-tetramethyl-hexadecane,

33B, 28
$C_{20}H_{42}$, n-Eicosane, 40B, 54
$C_{21}H_{28}O_2$, 1,1-Bis-(p-ethoxyphenyl)-2,2-dimethylpropane, 38B, 60
$C_{22}H_{46}$, n-Docosane, 40B, 54
$C_{23}H_{24}O_4$, Bis(p-acetoxyphenyl)cyclohexylidenemethane, 38B, 61
$C_{23}H_{46}O$, 12-Tricosanone, 39B, 30; 40B, 55
$C_{24}H_{49}I$, 1-Iodo-3-methyltricosane, 34B, 25
$C_{28}H_{22}$, 1,2,3,4-Tetraphenyl-cis,cis-butadiene, 30B, 37
$C_{28}H_{58}$, Octacosane, 42B, 31
$C_{29}H_{60}$, Nonacosane, 1, 620, 683
$C_{30}H_{30}O_4$, 1,1,2,2-Tetrakis(2-methoxyphenyl)ethane, 38B, 61
$C_{32}H_{66}O$, Dicetyl ether, 8, 267
$C_{36}H_{74}$, n-Hexatriacontane (monoclinic form), 20, 486
$C_{36}H_{74}$, n-Hexatriacontane (orthorhombic form), 23, 547
$C_{38}H_{30}O$, Bis(triphenylmethyl) ether, 44B, 42
$C_{41}H_{36}O_8$, Pentaerythritol tetracinnamate, 38B, 62

1. ALIPHATIC CARBOXYLIC ACID DERIVATIVES

CClFO, Fluoroformyl chloride (gas-mw), 26, 543
CCl_2O, Phosgene (gas-ed), 3, 705
CCl_2O, Phosgene (gas-mw), 17, 609
CCl_2O, Phosgene, 16, 417
CF_2O, Carbonyl fluoride (gas-ed), 16, 424
CF_2O, Carbonyl fluoride (gas-mw), 27, 729
CHClO, Formyl chloride (gas-mw), 42B, 1033
CHFO, Formyl fluoride (gas-ed), 19, 517
CHFO, Formyl fluoride (gas-mw), 23, 526; 26, 543
CHI_2NO, N,N-Diiodoformamide, 40B, 27
CH_2O_2, Formic acid, 17, 610; 44B, 3
CH_2O_2, Formic acid (gas-ed), 3, 716; 9, 312; 10, 185; 15, 382; 18, 638; 40B, 1140
CH_2O_2, Formic acid (gas-mw), 19, 519; 21, 498
CH_3NO, Formamide (gas-ed), 40B, 1136
CH_3NO, Formamide (gas-mw), 21, 498; 24, 541
CH_3NO, Formamide, 18, 651; 41B, 3; 44B, 4
$C_2Br_2O_2$, Oxalyl bromide (gas-ed), 39B, 888
$C_2Br_2O_2$, Oxalyl bromide, 26, 546; 27, 729
C_2Cl_2FNO, Dichlorofluoroacetamide, 46B, 3
$C_2Cl_2O_2$, Oxalyl chloride (gas-ed), 39B, 888
$C_2Cl_2O_2$, Oxalyl chloride, 27, 729
$C_2D_6O_6$, Oxalic acid dihydrate (deuterated), 32B, 3; 34B, 6
$C_2HCl_3O_2$, Trichloroacetic acid, 38B, 3
$C_2HF_3O_2$, Trifluoroacetic acid, 45B, 3; 44B, 5
$C_2HF_3O_2$, Trifluoroacetic acid (gas-ed), 9, 312
$C_2H_2BrF_2NO$, Bromodifluoroacetamide, 44B, 4
$C_2H_2ClF_2NO$, Chlorodifluoroacetamide, 42B, 3
$C_2H_2Cl_2O$, Monochloroacetyl chloride (gas-ed), 18, 634
$C_2H_2Cl_2O$, Monochloroacetyl chloride (gas-mw), 42B, 1036
$C_2H_2Cl_2O_2$, Monochloromethyl chloroformate (gas-ed), 16, 425
$C_2H_2D_2O_4$, Formic acid dimer (gas-ed), 35B, 890
$C_2H_2F_3NO$, Trifluoroacetamide, 44B, 4
$C_2H_2O_3$, Formic anhydride (gas-ed), 38B, 1066
$C_2H_2O_3$, Glyoxylic acid (gas-mw), 39B, 895; 42B, 1035
$C_2H_2O_4$, Oxalic acid (α form), 3, 671, 723; 16, 427; 40B, 3
$C_2H_2O_4$, Oxalic acid (β form), 3, 672, 723; 40B, 3

$C_2H_2O_4$, Oxalic acid (gas-ed), 18, 642; 35B, 889
$C_2H_2O_4 \cdot 2\ H_2O$, α-Oxalic acid dihydrate, 46B, 3
C_2H_3BrO, Acetyl bromide (gas-mw-ed), 15, 381; 38B, 1063, 1072
$C_2H_3BrO_2$, Bromoacetic acid (monoclinic), 41B, 4
$C_2H_3BrO_2$, Bromoacetic acid (orthorhombic), 41B, 4
C_2H_3ClO, Acetyl chloride (gas-mw-ed), 15, 381; 18, 634; 26, 546; 38B, 1063, 1072
$C_2H_3ClO_2$, Chloroacetic acid (α-form), 42B, 3
$C_2H_3ClO_2$, Chloroacetic acid (β-form), 42B, 3
$C_2H_3ClO_2$, Methyl chloroformate (gas-ed), 13, 446
C_2H_3FO, Acetyl fluoride (gas-mw-ed), 15, 381; 23, 526; 40B, 1138
$C_2H_3FO_2$, Fluoroacetic acid (gas-mw), 38B, 1071
$C_2H_3F_2NO$, Difluoroacetamide, 38B, 4
C_2H_3IO, Acetyl iodide (gas-ed), 15, 381; 40B, 1138
$C_2H_3NO_4$, Nitroacetic acid, 46B, 3
$C_2H_3O_2$, Methyl formate (gas-ed), 43B, 1484
C_2H_4BrNO, Bromoacetamide, 23, 582
C_2H_4BrNO, N-Bromoacetamide, 37B, 3
C_2H_4ClNO, Chloroacetamide (α form), 20, 498; 44B, 5
C_2H_4ClNO, Chloroacetamide (β form), 20, 498
C_2H_4ClNO, Chloroacetamide (unstable β form), 20, 498
C_2H_4ClNO, Monochloroacetamide (gas-ed), 18, 654
C_2H_4FNO, Monofluoroacetamide, 27, 756
$C_2H_4N_2O_2$, Oxamide, 8, 282; 18, 652; 43B, 3
$C_2H_4N_2O_3$, N-Hydroxyoxamide, 46B, 4
$C_2H_4O_2$, Acetic acid (gas-mw-ed), 9, 312; 37B, 691, 696
$C_2H_4O_2$, Acetic acid, 22, 561; 35B, 3; 37B, 3; 44B, 6
$C_2H_4O_2$, Methyl formate (gas-ed), 13, 446
$C_2H_4O_3$, Glycolic acid, 37B, 4
$C_2H_4O_4$, Formic acid dimer (gas-ed), 40B, 1140
$C_2H_5FO_5S$, Acetic acid - fluorosulfuric acid, 40B, 1076
C_2H_5NO, Acetamide (gas-ed), 17, 652; 39B, 888
C_2H_5NO, Acetamide (orthorhombic form), 30B, 12
C_2H_5NO, Acetamide (rhombohedral form), 8, 280; 37B, 5; 41B, 3; 46B, 5
C_2H_5NO, N-Methylformamide (gas-ed), 40B, 1136
$C_2H_5NO_2 \cdot 0.5\ H_2O$, Acetohydroxamic acid hemihydrate, 35B, 20
$C_2H_6N_2O_4$, Monoacetamide nitrate, 46B, 5
$C_2H_6O_6$, Oxalic acid dihydrate, 3, 673, 724; 4, 260, 290; 7, 250; 9, 318; 11, 573; 17, 617; 18, 642; 19, 519; 29, 459; 34B, 3, 4
$C_2H_7O_6P$, Acetic acid - phosphoric acid, 38B, 5
$C_2H_9Al_2NO_{10}Si_2$, Dickite:N-methylformamide, 45B, 3
$C_3H_2F_2O_4$, Difluoromalonic acid, 38B, 7
C_3H_3ClO, Acrylyl chloride (gas-ed), 21, 507
$C_3H_3FO_4$, Fluoromalonic acid, 37B, 6; 44B, 6
$C_3H_3NO_2$, α-Cyanoacetic acid, 44B, 7
$C_3H_3NO_2$, β-Cyanoacetic acid, 44B, 7
$C_3H_4Cl_2N_2O_2$, Dichloromalonamide, 37B, 7
$C_3H_4N_2O$, Cyanoacetamide, 35B, 21
$C_3H_4O_2$, Acrylic acid (gas-ed), 23, 539
$C_3H_4O_2$, Acrylic acid, 28, 365
$C_3H_4O_2$, β-Propiolactone (gas-ed), 19, 593
$C_3H_4O_2$, β-Propiolactone (gas-mw), 20, 601
$C_3H_4O_3$, Formic acetic anhydride (gas-ed), 37B, 693
$C_3H_4O_3$, Pyruvic acid (gas-mw), 40B, 1146
$C_3H_4O_3$, Pyruvic acid, 43B, 4
$C_3H_4O_4$, Malonic acid, 21, 507

$C_3H_4O_5$, Hydroxymalonic acid (tartronic acid), 30B, 3; 44B, 6
$C_3H_5BrN_2O_2$, Bromomalonamide, 40B, 3
C_3H_5ClO, Propionyl chloride (gas-mw), 42B, 1035
C_3H_5NO, Acrylamide, 31B, 3
$C_3H_5NO_4$, Aminomalonic acid, 31B, 3
$C_3H_5NO_4$, β-Nitropropionic acid, 18, 655
$C_3H_5N_3O_4$, Nitromalonamide, 45B, 4
C_3H_6ClNO, N-Methylchloroacetamide, 37B, 9
$C_3H_6N_2O_2$, Malonamide, 35B, 23
$C_3H_6O_2$, Methyl acetate (gas-ed), 13, 446
$C_3H_6O_2$, Propionic acid (gas-ed), 37B, 692
$C_3H_6O_2$, Propionic acid, 27, 736
C_3H_7NO, Dimethylformamide (gas), 27, 759
C_3H_7NO, N-Methylacetamide (gas-ed), 17, 652; 39B, 887
C_3H_7NO, N-Methylacetamide, 24, 548
$C_3H_7NO_2$, Ethyl carbamate, 32B, 5
$C_3H_7N_3O_2$, 2-(N-Nitrosomethylamino)acetamide, 39B, 3
$C_4F_6O_3$, Hexafluoroacetic anhydride (gas-ed), 37B, 690
$C_4H_2O_3$, Maleic anhydride (gas-ed), 38B, 1065
$C_4H_2O_3$, Maleic anhydride, 27, 945
$C_4H_2O_4$, Acetylenedicarboxylic acid, 38B, 7; 39B, 3
$C_4H_4BrNO_2$, N-Bromosuccinimide, 43B, 62
$C_4H_4ClNO_2$, N-Chlorosuccinimide, 26, 599
$C_4H_4Cl_2O_3$, Chloroacetic anhydride, 34B, 5
$C_4H_4O_2$, But-3-ynoic acid, 38B, 8
$C_4H_4O_2$, Ketene dimer (gas-ed), 19, 592
$C_4H_4O_2$, Tetrolic acid (α form), 38B, 9
$C_4H_4O_2$, Tetrolic acid (β form), 38B, 10
$C_4H_4O_3$, Succinic anhydride (gas-ed), 39B, 885
$C_4H_4O_3$, Succinic anhydride, 30B, 4
$C_4H_4O_4$, Fumaric acid (α form), 31B, 4
$C_4H_4O_4$, Fumaric acid (β form), 31B, 5
$C_4H_4O_4$, Maleic acid, 16, 431; 40B, 3; 41B, 4
$C_4H_5ClO_2$, cis-β-Chlorocrotonic acid, 33B, 3
$C_4H_5ClO_4$, Chlorosuccinic acid, 38B, 11
$C_4H_5Cl_3O_3$, 4,4,4-Trichloro-3-hydroxybutanoic acid, 42B, 1017
C_4H_5NO, N-Methylpropiolamide, 44B, 8
$C_4H_5NO_2$, Succinimide, 26, 597
$C_4H_5NO_2$, 4-Aminotetrolic acid, 42B, 4
$C_4H_5NO_3$, Fumaramic acid, 37B, 9; 38B, 11
$C_4H_6BrNO_3$, (-)-2-Bromosuccinamic acid, 40B, 1074
$C_4H_6O_2$, Crotonic acid, 40B, 4
$C_4H_6O_2$, Methyl acrylate (gas-ed), 23, 539
$C_4H_6O_2$, α-Methylacrylic acid (gas-ed), 23, 539
$C_4H_6O_3$, Acetic anhydride (gas-ed), 37B, 692
$C_4H_6O_4$, Methylmalonic acid, 35B, 4
$C_4H_6O_4$, Succinic acid (α), 12, 315
$C_4H_6O_4$, Succinic acid (β), 6, 236; 7, 223, 254; 12, 316; 23, 537
$C_4H_6O_4$, trans-trans-Dimethyl oxalate, 17, 620
$C_4H_6O_4S$, Thioglycollic acid, 32B, 6
$C_4H_6O_5$, Diglycollic acid, 39B, 4
$C_4H_6O_6$, Acetylenedicarboxylic acid dihydrate, 11, 575
$C_4H_6O_6$, D-Tartaric acid, 13, 452; 31B, 5; 45B, 5
$C_4H_6O_6$, meso-Tartaric acid (triclinic form), 32B, 7
C_4H_7NO, Crotonamide, 40B, 4
$C_4H_7NO_2$, Diacetamide (trans-cis form), 41B, 6
$C_4H_7NO_2$, Diacetamide (trans-trans form), 41B, 6

$C_4H_8N_2O_2$, Succinamide, 20, 505
$C_4H_8O_2$, n-Butyric acid, 27, 737
$C_4H_8O_3$, α-Hydroxyisobutyric acid, 44B, 9
$C_4H_8O_7$, DL-Tartaric acid monohydrate, 12, 317; 15, 391
$C_4H_8O_7$, meso-Tartaric acid monohydrate (monoclinic form), 32B, 9
$C_4H_8O_7$, meso-Tartaric acid monohydrate (triclinic form), 32B, 8
$C_4H_8O_8$, Dihydroxyfumaric acid dihydrate, 33B, 3
C_4H_9NO, Isobutyramide, 44B, 9
$C_4H_{10}ClNO_3 \cdot H_2O$, (RS)-2-Hydroxy-4-aminobutyric acid hydrochloride monohydrate, 45B, 5
$C_4H_{11}BrN_2O_2$, Acetamide hemihydrobromide, 32B, 9; 34B, 3; 43B, 5
$C_4H_{11}ClN_2O_2$, Bis(acetamide) hydrochloride, 45B, 6
$C_5H_6O_4$, Itaconic acid, 39B, 4
$C_5H_6O_4$, Mesaconic acid, 38B, 12
$C_5H_6O_5$, β-Ketoglutaric acid, 32B, 10
$C_5H_7BrN_4O_7$, N-(2,2,2-Trinitroisopropyl)bromoacetamide, 37B, 9
$C_5H_7ClN_4O_7$, N-(2,2,2-Trinitroisopropyl)chloroacetamide, 37B, 9
C_5H_7NO, N-Methyltetrolamide, 44B, 7
$C_5H_7NO_2$, Glutarimide, 37B, 10
C_5H_8ClNO, α-Chloro-δ-valerolactam, 32B, 11
$C_5H_8O_2$, Methyl α-methylacrylate (gas-ed), 23, 540
$C_5H_8O_2$, 2-Methyl-cis-but-2-enoic acid, 23, 540
$C_5H_8O_2$, 2-Methyl-trans-but-2-enoic acid, 23, 541
$C_5H_8O_4$, DL-Methylsuccinic acid, 39B, 5
$C_5H_8O_4$, Dimethylmalonic acid, 31B, 6
$C_5H_8O_4$, Glutaric acid (β form), 11, 582
$C_5H_8O_4$, Monomethyl succinate, 43B, 5
$C_5H_8O_4S_2$, Methylenedithioacetic acid, 44B, 10
$C_5H_9N_3O_3$, Hydroxyimino(N,N'-dimethyl)malonamide, 43B, 6
$C_5H_{10}N_2O_2$, Glutaramide, 31B, 7
$C_5H_{10}N_2O_2S$, S-Methyl N-methylcarbamoyloxythioacetimidate, 43B, 95
$C_5H_{10}O_2$, Valeric acid, 27, 739
$C_5H_{10}O_5$, Tris(hydroxymethyl)acetic acid, 45B, 7
$C_5H_{12}ClNO_3$, 2-Hydroxy-4-aminovaleric acid hydrochloride, 44B, 10
$C_6H_4Cl_2O_2$, Muconyl chloride, 44B, 10
$C_6H_6O_4$, trans,trans-Muconic acid, 38B, 13
$C_6H_6O_6$, Diacetylenedicarboxylic acid dihydrate, 11, 578
$C_6H_8O_2$, Methylketene dimer (gas-ed), 19, 593
$C_6H_8O_2$, Sorbic acid, 8, 266
$C_6H_8O_4$, Acrylic acid dimer (gas-ed), 23, 539
$C_6H_8O_4$, trans-2-Butene-1,4-dicarboxylic acid, 31B, 8
$C_6H_8O_5$, 3-Methylglutaconic acid, 44B, 11
$C_6H_8O_7$, Citric acid, 24, 521; 34B, 7
C_6H_9NO, trans,trans-Sorbamide, 32B, 12
$C_6H_9NO_3$, N-Methyl-citraconamic acid, 39B, 6
$C_6H_9NO_6$, Nitrilotriacetic acid, 32B, 13
$C_6H_{10}O_4$, (E)-3-Methoxymethoxybut-2-enoic acid, 44B, 94
$C_6H_{10}O_4$, Adipic acid, 12, 321; 13, 456; 30B, 5
$C_6H_{10}O_4$, meso-α,α'-Dimethylsuccinic acid, 32B, 14
$C_6H_{10}O_4$, racemic-α,α'-Dimethylsuccinic acid, 32B, 14
$C_6H_{10}O_6$, Dimethyl meso-tartrate, 39B, 6
$C_6H_{10}O_8$, Citric acid monohydrate, 38B, 14
$C_6H_{12}N_2O_2$, Adipamide, 31B, 9
$C_6H_{12}N_2O_2$, Ethylidene-N,N-diacetamide, 34B, 8
$C_6H_{12}N_2O_2$, N,N,N',N'-Tetramethyloxamide, 43B, 7
$C_6H_{12}N_2O_3$, Dilactylamide, 39B, 7
$C_6H_{13}NO$, Diethylacetamide, 45B, 8

$C_6H_{13}NO_2$, ϵ-Aminocaproic acid, 32B, 15
$C_6H_{19}CoN_6O_4S_2$, trans-Tetraamminebis(isothiocyanato)cobalt(III) acetate acetic acid, 46B, 23
$C_7H_7BrO_4$, 3-Bromo-4,4-dihydroxypimelic acid dilactone, 35B, 5
$C_7H_8O_3$, 5-Hydroxy-trans-2,trans-4-pentadienal acetate, 43B, 9
$C_7H_8O_4$, Monomethyl-trans,trans-muconate, 32B, 15
$C_7H_8O_4$, β-Methylmuconic acid γ-lactone, 23, 542
$C_7H_{11}NO$, N-Methylsorbamide, 44B, 8
$C_7H_{12}O_4$, Pimelic acid (α form), 22, 590; 31B, 10
$C_7H_{12}O_4$, Pimelic acid (β form), 11, 584
$C_7H_{12}O_4$, meso-α,α'-Dimethylglutaric acid, 29, 468; 32B, 16
$C_7H_{12}O_4$, racemic-α,α'-Dimethylglutaric acid, 33B, 4
$C_7H_{12}O_4$, 3,3-Dimethylglutaric acid, 38B, 15; 39B, 8
$C_7H_{14}N_2O_2S$, 2-Methyl-2-(methylthio)propionaldehyde O-(methylcarbamoyl)oxime, 43B, 100
$C_7H_{15}NO$, Diethylpropionamide, 41B, 7
$C_7H_{16}INO_2$, Methyl 3-(dimethylamino)propionate methiodide, 41B, 9
$C_8H_5Cl_3O_2$, 2,3,6-Trichlorophenylacetic acid, 45B, 9
$C_8H_6O_6$, Butane-1,2,3,4-tetracarboxylic dianhydride, 37B, 10
$C_8H_7ClO_3S$, 4-Chlorophenylsulfinyl-acetic acid, 41B, 10
$C_8H_7ClO_5$, Dilactophorbic acid chloride, 37B, 11
$C_8H_8Cl_2O_4$, Dimethyl trans,trans-2,5-dichloromuconate, 39B, 8
$C_8H_8N_2O_4$, N,N'-Bisuccinimidyl, 31B, 10
$C_8H_8O_3$, α-Hydroxyphenylacetic acid, 40B, 6; 43B, 10
$C_8H_9NO_2S$, Phenylsulfinyl-acetamide, 41B, 11
$C_8H_{10}O_4$, Dimethyl trans,trans-muconate, 32B, 17
$C_8H_{10}O_4$, Octa-cis-2,trans-6-diene-1,8-dioic acid, 34B, 9
$C_8H_{10}O_4$, Octa-trans-2,trans-6-diene-1,8-dioic acid (low temperature form), 34B, 9
$C_8H_{12}Cl_2O_4$, Dimethyl meso-β,β'-dichloroadipate, 40B, 7
$C_8H_{12}O_3$, Tetramethylsuccinic anhydride (gas-ed), 42B, 1027
$C_8H_{12}O_4$, α-Methylacrylic acid dimer (gas-ed), 23, 539
$C_8H_{13}NO_4$, Dimethyl (dimethylaminomethylene)malonate, 39B, 9
$C_8H_{14}O_2$, 2-Propyl-2-pentenoic acid, 43B, 11
$C_8H_{14}O_4$, Suberic acid, 29, 470; 30B, 6
$C_8H_{14}O_4$, meso-β,β'-Dimethyladipic acid, 31B, 12; 32B, 18
$C_8H_{16}N_2O_2$, Suberamide, 31B, 12
$C_8H_{16}N_2S_2$, N,N'-Diisopropyldithiooxalamide, 46B, 6
$C_8H_{20}Br_6N_4O_5Te$, Disuccinamide (protonated) hexabromotellurate monohydrate, 42B, 6
$C_9H_6O_2$, Phenylpropiolic acid, 19, 565
$C_9H_7ClO_2$, β-Chloro-cis-cinnamic acid, 38B, 158
$C_9H_7ClO_2$, β-Chloro-trans-cinnamic acid, 38B, 158
$C_9H_7ClO_2$, p-Chloro-trans-cinnamic acid, 41B, 12
$C_9H_7ClO_4$, 3-Chloro-4-hydroxyphenyl glyoxylic methyl ester, 46B, 492
C_9H_8ClNO, p-Chlorocinnamide, 40B, 1074
$C_9H_8O_2$, α-trans-Cinnamic acid, 41B, 14
$C_9H_9BrO_3$, L(-)-p-Bromo-β-phenylhydracrylic acid, 40B, 8
$C_9H_9ClO_2$, β-(p-Chlorophenyl)propionic acid, 41B, 12
$C_9H_9ClO_3$, D(+)-p-Chloro-β-phenylhydracrylic acid, 40B, 9
$C_9H_9FO_3$, D(+)-p-Fluoro-β-phenylhydracrylic acid, 40B, 9
$C_9H_9FO_3$, DL-p-Fluoro-β-phenylhydracrylic acid, 40B, 9
$C_9H_{10}O_3$, D(+)-β-Phenylhydracrylic acid, 40B, 10
$C_9H_{10}O_3$, S(-)-Phenyl-3-lactic acid, 41B, 15
$C_9H_{10}O_4$, (+)(2S,3S)-Phenylglyceric acid, 41B, 17
$C_9H_{10}O_4$, (-)(2S,3R)-Phenylglyceric acid, 41B, 16
$C_9H_{14}O_5$, Tetramethyl-β-oxoglutaric acid (monoclinic form), 37B, 12

$C_9H_{14}O_5$, α,α,α',α'-Tetramethyl-β-ketoglutaric acid (triclinic form), 34B, 11
$C_9H_{16}O_4$, Azelaic acid (form 1), 32B, 18
$C_9H_{16}O_4$, Azelaic acid (form 2), 32B, 18
$C_9H_{16}O_4$, DL-Pentane-2,4-diol diacetate, 31B, 14
$C_9H_{16}O_4$, n-Hexylmalonic acid, 42B, 6
$C_9H_{17}NO$, 3-Propyl-3-hexenamide, 43B, 11
$C_9H_{18}N_2O_2$, Azelamide, 31B, 15; 37B, 12
$C_9H_{18}O_3$, Peroxypelargonic acid, 30B, 7
$C_9H_{19}NO$, Dipropylpropionamide, 41B, 7
$C_9H_{19}NO$, N-Methyldipropylacetamide, 39B, 9
$C_9H_{19}NO_2$, 2-Propylpentyl carbamate, 43B, 12
$C_{10}H_7BrCl_2O_3$, p-Bromophenacyl dichloroacetate, 35B, 17
$C_{10}H_7BrF_2O_3$, p-Bromophenacyl difluoroacetate, 35B, 17
$C_{10}H_7Cl_4IO_4$, Phenyliodine(III) bis(dichloroacetate), 45B, 10
$C_{10}H_8BrNO_2$, N-p-Bromophenylsuccinimide, 28, 366
$C_{10}H_9BrO_2$, Methyl m-bromocinnamate, 30B, 8
$C_{10}H_9BrO_2$, Methyl p-bromocinnamate, 30B, 8
$C_{10}H_{10}O_2$, β-Methyl-cis-cinnamic acid, 38B, 158
$C_{10}H_{10}O_3$, p-Methoxy-cinnamic acid, 41B, 14
$C_{10}H_{10}O_4$, Benzylmalonic acid, 41B, 18
$C_{10}H_{11}IO_4$, Phenyliodine(III) diacetate, 45B, 10
$C_{10}H_{14}O_4$, Deca-trans-3,trans-7-diene-1,10-dioic acid (high temperature form), 34B, 13
$C_{10}H_{14}O_4$, Deca-trans-3,trans-7-dienedioic acid, 32B, 20
$C_{10}H_{14}O_8$, 1,1,2,2-Tetracarbomethoxyethane, 33B, 5
$C_{10}H_{15}NO_4$, Methyl trans-5-dimethylamino-2-carbomethoxypentadienecarboxylate, 43B, 15
$C_{10}H_{16}Br_2O_4$, Dimethyl (RRSS)-α,α'-dimethyl-β,β'-dibromoadipate, 43B, 15
$C_{10}H_{16}Br_2O_4$, Dimethyl (RSRS)-α,α'-dimethyl-β,β'-dibromoadipate, 43B, 15
$C_{10}H_{16}N_2O_8$, Ethylenediaminetetra-acetic acid, 27, 775; 38B, 15
$C_{10}H_{16}N_2O_8$, Ethylenediaminetetraacetic acid (β), 39B, 358
$C_{10}H_{16}O_3$, 9-Keto-trans-2-decenoic acid, 38B, 16
$C_{10}H_{18}O_4$, Sebacic acid, 12, 323; 31B, 17
$C_{10}H_{18}O_4$, Trimethylpimelic acid, 37B, 13
$C_{10}H_{18}O_4$, 2,2,4,4-Tetramethyladipic acid, 35B, 6
$C_{10}H_{20}N_2O_2$, N,N'-Diethyladipamide, 39B, 10
$C_{10}H_{20}N_2O_2$, Sebacamide, 31B, 18
$C_{10}H_{21}NO$, Decanamide, 22, 635
$C_{10}H_{21}NO$, Dibutylacetamide, 40B, 11
$C_{10}H_{26}N_2O_{13}$, N,N'-Ethylenediaminedisuccinic acid pentahydrate, 42B, 7
$C_{11}H_8Br_2O_3$, α-(2-Hydroxy-3,5-dibromobenzylidene)-γ-butyrolactone, 34B, 13
$C_{11}H_{10}O_2$, 2-Methyl-4-phenyl-2,3-butadiene carboxylic acid, 44B, 12
$C_{11}H_{11}NO_3$, Ethyl m-nitroso-trans-cinnamate, 44B, 12
$C_{11}H_{12}BrNO$, N,N-Dimethyl-p-bromocinnamamide, 37B, 52
$C_{11}H_{15}NO_2S$, 4-Methylthio-3,5-dimethylphenyl N-methylcarbamate, 43B, 16
$C_{11}H_{20}ClNOS$, N,N-Diisopropyl-3-chloro-4-methylthiobut-3-enoamide, 43B, 63
$C_{11}H_{20}Cl_2N_2O_2$, N,N'-Bis(β-chloroethyl)pimelamide, 43B, 16
$C_{11}H_{20}O_4$, Undecandioic acid (α form), 31B, 19
$C_{11}H_{21}BrO_2$, 11-Bromoundecanoic acid (form D), 28, 367
$C_{11}H_{21}BrO_2$, 11-Bromoundecanoic acid (form E), 28, 369

$C_{11}H_{22}O_2$, Tripropylacetic acid (low-temp), 42B, 7
$C_{11}H_{22}O_2$, Tripropylacetic acid, 42B, 7
$C_{11}H_{22}O_2S$, 3-Thiadodecanoic acid, 28, 371
$C_{11}H_{23}NO$, N-Propyl-dipropylacetamide, 40B, 11
$C_{11}H_{23}NO$, Tripropylacetamide, 40B, 11
$C_{11}H_{24}BrNO_2 \cdot 0.5\ H_2O$, 11-Aminoundecanoic acid hydrobromide hemihydrate, 19, 525
$C_{12}H_{15}NO_4S$, Ethyl N-methyl-N-(p-tolylsulphonylmethyl)carbamate, 39B, 10
$C_{12}H_{17}NO_4$, Methyl trans-7-dimethylamino-2-carbomethoxyheptatrienecarboxylate, 45B, 10
$C_{12}H_{22}Br_2N_2O_2$, N,N,N',N'-Tetraethyl-α,α'-dibromosuccinyldiamide (meso form), 16, 463
$C_{12}H_{22}O_4$, Dodecanedioic acid, 31B, 20
$C_{12}H_{24}N_2O_2$, N,N'-Hexamethylenebispropionamide, 21, 545; 27, 776
$C_{12}H_{24}N_2O_4$, N,N'-Dihydroxy-N,N'-diisopropylhexanediamide, 46B, 12
$C_{12}H_{24}O_2$, Lauric acid (A-super form), 44B, 13
$C_{12}H_{24}O_2$, Lauric acid (form A_1), 28, 373
$C_{12}H_{24}O_2$, Lauric acid (form C), 15, 385
$C_{12}H_{26}N_2O_2$, n-Dodecanoic acid hydrazide, 20, 483
$C_{13}H_{12}ClNO_2$, Ethyl p-chloro-α-cyano-β-methyl-cis-cinnamate, 38B, 165
$C_{13}H_{16}BF_4NO$, N,N-Dimethyl-(O-ethyl)phenylpropiolamidium tetrafluoroborate, 43B, 17
$C_{13}H_{18}O_2$, 2-(4-Isobutylphenyl)propionic acid, 40B, 12
$C_{13}H_{24}O_3$, DL-2-Methyl-7-oxododecanoic acid, 33B, 6
$C_{13}H_{24}O_4$, Brassylic acid, 33B, 7
$C_{13}H_{25}NO_6$, Methyl 8-t-butoxy-6-hydroxy-7-nitrooctanoate, 45B, 12
$C_{13}H_{26}O_2$, Tridecanoic acid (A' form), 46B, 13
$C_{14}H_{11}KO_3$, Potassium benzilate, 44B, 112
$C_{14}H_{13}NO$, N,N-Diphenylacetamide, 33B, 9
$C_{14}H_{14}O_3$, Kavaic acid, 40B, 13
$C_{14}H_{14}O_4$, Dimethyl p-phenylenediacrylate, 44B, 13
$C_{14}H_{14}O_4$, Dimethyl m-phenylenediacrylate, 43B, 18
$C_{14}H_{18}O_3$, Tetrahydrokavaic acid, 40B, 14
$C_{14}H_{27}BrO_4$, L-(11-Bromoundecanoic acid 1-monoglyceride) (β' form), 31B, 21
$C_{14}H_{28}O_3$, 2-DL-Hydroxytetradecanoic acid, 42B, 8
$C_{14}H_{29}NO$, N-Propyl-tripropylacetamide, 40B, 11
$C_{14}H_{29}NO$, Tetradecanamide, 19, 510
$C_{15}H_{13}FO_2$, DL-2-(2-Fluoro-4-biphenyl)propionic acid, 41B, 20
$C_{15}H_{13}NO_4$, Dimethyl 4-(2-carboxyvinyl)-α-cyanocinnamate, 44B, 14
$C_{15}H_{30}O_4$, 2-Monolaurin, 30B, 11
$C_{16}H_{14}O_4$, Diphenyl succinate, 45B, 14
$C_{16}H_{18}O_4$, Diethyl p-phenylenediacrylate, 44B, 14
$C_{16}H_{22}ClN$, 3'-(β-Methylaminoethyl)spiro(cyclopentane-1,1'-indene) hydrochloride, 42B, 24
$C_{16}H_{32}O_4$, (2R,5S)-5-Hydroxy-5-(hydroxymethyl)-2-methyltetradecanoic acid, 45B, 14
$C_{17}H_{24}O_2$, 9,10-trans-β-Ionylidene-γ-crotonic acid, 23, 637; 38B, 18
$C_{17}H_{27}NO_2$, 4-Propylheptyl N-phenylcarbamate, 43B, 12
$C_{17}H_{34}O_5$, 9,10,16-Trihydroxypalmitic acid, 42B, 8
$C_{18}H_{32}O_2$, Linoleic acid, 45B, 15
$C_{18}H_{32}O_2$, 4-Octadecynoic acid, 45B, 15
$C_{18}H_{34}O_2$, Oleic acid, 27, 740
$C_{18}H_{34}O_2$, cis-D,L-8,9-Methyleneheptadecanoic acid, 28, 375
$C_{18}H_{34}O_3$, 13-Oxoisostearic acid, 38B, 18
$C_{18}H_{35}BrO_2$, DL-3-Bromo-octadecanoic acid, 31B, 22

$C_{18}H_{36}O_2$, Isostearic acid, 38B, 19
$C_{18}H_{36}O_2$, Stearic acid (B form), 44B, 15
$C_{18}H_{36}O_2$, Stearic acid (C form), 37B, 13
$C_{19}H_{21}BrO_7$, (+)-m-Methoxyphenoxypropionic acid-(-)-m-bromophenoxypropionic acid, 31B, 16
$C_{19}H_{36}O_2$, cis-DL-11,12-Methyleneoctadecanoic acid, 23, 558
$C_{19}H_{38}O_2$, Methyl stearate, 24, 526; 35B, 7
$C_{19}H_{38}O_3$, 12-D-Hydroxyoctadecanoic acid methyl ester, 42B, 9
$C_{20}H_{20}N_2O_4$, p-Phenylenedi(α-cyanoacrylic acid) di-n-propyl ester, 42B, 9
$C_{20}H_{38}O_4$, Bis(di-t-butylmethyl) oxalate, 42B, 10
$C_{20}H_{40}O_2$, Ethyl stearate, 33B, 10
$C_{20}H_{41}NO_2$, N-(2-Hydroxyethyl)-octadecanamide, 43B, 19
$C_{21}H_{22}BrNO_3$, β-(p-Bromobenzoyloxy)-α-methyl-N,N-diethylcinnamamide, 37B, 53
$C_{21}H_{42}O_2$, n-Propyl stearate, 33B, 10
$C_{21}H_{42}O_4$, 1-Monostearin, 30B, 12
$C_{22}H_{29}NO_2$, Dextro-propoxyphene, 41B, 21
$C_{22}H_{42}O_5$, 2,4,6-Trimethyl-3-(3'-hydroxy-2',4',6'-trimethyloctanoyloxy)octanoic acid, 43B, 19
$C_{24}H_{18}O_4$, Diphenyl p-phenylenediacrylate, 44B, 15
$C_{25}H_{46}Br_2O_5$, 1,3-Diglyceride of 11-bromoundecanoic acid, 37B, 14
$C_{25}H_{48}O_5S_2$, 3-Thiadodecanoic acid 1,3-diglyceride, 28, 379
$C_{27}H_{26}BrNO_2$, 5-Methoxy-1,4-diphenyl-5-(p-bromophenyl)-penta-2-cis,4-trans-diene-1-(N,N-dimethylcarboxamide), 41B, 22
$C_{27}H_{41}Cl_2N_2O_6$, Chloramphenicol palmitate (β-form), 40B, 16
$C_{32}H_{64}O_2$, Cetyl palmitate, 8, 267
$C_{33}H_{62}O_6$, Tricaprin (β form), 31B, 24
$C_{39}H_{74}O_6$, Trilaurin (β form), 30B, 10
$C_{42}H_{85}NO_4$, N-Tetracosanoylphytosphingosine, 38B, 63

2. ALIPHATIC CARBOXYLIC ACID SALTS (AMMONIUM, IA, IIA METALS)

$CHKO_2$, Potassium formate, 46B, 16
$CHKS_2$, Potassium dithioformate, 40B, 17
$CHNaO_2$, Sodium formate, 6, 226; 8, 271; 41B, 23
CH_3LiO_3, Lithium formate monohydrate, 39B, 11; 43B, 20
CH_5NO_2, Ammonium formate, 33B, 11
$CH_6N_2O_2$, Ammonium carbamate, 39B, 11
$C_2CaO_4 \cdot H_2O$, Calcium oxalate monohydrate, 46B, 17
$C_2CaO_4 \cdot 2+x\ H_2O$, Calcium oxalate hydrate, 46B, 17
$C_2ClK_3OS_3$, Potassium trithiooxalate - potassium chloride, 42B, 11
$C_2D_2K_2O_6$, Potassium oxalate monoperdeuterate, 46B, 18
$C_2D_2O_6Rb_2$, Rubidium oxalate monoperdeuterate, 46B, 18
$C_2D_3LiO_5$, Lithium hydrogen oxalate monohydrate (deuterated), 38B, 20
$C_2K_2O_2S_2$, Potassium dithiooxalate, 41B, 1231
$C_2Li_2O_4$, Lithium oxalate, 29, 460
$C_2O_4Sr \cdot 0.17\ H_2O$, Strontium oxalate hydrate, 30B, 22
C_2HKO_4, Potassium hydrogen oxalate, 3, 674, 728; 33B, 11; 38B, 20
$C_2HK_2NO_4$, Dipotassium nitroacetate, 18, 657
C_2HNaO_4, Sodium hydrogen oxalate, 18, 643
$C_2H_2BaO_4$, Barium formate, 15, 382; 44B, 16
$C_2H_2CaO_4$, Calcium formate (α form), 11, 556; 43B, 20; 44B, 16; 46B, 19
$C_2H_2CaO_4$, Calcium formate (β form), 46B, 19
$C_2H_2CaO_5$, Calcium oxalate hydrate (whewellite), 26, 484; 27, 819

$C_2H_2ClCs_3O_3S_2$, Caesium 1,1-dithiooxalate - caesium chloride monohydrate, 42B, 11
$C_2H_2DLiO_3$, Lithium α-monodeuterioglycolate, 31B, 25
$C_2H_2FNaO_2$, Sodium fluoroacetate, 43B, 20
$C_2H_2K_2O_5$, Potassium oxalate monohydrate, 3, 675, 726; 29, 463; 34B, 16; 35B, 8
$C_2H_2K_2O_6$, Potassium oxalate monoperhydrate, 32B, 21
$C_2H_2Li_2O_6$, Lithium oxalate monoperhydrate, 34B, 15
$C_2H_2Na_2O_6$, Sodium oxalate perhydrate, 29, 462
$C_2H_2O_4Sr$, Strontium formate (α-form), 12, 333; 44B, 16
$C_2H_2O_5Rb_2$, Rubidium oxalate monohydrate, 30B, 24
$C_2H_2O_6Rb_2$, Rubidium oxalate monoperhydrate, 32B, 21
C_2H_3KOS, Potassium thioethanoate, 40B, 18
$C_2H_3KO_4$, Potassium hydrogen diformate, 33B, 13
$C_2H_3LiO_5$, Lithium hydrogen oxalate monohydrate, 35B, 9; 38B, 20
$C_2H_3NaO_5$, Sodium hydrogen oxalate monohydrate, 35B, 9; 37B, 16; 43B, 21
C_2H_3ORbS, Rubidium monothioacetate, 41B, 23
$C_2H_4CaO_6$, Calcium oxalate dihydrate, 30B, 22
$C_2H_4F_3NO_2$, Ammonium trifluoroacetate, 29, 456
$C_2H_5F_2NO_2$, Ammonium difluoroacetate, 42B, 12
$C_2H_5LiO_4$, Lithium glycollate monohydrate, 30B, 16
$C_2H_5NO_4 \cdot 0.5\ H_2O$, Ammonium hydrogen oxalate hemihydrate, 39B, 12
$C_2H_6FNO_2$, Ammonium fluoroacetate, 42B, 12
$C_2H_6MgO_6$, Magnesium formate dihydrate, 29, 455
$C_2H_6N_2O_3$, Ammonium oxamate, 28, 382
$C_2H_6N_2O_4$, Hydrazinium hydrogen oxalate, 33B, 36; 39B, 13; 44B, 16
$C_2H_6O_6Sr$, Strontium formate dihydrate, 22, 564; 26, 548; 37B, 17
$C_2H_7LiO_4$, Lithium acetate dihydrate, 22, 560; 35B, 11
$C_2H_7NO_2$, Ammonium acetate, 32B, 22
$C_2H_8N_2O_2$, Hydrazinium acetate, 40B, 1075
$C_2H_9NaO_5$, Sodium acetate trihydrate, 42B, 12; 43B, 3
$C_2H_{10}N_2O_2$, Ammonium acetate monoammine, 33B, 13
$C_2H_{10}N_2O_5$, Ammonium oxalate monohydrate, 4, 262, 291; 16, 429; 30B, 21; 38B, 22
$C_2H_{10}N_2O_6$, Ammonium oxalate monoperhydrate, 38B, 23
$C_2H_{13}N_3O_2$, Ammonium acetate diammine, 33B, 15
$C_2H_{15}CoN_6O_6$, trans-Dinitrotetramminecobalt(III) acetate, 44B, 17
$C_3H_2O_4Sr$, Strontium malonate, 43B, 22
$C_3H_3KO_3S$, Potassium S-methylmonothiooxalate, 44B, 17
$C_3H_3KO_4$, Potassium methyl oxalate, 44B, 17
$C_3H_3KO_4$, Potassium hydrogen malonate, 35B, 12
$C_3H_3NaO_3$, Sodium pyruvate, 26, 551
$C_3H_3NaO_4$, Sodium hydrogen malonate, 40B, 18
$C_3H_4FNaO_4$, Sodium β-fluoropyruvate, 45B, 16
$C_3H_4Na_2O_5$, Disodium malonate monohydrate, 44B, 18
$C_3H_6CaO_6$, Calcium malonate dihydrate (gel grown phase), 44B, 18
$C_3H_6CaO_6$, Calcium malonate dihydrate, 43B, 22, 23
$C_3H_7NO_4$, Ammonium hydrogen malonate, 41B, 24
$C_3H_7NO_4$, Monomethylammonium hydrogen oxalate, 41B, 24
$C_3H_9N_3O_5$, Guanidinium hydrogen oxalate monohydrate, 44B, 19
$C_4DF_6KO_4$, Potassium deuterium bis(trifluoroacetate), 38B, 25
$C_4HCl_6KO_4$, Potassium hydrogen bis-trichloroacetate, 40B, 19
$C_4HCl_6O_4Rb$, Rubidium hydrogen bis-trichloroacetate, 40B, 20
$C_4HCsF_6O_4$, Caesium hydrogen di-trifluoroacetate, 30B, 13
$C_4HF_2KO_4$, Potassium hydrogen difluorofumarate, 46B, 20
$C_4HF_2KO_4$, Potassium hydrogen difluoromaleate, 46B, 20

$C_4HF_6KO_4$, Potassium hydrogen bis(trifluoroacetate), 30B, 25; 38B, 25
C_4HKO_4, Potassium hydrogen acetylenedicarboxylate, 39B, 13
C_4HO_4Rb, Rubidium hydrogen acetylenedicarboxylate, 39B, 14
$C_4H_2ClKO_4$, Potassium hydrogen chloromaleate, 30B, 25
$C_4H_2Cl_4DKO_4$, Potassium deuterium bis-dichloroacetate, 45B, 18
$C_4H_3Br_4CsO_4$, Caesium hydrogen bis(dibromoacetate); 46B, 21
$C_4H_3Br_4KO_4$, Potassium hydrogen bis(dibromoacetate), 45B, 17
$C_4H_3Br_4O_4Rb$, Rubidium hydrogen bis(dibromoacetate), 46B, 21
$C_4H_3Cl_4KO_4$, Potassium hydrogen bis-dichloroacetate, 40B, 21; 45B, 19
$C_4H_3CsO_5$, Caesium hydrogen acetylenedicarboxylate monohydrate, 42B, 13
$C_4H_3KO_4$, Potassium hydrogenfumarate, 44B, 20
$C_4H_3KO_4$, Potassium hydrogen maleate, 22, 573; 26, 555
$C_4H_3NaO_4$, Sodium hydrogen fumarate, 35B, 13
$C_4H_3O_4Rb$, Rubidium hydrogen fumarate, 34B, 17
$C_4H_3O_8Tl \cdot 2\ H_2O$, Thallium oxalate dihydrogen-oxalate dihydrate, 45B, 19
$C_4H_4Li_2O_4$, Lithium succinate, 39B, 14
$C_4H_4Na_2O_5$, Disodium maleate monohydrate, 40B, 20
$C_4H_5Cl_6NO_4$, Ammonium hydrogen bis-trichloroacetate, 41B, 26
$C_4H_5KO_4$, Potassium hydrogen succinate, 37B, 18
$C_4H_5KO_4S$, Potassium hydrogen thiodiacetate, 42B, 14
$C_4H_5KO_5$, Potassium hydrogen oxydiacetate, 39B, 15
$C_4H_5KO_6$, Potassium hydrogen meso-tartrate, 38B, 26; 41B, 26
$C_4H_5LiO_5$, Lithium hydrogen (+)-1-malate, 46B, 22
$C_4H_5LiO_5$, Lithium hydrogen oxydiacetate, 42B, 13
$C_4H_5NO_4$, Ammonium hydrogen acetylenedicarboxylate, 40B, 25
$C_4H_5NaO_3$, Sodium α-ketobutyrate, 28, 384
$C_4H_5NaO_5$, Sodium hydrogen oxydiacetate, 39B, 15
$C_4H_5NaO_6$, Sodium hydrogen acetylenedicarboxylate dihydrate, 40B, 24
$C_4H_5O_4RbS$, Rubidium hydrogen thiodiacetate, 42B, 14
$C_4H_5O_5Rb$, Rubidium hydrogen oxydiacetate, 39B, 15
$C_4H_6BaO_{10}$, Barium hydrogen oxalate dihydrate, 35B, 15
$C_4H_6K_2O_6$, Dipotassium fumarate dihydrate, 35B, 14
$C_4H_6Li_2O_6$, Lithium maleate dihydrate, 39B, 16
$C_4H_7CsO_5$, Caesium hydrogen succinate monohydrate, 37B, 17
$C_4H_7KO_4$, Potassium hydrogen diacetate, 38B, 26
$C_4H_7KO_6$, Potassium hydrogen bis(glycollate), 33B, 18
$C_4H_7KO_{10}$, Potassium tetroxalate dihydrate, 29, 465; 32B, 25
$C_4H_7LiO_6$, Lithium hydrogen maleate dihydrate, 41B, 25
$C_4H_7NaO_4$, Sodium hydrogen diacetate, 26, 556; 41B, 28
$C_4H_7O_6Rb$, Rubidium hydrogen glycollate, 30B, 17
$C_4H_8CaO_6 \cdot H_2O$, Diaquasuccinatocalcium(II) monohydrate, 46B, 22
$C_4H_8CaO_7$, Calcium fumarate trihydrate, 38B, 27
$C_4H_8CaO_7$, Calcium malate dihydrate, 31B, 26
$C_4H_8K_2O_8$, Potassium mesotartrate dihydrate, 30B, 26
$C_4H_8Na_2O_8$, Sodium D-tartrate dihydrate, 33B, 15
$C_4H_8O_8Rb_2$, Rubidium mesotartrate dihydrate, 30B, 26
$C_4H_9Cl_2NO_4$, Ammonium hydrogen bis-chloroacetate (low-temperature), 40B, 25
$C_4H_9Cl_2NO_4$, Ammonium hydrogen bis-chloroacetate, 38B, 28
$C_4H_9NO_4$, Dimethylammonium hydrogen oxalate, 41B, 29; 43B, 24
$C_4H_9NO_5$, Ammonium hydrogen L-malate, 44B, 21
$C_4H_9NO_6$, Ammonium hydrogen D-tartrate, 22, 575
$C_4H_9NaO_7$, Sodium hydrogen maleate trihydrate, 40B, 26
$C_4H_{10}LiNO_7$, Lithium ammonium tartrate monohydrate, 20, 480; 38B, 29
$C_4H_{10}O_9Sr$, Strontium tartrate trihydrate, 33B, 16; 34B, 19

$C_4H_{11}NO_4$, Ammonium hydrogen diacetate, 34B, 16
$C_4H_{11}NO_{10}$, Ammonium tetroxalate dihydrate, 32B, 23
$C_4H_{12}CaO_6S$, Calcium acetate monothioacetate trihydrate, 41B, 27
$C_4H_{12}CaO_{10}$, Calcium tartrate tetrahydrate, 33B, 16
$C_4H_{12}KNaO_{10}$, Potassium sodium DL-tartrate tetrahydrate, 13, 454
$C_4H_{12}KNaO_{10}$, Rochelle salt, 8, 272
$C_4H_{12}N_2O_6$, Ammonium tartrate, 39B, 16
$C_4H_{12}NaO_{10}Rb$, Sodium rubidium D-tartrate tetrahydrate, 13, 589
$C_4H_{14}MgO_8$, Magnesium acetate tetrahydrate, 21, 504
$C_4H_{14}O_7SSr$, Strontium ethanoate thioethanoate tetrahydrate, 41B, 28
$C_4H_{16}CaO_{11}$, Calcium oxydiacetate hexahydrate, 39B, 17
$C_4H_{18}N_6O_8$, Guanidinium oxalate dihydrate monoperhydrate, 42B, 14
$C_5H_5KO_4$, Potassium hydrogen mesaconate, 41B, 30
$C_5H_7KO_4$, Potassium hydrogen DL-methylsuccinate, 38B, 31
$C_5H_7KO_4$, Potassium hydrogen glutarate, 38B, 30
$C_5H_7NaO_3$, Sodium 2-oxovalerate, 34B, 19
$C_5H_7O_4Rb$, Rubidium hydrogen glutarate, 37B, 19
$C_5H_9KOS_2$, Potassium butyl xanthate, 45B, 6
$C_5H_{11}NO_4$, Ammonium hydrogen glutarate, 37B, 19
$C_5H_{11}NO_4$, Trimethylammonium hydrogen oxalate, 41B, 30
$C_6H_2K_2O_8$, Dipotassium ethylenetetracarboxylate, 30B, 31
$C_6H_4K_2O_6$, Dipotassium cis-aconitate, 38B, 32
$C_6H_5KO_6$, Potassium dihydrogen trans-aconitate, 38B, 33
$C_6H_5KO_6$, Potassium hydrogen isocitrate lactone, 28, 385; 39B, 18
$C_6H_7KO_7$, Potassium dihydrogen isocitrate, 33B, 21
$C_6H_7LiO_7$, Lithium dihydrogen citrate, 30B, 28
$C_6H_7LiO_8$, Lithium trihydrogen dimalonate, 44B, 21
$C_6H_7NaO_7$, Sodium dihydrogen citrate, 30B, 28
$C_6H_7O_7Rb$, Rubidium dihydrogen citrate, 24, 519
$C_6H_8NNa_3O_7$, Trisodium nitriloacetate monohydrate, 32B, 25
$C_6H_9NaO_3$, Sodium 2-oxocaproate, 33B, 21
$C_6H_{10}MgO_{10}$, Magnesium bis(hydrogen malonate) dihydrate, 44B, 22
$C_6H_{11}CaNO_8$, Calcium nitrilotriacetate dihydrate, 38B, 688
$C_6H_{11}KO_2$, Potassium caproate, 16, 432
$C_6H_{12}CaO_{10}$, Calcium hydrogen citrate trihydrate, 40B, 27
$C_6H_{12}CaO_{11}$, Calcium (+)-allo-hydroxycitrate lactone tetrahydrate, 34B, 18; 38B, 34
$C_6H_{12}CaO_{11}$, Calcium (-)hydroxycitrate lactone tetrahydrate, 34B, 18; 37B, 20
$C_6H_{12}LiNO_8$, Lithium ammonium hydrogen citrate monohydrate, 32B, 26
$C_6H_{13}FNO_9Rb$, Rubidium ammonium hydrogen fluorocitrate dihydrate, 39B, 19
$C_6H_{14}CaO_9$, Calcium ethylenedioxydiacetate trihydrate, 43B, 25
$C_6H_{14}CaO_{10}$, Calcium di-DL-glycerate dihydrate, 45B, 19
$C_6H_{14}N_2O_6$, Ethylenediamine (+)-tartrate, 42B, 15
$C_6H_{15}Br_2CaN_3O_6$, Tris(glycine)calcium(II) dibromide, 46B, 23
$C_7H_5D_3N_2O_4$, Deuterated imidazolium hydrogen maleate, 46B, 24
$C_8H_7NaO_3 \cdot 0.5\ H_2O$, Sodium phenoxyacetate hemihydrate, 37B, 21
$C_8H_{10}BaO_4$, Barium methacrylate, 42B, 15
$C_8H_{11}KO_4$, Potassium hydrogen dicrotonate, 43B, 26
$C_8H_{11}KO_8$, Potassium trihydrogen disuccinate, 39B, 20
$C_8H_{12}BaO_5$, Barium methacrylate monohydrate, 38B, 691
$C_8H_{13}NaO_3$, Sodium 2-oxocaprylate, 29, 471
$C_8H_{16}CaO_{13}$, Calcium di(hydrogen maleate) pentahydrate, 44B, 22; 46B, 201
$C_8H_{18}N_2O_8 \cdot H_2O$, Ethylenediammonium (2S,3S)-2-hydroxycitrate monohydrate, 45B, 20

$C_8H_{18}N_2O_8$, Ethylenediammonium (4S)-4-hydroxycitrate, 45B, 20
$C_8H_{18}O_{16}Zn$, Hexaaquozinc hydrogenacetylenedicarboxylate dihydrate, 43B, 26
$C_8H_{22}CaO_{16}$, Calcium di(hydrogen 1-malate) hexahydrate, 46B, 24
$C_9H_{16}BrCaNO_5$, Calcium bromide D-pantothenate, 45B, 21
$C_{10}H_{16}BaN_6O_6$, Barium N,N'-dimethylisonitrosomalonamide, 42B, 15
$C_{10}H_{16}K_2N_2O_9$, Dipotassium ethylenediaminetetra-acetate monohydrate, 34B, 20
$C_{10}H_{18}N_2O_{10}Rb_2$, Dirubidium ethylenediaminetetra-acetate dihydrate, 31B, 28
$C_{10}H_{19}KO_2$, Potassium caprate (form A), 11, 558; 34B, 21
$C_{10}H_{20}N_2O_3$, Ammonium 2,3-di-isopropylmaleamate, 39B, 20
$C_{10}H_{30}CoN_8NaO_8 \cdot 3.5\ H_2O$, Hexaamminecobalt(III) sodium ethylenediaminetetraacetate 3.5-hydrate, 43B, 28
$C_{12}H_{10}K_2O_{12}$, Dipotassium tetrahydrogen trifumarate, 35B, 16; 38B, 36; 39B, 21
$C_{12}H_{13}K_3O_{14}$, Potassium trans-aconitate, 39B, 21
$C_{12}H_{18}Be_4O_{13}$, Basic beryllium acetate, 3, 666, 716; 9, 317; 23, 534
$C_{12}H_{26}N_2O_4$, Hexamethylenediammonium adipate, 18, 659; 31B, 27
$C_{12}H_{30}Ba_2N_{12}O_{19}$, Barium isonitrosomalonamide hydrate, 42B, 15
$C_{12}H_{30}Mg_3O_{24}$, Magnesium citrate decahydrate, 30B, 30
$C_{14}H_{20}CaO_{10}$, Calcium mandelate hydroxide 3-acetic acid, 43B, 27
$C_{16}H_{15}KO_4$, Potassium hydrogen bis(phenylacetate), 12, 381; 21, 591; 24, 634; 33B, 45; 43B, 28
$C_{16}H_{18}NO_3$, (1-Phenyl)ethylammonium mandelate, 45B, 21
$C_{16}H_{31}KO_2$, Potassium palmitate (form B), 30B, 19
$C_{17}H_{21}NO_2$, n-α-Phenylethylammonium α-phenyl-α-methylacetate, 42B, 16
$C_{17}H_{21}NO_2$, p-α-Phenylethylammonium α-phenyl-α-methylacetate, 42B, 16
$C_{17}H_{24}FNO_7$, Methylbenzylammonium (2S,3S)-diethyl-2-fluorocitrate, 46B, 25
$C_{18}H_{19}NO_4$, Ammonium hydrogen dicinnamate, 28, 387
$C_{18}H_{19}NaO_6$, Sodium hydrogen di(α-methoxy-α-phenylacetate), 46B, 26
$C_{18}H_{19}O_6Rb$, Rubidium hydrogen di(α-methoxy-α-phenylacetate), 46B, 26
$C_{18}H_{30}BaCa_2O_{12}$, Barium dicalcium propionate, 3, 670, 719; 46B, 26
$C_{20}H_{37}Cl_2NO_4$, Tetra-n-butylammonium hydrogen dichloromaleate, 46B, 27
$C_{25}H_{35}N_3O_7$, (+)-[(2R)-(-)-4-(Diisopropylamino)-2-(2-pyridyl)-2-phenylbutyramide (+)-(2R,3R)-bitartrate], 46B, 35
$C_{26}H_{28}ClNO_4$, (1S,4R)-3'-Chloro-10',11'-dihydro-N,N-dimethylspiro[2-cyclohexene-1,5'-[5H]dibenzo[a,d]cyclohepten]-4-amine hydrogen maleate, 46B, 36

3. ALIPHATIC AMINES

CH_3Cl_2N, Methyldichloramine (gas-ed), 8, 277
CH_3N, Methylenimine (gas-mw), 43B, 1485
CH_5N, Methylamine (gas-ed), 30B, 400
CH_5N, Methylamine (gas-mw), 19, 507; 20, 494
CH_5N, Methylamine, 17, 649
CH_6BrN, Methylammonium bromide, 26, 585
CH_6ClN, Methylammonium chloride, 10, 206
CH_6ClNO, N-Methylhydroxylammonium chloride, 32B, 28
CH_6ClNO_4, Methylammonium perchlorate, 33B, 21
CH_6Cl_3NNi, Nickel(II) chloride - methylammonium chloride complex, 31B, 29
CH_6FN, Methylammonium fluoride, 45B, 22

$CH_6IN_3O_3$, Guanidinium iodate, 43B, 76
$CH_9N_4O_4P$, Aminoguanidinium dihydrogen orthophosphate, 43B, 29
$C_2H_4F_3N$, Trifluoroethylamine (gas-mw), 38B, 1070
$C_2H_5Br_2N$, Acetonitrile dihydrobromide, 32B, 28
$C_2H_5Cl_2N$, Chloroacetiminium chloride, 33B, 23
C_2H_6ClN, Dimethylchloramine (gas-ed), 8, 277
C_2H_6IN, N-Iododimethylamine, 43B, 29
$C_2H_7ClN_2$, Acetamidinium chloride, 42B, 4
C_2H_7N, Dimethylamine (gas-ed), 6, 233; 33B, 530
C_2H_7N, Dimethylamine (gas-mw), 33B, 539
C_2H_7NO, 2-Aminoethanol (gas-mw), 37B, 697; 42B, 1034
$C_2H_7N_5$, Biguanide, 43B, 30
C_2H_8BrN, Monoethylamine hydrobromide, 22, 620
C_2H_8ClN, Dimethylammonium chloride, 33B, 24; 38B, 36
$C_2H_8ClN_5$, Biguanide hydrochloride, 43B, 30
$C_2H_8Cl_3CuN$, Copper(II) dimethylammonium chloride, 31B, 29
$C_2H_8N_2$, Ethylenediamine (gas-ed), 37B, 688
$C_2H_8N_2$, Ethylenediamine, 39B, 22
$C_2H_{10}Br_2N_2$, Ethylenediammonium dibromide, 42B, 18
$C_2H_{10}Cl_2N_2$, Ethylenediammonium dichloride, 28, 391
$C_2H_{10}N_2O_4S$, Ethylenediammonium sulfate, 26, 589
$C_2H_{18}N_6O_{13}S_2U$, Guanidinium uranyl sulfate trihydrate, 43B, 32
C_3F_9N, Perfluorotrimethylamine (gas-ed), 20, 494
C_3HF_6N, Hexafluoropropylimine (gas-ed), 35B, 891
C_3H_9ClIN, Trimethylamine - iodomonochloride, 24, 554
C_3H_9N, Trimethylamine (gas-ed), 4, 277; 33B, 530
$C_3H_9NO_4S$, 3-Amino-2-hydroxy-1-propanesulphonic acid, 45B, 23
$C_3H_{10}BrN$, n-Propylammonium bromide, 13, 484
$C_3H_{10}ClN$, Trimethylammonium chloride, 33B, 25
$C_3H_{10}ClN$, n-Propylammonium chloride, 13, 483
$C_3H_{10}IN$, Trimethylammonium iodide, 35B, 25
$C_3H_{10}IN$, n-Propylammonium iodide, 13, 484
$C_3H_{12}Cl_2N_2$, Trimethylenediamine dihydrochloride, 33B, 26
$C_3H_{12}N_2O_5S \cdot H_2O$, 1,3-Diammonio-2-propanol sulfate monohydrate, 46B, 29
$C_4H_8N_2O_2$, trans-N,N-Dimethyl-2-nitroethenamine, 46B, 29
$C_4H_{12}ClNO_4$, Tetramethylammonium perchlorate, 2, 804, 815; 29, 476
$C_4H_{12}Cl_2IN$, Tetramethylammonium dichloroiodide, 29, 477
$C_4H_{12}F_6NP$, Tetramethylammonium hexafluorophosphate, 46B, 31
$C_4H_{12}I_5N$, Tetramethylammonium pentaiodide, 13, 484; 15, 401
$C_4H_{14}Cl_2N_2$, Putrescine dihydrochloride, 28, 390; 46B, 31
$C_4H_{18}N_2O_8P_2$, Putrescine diphosphate, 44B, 24; 45B, 23, 24
$C_4H_{26}N_{12}O_8P_2$, Guanidinium pyrophosphate monohydrate, 43B, 35
$C_5H_{12}ClNO_4$, N,N-Dimethylisopropylideniminium perchlorate, 31B, 30
$C_6H_6F_3N_7O_{12}$, Tris(2-fluoro-2,2-dinitroethyl)amine, 46B, 31
$C_6H_{12}N_2$, Triethylenediamine (gas-ed), 37B, 687
$C_6H_{14}I_3N_3$, Dimethyl-(3-dimethylamino-2-aza-propen(2)-yliden)ammonium triiodide, 39B, 23
$C_6H_{14}N_2O_6$, Ethylenediamine D-tartrate, 39B, 23; 43B, 36
$C_6H_{15}Cl_{0.5}I_{1.5}N$, 3-Chloropropyltrimethylammonium iodide - 3-iodopropyltrimethylammonium iodide, 40B, 28
$C_6H_{16}ClN$, Triethylammonium chloride, 30B, 33
$C_6H_{16}N_2$, Hexamethylenediamine, 13, 491
$C_6H_{18}Cl_2I_2N_2$, Tetramethylammonium μ-dimethylamino-bis(chloroiodate)(1-), 45B, 25
$C_6H_{18}Cl_2N_2$, 1,6-Hexamethylenediamine dihydrochloride, 12, 352; 44B, 25

$C_6H_{18}I_2N_2$, Hexamethylenediamine dihydroiodide, 28, 396
$C_6H_{21}Cl_3N_4$, 2,2',2''-Triamino-triethylamine trihydrochloride, 33B, 27
$C_7H_{18}ClN$, N-t-Butylpropylamine hydrochloride, 28, 397
$C_8H_{11}Cl_2N$, p-Chlorophenylethylamine hydrochloride, 44B, 25
$C_8H_{12}ClN$, β-Phenylethylamine hydrochloride, 26, 702
$C_8H_{12}ClNO$, Tyramine hydrochloride, 40B, 29
$C_8H_{16}N_2O_7$, Ethylenediamine citrate, 43B, 38
$C_8H_{19}BrFN$, (2-Fluoroethyl)triethylammonium bromide, 46B, 32
$C_8H_{20}BN_5O_{12}$, Tetraethylammonium tetranitratoborate, 44B, 26
$C_8H_{20}BrCl_2N$, Tetraethylammonium dichlorobromate, 40B,29
$C_8H_{20}IN$, Tetraethylammonium iodide, 22, 630
$C_8H_{20}I_3N$, Tetraethylammonium tri-iodide (form 1), 32B, 33
$C_8H_{20}I_3N$, Tetraethylammonium tri-iodide (form 2), 32B, 33
$C_8H_{20}I_7N$, Tetraethylammonium heptaiodide, 22, 631
$C_8H_{22}ClNO$, Tetraethylammonium chloride monohydrate, 44B, 27
$C_8H_{24}Cl_6N_2Sn$, Tetramethylammonium hexachlorostannate(IV), 46B, 33
$C_8H_{32}N_2O_8S$, Tetramethylammonium sulfate tetrahydrate, 33B, 27
$C_9H_{14}ClNO$, Norephedrine hydrochloride, 45B, 26
$C_9H_{15}NS$, 4-Dimethylamino-4-t-butyl-1-thiabuta-1,2,3-triene, 44B, 27
$C_9H_{17}ClNO_4$, 1,5-Bis(dimethylamino)pentamethinium perchlorate, 42B, 19
$C_9H_{20}BClF_4N_2$, 2-t-Butylammonium-3-chloro-N,N-dimethylpropionamide tetrafluoroborate, 46B, 33
$C_9H_{21}ClN_2O_2$, Bis(dimethyl)-pentamethine-cyanine chloride, 41B, 36
$C_9H_{22}N_2S_2$, Diethylammonium diethyldithiocarbamate, 44B, 28
$C_9H_{24}Ag_{11}I_{13}N_2$, N,N,N,N',N',N'-Hexamethylisopropylenediamine diiodide - silver iodide, 42B, 20
$C_{10}H_{18}N_2O_{10}Rb_2$, Dirubidium ethylenediaminetetraacetate dihydrate, 35B, 26
$C_{10}H_{20}ClN$, Geranylamine hydrochloride, 10, 218
$C_{10}H_{20}N_2$, N,N'-Di-t-butyl-1,2-ethanediimine (gas-ed), 42B, 1024
$C_{10}H_{20}N_2O_2$, N,N'-Diacetylhexamethylenediamine, 19, 513
$C_{10}H_{24}N_2O_{14}$, Ethylenediamine ditartrate dihydrate, 43B, 43
$C_{10}H_{24}N_2S_6 \cdot 0.5\ CS_2$, Tetramethylammonium perthiocarbonate carbon disulfide solvate, 41B, 37
$C_{10}H_{30}Br_2N_2O_2$, Tetramethylene bis(trimethylammonium) dibromide dihydrate, 38B, 41
$C_{10}H_{33}Cl_5N_6$, N,N,N',N'-Tetrakis(2-aminoethyl)ethylenediamine pentahydrochloride, 32B, 37
$C_{11}H_{15}Cl_2NO$, Dichloroisoproterenol, 43B, 45
$C_{11}H_{16}Cl_3NO$, Dichloroisoproterenol hydrochloride, 41B, 38
$C_{11}H_{17}NO_3 \cdot 0.5\ H_2SO_4 \cdot 0.5\ C_2H_5OH$, DL-1-(3,5-Dihydroxyphenyl)-2-(isopropylamino)ethanol sulfate hemiethanolate, 38B, 41
$C_{11}H_{18}ClNO_3$, Methoxamine chloride, 40B, 32
$C_{11}H_{28}I_2N_2 \cdot 0.25\ H_2O$, Pentamethylenedi(trimethylammonium) iodide quarterhydrate, 28, 398
$C_{11}H_{32}I_3N_3O$, Heptamethylhydrodiethylenetriamine triiodide monohydrate, 43B, 46
$C_{12}H_{18}ClNO$, 3-Phenyl-4-dimethylammonio-2-butanone chloride, 43B, 47
$C_{12}H_{20}N_2O_2$, N,N'-Ethylenebis(acetylacetoneimine), 45B, 12
$C_{12}H_{21}ClN_2O_3S$, Sotalol hydrochloride, 42B, 21
$C_{12}H_{28}BrN$, Tetra-n-propylammonium bromide, 21, 523
$C_{12}H_{28}BrN$, n-Dodecylammonium bromide, 40B, 34
$C_{12}H_{34}Br_2N_2O_2$, Hexamethylenebis(trimethylammonium) dibromide dihydrate, 30B, 34
$C_{12}H_{36}N_3F_2PS_2Se$, Tris(diethylammonium) fluoroselenodithiophosphate

fluoride, 45B, 1400
$C_{13}H_{19}ClN_2O_4$, 2-Diethylaminoethyl-p-nitrobenzoate hydrochloride, 39B, 24
$C_{13}H_{21}ClN_2O_2$, Procaine hydrochloride, 38B, 44
$C_{13}H_{21}NO_2$, 4-Ethyl-2,5-dimethoxyamphetamine, 40B, 36
$C_{13}H_{21}NO_3$, DL-N-t-butyl-2-(4-hydroxy-3-hydroxymethylphenyl)-2-hydroxyethylamine, 38B, 45
$C_{13}H_{30}N_2S_2$, Diisopropylammonium diisopropyldithiocarbamate, 44B, 28
$C_{14}H_{30}I_2N_2O_4$, Diteline, 39B, 25
$C_{15}H_{14}N_2O_2$, Bis(p-methoxyphenyl)carbodiimide, 44B, 31
$C_{15}H_{24}ClNO_2$, Alprenolol hydrochloride, 39B, 24; 40B, 38
$C_{15}H_{24}ClNO_3$, Oxprenolol hydrochloride, 43B, 49
$C_{16}H_{21}NO_2$, Propranolol, 43B, 45
$C_{16}H_{22}ClNO_2$, Propanolol hydrochloride, 39B, 24; 40B, 38; 43B, 45
$C_{16}H_{36}INO_4$, Tetrabutylammonium periodate, 43B, 49
$C_{16}H_{36}N_4S_3$, n-Tetrabutylammonium trisulfurtrinitride, 45B, 28
$C_{16}H_{42}Br_2N_2O_2$, Decamethylenebis(trimethylammonium) dibromide dihydrate, 30B, 34
$C_{16}H_{49}B_{11}N_2Se_3$, Tetraethylammonium triselenylundecaborane, 45B, 1402
$C_{17}H_{16}N_2O_2$, N-(3-Amino-2,3-diphenyl-acrolyl)-acetamide, 42B, 24
$C_{17}H_{22}BrNO_4$, DL-N-(1-(4-Hydroxyphenyl)-prop-2-yl)-2-(3,5-dihydroxyphenyl)-2-hydroxyethylamine hydrobromide (isomer Th 1165), 38B, 45
$C_{17}H_{22}BrNO_4$, DL-N-(1-(4-Hydroxyphenyl)-prop-2-yl)-2-(3,5-dihydroxyphenyl)-2-hydroxyethylamine hydrobromide (isomer Th 1179), 38B, 46
$C_{17}H_{24}N_4O_2S$, t-Butyl-(2-t-butylimino-1-phenylsulphonylimino-2-cyanoethyl)amine, 45B, 216
$C_{17}H_{44}Br_2N_4O$, Tetrabutylammonium guanidinium bromide monohydrate, 43B, 53
$C_{18}H_{23}NO_2$, α-Phenylethylammonium α-phenylbutyrate, 44B, 29
$C_{18}H_{24}IN$, (3,3-Diphenylpropyl)trimethylammonium iodide, 44B, 29
$C_{18}H_{28}ClNO_2$, 2-Diethylaminoethyl-1-phenylcyclopentanecarboxylate hydrochloride, 38B, 47
$C_{18}H_{29}NO_6S$, 1-(2'-Cyclopentylphenoxy)-3-t-butylaminopropan-2-ol sulfate, 42B, 25
$C_{18}H_{30}BrNO_4$, (-)-erythro-1'-(2,5-Dimethoxyphenyl)-3'-diethylaminobutyl acetate hydrobromide, 40B, 41
$C_{18}H_{48}Ag_{21}I_{25}N_4$, N,N,N,N',N',N'-Hexamethyl-1,3-propylenediamine diiodide - silver iodide, 42B, 25
$C_{19}H_{30}NO_5S$, (-)-1-t-Butylamino-3-(2-cyclopentylphenoxy)-propan-2-olmethylsulphonate, 40B, 41
$C_{19}H_{42}Cl_2IN$, Hexadecyltrimethylammonium dichloroiodide, 44B, 30
$C_{20}H_{26}ClNO_3$, Difemerine hydrochloride, 45B, 29
$C_{20}H_{26}ClNO_3$, Benzactyzine hydrochloride, 40B, 44
$C_{21}H_{19}NO$, 8-(Diphenylmethylene)amino-3,5,7-octatrien-2-one, 44B, 30
$C_{21}H_{21}N$, Tribenzylamine, 38B, 49
$C_{21}H_{28}BrNO_2$, γ-Diethylaminopropyl diphenylacetate hydrobromide, 39B, 25; 40B, 45
$C_{22}H_{30}ClNO_2$, Dextropropoxyphene hydrochloride, 39B, 26
$C_{23}H_{32}BrNO_3$, γ-Diethylaminopropyl α-ethoxydiphenylacetate hydrobromide, 39B, 25
$C_{23}H_{33}IN_2O$, γ-(Aminocarbonyl)-N-methyl-N,N-bis-(1-methylethyl)-γ-phenylbenzene-propan-amine iodide, 42B, 28
$C_{24}H_{56}Cl_4N_2Zn$, Bis(n-dodecylammonium) tetrachlorozincate, 43B, 57
$C_{25}H_{62}Br_3N_5$, Tetrapropylammonium guanidinium bromide (2:1 Complex), 43B, 59
$C_{27}H_{22}N_2$, Bis(diphenylmethyl)carbodiimide, 44B, 31
$C_{28}H_{24}N_2$, N,N'-Dibenzylidene-1,2-diphenylethylenediamine, 42B, 28

$C_{28}H_{32}ClN_3O_5$, 1,5-Bis(dimethylamino)-3-ethoxy-1,2,5-triphenyl-4-aza-pentamethinium perchlorate, 46B, 36
$C_{32}H_{88}Ag_{31}I_{39}N_8$, Hexamethylethylenediamine diiodide - silver iodide, 42B, 29
$C_{33}H_{90}Ag_{44}I_{53}N_9$, Heptamethyl-hydrodiethylenetriamine triiodide silver iodide, 44B, 31

4. ALIPHATIC (N AND S) COMPOUNDS

$CFN_5O_3S_5$, Fluorosulphonyl isocyanate tetrasulfurtetranitride, 44B, 32
CH_3AsF_7NOS, N-Methyl-S,S-difluorosulfoximine arsenic pentafluoride, 44B, 32
CH_3KN_2OS, Potassium thiocarbazate, 46B, 50
CH_5NO_2S, Methanesulphonamide, 31B, 32
CH_6ClN_3S, Thiosemicarbazide hydrochloride, 42B, 29
$C_2H_4N_2S_2$, Dithio-oxamide, 18, 654; 30B, 36
C_2H_5NS, Thioacetamide, 24, 545
$C_2H_6ClNO_2S$, N,N'-Dimethylsulfamoyl chloride (gas-ed), 42B, 1025
$C_2H_6N_2OS$, O-Methyl-thiocarbazate, 46B, 50
$C_2H_6N_2OS$, S-Methyl-thiocarbazate, 46B, 50
$C_2H_6N_2S_2$, trans,cis-S-Methyl dithiocarbazate, 46B, 61
$C_2H_6N_2S_8$, N-(Dimethylaminomercapto)cyclo(nitrogenheptasulfide), 44B, 33
$C_2H_8Cl_2N_4O_8S_3$, *a,a*'-Trithiobisformamidinium perchlorate, 46B, 37
$C_2H_8N_2S$, Dimethyl sulphone diimine, 41B, 46
$C_2H_8N_2S$, Dimethylsulfodiimine (gas-ed), 40B, 1143
$C_2H_9NO_5S_2$, Bis(methanesulphonyl)imide monohydrate, 41B, 46
$C_3F_9NS_3$, Tris(trifluoromethylthio)amine (gas-ed), 43B, 1484
$C_3H_9NO_2S_2$, S,S-Dimethyl-N-methylsulphonylsulfimide, 46B, 38
$C_3H_9NO_2S_2$, S,S-Dimethyl-N-(methylsulphonyl)sulfilimine, 32B, 38
$C_4H_8N_2O_2$, N,N'-Dimethyloxalamide, 46B, 6
$C_4H_{11}NO_3S$, Isopropylammoniomethanesulphonate, 46B, 29
$C_4H_{12}F_2N_2S$, Bis(dimethylamino)difluorosulfurane, 45B, 23
$C_4H_{12}N_2O_2S$, Bis-dimethylamine sulphone, 28, 399
$C_4H_{14}BrNOS_2$, Bis(dimethylthionium)amino bromide monohydrate, 41B, 47
$C_5H_9NO_4S_2$, N-Methyl-2,2-dimethylsulphonylvinylideneamine, 18, 663
$C_6H_{11}NO_4S_2$, N-Ethyl-2,2'-dimethylsulphonylvinylideneamine, 26, 604
$C_7H_{16}ClO_2N_3S_2$, 1,3-Bis(thiocarbamoyl)-2-N,N-dimethylaminopropane hydrochloride, 46B, 63
$C_8H_{10}N_2S_2$, Methyl 3-phenyldithiocarbazate, 46B, 88
$C_8H_{16}N_2O_2S_4$, Dimorpholino-tetrasulfane, 46B, 64
$C_8H_{16}N_6OS_2$, 2-Keto-3-ethoxybutyraldehyde-bis(thiosemicarbazone), 31B, 454; 34B, 24
$C_9H_{12}N_2O_3S_2$, N-Methyl-N'-phenyl-S-sulfomethylisothiourea, 46B, 48
$C_9H_{21}BF_4N_2S$, N,N'-Di-t-butyl-N-methyl-sulfur-di-imidium tetrafluoroborate, 45B, 30
$C_{10}H_{11}NO_4S_2$, N-Methyl-2-methylsulphonyl-2-phenylsulphonyl-vinylideneamine, 21, 549
$C_{10}H_{20}N_2S_4$, Tetraethylthiuram disulfide, 32B, 39
$C_{10}H_{24}Cl_6N_4S_2Te$, Bis(N,N'-diethyl)formamidine disulfide hexachlorotellurate, 46B, 67
$C_{12}H_{18}N_2O_2S$, 2-Methyl-2-(N-phenylsulfinamoyl)-N,N-dimethylpropanamide, 46B, 38
$C_{12}H_{27}N_3S$, Tris(t-butylimino)sulfur(VI), 45B, 701
$C_{14}H_{14}ClN_3S_4$, 1,7-Di-4-tolyltetrasulfurtrinitrogen chloride, 44B, 34

$C_{14}H_{21}N_3O_4S_2$, 1,1,3-Trimethyl-S-methylisothiouronium-3-(N-(2,6-dimethylphenoxysulphonyl)carboxamidate), 44B, 34
$C_{16}H_{18}N_4S$, S-Isopropyldithizone, 44B, 365
$C_{17}H_{18}ClNO_3S$, L-(1-Tosylamino-2-phenyl)ethyl chloromethyl ketone, 46B, 38
$C_{19}H_{18}ClNO_4S_2$, Methylphenyl(diphenylsulfylimino)sulphonium perchlorate, 44B, 35
$C_{23}H_{23}NO_2S_3$, Z-(S-Phenyl-2,4,6-trimethyldithiobenzoate S'-tosylimide), 44B, 35
$C_{24}H_{44}N_2S$, Bis(dicyclohexylamino)sulfide, 44B, 33
$C_{26}H_{20}N_4S_3$, Bis(diphenylmethylene)trisulfurtetranitride, 40B, 49

6. ENOLATES (ALIPHATIC AND AROMATIC)

$C_{16}H_{26}O_4$, 5,6-Diacetyl-2,9-dimethyl-4,7-decanedione, 42B, 32

7. NITRILES (ALIPHATIC AND AROMATIC)

CLi_2N_2, Lithium cyanamide, 44B, 36
CN_2Sr, Strontium cyanamide, 31B, 39
CHN, Hydrogen cyanide (gas-ed), 40B, 1136
CHN, Hydrogen cyanide (gas-mw), 22, 602
CHN_2Na, Sodium cyanamide, 44B, 43
CH_2N_2, Cyanamide, 26, 586
C_2Cl_3N, Trichloroacetonitrile (gas-ed), 24, 547
C_2Cl_3N, Trichloroacetonitrile (gas-mw), 20, 468; 21, 535; 26, 588
$C_2D_4N_4$, Dicyanodiamide, perdeuterated, 31B, 41
C_2F_3N, Trifluoromethyl cyanide (gas-ed), 19, 502; 22, 554
C_2F_3N, Trifluoromethyl cyanide (gas-mw), 13, 445; 16, 417; 21, 500
C_2N_2, Cyanogen (gas-ed), 3, 731
C_2N_2, Cyanogen (gas-irr), 19, 503
C_2N_2, Cyanogen, 28, 401
C_2H_2BrN, Bromoacetonitrile (gas-mw), 42B, 1034
C_2H_2ClN, Chloroacetonitrile (gas-mw), 24, 546; 26, 588
C_2H_2FN, Fluoroacetonitrile (gas-mw), 26, 587
C_2H_3N, Methyl cyanide (gas-mw-ed), 4, 276; 7, 240; 19, 502; 22, 602; 40B, 1136
C_2H_3N, Methyl isocyanide (gas-ed), 4, 276; 9, 336
C_2H_3NO, Methylisocyanate (gas-ed), 8, 277; 38B, 1067
C_2H_3NS, Methylisothiocyanate (gas-ed), 38B, 1067
$C_2H_4N_4$, 2-Cyanoguanidine, 8, 282; 31B, 41; 46B, 41
C_3BrN, Bromocyanoacetylene (gas-mw), 40B, 1146
C_3BrN, Bromocyanoacetylene, 33B, 29
C_3ClN, Chlorocyanoacetylene (gas-ed), 42B, 1025
C_3ClN, Chlorocyanoacetylene (gas-mw), 40B, 1146
C_3ClN, Chlorocyanoacetylene, 33B, 29
C_3IN, Iodocyanoacetylene (gas-mw), 40B, 1146
C_3IN, Iodocyanoacetylene, 27, 785
C_3HBrN_2, Bromomalononitrile, 44B, 43
C_3HN, Cyanoacetylene (gas-mw), 22, 602
C_3HN, Cyanoacetylene, 22, 622; 23, 563
$C_3H_2N_2$, Malononitrile (gas-mw), 24, 551
C_3H_3N, Acrylonitrile (gas-ed), 21, 536; 24, 552; 35B, 892
C_3H_3N, Acrylonitrile (gas-mw), 23, 563
C_3H_3NO, Acetyl cyanide (gas-ed), 40B, 1138

C_3H_3NO, Acetyl cyanide (gas-mw), 23, 526
C_3H_5N, Ethyl cyanide (gas-mw), 21, 506; 40B, 1146
C_3H_5NO, Ethyl isocyanate (gas-mw), 42B, 1031
C_3H_5NO, 2-Cyanoethanol (gas-ed), 20, 471
$C_3H_5N_3O$, α-Cyanoacetohydrazide, 39B, 30
$C_3H_6N_2$, Dimethycyanamide (gas-mw), 39B, 895
C_4BrN_3, Bromotricyanomethane, 38B, 64
C_4ClN_3, Chlorotricyanomethane, 38B, 64
$C_4Cl_2N_2$, Dichlorofumaronitrile, 38B, 66
C_4FN_3, Fluorotricyanomethane, 43B, 69
C_4N_2, Dicyanoacetylene, 17, 653
C_4N_3Na, Sodium tricyanomethanide, 32A, 234; 32B, 49
$C_4H_2N_2ORb_2S_2$, Dirubidium dimercaptomaleonitrile monohydrate, 38B, 687
C_4H_3N, Cyanoallene (gas-mw), 39B, 895
C_4H_3N, Methylcyanoacetylene (gas-mw), 18, 657
$C_4H_4N_4$, Ammonium tricyanomethanide, 30A, 291; 30B, 39
$C_4H_4N_4$, 1,2-Diamino-1,2-dicyanoethene (hydrogen cyanide tetramer), 21, 259; 24, 551; 26, 600
C_4H_7N, Isopropyl cyanide (gas-mw), 40B, 1146
C_4H_7N, 2-Cyanopropane (gas-ed), 23, 564
C_4H_9N, Trimethyl acetonitrile (gas-ed), 22, 603
C_5N_4, Tetracyanomethane (gas-ed), 37B, 695
C_5N_4, Tetracyanomethane, 40B, 58
C_5HN, Cyanobutadiyne (gas-mw), 42B, 1034
$C_5H_3N_3$, 1,1,1-Tricyanoethane, 38B, 65
$C_5H_6N_2S$, 2-Dimethylsulfuranylidenemalononitrile, 34B, 27
C_5H_9N, t-Butylcyanide (gas-mw), 27, 735
C_6N_4, Tetracyanoethylene, 24, 560; 37B, 23; 39B, 31
$C_6H_2KN_3O$, Potassium 1,1,3-tricyanopropanone, 37B, 24
$C_6H_4N_2$, trans,trans-Muconodinitrile, 32B, 50
$C_6H_4N_4$, 2-Amino-1,1,3-tricyanopropene (form A), 37B, 24
$C_6H_4N_4$, 2-Amino-1,1,3-tricyanopropene (form B), 38B, 66
$C_6H_6Br_2N_2$, meso-β,β'-Dibromoadiponitrile, 44B, 44
$C_6H_6N_2S_2$, Bis(methylthio)maleonitrile, 39B, 32
$C_6H_6N_2S_2$, 3-Thiocyanato-2-thiocyanatomethyl-1-propene, 43B, 70
C_7F_5N, Pentafluorobenzonitrile (gas-mw), 33B, 541
$C_7H_2Br_2IN$, 2,6-Dibromo-4-iodobenzonitrile, 44B, 44
$C_7H_2Br_3N$, 2,4,6-Tribromobenzoisonitrile, 43B, 70
$C_7H_2Br_3N$, 2,4,6-Tribromobenzonitrile, 38B, 67
$C_7H_2Cl_3N$, 2,4,6-Trichlorobenzonitrile, 16, 510; 38B, 67
C_7H_4BrN, 4-Bromobenzoisonitrile, 44B, 45
C_7H_4BrN, 4-Bromobenzonitrile, 43B, 71
C_7H_4ClN, 4-Chlorobenzonitrile, 45B, 36
C_7H_4FN, p-Fluorobenzonitrile, 43B, 72
C_7H_4IN, o-Iodobenzonitrile, 40B, 58
C_7H_4IN, p-Iodobenzonitrile, 30B, 38
C_7H_4IN, 4-Iodobenzoisonitrile, 44B, 45
$C_7H_4N_2O_2$, p-Nitrobenzonitrile, 43B, 72
C_7H_5N, Benzonitrile, 44B, 46
C_7H_5N, Benzonitrile (gas-mw), 37B, 697
C_7H_5N, Phenylisocyanide (gas-mw), 39B, 895
C_7H_5NS, Phenyl isothiocyanate (gas-mw), 42B, 1036
$C_7H_{10}N_2$, 2,4-Dicyanopentane, 31B, 42
$C_8H_3Cl_4NS$, 2,3,5,6-Tetrachloro-4-(methylthio)benzonitrile, 38B, 157
$C_8H_4N_2$, 1,4-Benzenedicarbonitrile, 43B, 55; 44B, 46
$C_8H_5Br_2N$, 4-Cyano-3,5-dibromotoluene, 42B, 33

C_8H_6BrN, o-(Bromomethyl)benzonitrile, 45B, 37
$C_8H_{10}Br_2N_2$, meso-(R,R,S,S)-α,α'-Dimethyl-β,β'-dibromoadiponitrile (C2/c), 44B, 44
$C_8H_{10}Br_2N_2$, meso-(R,R,S,S)-α,α'-Dimethyl-β,β'-dibromoadiponitrile ($P2_1/n$), 44B, 44
$C_8H_{10}Br_2N_2$, meso-(R,S,R,S)-α,α'-Dimethyl-β,β'-dibromoadiponitrile, 44B, 44
$C_8H_{12}N_4$, Azobisisobutyronitrile, 37B, 25; 38B, 79
$C_9H_2KN_5$, Potassium 2-cyanomethyl-1,1,3,3-tetracyanopropenide, 37B, 26
$C_9H_3Cl_3N_2S$, 2,4,5-Trichloro-6-(methylthio)isophthalonitrile, 38B, 68
$C_9H_4KN_3O_2$, Potassium p-nitrophenyldicyanomethanide, 32B, 51
C_9H_8BrNO, 4-Bromo-2,6-dimethylbenzonitrile N-oxide, 45B, 38
C_9H_8ClNO, 4-Chloro-3,5-diphenylcyanate, 39B, 32
$C_{10}H_2N_4$, 1,2,4,5-Tetracyanobenzene, 39B, 33
$C_{10}H_5ClN_2$, o-Chlorobenzylidenemalononitrile, 39B, 33
$C_{10}H_5ClN_2$, 3-(4-Chlorophenyl)-2-cyanopropenonitrile, 41B, 54
$C_{10}H_6N_2$, Benzylidenemalononitrile, 38B, 69
$C_{10}H_6N_2$, α-Cyanocinnamonitrile, 34B, 28; 37B, 27
$C_{10}H_{11}N$, Mesitonitrile, 45B, 38
$C_{10}H_{11}NO$, 2,4,6-Trimethylbenzonitrile N-oxide, 45B, 38
$C_{10}H_{11}NO_2$, 4-Methoxy-2,6-dimethylbenzonitrile N-oxide, 45B, 38
$C_{10}H_{12}BF_4N$, N-(2,6-Dimethylphenyl)-acetonitrilium tetrafluoroborate, 43B, 134
$C_{10}H_{14}N_6$, α,α'-Dicyanofumaricdialdehydebis(dimethylhydrazone), 44B, 46
$C_{10}H_{20}N_2S_2$, Tetraethylammonium cyanodithioformate, 40B, 59
$C_{12}H_4KN_4$, Potassium 7,7,8,8-tetracyanoquinodimethanide (high-temp), 43B, 73
$C_{12}H_4KN_4$, Potassium 7,7,8,8-tetracyanoquinodimethanide, 43B, 73
$C_{12}H_4N_4$, 7,7,8,8-Tetracyanoquinodimethane, 30B, 40
$C_{12}H_4N_4Rb$, Rubidium 7,7,8,8-tetracyanoquinodimethane, 43B, 74
$C_{13}H_5ClN_6 \cdot 0.5\ C_6H_6$, α-[2-(2,2-Dicyanovinyl)-2-(4-chlorophenyl)-hydrazono·malononitrile benzene solvate, 45B, 39
$C_{13}H_7N_7$, Ammonium 1,1,2,6,7,7,-hexacyanoheptatrienide, 35B, 31
$C_{13}H_8BrN$, 4-Bromo-4'-cyanobiphenyl, 42B, 95
$C_{14}H_6N_4$, Terephthaldehydene bis(Δ(1-α))malononitrile, 40B, 60
$C_{14}H_{18}N_4O_2$, 3,3,4,4-Tetracyano-1-methoxy-2-methyl-butyl 2-butyl ether, 41B, 55
$C_{16}H_{10}N_2$, trans-α,β-Dicyanostilbene, 26, 730
$C_{16}H_{10}N_4O_3$, O-(Phenyl-cyano-nitromethyl)benzoylcyanoxime, 41B, 56
$C_{17}H_{10}Cl_2N_2$, 2,3-Dichloro-1,1-dicyano-3,3-diphenylpropene, 41B, 57
$C_{17}H_{13}N$, 2,2-Diphenylpent-4-ynenitrile, 43B, 74
$C_{17}H_{15}N_3O_2$, trans-4-Nitro-4'-dimethylàmino-α-cyanostilbene, 23, 667
$C_{18}H_{18}N_6$, Hexacrylonitrile, 29, 478
$C_{18}H_{25}BrN_2O_4$, (-)-β-Bromoallyl-isopropyl-cyanacetic acid D(-)-threo-1-phenyl-2-amino-1,3-propanediol, 42B, 34
$C_{18}H_{28}N_2O_4$, (+)-Isopropyl-propyl-cyanacetic D(-)-threo-1-phenyl-2-amino-1,3-propanediol, 42B, 34
$C_{19}H_{21}N$, (2S,3S)-2,3-Diphenyl-3-hexanecarbonitrile, 45B, 40
$C_{22}H_{10}Br_2N_4$, (E)-3,4-Bis(4-bromophenyl)-1,3,5-hexatriene-1,1,6,6-tetracarbonitrile, 43B, 75
$C_{22}H_{12}N_4$, (E)-3,4-Diphenyl-1,3,5-hexatriene-1,1,6,6-tetracarbonitrile, 43B, 75

8. UREA COMPOUNDS (ALIPHATIC AND AROMATIC)

CD_4N_2S, Thiourea, deuterated, 33B, 30
$CH_4F_3N_2OSb$, Urea trifluoroantimony, 45B, 40
CH_4N_2O, Urea, 1, 617, 652; 2, 805, 881; 3, 660, 731; 16, 440; 21, 524; 26, 568; 27, 748; 29, 473; 44B, 47; 45B, 40; 46B, 44
$CH_4N_2O_2$, N-Hydroxyurea, 31B, 45; 32B, 52; 42B, 36; 44B, 383
$CH_4N_2O_2S$, Thiourea dioxide, 27, 749
CH_4N_2S, Thiourea, 1, 619, 653; 2, 805, 882; 22, 604; 24, 531; 26, 572; 32B, 53; 33B, 30; 44B, 47
CH_4N_2Se, Selenourea, 34B, 29
$CH_4N_4O_2$, Nitroguanidine, 20, 497
$CH_5N_3O_3S$, Thiourea nitrate, 33B, 31
$CH_5N_3O_4$, Urea nitrate, 34B, 30
CH_5N_3S, Thiosemicarbazide, 34B, 29, 32; 35B, 28
CH_6BrN_3, Guanidinium bromide, 3, 661, 733
CH_6ClN_3, Guanidinium chloride, 30B, 44
CH_6ClN_3O, Semicarbazide hydrochloride, 30B, 46
CH_6IN_3, Guanidinium iodide, 3, 663, 734
CH_6N_4O, Carbonohydrazide, 38B, 70; 45B, 41
CH_6N_4S, Thiocarbohydrazide, 34A, 349
CH_7ClN_4, Aminoguanidine hydrochloride, 21, 519
$CH_7N_2O_5P$, Urea phosphate, 21, 524; 38B, 595
CH_8ClN_3O, Urea - ammonium chloride, 24, 529
CH_9ClN_6, Triaminoguanidinium chloride, 21, 520
$CH_{18}AlN_3O_{14}S_2$, Guanidinium aluminum sulfate hexahydrate, 27, 828; 32B, 57
$CH_{18}CrN_3O_{14}S_2$, Guanidinium chromium(III) sulfate hexahydrate, 32B, 57
$CH_{18}GaN_3O_{18}S_3$, Guanidinium gallium sulfate hexahydrate, 23, 584
$C_2D_5N_3O_2 \cdot 0.77\ D_2O$, Biuret hydrate (perdeuterated), 39B, 33
$C_2H_5N_3O_3 \cdot 0.5\ H_2O$, 3-Hydroxybiuret hemihydrate, 42B, 36
$C_2H_5N_3O_3$, 1-Hydroxybiuret, 43B, 76
$C_2H_5N_3O_3$, 3-Hydroxybiuret, 42B, 36
$C_2H_5N_3S_2$, Dithiobiuret, 38B, 71
$C_2H_6N_2O$, N-Methylurea, 3, 660, 732; 42B, 37
$C_2H_6N_4O_2$, Biurea, 42B, 38
$C_2H_6N_4S_2$, N,N'-Bis(thiocarbamoyl)hydrazine, 41B, 58
$C_2H_7ClN_2O$, O-Methylisourea hydrochloride, 35B, 34
$C_2H_7N_3O_3$, Biuret hydrate, 26, 570; 39B, 33
$C_2H_7N_3O_4$, N-Methylurea nitrate, 21, 525
$C_2H_7N_5$, Biguanide, 44B, 48
$C_2H_8Cl_2N_4Se_2$, *a,a'*-Diselenobisformamidinium dichloride, 35B, 38
$C_2H_8N_2O_3$, Methylguanidinium nitrate, 19, 515
$C_2H_9IN_4S$, S-Methylisothiocarbonohydrazide hydroiodide, 44B, 49
$C_2H_{10}Br_2N_4OS_2$, Formamidinium disulfide dibromide monohydrate, 22, 619
$C_2H_{10}I_2N_4OS_2$, Formamidinium disulfide diiodide monohydrate, 22, 617
$C_2H_{10}N_3O_4P$, Methylguanidinium dihydrogenorthophosphate, 39B, 34
$C_2H_{11}N_5O_5S$, (Biguanidinium) sulfate monohydrate, 44B, 48
$C_2H_{12}CrN_6O_4$, Guanidinium chromate, 42B, 38
$C_2H_{14}N_4O_8Te$, Orthotelluric acid urea adduct, 45B, 43
$C_2H_{14}N_6O_7S$, Hydroxyguanidinium sulfate monohydrate, 41B, 58
$C_2H_{15}N_6O_5P$, Bisguanidinium hydrogen phosphate monohydrate, 42B, 38
$C_2H_{24}N_{12}O_{19}V_2W_4$, Tetrakis(guanidinium) divanado(V)tetratungstate(IV), 42B, 38
$C_3H_6N_4O_3$, Triuret, 30B, 41; 31B, 45
$C_3H_8N_2O_3S$, N,N-Dimethylthiourea S-trioxide, 37B, 35
$C_3H_{10}N_4O_4S$, Malondiamidinium sulfate, 44B, 48

$C_3H_{12}N_6O_3$, Guanidinium carbonate, 40B, 92
$C_3H_{14}Br_2N_3OSe_3$, Tris(selenourea) dibromide hydrate, 41B, 59
$C_3H_{14}Cl_2N_3OSe_3$, Tris(selenourea) dichloride hydrate, 41B, 59
$C_3H_{18}F_6N_9NbO$, Guanidinium oxahexafluoroniobate, 40B, 93
$C_3H_{19}N_9O_7P_2$, Trisguanidinium hydrogen pyrophosphate, 42B, 337
$C_4D_6N_4O_4$, Perdeuterated parabanic acid and urea, 46B, 44
$C_4H_8N_2O_5$, Urea malonic acid, 46B, 45
$C_4H_8N_2S$, Allylthiourea, 30B, 42
$C_4H_8N_2S$, Trimethylenethiourea, 29, 516
$C_4H_9KN_6O_6$, Potassium hydrogen di-(3-hydroxybiuret), 40B, 61
$C_4H_9N_3O_2$, β-Guanidinopropionic acid, 40B, 61
$C_4H_{14}N_4O_4S_3$, S-Methyisothiourea sulfate, 27, 771
$C_4H_{16}N_8O_4SSe_4 \cdot 2\ H_2O$, Tris(selenourea) sulfate selenourea solvate dihydrate, 45B, 44
$C_4H_{17}N_6O_4P$, Bis(methylguanidinium) monohydrogen orthophosphate, 40B, 62
$C_4H_{20}N_{10}O_6S$, Di(biguanidinium) sulfate dihydrate, 44B, 48
$C_4H_{26}N_{12}O_9P_2 \cdot 1.5\ H_2O$, Guanidinium pyrophosphate monoperhydrate sesquihydrate, 44B, 50
$C_5H_6N_8O_{13}$, N,N'-Bis-(β,β,β-trinitroethyl)urea, 35B, 36
$C_5H_{16}Cl_2N_8O$, Methylglyoxal bisguanylhydrazone dihydrochloride monohydrate, 33B, 31
$C_5H_{16}N_{10}O_3$, Di(biguanidinium) carbonate, 44B, 48
$C_7H_8N_2O$, Phenylurea, 43B, 79
$C_9H_6N_2O_3$, Phthaloylurea, 17, 709
$C_9H_{10}Cl_2N_2O$, 3-(3,4-Dichlorophenyl)-1,1-dimethylurea, 46B, 46
$C_9H_{10}NOSe$, N-Acetyl-N'-phenylselenourea, 34B, 33
$C_9H_{11}ClN_2O$, 3-(p-Chlorophenyl)-1,1-dimethylurea, 46B, 47
$C_{10}H_{15}N_5S$, 1-(Pentamethylguanidinium)-2,2-dicyanoethylene-1-thiolate, 44B, 50
$C_{10}H_{16}ClN_5$, 1-(2-Phenyl-ethyl)biguanide hydrochloride, 45B, 272
$C_{11}H_{17}Cl_2N_5$, 1-(p-Chlorophenyl)-5-isopropylbiguanidine hydrochloride, 32B, 59
$C_{11}H_{44}CeN_{18}O_{19}$, Guanidinium pentacarbonatocerate tetrahydrate, 41B, 62
$C_{11}H_{44}N_{18}O_{19}Th$, Guanidinium pentacarbonatothorate tetrahydrate, 41B, 63
$C_{13}H_8Cl_4N_2O$, N,N'-Bis(3,4-dichlorophenyl)urea, 44B, 51
$C_{13}H_{10}N_{10}O_{12}$, Guanidinium dipicrylaminate, 41B, 63
$C_{13}H_{12}N_2O$, N,N'-Diphenylurea, 45B, 45
$C_{13}H_{12}N_4O$, 1,5-Diphenylcarbazone, 45B, 45
$C_{13}H_{14}N_4O$, 1,5-Diphenylcarbonohydrazide, 45B, 46
$C_{13}H_{24}N_2O$, N,N'-Dicyclohexylurea, 37B, 27
$C_{14}H_9ClF_2N_2O_2$, 1-(4-Chlorophenyl)-3-(2,6-difluorobenzoyl)urea, 44B, 51
$C_{14}H_{12}N_2OSe$, N-Phenyl-N'-benzoylselenourea, 30B, 43
$C_{15}H_{12}N_6O_9$, N,N'-Dimethyl-N,N'-di(2,4-dinitrophenyl)urea, 42B, 39
$C_{15}H_{14}N_4O_5$, N,N'-Dimethyl-N,N'-di(p-nitrophenyl)urea, 42B, 39
$C_{17}H_{17}N_5O_6$, O-Benzoyl-N,N-dimethyl-N'-(N-methyl-2,4-dinitroanilino)-isourea, 43B, 81; 46B, 49
$C_{17}H_{20}N_2O_5S_2$, N,N'-Bis(tosylmethyl)urea, 44B, 52
$C_{24}H_{30}BrN_3O_5S$, N-(4-(β-2-Methoxy-5-bromobenzamidoethyl)benzenesulphonyl)-N'-4-methylcyclohexylurea, 38B, 72
$C_{32}H_{34}N_2O_6S_2$, N,N'-(Bis-(α-tosylbenzyl))urea acetone solvate, 42B, 39

9. NITROGEN-NITROGEN COMPOUNDS (ALIPHATIC AND AROMATIC)

CH_2N_2, Diazomethane (gas-ed), 3, 715
$CH_3ClN_2O_2$, N-Chloro-N-methylnitramine (gas-ed), 43B, 1484
CH_3N_2OK, Potassium syn-methyldiazotate, 30B, 48
CH_3N_3, Methyl azide (gas-ed), 3, 712; 22, 603; 24, 541; 38B, 1067
$CH_3N_3O_4$, Methyldinitramine (gas-ed), 43B, 1484
CH_4ClN_5, Azidoformamidinium chloride, 38B, 72
$CH_4N_2O_2$, N-Methylnitramine (gas-ed), 43B, 1484
$CH_8N_4O_4S_2$, Thiocarbonohydrazide sulfate, 38B, 73
$CH_8N_4O_5S$, Carbonohydrazide sulfate, 40B, 63
$CH_8N_4S_2$, Hydrazinium hydrazinedithiocarboxylate, 34B, 34
$C_2F_6N_2$, Hexafluoroazomethane (gas-ed), 35B, 891
$C_2H_3F_3N_2$, 1,1,1-Trifluoroazomethane (gas-ed), 35B, 891
$C_2H_4Cl_2N_6$, Azo bis-N-chloroformamide, 22, 613
$C_2H_4N_2$, Formaldazine (gas-ed), 43B, 1484
$C_2H_4N_2O_2$, Diformylhydrazine, 22, 611; 40B, 63; 44B, 53
$C_2H_4N_4Na_2O_4$, Disodium ethane-1,2-bisnitramine, 17, 665
$C_2H_4N_4O_2$, Azodicarbonamide, 26, 589
$C_2H_4N_4O_2$, 1,1'-Azobiscarbamide, 40B, 64
$C_2H_6N_2$, Azomethane (gas-ed), 3, 715; 35B, 890, 891
$C_2H_6N_2O$, Dimethylnitrosamine (gas-ed), 33B, 529; 40B, 1139
$C_2H_6N_2O$, N-Nitrosodimethylamine, 41B, 64
$C_2H_6N_2O_2$, Dimethylnitramime (gas-ed), 40B, 1140
$C_2H_6N_2O_2$, N,N-Dimethylnitramine, 11, 597; 45B, 22; 46B, 28
$C_2H_6N_2O_2$, N,N-Dimethylnitramine (low temperature form), 46B, 28
$C_2H_6N_4O_4$, 1,2-Bisnitroaminoethane, 11, 598
$C_2H_7ClN_2O$, Acetylhydrazonium chloride, 44B, 53
$C_2H_8N_2$, N,N-Dimethylhydrazine (gas-ed), 11, 597
$C_2H_8N_2$, N,N'-Dimethylhydrazine (gas-ed), 11, 597
$C_2H_{10}ClN_3$, 2,2-Dimethyltriazinium chloride, 42B, 40
$C_2H_{11}N_2O_2$, Hydrazine bismethanol, 33B, 37
$C_2H_{13}ClN_8S_2$, Thiocarbonohydrazide hemihydrochloride, 38B, 74
$C_3H_8N_4O_2$, Malonic dihydrazide, 41B, 65
$C_3H_9N_3O_2$, Trimethylammonionitramidate, 38B, 75
$C_3H_{10}N_4O_3$, Malonic dihydrazide monohydrate, 39B, 35
$C_3H_{11}ClN_2$, 1,1,1-Trimethylhydrazinium chloride, 42B, 40
$C_4F_{12}N_2$, Tetrakis(trifluoromethyl)hydrazine (gas-ed), 30B, 402
$C_4H_2N_4O_2$, 1,4-Bis(diazo)-2,3-butanedione, 38B, 76
$C_4H_4F_2N_6O_{10}$, Bis(2-fluoro-2,2-dinitroethyl)nitramine, 46B, 54
$C_4H_4N_2O_4$, Tetraformylhydrazine, 40B, 64
$C_4H_4N_8O_{12}$, Bis(2,2,2-trinitroethyl)nitramine (monoclinic), 46B, 55
$C_4H_4N_8O_{12}$, Bis(2,2,2-trinitroethyl)nitramine (orthorhombic), 46B, 55
$C_4H_8N_2O_2$, Diacetylhydrazine, 24, 554
$C_4H_8N_2O_2$, 1,2-Dimethyl-1,2-diformylhydrazine, 44B, 54
$C_4H_8N_4O_6$, 2-Nitronitrosoethane dimer, 34B, 34
$C_4H_9N_3O$, Acetone semicarbazone, 40B, 65
$C_4H_9N_3S$, Acetone thiosemicarbazone, 40B, 67
$C_5H_{13}Br_3Cu_2N_4$, 1,1,5,5-Tetramethylformazanium copper(I) bromide, 41B, 66
$C_6H_4Cl_4N_4Zn$, p-Benzenebis(diazonium) tetrachlorozincate(II), 33B, 52
$C_6H_4N_2Na_2O_4S \cdot 3\ H_2O$, syn-Disodium 4-sulphonatobenzenediazotate trihydrate, 46B, 50
$C_6H_4N_2O_3S$, p-Benzenediazonium sulphonate, 35B, 78; 38B, 77
$C_6H_4N_4O_2$, p-Nitrophenylazide, 30B, 50
$C_6H_5Br_3Cu_2N_2$, Benzenediazonium copper(I) bromide complex, 30B, 357
$C_6H_5Br_3N_2$, Benzenediazonium tribromide, 27, 889

$C_6H_5ClN_2$, Benzenediazonium chloride, 28, 402
$C_6H_5ClN_2O_3$, N-Chloroacetylamino isomaleimide, 40B, 66
$C_6H_6N_2OS$, N-Phenyl-N'-sulfinylhydrazine, 43B, 81
$C_6H_6N_2O_2$, Phenyl-hydroxydiazenium-oxide, 45B, 46
$C_6H_6N_2O_5S$, 2-Diazonium-4-phenolsulphonate monohydrate, 27, 981; 34B, 36
$C_6H_7ON_3$, Isonicotinic acid hydrazide, 17, 742
$C_6H_8N_2$, Phenylhydrazine, 33B, 54
$C_6H_9ClN_2$, Phenylhydrazine hydrochloride, 30B, 47
$C_6H_{12}N_2O_4$, Diethyl hydrazine-1,2-dicarboxylate, 43B, 8
$C_6H_{12}N_2S_4$, 3,6-Bis(methylthio)-4,5-diaza-2,7-dithiaocta-3,5-diene, 41B, 76
$C_6H_{20}Cl_2N_8O_2$, Dimethylglyoxal bisguanylhydrazone dihydrochloride dihydrate, 38B, 78
$C_7H_3Br_2N_3$, 2,4-Dibromophenyldiazocyanide, 32B, 60
$C_7H_4BrN_3$, o-Bromobenzene-anti-diazocyanide, 37B, 28
$C_7H_6BrN_3O$, o-Bromophenylazocarbamide, 37B, 28
$C_7H_7Cl_4FeN_2O$, o-Methoxybenzenediazonium tetrachloroferrate(III), 30B, 51
$C_7H_9N_3S$, 1-Phenyl-thiosemicarbazide, 39B, 36
$C_7H_9N_3S$, 4-Phenyl-thiosemicarbazide, 38B, 79
$C_7H_{12}N_2O_4$, Dimethyl 2-(1-methylhydrazino)maleate, 42B, 41
$C_7H_{14}N_2O_2$, cis-4-Methylcyclohexyl-hydroxydiazenium-oxide, 45B, 46
$C_8H_5ClN_4$, p-Chlorobenzene-anti-diazoimidoglyoxynitrile, 33B, 57
$C_8H_9N_3O$, Phenylazoacetaldoxime, 46B, 51
$C_8H_9N_3O$, Benzaldehyde semicarbazone, 40B, 65
$C_8H_{14}N_6O_{10}$, 1,7-Diacetoxy-2,4,6-trinitro-2,4,6-triazaheptane, 39B, 36
$C_9H_{11}N_3O_2$, Acetone 3-nitrophenylhydrazone, 40B, 1076
$C_9H_{11}N_3O_2$, Acetone 4-nitrophenylhydrazone, 38B, 81
$C_9H_{12}N_4O$, 2-(3,3-Dimethyl-1-triazeno)phenyl-1-carboxamide, 43B, 82
$C_9H_{20}N_2O$, n-Nonanoic acid hydrazide, 26, 562
$C_{10}H_8N_4O_2$, trans-4,4'-Azopyridine N-oxide, 23, 701
$C_{10}H_{10}Cl_3N_3$, 1,4-Dichloro-1'-p-chlorophenyl-4'-dimethylamino-2,3-diazabutadiene, 43B, 42
$C_{10}H_{14}N_2O$, Trimethylammoniobenzamidate, 38B, 76
$C_{10}H_{15}BrN_2O$, 1,1-Dimethyl-1-phenacylhydrazinium bromide, 43B, 83
$C_{10}H_{18}N_2O_3S$, N-(Trimethylammonio)toluene-p-sulphonamidate monohydrate, 42B, 42
$C_{11}H_9ClN_2S$, N-(1-Pyridinio)chlorobenzene-p-sulphonamidate, 42B, 42
$C_{11}H_{17}N_3O_2$, Tetramethylammonium benzonitrosolate, 45B, 26
$C_{12}F_{10}N_2O_2$, cis-Azo(pentafluorobenzene) dioxide, 40B, 74
$C_{12}H_4N_8O_{12}$, 2,2',4,4',6,6'-Hexanitroazobenzene (form 1), 40B, 68
$C_{12}H_4N_8O_{12}$, 2,2',4,4',6,6'-Hexanitroazobenzene (form 2), 40B, 68
$C_{12}H_8Br_2N_2$, trans-p,p'-Dibromoazobenzene, 31B, 46
$C_{12}H_8Cl_2N_2$, trans-p,p'-Dichloroazobenzene, 34B, 38
$C_{12}H_8Cl_2N_2$, trans-2,2'-Dichloroazobenzene, 39B, 37
$C_{12}H_9BrN_2O$, 4-Bromo-4'-hydroxyazobenzene, 40B, 69
$C_{12}H_9Br_2N_3$, p-Dibromodiazoaminobenzene, 26, 731
$C_{12}H_9Br_2N_3$, 2,4-Dibromodiazoaminobenzene, 30B, 49
$C_{12}H_{10}BrN_3$, p-Bromodiazoaniline (α form), 32B, 61; 38B, 82
$C_{12}H_{10}BrN_3$, p-Bromodiazoaniline (β form), 32B, 61; 38B, 82
$C_{12}H_{10}N_2$, cis-Azobenzene, 7, 267; 8, 314; 37B, 29
$C_{12}H_{10}N_2$, trans-Azobenzene, 7, 226, 268; 31B, 47
$C_{12}H_{10}N_2O_2$, cis-Azobenzene dioxide, 40B, 69
$C_{12}H_{10}N_4O_2$, p-Nitrodiazoaminobenzene, 40B, 71
$C_{12}H_{11}N_3$, Diazoaminobenzene, 37B, 30

$C_{12}H_{11}N_3$, β-Diazoaminobenzene, 38B, 83
$C_{12}H_{15}BrN_4O_4$, N-(2,4-Dinitrophenyl)-N-methylpivalohydrazonyl bromide, 43B, 83
$C_{12}H_{20}N_4$, Azobis-3-cyano-3-pentane, 38B, 80
$C_{13}H_9N_3S$, 2-Thiocyanatoazobenzene, 33B, 63; 43B, 107
$C_{13}H_{11}N_3O_2$, N-Picolinylidene-N'-salicyloylhydrazine, 40B, 71
$C_{13}H_{12}N_2O$, cis-(4'-Methyl-N,N,O)-azoxybenzene, 43B, 84
$C_{13}H_{12}N_4$, Diphenylformazane, 39B, 37; 40B, 72
$C_{13}H_{12}N_4S$, Dithizone, 43B, 85
$C_{14}H_{10}Br_2N_2$, p,p-Dibromobenzalazine, 38B, 83
$C_{14}H_{10}N_2O_6$, trans-2,2'-Dicarboxyazobenzene dioxide, 40B, 69
$C_{14}H_{12}N_2$, Benzalazine, 26, 731; 35B, 63; 42B, 45; 44B, 54
$C_{14}H_{12}N_2O_2$, Salicylaldehyde azine, 34B, 35
$C_{14}H_{14}N_2$, p,p'-Azotoluene, 22, 701; 31B, 48
$C_{14}H_{14}N_2O_3$, p-Azoxyanisole, 35B, 69; 37B, 33
$C_{14}H_{15}N_3$, o-Aminoazotoluene, 42B, 45
$C_{14}H_{15}N_3$, p-Dimethyldiazoaminobenzene, 29, 662
$C_{14}H_{15}N_3O$, β-p-Dimethylaminoazoxybenzene, 40B, 72
$C_{15}H_{11}N_3O$, 1-Azido-1-benzoyl-2-phenylethylene, 44B, 55
$C_{15}H_{12}Br_2N_4O_3$, *a*,4-Dibromo-*a*-(4-methyl-2-nitrophenylazo)-acetanilide, 32B, 62
$C_{15}H_{12}N_4$, 1-p-Tolyl-3-(*a*-cyano)benzylidenetriazene, 38B, 84
$C_{15}H_{14}N_4O_2S$, 3-Carboxymethylthio-1,5-diphenylformazan, 45B, 13
$C_{15}H_{15}N_3O_2$, 4-Dimethylaminoazobenzene-2'-carboxylic acid, 46B, 51
$C_{15}H_{16}N_4O_2 \cdot 0.25\ C_6H_6$, p-Dimethylamino-benzaldehyde-p-nitrophenylhydrazone benzene solvate, 46B, 87
$C_{16}H_{10}ClN_3O_3$, 4-Chloro-2-nitrophenylazo-2-naphthol, 43B, 86
$C_{16}H_{11}N_3O_3$, 1-p-Nitrobenzeneazo-2-naphthol, 34B, 39
$C_{16}H_{16}N_2O_2$, Anisaldehyde-azine, 33B, 75
$C_{16}H_{18}N_2O_2$, 4,4'-Azodiphenetole, 35B, 70
$C_{16}H_{18}N_2O_4$, Azoxyphenetole, 44B, 101
$C_{16}H_{18}N_4S$, 3-Methylthio-1,5-di(o-tolyl)formazan, 46B, 52
$C_{16}H_{20}Cl_4N_6Zn$, Bis(p-N,N-dimethylaminophenyldiazonium) tetrachlorozincate, 31B, 46; 37B, 62
$C_{16}H_{22}N_3NaO_5S$, Sodium 4'-dimethylaminoazobenzene-4-sulphonate monohydrate monoethanolate, 39B, 38
$C_{18}H_{12}Cl_2N_2O_4S$, 5,5'-Dichloro-2-hydroxy-2'-(phenylsulphonyl)azoxybenzene, 40B, 72; 42B, 47
$C_{18}H_{12}N_5O_6$, 2,2-Diphenyl-1-picrylhydrazyl, 42B, 64
$C_{18}H_{14}Br_2N_2$, 4,4'-Dibromocinnamaldazine, 32B, 63
$C_{18}H_{14}N_4$, 4-Phenylazoazobenzene, 38B, 84
$C_{18}H_{18}N_2O_5$, Ethyl p-azoxybenzoate, 37B, 30
$C_{18}H_{18}N_4$, 1,4-Bis(phenylazo)-1,4-dimethyl-1,3-butadiene, 45B, 47
$C_{18}H_{20}Cl_2N_2$, meso-1,1'-Dichloro-1,1'-diphenyl-1,1'-azopropane, 35B, 32
$C_{20}H_{17}N_3O$, N-p-Methoxybenzylidene-p-phenylazoaniline, 43B, 86
$C_{20}H_{18}Br_2N_4O_6$, *a*,*a*'-Azobis(4-bromobenzaldehyde)bis(O-carboxyoxime) diethyl ester (triclinic), 41B, 67
$C_{20}H_{18}Br_2N_4O_6$, *a*,*a*'-Azobis(4-bromobenzaldehyde)bis(O-carboxyoxime) diethyl ester (monoclinic), 41B, 67
$C_{20}H_{18}N_2O_5$, Methyl m-azoxy-trans-cinnamate dimer, 44B, 55
$C_{20}H_{20}N_2$, β,β'-Dimethylcinnamaldazine, 39B, 39
$C_{20}H_{26}Cl_2N_4$, 1,4-Di-t-butyl-1,4-bis(4-chlorophenyl)-2-tetrazene, 43B, 87
$C_{20}H_{26}N_2$, trans-2,2'-Azo-p-cymene, 40B, 73
$C_{20}H_{28}N_4$, 1,4-Di-t-butyl-1,4-diphenyl-2-tetrazene, 43B, 87
$C_{21}H_{18}N_2$, cis-1,3-Diphenyl-2-(phenylazo)propene, 37B, 31

$C_{21}H_{18}N_2$, trans-1,3-Diphenyl-2-(phenylazo)propene, 37B, 31
$C_{22}H_{20}Cl_2N_4$, N,N'-Bis[2-(4-chlorophenylamino)-2-cyclopenten-1-ylidene]hydrazine (monoclinic), 46B, 52
$C_{22}H_{20}Cl_2N_4$, N,N'-Bis[2-(4-chlorophenylamino)-2-cyclopenten-1-ylidene]hydrazine (orthorhombic), 46B, 52
$C_{23}H_{15}Cl_2N_3O_2$, 1'-(2,5-Dichlorophenyl)azo-2'-hydroxy-3'-phenylamidonaphthalene, 43B, 88
$C_{24}H_{16}Cl_3N_3O_3$, 1-(2,5-Dichlorophenylazo)-2-hydroxy-3-naphthoic acid 4-chloro-2-methoxyanilide, 40B, 73
$C_{26}H_{20}N_4$, Bisphenylazostilbene, 39B, 39
$C_{28}H_{19}BrN_2O_3$, Diphenyl triketone sym-benzoyl-p-bromophenylhydrazone, 38B, 87
$C_{28}H_{19}BrN_2O_3$, α-p-Bromophenylazo-β-benzoyloxybenzalacetophenone, 38B, 88
$C_{28}H_{19}BrN_2O_3$, p-Bromobenzeneazotribenzoylmethane, 31B, 50
$C_{28}H_{20}N_2O_3$, Diphenyl triketone sym-benzoylphenylhydrazone, 38B, 87
$C_{28}H_{20}N_2O_3$, Phenylazotribenzoylmethane, 38B, 89
$C_{36}H_{58}N_2$, 2,2',4,4',6,6'-Hexa-t-butylazobenzene, 46B, 53

10. NITROGEN-OXYGEN COMPOUNDS (ALIPHATIC AND AROMATIC)

$CBrN_3O_6$, Bromotrinitromethane (gas-ed), 40B, 1139; 42B, 1030
CBr_3NO_2, Tribromonitromethane (gas), 27, 746
$CClN_3O_6$, Chlorotrinitromethane (gas-ed), 39B, 888; 40B, 1139; 42B, 1030
CCl_3NO_2, Chloropicrin (gas-ed), 21, 497
CF_3NO, Trifluoronitrosomethane (gas-ed), 30B, 400
CF_3NO_2, Trifluoronitromethane (gas), 27, 746
CIN_3O_6, Iodotrinitromethane, 31B, 51
CN_3O_6Rb, Rubidium trinitromethane, 31B, 52
CN_4O_8, Tetranitromethane (gas-ed), 7, 240; 40B, 1139; 42B, 1028
CN_4O_8, Tetranitromethane, 9, 314
CHN_3O_6, Trinitromethane (gas-ed), 39B, 888; 40B, 1139; 42B, 1030
CH_2ClNO_2, Chloronitromethane (gas-ed), 38B, 1068
$CH_2K_2N_4O_4$, Dipotassium methylene bis-nitrosohydroxylamine, 23, 580
CH_3NO, Formaldoxime (gas-mw), 28, 664
CH_3NO, Nitrosomethane (gas-mw), 33B, 538
CH_3NO_2, Methyl nitrite (gas-ed), 9, 313
CH_3NO_2, Nitromethane (gas-ed), 3, 705; 9, 313
CH_3NO_3, Methyl nitrate (gas-ed), 9, 313
CH_4N_2O, Formamidoxime, 20, 495; 30B, 52
$CH_4N_2O_2$, O-Carbamoylhydroxylamine, 33B, 35
CH_5NO, α-Methylhydroxylamine (gas-ed), 3, 705
CH_6ClNO, O-Methylhydroxylamine hydrochloride, 29, 474
C_2F_6NO, Bis(trifluoromethyl)nitroxyl (gas-ed), 37B, 691
$C_2K_2N_4O_8$, Dipotassium tetranitroethide, 33B, 33
C_2HF_6NO, Bis(trifluoromethyl)hydroxylamine (gas-ed), 38B, 1065
$C_2H_4N_2O_2$, Glyoxime, 31B, 53
$C_2H_4N_4O_2$, Azodiformaldehyde dioxime, 46B, 53
C_2H_5NO, anti-Acetaldehyde oxime-d (gas-mw), 43B, 1485
$C_2H_5NO_3$, Ethyl nitrate (gas-mw), 40B, 1145
$C_2H_6N_2O_2$, cis-Bis(nitrosomethane), 28, 404
$C_2H_6N_2O_2$, trans-Nitrosomethane dimer, 23, 579
$C_2H_6N_4O_4$, N,N'-Dinitroethylenediamine, 33B, 33
C_3H_7NO, Acetoxime, 15, 400
$C_3H_7NO_2$, Methyl Z-acetohydroximate, 43B, 88

C_3H_9NO, Trimethylamine oxide (gas-ed), 7, 244
C_3H_9NO, Trimethylamine oxide, 29, 473
$C_3H_{10}ClNO$, Trimethylamine oxide hydrochloride, 24, 553; 27, 762
$C_4H_4F_2N_4O_{10}$, 1,4-Difluoro-1,1,4,4-tetranitro-2,3-butanediol, 45B, 47
$C_4H_6ClN_3O_5$, 2-Chloro-5,5-dinitro-3-aza-4-oxa-2-hexene, 41B, 68
$C_4H_6N_2Na_2O_2 \cdot 8\ H_2O$, Sodium dimethylglyoximate octahydrate, 46B, 55
$C_4H_8N_2O_2$, Dimethylglyoxime, 16, 461; 26, 596; 45B, 48; 46B, 55
$C_4H_8N_4O_8$, 2,2'-Dinitroxydiethylnitramine, 44B, 56
$C_4H_9NO_4$, 2-Methyl-2-nitro-1,3-propanediol, 43B, 89
$C_4H_{10}N_2O$, N,N-Dimethylacetamidoxime, 39B, 40
$C_4H_{11}N_5O_2$, 2,2'-Iminobis(acetamide oxime), 46B, 30
$C_5H_{12}Cl_3NO$, Methyl-bis(β-chloroethyl)amine-N-oxide hydrochloride, 41B, 34
C_6F_5NO, Pentafluoronitrosobenzene, 40B, 74
C_6H_4INO, p-Iodonitrosobenzene, 20, 565
$C_6H_{12}NNaO_6$, Sodium ethyl 3-oxobutanoate-2-oximate dihydrate, 44B, 56
C_7H_6ClNO, Z-p-Chlorobenzaldoxime, 46B, 57
$C_7H_6N_2O_3$, α-p-Nitrobenzaldoxime, 38B, 133; 39B, 41
$C_7H_7NO_2$, Salicylaldoxime, 39B, 41
$C_7H_7NO_3$, 5-Methoxy-2-nitrosophenol (red form), 23, 658
$C_8H_7ClN_2O_3$, O-Methyl-p-nitrobenzohydroximoyl chloride, 43B, 89
$C_8H_7NO_2$, 2-Hydroximino-1-phenylethan-1-one, 44B, 56
C_8H_8BrNO, anti-α-Bromoacetophenone oxime, 39B, 42
$C_8H_{18}N_2O_2$, Nitrosoisobutane trans-dimer, 26, 608
$C_9H_7F_2NO_2$, anti-3',4'-Difluoro-2-hydroxyiminopropiophenone, 37B, 32
$C_9H_9NO_2$, 1-Phenyl-1,2-propanedione-2-oxime, 43B, 90
$C_9H_{11}NO_2$, anti-Ethyl benzohydroximate, 37B, 32
$C_9H_{11}NO_2$, syn-Ethyl benzohydroximate, 37B, 32
$C_{10}H_{12}ClNO$, anti-2,6-Dimethyl-4-chloro-N-methylbenzaldoxime, 34B, 40
$C_{10}H_{14}N_2O$, (E)-N,N-Dimethyl-p-tolylamidoxime, 43B, 121
$C_{10}H_{19}NO$, 3,3,5,5-Tetramethylcyclohexanone oxime, 46B, 57
$C_{11}H_{15}NO$, 2,4,6-Trimethylacetophenone oxime, 46B, 58
$C_{11}H_{17}N_3O_2$, Tetramethylammonium benzonitrosolate, 39B, 897
$C_{12}H_4Br_2N_2O_2$, 2,4,6-Tribromonitrosobenzene dimer, 13, 547
$C_{12}H_8Br_2N_2O_2$, p-Bromonitrosobenzene dimer, 13, 546
$C_{13}H_{11}NO_3$, p-Methoxyindophenol N-oxide, 31B, 54
$C_{14}H_9Cl_2NO_3$, N,O-Bis(2-chlorobenzoyl)hydroxylamine, 40B, 75
$C_{14}H_{11}NO_3$, N,O-Dibenzoylhydroxylamine, 40B, 74
$C_{14}H_{14}N_2O$, N-Benzyl-syn-benzamidoxime, 43B, 91, 121
$C_{14}H_{16}N_2O_3 \cdot 2\ H_2O$, Oxybis(2-methylene-6-methylpyridine N-oxide) dihydrate, 46B, 58
$C_{15}H_{13}NO_3$, N-Benzoyl-O-o-toluoylhydroxylamine, 40B, 76
$C_{15}H_{13}NO_3$, N-Benzoyl-O-p-toluoylhydroxylamine, 40B, 76
$C_{15}H_{24}Cl_2N_2O_2$, 4-Chloro-O-(3-t-butylamino-2-hydroxypropyl)acetophenone oxime, 46B, 59
$C_{16}H_{22}N_2O_2$, N-[1-(N,N-Diethylcarbamoyl)ethylidene]-1-phenyl-1-propen-2-amine-N-oxide, 46B, 59
$C_{19}H_{15}NO$, α,α,N-Triphenylnitrone, 39B, 43
$C_{21}H_{14}BrNO_3$, Benzil α-monoxime p-bromobenzoate, 32B, 114
$C_{24}H_{24}N_2O_2$, (1,1-Bis(1-phenylethylidenaminooxy)ethyl)benzene, 44B, 57

11. SULFUR AND SELENIUM COMPOUNDS

$CClFS$, Chlorofluoromethyl sulfide (gas-mw), 42B, 1036
CCl_2S, Thiophosgene (gas-ed), 3, 705

CCl_3F_3Se, Trifluoromethylselenium trichloride, 43B, 93
$CF_3N_2O_3S_4 \cdot 0.5\ C_2H_3N$, Trisulfurdinitrido trifluoromethyl sulphonate acetonitrile solvate, 46B, 60
CF_8S, Trifluoromethyl sulfur pentafluoride (gas-ed), 22, 554
$CHF_3O_3S \cdot 0.5\ H_2O$, Trifluoromethanesulphonic acid hemihydrate, 41B, 72
CHF_3S, Trifluoromethylthiol (gas-ed), 40B, 1138
$CHK_3O_9S_3 \cdot H_2O$, Tripotassium methanetrisulphonate hydrate, 46B, 60
$CH_2Ag_2O_6S_2$, Silver(I) methanesulphonate, 45B, 49
CH_2F_4S, Methylenesulfurtetrafluoride, 46B, 61
$CH_2K_2O_6S_2$, Potassium methylenedisulphonate, 27, 752
CH_2OS, Monothioformic acid (gas-mw), 42B, 1036
CH_2OS, Sulfine (gas-mw), 42B, 1034
CH_2S_3, Trithiocarbonic acid, 30B, 53
CH_3ClO_2S, Methane sulphonyl chloride (gas-ed), 39B, 889
CH_3ClO_3S, Methyl chlorosulfate (gas-ed), 43B, 1484
CH_3CsO_3S, Caesium methylsulphonate, 32B, 64
CH_3FO_2S, Methane sulphonyl fluoride (gas-ed), 39B, 890
$CH_3F_3O_4S$, Oxonium trifluoromethanesulphonate, 39B, 43; 44B, 57
$CH_3F_3O_4S$, Trifluoromethanesulphonic acid monohydrate, 39B, 43
CH_3KO_4S, Potassium hydroxymethanesulphonate, 45B, 49
CH_3KS, Potassium methanethiolate, 38B, 51
CH_3LiS, Lithium methanethiolate, 38B, 51
CH_3NaS, Sodium methanethiolate, 38B, 51
$CH_4F_3NO_3S$, Ammonium trifluoromethylsulphonate, 41B, 71
CH_4O_2S, Methanesulfinic acid, 34B, 42
CH_4S, Methanethiol (gas-mw), 19, 547; 24, 513
CH_4Se, Methane selenol (gas-mw), 39B, 894
$CH_5F_3O_5S$, Trifluoromethanesulphonic acid dihydrate, 41B, 70
$CH_5NaO_3S_2$, Sodium methanethiosulphonate monohydrate, 29, 484
CH_7NaO_5S, Sodium hydroxymethanesulfinate dihydrate, 19, 534; 27, 754
$CH_8CaO_9S_2$, Calcium methanedisulphonate trihydrate, 45B, 49
$CH_8CdO_9S_2$, Cadmium methanedisulphonate trihydrate, 45B, 49
$CH_9F_3O_7S$, Trifluoromethanesulphonic acid tetrahydrate, 44B, 58
$CH_{11}F_3O_8S$, Trifluoromethanesulphonic acid pentahydrate, 44B, 58
$CH_{12}Cl_2N_4O_2S$, Thiocarbohydrazide dihydrochloride dihydrate, 37B, 33
$C_2Cl_6S_3$, Perchlorodimethyl trisulfide, 26, 592
C_2F_3NSe, Trifluoromethylselenium cyanide (gas-ed), 37B, 693
$C_2F_3N_3OS_3$, Trifluoroacetyl-trisulfurtrinitride, 43B, 62
$C_2F_4Se_2$, Tetrafluoro-1,3-diselenetane (gas-ed), 42B, 1029
C_2F_6S, Perfluorodimethyl sulfide (gas-ed), 18, 678
$C_2F_6S_2$, Perfluorodimethyl disulfide (gas-ed), 18, 678
$C_2F_6S_3$, Perfluorodimethyl trisulfide (gas-ed), 18, 678
C_2F_6Se, Bis(trifluoromethyl)selenide (gas-ed), 18, 678; 37B, 693
$C_2F_6Se_2$, Bis(trifluoromethyl)diselenide (gas-ed), 18, 678; 37B, 694
$C_2K_2O_2Se_2$, Potassium 1,2-diselenooxalate, 44B, 59
$C_2H_2K_2O_4S_2$, Potassium sulfenothiooxoacetate monohydrate, 44B, 59
$C_2H_3KO_2S$, Potassium O-methyl monothiocarbonate, 41B, 72
$C_2H_3KS_2$, Potassium dithioacetate, 41B, 73
$C_2H_4K_2OS_4$, Potassium perthiocarbonate methanol solvate, 41B, 73
C_2H_4OS, Thioacetic acid (gas-ed), 10, 208
$C_2H_5KO_4S$, Potassium ethyl sulfate, 17, 616; 22, 615
C_2H_5NOS, Thioacetamide-S-oxide, 37B, 34
$C_2H_6CaO_6S_2$, Calcium methanesulphonate, 43B, 91
$C_2H_6N_2S_2$, Methyl dithiocarbazate, 43B, 92
$C_2H_6Na_2O_4S_3$, Sodium thiosulfate dimethylsulfoxide solvate, 44B, 60
C_2H_6OS, Dimethylsulfoxide (gas-ed), 11, 534

C_2H_6OS, Dimethylsulfoxide, 31B, 55, 56
C_2H_6OS, 2-Mercaptoethanol (gas-mw), 43B, 1485
$C_2H_6O_2S$, Dimethylsulphone (gas-ed), 7, 232; 35B, 890; 40B, 1137
$C_2H_6O_2S$, Dimethylsulphone (gas-mw), 38B, 1069
$C_2H_6O_2S$, Dimethylsulphone, 28, 406
$C_2H_6O_4S_4$, Dimethanesulphonyl disulfide, 17, 667
C_2H_6S, Dimethylsulfide (gas-ed), 4, 276; 43B, 1484
C_2H_6S, Dimethylsulfide (gas-mw), 24, 545; 26, 586
C_2H_6S, Ethyl mercaptan (gas-ed), 12, 313
C_2H_6S, Ethyl mercaptan (gas-mw), 33B, 539; 40B, 1144; 42B, 1031
$C_2H_6S_2$, Dimethyldisulfide (gas-ed), 6, 233; 37B, 695
$C_2H_6S_3$, Dimethyltrisulfide (gas-ed), 11, 600
C_2H_6Se, Dimethylselenide (gas-ed), 19, 548
$C_2H_6Se_2$, Dimethyldiselenide (gas-ed), 37B, 691
$C_2H_7NO_3S_2$, 2-Aminoethanethiosulfuric acid, 38B, 90
$C_2H_8Cl_2N_4S_2$, *a,a'*-Dithiobisformamidinium dichloride, 38B, 91
$C_2H_{10}O_8S_2$, Dioxonium ethan-1,2-disulphonate, 35B, 40
$C_3H_2N_2S_2$, Methylene dithiocyanate, 37B, 34
$C_3H_3KO_2S_2$, Potassium S-methyl 1,2-dithiooxalate, 43B, 22
$C_3H_5KOS_2$, Potassium xanthate, 28, 407
C_3H_5NS, Ethyl thiocyanate (gas-mw), 38B, 1071
$C_3H_6CsNS_2$, Caesium dimethyldithiocarbamate, 42B, 762
C_3H_6S, Methyl vinyl sulfide (gas-ed), 37B, 686; 39B, 890
$C_3H_6S_3$, Dimethyltrithiocarbonate (gas-ed), 40B, 1136
$C_3H_8N_2S_2$, Methyl 2-methyldithiocarbazate, 43B, 92
C_3H_8S, 1-Propanethiol (gas-mw), 43B, 1485
$C_3H_8S_2$, Methyl ethyl disulfide (gas-ed), 42B, 1030
$C_3H_9BF_4OS$, Trimethyloxosulphonium fluoroborate, 28, 409
$C_3H_9ClO_5S$, Trimethyloxosulphonium perchlorate, 28, 410
C_3H_9IS, Trimethylsulphonium iodide, 23, 609
C_3H_9ISe, Trimethylselenonium iodide, 31B, 58
$C_3H_9NO_3S$, Homotaurine, 38B, 92
$C_3H_9NaO_3S_2$, Sodium xanthate dihydrate, 34B, 41
$C_3H_{12}Br_2N_3S$, S,2-Aminoethylisothiouronium bromide hydrobromide, 43B, 78
$C_3H_{14}Na_2O_{11}S_2$, Acraldehyde - sodium bisulfite addition compound tetrahydrate, 20, 521
$C_4H_4N_2S_2$, Ethylene dithiocyanate, 22, 626
$C_4H_6Cl_3NOS$, S,S-Dimethyl-N-trichloroacetylsulfilimine, 39B, 44
$C_4H_6Cl_4O_4S_4$, Bis(methylsulphonyldichloromethyl) disulfide, 46B, 62
$C_4H_6OS_3$, Trithiooxalic acid S,S'-dimethyl ester, 44B, 60
$C_4H_6O_2S_4Se$, Selenium dimethyl-xanthate, 41B, 74
$C_4H_6O_2S_5$, Sulfur dimethyl-xanthate, 41B, 74
C_4H_6S, Methyl allenyl sulfide (gas-ed), 39B, 890
$C_4H_8Cl_2O_3Se$, Seleninyl dichloride dioxan, 46B, 63
$C_4H_8I_2S_3$, 2,2'-Di-iododiethyl trisulfide, 11, 601
$C_4H_8S_2$, 1,1-Bis(methylthio)ethylene (gas-ed), 42B, 1028
$C_4H_9Cl_4FeS_3$, Tris(methylthio)carbonium tetrachloroferrate, 41B, 75
$C_4H_{10}ClNO$, N,N'-Dimethylacetamidonium chloride, 46B, 7
$C_4H_{10}N_2S_2$, Methyl 3,3-dimethyldithiocarbazate, 43B, 92
$C_4H_{10}O_2S_2$, meso-Ethane-1,2-bis(methyl sulfoxide), 42B, 51
$C_4H_{10}O_3S_2$, 1-Methylsulphonyl-2-methylsulfoxyethane, 44B, 61
$C_4H_{10}O_6S_3$, Tris(methylsulphonyl)methane, 30B, 53
$C_4H_{12}Cl_{11}NO_5$, Tetramethylammonium chloride - penta-selenyl-dichloride complex, 32B, 29
$C_4H_{12}N_2O_2S$, N,N'-Tetramethylsulfamide (gas-ed), 39B, 891
$C_4H_{12}N_2S$, N,N'-Thio-bis(dimethylamine) (gas-ed), 35B, 895; 39B, 887

$C_4H_{12}N_4S_4$, 1,3,5,7-Tetrasulfur-2,4,6,8-tetramethylimide, 39B, 45
$C_4H_{16}B_5CsO_{12}S_2$, Caesium hexaoxotetrahydroxopentaborate bis(dimethyl sulfoxide), 43B, 94
$C_5H_8N_2OS$, S-Cyanoethylacetothiohydroximate, 37B, 35
$C_5H_{10}N_2O_2S$, S-Methyl O-(N-methylcarbamoyl)acetothiohydroximate, 37B, 36
$C_5H_{12}N_2S$, Tetramethylthiourea, 24, 531
$C_5H_{12}S_4$, Tetramethyl orthothiocarbonate, 9, 339; 10, 212
$C_5H_{14}N_2S_2$, Dimethylammonium dimethyldithiocarbamate, 44B, 61
C_6HCl_5S, Pentachlorothiophenol, 46B, 98
$C_6H_5Br_2NO_2S$, N,N-Dibromobenzenesulphonamide, 43B, 62
$C_6H_5ClO_2S$, Benzene sulphonyl chloride (gas-ed), 42B, 1027
$C_6H_5ClO_2Se$, p-Chlorobenzeneselenic acid, 20, 567
$C_6H_5NO_3S_2$, N-Sulfinylbenzenesulphonamide, 45B, 50
$C_6H_6O_2Se$, Benzeneseleninic acid, 18, 703
$C_6H_9ClN_2O_2S$, m-Aminobenzenesulphonamide hydrochloride, 39B, 45
$C_6H_{10}Br_2O_6S$, Diaquaoxonium 2,5-dibromobenzenesulphonate, 38B, 92; 45B, 51
$C_6H_{10}Cl_2O_6S$, 2,5-Dichlorobenzene sulphonic acid trihydrate, 38B, 94; 44B, 62
$C_6H_{10}O_2S_2$, S,S'-Diethyldithiooxalate, 40B, 77
$C_6H_{10}O_2S_4$, Diethyldixanthogen, 37B, 37
$C_6H_{10}O_3S_2$, O-Ethyl S-(2-carboxyethyl) dithiocarbonate, 43B, 7
$C_6H_{10}O_4S$, Thiodilactic acid, 39B, 46
$C_6H_{12}N_2OS$, N,N,N',N'-Tetramethylmonothiooxamide, 43B, 7
$C_6H_{12}N_2S_3$, N,N,N',N'-Tetramethylthiuram monosulfide, 42B, 53; 43B, 9
$C_6H_{12}S_4$, Tetrakis(methylthio)ethylene, 44B, 62
$C_6H_{16}Na_6O_{26}S_6$, Hexasodium benzenehexasulphonate octahydrate, 43B, 96
$C_6H_{18}HgI_4S_2$, Bis(trimethylsulphonium) tetraiodomercurate, 31B, 57
$C_6H_{18}O_4S_3$, Trimethylsulphonium sulfate, 44B, 63
$C_6H_{33}O_{21}PrS_3$, Nonaaquopraseodymium ethyl sulfate, 43B, 97
$C_6H_{33}O_{21}S_3Y$, Nonaaquoyttrium tris(ethylsulfate), 45B, 51
$C_6H_{33}O_{21}S_3Yb$, Nonaaquoytterbium ethyl sulfate, 43B, 97
C_7H_5BrOSe, o-Formylphenylselenenyl bromide, 41B, 77
C_7H_7KOS, Potassium *a*-hydroxybenzenesulphonate, 32B, 67
$C_7H_8O_2S$, Methyl phenyl sulphone, 30B, 83
C_7H_8S, Thioanisole (gas-ed), 42B, 1031
$C_7H_9NO_5S$, Ammonium hydrogen o-carboxybenzenesulphonate, 32B, 81
$C_7H_{10}O_4S$, p-Toluenesulphonic acid monohydrate, 37B, 69; 39B, 46
$C_7H_{10}O_8S$, 5-Sulfosalicylic acid dihydrate, 42B, 69
$C_7H_{10}S$, Cyclopentadienylidene-dimethylsulfur, 43B, 99
$C_7H_{11}NO_6Se$, Ammonium o-carboxybenzeneselenonate monohydrate, 43B, 98
$C_7H_{11}N_3O_2S$, S-Cyanoethyl O-(N-methylcarbamoyl)acetothiohydroximate, 37B, 36
$C_7H_{12}N_4O_3S$, Sulfaguanidine monohydrate, 42B, 812
$C_7H_{12}O_4S_2$, Methylenedithiodipropionic acid, 43B, 99
$C_7H_{12}O_8S$, o-Sulfobenzoic acid trihydrate, 42B, 68
$C_7H_{15}ClO_6S$, (Carboxymethyl)methyl(1-methylpropyl)sulphonium perchlorate, 43B, 101
$C_7H_{15}NS_3$, t-Butyl-N,N-dimethyltrithiopercarbamate, 34B, 45
$C_7H_{16}O_6S_3$, Tri(ethylsulphonyl)methane, 34B, 46
$C_8H_6FNO_2S$, 2-Fluorophenyl cyanomethyl sulphone, 42B, 52
$C_8H_8O_3S$, (-)-o-Carboxyphenyl methyl sulfoxide, 40B, 105
$C_8H_8O_3S$, DL-o-Carboxyphenyl methyl sulfoxide, 39B, 46
$C_8H_8O_3Se$, o-Carboxyphenyl methyl selenium oxide, 39B, 46
$C_8H_9BrO_2S$, S(-)-*a*-Bromobenzylmethylsulphone, 39B, 48
$C_8H_9ClO_6S$, p-Tolylsulphonylmethyl perchlorate, 44B, 63

$C_8H_{10}N_2O_2S$, N-(p-Nitrophenyl)-S,S-dimethyliminosulfurane, 42B, 52
$C_8H_{10}SO$, (+)-Methyl p-tolyl sulfoxide, 35B, 83
$C_8H_{11}AuCl_4N_2S$, S-Benzylisothiouronium tetrachloroaurate(III), 41B, 1233
$C_8H_{11}ClO_4S$, Dimethylphenylsulphonium perchlorate, 29, 485
$C_8H_{12}O_2S_2$, Bis[(methylthio)methylene·-pentane-2,4-dione, 45B, 52
$C_8H_{12}O_2S_2$, trans-3,3-Bis(methylthio)-3-hexene-2,5-dione, 44B, 323
$C_8H_{16}N_2O_2S_2$, Bis(morpholino)-disulfide, 39B, 48
$C_8H_{16}N_2O_2Se_2$, Dimorpholino-diselane, 46B, 64
$C_8H_{16}N_2O_2Se_3$, Dimorpholino-triselane, 46B, 64
$C_8H_{16}N_2O_2Se_4$, Dimorpholino-tetraselane, 46B, 65
$C_8H_{22}Cl_2N_2S_2$, Bis-(2-(N,N-dimethylamino)ethyl)disulfide dihydrochloride, 39B, 48
$C_9H_9IO_2S$, 1-Iodo-1-phenyl-2-methanesulphonylethylene, 41B, 1232
$C_9H_{11}NOS$, N-Benzoyliminodimethylsulfur(IV), 42B, 53
$C_9H_{11}NO_2S_2$, N-(Phenylsulphonyl)thiopropionamide, 41B, 1232
$C_9H_{12}N_2O_5S$, Ammonium 2-oxamoyl-3-methyl-benzene sulphonate, 40B, 79
$C_9H_{12}O_4S_2$, 1-Methylsulphonyl-2-phenylsulphonylethane, 46B, 66
$C_9H_{13}NO_2S_2$, N-(Dimethylsulphonio)toluene-p-sulphonamidate, 39B, 49
$C_9H_{13}NO_3S$, Methyl p-dimethylaminobenzenesulphonate, 43B, 102
$C_9H_{14}K_2O_8S_2$, Dipotassium mesitylene-disulphonate dihydrate, 38B, 161
$C_9H_{14}O_3S$, 5,5-Dimethyl-3-methylsulphonylcyclohex-2-enone, 46B, 66
$C_9H_{16}O_3S_2$, O-Ethyl S-(5-carboxypentyl)dithiocarbonate, 40B, 79
$C_{10}H_8N_2O_2S_2$, 3,3'-Dihydroxydi-2-pyridyl disulfide, 44B, 63
$C_{10}H_{10}Cl_4N_2OSe$, Dipyridinium(II) oxytetrachloroselenate(II), 35B, 493
$C_{10}H_{12}Cl_3Se$, 2-Chloropropyl-1,4'-tolyl selenide dichloride, 40B, 80
$C_{10}H_{12}O_2S$, Propenyl p-tolyl sulphone, 39B, 50
$C_{10}H_{14}O_2S$, Mesityl methyl sulphone, 32B, 68
$C_{10}H_{14}O_4S_2$, 3,3'-Dithiobis-(2,4-pentanedione) (enol form), 42B, 53
$C_{10}H_{14}O_4S_3$, 3,3-Trithiobis-(2,4-pentanedione), 41B, 79
$C_{10}H_{14}O_4S_3$, 3,3'-Trithiobis-(2,4-pentanedione), 37B, 38
$C_{10}H_{16}N_2O_2S_4Se$, Selenium di(morpholyldithiocarbamate), 35B, 499
$C_{10}H_{16}N_2Se_5$, Selenium bis-(1-pyrrolidinecarbodiselenoate), 41B, 254
$C_{10}H_{18}N_2O_4S_6$, Di(t-butyloxycarbonyl)cyclohexasulfurhydrazide, 43B, 103
$C_{10}H_{20}N_2S_4Se$, Diethyldithiocarbamatoselenium(II), 35B, 500; 38B, 857
$C_{10}H_{20}N_2Se_4$, Dipiperidino-tetraselane, 46B, 65
$C_{10}H_{20}N_2Se_5$, Bis(N,N-diethyldiselenocarbamato)selenium(II), 42B, 54
$C_{11}H_{12}OS_2$, 3-Phenyl-1-propene-1,3-dione-1-(dimethyl mercaptole), 37B, 40
$C_{11}H_{13}ClO_2S$, (E)-4-Chloro-2-methyl-but-2-enyl-sulphonylbenzene, 45B, 53
$C_{11}H_{13}ClO_2S$, (Z)-4-Chloro-2-methyl-but-2-enyl-sulphonylbenzene, 45B, 53
$C_{11}H_{15}ClOS$, (R)(-)-α-Chlorobenzyl-t-butyl-(S) sulfoxide, 43B, 104
$C_{11}H_{15}ClO_2S$, (R)(-)-α-Chlorobenzyl-t-butyl sulphone, 43B, 105
$C_{11}H_{16}ClNOSe$, 2-Dimethylaminoethyl selenolbenzoate hydrochloride, 38B, 95
$C_{11}H_{16}O_2S$, (2RS,SRS)-3-Methyl-3-phenylsulfinylbutan-2-ol, 42B, 54
$C_{12}F_{10}S_2$, Bis(pentafluorophenyl)disulfide, 42B, 1017
$C_{12}F_{10}Se_2$, Bis(pentafluorophenyl)diselenide, 42B, 1017
$C_{12}H_6F_2N_2O_6S$, 3,3'-Dinitro-4,4'-difluorodiphenyl sulphone, 40B, 81
$C_{12}H_8Br_2O_2S$, 4,4'-Dibromodiphenyl sulphone, 9, 384; 10, 267; 40B, 82
$C_{12}H_8Br_2O_2S_2$, Di(p-bromophenyl)thiosulphonate, 32B, 68
$C_{12}H_8Br_2S$, Bis(p-bromophenyl) sulfide, 9, 383; 10, 266
$C_{12}H_8Br_2S_2$, Di-p-bromophenyl disulfide, 10, 269

$C_{12}H_8Cl_2N_2S_3$, 1,5-Bis(p-chlorophenyl)-2,4-diaza-1,3,5-trithiapenta-2,3-diene, 39B, 50
$C_{12}H_8Cl_2OS_3$, 1,3-Bis(4-chlorophenyl)trisulfane 2-oxide, 43B, 105
$C_{12}H_8Cl_2O_2S$, 4,4'-Dichlorodiphenyl sulphone, 24, 655; 26, 728; 40B, 81
$C_{12}H_8Cl_2S_2$, Bis(4-chlorophenyl) disulfide, 45B, 53
$C_{12}H_8Cl_2Se_2$, 4,4'-Dichlorodiphenyl diselenide, 21, 595
$C_{12}H_8Cl_4S$, Bis(p-chlorophenyl)dichlorosulfide, 34B, 42
$C_{12}H_8F_2O_2S$, 4,4'-Difluorodiphenyl sulphone, 40B, 81
$C_{12}H_8I_2O_2S$, 4,4'-Diiododiphenyl sulphone, 19, 570; 40B, 82
$C_{12}H_8N_2O_4SSe_2$, Bis-(o-nitrobenzeneselenyl) sulfide, 41B, 111
$C_{12}H_8N_2O_4S_2$, Bis(o-nitrophenyl) disulfide, 35B, 58
$C_{12}H_8N_2O_4S_2$, Bis(p-nitrophenyl) disulfide (α-form), 34B, 44
$C_{12}H_8N_2O_4S_2$, 3,3'-Dithiodi-2-pyridinecarboxylic acid, 46B, 67
$C_{12}H_9NO_2S$, 4-Nitrodiphenyl sulfide, 41B, 111
$C_{12}H_{10}BrNOS_2$, 1,1-Bis-methylthio-2-p-bromobenzoyl-2-cyanoethylene, 40B, 53
$C_{12}H_{10}Br_2Se$, Diphenylselenium dibromide, 8, 311
$C_{12}H_{10}Cl_2Se$, Diphenylselenium dichloride, 9, 385
$C_{12}H_{10}NNaO_4S_2$, Sodium dibenzenesulphonamide, 35B, 84
$C_{12}H_{10}N_2S_3$, Diphenylthiosulfodiimide, 39B, 51
$C_{12}H_{10}OS$, Diphenyl sulfoxide, 21, 593
$C_{12}H_{10}OS_3$, S,S-Diphenyldithiosulfite, 44B, 65
$C_{12}H_{10}O_2S$, Diphenyl sulphone, 40B, 81
$C_{12}H_{10}O_4S_2$, Diphenyl disulphone, 38B, 96
$C_{12}H_{10}O_4S_2Se$, Dibenzenesulphonyl selenide, 18, 721
$C_{12}H_{10}O_4S_3$, Diphenylsulphonyl sulfide, 11, 682
$C_{12}H_{10}S_2$, Diphenyl disulfide, 34B, 48; 42B, 55
$C_{12}H_{10}Se_2$, Diphenyl diselenide, 16, 525
$C_{12}H_{11}NO_4S_2$, Dibenzenesulphonamide, 35B, 84
$C_{12}H_{12}N_2S$, 4,4'-Diaminodiphenylsulfide, 39B, 52
$C_{12}H_{12}N_2S_2$, 2,2'-Diaminodiphenyl disulfide, 32B, 70; 35B, 60
$C_{12}H_{16}ClNO_4S$, (2R)-2-(Benzyloxycarbonylamino)-3-hydroxypropyl chloromethyl (R)-sulfoxide, 45B, 54
$C_{12}H_{16}S_2$, Methyl 4-t-butyldithiobenzoate, 43B, 107
$C_{12}H_{20}N_2S_4$, Bis(N,N-cyclopentamethylenethiocarbamyl)disulfide, 39B, 52
$C_{12}H_{25}NaO_4S \cdot 0.1\ H_2O$, Sodium dodecylsulfate hydrate, 43B, 66
$C_{13}H_{10}Br_2O_4S_2$, Bis(4-bromophenylsulphonyl)methane, 34B, 45; 38B, 97
$C_{13}H_{10}OS$, Thiobenzophenone S-oxide, 45B, 54
$C_{13}H_{10}S$, Thiobenzophenone, 45B, 54
$C_{13}H_{13}IO_4S$, Phenylhydroxytosyloxyiodine, 42B, 56
$C_{14}H_8Cl_2O_2S_2$, 3-Chlorophenyl dithiooxalate, 40B, 82
$C_{14}H_8O_4S$, Bis(2-carboxyphenyl)sulfur dihydroxide dilactone, 39B, 52
$C_{14}H_8O_4Se$, 3,3'-Spirobi-(3-selenaphthalide), 40B, 83
$C_{14}H_{12}N_2O_6S$, 2,4-Dinitrobenzyl p-tolyl sulphone, 40B, 111
$C_{14}H_{12}N_2S_2$, Benzylidenimine disulfide, 40B, 83
$C_{14}H_{12}N_2S_3$, Benzylidenimine trisulfide, 39B, 53
$C_{14}H_{12}N_2S_4$, Benzylidenimine tetrasulfide, 39B, 53
$C_{14}H_{14}Br_2Se$, Di-p-tolylselenium dibromide, 13, 545
$C_{14}H_{14}Cl_2Se$, Di-p-tolylselenium dichloride, 13, 544
$C_{14}H_{14}I_2S$, Benzyl sulfide - iodine, 24, 657
$C_{14}H_{14}N_2O_2S$, 4-Nitro-4'-dimethylaminodiphenyl sulfide, 41B, 111
$C_{14}H_{14}N_2S_2$, 3-Diphenyl-dithiocarbazic-methylester, 41B, 80
$C_{14}H_{14}OS_2$, S-p-Tolyl p-toluene thiosulfinate, 44B, 65
$C_{14}H_{14}O_2S_2$, (R,S)meso-1,2-Bis(phenylsulfinyl)ethane, 42B, 57; 44B, 65; 45B, 55

$C_{14}H_{14}O_2S_2$, (S,S)rac-1,2-Bis(phenylsulfinyl)ethane, 42B, 56; 44B, 65; 45B, 55
$C_{14}H_{14}S$, Di-p-tolyl sulfide, 19, 575
$C_{14}H_{14}S$, 2,2'-Dimethyldiphenylsulfide, 42B, 56
$C_{14}H_{14}S_2$, Di-p-tolyldisulfide, 24, 656; 32B, 71
$C_{14}H_{14}S_2$, Dibenzyl disulfide, 34B, 49; 37B, 39
$C_{14}H_{14}Se$, Di-p-tolyl selenide, 19, 575
$C_{14}H_{15}NO_2S$, N-p-Tolylsulphonylbenzylamine, 41B, 327
$C_{14}H_{15}NO_4S$, p-Methoxybenzenesulphon-p-anisidide, 39B, 54
$C_{14}H_{15}NS$, 4-Dimethylaminophenyl phenyl sulfide, 39B, 54; 40B, 84
$C_{14}H_{17}NO_2S$, trans-1-Isocyano-1-tosyl-3,3-dimethyl-1-butene, 45B, 56
$C_{14}H_{22}N_2O_5S$, 2-Carboxamidomethyl-4,5-dimethoxyphenyldiethylsulphonamide, 43B, 108
$C_{15}H_{11}N_3O_9S_2$, 1-Phenyl-2-thiomethylethenyl 2,4,6-trinitrobenzenesulphonate,(E), 34B, 50
$C_{15}H_{12}OS$, 3-Mercapto-1,3-diphenylprop-2-en-1-one, 42B, 55
$C_{15}H_{14}Br_2OS$, (+)-1,2-Dibromo-2-phenylethyl p-tolyl sulfoxide, 41B, 80
$C_{15}H_{14}N_2O_3S$, N-(p-Nitrobenzoyl)-2-iminophenyldimethylsulfur(IV), 42B, 57
$C_{15}H_{14}O_2S$, Methyl 2,2-diphenylvinyl sulphone, 42B, 57
$C_{15}H_{15}BrSO_2$, (+)-1-Bromo-2-hydroxy-2-phenylethyl p-tolyl sulfoxide, 42B, 58
$C_{15}H_{16}O_2S$, Phenyl mesityl sulphone, 42B, 58
$C_{15}H_{21}BrO_2S$, (E)-4-Bromo-2-ethyl-but-2-enyl-sulphonylmesitylene, 45B, 53
$C_{15}H_{21}BrO_2S$, (Z)-4-Bromo-2-ethyl-but-2-enyl-sulphonylmesitylene, 45B, 53
$C_{15}H_{22}BrNSO_3$, Triethylprop-2-ynylammonium p-bromobenzenesulphonate, 35B, 50
$C_{15}H_{28}O_3S_2$, O-Ethyl S-(11-carboxyundecyl)dithiocarbonate, 42B, 64
$C_{16}H_{14}O_5S$, 2-Tosyloxy-5-methyl-isophthalaldehyde, 46B, 68
$C_{16}H_{16}Br_2S_2$, Bis(o-(bromomethyl)benzyl) disulfide, 44B, 66
$C_{16}H_{16}O_2S_2$, E-Mesityl(phenylsulfinyl)sulfine, 40B, 85
$C_{16}H_{16}O_2S_2$, Z-Mesityl(phenylsulfinyl)sulfine, 41B, 81
$C_{16}H_{16}O_3S_2$, E-Mesityl(phenylsulphonyl)sulfine, 42B, 59
$C_{16}H_{16}O_3S_2$, cis-Mesityl(phenylsulphonyl)sulfine, 40B, 86
$C_{16}H_{16}S_4$, trans-1,2-Bis(methylthio)-1,2-bis(phenylthio)ethylene, 44B, 66
$C_{16}H_{17}NO_2S_2$, S-(2-Methoxyphenyl)-N-(2,6-dimethylphenyl)dithiourethane S-oxide, 38B, 97
$C_{16}H_{17}NO_2S_2$, S-(2-Methoxyphenyl)-N-(2,6-dimethylphenyl)dithiourethane sulfenic acid, 38B, 97
$C_{16}H_{19}NO_2S_2$, S-Propyl,S-phenyl-N-p-tolylsulphonyl sulfilimine, 38B, 98
$C_{16}H_{22}Cl_6N_4PtS_2$, S-Benzylisothiouronium hexachloroplatinate(IV), 41B, 1233
$C_{16}H_{22}O_2S_2$, O,O'-Diethyl 5-t-butyldithioisophthalate, 43B, 108
$C_{17}H_{16}O_2Se$, Diacetylmethylenediphenylselenurane, 40B, 86
$C_{17}H_{18}OS$, Methyl 2-phenyl-(3',4'-dimethyl-2-phenyl)vinyl sulfoxide, 38B, 99
$C_{17}H_{20}O_2S$, Methyl (2-hydroxy-2-phenyl-2-(2',4'-dimethylphenyl))ethyl sulfoxide, 40B, 87
$C_{17}H_{20}O_2S_2$, meso-2,2-Bis(methylsulfinyl)-1,3-diphenylpropane, 46B, 68
$C_{17}H_{21}NO_4S$, p-Methoxybenzenesulphon-N-i-propyl-p-anisidide, 39B, 54
$C_{17}H_{26}O_3S$, trans-3-t-Butylcyclohexyl-p-toluene sulphonate, 39B, 55

$C_{18}H_{12}Cl_2O_6S_2$, 2,4-Hexadiynylene bis(p-chlorobenzenesulphonate), 44B, 42

$C_{18}H_{14}F_6N_2O_4S_4$, Bis(6-(ethylthio)-5-nitro-α,α,α-trifluoro-m-tolyl) disulfide, 43B, 109

$C_{18}H_{15}ClSe \cdot H_2O$, Triphenylselenonium chloride monohydrate, 45B, 56

$C_{18}H_{17}ClN_2S$, trans-((2-Chlorophenyl)-1-benzocyclobutenyl)-(1,4,5,6-tetrahydro-2-pyrimidyl)-sulfide, 43B, 109

$C_{18}H_{18}S_2$, Dicinnamyl disulfide, 37B, 41; 41B, 82

$C_{18}H_{19}ClO_2Se$, Triphenylselenonium chloride dihydrate, 42B, 1018

$C_{18}H_{21}NO_2S_3$, t-Butylimido-λ⁴-thio(phenylthio)methyl p-tolyl sulphone, 44B, 66

$C_{18}H_{22}O_2S$, Dimesityl sulphone, 37B, 41

$C_{18}H_{38}Se_3$, Bis(t-butyl)triselenide, 44B, 67

$C_{19}H_{15}NSSe$, Triphenylselenonium isothiocyanate, 43B, 110

$C_{19}H_{16}Cl_3NO_2S_2$, N-(1-α-Naphthylethyl)-N-(benzenesulphonyl)trichloromethanesulfenamide, 37B, 42

$C_{19}H_{17}NO_2S_2$, S,S-Diphenyl-N-p-tolylsulphonyl sulfilimine, 37B, 43

$C_{19}H_{29}N_3O_9S_2$, 2,3-Dimethyl-2-butenyl-1,1,2-trimethylpropyl-methylsulphonium 2,4,6-trinitrobenzenesulphonate, 39B, 55

$C_{20}H_{14}Cl_4O_6S_2$, Tetrachloro-p-phenylenedi(toluene-p-sulphonate), 44B, 67

$C_{20}H_{16}F_6N_4O_6S_4$, S,S'-(2,2'-Dithiobis(6-nitro-α,α,α-trifluoro-p-tolyl)) bis(N,N-dimethylcarbamothioate), 42B, 59

$C_{20}H_{18}N_4S_2$, Dithiodinicotyrine, 46B, 69

$C_{20}H_{18}O_4S_2$, 1,1'-Diphenyl-2,2'-dithiobis(butane-1,3-dione), 41B, 83

$C_{20}H_{18}O_6S_2$, Poly-(1,2-bis-(p-tolylsulphonyloxymethylen)-1-buten-3-inylen), 40B, 88

$C_{20}H_{26}O_2S_2$, 1,4-Dimesityl-1,4-dithiabutane 1,4-dioxide, 44B, 68

$C_{20}H_{30}O_4Se$, Bis(2-hydroxy-4,4,6,6-tetramethyl-1-cyclohexen-3-one)selenide, 44B, 68

$C_{20}H_{36}N_4O_8S_4$, Tetra(t-butyloxycarbonyl)cyclotetrasulfurdihydrazide, 43B, 103

$C_{21}H_{24}Cl_2N_2O_4S$, 2-(N-(3-Dimethylammoniopropyl)amino)-4,2'-dichlorodiphenylsulfur maleate, 43B, 111

$C_{21}H_{25}ClN_2O_4S$, N-(3-Dimethylammoniopropyl)-2-amino-2'-chlorodiphenylsulfide maleate, 42B, 27

$C_{21}H_{25}ClN_2O_4S$, N-(3-Dimethylammoniopropyl)-2-amino-4-chlorodiphenylsulfide maleate, 42B, 27

$C_{22}H_{18}N_2O_3S$, N-Phthalimido-dibenzylsulfoximide, 40B, 88

$C_{22}H_{22}O_6S_2$, 3,5-Octadiynylene bis(p-toluenesulphonate), 46B, 69

$C_{22}H_{28}O_2S$, 4,4'-Dihydroxydiphenylsulfide decamethylene ether, 13, 548

$C_{22}H_{28}O_2Se$, (2S,6aR,9R,10aR)-6,6,9-Trimethyl-1-oxo-2-phenylseleno-1,2,3,4,6a,7,8,9,10,10a-decahydro-6H-dibenzo[b,d·pyran, 46B, 339

$C_{23}H_{21}NO_3S$, 2-Phenyl-4-acetylphenoxy-2,6-dimethylphenyliminomethanesulfenic acid, 38B, 100

$C_{23}H_{27}BrN_2O_3S$, (+)-N-Phthalimido-2(R)-bromo-2-octyl-p-tolyl-(R)-sulfoximide, 43B, 112

$C_{23}H_{27}BrN_2O_3S$, (+)-N-Phthalimido-2-bromo-2(S)-octyl-p-tolyl-(R)-sulfoximine, 39B, 56

$C_{24}H_{24}N_2S_2$, 2,2'-Diisopropyl-8,8'-diquinoline disulfide, 46B, 70

$C_{24}H_{26}O_6S_2$, Tetramethyl-p-phenylene ditoluene-p-sulphonate, 41B, 84

$C_{25}H_{20}S_4$, Tetraphenyl orthothiocarbonate, 38B, 101

$C_{26}H_{20}S_4$, Tetrakis(phenylthio)ethylene, 44B, 69

$C_{26}H_{22}Se_2$, Bis(diphenylmethyl) diselenide, 34B, 52

$C_{26}H_{24}F_{12}O_3S$, Diaryldialkoxysulfurane oxide derivative, 40B, 89

$C_{30}H_{20}F_{12}O_2S$, Diphenyldi(phenyl-di(trifluoromethyl)methoxy)sul-

furane, 38B, 102
$C_{32}H_{58}CoO_{24}S_4$, Hexaaquocobalt(II) bis(hydrogen-2,2'-dithiobisbenzoate) hexahydrate-tetrakismethanol, 40B, 90
$C_{38}H_{30}S$, Bis(triphenylmethyl) sulfide, 46B, 70
$C_{38}H_{30}S_6$, Hexakis(phenylthio)ethane, 40B, 91
$C_{44}H_{58}Cl_2N_4O_8S_2$, N-(3-Diethylammoniopropyl)-2-methylamino-4-chlorodiphenylsulfide oxalate, 44B, 69

12. CARBONIUM IONS, CARBANIONS, RADICALS

CBr_2, Dibromomethylene radical (gas-ed), 40B, 1137
CKN_3O_6, Potassium trinitromethane, 32A, 347; 32B, 64
CHN_2O_4Rb, Rubidium dinitromethane, 33B, 30
CHO, Formyl radical (gas-mw), 40B, 1145
$CH_5N_5O_6$, Hydrazinium trinitromethide, 35B, 18
$CH_8N_3O_2P$, Guanidinium phosphinate, 45B, 41
$CH_9N_7O_3$, 1,2,3-Triaminoguanidinium nitrate, 45B, 42
$C_2KN_3O_4$, Potassium dinitroacetonitrile, 38B, 104; 40B, 1076
$C_2K_2N_4O_8$, Potassium tetranitroethide, 33A, 432; 33B, 33
$C_2N_3O_4Rb$, Rubidium cyanodinitromethanide, 32B, 48; 33B, 30; 38B, 105
$C_2N_4O_8Rb_2$, Dirubidium 1,1,2,2-tetranitroethanediide, 38B, 106
$C_2H_2N_3O_5Rb$, Rubidium α,α-dinitroacetamide, 35B, 19
$C_2H_3AlCl_4O$, Methyloxocarbonium tetrachloroaluminate, 38B, 102
$C_2H_3Cl_6OSb$, Methyloxocarbonium hexachloroantimonate, 38B, 103
$C_2H_3KN_2O_4$, Potassium 1,1-dinitroethane, 39B, 57
$C_2H_6O_6S$, Acetic acid - sulfuric acid, 33B, 38
$C_2H_{10}Cl_5N_4O_2ReS_2 \cdot H_2O$, Dithiobisformamidinium trans-aquatetrachlorooxorhenate(V) chloride monohydrate, 45B, 43
$C_3HF_6O_4RbS_2$, Rubidium bis(trifluoromethylsulphonyl)methane, 45B, 57
$C_3H_4KN_3O_5$, Potassium N-methyl-α,α-dinitroacetamide, 37B, 43
$C_3H_5Cl_4GaO$, Ethyloxocarbonium tetrachlorogallate, 38B, 107
$C_3H_7N_3O_4 \cdot H_2O$, Guanidinium hydrogen oxalate monohydrate, 45B, 17
C_4KN_3, Potassium tricyanomethanide, 28, 107; 37B, 44
$C_4H_4KN_3O_5$, Potassium 4,4-dinitro-2-butenamide, 33B, 39
$C_4H_6KN_3O_5$, Potassium β,β-dinitroethylacetamide, 40B, 92
$C_4H_6N_3O_5Rb$, Rubidium N,N-dimethyl-α,α-dinitroacetamide, 40B, 93
$C_4H_7Cl_6OSb$, Isopropyloxocarbonium hexachloroantimonate, 38B, 107
$C_5H_6KN_3O_4 \cdot 0.5\ H_2O$, Potassium 1,1-dinitro-2,2-dimethylpropionitrile hemihydrate, 40B, 94
$C_5H_{13}ClN_2O_4$, N,N,N',N'-Tetramethylformamidinium perchlorate, 45B, 7
$C_7H_2KN_5O_{10}$, Potassium 2,4,6-trinitrophenyldinitromethane, 39B, 57
$C_7H_2N_5O_{10}Rb$, Rubidium 2,4,6-trinitrophenyldinitromethane, 37B, 44
$C_7H_4ClKN_2O_4$, Potassium p-chlorophenyldinitromethanide, 38B, 108
$C_7H_5KN_2O_4$, Potassium phenyldinitromethane, 35B, 30
$C_7H_6KN_3O_7 \cdot 0.5\ H_2O$, Potassium methoxide s-trinitrobenzene hemihydrate, 45B, 82
$C_7H_9N_4O_9Rb$, Rubidium 1,1,3,3-tetranitroheptan-6-one, 34B, 54
$C_7H_{15}ClN_2O_4$, 1,3-Bis(dimethylamino)trimethinium perchlorate, 39B, 58; 40B, 94
$C_7H_{21}Cl_4MnN_3$, Dimethylammonium bis(dimethylamino)carbonium tetrachloromanganate, 42B, 61
$C_8H_5KN_2O_2$, Potassium phenylnitroacetonitrile, 40B, 95
$C_8H_7Cl_6OSb$, 2-Methylphenyloxocarbonium hexachloroantimonate, 38B, 109
$C_9H_{17}ClN_2O_4$, 1,5-Bis(dimethylamino)pentamethinium perchlorate (Form 1), 40B, 97

$C_9H_{17}ClN_2O_4$, 1,5-Bis(dimethylamino)pentamethinium perchlorate (Form 2), 43B, 41
$C_{10}K_2N_6$, Dipotassium cis-hexacyanobutenediide, 38B, 110
$C_{10}H_{26}Cl_6N_4Sn$, Bis(dimethylaminecarbonium) hexachlorostannate, 43B, 44
$C_{11}H_{25}ClN_4O_4$, 1,1,3,3-Tetrakis(dimethylamino)allyl perchlorate, 40B, 97
$C_{12}Cl_{10}NO$, Perchlorodiphenylnitroxide, 44B, 70
$C_{12}H_4KN_4$, Potassium 7,7,8,8-tetracyanoquinodimethane, 44B, 70
$C_{12}H_8N_5$, Ammonium 7,7,8,8-tetracyanoquinodimethane (phase II), 44B, 71
$C_{12}H_{10}ClI$, Diphenyliodinium chloride, 20, 584
$C_{13}Cl_{11}$, Perchlorodiphenylmethyl, 37B, 44
$C_{13}H_{21}AlCl_4$, Heptamethylbenzene tetrachloroaluminate, 33B, 41
$C_{13}H_{21}Cl_2N_3O_3$, 2-Chloro-2-phenoxymalonylamide-amidium hydrochloride monohydrate, 40B, 98
$C_{14}H_{40}Br_3N_7$, Tetrapropylammonium guanidinium bromide, 45B, 27
$C_{15}H_{13}N_7$, Tetramethylammonium 1,1,2,4,5,5-hexacyanopentadienide, 40B, 99
$C_{15}H_{14}IN_5$, Trimethylammonium 7,7,8,8-tetracyanoquinodimethanide tri-iodide, 46B, 43
$C_{15}H_{14}IN_5O$, Trimethylammonium 7,7,8,8-tetracyanoquinodimethanide triiodide, 45B, 27
$C_{16}H_{21}IN_3$, 4,4'-Bis(dimethylamino)diphenylamine iodide, 31B, 59
$C_{17}H_{17}BrN_2 \cdot 0.5\ H_2O$, N-Phenyl-(5-anilino-2,4-pentadienylidene)ammonium bromide hemihydrate, 40B, 100
$C_{17}H_{17}BrN_2$, 5-Anilinopentadienylidenanilinium bromide, 40B, 100
$C_{18}H_{19}N_7$, Tetramethylammonium 3-i-propyl-1,1,2,4,5,5-hexacyanopentadienide, 42B, 65
$C_{18}H_{40}Li_2N_4$, (Hexatriene dianion) dilithium tetramethylethylenediamine, 41B, 86
$C_{19}H_{12}N_3O_6$, Tri-p-nitrophenylmethyl, 32B, 74
$C_{19}H_{15}$, Triphenylmethyl (gas-ed), 30B, 403
$C_{19}H_{15}Br$, Triphenylmethyl bromide, 17, 719; 31B, 82
$C_{19}H_{15}ClO_4$, Triphenylmethyl perchlorate, 30B, 54
$C_{19}H_{15}Cl_{13}Te_3$, Triphenylmethyl tridecachloro-tri-tellurium, 45B, 57
$C_{19}H_{18}ClN_3O_4$, Tri-(p-aminophenyl)carbonium perchlorate, 37B, 45
$C_{20}H_{17}N_4$, 3,4-Dihydro-2,4,6-triphenyl-s-tetrazin-1(2H)-yl free radical, 34B, 56
$C_{21}H_{15}ClO_4$, sym-Triphenylcyclopropenium perchlorate, 31B, 60
$C_{22}H_{30}Br_2O_7$, Tri-p-methoxyphenylmethyl hydrogen dibromide tetrahydrate, 30B, 55
$C_{22}H_{30}Cl_2O_7$, Tri-p-methoxyphenylmethyl hydrogen dichloride tetrahydrate, 30B, 55
$C_{24}H_{18}N_5O_6$, 2,2-Diphenyl-1-picrylhydrazyl benzene solvate, 32B, 73
$C_{24}H_{20}N_{12}Pt$, Tetraammineplatinum(II) 7,7,8,8-tetracyanoquinodimethanide (1:2), 44B, 71
$C_{24}H_{32}N_8Pt \cdot H_2O$, Bis(N,N,N',N'-tetramethyl-1,4-diaminobenzeniumyl) tetracyanoplatinate(II) monohydrate, 45B, 80
$C_{25}H_{33}O_2$, Galvinoxyl, 34B, 55
$C_{28}H_{46}NaO_8$, Bis[1,2-bis(2-methoxyethoxy)ethane·sodium biphenylide, 45B, 58
$C_{30}H_{24}N_9$, Triethylammonium 7,7,8,8-tetracyano-p-quinodimethane, 46B, 613
$C_{30}H_{28}Cl_4I_2O_4$, Tetra-p-anisylethylene bis(dichloroiodate(I)), 32B, 75
$C_{32}H_{28}N_9$, Tetraethylammonium bis(7,7,8,8-tetracyanoquinodimethyl-

ide), 46B, 44
$C_{36}H_{12}N_{12}Rb_2$, Rubidium 7,7,8,8-tetracyanoquinodimethane, 44B, 72; 45B, 58
$C_{38}H_{28}Cl_{12}Sb_2$, Bi(triphenylcarbonium) tetrachloroantimonate, 37B, 45
$C_{56}H_{32}N_{16}Pt$, Bis(2,2'-dipyridyl)platinum(II)-7,7,8,8-tetracyanoquinodimethane 1:3 radical, 44B, 72

13. BENZOIC ACID DERIVATIVES

$C_7HF_5O_2$, Pentafluorobenzoic acid, 38B, 117
$C_7H_4ClNO_4$, 2-Chloro-5-nitrobenzoic acid, 27, 897
$C_7H_5BrO_2$, m-Bromobenzoic acid, 32B, 76
$C_7H_5BrO_2$, o-Bromobenzoic acid, 27, 895
$C_7H_5BrO_2$, p-Bromobenzoic acid, 38B, 116
$C_7H_5Br_2NO_2$, 3,5-Dibromo-p-aminobenzoic acid, 30B, 57
$C_7H_5ClO_2$, m-Chlorobenzoic acid, 42B, 68
$C_7H_5ClO_2$, o-Chlorobenzoic acid, 26, 694
$C_7H_5ClO_2$, p-Chlorobenzoic acid, 15, 463; 40B, 102
$C_7H_5Cl_2NO_2$, 3,5-Dichloroanthranilic acid, 34B, 57
$C_7H_5FO_2$, o-Fluorobenzoic acid, 41B, 87
$C_7H_5FO_2$, p-Fluorobenzoic acid, 45B, 59
$C_7H_5IO_2$, o-Iodobenzoic acid, 43B, 114
$C_7H_5NO_4$, m-Nitrobenzoic acid (unstable form), 41B, 86
$C_7H_5NO_4$, m-Nitrobenzoic acid, 40B, 103
$C_7H_5NO_4$, o-Nitrobenzoic acid, 32B, 76; 39B, 58
$C_7H_5NO_4$, p-Nitrobenzoic acid (form II), 46B, 71
$C_7H_5NO_4$, p-Nitrobenzoic acid, 31B, 61; 37B, 46; 43B, 115
$C_7H_5NO_4$, 5-Nitrososalicylic acid, 43B, 115
$C_7H_5NO_5$, o-Nitroperoxybenzoic acid, 30B, 58
$C_7H_5NO_5$, p-Nitroperoxybenzoic acid, 35B, 41
C_7H_6BrNO, N-Bromobenzamide, 43B, 62
C_7H_6ClNO, α-p-Chlorobenzamide, 44B, 73
C_7H_6ClNO, γ-p-Chlorobenzamide, 44B, 73
C_7H_6ClNO, o-Chlorobenzamide (α form), 40B, 103
C_7H_6ClNO, o-Chlorobenzamide (β form), 40B, 103
C_7H_6ClNO, p-Chlorobenzamide (β form), 46B, 71
C_7H_6ClNO, p-Chlorobenzamide (high-temp.), 40B, 1077
C_7H_6FNO, m-Fluorobenzamide, 30B, 61
C_7H_6FNO, p-Fluorobenzamide, 30B, 60
C_7H_6INO, p-Iodobenzamide, 38B, 117
$C_7H_6N_2O_3$, o-Nitrobenzamide, 38B, 117
$C_7H_6N_2O_3$, p-Nitrobenzamide, 43B, 116
$C_7H_6O_2$, Benzoic acid, 19, 563; 46B, 72
$C_7H_6O_3 \cdot H_2O$, p-Hydroxybenzoic acid monohydrate, 45B, 60
$C_7H_6O_3$, Salicylic acid, 17, 702; 30B, 56; 39B, 58
C_7H_7NO, Benzamide, 23, 645; 38B, 118
$C_7H_7NO_2$, m-Aminobenzoic acid, 46B, 72
$C_7H_7NO_2$, m-Hydroxybenzamide, 31B, 63
$C_7H_7NO_2$, o-Aminobenzoic acid (Anthranilic acid) (Form 1), 32B, 77
$C_7H_7NO_2$, o-Aminobenzoic acid (Anthranilic acid) (Form 2), 43B, 117
$C_7H_7NO_2$, p-Aminobenzoic acid, 31B, 62; 32B, 79; 33B, 42
$C_7H_7NO_3$, Salicylohydroxamic acid, 44B, 74
$C_7H_7NO_3$, p-Aminosalicylic acid, 18, 710; 44B, 75
$C_7H_8ClNO_2$, p-Aminobenzoic acid hydrochloride, 46B, 72
$C_7H_8ClNO_2$, 3-Aminobenzoic acid hydrochloride, 39B, 59
$C_7H_8ClNO_3$, p-Aminosalicylic acid hydrochloride, 44B, 75

$C_7H_8O_4$, p-Hydroxybenzoic acid monohydrate, 39B, 60
$C_7H_{10}O_8S$, 5-Sulfosalicylic acid dihydrate, 42B, 69; 43B, 97, 113
$C_7H_{12}O_9S$, Diaquo-oxonium salicylic acid 5-sulphonate (orthorhombic), 35B, 43
$C_7H_{12}O_9S$, Diaquo-oxonium salicylic acid 5-sulphonate (triclinic), 43B, 113
$C_8Br_4O_3$, Tetrabromophthalic anhydride, 41B, 88
$C_8Cl_4O_3$, Tetrachlorophthalic anhydride, 37B, 47
$C_8H_2Cl_4O_4 \cdot 0.5\ H_2O$, Tetrachlorophthalic acid hemihydrate, 41B, 88
$C_8H_4Br_2O_2$, Terephthaloyl bromide, 44B, 75
$C_8H_4Cl_2O_2$, Terephthaloyl chloride, 44B, 76
$C_8H_4O_3$, Phthalic anhydride, 43B, 117
$C_8H_6N_2O_6$, 3,5-Dinitro-4-methylbenzoic acid, 34B, 60
$C_8H_6O_3$, 4-Formylbenzoic acid (monoclinic), 42B, 70
$C_8H_6O_3$, 4-Formylbenzoic acid (triclinic), 42B, 70
$C_8H_6O_4$, Isophthalic acid, 38B, 119; 40B, 104
$C_8H_6O_4$, Phthalic acid, 21, 582
$C_8H_6O_4$, Terephthalic acid (form 1), 32B, 82
$C_8H_6O_4$, Terephthalic acid (form 2), 32B, 82
$C_8H_8N_2O_2$, Terephthalamide, 38B, 119
$C_8H_8O_2$, o-Toluic acid, 32B, 83
$C_8H_8O_2$, p-Toluic acid, 37B, 48
$C_8H_8O_3$, p-Methoxybenzoic acid (Anisic acid), 32B, 84; 44B, 77
$C_8H_8O_3$, 2-Hydroxy-3-methylbenzoic acid, 37B, 49
C_8H_9NO, N-Methylbenzamide, 44B, 8
C_8H_9NO, m-Methylbenzamide, 28, 412
$C_8H_9NO_2$, N-Methylanthranilic acid, 38B, 120
$C_8H_9NO_2$, 2-Amino-3-methylbenzoic acid, 28, 414
$C_8H_{12}O_8S$, 5-Methyl-3-sulfosalicylic acid dihydrate, 44B, 77
$C_8H_{16}O_{10}S$, 4-Methyl-5-sulfosalicylic acid tetrahydrate, 44B, 77
$C_9H_6O_6 \cdot 0.83\ H_2O$, Trimesic acid hydrate, 43B, 118
$C_9H_6O_6$, Benzene-1,2,4-tricarboxylic acid, 39B, 61
$C_9H_6O_6$, Benzene-1,3,5-tricarboxylic acid, 34B, 62
$C_9H_8KN_3O_9$, Methyl 2,4,6-trinitrobenzoate potassium methylate, 43B, 119
$C_9H_8N_2O_6$, Ethyl 3,5-dinitrobenzoate, 37B, 49; 43B, 120
$C_9H_8N_4O_7$, N-Methyl-2,4,6-trinitroacetanilide, 39B, 62
$C_9H_8O_3$, p-Coumaric acid, 35B, 44
$C_9H_8O_4$, Acetylbenzoyl peroxide, 41B, 90
$C_9H_8O_4$, Acetylsalicylic acid, 29, 653
$C_9H_8O_4$, Toluene-α,2-dicarboxylic acid, 37B, 50
$C_9H_9NO_3$, N-Acetylanthranilic acid, 46B, 73
$C_9H_9NO_4$, 3,5-Dimethyl-4-nitrobenzoic acid, 39B, 63
$C_9H_{10}BrNO$, p-Bromobenzoyldimethylamine, 32B, 34; 33B, 61
$C_9H_{10}N_4O_5$, N-Methyl-2-(N-methylamino)-3,5-dinitrobenzamide, 40B, 108
$C_9H_{10}O_2$, 2,3-Dimethylbenzoic acid, 37B, 51
$C_9H_{10}O_2$, 2,6-Dimethylbenzoic acid, 32B, 84
$C_9H_{10}O_3$, o-Ethoxybenzoic acid, 38B, 121
$C_9H_{10}O_3$, p-Ethoxybenzoic acid, 41B, 91
$C_9H_{10}O_4$, 2,3-Dimethoxybenzoic acid, 43B, 120
$C_9H_{10}O_8$, 1,2,3-Benzenetricarboxylic acid dihydrate (Hemimellitic acid dihydrate), 38B, 121; 39B, 64; 41B, 89
$C_9H_{11}NO_2$, N,N-Dimethylanthranilic acid, 39B, 64
$C_9H_{11}NO_2$, p-N,N-Dimethylaminobenzoic acid, 44B, 78
$C_9H_{11}NS$, N,N-Dimethylthiobenzamide, 42B, 70
$C_9H_{11}N_3O_2S$, 1-Methyl-4-salicyloyl thiosemicarbazide, 45B, 60
$C_{10}H_4O_4S_2$, Pyromellitic dithioanhydride, 38B, 162

$C_{10}H_6BrNO_4$, Propargyl 2-bromo-3-nitrobenzoate, 31B, 64
$C_{10}H_8Cl_2O_6$, Dimethyl 3,6-dichloro-2,5-dihydroxyterephthalate, 38B, 122
$C_{10}H_8O_2S$, 4,5-Dimethylphthalic thioanhydride, 20, 574
$C_{10}H_{10}O_4$, Dimethyl terephthalate, 42B, 70
$C_{10}H_{10}O_{10}$, Pyromellitic acid dihydrate, 37B, 52
$C_{10}H_{11}NO$, N-Methylcinnamide, 44B, 8
$C_{10}H_{11}N_3O_4S$, p-Carboxyphenylazocyanide dimethylsulfoxide, 41B, 92
$C_{10}H_{12}O_2$, 2,4,6-Trimethylbenzoic acid, 35B, 46; 38B, 124
$C_{10}H_{12}O_2$, 3,4,5-Trimethylbenzoic acid, 35B, 45
$C_{10}H_{12}O_3$, p-n-Propoxybenzoic acid, 46B, 74
$C_{11}H_{12}O_7S$, Trimesic acid dimethylsulfoxide solvate, 44B, 78
$C_{11}H_{14}O_3$, p-n-Butoxybenzoic acid, 41B, 93
$C_{11}H_{16}N_2O$, N,N-Diethylbenzamideoxime, 44B, 79
$C_{12}H_6O_{12}$, Mellitic acid, 24, 653; 26, 713
$C_{12}H_{14}O_4$, Diethyl terephthalate, 12, 384
$C_{12}H_{14}O_6$, Bis-(2-hydroxyethyl) terephthalate, 41B, 94
$C_{12}H_{16}N_2O_4$, Piperazinium terephthalate, 39B, 65
$C_{12}H_{16}O_3$, p-n-Pentoxybenzoic acid, 46B, 74
$C_{13}H_6Cl_5NO_3 \cdot 0.5\ C_6H_5Cl$, 3,3',5,5',6-Pentachloro-2'-hydroxysalicylanilide chlorobenzene solvate, 45B, 62
$C_{13}H_6Cl_5NO_3$, 3,3',5,5',6-Pentachloro-2'-hydroxysalicylanilide (monoclinic), 45B, 61
$C_{13}H_6Cl_5NO_3$, 3,3',5,5',6-Pentachloro-2'-hydroxysalicylanilide (triclinic), 45B, 61
$C_{13}H_7Cl_2I_2NO_2$, 2',4'-Dichloro-3,5-diiodosalicylanilide, 46B, 75
$C_{13}H_7Cl_2I_2NO_2$, 3,5-Diiodo-3',5'-dichlorosalicylanilide, 45B, 63
$C_{13}H_9ClO_2$, 2-Chlorobiphenyl-4-carboxylic acid, 34B, 63
$C_{13}H_9ClO_2$, 2'-Chlorobiphenyl-4-carboxylic acid, 34B, 64
$C_{13}H_9IO_2$, 2'-Iodobiphenyl-4-carboxylic acid, 35B, 90
$C_{13}H_9IO_2$, 3'-Iodobiphenyl-4-carboxylic acid, 38B, 124
$C_{13}H_{10}Cl_2N_2O$, N-Phenyl-2,6-dichlorobenzamide oxime, 42B, 50
$C_{13}H_{10}O_2$, Phenyl benzoate, 42B, 71
$C_{13}H_{11}NO$, Benzanilide, 45B, 64
$C_{13}H_{11}NO_2S_2$, 2-(2-Pyridylmethldithio)benzoic acid, 34B, 64
$C_{13}H_{15}ClF_3NO$, 2-Chloro-5-trifluoromethylacetanilide, 39B, 66
$C_{13}H_{16}N_2O_6$, 3,3-Dimethylbutyl 3,5-dinitrobenzoate, 42B, 71
$C_{13}H_{16}O_4$, Trimesic acid monohydrate 1,4-dioxane solvate, 44B, 79
$C_{13}H_{18}O_3$, p-n-Hexoxybenzoic acid, 46B, 75
$C_{13}H_{22}ClN_3O$, 4-Amino-N-[2-(diethylamino)ethyl]benzamide hydrochloride, 46B, 76
$C_{13}H_{23}N_2O_6P \cdot 0.05\ H_2O$, Procaine dihydrogen orthophosphate hydrate, 41B, 95
$C_{14}H_8BrIO_4$, 2-Iodo-3'-bromodibenzoyl peroxide, 45B, 64
$C_{14}H_8BrNO_2$, N-(4-Bromophenyl)-phthalimide, 42B, 71
$C_{14}H_8Br_2N_2O_7$, 2,4-Dibromo-o-cresyl 3',5'-dinitrosalicylate, 41B, 96
$C_{14}H_8Br_2O_3$, p-Bromobenzoic anhydride, 29, 655
$C_{14}H_8ClIO_4$, 2-Iodo-3'-chlorodibenzoyl peroxide, 39B, 66
$C_{14}H_8ClIO_4$, 3-Oxo-3H-2,1-benzoxiodol-1-yl m-chlorobenzoate (form 2 *α*), 40B, 106
$C_{14}H_8ClIO_4$, 3-Oxo-3H-2,1-benzoxiodol-1-yl m-chlorobenzoate (form 2 *β*), 40B, 106
$C_{14}H_8ClNO_2$, N-(p-Chlorophenyl)phthalimide, 40B, 106
$C_{14}H_8Cl_2O_2$, Biphenyl-2,2'-dicarbonyl chloride, 44B, 80
$C_{14}H_8Cl_2O_3$, p-Chlorobenzoic anhydride, 31B, 65
$C_{14}H_8Cl_2O_4$, 4,4'-Dichlorodibenzoyl peroxide, 33B, 43
$C_{14}H_9ClO_4$, 2-(4'-Chloro-2'-hydroxybenzoyl)benzoic acid, 46B, 77

$C_{14}H_9ClO_4$, 2-(5'-Chloro-2'-hydroxybenzoyl)benzoic acid, 46B, 77
$C_{14}H_{10}ClNO_3$, N-(4-Chlorophenyl)phthalamic acid, 35B, 48
$C_{14}H_{10}F_3NO_2$, 3'-Trifluoromethyldiphenylamine-2-carboxylic acid, 39B, 66
$C_{14}H_{10}O_3$, o-Benzoylbenzoic acid, 34B, 58
$C_{14}H_{10}O_4$, Dibenzoyl peroxide, 32B, 85
$C_{14}H_{11}NO_5$, (4'-Carbomethoxy-2'-nitrophenoxy)benzene, 46B, 78
$C_{14}H_{20}ClNO$, 4-Chlorodipropylacetanilide, 39B, 66
$C_{14}H_{20}O_3$, p-n-Heptoxybenzoic acid, 46B, 78
$C_{14}H_{22}ClNO_3$, 2-Diethylaminoethyl p-methoxybenzoate hydrochloride, 35B, 50
$C_{14}H_{23}Cl_2N_3O_2 \cdot H_2O$, 4-Amino-5-chloro-N-[(2-diethylamino)ethyl]-2-methoxybenzamide hydrochloride monohydrate, 46B, 79
$C_{15}H_{15}NO_2$, N-(2,3-Xylyl)anthranilic acid, 42B, 72
$C_{15}H_{20}O_3$, 2,4,6-Trimethyl-3-pivaloylbenzoic acid, 39B, 67
$C_{15}H_{22}O_2$, 3,5-Di-t-butylbenzoic acid, 42B, 72
$C_{15}H_{22}O_3$, p-n-Octoxybenzoic acid (form 1), 46B, 78
$C_{15}H_{22}O_3$, p-n-Octoxybenzoic acid (form 2), 46B, 75
$C_{16}H_8O_4$, trans-Biphthalyl, 42B, 73
$C_{16}H_{12}Cl_2O_4$, Ethylene glycol di(p-chlorobenzoate), 42B, 73
$C_{16}H_{13}Cl_4NO_4$, 2-Methyl-5-ethylpyridinium tetrachlorophthalate, 42B, 74
$C_{16}H_{14}O_2S$, trans-3-p-Tolylthiocinnamic acid, 35B, 51
$C_{16}H_{14}O_4$, Ethylene glycol dibenzoate, 42B, 74
$C_{16}H_{16}N_2O_2$, N,N'-Ethylenedibenzamide, 46B, 80
$C_{16}H_{22}CoO_{14}$, Cobalt dihydrogen diphthalate hexahydrate, 44B, 81
$C_{16}H_{24}O_3$, p-n-Nonoxybenzoic acid, 46B, 78
$C_{17}H_{14}Cl_2O_4$, Trimethylene glycol di-p-chlorobenzoate, 42B, 75
$C_{17}H_{16}O_4$, Trimethylene glycol dibenzoate, 43B, 122
$C_{17}H_{17}BrO_2$, (2'-Bromophenoxyacetyl)-4-propylbenzene, 44B, 82
$C_{17}H_{17}NO_4S$, 2-(4'-Carbomethoxy-2'-nitrothiophenyl)-1,3,5-trimethylbenzene, 41B, 97
$C_{17}H_{17}NO_5$, 2-(4'-Carbomethoxy-2'-nitrophenoxy)-1,3,5-trimethylbenzene, 41B, 98
$C_{17}H_{17}NO_5S$, 2-(4'-Carbomethoxy-2'-nitrophenylsulfinyl)-1,3,5-trimethylbenzene, 41B, 98
$C_{17}H_{17}NO_6S$, 2-(4'-Carbomethoxy-2'-nitrophenylsulphonyl)-1,3,5-trimethylbenzene, 41B, 99
$C_{17}H_{18}BrNO_2$, N-(p-Methoxyphenyl-3-propyl)-p-bromobenzamide, 34B, 58
$C_{17}H_{19}NO_3$, 2-(4'-Carboxymethoxy-2'-aminophenoxy)-1,3,5-trimethylbenzene, 45B, 65
$C_{17}H_{26}O_3$, p-n-Decyloxybenzoic acid, 45B, 66; 46B, 78
$C_{18}H_{15}NO_5$, (Z)-2-(p-Nitrobenzoyloxy)-3-phenyl-2-penten-4-one, 42B, 75
$C_{18}H_{17}NO_5 \cdot CHCl_3$, N-(3',4'-Dimethoxycinnamoyl)anthranilic acid chloroform solvate, 45B, 67
$C_{18}H_{17}NO_5$, 2-(4'-Carbomethoxy-2'-nitrobenzoyl)-1,3,5-trimethylbenzene, 41B, 100
$C_{18}H_{19}NO_4$, 2-(4'-Carbomethoxy-2'-nitrobenzyl)-1,3,5-trimethylbenzene, 41B, 100
$C_{18}H_{20}N_2O_2$, N,N'-Tetramethylenedibenzamide, 46B, 80
$C_{18}H_{22}N_2O_2$, N,N'-Dimethyl-N,N'-dibenzoylethylenediamine, 43B, 54
$C_{18}H_{24}O_7$, Diethyl 2-(α-methoxymethylene)-3-methyl-4,6-dimethoxyhomophthalate, 42B, 76
$C_{19}H_{11}Cl_2I_2NO_3$, 3'-Chloro-4'-(p-chlorophenoxy)-3,5-diiodosalicylanilide, 46B, 81
$C_{19}H_{19}NO_3$, Ethyl 4-(4'-methoxybenzylideneamino)cinnamate, 43B, 123

$C_{20}H_{14}O_4$, 2,4-Hexadiynylene dibenzoate (form I), 41B, 101
$C_{20}H_{14}O_4$, 2,4-Hexadiynylene dibenzoate (form II), 41B, 101
$C_{20}H_{16}N_2O_2$, N,N'-(p-Phenylene)dibenzamide, 43B, 124
$C_{20}H_{16}N_2O_2$, N,N'-Dibenzoyl-p-phenylenediamine, 44B, 82
$C_{20}H_{16}N_2O_2$, N,N'-Diphenylterephthalamide, 45B, 67
$C_{20}H_{22}O_4$, Hexamethylene glycol dibenzoate, 43B, 124
$C_{20}H_{31}NO_4$, Tri-t-butylmethyl p-nitrobenzoate, 44B, 83
$C_{21}H_{15}NO_3$, Tribenzamide, 43B, 124
$C_{21}H_{17}N_3OS$, Benzaldehyde 4-benzoyl-2-phenylthiosemicarbazone, 45B, 68
$C_{21}H_{20}N_2O_4$, 1,4-Bis(acylamido)butadiene, 45B, 69
$C_{21}H_{25}NO_5$, 4'-Nitrophenyl 4-octyloxybenzoate, 46B, 81
$C_{21}H_{28}I_4N_2O_8$, 3,3',5,5'-Tetraiodothyroformic acid N-diethanolamine 1:2 salt, 46B, 81
$C_{22}H_{36}O_4$, 6-n-Pentadecyl-2,4-dihydroxybenzoic acid, 41B, 102
$C_{23}H_{14}Cl_4N_2O_4$, 2-(Benzoylaminocarbonyl-methyliminomethyl-3,4,5,6-tetrachlorophenyl) benzoate, 46B, 82
$C_{23}H_{21}N_5O_8$, Ethyl N-(N'-methylbenzimidoyl)benzimidate picrate, 43B, 125
$C_{26}H_{26}O_4$, Di-n-propyl p-terphenyl-4,4''-dicarboxylate, 44B, 83
$C_{26}H_{35}N_3O_3$, (S)-N,N'-Di-t-butyl-2-(N-(1-phenylethyl)benzamido)malonamide, 43B, 126; 44B, 84
$C_{27}H_{39}NO$, N-Methyl-N-benzyl-2,4,6-tri-t-butylbenzamide, 37B, 54
$C_{28}H_{16}Br_2O_4$, meso-3,3'-Di-(p-bromophenyl)bi-3-phthalidyl, 32B, 86
$C_{28}H_{16}Cl_2O_4$, meso-3,3'-Di-(p-chlorophenyl)bi-3-phthalidyl, 34B, 65
$C_{28}H_{20}O_4$, cis-Bis[(benzoyl)oxy]stilbene, 45B, 70
$C_{30}H_{22}O_7 \cdot C_6H_6$, 2,4,6-Tribenzoyloxypropiophenone benzene solvate, 46B, 83
$C_{30}H_{22}O_7$, 2,4,6-Tribenzoyloxypropiophenone, 46B, 82

14. BENZOIC ACID SALTS (AMMONIUM, IA, IIA METALS)

$C_8F_4K_2O_4$, Potassium tetrafluorophthalate, 39B, 68
$C_8H_4K_2O_4 \cdot H_2O$, Potassium phthalate monohydrate, 46B, 83
$C_8H_5KO_4$, Potassium acid phthalate, 30B, 62
$C_8H_5NaO_4 \cdot 0.5\ H_2O$, Sodium biphthalate hemihydrate, 41B, 103; 42B, 77
$C_8H_5O_4Rb$, Rubidium acid phthalate, 41B, 104
$C_8H_6CaO_5$, Calcium phthalate monohydrate, 41B, 105; 44B, 85
$C_8H_7LiO_5$, Lithium hydrogen phthalate monohydrate, 44B, 86
$C_8H_9LiO_6$, Lithium hydrogen phthalate dihydrate, 41B, 103
$C_8H_9NO_4 \cdot 0.5\ H_2O$, Ammonium hydrogen phthalate hemihydrate, 43B, 126
$C_8H_9NO_4$, Ammonium hydrogen phthalate, 21, 583; 41B, 106
$C_8H_9NO_4$, Ammonium hydrogen terephthalate, 38B, 125
$C_8H_{10}CaO_7$, Calcium terephthalate trihydrate, 38B, 126
$C_8H_{12}N_2O_4$, Diammonium phthalate, 41B, 107
$C_9H_9LiO_5$, Lithium hydrogen phthalate methanol solvate, 44B, 86
$C_{12}H_4KN_7O_{12}$, Dipicrylamine potassium salt, 46B, 86
$C_{12}H_{20}Ca_2O_{21}$, Dicalcium dihydrogen mellitate nonahydrate, 46B, 84
$C_{14}H_9Cl_2KO_4$, Potassium hydrogen bis-m-chlorobenzoate, 38B, 127
$C_{14}H_9Cl_2KO_4$, Potassium hydrogen di-p-chlorobenzoate, 28, 416
$C_{14}H_9KN_2O_8$, Potassium hydrogen di-p-nitrobenzoate, 26, 707
$C_{14}H_9N_2O_8Rb$, Rubidium hydrogen di-o-nitrobenzoate, 26, 705
$C_{14}H_{10}CaO_6 \cdot 2\ H_2O$, Calcium salicylate dihydrate, 45B, 70
$C_{14}H_{10}O_6Sr \cdot 2\ H_2O$, Strontium salicylate dihydrate, 45B, 70
$C_{14}H_{11}KO_4$, Potassium hydrogen dibenzoate, 18, 706

$C_{14}H_{13}KO_7$, Potassium hydrogen di(p-hydroxybenzoate) monohydrate, 15, 465; 33B, 44
$C_{14}H_{15}NO_4$, Ammonium hydrogen bisbenzoate, 43B, 127
$C_{14}H_{17}NO_7$, Ammonium hydrogen disalicylate monohydrate, 18, 708
$C_{14}H_{26}N_{12}O_8 \cdot 3\ H_2O_2$, Guanidinium pyromellitate triperhydrate, 45B, 71
$C_{14}H_{34}N_{12}O_{13}$, Guanidinium pyromellitate trihydrate monoperhydrate, 44B, 87
$C_{16}H_{14}O_{10}Sr$, Strontium dihydrogen diphthalate dihydrate, 44B, 88
$C_{16}H_{15}KO_6$, Potassium hydrogen dianisate, 33B, 46
$C_{16}H_{15}O_{10}Rb$, Rubidium trihydrogen diphthalate dihydrate, 43B, 128
$C_{18}H_{15}KO_8$, Potassium hydrogen bis(acetylsalicylate), 32B, 87; 33B, 47
$C_{18}H_{15}KO_8$, Potassium hydrogen bis(homophthalate), 38B, 128
$C_{18}H_{15}O_8Rb$, Rubidium homophthalate, 38B, 129
$C_{18}H_{15}O_8Rb$, Rubidium hydrogen bis(acetylsalicylate), 32B, 89
$C_{18}H_{24}CaO_{13}$, Calcium hydrogen dihomophthalate pentahydrate, 42B, 77

15. BENZENE NITRO COMPOUNDS

$C_6Cl_5NO_2$, Pentachloronitrobenzene, 38B, 129; 40B, 107
$C_6N_6O_{12}$, Hexanitrobenzene, 31B, 66
$C_6HCl_2N_3O_6$, 1,3-Dichloro-2,4,6-trinitrobenzene, 32B, 90
$C_6H_2ClN_3O_6$, 2-Chloro-1,3,5-trinitrobenzene (Picryl chloride), 17, 706; 37B, 54
$C_6H_2F_2N_4O_6$, N,N-Difluoroamino-2,4,6-trinitrobenzene, 41B, 113
$C_6H_2IN_3O_6$, Iodo-s-trinitrobenzene, 8, 303
$C_6H_2KN_3O_7$, Potassium picrate, 18, 715; 34B, 67; 38B, 129
$C_6H_3BrN_2O_4$, 1-Bromo-2,4-dinitrobenzene, 24, 646
$C_6H_3ClN_2O_4$, 1-Chloro-2,4-dinitrobenzene, 24, 645, 646
$C_6H_3N_3O_6$, sym-Trinitrobenzene (gas-ed), 42B, 1031
$C_6H_3N_3O_6$, 1,3,5-Trinitrobenzene, 38B, 130
$C_6H_3N_3O_7$, Picric acid, 44B, 93; 46B, 98
$C_6H_4(Br,Cl)(Br,NO_2)$, p-Dibromobenzene - p-chloronitrobenzene, 40B, 1077
$C_6H_4BrNO_2$, m-Bromonitrobenzene, 28, 418
$C_6H_4ClNO_2$, m-Chloronitrobenzene, 30B, 63
$C_6H_4ClNO_2$, p-Chloronitrobenzene (gas-ed), 42B, 1031
$C_6H_4ClNO_2$, p-Chloronitrobenzene, 27, 892
$C_6H_4ClNO_4S$, 2-Nitrobenzenesulphonyl chloride, 35B, 52
$C_6H_4ClNO_5$, 3-Nitroperchlorylbenzene, 33B, 48
$C_6H_4N_2O_4$, m-Dinitrobenzene, 11, 654; 26, 695; 31B, 67
$C_6H_4N_2O_4$, o-Dinitrobenzene (gas-ed), 40B, 1139
$C_6H_4N_2O_4$, p-Dinitrobenzene (gas-ed), 40B, 1139
$C_6H_4N_2O_4$, p-Dinitrobenzene, 3, 676, 759; 11, 654; 13, 531; 46B, 84
$C_6H_5NO_2$, Nitrobenzene (gas-mw), 37B, 697
$C_6H_5NO_2$, Nitrobenzene, 23, 643
$C_6H_6N_2O_3$, 2-Amino-5-nitrophenol, 46B, 99
$C_6H_6N_4O_7$, Ammonium picrate, 34B, 68
$C_6H_{11}N_3O_{13}S$, Picrylsulphonic acid tetrahydrate, 38B, 132; 40B, 107
$C_7H_5ClN_2O_3$, p-(Chlorohydroxamoyl)nitrobenzene, 41B, 108
$C_7H_5NO_4$, o-Nitrobenzaldehyde, 29, 651
$C_7H_7NO_2$, p-Nitrotoluene, 37B, 56
$C_7H_{10}O_6S_3$, Methylidynetrithioacetic acid, 46B, 8
$C_8H_7ClN_2O_4$, 1-Chloro-3,6-dimethyl-2,4-dinitrobenzene, 45B, 71
$C_8H_7ClN_2O_4$, 2-Chloro-3,5-dinitro-p-xylene, 46B, 85

$C_8H_7NO_3$, p-Nitroacetophenone, 39B, 68
$C_8H_7NO_4$, p-Nitrophenyl acetate, 38B, 134
$C_8H_7N_3O_6$, 2,4,6-Trinitro-m-xylene, 38B, 135
$C_8H_8N_2O_3$, anti-4-Nitro-N-methylbenzaldoxime, 41B, 109
$C_8H_{12}KN_3O_{10}$, Potassium 1,1'-dimethoxy-2,4,6-trinitrobenzene dihydrate, 33B, 50
$C_9H_{11}NO_2$, Nitromesitylene, 23, 661
$C_{10}H_9NO_4$, 1-(4-Nitrophenyl)butane-1,3-dione, 42B, 78
$C_{11}H_{11}N_3O_9$, Ethyl 3-nitrato-2-nitro-3-(4-nitrophenyl)propionate, 45B, 72
$C_{11}H_{15}NO_2$, Pentamethylnitrobenzene, 41B, 110
$C_{12}F_9NO_2$, 2-Nitrononafluorobiphenyl, 44B, 88
$C_{12}H_2F_6N_2O_4S$, 3H,3'H-4,4'-Dinitrohexafluorodiphenyl sulfide, 45B, 72
$C_{12}H_6Cl_2INO_3$, 6-Chloro-2-(2-chlorophenyliodonio)-4-nitro-1-phenolate, 46B, 86
$C_{12}H_8N_2O_4$, 4,4'-Dinitrobiphenyl, 9, 375; 28, 419
$C_{12}H_9NO_2$, p-Nitrobiphenyl, 39B, 68
$C_{13}H_8N_2O_5$, Bis(4-nitrophenyl) ketone, 46B, 86
$C_{14}H_8N_2O_6$, 4,4'-Dinitrobenzil, 45B, 73
$C_{14}H_{10}N_2O_4$, 1,1-Di-(p-nitrophenyl)ethylene, 37B, 57
$C_{14}H_{11}NO_3$, 4-Acetyl-2'-nitrobiphenyl, 40B, 110
$C_{14}H_{14}ClN_3O_6$, Methyl (2E,4Z)-5-(dimethylamino)-2-nitro-4-(4-chloro-2-nitrophenyl)-2,4-pentadienoate, 45B, 74
$C_{15}H_{10}BrNO_3$, 3-Bromo-2'-nitrochalcone, 35B, 54
$C_{15}H_{10}BrNO_3$, 4-Bromo-2'-nitrochalcone, 35B, 54
$C_{15}H_{10}ClNO_3$, 3-Chloro-2'-nitrochalcone, 35B, 54
$C_{15}H_{12}N_6O_9$, N,N'-Dimethyl-N,N'-di(2,4-dinitrophenyl)urea, 41B, 112
$C_{15}H_{14}N_4O_5$, N,N'-Dimethyl-N,N'-di(p-nitrophenyl)urea, 41B, 112
$C_{16}H_8N_2O_4$, Bis(p-nitrophenyl)butadiyne, 45B, 74
$C_{18}H_{13}NO_2S_2$, 2,4-Bis(phenylthio)nitrobenzene (1-phenyl-6-(phenylthio)-3H-2,1,3-benzoxathiazolium-3-olate), 44B, 88
$C_{25}H_{23}N_5O_6$, 3,5-Dimethyl-1,7-diphenyl-4-(2,4,6-trinitrophenyl)-2,6-diazahepta-2,4-diene, 46B, 87
$C_{27}H_{22}N_2O_6$, 2,2-Bis[4-(p-nitrophenoxy)phenyl·propane, 46B, 109

16. ANILINES

$C_6H_2Cl_5N$, Pentachloroaniline, 34B, 69
$C_6H_3F_2N_3O_4$, (N,N-Difluoroamino)-2,4-dinitrobenzene, 42B, 78
$C_6H_3N_5O_8$, 2,4,3,6-Tetranitroaniline, 31B, 68
$C_6H_4Br_3N$, 2,4,6-Tribromoaniline, 34B, 70
$C_6H_4Cl_2N_2O_2$, 2,6-Dichloro-4-nitroaniline, 37B, 58
$C_6H_4Cl_2N_2O_2$, 3,5-Dichloro-4-nitroaniline, 19, 561
$C_6H_4Cl_3N$, 2,4,6-Trichloroaniline, 37B, 59
$C_6H_4N_4O_6$, 2,4,6-Trinitroaniline, 39B, 69
$C_6H_5ClN_2O_2$, 2-Chloro-4-nitroaniline, 30B, 68
$C_6H_5Cl_2N$, 2,5-Dichloroaniline, 28, 420
$C_6H_5N_5O_6$, 1,3-Diamino-2,4,6-trinitrobenzene (form 1), 32B, 91
C_6H_6ClN, p-Chloroaniline, 31B, 69
$C_6H_6N_2O_2$, γ-o-Nitroaniline, 44B, 89
$C_6H_6N_2O_2$, m-Nitroaniline, 39B, 69
$C_6H_6N_2O_2$, p-Nitroaniline (gas-ed), 42B, 1031
$C_6H_6N_2O_2$, p-Nitroaniline, 11, 658; 20, 563; 26, 697; 34B, 69
$C_6H_6N_6O_6$, 1,3,5-Triamino-2,4,6-trinitrobenzene, 30B, 69
C_6H_7N, Aniline (gas-mw), 40B, 1146
$C_6H_7NO_3S$, Aniline m-sulphonic acid, 30B, 65

$C_6H_7NO_3S$, Orthanilic acid, 32B, 91
C_6H_8BrN, Aniline hydrobromide, 11, 638; 26, 698
C_6H_8ClN, Aniline hydrochloride, 12, 386
$C_6H_8ClNO_2$, 2-Aminoresorcinol hydrochloride, 37B, 59
$C_6H_8N_2$, p-Phenylenediamine, 41B, 115
$C_6H_8N_2O_2S$, Sulfanilamide (α form), 30B, 66
$C_6H_8N_2O_2S$, Sulfanilamide (β form), 30B, 66; 32B, 93
$C_6H_8N_2O_2S$, Sulfanilamide (γ form), 30B, 66
$C_6H_9BrF_3NOSi_{0.5}$, p-Bromoaniline dihydrogen hexafluorosilicate monohydrate, 37B, 60
$C_6H_9ClN_2$, 1,2-Diaminobenzene monohydrochloride, 42B, 79
$C_6H_9NO_4S$, Sulfanilic acid monohydrate, 27, 894
$C_6H_{10}Br_2N_2$, o-Phenylenediamine dihydrobromide, 38B, 135
$C_6H_{10}Cl_2N_2$, o-Phenylenediamine dihydrochloride, 40B, 112
$C_6H_{10}Cl_2N_2$, p-Phenylenediamine dihydrochloride, 34B, 72; 43B, 129
$C_6H_{10}NNaO_5S$, Sodium sulfanilate dihydrate, 41B, 114; 43B, 130
$C_6H_{10}N_2O_3S$, Sulfanilamide monohydrate, 33B, 55
$C_7H_5N_5O_8$, N-Methyl-N,2,4,6-tetranitroaniline, 32B, 94
$C_7H_6N_2S$, 1-Amino-4-thiocyanatobenzene, 30B, 64; 42B, 79; 43B, 130
$C_7H_7BrN_2O_2$, 4-Bromo-3-nitro-N-methylaniline, 37B, 56
C_7H_8BrN, 2-Amino-5-bromotoluene, 38B, 157
$C_7H_8NNaO_3S \cdot H_2O$, Sodium anilinomethanesulphonate monohydrate, 45B, 75
C_7H_9N, p-Toluidine, 3, 678, 758; 28, 422
C_7H_9NO, p-Anisidine, 41B, 116
$C_7H_9NO_2S$, Methanesulphonanilide, 33B, 55
$C_7H_9NO_3S$, 2'-Hydroxymethanesulphonanilide, 35B, 80
$C_7H_{10}ClN$, o-Toluidine hydrochloride, 42B, 80
$C_7H_{10}N_2$, 2,4-Diaminotoluene, 46B, 88
$C_7H_{11}F_2N$, p-Toluidinium bifluoride, 37B, 60
$C_7H_{12}F_3NOSi_{0.5}$, p-Toluidine dihydrogen hexafluorosilicate monohydrate, 37B, 60
$C_8H_4F_3N_5O_8$, N-(β,β,β-Trifluoroethyl)-N,2,4,6-tetranitroaniline, 34B, 71
$C_8H_6F_3N_3O_4$, N-Methyl-2,6-dinitro-4-trifluoromethylaniline, 43B, 131
$C_8H_6IN_3$, (Z)-2-Cyano-2-phenylhydrazonoacetonitrile, 35B, 59
$C_8H_7ClN_2O_2$, 2-Chloroisonitrosoacetanilide, 40B, 112
C_8H_8BrNO, p-Bromoacetanilide, 33B, 59
$C_8H_8BrN_3O_4$, N,3-Dimethyl-4-bromo-2,6-dinitroaniline, 32B, 96
C_8H_8ClNO, p-Chloroacetanilide, 31B, 70
$C_8H_8Cl_2N_2O_2$, 2,6-Dichloro-4-nitro-N,N-dimethylaniline, 24, 647
$C_8H_8N_2O_2$, Isonitrosoacetanilide, 39B, 69
$C_8H_9FN_2O_2$, 2-Fluoro-4-nitro-N,N-dimethylaniline, 43B, 131
C_8H_9NO, Acetanilide, 11, 638; 18, 705; 31B, 70
C_8H_9NO, p-Aminoacetophenone, 42B, 80
C_8H_9NOS, Thioacetanilide-S-oxide, 34B, 73
C_8H_9NOS, 2-Methoxythioformanilide, 43B, 132
$C_8H_9NO_2$, p-Hydroxyacetanilide, 40B, 113; 42B, 88
$C_8H_{10}N_2O$, p-Aminoacetanilide, 43B, 133
$C_8H_{10}N_2O_2$, m-Nitro-N,N-dimethylaniline, 41B, 117
$C_8H_{10}N_2O_2$, p-Nitro-N,N-Dimethylaniline, 30B, 68
$C_8H_{10}N_2O_2$, 4-Methyl-3-nitro-N-methylaniline, 37B, 56
$C_8H_{11}N$, N,N-Dimethylaniline (gas-ed), 30B, 403
$C_8H_{12}BrN_2$, N,N-Dimethyl-p-phenylenediamine bromide, 33B, 60
$C_8H_{13}ClN_2O_2$, N,N-Dimethyl-p-nitrosoaniline hydrochloride hydrate, 40B, 114
$C_9H_9ClN_2O_3$, N-Ethyl-N-p-nitrophenylcarbamoyl chloride, 37B, 61

C_9H_9NO, Phenyl-trimethine-merocyanine, 45B, 75
$C_9H_{10}N_2O$, 2-Oxopropionaldehyde phenylhydrazone, 42B, 42
$C_9H_{10}N_2O_3$, 2-Methoxyisonitrosoacetanilide, 42B, 80
$C_9H_{10}N_2O_3$, 4-Methoxyisonitrosoacetanilide, 40B, 114
$C_9H_{11}NO$, N-Methylacetanilide, 32B, 96
$C_9H_{11}NO$, p-Methylacetanilide (monoclinic form), 42B, 81; 43B, 133
$C_9H_{11}NO$, p-Methylacetanilide (orthorhombic form), 43B, 133
$C_9H_{11}NO$, 4-Isopropylideneaminophenol, 17, 704
$C_9H_{11}NO_2$, N-Phenylurethane, 37B, 61
$C_9H_{11}NO_2$, p-Acetanisidide, 46B, 100
$C_9H_{12}BrNO$, p-Dimethylaminobenzaldehyde hydrobromide, 39B, 70
$C_9H_{12}N_2O$, α-p-Dimethylaminobenzaldoxime, 38B, 136
$C_9H_{13}NO_3S$, Methyl-p-dimethylaminobenzenesulphonate, 41B, 118
$C_{10}H_8ClN_3O$, 3-Amino-3-chloro-2-cyano-acrylic anilide, 41B, 118
$C_{10}H_{10}N_2OS$, β-Thiocyanatopropionanilide, 46B, 397
$C_{10}H_{11}BrN_2O_2$, p-Dimethylamino-β,β-bromonitrostyrene, 40B, 125
$C_{10}H_{11}ClN_2O_2$, 1-(2-Chloro-4-dimethylaminophenyl)-2-nitroethylene, 39B, 70
$C_{10}H_{11}Cl_2NO$, 3,4-Dichloroisobutyranilide, 42B, 81
$C_{10}H_{11}FN_4O_6$, 1-(N-Fluoro-N-t-butyl)amino-2,4,6-trinitrobenzene, 40B, 109
$C_{10}H_{12}N_2O_3$, 2-Ethoxyisonitrosoacetanilide, 39B, 70
$C_{10}H_{12}N_2O_3$, 4-Ethoxyisonitrosoacetanilide, 39B, 70
$C_{10}H_{13}ClN_2$, N'-(4-Chloro-o-tolyl)-N,N-dimethylformamidine, 46B, 89
$C_{10}H_{14}N_2O$, N,N-Diethyl-p-nitrosoaniline, 43B, 134
$C_{10}H_{15}N_3O$, N,N,N',N'-Tetramethyl-1,5-diamino-4-nitrosobenzene, 42B, 49
$C_{10}H_{16}IN_2$, N,N,N',N'-Tetramethyl-p-diaminobenzene iodide, 33B, 61
$C_{10}H_{16}N_2$, N,N,N',N'-Tetramethyl-p-phenylenediamine, 45B, 76
$C_{10}H_{16}N_2O_4S_2$, N,N,N',N'-Tetramethyl-p-phenylenediamine - bis(sulfur dioxide), 41B, 119
$C_{10}H_{17}ClN_2O_4$, N,N,N',N'-Tetramethyl-p-diaminobenzene perchlorate, 38B, 137, 138
$C_{11}H_{12}F_3N_3O_4$, N-t-Butyl-2,6-dinitro-4-trifluoromethylaniline, 41B, 120
$C_{11}H_{14}N_2O_2$, 3-Methyl-N-ethyl-isonitrosoacetanilide, 40B, 115
$C_{11}H_{16}BF_4NO_3$, (2'-(2-Formyl-4-nitroanilino)ethyl)dimethylammonium tetrafluoroborate, 43B, 135
$C_{12}H_7IN_4O_6$, N-Picryl-p-iodoaniline, 12, 395
$C_{12}H_7N_5$, 4-Ethyl-2,3,5,6-tetracyanoaniline, 41B, 121
$C_{12}H_8N_4O_6$, 2,4,6-Trinitrodiphenylamine, 39B, 71
$C_{12}H_9BrClNO_2S$, N-p-Bromophenyl-p-chlorobenzenesulphonamide, 34B, 620
$C_{12}H_9Cl_2N$, p,p'-Dichlorodiphenylamine, 24, 654; 26, 727
$C_{12}H_{10}BrNO_2S$, N-p-Bromophenylbenzenesulphonamide, 32B, 69; 34B, 622
$C_{12}H_{10}Cl_2N_2$, 2,2'-Dichlorobenzidine, 11, 663
$C_{12}H_{10}Cl_2N_2$, 3,3'-Dichlorobenzidine (gas-ed), 12, 391
$C_{12}H_{10}Cl_2N_2$, 4,4'-Diamino-3,3'-dichlorobiphenyl, 33B, 64
$C_{12}H_{10}N_2O_2$, 2'-Nitro-4-aminobiphenyl, 40B, 109
$C_{12}H_{11}ClN_2O_2S$, N(1)-2-Chlorophenylsulfanilamide, 42B, 81
$C_{12}H_{12}N_2$, 2,2'-Diaminobiphenyl, 43B, 135
$C_{12}H_{14}N_2O_4$, N-Acetylkynurenine, 45B, 77
$C_{12}H_{15}BrN_2O_2$, 2-(Isopropylideneaminooxy)propion-p-bromoanilide, 42B, 50
$C_{12}H_{16}ClN_3O_4$, N,N-Di-1-propyl-2,6-dinitro-4-chloroaniline, 40B, 115
$C_{12}H_{16}Cl_4CuN_2$, Di-anilinium tetrachlorocuprate(II), 40B, 116
$C_{12}H_{16}N_2O_2$, N-Isobutylisonitrosoacetanilide, 43B, 136
$C_{12}H_{19}Cl_3N_2O$, 1-(4-Amino-3,5-dichlorophenyl)-2-t-butylaminoethanol

hydrochloride, 46B, 89
$C_{13}H_7Cl_2I_2NO_2$, 3,5-Diiodo-3',4'-dichlorosalicylanilide, 44B, 90
$C_{13}H_8N_4O_4$, Bis-p-nitrophenylcarbodi-imide, 38B, 139
$C_{13}H_9Br_2N$, N-(p-Bromobenzylidene)-p-bromoaniline, 41B, 122
$C_{13}H_9Cl_2N$, N-(p-Chlorobenzylidene)-p-chloroaniline, 38B, 139; 42B, 82
$C_{13}H_9Cl_2N$, N-(2,4-Dichlorobenzylidine)aniline, 38B, 140
$C_{13}H_{10}BrNO$, 2-Amino-4'-bromobenzophenone, 32B, 99
$C_{13}H_{10}BrNO$, 2-Bromo-N-(salicylidene)aniline, 34B, 73
$C_{13}H_{10}ClNO$, Diphenylcarbamoyl chloride, 39B, 71
$C_{13}H_{10}ClNO$, N-(5-Chlorosalicylidene)aniline, 29, 660
$C_{13}H_{10}ClNO$, 2-Amino-4'-chlorobenzophenone, 32B, 99
$C_{13}H_{10}ClNO$, 2-Chloro-N-(salicylidene)aniline, 29, 660
$C_{13}H_{10}N_4$, p-Tricyanovinyl-N,N-dimethylaniline, 42B, 34
$C_{13}H_{11}ClN_2$, Benzoyl chloride phenylhydrazone, 42B, 43
$C_{13}H_{11}N$, Benzylideneaniline, 35B, 61
$C_{13}H_{11}NO$, Salicylidineaniline, 44B, 90
$C_{13}H_{11}N_3O_3S$, (E)-4-Nitrobenzenediazo-4'-methoxythiophenolate, 42B, 44
$C_{13}H_{12}BrNO_2S$, N-(4-Bromo-2-methylphenyl)benzenesulphonamide, 34B, 623
$C_{13}H_{12}N_2O$, 4,4'-Diaminobenzophenone, 46B, 90
$C_{13}H_{13}N_3$, Diphenyl guanidine, 46B, 90
$C_{13}H_{15}NO_2$, 6-Methyl-4-p-methylphenylamino-5,6-dihydro-2-pyrone, 46B, 91
$C_{14}H_{10}Cl_2N_2S_2$, p,p'-Dichlorodithiooxanilide, 45B, 77
$C_{14}H_{11}I_2NO_2$, 4-Aldehydo-4'-methoxy-2,6-diiododiphenylamine, 42B, 82
$C_{14}H_{11}I_2NO_3$, 4-Carboxy-4'-methoxy-2,6-diiododiphenylamine, 42B, 83
$C_{14}H_{11}NO_2$, Benzylideneaniline-p-carboxylic acid, 35B, 61
$C_{14}H_{12}N_2O_2$, p-Methylbenzylidene-p-nitroaniline, 35B, 61
$C_{14}H_{12}N_2O_2$, p-Nitro-p'-methylbenzylideneaniline (monoclinic form), 43B, 136
$C_{14}H_{12}N_2O_2$, p-Nitro-p'-methylbenzylideneaniline (triclinic form), 43B, 136
$C_{14}H_{12}N_2O_3$, 4-Nitro-4'-methoxy-benzalaniline, 38B, 141
$C_{14}H_{13}I_2NO_2$, 4-Methanol-4'-methoxy-2,6-diiododiphenylamine, 42B, 83
$C_{14}H_{14}BrNO_2S$, N-(4-Bromo-2,6-dimethylphenyl)benzenesulphonamide, 34B, 624
$C_{14}H_{14}N_4S$, S-Methyldithizone, 41B, 123
$C_{14}H_{16}N_2$, m-Tolidine, 16, 523
$C_{14}H_{16}N_2$, 4,4'-Diamino-3,3'-dimethylbiphenyl, 33B, 64
$C_{14}H_{18}Cl_2N_2$, m-Tolidine hydrochloride, 13, 542
$C_{14}H_{19}Cl_2NO_2$, p-(Di-2-chloroethyl)aminophenylbutyric acid, 41B, 123
$C_{14}H_{19}NO_2S_4$, N,N-Di(ethoxythiocarbonylthiomethyl)aniline, 44B, 90
$C_{14}H_{20}Cl_6N_2Re$, p-Toluidinium hexachlorohenate(IV), 32B, 98
$C_{14}H_{22}N_2O$, Lidocaine, 40B, 116
$C_{14}H_{23}AsF_6N_2O$, Lidocaine hydrohexafluoroarsenate, 38B, 141
$C_{14}H_{25}ClN_2O_2$, Lidocaine hydrochloride monohydrate, 38B, 142
$C_{15}H_{12}ClNO_2$, p-Tolyl-glyoxylic acid p-chloroanilide, 45B, 78
$C_{15}H_{13}I_2NO_3$, 4-Carbomethoxy-4'-methoxy-2,6-diiododiphenylamine, 42B, 83
$C_{15}H_{14}ClNO_4$, N,O-Diacetyl-5-p-chloroanilino-4-hydroxy-penta-2,4-dienal, 40B, 117
$C_{15}H_{14}N_2$, Di-p-tolylcarbodi-imide, 38B, 143
$C_{15}H_{15}N$, N-(p-Methylbenzylidene)-p-methylaniline (form II), 43B, 137
$C_{15}H_{15}N$, p-(N-Methylbenzylidene)-p-methylaniline (form III), 42B, 84
$C_{15}H_{15}N_3O_2$, p-Dimethylaminobenzylidene-p-nitroaniline, 42B, 84

$C_{15}H_{15}N_3O_2$, p-Nitrobenzylidene-p-dimethylaminoaniline, 42B, 84
$C_{15}H_{16}N_2O_4S$, p-Dimethylaminobenzyl-p-nitrophenyl-sulphone, 44B, 100
$C_{15}H_{17}ClN_4O_2$, p-Dimethylamino-benzaldehyde-p-nitrophenylhydrazone hydrochloride, 42B, 46
$C_{15}H_{23}NS_2$, 2,4-Di-t-butyl-6-methyl-N-thiosulfinylaniline, 45B, 78
$C_{15}H_{26}ClFN_2O_4S$, 2'-Hydroxy-5'-(1-hydroxy-2-(3"-pentylamino)propyl)-fluoromethanesulphonanilide hydrochloride, 44B, 91
$C_{16}H_{12}F_3NO_2$, p-Tolyl-glyoxylic acid m-trifluoroanilide, 46B, 91
$C_{16}H_{14}N_2O$, p-[(p'-Ethoxybenzylidene)amino]benzonitrile, 46B, 92
$C_{16}H_{15}NO_3$, p-[(p'-Methoxybenzylidene)amino]phenylacetate, 46B, 92
$C_{16}H_{18}N_4O_4$, 2,4-Dinitro-4'-diethylaminodiphenylamine, 40B, 118
$C_{16}H_{21}ClN_3O_3$, 4,4'-Bis(dimethylamino)diphenylamine chlorate, 32B, 71; 37B, 62
$C_{17}H_{10}N_4$, p-Tricyanovinyl diphenylamine, 44B, 91
$C_{17}H_{15}BrN_2$, 2-Bromo-4'-dimethylamino-α-cyanostilbene, 24, 659
$C_{17}H_{16}N_2S$, 3,3-Bis(N-methyl-N-phenylamino)propadiene-1-thione, 43B, 52
$C_{17}H_{18}N_2O_2S$, N,N'-Diacetyl-3-methylthiobenzidine, 46B, 92
$C_{17}H_{18}N_2O_4$, Dimethyl 4,4'-methylenebis(phenylcarbamate), 46B, 93
$C_{17}H_{19}N_3O_6S$, p-Dimethylaminophenyl N-methyl-N-(p-nitrophenylsulphonylmethyl)-carbamate, 44B, 92
$C_{17}H_{22}ClN_3O_4$, Auramine perchlorate, 32B, 100
$C_{17}H_{23}IN_2$, (2-Diphenylaminoethyl)trimethylammonium iodide, 42B, 24; 43B, 139
$C_{18}Cl_{15}N$, Perchlorotriphenylamine, 46B, 93
$C_{18}H_{15}N$, Triphenylamine (gas-ed), 23, 669
$C_{18}H_{16}IN_2$, N,N'-Diphenyl-p-phenylenediamine iodide, 32B, 72
$C_{18}H_{16}N_2$, N,N'-Diphenyl-m-phenylenediamine, 43B, 140
$C_{18}H_{16}N_2$, N,N'-Diphenyl-p-phenylenediamine, 42B, 85
$C_{18}H_{18}BrCl_3N_2O_2$, (O-1-Methyl-p-bromoacetanilido)-3,4-dimethyl-4-trichloromethyl-2,5-cyclohexadiene-1-oxime, 46B, 94
$C_{18}H_{29}NOS$, 2,4,6-Tri-t-butyl-n-sulfinylaniline, 46B, 94
$C_{19}H_{15}N$, N-(Diphenylmethylene)aniline, 41B, 124
$C_{19}H_{23}ClN_4O_3S$, 2'-Acetamido-2-chloro-4'-diethylamino-4-mesylazobenzene, 46B, 117
$C_{19}H_{23}NO$, p-Ethoxybenzylidene-p-n-butylaniline, 46B, 94
$C_{19}H_{28}BrNO_5S$, (-)-N-Ethyl-N-methylaniline oxide (+)-3-bromocamphorsulphonate, 38B, 143
$C_{20}H_{14}BrN$, N-p-Bromophenyl-2,2-diphenylvinylideneamine, 35B, 64
$C_{20}H_{25}ClN_4O_3S$, 2-Chloro-4'-diethylamino-4-mesyl-2'-propionamidoazobenzene, 46B, 117
$C_{20}H_{25}NO_2$, p-Propoxysalicylidene-p'-butylaniline, 45B, 79, 93
$C_{20}H_{27}BrN_2O$, (+)-N-((2-Benzylmethylamino)propyl) propionanilide hydrobromide, 34B, 75
$C_{21}H_{16}ClN_3O_6S$, cis-(p-Nitrobenzoyl)-(p-chloro-o-phenylsulphonylhydrazono)methyl formate, 41B, 125
$C_{21}H_{17}N$, N-p-Tolyl-2,2-diphenylvinylideneamine, 35B, 64
$C_{22}H_{20}N_2O_4$, 2,4-Hexadiyne-1,6-bis(m-tolylurethane) (orange phase), 46B, 95
$C_{22}H_{21}N$, N-Mesitylbenzophenoneimine, 45B, 79
$C_{22}H_{26}N_2O$, N-p-Cyanobenzylidene-p-n-octyloxyaniline, 43B, 151
$C_{23}H_{14}N_4$, p-Tricyanovinyltriphenylamine (monoclinic form), 42B, 35
$C_{23}H_{14}N_4$, p-Tricyanovinyltriphenylamine (orthorhombic form), 42B, 35
$C_{24}H_{20}N_2$, Tetraphenylhydrazine, 41B, 126
$C_{24}H_{32}N_2O_7$, 1,11-Bis(2-acetylaminophenoxy)-3,6,9-trioxaundecane, 46B, 108
$C_{25}H_{21}N_3$, 4,4-Bis(N-methyl-N-phenylamino)-1-phenylbutatriene-1-car-

bonitrile, 43B, 57
$C_{26}H_{24}N_6$, 1,1',3,3'-Tetraphenyloxaldiamidrazone, 41B, 127
$C_{26}H_{25}NO_2$, Isobutyl 4-(4'-phenylbenzylideneamino)cinnamate, 46B, 95
$C_{26}H_{28}N_2O_4$, Bis(phenylcarbamoyloxy-n-butyl)diacetylene, 46B, 95
$C_{26}H_{31}N_4O_9P$, Lidocaine bis-p-nitrophenylphosphate, 41B, 128
$C_{28}H_{26}N_2$, cis-α,β-Bis(N-methylanilino)stilbene, 44B, 92
$C_{28}H_{32}N_2$, Bis-p-butylanilino-terephthalate, 43B, 141
$C_{29}H_{34}Cl_4N_6$, γ,γ-Dicyano-bis-(N-t-butyl-N'-2,6-dichlorophenyl)-pimelamidine, 46B, 43
$C_{30}H_{23}NO_2$, Methyl N-(biphenyl-2-yl)-N-(1-naphthyl)-anthranilate (α isomer), 46B, 1344
$C_{30}H_{23}NO_2$, Methyl N-(biphenyl-2-yl)-N-(1-naphthyl)-anthranilate (β_1 isomer), 46B, 1344
$C_{30}H_{23}NO_2$, Methyl N-(biphenyl-2-yl)-N-(1-naphthyl)-anthranilate (β_2 isomer), 46B, 1344
$C_{31}H_{25}N$, Diphenylaminotriphenylmethane, 41B, 152
$C_{32}H_{40}N_2O_2$, Bis(4'-p-hexylhydroxybenzylidene)-1,4-phenylenediamine, 46B, 97
$C_{36}H_{27}ClNO_4$, Tri(p-biphenylyl)aminium perchlorate, 43B, 141
$C_{38}H_{44}Cl_4N_4O_{12}S_2$, Bis(2-(3-dimethylammoniopropyl)amino)-4,4'-dichlorodiphenylsulfide oxalate oxalic acid, 45B, 80

17. PHENOLS AND ETHERS

$C_6D_6O_2$, Resorcinol (α form), 46B, 97
$C_6D_6O_2$, Resorcinol (β form), 46B, 97
C_6HCl_5O, Pentachlorophenol, 27, 903
$C_6H_2Cl_4O_2$, Tetrachlorohydroquinone, 27, 901; 32B, 101
$C_6H_3BrN_2O_5$, 2-Bromo-4,6-dinitrophenol, 41B, 129
$C_6H_3ClN_2O_5$, 2-Chloro-4,6-dinitrophenol, 41B, 130
$C_6H_4ClNO_3$, 2-Nitro-4-chlorophenol, 42B, 85
$C_6H_4Cl_2O$, 2,3-Dichlorophenol, 42B, 85
$C_6H_4Cl_2O$, 2,4-Dichlorophenol, 45B, 81
$C_6H_4Cl_2O$, 2,5-Dichlorophenol, 39B, 71
$C_6H_4Cl_2O$, 2,6-Dichlorophenol, 40B, 119
$C_6H_4Cl_2O$, 3,4-Dichlorophenol, 46B, 99
$C_6H_4Cl_2O$, 3,5-Dichlorophenol, 39B, 72
$C_6H_4KNO_3 \cdot 0.5\ H_2O$, Potassium o-nitrophenolate hemihydrate, 24, 639; 26, 704; 41B, 107
$C_6H_4N_2O_5$, 2,4-Dinitrophenol, 42B, 85; 43B, 142
$C_6H_4N_2O_5$, 2,6-Dinitrophenol, 42B, 86
C_6H_5BrO, 4-Bromophenol, 44B, 93
C_6H_5ClO, p-Chlorophenol, 39B, 73
$C_6H_5NO_3$, m-Nitrophenol, 41B, 131
$C_6H_5NO_3$, o-Nitrophenol, 44B, 94
$C_6H_5NO_3$, p-Nitrophenol (α form), 30B, 71
$C_6H_5NO_3$, p-Nitrophenol (β form), 18, 701; 30B, 71
C_6H_6ClNO, 2-Amino-4-chlorophenol, 45B, 82
$C_6H_6O \cdot 0.5\ H_2O$, Phenol hemihydrate, 24, 632
C_6H_6O, Phenol, 24, 632; 28, 423; 32B, 102
$C_6H_6O_2$, Quinol (α-form), 46B, 111
$C_6H_6O_2$, Catechol, 31B, 71; 37B, 63
$C_6H_6O_2$, Hydroquinone (monoclinic, γ form), 31B, 72
$C_6H_6O_2$, Resorcinol (α), 4, 266, 297; 20, 570; 39B, 74; 46B, 97
$C_6H_6O_2$, Resorcinol (β), 6, 207, 243; 46B, 97
$C_6H_6O_2$, o-Cresol, 38B, 145

$C_6H_6O_3$, Phloroglucinol (gas-ed), 9, 370; 10, 259
$C_6H_6O_3$, Phloroglucinol, 30B, 77
C_6H_7NO, m-Aminophenol, 40B, 120
C_6H_7NO, p-Aminophenol, 15, 461
C_6H_7NO, 2-Aminophenol, 45B, 82
C_6H_8ClNO, o-Aminophenol hydrochloride, 30B, 70
$C_6H_{10}NNaO_5$, Sodium p-nitrosophenolate trihydrate, 41B, 131
$C_6H_{10}O_5$, Phloroglucinol dihydrate, 21, 578
$C_7H_5Cl_3O_2$, 3,4,5-Trichloroguaiacol, 46B, 99
C_7H_5NO, p-Cyanophenol, 43B, 142
C_7H_6O, Salicylaldehyde, 17, 701
$C_7H_6O_2$, p-Hydroxybenzaldehyde, 43B, 143
C_7H_7BrO, 2-Methyl-3-bromophenol, 31B, 73; 39B, 77
$C_7H_7NO_2$, p-Nitrosoanisole, 45B, 83
$C_7H_7NO_3$, p-Nitroanisole, 44B, 95
C_7H_8O, Anisole (gas-ed), 39B, 887
C_7H_8O, m-Cresol, 39B, 74
C_7H_8O, p-Cresol (metastable form), 40B, 120
C_7H_8O, p-Cresol, 31B, 74; 35B, 65
$C_8H_5Cl_3O_3$, 2,4,5-Trichlorophenoxyacetic acid, 42B, 5
$C_8H_5Cl_3O_3$, 2,4,6-Trichlorophenoxyacetic acid, 43B, 10
$C_8H_6Br_4O_2$, 1,4-Dimethoxy-2,3,5,6-tetrabromobenzene, 44B, 95
$C_8H_6Cl_2O_3$, 2,4-Dichlorophenoxyacetic acid, 42B, 87
$C_8H_6Cl_2O_3$, 2,5-Dichlorophenoxyacetic acid, 41B, 5
$C_8H_6Cl_4O_2$, 2,3,5,6-Tetrachloro-1,4-dimethoxybenzene, 46B, 100
$C_8H_7Br_2NO_3$, 2,6-Dibromohydroquinone-3-acetamide, 41B, 132
$C_8H_7ClO_3$, 2-Chlorophenoxyacetic acid, 43B, 143
$C_8H_7NO_5$, p-Nitrophenoxyacetic acid, 46B, 8
$C_8H_7N_3O_7$, 2,4,6-Trinitrophenetole, 33B, 49
$C_8H_8ClNO_2$, 2-Chlorophenyl-methyl-carbamate, 41B, 133
$C_8H_8O_2$, p-Hydroxyacetophenone, 40B, 124
$C_8H_8O_3$, 2-Hydroxy-3-methoxybenzaldehyde, 42B, 87
$C_8H_9NO_3$, 5-Methoxysalicylaldoxime, 44B, 96
$C_8H_{10}O$, 2,3-Dimethylphenol, 31B, 73; 32B, 102; 39B, 77
$C_8H_{10}O$, 2,5-Dimethylphenol, 30B, 74; 39B, 77
$C_8H_{10}O$, 2,6-Dimethylphenol, 39B, 75
$C_8H_{10}O$, 3,4-Dimethylphenol, 39B, 76
$C_8H_{10}O_2$, 1,4-Dimethoxybenzene, 13, 535
$C_8H_{11}NO_3$, Noradrenaline, 41B, 134
$C_8H_{12}ClNO$, Tyramine hydrochloride, 45B, 84
$C_8H_{12}ClNO_2$, Dopamine hydrochloride, 46B, 32
$C_8H_{12}ClNO_2$, β-(2-Hydroxyphenyl)ethanolamine hydrochloride, 45B, 84
$C_9H_7Cl_3O_3$, 2-(2,4,5-Trichlorophenoxy)propanoic acid, 43B, 145
$C_9H_8Cl_2O_3$, DL-2-(3,5-Dichlorophenoxy)propanoic acid, 44B, 96
$C_9H_8O_3$, p-Hydroxy-trans-cinnamic acid, 46B, 8
$C_9H_8O_3$, 2-Coumaric acid, 45B, 85
$C_9H_{12}O$, 4-Isopropylphenol, 43B, 145
$C_9H_{12}O_3$, 1,3,5-Trimethoxybenzene, 45B, 86
$C_9H_{13}NO_3$, Adrenaline, 41B, 134
$C_9H_{13}NO_3S \cdot 0.5\ H_2O$, (DL)-Epinine β-sulphonate hemihydrate, 46B, 33
$C_9H_{13}NO_5S \cdot 0.5\ H_2O$, 1-(3,4-Dihydroxyphenyl)-2-methylaminoethanesulphonic acid hemihydrate, 45B, 86
$C_{10}H_8Cl_3NO_5$, α-Chloro-α-chlorosulphenyl-4-nitro-2,5-dimethoxyphenylacetyl chloride, 32B, 103
$C_{10}H_9NO_5$, Methyl 4-hydroxy-3-nitro-trans-cinnamate, 41B, 135
$C_{10}H_{10}O_4$, Methyl 3-(3,4-dihydroxyphenyl)-2-propenoate, 45B, 87
$C_{10}H_{10}O_4$, 2,5-Diacetylhydroquinone, 44B, 97

$C_{10}H_{11}ClO_3$, (DL)-2-(4-Chloro-2-methylphenoxy)propionic acid, 46B, 9
$C_{10}H_{11}NO_4$, (E)-1-(4'-Hydroxy-3'-methoxyphenyl)-2-nitropropene, 46B, 101
$C_{10}H_{12}CsN_3O_8$, 2,4,6-Trinitrophenetole - caesium ethoxide, 33B, 52
$C_{10}H_{12}KN_3O_8$, 2,4,6-Trinitrophenetole - potassium ethoxide, 33B, 52
$C_{10}H_{12}O_4$, 3,6-Dihydro-2,5-diacetylhydroquinone, 44B, 97
$C_{10}H_{14}O$, 2-Isopropyl-5-methylphenol, 46B, 102
$C_{10}H_{14}O$, 3-Methyl-4-isopropylphenol, 43B, 145
$C_{10}H_{14}O_2$, 1,4-Diethoxybenzene, 43B, 146
$C_{10}H_{14}O_4$, 1,2,4,5-Tetramethoxybenzene, 45B, 87
$C_{11}H_{18}BrNO$, Phenylcholine ether bromide, 46B, 102
$C_{12}H_6Cl_4O$, Bis(3,4-dichlorophenyl)ether, 46B, 102
$C_{12}H_6N_4O_{10}$, 4,4'-Dihydroxy-3,3',5,5'-tetranitrobiphenyl, 45B, 88
$C_{12}H_8Br_2O$, Di-p-bromophenyl ether, 10, 266
$C_{12}H_{10}O_2$, 4,4'-Dihydroxybiphenyl, 15, 470; 24, 654; 37B, 64
$C_{12}H_{10}O_4$, Quinhydrone (monoclinic α form), 2, 807, 888; 22, 694; 33B, 65
$C_{12}H_{10}O_4$, Quinhydrone (triclinic β form), 2, 807, 888; 30B, 75
$C_{12}H_{16}O_5S_2$, 1,2-Bis(methylsulphonyl)-1-p-anisylprop-1-ene, 43B, 106
$C_{12}H_{18}O_2$, 1,4-Dimethoxy-2,3,5,6-tetramethylbenzene, 41B, 135
$C_{13}H_{10}Cl_2O_2$, 3,3'-Dichloro-4,4'-dihydroxydiphenylmethane, 17, 713
$C_{13}H_{10}O_3$, 2,4-Dihydroxybenzophenone, 45B, 88
$C_{13}H_{12}Cl_2O_4$, Ethacrynic acid, 44B, 98
$C_{13}H_{12}O_3S$, 4,4'-Dihydroxythiobenzophenone monohydrate, 30B, 82
$C_{13}H_{22}BrNO_3$, o-Dimethoxyphenylcholine ether bromide, 46B, 103
$C_{13}H_{22}NO_5S_{0.5}$, 2-t-Butylamino-1-(4-hydroxy-3-hydroxymethylphenyl)-ethanol hemi(dihydrogen sulfate), 44B, 98
$C_{14}H_8Cl_3NO_5$, 2-(2'-Carbomethoxy-4'-nitrophenoxy)-1,3,5-trichloro-benzene, 38B, 147
$C_{14}H_{10}Cl_2N_2O_6$, 2,2'-Ethylidenebis(4-chloro-6-nitrophenol), 45B, 89
$C_{14}H_{11}ClO_3$, 2-Hydroxy-4-methoxy-4'-chlorobenzophenone, 42B, 88
$C_{14}H_{12}Cl_2O_2$, p,p'-Dichloro-1,2-diphenoxyethane, 32B, 104
$C_{14}H_{12}O_3$, 2-Hydroxy-4-methoxybenzophenone, 40B, 127
$C_{14}H_{13}ClO_3S$, p-Chlorophenyl-p-methoxybenzyl-sulphone, 44B, 100
$C_{14}H_{13}I_2NO$, 4-Methyl-4'-methoxy-2,6-diiododiphenylamine, 41B, 136
$C_{14}H_{13}NO_2$, p-((p-Methoxybenzylidene)amino)phenol, 44B, 99
$C_{14}H_{14}Cl_2O_2$, 3,5-Dichlorophenol 2,6-dimethylphenol (1:1), 46B, 103
$C_{14}H_{14}NO_3$, Di-p-anisyl nitric oxide, 17, 722
$C_{14}H_{14}O_2$, 1,2-Diphenoxyethane, 32B, 104
$C_{14}H_{20}N_2O_6S$, p-Methylaminophenol sulfate, 35B, 67
$C_{14}H_{21}BrO$, 4-Bromo-2,6-di-t-butyl-phenol, 38B, 148
$C_{14}H_{22}O_2$, Pentamethyl-o-hydroxyhydrocinnamyl alcohol, 38B, 148
$C_{14}H_{23}Cl_2NO_2$, Bupranolol hydrochloride, 41B, 137
$C_{15}H_{12}Cl_2N_2O_6$, 2,2'-Isopropylidenebis(4-chloro-6-nitrophenol), 45B, 89
$C_{15}H_{12}I_2O_4$, 3,5-Diiodothyropropionic acid, 43B, 147
$C_{15}H_{14}N_4O_7$, N-2,4,6-Trinitrobenzyl-N-methyl-anisidine, 41B, 138
$C_{15}H_{14}O_3$, p,p'-Dimethoxybenzophenone, 27, 909
$C_{15}H_{14}O_6$, Butein monohydrate, 38B, 149
$C_{15}H_{21}IO_4$, (DL)-erythro-Methyl 3-t-butoxy-2-iodo-3-(p-methoxyphen-yl)propionate, 46B, 104
$C_{15}H_{23}NO_6S$, 4,5-Dimethoxy-2-(carbomethoxymethylbenzenediethylsul-phonamide), 45B, 90
$C_{15}H_{24}O$, 2,6-Di-t-butyl-4-methylphenol, 39B, 77
$C_{16}H_{14}O$, 4-Methoxy-4'-methyltolane, 43B, 148
$C_{16}H_{15}BrO_2$, (4-Bromophenoxyacetyl)-4-ethylbenzene, 44B, 100
$C_{16}H_{15}Cl_3O_2$, 1,1,1-Trichloro-2,2-bis(p-methoxyphenyl)ethane, 42B, 96

$C_{16}H_{16}Br_2O_2$, 1,4-Bis(p-bromophenoxy)butane, 37B, 64
$C_{16}H_{16}Cl_2O_2$, 1,4-Bis(5-chloropent-4-ynyloxy)benzene, 46B, 104
$C_{16}H_{16}N_2O_2$, N,N'-Ethylenebis(salicylideneimine), 44B, 81
$C_{16}H_{16}N_2O_5$, 2-(2',6'-Dinitrophenoxy)-t-butylbenzene, 41B, 139
$C_{16}H_{16}O_2$, Dimethylstilbestrol, 40B, 128
$C_{16}H_{16}O_3$, Deoxyanisoin, 27, 911
$C_{16}H_{16}O_4$, 2-[2-(Phenoxy)phenoxy]-2-methylpropionic acid, 45B, 91
$C_{16}H_{16}O_4$, 2,2'-Dimethoxybenzoin, 43B, 122
$C_{16}H_{22}Cl_2N_2O_6$, 6-Hydroxydopamine 6-hydroxydopamine-quinone hydrochloride, 41B, 140
$C_{17}H_{14}Cl_2O_4$, 3,3'-Diacetyl-5,5'-dichloro-2,2'-dihydroxydiphenylmethane, 42B, 88
$C_{17}H_{15}ClO_2$, 4-Chlorophenoxyacetyl-4-propenylbenzene, 43B, 162
$C_{17}H_{15}I_3O_4$, Methyl 3,5,3'-triiodo-4'-methoxythyropropionate, 44B, 101
$C_{17}H_{15}I_3O_4$, Triiodothyropropionic acid ethyl ester, 40B, 14
$C_{17}H_{16}O$, 4-Methoxy-4'-ethyltolane, 43B, 148
$C_{17}H_{20}O_6$, Mycophenolic acid, 38B, 17
$C_{18}H_{14}O$, 2,6-Diphenylphenol, 44B, 102
$C_{18}H_{14}O_2$, 1,6-Diphenoxy-2,4-hexadiyne (monoclinic), 43B, 149
$C_{18}H_{14}O_2$, 1,6-Diphenoxy-2,4-hexadiyne (orthorhombic), 43B, 149
$C_{18}H_{16}O_6$, Phenoquinone, 33B, 65
$C_{18}H_{18}O_2$, Dienestrol, 38B, 150
$C_{18}H_{20}O_2$, Diethylstilbestrol, 35B, 332
$C_{18}H_{22}O_2$, DL-Hexestrol, 39B, 78
$C_{18}H_{22}O_2$, 1,6-Diphenoxyhexane, 43B, 149
$C_{18}H_{22}O_3$, Diethylstilbestrol monohydrate, 34B, 77
$C_{18}H_{29}ClN_2O_4$, 2-(2-Hydroxy-3-isopropylaminopropoxy)-5-butyrylaminoacetophenone hydrochloride, 45B, 92
$C_{18}H_{30}O_4$, 2,2,4,4-Tetramethyl-3-(3,4,5-trimethoxyphenyl)pentan-3-ol, 43B, 149
$C_{19}H_{11}BrN_4O_7$, p-Bromobenzophenone oxime O-picryl ether (anti form), 38B, 151
$C_{19}H_{11}BrN_4O_7$, p-Bromobenzophenone oxime O-picryl ether (syn form), 38B, 151
$C_{19}H_{14}Br_2O_2$, 3,5-Dibromo-p-hydroxy-triphenylmethane carbinol, 31B, 75; 37B, 64
$C_{19}H_{16}O$, (4-Hydroxyphenyl)diphenylmethanol, 45B, 92
$C_{19}H_{16}O_2$, [1-(2-Hydroxyphenyl)]diphenylmethanol, 46B, 105
$C_{19}H_{20}O$, p-Methoxy-p'-butyltolane, 43B, 150
$C_{19}H_{21}F_3O_2$, 3,4-Di(p-methoxyphenyl)-1,1,1-trifluoropentane, 41B, 140
$C_{19}H_{22}N_2O_3$, 4-(4'-Ethoxyphenylazo)phenyl valerate, 45B, 93
$C_{19}H_{23}NO_2$, p-Ethoxy-p'-butylaniline, 45B, 93
$C_{19}H_{23}NO_3$, 2-(3'-Methyl-4'-nitrophenoxy)-1,3-diisopropylbenzene, 38B, 152
$C_{19}H_{24}O_2$, 1,1-Bis(p-methoxyphenyl)-2,2-dimethylpropane, 46B, 105
$C_{19}H_{26}O_4$, Diethylstilboestrol methanol hydrate, 34B, 76; 39B, 79
$C_{20}H_{14}Br_4O_6S_2$, Tetrabromo-p-phenylene bis(toluene-p-sulphonate), 46B, 106
$C_{20}H_{14}F_6O_2$, 1,6-Di-(p-anisyl)-perfluorohexa-1,3,5-diene, 44B, 102
$C_{20}H_{16}N_2O_2$, N,N'-(o-Phenylene)bis(salicylideneamine), 42B, 745
$C_{20}H_{18}O_2$, Bis(o-ethoxyphenyl)butadiyne, 37B, 64
$C_{20}H_{18}O_5$, 1,4-Dimethoxy-2,4',4''-trihydroxy-p-terphenyl, 40B, 130
$C_{20}H_{18}O_8S_2$, 2,4-Hexadiynylene bis(p-methoxybenzenesulphonate), 45B, 94
$C_{20}H_{19}KO_{12}$, Potassium O,O'-catecholdiacetate, 41B, 141
$C_{20}H_{24}N_2O_3$, 4-(4'-Ethoxyphenylazo)phenyl hexanoate, 45B, 95

$C_{20}H_{24}N_2O_4$, 5,5'-Diethoxy-α,α'-dimethyl-α,α'-azinodi-o-cresol, 46B, 106
$C_{20}H_{24}O_2$, 4,4'-Dimethoxy-α,β-diethylstilbene, 41B, 142
$C_{20}H_{26}O_3$, Diethylstilboestrol ethanol, 39B, 79
$C_{20}H_{26}O_3S$, Diethylstilboestrol dimethyl sulfoxide, 39B, 79
$C_{20}H_{27}I_5N_2O_4$, Diphenacetin tri-iodine di(iodine), 45B, 96
$C_{21}H_{20}O_2$, (3,5-Dimethyl-4-hydroxyphenyl)diphenylmethanol, 45B, 92
$C_{21}H_{23}NO_3$, Propyl 4-(4'-methoxybenzylideneamino)-α-methylcinnamate, 41B, 143
$C_{21}H_{26}Cl_2O_2$, 2,2'-Methylenebis(4-chloro-3-methyl-6-isopropylphenol), 44B, 103
$C_{21}H_{26}NNaO_6S$, Bis((O-methoxyphenoxy)ethoxy)ethane sodium isothiocyanate, 44B, 103
$C_{21}H_{28}O_2$, Bis(hydroxyluryl)methane, 20, 580
$C_{22}H_{21}ClO_3$, Tris-(p-methoxyphenyl)chloromethane, 42B, 89
$C_{22}H_{36}Cl_4CuN_2O_6 \cdot H_2O$, Bis(3,4,5-trimethoxyphenethylammonium) tetrachlorocuprate(II) monohydrate, 45B, 97
$C_{23}H_{19}NO_2$, 3,3-Bis(p-methoxyphenyl)-2-phenylacrylonitrile, 45B, 97
$C_{24}H_{12}N_6O_{15}$, 1,2,3-Tris(2,4-dinitrophenoxy)benzene, 44B, 104
$C_{24}H_{32}I_2O_6$, Iodophyllanthin, 43B, 152
$C_{24}H_{32}N_2O_5$, 2-(2',4'-dinitrophenoxy)-1,3,5-tri-t-butylbenzene, 38B, 153
$C_{25}H_{36}O_2$, 2,2-Bis-(2-hydroxy-5-methyl-3-t-butylphenyl)propane, 42B, 90
$C_{26}H_{29}Cl_2NO$, (E)-1-(p-(Diethylaminoethoxy)phenyl)-1,2-diphenyl-2-chloroethylene hydrochloride, 42B, 91
$C_{26}H_{29}NO$, 1-(p-(2-Dimethylamino-ethoxy)-phenyl)-1,2-trans-diphenylbutene, 45B, 98
$C_{26}H_{38}N_2O_3$, 4,4'-Di-n-heptyloxyazoxybenzene, 45B, 98, 99
$C_{28}H_{22}Br_2N_2O_2$, 2,2'-Dibromo-4,4'-bis(p-methoxybenzylideneamino)biphenyl, 41B, 144
$C_{28}H_{24}N_2O_2$, N,N'-Disalicylidene-meso-1,2-diphenylethylenediamine, 43B, 153
$C_{48}H_{32}Br_2O_2 \cdot 1.5\ C_6H_6$, 3-Bromo-2,4,6-triphenylphenoxyl dimer benzene solvate, 33B, 66

18. BENZOQUINONES

$C_6Br_4O_2$, Tetrabromo-p-benzoquinone, 26, 714
$C_6Cl_4KO_2 \cdot C_3H_6O$, p-Chloranil potassium salt acetone solvate, 46B, 111
$C_6Cl_4KO_2$, Potassium chloroanilate (α form), 39B, 80
$C_6Cl_4O_2$, Tetrachloro-o-benzoquinone, 44B, 104
$C_6Cl_4O_2$, Tetrachloro-p-benzoquinone (Chloranil), 26, 714; 27, 899; 38B, 154
$C_6F_4O_2$, Fluoranil, 40B, 121
$C_6I_4O_2$, Tetraiodo-p-benzoquinone, 40B, 122
$C_6H_2Cl_2O_2$, 2,3-Dichloro-1,4-benzoquinone, 35B, 71
$C_6H_2Cl_2O_2$, 2,5-Dichloro-1,4-benzoquinone, 35B, 71
$C_6H_2Cl_2O_2$, 2,6-Dichloro-1,4-benzoquinone, 35B, 71
$C_6H_2Cl_2O_4$, Chloranilic acid, 32B, 105
$C_6H_2F_2O_4$, Fluoranilic acid, 41B, 144
$C_6H_3ClO_2$, Chloro-1,4-benzoquinone, 35B, 71
$C_6H_4ClNO_2$, 3-Chlorobenzoquinone 4-oxime, 21, 579
$C_6H_4Cl_2N_2O_2$, 3,6-Dichloro-2,5-diamino-1,4-benzoquinone, 34B, 78
$C_6H_4O_2$, o-Benzoquinone, 39B, 81

$C_6H_4O_2$, p-Benzoquinone (gas-ed), 18, 729; 39B, 889
$C_6H_4O_2$, p-Benzoquinone, 3, 684, 769; 24, 639; 44B, 105
$C_6H_4O_4$, 2,5-Dihydroxy-1,4-benzoquinone, 43B, 153
$C_6H_5NO_2$, Quinone-4-oxime, 40B, 121
$C_6H_6Cl_2O_6$, Chloranilic acid dihydrate, 32B, 106
$C_6H_8N_4O_8$, Ammonium nitranilate, 29, 648; 32B, 47
$C_6H_8O_8$, Tetrahydroxy-p-benzoquinone, 30B, 77
$C_6H_{10}Cl_2N_2O_5$, Ammonium chloranilate monohydrate, 32B, 48
$C_6H_{14}N_2O_{14}$, Hydronium nitranilate hydrate, 32B, 107; 41B, 145
$C_7H_6ClNO_2$, α-2-Chloro-5-methyl-p-benzoquinone-4-oxime (syn form), 24, 640
$C_8Cl_2N_2O_2$, 2,3-Dichloro-5,6-dicyano-p-benzoquinone, 46B, 112
$C_8H_6ClNO_3$, α-2-Chloro-p-benzoquinone-4-oxime acetate, 26, 709
$C_8H_6ClNO_3$, β-2-Chloro-p-benzoquinone-4-oxime acetate, 26, 709
$C_8H_8ClNO_3$, α-5-(2'-Chloroethoxy)-o-benzoquinone-2-oxime (anti form), 31B, 76
$C_8H_8ClNO_3$, β-5-(2'-Chloroethoxy)-o-quinone-2-oxime (syn form), 24, 642
$C_8H_8Cl_2N_2O_2$, 3,6-Dichloro-2,5-bis(monomethylamino)-1,4-benzoquinone, 37B, 65
$C_8H_8INO_3$, 1-(β-Aminoethyl)-4-hydroxy-5-iodo-3,6-benzoquinone, 45B, 100
$C_8H_8O_2$, 2,3-Dimethyl-1,4-benzoquinone, 32B, 108
$C_8H_8O_2$, 2,5-Dimethyl-1,4-benzoquinone, 29, 645; 32B, 109
$C_8H_8O_2$, 2,6-Dimethyl-1,4-benzoquinone, 32B, 110
$C_8H_8O_4$, 2,6-Dimethoxy-1,4-benzoquinone, 43B, 154
$C_8H_9NO_5$, (E)-3,6-Dihydroxy-2-methyl-1,4-benzoquinone 4-methoxyimine N-oxide, 44B, 105
$C_8H_{10}N_2O_4S_2$, p-Benzoquinone-N,N'-bis(methylsulphonyldiimine), 44B, 105
$C_8H_{14}N_2O_{10}$, Hydronium cyananilate hydrate, 41B, 145
$C_9H_9ClO_3$, Shanorellin chloride, 35B, 73
$C_9H_{10}O_4$, Shanorellin, 46B, 113
$C_9H_{10}O_4$, 5-Methyl-2,3-dimethoxy-p-benzoquinone, 37B, 66
$C_9H_{11}NO_3$, 5-n-Propoxy-o-benzoquinone-2-oxime (β form), 29, 647
$C_{10}H_{10}N_2O_2$, 2,5-Bis(ethyleneimino)-1,4-benzoquinone, 39B, 81
$C_{10}H_{12}O_2$, 2,3,5,6-Tetramethyl-1,4-benzoquinone, 32B, 111
$C_{10}H_{14}N_2O_4$, 2,5-Bis(2'-hydroxyethylamino)-1,4-benzoquinone, 41B, 146
$C_{12}H_4N_4Na$, Sodium 7,7,8,8-tetracyanoquinodimethanide (monoclinic), 41B, 85
$C_{12}H_4N_4Na$, Sodium 7,7,8,8-tetracyanoquinodimethanide (triclinic), 40B, 654
$C_{12}H_4N_4Rb$, Rubidium 7,7,8,8-tetracyanoquinodimethanide (monoclinic), 38B, 111
$C_{12}H_4N_4Rb$, Rubidium 7,7,8,8-tetracyanoquinodimethanide (triclinic), 39B, 82
$C_{13}H_{10}O_3$, 2-(4'-Methoxyphenyl)-1,4-benzoquinone (red form), 43B, 154
$C_{13}H_{10}O_3$, 2-(4'-Methoxyphenyl)-1,4-benzoquinone (yellow form), 43B, 154
$C_{14}H_{12}O_2$, 2,6-Dimethyl-3,5-dipropynylcyclohexa-2,5-diene-1,4-dione, 44B, 106
$C_{14}H_{20}O_2$, 2,6-Di-t-butyl-1,4-benzoquinone, 39B, 83
$C_{16}H_{22}N_2O_2$, 2,5-Dipiperidino-1,4-benzoquinone, 40B, 123
$C_{16}H_{22}O_2$, 2,5-Pentamethyleneimino-1,4-benzoquinone, 37B, 67
$C_{17}H_{16}O_3$, O-Methylobtusaquinone, 39B, 83
$C_{21}H_{18}O$, 2,6-Dimethyl-4-(α,α-diphenylmethylene)-1,4-benzoquinone (α form), 46B, 114

$C_{21}H_{18}O$, 2,6-Dimethyl-4-(α,α-diphenylmethylene)-1,4-benzoquinone (β form), 46B, 114
$C_{21}H_{18}O$, 2,6-Dimethyl-4-(α,α-diphenylmethylene)-1,4-benzoquinone (γ form), 46B, 114
$C_{30}H_{24}N_9$, Triethylammonium bis-7,7,8,8-tetracyanoquinodimethanide, 35B, 76
$C_{36}H_{12}Cs_2N_{12}$, Caesium tetracyanoquinodimethanide, 29, 650; 31B, 43
$C_{38}H_{34}O_6$, Obtusaquinone dimer benzene solvate, 42B, 91

19. BENZENE MISCELLANEOUS

C_6BrCl_5, Pentachlorobromobenzene, 31B, 77
$C_6Br_2Cl_4$, 1,4-Dibromotetrachlorobenzene, 33B, 67
C_6Br_5Cl, Pentabromochlorobenzene, 31B, 77
C_6Br_6, Hexabromobenzene (gas-ed), 11, 650
C_6Cl_6, Hexachlorobenzene (gas-ed), 11, 650; 24, 653
C_6Cl_6, Hexachlorobenzene, 2, 808, 892; 22, 694; 26, 713; 40B, 123
C_6F_6, Hexafluorobenzene (gas-ed), 29, 642
C_6F_6, Hexafluorobenzene, 39B, 84
C_6I_6, Hexaiodobenzene, 30B, 79; 33B, 68; 35B, 77
C_6HF_5, Pentafluorobenzene (gas-mw), 33B, 540
$C_6H_2BrCl_3$, 2,4,6-Trichlorobromobenzene, 23, 660
$C_6H_2Br_4$, 1,2,3,5-Tetrabromobenzene (gas-ed), 11, 650
$C_6H_2Br_4$, 1,2,4,5-Tetrabromobenzene (β form), 22, 686; 24, 649
$C_6H_2Br_4$, 1,2,4,5-Tetrabromobenzene (γ form), 29, 642
$C_6H_2Cl_4$, 1,2,4,5-Tetrachlorobenzene, 22, 685
$C_6H_3Br_3$, 1,3,5-Tribromobenzene, 24, 644
$C_6H_3Cl_3$, 1,2,3-Trichlorobenzene, 38B, 155
$C_6H_3Cl_3$, 1,3,5-Trichlorobenzene, 24, 644
$C_6H_3F_3$, 1,2,4-Trifluorobenzene (gas-mw), 42B, 1035
C_6H_4BrCl, 1-Chloro-4-bromobenzene, 2, 887; 11, 652
C_6H_4BrF, p-Fluorobromobenzene (gas-ed), 17, 697
$C_6H_4Br_2$, o-Dibromobenzene (gas-ed), 11, 650
$C_6H_4Br_2$, p-Dibromobenzene (gas-ed), 11, 650
$C_6H_4Br_2$, p-Dibromobenzene, 2, 886; 9, 368; 12, 380
C_6H_4ClF, o-Fluorochlorobenzene (gas-ed), 17, 697
C_6H_4ClF, p-Fluorochlorobenzene (gas-ed), 17, 697
$C_6H_4ClIO_2$, p-Chloroiodoxybenzene, 11, 653
$C_6H_4Cl_2$, m-Dichlobenzene (gas-mw), 42B, 1035
$C_6H_4Cl_2$, o-Dichlorobenzene (gas-ed), 11, 650
$C_6H_4Cl_2$, p-Dichlorobenzene (α phase), 42B, 92
$C_6H_4Cl_2$, p-Dichlorobenzene (β phase), 42B, 92
$C_6H_4Cl_2$, p-Dichlorobenzene (γ phase), 41B, 147
$C_6H_4Cl_2$, p-Dichlorobenzene (monoclinic), 2, 806, 885; 12, 380; 16, 507; 23, 648; 28, 424
$C_6H_4Cl_2$, p-Dichlorobenzene (triclinic), 20, 560; 21, 575
$C_6H_4F_2$, m-Difluorobenzene (gas-ed), 39B, 890
$C_6H_4I_2$, p-Diiodobenzene, 23, 649
C_6H_5Br, Bromobenzene (gas-ed), 17, 697
C_6H_5Cl, Chlorobenzene (gas-mw), 42B, 1035
C_6H_5Cl, Chlorobenzene, 37B, 67
$C_6H_5Cl_2I$, Benzene iododichloride, 17, 699
C_6H_5F, Fluorobenzene (gas-ed), 10, 259; 17, 697
C_6H_5F, Fluorobenzene (gas-mw), 17, 697; 18, 700; 33B, 540
$C_6H_5IO_2$, Iodylbenzene, 46B, 63
C_6H_6, Benzene (gas-ed), 7, 270; 9, 368; 10, 259; 16, 504; 22, 688;

24, 631; 42B, 1027
C_6H_6, Benzene (gas-irr), 19, 559
C_6H_6, Benzene (high-pressure), 34B, 79; 37B, 67
C_6H_6, Benzene, 22, 687; 29, 636
$C_7H_3Br_5$, Pentabromotoluene, 37B, 68
C_7H_5ClO, 2-Chlorobenzaldehyde (gas-ed), 42B, 1027
C_7H_6ClNO, anti-p-Chlorobenzaldoxime, 29, 657; 35B, 79
C_7H_6ClNO, syn-p-Chlorobenzaldoxime, 29, 657
$C_7H_6ClNO_2$, 5-Chlorosalicylaldoxime, 26, 711
C_7H_7Br, Benzyl bromide (gas-ed), 42B, 1028
C_7H_7Cl, Benzyl chloride (gas-ed), 42B, 1028
C_7H_7I, p-Iodotoluene, 38B, 156
C_7H_8, Toluene (α-form), 43B, 155
C_7H_8, Toluene (gas-ed), 20, 557
C_8Cl_8, Perchloro-p-xylylene, 42B, 92
$C_8H_2Cl_8$, 2H,5H-Octachloro-p-xylene, 39B, 84
$C_8H_4N_2$, p-Di-isocyanobenzene, 16, 508
$C_8H_4N_2Se_2$, 1,4-Diselenocyanatobenzene, 35B, 81
$C_8H_6Br_4$, Tetrabromo-m-xylene, 26, 712
$C_8H_6Cl_4$, Tetrachloro-p-xylene, 38B, 158
C_8H_7Br, omega-Bromostyrene (gas-ed), 24, 638
C_8H_7BrO, α-Bromoacetophenone, 37B, 70
C_8H_7ClO, α-Chloroacetophenone, 31B, 78
C_8H_8, p-Xylylene (gas-ed), 43B, 1485
C_8H_8ClNO, N-Methyl-p-chlorobenzaldoxime, 29, 657
C_8H_8O, Acetophenone, 39B, 85
C_8H_{10}, Perchloro-p-xylene, 44B, 107
C_9H_5ClO, o-Chlorobenzoylacetylene, 31B, 79
$C_9H_8O_2$, Phenylmalondialdehyde, 43B, 66
$C_9H_9Cl_3$, 1,2,3-Trichloro-4,5,6-trimethylbenzene, 38B, 160, 161
$C_9H_9Cl_3$, 1,2,4-Trichloro-3,5,6-trimethylbenzene, 32B, 112
C_9H_{12}, Cumene (gas-ed), 33B, 536
$C_{10}H_6$, p-Diethynylbenzene, 38B, 163
$C_{10}H_9BrO$, p-Bromobenzylideneacetone, 35B, 84
$C_{10}H_9BrO_2$, 1-(4-Bromophenyl)-1,3-butanedione, 42B, 31
$C_{10}H_{10}O_2$, Benzoylacetone (enol form), 38B, 63; 42B, 93; 45B, 100
$C_{10}H_{11}IO_4$, Iodobenzene diacetate, 43B, 13
$C_{10}H_{12}Cl_2$, 1,2-Dichlorotetramethylbenzene, 22, 693
$C_{10}H_{13}Br$, 2,3,5,6-Tetramethylbromobenzene (α form), 30B, 79
$C_{10}H_{13}Br$, 2,3,5,6-Tetramethylbromobenzene (β form), 30B, 79
$C_{10}H_{13}Cl$, 1-Chloro-2,3,5,6-tetramethylbenzene, 33B, 69
$C_{10}H_{13}NO$, (-)-(S)-N-(1-Phenylethyl)acetamide, 46B, 9
$C_{10}H_{14}$, Durene, 3, 683, 752; 38B, 163; 39B, 85; 40B, 126; 43B, 155
$C_{11}H_{10}O_2$, 5-Phenyl-2,4-pentadienoic acid, 46B, 10
$C_{11}H_{11}ClO_4$, p-Chlorobenzaldehyde hydrate diacetate, 44B, 107
$C_{11}H_{15}Cl$, Chloropentamethylbenzene, 33B, 70
$C_{11}H_{16}$, 1-Methyl-4-t-butylbenzene, 46B, 115
$C_{12}Br_2F_8$, 2,2'-Dibromooctafluorobiphenyl, 46B, 115
$C_{12}Cl_{10}$, Decachlorobiphenyl, 41B, 148
$C_{12}D_{10}$, Biphenyl (deuterated, low temperature), 45B, 101
$C_{12}F_{10}$, Decafluorobiphenyl (gas-ed), 33B, 537
$C_{12}F_{10}$, Perfluorobiphenyl, 42B, 93
$C_{12}F_{18}$, Hexakis(trifluoromethyl)benzene, 42B, 94
$C_{12}HF_9$, 2H-Nonafluorobiphenyl, 44B, 108
$C_{12}H_4Cl_6$, 2,2',4,4',6,6'-Hexachlorobiphenyl, 45B, 101
$C_{12}H_5F_5$, 2,3,4,5,6-Pentafluorobiphenyl, 44B, 108
$C_{12}H_6Br_4$, 3,3',5,5'-Tetrabromodiphenyl (gas-ed), 13, 540

$C_{12}H_6F_4$, 2,3,5,6-Tetrafluorobiphenyl, 44B, 108
$C_{12}H_8Br_2$, 2,2'-Dibromobiphenyl (gas-ed), 13, 540
$C_{12}H_8Br_2$, 3,3'-Dibromobiphenyl (gas-ed), 12, 391
$C_{12}H_8Br_2$, 4,4'-Dibromobiphenyl, 42B, 94
$C_{12}H_8Cl_2$, 2,2'-Dichlorobiphenyl (gas-ed), 13, 540; 40B, 127
$C_{12}H_8Cl_2$, 2,2'-Dichlorobiphenyl, 40B, 127
$C_{12}H_8Cl_2$, 4,4'-Dichlorobiphenyl, 44B, 109
$C_{12}H_8I_2$, 2,2'-Diiododiphenyl (gas-ed), 13, 540
$C_{12}H_9Br$, 4-Bromobiphenyl, 46B, 115
$C_{12}H_{10}$, Biphenyl (gas-ed), 9, 375; 12, 390
$C_{12}H_{10}$, Biphenyl, 26, 726; 27, 907; 42B, 94; 43B, 156; 45B, 101
$C_{12}H_{10}BF_4I$, Diphenyliodonium tetrafluoroborate, 24, 661
$C_{12}H_{10}INO_3$, Diphenyliodinium nitrate, 38B, 164
$C_{12}H_{10}I_2N_2O_7$, μ-Oxo-bis[nitrato(phenyl)iodine(III)], 45B, 102
$C_{12}H_{12}Br_6$, Hexa(bromomethyl)benzene, 3, 679, 755; 30B, 80
$C_{12}H_{12}O_3$, 1,3,5-Triacetylbenzene, 39B, 86
$C_{12}H_{15}Cl_3$, Pentamethylbenzotrichloride, 39B, 87
$C_{12}H_{16}BrNO_2$, L-3-Benzylamino-4-hydroxypentanoic acid lactone hydrobromide, 35B, 86
$C_{12}H_{16}ClNO_2$, DL-3-Benzylamino-4-hydroxypentanoic acid lactone hydrochloride, 35B, 86
$C_{12}H_{17}NO$, N-(1-Phenylethyl)isobutyramide, 46B, 12
$C_{12}H_{18}$, Hexamethylbenzene, 1, 621, 717; 7, 225, 264; 12, 379
$C_{12}H_{18}$, 1-Methyl-4-(1'-ethyl)propylbenzene, 45B, 103
$C_{12}H_{18}$, 1,4-Diisopropylbenzene, 45B, 102
$C_{13}H_8Br_2O$, 3,3'-Dibromobenzophenone, 22, 706; 39B, 87
$C_{13}H_8Cl_2O$, 4,4'-Dichlorobenzophenone, 11, 676; 43B, 160
$C_{13}H_8I_2O$, 4,4'-Diiodobenzophenone, 45B, 103
$C_{13}H_{10}O$, Benzophenone, 32B, 113; 33B, 71
$C_{13}H_{13}BrO_2$, Ethyl 5-(4'-bromophenyl)penta-2,4-dienoate, 41B, 148
$C_{13}H_{15}NO_3$, Z-Ethyl 3-amino-2-benzoyl-2-butenoate, 46B, 12
$C_{13}H_{19}Cl_2NO_2 \cdot H_2O$, 2-(4-Chlorophenyl)-1,1-dimethylethyl 2-aminopropanoate hydrochloride hydrate, 46B, 13
$C_{13}H_{20}$, 1-Methyl-4-hexylbenzene, 44B, 109
$C_{14}F_{10}$, Perfluorodiphenylacetylene, 45B, 104
$C_{14}H_8BrIO_4$, 2-Iodo-2'-bromodibenzoyl peroxide, 35B, 91
$C_{14}H_8Br_2O_4$, 2,2'-Dibromodibenzoyl peroxide, 35B, 91
$C_{14}H_8ClIO_4$, 2-Iodo-2'-chlorodibenzoyl peroxide, 35B, 91
$C_{14}H_8Cl_2$, Bis(m-chlorophenyl)acetylene, 43B, 157
$C_{14}H_8Cl_2O_4$, 2,2'-Dichlorodibenzoyl peroxide, 35B, 91
$C_{14}H_8Cl_4$, cis-1,2-Bis(4-chlorophenyl)-1,2-dichloroethylene, 43B, 158
$C_{14}H_8Cl_4$, trans-1,2-Bis(4-chlorophenyl)-1,2-dichloroethylene, 43B, 158; 44B, 109
$C_{14}H_8Cl_4$, 1-Chloro-2-(4-chlorophenyl)-2-(2,4-dichlorophenyl)ethylene, 44B, 110
$C_{14}H_8Cl_4$, 1,1-Dichloro-2,2-bis-(p-chlorophenyl)ethylene, 43B, 158
$C_{14}H_8Cl_6$, 1,1,1,2-Tetrachloro-2,2-bis(p-chlorophenyl)ethane, 44B, 110
$C_{14}H_8I_2O_4$, 2,2'-Diiododibenzoyl peroxide, 35B, 91
$C_{14}H_9Br_3Cl_2$, 1,1,1-Tribromo-2,2-bis(p-chlorophenyl)ethane, 44B, 110
$C_{14}H_9Cl_5$, (-)-1-(o-Chlorophenyl)-1-(p-chlorophenyl)-2,2,2-trichloroethane, 43B, 159
$C_{14}H_9Cl_5O$, 1,1-Bis(p-chlorophenyl)-2,2,2-trichloroethanol, 44B, 111
$C_{14}H_{10}$, Diphenylacetylene (Tolane), 6, 204, 246; 33B, 72; 43B, 159; 44B, 112
$C_{14}H_{10}Cl_2F_2$, 1,1-Dichloro-2,2-bis(p-fluorophenyl)ethane, 43B, 160
$C_{14}H_{10}Cl_2O_2$, 1,1-Bis(p-chlorophenyl)acetic acid, 43B, 160

$C_{14}H_{10}Cl_2O_2$, 2,2-Bis(p-chlorophenyl)acetic acid, 43B, 17
$C_{14}H_{10}Cl_4$, 1-(o-Chlorophenyl)-1-(p-chlorophenyl)-2,2-dichloroethane, 42B, 95
$C_{14}H_{10}Cl_4$, 1,1-Dichloro-2,2-bis-(p-chlorophenyl)ethane, 43B, 158
$C_{14}H_{10}F_4$, Tetrafluoro-1,2-diphenylethane, 23, 665
$C_{14}H_{10}O_2$, Benzil, 6, 245; 30B, 81
$C_{14}H_{11}BrO$, 4-Acetyl-3'-bromobiphenyl, 34B, 81
$C_{14}H_{11}ClO$, 4-Acetyl-2'-chlorobiphenyl, 33B, 72
$C_{14}H_{11}FO$, 4-Acetyl-2'-fluorobiphenyl, 33B, 73
$C_{14}H_{12}$, trans-Stilbene, 5, 144, 171; 40B, 127; 41B, 149; 42B, 95
$C_{14}H_{12}Br_2$, o,o'-Dibromodibenzyl, 45B, 104
$C_{14}H_{12}O_2$, Benzoin, 46B, 116
$C_{14}H_{14}$, Dibenzyl, 3, 685, 787; 12, 391
$C_{14}H_{14}$, p,p'-Bitolyl, 34B, 82
$C_{14}H_{22}$, p-Di-t-butylbenzene, 15, 467
$C_{14}H_{22}$, 1-t-Butyl-4-n-butylbenzene, 45B, 105
$C_{14}H_{22}$, 1,4-Di-t-butylbenzene, 46B, 116
$C_{15}H_8Br_2O_3$, 1,3-Bis(4'-bromophenyl)propane-1,2,3-trione, 44B, 325
$C_{15}H_{10}Br_2O_2$, Bis(m-bromobenzoyl)methane, 27, 913
$C_{15}H_{10}Cl_2O_2$, Bis(m-chlorobenzoyl)methane, 29, 655
$C_{15}H_{11}BrO$, p'-Bromochalcone, 39B, 87
$C_{15}H_{11}F_3$, (E)-α-(Trifluoromethyl)stilbene, 46B, 116
$C_{15}H_{12}O$, Benzylideneacetophenone (Chalcone, form 1), 35B, 55
$C_{15}H_{12}O$, Benzylideneacetophenone (Chalcone, form 2), 39B, 88
$C_{15}H_{12}O_2$, 1,3-Diphenyl-1,3-propanedione enol (dibenzoylmethane), 31B, 80; 39B, 88; 42B, 96
$C_{15}H_{12}O_4$, 1,3-Diphenyl-2,2-dihydroxypropane-1,3-dione, 46B, 40
$C_{15}H_{15}BrO_2$, Ethyl 7-(4'-bromophenyl)hepta-2,4,6-trienoate, 41B, 148
$C_{15}H_{15}F$, 4-Fluoro-2',4',6'-trimethylbiphenyl (gas-ed), 43B, 1485
$C_{16}H_{10}$, Diphenyldiacetylene, 8, 315
$C_{16}H_{10}Cl_2O_2$, cis-1,2-Di-p-chlorobenzoylethylene, 35B, 57
$C_{16}H_{10}O_2$, Dibenzoylacetylene, 38B, 124
$C_{16}H_{12}$, 1,4-Diphenylbuta-1,2,3-triene, 45B, 105
$C_{16}H_{12}Cl_2$, 1-(2,6-Dichlorophenyl)-4-phenyl-trans,trans-1,3-butadiene, 41B, 150
$C_{16}H_{12}O_2$, cis-1,2-Dibenzoylethylene, 35B, 57
$C_{16}H_{12}O_2$, trans-1,2-Dibenzoylethylene, 44B, 112
$C_{16}H_{14}F_2$, p,p'-Dimethyl-α,α-difluorostilbene, 33B, 74
$C_{16}H_{14}O_2$, ω-(p-Toluoyl)-acetophenone enol, 34B, 26; 37B, 71
$C_{16}H_{14}O_2$, p-Methoxychalcone, 35B, 56
$C_{16}H_{14}O_2$, 1,2-Bis(p-formylphenyl)ethane, 44B, 113
$C_{16}H_{16}$, 1,1-Di-p-tolyethylene, 35B, 93
$C_{16}H_{18}$, meso-2,3-Diphenylbutane, 43B, 161
$C_{16}H_{18}$, 4,4'-Dimethyldibenzyl, 18, 717
$C_{16}H_{26}$, 1,4-Bis(α-ethylpropyl)benzene, 45B, 105
$C_{17}H_{12}Br_2O$, (Z,Z)-2,4-Dibromo-1,5-diphenylpenta-1,4-dien-3-one, 42B, 96
$C_{17}H_{13}BrO$, 1-p-Bromophenyl-4-benzoyl-1,3-butadiene, 39B, 898
$C_{17}H_{14}O$, 1,5-Diphenyl-2,4-pentadien-1-one, 46B, 10
$C_{17}H_{16}O$, 4,4'-Dimethylchalcone, 40B, 128
$C_{17}H_{16}O_3$, Dibenzylpyruvic acid, 46B, 14
$C_{17}H_{18}BrCl$, 2,3,5,6-Tetramethyl-4-bromo-2'-chlorodiphenylmethane, 42B, 97
$C_{17}H_{18}Br_2$, 2,2-Dibenzyl-1,3-dibromopropane, 45B, 106
$C_{18}BrF_{13}$, 1-Bromo-3,4,5-trifluoro-2,6-bis(pentafluorophenyl)benzene, 46B, 117
$C_{18}H_{14}$, o-Terphenyl, 5, 171; 44B, 113; 45B, 106

$C_{18}H_{14}$, o-Terphenyl (gas-ed), 9, 375
$C_{18}H_{14}$, p-Terphenyl (low-temp), 42B, 97; 43B, 162
$C_{18}H_{14}$, p-Terphenyl, 3, 681, 780; 26, 734; 34B, 83; 35B, 95
$C_{18}H_{16}F_5NO_2$, N-[(S)-α-Phenylethyl]-(R)-α-methyl-α-methoxy(pentafluorophenyl)acetamide, 46B, 15
$C_{18}H_{16}O_2$, trans-1,2-Di-p-toluoylethylene, 44B, 112
$C_{18}H_{18}Br_2O_2$, 1-Acetoxy-2,4-dibromo-1,1-diphenylbutane, 43B, 18
$C_{18}H_{20}Cl_2$, 1,1-Dichloro-2,2-bis(p-ethylphenyl)ethane, 44B, 114
$C_{18}H_{30}$, 1,3,5-Tri-t-butylbenzene, 44B, 115
$C_{19}H_{12}Br_2O$, 3,5-Dibromofuschone, 34B, 80
$C_{19}H_{14}BrF_2N$, p-Bromophenyldiphenylcarbinyl-difluoramine, 41B, 1234
$C_{19}H_{15}BrO$, 1-(p-Bromophenyl)-6-benzoylhexa-1,3,5-triene, 40B, 129
$C_{19}H_{16}$, Triphenylmethane (gas-ed), 30B, 403
$C_{19}H_{16}$, Triphenylmethane, 40B, 129
$C_{19}H_{21}NO_4$, (+)-(S)-Ethyl 3-{4-[2-sec-butoxycarbonyl-(E)-vinyl]-phenyl}-2-cyano-(E)-acrylate, 46B, 15
$C_{20}H_{10}F_8$, 1,8-Diphenyl-perfluoroocta-1,3,5,7-tetraene, 44B, 115
$C_{20}H_{18}$, 1,1,1-Triphenylethane, 46B, 118
$C_{20}H_{18}$, 1,8-Diphenyl-1,3,5,7-octatetraene, 19, 572
$C_{20}H_{18}$, 2,7-Diphenyl-2,3,5,6-octatetraene, 39B, 89
$C_{21}H_{33}Br_3$, 2,4,6-Tribromo-1,3,5-trineopentylbenzene, 44B, 116
$C_{22}H_{19}Br$, 1-Bromo-2(p-ethylphenyl)-1,2-diphenylethylene, 38B, 165
$C_{22}H_{19}NO$, α-(1-Phenylethylimino)benzyl phenyl ketone, 45B, 107
$C_{22}H_{20}$, 1,10-Diphenyl-1,3,5,7,9-decapentaene, 18, 719
$C_{22}H_{28}$, 1,1-Diphenyl-2,2-di-t-butylethylene, 42B, 31
$C_{22}H_{30}$, DL-2,2,5,5-Tetramethyl-3,4-diphenylhexane, 46B, 119
$C_{22}H_{30}$, meso-2,2,5,5-Tetramethyl-3,4-diphenylhexane, 46B, 119
$C_{22}H_{38}$, 1,2,4,5-Tetra-t-butylbenzene, 38B, 165
$C_{24}H_{18}$, Quaterphenyl, 4, 268, 307; 42B, 98
$C_{24}H_{18}$, p-Quaterphenyl, 44B, 116
$C_{24}H_{18}$, 1,3,5-Triphenylbenzene, 18, 728; 41B, 151
$C_{24}H_{20}Br_2I_2$, Diphenyliodonium bromide, 43B, 157, 163
$C_{24}H_{20}Cl_2I_2$, Diphenyliodonium chloride, 43B, 157
$C_{24}H_{20}I_4$, Diphenyliodonium iodide, 43B, 157
$C_{25}H_{20}$, Tetraphenylmethane, 13, 553; 41B, 152
$C_{26}H_{18}O_2$, 4,4'-Dibenzoylbiphenyl, 45B, 69
$C_{26}H_{20}$, Tetraphenylethylene, 41B, 152
$C_{26}H_{22}$, 1,1,1,2-Tetraphenylethane, 46B, 119
$C_{26}H_{22}$, 1,1,2,2-Tetraphenylethane, 43B, 163
$C_{26}H_{22}O$, Bis(diphenylmethyl)ether, 41B, 1235
$C_{26}H_{30}BrNO$, 1-p-(2-Dimethylaminoethoxyphenyl)-1,2-cis-diphenylbut-1-ene hydrobromide, 35B, 97
$C_{28}H_{20}$, Tetraphenylbutatriene, 41B, 155; 43B, 164
$C_{28}H_{30}O_8$, Dimethyl meso-3,4-bis(2-(2-methoxycarbonylvinyl)phenyl)-1,6-hexanoate, 44B, 165
$C_{28}H_{34}$, Trimesitylmethane, 41B, 155
$C_{30}H_{20}$, 1,1,6,6-Tetraphenylhexapentaene, 17, 642
$C_{30}H_{20}O_4$, Tetrabenzoylethylene, 44B, 85
$C_{31}H_{20}O_4S_2$, Tetrabenzoylethylene carbon disulfide solvate, 44B, 85
$C_{33}H_{36}O_6$, Tri-o-thymotide, 40B, 130
$C_{38}H_{30}O_2$, Bis(triphenylmethyl) peroxide, 45B, 108
$C_{42}H_{30}$, Hexaphenylbenzene, 33B, 76

20. MONOCYCLIC HYDROCARBONS (3,4,5-MEMBERED RINGS)

C_3Cl_4, Perchlorocyclopropene (gas-ed), 37B, 694

C_3Cl_6, Hexachlorocyclopropane (gas-ed), 38B, 1068
C_3F_6, Hexafluorocyclopropane (gas-ed), 37B, 694
$C_3H_2F_2$, 3,3-Difluorocyclopropene (gas-mw), 42B, 1033
C_3H_2O, Cyclopropenone (gas-mw), 39B, 893
$C_3H_3F_3$, cis-1,2,3-Trifluorocyclopropane (gas-mw), 42B, 1034
C_3H_4, Cyclopropene (gas-ed), 16, 497; 35B, 892
$C_3H_4Cl_2$, 1,1-Dichlorocyclopropane (gas-ed), 10, 242; 38B, 1068
C_3H_5Cl, Chlorocyclopropane (gas-ed), 10, 242
C_3H_5Cl, Cyclopropyl chloride (gas-mw), 27, 732
$C_3H_5NO_2$, Nitrocyclopropane (gas-mw), 39B, 894
C_3H_6, Cyclopropane (gas-ed), 10, 242; 29, 490
C_3H_6, Cyclopropane (gas-irr), 19, 556
$C_4Cl_2F_6$, 1,1-Dichlorohexafluorocyclobutane (gas-ed), 40B, 1139
$C_4Cl_2O_2$, 1,2-Dichloro-1-cyclobutene-dione, 38B, 166
C_4Cl_8, Octachlorocyclobutane, 15, 448; 30B, 84
C_4F_6, Perfluorocyclobutene (gas-ed), 37B, 692
C_4F_8, Octafluorocyclobutane (gas-ed), 13, 523; 16, 498; 33B, 531; 37B, 692; 40B, 1139
$C_4H_2K_2OS_4$, Potassium tetrathiosquarate monohydrate, 42B, 98
$C_4H_2K_2O_5$, Potassium squarate monohydrate, 29, 674
$C_4H_2O_4$, Squaric acid (high-temp.), 43B, 165
$C_4H_2O_4$, Squaric acid, 39B, 89; 40B, 131; 41B, 156; 43B, 165
$C_4H_3KO_5$, Potassium 2-hydroxy-3,4-dioxocyclobut-1-ene-1-olate monohydrate, 39B, 90
$C_4H_3LiO_5$, Lithium hydrogen squarate monohydrate, 42B, 564
C_4H_5ClO, Cyclopropanecarboxylic acid chloride (gas-ed), 30B, 403
C_4H_5ClO, Cyclopropanecarboxylic acid chloride (gas-mw), 42B, 1035
C_4H_6, Cyclobutene (gas-ed), 20, 548
C_4H_6, Methylenecyclopropane (gas-mw), 35B, 904
C_4H_6O, Cyclobutanone (gas-mw), 42B, 1034
C_4H_6O, Cyclopropyl carboxaldehyde (gas-ed), 30B, 402
C_4H_7NO, Cyclopropanecarboxamide, 34B, 84
C_4H_8, Cyclobutane (gas-ed), 16, 497; 26, 641
C_4H_8, Cyclobutane (gas-irr), 27, 733
$C_4H_8N_2O$, Cyclopropanecarbohydrazide, 22, 676
$C_4H_8O_8$, Octahydroxycyclobutane, 33B, 77
$C_5Cl_3KN_2O_4$, Potassium 1,2-dinitro-3,4,5-trichlorocyclopentadienide, 38B, 167
C_5Cl_6, Perchlorocyclopentadiene (gas-ed), 37B, 694
C_5F_8, Perfluorocyclopentene (gas-ed), 37B, 694
C_5HO_5Rb, Rubidium hydrogen croconate, 31B, 38
C_5H_4O, Methylenecyclobutenone (gas-mw), 40B, 1146
$C_5H_5NO_5$, Ammonium hydrogen croconate, 31B, 38
C_5H_6, Cyclopentadiene (gas-ed), 7, 270
C_5H_6, Cyclopentadiene, 30B, 87
C_5H_6, Cyclopropylacetylene (gas-mw), 38B, 1070
$C_5H_6O_3$, 2-Cyclopentene-2,3-diol-1-one, 43B, 165
$C_5H_6O_4$, Cyclopropane-1,1-dicarboxylic acid, 37B, 71
C_5H_7ClO, Cyclobutanecarboxylic acid chloride (gas-ed), 37B, 692
C_5H_8, Cyclopentene (gas-ed), 35B, 896
C_5H_8, Methylenecyclobutane (gas-ed), 9, 366
C_5H_8, Vinylcyclopropane (gas-ed), 40B, 1142
$C_5H_8N_2O_2$, Cyclopropane-1,1-dicarboxamide, 45B, 109
$C_5H_8N_2O_5$, Diammonium croconate, 29, 668
C_5H_8O, Cyclopentanone (gas-ed), 37B, 694
C_5H_8O, Cyclopropyl methyl ketone (gas-ed), 30B, 403
C_5H_{10}, Cyclopentane (gas-ed), 10, 243; 26, 641; 35B, 897

C_5H_{10}, Methylcyclobutane (gas-ed), 16, 498
$C_5H_{12}ClNO_3$, 1-Aminocyclobutanecarboxylic acid hydrochloride monohydrate, 41B, 157
C_6Cl_6, Hexachloro-3,4-dimethylenecyclobutene (gas-ed), 39B, 887
C_6Cl_6, Hexachlorofulvene (gas-ed), 39B, 886
$C_6H_3N_3$, cis-1,2,3-Tricyanocyclopropane, 31B, 85
C_6H_6, Fulvene (gas-mw), 39B, 894
C_6H_6, Trimethylenecyclopropane (gas-ed), 33B, 532
C_6H_6, 3,4-Dimethylenecyclobutene (gas-ed), 33B, 533
$C_6H_6Cl_6$, β-1,2,3,4,5,6-Hexachlorocyclohexane, 42B, 105
$C_6H_6Cl_6$, γ-1,2,3,4,5,6-Hexachlorocyclohexane, 42B, 105
$C_6H_6O_4$, Cyclobut-1-ene-1,2-dicarboxylic acid, 40B, 132
$C_6H_6O_4$, cis-Cyclobut-1-ene-3,4-dicarboxylic acid, 40B, 1078
$C_6H_6O_4$, 3-Methylenecyclopropane-trans-1,2-dicarboxylic acid, 20, 547
$C_6H_8Br_2$, anti,cis,cis-2,2'-Dibromobicyclopropyl, 38B, 168
$C_6H_8Br_2$, 2,2'-Dibromobicyclopropyl (anti-cis,cis) (gas-ed), 39B, 886
$C_6H_8Br_2$, 2,2'-Dibromobicyclopropyl (anti-trans,trans) (gas-ed), 39B, 886
$C_6H_8O_4$, cis-1,2-Cyclobutanedicarboxylic acid, 38B, 169
$C_6H_8O_4$, cis-1,3-Cyclobutanedicarboxylic acid, 34B, 86
$C_6H_8O_4$, trans-1,2-Cyclobutanedicarboxylic acid, 35B, 97
$C_6H_8O_4$, trans-1,3-Cyclobutanedicarboxylic acid, 32B, 115
$C_6H_8O_4$, 1,1-Cyclobutanedicarboxylic acid, 37B, 72
$C_6H_9NO_3$, Methyl 1-carbamoyl-cyclopropane-1-carboxylate, 39B, 91
C_6H_{10}, Bicyclopropyl (gas-ed), 38B, 1062
C_6H_{10}, Bicyclopropyl, 32B, 116
$C_6H_{10}O_2$, trans-2,trans-3-Dimethylcyclopropanecarboxylic acid, 38B, 170
$C_6H_{13}NO_3$, 1-Aminocyclopentane carboxylic acid monohydrate, 38B, 171
$C_6H_{14}ClNO_3$, 1-Aminocyclopentanecarboxylic acid hydrochloride monohydrate, 44B, 117
$C_7H_2N_4$, 1,1,2,2-Tetracyanocyclopropane, 39B, 91; 42B, 32
$C_7H_6O_2$, 6-Hydroxy-1-fulvenecarbaldehyde, 41B, 157
$C_7H_{10}O_4$, (+)-1,2-trans-Cyclopentanedicarboxylic acid, 38B, 172
$C_7H_{10}O_4$, DL-1,2-trans-Cyclopentanedicarboxylic acid, 38B, 172
$C_7H_{12}Cl_2O$, 1-(2,2-Dichloro-3,3-dimethylcyclopropyl)ethanol, 43B, 166
C_7H_{14}, 1,1,2,2-Tetramethylcyclopropane (gas-ed), 16, 497
C_8Cl_8, Perchloro(4)radialene, 35B, 99
$C_8H_4N_4$, cis-trans-cis-1,2,3,4-Tetracyanocyclobutane, 33B, 77; 42B, 99
$C_8H_6N_2$, 1,2-Dimethyl-4,4-triafulvenedicarbonitrile, 44B, 117
C_8H_{10}, Dimethylfulvene (gas-ed), 35B, 899
C_8H_{10}, Dimethylfulvene, 26, 642
$C_8H_{10}Br_2O_4$, cis-1,2-Dibromo-1,2-dicarbomethoxycyclobutane, 31B, 85
$C_8H_{10}Br_2O_4$, trans-1,2-Dibromo-1,2-dicarbomethoxycyclobutane, 31B, 85
$C_8H_{10}O_2$, Dicyclopropylethanedione, 42B, 100
$C_8H_{11}N$, 6-(N,N-Dimethylamino)pentafulvene, 40B, 133
$C_8H_{11}NO_3$, Cyclopentene-1,2-dicarboxylic acid N-methyl-half-amide, 39B, 92
$C_8H_{12}OS$, 2,2,4,4-Tetramethyl-3-thio-1,3-cyclobutanedione, 40B, 134
$C_8H_{12}O_2$, Dimethylketene dimer (gas-ed), 10, 251; 19, 593
$C_8H_{12}O_2$, 2,2,4,4-Tetramethylcyclobutane-1,3-dione (gas-ed), 10, 251; 19, 593
$C_8H_{12}O_2$, 2,2,4,4-Tetramethylcyclobutane-1,3-dione, 20, 549; 40B, 135
$C_8H_{12}O_4$, 2-Methoxymethoxycyclopentene-1-carboxylic acid, 44B, 118
$C_8H_{12}S_2$, 2,2,4,4-Tetramethyl-1,3-cyclobutanedithione, 39B, 92
$C_8H_{13}NO_3$, 2-Nitro-3,5,5-trimethylcyclopentanone, 43B, 167

$C_8H_{16}INO_2$, (+)-(1S,2S)-trans-Acetoxycyclopropyltrimethylammonium iodide, 44B, 118
$C_8H_{16}O_2$, cis-2,2,4,4-Tetramethyl-1,3-cyclobutanediol, 42B, 100
$C_9H_9NO_4$, 2,3-Diacetyl-5-nitrocyclopentadiene, 41B, 159
C_9H_{10}, Phenylcyclopropane (gas-ed), 30B, 404
$C_9H_{10}O_2Cl_4$, cis-1,2-Diacetonyl-1,2,3,3-tetrachlorocyclopropane, 35B, 99
$C_9H_{11}NO$, 2-Formyl-6-(N,N-dimethylamino)pentafulvene, 40B, 133
$C_9H_{12}O_4$, 3-(2-Carboxy-1-cyclopentenyl)propionic acid, 41B, 161
$C_9H_{14}O$, 2,2,4,4-Tetramethyl-3-methylenecyclobutanone, 41B, 162
$C_9H_{14}O_4$, cis-2,5-Dimethylcyclopentane-1,1-dicarboxylic acid, 40B, 135
$C_9H_{14}O_4$, trans-2,5-Dimethylcyclopentane-1,1,-dicarboxylic acid, 44B, 119
$C_9H_{14}O_4$, 1,1-Cyclopentane diacetic acid, 39B, 93
$C_9H_{15}BrN_2O_3$, 1-Methyl-3-imino-4-α-hydroxyethylidenecyclopent-1-ene-2-carbonamide hydrobromide monohydrate, 27, 945
$C_9H_{15}NO_6$, trans-1-Amino-1,3-dicarboxycyclopentane (acetic acid solvate), 41B, 163
$C_9H_{18}ClN_3O_4$, 1,2,3-Trisdimethylaminocyclopropenium perchlorate, 38B, 173
$C_9H_{20}ClNO$, [(1RS,2RS,3SR)-3-Hydroxy-2-methylcyclopentyl]trimethylammonium chloride, 46B, 120
$C_{10}Br_8$, Octabromopentafulvalene, 40B, 136
$C_{10}Cl_8$, Perchlorofulvalene, 26, 643; 39B, 93
$C_{10}Cl_{10}$, synclinal-1,1',2,2',3,3',4,4',5,5'-Decachloro-1,1'-bis(cyclopenta-2,4-dienyl), 42B, 101
$C_{10}H_6Cl_2O$, 2,2-Dichloro-3-phenylcyclobutenone, 30B, 86
$C_{10}H_6F_{12}$, Tetrafluoro-1,2-(RS)-bis(2,2,3,3-tetrafluorocyclobutyl)-ethane, 46B, 120
$C_{10}H_6O_2$, Phenylcyclobutenedione, 29, 665
$C_{10}H_8Cl_2O$, 2,2-Dichloro-3-phenylcyclobutanone, 43B, 168
$C_{10}H_{10}O_2$, 1-Cyclohexenyl-1-cyclobutenedione, 29, 667
$C_{10}H_{11}NO_8$, Dimethylammonium hydrobis(hydrogen squarate), 40B, 136
$C_{10}H_{12}$, Phenylcyclobutane (gas-ed), 33B, 536
$C_{10}H_{14}Cl_4$, meso-2,2,2',2'-Tetrachloro-3,3,3',3'-tetramethyl-bicyclopropyl, 42B, 101
$C_{10}H_{16}N_2$, 6,6-Bis(dimethylamino)fulvene, 40B, 137
$C_{10}H_{18}ClNO_4$, N,N-Dimethyl-2,3,4,4-tetramethylcyclobutenylideneammonium perchlorate, 43B, 168
$C_{11}H_9ClO$, 4-Chloro-2-methyl-3-phenylcyclobut-2-enone, 34B, 87
$C_{11}H_{11}NO_3$, E-2-(p-Nitrophenyl)-cyclopropyl methyl ketone, 38B, 174
$C_{11}H_{12}O_2$, trans-3-Phenylcyclobutanecarboxylic acid, 46B, 10
$C_{11}H_{13}ClO_2S$, 1-Chloro-1-phenylsulphonyl-2,3-dimethylcyclopropane, 37B, 72
$C_{11}H_{14}O_2S$, cis-2-Ethoxy-1-phenylsulfinylcyclopropane, 45B, 109
$C_{11}H_{16}ClN$, N,N-Dimethyl-2-phenylcyclopropylamine hydrochloride, 41B, 163
$C_{11}H_{18}O_2$, cis-3-Cyclohexylcyclobutanecarboxylic acid, 46B, 10
$C_{11}H_{18}O_2$, trans-3-Cyclohexylcyclobutanecarboxylic acid, 46B, 10
$C_{12}H_{14}O_2$, 1-Phenylcyclopentanecarboxylic acid, 41B, 164
$C_{12}H_{16}O_8$, Tetramethyl cyclobutane-1,2,3,4-tetracarboxylate, 37B, 73
$C_{12}H_{18}$, Triisopropylidenecyclopropane, 35B, 100
$C_{12}H_{18}O_4$, Dimethyl 1,1'-dimethylbicyclopropyl-2,2'-dicarboxylate, 37B, 73
$C_{12}H_{20}Br_2O$, cis-2,4-Dibromo-2,4-di-t-butylcyclobutanone, 31B, 84; 40B, 138

$C_{12}H_{20}O_2$, trans-3,4-Di-t-butylcyclobutanedione, 42B, 102
$C_{12}H_{22}BF_4N$, N,N-Dimethyl-2-t-butyl-4,4-dimethylcyclobutenylideneammonium tetrafluoroborate, 43B, 168
$C_{12}H_{23}ClN_2O_4$, N,N-Dimethyl-3-dimethylamino-2,4-diethylcyclobutenylideneammonium perchlorate, 43B, 168
$C_{13}H_9Br_3O$, 1,2,3-Tribromo-6-(o-methoxyphenyl)fulvene, 30B, 88
$C_{13}H_{18}ClNOS$, 2-((p-Chlorophenyl)sulfinyl)-N,N,3,3-tetramethylcyclopropylamine, 43B, 170
$C_{13}H_{18}O_4$, trans-1,2-Bis-(β-carboethoxyvinyl)cyclopropane, 41B, 165
$C_{14}H_8N_4O$, 3-(2-Methoxyphenyl)-1,1,2,2-cyclopropanetetracarbonitrile, 46B, 42
$C_{14}H_{14}Cl_4$, 1,2,3,4-Tetrachloro-5,6-di-n-propylcalicene, 34B, 88
$C_{14}H_{14}O_4$, 3,3-Dimethoxycarbonyl-1-methyl-2-phenylcyclopropene, 45B, 110
$C_{14}H_{16}O_5S$, (SR,RS)-Dimethyl 2-(p-tolylsulfinyl)-1,1-cyclopropanedicarboxylate, 44B, 119
$C_{14}H_{16}O_5S$, (SS,RR)-Dimethyl 2-(p-tolylsulfinyl)-1,1-cyclopropanedicarboxylate, 44B, 119
$C_{14}H_{24}O_3$, 2,5-Di-t-butyl-4,5-dihydroxy-4-methyl-2-cyclopenten-1-one, 45B, 110
$C_{15}H_8Cl_2O$, Bis(p-chlorophenyl)cyclopropenone, 39B, 94
$C_{15}H_{10}Br_2N_2O_4$, 1,1-Dibromo-trans-2,3-bis(4'-nitrophenyl)cyclopropane, 43B, 170
$C_{15}H_{10}Cl_4$, 1,1-Bis-(p-chlorophenyl)-2,2-dichlorocyclopropane, 38B, 174
$C_{15}H_{12}Br_2$, 1,1-Dibromo-trans-2,3-diphenylcyclopropane, 43B, 170
$C_{15}H_{12}Br_2$, 1,1-Dibromo-2,2-diphenylcyclopropane, 41B, 165
$C_{15}H_{12}Cl_2$, 1,1-Dichloro-2,2-diphenylcyclopropane, 41B, 165
$C_{15}H_{12}O_2$, 2,3-Diphenylcyclopropenone monohydrate, 39B, 94
$C_{16}H_{12}Cl_8$, 1,1-Bis(2,3,4,5-tetrachloro-1,3-cyclopentadienyl)cyclohexane, 42B, 102
$C_{16}H_{14}O_2$, S-(+)-2,2-Diphenylcyclopropanecarboxylic acid, 43B, 171
$C_{16}H_{14}O_2$, 2,3-Dicyclopropyl-1,4-naphthoquinone, 45B, 111
$C_{16}H_{20}BrNO$, (+)-trans-Chrysanthemic acid (p-bromoanilide derivative), 41B, 166
$C_{16}H_{27}F_6N$, Tri-n-butylammonio-2,2,3,3,4,4-hexafluorocyclobutanide ylide, 43B, 172
$C_{16}H_{31}ClN_2$, N,N-Dimethyl-3-dimethylamino-2,4-di-t-butylcyclobutenylideneammonium chloride, 43B, 168
$C_{17}H_{16}$, 1,2-Diphenylcyclopentene, 41B, 167
$C_{17}H_{16}O_2$, (R)-(+)-2,2-Diphenyl-1-methylcyclopropanecarboxylic acid, 40B, 139; 43B, 171
$C_{17}H_{29}BrO$, 3-(4-Bromocyclohexyl)-4-(3-oxocyclopentyl)hexane, 37B, 74
$C_{18}H_{10}N_2$, 4,4-Dicyano-2,3-diphenyltriafulvene, 39B, 94; 45B, 112
$C_{18}H_{14}O_4$, 2,4-Di-p-methoxyphenylcyclobutadiene-1,3-quinone, 40B, 139
$C_{18}H_{15}BrO$, trans-1-(2'-(p-Bromophenyl)vinyl)-2-benzoylcyclopropane, 41B, 168
$C_{18}H_{15}NO$, 4-Oxo-1,2-diphenyl-1-cyclopentanecarbonitrile, 45B, 112
$C_{18}H_{17}ClN_2$, 1-(p-Chlorophenyl)-1-cyano-2-(p-dimethylaminophenyl)-cyclopropane, 37B, 75
$C_{18}H_{18}O_2S_2$, 3-Bis(phenylsulfinyl)methyl-1,2-dimethylcyclopropene, 45B, 113
$C_{18}H_{22}Na_2O_{12}$, trans-1,3-Cyclobutanedicarboxylic acid (disodium acid salt), 33B, 78
$C_{18}H_{25}NO_6S$, Penbutolol sulfate, 41B, 169
$C_{18}H_{26}O_6$, 1,2-Bis(1-ethoxycarbonyl-2-oxocyclopentyl)ethane, 46B, 121
$C_{18}H_{30}N_2O_4$, Diethyl 2,4-bis(diethylamino)-1,3-cyclobutadienedicarb-

oxylate, 40B, 140
$C_{18}H_{30}O_2$, Methyl tri-t-butyl(4)annulenecarboxylate, 41B, 169
$C_{18}H_{36}Cl_6N_6Pt_2$, Bis(1,2,3-tris(dimethylamino)cyclopropenium) hexachlorodiplatinate, 40B, 101; 42B, 65
$C_{19}H_{12}Br_2O$, 1,3-Bis(p-bromobenzylidene)-4-cyclopentene-2-one, 23, 634
$C_{19}H_{13}NO_4$, 2,3-Dibenzoyl-5-nitrocyclopentadiene, 41B, 159
$C_{19}H_{14}Br_2O$, 2,5-Di-p-bromobenzilidene-cyclopentanone, 21, 599
$C_{19}H_{14}I_2$, 2,5-Di(p-iodobenzilidene)cyclopentanone, 26, 734
$C_{19}H_{15}IN_2O_4$, Phenyliodonio-2,5-diethoxycarbonyl-3,4-dicyanocyclopentadienide, 44B, 120
$C_{19}H_{16}N_2$, N,N'-Diphenyl-6-aminopentafulvene-2-aldimine, 40B, 141
$C_{19}H_{23}ClO$, 2,5-Di-t-butyl-3-(4-chlorophenyl)cyclopentadienone, 44B, 121
$C_{19}H_{26}O_2$, 5-Vitacid, 37B, 75
$C_{19}H_{28}O_6$, 1,3-Bis(1-ethoxycarbonyl-2-oxocyclopentyl)propane, 46B, 121
$C_{19}H_{36}O_2$, Lactobacillic acid, 24, 527
$C_{20}H_{10}Cl_4$, 1,2,3,4-Tetrachloro-5,6-diphenylcalicene, 35B, 103
$C_{20}H_{14}F_6$, 2,3-Di-p-tolyl-4,4-bis(trifluoromethyl)triafulvene, 43B, 172
$C_{20}H_{18}N_2O_8$, cis-1,2-Dimethoxycarbonyl-1,2-bis(4-nitrophenyl)cyclobutane, 43B, 173
$C_{20}H_{20}Br_2N_2$, N,N'-Bis-p-bromophenyl-2,2,4,4-tetramethyl-cyclobutane-1,3-di-imine, 42B, 103
$C_{21}H_{20}Br_2O_3$, cis-3-Phenoxybenzyl 3-(2,2-dibromovinyl)-2,2-dimethylcyclopropanecarboxylate, 42B, 104
$C_{21}H_{20}Cl_2O_3$, cis-3-Phenoxybenzyl 3-(2,2-dichlorovinyl)-2,2-dimethylcyclopropanecarboxylate, 42B, 104
$C_{21}H_{20}O_4$, Bis(p-acetoxyphenyl)cyclobutylidenemethane, 44B, 122
$C_{22}H_{19}Br_2NO_3$, α-Cyano-3-phenoxybenzyl cis-3-(2,2-dibromovinyl)-2,2-dimethylcyclopropanecarboxylate, 42B, 104
$C_{22}H_{21}NO$, 3-E-Benzylidene-1-t-butyl-2-oxo-4-trans-phenylcyclobutane carbonitrile, 41B, 170
$C_{22}H_{22}O_4$, Bis(p-acetoxyphenyl)cyclopentylidenemethane, 44B, 122
$C_{22}H_{24}O_2$, cis-2,4-Diisopropyl-2,4-diphenyl-1,3-cyclobutandione, 45B, 113
$C_{22}H_{34}O_6$, 1,6-Bis(1-ethoxycarbonyl-2-oxocyclopentyl)hexane, 46B, 121
$C_{22}H_{36}N_2O$, N-Cyclohexyl-1-(N-cyclohexyl-1-(2,3,4,5-tetramethyl-2,4-cyclopentadienyl)-amino)carboxamide, 44B, 122
$C_{23}H_{15}Br_3O$, 4-Bromo-4-(α,2-dibromobenzyl)-2,3-diphenyl-2-cyclobuten-1-one, 44B, 123
$C_{24}H_{30}$, Hexacyclopropylbenzene, 43B, 174
$C_{24}H_{31}ClN_2O_4$, N,N-Diethyl-3-diethylamino-2,4-diphenylcyclobutenylideneammonium perchlorate, 43B, 168
$C_{25}H_{33}ClN_2O_4S_2$, N,N-Diethyl-3-diethylamino-4-methyl-2,4-bis(phenylthio)cyclobutenylideneammonium perchlorate, 43B, 168
$C_{26}H_{20}Cl_2N_2$, 1,3-trans-Bis(4-chlorophenyl)-2,4-trans-di(4-pyridyl)-cyclobutane, 46B, 122
$C_{28}H_{24}$, 1,2,3,4-Tetraphenylcyclobutane, 12, 402; 30B, 85
$C_{29}H_{21}Br$, 5-Bromo-1,2,3,4-tetraphenylcyclopentadiene, 37B, 76
$C_{29}H_{22}$, 1,2,3,4-Tetraphenylcyclopentadiene, 37B, 77
$C_{29}H_{22}O$, trans-2,3,4,5-Tetraphenylcyclopent-2-enone, 43B, 174
$C_{30}H_{24}Cl_2$, 2,2',3,3'-Tetraphenyl-3,3'-dichlorodicyclopropane, 41B, 171
$C_{30}H_{24}O_2$, cis-2,4-Dibenzyl-2,4-diphenyl-1,3-cyclobutandione, 45B, 113

$C_{30}H_{24}O_2$, 1,2-Dibenzoyl-3,4-diphenylcyclobutane, 44B, 123
$C_{30}H_{26}O_2$, 2-Benzoyl-1,3,4-triphenyl-1-cyclopentanol, 44B, 124
$C_{56}H_{40}$, 1,3-Bis(diphenylvinylidene)-2,2,4,4-tetraphenylcyclobutane, 40B, 142

21. MONOCYCLIC HYDROCARBONS (6-MEMBERED RINGS)

$C_6Cl_2F_5N$, N,4-Dichloropentafluorocyclohexa-2,5-dienylideneamine (gas-ed), 42B, 1029
C_6Cl_6O, Hexachlorocyclohexa-2,5-dien-1-one, 41B, 172
C_6F_{12}, Dodecafluoro-cyclohexane (gas-ed), 33B, 532
$C_6N_4O_4$, 3,6-Bis(diazo)cyclohexantetraone, 34B, 89
$C_6H_2Br_2Cl_2O$, 2,6-Dibromo-4,4-dichlorocyclohexa-2,5-diene-1-one, 33B, 79
$C_6H_2Cl_2N_2O$, 2,6-Dichloro-4-diazo-2,5-cyclohexadiene-1-one, 35B, 104
$C_6H_2Cl_4O$, 2,4,4,6-Tetrachlorocyclohexa-2,5-diene-1-one, 33B, 79
$C_6H_4F_8$, Octafluorocyclohexane, 44B, 125
$C_6H_5Cl_5$, 2,3,4,5,6-Pentachlorocyclohexene, 15, 456
$C_6H_6Br_6$, Hexabromocyclohexane, 1, 622, 715
$C_6H_6Cl_6$, Hexachlorocyclohexane (cubic), 1, 622, 715
$C_6H_6Cl_6$, Hexachlorocyclohexane (gas-ed), 9, 367
$C_6H_6Cl_6$, racemic-α-1,2,3,4,5,6-Hexachlorocyclohexane, 42B, 106
$C_6H_6Cl_6$, δ-1,2,3,4,5,6-Hexachlorocyclohexane, 42B, 106
$C_6H_6Cl_6$, ϵ-1,2,3,4,5,6-Hexachlorocyclohexane, 13, 526
$C_6H_6Cl_6$, γ-1,2,3,4,5,6-Hexachlorocyclohexane, 13, 524
$C_6H_7Cl_3O$, 2,2,4-Trichlorocyclohexanone, 43B, 166
$C_6H_7Cl_3O$, 2,3,6-Trichlorocyclohexanone, 43B, 175
$C_6H_7Cl_3O$, 2,4,6-Trichlorocyclohexanone, 38B, 175
$C_6H_7Cl_5$, Pentachlorocyclohexane, 44B, 125
C_6H_8, Cyclohexa-1,3-diene (gas-ed), 33B, 534; 40B, 1141
C_6H_8, Cyclohexa-1,4-diene (gas-ed), 40B, 1141
$C_6H_8Br_2Cl_2$, 1,2-Dichloro-4,5-dibromocyclohexane, 18, 697
$C_6H_8Br_4$, 1,2,3,4-Tetrabromocyclohexane, 13, 523
$C_6H_8Br_4$, 1,2,4,5-Tetrabromocyclohexane, 7, 261; 16, 501
$C_6H_8Cl_4$, 1,1,4,4-Tetrachlorocyclohexane, 18, 696
$C_6H_8Cl_4$, 1,2,4,5-Tetrachlorocyclohexane, 12, 374
$C_6H_8O_2$, Cyclohexane-1,4-dione, 29, 491
$C_6H_9Cl_3$, cis-1,3,5-Trichlorocyclohexane, 38B, 175
C_6H_{10}, Cyclohexene (gas-ed), 35B, 897; 40B, 1141
$C_6H_{10}BrCl$, cis-1-Bromo-4-chlorocyclohexane (gas), 24, 582
$C_6H_{10}BrCl$, trans-1-Bromo-4-chlorocyclohexane, 17, 695
$C_6H_{10}Br_2$, trans-1,4-Dibromocyclohexane (gas-ed), 23, 635
$C_6H_{10}Cl_2$, trans-1,4-Dichlorocyclohexane (gas-ed), 23, 635
$C_6H_{10}Cl_2$, trans-1,4-Dichlorocyclohexane, 29, 490
$C_6H_{10}N_2O_2$, Cyclohexane-1,4-dioxime, 33B, 80
$C_6H_{10}O$, Cyclohexanone (gas-ed), 20, 553
$C_6H_{10}O$, Cyclohexanone (gas-mw), 33B, 540
$C_6H_{11}Cl$, Chlorocyclohexane (gas-ed), 9, 367
C_6H_{12}, Cyclohexane (gas-ed), 9, 366; 28, 666; 37B, 693; 39B, 891; 42B, 1027
C_6H_{12}, Cyclohexane, 6, 242; 39B, 95
$C_6H_{12}O_6$, epi-Inositol, 37B, 239
$C_6H_{12}O_6$, myo-Inositol, 29, 487
$C_6H_{14}ClN$, Cyclohexylamine hydrochloride, 19, 558; 34B, 92
$C_6H_{15}BrN_2$, 1,2-trans-Diaminocyclohexane hydrobromide, 42B, 106
$C_6H_{16}Cl_2N_2$, 1,3-trans-Diaminocyclohexane dihydrochloride, 30B, 89

$C_6H_{16}Cl_2N_2$, 1,4-trans-Diaminocyclohexane dihydrochloride, 31B, 87
$C_6H_{16}Cl_2N_2O_3$, 2-Deoxy-cis-inosa-1,3-diamine dihydrochloride, 32B, 117
$C_6H_{16}O_5$, α-Phloroglucitol dihydrate, 11, 614
$C_6H_{16}O_8$, myo-Inositol dihydrate, 28, 425
$C_6H_{19}N_3O_2$, 1,3,5-Triaminocyclohexane dihydrate, 15, 452
$C_7H_9Cl_3O$, 2,3,6-Trichloro-4-methylcyclohexanone, 44B, 126
$C_7H_{12}O$, 1-Methoxycyclohexene (gas-ed), 39B, 889
$C_7H_{12}O$, 2-Methylcyclohexanone (gas-ed), 20, 553
$C_7H_{13}NO_2$, 1-Aminocyclohexanecarboxylic acid, 41B, 534
C_7H_{14}, Methylcyclohexane (gas-ed), 37B, 693
$C_7H_{14}ClNO_2$, 1-Aminocyclohexane-1-carboxylic acid hydrochloride, 37B, 77; 41B, 535
$C_7H_{14}N_2O$, cis-N-Cyclohexyl-N'-methyldiimide N-oxide, 43B, 175
$C_7H_{14}N_2O$, cis-N-Methyl-N'-cyclohexyldiimide N-oxide, 43B, 175
$C_8H_{11}Cl_3O$, 2,2,3-Trichloro-4,4-dimethylcyclohexanone, 44B, 126
$C_8H_{11}O_5RbS$, Rubidium 5,5-dimethylcyclohexan-1,3-dione-2-sulphonate, 40B, 1078
$C_8H_{12}Cl_2O$, 2,2-Dichloro-4,4-dimethylcyclohexanone, 40B, 143
$C_8H_{12}O_2$, Dimedone, 41B, 172
$C_8H_{12}O_2$, 5,5-Dimethyl-1,3-cyclohexanedione, 40B, 143
$C_8H_{12}O_4$, (+)-trans-1,2-Cyclohexane dicarboxylic acid, 34B, 90
$C_8H_{12}O_4$, DL-trans-1,2-Cyclohexane dicarboxylic acid, 34B, 91
$C_8H_{12}O_4$, cis-1,2-Cyclohexanedicarboxylic acid, 35B, 106
$C_8H_{12}O_4$, trans-Cyclohexane-1,4-dicarboxylic acid, 31B, 87; 38B, 176
$C_8H_{14}O$, 4,4-Dimethylcyclohexanone, 43B, 176
$C_8H_{15}NO_2$, trans-4-Aminomethylcyclohexanecarboxylic acid, 33B, 82
C_8H_{16}, 1,1-Dimethylcyclohexane (gas-ed), 38B, 1066
$C_8H_{16}BrNO_2$, cis-4-Aminomethylcyclohexanecarboxylic acid hydrobromide, 30B, 90
$C_8H_{16}BrNO_2$, trans-4-Aminomethylcyclohexane-1-carboxylic acid hydrobromide, 31B, 88
$C_8H_{16}ClNO_2$, cis-4-Aminomethylcyclohexane-1-carboxylic acid hydrochloride, 31B, 88
$C_8H_{16}O_2$, rel(1S,2R)-2-Hydroxymethyl-2-methylcyclohexanol, 46B, 122
$C_8H_{20}NO_4P$, Cyclohexylammonium ethyl hydrogen phosphate, 45B, 114
$C_9H_8KN_3O_7$, Potassium 1-acetonyl-2,4,6-trinitrocyclohexa-2,5-dienate, 42B, 62
$C_9H_9Br_2NO_3$, (+)-Aeroplysinin-I, 38B, 177
$C_9H_{12}O_2$, 2,6-Dimethyl-1,4-dihydrobenzoic acid, 46B, 73
$C_9H_{12}O_2$, 3,5-Dimethyl-1,4-dihydrobenzoic acid, 46B, 73
$C_9H_{16}O$, 2,2,6-Trimethylcyclohexanone (gas-ed), 42B, 1028
$C_{10}H_{12}Cl_2O$, 2,4-Dimethyl-4-(1,2-dichlorovinyl)-cyclohexene-3-one, 45B, 115
$C_{10}H_{12}I_2O$, 4-Diiodomethyl-3,4,5-trimethylcyclohexa-2,5-dien-1-one, 41B, 173
$C_{10}H_{12}O_6$, Dimethyl 2,5-dihydroxy-1,4-cyclohexadiene-1,4-dicarboxylate (form 1), 28, 427
$C_{10}H_{12}O_6$, Dimethyl 2,5-dihydroxy-1,4-dydlohexadiene-1,4-dicarboxylate (form 2), 28, 427
$C_{10}H_{16}Br_2O$, 2,6-Dibromo-3,3,5,5-tetramethylcyclohexanone, 29, 493
$C_{10}H_{16}Cl_2O$, 2,2-Dichloro-4-t-butylcyclohexanone, 39B, 96
$C_{10}H_{16}O_2$, 3-Acetyl-2-vinyl-cyclohexan-1-ol, 45B, 115
$C_{10}H_{16}O_3$, trans-2,4-Dihydroxy-2,4-dimethylcyclohexane-trans-1-acetic acid γ-lactone, 37B, 78
$C_{10}H_{16}O_4$, Cyclohexane-1,1'-diacetic acid, 35B, 107
$C_{10}H_{16}O_4$, 3,3,6,6-Tetramethoxy-1,4-cyclohexadiene, 42B, 107

$C_{10}H_{17}BrO$, 2-Bromo-3,3,5,5-tetramethylcyclohexanone, 29, 493
$C_{10}H_{17}ClO$, cis-2-Chloro-4-t-butylcyclohexanone, 38B, 178
$C_{10}H_{18}ClN_3O_2$, 1-(2-Chloroethyl)-3-(trans-4-methylcyclohexyl)-1-nitrosourea, 44B, 127
$C_{10}H_{18}O$, 4-t-Butylcyclohexanone, 41B, 174
$C_{10}H_{19}Cl$, cis-4-t-Butyl-1-chlorocyclohexane (gas-ed), 40B, 1142
$C_{10}H_{19}Cl$, trans-4-t-Butyl-1-chlorocyclohexane (gas-ed), 40B, 1142
$C_{10}H_{20}O_2$, 4α-t-Butylcyclohexane-1β,2β-diol, 38B, 178
$C_{11}H_{16}O_5$, 2β,4β-Diacetoxy-3β-methylcyclohexanone, 44B, 127
$C_{11}H_{17}NO$, trans-3-t-Butyl-4-cyanocyclohexane, 44B, 128
$C_{11}H_{18}O_2$, 4-t-Butyl-1-cyclohexene-1-carboxylic acid, 44B, 128
$C_{11}H_{20}O_2$, cis-4-t-Butylcyclohexane-1-carboxylic acid, 38B, 179
$C_{12}H_8N_4$, 1,4-Bis(dicyanomethylene)cyclohexane, 35B, 108
$C_{12}H_{10}Cl_6$, 1,2,3,4,5,6-Hexachloro-1-phenylcyclohexane, 38B, 180
$C_{12}H_{16}O_4$, 1,2-Dicarbomethoxy-4,5-dimethylcyclohexa-1,4-diene, 44B, 128
$C_{12}H_{16}O_6$, Diethyl succinylsuccinate, 39B, 96
$C_{12}H_{17}NO_2$, 2-(β-Cyanoethyl)-2,5,5-trimethylcyclohexane-1,3-dione, 42B, 107
$C_{12}H_{18}O_2$, 3-(1,1,5-Trimethyl-5-cyclohexene-6-yl)-propenoic acid, 38B, 180
$C_{12}H_{19}NO_3$, 4-Diethylcarbamoyl-1-cyclohexene-5-carboxylic acid, 35B, 111
$C_{12}H_{20}$, Bicyclohexylidene, 30B, 91
$C_{12}H_{22}N_2O_2$, Bis(nitrosocyclohexane), 35B, 111
$C_{12}H_{22}O_6$, 1,1'-Dihydroperoxycyclohexanylperoxide-1,1', 40B, 1079
$C_{13}Cl_{12}$, Perchlorobenzylidenecyclohexa-2,5-diene, 44B, 129
$C_{13}H_{18}O$, Δ(1,1')-Dicyclohexenyl ketone, 41B, 174
$C_{13}H_{18}O_3$, Methyl 6-isobutenyl-2-methyl-4-oxocyclohex-2-enecarboxylate, 45B, 116
$C_{13}H_{18}O_3S$, Cyclohexyltosylate, 37B, 78
$C_{13}H_{19}NO_2$, cis-1-Acetoxy-3-t-butyl-4-cyanocyclohexene, 44B, 130
$C_{13}H_{20}O_2$, 4-(1-Hydroxy-2,6,6-trimethyl-2-cyclohexenyl)-3-buten-2-one, 44B, 130
$C_{14}H_{16}BrNO_3$, 2-(α-p-Bromophenyl-β-nitroethyl)cyclohexanone, 31B, 89
$C_{14}H_{16}ClFO_3$, (1S,2S)-2-((S)-Chlorofluoroacetoxy)-1-phenylcyclohexanol, 39B, 97
$C_{14}H_{20}O_2S$, (1SR,SRS)-1-(1-Phenylsulfinylcyclohexyl)ethanol, 43B, 176
$C_{14}H_{24}O_2$, Octamethyl-1,4-cyclohexanedione, 42B, 108
$C_{14}H_{28}$, cis-1,4-Di-t-butylcyclohexane (gas-ed), 39B, 890
$C_{15}H_{18}BrNO_3$, 2-(α-p-Bromophenyl-β-nitro)ethyl-5-methylcyclohexanone, 40B, 144
$C_{15}H_{19}ClO$, 3-(p-Chlorophenyl)-3,5,5-trimethylcyclohexanone, 38B, 180
$C_{15}H_{19}IO_2S$, 1-Iodo-1-cyclohexyl-2-p-toluenesulphonylethylene, 41B, 175
$C_{15}H_{20}O_3$, 1-Hydroxy-α-methyl-2-phenylcyclohexaneacetic acid, 45B, 116
$C_{15}H_{21}NO$, 3,3,5-Trimethyl-5-phenyl-cyclohexanone-oxime, 45B, 117
$C_{15}H_{26}O$, cis-1,3-Di-t-butyl-2-carbonylcyclohexane, 45B, 118
$C_{15}H_{26}OS$, cis-1,3-Di-t-butyl-2-thiocarbonylcyclohexane syn-S-oxide, 45B, 118
$C_{15}H_{26}S$, cis-1,3-Di-t-butyl-2-thiocarbonylcyclohexane, 45B, 118
$C_{16}H_{18}BrNO_2$, Cyclohexanonecyanohydrin 3-p-bromophenylpropionate, 37B, 78
$C_{16}H_{19}BrO_5S$, 2-p-Bromophenylthiomethyl-4,5,6-trihydroxy-cyclohex-2-enone acetone solvate, 41B, 176
$C_{16}H_{22}N_2S_2$, 4,4'-R,R-1,2-Cyclohexanediimine-di(3-penten-2-thione),

44B, 131
$C_{16}H_{28}O_4$, trans-1,2-Bis-(2-carboxymethyl-2-propyl)cyclohexane, 39B, 97
$C_{17}H_{19}NO_6$, 2,trans-4-Diacetyl-cis-5-methyl-r-3-(p-nitrophenyl)cyclohex-1-en-1,trans-5-diol, 43B, 178
$C_{17}H_{21}BrO_2$, trans-4-t-Butylcyclohex-2-enyl p-bromobenzoate, 44B, 131
$C_{17}H_{21}Br_2NO_4$, cis-2,trans-3-Dibromo-cis-4-t-butylcyclohexyl p-nitrobenzoate, 42B, 109
$C_{17}H_{22}BrNO_4$, trans-2-Bromo-trans-6-t-butylcyclohexyl p-nitrobenzoate, 42B, 109
$C_{17}H_{23}BrO_2$, trans-4-t-Butylcyclohexanol p-bromobenzoate, 38B, 181
$C_{17}H_{24}O_2$, 1-Phenyl-4-t-butylcyclohexane-1-carboxylic acid, 40B, 145
$C_{17}H_{24}O_{12}$, DL-2-Acetoxymethyl-1,3,4,6-tetra-O-acetylepiinositol, 39B, 98
$C_{17}H_{26}O_2$, 1-(4-Methoxyphenyl)-2,2,6,6-tetramethylcyclohexanol, 39B, 101
$C_{17}H_{26}O_3S$, cis-4-t-Butylcyclohexyl p-toluenesulphonate (form A), 39B, 99
$C_{17}H_{26}O_3S$, cis-4-t-Butylcyclohexyl p-toluenesulphonate (form B), 38B, 182
$C_{17}H_{26}O_3S$, trans-4-t-Butylcyclohexyl toluene-p-sulphonate, 39B, 100; 41B, 177
$C_{18}H_{18}O$, 4,4-Diphenylcyclohexanone, 34B, 93
$C_{18}H_{23}BrO_3$, (-)Menthyl-p-bromophenylglyoxylate, 41B, 178
$C_{18}H_{24}$, Hexaethylidenecyclohexane, 41B, 1236
$C_{18}H_{25}NO_3$, 2-(α-Phenyl-β-nitro)ethyl-4-t-butyl-cyclohexanone, 41B, 179
$C_{18}H_{26}O_2$, 1,1',5,5,5',5'-Hexamethyl-(4,4'-bicyclohexenyl)-3,3'-dione, 46B, 123
$C_{18}H_{27}IO_2S$, 2-(Cyclohexyl(phenyl)acetoxy)ethyldimethylsulphonium iodide, 41B, 180
$C_{18}H_{27}N$, 1-Piperidino-1-benzylcyclohexane, 40B, 145
$C_{18}H_{30}BrNO$, 2-(cis-4-Bromo-1-methylcyclohexyl)-3-(4-hydroxylimino-1-methylcyclohexyl)trans-2-butene, 34B, 93
$C_{18}H_{30}Cl_2$, 2,3-Bis-(cis-4-chloro-1-methylcyclohexyl)-trans-2-butene, 33B, 82
$C_{18}H_{30}O_2$, rac-Bis(4-oxocyclohexyl)-3,4-hexane, 45B, 118
$C_{18}H_{32}$, p-Tercyclohexane (form I), 42B, 109
$C_{18}H_{32}$, p-Tercyclohexane-2, 38B, 183
$C_{18}H_{34}O_2$, meso-Bis(trans-4-hydroxycyclohexyl)-3,4-hexane, 45B, 119
$C_{18}H_{34}O_2$, rac-Bis(trans-4-hydroxycyclohexyl)-3,4-hexane, 45B, 119
$C_{18}H_{36}$, trans-Hexaethylcyclohexane, 33B, 83
$C_{19}H_{19}BrO$, 1-Bromo-1-benzoyl-2-phenylcyclohexane, 38B, 184
$C_{19}H_{22}N_2O_3$, 1-Hydroperoxy-1-(o-methoxyphenylazo)-2-phenylcyclohexane, 43B, 178
$C_{19}H_{23}BrO$, trans-3-(4'-Bromo-1-naphthyl)-1,5,5-trimethylcyclohexanol, 40B, 146
$C_{20}H_{18}Br_2$, cis,trans-1,2-Di(4-bromobenzylidene)cyclohexane, 40B, 1079
$C_{20}H_{24}BrNO$, 4-(p-Bromophenyl)imino-2,6-di-t-butylcyclohexa-2,5-diene-1-one, 37B, 79
$C_{20}H_{28}ClNO_3$, 4-Dimethylamino-2-butynyl 2-cyclohexyl-2-hydroxy-2-phenylacetate hydrochloride, 45B, 120
$C_{20}H_{38}$, 3,4-Dicyclohexyl-3,4-dimethylhexane, 46B, 41
$C_{21}H_{22}O_2$, α-Acetoxy-α,2-anti-diphenylmethylenecyclohexane, 40B, 147
$C_{21}H_{22}O_2$, α-Acetoxy-α,2-syn-diphenylmethylenecyclohexane, 42B, 110
$C_{21}H_{30}BrNO_2$, 1-Hydroxymethyl-2-(1-methylpentyl)-3-methylcyclohex-1-

ene p-bromophenylurethane, 45B, 120
$C_{21}H_{34}O_3S$, cis,trans-2,5-Di-t-butylcyclohexanol toluene-p-sulphonate, 40B, 148
$C_{22}H_{27}NS$, N-Cyclohexyl-2,3,3-triphenyl-thioacrylamide, 44B, 133
$C_{22}H_{38}$, meso-3,4-Di(1-cyclohexen-1-yl)-2,2,5,5-tetramethylhexane, 44B, 132
$C_{22}H_{42}$, DL-3,4-Dicyclohexyl-2,2,5,5-tetramethylhexane, 44B, 133
$C_{22}H_{42}$, meso-3,4-Dicyclohexyl-2,2,5,5-tetramethylhexane, 44B, 133
$C_{22}H_{42}INO_2$, (Dicyclohexylacetoxyethyl)triethylammonium iodide, 46B, 27
$C_{23}H_{25}NO_4$, Benzyl trans-6-(ethoxycarbonyl)-cis-6-phenyl-2-cyclohexene-1-carbamate, 45B, 121
$C_{24}H_{31}BrO_4$, p-Bromophenacyl 1-(1',5'-dimethyl-1'-hydroxy-4'-hexene)-4-methyl-3-cyclohexenecarboxylate, 41B, 110
$C_{24}H_{34}K_2O_{12}$, Monopotassium sesqui(cyclohexane-1,4-dicarboxylate), 35B, 112; 38B, 184
$C_{24}H_{36}BrNO_2$, (+)-trans-2-(o-Bromophenyl)cyclohexylamine menthoxyacetamide, 42B, 111
$C_{26}H_{26}Br_2O_6$, trans-2,5-Di-p-bromobenzyl-2,5-diethoxycarbonylcyclohexane-1,4-dione, 42B, 111
$C_{26}H_{46}$, 1,1,2,2-Tetracyclohexylethane, 45B, 121
$C_{34}H_{28}O_4$, 1,2,4,5-Tetraphenyl-3,6-dicarbomethoxycyclohexa-1,4-diene, 44B, 128
$C_{36}H_{46}N_2$, 1,14-Bis-(2',6',6'-trimethylcyclohex-1'-enyl)-3,12-dimethyl-tetradeca-1,3,5,7,9,11,13-heptaene-6,9-nitrile, 37B, 80
$C_{42}H_{38}O_2$, 1-Diphenylmethylene-4-triphenylmethyl-2,5-cyclohexadiene ethyl acetate solvate, 44B, 134

22. MONOCYCLIC HYDROCARBONS (7,8-MEMBERED RINGS)

C_7F_6O, Hexafluorotropone, 41B, 181
C_7H_5ClO, 2-Chlorotropone, 37B, 81
$C_7H_5NaO_2$, Sodium tropolonate, 20, 554; 40B, 1079
C_7H_6O, Tropone (gas-ed), 21, 569; 22, 678; 38B, 1063
$C_7H_6O_2$, Tropolone (gas-ed), 17, 710
$C_7H_6O_2$, Tropolone, 39B, 102
$C_7H_7ClO_2$, Tropolone hydrochloride, 18, 743; 21, 569
$C_7H_7ClO_4$, Tropylium perchlorate, 24, 583
C_7H_7I, Tropylium iodide, 24, 583
C_7H_8, 1,3,5-Cycloheptatriene (gas-ed), 29, 503
C_7H_{10}, 1,3-Cycloheptadiene (gas-ed), 31B, 93; 38B, 1062
$C_8H_3F_5O_2$, 3-Methoxypentafluorotropone, 42B, 111
$C_8H_6N_2O_4$, 1,4-Dinitrocyclooctatetraene, 44B, 134
$C_8H_7BrO_2$, 3-Bromo-2-methoxytropone, 29, 675
$C_8H_7BrO_2$, 7-Bromo-2-methoxytropone, 29, 675
C_8H_8, 1,3,5,7-Cyclo-octatetraene (gas-ed), 16, 505; 21, 571
$C_8H_{10}Br_2$, syn-3,7-Dibromo-cis,cis-cycloocta-1,5-diene, 38B, 185
C_8H_{12}, Cyclooctyne (gas-ed), 37B, 695
C_8H_{12}, 1,3-Cyclooctadiene (gas-ed), 35B, 899
$C_8H_{12}O_2$, Cyclooctane-1,5-dione, 45B, 123
C_8H_{14}, trans-Cyclooctene (gas-ed), 39B, 888
$C_8H_{14}Cl_2$, trans-1,4-Dichlorocyclooctane, 34B, 95
$C_8H_{14}N_2O_2$, Cyclooctane-1,5-dione dioxime, 42B, 112; 45B, 122
$C_8H_{16}O_2$, cis-Cyclooctane-1,5-diol, 45B, 123
$C_8H_{18}BrNO_3$, 1-Aminocycloheptane carboxylic acid hydrobromide monohydrate, 37B, 81

$C_9H_8O_2$, 1,3,5,7-Cyclo-octatetraenecarboxylic acid, 30B, 92
$C_9H_9BrO_2$, Acetoxytropylium bromide, 42B, 63
$C_9H_{12}N_2$, 1-Methylamino-7-methylimino-1,3,5-cycloheptatriene, 32B, 118
$C_9H_{18}BrNO_2$, 1-Aminocyclooctane carboxylic acid hydrobromide, 37B, 82
$C_{10}H_6N_2$, 8,8-Dicyanoheptafulvene, 31B, 91
$C_{10}H_{10}CaO_6$, Calcium 2,4,6,8-cyclooctatetraene-1,2-dicarboxylate dihydrate, 38B, 186
$C_{10}H_{12}O_2$, 4-Isopropyltropolone, 38B, 187
$C_{10}H_{16}O_4$, Cyclooctane-cis-1,2-dicarboxylic acid, 33B, 85
$C_{10}H_{16}O_4$, Cyclooctane-1,2-trans-dicarboxylic acid, 31B, 92
$C_{11}H_{13}BrO_5$, 4-Bromo-2,3-dicarbomethoxy-2-cyclohepten-1-one, 40B, 148
$C_{11}H_{14}O_5$, 1-Hydroxy-2,3-dicarbomethoxy-1,3-cycloheptadiene, 42B, 113
$C_{12}Cl_{14}$, Perchloro-4,7-dimethyl-3,8-dimethylene-cyclo-octa-1,4,6-triene, 33B, 86
$C_{12}H_{10}N_2$, 1,6-Dimethyl-8,8-dicyanoheptafulvene, 40B, 149
$C_{12}H_{16}$, 1,3,5,7-Tetramethylcycloocta-cis,cis,cis,cis-1,3,5,7-tetraene, 34B, 95
$C_{12}H_{16}Br_2O_4$, cis-5,8-Dibromo-trans-cyclooct-2-ene-1,4-dione bisethyleneketal, 42B, 113
$C_{12}H_{18}O_5$, 1,2-Epoxycyclooctane-trans-1,5-diol diacetate, 42B, 113
$C_{13}H_{12}N_2$, 1-Isopropyl-8,8-dicyanoheptafulvene, 40B, 150
$C_{14}H_9ClO_3$, Tropolonyl p-chlorobenzoate, 37B, 83
$C_{14}H_{12}$, Heptafulvalene, 38B, 187
$C_{14}H_{17}BrO_2$, Cycloheptyl p-bromobenzoate, 46B, 123
$C_{16}H_{24}$, Octamethylcyclooctatetraene, 38B, 188
$C_{16}H_{36}CaO_9$, Calcium cycloheptanecarboxylate pentahydrate, 43B, 179
$C_{19}H_{15}BF_4$, o-Tropylbiphenyl tetrafluoroborate, 45B, 123
$C_{21}H_{20}$, 2,5-Dimethyl-3,4-diphenyl-1,3,5-cycloheptatriene, 45B, 124
$C_{23}H_{35}BrO_2$, cis-3,7-Di-t-butylcyclooctyl cis-p-bromobenzoate, 43B, 180
$C_{25}H_{28}O_4$, Bis(p-acetoxyphenyl)cyclooctylidenemethane, 44B, 122
$C_{31}H_{24}$, 2,3,4,5-Tetraphenyl-1,3,5-cycloheptatriene, 46B, 124
$C_{38}H_{24}Cl_4$, (E)-2,2',3,3'-Tetrachloro-4,4',7,7'-tetraphenyl-1,1'-bicycloheptatrienylidene, 46B, 124
$C_{56}H_{40}$, Octaphenylcyclo-octatetraene, 30B, 93

23. MONOCYCLIC HYDROCARBONS (9- AND HIGHER-MEMBERED RINGS)

$C_9H_{17}NO$, Cyclononane oxime, 46B, 125
$C_9H_{20}BrN$, Cyclononylamine hydrobromide, 24, 587
$C_{10}H_{10}Cl_2$, cis-1,6-Dichlorocyclodeca-1,3,6,8-tetraene, 35B, 113
$C_{10}H_{10}O_2$, cis-trans-cis-Cyclodeca-2,4,8-triene-1,6-dione, 37B, 83
$C_{10}H_{14}Br_2O_2$, trans-3,10-Dibromocyclodeca-1,2-dione, 35B, 114
$C_{10}H_{16}$, cis,cis-Cyclodeca-1,6-diene (gas-ed), 40B, 1140
$C_{10}H_{18}Br_2$, 1,6-trans-Dibromocyclodecane, 29, 509
$C_{10}H_{18}O$, Cyclodecanone, 42B, 114
$C_{10}H_{19}NO$, Cyclodecaneoxime, 45B, 125
$C_{10}H_{20}$, Cyclodecane (gas-ed), 39B, 889
$C_{10}H_{20}O_2$, Cyclodecane-1,6-trans-diol, 39B, 103
$C_{10}H_{22}ClN \cdot 1.5\ H_2O$, Cyclodecylamine hydrochloride sesquihydrate, 29, 511
$C_{10}H_{24}Cl_2N_2$, 1,6-trans-Diaminocyclodecane dihydrochloride (monoclinic), 26, 645; 31B, 93
$C_{10}H_{24}Cl_2N_2$, 1,6-trans-Diaminocyclodecane dihydrochloride (triclinic), 26, 645

$C_{10}H_{28}Cl_2N_2O_2$, 1,6-cis-Diaminocyclodecane dihydrochloride dihydrate, 26, 645
$C_{11}H_{20}O$, Cycloundecanone, 40B, 150
$C_{11}H_{21}NO$, Cycloundecaneoxime, 45B, 125
$C_{12}H_{18}$, 1,5,9-trans-trans-trans-Cyclododecatriene, 32B, 119
$C_{12}H_{18}Br_2O_2$, 2,7-Dibromo-3,8-dimethoxy-trans,trans-cyclodeca-1,6-diene, 39B, 103
$C_{12}H_{20}Br_2O$, 2,12-Dibromocyclododecanone, 34B, 96
$C_{12}H_{22}O$, Cyclododecanone, 45B, 126
$C_{12}H_{23}NO$, Cyclododecaneoxime, 45B, 125
$C_{12}H_{24}$, Cyclododecane, 22, 681; 24, 590
$C_{12}H_{24}O_4$, 1,1-Dihydroperoxycyclododecane, 41B, 183
$C_{13}H_{24}O$, 4,4,7,7-Tetramethylcyclononanone, 42B, 115
$C_{13}H_{25}NO$, Cyclotridecaneoxime, 45B, 125
$C_{14}H_{10}$, 1,8-bis-Dehydro-[14]annulene, 31B, 94
$C_{14}H_{14}$, [14]Annulene, 38B, 189
$C_{14}H_{26}O$, Cyclotetradecanone, 41B, 183
$C_{14}H_{27}NO$, Cyclotetradecaneoxime, 45B, 125
$C_{14}H_{28}$, Cyclotetradecane, 42B, 115
$C_{15}H_{28}O_2$, 1,1,5,5-Tetramethylcyclodecane-8-carboxylic acid, 32B, 484
$C_{16}H_{16}$, [16]Annulene, 35B, 115
$C_{16}H_{19}BrN_4O_4$, 5-Bromomethylenecyclononanone-2,4-dinitrophenylhydrazone, 43B, 180
$C_{16}H_{19}NO_2$, 1-N-Phenylcarbamatocyclonona-2,3-diene, 43B, 181; 46B, 126
$C_{16}H_{23}N_3O$, Cyclononane phenylsemicarbazone, 46B, 125
$C_{17}H_{21}NO_2$, 1-N-Phenylcarbamatocyclodeca-2,3-diene, 46B, 126
$C_{17}H_{25}N_3O$, Cyclodecane phenylsemicarbazone, 46B, 127
$C_{18}H_{18}$, [18]Annulene, 30B, 94
$C_{18}H_{23}NO_2$, 1-N-Phenylcarbamatocycloundeca-2,3-diene, 46B, 126
$C_{18}H_{27}N_3O$, Cycloundecane phenylsemicarbazone, 46B, 127
$C_{19}H_{29}N_3O$, Cyclododecane phenylsemicarbazone, 46B, 127
$C_{20}H_{31}N_3O$, Cyclotridecane phenylsemicarbazone, 46B, 127
$C_{20}H_{40}$, 1,1,9,9-Tetramethylcyclohexadecane, 40B, 151
$C_{22}H_{35}N_3O$, Cyclopentadecanone phenylsemicarbazone, 45B, 126; 46B, 129
$C_{23}H_{37}N_3O$, Cyclohexadecane phenylsemicarbazone, 46B, 127
$C_{24}H_{46}O_6$, 1,1'-Dihydroperoxycyclododecanyl-peroxide-1,1', 41B, 184
$C_{24}H_{48}$, Cyclotetraeicosane, 45B, 127
$C_{26}H_{52}$, Cyclohexaeicosane, 45B, 127
$C_{32}H_{32}$, [2_4]Metacyclophane I, 40B, 152
$C_{34}H_{68}$, Cyclotetratriacontane, 33B, 87
$C_{48}H_{48} \cdot 0.33\ C_6H_{14}$, [$2_6$]Metacyclophane hexane solvate, 40B, 152

24. NAPHTHALENE COMPOUNDS

$C_{10}Cl_8$, Octachloronaphthalene, 28, 429; 45B, 128
$C_{10}D_8$, Naphthalene, 34B, 97
$C_{10}F_8$, Octafluoronaphthalene, 44B, 135
$C_{10}H_2Br_6$, 1,2,3,5,6,7-Hexabromonaphthalene, 43B, 182
$C_{10}H_4Br_2Cl_2$, 1,5-Dibromo-4,8-dichloronaphthalene, 30B, 99
$C_{10}H_4Br_4$, 1,4,5,8-Tetrabromonaphthalene, 33B, 87
$C_{10}H_4Br_4$, 2,3,6,7-Tetrabromonaphthalene, 46B, 129
$C_{10}H_4Cl_4$, 1,4,5,8-Tetrachloronaphthalene, 27, 921
$C_{10}H_4Cl_4O$, 2,3,4,4-Tetrachloro-1-keto-naphthalenedihydride, 37B, 85
$C_{10}H_6Br_2$, 1,4-Dibromonaphthalene, 26, 741; 44B, 135

$C_{10}H_6Br_2$, 2,6-Dibromonaphthalene, 42B, 116
$C_{10}H_6Cl_2$, 1,3-Dichloronaphthalene, 26, 740
$C_{10}H_6Cl_2$, 1,4-Dichloronaphthalene, 44B, 135
$C_{10}H_6Cl_2$, 1,5-Dichloronaphthalene, 15, 477
$C_{10}H_6F_2$, 1,5-Difluoronaphthalene, 39B, 104; 41B, 185
$C_{10}H_6F_2$, 1,8-Difluoronaphthalene, 41B, 185
$C_{10}H_6N_2O_2$, 1,8-Dinitrosonaphthalene, 37B, 85
$C_{10}H_6N_2O_4$, 1,5-Dinitronaphthalene, 11, 694; 24, 663
$C_{10}H_6N_2O_4$, 1,8-Dinitronaphthalene (orthorhombic), 30B, 98; 43B, 183
$C_{10}H_7Br$, 2-Bromonaphthalene, 39B, 104
$C_{10}H_7BrO$, 8-Bromo-2-naphthol, 33B, 88
$C_{10}H_7Cl$, 2-Chloronaphthalene, 46B, 129
$C_{10}H_7F$, β-Fluoronaphthalene, 38B, 193
$C_{10}H_8$, Naphthalene (gas-ed), 8, 318; 26, 739
$C_{10}H_8$, Naphthalene, 12, 405; 16, 533; 17, 730; 19, 580; 21, 603; 34B, 97; 42B, 116
$C_{10}H_8O$, α-Naphthol, 29, 677
$C_{10}H_8O$, β-Naphthol, 22, 734
$C_{10}H_8O_2$, Naphthalene-1,4-diol, 32B, 121
$C_{10}H_{16}NNaO_7S$, Sodium naphthionate tetrahydrate, 31B, 95
$C_{11}H_7IN_2O_2 \cdot H_2O$, 3-Carboxy-2-naphthalenediazonium iodide hydrate, 45B, 129
$C_{11}H_7IO_2$, 3-Iodo-2-naphthoic acid, 43B, 184
$C_{11}H_8O_2$, 1-Naphthoic acid, 24, 662
$C_{11}H_8O_2$, 2-Naphthoic acid, 26, 739
$C_{11}H_8O_3$, 1-Hydroxy-2-naphthoic acid, 42B, 117
$C_{11}H_8O_3$, 3-Hydroxy-2-naphthoic acid, 41B, 186
$C_{11}H_9BrN_2O_3$, 3-Carboxy-2-naphthalenediazonium bromide monohydrate, 44B, 136
$C_{11}H_{10}N_2O_7S$, 3-Carboxy-2-naphthalenediazonium hydrogensulfate monohydrate, 44B, 136
$C_{11}H_{10}O_2$, 2-Methyl-1,4-dihydroxynaphthalene, 34B, 97
$C_{12}H_4Cl_2O_3$, 4,5-Dichloronaphthalic anhydride, 37B, 86
$C_{12}H_4N_2O_7$, 4,5-Dinitronaphthalic anhydride, 38B, 194
$C_{12}H_6O_3$, Naphthalic anhydride, 37B, 86
$C_{12}H_8O_4$, 1,4-Naphthalenedicarboxylic acid, 45B, 129
$C_{12}H_{10}Br_2$, 1,8-Di(bromomethyl)naphthalene, 40B, 153
$C_{12}H_{10}O_2$, 1-Naphthaleneacetic acid, 44B, 137
$C_{12}H_{10}O_2$, 2-Hydroxy-6-acetylnaphthalene, 42B, 117
$C_{12}H_{10}O_3$, 2-Naphthyloxyacetic acid, 44B, 137
$C_{12}H_{10}O_3$, 8-Methoxynaphthalene-1-carboxylic acid, 44B, 137
$C_{12}H_{10}O_3S$, 1-Naphthylsulfinylacetic acid, 41B, 187
$C_{12}H_{11}Br$, 3-Bromo-1,8-dimethylnaphthalene, 30B, 97
$C_{12}H_{11}NO$, N-2-Naphthylacetamide, 46B, 100
$C_{12}H_{11}NO_2$, Methyl N-(1-naphthyl)carbamate, 42B, 117
$C_{12}H_{12}$, 1,5-Dimethylnaphthalene, 30B, 96; 38B, 194
$C_{12}H_{12}$, 1,8-Dimethylnaphthalene, 39B, 104
$C_{12}H_{12}$, 2,6-Dimethylnaphthalene, 10, 280
$C_{13}H_{10}O_3$, 3-Methoxynaphthalide, 42B, 117
$C_{13}H_{12}O_2$, 1-Methyl-2-naphthyl acetate, 43B, 184
$C_{13}H_{12}O_2$, 8-Methoxy-1-naphthyl methyl ketone, 44B, 137
$C_{13}H_{13}NO_2$, N,N-Dimethyl-8-hydroxynaphthalene-1-carboxamide, 44B, 138
$C_{13}H_{13}NO_2$, 8-(N,N-Dimethylamino)naphthalene-1-carboxylic acid, 44B, 137
$C_{13}H_{14}O_3$, 3,4-Epoxy-3,4-dihydro-1-hydroxy-1-isopropylnaphthalen-2-one, 46B, 130
$C_{14}H_8$, 1,4-Diethynylnaphthalene, 43B, 185

$C_{14}H_{12}F_3NO$, (R)-N-Trifluoroacetyl-1-(1-naphthyl)ethylamine, 46B, 34
$C_{14}H_{14}O_2$, 1-Acetyl-2-ethoxynaphthalene, 38B, 196
$C_{14}H_{14}O_2$, 1,8-Dimethyl-2-naphthyl acetate, 43B, 185
$C_{14}H_{14}O_2$, 2-Ethoxy-6-acetylnaphthalene, 40B, 154
$C_{14}H_{15}NO$, (R)-N-Acetyl-1-(1-naphthyl)ethylamine, 46B, 34
$C_{14}H_{15}NO$, (RS)-N-Acetyl-1-(1-naphthyl)ethylamine, 46B, 34
$C_{14}H_{15}NO$, 8-(N,N-Dimethylamino)-1-naphthyl methyl ketone, 44B, 137
$C_{14}H_{15}NO_2$, Methyl 8-(N,N-dimethylamino)naphthalene-1-carboxylate, 44B, 137
$C_{14}H_{15}NO_2$, N,N-Dimethyl-8-methoxynaphthalene-1-carboxamide, 44B, 138
$C_{14}H_{18}N_2$, 1,8-Bis(dimethylamino)naphthalene, 39B, 105
$C_{15}H_{15}IO_5$, Methyl 8-iodo-4,5,7-trimethoxy-2-naphthoate, 43B, 186
$C_{15}H_{16}O_2$, 1-Isopropyl-2-naphthyl acetate, 43B, 187
$C_{15}H_{18}O_2$, 1-(2-Ethoxy-1-naphthyl)ethyl methyl ether, 42B, 118
$C_{15}H_{20}ClNO$, Pronethanol hydrochloride, 41B, 187
$C_{16}H_9Cl_3$, 1-(p-Chlorophenyl)-2,6-dichloronaphthalene, 42B, 119
$C_{16}H_{12}$, 1,8-Bis(prop-1-ynyl)naphthalene, 38B, 195
$C_{16}H_{16}N_2O_4S \cdot H_2O$, Ammonium 1-anilino-8-naphthalenesulphonate monohydrate, 46B, 130
$C_{16}H_{17}IN_2O$, α-(4'-Iodo-1'-diazocyclohexane)-β-naphthol, 42B, 46
$C_{16}H_{22}ClNO_2$, Propranolol hydrochloride, 41B, 188
$C_{17}H_{13}N_3O_3$, 4-Methyl-2-nitrophenylazo-2-naphthol, 44B, 138
$C_{17}H_{14}KNO_3S$, Potassium 2-p-toluidinyl-6-naphthalenesulphonate, 35B, 116
$C_{17}H_{20}O_3$, 3-(6'-Methoxy-2'-naphthyl)-2,2-dimethyl-butyric acid, 40B, 155
$C_{18}H_{11}NO_5S$, N-Benzenesulfoxynaphthalimide, 43B, 364
$C_{18}H_{12}ClNO_3$, α-Naphthyl-4-chlorophthalamic acid, 31B, 96
$C_{18}H_{14}N_2O_5$, 2,4-Dinitronaphthyl-2',6'-dimethylphenyl ether, 42B, 89
$C_{18}H_{17}NO_3$, N-α-Naphthyl-1,2,3,6-tetrahydrophthalamic acid, 32B, 122
$C_{18}H_{24}$, Octamethylnaphthalene, 17, 732
$C_{18}H_{25}NOS$, 2-t-Butylimino-λ^4-thio-3,4-dihydro-3,3,5,8-tetramethyl-1(2H)-naphthalenone, 44B, 139
$C_{19}H_{16}O_3$, 8-Benzoyl-5-ethoxy-1-naphthol, 44B, 139
$C_{19}H_{20}O_2$, Di-O-methylsequirin-D, 44B, 140
$C_{19}H_{24}O_3$, 2-(2-Hydroxyethyl)-2-methyl-3-(6-methoxy-2-naphthyl)cyclopentanol, 44B, 140
$C_{20}H_{14}$, cis-1,1'-Binaphthyl (racemic B form), 40B, 1080; 46B, 131
$C_{20}H_{14}$, trans-1,1'-Binaphthyl (chiral A form), 46B, 131
$C_{20}H_{16}F_6P$, Bis(naphthalene) hexafluorophosphate, 44B, 141
$C_{21}H_{27}BrO_2$, 1,2,3,4,4a,7,8,8a-Octahydro-2α,4aβ,5,8aβ-tetramethylnaphthalen-1β-yl 4-bromobenzoate, 37B, 87
$C_{22}H_{16}Br_2$, (-)-2,2'-Bisbromomethyl-1,1'-binaphthyl, 39B, 106
$C_{22}H_{23}NO_2$, (+)-(R)-N-Methyl-1-((1-naphthyl)ethyl)-(R)-o-methylmandelamide, 34B, 99
$C_{23}H_{18}O$, Bis(2-methyl-1-naphthyl) ketone, 46B, 132
$C_{23}H_{18}O_2$, [1-(2-Hydroxynaphthyl)]diphenylmethanol, 46B, 105
$C_{24}H_{16}O_2$, 1,5-Dibenzoylnaphthalene, 44B, 141
$C_{25}H_{18}Cl_3N_3O_4$, 1-(2,5-Dichlorophenylazo)-2-hydroxy-3-naphthoic acid 4-chloro-2,5-dimethoxyanilide, 38B, 196
$C_{25}H_{22}O_2$, Bis(2-methyl-1-naphthyl)methyl acetate, 46B, 133
$C_{26}H_{16}$, 1,8-Bis(phenylethynyl)naphthalene, 37B, 87
$C_{26}H_{18}N_2$, N,N'-(Di-β-naphthyl)-p-quinone-diimine, 46B, 133
$C_{26}H_{18}O_6S_2$, 2,4-Hexadiynylene bis(β-naphthalenesulphonate), 45B, 129
$C_{26}H_{20}N_2$, N,N'-Di-β-naphthyl-p-phenylenediamine, 43B, 187
$C_{26}H_{20}O_2$, 1,8-Dibenzoyl-2,7-dimethylnaphthalene, 46B, 134
$C_{26}H_{24}$, 1-Isopropenylnaphthalene dimer, 44B, 142

$C_{26}H_{40}$, 1,3,6,8-Tetra-t-butylnaphthalene, 43B, 188
$C_{28}Cl_{20}$, Perchloro-1,2,3-triphenylnaphthalene, 44B, 142
$C_{28}H_{16}Cl_4O_4$, 2,6-Naphthalenediacrylic acid bis(2,4-dichlorophenyl) ester, 44B, 143
$C_{28}H_{30}O_8$, Trimethyl 2-carboxylato-trans-3-(2-(2-methoxycarbonylvinyl)phenyl)-1,2,3,4-tetrahydronaphthalene-cis-1,4-diacetate, 44B, 165
$C_{29}H_{19}IO_2$, 1-p-(1,2,3,4-Tetrahydro-1-naphthyl)-phenol p-iodobenzoate, 35B, 117
$C_{29}H_{32}ClNO_2 \cdot C_2H_6O$, 1-(p-2-Pyrrolidinethoxyphenyl)-2-phenyl-3,4-dihydro-6-methoxynaphthalene hydrochloride ethanol solvate, 46B, 134
$C_{30}H_{23}BrO_6$, (+)-2,2'-Dihydroxy-1,1'-binaphthalene-3,3'-dicarboxylic acid dimethyl ester bromobenzene solvate, 34B, 100
$C_{32}H_{48}MgN_2O_{18}S_2$, Hexaaquomagnesium bis(8-anilino-1-naphthalenesulphonate) hexahydrate, 43B, 189
$C_{34}H_{24}$, 1,4,5,8-Tetraphenylnaphthalene, 38B, 197
$C_{40}H_{42}K_2N_2O_8S_2$, Potassium 4,4'-bis-1-phenylaminonaphthalene-8-sulphonate bis(1-butanol), 44B, 143

25. NAPHTHOQUINONES

$C_{10}H_4Br_2O_2$, 2,3-Dibromo-1,4-naphthoquinone, 31B, 97; 32B, 124
$C_{10}H_4Cl_2O_2$, 2,3-Dichloro-1,4-naphthoquinone (orthorhombic), 43B, 189
$C_{10}H_4Cl_2O_2$, 2,3-Dichloro-1,4-naphthoquinone (triclinic), 26, 742
$C_{10}H_4Cl_2O_4$, 2,3-Dichloro-5,8-dihydroxynaphtho-1,4-quinone, 44B, 144
$C_{10}H_5BrO_2$, 2-Bromo-1,4-naphthoquinone, 30B, 100
$C_{10}H_5BrO_2$, 3-Bromo-1,2-naphthoquinone, 41B, 189
$C_{10}H_5ClO_2$, 5-Chloro-1,4-naphthoquinone, 39B, 107
$C_{10}H_5ClO_3$, 2-Chloro-3-hydroxy-1,4-naphthoquinone, 30B, 100
$C_{10}H_6BrNO_2$, 3-Bromo-2-amino-1,4-naphthoquinone, 31B, 98
$C_{10}H_6BrNO_2$, 4-Amino-3-bromo-1,2-naphthoquinone, 42B, 119
$C_{10}H_6ClNO_2$, 3-Chloro-2-amino-1,4-naphthoquinone, 30B, 100
$C_{10}H_6NNaO_5S \cdot 1.5\ H_2O$, Sodium 1,2-naphthoquinone-2-oxime-5-sulphonate sesquihydrate, 43B, 191
$C_{10}H_6O_2$, 1,4-Naphthoquinone, 30B, 100
$C_{10}H_6O_3$, Juglone, 37B, 88
$C_{10}H_6O_4$, Naphthazarin (form A), 21, 606; 22, 730; 27, 927; 37B, 88
$C_{10}H_6O_4$, Naphthazarin (form B), 19, 582; 22, 730; 27, 927
$C_{10}H_6O_4$, Naphthazarin (form C), 22, 730; 23, 672; 24, 666; 26, 743, 744; 27, 927; 37B, 89
$C_{10}H_7NO_2 \cdot 0.5\ H_2O$, 4-Amino-1,2-naphthoquinone hemihydrate, 35B, 118
$C_{10}H_7NO_2$, 2-Amino-1,4-naphthoquinone, 31B, 98; 34B, 101
$C_{10}H_8BrNO_3$, 4-Amino-3-bromo-1,2-naphthoquinone monohydrate, 42B, 119
$C_{10}H_8KNO_6S$, Potassium 1,2-naphthoquinone-1-oxime-7-sulphonate monohydrate, 43B, 190
$C_{10}H_8N_2O_2$, (1Z,2E)-1,2-Naphthoquinone dioxime (form A), 43B, 193
$C_{10}H_8N_2O_2$, (1Z,2E)-1,2-Naphthoquinone dioxime (form B), 43B, 193
$C_{10}H_9NO_6S$, 1,2-Naphthoquinone-2-oxime-5-sulphonic acid monohydrate, 43B, 192
$C_{10}H_{10}KNO_7S$, Potassium 1,2-naphthoquinone-2-oxime-8-sulphonate dihydrate, 43B, 191
$C_{10}H_{10}O_2$, cis-4a,5,8,8a-Tetrahydro-1,4-naphthoquinone, 43B, 193
$C_{10}H_{13}ClN_2O_3$, 2-Amino-1,4-naphthoquinone iminium chloride dihydrate, 40B, 155
$C_{11}H_8O_2$, 2,3-Dihydro-2,3-methylene-1,4-naphthoquinone, 22, 736; 27, 906

$C_{11}H_8O_3$, 2-Hydroxy-3-methyl-1,4-naphthoquinone, 30B, 100
$C_{11}H_8O_4$, 2-Methyl-5,8-dihydroxy-1,4-naphthoquinone, 43B, 194
$C_{11}H_9NO_2 \cdot 0.5\ H_2O$, 3-Methyl-2-amino-1,4-naphthoquinone hydrate, 31B, 101
$C_{11}H_9NO_2$, 3-Methyl-2-amino-1,4-naphthoquinone, 31B, 100
$C_{11}H_{10}BrNO_3$, 4-Amino-3-bromo-1,2-naphthoquinone methanol solvate, 42B, 119
$C_{11}H_{11}NO_3$, 4-Amino-3-methyl-1,2-naphthoquinone monohydrate, 42B, 119
$C_{12}H_{10}O_2$, 2,3-Dimethyl-1,4-naphthoquinone, 31B, 102; 32B, 123
$C_{12}H_{11}NO_2$, 4-Dimethylamino-1,2-naphthoquinone, 41B, 189; 42B, 119
$C_{12}H_{14}O_2$, 5α,8α-Dimethyl-4aβ,5,8,8aβ-tetrahydro-1,4-naphthoquinone, 43B, 195
$C_{13}H_{12}O_6$, 2,5-Dihydroxy-3,8-dimethoxy-7-methyl-1,4-naphthoquinone, 46B, 135
$C_{14}H_{10}O_6$, Naphthazarin diacetate, 43B, 195
$C_{14}H_{12}N_2O_2$, 5α,8α-Dimethyl-4aβ,8aβ-dicyano-4a,5,8,8a-tetrahydro-1,4-naphthoquinone, 42B, 121
$C_{14}H_{12}N_2O_2$, 6,7-Dimethyl-cis-4a,8a-dicyano-4a,5,8,8a-tetrahydro-1,4-naphthoquinone, 42B, 120
$C_{14}H_{12}O_7$, 7-Acetyl-2,5,8-trihydroxy-3-methoxy-6-methyl-1,4-naphthoquinone, 30B, 104
$C_{14}H_{15}NO_2$, 4-Diethylamino-1,2-naphthoquinone, 41B, 191; 42B, 119
$C_{14}H_{15}NO_2$, 6-t-Butyl-1,2-naphthoquinone-1-oxime, 42B, 121
$C_{14}H_{18}O_2$, 2,3,6,7-Tetramethyl-cis-4a,5,8,8a-tetrahydro-1,4-naphthoquinone, 42B, 121
$C_{16}H_6N_4$, 11,11,12,12-Tetracyano-1,4-naphthoquinodimethane, 37B, 89
$C_{16}H_{10}ClN_3O_3$, 1,2-Naphthoquinone 1-(2-nitro-4-chloro-phenylhydrazone), 41B, 191
$C_{16}H_{10}O_2$, 3-Phenyl-1,4-naphthoquinone, 34B, 102
$C_{16}H_{11}N_3O_3$, 1-[(4-Nitrophenyl)azo]-2-naphthol (β-form), 46B, 136
$C_{16}H_{22}O_2$, cis-2,3,4a,6,7,8a-Hexamethyl-4a,5,8,8a-tetrahydro-1,4-naphthoquinone, 42B, 122
$C_{16}H_{22}O_2$, 2,3,4aβ,5β,8β,8aβ-Hexamethyl-4a,5,8,8a-tetrahydro-1,4-naphthoquinone, 42B, 123
$C_{18}H_{18}O_6$, Tetramethylnaphthazarin diacetate, 45B, 130
$C_{19}H_{16}ClNO_2$, 2-Chloro-3-N-methyl-N-p-tolylaminomethyl-1,4-naphthoquinone, 41B, 192
$C_{19}H_{17}NO_2$, 2-Methyl-3-N-methylanilinomethyl-1,4-naphthoquinone, 35B, 120
$C_{20}H_{10}O_4$, 2,2'-Di(1,4-naphthoquinone), 34B, 103
$C_{20}H_{18}KN_2NaO_{15}S_2$, Potassium sodium bis(1,2-naphthoquinone-2-oxime-5-sulphonate) trihydrate, 43B, 197
$C_{20}H_{20}O_4$, Pentacyclo[$10.8.0.0^{2-11}.0^{4-9}.0^{14-19}$]eicosa-6,16-diene-3,10,13,20-tetrone, 43B, 197
$C_{20}H_{32}N_2NiO_{20}S_2$, Hexaaquonickel(II) bis(1,2-naphthoquinone-2-oxime-4-sulphonate) tetrahydrate, 43B, 198
$C_{21}H_{18}ClNO_4$, 2-Chloro-3-N-methyl-p-ethoxycarbonylphenylaminomethyl-1,4-naphthoquinone, 45B, 131
$C_{21}H_{20}ClNO_6S$, 3-Chloro-6-methoxy-7-methyl-2-(p-tosyloxyethylamine)-1,4-naphthoquinone, 46B, 136
$C_{22}H_{18}O_2$, 6,7-Diphenyl-cis-4a,5,8,8a-tetrahydro-1,4-naphthoquinone, 42B, 123
$C_{22}H_{21}ClN_2O_3$, 2-(N-Acetyl-2-(N'-ethyl-N'-phenylamino)ethylamino)-3-chloro-1,4-naphthoquinone, 41B, 193
$C_{23}H_{16}O$, 1-(α,α-Diphenylmethylene)-1,2-naphthoquinone, 44B, 144
$C_{24}H_{28}O_4$, 5,8,15,18-Tetramethylpentacyclo[$10.8.0.0^{2-11}.0^{4-9}.0^{14-19}$]-eicosa-6,16-diene-3,10,13,20-tetrone, 43B, 195

$C_{24}H_{28}O_4$, 6,7,16,17-Tetramethylpentacyclo[10.8.0.0$^{2-11}$.0$^{4-9}$.0$^{14-19}$]-eicosa-6,16-diene-3,10,13,20-tetrone, 43B, 197

$C_{26}H_{21}NO_4$, 2-Phenyl-3-(N-methyl-p-anisidino)acetyl-1,4-naphthoquinone, 43B, 198

26. ANTHRACENE COMPOUNDS

$C_{14}F_8O_2$, Octafluoroanthraquinone, 39B, 897

$C_{14}H_6Br_2O_2$, 1,5-Dibromoanthraquinone, 26, 751; 28, 431; 32B, 125

$C_{14}H_6Cl_2O_2$, 1,5-Dichloroanthraquinone, 22, 737

$C_{14}H_6Cl_2O_2$, 1,8-Dichloroanthraquinone, 38B, 199

$C_{14}H_6F_2O_2$, 1,5-Difluoroanthraquinone, 32B, 126

$C_{14}H_6I_2O_2$, 1,5-Diiodoanthraquinone, 28, 432; 32B, 125

$C_{14}H_6I_2O_2$, 1,8-Diiodoanthraquinone, 39B, 107

$C_{14}H_6N_2O_8$, 1,5-Dinitro-4,8-dihydroxyanthraquinone, 32B, 127

$C_{14}H_7BrO_2$, 1-Bromoanthraquinone, 34B, 103

$C_{14}H_7BrO_2$, 2-Bromo-1,4-anthraquinone, 41B, 194

$C_{14}H_7ClO_2$, 1-Chloroanthraquinone, 34B, 105

$C_{14}H_7IO_2$, 1-Iodoanthraquinone, 39B, 108

$C_{14}H_8BrCl$, 9-Chloro-10-bromoanthracene, 26, 748

$C_{14}H_8Br_2$, 9,10-Dibromoanthracene, 10, 282; 22, 739

$C_{14}H_8Br_2O_2$, 10,10-Dibromoanthrone, 28, 433

$C_{14}H_8Cl_2$, 1,5-Dichloroanthracene, 15, 479

$C_{14}H_8Cl_2$, 9,10-Dichloroanthracene (α form), 23, 679

$C_{14}H_8Cl_2$, 9,10-Dichloroanthracene (β form), 44B, 144; 45B, 131

$C_{14}H_8Cl_4$, 9,9,10,10-Tetrachloro-9,10-dihydroanthracene, 31B, 103

$C_{14}H_8N_2O_4$, 9,10-Dinitroanthracene, 23, 678

$C_{14}H_8O_2$, Anthraquinone, 6, 208, 247; 11, 697; 24, 667; 32B, 128

$C_{14}H_8O_4$, 1,2-Dihydroxy-9,10-anthraquinone, 32B, 129

$C_{14}H_8O_4$, 1,4-Dihydroxyanthraquinone, 46B, 136

$C_{14}H_8O_4$, 1,5-Dihydroxy-9,10-anthraquinone, 31B, 103; 32B, 129

$C_{14}H_8O_4$, 1,8-Dihydroxyanthraquinone, 30B, 104

$C_{14}H_9BrO$, 10-Bromoanthrone, 39B, 108

$C_{14}H_9NO_2$, 9-Nitroanthracene, 23, 677

$C_{14}H_{10}$, Anthracene (gas-ed), 26, 739

$C_{14}H_{10}$, Anthracene, 3, 686, 790; 13, 555; 16, 533; 20, 591; 29, 682; 38B, 198

$C_{14}H_{10}O$, Anthrone, 29, 683; 35B, 121

$C_{14}H_{10}O_3$, Anthralin, 46B, 137

$C_{15}H_9N$, 9-Cyanoanthracene, 23, 673

$C_{15}H_{10}Cl_2$, 1,8-Dichloro-9-methylanthracene, 34B, 107

$C_{15}H_{10}O$, 9-Anthraldehyde, 23, 675

$C_{15}H_{11}Br$, 9-Bromo-10-methylanthracene, 26, 749; 46B, 137

$C_{15}H_{12}$, 9-Methylanthracene, 34B, 106; 37B, 90; 45B, 132

$C_{15}H_{12}O$, 9-Methoxyanthracene, 37B, 90

$C_{16}H_{12}Cl_2$, 9,10-Bis(chloromethyl)anthracene, 37B, 90

$C_{16}H_{12}O$, 9-Methoxycarbonylanthracene, 37B, 90

$C_{16}H_{13}Br$, 9-Ethyl-10-bromoanthracene, 24, 667

$C_{16}H_{13}Cl$, 9-Methyl-10-chloromethylanthracene, 43B, 200

$C_{16}H_{14}$, 9,10-Dimethylanthracene, 40B, 156

$C_{16}H_{16}O_2$, 2,3-Dimethyl-1,4,4a,9a-tetrahydro-9,10-anthraquinone, 43B, 201

$C_{17}H_8N_2O$, 10-(Dicyanomethylene)anthrone, 32B, 131

$C_{17}H_{15}ClS$, 9-(((2-Chloroethyl)thio)methyl)anthracene, 42B, 124

$C_{17}H_{16}O_4$, Methyl 1'β,4',4'$a\beta$,9',9'$a\beta$,10'-hexahydro-9',10'-dioxoanthryl acetate, 37B, 91

$C_{17}H_{18}$, cis-9-Methyl-10-ethyl-9,10-dihydroanthracene, 39B, 109
$C_{18}H_{17}Cl$, 10-Chloromethyl-2,3,9-trimethylanthracene, 38B, 199
$C_{18}H_{17}ClS$, 10-Methyl-9(((2-chloroethyl)thio)methyl)anthracene, 38B, 200
$C_{18}H_{20}$, trans-9-Isopropyl-10-methyl-9,10-dihydroanthracene, 39B, 110
$C_{19}H_{13}BrN_2O_5S$, Pyridinium 1-amino-4-bromo-9,10-dioxoanthracene-2-sulphonate, 44B, 145
$C_{23}H_{16}O$, 9-Anthryl styryl ketone, 41B, 194
$C_{24}H_{17}Br$, 10-Bromo-1,8-diphenylanthracene, 46B, 138
$C_{24}H_{33}ClF_3NO_2$, (+)-cis-9-(3-Dimethylaminopropyl)-10-methyl-2-(trifluoromethyl)-9,10-dihydroanthracene hydrochloride monohydrate acetone solvate, 42B, 125
$C_{26}H_{14}Cl_4O_2$, Anthracene-1,1'-bi-3,5-dichloro-4-oxo-hexa-2,5-dienylidene (1:1), 46B, 139
$C_{26}H_{18}$, 9,10-Diphenylanthracene, 45B, 133
$C_{26}H_{18}N_2O_2$, N,N'-Diphenyl-1,5-diaminoanthraquinone, 32B, 132
$C_{26}H_{18}N_2O_2$, N,N'-Diphenyl-1,8-diaminoanthraquinone, 32B, 133
$C_{28}H_{16}N_2O_4$, 4,4'-Diamino-1,1'-bianthraquinone, 40B, 1080
$C_{28}H_{18}O_4$, 9,10-Anthrahydroquinone dibenzoate, 27, 936
$C_{30}H_{26}O_5$, 10,10-Divanillyl-9(10H)-anthracenone, 46B, 139
$C_{32}H_{22}O_2$, 9,10-Anthryl bis(styryl ketone), 41B, 194

27. HYDROCARBONS (2 FUSED RINGS)

C_5H_6, Bicyclo[2.1.0]pent-2-ene (gas-mw), 35B, 905
C_5H_6, Bicyclo[2.1.0]pentene (gas-ed), 38B, 1065
C_5H_8, Bicyclo[2.1.0]pentane (gas-ed), 35B, 896
C_5H_8, Bicyclo[2.1.0]pentane (gas-mw), 38B, 1070; 42B, 1033
C_6F_6, Hexafluorobicyclo[2.2.0]hexa-2,5-diene (gas-ed), 40B, 1138
$C_6H_4N_2$, 1,3-Dicyanobicyclo[1.1.0]butane, 38B, 201
C_6H_6, Dewar-benzene (gas-ed), 42B, 1027
C_6H_{10}, Bicyclo[2.2.0]hexane (gas-ed), 38B, 1062
C_6H_{10}, Bicyclo[3.1.0]hexane (gas-ed), 43B, 1485
C_7H_8, Spiro[2.4]hepta-4,6-diene (gas-ed), 39B, 889
$C_7H_8Cl_2O$, cis-6-Chloro-trans-7-chloro-cis-bicyclo[3.2.0]heptan-2-one, 38B, 202
C_7H_{10}, Bicyclo[3.2.0]hept-6-ene (gas-ed), 31B, 93
C_7H_{10}, Bicyclo[4.1.0]hept-2-ene (gas-ed), 38B, 1062
$C_7H_{10}Cl_2$, 7,7-Dichlorobicyclo[4.1.0]heptane (gas-ed), 35B, 898
C_8Cl_8, Octachloro-2,4-dihydropentalene, 39B, 110
$C_8H_4O_2$, Benzocyclobutene-1,2-dione, 35B, 121
$C_8H_6Br_2$, cis-1,2-Dibromobenzocyclobutene, 33B, 90
$C_8H_6Cl_2$, cis-1,2-Dichlorobenzocyclobutene, 33B, 90
$C_8H_6I_2$, cis-1,2-Diiodobenzocyclobutene, 33B, 90
$C_8H_6N_2O_6$, cis-Benzocyclobutene-1,2-diol dinitrate, 35B, 123
$C_9H_4O_3$, Triketoindane, 30B, 105
$C_9H_6O_2$, 1,3-Indandione, 42B, 126
$C_9H_6O_4$, Ninhydrin, 34B, 108
C_9H_8, Indene (gas-ed), 37B, 691
$C_9H_9NO_6$, 2-Nitro-1,3-indandione dihydrate, 43B, 201; 46B, 140
$C_9H_{10}O_4$, Methyl 2*a*,3*a*:4*a*,5*a*-diepoxy-cis-(1*a*H,2*a*H)-bicyclo[4.1.0]heptane-7*a*-carboxylate, 42B, 126
$C_9H_{14}O_4$, Methyl 2,5-dihydroxybicyclo[4.1.0]heptane-7-carboxylate, 42B, 126
$C_{10}H_2F_{16}O_2$, cis-Perfluorobicyclo[4.4.0]decane-1,6-diol, 42B, 127
$C_{10}H_4Cl_4O$, 2,2,3,4-Tetrachloro-1-oxo-1,2-dihydronaphthalene (*a*

isomer), 45B, 133
$C_{10}H_4Cl_4O$, 2,3,4,4-Tetrachloro-1-oxo-1,4-dihydronaphthalene (monoclinic β isomer), 45B, 134
$C_{10}H_4Cl_4O$, 2,3,4,4-Tetrachloro-1-oxo-1,4-dihydronaphthalene (orthorhombic β isomer), 45B, 134
$C_{10}H_4F_{12}N_2O$, 6,10-Diaminododecafluorobicyclo[4.4.0]dec-1(10)-en-2-one, 42B, 128
$C_{10}H_6Cl_6$, r-1,t-2,c-3,t-4,5,6-Hexachlorotetralin, 46B, 140
$C_{10}H_8$, Azulene, 20, 590; 27, 926
$C_{10}H_8Cl_4$, 1,2,3,4-Tetrachloro-1,2,3,4-tetrahydronaphthalene (monoclinic), 16, 534; 20, 588
$C_{10}H_8Cl_4$, 1,2,3,4-Tetrachlorotetralin (orthorhombic), 39B, 111
$C_{10}H_9N$, 2-Aminoazulene, 23, 672
$C_{10}H_{13}BrO_3$, 4-Bromo-3a-hydroxy-7a-methyl-octahydroindene-1,5-dione, 37B, 92
$C_{10}H_{14}O_2$, 1-Hydroxy-cis-bicyclo[5.3.0]dec-9-en-8-one, 44B, 145
$C_{10}H_{15}BrO$, 3-Bromo-2-decalone, 37B, 93
$C_{10}H_{15}ClO$, 3-Chloro-2-decalone, 34B, 109; 37B, 93
$C_{10}H_{16}N_2O$, N'-Isopropylidenebicyclo[3.1.0]hexane-6-exo-carbohydrazide, 43B, 202
$C_{10}H_{16}O$, trans-2-Decalone (gas-ed), 33B, 537
$C_{10}H_{18}$, Decalin (gas-ed), 10, 246
$C_{10}H_{18}N_2O_2$, exo-5-Ethoxy-5-(N,N-dimethylamido)aminobicyclo[2.1.0]-pentane, 44B, 147
$C_{10}H_{18}S_2$, 1,3-Di(methylthio)-2,2,4,4-tetramethylbicyclo[1.1.0]butane, 45B, 135
$C_{11}H_8O$, 4,5-Benzotropone, 41B, 196
$C_{11}H_8O_3$, 2-Acetyl-1,3-indandione, 46B, 143
$C_{11}H_{10}N_2$, 2,5-Dimethyl-7,7-dicyanonorcaradiene, 31B, 104
$C_{11}H_{10}O_2S$, 2-Dimethylsulfuranylidene-1,3-indanedione, 37B, 94
$C_{11}H_{10}O_4$, 1,4-Dihydroxybenzocycloheptene-5,9(6H,8H)-dione, 46B, 141
$C_{11}H_{14}O_2$, 9-Methyl-5(10)-octalin-1,6-dione, 39B, 112
$C_{11}H_{15}BrO_2$, 5α-Bromo-8β,9α-dimethyl-hydrindane-1,4-dione, 40B, 157
$C_{11}H_{15}BrO_3$, 1-Bromo-5-acetoxy-trans-hydrinan-4-one, 41B, 197
$C_{11}H_{16}BrNO$, 4-Amino-3-hydroxybenzo-1-cycloheptene hydrobromide, 44B, 146
$C_{11}H_{16}O_3$, 3-Oxo-cis-bicyclo[4.4.0]decane-1-carboxylic acid, 45B, 136
$C_{11}H_{18}N_3O$, 3α-Methyl-trans-4-hydrindanone semicarbazone, 44B, 146
$C_{11}H_{18}O$, 1,4,5,6,7,7a-Hexahydro-3,4β-dimethyl-2H-4α-hydroxyindene, 46B, 141
$C_{11}H_{18}O$, 10-Methyl-trans-2-decalone (gas-ed), 42B, 1028
$C_{11}H_{18}O_2$, 7α-Hydroxy-bicyclo[5.4.0]undecan-9-one, 45B, 137
$C_{11}H_{20}N_2O_2$, exo-6-Ethoxy-6-(N,N-dimethylamido)aminobicyclo[3.1.0]-hexane, 44B, 147
$C_{12}H_4N_2O_2$, 2-Dicyanomethyleneindane-1,3-dione, 40B, 157
$C_{12}H_{12}O_2$, 1-exo-Phenylbicyclo[2.1.0]pentane-5-carboxylic acid, 42B, 129
$C_{12}H_{14}O_3S$, exo-Bicyclo[2.1.0]pentyl tosylate, 39B, 112
$C_{12}H_{15}ClO_2$, trans-4a-Acetoxy-8a-chloro-1,4,4a,5,8,8a-hexahydronaphthalene, 42B, 129
$C_{12}H_{16}O_2$, 5α,8α-Dimethyl-4aβ,5,8,8aβ-tetrahydro-1-naphthoquin-4α-ol, 46B, 142
$C_{12}H_{16}O_4S$, Hydroazulene derivative, 43B, 203
$C_{12}H_{18}$, Hexamethyl(Dewar benzene) (gas-ed), 35B, 902
$C_{12}H_{18}O_3$, Methyl 3-oxo-trans-bicyclo[4.4.0]decane-1-carboxylate, 45B, 136
$C_{12}H_{23}NO$, (+)-2(S)-N,N-Dimethylamino-3(S)-hydroxy-9(S),10(S)-deca-

lin, 39B, 112
$C_{13}H_{12}O$, 2,7-Dimethyl-4,5-benzotropone, 42B, 130
$C_{13}H_{13}N_3O_2$, cis-Bicyclo[3.2.0]hept-2-en-6-one p-nitrophenylhydrazone, 45B, 138
$C_{13}H_{14}O_2$, 1,3,4-Trimethylbicyclo[4.4.0]dec-3,6,8-triene-2,5-dione, 39B, 113
$C_{13}H_{14}O_4$, 1,4-Dimethoxybenzocycloheptene-5,9(6H,8H)-dione, 46B, 142
$C_{13}H_{19}BrO_2$, 7-Bromo-4,4-dimethylbicyclo[6.3.0]undecane-2,6-dione, 44B, 147
$C_{13}H_{20}O_3$, 1,7-Diacetyl-5-hydroxy-2,2-dimethylbicyclo[4.1.0]heptane, 46B, 143
$C_{14}H_9NO_2$, 2-N-Pyridinium-indan-1,3-dione (monoclinic form), 42B, 131
$C_{14}H_{12}$, 3,4-Benzocyclodeca-1,5-diyne, 43B, 204
$C_{14}H_{14}O_3$, 2-Pivaloyl-1,3-indandione, 42B, 131; 46B, 143
$C_{14}H_{20}O_2$, 2,3,4aβ,8aβ-Tetramethyl-4a,5,8,8a-tetrahydro-1-naphthoquin-4β-ol, 46B, 144
$C_{14}H_{20}O_2$, 4α-Hydroxy-2,3,4aβ,8aβ-tetramethyl-4a,5,8,8a-tetrahydro-1(4H)-naphthalenone, 46B, 144
$C_{14}H_{20}O_3$, 1,6,6-Trimethyl-8-methoxycarbonylmethylspiro[2.5]oct-1-ene-7-one, 45B, 138
$C_{14}H_{20}O_3$, 2-(1α,8β,10β-Bicyclo[6.3.0]undec-4-en-11-one)propionic acid, 39B, 113
$C_{14}H_{24}O_3$, 1-Ethoxycarbonyl-2-hydroxy-1-methyl-cis-perhydronaphthalene, 44B, 148
$C_{15}H_8F_6$, 1,2,3,4,5,6-Hexafluoro-7-methyl-8-phenyl-bicyclo[4.2.0]-octa-2,4,7-triene, 44B, 149
$C_{15}H_9BrO_2$, 2-p-Bromophenylindane-1,3-dione (orthorhombic form), 40B, 158
$C_{15}H_9BrO_2 \cdot 0.25\ C_6H_6$, 2-p-Bromophenylindane-1,3-dione benzene solvate (triclinic form), 38B, 204; 40B, 158
$C_{15}H_9ClN_4$, trans-6-Chloro-10,10,11,11-tetracyanobicyclo[7.2.0]-undeca-2,4,7-triene, 39B, 114
$C_{15}H_{10}ClNO$, 1-Amino-2-p-chlorophenyl-3-indenone, 38B, 205
$C_{15}H_{10}N_4$, trans-10,10,11,11-Tetracyanobicyclo[7.2.0]undeca-2,4,7-triene, 39B, 114
$C_{15}H_{10}O_2$, 2-Phenyl-1,3-indandione, 40B, 159
$C_{15}H_{15}Br_2O_3P$, Dimethyl 2,5-dibromo-7-phenylnorcaradiene-7-phosphonate, 40B, 409
$C_{15}H_{15}Cl_2O_3P$, Dimethyl 2,5-dichloro-7-phenylnorcaradiene-7-phosphonate, 40B, 410
$C_{15}H_{15}Cl_7O_3$, 2,2,3,3,4,6,7-Heptachloro-1,1,5-triethoxyindan, 45B, 139
$C_{15}H_{16}N_2O_6$, 3-Ethylidenecyclohexanol 3,5-dinitrobenzoate, 45B, 140
$C_{15}H_{17}O_3P$, 7-Methoxyphosphoryl-7-phenylnorcaradiene, 40B, 410
$C_{15}H_{20}O_2$, 2-Isopropyl-α-methyl-5-indanacetic acid, 45B, 140
$C_{15}H_{22}O_3$, 6-Carboxy-7-hydroxymethyl-2,2,4-trimethyl-1,2,3,3a,4,8a-hexahydro-azulene, 44B, 149
$C_{15}H_{24}O_2$, (DL)-5α-Ethenyl-5β-hydroxy-1,1,4aβ-trimethyl-1,4,4a,5,6,7,8,8aα-octahydronaphthalene-2(3H)-one, 46B, 145
$C_{15}H_{24}O_3$, 9-Carbomethoxy-4,6,6-trimethyl-trans-decal-3-one, 40B, 159
$C_{15}H_{24}O_4$, cis-1,1-Bismethoxycarbonyl-4a-methyl-decahydronaphthalene, 38B, 206
$C_{15}H_{27}NO$, 1-Decalone enolate, 45B, 141
$C_{15}H_{28}INO_2$, (-)-3(a)-Triethylammonio-2(a)-acetoxy-trans-decalin iodide, 42B, 22
$C_{15}H_{28}INO_2$, 3(a)-Dimethylamino-2(a)-acetoxy-trans-decalin methiodide, 37B, 95

$C_{16}H_{8}Cl_{4}$, 1,2,3,4-Tetrachlorobenzo[g]sesquifulvalene, 31B, 106
$C_{16}H_{12}Cl_{4}$, 1-Phenyl-1,2,3,4-tetrachlorotetralin, 41B, 199
$C_{16}H_{12}O$, 2-Benzylidene-1-indanone, 46B, 145
$C_{16}H_{12}O_{2}$, 3-Hydroxy-9-phenyl-1,9-dihydroazulen-1-one, 37B, 95
$C_{16}H_{14}O_{4}$, Dimethyl heptalene-3,8-dicarboxylate, 43B, 204
$C_{16}H_{16}O_{4}$, Azulene-1,3-dipropionic acid, 31B, 107
$C_{16}H_{19}ClO_{2}$, D-6-Chloro-5-cyclohexylindan-1-carboxylic acid, 41B, 200
$C_{16}H_{19}NO_{5}S$, 7-exo-Isopropenylbicyclo[4.1.0]hept-7-endo-yl p-nitrobenzosulphonate, 45B, 142
$C_{16}H_{22}BrNO_{3}$, 6α-Bromo-4α-(3,5-dimethyl-4-isoxazoylmethyl)-1β-hydroxy-7aβ-methyl-3aα,6,7,7a-tetrahydro-5(4H)-indanone, 40B, 160
$C_{16}H_{22}O_{2}$, 2,3,4aβ,6,7β,8aβ-Hexamethyl-4a,7,8,8a-tetrahydro-1,4-naphthoquinone, 46B, 146
$C_{16}H_{24}O_{2}$, 2,3,4aβ,6,7,8aβ-Hexamethyl-4a,5,8,8a-tetrahydro-1-naphthoquin-4α-ol, 46B, 147
$C_{16}H_{24}O_{2}$, 2,3,4aβ,6,7,8aβ-Hexamethyl-4a,5,8,8a-tetrahydro-1-naphthoquin-4β-ol, 46B, 147
$C_{16}H_{25}NO_{5}$, Dimethyl r-1-morpholino-c-6H-bicyclo[4.2.0]octane-c-7,t-8-dicarboxylate, 43B, 205
$C_{16}H_{26}ClNO$, (+)-2-Dipropylamino-5-hydroxytetralin hydrochloride, 46B, 34
$C_{16}H_{26}O_{3}$, trans-1β,6α-Dimethyl-8(E)-(1-ethoxyethylidene)-2α-methoxybicyclo[4.3.0]nonan-7-one, 42B, 132
$C_{16}H_{27}ClN$, 11-Methyl-10-epieudesm-4-en-3-one 3-oxime, 46B, 148
$C_{17}H_{12}O_{2}$, DL-2,2'-Spirobi(indan)-1,1'-dione, 40B, 160
$C_{17}H_{12}O_{2}$, 2-p-Methylbenzilidene-1,3-indandione, 35B, 124
$C_{17}H_{14}N_{2}O_{2}$, 2-(4'-Dimethylaminophenylimino)indan-1,3-dione, 46B, 148
$C_{17}H_{14}O$, 2-Benzylidene-1-tetralone, 44B, 150
$C_{17}H_{14}O_{2}$, 2-Ethyl-2-phenyl-1,3-indandione, 40B, 161
$C_{17}H_{20}BrNO \cdot H_{2}O$, 1-Amino-2-phenyl-benzocycloheptanol hydrobromide monohydrate, 46B, 148
$C_{17}H_{21}BrO_{2}$, 1-(7'-Bromobicyclo[4.1.0]hept-7'-yl)-methylethyl benzoate, 45B, 142
$C_{17}H_{24}O_{3}S$, trans,trans-Decalyl-2-tosylate, 39B, 115
$C_{18}H_{10}O_{3}$, 2'-Dionylidene-indane-3-indanone, 42B, 132
$C_{18}H_{14}$, 1,1'-Biindenyl, 45B, 143
$C_{18}H_{15}NO_{2}$, 2-(p-Dimethylaminobenzylidene)-1,3-indandione (α-form), 44B, 150
$C_{18}H_{15}NO_{2}$, 2-(p-Dimethylaminobenzylidene)-1,3-indanedione (β-form), 46B, 149
$C_{18}H_{15}NO_{2}$, 2-(p-Dimethylaminobenzylidene)-1,3-indanedione (γ-form), 46B, 149
$C_{18}H_{17}Cl$, 3-Chloro-1-ethyl-2-methyl-1-phenylindene, 46B, 150
$C_{18}H_{21}NO_{2}$, 4-Methyl-6-(N-methyl-N-phenylamino)-spiro[4.5]dec-6-ene-1,8-dione, 46B, 150
$C_{19}H_{12}Cl_{2}$, 1,1-Dichloro-2,5-diphenylcyclopropabenzene, 42B, 132
$C_{19}H_{12}O_{2}$, 2-(1-Naphthyl)-1,3-indandione, 45B, 144
$C_{19}H_{12}O_{2}$, 2-(2-Naphthyl)-1,3-indandione, 45B, 144
$C_{19}H_{16}O_{2}$, 5-Methoxy-6,6-diphenylbicyclo[3.1.0]hex-3-en-2-one, 46B, 151
$C_{19}H_{16}O_{4}$, Dimethyl 3-phenyl-1H-indene-1,2-dicarboxylate, 43B, 205
$C_{19}H_{24}IN$, (+)-N,N-Dimethyl-4-methyl-4-phenyl-1,2,3,4-tetrahydro-2-naphthylamine hydroiodide, 45B, 144
$C_{19}H_{25}BrO_{2}$, Decalyl p-bromobenzoate, 41B, 201
$C_{19}H_{25}BrO_{2}$, 2-p-Bromobenzoato-5-t-butylbicyclo[4.2.0]octane, 46B, 151
$C_{19}H_{28}O_{4}$, 6-Methoxycarbonyl-5-(2-methoxycarbonylethyl)-2,6-dimethyl-

bicyclo[6.3.0]undeca-1(11),2-diene, 43B, 206
$C_{20}H_{17}NO_3$, 2-(4'-Dimethylaminocinnamoyl)indan-1,3-dione, 46B, 152
$C_{20}H_{18}$, 2,6-Diphenylhomotropylidene, 46B, 152
$C_{20}H_{18}O_4$, endo,endo-2,4-Bis(methyoxycarbonyl)-1,3-diphenylbicyclo-[1.1.0]butane, 45B, 145
$C_{20}H_{18}O_4$, endo,exo-2,4-Bis(methoxycarbonyl)-1,3-diphenylbicyclo-[1.1.0]butane, 45B, 145
$C_{20}H_{22}O$, 1,1-Bis(bicyclo[6.1.0]nona-2,4,6-trien-9-anti-yl)dimethyl ether, 45B, 146
$C_{20}H_{26}O_4$, Dimethyl 4,6-di-t-butylpentalene-1,2-dicarboxylate, 43B, 207
$C_{20}H_{29}BrN_2O_2S$, (p-Bromophenyl)sulphonylhydrazone, 45B, 147
$C_{20}H_{30}$, 1,3,5-Tri-t-butylpentalene, 43B, 207
$C_{21}H_{15}NO$, 1-Phenylamino-2-phenyl-3-indenone, 39B, 116
$C_{21}H_{21}NO_4$, 4-Anilino-3-methoxycarbonyl-1-(methoxycarbonylmethylidene)-1,2,3,4-tetrahydronaphthalene, 45B, 146
$C_{21}H_{27}NO_4$, 1-Decalone enol p-nitrobenzoate, 45B, 141
$C_{21}H_{31}NO_5$, 6,7-Bismethoxycarbonyl-2,2,4-trimethyl-8-(N-morpholino)-1,2,3,3a,4,8a-hexahydro-azulene, 44B, 151
$C_{21}H_{32}N_2O_2S$, Tosylhydrazone, 45B, 147
$C_{22}H_{13}BrN_2O_3$, Indane-1,2,3-trione 2-(N-bromobenzoyl-N-phenylhydrazone), 43B, 207
$C_{22}H_{14}N_2O_3$, Indane-1,2,3-trione 2-(N-benzoyl-N-phenylhydrazone), 43B, 207
$C_{22}H_{16}N_2O_4$, 2-Hydroxy-2-(β-benzoyl-β-phenylhydrazyl)indan-1,3-dione, 42B, 47
$C_{22}H_{23}BrO$, [2-(7-exo-Bromobicyclo[4.1.0]hept-7-endo-yl)-2-phenylethyl]phenylketone, 45B, 142
$C_{22}H_{24}N_2O_4$, endo-endo-2,6-Bis(phenylcarbamoyloxy)-cis-bicyclo-[3.3.0]octane, 41B, 201
$C_{22}H_{28}O_7S$, Δ(3a)-4-(m-Methoxybenzylsulphonyl)-1,5-diacetoxy-7a-methylhexahydroindene, 38B, 206
$C_{22}H_{30}O_3$, trans-1-(2-o-Methoxyphenylethyl)-4β-hydroxy-2,5α,9β-trimethyl-1-octal-8-one, 40B, 162
$C_{23}H_{16}O$, 2,7-Diphenyl-4,5-benzotropone, 43B, 208
$C_{23}H_{35}BrO_9$, Tetraethyl 4-bromo-8-methoxydecahydronaphthalene-2,2,6,6-tetracarboxylate, 38B, 207
$C_{23}H_{36}O_3S$, (12β-H)-13β-(p-Tolylsulphonyl)-14α-methylbicyclo[10.3.0]-pentadecan-1β-ol, 42B, 133
$C_{24}H_{21}NO_2$, 1-Phenylimino-2-p-methoxyphenyl-3-ethoxy-ind-2-ene, 40B, 1081
$C_{24}H_{22}O$, 3-Methoxy-1,4-dimethyl-1,2-diphenyl-indene, 45B, 148
$C_{25}H_{20}O_2$, endo-6-Methoxy-1,3,6-triphenylbicyclo[3.1.0]hex-3-en-2-one, 41B, 203
$C_{26}H_{20}$, cis,cis-1,3-Bis(styryl)azulene, 40B, 163
$C_{26}H_{22}N_2O_3$, Indan-1,2,3-trione 2-(N-p-t-butylbenzoyl-N-phenylhydrazone), 44B, 152
$C_{26}H_{25}BrO_5$, (1R,S)-2-(5,6,7,7a-Tetrahydro-(7aS,R)-methyl-1,5-dioxo-4-indanyl)-1-(3-methoxyphenyl)ethyl 4-bromobenzoate, 39B, 116
$C_{29}H_{28}N_2O_4$, Indan-1,2,3-trione 2-(N-p-t-butylbenzoyl-N-phenylhydrazone) acetone solvate, 44B, 152
$C_{30}H_{22}O_2$, meso-2,2'-Dioxo-1,1'-diphenyl-1,1'-biindane, 44B, 153
$C_{37}H_{24}N_4O_5$, Indan-1,2,3-trione 2-(N-benzoyl-N-phenylhydrazone) indan-1,2,3-trione 2-(N-phenylhydrazone), 44B, 153
$C_{43}H_{30}O_2$, 1-(Diphenylvinylidene)-3-phenyl-2-indenyl diphenylacetate, 45B, 148
$C_{48}H_{36}$, Tetramethylverdene, 37B, 96

$C_{51}H_{44}N_4O$, 1,4-Di(t-butyl)-2,3-diphenyl-5,6-bis(1-phenyl-2,2-dicyanovinyl)benzocyclobutene - acetone, 43B, 211

28. HYDROCARBONS (3 FUSED RINGS)

C_6H_8, Tricyclo[3.1.0.0$^{2-4}$]hexane (gas-ed), 42B, 1028
C_7H_{10}, Tricyclo[4.1.0.0$^{1-3}$]heptane (gas-ed), 43B, 1485
$C_8H_6Br_2N_2O_5$, 6,6-Dibromo-2,3:4,5-dimethano-2,4-dinitrocyclohexanone, 30B, 106
C_8H_8, Semibullvalene (gas-ed), 38B, 1064
C_8H_{12}, Tricyclo[4.2.0.0$^{2-5}$]octane, 35B, 899
$C_9H_{10}N_2O_7$, 2,3:4,5-Dimethano-2,4-dinitro-1-hydroxycyclohexane-1-carboxylic acid, 42B, 135
$C_{10}Cl_{10}$, Perchloro-all-cis-tricyclo[5.2.1.0$^{4-10}$]deca-2,5,8-triene, 39B, 117
$C_{10}H_{10}$, Benzo(1,2:4,5)dicyclobutene, 34B, 110
$C_{10}H_{10}$, Triquinacene, 42B, 153
$C_{10}H_{12}O_2$, Tricyclo[5.3.0.0$^{2-6}$]decan-3,8-dione, 30B, 108
$C_{10}H_{12}O_2$, Tricyclo[5.3.0.0$^{2-6}$]decane-4,9-dione, 40B, 164
$C_{11}H_7Br$, 1-Bromo-1H-cyclobuta[de]naphthalene, 43B, 183
$C_{11}H_8$, Naphtho[b]cyclopropene, 39B, 118
$C_{11}H_{16}Br_2$, trans-8,8-Dibromo-1,4,4-trimethyl-tricyclo[5.1.0.0$^{3-5}$]-octane, 35B, 125
$C_{12}Cl_{12}$, Perchloro-(3,4,7,8-tetramethylene-tricyclo[4.2.0.0$^{2-5}$]-octane), 32B, 135
$C_{12}H_6O_2$, Acenaphthenequinone, 28, 434
$C_{12}H_8$, Acenaphthylene, 39B, 118; 40B, 164
$C_{12}H_8$, Biphenylene (gas-ed), 40B, 1137
$C_{12}H_8$, Biphenylene, 9, 375; 27, 935; 31B, 109
$C_{12}H_8BrCl$, 5-Bromo-6-chloroacenaphthene, 29, 679
$C_{12}H_8Br_2$, trans-1,2-Dibromoacenaphthene, 38B, 208
$C_{12}H_8Br_2$, 3,5-Dibromoacenaphthene, 27, 933
$C_{12}H_8Cl_2$, cis-1,2-Dichloroacenaphthene, 40B, 165
$C_{12}H_8Cl_2$, trans-1,2-Dichloroacenaphthene, 38B, 208
$C_{12}H_8Cl_2$, 5,6-Dichloroacenaphthene, 26, 746
$C_{12}H_8I_2$, 3,5-Diiodoacenaphthene, 28, 438
$C_{12}H_8N_2O_6$, cis-1,2-Acenaphthenediol dinitrate, 29, 681
$C_{12}H_8N_2O_6$, trans-1,2-Acenaphthenediol dinitrate, 39B, 119
$C_{12}H_8N_4$, 1,2,5,6-Tetracyano-anti-tricyclo[4.2.0.0$^{2-5}$]octane, 40B, 166
$C_{12}H_8N_4O$, 1,2,5,6-Tetracyano-cycloocta-1-ene-5-epoxide, 40B, 166
$C_{12}H_9I$, 5-Iodoacenaphthene, 28, 438
$C_{12}H_{10}$, Acenaphthene, 5, 143, 172; 12, 408; 21, 604
$C_{12}H_{10}$, Naphtho[b]cyclobutene, 39B, 119
$C_{12}H_{10}O$, 7-Acenaphthenol, 41B, 204
$C_{12}H_{10}O_2$, cis-1,2-Acenaphthenediol, 28, 435
$C_{12}H_{12}Cl_4$, 1,2,5,6-Tetrachloro-3,4,7,8-tetramethyl-tricyclo-[4.2.0.0$^{2-5}$]octa-3,7-diene, 32B, 134
$C_{12}H_{12}O_2$, 2,2a,3,4-Tetrahydro-1H-cyclopent[cd]indene-1-carboxylic acid, 39B, 120
$C_{12}H_{16}O_2$, 6,7-Dimethyl-cis-anti-cis-tricyclo[5.3.0.0$^{2-6}$]decane-3,10-dione, 40B, 167
$C_{12}H_{20}$, trans-anti-trans-Tricyclo[6.4.0.0$^{2-7}$]dodecane, 40B, 167
$C_{13}H_5N_3O_7$, 2,4,7-Trinitro-9-fluorenone, 38B, 209
$C_{13}H_7BrN_2$, 2-Bromodiazofluorene, 35B, 126
$C_{13}H_7BrO$, 2-Bromoketofluorene, 35B, 125

$C_{13}H_8N_2$, 9-Diazafluorene, 44B, 154
$C_{13}H_8O$, 9-Fluorenone, 38B, 211
$C_{13}H_8O_2$, Acenaphthylene-1-carboxylic acid, 39B, 120
$C_{13}H_8O_2$, 9-Hydroxyphenalenone, 45B, 149
$C_{13}H_{10}$, Fluorene, 19, 583
$C_{13}H_{10}N_2O$, 2,7-Diamino-9-fluorenone, 43B, 212
$C_{13}H_{10}N_2O_2$, 2-Amino-7-nitrofluorene, 40B, 168
$C_{13}H_{12}O_2$, cis-2a,5-Dihydro-5-acenaphthoic acid, 43B, 212
$C_{13}H_{18}O$, trans,trans-Tricyclo[7.3.1.0$^{5-13}$]tridec-1-en-3-one, 45B, 150
$C_{14}Cl_{10}$, Decachlorophenanthrene, 42B, 135
$C_{14}H_{10}$, Phenanthrene, 26, 752; 28, 439; 37B, 97
$C_{14}H_{12}N_2S$, S-(9-Fluorenyl)isothiourea (monoclinic), 43B, 213
$C_{14}H_{12}N_2S$, S-(9-Fluorenyl)isothiourea (triclinic), 43B, 213
$C_{14}H_{20}O_2$, 9b-Methyl-dodecahydro-1H-cyclopenta[a]naphthalene-1,5-dione, 45B, 150
$C_{14}H_{22}O$, Norcubebanone, 43B, 214
$C_{14}H_{22}O$, 2-Hydroxy-8-methyl-12-methylenetricyclo[8.2.0.0$^{3-8}$]dodecane, 44B, 154
$C_{14}H_{22}O$, 6,6,9-Trimethyltricyclo[6.3.0.0$^{5-7}$]undecen-2-one, 45B, 151
$C_{14}H_{22}O_2$, 3,3-Dimethyltricyclo[6.4.0.0$^{2-7}$]dodec-6-en-1,2-diol, 43B, 214
$C_{14}H_{24}$, cis-anti-cis-Perhydroanthracene, 24, 666; 32B, 136
$C_{14}H_{24}O$, (1RS,2SR,9SR)-Tricyclo[7.5.0.0$^{2-8}$]tetradeca-7-en-1-ol, 41B, 205
$C_{15}H_{13}N$, 10-(N-Methylamino)benz[f]azulene, 44B, 155
$C_{15}H_{13}NO$, N-2-Fluorenylacetamide, 46B, 100
$C_{15}H_{13}NO$, 2-Acetylaminofluorene, 46B, 153
$C_{15}H_{13}NO$, 4-Acetylaminofluorene, 45B, 151
$C_{15}H_{16}BrNO$, endo-N-(p-Bromophenyl)tricyclo[4.2.0.0$^{1-4}$]octane-3-carboxamide, 46B, 153
$C_{15}H_{22}O$, (+)-6,7-Dimethyl-4-isopropyl-tricyclo[4.4.0.0$^{2-4}$]dec-1-ene-9-one, 44B, 156
$C_{15}H_{24}Br_2$, 8,9-Dibromo-1,4,4,8-tetramethyltricyclo[5.4.0.0$^{2-5}$]undecane, 39B, 121
$C_{15}H_{24}Br_2$, 8,9-Dibromo-2,2,4,8-tetramethyltricyclo[5.3.1.0$^{4-11}$]undecane, 39B, 121
$C_{15}H_{26}O$, 7β-Hydroxy-8α-isopropyl-1α,5β-dimethyltricyclo[4.4.0.0$^{2-5}$]-decane, 45B, 152
$C_{16}H_4N_6O_8$, 9-Dicyanomethylene-2,4,5,7-tetranitrofluorene, 40B, 169
$C_{16}H_5N_5O_6$, 9-Dicyanomethylene-2,4,7-trinitrofluorene, 32B, 137
$C_{16}H_6Br_2N_2$, 9-Dicyanomethylene-2,7-dibromofluorene, 39B, 122
$C_{16}H_6N_4O_4$, 9-Dicyanomethylene-2,7-dinitrofluorene, 33B, 91
$C_{16}H_8$, sym-Dibenzo-1,5-cyclooctadiene-3,7-diyne, 41B, 206; 43B, 215
$C_{16}H_8$, 5,6,11,12-Tetradehydrodibenzo[a,e]cyclooctene, 41B, 206
$C_{16}H_{10}O_2$, Benzoditropone, 42B, 136
$C_{16}H_{13}Br$, 5-(Bromomethylene)-10,11-dihydro-5H-dibenzo[a,d]-cycloheptene, 35B, 127
$C_{16}H_{20}$, 4,4,8,8-Tetramethyl-1,4,5,8-tetrahydro-3-indacene 4,4,8,8-tetramethyl-1,4,7,8-tetrahydro-3-indacene (1:1), 45B, 152
$C_{16}H_{24}ClNO \cdot 1.5\ H_2O$, 1-Hydroxy-13-isopropylamino-tricyclo[6.5.0.-0$^{2-7}$]trideca-2,4,6-triene hydrochloride sesquihydrate, 46B, 154
$C_{16}H_{24}Cl_4O_4$, 6,6,12,12-Tetrachloro-3,3,9,9-tetramethoxytricyclo-[9.1.0.0$^{5-7}$]dodecane, 38B, 212
$C_{16}H_{24}O_2$, 3,14-Dimethoxy-tricyclo[7.5.0.0$^{2-8}$]tetradecane-5,11-diene, 46B, 155
$C_{17}H_{12}$, 7,8,12,13-Tetradehydro-10,11-dihydro-9H-cyclodeca[1,2,3-de]-

naphthalene, 43B, 216
$C_{17}H_{12}F_8$, 2a,3,4,5,6,7,8,8b-Octafluoro-2-methyl-2-(2-methylprop-1-enyl)-1,2,2a,8b-tetrahydrocyclobuta[a]naphthalene, 43B, 217
$C_{17}H_{16}O_3$, 2',3'-Dimethoxy-2,3,6,7-dibenzo-1-suberone, 40B, 170
$C_{17}H_{30}O_4$, 2-Hydroxymethyl-3,3-dimethoxy-6,10,10-trimethyltricyclo-[6.3.0.0$^{2-5}$]undecan-6-ol, 46B, 155
$C_{18}H_{14}$, 7,8,13,14-Tetradehydro-9,10,11,12-tetrahydrocycloundeca-[1,2,3-de]naphthalene, 43B, 216
$C_{18}H_{16}O_2$, 7-Methoxycarbonyl-anti-1,6:8,13-dimethano[14]annulene, 39B, 123
$C_{18}H_{18}$, Retene, 8, 318
$C_{18}H_{18}N_4$, 7,7,8,8-Tetracyano-2-endo-3,4-trans-5,6-pentamethyltricyclo[4.3.0.0$^{1-5}$]non-3-ene, 42B, 136
$C_{18}H_{28}O_2$, dl-1,1,4aα-Trimethyl-2α-hydroxy-8β-methoxy-1,2,3,4,4a,5,6,7,8,9-decahydrophenanthrene, 40B, 171
$C_{18}H_{30}O_2$, Isophorone photodimer, 35B, 128
$C_{19}H_{14}$, 9-Phenylfluorene, 45B, 153
$C_{19}H_{15}N$, 1-Methyl-2-(9'-fluorenylidene)-1,2-dihydropyridine, 34B, 111; 42B, 137
$C_{19}H_{15}NO$, 1-Phenoxy-1,2,2a,8b-tetrahydro-cyclobuta[a]naphthalene-8b-carbonitrile, 46B, 156
$C_{19}H_{17}BrO_2$, Tricyclo[6.4.0.0$^{2-7}$]dodeca-2,4,6-triene-1-yl p-bromobenzoate, 39B, 124
$C_{19}H_{21}NO_5$, 2-N,N-Diethylamino-1-carbomethoxy-2a,8a-dihydro-2a-methoxycyclobuta[b]naphthalene-3,8-dione, 37B, 97
$C_{19}H_{22}O_7$, 4,6-Dicarbomethoxy-7,9-dimethoxy-2,3,3a,4,5,9b-hexahydro-1H-benz[e]inden-2-one, 43B, 218
$C_{19}H_{23}ClO_2$, 3a,6-Dimethyl-6-(trans-3-chlorobut-2-ene-1-yl)-2,4,5,6,8,9-hexahydro-3H-benz[e]indene-3,7-(3aH)-dione, 38B, 213
$C_{19}H_{23}IO_3$, 9-Hydroxy-1-methoxy-2-methyltricyclo[5.2.1.0$^{2-10}$]decane p-iodobenzoate, 44B, 157
$C_{19}H_{23}N$, 4-(Diethylamino)-1,10-ethano-5-methylcyclopentacyclononene, 46B, 156
$C_{19}H_{23}NO_2$, 7-Methyl-3-methylenetricyclo[5.3.0.0$^{1-4}$]dec-5-yl phenylcarbamate, 46B, 157
$C_{19}H_{24}$, 1,1-Dimethyl-7-isopropyl-1,2,3,4-tetrahydrophenanthrene, 40B, 1081
$C_{19}H_{28}O_3$, 1,1,4aα-Trimethyl-2α-hydroxy-8α-acetoxy-1,2,3,4,4a,5,6,7,8,9-decahydrophenanthrene, 39B, 124
$C_{20}H_{12}IN_3$, C-Biphenylene-N(1)-(4-iodophenyl)-N(2)-cyano-azomethinimine, 32B, 138
$C_{20}H_{16}$, 5,7-Dimethyl-2-phenylcyclopent[cd]azulene, 35B, 130
$C_{20}H_{16}$, 9,10,15,16-Tetradehydro-11,12,13,14-tetrahydrodibenzo[a,c]-cyclododecene, 38B, 190
$C_{20}H_{16}Cl_2O_7$, 5-Acetoxy-2,6-dichloro-4aα,10aα-dihydro-8-hydroxy-10β-methyl-4aα-propionyl-phenanthrene-1,4,9(10H)-trione, 45B, 153
$C_{20}H_{18}$, 1,4,7,10-Tetramethyl-5,6-didehydrodibenzo[a,e]cyclooctene, 45B, 153
$C_{20}H_{18}BrNO_2$, 2,3-Benzobicyclo[4.2.0]octa-2,4-dienyl-endo-7-carbinol p-bromophenylurethane, 39B, 126
$C_{20}H_{18}O_2$, 4-Hydroxy-4-(tetrahydronaphthalene-1-one-4-yl)-2,3-benzobicyclo[3.1.0]hex-2-ene, 44B, 158
$C_{20}H_{19}BrO_2$, Tricyclo[6.5.0.0$^{2-7}$]trideca-2,4,6-triene-1-yl p-bromobenzoate, 39B, 126
$C_{20}H_{24}N_2O_2$, 9-(3-t-Butylamino-2-hydroxy-propoxyimino)fluorene, 45B, 154
$C_{20}H_{24}O_8$, Tetramethyl 2β-tricyclo[6.4.0.0$^{2-7}$]dodeca-4,10-diene-

1,4,5,8-tetracarboxylate, 44B, 159
$C_{20}H_{26}O_3$, (9R,10R)-20(10-9)abeo-12-Hydroxy-5,7,12-abietatriene-11,14-dione, 41B, 383
$C_{20}H_{27}N$, 8-Dimethylamino-1,3,4a,8a-tetramethyl-4a,8a-dihydro-cyclopenta[1,2-a]indene, 46B, 157
$C_{20}H_{27}NO_3$, 10aα-Cyano-1α,4aβ-dimethyl-7-ethoxy-2β-methoxy-1,2,3,4,4a,9,10,10a-octahydro-1β-phenanthrenol, 40B, 172
$C_{20}H_{30}$, DL-Bivalvane, 45B, 154
$C_{20}H_{30}$, meso-Bivalvane, 45B, 154
$C_{20}H_{30}O_2$, Levopimaric acid, 38B, 214
$C_{20}H_{30}O_3$, 12β-Hydroxysandaracopimaric acid, 38B, 215
$C_{21}H_{18}$, 5,6,12,13-Tetradehydro-8,9,10,11-tetrahydro-7H-dibenzo[a,c]-cyclotridecene, 43B, 218
$C_{21}H_{23}NO_4S$, 8,12-Dimethyltricyclo[6.4.0.0$^{2-7}$]dodeca-2,4,6-trien-1-ol benzosulphonylurethane, 41B, 207
$C_{21}H_{26}BrN$, 9-(3-Dimethylaminopropylidene)-10,10-dimethyl-9,10-dihydroanthracene hydrobromide, 45B, 155
$C_{22}H_{12}O_2$, E,E-(1,2,8,9)-Dibenzo-3,10-bisdehydro[14]annulene-5,12-dione, 45B, 156
$C_{22}H_{12}O_2$, Z,Z-(1,2,8,9)-Dibenzo-3,10-bisdehydro[14]annulene-5,12-dione, 45B, 156
$C_{22}H_{16}$, α-(6-Fulvenyl)dibenzo[a,e]heptafulvene, 42B, 137
$C_{22}H_{17}Br$, 5-(2'-Bromo-3',4'-dihydronaphthyl)acenaphthene, 38B, 216
$C_{22}H_{17}Br_2N$, 2,7-Dibromo-9-[4-(dimethylamino)benzylidene]fluorene, 45B, 157
$C_{22}H_{20}O_2$, cis-anti-cis-3-Phenyl-2-cyclopentenone photodimer, 40B, 173
$C_{22}H_{22}O$, Phenyl[3'-exo-phenyl-bicyclo[4.1.0]heptan-7-spiro-1'-cycloprop-2'-endo-yl]ketone, 45B, 142
$C_{22}H_{23}Cl_2N$, (1S,4R)-3'-Chloro-N,N-dimethylspiro[2-cyclohexene-1,5'-[5H]dibenzo[a,d]cyclohepten]-4-amine hydrochloride, 46B, 35
$C_{22}H_{26}BrN$, 3-(10,11-Dihydro-5H-dibenzo[a,d]cyclohepten-5-ylidene)-1-ethyl-2-methylpyrrolidine hydrobromide, 37B, 98
$C_{22}H_{29}BrO_2$, 9-p-Bromobenzoyloxy-1,4,4,8-tetramethyltricyclo-[5.4.0.0$^{3-5}$]undecane, 42B, 138
$C_{22}H_{29}O_2$, 5-(p-Methoxybenzylidene)-2,2,9-trimethyl-tricyclo-[6.3.0.0$^{3-8}$]undecan-4-one, 45B, 157
$C_{22}H_{30}O_5$, 3"-(3-Methyl-3-butenyl)-2,2',2"-trioxo-1,1':3',1"-tercyclopent-3-ylacetic acid, 42B, 1019
$C_{23}H_{22}O_3$, 1-Phthaloyl-2-p-amylbenzoylcyclopropane, 45B, 158
$C_{23}H_{25}BrO_2$, cis-10-t-Butyltricyclo[6.4.0.0$^{2-7}$]dodeca-2,4,6-trien-1-ol p-Bromobenzoate, 43B, 219
$C_{23}H_{28}O$, 2-Butyl-7-(1-oxopentyl)-9,10-dihydrophenanthrene, 42B, 139
$C_{24}H_{14}Br_2N_2O_2$, 2,7-Dibromo-4,5-bis-(2-pyridyl)phenanthrene-3,6-diol, 37B, 99
$C_{24}H_{18}$, peri-Diphenylacenaphthene, 42B, 139
$C_{24}H_{23}NO_4$, 5-(3-Dimethylammonioprop-1-enylidene)-5H-dibenzo[a,d]cycloheptene maleate, 44B, 159
$C_{24}H_{30}O$, 2-(2-Methylbutyl)-7-(1-oxopentyl)-9,10-dihydrophenanthrene, 41B, 207
$C_{24}H_{32}O_{12}$, anti-Tricyclo[4.4.0$^{1-6}$.0$^{7-12}$]-1,4,7,10-tetraethoxycarbonyl-3,6,9,12-tetrahydroxy-dodeca-3,9-diene, 39B, 127
$C_{25}H_{20}O$, 8b-Methoxy-1,3-diphenyl-2a,8b-dihydrocyclobuta[a]naphthalene, 46B, 158
$C_{25}H_{20}O$, 8b-Methoxy-2,3-diphenyl-2a,8b-dihydrocyclobuta[a]naphthalene, 46B, 158
$C_{25}H_{31}BrO_3$, 1,1,4aα-Trimethyl-2α-p-bromobenzoyloxy-8α-methoxy-

1,2,3,4,4a,5,6,7,8,9-decahydrophenanthrene, 39B, 125
$C_{26}HCl_{17}$, 9H-Heptadecachloro-9,9'-bifluorenyl, 46B, 159
$C_{26}H_2Cl_{16}$, 9H,9'H-Hexadecachloro-9,9'-bifluorenyl, 46B, 159
$C_{26}H_{16}$, Bis(biphenylene)ethylene, 11, 666; 18, 738
$C_{26}H_{16}$, 9,9'-Bifluorenylidene, 44B, 160
$C_{26}H_{18}$, 9,9'-Bifluorenyl, 44B, 160
$C_{26}H_{36}O_6$, Methyl 6-isobutenyl-2-methyl-4-oxocyclohex-2-enecarboxylate photodimer, 45B, 116
$C_{26}H_{40}$, 1,3,6,8-Tetra-t-butyl-hemi-Dewar-naphthalene, 43B, 220
$C_{27}H_{18}$, 1,1,3,3-Bis(2,2'-biphenylylene)propene, 43B, 221
$C_{28}H_{16}O_2$, Dianthronylidene, 18, 740
$C_{28}H_{18}O_2$, 10,10'-Dianthronyl, 32B, 139
$C_{28}H_{24}$, trans-1,2,3,4,1',2',3',4'-Octahydro-4,4'-biphenanthrylidene, 45B, 159
$C_{28}H_{30}O_2$, (+)-trans-threo-trans-3,3'-Bi-1,2,3,9,10,10a-hexahydro-3-phenanthrol, 45B, 159
$C_{28}H_{36}O_2$, 1,10-Dimethyl-2-(2-(3-keto-1,4,4-trimethylcyclohexyl)-ethyl)-7-methoxy-3,4-dihydroanthracene, 44B, 161
$C_{29}H_{40}O$, 7-Hexanoyl-2-nonyl-9,10-dihydrophenanthrene, 45B, 160
$C_{30}H_{20}$, syn-Δ(5,5')-Bi-5H-dibenzo[a,d]cycloheptene, 40B, 173
$C_{30}H_{20}$, trans-Δ(5,5')-Bi-5H-dibenzo[a,d]cycloheptene, 40B, 173
$C_{30}H_{26}$, 9-t-Butyl-9-(9-fluorenyl)fluorene, 46B, 159
$C_{31}H_{27}LiO$, Lithium 1,1,3,3-bis(2,2'-biphenylylene)propenide diethyl ether, 43B, 221
$C_{34}H_{22}$, 9,9'-(1,4-Phenylenedimethylidene)difluorene, 45B, 157
$C_{34}H_{28}O_4$, Diisopropyl 9,9'-bifluorenylidene-1,1'-dicarboxylate, 44B, 160
$C_{34}H_{28}O_4$, anti-1,2,4,5-Tetraphenyl-3,6-dicarbomethoxytricyclo-[3.1.0.0$^{2-4}$]hexane, 44B, 161
$C_{37}H_{30}$, 7-Methyl-2,3,5,6-tetraphenyltricyclo[6.4.0.0$^{4-8}$]dodeca-3,5,9,11-tetraene, 45B, 160
$C_{38}H_{34}Br_2O_2$, 2-Benzyl-5-p-bromobenzylidenecyclopentanone dimer, 37B, 99
$C_{42}H_{28}$, 1-Phenyl-3,3-biphenylene-allene dimer, 40B, 142
$C_{44}H_{30}F_2$, anti-1,2-Difluoro-3,4,5,6,7,8-hexaphenyltricyclo-[4.2.0.0$^{2-5}$]octa-3,7-diene, 27, 786
$C_{60}H_{80}O_4 \cdot 0.5\ C_4H_{10}O$, 3,3a',5,5'-Tetra-t-butyl-2'-(3,5-di-t-butyl-4-hydroxyphenyl)-7'-[1-(3,5-di-t-butyl-4-hydroxyphenyl)ethylidene]-3a',4',7'7a'-tetrahydrospiro[cyclohexa-2,5-diene-1,1'-[1H]indene]-4,4'-dione diethyl ether solvate, 46B, 110

29. HYDROCARBONS (4 FUSED RINGS)

C_9H_{12}, Trispiro[2.0.2.0.2.0]nonane, 46B, 163
$C_{12}F_{12}$, Perfluoro-1,2-3,4-5,6-triethanobenzene, 42B, 141
$C_{12}F_{12}$, Perfluorobenzo[1,2:3,4:5,6]tricyclobutene, 42B, 141; 43B, 222
$C_{13}H_{14}O_3$, endo-Tetracyclo[5.5.1.0$^{2-6}$.0$^{10-13}$]tridecane-4,8,12-trione, 41B, 209
$C_{14}H_8Cl_2F_6$, exo-1,7-Dichloro-8,8,9,9,10,10-hexafluoro-3,4-benzotricyclo[5.3.0.0$^{2-6}$]dec-3-ene, 46B, 160
$C_{14}H_{10}$, 4,8-Dihydrodibenzo[cd,gh]pentalene, 37B, 100
$C_{14}H_{10}$, 6b,8a-Dihydrocyclobut[a]acenaphthylene, 42B, 142
$C_{14}H_{12}$, Naphtho[b,e]dicyclobutene, 45B, 161
$C_{14}H_{12}$, Pyracene, 26, 753
$C_{14}H_{12}Cl_4$, anti-1',1',2,2-Tetrachloro-3,3-dimethyl-1a',6b'-dihydro-

spiro[cyclopropane-1,2'-[1H]cycloprop[a]indene], 46B, 177
$C_{14}H_{16}$, Dicyclopentenobenzocyclobutene, 43B, 222
$C_{14}H_{16}N_2O_2S$, 5-N,N-Dimethylaminonaphthylsulphonylaziridine, 43B, 223
$C_{15}H_{10}$, 2H-Cyclopenta[jk]fluorene, 42B, 142
$C_{15}H_{12}O_2$, exo-6b,7,8,8a-Tetrahydrocyclobut[a]acenaphthylene-7-carboxylic acid, 44B, 162
$C_{15}H_{14}O_3S$, endo-7-6b,7,8,8a-Tetrahydrocyclobut[a]acenaphthylenyl methanesulphonate, 44B, 162
$C_{15}H_{18}$, Tris(trimethylene)benzene, 29, 693
$C_{15}H_{24}$, Hexamethyl-trans-σ-tris-homobenzene, 40B, 175
$C_{16}Cl_6F_4$, 1,3,6,8-Tetrafluoro-2,4,5,7,9,10-hexachloropyrene, 41B, 210
$C_{16}Cl_{10}$, Decachloropyrene, 42B, 143
$C_{16}H_{10}$, Acepleiadylene, 24, 679
$C_{16}H_{10}$, Fluoranthene, 43B, 224
$C_{16}H_{10}$, Pyrene, 11, 700; 30B, 109; 35B, 131; 38B, 217; 44B, 186
$C_{16}H_{12}$, 6,7-Benzoelassovalene, 43B, 224
$C_{16}H_{12}$, 9,10-Dihydroindeno[1,2-a]indene, 40B, 175
$C_{16}H_{12}Br_2$, 4b,9a-Dibromo-9,10-dihydroindeno[1,2-a]indene, 37B, 101
$C_{16}H_{13}Cl$, 2-Chloro-4,5,9,10-tetrahydropyrene, 39B, 127
$C_{16}H_{16}$, 4,8-Dimethylnaphtho[1,2:5,6]dicyclobutene, 46B, 161
$C_{17}H_{10}O$, Benzo[b]fluorenone, 46B, 161
$C_{17}H_{14}$, 1,2-Cyclopentenophenanthrene, 23, 681; 26, 752
$C_{18}F_{12}$, Dodecafluorotriphenylene, 43B, 225
$C_{18}H_9ClO_2$, 2-Chloro-1,8-phthaloylnaphthalene, 32B, 141
$C_{18}H_{11}Br$, 4-Bromobenz[a]anthracene, 43B, 225
$C_{18}H_{12}$, Benzo[4,5]cyclohepta[1,2,3-de]naphthalene, 44B, 163
$C_{18}H_{12}$, Benzo[c]phenanthrene, 28, 444
$C_{18}H_{12}$, Chrysene, 3, 687, 792; 24, 670
$C_{18}H_{12}$, Tetracene, 26, 754; 27, 938
$C_{18}H_{12}$, Triphenylene, 13, 559; 20, 595; 28, 440; 39B, 128
$C_{18}H_{12}$, 1,2-Benzanthracene, 20, 596
$C_{18}H_{12}$, 3,4-Benzophenanthrene, 18, 735
$C_{18}H_{14}$, 2,7-Dimethylpyrene, 43B, 228
$C_{18}H_{14}O_2$, trans-1,2-Dihydro-1,2-dihydroxybenz[a]anthracene, 45B, 162
$C_{18}H_{14}O_2$, trans-10,11-Dihydro-10,11-dihydroxybenz[a]anthracene, 45B, 162
$C_{18}H_{18}O_2$, 2,8-Dihydroxy-5,6,11,12,4b,10b-hexahydrochrysene, 28, 442
$C_{18}H_{20}$, 3,4,5,6,9,10,11,12-Octahydrochrysene, 22, 744
$C_{18}H_{30}$, trans-Perhydrotriphenylene (form 1), 32B, 142
$C_{18}H_{30}$, trans-Perhydrotriphenylene (form 2), 32B, 143
$C_{18}H_{30}$, 2,7-Dimethylperhydropyrene, 29, 502
$C_{19}H_{12}O_2$, 2'-Methyl-1,2-benzanthraquinone (form 1), 28, 447
$C_{19}H_{12}O_2$, 5-Methyl-1,2-benzanthraquinone, 28, 449
$C_{19}H_{13}Cl$, 7-Chloromethylbenz[a]anthracene, 43B, 226
$C_{19}H_{14}$, 1-Methylbenz[a]anthracene, 45B, 162
$C_{19}H_{14}$, 5-Methylnaphthacene, 45B, 132
$C_{19}H_{16}O_3$, 11-Hydroxy-3-methyl-1,4,4a,12a-tetrahydronaphthalene-5,12-quinone, 45B, 163
$C_{20}H_{16}$, 1,12-Dimethylbenz[a]anthracene, 44B, 164
$C_{20}H_{16}$, 1,12-Dimethylbenzo[c]phenanthrene, 28, 446
$C_{20}H_{16}$, 9,10-Dimethyl-1,2-benzanthracene, 24, 672; 29, 696
$C_{20}H_{16}O_5$, 7,8-Dihydro-1,6-dimethoxy-5,10,12(9H)-naphthacenetrione, 45B, 164
$C_{20}H_{18}O_2$, DL-cis-5,6-Dihydro-5,6-dihydroxy-7,12-dimethylbenz[a]anthracene, 43B, 226
$C_{20}H_{18}O_7$, γ-Rhodomycinone, 35B, 132

$C_{20}H_{20}$, trans-15,16-Diethyldihydropyrene, 32B, 144
$C_{21}H_{18}O$, 1,4-Dimethyl-9H-tribenzo[a,c,e]cycloheptene-9e-ol, 45B, 164
$C_{22}H_6Cl_{10}$, Decachloropyrene benzene solvate, 44B, 164
$C_{22}H_{20}O_4$, 2,7-Diacetoxy-trans-15,16-dimethyl-15,16-dihydropyrene, 30B, 110
$C_{22}H_{30}O_2$, 8-Methoxy-4aβ,10bβ,12aα-trimethyl-3,4,4a,4bα,5,6,10b,11,-12,12a-decahydrochrysen-1(2H)-one, 41B, 211
$C_{22}H_{32}O$, 9β,14β,18α,18β-Tetramethyl-17β-ol-dodecahydrochrysene, 40B, 522
$C_{24}H_{12}$, 5,6,11,12,17,18-Hexadehydro-tribenzo[a,e,2']cyclododecene, 35B, 133
$C_{24}H_{26}$, 2,7-Di-t-butylpyrene, 38B, 218
$C_{24}H_{28}$, 8a-Phenyl-dodecahydrotriphenylene, 45B, 165
$C_{24}H_{29}BrO_4$, Decahydro-3-hydroxy-9a-methyl-1H-cyclopenta[1,3]cyclopropa[1,2-a]naphthalene-7(6H)-one cyclic ethylene acetal p-bromobenzoate, 42B, 145
$C_{28}Cl_{18}$ · 0.5 CCl_4, Perchloro-5,10-diphenyldibenzo[a,e]pentalene carbon tetrachloride, 45B, 165
$C_{28}H_{30}O_8$, Tetramethyl cis-5,11-dicarboxylato-cis-4b,5,6,10b,11,12-hexahydrochrysene-cis-6,12-diacetate, 44B, 165
$C_{29}H_{19}Cl_2NO_3S$, Pyrene 2,6-dichloro-n-tosyl-p-benzoquinone-imine (1:1), 46B, 161
$C_{30}H_{18}Cl_2$, 5,6-Dichloro-11,12-diphenylnaphthacene, 29, 697
$C_{50}H_{41}N_5$, 2,7-Di-t-butyl-10,10,11,11-tetracyano-4,5,9,12-tetraphenyltetracyclo[6.4.0.0$^{3-6}$.0$^{9-12}$]dodeca-1(8),2,4,6-tetraene - acetonitrile, 44B, 162
$C_{58}H_{40}$, 1,1,2,3,4,5,6-Heptaphenyl-1,4-dihydrobenz[e]-as-indacene, 46B, 162

30. HYDROCARBONS (5 OR MORE FUSED RINGS)

$C_{12}H_{16}$, Tetraspiro[2.0.2.0.2.0.2.0]dodecane, 46B, 163
$C_{15}H_{20}$, Pentaspiro[2.0.2.0.2.0.2.0.2.0]pentadecane, 46B, 163
$C_{16}H_{16}O_2$, 2,2a,3,4,4a,4b,5,7a,7b,7c-Dodecahydro-3,4,5-(1)-propanyl-(3)ylidene-1H-dicyclopenta-[a,cd]-pentalene-1,9-dione, 46B, 164
$C_{17}H_{18}O_4$, Peristylane diketo-acetate derivative, 40B, 188
$C_{18}H_{10}$, 1,10-Benzofluoranthene, 20, 592
$C_{18}H_{12}$, 3,4-Benzopyrene, 20, 599
$C_{18}H_{12}$, 6b,10b-Dihydrobenzo[j]cyclobut[a]acenaphthylene, 43B, 230
$C_{18}H_{16}$, Naphtho[2,3:6,7]dicyclobutene[1,8:4,5]dicyclopentene, 46B, 164
$C_{18}H_{24}$, Hexaspiro[2.0.2.0.2.0.2.0.2.0.2.0]octadecane, 46B, 163
$C_{19}H_{10}O$, Naphthanthrone, 42B, 145
$C_{19}H_{12}$, 7bH-Indeno[1,2,3-jk]fluorene, 41B, 212
$C_{20}F_{24}$, Perfluorododecahydrotetra(cyclopenta)cyclooctene, 43B, 227
$C_{20}H_8Br_4O_4$, 1,2,3,4-Tetrabromo-1,2,3,4-diphthaloylcyclobutane, 33B, 93
$C_{20}H_{10}$, Corannulene, 42B, 146
$C_{20}H_{10}Br_2$, 3,9-Dibromoperylene, 45B, 167
$C_{20}H_{12}$, Cyclo[1,2,3,4-def]benzo[3,4]cyclobuta[6,7]biphenylene, 42B, 147
$C_{20}H_{12}$, Perylene (β-form), 28, 451
$C_{20}H_{12}$, Perylene (gas-ed), 30B, 404
$C_{20}H_{12}$, Perylene, 17, 735; 29, 694
$C_{20}H_{12}$, 3,4-Benzopyrene, 42B, 146
$C_{20}H_{14}Br_2O_2$, 4,8-Dibromo-3,7-diketo-3,4,4a,4b,7,8,8a,8b-octahydrodi-

benzo-[a,g]-biphenylene, 37B, 102
$C_{20}H_{14}O_3$, (DL)-7α,8β-Dihydroxy-9β,10β-epoxy-7,8,9,10-tetrahydrobenzo[a]pyrene, 46B, 165
$C_{20}H_{20}$, Octahydrodibenzo[a,i]biphenylene, 44B, 165
$C_{20}H_{20}$, (DL)-cis-anti-4,5,6,6a,6b,7,8,12b-Octahydrobenzo[j]fluoranthene, 46B, 165
$C_{20}H_{20}$, (DL)-cis-syn-4,5,6,6a,6b,7,8,12b-Octahydrobenzo[j]fluoranthene, 46B, 165
$C_{21}H_{16}$, 20-Methylcholanthrene, 19, 585; 24, 675; 41B, 213
$C_{22}H_{10}O_2$, Anthanthrone, 37B, 102
$C_{22}H_{12}$, 1,12-Benzoperylene, 11, 704; 23, 682
$C_{22}H_{12}O_2$, 1:2,5:6-Dibenzanthraquinone, 34B, 111
$C_{22}H_{12}O_2$, 6,13-Pentacenequinone, 45B, 167
$C_{22}H_{14}$, Dibenz[a,h]anthracene, 41B, 214
$C_{22}H_{14}$, Pentacene, 26, 756; 27, 938
$C_{22}H_{14}$, 1,2:5,6-Dibenzanthracene (monoclinic), 20, 598
$C_{22}H_{14}$, 1,2:5,6-Dibenzanthracene (orthorhombic), 11, 707
$C_{22}H_{14}$, 3,4:5,6-Dibenzophenanthrene, 18, 739
$C_{22}H_{14}$, 6b,12b-Dihydronaphtho[2,3-j]cyclobut[a]acenaphthylene, 42B, 147
$C_{22}H_{16}$, 5,6-Dihydrodibenz[a,h]anthracene, 38B, 219
$C_{22}H_{16}$, 5,6-Dihydrodibenz[a,j]anthracene, 38B, 220
$C_{22}H_{16}$, 9,10-Dihydro-1,2:5,6-dibenzanthracene, 22, 747
$C_{24}H_{12}$, Coronene (gas-ed), 26, 739
$C_{24}H_{12}$, Coronene, 11, 709; 30B, 111
$C_{24}H_{16}$, Tetraphenylene (gas-ed), 9, 375
$C_{24}H_{16}$, cis-Dimer of acenaphthylene, 37B, 103
$C_{24}H_{16}$, 6,12-Dimethylanthanthrene, 40B, 176
$C_{24}H_{24}$, syn-5,12:6,11-Dibutanodibenzo[a,e]cyclooctene, 46B, 166
$C_{24}H_{24}O_2$, 6aα,6bβ,12bβ,12cα-Tetrahydro-5,5,8,8-tetramethyldinaphtho-[1,2-a:1',2'-c]cyclobutene-6(5H),7(8H)-dione, 40B, 177
$C_{25}H_{16}$, Bis-(2,2'-biphenylene)methane, 38B, 220
$C_{26}H_{14}N_2$, 1-Cyanoacenaphthylene cis-photodimer, 38B, 221
$C_{26}H_{15}Br$, (-)-2-Bromohexahelicene, 38B, 222
$C_{26}H_{16}$, Hexahelicene, 39B, 128
$C_{26}H_{16}$, Tetrabenzonaphthalene, 18, 737; 45B, 128
$C_{26}H_{16}O_2$, Tetrabenzocyclodecan-1,16-dione, 46B, 167
$C_{26}H_{28}O_2$, 3,4:13,14-Dibenzo-6,6,11,11-tetramethyl-1,8,9-cis-tricyclo[7.5.0.0$^{2-8}$]tetradecane-5,12-dione, 45B, 168
$C_{26}H_{28}O_2$, 6,6-Dimethyl-2,3-benzo-2,4-cycloheptadienone photodimer A, 40B, 178
$C_{26}H_{28}O_2$, 6,6-Dimethyl-2,3-benzo-2,4-cycloheptadienone photodimer B, 40B, 178
$C_{27}H_{16}$, 5-H-Benzo[c,d]pyrene-5-spiro-1'-indene, 41B, 214
$C_{27}H_{16}O$, 1-Formylhexahelicene, 45B, 168
$C_{27}H_{18}$, 1-Methylhexahelicene, 46B, 167
$C_{27}H_{18}$, 2-Methylhexahelicene, 39B, 129
$C_{28}H_{16}$, β-Tribenzopyrene, 43B, 230
$C_{28}H_{16}$, 2,3:8,9-Dibenzoperylene, 23, 683
$C_{28}H_{18}O$, 1-Acetylhexahelicene, 45B, 168
$C_{28}H_{20}$, 1,16-Dimethylhexahelicene, 39B, 129
$C_{28}H_{20}$, 5-cis-15-cis-Tetrabenzo[a,c,g,i]cyclododecene, 39B, 130
$C_{28}H_{20}$, 5-cis-15-trans-Tetrabenzo[a,c,g,i]cyclododecene, 38B, 191
$C_{28}H_{20}$, 5-trans-15-trans-Tetrabenzo[a,c,g,i]cyclododecene, 39B, 130
$C_{28}H_{24}$, meso-9,10,19,20-Tetrahydrotetrabenzo[a,c,g,i]cyclodecene, 42B, 148
$C_{28}H_{24}$, racemic-9,10,19,20-Tetrahydrotetrabenzo[a,c,g,i]cyclodecene,

42B, 148
$C_{28}H_{36}O_2$, 10-Ethoxy-3-methoxy-6bα,12bα,14aα-trimethyl-5,6,6aβ,6b,-7,8,12b,13,14,14a-decahydropicene, 39B, 131
$C_{28}H_{36}O_2$, 10-Ethoxy-3-methoxy-6bβ,12bα,14aβ-trimethyl-5,6,6aα,6b,-7,8,12b,13,14,14a-decahydropicene, 39B, 131
$C_{28}H_{38}O_2$, 10-Methoxy-2,2,4aβ,6aβ,12bβ-pentamethyl-1,2,3,4,4a,5,-6,6a,6bα,7,8,12β,13,14-tetradecahydro-1-picenone, 44B, 166
$C_{30}H_{14}$, 1,14-Benzobisanthrene, 22, 751
$C_{30}H_{14}O_2$, Acedianthrone, 18, 742
$C_{30}H_{16}$, Pyreno-[1',2':1,2]-pyrene, 38B, 223
$C_{30}H_{16}$, 5,6:11,12-Diphenylenenaphthacene, 17, 738
$C_{30}H_{18}$, Dinaphtho[1,2-a;1',2'-h]anthracene, 41B, 215
$C_{30}H_{18}$, Heptahelicene (form 1), 42B, 149
$C_{30}H_{18}$, Heptahelicene (form 2), 42B, 149; 43B, 231
$C_{30}H_{18}O_2$, 14c-Hydro-5a-phenylbenz[a]indeno[2,1-c]fluorene-5,10-dione, 33B, 93
$C_{30}H_{20}$, Penta-m-phenylene, 35B, 134
$C_{31}H_{19}$, 9,10,19,20-Tetradehydro-tetrabenzo[a,c,g,i]cyclododecene - benzene, 39B, 132
$C_{32}H_{14}$, Ovalene, 12, 411; 17, 739; 39B, 133
$C_{32}H_{16}$, 1,2:7,8-Dibenzocoronene, 26, 757
$C_{32}H_{20}$, 3,15-Ethano-[7]helicene, 43B, 231
$C_{32}H_{20}O \cdot 0.5\ C_6H_6$, 3,15-Oxapropano-[7]helicene hemibenzene, 43B, 231
$C_{32}H_{22}$, 3,15-Dimethyl-[7]helicene, 43B, 231
$C_{32}H_{22}N_2O_2$, 9-Cyano-10-methoxyphenanthrene head-to-tail photodimer, 41B, 216
$C_{32}H_{24}O_4$, Dimethyl 8b,8c,16b,16c-tetrahydrocyclobuta[1,2-l:3,4-l']diphenanthrene-8b,16b-cis-dicarboxylate, 40B, 179
$C_{32}H_{35}BrO_3$, trans-8-Bromo-7-(menthyloxyacetoxy)-7,8,9,10-tetrahydrobenz[a]pyrene, 46B, 168
$C_{34}H_{16}O_2$, Dibenzanthrone, 29, 698
$C_{34}H_{16}O_2$, Isodibenzanthrone, 29, 699
$C_{34}H_{18}$, Tetrabenzo[a,cd,j,lm]perylene, 41B, 217
$C_{34}H_{18}$, 1,9:5,10-Diperinaphthyleneanthracene, 23, 685
$C_{36}H_{24}$, Hexa-m-phenylene, 35B, 134
$C_{36}H_{24}$, Hexa-o-phenylene, 38B, 192; 39B, 133
$C_{36}H_{24}$, Hexa-o,m,o,o,m,o-phenylene, 42B, 115
$C_{38}H_{18}$, Diphenanthro[5,4,3-abcd:5',4',3'-jklm]perylene, 44B, 167
$C_{38}H_{18}$, Dinaphtho[7',1':1,13][1",7":6,8]peropyrene (*a* form), 23, 686
$C_{40}H_{20}$, Quaterrylene, 24, 680; 41B, 218
$C_{42}H_{18}$, 1,12:2,3:4,5:6,7:8,9:10,11-Hexabenzcoronene, 26, 758
$C_{42}H_{22}$, Tetrabenzoheptacene, 45B, 169
$C_{42}H_{24}$, Tribenzo[f,l,r]heptahelicene, 40B, 180
$C_{42}H_{24}$, 10-Helicene, 42B, 150
$C_{44}H_{36}$, o-Distyrylbenzene photodimer, 40B, 181
$C_{46}H_{26}$, Diphenanthro[4,3-a;3',4'-o]picene, 45B, 169
$C_{46}H_{26}$, 11-Helicene, 42B, 151
$C_{84}H_{84}Br_4O_{16}$, Resorcinol - p-bromobenzaldehyde condensation product, 33B, 95

31. BRIDGED-RING HYDROCARBONS

C_5H_8, Bicyclo[1.1.1]pentane (gas-ed), 35B, 896; 37B, 686
C_5H_8, Spiropentane (gas-ed), 10, 242
C_6H_6, Tricyclo[3.1.0.0$^{2-6}$]hex-3-ene (gas-mw), 39B, 894

C_6H_8, Bicyclo[2.1.1]hex-2-ene (gas-mw), 42B, 1032
C_6H_{10}, Bicyclo[2.1.1]hexane (gas-ed), 37B, 690
C_7H_8, Norbornadiene (gas-ed), 37B, 688
C_7H_9Cl, 4-Chloronortricyclene (gas-ed), 40B, 1142
C_7H_9Cl, 4-Chloronortricyclene (gas-mw), 39B, 893
C_7H_{10}, Nortricyclene (gas-ed), 16, 504; 39B, 888
$C_7H_{10}Cl_2$, 1,4-Dichloronorbornane (gas-ed), 33B, 534
C_7H_{12}, Bicyclo[3.1.1]heptane (gas-ed), 40B, 1142
C_7H_{12}, Norbornane (gas-ed), 33B, 534; 37B, 688
C_8F_{12}, Dodecafluoro-tricyclo[3.3.0.0$^{2-6}$]octane, 30B, 112
$C_8H_6Cl_4$, 1,2,3,3-Tetrachloro-4,5-dimethylspiro[2.3]hexa-1,4-diene, 39B, 134
$C_8H_8Br_2O_2$, 3-endo,5-endo-Dibromotricyclo[2.2.1.0$^{2-6}$]heptane-7-carboxylic acid, 45B, 170
$C_8H_8O_3$, 5-Oxotricyclo[2.2.1.0$^{2-6}$]heptane-3-carboxylic acid, 45B, 170
$C_8H_9BrO_2$, 5-exo-Bromo-6-endo-norbornanol-2-endo-carboxylic acid lactone, 40B, 182
$C_8H_9IO_2$, 5-exo-Iodo-6-endo-norbornanol-2-endo-carboxylic acid lactone, 40B, 182
C_8H_{10}, Bicyclo[2.2.2]octadiene (gas-ed), 37B, 687
C_8H_{10}, Tricyclo[3.3.0.0$^{2-6}$]oct-3-ene (gas-ed), 38B, 1067
$C_8H_{10}Br_2O$, cis-2,4-Dibromobicyclo[3.2.1]octan-3-one, 45B, 171
$C_8H_{10}Cl_2$, 8,8-Dichlorotricyclo[3.2.1.0]octane, 38B, 224
C_8H_{12}, Bicyclo[2.2.2]octene (gas-ed), 37B, 687
C_8H_{12}, Tricyclo[3.3.0.0$^{2-6}$]octane (gas-ed), 33B, 535
$C_8H_{12}I_2$, 1,4-Iodobicyclo[2.2.2]octane, 40B, 183
$C_8H_{13}Br$, 1-Bromobicyclo[2.2.2]octane (gas-mw), 16, 505; 17, 696
$C_8H_{13}Cl$, 1-Chloro-bicyclo[2.2.2]octane (gas-mw), 17, 696
C_8H_{14}, Bicyclo[2.2.2]octane (gas-ed), 35B, 900
$C_8H_{14}BrNO_2$, (-)-2-exo-Aminonorbornane-2-carboxylic acid hydrobromide, 38B, 225
$C_8H_{16}ClNO$, α-Chlorotropane monohydrate, 35B, 135
$C_9H_5F_9O_2$, endo-1-Methoxycarbonyl-3H,4H-nonafluoronorbornane, 44B, 167
C_9H_7N, 1(5)-Cyanosemibullvalene, 41B, 218
C_9H_7N, 1-Cyanotricyclo[3.3.0.0$^{2-8}$]octa-3,6-diene, 44B, 168
$C_9H_8O_3$, Bicyclo[2.2.1]hept-5-ene-2,3-exo-dicarboxylic anhydride, 38B, 226
$C_9H_8O_3$, 5-Norbornene-2,3-endo-dicarboxylic anhydride, 34B, 304
$C_9H_9KO_4 \cdot 1.5\ H_2O$, Potassium hydrogen norborn-5-ene-2-endo,3-endo-dicarboxylate sesquihydrate, 42B, 151
$C_9H_{10}Br_2O$, 2,7-Dibromotricyclo[4.2.1.0$^{3-7}$]nonan-4-one, 46B, 168
$C_9H_{10}O_4$, Norborn-5-ene-2,3-endo-dicarboxylic acid, 39B, 135
$C_9H_{10}O_4$, 5-Hydroxy-2,3-norbornane dicarboxylic acid γ-lactone, 39B, 135
$C_9H_{12}O_4$, D-Spiro[3.3]heptane-2,6-dicarboxylic acid, 38B, 203
$C_9H_{13}ClO$, 2-Chlorobicyclo[3.3.1]nonan-9-one, 32B, 144
$C_9H_{13}NO_2$, 6-endo-Hydroxy-3-endo-aminomethylbicyclo[2.2.1]heptane-2-endo-carboxylic acid lactam, 43B, 232
$C_9H_{14}O_2$, exo-7-Hydroxybicyclo[3.3.1]nonan-3-one, 45B, 172
$C_9H_{14}O_2S$, 6-endo-(Methylthio)bicyclo[2.2.1]heptane-2-endo-carboxylic acid, 46B, 169
$C_9H_{14}O_2S$, 6-exo-(Methylthio)bicyclo[2.2.1]heptane-2-endo-carboxylic acid, 46B, 169
$C_9H_{14}O_3$, 2-exo-Hydroxy-7-methylbicyclo[2.2.1]heptane-7-syn-carboxylic acid, 43B, 232
$C_9H_{17}NO_3$, (-)-2-exo-Aminobicyclo[3.2.1]octane-2-carboxylic acid

monohydrate, 44B, 168
$C_9H_{19}NaO_6$, Sodium 2-methyl-6-endo-hydroxybicyclo[2.2.1]heptane-2-endo-carboxylate trihydrate, 43B, 233
$C_{10}Cl_{12}O_3S$, Undecachloro-pentacyclo-[5.3.0.0$^{2-6}$.0$^{3-9}$.0$^{4-8}$]-decan-5-chlorosulphonate, 32B, 146
$C_{10}H_5Cl_7$, Heptachlor, 39B, 136
$C_{10}H_6Cl_6O$, 1-Hydroxy-4,5,6,7,8,8-hexachlorotetrahydro-4,7-methanoindene, 44B, 169
$C_{10}H_6Cl_8$, Photo-cis-chlordane, 45B, 172
$C_{10}H_6Cl_8$, Photo-trans-chlordane, 45B, 172
$C_{10}H_6Cl_8$, cis-Chlordane, 45B, 172
$C_{10}H_6Cl_8$, trans-Chlordane, 45B, 172
$C_{10}H_8Cl_4$, 2,3,4,5-Tetrachlorotricyclo[4.2.2.0$^{2-5}$]deca-3,7-diene, 42B, 152
$C_{10}H_9Cl_3O$, 3,4,5-Trichlorotetracyclo[4.4.0.0$^{3-9}$.0$^{4-8}$]decan-2-one, 33B, 95
$C_{10}H_9Cl_3O_3$, Methyl 2,4,5-trichloro-6-methyltricyclo[2.2.1.0$^{2-6}$]heptan-3-one-syn-5-carboxylate, 42B, 152
$C_{10}H_{10}$, Bullvalene (gas-ed), 37B, 686
$C_{10}H_{10}$, Bullvalene, 33B, 96
$C_{10}H_{10}BrClO_2$, 7-Bromo-2-chlorotricyclo[4.3.1.0$^{3-7}$]decan-4,8-dione, 46B, 169
$C_{10}H_{10}O_2$, Tricyclo[4.4.0.0$^{2-8}$]dec-3-ene-7,10-dione, 38B, 226
$C_{10}H_{10}O_3$, Bicyclo[2.2.2]octene-2,3-endo-dicarboxylic anhydride, 34B, 113; 37B, 104
$C_{10}H_{10}O_4$, Bicyclo[2.2.2]octa-2,5-diene-2,3-dicarboxylic acid, 35B, 137
$C_{10}H_{11}Cl_7$, 2,2,5-endo,6-exo,8,9,10-Heptachlorobornane, 41B, 219
$C_{10}H_{12}BrCl_3O_3$, 1,2-endo,4-Trichloro-7,7-dimethoxy-5-endo-bromomethyl-3-norbornanone, 44B, 170
$C_{10}H_{12}Br_4O$, 1,7-Dibromo-4-dibromomethyl-3,3-dimethylnorbornan-2-one, 42B, 153
$C_{10}H_{12}O_2$, 1,2,5,6-Tetramethyltricyclo[3.1.0.0$^{2-6}$]hexane-3,4-dione, 43B, 233
$C_{10}H_{12}O_4$, 7-syn-Acetoxy-6-endo-dehydroxybicyclo[2.2.1]heptane-2-endo-carboxylic acid α-lactone, 38B, 227
$C_{10}H_{14}Br_2O$, (-)-3,3,4-Trimethyl-1,7-dibromonorbornan-2-one, 41B, 220
$C_{10}H_{14}Br_2O$, 3,9-Dibromobornan-2-one, 39B, 137
$C_{10}H_{14}I_2$, 1,8-Diiodotricyclo[5.3.0.0$^{4-8}$]decane, 46B, 170
$C_{10}H_{14}O_2$, 1,7,7-Trimethylbicyclo[2.2.1]hepta-2,3-dione, 46B, 170
$C_{10}H_{14}O_4$, Bicyclo[2.2.2]octane-1,4-dicarboxylic acid, 40B, 1082
$C_{10}H_{15}Br$, 1-Bromoadamantane (gas-mw), 33B, 541
$C_{10}H_{15}Cl$, 1-Chloroadamantane (gas-mw), 33B, 541
$C_{10}H_{15}F$, 1-Fluoroadamantane (gas-mw), 33B, 541
$C_{10}H_{16}$, Adamantane (disordered cubic phase), 46B, 170
$C_{10}H_{16}$, Adamantane (gas-ed), 11, 618
$C_{10}H_{16}$, Adamantane, 10, 247; 30B, 115; 32B, 148
$C_{10}H_{16}O$, 1-Adamantol, 45B, 174
$C_{10}H_{16}O_2$, exo-7-Methoxybicyclo[3.3.1]nonan-3-one, 45B, 172
$C_{10}H_{18}BrNO_2 \cdot H_2O$, 9-Aminobicyclo[3.3.1]nonane-9-carboxylic acid hydrobromide monohydrate, 45B, 520
$C_{11}H_8F_2$, 11,11-Difluoro-1,6-methano[10]annulene, 37B, 105; 42B, 154
$C_{11}H_8O_2$, Spiro[5.5]undeca-1,4,7,10-tetraene-3,9-dione, 42B, 128
$C_{11}H_{10}$, 1,6-Methano[10]annulene, 46B, 171
$C_{11}H_{10}O_3$, 7-Spirocyclopropylbicyclo[2.2.1]heptene anhydride, 42B, 154
$C_{11}H_{12}Br_2O_3$, 3,9-Dibromotricyclo[4.2.1.0$^{3-7}$]nonan-2,5-dione 2-ethyl-

ene acetal, 44B, 170
$C_{11}H_{12}Br_4O$, 3exo,4endo,6exo-Tribromo-7-bromomethyl-1,5-dimethyltricyclo[3.2.1.0$^{2-7}$]octan-8-one, 41B, 221
$C_{11}H_{12}Cl_2N_2$, 7,7-Dichloro-1,6-dimethylbicyclo[4.1.0]heptane-trans-3,trans-4-dicarbonitrile, 46B, 171
$C_{11}H_{12}I_2$, syn-4-syn-7-Diiodopentacyclo[6.3.0.0$^{2-6}$.0$^{3-10}$.0$^{5-9}$]undecane, 42B, 155
$C_{11}H_{12}N_2O$, 1-Methoxybicyclo[2.2.2]oct-5-ene-2,3-dicarbonitrile, 45B, 175
$C_{11}H_{12}O_2$, 3,3'-Spirobi(bicyclo[3.1.0]hexane)-2,2'-dione, 37B, 105
$C_{11}H_{13}Cl$, (+)-1-Chlorotrishomobarrelene, 43B, 234
$C_{11}H_{13}N$, 1-Cyanotetracyclo[3.3.1.1$^{3-7}$.0$^{3-7}$]decane, 39B, 137
$C_{11}H_{14}ClNO$, 1-Methoxy-3-methyl-6-chloro-6-cyanobicyclo[2.2.2]octene, 43B, 235
$C_{11}H_{14}O$, Spirodienone II, 38B, 204
$C_{11}H_{14}O_2$, Tricyclo[4.3.1.1$^{3-8}$]undecane-4,5-dione, 35B, 138
$C_{11}H_{15}N$, Tricyclo[3.3.1.1$^{3-7}$]decane-1-carbonitrile, 45B, 175
$C_{11}H_{15}NO$, 3-Cyano-4-exo-protoadamantanol, 46B, 172
$C_{11}H_{16}O$, Tricyclo[5.3.1.0$^{3-8}$]undecan-5-one, 43B, 235
$C_{11}H_{16}O_3$, 2-Methyl-3-(3'-hydroxy-bicyclo[2.2.1]-2'-heptylidenyl)propionic acid, 45B, 176
$C_{11}H_{18}BrNO_2$, 2-Aminoadamantane-2-carboxylic acid hydrobromide, 39B, 138
$C_{11}H_{20}O_2$, Bicyclo[3.3.3]undecane-1,5-diol, 46B, 172
$C_{11}H_{20}O_2$, Bicyclo[4.4.1]undecan-1,6-diol, 43B, 236
$C_{12}Cl_{12}$, Dodecachlorotetracyclo[7.2.1.0$^{2-8}$.0$^{5-12}$]dodeca-3,6,10-triene, 41B, 223
$C_{12}Cl_{12}$, Perchloro-(4,8-dimethylene-tricyclo[3.3.2.0$^{1-5}$]-deca-2,6-diene), 32B, 148
$C_{12}H_5Cl_7O_4$, 6,6-Ethylenedioxyheptachloropentacyclo-[5.2.0.0$^{2-5}$.0$^{3-9}$.0$^{4-8}$]-nonane-3-carboxylic acid, 34B, 114
$C_{12}H_6F_4$, 2,3-(Tetrafluorobenzo)bicyclo[2.2.2]octatriene, 45B, 176
$C_{12}H_8Cl_6$, 1,2,3,4,10,10-Hexachloro-1,4,4a,5,8,8a-hexahydro-endo-1,4-endo-5,8-dimethanonaphthalene, 45B, 177
$C_{12}H_8Cl_6$, 1,2,3,4,10,10-Hexachloro-1,4,4a,5,8,8a-hexahydro-endo-1,4-exo-5,8-dimethanonaphthalene, 38B, 230
$C_{12}H_8Cl_6$, 3,4,5,6,6,7-Hexachloropentacyclo[6.4.0$^{2-10}$.0$^{3-7}$.0$^{5-9}$]-dodec-11-ene, 38B, 228
$C_{12}H_8Cl_6O$, 1,2,3,4,10,10-Hexachloro-6,7-epoxy-1,4,4a,5,6,7,8,8a-octahydro-endo-1,4-endo-5,8-dimethanonaphthalene, 38B, 229
$C_{12}H_8F_6$, 4,5,9,10,11,12-Hexafluoropentacyclo[6.4.0.0$^{3-6}$.0$^{4-12}$.0$^{5-9}$]-dodec-10-ene, 46B, 173
$C_{12}H_9ClO$, 11-Chloro-3,8-methano-[11]annulenone, 38B, 231
$C_{12}H_9Cl_3$, 2,3,4-Trichloropentacyclo[6.4.0.0$^{2-4}$.0$^{5-9}$]dodeca-6,11-diene, 38B, 232
$C_{12}H_{10}Br_4$, 3,5,8,11-Tetrabromo-tricyclo[5.4.1.0$^{3-5}$]dodeca-1,6,9-triene, 41B, 221
$C_{12}H_{10}O_2$, 1,6-Methano-cyclodecapentaene-2-carboxylic acid, 30B, 117
$C_{12}H_{10}O_3$, anti-Tricyclo[4.2.2.0$^{2-5}$]deca-3,9-diene-7,8-endo-dicarboxylic anhydride, 39B, 138
$C_{12}H_{11}BF_4$, Bicyclo[5.4.1]dodecapentaenylium tetrafluoroborate, 45B, 177
$C_{12}H_{11}BrO_2$, Methyl 8-exo-bromo-1,3-methanoindane-2-endo-carboxylate, 40B, 1082
$C_{12}H_{11}F_6P$, Bicyclo[5.4.1]dodecapentaenylium hexafluorophosphate, 45B, 178
$C_{12}H_{11}F_6Sb$, Bicyclo[5.4.1]dodecapentaenylium hexafluoroantimonate,

45B, 177
$C_{12}H_{12}O_2$, [4.4.2]Propella-3,8-diene-11,12-dione, 41B, 222
$C_{12}H_{12}O_2$, Tetracyclo[6.4.0.0$^{2-11}$.0$^{4-9}$]dodec-6-ene-3,10-dione, 41B, 222
$C_{12}H_{12}O_2$, 4a,5,8,8a-Tetrahydro-5,8-ethano-1,4-naphthoquinone, 46B, 173
$C_{12}H_{12}O_4$, trans-9,10-Pentacyclo[4.4.0.0$^{2-5}$.0$^{3-8}$.0$^{4-7}$]decandioic acid, 37B, 106
$C_{12}H_{13}NO_4S$, 5-Cyano-hydroxytetracyclo[4.3.1.0$^{2-9}$.0$^{4-8}$]decan-3-one methylsulphonate, 43B, 237
$C_{12}H_{14}BrNO_2S$, N-exo-6-Bicyclo[3.1.0]hexyl-p-bromosulphonamide, 37B, 107
$C_{12}H_{14}Cl_2O_2$, Adamantane-1,3-dicarbonyl chloride, 44B, 170
$C_{12}H_{16}Br_2O$, 9,11-Dibromotricyclo[6.3.1.0$^{1-5}$]dodecan-10-one, 43B, 237
$C_{12}H_{16}ClNO$, DL-9-exo-Amino-5,6,7,8-tetrahydro-5,8-methano-9H-benzocyclohepten-8-ol hydrochloride, 44B, 171
$C_{12}H_{16}O_2$, (3aα,6β,8aβ)-Hexahydro-6-hydroxy-5-methylene-1H-3a,6-methanoazulen-2(3H)-one, 43B, 238
$C_{12}H_{16}O_4$, 1,4-Dihydroxytricyclo[6.4.0.0$^{4-9}$]dodecane-7,10-dione, 40B, 183
$C_{12}H_{17}BrO_3$, 1-Hydroxy-6-acetoxy-10α-bromobicyclo[4.3.1]dec-3-ene, 43B, 180
$C_{12}H_{18}$, Tetracyclo[5.3.1.1$^{2-6}$.0$^{4-9}$]dodecane, 43B, 239
$C_{12}H_{18}O_2$, Norbornane-2-exo-methallyl-3-exo-carboxylic acid, 39B, 139
$C_{12}H_{24}BrN$, 3-N-Methylaminomethylpinane hydrobromide, 35B, 139
$C_{13}H_6Cl_2N_2O_2$, exo-2,3-Dichloro-4a,8a-dicyano-4a,5,8,8a-tetrahydro-5,8-methano-1,4-naphthaquinone, 37B, 110
$C_{13}H_8F_6$, 2,3,4,5,6,7-Hexafluoropentacyclo[6.4.0.1$^{9-12}$.0$^{2-7}$.0$^{3-6}$]-trideca-4,10-diene, 46B, 174
$C_{13}H_{10}Br_2$, 4,10-Dibromo-1,7-methano[12]annulene, 42B, 155
$C_{13}H_{10}F_6$, 3,4,5,6,7,8-Hexafluoropentacyclo[8.2.1.0$^{2-9}$.0$^{3-8}$.0$^{4-7}$]-tridec-5-ene, 44B, 171
$C_{13}H_{11}ClN_2O_4S$, 3-Chloro-5-(2',4'-dinitrophenylthio)nortricyclene, 46B, 174
$C_{13}H_{11}N$, 11-Methyltricyclo[4.4.1.0$^{1-6}$]undeca-2,4,7,9-tetraene-11-carbonitrile, 44B, 172
$C_{13}H_{12}ClNO_2S$, trans-5-Chloro-6-(2-nitrophenylthio)bicyclo[2.2.1]-hept-2-ene, 46B, 175
$C_{13}H_{12}Cl_2$, exo-2,exo-3-Dichloro-endo-5,6-(o-phenylene)norbornane, 44B, 172
$C_{13}H_{13}BrO_4$, 10-Bromo-12-methoxycarbonyl-4-oxapentacyclo-[7.1.1.1$^{3-11}$.0$^{2-8}$.0$^{6-12}$]dodecan-5-one, 44B, 173
$C_{13}H_{14}$, 11,11-Dimethyltricyclo[4.4.1.0$^{1-6}$]undeca-2,4,7,9-tetraene, 39B, 140
$C_{13}H_{14}N_2$, 2-Adamantylidene-malondinitrile, 41B, 224
$C_{13}H_{14}O_3$, 7-Spirocyclopentylbicyclo[2.2.1]hept-2-ene-5,6-dicarboxylic anhydride, 43B, 395
$C_{13}H_{15}ClN \cdot 0.5\ CHO_2 \cdot 0.5\ ClO_4$, 11-anti-Ammonio-8-chlorobenzo[b]bicyclo[3.3.1]nona-3,6a(10a)-diene hemiformate hemiperchlorate, 45B, 178
$C_{13}H_{16}O_2$, Tetracyclo[7.2.1.1$^{5-8}$.0$^{4-9}$]tridecan-2,10-dione, 38B, 234
$C_{13}H_{17}BrCl_3F_3O_4$, 1,3-endo-4-Trichloro-2-endo-7,7-trimethoxy-5-endo-bromoethyl-2-exo-(2',2',2'-trifluoroethoxy)-bicyclo[2.2.1]heptane, 45B, 179
$C_{13}H_{18}ClN$, DL-2-endo-Methylamino-1,2,3,4-tetrahydro-1,4-ethanonaphthalene hydrochloride, 44B, 174
$C_{13}H_{18}ClN$, DL-2-exo-Methylamino-1,2,3,4-tetrahydro-1,4-ethanonaphth-

alene hydrochloride, 44B, 174
$C_{13}H_{18}O_2$, (3aRS,4RS,6RS,8aSR)-5-Methyleneoctahydro-4H-3a,6-methanoazulene-4-carboxylic acid, 46B, 176
$C_{13}H_{18}O_2$, 3,4-Epoxy-5-acetyl-5-methyltricyclo[5.2.1.0$^{2-6}$]decane, 46B, 176
$C_{13}H_{20}BrCl_3O_4$, 6-Bromomethyl-1,2,4-trichloro-3-ethoxy-3,7,7-trimethoxy[2.2.1]heptane, 46B, 176
$C_{13}H_{21}NO$, N,N-Dimethyltricyclo[3.3.1.1$^{3-7}$]decane-1-carboxamide, 43B, 239
$C_{14}H_{10}$, 1,8-Naphthotricyclo[4.1.0.0$^{2-7}$]heptene, 43B, 229
$C_{14}H_{10}N_4$, 5,5,6,6-Tetracyanotricyclo[5.2.1.0$^{4-8}$]dec-2-ene, 41B, 225
$C_{14}H_{10}O$, 4,10b-Methano-8H-benzo[ab]cyclodecen-8-one, 39B, 140
$C_{14}H_{12}N_2O_2$, 1,6-Dicyano-2-hydroxy-8,9-dimethyltricyclo[4.4.0.0$^{2-8}$]-dec-3,9-dien-5-one, 43B, 240
$C_{14}H_{12}O_2$, Decahydro-1,2,4:5,6,8-dimetheno-s-indacenedione, 37B, 107
$C_{14}H_{12}O_2$, 3,5-Diformylbicyclo[5.4.1]dodeca-2,5,7,9,11-pentaene, 46B, 178
$C_{14}H_{13}BrO_2$, anti-7-Norbornenyl p-bromobenzoate, 30B, 224
$C_{14}H_{13}BrO_2$, exo-anti-Tricyclo[3.1.1.0$^{2-4}$]heptan-6-yl p-bromobenzoate, 39B, 141
$C_{14}H_{13}ClN_2O_4S$, 2-Chloromethyl-5-(2'-4'-dinitrophenylthio)nortricyclene, 46B, 178
$C_{14}H_{13}ClO$, 7-Chloropentacyclo[7.5.0.0$^{2-7}$.0$^{5-13}$.0$^{6-12}$]tetradeca-3,10-diene-8-one, 34B, 115
$C_{14}H_{14}Br_2O_4 \cdot 0.5\ C_4H_8O_2$, 1,2,3,4,4a,5,6,8a-Octahydro-3,5-bisbromomethyl-3,5-dihydroxy-2,6-dioxo-1,4-ethenonaphthalene dioxane solvate, 35B, 141
$C_{14}H_{14}Cl_2O_4$, 4(RS),9(RS)-Dichloro-5,6-dimethoxycarbonyltetracyclo-[5.3.0.0$^{2-10}$.0$^{3-8}$]dec-5-ene, 46B, 179
$C_{14}H_{14}F_4$, 5,6,7,8-Tetrafluoro-1,2,3,4-tetrahydro-9-isopropyl-1,4-methanonaphthalene, 44B, 174
$C_{14}H_{15}BrO_3S$, anti-8-Tricyclo[3.2.1.0$^{2-4}$]octyl-p-bromobenzenesulphonate, 30B, 113
$C_{14}H_{16}Br_2O_4$, 3,4-Dibromo-9,10-bis(methoxycarbonyl)-tricyclo-[4.2.2.0$^{2-5}$]dec-7-ene, 44B, 173
$C_{14}H_{16}O_6$, 2,3-Bis(methoxycarboxyl)-5-hydroxy-1,5-dimethylbicyclo-[2.2.2]octa-2,7-diene-6-one, 45B, 179
$C_{14}H_{17}IO_4$, 5-exo-Iodo-2,3-endo-biscarbomethoxy-7-isopropylidenetricyclene, 39B, 141
$C_{14}H_{17}NO$, exo-N-(2-Norbornyl)benzamide, 44B, 174
$C_{14}H_{18}$, [4.4.4]Propellatriene, 37B, 108
$C_{14}H_{18}O_2$, 3-Carboxy[7]paracyclophane, 40B, 184
$C_{14}H_{18}O_3S$, 2-exo-Norbornanol p-toluenesulphonate, 38B, 235
$C_{14}H_{18}O_4$, Dimethyl tetracyclo[4.4.0.0$^{2-4}$.0$^{3-8}$]decane-9-endo-10-endo-dicarboxylate, 42B, 156
$C_{14}H_{18}O_6$, 2,3-Bis(methoxycarboxyl)-5-hydroxy-1,5-dimethylbicyclo-[2.2.2]oct-2-ene-6-one, 45B, 179
$C_{14}H_{18}O_6$, 2,3-Bis(methoxycarboxyl)-5-hydroxy-1,5-dimethylbicyclo-[2.2.2]oct-7-ene-6-one, 45B, 179
$C_{14}H_{19}BrO_4$, 1,6-Diacetoxy-10α-bromobicyclo[4.3.1]dec-3-ene, 43B, 180
$C_{14}H_{19}NO$, Diamantane-3-one oxime (E isomer), 45B, 180
$C_{14}H_{19}NO$, Diamantane-3-one oxime (Z isomer), 45B, 180
$C_{14}H_{20}$, Congressane, 30B, 119
$C_{14}H_{20}O_2$, syn(1R,2R,4S)-4-Methyl-spiro(bicyclo[2.2.2]octane-2,1'-cyclohexane)-3',6-dione, 46B, 179
$C_{14}H_{20}O_2$, 1β,5β,6β,11β-Tetramethyl-3-oxatetracyclo[5.4.0.0$^{2-6}$.-0$^{8-11}$]undec-8-en-2β-ol, 46B, 144

$C_{14}H_{20}O_2$, 6-Hydroxy-2-isopropenyl-cis-perhydro-3a,6-methanoazulen-8-one, 45B, 181
$C_{14}H_{20}O_3$, 1,5,5-Trimethyl-2-methylidene-7-methoxycarbonylbicyclo-[4.1.1]octan-8-one, 45B, 138
$C_{14}H_{20}O_3$, 7-Methoxycarbonyl-4,4,8-trimethyltricyclo[4.3.0.0^{2-8}]-nonan-5-one, 45B, 138
$C_{14}H_{22}O_2$, 1-Methyl-7-exo-t-butylbicyclo[3.3.1]nonane-2,9-dione, 40B, 184
$C_{14}H_{22}O_2$, 10-Hydroxy-3-methoxy-2,4a-ethano-5,8-methanoperhydronaphthalene, 43B, 241
$C_{14}H_{24}$, [4.4.4]Propellane, 37B, 108
$C_{15}H_{12}$, Spiro(indene-1,7'-norcaradiene), 39B, 142
$C_{15}H_{14}BrNO_4$, exo-6-Bromo-exo-tricyclo[3.2.1.0^{2-4}]oct-anti-8-yl p-nitrobenzoate, 43B, 242
$C_{15}H_{15}BrO$, 11-anti-Bromo-2,3-benzotricyclo[4.4.1.0]undecen-4-one, 42B, 158
$C_{15}H_{15}NO_4S$, 3-Acetoxy-5-(2-nitrophenylthio)nortricyclene, 46B, 179
$C_{15}H_{16}N_2O_6$, trans-Bicyclo[4.2.0]oct-1-yl 3,5-dinitrobenzoate, 35B, 142
$C_{15}H_{16}N_2O_6S$, 2-exo-Acetoxy-7-anti-dinitrophenylthionorbornane, 45B, 181
$C_{15}H_{16}O_3S$, syn-Bicyclo[3.2.1]octa-2,6-dien-8-yl tosylate, 44B, 175
$C_{15}H_{17}BrO_3$, 1-Methyl-2-hydroxy-tetracyclo[4.4.2^{2-5}]dodeca-6,11-diene-8-one monobromoacetate, 31B, 111
$C_{15}H_{18}O_3S$, endo-3-Benzylsulfinyl-bicyclo[2.2.1]heptane-endo-2-carboxylic acid, 42B, 156
$C_{15}H_{19}BrO_3S$, 1-p-Bromobenzenesulphonyloxymethylbicyclo[2.2.2]octane, 33B, 98
$C_{15}H_{20}ClNO$, 3-N-Methylamino-3-phenyl-2-bicyclo[3.2.1]octanone hydrochloride, 39B, 143
$C_{15}H_{20}O$, Tetracyclo[6.4.3.0.0^{2-7}]pentadec-4-en-13-one, 46B, 180
$C_{15}H_{20}O_2$, 4-Carboxy[8]paracyclophane, 39B, 143
$C_{15}H_{22}O$, Tetracyclo[6.4.3.0.0^{2-7}]pentadecan-13-one, 42B, 157
$C_{15}H_{22}O_2$, 3,7-Dimethyl-1,1-(2-methyl-1-oxopropyl)tricyclo-[4.2.1.0^{3-7}]nonan-2-one, 46B, 180
$C_{15}H_{22}O_4$, 1,8-Dicarboxymethyl-tricyclo[4.3.1.1^{3-8}]undecane, 41B, 226
$C_{15}H_{23}N_3O_3$, 6-(2-Butenyl)-9-semicarbazonobicyclo[3.3.1]nonan-2-carboxylic acid, 39B, 144
$C_{15}H_{25}NO$, DL-7,7-(2',2'-Dimethyl)pentamethylene-1-methyl-norbornane-2-oxime, 41B, 227
$C_{16}H_7BrO$, 3-(p-Bromobenzylidene)nopinone, 46B, 181
$C_{16}H_8F_8$, 1,1,2,2,9,9,10,10-Octafluoro-[2.2]paracyclophane, 38B, 237
$C_{16}H_{10}F_2$, anti-8,16-Difluorometacyclophane-1,9-diene, 41B, 228
$C_{16}H_{10}O$, 8,16-Oxido-cis[2.2]metacyclophane-1,9-diene, 34B, 117
$C_{16}H_{10}O_2$, anti-1,6:8,13-Biscarbonyl[14]annulene, 46B, 181
$C_{16}H_{10}O_2$, syn-1,6:8,13-Biscarbonyl[14]annulene, 43B, 242
$C_{16}H_{12}$, [2.2]Metacyclophane-1,9-diene, 38B, 236
$C_{16}H_{12}$, [2.2]Metaparacyclophane-1,9-diene, 37B, 109
$C_{16}H_{12}$, [2.2]Paracyclophane diene, 28, 452
$C_{16}H_{12}$, 3,4:7,8-dibenzotricyclo[4.2.0.0^{2-5}]octa-3,7-diene, 35B, 144
$C_{16}H_{14}$, [2.2]Metacyclophan-1-ene, 44B, 176
$C_{16}H_{14}$, syn-1,6:8,13-Bismethano[14]annulene, 43B, 243
$C_{16}H_{14}Cl_2O$, 3,6-Dichloro-11,12-benzotetracyclo[5.3.2.0^{2-6}.0^{3-8}]-dodeca-9-one, 35B, 145
$C_{16}H_{16}$, [2.2]Metacyclophane, 43B, 243
$C_{16}H_{16}$, [2.2]Paracyclophane, 38B, 237
$C_{16}H_{16}$, [3.3]Paracyclophadiyne, 41B, 228

$C_{16}H_{16}$, Di-m-xylylene, 17, 716
$C_{16}H_{16}$, Di-p-xylylene, 17, 715; 24, 676
$C_{16}H_{16}N_4O_4$, Tetracyclo[5.2.1.0^{2-6}.0^{4-9}]decan-3-one 2,4-dinitrophenylhydrazone, 41B, 229
$C_{16}H_{16}O$, 2,7-Pentamethylene-4,5-benzotropone, 41B, 230
$C_{16}H_{16}O_2$, 1,4,4a,5,8,8a,9a,10a-Octahydro-1,4:5,8-dimethano-9,10-anthraquinone, 40B, 185
$C_{16}H_{17}ClO_2$, Tricyclo[3.3.1.0^{2-4}]nonyl p-chlorobenzoate, 43B, 244
$C_{16}H_{18}Br_2$, 6,12-Dibromoheptacyclo[7.7.0.0^{2-6}.0^{3-15}.0^{4-12}.0^{5-10}.-0^{11-16}]hexadecane, 46B, 182
$C_{16}H_{18}Cl_2O_2$, 1,4-Ethyleno-2,8-dichloro-2,4,6,8-tetramethyloctahydronaphthal-5-ene-3,7-dione, 41B, 232
$C_{16}H_{18}Cl_2O_4 \cdot H_2O$, 7,12-Dichloro-6,13-dihydroxy-5,14-dimethylpentacyclo[7.5.0.0^{2-6}.0^{3-13}.0^{4-10}]-tetradecane-8,11-dione monohydrate, 45B, 182
$C_{16}H_{18}Cl_4$, 2,3-Dihydro-1,3-(11,11-dimethylpropano)-1,3-dimethyl-4,5,6,7-tetrachloroindene, 39B, 145
$C_{16}H_{18}O_2S$, (+)-2-Z-Ethylidene-3-exo-p-tolylsuphonyl-5-norbornene, 45B, 182
$C_{16}H_{18}O_3$, 1,4-Etheno-2,7-dihydroxy-2,4,6,8-tetramethyl-1,2,3,4-tetrahydronaphthalen-3-one, 43B, 245
$C_{16}H_{18}O_4$, 1,4-Etheno-2-hydroxy-2,4a,6,9-tetramethyloctahydronaphthal-5-ene-3,7,8-trione, 43B, 245
$C_{16}H_{19}Br$, 5-Bromoheptacyclo[8.6.0.0^{2-8}.0^{3-13}.0^{4-11}.0^{5-9}.0^{12-16}]hexadecane, 46B, 182
$C_{16}H_{20}BrN$, 2-Anilino-3-bromo-tetrahydro-exo-dicyclopentadiene, 32B, 149
$C_{16}H_{20}O_2$, trans-1,4:5,8-Dimethylene-cis,anti,cis-perhydroanthraquinone, 30B, 119
$C_{16}H_{20}O_3$, 2-(Tetracyclo[5.5.1.0^{2-6}.0^{8-12}]tridec-9-en-4-yl)propionic acid, 43B, 246
$C_{16}H_{20}O_4$, 1,4-Ethyleno-2,8-dihydroxy-2,4,6,8-tetramethyl-octahydronaphthal-5-en-3,7-dione, 39B, 146
$C_{16}H_{20}O_4$, 1,4-Ethyleno-2,8-dihydroxy-2,4a,8,9-tetramethyl-octahydronaphthal-5-en-3,7-dione, 39B, 146
$C_{16}H_{20}O_4$, 4-Acetyl-11-hydroxy-2,7,11-trimethyltricyclo[4.3.0.2^{2-5}]-undec-3-ene-8,10-dione, 42B, 157
$C_{16}H_{20}O_6$, DL-1,7-Dicarbomethoxy-3α,7-methano-3αH-decahydrocyclopentacyclooctene-2,10-dione, 40B, 186
$C_{16}H_{21}Br$, 4-Bromo-1,7,7-trimethyl-2,3-benzobicyclo[3.3.1]non-2-ene, 44B, 176
$C_{16}H_{21}NO$, (10RS,5RS,9SR)-10β-Ethyl-5,6,7,8,9,10-hexahydro-5α,9α-methanobenzocyclooctene-10α-carboxamide, 46B, 154
$C_{16}H_{21}NO$, (10SR,5RS,9SR)-10α-Ethyl-5,6,7,8,9,10-hexahydro-5α,9α-methanobenzocyclooctene-10β-carboxamide, 46B, 146
$C_{16}H_{22}$, Hexacyclo[10.3.1.0^{2-10}.0^{3-7}.0^{6-15}.0^{9-14}]hexadecane, 38B, 238
$C_{16}H_{22}ClNO$, 2-Methylamino-1-(spiro(cyclopentane-1,1'-inden)-3'-yl)-ethanol hydrochloride, 44B, 156
$C_{16}H_{22}O_2$, 1β,3,4β,6β,8-Pentamethyl-9-exo-methylenetricyclo-[4.4.0.0^{3-8}]decane-2,5-dione, 46B, 183
$C_{16}H_{22}O_2$, 2-Hydroxy-1,3,4,6,8,9-hexamethyltricyclo[4.4.0.0^{2-8}]deca-3,9-dien-5-one, 40B, 187
$C_{16}H_{22}O_2$, 3,4,6,8β,9,10-Hexamethyltetracyclo[4.3.1.0^{3-10}.0^{4-9}]-decane-2,5-dione, 45B, 183
$C_{16}H_{22}O_4$, 6,12-Dihydroxy-2,6,9,12-tetramethyltetracyclo-[6.2.2.0^{2-7}.0^{4-9}]dodecane-5,11-dione, 46B, 183
$C_{16}H_{24}$, Tetramethyltetraasteranene, 39B, 147

$C_{17}H_{12}$, 1,6:8,13-Cyclopropanediylidene[14]annulene, 40B, 187
$C_{17}H_{14}$, 1,6:8,13-Propane-1,3-diylidene[14]annulene, 38B, 238
$C_{17}H_{15}BF_4$, Tricyclo[9.4.1.1$^{3-9}$]heptadecaheptaenylium tetrafluoroborate, 45B, 183
$C_{17}H_{15}BrO_2$, Tricyclo[5.2.1.0$^{2-6}$]deca-4,8-dienyl p-bromobenzoate, 34B, 117
$C_{17}H_{15}BrO_3S$, Benzonorbornen-anti-9-yl p-bromobenzenesulphonate, 34B, 118
$C_{17}H_{15}BrO_3S$, syn-9-Benzonorbornenyl p-bromobenzenesulphonate, 33B, 205
$C_{17}H_{16}O_2$, [2]Paracyclo[2](3,7)tropolonophane, 45B, 184
$C_{17}H_{18}O_3$, 11-anti-Acetoxy-2,3-benzotricyclo[4.4.1.0]undecen-4-one, 41B, 158
$C_{17}H_{19}BrO_2$, 7-p-Bromobenzoyloxy-1,2-dimethyltricyclo[3.3.0.0$^{2-7}$]-octane, 46B, 184
$C_{17}H_{19}NO_4$, 2-Hydroxy[4,2,2]propellane p-nitrobenzoate, 40B, 190
$C_{17}H_{22}O_2$, (1R,3S,αR)-3-(α-Hydroxybenzyl)-1,7,7-trimethylbicyclo-[2.2.1]heptan-2-one, 42B, 159
$C_{17}H_{22}O_3$, 2,2,8-Trimethyltricyclo[6.2.2.0$^{1-6}$]dodec-5-ene-9,10-dicarboxylic anhydride, 43B, 246
$C_{17}H_{23}BrO_4S$, 1-p-Bromobenzenesulphonyloxymethyl-5-methyl-bicyclo-[3.3.1]nonan-9-ol, 30B, 116
$C_{17}H_{24}O$, Tetracyclo[8.4.3.0.0$^{2-9}$]heptadec-5-en-15-one, 43B, 247
$C_{17}H_{24}O_2$, 3-(α-Hydroxybenzyl)-1,7,7-trimethylbicyclo[2.2.1]heptan-2-ol, 42B, 159
$C_{18}H_{12}$, [2.2.2](1,3,5)Cyclophane-1,9,17-triene, 38B, 239
$C_{18}H_{12}$, 1,10-Phenanthrotricyclo[4.1.1.0$^{2-7}$]heptene, 43B, 248
$C_{18}H_{15}BrO_5$, syn-3-exo-p-Bromobenzoyloxybicyclo[3.2.2]non-6-ene-8,9-endo-cis-dicarboxylic acid anhydride, 35B, 147
$C_{18}H_{16}$, 1,6:8,13-Butane-1,4-diylidene[14]annulene, 38B, 240
$C_{18}H_{16}$, 15,16-Dimethyl-1,6:8,13-ethanediylidene[14]annulene, 41B, 232
$C_{18}H_{16}BrNO_2$, 2-Hydroxy-2-phenylbicyclo[1.1.1]pentane p-bromophenylurethane, 33B, 99
$C_{18}H_{16}Cl_2N_2O_2$, endo-2,3-Dichloro-4a,8a-dicyano-4a,5,8,8a-tetrahydro-5,7-dimethyl-5,8-(2,2-dimethylethano)-1,4-naphthaquinone, 37B, 110
$C_{18}H_{18}$, [2.2.2](1,2,3)Cyclophane, 46B, 191
$C_{18}H_{18}$, [2.2.2](1,3,5)Cyclophane, 46B, 184
$C_{18}H_{18}O_2$, 1-Carbomethoxytetracyclo[9.2.2.1$^{4-11}$.0$^{8-16}$]hexadeca-5,7,12,14,16(4)-pentaene, 41B, 233
$C_{18}H_{19}BrO_3$, 2-p-Bromobenzoyl-1,5,5-trimethylbicyclo[2.2.2]octa-6,8-dione, 34B, 120
$C_{18}H_{19}Cl$, (E)-2-Chloro-2-ethynyl-5-phenyladamantane, 46B, 185
$C_{18}H_{19}Cl$, (E)-2-Chlorovinylidene-5-phenyladamantane, 44B, 177
$C_{18}H_{19}Cl$, (Z)-2-Chloro-2-ethynyl-5-phenyladamantane, 46B, 185
$C_{18}H_{19}ClO_6$, 1,9-Diacetoxy-5-chloro-12,12-dimethyltricyclo-[6.2.2.0$^{2-7}$]dodeca-4,9-diene-3,6-dione, 44B, 177
$C_{18}H_{20}$, [3.3]Paracyclophane, 30B, 121
$C_{18}H_{20}$, 4,12-Dimethyl[2.2]metacyclophane, 27, 918
$C_{18}H_{20}BrNO_2S$, trans-3-endo-Bromo-4-(2-nitrophenylthio)-9,10-cis-endo-dimethyltricyclo[4.2.2.0$^{2-5}$]dec-7-ene, 44B, 178
$C_{18}H_{20}Cl_4$, (1R,2S,7S,8R,9S,10R)-7,8,9,10-Tetrachloropentacyclo-[8.4.2.2$^{2-7}$.0$^{1-9}$.0$^{2-8}$]octadeca-4,12-diene, 41B, 234
$C_{18}H_{20}O$, (Z)-2-Ethynyl-5-phenyl-2-adamantanol, 45B, 184
$C_{18}H_{20}O_2$, 1,4,4a,5,8,8a,9a,10a-Octahydro-1,4,5,8-diethanoanthracene-9,10-dione, 46B, 186
$C_{18}H_{20}O_6$, Catechinic acid (trimethylated), 42B, 165

$C_{18}H_{21}BrO_4S$, (1R,6S,8R,10R)-8-(p-Bromobenzenesulphonoxy)-1,10-dimethyltricyclo[4.4.0.0^{3-8}]decan-2-one, 41B, 235
$C_{18}H_{21}ClS$, trans-3-endo-Chloro-4-phenylthio-9,10-cis-endo-dimethyltricyclo[4.2.2.0^{2-5}]dec-7-ene, 46B, 186
$C_{18}H_{24}O_2$, Bicyclo[3.3.1]non-1-ene-3-one cis-anti-dimer, 45B, 186
$C_{18}H_{24}O_2$, Bicyclo[3.3.1]non-1-ene-3-one trans-anti-dimer, 45B, 186
$C_{18}H_{24}O_2$, Bicyclo[3.3.1]non-1-ene-3-one trans-syn-dimer, 45B, 186
$C_{18}H_{24}O_2$, 1,5,8,8,11,11-Hexamethylpentacyclo[4.2.2.2.0^{3-10}.0^{7-12}]-dodeca-4,9-dione, 42B, 157
$C_{18}H_{26}O_4S$, 5-Methyl-1-p-toluenesulphonyloxymethylbicyclo[3.3.1]-nonan-9-ol, 45B, 185
$C_{18}H_{27}BrO_3$, 2-Bromo-11-ethyl-5,9-dimethoxytetracyclo-[5.4.1.1^{4-12}.1^{8-11}]tetradecan-3-one, 37B, 112
$C_{19}H_{16}O_2$, syn-5,7-Diformyltricyclo[9.4.1.1^{3-9}]heptadeca-2,4,7,9,11,13,15-heptaene, 43B, 248
$C_{19}H_{18}$, 1,6:8,13-Pentane-1,5-diylidene[14]annulene, 41B, 235
$C_{19}H_{19}NO_4$, 1endo,4endo:5exo,8exo-Dimethano-1,2,3,4,4a,5,8,8a-octahydronaphthalene 10-syn-p-nitrobenzoate, 46B, 187
$C_{19}H_{21}Br$, 1-Bromo-3,3,5-trimethyl-6-(1',2'-naphtho)-bicyclo[3.2.1]-octene, 40B, 190
$C_{19}H_{22}O_2$, 1,1,2,2-Tetramethyl-6-phenylspiro[2.5]octa-4,7-diene-6-carboxylic acid, 42B, 133
$C_{19}H_{23}Cl_3O_3S$, (1S,5S)-6,6-Dimethyl-2-((2S)-3,3,3-trichloro-2-tosyloxypropyl)bicyclo[3.1.1]hept-2-ene, 44B, 178
$C_{19}H_{24}BrNO_3$, 2-Bromo-3-methoxy-5-hydroxy-5,6,7,8,9,10,13,14-octahydro-8-methyl-10,13-N-acetylaminomethano-phenanthrene, 39B, 148
$C_{19}H_{24}S$, 4-Benzylthio-2,6-dimethyl-endo-tricyclo[5.2.1.0^{2-6}]dec-3-ene, 45B, 186
$C_{20}H_{14}$, Triptycene, 35B, 149; 37B, 112
$C_{20}H_{16}F_6N_2O_2$, anti-10-Cyano-syn-10-[(E)-(2,2,2-trifluoroethyl)vinyl]-endo-5-cyano-exo-5-[(E)-(2,2,2-trifluoroethyl)vinyl]-endo-tricyclo[5.2.1.0^{2-6}]deca-3,8-diene, 46B, 188
$C_{20}H_{16}O_3$, 1-Phthaloyl-2-p-ethylbenzoylcyclopropane, 44B, 158
$C_{20}H_{19}ClN_2O_8S$, trans-3-endo-Chloro-4-(2,4-dinitrophenylthio)-9,10-cis-endo-dimethoxycarbonyltricyclo[4.2.2.0^{2-5}]dec-7-ene, 44B, 179
$C_{20}H_{20}$, [2.2.2.2](1,2,4,5)Cyclophane, 43B, 249
$C_{20}H_{20}$, 5,6,7,12,13,14-Hexahydro-5,13:6,12-dimethanodibenz[a,f]cyclodecene, 42B, 159
$C_{20}H_{21}ClO_4S$, trans-3-endo-Chloro-4-phenylthio-9,10-cis-endo-di(methoxycarbonyl)tricyclo[4.2.2.0^{2-5}]dec-7-ene, 44B, 180
$C_{20}H_{22}O_4$, syn-15,18-Dimethoxy[3](2,6)-p-benzoquinone[3.3]metacyclophane, 46B, 188
$C_{20}H_{24}$, 1,3,3,5-Tetramethyl-6-(1',2'-naphtho)-bicyclo[3.2.1]octene, 42B, 160
$C_{20}H_{24}$, 3,3',6,6'-Tetrahydro[2.2.2.2](1,2,4,5)cyclophane, 43B, 249
$C_{20}H_{24}O_6$, 1,4-Etheno-2,8-diacetoxy-2,4,6,8-tetramethyloctahydronaphthal-5-ene-3,7-dione, 41B, 236
$C_{20}H_{26}$, Dodecahydrotriptycene (A-C'-C isomer), 43B, 250
$C_{20}H_{26}BrNO_2$, exo-1α,7α,2β,6β-11,11-Dimethyltricyclo[5.4.0.0^{2-6}]-undecan-8-ol p-bromophenylurethane, 38B, 240
$C_{20}H_{28}ClNO$, 2-(α-(Tricyclo[2.2.1.0^{2-6}]hept-3-ylidene)benzyloxy)triethylamine hydrochloride, 46B, 189
$C_{20}H_{30}S_2$, Di-t-adamantyl disulfide, 46B, 189
$C_{20}H_{32}$, cis,cis,cis-Perhydrotriptycene (hexagonal), 44B, 180
$C_{20}H_{32}$, cis-cis-cis-Perhydrotriptycene (orthorhombic), 42B, 160
$C_{20}H_{32}$, syn-2,2'-Bifenchylidene E, 43B, 251
$C_{20}H_{32}$, trans,cis,cis-Perhydrotriptycene, 44B, 181

$C_{20}H_{32}$, trans,trans,trans-Perhydrotriptycene, 44B, 180
$C_{21}H_{10}F_{10}O_2$, 1,4-Dibenzoyldecafluoronorbornane, 40B, 191
$C_{21}H_{12}BrIO_2$, 10-Bromo-9-triptycyl iodoformate, 44B, 181
$C_{21}H_{18}$, 9,10-Dihydro-9,10-(tricyclo[4.1.0.0^{2-7}]heptan-1,7-diyl)anthracene, 44B, 182
$C_{21}H_{20}Br_2O_8$, Indene - dimethyl acetylenedicarboxylate adduct, 33B, 103
$C_{21}H_{23}BrO_2$, Tetracyclic bromobenzoate, 40B, 192
$C_{21}H_{25}NO_6$, (1R,3S,4S,5S,9R)-3,6,6-Trimethyl-4-methoxycarbonyl-9-(p-nitrobenzoyloxy)tricyclo-[3.3.1.0^{1-3}]nonane, 46B, 190
$C_{21}H_{26}ClN$, N-Isopropyl-7-endo-phenyl-5,9-methano-6,7,8,9-tetrahydrobenzocycloheptene-6-exo-yl-amine hydrochloride, 44B, 182
$C_{21}H_{28}$, exo,exo-Octacycloheneicosane, 45B, 187
$C_{21}H_{28}O_3$, 1-Phthaloyl-2-p-propylbenzoylcyclopropane, 44B, 158
$C_{22}H_{17}Cl$, β-Chloroethyltriptycene, 34B, 121
$C_{22}H_{17}ClO_2$, (+)-2,5-Dimethoxy-8-chlorotriptycene, 40B, 193
$C_{22}H_{20}N_2O_{10}S$, 9(SR)-Acetoxy-10(RS)-(2,4-dinitrophenylthio)-7,8-dimethoxycarbonyltricyclo[4.3.1.0^{2-5}]deca-3,7-diene, 46B, 190
$C_{22}H_{21}NO_8S$, 5(SR)-(2-Nitrophenylthio)-10(SR)-acetoxy-8,9-dimethoxycarbonyltetracyclo[4.4.0.0^{2-4}.0^{3-7}]dec-8-ene, 45B, 187
$C_{22}H_{22}$, [2.2.2.2.2](1,2,3,4,5)Cyclophane, 46B, 191
$C_{22}H_{24}$, 4,13-Dimethyl[2.2.2.2](1,2,3,4)cyclophane, 46B, 191
$C_{22}H_{24}O_2$, 6,6a,7,12,13,13a-Hexahydro-2,9-dimethoxy-7,13-methano-5H-benzo[4,5]cyclohepta[1,2-a]naphthalene, 43B, 251
$C_{22}H_{25}N$, N,N-Dimethylspiro(5H-dibenzo[a,d]cycloheptene-5,1'-cyclohexane)-4'-amine, 42B, 144
$C_{22}H_{28}$, Nonacyclo[11.7.1.1^{2-18}.0^{3-16}.0^{4-13}.0^{5-10}.0^{6-14}.0^{7-11}.-0^{15-20}]docosane, 46B, 182
$C_{22}H_{28}$, anti-Tetramantane, 43B, 252
$C_{22}H_{28}BrNO_7$, 16-Bromomyricanol-nitromethane, 37B, 113
$C_{22}H_{32}O_2$, Pentacyclo[8.6.3.0.0^{2-9}.3^{2-9}]docosane-17,20-dione, 45B, 187
$C_{22}H_{32}O_2$, 13,14-Di-t-butoxyheptacyclo[5.5.1.1^{4-10}.0^{2-6}.0^{3-11}.0^{5-9}.-0^{8-12}]tetradecane, 42B, 161
$C_{22}H_{32}O_{10}$, 1,7-Bis(dimethoxymethyl)-5,5,9,9-tetramethoxy-1,4a,5,8a-tetrahydro-1,4-ethanonaphthalene-6,10(4H)-dione, 44B, 183
$C_{22}H_{32}O_{10}$, 2,8-Bis(dimethoxymethyl)-5,5,9,9-tetramethoxy-1,4a,5,8a-tetrahydro-1,4-ethanonaphthalene-6,10(4H)-dione, 44B, 183
$C_{22}H_{34}$, Tricyclo[3.3.3.0^{2-6}]undec-2(6)-ene dimer, 43B, 252
$C_{23}H_{18}O$, endo-1,2,3,4,4a,9a-Hexahydro-1,4-(peri-naphthaleno)fluoren-9-one, 43B, 253
$C_{23}H_{26}Br_2O_2$, (-)-7,7'-Dibromo-3,3,3',3',5,5'-hexamethyl-1,1'-spirobi-indane-6,6'-diol, 42B, 144
$C_{23}H_{28}$, (DL)-3,3,3',3',5,5'-Hexamethyl-1,1'-spirobi-indane, 42B, 144
$C_{23}H_{32}O_3$, 1,4,4aα,4bβ,5,8,8aβ,9aα-Octahydro-anti,anti-10,11-di-t-butoxy-1,4:5,8-dimethanofluorene-9-one, 41B, 237
$C_{24}H_{17}BrO_2$, 1,5-Diphenyltricyclo[2.1.0.0^{2-5}]pent-3-yl p-bromobenzoate, 32B, 151
$C_{24}H_{18}$, 1,8-Diphenyl-1a,2,7,7a-tetrahydro-1,2,7-metheno-1H-cyclopropa[b]naphthalene, 44B, 183
$C_{24}H_{18}$, 2,3:6,7:10,11-Tribenzotetracyclo[6.4.0.0^{4-12}.0^{5-9}]dodeca-2,6,10-triene, 41B, 237
$C_{24}H_{18}Cl_4$, 1,2,3,4-Tetrachloro-9-t-butyltriptycene, 41B, 239
$C_{24}H_{20}$, [2.2](1,3)Azulenophane, 45B, 188
$C_{24}H_{24}$, [2.2.2.2.2.2](1,2,3,4,5,6)Cyclophane, 46B, 191
$C_{24}H_{24}$, exo-7,7-Dimethyl-11-dimethylmethylene(3,4:9,10)tricyclo-[6.2.1.0^{2-6}]undeca-3,5,9-triene, 45B, 189

$C_{24}H_{24}BrNO_2$, (+)-2,5-Dimethoxy-7-dimethylaminotriptycene hydrobromide, 38B, 241
$C_{24}H_{25}Br_3O_3$, 2,2,10-Tribromo-7,11-diphenylspiro[5.5]undecane-1,9-dione methanol solvate, 44B, 151
$C_{24}H_{28}N_4O_4S_2$, 1,2,5,6-Tetramethyl-3,4-bis(tosylhydrazone)tricyclo-[3.1.0.0$^{2-6}$]hexane, 45B, 189
$C_{24}H_{34}N_2O_8$, Cedrone dimethylformamide solvate, 38B, 242
$C_{24}H_{36}$, trans-1-Ethyl-2(1-ethyl-2-adamantanylidene)adamantane, 46B, 192
$C_{24}H_{36}O_8$, Dihydroxyisoasatone, 40B, 193
$C_{25}H_{26}N_2O_8$, 1,8,8-Trimethylbicyclo[3.2.1]octane-2β,4β-diol bis-p-nitrobenzoate, 43B, 253
$C_{26}H_{24}$, 2,3,4-Triphenylbicyclo[3.2.1]oct-2-ene, 43B, 254
$C_{26}H_{26}$, [2.2]Metacyclophane (triple-layered ud isomer), 43B, 243
$C_{26}H_{26}Cl_2$, 13'-Chloro-7'-(4"-chlorophenyl)spiro(bicyclo[2.2.1]heptane-2,9'-tetracyclo[8.4.0.0$^{2-7}$.0$^{4-8}$])tetradeca-10',12',14'-triene), 44B, 184
$C_{26}H_{26}Cl_2$, 5'-Chloro-2'-(4"-chlorophenyl)spiro[bicyclo[2.2.1]-heptane-2,9'-tetracyclo[9.2.1.0$^{2-10}$.0$^{3-8}$]tetradeca-3',5',7'-triene], 45B, 190
$C_{26}H_{26}O_8$, Tetramethyl 3,3'-bitricyclo[3.2.2.0$^{2-4}$]nona-6,8-diene-6,6',7,7'-tetracarboxylate, 37B, 113
$C_{26}H_{28}O$, 2,3,6,7,7,8-Hexamethyl-1,5-diphenyl-tetracyclo-[3.3.0.0$^{2-8}$.0$^{3-6}$]octan-4-one, 40B, 194
$C_{26}H_{28}O_{12}$, 3,3-Dimethyl-4,5,9,10,11,12-hexacarboxymethyl-tetracyclo-[7.2.1.0$^{2-4}$.0$^{2-8}$]dodeca-5,7,10-triene, 39B, 149
$C_{26}H_{30}O_4$, 5,6,6a,7,7a,11b-Hexahydro-2,3,5,5,7,7,9,10-octamethyl-6,-11a,11c-metheno-1H-benzo[c]fluorene-1,4,8,11(4aH)-tetrone, 39B, 149
$C_{26}H_{32}O_4$, Di-t-butylsesquifulvalene - dimethyl acetylenedicarboxylate addition product, 41B, 239
$C_{27}H_{22}$, 1-(9-Anthryl)-3-(1-naphthyl)propane photoisomer, 45B, 190
$C_{27}H_{28}$, [2.2]Metacyclophane (methyl triple-layered ud-isomer), 44B, 185
$C_{27}H_{28}$, [2.2]Metacyclophane (triple-layered, methylated, uu-isomer), 43B, 255
$C_{27}H_{36}BaO_{13}$, Barium hydrogen bis(d-spiro[3.3]heptane-2,6-dicarboxylate) d-spiro[3.3]heptane-2,6-dicarboxylic acid monohydrate, 42B, 17
$C_{27}H_{38}O$, Isophorone condensation trienone, 43B, 256
$C_{28}H_{20}$, Di-p-anthracene, 31B, 112; 46B, 192
$C_{28}H_{26}BrO_4$, Norbornadiene dimer di-p-bromobenzoate, 38B, 242
$C_{28}H_{28}$, 9,10-Benzotricyclo[3.3.2.0$^{3-7}$]deca-3(7),9(10)-diene dimer, 43B, 256
$C_{28}H_{30}$, 4b-Phenylmariontetraene, 46B, 193
$C_{28}H_{31}BrO_3$, 1,2,3,4,4b,5,6,10b,11,12-Decahydro-8-methoxy-4a,12a-ethanochrysen-1-yl p-bromobenzoate, 41B, 163
$C_{28}H_{39}BrO_4$, Heptacyclic cage system, 40B, 195
$C_{30}H_{18}Br_2$, 2-Bromo-4-bromomethylene-3-fluorenylidene-spiro(cyclobutane-1,9'-fluorene), 42B, 140
$C_{30}H_{20}$, Tetrabenzopentacyclo[6.2.2.2$^{2-6}$.0$^{2-7}$.0$^{3-7}$]tetradeca-4,9,11,13-tetraene, 45B, 191
$C_{30}H_{20}Br_2$, 5a,11a-Dibromojanusene, 39B, 150
$C_{30}H_{22}$, Lepidopterene, 42B, 163
$C_{30}H_{22}$, 9,9'-Ethano-9,9',10,10'-tetrahydro-(9,9':10,10'-bianthryl), 45B, 192
$C_{30}H_{24}O_4$, Dodecacyclotriaconta-3,5,27-triene-14,18,29,30-tetraone, 45B, 193

$C_{31}H_{22}Br_2O_4$, (+)-2,3:6,7-Dibenzobicyclo[3.3.1]nona-2,6-diene-4,8-diyl bis(p-bromobenzoate), 42B, 164
$C_{31}H_{22}O_2$, 1-Formyl[6]helicene cycloaddition product, 42B, 164
$C_{32}H_{24}$, [2_4]Paracyclophanetetraene, 44B, 185
$C_{32}H_{24}$, Bi(anthracene-9,10-dimethylene) (photo form), 31B, 114
$C_{32}H_{24}$, Bi(anthracene-9,10-dimethylene), 43B, 257
$C_{33}H_{42}O_4$, Clusianone, 42B, 165
$C_{34}H_{26}$, 9,11-Dihydro-9,10-dimethylanthracene-(9,5':10,12'-diyl)-5',12'-dihydrotetracene, 45B, 193
$C_{34}H_{28}Br_2O_4$, 8-(1,2,3,4,5,6-Hexahydro-5-hydroxy-3a,6-methano-3aH-inden-8-ylidene)-1,2,3,4,5,6-hexahydro-3a,6-methano-3aH-inden-5-ol bis(p-bromobenzoate), 46B, 194
$C_{34}H_{30}$, Anthracenophane, 41B, 242
$C_{34}H_{30}$, Polycyclic compound, 42B, 166
$C_{34}H_{42}N_2$, 5,22-Bis(diethylamino)-6,21-dimethylheptacyclo[12.8.2.-$0^{1-15}.0^{2-13}.0^{4-12}.0^{7-12}.0^{15-20}$]tetracosa-3,5,8,10,16,18,21,23-octene, 46B, 195
$C_{35}H_{48}O_5$, 7-t-Butoxynorbornadiene trimer ketone, 44B, 186
$C_{36}H_{20}$, [2.2](2,7)Pyrenophane-1,1'-diene, 44B, 186
$C_{36}H_{24}$, [2.2](2,7)Pyrenophane, 43B, 228
$C_{36}H_{24}$, Tetracene photodimer, 41B, 243
$C_{37}H_{25}BrO \cdot C_6H_6 \cdot CH_4O$, 1,4-Diphenyl-5-(p-bromophenyl)-7-oxo-[2,3-l]phenanthrenobicyclo[2.2.1]hept-2-ene benzene methanol solvate, 46B, 195
$C_{37}H_{29}BrO_3$, 1,3,4,5,6,11-Hexahydro-6,11-diphenyl-6,11-epoxy-1,5a-ethano-5aH-cyclobut[d]anthracen-14-ol p-bromobenzoate (form 1), 46B, 194
$C_{39}H_{30}Cl_4O_2$, endo-exo-1,10,11,12-Tetrachloro-13,13-dimethoxy-4,5,6,7-tetraphenylpentacyclo[8.2.1.$0^{2-9}.0^{3-7}.0^{6-8}$]trideca-4,11-diene, 46B, 196
$C_{40}H_{26}$, 9,9'-Bitriptycyl, 44B, 187
$C_{40}H_{44}$, Tetramethylcyclophane (quadruple-layered), 43B, 257
$C_{46}H_{46}N_6$, 2,3,9,10-Tetrakis(2,4-dimethylphenylamino-1,2,3,4-tetrahydro-1,4-ethanonaphthalene-1,4-dinitrile, 46B, 197
$C_{48}H_{38}N_4$, 2,7-Di-t-butyl-11,11,12,12-tetracyano-4,5,9,10-tetraphenyltetracyclo[4.4.2.$0^{1-8}.0^{3-6}$]dodeca-2,4,7,9-tetraene, 46B, 197
$C_{68}H_{52}Br_4O_{16}$, Resorcinol p-bromobenzoate condensation tetramer octaacetate (cis form), 42B, 166
$C_{68}H_{52}Br_4O_{16}$, Resorcinol p-bromobenzoate condensation tetramer octaacetate (trans form), 42B, 166

32. HETERO-NITROGEN (3,4,5-MEMBERED MONOCYCLIC)

$CHBrN_4$, 5-Bromotetrazole, 39B, 150
CH_2N_4, Tetrazole, 40B, 196
CH_3N_4NaO, Sodium tetrazolate monohydrate, 28, 56, 453
$CH_3N_5 \cdot H_2O$, 5-Aminotetrazole monohydrate, 32B, 152; 45B, 194
CH_7N_7, Hydrazinium 5-aminotetrazolate, 22, 755
$C_2H_2ClN_3$, 3-Chloro-1,2,4-triazole, 46B, 198
$C_2H_3BrN_2$, Methylbromodiazirine (gas-mw), 35B, 903
$C_2H_3N_3$, 1,2,4-Triazole, 30B, 127; 34B, 122
C_2H_4ClN, 1-Chloroaziridine (gas-mw), 38B, 1071
$C_2H_4N_4$, 5-Amino-1H-1,2,4-triazole, 44B, 187
C_2H_5N, Ethyleneimine (gas-ed), 26, 590
C_2H_5N, Ethyleneimine (gas-mw), 37B, 697
$C_2H_5N_5$, 3,5-Diamino-1H-1,2,4-triazole, 45B, 194; 46B, 198

$C_2H_5N_5$, 5-Amino-2-methyltetrazole, 20, 636
$C_2H_5N_5S$, 3-Hydrazino-5-mercapto-1,2,4-triazole, 22, 758
$C_2H_6N_2$, 3-Methyldiaziridine (gas-ed), 42B, 1028
$C_2H_6N_6S$, 4-Amino-3-hydrazino-5-mercapto-1,2,4-triazole, 37B, 114
$C_2H_7BrN_6$, 3,4,5-Triamino-1,2,4-triazole hydrobromide, 39B, 151
$C_2H_9N_4NaO_3S$, Sodium 4-amino-5-mercapto-1,2,4-triazole trihydrate, 39B, 152
$C_3H_2N_2O_3$, Parabanic acid, 19, 599; 45B, 195
$C_3H_3N_3O_2$, N-Nitropyrazole, 43B, 258
$C_3H_4N_2$, Imidazole, 31B, 115; 34B, 123; 37B, 115; 43B, 259; 45B, 195
$C_3H_4N_2$, Pyrazole (gas-mw), 40B, 1146
$C_3H_4N_2$, Pyrazole, 24, 682; 35B, 152; 39B, 153
$C_3H_4N_2OS$, 2-Thiohydantoin, 34B, 124
$C_3H_4N_4OS$, 4-Formylamino-Δ^2-1,2,4-triazoline-5-thione, 45B, 196
$C_3H_5N_5O$, 5-Amino-1H-1,2,3-triazole-4-carboxamide, 40B, 196
$C_3H_6N_2S$, Ethylenethiourea, 17, 745
$C_3H_7ClN_2$, Pyrazoline hydrochloride, 24, 683; 27, 951
C_3H_7N, Azetidine (gas-ed), 39B, 888; 42B, 1029
$C_3H_8ClN_5$, 5-Imino-1,3-dimethyltetrazole hydrochloride, 19, 605
$C_3H_8N_2$, 1,2-Dimethyldiaziridine (gas-ed), 42B, 1028
$C_3H_{15}Ca_{1.5}N_2O_{12}P_2$, Calcium 1,3-diphosphorylimidazole hexahydrate, 33B, 107
$C_4H_4N_2OS$, 4-Formylimidazoline-2-thione, 43B, 260
$C_4H_4N_6 \cdot 0.5\ H_2O$, 3-(1-Tetrazolyl)pyrazole hydrate, 35B, 154
$C_4H_4N_6$, 4-Triazolyl-4H-1,2,4-triazole, 43B, 260
C_4H_5N, Pyrrole (gas-ed), 7, 270
C_4H_5N, Pyrrole (gas-mw), 20, 602
$C_4H_5N_3O_2$, 2-Methyl-4-nitro-imidazole, 46B, 199
$C_4H_6Cl_3NO$, α-Trichloro-N-ethyleniminocarbinol, 30B, 122; 32B, 153
$C_4H_6N_2O$, 3-Methyl-3-pyrazolin-5-one, 37B, 115
$C_4H_6N_2O_2$, 1,3-Dimethyldiazetidin-2,4-dione, 43B, 261
$C_4H_6N_4O_3$, Allantoin, 30B, 125
$C_4H_6N_4O_4 \cdot C_3H_8O$, 5-Amino-1H-imidazole-4-carboxamide 2-propanol solvate, 46B, 199
$C_4H_6N_4O_4 \cdot H_2O$, 4-Amino-1H-imidazole-5-carboxamide hydrate, 46B, 199
C_4H_7NO, N-Acetylethylenimine (gas-ed), 33B, 531
$C_4H_7NO_2$, L-Azetidine-2-carboxylic acid, 34B, 124
$C_4H_7N_3S$, 1,3-Dimethyl-4-(1,2,3-triazolio)sulfide, 41B, 244
$C_4H_7N_3S$, 2,3-Dimethyl-4-(1,2,3-triazolio)sulfide, 45B, 196
$C_4H_7N_3S_2$, 1-Thiocarbamoylimidazolidine-2-thione, 35B, 155
$C_4H_7N_5O_3S$, 5-Mesylamino-2H-1,2,3-triazole-4-carbonitrile monohydrate, 42B, 167
$C_4H_8Br_2ClN$, cis-3,4-Dibromopyrrolidine hydrochloride, 46B, 200
$C_4H_8N_2$, 1,1'-Biaziridyl (gas-ed), 38B, 1061
$C_4H_8N_2O_2 \cdot 3\ H_2O$, DL-3-Amino-1-hydroxy-2-pyrrolidone trihydrate, 46B, 200
C_4H_9N, 2,2-Dimethylethyleneimine (gas-ed), 40B, 1143
$C_4H_9N_4O_5P$, 5-Amino-1H,3H-imidazolium-4-carboxamide dihydrogen phosphate, 46B, 199
$C_5H_5BrN_2O$, 1-Acetyl-4-bromopyrazole, 38B, 243
$C_5H_6BrN_3O$, 3-Methyl-4-bromo-5-carboxamide-pyrazole, 35B, 157
$C_5H_6N_4O_5 \cdot 0.5\ H_2O$, 4-Hydroxy-2,5-dioxo-4-imidazolidinecarboxyureide hemihydrate, 46B, 201
$C_5H_7ClN_2O_2$, Imidazole-4-acetic acid hydrochloride, 42B, 168
C_5H_7N, N-Methylpyrrole (gas), 27, 759
$C_5H_7N_3OS$, 1,3-Dimethyl-4-imino-5-oxoimidazolidine-2-thione, 42B, 168
$C_5H_8Br_2N_2S$, S,S-Dibromo-1,3-dimethylimidazoline-2-thione, 43B, 262

C_5H_8INO, 4-Iodomethylpyrrolid-2-one, 39B, 154

$C_5H_8Li_2N_3O_5P \cdot 2\ H_2O$, Lithium 1-carboxymethyl-2-imino-3-phosphonoimidazolidene dihydrate, 45B, 197

$C_5H_8N_2S$, 1,3-Dimethylimidazole-2(3H)-thione, 38B, 244

$C_5H_8N_4O_2$, 5-Amino-4-ethoxycarbonyl-1H-1,2,3-triazole, 43B, 263

$C_5H_9Cl_3N_2O$, 1-(α-Hydroxy-β-trichloroethyl)-3,3-dimethyldiaziridine, 37B, 116

$C_5H_9N_3$, Histamine, 39B, 154

$C_5H_{10}BrN_3$, Histamine monohydrobromide, 40B, 197

$C_5H_{11}Cl_2N_3$, Histamine dihydrochloride, 27, 992

$C_5H_{11}Cl_4CoN_3$, Histamine tetrachlorocobaltate(II), 38B, 245

$C_5H_{13}N_3O_5S$, Histamine sulfate monohydrate, 39B, 155

$C_5H_{17}N_3O_9P_2$, Histamine diphosphate monohydrate, 40B, 1083

$C_6H_5N_5$, 3-Amino-4,5-dicyano-1-methylpyrazole, 42B, 169

$C_6H_6N_2O_2 \cdot 2\ H_2O$, Urocanic acid dihydrate, 45B, 198

$C_6H_8BrN_3S$, S-Cyano-1,3-dimethylimidazoline-2-thionium bromide, 43B, 262

$C_6H_8N_4O_4$, 1,4-Dinitro-2-isopropylimidazole, 38B, 245

$C_6H_9ClN_4O_4$, Imidazole imidazolium perchlorate, 42B, 169

C_6H_9NS, N-Vinyl-2-thiopyrrolidone, 42B, 170

$C_6H_9N_3O_3$, 2-(2-Methyl-5-nitro-1-imidazolyl)ethanol, 45B, 198

$C_6H_9N_5O_2S$, 2-Methyl-5-(methyl(mesyl)amino)-2H-1,2,3-triazole-4-carbonitrile, 43B, 263

$C_6H_{10}N_2O_2S$, Methyl 3-methyl-3-thioimidazolidine-1-carboxylate, 44B, 188

$C_6H_{10}N_2O_4$, 4-Imidazoleacrylic acid dihydrate, 43B, 264

$C_6H_{10}N_6O$, 5-(3,3-Dimethyl-1-triazenyl)imidazole-4-carboxamide, 45B, 199

$C_6H_{12}N_2O_2$, 3,3,4,4-Tetramethyldiazetine 1,2-dioxide, 44B, 188

$C_6H_{13}ClN_6O_2$, 5-(3,3-Dimethyl-1-triazeno)imidazole-4-carboxamide hydrochloride monohydrate, 40B, 198

$C_6H_{13}NO_4$, 2,5-Dihydroxymethyl-3,4-dihydroxypyrrolidine, 43B, 264

$C_6H_{13}N_3O$, 1,4-Dimethyl-5-ethyl-5-hydroxy-δ2-1,2,3-triazoline, 39B, 156

$C_6H_{14}N_4O_6S$, Imidazolium sulfate dihydrate, 41B, 244

$C_7H_3F_{12}N_3$, 4-Amino-2,2,5,5-tetrakis(trifluoromethyl)-3-imidazoline, 39B, 157

$C_7H_6N_4O$, 1,1'-Carbonyldipyrazole, 43B, 265

$C_7H_8N_2O_4$, Imidazolium hydrogen maleate, 42B, 170; 46B, 201

$C_7H_8N_2O_4$, Imidazolium maleate, 42B, 170

$C_7H_8N_4O$, Cyano(2,2-dimethyl-1,2,4-triazole-3-ylio)formylmethanide, 44B, 189

$C_7H_9NO_2 \cdot 0.33\ H_2O$, 4-Acetyl-3-hydroxymethylpyrrole hydrate, 44B, 189

$C_7H_9NO_2$, 4-Acetyl-3-hydroxymethylpyrrole, 44B, 189

$C_7H_{10}N_4O_4$, 1-Methylimidazolium oxalurate, 45B, 199

$C_7H_{12}N_2O_2$, 4,4-Diethyl-3,5-dioxo-1,2-pyrazolidine, 40B, 199

$C_7H_{13}BrN_2O_2$, 3-Methoxycarbonyl-trans-3,5-dimethyl-Δ^1-pyrazoline hydrobromide, 30B, 123

$C_7H_{13}NO_2$, 5-Hydroxy-2,4,4-trimethyl-1-pyrroline-1-oxide, 42B, 171

$C_7H_{14}N_2O$, 3-Isopropyl-5,5-dimethyl-2-thiohydantoin, 43B, 266

$C_8H_6N_4O_6$, 1-Ethyleneimino-2,4,6-trinitrobenzene, 43B, 266

$C_8H_7N_3O_2$, 3-(2-Aminophenyl)sydnone, 45B, 200

$C_8H_8N_2O_4S_2$, N,N'-Dithiodisuccinimide, 40B, 200

$C_8H_9F_6N_5$, 1-(4-Methyl-1-pyrazolin-3-yl)-5,5-bis(trifluoromethyl)-Δ^2-1,2,3-triazoline, 39B, 158

C_8H_9NO, (E)-2-Methyl-3-(pyrrol-2-yl)prop-2-enal, 43B, 267

$C_8H_9N_5$, 4-Amino-3-(2-aminophenyl)-4H-1,2,4-triazole, 42B, 172
$C_8H_{10}N_4O$, 3-Hydroxy-5,5'-dimethyl-4,4'-bi-1H-pyrazole, 40B, 201
$C_8H_{12}Cl_2N_6O$, 1-(2-Chloroethyl)-3-(4-carbamoylpyrazol-3-yl)-Δ²-1,2,3-triazolium chloride, 38B, 246
$C_8H_{12}N_4O$, 3-Methyl-3-(5'-amino-3'-methyl-pyrazol-1-yl)acrylonitrile monohydrate, 41B, 246
$C_8H_{12}N_4O_3S \cdot H_2O$, O-Methyl[2-(2-methyl-5-nitro-1H-imidazol-1-yl)ethyl]thiocarbamate monohydrate, 45B, 202
$C_8H_{13}NO_2S_2$, N-(t-Butyldithio)succinimide, 46B, 202
$C_8H_{14}N_4OS$, 4-(2-Morpholinoethyl)-1,2,4-triazoline-5-thione, 43B, 267
$C_8H_{14}N_4S_2$, N-(2-((Imidazol-4-yl)methylthio)ethyl)-N'-methylthiourea, 43B, 271
$C_8H_{15}N_2O_2$, 2,2,5,5-Tetramethyl-1-aza-3-cyclopentanone-3-oxime-1-oxyl, 41B, 246
$C_8H_{16}Cl_2N_4O$, 1-(2-Morpholinoethyl)-1,2,4-triazole dihydrochloride, 43B, 268
$C_8H_{16}Cl_4CuN_6O_2$, Creatininium tetrachlorocuprate(II), 45B, 202
$C_8H_{16}NO_2$, (+)-2,2,5,5-Tetramethyl-3-hydroxy-pyrrolidine N-oxide, 42B, 172
$C_8H_{16}NO_2$, DL-2,2,5,5-Tetramethyl-3-hydroxy-pyrrolidine N-oxide, 42B, 172
$C_8H_{16}N_2O_2$, meso-1,4-Diaziridinyl-2,3-butanediol, 26, 784
$C_9H_5BrN_4O_4$, 1-(2',4'-Dinitrophenyl)-4-bromopyrazole, 34B, 126
$C_9H_5ClN_4O_4$, 1-(2',4'-Dinitrophenyl)-4-chloropyrazole, 35B, 158
$C_9H_7N_5$, 4-Phenyl-3(5)azidopyrazole, 40B, 201
C_9H_8BrN, p-Bromobenzoylethylenimine, 32B, 154
C_9H_8BrNO, 1-(2-Bromophenyl)azetidin-2-one, 43B, 268
C_9H_8BrNO, 1-(4-Bromophenyl)azetidin-2-one, 39B, 158
$C_9H_8BrN_3O$, 3-p-Bromophenyl-1-nitroso-2-pyrazoline, 37B, 117
$C_9H_8N_4O_5$, 2,4-Dinitro-5-ethyleneiminobenzamide, 41B, 247
$C_9H_8N_4S$, 4-Benzylideneamino-5-mercapto-1,2,4-triazole, 39B, 159
$C_9H_9ClN_4O \cdot CH_4O$, 5-(4-Chlorophenylamino)-2,4-dihydro-4-methyl-3H-1,2,4-triazol-3-one methanol solvate, 46B, 203
$C_9H_9Cl_3N_6 \cdot 3\ H_2O$, 4-Amino-3-(2,6-dichlorobenzylidenehydrazino)-1,2,4-triazole hydrochloride trihydrate, 45B, 203
$C_9H_9N_3$, 1-(2-Pyridyl)-5-methylimidazole, 43B, 269
$C_9H_9N_3OS$, 3-(2-Hydroxyphenyl)-4-methyl-1,2,4-Δ²-triazoline-5-thione, 45B, 204
$C_9H_9N_3S$, 1-Methyl-3-phenyl-4-(1,2,3-triazolio)sulfide, 45B, 204
$C_9H_9N_3S$, 1-Phenyl-3-methyl-4-(1,2,3-triazolio)sulfide, 45B, 204
$C_9H_9N_3S$, 2-Phenyl-3-methyl-4-(1,2,3-triazolio)sulfide, 45B, 196
$C_9H_{10}Cl_3N_3$, 2-(2,6-Dichlorophenylimino)-2-imidazole hydrochloride, 45B, 204
$C_9H_{10}N_4O_2$, 5-Amino-4-(2-methoxyphenyl)-2,4-dihydro-3H-1,2,4-triazol-3-one, 46B, 203
$C_9H_{10}N_4S$, 4,5-Dicyano-2-imidazolyl(diethyl)sulphonium ylide, 44B, 190
$C_9H_{10}N_6OS$, 4-Amino-3-(β-benzoylhydrazino)-5-mercapto-1,2,4-triazole, 39B, 160
$C_9H_{12}Cl_2N_3O_4P$, 2-(2,6-Dichlorophenylamino)imidazolin phosphate, 45B, 205
$C_9H_{13}KNO_3 \cdot 0.3\ H_2O$, Potassium 2,2,5,5-tetramethyl-3-carboxypyrroline-1-oxyl hydrate, 35B, 159
$C_9H_{13}NO_3$, 3-Acetyl-5-isopropylpyrrolidine-2,4-dione, 46B, 203
$C_9H_{14}BrN_3O$, 4-Acetylamino-2-bromo-5-isopropyl-1-methylimidazole, 37B, 117
$C_9H_{14}N_4O_3S$, O-Methyl(2-(2-ethyl-5-nitroimidazol-1-yl)ethyl)thiocar-

bamate, 43B, 270
$C_9H_{15}BrN_4O_4$, 1,2-Bis(methoxycarbonylamino)-3,5-dimethylpyrazolium bromide, 41B, 248
$C_9H_{15}NO_4$, N-(t-Butyloxycarbonyl)-L-azetidine-2-carboxylic acid, 41B, 249
$C_9H_{15}N_2O_2$, 2,2,5,5-Tetramethyl-3-carbamidopyrroline-1-oxyl, 38B, 109
$C_9H_{16}NO_3$, R-(+)-3-Carboxy-2,2,5,5-tetramethyl-1-pyrrolidinyloxy, 40B, 96
$C_9H_{16}N_4O_4$, 1,1,1-Trimethylhydrazinium 3-carbomethoxy-5-pyrazolecarboxylate, 40B, 68
$C_9H_{16}N_4S$, N-(4-Imidazol-4-ylbutyl)-N'-methylthiourea, 39B, 160
$C_9H_{16}N_4S_2$, N-Methyl-N'-(2-((5-methylimidazol-4-yl)methylthio)ethyl)-thiourea, 43B, 271
$C_9H_{16}N_4S_2$, 1,1'-Trimethylenedi-2-imidazolidinethione, 46B, 204
$C_9H_{17}N_2O_2$, 2,2,5,5-Tetramethylpyrrolidine-3-carboxamide-1-oxyl (optically active form), 41B, 250
$C_9H_{17}N_2O_2$, 2,2,5,5-Tetramethylpyrrolidine-3-carboxamide-1-oxyl (racemic form), 41B, 250
$C_9H_{18}NO_2$, D(+)-2,2,5,5-Tetramethyl-3-hydroxymethyl-pyrrolidine-1-oxyl, 44B, 191
$C_9H_{18}N_2S_2$, Pyrrolidinium 1-pyrrolidinecarbodithioate, 45B, 226
$C_9H_{21}NO_4S$, N-Methyl-N-t-butyl-3-hydroxyazetidinium methanesulphonate, 34B, 125
$C_{10}H_4Br_5NO$, 2-(2'-Hydroxy-3',5'-dibromophenyl)-3,4,5-tribromopyrrole (form 2), 31B, 117
$C_{10}H_5Cl_2FN_2O_2$, 4'-Fluoropyrrolnitrin, 33B, 118
$C_{10}H_8BrNO_2$, p-Bromophenyl-N-succinimide, 26, 720
$C_{10}H_8BrN_3O_2$, 1-p-Nitrophenyl-3-methyl-4-bromopyrazole, 38B, 247
$C_{10}H_9NO_2$, 3-Phenylpyrrolidine-2,5-dione, 39B, 161
$C_{10}H_9N_3O$, 1-(2-Formamidophenyl)imidazole, 44B, 191
$C_{10}H_{10}BrNO$, 1-(2-Bromophenyl)pyrrolidin-2-one, 43B, 268
$C_{10}H_{10}N_2$, 3-Methyl-5-phenylpyrazole, 40B, 202; 41B, 251
$C_{10}H_{10}N_2O$, 1-Phenyl-3-methyl-5-pyrazolone, 39B, 162
$C_{10}H_{10}N_2O$, 1-Phenyl-5-methyl-3-pyrazolone, 39B, 162
$C_{10}H_{10}N_2O_2$, 3-(N-Phenyl)aminopyrrolidine-2,5-dione, 39B, 161
$C_{10}H_{10}N_2O_2S$, N-Tosylimidazole, 43B, 271
$C_{10}H_{11}Cl_3N_6$, 4-Amino-3-(2,6-dichlorobenzylidenehydrazino)-5-methyl-1,2,4-triazole hydrochloride, 43B, 272
$C_{10}H_{11}N_3S$, 1-Methyl-3-benzyl-4-(1,2,3-triazolio)sulfide, 45B, 205
$C_{10}H_{12}BrNO_3$, α-(3-Bromo-4-ethoxyphenyl)succinimide, 46B, 205
$C_{10}H_{12}ClN_5$, 5-((3-Chlorobenzyl)dimethylammonio)tetrazolide, 41B, 252
$C_{10}H_{13}NO$, 2-Hydroxy-2-phenyl-1-aziridinoethane, 41B, 253
$C_{10}H_{13}NO_2$, 2,5-dimethyl-3,4-diacetylpyrrole, 39B, 163
$C_{10}H_{13}NO_3$, 3-Ethoxycarbonyl-1,2-dimethyl-4-pyrrolecarbaldehyde, 45B, 206
$C_{10}H_{14}ClN_4O_3$, 2-(2-Chloro-4-methylphenylamino)imidazoline nitrate, 45B, 206
$C_{10}H_{14}ClN_5O_2$, 1-(4-Chlorobenzyl)-1-nitroso-2-(4,5-dihydro-2-imidazolyl)-hydrazine monohydrate, 30B, 124
$C_{10}H_{15}Cl_3N_5O_2$, 4-Amino-3-(2,6-dichlorobenzylidenehydrazono)-2-methyl-1,2,4-triazoline hydrochloride dihydrate, 43B, 272
$C_{10}H_{16}NO_3$, 2,2,5,5-Tetramethyl-4-methoxycarbonylpyrroline-1-oxyl, 44B, 192
$C_{10}H_{16}N_6S \cdot H_2O$, 2-Cyano-1-methyl-3-{2-[(5-methyl-1H-imidazol-4-yl)-methylthio]ethyl}guanidine monohydrate, 46B, 205
$C_{10}H_{16}N_6S$, N''-Cyano-N-methyl-N'-(2-((5-methyl-1H-imidazol-4-yl)-methylthio)ethyl)guanidine, 44B, 192

$C_{10}H_{18}NO_2$, Spiro(cyclohexane-1,2'-(dimethyl-4',4'-oxazolidine)), 40B, 202
$C_{10}H_{19}N_3S \cdot CHCl_3$, 5-Dimethylimonio-1-isopropyl-4,4-dimethyl-2-imidazolin-2-thiolate chloroform solvate, 45B, 207
$C_{10}H_{20}N_6O$, 5-Amino-n-octyl-1H-tetrazole-1-carboxamide, 46B, 389
$C_{11}H_5BrN_4$, 4,5-Dicyano-2-imidazolyl(phenyl)bromonium ylide, 41B, 255
$C_{11}H_8ClNO$, 2-(4-Chlorobenzoyl)pyrrole, 46B, 206
$C_{11}H_9NO$, 2-Benzoylpyrrole, 46B, 206
$C_{11}H_{10}N_2O_4$, 1-Methyl-3-(2'-nitrophenyl)pyrrolidine-2,5-dione, 44B, 199
$C_{11}H_{10}N_6$, 3'-Methyl-3-phenylbistriazolyl, 43B, 273
$C_{11}H_{11}BrNO_2$, 4-Bromo-2,3-dimethyl-1-phenyl-5-pyrazolone, 22, 785
$C_{11}H_{12}BrNO$, cis-4-(p-Bromophenyl)-5-methyl-2-pyrrolidinone, 44B, 192
$C_{11}H_{12}Cl_2N_2O$, α-2-(2,6-Dichlorophenoxyethyl)imidazolidine, 46B, 206
$C_{11}H_{12}N_2O$, Antipyrine, 39B, 164
$C_{11}H_{12}N_2O$, 1,1-Dimethyl-3-phenylpyrazolium-5-oxide, 35B, 161
$C_{11}H_{12}N_2O$, 3,4-Dimethyl-2,2'-pyrromethen-5(1H)-one, 45B, 208
$C_{11}H_{12}N_2OS$, 2-Oxo-1-(phenylaminothiocarbonyl)pyrrolidine, 39B, 165
$C_{11}H_{13}ClN_2O_3$, p-Chlorobenzylammonium N-hydroxysuccinimide, 43B, 273
$C_{11}H_{13}Cl_3N_2O$, α-2-(2,6-Dichlorophenoxyethyl)imidazolidine hydrochloride, 46B, 207
$C_{11}H_{14}ClNO_2$, Pyrrolidinium p-chlorobenzoate, 44B, 193
$C_{11}H_{14}N_2O_3$, 3-(N-Benzil)aminopyrrolidine-2,5-dione monohydrate, 40B, 203
$C_{11}H_{14}N_2O_6$, DL-Methylene-bis(N-pyrrolid-2-one-5-carboxylic) acid, 41B, 256
$C_{11}H_{14}N_4O_2$, 4-(5-Hydroxy-1,3-dimethyl-4-pyrazolylmethylene)-1,3-dimethyl-2-pyrazolin-5-one, 46B, 207
$C_{11}H_{14}N_4O_2$, 5-(1-Perhydroxy-1-isopropyl)-1-p-tolyl-1,2,3,4-tetrazole, 44B, 194
$C_{11}H_{15}N_5$, 5-(1-Amino-1-methylethyl)-1-p-tolyl-1,2,3,4-tetrazole, 44B, 194
$C_{11}H_{19}ClN_2O_4$, Dipyrrolidyl-trimethinecyanine perchlorate, 41B, 257
$C_{11}H_{19}IN_2O$, Trimethyl-(4-2-oxopyrrolidin-1-yl)but-2-ynyl)ammonium iodide, 39B, 165
$C_{11}H_{20}N_2$, 4,5-Di-t-butylimidazole, 37B, 118
$C_{12}H_9BrN_2O_3S$, Methyl (1-(p-bromophenyl)-5-oxo-2-thioxoimidazolidine-4-ylidene)acetate, 44B, 194
$C_{12}H_9ClN_2O_2$, α-p-Chlorophenyl-α-methyl-α'-cyanosuccinimide, 37B, 119
$C_{12}H_{10}N_2O_3$, 3-Cyano-2-phenylglyoxyl-N-methoxy-aziridine(E), 41B, 257
$C_{12}H_{10}N_4O_2S$, 1-(4-Imidazolylsulphonyl)-4-phenylimidazole, 41B, 258
$C_{12}H_{11}NO_2$, 2-(4-Methoxybenzoyl)pyrrole, 46B, 206
$C_{12}H_{12}N_2O$, 4-Acetyl-5-methyl-2-phenylimidazole, 44B, 195
$C_{12}H_{12}N_2O_4$, 1,3-Dimethyl-3-(2'-nitrophenyl)pyrrolidine-2,5-dione, 44B, 199
$C_{12}H_{13}N_5$, 1-Phenyl-3-dimethylamino-4-cyano-5-aminopyrazole, 43B, 274
$C_{12}H_{14}N_2O$, 1',3,4-Trimethyl-2,2'-pyrromethen-5(1H)-one, 45B, 208
$C_{12}H_{14}N_2O_2$, Methoin, 46B, 208
$C_{12}H_{14}N_2O_2S$, 3-Ethyl-5-p-hydroxybenzyl-2-thiohydantoin, 45B, 214
$C_{12}H_{16}BrNO_2S$, trans-(2R,5R)-1-(p-Bromophenylsulphonyl)-2,5-dimethylpyrrolidine, 46B, 208
$C_{12}H_{17}NO_2$, Pyrrolidinium p-toluate, 44B, 193
$C_{12}H_{17}NO_3$, Ethyl 4-acetyl-3-ethyl-5-methylpyrrole-2-carboxylate, 38B, 247
$C_{12}H_{17}N_3O_3$, 2,5-Bis(pyrrolidin-2-on-5-yl)pyrrole monohydrate, 40B, 203
$C_{12}H_{20}N_4$, 1,4-Bis(3,3-dimethylazirinyl)piperazine, 46B, 209

$C_{12}H_{22}N_2O_2$, 2,2-Pentamethylene-4,4,5,5-tetramethylimidazolidine-1,3-dioxyl, 44B, 195
$C_{13}H_{10}N_4O$, 2,3-Diphenyltetrazolium-5-olate, 45B, 200
$C_{13}H_{10}N_4O$, 3-Oxo-1,5-diphenyltetrazolium, 43B, 275
$C_{13}H_{10}N_4S$, Anhydro-5-mercapto-2,3-diphenyltetrazolium hydroxide, 35B, 162
$C_{13}H_{10}N_4S$, 1,3-Diphenyltetrazolium-5-thiolate, 45B, 200
$C_{13}H_{11}N_3O_2$, N-(2-Imidazol-4-ylethyl)phthalimide, 43B, 275
$C_{13}H_{12}Cl_2N_6O_2$, 4-Acetoamido-3-(1-acetyl-2-(2,6-dichlorobenzylidene)-hydrazino)-1,2,4-triazole, 42B, 44
$C_{13}H_{12}N_2O_4$, 1-(2'-Carboxyphenyl)-2-acetyl-5-methyl-3-pyrazolone, 38B, 248
$C_{13}H_{14}N_2O_3$, N-2-(6'-Methyl)pyridyl-4-ethoxycarbonyl-2H,3H-pyrrole-2-one, 42B, 173
$C_{13}H_{15}NO$, Cinnamic acid pyrrolidid, 40B, 204
$C_{13}H_{16}BrN_3O$, 1-(4-Bromophenyl)-4-dimethylamino-2,3-dimethyl-3-pyrazolin-5-one, 46B, 209
$C_{13}H_{16}N_3NaO_4S \cdot H_2O$, Metamizol monohydrate, 45B, 208
$C_{13}H_{17}BrN_2$, 1-Cyclohexyl-3-(p-bromophenyl)diaziridine, 39B, 166
$C_{13}H_{17}N_2O_2$, 2,2,5,5-Tetramethyl-4-phenyl-3-imidazoline-3-oxide-1-oxyl, 45B, 209
$C_{13}H_{17}N_3O$, Amidopyrine, 41B, 259; 42B, 173
$C_{13}H_{17}N_3O_2$, 3-Phenoxy-3-dimethylcarbamoyldimethylamino-2-azirine, 40B, 205
$C_{13}H_{18}N_2O_2$, 2-Hydroxy-3-phenyl-3-pyrrolidinylpropionamide, 44B, 196
$C_{13}H_{20}BrN$, 1-Benzyl-1,2,2-trimethylazetidinium bromide, 33B, 109
$C_{13}H_{20}ClNO$, 1-(1-Methyl-2-phenylethyl)-2-methyl-3-hydroxyazetidinium hydrochloride, 40B, 205
$C_{13}H_{20}ClN_5O$, 3,5-Bis(dimethylamino)-1-methyl-4-(o-hydroxyphenyl)-1,2,4-triazolium chloride, 41B, 259
$C_{13}H_{20}NI$, 1-Benzyl-1,3,3-trimethylazetidinium iodide, 37B, 121
$C_{13}H_{20}N_4O_2$, 4,4'-Methylenebis(1,3,5-trimethyl-4-imidazolin-2-one), 46B, 210
$C_{13}H_{24}N_4O$, 2-Ethyl-2-(5'-dimethylamino-4',4'-dimethyl-4'H-imidazol-2'-yl)butyramide, 45B, 210
$C_{14}H_{10}N_2O_2$, Phenyl isocyanate dimer, 40B, 206
$C_{14}H_{13}O_3NCl_2$, O,O',N-Trimethylpyoluteorin, 35B, 163
$C_{14}H_{14}BrN_3O$, 1,3-Dimethyl-2-(p-bromobenzoyl-cyanomethylene)-imidazolidine, 40B, 53
$C_{14}H_{18}N_2O_2$, 2,5-Bis(pyrrolidino)-1,4-benzoquinone, 45B, 210
$C_{14}H_{18}N_2O_3$, E-5'-Ethoxycarbonyl-3,4-dihydro-3',4'-dimethyl-5(1H)-2,2'-pyrromethenone, 43B, 276
$C_{14}H_{18}N_2O_3$, Z-5'-Ethoxycarbonyl-3,4-dihydro-3',4'-dimethyl-5(1H)-2,2'-pyrromethenone, 43B, 276
$C_{14}H_{19}N_3O_2S$, N-(1,2,3,5-Tetramethyl-4-pyrazolio)toluene-p-sulphonamidate, 43B, 278
$C_{14}H_{20}IN_3O$, 1-Phenyl-2,3-dimethyl-4-trimethylammonio-pyrazole-5-one iodide, 46B, 210
$C_{14}H_{21}NO_3$, trans-(2R,5R)-2,5-Dimethylpyrrolidinium (S)-mandelate, 46B, 211
$C_{15}H_6F_{12}N_6$, Phenylbis[3,5-bis(trifluoromethyl)-1,2,4-triazol-1-yl]-methane, 46B, 212
$C_{15}H_{12}N_2O_2$, Diphenylhydantoin, 37B, 121
$C_{15}H_{13}ClN_2O_4$, 1,3-Diphenyl-imidazole perchlorate, 42B, 174
$C_{15}H_{14}N_2$, Diphenyl-Δ^2-pyrazole, 33B, 110
$C_{15}H_{16}N_6O_4$, 4-(1'-Methyl-4'-nitropyrrole-2'-carbamoyl)pyrrole-2-(N-2"-cyanoethylcarbamide), 44B, 196

$C_{15}H_{18}N_2O_4$, 1-(p-Nitrophenyl)-2-(2,5-dimethylpyrryl)-1,3-propanediol, 45B, 211
$C_{15}H_{19}Br_3N_2$, 5,5'-Dibromo-3,3'-diethyl-4,4'-dimethyl-2,2'-pyrromethene hydrobromide, 44B, 197
$C_{15}H_{20}BrN_2O_2$, 1-Oxyl-2,2,5,5-tetramethylpyrrolidine-2-hydroxy-5-bromosalicylidene-imine, 44B, 198
$C_{15}H_{20}N_2O_5S$, 1-p-Methoxyphenyl-3-methyl-4-(D-arabinotetrahydroxybutyl)imidazoline-2-thione, 42B, 175
$C_{15}H_{21}N_2O_7$, Oxotremorine sesquioxalate, 41B, 260
$C_{15}H_{22}ClNO_2$, 1,2-Dimethyl-3-phenyl-3-pyrrolidynyl propionate hydrochloride, 43B, 278
$C_{15}H_{22}N_2O_9$, (3SR,4RS,5SR)-2-Acetyl-5-(methoxycarbonylmethyl)-1-methyl-3,4,5-tris(methoxycarbonyl)pyrazoline, 45B, 211
$C_{16}H_{12}N_2O$, 4-Benzylidene-2-phenyl-2-imidazoline-5-one, 46B, 212
$C_{16}H_{13}Cl_2N_3O \cdot CH_4O$, rac-2-Methylamino-4,5-bis(p-chlorophenyl)-4-hydroxy-4H-imidazole methanol solvate, 45B, 212
$C_{16}H_{14}N_2O_2$, 5-Benzyl-5-hydroxy-2-phenyl-imidazolin-4-one, 45B, 213
$C_{16}H_{14}N_2S_2$, 1,3-Diphenylimidazolinium-2-dithiocarboxylate, 46B, 212
$C_{16}H_{15}BrN_2O$, 2-(1-Imidazolin-2-yl)benzophenone hydrobromide, 33B, 111
$C_{16}H_{16}N_6$, 5-Dicyanomethylene-2,4-di(1-pyrrolidinyl)-3-pyrrolecarbonitrile, 45B, 213
$C_{16}H_{17}ClN_2OS$, 3-Chloro-3-cyano-1-cyclohexyl-4-(phenylthio)-2-azetidinone, 46B, 213
$C_{16}H_{17}NO$, 1-(Diphenylmethyl)azetidin-3-ol, 43B, 279
$C_{16}H_{19}NO_2$, Enamine 4, 44B, 198
$C_{16}H_{20}N_2O_3$, 1-Morpholinomethyl-3-methyl-3-phenylpyrrolidin-2,5-dione, 39B, 166
$C_{16}H_{21}N_2O_3$, 1-Ethyl-5'-ethoxycarbonyl-3,4-dihydro-3',4'-dimethyl-5(1H)-2,2'-pyrromethenone, 43B, 276
$C_{16}H_{21}N_2O_3$, 5-Ethoxy-5'-ethoxycarbonyl-3,4-dihydro-3',4'-dimethyl-2,2'-pyrromethene, 43B, 276
$C_{16}H_{22}N_2O_2$, 2,5-Bis(piperidino)-1,4-benzoquinone, 45B, 210
$C_{16}H_{26}N_2$, N-(1'-t-Butyl-spiro(adamantane-2,2'-aziridine)-3'-ylidene)methylamine, 46B, 213
$C_{16}H_{28}N_4O_2$, Tetramethyl-2,2,5,5-aza-1-cyclopentanone-3-azine-3-oxyl, 38B, 112
$C_{17}H_{14}Cl_2N_2O$, 1-(Bis-p-chlorophenyl)methylene-3-oxo-4,4-dimethyl-1,2-diazetidinium hydroxide inner salt, 40B, 207
$C_{17}H_{14}N_2OS$, 5-Benzylidene-3-methyl-1-phenyl-1,3-diaza-2-cyclopentanone-4-thione, 46B, 214
$C_{17}H_{15}ClN_2O_2$, 5-p-Chlorobenzyl-3-(o-tolyl)hydantoin, 45B, 214
$C_{17}H_{16}Br_2N_4O_3$, 1,3-Bis(4-(2-bromoethoxy)phenyl)-5-tetrazolone, 38B, 249
$C_{17}H_{16}N_2O_2$, cis-3,4-Diphenyl-3-(methoxycarbonyl)-1-pyrazoline, 41B, 261
$C_{17}H_{16}N_2O_2$, 4,5-Diphenyl-5-methoxycarbonyl-2(E)-pyrazoline, 43B, 279
$C_{17}H_{17}BrN_2O_2$, 3-Bromo-4,5-dihydro-5-hydroperoxy-4,4-dimethyl-3,5-diphenyl-3H-pyrazole, 46B, 214
$C_{17}H_{20}N_2O_8$, Diethyl cis-1-formyl-5-hydroxy-4-(2-nitrophenyl)pyrrolidine-2,2-dicarboxylate, 45B, 74
$C_{17}H_{21}NO_2$, Enamine 6, 44B, 198
$C_{17}H_{22}Br_2N_2$, (3S,5R)-2,2-Dimethyl-5-phenyl-3-(4-pyridyl)pyrrolidine dihydrobromide, 46B, 215
$C_{17}H_{22}N_2O_2$, 4-Ethyl-1-isopropyl-3-[(phenylcarbamoyl)methyl]-3-pyrrolin-2-one, 45B, 215
$C_{17}H_{24}N_2O$, 5'-Oxo-3',4,4'-triethyl-3,5-dimethyl-1',5'-dihydro-

(2,2')dipyrromethene, 43B, 280
$C_{18}H_{14}Cl_4N_2O \cdot 0.5\ H_2O$, 1-{2,4-Dichloro-β-[(2,4-dichlorobenzyl)oxy]-phenethyl}imidazole hemihydrate, 45B, 215
$C_{18}H_{16}N_4O_2$, 1-(α-Benzoyloxy-benzylideneamino)-4,5-dimethyl-1,2,3-triazole, 38B, 250
$C_{18}H_{17}NO_2$, 1-Methyl-3-phenyl-3-benzylpyrrolidine-2,5-dione, 44B, 199
$C_{18}H_{17}N_3OS$, 4-(2-Methylmercapto-anilinomethylene)-3-methyl-1-phenyl-5-pyrazolone, 43B, 281
$C_{18}H_{17}N_3O_4$, trans-3-Phenyl-4-p-nitrophenyl-3-carbomethoxy-anti-5-methyl-1-pyrazoline, 41B, 262
$C_{18}H_{18}ClNO$, N-(p-Chlorophenyl)-α-isopropyl-β-phenyl-β-lactam, 35B, 149
$C_{18}H_{18}N_2O_2$, cis-3,4-Diphenyl-3-(methoxycarbonyl)-syn-5-methyl-1-pyrazoline, 41B, 262
$C_{18}H_{18}N_2O_3$, trans-3-Phenyl-3-methoxycarbonyl-4-p-methoxyphenyl-1-pyrazoline, 38B, 250
$C_{18}H_{19}NO_2$, 1-Methyl-4-trans-phenyl-4-benzyl-5-hydroxy-pyrrolidine-2-one, 44B, 199
$C_{18}H_{19}NO_2$, 4-Hydroxy-4,5-dimethyl-3,5-diphenylpyrrolidin-2-one, 40B, 207
$C_{18}H_{22}N_2O_6S_2$, (DL)-4β-(2,2'-Trimethylenedithioethyl)-3α-((1'S*)-p-nitrobenzyloxycarbonyloxyethyl)-2-azetidinone, 46B, 215
$C_{18}H_{25}N_2O$, Enamine 7, 44B, 198
$C_{18}H_{25}N_6O_2S_2$, 1-t-Butyl-5-(2-(p-tolylsulphonylamino)-1-(t-butylimino)-2-thioxo-ethyl)-1H-tetrazole, 45B, 216
$C_{18}H_{29}N_3$, N-(1,2-Di-t-butyldiaziridin-3-ylidene)-2,4,6-trimethylaniline, 42B, 175
$C_{19}H_{13}N_5O_2$, 4-(5-Hydroxy-1,3-dimethyl-4-pyrazolylimino)-1,3-dimethyl-2-pyrazolin-5-one, 46B, 207
$C_{19}H_{19}Cl_3N_2O_4$, 2,2,2-Trichloroethyl 2-(2-benzyl-4-methoxycarbonyl-1-imidazolyl)-3-methylisocrotonate, 46B, 216
$C_{19}H_{20}Br_2N_2O$, Bis(p-bromo-α,α-dimethylbenzyl)diaziridinone, 44B, 200
$C_{19}H_{20}N_2O_2$, Phenylbutazone, 41B, 273; 43B, 281
$C_{19}H_{23}N_3O$, 1-n-Butyl-4,5-diphenyl-5-methyl-1,2,4-triazolidin-3-one, 45B, 217
$C_{19}H_{24}BrN$, 1,1-Dibenzyl-3,3-dimethylazetidinium bromide, 34B, 127
$C_{19}H_{25}ClN_2O$, Triprolidine hydrochloride monohydrate, 40B, 208
$C_{19}H_{26}N_2O_3$, 5-Ethoxycarbonyl-3,4'-diethyl-4,3',5'-trimethyl-2,2'-dipyrrolyl ketone, 44B, 200
$C_{19}H_{28}BrNO_3$, Glycopyrronium bromide, 39B, 167
$C_{19}H_{30}ClNO$, Procyclidine hydrochloride, 37B, 122
$C_{20}H_{16}BrNO_2$, cis-1-Acetyl-4-(1-(4-bromophenyl)-2-phenyl)-vinyl-3-pyrrolin-2-one, 40B, 209
$C_{20}H_{16}N_4 \cdot 0.5\ C_2H_6O$, N-[1,4-Diphenyl-3-(1,2,4-triazolio)]anilide ethanol solvate, 46B, 216
$C_{20}H_{17}ClN_4 \cdot 0.7\ HCl \cdot 3.3\ H_2O$, N-[1,4-Diphenyl-3-(1,2,4-triazolio)]anilide hydrochloride hydrate, 46B, 216
$C_{20}H_{17}N_5$, 2-Phenyl-4,5-dianilino-2H-1,2,3-triazole, 43B, 282
$C_{20}H_{18}N_2O_4$, cis-3-Acetamido-1-acetyl-3,5-diphenylpyrrolidin-2,4-dione, 42B, 176
$C_{20}H_{20}N_2OS$, 1,3-Diphenyl-4-(1'-t-butylthiomethylene)pyrazolin-5-one, 43B, 276
$C_{20}H_{21}N_3O_9$, 4,5-Dimethoxycarbonyl-3-(2,3-O-isopropylidene-β-D-erythrofuranosyl)-1-p-nitrophenylpyrazole, 42B, 176
$C_{20}H_{22}N_4O$, 4-(4-N,N-Diethylaminophenylimino)-3-methyl-1-phenyl-2-pyrazolin-5-one, 37B, 122
$C_{20}H_{23}N_5$, 5-Diethylamino-4-methyl-1-phenyl-3-phenylazopyrazole, 42B,

295
$C_{20}H_{26}ClN_3O$, Pentamethyltripyrro-methene-ethene hydrochloride hydrate, 44B, 1104
$C_{20}H_{30}BrNO_3$, Hexapyrronium bromide, 39B, 167
$C_{21}H_{16}ClNO$, 3-Chloro-1,3,4-triphenyl-azetidin-2-one, 40B, 210
$C_{21}H_{17}ClN_2$, 1,3-Diphenyl-5-(p-chlorophenyl)-2-pyrazoline (α form), 43B, 283
$C_{21}H_{17}ClN_2$, 1,3-Diphenyl-5-(p-chlorophenyl)-2-pyrazoline (β form), 43B, 283
$C_{21}H_{18}N_2O_9$, 3,4,5-Tri-(3-methoxycarbonylfuran-2-yl)-Δ^2-imidazoline, 46B, 217
$C_{21}H_{19}N_5O_2$, 5-Methoxycarbonyl-4-phenyl-4-(4-phenyl-3-Δ^1-pyrazolinyl)-5-Δ^2-pyrazolinecarbonitrile, 44B, 201
$C_{21}H_{22}N_2O$, 5-Dimethylamino-4-methyl-3,3-diphenyl-4-vinyl-5-pyrroline-2-one, 46B, 218
$C_{21}H_{22}N_2O_3$, 1-Ethyl-2-dimethylamino-3-phenyl-5-(2-carboxyphenyl)-2-pyrrolin-4-one, 42B, 177
$C_{21}H_{28}N_2O_3$, Enamine 8, 44B, 198
$C_{21}H_{28}N_2O_4$, 5,5'-Diethoxycarbonyl-3,3'-diethyl-4,4'-dimethyl-2,2'-pyrromethene, 44B, 201
$C_{21}H_{29}BrN_2O_4$, 5,5'-Diethoxycarbonyl-3,3'-diethyl-4,4'-dimethylpyrromethene hydrobromide, 44B, 202
$C_{22}H_{15}ClN_2O$, 1,3-Diphenyl-4-p-chlorobenzal-5-pyrazolone, 38B, 251; 40B, 210
$C_{22}H_{18}ClN_3O_2$, 1-(p-Chlorophenyl)-2-Z-(α-hydroxyiminobenzyl)-4-phenyl-Δ^3-imidazoline-3-oxide, 40B, 210
$C_{22}H_{26}N_4O$, 4-(2,6-Dimethyl-4-N,N-diethylaminophenylimino)-3-methyl-1-phenyl-2-pyrazolin-5-one, 37B, 122
$C_{22}H_{27}N_3O_2S$, 2-Dimethylamino-4-benzenesulphonylamino-1-t-butyl-3-phenylpyrrole, 44B, 203
$C_{22}H_{32}N_2O_5$, 5'-t-Butoxycarbonyl-4'-methoxycarbonylethyl-3',4-dimethyl-3-ethyl-2,2'-methylenedipyrrol-5(2H)-one, 43B, 283
$C_{22}H_{36}N_6O_7$, 1,3-Dinitro-4,6-di(3-(2,2,5,5-tetramethyl)-pyrrolidinyl-N-oxide)aminobenzene monohydrate, 42B, 177
$C_{23}H_{20}N_2O_3$, Ethyl 2,4,4-triphenyl-1,2-diazetidine-3-one-1-carboxylate, 40B, 211
$C_{23}H_{24}N_4O_5$, 4-((N-Benzoyloxycarbonylvalyl)oxyimino)-3-methyl-1-phenyl-2-pyrazolin-5-one, 44B, 203
$C_{23}H_{34}N_2O_4$, 5,5'-Diethoxycarbonyl-3,3',4,4'-tetraethyldipyrrol-2-ylmethane, 38B, 251
$C_{24}H_{19}N_3O_3$, Methyl (Z)-4-oxo-1,3-diphenyl-2-phenylimino-5-imidazolidinylideneacetate, 46B, 218
$C_{24}H_{25}BrN_2$, 2,4,5-Tritolyl-imidazolinium bromide, 45B, 217
$C_{24}H_{26}BrN_3O$, 1-Phenyl-4-(1-diethylamino-2-methyl-3-p-bromophenylpropenylidene)pyrazol-5-one, 45B, 218
$C_{24}H_{29}N_3O_2$, 3,8,12-Triethyl-14-formyl-2,7,13-trimethyl-1(15H)-tripyrrinone, 44B, 204
$C_{24}H_{32}ClN_3 \cdot 1.5\ H_2O$, 5-Dimethylamino-3,4-diethyl-3,4-diphenylpyrroline-2-dimethyliminium hydrochloride hydrate, 46B, 219
$C_{24}H_{32}N_2O_2$, Dextromoramide, 42B, 178
$C_{24}H_{32}N_4O_2$, N-Methylpyrazinobutazone, 40B, 212
$C_{25}H_{19}INO_4$, 1-p-Iodophenyl-3-carbomethoxy-4-phenacyliden-5-hydroxy-5-phenylpyrazoline, 38B, 252
$C_{25}H_{30}ClNO_5$, (+)2-(2-((p-Chloro-α-methyl-α-phenylbenzyl)oxy)ethyl)-1-methylpyrrolidinium fumarate, 44B, 205
$C_{26}H_{21}ClN_8$, 5-(1,5-Diphenyl-3-formazanyl)-1,3-diphenyl-tetrazolium chloride, 45B, 218

$C_{26}H_{32}Cl_8N_4Sn$, Bis(1-m-chlorobenzylidene-3,5,5-trimethylpyrazolinium) hexachlorostannate, 45B, 219
$C_{26}H_{34}Cl_6N_4O_2Sn$, Bis(1-p-hydroxybenzylidene-3,5,5-trimethylpyrazolinium) hexachlorostannate, 45B, 219
$C_{27}H_{22}BrN_5O_5$, 1-Phenyl-3-methyl-4-(p-bromobenzoyloximino)-5-(N-benzyl-oxycarbonylglycyl-imino)-2-pyrazoline, 40B, 213
$C_{28}H_{18}Br_2N_4O_2$, 1-(α-o-Bromobenzoyloxy-o'-bromobenzylideneamino)-4,5-diphenyl-1,2,3-triazole, 41B, 264
$C_{28}H_{18}Cl_2N_4O_2$, 1-(N,N-Bis-p-chlorobenzoylamino)-4,5-diphenyl-1,2,3-triazole, 41B, 265
$C_{28}H_{21}NO$, 1-(2-Hydroxyphenyl)-2,3,5-triphenylpyrrole, 46B, 219
$C_{28}H_{37}ClN_2O_4$, [S(R,R)]-Viminol p-hydroxybenzoate, 41B, 266
$C_{29}H_{22}N_2O_2$, 4-(o-Hydroxybenzoyl)-2,5-diphenyl-1-(p-tolyl)-imidazole, 46B, 220
$C_{29}H_{34}N_2O_8S$, (+)-5-p-Hydroxyphenyl-5-phenylhydantoin camphor-10-sulphonate ethyl acetate, 41B, 267
$C_{29}H_{38}Br_2N_4 \cdot CHCl_3$, 1,2,3,7,8,12,13,17,18,19-Decamethylbiladiene-a,c,d dihydrobromide chloroform, 45B, 221
$C_{30}H_{28}N_4$, N,N',N'',N'''-Tetraphenylbis(1,3-imidazolidin-2-ylidene), 45B, 220
$C_{31}H_{35}ClN_2O_2$, 1-(4-(4-(p-Chlorophenyl)-4-hydroxypiperidino)-2,2-diphenylbutyryl)pyrrolidine, 43B, 316
$C_{32}H_{26}N_4O_2$, 4,4'-Dibenzyl-2,2'-diphenyl-[4,4'-biimidazoline]-5,5'-dione, 45B, 220
$C_{32}H_{28}N_2O_8$, Methyl 2-(3,4-dicarbomethoxy-5-phenylpyrazolyl)-3-carbomethoxy-4,4-diphenylbut-3-enoate, 43B, 284
$C_{32}H_{37}ClN_2O_2$, 1-(4-(4-(p-Chlorophenyl)-4-hydroxypiperidino)-2,2-diphenylvaleryl)pyrrolidine, 43B, 317
$C_{33}H_{21}N_5$, 3-(N-Phenyl-N-tricyanovinyl)amino-1,2,5-triphenylpyrrole, 43B, 284
$C_{33}H_{40}N_4O_4$, 1,19-Di(ethoxycarbonyl)octamethylbilatriene-abc, 44B, 205
$C_{33}H_{41}BrN_4O_4 \cdot C_2H_6O$, 1,19-Bis(ethoxycarbonyl)-2,3,7,8,12,13,17,18-octamethylbilatriene-a,b,c hydrobromide ethanol, 45B, 221
$C_{33}H_{42}Br_2N_4O_4$, 1,19-Bis(ethoxycarbonyl)-2,3,7,8,12,13,17,18-octamethyl-L-biladiene-a,c dihydrobromide, 45B, 221
$C_{39}H_{52}BrN_3O_{12}$, 1-t-Butoxycarbonyl-2,7,12-tris(methoxycarbonylethyl)-8,13-bis(methoxycarbonylmethyl)-3,14-dimethyl-5,16-dihydro-15H-tripyrrin hydrobromide, 44B, 206
$C_{42}H_{30}N_2O$, 1,2,3-Triphenyl-5-(1,2-diphenyl-2-phenyliminoethylidene)-2-pyrrolin-4-one, 45B, 222
$C_{44}H_{34}CoN_{12}S_4$, 1,4-Diphenyl-3-phenylamino-1,2,4-triazolium tetrakis-(isothiocyanato)cobaltate(II), 37B, 124
$C_{50}H_{50}N_4$, (1Z,3E)-1,4-Bis(1,3-dibenzyl-2-imidazolidinyl)-1,4-diphenylbutadiene, 46B, 241

33. HETERO-NITROGEN (6-MEMBERED MONOCYCLIC)

$C_2H_2N_4$, s-Tetrazine, 20, 637
C_2H_7NO, Acetaldehyde - ammonia, 6, 224
$C_3N_3Cl_3$, Cyanuric chloride (gas-ed), 19, 605
C_3N_{12}, Cyanuric triazide, 3, 688, 799
$C_3H_3N_3$, s-Triazine, 19, 603; 32B, 155
$C_3H_3N_3O_3$, Cyanuric acid, 6, 213, 226; 37B, 124; 44B, 206
$C_3H_6N_6$, Melamine (gas-ed), 19, 605
$C_3H_6N_6$, Melamine, 8, 320; 40B, 21543B, 285

$C_3H_6N_6O_6$, Cyclotrimethylene-trinitramine, 38B, 253
$C_3H_9N_9$, Trihydrazinotriazine, 42B, 178
$C_4H_2Cl_2N_2$, 2,6-Dichloropyrazine, 42B, 179
$C_4H_2Cl_2N_2$, 3,6-Dichloropyridazine (gas-ed), 43B, 1484
$C_4H_2Cl_2N_2O_2$, 4,5-Dichloro-3,6-pyridazinedione, 39B, 168
$C_4H_4ClN_3$, 2-Amino-3-chloropyrazine, 38B, 254
$C_4H_4N_2$, Pyrazine (gas-ed), 7, 270
$C_4H_4N_2$, Pyrazine, 21, 610; 42B, 179
$C_4H_4N_2$, Pyridazine (gas-ed), 43B, 1484
$C_4H_4N_2O_2$, Maleic hydrazide, 42B, 179
$C_4H_5ClN_2$, Pyridazine hydrochloride, 41B, 268
$C_4H_5N_3$, Aminopyrazine, 42B, 180
$C_4H_6N_2O_2$, Hexahydro-3,6-pyridazinedione, 44B, 207
$C_4H_6N_2O_2$, Diketopiperazine (gas-ed), 19, 601
$C_4H_6N_2O_2$, Diketopiperazine, 6, 212, 253; 23, 709; 26, 767
$C_4H_6N_2O_2$, Hexahydro-3,6-pyridazine-dione, 41B, 270
$C_4H_6N_2S$, 2H-Pyridaz-3-thione, 31B, 117
$C_4H_8N_2O$, Perhydropyrimidin-2-one, 46B, 1153
$C_4H_8N_2O_2$, 2,6-Bishydroxyiminopiperazine, 46B, 221
$C_4H_{10}N_2$, Piperazine (gas-ed), 28, 666; 37B, 687
$C_4H_{10}N_4O$, 1,4-Dimethyl-1,4,5,6-tetrahydro-1,2,3,4-tetrazine-2-oxide, 44B, 207
$C_4H_{12}Cl_4I_2N$, Piperazinium bis(dichloroiodide), 22, 768
$C_4H_{12}MoN_2S_4$, Piperazine thiomolybdate, 37B, 125
$C_4H_{12}N_4$, Hexahydro-1,4-dimethyl-s-tetrazine, 41B, 270
$C_4H_{14}Cl_2N_2O$, Piperazine dihydrochloride monohydrate, 24, 693
$C_4H_{16}N_{12}$, N,N,N',N'-Tetraaminopiperazindiium bis(azide), 42B, 180
$C_4H_{22}N_2O_6$, Piperazine hexahydrate, 33B, 113
C_5Cl_5N, Pentachloropyridine, 39B, 170
$C_5H_2N_2O_3$, p-Nitropyridine N-oxide, 42B, 181
$C_5H_3N_3O_4$, 3,5-Dinitropyridine, 40B, 216
C_5H_4BrNO, 6-Bromo-2-hydroxypyridine, 42B, 182
C_5H_4ClNO, 5-Chloro-2-pyridone, 38B, 255
C_5H_4ClNO, 6-Chloro-2-hydroxypyridine, 34B, 128
$C_5H_4Cl_2N_2O$, 2,5-Dichloro-3-methoxypyrazine, 38B, 256
$C_5H_4N_2O_2$, Pyrazinic acid, 40B, 216
$C_5H_4N_2O_3$, 4-Nitropyridine N-oxide, 20, 608
$C_5H_5BrClNO_4 \cdot 0.5\ H_2O$, 3-Bromopyridinium perchlorate hemihydrate, 46B, 221
$C_5H_5ClN_2$, 2-Amino-5-chloropyridine, 40B, 217; 42B, 183
C_5H_5N, Pyridine (gas-mw-ed), 7, 270; 18, 749; 19, 596; 22, 772
C_5H_5N, Pyridine, 17, 741
C_5H_5NO, Pyridine 1-oxide, 37B, 126
C_5H_5NO, Pyridine-N-oxide (gas-ed), 40B, 1137
C_5H_5NO, 2-Pyridone, 17, 748
$C_5H_5NO_3S$, 3-Pyridinesulphonic acid, 43B, 286
C_5H_5NS, 2-Pyridthione, 17, 750
$C_5H_5N_3O$, Pyrazinamide (α form), 24, 684
$C_5H_5N_3O$, β-Pyrazine-2-carboxamide, 38B, 257
$C_5H_5N_3O$, δ-Pyrazine-2-carboxamide, 38B, 258
$C_5H_5N_3O$, syn-4-Pyrimidinecarboxaldehyde oxime, 39B, 170
$C_5H_5N_3O_2$, 2-Amino-3-nitropyridine, 41B, 271
$C_5H_5N_3O_2$, 6-Amido-3-pyridazone, 18, 754; 28, 458
$C_5H_6AsF_6N$, Pyridinium hexafluoroarsenate(V), 31B, 119
C_5H_6ClN, Pyridine hydrochloride, 27, 959
C_5H_6ClNO, Pyridine-N-oxide hydrochloride, 24, 690; 26, 703
$C_5H_6F_6NP$, Pyridinium hexafluorophosphate(V), 31B, 120

$C_5H_6F_6NSb$, Pyridinium hexafluoroantimonate(V), 31B, 119
C_5H_6IN, Pyridinium iodide, 41B, 273
$C_5H_6N_2$, 2-Aminopyridine, 41B, 272
$C_5H_6N_2$, 3-Aminopyridine, 41B, 272
$C_5H_6N_2$, 4-Aminopyridine, 43B, 286
$C_5H_6N_2O_2$, 1-Methyl-3,6-pyridazinedione, 40B, 218
$C_5H_6N_2O_3$, Pyridine hydrogen nitrate, 30B, 129
$C_5H_7N_3O$, 2-Amino-6-methoxypyrazine, 45B, 223
$C_5H_8Cl_3MnNO$, Pyridinium trichloromanganate(II) monohydrate, 44B, 208
$C_5H_9N_3O_3S$, 2-Amino-1,6-dihydro-1-methylpyrimidinium-6-sulphonate, 35B, 164
$C_5H_{10}N_2O_3$, 1-Methyl-2,5-piperazinedione monohydrate, 40B, 219
$C_5H_{12}BrN$, Piperidine hydrobromide, 39B, 171
$C_5H_{12}ClN$, Piperidine hydrochloride, 24, 691; 41B, 273
$C_5H_{13}NS$, Piperidinium hydrogen sulfide, 39B, 171
$C_6H_4N_2$, 4-Cyanopyridine, 37B, 126
$C_6H_4N_2O$, 4-Cyanopyridine-N-oxide, 40B, 219
$C_6H_5Br_2NO$, N-Methyl-3,5-dibromopyridone, 31B, 120
$C_6H_5NO_2$, Isonicotinic acid, 42B, 183
$C_6H_5NO_2$, Nicotinic acid, 17, 751; 41B, 274
$C_6H_5NO_2S \cdot 0.5\ H_2O$, 2-Mercaptopyridine-6-carboxylic acid hemihydrate, 43B, 287
$C_6H_5NO_2S$, 2-Mercaptopyridine-4-carboxylic acid, 43B, 287
$C_6H_6ClNO_2$, Picolinic acid hydrochloride, 30B, 130
$C_6H_6N_2O$, Nicotinamide, 18, 752
$C_6H_6N_2O$, α-Picolinamide, 31B, 121
$C_6H_6N_2O$, anti-4-Pyridinecarboxaldehyde oxime, 42B, 48
$C_6H_6N_2O$, syn-4-Pyridinecarboxaldehyde oxime, 42B, 48
$C_6H_6N_2O_3$, 3-Methyl-4-nitropyridine N-oxide, 43B, 288
$C_6H_6N_2S$, 2-Thioamidopyridine, 38B, 260
$C_6H_6N_2S$, 3-Thioamidopyridine, 39B, 172
$C_6H_6N_2S$, 4-Thiocarbamoylpyridine, 31B, 121; 32B, 156
$C_6H_6N_9Na_3O_3$, Trisodium tricyanmelamine trihydrate, 6, 211, 252
$C_6H_7ClN_2O_4$, 4-Chloro-5-methylamino-2,3,6-pyridinetrione monohydrate, 39B, 173
C_6H_7NOS, 3-Hydroxy-6-methylpyrid-2-thione, 43B, 289
$C_6H_7NO_2$, 2-Hydroxymethylpyridine N-oxide, 37B, 127
$C_6H_7N_3O$, Isonicotinic acid hydrazide, 40B, 220
$C_6H_8BiBr_4N$, 2-Picolinium tetrabromobismuthate(III), 32B, 157
$C_6H_8BiI_4N$, 2-Picolinium tetraiodobismuthate(III), 32B, 157
C_6H_8ClNO, 3-Pyridylcarbinol hydrochloride, 26, 764; 29, 725
C_6H_8IN, N-Methylpyridinium iodide, 44B, 208
$C_6H_8I_4NSb$, 2-Picolinium tetraiodoantimonate(III), 39B, 878
$C_6H_8N_2$, 2-Amino-4-methylpyridine, 43B, 290
$C_6H_8N_2$, 2-Amino-5-methylpyridine, 43B, 291
$C_6H_8N_2O_2$, 1-Methyl-3-methoxy-6-pyridazone, 39B, 173; 40B, 221
$C_6H_8N_2O_2$, 1,2-Dimethyl-3,6-pyridazinedione, 39B, 169
$C_6H_9ClN_2$, 2-Amino-5-methylpyridine hydrochloride, 41B, 275
$C_6H_9Cl_2N_3O$, Isoniazide dihydrochloride, 24, 688; 29, 724
$C_6H_{10}Cl_2N_2$, 3-Picolylamine dihydrochloride, 30B, 129
$C_6H_{10}N_2O_2$, Hexahydro-1,2-dimethyl-3,6-pyridazinedione, 43B, 291
$C_6H_{10}N_2O_2$, 2-(Nitromethylene)piperidine, 41B, 276
$C_6H_{11}NO_2$, Nipecotic acid, 42B, 184
$C_6H_{11}NO_3$, (3RS,4SR)-4-Hydroxypiperidine-3-carboxylic acid, 45B, 223
$C_6H_{12}ClNO_4$, 2(S)-Carboxy-4(R),5(S)-dihydroxypiperidine hydrochloride (form 1), 38B, 488
$C_6H_{12}ClNO_4$, 2(S)-Carboxy-4(R),5(S)-dihydroxypiperidine hydrochloride

(form 2), 40B, 221
$C_6H_{13}N_3S$, Piperidinothiosemicarbazide, 41B, 277
$C_6H_{14}N_2$, N,N'-Dimethylpiperazine (gas-ed), 28, 666
$C_6H_{14}N_2O_2S$, N-Methylpiperazine methanesulphonamide, 43B, 292
$C_6H_{14}N_6O_4Pd$, Tetraamminepalladium(II) pyrazine-2,5-dicarboxylate, 40B, 222
$C_6H_{15}N_3O_6S_3$, Tri-N-(methylsulphonyl)hexahydro-s-triazine, 44B, 209
$C_6H_{16}Cl_2N_2$, trans-2,5-Dimethylpiperazine dihydrochloride, 44B, 209
$C_6H_{21}N_3O_3$, Acetaldehyde-ammonia trihydrate, 22, 589
$C_7H_3D_2NO_4$, 2,3-Pyridinedicarboxylic acid, 45B, 224
$C_7H_5NO_4$, Quinolinic acid, 44B, 210; 45B, 224
$C_7H_5NO_4$, Cinchomeronic acid, 39B, 176
$C_7H_5NO_4$, Dinicotinic acid, 39B, 174
$C_7H_5NO_4$, Pyridine-2,3-dicarboxylic acid, 39B, 175; 40B, 222
$C_7H_5NO_5$, 2,6-Dicarboxypyridine N-oxide, 43B, 293
$C_7H_7Cl_2NO$, 3,5-Dichloro-2,6-dimethyl-4-pyridinol, 38B, 260
$C_7H_7NO_3$, 2-Carboxy-6-methylpyridine N-oxide, 41B, 277
$C_7H_7NO_5$, Dipicolinic acid monohydrate, 39B, 176
C_7H_8ClNO, 2-Acetylpyridine hydrochloride, 31B, 123
$C_7H_8N_2O_3$, 3,5-Dimethyl-4-nitropyridine N-oxide (orthorhombic), 43B, 288
$C_7H_8N_2O_3$, 3,5-Dimethyl-4-nitropyridine N-oxide (tetragonal), 43B, 288
$C_7H_8N_2S$, 2-Methyl-4-thiocarbamoylpyridine, 34B, 130, 136
$C_7H_8N_4OS \cdot 1.5\ H_2O$, 5-Hydroxy-2-formylpyridine thiosemicarbazone sesquihydrate, 40B, 67
$C_7H_8N_4O_4 \cdot H_2O$, 2-Methylaminopyridinium dinitromethylide monohydrate, 46B, 222
$C_7H_8N_4S$, 4-Formylpyridine thiosemicarbazone, 35B, 166
$C_7H_8N_4Se$, 2-Formylpyridine selenosemicarbazone, 38B, 261, 354
$C_7H_9CaNO_7$, Calcium dipicolinate trihydrate, 33B, 113
$C_7H_9ClN_2O$, 1-Methylnicotinamide chloride, 40B, 545
$C_7H_9ClN_4$, N,N-Dimethyl-N'-(6-chloropyridazinyl-3) formamidine, 40B, 223
$C_7H_9Cl_2N_3O$, 2,3-Dichloro-5-ethylamino-6-methoxypyrazine, 40B, 224
$C_7H_9IN_2O$, N-Methylpyridine-2-aldoxime iodide, 31B, 123
$C_7H_9IN_2O$, 1-Methylnicotinamide iodide, 40B, 545
$C_7H_9NO_3S$, 2-[[(Hydroxymethoxy)methyl]thio]pyridine 1-oxide, 46B, 222
$C_7H_{10}Br_4FeN$, 4-Ethylpyridinium tetrabromoferrate(III), 37B, 128
$C_7H_{10}N_2O_2S$, 1-Methyl-3-methylsulphonylimidopyridinium, 43B, 294
$C_7H_{10}N_4O_3$, 1,4,5,6-Tetrahydro-1,5,N-trimethyl-4,6-dioxo-1,3,5-triazine-2-carboxamide, 45B, 224
$C_7H_{11}NO_8Sr$, Strontium dipicolinate tetrahydrate, 38B, 690
$C_7H_{12}N_2O_2$, 1-iso-Propyl-2,5-piperazinedione, 40B, 225
$C_7H_{15}ClN_2O_2$, 4-Dimethylaminopyridine hydrochloride dihydrate, 43B, 294
$C_8D_5N_3$, Pyridinium-1-dicyanomethylide, 46B, 223
$C_8H_5N_3$, Pyridinium dicyanomethylide, 30B, 131
$C_8H_8Cl_6N_2O_2$, 5-Chloro-6-dichloromethylene-4-methoxy-1-methyl-4-trichloromethylhexahydro-2-pyrimidinone, 46B, 223
$C_8H_8N_2O_2S$, Cyanomethyl 2-picolyl sulphone, 40B, 225
$C_8H_9NO_2S$, 2-((Methylsulfinyl)acetyl)pyridine, 40B, 226
$C_8H_{10}N_2S$, 2-Ethyl-4-thiocarbamoylpyridine, 39B, 177
$C_8H_{11}BrN_2S$, 2-Ethyl-4-thiocarbamoylpyridine hydrobromide, 31B, 125; 33B, 114
$C_8H_{11}ClN_2S$, 2-Ethyl-4-thiocarbamoylpyridine hydrochloride, 31B, 126; 33B, 114

$C_8H_{11}IN_2O_2$, 1-Ethyl-2-methyl-3-nitropyridinium iodide, 46B, 224
$C_8H_{12}ClNO_3$, Pyridoxinium chloride, 46B, 223
$C_8H_{12}IN$, 1,2,6-Trimethylpyridinium iodide, 46B, 224
$C_8H_{12}NO_7P$, Pyridoxal phosphate hydrate, 39B, 178
$C_8H_{12}N_2$, Tetramethylpyrazine, 15, 484; 21, 611
$C_8H_{13}ClN_2O_2$, Pyridoxamine monohydrochloride, 46B, 225
$C_8H_{13}NO_3$, (DL)-(E)-N-Acetylpiperidine-2-carbocylic acid, 46B, 225
$C_8H_{15}N_5O$, 2,4-Bis(ethylamino)-6-methoxy-s-triazine, 44B, 210
$C_8H_{15}N_5S$, 2,4-Bis(ethylamino)-6-methylthio-s-triazine, 44B, 211
$C_8H_{18}ClN$, 1,(2,6)-cis-Trimethyl-piperidinium hydrochloride, 41B, 278
$C_9H_6ClN_3$, 5-(p-Chlorophenyl)-1,2,4-triazine, 40B, 227
$C_9H_6N_6$, 2,4-Diamino-3,5-dicyano-6-cyanomethyl-pyridine, 41B, 279
$C_9H_9NO_2S$, trans-3-(6-Methyl-2-pyridylthio)-propenic acid, 38B, 261
$C_9H_9NO_4$, N-Succinopyridine, 42B, 184
$C_9H_{12}CrN_7S_4$, Pyridinium reineckate, 21, 405, 629
$C_9H_{12}INO$, 1,2-Dimethyl-3-acetylpyridinium iodide, 46B, 224
$C_9H_{12}N_2S$, 2-Propyl-thioisonicotinamide, 34B, 137, 145; 35B, 168
$C_9H_{13}IN_2O_2$, 1,2,4,6-Tetramethyl-3-nitropyridinium iodide, 46B, 224
$C_9H_{14}Br_2NO_2$, cis-3,5-Dibromo-4-oxo-2,2,6,6-tetramethylpiperidin-1-yloxy, 43B, 295
$C_9H_{14}NO_7P$, Pyridoxal phosphate methyl hemiacetal, 39B, 178
$C_9H_{14}N_2O$, n-Propyldihydronicotinamide, 28, 461
$C_9H_{14}N_2O_6$, 2,5-Pyridinedicarboxylic acid N,N-dimethylformamide monohydrate, 42B, 185
$C_9H_{15}N_3O_3$, 1,3,5-Triacetyl-2,4,6-hexahydro-s-triazine, 41B, 279
$C_9H_{15}N_3O_6$, 1,3,5-Triacetoxyhexahydro-1,3,5-triazine, 43B, 297
$C_9H_{16}NO_2$, 2,2,6,6-Tetramethyl-4-piperidinone-1-oxyl nitroxide (orthorhombic), 38B, 262; 40B, 227
$C_9H_{16}NO_2$, 2,2,6,6-Tetramethyl-4-piperidone-1-oxyl (gas-ed), 40B, 1135
$C_9H_{16}N_2$, Dimethylpiperidinoacetonitrile, 40B, 1084
$C_9H_{16}N_3O_5P$, N'-Isopropylpyridine-4-carbohydrazinium phosphate, 37B, 128
$C_9H_{17}NO$, 4-t-Butyl-1-aza-2-cyclohexanone, 45B, 225
$C_9H_{17}NO_2$, 2,2,6,6-Tetramethyl-4-piperidone-1-hydroxyl (gas-ed), 40B, 1135
$C_9H_{17}N_2O_2$, 2,2,6,6-Tetramethyl-4-(hydroxyimino)-piperidine-1-oxyl, 43B, 297
$C_9H_{18}ClNO \cdot 0.4\ H_2O$, 2,2,6,6-Tetramethyl-4-piperidone hydrochloride, 37B, 129
$C_9H_{18}ClNO_5$, 2,2,6,6-Tetramethyl-1-oxopiperidinium perchlorate, 41B, 280
$C_9H_{18}NO$, 2,2,6,6-Tetramethyl-1-piperidine-1-oxyl nitroxide radical (tetragonal), 40B, 227
$C_9H_{18}NO_2$, 2,2,6,6-Tetramethyl-4-piperidinol-1-oxyl, 33B, 117
$C_9H_{18}N_6$, Hexamethylmelamine, 38B, 262
$C_9H_{19}NO_2$, 2,2,6,6-Tetramethyl-4-piperidone monohydrate, 40B, 228
$C_{10}H_8ClN_3O$, 5-Amino-4-chloro-2-phenylpyridazin-3-one, 44B, 211
$C_{10}H_8N_2$, 2,2-Bipyridyl, 20, 609
$C_{10}H_8N_2S_2$, Di-2-pyridyl disulfide, 43B, 298
$C_{10}H_9ClN_2O_4$, 2,2'-Pyridyl-pyridinium perchlorate, 42B, 185
$C_{10}H_9N_3$, Di-(2-pyridyl)amine, 39B, 179
$C_{10}H_{10}Br_4CoN_2$, 2,2'-Bipyridinium tetrabromocobaltate(II), 37B, 130
$C_{10}H_{10}Br_5Cl_2N_2Sb$, 2-Chloropyridinium pentabromoantimonate(III), 42B, 186
$C_{10}H_{10}Cl_8N_2Sn$, 4-Chloropyridinium hexachlorostannate(IV), 39B, 179
$C_{10}H_{12}As_2F_8N_2O_2$, Dipyridinium octafluoro-di-μ-oxo-diarsenate, 39B,

880
$C_{10}H_{12}Br_7Cu_5N_2$, Pyridinium copper(I) bromide, 44B, 211
$C_{10}H_{12}Br_{12}Cl_6N_2Nb_6$, Dipyridinium dodecabromohexaniobate hexachloride, 37B, 674
$C_{10}H_{12}Br_{12}Cl_6N_2Nb_6$, Dipyridinium hexabromohexachlorohexaniobate hexabromide, 37B, 674
$C_{10}H_{12}Cl_{18}N_2Nb_6$, Dipyridinium dodecachlorohexaniobate hexachloride, 37B, 674
$C_{10}H_{12}I_6N_2Pt$, Pyridinium hexaiodoplatinate(IV), 44B, 212
$C_{10}H_{13}ClN_4O_4$, μ_2-Hydrobis(4-aminopyridine) perchlorate, 46B, 225
$C_{10}H_{14}Cl_8N_2Re_2$, Pyridinium rhenium(II) tetrachloride, 28, 459
$C_{10}H_{14}N_2S$, 2-Butyl-4-thiocarbamoylpyridine, 39B, 180
$C_{10}H_{14}O_8S_4$, Ethanediylidenetetrathiotetraacetic acid, 44B, 64
$C_{10}H_{16}Br_2N_2O_2$, N,N'-Bis-(3-bromopropionyl)piperazine, 39B, 181
$C_{10}H_{16}Cl_2N_2O_2$, N,N'-Bis-(3-chloropropionyl)piperazine, 41B, 281
$C_{10}H_{16}Cl_2N_2O_2$, 1,4-Bis(chloroacetyl)-trans-2,5-dimethylpiperazine, 43B, 298
$C_{10}H_{18}Cl_4Mo_2N_2O_7$, Pyridinium di-$\mu$-oxo-bis(oxodichloroaquomolybdate(V)) monohydrate, 41B, 282
$C_{10}H_{19}N_5O$, 2,4-Bis(isopropylamino)-6-methoxy-s-triazine, 43B, 299
$C_{10}H_{20}N_2S_2$, Dipiperidine disulfide, 43B, 299
$C_{10}H_{22}BiBr_5N_2$, Bispiperidinium pentabromobismuthate(III), 33B, 294
$C_{10}H_{24}As_4N_2S_6$, Piperidinium hexathiotetraarsenate, 37B, 675
$C_{10}H_{24}Br_5N_2Sb$, Bispiperidinium pentabromoantimonate(III), 39B, 880
$C_{11}H_8Br_2N_2O_2$, 1-(p-Bromophenyl)-2-methyl-5-bromopyridazine-3,6-dione, 45B, 226
$C_{11}H_9NO_2S$, 2-Pyridyl-phenyl-sulphone, 45B, 227
$C_{11}H_{10}N_6O_4$, N,N'-Bis(3-nitro-2-pyridinylimino)methanediamine, 46B, 226
$C_{11}H_{12}BrNO_3$, 2,5-Dihydroxyphenylpyridinium bromide monohydrate, 39B, 181
$C_{11}H_{13}NO_2$, 3,5-Diacetyl-2,6-dimethylpyridine, 46B, 226
$C_{11}H_{15}BrClN_5$, 4,6-Diamino-1-p-chlorophenyl-2,2-dimethyl-1,3,5-triazine hydrobromide, 18, 756
$C_{11}H_{15}Cl_2N_5$, 4,6-Diamino-1-p-chlorophenyl-2,2-dimethyl-1,3,5-triazine hydrochloride, 18, 756
$C_{11}H_{18}NO_2$, 4-Ethynyl-4-hydroxy-2,2,6,6-tetramethylpiperidine-1-oxyl, 45B, 228
$C_{11}H_{19}NO$, 4-Ethynyl-2,2,6,6-tetramethylpiperidin-4-ol, 44B, 212
$C_{11}H_{22}N_2S_2$, Piperidinium 1-piperidinecarbodithioate (β form), 46B, 227
$C_{11}H_{23}NO$, 1,2,2,4,6,6-Hexamethyl-4-piperidinol, 46B, 227
$C_{12}H_8N_2O_2$, 2,2'-Pyridil, 26, 788; 35B, 169
$C_{12}H_9BrN_2O$, N-(5-Bromosalicylidene)-2-aminopyridine, 44B, 213
$C_{12}H_9Cl_2NO$, 1-(2',6'-Dichlorobenzyl)-2-pyridone, 40B, 228
$C_{12}H_9Cl_4FeN_4$, 3-Cyanopyridinium tetrachloroferrate(III) 3-cyano-pyridine, 45B, 228
$C_{12}H_9N_3O_4$, 2-(2',4'-Dinitrobenzyl)pyridine, 33B, 119
$C_{12}H_9N_3O_4$, 4-(2',4'-Dinitrobenzyl)pyridine, 40B, 229
$C_{12}H_{10}N_2O$, N-Phenyl-nicotinamide, 45B, 229
$C_{12}H_{10}N_2O$, N-Salicylidene-2-aminopyridine, 44B, 213
$C_{12}H_{10}N_2O$, N-Salicylidene-3-aminopyridine, 46B, 230
$C_{12}H_{10}N_2O_2$, α-Pyridoin, 30B, 132
$C_{12}H_{10}N_6$, 3,6-Di-(2-pyridyl)-1,4-dihydro-1,2,4,5-tetrazine, 42B, 187
$C_{12}H_{12}N_2O_2$, 2,2'-Bis(6-methyl-3-pyridinol), 37B, 131
$C_{12}H_{12}N_2O_2S_2 \cdot 0.5\ C_4H_8O$, Bis-3-hydroxy-6-methylpyridyl-2,2'-disulfide hemi(tetrahydrofuran), 43B, 289

$C_{12}H_{14}Br_2N_2$, N,N'-Dimethyl-4,4'-bipyridylium dibromide, 38B, 264
$C_{12}H_{14}Cl_2N_2$, N,N'-Dimethyl-4,4'-bipyridylium dichloride, 38B, 263
$C_{12}H_{14}Cl_4CoN_2$, N,N'-Dimethyl-4,4'-dipyridylium tetrachlorocobaltate(II), 34B, 131
$C_{12}H_{14}Cl_4CuN_2$, N,N'-Dimethyl-4,4'-bipyridinium tetrachlorocuprate, 34B, 138
$C_{12}H_{14}Cl_4Cu_2N_2$, N,N'-Dimethyl-4,4'-dipyridylium tetrachlorodicuprate(I), 34B, 131
$C_{12}H_{14}Cl_4N_2Pd$, N,N'-Dimethyl-4,4'-dipyridylium tetrachloropalladate(II), 34B, 131
$C_{12}H_{14}Cl_6Cu_2N_2$, N,N'-Dimethyl-4,4'-bipyridylium hexachlorodicuprate(II), 41B, 282
$C_{12}H_{14}I_2N_2$, N,N'-Dimethyl-4,4'-bipyridylium diiodide, 38B, 264
$C_{12}H_{14}N_2O_3$, p-Nitrobenzoylpiperidine, 43B, 300
$C_{12}H_{14}N_4O_4S$, Sulfadimethoxine, 38B, 265
$C_{12}H_{14}N_4O_4S$, Sulfadoxine, 38B, 265
$C_{12}H_{16}Br_9N_2Sb$, α-Picolinium nonabromoantimonate(V), 33B, 289
$C_{12}H_{16}Br_9N_2Sb$, 4-Methylpyridinium nonabromoantimonate(V), 39B, 182
$C_{12}H_{16}Cl_3N_5O \cdot 0.29\ H_2O$, 1-[(3,4-Dichlorophenyl)methoxy]-1,6-dihydro-6,6-dimethyl-1,3,5-triazine-2,4-diamine hydrochloride hydrate, 45B, 229
$C_{12}H_{17}ClN_2O_5S_2$, N-(3-Chloro-4-sulfamyl)-benzsulphonyl-4-methoxypiperidine, 45B, 230
$C_{12}H_{17}Cl_2NO_2$, 4-Chloro-5-(chloromethyl)-1-cyclohexyl-5,6-dihydro-3-hydroxy-2(1H)pyridinone, 43B, 300
$C_{12}H_{17}N_5S_3$, Methyl (E,Z,E)-[4-(2-methyl-3-(methylthio)-1,2,4-triazin-5(2H)-ylidene-2-butenylidene]methylhydrazinecarbodithioate, 46B, 231
$C_{12}H_{19}IN_2$, 1,1-Dimethyl-4-phenylpiperazinium iodide, 44B, 213
$C_{12}H_{20}Cl_2N_2$, N,N'-Bis(1-chloro-2-methylpropenyl)piperazine, 45B, 231
$C_{12}H_{20}I_2N_2$, 1,4-Bis(1-iodo-2,2-dimethylvinyl)piperazine, 45B, 238
$C_{12}H_{20}N_2S_3$, Dicyclopentamethylenethiuram monosulfide, 35B, 170
$C_{12}H_{21}N_5S_3$, 2-(1-Dimethylthiocarbamoyl-1-methyl-ethylimino)-1,3,5-trimethyl-4,6-dithioxo-hexahydro-1,3,5-triazine, 42B, 187
$C_{12}H_{22}N_2O$, 1-(Piperidinoacetyl)piperidine, 44B, 214
$C_{12}H_{22}N_2O_4$, 1,4-Piperazine-γ,γ'-dibutyric acid, 31B, 127
$C_{12}H_{23}ClN_2O_5$, N-(N-Piperidylacetyl)piperidinium perchlorate, 43B, 301
$C_{12}H_{23}NO$, 1,2,2,6,6-Pentamethyl-4-vinyl-4-piperidinol, 46B, 231
$C_{12}H_{24}N_2O_2 \cdot 4\ H_2O$, 1,2-Di(N-piperidyl)ethane bis-N-oxide tetrahydrate, 45B, 231
$C_{12}H_{36}Ag_4I_6N_2O_4S_4$, Silver iodide piperazinium tetrakis(dimethylsulfoxide), 41B, 284
$C_{13}H_9F_3N_2O_2$, 2-{[3-(Trifluoromethyl)phenyl]amino}-3-pyridinecarboxylic acid, 45B, 233
$C_{13}H_9F_3N_2O_2$, 4-[(m-Trifluoromethyl)phenylamino]nicotinic acid, 45B, 232
$C_{13}H_9NO_2$, 1-Phenyl-2-(2-pyridyl)ethane-1,2-dione, 31B, 128
$C_{13}H_{10}BrN$, cis-β-Bromo-β-(2-pyridyl)styrene, 35B, 171
$C_{13}H_{10}BrNO$, (Z)-N-(2-Bromo-1-phenylvinyl)-pyridin-2-one, 45B, 233
$C_{13}H_{11}ClN_2O_2$, 5-Chloro-N-(5-methoxysalicylidene)-2-aminopyridine, 44B, 213
$C_{13}H_{12}BrNO_2$, α-Bromo-γ-phenyl-γ-ethylglutaconimide, 27, 967
$C_{13}H_{12}N_2O_2$, N-(5-Methoxysalicylidene)-3-aminopyridine, 46B, 230
$C_{13}H_{13}Cl_2IN_2O_2$, N(1)-(2,6-Dichlorobenzyl)-3-carbamidopyridinium iodide monohydrate, 34B, 134
$C_{13}H_{13}IN_2$, Methylbenzylideneiminopyridinium iodide, 42B, 44

$C_{13}H_{13}N_5S$, 4-m-Aminophenyl-2-formylpyridine thiosemicarbazone, 45B, 234
$C_{13}H_{14}INO_2$, α-Ethyl-α-phenyl-α'-iodoglutarimide, 26, 768
$C_{13}H_{14}N_2O$, N-Benzyl-1,4-dihydronicotinamide, 26, 787
$C_{13}H_{17}BrN_2O$, Piperidino-acet-m-bromoanilide, 40B, 230
$C_{13}H_{17}ClN_2O$, Piperidino-acet-o-chloroanilide, 39B, 183
$C_{13}H_{17}N_3$, 1-Ethyl-2,4,4,6-tetramethyl-1,4-dihydro-3,5-pyridinedicarbonitrile, 45B, 234
$C_{13}H_{18}Cl_2N_2O \cdot H_2O$, o-Chloro-N-(piperid-1-yl-acetyl)-aniline hydrochloride monohydrate, 45B, 236
$C_{13}H_{18}Cl_2N_2O$, Piperidinium-acetyl-(m-chloroanilide) chloride, 45B, 235
$C_{13}H_{18}Cl_2N_2O$, Piperidinium-acetyl-(p-chloroanilide) chloride, 45B, 235
$C_{13}H_{18}Cl_2N_2O$, o-Chloro-N-(piperid-1-yl-acetyl)-aniline hydrochloride, 45B, 236
$C_{13}H_{18}Cl_2N_2O_4$, 1-(2,6-Dichlorobenzyl)-6-hydroxy-1,4,5,6-tetrahydronicotinamide dihydrate, 34B, 140
$C_{13}H_{19}NO_4$, 3,5-Diethoxycarbonyl-2,6-dimethyl-1,4-dihydropyridine, 45B, 237
$C_{13}H_{19}N_3$, (2,6-cis-Dimethylpiperidyl)diazobenzene, 44B, 214
$C_{13}H_{20}ClNO$, 1,2-Dimethyl-4-hydroxy-4-phenylpiperidine hydrochloride, 40B, 230
$C_{13}H_{20}N_2$, 6-Dimethylamino-6-piperidine-fulvene, 46B, 232
$C_{14}H_9NO_2$, 2-(2'-Pyridyl)indan-1,3-dione, 44B, 214
$C_{14}H_{10}N_4$, 3,6-Diphenyl-s-tetrazine, 38B, 267
$C_{14}H_{14}BCl_2F_4NO$, 1-(2',6'-Dichlorobenzyl)-2-ethoxypyridinium tetrafluoroborate, 40B, 229
$C_{14}H_{14}N_2O_2$, 3-Methyl-N-(3-methoxysalicylidene)-2-aminopyridine, 44B, 213
$C_{14}H_{14}N_4S$, 2-Formyl-5-benzylpyridine thiosemicarbazone, 44B, 215
$C_{14}H_{15}NO_2S$, 2-Hydroxy-2-phenyl-2-(2-pyridyl)ethyl(methyl)sulfoxide, 45B, 237
$C_{14}H_{16}ClNO_4$, N-Phenyl-2,4,6-trimethylpyridinium perchlorate, 34B, 141
$C_{14}H_{16}N_2O_2$, 2,3-Di(2-pyridyl)-2,3-butanediol, 38B, 267
$C_{14}H_{17}ClN_4O_4$, N-Pyridoxylidene-N'-picolinoylhydrazine hydrochloride monohydrate, 44B, 215
$C_{14}H_{19}NO$, N-Benzoyl-2,6-dimethylpiperidine, 43B, 302
$C_{14}H_{20}ClIN_2O$, N-Methyl-piperidinium-m-chloracetanilide iodide, 40B, 231
$C_{14}H_{20}ClIN_2O$, 2-(1-Methyl-piperidinio-(1))-2'-chloroacetanilide iodide, 39B, 183
$C_{14}H_{20}ClN_3S$, 2-((2)-Dimethylaminoethyl-2-thenylamino)pyridine hydrochloride, 38B, 268
$C_{14}H_{20}IN_3O_3$, N-Methyl-piperidinium p-nitroacetanilide iodide, 42B, 188
$C_{14}H_{20}N_4$, 1,4-Bis(1-cyano-2,2-dimethylvinyl)piperazine, 45B, 238
$C_{14}H_{21}NO$, DL-α-Promedol alcohol, 38B, 269
$C_{14}H_{21}NO$, DL-β-Promedol alcohol, 38B, 270
$C_{14}H_{21}NO$, DL-γ-Promedol alcohol, 38B, 271
$C_{14}H_{22}ClIN_2O_2$, N-Methylpiperidiniumaceto-m-chloroanilide iodide monohydrate, 44B, 216
$C_{14}H_{22}ClNO_4$, 1-Ethyl-1-methyl-4-phenylpiperidinium perchlorate ($P2_1$), 42B, 188
$C_{14}H_{22}ClNO_4$, 1-Ethyl-1-methyl-4-phenylpiperidinium perchlorate, 35B, 173

$C_{14}H_{25}N_5S_3$, 2-(1-Isopropyl-3,5-dimethyl-4,6-dithioxo-hexahydro-1,3,-5-triazine-2-ylideneamino)-N,N-dimethylthioisobutyramide, 43B, 302
$C_{14}H_{26}N_2$, 1,4-Bis(1,2,2-trimethylvinyl)piperazine, 45B, 238
$C_{14}H_{29}NO$, 4-t-Butyl-4-hydroxy-2,2,6,6-tetramethylpiperidine, 46B, 232
$C_{15}H_{13}NOS_2$, 1-(1-Pyridinio)-1-benzoyl-2-methylthio-2-thioxo-1-ethanide, 45B, 239
$C_{15}H_{14}F_3N_3O_3S$, Galosemide, 44B, 217
$C_{15}H_{15}BrN_2O$, N-(2-Pyridylacetomethylene)-4-bromoaniline, 39B, 184
$C_{15}H_{15}IN_2O$, 4-Cyano-N-(2-(p-methoxyphenyl)ethyl)pyridinium iodide, 34B, 135
$C_{15}H_{15}N_3O_3$, Ethyl 3-cyano-6-methyl-2-oxo-4-(3-pyridyl)-3,4-dihydro-5-pyridinecarboxylate, 44B, 217
$C_{15}H_{16}ClN_3O_3S_2$, 1-Isopropyl-3-((4-(3-chlorophenylthio)-pyrid-3-yl)-sulphonyl)urea, 45B, 240
$C_{15}H_{16}Cl_2N_4O_3S$, 1-Isopropyl-3-{[4-(3,4-dichlorophenylamino)-pyrid-3-yl]-sulphonyl}-urea, 46B, 233
$C_{15}H_{17}Cl_6Fe_2N_3O$, Pyridinium μ-oxo-bis(trichloroferrate(III)) pyridine solvate, 44B, 217
$C_{15}H_{18}BrN_3O_2$, 1-(p-Bromophenyl)-2-methyl-4-diethylamino-pyridazine-3,6-dione, 45B, 226
$C_{15}H_{18}N_4$, N,N,1-Trimethyl-2-(6-methyl-3-phenyl-1,2,4-triazin-5-yl)-vinylamine, 40B, 233
$C_{15}H_{19}Br_2NO_3$, (+)-1-Methyl-3-benzoyl-3-bromoacetoxypiperidine hydrobromide, 39B, 184
$C_{15}H_{19}ClN_4O_7S_2$, 1-{[4-(3-Chlorophenylamino)-3-pyridyl]sulphonyl}-3-ethyl-1-methylurea hydrogen sulfate, 46B, 48
$C_{15}H_{20}Cl_2N_4O_3$, 1,1'-Trimethylenebisnicotinamide dichloride dihydrate, 39B, 185
$C_{15}H_{20}N_2O_3$, 1-p-(Oximinoethyl)phenoxyacetylpiperidine, 40B, 233
$C_{15}H_{21}NOS$, 1-Benzenesulphenyl-2,2,6,6-tetramethyl-4-oxopiperidine, 41B, 284
$C_{15}H_{21}NO_2S$, 1-Benzenesulfinyl-2,2,6,6-tetramethyl-4-oxopiperidine, 41B, 284
$C_{15}H_{21}NO_3S$, 1-Benzenesulphonyl-2,2,6,6-tetramethyl-4-oxopiperidine, 41B, 284
$C_{15}H_{22}BrNO_2$, Ethyl 1-methyl-4-phenylpiperidine-4-carboxylate hydrobromide (Pethidine hydrobromide), 26, 769; 35B, 175
$C_{15}H_{22}ClNO_2$, Pethidine hydrochloride, 40B, 234
$C_{15}H_{22}INO$, (+)-1,3-Dimethyl-3-benzoylpiperidine methiodide, 42B, 188
$C_{15}H_{22}N_2$, 2,6-cis-Dimethylpiperidyl-N-phenylacetamidine, 45B, 240
$C_{15}H_{23}IN_2O$, N-Methyl-piperidinium-acet-o-methyl-anilide iodide, 41B, 285
$C_{16}H_{12}N_2$, 2,5-Diphenylpyrazine, 42B, 189
$C_{16}H_{14}N_6Ni$, N,N'-Dimethyl-4,4'-dipyridinium tetracyanonickelate(II), 34B, 138
$C_{16}H_{18}N_2O_2$, 2,6-Dimethyl-3,5-diacetyl-4-(β-pyridyl)-1,4-dihydropyridine, 43B, 303
$C_{16}H_{18}N_2O_6S_2$, trans-Dibenzenesulphonyl-2,5-dihydroxypiperazine, 39B, 186
$C_{16}H_{19}BrN_2O_2S$, 1-Methyl-2-oxo-3-(3'-p-bromobenzylthiocrotonyl)perhydropyrimidine (low-melting isomer), 43B, 304
$C_{16}H_{19}ClN_4O_2$, N-Benzyl-β-(N^2-isonicotinoylhydrazino)propionamide hydrochloride, 43B, 304
$C_{16}H_{19}N_5OS$, 2-Formyl-4-phenylpyridine thiosemicarbazone - dimethylformamide, 43B, 305
$C_{16}H_{20}N_4O_3S$, Torasemide, 44B, 218

$C_{16}H_{21}BrCl_2N_2O$, 3-(4-Bromophenyl)-N,N-dimethyl-3-(3-pyridyl)-allylamine dichloride, 42B, 23
$C_{16}H_{23}Cl_3INO_2$, (-)-1-Methyl-3-methoxy-3-benzoylpiperidine methiodide chloroform solvate, 42B, 189
$C_{16}H_{24}BrNO_2$, DL-β-Prodine hydrobromide, 28, 462
$C_{16}H_{24}BrNO_2$, 1e,2a,6e-Trimethyl-4e-phenyl-4a-acetoxypiperidine hydrobromide, 39B, 187
$C_{16}H_{24}ClNO_2$, DL-α-Prodine hydrochloride, 24, 709
$C_{16}H_{24}ClNO_2$, DL-β-Prodine hydrochloride, 28, 464
$C_{16}H_{28}N_2O$, 3,5,6-Tri-t-butyl-1,2-dihydro-pyrazine-2-one, 41B, 287
$C_{16}H_{31}NO$, Tripropylacetylpiperidine, 42B, 190
$C_{16}H_{40}Ag_{10}I_{12}N_6O_4$, Silver iodide piperazinium tetrakis(dimethylformamide), 41B, 288
$C_{17}H_{13}Cl_2N$, 1-(2',6'-Dichlorobenzyl)-4-cyclopentadienylidene-1,4-dihydropyridine, 41B, 289
$C_{17}H_{15}N$, 1-Benzyl-2-cyclopentadienylidene-1,2-dihydropyridine, 41B, 289
$C_{17}H_{17}BrN_2 \cdot 0.5\ H_2O$, 1-(1'-Methyl-2'-pyridinium)-3-(1"-methyl-4"-pyridinium)cyclopentadienide bromide hemihydrate, 46B, 235
$C_{17}H_{21}NO_2$, Enamine 5, 44B, 198
$C_{17}H_{22}N_4O_3S$, S-[1-Methyl-4-(3-methylphenylamino)-3-pyridinio]isopropylcarbamoylsulfamoylate, 45B, 241
$C_{17}H_{25}NO_8$, (+)-((-)-1-Methyl-3-benzoylpiperidine R:R-(+)-bitartrate) monohydrate, 39B, 187
$C_{17}H_{26}ClN$, 1-(1-Phenylcyclohexyl)piperidine hydrochloride, 35B, 176
$C_{18}H_{16}N_6Ni$, 1,2-Bis(N-methyl-4-pyridinium)ethylene tetracyanonickelate(II), 35B, 177
$C_{18}H_{17}Cl_6MnN_6O$, 2,4,6-Tris(2'-pyridinio)-1,3,5-triazine pentachloromanganate(III) chloride monohydrate, 43B, 305
$C_{18}H_{20}Br_2N_2$, 2,6-Bis(bromomethyl)-1,4-diphenylpiperazine, 38B, 272
$C_{18}H_{22}N_2O_2$, N,N'-Di(O-benzyl alcohol)piperazine, 43B, 306
$C_{18}H_{22}N_2O_4$, 2,6-Dimethyl-3,5-dicarboethoxy-4-(β-pyridyl)-1,4-dihydropyridine, 43B, 307
$C_{18}H_{23}ClN_2$, Cyclizine hydrochloride, 46B, 233
$C_{18}H_{23}NO_3$, 2,6-Diacetonyl-N-benzoylpiperidine, 44B, 219
$C_{18}H_{24}N_4$, 2,5-Dibenzyl-1,4-dimethyl-1,2,4,5-tetraazacyclohexane, 46B, 278
$C_{18}H_{24}N_6NiO_3$, 1,2-Bis-(2-methylpyridinium)ethane tetracyanonickelate(II) trihydrate, 37B, 132
$C_{18}H_{26}Br_2N_2O_4$, 4-((2',2',5',5'-Tetramethylpyrrolin-1-yloxy)-3'-carbonyloxy)-2,2,6,6-tetramethyl-3,5-dibromo-3,4-dehydropiperidin-1-yloxy, 43B, 295
$C_{18}H_{28}F_2N_2$, 1,4-Bis(fluoro-(cyclohexylidene)-methyl)piperazine, 45B, 238
$C_{18}H_{28}N_2$, 2,6-cis-Dimethylpiperidyl-N-phenyl-2,2-dimethylpropionamidine, 45B, 240
$C_{19}H_{18}N_4O_2$, Ethyl 2-amino-3-cyano-4,5-di(4-pyridyl)cyclopent-2-en-1-carboxylate, 43B, 307
$C_{19}H_{22}FN_3O$, 4'-Fluoro-4-(4-(2-pyridyl)-1-piperazinyl)butyrophenone, 43B, 307
$C_{19}H_{22}NO_4$, 2,6-Dimethyl-4-phenyl-3,5-diethoxycarbonyl-1,4-dihydropyridine, 43B, 1479
$C_{19}H_{22}N_4O_2$, 1-Propyl-3,5-dicyano-4-phenyl-6-hydroxy-2-pyridone propylamine, 41B, 290; 42B, 190
$C_{19}H_{23}NO_4$, Diethyl 2,6-dimethyl-4-phenyl-1,4-dihydro-3,5-pyridinedicarboxylate, 44B, 219
$C_{19}H_{24}N_2O_6S$, Ethyl 3-(4,5-dimethoxy-2-(5-methyl-2-pyridylsulfamoyl)-

phenyl)propionate, 44B, 219
$C_{19}H_{25}ClNO_2$, Diphenylpyraline hydrochloride monohydrate, 41B, 291
$C_{19}H_{25}IN_2$, 1-Phenylamino-4-N-piperidiniomethylbenzene iodide, 43B, 308
$C_{19}H_{27}NO_7$, (-)-((-)-1-Methyl-3-ethyl-3-benzoylpiperidine R:R-(+)-bitartrate), 42B, 191
$C_{19}H_{34}N_2O_5$, Di(4-(2,2,6,6-tetramethylpiperidine-1-oxyl)) carbonate, 42B, 66
$C_{20}H_{14}N_2O_2$, trans-2,5-Dimethyl-1,4-dibenzoylpiperazine, 43B, 54
$C_{20}H_{16}N_2$, 2,5-Distyrylpyrazine, 37B, 132
$C_{20}H_{16}N_2$, 2,5-Distyrylpyrazine (γ form), 42B, 191
$C_{20}H_{16}N_8$, r-1,c-2,t-3,t-4-Tetra(2-pyrazinyl)cyclobutane, 46B, 238
$C_{20}H_{17}N_4$, 2,4,6-Triphenylverdazyl, 39B, 188
$C_{20}H_{18}Cl_4N_4O_2S_2U$, Bis(2-pyridylthio-2-pyridinium) tetrachlorodioxouranate(VI), 44B, 220
$C_{20}H_{22}N_2O_2$, (RS)-1,4-Dibenzoyl-cis-2,5-dimethylpiperazine, 43B, 309
$C_{20}H_{22}N_2O_2$, (S)-1,4-Dibenzoyl-cis-2,5-dimethylpiperazine, 44B, 221
$C_{20}H_{22}N_2O_2$, trans-1,4-Dibenzoyl-2,5-dimethylpiperazine, 45B, 242
$C_{20}H_{22}N_4P_2S_8$, Pyridinium 2,2,5,5-tetrathio-cyclo-di(phosphadithianate), 44B, 221
$C_{20}H_{23}ClN_2O_5$, Carbinoxamine maleate, 46B, 234
$C_{20}H_{25}NO$, 1,1-Diphenyl-3-piperidino-1-propanol, 46B, 234
$C_{20}H_{25}N_2O_2$, 2,2,6,6-Tetramethylpiperidine-1-iminoxyl-4-(N-2-hydroxy-1-naphthaldehydeimine), 40B, 234
$C_{20}H_{26}ClN$, N-Benzyl-N-ethyl-4-phenylpiperidinium chloride, 39B, 189
$C_{20}H_{26}ClNO_2$, (1S,3R,4S)-1-Methyl-3-(4-methoxyphenoxymethyl)-4-phenylpiperidinium chloride, 45B, 243
$C_{20}H_{27}BrN_2$, N-(2-Diphenylaminoethyl)-N-methylpiperidinium bromide, 43B, 310; 44B, 220
$C_{20}H_{31}NO$, *a*-Cyclohexyl-*a*-phenyl-1-piperidinepropanol, 38B, 273
$C_{21}H_9BrN_2$, 1,3-Bis(1'-methyl-2'-pyridinium)indenide bromide, 46B, 235
$C_{21}H_{12}Cl_3N_3$, 2,4,6-Trichloro(p-chlorophenyl)-s-triazine, 30B, 135
$C_{21}H_{15}N_3$, s-Triphenyltriazine, 30B, 134
$C_{21}H_{15}N_3O_3$, 1,3,5-Triphenyl-1,3,5-perhydrotriazine-2,4,6-trione, 45B, 243
$C_{21}H_{19}BrN_2$, 1,3-Bis(1'-methylpyridinium)indenide bromide, 41B, 292
$C_{21}H_{19}N_4$, 1,3,5-Triphenyl-6-methylverdazyl, 41B, 293
$C_{21}H_{19}N_7O_2 \cdot 0.5\ H_2O$, 2,6-Diacetylpyridine bis(picolinoylhydrazone) hemihydrate, 45B, 243
$C_{21}H_{20}N_2O_3$, N-Carbomethoxyamino-3,6-dimethyl-4,5-diphenyl-2-pyridone, 41B, 294
$C_{21}H_{21}N_3O_6S_3$, Tri-N-(phenylsulphonyl)hexahydro-s-triazine, 44B, 209
$C_{21}H_{22}ClN \cdot 1.5\ H_2O$, 5-(1-Methylpiperidylidene-4)-5H-dibenzo[a,d]-cycloheptene hydrochloride sesquihydrate, 43B, 311; 44B, 222
$C_{21}H_{22}N_2O_3$, 1-Benzoyl-2-ethoxymethyl-4-methyl-5-phenyl-1,2,3,4-tetrahydropyridazin-3-one, 42B, 192
$C_{21}H_{23}ClFNO_2$, Haloperidol, 39B, 190
$C_{21}H_{23}F_2NO_2$, 4'-Fluoro-4-(1-(4-hydroxy-4-(4'-fluoro)-phenylpiperidino))-butyrophenone, 38B, 274
$C_{21}H_{24}BrClFNO_2$, 4-[4-(4-Chlorophenyl)-4-hydroxypiperidino]-4'-fluorobutyrophenone hydrobromide, 45B, 244
$C_{21}H_{24}ClF_2NO_2$, 4'-Fluoro-4-(1-(4-hydroxy-4-(4'-fluoro)-phenylpiperidino))-butyrophenone hydrochloride, 38B, 274
$C_{21}H_{26}BrNO_3 \cdot H_2O$, 3-(Hydroxy(diphenyl)acetyloxy)-1,1-dimethylpiperidinium bromide monohydrate, 45B, 245
$C_{21}H_{26}ClNO_2$, N-Ethyl-3-piperidyl diphenylacetate hydrochloride, 40B,

15
$C_{21}H_{28}IN$, N-(3,3-Diphenylpropyl)-N-methylpiperidinium iodide, 43B, 311
$C_{21}H_{30}ClNO$, N-Benzyl-N-isopropyl-4-phenylpiperidinium chloride monohydrate, 39B, 189
$C_{21}H_{30}FN_3O_2$, 1'-(4-(4-Fluorophenyl)-4-oxobutyl)-(1,4'-bipiperidine)-4'-carboxamide, 41B, 295
$C_{21}H_{33}ClN_2O_2$, Soventol hydrochloride monohydrate ethanolate, 43B, 312
$C_{22}H_{18}N_6$, r-1,c-2-Di(3-pyridyl)-t-3,t-4-di(2-pyrazinyl)cyclobutane, 46B, 238
$C_{22}H_{18}N_6$, r-1,t-3-Di(3-pyridyl)-c-2,t-4-di(2-pyrazinyl)cyclobutane, 46B, 238
$C_{22}H_{21}NO_4 \cdot 0.5\ CH_2Cl_2$, 4-(o-Carboxyphenyl)-1-oxo-3-oxy-2-phenyl-5-azoniaspiro[4.5]decane inner salt methylene chloride solvate, 42B, 192
$C_{22}H_{21}N_4$, 1,3-Di-p-tolyl-5-phenylverdazyl radical, 46B, 237
$C_{22}H_{22}Br_2N_2O_2$, 1-Ethoxycarbonyl-2-(4-methylphenyl)vinylenedipyridinium dibromide, 46B, 237
$C_{22}H_{26}N_2O_3S$, Z-4-Phenyl-4-piperidino-3-(p-tolylsulphonylamino)but-3-en-2-one, 45B, 245
$C_{22}H_{30}INO$, 3-(2-Methylpiperidino)-1-phenylpropyl phenyl ether methiodide, 41B, 295
$C_{22}H_{30}N_2O_2$, 1,4-Di-(4-aza-N-oxide-3,3,5,5-tetramethyl-cyclohex-1-enyl)buta-1,3-diyne, 41B, 296
$C_{22}H_{32}N_4OS_4$, 2,2',6,6'-Tetrakis(propylthio)-3,3'-azoxypyridine, 46B, 238
$C_{22}H_{42}N_2O_2$, 1,4-Bis(2,2,6,6-tetramethyl-1-oxyl-4-piperidyl)butane, 40B, 76
$C_{22}H_{46}N_4O_4$, N,N'-Bis-(2,2,6,6-tetramethylpiperidyl-4)-succinic acid diamide dihydrate, 40B, 16
$C_{23}H_{23}N_4$, 1,3,5-Tri-p-tolylverdazyl, 41B, 297
$C_{23}H_{26}N_2O_2$, 1-Benzyl-4-(2,6-dioxo-3-phenyl-3-piperidyl)piperidine, 39B, 190
$C_{23}H_{27}BrN_2O_2 \cdot 0.5\ H_2O$, (+)-1-Benzyl-4-(2,6-dioxo-3-phenyl-3-piperidyl)piperidine hydrobromide hemihydrate, 42B, 193
$C_{23}H_{29}NO_3$, DL-4-Phenyl-4-ethoxycarbonyl-1-(3-hydroxy-3-phenylpropyl)piperidine, 44B, 223
$C_{23}H_{31}ClN_6O_5S$, 4,6-Diamino-1-[3-chloro-4-(m-dimethylcarbamoylbenzyloxy)phenyl]-1,2-dihydro-2,2-dimethyl-s-triazine ethanesulphonate, 45B, 246
$C_{23}H_{31}FN_6O_6S_2 \cdot 2\ H_2O$, 4,6-Diamino-1-[4-(4'-fluorosulphonyl-3'-methylanilinocarbonylethyl)phenyl]-1,2-dihydro-2,2-dimethyl-s-triazine ethanesulphonate dihydrate, 45B, 246
$C_{23}H_{32}INO$, 3-(2-Methylpiperidino)-1-phenylpropyl 2-tolyl ether methiodide, 42B, 90
$C_{23}H_{32}INO$, 3-(2-Methylpiperidino)-1-phenylpropyl 3-tolyl ether methiodide, 43B, 313
$C_{24}H_{20}N_4$, r-1,c-2,t-3,t-4-Tetra(2-pyridyl)cyclobutane, 46B, 238
$C_{24}H_{20}N_4$, r-1,c-2,t-3,t-4-Tetra(4-pyridyl)cyclobutane, 46B, 238
$C_{24}H_{21}N_3O_3$, 4,5,6-Tris-(p-methoxyphenyl)-1,2,3-triazine, 38B, 275
$C_{24}H_{22}I_2N_2$, N,N'-Dibenzyl-4,4'-bipyridylium diiodide, 37B, 133
$C_{24}H_{23}N_3O_4S$, Ethyl 2-benzamido-5-benzoyl-4-dimethylamino-6-thioxonicotinate, 37B, 134
$C_{24}H_{29}FN_2O_2$, N-(1-(3-(p-Fluorobenzoyl)propyl)-4-piperidyl)propionanilide, 40B, 119
$C_{24}H_{29}FN_2O_2$, N-(1-(4-Fluorophenyl-4-oxobutyl)-4-phenyl-4-

piperidinylmethyl)acetamide, 43B, 314

$C_{24}H_{32}N_2O_2$, N-(4-(Methoxymethyl)-1-(2-phenylethyl)-4-piperidinyl)-N-phenylpropanamide, 42B, 193

$C_{26}H_{28}N_2$, trans-1-Cinnamyl-4-diphenylmethyl-piperazine, 41B, 298

$C_{26}H_{30}ClFN_2O_2$, 1-(4-(3-Chlorophenyl)-1-(4-fluorophenyl-4-oxobutyl)-4-piperidinylcarbonyl)-pyrrolidine, 42B, 194

$C_{26}H_{46}N_2O_6$, Di-(2,2,6,6-tetramethyl-4-piperidinyl-1-oxyl) suberate, 38B, 115

$C_{27}H_{38}N_4O_3$, 1'-(3-Cyano-3,3-diphenylpropyl)(1,4'-bipiperidine)-4'-carboxamide dihydrate, 43B, 314

$C_{28}H_{20}Cl_2N_2$, 1,4-Di-p-chlorophenyl-2,6-diphenyl-1,4-dihydropyrazine, 41B, 299

$C_{28}H_{23}N_3$, 2-(1,5-Dimethyl-3,4,6-triphenyl-1,2,3,4-tetrahydropyridine-2-ylidene)malononitrile, 44B, 223

$C_{28}H_{27}ClF_5NO$, 4-(4-Chloro-α,α,α-trifluoro-m-tolyl)-1-(4,4-bis(p-fluorophenyl)butyl)-4-piperidinol, 39B, 191

$C_{29}H_{21}NO_2 \cdot 0.5\ H_2O$, 1-(4-Hydroxyphenyl)-2,4,6-triphenylpyridinium-3-olate hemihydrate, 46B, 219

$C_{29}H_{35}ClN_2O_3$, 4-(p-Chlorophenyl)-4-hydroxy-N,N-dimethyl-α,α-diphenylpiperidine-1-butyramide hydrate, 43B, 315

$C_{30}H_{32}N_4$, 1,2,4,5-Tetrabenzylhexahydro-s-tetrazine, 44B, 223

$C_{30}H_{35}ClN_2O_2$, 4-(p-Chlorophenyl)-4-hydroxy-N,N,γ-trimethyl-α,α-diphenyl-1-piperidinebutyramide, 43B, 315

$C_{30}H_{35}F_2N_3O$, 4-(4,4-Bis(p-fluorophenyl)butyl)-1-piperazineacetyl-2',6'-xylidide, 43B, 316

$C_{35}H_{44}N_2O_8$, N-[1-(2-Phenylethyl)-4-piperidinylium]-N-phenylpropanamide citrate toluene solvate, 45B, 247

$C_{40}H_{32}N_4$, 2,5-Distyrylpyrazine 1,4-bis[2-(2-pyridyl)vinyl]benzene, 45B, 248

$C_{43}H_{34}BrNO_2$, 2,6-Diphenyl-4-(4-bromophenyl)-N-(p-oxy-m,m'-diphenyl)-phenyl-pyridinium betaine monoethanol solvate, 34B, 143

$C_{44}H_{47}Cl_4N_{11}O_2U$, Bis(protonated 2,6-diacetylpyridine-bis(phenylhydrazone)) uranyl tetrachloride acetonitrile, 41B, 300

$C_{52}H_{48}N_8O_4$, 1-Benzyl-3-carbamoylpyridine cyclotetracondensate, 43B, 317

34. HETERO-NITROGEN (7- AND HIGHER-MEMBERED MONOCYCLIC)

$C_4H_8N_8O_8$, Octahydro-1,3,5,7-tetranitro-1,3,5,7-tetrazocine (α form), 28, 467

$C_4H_8N_8O_8$, Octahydro-1,3,5,7-tetranitro-1,3,5,7-tetrazocine (β form), 19, 608; 28, 469; 35B, 179

$C_4H_8N_8O_8$, 1,3,5,7-Tetranitro-1,3,5,7-tetraazacyclooctane (δ form), 40B, 235

$C_6H_{11}NO$, ϵ-Caprolactam, 40B, 236; 41B, 301

$C_6H_{12}ClNO$, Caprolactam hydrochloride, 41B, 302

$C_7H_{13}ClN_2O_4$, 2,3-Dihydro-5,7-dimethyl-1,4-diazepinium perchlorate, 46B, 239

$C_7H_{13}NO$, 7-Heptanelactam, 45B, 249

$C_7H_{14}ClNO$, Enantholactam hydrochloride, 41B, 303

$C_8H_{14}N_2O_2$, 1,5-Diazacyclodeca-6,10-dione, 41B, 304

$C_8H_{14}N_6O_6$, 1,5-Diacetyl-3,7-dinitro-1,3,5,7-tetraazacyclooctane, 41B, 305

$C_8H_{15}NO$, Caprylolactam, 41B, 305

$C_8H_{16}ClNO$, Caprylolactam hydrochloride, 41B, 306

$C_9H_{10}N_2O$, 1,2,3,4-Tetrahydro-1-methyl-2-oxo-3-azocinecarbonitrile,

45B, 249
$C_9H_{17}NO$, Pelargolactam, 41B, 307
$C_9H_{17}NS$, syn-N-Methylthiocapryllactam, 38B, 277
$C_9H_{17}N_2O_2$, 2,2,7,7-Tetramethyl-5-oxo-homopiperazin-1-oxyl, 45B, 250
$C_9H_{18}N_2O$, 2,2,7,7-Tetramethyl-5-homopiperazinone, 45B, 250
$C_{10}H_{16}N_4O_4$, 1,4,7,10-Tetraazacyclotetradecane-3,8,11,14-tetraone, 46B, 240
$C_{10}H_{26}Cl_2N_4O_8$, 1,4,8,11-Tetra-azacyclotetradecane di(hydroperchlorate) (α-form), 40B, 236
$C_{11}H_{14}N_2O$, N-(N',N'-Dimethylcarbamyl)azonine, 40B, 237
$C_{11}H_{21}N_2O_2$, 4-Acetyl-1-oxyl-2,2,7,7-tetramethylhomopiperazine, 45B, 251
$C_{11}H_{24}ClN$, Azacyclododecane hydrochloride, 29, 513
$C_{12}H_{12}N_2O_2S$, 1-Tosyl-1,2-diazepine, 38B, 275
$C_{12}H_{20}N_4O_4$, Cyclotetrasarcosyl, 35B, 181
$C_{12}H_{20}N_4O_4$, 1,3,5,7-Tetraaceto-1,3,5,7-tetraazacyclooctane, 39B, 191
$C_{12}H_{22}N_2O_2$, 1,8-Diazacyclotetradecane-2,7-dione, 33B, 120
$C_{12}H_{22}N_2O_2$, 1,8-Diazacyclotetradecane-2,9-dione, 37B, 134
$C_{12}H_{23}NO$, 5,5,8,8-Tetramethylazacyclononan-2-one, 45B, 251
$C_{12}H_{26}N_2O_2$, 1,8-Dihydroxy-1,8-diazacyclotetradecane, 31B, 129
$C_{12}H_{28}Br_2N_2$, 1,8-Diazacyclotetradecane dihydrobromide, 30B, 136
$C_{12}H_{28}N_4$, 1,5,9,13-Tetraazacyclohexadecane, 44B, 224
$C_{13}H_{11}NO_2$, N-(Phenoxycarbonyl)azepine, 38B, 276
$C_{13}H_{23}N_3O_2$, 6,6-Diethyl-3-dimethylamino-2,2-dimethyl-1,5,6,7-tetrahydro-2H-(1,4)-diazepin-5,7-dione, 44B, 224
$C_{13}H_{25}NS$, syn-N-Methylthiolauryllactam, 38B, 277
$C_{14}H_{19}NO$, 1-p-Tolyl-1-azacyclooctan-5-one, 41B, 307
$C_{14}H_{24}N_4O_4$, 1,2,8,9-Tetraaza-1,8-cyclo-tetradecadiene-3,10-ylene diacetate, 40B, 238
$C_{16}H_{36}N_4 \cdot 2\ H_2O$, meso-5,5,7,12,12,14-Hexamethyl-1,4,8,11-tetraazacyclotetradecane dihydrate, 46B, 240
$C_{16}H_{36}N_4$, 2,5,8,11-Tetraethyl-1,4,7,10-tetraazacyclododecane, 44B, 225
$C_{16}H_{38}N_4O$, rac-5,5,7,12,12,14-Hexamethyl-1,4,8,11-tetraazacyclotetradecane monohydrate, 43B, 318
$C_{18}H_{19}N_3OS$, 2,5-N,N'-Dibenzyl-2,5,7-triazacycloheptane-1-thione-6-one, 44B, 225
$C_{18}H_{20}N_2$, trans-3,8-Diphenyl-1,2-diaza-(E)-1-cyclooctene, 42B, 195
$C_{18}H_{35}ClN_2O_2$, Pelargolactam hemihydrochloride, 41B, 308
$C_{19}H_{17}NO$, 3-(Diphenylhydroxymethyl)-3H-azepine, 39B, 192
$C_{19}H_{22}ClNO$, DL-1-Benzyl-5-phenyl-1-azacycloheptan-4-one hydrochloride, 40B, 240
$C_{20}H_{21}ClN_2O_4$, 6-Cyclopropyl-2,3-dihydro-1,4-diphenyl-1,4-diazepinium perchlorate, 45B, 252
$C_{20}H_{26}Br_2N_2O_3$, 3,7-Dihydroxy-1,5-bis(p-bromophenyl)octahydro-1,5-diazocine ethanolate, 42B, 195; 43B, 319
$C_{20}H_{26}N_2O_6S_2$, cis-1,5-Bis(toluene-p-sulphonyl)-3,7-dihydroxy-octahydro-1,5-diazocine, 41B, 309
$C_{20}H_{39}ClN_2O_2$, Caprinolactam hemihydrochloride, 41B, 310
$C_{21}H_{18}Cl_2N_2O_2$, 1,4-Bis(p-chlorobenzoyl)-5-methylene-7-methyl-2,3,4,5-tetrahydro-1H-1,4-diazepine, 43B, 320
$C_{21}H_{18}N_2O_3$, 5-Acetyloxy-1-benzoyl-6-methyl-7-phenyl-1,3-diazepine, 44B, 226
$C_{21}H_{25}N_5$, 3,7-Bis(dimethylamino)-6-methyl-2,5-diphenyl-1,2,4-triazepine, 45B, 252
$C_{24}H_{36}N_4$, 5,12-Dimethyl-7,14-diphenyl-1,4,8,11-tetraazatetradecane, 44B, 820

$C_{24}H_{44}N_4O_4$, 1,8,15,22-Tetra-aza-2,7,16,21-tetra-oxocyclo-octacosane, 35B, 181
$C_{26}H_{23}NO_8$, 3,5-Diphenyl-2,4,6,7-tetrakis-methoxycarbonyl-3H-azepine, 38B, 279
$C_{26}H_{23}NO_8$, 4,6-Diphenyl-2,3,5,7-tetrakis-methoxycarbonyl-3H-azepine, 38B, 279
$C_{27}H_{54}ClN_3O_4$, Tripelargolactam oxonium chloride, 41B, 310
$C_{29}H_{45}NO_2 \cdot C_2H_6O$, 3,5-Di-t-butyl-7-(3,5-di-t-butyl-2-hydroxyphenyl)-1-methyl-2,3-dihydro-1H-azepin-2-one ethanol, 45B, 253
$C_{35}H_{24}I_2N_2$, 3,7-Bis(p-iodophenyl)-4,5,6-triphenyl-4-H-1,2-diazepine, 35B, 184
$C_{44}H_{60}N_4$, (2R,5R,8S,11S)-1,4,7,10-Tetrabenzyl-2,5,8,11-tetraethyl-1,4,7,10-tetraazacyclododecane, 45B, 253
$C_{44}H_{60}N_4$, (2RS,5RS,8RS,11SR)-1,4,7,10-Tetrabenzyl-2,5,8,11-tetraethyl-1,4,7,10-tetraazacyclododecane, 46B, 241
$C_{44}H_{60}N_4$, (2RS,5RS,8RS,11RS)-1,4,7,10-Tetrabenzyl-2,5,8,11-tetraethyl-1,4,7,10-tetraazacyclododecane, 44B, 226
$C_{50}H_{50}N_4$, meso-7,7'-Bi(1,4-dibenzyl-6-phenyl-1,2,3,4-tetrahydro-1,4-diazepine), 46B, 241
$C_{50}H_{50}N_4$, racemic-7,7'-Bi(1,4-dibenzyl-6-phenyl-1,2,3,4-tetrahydro-1,4-diazepine), 46B, 241

35. HETERO-NITROGEN (2 FUSED RINGS)

$C_4H_3N_5$, Tetrazolo[1,5-b]pyridazine, 44B, 227
$C_4H_6N_4$, Trimethylenetetrazole, 45B, 254
$C_5H_4N_4$, Imidazo[1,2-b]-as-triazine, 43B, 320
$C_5H_4N_4$, s-Triazolo[1,5-b]pyridazine, 44B, 227
$C_5H_4N_4$, s-Triazolo[4,3-b]pyridazine, 44B, 227
$C_5H_4N_4O$, 6-Hydroxy-s-triazolo[4,3-b]pyridazine, 43B, 321
$C_5H_5N_5O$, 3-Methoxy-1H-pyrazolo[4,3-e]as-triazine, 38B, 281
$C_6H_4N_4$, Pteridine, 20, 627; 41B, 312
$C_6H_4N_4$, Pyridazino[4,5-d]pyridazine, 34B, 147
$C_6H_4N_4$, 1,4,5,8-Tetraazanaphthalene, 37B, 136
$C_6H_5N_3$, Benzotriazole, 40B, 241
$C_6H_5N_3O$, 1,2-Dihydro-3H-pyrazolo-[3,4-b]pyridin-3-one, 45B, 256
$C_6H_6Br_2N_4$, 2,5-Dimethyl-3,6-dibromo-1,3a,4,6a-tetra-azapentalene, 28, 457
$C_6H_6ClN_5O_2$, Xanthopterine hydrochloride, 42B, 197
$C_6H_7N_3O_4S$, Benzotriazolium hydrogensulfate, 46B, 241
$C_6H_{10}N_4$, Pentamethylenetetrazole, 45B, 254
$C_6H_{10}N_4Na_2O_8$, 2,4,6,8-Tetrahydroxypyrimido[5,4-d]pyrimidine disodium salt, 31B, 131
$C_6H_{10}N_4O_2$, 3a,6a-Dimethylglycouril, 44B, 227
$C_6H_{13}N_3$, 2,4,6-Trimethyl-1,3,5-triazabicyclo[3.1.0]hexane, 45B, 256
$C_6H_{14}N_4$, 1,4,6,9-Tetraazabicyclo[4.4.0]decane, 44B, 228
$C_7H_3Cl_2N_3$, 1,4-Dichloro-7-methylpyrrolo[3,2-d]pyridazine, 44B, 228
$C_7H_4N_4$, 3-Diazoindazole, 44B, 229
$C_7H_5N_3O$, 1,2,3-Benzotriazin-4(3H)-one, 39B, 193
$C_7H_6N_2$, Benzimidazole, 39B, 194; 40B, 264
$C_7H_6N_2$, Indazole, 40B, 243
$C_7H_6N_2S$, 2-Mercaptobenzimidazole, 42B, 197
$C_7H_8N_4O$, 6,8-Dimethylimidazo[1,5-a]-1,3,5-triazin-4(3H)-one, 45B, 257
$C_7H_9N_5O$, 3,6-Dimethyl-7,8-dihydro-9H-s-triazola[4,3-b](1,2,4)-triazepin-8-one, 41B, 312

$C_7H_{10}N_2O_2$, 1,7-Diazaspiro[4.4]nonane-2,6-dione, 45B, 257
$C_7H_{10}N_2O_2$, 2,7-Diazaspiro[4.4]nonane-1,6-dione, 46B, 242
$C_7H_{12}ClN_5O_2$, 6-Methyl-7,8-dihydropterine monohydrochloride monohydrate, 43B, 321
$C_7H_{13}Br_2N$, trans-1-Bromopyrrolizidine hydrobromide, 45B, 258
$C_7H_{13}N_3O_4$, Spinacine dihydrate, 37B, 136
$C_8Cl_2F_4N_2$, 4,5-Dichloro-1,3,6,8-tetrafluoro-2,7-naphthyridine, 43B, 326
$C_8Cl_6N_2$, Hexachloroquinoxaline, 45B, 258
C_8Cl_7N, Heptachloro-5H-1-pyrindine, 40B, 243
$C_8H_4ClN_3O_4$, 5-Chloro-7-nitro-2,3-dihydroxyquinoxaline, 43B, 322
$C_8H_4N_2O_4$, 1,4,5,8-Pyridazino[1,2-a]pyridazinetetrone, 43B, 322
$C_8H_5NO_2$, Isatin, 13, 570
$C_8H_5NO_2$, Phthalimide, 38B, 282
$C_8H_6ClN_3O_2 \cdot 0.5\ HCl$, 5-Chloro-7-amino-2,3-dihydroxyquinoxaline hydrochloride, 43B, 323
$C_8H_6Cl_2N_4O_2$, 2,6-Dimethyl-4,8-dichloro-2H,6H-pyridazino[4,5-d]pyridazin-1,5-dione, 38B, 283
$C_8H_6N_2$, 1,3-Diazanaphthalene, 42B, 198
$C_8H_6N_2$, 1,5-Diazanaphthalene, 44B, 229
$C_8H_6N_2$, 1,8-Naphthyridine, 38B, 284
$C_8H_6N_2$, 2,3-Diazanaphthalene, 38B, 284
$C_8H_6N_2$, 2,6-Diazanaphthalene, 44B, 229
$C_8H_6N_2$, 2,7-Diazanaphthalene, 43B, 324
$C_8H_6N_2O$, 2-Hydroxyquinoxaline, 42B, 199
$C_8H_6N_2O_2$, 1,4-Dihydro-2,3-quinoxalinedione, 42B, 199
$C_8H_7ClN_2O_4$, Quinoxalinium monoperchlorate, 43B, 324
C_8H_7N, Indole, 41B, 313
$C_8H_7NO_2$, 3-Hydroxyphthalimidine, 44B, 230
$C_8H_7N_3O_2$, Benzotriazol-2-ylacetic acid, 44B, 230
$C_8H_7N_3O_2$, 1-Benzotriazoleacetic acid, 43B, 325
$C_8H_8BrN_3 \cdot H_2O$, 2-Amino-1,3-diazaazulene hydrobromide monohydrate, 45B, 259
$C_8H_8ClN_5 \cdot H_2O$, 2,4,6-Triamino-5-chloroquinazoline hydrate, 46B, 242
$C_8H_9ClN_4$, 1-Hydrazinophthalazine hydrochloride, 44B, 231
$C_8H_9NO_2$, 1,2,3,6-Tetrahydrophthalimide, 42B, 199
$C_8H_9NO_2$, 3,4,5,6-Tetrahydrophthalimide, 41B, 313
$C_8H_{10}BrNO_3$, 7α-Bromo-7β-carboxy-8-oxo-6αH-1-azabicyclo[4.2.0]octane, 41B, 314
$C_8H_{10}N_2O_2$, 1,5-Naphthyridine dihydrate, 26, 779
$C_8H_{10}N_4 \cdot 0.5\ H_2O$, 7-Amino-2,5-dimethyl-pyrazolo[1,5-a]pyrimidine hemihydrate, 41B, 315
$C_8H_{10}N_4O_6Rb_2$, Rubidium 2,5-dimethyl-1,3a,4,6a-tetra-azapentalene-3,6-dicarboxylate dihydrate, 28, 456
$C_8H_{11}N_3O_3$, 6,7-Dihydro-6,6-dimethylbenzofurazan-4(5H)-one 3-oxide oxime, 43B, 325
$C_8H_{11}N_5O_2$, 1,6-Dimethyl-7-ethoxycarbonylpyrazolo[1,5-d]tetrazole, 44B, 231
$C_8H_{12}BrNO_4$, 7α-Carboxy-7β-bromo-8-oxo-6αH-1-azabicyclo[4.2.0]octane monohydrate, 41B, 314
$C_8H_{12}BrN_5O$, 2-Amino-4-ethoxy-3,4-dihydropteridin-1-ium bromide, 34B, 146
$C_8H_{12}N_2O$, (DL)-8,8-Dimethyl-6,7-diazabicyclo[3.3.0]octa-1,6-diene 7-oxide, 46B, 243
$C_8H_{12}N_6O_4S \cdot 2.64\ H_2O$, 1,4-Dihydrazinophthalazinium sulfate 2.64-hydrate, 45B, 260
$C_8H_{12}N_6O_4S \cdot 3\ H_2O$, 1,4-Dihydrazinophthalazine hydrosulfate trihyd-

rate, 45B, 272
$C_9H_5Br_2NO$, 5,7-Dibromo-8-hydroxyquinoline, 42B, 200
C_9H_5ClINO, 5-Chloro-7-iodo-8-quinolinol, 39B, 194
$C_9H_5ClN_6$, 5-p-Chlorophenyl-tetrazolo[1,5-b]-1,2,4-triazine, 42B, 200
C_9H_6ClN, 2-Chloro-1-aza-azulene, 23, 704
C_9H_6ClN, 2-Chloroquinoline, 28, 470; 33B, 123
C_9H_6ClN, 6-Chloroquinoline, 33B, 124
C_9H_6ClNO, 1-Chloro-3-hydroxyisoquinoline, 40B, 244
$C_9H_6Cl_2IN$, 2-Chloroquinoline iodomonochloride, 42B, 200
$C_9H_6IO_4NS$, 7-Iodo-8-hydroxyquinoline-5-sulphonic acid, 35B, 185
$C_9H_6KNO_3S_2$, Potassium 5-sulphonato-8-mercaptoquinoline, 46B, 243
$C_9H_6NNaS \cdot 2\ H_2O$, Sodium quinoline-8-thiolate dihydrate, 45B, 260
$C_9H_6NO_3RbS_2$, Rubidium 8-mercaptoquinoline-5-sulphonate, 45B, 261
$C_9H_6N_2NaO_6S$, Disodium 7-oximato-5-sulphonato-quinoline-8-one monohydrate, 44B, 232
$C_9H_6N_2O_5S$, 7-Hydroximino-5-sulphonato-quinolinium-8-one, 44B, 232
C_9H_7NO, 2-Hydroxyquinoline, 43B, 327
C_9H_7NO, 8-Hydroxyquinoline, 44B, 233
$C_9H_7NO_2$, 8-Hydroxyquinoline-N-oxide, 37B, 137
$C_9H_8BrN_3O_2$, 3-Amino-6-bromo-1-methyl-2,4(1H,3H)-quinazolinedione, 44B, 233
C_9H_8ClN, Isoquinoline hydrochloride, 30B, 137
$C_9H_8Cl_3NO_2Se$, 8-Hydroxyquinolinium trichlorooxyselenate, 32B, 158
$C_9H_8N_2$, 4-Methylcinnoline, 45B, 262
$C_9H_8N_2O$, 2-Methyl-quinazoline-3-oxide, 41B, 317
$C_9H_8N_2O_2$, 2-Acetyl-3-indazoline, 34B, 151
$C_9H_8INO_3$, 7-Iodoadrenochrome, 45B, 262
$C_9H_9NO_5S$, 8-Quinolinol-7-sulphonic acid monohydrate, 42B, 201
$C_9H_{10}ClNO$, Isoquinoline hydrochloride monohydrate, 30B, 137
$C_9H_{10}Cl_3MnNO$, Quinolinium trichloromanganate(II) monohydrate, 44B, 208
$C_9H_{11}NO_2$, Octahydroquinoline-2,5-dione, 42B, 201
$C_9H_{11}NO_2$, 5-Hydroxy-5,6,7,8-tetrahydroquinoline-1-oxide, 41B, 319
$C_9H_{13}AuCl_4N_2O_3$, Hydronium tetrachloroaurate(III) - 2,2-dimethyl-2H-benzimidazole 1,3-dioxide, 43B, 327
$C_9H_{13}NO_3$, 7α-Carboxy-7β-methyl-8-oxo-6αH-1-azabicyclo[4.2.0]octane, 41B, 314
$C_9H_{14}ClN_5$, 2-Amino-5-propyl-7-methyl-s-triazolo-[2,3-c]-pyrimidine hydrochloride, 27, 978
$C_9H_{15}N_7S$, 5,7-Bis(dimethylamino)-2-(methylthio)-s-triazolo-[1,5-a]-s-triazine, 39B, 195
$C_9H_{18}BrN$, Decahydroquinoline hydrobromide, 43B, 328
$C_9H_{18}IN$, 8,8-Dimethyl-8-azoniabicyclo[5.1.0]octane iodide, 29, 506
$C_9H_{19}Cl_2N_5O_2$, 5,6,7-Trimethyl-5,6,7,8-tetrahydropterine dihydrochloride monohydrate, 43B, 328
$C_{10}H_6ClNO_3$, 4-Chloro-2-carboxyquinoline N-oxide, 39B, 896
$C_{10}H_8N_2O \cdot 0.5\ H_2O$, Quinoline-2-carboxamide hemihydrate, 43B, 329
$C_{10}H_8N_2O$, Quinoline-2-carboxamide, 43B, 329
$C_{10}H_8N_2O_4$, N-Phthaloylglycinehydroxamic acid, 46B, 244
$C_{10}H_8N_2O_4$, 6-Methoxy-8-nitro-5(1H)-quinoline, 34B, 151
$C_{10}H_8N_4$, (E)-2,2',5,5'-Tetrazastilbene, 46B, 229
$C_{10}H_9BrN_2O_2$, 8a-Bromo-1,2,3,5,6,7,8,8a-octahydro-1,3-dioxoisoquinoline-4-carbonitrile, 41B, 319
$C_{10}H_9N$, 3-Methylisoquinoline, 40B, 245
$C_{10}H_9NO_2$, 3-Indolylacetic acid, 29, 726
$C_{10}H_{10}N_2O_2$, 4,8-Dimethoxy-1,5-naphthyridine, 43B, 330
$C_{10}H_{10}N_2O_3$, N-Ethanol-β-isatoxime, 39B, 195

$C_{10}H_{10}N_4O_2$, (1E,2Z)-2,3-Dimethyl-1-phthalimido-azimine, 45B, 263
$C_{10}H_{10}N_4O_2$, (1Z,2E)-2,3-Dimethyl-1-phthalimido-azimine, 45B, 263
$C_{10}H_{11}BrN_2$, 2-p-Bromophenyl-1,3-diazabicyclo[3.1.0]hexane, 41B, 320
$C_{10}H_{11}NS$, N-Thioacetylindolin, 44B, 233
$C_{10}H_{11}N_3$, 6-Dimethylamino-5,7-diaza-azulene, 35B, 187
$C_{10}H_{11}O_5NS$, 2-Methyl-8-hydroxyquinoline-5-sulphonic acid monohydrate, 35B, 185
$C_{10}H_{12}AsNO_5$, 1-Methyl-2-quinolinium dihydrogen arsenate, 34B, 148
$C_{10}H_{12}N_2O_2$, 2,3,6,7-Tetramethyl-1,5-diazabicyclo[3.3.0]octa-2,6-diene-4,8-dione, 46B, 244
$C_{10}H_{12}N_2O_2$, 3,4,6,7-Tetramethyl-1,5-diazabicyclo[3.3.0]octa-3,6-diene-2,8-dione, 46B, 244
$C_{10}H_{12}N_2O_2S$, S,S-Dimethyl-N-(2-oxo-1-indolinyl)sulfoximide, 43B, 331
$C_{10}H_{12}N_4O$, 2,7-Dimethyl-5-acetylaminopyrazolo[1,5-a]pyrimidine, 41B, 321
$C_{10}H_{12}N_4O_4$, 1,4,5,8-Tetramethoxypyridazino[4,5-d]pyridazine, 38B, 285
$C_{10}H_{12}N_4S$, (Z)-5,6-Dihydro-8(7H)quinolinone thiosemicarbazone, 43B, 331; 44B, 234
$C_{10}H_{13}ClN_2$, Tryptamine hydrochloride, 39B, 196
$C_{10}H_{13}ClN_4O$, 2-[1-(4-Quinazolinyl)hydrazino]ethanol hydrochloride, 45B, 263
$C_{10}H_{14}N_2O_4$, 2,4-Diazabicyclo[4.2.0]-octane-3,5-dione-8-acetate, 40B, 245
$C_{10}H_{15}NO_2$, 1,9-Dimethyl-8-azabicyclo[4.3.0]nonane-3,7-dione, 46B, 245
$C_{10}H_{16}N_2O_2$, Cyclo-octane-spiro-5'-hydantoin, 45B, 264
$C_{10}H_{16}N_2O_2$, 7-(Aminoethylidene)-6-methoxy-8-oxo-1-azabicyclo-[4.2.0]octane, 45B, 264
$C_{10}H_{17}N_5O_3$, 5-Formyl-6,7-dimethyl-5,6,7,8-tetrahydropterine methanolate, 43B, 332
$C_{10}H_{17}N_5O_4S \cdot H_2O$, 2-Amino-5-methyl-7-propylimidazo-[5,1-f][1,2,4]-triazin-4(3H)-one methanesulphonate monohydrate, 45B, 265
$C_{10}H_{18}IN$, cis-8-Azabicyclo[4.3.0]non-3-ene methiodide, 40B, 246
$C_{10}H_{18}N_4$, Butylpentamethylenetetrazole, 45B, 254
$C_{10}H_{19}ClINO_4$, (-)-Iodolupinane perchlorate, 46B, 245
$C_{10}H_{20}ClN_3O_4$, DL-Azabiotin hydrochloride monohydrate, 43B, 333
$C_{10}H_{20}IN$, 9,9-Dimethyl-9-azoniabicyclo[6.1.0]nonane iodide, 28, 471
$C_{10}H_{20}N_2$, 3,4-Dimethyl-3,4-diazabicyclo[4.4.0]decane, 43B, 333
$C_{11}H_8Cl_3NO_3$, Methyl 3,3,4-trichloro-5-methoxyindolenine-2-carboxylate, 40B, 246
$C_{11}H_8N_2O_2S$, 8-Quinolyl cyanomethyl sulphone, 42B, 202
$C_{11}H_9BF_4N_4$, 3-Phenyltetrazolo[4,5-a]pyridinium tetrafluoroborate, 45B, 266
$C_{11}H_9NO_3$, 2-Methoxy-4-hydroxy-5-oxo-benz[f]azepine, 38B, 286
$C_{11}H_9N_3$, (E)-2',5,5'-Triazastilbene, 46B, 229
$C_{11}H_9N_3O_2$, 7-Phenyl-2,3,7-triazabicyclo[3.3.0]oct-2-ene-6,8-dione, 45B, 266
$C_{11}H_9N_5O$, 2-Phenyl-7-methyl-8-azahypoxanthine, 46B, 246
$C_{11}H_9N_5O$, 2-Phenyl-8-methyl-8-azahypoxanthine, 46B, 246
$C_{11}H_{11}NO$, 1-Acetylskatole, 44B, 235
$C_{11}H_{11}NO_2$, 3-Acetyl-1-methoxyindole, 46B, 246
$C_{11}H_{11}NO_3$, D-Methyl 3,4-dihydroisocarbostyril-3-carboxylate, 42B, 202
$C_{11}H_{11}NO_3$, 5-Methoxyindole-3-acetic acid, 41B, 322
$C_{11}H_{12}N_2$, 6-Dimethylamino-5-aza-azulene, 34B, 149
$C_{11}H_{12}N_2O_3$, 5-Hydroxy-DL-tryptophan, 39B, 197

$C_{11}H_{13}ClN_4O_2 \cdot H_2O$, 1-(N'-Ethoxycarbonylhydrazino)phthalazinium chloride monohydrate, 45B, 267
$C_{11}H_{13}NO$, trans-4-Methyl-10-azabicyclo[7.2.0]undeca-2,5,7-trien-11-one, 39B, 198
$C_{11}H_{13}N_3$, 2,3,5,6-Tetrahydro-5-phenyl-1H-imidazo[1,2-a]imidazole, 44B, 235
$C_{11}H_{14}N_2O$, 5-Methoxytryptamine, 40B, 247
$C_{11}H_{14}N_2O_3$, 1,6-Dimethyl-4-oxo-1,6,7,8-tetrahydro-3-homo-pyrimidazolecarboxylic acid, 41B, 323
$C_{11}H_{15}NO$, α-2,6-Dimethyl-4-hydroxy-1,2,3,4-tetrahydroquinoline, 42B, 203
$C_{11}H_{15}NO$, 2,6-Dimethyl-4-hydroxy-1,2,3,4-tetrahydroquinoline (β-isomer), 43B, 335
$C_{11}H_{17}ClN_2O_2$, 2,4-Dimethyl-1,5-benzodiazepinium chloride dihydrate, 42B, 203
$C_{12}H_7N_5O_4$, 1-(2,4-Dinitrophenyl)-imidazo[4,5-b]pyridine, 46B, 247
$C_{12}H_9N_3$, Dicyano-1-methyl-3H-isoindoleninium methylide, 44B, 236
$C_{12}H_{10}N_2$, (E)-2,2'-Diazastilbene, 46B, 229
$C_{12}H_{10}N_2$, (E)-4,4'-Diazastilbene, 46B, 229
$C_{12}H_{10}N_2O_5$, 5-Acetoxy-6-methoxy-8-nitroquinoline, 32B, 160
$C_{12}H_{12}ClN_3$, N',N'-(2-Chlorobenziliden)histamine, 37B, 138
$C_{12}H_{12}N_2O_3$, 1-Ethyl-1,4-dihydro-7-methyl-4-oxo-1,8-naphthyridine-3-carboxylic acid, 42B, 205; 46B, 248
$C_{12}H_{12}N_2O_3$, 3-Carbethoxy-4-oxo-6-methylhomopyrimidazole, 38B, 287
$C_{12}H_{13}ClN_2O_3$, 3-Carboethoxy-4-oxo-6-methylhomopyrimidazole hydrochloride, 42B, 205
$C_{12}H_{13}NO_2$, 4-(3-Indolyl)butyric acid, 46B, 11
$C_{12}H_{13}NO_2S_2$, N-t-Butyldithiophthalimide, 45B, 267
$C_{12}H_{13}NO_4$, L-5-Methoxycarbonyl-7-formyl-1,2,5,6-tetrahydro-3H-pyrrolo[1,2-a]azepin-3-one, 32B, 162
$C_{12}H_{14}IN$, 1-Ethyl-2-methylquinolinium iodide, 35B, 188
$C_{12}H_{14}INO_2S$, 7-(p-Iodobenzenesulphonyl)-7-azabicyclo[4.1.0]heptane, 30B, 144
$C_{12}H_{14}N_2O_5$, Serotonin hydrogen oxalate, 44B, 236
$C_{12}H_{14}N_4O_4$, 5,8-Diacetyl-1,3-dimethyl-5,8-dihydrolumazine, 43B, 335
$C_{12}H_{14}N_8O_7$, Dilumazine trihydrate, 38B, 288
$C_{12}H_{16}N_2$, N,N-Dimethyltryptamine, 38B, 301
$C_{12}H_{16}N_2O$, Psilocin, 40B, 248
$C_{12}H_{16}N_2O$, 5-Hydroxy-(N,N)-dimethyltryptamine (bufotenine), 38B, 289
$C_{12}H_{16}N_2O_3$, 3-Carbethoxy-4-oxo-6-methyl-6,7,8,9-tetrahydrohomopyrimidazole, 39B, 198
$C_{12}H_{17}NO$, 2,2,4-Trimethyl-8-hydroxy-1,2,3,4-tetrahydroquinoline, 46B, 248
$C_{13}H_7Br_2N_3O$, 2-(2,4-Dibromophenyl)-4-oxo-1,2,3-benzotriazin-2-ium-3-ide, 42B, 206
$C_{13}H_9BrN_2O_2 \cdot 0.5\ C_4H_8O_2$, N-Methyl-2-bromo-3-(2-indolyl)maleimide - dioxan, 43B, 339
$C_{13}H_9BrN_2O_4$, N-(α-Glutarimido)-4-bromophthalimide, 40B, 1084
$C_{13}H_9N_3O_2$, 1-Benzoyloxybenzotriazole, 43B, 336
$C_{13}H_9N_3O_2$, 2-Phenylbenzo-1,2,3-triaziniumbetaine-1-oxide, 41B, 325
$C_{13}H_{10}N_2O_4$, Thalidomide, 37B, 139
$C_{13}H_{11}N_3$, 3-Methyl-2-phenylimidazo[1,2-a]pyrimidine, 41B, 326
$C_{13}H_{13}IN_2$, (E)-6-Methyl-6-azonia-2'-azastilbene iodide, 46B, 229
$C_{13}H_{14}ClNO_2$, Methyl 2-chloro-3-(indol-3-yl)butyrate, 44B, 237
$C_{13}H_{15}NO_3$, Methyl 2-hydroxy-3-(indol-3-yl)butyrate, 44B, 237
$C_{13}H_{16}N_2O_2$, Melatonin, 40B, 249
$C_{13}H_{19}ClN_2O$, 5-Methoxy-(N,N)-dimethyltryptamine hydrochloride, 37B,

140
$C_{13}H_{20}N_2O_3$, 1,6-Dimethyl-3-carbethoxy-4-oxo-1,6,7,8,9,10-hexahydrohomopyrimidazole, 38B, 290
$C_{13}H_{21}N_2O_5P$, Psilocybin monomethanolate, 40B, 408
$C_{13}H_{26}ClN$, 2-n-Propyl-7-methyl-trans-decahydroquinoline hydrochloride, 40B, 250
$C_{14}H_8INO_2$, N-(4-Iodophenyl)-phthalimide, 40B, 250
$C_{14}H_{10}N_2O$, 2-Phenyl-4(3H)-quinazolinone, 43B, 336
$C_{14}H_{10}N_2O_2$, 2-Phenyl-6-nitroindolizine, 46B, 249
$C_{14}H_{10}N_2O_2$, 2-Phenyl-8-nitroindolizine, 46B, 249
$C_{14}H_{10}N_2O_2$, 2-Phenylquinazoline 1,3-dioxide, 45B, 268
$C_{14}H_{10}N_2O_2$, 3-Phenyl-5-nitroindole, 46B, 249
$C_{14}H_{10}N_2O_2$, 3-Phenyl-7-nitroindole, 46B, 249
$C_{14}H_{10}N_2O_2$, 3-Phenyl-2,4-(1H,3H)-quinazolinedione, 38B, 290
$C_{14}H_{11}ClN_2O_2$, 2-Benzoyl-8β-chloro-1,2-diazabicyclo[5.2.0]-3,5-nonadiene-9-one, 40B, 251
$C_{14}H_{11}N_3O$, 2-(2-Pyridylamino)-8-hydroxyquinoline, 46B, 249
$C_{14}H_{11}N_5O_2$, Anhydro-2-methyl-4-o-nitroanilino-1,2,3-benzotriazinium hydroxide, 44B, 238
$C_{14}H_{13}BF_4N_4$, Benzimidazole benzimidazolium fluoroborate, 42B, 206
$C_{14}H_{15}BrN_2O$, 2-(p-Bromophenyl)-trans-4a,5,6,7,8,8a-hexahydroquinazolin-4(3H)-one, 45B, 269
$C_{14}H_{15}BrN_6O$, 2,6-Dimethyl-3,4-dioxo-2,3,4,6,7,8-hexahydropyridazino[4,3-c]pyridazine-4-[p-bromophenylhydrazone], 46B, 250
$C_{14}H_{16}BrNO$, 3-(p-Bromobenzyl)-1-aza-bicyclo[3.3.0]octan-2-one, 45B, 269
$C_{14}H_{16}I_2N_2$, (E)-4,4'-Dimethyl-4,4'-diazoniastilbene diiodide, 46B, 229
$C_{14}H_{16}I_2N_2$, (E)-6,6'-Dimethyl-6,6'-diazoniastilbene diiodide, 46B, 229
$C_{14}H_{16}N_2O$, 2-Phenyl-cis-5,6-tetramethylene-5,6-dihydropyrimidin-4(3H)-one, 46B, 250
$C_{14}H_{16}N_2O$, 2-Phenyl-trans-5,6-tetramethylene-5,6-dihydropyrimidin-4(3H)-one, 46B, 250
$C_{14}H_{17}BrN_4$, 1-(3,5,5-Trimethyl-1-pyrazolinyl)phthalazine hydrobromide, 43B, 337
$C_{14}H_{17}NO_2$, 6-Acetoxy-1,2-dihydro-2,2,4-trimethylquinoline, 45B, 270
$C_{14}H_{17}NO_3$, 2-Ethoxy-1-ethoxycarbonyl-1,2-dihydroquinoline, 46B, 251
$C_{14}H_{18}N_2O_3$, 8-Nitroethoxyquin, 45B, 270
$C_{14}H_{20}N_2O_2$, 7-Acetyl-3,5,5,9,9-pentamethyl-1,6-diazabicyclo-[4.3.0]nona-3,7-dien-2-one, 43B, 338
$C_{14}H_{22}N_2O_7S$, 3-Carbethoxy-1,6-dimethyl-4-oxo-6,7,8,9-tetrahydrohomopyrimidazolium methyl sulfate, 41B, 330
$C_{14}H_{23}NO$, cis-1,2,3,4,4a,5,6,8a-Octahydro-3,3-dimethyl-1-(1-propyl)-quinolin-2-one, 43B, 338
$C_{14}H_{28}IN$, cis-13,13-Dimethyl-13-azoniabicyclo[10.1.0]tridecane iodide, 32B, 163
$C_{14}H_{28}IN$, trans-13,13-Dimethyl-13-azoniabicyclo[10.1.0]tridecane iodide, 28, 472
$C_{15}H_7Cl_4NO_2$, N-(p-Tolyl)tetrachlorophthalimide (α phase), 44B, 238
$C_{15}H_{10}BrNO_3$, 1-(5'-Bromo-2'-hydroxy-4'-methoxyphenyl)-3-oxo-isoindolo-1-ene, 26, 722
$C_{15}H_{10}ClN_3O_3$, 5-(2-Chlorophenyl)-1,3-dihydro-7-nitro-2H-1,4-benzodiazepin-2-one, 45B, 270
$C_{15}H_{10}INO_5$, 5-Iodo-2-phthalimidobenzoic acid monohydrate, 28, 474
$C_{15}H_{11}ClN_2O_2$, 7-Chloro-1,3-dihydro-3-hydroxy-5-phenyl-2H-1,4-benzodiazepin-2-one, 44B, 239

$C_{15}H_{11}N_3O_3$, 7-Nitro-1,3-dihydro-5-phenyl-2H-1,4-benzodiazepin-2-one, 43B, 339
$C_{15}H_{11}N_5O_6$, 1-Methyl-2-picryliminoindoline, 35B, 190
$C_{15}H_{12}N_2O_2$, N-(N'-Methyl-anilino)-phthalimide, 45B, 271
$C_{15}H_{16}BrN$, 2,4-Diamino-5-methyl-6-benzylpyrido[2,3-d]pyrimidine hydrobromide, 39B, 199
$C_{15}H_{16}N_4O_3$, 2,5-Diethyl-7-(1-methyl-2-imidazolyl)-1H-pyrrolo[3,4-c]-pyridin-1,3,6(2H,5H)-trione, 41B, 331
$C_{15}H_{17}BrN_2O_4$, Dimethyl 6-bromo-1-(1-pyrrolidinyl)-3H-pyrroline-2,3-dicarboxylate, 40B, 251
$C_{15}H_{18}BrN_3$, 10-(4-Bromophenyl)-10,11,12-triaza-Δ^{11}-bicyclo[7.3.0]-dodec-1-ene, 40B, 252
$C_{15}H_{18}N_2O_3$, L-3-(2-Acetamido-2-ethoxycarbonyl)ethylindole, 43B, 340
$C_{15}H_{21}ClN_2 \cdot H_2O$, 3-[2-(4-Piperidyl)ethyl]indole hydrochloride monohydrate, 46B, 252
$C_{16}H_8Br_2N_2O_2$, 6,6'-Dibromoindigotin, 46B, 252
$C_{16}H_8N_2O_4$, N,N'-Biphthalimide, 44B, 240
$C_{16}H_{10}N_2O_2$, Indigo (form A), 18, 763; 19, 594; 20, 612
$C_{16}H_{10}N_2O_2$, Indigo (form B), 46B, 253
$C_{16}H_{10}N_2O_2$, Indirubine, 26, 783
$C_{16}H_{10}N_2O_2$, Isoindigo, 23, 718
$C_{16}H_{11}Cl_2NO$, 2-(2',6'-Dichlorobenzyl)-1-isoquinolone, 40B, 253
$C_{16}H_{11}N_3$, 6-Cyano-7:2'-(1'-aminophenyl)isoquinoline, 42B, 206
$C_{16}H_{11}N_5$, 6,7-Diphenyl(1,2,4)triazolo[5,1-c](1,2,4)triazine, 44B, 240
$C_{16}H_{13}ClN_2O$, 7-Chloro-1,3-dihydro-1-methyl-5-phenyl-2H-1,4-benzodiazepin-2-one, 38B, 291
$C_{16}H_{13}ClN_2O_2$, Anhydro(1-benzyl-3-chloroacetyl-2-hydroxyimidazo[1,2-a]pyridinium hydroxide), 44B, 240
$C_{16}H_{13}ClN_2O_2$, 7-Chloro-3-hydroxy-1-methyl-5-phenyl-1,3-dihydro-2H-1,4-benzodiazepin-2-one, 46B, 253
$C_{16}H_{13}NO$, 1-Phenyl-2-methyl-3-isoquinolone, 40B, 254
$C_{16}H_{14}BrN$, 1-Phenyl-3-methylisoquinoline hydrobromide, 41B, 331
$C_{16}H_{14}BrNO_2$, 2-(4'-Bromophenyl)-4,6-dimethoxyindole, 46B, 253
$C_{16}H_{14}BrN_3O_3$, (5-Benzoyl-1H-benzimidazol-2-yl)-carbamic acid methyl ester hydrobromide, 46B, 254
$C_{16}H_{14}ClN_3O$, 7-Chloro-2-methylamino-5-phenyl-3H-1,4-benzodiazepin-3-ol, 46B, 255
$C_{16}H_{14}Cl_2N_2$, 1-Methyl-3(N-methyl-4'-chloroanilino)-5-chloroindole, 38B, 292
$C_{16}H_{14}N_2$, 1-Methyl-4-phenyl-1H-2,3-benzodiazepine, 40B, 255
$C_{16}H_{14}N_2O_2$, N-(4-Dimethylaminophenyl)-phthalimide, 46B, 255
$C_{16}H_{15}ClN \cdot 0.5\ H_2O$, 1-Benzyl-3,4-dihydro-isoquinoline hydrochloride hemihydrate, 41B, 332
$C_{16}H_{15}ClN_2$, 7-Chloro-2,3-dihydro-1-methyl-5-phenyl-1H-1,4-benzodiazepine, 44B, 241
$C_{16}H_{15}ClN_2O$, (-)-7-Chloro-1,3,4,5-tetrahydro-1-methyl-5-phenyl-2H-1,4-benzodiazepin-2-one, 43B, 340
$C_{16}H_{15}Cl_2N_3O$, 7-Chloro-2-methylamino-5-phenyl-3H-1,4-benzodiazepine 4-oxide hydrochloride, 45B, 272
$C_{16}H_{16}ClN$, cis-1-(4-Chlorophenyl)-3-methyl-2-methylisoindoline, 40B, 269
$C_{16}H_{16}N_2O_2S$, 1-Methyl-3-p-tolylsulphonylaminoindole, 41B, 327
$C_{16}H_{18}ClN_3O_2$, 1-(2-Indol-3-ylethyl)-3-carbamidopyridinium chloride monohydrate, 40B, 255
$C_{16}H_{18}Cl_2N_2O$, Medazepam hydrochloride, 46B, 256
$C_{16}H_{21}NO$, 1,3,3-Trimethyl-2-[2-oxo-3-methyl-butylidene]indole, 46B,

256
$C_{16}H_{22}INO_2S$, cis-11-(p-Iodobenzenesulphonyl)-11-azabicyclo[8.1.0]-undecane, 35B, 190
$C_{16}H_{22}N_2$, 2,3-Di-t-butylquinoxaline, 37B, 141
$C_{17}H_{12}BrNO_2$, N-Phenyl-2-(p-bromophenyl)-cyclopropane-1,3-dicarboximide, 38B, 249
$C_{17}H_{13}ClN_2O_4$, Ethyl 5-chloro-3-nitro-3-phenyl-3H-indole-2-carboxylate, 39B, 200
$C_{17}H_{14}ClN_3$, 5-Chloro-3-cyano-2-dimethylamino-3-phenyl-3H-indole, 38B, 293
$C_{17}H_{15}BrN_2O$, 8-Bromo-6-phenyl-1-methyl-1,2,3,4-tetrahydro-1,5-benzdiazocin-2-one, 45B, 274
$C_{17}H_{15}Cl_3N_2O$, 7-Chloro-(2,4-dichlorophenyl)-4,5-dihydro-1,4-dimethyl-3H-1,4-benzodiazepin-2-one, 32B, 164
$C_{17}H_{16}BrNO$, 1-(p-Bromobenzoyl)-2-methyl-1,2,3,4-tetrahydroquinoline, 43B, 341
$C_{17}H_{16}ClFN_2O \cdot H_2O$, 7-Chloro-5-(2-fluorophenyl)-1,3,4,5-tetrahydro-1,4-dimethyl-2H-1,4-benzodiazepin-2-one monohydrate, 46B, 256
$C_{17}H_{16}ClN_3O_2$, (+)-4-Carbamoyl-7-chloro-1,3,4,5-tetrahydro-1-methyl-5-phenyl-2H-1,4-benzodiazepin-2-one, 43B, 340
$C_{17}H_{17}NO_2$, 2-Acetyl-1-(4-methoxyphenyl)-trans-1,8a-dihydroindolizine, 45B, 273
$C_{17}H_{20}N_4O_4$, anti-6-[(Hydroxyiminophenyl)methyl]-1-[(1-methylethyl)-sulphonyl]-1H-benzimidazol-2-amine monohydrate, 46B, 257
$C_{17}H_{24}BrNO$, 1,3,4,7,8,8a-Hexahydro-2-methyl-4a-phenylisoquinolin-6(8H)-one methobromide, 40B, 256
$C_{17}H_{24}NO_3$, Mesembranol, 39B, 200
$C_{17}H_{25}NO_3$, 1β-(p-Methoxybenzyl)-9α,10β-dihydroxydecahydroisoquinoline, 44B, 241
$C_{17}H_{25}NO_3$, 1β-(p-Methoxybenzyl)-9β,10α-dihydroxydecahydroisoquinoline, 44B, 241
$C_{18}H_{10}Cl_2N_2O_2$, 3,7-Dichloro-4,6-diphenyl-1,5-diazabicyclo[3.3.0]-octa-3,6-diene-2,8-dione, 46B, 257
$C_{18}H_{10}Cl_2N_2O_2$, 3,7-Dichloro-4,8-diphenyl-1,5-diazabicyclo[3.3.0]-octa-3,7-diene-2,6-dione, 46B, 257
$C_{18}H_{12}N_2$, 2,2'-Biquinolyl, 21, 623; 43B, 342
$C_{18}H_{12}N_2$, 8,8'-Biquinolyl, 42B, 207
$C_{18}H_{13}Cl_3I_2N_2$, 2,2'-Biquinolinium diiodotrichloride, 45B, 276
$C_{18}H_{14}BrClN_2O_4$, Bis(quinoline)bromine perchlorate, 41B, 333
$C_{18}H_{14}F_{12}N_2O_{10}$, 1,5-Dimethyl-1,5-naphthyridinium-4(1H),8(5H)-dione hydrogen trifluoroacetate, 44B, 242
$C_{18}H_{14}N_2O_5$, 2,3-Dicarbomethoxy-3,4-dihydro-3-(2-pyridyl)quinolin-4-one, 41B, 323
$C_{18}H_{15}Cl_2N_3O_3$, 3-trans-Bis(4-chlorophenyl)-2-cis-nitro-5-oxoperhydropyrazolo[1,2-a]pyrazole, 45B, 277
$C_{18}H_{15}Cl_2N_3O_3$, 3-trans-Bis(4-chlorophenyl)-2-trans-nitro-5-oxoperhydropyrazolo[1,2-a]pyrazole, 45B, 278
$C_{18}H_{15}N_5O$, 3-Benzyl-7-methoxy-6-phenylimidazo[1,2-b]-s-tetrazine, 45B, 278
$C_{18}H_{16}BrN_2O$, 2-Methyl-4-(N-acetanylino)-6-bromo-1,2,3,4-tetrahydroquinoline, 41B, 334
$C_{18}H_{16}Br_9N_2Sb$, Quinolinium hexabromoantimonate(V) tribromide, 39B, 201
$C_{18}H_{16}CdCl_4N_2$, Diquinolinium tetrachlorocadmate, 40B, 1084
$C_{18}H_{16}ClN_3O_2$, 7-(4-Chlorophenyl)-8-phenyl-2,3-dihydro-imidazo[1,2-a]pyrimidin-5(8H)-one hydrate, 42B, 207
$C_{18}H_{16}ClN_3O_2$, 7-Chloro-1,3-dihydro-1-(N-methylacetamido)-5-phenyl-

2H-1,4-benzodiazepin-2-one, 46B, 258
$C_{18}H_{16}Cl_8N_4O_2Pt_2$, Bisquinolinium di-μ-chloro-bis(trichloronitrosoplatinate), 40B, 257
$C_{18}H_{16}N_2O_4$, 2,2'-Dimethoxyindigotin, 46B, 252
$C_{18}H_{18}ClN_3O$, 5-Chloro-2-dimethylamino-3-methoxyiminomethyl-3-phenyl-3H-indole, 38B, 293
$C_{18}H_{18}N_2O_6$, 6-Acetamino-2-ethoxycarbonyl-1-phenyl-pyrrolizidin-3,5,7-trione, 42B, 208
$C_{18}H_{19}IN_2$, 1-Methyl-6-(5-(1-methyl-1H-1-pyrindinyl))-1-azoniaindan iodide, 31B, 133
$C_{18}H_{19}N_3O_2$, 2-Phenyl-5-t-butylamino-8-nitroindolizine, 46B, 249
$C_{18}H_{23}Cl_2N_3O_2$, 5-Chloro-1-(3-(dimethylamino)propyl)-1,3-dihydro-3-phenyl-2H-benzimidazol-2-one monohydrochloride monohydrate, 44B, 243
$C_{18}H_{26}ClN_3$, 7-Chloro-4-(4-diethylamino-1-methylbutylamino)quinoline, 39B, 202
$C_{18}H_{29}NO_3S$, 8α,Nβ-Dimethyl-trans-decahydroquinoline p-toluenesulphonate, 46B, 260
$C_{19}H_{12}Br_2N_2$, 8,8'-Dibromo-2,2'-methylenediquinoline (red form), 31B, 134
$C_{19}H_{13}BrN_2$, 8-Bromo-2,2'-methylenediquinoline, 28, 476
$C_{19}H_{13}N$, 1-(1-Naphthyl)isoquinoline, 44B, 243
$C_{19}H_{15}N_3O$, 5-Hydroxy-3-phenyl-1-(3-methyl-1-isoquinolyl)pyrazole, 37B, 142
$C_{19}H_{15}N_3O_2$, Methyl-bis-(8-hydroxy-2-quinolyl)-amine, 46B, 258
$C_{19}H_{16}ClNO_4$, 1-(p-Chlorobenzoyl)-5-methoxy-2-methylindole-3-acetic acid, 38B, 294
$C_{19}H_{17}ClN_2OS$, N-(1,2-Dihydro-2-oxoquinolin-1-yl)-N-(1-methylallyl)-p-chlorobenzenesulfenamide, 44B, 366
$C_{19}H_{17}ClN_4O_3$, 2-Acetoxyamino-4-acetyl-8-chloro-3,4-dihydro-6-phenyl-1,4,5-benzotriazocine, 39B, 212
$C_{19}H_{17}NO_2$, 1-Isoquinolylisopropanyl benzoate, 45B, 285
$C_{19}H_{17}NO_3$, 1-Isoquinolyl(phenyl)methanol ethyl carbonate, 43B, 343
$C_{19}H_{18}ClNO_5S$, 1-Benzazepinone derivative, 42B, 208
$C_{19}H_{18}N_2O_2$, 1-(Cyclopropylmethyl)-4-phenyl-6-methoxy-2(1H)-quinazolinone, 42B, 209
$C_{19}H_{19}N_3O_2$, 2-Phenyl-5-piperidino-8-nitroindolizine, 46B, 249
$C_{19}H_{20}N_2O_5$, 2,3,4,7-Tetrahydro-3a,4-bis(methoxycarbonyl)-2,6-dimethyl-5-phenylindazol-7-one, 40B, 258
$C_{19}H_{22}BrN_3O_2$, 3-p-Bromophenyl-6,7-bisisopropylidene-8,8-dimethyl-1,3,5-triazabicyclo[3.3.0]octan-2,4-dione, 41B, 335
$C_{19}H_{22}N_2O_4S$, 9-(2-Aminophenylthio)-2,3-dimethyl-1,7-dioxo-2-azaspiro[4.4]non-3-en-4-carboxylic acid ethyl ester, 46B, 259
$C_{19}H_{23}ClN_2O$, 3-Methyl-3-(3-methylaminopropyl)-1-phenyl-2-indolinone hydrochloride, 46B, 260
$C_{19}H_{23}Cl_2N_5O$, 2-{3-[4-(m-Chlorophenyl)-1-piperazinyl]propyl}-s-triazolo-[4,3-a]-pyridin-3(2H)-one hydrochloride, 45B, 279
$C_{19}H_{24}ClNO_5 \cdot H_2O$, 1-(3,4,5-Trimethoxybenzyl)-6,7-dihydroxy-1,2,3,4-tetrahydroisoquinoline hydrochloride hydrate, 45B, 279
$C_{19}H_{28}ClNO$, 7-(cis-1-Buten-3-ynyl)-8-hydroxy-2-(3,4-pentadienyl)-1-azaspiro[5.5]undecane hydrochloride, 39B, 203
$C_{19}H_{28}N_4O_7$, 8α-t-Butyl-trans-decahydroquinoline picrate, 46B, 260
$C_{19}H_{28}N_4O_7$, 8β-t-Butyl-trans-decahydroquinoline hydrogen picrate, 45B, 280; 46B, 260
$C_{20}H_{13}BrN_2O$, 1-(Pyridyl-2)-3-benzoyl-6-bromoindolizine, 42B, 209
$C_{20}H_{14}NO_2$, 1,2-Dihydro-3-oxo-2,2-diphenyl-3H-indole 1-oxyl, 46B, 261
$C_{20}H_{14}N_2$, (E)-Dibenzo[e,e']-3,3'-diazastilbene, 46B, 229

$C_{20}H_{14}N_2$, (Z)-Dibenzo[e,e']-3,3'-diazastilbene, 46B, 229
$C_{20}H_{14}N_2O_2$, N-(4-Phenylaminophenyl)phthalimide, 44B, 244
$C_{20}H_{14}N_6 \cdot 0.67\ CH_4O$, 2-(2-2H-Benzotriazolyl)-N-(6-quinoxalinyl)aniline methanolate, 43B, 344
$C_{20}H_{15}N$, 2,3-Diphenylindole, 39B, 204
$C_{20}H_{16}N_4O_4$, Dimethyl 2,5-diphenyl-1,3a,4,6a-tetraazapentalene-3,6-dicarboxylate, 37B, 142
$C_{20}H_{17}NO$, 3-Hydroxy-2,3-diphenylindoline, 46B, 261
$C_{20}H_{18}N_2$, 4,5-Dihydro-1,8-diphenyl-3H-pyrrolo[1,2-d][1,4]-diazepine, 45B, 281
$C_{20}H_{19}AsF_6N_2O_2$, Bis(1-methyl-2-quinoline) hydrogen hexafluoroarsenate(V), 34B, 150
$C_{20}H_{19}BrN_2OS_2$, N-(4-Bromobenzyl)isoquinolinium 4-dithiocarboxylate dimethylformamide, 39B, 204
$C_{20}H_{19}ClN_2O_4$, 4,5-Dihydro-1,8-diphenyl-7H-pyrrolo[1,2-d][1,4]-diazepinium perchlorate, 45B, 281
$C_{20}H_{21}NO_4$, 6,7-Dimethoxy-1-veratrylisoquinoline, 39B, 205
$C_{20}H_{21}NS_2$, 2,2-Dimethyl-1-methylthio-4,5-diphenyl-2-azoniabicyclo-[3.1.0]hex-3-en-3-thiolate, 42B, 210
$C_{20}H_{21}N_4O_5S_2 \cdot 4\ H_2O$, 3-Hydroxy-2,2'-dioxo[Δ(3-3')-biindoline]-N,N'-di(S,S-dimethylsulfoximide) tetrahydrate, 45B, 282
$C_{20}H_{22}N_2O_2S$, 1-Methyl-3-p-tolylsulphonyliminoindoline-2-spirocyclopentane, 41B, 327
$C_{20}H_{22}N_{10}O_8$, 1,4-Dihydrazinophthalazine bis(2'-pyridiniumcarboxaldimine) nitrate dihydrate, 43B, 345
$C_{21}H_{15}Cl_2N$, 2-(2',6'-Dichlorobenzyl)-1-cyclopentadienylidene-1,2-dihydroisoquinoline, 41B, 289
$C_{21}H_{16}NO$, 1,2-Dihydro-2,2-diphenylquinoline 1-oxyl, 46B, 261
$C_{21}H_{16}N_2O_2$, 2-(o-Hydroxybenzoyl)-7-methyl-3-phenylimidazo[1,2-a]pyridine, 46B, 220
$C_{21}H_{18}N_2O$, 2-Phenyl-3-(N-p-methoxyphenyl)amine-indole, 46B, 262
$C_{21}H_{21}NO_6$, Ethyl 1-(p-methoxyphenyl)-2-methyl-4-hydroxy-5-acetoxyindole-3-carboxylate, 40B, 258
$C_{21}H_{22}BrN_2O_8$, Tetramethyl 2-bromo-9-(1-pyrrolidinyl)-5H-pyrrolo[1,2-a]azepine-5,6,7,8-tetracarboxylate, 40B, 259
$C_{21}H_{22}ClFN_4O_2$, Halopemide, 46B, 236
$C_{21}H_{23}NO_6$, 1-Benzylidene-N-formyl-tetrahydroisoquinoline, 43B, 140
$C_{21}H_{23}N_5O_5$, 2-Acetamido-8-methyl-4,9-dioxo-6,7-diphenyl-6,7,8,9-tetrahydro-4H-pyrazino[1,2-a]-s-triazine dihydrate, 40B, 260
$C_{21}H_{25}N_5O$, 2-Cyano-2-(3-cyano-4-diethylamino-1H-quinolylidene-2)-N,N-diethylacetamide, 38B, 295
$C_{21}H_{26}N_2O_3$, Hydroxybenzylpindolol, 43B, 346
$C_{21}H_{26}N_4$, 2,4-Bis(dimethylamino)-3-ethyl-1-phenyl-1,5-benzodiazepine, 45B, 283
$C_{21}H_{28}BrNO$, 6,6-Diphenyl-3,3-diethyl-3-azabicyclo[3.1.0]hexane bromide monohydrate, 29, 503
$C_{22}H_{14}N_6$, 2,3-Bisbenzimidazol-2-yl-quinoxaline, 46B, 263
$C_{22}H_{16}N_2O$, 2,4-Diphenyl-2,3-benzodiazocin-1(2H)-one, 42B, 211
$C_{22}H_{18}BrNO_4$, E,E-1-Acetyl-3-(1'-p-bromophenyl)-3'-ethoxycarbonylallylidene)-2-oxo-indoline, 45B, 283
$C_{22}H_{19}N_3O_5$, 4-Ethoxycarbonyl-5-hydroxy-1-phenyl-3-(α-phthalimidoethyl)pyrazole, 46B, 263
$C_{22}H_{20}N_2O_2$, N,N'-Diacetyl-1,2,1',2'-tetrahydro-1,1'-diisoquinoline, 42B, 212; 43B, 347
$C_{22}H_{20}N_2O_4$, Dimethyl 2,3-dihydro-2-indol-3-ylbenz(b)azepine-3,4-dicarboxylate, 38B, 296
$C_{22}H_{22}FN_3O_2 \cdot 2\ H_2O$, 1-{1-[4-(4-Fluorophenyl)-4-oxobutyl]-1,2,3,6-

tetrahydro-4-pyridyl}-1,3-dihydro-2H-benzimidazol-2-one dihydrate, 46B, 264
$C_{22}H_{23}ClFN_3O_2$, 5-Chloro-1-(3-(4-(4-fluorobenzoyl)piperidino)propyl)-1,3-dihydro-2H-benzimidazol-2-one, 43B, 348
$C_{22}H_{23}NO$, 1,3,3-Trimethyl-2-[3'-(p-tolyl)-protenylidene-1']indoline, 46B, 264
$C_{22}H_{24}FN_3O_2$, 1-(1-(3-(p-Fluorobenzoyl)propyl)-4-piperidyl)-2-benzimidazolinone, 39B, 206
$C_{22}H_{26}I_4N_4Zn$, 2,4-Dimethyl-1H-1,5-benzodiazepinium tetraiodozincate, 40B, 261
$C_{22}H_{30}INO_2$, 2-Benzyl-1,2,3,4-tetrahydro-6,7-dimethoxy-2-methyl-1-isopropyl-isoquinolinium iodide, 45B, 284
$C_{23}H_{17}NO_2$, 1-Isoquinolyl(phenyl)methyl benzoate, 45B, 285
$C_{23}H_{20}Br_2N_2O_5$, Dimethyl 1,2-bis-(4-bromophenyl)-3,5-dimethyl-4-oxo-6,7-diazabicyclo[3.2.0]hept-2-ene-6,7-dicarboxylate, 38B, 297
$C_{23}H_{21}N_3O_5$, 4-Ethoxycarbonyl-5-methoxy-1-phenyl-3-(α-phthalimidoethyl)pyrazole, 46B, 265
$C_{23}H_{23}IN_2$, 1,1'-Diethyl-2,2'-cyanine iodide, 43B, 348
$C_{23}H_{25}ClN_2O$, N,N'-Diethylpseudoisocyanin chloride monohydrate, 37B, 143
$C_{23}H_{27}BF_4N_2$, Bis(1,3,3-trimethyl-indolenine-2-yl)monomethinium tetrafluoroborate, 42B, 212
$C_{23}H_{31}N_3O_2$, cis-5,6-Dimethoxy-2-methyl-3-[2-(4-phenyl-1-piperazinyl)ethyl]indoline, 45B, 286
$C_{24}H_{19}BrN_2O_5$, 5,6-Bis(methoxycarbonyl)-7-p-bromophenyl-3-methyl-4-phenylpyrrolo[1,2-b]pyridazin-2-one, 39B, 207
$C_{24}H_{21}NO$, 3-Methyl-α,α-diphenyl-1-isoquinolineethanol, 44B, 245
$C_{24}H_{21}N_3O_2$, 1,3-Diphenyl-5-carbethoxymethylene-1H-4,5-dihydro-1,2,4-benzotriazepine, 45B, 286
$C_{24}H_{22}N_6O_{10}Sr_2$, Strontium bis(quinoxaline-2-one-3-olate) quinoxaline-2,3-diolate tetrahydrate, 43B, 349
$C_{24}H_{33}ClN_4O_5$, 2-(2-(4-Ethoxyphenyl)methyl-5-nitro-1H-benzimidazolyl)-N,N-diethylethanaminium chloride acetic acid solvate, 44B, 245
$C_{24}H_{34}BrNO_4$, Indolizidine hydrobromide ethanolate, 43B, 424
$C_{25}H_{18}N_2O_3S$, N-Phthalimido-p-tolyl-α-naphthylsulfoximide, 41B, 337
$C_{25}H_{18}N_2O_3S$, S(+)-N-Phthalimido-p-tolyl-α-naphthylsulfoximide, 41B, 338
$C_{25}H_{23}N_5O_5$, 2-Acetamido-8-methyl-4,9-dioxo-6,7-diphenyl-6,7,8,9-tetrahydro-4H-pyrazino[1,2-a]-s-triazine dihydrate, 41B, 336
$C_{25}H_{28}INO_2$, 2-Benzyl-1,2,3,4-tetrahydro-6,7-dimethoxy-2-methyl-1-phenylisoquinolinium iodide, 43B, 350
$C_{26}H_{19}N_3$, 1,3,5-Triphenyl-1H-1,2,4-benzotriazepine, 44B, 246
$C_{26}H_{20}Br_2N_6$, 1,1'-Azo-2-phenylimidazo[1,2-a]pyridinium dibromide, 38B, 86
$C_{26}H_{20}ClN_3$, 4-(4-Chlorophenyl)-4,4a-dihydro-1,3-diphenyl-1H-cyclohepta[e]-1,2,4-triazine, 45B, 287
$C_{26}H_{26}ClNO_9S$, 1H-1-Benzazonine derivative, 42B, 208
$C_{26}H_{28}N_2O_4$, Dimethyl 2-(1,3-dimethylindol-2-yl)-3-(trans-2,3-dihydro-1,3-dimethylindol-2-yl)-maleate, 46B, 265
$C_{26}H_{28}N_3O_4S$, 1,2,3,4-Tetrahydro-1,2,4-trimethyl-4-p-tolylsulphonylamino-3-p-tolylsulphonylimino-quinoline, 41B, 327
$C_{26}H_{31}Cl_4IN_4O$, 5,5',6,6'-Tetrachloro-1,1',3,3'-tetraethylbenzimidazolocarbocyanine iodide methanol, 43B, 350
$C_{27}H_{26}Cl_2N_2O_4$, (αS,1S)-(+)-α,1-Bis(4-chlorophenyl)isoindoline-1-ethanol S-(-)-5-carboxylato-2-pyrrolidone, 46B, 266
$C_{27}H_{30}Cl_4IN_5$, 5,5',6,6'-Tetrachloro-1,1',3,3'-tetraethylbenzimidazolo-carbocyanine iodide (acetonitrile solvate), 38B, 297

$C_{27}H_{31}Cl_8IN_4$, 5,5',6,6'-Tetrachloro-1,1',3,3'-tetraethylbenzimidazolocarbocyanine iodide bis(dichloromethane), 43B, 351
$C_{27}H_{32}N_4O_5S$, Bis[(8-quinolyloxy)ethoxyethyl]ether thiourea, 46B, 109
$C_{27}H_{33}Cl_4IN_4O$, 5,5',6,6'-Tetrachloro-1,1',3,3'-tetraethylbenzimidazolocarbocyanine iodide ethanol, 43B, 350
$C_{27}H_{35}Cl_4IN_4O_2$, 5,5',6,6'-Tetrachloro-1,1',3,3'-tetraethylbenzimidazolo-carbocyanine iodide (methanol solvate), 38B, 297
$C_{28}H_{18}N_2$, 3,3'-Diphenyl-1,1'-biisoindolylidene, 37B, 144
$C_{28}H_{20}Cl_3N_3O_2$, 4,6-Dianilino-2-phenylquinoline-5,8-quinone chloroform, 43B, 352
$C_{28}H_{24}B_2F_8N_4S$, 3,3'-Thiobis-(2-methyl-1-phenylimidazo-[1,5-a]pyridinium) bistetrafluoroborate, 46B, 266
$C_{28}H_{28}ClF_2N_3O$, 1-(1-(4,4-Bis(4-fluorophenyl)butyl)-4-piperidinyl)-5-chloro-1,3-dihydro-2H-benzimidazol-2-one, 43B, 353
$C_{28}H_{30}ClF_2N_3O$, 1-{1-[4,4-Bis(4-fluorophenyl)butyl]-4-piperidyl}-2-benzimidazolinone hydrochloride, 45B, 288
$C_{28}H_{30}ClF_2N_3O_5$, Pimozide perchlorate, 45B, 288
$C_{28}H_{56}Br_{16}N_4Sb_2$, Quinuclidinium dodecabromoantimon(III)antimon-(V)ate-2-dibromine, 37B, 678
$C_{29}H_{22}N_2O$, 2-Phenyl-2-(1-methyl-2-phenyl-3-indolyl)-3-indolinone, 40B, 262
$C_{29}H_{45}NO_2$, 4,6-Di-t-butyl-1-(3,5-di-t-butyl-2-hydroxyphenyl)-2-methyl-2-azabicyclo[3.2.0]hept-6-en-3-one, 45B, 253
$C_{30}H_{35}BF_4N_2$, γ-Cyclopropylbis(1,3,3-trimethylindolenin-2-yl)pentamethinium tetrafluoroborate, 42B, 66
$C_{30}H_{35}N_5O_3$, N-Acetyl-3-(benzoyl(2-piperidyl-2-piperidyleneium-ethyl)methylene)-indol-2-olate, 45B, 289
$C_{32}H_{26}Cl_4N_4O_5$, 7-Chloro-5-(2-chlorophenyl)-1,3-dihydro-3-hydroxy-1,4-benzodiazepin-2-one ethanol solvate, 42B, 212
$C_{32}H_{32}Cl_4N_4O_3S \cdot 0.2\ H_2O$, 5,6-Dichloro-1,3-diethyl-2-((5,6-dichloro-1,3-diethyl-2-benzimidazolinylidene)-1-propynyl)benzimidazolium toluene-p-sulphonate hydrate, 41B, 339
$C_{32}H_{33}Cl_3N_4O_2$, 1-(1-(3-Cyano-3,3-diphenylpropyl)-4-piperidinyl)-1,3-dihydro-3-(1-oxopropyl)-2H-benzimidazol-2-one chloroform, 43B, 353
$C_{34}H_{24}BrN_3$, 2-Phenyl-2-(2-phenyl-3-indolyl)-3-(p-bromophenyl)imino-indoline, 40B, 262
$C_{35}H_{27}N_3$, 2-(1-Methyl-2-phenyl-3-indolyl)-3-phenyl-3-phenylamino-3H-indole, 41B, 340
$C_{35}H_{29}N_3O$, 2-Anilino-3-(1-methyl-2-phenyl-indol-3-yl)-3-phenyl-3H-indole monohydrate, 42B, 213
$C_{36}H_{34}Br_2N_2O_{12} \cdot C_3H_6O$, 1-(4-Bromo-2,6-dimethylphenyl)-2-(4-bromo-2,6-dimethylphenyl)imino-1,2-dihydro-3,4,5,5,6,7-hexacarbomethoxy-5h-1-pyrindine (acetone solvate), 37B, 144
$C_{36}H_{36}N_4O_3 \cdot 2\ H_2O$, Tris[(2-methyl-8-quinolyloxy)ethyl]amine dihydrate, 46B, 267
$C_{38}H_{31}N_3O_2$, 1-Ethyl-2-phenyl-2-(1-methyl-2-phenyl-indolyl-3)-3-indolinon-O-benzoyl oxime, 40B, 263
$C_{38}H_{33}N_3$, 1-Ethyl-2-(1-ethyl-2-phenylindol-3-yl)-2-phenyl-3-phenyliminoindoline, 42B, 214
$C_{38}H_{33}N_3$, 1-Ethyl-3-(1-ethyl-2-phenylindol-3-yl)-3-phenyl-2-phenyliminoindoline, 42B, 214
$C_{46}H_{34}N_2O_2$, 1,1-Bis(5-methyl-2-phenyl-1-benzoyl-3-indolizinyl)ethylene, 38B, 298

36. HETERO-NITROGEN (MORE THAN 2 FUSED RINGS)

$C_6H_9N_3$, cis-[1,2;3,4;5,6]-Triiminocyclohexane, 45B, 291
$C_6H_{10}N_6$, Allyl azide dimer, 39B, 207
$C_7H_9N_5O_2 \cdot H_2O$, 3a,4,5,5a,6,7,8,9-Octahydro-3-H-pyrazolo[3',4';3,4]-pyrrolo[2,3-d]pyrimidin-7,9-dione monohydrate, 45B, 291
$C_8H_5N_5$, 1,3,4,6-Tetraazacycl[3.3.3]azine, 44B, 247
$C_8H_8N_6$, 3,3'-Dimethyl-bis-s-triazolo[4,3-b;3',4'-f]pyridazine, 43B, 354
$C_9H_9N_5O$, 1H-4,6-Dimethylimidazo[1,2-a]purine-9-one, 41B, 341
$C_9H_9N_7O_3$, 1,5-Diamino-1H-1,2,4-triazolo[1,5-c]quinazolinium nitrate, 45B, 292
$C_9H_{12}N_2O_2$, 5,8-Diaza-4,9-dioxotricyclo[6.3.0.0^{1-5}]undecane, 41B, 342
$C_9H_{21}Cl_3N_4 \cdot 0.5\ H_2O$, Dodecahydro-1,4,7,9b-tetra-azaphenalene trihydrochloride hemihydrate, 30B, 142
$C_{10}H_4N_2O_4$, Pyromellitic acid diimide, 42B, 215
$C_{10}H_5Br_2N$, 1,4-Dibromo-cycl[3.2.2]azine, 26, 790
$C_{10}H_7N$, Cycl[3.2.2]azine, 26, 789
$C_{10}H_9ClN_4$, 8-Chloro-6,7-dihydro-3-methyl-dipyridazino[2,3-a.4,3-d]-pyrrole, 33B, 124
$C_{10}H_{10}N_2O_2$, 3,4-Dehydroproline anhydride, 40B, 264
$C_{10}H_{14}N_2O_2$, cis-Bicyclo[3.3.0]octane-3-spiro-5'-hydantoin, 45B, 292
$C_{10}H_{16}N_4 \cdot H_2O$, 6H,13H-1,4,8,11-Tetrahydrobis(pyridazino-[1,2-*a*:1',2'-d]-s-tetrazine monohydrate, 46B, 278
$C_{10}H_{20}N_4 \cdot 0.5\ C_6H_6\ O_2 \cdot H_2O$, 6H,13H-Octahydrobis(pyridazino[1,2-a:1',2'-d]-s-tetrazine hemihydroquinone monohydrate, 46B, 278
$C_{10}H_{20}N_4$, 6H,13H-Octahydrobis(pyridazino[1,2-a:1',2'-d]-s-tetrazine, 46B, 278
$C_{11}H_7BrN_2O_2$, 2-Bromo-1-methylbenzo[c]pyrazolo[1,2-a]pyrazole-3,9-dione, 39B, 208
$C_{11}H_7NO$, 5-Keto-1,5-dihydrobenz[cd]indole, 38B, 299
$C_{11}H_8N_4O_2$, 10-Methylisoalloxazine, 39B, 208
$C_{11}H_9N_3$, 2-Methyl-2H-naphtho[1,8-de]triazine, 46B, 267
$C_{11}H_{10}N_4$, 2-Amino-6-methyldipyrido[1,2-a:3',2'-d]imidazole, 45B, 293
$C_{11}H_{11}BrN_4 \cdot 2\ H_2O$, 2-Amino-6-methyldipyrido[1,2-a:3',2'-d]imidazole hydrobromide dihydrate, 45B, 293
$C_{11}H_{13}BrN_4O_4$, 10-Methylisoalloxazinium bromide dihydrate, 34B, 153; 35B, 191
$C_{11}H_{14}N_2O_2$, 1,4-Diaza-5,12-dioxotetracyclo[5.5.1.0^{4-13}.0^{10-13}]tridecane, 42B, 215
$C_{11}H_{17}BrN_2$, Octahydrodipyrido[1,2-a:1',2'-c]imidazol-10-ium bromide, 35B, 192
$C_{12}H_4Cl_4N_2$, 1,4,6,9-Tetrachlorophenazine, 30B, 141
$C_{12}H_4Cl_4N_2$, 2,3,7,8-Tetrachlorophenazine, 30B, 139
$C_{12}H_6BrNO_3$, 5-Bromo-N-hydroxynaphthaloimide, 37B, 145
$C_{12}H_7NO_2$, Naphthaloimide, 39B, 897
$C_{12}H_8N_2$, Phenazine (*a* form), 19, 613; 21, 619; 35B, 121
$C_{12}H_8N_2$, o-Phenanthroline, 44B, 247
$C_{12}H_8N_2$, 8b,8c-Diazacyclopent[fg]acenaphthylene, 40B, 265
$C_{12}H_8N_2O$, N-Oxyphenazine, 26, 781; 35B, 121
$C_{12}H_8N_2O_2$, Phenazine-5,10-dioxide, 24, 701; 26, 781; 28, 477
$C_{12}H_8N_2O_4 \cdot 2\ C_2H_4O_2$, 1,2,3,4-Tetrahydro-1,4-dioxo-5,10-dihydroxybenzo[g]phtalazine acetic acid, 46B, 268
$C_{12}H_8N_2O_4$, Iodinin, 34B, 155
$C_{12}H_8N_4$, Dibenzo-1,3a,4,6a-tetraazapentalene (metastable form), 42B, 216
$C_{12}H_8N_4$, Dibenzo-1,3a,4,6a-tetraazapentalene, 42B, 216
$C_{12}H_9ClN_2O_4$, o-Phenanthroline perchlorate, 44B, 248
$C_{12}H_9N$, Carbazole, 34B, 157; 40B, 1085

$C_{12}H_9N_3O_2$, 3-Methylnaphtho[1,2-e](1,2,4)triazin-2(3H)-one 1-oxide, 44B, 248
$C_{12}H_{10}N_2O_4S$, o-Phenanthrolinium hydrogensulfate, 44B, 249
$C_{12}H_{10}N_4O$, 3-Acetyl-4-methyl-as-triazino[4,3-b]indazole, 40B, 266
$C_{12}H_{10}N_4O_6$, o-Phenanthroline nitrate nitric acid, 44B, 249
$C_{12}H_{11}BrN_2O$, 1,10-Phenanthroline hydrobromide monohydrate, 43B, 356
$C_{12}H_{11}ClN_2O$, 1,10-Phenanthroline hydrochloride monohydrate, 43B, 356
$C_{12}H_{12}Br_2N_2$, 6,7-Dihydrodipyrido[1,2-a:2',1'-c]pyrazinediium dibromide, 42B, 217
$C_{12}H_{12}Br_2N_2O$, Dipyrido[1,2-a:2',1'-c]pyrazinium dibromide monohydrate, 38B, 301
$C_{12}H_{12}ClNO$, cis-1-Chloro-spiro[indan-2,2'-pyrrolidin]-5'-one, 45B, 311
$C_{12}H_{12}Cl_2N_2O_9$, o-Phenanthroline diperchlorate monohydrate, 44B, 250
$C_{12}H_{12}Cl_3N_2O_4Re$, 1,10-Phenanthrolinium aquadichlorotrioxorhenate(VII) chloride, 45B, 294
$C_{12}H_{12}N_2O_2 \cdot H_2O$, 1,2,4,5-Tetrahydro-7-methoxy-3H-benz[g]indazol-3-one monohydrate, 45B, 294
$C_{12}H_{12}N_2O_4$, 1,4-trans-Phenazinediol-1,2,3,4-tetrahydro-5,10-dioxide, 40B, 267
$C_{12}H_{14}Br_2N_2O$, 1,1'-Ethylene-2,2'-bipyridylium dibromide monohydrate, 34B, 158
$C_{12}H_{16}BaN_4O_{11}$, Barium 5-methylorotate trihydrate, 42B, 217
$C_{12}H_{16}N_2O_2$, 1,5-Diaza-6,13-dioxotetracyclo[6.5.1.0^{4-14}.0^{11-14}]tetradecane, 42B, 215
$C_{12}H_{16}N_4O_{10}$, Methyl orotate trans-syn-photodimer dihydrate, 38B, 302
$C_{12}H_{17}BrN_2O$, 1,2,3,4,6,7,8,9-Octahydro-11H-pyrido[2,1-b]quinazolin-11-one hydrobromide, 41B, 323
$C_{12}H_{18}N_2$, 9,10-Diazaphenanthrene, 38B, 300
$C_{12}H_{18}N_6$, 4,4a,9,9a,14,14a-Hexahydro-3H,8H,13H-tripyridazino[1,6-a:1',6'-c:1'',6''-e]-s-triazine, 39B, 210
$C_{12}H_{22}K_2N_4O_4$, Potassium trans-anti-bis(5-methylorotate) hexahydrate, 44B, 251
$C_{13}H_8ClN$, 9-Chloroacridine, 43B, 357
$C_{13}H_9N$, Acridine (form 2), 24, 703
$C_{13}H_9N$, Acridine (form 3), 20, 614
$C_{13}H_9N$, Phenanthridine, 39B, 210
$C_{13}H_9NO$, Phenanthridone, 35B, 193; 42B, 218
$C_{13}H_{10}N_6$, 3-Phenyl-3'-methyl-bis-s-triazolo[4,3-b;3',4'-f]pyridazine, 43B, 357
$C_{13}H_{11}BrN_2$, 2-Allyl-2-azonia-7-azabiphenylene bromide, 46B, 268
$C_{13}H_{11}ClN_2 \cdot 2\ H_2O$, 9-Aminoacridinium chloride dihydrate, 46B, 269
$C_{13}H_{11}N$, N-Methylcarbazole, 45B, 295
$C_{13}H_{12}ClN_2O_4$, N-Hydro-N'-methylphenaziniumyl perchlorate, 44B, 252
$C_{13}H_{12}N_3 \cdot 0.5\ SO_4 \cdot 1.75\ H_2O$, Proflavine hemisulfate hydrate, 41B, 343
$C_{13}H_{13}ClN_2O$, 9-Aminoacridine hydrochloride monohydrate, 40B, 268
$C_{13}H_{13}IN_4O_2$, 1,3,10-Trimethylisoalloxazinium iodide, 35B, 195
$C_{13}H_{13}N_3O$, Proflavine monohydrate, 42B, 218
$C_{13}H_{13}N_3O_3$, Methyl 1,2,3,4-tetrahydro-1-methyl-2-oxo-pyrimido[1,2-a]benzimidazol-4-carboxylate, 40B, 249
$C_{13}H_{14}N_4OS$, 6,6-Dimethyl-3-methylthio-6,7-dihydro-as-triazino[1,6-c]quinazolin-5-ium-1-olate, 41B, 343
$C_{13}H_{15}NO_2$, trans-1'-Hydroxy-spiro[pyrrolidine-2,2'-tetralin]-5-one, 45B, 311
$C_{13}H_{16}ClN_3O_5$, Azotobacter vinelandii chromophore, 39B, 210
$C_{13}H_{16}N_2O_3$, 4,5-Dihydro-7,8-dimethoxy-2H-benz[g]indazole monohyd-

rate, 43B, 358
$C_{13}H_{16}N_4O_4$, 7,9a,10-Trimethyl-8-oxo-8,9,9a,10-tetrahydroalloxazine monohydrate, 40B, 270
$C_{13}H_{16}N_6 \cdot H_2O$, 2-Amino-3,9-dimethyl-5-dimethylamino-3H-1,3,4,6-tetrazacyclopent[e]azulene monohydrate, 45B, 295
$C_{13}H_{17}Cl_2N_3O_2$, 3,6-Diaminoacridine dihydrochloride dihydrate, 40B, 270
$C_{13}H_{23}ClN_2O_6$, Nitropolyzonammonium perchlorate, 41B, 345
$C_{13}H_{24}Br_2N_6O_3$, Paragracine dihydrobromide trihydrate, 41B, 344
$C_{13}H_{26}I_2N_2$, N,N'-Dimethyl-trans-perhydrocyclopenta[1,2-c:3,4-c']dipyrrol-bis-methiodide, 33B, 125
$C_{14}H_9N$, 4,5-Iminophenanthrene, 37B, 146
$C_{14}H_9NO_2$, 2,2'-Biphenyldicarboximide, 44B, 252
$C_{14}H_9NO_2$, 5H-10,11-Dioxodihydrodibenzo[b,f]azepine, 39B, 211
$C_{14}H_9N_3$, 1-Phenyl-2,3-diazacycl[3.2.2]azine, 46B, 269
$C_{14}H_{10}Cl_2N_2$, 1,10-Dichloro-3,8-dimethyl-4,7-phenanthroline, 38B, 303
$C_{14}H_{11}Cl_2N$, 9-Chloromethylacridine hydrochloride, 40B, 271
$C_{14}H_{11}N$, N-Vinylcarbazole, 42B, 218
$C_{14}H_{11}N$, 5H-Dibenzo[b,f]azepine, 46B, 270
$C_{14}H_{11}NO$, N-Methylacridone, 46B, 270
$C_{14}H_{12}N_2O_2$, 1,6-Dimethoxyphenazine, 32B, 166
$C_{14}H_{13}N$, Iminodibenzyl, 46B, 271
$C_{14}H_{14}I_3N_2$, 5,10-Dihydro-5,10-dimethylphenazinium triiodide, 44B, 253
$C_{14}H_{14}N_2$, 1,3,5-Trimethylazuleno[1,8-cd]pyridazine, 44B, 254
$C_{14}H_{14}N_2O_2$, 2,3-Dimethyl-4a,9a-diaza-1,4,4a,9,9a,10-hexahydroanthracene-9,10-dione, 43B, 358
$C_{14}H_{14}N_2O_3$, 2,3-Epoxy-cis-1,3-dimethyl-4a,9a-diaza-1,2,3,4,4a,9,9a,10-octahydroanthracene-9,10-dione, 43B, 359
$C_{14}H_{14}N_2O_3$, 8-Amino-5,7-dimethoxy-6-methylpyrrolo[1,2-a]indol-9-one, 43B, 360; 44B, 254
$C_{14}H_{14}N_4O_2$, 3,7,8,10-Tetramethylisoalloxazine, 38B, 305
$C_{14}H_{15}BrN_4O_3$, 9-Bromo-3,7,8,10-tetramethylisoalloxazine monohydrate, 38B, 306
$C_{14}H_{15}Cl_2N_3 \cdot 2\ H_2O$, 7-Chloro-2,6-dihydro-4-methyl-5,6-ethano-1H-(1,4)diazepino[1,7-a]benzimidazole hydrochloride dihydrate, 46B, 271
$C_{14}H_{15}IN_4O_3$, 1,10-Ethylene-7,8-dimethylisoalloxazinium iodide monohydrate, 34B, 156; 35B, 196
$C_{14}H_{15}N_3O_5 \cdot 0.5\ CH_4O$, 3-Carbamoyl-1,2-dihydro-4-hydroxy-5-methoxy-3H-pyrrolo[3,2-e]indole-7-carboxylic acid methyl ester methanol solvate, 44B, 255
$C_{14}H_{16}N_2O_3 \cdot 0.5\ H_2O$, 2,3a,4,5-Tetrahydro-7,8-dimethoxy-3a-methyl-3H-benz[g]indazol-3-one hemihydrate, 46B, 272
$C_{14}H_{16}N_2O_4$, 2,3-Hydroxy-2,3-dimethyl-4a,9a-diaza-1,2,3,4,4a,9,9a,10-octahydroanthracene-9,10-dione, 44B, 256
$C_{14}H_{16}N_2O_5$, 1,2β-Dimethoxycarbonyl-3aα-hydroxy-2α,3,3,8,8aα-pentahydropyrrolo[2,3-b]indole, 44B, 256
$C_{14}H_{17}Br_2N$, trans-6,8-Dibromo-1,2,3,4,4a,9a-hexahydro-4a,9-dimethylcarbazole, 37B, 146
$C_{14}H_{17}N_3O$, 3a,9a-Dihydro-1,3,3a,9a-tetramethyl-4H-pyrazolo[3,4-b]quinolin-4-one, 46B, 272
$C_{14}H_{21}ClN_2O$, 3a,9-Dimethylpyrroloquinoline hydrochloride, 45B, 296
$C_{14}H_{24}N_2$, 2,3-Dimethyl-2,3-diazatricyclo[8.4.0.0$^{4-9}$]tetradec-9-ene, 43B, 333
$C_{14}H_{26}N_4$, 1,6-Dimethyl-3a,5a,8a,10a-tetraaza-cis-10b,10c-perhydropyrene, 43B, 360

$C_{15}H_8Cl_3NO_2S$, 9-Chloroacridinium 2-chloro-1-(chlorosulfinyl)-2-oxoethylide, 42B, 63
$C_{15}H_{13}NO$, N-Ethylacridone, 45B, 296
$C_{15}H_{13}NO$, 6,7-Dihydro-6-methyl-5H-dibenz[c,e]azepin-5-one, 39B, 211
$C_{15}H_{14}N_4O_2$, 4,5-Benzo-8,11,11-trimethyl-1,2,7,10-tetraazatricyclo-[7.3.0.0$^{3-7}$]dodeca-2,9-diene-6,12-dione, 46B, 273
$C_{15}H_{15}Br_2N$, 1,1-Bis(bromoethyl)-1,2,3,4-tetrahydroacridine, 46B, 273
$C_{15}H_{15}N_2O_5$, Methyl 8-methoxycarbonylmethyl-2-oxo-2,3,3a,8-tetrahydro-1H-pyrazolo[5,1-a]isoindol-3-carboxylate, 41B, 346
$C_{15}H_{16}N_2O_2$, 5,7-Dihydroxy-6,6-dimethyl-6,7-dihydrodibenzo[d,f](1,3)-diazepine, 44B, 256
$C_{15}H_{16}N_2O_2S$, 9,9a-Dihydro-1,9,9-trimethyl-2-methylene-3,10-dioxo-9a-mercaptopiperazino[1,2-a]indole, 44B, 257
$C_{15}H_{16}N_2O_5 \cdot 0.5\ C_3H_6O$, 3-Acetyl-1,2-dihydro-4-hydroxy-5-methoxy-3H-pyrrolo[3,2-e]indole-7-carboxylic acid methyl ester acetone solvate, 44B, 255
$C_{15}H_{17}BrN_4O_2$, 9-Bromo-1,3,7,8,10-pentamethyl-1,5-dihydroisoalloxazine, 34B, 154; 38B, 306
$C_{15}H_{17}NO_2$, 4-Methyl-4-hydroxy-1,2,3,4-tetrahydro-6-methoxyacridine, 42B, 219
$C_{15}H_{17}N_3O_2$, 3-Amino-1,4-dimethyl-5H-pyrido[4,3-b]indole acetate, 44B, 258
$C_{15}H_{18}N_2O_2$, trans-1-Ethyl-3-(methoxycarbonyl)-1,2,3,4-tetrahydro-β-carboline, 46B, 274
$C_{15}H_{18}N_4O_4$, 4a,10a-Ethylenedioxy-1,3,10-trimethyl-4a,5,10,10a-tetrahydroalloxazine, 43B, 361
$C_{15}H_{19}IN_4O_3$, 1,3,7,8,10-Pentamethylisoalloxazinium iodide monohydrate, 34B, 156; 38B, 307
$C_{15}H_{20}BrN$, 1,2,3,4,4a,9a-Hexahydro-4a,9-propanocarbazolium hydrobromide, 35B, 197
$C_{15}H_{20}N_4O_4$, 3a-Methoxy-9a-methoxycarbonyl-1,3,4-trimethyl-2-oxo-2,3,3a,4,9,9a-hexahydro-1H-imidazo[4,5-b]quinoxaline, 43B, 361
$C_{16}H_8N_4$, Dichinoxalylene, 40B, 271
$C_{16}H_9ClN_2O$, 1-Chlorobenz[a]phenazine-7N-oxide, 37B, 146
$C_{16}H_{10}N_2$, Naphtho[2,1-c]cinnoline, 38B, 308
$C_{16}H_{11}ClN_4$, 8-Chloro-6-phenyl-4H-s-triazolo[4,3-a](1,4)benzodiazepine, 39B, 212
$C_{16}H_{12}N_6O_{11}S$, 6-Hydroxy-1,3,7,9-tetranitroindazolo[2,1-a]indazol-12-one dimethylsulfoxide, 44B, 258
$C_{16}H_{16}BrN$, 3-Bromo-N-methyl-5,6-dihydro-7H,12H-dibenz[c,f]azocine, 41B, 346
$C_{16}H_{17}BrN_4O_3$, 5-Acetyl-9-bromo-1,3,7,8-tetramethyl-1,5-dihydroalloxazine, 37B, 147
$C_{16}H_{17}N$, N-Methyl-5,6-dihydro-7H,12H-dibenz[c,f]azocine, 40B, 272
$C_{16}H_{17}N$, 9-Isopropyl-9,10-dihydroacridine, 46B, 274
$C_{16}H_{18}ClN_3O$, 10-Chloro-3,4-dihydro-4,4,7,8-tetramethyl-2-methoxypyrimido[5,4-b]quinoline, 38B, 309
$C_{16}H_{18}I_3N_2$, 5,10-Dihydro-5,10-diethylphenazinium triiodide, 44B, 253
$C_{16}H_{18}N_4O_3$, 4a,5-Epoxyethano-3-methyl-4a,5-dihydrolumiflavin, 44B, 260
$C_{16}H_{19}BrClN_5O_3 \cdot C_2H_3N$, 11-Chloro-2-[2-(dimethylamino)ethyl]-2,8-dihydro-8-methyl[1,2,4]triazino[4,3-d][1,4]benzodiazepine-3,4,7(6H)-trione hydrobomide acetonitrile solvate, 45B, 297
$C_{16}H_{19}ClN_4O_6$, 5-Ethyl-3,7,8,10-tetramethylisoalloxazinium perchlorate, 38B, 309
$C_{16}H_{19}NO_3$, 1,2,3,4,4aα,5,11aα-Heptahydroacetoxy-11βH-dibenz[b,e]-azepine-6-one, 41B, 347

$C_{16}H_{19}N_3O$, Spiro(N(1)-phenyl-1,2,3-triazole-5-one-4,9'-bicyclo-[6.1.0]nonane), 40B, 273
$C_{16}H_{21}ClIN_5O_2 \cdot 0.5\ H_2O$, 10-Chloro-2-[2-(dimethylamino)ethyl]-2,7-dihydro-7-methyl-3H-1,2,4-triazolo[4,3-d][1,4]benzodiazepine-3,6(5H)-dione methiodide hemihydrate, 45B, 297
$C_{16}H_{24}BrNO_3$, (-)-trans-4-Methyl-10b-methoxycarbonyl-1,2,3,4,4a,5,6,-10b-octahydrobenzo(f)quinoline hydrobromide monohydrate, 44B, 259
$C_{16}H_{24}N_4O_6$, 3-Carboethoxy-5-methyl-5-formyl-Δ^2-pyrazoline aminohemiacetal dimer, 42B, 219
$C_{16}H_{24}N_4O_6$, 3-Carbomethoxy-5-methyl-5-formyl-Δ^2-pyrazoline dimer, 41B, 348
$C_{17}H_8ClNO_2$, 6-Chloro-α-pyridineanthraquinone, 42B, 220
$C_{17}H_{12}N_2O_9 \cdot 2\ H_2O$, 4,5-Dihydro-5-hydroxy-4-oxo-5-(2-oxopropyl)-1H-pyrrolo[2,3-f]quinoline-2,7,9-tricarboxylic acid dihydrate, 46B, 274
$C_{17}H_{13}ClN_4$, 9-Chloro-2-methyl-7-phenyl-5(H)-[1,2,4]triazolo[5,1-a]-[2,4]benzodiazepine, 45B, 298
$C_{17}H_{14}N_2$, 5,11-Dimethyl-6H-pyrido[4,3-b]carbazole, 40B, 273
$C_{17}H_{14}N_2O_2$, 1-Phenyl-4,5-dihydro-7,8-dihydroxy-1H-benz[g]indazole, 43B, 363
$C_{17}H_{14}N_2O_2$, 1H,4H-1-Methylpyridazino[1,2-b]benzo[g]phthalazine-6,13-dione, 44B, 259
$C_{17}H_{14}N_2O_2$, 12,12-Dimethyl-2,4-dioxo-2,3,4,12-tetrahydrobenzo-[e]pyrimido[3,4-a]indole, 42B, 220
$C_{17}H_{14}N_2O_2$, 7a-Methoxycarbonyl-1,2-diaza-4,5:6,7-dibenzobicyclo-[4.3.0]non-1-ene, 43B, 362
$C_{17}H_{17}Br_5N_4O$, 2,2-Di(bromomethyl)-4,11-dibromo-5H-imidazolidino[5,4-b]phenazinium bromide monoethanolate, 43B, 364
$C_{17}H_{19}BrN_4O_3$, 5-Acetyl-9-bromo-1,3,7,8,10-pentamethyl-1,5-dihydroisoalloxazine, 35B, 199
$C_{17}H_{19}N$, 9-t-Butyl-9,10-dihydroacridine, 45B, 298
$C_{17}H_{20}N_2O_4$, 6-Deoxy-6-azidodihydroisomorphine, 40B, 274
$C_{17}H_{21}ClN_4O_6$, 2-O,3-Diethyl-7,8,10-trimethylisoalloxazinium perchlorate, 45B, 299
$C_{17}H_{22}N_4O_2$, 4a-Isopropyl-3-methyl-4a,5-dihydrolumiflavin, 44B, 260
$C_{17}H_{25}BrN_2O_3$, 4-Ethyl-1,4,5,7,8,12b-hexahydro-10,11-dimethoxy-(1,4)diazepino[7,1-a]isoquinolin-2(3H)-one hydrobromide, 39B, 898
$C_{17}H_{25}N_3O_9$, Chromophore from azotobacter vinelandii, 37B, 148
$C_{17}H_{27}NO$, 4-Methyl-1-azatetracyclo[12.3.0.0$^{2-11}$.0$^{4-9}$]heptadecan-12-one, 42B, 221
$C_{18}H_{12}N_2O_2$, N-Phenylaminonaphthaloimide, 41B, 349
$C_{18}H_{12}N_6$, 3,3'-Diphenyl-bis-s-triazolo[4,3-b;3',4'-f]pyridazine, 43B, 365
$C_{18}H_{14}N_2$, 5,6-Cyclopentenopyrido[3,2-a]carbazole, 46B, 276
$C_{18}H_{14}N_2OS$, Anhydro-2-benzyl-1-mercapto-9-methyl-9-oxo-(9H)-imidazo-[1,5-a]indolium hydroxide, 39B, 213
$C_{18}H_{15}NO_3$, 4,5,10,11,12,13-Hexahydro-10,13a-methanocycloocta[c]pyrrolo[3,2,1-ij]quinoline-7,9,14-trione, 44B, 260
$C_{18}H_{16}BrNO$, trans-11-Bromo-5,6-dihydro-4,6-dimethyl-4H,8H-pyrido-[3,2,1-de]phenanthridin-8-one, 44B, 261
$C_{18}H_{16}N_2O_2$, 1H,4H-2,3-Dimethylpyridazino[1,2-b]benzo[g]phthalazine-6,13-dione, 46B, 276
$C_{18}H_{16}N_2O_2$, 3H,4H-2,3-Dimethylpyridazino[1,2-b]benzo[g]phthalazine-6,13-dione, 46B, 276
$C_{18}H_{16}N_4$, 6,7,14,15-Tetrahydrobisbenzimidazo[1,2-a:1',2'-e]-(1,5)diazocine, 42B, 222
$C_{18}H_{17}NO$, cis-5,6-Dihydro-4,6-dimethyl-4H,8H-pyrido[3,2,1-de]-

phenanthridin-8-one, 44B, 261
$C_{18}H_{17}NO_3$, 1,2,3,9b-Tetrahydro-9bβ-hydroxy-2β-methoxy-1α-phenyl-5H-pyrrolo-[2,1-a]isoindol-5-one, 46B, 277
$C_{18}H_{19}BrN_2O$, 1-Methyl-2-phenyl-5,10-dihydro-1H-imidazo[2,1-b]isoquinolin-4-ium bromide monohydrate, 43B, 366
$C_{18}H_{19}BrN_4O_4$, 4-Ethoxycarbonyl-5-p-bromooxobenzoyl-8,8-dimethyl-3,4,9,10-tetraazatricyclo[5.3.0.0$^{1-6}$]deca-2,9-diene, 40B, 275
$C_{18}H_{19}Br_3N_2O_3$, 11,12,13-Tribromo-3,6-diethyl-10-ethoxycarbonyl-3,6-diazatricyclo[7.4.0.0$^{2-7}$]trideca-2(7),9,11,13-tetraene-8-one, 46B, 277
$C_{18}H_{19}ClN_4$, 8-Chloro-11-(4-methylpiperazin-1-yl)dibenzo[b,e](1,4)-diazepine, 42B, 223
$C_{18}H_{19}N$, 4-Phenyl-tetrahydro-benz[f]isoindoline, 42B, 223
$C_{18}H_{20}ClN_3O_2$, 2-(2-(6-Chloro-2-methoxy-9-acridinylamino)ethylamino)-ethanol, 38B, 310
$C_{18}H_{20}N_4O_2$, 9-(3-Dimethylaminopropylamino)-2-nitroacridine, 45B, 299
$C_{18}H_{20}N_4O_3 \cdot H_2O$, 9-[3-(Dimethyloxyamino)propylimino]-1-nitro-9,10-dihydroacridine monohydrate, 45B, 300
$C_{18}H_{21}ClN_4O$, 2-Chloro-11-(4-methylpiperazin-1-yl)dibenzo[b,e](1,4)-diazepine monohydrate, 42B, 223
$C_{18}H_{21}ClN_4O_2$, 2-(3-(7-Chloro-2-methoxy-10-(benzo[b]-1,5-naphthyridinyl)amino)propylamino)ethanol, 41B, 350
$C_{18}H_{21}IN_4O_2$, 1-Nitro-9-(3-dimethylaminopropylamino)-acridine monoiodide, 41B, 351
$C_{18}H_{21}N_3O$, 2-(3-(9-Acridinylamino)-propylamino)ethanol, 39B, 214
$C_{18}H_{22}N_4O_2$, 4a-Allyl-3,5,7,8,10-pentamethyl-4a,5-dihydroisoalloxazine, 38B, 311
$C_{18}H_{24}BrNO_4$, trans-5α,10bβ-Dimethoxycarbonyl-4-methyl-1,2,3,4,4a,5,6,10b-octahydrobenzo[f]quinoline hydrobromide, 45B, 300
$C_{18}H_{29}NO$, 4-Methyl-1-azatetracyclo[12.4.0.0$^{1-11}$.0$^{4-9}$]octadecan-12-one, 42B, 222
$C_{18}H_{30}N_4O_5$, 5,5-Diethyl-3,7,8,10-tetramethyl-1,5-dihydroisoalloxazine trihydrate, 37B, 148
$C_{18}H_{34}N_4$, meso-1,1,3,6,6,8-Hexamethyl-3a,5a,8a,10a-tetraazaperhydropyrene, 46B, 279
$C_{18}H_{34}N_4$, racemic-1,1,3,6,6,8-Hexamethyl-3a,5a,8a,10a-tetraazaperhydropyrene, 46B, 279
$C_{18}H_{36}N_4 \cdot 0.25\ H_2O$, 5,5,7,12,12,14-Hexamethyl-1,4,8,11-tetraazatricyclo[9.3.1.1$^{4-8}$]hexadecane hydrate, 46B, 279
$C_{19}H_{16}Cl_2N_4O_4$, 9,10-Dichloro-8-(2,4-dinitrophenyl)-5,8a,9,10,10a,-10b-hexahydro-6H-cyclobuta(4,5)pyrazolo[3,2-a]isoquinoline, 44B, 261
$C_{19}H_{16}N_6O_5$, 6-Acetyl-8-(acetyloxyimino)-2-phenyl-4-oxo-4,8-dihydro-2H,6H-pyrazolo[3,4-f]-1,2,3-benzotriazole - dioxane (2:1), 43B, 366
$C_{19}H_{19}ClN_2O_2S$, 1,2,3,4-Tetrahydro-10-methyl-4a-(p-chlorophenylsulphonylamino)pyrido[1,2-a]indole, 45B, 301
$C_{19}H_{21}N$, 3a,4,9,9a-Tetrahydro-2-methyl-4β-phenylbenz[f]isoindoline, 46B, 280
$C_{19}H_{21}NO_2$, DL-4β,10β-Dimethyl-6-ethylamino-4-hydroxycarbonyl-2,3,5β,10-tetrahydrophenanthr-1-one lactam, 35B, 351
$C_{19}H_{22}BrN$, 2,3,4,4a,9,9a-Hexahydro-2-methyl-9-phenyl-1H-indeno[2,1-c]pyridine hydrobromide, 37B, 150
$C_{19}H_{22}ClN_3$, 2-Methoxy-6-chloro-9-(3-dimethylaminopropylamino)acridine, 43B, 367
$C_{19}H_{22}ClN_3O_2$, 2-(3-(6-chloro-2-methoxy-9-acridinylamino)propylamino)ethanol, 38B, 311

$C_{19}H_{23}BrN_3O_6$, Diethyl (2-nitrophenazine-1)-bromomalonate, 37B, 149
$C_{19}H_{23}N$, N-t-Butyl-5,6-dihydro-7H,12H-dibenz[c,f]azocine, 40B, 276
$C_{19}H_{25}BrN_2$, Imipramine hydrobromide, 44B, 262
$C_{19}H_{25}ClN_2$, Imipramine hydrochloride, 41B, 351
$C_{19}H_{28}N_4O_4$, 1,1'-Trimethylene-3,3'-dipropylbisthymine photodimer, 43B, 368
$C_{19}H_{31}ClN_2O$, Iprindole hydrochloride monohydrate, 40B, 276
$C_{20}H_{12}N_2$, asym-α,β-Naphthazine, 30B, 143
$C_{20}H_{12}N_4O_4$, 5-Cyano-1,3-dioxo-2-phenyl-2,10b-dihydro-s-triazino[1,2-a]quinoline-6-carboxylic acid methyl ester, 42B, 224
$C_{20}H_{18}Br_2N$, 12-Bromo-1-(4'-bromophenyl)-1,4a,4b,5,6,13b-hexahydro-4H-dipyridazino[1,6-a:4,3-c]quinoline (two isomers), 37B, 151
$C_{20}H_{18}Br_2N_4$, (4aRS,4bRS,13bRS)-12-Bromo-1-(p-bromophenyl)-1,4a,4b,5,6,13b-hexahydro-4H-dipyridazino-[1,6-a:4,3-c]quinoline, 38B, 313
$C_{20}H_{18}Br_2N_4$, (4aRS,4bSR,13bRS)-12-Bromo-1-(p-bromophenyl)-1,4a,4b,5,6,13b-hexahydro-4H-dipyridazino-[1,6-a:4,3-c]quinoline, 38B, 312
$C_{20}H_{18}ClN_3O_2S$, 4'-(Acridin-9-ylamino)methanesulphonanilide hydrochloride, 40B, 277
$C_{20}H_{21}ClN_2O_4$, Dimethyl-1-p-chlorophenyl-3,4-propano-4,5,6,7-tetrahydroindazol 5,5-dicarboxylate, 46B, 280
$C_{20}H_{21}NO_4$, Dimethyl 4,5,10,11,12,12a-hexahydroindolo[1,7-cd]benzazepine-7,8-dicarboxylate, 44B, 260
$C_{20}H_{22}BrN_3O$, Dimidium bromide, 40B, 278
$C_{20}H_{22}N_2$, 3,4,10,11-Dibenzo-1,8-diazacyclotetradeca-1,3,8,10-tetraene, 40B, 240
$C_{20}H_{24}N_4$, 5,6,7,8,15,16,17,18-Octahydrodibenzo[e,o](1,4,8,13)tetraazacyclohexadecene, 44B, 263
$C_{20}H_{26}BrNO_6$, cis-5,6,13,13a-Tetrahydro-3,9-dihydroxy-1,2,10-trimethoxy-8H-dibenzo[a,f]quinolizine hydrobromide monohydrate, 35B, 199
$C_{21}H_{12}N_2O_3$, 6-Methylbenzo[g]naphtho[1,2-c]-cinnoline-7,12,14-trione, 35B, 201
$C_{21}H_{12}N_4$, Tricycloquinazoline, 34B, 159
$C_{21}H_{13}N$, 1,2:8,9-Dibenzacridine, 21, 608; 24, 705
$C_{21}H_{13}N_3O_2$, 4b,9a,13b-Triazadibenzo[a,e]acephenanthrylene-9,14-dione, 42B, 224
$C_{21}H_{15}N_5O_4$, 1,2-b-[2-Methoxycarbonylpyrimido[1,2-a]-3-methoxycarbonylpyrazine]phenazine, 46B, 281
$C_{21}H_{16}ClN_3O_4$, Methyl 8-chloro-3-hydroxy-5-phenyl-1-methyl-2-oxo-2,3,3a,6-tetrahydropyrrolo[1,2,3-cd]pyrrolo[2,3-b]-1H-quinoxaline-4-carboxylate, 43B, 1479
$C_{21}H_{20}ClN_3O_3S$, 4'-(9-Acridinylamino)-3'-methoxymethanesulphonanilide hydrochloride, 46B, 282
$C_{21}H_{20}ClN_5O$, 8-Azidoethidium chloride monohydrate, 44B, 263
$C_{21}H_{21}ClN_2O_3S$, 2'-p-Chlorophenylsulphonylaminomethylene-N,N'-dimethyloxindole-3-spiro-cyclopentane, 45B, 302
$C_{21}H_{22}BrN_3O$, Ethidium bromide monohydrate, 37B, 152
$C_{21}H_{22}BrN_5$, 5-(2-Bromo-4-methylphenyl)-3,7-dimethyl-1-propyl-1,5-dihydrobenzo[f]pyrazolo[3,4-c][1,2,5]triazepine, 45B, 302
$C_{21}H_{26}ClN_3O_2$, 2-Methoxy-6-chloro-9-(3-(ethyl-2-hydroxyethyl)aminopropylamino)acridine, 38B, 314
$C_{21}H_{28}NNaO_5 \cdot 0.5\ H_2O \cdot 0.5\ CH_4O$, Sodium dimethoxy(depyrrolo)dimethylcorynantheidinate hemihydrate methanol solvate, 45B, 303
$C_{21}H_{31}N_7O_{10}$, 10-(3-(3-Carbamoyl-1-pyridinium)propyl)-1,5-dihydro-7,8-dimethyl-isoalloxazine nitrate tetrahydrate, 44B, 264
$C_{21}H_{34}N_6O_{10}$, 10-(3-(3-Carbamoyl-1-pyridinium)-propyl)-7,8-dimethyl-

isoalloxazine heptahydrate, 44B, 264
$C_{22}H_{16}N_2O$, 3a,8b-Dihydro-1,3-diphenylindeno[1,2-c]pyrazol-4(1H)-one, 42B, 211
$C_{22}H_{19}Br_3N_2O_3$, 10,11,12-Tribromo-3,6-diethyl-13-(phenoxycarbonyl)-3,6-diazatricyclo[7.4.0.0$^{2-7}$]trideca-2(7),9,11,13-tetraene-8-one, 46B, 277
$C_{22}H_{20}BrNO_8$, Tetramethyl 3-bromo-7a,8,9,9a-tetrahydrocyclobuta(4,5)-pyrrolo[1,2-a]quinoline-7,r-7a,t-9,c-9a-tetracarboxylate, 43B, 369
$C_{22}H_{22}Cl_3NO_3S$, 8-Methoxy-N-tosyl-3-trichloromethyl-2,3,4,4a,5,6-hexahydrobenzo[f]quinoline, 39B, 215
$C_{22}H_{23}BrN_4O_4$, [[5,6-c]-(1-Bromobenzo)-1-ethyl-4-(3-methoxyphenylethyl)-1,4-diazocyclohex-5-en-2-one]-3-spiro-5'-(3'-methylhydantoin), 46B, 282
$C_{22}H_{23}N_3O_6S_2$, 4'-(2-Methoxy-9-acridinylamino)methanesulphonanilide methanesulphonate, 46B, 282
$C_{22}H_{24}BrN_3O$, cis-5-Acetyl-5,5a,6,7,8,10,11,11a-octahydro-9H-cyclooct[b]indol-9-one p-bromophenylhydrazone, 38B, 315
$C_{22}H_{24}BrN_3O$, 2,7-Diamino-9-phenyl-10-ethylphenanthridinium bromide methanol solvate, 34B, 160
$C_{22}H_{24}ClN_3O_2$, 7-Chloro-3,3-diethyl-2-oxo-9-phenylpyrazolo[5,1-b]-quinazolinium betaine ethanolate, 43B, 370
$C_{22}H_{24}N_4$, 7,16-Dihydro-6,8,15,17-tetramethyldibenzo[b,i](1,4,8,11)-tetraazacyclotetradecene, 42B, 853
$C_{22}H_{26}Cl_2N_2O_4$, 10,12-Dichloro-2,16-dihydro-16-hydroxy-2-methoxytabersonine, 46B, 283
$C_{22}H_{26}N_4$, 5,6-Diethyl-11a-methyl-5,6,11,13-tetraazadibenzo[b,h]-5,5a,5b,6,11,11a-hexahydrofluorene, 39B, 215
$C_{22}H_{26}N_8O_6$, 1,2-Bis(1,3,7-trimethyl-6-lumazinyl)-threo-1,2-butanediol, 45B, 303
$C_{23}H_{19}N_5O_4$, 1,2-c-[5,6-Bis(methoxycarbonyl)-2,4-dimethyl-1,9,10-triazatricyclo[5.2.1.0$^{4-10}$]deca-2,5,8-triene]-1,2-dihydrophenazine, 46B, 281
$C_{23}H_{25}BrN_3O$, Ethidium bromide ethanol solvate, 40B, 278
$C_{23}H_{25}ClN_3O$, Ethidium chloride ethanol solvate, 40B, 278
$C_{23}H_{36}Cl_3N_3O$, 6-Chloro-9-((4-(diethylamino)-1-methylbutyl)amino)-2-methoxyacridine dihydrochloride dihydrate, 39B, 216
$C_{24}H_{14}N_6$, 1,14:7,8-Diethenotetrapyrido[1,3,5,8,10,12]hexaazacyclotetradecine, 45B, 304
$C_{24}H_{16}N_2$, 4,7-Diphenyl-1,10-phenanthroline, 45B, 305
$C_{24}H_{16}N_2O_2$, 2-(o-Hydroxybenzoyl)-3-phenylimidazo[2,1-a]isoquinoline, 46B, 220
$C_{24}H_{16}N_4O_2$, Cyclobuta[1,2-h:3,4-j]bis(pyrido[2,1-h]quinazolin)-5,6-dione, 44B, 265
$C_{24}H_{18}N_4O_3$, 1,3-Dimethyl-5-(10-phenyl-2,10-dihydrophenazin-2-ylidene)hexahydropyrimidine-2,4,6-trione, 45B, 305
$C_{24}H_{27}N_3O$, 7-(3-(Ethyl-2-hydroxyethylamino)propylamino)benz[c]acridine, 43B, 370
$C_{24}H_{34}N_8O_9$, Pyrimidine tetramer trihydrate, 38B, 326
$C_{25}H_{23}ClFN_3O_2S$, 8-Chloro-cis-6-(2-fluorophenyl)-1-(((4-methylphenyl)sulphonyl)methylene)-2,3,3a,4,5,6-hexahydro-1H-imidazo[1,5-a](1,4)benzodiazepine, 44B, 266
$C_{25}H_{23}N_3O_8$, 3-Ethyl-1,2,5-trimethyl-3H-benzo[c](1,2,5)triazepino1,2-a]cinnoline-1,2,3,5-tetracarboxylate, 40B, 280
$C_{25}H_{26}BrN_3O_5$, 7-Benzyl-3-(p-bromobenzyloxycarbonyl)-2-hydroxy-3,6,9-triazatricyclo[7.3.0.0$^{2-6}$]dodecane-5,8-dione, 46B, 284
$C_{25}H_{26}N_2$, 6,8,8-Trimethyl-6-(3-indolyl)-4-(isopropenyl)-(2,3)-benzo-1-azabicyclo[3.3.0]octa-2,4-diene, 46B, 285

$C_{25}H_{27}N_3O_5$, 7-Benzyl-3-(benzyloxycarbonyl)-2-hydroxy-3,6,9-triazatricyclo[7.3.0.0$^{2-6}$]dodecane-5,8-dione, 46B, 284
$C_{25}H_{31}N$, (DL)-Deoxybutaclamol, 46B, 285
$C_{25}H_{32}BrNO$, (+)-Isobutaclamol hydrobromide, 45B, 306
$C_{26}H_{17}N_3O$, N'-Acetyl-di-indolo[2,3-a:2',3'-c]carbazole, 45B, 306
$C_{26}H_{18}N_2$, 6,7-Diphenyldibenzo[e,g])(1,4)diazocine, 39B, 217
$C_{26}H_{23}NO_{12}$, 1,2,3,3a,4,5-Hexacarbomethoxy-3a,4-isoindolo[3,2,1-cd]-indolizin, 40B, 281
$C_{26}H_{28}ClN_3O_2$, Ethyl 7-chloro-2-methyl-9-phenyl-1-(3-dimethylamino)-propyl-1H-pyrrolo[3,2-b]quinoline-3-carboxylate, 42B, 225
$C_{26}H_{32}ClN_5O_7$, 5-(N-Benzyl-N-methyliminioprop-2-enyl)-3,7,8,10-tetramethyl-1,5-dihydroisoalloxazine perchlorate methanol, 43B, 372
$C_{27}H_{18}N_2$, 6,6'-Methylenediphenanthridine, 33B, 127
$C_{27}H_{23}NO_{11}$, Pentamethyl 11,11a-dihydro-9-oxo-9H,10H-cyclobuta(4,5)-cyclopenta(3,4)pyrrolo[1,2-a]quinoline-7,8,10,11,11a-pentacarboxylate, 42B, 226
$C_{27}H_{24}Cl_3FN_4O_3S$, 8-Chloro-trans-6-(2-fluorophenyl)-5-nitroso-1-(((4-methylphenyl)sulphonyl)methylene)imidazo[1,5-a](1,4)benzodiazepine dichloromethane solvate, 44B, 266
$C_{27}H_{25}F_3N_2O$, dl-8-Fluoro-5-(4-fluorophenyl)-2-[4-hydroxy-4-(4-fluorophenyl)butyl]-2,3,4,5-tetrahydro-1H-pyrido[4,3-b]indole, 46B, 286
$C_{27}H_{29}N_3O_4S_2$, 1-Methyl-2'-(p-tolylsulphonamido)-2-(p-tolylsulphonylimino)indoline-3-spirocyclopentane, 37B, 153
$C_{27}H_{48}N_6O_3$, 3,7,11-Tri-t-butyl-4,4,8,8,12,12-hexamethyl-1H,2H,5H,6H,9H,10H-triimidazo[3,4-a:3',4'-c:3'',4''-e](1,3,5)-triazine-2,6,10-trione, 44B, 267
$C_{28}H_{12}N_2O_2$, Flavanthrone, 17, 758
$C_{28}H_{14}N_2O_4$, Indanthrone (α form), 19, 587
$C_{28}H_{16}N_2$, Bis(1-carbazolyl)butadiyne, 44B, 267
$C_{28}H_{16}N_2$, Tetrabenzo[a,c,h,j]phenazine, 46B, 286
$C_{28}H_{22}ClN_3O_4$, Bis(10-methyl-9-acridine)monoazamonomethinecyanine perchlorate, 38B, 317; 42B, 226
$C_{28}H_{24}N_2$, 10,10'-Dimethyl-9,10,9',10'-tetrahydro-9,9'-biacridinyl, 39B, 217; 43B, 373
$C_{28}H_{31}N_3O_4S_2$, cis-5,6,6a,7,8,9,10,10a-Octahydro-5-methyl-10a-p-tolylsulphonylamino-6-p-tolylsulphonyliminophenanthridine, 43B, 373
$C_{30}H_{20}N_2$, 1,6-Di-(N-carbazolyl)-2,4-hexadiyne, 43B, 1480
$C_{30}H_{28}NO_{12}$, Tetramethyl 7-((E)-1,2-dimethoxycarbonylvinyl)-1,3-dimethyl-7a,8,9,9a-tetrahydrocyclobuta(4,5)pyrrolo[1,2-a]quinoline-7a,8,9,9a-tetracarboxylate, 43B, 375
$C_{30}H_{35}NO_2$, 3,4-Dibenzyl-6,7,10,10-tetramethyl-3-azatricyclo-[7.4.0.0$^{1-5}$]tridec-6-ene-2,13-dione, 46B, 287
$C_{30}H_{37}IN_8O$, 1,1',3,3'-Tetraethylimidazo[4,5-b]quinoxalinocyanine iodide-2-propanol, 37B, 154
$C_{32}H_{20}N_4$, 4,4''',6,6'''-Tetraazahexa-m-phenylene, 45B, 316
$C_{37}H_{37}Cl_8N_5O_2$, 1,10-Di(2-methoxy-6-chloro-9-acridinyl)-1,5,10-triazadecane bis(chloroform), 43B, 376
$C_{40}H_{44}Cl_8N_6O_2$, 1,14-Di(2-methoxy-6-chloro-9-acridinyl)-1,5,10,14-tetraazatetradecane bis(chloroform), 43B, 376
$C_{40}H_{51}Cl_2N_5O_5$, 1,9-Di(2-methoxy-6-chloro-9-acridinyl)-1,5,9-triazanonane tris(ethanol), 43B, 377
$C_{47}H_{30}Br_2N_6O_4$, 4'-Methylene-1,2-di-m-bromophenyl-1',2',6',7'-tetraphenylspiro(pyrazolidine-4,8'-(8'H,4'H)-benzo[1,2-c:4,5-c']dipyrazoline)-3,5,3',5'-tetraone, 40B, 282

37. HETERO-NITROGEN (BRIDGED-RING SYSTEMS)

$C_5H_8N_2$, 2,3-Diazabicyclo[2.2.1]hept-2-ene (gas-mw), 42B, 1035
$C_5H_8N_2O_2$, 2,3-Diazabicyclo[2.2.1]hept-2-ene 2,3-dioxide, 44B, 188
$C_5H_{10}N_6O_4$, Dinitropentamethylenetetramine, 40B, 283
$C_6H_3Cl_6N$, 1,3,4,5-endo,7,7-Hexachloro-2-azabicyclo[2.2.1]hept-2-ene, 45B, 307
$C_6H_{10}N_2$, 2,3-Diazabicyclo[2.2.2]oct-2-ene, 42B, 226
$C_6H_{10}N_2O$, 2,3-Diazabicyclo[2.2.2]oct-2-ene N-oxide, 45B, 308
$C_6H_{12}I_2N_4$, Hexamethylenetetramine-1-diiodine, 41B, 352
$C_6H_{12}I_4N_4$, Hexamethylenetetramine-2-diiodine, 41B, 352
$C_6H_{12}N_2$, Triethylenediamine (phase I), 42B, 227
$C_6H_{12}N_2$, Triethylenediamine (phase II), 24, 697; 29, 515; 42B, 227; 46B, 288
$C_6H_{12}N_4$, Hexamethylenetetramine (gas-ed), 6, 232; 10, 257
$C_6H_{12}N_4$, Hexamethylenetetramine, 1, 624, 734; 3, 688, 715; 7, 244; 18, 662; 21, 544; 26, 603; 28, 478
$C_6H_{12}N_4O$, Hexamethylenetetramine oxide, 45B, 308
$C_6H_{16}BF_4N_5$, Hexamethylenetetramine ammonium tetrafluoroborate, 44B, 267
$C_6H_{16}N_4O_4$, Hexamethylenetetramine oxide - hydrogen peroxide - water (1:1:1), 44B, 268
$C_7H_8N_2O_2$, 6,7-Diazatetracyclo[3.2.1.1$^{3-8}$.0$^{2-4}$]non-6-ene 6,7-dioxide, 44B, 188
$C_7H_9NO_3$, 1,S-4,S-5-(N-Acetyl)-5-aza-2-oxa-3-oxo-bicyclo[2.2.1]heptane, 45B, 309
$C_7H_{10}N_2O_2$, 1,6-Diazaspiro[4.4]nonane-2,7-dione, 42B, 171
$C_7H_{11}NO$, 3-Isoquinuclidone, 39B, 217
$C_7H_{13}BrClN$, 4-Bromoquinuclidium chloride, 46B, 289
$C_7H_{13}BrClNO_4$, 4-Bromoquinuclidium perchlorate, 46B, 288
$C_7H_{13}ClFN$, 4-Fluoroquinuclidium chloride, 46B, 289
$C_7H_{13}ClFNO_4$, 4-Fluoroquinuclidium perchlorate, 46B, 288
$C_7H_{13}ClINO_4$, 4-Iodoquinuclidium perchlorate, 46B, 288
$C_7H_{13}Cl_2N$, 4-Chloroquinuclidium chloride, 46B, 289
$C_7H_{13}Cl_2NO_4$, 4-Chloroquinuclidium perchlorate, 46B, 288
$C_7H_{13}NO_3S$, Quinuclidine sulfur trioxide, 45B, 309
$C_7H_{14}ClN$, Quinuclidium chloride, 46B, 289
$C_7H_{14}ClN$, 1-Azabicyclo[3.2.0]heptane-1-methyl chloride, 33B, 127
$C_7H_{14}I_5N$, Quinuclidinium tri-iodide iodine, 42B, 227
$C_7H_{14}N_4O_3$, Hexamethylenetetramine oxide formic acid, 44B, 268
$C_7H_{16}N_6OS$, Hexamethylenetetramine oxide - thiourea (1:1), 44B, 269
$C_8H_8N_2$, 7,8-Diazapentacyclo[4.2.2.0$^{2-5}$.0$^{3-9}$.0$^{4-10}$]dec-7-ene, 42B, 226
$C_8H_{10}N_2O$, 5,6-Diazatricyclo[5.3.0.0$^{4-8}$]deca-2,5-diene 5-oxide, 46B, 290
$C_8H_{12}NO$, 2-Azanoradamantane N-oxyl (cubic plastic phase), 43B, 380
$C_8H_{12}NO$, 2-Azanoradamantane N-oxyl (orthorhombic), 43B, 380
$C_8H_{12}NO_2$, 9-Azabicyclo[3.3.1]nonan-3-one-9-oxyl, 37B, 155
$C_8H_{12}N_2$, 2,3-Diaza-7-isopropylidene-bicyclo[2.2.1]hept-2-ene, 46B, 290
$C_8H_{12}N_2$, 7,8-Diazatricyclo[4.2.2.0$^{2-5}$]dec-7-ene, 42B, 226
$C_8H_{15}N$, 3-Azabicyclo[3.2.2]nonane, 40B, 284
$C_8H_{16}BrN$, 3-Azabicyclo[3.3.1]nonane hydrobromide, 29, 507
$C_8H_{16}BrNO$, Tropine hydrobromide, 18, 698
$C_8H_{16}ClNO$, Tropine hydrochloride, 18, 698
$C_8H_{16}N_4$, 1,3,6,8-Tetra-azatricyclo[4.4.1.4$^{3-8}$]dodecane, 40B, 285
$C_9H_{10}N_2O_4$, 2,7-Diazo-spiro[5.5]undecane-1,3,6,8-tetraone, 41B, 318

$C_9H_{12}Cl_2N_2O_4$, N,N'-Dicarbomethoxy-2,3-diaza-5(exo)-7(anti)-dichlorobicyclo[2.2.1]heptane, 39B, 218
$C_9H_{13}NO$, DL-(3S)-4-Azatricyclo[4.3.1.0$^{3-7}$]decan-5-one, 44B, 269
$C_9H_{13}NO$, 4-Aza-5-oxo-tri[4.4.0.0$^{3-8}$]decane, 43B, 381
$C_9H_{13}N_3O_5$, 1,3-Dinitro-6-hydroxymethylene-3,4-methanonortropane, 38B, 320
$C_9H_{13}N_3O_5$, 1,5-Dinitro-3-methyl-3-azabicyclo[3.3.1]nonan-7-one (A-form), 42B, 228
$C_9H_{13}N_3O_5$, 1,5-Dinitro-3-methyl-3-azabicyclo[3.3.1]nonan-7-one (B-form), 42B, 228
$C_9H_{13}N_3O_6$, 1,3-Dinitro-6-hydroxymethylene-N-hydroxy-3,4-methanonortropane, 38B, 320
$C_9H_{14}NO_2$, 1,5-Dimethyl-8-azabicyclo[3.2.1]octan-3-one-8-oxyl, 39B, 219
$C_9H_{16}ClNO_4$, 4-azoniaspiro[3.5]nonane perchlorate, 37B, 156
$C_9H_{16}N_4O_2$, 3,7-Diacetyl-1,3,5,7-tetraazabicyclo[3.3.1]nonane, 42B, 228
$C_9H_{18}Cl_4N_2O_8$, 1,4-Bis(2-chloroethyl)-1,4-diazabicyclo[2.2.1]heptane diperchlorate, 37B, 156
$C_{10}H_{12}N_2$, Benzo[b][1,4]diazabicyclo[3.2.1]octane, 45B, 309
$C_{10}H_{16}N_2O_3$, 6,9-Diaza-5,10-dioxotricyclo[7.3.0.0$^{1-6}$]dodecane monohydrate, 43B, 355
$C_{10}H_{18}INO_2$, (R)-(-)-3-Acetoxyquinuclidine methiodide, 38B, 321
$C_{10}H_{20}ClN$, 1-Azabicyclo[3.3.3]undecane hydrochloride, 38B, 321
$C_{11}H_{15}N_3$, 4-(1,5-Diazabicyclo[3.2.1]oct-8-yl)pyridine, 31B, 137
$C_{11}H_{16}ClNO_3S$, 4-Chlorosulphonyl-7,8,8-trimethyl-4-azatricyclo-[4.2.1.0$^{3-7}$]nonan-5-one, 39B, 219
$C_{11}H_{17}NO_2$, 3-Amino-3-hydroxy-trans-bicyclo[4.4.0]decane-1-carboxylic acid lactam, 44B, 270
$C_{11}H_{17}N_3O_2$, N-Methylgranatanine-3-spiro-5'-hydantoin, 44B, 270
$C_{11}H_{17}N_3O_2$, 3-Ethyl-3-azabicyclo[3.2.1]octane-8-spiro-5'-hydantoin, 45B, 310
$C_{11}H_{18}N_4O_3$, 1,5,6,1',3'-Pentamethyl-3-oxopiperazine-2-spiro-5'-hydantoin, 42B, 204
$C_{11}H_{19}NO$, 11-Methyl-11-azabicyclo[5.3.1]undecan-4-one, 41B, 354; 44B, 271
$C_{12}H_8Cl_5N$, 1,3,4,7,7-Pentachloro-5-phenyl-2-azabicyclo[2.2.1]hept-2-ene, 46B, 291
$C_{12}H_{12}BrNO$, 6-(p-Bromobenzoyl)-6-azabicyclo[3.1.0]hexane, 33B, 128
$C_{12}H_{13}N_3O_2$, ([4.2.1]Propella-2,4-diene)-(4-methyl-1,2,4-triazoline-3,5-dione), 46B, 291
$C_{12}H_{15}NO_3$, 1,2,5a,7b-Tetrahydro-5a,5b-dimethoxy-5bH-cyclobuta(1,4)-cyclobuta[1,2,3-gh]-pyrrolizin-4(5H)-one, 34B, 165
$C_{12}H_{16}N_2$, 2,3-Dihydrobenzimidazole-2-spirocyclohexane, 44B, 251
$C_{12}H_{17}NO_2$, 2,9-Diacetyl-9-azabicyclo[4.2.1]non-2,3-ene, 38B, 322
$C_{12}H_{18}N_4O_3$, 1,3,5,7-Tetraazaadamantane N-oxide hydroquinone, 45B, 310
$C_{12}H_{19}N_3O_2$, 8-Methyl-8-azabicyclo[4.3.1]decane-10-spiro-5'-hydantoin, 45B, 311
$C_{12}H_{21}INO_2$, exo-2-Iodomethyl-2,4,4-exo-6-tetramethyl-3-azabicyclo-[3.3.1]nonan-7-on-3-oxyl, 40B, 286
$C_{12}H_{24}I_4N_8$, Bis(hexamethylenetetramine)iodinium triiodide, 41B, 355
$C_{12}H_{38}CaCr_2N_8O_{14}$, Calcium dichromate bis(hexamethylenetetramine) heptahydrate, 41B, 354
$C_{13}H_{11}NO_2$, 12-Methyl-11,13-dioxo-12-aza-pentacyclo-[4.4.3.0$^{1-6}$.0$^{2-10}$.0$^{5-7}$]trideca-3,8-diene, 38B, 233
$C_{13}H_{13}N_3O$, 11-Methyl-3,4,11-triaza-8,9-benzotricyclo-

[5.2.2.0$^{2-6}$]undec-3-ene-10-one, 43B, 382
$C_{13}H_{15}NO_2$, 11,13-Dioxo-12-methyl-12-aza[4.4.3]propella-3,8-diene, 42B, 229
$C_{13}H_{15}NO_3$, exo-2-Methoxy-3-aza-4-keto-7,8-benzobicyclo[4.2.1]nonene, 39B, 220
$C_{13}H_{15}NO_3$, syn-8,9-Epoxy-11,13-dioxo-12-methyl-12-aza[4.4.3]propell-3-ene, 42B, 230
$C_{13}H_{15}NO_4$, syn-2,3;4,5-Diepoxy-12-methyl-12-aza[4.4.3]propellane-11,13-dione, 44B, 271
$C_{13}H_{16}INO_2S$, 1-(p-Iodobenzenesulphonyl)-1-azaspiro[2.5]octane, 35B, 203
$C_{13}H_{19}NO_3$, 3-Aza-11-formyloxy-tricyclo[7.3.1.0$^{3-8}$]tridecan-4-one, 44B, 271
$C_{13}H_{19}NO_5 \cdot 0.5\ H_2O$, 7-Hydroxy-1,9,10-trimethoxy-4-azabicyclo-[5.2.2]undeca-8,10-dien-3-one, 34B, 161; 35B, 204
$C_{13}H_{19}N_3O_3$, N-(β-Hydroxyethyl)granatanine-3-spiro-5'-hydantoin, 44B, 272
$C_{13}H_{20}ClNO_9$, 1,8,8-Trimethoxy-2,6-dioxo-4-(2'-aminoethyl)bicyclo-[3.2.1]oct-3-ene perchlorate, 35B, 206
$C_{14}H_8BrN_5O_2 \cdot 0.5\ C_7H_8$, 6-Bromo-8,8,9,9-tetracyano-2-methoxycarbonyl-2-azabicyclo[3.2.2]nona-3,6-diene toluene solvate, 31B, 138; 32B, 167
$C_{14}H_{14}BrNO$, N-p-Bromobenzoyl-exo-2,3-aziridinobicyclo[2.2.1]heptane, 38B, 323
$C_{14}H_{14}N_2$, [2.2](2,6)pyridinophane, 44B, 272
$C_{14}H_{15}Br_2N_2O_2$, N-Methylazepine dimer dihydrobromide dihydrate, 37B, 157
$C_{14}H_{15}N_3O_2$, 4-Phenyl-2,4,6-triazatricyclo[5.2.2.0$^{2-6}$]undecan-3,5-dione, 40B, 286
$C_{14}H_{15}N_3O_2$, 8-Methyl-4,6,8-triazapentacyclo[9.2.2.1$^{3-9}$.0$^{2-10}$.$^{4-8}$]-hexadeca-2(10),12-diene-5,7-dione, 46B, 295
$C_{14}H_{17}NO_2$, 2-(α-Hydroxybenzyl)-1-azabicyclo[2.2.2]octan-3-one, 42B, 230
$C_{14}H_{19}NO_2$, 2-(α-Hydroxybenzyl)-1-azabicyclo[2.2.2]octan-3-ol, 42B, 230
$C_{14}H_{19}N_3O_2$, ([4.4.1]Propella-3,4-diene)-(4-methyl-1,2,4-triazoline-3,5-dione), 46B, 291
$C_{14}H_{20}N_4O_2$, 1-Methyl-1,4-dihydronicotinamide dimer, 32B, 168
$C_{14}H_{22}N_4O$, 11-Acetyl-1,4,8,9-tetramethyl-2,3,10,11-tetraazatricyclo-[6.3.1.0$^{2-7}$]dodeca-3,9-diene, 44B, 273
$C_{14}H_{30}I_8N_8$, 1-Methyl-1,3,5,7-tetraazaadamantan-1-ium octaiodide, 45B, 312
$C_{14}H_{32}Cl_8N_4Ni_2$, Bis-N-methyl-N'-diazabicyclo[2.2.2]octonium di-μ-chloro-hexachlorodinickelate(II), 35B, 878
$C_{15}H_{11}N$, [2.2](2,6)Pyridinoparacyclophane-1,9-diene, 40B, 287
$C_{15}H_{16}BrNO_2$, 9-Benzoyl-3α-bromo-9-azabicyclo[3.3.1]nonan-2-one, 37B, 158
$C_{15}H_{18}BrNO_2$, 9-Benzoyl-3α-bromo-2β-hydroxy-9-azabicyclo[3.3.1]-nonane, 33B, 129
$C_{15}H_{22}BrNO$, (+)-1-(3-Hydroxyphenyl)-6,7-dimethyl-6-azabicyclo-[3.2.1]octane hydrobromide, 43B, 1481
$C_{15}H_{25}ClN_2O_6$, Lupanine-N-oxide perchlorate, 38B, 323
$C_{15}H_{29}IN_2$, N-Methylazepine dimer perhydromonomethiodide, 40B, 1085
$C_{16}H_9NO_2$, Spiro(3-phenylazirine-2,2'-indan-1',3'-dione), 43B, 362
$C_{16}H_{15}N$, 8,16-Imino-cis-[2.2]metacyclophane, 37B, 158
$C_{16}H_{16}IN$, 9-(p-Iodophenyl)-9-azatetracyclo[5.3.1.0$^{2-6}$.0$^{8-10}$]undec-4-ene, 35B, 207

$C_{16}H_{16}N_4$, 1,4-Dimethanodibenzo[d,i]-1,3,6,8-tetrazecine, 41B, 356
$C_{16}H_{18}N_2$, 7,8,9,10-Tetrahydro-6,10-propano-6H-cyclohepta[b]quinoxaline, 41B, 356
$C_{16}H_{18}N_2O_5$, 3-Oxo-4,5-diaza-6,7-dicarbomethoxy-1,2,9,10-tetramethyltetracyclo[7.1.0.0$^{2-10}$.0$^{4-8}$]deca-5,7-diene, 44B, 273
$C_{16}H_{20}BrNO$, 10-Hydroxy-2-methyl-1,11-propano-9H-indeno-3,4,10,11-tetrahydro[2,1-c]pyridinium bromide, 41B, 357
$C_{16}H_{28}F_6N_2P$, 9,9'-Bis-9-azabicyclo[3.3.1]nonenium hexafluorophosphate, 44B, 274
$C_{16}H_{28}N_2$, 9,9'-Bis-9-azabicyclo[3.3.1]nonane, 44B, 274
$C_{16}H_{29}BrN_2O$, 1,2,4,4,5,8-Hexamethyl-8-N-acetamido-3-azabicyclo-[3.3.1]non-2-ene hydrobromide, 42B, 231
$C_{16}H_{30}N_2O$, 1,2,4,4,5,8-Hexamethyl-8-N-acetamido-3-azabicyclo[3.3.1]-nonane, 42B, 231
$C_{17}H_{17}N_3$, Spiro(2H-indole-3,7'-(1'-methyl-6'-cyano-1',8',2',3'-tetrahydro-7'H-1'-pyrindine)), 42B, 221
$C_{17}H_{19}NO_3$, 1α,2,2aα,3α,4,9bα-Hexahydro-9b-acetoxy-1,3-propano-5H-cyclobuta[d](2)benzazapin-5-one, 40B, 283
$C_{17}H_{19}NO_3S_2$, Quinuclidinyl di-α,α-thienylglycollate, 35B, 208
$C_{17}H_{22}N_2O_2S$, 2-(2'-Propenyl)-8-toluenesulphonyl-7,8-diazatricyclo-[4.2.1.0$^{3-7}$]nonane, 46B, 295
$C_{17}H_{25}NO_2$, 13-Acetyl-11,11-dimethyl-13-azatetracyclo-[5.5.2.0$^{2-8}$.0$^{2-7}$]tetradecan-9-one, 45B, 312
$C_{17}H_{25}NO_3S$, 1,5-endo-Methylene-quinolizidinium p-toluenesulphonate, 34B, 163
$C_{17}H_{26}BrNO \cdot 2\,H_2O$, (-)-5-Ethyl-2'-hydroxy-2,9,9-trimethyl-6,7-benzomorphan hydrobromide dihydrate, 45B, 313
$C_{18}H_{15}N_3O_5$, 14-Methyl-6-oxa-12,14,16-triazaheptacyclo-[9.5.2.2$^{3-9}$.1$^{2-10}$.0$^{2-10}$.0$^{4-8}$.0$^{12-16}$]heneicosa-17,19-diene-5,7,13,15-tetrone, 44B, 275
$C_{18}H_{17}N_3O_2$, ([4.3.1]Propella-2,4-diene)-(4-phenyl-1,2,4-triazoline-3,5-dione), 46B, 291
$C_{18}H_{20}ClNO_2$, Bicyclo[5.3.1]undec-7-en-11-one-1-carboxylic acid p-chloroanilide, 37B, 160
$C_{18}H_{20}Cl_2N_2O_4$, 1α,9β-Dichloro-2,2,10,10-tetramethoxy[2.2](2,6)pyridinophane, 44B, 275
$C_{18}H_{21}NO_3$, (+)-4-Hydroxy-7-oxo-3-methoxy-17-methyl-5,6-dehydromorphinan, 46B, 295
$C_{18}H_{22}N_2O_2$, 1,3,4,7,8,9,10,11-Octahydro-2H,5H-4a,11:10a,5-bis(iminomethano)dibenzo[a,e]cyclooctene-13,16-dione, 37B, 161
$C_{18}H_{23}N_3O_2$, N-Phenethylgranatanine-3-spiro-5'-hydantoin, 46B, 296
$C_{19}H_{17}N_3O_2$, Bishomocubane photoproduct, 43B, 383
$C_{19}H_{24}N_2O_4$, 1,8,9-Trimethylbicyclo[4.3.0]nona-2,8-diene-7,4'-spiro-(1',4'-dihydro-3',6'-bis(methoxycarbonyl)pyridazine), 44B, 262
$C_{19}H_{30}ClNO$, dl-1,2,3,4,5,6-Hexahydro-6,11,12,12-tetramethyl-2,6-methano-3,11-propano-3-benzazocine hydrochloride monohydrate, 40B, 288
$C_{20}H_{15}N_3O_4$, Phenyl-dioxa-triaza-nonacyclononadecadione, 42B, 232
$C_{20}H_{18}N_2O_2$, 4,10-Diphenyl-2,6-dioxo-1,7-diazatricyclo[5.2.1.0$^{4-10}$]-decane, 42B, 233
$C_{20}H_{19}BrN_2O$, 3-p-Bromophenyl-10-phenyl-3,10-diazatricyclo-[4.2.1.1$^{2-5}$]decan-4-one, 32B, 170
$C_{20}H_{19}ClN_2O_3S$, 12-(p-Chlorophenylsulphonylamino)-6,7,10,11-tetrahydroazocino[1,2-a]indol-8(9H)-one, 46B, 297
$C_{20}H_{19}ClN_2O_3S$, 6-(p-Chlorophenylsulphonylamino)-2,2a,3,4,5,6-hexahydro-2a,6-methano-1H-azeto[1.2-a]-[1]benzazocin-12-one, 46B, 297
$C_{20}H_{20}ClN_3O_4S \cdot C_5H_5N$, 8-(p-Chlorophenylsulphonylamino)-4-hydroxyim-

ino-1,2,3,4,5,6,7,8-octahydro-1,8-methano[1]benzazecin-13-one pyridine solvate, 46B, 297
$C_{20}H_{20}N_2$, cis-Dihydroquinaldine dimer, 40B, 289
$C_{20}H_{23}N_3O_2$, 13-Phenyl-11,13,15-triazapentacyclo-[8.5.2.0^{2-6}.0^{2-8}.0^{11-15}]heptadeca-12,14-dione, 46B, 298
$C_{20}H_{24}Cl_5NO_3$, Pentachloroethoxycodide, 34B, 166
$C_{20}H_{28}ClN$, N-n-Butyl-2,3-dimethyl-1,4-endo-(3',4'-pyrrolidino)-1,4-dihydronaphthalene hydrochloride, 39B, 221
$C_{20}H_{28}N_4O_2$, Hexamethylenetetramine m-cresol (1:2), 44B, 276
$C_{21}H_{23}IN_2O$, 1-Methyl-5,7-diphenyl-1,3-diazaadamantan-6-one iodide, 42B, 233
$C_{21}H_{23}N_3O_2$, 5-Phenyl-3,5,7-triazapentacyclo[7.4.3.2^{2-8}.0.0^{3-7}]-octadec-17-ene-4,6-dione, 44B, 276
$C_{21}H_{24}BrNO_3$, Quinuclidinyl benzilate hydrobromide, 34B, 169
$C_{21}H_{25}N_3O$, 8-((4-Methylphenyl)methyl)-1-phenyl-1,3,8-triazaspiro-[4.5]decan-4-one, 42B, 210
$C_{21}H_{25}N_3O_2$, 8-(2-Phenoxyethyl)-1-phenyl-1,3,8-triazaspiro[4.5]decan-4-one, 39B, 222
$C_{21}H_{27}N_3O_2$, (+)-8-Methyl-4-(2'-(-)-endo-bornyl)-2,4,6-triazapentacyclo[5.3.3.0^{2-6}.0^{8-10}.0^{11-13}]-trideca-3,5-dione, 46B, 299
$C_{21}H_{32}BrNO$, (-)-2-Cyclobutylmethyl-5-ethyl-2'-hydroxy-9,9-dimethyl-6,7-benzomorphan hydrobromide, 45B, 313
C??H??INO? ? 0.5 CH?O, 3,13,17-Triaza-15,21-ethanopentacyclo-[12.6.1^{1-17}.0^{1-1}@^{4}.0^{2-7}.0^{8-13}]eicosane 17-methiodide methanol solvate, 32B, 171
$C_{22}H_{18}N_2O_4$, 3-Acetonylideneindolin-2-one dimer, 44B, 265
$C_{22}H_{21}N_5O_3 \cdot 0.5\ C_6H_6$, Ethyl 3-methyl-4-oxo-1-phenyl-2-phenylimino-1,3,7,8-tetraazaspiro[4.5]deca-6,9-diene-10-carboxylate hemi-benzene solvate, 44B, 244
$C_{22}H_{23}N_3O_2$, 8,9-Dimethyl-14-phenyl-1,12,14-triazapentacyclo-[10.3.2^{2-11}.0.0^{2-7}.0^{7-10}]pentadeca-8,16-diene-13,15-dione, 42B, 234
$C_{22}H_{23}N_3O_2$, 9,10-Dimethyl-14-phenyl-1,12,14-triazapentacyclo-[10.3.2^{2-11}.0.0^{3-10}.0^{3-8}]pentadeca-8,16-diene-13,15-dione, 42B, 234
$C_{22}H_{27}N_3O \cdot 0.5\ CH_4O$, 8-(1-(4-Methylphenyl)ethyl)-1-phenyl-1,3,8-triazaspiro[4.5]decan-4-one hemi-methanol solvate, 44B, 245
$C_{22}H_{28}BrNO \cdot 0.5\ H_2O$, (DL)-2'-Hydroxy-5,9-dimethyl-2-phenethyl-6,7-benzomorphan hydrobromide hemihydrate, 46B, 299
$C_{22}H_{30}N_2O_3$, Quinidine ethanol solvate, 44B, 277
$C_{22}H_{30}N_2O_6$, N,N'-Dimethyldecahydro-7,14a,7a,14-ethanediylidenenaphtho[1,8-de:4,5-d'e']bisazocine-4,6,11,13(1H,7H,8H,14H)-tetrone dihydrate, 40B, 290
$C_{22}H_{30}N_2O_8$, Tetraethyl 3,9-diazahexacyclo[6.4.0^{2-7}.0^{4-11}.0^{5-10}]-dodecane-1,5,7,11-tetracarboxylate, 40B, 291
$C_{23}H_{18}N_2O_4$, 15-Methyl-5-phenyl-5,15-diazapentacyclo[7.4.3.2^{2-8}0.-0^{3-7}]octadeca-10,12,17-triene-4,6,14,16-tetrone, 46B, 292
$C_{23}H_{24}N_4O_5$, 18-Ethoxy-15-methyl-5-phenyl-3,5,7,15-tetraazapentacyclo[7.4.3.2^{2-8}.0.0^{3-7}]octadec-17-ene-4,6,14,16-tetraone, 46B, 292
$C_{23}H_{26}FN_3O_2$, 8-(3-(p-Fluorobenzoyl)propyl)-1-phenyl-1,3,8-triazaspiro[4.5]decan-4-one, 39B, 222
$C_{24}H_{16}N_2$, (3-Pyrida-3-(2,6))(6-benza-6-(1,4;2,5))(9-pyrida-9-(2,6))-spiro[5.5]undecaphane-1,4,7,10-tetraene, 43B, 384
$C_{24}H_{21}N_3O_3$, 3aα,4β,8β-3a,4-Dihydro-3-oxo-2,3a-diphenyl-4,8-etheno-3H-pyrazolo[1,5-c](1,3)diazepine-7(8H)-carboxylic acid ethyl ester, 46B, 300
$C_{24}H_{21}N_3O_3$, 4,6-Diphenyl-10-ethoxycarbonyl-2,3,10-triazatricyclo-[5.3.2.0^{2-6}]dodeca-3,8,11-triene-5-one, 45B, 314
$C_{24}H_{26}N_2O_6$, 4,11-Diacetyldodecahydro-7H-1,7,8a-ethanylylidene-8,14-

methanocyclopropa(1,6)benzo[1,2-d:4,3-d']-bisazocine-3,12,15,17(4H,9H)-tetrone, 40B, 290

$C_{24}H_{28}FN_3O$, 8-(4-(4-Fluorophenyl)-3-pentenyl)-1-phenyl-1,3,8-triazaspiro[4.5]decan-4-one, 39B, 223

$C_{24}H_{31}NO_9$, Delphinine intermediate oxalate complex, 38B, 324

$C_{24}H_{32}BrNO$, (DL)-5-Ethyl-2'-hydroxy-9,9-dimethyl-2-phenethyl-6,7-benzomorphan hydrobromide, 46B, 299

$C_{26}H_{46}N_8$, 1,3,7,9,13,15,19,21-Octaazapentacyclo[19.3.1.1^{3-7}.-1^{9-13}.1^{15-19}]octeicosane benzene solvate, 41B, 358

$C_{27}H_{19}N_3O_2$, 3,4;5,6-Dibenzo-12-phenyl-10,12,14-triazatricyclo-[7.5.2.0^{2-7}.0^{2-8}.0^{10-14}]hexadeca-3,5,15-triene-11,13-dione, 45B, 315

$C_{27}H_{25}NO_8$, 2-Phenyl-3-methylpyrrocoline acetylene dicarboxylate adduct, 42B, 235

$C_{27}H_{29}NO_5S$, N-Tosyl-2,12-ethano-2-ethyl-8-methoxy-1,4-methylene-1,2,3,4,5,6,12,13-octahydrophenanthridin-3-one, 40B, 292

$C_{27}H_{33}N$, 1,2,3,4,5,6,7,8-Octahydro-1,4:5,8-diisopropano-4,5-dimethyl-9-phenyl-acridine, 45B, 315

$C_{28}H_{27}N_5O_3$, 6-Dimethylamino-1,4-etheno-5,5-dimethyl-8-oxo-N,9-diphenyl-2,3,4,4a,5,8-hexahydro-1H-pyridazino[4,5-d]azepine-2,3-dicarboximide, 46B, 300

$C_{28}H_{32}N_4O_2$, Ethyl 9,10,10-tricyano-4,7-dimethyl-2-(2,2,6-trimethylcyclohexen-1-yl)-3,5,8,9-tetrahydro-5,9-methano-2H-cyclohepta[b]-pyridine-8-carboxylate, 41B, 358

$C_{30}H_{20}N_2$, 3-Phenyl-1-(spiro(indazolium-3,1'-inden)-2-yl)indenide, 43B, 374

$C_{30}H_{27}NO_7$, (1α,2α,5α,6α)-11-Ethoxycarbonyl-2,5-dimethoxycarbonyl-3,4-diphenyl-exo-11-azatricyclo[4.4.1.1^{2-5}]dodeca-3,7,9-trien-12-one, 46B, 300

$C_{30}H_{27}NO_7$, (1α,5α,5aβ,8aβ)-2-Ethoxycarbonyl-5a,7-dimethoxycarbonyl-6,8a-dihydro-6-oxo-8,8a-diphenyl-(1H,5H)-1,5-ethenocyclopent[c]azepine, 46B, 300

$C_{32}H_{41}BrN_2O_{12}$, Dimethyl acetylenedicarboxylate-cyclohexyl isocyanide adduct, 37B, 161

$C_{33}H_{47}NO_9S$, Talatisamine intermediate, 40B, 293

$C_{34}H_{48}N_6$, 2,3-Dihydrobenzimidazole-2-spirocyclohexane 5,6-(N,N'-dipiperidino)isobenzimidazole-2-spirocyclohexane (1:1), 44B, 251

$C_{38}H_{29}NO_3$, 3,4:5,6-Dibenzo-14-ethoxycarbonyl-8,10-diphenyl-14-azatetracyclo[9.3.2.0^{2-7}.0^{2-10}]hexadeca-7,12,15-triene-9-one, 46B, 301

38. HETERO-OXYGEN

C_2H_4O, Ethylene oxide (gas-ed), 4, 277; 17, 615

C_2H_4O, Ethylene oxide (gas-mw), 15, 413; 40B, 1143

$C_2H_4O_3$, Ethylene ozonide (gas-ed), 40B, 1140

$C_2H_4O_3$, Ethylene ozonide (gas-mw), 38B, 1069; 42B, 1034

$C_3H_2O_3$, Vinylene carbonate (gas-mw), 37B, 696

$C_3H_2O_3$, Vinylene carbonate, 40B, 294

$C_3H_2O_4$, Methylene oxalate, 46B, 302

$C_3H_4O_2$, 3-Oxetanone (gas-mw), 38B, 1070

$C_3H_4O_3$, Ethylene carbonate, 18, 635

C_3H_5BrO, Epibromohydrin (gas-ed), 26, 594

C_3H_5ClO, Epichlorhydrin (gas-ed), 19, 592

C_3H_6O, Propylene oxide (gas-ed), 19, 592

$C_3H_6O_2$, Glycidol (gas-ed), 26, 594

$C_3H_6O_3$, Propylene ozonide (gas-mw), 40B, 1144

$C_3H_6O_3$, Trioxan, 5, 139, 159; 28, 480; 34B, 169
$C_3H_6O_3$, 1,3,5-Trioxan (gas-ed), 15, 548; 37B, 693
$C_4H_4Cl_4O_2$, trans-syn-trans-2,3,5,6-Tetrachloro-1,4-dioxane, 33B, 130
C_4H_4O, Epoxybutyne (gas-mw), 38B, 1070
C_4H_4O, Furan (gas-ed), 7, 270; 19, 594
C_4H_4O, Furan, 38B, 327
$C_4H_4O_2$, Diketene, 16, 539; 22, 760
$C_4H_4O_3$, Tetrahydro-3,4-furandione, 46B, 302
$C_4H_4O_4$, Diglycollic anhydride, 41B, 359
$C_4H_6Br_2O_2$, trans-2,3-Dibromo-1,4-dioxane, 28, 485
$C_4H_6Cl_2O_2$, cis-2,3-Dichloro-1,4-dioxane, 28, 482
$C_4H_6Cl_2O_2$, trans-2,3-Dichloro-1,4-dioxane, 28, 483
$C_4H_6Cl_2O_2$, trans-2,5-Dichloro-1,4-dioxane, 28, 487
$C_4H_6O_2$, 1,2,3,4-Diepoxybutane (gas-ed), 38B, 1064
$C_4H_6O_2$, 2,3-Dihydro-p-dioxin (gas-mw), 40B, 1145
$C_4H_7ClO_6$, 2-Methyl-1,3-dioxolane-2-ylium perchlorate, 42B, 60
C_4H_8O, Tetrahydrofuran (gas-ed), 40B, 1140, 1143
C_4H_8O, 2,3-Epoxybutane (gas-ed), 4, 277
$C_4H_8O_2$, 1,4-Dioxan (gas-ed), 3, 705; 10, 284; 15, 458; 28, 666
$C_4H_8O_4$, p-Dioxanyl hydroperoxide, 39B, 223
$C_4H_8O_4$, 1,3,5,7-Tetroxocane, 40B, 295
$C_4H_8O_5$, Tetrahydrofuran-3,3,4,4-tetrol, 29, 518
$C_5H_2Cl_6O_3$, Trichloroethylidene trichlorolactic ester, 46B, 303
$C_5H_3BrO_2$, 2-Formyl-4-bromofuran, 39B, 224
C_5H_3NO, 2-Cyanofuran (gas-mw), 42B, 1036
$C_5H_4O_2$, 2-Pyrone (gas-mw), 39B, 893
$C_5H_4O_3$, α-Furoic acid, 18, 746; 27, 947
$C_5H_5ClO_3$, β-Chloroglutaric anhydride, 38B, 327
$C_5H_5NO_2$, anti-Furfuraldoxime, 32B, 172
C_5H_6O, 3-Methylfuran (gas-mw), 37B, 695
$C_5H_6O_3$, α-Methyltetronic acid, 28, 489; 41B, 360
C_5H_8O, Cyclopentene oxide (gas-ed), 40B, 1139
C_5H_8O, 3,6-Dihydro-2H-pyran (gas-mw), 40B, 1145
$C_5H_{10}O_5$, 1,3,5,7,9-Pentoxecane, 39B, 225
C_6N_4O, Tetracyanoethylene oxide, 37B, 162
$C_6H_3Cl_9O_3$, α-[2,4,6-Tris(trichloromethyl)-1,3,5-trioxane], 46B, 303
$C_6H_3Cl_9O_3$, β-[2,4,6-Tris(trichloromethyl)-1,3,5-trioxane], 46B, 303
$C_6H_3KO_5$, Potassium hydrogen furan-3,4-dicarboxylate, 44B, 278
$C_6H_4Cl_4O_4$, 2,2,4a,5a-Tetrachloro-1a,3a-dihydroxycyclopentane-1,4-carbolactone, 43B, 384
$C_6H_4O_5$, Furan-α,α'-dicarboxylic acid, 33B, 131
$C_6H_4O_5$, Furan-3,4-dicarboxylic acid, 29, 729
$C_6H_5NO_3$, α-Aminomethyleneglutaconic anhydride, 44B, 278
$C_6H_6O_4$, Kojic acid, 16, 478
$C_6H_7ClO_5 \cdot 0.5\ CH_3NO_2$, 6-Deoxy-6-chloro-L-ascorbic acid nitromeothane solvate, 46B, 304
$C_6H_7O_3$, α,γ-Dimethyltetronic acid, 34B, 172
$C_6H_8Br_4O_4$, 3,3,6,6-Tetra(bromoethyl)-1,2,4,5-tetroxane, 32B, 173
$C_6H_8O_4$, cis-1,4,6-Trioxabicyclo[4.3.0]nonan-7-one, 43B, 385
$C_6H_8O_5$, Cortalcerone, 42B, 235
$C_6H_{10}N_2O_3$, E-1-(5-Nitro-2-furyl)-2-dimethylaminoethylene (purple form), 45B, 316
$C_6H_{10}N_2O_3$, E-1-(5-Nitro-2-furyl)-2-dimethylaminoethylene (red form), 45B, 316
$C_6H_{10}O$, 1,2-Epoxycyclohexane (gas-ed), 11, 613
$C_6H_{10}O_4$, bis-1,3-Dioxa-2-cyclopentyl, 13, 567
$C_6H_{10}O_4$, cis-1,4,5,8-Tetraoxadecalin, 38B, 328

$C_6H_{12}O_3$, 2,4,6-Trimethyltrioxan (gas-ed), 4, 280; 39B, 886
$C_6H_{12}O_4$, 1,3,6,8-Tetraoxacyclodecane, 41B, 360
$C_6H_{12}O_6$, 1,3,5,7,9,11-Hexoxecane, 39B, 225
$C_7H_5ClO_4$, 5Z-Carboxymethylene-3-chloro-4-methyl-2(5H)-furanone, 46B, 304
$C_7H_6O_4$, 4-Hydroxy-4H-furo[3,2-c]pyran-2(6H)-one, 43B, 385
$C_7H_7NO_2$, trans-β-2-Furylacrylamide, 40B, 1085
$C_7H_8Br_2O$, anti-7,7-Dibromonorcar-3-ene oxide, 45B, 317
C_7H_8OS, 2,6-Dimethylpyran-4-thione, 20, 617
$C_7H_8O_3$, 6,7-endo-Epoxy-2-oxabicyclo[3.3.0]octan-3-one, 45B, 318
$C_7H_{10}O_3$, meso-2,4-Dimethylglutaric anhydride, 46B, 305
$C_7H_{11}BF_4O_2$, endo-2-Methyl-cis-4,5-trimethylene-1,3-dioxolan-2-ylium tetrafluoroborate, 45B, 318
$C_7H_{11}BrO_3$, 2,6-Dimethyl-γ-pyrone hydrobromide monohydrate, 30B, 145
$C_7H_{12}O_4$, 7-Methoxy-3,5,9-trioxabicyclo[4.3.0]nonane, 45B, 317
$C_8H_3O_9Rb$, Monorubidium furanetetracarboxylate, 32B, 174
$C_8H_4N_2O_5$, 5-Nitro-2,3-benzofurandione-(Z)-2-oxime, 45B, 319
$C_8H_4O_4$, cis-Octa-2,4,6-triene-1:4,5:8-diolide, 24, 584; 31B, 140
$C_8H_4O_4$, trans-Octa-2,4,6-triene-1:4,5:8-diolide, 24, 584; 31B, 140
$C_8H_6O_3$, trans-β-2-Furylacrylic acid, 32B, 174
$C_8H_6O_4$, 7-Oxabicyclo[2.2.1]hept-5-ene-2,3-exo-dicarboxylic anhydride, 38B, 328
$C_8H_7NO_3$, p-Nitrostyrene oxide, 41B, 361
$C_8H_8N_2O_2$, 2-Amino-3-acetyl-4-cyano-5-methylfuran, 44B, 279
$C_8H_8O_5$, Dimethyl 3,4-furandicarboxylate, 37B, 163
$C_8H_{10}Br_2O$, cis-6,8-Dibromo-exo-3-oxatricyclo[3.3.1.0^{2-4}]nonane, 43B, 387
$C_8H_{11}NO_7$, Methyl cis-2-acetoxy-5-nitro-2,5-dihydro-2-furoate, 46B, 306
$C_8H_{12}O_3$, 3-Methyl-2,4,10-trioxa-adamantane, 46B, 306
$C_8H_{12}O_4$, 5-Ethyl-2,2-dimethyl-1,3-dioxane-4,6-dione, 45B, 327
$C_8H_{13}ClO_6$, 2-Methyl-4,5-tetramethylene-1,3-dioxolan-2-ylium perchlorate, 42B, 62
$C_8H_{14}Cl_{12}O_4$, 2(e),4(e),6(e),8(e)-Tetrakis(trichloromethyl)-1,3,5,7-tetraoxocane, 46B, 305
$C_8H_{14}O_2$, 1,2,7,8-Diepoxyoctane (gas-ed), 40B, 1137
$C_8H_{14}O_4$, dl-Bi-1,4-dioxanyl, 40B, 295
$C_8H_{15}ClO_6$, 2,4,4,5,5-Pentamethyl-1,3-dioxolan-2-ylium perchlorate, 39B, 226
$C_8H_{15}NO_2S$, cis-2-Isopropyl-1,3-dioxane-5-carbothioamide, 45B, 319
$C_8H_{16}O_4$, Metaldehyde, 4, 253, 281
$C_8H_{16}O_4$, 1,4,7,10-Tetraoxacyclododecane, 44B, 279
$C_8H_{28}Cl_2MgO_{10}$, Magnesium chloride hexahydrate - 1,4,7,10-tetraoxacyclodecane, 41B, 362
$C_9H_4Cl_8O$, exo-1,exo-3,4,5,6,7,8,8-Octachloro-1,3,3a,4,7,7a-hexahydro-endo-4,7-methanoisobenzofuran, 43B, 387
$C_9H_5NO_4$, 3-Nitromethylenephthalide, 46B, 306
$C_9H_6O_2$, Coumarin, 39B, 226; 40B, 296
$C_9H_6O_2$, 1-Oxa-azulen-2-one, 23, 703
$C_9H_6O_4 \cdot 0.5\ H_2O$, Daphnetin hemihydrate, 42B, 236
$C_9H_6O_4$, 6,7-Dihydroxycoumarin, 43B, 388
$C_9H_7BrO_4$, 3-Bromo-4-hydroxycoumarin monohydrate, 30B, 146
$C_9H_8F_6O_4$, 2H,4H-Hexafluoro-1,5-dimethoxy-8-oxabicyclo[3.2.1]octan-3-one, 43B, 389
$C_9H_8O_4$, 4-Hydroxycoumarin monohydrate, 31B, 141
$C_9H_8O_6$, 3-Acetyl-2,6-dioxo-4-hydroxy-5-hydroxyethylidenepyran, 35B, 210

$C_9H_9Cl_3O_3$, Glycolic acid-2-methyl-4,5,6-trichlorocyclohex-2-en-1-one ester acetal, 29, 495

$C_9H_9NO_5$, Methyl 4-amino-5-oxaspiro[5.5]oxirano-7-oxabicyclo[4.1.0]-hept-3-ene-2-one-3-carboxylate, 44B, 280

$C_9H_{10}N_2O_4$, 3',4'-Dioxymethylene-2-phenyl-2-hydroxyacetamide oxime, 42B, 237

$C_9H_{10}O_4$, (E,R)-5-Methyl-3-(2'-tetrahydrofurylidene)tetrahydrofuran-2,4-dione, 46B, 307

$C_9H_{11}Br_2O$, anti-1,6-Dimethyl-7,7-dibromonorcar-3-ene oxide, 45B, 317

$C_9H_{12}Cl_2O$, 9,9-Dichloro-trans,trans-bicyclo[6.1.0]non-4-ene oxide, 39B, 227

$C_9H_{12}O_6$, 2,3,-di-O-Acetyl-2-C-methylerythrono-1,4-lactone, 46B, 307

$C_9H_{13}ClO_4$, 3-Chloromethyl-1,5,7-trimethyl-2,4,6,8-tetraoxatricyclo-[3.3.1.0$^{3-7}$]nonane, 38B, 329

$C_9H_{14}O_3$, Oxacyclodeca-2,7-dione, 33B, 132

$C_9H_{14}O_5$, cis-(2S,5S)-2-t-Butyl-5-carboxymethyl-1,3-dioxolan-4-one, 42B, 238

$C_9H_{18}O_6$, 3,3,6,6,9,9-Hexamethyl-1,2,4,5,7,8-hexaoxacyclononane, 34B, 170

$C_9H_{20}INO_2$, Trimethyl(tetrahydro-3-hydroxy-2-methyl-5-furyl)methylammonium iodide, 21, 557

$C_{10}H_2O_6$, Pyromellitic acid dianhydride, 45B, 320

$C_{10}H_5Cl_7O$, racemic-1-exo-4,5,6,7,8,8-Heptachloro-exo-2,3-epoxy-3a,4,7,7a-tetrahydro-4,7-methanoindane, 44B, 280

$C_{10}H_6Cl_6O_3$, 1,3,7,8,8a,9-Hexachloro-4a,5,6,8a-tetrahydro-1,3-epidioxy-4,6-methanoisochroman, 44B, 281

$C_{10}H_8O_3$, 4-Methylumbelliferone, 41B, 362

$C_{10}H_8O_4$, Anemonin, 31B, 142

$C_{10}H_8O_4$, trans-2,7-Dimethylocta-2,4,6-triene-1:4,5:8-diolide, 31B, 140

$C_{10}H_8O_4$, 7-Hydroxy-6-methoxycoumarin, 46B, 308

$C_{10}H_9BrO_2$, 2-Bromomethyl-2,3-dihydrofuro[2,3-b]tropone, 32B, 176

$C_{10}H_9FO_2$, α-(2-Hydroxy-5-fluorophenyl)-α,α-dimethyl acetic acid lactone, 41B, 363

$C_{10}H_{10}O_2$, 2,3.5,6-di-Epoxytricyclo[5.2.1.0$^{2-6}$]decane, 46B, 308

$C_{10}H_{10}O_4S$, 2,2-Dimethyl-5-thienyl-1,3-dioxan-4,6-dione, 43B, 390

$C_{10}H_{11}BrO_3$, 2-Bromomethyl-3,4-diacetyl-5-methylfurane, 33B, 133

$C_{10}H_{11}ClO_2$, 2-p-Chlorophenyl-1,3-dioxane, 35B, 211

$C_{10}H_{12}O_3$, (ps,3R,5S,9S)-9-Methoxy-4,11-dioxatricyclo[5.3.1.0$^{3-5}$]-undeca-1,6-diene, 45B, 321

$C_{10}H_{12}O_3$, 2,5-Dimethyl-3,4-diacetylfuran, 39B, 227

$C_{10}H_{12}O_4$, Isotetrahydroanemonin, 43B, 390

$C_{10}H_{12}O_8$, Dimethyl 2,5-dimethoxy-3-oxo-2,3-dihydro-2,4-furandicarboxylate, 43B, 391

$C_{10}H_{13}ClO_4$, Sceleratinic acid, 39B, 228

$C_{10}H_{13}NO_2$, anti-3-Methyl-6-(2-furyl)-piperidin-2-one, 45B, 321

$C_{10}H_{13}NO_2$, syn-3-Methyl-6-(2-furyl)-piperidin-2-one, 45B, 321

$C_{10}H_{14}O_3$, cis-10-Methyl-1-oxadecalin-2,5-dione, 45B, 322

$C_{10}H_{14}O_3$, 1-Hydroxymethyl-4-methyl-7-oxabicyclo[4.3.0]non-4-ene-8-one, 39B, 228

$C_{10}H_{16}O_4$, (5Z,12Z)-1,3,8,10-Tetraoxacyclotetradeca-5,12-diene, 45B, 323

$C_{10}H_{16}O_4$, 1,4,9,12-Tetraoxadispiro(4,2,4,2)tetradecane, 43B, 391

$C_{10}H_{16}O_4$, 3,4,7,8-Tetramethyl-trans-3,8:4,7-diepoxy-perhydro-1,5-dioxocine, 42B, 238

$C_{10}H_{16}O_5$, (2S,4R,5R)-2-Carboxymethyl-5-carboxy-2,4,5-trimethyl-2,3,4,5-tetrahydrofurane, 46B, 309

$C_{10}H_{20}O_4$, 1,3,8,10-Tetraoxacyclotetradecane, 38B, 329
$C_{10}H_{20}O_4$, 1,4,8,11-Tetraoxacyclotetradecane, 44B, 281
$C_{10}H_{20}O_7$, rac-2,3,3,4,4,5-Hexamethoxy-tetrahydrofuran, 43B, 392
$C_{10}H_{24}Br_2CuO_7$, 1,4,7,10,13-Pentaoxacyclopentadecane dibromodiaquo-copper(II), 45B, 323
$C_{11}H_5BrCl_3NO_6S$, 1-(5-Nitro-2-furyl)-1-trichloromethylsulphonyl-2-(5-bromo-2-furyl)-ethylene, 45B, 324
$C_{11}H_6O_3$, Furo[2,3-f]coumarin, 46B, 309
$C_{11}H_6O_3$, Furo[2,3-h]coumarin, 46B, 310
$C_{11}H_6O_3$, Furo[3,2-g]coumarin, 45B, 324
$C_{11}H_8O_3$, trans-1a,7b-Dihydro-oxireno[a]naphthalene-3-spiro-2'-oxiran-2(3H)-one, 38B, 193
$C_{11}H_9BrO_3$, 6-Bromo-1,2,3,4,4a,9a-hexahydro-4,9-dioxafluoren-2-one, 45B, 325
$C_{11}H_{10}O_4$, Methyl 3,4-dihydroisocoumarin-3-carboxylate, 42B, 238
$C_{11}H_{10}O_4$, α-Benzyloxy-γ-butyrolactone, 46B, 310
$C_{11}H_{10}O_6$, 8-Carboxy-1-hydroxy-2-oxobicyclo[3.2.2]non-6-ene-9,4-carbolactone, 37B, 164
$C_{11}H_{11}BrO_4$, 5-Bromo-austdiol, 42B, 239
$C_{11}H_{11}ClO_2$, exo-7-Chloro-7-phenyl-2,5-dioxabicyclo[4.1.0]heptane, 42B, 239
$C_{11}H_{11}IO_5$, Gibberellic acid intermediate, 37B, 165
$C_{11}H_{12}O_4$, 4,7-Dimethoxy-5-methylphthalide, 46B, 311
$C_{11}H_{12}O_6$, 3-Carbomethoxy-5-anisyl-1,2,4-trioxacyclopentane, 35B, 212
$C_{11}H_{13}NO_9$, trans-2-Acetoxy-5-nitro-2,5-dihydro-2-furfural diacetate, 46B, 311
$C_{11}H_{14}O_5$, Methyl (1S,5S,6S,7R)-6-formyl-7-methyl-2,8-dioxabicyclo-[3.3.1]non-3-ene-4-carboxylate, 46B, 312
$C_{11}H_{16}O_3$, 3-Hydroxy-trans-bicyclo[4.4.0]decane-1,3-carbolactone, 45B, 136
$C_{11}H_{17}ClO_2$, 5-Chloromethyl-4-oxahomoadamantan-5-ol, 41B, 364
$C_{11}H_{18}O_3$, 2,10,11-Trioxatricyclo[4.4.4.0$^{1-6}$]tetradecane, 42B, 240
$C_{11}H_{18}O_7$, 1,4,7,10,13-Pentaoxa-14,16-cyclohexadecanedione, 45B, 325
$C_{11}H_{20}O_5$, (4S,5R,8R,9R)-3,3'-[(1,1',2,2'-Tetraoxa)-6-(1-hydroxyprop-1-yl)]bicyclohexane, 46B, 313
$C_{11}H_{20}O_5$, (4S,5R,8R,9S)-3,3'-[(1,1',2,2'-Tetraoxa)-6-(1-hydroxyprop-1-yl)]bicyclohexane, 46B, 313
$C_{12}Cl_8O_2$, Octachlorodibenzo-p-dioxin, 38B, 334
$C_{12}F_8O_2$, Octafluorodibenzo-1,4-dioxane, 46B, 313
$C_{12}H_4Cl_4O$, 2,3,7,8-Tetrachlorodibenzofuran, 44B, 282
$C_{12}H_4Cl_4O_2$, 2,3,7,8-Tetrachlorodibenzo-p-dioxin, 38B, 330
$C_{12}H_5ClO_6$, α-Chloro-3,4-methylenedioxybenzylidenetetrahydrofuran-2,4,5-trione, 43B, 393
$C_{12}H_6Cl_2O_2$, 2,7-Dichlorodibenzo-p-dioxin, 38B, 331
$C_{12}H_6Cl_2O_2$, 2,8-Dichlorodibenzo-p-dioxin, 38B, 331
$C_{12}H_8ClNO_5S$, E-1-(5-Nitro-2-furyl)-2-p-chlorophenylsulfoethylene, 45B, 330
$C_{12}H_8Cl_6O$, 1,2,3,4,10,10-Hexachloro-6,7-exo-epoxy-1,4,4a,5,6,7,8,8a-octahydro-endo-ex0-1,4:5,8-dimethanonaphthalene, 39B, 229
$C_{12}H_8O$, Dibenzofuran, 38B, 332; 39B, 230; 40B, 296
$C_{12}H_8O_2$, Dibenzo-p-dioxin, 39B, 230; 40B, 297; 44B, 282
$C_{12}H_8O_3$, 2-Acetyldifuro[2,3-a;3,2-d]benzene, 44B, 283
$C_{12}H_8O_3$, 3-Formylfuro[3,2-f]chromene, 45B, 326
$C_{12}H_8O_4$, Xanthotoxin, 38B, 333
$C_{12}H_{11}BrO_4$, 5-Bromo-12-carboxyl-10-oxa-pentacyclo-[5.3.1.1$^{2-8}$.0$^{3-6}$.0$^{4-11}$]dodeca-9-one, 46B, 314
$C_{12}H_{11}ClN_2O_5S$, 2-Furfurylamino-4-chloro-5-sulfamoylbenzoic acid,

44B, 283
$C_{12}H_{12}Cl_2O_2$, syn-8,8-Dichloro-4-phenyl-3,5-dioxabicyclo[5.1.0]octane, 39B, 231
$C_{12}H_{12}N_2O_6$, 3,4,4a*α*,10a*α*-Tetrahydro-7,9-dinitro-2H,5H-[1]benzopyrano[2,3-b]pyran, 45B, 326
$C_{12}H_{12}N_4O_2$, 5,5'-Diamino-3,3'-dimethyl-2',3'-dihydro-2,3'-bifuryl-4,4'-dicarbonitrile, 46B, 315
$C_{12}H_{12}O_2$, [2.2](2,5)Furanophane, 44B, 284
$C_{12}H_{12}O_3$, anti-2,3;4,5-Diepoxy-12-oxa[4.4.3]propella-7,9-diene, 45B, 327
$C_{12}H_{12}O_4$, Di-*α*-methyleneanemonin, 43B, 390
$C_{12}H_{12}O_4$, 2,2-Dimethyl-5-phenyl-1,3-dioxane-4,6-dione, 45B, 327
$C_{12}H_{12}O_5$, anti-2,3;4,5-syn-7,8;9,10-Tetraepoxy-12-oxa[4.4.3]propellane, 45B, 327
$C_{12}H_{12}O_5$, syn-2,3;4,5-Diepoxy-12-oxa[4.4.3]propellane-11,13-dione, 45B, 328
$C_{12}H_{12}O_5$, 5-Carboxyl-11-hydroxy-2-oxa-tetracyclo[5.5.0.0^{4-12}.0^{6-10}]-dodeca-8-ene-3-one, 46B, 314
$C_{12}H_{12}O_5$, 5-Carboxyl-8-hydroxy-2-oxa-pentacyclo-[5.5.0.0^{4-12}.0^{6-10}.0^{9-11}]dodeca-3-one, 46B, 314
$C_{12}H_{12}O_5Br_2$, Dibromoradicinine, 35B, 213
$C_{12}H_{13}BrO$, 1-(p-Bromophenyl)-1,2-epoxycyclohexane, 38B, 334
$C_{12}H_{13}ClO$, endo-7-Chloro-7-phenyl-2-oxabicyclo[4.1.0]heptane, 42B, 240
$C_{12}H_{13}NO_4$, 3,4,4a*α*,10a*α*-Tetrahydro-7-nitro-2H,5H-[1]benzopyrano[2,3-b]pyran, 45B, 329
$C_{12}H_{14}N_4O_2S \cdot C_4H_8O_2$, 5,5'-Diamino-4-cyano-3,3'-dimethyl-2',3'-dihydro-2,3'-bifuryl-4'-thiocarbamide, 46B, 315
$C_{12}H_{14}O$, 3,6-Epoxypentacyclo[6.2.2.0^{2-7}.0^{4-10}.0^{5-9}]dodecane, 44B, 284
$C_{12}H_{14}O_4$, Hydroquinone diglycidyl ether, 43B, 394
$C_{12}H_{14}O_6$, 8-Hydroxy-3-methoxy-3-methoxycarbonyl-2-oxaspiro[4.5]deca-6,9-diene, 45B, 329
$C_{12}H_{15}BrO_2$, r-2-(p-Bromophenyl)-cis-4,cis-6-dimethyl-1,3-dioxane, 41B, 365; 42B, 241
$C_{12}H_{16}BrO_5$, 3-(1-Acetoxy-n-butyl)-4-bromo-5-methoxy-5-bromoethyl-2,5-dihydrofuran-2-one, 43B, 767
$C_{12}H_{16}O_3$, anti-2,3:4,5-Diepoxy-12-oxa[4.4.3]propellane, 46B, 315
$C_{12}H_{16}O_3$, syn-2,3:4,5-Diepoxy-12-oxa[4.4.3]propellane, 46B, 315
$C_{12}H_{16}O_6$, Methyl-2-acetonyl-3-ethyl-4-methoxy-5-oxo-dihydro-2H-furan-2-carboxylate, 37B, 165
$C_{12}H_{16}O_6$, cyclo-Bis(2-carboxy-tetrahydropyran-6-yl), 46B, 316
$C_{12}H_{18}BrNO$, 2,3-Dihydrobenzofuran-2-ylmethyl trimethylammonium bromide, 42B, 241
$C_{12}H_{18}O_2$, Hexamethyl-2,3:5,6-diepoxybicyclo[2.2.0]hexane, 46B, 316
$C_{12}H_{18}O_2$, Pentacyclo[6.2.2.0^{2-7}.0^{4-10}.0^{5-9}]dodecane-4,5-diol, 44B, 284
$C_{12}H_{18}O_3$, cis-1-Oxa-7,7,10-trimethyldecalin-2,5-dione, 42B, 242
$C_{12}H_{18}O_3$, 3-Methoxy-trans-bicyclo[4,4,0]decane-1,3-carbolactone, 45B, 136
$C_{12}H_{18}O_5$, 8-Acetyl-1,3,6,7,8-pentamethyl-2,4,5,9-tetraoxa-tricyclo-[4.2.1.0^{3-7}]nonane, 41B, 366
$C_{12}H_{18}O_8$, Methyl o-acetyleurekanate, 46B, 317
$C_{12}H_{18}O_8$, 2*β*,3*α*-Di(ethoxycarbonyl)-2,3,4a*α*,6,7,8a*α*-hexahydro-p-dioxino[2,3-e]-p-dioxin, 39B, 231
$C_{12}H_{20}O_4$, 3,6-Spiro-dicyclohexylidene-1,2,4,5-tetraoxacyclohexane, 32B, 177

$C_{12}H_{20}O_6$, 1,3,5,7-Tetramethoxy-2,6-dioxa-adamantane, 33B, 135
$C_{12}H_{20}O_8$, 1,4,7,10,13,16-Hexaoxa-2,6-cyclooctadecanedione, 45B, 330
$C_{12}H_{22}N_2O_4$, Cyclobisurethane, 42B, 242
$C_{12}H_{24}ClNaO_{10}$, Tris(1,4-dioxane) sodium perchlorate, 44B, 285
$C_{12}H_{24}O_4$, 1,3,9,11-Tetraoxacyclohexadecane, 41B, 367
$C_{12}H_{24}O_4$, 1,5,9,13-Tetraoxocyclohexadecane, 37B, 167
$C_{12}H_{24}O_6$, 1,4,7,10,13,16-Hexaoxacyclooctadecane, 40B, 297; 46B, 317
$C_{12}H_{28}N_2O_{16}U$, 1,4,7,10,13,16-Hexaoxacyclooctadecane diaquo-uranylbis(nitrate), 44B, 286
$C_{12}H_{32}BrNO_8$, Ammonium bromide - 1,4,7,10,13,16-hexaoxacyclooctadecane dihydrate, 44B, 286
$C_{12}H_{36}Cl_2MnO_{20}$, Hexaaquamanganese(II) perchlorate 18-crown-6, 46B, 318
$C_{13}H_2Cl_6O_2$, 1,3,4,6,7,9-Hexachloroxanthone, 42B, 243
$C_{13}H_4Cl_6O$, 1,3,4,6,7,9-Hexachloroxanthene, 42B, 243
$C_{13}H_8ClNO_4$, 1-(o-Chlorophenyl)-3-(5-nitro-2-furanyl)-2-propen-1-one, 44B, 287
$C_{13}H_8N_2O_6$, 1-(2-Nitrophenyl)-3-(5-nitro-2-furanyl)-2-propen-1-one, 44B, 287
$C_{13}H_8N_2O_6$, 1-(4-Nitrophenyl)-3-(5-nitro-2-furanyl)-2-propen-1-one, 44B, 287
$C_{13}H_9NO_4$, 1-Phenyl-3-(5-nitro-2-furanyl)-2-propen-1-one, 44B, 287
$C_{13}H_{10}O_3$, 3-Acetylfuro(2,3-h)chromene, 44B, 288
$C_{13}H_{10}O_3$, 3-Formyl-8-methyl-furo[3,2-g]chromene, 46B, 318
$C_{13}H_{10}O_5$, Isopimpinellin, 43B, 394
$C_{13}H_{11}NO_5S$, E-1-(5-Nitro-2-furyl)-2-p-tolylsulfoethylene, 45B, 330
$C_{13}H_{11}O_4Rb$, Rubidium D-1,2-dihydronaphtho[2,1-b]furan-2-carboxylate monohydrate, 42B, 243
$C_{13}H_{12}O_3$, (+)-Tetralin-1-spiro(succinic anhydride), 44B, 288
$C_{13}H_{12}O_4$, 5-(p-Methoxyphenyl)-3-methoxy-2,4-pentadien-4-olide, 44B, 289
$C_{13}H_{12}O_5$, Methyl 1,6-dihydro-4-methyl-6-oxo-2,5-benzodioxocin-3-carboxylate, 42B, 244
$C_{13}H_{14}F_3O_2$, r-2-(p-Trifluoromethylphenyl)-trans-4,trans-6-dimethyl-1,3-dioxane, 41B, 367
$C_{13}H_{15}F_3O_2$, 2-(p-Trifluoromethylphenyl)-4,6-dimethyl-1,3-dioxan, 42B, 241
$C_{13}H_{16}O_3$, 2,2-Dimethyl-3,4-dihydroxy-5-phenylvaleric acid γ-lactone, 41B, 368
$C_{13}H_{16}O_5$, 1-Hydroxy-12-methoxycarbonyl-2-oxatricyclo[$6.3.1.0^{4-12}$]-dodec-5-ene-9-one, 44B, 289
$C_{13}H_{17}BrO_7$, 6-Bromomethyl-7,7a-epoxy-4,6-dihydroxy-7-methoxycarbonyl-3,4-dimethyl-octahydrobenzo[c]furan-1-one, 34B, 171
$C_{13}H_{18}O_3$, (Z)-5-Isopropyl-2-phenyl-1,3-dioxan-5-ol, 43B, 396
$C_{13}H_{18}O_3$, endo-1,3,5,6,7,9-Hexamethyl-4,8-dioxatetracyclo-[$4.3.0.0^{3-5}.0^{7-9}$]nonan-2-one, 46B, 319
$C_{13}H_{18}O_3$, exo-1,3,5,6,7,9-Hexamethyl-4,8-dioxatetracyclo-[$4.3.0.0^{3-5}.0^{7-9}$]nonan-2-one, 46B, 319
$C_{13}H_{18}O_3$, 3-(5α-Hydroxy-7aβ-methyl-1-oxo-3aαH-hexahydroindan-4α-yl)-propionic acid δ-lactone, 44B, 290
$C_{13}H_{18}O_5$, 1-Hydroxy-12-methoxycarbonyl-2-oxatricyclo[$6.3.1.0^{4-12}$]-dodecan-9-one, 44B, 289
$C_{13}H_{20}O_4$, 1-Carboxy-trans-bicyclo[4.4.0]decan-3-one ethylene acetal, 44B, 290
$C_{13}H_{20}O_4$, 6-Acetoxymethyl-1,2,5-trimethyl-4-oxabicyclo[3.3.0]octan-3-one, 46B, 320
$C_{13}H_{22}O_2$, trans-8-t-Butyl-1-oxaspiro[4.5]decan-2-one, 41B, 369

$C_{14}H_9BrO_2$, 3-(p-Bromophenyl)phthalide, 34B, 173
$C_{14}H_{10}O_2$, cis-5aH,10bH-Benzofuro[2,3-b]benzofuran (polymorph 1), 40B, 302
$C_{14}H_{10}O_2$, cis-5aH,10bH-Benzofuro[2,3-b]benzofuran (polymorph 2), 40B, 302
$C_{14}H_{10}O_2$, 1,6:8,13-Bis-oxido[14]annulene, 32B, 178
$C_{14}H_{10}O_5$, 3-Carboethoxyfuro[3,2-f]coumarin, 44B, 291
$C_{14}H_{11}BrO_4$, 3-Bromo-4'a,10'a-dihydrospiro(2,5-cyclohexadiene-1,3-cycloocta-as-trioxin)-4-one, 40B, 303
$C_{14}H_{11}NO_4$, 1-(4-Methylphenyl)-3-(5-nitro-2-furanyl)-2-propen-1-one, 43B, 396
$C_{14}H_{11}NO_6$, 1-(3-Methoxy-4-hydroxyphenyl)-3-(5-nitro-2-furanyl)-2-propen-1-one, 44B, 287
$C_{14}H_{12}BrNO_3$, 3-(4-Bromophenyl)-6,7-dihydro-2-hydroxyiminobenzofuran-4(5H)-one, 37B, 168
$C_{14}H_{12}N_2O_2$, 11-Oxa-5,10-dicyano-6,9-dimethyltetracyclo-[6.2.1.0^{1-7}.0($^{5-10}$]undec-2-en-4-one, 43B, 240
$C_{14}H_{12}O_3$, D-Methyl 1,2-dihydronaphtho[2,1-b]furan-2-carboxylate, 42B, 118
$C_{14}H_{12}O_3$, 2,8-Dimethoxydibenzofuran, 44B, 291
$C_{14}H_{12}O_4$, 3,6-Diphenyl-1,2,4,5-tetraoxacyclohexane, 32B, 179
$C_{14}H_{12}O_5$, Phenacylkojate, 42B, 245
$C_{14}H_{12}O_5$, 4,9-Dimethoxy-7-methyl-5H-furo[3,2-g](1)benzopyran-5-one, 39B, 232
$C_{14}H_{12}O_5$, 5,8-Dimethoxy-2-methyl-furo[2,3-e]chromone, 45B, 331
$C_{14}H_{13}NO_3$, 3-(1-(Phenylamino)ethylidene)-6-methyl-2,4-dioxo-2,3-dihydro-4H-pyran, 44B, 292
$C_{14}H_{14}O_3$, DL-4-Methoxy-6-styryl-5,6-dihydro-α-pyrone, 38B, 335
$C_{14}H_{14}O_3$, 1,2,3,4,4a,5,8,8a-Octahydro-1,4,5,8-exo,exo-dimethanonaphthalene-4a,8a-dicarboxylic anhydride, 46B, 320
$C_{14}H_{14}O_4$, cis-2-Hydroxy-2,4-dimethyl-3,4-dihydro-2H,5H-pyrano[3.2-c][1]benzopyran-5-one, 45B, 332
$C_{14}H_{14}O_4$, exo-4-Benzoyloxy-5,9-dioxatricyclo[4.2.1.0^{3-8}]nonane, 45B, 332
$C_{14}H_{14}O_4$, 3-O-Benzoyl-1,7,4,5-dianhydro-1-C-hydroxymethylcyclohexane-1,3,4,5-tetrol, 39B, 233
$C_{14}H_{14}O_6$, Phenacylkojate monohydrate, 41B, 370
$C_{14}H_{16}O_3$, 1,2,3,4,4a,5,6,7,8,8a-Decahydro-1,4,5,8-exo,endo-dimethanonaphthalene-4a,8a-dicarboxylic anhydride, 46B, 320
$C_{14}H_{16}O_3$, 1,2,3,4,4a,5,6,7,8,8a-Decahydro-1,4,5,8-exo,exo-dimethanonaphthalene-4a,8a-dicarboxylic anhydride, 46B, 320
$C_{14}H_{16}O_3$, 4-Methoxy-6-(1',2'-dihydrostyryl)-5,6-dihydro-α-pyrone, 38B, 335
$C_{14}H_{16}O_5$, 2,5-Dimethoxy-2,5-dihydro-2-furylmethyl benzoate, 44B, 292
$C_{14}H_{17}BrO_2$, 2-((1'-Bromo-2'-phenyl)-ethenyl)-4,6-dimethyl-1,3-dioxan, 43B, 397
$C_{14}H_{17}BrO_2$, 2-((1'-Bromo-2'-phenyl)-ethenyl)-5,5-dimethyl-1,3-dioxan, 43B, 397
$C_{14}H_{17}NO_2$, 7-Diethylamino-4-methylcoumarin, 40B, 303
$C_{14}H_{18}O_2$, Pentamethylhydrocoumarin, 38B, 336
$C_{14}H_{18}O_2$, exo-7-Phenyl-5,7-dimethyl-6,8-dioxabicyclo[3.2.1]octane, 44B, 292
$C_{14}H_{18}O_2$, 2-Methylbicyclo[3.3.1]non-1-en-3-one furan adduct, 46B, 321
$C_{14}H_{18}O_5$, Gilmicolin, 45B, 333
$C_{14}H_{18}O_6$, 5-Methyl-2-oxo-3-acetyl-4-(2'-oxo-3'-acetyl-2',5'-dihydrofuran-5'-yl)tetrahydrofuran, 43B, 398

$C_{14}H_{19}BrO_2$, 2-(4-Bromophenyl)-r-2,4,4,c-6-tetramethyl-1,3-dioxan, 38B, 337
$C_{14}H_{20}F_2O_2$, 8β,9β-Difluoromethylene-10-methyl-2-decalone-2-ethylene acetal, 41B, 371
$C_{14}H_{20}O_3$, (+)-trans-1β-Carboxy-8β-hydroxy-1α,4a,6β-trimethyl-5-oxo-decahydronaphthalene lactone, 41B, 373
$C_{14}H_{20}O_3$, DL-trans-1β-Carboxy-8β-hydroxy-1α,4a,6β-trimethyl-5-oxo-decahydronaphthalene lactone, 41B, 372
$C_{14}H_{20}O_4$, (DL)-Decahydro-4-hydroxy-4a,8-dimethylazuleno[6,5-b]furan-2,5(3H)-dione, 46B, 321
$C_{14}H_{20}O_4$, cis,cis-8,8-Dimethyl-5,11-dioxatricyclo[8.4.0.0$^{1-6}$]tetra-deca-4,12-dione, 45B, 334
$C_{14}H_{20}O_5$, Benzo-15-crown-5, 44B, 293
$C_{14}H_{21}ClO_6$, Methyl (5S)-4-chloro-3,4-didesoxy-1,2:6,7-di-O-isopro-pylidene-α-D-erythro-3-hepten-5-ulo-(1,5)-pyrannoside, 46B, 322
$C_{14}H_{21}NO_5$, 4-Carboxy-1,2,4,8-tetramethyl-3,9-dioxatricyclo-[5.2.1.0$^{1-6}$]decane-8-carboxamide, 43B, 399
$C_{14}H_{22}O_2$, cis-8-t-Butyl-3-methylene-1-oxaspiro[4.5]decan-2-one, 44B, 293
$C_{14}H_{22}O_4$, (DL)-Decahydro-4,5-dihydroxy-4a,8-dimethylazuleno[6,5-b]furan-2(3H)-one(3aα,4α,4aβ,5α,7aα,8α,9aβ), 46B, 322
$C_{14}H_{24}O_2$, trans-9-t-Butyl-1-oxaspiro[5.5]undecan-2-one, 41B, 373
$C_{14}H_{24}O_4$, trans-anti-trans-4,5:9,10-Bis(cyclohexano)-1,3,6,8-tetra-oxecane, 45B, 334
$C_{14}H_{24}O_4$, trans-syn-trans-4,5:9,10-Bis(cyclohexano)-1,3,6,8-tetra-oxecane, 43B, 399
$C_{14}H_{24}O_4$, 3,6-Spiro-dicycloheptylidene-1,2,4,5-tetraoxacyclohexane, 32B, 180
$C_{15}H_9BrO_3$, 6-Bromo-1,4,4a,9a-tetrahydro-endo-4a,9a-epoxy-1,4-meth-anoanthraquinone, 45B, 335
$C_{15}H_{11}BrO_2$, 4'-Bromoflavanone, 40B, 305
$C_{15}H_{11}NO$, 1-Cyano-7,8-benzo-11-oxatricyclo[4.2.2.1$^{2-5}$]undeca-3,7,9-triene, 45B, 336
$C_{15}H_{11}NO_4$, 3,4-Dihydro-2-((5-nitro-2-furanyl)methylene)-1(2H)-naph-thalenone, 44B, 287
$C_{15}H_{12}O_8$, DL-1-Methoxycarbonyl-1-hydroxy-2-oxa-3,4-dioxo-7-methoxy-9-methylcyclopenteno[3,4-c]chromene, 41B, 374
$C_{15}H_{13}BrO_7$, Cyanidin bromide monohydrate, 43B, 400
$C_{15}H_{13}ClO_5$, Apigeninidin chloride monohydrate, 40B, 304
$C_{15}H_{13}ClO_5$, 4',6,7-Trihydroxyflavylium chloride monohydrate, 43B, 401
$C_{15}H_{14}O_4$, Piperolide, 43B, 402
$C_{15}H_{14}O_4$, trans-δ-(p-Methoxystyryl)-4-methoxy-α-pyrone (yangonin), 37B, 169
$C_{15}H_{14}O_5$, Epoxypiperolide, 43B, 402
$C_{15}H_{14}O_5$, 4-Methoxy-6-(5',6'-dioxymethylene-styryl)-5,6-dihydro-α-pyrone, 38B, 339
$C_{15}H_{16}O_4$, trans-2-Methoxy-2,4-dimethyl-3,4-dihydro-2H,5H-pyrano[3.2-c][1]benzopyran-5-one, 45B, 332
$C_{15}H_{16}O_6$, Dimethyl 6',7'-dimethyl-3'-oxospiro(oxirane-2,4'-tricyclo-[3.3.0.0$^{2-8}$]oct(6)ene)-1',8'-dicarboxylate, 44B, 293
$C_{15}H_{16}O_6$, Allamdin, 40B, 306
$C_{15}H_{18}O_2$, Lindenenol, 40B, 307
$C_{15}H_{18}O_4$, 3,3a-Diacetyl-2,7a-dimethyl-3aα,4,6aα,7aα-tetrahydro-3bαH-cyclopenta[d]furo-[2,3-b]-furan, 45B, 337
$C_{15}H_{18}O_5$, 5,8-Etheno-3,4a,7,9-tetramethyl-4a,5,6,7,8,8a-hexahydro-chromene-7,8a-diol-2,6-dione, 43B, 404

$C_{15}H_{19}NO_4$, 2e-(4-Nitrophenoxy)-trans-1-oxadecalin, 45B, 337
$C_{15}H_{20}BrClO_2$, Chlorofucin, 46B, 323
$C_{15}H_{20}BrClO_3$, Poiteol, 46B, 323
$C_{15}H_{20}O_2$, 2-Phenoxy-trans-1-oxadecalin (equatorial isomer), 44B, 295
$C_{15}H_{20}O_2$, 2a-Phenoxy-trans-1-oxadecalin, 44B, 294
$C_{15}H_{22}O_4$, (1,5,5-Trimethyl-bicyclo[4.2.0]octan-7-one)-8-spiro-4-(2,2-dimethyl-5-oxo-1,3-dioxolane), 46B, 323
$C_{15}H_{22}O_5$, 4-Ethyl-1-hydroxy-4,8,8,10,10-pentamethyl-7,9-dioxo-2,3-dioxabicyclo[4.4.0]decene-5, 37B, 94
$C_{15}H_{24}O_2$, Tricyclo[7.6.0.0$^{2-8}$]pentadec-7-ene-1-ol epoxide, 45B, 338
$C_{15}H_{24}O_3$, Tricyclic photoproduct, 40B, 307
$C_{15}H_{24}O_4$, 5,6-Epoxy-4-hydroxy-2-methoxy-4,6-di-t-butylcyclohex-2-enone, 41B, 375
$C_{15}H_{24}O_5$, 2,6-Di-t-butyl-5,6-epoxy-2,4-dihydroxy-4-methyl-1,3-cyclohexanedione, 45B, 338
$C_{15}H_{26}O_6$, 1R,6S-4S,10S-Dihydroxy-3R-(2S-hydroxy-1S-methyl-propyl)-2,8-dioxabicyclo[4.4.0]decan-9S-yl-acetone, 45B, 339
$C_{15}H_{26}O_6$, 1S,6R-8R-(1S,3S-Dihydroxy-2S-methylbutyl)-5S-hydroxy-3,7-dioxabicyclo[4.3.0]nonan-4S-yl-acetone, 45B, 339
$C_{15}H_{26}O_{10}$, Euonyminol, 43B, 404
$C_{15}H_{42}Cl_4Co_2O_{13}$, Hexaaquacobalt tetrachlorocobaltate 18-crown-6 acetone, 46B, 324
$C_{16}H_8O_4$, Oxindigo, 27, 983
$C_{16}H_9BrO$, 2-Bromobenzo[b]indeno[1,2-e]pyran, 38B, 339
$C_{16}H_{10}Br_4O_5$, 5,7-Dibromo-2-(3,5-dibromo-2-hydroxyphenyl)-2-methoxymethoxy-3(2H)-benzofuranone, 37B, 170
$C_{16}H_{10}N_2O$, trans-α,β-Dicyanostilbene oxide, 44B, 295
$C_{16}H_{10}O$, Benzo[b]indeno[1,2-e]pyran, 45B, 340
$C_{16}H_{11}ClO_3$, 1a-(p-Chlorophenoxy)1a,7b-dihydrobenzo[d]cyclopropa[b]-pyran-3(1H)one, 34B, 174
$C_{16}H_{12}O_3$, (Z)-2-p-Methoxyphenylmethylenebenzofuran-3(2H)-one, 42B, 245
$C_{16}H_{12}O_3$, 4'-Oxo-spiro[isobenzofuran-1(3H)-3'-isochroman], 45B, 341
$C_{16}H_{12}O_5$, 1-(2-Hydroxy-6-methoxyphenyl)-5-methyl-4,7-dioxo-4,7-dihydroisobenzofuran, 46B, 324
$C_{16}H_{13}N_2O_8$, 3,5,7-Trimethyltropylium-O,O-(2',4',6'-trinitrophenylide), 46B, 325
$C_{16}H_{14}N_2O_6$, O-(2',6'-Dinitrophenyl)-3,5,7-trimethyltropolone, 46B, 325
$C_{16}H_{14}O$, 2,3,6,7-Dibenzo-9-oxa[3.3.1]bicyclonona-2,6-diene, 44B, 295
$C_{16}H_{14}O$, 8,16-Oxido-cis[2.2]metacyclophane, 33B, 135
$C_{16}H_{14}O_3$, Ruscodibenzofuran, 43B, 405
$C_{16}H_{14}O_4$, 2-(4-Dibenzofuranyloxy)-2-methylpropionic acid, 45B, 341
$C_{16}H_{14}O_4$, 3,8-Dimethoxy-4,5,6,7-dibenzo-1,2-dioxacyclooctane, 42B, 246
$C_{16}H_{14}O_4$, 5,15-Dioxapentacyclo[7.4.3.2$^{2-8}$.0.0$^{3-7}$]octadeca-10,12,17-triene-4,6-dione, 46B, 292
$C_{16}H_{15}ClO_5$, 2-Phenyl-5,5-dimethyl-(3,4-benzo)-2,5-dihydrofurylium perchlorate, 45B, 342
$C_{16}H_{16}O$, 9-Isopropylxanthene, 43B, 406
$C_{16}H_{16}O_2$, 7,8-Dihydrodibenzo[f,h](1,4)dioxecin, 43B, 407
$C_{16}H_{16}O_4$, 6,7,14,15-Tetrahydrodibenzo[b,h](1,4,7,10)tetraoxacyclododecin, 44B, 296
$C_{16}H_{16}O_4Se$, 2,3-Epoxy-5-(1-hydroxy-2,3-epoxy-5-phenyl-selenocyclopent-1-yl)cyclopentan-1-one, 45B, 343
$C_{16}H_{16}O_5$, 7,10-Dihydroxy-1,3,8-trimethyl-3,4,6,9-tetrahydro-1H-naphtho[2,3-c]pyran-6,9-dione, 46B, 326

$C_{16}H_{17}NO_3$, 6,7-Dihydro-6,6-dimethyl-3-phenylbenzofuran-2,4(3H,5H)-dione-2-oxime, 38B, 340
$C_{16}H_{20}MgN_2O_5S_2$, Diisothiocyanato(2,3,5,6,8,9,11,12-octahydro-1,4,7,10,13-pentaoxabenzocyclopentadecine)magnesium, 44B, 296
$C_{16}H_{20}N_2O_7$, threo-3,3,4,4,α-Pentamethyl-2-oxetanemethanol 3,5-dinitrobenzoate, 44B, 297
$C_{16}H_{20}N_2O_7$, 5,5'-Dinitro-2-(1-adamantyl)-2'-carbomethoxy-(2H,5H)-furane, 46B, 327
$C_{16}H_{20}O_8$, meso-2,10-Dimethyl-3,11-dimethoxycarbonyl-1,6,9,13-tetraoxadispiro-[4.2.4.2]-tetradeca-2,10-diene, 45B, 343
$C_{16}H_{22}CaN_2O_6S_2$, Diisothiocyanato(2,3,5,6,8,9,11,12-octahydro-1,4,7,10,13-pentaoxabenzocyclopentadecine)calcium hydrate, 44B, 296
$C_{16}H_{22}O$, cis-4-t-Butyl-1-phenyl-1,2-epoxycyclohexane, 46B, 327
$C_{16}H_{22}O$, trans-4-t-Butyl-1-phenyl-1,2-epoxycyclohexane, 46B, 327
$C_{16}H_{22}O_6$, 5-Methyl-2-oxo-3-propionyl-4-(2'-oxo-3'-propionyl-2',5'-dihydrofuran-5'-yl)tetrahydrofuran, 43B, 398
$C_{16}H_{24}O_4$, Brefeldin A, 37B, 169
$C_{16}H_{25}NO_3$, 2,6-Bis(t-butyl)-3-dimethylcarbamoyl-4-pyranone, 45B, 344
$C_{16}H_{26}O_2$, cis-4,4,8,8-Pentamethylene-2,6-dioxa-bicyclo[3.3.0]octane, 42B, 246
$C_{16}H_{26}O_2$, trans-4,4,8,8-Pentamethylene-2,6-dioxa-bicyclo-[3.3.0]octane, 42B, 246
$C_{16}H_{28}BaCl_2O_{16}$, 6,7,9,10,12,13,15,16,18,19-Decahydrobenzo[b]-1,4,7,-10,13,16-hexaoxacyclooctadecin barium perchlorate dihydrate, 44B, 298
$C_{16}H_{28}O_4$, 3,6-Spiro-dicyclo-octylidene-1,2,4,5-tetraoxacyclohexane, 32B, 180
$C_{16}H_{30}Cl_2O_{17}Sr$, 6,7,9,10,12,13,15,16,18,19-Decahydrobenzo[b]-1,4,7,-10,13,16-hexaoxacyclooctadecin strontium perchlorate trihydrate, 44B, 298
$C_{17}H_{10}Cl_2O_3$, 8c,11a-Dichloro-8b,8c,11a,11b-tetrahydroexophenanthro-(9',10':3,4)cyclobuta[1,2-d]-(1,3)dioxol-10-one, 46B, 328
$C_{17}H_{11}NO_4$, 1-(2-Naphthyl)-3-(5-nitro-2-furanyl)-2-propen-1-one, 44B, 287
$C_{17}H_{12}Br_2O_3$, 3-(3,5-Dibromo-4-hydroxybenzoyl)-2-ethyl-benzo[b]furan, 41B, 376
$C_{17}H_{16}O_2$, Spiro(2H-dibenzo[f,h]-3,4-dihydro-1,5-dioxacyclononene-3,1'-cyclopropane, 45B, 344
$C_{17}H_{16}O_2$, 1,4-Diphenyl-2,3-dioxabicyclo[2.2.1]heptane, 44B, 299
$C_{17}H_{16}O_4$, 3,3'-Spirobi[3H-2,4-dihydrobenzo[1,4]dioxepin], 45B, 345
$C_{17}H_{16}O_5$, rel-(2S,3S)-3-Hydroxy-7-methoxy-3',4'-methylenedioxyflavin, 45B, 345
$C_{17}H_{17}ClO_6$, (2S,6'R)-7-Chloro-2',4,6-trimethoxy-6'-methylspiro(benzofuran-2(3H),2-(2')-cyclohexene)-3,4'-dione, 43B, 409
$C_{17}H_{18}O_2 \cdot 0.1667\ CCl_4$, 4-p-Hydroxyphenyl-cis-2,4-dimethylchroman carbon tetrachloride solvate, 45B, 346
$C_{17}H_{18}O_2$, 4-p-Hydroxyphenyl-2,2-dimethylchroman, 45B, 346
$C_{17}H_{20}O_4$, Formylcitran, 43B, 410
$C_{17}H_{20}O_7$, Micordilin, 43B, 409
$C_{17}H_{20}O_7$, 1a,5,6,7,8,9-Hexahydrospiro(cyclohepta[2,3]benzo[1,2-b]oxiren-5,2'-[1,3]dioxolane)-3,4-dicarboxylic acid dimethyl ester, 46B, 328
$C_{17}H_{23}BrO_2$, (6aRS,9RS,10aRS)-4-Bromo-1-methoxy-6,6,9-trimethyl-6a,7,8,9,10,10a-hexahydro-6H-dibenzo[b,d]pyran, 46B, 329
$C_{17}H_{23}N_2O_8RbS$, 4-Nitrobenzo-1,4,7,10,13,16-hexaoxacyclooctadecane rubidium thiocyanate, 44B, 299
$C_{17}H_{24}CaN_2O_6S_2$, Diisothiocyanato(2,3,5,6,8,9,11,12-octahydro-1,4,7,-

10,13-pentaoxabenzocyclopentadecine)calcium methanol, 44B, 296
$C_{17}H_{24}NO_6RbS$, Rubidium thiocyanate benzo-18-crown-6, 44B, 300
$C_{17}H_{24}O_2$, 3,4,5,6-Tetrahydro-2,9-dimethyl-7-hydroxy-5-isopropyl-2,6-methano-2H-1-benzoxocin, 43B, 411
$C_{17}H_{24}O_2$, 8,5a-trans-5a,9a-cis-1,8-Dimethyl-5a-iso-propyl-5a,6,7,8,9,9a-hexahydrodibenzofuran-3-ol, 40B, 308
$C_{17}H_{24}O_5$, Hysterin, 46B, 330
$C_{17}H_{24}O_5$, 1,13,16,19-Tetraoxatetraspiro[4.0.0.0.4.3.3.3]heneicosan-4-one, 46B, 330
$C_{17}H_{24}O_7$, 3,6,9,12,15-Pentaoxa-21-carboxybicyclo[15.3.1]-heneicosa-1(21),17,19-triene, 42B, 248
$C_{17}H_{26}O_3$, 1-α,4,5-α,7-α,9,10-Hexamethyl-2-β-methoxy-3-oxatricyclo-[5.4.0.0$^{4-11}$]undec-9-ene-6-one, 45B, 347
$C_{17}H_{28}O_4$, Dihydrophytuberin, 42B, 248
$C_{17}H_{28}O_5$, 2α-Hydroxydihydrophytuberin, 42B, 248
$C_{18}H_5BrN_2O$, 2-Cyano-5-bromobenz[f](1,3)oxazepine, 37B, 163
$C_{18}H_{10}O_4$, Bis-(enol-lactone) of α,α'-dibenzoylsuccinic acid, 46B, 331
$C_{18}H_{12}O_3$, 5,6:7,8-Dibenzo-bicyclo[2.2.2]octa-5,7-diene-(2,3-dicarboxylic acid anhydride), 45B, 347
$C_{18}H_{12}O_5$, 12-Oxapentacyclo[8.4.3.2$^{2-9}$.0.0$^{3-8}$]nonadeca-5,14,16,18-tetraene-4,7,11,13-tetrone, 46B, 292
$C_{18}H_{16}O$, [2.2](2,5)Furano(1,4)naphthalenophane, 44B, 301
$C_{18}H_{16}O_3$, (-)-3-(1-Phenylpropyl)-4-hydroxycoumarin, 42B, 249
$C_{18}H_{16}O_7$, Usnic acid, 40B, 308; 42B, 250
$C_{18}H_{17}ClN_2O_5S$, 9-syn-Chloro-11-exo-(2,4-(dinitrophenylthio)-4-oxapentacyclo[6.5.0.0$^{2-6}$.0$^{7-12}$.0$^{10-13}$]-tridecane, 46B, 331
$C_{18}H_{17}ClO_6$, 1-(2-Hydroxystyryl)-3,3-dimethyl-3H-isobenzofurylium perchlorate, 44B, 301
$C_{18}H_{18}Cr_2N_2O_{12}$, μ-Diiminobis(pentacarbonylchromium) tetrahydrofuran complex, 41B, 376
$C_{18}H_{18}N_2O_4$, 1-(4-Piperidylphenyl)-3-(5-nitro-2-furanyl)-2-propen-1-one, 44B, 287
$C_{18}H_{18}O_5$, 5,7,4'-Trimethoxyflavanone, 44B, 302
$C_{18}H_{18}O_5$, 5,9-Dimethoxy-3,3,8-trimethyl-7,10-dihydro-3H-naphtho[2,1-b]pyran-7,10-dione, 44B, 303
$C_{18}H_{19}F_6NO_3$, 7-Diethylamino-6-methyl-4-phenyl-2,2-bis(trifluoromethyl)-4,5-dihydro-1,3-dioxepin-5-one, 41B, 378
$C_{18}H_{19}NaO_5 \cdot H_2O$, Sodium 8-allyl-5-(3-methylbutoxy)-4-oxo-4H-1-benzopyran-2-carboxylate monhydrate, 45B, 349
$C_{18}H_{20}O_2$, 4-Hydroxyphenyl-2,2,4-trimethylchroman, 35B, 213
$C_{18}H_{20}O_3$, 4-(1-Oxo-2-phenylethyl)-5,6-dimethyl-1,4,5,7a-tetrahydro-isobenzofuran-3(3aH)-one, 44B, 303
$C_{18}H_{20}O_4$, 2-(Di-p-anisylmethyl)-1,3-dioxolane, 40B, 309
$C_{18}H_{22}O_4$, Acetylcitran, 43B, 410
$C_{18}H_{22}O_6$, Dimethyl 6'-t-butyl-2'-methyl-3'-oxospiro[oxirane-2,4'-tricyclo[3.3.0.0$^{2-8}$]oct-6'-ene]-1,8-dicarboxylate, 45B, 349
$C_{18}H_{22}O_6$, Dimethyl 8-t-butyl-1-methyl-2-oxospiro[bicyclo[2.2.2]octa-5,7-diene-3,2'-oxirane]-5,6-dicarboxylate, 45B, 349
$C_{18}H_{22}O_6$, Dimethyl 8-t-butyl-5-methyl-2-oxospiro[bicyclo[2.2.2]octa-5,7-diene-3,2'-oxirane]-1,6-dicarboxylate, 45B, 349
$C_{18}H_{23}NO_3$, 3-Isopropyl-6,6-dimethyl-5-(1-naphthylamino)-1,2,4-trioxane, 43B, 412
$C_{18}H_{24}O_5$, 2,4-Octano-1,5-dicarbomethoxy-3-oxa-quadricyclane, 44B, 303
$C_{18}H_{24}O_7$, 8-Hydroxyzearalenone monohydrate, 42B, 250
$C_{18}H_{30}O_4$, 2,2'-(Butane-1,4-diyl)bis(hexahydro-1,3-benzodioxole),

46B, 332
$C_{18}H_{30}O_{10}$, 1,4,7,10,13,16-Hexaoxacyclooctadecane - dimethyl acetylenedicarboxylate, 41B, 379
$C_{18}H_{31}NO_2$, 2,4,6-Tri-t-butyl-4,5-epoxy-6-hydroxycyclohex-2-ene-1-imine, 44B, 304
$C_{18}H_{33}KO_9$, 1,4,7,10,13,16-Hexaoxacyclooctadecanepotassium ethyl acetoacetate, 44B, 304
$C_{19}H_{10}Br_2O_6$, 3,3'-Methylene(bis-6-bromo-4-hydroxycoumarin), 38B, 342
$C_{19}H_{16}O_4$, (-)-(S)-Warfarin, 41B, 380
$C_{19}H_{17}NO_6S$, 12-Methoxycarbonyl-5-(2-nitrophenylthio)-10-oxapentacyclo[5.3.1.1$^{2-8}$.0$^{3-6}$.0$^{4-11}$]dodecan-9-one, 44B, 305
$C_{19}H_{18}Br_2O_4$, Dibromodeoxybruceol, 43B, 413
$C_{19}H_{18}O_3$, Hexacyclo[9.3.2.2$^{4-7}$.0$^{2-9}$.0$^{3-8}$.0$^{10-12}$]octadeca-13,15,17-triene-5,6-carbonate, 41B, 381
$C_{19}H_{18}O_6$, 3',5,5',6-Tetramethoxyflavone, 38B, 342
$C_{19}H_{20}O_5$, (-)-Bruceol, 43B, 413
$C_{19}H_{21}BrN_2O_3S$, 1α,6β,9α,10α-13-Oxatetracyclo[8.2.1$^{2-6}$.0$^{2-9}$]tetradec-11-en-8-one p-bromophenylsulphonyl hydrazone, 45B, 185
$C_{19}H_{21}BrN_2O_3S$, 1β,6β,9α,10β-13-Oxatetracyclo[8.2.1$^{2-6}$.0$^{2-9}$]tetradec-11-en-8-one p-bromophenylsulphonyl hydrazone, 45B, 185
$C_{19}H_{22}O$, α-2-Ethyl-5-methyl-3,3-diphenyltetrahydrofuran, 34B, 176
$C_{19}H_{22}O_4$, 4-Cyclohexyl-3,4-dihydro-2-hyroxy-2-methyl-2H,5H-pyrano[3,2-c][1]benzopyran-5-one, 45B, 350
$C_{19}H_{22}O_5$, 6-t-Butyl-4,4,8-trimethoxy-1-oxo-1,4-dihydrodibenzofuran, 44B, 306
$C_{19}H_{28}O_2$, 10,10-Dimethyl-3,4-dioxatricyclo-[5.2.1.0$^{1-5}$]decane-2-spiro-2'-adamantane, 45B, 350
$C_{19}H_{38}O_3$, (6S,7S,9R,10R)-6,9-Epoxynonadecane-7,10-diol, 46B, 332
$C_{20}H_{14}O_4$, 2,11-Dimethoxybenzo[1,2-b;4,3-b']bis(benzofuran), 46B, 333
$C_{20}H_{14}O_9$, Variolaric acid diacetate, 43B, 414
$C_{20}H_{16}O_2$, 1,4-Epoxy-1,4-dihydronaphthalene photodimer, 35B, 215
$C_{20}H_{16}O_7$, Averufin, 38B, 343
$C_{20}H_{17}ClO_7$, 8-(3'-Chloro-2'-hydroxy-6'-methoxybenzoyl)-5,7-dimethoxy-4-methylcoumarin, 44B, 307
$C_{20}H_{20}O_2$, 2-Isopropyl-4,4,7-trimethyl-1H-phenaleno[1,9-bc]fura-1-one, 41B, 383
$C_{20}H_{20}O_4$, Diacetonephenanthroquinone, 46B, 333
$C_{20}H_{20}O_4$, Avicennin, 40B, 309
$C_{20}H_{20}O_6$, 2,10-Diphenyl-1,3,6,9,11,13-hexaoxadispiro[4.1.5.2]tetradecane, 40B, 310
$C_{20}H_{22}O_2$, 7,7a,8,9,11,11a-Hexahydro-7,7,10-trimethyl-1,10-epoxy-10H-benz[de]anthracen-6-ol, 41B, 381
$C_{20}H_{22}O_6$, O-Tetramethylhaematoxylin, 46B, 333
$C_{20}H_{24}IKO_6 \cdot 0.5\ H_2O$, Dibenzo-18-crown-6 potassium iodide hemihydrate, 46B, 334
$C_{20}H_{24}O_8$, (2R,3'aR,5R,6'aR)-3,3'a,4,4'-Tetraacetyl-5-hydroxy-5,5',6'a-trimethylspiro[furan-2(5H),2'(3'H)-furo[2,3-b]furan], 46B, 334
$C_{20}H_{26}O_2$, 2,6-Di-cis-4-hydroxyretinoic acid γ-lactone, 41B, 382
$C_{20}H_{28}O_2$, Adamantylideneadamantane peroxide, 43B, 415
$C_{20}H_{28}O_2$, 2-Methylene-bicyclo[3.3.1]nonan-3-one-1-yl-1-(2-methylbicyclo[3.3.1]nonan-3-one), 46B, 321
$C_{20}H_{28}O_5$, (1RS,14SR,17RS,18RS)-2,20-Dioxa-17-methyltricyclo-[16.3.0$^{1-14}$.0$^{1-18}$]heneicosa-15-ene-3,9,21-trione, 46B, 335
$C_{20}H_{32}O_2$, Dihydroisopimaric acid γ-lactone, 45B, 351
$C_{20}H_{34}N_2O_6S$, 1,4,7,10,13,16-Hexaoxacyclooctadecane benzylammonium thiocyanate, 46B, 335

$C_{21}H_{14}O$, 14-[aj]-Dibenzoxanthene, 40B, 311
$C_{21}H_{14}O_4$, O,O'-Diacetylenyldiphenyl glutarate, 44B, 307
$C_{21}H_{18}O_9$, DL-4,9,10-Triacetoxy-2-methoxydibenzo[c,e]oxepine-5-one, 42B, 251
$C_{21}H_{22}O_5$, 7,11-Bis(5-methyl-2-furyl)spiro[5.5]undecane-1,5,9-trione, 45B, 352
$C_{21}H_{26}O_4 \cdot CHCl_3$, 4,6-Di-t-butyl-4a-hydroxy-8-methoxydibenzofuran-2(4aH)-one chloroform solvate, 45B, 352
$C_{21}H_{30}O_2$, 2,3,4,4a-Tetrahydro-3-methyl-6-pentyl-8-hydroxy-9,9-dimethylxanthene, 43B, 416
$C_{21}H_{32}O_4$, 2-(2-Oxo-3,3,5,5-tetramethylcyclopentyl)-1-oxa-7,7,9,9-tetramethylspiro[4.4]nonane-3,6-dione, 45B, 353
$C_{21}H_{35}NO_7$, 3,6,9,12,15-Pentaoxa-21-carboxybicyclo[15.3.1]heneicosa-1(21),17,19-triene t-butylamine complex, 41B, 384
$C_{21}H_{36}O_3$, 2,4,6-Tricyclohexyltrioxane, 28, 490
$C_{22}H_{15}BrO_4$, 4-[(4-Bromophenyl)methyl]-1,4-epoxy-1-phenyl-1H-2,3-benzodioxepin-5(4H)-one, 45B, 354
$C_{22}H_{15}NO_6$, 1,4-Epoxy-4-(4-nitrophenylmethyl)-1-phenyl-1H-2,3-benzodioxepin-5(4H)-one, 45B, 354
$C_{22}H_{16}Cl_6O_8$, Bikaverin (chloroform solvate), 37B, 171
$C_{22}H_{16}O_5$, cis-1,3-Diphenyl-1,3-dihydroisobenzofuran-1,3-dicarboxylic acid, 45B, 355
$C_{22}H_{18}Br_2O_2$, 2,6-Dibromo-4-[3(S),4(S)-diphenyl-2(R)-tetrahydrofuryl]phenol, 45B, 356
$C_{22}H_{18}O_2$, 3,3-Di(o-tolyl)phthalide, 46B, 336
$C_{22}H_{22}O_8$, (+)-Tetra-O-methyldehydrodicaffeic acid dilactone, 43B, 417
$C_{22}H_{22}O_8$, (-)-Tetra-O-methyldehydrodicaffeic acid dilactone, 43B, 416
$C_{22}H_{22}O_8$, 8a-Hydroxyisopicrostegane, 46B, 338
$C_{22}H_{23}NO_5$, 3-o-Acetamidophenyl-4,4-dimethyl-5-hydroxy-6a-phenyl-3a,4,5,6a-tetrahydrofuro[2,3-b]-furan-2(3H)-one, 46B, 338
$C_{22}H_{25}BrO_2$, 1-Tocoquinone-p-bromophenylethylene condensation product, 37B, 172
$C_{22}H_{26}O_2$, 5-Allyl-5-methoxy-2-(3',4',5'-trimethoxyphenyl)-3-methyl-2,3,5,6-tetrahydro-6-oxobenzofuran, 43B, 418
$C_{22}H_{26}O_6$, 5-Allyl-3a-methoxy-2-(3',4',5'-trimethoxyphenyl)-3-methyl-2,3,3a,6-tetrahydro-6-oxobenzofuran, 43B, 417
$C_{22}H_{28}O_3$, 7-t-Butyl-3-(5'-t-butyl-3'-methylfuran-2'-yl)-5-methylbenzofuran-2(3H)-one, 44B, 308
$C_{22}H_{28}O_5$, (E)-7-t-Butyl-5-methoxy-3-((E)-2'-methoxy-5',5'-dimethyl-4'-oxohex-2'-enylidene)benzofuran-2(3H)-one, 43B, 418
$C_{22}H_{28}O_5$, (Z)-7-t-Butyl-5-methoxy-3-((Z)-2'-methoxy-5',5'-dimethyl-4'-oxohex-2'-enylidene)benzofuran-2(3H)-one, 43B, 419
$C_{22}H_{32}O_3$, (1RS,2RS,1''SR,3''RS)-1,2-Diethyl-1-(4'-cyclohexanone-yl)-2-(3''-(a-pyrone-5-yl)cyclopentane)ethane, 45B, 356
$C_{22}H_{50}N_4O_{14}S_4U$, Tetrathiocyanatotetraaquouranium(IV) (18-crown-6) (methyl isobutyl ketone) water (1:1.5:1:3), 43B, 421
$C_{23}H_{14}Cl_4O$, 1,1,1a,7a-Tetrachloro-1a,2,7,7a-tetrahydro-2,7-diphenyl-2,7-epoxy-1H-cyclopropa[b]naphthalene, 41B, 385
$C_{23}H_{17}ClO_3$, 14-[aj]-Dibenzoxanthenium chloride (acetic acid solvate), 39B, 233
$C_{23}H_{17}Cl_4FeO$, 2,4,6-Triphenylpyran tetrachloroferrate, 43B, 422
$C_{23}H_{18}O_6$, Fluorescein (lactoid form) - acetone, 41B, 386
$C_{23}H_{26}O_9S$, Methyl-dimethyl-a-conidendrin sulphonate, 42B, 252
$C_{23}H_{27}NO_3$, (-)-2(R)-[2,3-Dihydro-2(R)-isopropyl-4-oxo-4H-1-benzopyran-6-yl]-N-[1(R)-phenylethyl]-propionamide, 46B, 337

$C_{24}H_{12}O_3$, Dibenzo[b,b']furo[3,2-e;4,5-e']bis(benzofuran), 46B, 339
$C_{24}H_{16}I_3KO_8$, Bis(8-methoxy-3',2':6,7-furocoumarin) potassium triiodide, 46B, 491
$C_{24}H_{16}O_2$, 6-Phenyl-4-oxa-8,9:10,11-dibenzotricyclo[5.4.0.0^{1-6}]-undeca-2,8,10-triene-5-one, 44B, 308
$C_{24}H_{18}O_7$, 7,10-Dihydroxy-9-methoxy-6-(4-methoxyphenyl)-6H,11H-(2)-benzopyrano[4,3-c]-(1)-benzopyran-11-one, 46B, 340
$C_{24}H_{20}Cl_4O_6$, (13-anti-Acetoxy-1,8,11,12-tetrachlorotetracyclo-[6.2.2.1^{3-6}.0^{2-7}]trideca-4,11-diene-9-one)spiro-10,4'-(9-anti-acetoxy-3-oxatricyclo[4.2.1.0^{2-5}])non-7'-ene, 46B, 337
$C_{24}H_{20}Cl_4O_6$, (13-syn-Acetoxy-1,8,11,12-tetrachlorotetracyclo-[6.2.2.1^{3-6}.0^{2-7}]trideca-4,11-diene-9-one)spiro-10,4'-(9-anti-acetoxy-3-oxatricyclo[4.2.1.0^{2-5}])non-7'-ene, 46B, 337
$C_{24}H_{22}O_8$, 2,3,6,8,9,12-Hexamethoxybenzobis[1-2-b,4-5-b']benzofuran, 39B, 234
$C_{24}H_{23}Br_3O_2$, 1,9-Epoxy-2,8,10-tribromo-9-methoxy-7,11-diphenylspiro-[5.5]undec-1(2)-ene, 44B, 309
$C_{24}H_{24}O$, [2.2](2,5)Furano(4,7)[2.2]paracyclophane, 46B, 341
$C_{24}H_{24}O_4$, (E)-Bis-3,3'-(7-t-butyl-3-oxo-2,3-dihydrobenzo[b]furylidene), 45B, 358
$C_{24}H_{26}BrO_2$, 6-Bromo-7-hydroxy-1,3,4,1',3',4'-hexamethyl-2,1'-diindanyl-1-7'-ether, 40B, 312
$C_{24}H_{26}O_5$, Diethyl 4,8-dimethyl-2,3;6,7-dibenzo-9-oxabicyclo[3.3.1]-nona-2,6-diene-4,8-dicarboxylate, 44B, 309
$C_{24}H_{32}BaCl_2O_{16}$, Bisperchlorato(dodecahydrodibenz[b,n]octaoxacyclotetracosin)barium, 44B, 309
$C_{24}H_{32}O_6$, (7R,9R,18S,20S)-6,7,9,10,17,18,20,21-Octahydro-7,9,18,20-tetramethyldibenzo[b,k](1,4,7,10,13,16)hexaoxacyclooctadecin (isomer F), 41B, 386
$C_{24}H_{32}O_7$, Dimethyldibenzo-21-crown-7, 44B, 310
$C_{24}H_{32}O_8$, 6,7,9,10,12,13,20,21,23,24,26,27-Dodecahydrodibenzo-[b,n](1,4,7,10,13,16,19,22)octaoxacyclotetracosin, 42B, 252
$C_{24}H_{34}BrNO_3$, (4R-Bromo-4,5-dimethyl-6-oxatricyclo[3.2.1.1^{3-8}]non-2-endo-yl)methyl N-((1R)-2-exo-bornyl)carbamate, 41B, 1236
$C_{24}H_{36}O_7$, Erythronolide A anhydride, 46B, 341
$C_{24}H_{44}N_4O_8$, Cyclotetraurethane, 42B, 242
$C_{24}H_{48}K_2MoO_{16} \cdot 5\ H_2O$, Potassium molybdate 1,4,7,10,13,16-hexaoxacyclooctadecane pentahydrate, 45B, 358
$C_{24}H_{50}K_2Mo_6O_{32}$, Bis(1,4,7,10,13,16-hexaoxacyclo-octadecane) potassium hexamolybdate monhydrate, 45B, 359
$C_{25}H_{20}O_2$, Spiro[3,4-dihydro-2H-1,5-dioxadinaphtho[2,1-f:1,2-h]cyclononene-3,1'-cyclopropane], 45B, 360
$C_{25}H_{20}O_4$, 3,3'-Spirobi(3H-2,4-dihydronaphtho[2,3-b][1,4]dioxepin), 45B, 360
$C_{25}H_{20}O_8$, Penioflavin diacetate, 44B, 310
$C_{25}H_{21}NO_2$, 9-Cyano-1,10-dimethyl-1,6-ethylenedioxy-1-octalin, 41B, 387
$C_{25}H_{24}Br_2O_2$, 1,10-Epoxy-2,8-dibromo-9-ethoxy-7,11-diphenylspiro-[5.5]undeca-1(2),8(9)-diene, 44B, 311
$C_{25}H_{25}Br_3O_2$, 1,9-Epoxy-2,8,10-tribromo-9-ethoxy-7,11-diphenylspiro-[5.5]undec-1(2)-ene, 44B, 311
$C_{25}H_{30}N_2O_2$, 2,6-Bis(diethylamino)-3,5-diphenyl-γ-pyrone, 42B, 253
$C_{26}H_{10}Cl_6O_5$, 2,2'-Epoxy-1,1'-bis(3,6,8-trichloro-1,4-dihydro-5-hydroxy-4-oxonaphthylidene) benzene solvate, 44B, 312
$C_{26}H_{14}N_2O_4$, N-(2'-Pyridyl)-8,13-dioxodinaphtho[2,1-b:2',3'-d]-furan-6-carboxamide, 41B, 390
$C_{26}H_{16}O_2$, 9,9'-Bixanthenylidene (α-form), 28, 492

$C_{26}H_{16}O_2$, 9,9'-Bixanthenylidene (β-form), 28, 491
$C_{26}H_{22}O_2$, 16endo-6,10b-Ethano-4a(exo),5,6,10b-tetrahydro-4H-dinaphtho[2,1-b:1',2'-d]pyran-15-one, 41B, 391
$C_{26}H_{26}O_8$, Polystachin, 45B, 361
$C_{26}H_{28}O_4$, meso-Bi-(2,γ')-(2-methyl-6-propoxyindan-1-one), 45B, 362
$C_{26}H_{28}O_4$, racemic-Bi-(2,γ')-(2-methyl-6-propoxyindan-1-one), 45B, 362
$C_{26}H_{34}O_4$, Tridentoquinone, 41B, 240
$C_{26}H_{35}NO_{12}$, cyclo-Tetrakis(2-carboxy-tetrahydropyran-6-yl)-acetonitrile (1:1), 46B, 316
$C_{27}H_{22}O$, 1,9-Diphenyl-12-oxa-benzo(10,11)pentacyclo-[7.2.1.0$^{2-7}$.0$^{2-8}$.0$^{3-8}$]dodec-10-ene, 45B, 191
$C_{27}H_{24}O_2$, Spiro[3,4-dihydro-2H-1,5-dioxadinaphtho[2,1-f:1,2-h]cyclononene-3,1'-cyclopentane], 45B, 363
$C_{27}H_{28}N_2O_9$, 3-(5-Methoxy-2,2,8,8-tetramethyl-2H,8H-benzo[1,2-b:3,4-b']dipyran-6-yl)propyl 3,5-dinitrobenzoate, 46B, 342
$C_{27}H_{28}N_2O_9$, 3-(5-Methoxy-2,2,8,8-tetramethyl-2H,8H-benzo[1,2-b:5,4-b']dipyran-10-yl)propyl 3,5-dinitrobenzoate, 46B, 342
$C_{27}H_{29}NO_7$, 3-(5-Methoxy-2,2,8,8-tetramethyl-2H,8H-benzo[1,2-b:5,4-b']dipyran-10-yl)propyl p-nitrobenzoate, 46B, 342
$C_{27}H_{32}N_2O_9$, Tetrahydro-3-(5-methoxy-2,2,8,8-tetramethyl-2H,8H-benzo-[1,2-b:5,4-b']dipyran-10-yl)-propyl 3,5-dinitrobenzoate, 46B, 342
$C_{28}H_{20}Cl_8O_4$, Tetrachlorobenzoquinone - fulvene cycloaddition product (monoclinic), 42B, 162
$C_{28}H_{20}Cl_8O_4$, Tetrachlorobenzoquinone - fulvene cycloaddition product (triclinic), 42B, 162
$C_{28}H_{22}O_9$, 8,11-Diacetoxy-10-methoxy-7-(4-methoxyphenyl)-6H,7H-(1)-benzopyrano[4,3-b]-(1)-benzopyran-6-one, 46B, 340
$C_{28}H_{28}INaO_{12}$, Phenacylkojate sodium iodide dihydrate, 41B, 392
$C_{28}H_{30}O_6$, 4,10a-Dihydro-7-methoxy-4-oxo-2,9-di-t-butyloxepino[2,3-b]benzofuran-10a-yl benzoate, 45B, 364
$C_{28}H_{34}O_4$, 2,2'-Dihydroxy-7,7,7',7'-tetramethyl-exo-2,2'-bi(3,4-benzo-9-oxabicyclo[3.3.1]non-3-enyl), 44B, 313
$C_{28}H_{40}O_{10}$, Dibenzo-30-crown-10, 38B, 345
$C_{29}H_{18}O_6 \cdot C_3H_6O$, 3,3'-(1-Naphthylmethylene)bis(4-hydroxycoumarin) acetone solvate (orthorhombic), 45B, 365
$C_{29}H_{18}O_6 \cdot C_3H_6O$, 3,3'-(1-Naphthylmethylene)bis(4-hydroxycoumarin) acetone solvate (triclinic), 45B, 365
$C_{29}H_{22}O_6$, Peniophorin trimethyl ether, 46B, 342
$C_{29}H_{40}KNO_{10}S$, Dibenzo-30-crown-10 potassium thiocyanate, 46B, 343
$C_{29}H_{42}O_4$, (E)-5,7-Di-t-butyl-3-(3',5'-di-t-butyl-5'-methoxy-2',5'-dihydrofuran-2'-ylidene)benzofuran-2(3H)-one, 44B, 313
$C_{30}H_{20}O$, 2,5-Diphenylphenanthro[9,10-b]oxepin, 46B, 343
$C_{30}H_{20}O_4$, 4-Phenoxy-3,4,6-triphenyl-1H,4H-furo[3,4-c]furan-1-one, 44B, 85
$C_{30}H_{24}O_{12}$, Poly(ethylene terephthalate) cyclic trimer, 46B, 344
$C_{30}H_{34}O_8$, 3,3'-(1,1'-Bi-2-naphthol)-21-crown-5 hydrate, 44B, 313
$C_{30}H_{36}O_6$, 2,3,4,5,6,7,8,9,10-Tri[1,3-(2-methoxy-5-methylbenzo)]-12,15,18-trioxacyclooctadeca-2,5,8-triene, 46B, 345
$C_{30}H_{40}N_2Na_2O_{10}S_2 \cdot H_2O$, Hexadecahydrodibenzo[b,q]-decaoxocyclotriacontin bis(sodium isothiocyanate) monohydrate, 45B, 366
$C_{30}H_{42}N_2O_8$, 9,19-Dimethyl-9,19-bis(benzoylamino)-1,4,7,11,14,17-hexaoxacycloeicosane, 44B, 314
$C_{30}H_{48}O_{12}$, Cyclohexahydroxyisovaleryl, 42B, 253
$C_{31}H_{16}Br_4O_7$, Bis(p-bromophenyl) tetraketone photoproduct, 45B, 367
$C_{31}H_{21}NO_4$, (E)-1-Phenyl-1-(3,5-diphenyl-2-furyl)-2-(3-nitrobenzoyl)-ethylene, 41B, 392

$C_{31}H_{22}O$, 1,6-Methano(10)annulen-11-ylidene - 1,3-diphenylisobenzofuran (1:1 adduct), 42B, 254
$C_{31}H_{28}O_2$, 2,2,7,7-Tetraphenyl-6-oxabicyclo[3.2.1]octan-5-ol, 44B, 315
$C_{32}H_{18}O_6$, 6,12-Disalicyloylbenzo[1,2-b:4,5-b']bisbenzofuran, 45B, 367
$C_{32}H_{30}O_2$, 1,3,7,9-Tetramethyl-5,11-diphenyl-(1)-benzopyrano[4,3-c]-(1)-benzopyran, 40B, 313
$C_{32}H_{36}O_6$, 2,3:4,5-Bis[1,2-(3-methylnaphtho)]-1,6,9,12,15,18-hexaoxacycloeicosa-2,3-diene, 46B, 345
$C_{32}H_{36}O_{10}$, 1,5,12,16,23,26,29-Heptaoxa[7^{3-14}][5,5]orthocyclophane naphthalene-2,3-diol monohydrate, 46B, 346
$C_{33}H_{42}O_6$, 3',4,9'-Tri-t-butyl-2',5,10'-trimethoxyspiro[cyclohexa-3,5-diene-1,6'-dibenzo[d,f][1,3]dioxepin]-2-one, 45B, 352
$C_{34}H_{24}O_3$, 2-(2',6'-Diphenylpyranilidene)-1,5-diphenylpent-3-ene-1,5-dione, 43B, 424
$C_{34}H_{28}Cl_8O_4$, 5,6,7,8-Tetrachloro-10-cyclohexylidene-3a,10-dihydrobenzo[b]-cyclopenta[e](1,4)dioxepine dimer, 42B, 149
$C_{36}H_{40}N_2Na_2O_{14}$, Dibenzo-24-crown-8-bis(μ-o-nitrophenolato)disodium, 41B, 393
$C_{36}H_{44}O_{16}$, 3,4',11,12-Tetra-O-acetylbruceine C, 46B, 347
$C_{36}H_{38}ClO_{10} \cdot C_6H_6$, 2,3:4,5-Bis[1,2-(3-methylnaphtho)]-1,6,9,12,15,18-hexaoxacycloeicosa-2,3-diene t-butylammonium perchlorate benzene solvate, 46B, 345
$C_{40}H_{36}O_6$, E-2,2,2',2'-Tetraethoxy-Δ[3,3'(2H,2'H)]-bis(phenanthro[9,10-b]furan), 46B, 347
$C_{41}H_{57}Br_2NO_{12}$, Di-(p-bromobenzoyl)pederin monoethanolate, 37B, 172
$C_{42}H_{36}O_4$, 6-Ethoxycarbonyl-11,12,13,14-tetraphenyl-3-oxahexacyclo-[7.6.1.0^{2-7}.0^{5-16}.0^{8-14}.0^{10-15}]-hexadec-12-en-4-one, 46B, 348
$C_{42}H_{60}O_4$, Furo[3,4-c]pyran derivative, 43B, 425
$C_{56}H_{68}O_{10}$, Bis[1,1'-bi(5,6,7,8-tetrahydronaphthyl)]-crown-polyether, 46B, 349

39. HETERO-SULFUR AND HETERO-SELENIUM

$C_2Br_4O_4S_2$, 2,2,4,4-Tetrabromo-1,3-dithietane 1,1,3,3-tetroxide, 46B, 349
$C_2Cl_4O_2S_2$, 2,2,4,4-Tetrachloro-1,3-dithietane 1,1-dioxide, 46B, 349
$C_2Cl_4O_4S_2$, 2,2,4,4-Tetrachloro-1,3-dithietane 1,1,3,3-tetroxide, 46B, 349
$C_2Cl_4S_2$, 2,2,4,4-Tetrachloro-1,3-dithiacyclobutane, 34B, 178
$C_2F_4O_4S_2$, 2,2,4,4-Tetrafluoro-1,3-dithietane 1,1,3,3-tetroxide, 46B, 349
$C_2F_4S_2$, Tetrafluoro-1,3-dithietane (gas-ed), 42B, 1025
C_2H_4S, Ethylene sulfide (gas-mw), 15, 414
$C_2H_4SO_2$, Ethylene sulphone (gas-mw), 38B, 1070
$C_3O_2S_3$, 4,5-Dioxo-2-thioxo-1,3-dithiolane, 34B, 179
$C_3H_4S_3$, 1,3-Dithiolane-2-thione, 38B, 347
$C_3H_5ClO_2S$, 3-Chlorothietane 1,1'-dioxide, 39B, 235
$C_3H_5IN_2S_2$, 3,5-Diamino-1,2-dithiolium iodide, 30B, 148
$C_3H_6OS_3$, 1,3,5-Trithiane 1-oxide, 46B, 350
$C_3H_6O_3S$, 3-Hydroxythietane 1,1-dioxide, 39B, 235
$C_3H_6S_3$, 1,3,5-Trithiane (gas-ed), 10, 284
$C_3H_6S_3$, 1,3,5-Trithiane, 5, 141, 157; 30B, 152; 32B, 181; 34B, 181
$C_3H_6Se_3$, 1,3,5-Triselenane, 33B, 136
$C_4Br_2O_2S$, Dibromo-maleic acid thioanhydride, 46B, 350

$C_4F_6S_2$, 1,2-Bis(trifluoromethyl)dithiete (gas-ed), 42B, 1026
$C_4I_2O_2S$, Diiodo-maleic acid thioanhydride, 46B, 350
C_4H_3BrS, 2-Bromothiophene (gas-ed), 35B, 893; 37B, 686
C_4H_3ClS, 2-Chlorothiophene (gas-ed), 35B, 893; 37B, 686
$C_4H_3NO_2S$, 2-Nitrothiophene (gas-mw), 43B, 1485
$C_4H_4O_2S_2$, 1,4-Dithan-1,4-dioxide, 23, 716
C_4H_4S, Thiophene (gas-ed), 7, 270; 35B, 893; 37B, 686
C_4H_4S, Thiophene (gas-mw), 22, 763; 26, 760
C_4H_4S, Thiophene, 16, 541
$C_4H_4S_2$, 1,4-Dithiadiene, 18, 759
$C_4H_4S_4$, 4-Methyl-1,2-dithia-4-cyclopentene-3-thione, 22, 764
$C_4H_4S_6$, Hexathia-adamantane, 20, 633
C_4H_4Se, Selenophene (gas-mw), 33B, 540; 35B, 904
$C_4H_6Br_2S_2$, trans-2,5-Dibromo-1,4-dithiane, 31B, 143
$C_4H_6Cl_2S_2$, trans-2,3-Dichloro-1,4-dithiane, 30B, 150
$C_4H_6O_2S$, 2,3-Dimethylthiirene 1,1-dioxide, 42B, 262
$C_4H_6O_2S$, 2,5-Dihydrothiophene-1,1-dioxide, 32B, 65
$C_4H_6O_2S_2$, 1,2-Dithiolane-4-carboxylic acid, 22, 766; 31B, 143
$C_4H_6O_3S$, 1-Thiacyclobutane-3-carboxylic acid-1-oxide, 38B, 348
$C_4H_6O_3S$, 3,4-Epoxysulfolane, 38B, 347
$C_4H_8Br_2I_2S_2$, 1,4-Dithiane-iodine monobromide, 37B, 173
$C_4H_8Br_2S$, 1-Bromothiophanium bromide, 35B, 217
$C_4H_8Cl_4Se_2$, 1,4-Diselenane tetrachloride, 29, 521
$C_4H_8I_4S_2$, 1,4-Dithiane - iodine, 24, 556
$C_4H_8I_4Se_2$, 1,4-Diselenane iodine, 26, 771
$C_4H_8O_2S$, Tetramethylene sulphone (gas-ed), 39B, 892
$C_4H_8O_2S$, cis-2-Butene episulphone, 32B, 66
$C_4H_8O_2S_2$, 1,2-Dithiane-4,5-diol, 41B, 394
$C_4H_8O_2S_2$, 1,4-Dithiane-disulfoxide, 24, 557
C_4H_8S, Tetrahydrothiophene (gas-ed), 40B, 1140
$C_4H_8S_2$, 1,4-Dithiane (gas-ed), 10, 284
$C_4H_8S_2$, 1,4-Dithiane, 19, 616
$C_4H_8S_4$, 1,3,5,7-Tetrathiocane, 39B, 236
C_4H_8Se, Tetrahydroselenophene (gas-ed), 35B, 894
C_4H_8Se, Tetrahydroselenophene (gas-mw), 42B, 1035
$C_4H_8Se_2$, 1,4-Diselenane, 15, 506
$C_4H_8Se_4$, 1,3,5,7-Tetraselenocane, 44B, 316
$C_5H_4O_2S$, Thiophene-α-carboxylic acid, 22, 770; 27, 948
$C_5H_4O_2S$, β-Thiophenic acid, 29, 731
$C_5H_4O_2Se$, Selenophene-α-carboxylic acid, 22, 770; 27, 948
$C_5H_4S_3$, 6a-Thiathiophthene, 39B, 236
$C_5H_4Se_3$, 6a-Selenaselenophthene, 37B, 181
$C_5H_6O_2S$, 2H-Thiopyran-1,1-dioxide, 32B, 182
$C_5H_6S_3$, 4,5-Dimethyl-1,3-dithiole-2-thione, 46B, 351
$C_5H_8O_2S$, β-Isoprene sulphone, 9, 395; 15, 502; 35B, 217
$C_5H_8O_4S$, 3-Acetoxythietane 1,1-dioxide, 38B, 348
$C_5H_8O_4S_2$, 2,3-Dihydro-5-methyl-1,4-dithiin 1,1,4,4-tetroxide, 46B, 351
C_5H_8S, 5-Thiabicyclo[2.1.1]hexane (gas-ed), 37B, 690; 42B, 1025
C_5H_8SO, 4-Thiacyclohexanone (gas-ed), 42B, 1028
$C_5H_9NO_2S_3$, 1,2,3-Trithian-5-yl N-methylcarbamate, 44B, 316
$C_5H_{10}O_2S$, 2,2-Dimethylthietane 1,1-dioxide, 39B, 237
$C_5H_{10}O_3S$, 4,6-Dimethyl trimethylene sulfite (gas-ed), 38B, 1063
$C_5H_{10}S_5$, 1,3,5,7,9-Pentathio-cyclodecane (monoclinic), 38B, 349, 350
$C_5H_{10}S_5$, 1,3,5,7,9-Pentathio-cyclodecane (orthorhombic), 38B, 350
$C_6Cl_2N_4S_3$, 2,5-Bis(N-chlorothioimino)-3,4-dicyanothiophene, 46B, 352
$C_6Cl_4N_4S_2Se$, 1,1-Dichloro-2,5-bis(n-chlorothioimino)-3,4-dicyanose-

lenophene, 46B, 352
$C_6H_4ClO_4S_4$, Tetrathiafulvalenium perchlorate, 46B, 353
$C_6H_4ClS_4$, Δ(2,2')-Bis-1,3-dithiolidene chloride, 41B, 395
$C_6H_4Cl_{0.67}S_4$, Tetrathiafulvalenium chloride, 46B, 353
$C_6H_4Cl_3HgS_4$, Tetrathiafulvalene trichloromercurate(II), 44B, 317
$C_6H_4I_3S_4$, Tetrathiafulvalenium triiodide, 46B, 354
$C_6H_4S_2$, 1,4-Thiophthen, 12, 420
$C_6H_4S_2Se_2$, Diselenadithiafulvalene, 44B, 317
$C_6H_4S_4 \cdot 0.57$ SCN, Tetrathiafulvalene thiocyanate, 43B, 427
$C_6H_4S_4$, 2-2'-Bi-1,3-dithiole, 40B, 314
$C_6H_4S_5$, Benzopentathiepine, 45B, 368
$C_6H_4Se_2$, trans-Selenophthene, 34B, 186
C_6H_5BrSO, 2-Acetyl-5-bromothiophene, 35B, 218
$C_6H_6N_2O_4S$, 2,5-Dimethyl-3,4-dinitrothiophene, 44B, 318
$C_6H_6O_2S$, Thiepin 1,1-Dioxide, 35B, 219
$C_6H_6O_2S$, 2-Acetyl-3-hydroxythiophene, 34B, 180
$C_6H_6O_3S_2$, 1,4-Dithiane-2,3-dicarboxylic acid anhydride, 34B, 181
$C_6H_6S_3$, 2-Methyl-6a-thiathiophthene, 41B, 396
$C_6H_7N_3S_2$, 2-Formylthiophene thiosemecarbazone, 37B, 174
$C_6H_8O_2S$, 4,5-Dihydrothiepin 1,1-dioxide, 38B, 352
$C_6H_8O_4S_2$, DL-1,2-Dithiane-3,6-dicarboxylic acid, 29, 520
$C_6H_8O_4Se_2$, DL-1,2-Diselenane-3,6-dicarboxylic acid, 29, 520
$C_6H_9IS_5$, 3,4,5-Tris(methylthio)-1,2-dithiolium iodide, 40B, 314
$C_6H_{10}O_4S_2$, 2,3-Dihydro-5,6-dimethyl-1,4-dithiin 1,1,4,4-tetroxide, 44B, 318
$C_6H_{10}S$, 7-Thianorbornane (gas-ed), 37B, 690; 42B, 1025
$C_6H_{10}S_3$, 2,5-Dimethyl-2,5-endo-thio-1,4-dithiane, 32B, 183
$C_6H_{10}S_4$, Bis-1,3-dithia-2-cyclopentyl, 18, 694, 748
$C_6H_{10}S_4$, 1,2,3,4-Tetrathiadecalin, 38B, 353
$C_6H_{11}NO_2S$, 4-Amino-4-carboxylthiapyran, 41B, 1236
$C_6H_{12}BrNO_2S$, 4-Amino-4-carboxylthiapyran hydrobromide, 41B, 529
$C_6H_{12}Cl_4S_2Zn$, 1,4-Dithionabicyclo[2.2.2]octane tetrachlorozincate, 38B, 353
$C_6H_{12}OS_2$, r-4,c-6-Dimethyl-1,3-dithian t-1-oxide, 42B, 255
$C_6H_{12}OS_3$, t-2,c-4,t-6-Trimethyl-1λ^4,3,5-trithiane r-1-oxide, 46B, 355
$C_6H_{12}O_2S_2$, r-4,c-6-Dimethyl-1,3-dithian t-1,t-3-dioxide, 42B, 255
$C_6H_{12}S_2$, r-4,c-6-Dimethyl-1,3-dithian, 42B, 255
$C_6H_{12}S_3$, 1,4,7-Trithiacyclononane, 46B, 355
$C_6H_{12}S_3$, 2,4,6-Trimethyl-1,3,5-trithiane (α-form), 43B, 428
$C_6H_{12}S_3$, 2,4,6-Trimethyl-1,3,5-trithiane (β-form), 34B, 188
$C_6H_{12}S_3$, 2,4,6-Trimethyl-1,3,5-trithiane (gas-ed), 10, 285
$C_6H_{12}Se_3$, 2,4,6-Trimethyl-1,3,5-triselenane, 39B, 237
$C_6H_{13}IS$, 1-Methyl-1-thiona-cyclohexane iodide, 40B, 315
$C_7H_5NOS_2$, 1,2-Benzodithiol-3-one oxime, 34B, 189
$C_7H_6O_2S$, trans-β-2-Thienylacrylic acid, 32B, 185
$C_7H_6O_3S$, 2-Carboxy-3-acetylthiophene, 38B, 351
$C_7H_6O_3S$, 3-Carboxy-2-acetylthiophene, 38B, 351
$C_7H_6O_3S$, 3-Carboxy-4-acetylthiophene, 38B, 351
$C_7H_6S_4$, 1,2,4,5-Benzotetrathiepine, 45B, 369
$C_7H_7N_2NaO_2S_2 \cdot 2.5\ H_2O$, Sodium 3,5-diacetylimino-1,2-dithiole hydrate, 31B, 145
$C_7H_8OS_2$, 2,3-Dimethyl-dithiofurophthene, 26, 761
$C_7H_8OS_2$, 2,5-Dimethyl[1,2]dithiolo[1,5-b][1,2]oxathiole, 45B, 370
$C_7H_8S_3$, 2,5-Dimethyl-thiothiophthene, 22, 776; 27, 973
$C_7H_9BrN_2O_2S_2$, 3,5-Diacetamido-1,2-dithiolium bromide, 31B, 146
$C_7H_{14}O_3S$, cis-5-t-Butyl-1,3,2-dioxathiane 2-oxide, 46B, 355

C_8N_4S, Tetracyanothiophene, 33B, 138
$C_8N_4S_2$, Tetracyano-1,4-dithiin, 30B, 151
$C_8HBrN_4O_5S$, 3-Bromo-2,7-dinitrothionaphthen-5-diazo-4-oxide, 27, 982
$C_8H_4CsF_3O_2S$, Caesium 1-(2-thienyl)-4,4,4-trifluoro-1,3-butanedione, 41B, 397
$C_8H_4O_2S$, Thiophthalic anhydride, 41B, 397
$C_8H_4S_4$, Dithieno[2,3-b;3',2'-e]-[1,4]-dithiin, 45B, 370
$C_8H_4S_4$, Dithieno[3,4-b;3',4'-e]-[1,4]-dithiin, 45B, 370
$C_8H_5F_3O_2S$, Thenoyltrifluoroacetone (enol form), 42B, 255
C_8H_6OS, Thiolphthalide, 33B, 138
$C_8H_6O_2S$, Benzothiophene-1,1-dioxide, 40B, 316
$C_8H_6O_2S_2$, Thieno[3,4-d]thiepin 6,6-dioxide, 34B, 182
$C_8H_6S_2$, Thieno[3,4-d]thiepin, 34B, 182
$C_8H_6S_2$, 2,2'-Dithienyl (gas-ed), 22, 781
$C_8H_6S_2$, 2,2'-Dithienyl, 33B, 140
$C_8H_6S_2$, 2,3'-Dithienyl, 33B, 140
$C_8H_6S_2$, 3,3'-Dithienyl, 33B, 140
$C_8H_6S_2$, 6,6-(Ethylenedithio)fulvene, 35B, 221
$C_8H_7NO_3S$, 3-Nitro-4-(2-thienyl)-3-buten-2-one, 45B, 371
$C_8H_7NO_3S$, 4-(5-Nitro-2-thienyl)-3-buten-2-one, 45B, 371
$C_8H_7NO_3S_2$, 3-Methoxyimino-1,2-benzodithiol-1,1-dioxide, 39B, 237
$C_8H_8Cl_2S$, 2,6-Dichloro-9-thiabicyclo[3.3.1]nona-3,7-diene, 42B, 256
$C_8H_8Cl_4S_2$, 4,8,9,10-Tetrachloro-2,6-dithiaadamantane, 38B, 354
$C_8H_8Cl_4S_6$, Tetrakis(chloromethyl)hexathiaadamantane, 40B, 316
$C_8H_8O_2S$, 9-Thiabicyclo[4.2.1]nona-2,4,7-triene 9,9-dioxide, 42B, 256
$C_8H_8O_2S_2$, trans-3,6-Dimethyl-thieno[3,2-b]thiophen-2,5(3H,8H)-dione, 42B, 257
$C_8H_8O_2S_2$, 6-Acetoxy-1,4-dithiocin, 44B, 319
$C_8H_8S_4$, 3-Methyl-1,2,4,5-benzotetrathiepine, 45B, 369
$C_8H_8Se_3$, 3,4-Trimethylene-6a-selenaselenophthene, 39B, 238
$C_8H_{10}ClN_3S$, 2-(2-Chloro-4-methyl-thiophene-3-yl-imino)-tetrahydro-imidazole, 45B, 371
$C_8H_{10}O_2S$, 9-Thiabicyclo[3.3.1]nonane-2,6-dione, 45B, 372
$C_8H_{10}O_2S_2$, 5,6-Dihydro-6-acetoxy-1,4-dithiocin, 41B, 398
$C_8H_{10}S_6$, 1-(2-(1,3-Dithiolanylidene))-2,3,6,9-tetrathiaspiro[4.4]-nonane, 40B, 317
$C_8H_{12}O_4S$, 3,8-Thionanedione 1,1-dioxide, 45B, 372
$C_8H_{12}S_6$, trans-2,4-Dimethyl-2,4-bis(thioacetylthio)-1,3-dithietane, 43B, 428
$C_8H_{14}N_2O_4S_2$, Diethyl 1,4-dithia-2,3-diazacyclohexane-2,3-dicarboxylate, 43B, 429
$C_8H_{14}N_2O_4S_4$, Diethyl 1,2,5,6-tetrathia-3,4-diazacyclooctane-3,4-dicarboxylate, 43B, 430
$C_8H_{14}O_2S_2$, DL-6-Thioctic acid, 32B, 186; 38B, 355
$C_8H_{15}IO_2S$, cis-4-Acetoxy-1-methylthiacyclohexane iodide, 45B, 373
$C_8H_{15}IO_2S$, trans-4-Acetoxy-1-methylthiacyclohexane iodide, 45B, 373
$C_9H_4BrS_2$, 4-Phenyl-1,2-dithiolium bromide, 34B, 184
$C_9H_6OS_2$, 3,3'-dithienyl ketone, 44B, 319
$C_9H_6S_2$, 4H-Cyclopenta[2,1-b:3,4-b']dithiophene, 35B, 221
$C_9H_7IS_2$, 3-Phenyl-1,2-dithiolium iodide, 30B, 147
$C_9H_7IS_2$, 4-Phenyl-1,2-dithiolium iodide, 31B, 147
C_9H_7NOS, 2-Pyrrolyl 3-thienyl ketone, 45B, 374
$C_9H_8O_3S$, Thiochroman-4-one 1,1-dioxide, 45B, 374
$C_9H_8O_3S$, 7-Thiapentacyclo[4.4.0.0$^{2-5}$.0$^{3-9}$.0$^{4-8}$]decan-10-one-7,7-dioxide, 41B, 398
$C_9H_8S_2$, Bis(2-thienyl)methane, 41B, 399
$C_9H_9BrO_2S$, 4-Bromo-thiochroman 1,1-dioxide, 44B, 319

$C_9H_9ClOS_2$, 4-Phenyl-1,2-dithiolium chloride monohydrate, 37B, 175
$C_9H_9NO_3S_2$, (S)-α-(Benzenesulphonamido)-β-propiothiolactone, 43B, 430
$C_9H_{10}S_3$, 9-Thiono-8,10-dithiabicyclo[5.3.1]undeca-2,5-diene (monoclinic), 41B, 400
$C_9H_{12}O_2S_2$, 2-(Methyl-acetyl-methylene)-4-methyl-4-acetyl-1,3-dithia-cyclobutane, 33B, 142
$C_9H_{12}O_2Se$, 1,5,5-Trimethyl-7-selenabicyclo[2.2.1]heptane-2,3-dione, 45B, 375
$C_9H_{12}O_4S$, 7a-Methyl-4,7-dioxocyclopenta[b]thiopyran-1,1-dioxide, 42B, 257
$C_9H_{14}O_4S_2$, cis-Perhydrocyclopenta[1,2-c:3,4-c']dithiophene-S,S,S',S'-tetroxide, 33B, 143
$C_9H_{16}ClNO_3S$, cis-2-Chloro-3-morpholino-4,4-dimethylthietane 1,1-dioxide, 40B, 317
$C_9H_{17}BrS$, trans-1-Thioniabicyclo[4.4.0]decane bromide, 42B, 258
$C_9H_{17}ClOS$, 2-Chloro-4-t-butylthiacyclohexane sulfoxide, 43B, 431
$C_9H_{17}ClO_5S_2$, 1-Acetonyl-1-thionia-5-thiacyclo-octane perchlorate, 35B, 223
$C_{10}H_2N_4S$, 2,5-Bis-(dicyanomethylene)-2,5-dihydrothiophene, 38B, 355
$C_{10}H_6BrNO_2S_2$, 1-Bromo-2,4-epidithio-1-nitro-4-phenylbutadiene, 37B, 176
$C_{10}H_6N_2O_3S_2$, 2,4-Epidithio-1-nitro-1-nitroso-4-phenylbutadiene, 37B, 186
$C_{10}H_6O_2S$, Naphtho[1,8-bc]thiete 1,1-dioxide, 42B, 258
$C_{10}H_6SSe$, Naphtho[1,8-cd]-1,2-selenathiole, 43B, 432
$C_{10}H_6SSe$, Selenolo[2,3-b]benzothiophene, 43B, 432
$C_{10}H_6SSe$, Selenolo[3,2-b]benzothiophene, 43B, 432
$C_{10}H_7Cl_2NO_4S_3$, 5-(2,5-Dichloro)-benzsulphonyl-2-sulfamyl-thiophene, 45B, 375
$C_{10}H_7NS_3$, 4-Phenyl-1,2-dithiolium thiocyanate, 34B, 185
$C_{10}H_8BrNO_4S_3$, 5-Benzsulphonyl-4-bromo-2-sulfamyl-thiophene, 45B, 376
$C_{10}H_8O_2S$, 1-Benzothiepin 1,1-dioxide, 41B, 401
$C_{10}H_8S_2$, trans-1,2-Di-2-thienylethene, 41B, 401
$C_{10}H_9BrS$, 5-Bromo-2,3-dimethylbenzo[b]thiophen, 40B, 318
$C_{10}H_{10}OS_2$, 2-Phenyl-1,3-dithian-5-one, 43B, 432
$C_{10}H_{10}O_2S$, cis-2-Phenyl-4-thiolanone 1-oxide, 43B, 433
$C_{10}H_{10}O_2S$, 3-Phenoxy-2-thiolanone, 46B, 356
$C_{10}H_{10}O_3S$, Homothiochromanone 1,1-dioxide, 37B, 176
$C_{10}H_{10}O_3S$, 2,2-Dimethylthioindoxyl 1,1-dioxide, 37B, 176
$C_{10}H_{12}OS_2$, cis-2-Phenyl-1,3-dithiane 1-oxide, 43B, 434
$C_{10}H_{12}OS_2$, trans-2-Phenyl-1,3-dithiane 1-oxide, 43B, 434
$C_{10}H_{12}O_2S_2$, 2-Phenyl-1,3-dithiane trans-1,trans-3-dioxide, 43B, 434
$C_{10}H_{12}O_2S_2$, 2,2,5,5-Tetramethyl-2,3,5,6-tetrahydrothieno[3,2-b]thiophene-3,6-dione, 43B, 434
$C_{10}H_{12}O_2S_4$, 3,6-Bis(2'-oxobut-3'-ylidene)-1,2,4,5-tetrathiin, 31B, 148
$C_{10}H_{12}S_2$, 2-Phenyl-1,3-dithiane, 31B, 149
$C_{10}H_{12}Se_4$, 4,4',5,5'-Tetramethyl-Δ(2,2')-bis-1,3-diselenole, 45B, 377
$C_{10}H_{14}Cl_4CoS_4$, Bis(3,5-dimethyl-1,2-dithiolium) tetrachlorocobaltate(II), 43B, 435
$C_{10}H_{14}Cl_4FeS_4$, Bis-(3,5-dimethyl-1,2-dithiolium) tetrachloroferrate(II), 40B, 318
$C_{10}H_{16}S$, 3,3,6,6-Tetramethyl-1-thiacycloheptyne (gas-ed), 38B, 1068
$C_{10}H_{19}ClOS$, 2-Chloro-4-t-butyl-6-methylthiacyclohexane sulfoxide, 43B, 437
$C_{10}H_{20}Hg_2I_6N_2S_3$, 3,5-Bis(N,N-diethylimonium)-1,2,4-trithiolane tet-

raiododi-μ-iododimercurate(II), 38B, 356
$C_{10}H_{20}S_4$, 1,4,8,11-Tetrathiacyclotetradecane (α form), 42B, 259
$C_{10}H_{20}S_4$, 1,4,8,11-Tetrathiacyclotetradecane (β form), 42B, 259
$C_{10}H_{21}ClO_4S$, 4E-t-Butylthiacyclohexyl-A-methylsulphonium perchlorate, 40B, 319
$C_{10}H_{21}ClO_4S$, 4E-t-Butylthiacyclohexyl-E-methylsulphonium perchlorate, 40B, 319
$C_{11}H_6OS_2$, 4H-Cyclohepta[1,2-b;4,5-c']dithiophene-4-one, 44B, 320
$C_{11}H_6OS_2$, 4H-Cyclohepta[1,2-b;5,4-b']dithiophene-4-one, 44B, 320
$C_{11}H_6OS_2$, 9H-Cyclohepta[2,1-b:5,6-c']dithiophen-9-one, 42B, 260
$C_{11}H_7BF_4S_2$, Dithieno[1,2-b;5,4-b']tropylium tetrafluoroborate, 45B, 377
$C_{11}H_7ClO_4S_2$, Dithieno-[2,1-b:4,5-b']-tropylium perchlorate, 40B, 320
$C_{11}H_8OS_2$, 8,9-Dihydro-4H-cyclohepta[1,2-b:5,4-b']dithiophene-4-one, 41B, 402
$C_{11}H_8S_3$, 2-Phenylthiothiophen, 44B, 321
$C_{11}H_{10}O_3SSe$, 2-Selena-1-dioxothia-3-formyl-5,8-dimethyl-1,2-dihydronaphthalene, 46B, 356
$C_{11}H_{11}NOS$, N-Phenyl-2,5-dihydrothiophene-3-carboxamide, 46B, 357
$C_{11}H_{11}NOS$, 4-(Anilinomethylene)tetrahydrothipophene-3-one, 46B, 357
$C_{11}H_{12}O_3S$, 2,2-Dimethylthiochromanone 1,1-dioxide, 37B, 176
$C_{11}H_{15}ClO_5S$, 1-p-Hydroxyphenylthianium perchlorate, 46B, 358
$C_{11}H_{15}NS_2$, 2-Mercapto-3-(N-cyclohexyliminomethyl)thiophene, 44B, 321
$C_{11}H_{19}IO_2S$, 6-Methoxycarbonyl-1-thioniabicyclo[4.4.0]decane iodide, 45B, 378
$C_{11}H_{21}BF_4S$, 1-Methyl-2,3-di-t-butylthiirenium tetrafluoroborate, 45B, 378
$C_{12}F_8S_2$, Perfluorothianthrene, 46B, 358
$C_{12}F_8Se_2$, Perfluoroseleneanthrene, 46B, 358
$C_{12}H_8Br_2N_2O_4S_2$, trans-1,2-Bis(5-bromo-3-nitro-2-thienyl)cyclobutane, 43B, 437
$C_{12}H_8Br_2O_4S_2$, (R)-(+)-4,4'-Dibromo-2,2'-dicarbomethoxy-3,3'-bithienyl, 41B, 403
$C_{12}H_8O_2S$, Dibenzothiophene sulphone, 33B, 144
$C_{12}H_8S$, Dibenzothiophen, 35B, 225
$C_{12}H_8S_2$, Thianthrene, 20, 628
$C_{12}H_8Se$, Dibenzoselenophene, 35B, 226
$C_{12}H_8Se_2$, Selenanthrene, 9, 398
$C_{12}H_{10}Cl_2O_2Se$, 4-Acetoxy-2,3-dichloro-5,6-benzobicyclo[3.2.0]-7-selenaheptene, 43B, 438
$C_{12}H_{10}S$, 1H,3H-Naphtho(1,8)thiopyrane, 39B, 239
$C_{12}H_{10}S_2$, trans-1,4-Di(2-thienyl)-1,3-butadiene, 44B, 322
$C_{12}H_{10}S_3$, 2-Methyl-4-phenyl-thiathiophthene, 37B, 178
$C_{12}H_{11}NOS_3$, 2-Benzylsulfinyl-2-(1',3'-dithiolan-2'-ylidene)ethanonitrile, 46B, 359
$C_{12}H_{11}NS_4$, 3-Amino-2-methylthio-5-phenyl(1,2)dithiolo[1,5-b](1,2)dithiole-7-S(IV), 42B, 261
$C_{12}H_{12}O_2S_2$, 5-Methylthiopyrylium-3-oxide anti-dimer, 41B, 404
$C_{12}H_{12}O_2S_2$, 5-Methylthiopyrylium-3-oxide syn-dimer, 41B, 404
$C_{12}H_{12}S_2$, [2.2](2,5)thiophenophane, 44B, 284
$C_{12}H_{12}S_2$, 8,9-Dimethyl-1,6-benzodithiocin, 44B, 322
$C_{12}H_{14}N_2S$, trans-6-Thiaperhydronaphthylidene-2-malonitrile, 44B, 323
$C_{12}H_{16}O_2S_2$, trans-Δ(2,2)-Bis(4,4-dimethylthiolan-3-one), 44B, 323
$C_{12}H_{16}S_3$, 2,6,10-Trithiabicyclo[9.4.0]pentadeca-11(1),12,14-triene, 46B, 694
$C_{12}H_{18}Cl_2OS$, 2,9-Dichloro-13-thiabicyclo[8.2.1]-cis-tridec-5-ene sulfoxide, 43B, 439

$C_{12}H_{18}O_2S$, 1,3,3,4,5,6-Hexamethyl-7-thiabicyclo[2.2.1]hept-5-en-2-one 7-anti-oxide, 43B, 440
$C_{12}H_{18}O_2S$, 8-Hydroxy-2-thiatricyclo[7.3.1.0$^{3-8}$]tridecan-13-one (α-isomer), 44B, 324
$C_{12}H_{18}O_2Se$, Tetramethyleneselenium dimedonylide, 38B, 356
$C_{12}H_{20}O_2S$, 2,5-Di-t-butylthiophene-1,1-dioxide, 31B, 150
$C_{12}H_{20}S_4$, 1,3-Dispiro-[2'-(1',3'-dithiolane)]-5,5-dimethylcyclohexane, 45B, 379
$C_{12}H_{20}S_4$, 3,3:6,6-Bis(pentamethylene)-s-tetrathiane, 41B, 404
$C_{12}H_{20}S_6$, trans,trans-Perhydrodibenzo[d,i](1,2,3,6,7,8)hexathiecine, 39B, 239
$C_{12}H_{20}S_6$, 7,8,9,16,17,18-Hexathiadispiro[5.3.5.3]octadecane, 45B, 380
$C_{13}H_7ClOS$, 2-Chlorothioxanthone, 42B, 261
$C_{13}H_8Cl_2O_4S$, 2,3-Dichloro-4-(2-thionyl)-phenoxyacetic acid, 46B, 360
$C_{13}H_8O_2S$, Thioxanthone 10-oxide, 42B, 262
$C_{13}H_{10}O_2S$, Thioxanthen-9-ol-10-oxide, 32B, 187
$C_{13}H_{10}S$, Thioxanthene, 39B, 240
$C_{13}H_{11}NS$, 2,6-Dimethyl-4-(3'-cyanocyclopentadienylidene)thiopyran, 34B, 186
$C_{13}H_{14}Br_2O_2S$, trans-2,5-Dibromo-7-thiabicyclo[4.2.0]-1(6)-octene-7,7-dioxide benzene solvate, 41B, 405
$C_{13}H_{18}O_2S$, (Z)-2-(2-(Thiacyclohexylidene-3)ethyl)-2-methyl-cyclopentane-1,3-dione, 39B, 241
$C_{13}H_{19}ClO_2S$, 10-Chloro-1,4,6,7,8,9-hexahydro-4a,9a-(methanothiomethano)-5H-benzo-cycloheptene 11,11-dioxide, 37B, 182
$C_{13}H_{22}S_3$, 4-t-Butyl-2-(2,2-dimethyl-1-methylthiopropylidene)-1,3-dithiole, 39B, 241
$C_{14}H_4F_6N_2O_4S_2$, 1,6-Dinitro-3,8-bis(trifluoromethyl)thianthrene, 43B, 440; 44B, 324
$C_{14}H_8I_3S_4$, Dibenzotetrathiofulvalene - iodine (1:3), 45B, 381
$C_{14}H_8O_4S_2$, (1)Benzothieno[2,3-b](1)benzothiophene disulphone, 37B, 182
$C_{14}H_8S_2$, (1)Benzothieno[2,3-b](1)benzothiophene, 37B, 182
$C_{14}H_{10}OS$, 2,3-Diphenylthiirene 1-oxide, 42B, 262
$C_{14}H_{10}O_2S$, 2,3-Diphenylthiirene 1,1-dioxide, 42B, 262
$C_{14}H_{11}NO_2S_2$, 8-Nitro-1,10-dithia[2.2]metacyclophane, 41B, 406
$C_{14}H_{12}N_4S_3$, 1,2,4-Trithiolane-3,5-dione diphenylhydrazone, 37B, 184
$C_{14}H_{12}OS$, cis-9-Methylthioxanthene 10-oxide, 34B, 191
$C_{14}H_{12}O_2S_2$, 5H,8H-Dibenzo[d,f](1,2)dithiocin-1,1-dioxide, 39B, 242
$C_{14}H_{12}O_8S_4$, Tetracarbomethoxytetrathiofulvalene, 42B, 263
$C_{14}H_{12}S_2$, trans-1,6-Di(2-thienyl)-1,3,5-hexatriene, 44B, 322
$C_{14}H_{12}S_2$, 1,10-Dithia[2.2]metacyclophane, 41B, 407
$C_{14}H_{12}S_2$, 2,7-Dimethylthianthrene, 37B, 184
$C_{14}H_{12}S_4$, 2,8-Dimethyldibenzo[c,g][1,2,5,6]tetrathiacyclooctadiene, 45B, 381
$C_{14}H_{14}O_4Se_2$, 4,4'-Dicarboxy-2,2',5,5'-tetramethyl-3,3'-biselenienyl, 37B, 185
$C_{14}H_{14}S_7$, 5-Methyl-3-(4-methyl-5-(methylthio)thiocarbonyl-2-thienyl)-2-methylthio(1,2)dithiolo[1,5-b](1,2)dithiole-7-S(IV), 42B, 264
$C_{14}H_{15}ClO_5S$, 1-(4-Hydroxy-1-napthyl)thiophanium perchlorate, 46B, 360
$C_{14}H_{16}S_5$, 7-(5-t-Butyl-1,2-Dithiole-3-ylidene)-4,5,6,7-tetrahydro-1,2-benzodithiole-3-thione, 38B, 357
$C_{14}H_{18}O_3S$, 8-Methoxy-1,2,3,4,5,6-cis-4a,10b-octahydro-6-thiaphenanthrene-6,6-dioxide, 41B, 408

$C_{14}H_{20}OS$, (10)-α-Cyclothien-1-one, 43B, 441
$C_{14}H_{20}O_2S_2$, 2-Pivaloylmethylene-4-pivaloyldithiolene, 44B, 324
$C_{14}H_{20}O_2S_3$, 3,5-Bis(pivaloylmethylene)-1,2,4-trithiolane, 37B, 40
$C_{14}H_{20}S_3$, 3,5-Biscyclohexylidene-1,2,4-trithiolane, 46B, 360
$C_{14}H_{23}N_3O_4S$, 2,5-Dimethyl-trans-2,3-dimorpholino-4-nitro-2,3-dihydrothiophen, 46B, 361
$C_{14}H_{24}O_2S_2$, 2-(1,3-Dithian-2-yl)isoborneol 1'-oxide, 44B, 325
$C_{15}H_{10}OS_2$, 3,5-Diphenyl-1,2-dithiolylium-4-olate, 44B, 325
$C_{15}H_{11}ClO_5S_2$, 4-Hydroxy-3,5-diphenyl-1,2-dithiolium perchlorate, 44B, 325
$C_{15}H_{12}Cl_2O_2S$, cis-2,2-Diphenyl-3,4-dichlorothietane 1,1-dioxide, 41B, 408
$C_{15}H_{12}Cl_2S$, cis-2,2-Diphenyl-3,4-dichlorothietane, 41B, 409
$C_{15}H_{12}O_3S_6$, 2,4,6-Tris(1,3-dithiolan-2-ylidene)-cyclohexane-1,3,5-trione, 46B, 361
$C_{15}H_{14}OS$, cis-2,4-Diphenylthietane trans-1-monoxide, 40B, 321
$C_{15}H_{14}OS$, 1,4-Dimethylthioxanthene 10-oxide, 40B, 320
$C_{15}H_{18}S_3$, 5,6,7-Trithiahexacyclo[9.5.1.1^{3-9}.0^{2-10}.0^{4-8}.0^{12-16}]-octadec-13-ene, 45B, 382
$C_{15}H_{22}OS$, 2,2,6,6-Tetramethyl-4(e)-phenylthian-4(a)-ol, 46B, 362
$C_{16}H_8O_2S_2$, Thioindigo, 18, 764
$C_{16}H_8O_2Se_2$, Selenindigo, 18, 765
$C_{16}H_{11}BrO_2S_2$, 3-Bromo-2,5-diphenyl-1,4-dithiin 1,1-dioxide, 46B, 362
$C_{16}H_{12}OS_2$, 2,5-Diphenyl-1,4-dithiin-1-oxide, 37B, 185
$C_{16}H_{12}S_3$, 2,5-Bis(2-thienylvinyl)thiophene, 44B, 328
$C_{16}H_{14}N_2O_2S$, 2S,3S-1-Cyano-2-hydroxy-3,4-epithiobutane-α-naphthylurethane, 38B, 358
$C_{16}H_{14}S_3$, 4,7-Di-(2-thienyl)-4,5,6,7-tetrahydrobenzo[b]thiophen, 37B, 186
$C_{16}H_{16}OS$, cis-2,4,9-Trimethylthioxanthene 10-oxide, 45B, 382
$C_{16}H_{16}OS$, cis-9-Ethyl-9-methylthioxanthene 10-oxide, 46B, 363
$C_{16}H_{16}OS$, cis-9-Isopropylthioxanthene 10-oxide, 41B, 411
$C_{16}H_{16}O_2S_2$, 2,2-Diphenyl-1,3-dithiane cis-1,3-dioxide, 45B, 383
$C_{16}H_{16}O_4S_6$, 5,7,14,16-Tetramethoxy-1,2,3,10,11,12-hexathia[3.3]metacyclophane, 46B, 363
$C_{16}H_{16}S_2$, syn-2,11-Dithia[3.3]metacyclophane, 45B, 384
$C_{16}H_{16}S_2$, 10,11-(4',5'-Dimethylbenzo)-9,12-dithia-trans-bicyclo-[6.4.0]dodeca-2,4,6,10-tetraene, 45B, 383
$C_{16}H_{20}S_5$, 2-t-Butyl-4,5-(1-(1,3-dithiolane-2-ylidene)-tetramethylene)-1,6,6a-thiathiophthene, 37B, 179
$C_{16}H_{24}O_4S$, 2,10-Diacetoxy-13-thiabicyclo[7.3.1]-cis-tridec-5-ene, 43B, 439
$C_{16}H_{24}O_4S_3$, 1,1,3,3,8,8,10,10-Octamethyl-4,11-dioxo-2,9-dioxa-6,12,13-trithia-dispiro[4.1.4.2]tridecane, 41B, 1237
$C_{16}H_{24}S_5$, 2-Isopropylidene-1,1,7,7,9,9-hexamethyl-3,5,10,11-tetrathiadispiro[3.1.3.2]undecane-8-thione, 39B, 243
$C_{17}H_{11}NO_2S_2$, 3,5-Epidithio-2-nitroso-1,5-diphenylpenta-2,4-dien-1-one, 34B, 193; 37B, 186
$C_{17}H_{12}OS_2$, 2,4-Diphenyldithiofurophthene, 34B, 189
$C_{17}H_{12}S_3$, 2,4-Diphenylthiothiophthene, 31B, 151; 34B, 191
$C_{17}H_{12}S_3$, 2,5-Diphenylthiothiophthene, 37B, 180
$C_{17}H_{12}S_3$, 3,4-Diphenyl(1,2)dithiolo[1,5-b](1,2)dithiole-7-S(IV), 42B, 264
$C_{17}H_{14}OS$, 1-Benzoyl-2-methyl-2-thianaphthalene, 45B, 386
$C_{17}H_{16}O_3S$, 2,3-Benzo-5-oxa-(10)-α-cyclothiene-1,4-dione, 42B, 265
$C_{17}H_{18}O_2S$, 9-Isobutylthioxanthene 10,10-dioxide, 39B, 243
$C_{17}H_{18}S$, 9-Isobutylthioxanthene, 39B, 244

$C_{17}H_{18}S$, 9-Methyl-9-isopropylthioxanthene, 41B, 411
$C_{17}H_{18}S$, 9-t-Butylthioxanthene, 44B, 326
$C_{17}H_{27}NO_5S_2$, 3(S)-(tert-Butoxycarbonylamino)-1(S)-methyltetrahydrothiophenium tosylate, 46B, 364
$C_{18}H_8BrS_4$, Tetrathiotetracene bromide, 46B, 365
$C_{18}H_8S_4$, Naphthaceno[5,6-cd:11,12-c'd']bis-(1,2-dithiolane), 40B, 322
$C_{18}H_8S_4$, Tetrathiotetracene, 32B, 188
$C_{18}H_8Se_4$, Tetraselenotetracene, 44B, 567
$C_{18}H_{12}O_2S_2$, Benzo[c]thiopyrylium-4-oxide syn-dimer, 41B, 412
$C_{18}H_{12}O_2S_2$, Desaurin from acetophenone, 37B, 39
$C_{18}H_{12}S_2$, 5,6,11,12-Tetradehydro-7,10-dihydro-8,9-dithiadibenzo-[a,c]cyclododecene, 39B, 245
$C_{18}H_{12}S_4$, Diphenyl-tetrathiofulvalene, 45B, 387
$C_{18}H_{13}N_3O_2S_2$, 4-Phenyl-1-[3-(4-methyl-5-phenyl-1,2-dithiolio)]urazolide, 45B, 387
$C_{18}H_{14}S_2$, 1,4-Bis(2-thienylvinyl)benzene, 42B, 266
$C_{18}H_{14}S_4$, Bis(2-methyl-benzo[b]thien-3-yl) disulfide, 45B, 348
$C_{18}H_{14}S_4$, 4-(Thio-p-toluoyl)-5-p-tolyl-1,2-dithiole-3-thione, 43B, 443
$C_{18}H_{16}O_2S_2 \cdot 0.5\ C_6H_6$, 2,6-Dimethyl-5'-p-methoxyphenyl-1',2'-dithiole-3',4-ylidene-2,5-cyclohexadienone hemibenzenate, 43B, 443
$C_{18}H_{16}O_3S$, 3,5-Dimethoxy-4-phenyl-1-benzothiepin-1-oxide, 44B, 326
$C_{18}H_{17}NO_4S$, 2-N,N-Diethylamino-3-carbomethoxynaphtho[2,3-b]thiophene-4,9-dione, 38B, 358
$C_{18}H_{18}ClNS$, 2-Chloro-9-(ω-dimethylaminopropylidene)thioxanthene (α-Chlorprothixene), 29, 735; 40B, 323
$C_{18}H_{18}OS$, r-2,trans-6-Diphenyl-cis-3-methyl-4-thianone, 45B, 388
$C_{18}H_{18}S_3$, 2,11,20-Trithia[3.3.3](1,3,5)cyclophane, 38B, 359
$C_{18}H_{20}S_2$, syn-2,11-Dithia-9,18-dimethyl[3.3]metacyclophane, 37B, 111
$C_{18}H_{20}S_2$, 2,11-Dithia-6,15-dimethyl[3.3]metacyclophane, 43B, 444
$C_{18}H_{20}S_6$, 6,8,10,14,16,18-Hexamethyl-1,2,3,4,11,12-hexathia[4.2]-metacyclophane, 46B, 363
$C_{18}H_{22}ClNS$, trans-9,10-Dihydro-4-(3-dimethylamino-propylidene)-4H-benzo(4,5)cyclohepta[1,2-b]thiophene hydrochloride, 37B, 188
$C_{18}H_{22}N_2O_6S_2$, β-Hydroxyethylammonium 2-(3'-β-hydroxyethylamino-1',1'-dioxo-2'-benzo[b]thienyl)benzenesulfinate, 46B, 365
$C_{18}H_{30}BF_4NO_2S_2$, trans-4-t-Butyl-1-(N-ethyl-N-p-toluenesulphonylamino)-1-thioniacyclohexane fluoroborate, 37B, 188
$C_{18}H_{30}BrNO_3S$, Penthienate bromide, 40B, 40
$C_{19}H_{13}BrOS_4$, 3-Benzoyl-5-p-bromophenyl-2-methylthio-6a-thiathiophthen, 34B, 193
$C_{19}H_{14}O_2S_2$, syn-5,12-Dihydro-5,14(epithiomethylenoxy)-6,12-epithiodibenzo[a,f]cyclodecen-7(6H)-one, 46B, 366
$C_{19}H_{17}NOS_2$, 2-(p-Dimethylanilino)-4-phenyl-6,6a-dithiafurophthene, 41B, 414
$C_{19}H_{17}NS_3$, 2-(p-Dimethylanilino)-4-phenyl-6a-thiathiophthene, 38B, 360
$C_{19}H_{18}OS_5$, 2-p-Methoxyphenyl-4,5-(1-(1,3-dithiolane-2-ylidene)-tetramethylene)-1,6,6a-thiathiophthene, 40B, 324
$C_{19}H_{20}BrN_5O_2S$, N-(p-Bromophenylcarbamoyl)thiamine anhydride (N,S-cis-Type), 38B, 361
$C_{19}H_{20}OS$, r-2,trans-6-Diphenyl-cis-3-ethyl-4-thianone, 45B, 388
$C_{19}H_{20}OS_5$, 2-(2-p-Methoxyphenyl-2-methylthiovinyl)-3,4-trimethylene-5-methylthio-1,6,6a-thiathiophthene, 39B, 245
$C_{19}H_{22}OS$, cis-2,trans-6-Diphenyl-cis-3-ethylthian-r-4-ol, 46B, 366
$C_{19}H_{22}OS$, 4-p-Hydroxyphenyl-2,2,4,7-tetramethylthiochroman, 43B, 446

$C_{19}H_{22}O_2S$, 1,4-Dimethyl-9-isobutylthioxanthene 10,10-dioxide, 44B, 327
$C_{20}H_{12}N_2S_2$, 2-Cyano-3-phenyl-4-(5-phenyl-1,2-dithiole-3-ylidene)-Δ^2-butyronitrile, 44B, 327
$C_{20}H_{16}N_2O_4S_2$, 5-Benzamido-2-benzoyl-imino-1,3-dithiol 4-carbonic acid ethyl ester, 46B, 367
$C_{20}H_{16}O_2S_2$, 3,4-Dihydro-2H,6H-2-benzothiopyrano[4,3-b]pyran-2-spiro-3'-[1H-2]benzothiopyran-4'(3'H)-one, 45B, 389
$C_{20}H_{16}S$, 2,5-Distyrylthiophene, 44B, 328
$C_{20}H_{16}S_3$, 2,5-Diphenyl-3,4-trimethylene-6a-thiathiophthene, 41B, 414
$C_{20}H_{17}IOS_4$, 5-Phenyl-3-(5-phenyl-1,2-dithiol-3-ylidenmethyl)-1,2-dithiolium iodide methanol, 30B, 149
$C_{20}H_{20}OS_2$, Xanthene-9-spiro-2'-[4'-t-butyl-3'-(methylthio)thiete], 46B, 369
$C_{20}H_{22}O_2S$, 9-(Cyclohexylmethyl)thioxanthene 10,10-dioxide, 41B, 415
$C_{20}H_{24}BF_4S_8$, Bis(tetramethyltetrathiofulvalene) tetrafluoroborate, 43B, 783; 45B, 389
$C_{20}H_{24}BrS_8$, Tetramethyltetrathiafulvalene bromide, 44B, 328
$C_{20}H_{24}IS_8$, Tetramethyltetrathiofulvalene iodide, 46B, 607
$C_{20}H_{24}O_4S_8$, 7,15,17,19-Tetraethoxy-2,3,4,5,10,11,12,13-octathiatricyclo[12.2.2.2$^{6-9}$]eicosa-6,8,14,16,17,19-hexaene, 37B, 189
$C_{20}H_{24}S$, Benzo[p]-2-thiatetracyclo[7.5.3.0$^{3-8}$.0$^{1-10}$]heptadeca-3(8),15-diene, 46B, 369
$C_{20}H_{24}S$, 4-Benzyl-2,3-5,6-bis(tetramethylene)-4H-thiopyran, 46B, 370
$C_{20}H_{26}ClNOS$, Methixene hydrochloride monohydrate, 38B, 363
$C_{20}H_{26}S_5$, 2,6-Bis(5-t-butyl-1,2-dithiole-3-ylidene)cyclohexanethione, 39B, 246
$C_{20}H_{29}N_3O_2S_2$, 2-Diethylamino-5-(t-butylimino)-4-(p-tolylsulphonylimino)-3-methyl-4,5-dihydrothiophene, 44B, 203
$C_{20}H_{32}O_2S_2$, 3,3,3',3',6,6,6',6'-Octamethyl-4,4'-bi(1-thiacycloheptylidene)-5,5'-dione, 44B, 329
$C_{20}H_{32}O_2S_2$, 3,3,7,7,10,10,14,14-Octamethyl-2,12-dioxo-5,12-dithiatricyclo[7.5.0.0$^{1-8}$]tetradec-8-ene, 45B, 390
$C_{21}H_{15}NS_2$, N-(3-Phenyl-2-benzo[b]thienyl)thiobenzamide, 39B, 248
$C_{21}H_{18}S_3$, 6H,12H,18H-5,11,17-Trithiatribenzo[a,e,i]cyclododecene, 41B, 415
$C_{21}H_{24}NS_9$, Tetramethyltetrathiafulvalene thiocyanate, 45B, 390
$C_{21}H_{28}N_2O_5S$, Hycanthone methanesulphonate, 44B, 329
$C_{22}H_{16}N_2OS_2$, 2-(α-(5-p-Methoxyphenyl-3-methylthio-2-thienyl)benzylidene)propanedinitrile, 46B, 370
$C_{22}H_{19}BrOS$, 2-(p-Bromobenzoyl)-5-isopropyl-7-methyl-8H-azuleno[1,8-bc]thiophene, 37B, 190
$C_{22}H_{22}OS_4$, α-(7-(5-t-Butyl-1,2-dithiole-3-ylidene)-4,5,6,7-tetrahydro-1,2-benzodithiole-3-ylidene) - acetophenone, 39B, 247
$C_{22}H_{22}S_5$, 2-(5-Phenyl-1,2-dithiole-3-ylidene)-6-(5-t-butyl-1,2-dithiole-3-ylidene)-cyclohexanethione (modification 2), 42B, 266
$C_{22}H_{28}S$, 2,2-Di-t-butyl-3,3-diphenylthiirane (monoclinic), 42B, 266
$C_{22}H_{28}S$, 2,2-Di-t-butyl-3,3-diphenylthiirane (orthorhombic), 42B, 266
$C_{22}H_{29}ClO_6S_2$, 1-p-Hydroxyphenyl-(1'-p-phenolate)-dithianium perchlorate, 46B, 371
$C_{22}H_{30}N_2O_2S$, N-{4-(Methoxymethyl)-1-[2-(2-thienyl)ethyl]-piperidinyl}-N-phenylpropanamide, 45B, 391
$C_{23}H_{16}S_3$, 2,3,4-Triphenyl-thiathiophthene, 37B, 180
$C_{23}H_{26}O_2S$, 1-Methoxy-3-(4-methoxyphenyl)-5-phenyl-2-thiabicyclo-[4.4.0]dec-3-ene, 44B, 330
$C_{24}H_{16}S_6$, 3,5-Diphenyl-1,2-dithiolylium 4-phenyl-5-thioxo-1,2-di-

thiole-3-thiolate, 43B, 448

$C_{24}H_{20}Se_4$, 1H,4H-Naphtho(1,8)diselenepine dimer, 38B, 364

$C_{24}H_{24}O_6S_3 \cdot 0.5\ C_6H_6$, 2,3,7,8,12,13-Hexamethoxy-10,15-dihydro-5H-5,10,15-trithiatribenzo[a,d,g]cyclononene hemibenzene solvate, 45B, 392

$C_{24}H_{40}S_2$, 2,5-Di-t-butyl-3,6-dineopentylthioeno[3,2-b]thiophene, 41B, 417

$C_{25}H_{20}OS$, 2,6-Diphenyl-3-benzyl-2H-thiopyran-5-carboxaldehyde, 32B, 189; 42B, 267

$C_{25}H_{25}NS$, 2-(2,6-Dimethylphenylimino)-3,3-dimethyl-4,4-diphenylthietane, 44B, 330

$C_{25}H_{38}N_4O_2S_2$, 2-Diethylamino-5,6-bis(t-butylimino)-4-(p-tolylsulphonylimino)-3-methyl-1-thiacyclohex-ene, 44B, 203

$C_{26}H_{18}N_2S_2$, 3,5-Epidithio-2,5-diphenyl-2,4-pentadienylidene-3-aminoquinoline, 37B, 191

$C_{26}H_{20}S_2$, 1-t-Butylbenzo[d]naphtho[1,2-d']benzo[1,2-b:4,3-b']dithiophene, 44B, 331

$C_{26}H_{22}O_2S_5 \cdot 0.5\ CS_2$, 2,6-Bis(p-methoxyphenyl-1,2-dithiole-3-ylidene)-cyclohexanethione carbon disulfide solvate, 41B, 418

$C_{26}H_{28}S$, 12,15-Dimethyl[2.2](2,5)thiopheno(4,7)[2.2]paracyclophane, 46B, 341

$C_{26}H_{32}O_6S_2$, 5,19-Dioxa-(10,10)-*a*-cyclodithiene-1,4,15,18-tetraone, 42B, 267

$C_{30}H_{20}S_2$, Tetraphenylthieno[3,4-c]thiophene, 38B, 365

$C_{30}H_{22}Cl_4FeS_4$, Bis-(3,5-diphenyl-1,2-dithiolium) tetrachloroferrate(II), 40B, 324

$C_{30}H_{22}Cl_4HgS_4$, Bis-(3,5-diphenyl-1,2-dithiolium) tetrachloromercurate(II), 40B, 324

$C_{30}H_{22}Cl_5FeS_4$, Bis-(3,5-diphenyl-1,2-dithiolium) tetrachloroferrate(III) chloride, 40B, 325

$C_{30}H_{38}Cl_2O_{11}S_3$, 1,1'-Di-p-hydroxyphenyl-(1"-p-phenolate)-trithiophanium perchlorate, 46B, 371

$C_{32}H_{22}S_4$, Bis(2,4-diphenyl-3-thienyl)-disulfide, 45B, 393

$C_{32}H_{28}S_4$, 2,17,32,35-Tetrathia[3.3.3.3](3,5',5,3')biphenylo[4]phane, 42B, 268

$C_{32}H_{32}ClO_4S_8$, Bis(4,5-dimethyl-2H-1,3-dithiol-2-ylidene)-1,4-cyclohexa-2,5-diene perchlorate, 46B, 372

$C_{32}H_{32}S_8$, 4,6,12,14,20,22,28,30-Octamethyl-1,2,9,10,17,18,25,26-octathia[2.2.2.2]metacyclophane, 46B, 372

$C_{34}H_{24}I_{2.15}S_2$, Tetraphenyldithiapyranylidene polyiodide, 46B, 373

$C_{34}H_{24}I_{3.09}S_2$, Tetraphenyldithiapyranylidene polyiodide, 46B, 373

$C_{34}H_{24}S_2$, Tetraphenyldithiapyranylidene, 46B, 373

$C_{36}H_{16}ClSe_8$, Bis(tetraselenotetracene) chloride, 44B, 331

$C_{36}H_{16}ISe_8$, Bis(tetraselenotetracene) iodide, 44B, 567

$C_{36}H_{16}I_3S_8$, Bis(tetrathiotetracene) triiodide, 43B, 449; 44B, 332

$C_{37}H_{16}NSSe_8$, Tetraselenotetracene thiocyanate, 45B, 682

$C_{42}H_{28}I_5S_{28}$, Hepta(tetrathiafulvalene) pentaiodide, 41B, 419; 42B, 1020

$C_{54}H_{40}Cl_7S$, 1-(2,6-Dichlorophenyl)-4-phenyl-trans,trans-1,3-butadiene 1-(2,6-dichlorophenyl)-4-(2-thienyl)-trans,trans-1,3-butadiene (2.5:1 mixed crystal), 45B, 108

$C_{79}H_{68}Cl_2N_2O_{14}S_2$, Isopropylidene-bis-4,4'-(diphenylcarbonate) bis(2,5-diphenyl-4-p-dimethylaminophenylthiapyrylium perchlorate), 44B, 333

40. HETERO-(NITROGEN AND OXYGEN)

$C_2H_2N_2O$, 1,3,4-Oxadiazole (gas-mw), 38B, 1071
$C_3H_3N_3O_4$, 3-Methyl-4-nitrofuroxan, 40B, 326
$C_3H_5NO_2$, 2-Oxazolidinone, 38B, 365
$C_3H_5N_3O$, 3-Amino-4-methylfurazan, 44B, 333
$C_3H_7ClN_2O_2$, 4-Amino-3-isoxazolidone hydrochloride, 20, 639
$C_4H_4N_2O_2$, 3a,6a-Dihydroisoxazolo[5,4-d]isoxazole, 45B, 393
$C_4H_5N_3O_3$, 3-Methyl-4-furoxancarboxamide, 41B, 419
$C_4H_5N_3O_3$, 4-Methyl-3-furoxancarboxamide, 41B, 419
$C_4H_6N_4O_2$, 4-Methyl-3-furazancarbohydrazide, 42B, 269
$C_4H_6N_4O_3$, 3-Methyl-4-furoxancarbohydrazide, 38B, 366; 39B, 248
$C_4H_6N_4O_3$, 4-Methyl-3-furoxancarbohydrazide, 38B, 366; 39B, 248
$C_4H_8N_2O_3$, 3-Aminomethyl-5-isoxazolol monohydrate, 44B, 334
$C_4H_{10}N_2O_4$, Morpholinium nitrate, 42B, 270
$C_5H_5N_3O_3$, 2,3-Dihydro-5-nitroimidazo[2,1-b]oxazole, 45B, 394
$C_5H_6N_2O_2$, 5,6-Dihydro-4H-cyclopenta[c]furazan-2-oxide, 45B, 394
$C_5H_8N_2O_2 \cdot 0.5\ H_2O$, 4-Aminomethyl-5-methyl-3-isoxazolol hemihydrate, 43B, 450
$C_5H_8N_2O_2$, 3-Hydroxy-5-(2-aminoethyl)isoxazole zwitterion, 40B, 327
$C_5H_8N_2O_4$, Nitroisoxazolisidine, 40B, 1086
$C_5H_9NO_5$, 1,3,5,7-Tetraoxa-9-azacyclodecan-10-one, 39B, 249
$C_5H_9N_3O_2$, 2-Methylamino-5-ethoxy-1,3,4-oxadiazole, 44B, 334
$C_5H_{11}NO_3S$, Morpholine methanesulphonamide, 43B, 451
$C_6N_6O_6$, Benzotrifuroxan, 31B, 152
$C_6H_2Cl_2N_2O_2$, 5,6-Dichlorobenzfurazan-1-oxide, 38B, 369
$C_6H_2N_4O_6$, 4,6-Dinitrobenzfuroxan, 38B, 368
$C_6H_3BrN_2O_2$, 5-Bromobenzfurazan-1-oxide, 38B, 369
$C_6H_3ClN_2O_2$, 5-Chlorobenzfurazan-1-oxide, 27, 973
$C_6H_3IN_2O_2$, 5-Iodobenzfurazan-1-oxide, 38B, 370
$C_6H_3KN_4O_7$, Potassium 4-hydroxy-5,7-dinitrobenzfurazan monohydrate, 37B, 191
$C_6H_4Cl_2N_4O_4$, 4,4'-Dichloro-3,3'-ethylenebis(sydnone), 34B, 194
$C_6H_4N_2O$, Benzo-2,1,3-oxadiazole, 15, 505
$C_6H_4N_2O_2$, Benzfurazan 1-oxide, 45B, 395
$C_6H_4N_2O_2$, 3,3'-Bi-isoxazole, 32B, 190
$C_6H_4N_2O_2$, 3,4'-Bi-isoxazole, 34B, 195
$C_6H_4N_2O_2$, 5,5'-Bi-isoxazole, 33B, 145
$C_6H_5N_5O_3$, cis,trans-5,6,7,8-Diepoxy-8-carboxamido-5,6,7,8-tetrahydrotetrazolo[1,5-a]pyridine, 39B, 249
$C_6H_6N_4O_4$, 3,3'-Ethylenebis(sydnone), 34B, 196
$C_6H_7Cl_2NO_3$, 2,4-Dimethyl-2-(dichloromethyl)-5-oxazolinone N-oxide, 44B, 334
$C_6H_8N_2O_2$, 3,3'-Bi-2-isoxazoline, 31B, 153
$C_6H_9NO_3$, N-Methyl-5,5-dimethyloxazolidine-2,4-dione, 35B, 228
$C_6H_9NO_5$, α-D-Ribofuro(1',2':4,5)-2-oxazolidone, 42B, 271
$C_6H_9N_3O_2$, 4-Methyl-3-furazancarbodimethylamide, 41B, 421
$C_6H_9N_3O_3$, 3-Methyl-4-furoxancarbodimethylamide, 40B, 328
$C_6H_9N_3O_3$, 4-Methyl-3-furoxancarbodimethylamide, 40B, 328
$C_6H_{12}N_2O_3$, 3-Hydroxy-5-(3-aminopropyl)isoxazole zwitterion hydrate, 40B, 329
$C_6H_{14}BrN_5O$, Morpholine biguanide hydrobromide, 39B, 253
$C_6H_{14}ClN_5O$, Morpholine biguanide hydrochloride, 37B, 193
$C_7H_4BrN_3O_2$, 4-Bromo-3-(3-pyridyl)sydnone, 45B, 395
C_7H_5NOS, Benzoxazoline-2-thione, 39B, 250
$C_7H_5NO_2$, Benzoxazoline-2-one, 39B, 250
$C_7H_5N_3O_2$, O-(3-Phenyl-5-(1,2,3,4-oxatriazolio)) oxide, 41B, 245
$C_7H_6N_2O_2$, 5-Methylbenzfurazan-1-oxide, 38B, 371
$C_7H_7Cl_2NO$, N-(2,2-Dichlorovinyl)-1,2-epimino-4,5-epoxycyclopentadi-

ene, 43B, 386
$C_7H_{10}N_2O_4 \cdot H_2O$, (RS)-α-Amino-3-hydroxy-5-methylisoxazole-4-propionic acid monohydrate, 46B, 374
$C_7H_{11}NO_2$, 4,4,6-Trimethyl-2-oxo-3,4-dihydro-2H-1,3-oxazine, 45B, 396
$C_7H_{11}N_3O_4$, Isopropyl 3-methyl-4-furoxancarbamate, 43B, 452
$C_7H_{11}N_3O_4$, Isopropyl 4-methyl-3-furoxancarbamate, 43B, 452
$C_7H_{12}N_2O_3$, 5,6,7,8-Tetrahydro-4H-isoxazolo[4,5-d]azepin-3-ol monohydrate, 42B, 271
$C_8H_4N_4O_2$, [1,2,3]Oxadiazolo[4,3-c][1,2,4]benzotriazinium-3-olate, 45B, 200
$C_8H_5BrN_2O_2$, N-(p-Bromophenyl)sydnone, 28, 494
$C_8H_5NO_3$, 2H-1,3-Benzoxazine-2,4(3H)-dione, 44B, 335
$C_8H_5NO_3$, 2H-3,1-Benzoxazine-2,4(1H)-dione, 44B, 335
$C_8H_6N_2O_3$, 3-(p-Hydroxyphenyl)-1,2,5-oxadiazole 2-oxide, 46B, 374
$C_8H_7BrN_2O_2$, 5-Bromo-4,6-dimethylisoxazolo[3,4-b]pyridin-3(1H)-one, 45B, 397
$C_8H_7KN_4O_7$, 4-Methoxy-5,7-dinitrobenzfurazan potassium methoxide adduct, 37B, 192
C_8H_7NOS, 3-Methyl-benzoxazoline-2-thione, 38B, 372
$C_8H_7NO_2$, 3-Methyl-benzoxazoline-2-one, 39B, 250
$C_8H_8N_2O_2 \cdot H_2O$, 4,6-Dimethylisoxazolo[3,4-b]pyridin-3(7H)-one monohydrate, 45B, 397
$C_8H_8N_4O$, 2-Amino-3-(5-methyl-1,2,4-oxadiazol-3-yl)pyridine, 45B, 397
$C_8H_{10}ClNO_2$, 2-Chloroethanoyl-4-methyl-5-methylene-6-hydro-1,2-oxazine, 46B, 375
$C_8H_{10}N_2O_2$, syn-5,10-Dimethyl-3,8-oxa-4,9-aza-tricyclo[5.3.0.0$^{2-6}$]-deca-4,9-diene, 39B, 252
$C_8H_{10}N_2O_3$, 6,7-Dihydro-6,6-dimethylbenzofurazan-4(5H)-one 3-oxide, 42B, 273
$C_8H_{11}N_3O_2$, 6,7-Dihydro-6,6,-dimethylbenzofurazan-4(5H)-one oxime, 44B, 335
$C_8H_{11}N_3O_3$, 6,7-Dihydro-6,6-dimethylbenzofurazan-4(5H)-one 1-oxide oxime, 42B, 272
$C_8H_{13}NO_2$, 3-Oxa-9-methyl-9-azabicyclo[3.3.1]nonan-7-one, 41B, 421
$C_8H_{14}ClNO$, Perkinamine hydrochloride, 29, 735
$C_8H_{14}N_2O_4S$, 1-Methyl-4,5-(D-glucofurano)imidazolidine-2-thione, 40B, 329
$C_8H_{15}NO_4S$, Piperidine-4-spiro-1'-(2',5'-dioxolane) methanesulphonamide, 42B, 273
$C_8H_{16}N_2O_2S_2$, Dithiobismorpholine, 44B, 336
C_9H_6ClNO, 3-Chloro-5-phenylisoxazole, 45B, 398
$C_9H_6N_2O_2S$, 7-Methylbenzothieno[3,2-c]furoxan, 41B, 422
$C_9H_7BrN_2O_2$, 4-Methyl-3-(p-bromophenyl)-1,2,5-oxadiazole 2-oxide, 34B, 197
$C_9H_7BrN_2O_2$, 4-Methyl-3-(p-bromophenyl)-1,2,5-oxadiazole 5-oxide, 34B, 198
$C_9H_7NO_2$, 3-Hydroxy-5-phenylisoxazole (α form), 40B, 1086
$C_9H_7NO_2$, 3-Hydroxy-5-phenylisoxazole (β form), 34B, 199
$C_9H_7NO_2$, 3-Phenylisoxazoline-5-one, 34B, 201
$C_9H_7N_3O_2$, 5-Phenyl-1,2,4-oxadiazole-3-carboxamide, 46B, 375
$C_9H_8N_2O_4S$, 3-Methyl-4-phenylsulphonylfuroxan, 39B, 254
$C_9H_8N_2O_4S$, 4-Methyl-3-(phenylsulphonyl)furoxan, 42B, 273
$C_9H_9NO_3$, N-Acetyl-5,6-dihydrofuro[3,2-b]pyrid-2-one, 42B, 274
$C_9H_{10}N_2O_5$, Dimethyluracil - vinyl carbonate photolysis product, 39B, 254
$C_9H_{13}BrN_2O_4$, 2-Amino-3-hydroxy-4-hydroxymethyl-7-methyl-Δ^2-furo[2,3-c]pyridine hydrobromide, 35B, 229

$C_9H_{13}NO_2$, Morpholylpentamethinemerocyanine, 42B, 274
$C_9H_{15}NO_3$, 2,6,10-Trioxa-13-azatricyclo[7.3.1.0$^{5-13}$]tridecane, 42B,275
$C_9H_{17}NO_3$, 1,3-Dioxole-2-spiro-4'-(3',3'-diethylpyrrolidin-2'-one), 44B, 280
$C_{10}H_6BrN_2O$, 2-Cyano-5-bromobenz[f](1,3)oxazepine, 35B, 232
$C_{10}H_8BrNO_2$, N-Methyl-3-phenyl-4-bromoisoxazolin-5-one, 32B, 191; 34B, 201
$C_{10}H_9NO_2$, N-Methyl-4-phenylisoxazolin-5-one, 34B, 203
$C_{10}H_{10}BrN_3O$, 5,5-Dimethyl-2-p-bromophenylimino-Δ^3-1,3,4-oxadiazoline, 39B, 255
$C_{10}H_{10}INO_3S$, 6-(p-Iodobenzenesulphonyl)-3-oxa-6-azabicyclo[3.1.0]-hexane, 31B, 154
$C_{10}H_{10}N_2O_3$, 1-Acetyl-4,6-dimethylisoxazolo[3,4-b]pyridin-3(1H)-one, 45B, 399
$C_{10}H_{10}N_2O_3S$, 4-Methyl-3-tolylsulphonylfurazan, 42B, 275
$C_{10}H_{12}BF_4N_3O$, 4-Morpholinobenzenediazonium tetrafluoroborate, 46B, 376
$C_{10}H_{12}N_2O_3$, cis-2-Isopropyl-3-(4-nitrophenyl)oxaziridine, 38B, 373
$C_{10}H_{12}N_2O_6$, 6a-Methoxycarbonyl-3-ethoxycarbonyl-4-oxofuro[3,4-c]-2-pyrazoline, 41B, 363
$C_{10}H_{13}ClNO$, racemic-trans-2-Methyl-3-(2,6-dimethyl-4-chlorophenyl)-oxaziridine, 32B, 192
$C_{10}H_{14}N_2O_5$, 3-Amino-3-(3-ethoxy-5-oxo-2-isoxazoline-4-ylidene)propionate, 43B, 14
$C_{10}H_{14}N_2O_6$, 2-Amino-3-hydroxy-5-(hydroxymethyl)-8-methyl-7-azachromone dihydrate, 39B, 256
$C_{10}H_{14}N_4O_2$, Decahydropyrazino[2,3-b]pyrazino-1,6'.4,2'.5,3'.8,5-dioxane, 38B, 373
$C_{10}H_{16}N_2O_2$, Perhydrodipyrido[1,2-b;1',2'-e]-1,4,2,5-dioxadiazone, 44B, 336
$C_{10}H_{16}N_2O_2S_5$, Bis(4-morpholinethiocarbonyl) trisulfide, 39B, 257
$C_{10}H_{18}N_2O_2$, 3-Isopropyl-2-isopropylimino-5-methoxy-Δ^4-oxazoline, 45B, 399
$C_{11}H_5BrClN_3O$, 3-Phenyl-5-chloro-7-bromoisoxazolo[4,5-d]pyrimidine, 38B, 374
$C_{11}H_6BrN_3O_2$, 3-Phenyl-7-bromoisoxazolo[4,5-d]pyridazin-4(5H)-one, 38B, 375
$C_{11}H_7BrN_2O_2$, 3-Bromo-2-methylpyrazolo[a](3,1)benzoxazin-5-one, 39B, 208
$C_{11}H_8NO_2$, Spiro(2,3-norbornane(5,5'-dimethyl-oxazolidine-4-oxyl)), 42B, 49
$C_{11}H_9Br_2NO$, 4-Phenyl-6α,7β-dibromo-2,3-oxazabicyclo[3.2.0]hept-1β,5β,6β,7α-3-ene, 39B, 257
$C_{11}H_9Br_2NO$, 4-Phenyl-6β,7α-dibromo-2,3-oxazabicyclo[3.2.0]hept-1β,5β,6α,7β-3-ene, 39B, 257
$C_{11}H_9ClN_2O_3$, 2-Methyl-4-chloromethyl-5-p-nitrophenyloxazole, 28, 495
$C_{11}H_9Cl_2NO$, 4-Phenyl-6β,7α-dichloro-2,3-oxazabicyclo[3.2.0]hept-1β,5β,6α,7β-3-ene, 39B, 258
$C_{11}H_9NO_2$, 3-Methyl-4-benzylideneisoxazoline-5-one, 38B, 376
$C_{11}H_{11}BrN_2O_2$, 3-Cyano-5-p-bromophenyl-N-methoxyisoxazolidine, 40B, 331
$C_{11}H_{13}NOS$, 4-Thiobenzoylmorpholine, 45B, 399
$C_{11}H_{13}NO_2$, N-Benzoylmorpholine, 43B, 453
$C_{11}H_{13}NO_2S$, 4-(2'-Hydroxythiobenzoyl)morpholine, 45B, 400; 46B, 376
$C_{11}H_{13}N_3O_3S$, Sulfisoxazole, 38B, 265
$C_{11}H_{13}N_3O_4$, (E)-Morpholino-p-nitrobenzamidoxime, 46B, 377

$C_{11}H_{15}N_5OS$, 2-Formyl-4-morpholinopyridine thiosemicarbazone, 44B, 337

$C_{11}H_{16}ClNO$, 3-Methyl-2-phenylmorpholine hydrochloride, 40B, 331

$C_{11}H_{20}N_2O_2S_3$, 7,7,9,9-Tetramethyl-8-thiosulfinyliminothio-1,4-dioxa-8-azaspiro[4.5]decane, 43B, 393

$C_{12}H_6N_2O_2$, Acenaphthofuroxan, 39B, 209

$C_{12}H_7BrF_3NO_3$, 2-(m-Bromophenyl)-3-methyl-4-(trifluoroacetyl)oxazolium 5-oxide, 41B, 423

$C_{12}H_8BrNO_4$, 2-(3'-Bromophenyl)-4-acetoxy-6H-1,3-oxazin-6-one, 39B, 259

$C_{12}H_{10}N_2O_3$, 2-(Pyrrol-1-yl)-4,7-dihydro-4,7-epoxyisoindole-1,3(7aH,3aH)-dione, 44B, 337

$C_{12}H_{10}N_2O_6$, 3,7-Diethoxy-4H,8H-benzo[1,2-c:4,5-c']diisoxazole-4,8-dione, 44B, 338

$C_{12}H_{12}ClNO_2S$, 1-(4-Chlorophenyl)-2-morpholino-2-thioxoethanone, 45B, 401

$C_{12}H_{13}N_3O$, 8-Phenyl-3,4,8-triaza-9-oxatricyclo[5.2.1.0$^{1-6}$]dec-3-ene, 39B, 259

$C_{12}H_{13}N_3O_3S$, 3-Phenylthio-8-oxa-1-aza-2,5-diazoniaspiro[4.5]deca-1,3-diene-2,4-diolate, 44B, 338

$C_{12}H_{14}BrN_3O_2$, 4-(4-Bromophenylamino)-3-butylfuroxan, 44B, 339

$C_{12}H_{14}BrN_3O_3$, 2-Methyl-3-phenyl-4-(N-methyl-N-hydroxyamidin)-isoxazolin-5-one hydrobromide, 38B, 377

$C_{12}H_{14}Br_2N_2O$, (E)-6-(Bromomethylene)-5,6-dihydro-4,4-dimethyl-2-phenyl-4H-1,3,4-oxadiazinium bromide, 45B, 401

$C_{12}H_{14}NO_4$, DL-Flavipucine, 44B, 339

$C_{12}H_{14}N_2O_4$, 2-Acetyl-1-methyl-8-nitro-1,2,4,5-tetrahydro-3,2-benzoxazepine, 40B, 332

$C_{12}H_{15}N_3O_2S$, Salicylaldehyde-4-morpholinothiosemicarbazone, 43B, 454

$C_{12}H_{16}INO_2$, 1,2,3,4,4a,10a-Hexahydro-4a-hydroxyphenoxazine hydroiodide, 42B, 276

$C_{12}H_{16}N_4OS$, 1-Methylthio-3-(α-morpholino)benzylidenetriazene, 45B, 402

$C_{12}H_{18}N_2O_3S$, 3-Cyanomethylsulphonyl-2-morpholinocyclohexene, 42B, 277

$C_{12}H_{21}N_3O_3$, 2,4,7-Trimethylperhydroisoxazolo[2,3-a]pyridine-2,7-dicarbaldehyde dioxime (α form), 46B, 378

$C_{12}H_{22}N_4O_2S$, N^1,N^2-Dimethyl-N^1-morpholinothiocarbonyl-4-morpholinecarboxamidine, 42B, 294

$C_{12}H_{24}N_2O_4$, 1,4-Bis-(4',4'-dimethyloxazolidine-N-oxyl)cyclohexane, 39B, 260

$C_{12}H_{26}N_2O_4$, 1,7,10,16-Tetraoxa-4,13-diazacyclooctadecane, 38B, 377

$C_{12}H_{28}ClNO_6$, 1,4,7,10,13-Pentaoxa-16-azacyclooctadecane hydrochloride monohydrate, 43B, 454

$C_{13}H_{10}N_4O$, N-(3-Phenyl-5-(1,2,3,4-oxatriazolio))-phenylamide, 41B, 423

$C_{13}H_{11}ClN_2O_2$, 3-(p-Chlorophenyl)-6-methyl-2-oxo-2,3,3a,7a-tetrahydro-oxazolo[4,5-b]pyridine, 45B, 403

$C_{13}H_{11}ClN_2O_2$, 3-(p-Chlorophenyl)-7a-methyl-2-oxo-2,3,3a,7a-tetrahydro-oxazolo[4,5-b]pyridine, 45B, 403

$C_{13}H_{12}N_2O_5$, 1-Ethyl-1,4-dihydro-4-oxo-5-amino-6,7-methylenedioxy-3-quinoline-carboxylic acid, 42B, 277

$C_{13}H_{13}ClN_2O_3S$, 1-p-Chlorophenyl-4-(α-D-erythrofuranosyl)-4-imidazoline-2-thione, 40B, 332

$C_{13}H_{13}NO$, [2.2](2,5)Furano(2,5)pyridinophane, 41B, 424

$C_{13}H_{13}NO_2S$, 3-Phenyl-4-(1'-isopropylthiomethylene)isoxazolin-5-one, 43B, 276

$C_{13}H_{14}ClNO_2$, 2-(p-Chlorophenyl)-cis-5,6-trimethylene-2,3,5,6-tetrahydro-1,3-oxazin-4-one, 46B, 378
$C_{13}H_{14}Cl_3NO_3S$, N-Tosyl-2-trichloromethyl-4,5-epoxy-piperidine, 39B, 261
$C_{13}H_{14}N_2O_2$, 3-Phenyl-4-(3-methyl-1-buten-2-yl)sydnone, 44B, 340
$C_{13}H_{15}NOS$, N-Thiocinnamoylmorpholine, 45B, 403
$C_{13}H_{15}NO_2$, 6,7,8,8aα-Tetrahydro-2α-phenyl-5H-oxazolo[3,2-a]pyridin-3(2H)-one, 43B, 455
$C_{14}H_4Cl_2N_2O_3$, 7,10-Dichloroanthraquinoneoxadiazole, 38B, 198
$C_{14}H_5ClN_2O_3$, 4-Chloroanthraquinonoxadiazole, 38B, 379
$C_{14}H_6N_2O_3$, Anthra[1,2-c](1,2,5)oxadiazole-6,11-dione, 38B, 379
$C_{14}H_8Cl_5NO$, 1,7,8,9,10-Pentachloro-3-phenyl-4-aza-5-oxa-tricyclo-[5.2.1.0^{2,6}]deca-3,8-diene, 46B, 379
$C_{14}H_9Cl_4NO$, 1,7,8,9-Tetrachloro-3-phenyl-4-aza-5-oxa-tricyclo-[5.2.1.0^{2,6}]deca-3,8-diene, 46B, 379
$C_{14}H_9NO_2$, 3-Benzoylanthranil, 27, 975
$C_{14}H_{10}N_2O$, 3,4-Diphenyl-1,2,5-oxadiazole, 42B, 301
$C_{14}H_{10}N_2O_2 \cdot 0.25\ C_6H_6$, 4,5-Diphenylisosydnone benzene solvate, 45B, 200
$C_{14}H_{10}N_2O_2$, 3,4-Diphenylfurazan N-oxide, 44B, 340
$C_{14}H_{11}ClN_2O$, 7-Chloro-3-p-tolylbenz-1,2,4-oxadiazine, 42B, 279
$C_{14}H_{11}N_3O$, 3-(p-Aminophenyl)-5-phenyl-1,2,4-oxadiazole, 44B, 341
$C_{14}H_{12}ClNO_3S$, 2-(p-Toluenesulphonyl)-3-(p-chlorophenyl)oxaziridine, 45B, 404
$C_{14}H_{12}N_2O$, 2-Benzylimino-benzoxazoline, 39B, 252
$C_{14}H_{12}N_2O_5S$, trans-2-(p-Toluenesulphonyl)-3-(m-nitrophenyl)oxaziridine, 44B, 341
$C_{14}H_{13}NO_5$, (2S,4R)-4-Ethoxy-2-phthalimido-γ-butyrolactone, 45B, 331
$C_{14}H_{13}N_3O_4$, 5-Methyl-14,16-dioxa-3,5,7-triazapentacyclo-[7.4.3.2^{2,8}.0.0^{3,7}]octadeca-10,12,17-triene-4,6-dione, 46B, 292
$C_{14}H_{14}N_2O_2$, 4-(1-Cyclohexenyl)-3-phenylsydnone, 45B, 404
$C_{14}H_{14}N_2O_3$, 2,3-Epoxy-trans-1,3-dimethyl-4a,9a-diaza-1,2,3,4,4a,9,9a,10-octahydroanthracene-9,10-dione, 44B, 342
$C_{14}H_{16}ClNO$, 2-(p-Chlorophenyl)-cis-4,5-tetramethylene-4,5-dihydro-6H-1,3-oxazine, 45B, 405
$C_{14}H_{16}ClNO$, 2-(p-Chlorophenyl)-cis-5,6-tetramethylene-5,6-dihydro-4H-1,3-oxazine, 45B, 405
$C_{14}H_{18}BrNO$, Hexahydro-cis-(1-H,4a-H)-1-p-bromophenyl-1H,3H-pyrido-[1,2-c](1,3)oxazine, 39B, 261
$C_{14}H_{18}ClNO$, rac-cis-5-Chloro-2,3,3a,9-tetrahydro-6,8,9-trimethyl-1H-pyrrolo[2,1-b]benzoxazine, 45B, 406
$C_{14}H_{18}ClNO_3S$, β-Chloromethylsulphonyl-β-methyl-α-morpholinostyrene (synclinal-5 and 5'), 41B, 425
$C_{14}H_{18}ClNO_3S$, β-Chloromethylsulphonyl-β-methyl-α-morpholinostyrene (antiperiplanar-4), 41B, 425
$C_{14}H_{19}NO_7S$, N-Morpholine-4,5-dimethoxy-2-carboxymethylbenzenesulphonamide, 43B, 456
$C_{14}H_{19}N_3O_3$, 1-(1-Morpholino-1-cyclohexen-6-ylideneammonio)-1-cyano-2-methoxy-2-oxoethanide, 45B, 406
$C_{14}H_{20}N_2O_2$, 4,8-Dimethyl-4a-phenylperhydro[1,4]oxazino[3,2-b]-1,4-oxazine, 46B, 379
$C_{14}H_{20}N_4O_2$, 4-Dimethylamino-5-(dimethylamino)methylene-2-phenylimino-oxazoline monohydrate, 43B, 456
$C_{14}H_{21}NO_7$, 1-Acetyl-trans-3,trans-4-isopropylidenedioxy-cis-4-acetoxymethyl-2-acetoxypyrrolidine, 40B, 333
$C_{14}H_{22}N_2O_2$, 3,7-Dimethyl-1,5-dioxa-3,7-diazacyclo-octane-2,4,6,8-tetraspirocyclopropane, 37B, 193

$C_{14}H_{23}NO$, (1RS,8SR,10SR,4(15)Z)-4-Ethylidene-5-oxa-3-azatricyclo-[8.4.0.0$^{3-8}$]tetradecane, 46B, 380
$C_{14}H_{24}N_2O_4$, 5,6-Dihydro-3,5,5-trimethyl-1,4-oxazin-2-one photoreductive dimer, 43B, 457
$C_{14}H_{25}N_3O_3$, 1,1,2-Trimorpholinoethene, 40B, 232
$C_{15}H_{10}BrNO$, 2-Phenyl-7-bromo-benz[d](1,3)oxazepine, 38B, 380
$C_{15}H_{11}NO$, 3,5-Diphenylisoxazole, 45B, 406
$C_{15}H_{13}NO_6$, Dimethyl 3a,4,9,9a-tetrahydro-9-oxo-cis-furo[3,2-b]quinoline-2,3-dicarboxylate, 46B, 381
$C_{15}H_{14}BrNO$, (+)-2R,3R-2-(S-1-Phenylethyl)-3-p-bromophenyloxaziridine, 45B, 407
$C_{15}H_{14}N_2O$, 2-Benzylimino-3-methyl-benzoxazoline, 39B, 251
$C_{15}H_{16}N_2O_3$, Ethyl 8-phenyl-7-oxa-8,9-diazabicyclo[4.2.1]nona-2,4-diene-9-carboxylate, 43B, 457
$C_{15}H_{18}ClNO$, 2-p-Chlorophenyl-cis-4,5-pentamethylene-4,5-dihydro-1,3-oxazine, 43B, 458
$C_{15}H_{18}ClNO_2$, 2-(p-Chlorophenyl)-cis-5,6-pentamethylene-2,3,5,6-tetrahydro-1,3-oxazin-4-one, 46B, 381
$C_{15}H_{18}ClN_3O$, N-(6-Chlorophthalan-1-ylidene)-1-(methylazo)cyclohexylamine, 37B, 195
$C_{15}H_{19}NO_{10}$, [3aR-(3aα,5α,6α,6aα)]-5-(1,2-Diacetyloxyethyl)-3a,5,6,6a-tetrahydro-6-(acetyloxy)furo[3,2-d]-1-acetyloxazol-2(1H)-one, 45B, 408
$C_{15}H_{20}N_2O_3$, 1-p-Nitrophenyl-3-methylperhydro-2,9-pyridoxazine, 38B, 381
$C_{15}H_{20}N_4OS$, 3-(2-Morpholinoethyl)-4-benzyl-1,2,4-Δ^2-triazoline-5-thione, 45B, 408
$C_{15}H_{23}N_3O_5$, S-(α)-α-Phenylethlyammonium 2-methoxyisoxazolidine-3,3-dicarboxylic acid trans-monomethylamide, 45B, 409
$C_{16}H_8Cl_2N_2O_4$, 3,4-(p-Chlorobenzoyl)-1,2,5-oxadiazole N-oxide, 35B, 233
$C_{16}H_9N_3O_6$, 4-(2,4-Dinitrobenzylidene)-2-phenyloxazolin-5-one, 43B, 459
$C_{16}H_{10}N_2O$, (1)Benzoxepino[2,3-b]quinoxaline, 44B, 342
$C_{16}H_{10}N_2O$, (3)Benzoxepino[1,2-b]quinoxaline, 44B, 342
$C_{16}H_{15}ClN_2O_3$, 8-Chloro-4a,14a-epoxy-1,2,3,4,5,14-hexahydrophthalazino[2,3-b]-phthalazine-7,12-dione, 45B, 342
$C_{16}H_{15}NO_3$, cis-4,5-Dihydroxy-3,4-diphenyl-5-methyl-2-isoxazoline, 46B, 382
$C_{16}H_{15}N_3O$, 5-Cyano-2,3,4,6,7,12-hexahydropyrano[2',3':2,3]azepino-[4,5-b]indole, 43B, 406
$C_{16}H_{17}BrN_2O_4$, Perlolyrine hydrobromide dihydrate, 35B, 233
$C_{16}H_{18}N_2O_5 \cdot 0.32\ H_2O$, 8a-Hydroxy-4-carboxy-perhydrophthalazino[8a,2-bc]benzo[f]-2,3,4,5-tetrahydro(1,4)oxazepin-5-one hydrate, 44B, 342
$C_{16}H_{18}N_2O_5$, 8a-Hydroxy-4-carboxyperhydrophthalazino[8a,2-bc]benzo-[f]-2,3,4,5-tetrahydro(1,4)oxazepin-5-one, 43B, 459
$C_{16}H_{19}KN_2O_7S$, 3,6,9,12,15-Pentaoxa-21-azabicyclo[15.3.1]henicosa-1(21),17,19-triene-2,16-dione potassium thiocyanate, 46B, 326
$C_{16}H_{20}N_2O_4$, 2-(1'-Benzyloxycarbonylamino-1'-methylethyl)-4,4-dimethyl-5-oxazolone, 46B, 382
$C_{16}H_{20}N_4O_6$, 10,22,25,26-Tetraazatricyclo-2,5,8,14,17,20-hexaoxatricyclo-[19.3.1.1$^{9-13}$]hexacosa-1(25),9,11,13(26),21,23-hexaene, 45B, 410
$C_{16}H_{21}NO$, 8,9,10,11,11a,11b,12,13-Octahydro-6H,7aH-benzo-(5,6)(1,3)oxazino[3,4-a]quinoline, 43B, 460
$C_{16}H_{21}NO_5$, (2R,4S,5R)-2-Methoxycarbonyl-2-methoxycarbonylmethyl-3,4-

dimethyl-5-phenyl-1,3-oxazolidine, 45B, 410
$C_{16}H_{21}NO_5$, (2S,4R,5R)-2-Methoxycarbonyl-2-methoxycarbonylmethyl-3,4-dimethyl-5-phenyl-1,3-oxazolidine, 44B, 343
$C_{16}H_{24}N_8O_8$, Acetonitrile oxide octamer, 43B, 461
$C_{16}H_{30}N_2O_2$, Spiro((1'-ethyl-3'-hydroxy-3'-methylpyrrolidine)-2,2'-(1a,3a-dimethyl-4-ethyl-perhydrofuro[3,2-b]pyrrole)), 44B, 343
$C_{17}H_{12}ClN_3O_3$, N-((6-Chloro-4-oxo-4H-3,1-benzoxazin-2-yl)methyleneamino)-N-phenylacetamide, 42B, 279
$C_{17}H_{13}NO_2$, cis-1,5-Diphenyl-6-oxa-4-aza-spiro[2.4]hept-4-en-7-one, 44B, 344
$C_{17}H_{13}NO_2$, trans-1,5-Diphenyl-6-oxa-4-aza-spiro[2.4]hept-4-en-7-one, 44B, 344
$C_{17}H_{14}N_2O_2$, 2-Oxo-1-phenyl-1,5-methano-1,2,4,5,6,11-hexahydro-(1,2,3)oxadiazolo[3,2-a]cinnoline, 44B, 344
$C_{17}H_{15}ClN_2O_2$, 7-Chloro-1,2-dimethyl-2,5-epoxy-5-phenyl-1,2,4,5-tetrahydro-3H-1,4-benzodiazepin-3-one, 37B, 159
$C_{17}H_{15}NO_6$, 7,8-Benzo-3-ethoxycarbonyl-2-methoxycarbonyl-9-methyl-9-aza-5-oxabicyclo[4.3.0]nona-(1(6),2)-dien-4-one, 46B, 275
$C_{17}H_{16}ClN_3O$, 2-Chloro-11-(1-piperazinyl)dibenz[b,f](1,4)oxazepine, 43B, 461
$C_{17}H_{17}NO_2$, (3S,5R)-3-Methyl-5-(4'-biphenylyl)-2,3,5,6-tetrahydro-1,4-oxazin-2-one, 45B, 411
$C_{17}H_{17}NO_3$, 5-Methyl-15-oxa-5-azapentacyclo[$7.4.3.2^{2-8}.0.0^{3-7}$]octadeca-10,12,17-triene-4,6-dione, 46B, 292
$C_{17}H_{18}BrNO$, 2-p-Bromophenyl-3,4-dimethyl-5-phenyloxazolidine, 37B, 116
$C_{17}H_{24}BrNO_6$, Scopolamine N-oxide hydrobromide monohydrate, 37B, 196
$C_{17}H_{32}KN_3O_5S$, 4,7,13,16,21-Pentaoxa-1,10-diazabicyclo-[8.8.5]tricosane-potassium thiocyanate, 44B, 300
$C_{17}H_{32}N_3NaO_5S$, 4,7,13,16,21-Pentaoxa-1,10-diazabicyclo-[8.8.5]tricosane-sodium thiocyanate, 44B, 300
$C_{18}H_{12}N_2O_4$, 3-Phenylisoxazolin-4,5-dione-4-(3'-phenyl-(2'H)Δ(3')-isoxazolin-5'-on-4'-yl)hydrazone, 44B, 345
$C_{18}H_{16}N_2O_3$, 7-Methoxy-6-(2-methyl-3-indolyl)-2H-1,4-benzoxazin-3(4H)-one, 46B, 383
$C_{18}H_{17}N_3O_4$, syn-Methyl-3,7a-diphenyl-7,7a-dihydro-6H-isoxazolo[2,3-d][1,2,4]oxadiazol-6-yl, 46B, 383
$C_{18}H_{18}BrNO$, 2-Phenyl-4-(4-bromophenyl)-5,6-dimethyl-5,6-dihydro-4H-1,3-oxazine, 39B, 262
$C_{18}H_{18}ClN_3O$, 2-Chloro-11-(4-methyl-1-piperazinyl)dibenz[b,f]-(1,4)oxazepine (monoclinic), 43B, 461
$C_{18}H_{18}ClN_3O$, 2-Chloro-11-(4-methyl-1-piperazinyl)dibenz[b,f]-(1,4)oxazepine (orthorhombic), 42B, 223
$C_{18}H_{18}ClN_3O_2$, 10-Chloro-2,3,5,6,7,11b-hexahydro-3-hydroximino-7-methyl-11b-phenyl-oxazolo[3,2-d]benzo-1,4-diazepine, 38B, 382
$C_{18}H_{21}NO_7S$, 7,8-Dicarbomethoxy-9-oxa-2-azabicyclo[4.2.1]non-7-ene 2-p-toluenesulphonamide, 43B, 462
$C_{18}H_{22}N_2O_2$, 5-(Hydroxy(phenyl)amino)-3,3,5-trimethyl-2-phenylisoxazolidine (form A), 40B, 334
$C_{18}H_{22}N_2O_2$, 5-(Hydroxy(phenyl)amino)-3,3,5-trimethyl-2-phenylisoxazolidine (form B), 40B, 334
$C_{18}H_{22}N_2O_7$, 7-(1-p-Nitrobenzyloxycarbonyl)oxy-ethyl)-8-oxo-2,2-dimethyl-3-oxa-1-azabicyclo[4.2.0]octane, 46B, 383
$C_{18}H_{25}Cl_2NO_3$, 1β-Acetoxy-5α,6α-(dichloromethano)-6β-morpholino-8a-β-methyl-1,2,3,5,6,7,8,8a-octahydronaphthalene, 39B, 263
$C_{18}H_{32}N_2O_3$, 5,8-Di-t-butyl-3,3-dimethyl-9-isopropylidene-5,8-diaza-4,7-dioxabicyclo[4.2.1]nonan-2-one, 41B, 427

$C_{18}H_{36}N_2O_6$, 4,7,13,16,21,24-Hexaoxa-1,10-diazabicyclo[8.8.8]hexacosane, 42B, 280
$C_{19}H_{16}N_2O_3$, 5-Diacetylamino-3,4-diphenylisoxazole, 40B, 335
$C_{19}H_{18}BrN_3O_2$, 2-(4-Methyl-2-morpholino-1H-5-imidazolylmethylidene)-4-bromo-1,2-dihydro-1-oxonaphthalene, 40B, 336
$C_{19}H_{20}N_2O$, 2-(N-Morpholinomethyl)-5H-dibenzo[b,f]azepine, 45B, 411
$C_{19}H_{20}N_2O_2$, 2,3:11,12-Dibenzo-1,13-dioxa-5,9-diaza-2,4,9,11-cyclopentadecatetraene, 45B, 412
$C_{19}H_{21}NO_3$, 7-Benzyl-5,6-dimethyl-3,3a,6,6a,7,8-hexahydrofuro[3,4-h]-isoindole-1,9(9aH)-dione, 44B, 306
$C_{19}H_{21}NO_5$, (-)-O-Methyllaurepukin monohydrate, 37B, 170
$C_{19}H_{23}NO_3$, 8,9-Dimethoxy-3-methyl-1-phenyl-3,4,5,6-tetrahydro-1H-2,3-benzoxazocine, 46B, 384
$C_{19}H_{26}N_4O_8$, (2,3:11,12)-Dibenzo-1,13-dioxa-5,9-diazonia-2,11-cyclopentadecadiene dinitrate, 45B, 413
$C_{19}H_{35}ClN_2O_9$, 3,6,9,12,15-Pentaoxa-21-azabicyclo[15.3.1]heneicosa-1(20),17,19-triene t-butylammonium perchlorate, 45B, 413
$C_{20}H_{12}Br_2N_2O$, 2,4-Bis-(4-bromophenyl)benz[d](1,3,6)oxadiazepine, 33B, 146
$C_{20}H_{15}N_3O$, 2,4,6-Triphenyl-4H-1,3,4,5-oxatriazine, 46B, 384
$C_{20}H_{18}N_2O_3$, 9,10-Carbonyldioxy-8-phenyl-5,8a,9,10,10a,10b-hexahydro-6H-cyclobuta(4,5)pyrazolo[3,2-a]isoquinoline, 44B, 345
$C_{20}H_{18}N_2O_4$, 2,4-Dimethyl-9-phenyl-11,13-dioxa-1,9-diaza-1,2,3,4-tetrahydro-1,2-propan-3,4-propanquinoline-10,14-dione, 43B, 463
$C_{20}H_{18}N_2O_6$, cis-3,4-Dihydro-4-morpholinocarbonyl-3-p-nitrophenyl-1H-p-benzopyran-1-one, 44B, 345
$C_{20}H_{19}NOS_2$, 2-Phenylcyclopropane-1-spiro-4'-(2'-benzylthio-4',5'-dihydro-6'H-1',3'-thiazine)-5'-spiro-2"-oxirane, 43B, 463
$C_{20}H_{22}N_2O_8$, 2,2'-(μ_2-(1,4,7-Trioxahepta-1,7-diyl)-5,5'-(μ_2-(1,9-dioxo-2,5,8-trioxanona-1,9-diyl))bis(pyridine), 46B, 385
$C_{20}H_{23}BrN_2O_2$, 2-Benzylamino-2-(1-bromoethyl)-3-benzyl-5-methyloxazolidin-4-one, 46B, 385
$C_{20}H_{23}BrN_2O_3$, 10-Bromo-2,3,5,6,7,11b-hexahydro-2-methyl-11b-phenylbenzo(6,7)-1,4-diazedino[5,4-b]oxazol-6-one-ethanol, 37B, 197
$C_{20}H_{23}NO_6$, 2,6-Dimethylylbenzoic acid-2,6-dimethylylpyridine-18-crown-5, 44B, 346
$C_{20}H_{24}N_2O_3 \cdot 0.5\ H_2O$, 7-Methyl-5,6-dibenzo-1,7-diazatricyclo-[9.3.1.0$^{4-8}$]pentadeca-4(8),5-diene-2-one-spiro-9,2'-dioxolane hemihydrate, 44B, 347
$C_{20}H_{24}N_2O_4$, 5-Methoxyindole-2-yl-(1'-ethylenedioxy-2'-((1"-acetyl-a,7-yl)-piperidine-3"-yl)ethane), 44B, 347
$C_{20}H_{25}N_3O_6$, 4-Benzoyl-5-ethoxy-2-methyl-7,8-methylenedioxy-4,5-dihydro-1H-1,3,4-benzotriazepine hydrate methanol solvate, 42B, 281
$C_{20}H_{26}N_2O_6$, 6,6'-(3,6,9,12-Tetraoxa-tetradeca-1,14-dioxy)-2,2'-bipyridine, 45B, 414
$C_{20}H_{31}N_3O_4S$, (+)-trans-N',O-Cyclohexylidene-N-(2-hydroxycyclohexyl)guanidinium p-toluenesulphonate, 44B, 347
$C_{21}H_{13}BrN_2O_3$, O-(4-Bromobenzoyl)-3-benzoyl-2,1-benzisoxazole oxime(10), 37B, 198
$C_{21}H_{16}N_2O$, 2-Phenyl-4-p-tolylbenz[f]-1,3,5-oxadiazepine, 42B, 281
$C_{21}H_{16}N_2O_4S$, (2-(2-Tosylaminophenyl)-4H-3,1-benzoxazine-4-one, 45B, 415
$C_{21}H_{17}ClN_4O_3$, 8-Chloro-4,11-diacetyl-4,11-dihydro-2-methyl-6-phenyloxazolo[4,5-b](1,4,5)benzotriazocine, 39B, 212
$C_{21}H_{19}NO$, (+)-(2R)-2-((R)-a-Methylbenzyl)-3,3-diphenyloxaziridine, 44B, 348
$C_{21}H_{19}NO$, (-)-(2S)-2-((R)-a-Methylbenzyl)-3,3-diphenyloxaziridine,

44B, 348
$C_{21}H_{24}N_4O_3$, 6-Dimethylamino-5-(4-nitrophenylamino)-3-(2,4,6-trimethylphenyl)-6H-1,2-oxazine, 46B, 386
$C_{21}H_{27}NO_3S$, (+)-(2R)-(3S)-(5R)-2-Ethyl-p-tolylsulphonylmethyl-3,4-dimethyl-5-phenyl-1,3-oxazolidine, 45B, 415
$C_{21}H_{28}BrNO_4$, N-Cyclopropylmethyl-scopolammonium bromide, 45B, 416
$C_{22}H_{21}Br_2N_3O_2$, 4(RS),5'(SR)-trans-3-p-Bromophenyl-4-(3'-p-bromophenyl-2'-isoxazolin-5'-yl)-5-(1-pyrrolidinyl)-2-isoxazoline, 39B, 263
$C_{22}H_{22}N_2O_4$, 4H-5,11,13-Trimethyl-1-p-tolyl-1-aza-3,7-dioxa-dicyclopenta[a,c]quinolin-2,8-dione, 44B, 349
$C_{22}H_{24}BrN_3O_2$, 1-(4-Bromobenzoyl)-2,5-dimethyl-4-morpholino-5-phenyl-4,5-dihydroimidazole, 45B, 417
$C_{22}H_{25}N_3O_2$, 4-Dimethylamino-5,5-dimethyl-2-(α-(N-methylbenzamido)-benzylidene)-Δ^3-1,3-oxazoline, 43B, 464
$C_{22}H_{32}N_2O_4$, 2,2,6,6-Tetramethyl-4-oxopiperidin-1-oxyl formaldehyde condensation product, 43B, 420
$C_{22}H_{33}NO_7S$, Cyclostachine A amine sulfate monohydrate, 41B, 202
$C_{22}H_{37}NO_5$, Diethyl r-1-morpholino-c-4-t-butyl-c-6H-bicyclo-[4.2.0]octane-c-7,t-8-dicarboxylate, 41B, 427
$C_{22}H_{42}N_2O_2$, trans-(8-H,5a-H)-8-Ethyl-cis(7-H,5a-H)-7-methylperhydropyrido[1,2-c](1,3)oxazepine dimer, 42B, 282
$C_{23}H_{21}ClFN_5O_5$, N,N-Dimethyl-(2-chloro-13a-(o-fluorophenyl)-9,12-dimethoxy-11-oxo-10,11,12,13a-tetrahydro-5H,9H-(1,3)oxazolo[3,2-d](1,2,4)triazolo[1,5-a](1,4)benzodiazepin)-7-carboxamide, 44B, 350
$C_{23}H_{33}BrNO_3S$, 1α-(2-p-Bromophenylsulphonylethyl)-4a-methyl-2-morpholino-Δ^2-trans-octalin, 43B, 464
$C_{24}H_{16}N_2O_2$, 2,2'-p-Phenylenebis(5-phenyloxazole), 30B, 152
$C_{24}H_{19}ClN_2O_3$, Spinopyran, 45B, 357
$C_{24}H_{20}N_2O_5$, Methyl 4-(2-furyl)-6,8-dioxo-2,7-diphenyl-3,7-diazabicyclo[3.3.0]octane-2-carboxylate, 45B, 357
$C_{24}H_{22}N_6O_8$, N-(6,7-Methylenedioxy-3-quinazolinio)ethoxy formamidate dimer, 43B, 423
$C_{24}H_{26}N_2O_4$, 2,11-Dimethoxy-5,13-diphenyl-3,10-dioxa-5,13-diazadiamantane, 43B, 465
$C_{24}H_{32}BrNO_2$, 2-p-Bromophenyl-4a,5,5a,6,7,8,9,9a,10,10a-decahydro-5a-methyl-10a-morpholino-4H-naphtho[2,3-b]pyran, 40B, 336
$C_{24}H_{36}N_8O_4$, N,N'-Bis-(2-cyanoethyl)-1,10-diaza-4,7,13,16-tetraoxacyclo-octadecane malononitrile complex, 45B, 417
$C_{25}H_{20}ClNO_{13}$, 6,7-Bis(methoxycarbonyloxy)-1,2,3,4-tetrahydro-isoquinoline-[1,2-c]-oxazol-2-one-[3,4-b]-1-chloro-3-methoxycarbonyloxy-6,7-methylenedioxyindane, 45B, 418
$C_{25}H_{21}Cl_2NO$, 3-(3,5-Dichloro-2,4,6-trimethylphenyl)-4-diphenylmethylene-2-isoxazoline, 42B, 283
$C_{25}H_{21}Cl_2NO$, 3-(3,5-Dichloro-2,4,6-trimethylphenyl)-4-methylene-5,5-diphenyl-2-isoxazoline (form I), 42B, 283
$C_{25}H_{21}Cl_2NO$, 3-(3,5-Dichloro-2,4,6-trimethylphenyl)-4-methylene-5,5-diphenyl-2-isoxazoline (form II), 42B, 283
$C_{25}H_{29}N_3O_2$, 2-Phenyl-4-benzoyl-4a,5,6,7,8,8a-hexahydro-8a-piperidino-4H-1,3,4-benzoxadiazine, 39B, 264
$C_{26}H_{22}BrNO_2$, cis-3-(4-Bromophenyl)-4a,7a-dihydro-6,7a-dimethyl-4a,5-diphenyl-7H-cyclopenta-1,4,2-dioxazine, 45B, 419
$C_{26}H_{23}Cl_2N_3O_4S_2$, 14-(p-Chlorophenyl)-8-(p-chlorophenylsulphonylamino)-13-oxa-14-thia-1,15-diazapentacyclo[10.4.2.0^{2-7}.0^{8-16}.-0^{12-16}]octadeca-2,4,6,14-tetraene s-oxide, 46B, 297
$C_{26}H_{28}Cl_2N_4O_4$, cis-1-Acetyl-4-(4-{[2,4-(dichlorophenyl)-2-(1H-1-imidazolylmethyl)-1,3-dioxolan-4-yl]methoxy}phenyl)piperazine, 45B, 361

$C_{26}H_{30}Br_2N_2O_3$, 4-p-Bromobenzylidene-9-p-bromophenyl-10a-hydroxy-2,7-dimethyl-2,7-diaza-10-oxa-1,2,3,4,5,6,7,8,8a,10a-decahydroanthracene monohydrate, 39B, 265
$C_{28}H_{18}Cl_4N_2O_3$, 7,8,9,10-Tetrachloro-2,5-imino-3-methyl-2,5,11-triphenyl-2,3,4,5-tetrahydro-1,6,3-benzodioxazocine-4-one, 46B, 387
$C_{28}H_{20}N_2O_8$, 3a,5-Bis(methoxycarbonyl)-3-(p-nitrophenyl)-4-oxo-6,6a-diphenyl-3a,4-dihydrocyclopenta[2,3-d]isoxazoline, 44B, 350
$C_{29}H_{38}N_2O_8$, Dextro-moramidinium hydrogen tartrate, 41B, 428
$C_{29}H_{43}NO_4$, 4',5,7-Tri-t-butyl-3'-(2,2-dimethylpropionyl)-1'-methyl-spiro[benzofuran-3(2H)-2'-pyrrolidine]-2,5'-dione, 45B, 253
$C_{30}H_{28}BrN_3O_5$, 1-(Benzyl-(2-p-bromobenzyloxy-carbonyl-amino-propion-amido-methyl)-1H,3H,5H-oxazolo[3,4-a]quinolin-3, 45B, 419
$C_{30}H_{36}IN_3O_3$, 3-Ethoxy-14-(2-quinolyl)-3-azabenzo[d]tricyclo-[5.3.1.1$^{2-8}$]dodecan-13-ylidenemorpholinium iodide monohydrate (triclinic form), 43B, 466
$C_{30}H_{36}IN_3O_3$, 3-Ethoxy-14-(2-quinolyl)-3-azabenzo[d]tricyclo-[5.3.1.1$^{2-8}$]dodecan-13-ylidenemorpholinium iodide monohydrate (orthorhombic form), 43B, 466
$C_{31}H_{22}ClNO_2$, 3-(p-Chlorophenyl)-3a-methyl-4-oxo-5,6,6a-triphenyl-3a,4-dihydrocyclopenta[2,3-d]isoxazoline, 46B, 387
$C_{31}H_{22}N_2O_4$, 1-(4-Methoxyphenyl)-2,3-dioxo-4,6a-diphenyl-6-phenylimino-2,3,6,6a-tetrahydro-1H-furo[3,4-b]pyrrole, 46B, 388
$C_{31}H_{31}NO_2$, 3a,5-Diethyl-4-oxo-6,6a-diphenyl-3-(2,4,6-trimethylphenyl-3a,4-dihydrocyclopenta[2,3-d]isoxazoline, 45B, 420
$C_{32}H_{25}N_3O_6 \cdot 0.5\ C_6H_6$, 5'-(4-Tolylimino)-6,6",7,7"-tetrahydrodispiro(3H,5H-benzo[ij]chinolizine-2(1H),2'-(1',3'-dioxolan)-4',2"(1"H)-3"H,5"h-benzo[ij]chinolizine)-1,1",3,3"-tetrone benzene solvate, 46B, 287
$C_{32}H_{49}NO_3$, 2,4-Di-t-butyl-6-(3,5-di-t-butyl-2-hydroxyphenyl)-6-morpholinocyclohexa-2,4-dienone, 44B, 351
$C_{35}H_{24}BrNO$, 2,4,5,7-Tetraphenyl-6-(4-bromophenyl)-1,3-oxazepine, 38B, 383
$C_{36}H_{24}ClNO_2$, 3-(p-Chlorophenyl)-4-oxo-3a,5,6,6a-tetraphenyl-3a,4-dihydrocyclopenta[2,3-d]isoxazoline, 44B, 351
$C_{54}H_{108}N_6Na_3O_{18}Sb_7$, Tris(sodium-4,7,13,16,21,24-hexaoxa-1,10-diazobicyclo[8.8.8]hexacosane) heptaantimonide, 41B, 429

41. HETERO-(NITROGEN AND SULFUR)

$CAsF_5N_2OS_2$, (5-Oxo-1,3λ^4,2,4-dithiadiazolyl) arsenic pentafluoride, 46B, 388
CN_2OS_2, 5-Oxo-1,3,2,4-dithiadiazole, 44B, 352
CH_2N_4S, 5-Amino-1,2,3,4-thiatriazole, 45B, 420
$C_2Cl_4N_2S_2$, 4-Trichloromethyl-1,2-dithia-3,5-diazolium chloride, 43B, 470
$C_2H_2N_2OS_2$, Rhodan hydrate, 31B, 155
$C_2H_2N_2S$, 1,2,4-Thiadiazole (gas-mw), 42B, 1036
$C_2H_2N_2S$, 1,2,5-Thiadiazole (gas-ed), 26, 760; 29, 733
$C_2H_2N_2S$, 1,2,5-Thiadiazole (gas-mw), 28, 664
$C_2H_2N_2S$, 1,3,4-Thiadiazole (gas-ed), 35B, 889
$C_2H_2N_2S$, 1,3,4-Thiadiazole (gas-mw), 37B, 697
$C_2H_2N_2S$, 1,3,4-Thiadiazole, 40B, 337
$C_2H_2N_2S_3$, Xanthan hydride, 28, 497
$C_2H_2N_2S_3$, 2,5-Dimercaptothiadiazole, 42B, 284
$C_2H_2N_6O_3S$, 5-Azido-2H-1,2,4,6-thiatriazin-3(4H)-one 1,1-dioxide, 46B, 389

$C_2H_3NOS_7$, N-Acetyl-octacycloheptathioimide, 44B, 352
$C_2H_3N_3S_2$, 5-Amino-2-thiol-1,3,4-thiadiazole, 38B, 384
$C_2H_4BrN_3S_2$, Thiuret hydrobromide, 30B, 157
$C_2H_4ClN_3S_2 \cdot 0.5\ H_2O$, Thiuret hydrochloride hemihydrate, 31B, 156
$C_2H_4IN_3S_2$, 3,5-Diamino-1,2,4-dithiazolium iodide, 22, 757; 37B, 199
$C_2H_6N_6S_5$, S,S-Dimethylpentasulfur hexanitride, 46B, 389
$C_3Cl_2N_2OS$, 3,5-Dichloro-4H-1,2,6-thiadiazin-4-one, 44B, 352
$C_3H_3NOS_2$, Rhodanine, 27, 952
$C_3H_3NO_2S$, Thiazolidine-2,4-dione, 41B, 430
C_3H_3NS, Thiazole (gas-mw), 37B, 697
$C_3H_3N_5O_3S$, 7-Amino-4H-furazo[3,4-d]-1,2,6-thiadiazine-1,1-dioxide, 41B, 431
$C_3H_4N_2OS$, 2-Imino-4-thiazolidinone, 38B, 385
$C_3H_4N_6O_2S$, 7-Amino-2H,4H-vic-triazolo[4,5-c]-1,2,6-thiadiazine-1,1-dioxide, 41B, 432
$C_3H_5ClN_2OS$, 2-Imino-4-oxo-1,3-thiazolidine hydrochloride, 41B, 433; 42B, 284
$C_3H_5KN_2O_5S$, Potassium 2H-1,2,6-thiadiazine-3,5(4H,6H)-dione 1,1-dioxide monohydrate, 44B, 353
$C_3H_5N_3O_2S$, 5-Amino-2H-1,2,6-thiadiazine 1,1-dioxide, 45B, 421
$C_3H_5N_5O_3S$, 3,5-Diamino-4-hydroxyimino-4H-1,2,6-thiadiazine 1,1-dioxide, 43B, 470
$C_3H_6AsF_6N_5OS_3$, 6,8-Dimethyl-7-oxo-1λ³-thionia-3λ⁴,5λ⁴-dithia-2,4,6,8,9-pentaazabicyclo[3.3.1]nona-2,3,5(9)-triene hexafluroarsenate, 46B, 390
$C_3H_6N_6O_3S$, 7-Amino-2H,4H-vic-triazolo[4,5-c]-1,2,6-thiadiazine-1,1-dioxide monohydrate, 41B, 433
$C_3H_8N_2O_4S_2$, 2-Amino-1,3-dithia-2-aza-1,1,3,3-cyclohexane tetrone, 41B, 434
$C_3H_9N_3O_6S_3$, Methyl sulfimide, 40B, 338
$C_4Cl_4N_4S_4$, 2,3,7,8-Tetrachloro-5,10,11,12-tetrathia-1,4,6,9-tetraazatricyclo[5.3.1.1²⁻⁶]dodeca-3,8-diene, 32B, 193
$C_4F_{12}N_4O_8S_{12}$, Cyclo-tetra(N-trifluoromethylsulphonyl-iminodisulfide), 45B, 421
$C_4H_3N_3O_3S$, 4-Cyano-3-hydroxy-6H-1,2,6-thiadiazine 1,1-dioxide, 45B, 422
$C_4H_4N_4O_2S$, 1-Thia-2,5-diazole-3,4-dicarboxamide, 27, 954
$C_4H_6N_2S$, 2,5-Dimethylthiadiazole, 37B, 200
$C_4H_6N_2S_3$, 2,4-Dimethyl-1,2,4-thiadiazolidine-3,5-dithione, 40B, 338
$C_4H_6N_4O_3S_2$, 5-Acetamido-1,3,4-thiadiazole-2-sulphonamide, 40B, 338
$C_4H_7NO_2S$, (R)-Thiazolidene-4-carboxylic acid, 45B, 368
$C_5H_5NO_3S$, 4-Hydroxymethylisothiazole-3-carboxylic acid, 44B, 353
$C_5H_5N_3OS$, [1,2,4]Triazolo[3,2-b][1,3]thiazin-5-one, 46B, 390
$C_5H_5N_3OS_2$, 2,3,6,7-Tetrahydro-4H-thiazolo[3,2-a]s-triazin-2-on-4-thione, 40B, 263
$C_5H_6N_2O_2S$, 4-Carboxylato-L-thiazolidinehydantoin, 46B, 391
$C_5H_6N_2O_3S$, 2-Amino-4-thiazolidinone-5-acetic acid, 38B, 386
$C_5H_7NOS_2$, 3-Methyl-4-oxo-1,3-thiazine-2-thione, 38B, 386
$C_5H_7N_5O_2S$, 7-Amino-3-methyl-4H-imidazo[2,3-c](1,2,6)thiadiazine 5,5-dioxide, 43B, 471
$C_5H_8N_2O_2S$, 5-Ethyl-2-hydroximino-1,3-thiazolidine-4-one, 44B, 354
$C_5H_8N_2O_4S_4$, Dimethyl 1,2,5,6-tetrathia-3,4-diazacycloheptane-3,4-dicarboxylate, 42B, 285
$C_5H_9NO_3S$, Chondrine, 38B, 387
$C_5H_{10}N_2S_2$, 1,5-Diaza-3,7-dithiabicyclo[3.3.1]nonane, 43B, 471
$C_5H_{11}NO_2S$, 4-Methylthiomorpholine-1,1-dioxide, 34B, 203
$C_5H_{11}NO_3S$, DL-1-Aza-3-thiacyclohexane-6-carboxylic acid monohydrate,

42B, 286
$C_5H_{15}N_7S_5Si$, 1-[S,S-Dimethyl-N-(trimethylsilyl)sulfodiimide]bicyclo-[3.3.1]pentaazatetrathiane, 46B, 389
$C_6H_2N_4O_2S$, (1,2,5)Thiadiazolo[3,4-g]benzofurazan 1-oxide, 43B, 472
$C_6H_4N_2S$, Benzo-2,1,3-thiadiazole, 11, 722; 15, 504
$C_6H_5F_2N_3O_3S_3$, 3,5-Difluoro-1,3,5-trioxo-1-phenylcyclotriazathiane, 35B, 234
$C_6H_5NO_2S_2$, 3,6-Dithia-3,4,5,6-tetrahydrophthalimide, 41B, 396
$C_6H_6N_2O_2S$, 2,3-Dihydro-3-hydroxy-5H-thiazolo[3,2-a]pyrimidin-5-one, 46B, 391
$C_6H_6N_4S_6$, 5,5'-Dithiobis(3-methyl-1,3,4-thiadiazoline-2-thione), 44B, 354
$C_6H_7N_3S$, 5,6-Dimethylimidazo[2,1-b]-1,3,4-thiadiazole, 46B, 392
$C_6H_8BrN_3S \cdot H_2O$, 5,6-Dimethylimidazo[2,1-b]-1,3,4-thiadiazole hydrobromide monohydrate, 46B, 392
$C_6H_8N_2O_2S$, L-Hydantoino(C)-1,4-perhydrothiazine, 46B, 393
$C_6H_8N_2O_2S_4$, N,N'-Dimethyl-3,6-epitetrathio-2,5-piperazinedione, 39B, 266
$C_6H_9NO_3S$, N-Acetyl-4-carboxylato-L-thiazolidine, 46B, 393
$C_6H_9N_5O_2S$, 4-Amino-N^1,N^7-dimethylimidazo[4,5-c]-1,2,6-thiadiazine 1,1-dioxide, 43B, 472
$C_6H_{10}BrNS$, 2,3,4-Trimethylthiazolium bromide, 38B, 388
$C_6H_{14}ClNO_3S$, Cycloalliin hydrochloride monohydrate, 31B, 157
C_7H_4ClNOS, 7-Chloro-1,2-benzoisothiazolin-3-one, 34B, 206
$C_7H_4N_4S$, Tetrazole[1,5-b]benzothiazole, 40B, 339
C_7H_5ClNS, 5-Chloro-2,1-benzisothiazole, 38B, 389
$C_7H_5ClN_2S$, 6-Chloro-4-methyl-1,2,3-benzothiadiazole, 45B, 422
C_7H_5NOS, 1,2-Benzoisothiazolin-3-one, 35B, 235
$C_7H_5NO_3S$, Saccharin (o-sulfobenzoic imide), 33B, 147; 34B, 207
$C_7H_5NS_2$, 2-Mercaptobenzothiazole, 21, 586; 37B, 200
$C_7H_5N_3OS$, 5-Phenyl-1,2,3,4-thiatriazolio-3-oxide, 42B, 286
$C_7H_5N_3S$, 5-Phenyl-1,2,3,4-thiatriazole, 42B, 286
$C_7H_6N_2O_2S_2 \cdot 0.75\ H_2O$, 5-Oxo-6-N-methylformylamino-4,5-dihydro-1,2-dithiolo[4,3-b]pyrrole hydrate, 37B, 201
C_7H_7NSe, 1,2-Benzisoselenazole, 44B, 355
$C_7H_7N_3OS$, trans-3-(7-Dihydrothiazolo[2,3-e]-1,2,3-triazolyl)propenal, 41B, 434
$C_7H_8ClN_3O_4S_2$, Hydrochlorothiazide, 38B, 389
$C_7H_8N_2O_3S$, 2-Imino-3-methyl-5-methoxycarbonylmethylene-1,3-thiazolidin-4-one, 41B, 435
$C_7H_9NO_2S_2$, N-Methyl-1,4-dithiane-2,3-dicarboximide, 46B, 395
$C_7H_9NO_2S_2$, 7-Methyl-1,4-dithia-7-azaspiro[4.4]nonane-6,8-dione, 46B, 394
$C_7H_9N_3O_2S$, 7,8-Dihydro-8-hydroxy-2-methyl-3H,6H-(1,3)thiazino[3,2-b](1,2,4)triazine-3-one, 40B, 340
$C_7H_{10}N_2O_2S$, L-N-Methylhydantoino(C)-1,4-perhydrothiazine, 46B, 393
$C_7H_{11}NO_3S$, 3-Oxo-2-(methoxycarbonylmethyl)-tetrahydro-1,4-thiazine, 45B, 423
$C_7H_{12}BrNOS$, 2-(α-Hydroxyethyl)-3,4-dimethylthiazolium bromide, 40B, 341
$C_7H_{12}N_2O_3S$, Perhydro-trans-thieno[3,4-d]imidazol-2-one S,S-dioxide, 38B, 390
$C_7H_{12}N_2S_2$, 5-Dimethyliminio-4,4-dimethyl-Δ^2-thiazoline-2-thiolate, 43B, 473
$C_7H_{12}N_4S_3$, 2,5-Bis(dimethylamino)-3,4-diazatrithiapentalene, 43B, 474
$C_7H_{15}N_5S_4$, 3,5-Bis(N,N-dimethylthiocarbamoyl)-4-methyl-1,2,4-dithia-

zolidine, 39B, 266
$C_7H_{16}ClNO_5S$, 5-Methyl-1-thia-5-azacyclo-octane 1-oxide perchlorate, 34B, 208
$C_8F_{24}N_8O_4S_4$, Tetrathiazyl tetra-bis-(trifluoromethyl)nitroxide, 37B, 203
$C_8H_6N_2OS$, 4-Phenyl-1,2,3-thiadiazole-3-oxide, 44B, 355
$C_8H_6N_2O_4S_2$, cis-syn-cis-Tetrahydrocyclobuta[1,2-e:4,3-e']bis(1,3)-thiazine-2,4,5,7-(3H,6H)-tetrone, 37B, 203
$C_8H_7ClN_2O_2S$, 7-Chloro-3-methyl-4H-1,2,4-benzothiadiazine-1,1-dioxide, 43B, 474
$C_8H_7NO_2S$, 4H-1,4-Benzothiazine 1,1-dioxide, 40B, 343
$C_8H_7NS_2$, 2-Methylthiobenzothiazole, 27, 980
$C_8H_7NS_2$, 3-Methylbenzothiazoline-2-thione, 26, 719
$C_8H_7NS_2$, 6-Methyl-5-thioformylpyrrolo[2,1-b]thiazole, 41B, 435
$C_8H_7N_3O_3S$, 6a-Hydroxy-5,6,6a,7-tetrahydro-8-thia-1,4-diazacycl-[3.3.2]azin-2,5-dione, 45B, 424
$C_8H_8N_2OS$, (Z)-2-Hydroxylimino-4-methyl-2H-1,3-benzo[e]thiazine, 43B, 476
$C_8H_8N_2S$, 2-Methylaminobenzothiazole, 35B, 236
$C_8H_9NOS_4$, 3-Ethyl-5-(2'-(1',3'-dithiolanylidene))rhodanine, 39B, 267
$C_8H_{12}N_2O_2S$, 3-Hydroxy-1,2,3,6,6aα,7,8aα,8bα-Octahydro-5H-pyrido-[1,2,3-cd]thieno[3,4-d]imidazol-5-one, 42B, 287
$C_8H_{12}N_2S_2$, 4-Dimethylamino-5-methyl-5-vinyl-2-thiazolinethione, 45B, 430
$C_8H_{14}Cl_2N_4S$, 3,5-Dichloro-1,1-diisopropylamino-1H-1λ^4,2,4,6-thiatriazine, 45B, 424
$C_8H_{15}NO_4S$, 8-Methyl-dihydrothiazolo[3,2-a]pyridin-5-one trihydrate, 43B, 475
$C_8H_{20}BrNS$, 3-Methyl-2-t-butylthiazolium bromide, 42B, 287
$C_8H_{20}BrNS$, 3-Methyl-4-t-butylthiazolium bromide, 42B, 287
$C_9H_7Cl_2N_3O_5S_2$, N-Chloroacetylchlorothiazide, 38B, 391
$C_9H_7NO_2S$, 3-Phenyl-1,3-thiazolidine-2,4-dione, 45B, 425
$C_9H_7N_3OS_2$, 5-(4-Methylbenzoylimino)-5H-1,3,2,4-dithiadiazole, 44B, 356
$C_9H_7N_3O_2S$, 4-(2'-Thiazolylazo)pyrocatechol, 44B, 356
$C_9H_8N_2OS$, 1,2-Benzisothiazol-3-yl methyl ketoxime, 39B, 268
$C_9H_8N_2OS$, 2-Amino-5-phenyl-4-thiazolinone, 37B, 204; 38B, 392
$C_9H_9NO_2S$, 3-Methyl-4H-1,4-benzothiazine 1,1-dioxide, 42B, 288
$C_9H_9NS_2$, 3,6-Dimethyl-5-thioformylpyrrolo[2,1-b]thiazole, 40B, 343
$C_9H_9N_3O_2S_2$, Sulfathiazole (polymorph 1), 38B, 393
$C_9H_9N_3O_2S_2$, Sulfathiazole (polymorph 2), 37B, 205
$C_9H_9N_3O_2S_2$, Sulfathiazole (polymorph 3), 38B, 394
$C_9H_{10}BF_4N_3S$, 3-Ethyl-5-phenyl-1,2,3,4-thiatriazolium tetrafluoroborate, 42B, 286
$C_9H_{10}N_2O_2S$, 2-Methyl-1-oxo-4-phenyl-1-λ^4-1,2,4-thiadiazolidin-3-one, 44B, 357
$C_9H_{10}N_2S$, 2-Phenyliminothiazolidine, 43B, 476
$C_9H_{11}BrN_2OS$, 2-Amino-4-phenylthiazole hydrobromide monohydrate, 40B, 344
$C_9H_{11}N_3OS_4$, Merocyanine, 29, 738
$C_9H_{11}N_3O_3S$, 2-Ethoxycarbonylimino-2H-1,2,4-thiadiazolo[2,3-a]pyridine monohydrate, 42B, 288
$C_9H_{12}N_2OS_3$, 2-Thio-3-ethyl-5-(2'-(3'-methylthiazolidinylidene))-thiazolidine-2,4-dione, 26, 785
$C_9H_{13}NO_4S_2$, 5,7-Dimethyl-1,1,3,3-tetraoxo-4,4a-dihydropyrido[1,2-d]dithiazine, 44B, 357
$C_9H_{15}NO_4S$, N-t-Butyloxycarbonyl-L-thiazolidine-4-carboxylic acid,

42B, 359
$C_9H_{15}NS_2$, 3,4-Diisopropyl-Δ⁴-thiazoline-2-thione, 42B, 288
$C_{10}Cl_6N_2S$, 1,3,4,5,6,8-Hexachlorothieno[2,3-c:5,4-c']dipyridine, 40B, 344
$C_{10}H_4N_4S_2$, Naphtho[1,8-cd:4,5-c'd']bis[1,2,6]thiadiazine, 45B, 425
$C_{10}H_7N_3OS$, 1-Oxo-1,2-dihydro-2,3-diazaphenothiazine, 46B, 396
$C_{10}H_7N_3S$, 2-(4'-Thiazolyl)benzimidazole, 39B, 269
$C_{10}H_8BrNOS_2$, 3-(p-Bromobenzoyl)-1,3-thiazolidine-2-thione, 46B, 396
$C_{10}H_8N_2O_2S$, 1,9-Dihydro(1)benzothiopyrano[4,3-c]pyrazole 5,5-dioxide, 43B, 477
$C_{10}H_8N_4S$, 2-Methyl-5-phenyl-s-triazolo[3,4-b]-1,3,4-thiadiazole, 40B, 345
$C_{10}H_9ClN_2OS$, 2-(2-Chlorobenzoyl)iminothiazolidine, 44B, 358; 45B, 427
$C_{10}H_9NO_3S_2$, 3-Hydroxy-5-(methylsulphonyl)-4-phenylisothiazole, 43B, 477
$C_{10}H_9NO_4S$, trans-2-Carboxy-5-methyl-dihydrothiazolo[3,2-a]pyridinium-3-carboxylate, 37B, 206
$C_{10}H_9N_3O_2S$, 3,4-Dihydro-4-oxo-2,3-diazaphenothiazine monohydrate, 40B, 346
$C_{10}H_{10}BrNO_4S$, 7-Bromo-3-carboxy-2,5-dimethyl-1-oxo-dihydrothiazolo[3,2-a]pyridinium-8-hydroxylate, 37B, 207
$C_{10}H_{10}N_2OS$, 4-Oxo-2-phenyliminoperhydro-1,3-thiazene, 46B, 397
$C_{10}H_{10}N_2OS_2$, Anhydro-2-methylthio-3-amino-4-hydroxy-5-phenyl-thiazolium hydroxide, 32B, 194
$C_{10}H_{10}N_2O_2S$, μ-(S)-syn-(CH_2,CH_3)-9,10-Dioxabimane, 46B, 398
$C_{10}H_{10}N_2O_4S$, μ-(SO_2)-syn-(CH_2,CH_3)-9,10-Dioxabimane, 46B, 398
$C_{10}H_{11}BF_4N_2O_2S$, 3-Ethyl-2-methyl-6-nitrobenzothiazolium tetrafluoroborate, 44B, 358
$C_{10}H_{11}Cl_2N_3S$, 5-Methyl-1,3-thiazolidin-2-one (2,6-dichlorophenyl)-hydrazone, 46B, 398
$C_{10}H_{11}NO_2S$, trans-3,5-Dihydro-3-methyl-4,1-benzothiazepin-2(1H)-one 4-oxide, 34B, 211
$C_{10}H_{11}N_3S_2$, 1-Phenyl-3-(2-thiazolin-2-yl)-2-thiourea, 35B, 237
$C_{10}H_{11}N_3S_2$, 5-Dimethylamino-3-phenylimino-1,2,4-dithiazole decomposition product (1:1 isomeric mixture), 43B, 479
$C_{10}H_{12}BF_4NS$, 3-Ethyl-2-methylbenzothiazolium tetrafluoroborate, 40B, 346
$C_{10}H_{12}N_2OS_3$, 2-Thio-3-allyl-5-(2-(3'-methylthiazolidinylidene))-thiazolidine-2,4-dione, 27, 954
$C_{10}H_{12}N_2S$, 2-Phenylimino-perhydro-1,3-thiazine, 43B, 478
$C_{10}H_{14}N_2OS$, 2-Phenylmethylamino-5-phenyl-thiazolin-4-one, 39B, 270
$C_{10}H_{14}N_2O_4S_2$, Benzenesulphonamide-p-(tetrahydro-2H-1,2-thiazine-2-yl)-S,S-dioxide, 40B, 330
$C_{10}H_{14}N_2O_4S_2$, N-(4'-Sulfamylphenyl)-1,4-butan-sultam, 41B, 436
$C_{10}H_{14}N_2S$, 1,6-Dimethyl-3,4-trimethylene-6a-thia-azophthene, 38B, 394
$C_{10}H_{15}N_3OS$, 5,7-Diaza-3-cyano-3,5,7-trimethyl-6-oxo-1-thia-spiro-[3.5]-nonane, 43B, 436
$C_{10}H_{16}N_2O_2S$, 3,6-Diethyl-1,4-dimethyl-3,6-epithio-2,5-piperazinedione, 46B, 399
$C_{10}H_{16}N_2O_3S$, Biotin, 20, 638
$C_{10}H_{17}NS_2$, 3,4-Diisopropyl-5-methyl-Δ⁴-thiazoline-2-thione, 42B, 289
$C_{10}H_{18}ClNO_3S$, trans-2-Chloro-3-morpholino-2,4,4-trimethylthietane-1,1-dioxide, 37B, 208
$C_{10}H_{18}N_2O_2S$, 2,5-Di-t-butyl-1,2,5-thiadiazolidine-3,4-dione, 45B, 426

$C_{10}H_{18}N_6S_3$, 5-Dimethylamino-3-methylimino-1,2,4-dithiazole decomposition product, 43B, 479
$C_{11}H_7ClN_4O_2S$, 5-(Pyridyl-(2))-3-(5-chloropyridyl-(2))-3H-1,2,3,4-oxathiadiazole-S-oxide, 35B, 232
$C_{11}H_8ClN_3S$, 1-Chloro-10-methyl-2,3-diazaphenothiazine, 46B, 396
$C_{11}H_8N_2O_3S_2$, D-(-)-Luciferin, 37B, 208; 39B, 270
$C_{11}H_8N_2S$, 6-Phenyl-imidazo[2,1-b]thiazole, 38B, 395
$C_{11}H_9N_3S$, 10-Methyl-2,3-diazaphenothiazine, 40B, 347
$C_{11}H_{10}N_2OS_2$, 1-Methyl-3-phenyl-2,7-dithia-4,5-diazabicyclo-[3.3.0]oct-3-en-6-one, 44B, 359
$C_{11}H_{10}N_2O_2S$, 2-Benzoylimino-3-methylthiazolid-5-one, 26, 717
$C_{11}H_{11}NO_2S_3$, 4-Aza-2,7-dithio-3-thioxo-4-phenyl-bicyclo-[3.3.0]octane-7,7-dioxide, 45B, 426
$C_{11}H_{12}BrNS$, N-Benzyl-4-methylthiazolium bromide, 35B, 239
$C_{11}H_{12}Cl_2N_2S$, 2-(2,6-Dichlorophenyl)imino-3-methylperhydro-1,3-thiazine, 46B, 400
$C_{11}H_{12}N_2OS$, 2-(2-Methylbenzoyl)aminothiazoline, 45B, 427
$C_{11}H_{12}N_2OS$, 2-Benzoylimino-3-methyl-1,3-thiazolidine, 46B, 400
$C_{11}H_{12}N_2S$, 2,3,5,6-Tetrahydro-6-phenyl-imidazo[2,1-b]thiazole, 38B, 396
$C_{11}H_{13}ClN_2S$, 2,3,6,7-Tetrahydro-6-phenyl-5H-imidazo[2,1-b]thiazolium chloride, 39B, 271
$C_{11}H_{13}NO_5S$, N-Methyl-4H-6,7-dimethoxy-benzo-1,2-thiazine-3-one 1,1-dioxide, 46B, 401
$C_{11}H_{13}NO_5S$, 3-Methyl-2,4-dicarbomethoxy-Δ^3-cephem, 40B, 348
$C_{11}H_{13}N_3O_2S_2$, 2,4-Dimethyl-5-tosylimino-Δ^3-1,2,3-thiadiazoline, 44B, 359
$C_{11}H_{14}BF_4NOS$, 3-Ethyl-6-methoxy-2-methylbenzothiazolium tetrafluoroborate, 44B, 360
$C_{11}H_{14}N_2O_3S$, 5-Methoxycarbonylmethylene-2-piperidino-Δ^2-thiazolin-4-one, 37B, 209
$C_{11}H_{14}N_2S$, 2-(2,6-Dimethylphenyl)iminothiazolidine, 43B, 479
$C_{11}H_{15}N_3S$, 1,3-Thiazolidin-2-one (2,6-dimethylphenyl)hydrazone, 46B, 398
$C_{11}H_{16}ClN_3OS$, 3,4-Dihydro-7-chloro-6-diethylamino-2H,8H-pyrimido-[2,1-b]thiazine-8-one, 40B, 348
$C_{11}H_{18}N_2O_3S \cdot 0.5\ HI$, 6$\beta$-Trimethylammoniopenicillanate hemihydriodide, 40B, 349
$C_{12}H_8ClNO_2S_2$, N-(2-Chlorophenyl)-3,6-dithiacyclohexene-1,2-dicarboximide, 45B, 428
$C_{12}H_8ClNO_2S_2$, N-(4-Chlorophenyl)-3,6-dithiacyclohexene-1,2-dicarboximide, 45B, 428
$C_{12}H_8N_2O_3S$, Methyl 2-oxopyrimido[2,1-b]benzothiazole-4-carboxylate, 43B, 480
$C_{12}H_9NS$, Phenothiazine, 42B, 289
$C_{12}H_9NSe$, Phenoselenazine, 42B, 260
$C_{12}H_{10}N_2O_7S$, Berninamycinic acid dihydrate, 42B, 290
$C_{12}H_{10}N_2S_3$, 5,6-Dihydro-3-thiobenzoylmethylene-3H-thiazolo[2,3-c]-[1,2,4]thiadiazole, 46B, 401
$C_{12}H_{11}N_3O_2S$, 3-Hydroxy-N-(5-methyl-1,3,4-thiadiazol-2-yl)-3-phenyl-2-propenamide, 46B, 402
$C_{12}H_{12}N_2O_4S_2$, 4,4'-Diacetoxy-5,5'-dimethyl-2,2'-bithiazolyl, 37B, 210
$C_{12}H_{12}N_2S_2$, 3,4-Dimethyl-6-phenyl-5,6-diaza-1,6a-dithiapentalene, 43B, 481
$C_{12}H_{15}NO_2S_2$, 2,9-Bis(p-methoxyphenyl)-5,12-dimethyl-5,12-diaza-1,3,8,10-tetrathiacyclotetradecane-6,13-dione, 45B, 429

$C_{12}H_{16}N_2O_2S_2$, 2-(2,6-Dimethylphenyl)imino-3-mesyl-thiazolidine, 44B, 361
$C_{12}H_{16}N_2S$, 2-(2,6-Dimethylphenyl)imino-3-methyl-thiazolidine, 46B, 402
$C_{12}H_{16}N_2S$, 2-(2,6-Dimethylphenyl)imino-1,3-thiazine, 43B, 481
$C_{12}H_{16}N_4OS$, Thiamine, 44B, 361
$C_{12}H_{17}NO_4S$, Dimethyl 1,3-dimethyl-8-thia-2-azabicyclo[3.2.1]oct-3-ene-4,7-dicarboxylate, 38B, 397
$C_{12}H_{17}N_5S_4$, 3,5-Bis(N,N-dimethylthiocarbamoyl)-4-phenyl-1,2,4-dithiazolidine, 39B, 266
$C_{12}H_{18}Br_2N_4OS \cdot 0.5\ H_2O$, Thiamine bromide hydrobromide hemihydrate, 43B, 482
$C_{12}H_{18}Cl_4CuN_4OS$, Thiamine tetrachlorocuprate, 40B, 553
$C_{12}H_{18}N_2O_3S$, 2,3-Dihydro-2-acetyl-3,3-dimethyl-4-[(1E)-4-methyl-1,3-pentadienyl]-1,2,5-thiadiazole 1,1-dioxide, 46B, 403
$C_{12}H_{18}N_4O_3S$, 2,7-Dimethylthiachromine-8-ethanol dihydrate, 40B, 350
$C_{12}H_{19}N_2O_4PS$, N-(2,6-Dimethylphenyl)-2-amino-5,6-dihydro-4H-1,3-thiazine phosphate, 45B, 429
$C_{12}H_{20}Cl_2N_4O_2S$, Thiamine chloride hydrochloride monohydrate, 27, 994
$C_{13}H_8BrN_3OS$, 1-(2-Thiazolylazo)-6-bromo-2-naphthol, 42B, 290
$C_{13}H_8F_3NS$, 2-(Trifluoromethyl)phenothiazine, 42B, 291
$C_{13}H_9BBrF_4NS$, 3-(p-Bromophenyl)thiazolo[3,2-a]pyridinium tetrafluoroborate, 46B, 403
$C_{13}H_9NOS$, 2-(o-Hydroxyphenyl)benzothiazole, 35B, 241
$C_{13}H_9N_3OS$, 1-(2-Thiazolylazo)-2-naphthol, 42B, 291
$C_{13}H_{11}NOS$, 2-Methoxyphenothiazine, 41B, 437
$C_{13}H_{11}NO_2S_2$, N-(2-Methylphenyl)-3,6-dithiacyclohexene-1,2-dicarboximide, 46B, 404
$C_{13}H_{11}NO_3S_2$, N-(4-Methoxyphenyl)-3,6-dithiacyclohexene-1,2-dicarboximide, 46B, 404
$C_{13}H_{11}NS$, N-Methylphenothiazine, 37B, 211; 40B, 350
$C_{13}H_{12}ClNO_3S$, 2-Oxoindoline-3-spiro-2'-(3'-chloro-5',5'-dimethyl-4'-oxothiolane-1'-oxide), 37B, 181
$C_{13}H_{12}ClN_3O_2S$, 7-Chloro-1,3,5-trimethyl-5H-pyrimido[5,4-b](1,4)benzothiazine-2,4(1H,3H)-dione, 38B, 398
$C_{13}H_{12}N_2O_2S$, 5,6-Dihydro-6-(2-oxo-propyl)-4H-pyrrolo[1,2-a]thieno-[3,2-f][1,4-diazepine]-4-one, 45B, 380
$C_{13}H_{14}N_2O_2S$, 2-Amino-4,5-dihydro-7,8-dimethoxynaphtho[1,2-d]thiazole, 43B, 483
$C_{13}H_{15}NO_3S$, N-Acetyl-2-(p-tolyl)thiazolidine-4-carboxylic acid, 42B, 292
$C_{13}H_{15}N_3O_6S_2$, p-(N-(1,3-Thiazol-2-ylidene)sulfamoyl)succinanilic acid monohydrate, 44B, 361
$C_{13}H_{16}N_2O_2S$, 1-Benzyl-1,4,4-trimethyl-1,2,6-thia(IV)diazine-3,5(4H)-dione, 39B, 272
$C_{13}H_{16}N_2S_2$, 4-Dimethylamino-5-ethyl-5-phenyl-2-thiazolinethione, 45B, 430
$C_{13}H_{17}N_3OS$, 2-Methyl-4-phenyl-5-morpholino-Δ^3-1,2,3-thiadiazoline, 46B, 404
$C_{13}H_{17}N_3O_3S$, 3-Dimethylamino-4,4-dimethyl-5,6-dihydro-4H-1,2,5-benzothiadiazocin-6-one-1,1-dioxide, 43B, 483
$C_{13}H_{18}N_2O_2S$, 2-(N-2,6-Dimethylphenyl-N-mesyl-amino)-dihydro-Δ^2-1,3-thiazine, 43B, 484
$C_{13}H_{18}N_2O_2S$, 3-Diisopropylamino-benzisothiazol-1,1-dioxide, 34B, 204
$C_{13}H_{18}N_2O_2S_2$, 2-(2,6-Dimethylphenyl)imino-3-mesyl-1,3-thiazine, 44B, 361
$C_{13}H_{18}N_2S$, 2-(2,6-Dimethylphenyl)imino-3-methylperhydro-1,3-thia-

zine, 46B, 400
$C_{13}H_{18}N_2S$, 2-[N-(2,6-Dimethylphenyl)-N-methylamino]-4,5-dihydro-6H-1,3-thiazine, 46B, 400
$C_{13}H_{19}NO_2S_2$, (-)-14-Hydroxy-2,8-dithial[9](2,5)-pyridinophane-15-methanol, 45B, 430
$C_{13}H_{20}N_2O_5S$, N1'-Methoxycarbonylbiotin methyl ester, 46B, 405
$C_{13}H_{21}IN_2S_2$, 1,1'-Diethyl-2,2'-thiazolinecarbocyanine iodide, 38B, 398
$C_{14}H_5ClN_2O_2S$, 4-Chloroanthraquinonethiadiazole, 39B, 272
$C_{14}H_6N_2O_2S$, Anthraquinonethiadiazole, 39B, 273
$C_{14}H_8N_2S_4$, Bis(1,3-benzothiazole)-2,2'-disulfide, 46B, 406
$C_{14}H_{10}N_2O_2S$, Phenanthro[9,10-c]-1,2,5-thiadiazole 1-oxide hydrate, 40B, 351
$C_{14}H_{10}N_2S$, 2,5-Diphenylthiadiazole, 30B, 156
$C_{14}H_{10}N_2S$, 3,4-Diphenyl-1,2,5-thiadiazole, 42B, 301
$C_{14}H_{10}N_4S_4$, 4-Phenyl-1,2-dithia-3,5-diazole dimer, 46B, 394
$C_{14}H_{11}N_3OS$, 4-Phenyl-3-phenylamino-1,2,4-thiadiazolin-5-one, 45B, 431
$C_{14}H_{12}N_2S$, 3-Phenyl-2-phenylimino-1,3-thiazetidine, 44B, 362
$C_{14}H_{12}N_4O_2S_3$, 4-(6-Methoxy-3-pyridinyl)-5-[(6-methoxy-3-pyridinyl)-imino]-1,2,4-dithiazolidine-3-thione, 46B, 406
$C_{14}H_{12}N_4S$, 5-Imino-4-phenyl-3-phenylamino-4H-1,2,4-thiadiazoline, 44B, 363
$C_{14}H_{13}NO_2S_2$, N-(2,6-Dimethylphenyl)-3,6-dithiacyclohexene-1,2-dicarboximide, 46B, 406
$C_{14}H_{13}NS$, N-Ethylphenothiazine, 41B, 438
$C_{14}H_{13}N_8NaO_4S_3 \cdot NaCl \cdot H_2O$, Sodium 3-[(5-methyl-1,3,4-thiadiazol-2-yl)thiomethyl]-8-oxo-7-(1H-tetrazol-1-ylacetamido)-1-aza-5-thiabicyclo[4.2.0]oct-2-en-2-ylcarboxylate sodium chloride hydrate, 45B, 432
$C_{14}H_{15}N_3OS$, 5-(1-Indol-3-ylethyl)-2-methylamino-Δ^2-thiazolin-4-one, 44B, 237
$C_{14}H_{15}N_3O_3S$, (2Z)-3-(2'-Acetylaminoethyl)-2-nitromethylene-4-phenyl-2,3-dihydrothiazole, 43B, 484
$C_{14}H_{16}N_2OS_2$, p-Diethylaminobenzylidene rhodanine, 46B, 407
$C_{14}H_{16}N_2O_3S_4$, Hyalodendrin tetrasulfide, 43B, 485
$C_{14}H_{16}N_2S_2$, 2-t-Butyl-6-phenyl-5,6-diaza-1,6a-dithiapentalene, 43B, 486
$C_{14}H_{16}N_4S_4$, Tetrasulfur tetranitride bis(norbornadiene), 41B, 439
$C_{14}H_{16}N_4S_4$, 1,4,4a,7a,8,11,11a,14a-Octahydro-1,4:8,11-dimethano-5,14:12,7-dinitrilodibenzo[d,i](1,3,6,8,2,7)tetrathiadiazecine-5,7,12,14-S(IV), 42B, 1021
$C_{14}H_{17}NO_7S$, (5RS,6RS,7SR)-3,7-Dimethyl-4,4,7-tricarbomethoxy-Δ^2-cephem, 40B, 352
$C_{14}H_{18}BrNO_2S$, cis-2,3-(3'-Cyclohexanon-1',2'-ylene)-5-methyl-8-ethoxydihydrothiazolo[3,2-a]pyridinium bromide, 38B, 399
$C_{14}H_{18}BrNS$, Hexahydro-cis-(1-H,4a-H)-1-p-bromophenyl-1H,3H-pyrido-[1,2-c](1,3)thiazine, 39B, 261
$C_{14}H_{18}ClNO_3S$, 3-(2-(2-Chloroethoxy)ethyl)-5-methyl-4-phenyl-Δ^4-thiazoline 1,1-dioxide, 40B, 352
$C_{14}H_{18}N_4S_2$, 1,5-Dimethyl-3-isopropyl-6-phenyl-1,2,3,5-tetrahydro-[1,2,4]thiadiazolo[5,1-e][1,2,4]-thiadiazol-4-S-2-thione, 46B, 407
$C_{14}H_{22}Cl_2N_4O_2S$, DL-2-(α-Hydroxyethyl)thiamine chloride hydrochloride, 40B, 341
$C_{14}H_{26}ClN_5S$, 3-Chloro-5-cyclohexylamino-1-diisopropylamino-1H-1,2,4,6-thia(IV)triazine, 43B, 486
$C_{15}H_9N_3S_3$, 6-Chloropyrid-2-thione cyclization product, 39B, 274

$C_{15}H_{10}N_2S_2$, 3-(3-Aminobenzo[b]thiophen-2-yl)-1,2-benzisothiazole, 45B, 433
$C_{15}H_{10}N_2S_3$, 2,5-Diphenyl-3,4-diaza-6a-thiathiophthene, 39B, 274
$C_{15}H_{11}NO_3S$, cis-6a,11b-Dihydrobenzo[b]thieno[2,3-c]quinolin-6-one 7,7-dioxide, 43B, 441
$C_{15}H_{12}N_2OS \cdot 0.5\ C_2H_6O$, 2-Phenylamino-5-phenylthiazolin-4-one (ethanol solvate), 39B, 276
$C_{15}H_{12}N_2OS$, 2-Phenylamino-5-phenylthiazolin-4-one, 39B, 275
$C_{15}H_{12}N_2OS$, 3-p-Tolyl-4-oxo-5-phenyl-1-thia-2,3-diazoline, 40B, 353
$C_{15}H_{12}N_2S$, Phenothiazine-10-propionitrile, 39B, 276
$C_{15}H_{12}N_4S_3 \cdot 0.7\ H_2O$, 5-(1,2-Diphenylguanidino)-3H-1,2,4-dithiazole-3-thione hydrate, 44B, 363
$C_{15}H_{13}FN_2O_3S_2$, 3-Methyl-4,5-diphenyl-1,2,3-thiadiazolium fluorosulfate, 44B, 364
$C_{15}H_{13}NO_2S$, Phenothiazine-10-propionic acid, 38B, 400
$C_{15}H_{14}N_2O_2S$, 2-Benzyl-1-oxo-4-phenyl-1-λ^4-1,2,4-thiadiazolidin-3-one, 44B, 357
$C_{15}H_{14}N_4S$, 2-Methyl-3-phenyl-5-phenylazo-1,3,4-thiadiazoline, 46B, 408
$C_{15}H_{14}N_4S$, 5,6-Dihydro-4-phenyl-2-phenylazo-4H-1,3,4-thiadiazine, 46B, 408
$C_{15}H_{15}NS$, N-Isopropylphenothiazine, 42B, 293
$C_{15}H_{16}N_2O_2S$, 9,9a-Dihydro-1,2,9,9-tetramethyl-2,9a-epithio-3,10-diketopiperazino[1,2-a]indole, 44B, 364
$C_{15}H_{16}N_2O_2S_2$, 2,2,7,8-Tetramethyl-(3,4)benzo-10,11-dithia-5,8-diazatricyclo[5.2.2.0$^{1-5}$]undec-3-ene-6,9-dione, 44B, 364
$C_{15}H_{16}N_2O_2S_4$, 9,9a-Dihydro-1,2,9,9-tetramethyl-2,9a-epitetrathio-3,10-diketopiperazino[1,2-a]indole, 46B, 408
$C_{15}H_{17}BrN_2O_2S$, DL-2-Amino-7-oxa-3-thia-1-azaspiro[5.5]undec-1-ene p-bromobenzoyl derivative, 40B, 354
$C_{16}H_8N_4S_2$, Bis(quinoxaline)-2,2',3,3'-disulfide, 43B, 487
$C_{16}H_{12}ClNO_2S_2$, 6-Chloro-2-(2-acetoxy-1-ethylidene)-3-(2-thienyl)-4H-1,4-benzothiazine, 46B, 409
$C_{16}H_{12}N_2O_3S$, (Z)-2,3-Dihydro-N-methyl-2-(4-nitrobenzoylmethylene)-benzothiazole, 46B, 409
$C_{16}H_{12}N_4OS$, 5-Anilino-3-oxo-2-phenyl-2,3-dihydro-1H-pyrazolo[3,4-d]-thiazole, 34B, 212
$C_{16}H_{13}NOS$, trans-6a,11b-Dihydro-5-methylbenzo[b]thieno[2,3-c]quinolin-6-one, 43B, 441
$C_{16}H_{13}N_3O_4S$, 1,2-Bis(methoxycarbonyl)-4-cyano-5-methyl-5H(1,4)thiazino[4,3-a]benzimidazole, 45B, 434
$C_{16}H_{14}N_2OS$, 2-Phenylimino-3-methyl-5-phenylthiazolidin-4-one, 39B, 276
$C_{16}H_{15}N_2NaO_6S_2$, Cephalothin sodium salt, 45B, 433
$C_{16}H_{15}N_3O_3S_2$, 2-N-Phenyl-4-N-methyl-5-(N-(p-toluolsulphonyl)imino)-1,2,4-thiadiazolidin-3-one, 42B, 293
$C_{16}H_{15}N_3O_4S$, 1,2-Bis(methoxycarbonyl)-4-cyano-4,4a-dihydro-5-methyl-5H-(1,4)thiazino[4,3-a]benzimidazole, 45B, 434
$C_{16}H_{16}N_2O_4S_2$, 3-(4-Toluenesulphonyl)-2-(4-toluenesulphonylimino)-1,3-thiazetidine, 43B, 488
$C_{16}H_{16}N_4S$, 2,4-Dimethyl-3,5-bis(phenylimino)-1,2,4-thiadiazolidine, 41B, 439; 42B, 294
$C_{16}H_{17}NOS$, exo-7,11-Methano-5,6a,7,7a,10,11,11a-octahydro-benz[c]indeno[5,6-e]thiazine 6-oxide, 38B, 400
$C_{16}H_{18}IN_3S$, Methylene blue iodide, 3, 796
$C_{16}H_{18}I_3N_3S$, 3,3'-Bis(dimethylamino)phenothiazine triiodide, 43B, 488

$C_{16}H_{18}N_2O_2S$, 2-Allylamino-4,5-dihydro-7,8-dimethoxynaphtho[1,2-d]-thiazole, 46B, 410
$C_{16}H_{19}N_3OS \cdot 0.5\ H_2O$, [3aS-(3aα,4β,6aα)]-4-[3-(Indol-3-yl)propyl]-hexahydro-2-oxo-1H-thieno[3,4-d]imidazole hemihydrate, 45B, 385
$C_{16}H_{19}N_3O_2S$, [3aS-(3aα,4β,6aα)]-4-[3-(Indol-3-yl)propyl]hexahydro-2,5-α-dioxo-1H-thieno[3,4-d]-imidazole, 45B, 385
$C_{16}H_{22}N_4S_3$, 2-Diisopropylamino-4-methyl-6-phenyl-3,4,6-triaza-1,6a-dithiapentalenylium-5-thiolate, 43B, 489
$C_{16}H_{28}ClN_3O_5S$, 3,7-Bisdimethylaminophenazothionium chloride pentahydrate, 39B, 277
$C_{17}H_{10}BrNS_2$, 3-(p-Bromophenyl)thiazolo[2,3-a]isoquinolinium-2-thione betaine, 32B, 196
$C_{17}H_{12}ClNO_2S$, (2-Phenyl-4-(p-chlorophenyl)-5-thiazolyl)acetic acid, 44B, 366
$C_{17}H_{13}N_3OS$, 3-(1-Phenyl-5-methoxy-pyrazol-3-yl)-1,2-benzisothiazole, 40B, 354
$C_{17}H_{13}N_3S$, 4,6-Diphenyl-2-methylthieno[3,4-d]-1,2,3-triazole, 45B, 386
$C_{17}H_{14}N_4S_2$, 2-Phenylimino-3-phenyl-4-thioxothiazolo[3,2-a]tetrahydro-s-triazine, 32B, 197
$C_{17}H_{15}NO_2S$, 1,2-Dihydro-1-methyl-2-(5-phenylthiopyran-2-ylidene)pyridine SS-dioxide, 39B, 279
$C_{17}H_{16}BrNO_5S$, N-(p-Bromophenyl)-4,5-dihydro-7,8-dimethoxybenzothiazepine-3-one 1,1-dioxide, 46B, 410
$C_{17}H_{17}N_3OS$, 5-Benzoylimino-2,2-dimethyl-4-phenyl-1,3,4-thiadiazolidine, 45B, 68
$C_{17}H_{18}N_2O_5S$, 5,5-Dimethyl-2-(2-phenoxymethyl-5-oxo-1,3-oxazolin-4-ylidene)-1,3-thiazolidine-4-carboxylic acid methyl ester, 45B, 434
$C_{17}H_{18}N_2O_5S$, 6-Methoxyphenoxymethylanhydropenicillin, 39B, 280
$C_{17}H_{18}N_4S_2$, Methylene blue thiocyanate, 39B, 278
$C_{17}H_{19}ClN_2OS$, 7-Hydroxy-2-chloro-10-(3'-dimethylamino-n-propyl)-phenothiazine, 43B, 490
$C_{17}H_{19}ClN_2S$, 3-Chloro-10-(3'-dimethylamino-n-propyl)phenothiazine, 34B, 213
$C_{17}H_{20}N_2O_3S$, Tricyclic thia-cyclol from [(RS)-2-tritylthiopropionyl]-L-phenylalanyl-L-proline, 46B, 411
$C_{17}H_{21}BrN_2S$, 10-(2'-Dimethylaminopropyl)phenothiazine hydrobromide, 39B, 282
$C_{17}H_{22}N_4S_4$, 3-(N-Diisopropylthiocarbamoylimino)-5-(thiobenzimino)-4-methyl-1,2,4-dithiazolidine, 43B, 442
$C_{17}H_{25}N_3O_4S_2$, 2,5-Di-t-butyl-1-(p-tolylsulphonylimino)-1λ^4,2,5-thiadiazolidine-3,4-dione, 46B, 412
$C_{17}H_{31}N_5S_4$, 3,5-Bis(N,N-diisopropylthiocarbamoylimino)-4-methyl-1,2,4-dithiazolidine, 40B, 322
$C_{18}H_8N_2O_4S_3$, 6a,13a-Epitrithio-6a,7,13a,14-tetrahydro-6H,13H-pyrazino[1,2-a:4,5-a']diindole-6,7,13,14-tetraone, 43B, 491
$C_{18}H_{12}BrNS$, N-(p-Bromophenyl)phenothiazine, 43B, 492
$C_{18}H_{13}NO_2S$, 3-Phenyl-2H-thiapyran[3,2-b]quinoline 1,1-dioxide, 40B, 355
$C_{18}H_{15}N_4PS_3$, 1-((Triphenylphosphoranylidene)amino)-1,3,5-trithia-2,4,6-triazine, 40B, 357
$C_{18}H_{16}Cl_6N_2O_2S_4Te \cdot C_4H_8O_2$, Bis[2,3-dihydro-3-hydroxy-thiazolo[2,3-b]benzothiazolium] hexachlorotellurate(IV) dioxane, 46B, 412
$C_{18}H_{16}N_2OS$, 2,3,4,5-Tetrahydro-6,6-diphenylimidazo[2,1-b]thiazine-7(6H)-one, 46B, 413
$C_{18}H_{17}ClN_2OS$, 2-(2,6-Dimethylphenyl)imino-3-(2-chlorobenzoyl)thiazolidine, 41B, 440

$C_{18}H_{17}Cl_2N_3O_2S$, N-Benzyloxycarbonyl-N'-(2,6-dichlorophenyl)-N-(5-methyl-1,3-thiazolin-2-yl)hydrazine, 46B, 414

$C_{18}H_{18}N_2S$, 6-Phenyl-3,4,6,11-tetrahydro-2H-pyrimido[2,1-c](2,4)benzothiazepine, 43B, 492

$C_{18}H_{18}N_2S_2$, 2-Methylene-3-methylbenzothiazoline dimer, 40B, 1087

$C_{18}H_{18}N_6S_2$, 1,4-Bis(N-ethyl-1,2-dihydrobenzthiazol-2-ylidene)tetrazene, 32B, 198

$C_{18}H_{19}BrN_2S \cdot CH_4O$, 11-Phenyl-3,4,6,11-tetrahydro-2H-pyrimido[2,1-c](2,4)benzothiazepine hydrobromide methanolate, 43B, 492

$C_{18}H_{20}ClF_3N_2S$, Triflupromazine, 40B, 357

$C_{18}H_{20}N_2S_2$, Bis(2,3-dimethylbenzothiazoline), 39B, 281

$C_{18}H_{20}N_4S$, 4-Ethyl-5-ethylimino-2-phenyl-3-phenylimino-1,2,4-thiadiazolidine, 46B, 414

$C_{18}H_{22}BrNO_3S$, α-Heteronium bromide, 41B, 413

$C_{18}H_{23}ClN_2S$, Diethazine, 37B, 211

$C_{18}H_{29}NOS_2$, 2,4,6-Tri-t-butyl-7,8,9-dithiazabicyclo[4.3.0]nona-1(9),2,4-triene 7-oxide, 46B, 415

$C_{18}H_{29}NS_2$, 2,4,6-Tri-t-butyl-7,8,9-dithiazabicyclo[4.3.0]nona-1(9),2,4-triene, 46B, 415

$C_{19}H_{12}N_4S_2$, 5,11-(p-Cyanophenylimino)-5H,11H-dipyrido[2,3-b:2',3'-f][1,5]-dithiocine, 45B, 435

$C_{19}H_{15}NOS$, N-(o-Methoxyphenyl)phenothiazine, 42B, 295

$C_{19}H_{15}NO_2S$, 3-Phenyl-5-methyl-5H-thiapyran[3,2-b]quinoline 1,1-dioxide, 40B, 356

$C_{19}H_{15}NS$, N-Benzylphenothiazine, 43B, 494

$C_{19}H_{16}N_4O_4S$, Azobenzene N-(4,5-bismethoxycarbonylthiazol-2-yl)imide, 43B, 495

$C_{19}H_{17}N_3S_2$, 3-Phenyl-8-phenylthiocarbamoyl-6,7-dihydro-5H-thiazolo-[3,2-a]pyrimidinium-betaine, 41B, 441

$C_{19}H_{18}Cl_3NO_2S$, 1,4-Dihydro-1-methyl-4-(2-methyl-5-phenylthiopyran-4-ylidene)pyridine SS-dioxide chloroform solvate, 39B, 280

$C_{19}H_{20}BrN_5O_2S$, N-(p-Bromophenylcarbamoyl)thiamine anhydride, 37B, 212

$C_{19}H_{20}N_2OS$, 2-(2,6-Dimethylphenyl)imino-3-(2-methylbenzoyl)thiazolidine, 41B, 442

$C_{19}H_{21}ClN_2S$, (+)-Octoclothepin, 43B, 445

$C_{19}H_{21}ClN_2S$, Octoclothepin, 43B, 445

$C_{19}H_{24}N_2OS$, 3-(2-Methoxy-10-phenothiazinyl)-N,N,2-trimethylpropanamine, 46B, 415

$C_{19}H_{25}ClN_2S$, 10-(2'-Methyl-2'-diethylaminoethyl)phenothiazine hydrochloride (Isothazine hydrochloride), 34B, 210; 37B, 212

$C_{19}H_{26}N_2O_4S_2$, (1-(10'-Phenothiazinyl)-prop-2-yl)trimethylammonium methylsulfate, 39B, 282

$C_{19}H_{30}Cl_2N_4O_5S$, DL-2-(α-Hydroxybenzyl)thiamine chloride hydrochloride trihydrate, 43B, 497

$C_{19}H_{35}N_5S_4$, 3,5-Bis(N,N-diisopropylthiocarbamoylimino)-4-isopropyl-1,2,4-dithiazolidine, 43B, 447

$C_{20}H_{15}N_3O_2S$, 2,5-Diphenyl-1-(phenylimino)-1λ^4,2,5-thiadiazolidin-3,4-dione, 43B, 497

$C_{20}H_{19}IN_2S$, (2-(1-Methylquinoline))-(2-(3-methylbenzothiazole))methylmonomethine cyanine iodide, 24, 707

$C_{20}H_{19}N_3O_2S_2$, 3-Benzoylsulfinyl-5-(N-n-butylbenzamido)-1,2,4-thiadiazole, 45B, 435

$C_{20}H_{21}N_3O_3S$, 5-Phenyl-15-thia-3,5,7-triazapentacyclo-[7.4.3.2$^{2-8}$.0.0$^{3-7}$]octadeca-17-ene-4,6,15-trione, 44B, 367

$C_{20}H_{22}N_2O_2S$, 2,2'-Bis(isopropyl)-3,3'-dioxo-1,1'-spiro(3H-2,1-benzazathiole), 44B, 367

$C_{20}H_{24}N_4O_2S$, 2-Ethyl-5-ethylimino-3-p-methoxyphenylimino-4-p-methoxyphenyl-1,2,4-thiadiazolidine, 43B, 498
$C_{20}H_{24}N_4S_3$, 3,5-Bis(phenylthio)-1,1-diisopropylamino-1H-1λ^4,2,4,6-thiatriazine, 45B, 424
$C_{20}H_{25}ClN_2S$, 10-(1,3-Dimethyl-3-piperidylmethyl)phenothiazine hydrochloride, 44B, 368
$C_{20}H_{28}N_4O_2S$, 8-Dimethylamino-5α-t-butyl-5β-cyano-2-(cyano(t-butyl)-methylene)-1β-methyl-3-oxa-9-thia-7-azabicyclo[4.3.0]non-7-ene-4-one, 46B, 416
$C_{20}H_{28}N_4O_2S$, 8-Dimethylamino-5β-t-butyl-5α-cyano-2-(cyano(t-butyl)-methylene)-1β-methyl-3-oxa-9-thia-7-azabicyclo[4.3.0]non-7-ene-4-one, 46B, 416
$C_{21}H_{14}BrN_3OS$, 5-Benzoylimino-2-(4-bromophenyl)-3-phenyl-2,5-dihydro-1,2,4-thiazole, 38B, 402
$C_{21}H_{15}N_3OS$, 2-Benzoylimino-3,5-diphenyl-2,3-dihydro-1,3,4-thiadiazole, 45B, 68
$C_{21}H_{17}N_3S_2$, 4-Benzyl-3,5-bis(phenylimino)-1,2,4-dithiazolidine, 45B, 436
$C_{21}H_{19}N_5OS_2$, 2,4-Bis-(1,2-benzisothiazol-3-yl)3-aminocrotonitrile dimethylformamide solvate, 38B, 402
$C_{21}H_{21}AuCl_2N_2S_2$, 3,3'-Diethylthiacarbocyanine dichloroaurate, 39B, 284
$C_{21}H_{21}BrN_2S_2$, 3,3-Diethylthiacarbocyanine bromide, 23, 720
$C_{21}H_{22}N_2O_8S$, (5RS,6RS,7RS)-7-Phenylacetamino-3-methyl-4,4,7-tricarbomethoxy-Δ^2-cephem, 40B, 358
$C_{21}H_{22}N_2O_8S$, (5SR,6SR,7RS)-7-Phenylacetamino-3-methyl-4,4,7-trimethoxycarbonyl-Δ^2-cephem, 43B, 499
$C_{21}H_{26}ClN_3OS$, 2-Chloro-10-(3-(4-(2-hydroxyethyl)piperazin-1-yl)propyl)phenothiazine, 44B, 369
$C_{21}H_{26}Cl_2F_3N_3S$, 10-[3-(4-Methyl-1-piperazinyl)propyl]-2-trifluoromethylphenothiazine dihydrochloride, 46B, 416
$C_{21}H_{26}N_2O_4S_2$, t-Butyl 7α-methylthio-7-phenylacetamidodeacetoxycephalosporanate, 39B, 285
$C_{21}H_{26}N_2S_2$, Thioridazine, 41B, 442
$C_{22}H_{16}N_2O_3S$, (3-Benzyl-5-(4-nitrophenyl)-2-phenyl-1,3-thiazolio)-4-oxide, 44B, 369
$C_{22}H_{19}N_3O_2S_2$, 3-Phenyl-4-benzyl-5-tosylimino-1,2,4-thiadiazoline, 42B, 296
$C_{22}H_{20}BF_4N_3O_2S$, 1-(Ethylphenylamino)-3,4-dioxo-2,5-diphenyl-1,2,5-thiadiazolidinium tetrafluoroborate, 46B, 412
$C_{22}H_{21}N_3O_5S_2$, [2S-(2α,5α)]-3,3-Dimethyl-6-[[(4-methylphenyl)thio]imino]-7-oxo-4-thia-1-azabicyclo[3.2.0]heptane-2-carboxylic acid p-nitrobenzyl ester, 46B, 417
$C_{22}H_{21}N_3O_5S_4$, 4-Benzyl-3-oxo-2-p-toluenesulphonyl-5-p-toluenesulphonylimino-1,3,2,4-dithiadiazolidine, 44B, 370
$C_{22}H_{24}N_2S$, 2,3,3a,4,6,7-Hexahydro-3-phenyl-7-(phenylmethylene)-2-propylthiopyrano[4,3-c]pyrazole, 45B, 391
$C_{22}H_{26}N_2O_5S$, 3-Methoxy-10-(3'-dimethylaminopropyl)phenothiazine maleate, 39B, 283
$C_{22}H_{28}N_2OS_2 \cdot 0.5\ H_2O$, Oxyprothepine, 40B, 359
$C_{22}H_{29}N_3S_2$, Thiethylperazine, 35B, 242
$C_{22}H_{32}ClN_3O_6S_3$, 2-Chloro-10-[3-(4-methyl-1-piperazinyl)propyl]phenothiazine-methanesulphonic acid (1:2), 45B, 437
$C_{22}H_{32}N_6O_2S_4$, 3,5-Bis(N,N-diisopropylthiocarbamoylimino)-4-(4-nitrophenyl)-1,2,4-dithiazolidine, 41B, 443
$C_{23}H_{17}BrN_2OS$, 9-Phenacyl-3-phenylbenzimidazo[2,1-b]thiazole bromide, 45B, 437

$C_{23}H_{17}NS_2$, 2,5,6-Triphenyl-1,6a-dithia(S(IV))-6-azapentalene, 43B, 499

$C_{23}H_{17}NS_2$, 3,5-Epidithio-2,5-diphenyl-2,4-pentadienylideneaniline, 38B, 403

$C_{23}H_{19}NS$, 5-Benzyl-2-phenyl-4-tolylthiazole, 44B, 370

$C_{23}H_{22}N_2OS_2$, 2-p-Methoxyphenyl-3,4-dibenzyl-1,3,4-thiadiazolidine-5-thione, 30B, 154

$C_{23}H_{25}F_3N_2OS$, α-Flupenthixol, 41B, 416, 1238

$C_{23}H_{25}F_3N_2OS$, β-Flupenthixol, 41B, 417, 1238

$C_{23}H_{25}N_3O_7S$, 4-(p-Nitrobenzyloxycarbonyl)-5,5-dimethyl-2-(methoxy-(phenoxyacetamido)methyl)-Δ^3-thiazoline, 44B, 370

$C_{23}H_{27}BrN_2OS_2$, 3-3-Diethylthiacarbocyanine bromide ethanol solvate, 23, 720

$C_{24}H_{19}BrN_2OS$, 9-(p-Methylphenacyl)-3-phenylbenzimidazo[2,1-b]thiazole bromide, 45B, 437

$C_{24}H_{20}INS_2$, 2,4-Diphenyl-5-(3'-phenyl-2'-thiabutene-4'-yl)-isothiazolium iodide 6, 38B, 404

$C_{24}H_{20}N_2S_2$, Mesionic thiadiazole derivative, 38B, 404

$C_{25}H_{25}IN_2S_2$, 3,3'-Diethylthiatricarbocyanine iodide, 44B, 371

$C_{25}H_{27}BrCl_2N_2O_2S_2$, 5,5'-Dichloro-3,3',9-triethylthiacarbocyanine bromide acetic acid solvate, 40B, 360

$C_{25}H_{27}N_2O_6S$, (+)-7-Benzylidenamino-7-carbomethoxy-4-carbo(p-methoxybenzyloxy)-Δ^3-cephem, 40B, 361

$C_{26}H_{23}N_5S$, 3'-Phenyl-5'-phenylazo-2-pyrrolidinospiro-(1H-indene-1,2'(3'H)-(1,3,4)thiadiazole), 42B, 295

$C_{26}H_{28}N_6OS$, 4a,5,6,7,8,8a-Hexahydro-6-methyl-8a-morpholino-1-phenyl-3-phenyl-azo-1H-pyrido[4,3-e](1,3,4)thiadiazine, 42B, 295

$C_{26}H_{38}Cl_4CoN_4O_2S_2$, 3-(2-Diethylammoniumethoxy)-1,2-benzisothiazole tetrachlorocobaltate, 38B, 406

$C_{26}H_{38}Cl_4CuN_4O_2S_2$, 3-(2-Diethylammoniumethoxy)-1,2-benzisothiazole tetrachlorocuprate(II), 37B, 213

$C_{27}H_{16}F_6N_2S_2$, 10-Methyl-2,2'-bis(trifluoromethyl)-7,10'-biphenothiazine, 43B, 500

$C_{27}H_{22}ClNO_5S$, 4,5-Bis-exo-(methoxycarbonyl)-1,6-diphenyl-3-p-chlorophenyl-3,6-epithiopiperidine-2-one, 44B, 371

$C_{27}H_{25}IN_2S_2$ · X CH_4O, 3,3'-Diethyl-9-phenylthiacarbocyanine iodide methanol solvate, 45B, 438

$C_{27}H_{33}ClN_2O_4S_2$, 5,5',7,7'-Tetramethyl-3,3',9-triethylthiacarbocyanine perchlorate, 40B, 361

$C_{28}H_{16}Br_2N_2S_2$, cis-Δ(2,2')-Bi-(3-p-bromophenyl-2H-1,4-benzothiazine), 37B, 214

$C_{28}H_{16}Br_2N_2S_2$, trans-Δ(2,2')-Bi-(3-p-bromophenyl-2H-1,4-benzothiazine), 37B, 214

$C_{28}H_{21}NS$, 2,2,4,5-Tetraphenyl-2H-1,3-thiazine, 45B, 439

$C_{28}H_{21}N_3O_2S$, 4-[(E)-1,2-Diphenylvinyl]-2,6-diphenyl-1,3,4,5-thiatriazine 1,1-dioxide, 46B, 417

$C_{28}H_{27}F_3N_4OS$, 1-(1-(3-(2-Trifluoromethyl-10-phenothiazinyl)propyl)-4-piperidinyl)-2-benzimidazolinone, 43B, 501

$C_{28}H_{28}N_2O_3S$ · 0.5 H_2O, 3,3'-Diethylbenzthiacarbocyanin p-toluenesulphonate hemihydrate, 41B, 444

$C_{28}H_{29}NOS$, 7,11b-Dihydro-11b-isopropyl-2-methoxy-3,4-diphenyl-2H,6H-[1,3]-thiazino[2,3-a]isoquinoline, 45B, 439

$C_{29}H_{21}NO_2S$ · 0.5 C_6H_6, 1,2-Dihydronaphtho[2,1-d]-3H-indano(2,1-f)-N-phenylsuccinimido[4,3-b]-7α-thia-2β,3β,5α,6β-tetrahydrobicyclo-[2.2.1]hepta-2,5-diene benzene solvate, 42B, 297

$C_{30}H_{22}N_2O_2S_2$ · CH_4O, 8-Benzoyl-15-hydroxy-15-phenyl-6aH,14aH-benzothiazolo[2,3-a]benzothiazino[4,3-c]piperazine methanol (1:1), 46B,

418
$C_{31}H_{33}BrClN_3O_2S$, N,N-Dimethyl-(9-acridinylmethyl)-(3-(2-chlorophenothiazine-10-yl)propyl)ammonium bromide dihydrate, 44B, 372
$C_{31}H_{34}N_4O_8S_3$, Methyl 6β-phenylacetamidopenicillanate β-lactam fused ylide, 42B, 298
$C_{32}H_{23}N_3O_7S$, 3-[4,5-Bis(methoxycarbonyl)isoxazol-3-yl]-4,6-dioxo-syn-cis-syn-1,3,5-triphenylperhydrothieno[3,4-c]pyrrole-1-carbonitrile, 46B, 419
$C_{34}H_{27}NS$, 2-Diphenylmethyl-4,4-diphenyl-6-methyl-4H-3,1-benzothiazine, 43B, 502
$C_{36}H_{25}N_3OS$, cis-3a,6a-Dihydro-3a,5,6,6a-tetraphenyl-2-phenylazo-4H-cyclopentathiazol-4-one, 43B, 503
$C_{36}H_{28}N_2O_2S_2$, 2,2'-Diphenyl-4,4'-di(2-phenyl-2-propenyl)-4,4'-bi-5(4H)thiazolone, 44B, 372
$C_{38}H_{32}Br_2N_{10}O_2S \cdot 0.5\ CHCl_3$, 2,4-Bis(p-bromophenyl)-3,5-bis(2-morpholino-4-quinazolinylimino)-1,2,4-thiadiazolidine chloroform solvate, 44B, 373
$C_{42}H_{30}N_4OS$, Azobenzene N-(cis-3a,6a-dihydro-4-oxo-3a,5,6,6a-tetraphenyl-4H-cyclopentathiazol-2-yl)imide, 43B, 503

42. HETERO-MIXED MISCELLANEOUS

$C_2N_2O_6S_2$, 1,2,3-Oxathiazolo[5,4-d][1,2,3]oxathiazole 2,2,5,5-tetraoxide, 46B, 419
$C_2H_4O_3Se_2$, trans-Ethanediseleninic anhydride, 20, 524
$C_3H_4Cl_2O_3S$, 2,2'-Dichlorotrimethylene sulfite, 33B, 147
$C_3H_6OSe_2$, 1-Oxa-3,5-diselenane, 41B, 445
$C_3H_6O_3S$, Trimethylene sulfite (gas-ed), 35B, 893
$C_3H_6O_3S$, Trimethylene sulfite, 31B, 159
$C_4H_4KNO_4S$, Potassium 6-methyl-1,2,3-oxathiazin-4-on-2,2-dioxide, 41B, 446
$C_4H_6Cl_2OS$, trans-2,3-Dichloro-1,4-thioxane, 32B, 199
$C_4H_7NO_5S$, cis-5-Methyl-trans-5-nitro-1,3,2-dioxathian-2-oxide, 46B, 420
$C_4H_8Br_2SSe$, 1-Thia-4-selenacyclohexane 4,4-dibromide, 32B, 200
$C_4H_8ClIOSe$, 1-Oxa-4-selenacyclohexane - iodine monochloride, 33B, 149
C_4H_8OS, 1,4-Thioxane (gas-ed), 38B, 1066
$C_5H_{10}OS_4$, 1-Oxa-3,5,7,9-tetrathiacyclodecane, 43B, 504
$C_5H_{10}O_3S$, cis,cis-4,6-Dimethyltrimethylene sulfite, 44B, 374
$C_6H_4N_2Se$, 3,4-Benzo-1,2,5-selenodiazole, 11, 721; 15, 503
$C_6H_4O_4S$, Catechol sulfate, 40B, 1087
$C_6H_8O_2S$, 6-Oxa-7-thia-bicyclo[3.2.1]oct-2-ene-7-exo-oxide, 45B, 440
$C_6H_8O_4Se$, 4,4'-Spirobi(4-selena-4-butanolide), 45B, 440
$C_6H_{10}OSSe$, 9-Oxo-3-seleno-7-thiobicyclo[3.3.1]nonane, 43B, 505
$C_6H_{10}O_2S$, 3,9-Dioxa-7-thiabicyclo[3.3.1]nonane, 41B, 448
$C_6H_{10}O_4S_2$, 1,5-Dihydroxy-9-oxa-3,7-dithiabicyclo[3.3.1]nonane-3-oxide, 38B, 407
$C_6H_{12}O_2Se$, 2,4,6-Trimethyl-1,3,5-dioxaselenan, 39B, 285
$C_6H_{12}O_3S$, 1,1,3-Trimethyltrimethylene sulfite, 44B, 374
$C_6H_{12}O_4S$, (2R,4S,6S)-2-Hydroxymethyl-6-methoxy-1,4-oxathian S-oxide, 37B, 215
$C_7H_4Cl_2INO$, N-Chloro-3-aza-3H,2,1-benzoxiodol-1-yl chloride, 41B, 449
$C_7H_5IO_3$, 1,3-Dihydro-1-hydroxy-3-oxo-1,2-benziodoxole, 30B, 157
$C_7H_8OSe_2$, 2,5-Dimethyl-1,2-diselenolo[1,5-b](1,2)oxaselenole, 44B,

375
$C_7H_{10}ClNO_3S$, 6-Chloro-4,5-diethyl-1,2,3-oxathiazine 2,2-dioxide, 44B, 375
$C_8F_{16}O_4S$, 2,2'-Spiro-bis(4,5-difluoro-4,5-bis(trifluoromethyl)-1,3,2-dioxasulfolane), 43B, 101
$C_8H_5NO_2S$, 4-Phenyl-1,3,2-oxathiazol-5-one, 38B, 408
$C_8H_7IO_3$, 1-Methoxy-1,2-benziodoxolin-3-one (α form), 42B, 299
$C_8H_7IO_3$, 1-Methoxy-1,2-benziodoxolin-3-one (β form), 42B, 299
$C_8H_{10}N_2O_2Se$, (1,2,5)Oxaselenazolo[2,3-b](1,2,5)oxaselenazole-7-Se, 38B, 409
$C_8H_{17}NO_2S$, 3-t-Butyl-4c-methyl-2r-oxo-1-oxa-2-thia-3-azacyclohexane, 46B, 420
$C_9H_6Cl_6O_3S$, β-Thiodan, 43B, 506
$C_9H_6Cl_6O_3S$, β-6,7,8,9,10,10-Hexachloro-1,5,5a,6,9,9a-hexahydro-endo-6,9-methano-2,4,3-benzodioxathiepin 3-oxide, 43B, 506
$C_9H_6O_4S$, 3-Oxo-3H-2,1-benzoxathiole-7-carboxylic acid methyl ester, 44B, 376
$C_9H_6O_5S$, 7-Methoxycarbonyl-1,3-dioxo-2,1-benzoxathiole, 46B, 421
$C_9H_7IO_4$, 1-Acetoxy-1,2-benziodoxolin-3-one, 38B, 409
$C_9H_7NO_2Se$, 5-Phenylselenazolidine-2,4-dione, 40B, 364
$C_9H_9NO_4S$, trans-2-p-Nitrophenyl-1,3-oxathiolane-S-oxide, 41B, 451
$C_9H_{10}N_2O_4S$, 2,2'-Anhydro-1-β-D-arabinofuranosyl-2-thiouracil, 46B, 421
$C_{10}H_4N_4Se_2$, Naphtho[1,8-cd:4,5-c'd']bis[1,2,6]selenadiazine, 46B, 422
$C_{10}H_6Cl_2F_6OS$, 1,1-Dichloro-3,3-bis(trifluoromethyl)-5-methyl-[3H-2,1]-benzoxathiole, 45B, 441
$C_{10}H_9ClINO_3$, 1-Chloro-2-carbomethoxymethyl-1,3-dihydro-3-oxo-1,2-benziodazole, 45B, 442
$C_{10}H_{11}NO_3S$, 2-p-Nitrophenyl-1,3-oxathiane, 38B, 413
$C_{10}H_{12}O_4S$, 5-Methyl-2,2,4-triacetyl-1,3-oxathiole, 40B, 364
$C_{10}H_{15}NO_4S$, (4R,6R)-3-Methoxycarbonyl-9,9-dimethyl-8-oxa-4-thia-1-azabicyclo[4.3.0]non-2-ene 4-oxide, 40B, 365
$C_{11}H_6ClNOS$, 7-Chloro-1-azaphenoxathiin, 44B, 376
$C_{11}H_6ClNOS$, 8-Chloro-1-azaphenoxathiin, 44B, 376
$C_{11}H_{12}N_2O_2S$, 3-Benzoylimino-4-methyl-perhydro-1,2,4-oxathiazine, 43B, 507
$C_{11}H_{21}ClO_3S$, 5-trans-Chloro-4,6-di-t-butyl-2-oxo-1,3,2-dioxathiane, 44B, 377
$C_{12}H_7Cl_2NSe$, 3,7-Dichlorophenoselenazine, 40B, 366
$C_{12}H_8OS$, Phenoxthionine, 31B, 159
$C_{12}H_8O_2S$, Phenoxathiin S-oxide, 45B, 442
$C_{12}H_{14}N_2O_4S$, 2,5'-Anhydro-1-(2',3'-O-isopropylidene-β-D-ribofuranosyl)-2-thiouracil, 46B, 422
$C_{12}H_{16}O_6S_8$, Diethylene tetraxanthogen, 40B, 367
$C_{13}H_{10}O_2S$, Thioxanthene 10,10-dioxide, 40B, 367
$C_{14}H_6N_2O_2Se$, Anthraquinoneselenadiazole, 39B, 273
$C_{14}H_8BrIO_4$, 3-Oxo-3H-2,1-benzoxiodol-1-yl o-bromobenzoate (α-form), 42B, 300
$C_{14}H_8BrIO_4$, 3-Oxo-3H-2,1-benzoxiodol-1-yl o-bromobenzoate (β-form), 42B, 300
$C_{14}H_8ClIO_4$, 3-Oxo-3H-2,1-benzoxiodol-1-yl o-chlorobenzoate, 42B, 300
$C_{14}H_8FIO_4$, 3-Oxo-3H-2,1-benzoxiodol-1-yl o-fluorobenzoate, 42B, 300
$C_{14}H_8I_2O_4$, 1-(2'-Iodobenzoyloxy)-1,2-benziodoxolin-3-one, 38B, 410
$C_{14}H_{10}N_2Se$, 3,4-Diphenyl-1,2,5-selenadiazole, 42B, 301
$C_{14}H_{14}N_2O_2S$, 2-Methyl-5-(3-hydroxy-6-methylpyridin-2-yl)-methyl-[1,3]oxathiolo[4,5-b]pyridine, 45B, 442

$C_{14}H_{22}O_3S$, 5,8-Di-t-butyl-3-oxa-4-thiabicyclo[4.2.0]octa-1(8),5-diene 4,4-dioxide, 41B, 453
$C_{15}H_{11}NO_3S$, 1-Benzothiophene[2,3-c]-cis-14,15-dihydroquinolin-6-one sulphone, 41B, 454
$C_{15}H_{12}O_4S$, Thiathiophthene oxygen analogue, 37B, 216
$C_{15}H_{14}O_3S$, 2-(o-Hydroxyphenyl)-1-phenylpropane sulphonic acid sultone, 32B, 201
$C_{15}H_{14}O_3S$, 3,3-Diphenyl-1,2-oxathiolane 2,2-dioxide, 46B, 423
$C_{16}H_{12}F_{12}OS$, 1,7,8,9-Tetramethyl-2,3,5,6-tetrakis(trifluoromethyl)-4-thia-10-oxatetracyclo[5.2.1.0$^{2-6}$.0$^{3-5}$]dec-8-ene, 41B, 410
$C_{16}H_{24}O_4S$, 2',2',4,4,5',5',6,6-Octamethyl-2',4,5',6-tetrahydrofuro[3,4-d]-1,3-oxathiol-2-spiro-3'-furan-4'-(3'H)-one, 43B, 507
$C_{18}H_{14}F_6O_2S$, (2-Isopropyloxyphenyl)(2-hexafluoroisopropyloxyphenyl)-spirosulfurane, 43B, 508
$C_{18}H_{14}IOS$, Phenylphenoxtinium iodide, 39B, 286
$C_{19}H_{26}O_5S_2$, 11,13-Dioxa-1,4-dithia-12,12,18-trimethyl-10,14,17-trioxotrispiro[4.2.0.5.4.2]eicosane, 43B, 447
$C_{20}H_{12}O_2S$, Spiro(2(1H)-naphthalenone-1,2'-naphtho[1,2-d](1,3)oxathiole), 44B, 377
$C_{20}H_{12}O_2S_2$, Spiro(2(1H)-naphthalenone-1,3'-naphtho[1,2-e](1,3,4)oxadithiin), 44B, 378
$C_{20}H_{22}BrNO_4S_2$, 2H-Thiopyran p-bromobenzylester, 38B, 413
$C_{20}H_{24}N_2OS$, 2,2,4,4-Tetramethyl-1,5-diphenyl-8-oxa-3-thia-6,7-diazabicyclo[3.2.1]octane, 41B, 455
$C_{20}H_{28}O_4Se$, 4-Oxo-5,5,7,7-tetramethyltetrahydro-2H-3,1-benzoxaselenole-2-spiro-4',4',6',6'-tetramethylcyclohexane-2',3'-dione, 44B, 379
$C_{26}H_{24}F_{12}O_2S$, 2,2'-Bis(1,1,1,3,3,3-hexafluoro-2-hydroxy-2-propyl)-4,4'-di-t-butyl-diphenyl spirosulfurane, 40B, 369
$C_{26}H_{24}F_{12}O_3S$, 2,2'-Bis(1,1,1,3,3,3-hexafluoro-2-hydroxy-2-propyl)-4,4'-di-t-butyl-diphenyl spirosulfurane oxide, 40B, 369
$C_{28}H_{16}F_{18}O_3S$, 1,1-Bis[1,1,1,3,3,3-hexafluoro-2-phenyl-2-propanolato]-5-methyl-3,3-bis(trifluoromethyl)[3H-2,1-benzoxathiole], 45B, 443
$C_{28}H_{20}O_4S$, 3,3,5,6-Tetraphenyl-2,3-dihydro-1,4-oxathiin-2-one-4,4-dioxide, 38B, 414

43. BARBITURATES

$C_4D_5N_3O_5$, Violuric acid monohydrate, perdeuterated, 29, 721
$C_4H_2N_3O_4Rb$, Rubidium violurate, 21, 612; 30B, 164
$C_4H_3N_3O_5$, 5-Nitrobarbituric acid, 28, 499
$C_4H_4N_2O_3$, Barbituric acid, 28, 501
$C_4H_4N_2O_5$, 5,5-Dihydroxybarbituric acid, 30B, 159
$C_4H_5N_3O_5$, Violuric acid monohydrate, 29, 721
$C_4H_6KN_3O_6$, Potassium violurate dihydrate, 24, 694; 28, 454; 30B, 164
$C_4H_6N_2O_5$, Dialuric acid monohydrate, 20, 625; 30B, 167; 40B, 1088
$C_4H_7N_3O_3$, Ammonium barbiturate, 29, 714
$C_4H_8N_2O_5$, Barbituric acid dihydrate, 26, 764; 43B, 509
$C_4H_9N_3O_8$, 5-Nitrobarbituric acid trihydrate, 29, 719
$C_4H_{10}N_2O_8$, 5,5-Dihydroxybarbituric acid trihydrate, 30B, 160
$C_6H_7KN_2O_3 \cdot 1.67\ H_2O$, Potassium 5-ethylbarbiturate hydrate, 41B, 456
$C_6H_8N_2O_3$, 5-Ethylbarbituric acid, 37B, 217
$C_6H_8N_2O_4$, 5-Hydroxy-5-ethylbarbituric acid, 37B, 217
$C_8H_8LiN_5O_8$, Lithium purpurate dihydrate, 38B, 415
$C_8H_{10}KN_5O_9$, Potassium purpurate trihydrate, 43B, 509

$C_8H_{10}N_4O_{10}$, Alloxantin dihydrate, 30B, 168
$C_8H_{10}N_6O_7$, Ammonium purpurate monohydrate, 43B, 509
$C_8H_{11}KN_2O_3$, Potassium 5,5-diethylbarbiturate, 30B, 163
$C_8H_{11}N_2NaO_3$, Sodium 5,5-diethylbarbiturate, 27, 973; 37B, 217
$C_8H_{12}N_2O_2S$, 1,3-Diethyl-2-thio-barbituric acid, 42B, 302; 43B, 510
$C_8H_{12}N_2O_3$, 1,3-Diethyl-2-oxo-barbituric acid, 41B, 457; 42B, 302
$C_8H_{12}N_2O_3$, 5,5-Diethylbarbituric acid (polymorph I), 34B, 216
$C_8H_{12}N_2O_3$, 5,5-Diethylbarbituric acid (polymorph II), 34B, 216; 44B, 379
$C_8H_{12}N_2O_3$, 5,5-Diethylbarbituric acid (polymorph IV), 37B, 219
$C_8H_{12}N_6O_{12}Sr$, Strontium violurate tetrahydrate, 42B, 567
$C_9H_6NNaO_2S$, Sodium phenylthiazolidinedione, 29, 715
$C_9H_{13}N_2NaO_3$, Sodium 1-methyl-5,5-diethylbarbiturate, 38B, 416
$C_9H_{14}N_2O_3$, 1-Methyl-5,5-diethylbarbituric acid, 39B, 286
$C_9H_{21}N_5O_5$, Guanidinium 5,5-diethylbarbiturate dihydrate, 39B, 287
$C_{10}H_{12}N_2O_3$, 5,5-Diallyl-barbituric acid, 41B, 458
$C_{10}H_{14}N_2O_3$, 5-Allyl-5-isopropylbarbituric acid, 41B, 459
$C_{11}H_{15}BrN_2O_3$, (+)-1-Methyl-5-isopropyl-5-β-bromoallyl-barbituric acid, 42B, 302
$C_{11}H_{16}N_2O_3$, 5-Ethyl-5-(1-methylbutenyl)barbituric acid, 34B, 218
$C_{11}H_{18}N_2O_3$, 5-Ethyl-5-isoamylbarbituric acid, 34B, 220
$C_{11}H_{18}N_2O_3$, 5-Ethyl-5-n-pentylbarbituric acid, 40B, 371
$C_{12}H_{12}N_2O_3$, 5-Ethyl-5-phenylbarbituric acid (form 3), 40B, 371
$C_{12}H_{13}N_3O_6$, 5-Hydroxy-5-(3'-indole)-barbituric acid dihydrate, 40B, 371
$C_{12}H_{14}N_2O_4$, 3-Oxocyclobarbital, 43B, 511
$C_{12}H_{14}N_2O_4$, 5-Ethyl-5-phenylbarbituric acid monohydrate, 39B, 288
$C_{12}H_{14}N_2O_4$, 6-Oxocyclobarbital, 43B, 511
$C_{12}H_{20}N_2O_3$, 5-Ethyl-5-(1,3-dimethylbutyl)barbituric acid, 40B, 372
$C_{12}H_{20}N_2O_3$, 5-Ethyl-5-(3,3-dimethylbutyl)barbituric acid, 37B, 220
$C_{12}H_{23}N_3O_4$, 2-Ethoxyethylammonium 5,5-diethylbarbiturate, 41B, 459
$C_{12}H_{24}N_4O_3$, 2-Dimethylaminoethylammonium 5,5-diethylbarbiturate, 41B, 459
$C_{13}H_{14}N_2O_3$, 5-Ethyl-1-methyl-5-phenylbarbituric acid, 34B, 214
$C_{13}H_{18}N_2O_3$, 5-(1'-Cyclohepten-1'-yl)-5-ethylbarbituric acid, 34B, 222
$C_{14}H_{15}BrN_4O_4$, 5-(6'-Bromo-3'-ethyl-2'-methylbenzimidazolium)barbiturate monohydrate, 30B, 161
$C_{16}H_{15}BrN_2O_3$, 5,5-Diallyl-1-(p-bromophenyl)barbituric acid, 46B, 424
$C_{16}H_{22}N_2O_3$, N-Cyclohexyl-5,5-diallyl-barbituric acid, 41B, 460
$C_{16}H_{28}CaN_4O_9$, Calcium 5,5-diethylbarbiturate trihydrate, 38B, 417
$C_{18}H_{18}N_6O_9$, 5,6-Dihydro-1,3-dimethyl-5,6-di(1',3'-dimethyl-2',4',6'-trioxopyrimid-(5',5')yl)furo[2,3-d]-uracil, 40B, 373; 42B, 303
$C_{22}H_{32}N_2O_3$, 2,4,6-Trioxo-5,5-diallyl-N,N-dicyclohexyl-1,3-diazine, 40B, 375

44. PYRIMIDINES AND PURINES

$C_3H_3N_3O_2$, 6-Azauracil, 40B, 214, 375
C_4HIN_8, 2,4-Diazido-5-iodopyrimidine, 42B, 304
$C_4H_2D_3N_3O \cdot D_2O$, Deuterated cytosine monohydrate, 46B, 424
$C_4H_3BrN_2O_2$, 5-Bromouracil, 41B, 461
$C_4H_3ClN_2$, 2-Chloropyrimidine, 45B, 444
$C_4H_3ClN_2O$, 5-Chloropyrimidin-2-one, 40B, 215
$C_4H_3ClN_2O_2$, 5-Chlorouracil, 41B, 461
$C_4H_3Cl_2N_3$, 2-Amino-4,6-dichloropyrimidine, 11, 716

$C_4H_3Cl_2N_3$, 4-Amino-2,6-dichloropyrimidine, 12, 414
$C_4H_3FN_2O$, 5-Fluoropyrimidine-2-one, 38B, 418
$C_4H_3FN_2O_2$, 5-Fluorouracil, 39B, 289
$C_4H_3IN_2O_2$, 5-Iodouracil, 41B, 462
$C_4H_3N_5O$, 7-Aminofurazano[3,4-d]pyrimidine, 37B, 221
$C_4H_3N_5S$, 7-Amino-1,2,5-thiadiazolo[3,4-d]pyrimidine, 37B, 221
$C_4H_4BrN_3O$, 5-Bromocytosine, 45B, 444
$C_4H_4N_2$, Pyrimidine, 24, 689; 45B, 444
$C_4H_4N_2O$, Pyrimidine-2-one, 35B, 248
$C_4H_4N_2O_2$, Uracil, 18, 758; 32B, 203
$C_4H_4N_2S_2$, 2,4-Dithiouracil, 32B, 204
$C_4H_5BrN_4$, 4,6-Diamino-5-bromopyrimidine, 12, 416
$C_4H_5ClN_2$, Pyrimidine hydrochloride, 41B, 463
$C_4H_5ClN_2O$, Pyrimidin-2-one hydrochloride, 41B, 463
$C_4H_5ClN_4$, 4,5-Diamino-2-chloropyrimidine, 20, 623
$C_4H_5ClN_6$, 8-Azaadenine hydrochloride, 42B, 305
$C_4H_5N_2NaO_5S \cdot H_2O$, Sodium 5,6-dihydrouracicl-6-sulphonate monohydrate, 46B, 425
$C_4H_5N_3$, 2-Aminopyrimidine, 42B, 305; 45B, 444
$C_4H_5N_3O$, Cytosine, 29, 703; 39B, 289
$C_4H_5N_3O$, Isocytosine, 30B, 172
$C_4H_5N_3O_2$, 6-Azathymine, 41B, 269
$C_4H_5N_3O_5$, 5-Nitrouracil monohydrate, 32B, 205
$C_4H_5N_3S$, Thiocytosine, 35B, 244
$C_4H_5N_5O_3$, Xanthazole monohydrate, 19, 610; 40B, 1088
$C_4H_6ClN_3 \cdot 0.5\ H_2O$, 2-Aminopyrimidine hydrochloride hemihydrate, 46B, 425
$C_4H_6ClN_3O$, Cytosine hydrochloride, 43B, 513
$C_4H_6N_2OS$, 5,6-Dihydro-2-thiouracil, 42B, 306
$C_4H_6N_2O_2$, Dihydrouracil, 35B, 249
$C_4H_6N_4O$, N^4-Aminocytosine, 46B, 426
$C_4H_6N_6O_2$, 8-Azaguanine monohydrate, 30B, 138; 33B, 150
$C_4H_7BrN_6O_2$, 8-Azaguanine hydrobromide monohydrate, 41B, 311
$C_4H_7CaCl_2N_3O_2$, Cytosine calcium chloride monohydrate, 46B, 426
$C_4H_7ClN_6O_2$, 8-Azaguanine hydrochloride monohydrate, 40B, 376
$C_4H_7N_2NaO_5S_2$, Sodium 5,6-dihydro-2-thiouracil-6-sulphonate monohydrate, 44B, 380
$C_4H_7N_3O_2$, Cytosine monohydrate, 28, 503; 39B, 289; 42B, 307
$C_4H_7N_3O_2S$, 6-Amino-2-thiouracil monohydrate, 44B, 380
$C_4H_7N_3O_4$, trans-1-Carbamoyl-imidazolidine-4,5-diol, 39B, 291
$C_4H_{14}N_2Na_2O_9S$, Disodium 4-oxopyrimidine-2-sulfinate hexahydrate, 34B, 224
$C_5H_4FN_2O_5Rb$, Rubidium 5-fluoro-orotate monohydrate, 29, 712
$C_5H_4N_4$, Purine, 30B, 173
$C_5H_4N_4O$, Allopurinol, 38B, 420
$C_5H_4N_4O_3$, Uric acid, 31B, 162
$C_5H_4N_4S$, 5H-Pyrazolo[3,4-d]pyrimidine-4-thione (Thiopurinol), 39B, 290; 40B, 377
$C_5H_5ClN_2O$, 6-Chloro-2-methyl-4(3H)-pyrimidone, 39B, 290
$C_5H_5KN_2O_2 \cdot 3\ H_2O$, Potassium thyminate trihydrate, 45B, 445
$C_5H_5N_3O$, anti-4-Pyrimidinecarboxaldehyde oxime, 40B, 378
$C_5H_5N_3O_4$, 5-Nitro-6-methyluracil, 43B, 514
$C_5H_5N_4NaO_4$, Sodium urate monohydrate, 42B, 307
$C_5H_5N_5 \cdot 0.5\ H_2SO_4 \cdot H_2O$, Adeninium hemisulfate hydrate, 44B, 381
$C_5H_5N_5S$, 6-Thioguanine, 35B, 251
$C_5H_5N_7O$, Azidopurine monohydrate, 33B, 151
$C_5H_6BrN_5 \cdot 0.5\ H_2O$, Adenine hydrobromide hemihydrate, 44B, 381

$C_5H_6ClN_3$, 2-Amino-6-chloro-4-methylpyrimidine, 11, 718
$C_5H_6ClN_5 \cdot 0.5\ H_2O$, Adenine hydrochloride hemihydrate, 40B, 379
$C_5H_6Cl_3HgN_5 \cdot 1.5\ H_2O$, Adeninium trichloromercurate(II) sesquihydrate, 41B, 464
$C_5H_6N_2$, 5-Methylpyrimidine, 45B, 444
$C_5H_6N_2OS$, 1-Methyl-4-thiouracil, 41B, 465
$C_5H_6N_2O_2$, Thymine, 34B, 237
$C_5H_6N_2O_5$, Orotic acid monohydrate, 39B, 291
$C_5H_6N_4OS$, 6-Mercaptopurine monohydrate, 34B, 234
$C_5H_6N_6$, 7-Methyl-8-azaadenine, 43B, 354
$C_5H_7BrN_2O_2$, 1-Methyluracil hydrobromide, 29, 705
$C_5H_7ClN_4O_2$, Hypoxanthine hydrochloride monohydrate, 34B, 236
$C_5H_7Cl_2N_5$, Adenine dihydrochloride, 40B, 379
$C_5H_7NO_3 \cdot 0.5\ H_2O$, 3-Methylcytosine hemihydrate, 44B, 381
$C_5H_7N_3O \cdot 0.5\ H_2O$, 5-Methylcytosine hemihydrate, 46B, 427
$C_5H_7N_3O$, 1-Methylcytosine, 43B, 515
$C_5H_7N_5O_2$, Guanine monohydrate, 37B, 222
$C_5H_7N_5O_4S$, Adeninium sulfate, 44B, 382
$C_5H_7N_5O_5S$, Adenine N^1-oxide-sulfuric acid, 38B, 421
$C_5H_8BrN_3O$, 1-Methylcytosine hydrobromide, 27, 971
$C_5H_8BrN_5O_2$, Guanine hydrobromide monohydrate, 43B, 515
$C_5H_8ClN_3O$, 1-Methylcytosine hydrochloride, 38B, 423
$C_5H_8ClN_3O_2$, 1-Methyl-N^4-hydroxycytosine hydrochloride, 45B, 445
$C_5H_8ClN_3O_5$, 1-Methyl-cytosine perchlorate, 45B, 450
$C_5H_8ClN_5O_2$, Guanine hydrochloride monohydrate, 15, 494
$C_5H_8I_3N_3O$, Methylcytosinium triiodide, 44B, 382
$C_5H_8N_2OS$, 5,6-Dihydro-1-methyl-4-thiouracil, 38B, 424
$C_5H_8N_2O_2$, Dihydrothymine, 33B, 154
$C_5H_8N_2O_2$, 6-Methyl-5,6-dihydrouracil, 43B, 516
$C_5H_8N_2O_3$, Thymine monohydrate, 26, 765
$C_5H_8N_2O_4$, cis-Thymine glycol, 39B, 291
$C_5H_8N_4O_5$, 3-Hydroxyxanthine dihydrate, 44B, 383
$C_5H_8N_4S$, 2,5-Diamino-4-mercapto-6-methylpyrimidine, 22, 774
$C_5H_8N_5O_4P$, Adeninium phosphate, 45B, 481
$C_5H_9AuCl_4N_4O_3$, Hypoxanthine tetrachloroaurate(III) dihydrate, 41B, 463
$C_5H_9BrN_6O_2$, 3-Methyl-8-azaguanine hydrobromide monohydrate, 44B, 384
$C_5H_9N_3O_5$, Ammonium ororate monohydrate, 37B, 224
$C_5H_{10}ClN_5O_3$, Guanine hydrochloride dihydrate, 30B, 175
$C_5H_{11}N_4NaO_6$, Sodium xanthine tetrahydrate, 40B, 1088
$C_6H_7BrN_2O_2$, 1-Ethyl-5-bromouracil, 38B, 425
$C_6H_7N_3O_3$, Cytosine-5-acetic acid, 27, 965
$C_6H_7N_5$, N^6-Methyladenine, 39B, 292
$C_6H_7N_5$, 9-Methyladenine, 29, 708; 43B, 517; 46B, 427
$C_6H_8BrN_5O$, 9-Methylguanine hydrobromide, 29, 706
$C_6H_8ClN_5$, N^6-Methyladenine hydrochloride, 44B, 384
$C_6H_8N_2O_2$, 1-Methylthymine, 28, 505; 40B, 380
$C_6H_8N_2O_2$, 1,3-Dimethyluracil, 43B, 517
$C_6H_8N_4OS$, 2-Mercapto-6-methylpurine monohydrate, 33B, 156
$C_6H_9Br_2N_5$, 9-Methyladenine dihydrobromide, 27, 977
$C_6H_9ClN_4O_3$, 7-Methylxanthine hydrochloride monohydrate, 41B, 466
$C_6H_9Cl_2N_5$, 7-Methyladenine dihydrochloride, 41B, 467
$C_6H_9FN_2O_3$, 1-Methyl-5-fluoro-6-methoxy-5,6-dihydrouracil, 42B, 308
$C_6H_{10}ClN_3S$, 4-Amino-1-methyl-2-(methylthio)pyrimidinium chloride, 43B, 518
$C_6H_{12}ClN_5O_3$, 9-Methylisoguanine hydrochloride dihydrate, 44B, 385
$C_6H_{12}N_2O_3$, 2-Hydroxy-4,6-dimethylpyrimidine dihydrate, 11, 719

$C_6H_{12}N_4O_3S$, 6-Methylmercaptopurine trihydrate, 42B, 309
$C_7H_8N_2O_3$, Cyclobutane-1,5-spiro-2,4,6-triketo-hexahydropyrimidine, 30B, 169
$C_7H_8N_2O_4$, 6-Methyluracil-5-acetic acid, 38B, 426
$C_7H_9ClN_4O_2$, Theophylline hydrochloride, 44B, 385
$C_7H_9N_5$, N^6,N^9-Dimethyladenine, 44B, 384
$C_7H_9N_5$, 3-Ethyladenine, 41B, 467
$C_7H_9N_5O \cdot 0.5\ HCl$, 9-Ethylguanine hemihydrochloride, 41B, 468
$C_7H_9N_5O$, 9-Ethylguanine, 40B, 381
$C_7H_{10}ClN_5$, 1,9-Dimethyladeninium chloride, 45B, 446
$C_7H_{10}N_2O_2$, 5-Ethyl-6-methyluracil, 31B, 166
$C_7H_{10}N_4O_3$, Theophylline monohydrate, 22, 777
$C_7H_{10}N_4O_4$, Cytosine - N-formylglycine, 41B, 469
$C_7H_{11}N_3O_3$, N'-(N-Methylcarbamoyl)-N^3-methyl-5,6-dihydrouracil, 39B, 293
$C_7H_{11}N_3O_5S$, 5-S-Cysteinyluracil monohydrate, 43B, 518
$C_8H_6N_4S_2$, Bis(2-pyrimidyl)disulfide, 45B, 1152
$C_8H_7BrN_4O_5$, Bromo-meso-sarcosinuric acid, 35B, 252
$C_8H_7Cl_3N_4$, 4,6-Dichloro-5-cyano-N-methyl-N-β-chloroethyl-2-aminopyrimidine, 45B, 446
$C_8H_7N_5O$, 3-(7-Adeninyl)propionic acid lactam, 45B, 447
$C_8H_7N_5O$, 3-Methyl-3H-imidazo[2,1-i]purine-8(7H)-one, 35B, 254
$C_8H_8Cl_2HgN_4O_4$, Uracil-mercuric chloride, 37B, 225
$C_8H_8N_4O_4$, cis-anti Uracil photodimer, 37B, 226
$C_8H_8N_4O_4$, cis-syn Uracil photodimer, 35B, 256
$C_8H_9KN_6O_4$, Potassium N-(purin-6-ylcarbamoyl)glycine monohydrate, 42B, 309
$C_8H_{10}FN_5O_4$, 5-Fluorouracil - cytosine monohydrate, 34B, 226
$C_8H_{10}N_4O_2S_2$, Bis-pyrimidyl-2,2'-disulfide dihydrate, 39B, 294
$C_8H_{11}Cl_2N_6O_2Zn_{0.5}$, Cytosinium hemitetrachlorozincate-cytosine, 45B, 448
$C_8H_{12}Cl_2HgN_4O_4$, Dihydrouracil-mercuric chloride, 37B, 225
$C_8H_{12}Cl_4CuN_6O_2$, Dicytosinium tetrachlorocuprate, 45B, 448
$C_8H_{12}Cl_4N_6O_2Pd$, Cytosine tetrachloropalladate, 41B, 470
$C_8H_{12}N_4O_3$, Caffeine monohydrate, 22, 779
$C_8H_{13}N_3O$, 1,5,N^4,N^4-Tetramethylcytosine, 43B, 519
$C_8H_{14}N_{14}O_5S$, 8-Aza-2,6-diaminopurine sulfate monohydrate, 41B, 316
$C_8H_{15}ClN_4O_4$, Caffeine hydrochloride dihydrate, 44B, 386
$C_9H_{10}FN_5O_3$, 5-Fluorouracil - 1-methylcytosine, 34B, 224
$C_9H_{10}N_4O_2$, 1,3,9-Trimethyl-2,6-dioxypurine, 39B, 294
$C_9H_{11}N_7O_4S$, Azathioprine dihydrate, 42B, 309
$C_9H_{12}N_4O$, 2-(4'-Amino-5'-aminopyrimidyl)pent-2-ene-4-one, 30B, 171
$C_9H_{12}N_4O_3$, 1,3,7,9-Tetramethyluric acid, 24, 700; 28, 510
$C_9H_{13}N_3O_5$, Cytidine, 13, 571
$C_{10}H_8N_4O_3$, 6-(p-Hydroxyphenylazo)uracil, 40B, 383
$C_{10}H_9AgN_4O_2S$, Silver sulfadiazine, 41B, 471
$C_{10}H_9BrN_4O_2S$, 5-Bromo-2-metanilamidopyrimidine, 16, 543
$C_{10}H_9Br_2KN_4O_6$, (5-Bromo-3-hydroxy-6-methyluracil)(5-bromo-3-hydroxy-6-methyl-uricilato)potassium, 37B, 228
$C_{10}H_9Br_2N_4O_6Rb$, (5-Bromo-3-hydroxy-6-methyluracil)(5-bromo-3-hydroxy-6-methyl-uricilato)rubidium, 37B, 228
$C_{10}H_9N_5O$, 6-Furfurylaminopurine, 43B, 520
$C_{10}H_{12}N_4O_4$, trans-(5,6':5',6)-Thymine photodimer, 34B, 239
$C_{10}H_{12}N_4O_4$, 6-(p-Hydroxyphenylhydrazino)-uracil monohydrate, 41B, 472
$C_{10}H_{13}ClIN_5$, 9-β-Chloroethyl-7,8-dihydro-9H-imidazo[2,1-i]purine methiodide, 28, 512

$C_{10}H_{13}N_5$, N^6-(Δ^2-Isopentenyl)adenine, 38B, 427
$C_{10}H_{13}N_5$, 6-Amino-3-dimethylallylpurine, 43B, 521
$C_{10}H_{14}Cl_2N_{10}O$, Adenine hydrochloride hemihydrate, 11, 723; 15, 492
$C_{10}H_{14}N_4O_5$, cis-syn 6-Methyluracil photodimer monohydrate, 37B, 227
$C_{10}H_{14}N_4O_5$, 5α-Hydroxy-6,4'-(5'-methylpyrimidin-2'-one)-dihydrothymine monohydrate, 34B, 240
$C_{10}H_{15}IN_6O_2 \cdot H_2O$, Bis(1-methyl-cytosine) hydroiodide monohydrate, 45B, 449
$C_{10}H_{15}N_7O_2$, 6-Histaminopurine dihydrate, 38B, 428
$C_{10}H_{17}BF_4N_2O_2$, 4,6-Diethoxy-1-ethylpyrimidinium tetrafluoroborate, 43B, 522
$C_{10}H_{17}BF_4N_2O_2$, 6-Ethoxy-1,3-diethyl-1,4(3,4)-dihydro-4-oxopyrimidinium tetrafluoroborate, 43B, 522
$C_{10}H_{19}KN_6O_8$, Potassium N-(purin-6-ylcarbamoyl-L-threoninate tetrahydrate, 40B, 383
$C_{10}H_{24}Cl_4N_{10}O_4Pt$, 9-Ethylguanidinium tetrachloroplatinate(II) dihydrate, 42B, 310
$C_{11}H_{10}N_8O_8S$, 6-Thioguanine picrate monohydrate, 41B, 473
$C_{11}H_{10}N_8O_9$, Guanine picrate monohydrate, 41B, 473
$C_{11}H_{11}FN_6O_3$, 5-Fluorouracil - 9-ethylhypoxanthine, 32B, 206
$C_{11}H_{12}BrN_7O_2$, 9-Methyladenine - 1-methyl-5-bromouracil, 31B, 161
$C_{11}H_{12}N_4O_3S$, 2-Sulfanilamido-5-methoxypyrimidine (form I), 43B, 104
$C_{11}H_{12}N_4O_3S$, 2-Sulfanilamido-5-methoxypyrimidine (form III), 43B, 104
$C_{11}H_{12}N_4O_4$, 1-Benzyl-cytosine nitrate, 45B, 450
$C_{11}H_{13}ClN_4O_3$, trans-6-Chloro-9-(2-ethoxy-1,3-dioxan-5-yl)purine, 45B, 450
$C_{11}H_{13}N_3O_3$, 6-Ethoxycarbonyl-4-ethylpyrazolo[1,5-a]pyrimidin-7(4H)-one, 46B, 428
$C_{11}H_{15}N_5O_3$, trans-9-(2-Ethoxy-1,3-dioxan-5-yl)adenine, 45B, 451
$C_{11}H_{15}N_5S$, N^6-(Δ^2-Isopentenyl)-2-methylthioadenine, 37B, 229
$C_{11}H_{17}Cl_2N_3OS$, 2-(3-Diethylaminopropylthio)-4,5-dichloropyrimidin-6-one, 40B, 384
$C_{12}H_{10}N_4S \cdot 0.5\ H_2O$, 9-Methyl-8-phenyl-6-thiopurine hemihydrate, 39B, 295
$C_{12}H_{12}N_4O_3$, 6-Ethoxycarbonyl-4-ethyl-1,2,4-triazolo[1,5-a]pyrimidin-7(4H)-one, 46B, 428
$C_{12}H_{13}Br_2N_7O_2$, 1-Methyl-5-bromouracil - 9-ethyl-8-bromoadenine, 34B, 227
$C_{12}H_{14}BrN_7O_2$, 1-Methyl-5-bromouracil - 9-ethyl-2-aminopurine, 34B, 228
$C_{12}H_{14}BrN_7O_2$, 9-Ethyladenine - 1-methyl-5-bromouracil complex, 31B, 163
$C_{12}H_{14}FN_7O_2$, 9-Ethyl-2-aminopurine - 1-methyl-5-fluorouracil, 31B, 164; 34B, 230
$C_{12}H_{14}FN_7O_2$, 9-Ethyladenine - 1-methyl-5-fluorouracil, 32B, 207
$C_{12}H_{14}N_2O_2$, Primidone, 41B, 474
$C_{12}H_{14}N_4O_4S$, Sulfadimethoxine, 41B, 474
$C_{12}H_{15}BrN_8O_2$, 1-Methyluracil - 9-ethyl-8-bromo-2,6-diaminopurine, 35B, 275
$C_{12}H_{15}FN_8O_2$, 9-Ethylguanine - 1-methyl-5-fluorocytosine, 32B, 210
$C_{12}H_{15}N_7O_2$, 9-Methyladenine - 1-methylthymine, 28, 507; 39B, 296
$C_{12}H_{16}N_8O_2$, 9-Ethylguanine - 1-methylcytosine, 32B, 209
$C_{12}H_{17}Cl_2N_3O_2S \cdot H_2O$, Oxythiamin chloride hydrochloride monohydrate, 45B, 451
$C_{12}H_{18}Cl_2N_4OS \cdot 0.5\ H_2O$, Thiamine chloride hydrochloride hemihydrate, 45B, 452

$C_{12}H_{18}I_2N_4OS$, Thiamine iodide hydroiodide, 42B, 311
$C_{12}H_{19}ClN_4O_2S$, Thiamine chloride monohydrate, 38B, 429
$C_{12}H_{20}CdCl_4N_4O_2S$, Thiaminium tetrachlorocadmate monohydrate, 41B, 475
$C_{12}H_{24}N_4O_5$, 6,6'-Bis(3,6-dihydro-4,6-dimethylpyrimidone-2) cyclization product hydrate, 43B, 525
$C_{12}H_{28}Cl_3Mg_{0.5}N_4O_6S$, Thiamine chloride hydrochloride - magnesium chloride hydrate, 42B, 312
$C_{13}H_{11}N_5O_2$, Benzyl 6-aminopurine-9-carboxylate, 45B, 453
$C_{13}H_{13}N_5O_5$, Theophylline p-nitrophenol, 44B, 386
$C_{13}H_{16}N_4O_2$, 2,4-Diamino-5-(3',4'-dimethoxybenzyl)pyrimidine, 44B, 387
$C_{13}H_{16}N_4O_4$, cis-syn-1,1'-Trimethylenebisthymine photodimer, 40B, 1089
$C_{13}H_{16}N_4O_4$, 1,1'-Trimethylenebisthymine, 39B, 297
$C_{13}H_{17}N_9O_8$, 9-Ethyladenine - parabanic acid - oxaluric acid monohydrate, 41B, 476
$C_{13}H_{20}N_4O_3$, 2-Ethoxy-1,7,9-triethyl-7,9-dihydro-1H-purine-6,8-dione, 45B, 453
$C_{13}H_{20}N_6O_2$, 1,3-Dimethyl-4-imino-5,5-(spiro-2',4'-bisdimethylaminopyrrolyl)uracil, 46B, 428
$C_{14}H_{10}BrNO_5$, Purine-6-malonodialdehyde mono(p-bromoanil), 37B, 230
$C_{14}H_{12}CaN_4O_8 \cdot 1.5\ H_2O$, Calcium 5-ethylidenehydroorotate sesquihydrate, 42B, 572
$C_{14}H_{14}Br_2N_9O$, 9-Ethyl-8-bromoadenine - 9-ethyl-8-bromohypoxanthine, 34B, 235
$C_{14}H_{18}I_8N_8O_4$, Bis(theobromine) bis(tri-iodine) iodine, 45B, 96
$C_{14}H_{18}N_4O_3$, 2,4-Diamino-5-(3,4,5-trimethoxybenzyl)pyrimidine, 42B, 312
$C_{14}H_{19}BrN_8O_2$, 1-Ethylthymine - 9-ethyl-8-bromo-2,6-diaminopurine, 35B, 275
$C_{14}H_{20}N_4O_4$, 1,3-Dimethylthymine photodimer A, 35B, 257
$C_{14}H_{20}N_4O_4$, 1,3-Dimethylthymine photodimer C, 34B, 241
$C_{14}H_{22}N_4O_6$, 2,4-Diamino-5-(3,4,5-trimethoxybenzyl)pyrimidine-1-oxide dihydrate, 35B, 172
$C_{15}H_{15}N_3O_3$, 6-Amino-5-cinnamoyl-1,3-dimethyluracil, 46B, 429
$C_{15}H_{18}N_4O_2$, 6-Benzyl-1,2,3,4,5,6,7,8-octahydro-1,3-dimethylpyrimido[4,5-d]pyrimidine-2,4-dione, 45B, 271
$C_{15}H_{22}N_6O_8$, Thymine trimer monohydrate, 37B, 231
$C_{15}H_{24}N_4O_2S_2$, Thiamine n-propyl disulfide, 34B, 238
$C_{16}H_{17}N_3O_2$, 1-[3-(Indol-3-yl)propyl]thymine, 46B, 429
$C_{16}H_{22}N_6O_4$, 3-(9-Adenyl)propionyltyramine dihydrate, 43B, 524
$C_{16}H_{30}N_4O_4$, 6,6'-Bis(3,6-dihydro-4,6-dimethylpyrimid-2-one) bis(ethanol), 43B, 525
$C_{17}H_{19}I_2N_9O_4$, 1-Methyl-5-iodouracil - 9-ethyladenine, 34B, 231
$C_{17}H_{19}N_7O_2 \cdot 0.5\ H_2O$, Tryptamine adenin-9-ylacetic acid (1:1) hemihydrate, 45B, 454
$C_{17}H_{20}I_2N_{10}O_4$, Bis(1-methyl-5-iodouracil) - 9-ethyl-2,6-diaminopurine, 34B, 233
$C_{17}H_{20}N_4O_4$, Tryptamine:1-thyminylacetic acid, 45B, 275
$C_{17}H_{22}N_8O_5$, 1,3-Bis-(8-theophylline)propane monohydrate, 37B, 232
$C_{18}H_{19}N_7O$, 3-(Adenin-9-yl)propiontryptamide, 43B, 527
$C_{18}H_{21}N_7O_2$, 3-(9-Adenyl)propionyltryptamine monohydrate, 43B, 528
$C_{18}H_{22}N_6O_3$, 9-Diethylcarbamoyl-2-(2-propoxyphenyl)-8-azahypoxanthine, 44B, 388
$C_{19}H_{16}N_6O_4S$, 2-Methylthio-6-benzamidopurine p-nitrophenol, 44B, 388
$C_{19}H_{23}Cl_2N_3O_3S \cdot 3\ H_2O$, DL-2-($\alpha$-Hydroxybenzyl)oxythiamin chloride

hydrochloride trihydrate, 45B, 454
$C_{19}H_{28}N_{10}O_5$, 1-Methylthymine - 9-ethyl-2,6-diaminopurine hydrate, 34B, 232
$C_{22}H_{19}N_5O_4$, 2-Benzoyl-2-(1,3-dimethyl-2,6-dioxo-7-purinyl)-acetanilide, 46B, 430
$C_{23}H_{25}N_5O_3$, N-Methyl-N-(2-benzhydryloxyethyl)-1,3-dimethyl-2,6-dioxo-1,2,3,6-tetrahydro-7H-purin-8-amine, 43B, 530
$C_{24}H_{27}N_5O_3$, N,N-Dimethyl-N-(2-benzhydryloxyethyl)-1,3-dimethyl-2,6-dioxo-1,2,3,6-tetrahydro-9H-purin-8-yl ammonium hydroxide inner salt, 43B, 531
$C_{29}H_{30}AsN_5O_3$, Tetraphenylarsonium adenine trihydrate, 39B, 298

45. CARBOHYDRATES

$C_3H_8O_3$, Glycerol, 33B, 157
$C_4H_8ClNaO_7$, 1,4-Anhydroerythritol sodium perchlorate, 40B, 649
$C_4H_{10}O_4$, meso-Erythritol, 23, 544
$C_5H_8N_4O_{12}$, Pentaerythritol tetranitrate (high-temp.), 41B, 69
$C_5H_8N_4O_{12}$, Pentaerythritol tetranitrate, 11, 604; 28, 405; 41B, 69
$C_5H_8O_5$, D-Arabono-γ-lactone, 45B, 455
$C_5H_{10}O_4$, 2-Deoxyribose, 24, 565
$C_5H_{10}O_5$, β-L-Arabinopyranose, 46B, 431
$C_5H_{10}O_5$, α-L-Xylopyranose, 37B, 233; 45B, 456; 46B, 431
$C_5H_{10}O_5$, β-D,L-Arabinopyranose, 32B, 211; 45B, 457
$C_5H_{10}O_5$, β-L-Lyxopyranose, 44B, 389
$C_5H_{10}O_5$, β-L-Arabinose, 21, 551; 26, 622; 43B, 532
$C_5H_{10}O_5$, β-Lyxose, 26, 621; 31B, 168
$C_5H_{11}NO_5$, anti-D-Arabinose oxime, 44B, 394
$C_5H_{11}NO_5$, syn-D-Arabinose oxime, 44B, 394
$C_5H_{12}O_4$, Pentaerythritol, 5, 140, 155; 22, 581
$C_5H_{12}O_5$, DL-Arabinitol, 33B, 157
$C_5H_{12}O_5$, Ribitol, 34B, 245
$C_5H_{12}O_5$, Xylitol, 34B, 246
$C_5H_{18}CaCl_2O_9$, α-L-Arabinose calcium chloride tetrahydrate, 44B, 389
$C_5H_{19}BaO_{13}P$, Barium ribose-5-phosphate pentahydrate, 27, 792
$C_6H_7NaO_6$, Sodium ascorbate, 34B, 247
$C_6H_7O_6Tl$, Thallium(I) L-ascorbate, 39B, 298
$C_6H_8F_2O_3$, 1,6-Anhydro-2,4-deoxy-2,4-difluoro-β-D-glucopyranose, 45B, 457
$C_6H_8N_6O_3$, 2,4-Diazido-2,4-dideoxy-1,6-anhydro-β-D-glucopyranose, 46B, 431
$C_6H_8O_3$, 1,4:2,5:3,6-Trianhydro-D-mannitol, 41B, 477
$C_6H_8O_4$, 1,4:3,6-Dianhydro-α-D-glucose, 44B, 390
$C_6H_8O_4$, 1,6:3,5-Dianhydro-α-D-gulofuranose, 45B, 458
$C_6H_8O_4$, 1,6:2,3-Dianhydro-β-D-gulopyranose, 37B, 234
$C_6H_8O_6$, D-iso-Ascorbic acid, 38B, 430
$C_6H_8O_6$, L-Ascorbic acid, 4, 268, 317; 33B, 158
$C_6H_8O_6$, β-D-Glucurono-1,4-lactone, 32B, 212
$C_6H_9KO_8$, Potassium D-glucarate, 43B, 24
$C_6H_9NaO_7$, Sodium D-isoascorbate monohydrate, 43B, 533
$C_6H_{10}BaO_{11}S$, Barium 2-O-sulphonato-L-ascorbate dihydrate, 40B, 652
$C_6H_{10}O_4S$, 1-Deoxy-1-thio-1,6-anhydro-β-D-glucopyranose, 44B, 390
$C_6H_{10}O_5$, 1,6-Anhydro-β-D-galactopyranose, 46B, 432
$C_6H_{10}O_5$, 1,6-Anhydro-α-L-gulofuranose, 45B, 458
$C_6H_{10}O_5$, 1,6-Anhydro-β-D-glucopyranose, 37B, 235; 40B, 386
$C_6H_{10}O_5$, 1,6-Anhydro-β-D-mannofuranose, 38B, 431

$C_6H_{10}O_5$, 2,6-Anhydro-β-D-fructofuranose, 39B, 299
$C_6H_{10}O_6$, D-Galactono-1,4-lactone, 32B, 214
$C_6H_{10}O_6$, D-Glucono-(1,5)-lactone, 37B, 237
$C_6H_{10}O_6$, γ-D-Gulonolactone, 37B, 236
$C_6H_{10}O_8$, D-Glucaro-1,4-lactone monohydrate, 42B, 313
$C_6H_{11}FO_5$, 2-Deoxy-2-fluoro-β-D-mannopyranose, 41B, 480
$C_6H_{11}KO_7$, Potassium gluconate, 17, 636
$C_6H_{11}NO_4$, 3-Amino-1,6-anhydro-3-deoxy-β-d-glucopyranose, 43B, 534
$C_6H_{11}NO_6$, α-Glucuronamide, 40B, 386
$C_6H_{11}NaO_8$, Sodium β-D-glucuronate monohydrate, 44B, 391
$C_6H_{12}Br_2O_4$, D-Dibromomannitol, 39B, 300
$C_6H_{12}O_3S_2 \cdot 0.25\ H_2O$, Methyl 1,5-dithio-α-D-ribopyranoside quarter-hydrate, 40B, 387
$C_6H_{12}O_3S_2$, Methyl 1,5-dithio-β-D-ribopyranoside, 40B, 387
$C_6H_{12}O_4$, β-D-Digitoxose, 44B, 391
$C_6H_{12}O_4S$, Methyl 1-thio-α-D-ribopyranoside, 39B, 300
$C_6H_{12}O_4S$, Methyl 1-thio-β-D-xylopyranoside, 31B, 169
$C_6H_{12}O_4S$, Methyl 5-thio-α-D-ribopyranoside, 39B, 302
$C_6H_{12}O_4S$, Methyl 5-thio-β-D-ribopyranoside, 39B, 302
$C_6H_{12}O_5$, Methyl α-D-xylopyranoside, 44B, 393
$C_6H_{12}O_5$, Methyl α-L-arabinopyranoside, 44B, 392
$C_6H_{12}O_5$, Methyl β-D-arabinopyranoside, 45B, 459
$C_6H_{12}O_5$, Methyl β-D-ribopyranoside, 40B, 387; 44B, 393
$C_6H_{12}O_5$, Methyl β-L-arabinopyranoside, 44B, 392
$C_6H_{12}O_5$, Methyl α-D-lyxofuranoside, 33B, 160
$C_6H_{12}O_5$, Methyl β-D-xylopyranoside, 31B, 170; 43B, 532
$C_6H_{12}O_5$, α-DL-Fucopyranose, 43B, 535
$C_6H_{12}O_5$, α-L-Fucose, 41B, 484
$C_6H_{12}O_5$, 1,6-Anhydro-L-iditol, 41B, 483
$C_6H_{12}O_5$, 6-Deoxy-α-L-sorbofuranose, 45B, 458
$C_6H_{12}O_6$, α-D-Galactose, 42B, 314
$C_6H_{12}O_6$, α-D-Glucose, 16, 479; 30B, 179; 45B, 460
$C_6H_{12}O_6$, α-D-Mannopyranose, 42B, 314
$C_6H_{12}O_6$, α-D-Tagatose, 34B, 242
$C_6H_{12}O_6$, α-D-Talose, 43B, 536
$C_6H_{12}O_6$, α-L-Sorbopyranose, 32B, 215; 45B, 461
$C_6H_{12}O_6$, β-D-Fructose, 43B, 536
$C_6H_{12}O_6$, β-D-Galactose, 41B, 484; 42B, 314
$C_6H_{12}O_6$, β-D-Glucose, 28, 516; 33B, 161
$C_6H_{12}O_6$, 1,2,4,5/3,6-Cyclohexanehexol, 45B, 460
$C_6H_{12}O_8$, 2-Keto-L-gulonic acid monohydrate, 41B, 479
$C_6H_{13}KO_8$, Potassium D-gluconate monohydrate (form A), 38B, 433; 40B, 388
$C_6H_{13}KO_8$, Potassium D-gluconate monohydrate (form B), 40B, 388
$C_6H_{13}KO_9$, Potassium β-D-glucuronate dihydrate, 28, 513
$C_6H_{13}NO_6$, Glucopyranosylhydroxylamine, 44B, 394
$C_6H_{13}O_9Rb$, Rubidium β-D-glucuronate dihydrate, 28, 513
$C_6H_{14}BrNO_5$, α-D-Glucosamine hydrobromide, 7, 272; 30B, 180; 34B, 243
$C_6H_{14}CaO_{10}$, Calcium DL-glycerate dihydrate, 44B, 396
$C_6H_{14}ClNO_4$, Mycosamine hydrochloride, 27, 796
$C_6H_{14}ClNO_5$, α-D-Glucosamine hydrochloride, 7, 272; 30B, 180; 34B, 243
$C_6H_{14}ClNO_5$, β-D-Galactosamine hydrochloride, 38B, 434
$C_6H_{14}ClNO_5$, 3-Ammonio-1,6-anhydro-3-deoxy-β-D-glucopyranose chloride monohydrate, 43B, 537
$C_6H_{14}O_6$, Allitol, 38B, 435
$C_6H_{14}O_6$, D-Iditol, 38B, 436

$C_6H_{14}O_6$, D-Mannitol, 33B, 163
$C_6H_{14}O_6$, DL-Mannitol, 43B, 537
$C_6H_{14}O_6$, Galactitol, 33B, 162
$C_6H_{14}O_6$, α-l-Rhamnose monohydrate, 21, 553; 37B, 239; 44B, 396
$C_6H_{14}O_7$, α-D-Glucose monohydrate, 27, 795; 39B, 303
$C_6H_{15}BrCaO_{10}$, α-D-Glucuronate calcium bromide trihydrate, 41B, 478
$C_6H_{15}K_2O_{11}P$, Dipotassium glucose-1-phosphate dihydrate, 30B, 191
$C_6H_{16}Br_2CaO_8$, β-D-Fructose - calcium bromide dihydrate, 42B, 566
$C_6H_{16}CaCl_2O_8$, β-D-Fructopyranose calcium chloride dihydrate, 40B, 653
$C_6H_{16}CaO_{12}$, Calcium D-glucarate tetrahydrate, 42B, 566
$C_6H_{18}Br_2CaO_8$, α-Fucose calcium bromide trihydrate, 41B, 481
$C_6H_{18}Br_2CaO_9$, α-Galactose - calcium bromide trihydrate, 39B, 304
$C_6H_{22}Cl_2O_{11}Sr$, epi-Inositol - strontium chloride pentahydrate, 43B, 538
$C_7H_{10}N_2O_3S$, 4-(β-D-Erythrofuranosyl)imidazoline-2-thione, 39B, 896
$C_7H_{12}Cl_2O_4$, Methyl 4,6-dichloro-4,6-dideoxy-α-D-galactopyranoside, 34B, 249
$C_7H_{12}Cl_2O_4$, Methyl 4,6-dichloro-4,6-dideoxy-α-D-glucopyranoside, 33B, 165
$C_7H_{12}O_4$, Methyl 3,4-dideoxy-α-DL-threo-hex-3-enopyranoside, 45B, 461
$C_7H_{12}O_5$, Methyl 3,6-anhydro-α-D-galactoside, 38B, 436
$C_7H_{13}BrO_5$, α-Methyl-D-galactoside 6-bromohydrin, 30B, 177
$C_7H_{13}ClO_5$, Methyl 2-chloro-2-deoxy-D-galactopyranoside, 34B, 249
$C_7H_{13}FO_5$, Methyl 4-deoxy-4-fluoro-α-D-glucopyranoside, 41B, 485
$C_7H_{14}ClNaO_6$, 2,5-O-Methylene-D-mannitol sodium chloride, 42B, 567
$C_7H_{14}N_2O_5$, N'-Methyl-N-(β-D-xylosyl)urea, 44B, 397
$C_7H_{14}O_6$, 5-O-Methyl-myo-inositol, 46B, 438
$C_7H_{14}O_6 \cdot H_2O$, methyl α-D-Galactopyranoside monohydrate, 37B, 243; 45B, 462
$C_7H_{14}O_6 \cdot 0.5\ H_2O$, Methyl β-D-glucopyranoside hemihydrate, 43B, 541
$C_7H_{14}O_6$, Methyl α-D-altropyranoside, 37B, 242; 41B, 486
$C_7H_{14}O_6$, Methyl α-D-galactofuranoside, 42B, 315
$C_7H_{14}O_6$, Methyl α-D-glucopyranoside, 33B, 168; 43B, 539
$C_7H_{14}O_6$, Methyl α-D-mannopyranoside, 35B, 258; 43B, 539
$C_7H_{14}O_6$, Methyl β-D-galactopyranoside, 43B, 540; 44B, 397; 45B, 462
$C_7H_{14}O_6$, Methyl β-D-hamameloside, 39B, 305
$C_7H_{14}O_7$, Coriose, 35B, 259
$C_7H_{14}O_7$, α-D-Manno-2-heptulose, 45B, 463
$C_7H_{14}O_7$, β-D-Glucoheptose, 44B, 398
$C_7H_{16}O_7$, meso-L-Glycero-L-gulo-heptitol, 42B, 315
$C_8H_{10}O_5$, 2-O-Acetyl-1,6:3,5-dianhydro-α-L-idofuranose, 45B, 464
$C_8H_{11}IO_5S$, 1,6-Anhydro-3-deoxy-3-iodo-2-O-(methylthio)carbonyl-β-D-altropyranose, 46B, 432
$C_8H_{11}NO_5$, 2-Acetamido-2,3-dideoxy-D-erthyro-hex-2-enono-1,4-lactone, 44B, 398
$C_8H_{11}NO_5$, 2-Acetamido-2,3-dideoxy-D-threo-hex-2-enono-1,4-lactone, 42B, 236
$C_8H_{12}N_2O_3S$, 1-Methyl-4-(β-D-erythrofuranosyl)-4-imidazoline-2-thione, 41B, 487
$C_8H_{13}NO_3$, Ethyl 3-cyano-3,4-dideoxy-α-DL-threo-pentapyranoside, 39B, 305
$C_8H_{15}NO_6$, N-Acetyl-α-D-galactosamine, 40B, 389; 41B, 487
$C_8H_{15}NO_6$, N-Acetyl-α-D-glucosamine, 31B, 171; 41B, 488
$C_8H_{16}ClNO_6$, N-(2-Chloroethyl)-D-gluconamide, 41B, 490; 43B, 541
$C_8H_{16}INaO_6$, Bis(1,4-anhydroerythritol) sodium iodide, 42B, 568
$C_8H_{16}O_5S$, Ethyl-1-thio-α-D-glucofuranoside, 32B, 216

$C_8H_{16}O_8$, 2,7-Anhydro-L-glycero-β-D-manno-octulopyranose monohydrate, 43B, 542
$C_8H_{17}NO_7$, N-Acetyl-β-D-mannosamine monohydrate, 41B, 489
$C_8H_{18}CaCl_2O_8$, Methyl D-glycero-α-D-gulo-heptopyranoside calcium chloride hydrate, 45B, 464
$C_8H_{18}O_3$, Pentamethylglycerol, 40B, 52
$C_8H_{18}O_7$, 2,4-Di-O-methyl α-D-galactopyranose monohydrate, 40B, 390
$C_9H_{11}NO_6$, 3-(β-D-Ribofuranosyl)-3-pyrroline-2,5-dione, 35B, 260
$C_9H_{14}IN_3O_4$, 6-Azido-5,6-dideoxy-5-iodo-1,2-O-isopropylidene-β-L-idofuranose, 35B, 262
$C_9H_{14}O_5$, Methyl 3,4-O-isopropylidene-β-L-erythro-pentopyranosid-2-ulose, 42B, 316
$C_9H_{14}O_5$, 1,6-Anhydro-3,4-O-isopropylidine-β-D-talopyranose, 40B, 394
$C_9H_{15}ClO_5S_2$, 5-Chloro-5-deoxy-1,2-O-isopropylidene-3-methanesulphonyl-4-thio-β-L-arabinofuranose, 41B, 490
$C_9H_{16}O_4S_2$, 5-Desoxy-3-C-formyl-β-L-lyxofuranose-trimethylenedithioacetal, 41B, 491
$C_9H_{16}O_6$, exo-Methyl 3,4-O-ethylidene-β-D-galactopyranoside, 42B, 317
$C_9H_{16}O_6$, 1,2-O-Isopropylidene-D-glucofuranose, 45B, 465
$C_9H_{16}O_6$, 2,3,6-Trimethylgalactono-γ-lactone, 39B, 306
$C_9H_{16}O_7$, Methyl 6-O-acetyl-β-D-galactopyranoside, 42B, 318
$C_9H_{16}O_7$, Methyl 6-O-acetyl-β-D-glucopyranoside, 42B, 317
$C_9H_{16}O_8$, Methyl D-threo-2,5-hexodiulosonate 5-(dimethyl acetal), 45B, 466
$C_9H_{17}O_{10}RbS$, 6-Sulfo-6-deoxy-α-D-glucopyranosyl-(1,1')-D-glycerol rubidium salt, 29, 740
$C_9H_{18}INO_6$, 1,2-O-Aminoisopropylidene-α-D-glucopyranose hydroiodide, 31B, 172
$C_9H_{20}O_4S_2$, D-Ribose diethyldithioacetal, 44B, 399
$C_{10}H_{14}O_6S_2$, 1,4:3,6-Bis(thioanhydro)-2,5-O-acetyl-D-iditol (R,R)-disulfoxide, 43B, 436
$C_{10}H_{14}O_7$, 2,3-Di-O-acetyl-1,6-anhydro-β-D-galactopyranose, 42B, 318
$C_{10}H_{16}Cl_2O_4$, Methyl 2,6-dichloro-2,6-dideoxy-3,4-O-isopropylidene-α-D-altropyranoside, 39B, 306
$C_{10}H_{16}N_2O_4S$, 1-Allyl-4,5-(D-glucofurano)imidazolidine-2-thione, 42B, 276
$C_{10}H_{16}O_5$, Methyl-2,3-O-isopropylidene-β-D-allohept-6-ynofuranoside, 41B, 492
$C_{10}H_{18}KNO_{10}S_2$, Potassium allylglucosinate monohydrate, 43B, 543
$C_{10}H_{18}KNO_{10}S_2$, Potassium myronate monohydrate (Sinigrin), 28, 518; 35B, 263
$C_{10}H_{20}O_4S_2$, Ethyl 2-S-ethyl-1,2-dithio-α-D-mannofuranoside, 38B, 437
$C_{10}H_{22}INO_4$, 4,6-Dideoxy-4-(N,N-dimethylamino)-α-D-talopyranoside methiodide, 35B, 264
$C_{10}H_{28}CaO_{17}$, Calcium arabonate, 27, 788
$C_{10}H_{28}O_{17}Sr$, Strontium arabonate, 27, 788
$C_{11}H_{12}N_8O_{14}$, 2-Methyl-5-(N-nitrocarboxamido)-1-(2',3',5'-tri-O-nitro-β-D-ribofuranosyl)imidazol-4-carboxamide, 46B, 433
$C_{11}H_{13}NO_7$, p-Nitrophenyl-β-D-xylopyranoside, 42B, 319
$C_{11}H_{15}BrN_2O_4$, Arabinose p-bromophenylhydrazone, 27, 790
$C_{11}H_{15}BrN_2O_4$, Ribose-p-bromophenylhydrazone, 29, 489
$C_{11}H_{15}ClO_7$, 2,3,4-Tri-O-acetyl-β-D-xylopyranosyl chloride, 40B, 391
$C_{11}H_{15}FO_7$, 1,3,4-Tri-O-acetyl-2-desoxy-2-fluoro-α-D-xylopyranose, 42B, 319
$C_{11}H_{15}FO_7$, 2,3,4-Tri-O-acetyl-β-D-xylopyranosyl fluoride, 45B, 478
$C_{11}H_{15}NO_5S \cdot H_2O$, Pyridyl 1-thio-β-D-glucopyranoside monohydrate, 46B, 433

C??H??NO?, 2-Acetamido-2,3,-dideoxy-5,6-O-isopropylidene-D-threo-hex-2-enono-1,4-lactone, 44B, 399

$C_{11}H_{15}N_3O_7$, Tri-O-acetyl-α-D-arabinopyranosyl azide, 40B, 391

$C_{11}H_{15}N_3O_7$, Tri-O-acetyl-β-D-xylopyranosyl azide, 42B, 320

$C_{11}H_{16}O_4$, 1-Deoxy-2-C-phenyl-D-arabinitol, 44B, 400

$C_{11}H_{16}O_6$, Methyl 4,6-di-O-acetyl-2,3-dideoxy-α-D-threo-hex-2-enopyranoside, 45B, 466

$C_{11}H_{16}O_7$, Methyl 3,4-di-O-acetyl-2,6-anhydro-α-D-altropyranoside, 46B, 434

$C_{11}H_{16}O_7$, Methyl 3,4-di-O-acetyl-2,6-anhydro-β-D-talopyranoside, 43B, 543

$C_{11}H_{17}NO_5$, 2-Acetamido-3-deoxy-5,6-O-isopropylidene-α-D-hex-2-enofuranose, 42B, 320

$C_{11}H_{17}NO_5$, 3,6-(Acetylepimino)-3,6-dideoxy-1,2-O-isopropylidene-β-L-idofuranose, 38B, 437

$C_{11}H_{17}NO_8$, Methyl 2,4-di-O-acetyl-3-deoxy-3-C-methyl-3-nitro-β-D-xylopyranoside, 46B, 435

$C_{11}H_{17}NO_8$, Methyl 2,4-di-O-acetyl-3-deoxy-3-C-methyl-3-nitro-β-DL-arabinopyranoside, 46B, 434

$C_{11}H_{18}O_4S_2$, Methyl 2,3,6-trideoxy-2-C-(2-hydroxy-1,1-(ethylenedithio)ethyl)-α-L-threo-hexopyranosid-4-ulo-2(2),4-pyranose, 43B, 544

$C_{11}H_{18}O_4S_2$, 1,2-SS'-Ethylene-5,6-O-isopropylidene-1,2-dithio-α-D-mannofuranoside, 45B, 467

$C_{11}H_{18}O_8$, trans-O-β-D-Glucopyranosyl methyl acetoacetate, 40B, 392

$C_{11}H_{18}O_9$, γ-Lactone glucoside, 39B, 306

$C_{11}H_{21}NO_9$, N-Acetyl-α-D-muramic acid monohydrate, 40B, 393

$C_{11}H_{23}NO_{11}$, β-D-N-Acetylneuraminic acid dihydrate, 39B, 307

$C_{12}H_{11}BrN_2O_5$, Dehydroascorbic acid p-bromophenylhydrazone, 42B, 321

$C_{12}H_{11}O_5$, α-D-Xylopyranose 1,2,4-orthobenzoate, 41B, 494

$C_{12}H_{12}BrNO_8S$, 2-O-(p-Bromobenzenesulphonyl)-1,4:3,6-dianhydro-D-glucitol-5-nitrate, 30B, 184

$C_{12}H_{12}O_{12}$, Dehydro-L-ascorbic acid dimer, 38B, 438

$C_{12}H_{14}ClN_3O_6$, 1-(3"-Chloro-2"-oxo-5"-cyanopyrrolidinyl)methylene-2-oxo-β-D-arabinofurano[1',2':4,5]oxazoline, 46B, 435

$C_{12}H_{14}O_{12}Sr \cdot 4.5\ H_2O$, Strontium 4-O-(4-deoxy-β-L-threo-hex-4-enosyl)-α-D-galacturonate - 4.5 water, 42B, 571

$C_{12}H_{15}NO_7$, 1-(4-Acetyl-5-methyl-2-furyl)-1,3-dideoxy-3-nitro-β-D-xylopyranose, 46B, 436

$C_{12}H_{16}BrFO_7$, 3,4,6-Tri-O-acetyl-2-bromo-2-deoxy-α-D-mannopyranosyl fluoride, 34B, 252

$C_{12}H_{16}O_7$, 3,4,6-Tri-O-acetyl-1,2-dideoxy-D-hex-1-enopyranose, 45B, 469

$C_{12}H_{16}O_7$, 3,4,6-Tri-O-acetyl-1,5-anhydro-2-desoxy-D-arabinohex-1-enitol, 45B, 468

$C_{12}H_{16}O_8$, Levoglucosan triacetate, 40B, 394

$C_{12}H_{16}O_8$, 3,6-Anhydro-α-D-glucosyl-1,4:3,6-dianhydro-β-D-fructoside, 38B, 439

$C_{12}H_{17}BrN_2O_5$, D-β-Glucose-p-bromophenylhydrazone, 30B, 181

$C_{12}H_{17}BrN_2O_5$, Mannose p-bromophenylhydrazone, 40B, 1090

$C_{12}H_{17}ClO_7$, Tri-O-acetyl-6-deoxy-α-L-mannopyranosyl chloride, 42B, 322

$C_{12}H_{18}CaO_{14}$, Calcium L-ascorbate dihydrate, 40B, 655; 44B, 401

$C_{12}H_{18}O_6$, 4-Ethyl-6-di-O-acetyl-2,3-dideoxy-α-D-erythro-hex-2-enopyranoside, 45B, 466, 479

$C_{12}H_{18}O_8$, Methyl 2,3,4-tri-O-acetyl-α-D-xylopyranoside, 40B, 395

$C_{12}H_{19}ClO_5$, 3-Chloro-3-deoxy-1,2:5,6-di-O-isopropylidene-β-D-idose, 38B, 440

$C_{12}H_{20}O_4S_2$, 5-Desoxy-3-C-formyl-1,2-O-isopropylidene-β-L-lyxofuranose trimethylenedithioacetal, 41B, 495
$C_{12}H_{20}O_6$, 1,2:4,5-Di-O-isopropylidene-β-D-fructopyranose, 39B, 308
$C_{12}H_{20}O_8$, 2,3:4,6-di-O-isopropylidene-2-keto-L-gulonic acid monohydrate, 44B, 401
$C_{12}H_{21}N_3O_8 \cdot 1.5\ H_2O$, 2-Acetamido-1-N-(L-aspart-4-oyl)-2-deoxy-β-D-glucopyranosylamine hydrate, 46B, 436
$C_{12}H_{22}CaO_{12}$, Calcium α-D-glucoisosaccharate, 33B, 168
$C_{12}H_{22}CaO_{16}$, Calcium 5-keto-D-glucuronate dihydrate, 30B, 182
$C_{12}H_{22}O_{11}$, Turanose, 44B, 402
$C_{12}H_{22}O_{11}$, Sucrose, 16, 483; 28, 519; 39B, 309
$C_{12}H_{22}O_{11}$, β-D-Glucopyranosyl-β-D-glucopyranose, 46B, 437
$C_{12}H_{22}O_{11}$, α-Maltose, 44B, 402
$C_{12}H_{22}O_{11}$, β-Cellobiose, 26, 625; 31B, 174; 33B, 161
$C_{12}H_{22}O_{11}$, β-Lactose, 40B, 395
$C_{12}H_{22}O_{11} \cdot 0.19\ H_2O$, Laminarabiose - O-α-D-glucopyranosyl-(1-3)-β-D-glucopyranose (3:2) hydrate, 43B, 544
$C_{12}H_{22}O_{12}Sr$, Strontium 3-deoxy-2C-hydroxymethyl-D-erythro-pentoate, 34B, 250
$C_{12}H_{23}NO_{10}$, N-Acetylneuraminic acid methyl ester monohydrate, 39B, 309
$C_{12}H_{24}Br_2CaO_{12}$, α,α-Trehalose calcium bromide monohydrate, 40B, 397
$C_{12}H_{24}CaO_{17}$, Calcium 2-keto-D-gluconate trihydrate, 42B, 322
$C_{12}H_{24}O_{10}$, 4-O-β-D-Galactopyranosyl-L-rhamnitol, 43B, 546
$C_{12}H_{24}O_{11}$, Gentiobiose, 45B, 469
$C_{12}H_{24}O_{11}$, 4-O-β-D-Glucopyranosyl-D-glucitol, 45B, 470
$C_{12}H_{24}O_{12}$, Sophorose monohydrate, 44B, 403
$C_{12}H_{24}O_{12}$, Isomaltulose, 39B, 310
$C_{12}H_{24}O_{12}$, α-Lactose monohydrate, 37B, 244
$C_{12}H_{24}O_{12}$, α-Melibiose monohydrate, 42B, 323
$C_{12}H_{24}O_{12}$, α,β-Melibiose monohydrate, 44B, 403
$C_{12}H_{24}O_{12}$, β-Maltose monohydrate, 35B, 266; 43B, 546
$C_{12}H_{26}BrNaO_{13}$, Sucrose-sodium bromide dihydrate, 11, 624
$C_{12}H_{26}O_4S_3$, 2-S-Ethyl-2-thio-D-mannose diethyl dithioacetal, 40B, 397
$C_{12}H_{26}O_{13}$, α,α-Trehalose dihydrate, 38B, 441
$C_{12}H_{29}BrCaO_{16}$, Calcium lactobionate bromide hydrate, 39B, 311
$C_{12}H_{30}CaCl_2O_{15}$, Bis-(β-D-fructopyranose) calcium chloride trihydrate, 40B, 657
$C_{12}H_{32}CaCl_2O_{16}$, Trehalose calcium chloride pentahydrate, 44B, 404
$C_{12}H_{36}Br_2CaO_{18}$, Lactose - calcium bromide heptahydrate, 39B, 312
$C_{12}H_{36}CaCl_2O_{18}$, Lactose - calcium chloride heptahydrate, 39B, 313
$C_{13}H_{15}BrN_2O_4S$, 1-p-Bromophenyl-4,5-(1,2-cis-D-glucofurano)imidazolidine-2-thione, 42B, 278
$C_{13}H_{15}ClN_2O_4S$, 1-p-Chlorophenyl-4,5-(D-glucofurano)imidazolidine-2-thione, 42B, 279
$C_{13}H_{15}NO_6$, 1-Carboxyamido-5-O-benzoyl-α-D-arabinofuranose, 42B, 324
$C_{13}H_{16}N_2O_4S$, 1-Phenyl-4,5-(D-glucofurano)imidazolidine-2-thione, 42B, 279
$C_{13}H_{16}N_2O_5$, 1-Phenyl-4,5-(1,2-D-glucofurano)imidazolidin-2-one, 46B, 438
$C_{13}H_{17}NO_4$, Methyl 3,4-dideoxy-3-(salicylidenylamino)-α-D-erythro-pentopyranoside, 42B, 324
$C_{13}H_{18}Cl_2O_5$, 3-Deoxy-3,4-C-(dichloromethylene)-1,2,5,6-di-O-isopropylidene-α-D-galactofuranose, 33B, 170
$C_{13}H_{18}O_9$, Tetra-O-acetyl-α-D-lyxopyranose, 42B, 325
$C_{13}H_{18}O_9$, 1,2,3,4-Tetra-O-acetyl-α-D-arabinopyranose, 40B, 398

$C_{13}H_{18}O_9$, 1,2,3,4-Tetra-O-acetyl-α-D-ribopyranose, 43B, 547
$C_{13}H_{18}O_9$, 1,2,3,4-Tetra-O-acetyl-β-D-arabinopyranose, 40B, 398
$C_{13}H_{18}O_9$, 1,2,3,4-Tetra-O-acetyl-β-D-xylopyranose, 42B, 324
$C_{13}H_{18}O_9$, 1,2,3,5-Tetra-O-acetyl-β-D-ribofuranose, 39B, 313; 42B, 325
$C_{13}H_{19}NO_5$, Methyl hexopyranoside derivative, 40B, 301
$C_{13}H_{20}O_6$, 1-O-Acetyl-2,3:4,5-di-O-isopropylidene-δ-erythro-pent-1-enitol, 43B, 548
$C_{13}H_{20}O_6$, 1-O-Acetyl-2,3:4,5-di-O-isopropylidene-δ-threo-pent-1-enitol, 43B, 548
$C_{13}H_{20}O_7$, 2,2':5,6-Di-O-isopropylidene-2-C-hydroxymethyl-L-gulonolactone, 43B, 549
$C_{13}H_{20}O_8$, Pentaerythritol tetraacetate, 6, 203, 223
$C_{13}H_{22}O_6$, Methyl 2,3:4,5-di-O-isopropylidene-α-D-glucoseptanoside, 45B, 470
$C_{13}H_{22}O_6$, Methyl 2,3:4,5-di-O-isopropylidene-α-D-alloseptanoside, 40B, 399
$C_{13}H_{24}N_2O_{11}$, (Z)-O-β-D-Xylopyranosyl(1,6)-β-D-glucopyranosyloxy-N,N,O-azoxymethane, 46B, 438
$C_{13}H_{24}O_{11}$, Methyl 3-O-α-D-glucopyranosyl-α-D-glucopyranoside, 46B, 438
$C_{13}H_{26}NO_8P$, Cyclohexylammonium methyl α-D-glucopyranoside cyclic 4,6-phosphate, 42B, 326
$C_{13}H_{26}O_{12}$, Methyl β-maltoside monohydrate, 32B, 218
$C_{14}H_{16}N_2O_3S$, 1-p-Tolyl-4-(β-D-erthyrofuranosyl)imidazoline-2-thione, 44B, 404
$C_{14}H_{16}O_5$, Methyl 2,3-anhydro-4,6-benzylidene-α-D-mannopyranoside, 38B, 447
$C_{14}H_{16}O_6$, 2-C-Benzyl-3-keto-L-lyxo-hexulosonic acid lactone methyl glycoside, 42B, 326
$C_{14}H_{18}N_2O_5S$, 1-Phenyl-4,5-(D-glycero-L-gulo-heptofurano)imidazolidine-2-thione, 44B, 405
$C_{14}H_{18}N_2O_7$, 1-(Tri-O-acetyl-α-D-xylopyranosyl)-imidazole, 40B, 399
$C_{14}H_{18}O_9$, 2,3,4,6-Tetra-O-acetyl-1,5-anhydro-D-arabinohex-1-enitol, 45B, 468
$C_{14}H_{19}Br_2ClO_8$, 3,4,5-Tri-O-acetyl-1,7-dibromo-6-chloromethyl-1,7-dideoxy-α-DL-ido-heptopyranos-2-ulose, 43B, 550
$C_{14}H_{19}ClO_9$, Tetra-O-acetyl-α-D-mannopyranosyl chloride, 42B, 322
$C_{14}H_{20}N_2O_9$, p-Nitrophenyl-β-D-N-acetylglucosaminide monohydrate, 41B, 495
$C_{14}H_{20}O_9$, 3,4,6-Tri-O-acetyl-1,2-O-(R)-ethylidene-α-D-allopyranose, 46B, 439
$C_{14}H_{21}ClO_7$, 5-O-(Chloroacetyl)-1,2:3,4-di-O-isopropylidene-α-D-glucoseptanose, 39B, 314
$C_{14}H_{21}NO_5$, N-Benzyl-1-methylamino-1-deoxy-β-D-arabino-2-hexulopyranose, 44B, 406
$C_{14}H_{21}NO_5$, 1-Benzyl(methyl)amino-1-deoxy-α-D-lyxo-hexulopyranose, 44B, 406
$C_{14}H_{22}O_7$, 5-O-Acetyl-1,2:3,4-di-O-isopropylidene-α-D-galactoseptanose, 46B, 440
$C_{14}H_{22}O_8$, 4,5-Di-O-acetyl-1,2-O-isopropylidene-3-O-methyl-α-D-guloseptanose, 39B, 314
$C_{14}H_{22}O_9S$, Methyl tri-O-acetyl-6-deoxy-6-methylsulfinyl(S)-α-D-glucopyranoside, 42B, 327
$C_{14}H_{24}N_4O_7S$, 5-Azido-5-desoxy-1,2-O-isopropylidene-3-O-mesyl-β-L-iduronic acid diethylamide, 44B, 407
$C_{14}H_{24}O_5S$, Ethyl 2,3:4,5-di-O-isopropylidene-1-thio-β-D-glucosep-

tanoside, 38B, 442
$C_{14}H_{28}O_{12}$, Methyl β-cellobioside - methanol, 35B, 268
$C_{15}H_{17}IO_4$, Methyl 6-O-benzoyl-4-iodo-2,3,4-trideoxy-α-D-hex-2-enopyranoside, 42B, 328
$C_{15}H_{19}NO_9$, 2,3,4,6-Tetra-O-acetyl-1-cyano-β-D-galactopyranose, 42B, 328
$C_{15}H_{19}NO_9$, 3,4,6-Tri-O-acetyl-1,2-O-(1-cyanoethylidene)-α-D-glucopyranose, 42B, 329
$C_{15}H_{20}Cl_2O_9$, DL-1,4,5,6-Tetra-O-acetyl-3-chloro-2-C-(chloromethyl)-epi-inositol, 46B, 440
$C_{15}H_{20}O_{10}$, 1,4,6-Tri-O-acetyl-3-O-1'(R)-C-carboxyethyl-β-D-glucose 2,2'-lactone, 45B, 472
$C_{15}H_{20}O_{10}$, 1,6-Anhydro-isopropylidene-β-D-arabino-hexopyranos-3-ulose dimer, 45B, 471
$C_{15}H_{20}O_{11}$, Daphnetin 8-β-D-glucopyranoside dihydrate, 44B, 407
$C_{15}H_{20}O_{11}$, Methyl 1,2,3,4-tetra-O-acetyl-β-D-galactopyranuronate, 41B, 497
$C_{15}H_{22}O_9$, 1,4,5-Tri-O-acetyl-2,3-O-isopropylidene-β-D-fructopyranose, 42B, 329
$C_{15}H_{22}O_{10}$, Methyl 2,3,4,6-tetra-O-acetyl-β-D-glucopyranoside, 44B, 408
$C_{15}H_{22}O_{10}$, Methyl 2,3,4,5-tetra-O-acetyl-α-D-galactoseptanoside, 46B, 441
$C_{15}H_{22}O_{10}$, Methyl 2,3,4,5-tetra-O-acetyl-α-D-glucoseptanoside, 40B, 400
$C_{15}H_{22}O_{10}$, 1,3,4,6-Tetra-O-acetyl-2,5-O-methylene-D-mannitol, 43B, 551
$C_{15}H_{22}O_{10}$, 3,4,6-Tri-O-acetyl-β-D-mannose 1,2-(methyl orthoacetate), 42B, 330
$C_{15}H_{24}O_7$, (1S)-5,7-Anhydro-8-deoxy-1,2:3,4-di-O-isopropylidene-1-O-methyl-D-glycero-D-galacto-octos-6-ulose, 42B, 330
$C_{16}H_{20}O_5S_2$, Methyl 4,6-O-benzylidene-2-deoxy-3-O-[(methylthio)thiocarbonyl]-α-D-ribopyranoside, 45B, 472
$C_{16}H_{20}O_5S_2$, Methyl 4,6-O-benzylidene-2-deoxy-3-O-[(methylthio)thiocarbonyl]-α-D-arabinopyranoside, 45B, 472
$C_{16}H_{20}O_5S_2$, Methyl 4,6-O-benzylidene-3-deoxy-2-O-[(methylthio)thiocarbonyl]-α-D-arabinopyranoside, 45B, 472
$C_{16}H_{20}O_6$, Methyl 3-C-acetyl-4,6-O-benzylidene-2-deoxy-α-D-ribohexopyranoside, 46B, 441
$C_{16}H_{21}NO_9$, 1,4,6-Tri-O-acetyl-2-(N-acetylacetamido)-2,3-dideoxy-α-D-erythro-hex-2-enopyranose, 46B, 442
$C_{16}H_{21}NO_9$, 1,4,6-Tri-O-acetyl-2-(N-acetylacetamido)-2,3-dideoxy-α-D-threo-hex-2-enopyranose, 42B, 331
$C_{16}H_{21}NO_9$, 3,4,6-Tri-O-acetyl-2-(N-acetylacetamido)-1,2-dideoxy-D-arabino-hex-1-enopyranose, 44B, 408
$C_{16}H_{21}NO_9$, 3,4,6-Tri-O-acetyl-2-(N-acetylacetamido)-1,2-dideoxy-D-ribo-hex-1-enopyranose, 43B, 552
$C_{16}H_{21}NO_9$, 3,4,6-Tri-O-acetyl-2-(N-acetylacetamido)-1,2-dideoxy-D-xylo-hex-1-enopyranose, 43B, 552
$C_{16}H_{22}O_6$, Methyl 4,6-O-(R)-benzylidene-2,3-di-O-methyl-α-D-glucopyranoside, 45B, 473
$C_{16}H_{22}O_6$, Methyl 4,6-O-(R)-benzylidene-2,3-di-O-methyl-β-D-galactopyranoside, 45B, 473
$C_{16}H_{22}O_{11}$, 1,2,3,4,6-Penta-O-acetyl-α-D-altopyranose, 41B, 497
$C_{16}H_{22}O_{11}$, 1,2,3,4,6-Penta-O-acetyl-α-D-gulopyranose, 43B, 553
$C_{16}H_{22}O_{11}$, 1,2,3,4,6-Penta-O-acetyl-α-D-idopyranose, 42B, 332
$C_{16}H_{24}O_{10}$, 3,4,6-Tri-O-acetyl-1,2-O-(1-(exo-ethoxy)ethylidene)-α-D-

glucopyranose, 40B, 400
$C_{16}H_{27}NO_6$, 7-Acetamido-6,7,8-trideoxy-1,2:3,4-di-O-isopropylidene-α-D-glycero-D-galactooctopyranose, 44B, 409
$C_{16}H_{27}NO_7 \cdot 0.5\ H_2O$, 7-Acetamido-7,8-dideoxy-1,2:3,4-di-O-isopropylidene-β-L-threo-D-galacto-octopyranose hemihydrate, 45B, 474
$C_{16}H_{28}N_2O_{11} \cdot 3\ H_2O$, β-N,N'-Diacetylchitobiose trihydrate, 45B, 475
$C_{16}H_{30}N_2O_{12}$, α-N,N'-Diacetylchitobiose monohydrate, 44B, 409
$C_{16}H_{30}O_5S_2$, 2,3:4,5-Di-O-isopropylidene-D-gulose diethyl dithioacetal, 45B, 475
$C_{16}H_{32}O_6$, 1-Decyl α-D-glucopyranoside, 42B, 332
$C_{17}H_{17}BBrNO_4$, N-(p-Bromophenyl)-α-D-ribopyranosylamine 2,4-diphenylboronate, 34B, 253
$C_{17}H_{18}BrNO_5$, endo-3,6-Anhydro-2-deoxy-4,5-O-isopropylidene-7-O-p-bromobenzoyl-D-ribo-heptononitrile, 43B, 554
$C_{17}H_{18}BrNO_5$, 3,6-Anhydro-7-O-p-bromobenzoyl-2-deoxy-4,5-O-isopropylidene-D-altro-heptononitrile, 41B, 498
$C_{17}H_{20}O_4S_2$, D-Ribose diphenyldithioacetal, 44B, 399
$C_{17}H_{21}NO_5$, Spiro-[(1-acetylaziridine)-2,3'-(methyl 4,6-O-benzylidene-2,3-dideoxy-α-D-arabinohexapyranoside)], 46B, 442
$C_{17}H_{23}BrO_9$, Methyl 2,3-anhydro-6-bromo-6-deoxy-4O-(4,6-di-O-acetyl-2,3-dideoxy-α-D-erythro-hex-2-eno-pyranosyl)-α-D-allopyranoside, 45B, 476
$C_{17}H_{23}NO_{10}$, 1,5-Di-O-acetyl-3-C-(R)-ethoxycarbonylmethyl-5(R),1'(R)-N-formylepimino-2,3-O-isopropylidene-β-D-ribofuranose, 42B, 333
$C_{17}H_{24}O_{10}$, 1-(2,3,4,6-Tetra-O-acetyl-β-D-glucopyranosyl)-2,3-(2R)-epoxypropane, 43B, 554
$C_{17}H_{26}BrNO_5$, N-(p-Bromobenzyl)nogalonamide, 37B, 245
$C_{17}H_{26}N_2O_7$, Spiro-3,4'-R-(3,3-dideoxy-1,2:5,6-di-O-isopropylidene-α-D-ribo-hexafuranose)-3'-S-acetamido-2'-pyrrolidone, 43B, 555
$C_{17}H_{26}N_2O_7$, Spiro-3,4'-R-(3,3-dideoxy-1,2:5,6-di-O-isopropylidene-α-D-ribo-hexafuranose)-3'-R-acetamido-2'-pyrrolidone, 43B, 555
$C_{17}H_{27}N_3O_5S_2$, 7-Azido-8-deoxy-1,2,3,4-di-O-isopropylidene-6,7-S,S-trimethylene-6,7-dithio-α-D-erythro-D-galacto-octapyranose, 41B, 499
$C_{17}H_{34}O_{18}$, O-(4-O-Methyl-α-D-glucopyranosyluronic acid)-(1-2)-O-β-D-xylopyranosyl-(1-4)-D-xylopyranose trihydrate, 39B, 315
$C_{18}H_{19}BrN_4O_4$, 2,3-Diphenyl-5-(α-D-lyxofuranosyl)tetrazolium bromide, 44B, 410
$C_{18}H_{21}BrO_7S$, 2-C-Hydroxymethyl-2,2'-O-cyclohexylidene-3-deoxy-5-O-(p-bromobenzenesulphonyl)-D-erythro-pentono-1,4-lactone, 35B, 270
$C_{18}H_{23}IO_9S$, Methyl 3,4-di-O-acetyl-6-deoxy-6-iodo-2-O-p-tolylsulphonyl-α-D-mannopyranoside, 43B, 556
$C_{18}H_{25}IO_{10}$, 6'-Iodo phenyl α-maltoside, 42B, 333
$C_{18}H_{25}NO_{11}$, 1,3,4,6-Tetra-O-acetyl-2-(N-acetylacetamido)-2-deoxy-β-D-galactopyranose, 46B, 443
$C_{18}H_{26}O_{11}$, Phenyl α-maltoside, 42B, 333
$C_{18}H_{26}O_{11}$, cis-2,3,4,6-Tetra-O-acetyl-1-deoxy-D-glucopyranoside-1,2'-spiro(3'-methyl-3'-hydroxy-tetrahydrofuran), 46B, 443
$C_{18}H_{26}O_{11}$, trans-2,3,4,6-Tetra-O-acetyl-1-deoxy-D-glycopyranoside-1,2'-spiro(3'-methyl-3'-hydroxy-tetrahydrofuran), 46B, 443
$C_{18}H_{28}O_{10}$, 3,4,6-Tri-O-acetyl-1,2-O-(R)-(1-t-butoxyethylidene)-α-D-galactopyranose, 46B, 439
$C_{18}H_{29}NO_8$, 7-Acetamido-6-O-acetyl-7,8-dideoxy-1,2:3,4-di-O-isopropylidene-α-D-erythro-D-galacto-octopyranose, 45B, 474
$C_{18}H_{30}O_{10}$, Methyl 3,5-O-isopropylidene-2-O-(methyl 3,5-O-isopropylidene-α-D-xylofuranosid-3-yl)-α-D-xylofuranoside, 44B, 410
$C_{18}H_{32}O_{16}$, 1-Kestose, 38B, 443

$C_{18}H_{34}O_{17}$, Melezitose monohydrate, 42B, 334

$C_{18}H_{36}O_{18}$, Planteose dihydrate, 38B, 446

$C_{18}H_{39}CaNaO_{27}$, Calcium sodium α-D-galacturonate hexahydrate, 41B, 500; 42B, 334

$C_{18}H_{39}NaO_{27}Sr$, Strontium sodium galacturonate hexahydrate, 41B, 500

$C_{18}H_{42}O_{21}$, Raffinose pentahydrate, 35B, 270

$C_{19}H_{22}O_{10}S$, 3-O-Acetyl-1,2-O-isopropylidene-5-O-tosyl-α-D-gulofuranose-4,6-carbolactone, 43B, 556

$C_{19}H_{23}BrO_{10}S$, 1-O-(p-Bromobenzenesulphonyl)-4,5,7-tri-O-acetyl-2,6-anhydro-3-deoxy-D-glucoheptitol, 30B, 183

$C_{19}H_{24}BrNO_9S$, 5-Brosyl-3-deoxy-3-C-(R)-(ethoxycarbonylformamido)-methyl-2-O-isopropylidene-α-D-ribofuranose, 38B, 448

$C_{19}H_{24}O_6S_2$, Ethyl 3,7-anhydro-6,8-O-benzylidene-4-deoxy-2-ethyl-enedithio-D-talo-2-octulosonate, 44B, 411

$C_{19}H_{25}N_3O_{12}$, 3,5-Dimethyl-1-(2,3,4,6-tetra-O-acetyl-α-D-mannopyranosyl)isocyanuric acid, 44B, 411

$C_{19}H_{31}ClN_2O_7 \cdot H_2O$, Procaine N-D-glucoside hydrochloride monohyd-rate, 46B, 444

$C_{19}H_{34}O_7S_2$, 1-(1,3-Dithian-2-yl)-2,3:5,6-di-O-isopropylidene-β-L-gulofuranose isopropanol solvate, 44B, 412

$C_{20}H_{22}N_2O_{14}$, 1-O-(2,4-Dinitrophenyl)-2,3,4,6-tetra-O-acetyl-α-D-glu-copyranoside, 46B, 444

$C_{20}H_{22}O_4$, 2(S)-4-O-Acetyl-3,5-O-benzylidene-1,2-dideoxy-2-C-phenyl-D-erthyro-pentitol, 44B, 400

$C_{20}H_{23}NO_7$, 1-Phthalimido-1-deoxy-2,3:4,6-di-O-isopropylidene-α-L-sorbofuranose, 42B, 335

$C_{20}H_{24}NO_{11}P$, 1,2,3,4-Tetra-O-acetyl-5,6-dideoxy-6-C-nitro-5-(phenyl-phosphinyl)-L-idopyranose, 46B, 445

$C_{20}H_{26}O_{11}$, Apterin monohydrate, 44B, 412

$C_{20}H_{35}NO_{16}$, O-α-D-Mannopyranosyl-(1-3)-O-β-D-mannopyranosyl-(1-4)-2-acetamido-2-deoxy-α-D-glucopyranose, 45B, 476

$C_{21}H_{20}BrNO_7S$, Methyl 4,6-O-benzylidene-2-O-p-bromobenzenesulphonyl-3-cyano-3-deoxy-α-D-altropyranoside, 35B, 273

$C_{22}H_{29}NO_9$, Ethyl 5-O-acetyl-3-O-benzyl-6-deoxy-6-formylamino-1,2-O-isopropylidene-D-glycero-L-talo-heptafuranuronate, 43B, 557

$C_{22}H_{29}NO_9$, Ethyl 5-O-acetyl-3-O-benzyl-6-deoxy-6-formylamino-1,2-O-isopropylidene-L-glycero-D-allo-heptofuranuronate, 43B, 557

$C_{22}H_{30}O_{15}$, β-D,1-4 Xylobiose hexaacetate, 39B, 315

$C_{24}H_{26}O_{10}$, 1-Naphthyl 2',3',4',6'-tetra-O-acetyl-β-D-glucopyrano-side, 44B, 413

$C_{24}H_{52}ClNO_9$, Glucosylphytosphingosine hydrochloride monohydrate, 43B, 558

$C_{24}H_{52}O_{26}$, Stachyose pentahydrate, 41B, 501

$C_{25}H_{33}Br_2Cl_3O_{15}$, 6,6'-Dibromo-6,6'-dideoxy-α,α-trehalose hexaacetate chloroform adduct, 45B, 477

$C_{26}H_{20}Cl_2O_7$, 2,3,4-Tri-O-benzoyl-2-C-chloro-α-D-xylopyranosyl chlor-ide, 46B, 446

$C_{26}H_{21}BrO_7$, Tri-O-benzoyl-β-D-xylopyranosyl bromide, 40B, 402

$C_{26}H_{21}ClO_7$, 2,3,4-Tri-O-benzoyl-β-D-xylopyranosyl chloride, 42B, 336

$C_{26}H_{21}FO_7$, 2,3,4-Tri-O-benzoyl-β-D-xylopyranosyl fluoride, 45B, 478

$C_{26}H_{22}O_7$, 1,5-Anhydro-2,3,4-tri-O-benzoylribitol, 45B, 480

$C_{26}H_{22}O_7$, 1,5-Anhydro-2,3,4-tri-O-benzoylxylitol, 45B, 480

$C_{26}H_{26}N_4O_4$, Methyl 4,6-O-benzylidene-2,3-dideoxy-2,3-diphenylazo-α-D-mannoside, 44B, 413

$C_{26}H_{26}O_8$, (+)-2,4:3,5-Di-O-methylene-D-mannitol 1,6-di-trans-cinna-mate, 46B, 446

$C_{26}H_{32}O_{10}S_2$, 1,2:5,6-Di-O-isopropylidene-3,4-di-O-tosyl-L-chiro-ino-

sitol, 38B, 449
$C_{26}H_{36}O_{11} \cdot 2\ Ch_4O$, Mascaroside methanol solvate, 43B, 559
$C_{26}H_{38}O_{15}$, 1',2:4,6-Di-O-isopropylidenesucrose tetraacetate, 45B, 478
$C_{27}H_{24}O_8$, Methyl 2,3,4-tri-O-benzoyl-β-D-xylopyranoside, 46B, 446
$C_{27}H_{28}O_4$, 1,5-Anhydro-3,4,6-tri-O-benzyl-2-deoxy-D-arabino-hex-1-enitol, 45B, 479
$C_{27}H_{30}N_2O_{13}$, 1-Acetyl-3-benzamido-4-(2,3,4,6-tetra-O-acetyl-β-D-glucopyranosyloxy)-Δ^3-2-pyrrolinone, 44B, 414
$C_{27}H_{38}O_{18}$, Methyl 2,3,4,6,2',4',6'-hepta-O-acetyl-β-L-laminarabioside, 44B, 414
$C_{28}H_{38}O_{19}$, β-D-Acetyl-cellobiose, 42B, 336
$C_{28}H_{38}O_{19}$, 1,2,4,6-Tetra-O-acetyl-3-O-(2,3,4,6-tetra-O-acetyl-β-D-galactopyranosyl)-α-D-galactopyranose, 46B, 447
$C_{30}H_{14}O_{10}$, 2α,3β,11-Trihydroxy-12-oxo-5β-carda-9,20-dienolide-3-O-methyl-4,6-dideoxy-D-allo-2-hexosulose, 45B, 480
$C_{33}H_{26}O_9$, 1,2,3,4-Tetra-O-benzoyl-β-D-xylopyranose, 45B, 480
$C_{33}H_{38}O_{14}$, Methyl 2,3',6-tri-O-acetyl-2',3;4',6'-di-O-benzylidene-7(R)-β-D-cellobioside, 46B, 447
$C_{34}H_{26}Cl_2O_9$, 2,3,4,6-Tetra-O-benzoyl-2-chloro-α-D-mannopyranosyl chloride, 43B, 560
$C_{35}H_{42}O_6S_4$, 4,5,6-Tri-O-benzoyl-2,3-di-S-ethyl-2,3-dithio-D-allose diethyl dithioacetal, 43B, 561
$C_{36}H_{60}O_{30} \cdot 6\ H_2O$, α-Cyclodextrin hexahydrate, 46B, 448
$C_{36}H_{68}I_2O_{34}$, Cyclohexa-amylose iodine tetrahydrate, 39B, 899
$C_{36}H_{72}O_{36}$, Cyclohexaamylose hexahydrate, 40B, 402
$C_{39}H_{36}N_2O_{10}$, Methyl 2,4-bis(N-acetyl-N-benzoylamino)-3,6-di-O-benzoyl-2,4-didesoxy-α-D-idopyranoside, 44B, 415
$C_{40}H_{54}O_{27}$, β-Cellotriose undecaacetate, 43B, 562
$C_{42}H_{72}INO_{33}$, α-Cyclodextrin-p-iodoaniline complex, 41B, 501
$C_{42}H_{83}NO_9 \cdot 0.5\ C_2H_6O$, β-D-Galactosyl-N-(2-D-hydroxyoctadecanoyl)-D-dihydrosphingosine ethanol solvate, 43B, 564
$C_{72}H_{120}O_{60}$, Cyclohexa-amylose-potassium acetate complex, 30B, 186
$C_{72}H_{136}I_5LiO_{68}$, bis(α-Cyclodextrin)-lithium triiodide-iodine-8 water, 46B, 449
$C_{72}H_{174}Cd_{0.5}I_5O_{87}$, bis(α-Cyclodextrin)-hemi(cadmium pentaiodide)-27 water, 46B, 449

46. PHOSPHATES

$CH_{15}N_2O_6P$, Methyl diammonium phosphate dihydrate, 38B, 449
$C_2H_8NO_4P$, 2-Amino-ethanol phosphate, 26, 591
$C_2H_{10}NO_4P$, Dimethyl ammonium phosphate, 39B, 316
$C_2H_{13}K_2O_8P$, Dipotassium ethyl phosphate tetrahydrate, 37B, 246
$C_3H_7Na_2O_6P \cdot 6\ H_2O$, Disodium DL-α-glycerophosphate hexahydrate, 46B, 450
$C_3H_7O_4P$, Methyl ethylene phosphate, 30B, 194
$C_3H_9O_4P$, Trimethylphosphate (gas-ed), 39B, 892
$C_3H_{11}CdO_{10}P$, Cadmium D(-)-phosphoglycerate trihydrate, 37B, 246
$C_3H_{17}Na_2O_{11}P$, Disodium β-glycerolphosphate pentahydrate, 31B, 176
$C_3H_{18}Na_2O_{12}P$, Disodium DL-glycerol 3-phosphate hexahydrate, 38B, 450
$C_4H_{10}AgO_4P$, Silver diethyl phosphate, 38B, 451
$C_5H_{13}CaClNO_4P \cdot 4\ H_2O$, Choline phosphate calcium chloride tetrahydrate, 45B, 482
$C_5H_{13}N_2O_3P$, Tetramethylformamidiniumphosphonate, 38B, 652
$C_5H_{16}NO_7P$, L-α-Glycerylphosphorylethanolamine monohydrate, 39B, 317

$C_6H_5K_2O_4P \cdot 1.5\ H_2O$, Dipotassium phenylphosphate sesquihydrate, 32B, 220
$C_6H_5O_4P$, Catechol cyclic phosphate, 38B, 451
$C_6H_9Cl_6O_4PS_2$, O,O-Dimethyl-S-(2'-trichloro-1'-hydroxyethyl-1-mercapto-2-trichloroethyl)-thiolphosphate (triclinic), 45B, 483
$C_6H_9Cl_6O_4PS_2$, O,O-Dimethyl-S-(2'-trichloro-1'-hydroxyethyl-1-mercapto-2-trichloroethyl)-thiolphosphate (monoclinic), 45B, 483
$C_6H_{14}NO_8P \cdot H_2O$, α-D-Galactosamine-1-phosphate monohydrate, 46B, 451
$C_6H_{14}Na_3O_{12}P$, Trisodium 6-phospho-D-gluconate dihydrate, 40B, 46
$C_6H_{15}O_{10}P$, myo-Inositol-2-phosphate monohydrate, 40B, 405
$C_6H_{17}Na_2O_8P$, Disodium trans-1,2-hydroxycyclohexane monophosphate trihydrate, 39B, 318
$C_6H_{82}Na_{12}O_{62}P_6$, Dodecasodium myoinositol hexaphosphate octatriacontahydrate, 41B, 504
$C_7H_{15}O_4P$, Methyl pinacol phosphate, 31B, 179
$C_8H_{12}O_7P_2$, Acetoinenediol cyclopyrophosphate, 41B, 505
$C_8H_{14}ClN_2O_5P$, Pyridoxamime-5'-phosphate hydrochloride, 37B, 247
$C_8H_{15}N_2O_8P$, Pyridoxal phosphate oxime dihydrate, 34B, 254
$C_8H_{20}BaO_8P_2$, Barium diethyl phosphate, 31B, 177
$C_8H_{20}MgO_8P_2$, Magnesium diethyl phosphate, 39B, 318
$C_8H_{20}NO_6P$, L-α-Glycerylphosphorylcholine, 31B, 180
$C_8H_{22}N_3O_4P$, Propylguanidinium diethylphosphate, 38B, 452
$C_8H_{26}CdCl_2NO_9P$, L-α-Glycerophosphorylcholine cadmium chloride trihydrate, 30B, 192
$C_9H_{11}N_2Na_2O_8P \cdot 5\ H_2O$, Disodium deoxyuridine 5'-phosphate pentahydrate, 46B, 452
$C_9H_{16}O_{11}P_2$, Bis(dimethylphosphatovinyl) carbonate, 42B, 338
$C_9H_{18}NO_6P$, Cyclohexylammonium phosphoenolpyruvate, 39B, 319
$C_{10}H_9Cl_4O_4P$, 2-Chloro-1-(2,4,5-trichlorophenyl)vinyl dimethylphosphate, 44B, 416
$C_{10}H_{10}BrCl_2O_4P$, O-[1-(2,4-Dichlorophenyl)-2-bromovinyl] O,O'-dimethyl phosphate, 45B, 484
$C_{10}H_{20}NO_5P$, 2-O-Methylxylitanylidene phosphorodiethylamidate, 45B, 484
$C_{10}H_{24}N_4O_5P_2$, Tetramethylformamidinium phosphonic anhydride, 38B, 453
$C_{10}H_{25}N_4O_6P$, Arginine diethyl phosphate, 39B, 320
$C_{10}H_{44}N_4O_{14}P_2$, Spermine phosphate hexahydrate, 30B, 187
$C_{12}H_9Cl_2O_4P$, Di-p-chlorophenyl hydrogen phosphate, 29, 663
$C_{12}H_{11}O_2P$, Diphenylphosphinic acid, 30B, 195; 39B, 321
$C_{12}H_{18}N_4O_7P_2S \cdot 4.5\ H_2O$, Thiamin pyrophosphate hydrate, 45B, 485
$C_{12}H_{19}ClN_4O_7P_2S \cdot 0.5\ H_2O$, Thiamine pyrophosphate hydrochloride hemihydrate, 31B, 181
$C_{12}H_{19}ClN_4O_7P_2S$, Thiamine pyrophosphate hydrochloride, 38B, 396
$C_{12}H_{26}N_4O_{11}P_2S$, Thiamine pyrophosphate tetrahydrate, 34B, 255; 43B, 566
$C_{12}H_{36}N_4O_{11}P_2S$, Hydrolysed cocarboxylase trihydrate, 31B, 182
$C_{14}H_{15}O_4P$, Dibenzyl hydrogen phosphate, 20, 586
$C_{18}H_{12}N_3O_{10}P$, Tri-(p-nitrophenyl)phosphate, 35B, 274
$C_{18}H_{15}O_4P$, Triphenylphosphate, 27, 916; 30B, 193
$C_{18}H_{39}O_2P$, Di-n-nonylphosphinic acid, 39B, 321
$C_{20}H_{22}CaO_{11}P_2$, Calcium 1-naphthylphosphate trihydrate, 30B, 188
$C_{20}H_{42}NO_6P \cdot H_2O$, Deoxylysophosphatidylcholine monohydrate, 46B, 452
$C_{31}H_{62}NO_{10}P$, 1,2-Dilauroyl-DL-phosphatidylethanolamine - acetic acid, 40B, 411; 43B, 567
$C_{32}H_{42}N_2O_7P_2$, Bis(cyclohexylammonium) P(1),P(2)-di-β-naphthyl pyrophosphate, 41B, 506

$C_{35}H_{35}O_4P$, Diisopropyl-2,3,4,5-tetraphenylcyclopenta-1,4-dienyl phosphate, 45B, 485
$C_{72}H_{70}Mg_3O_{29}P_6$, Pentaaquohexa(diphenyl phosphato)trimagnesium(II), 44B, 416

47. NUCLEOSIDES AND NUCLEOTIDES

$C_8H_{10}N_2O_3$, Tetrahydrofuranyluracil, 45B, 486
$C_8H_{11}N_3O_6$, 6-Azauridine, 39B, 321
$C_8H_{12}N_2O_4$, N-(β-D-ribofuranosyl)imidazole, 39B, 322
$C_8H_{12}N_4O_5$, 6-Azacytidine, 40B, 382
$C_8H_{12}O_9P \cdot 3\ H_2O$, 6-Azauridine-5'-phosphoric acid trihydrate, 45B, 486
$C_9H_9N_3O_9 \cdot H_2O$, 5-Nitro-1-(β-D-ribosyluronic acid)-uracil monohydrate, 45B, 487
$C_9H_{10}N_2O_5$, O(2),2'-Anhydro-1-α-D-xylofuranosyluracil, 42B, 338
$C_9H_{10}N_2O_5$, O(2),2'-Cyclouridine, 39B, 322
$C_9H_{11}BrN_2O_5$, 5-Bromo-2'-deoxyuridine, 31B, 183
$C_9H_{11}BrN_2O_6$, 1-β-D-Arabinofuranosyl-5-bromouracil, 39B, 324
$C_9H_{11}BrN_2O_6$, 5-Bromouridine, 31B, 183
$C_9H_{11}ClN_2O_5$, 2'-Chloro-2'-deoxyuridine, 38B, 455
$C_9H_{11}ClN_2O_5$, 5-Chloro-2'-deoxyuridine, 39B, 324
$C_9H_{11}ClN_2O_6$, 5-Chlorouridine, 37B, 248
$C_9H_{11}FN_2O_5$, 5-Fluoro-2'-deoxy-β-uridine, 29, 710
$C_9H_{11}IN_2O_5$, 5-Iodo-2'-deoxyuridine, 30B, 204
$C_9H_{11}IN_2O_6$, 5-Iodouridine, 35B, 276
$C_9H_{11}N_2NaO_8P$, Disodium deoxyuridine-5'-phosphate, 41B, 507
$C_9H_{11}N_2Na_2O_9P \cdot 7\ H_2O$, Disodium uridine 5'-phosphate heptahydrate, 46B, 453
$C_9H_{11}N_3O_5$, 6,2'-Anhydro-1-β-D-arabinofuranosyl-6-hydroxycytosine, 45B, 487
$C_9H_{11}N_3O_{10}$, 1-(5-Nitro-2,4-dioxopyrimidinyl)-β-D-ribofuranuronic acid monohydrate, 44B, 417
$C_9H_{12}BaN_3O_8P \cdot 8.5\ H_2O$, Barium cytidine 5'-monophosphate hydrate, 46B, 453
$C_9H_{12}IN_3O_4$, 5-Iodo-5'-amino-2',5'-dideoxyuridine, 45B, 488
$C_9H_{12}IN_3O_5$, 5-Iodocytidine, 45B, 489
$C_9H_{12}K_2N_2O_{12}P_2 \cdot 3\ H_2O$, Dipotassium uridine 5'-phosphate trihydrate, 45B, 489
$C_9H_{12}N_2O_5$, 2'-Deoxyuridine, 38B, 456
$C_9H_{12}N_2O_5S \cdot 1.5\ H_2O$, 4-Thiouridine hydrate, 35B, 278
$C_9H_{12}N_2O_5S$, 2-Thiouridine, 43B, 569
$C_9H_{12}N_2O_5S$, 4-Thiopseudouridine, 46B, 454
$C_9H_{12}N_2O_6$, Uridine, 41B, 508
$C_9H_{12}N_2O_6$, 1-β-D-Lyxofuranosyluracil, 45B, 490
$C_9H_{12}N_2O_6$, 1-β-D-Arabinofuranosyluracil, 39B, 325; 40B, 412
$C_9H_{12}N_2O_7$, 5-Hydroxyuridine, 39B, 325
$C_9H_{12}N_3O_7P$, 2',5'-Arabinosylcytidine monophosphate, 43B, 570
$C_9H_{12}N_6O_4 \cdot 0.5\ H_2O$, 2-Azaadenosine hemihydrate, 45B, 490
$C_9H_{13}N_2O_9P \cdot H_2O$, 3'-Uridine monophosphate monohydrate, 45B, 491
$C_9H_{13}N_3O_4$, 2'-Deoxycytidine, 41B, 509
$C_9H_{13}N_3O_5$, Cytidine, 30B, 198
$C_9H_{13}N_3O_5$, α-Cytidine, 43B, 571
$C_9H_{13}N_3O_5$, 1-β-D-Arabinofuranosyl-cytosine, 40B, 413
$C_9H_{13}N_3O_5$, 6-Azathymidine, 44B, 417
$C_9H_{13}N_3O_6$, 5-Aminouridine, 44B, 418

$C_9H_{13}N_3O_9$, 5-Nitrouridine monohydrate, 43B, 568
$C_9H_{13}N_3O_{10}P_2 \cdot H_2O$, 2,2'-Anhydro-1-β-D-arabinofuranosylcytosine-3',5'-diphosphate monohydrate, 45B, 491
$C_9H_{14}ClN_3O_4$, 2'-Deoxycytidine hydrochloride, 35B, 279
$C_9H_{14}ClN_3O_5$, Cytidinium chloride, 45B, 492
$C_9H_{14}ClN_3O_5$, Pseudoisocytidine hydrochloride, 46B, 454
$C_9H_{14}ClN_3O_5$, 1-(β-D-Arabinofuranosyl)cytosine hydrochloride, 39B, 326
$C_9H_{14}KN_2O_9P \cdot 0.5\ H_2O$, Potassium dihydrouridine 3'-monophosphate hemihydrate, 46B, 455
$C_9H_{14}N_2O_4S_2$, 5,6-Dihydro-2,4-dithiouridine, 42B, 339
$C_9H_{14}N_2O_5S$, 5,6-Dihydro-2-thiouridine, 40B, 413
$C_9H_{14}N_2O_5S_2$, 2,4-Dithiouridine monohydrate, 37B, 249
$C_9H_{14}N_2O_6 \cdot 0.5\ H_2O$, Dihydrouridine hemihydrate, 37B, 252; 38B, 458
$C_9H_{14}N_2O_6S$, 1-β-D-Arabinofuranosyl-4-thiouracil monohydrate, 38B, 459
$C_9H_{14}N_2O_7$, α-Pseudouridine monohydrate, 35B, 294
$C_9H_{14}N_3O_8P \cdot 3\ H_2O$, 1-β-D-Arabinofuranosylcytosine 5'-monophosphate trihydrate, 45B, 493
$C_9H_{14}N_3O_8P$, Cytidine-3'-phosphate (monoclinic), 32B, 223
$C_9H_{14}N_3O_8P$, Cytidine-3'-phosphate (orthorhombic), 23, 711; 30B, 202
$C_9H_{14}N_4O_5$, 5-Amino-1-β-D-ribofuranosylimidazole-4-carboxamide, 45B, 493
$C_9H_{14}N_4O_8$, Cytidinium nitrate, 42B, 340
$C_9H_{14}N_6O_5$, 8-Azaadenosine monohydrate, 43B, 570
$C_9H_{15}N_3NaO_9P$, Sodium β-cytidine 2',3'-cyclic phosphate dihydrate, 39B, 319
$C_9H_{15}N_3O_7$, Bredinin monohydrate, 41B, 510
$C_9H_{16}N_2O_7$, Ribosylpyrimidin-2-one dihydrate, 44B, 418
$C_9H_{16}N_3O_8P$, Deoxycytidine-5'-phosphate monohydrate, 43B, 572
$C_9H_{17}N_3O_6$, 5,6-Dihydroisocytidine monohydrate, 42B, 340
$C_9H_{17}N_3O_6P$, Deoxycytidine 5'-phosphate monohydrate, 37B, 254
$C_9H_{17}N_3O_6S$, 2-Thiocytidine dihydrate, 37B, 251
$C_9H_{17}N_3O_{12}P_2$, Cytidine-5'-diphosphate monohydrate, 41B, 511
$C_9H_{19}N_2Na_2O_{13}P$, Disodium uridine 3'-phosphate tetrahydrate, 38B, 460
$C_9H_{25}BaN_2O_{16}P$, D(+)-Barium uridine-5'-phosphate, 30B, 206
$C_{10}H_{10}N_2O_7$, 7,2'-Anhydro-β-D-arabinosylorotidine, 46B, 463
$C_{10}H_{10}N_4O_4S_2$, 1-Methyl-5-mercaptouracil disulfide, 35B, 280
$C_{10}H_{11}BrN_4O_5$, 8-Bromoinosine, 43B, 572
$C_{10}H_{11}CaN_4O_8P \cdot 6.5\ H_2O$, Calcium inosine-5'-monophosphate hydrate, 46B, 455
$C_{10}H_{11}ClN_4O_4$, 6-Chloropurine riboside, 41B, 512
$C_{10}H_{11}NO_5$, 2,2'-Anhydro-2-hydroxy-1-(β-D-arabino-pentofuranosyl)-4-pyridone, 43B, 573
$C_{10}H_{11}N_2O_7P$, Cytidine 2',3'-cyclophosphate, 44B, 419
$C_{10}H_{11}N_4O_8PSr \cdot 6.5\ H_2O$, Strontium inosine-5'-monophosphate hydrate, 46B, 456
$C_{10}H_{11}N_5O_2$, 2',3'-Dideoxy-2',3'-didehydroadenosine, 40B, 414
$C_{10}H_{11}N_5O_3S \cdot H_2O$, 8,3'-Anhydro-8-mercapto-9-β-D-xylofuranosyladenine monohydrate, 45B, 494
$C_{10}H_{11}N_5O_4 \cdot H_2O$, 8,5'-Cycloadenosine monohydrate, 46B, 457
$C_{10}H_{11}N_5O_4 \cdot 3\ H_2O$, 8,2'-O-Cyclo-9-β-D-arabinofuranosyladenine trihydrate, 45B, 495
$C_{10}H_{12}N_2O_6$, Thymidine 5'-carboxylic acid, 40B, 415
$C_{10}H_{12}N_4O_4$, Hypoxanthine-8-(2'-desoxy-riboside), 41B, 513
$C_{10}H_{12}N_4O_4$, Nebularine, 40B, 416
$C_{10}H_{12}N_4O_4S$, 6-Thiopurine riboside, 33B, 170

$C_{10}H_{12}N_4O_5$, Inosine (monoclinic), 35B, 288
$C_{10}H_{12}N_4O_5$, Inosine (orthorhombic), 45B, 496
$C_{10}H_{12}N_5O_6PS \cdot 3\ H_2O$, 8,2'-Anhydro-8-mercapto-9-β-D-arabinofuranosyl-adenine 5'-monophosphate trihydrate, 45B, 496
$C_{10}H_{13}ClN_4O_5$, Formycin B hydrochloride, 43B, 574
$C_{10}H_{13}NO_5$, 3-Deaza-4-deoxyuridine, 43B, 575
$C_{10}H_{13}NO_6$, 3-Deazauridine, 39B, 327
$C_{10}H_{13}N_5O_3$, Cordycepin, 46B, 457
$C_{10}H_{13}N_5O_4$, Adenosine, 38B, 461
$C_{10}H_{13}N_5O_4$, 8β-D-Ribofuranosyl-adenine, 42B, 341
$C_{10}H_{13}N_5O_4$, 9α-D-Arabinofuranosyladenine, 45B, 497; 46B, 458
$C_{10}H_{13}N_5O_4$, 9β-D-Arabinofuranosyladenine, 40B, 417
$C_{10}H_{13}N_5O_4S \cdot H_2O$, 8-Thioxoadenosine monohydrate, 46B, 458
$C_{10}H_{14}BrN_2O_4$, 5-Bromo-5-deoxythymidine, 21, 555
$C_{10}H_{14}BrN_5 \cdot 0.5\ H_2O$, Guanosine hydrobromide hemihydrate, 40B, 418
$C_{10}H_{14}ClN_5O_4$, Adenosine hydrochloride, 39B, 328
$C_{10}H_{14}ClN_5O_4$, 9-β-D-Arabinofuranosyladenine hydrochloride, 40B, 419
$C_{10}H_{14}IN_5O_6$, 8-Iodoguanosine monohydrate, 44B, 420
$C_{10}H_{14}N_2O_5$, Thymidine, 34B, 262
$C_{10}H_{14}N_2O_5$, 3-Deazacytidine, 43B, 576
$C_{10}H_{14}N_2O_5$, 6-Methyl-2'-deoxyuridine, 46B, 459
$C_{10}H_{14}N_2O_6 \cdot 0.5\ H_2O$, 5-Methyluridine hemihydrate, 34B, 263
$C_{10}H_{14}N_2O_6$, 1-β-D-Arabinofuranosylthymine, 39B, 329
$C_{10}H_{14}N_2O_6$, 5-Hydroxymethyl-2'-deoxyuridine, 46B, 459
$C_{10}H_{14}N_2O_6$, 6-Methyluridine, 38B, 462
$C_{10}H_{14}N_4O_6 \cdot H_2O$, 4-Amino-1-[4-amino-2-oxo-1(2H)-pyrimidinyl]-1,4-dideoxy-β-D-glucopyranuronic acid monohydrate, 46B, 460
$C_{10}H_{14}N_5O_8P \cdot 3\ H_2O$, Guanosine 5'-monophosphate trihydrate, 46B, 461
$C_{10}H_{14}N_6O_3$, 3'-Amino-3'-deoxyadenosine, 46B, 461
$C_{10}H_{14}N_{10}O_6S$, Isoguanine sulfate monohydrate, 37B, 258
$C_{10}H_{15}N_3O_5$, 2'-O-Methylcytidine, 43B, 577
$C_{10}H_{15}N_3O_5$, 3'-O-Methyl-1-β-D-arabinofuranosylcytosine, 41B, 514
$C_{10}H_{15}N_5O_4$, Deoxyadenosine monohydrate, 30B, 197
$C_{10}H_{15}N_5O_4S$, 2'-Deoxy-6-thioguanosine monohydrate, 43B, 578
$C_{10}H_{15}N_5O_5$, Formycin monohydrate, 39B, 330
$C_{10}H_{15}N_5O_5S$, 6-Thioguanosine monohydrate, 38B, 464
$C_{10}H_{16}BrN_5O_5$, Formycin hydrobromide monohydrate, 40B, 420
$C_{10}H_{16}BrN_5O_7$, 8-Bromoguanosine dihydrate, 34B, 259
$C_{10}H_{16}N_2O_5$, Dihydrothymidine, 35B, 290
$C_{10}H_{16}N_2O_6$, 5-Hydroxy-5,6-dihydrothymidine, 44B, 420
$C_{10}H_{16}N_4O_7$, Inosine dihydrate, 35B, 283
$C_{10}H_{16}N_4O_8$, Xanthosine dihydrate, 42B, 342
$C_{10}H_{16}N_5O_5$, 8-Azatubercidin monohydrate, 44B, 421
$C_{10}H_{16}N_5O_8P$, Adenosine-5'-monophosphate monohydrate (orthorhombic form), 42B, 342
$C_{10}H_{16}N_5O_8P$, Adenosine-5'-phosphate monohydrate, 28, 521
$C_{10}H_{16}N_5O_{11}P_2Rb$, Rubidium adenosine-5'-diphosphate monohydrate, 42B, 343
$C_{10}H_{16}N_6O_4$, α-D-2'-Amino-2'-deoxyadenosine monohydrate, 35B, 296
$C_{10}H_{17}N_5O_7$, Guanosine dihydrate, 35B, 283
$C_{10}H_{18}KN_5O_{12}P_2$, Potassium adenosine-5'-diphosphate dihydrate, 46B, 462
$C_{10}H_{18}N_5O_9P$, Adenosine-3'-phosphate dihydrate, 31B, 185
$C_{10}H_{19}N_4O_{10}P$, Guanosine-5'-phosphate trihydrate, 34B, 260
$C_{10}H_{19}N_5NaO_{11}P$, Sodium guanosine 3',5'-cyclic monophosphate tetrahydrate, 40B, 406
$C_{10}H_{20}N_5Na_2O_{11}P$, Disodium deoxyguanosine-5'-phosphate tetrahydrate,

40B, 422
$C_{10}H_{20}N_5Na_2O_{16}P_3$, Disodium adenosine 5'-triphosphate trihydrate, 37B, 255; 44B, 421
$C_{10}H_{25}CaN_2O_{14}P$, Calcium thymidylate hexahydrate, 26, 803
$C_{10}H_{25}N_5NaO_{12}P$, Sodium 2'-deoxyadenosine-5'-phosphate hexahydrate, 39B, 328; 41B, 514
$C_{10}H_{28}N_4NaO_{16}P$, Monosodium inosine-5'-phosphate octahydrate, 34B, 261
$C_{11}H_{12}N_2O_5$, 5-Ethynyl-2'-deoxyuridine, 44B, 422
$C_{11}H_{13}Cl_2N_3O_3$, 2-(4-O-Acetyl-2,3-dideoxy-β-L-glycero-pent-2-enopyranosyl)-5,6-dichlorobenzotriazole, 39B, 335
$C_{11}H_{14}ClN_3O_5$, 3,N[4]-Ethenocytidine hydrochloride, 42B, 344
$C_{11}H_{14}N_2O_5$, 5-Vinyl-2'-deoxyuridine, 44B, 422
$C_{11}H_{14}N_2O_6$, β-5-Acetyl-2'-deoxyuridine, 46B, 463
$C_{11}H_{14}N_2O_6$, α-5-Acetyl-2'-deoxyuridine, 43B, 578
$C_{11}H_{14}N_2O_7$, 2',3'-O-Methoxymethyleneuridine, 43B, 579
$C_{11}H_{14}N_2O_8$, 5-Carboxymethyluridine, 44B, 423
$C_{11}H_{14}N_4O_4$, Tubercidin, 39B, 331
$C_{11}H_{14}N_4O_4$, 3-Deazaadenosine, 42B, 344
$C_{11}H_{14}N_4O_4$, 6-Methyl-9-β-D-ribofuranosylpurine, 41B, 515
$C_{11}H_{14}N_4O_5 \cdot 0.5\ H_2O$, 6-Methoxypurine riboside hemihydrate, 46B, 464
$C_{11}H_{14}N_6O_4$, 6-Amino-10-(β-D-ribofuranosylamino)pyrimido[5,4-d]pyrimidine, 41B, 493
$C_{11}H_{15}N_3O_7$, 5-Carbamoylmethyluridine, 44B, 423
$C_{11}H_{15}N_5O_3S$, 5'-Methylthioadenosine, 44B, 423
$C_{11}H_{15}N_5O_4$, 2'-O-Methyladenosine, 42B, 345
$C_{11}H_{16}ClN_3O_5S$, 2,2'-Anhydro-1-β-D-arabinofuranosyl-5-dimethylsulphonio-6-oxocytosine chloride, 45B, 497
$C_{11}H_{16}N_2O_8S$, 2-Thio-5-carboxymethyluridine monohydrate, 44B, 424
$C_{11}H_{16}N_4O_5S$, 6-Methylmercaptopurine-riboside monohydrate, 42B, 346
$C_{11}H_{16}N_5O_6P \cdot 0.5\ H_2O$, Adenosine 5'-methylphosphonate hemihydrate, 45B, 498
$C_{11}H_{16}N_5O_6P$, 5-Methyleneadenosine 3',5'-cyclic monophosphonate monohydrate, 38B, 466
$C_{11}H_{16}N_5O_7P$, Adenosine-5'-O-methylphosphate, 44B, 424
$C_{11}H_{17}FN_2O_6S$, 1-Deoxy-1-ethylthio-1-(5-fluorouracil-1-yl)-D-arabinose aldehydrol, 43B, 580
$C_{11}H_{17}N_3O_5S \cdot 2\ H_2O$, 2-Thio-5-methylaminomethyluridine dihydrate, 45B, 499
$C_{11}H_{17}N_3O_6$, 5-Dimethylaminouridine, 45B, 500
$C_{11}H_{17}N_5O_6$, 2-N-Methylguanosine monohydrate, 44B, 425
$C_{11}H_{18}N_2NaO_{10}P$, Sodium uridine-5'-O-methylphosphate methanol solvate, 44B, 426
$C_{11}H_{19}IN_6O_4$, 5'-Methylammonium-5'-deoxyadenosine iodide monohydrate, 37B, 256
$C_{11}H_{19}N_4O_{10}P$, Inosine-5'-phosphate, 40B, 407
$C_{11}H_{20}N_3O_{10}S$, 3-Methylcytidine methosulfate monohydrate, 40B, 422
$C_{12}H_{13}ClN_2O_4$, 2-Chloro-1-(β-D-ribofuranosyl)benzimidazole, 39B, 332
$C_{12}H_{14}N_2O_5$, 2,5'-Anhydro-2',3'-isopropylidene cyclouridine, 40B, 385, 424
$C_{12}H_{15}N_2O_5S$, 3'-O-Acetyl-4-thiothymidine, 37B, 257
$C_{12}H_{15}N_5O_5$, 3-O-Acetyladenosine, 35B, 297
$C_{12}H_{16}N_2O_5S$, 2-Thio-1-(β-D-ribofuranosyl)-3H-benzimidazole monohydrate, 39B, 334
$C_{12}H_{16}N_2O_7S$, 2-Thio-5-(methoxycarbonylmethyl)uridine, 44B, 424
$C_{12}H_{17}N_5O_5$, N[2]-Dimethylguanosine, 38B, 467
$C_{12}H_{18}N_2O_5$, β-5-Isopropyl-2'-deoxyuridine, 45B, 500

$C_{12}H_{18}N_2O_{10}$, Methyl uridine-5-oxyacetate monohydrate, 41B, 516
$C_{12}H_{18}N_3O_6P$, Thymidine 3',5'-cyclic N,N-dimethylphosphoramidate, 43B, 856
$C_{12}H_{19}ClN_4O_5S$, 1-(6-Chloropurin-9-yl)-1-deoxy-1-ethylthio-aldehydo-D-glucose aldehydrol, 43B, 580
$C_{12}H_{26}N_7O_{10}P$, 8-((2-Aminoethyl)amino)adenosine cyclic 3',5'-monophosphate tetrahydrate, 44B, 426
$C_{13}H_{14}ClN_3O_7$, 2,2'-Anhydro-1-{3',5'-di-O-acetyl-β-D-arabinofuranosyl}-5-chloro-6-oxocytosine, 45B, 497
$C_{13}H_{15}FN_2O_7$, 3',5'-Diacetyl-2'-deoxy-2'-fluorouridine, 40B, 424
$C_{13}H_{16}IN_5O_3$, 2,3-iso-Propylidene-3,5-cycloadenosine iodide, 17, 765
$C_{13}H_{16}N_2O_8$, 3',5'-Di-O-acetyluridine, 43B, 581
$C_{13}H_{16}N_4O_6$, 6,7-Dimethyl-N-1-β-D-ribofuranosyllumazine, 43B, 582
$C_{13}H_{16}N_6O_7$, N6-(N-Glycylcarbonyl)adenosine, 43B, 582
$C_{13}H_{17}N_5O_4$, 2'-3'-O-Isopropylideneadenosine, 44B, 427
$C_{13}H_{20}N_4O_6S$, 2-Ethylthio-8-methylinosine monohydrate, 40B, 425
$C_{13}H_{22}N_5O_7P$, 3'-Deoxy-3'-(dihydroxyphosphinylmethyl)adenosine ethanol solvate, 38B, 469
$C_{13}H_{23}N_5O_7$, 8-(α-Hydroxyisopropyl)adenosine dihydrate, 44B, 427
$C_{14}H_{14}BrN_7O_6$, Adenosine 5-bromouracil, 44B, 428
$C_{14}H_{15}Cl_2N_3O_3$, cis-1-(6-Acetoxymethyltetrahydro-2-pyranyl)-5,6-dichlorobenzotriazole, 38B, 470
$C_{14}H_{16}N_2O_3S$, 1-p-Tolyl-4-(α-D-erythrofuranosyl)imidazoline-2-thione, 43B, 549
$C_{14}H_{18}N_2O_4$, 5,6-Dimethyl-1-(α-D-ribofuranosyl)benzimidazole, 45B, 501
$C_{14}H_{18}N_2O_8S$, 3,1'-Anhydro-2-(4',6'-di-O-acetyl-2',3'-dideoxy-α-D-ribo-hexopyranose)-3-hydroxy-5-methyl-2H-1,2,6-thiadiazine-1,1-dioxide, 41B, 517
$C_{14}H_{19}N_3O_6 \cdot CHCl_3$, 5'-Acetamido-3'-acetyl-5'-deoxythyimidine chloroform solvate, 46B, 464
$C_{14}H_{19}N_7O_6$, (1-Methylnicotinamide)(+) aden-9-ylacetate(-) dihydrate, 39B, 335
$C_{14}H_{20}ClN_5O_5$, 7-Ethyl-3-β-D-ribofuranosylimidazo[2,1-i]purine hydrochloride monohydrate, 40B, 426
$C_{14}H_{20}N_2O_7$, 2'-O-Tetrahydropyranyluridine, 39B, 337
$C_{14}H_{20}N_6O_3 \cdot CH_4O$, 3'-Cyclobutylamino-3'-deoxyadenosine, 46B, 461
$C_{14}H_{20}N_6O_5 \cdot 2\ H_2O$, 9-β-D-Arabinofuranosyl-8-morpholinoadenine dihydrate, 46B, 465
$C_{14}H_{21}Br_2N_7O_3$, N-(3-(Aden-9-yl)propyl)-3-carbamoylpyridinium bromide hydrobromide dihydrate, 39B, 337
$C_{14}H_{22}BrN_7O_4$, N-(3-(Aden-9-yl)propyl)-3-carbamoylpyridinium bromide trihydrate, 39B, 338
$C_{14}H_{22}N_6O_4$, 9-β-D-Arabinofuranosyl-8-n-butylaminoadenine, 45B, 501
$C_{14}H_{26}N_6O_{13}P \cdot 2\ H_2O$, Tris(hydroxymethyl)methylammonium adenosine-5'-diphosphate dihydrate, 46B, 466
$C_{14}H_{35}N_4NaO_{16}P_2$, Sodium cytidine-5'-diphospho-choline pentahydrate, 41B, 511
$C_{15}H_{19}N_5O_6$, 2',3'-O-(2-Carboxyethyl)ethylideneadenosine, 44B, 428
$C_{15}H_{20}N_6O_7 \cdot 1.5\ H_2O$, N-(9-β-D-Ribofuranosylpurin-6-yl)glycyl-L-alanine sesquihydrate, 42B, 346
$C_{15}H_{20}N_6O_8$, N6-(N-Threonylcarbonyl)adenosine, 43B, 582
$C_{15}H_{21}N_5O_5$, 2'-O-Tetrahydropyranyladenosine, 37B, 260
$C_{15}H_{24}N_4O_7 \cdot 2.5\ H_2O$, 5-(N-(L-Leucyl)amino)uridine hydrate, 43B, 584
$C_{15}H_{26}N_3O_7PS$, Triethylammonium uridine-2',3'-O,O-cyclothiophosphate, 34B, 258; 35B, 292
$C_{15}H_{26}N_3O_8P$, Triethylammonium cyclic-uridine-3',5'-phosphate, 34B,

256
$C_{16}H_{30}N_3O_8PS$, Triethylammonium uridine 3'-O-thiophosphate methyl ester, 40B, 427
$C_{17}H_{19}N_5O_4$, 6-Benzylamino-9-β-D-ribofuranosylpurine, 42B, 347
$C_{17}H_{21}N_3O_5$, 2-(3',4'-Di-O-acetyl-2'-deoxy-β-L-erythro-pentapyranosyl)-5,6-dimethylbenzotriazole, 42B, 348
$C_{17}H_{23}BrN_4O_7$, Riboflavin hydrobromide monohydrate, 40B, 1089
$C_{18}H_{19}N_2O_8P$, 2'-Acetyluridine-3',5'-cyclophosphate benzyl ester, 43B, 584
$C_{18}H_{20}Cl_2N_2O_7$, 5'-Acetyl-7,7-dichloro-2',3'-isopropylidene-3-methylcyclothymidine, 43B, 413
$C_{18}H_{20}N_6O_3$, 3'-(N-Benzyl-N-methylamino)-3'-deoxyadenosine, 46B, 461
$C_{18}H_{22}N_4O_7$, 5-(N-(L-Phenylalanyl)amino)uridine, 39B, 339
$C_{18}H_{22}N_4O_{10}S_2$, 5-(1-(2'-Deoxy-α-D-ribofuranosyl)uracilyl)disulfide, 32B, 224
$C_{18}H_{23}N_5O_7$, 7-(Methyl 2-acetamido-6-O-acetyl-2,3,4-trideoxy-α-D-threo-hex-2-enopyranosid-4-yl)-theophylline, 45B, 502
$C_{18}H_{23}N_5O_7$, 7-(Methyl 2-acetamido-6-O-acetyl-2,3,4-trideoxy-β-D-erythro-hex-2-enopyranosid-4-yl)-theophylline, 45B, 503
$C_{19}H_{24}N_7O_{12} \cdot 0.5\ H_2O$, Uridylyl-(3'-5')-adenosine hemihydrate, 38B, 471
$C_{19}H_{26}BrN_7O_{11}$, Adenosine 5-bromouridine monohydrate, 30B, 196
$C_{19}H_{27}BrN_8O_8$, Deoxyguanosine 5-bromodeoxycytidine complex, 30B, 200
$C_{19}H_{32}N_7O_{16}P$, β-Adenosine-2'-β-uridine-5'-phosphoric acid tetrahydrate, 34B, 265
$C_{19}H_{42}CaN_8O_{21}P$, Calcium guanosine-3',5'-cytidine monophosphate hydrate, 41B, 518
$C_{20}H_{22}N_4O_5$, (5'-Tryptaminocarbonyl)-2'-(deoxyribofuranosyl)thymine, 46B, 466
$C_{20}H_{24}N_{10}O_{10} \cdot 4\ H_2O$, 8-(8-Guanosyl)guanosine tetrahydrate, 45B, 504
$C_{20}H_{26}N_4O_8S_2$, (1R)-2,3,4,5-Tetra-O-acetyl-1-deoxy-1-S-ethyl-1-(purin-9-yl-6(1H)-thione)-1-thio-aldehydo-D-arabinose aldehydrol, 41B, 519
$C_{20}H_{44}N_7NaO_{21}P$, Sodium guanylyl-3',5'-cytidine hydrate, 39B, 899
$C_{21}H_{31}FN_2O_{11}S$, 2,3,4,5-Tetraacetyl-1-deoxy-1-ethylthio-1-(5-fluorouracil-1-yl)-D-arabinose aldehydrol ethanol solvate, 43B, 580
$C_{22}H_{32}N_3O_{13}$, Cytidine N-carbobenzoxyglutamic acid (1:1) dihydrate, 41B, 519
$C_{22}H_{32}N_7O_8PS$, Triethylammonium adenosine-5'-(O-p-nitrophenyl-O-phosphorothioate), 45B, 504
$C_{22}H_{41}Cl_2N_7O_{10}$, Puromycin dihydrochloride pentahydrate, 38B, 518
$C_{24}H_{29}N_5O_{10}S$, 2',3',5'-Tri-O-acetyl-6-O-(mesitylenesulphonyl)guanosine, 44B, 429
$C_{24}H_{36}Cl_6N_8O_4S_2U$, Bis(protonated thiamine chloride) tetrachlorodioxouranate(VI), 40B, 428
$C_{24}H_{36}N_4O_9$, Cytosamine triacetate, 40B, 401
$C_{30}H_{49}N_{15}O_{22}P_2$, Adenylyl-(3',5')-adenylyl-(3',5')-adenosine hexahydrate, 42B, 349
$C_{31}H_{63}N_9O_{27}$, Adenylyl-3',5'-uridine 9-aminoacridine hydrate, 41B, 520
$C_{38}H_{84}CaN_{16}O_{42}P_2$, Calcium guanylyl-3',5'-cytidine hydrate, 42B, 349
$C_{40}H_{42}IN_{10}O_{12}P \cdot 13.5\ H_2O$, Ethidium - 5-iodouridylyl(3'-5')adenosine hydrate, 43B, 585
$C_{42}H_{51}IN_{11}O_{14}P \cdot 13.5\ H_2O$, Ethidium - 5-iodocytidylyl(3'-5')guanosine hydrate methanolate, 43B, 586
$C_{64}H_{122}N_{22}O_{45}P_2$, Proflavine deoxycytidyl-(3',5')-guanosine hydrate, 46B, 467

48. AMINO-ACIDS AND PEPTIDES

$C_2H_5NO_2$, α-Glycine, 2, 805, 877; 4, 264, 287; 7, 221, 248; 22, 644; 38B, 473; 39B, 340; 42B, 350; 46B, 468
$C_2H_5NO_2$, β-Glycine, 24, 542
$C_2H_5NO_2$, γ-Glycine, 26, 588; 46B, 468
$C_2H_6ClNO_2$, Glycine hydrochloride, 41B, 521; 43B, 587
$C_2H_6N_2O_5$, Monoglycine nitrate, 41B, 522
$C_2H_7NO_3S$, Taurine, 28, 524; 29, 484; 31B, 187
$C_2H_{10}N_2O_6S$, Ammonium glycinium sulfate, 40B, 429; 42B, 350
$C_3H_3NO_3$, Glycine N-carboxyanhydride, 42B, 269
$C_3H_7NO_2$, DL-Alanine, 13, 480
$C_3H_7NO_2$, L-Alanine, 31B, 187; 38B, 474
$C_3H_7NO_2$, β-Alanine, 30B, 207
$C_3H_7NO_2S$, L-Cysteine, 33B, 172; 39B, 341; 41B, 523
$C_3H_7NO_3$, DL-Serine, 17, 655; 39B, 341; 40B, 430
$C_3H_7NO_3$, L-(-)-Serine, 38B, 475; 39B, 340; 40B, 429, 430
$C_3H_7NO_5S$, L-Cysteic acid, 33B, 174
$C_3H_7N_3O_2$, Guanidinoacetic acid, 45B, 505
$C_3H_7N_3O_2$, Glycocyamine, 39B, 340
$C_3H_8ClNO_2$, Sarcosine hydrochloride, 44B, 430
$C_3H_8ClNO_2$, L-Alanine hydrochloride, 43B, 587
$C_3H_8Cl_4FeNO_2$, β-Alaninium tetrachloroferrate(III), 45B, 505
$C_3H_8NO_6P$, L-O-serine phosphate, 35B, 302
$C_3H_9NO_4$, L-Serine monohydrate, 39B, 341
$C_3H_9NO_6S$, L-Cysteic acid monohydrate, 37B, 260; 39B, 342
$C_3H_{10}ClNO_3S$, L(+)-Cysteine hydrochloride monohydrate, 33B, 174
$C_3H_{10}NO_7P$, DL-O-serine phosphate monohydrate, 35B, 299
$C_3H_{11}N_2O_5P$, Glycyl aminomethylphosphonic acid monohydrate, 42B, 351
$C_4D_8N_2O_3$, Perdeutero-α-glycylglycine, 35B, 304; 41B, 524
$C_4H_5NO_3$, N-Carboxy-L-alanine anhydride, 42B, 351
$C_4H_7KN_4O_4$, Potassium allantoinate, 33B, 175
$C_4H_7NO_2$, trans-4-Aminocrotonic acid, 41B, 5
$C_4H_7NO_2S$, L-Thioproline, 40B, 431
$C_4H_7NO_3$, N-Acetylglycine, 13, 486; 27, 764; 41B, 524
$C_4H_7NO_4$, Iminodiacetic acid (form i), 40B, 432
$C_4H_7NO_4$, Iminodiacetic acid (form ii), 45B, 506
$C_4H_7NO_4$, Iminodiacetic acid (form iii), 45B, 506
$C_4H_7NO_4$, DL-Aspartic acid, 33B, 177; 39B, 343
$C_4H_7NO_4$, L-Aspartic acid, 33B, 176
$C_4H_7N_3O$, Creatinine, 19, 597
$C_4H_8BrLiN_2O_3$, Glycylglycine lithium bromide, 37B, 261
$C_4H_8BrNO_4$, Iminodiacetic acid hydrobromide, 39B, 343; 42B, 5
C_4H_8ClNOS, DL-Homocysteine thiolactone hydrochloride, 30B, 210
$C_4H_8ClNO_4$, Aspartic acid hydrochloride, 43B, 588
$C_4H_8ClNO_4$, Iminodiacetic acid hydrochloride, 39B, 343; 40B, 431
$C_4H_8INO_4$, Iminodiacetic acid hydroiodide, 40B, 433
$C_4H_8N_2O_3$, Glycylglycine (α form), 33B, 179; 35B, 304; 41B, 524; 43B, 588
$C_4H_8N_2O_3$, Glycylglycine (β form), 12, 349
$C_4H_9NO_2$, DL-α-Amino-n-butyric acid (form A), 33B, 180; 46B, 469
$C_4H_9NO_2$, DL-α-Amino-n-butyric acid (form B), 33B, 180; 46B, 469
$C_4H_9NO_2$, DL-α-Amino-n-butyric acid (form C), 38B, 475
$C_4H_9NO_2$, DL-α-Amino-n-butyric acid (form D), 46B, 469
$C_4H_9NO_2$, α-Amino-α-methylpropionic acid, 16, 443
$C_4H_9NO_2$, γ-Aminobutyric acid, 39B, 344
$C_4H_9NO_3$, DL-γ-Amino-β-hydroxybutyric acid, 39B, 344

$C_4H_9NO_3$, L(S)-Threonine, 13, 488; 40B, 434
$C_4H_9NO_3$, L-Allothreonine, 41B, 525
$C_4H_9NO_3$, L-Threonine - L-allothreonine, 41B, 526
$C_4H_9NO_3S$, (+)-S-Methyl-L-cysteine sulfoxide, 27, 765
$C_4H_9NO_5S$, DL-Homocysteic acid, 43B, 589
$C_4H_9N_3O_2 \cdot H_2O$, Creatine monohydrate, 18, 655; 19, 531; 45B, 507
$C_4H_9N_3O_6$, Glycylglycine nitrate, 39B, 346
$C_4H_{10}ClNO_2$, γ-Aminobutyric acid hydrochloride, 39B, 345
$C_4H_{10}FNO_5$, Iminodiacetic acid hydrofluoride monohydrate, 40B, 433
$C_4H_{10}KNO_4$, Monopotassium L-aspartate dihydrate, 37B, 262
$C_4H_{10}N_2O_4$, L-Asparagine monohydrate, 26, 602; 38B, 476
$C_4H_{10}N_2O_4S$, Glycyltaurine, 43B, 590
$C_4H_{11}BrN_2O_4$, Diglycine hydrobromide, 10, 217; 20, 509; 21, 538
$C_4H_{11}ClN_2O_2$, L-α,γ-Diaminobutyric acid monochloride, 37B, 262
$C_4H_{11}ClN_2O_4$, Diglycine hydrochloride, 21, 538; 24, 544
$C_4H_{11}ClN_2O_4$, Glycylglycine monohydrochloride monohydrate, 34B, 267; 38B, 477
$C_4H_{11}IN_2O_4$, Diglycine hydroiodide, 38B, 479
$C_4H_{11}N_3O_3S$, γ-Guanidino-propane sulphonic acid, 39B, 347
$C_4H_{11}N_3O_4S$, γ-Guanidino-β-hydroxypropane sulphonic acid, 39B, 348
$C_4H_{11}N_3O_7$, Diglycine nitrate, 33B, 183
$C_4H_{12}BaCl_2N_2O_5$, Bisglycinebarium(II) dichloride monohydrate, 41B, 527
$C_4H_{12}IN_2NaO_5$, Sodium iodide diglycine monohydrate, 37B, 263
$C_4H_{12}N_2O_8Se$, Diglycine selenate, 41B, 528
$C_4H_{13}N_2O_7P$, Glycylglycine phosphate monohydrate, 38B, 479
$C_4H_{14}N_2O_9S$, Diglycine sulfate monohydrate, 40B, 434
$C_4H_{16}Cl_2N_2O_7Sr$, Bisglycinestrontium(II) dichloride trihydrate, 41B, 527
$C_4H_{18}CaCl_2N_2O_8$, Bis(glycine)calcium dichloride tetrahydrate, 46B, 470
$C_5H_7NO_3$, L-Pyroglutamic acid, 44B, 430
$C_5H_7NO_3$, Pyroglutamic acid, 38B, 480; 40B, 197; 43B, 604
$C_5H_7NO_3$, 2-(Acetylamino)prop-2-enoic acid, 45B, 507
$C_5H_7NO_3$, 2-Amino-3-hydroxypent-4-ynoic acid, 43B, 591
$C_5H_7N_3O_2$, Glycocyamine, 42B, 352
$C_5H_7N_3O_4$, O-Diazoacetyl-L-serine, 44B, 431
$C_5H_7N_3O_5$, Quisqualic acid, 42B, 352
$C_5H_8ClNO_3$, DL-N-Chloroacetylalanine, 35B, 305
$C_5H_9NO_2$, L-Proline, 30B, 209
$C_5H_9NO_2S$, Dehydromethionine, 42B, 285
$C_5H_9NO_3$, 4-Hydroxy-L-proline, 15, 493; 16, 445; 39B, 349
$C_5H_9NO_3S$, N-Acetyl-L-cysteine, 46B, 471
$C_5H_9NO_4$, L-Glutamic acid (α form), 46B, 471
$C_5H_9NO_4$, L-Glutamic acid (β form), 19, 529; 38B, 481
$C_5H_9NO_4$, 2,3-cis-3,4-trans-3,4-Dihydroxy-L-proline, 34B, 278; 35B, 307
$C_5H_9NO_4S$, S-Carboxymethyl-L-cysteine, 45B, 507
$C_5H_9NO_5$, 2-(Hydroxymethyl)aspartic acid, 46B, 472
$C_5H_9NO_5S$, S-Carboxymethyl-L-cysteine sulfoxide, 42B, 353
$C_5H_9NO_6S$, S-Carboxymethyl-L-cysteine sulphone, 42B, 353
$C_5H_{10}CaClNO_5$, Calcium L-glutamate chloride monohydrate, 43B, 591
$C_5H_{10}ClNO_2$, DL-Proline hydrochloride, 34B, 268
$C_5H_{10}ClNO_2$, DL-4-Hydroxyvaline lactone hydrochloride, 44B, 431
$C_5H_{10}ClNO_4$, DL-Glutamic acid hydrochloride, 17, 633
$C_5H_{10}ClNO_4$, L-Glutamic acid hydrochloride, 38B, 482
$C_5H_{10}N_2O_2$, Acetylglycine-N-methylamide, 40B, 435

$C_5H_{10}N_2O_3$, Glycyl-D,L-alanine, 45B, 508
$C_5H_{10}N_2O_3$, Glycyl-L-alanine, 45B, 508
$C_5H_{10}N_2O_3$, L-Alanylglycine, 35B, 308
$C_5H_{10}N_2O_3$, L-Glutamine, 16, 449; 39B, 349
$C_5H_{10}N_2O_4$, L-Serylglycine, 44B, 431
$C_5H_{11}ClN_2O_3$, Glycyl-L-alanine hydrochloride, 38B, 483
$C_5H_{11}ClN_2O_3$, L-Glutamine hydrochloride, 38B, 483
$C_5H_{11}ClN_2O_6$, Cucurbitine perchlorate, 30B, 212
$C_5H_{11}NO_2$, DL-Valine, 34B, 269
$C_5H_{11}NO_2$, L-Valine, 35B, 310
$C_5H_{11}NO_2S$, D,L-Methionine (α form), 16, 464; 46B, 472
$C_5H_{11}NO_2S$, D,L-Methionine (β form), 16, 464; 46B, 472
$C_5H_{11}NO_2S$, L-Methionine, 39B, 349
$C_5H_{12}BrNO_2$, L-Valine hydrobromide, 31B, 193
$C_5H_{12}BrN_3O_2$, γ-Guanidinobutyric hydrobromide, 38B, 484
$C_5H_{12}ClNO_2$, Betaine hydrochloride, 35B, 313
$C_5H_{12}ClNO_2$, DL-Valine hydrochloride, 43B, 592
$C_5H_{12}ClNO_2$, L-Valine hydrochloride, 31B, 188; 32B, 226; 40B, 435
$C_5H_{12}ClNO_2S$, L-Methionine hydrochloride, 43B, 593
$C_5H_{12}ClN_3O_2$, γ-Guanidinobutyric acid hydrochloride, 38B, 484
$C_5H_{12}N_2O_3S$, (2S,SR)-Methionine sulfoximine, 35B, 312
$C_5H_{13}BrN_2O_2$, DL-Ornithine hydrobromide, 35B, 314; 37B, 264
$C_5H_{13}CaNO_7$, Calcium L-glutamate trihydrate, 40B, 651
$C_5H_{13}ClN_2O_2$, L-Ornithine hydrochloride, 32B, 227; 34B, 266
$C_5H_{13}ClN_2O_4$, Glycyl-L-alanine hydrochloride monohydrate, 20, 515
$C_5H_{13}NO_5$, allo-4-Hydroxy-L-proline dihydrate, 42B, 354
$C_5H_{14}BrLiN_2O_5$, Lithium bromide L-alanylglycine dihydrate, 37B, 264
$C_5H_{14}ClNO_3$, L-Valine hydrochloride monohydrate, 34B, 265
$C_5H_{14}ClNO_3S$, L-β,β'-Dimethylcysteine hydrochloride monohydrate, 39B, 351
$C_5H_{15}N_2O_5P$, β-Alanyl-ciliatine monohydrate, 43B, 593
$C_6H_7N_2O_2S$, Cyclo-glycyl-4-thiopropyl, 42B, 354
$C_6H_8N_2O_2 \cdot 2\,H_2O$, Glycyl-D-threonine dihydrate, 45B, 509
$C_6H_8N_2O_2$, L-Prolinehydantoin, 46B, 473
$C_6H_8N_2O_3$, D-Allohydroxyprolinehydantoin, 46B, 473
$C_6H_9ClN_2O_4$, Chloroacetylglycylglycine, 39B, 351
$C_6H_9NO_2 \cdot H_2O$, 2,4-Methanoproline monohydrate, 46B, 474
$C_6H_9NO_4 \cdot H_2O$, 2,4-Methanoglutamic acid monohydrate, 46B, 474
$C_6H_9NO_4$, trans-4-Carboxy-L-proline, 44B, 432
$C_6H_9NO_6$, L-γ-Carboxyglutamic acid, 45B, 510
$C_6H_9N_3O_2$, DL-Histidine, 40B, 436
$C_6H_9N_3O_2$, L-Histidine, 38B, 485
$C_6H_9N_3O_2$, β-(Pyrazolyl-3)-L-alanine, 38B, 487
$C_6H_{10}N_2O_2$, Cyclo-D-alanyl-L-alanyl, 34B, 129; 35B, 318
$C_6H_{10}N_2O_2$, Cyclo-L-alanyl-L-alanyl, 35B, 318
$C_6H_{10}N_2O_2$, Cyclo-di-β-alanyl, 38B, 487
$C_6H_{10}N_2O_2$, L-Alanyl-L-alanyl-2,5-diketopiperazine, 40B, 1091
$C_6H_{10}N_2O_2$, N,N'-Dimethyldiketopiperazine, 40B, 1083
$C_6H_{10}N_2O_2$, trans-3,6-Dimethylpiperazine-2,5-dione, 46B, 473
$C_6H_{11}BrLiN_3O_4$, Glycylglycylglycine lithium bromide, 37B, 267
$C_6H_{11}Cl_2N_3O_2$, Histidine dihydrochloride, 39B, 352; 40B, 437
$C_6H_{11}NO_2 \cdot 4\,H_2O$, DL-Homoproline tetrahydrate, 45B, 510
$C_6H_{11}NO_3$, trans-3,4-Methylene-L-proline monohydrate, 37B, 265
$C_6H_{11}NO_3$, 3-Hydroxy-4-methylproline, 40B, 441
$C_6H_{11}NO_3S$, N-Formyl-L-methionine, 43B, 594
$C_6H_{11}NO_4 \cdot 0.5\,H_2O$, D,L-$\alpha$-Methylglutamic acid hemihydrate, 45B, 511
$C_6H_{11}NO_5$, (2S,3R,4S)-β-Hydroxy-α-methylglutamic acid, 40B, 442

$C_6H_{11}N_3O_4$, Glycyl-L-asparagine, 16, 451; 18, 670
$C_6H_{12}ClNO_2$, 2-Butenylglycine hydrochloride, 45B, 511
$C_6H_{12}ClN_3O_3$, L-Histidine hydrochloride monohydrate, 20, 510; 29, 701; 38B, 489; 43B, 595
$C_6H_{12}ClO_3$, cis-3,4-Methylene-L-proline hydrochloride monohydrate, 37B, 266
$C_6H_{12}N_2O_2$, N-Acetyl-α-aminoisobutyric acid methyl amide, 44B, 433
$C_6H_{12}N_2O_2$, N-Acetyl-DL-alanine-N-methylamide, 40B, 438
$C_6H_{12}N_2O_2$, N-Acetyl-L-alanine-N-methylamide, 40B, 438
$C_6H_{12}N_2O_3$, L-Alanyl-L-alanine, 37B, 267
$C_6H_{12}N_2O_4$, L-Alanyl-L-serine, 44B, 433
$C_6H_{12}N_2O_4S_2$, L-Cysteinyl-L-cysteine monohydrate, 43B, 596
$C_6H_{12}N_2O_4S_2$, L-Cystine (hexagonal form), 23, 593
$C_6H_{12}N_2O_4S_2$, L-Cystine (tetragonal form), 40B, 439
$C_6H_{12}N_2O_6$, Ammonium DL-γ-carboxyglutamate, 45B, 512
$C_6H_{13}ClN_2O_3$, L-Alanyl-L-alanine hydrochloride, 34B, 270
$C_6H_{13}NO_2$, D-Alloisoleucine, 41B, 531
$C_6H_{13}NO_2$, DL-Isoleucine, 39B, 352
$C_6H_{13}NO_2$, DL-Leucine, 41B, 530
$C_6H_{13}NO_2$, DL-Norleucine, 17, 659
$C_6H_{13}NO_2$, L-Isoleucune, 37B, 268
$C_6H_{13}NO_2$, L-Leucine, 42B, 355
$C_6H_{13}NO_2$, L-Norleucine, 39B, 349
$C_6H_{14}BrNO_2$, L-Leucine hydrobromide, 32B, 229
$C_6H_{14}Br_2N_2O_4S_2$, L-Cystine dihydrobromide, 24, 561; 29, 484
$C_6H_{14}ClN_3O_3$, L-Citrulline hydrochloride, 37B, 269; 38B, 489
$C_6H_{14}ClN_3O_4$, DL-Histidine hydrochloride dihydrate, 35B, 316
$C_6H_{14}Cl_2N_2O_4S$, meso-Lanthionine dihydrochloride, 40B, 439
$C_6H_{14}Cl_2N_2O_4S_2$, L-Cystine dihydrochloride dihydrate (copper(II) doped), 42B, 356
$C_6H_{14}Cl_2N_2O_4S_2$, L-Cystine dihydrochloride, 22, 647; 40B, 440
$C_6H_{14}INO_2$, L-Leucine hydroiodide, 37B, 271
$C_6H_{14}N_2O_5$, Glycyl-DL-threonine monohydrate, 41B, 530
$C_6H_{15}BrN_2O_4$, Disarcosine hydrobromide, 44B, 434
$C_6H_{15}ClN_2O_2$, DL-Lysine hydrochloride, 45B, 513
$C_6H_{15}ClN_4O_2$, L-Arginine hydrochloride, 40B, 1091
$C_6H_{15}Cl_2NO_2S$, Vitamin U hydrochloride, 43B, 597
$C_6H_{16}BrNO_3$, D-iso-Leucine hydrobromide monohydrate, 18, 666
$C_6H_{16}ClNO_3$, D-Alloisoleucine hydrochloride monohydrate, 42B, 356
$C_6H_{16}ClNO_3$, D-iso-Leucine hydrochloride monohydrate, 18, 666
$C_6H_{16}Cl_2N_4O_2S_2$, L-Cystinediamide dihydrochloride, 33B, 184
$C_6H_{16}Cl_6N_2O_2Pt$, L-Lysine hexachloroplatinate(IV), 44B, 434
$C_6H_{16}N_2O_6$, Glycyl-L-threonine dihydrate, 39B, 352
$C_6H_{17}BeF_4N_3O_6$, Tri-glycine fluoroberyllate, 39B, 353; 45B, 513
$C_6H_{17}BrN_4O_3$, L-Arginine hydrobromide monohydrate, 31B, 189
$C_6H_{17}CaCl_2N_3O_7$, Glycylglycylglycine-calcium chloride complex trihydrate, 34B, 355
$C_6H_{17}ClN_4O_3$, L-Arginine hydrochloride monohydrate, 35B, 319
$C_6H_{17}N_3O_{10}S$, Triglycine sulfate, 23, 590; 39B, 354; 40B, 441
$C_6H_{17}N_3O_{10}Se$, Triglycine selenate, 40B, 597
$C_6H_{18}Br_2N_2O_6S_2$, L-Cystine dihydrobromide dihydrate, 41B, 532
$C_6H_{18}N_4O_4$, L-Arginine dihydrate, 29, 479; 39B, 354
$C_6H_{19}ClN_2O_4$, L-Lysine hydrochloride dihydrate, 23, 596; 27, 767; 38B, 492
$C_6H_{19}N_4O_7P$, L-Arginine phosphate monohydrate, 37B, 271; 38B, 490
$C_7H_9NO_3S \cdot H_2O$, Thienyl-DL-serine monohydrate, 45B, 513
$C_7H_{10}N_2O_2$, Cyclo-L-prolyl-glycyl, 41B, 533

$C_7H_{10}N_2O_5$, 3R-(1'S-Aminocarboxymethyl)-2-pyrrolidone-5S-carboxylic acid, 41B, 533
$C_7H_{11}NO_3$, N-Carboxy-L-leucine anhydride, 44B, 434
$C_7H_{11}NO_4 \cdot H_2O$, N-Acetyl-L-4-hydroxyproline monohydrate, 45B, 514
$C_7H_{12}ClNO_3$, N-Chloroacetyl-L-norvaline, 42B, 357
$C_7H_{12}N_2O_2$, 1-Acetylprolinamide, 45B, 515
$C_7H_{12}N_2O_4$, Glycyl-L-4-hydroxyproline, 46B, 475
$C_7H_{12}N_2O_4$, N-Acetyl-L-glutamine, 40B, 442; 42B, 357
$C_7H_{13}NO_2 \cdot 0.5\ H_2O$, 1-Amino-3-methylcyclopentanecarboxylic acid hemihydrate, 39B, 354
$C_7H_{13}NO_3$, N-Acetyl-L-norvaline, 40B, 443
$C_7H_{13}N_3O_4$, Sarcosylglycylglycine, 43B, 200
$C_7H_{15}ClN_2O_4$, LL-Diaminopimelic acid hydrochloride, 43B, 598
$C_7H_{15}ClN_2O_4$, meso-Diaminopimelic acid hydrochloride, 43B, 598
$C_7H_{15}NO_2 \cdot C_6H_6$, N,N-Diethyl-β-alanine benzene solvate, 45B, 516
$C_7H_{16}ClNO_3$, DL-Carnitine hydrochloride, 40B, 444
$C_7H_{16}ClN_3O_3$, L-Homocitrulline hydrochloride, 38B, 489
$C_7H_{18}N_2O_6S$, β-(S)-Methyllanthionine dihydrate, 41B, 536
$C_8H_{10}N_2O_4$, S-β-N-(3-Hydroxy-4-pyridone)-α-aminopropionic acid, 39B, 355
$C_8H_{11}NO_2$, 1,4-Cyclohexadiene-1-glycine, 40B, 1092
$C_8H_{11}NO_6$, Lycoperdic acid, 44B, 435
$C_8H_{12}N_2O_2$, L-Prolyl-L-alanyl cyclodipeptide, 45B, 517
$C_8H_{12}N_2O_8S \cdot 1.5\ H_2O$, L-Mimosine sulfate hydrate, 40B, 444
$C_8H_{13}N_2O_8Rb$, Rubidium hydrogen iminodiacetate iminodiacetic acid, 43B, 599
$C_8H_{13}N_3O_4$, L-N-Acetylhistidine monohydrate, 38B, 493
$C_8H_{14}BrNO_3$, N-(Bromoacetyl)-L-leucine, 41B, 536
$C_8H_{14}N_2O_2$, Acetyl-L-proline-N-methylamide, 37B, 272
$C_8H_{14}N_2O_2$, Cyclobis(N-methyl-L-alanyl), 42B, 358
$C_8H_{14}N_2O_2$, Cyclosarcosyl-L-valyl, 41B, 537
$C_8H_{14}N_2O_2$, rac-Cyclobis(N-methylalanyl), 42B, 358
$C_8H_{14}N_2O_3 \cdot H_2O$, L-Prolylsarcosine monohydrate, 46B, 476
$C_8H_{15}NO_3$, N-Acetyl-L-leucine, 41B, 539
$C_8H_{15}NO_3S$, N-Acetylmethionine methyl ester, 40B, 445
$C_8H_{15}NO_7$, O-(β-D-Xylopyranosyl)-L-serine, 41B, 538
$C_8H_{16}BrNO_2 \cdot H_2O$, 1-Aminocycloheptanecarboxylic acid hydrobromide monohydrate, 46B, 477
$C_8H_{16}BrN_3O_4$, Diglycylglycine ethyl ester hydrobromide, 16, 451
$C_8H_{16}CaCl_2N_4O_6$, Glycylglycine calcium chloride, 37B, 261
$C_8H_{16}ClN_3O_4$, Diglycylglycine ethyl ester hydrochloride, 16, 451
$C_8H_{16}N_2O_3$, Glycyl-L-leucine, 39B, 356; 40B, 446
$C_8H_{16}N_2O_3S$, DL-Alanyl-LD-methionine, 40B, 447
$C_8H_{17}BrN_2O_3$, D-Leucylglycine hydrobromide, 31B, 190; 34B, 267
$C_8H_{17}N_3O_2$, N(α)-Acetyl-aza-α'-homo-L-valine methylamide, 42B, 359
$C_8H_{19}BrN_2O_6$, DL-Allothreonine hydrobromide, 41B, 541
$C_8H_{20}Cl_2N_2O_5S_2$, L-Cystine dimethyl ester dihydrochloride monohydrate, 41B, 540
$C_8H_{20}I_3KN_4O_8$, Tetrakis(glycine) potassium triiodide, 46B, 491
$C_8H_{22}CaCl_2N_2O_6$, Bis(DL-2-aminobutyric acid) calcium chloride dihydrate, 44B, 438
$C_9H_9NO_3$, Hippuric acid, 37B, 273; 38B, 494; 40B, 448
$C_9H_9N_3O_7 \cdot H_2O$, 3,5-Dinitro-L-tyrosine monohydrate, 45B, 518
$C_9H_{11}ClN_2O_4$, DL-p-Nitrophenylalanine hydrochloride, 44B, 436
$C_9H_{11}NO_3$, DL-Tyrosine, 39B, 356
$C_9H_{11}NO_3$, L-Tyrosine, 38B, 495; 39B, 357
$C_9H_{11}NO_3$, 2-Hydroxyphenylalanine (o-Tyrosine), 41B, 542

$C_9H_{11}NO_3$, 3-Hydroxyphenylalanine, 40B, 448
$C_9H_{11}NO_4$, 2S-3-(3,4-Dihydroxyphenyl)alanine (L-dopa), 37B, 276; 39B, 358
$C_9H_{11}NO_5$, DL-Threo-3,4-dihydroxyphenylserine, 45B, 517
$C_9H_{12}BrNO_3$, D-β-Tyrosine hydrobromide, 40B, 449
$C_9H_{12}BrNO_3$, L-Tyrosine hydrobromide, 22, 649; 23, 597
$C_9H_{12}ClNO_2$, L-Phenylalanine hydrochloride, 29, 700; 41B, 543
$C_9H_{12}ClNO_3$, D-β-Tyrosine hydrochloride, 40B, 449
$C_9H_{12}ClNO_3$, L-Tyrosine hydrochloride, 23, 597; 39B, 357
$C_9H_{12}ClNO_3$, o-Tyrosine hydrochloride, 43B, 600
$C_9H_{12}ClNO_4$, 3-(3,4-Dihydroxyphenyl)-L-alanine hydrochloride (L-dopa hydrochloride), 37B, 277; 40B, 450
$C_9H_{12}ClNO_5 \cdot 3\ H_2O$, DL-Threo-3,4-dihydroxyphenylserine hydrochloride trihydrate, 45B, 517
$C_9H_{12}N_3NaO_9 \cdot H_2O$, Diaquasodium 3,5-dinitro-L-tyrosinate monohydrate, 45B, 518
$C_9H_{13}I_2NO_5$, Di-iodo-L-tyrosine dihydrate, 32B, 230
$C_9H_{13}NO_4$, D-β-Tyrosine monohydrate, 40B, 450
$C_9H_{14}KNO_8S$, Potassium L-tyrosine-O-sulfate dihydrate, 37B, 275
$C_9H_{14}N_4O_2$, N-Acetyl-L-histidine-N-methylamide, 43B, 600
$C_9H_{14}N_4O_3$, β-Alanyl-L-histidine, 42B, 359; 43B, 601
$C_9H_{14}N_4O_4$, Cyclo(L-seryl-L-histidyl) monohydrate, 44B, 437
$C_9H_{15}NO_3S$, 1-(D-3-Mercapto-2-methylpropionyl)-L-proline, 46B, 478
$C_9H_{15}NO_4$, 3-Hydroxy-3-isobutyl-2-pyrrolidone-5-carboxylic acid, 43B, 270
$C_9H_{15}NO_9$, L-Serine L-ascorbic acid, 46B, 478
$C_9H_{15}N_3O_3$, Cyclotrisarcosyl, 42B, 194
$C_9H_{17}NO_4$, N-(t-Butyloxycarbonyl)-2-methylalanine, 46B, 479
$C_9H_{17}N_3O_4 \cdot 0.5\ H_2O$, L-Alanyl-L-alanyl-L-alanine hemihydrate, 41B, 544
$C_9H_{18}N_2O_2$, DL-Acetylleucine N-methylamide, 34B, 272
$C_9H_{18}N_2O_3$, N-Pivalyl-L-seryl-methylamide, 42B, 360
$C_9H_{19}BrN_2O_3$, N-Methyl-DL-leucylglycine hydrobromide, 34B, 272
$C_9H_{19}N_3O_4S$, L-Ergothioneine dihydrate, 42B, 360
$C_9H_{21}CaCl_2N_3O_6$, Tris-sarcosine calcium chloride, 38B, 889
$C_{10}H_{10}N_2O_5$, p-Nitrophenaceturic acid, 43B, 602
$C_{10}H_{10}N_4O_7$, N-(2,4-Dinitrophenyl)asparagine, 42B, 361
$C_{10}H_{11}N_3O_6$, Methyl N-(2,4-dinitrophenyl)alaninate, 42B, 361
$C_{10}H_{12}BrNO_4$, L-(3-Bromo-4-hydroxy-5-methoxy)phenylalanine, 40B, 451
$C_{10}H_{13}NO_3$, DL-α-Methyltyrosine, 42B, 362
$C_{10}H_{14}N_2O_2$, Cyclo-L-prolyl-L-prolyl, 41B, 1239
$C_{10}H_{14}N_2O_3 \cdot H_2O$, cyclo-(L-Prolyl-L-4-hydroxyproline) monohydrate, 46B, 475
$C_{10}H_{14}N_2O_4 \cdot H_2O$, 2-Amino-4-hydroxy-4-(5-hydroxy-2-pyridyl)-3-methylbutyric acid hydrate, 46B, 479
$C_{10}H_{15}N_3O_2$, N-Methyltropane-3-spiro-5'-hydantoin, 43B, 381
$C_{10}H_{16}N_2O_4 \cdot H_2O$, L-Prolyl-L-4-hydroxyproline monohydrate, 46B, 475
$C_{10}H_{16}N_2O_8 \cdot 0.39\ H_2O$, Ethylenediaminetetraacetic acid hydrate (α form), 40B, 451
$C_{10}H_{16}N_4O_4$, Cyclo(glycyl-sarcosyl-glycyl-sarcosyl), 41B, 546
$C_{10}H_{17}NO_4$, N-(t-Butyloxycarbonyl)-L-proline, 40B, 452
$C_{10}H_{17}N_3O_6S$, γ-L-Glutamyl-L-cysteinylglycine, 22, 651
$C_{10}H_{18}N_2O_5$, L-Proline L-hydroxyproline monohydrate, 43B, 603
$C_{10}H_{18}N_2O_5$, t-Butyloxycarbonyl-glycyl-L-alanine, 43B, 602
$C_{10}H_{18}N_2O_8$, L-Glutamic acid L-pyroglutamic acid monohydrate, 43B, 604
$C_{10}H_{18}N_4O_5$, Cyclo-(L-threonyl-L-histidyl) dihydrate, 42B, 362

$C_{10}H_{18}N_4O_7$, Cyclo(L-histidyl-L-aspartyl) trihydrate, 44B, 438
$C_{10}H_{18}N_4O_7$, L-Histidine L-aspartic acid monohydrate, 44B, 437
$C_{10}H_{19}N_3O_4$, DL-Leucylglycylglycine, 43B, 605
$C_{10}H_{19}N_6O_8Rb$, Rubidium N-(purin-6-ylcarbamoyl)-L-threonine tetrahydrate, 41B, 545
$C_{10}H_{20}IN_4NaO_6S_2$, Cystylglycine-sodium iodide, 15, 437
$C_{10}H_{20}N_2O_2$, N-Acetyl-DL-pseudoleucyl-dimethylamide, 39B, 359
$C_{10}H_{20}N_2O_2$, N-Pivaloyl-glycyl-isopropylamide, 39B, 359
$C_{10}H_{20}N_2O_3S_2$, L-Methionyl-L-methionine, 41B, 547
$C_{10}H_{21}N_3O_6$, L-Lysine L-aspartate, 42B, 363
$C_{10}H_{22}Cl_2N_2O_4S_2$, 3,3,3',3'-Tetramethyl-D-cystine dihydrochloride, 41B, 548
$C_{10}H_{22}N_2O_8$, Palythine trihydrate, 46B, 480
$C_{10}H_{23}N_4O_8S_2$, N,N'-Diglycyl-L-cystine dihydrate, 18, 672; 42B, 363
$C_{10}H_{24}N_2O_6S_2$, meso-3,3'-Dithiobisvaline dihydrate, 40B, 453
$C_{10}H_{26}CaCl_2N_2O_6$, Bis(DL-valine) calcium chloride dihydrate, 44B, 438
$C_{11}H_{12}N_2O_2$, DL-Tryptophan, 46B, 480
$C_{11}H_{12}N_2O_3$, Cyclo(glycyl-L-tyrosyl), 39B, 360
$C_{11}H_{13}BrN_2O_2$, L-Tryptophan hydrobromide, 31B, 191; 32B, 231
$C_{11}H_{13}ClN_2O_2$, L-Tryptophan hydrochloride, 31B, 191
$C_{11}H_{14}N_2O_3$, Glycyl-DL-phenylalanine, 42B, 364
$C_{11}H_{15}NO_3$, Tyrosine ethyl ester, 35B, 325
$C_{11}H_{17}ClN_2O_4$, Glycyl-L-phenylalanine hydrochloride monohydrate, 40B, 454
$C_{11}H_{17}ClN_2O_5$, Glycyl-L-tyrosine hydrochloride monohydrate, 17, 661
$C_{11}H_{18}N_2O_2$, Cyclo(L-prolyl-D-t-leucyl-), 45B, 521
$C_{11}H_{18}N_2O_2$, Cyclo-L-propyl-L-leucyl, 38B, 496
$C_{11}H_{18}N_2O_4$, N-Acetyl-L-prolyl-D-lactylmethylamide, 40B, 456
$C_{11}H_{18}N_2O_4$, N-Acetyl-L-prolyl-L-lactylmethylamide, 40B, 455
$C_{11}H_{18}N_2O_6$, Glycyl-L-tyrosine dihydrate, 40B, 456
$C_{11}H_{19}NO_3$, Cyclo-D-methylvalyl-D-α-hydroxyisovaleryl, 43B, 453; 46B 481
$C_{11}H_{20}N_4O_5$, Cyclo(glycyl-sarcosyl-sarcosyl-sarcosyl) hydrate, 41B, 546
$C_{11}H_{22}BrNO_4$, D-N-Methylvalyl-D-α-hydroxyvaleric acid hydrobromide, 43B, 607
$C_{11}H_{22}N_2O_2S$, N-Acetyl-DL-methionine-diethylamide, 45B, 522
$C_{11}H_{24}N_2O_2$, N-Acetyl-L-leucyl-isopropylamide, 41B, 548
$C_{11}H_{25}N_5O_7$, L-Arginine L-glutamate monohydrate, 43B, 607
$C_{12}H_{14}N_2O_4$, DL-Tryptophan formate, 39B, 361
$C_{12}H_{16}N_2O_2$, N-Acetyl-DL-phenylalanine-N-methylamide, 40B, 457
$C_{12}H_{16}N_2O_2$, N-Acetyl-L-phenylalanine-N-methylamide, 43B, 608
$C_{12}H_{16}N_2O_3$, N-Acetyl-L-tyrosinemethylamide, 39B, 362; 40B, 456
$C_{12}H_{16}N_2O_5$, Cyclo(L-seryl-L-tyrosyl) monohydrate, 39B, 360
$C_{12}H_{16}N_4O_4$, Methyl L-pyroglutamyl-L-histidine, 39B, 362
$C_{12}H_{17}N_3O_2$, N(α)-Acetyl-aza-α-homo-phenylalanine methylamide, 45B, 11
$C_{12}H_{18}N_2O_5$, N-Acetyl-L-prolyl-L-4-hydroxyproline, 46B, 475
$C_{12}H_{18}N_6O_6 \cdot 0.5\ H_2O$, Cyclo(hexaglycyl) hemihydrate, 28, 525
$C_{12}H_{20}BrNO_6$, (+)-2-Benzylglutamic acid hydrobromide dihydrate, 32B, 232
$C_{12}H_{20}ClLiN_4O_8$, Bis(cyclodisarcosyl)lithium perchlorate, 43B, 610
$C_{12}H_{20}N_2O_5$, t-Butyloxycarbonylglycyl-L-proline, 43B, 610
$C_{12}H_{20}N_2O_5S_2$, t-Butyloxycarbonyl-L-cysteinyl-L-cysteine disulfide methyl ester, 43B, 612
$C_{12}H_{20}N_4O_3$, Cyclo-L-leucyl-L-histidyl monohydrate, 43B, 609
$C_{12}H_{20}N_4O_4$, Cyclo(alanyl-sarcosyl-sarcosyl-sarcosyl), 41B, 546

$C_{12}H_{21}ClN_2O_5$, L-Alanyl-L-phenylalanine hydrochloride dihydrate, 39B, 362; 40B, 454
$C_{12}H_{22}N_2O_2$, Cyclo(N-methyl-L-valyl-n-methyl-D-valyl), 43B, 358
$C_{12}H_{22}N_2O_2$, Cyclobis(N-methyl-L-valyl), 42B, 358
$C_{12}H_{22}N_4O_8$, L-Arginine L-ascorbate, 46B, 481
$C_{12}H_{23}N_3O_3$, N-Pivalyl-N'-methyl-L-glutaminyl-methylamide, 43B, 612
$C_{12}H_{27}ClN_2O_4$, Di-L-leucine hydrochloride, 38B, 497
$C_{12}H_{28}I_6N_2O_4Te$, DL-Norleucine hexaiodotellurate, 45B, 522
$C_{13}H_{13}NO_5$, N-Carboxy-γ-benzyl-L-glutamate anhydride, 44B, 439
$C_{13}H_{13}N_3O_2$, Cyclo-glycyl-tryptophyl, 40B, 458
$C_{13}H_{14}N_2O_3$, N-Acetyl-L-tryptophan, 43B, 613
$C_{13}H_{14}N_2O_6$, D-Tryptophan hydrogen oxalate, 46B, 480
$C_{13}H_{16}ClNO_4$, 1-Chloro-2-ethoxy-5-(α-(methyl N-acetylglycyl))benzene, 46B, 492
$C_{13}H_{16}N_2O_2$, 1-Amino-2-(1-hydroxyethylindoline)hydrazino lactone, 35B, 326
$C_{13}H_{16}N_2O_5S$, (R,R)-N-Acetyl-S-(2-nitro-1-phenylethyl)-L-cysteine, 45B, 523
$C_{13}H_{16}N_4O_{13}S_2$, (R)-(((3S)-3-Amino-3-carboxypropyl)(carboxymethyl)-methylsulphonium) 2,4,6-trinitrobenzenesulphonate, 43B, 614
$C_{13}H_{17}ClN_2O_2$, DL-Tryptophan ethyl ester hydrochloride, 41B, 549
$C_{13}H_{19}NO_6$, N-Acetyl-L-tyrosine ethyl ester monohydrate, 38B, 498
$C_{13}H_{19}N_3O_5$, DL-Glycyl-phenylalanyl-glycine monohydrate, 26, 724
$C_{13}H_{19}N_3O_5$, Glycyl-L-tryptophan dihydrate, 20, 515
$C_{13}H_{19}N_3O_6$, L-Tyrosyl-glycyl-glycine monohydrate, 44B, 440
$C_{13}H_{22}N_2O_5$, t-Butyloxycarbonyl-L-prolylsarcosine, 44B, 440
$C_{13}H_{22}N_3O_4 \cdot 0.84\ H_2O$, L-Leucyl-L-prolylglycine, 22, 654
$C_{13}H_{23}N_3O_4 \cdot H_2O$, L-Leucyl-L-prolylglycine monohydrate, 46B, 482
$C_{14}H_{16}N_2O_2$, Cyclo-L-prolyl-D-phenylalanyl, 42B, 365
$C_{14}H_{16}N_2O_3$, N-Acetyl-L-tryptophan methyl ester, 40B, 458
$C_{14}H_{17}BrN_2O_2$, L-6-Bromohypaphorine, 43B, 615
$C_{14}H_{17}N_3O_2$, N-Acetyl-DL-tryptophan-N-methylamide, 43B, 615
$C_{14}H_{19}NO_4$, t-Butoxycarbonyl-L-phenylalanine, 46B, 483
$C_{14}H_{23}ClN_4O_4S$, Tosyl-L-arginine methyl ester chloride, 42B, 365
$C_{14}H_{25}N_3O_3$, N-Pivalyl-L-prolyl-glycyl-dimethylamide, 41B, 550
$C_{14}H_{28}N_6O_9$, Glycylglycyl-D-alanyl-D-alanylglycylglycyl cyclic poly-peptide trihydrate, 35B, 328
$C_{15}H_{12}I_3NO_4$, 3,5,3'-Triiodo-L-thyronine, 40B, 459
$C_{15}H_{14}ClI_4NO_5$, L-Thyroxine hydrochloride monohydrate, 40B, 460
$C_{15}H_{15}N_3O_8 \cdot 0.33\ C_3H_6O$, DL-Histidinium trimesate acetone solvate, 45B, 523
$C_{15}H_{15}N_3O_8 \cdot 0.33\ C_3H_6O$, L-Histidinium trimesate acetone solvate, 45B, 523
$C_{15}H_{18}N_2O_5$, Benzyloxycarbonylglycyl-DL-proline, 44B, 441
$C_{15}H_{18}N_2O_5$, Benzyloxycarbonylglycyl-L-proline, 43B, 616
$C_{15}H_{19}ClI_3NO_7$, 3,5,3'-Triiodo-L-thyronine hydrochloride trihydrate, 40B, 460
$C_{15}H_{19}N_3O_5 \cdot H_2O$, 1-Thyminylacetic acid:tyramine (1:1) monohydrate, 46B, 483
$C_{15}H_{20}BrNO_4$, 4-Amino-3-hydroxy-6-methylheptanoic acid (p-bromobenzo-yl derivative), 39B, 363
$C_{15}H_{21}BrN_2O_3$, L-Prolyl-L-phenylalanine-O-methoxy hydrobromide, 37B, 278
$C_{15}H_{21}N_3O_3$, Cyclo(tri-L-prolyl), 42B, 366
$C_{15}H_{21}N_3O_3$, cyclo-(Di-L-prolyl-D-proline), 46B, 490
$C_{15}H_{21}N_3O_4$, Cyclo-L-prolyl-L-prolyl-L-hydroxyproline, 41B, 550
$C_{15}H_{25}N_5O_5 \cdot 0.5\ H_2O$, Cycloalanyltetrasarcosyl hemihydrate, 40B, 239

$C_{15}H_{26}N_6O_7$, Cyclo(L-alanyl-L-alanylglycylglycyl-L-alanylglycyl) monohydrate, 44B, 441
$C_{15}H_{27}N_3O_3$, Isobutyryl-L-alanyl-N'-isopropyl-L-prolinamide, 46B, 484
$C_{15}H_{27}N_3O_3$, Pivaloyl-L-prolyl-N'-isopropyl-glycinamide, 46B, 484
$C_{15}H_{27}N_3O_3$, N-Isobutyryl-L-prolyl-D-alanine-isopropylamide, 43B, 617
$C_{15}H_{27}N_3O_3$, N-Isobutyryl-L-prolyl-L-alanine-isopropylamide, 43B, 617
$C_{15}H_{28}N_6O_8$, Cyclo(L-alanyl-L-alanylglycyl-L-alanylglycylglycyl) dihydrate, 44B, 441
$C_{15}H_{29}N_5O_7$, Cyclopentasarcosyl dihydrate, 39B, 363
$C_{16}H_{13}NO_4$, N,N-Bis(2-hydroxyethyl)glycine, 43B, 617
$C_{16}H_{14}I_3NO_4$, 3,5,3'-Triiodo-L-thyronine methyl ester, 41B, 551
$C_{16}H_{18}N_2O_2$, Clavicipitic acid, 46B, 485
$C_{16}H_{21}NO_5$, Dimethyl 2-[N-(2-hydroxy-1-methyl-2-phenylethyl)-N-methylamino]maleate, 46B, 14
$C_{16}H_{29}N_3O_3 \cdot H_2O$, Pivaloyl-D-alanyl-N-isopropyl-D-prolinamide monohydrate, 46B, 485
$C_{16}H_{29}N_3O_3$, N-Pivaloyl-D-alanyl-L-proline-N-isopropylamide, 45B, 524
$C_{16}H_{30}N_2O_5S_2$, t-Butoxycarbonyl-D-methionyl-L-methionine methyl ester, 44B, 442
$C_{17}H_{12}N_2O_2$, Cyclo-N-methyl-L-alanyl-L-alanyl, 42B, 366
$C_{17}H_{13}NO_4$, Phthalylphenylalanine, 41B, 552
$C_{17}H_{17}N_3O_5$, N-Acetyl-L-tyrosine-p-nitroanilide, 42B, 367
$C_{17}H_{19}BrN_2O_6S$, Glycyl-L-phenylalanine p-bromobenzenesulphonate, 38B, 498
$C_{17}H_{20}ClI_2NO_5$, 2',3'-Dimethyl-3,5-diiodo-D,L-thyronine hydrochloride hydrate, 42B, 367
$C_{17}H_{20}N_4O_9$, Cytosine N,N-phthaloyl-DL-glutamic acid dihydrate, 46B, 486
$C_{17}H_{22}ClNO_5$, L-Thyronine hydrochloride ethyl ester monohydrate, 40B, 461
$C_{17}H_{24}N_2O_4$, Benzoyl-DL-leucylglycine ethyl ester, 41B, 553
$C_{17}H_{24}N_2O_5$, t-Butyloxycarbonylsarcosylglycine benzyl ester, 42B, 368
$C_{17}H_{24}N_2O_7S$, Tosyl-L-prolyl-L-hydroxyproline monohydrate, 27, 1051; 37B, 278
$C_{17}H_{25}N_5O_5$, Cyclo(glycylprolylglycyl-D-alanylprolyl), 44B, 443
$C_{17}H_{28}N_4O_5$, t-Butyloxycarbonyl-L-prolyl-L-prolylglycinamide, 45B, 525
$C_{17}H_{28}N_4O_7$, Cyclo-L-(O-t-butylseryl)-β-alanyl-glycyl-L-β-(O-methyl-aspartate), 41B, 554
$C_{17}H_{29}N_3O_6 \cdot 0.5\ H_2O$, N-(t-Butoxycarbonyl)-L-prolyl-L-valylglycine hemihydrate, 46B, 487
$C_{17}H_{35}N_2O_{15}P$, N-(5-O-Phosphopyridoxyl)-L-tyrosine heptahydrate, 41B, 552
$C_{18}H_{19}NO_4$, N-Benzoyl-L-tyrosine ethyl ester, 38B, 500
$C_{18}H_{21}I_2N_2O_5$, 3,5-Iodo-L-thyronine N-methylacetamide, 38B, 500
$C_{18}H_{22}N_2O_6S$, Glycyl-L-phenylalanine p-toluenesulphonate, 38B, 499
$C_{18}H_{25}N_3O_4$, N-Benzyloxycarbonyl-α-aminoisobutyryl-L-prolyl methylamide, 45B, 525
$C_{18}H_{26}ClI_2NO_7$, 3'-Isopropyl-3,5-diiodo-L-thyronine hydrochloride trihydrate, 42B, 368
$C_{18}H_{26}N_2O_5S$, N-t-Butyloxycarbonyl-S-benzylcysteinylglycine methyl ester, 40B, 239
$C_{18}H_{26}N_6O_6 \cdot 4\ H_2O$, Cyclo-bis(glycyl-L-prolyl-glycyl) tetrahydrate, 45B, 526
$C_{18}H_{27}N_3O_5$, Methyl L-alanyl-O,N-dimethyl-L-tyrosyl-L-alanate, 45B, 527
$C_{18}H_{27}N_5O_6 \cdot CH_2Cl_2$, Cyclo-glycyl-L-prolyl-L-seryl-D-alanyl-L-prolyl

dichloromethane solvate, 45B, 527
$C_{18}H_{29}N_3O_3$, L-Pyroglutamyl-N,N'-dicyclohexylurea, 45B, 528
$C_{18}H_{31}N_3O_6 \cdot 0.25\ H_2O$, t-Butoxycarbonyl-L-prolyl-L-isoleucylglycine hydrate, 46B, 487
$C_{18}H_{31}N_3O_7$, t-Butyloxycarbonyl-L-prolyl-L-leucylglycine hydrate, 43B, 619
$C_{19}H_{20}N_2O_4$, N-Phenylacetyl-glycyl-DL-phenylalanine, 42B, 369
$C_{19}H_{26}N_2O_5$, t-Butyloxycarbonylglycyl-L-proline benzyl ester, 43B, 610
$C_{19}H_{28}N_2O_5S$, N-(t-Butoxycarbonyl)-L-methionylglycine benzyl ester, 46B, 488
$C_{19}H_{28}N_4O_5 \cdot H_2O$, Glycyl-glycyl-phenylalanyl-leucine monohydrate, 45B, 529
$C_{19}H_{32}N_4O_4$, Pivaloyl-D-prolyl-L-prolyl-L-alanyl-N-methylamide, 46B, 488
$C_{19}H_{33}N_3O_7$, D-Valyl-L-tyrosyl-L-valine dihydrate, 44B, 443
$C_{20}H_{22}N_2O_2$, Cyclo(N-methyl-L-phenylalanyl-N-methyl-D-phenylalanyl), 42B, 358
$C_{20}H_{22}N_2O_2$, Cyclobis(N-methyl-L-phenylalanyl), 42B, 358
$C_{20}H_{22}N_2O_5$, N-Acetyl-L-phenylalanyl-L-tyrosine, 39B, 365
$C_{20}H_{22}N_2O_6$, Carbobenzoxy-L-leucyl-p-nitrophenyl ester, 40B, 462
$C_{20}H_{24}BrN_3O_6$, L-Threonyl-L-phenylalanine-p-nitrobenzyl ester hydrobromide, 34B, 274
$C_{20}H_{24}N_2O_3S_2$, S-Benzyl-L-cysteinyl-S-benzyl-L-cysteine, 41B, 555
$C_{20}H_{28}N_2O_5$, t-Butoxycarbonyl-L-prolylsarcosine benzyl ester, 46B, 476
$C_{20}H_{29}N_3O_6$, Benzyloxycarbonyl-bis(α-aminoisobutyryl)-L-alanyl methyl ester, 46B, 489
$C_{20}H_{30}N_6O_6$, Cyclo-bis(glycyl-L-prolyl-D-alanyl), 45B, 528
$C_{20}H_{38}N_6O_8$, Cyclohexasarcosyl bis(methanol), 43B, 619
$C_{21}H_{24}N_2O_7$, N-Benzyloxycarbonyl-L-isoleucine 4-nitroguaiacyl ester, 46B, 489
$C_{21}H_{29}N_3O_6$, Benzyloxycarbonyl-glycyl-prolyl-leucine, 42B, 370
$C_{21}H_{30}N_2O_5$, N-(Benzyloxycarbonyl)prolylleucine ethyl ester, 44B, 444
$C_{21}H_{33}N_3O_6$, t-Amyloxycarbonyl-L-prolyl-L-prolyl-L-proline, 40B, 462
$C_{21}H_{37}N_7O_8$, Cycloheptasarcosyl monohydrate, 41B, 556
$C_{22}H_{21}N_3O_5$, N-Acetyl-bis(dehydrophenylalanine)-glycine, 41B, 556
$C_{22}H_{25}BrN_2O_4$, N-(Bromoacetyl)-L-phenylalanyl-L-phenylalanine ethyl ester, 38B, 501
$C_{22}H_{25}ClN_2O_4$, N-(Chloroacetyl)-L-phenylalanyl-L-phenylalanine ethyl ester, 38B, 502
$C_{22}H_{26}N_4O_6 \cdot C_2H_6OS \cdot 3\ H_2O$, Tyrosyl-glycyl-glycyl-phenylalanine dimethylsulfoxide trihydrate, 45B, 529
$C_{22}H_{38}N_2O_6$, Cyclo-(N-methylvalyl-α-hydroxyisovaleryl-N-methylvalyl-α-hydroxyisovaleryl), 45B, 530
$C_{22}H_{38}N_2O_6$, Tetraenniatin, 46B, 490
$C_{23}H_{23}N_2O_5$, N-(t-Butoxycarbonyl)prolylleucine benzyl ester, 44B, 444
$C_{23}H_{25}N_3O_3 \cdot H_2O$, cyclo-(Dibenzylglycyl-L-proline) monohydrate, 46B, 490
$C_{23}H_{29}NO_5$, N-Acetyl-4'-methoxy-3,5,3'-trimethyl-L-thyronine ethyl ester, 46B, 491
$C_{23}H_{29}N_3O_6$, 9-Methyl-10-methylanthrylsarcosylglycylglycine dihydrate, 43B, 200
$C_{23}H_{31}BrN_4O_7$, p-Bromocarbobenzoxy-glycyl-L-prolyl-L-leucyl-glycine, 34B, 276
$C_{23}H_{35}N_5O_4S$, S-Benzyl-L-cysteinyl-L-prolyl-L-leucylglycinamide, 41B, 557

$C_{23}H_{35}N_5O_4Se$, S-Benzyl-L-cysteinyl-L-prolyl-L-leucylglycinamide (selenium analogue), 41B, 557
$C_{24}H_{34}N_2O_9$, N-Benzyloxycarbonyl-(γ-ethyl)-L-glutamyl-(γ-ethyl)-L-glutamic acid ethyl ester, 45B, 530
$C_{24}H_{37}NO_6$, Benzyloxycarbonyl-L-α-hydroxyisovaleryl-D-N-methyl-leucine t-butyl ester, 43B, 620
$C_{24}H_{42}N_2O_6$, Cyclo(D-α-hydroxyisovaleryl-L-methylisoleucyl-D-α-hydroxyisovaleryl-L-methylleucyl), 34B, 270
$C_{24}H_{42}N_2O_6$, Cyclo-bis(L-N-methylisoleucyl-D-α-hydroxyisovaleryl), 43B, 621
$C_{24}H_{48}N_8O_{12}$, Cyclooctasarcosyl tetrahydrate, 39B, 364
$C_{25}H_{30}N_2O_9$, 1-[2'-Methoxy-5'-(α-(methyl N-acetylglycyl))phenyl]-2,4-dimethoxy-6-(α-(methyl N-acetylglycyl))benzene, 46B, 492
$C_{25}H_{39}N_4O_5 \cdot 3.6\ H_2O$, Cyclo(L-leucyl-L-tyrosyl-δ-aminovaleryl-δ-aminovaleryl) hydrate, 44B, 444
$C_{26}H_{38}N_2O_9$, (±)-N-Carbobenzoxy-(γ,γ'-di-t-butyl)-γ-carboxyglutamylglycine ethyl ester, 45B, 531
$C_{26}H_{40}N_4O_6$, L-Prolyl-L-tyrosyl-L-isoleucyl-L-leucine, 45B, 532
$C_{28}H_{43}N_5O_{10}$, N-Benzyloxycarbonyl-glycyl-L-prolyl-L-leucyl-glycyl-L-proline dihydrate, 44B, 445
$C_{28}H_{47}N_3O_6$, Majusculamide B hydrate, 43B, 621
$C_{29}H_{46}N_9O_{11}RbS$, Cyclo-tetra(L-prolylglycyl)rubidium thiocyanate trihydrate, 43B, 622
$C_{31}H_{36}BNO_2$, Acetylcholine tetraphenylborate, 46B, 37
$C_{32}H_{38}N_6O_6 \cdot C_2H_6OS$, Cyclo-bis(glycyl-L-prolyl-D-phenylalanyl) dimethylsulfoxide, 45B, 532
$C_{32}H_{46}N_4O_8$, t-Butyloxycarbonyltetra-L-prolinebenzyl ester monohydrate, 40B, 213
$C_{32}H_{48}BrN_5O_{11}$, o-Bromocarbobenzoxy-glycyl-L-prolyl-L-leucyl-glycyl-L-proline ethyl acetate monohydrate, 37B, 279
$C_{33}H_{57}N_3O_9$, Enniatin B, 43B, 622
$C_{34}H_{46}N_6O_6 \cdot 2\ H_2O$, Cyclo(-L-leucyl-L-phenylalanyl-glycyl-D-leucyl-D-phenylalanyl-glycyl-) dihydrate, 45B, 533
$C_{34}H_{50}N_6O_{10}$, Cyclo-bis(L-alanyl-L-prolyl-L-phenylalanyl) tetrahydrate, 42B, 370
$C_{34}H_{57}N_4O_9RbS$, Cyclo(L-methylvalyl-D-α-hydroxyisovaleryl-L-methylvalyl-L-α-hydroxyisovaleryl-D-methylvalyl-L-α-hydroxyisovaleryl)-rubidium isothiocyanate, 43B, 623
$C_{34}H_{60}N_4O_8$, Sporidesmolide, 42B, 371
$C_{34}H_{66}N_{10}O_{14}$, Cyclodecasarcosyl tetramethanolate, 42B, 196
$C_{39}H_{53}N_9O_{15}S$, β-Amanitin, 43B, 624
$C_{40}H_{48}N_6O_{10}CH_4O \cdot 0.6\ H_2O$, Bouvardin methanolate hydrate, 43B, 625
$C_{40}H_{68}N_4O_{12}$, cyclo-Bis(D-isoleucyl-L-lactyl-L-isoleucyl-D-hydroxyvaleryl), 46B, 493
$C_{45}H_{85}N_9O_{25}S$, β-Amanitin ethanol solvate heptahydrate, 44B, 446
$C_{46}H_{61}N_7O_{13}$, Virginiamycin factor S methanol solvate, 44B, 446
$C_{57}H_{93}N_{15}O_{15} \cdot 1.2\ C_2H_6OS \cdot 5.7\ H_2O$, cyclo-Tris(L-valyl-L-prolylglycyl-L-valylglycyl) dimethylsulfoxide solvate hydrate, 46B, 493
$C_{62}H_{110}IN_{11}O_{12}$, Iodocyclosporin A, 42B, 371
$C_{66}H_{82}N_{10}O_{10} \cdot 0.3\ H_2O$, (Phe(4),Val(6))Antamanide hydrate, 43B, 626

49. PORPHYRINS AND CORRINS

$C_{20}H_{14}N_4$, Porphine, 30B, 216; 38B, 503; 39B, 371
$C_{22}H_{18}GeN_4O_2$, Dimethoxyporphinatogermanium(IV), 42B, 372

$C_{24}H_{14}N_8Ni$, Nickel phthalocyanine analogue, 17, 762
$C_{25}H_{30}ClN_5Ni \cdot X\ CH_4O$, Nickel(II) 1,8,8,13,13-pentamethyl-5-cyano-trans-corrin chloride methanolate, 37B, 280
$C_{27}H_{34}ClN_5NiO_4$, Nickel A,D-seco-corrinoid perchlorate, 37B, 281
$C_{27}H_{34}ClN_5NiO_5$, (2,3,4,6,7,8,12,13,17,18-Decahydro-3,3,4,13,13,18,18-heptamethyl-21H-5-oxaporphine-10-carbonitrilato)-nickel perchlorate, 42B, 862
$C_{27}H_{34}ClN_5O_4Pd$, Palladium A,D-seco-corrinoid perchlorate, 37B, 281
$C_{27}H_{34}ClN_5O_4Pt$, Platinum A,D-seco-corrinoid perchlorate, 37B, 281
$C_{29}H_{34}N_4$, 8,12-Diethyl-2,3,7,13,17,18-hexamethylcorrole, 37B, 282
$C_{29}H_{35}BrN_4 \cdot 1.5\ CHCl_3$, 8,12-Diethyl-2,3,7,13,17,18-hexamethylcorrole hydrobromide, 40B, 463
$C_{29}H_{42}ClN_5O$, rac-15-Cyano-1,2,2,7,7,12,12-heptamethylcorrin hydrochloride ethanol solvate, 41B, 558
$C_{30}H_{38}ClCoN_7O_5$, pseudo-Corrin perchlorate, 32B, 233
$C_{32}H_{16}CoN_8$, Phthalocyaninatocobalt(II), 45B, 534; 46B, 494
$C_{32}H_{16}CuN_8$, Copper phthalocyanine, 33B, 431
$C_{32}H_{16}FeN_8$, Iron(II) phthalocyanine (β), 42B, 372
$C_{32}H_{16}IN_8Ni$, Nickel phthalocyanine iodide, 46B, 495
$C_{32}H_{16}MnN_8$, Phthalocyaninatomanganese(II) (β form), 42B, 372; 45B, 534; 46B, 495, 496
$C_{32}H_{16}N_8Ni$, Nickel phthalocyanine, 5, 175
$C_{32}H_{16}N_8OV$, Vanadyl phthalocyanine (phase II), 46B, 496
$C_{32}H_{16}N_8Pb$, Lead phthalocyanine, 39B, 365
$C_{32}H_{16}N_8Pt$, Platinum phthalocyanine, 8, 329; 33B, 432
$C_{32}H_{16}N_8Si_{0.17}$, Tetraphenylporphinatosilicon, 46B, 497
$C_{32}H_{16}N_8Zn$, Phthalocyanatozinc(II), 43B, 626
$C_{32}H_{18}N_8$, Phthalocyanine, 4, 270, 320
$C_{32}H_{35}N_4O_2Rh$, Dicarbonyl-(2,3,7,13,17,18,21-heptamethyl-8,12-diethyl-corrolato)rhodium(I), 45B, 534
$C_{32}H_{35}N_4O_2Rh$, Dicarbonyl-(2,3,7,13,17,18,22-heptamethyl-8,12-diethyl-corrolato)rhodium(I), 45B, 534
$C_{32}H_{36}CuN_4$, Copper(II) α,β,γ,δ-n-propylporphine, 39B, 366
$C_{32}H_{36}N_4Ni$, Nickel(II) etioporphyrin-I, 28, 528
$C_{32}H_{36}N_4Pb$, (5,10,15,20-Tetra-n-propylporphinato)lead(II), 46B, 497
$C_{32}H_{38}N_4$, α,β,γ,δ-Tetra-n-propylporphine, 38B, 504; 39B, 371
$C_{33}H_{38}N_4O_2$, Phyllochlorine ester, 34B, 281
$C_{34}H_{32}ClFeN_4O_4$, Chlorohemin, 30B, 217
$C_{34}H_{32}GeN_8O_4$, Hemiporphyrazine-germanium-diethylene glycol monoethyl ether, 40B, 464
$C_{34}H_{38}Cl_2N_4OV$, Vanadyl deoxophylloerythro-etioporphyrin 1,2-dichloroethane solvate, 34B, 279
$C_{34}H_{43}I_3N_4$, trans-N,N'-Dimethyletioporphyrin triiodide, 46B, 498
$C_{35}H_{21}N_5$, Tetrabenzmonazaporphin, 38B, 505
$C_{35}H_{38}ClN_4NiO_2$, Nickel(II) deoxophylloerythrin methyl ester 1,2-dichloroethane solvate, 37B, 284
$C_{35}H_{39}ClN_4O_5Zn$, Zinc 5-oxoniaporphin dimethyl ester chloride, 43B, 627
$C_{36}H_{23}FeN_9O_2 \cdot C_3H_7NO$, Carbonyl-(N,N-dimethylformamide)-phthalocyaninatoiron(II) dimethylformamide solvate, 46B, 498
$C_{36}H_{28}FeN_8O_2S_2 \cdot 2\ C_2H_6OS$, Bis(dimethyl sulfoxide-S)phthalocyaninatoiron(II) dimethyl sulfoxide solvate, 46B, 499
$C_{36}H_{36}N_4NiO \cdot 0.5\ C_6H_6$, Nickel(II) 2,4-diacetyldeuteroporphyrin-IX dimethyl ester benzene solvate, 30B, 220
$C_{36}H_{37}N_5O_6 \cdot C_2H_4Cl_2$, 8^1(E)-8^2-Nitroprotoporphyrin dimethyl ester dichloroethane solvate, 46B, 499
$C_{36}H_{38}N_4O_4$, Protoporphyrin IX dimethyl ester, 43B, 628

$C_{36}H_{38}N_4O_5$, Methyl pheophorbide a, 38B, 506
$C_{36}H_{40}MgN_4O_7$, Methylchlorophyllide A dihydrate, 43B, 629
$C_{36}H_{43}N_4O_4$, Mesoporphyrin-IX dimethyl ester, 41B, 559
$C_{36}H_{44}ClFeN_4O_4$, (Octaethylporphinato)iron(III) perchlorate, 46B, 500
$C_{36}H_{44}N_4Ni$, 1,2,3,4,5,6,7,8-Octaethylporphinatonickel(II) (tetragonal form), 38B, 507
$C_{36}H_{44}N_4Ni$, 1,2,3,4,5,6,7,8-Octaethylporphinatonickel(II) (triclinic form), 40B, 465
$C_{36}H_{44}N_4OTi$, (Octaethylporphinato)oxotitanium(IV) (monoclinic), 44B, 447
$C_{36}H_{44}N_4OTi$, (Octaethylporphinato)oxotitanium(IV) (orthorhombic), 41B, 560
$C_{36}H_{44}N_4OV$, 2,3,7,8,12,13,17,18-Octaethylporphinatooxovanadium(IV), 42B, 373
$C_{36}H_{44}N_4O_2Ti$, 2,3,7,8,12,13,17,18-Octaethylporphinatoperoxotitanium-(IV), 44B, 447
$C_{36}H_{46}N_4$, 1,2,3,4,5,6,7,8-Octaethylporphyrin, 39B, 367
$C_{36}H_{48}N_4O_6$, Octaethylxanthoporphinogen dihydrate, 42B, 374
$C_{36}H_{54}ClN_6O_2Rh$, Bis(dimethylamine)etio(I)porphinatorhodium(III) chloride dihydrate, 39B, 368
$C_{37}H_{26}CoN_4P$, Corrole(triphenylphosphine)cobalt(III), 42B, 374
$C_{37}H_{40}MgN_4O_8$, Ethyl chlorophyllide B dihydrate, 41B, 561
$C_{37}H_{42}MgN_4O_7$, Ethyl chlorophyllide A dihydrate, 41B, 562
$C_{37}H_{43}FeN_4O_5$, Methoxyiron(III) mesoporphyrin-IX dimethyl ester, 30B, 218
$C_{37}H_{47}N_4Rh$, Octaethylporphinato(methyl)rhodium(III), 42B, 375
$C_{37}H_{48}Cl_3N_4OTl$, 2,3,7,8,12,13,17,18-Octaethylporphinatochlorothallium(III) dichloromethane hydrate, 43B, 629
$C_{38}H_{21}N_8Os$, Carbonyl(phthalocyanato)(pyridine)osmium(II), 46B, 501
$C_{38}H_{32}GeN_8$, trans-Bis(3,3-dimethyl-1-butynyl)hemiporphyracinegermanium, 46B, 501
$C_{38}H_{46}MgN_4O_5$, Methylpyrochlorophyllide A monohydrate monoetherate, 43B, 630
$C_{38}H_{50}Cl_2N_6O_4Sn$, 1,2,3,4,5,6,7,8-Octaethylporphinatodichlorotin(IV) nitromethane solvate, 39B, 369
$C_{38}H_{50}N_4Ni$, α,γ-Dimethyl-α,γ-dihydrooctaethylporphinatonickel(II), 40B, 466
$C_{38}H_{50}N_4OTi$, (α,γ-Dimethyl-α,γ-dihydrooctaethylporphinato)oxotitanium(IV), 41B, 563
$C_{38}H_{52}ClN_4O_7Sb$, Dihydroxo(1,2,3,4,5,6,7,8-octaethylporphinato)antimony(V) perchlorate monoethanol solvate, 43B, 631
$C_{39}H_{46}N_4Ni$, 2,2,7,8,12,13,17,18-Octaethyl-benzo(3,4,5)porphinatonickel, 44B, 447
$C_{40}H_{34}Cl_6N_8O$, $\alpha,\beta,\gamma,\delta$-Tetra(4-pyridyl)porphine hexa(hydrogen chloride) monohydrate, 33B, 185
$C_{40}H_{44}N_4O_4Rh_2$, μ-1,2,3,4,5,6,7,8-Octaethylporphinatobis(dicarbonylrhodium(I)), 41B, 565
$C_{40}H_{48}Cl_4N_4O_4Rh_2$, Octaethylporphin bis(cis-dicarbonyldichlororhodate(I)), 40B, 467
$C_{40}H_{50}CoN_6$, 2,3,7,8,12,13,17,18-Octaethylporphinato-(1-methylimidazole)cobalt(II), 40B, 468
$C_{40}H_{51}ClCoN_4O_2$, Chloro-(2,3,7,8,12,13,17,18-octaethyl-N-ethylacetatoporphine)cobalt(II), 42B, 376
$C_{40}H_{88}CoN_{11}O_{20}$, Factor V(1A), 27, 1040
$C_{41}H_{49}N_5Zn$, 2,3,7,8,12,13,17,18-Octaethylporphinatomonopyridinezinc(II), 42B, 376
$C_{42}H_{28}MgN_{10}O$, Magnesium phthalocyanin monohydrate dipyridinate, 37B,

283

$C_{42}H_{40}FeN_5O_6S$, Iron(III) protoporphyrin IX dimethyl ester p-nitrobenzenethiolate, 42B, 377

$C_{42}H_{49}Cl_3N_4O_7Re_2$, Octaethylporphinium tri-μ-chloro-hexacarbonyldirhenate(I) monohydrate, 44B, 448

$C_{42}H_{62}ClFeN_4O_7$, Bis(ethanol)(octaethylphorphinato)iron(III) perchlorate ethanol, 44B, 449

$C_{43}H_{49}FeN_8O_6$, Bis(1-methylimidazole)-(protoporphyrin IX)iron-methanol-water, 41B, 565

$C_{43}H_{50}N_4O_2$, 5-Benzoyloxyoctaethylporphyrin, 44B, 449

$C_{43}H_{51}Cl_2Hg_2N_5O_2S$, N-Tosylaminooctaethylporphyrin-bis(chloromercury-(II)), 45B, 535

$C_{43}H_{59}IN_4O_3$, 21-Ethoxycarbonylmethyl-2,3,7,8,12,13,17,18-octaethylporphyrin iodide acetone solvate, 40B, 468

$C_{44}H_{28}AuClN_4 \cdot 0.7\ CHCl_3$, Chloro(α,β,γ,δ-tetraphenylporphinato)gold-(III) chloroform, 43B, 631

$C_{44}H_{28}BrFeN_4$, Bromo(tetraphenylporphyrinato)iron(III), 43B, 632

$C_{44}H_{28}Br_2N_4Ti$, trans-Dibromo(tetraphenylporphinato)titanium(IV), 44B, 450

$C_{44}H_{28}ClCoN_4$, Chloro-α,β,γ,δ-tetraphenylporphinatocobalt(III), 42B, 378

$C_{44}H_{28}ClFeN_4$, Chloroiron(III) tetraphenylporphine, 32B, 234

$C_{44}H_{28}ClFeN_4O_4 \cdot 0.5\ C_4H_8O$, Perchlorato-tetraphenylporphinatoiron-(III) tetrahydrofuran solvate, 44B, 450

$C_{44}H_{28}ClInN_4$, Chloro(5,10,15,20-tetraphenylporphinato)indium(III), 46B, 502

$C_{44}H_{28}ClN_4O_4Zn$, Perchloratotetraphenylporphinatozinc(II), 40B, 469

$C_{44}H_{28}ClN_4Tl$, Chloro-5,10,15,20-tetraphenylporphinatothallium(III), 43B, 635

$C_{44}H_{28}Cl_2N_4Sn$, α,β,γ,δ-Tetraphenylporphinatodichlorotin(IV), 38B, 509

$C_{44}H_{28}CoN_4$, α,β,γ,δ-Tetraphenylporphinatocobalt(II), 42B, 378

$C_{44}H_{28}CoN_5O$, Nitrosyl-α,β,γ,δ-tetraphenylporphinatocobalt(II), 39B, 370

$C_{44}H_{28}CuN_4$, Copper(II) tetraphenylporphine, 29, 617

$C_{44}H_{28}FeIN_4$, Iodo(meso-tetraphenylporphinato)iron(III), 45B, 536

$C_{44}H_{28}FeN_4$, α,β,γ,δ-Tetraphenylporphinatoiron(II), 41B, 566

$C_{44}H_{28}FeN_5O$, Nitrosyl-α,β,γ,δ-tetraphenylporphinatoiron(II), 41B, 567

$C_{44}H_{28}FeN_7 \cdot 0.5\ C_5H_5N$, Azido(pyridine)(tetraphenylporphinato)iron-(III) pyridine solvate, 45B, 536

$C_{44}H_{28}N_4Pd$, Palladium(II) tetraphenylporphine, 29, 617

$C_{44}H_{29}Ag_{0.5}N_4$, (Tetraphenylporphine)silver-tetraphenylporphine, 38B, 508

$C_{44}H_{30}Cl_5FeN_4$, α,β,γ,δ-Tetraphenylporphine diacid chloride ferrichloride, 33B, 185

$C_{44}H_{30}MgN_4O$, Aquomagnesium tetraphenylporphyrin, 34B, 278

$C_{44}H_{30}N_4$, Tetraphenylporphine (tetragonal), 29, 728

$C_{44}H_{30}N_4$, Tetraphenylporphine (triclinic), 32B, 235; 39B, 371

$C_{44}H_{30}N_4OZn$, Aquozinc(II) tetraphenylporphine, 32B, 237

$C_{44}H_{30}N_4O_2Sn \cdot 2\ CHCl_3 \cdot 2\ CCl_4$, α,β,γ,δ-Tetraphenylporphinatodihydroxytin(IV) chloroform solvate carbon tetrachloride solvate, 45B, 537

$C_{44}H_{31}FeN_4O_2$, Aquohydroxyiron(III) tetraphenylporphine, 29, 617

$C_{44}H_{32}N_4O_2Zn$, Zinc tetraphenylporphine dihydrate, 29, 617

$C_{44}H_{36}IN_4Ni \cdot 0.08\ I$, 1,4,5,8,9,12,13,16-Octamethyl-tetrabenzo-porphinato-nickel(II) iodide (partially oxidized), 46B, 502

$C_{44}H_{55}N_5OOs$, Carbonyl(α,γ-dimethyl-α,γ-dihydrooctaethylporphinato)-pyridineosmium(II), 42B, 872
$C_{44}H_{56}CoN_5O_7$, (trans-21,22-Bis(ethoxycarbonylmethyl)octaethyl-porphyrinato(2-))nitratocobalt(III), 43B, 633
$C_{44}H_{56}N_4O_4$, 2,3,7,8,12,13,17,18-Octaethyl-5-(2,2-bis(ethoxycarbonyl)vinyl)-22H,24H-porphine, 43B, 633
$C_{45}H_{29}N_9Zn$, $\alpha,\beta,\gamma,\delta$-Tetra(4-pyridyl)porphinatomonopyridinezinc(II), 35B, 717
$C_{45}H_{31}ClCoN_4$, Chloro-N-methyl-$\alpha,\beta,\gamma,\delta$-tetraphenylporphinatocobalt-(II), 43B, 634
$C_{45}H_{31}ClFeN_4$, Chloro(N-methyl-5,10,15,20-tetraphenylporphinato)iron-(II), 46B, 503
$C_{45}H_{31}ClMnN_4$, Chloro-N-methyl-$\alpha,\beta,\gamma,\delta$-tetraphenylporphinatomanganese(II), 43B, 635
$C_{45}H_{31}ClN_4Zn \cdot 0.67\ CH_2Cl_2$, Chloro-N-methyl-tetraphenylporphinato-zinc(II) - dichloromethane, 44B, 451
$C_{45}H_{31}InN_4$, Methyl(tetraphenylporphinato)indium(III), 46B, 503
$C_{45}H_{31}InN_4O_3S \cdot 2\ C_2H_4Cl_2$, Methanesulphonato(5,10,15,20-tetraphenyl-porphyrinato)indium(III) 1,2-dichloroethane solvate, 46B, 504
$C_{45}H_{31}N_4Tl$, Methyl-5,10,15,20-tetraphenylporphinatothallium(III), 43B, 635
$C_{46}H_{28}KFeN_6 \cdot 2\ C_3H_6O$, Potassium dicyano(meso-tetraphenylporphinato)-iron(III) bis(acetone), 46B, 504
$C_{46}H_{31}N_4NbO_3 \cdot C_2H_4O_2$, Acetato-oxo(5,10,15,20-tetraphenyl-porphyrinato)niobium(V) acetic acid solvate, 45B, 537
$C_{46}H_{54}MgN_6$, Di(pyridine)magnesium(II) octaethylporphyrinate, 43B, 636
$C_{46}H_{54}N_6Ru$, Ruthenium(II) octaethylporphyrin dipyridinate, 41B, 568
$C_{46}H_{88}CoN_{11}O_{20}$, Cobyric acid, 37B, 285
$C_{47}H_{31}MnN_6$, (Imidazolato)tetraphenylporphinatomanganese(III), 44B, 451; 46B, 505
$C_{47}H_{33}CoN_4O$, Acetonyl-tetraphenylporphinatocobalt(III), 44B, 1105
$C_{47}H_{34}ClMnN_4O$, Chloro-$\alpha,\beta,\gamma,\delta$-tetraphenylporphinatomanganese(III) acetone solvate, 43B, 637
$C_{47}H_{34}N_4O_2Ru$, Ruthenium(II) carbonyl - tetraphenylporphine - ethanol, 39B, 372
$C_{47}H_{61}N_8O_{12}Rh \cdot 3\ H_2O$, Dicyanorhodibyrinic acid A,C-diamide trihydrate, 46B, 506
$C_{48}H_{33}ClFeN_4O_4$, Perchlorato(m-tetraphenylporphinato)iron(III) hemi-m-xylene solvate, 45B, 538
$C_{48}H_{34}N_4NiO_2 \cdot 0.5\ CH_2Cl_2$, 21-Ethoxycarbonyl-5,10,15,20-tetraphenyl-21H-21-homoporphine-nickel(II) hemi(dichloromethane), 41B, 568
$C_{48}H_{34}N_4NiO_3$, (10H-10-Hydroxyl-21-ethoxycarbonyl-5,10,15,20-tetraphenyl-21-homoporphinato)nickel(II), 42B, 379
$C_{48}H_{36}MnN_5O \cdot C_6H_6$, Nitrosyl(5,10,15,20-tetratolyl-porphinato)manganese(II) benzene solvate, 45B, 539
$C_{48}H_{36}MoN_4O_2 \cdot 1.5\ C_7H_8$, cis-Dioxo(5,10,15,20-tetra-p-tolylporphinato)molybdenum(VI) toluene solvate, 46B, 506
$C_{48}H_{58}CoN_6$, 2,3,7,8,12,13,17,18-Octaethylporphinatobis-(3-methylpyridine)cobalt(II), 40B, 470
$C_{49}H_{29}Cl_3FeN_7$, 5,10,15,20-Tetraphenylporphinatoiron(III) tricyano-methanide - chloroform, 44B, 452
$C_{49}H_{32}N_5ORu \cdot 1.5\ C_7H_8$, (Tetraphenylporphinato)(carbonyl)(pyridine)-ruthenium(II) - 1.5-toluene, 39B, 372
$C_{49}H_{35}Cl_3FeN_7O$, Nitrosyl-$\alpha,\beta,\gamma,\delta$-tetraphenylporphinato(1-methylimidazole)iron(II) chloroform solvate, 40B, 471; 42B, 380
$C_{49}H_{35}Cl_3N_4NiO_2$, Nickel(II) meso-tetraphenylporphine ethoxycarbonyl-

carbene insertion complex chloroform solvate, 42B, 380
$C_{49}H_{68}ClCoN_6O_{16}$, Vitamin B_{12} hexacarboxylic acid degradation product, 23, 729
$C_{50}H_{28}N_4O_6Re_2$, μ-(Meso-tetraphenylporphinato)-bis(tricarbonylrhenium(I)), 41B, 569
$C_{50}H_{28}N_4O_6Tc_2$, μ-(Meso-tetraphenylporphinato)-bis(tricarbonyltechnetium(I)), 41B, 569
$C_{50}H_{34}MnN_7$, Azido-(α,β,γ,δ-tetraphenylporphinato)manganese(III) - benzene, 41B, 571
$C_{50}H_{36}ClFeN_8$, Bis(imidazole)-α,β,γ,δ-tetraphenylporphinatoiron(III) chloride, 34B, 279
$C_{50}H_{41}FeN_6O$, Nitrosyl-α,β,γ,δ-tetraphenylporphinato(4-methylpiperidine)iron(II), 43B, 638
$C_{50}H_{41}MnN_6O \cdot CHCl_3$, Nitrosyl(4-methylpiperidine)(5,10,15,20-tetraphenylporphinato)manganese(II) chloroform solvate, 45B, 539
$C_{51}H_{37}CoN_5$, (3,5-Dimethylpyridine)-α,β,γ,δ-tetraphenylporphinatocobalt(II), 41B, 573
$C_{51}H_{37}CoN_6O_2$, Nitro-α,β,γ,δ-tetraphenylporphinato(3,5-lutidine)cobalt(III), 40B, 471
$C_{51}H_{40}ClFeN_8O$, Bis(imidazole)-α,β,γ,δ-tetraphenylporphinatoiron(III) chloride methanol solvate, 38B, 511
$C_{51}H_{42}Cl_3FeN_6O$, Nitrosyl-α,β,γ,δ-tetraphenylporphinato(4-methylpiperidine)iron(II) chloroform solvate, 43B, 638
$C_{51}H_{42}Cl_3MnN_6O$, Nitrosyl-α,β,γ,δ-tetraphenylporphinato(4-methylpiperidine)manganese chloroform solvate, 40B, 471
$C_{52}H_{32}Cl_{10}N_4O_6Re_2Sb$, Bis(tricarbonylrhenium)-μ-tetraphenylporphyrin hexachloroantimonate bis(dichloromethane) solvate, 43B, 638
$C_{52}H_{36}ClCoN_5$, Chloro-α,β,γ,δ-tetraphenylporphinato(pyridine)cobalt(III) benzene solvate, 41B, 574
$C_{52}H_{36}N_6OZn \cdot C_6h_6 \cdot 0.5\ C_2H_6O$, (5-(2-((2-(3-Pyridyl)ethyl)carbonylamino)phenyl)-10,15,20-triphenylporphinato)zinc(II) benzene hemiethanol solvate, 46B, 507
$C_{52}H_{37}Cl_2N_2NiO_2S$, N-Tosylamino-5,10,15,20-tetraphenylporphinatonickel(II) dichloromethane solvate, 44B, 453
$C_{52}H_{42}MnN_6O$, α,β,γ,δ-Tetraphenylporphinato(1-methylimidazole)manganese(II) tetrahydrofuran solvate, 41B, 576; 43B, 639
$C_{52}H_{44}ClFeN_4O_6S_2$, Bis(tetramethylene sulfoxide)(tetraphenylporphinato)iron(III) perchlorate, 44B, 453
$C_{52}H_{48}ClFeN_4O_8$, Diaquo-tetraphenylporphinatoiron(III) perchlorate tetrahydrofuran solvate, 44B, 450
$C_{53}H_{42}Cl_3CoN_8O_3$, meso-Tetraphenylporphinatobis(imidazole)cobalt(III) acetate monohydrate monochloroformate, 40B, 473
$C_{53}H_{48}CoN_5O_4$, Methoxy-tetraphenylporphinato(pyridine)cobalt(III) methanol solvate, 44B, 454
$C_{54}H_{28}FeN_4O_2$, (meso-Tetraphenylporphinato)bis(tetrahydrofuran)iron(II), 46B, 507
$C_{54}H_{38}CrN_6$, Bis(pyridine)(meso-tetraphenylporphinato)chromium(III), 45B, 539
$C_{54}H_{40}CoN_6$, (1-Methylimidazole)-α,β,γ,δ-tetraphenylporphinatocobalt(II) benzene solvate, 40B, 473
$C_{54}H_{42}Cl_2MoN_4 \cdot C_6H_6$, Dichloro(5,10,15,20-tetra-p-tolylporphyrinato)molybdenum(IV) benzene solvate, 45B, 540
$C_{54}H_{42}MoN_4O \cdot C_6H_6$, Oxo(5,10,15,20-tetra-p-tolylporphyrinato)molybdenum(IV) benzene solvate, 45B, 540
$C_{54}H_{46}FeN_6 \cdot 2\ C_7H_8$, Bis(t-butyl isocyanide)-(meso-tetraphenylporphyrinato)iron(II) toluene solvate, 45B, 540
$C_{54}H_{48}CoN_6$, Bis(piperidine)-α,β,γ,δ-tetraphenylporphinatocobalt(II),

40B, 474
$C_{54}H_{50}FeN_6$, Bis(piperidine)-$\alpha,\beta,\gamma,\delta$-tetraphenylporphinatoiron(III), 38B, 511
$C_{54}H_{73}CoN_6O_{14} \cdot H_2O$, Heptamethyl dicyanocobyrinate(III) monohydrate, 46B, 508
$C_{55}H_{39}ClMnN_5$, Chloro-tetraphenylporphinato(pyridine)manganese(III) benzene solvate, 41B, 575
$C_{55}H_{41}N_5Zn$, 2,3-Dihydro-$\alpha,\beta,\gamma,\delta$-tetraphenylporphyrinatopyridine-zinc(II) - benzene solvate, 43B, 640
$C_{55}H_{60}CuN_4O_5$, Octaethyl-5-formyl-10-(2,2-bis(benzyloxycarbonyl)vinyl)porphinatocopper(II), 44B, 454
$C_{56}H_{44}CoN_{12}$, Bis(4-methylpyridine)phthalocyaninatocobalt(II) 4-methylpyridine solvate, 41B, 576; 44B, 455
$C_{56}H_{44}FeN_{12}$, Bis(4-methylpyridine)phthalocyaninatoiron(II) 4-methylpyridine solvate, 44B, 455
$C_{56}H_{56}Cl_4N_{12}NiO_{18}$, Bis(imidazole)nickel(II)-tetra(4-N-methylpyridyl)porphine perchlorate acetone solvate, 41B, 572
$C_{58}H_{44}CrN_4$, Tetraphenylporphinatochromium(II) - toluene, 44B, 456
$C_{58}H_{44}MnN_4$, $\alpha,\beta,\gamma,\delta$-Tetraphenylporphinatomanganese(II) toluene solvate, 41B, 576; 43B, 640
$C_{58}H_{44}N_4Zn$, Tetraphenylporphinatozinc(II) - toluene, 44B, 456
$C_{61}H_{48}CoN_6$, (1,2-Dimethylimidazole)-$\alpha,\beta,\gamma,\delta$-tetraphenylporphinatocobalt(II) benzene solvate, 40B, 475
$C_{62}H_{44}N_4O_{12} \cdot 5\ CHCl_3 \cdot CH_4O$, 5,10,15,20-(Pyrromellitoyl-(tetrakis-o-oxyethoxyphenyl))-porphyrin chloroform methanol solvate, 46B, 508
$C_{62}H_{45}FeN_5O$, (Pyridine)(carbonyl)(5,10,15,20-tetraphenylporphinato)-iron(II) - di(benzene), 42B, 381
$C_{62}H_{52}BrCl_6CoN_6$, Bis-((RS)-1-phenylethylamine)tetraphenylporphinato-cobalt(III) bromide chloroform, 44B, 457
$C_{63}H_{88}CoN_{14}O_{14} \cdot 18\ H_2O$, Vitamin B_{12}, 27, 1042
$C_{63}H_{88}CoN_{14}O_{14} \cdot 20\ H_2O$, Neovitamin B_{12}, 38B, 512
$C_{63}H_{88}CoN_{14}O_{14} \cdot 22\ H_2O$, Vitamin B_{12}, 29, 751
$C_{64}H_{32}N_{16}Sn$, Tin(IV) phthalocyanine, 39B, 374
$C_{64}H_{32}N_{16}Th$, Bis(phthalocyaninato)thorium(IV), 44B, 457
$C_{64}H_{72}CoN_8 \cdot 0.5\ NO_3 \cdot 0.5\ HCO_3$, Bis(piperidine)-$\alpha,\beta,\gamma,\delta$-tetraphenylporphinatocobalt(III) (nitrate, bicarbonate) piperidine solvate, 39B, 375
$C_{64}H_{72}N_{16}Th$, Thorium diphthalocyanine, 40B, 930
$C_{64}H_{72}N_{16}U$, Uranium diphthalocyanine, 40B, 930
$C_{66}H_{54}MoN_{10}$, Bis(phenyldiazo)(meso-tetra-p-tolylporphyrinato)molybdenum(IV)-phenylhydrazine (1:1), 46B, 509
$C_{68}H_{70}FeN_{10}O_4 \cdot C_2H_6O$, (2-Methylimidazole)-meso-tetra($\alpha,\alpha,\alpha,\alpha$-o-pivalamidophenyl)porphyrinatoiron(II) ethanol, 46B, 510
$C_{68}H_{70}FeN_{10}O_6 \cdot C_2H_6O$, (2-Methylimidazole)-dioxygen-meso-tetra-($\alpha,\alpha,\alpha,\alpha$-o-pivalamidophenyl)porphyrinatoiron ethanol, 46B, 510
$C_{68}H_{74}FeN_8O_5S$, Tetrakis(pivalamidophenyl)porphinato-aquoiron(II) - tetrahydrothiophene, 44B, 458
$C_{72}H_{134}CoN_{18}O_{24}P$, Vitamin B_{12} coenzyme, 33B, 186
$C_{73}H_{76}FeN_{11}O_6$, (1-Methylimidazole)(dioxygen)-meso-tetrakis(pivalamidophenyl)porphinatoiron(II) hemibenzene hemi(1-methylimidazole), 44B, 459
$C_{80}H_{90}N_8O_{16}S_2$, 10,10'-Dithiobis(coproporphyrin II tetramethyl ester), 45B, 541
$C_{82}H_{77}N_{19}Zn$, Zinc phthalocyanine - n-hexylamine, 37B, 286
$C_{84}H_{52}Mn_2N_{20}O$, μ-Oxo-bis(phthalocyanatopyridinemanganese(III)) dipyridinate, 32B, 238
$C_{88}H_{56}Fe_2N_8O$, μ-Oxo-bis($\alpha,\beta,\gamma,\delta$-tetraphenylporphinatoiron(III)),

38B, 516
$C_{88}H_{56}N_8Nb_2O_3 \cdot 4\ C_2H_4Cl_2$, Tri-μ-oxo-bis[5,10,15,20-tetraphenylporphyrinatoniobium(V)] 1,2-dichloroethane solvate, 45B, 537
$C_{89}H_{57}Cl_3N_8Nb_2O_3$, Tri-μ-oxo-bis(tetraphenylporphinatoniobium(V)) - chloroform, 43B, 641; 44B, 459
$C_{90}H_{58}Cl_6Fe_2N_9$, μ-Nitrido-bis(α,β,γ,δ-tetraphenylporphinatoiron) chloroform solvate, 42B, 382
$C_{96}H_{66}Fe_2N_8O$, μ-Oxobis(tetraphenylporphineiron(III)) - p-xylene, 34B, 283
$C_{100}H_{70}Cl_{12}Mo_2N_8O_3$, μ-Oxo-bis(oxotetraphenylporphinatomolybdenum(V)) - chloroform - m-xylene, 43B, 641; 44B, 459
$C_{104}H_{76}Fe_2N_9$, μ-Nitrido-bis(α,β,γ,δ-tetraphenylporphinatoiron) xylene solvate, 42B, 382

50. ANTIBIOTICS

$C_4H_3BrN_2O_2$, Bromoemimycin, 28, 529
$C_5H_7NO_2$, trans-2-Azabicyclo[2.1.0]pentane-3-(S)-carboxylic acid, 46B, 510
$C_5H_{13}NO_4S$, Choline O-sulfate, 44B, 24
$C_7H_{12}N_2O_4S$, S-Nitroso-N-acetyl-DL-pencillamine, 44B, 435
$C_8H_9NO_4$, Trichoviridin, 46B, 511
$C_8H_{12}N_4O_5$, 1-β-D-Ribofuranosyl-1,2,4-triazole-3-carboxamide (polymorph V1), 42B, 383
$C_8H_{12}N_4O_5$, 1-β-D-Ribofuranosyl-1,2,4-triazole-3-carboxamide (polymorph V2), 42B, 383
$C_8H_{18}INO_2$, L-Acetyl-β-methylcholine iodide, 44B, 26
$C_8H_{18}N_2O_4$, Fortamine, 46B, 511
$C_8H_{20}Cl_2N_2O_4$, Fortamine dihydrochloride, 46B, 512
$C_9H_{10}O_4$, DL-Cyclomethylenomycin A, 44B, 460
$C_9H_{20}N_4O_5S$, N-(2'-Amidinoethyl)-3-aminocyclopentanecarboxamide hydrogen sulfate, 46B, 512
$C_{10}H_{12}N_4O_5$, Formycin B, 42B, 1021
$C_{10}H_{12}N_4O_6$, Oxoformycin B, 42B, 1021
$C_{10}H_{13}NO_5Se$, Latumcidin selenate, 37B, 287
$C_{10}H_{16}BrN_5O_5$, Formycin hydrobromide monohydrate, 31B, 194
$C_{11}H_{12}Cl_2N_2O_5$, D-(-)-threo-2,2-Dichloro-N-[β-hydroxy-α-(hydroxymethyl)-p-nitrophenethyl]acetamide, 45B, 542
$C_{11}H_{12}N_2O_5Br_2$, Bromomycetin, 16, 551
$C_{11}H_{16}N_4O_5 \cdot 1.5\ H_2O$, Coformycin sesquihydrate, 42B, 384
$C_{12}H_{11}NO_4$, Mimosamycin, 44B, 460
$C_{12}H_{15}N_5O_5$, Toyocamycin, 44B, 461
$C_{12}H_{18}N_2O_7$, Bicyclomycin, 40B, 476
$C_{13}H_{10}N_2O_4$, Myxin, 33B, 187
$C_{13}H_{14}N_2O_4S_2$, Gliotoxin, 32B, 239
$C_{13}H_{22}BrNO_4$, N-Methylbisdeoxocycloshowdomycin acetonide hydrobromide, 34B, 283
$C_{13}H_{23}IN_2O_7 \cdot 0.25\ C_2H_3N$, Actinobolin hydrogen iodide monohydrate acetonitrile, 41B, 577
$C_{14}H_7Cl_5N_2O_6$, 2,2'-(2,2,2-Trichlorethylidene)bis(4-chloro-6-nitrophenol), 46B, 512
$C_{14}H_{15}NO_6$, Reductiomycin, 46B, 513
$C_{14}H_{16}N_2O_3S_2$, Epidithiopiperazinedione antibiotic A26771A, 40B, 477
$C_{14}H_{20}N_2O_2S$, N-Acetylthienamycin methyl ester, 44B, 461
$C_{14}H_{21}BrCl_2N_2O_6 \cdot 2\ H_2O$, 2-Amino-N-(3-dichloromethyl-3,4,4a,5,6,7-hexahydro-5,6,8-trihydroxy-3-methyl-1-oxo-1H-2-benzopyran-4-yl)pro-

panamide hydrobromide dihydrate, 46B, 513
$C_{15}H_8N_2O_2$, Indolo[2,1-b]quinazoline-6,12-dione, 40B, 478
$C_{15}H_{18}N_4O_5 \cdot 2\ H_2O$, Mitomycin C dihydrate, 45B, 542
$C_{15}H_{21}BrO_5 \cdot 0.5\ C_3H_6O$, Tetrahydropentalenolactone bromohydrin hemiacetonate, 38B, 517
$C_{15}H_{22}N_2O_4S$, D-α-Benzylpenilloic acid monohydrate, 44B, 462
$C_{15}H_{22}N_2O_7$, N-Acetylactinobolin, 42B, 384
$C_{15}H_{23}NO_4$, Cycloheximide, 43B, 642
$C_{15}H_{31}N_4O_5 \cdot 3.5\ H_2O$, Fortimicin B hydrate, 44B, 462
$C_{15}H_{36}BrN_3O_{11}$, 3-O-(4-Deoxy-4-propionamido-α-D-glucopyranosyl)-1,4-diamino-1,4-dideoxyhexytol hydrobromide dihydrate, 44B, 463
$C_{16}H_{17}KN_2O_4S$, Potassium benzylpenicillin, 12, 430; 16, 553; 44B, 463
$C_{16}H_{17}N_2NaO_4S$, Sodium benzylpenicillin, 12, 424
$C_{16}H_{17}N_2O_4RbS$, Rubidium benzylpenicillin, 12, 431
$C_{16}H_{18}N_2O_5S$, Phenoxymethylpenicillin, 28, 530
$C_{16}H_{19}N_3O_4S$, Ampicillin, 42B, 385
$C_{16}H_{20}BrNO_5$, N-Acetylbromoanisomycin, 33B, 188
$C_{16}H_{25}N_3O_8S$, D(-)-α-Amino-p-hydroxybenzylpenicillin trihydrate, 44B, 464
$C_{17}H_{15}ClN_2O_5S$, Methyl 6β-phthalimido-2β-chloromethyl-2α-methylpenam-3α-carboxylate, 41B, 578
$C_{17}H_{16}BrClO_6$, 5-Bromogriseofulvin, 28, 533
$C_{17}H_{18}N_2O_4S$, 4-Acetyl-3-methyl-7β-phenoxyacetamido-Δ^3-cephem, 44B, 465
$C_{17}H_{19}N_3O_4 \cdot H_2O$, Anthramycin methyl ether monohydrate, 45B, 543
$C_{17}H_{19}N_3O_{10}S_3$, Cephapyrine hydrogen sulfate, 43B, 490
$C_{17}H_{20}N_2O_4S$, Methyl benzylpenillonate, 44B, 466
$C_{17}H_{21}N_3O_5$, Anthramycin methyl ether monohydrate, 44B, 466
$C_{17}H_{22}N_2O_7S$, Phenoxymethylpenicillin sulfoxide methanolate, 34B, 285
$C_{17}H_{28}O_5$, Dihydrobotrydial, 40B, 478
$C_{18}H_{16}N_2O_6S_2 \cdot 0.5\ C_2H_6O$, Epicorazine B ethanol solvate, 45B, 543
$C_{18}H_{22}N_2O_5S$, Phenoxymethylanhydropenicillin ethanol solvate, 38B, 401
$C_{18}H_{22}N_2O_6S$, (2,6-Dimethoxyphenyl)penicillin methyl ester, 43B, 493
$C_{18}H_{27}N_{10}O_7S \cdot 5\ H_2O$, Netropsin sulfate pentahydrate, 45B, 544
$C_{19}H_{16}ClFN_3NaO_5S \cdot H_2O$, Sodium flucloxacillin monohydrate, 46B, 514
$C_{19}H_{19}N_3O_4S$, 7-(2,2-Dimethyl-5-oxo-4-phenyl-1-imidazolidinyl)-desacetylcephalosporanic acid lactam, 43B, 496
$C_{19}H_{20}N_3NaO_6S$, Sodium (5-methyl-3-phenyl-4-isoxazolyl)penicillin monohydrate, 43B, 495
$C_{20}H_{19}Cl_2N_3O_5S$, 3-(2,6-Dichlorophenyl)-5-methyl-4-isoxazolyl penicillin methyl ester, 43B, 643
$C_{20}H_{20}ClN_3O_5S$, 3-(2-Chlorophenyl)-5-methyl-4-isoxazolyl-penicillin methyl ester, 42B, 386
$C_{20}H_{20}N_2O_7S$, 6-Thiatetracycline, 46B, 367
$C_{20}H_{23}ClN_2O_4$, (+)-Chlorpheniramine maleate, 40B, 42
$C_{20}H_{24}O_8$, Vermiculine, 40B, 479
$C_{20}H_{25}N_3O_7S$, Methyl 6α-ethoxyformamido-6β-phenoxyacetamidopenicillanate, 45B, 545
$C_{20}H_{30}O_5$, Kromycin, 37B, 288
$C_{20}H_{32}ClN_3O_6S$, (S)-1'-Ethoxycarbonyloxyethyl 6β-((hexahydro-1H-azepin-1-yl)methyleneamino)penicillanate hydrochloride, 44B, 467
$C_{21}H_{18}N_2O_5S_2$, 4-Nitrobenzyl-(3SR,5RS,Z)-2-(2-phenylthioethylidene)-penam-3-carboxylate, 46B, 515
$C_{21}H_{25}N_3O_6S$, Dimethyl (5-methyl-3-phenyl-4-isoxazolyl)penicilloate, 43B, 496
$C_{21}H_{26}N_2O_5S$, t-Butyl 7α-methoxy-7-phenylacetamidodeacetoxycephalos-

poranate, 40B, 480
$C_{21}H_{26}N_2O_7S$, Benzylpenicillin 1'-diethyl carbonate ester, 43B, 643
$C_{21}H_{27}N_3O_6$, Naphthyridinomycin, 42B, 386
$C_{21}H_{28}Cl_2N_2O_{11}$, 6-Demethyl-7-chlorotetracycline hydrochloride trihydrate, 44B, 469
$C_{21}H_{30}O_2$, 9-Ethyl-vitamin A acid, 44B, 132
$C_{21}H_{31}BrN_2O_5$, 3R,4R-Dihydroxy-5S-(2S,3S-epoxy-5S-hydroxy-4S-methylhexyl)tetrahydropyran-2S-ylacetone o-bromophenylhydrazone, 44B, 1105
$C_{21}H_{51}N_8O_{22}Se_{1.5}$, Streptomycin oxime selenate tetrahydrate, 44B, 467
$C_{22}H_{21}ClN_2O_8$, 5a,11a-Dehydro-7-chlorotetracycline, 46B, 516
$C_{22}H_{22}BrN_3O_8S \cdot 0.5\ C_6H_6$, N-Brosylmitomycin A benzene solvate, 32B, 240
$C_{22}H_{24}Cl_2N_2O_8$, Aureomycin hydrochloride, 28, 534; 44B, 468
$C_{22}H_{24}Cl_2N_2O_8$, 5a-epi-7-Chlorotetracycline hydrochloride, 46B, 516
$C_{22}H_{24}N_2O_5S \cdot C_2H_6O$, Nafcillin methyl ester ethanol solvate, 46B, 514
$C_{22}H_{24}N_2O_9$, Oxytetracycline (form 1), 42B, 388
$C_{22}H_{24}N_2O_9$, Oxytetracycline (form 2), 43B, 645
$C_{22}H_{25}BrN_2O_8$, Anhydrotetracycline hydrobromide monohydrate, 44B, 469
$C_{22}H_{25}ClN_2O_8$, β-6-Deoxyoxytetracycline hydrochloride, 45B, 545
$C_{22}H_{25}ClN_2O_9$, Terramycin hydrochloride, 30B, 221
$C_{22}H_{28}Cl_2HgN_2O_{11}$, Oxytetracycline - mercury(II) chloride dihydrate, 42B, 387
$C_{22}H_{28}N_2O_{11}$, Oxytetracycline dihydrate, 42B, 388
$C_{22}H_{29}BrN_2O_{11}$, Oxytetracycline hydrobromide dihydrate, 42B, 144
$C_{22}H_{34}INO_3$, (R)-N-(2-(2-Cyclohexylmandeloyloxy)ethyl)-N-methylpiperidinium iodide, 43B, 312
$C_{22}H_{36}N_2O_{14}$, Tetracycline hexahydrate, 42B, 388; 43B, 644
$C_{22}H_{40}N_2O_{12}S$, DL-Isoproterenol sulfate dihydrate, 37B, 288
$C_{23}H_{20}BrN_3O_5S$, 6β-Phthalamido-6α-methylpenam-3α-p-bromocarboxanilide 1β-oxide, 44B, 469
$C_{23}H_{22}N_2O_3S$, Benzyl 6α-benzyl-6β-isocyanopencillanate, 45B, 546
$C_{23}H_{25}N_2O_{10} \cdot 4\ H_2O \cdot CH_2Cl_2$, 4-Epioxytetracycline tetrahydrate dichloromethane solvate, 45B, 547
$C_{23}H_{26}N_2O_8S$, 11a-Hydroxy-12a-dehydroxy-6-thiatetracycline-acetone (1:1), 46B, 367
$C_{23}H_{27}N_3O_8S$, 5a-epi-6-Thiatetracycline-N,N-dimethylformamide (1:1), 46B, 367
$C_{23}H_{28}ClN_3O_9S$, 3-(2-Chlorophenyl)-5-methyl-4-isoxazolyl-penicillin sulfoxide dioxan monohydrate, 42B, 388
$C_{23}H_{29}BrN_2O_{10}$, α-6-Deoxyoxytetracycline hydrobromide hemihydrate hemiethanolate, 43B, 655
$C_{23}H_{29}ClN_2O_{10}$, α-6-Deoxyoxytetracycline hydrochloride hemihydrate hemiethanolate, 43B, 655
$C_{23}H_{29}N_5O_8S$, Piperacillin hydrate, 44B, 470
$C_{23}H_{31}N_3O_8S$, Griseoviridin methanolate, 42B, 389
$C_{23}H_{33}IN_2O$, Isopropamide iodide, 43B, 646
$C_{23}H_{36}N_2O_{13}$, Tetracycline methyl betaine pentahydrate, 44B, 470
$C_{23}H_{36}N_4O_3$, Tetracycline - urea tetrahydrate, 44B, 471
$C_{23}H_{38}O_5$, Protylonolide, 46B, 517
$C_{23}H_{39}NO_6$, Thermozymocidin N-acetyl-γ-lactone, 45B, 547
$C_{24}H_{25}F_4NOS$, Piflutixol, 43B, 448
$C_{24}H_{34}K_2N_2O_{13}$, Potassium oxytetracycline dihydrate dimethanolate, 42B, 387
$C_{24}H_{34}O_8$, 5,9-Diacetyl-(3,6)bicycloleuconolide A(3), 43B, 646
$C_{25}H_{30}ClNO_{10} \cdot C_4H_{10}O$, Daunomycin chloride butanol solvate, 45B, 548

$C_{25}H_{33}BrO_8$, Bromouliginosin B, 33B, 234
$C_{25}H_{33}ClO_5$, 4-O-Ethyl ascofuranone, 41B, 579
$C_{25}H_{34}Br_2Cl_2O_8S_2$, Caldariomycin bis((+)-3-bromocamphor-9-sulphonate), 33B, 190
$C_{26}H_{30}ClNO_{11}$, Carcinomycin I hydrochloride monohydrate, 41B, 579; 43B, 647
$C_{26}H_{32}N_2O_8$, N-t-Butyl-4-(dimethylamino)-1,4,4a,5,5a,6,11,12a-octahydro-3,10,12,12a-tetrahyroxy-8-methoxy-1,11-dioxo-2-naphthacenecarboxamide, 45B, 548
$C_{27}H_{24}O_{10}$, 11-Deoxydaunorubicin aglycone triacetate, 46B, 517
$C_{27}H_{30}N_2O_9S$, p-Methoxybenzyl 2α-methyl-2β-[(R)-acetoxy(methoxy)methyl]-6β-phenoxyacetamidopenam-3α-carboxylate, 45B, 549
$C_{28}H_{27}BrO_6$, p-Bromobenzoyl-leukopleurotin, 41B, 580
$C_{28}H_{33}BrO_5S$, Siccanin p-bromobenzenesulphonate, 40B, 1093
$C_{29}H_{30}N_4O_{10}$, Streptonigrin ethyl acetate, 41B, 581
$C_{29}H_{32}BrClO_6S$, Ascochlorin p-bromobenzenesulphonate, 34B, 286
$C_{29}H_{34}N_2O_4$, 6,7-Epoxy-10(indol-3-yl)-16-methyl(11)cytochalas-13'-ene-1,18,21-trione, 40B, 480
$C_{29}H_{37}N_3O_6$, Carboxylic acid antibiotic A23187, 40B, 481
$C_{29}H_{40}N_2O_4S$, Procaine penicillin G monohydrate, 44B, 463
$C_{30}H_{27}IO_{14}$, Tri-O-acetyl-O-iodoacetyl-granaticin, 33B, 190
$C_{30}H_{42}N_2O_9$, Herbimycin A, 46B, 517
$C_{30}H_{43}CsO_{11}$, Caesium chlorothricolide methyl ester trihydrate, 38B, 519
$C_{30}H_{44}BrN_3O_3$, Dihydroteleocidin B monobromoacetate, 31B, 194
$C_{31}H_{38}BrN_3O_8S$, Bundlin A p-bromophenylsulphonylhydrazone, 37B, 291
$C_{31}H_{38}N_2O_{12}$, Novobiocin monohydrate, 41B, 582
$C_{31}H_{44}O_8$, Polyangi-1,5,6-triol triformate, 43B, 648
$C_{32}H_{24}O_{12} \cdot CHCl_3$, Thermorubin, 46B, 518
$C_{32}H_{30}N_2O_4$, Cochliodinol, 43B, 649
$C_{32}H_{30}N_6O_6S$, 7α-Methoxy-1-oxacephem, 46B, 518
$C_{32}H_{37}ClN_2O_{11}$, Daunomycin hydrochloride monohydrate pyridinate, 43B, 649
$C_{32}H_{40}O_{12}$, Pillaromycin A diethylether solvate, 41B, 582
$C_{32}H_{43}N_3O_9$, Virginiamycin factor-M dioxane solvate, 40B, 482
$C_{33}H_{40}BrN_3O_7$, Bundlin B p-bromophenylhydrazone, 34B, 287
$C_{33}H_{57}IKN_3O_9$, Enniatin B potassium complex, 34B, 288
$C_{34}H_{21}Br_2N_7O_3S_4 \cdot 0.5\ C_7H_8O$, Micrococcinic acid bis-4-bromoanilide (anisole solvate), 31B, 198
$C_{34}H_{57}N_4O_9RbS$, Enniatin B rubidium isothiocyanate, 44B, 472
$C_{34}H_{59}AgO_{10} \cdot 0.5\ H_2O$, Silver lysocellin hemihydrate, 42B, 390
$C_{35}H_{52}BrNO_{10}$, p-Bromobenzoylpikromycin monohydrate, 43B, 1481
$C_{35}H_{59}N_3O_{10}S \cdot H_2O$, (4,4-Dioxo-1,4-perhydrothiazine-1-imino)-rosaramicin monohydrate, 46B, 519
$C_{36}H_{43}IO_{12}S$, Verrucarin A p-iodobenzenesulphonate acetone solvate, 31B, 196
$C_{36}H_{62}BrNaO_{11}$, Monensin sodium bromide, 44B, 472
$C_{36}H_{64}O_{12}$, Monensin monohydrate, 37B, 292
$C_{37}H_{29}N_2O_{11}$, Kinamycin C p-bromobenzoate benzene solvate, 38B, 520
$C_{37}H_{45}NO_{14}$, Tolypomycinone monohydrate, 44B, 473
$C_{37}H_{51}BrClN_3O_{10}$, Maytansine (3-bromopropyl) ether, 39B, 376
$C_{38}H_{36}ClNO_{15}$, Lysolipin triacetate acetone solvate, 43B, 650
$C_{38}H_{65}NO_{14}$, Erythromycin A cyclic carbonate, 44B, 474
$C_{40}H_{48}BrNO_8$, Isozygosporin A p-bromobenzoate isopropyl alcohol solvate, 38B, 521
$C_{40}H_{64}AgBO_{16}$, Silver asplasmomycin dihydrate, 43B, 651
$C_{40}H_{64}O_{12}$, Nonactin, 38B, 522

$C_{40}H_{70}INO_{14}$, Anhydroerythromycin A cyclic carbonate N-methyl iodide methanolate, 41B, 583
$C_{40}H_{70}O_{15}$, des-Boron-des-valine-boromycin monohydrate, 40B, 483
$C_{41}H_{50}N_2O_{11}$, Hedamycin, 45B, 550
$C_{41}H_{64}CsNO_{12}S$, Caesium nonactin, 43B, 652
$C_{41}H_{64}KNO_{12}S$, Nonactinpotassium thiocyanate complex, 32B, 242
$C_{41}H_{68}N_2O_{12}S$, Nonactin - ammonium thiocyanate, 42B, 390
$C_{41}H_{70}CaO_9 \cdot 0.5\ C_7H_{16}$, Ionomycin calcium salt n-heptane solvate, 45B, 550
$C_{41}H_{70}CdO_9 \cdot 0.5\ C_6H_{14}$, Ionomycin cadmium salt n-hexane solvate, 45B, 550
$C_{41}H_{70}CdO_9 \cdot 0.5\ C_7H_{16}$, Ionomycin cadmium salt n-heptane solvate, 45B, 550
$C_{42}H_{64}BrNO_8 \cdot CH_2Cl_2 \cdot C_6H_{14}$, 1-Amino-1-(4-bromophenyl)ethane lasalocid salt, 43B, 652
$C_{42}H_{78}N_2O_{14} \cdot C_2H_3N \cdot 3\ H_2O$, 9,11-Dideoxy-9-11-(imino(2-(2-methoxyethoxy)-ethylidene)oxy)-(9S)-erythromycin acetonitrile water solvate, 45B, 551
$C_{43}H_{68}N_4O_{17}$, Rifampicine pentahydrate, 41B, 585
$C_{44}H_{72}O_{12}$, Tetranactin, 40B, 484
$C_{44}H_{75}O_{14}Tl$, Thallium(I) lonomycin, 42B, 391; 43B, 653
$C_{45}H_{57}N_3O_9 \cdot X\ H_2O$, Beauvericin hydrate, 42B, 391
$C_{45}H_{72}CsNO_{12}S$, Caesium tetranactin, 43B, 652
$C_{45}H_{72}KNO_{12}S$, Potassium tetranactin (form I), 42B, 392
$C_{45}H_{72}KNO_{12}S$, Potassium tetranactin (form II), 42B, 392
$C_{45}H_{72}NNaO_{12}S$, Sodium tetranactin, 42B, 392
$C_{45}H_{72}NO_{12}RbS$, Rubidium tetranactin, 42B, 392
$C_{45}H_{76}N_2O_{12}S$, Tetranactin ammonium thiocyanate, 41B, 586; 43B, 654
$C_{47}H_{76}Cl_6I_3NO_{17}$, 9-Propionylmaridomycin triiodide tetrachloromethane 1,2-dichloroethane solvate, 44B, 474
$C_{47}H_{77}AgO_{13}$, Antibiotic A-130A silver salt, 43B, 656
$C_{47}H_{79}O_{15}Tl$, Carriomycin thallium salt, 45B, 552
$C_{48}H_{64}I_2N_2O_{13} \cdot 4\ CH_4O \cdot H_2O$, Triacetylmethoxykidamycin bis(trimethylammonium) iodide tetramethanol hydrate, 46B, 519
$C_{48}H_{82}O_{20}$, 3-(β-L-Mycarose)-5-(β-D-4,6-dideoxy-3-ketoallose)-13-(β-D-mycinose)-lankamycin-11-α-hydroxyisovalerate ester, 46B, 520
$C_{50}H_{75}IO_{12}$, Salinomycin p-iodophenacyl ester, 41B, 587
$C_{51}H_{43}N_{13}O_{12}S_6$, Nosiheptide, 43B, 656
$C_{51}H_{83}IN_2O_{16} \cdot 2\ H_2O$, 4"-O-(4-Iodobenzoyl)megalomicin A dihydrate, 45B, 552
$C_{52}H_{79}Br_2NO_{17}$, 11,4"-Bis(O-(p-bromobenzoyl))oleandomycin monohydrate methanolate ethanolate, 44B, 475
$C_{52}H_{89}AgO_{18}$, Antibiotic A204A (silver salt), 39B, 377
$C_{52}H_{89}NaO_{18}$, Antibiotic A204A (sodium salt), 39B, 377
$C_{52}H_{92}O_{19}$, Antibiotic A204A acetone solvate, 44B, 475
$C_{53}H_{54}Br_2N_2O_{11} \cdot 0.5\ C_6H_6$, Isokidamycin bis(m-bromobenzoate) hemibenzene solvate, 46B, 519
$C_{53}H_{61}BBrCl_2NO_{17}$, Streptovaricin C triacetate (cyclic p-bromobenzeneboronate atropisomer) - methylene dichloride (1:1), 42B, 393
$C_{54}H_{90}I_4KN_6O_{18}$, Valinomycin potassium iodide, 41B, 1240
$C_{54}H_{90}N_6O_{18}$, Valinomycin, 41B, 588
$C_{54}H_{98}Ba_2Cl_4N_6O_{32}$, Valinomycin barium perchlorate tetrahydrate, 46B, 520
$C_{57}H_{61}BaN_9O_{23} \cdot 2\ C_7H_8$, Beauvericin-barium picrate toluene solvate (form B), 46B, 522
$C_{58}H_{52}Br_3NO_{16}$, Tolypomycinone tri-m-bromobenzoate, 34B, 287
$C_{58}H_{76}CaN_6O_{14}$, Calcium complex of A23187, 42B, 394

$C_{58}H_{99}N_6O_{18}$, Valinomycin hemi(n-octane) solvate, 41B, 588
$C_{60}H_{102}N_6O_{18}$, Cyclo-tris(D-valyl-L-α-hydroxyisovaleryl-L-valyl-D-α-hydroxyisovaleryl), 45B, 553
$C_{60}H_{102}N_6O_{18}$, Isoleucinomycin, 46B, 521
$C_{61}H_{80}BrN_9O_{11}$, Ilamycin B_1 p-bromobenzoate, 40B, 485
$C_{61}H_{100}INO_{22}$, N-Iodoacetylamphotericin B tritetrahydrofuran monohydrate, 37B, 290
$C_{66}H_{98}N_{15}O_{19}Rb \cdot 1.5\ C_7H_8 \cdot CHCl_3$, Prolinomycin rubidium picrate toluene chloroform, 46B, 522
$C_{68}H_{108}BaO_{17}$, Antibiotic X-537A, 35B, 406
$C_{68}H_{110}Na_2O_{18}$, Sodium lasalocid A hydrate, 44B, 476
$C_{70}H_{96}BrN_3O_8$, Antibiotic X-14547A (R-(+)-1-amino-1-(4-bromophenyl)-ethane salt), 44B, 476

51. STERIODS

$C_{16}H_{20}BrN_3O_2$, 12-Amino-11,13-diaza-9β,14β-11,12-dehydroestrone methyl ether hydrobromide, 35B, 331
$C_{17}H_{22}BrNO_2$, 8-Azaestrone hydrobromide, 34B, 289
$C_{17}H_{22}N_2O_3$, DL-2,3-Dimethoxy-18-nor-8,13-diaza-1,3,5(10)-estratrien-17-one, 41B, 589
$C_{17}H_{23}NO_2$, 10-Aza-19-nor-5β,9β-androst-8(14)-en-3,17-dione, 39B, 377
$C_{17}H_{23}NO_2$, 8-Azaestradiol, 38B, 523
$C_{17}H_{26}O_3$, 17β-Hydroxy-1-oxa-A-nor-5β-androstan-2-one, 46B, 523
$C_{18}H_{21}BrO_2$, 4-Bromoestrone, 28, 536
$C_{18}H_{22}Br_2O_2$, 2,4-Dibromoestradiol, 37B, 293
$C_{18}H_{22}O_2 \cdot 0.5\ H_2O$, 1,3,5(10),14-Estratetraene-3,17β-diol hemihydrate, 45B, 554
$C_{18}H_{22}O_2$, DL-3-Methoxy-B-nor-9β-estra-1,3,5(10)-trien-17-one, 41B, 589
$C_{18}H_{22}O_2$, Estrone, 38B, 524; 39B, 378
$C_{18}H_{22}O_2$, 17β-Hydroxy-4,9,11-estratriene-3-one, 45B, 554
$C_{18}H_{22}O_3$, 16α-Hydroxyestrone, 42B, 395
$C_{18}H_{22}O_3$, 5(10)-Secoestra-5,6-yne-3,10,17-trione, 44B, 478
$C_{18}H_{22}O_3$, 5,10-Secoestra-4,5-diene-3,10,17-trione, 44B, 478
$C_{18}H_{22}O_4$, 7-Hydroxy-3-methoxy-6-oxaestra-1,3,5(10)-trien-17-one, 46B, 523
$C_{18}H_{24}BrNO_2$, 12-Keto-17-deoxo-8-azaestrone methyl ether hydrobromide, 37B, 293
$C_{18}H_{24}O_2 \cdot 0.5\ H_2O$, Estradiol hemihydrate, 38B, 524
$C_{18}H_{24}O_2$, 1,3,5(10)-Estratrien-3,17α-diol, 42B, 395
$C_{18}H_{24}O_2$, 17β-Hydroxy-4,14-estradien-3-one, 44B, 477
$C_{18}H_{24}O_3$, Estriol, 34B, 291
$C_{18}H_{24}O_3$, 17β-Hydroxy-8(9-10β)abeo-estr-4-en-3,10-dione, 38B, 526
$C_{18}H_{26}O_2$, 17β-Hydroxyestr-5(10)-en-3-one, 38B, 526
$C_{18}H_{26}O_2$, 19-Nortestosterone, 41B, 590
$C_{18}H_{26}O_3$, Estradiol monohydrate, 34B, 290
$C_{18}H_{28}O_3$, 5,7-Estradiene-3β,17β-diol monohydrate, 43B, 657
$C_{18}H_{28}O_5S$, 17β-Acetoxy-2,4-dioxa-3-thia-5α-androstan-3-one, 42B, 396
$C_{18}H_{31}NO_4$, 17β-Hydroxyestr-5(10)-en-3-one oxime dihydrate, 42B, 396
$C_{19}H_{22}O_2$, 3-Methoxy-1,3,5(10),14-estratetraen-17-one, 45B, 554
$C_{19}H_{24}O_2$, DL-14-β-androsta-4,8-diene-3,17-dione, 39B, 379
$C_{19}H_{24}O_2$, 3-Methoxy-1,3,5(10)-estratrien-17-one, 45B, 554
$C_{19}H_{24}O_2$, 3-Methoxy-1,3,5(10),14-estratetraen-17β-ol, 45B, 554
$C_{19}H_{24}O_2$, 9β,10α-Androsta-4,6-diene-3,17-dione, 40B, 1093
$C_{19}H_{25}BrO_2$, 3-Methoxy-16α-bromo-17α-hydroxy-estra-1,3,5(10)-triene,

45B, 556
$C_{19}H_{25}BrO_2$, 3-Methoxy-16α-bromo-17β-hydroxy-estra-1,3,5(10)-triene, 45B, 557
$C_{19}H_{25}BrO_2$, 4α-Bromo-5α-androst-2-ene-1,17-dione, 35B, 333
$C_{19}H_{25}BrO_2$, 6α-Bromo-4-androsten-3,17-dione, 43B, 658
$C_{19}H_{25}BrO_2$, 6β-Bromo-4-androstene-3,17-dione, 42B, 396
$C_{19}H_{26}O \cdot H_2O$, 17β-Hydroxy-5α-androst-1-en-3-one hydrate, 45B, 557
$C_{19}H_{26}O_2 \cdot H_2O$, 17β-Hydroxymethyl-1,3,5(10)-estratrien-3-ol monohydrate, 46B, 525
$C_{19}H_{26}O_2$, Androst-4-en-3,17-dione, 38B, 527
$C_{19}H_{26}O_2$, 17β-Hydroxy-9α-methyl-4,14-estradien-3-one, 46B, 523
$C_{19}H_{26}O_2$, 17β-Hydroxy-4,14-androstadien-3-one, 44B, 477
$C_{19}H_{26}O_2$, 17β-Hydroxy-7β-methyl-4,14-estradien-3-one, 45B, 558
$C_{19}H_{26}O_2$, 5-Androstene-3,17-dione, 44B, 478
$C_{19}H_{26}O_3$, 19-Hydroxyandrost-4-ene-3,17-dione, 46B, 524
$C_{19}H_{26}O_3$, 3β-Hydroxy-17-oxo-5-androsten-19-al, 41B, 591
$C_{19}H_{26}O_3$, 4-Hydroxy-4-androstene-3,17-dione, 46B, 524
$C_{19}H_{26}O_3$, 9β-Hydroxy-10α-androst-4-ene-3,17-dione, 40B, 1093
$C_{19}H_{27}BrO_3$, 4-Bromoestradiol methanolate, 29, 750
$C_{19}H_{28}O_2$, Testosterone, 39B, 381
$C_{19}H_{28}O_2$, 10β-Methyl-1(10-5)abeo-estran-3,17-dione, 44B, 479
$C_{19}H_{28}O_2$, 17α-Hydroxyandrost-4-en-3-one, 38B, 528
$C_{19}H_{28}O_2$, 5α-Androstan-3,17-dione, 39B, 380
$C_{19}H_{28}O_6$, Tetradecahydro-3,7-dihydroxy-3a-methyl-1H-5b,7-methano-8H-as-indaceno(3',2':4,5)furo[2,3-b]pyran-8-one monohydrate, 41B, 591
$C_{19}H_{29}ClO \cdot 0.5\ CH_3OH$, 3β-Chloro-5-androsten-17β-ol methanol solvate, 37B, 294
$C_{19}H_{29}O_3Br$, 6α-Bromo-17β-hydroxy-17α-methyl-4-oxa-5α-androstan-3-one, 35B, 334
$C_{19}H_{30}Br_2$, 16β,17β-Dibromoandrostane, 38B, 529
$C_{19}H_{30}O$, Androstan-17-one, 44B, 479
$C_{19}H_{30}O_2$, 17β-Hydroxyandrostan-3-one, 39B, 381
$C_{19}H_{30}O_2$, 5α-Androstan-3α-ol-17-one, 31B, 201
$C_{19}H_{30}O_2$, 5α-Androstan-3β-ol-17-one, 37B, 295
$C_{19}H_{30}O_2S$, 2α,3α-Epithio-5α-androstan-17β-ol (R)-S-oxide, 43B, 658
$C_{19}H_{30}O_2S_2$, 1α,5α-Epidithioandrostane-3α,17β-diol, 43B, 659
$C_{19}H_{30}O_3$, Testosterone monohydrate, 38B, 530; 39B, 382
$C_{19}H_{30}O_3$, 17β-Hydroxy-17α-methyl-2-oxa-5α-androstan-3-one, 41B, 592
$C_{19}H_{30}O_4$, 2,4-Dioxa-5α-androstan-17β-ol acetate, 41B, 593
$C_{19}H_{31}N_3$, 14β-Azido-5α-androstane, 40B, 486
$C_{19}H_{32}O_2$, 3α,17β-Dihydroxy-5α-androstane, 38B, 531
$C_{19}H_{32}O_2$, 5β-Androstane-3α,17β-diol, 37B, 295
$C_{19}H_{32}O_3$, 17β-Hydroxyandrostan-3-one monohydrate, 38B, 532
$C_{19}H_{34}O_3$, 3β,17β-Dihydroxy-5α-androstane monohydrate, 39B, 382
$C_{20}H_{22}O_2$, 17β-Hydroxy-19-nor-4,9,11-pregnatriene-20-yne-3-one, 40B, 487
$C_{20}H_{22}O_3$, 17β-Hydroxy-19-norpregna-4,9-diene-20-yne-3,11-dione, 39B, 383
$C_{20}H_{24}O_2$, 17β-Hydroxy-19-norpregna-4,9-diene-20-yne-3-one, 46B, 525
$C_{20}H_{24}O_2$, 19-Norpregna-4,9,11-triene-3,20-dione, 43B, 660
$C_{20}H_{24}O_3S$, 9β-3-Methoxy-17-acetoxy-7-thiaoestra-1,3,5(10),8(14)-tetraene, 38B, 362
$C_{20}H_{24}O_5$, 3-Methoxy-6-oxaestra-1,3,5(10)-triene-7,17-dione 17-(ethylene acetal), 46B, 526
$C_{20}H_{24}O_5S$, 17-Homo-3-methoxy-15-ethylenedioxy-17-thiaestra-1,3,5(10),8-tetraene S,S-dioxide, 43B, 661
$C_{20}H_{25}O_4$, 17β-Acetoxy-4-oxa-6β,10-cyclo-1(10-5)abeo-5(R),10α-1-an-

drosten-3-one, 44B, 480
$C_{20}H_{26}O$, 19-Nor-17β-ethynyl-Δ⁴-5-androstene-3-one, 40B, 488
$C_{20}H_{26}O_2$, 17α-Ethynyl-17β-hydroxyestr-5(10)-en-3-one, 43B, 661
$C_{20}H_{26}O_2$, 3-Methoxy-14-methyl-14β-estra-1,3,5(10)-trien-15-one, 45B, 559
$C_{20}H_{26}O_3$, (10S)-17β-Acetoxy-3,10-cyclo-3,4-seco-4,9(11)-estradien-1-one, 42B, 397
$C_{20}H_{26}O_3$, 1α-Hydroxynorethisterone, 43B, 662
$C_{20}H_{27}BrO_3$, 9α-Bromo-17β-hydroxy-17α-methylandrost-4-ene-3,11-dione, 33B, 192
$C_{20}H_{27}FO$, 11β-Fluoro-19-nor-17α-pregn-4-en-20-yn-17β-ol, 44B, 480
$C_{20}H_{27}NO_3$, 3-Methoxy-2-aza-1,3,5(10)-estratrien-17β-yl acetate, 44B, 481
$C_{20}H_{28}O$, 19-Nor-17α-pregn-4-en-20-yn-17β-ol, 42B, 397
$C_{20}H_{28}O_2$, 19-Norpregn-4-ene-3,20-dione, 45B, 559
$C_{20}H_{28}O_2$, 3-Methoxy-17β-hydroxymethyl-estra-1,3,5(10)-triene, 46B, 526
$C_{20}H_{28}O_3$, 19-Nor-17β-acetoxy-4-androsten-3-one, 46B, 527
$C_{20}H_{28}O_3$, 19-Nor-retrotestosterone acetate, 38B, 533
$C_{20}H_{28}O_3$, 3,3-Dimethoxy-19-norandrosta-5(10),6-dien-17-one, 44B, 481
$C_{20}H_{29}BrO_3$, 17β-Bromoacetoxy-19-nor-5α-androstan-3-one, 44B, 482
$C_{20}H_{29}FO_3$, 9α-Fluoro-11β,17β-dihydroxy-17α-methyl-4-androsten-3-one, 45B, 560
$C_{20}H_{29}IO_2$, 17β-Hydroxyestr-5(10)-ene iodoacetate, 39B, 383
$C_{20}H_{30}O_2 \cdot 0.5\ H_2O$, 17β-Hydroxy-17α-methyl-5α-androst-1-en-3-one hemihydrate, 40B, 488
$C_{20}H_{31}BrO_3$, 16-Bromo-3β-hydroxy-5-androsten-17-one methanol (1:1), 41B, 594
$C_{20}H_{32}$, 3-Methylene-5α-androstane, 41B, 595
$C_{20}H_{33}BrO$, 2α-Bromo-17α-methyl-5α,14β-androstan-3α-ol, 38B, 533
$C_{20}H_{36}O_5$, (19S)-19-Methyl-5-androstene-3β,17β,19-triol dihydrate, 42B, 397
$C_{21}H_{22}O_2$, 3-Methoxy-1,11-ethenoestra-1,3,5(10),9(11)-tetraen-17-one, 43B, 662
$C_{21}H_{22}O_3$, 3α-Hydroxy-5α-pregnane-11,20-one, 44B, 482
$C_{21}H_{22}O_4$, 6α,7α-Epoxy-5-hydroxy-17(13-18)-abeo-5α-pregna-2,13,15,17-tetraene-1,20-dione, 42B, 398
$C_{21}H_{24}O_2$, 17α-Ethynyl-17β-hydroxy-12-methyl-4,9,11-estratrien-3-one, 46B, 527
$C_{21}H_{24}O_2$, 17β-Hydroxy-18-methyl-19-nor-pregna-4,9,11-triene-20-yne-3-one, 41B, 595
$C_{21}H_{26}O_2$, 17β-Hydroxy-18-methyl-19-norpregna-4,9-diene-20-yne-3-one, 46B, 528
$C_{21}H_{26}O_3$, 11-Methoxy-17β-hydroxy-19-nor-4,9-pregnadiene-20-yne-3-one, 41B, 596
$C_{21}H_{26}O_3$, 11β,17β-Dihyroxy-18-methyl-19-nor-pregna-4,9-diene-20-yne-3-one, 43B, 663
$C_{21}H_{26}O_3$, 21-Hydroxy-4,9(11),16-pregnatriene-3,20-dione, 45B, 561
$C_{21}H_{27}BrO_2$, 4-Bromo-9β,10α-pregna-4,6-diene-3,20-dione, 31B, 202
$C_{21}H_{27}ClO_5$, 4-Chlorocortisone, 37B, 296
$C_{21}H_{28}BrO_3$, 12α-Bromo-11β-hydroxyprogesterone, 33B, 193
$C_{21}H_{28}NO_3$, 11β-Methoxyestradiol acetonitrile, 39B, 384
$C_{21}H_{28}N_2O_3$, 2-Methoxy-8,13-diaza-cyclohexa(15,16-α)estrone methyl ether, 42B, 398
$C_{21}H_{28}O \cdot 0.5\ H_2O$, 4,6β-Ethano-3-methoxy-8α-estra-1,3,5(10)-trien-17β-ol hemihydrate, 45B, 561
$C_{21}H_{28}O_2 \cdot 0.25\ H_2O$, rac-4,6β-Ethano-3-methoxy-8α-estra-1,3,5(10)-

trien-17β-ol hydrate, 45B, 562
$C_{21}H_{28}O_2$, d-Norgestrel, 41B, 597
$C_{21}H_{28}O_2$, 17α-Methyl-19-nor-Δ(9-10)-progesterone, 41B, 598
$C_{21}H_{28}O_2$, 3-Methoxy-1'β-methyl-1,11α-methano-9β-estra-1,3,5(10)-trien-17β-ol, 42B, 399
$C_{21}H_{28}O_2$, 9β,10α-Pregna-4,6-diene-3,20-dione, 40B, 1093
$C_{21}H_{28}O_3$, 17β-Hydroxy-11β-methoxy-19-norpregn-4-ene-20-yne-3-one, 43B, 664
$C_{21}H_{28}O_5$, Cortisone, 38B, 534
$C_{21}H_{28}O_6$, 16α-Hydroxyprednisolone, 40B, 489
$C_{21}H_{29}BrO_2$, 6β-Bromoprogesterone, 34B, 293
$C_{21}H_{29}BrO_3$, 17β-Bromoacetoxy-9β,10α-androst-4-ene-3-one, 32B, 243
$C_{21}H_{29}BrO_5$, 9α-Bromocortisol, 39B, 384
$C_{21}H_{29}ClO_5$, 9α-Chlorocortisol, 40B, 490
$C_{21}H_{29}FO_5$, 9α-Fluorocortisol, 38B, 535
$C_{21}H_{30}O$, 11β-Methyl-19-nor-17α-pregn-4-en-20-yn-17β-ol, 42B, 400
$C_{21}H_{30}O_2$, DL-4α,8α,14β-Trimethyl-18-nor-5α,13β-androst-9(11)-en-3,17-dione, 39B, 385
$C_{21}H_{30}O_2$, Progesterone (form I), 38B, 536
$C_{21}H_{30}O_2$, Progesterone (form II), 41B, 599
$C_{21}H_{30}O_2$, ψ-Retroprogesterone, 40B, 491
$C_{21}H_{30}O_2$, 21-Methyl-Δ(9,10)-norprogesterone, 40B, 490
$C_{21}H_{30}O_3$, Desoxycorticosterone, 38B, 537
$C_{21}H_{30}O_3$, 11α-Hydroxy-9β,10α-pregn-4-ene-3,20-dione, 40B, 1093
$C_{21}H_{30}O_3$, 17α-Hydroxyprogesterone, 38B, 537
$C_{21}H_{30}O_3$, 2β-Ethynyl-5β,17β-dihydroxy-3,4-bisnorandrostane 17-acetate, 41B, 599
$C_{21}H_{30}O_3$, 21-Hydroxy-4-pregnen-3,20-dione, 39B, 385
$C_{21}H_{30}O_3$, 3-Oxo-17β-acetoxy-Δ^4-14α-methyl-8α,9β,10α,13α-estrene, 40B, 492
$C_{21}H_{30}O_3$, 6-Oxo-3β,5-cycloandrostan-17-yl acetate, 38B, 552
$C_{21}H_{30}O_3$, 6α-Hydroxy-4-pregnene-3,20-dione, 45B, 562
$C_{21}H_{30}O_3S$, (17S)-Spiro((androst-4-ene)-17:5'-(1',2'-oxathiolane))-3-one 2-oxide (isomer B), 45B, 563
$C_{21}H_{30}O_3S$, (17S)-Spiro((androst-4-ene)-17:5'-(1',2'-oxathiolane))-3-one 2-oxide (isomer A), 45B, 563
$C_{21}H_{30}O_4$, Corticosterone, 39B, 386
$C_{21}H_{30}O_4$, 4-Pregnene-17α,21-diol-3,20-dione, 38B, 539; 39B, 386
$C_{21}H_{30}O_6$, Aldosterone monohydrate, 38B, 540
$C_{21}H_{31}BrO_3$, 17β-Bromoacetoxy-5α-androstan-3-one, 44B, 482
$C_{21}H_{31}BrO_3$, 2β-Bromo-3α-hydroxy-5α-pregnane-11,20-dione, 46B, 528
$C_{21}H_{31}BrO_3$, 2α-Bromo-17β-acetoxy-9-methyl-5α,9β,10α-estran-3-one, 44B, 483
$C_{21}H_{31}BrO_3$, 2β-Bromo-17β-acetoxy-9-methyl-5α,9β,10α-estran-3-one, 44B, 483
$C_{21}H_{31}BrO_3$, 5α-Bromo-6β,19-oxido-pregnan-3β-ol-20-one, 34B, 294
$C_{21}H_{31}BrO_4$, 3β,17α-Dihydroxy-16β-bromo-5α-pregnan-11,20-dione, 34B, 296
$C_{21}H_{31}BrO_4$, 3β,17α-Dihydroxy-21-bromo-5α-pregnan-11,20-dione, 33B, 194
$C_{21}H_{31}ClO_2$, 5α-Androst-2-en-17β-yl chloroacetate, 42B, 400
$C_{21}H_{31}IO_2$, 3β-Acetoxy-17α-iodo-Δ^5-androstene, 38B, 541
$C_{21}H_{32}O_2$, 20(S)-Hydroxyprogesterone, 38B, 542
$C_{21}H_{32}O_2$, 3β-Hydroxypregn-5-ene-20-one, 44B, 483
$C_{21}H_{32}O_2S$, 3α-Acetylthio-5α-androstan-17-one, 45B, 564
$C_{21}H_{32}O_3$, Estradiol - propanol solvate, 38B, 543
$C_{21}H_{32}O_3$, 17β-Acetoxy-5-methyl-5α,10α-estran-3-one, 46B, 529

$C_{21}H_{32}O_3$, 17β-Acetoxy-5-methyl-5α,10β-estran-3-one, 46B, 529
$C_{21}H_{32}O_3$, 17β-Acetoxy-5-methyl-5β,10β-estran-3-one, 46B, 529
$C_{21}H_{32}O_3$, 17β-Acetoxy-9-methyl-5α,9β,10α-estran-3-one, 44B, 483
$C_{21}H_{32}O_4$, 16α,17α-Epoxypregnenolone monohydrate, 42B, 402
$C_{21}H_{34}O_2$, 6α-Hydroxy-4,4-dimethylandrostan-3-one, 46B, 530
$C_{21}H_{35}ClO$, 14β-Chloro-4α-hydroxymethyl-4β,5β,13β-trimethylperhydrocyclopenta[a]phenanthrene, 43B, 664
$C_{21}H_{35}N_3$, 5α-Azido-pregnane, 41B, 600
$C_{21}H_{36}O_2 \cdot 0.5\ H_2O$, 5α,17α-Pregnane-3β,20α-diol hemihydrate, 40B, 492
$C_{21}H_{37}N$, 18,19-Bis-nor-5β,14β-dimethyl-13β-aminopregnane, 40B, 493
$C_{22}H_{23}O_3I$, 17β-Iodoacetoxy-4,4-dimethyl-19-nor-5α-androstan-3-one, 46B, 530
$C_{22}H_{24}Cl_4O_3$, 6,21,21-Trichloro-16α-chloromethyl-16β,20-oxido-17α-hydroxy-4,6,20-pregnatrien-3-one, 42B, 401
$C_{22}H_{25}Cl_3O_4$, 6-Chloro-15α-acetyl-16β-dichloromethyl-16α-hydroxy-4,6-androstadiene-3,17-dione, 42B, 401
$C_{22}H_{26}O_3$, 17-Ethylenedioxy-3-methoxy-6,7,8-methylidyne-1,3,5(10)-estratriene, 38B, 544
$C_{22}H_{27}BrO_3$, 17β-Bromoacetoxy-3-methoxy-8α-methyl-1,3,5(10),6-estratetraene, 38B, 545
$C_{22}H_{27}BrO_3$, 3-Methoxy-7α,8α-methylene-1,3,5(10)-oestratrien-17β-yl bromoacetate, 35B, 335
$C_{22}H_{28}O_3$, 17β-Hydroxy-18-methyl-11-methoxy-19-nor-4,9-pregnadiene-20-yne-3-one, 40B, 494
$C_{22}H_{28}O_3$, 17β-Hydroxy-3-oxo-17α-pregna-4,6-diene-21-carboxylic acid γ-lactone, 42B, 403
$C_{22}H_{29}BrO_3$, 3-Methoxy-8β-methylestradiol 17-monobromoacetate, 34B, 299
$C_{22}H_{29}FO_4$, 17-Desoxymethasone, 40B, 494
$C_{22}H_{29}FO_5$, 9α-Fluoro-16α-methyl-11β,17α,21-trihydroxy-1,4-pregnadiene-3,20-dione, 43B, 665
$C_{22}H_{29}FO_5$, 9α-Fluoro-6α-methylprednisolone, 40B, 495
$C_{22}H_{29}KO_4$, Potassium 17β-hydroxy-3-oxo-4,6-pregnadiene-21-carboxylate, 42B, 402
$C_{22}H_{30}O_2$, 18-Ethyl-17β-hydroxy-19-nor-pregn-4-ene-20-yne-3-one, 46B, 531
$C_{22}H_{30}O_2$, 6α-Methyl-4,16-pregnadiene-3,20-dione, 45B, 565
$C_{22}H_{30}O_3$, 17α-Ethynyl-17β-hydroxy-13β-(3-hydroxypropyl)gon-4-en-3-one, 43B, 665
$C_{22}H_{30}O_3$, 17β-Acetoxy-3-methoxy-C-homo-9β-1,3,5(10)-estratriene, 43B, 666
$C_{22}H_{30}O_3$, 5,6-Dihydroxycanrenone, 43B, 756
$C_{22}H_{30}O_3$, 6β,7β-Methylene-17β-hydroxyandrost-4-en-3-one 17-acetate, 35B, 336
$C_{22}H_{30}O_4$, 3,17-Dioxo-19-nor-androst-4-en-7α-butyric acid, 43B, 667
$C_{22}H_{30}O_4$, 3,17-Dioxo-19-nor-androst-4-en-7β-butyric acid, 43B, 667
$C_{22}H_{30}O_5$, 6α-Methyl-11β,17α,21β-trihydroxy-1,4-pregnadiene-3,20-dione, 38B, 546
$C_{22}H_{31}BrO_3$, 8β-Methyltestosterone 17β-monobromoacetate, 35B, 337
$C_{22}H_{31}FO_3$, 6α-Fluoro-6β-methyl-17β-hydroxy-9β,10α-androst-4-en-3-one 17-acetate, 40B, 1093
$C_{22}H_{31}FO_3$, 6β-Fluoro-6α-methyl-17β-hydroxy-9β,10α-androst-4-en-3-one 17-acetate, 40B, 1093
$C_{22}H_{31}FO_3$, 9α-Fluoro-11β-hydroxy-2α-methylprogesterone, 42B, 403
$C_{22}H_{31}FO_5$, 9α-Fluoro-2α-methylcortisol, 42B, 403
$C_{22}H_{31}NO_2$, 3β-Hydroxy-20-oxo-5-pregnene-16α-carbonitrile, 41B, 601

$C_{22}H_{31}NO_3$, 17β-Acetoxy-5α-cyanoandrostan-3-one, 44B, 484
$C_{22}H_{32}O$, 13-Ethyl-11-methylene-18,19-dinor-17α-pregn-4-en-20-yn-17-ol, 46B, 531
$C_{22}H_{32}O_2$, 16β-Methylprogesterone, 42B, 402
$C_{22}H_{32}O_2$, 3β-Hydroxy-16-methyl-5,16-pregnadien-20-one, 45B, 565
$C_{22}H_{32}O_2$, 6α-Methyl-9β,10α-pregn-4-ene-3,20-dione, 40B, 1093
$C_{22}H_{32}O_6$, 9α-Methoxycortisol, 42B, 404
$C_{22}H_{34}O$, 20-Methyl-14β,17α-pregn-4-en-3-one, 43B, 668
$C_{22}H_{34}O_3$, 17β-Acetoxy-7α-methyl-5α-androstan-3-one, 42B, 405
$C_{22}H_{34}O_3$, 3α-Hydroxy-2α-methyl-5α-pregnane-11,20-dione, 46B, 528
$C_{22}H_{34}O_4 \cdot H_2O$, 3α-Hydroxy-2β-methoxy-5α-pregnane-11,20-dione monohydrate, 46B, 528
$C_{22}H_{34}O_6$, Cortisol methanol solvate, 39B, 386
$C_{22}H_{36}O_2 \cdot CH_4O$, 20-Methyl-5-pregnene-3β,20-diol methanol solvate, 46B, 532
$C_{22}H_{36}O_2$, 14β,22-Epoxy-23,24-bisnor-5α-cholan-3β-ol, 45B, 566
$C_{22}H_{36}O_2Si$, 17β-Trimethylsiloxy-4-androsten-3-one, 38B, 547
$C_{23}H_{25}O_3I$, 17β-Iodoacetoxy-4,4-dimethyl-5α-androstan-3-one, 46B, 530
$C_{23}H_{27}BrO_3$, Fusidic acid intermediate, 35B, 338
$C_{23}H_{27}ClO_4$, 6-Chloro-17-hydroxypregna-1,4,6-triene-3,20-dione acetate, 44B, 484
$C_{23}H_{28}O_4$, 17-Hydroxypregna-1,4,6-triene-3,20-dione acetate, 44B, 484
$C_{23}H_{29}ClO_4$, Chlormadinone acetate, 41B, 602
$C_{23}H_{30}O_2$, 1,1-Dimethyl-14-anthrapregna-5,7,9(10)-triene-2,20-dione, 43B, 669
$C_{23}H_{30}O_6$, Cortisone acetate, 38B, 547
$C_{23}H_{31}BrO_3$, 4α,1α,5-Ethanylylidene-17β-hydroxy-5α-androstan-3-one bromoacetate, 42B, 405
$C_{23}H_{31}ClO_5$, 17β-Chloroacetoxy-2β-acetoxy-4-androsten-3-one, 41B, 602
$C_{23}H_{31}NO$, 3β-Methoxy-4,4,8α-trimethyl-13,17-seco-7,16-cycloandrosta-5,7(16),13(18)-triene-17-nitrile, 43B, 669
$C_{23}H_{31}O_4$, 11β,17β-Dihydroxy-18-methyl-19-norpregna-4,9-diene-20-yne-3-one ethanolate, 43B, 669
$C_{23}H_{32}O_3$, Δ-8,14-Anhydrodigitoxigenin, 35B, 338
$C_{23}H_{32}O_5$, 2α-Hydroxytestosterone diacetate, 41B, 603
$C_{23}H_{32}O_5$, 2β,17β-Diacetoxy-4-androsten-3-one, 41B, 604
$C_{23}H_{32}O_6 \cdot 0.5\,H_2O$, Strophanthidin, 39B, 387
$C_{23}H_{32}O_7 \cdot C_4H_8O_2$, Spirohydroxyprednisolone acetate ethyl acetate solvate, 46B, 532
$C_{23}H_{33}BrO_5$, 3β-Acetoxy-17α-hydroxy-16β-bromo-5α-pregnan-11,20-dione, 33B, 194
$C_{23}H_{33}IO_3$, 17β-Iodoacetoxy-4,4-dimethyl-5α-androst-7-en-3-one, 44B, 486
$C_{23}H_{33}IO_3$, 17β-Iodoacetoxy-4,4-dimethylandrostan-5-en-3-one, 44B, 485
$C_{23}H_{34}NO_3$, (20R,21S)-3β-Hydroxy-14,21-epoxy-5β,14β,20-cardanolactam, 42B, 406
$C_{23}H_{34}O_4$, Digitoxigenin, 34B, 298
$C_{23}H_{34}O_4$, 3α,17α-Dihydroxy-4,4,14α-trimethyl-19-nor-10α-pregn-5-ene-11,20-dione, 43B, 670
$C_{23}H_{34}O_5$, 19R-Methoxy-5,19-methyleneoxido-17β-acetoxy-5β-androstan-3-one, 40B, 496
$C_{23}H_{34}O_6$, 3β,17β-Diacetoxy-14α,15α-oxido-19-hydroxy-5α,8β-androstane, 39B, 388
$C_{23}H_{34}O_6$, 5β-Hydroxygitoxigenin, 45B, 566
$C_{23}H_{35}ClO_3$, 11β-Chloro-19-nor-17α-pregn-4-en-20-yn-17β-ol monoacetonate monohydrate, 43B, 671

$C_{23}H_{35}NO_3$, 3β-Hydroxy-16β-morpholino-5-androsten-17-one, 46B, 533
$C_{23}H_{35}O_4$, 18-Methylestradiol ethyl acetate solvate, 42B, 405
$C_{23}H_{35}O_5 \cdot 2\ H_2O$, Digoxigenin dihydrate, 46B, 533
$C_{23}H_{36}NO_4$, (20S,21R)-3β-Hydroxy-14,21-epoxy-5β,14β,20-cardanolactam hydrate, 42B, 406
$C_{23}H_{36}O_3$, 3α-Acetoxy-4α,8α,14β-trimethyl-18-nor-5α,9β,13β-androstan-17-one, 45B, 567
$C_{23}H_{36}O_4$, Methyl 3β-acetoxy-17α-methyl-18-nor-5α-androstane-17β-carboxylate, 46B, 534
$C_{23}H_{37}BrN_2$, N-Cyano-N-methyl-18-amino-20-bromo-C-nor-D-homo-pregnane, 38B, 548
$C_{23}H_{40}O_4$, 6β,17β-Dihydroxy-6α-pentyl-4-nor-3,5-secoandrostan-3-oic acid, 45B, 568
$C_{23}H_{44}ClN_2O_2$, 3β-Amino-20α-dimethylamino-5-pregnene hydrochloride dihydrate, 43B, 671
$C_{24}H_{28}BrO_5$, 8-Acetoxy-6-(2,4-dimethoxy-5-bromophenyl)-3-methyl-tricyclo[5.2.1.0$^{3-8}$]decan-2-one, 39B, 388
$C_{24}H_{28}ClFO_4$, 6-Chloro-21-fluoro-17α-acetoxy-16-methylene-4,6-pregnadiene-3,20-dione, 40B, 496
$C_{24}H_{29}ClO_4$, 6-Chloro-17-hydroxy-1α,2α-methylenepregna-4,6-diene-3,20-dione acetate, 40B, 497
$C_{24}H_{31}FO_6 \cdot 0.67\ CH_4O$, 9α-Fluoro-11β,21-dihydroxy-16α,17α-dioxyisopropylidene-1,4-pregnadiene-3,20-dione methanol, 45B, 568
$C_{24}H_{32}O_4S$, 7α-Acetylthio-3-oxo-17β-4-pregnene-21,17β-carbolactone, 38B, 549
$C_{24}H_{33}FO_7$, 9α-Fluoro-16α-methyl-11β,17,21-trihydroxy-1,4-pregnadiene-3,20-dione-21-acetate monohydrate, 41B, 604
$C_{24}H_{34}O_3$, (20R)-3β-Hydroxy-22-methylene-5β-card-14-enolide, 42B, 406
$C_{24}H_{34}O_3S$, 7α-Thioacetyl-(17R)-spiro(androst-4-en-17,2(3H)furan), 44B, 487
$C_{24}H_{34}O_4$, Medroxyprogesterone acetate, 44B, 486
$C_{24}H_{35}ClO_6$, 2β-Acetoxy-17β-chloroacetoxy-4-androsten-3-one - methanol, 37B, 297
$C_{24}H_{36}O_3$, 3β-Acetoxy-16β-methyl-5-pregnen-20-one, 45B, 569
$C_{24}H_{37}NO_2$, 3β-Hydroxy-6β-piperidine-5β,19-cycloandrostan-17-one, 44B, 487
$C_{24}H_{39}NaO_5 \cdot H_2O$, Sodium 3α,7α,12α-trihydroxy-5β-cholan-24-oate monohydrate, 46B, 534
$C_{24}H_{39}O_4Rb \cdot H_2O$, Deoxycholic acid rubidium salt, 46B, 535
$C_{24}H_{40}N_2$, 17α-Methyl-3β-pyrrolidinyl-17a-aza-D-homo-5-androstene, 43B, 371
$C_{24}H_{40}O_3$, Lithocholic acid, 42B, 407
$C_{24}H_{40}O_4 \cdot 0.125\ C_{16}H_{32}O_2 \cdot 0.125\ C_2H_6O$, 3α,12α-Dihydroxy-5β-cholan-24-oic acid palmitic acid ethanol, 46B, 536
$C_{24}H_{40}O_4 \cdot 0.5\ C_2H_6O \cdot 0.5\ H_2O$, Deoxycholic acid ethanol water, 45B, 571
$C_{24}H_{40}O_4 \cdot 0.5\ C_2H_6OS \cdot 0.5\ H_2O$, Deoxycholic acid dimethylsulfoxide hemihydrate, 45B, 571
$C_{24}H_{40}O_4 \cdot 0.5\ C_3H_6O$, Deoxycholic acid - acetone (2:1), 44B, 487
$C_{24}H_{40}O_4 \cdot 0.67\ C_2H_6O \cdot 0.33\ H_2O$, Deoxycholic acid - ethanol - water (3:2:1), 44B, 488
$C_{24}H_{40}O_4 \cdot 1.5\ H_2O$, 2,3-Deoxycholic acid hydrate, 45B, 570
$C_{24}H_{40}O_4$, Chenodeoxycholic acid, 46B, 535
$C_{24}H_{40}O_4$, 3α,6α-Dihydroxy-5β-cholan-24-oic acid, 40B, 497
$C_{24}H_{42}O_2 \cdot 0.5\ C_4H_8O_2$, 4,4,14α-Trimethyl-19(10-9β)abeo-5β,10α-pregnane-6β,11α-diol dioxan solvate, 45B, 572
$C_{24}H_{42}O_2$, 4,4,14α-Trimethyl-19(10-9β)abeo-5β,10α-pregnane-6α,11β-

diol, 45B, 572
$C_{24}H_{42}O_2$, 4,4,14α-Trimethyl-19(10-9β)abeo-5β,10α-pregnane-6β,11β-diol, 45B, 572
$C_{24}H_{42}O_3$, (20S)-20-Ethyl-5-pregnene-3β,20-diol methanol solvate, 44B, 489
$C_{25}H_{26}F_6O_2$, 14α,17α-Etheno-15,16-di(trifluoromethyl)-4,15-pregnadiene-3,20-dione, 39B, 389
$C_{25}H_{27}BrO_3$, Estradiol 3-p-bromobenzoate, 33B, 195
$C_{25}H_{27}BrO_3$, 1,4,6-Androstatriene-3,17-dione - p-bromophenol (1:1), 40B, 498
$C_{25}H_{31}BrO_3$, 17β-Hydroxy-1,4-androstadien-3-one - p-bromophenol, 37B, 298
$C_{25}H_{31}BrO_4S$, 2β-Methyl-19-nortestosterone p-bromobenzenesulphonate, 38B, 549
$C_{25}H_{33}BrO_3$, 5α-Androstane-3,17-dione - p-bromophenol (1:1), 40B, 499
$C_{25}H_{33}BrO_3S$, Androst-4-en-17β-ol brosylate, 39B, 390
$C_{25}H_{34}O_6$, (22R)-11β,21-Dihydroxy-16α,17α-propylmethylenedioxy-1,4-pregnadiene-3,20-dione, 44B, 489
$C_{25}H_{34}O_6$, (22S)-11β,21-Dihydroxy-16α,17α-propylmethylenedioxy-1,4-pregnadiene-3,20-dione, 44B, 489
$C_{25}H_{35}NO_6$, N(β)-Methoxy-(progesterone[16α,17α-d]tetrahydro-1',2'-oxazole) (isomer I), 45B, 573
$C_{25}H_{35}NO_6$, N(β)-Methoxy-(progesterone[16α,17α-d]tetrahydro-1',2'-oxazole) (isomer II), 45B, 573
$C_{25}H_{36}O_4$, 3,20-Bis(ethylenedioxy)-pregna-5,7-diene, 39B, 390
$C_{25}H_{38}Br_2O_3$, 11β,12α-Dibromo-3α,9-oxidocholanic acid methyl ester, 34B, 300
$C_{25}H_{38}Br_2O_3$, 11β,12β-Dibromo-3α,9-oxidocholanic acid methyl ester, 34B, 301
$C_{26}H_{25}BrO_3$, 9-Methyl-9β,10α-estra-4-en-3-one-17β-ol p-bromobenzoate, 42B, 407
$C_{26}H_{25}BrO_4$, 3-Methoxy-16-hydroxy-13α-estra-1,3,5(10),15-tetraen-17-one p-bromobenzoate, 41B, 605
$C_{26}H_{30}O_4$, 17β-Benzoyloxy-3-oxo-4-androsten-19-al, 42B, 408
$C_{26}H_{31}BrO_2$, 3-Methylene-17β-hydroxy-5(10)-estrene p-bromobenzoate, 42B, 408
$C_{26}H_{31}BrO_2S$, 2α,3α-Epithio-5α-androst-6-en-17β-yl p-bromobenzoate, 41B, 606
$C_{26}H_{31}BrO_3$, Testosterone 17β-p-bromobenzoate, 35B, 337
$C_{26}H_{31}BrO_3$, 3-Methoxyestra-2,5(10)-dien-17β-ol p-bromobenzoate, 42B, 408
$C_{26}H_{31}BrO_3$, 3β-p-Bromobenzoyloxy-13α-androst-5-en-17-one, 35B, 341
$C_{26}H_{31}BrO_3$, 3β-p-Bromobenzoyloxyandrost-5-en-17-one, 38B, 550
$C_{26}H_{31}BrO_3$, 6-Oxo-3α,5-cycloandrostan-17-yl p-bromobenzoate, 38B, 552
$C_{26}H_{31}BrO_5$, 3α,11α,17β-Trihydroxy-13α-C-nor-5β-androstane-11β-carboxylic acid 11a,17-lactone 3-p-bromobenzoyl ester, 33B, 196
$C_{26}H_{32}BrO_4$, 17β-Hydroxy-10β-methoxy-estr-4(5)-en-3-one p-bromobenzoate, 41B, 607
$C_{26}H_{32}O_4$, 19-Hydroxytestosterone 17-benzoate, 42B, 409
$C_{26}H_{33}BrO_2S$, 2α,3α-Epithio-5α-androstan-17β-yl p-bromobenzoate, 39B, 391
$C_{26}H_{33}BrO_3$, A-Homo-19-nor-5β-androst-9(10)-ene-4β,17β-diol 4-p-bromobenzoate, 43B, 673
$C_{26}H_{33}BrO_3$, 3-Methoxy-estr-5(10)-ene-3α,17β-diol 17-p-bromobenzoate, 42B, 409
$C_{26}H_{33}BrO_4$, 5-Androsten-3β,17β,19-triol 17-p-bromobenzoate, 43B, 673

$C_{26}H_{33}BrO_4S$, 17aβ-p-Bromobenzenesulphonyloxy-17α-methyl-19-nor-9β,10α-D-homoandrost-4-en-3-one, 32B, 245
$C_{26}H_{34}BrNO_2$, 3β-Bromoacetoxy-16α-ethyl-16(2)-cyano-16(2),21-cyclo-5α-pregna-17,21-diene, 37B, 299
$C_{26}H_{36}O_4S$, 3-Oxo-5α-androstan-17β-ol toluene-p-sulphonate, 40B, 500
$C_{26}H_{46}O_6$, Cholic acid - ethanol addition compound, 38B, 554
$C_{27}H_{29}BrO_3$, 4,6β-Ethanoestradiol 17-p-bromobenzoate, 43B, 674
$C_{27}H_{31}BrO_3$, 3-Methoxy-5β,19-cyclo-5,10-secoandrosta-1(10),2,4-trien-17β-ol p-bromobenzoate, 33B, 198
$C_{27}H_{33}BrO_3$, 17α-Hydroxy-7,7-dimethyl-8α,14β-estr-4-en-3-one p-bromobenzoate, 37B, 299
$C_{27}H_{33}BrO_3$, 3β-Hydroxy-8-methyl-5α,13α-androst-9(11)-en-15-one, 41B, 609
$C_{27}H_{34}BBrO_4$, 20β-Hydroxy-3-oxopregn-4-ene-17α,21-diyl p-bromophenylboronate, 42B, 518
$C_{27}H_{34}FN_3O_7$, Δ^6-6-Azido-betamethasone-21-acetate acetone solvate, 40B, 501
$C_{27}H_{37}IO_4$, Methyl 3α-iodoacetoxy-12-methyl-18-nor-5β,17α-chola-8,11,13-triene-24-oate, 39B, 391
$C_{27}H_{38}O_5$, 3α-Acetoxy-chola-7,9-diene-12-one-24-oic-acid methyl ester, 46B, 536
$C_{27}H_{39}O_5S$, 3α,3β-Dimethoxy-5α-oestran-17β-ol toluene-p-sulphonate, 40B, 502
$C_{27}H_{40}O_7$, 3,20-Diethylenedioxy-9β,11β-oxido-11α-acetoxy-9,11-seco-11,19-cyclo-5α,14β,17α-pregnane, 41B, 609
$C_{27}H_{42}Cl_2O$, 3',3'-Dichloro-2β,3β-dihydrocyclobuta(2,3)-5α-cholestan-4'(3'H)-one, 43B, 675
$C_{27}H_{42}Cl_2O$, 4',4'-Dichloro-2β,3β-dihydrocyclobuta(2,3)-5α-cholestan-3'(4'H)-one, 43B, 675
$C_{27}H_{44}Br_2$, 7α-Bromocholesteryl bromide, 20, 641
$C_{27}H_{44}O$, Cholest-4-en-3-one, 42B, 409
$C_{27}H_{44}O$, Cholest-4-en-6-one, 43B, 676
$C_{27}H_{44}O$, Δ^6-Cholesten-3-one, 43B, 675
$C_{27}H_{44}O$, Vitamin D_3, 42B, 134
$C_{27}H_{44}O_2$, 4β,5β-Epoxycholestan-3-one, 43B, 677
$C_{27}H_{44}O_2$, 4β,5β-Epoxycholestan-6-one, 43B, 677
$C_{27}H_{44}O_5$, Apocholic acid - acetone (1:1), 44B, 487
$C_{27}H_{44}O_6$, Ecdyson, 30B, 222
$C_{27}H_{44}O_7$, 20-Hydroxy-ecdysone, 37B, 300
$C_{27}H_{45}Br$, Cholesteryl bromide, 45B, 574
$C_{27}H_{45}Cl$, Cholesteryl chloride, 45B, 574
$C_{27}H_{45}Cl_2NO$, 4,4-Dichloro-2a-aza-A-homocholestan-3-one, 35B, 342
$C_{27}H_{45}I$, Cholesteryl iodide B, 10, 287
$C_{27}H_{46}$, 5α-Cholest-2-ene, 45B, 574
$C_{27}H_{46}Br_2$, 2α,3β-Dibromo-5α-cholestane, 31B, 203
$C_{27}H_{46}Br_2$, 3α,5β-Dibromocholestane, 42B, 409
$C_{27}H_{46}Cl_2$, 2α,3β-Dichloro-5α-cholestane, 31B, 203
$C_{27}H_{46}Cl_2$, 2β,3α-Dichloro-5α-cholestane, 31B, 206
$C_{27}H_{46}O_2$, (E)-3α-Acetoxy-5,10-seco-1(10)-cholesten-5-one, 45B, 575
$C_{27}H_{46}O_2$, (25R)-Cholest-5-ene-3β,26-diol, 43B, 679
$C_{27}H_{46}O_3$, 1α,25-Dihydroxycholesterol, 40B, 503
$C_{27}H_{46}O_3$, 25-Hydroxy-vitamin D_3 monohydrate, 43B, 210
$C_{27}H_{46}O_5$, 3α,7α,12α-Trihydroxy-5β-cholestan-26-oic acid, 45B, 575
$C_{27}H_{47}Cl$, 5α-Chlorocholestane, 41B, 610
$C_{27}H_{47}O \cdot H_2O$, Cholesterol monohydrate, 45B, 576
$C_{27}H_{48}INO_2$, Tetrahydroveralkamine hydroiodide, 38B, 555
$C_{27}H_{48}INO_2$, 22,26-Epimino-5α-cholestan-3β,20-diol hydroiodide, 39B,

392
$C_{27}H_{48}O_4$, 3β,16β,23(R),26-Tetrahydroxy-5β-cholestane, 46B, 537
$C_{27}H_{49}NaO_6S$, Sodium cholesteryl sulfate dihydrate, 43B, 678
$C_{27}H_{50}BrNO_4$, 22,26-Epimino-5α-cholestan-3β,16β,23-triol hydrobromide monohydrate, 39B, 393
$C_{28}H_{36}O_5$, 3-Ethylenedioxy-5-androstene-17β,19-diol 17-benzoate, 41B, 611
$C_{28}H_{38}O_3$, 4,4-Dimethylandrostan-3-one-17β-yl benzoate, 43B, 679
$C_{28}H_{44}Br_2O$, 22,23-Dibromo-9β-ergost-4-en-2-one, 34B, 303
$C_{28}H_{44}O$, Vitamin D_2, 42B, 410
$C_{28}H_{44}O_2$, Toxisterol(2)-D epoxide, 44B, 554
$C_{28}H_{44}O_2$, 23-Hydroxy-3α,5α-cycloergost-7-en-6-one, 39B, 393
$C_{28}H_{44}O_6$, 3β-Acetoxy-17aα-(2-acetoxyethoxy)-17α,17aβ-dimethyl-D-homo-5-androsten-17β-ol, 43B, 680
$C_{28}H_{45}ClO_2$, Cholesteryl chloroformate, 44B, 490
$C_{28}H_{46}O$, 5α-Vinyl-A-norcholestan-3-one, 41B, 612
$C_{28}H_{46}O_2$, Ergosterol monohydrate, 42B, 410
$C_{28}H_{46}O_2$, 24-Norcholesterol acetate, 45B, 576
$C_{28}H_{46}O_3$, 24-Methylenecholest-5-en-3β,7β,19-triol, 44B, 490
$C_{28}H_{52}INO_4$, (20R:22S:25S)-22,26-Epimino-5α-cholestan-3β,16α,20-triol hydriodide (methanol solvate), 37B, 301
$C_{29}H_{37}BrO_6$, 2β,17β-Diacetoxy-4-androsten-3-one - p-bromophenol, 37B, 297
$C_{29}H_{39}BrN_2O_5S$, 3β,6β-Dimethoxy-5β,19-cycloandrostan-17-one N-acetyl-p-bromobenzenesulphonylhydrazone, 33B, 199
$C_{29}H_{43}IO_4$, Diosgenin iodoacetate, 32B, 246
$C_{29}H_{44}O_5$, Dendrosterone, 43B, 680
$C_{29}H_{44}O_9$, Actodigin, 46B, 537
$C_{29}H_{44}O_{12} \cdot 8\ H_2O$, Ouabain octahydrate, 46B, 538
$C_{29}H_{46}BrO_2$, 6β-Bromoacetyl-3,5α-cyclo-5α-cholestane, 37B, 301
$C_{29}H_{46}ClO_2$, 6β-Chloroacetyl-3,5α-cyclo-5α-cholestane, 37B, 301
$C_{29}H_{46}O$, 22,23-Methylene-5,24(28)-ergostadien-3β-ol, 41B, 613
$C_{29}H_{48}N_2O_4$, 3β-Acetoxy-5β,6β-N-nitroaziridinylcholestene, 46B, 538
$C_{29}H_{48}O_2$, Cholesteryl acetate, 45B, 577
$C_{29}H_{48}O_2S$, Cholestan-4-one-3-spiro(2,5-oxathiolane), 33B, 201
$C_{29}H_{49}ClO_2$, 3β-Acetyloxy-14-chloro-5α,14β,17α-cholestane, 41B, 614
$C_{29}H_{50}O_2$, 9β,10α-Cholesta-5,7-diene-3β-ol ethanol solvate, 40B, 503
$C_{30}H_{33}BrF_4O_3$, 3β-(p-Bromobenzoyloxy)-androst-5-eno-[16α,17-d]-2',2',3',3'-tetrafluoro-2',3'-dihydro-6-methylpyran, 37B, 302
$C_{30}H_{42}O_7$, 6α,7α:22,26:24,25-Triepoxy-5,26-dihydroxy-5α-ergost-2-en-1-one acetate, 42B, 411
$C_{30}H_{44}BrNO_3$, Desoxycholic acid p-bromoanilide, 38B, 554
$C_{30}H_{44}O_7$, 6α,7α:17,24:22,26-Triepoxy-5,25,26-trihydroxy-5α,17α-ergost-2-en-1-one ethyl ether, 42B, 411
$C_{30}H_{49}O_4$, 24(R),25-Dihydroxyvitamin d_2, 45B, 578
$C_{31}H_{44}O_2S_2$, 3β-Acetoxy-6,7-epidithio-19-norlanosta-5,7,9,11-tetraene, 42B, 412
$C_{31}H_{45}ClO_5$, 17-Hydroxyprogesterone 17-(10-chloro-9-ketodecanoate), 39B, 394
$C_{31}H_{46}O_4$, Methyl (13α,14β,17α,20S,24Z)-3,21-dioxo-lanosta-8,24-dien-26-oate, 41B, 615
$C_{31}H_{46}O_6$, 17-Hydroxyprogesterone 17-(10-hydroxy-9-ketodecanoate), 39B, 394
$C_{31}H_{48}O_3$, Abieslactone, 39B, 395
$C_{31}H_{48}O_3$, Methyl (13α,14β,17α,20S,24Z)-3-oxo-lanosta-8,24-dien-26-oate, 41B, 615
$C_{32}H_{50}O_2$, 14α-Methyl-9β,19-cyclo-5α-cholestan-3β-yl acetate, 42B,

412
$C_{32}H_{52}Br_2O_2$, 24,25-Dibromolanost-8-en-3β-yl acetate A, 40B, 504
$C_{32}H_{52}Br_2O_3$, 3β-Acetoxy-7α,11α-dibromolanostane-8α,9α-epoxide, 31B, 208
$C_{32}H_{52}O_{17}$, Scillicyanoside pentahydrate, 39B, 395
$C_{32}H_{53}ClO_2$, Euphenyl chloroacetate, 44B, 491
$C_{32}H_{53}IO_2$, Lanostenyl iodoacetate, 17, 774
$C_{34}H_{45}BrO_2$, 1(10-6)abeo-Cholesta-5,7,9-trien-3-yl p-bromobenzoate, 43B, 681
$C_{34}H_{46}N_2O_6$, Toxisterol C_1 3,5-dinitrobenzoate, 42B, 413
$C_{34}H_{47}BrCl_2O_2$, 3β-p-Bromobenzoyloxy-7α,15β-dichloro-5α-cholest-8(14)-ene, 46B, 538
$C_{34}H_{47}BrO_3$, 3β-p-Bromobenzoyloxy-14α,15α-epoxy-5α-cholest-7-ene, 43B, 682
$C_{34}H_{48}N_4O_4$, 5α-Ergosta-7,22-dien-3-one 2,4-dinitrophenylhydrazone, 45B, 578
$C_{34}H_{49}BrO_3$, (E)-3β-(p-Bromobenzoyloxy)-5,10-seco-1(10)-cholesten-5-one, 42B, 140
$C_{34}H_{49}BrO_3$, 5(10-1βH)abeo-Cholest-10(19)-ene-3β,5α-diol 3-p-bromobenzoate, 45B, 579
$C_{34}H_{52}O_3S$, Cholesteryl p-toluenesulphonate, 43B, 683
$C_{34}H_{54}N_4O_2S$, 3β-Azido-2α-(toluene-p-sulphonamido)-5α-cholestane, 45B, 579
$C_{34}H_{55}NO_3S$, 3β-(Toluene-p-sulphonamido)-5α-cholestan-2β-ol, 45B, 579
$C_{35}H_{41}Br_2ClO_4$, (20S)-20-Chloro-3β,16-di-(p-bromobenzoyloxy)-5α-pregnane, 37B, 303
$C_{35}H_{46}INO_4$, Calciferyl 4-iodo-3-nitrobenzoate, 28, 537
$C_{35}H_{46}INO_4$, Lumisteryl 4-iodo-3-nitrobenzoate, 16, 560
$C_{35}H_{46}INO_4$, Suprasteryl II 4-iodo-3-nitrobenzoate, 30B, 223
$C_{35}H_{46}N_2O_6$, Isopyrocalciferol 3,5-dinitrobenzoate, 41B, 616
$C_{35}H_{46}N_2O_6$, Pyrocalciferol 3,5-dinitrobenzoate, 41B, 616
$C_{35}H_{46}N_2O_6$, Toxisterol(2)-A 3,5-dinitrobenzoate, 42B, 413
$C_{35}H_{47}BrO_2$, Photoisopyrocalciferyl m-bromobenzoate, 33B, 202
$C_{35}H_{51}BrO_3$, 3β-p-Bromobenzoyloxy-14α-methyl-5α-cholest-7-en-15β-ol, 44B, 491
$C_{35}H_{52}O_3S$, 7,22-Ergostadiene-3β-ol tosylate, 42B, 413
$C_{35}H_{53}BrO_7S$, Eucosterol-p-bromobenzenesulphonate, 41B, 617
$C_{35}H_{60}O_2$, Cholesteryl octanoate, 45B, 580
$C_{36}H_{44}N_2O_4$, 3(R)-Spiro-8'-(3',6'(R)-diphenyl-1',5'-dioxa-2',4'-diazabicyclo[3.3.0]oct-3'-ene)-5α-androstan-17β-yl acetate, 44B, 492
$C_{36}H_{51}IO_2$, 23-Demethylgorgosterol p-iodobenzoate, 39B, 396
$C_{36}H_{54}O_2$, Cholest-5-en-3-ol dihydrocinnamate, 46B, 539
$C_{36}H_{62}O_2$, Cholesteryl nonanoate, 45B, 581
$C_{36}H_{64}Br_2Cl_2N_2O_5$, 3α,17β-Diacetoxy-2β,16β-piperidino-5α-androstane dimethobromide, 37B, 304
$C_{37}H_{37}F_2N_2O_5Br$, Pregnene derivative, 35B, 344
$C_{37}H_{43}BrO_8 \cdot 0.5\ C_4H_8O_2$, Withaferin A acetate p-bromobenzoate ethyl acetate solvate, 33B, 202
$C_{37}H_{53}BrO_7$, Lobosterol 4-p-bromobenzoate, 42B, 414
$C_{37}H_{56}IO_2$, Dinosterol p-iodobenzoate, 44B, 492
$C_{37}H_{64}O_2$, Cholesteryl decanoate, 45B, 581
$C_{38}H_{56}Cl_2HgO_4$, Testosterone - mercuric chloride, 33B, 203
$C_{38}H_{62}INO_{10}$, Zygacine acetonide hydroiodide acetone solvate, 39B, 397
$C_{38}H_{66}O_2$, Cholesteryl undecanoate, 46B, 540
$C_{39}H_{53}BrO_7$, Fusidic acid methyl ester O-p-bromobenzoate, 33B, 204

$C_{39}H_{68}O_2$, Cholesteryl dodecanoate, 45B, 582
$C_{39}H_{68}O_2$, Cholesteryl laurate, 45B, 583; 46B, 539
$C_{40}H_{47}BrSO_{14}$, Trillenogenin tetraacetyl monobrosylate, 46B, 540
$C_{41}H_{52}Br_2O_4$, 5α,14β-Cholest-7-ene-3β,15β-diol di-p-bromobenzoate, 43B, 683
$C_{41}H_{64}O_{14}$, Digoxin, 45B, 584; 46B, 541
$C_{41}H_{64}O_{14}$, Gitoxin, 46B, 541
$C_{41}H_{72}O_2$, Cholesteryl myristate, 42B, 414
$C_{43}H_{49}BrF_2N_2O_7$, 6α,7α-Difluoromethylene-11β-hydroxy-16α,17α-isopropylidenedioxy-21-p-bromobenzoyloxypregn-4-en-20-one[3,2-c]-2'-phenylpyrazole, 37B, 305
$C_{43}H_{56}Br_2O_4$, 14α-Ethyl-5α-cholest-7-ene-3β,15α-diol-di-p-bromobenzoate, 46B, 542
$C_{43}H_{66}O_4$, 16-Methoxy-16'-oxo-21,20'-di(20,18-epoxy-pregnane), 46B, 542
$C_{44}H_{77}BrO_2$, Cholesteryl 17-bromoheptadecanoate, 43B, 684
$C_{45}H_{78}O_2$, Cholesteryl oleate, 45B, 584
$C_{52}H_{64}I_2O_{16}$, Datiscoside bis(p-iodobenzoate) dihydrate, 39B, 398
$C_{57}H_{96}Cl_4N_2O_3$, 4,4-Dichloro-2a-aza-A-homo-coprostane-3-one:acetone (2:1 adduct), 38B, 318

52. MONOTERPENES

$C_{10}H_{11}Cl_7$, 2,5,6-exo,8,8,9,10-Heptachlorodihydrocamphene, 44B, 493
$C_{10}H_{12}O_2$, 5-Isopropyltropolone, 42B, 415
$C_{10}H_{13}Br_2Cl_3$, 8-Bromo-2-bromomethyl-6-methyl-2,5,6-trichloro-3,7-octadiene, 45B, 585
$C_{10}H_{13}Br_3O$, endo-3,9,9-Tribromocamphor, 43B, 685
$C_{10}H_{13}Br_3O$, exo-3,9,9-Tribromocamphor, 41B, 220
$C_{10}H_{14}BrCl_3$, (1R,2S,4S,5R)-1-Bromo-trans-2-chlorovinyl-4,5-dichloro-1,5-dimethyl-cyclohexane, 41B, 618
$C_{10}H_{14}BrNO$, Anhydrobromonitrocamphane, 38B, 556
$C_{10}H_{14}Br_2O$, 2α,4α-Dibromo-10β-pinan-3-one, 39B, 398
$C_{10}H_{14}N_2O$, (+)-3-Diazocamphor, 38B, 82
$C_{10}H_{14}O_2$, (1S)-cis,cis-Iridolactone, 45B, 586
$C_{10}H_{14}O_2$, Carbocamphenilone, 43B, 686
$C_{10}H_{15}BrClNO$, (+)-10-Bromo-2-chloro-2-nitrosocamphane, 26, 649
$C_{10}H_{15}BrO$, (+)-3-Bromocamphor, 37B, 306
$C_{10}H_{15}BrO$, (+)-8-Bromocamphor, 41B, 619
$C_{10}H_{15}BrO$, (-)-Bromodihydroumbellulone, 34B, 305
$C_{10}H_{15}BrO$, 6-Bromoisofenchone, 34B, 306
$C_{10}H_{15}NO$, Carvoxime mixed crystal, 44B, 493
$C_{10}H_{15}NO$, dl-Carvoxime, 41B, 619; 42B, 416
$C_{10}H_{15}NO$, l-Carvoxime, 42B, 417
$C_{10}H_{15}N_3O_4$, 4,N-Dinitrobornan-2-imine, 45B, 174
$C_{10}H_{16}$, β-Pinene (gas-ed), 38B, 1069
$C_{10}H_{16}BrNO_2$, (-)-2-Bromo-2-nitrocamphane, 27, 1000
$C_{10}H_{16}Br_2O$, 2,4-Dibromomenthone, 24, 717
$C_{10}H_{16}Cl_2$, 3α,4β-Dichlorocarane (gas-ed), 38B, 1064
$C_{10}H_{16}O_2$, (-)-5-endo,6-exo-Dihydroxycamphene, 45B, 174
$C_{10}H_{16}O_2$, (1S)-cis,cis-Isoiridolactone, 45B, 586
$C_{10}H_{16}O_2$, Iridomyrmecin, 29, 742
$C_{10}H_{16}O_2$, Isoiridomyrmecin, 27, 997
$C_{10}H_{17}NO$, (-)-Camphoroxime, 44B, 494
$C_{10}H_{18}O_2$, trans-2,8-Dihydroxy-1(7)-p-menthene, 37B, 306
$C_{11}H_{14}O_5$, Sarracenin, 42B, 417

$C_{11}H_{15}NO$, D-α-Cyanocamphor, 10, 249
$C_{11}H_{16}O_2$, (-)-Camphene-8-carboxylic acid, 43B, 687; 44B, 494
$C_{13}H_{21}NOS$, (1R,3S,4R)-N,N-Dimethyl-3-camphorcarbothioamide, 46B, 177
$C_{13}H_{24}BrN$, 3-N-Dimethylaminomethylpinene-2(10) hydrobromide, 37B, 307
$C_{13}H_{25}Br_2N$, 2-Bromo-6-N-dimethylaminomethylfenchane hydrobromide, 37B, 307
$C_{13}H_{28}IN$, Menthyl trimethylammonium iodide, 27, 999
$C_{14}H_{22}$, 1-Bi(norbornane), 33B, 206
$C_{15}H_{16}Br_2O_2$, Isomaneonene-B, 44B, 495
$C_{15}H_{22}O_5S$, Dihydrofukinolidol sulfite, 38B, 557
$C_{16}H_{18}ClNO$, o-Chlorophenyliminocamphor, 41B, 620
$C_{16}H_{23}NO$, 3-(N-Benzyl-N-methylaminomethyl)-2-norbornanol, 32B, 247
$C_{17}H_{19}BrO$, (+)-3-p-Bromobenzylidenecamphor, 39B, 399
$C_{17}H_{19}NO_4$, cis-Pinocarvyl p-nitrobenzoate, 40B, 504
$C_{17}H_{22}BrNO_4$, 1-(-)-Menthyl 4-bromo-2-nitrobenzoate, 42B, 418
$C_{17}H_{24}BrNO_2$, (+)-Isomenthyl p-bromophenylcarbamate, 42B, 418
$C_{17}H_{26}O_{10}$, Loganin, 46B, 543
$C_{18}H_{21}BrO_3$, p-Bromophenacyl (+)-isocamphenilate, 44B, 495
$C_{18}H_{30}$, 1-Bi(apocamphane), 33B, 206
$C_{20}H_{30}$, 1-Biadamantane, 33B, 206
$C_{21}H_{20}Br_2$, 2,syn-7-Dibromo-5,5-dimethyl-3-exo-6-diphenyl-norbornene, 40B, 505
$C_{21}H_{26}O_2$, Cannabinol, 43B, 687
$C_{21}H_{30}O_2$, Cannabidiol, 43B, 688
$C_{24}H_{37}NO_4$, 8β-Hydroxy-Δ^9-tetrahydrocannabinol - N,N-dimethylformamide, 43B, 689
$C_{40}H_{30}BrN_3O_3$, p-Bromobenzyl-norbormide (active isomer), 31B, 210
$C_{40}H_{30}BrN_3O_3$, p-Bromobenzyl-norbormide (inactive isomer), 33B, 208

53. SESQUITERPENES

$C_{14}H_{15}BrO_3$, Bromomexicanin-E, 32B, 248
$C_{14}H_{20}O_3$, Guaianolide, 46B, 543
$C_{14}H_{20}O_4$, Herbasolide, 44B, 498
$C_{14}H_{21}BrO$, (+)-2,5-Diepi-β-cedrene α-bromonorketone, 39B, 399
$C_{14}H_{21}BrO$, Photocaryophyllene A bromoketone, 34B, 308
$C_{14}H_{21}BrO$, Photocaryophyllene D bromoketone, 34B, 308
$C_{14}H_{23}ClO_2$, Caryophyllene chlorohydrin, 40B, 505
$C_{14}H_{26}O_4$, 4,11-Dihydroxy-10-oxo-10-norguaiane monohydrate, 42B, 419
$C_{15}H_{12}O_3$, Freelingyne, 41B, 621
$C_{15}H_{14}O_5$, Miscandenin, 40B, 506
$C_{15}H_{15}NO_5$, 4,9-Dimethoxy-7-methyl-5H-furo[3,2-g](1)benzopyran-5-one oxime methyl ether, 43B, 403
$C_{15}H_{16}BrO_4$, Bromoambrosin, 31B, 213
$C_{15}H_{16}Br_2O_3$, 2,7-Dibromo-(-)-β-desmotroposantonin, 43B, 690
$C_{15}H_{16}O_3$, Linderalactone, 43B, 691
$C_{15}H_{16}O_5$, Lactucin, 44B, 496
$C_{15}H_{16}O_6$, 1,10:2,3-Diepoxy-6,8-dihydroxygermacr-4-ene-12,15-dioic acid di-γ-lactone, 40B, 507
$C_{15}H_{17}BrO_3$, 2-Bromo-(-)-α-desmotroposantonin, 43B, 691
$C_{15}H_{17}BrO_3$, 2-Bromo-(-)-β-desmotroposantonin, 32B, 249
$C_{15}H_{17}BrO_3$, 2-Bromo-α-santonin, 30B, 226
$C_{15}H_{17}BrO_3$, 2-Bromo-β-santonin, 34B, 309
$C_{15}H_{17}BrO_3$, 2-Bromo-6-epi-α-santonin, 41B, 621
$C_{15}H_{17}BrO_3$, 2-Bromo-6-epi-β-santonin, 41B, 621

$C_{15}H_{17}BrO_3$, 2-Bromolumisantonin, 33B, 209
$C_{15}H_{17}BrO_4$, Bromohelenalin, 40B, 1094
$C_{15}H_{17}ClO_3$, 14-Chlorosantonin, 44B, 497
$C_{15}H_{18}BrClO_2$, Epoxyrhodophytin, 46B, 544
$C_{15}H_{18}O_3$, Ergoyazin, 46B, 545
$C_{15}H_{18}O_3$, Hibiscone C, 46B, 544
$C_{15}H_{18}O_3$, Thieleanine, 46B, 545
$C_{15}H_{18}O_3$, Cannabispiran, 43B, 692
$C_{15}H_{18}O_3$, Chanootin, 39B, 400
$C_{15}H_{18}O_3$, γ-Metasantonin, 46B, 545
$C_{15}H_{18}O_4$, Bahia I, 46B, 546
$C_{15}H_{18}O_4$, Mikanokryptin, 45B, 586
$C_{15}H_{18}O_4$, Stramonin-B, 45B, 587
$C_{15}H_{18}O_4$, Helenalin oxide, 41B, 622
$C_{15}H_{19}BrO_6$, Shellolic bromolactone monohydrate, 27, 1019
$C_{15}H_{20}O_2$, Grilactone, 46B, 546
$C_{15}H_{20}O_2$, Costunolide, 42B, 420
$C_{15}H_{20}O_3$, Euryopsonol, 45B, 587
$C_{15}H_{20}O_3$, Istanlbulin-B, 46B, 547
$C_{15}H_{20}O_3$, Tamaulipin-A, 44B, 497
$C_{15}H_{20}O_3$, Eupatolide, 41B, 623
$C_{15}H_{20}O_3$, 4,5-Epoxygermacra-1(10),11(13)-dien-12,6-olactone, 42B, 420
$C_{15}H_{20}O_3$, 5α-Hydroxy-4αH,1,6,11βH-guai-2,10(15)-dien-6,12-olide, 44B, 498
$C_{15}H_{20}O_4$, (R,S)-Abscisic acid, 42B, 421
$C_{15}H_{20}O_4$, Crotocol, 41B, 624
$C_{15}H_{20}O_4$, DL-2-cis-4-trans-Abscisic acid, 43B, 177, 693
$C_{15}H_{20}O_4$, Deacetylneotenulin, 42B, 422
$C_{15}H_{20}O_4$, Parthemollin, 41B, 624
$C_{15}H_{20}O_5$, Autumnolide, 41B, 625
$C_{15}H_{20}O_5$, Carolenalone, 42B, 422
$C_{15}H_{20}O_5$, Solstitialin, 35B, 345
$C_{15}H_{20}O_5S$, 3β,8β-Dihydroxy-2βH,9βH-lactar-6-en-5,13-olide sulfite, 43B, 694
$C_{15}H_{21}BrO_2$, 15-Bromolongibornane-8,9-dione, 42B, 423
$C_{15}H_{21}BrO_3$, Aplysistatin, 43B, 690; 46B, 548
$C_{15}H_{21}BrO_3$, 2α-Bromo-α-tetrahydrosantonin, 46B, 548
$C_{15}H_{21}Br_2ClO$, 3,4'-Dibromo-4-chloro-1',3',3',4-tetramethylspirocyclohexane-1,2'-(7)oxabicyclo[4.1.0]hept-4'-ene, 42B, 424
$C_{15}H_{22}BrClO$, Nidofocene, 43B, 694
$C_{15}H_{22}Br_2O_4 \cdot 0.5 H_2O$, 11,13-Dibromopulchellin hemihydrate, 37B, 308
$C_{15}H_{22}ClHgO_2$, 2-Chloromercury-tetradymol, 40B, 508
$C_{15}H_{22}O$, (-)-Aristolone, 39B, 401
$C_{15}H_{22}O_2$, Herbadysidolide, 44B, 498
$C_{15}H_{22}O_3$, Lychnopholic acid, 44B, 499
$C_{15}H_{22}O_3$, Periplanol-B, 45B, 588
$C_{15}H_{22}O_3$, Valerenolic acid, 44B, 499
$C_{15}H_{22}O_3$, Tetradymodiol, 42B, 424
$C_{15}H_{22}O_3$, 2-Deoxy-12-oxolemnacarnol, 43B, 695
$C_{15}H_{22}O_4 \cdot 0.37 H_2O$, Hypochaerin hydrate, 43B,
$C_{15}H_{22}O_4$, Ivalbin, 45B, 588
$C_{15}H_{22}O_5$, Hymenoxon, 42B, 425
$C_{15}H_{22}O_7S$, Breynolide, 40B, 508
$C_{15}H_{23}Br$, ω-Bromolongifolene, 38B, 559
$C_{15}H_{23}Br$, 3α-Bromolongifolene, 38B, 558
$C_{15}H_{23}Br$, 7-Bromocyclo(3:15)longifolane, 38B, 558

$C_{15}H_{23}BrO$, 10-Bromo-3,4-epoxy-α-chamigrene, 45B, 589
$C_{15}H_{23}FO_4$, Fluoroguaianolide, 45B, 589
$C_{15}H_{24}Ag_2N_2O_6$, Humulenebis(silver nitrate), 31B, 213
$C_{15}H_{24}IN_2O_2$, Caryophyllene iodonitrosite, 33B, 210
$C_{15}H_{24}O$, Dehydro-10-epi-caparrapioxide, 42B, 425
$C_{15}H_{24}O_2$, (-)-α-Muurolene diepoxide (isomer I), 45B, 590
$C_{15}H_{24}O_2$, (-)-α-Muurolene diepoxide (isomer II), 45B, 590
$C_{15}H_{24}O_2$, Epoxyisoacoragermacrone, 46B, 548
$C_{15}H_{24}O_2$, Capsidiol, 40B, 509
$C_{15}H_{24}O_2$, Humulene diepoxide, 39B, 402
$C_{15}H_{24}O_2$, Hydroxydihydroeremophilone, 21, 647
$C_{15}H_{24}O_2$, 7-Epi-dihydrolemnalactone, 45B, 590
$C_{15}H_{24}O_3 \cdot 0.5\ H_2O$, Δ9(12)-Capnellene-3β,8β,10α-triol hemihydrate, 43B, 697
$C_{15}H_{24}O_3$, Ageratriol, 44B, 500
$C_{15}H_{24}O_3$, Agerol diepoxide, 44B, 500
$C_{15}H_{24}O_3$, Cuauhtemone, 40B, 510
$C_{15}H_{24}O_3$, Humulene triepoxide, 43B, 696
$C_{15}H_{24}O_3$, Lemnacarnol, 42B, 426
$C_{15}H_{24}O_3$, 3-Hydroxylubimin, 43B, 696
$C_{15}H_{24}O_4$, Δ[9(12)]-Capnellene-2β,5α,8β,10α-tetrol, 45B, 591
$C_{15}H_{24}O_4$, 2β-Carbomethoxy-5α,6α-dihydroxy-7,7-dimethyltricyclo-[6.2.1.0$^{1-6}$]undecane, 40B, 511
$C_{15}H_{25}Br$, 3α-Bromo-7βH-longifolane, 38B, 560
$C_{15}H_{25}BrO$, Humulene bromohydrin, 33B, 211
$C_{15}H_{25}BrO$, Oppositol, 39B, 402
$C_{15}H_{25}Br_2ClO_2$, Deodactol, 45B, 592
$C_{15}H_{25}Br_2ClO_2$, Isocaespitol, 41B, 626
$C_{15}H_{25}Cl$, Himachalene monohydrochloride, 33B, 212
$C_{15}H_{25}Cl$, Isoclovene hydrochloride, 26, 798
$C_{15}H_{25}Cl$, Longifolene hydrochloride, 30B, 225
$C_{15}H_{25}Cl$, β-Caryophyllene chloride, 19, 628
$C_{15}H_{26}BrClO$, Heterocladol, 43B, 698
$C_{15}H_{26}Br_2$, 1,6-Dibromodihydrocadinene, 22, 799
$C_{15}H_{26}Cl_2$, DL-Candinene dihydrochloride, 28, 539
$C_{15}H_{26}O \cdot 0.5\ C_5H_5N$, Koraiol pyridine solvate, 45B, 592
$C_{15}H_{26}O$, Africanol, 42B, 426
$C_{15}H_{26}O_2$, (1R,4R,5R,7R)-4,11-Dihydroxy-10,15-dehydroguaiane, 43B, 698
$C_{15}H_{26}O_2$, Gleenol, 45B, 593
$C_{15}H_{26}O_2$, Murol-4-one-9α-ol, 45B, 593
$C_{15}H_{26}O_2$, Poitediol, 44B, 500
$C_{15}H_{26}O_2$, Sibirin, 45B, 594
$C_{15}H_{26}O_2$, Clausantalene, 42B, 427
$C_{15}H_{26}O_2$, Flourensadiol, 41B, 626
$C_{15}H_{26}O_4$, Isocelorbicol, 42B, 429
$C_{15}H_{28}O_2$, Muurolane-4α,9β-diol, 45B, 595
$C_{15}H_{28}O_2$, Muurolane-4β,9β-diol, 45B, 595
$C_{16}H_{18}ClNO$, (±)-3-(o-Chlorophenylimino)camphor, 45B, 595
$C_{16}H_{22}Cl_4O$, 8,9,10,11-Tetrachloro-2,2,6-trimethyl-7-methoxymethyl-tricyclo[6.2.1.0$^{1-6}$]undec-9-ene, 44B, 501
$C_{16}H_{22}O_3$, Onitin monomethyl ether, 43B, 699
$C_{16}H_{24}Br_2O_5$, Ovalicine dibromide, 39B, 402
$C_{16}H_{25}N$, 2-Isocyanopupukeanane, 45B, 596
$C_{16}H_{26}O_4$, Methyl cis-tetrahydro-α-santoninate, 45B, 141
$C_{16}H_{26}O_4$, Methyl cis-tetrahydro-β-santoninate, 45B, 141
$C_{17}H_{18}O_7$, Melampodin B, 46B, 549

$C_{17}H_{20}O_4$, Chloranthalactone C, 46B, 550
$C_{17}H_{20}O_5$, Arteglasin A, 43B, 700
$C_{17}H_{21}BrO_2$, Laurinterol acetate, 34B, 310
$C_{17}H_{21}BrO_4$, Pseudoivalin bromoacetate, 40B, 512
$C_{17}H_{21}BrO_5$, Bromogaillardin, 37B, 309
$C_{17}H_{21}BrO_5$, Bromogeigerin acetate, 27, 1026
$C_{17}H_{21}BrO_5$, Bromoisotenulin, 40B, 513
$C_{17}H_{21}NO_6$, Eremofortin E, 46B, 550
$C_{17}H_{22}Br_2O_2$, Bromochamigrene derivative acetate, 45B, 596
$C_{17}H_{22}O_3$, 3β-Acetoxyatractylon, 43B, 699
$C_{17}H_{22}O_5$, Pleniradin acetate, 45B, 597
$C_{17}H_{22}O_5$, Eupaformonin, 42B, 428
$C_{17}H_{22}O_5$, Ovatifolin, 43B, 701
$C_{17}H_{22}O_5$, 2a,4a,5,6,8,9a,9b,9c-Octahydro-2-hydroxy-2,2a,6,9a-tetramethyl-2H-1,4-dioxadicyclopent[cd,f]azulene-3,9-dione, 44B, 501
$C_{17}H_{23}BrCl_2O_3$, Laurencienyne, 46B, 550
$C_{17}H_{23}BrO_5$, 2-Bromodihydroisophoto-α-santonic lactone acetate, 30B, 227
$C_{17}H_{24}O_4$, (9S,11S)-9-Acetoxydihydrocostunolide, 45B, 597
$C_{17}H_{24}O_5$, 9β-Acetoxy-6(βH),11(βH)-germacra-1β,10α-epoxy-4(5)-transene-6,12-olide, 44B, 502
$C_{17}H_{24}O_6$, Axivalin hydrate, 39B, 403
$C_{17}H_{24}O_6$, Eremofortin D, 43B, 701
$C_{17}H_{25}BrO_4$, Buddledin-A bromohydrin, 44B, 502
$C_{17}H_{25}Br_2ClO_2$, Obtusol acetate, 45B, 598
$C_{17}H_{25}ClO_5$, 2α-Carbomethoxy-5β-chloroacetoxy-7,7-dimethyltricyclo-[6.2.1.0$^{1-6}$]undecane-6β-ol, 40B, 511
$C_{17}H_{28}O_3$, 3β-Acetoxy-2β-hydroxy-thujopsane, 45B, 598
$C_{17}H_{28}O_4$, 5-Acetoxyisomarasman-7α,13-diol, 46B, 551
$C_{19}H_{24}Cl_2O_7$, Centaurepensin, 38B, 561; 44B, 503
$C_{19}H_{24}O_6 \cdot C_{20}H_{26}O_6$, Dehydroeriolanin-dehydroeriolangin, 41B, 627
$C_{19}H_{24}O_6$, Acantholide, 45B, 599
$C_{19}H_{24}O_6$, Eremantholide A, 44B, 504
$C_{19}H_{24}O_6$, Radiatin, 44B, 504
$C_{19}H_{24}O_6$, Tulirinol acetate, 46B, 551
$C_{19}H_{24}O_6$, Dehydroeriolanin, 41B, 627
$C_{19}H_{26}O_3S$, β-Methylmercaptoacrylol-petasol, 45B, 600
$C_{19}H_{26}O_7$, Hymenograndin, 46B, 552
$C_{19}H_{28}O_7$, Alatolide monohydrate, 43B, 702
$C_{19}H_{30}O_4$, Voleneol diacetate, 44B, 505
$C_{19}H_{32}O_5$, 7α-Acetoxy-8β-hydroxy-11-acetoxydrimane, 46B, 553
$C_{20}H_{24}BrNO_2$, Hinesol intermediate, 35B, 348
$C_{20}H_{24}O_7$, Eufoliatorin, 43B, 703
$C_{20}H_{26}O_4$, 3α-Angeloyloxy-9-oxy-10αH-furanoeremophilane, 45B, 600
$C_{20}H_{26}O_6$, Eremantholide B, 44B, 504
$C_{20}H_{26}O_6$, Dehydroeriolangin, 41B, 627
$C_{20}H_{28}O_6$, Neurolenin A, 44B, 505
$C_{20}H_{28}O_7 \cdot 0.25\ C_{21}H_{30}O_7$, Melnerin A - mernerin B (4:1), 44B, 506
$C_{20}H_{28}O_7$, Epoxyineupatorolide B, 46B, 557
$C_{20}H_{28}O_7S$, 2-Hydroxy-8-(2-hydroxy-2-hydroxymethyl-3-mercapto-butyryloxy)-trans,trans-1(10)-4-germacradienolide, 46B, 553
$C_{20}H_{30}O_5$, Hymenosignin, 46B, 552
$C_{21}H_{19}BrO_7S$, Vernolepin p-bromobenzenesulphonate, 37B, 310
$C_{21}H_{22}BrNO_4$, Chamaecynenol 4-bromo-3-nitrobenzoate, 34B, 311
$C_{21}H_{24}O_9$, Melampodin, 40B, 513
$C_{21}H_{26}O_6$, Euserotin, 45B, 600
$C_{21}H_{28}O_8$, Dihydroglaucolide-C, 46B, 554

$C_{21}H_{28}O_8$, Dihydrodesacetoxyglaucolide-A, 40B, 514
$C_{21}H_{29}BrN_2O_3S$, Acorone p-bromophenylsulphonylhydrazone, 31B, 216
$C_{21}H_{29}BrO_3S$, α-Caryophyllene alcohol p-bromobenzenesulphonate, 35B, 386
$C_{21}H_{29}BrO_4S$, Pseudoclovene A-diol mono-p-bromobenzenesulphonate, 34B, 312
$C_{21}H_{30}O_8$, Hymenolane, 42B, 425
$C_{21}H_{32}O_6$, O-Ethyl-tirotundin, 44B, 506
$C_{22}H_{19}BrO_7$, Elephantol p-bromobenzoate, 38B, 561
$C_{22}H_{20}O_7$, Isocollybolide, 38B, 562
$C_{22}H_{23}BrO_5$, Lactarorufin B 3,8-ether 14-p-bromobenzoate, 45B, 601
$C_{22}H_{23}IO_5$, Plenolin-p-iodobenzoate, 41B, 628
$C_{22}H_{24}O_4$, α-Pipitzol benzoate, 43B, 703
$C_{22}H_{25}BrO_2$, (-)-Myliol p-bromobenzoate, 45B, 602
$C_{22}H_{25}BrO_2$, 7-Hydroxycalamenene p-bromobenzoate, 44B, 507
$C_{22}H_{25}BrO_4$, Trichodermol p-bromobenzoate, 31B, 218
$C_{22}H_{25}BrO_5$, Deacetyldihydrogaillardin p-bromobenzoate, 37B, 309
$C_{22}H_{25}BrO_5$, Isolactarorufin p-bromobenzoate, 43B, 704
$C_{22}H_{26}O_7$, Berlandin, 41B, 628
$C_{22}H_{28}O_7$, 9β-Acetoxy-8α-epoxy-angelyloxy-trans,trans-germacra-1(10),4-diene-cis-6,12-olide, 46B, 554
$C_{22}H_{28}O_8$, 9β-Acetoxy-8α-epoxy-angelyloxy-7α-hydroxy-trans,trans-germacra-1(10),4-diene-cis-6,12-olide, 46B, 554
$C_{22}H_{29}BrO_2$, (+)-Allohimachalyl p-bromobenzoate, 46B, 555
$C_{22}H_{29}BrO_2$, 15-7αH-Longifolyl bromobenzoate, 38B, 563
$C_{22}H_{29}BrO_2$, 2α,6,6,8-Tetramethyltricyclo[6.2.1.0^{1,5}]undecan-7β-yl p-bromobenzoate, 42B, 428
$C_{22}H_{29}BrO_3$, 9-p-Bromobenzoxy-8-hydroxy-1,2,6,8-tetramethyltricyclo-[5.3.1.0^{2,6}]undecane, 44B, 507
$C_{22}H_{29}BrO_5$, Celorbicol p-bromobenzoate, 42B, 429
$C_{22}H_{29}BrO_6$, p-Bromobenzoyl-laserol, 32B, 251
$C_{22}H_{30}O_4$, Ilimaquinone, 45B, 602
$C_{22}H_{30}O_8$, Neurolenin B, 44B, 505
$C_{23}H_{24}BrNO_6$, Dihydrofomannosin p-bromobenzoylurethane, 33B, 213
$C_{23}H_{25}BrO_5$, Hirsutic acid p-bromophenacyl ester, 31B, 220
$C_{23}H_{28}O_{9.3}$, Glaucolide (D and E), 43B, 704
$C_{23}H_{28}O_{10}$, Glaucolide A, 41B, 629
$C_{23}H_{28}O_{10}$, Glaucolide-D, 42B, 430
$C_{23}H_{29}BrO_{10}$, Enhydrin bromohydrin, 42B, 431
$C_{23}H_{30}O_{10}$, Diacetylspathulin, 43B, 705
$C_{23}H_{31}BrO_3$, Bromoaureol acetate, 46B, 556
$C_{23}H_{33}N_2S$, 9-Isocyanopupukeanane phenylthiourea, 41B, 630
$C_{23}H_{34}O_{11}$, Paucin monohydrate, 43B, 706
$C_{24}H_{25}IO_6$, 4-O-Acetyl-2-O-p-iodobenzoylflorilenalin, 41B, 631
$C_{24}H_{30}O_4$, Gummosine, 43B, 707
$C_{24}H_{30}O_9$, Eupahyssopin diacetate, 44B, 508
$C_{24}H_{34}O_2$, 2,13-Dimethyl-10-(2-methylhept-2-ene-6-yl)tricyclo-[9.3.0.0^{9,13}]tetradeca-2,8-diene-4,7-dione, 46B, 556
$C_{25}H_{28}BrO_8$, Euparotin bromoacetate, 39B, 404
$C_{25}H_{29}BrO_6S$, 13β-p-Bromophenylthio-11α,13-dihydropulchellin-C diacetate, 39B, 405
$C_{25}H_{36}O_9$, Ineupatolide, 46B, 557
$C_{26}H_{33}BrO_6$, (2R,7S,11R)-2,7-Diacetoxynardosin-1(10)-en-12-ol p-bromobenzoate, 46B, 557
$C_{27}H_{33}BrO_{10}$, Spicatine hydrobromide, 45B, 603
$C_{28}H_{35}N_3O_2$, (-)-α-Bisabolol-8-p-phenylazophenylmethane, 45B, 603
$C_{29}H_{28}Br_2O_6$, Iresin di-p-bromobenzoate, 22, 808

$C_{30}H_{41}BrO_4S$, (13-p-Bromobenzenesulphonyl)-10-hydroxy-5,8-dimethyl-2-(2-methylhept-2-ene-6-yl)tetracyclo[9.3.0.0$^{1-5}$.0$^{7-11}$]tetradecan-8-ene, 46B, 556
$C_{30}H_{44}O_2$, cis,trans-Tetrahydromitchelladione, 45B, 604
$C_{30}H_{50}CrO_4$, Cedryl chromate, 38B, 563
$C_{31}H_{40}O_4$, (αR)-4-O-α-Cadinylangolensin, 46B, 558

54. DITERPENES

$C_{17}H_{20}O$, DL-19,20-Cyclopodocarpa-8,11,13-trien-19-one, 44B, 508
$C_{18}H_{20}O_5$, Salignone-D, 46B, 558
$C_{18}H_{20}O_8$, Inumakilactone, 41B, 631
$C_{18}H_{22}O_8$, Inumakilactone D, 43B, 707
$C_{18}H_{28}O_3$, (13R)-8α,13:13,17-Diepoxy-14,15-bisnorlabdan-17-one, 45B, 605
$C_{18}H_{30}O_2$, 14,15-Bis-nor-8α-hydroxylabd-11-en-13-one, 45B, 605
$C_{19}H_{20}O_5 \cdot CH_4O$, Teuflin methanol solvate, 45B, 606
$C_{19}H_{20}O_5$, Teucvidin, 40B, 515
$C_{19}H_{20}O_6$, 3-Dehydro-gibberellin A_3, 41B, 632
$C_{19}H_{20}O_6$, 3β-Hydroxyteucvidin, 44B, 509
$C_{19}H_{22}O_5$, Podolide, 41B, 633
$C_{19}H_{22}O_6$, Diosbulbin-G, 44B, 509
$C_{19}H_{22}O_7$, 1-Hydroxy-3-keto-gibberellin, 41B, 669
$C_{19}H_{23}BrO_4$, Methyl 6α-bromo-12-methoxy-7-oxopodocarpate, 35B, 129
$C_{19}H_{27}BrO_2$, Bromo-epoxynorcafestanone, 27, 1031
$C_{19}H_{28}O$, ent-9(8-15αH)abeo-17-Norkaur-8(14)-en-16-one, 43B, 708
$C_{20}H_{20}O_6$, Annonalide, 44B, 510
$C_{20}H_{21}BrO_7$, Sellowin B bromohydrin acetate, 42B, 431
$C_{20}H_{22}O_5 \cdot H_2O$, Crotofolin E monohydrate, 45B, 606
$C_{20}H_{22}O_5$, Bacchotricuneatin A, 44B, 511
$C_{20}H_{22}O_5$, Bacchotricuneatin B, 44B, 511
$C_{20}H_{22}O_5$, Icetexone, 42B, 432
$C_{20}H_{22}O_5$, 15,16-Epoxy-trans-cleroda-2,13(16),14-trieno-12,17:19,18-dilactone, 44B, 510
$C_{20}H_{23}BrO_6$, Methyl bromogibberellate, 31B, 223
$C_{20}H_{24}O_2$, 2,6,11,15-Tetramethylhexadeca-2,4,6,8,10,12,14-heptaen-1,16-dial, 38B, 564
$C_{20}H_{24}O_3$, Jatrophone, 42B, 432
$C_{20}H_{24}O_4$, Isoovatodiolide, 43B, 709
$C_{20}H_{24}O_4$, Ovatodiolic acid, 42B, 433
$C_{20}H_{24}O_4$, Ovatodiolide transannular cyclization product, 43B, 709
$C_{20}H_{24}O_4$, Ovatodiolide, 42B, 433; 43B, 709
$C_{20}H_{24}O_5$, Crotofolin A, 41B, 634
$C_{20}H_{24}O_6$, Triptolide, 39B, 405
$C_{20}H_{24}O_7$, Tripdiolide, 39B, 405
$C_{20}H_{25}BrO_7$, 3-Bromo-2-oxo-tetrahydrodiosbulbin-A, 43B, 710
$C_{20}H_{26}O_2$, (E)-4-Methyl-5-(5-(2,6,6-trimethylcyclohexen-1-yl)-3-methyl-2(E),4(E)-pentadienylidene)-2(5H)-furanone, 42B, 434
$C_{20}H_{26}O_2$, Iso-eremolactone, 33B, 214
$C_{20}H_{26}O_2$, 2-cis-4-Hydroxyretinoic acid γ-lactone, 40B, 515
$C_{20}H_{26}O_3$, Bertyadional photoproduct, 44B, 512
$C_{20}H_{26}O_3$, Jolkinolide A, 46B, 559
$C_{20}H_{26}O_3$, Jatrophatrione, 42B, 435
$C_{20}H_{26}O_3$, cis-Clerodanefuranolactone, 42B, 434
$C_{20}H_{26}O_5$, Baccharis trimera diterpene, 43B, 710
$C_{20}H_{26}O_6$, Conacytone, 42B, 436

$C_{20}H_{26}O_6$, cis-Coleon D, 42B, 435
$C_{20}H_{27}$, Cembrene cyclisation product, 45B, 607
$C_{20}H_{27}BrO$, Bromodehydrobispulegone, 40B, 515
$C_{20}H_{27}BrO_2$, 6α-Bromo-13-hydroxy-14-isopropyl-podocarpa-8,11,13-trien-7-one, 40B, 516
$C_{20}H_{28}O$, all-trans-Retinal, 38B, 567
$C_{20}H_{28}O$, 11-cis-Retinal, 38B, 566
$C_{20}H_{28}O_2$, Tetrachyrin, 46B, 559
$C_{20}H_{28}O_2$, Retinoic acid, 38B, 568
$C_{20}H_{28}O_3$, Cleomeolide lactone, 46B, 560
$C_{20}H_{28}O_3$, Lobophytolide, 43B, 711
$C_{20}H_{28}O_4$, Dihydroxyserrulatic acid, 43B, 712
$C_{20}H_{28}O_4$, 5β,6β-Isopropylidenedioxy-15,16,17-trinorgrayan-10(20)-ene-3,14-dione, 45B, 607
$C_{20}H_{28}O_6$, Nepetaefolinol, 46B, 566
$C_{20}H_{29}BrO_3$, Methyl (4R)-2β-bromo-3-oxo-19-nor-16α-(-)-kauran-17-oate, 42B, 436
$C_{20}H_{30}Br_2O_3$, Angasiol, 44B, 512
$C_{20}H_{30}O$, (-)-Kaur-15-en-19-al, 38B, 570
$C_{20}H_{30}O_2$, Dictyolactone, 45B, 612
$C_{20}H_{30}O_2$, Vitamin-A acid, 28, 377
$C_{20}H_{30}O_3$, Cleomeolide, 45B, 608; 46B, 560
$C_{20}H_{30}O_3$, 18-Hydroxydecipia-2(4),14-dien-1-oic acid, 42B, 437
$C_{20}H_{30}O_3$, 3β,5β,6β-Trihydroxy(1βH)grayana-10(20),15-diene, 46B, 560
$C_{20}H_{30}O_4$, (4R*,5S*,6R*,9S*,10R*,13E)-3,6,10,14-Tetramethyl-3,4,5,6,7,8,9,10,11,12-decahydro-6,9-epoxycyclotetradeca[b]furan-4,5-diol, 45B, 610
$C_{20}H_{30}O_4$, Sinularin, 45B, 609
$C_{20}H_{30}O_4$, Sinulariolide, 43B, 712
$C_{20}H_{30}O_4$, ent-7β-Hydroxy-9,10-friedokauran-19,10β-olide, 45B, 608
$C_{20}H_{31}Br$, 2-Bromo-laurenene, 45B, 611
$C_{20}H_{31}O_2$, 8β-Hydroxymethylpodocarpane-13β-carboxylic acid lactone, 45B, 611
$C_{20}H_{32}$, Cubitene, 44B, 512
$C_{20}H_{32}$, Cembrene, 34B, 315
$C_{20}H_{32}$, Rimuene, 35B, 349
$C_{20}H_{32}O$, 1-Isopropyl-6,10,12a-trimethyl-octahydrocyclopentacycloundecen-7-ol, 44B, 513
$C_{20}H_{32}O_2$, Isovirescenol B, 41B, 635
$C_{20}H_{32}O_2$, ent-Kaurenediol, 43B, 713
$C_{20}H_{32}O_3$, [1aS,3aS-(1β,1aα,3aβ,5α,5aα,6α,10aβ)]-1-[1-(Hydroxymethyl)vinyl]-3a,5a,8-trimethyl-2,3,3a,4,5,5a,9,10,10a,10b-decahydro-1H,6H-cyclohept[e]indene-5,6-diol, 46B, 560
$C_{20}H_{32}O_4$, Abestinin-1-diol, 46B, 561
$C_{20}H_{32}O_4$, Dihydrosinularin, 45B, 609
$C_{20}H_{32}O_4$, 1,18-Dihydroxydeciπ-14-en-19-oic acid, 46B, 561
$C_{20}H_{32}O_5$, Grayanotoxin II, 46B, 562
$C_{20}H_{33}BrO_2$, (12S,13S)-8,12;12,15-Diepoxy-13-bromolabdane, 44B, 513
$C_{20}H_{34}$, α-Dihydrophyllocladene, 44B, 514
$C_{20}H_{34}O_2$, Dictyodiol, 45B, 612
$C_{20}H_{34}O_2$, 1α,4β-Dihydroxyclavular-17-ene, 44B, 514
$C_{20}H_{34}O_2$, 3,15-Epoxy-4-hydroxycembra-7(Z),11(Z)-diene, 43B, 714
$C_{20}H_{34}O_4$, Neoconcinndiol hydroperoxide, 43B, 715
$C_{20}H_{34}O_6$, Thromboxane B2, 46B, 562
$C_{20}H_{35}BrO_2$, 14-Bromoobtus-1-ene-3,11-diol, 45B, 612
$C_{20}H_{36}O_4$, Verticillol diepoxide monohydrate, 44B, 515
$C_{20}H_{38}O_3$, (Z)-Cembr-4-ene-15,19,20-triol, 43B, 716

$C_{21}H_{24}Br_2O_7$, Gibberellic acid (3-acetyl 16α,17-dibromide), 37B, 311
$C_{21}H_{24}O_6$, Tinophyllone, 37B, 312
$C_{21}H_{26}O_4$, Ovatodiolide transannular cyclization product, 43B, 709
$C_{21}H_{26}O_5$, Methyl barbascoate, 42B, 438
$C_{21}H_{27}BrO_3$, Methyl 6α-bromo-13-isopropyl-7-oxo-podocarpa-8,11,13-trien-15-oate, 40B, 517
$C_{21}H_{28}O_6$, O-Methylshikodonin, 44B, 515
$C_{21}H_{28}O_7$, Gibberelin (cyclobutane annelated derivative) monohydrate, 42B, 439
$C_{21}H_{30}O_2$, Methyl dehydroabietate, 44B, 516
$C_{21}H_{30}O_3$, Oxetanol (from royleanone methyl ether), 44B, 517
$C_{21}H_{30}O_4$, Sarcoglaucol, 44B, 517
$C_{21}H_{31}N$, 7-Isocyano-11(20),14-epiamphilectadiene, 46B, 563
$C_{21}H_{31}N$, 8-Isocyano-10-cycloamphilectene, 46B, 563
$C_{21}H_{32}O_2$, Helifulvanic acid methyl ester, 46B, 564
$C_{21}H_{32}O_4$, 3-Acetoxy-9-hydroxy-17-norkauran-16-one, 41B, 635
$C_{21}H_{32}O_6$, ent-6α,18-Diacetoxy-14,15,16-trinorlabdan-13,8α-olide, 43B, 716
$C_{21}H_{34}O_4$, 5-Methoxycarbonyl-11,13-dihydroxy-5,9,13-trimethyltetracyclo[10.2.2.0$^{1-10}$.0$^{4-9}$]hexadecane, 42B, 439
$C_{22}H_{26}O_8$, Corylifuran, 42B, 440
$C_{22}H_{28}O_7$, Nepetaefolin, 41B, 636
$C_{22}H_{29}IO_6$, Deacetylcascarillin acetal iodoacetate, 31B, 225
$C_{22}H_{30}O_3$, Kempene-2, 43B, 717
$C_{22}H_{30}O_4$, (5E,12E)-7β-Acetoxybertya-5,12-diene-3,14-dione, 41B, 637
$C_{22}H_{30}O_5$, (5E,12E)-7β-Acetoxy-15β-hydroxybertya-5,12-diene-3,14-dione, 41B, 637
$C_{22}H_{30}O_5$, Diterpene D photoproduct, 44B, 512
$C_{22}H_{30}O_7$, Leonitin, 45B, 613
$C_{22}H_{30}O_7$, trans-Δ(1,10)-cis-Δ(4,5)-Heliangolide, 44B, 518
$C_{22}H_{30}O_7$, 15-Deoxyeurocurvin, 44B, 518
$C_{22}H_{31}BrO_4$, ent-3β-Acetoxy-11α-bromobeyer-2,12-dione, 42B, 440
$C_{22}H_{32}Br_2O_5$, Eupalmerin acetate dibromide, 41B, 638
$C_{22}H_{32}Br_2O_5$, 15(R)-Acetoxy-6(S),10(S)-dibromo-3a(S),4,7,8,11,12,13,-14(S),15,15a(R)-decahydro-6,10,14-trimethyl-3-methylene-5(R),9(R)-epoxycyclotetradeca[b]2-furanone, 40B, 518
$C_{22}H_{32}N_2$, 7,15-Diisocyanoadociane, 46B, 563
$C_{22}H_{32}O_2$, Vitamin-A acetate, 40B, 519
$C_{22}H_{32}O_4$, 15β-Acetoxy-(-)-kaur-16-en-19-oic acid, 45B, 613
$C_{22}H_{32}O_5$, Eupalmerin acetate, 41B, 638
$C_{22}H_{32}O_6$, O-Ethyldihydroshikodonin, 44B, 515
$C_{22}H_{33}BrO_3$, ent-3β-Acetoxy-11α-bromoisopimar-8(14)-en-12-one, 42B, 441
$C_{22}H_{33}BrO_4$, O-Bromoacetyl-pleuromutilin, 41B, 639
$C_{22}H_{33}NO_2$, Isoatisine, 44B, 519
$C_{22}H_{33}NO_2$, Veatchine, 44B, 519
$C_{22}H_{33}O_3$, 3α-Acetoxy-15-epoxyrippertane, 46B, 564
$C_{22}H_{34}ClNO_2$, Atisinium chloride, 44B, 519
$C_{22}H_{34}O_4$, Trinervi-2β,3α,9α-triol 9-O-acetate, 42B, 441
$C_{22}H_{34}O_4$, 6-Acetyldolatriol, 43B, 718
$C_{22}H_{34}O_5$, Evillosin, 45B, 614
$C_{22}H_{34}O_7$, Forskolin, 46B, 564
$C_{22}H_{34}O_7$, Phorbol (ethanol solvate), 37B, 312
$C_{22}H_{34}O_8$, Cinnzeylanine, 43B, 719
$C_{22}H_{35}BrO_2$, Phyllocladan-15-yl bromoacetate, 42B, 442
$C_{22}H_{35}BrO_3$, ent-3β-Acetoxy-11α-bromoisopimar-12-one, 42B, 443
$C_{22}H_{35}NO_2$, Dihydroatisine, 44B, 519

$C_{22}H_{36}O_3$, Dollabella californica diterpene, 42B, 443
$C_{22}H_{36}O_3$, ent-1β-Hydroxy-16S-atis-13-en-17-al dimethyl acetal, 42B, 444
$C_{22}H_{36}O_3$, 3α-Acetoxy-15β-hydroxy-7,16-secotrinervita-7,11-diene, 46B, 565
$C_{22}H_{36}O_4$, Plexaurolone acetate, 46B, 565
$C_{22}H_{36}O_5$, 13-Labden-6α-acetoxy-8α-hydroxy-15-oic acid, 45B, 614
$C_{23}H_{32}O_8$, Methoxynepetaefolin, 46B, 566
$C_{23}H_{33}BrO_6$, 16α-Methyl-15β-bromoacetatoenmein, 34B, 314
$C_{24}H_{30}O_8$, Dehydrocyclobutatusin, 43B, 720
$C_{24}H_{30}O_9$, Eriocephalin, 45B, 615
$C_{24}H_{31}BrO_8$, Acetyl-bromoacetyl-dihydroenmein, 31B, 227
$C_{24}H_{32}O_6$, Oxidopanamensin 2,6-diacetate, 46B, 566
$C_{24}H_{33}BrO_8$, Clerodin bromolactone, 27, 1029
$C_{24}H_{35}IO_4$, Beyerol monoethylidene iodoacetate, 31B, 229
$C_{24}H_{36}O_4$, 3,20-Bis(ethylenedioxy)-9,10-seco-pregna-5,7,10(19)-triene, 38B, 571
$C_{25}H_{29}BrN_4O_6$, (4S)-2-(Prop-2-enyl)rethron-4-yl (1R,3R)-chrysanthemate 6-bromo-2,4-dinitrophenylhydrazone, 40B, 519
$C_{25}H_{29}BrO_8 \cdot 0.8\ CHCl_3$, Phorbol bromofuroate chloroform solvate, 33B, 215
$C_{25}H_{36}O_5$, Vibsanine E, 46B, 567
$C_{26}H_{31}BrO_5S$, 7β-Hydroxykaurenolide p-bromobenzenesulphonyl derivative, 38B, 572
$C_{26}H_{34}O_8$, 3α,17,19-Triacetoxyspongia-13(16),14-dien-2-one, 45B, 615
$C_{26}H_{35}ClO_{10}$, Stylatulide, 43B, 720
$C_{26}H_{36}O_8$, 13-Epi-9-desacetylxenicin, 45B, 616
$C_{26}H_{37}BrN_2O_5S$, Portulal p-bromophenylsulphonylhydrazone, 37B, 313
$C_{26}H_{37}BrO_3S$, Beyran-3α-ol p-bromobenzenesulphonyl derivative, 38B, 573
$C_{26}H_{38}Br_2O_8$, Dibromowightionolide triacetate, 40B, 520
$C_{26}H_{38}O_8$, Tetrahydrofruticolone triacetate, 44B, 520
$C_{26}H_{40}O_7$, 12α,15α,16α-Triacetoxyspongan, 45B, 617
$C_{27}H_{33}BrO_3$, 3β-(p-Bromobenzoyloxy)-7β-kemp-8(9)-en-6-one, 45B, 617
$C_{27}H_{33}BrO_4$, ent-(1aR)-1-α-(1a-Hydroxy-1a-phenyl)-3β-hydroxy-11α-bromo-12-oxo-2-norbeyeran-3β-carboxylic acid 2,1a:lactone, 44B, 520
$C_{27}H_{33}BrO_6$, 1,2-Desacetyl-ε-caesalpin 2-p-bromobenzoate, 34B, 317
$C_{27}H_{33}IO_5$, Crassin p-iodobenzoate, 40B, 1094
$C_{27}H_{33}IO_5$, Jeunicin p-iodobenzoate, 42B, 444
$C_{27}H_{34}BrClO_4$, 17-Chloro-(C17,O4)-secospatol p-bromobenzoate, 46B, 567
$C_{27}H_{36}BrNO_3$, Dictyoxepin p-bromophenylurethane, 43B, 721
$C_{27}H_{37}BrO_2$, 19-(p-Bromobenzoyloxy)-biflora-4,15-diene, 46B, 568
$C_{27}H_{37}BrO_3$, (-)-2β,9α-Dihydroxyverrucosane mono-p-bromo-benzoate, 46B, 568
$C_{28}H_{36}O_5$, Acetylstrongylophorine-2, 44B, 521
$C_{28}H_{37}BrO_4$, Methyl ent-16β-p-bromobenzyloxy-17(16-12)abeo-atisan-19-oate, 42B, 445
$C_{28}H_{37}BrO_4$, 12α-Beyeranol derivative, 41B, 639
$C_{28}H_{38}O_9$, Xenicin, 43B, 722
$C_{28}H_{38}O_{10}$, Ingol-3,7,8,12-tetraacetate, 45B, 618
$C_{28}H_{39}BrO_9$, 2,5,9,10-Tetra-O-acetyl-14-bromotaxinol, 37B, 314
$C_{28}H_{41}BrO_4$, p-Bromophenacyl labdanolate, 33B, 216
$C_{28}H_{42}Br_2O_9$, Eunicellin dibromide, 35B, 396
$C_{29}H_{37}BrO_5$, ent-1α-Acetoxy-16β-hydroxy-17-p-bromobenzoyloxy-atis-13-ene, 46B, 569
$C_{30}H_{39}BrO_4$, Taxa-4(16),11-diene-5α,9α,10β,13α-tetraol p-bromobenzo-

ate dihydroanhydroacetonide, 32B, 252
$C_{30}H_{39}ClO_{13}$, Briarein A, 43B, 722
$C_{30}H_{42}O_{11} \cdot CDCl_3$, 1-Epigrayanol A pentaacetate chloroform solvate, 46B, 570
$C_{31}H_{35}BrO_9$, Neophorbol-13,20-diacetate-3-p-bromobenzoate, 37B, 312
$C_{31}H_{35}BrO_{10}$, Cyclobutatusin p-bromobenzoate, 40B, 521; 43B, 723
$C_{31}H_{38}O_{11}$, Baccatin V, 39B, 406
$C_{31}H_{43}BrO_7$, 1-Benzyl-7-desacetyl-7-bromoisobutyryl-forskolin, 46B, 570
$C_{33}H_{30}Br_2O_9$, Nagilactone B 2,7-bis-p-bromobenzoate, 45B, 618
$C_{33}H_{33}Br_3O_{12}$, 12-Hydroxydaphnetoxin tribromoacetate, 38B, 573
$C_{34}H_{30}Br_2O_8$, Methyl gibberellate di-p-bromobenzoate, 28, 540
$C_{36}H_{37}INO_7$, Isocolumbin 1-p-iodophenyl-3-phenylpyrazoline adduct acetone solvate, 31B, 230
$C_{36}H_{47}BrO_{11}$, Ajugareptansin p-bromobenzoate, 45B, 619
$C_{37}H_{39}BrO_9$, Barbatusin p-bromobenzoate benzene solvate, 39B, 406; 43B, 723
$C_{39}H_{38}O$, Mezerein, 41B, 640

55. SESTERPENES

$C_{25}H_{28}Br_2O_5$, Heliocide H_2 dibromide, 43B, 724
$C_{25}H_{30}O_6$, Andibenin, 45B, 619
$C_{25}H_{32}O_4$, Ircinianin, 43B, 725
$C_{25}H_{32}O_6 \cdot C_3H_8O$, Andilesin-A 2-propanol solvate, 45B, 619
$C_{26}H_{39}BrO_5$, Ophiobolin methoxybromide, 30B, 244; 33B, 217
$C_{28}H_{36}O_9 \cdot H_2O$, Amoorastatine monohydrate, 46B, 572
$C_{28}H_{36}O_{10} \cdot H_2O$, 12-Hydroxyamoorastatine monohydrate, 46B, 572
$C_{28}H_{39}BrO_5$, Methylcephalonate bromoacetate, 34B, 318
$C_{28}H_{40}O_5$, 22-Acetoxy-24-methyl-12,24-dioxoscalar-16-en-25-al, 46B, 570
$C_{28}H_{48}$, 28,30-Dinor-17α(H),18α(H),21β(H)-hopane, 45B, 621
$C_{30}H_{38}O_9$, Dihydronimbin, 46B, 571
$C_{30}H_{48}O_2$, 3-Oxofriedelan-12-ol, 45B, 621
$C_{31}H_{42}BrNO$, Retigeranic acid p-bromoanilide, 40B, 521
$C_{32}H_{43}BrO_2$, Ceroplastol I p-bromobenzoate, 34B, 320
$C_{37}H_{61}NO_2$, Dicyclohexylammonium gascardate, 45B, 620

56. TRITERPENES

$C_{27}H_{31}BrO_3$, 13β-Bromo-3,10-dimethoxy-6aβ,12bβ,14aβ-trimethyl-5,6,6a,6bβ,7,8,12b,13-octahydro-14(14aH)-picenone, 40B, 1095
$C_{27}H_{46}$, 22,29,30-Trisnorhopane II, 41B, 641
$C_{28}H_{31}IO_6$, Cedrelone iodoacetate, 28, 542
$C_{28}H_{33}IO_9$, Epilimonol iodoacetate, 26, 801
$C_{29}H_{50}$, 29-Nor-17αH-hopane, 41B, 642
$C_{30}H_{38}O_{10}$, 7α-Acetoxydihydronomilin, 44B, 521
$C_{30}H_{39}IO_8$, Dihydrogedun-3β-yl iodoacetate, 27, 1033
$C_{30}H_{42}O_4$, Papyriogenin-A, 43B, 725
$C_{30}H_{42}O_4$, 13α,14β,17α-Lanosta-7,24-diene-3,6-dione-21,16β-oide, 44B, 522
$C_{30}H_{46}O_4$, Benulin, 43B, 726
$C_{30}H_{47}BrO$, 2α-Bromoarborinone, 32B, 252
$C_{30}H_{48}O$, Lup-20(29)-en-3-one, 43B, 727
$C_{30}H_{48}O_4$, Hederagenine, 44B, 523

$C_{30}H_{50}$, Squalene, 45B, 622
$C_{30}H_{50}$, D:C-friedo-B':A'-neo-Gammacer-9(11)-ene, 46B, 572
$C_{30}H_{50}O$, Baccharis oxide, 39B, 407
$C_{30}H_{50}O$, Campanulin, 43B, 728
$C_{30}H_{50}O$, 3α,4α-Epoxy-D:A-friedo-18β,19αH-lupane, 46B, 571
$C_{30}H_{50}O_3$, (+)-Malabaricol, 45B, 622
$C_{30}H_{50}O_6$, Gymnemagenin, 40B, 522
$C_{30}H_{52}$, 1(10-5)-abeo-3β-Methyl-24β-nor-25α-18-α-oleanane, 35B, 350
$C_{30}H_{52}$, 18-α(H)-Oleanane, 44B, 523
$C_{30}H_{52}O \cdot 0.5\ H_2O$, Tetrahymanol hemihydrate, 43B, 729
$C_{30}H_{52}O$, Epifriedelinol, 43B, 729
$C_{30}H_{52}O_5 \cdot H_2O$, Cucurbit-5-ene-3β,22(S),23(R),24(R),25-pentaol monohydrate, 46B, 572
$C_{31}H_{46}O_4 \cdot 0.5\ H_2O$, Meristotropic acid methyl ester hemihydrate, 44B, 523
$C_{31}H_{47}O_5$, Holotoxinogenin 25-methyl ether, 41B, 642
$C_{31}H_{51}BrO_2$, Adiantol B bromoacetate, 35B, 403
$C_{31}H_{52}O_2$, 21α-Methoxy-Δ^{13}-serraten-3β-ol, 43B, 730
$C_{31}H_{52}O_3$, (11R,20S,24R)-20,24-Epoxy-11-hydroxy-24-methyldammaran-3-one, 43B, 730; 44B, 524
$C_{32}H_{39}BrO_{11}$, Glaucarubin p-bromobenzoate, 29, 753
$C_{32}H_{48}O_6$, 23-Hydroxy-2,3-secours-12-ene-2,3,28-trioic acid (2-23)-lactone 3,28-dimethyl ester, 37B, 316
$C_{32}H_{50}O_4$, 3β-e-Acetoxyolean-12-ene-28β-carboxylic acid, 44B, 524
$C_{32}H_{50}O_4$, 3β,16β-Dimethoxyolean-12(13)-ene-28-21β-olide, 43B, 731
$C_{32}H_{51}IO_2$, Davallol iodoacetate, 31B, 232
$C_{32}H_{52}IO_2$, Motiol iodoacetate, 33B, 218
$C_{32}H_{52}O_2$, 3-β-Acetoxy-α-amyrin, 45B, 622
$C_{32}H_{52}O_6$, Dimethyl tetrahydro-officinilate, 45B, 623
$C_{32}H_{52}O_7$, Passifloric acid methyl ester, 43B, 732
$C_{32}H_{53}IO_2$, Euphenyl iodoacetate, 31B, 234
$C_{32}H_{54}O_3$, 3β-Acetoxy-20-hydroxylupane, 38B, 574
$C_{33}H_{51}BrO_4$, Methylmicromerol bromoacetate, 28, 543
$C_{33}H_{51}IO_4$, Methyl oleanolate iodoacetate, 20, 645
$C_{33}H_{54}O_6$, Glycyrrhetinic acid acetone monohydrate, 43B, 733
$C_{34}H_{37}IO_9$, Detigloylswietenine p-iodobenzoate, 31B, 236
$C_{34}H_{50}Br_2O_6$, Bacogenin-A_1 dibromoacetate, 39B, 407
$C_{34}H_{50}Cl_2O_7$, Chloroacetoxy-glabretal-7-lactone chlorohydrin, 41B, 643
$C_{34}H_{51}BrO_6$, Echinocystic acid diacetate bromolactone, 42B, 446
$C_{34}H_{51}BrO_6$, Oleanolic acid diacetate bromolactone, 40B, 523
$C_{34}H_{51}IO_6$, Methyl melaleucate p-iodoacetate, 30B, 228
$C_{34}H_{54}O_5$, 3β,28-Diacetoxy-18β,19β-epoxylupane, 45B, 624
$C_{34}H_{56}O_6$, 12-O-Acetylspergulagenin A ethylene ketal, 42B, 445
$C_{35}H_{46}O_{13}$, Aphanastatine, 44B, 525; 46B, 572
$C_{37}H_{51}BrO_5 \cdot 0.5\ C_4H_8O_2$, Jujubogenin p-bromobenzoate hemi(ethyl acetate), 40B, 524
$C_{37}H_{53}IO_2$, Hortensenyl p-iodobenzoate, 42B, 447
$C_{37}H_{54}O_2$, 3-β-Benzoxy-α-amyrin, 45B, 622
$C_{38}H_{53}BrO_4$, Methyl ursolate-3-p-bromobenzoate, 45B, 624
$C_{38}H_{53}BrO_5$, p-Bromophenacyl retigerate A, 38B, 574
$C_{39}H_{55}IO_5$, Secogorgosterol 3-p-iodobenzoate-11-acetate, 40B, 525
$C_{39}H_{57}BrO_4$, 3-O-Acetyl-16-O-p-bromobenzoylpachysandiol B, 41B, 644
$C_{42}H_{59}IO_{14}S$, Fusicoccin A p-iodobenzenesulphonate, 37B, 315
$C_{44}H_{53}BrO_{18}S \cdot 0.75\ CH_2Cl_2$, Prieurianin p-bromobenzenesulphonate methylene chloride solvate, 42B, 447
$C_{45}H_{60}Br_2O_5$, Cyclograndisolide bis-p-bromobenzoate, 37B, 317

$C_{58}H_{84}O_{18}$, Acetylated napoleogenin, 46B, 573

57. TETRATERPENES

$C_{40}H_{50}O_2$, 15,15'-Dehydrocanthaxanthin, 33B, 84
$C_{40}H_{54}$, 15,15'-Dehydro-β-carotene, 29, 499
$C_{40}H_{56}$, β-Carotene, 29, 496
$C_{40}H_{58}$, 7,7-Dihydro-β-carotene, 29, 497
$C_{54}H_{62}Br_2O_5$, Capsanthin bis(p-bromobenzoate), 40B, 525

58. ALKALOIDS

$C_7H_{12}ClN_5O_2$, 1-Methyl-3-guanidino-6-hydroxymethylpyrazin-2-one chloride, 44B, 525
$C_8H_{11}NO \cdot 0.5\ H_2O$, Tyramine hemihydrate, 43B, 144
$C_8H_{12}ClNO_2$, Octopamine hydrochloride, 43B, 144
$C_8H_{15}Br_2N \cdot H_2O$, 3α-Bromotropane hydrobromide monohydrate, 46B, 574
$C_8H_{15}NO$, Pseudotropine, 32B, 253
$C_8H_{18}BrNO$, ψ-Conhydrine hydrobromide, 23, 697
$C_9H_{10}Cl_3N_3$, Clonidine hydrochloride, 42B, 172
$C_9H_{13}NO_2$, (-)-Phenylephrine, 42B, 18
$C_9H_{16}INO_2$, Arecoline methiodide, 41B, 644
$C_9H_{16}NO_4P$, (+)-Amphetamine dihydrogen phosphate, 44B, 27
$C_9H_{20}ClNO_5$, Acetylcarnitine hydrochloride monohydrate, 43B, 13
$C_{10}H_{15}NO$, (+)-Pseudoephedrine, 43B, 43
$C_{10}H_{16}ClNO$, (+)-Pseudoephedrine hydrochloride, 43B, 43
$C_{10}H_{16}ClNO$, (-)-Ephedrine hydrochloride, 18, 769; 37B, 318
$C_{10}H_{18}NO_5P$, (-)-Ephedrine dihydrogen phosphate, 38B, 452
$C_{10}H_{19}NO$, (+)-Epilupinine, 46B, 574
$C_{10}H_{19}NO$, Lupinine, 44B, 526
$C_{10}H_{20}ClNO$, (-)-Lupinine hydrochloride, 46B, 574
$C_{10}H_{23}N_3O$, N-Cyanomethylangustifoline, 45B, 625
$C_{11}H_{15}BrNO_2$, (S)-(-)-Anhalonine hydrobromide, 37B, 319
$C_{11}H_{17}Cl_3GeN_2O_2 \cdot 0.5\ H_2O$, Pilocarpine trichlorogermanate(II) hemihydrate, 33B, 219
$C_{11}H_{18}BrN_3O_8$, Tetrodonic acid hydrobromide, 31B, 239
$C_{11}H_{18}ClNO$, 6,9-endo-Methylenehomopseudopelletierine hydrochloride, 41B, 645
$C_{11}H_{19}NO_2S_4$, Gerrardine, 37B, 320
$C_{12}H_{12}N_2O_3$, trans-3-Hydroxyspiro[cyclopentane-1,4'(1'H)-[2,7]naphthyridine]-1',3'(2'H)-dione, 45B, 625
$C_{12}H_{14}N_2O$, 9-Nitroso-2,3,6,7-tetrahydro-1H,5H-benzo[ij]quinolizine, 45B, 626
$C_{12}H_{18}BrNO_3$, (+)-O-Methylanhalonidine hydrobromide, 37B, 319
$C_{12}H_{19}NO_5$, Swainsonine diacetate, 46B, 575
$C_{12}H_{20}ClNO_5$, Tecomanine methoperchlorate, 41B, 646
$C_{12}H_{20}IN$, (-)-N,N-Dimethylamphetamine methiodide, 43B, 47
$C_{12}H_{23}Cl_2N_3O_5$, Casimidine dihydrochloride, 28, 545
$C_{12}H_{24}INO$, Alkaloid C methiodide, 41B, 646
$C_{12}H_{25}Cl_2N_7O_5$, Saxitoxin ethyl hemiketal dihydrochloride monohydrate, 41B, 647
$C_{12}H_{34}N_2O_8P_2$, Putrescinium di-(diethyl phosphate), 38B, 454
$C_{13}H_{13}Br_2N_5O_2$, Monoacetyldibromophakellin, 43B, 733
$C_{13}H_{15}NO_2$, Phyllochrysine, 39B, 408
$C_{13}H_{20}BrNO_4$, Securinine hydrobromide dihydrate, 30B, 239

$C_{13}H_{22}BrN_3O$, N-Methyl-alchorneine bromide, 38B, 575
$C_{13}H_{22}ClNO$, Hippocasine oxide hydrochloride, 42B, 448
$C_{13}H_{23}ClN_2O_6$, Nitropolyzonammonium perchlorate, 44B, 526
$C_{13}H_{23}NO_3$, Coccutrine, 43B, 740
$C_{13}H_{26}ClN$, Pumiliotoxin C hydrochloride, 43B, 746
$C_{14}H_8N_2O_2 \cdot H_2O$, Amarorine monohydrate, 46B, 575
$C_{14}H_{18}INO_2$, Phyllochrysine methiodide, 31B, 240
$C_{14}H_{18}N_2O_6$, Pukeleimide C, 45B, 627
$C_{14}H_{20}ClN$, 2,9β-Dimethyl-6,7-benzomorphan hydrochloride, 41B, 648
$C_{14}H_{20}N_2O_2$, Pindolol, 42B, 448
$C_{14}H_{21}ClN_2O_5$, Albine perchlorate, 46B, 576
$C_{14}H_{22}N_2S_4$, Cassipourine, 34B, 321
$C_{15}H_{20}ClNO_2$, O-Benzoyl-ψ-tropine hydrochloride, 44B, 527
$C_{15}H_{20}ClNO_2$, O-Benzoyltropine hydrochloride, 42B, 449
$C_{15}H_{20}IN_3O_9 \cdot 0.5\ CH_4O$, 6,11-Diacetylanhydrotetrodotoxin hydroiodide methanol solvate, 31B, 241
$C_{15}H_{20}N_2O_2$, 6-Hydroxy-2'-(2-methylpropyl)-3,3'-spirotetrahydropyrrolidino-oxindole, 38B, 576
$C_{15}H_{21}BrN_2O$, Tsukushinamine-A hydrobromide, 46B, 576
$C_{15}H_{21}ClN_2O_5$, Dehydromultiflorine perchlorate, 46B, 576
$C_{15}H_{21}N_3O_2$, Eserine, 39B, 408
$C_{15}H_{23}ClN_2O_5 \cdot 0.5\ H_2O$, Multiflorine perchlorate hemihydrate, 46B, 577
$C_{15}H_{24}BrN$, Porantherine hydrobromide, 39B, 410
$C_{15}H_{24}BrNO$, Deoxynupharidine hydrobromide, 35B, 373
$C_{15}H_{24}BrNO_2$, Nupharidine hydrobromide, 39B, 409
$C_{15}H_{24}N_2O$, Allomatrine, 45B, 627
$C_{15}H_{24}N_2O$, Isosoforidine, 45B, 628
$C_{15}H_{24}N_2O$, Matrina alkaloid, 44B, 527
$C_{15}H_{24}N_2O$, Soforidine, 45B, 628
$C_{15}H_{24}N_2O$, Tetrahydroneosoforamine, 45B, 629
$C_{15}H_{24}N_2O$, DL-Lupanine, 42B, 449
$C_{15}H_{24}N_2O$, 17-Oxosparteine, 46B, 577
$C_{15}H_{24}N_2O_2 \cdot H_2O$, Matrine N-oxide monohydrate, 45B, 629
$C_{15}H_{24}N_2O_2 \cdot H_2O$, N-Oxy-allomatrine monohydrate, 45B, 627
$C_{15}H_{24}N_2O_2$, (5R,6S,11R)Sophocarpine monohydrate, 44B, 528
$C_{15}H_{24}N_2O_2$, 13α-Hydroxylupanine, 43B, 734
$C_{15}H_{25}ClN_2O_5 \cdot H_2O$, (+)-Lupanine perchlorate monohydrate, 45B, 630
$C_{15}H_{25}ClN_2O_5$, (+)-Lupanine perchlorate, 46B, 577
$C_{15}H_{26}BrNO_2$, DL-3-Epinupharamine hydrobromide, 43B, 734
$C_{15}H_{27}ClN_2O_5$, Sparteine N(16)-oxide monoperchlorate, 44B, 528
$C_{15}H_{27}ClN_2O_5$, 7-Hydroxy-β-isosparteine perchlorate, 32B, 254
$C_{15}H_{28}BrNO$, Poranthericine hydrobromide, 39B, 411
$C_{15}H_{28}BrNO$, Porantheridene hydrobromide, 39B, 411
$C_{15}H_{28}Cl_2N_2O_8$, α-Isosparteine diperchlorate, 40B, 526
$C_{15}H_{28}N_2O$, α-Isosparteine monohydrate, 17, 768
$C_{15}H_{29}ClN_2O_3$, (+)-Lupanine hydrochloride dihydrate, 44B, 529
$C_{16}H_{12}N_2O_2$, Eupolauramine, 42B, 450
$C_{16}H_{15}NO_5$, 1,2-Epoxy-3-hydroxy-α-lycoran-7-one, 44B, 529
$C_{16}H_{17}NO_4$, Lycorine, 40B, 528; 42B, 450
$C_{16}H_{18}BrNO_3$, Higenamine hydrobromide, 43B, 735
$C_{16}H_{18}BrNO_4$, Lycorine hydrobromide, 40B, 527
$C_{16}H_{18}N_2O_3$, (+)-Isopilosine, 38B, 577
$C_{16}H_{19}NO_3$, De-N-methyl-DL-galanthamine, 42B, 451
$C_{16}H_{20}BrNO_2$, Elaeocarpine hydrobromide, 34B, 325
$C_{16}H_{20}BrNO_4$, Dihydrolycorine hydrobromide, 33B, 220
$C_{16}H_{20}ClNO_3 \cdot 2\ H_2O$, Norgalanthamine hydrochloride dihydrate, 46B,

578
$C_{16}H_{21}BrNO_4$, Norcocaine hydrobromide, 43B, 736
$C_{16}H_{21}NO_3$, Phellibiline, 40B, 528
$C_{16}H_{22}BrN$, N-Methyl-D-normorphinan hydrobromide, 41B, 648
$C_{16}H_{22}BrNO_3$, Annotinine bromohydrin, 22, 801
$C_{16}H_{22}BrNO_3$, Dihydro-β-erythroidine hydrobromide, 28, 546
$C_{16}H_{22}BrNO_6$, (-)Isolunine hydrobromide dihydrate, 34B, 324
$C_{16}H_{22}Cl_3NO_4$, Dysidin, 43B, 737
$C_{16}H_{23}NO_5$, Fulvine, 39B, 412
$C_{16}H_{23}NO_6$, Monocrotaline, 45B, 631
$C_{16}H_{26}ClNO$, Lycopodine hydrochloride, 41B, 649
$C_{16}H_{27}BrN_2O_2$, Haloxine hydrobromide, 33B, 221
$C_{16}H_{27}ClN_2O_5$, 17β-Methyl-α-isolupanine perchlorate, 46B, 579
$C_{16}H_{27}ClN_2O_5$, 17β-Methyllupanine perchlorate, 46B, 579
$C_{16}H_{27}NO_5$, Heliotrine, 41B, 650
$C_{16}H_{29}ClN_2O_4$, 2-Methylsparteine perchlorate, 46B, 580
$C_{16}H_{30}ClNO$, 8-Hydroxy-8-methyl-6-(2'-methylhexylidene)-1-azabicyclo-[4.3.0]nonane hydrochloride, 46B, 580
$C_{16}H_{30}Cl_2N_2O_8$, 17β-Methylsparteine diperchlorate, 45B, 631
$C_{17}H_{17}NO_4$, 14-Hydroxymorphinone, 44B, 530
$C_{17}H_{17}NO_5$, Renierone, 45B, 632
$C_{17}H_{18}BrNO_2 \cdot 0.75\ H_2O$, Apomorphine hydrobromide hydrate, 43B, 737
$C_{17}H_{18}ClNO_2 \cdot 0.75\ H_2O$, Apomorphine hydrochloride hydrate, 39B, 412
$C_{17}H_{18}N_2O$, Borreline, 43B, 738
$C_{17}H_{18}N_2O$, Desethyleburnamonine, 39B, 414
$C_{17}H_{19}NO_3$, 1-Piperoylpiperidine, 41B, 651
$C_{17}H_{19}NO_4$, Crinamine, 43B, 738
$C_{17}H_{19}NO_4$, N-Demethyl-N-formylmesembrenone, 43B, 739
$C_{17}H_{19}NO_5$, 6-Hydroxycrinamine, 32B, 256
$C_{17}H_{20}N_4O_3$, 6-Deoxy-6-azido-14-hydroxydihydroisomorphine, 42B, 232
$C_{17}H_{21}NO_2$, Cocculine, 43B, 740
$C_{17}H_{21}NO_3$, Maritidine, 45B, 632
$C_{17}H_{21}NO_4$, (-)-Morphine monohydrate, 42B, 452
$C_{17}H_{21}NO_5$, Oxymorphone hydrate, 42B, 452
$C_{17}H_{22}BrNO_4$, Coclaurine hydrobromide monohydrate, 33B, 223
$C_{17}H_{22}ClNO_4$, L-Cocaine hydrochloride, 28, 548
$C_{17}H_{24}BrNO \cdot 0.7\ CH_4O$, 3-Hydroxy-N-methyl-hasubanan hydrobromide methanolate, 43B, 741
$C_{17}H_{24}BrNO_3$, (-)-Hyoscyamine hydrobromide, 38B, 578
$C_{17}H_{24}INO_5$, Morphine hydroiodide dihydrate, 19, 624
$C_{17}H_{25}NO_2$, (2R,6S,8R,10S)-8-Methyl-10-phenyl-3,4-dehydrolobelidiol, 46B, 581
$C_{17}H_{25}NO_3$, Sedacryptine, 46B, 581
$C_{17}H_{26}BrNO_2$, 2-Allyl-2-hydroxy-5,9-dimethyl-6,7-benzomorphan hydrobromide monohydrate, 35B, 379
$C_{17}H_{26}BrNO_3$, Annopodine hydrobromide, 39B, 414
$C_{17}H_{26}BrNO_4 \cdot 0.5\ H_2O$, Stemonine hydrobromide hemihydrate, 35B, 382
$C_{17}H_{26}ClNO_2$, (2R,6S,8R,10S)-8-Methyl-10-phenyl-3,4-dehydrolobelidiol hydrochloride, 46B, 581
$C_{17}H_{26}ClNO_6$, Morphine hydrochloride trihydrate, 39B, 413
$C_{17}H_{27}ClN_2O_2$, Dimethisoquin hydrochloride monohydrate, 43B, 346
$C_{17}H_{28}BrNO_2$, Paniculatine hydrobromide, 41B, 652; 42B, 453
$C_{17}H_{28}ClNO_3$, (-)-Mesembrane hydrochloride monohydrate, 43B, 741
$C_{18}H_{14}BrNO_2$, 1-endo-Carboxypyrrolizidine hydrobromide, 37B, 317
$C_{18}H_{15}NO_4$, Norrufescine, 44B, 531
$C_{18}H_{19}NO_4$, Laurelliptine, 44B, 531
$C_{18}H_{20}BrNO_3$, Erythraline hydrobromide, 22, 803

$C_{18}H_{20}ClN_3O_6S_2 \cdot 0.65\ CH_2Br_2$, Sporidesmin-methylene dibromide adduct, 30B, 240
$C_{18}H_{21}BrN_2$, (+)-Mianserin hydrobromide, 39B, 415
$C_{18}H_{21}NO_4$, 17-Epihomolycorine, 44B, 532
$C_{18}H_{21}NO_5$, Ambelline, 42B, 453
$C_{18}H_{22}BrNO_3$, DL-11,12-Dihydroglaziovine hydrobromide, 42B, 454
$C_{18}H_{22}ClNO_3$, Dihydrometacodeinone hydrochloride, 39B, 416
$C_{18}H_{22}NO_4$, Anhydroperforine, 45B, 633
$C_{18}H_{22}N_2O_2$, Isofumigaclavine A, 44B, 532
$C_{18}H_{24}INO_4$, Cocaine methiodide, 41B, 652
$C_{18}H_{25}NO$, DL-Cyclazocine, 34B, 326
$C_{18}H_{25}NO_5$, (1aR,6bR,10R,11R)-9,15-Dioxo-10-hydroxy-10,11,13-trimethyl-1a,2,3,6b-tetrahydro-5H-pyrrolizino-[1a,6b,6a-b,c]-1,8-dioxa-15-cis-tridecene, 46B, 581
$C_{18}H_{25}NO_6$, Jacobine, 44B, 532
$C_{18}H_{25}NO_6$, Retrorsine, 46B, 582
$C_{18}H_{26}BrNO_5$, Codeine hydrobromide dihydrate, 19, 623; 27, 1023
$C_{18}H_{26}BrNO_6 \cdot C_2H_6O$, Retrorsine hydrobromide ethanol solvate, 45B, 634
$C_{18}H_{26}BrNO_6 \cdot 0.5\ C_2H_6O$, Jacobine bromohydrin, 28, 550
$C_{18}H_{27}NO_5 \cdot C_6H_6$, Inkanine benzene solvate, 45B, 634
$C_{18}H_{27}NO_6$, Tricodesime, 45B, 635
$C_{18}H_{28}BrNO_2$, (+)-3-Methoxy-N-methylmorphinan hydrobromide monohydrate, 43B, 742
$C_{18}H_{28}BrNO_2$, L-Cyclazocine hydrobromide monohydrate, 34B, 326
$C_{18}H_{28}BrNO_2$, Magellanine methobromide, 42B, 454
$C_{18}H_{28}ClNO_6$, Stemonamine hydrochloride dihydrate, 39B, 416
$C_{18}H_{28}INO_3$, 6-Epimesembranol methiodide, 35B, 383
$C_{18}H_{36}ClNO_2$, 2-Methyl-3-hydroxy-6-(11-oxodecyl)piperidinium hydrochloride, 42B, 455
$C_{19}H_{20}N_2O \cdot CH_4O$, Ervistine methanol solvate, 45B, 636
$C_{19}H_{22}BrNO_3$, 7,8-Didehydro-4,5-epoxy-17-(2-propenyl)morphinan-3,6-diol hydrobromide, 45B, 636
$C_{19}H_{22}BrNO_4$, Isoboldine hydrobromide, 43B, 743
$C_{19}H_{22}INO_5$, Narcissidine methiodide, 37B, 320
$C_{19}H_{22}N_2O$, Cinchonine, 45B, 637
$C_{19}H_{22}N_2O$, Vincamone, 39B, 417
$C_{19}H_{24}BrNO_3$, (-)-4-Bromo-2,3-dimethoxy-N-methyl-6-oxomorphinan, 46B, 582
$C_{19}H_{24}BrNO_4$, Schelhammerine hydrobromide, 40B, 1095
$C_{19}H_{24}ClNO_6$, Acutumine, 33B, 223
$C_{19}H_{24}Cl_2N_2$, Chlorimipramine hydrochloride, 43B, 368
$C_{19}H_{24}INO_4$, Lycorenine methiodide, 38B, 579
$C_{19}H_{24}N_2O_2$, 16,17-Dihydrogoniomine, 46B, 583
$C_{19}H_{25}BrN_2O$, 16-Demethoxycarbonyl-20-epiervatamine hydrobromide, 43B, 744
$C_{19}H_{25}IN_2O_2$, Voaphylline hydroxyindolenine hydroiodide, 35B, 353
$C_{19}H_{25}N_2Br$, Yohimbane hydrobromide, 35B, 354
$C_{19}H_{26}BrNO_2$, (-)-N-Allyl-3,6β-dihydroxymorphinan hydrobromide, 39B, 418
$C_{19}H_{26}ClNO_6$, Naloxone hydrochloride dihydrate, 40B, 288; 41B, 653
$C_{19}H_{26}N_2$, Quebrachamine, 42B, 455
$C_{19}H_{27}NO_6$, Senkirkine, 40B, 529
$C_{19}H_{27}NO_7$, Otosenine, 43B, 745
$C_{19}H_{28}CdCl_4N_2O_3$, Cinchoninium tetrachlorocadmate(II) dihydrate, 44B, 533
$C_{19}H_{30}BrNO$, Gephyrotoxin hydrobromide, 43B, 746

$C_{20}H_{16}ClNO$, 6-Chloro-hyellazole, 45B, 638
$C_{20}H_{17}NO_5$, Imenine, 35B, 356
$C_{20}H_{17}NO_5$, Imerubrine, 43B, 746
$C_{20}H_{17}NO_6$, Bicuculline, 39B, 418
$C_{20}H_{19}NO_5$, Protopine, 33B, 224
$C_{20}H_{20}NO_3Br$, Bromodihydroacronycine, 35B, 357
$C_{20}H_{20}N_2O_2$, Meloscandonine, 44B, 534
$C_{20}H_{20}N_2O_4$, Nareline, 43B, 747
$C_{20}H_{21}NO_6$, (-)-Ophiocarpine N-oxide, 44B, 534
$C_{20}H_{21}N_2O_3$, Catharine derivative, 45B, 638
$C_{20}H_{22}BrNO_5$, Leucoxine hydrobromide, 43B, 743
$C_{20}H_{22}ClNO_4$, Papaverine hydrochloride, 40B, 530
$C_{20}H_{22}INO_4$, 6-Acetyl-1-iodocodeine, 44B, 535
$C_{20}H_{24}INO_4$, Cryptostyline I methiodide, 38B, 579
$C_{20}H_{24}N_2O$, Aristone, 46B, 584
$C_{20}H_{24}N_2O_2$, Akagerine, 41B, 654
$C_{20}H_{24}N_2O_2$, Sceletium alkaloid A_4, 38B, 581
$C_{20}H_{24}N_2O_3$, Strictamine monohydrate, 43B, 748
$C_{20}H_{25}BN_2O_2$, Condylocarpine-borine, 43B, 750
$C_{20}H_{25}BrN_2O_2$, (+)-10-Bromo-10,11-dihydroepiquinidine, 40B, 531
$C_{20}H_{26}BrNO_2$, p-Bromobenzoyldihydroluciduline, 39B, 419
$C_{20}H_{26}BrNO_4$, Cryptostyline II hydrobromide, 39B, 419
$C_{20}H_{26}ClNO$, Normethadone hydrochloride, 42B, 26
$C_{20}H_{26}ClNO_3$, Delnudine hydrochloride, 37B, 321
$C_{20}H_{26}N_2O_2$, Aristotelinine, 44B, 535
$C_{20}H_{26}N_2O_3 \cdot 0.5\ H_2O_4S$, Quinine sulfate monohydrate, 19, 625
$C_{20}H_{26}N_2O_3 \cdot 0.5\ H_2O_4Se$, Quinine selenate monohydrate, 19, 625
$C_{20}H_{27}BrN_2O$, Ibogaine hydrobromide, 24, 711
$C_{20}H_{27}IN_2$, Cleavamine methiodide, 29, 746
$C_{20}H_{27}IN_2O_2$, Hunterburnine α-methiodide, 27, 1008
$C_{20}H_{27}IN_2O_2$, Hunterburnine β-methiodide, 30B, 235
$C_{20}H_{27}N_3O_4$, Calpurmenin 13α-(2'-pyrrolecarboxylic acid) ester, 45B, 639
$C_{20}H_{28}BrNO_3$, (+)-Hetisine hydrobromide, 28, 552
$C_{20}H_{28}ClNO_6S$, Deacetylthiocolchicine hydrochloride dihydrate, 43B, 748
$C_{20}H_{28}INO_6$, Tazettine methiodide, 37B, 322
$C_{20}H_{29}IN_2O$, Deacetyl-aspidospermine N(a)-hydroiodide, 34B, 328
$C_{20}H_{30}BrNO$, (-)-2-Cyclopropylmethyl-2'-hydroxy-5-ethyl-9,9-dimethyl-6,7-benzo-morphan hydrobromide, 45B, 639
$C_{20}H_{32}BrNO$, (-)-5-Ethyl-2'-hydroxy-2-isobutyl-9,9-dimethyl-6,7-benzomorphan hydrobromide, 46B, 584
$C_{20}H_{32}ClN_3O_3$, Dibucaine hydrochloride monohydrate, 43B, 346
$C_{20}H_{34}BrNO_8$, Axillarine hydrobromide ethanol solvate, 42B, 455
$C_{20}H_{34}BrN_3$, Podopetaline hydrobromide, 41B, 654
$C_{20}H_{35}Cl_2N_3O_8$, Panamine diperchlorate, 31B, 245
$C_{20}H_{35}N_2O_7P$, Ephedrine hydrogen phosphate monohydrate, 39B, 435
$C_{20}H_{35}N_3$, 16-Epiormosanine, 42B, 456
$C_{20}H_{36}BrNO_3$, Samandarine hydrobromide methanolate, 26, 798; 27, 1034
$C_{20}H_{36}Br_3N_3 \cdot 1.75\ H_2O$, 6-Epipodopetaline hydrobromide hydrate, 46B, 585
$C_{21}H_{22}INO_6$, N-Methylrhoeagenine iodide, 35B, 359
$C_{21}H_{22}N_2O$, Andranginine, 45B, 640
$C_{21}H_{23}NO_5$, Cryptopine, 33B, 226
$C_{21}H_{23}NO_5$, 7,8-Didehydro-4,5α-epoxy-17-methylmorphinan-3,6α-diol diacetate, 45B, 641
$C_{21}H_{23}NO_6 \cdot C_4H_8O_2 \cdot H_2O$, Colchiceine ethyl acetate water solvate,

45B, 641
$C_{21}H_{23}NO_6$, Polycarpine, 43B, 751
$C_{21}H_{23}NO_7$, Ochrobirine methanolate, 41B, 655
$C_{21}H_{24}N_2O_3$, Pandine, 43B, 752
$C_{21}H_{24}N_2O_4$, Cathovalinine, 42B, 457
$C_{21}H_{25}ClN_2O_4$, Vindolinine hemihydrochloride hemiperchlorate, 42B, 457
$C_{21}H_{25}IN_2$, (-)-Kopsanone N[b]-methiodide, 34B, 330
$C_{21}H_{25}NO_5$, N-Desacetyl-N-methylcolchicine, 40B, 531
$C_{21}H_{25}NO_6S$, Demethylisothiocolchicine hydrate, 43B, 750
$C_{21}H_{25}N_2O$, 19-Epi-N(*a*)-methylvindolininol, 43B, 753
$C_{21}H_{26}BrNO_5$, Capaurine hydrobromide, 34B, 332
$C_{21}H_{26}ClNO_7$, Acutumine acetate, 33B, 223
$C_{21}H_{26}INO_4$, Cularine methiodide, 39B, 420
$C_{21}H_{26}N_2O_3$, Vincamene, 39B, 421
$C_{21}H_{26}N_2O_3$, 11-Ethyl-8,11,11a,12-tetrahydro-6,11-methanoindolo[3,2-b]quinolizine-6(5H)-carboxylic acid methyl ester methanol solvate, 42B, 234
$C_{21}H_{26}N_2O_4$, Oxymetavincadifformine, 44B, 535
$C_{21}H_{26}N_2O_4$, 14R,16S-De(vindolinyl)catharinine, 44B, 536
$C_{21}H_{27}BrN_2O_4$, Strychnine hydrobromide dihydrate, 15, 497
$C_{21}H_{27}ClN_2O_3$, Yohimbine hydrochloride, 39B, 422
$C_{21}H_{27}NO$, L-Methadone, 40B, 44
$C_{21}H_{27}NO$, Methadone, 39B, 422
$C_{21}H_{27}NO_4$, 5,6:12,20-Diepoxy-19,20-methylimino-14,20-cyclo-ent-kaur-16-ene-2,11-diol, 42B, 458
$C_{21}H_{28}BrNO$, D-Methadone hydrobromide, 22, 720
$C_{21}H_{28}Cl_2N_2O_2$, Ajmaline dichloromethane solvate, 44B, 536
$C_{21}H_{28}INO_2$, Spiradine A methiodide, 37B, 323
$C_{21}H_{28}N_2O_2$, Capuronine acetate, 46B, 585
$C_{21}H_{28}N_2O_4$, Iboxyphylline monohydrate, 42B, 458
$C_{21}H_{29}IN_2O_4$, (-)-N-Methyl gelsemicine hydroiodide, 27, 1021
$C_{21}H_{29}NO_8$, Clivorine hydrate, 38B, 581
$C_{21}H_{30}N_2O_9S$, Strychnine sulphonic acid tetrahydrate, 39B, 423
$C_{21}H_{31}ClN_2O_5$, 2-Phenylsparteine N(16)-oxide monoperchlorate, 43B, 753
$C_{21}H_{31}NO_2$, Lycoctonine alkaloid derivative, 42B, 460
$C_{21}H_{31}NO_5$, Oxotuberostemonine, 38B, 583
$C_{21}H_{32}ClNO_6$, Nalbuphine hydrochloride dihydrate, 42B, 459
$C_{21}H_{32}INO_3$, Megastachine methiodide, 45B, 642
$C_{21}H_{34}INO$, Thelepogine methiodide, 28, 553
$C_{21}H_{34}N_2O_{11}S$, Strychnine sulfate pentahydrate, 15, 500
$C_{21}H_{35}N_3$, Homodasycarpine, 44B, 537
$C_{21}H_{35}N_3$, Jamine (optically active form), 44B, 537
$C_{21}H_{35}N_3$, Jamine, 29, 747
$C_{22}H_{17}IN_2O_5$, Camptothecin iodoacetate, 33B, 226
$C_{22}H_{21}NO_2$, Melochinone, 41B, 656
$C_{22}H_{21}NO_{11}$, Narciclasine tetraacetate, 39B, 424
$C_{22}H_{22}BrNO_8$, Bromo-anhydro-N-oxy-nornarceine, 40B, 532
$C_{22}H_{23}NO_6$, 1-Acetyl-6-O-acetyl-laurelliptine, 44B, 538
$C_{22}H_{24}INO_4$, Ochotensine methiodide, 31B, 249
$C_{22}H_{25}NO_7$, (-)-10,11-Oxy-10,12a-cyclo-10,11-secocolchicine, 46B, 585
$C_{22}H_{26}BrNO$, Piroheptine metabolite hydrobromide, 41B, 657
$C_{22}H_{27}IN_2O_3$, Macusine-A iodide, 28, 558
$C_{22}H_{27}NO_4$, (+)-Coralydin, 42B, 461
$C_{22}H_{27}NO_4$, (+)-O-Methylcorytenchirin, 42B, 461
$C_{22}H_{28}Br_2N_4$, Chimonanthine dihydrobromide, 30B, 232

$C_{22}H_{28}Br_2N_4O_2$, Calycanthine dihydrobromide dihydrate, 27, 1013
$C_{22}H_{29}BrN_2O_2$, (+)-Meloscin N[b]-methobromide ethanolate, 40B, 1096
$C_{22}H_{29}BrN_2O_5$, Corymine hydrobromide monohydrate, 30B, 233
$C_{22}H_{29}ClINO$, Clemastine methiodide, 42B, 475
$C_{22}H_{29}IN_2O$, Isoaffinisine methiodide, 43B, 754
$C_{22}H_{29}IN_2O_4$, Akuammidine methiodide monohydrate, 28, 557
$C_{22}H_{29}IN_2O_4$, Echitamine iodide, 27, 1011
$C_{22}H_{29}NO_7$, Nokoensine methanolate, 43B, 755
$C_{22}H_{30}N_2O_4$, Ervatamine methanol solvate, 39B, 424; 40B, 533
$C_{22}H_{30}N_2O_5$, Powerine methanolate, 43B, 755
$C_{22}H_{31}BrN_2O_4$, Vincamene hydrobromide methanolate, 39B, 421
$C_{22}H_{31}NO_5S$, Benztropine methanesulphonate monohydrate, 44B, 538
$C_{22}H_{31}NO_6$, Icacine, 43B, 718
$C_{22}H_{32}INO_3$, Crepidine methiodide, 37B, 324
$C_{22}H_{32}INO_6$, (-)-Argemonine methiodide dihydrate, 42B, 460
$C_{22}H_{32}IN_2O$, N(a)-Acetyl-7-ethyl-5-desethyl-aspidospermidine N(b)-methiodide, 30B, 229
$C_{22}H_{33}BrN_2O_2$, Astrocasine methobromide methanol solvate, 42B, 461
$C_{22}H_{33}NO_2$, Denudatine, 37B, 325
$C_{22}H_{33}NO_3$, Atidine, 45B, 642
$C_{22}H_{33}NO_8$, Parsonsine (form I), 45B, 643
$C_{22}H_{33}NO_8$, Parsonsine (form II), 46B, 586
$C_{22}H_{33}NO_{11}$, Bulgarsenine (R:R)-(+)-bitartrate, 46B, 587
$C_{22}H_{34}BrNO_2$, 4-Methyl-6-(2'-benzoylpentyl)-quinolizidine hydrobromide, 39B, 425
$C_{22}H_{34}BrNO_5$, N-Butylhyoscine bromide methanol solvate, 44B, 538
$C_{22}H_{36}BrNO_4$, Samandaridine hydrobromide methanol solvate, 28, 555
$C_{22}H_{36}BrNO_6$, Heteratisine hydrobromide monohydrate, 30B, 234
$C_{22}H_{37}NO_{11}S$, Thiocolchicine hexahydrate, 43B, 749
$C_{22}H_{38}BrNO_3$, Himbacine hydrobromide monohydrate, 27, 1003
$C_{22}H_{38}ClNO_5$, N-α-Oxyconanine perchlorate, 43B, 756
$C_{22}H_{42}I_2N_2$, 4,17a-Dimethyl-4,17a-diaza-D-homo-5α-androstane dimethiodide, 46B, 589
$C_{23}H_{22}BrNO_6$, (+)-14-Epicorynolin bromoacetate, 45B, 644
$C_{23}H_{23}NO_7$, N,O-Diacetyl-4-hydroxynornantenine, 45B, 645
$C_{23}H_{24}BrN_3O_2$, 8β-(5-Bromonicotinoyloxymethyl)-1,6-dimethyl-10α-ergoline, 46B, 587
$C_{23}H_{24}N_2O_8$, N-Cyano-1-hydroxy-1,2-seconarcotine, 45B, 645
$C_{23}H_{28}INO_5$, Acetyl dienone II methiodide, 44B, 539
$C_{23}H_{31}BrN_3O_2 \cdot 1.5\ H_2O$, Oxypertine hydrobromide sesquihydrate, 44B, 539
$C_{23}H_{31}IN_2O_2$, 3,4-Seco-3,5-cyclo-1,2-dihydro-N^1,N^4-dimethylstrictamine methiodide, 45B, 646
$C_{23}H_{31}IN_2O_4$, Mitragynine hydroiodide, 30B, 237
$C_{23}H_{33}BrN_2O_5$, Echitamine bromide methanolate, 27, 1010
$C_{23}H_{33}IN_2O$, 1-Acetyl-3-methylaspidospermidine 9-methiodide, 35B, 389
$C_{23}H_{33}IN_2O_2$, (-)-Aspidospermine N(b)-methiodide, 24, 714
$C_{23}H_{34}BrNO_8$, Miyaconitine hydrobromide dihydrate, 37B, 326
$C_{23}H_{35}NO_2$, Daphnilactone A, 38B, 585
$C_{23}H_{35}NO_6$, Gadesine, 45B, 646
$C_{23}H_{38}BrNO_6$, Lappaconine hydrobromide, 34B, 330; 35B, 361
$C_{23}H_{44}I_2N_2$, 4-Methyl-17β-dimethylamino-4-aza-5α-androstane dimethiodide, 46B, 589
$C_{24}H_{27}N_3O_2 \cdot H_2O$, 8β-[(Benzyloxycarbonyl)aminomethyl]-6-methyl-10α-ergoline monohydrate, 46B, 588
$C_{24}H_{30}BrNO$, Dexclamol hydrobromide, 42B, 462
$C_{24}H_{31}IN_2O_6$, Rauvoxinine methiodide, 33B, 228

$C_{24}H_{32}INO_6$, Colchicum cornigerum alkaloid (methiodide derivative), 39B, 425
$C_{24}H_{32}N_2O_5$, 18(17-16)abeo-Yohimbine acetate methanol solvate, 43B, 757
$C_{24}H_{35}NO_6$, Chasmanine intermediate, 46B, 588
$C_{24}H_{36}INO_4$, Lucidusculine hydroiodide, 31B, 254
$C_{24}H_{42}INO_7$, Des(oxymethylene)-lycoctonine hydroiodide monohydrate, 26, 795
$C_{24}H_{42}INO_{11}$, (+)-Demethanolaconinone hydroiodide trihydrate, 26, 793; 29, 749
$C_{25}H_{29}N_3O_2$, 1,6-Dimethyl-8β-((benzyloxycarbonyl)aminomethyl)-10α-ergoline, 43B, 758
$C_{25}H_{30}INO_3$, DL-iso-Cryptopleurine methiodide, 19, 620
$C_{25}H_{31}NO_3$, 8,14-But-1-eno-18-ethyl-7,8-dihydro-19-methylcodeinone, 45B, 647
$C_{25}H_{31}NO_6$, Norpropoxyphene maleate, 43B, 58
$C_{25}H_{32}BrNO$, Butaclamol hydrobromide, 42B, 462
$C_{25}H_{34}BrN_3O_5$, Lunarine hydrobromide monohydrate, 30B, 236
$C_{25}H_{34}IN_3O_5$, Lunarine hydroiodide monohydrate, 30B, 236
$C_{25}H_{34}NO_2$, Daphnilactone B, 39B, 426
$C_{25}H_{38}BrNO_3$, Methyl N-bromoacetylhomosecodaphniphyllate, 37B, 326
$C_{25}H_{40}INO_6$, Condelphine hydroiodide, 43B, 759
$C_{25}H_{44}Br_2N_2$, Buxenine-G dihydrobromide, 37B, 327
$C_{25}H_{44}I_2N_2$, Buxenine-G dihydroiodide, 37B, 327
$C_{26}H_{27}NO_5Se$, N-Carbethoxy-7-phenylselenonordihydrocodeinone, 43B, 760
$C_{26}H_{28}BrNO_7S$, 4-Demethylhasubanonine p-bromobenzenesulphonate, 38B, 585
$C_{26}H_{28}BrN_3O_9$, Erythristemine 2-bromo-4,6-dinitrophenolate, 42B, 463
$C_{26}H_{29}NO_8$, Colchicine O,N-diacetate, 45B, 647
$C_{26}H_{30}IN_3O_6$, Aci-ergotamine p-iodobenzoylamide (ethanol solvate), 31B, 250
$C_{26}H_{32}BrNO_4$, Lythrumine hydrobromide, 39B, 426
$C_{26}H_{32}BrNO_5$, Vertaline hydrobromide, 37B, 328
$C_{26}H_{33}N_3O_5 \cdot CH_4O$, Cyclopiamine B methanol solvate, 45B, 648
$C_{26}H_{42}N_2O_2$, Diethylaminodihydroveatchinone, 46B, 589
$C_{26}H_{46}I_2N_2$, 17a-Methyl-3β-pyrrolidino-17a-aza-D-homo-5-androstene dimethiodide, 46B, 589
$C_{27}H_{26}BrNO_2$, 2,9-Dimethyl-5-phenyl-6,7-benzomorphan-3'-p-bromobenzoate, 44B, 540
$C_{27}H_{26}BrNO_6$, Capaurimine mono-p-bromobenzoate, 37B, 329
$C_{27}H_{30}INO_6$, Diacetyltylophorinidine methiodide, 42B, 463
$C_{27}H_{31}NO_3$, Paspalicine, 46B, 591
$C_{27}H_{31}NO_4$, Paspalinine, 46B, 590
$C_{27}H_{38}BrNO_7$, Yuzurimine hydrobromide, 31B, 255
$C_{27}H_{46}BrNO_2$, Tomatidine hydrobromide, 32B, 258
$C_{27}H_{46}ClNO_2$, 25-Isosolafloridine hydrochloride, 43B, 761
$C_{27}H_{46}INO \cdot 0.5\ C_2H_6O$, Demissidin hydroiodide hemiethanolate, 38B, 586
$C_{27}H_{46}INO_2$, Tomatidine hydroiodide, 32B, 260
$C_{27}H_{48}INO_{10}$, Cevine hydroiodide, 24, 722
$C_{28}H_{18}BrNO_6$, Corynoline p-bromobenzoate, 39B, 427
$C_{28}H_{28}N_4O_7$, Nortryptoquivaline, 45B, 649
$C_{28}H_{36}BrNO_4$, 7α-(1-(R)-Hydroxy-1-methylbutyl)-6,14-endo-ethenotetrahydrothebaine hydrobromide, 32B, 260
$C_{28}H_{36}BrNO_6$, O-Methyl-lythrine hydrobromide methanol solvate, 30B, 237

$C_{28}H_{38}Br_2N_4O_4 \cdot H_2O$, Ephedradine A dihydrobromide monohydrate, 45B, 650
$C_{28}H_{39}NO \cdot C_4H_8O_2$, Alflavinine ethyl acetate solvate, 46B, 590
$C_{28}H_{39}NO_2 \cdot 0.5\ CH_4O$, Paspaline methanol solvate, 46B, 591
$C_{28}H_{44}ClNO_8$, Delphisine hydrochloride, 42B, 464
$C_{28}H_{46}BrNO_3$, Verticinone methyl bromide, 35B, 394
$C_{28}H_{48}INO_3$, Veralkamine hydroiodide methanol solvate, 33B, 228
$C_{29}H_{34}Br_2N_4O_2$, Usambarensine dihydrobromide dihydrate, 41B, 658
$C_{29}H_{42}BrNO_{12}S$, Retusamine α'-bromo-D-camphor-trans-π-sulphonate monohydrate, 32B, 256
$C_{29}H_{45}NO_4$, O-Acetylveramarine, 45B, 650
$C_{30}H_{37}NO_5$, 3-Methyl-mono-o-benzylautumnaline, 43B, 761
$C_{30}H_{38}N_2O_4$, Lophocine, 46B, 591
$C_{30}H_{39}NO_5$, Kodo-cytochalasin-1, 43B, 762
$C_{30}H_{39}NO_5$, Paxilline acetone solvate, 41B, 659
$C_{30}H_{42}N_4O_2$, Homaline, 42B, 465
$C_{30}H_{43}Br_2NO_6$, Bromolythranine hydrobromide ethanol solvate, 38B, 586
$C_{30}H_{44}Br_2N_2O_2$, Isoalfileramine dihydrobromide, 45B, 651
$C_{30}H_{48}Br_2N_2O_4S$, Thiobinupharidine dihydrobromide dihydrate, 39B, 428
$C_{30}H_{51}NO_6$, Solaphyllidine, 35B, 362
$C_{30}H_{52}Br_2N_2O_6S$, neothiobinupharidine dihydrobromide tetrahydrate, 32B, 261
$C_{30}H_{55}Cl_3N_4O_{14}$, Episparteine N(16)-oxide sesquiperchlorate, 43B, 763
$C_{30}H_{55}Cl_3N_4O_{14}$, Sparteine-N(16) oxide sesquiperchlorate, 34B, 322
$C_{31}H_{38}BrNO_6$, Batrachotoxinin A O-p-bromobenzoate, 34B, 334
$C_{31}H_{44}INO_{14}$, Maytoline methiodide methanolate, 37B, 330
$C_{32}H_{38}N_2O_6$, Chaetoglobosin A monohydrate, 44B, 540
$C_{32}H_{41}N_3O_7$, Fumitremorgin A, 41B, 660
$C_{32}H_{46}ClNO_7$, Chasmanine 14-α-benzoate hydrochloride, 43B, 764
$C_{32}H_{50}BrNO_5$, Daphniphylline hydrobromide, 31B, 257
$C_{32}H_{52}INO_3$, Alkaloid II methiodide (from daphniphyllum macropodum), 31B, 258
$C_{33}H_{40}N_2O_9$, Reserpine, 33B, 230
$C_{33}H_{41}NO_8$, Lythrancine-IV, 40B, 535
$C_{33}H_{44}BrN_5O_8S \cdot 0.5\ C_3H_8O$, Bromocriptine methanesulphonate hemi-isopropanol solvate, 45B, 651
$C_{33}H_{44}N_4O_4$, Isocinchophyllamine methanol solvate, 40B, 534
$C_{33}H_{48}Br_2N_2O_4$, Dendocrepine hydrobromide monohydrate, 39B, 429
$C_{34}H_{32}BrN_5O_7$, Tryptoquivaline-p-bromophenylurethane, 41B, 661
$C_{34}H_{40}N_2O_5 \cdot C_2H_6O$, Chaetoglobosin K acetone solvate, 46B, 592
$C_{34}H_{41}N_5O_8S \cdot H_2O$, (-)-Dihydroergotamine methanesulphonate monohydrate, 45B, 652
$C_{34}H_{46}N_4O_4$, Strychnofoline - ethanol (1:2), 43B, 764
$C_{35}H_{40}BrNO_8S$, Lythrancine-II O-p-bromobenzenesulphonate, 40B, 534
$C_{35}H_{43}N_5O \cdot 3.5\ H_2O$, Strychnopentamine hydrate, 43B, 765
$C_{35}H_{50}ClNO_{16}$, Jesaconitine perchlorate, 45B, 653
$C_{36}H_{44}BrNO_{18}$, Bromoacetylneoevonine monohydrate, 38B, 587
$C_{36}H_{58}INO_5$, Daphmacrine methiodide acetonate, 34B, 333
$C_{36}H_{60}Cl_6N_2O_{14}Pt$, Alkaloid chloroplatinate dihydrate (from senecio kirkii), 31B, 243
$C_{37}H_{38}N_2O_6$, Methylwarifteine, 44B, 541
$C_{37}H_{42}Br_2N_4O_7$, (+)-Haplophytine dihydrobromide, 35B, 365
$C_{37}H_{52}Cl_2N_2O_{11}$, (+)-Tubocurarine dichloride pentahydrate, 39B, 429
$C_{38}H_{40}N_2O_6$, Dimethylwarifteine, 44B, 541
$C_{38}H_{42}N_2O_6$, Tetrandrine, 42B, 465
$C_{38}H_{43}BrN_2O_7$, Dihydro-O-methylcancentrinemethine hydrobromide, 38B, 588

$C_{38}H_{44}N_4O_8$, Haplophytine methanol solvate, 42B, 466
$C_{40}H_{44}I_2N_4O_2$, Caracurine-II dimethiodide, 30B, 230
$C_{40}H_{48}N_4O_3 \cdot 2\ H_2O$, Geissospermine dihydrate, 45B, 653
$C_{41}H_{50}I_2N_4O_2$, Anhydroisocalebassin methyl ether diiodide monohydrate, 34B, 333
$C_{41}H_{58}Br_2N_2O_{10}$, (+)-Tubocurarine dibromide methanol solvate, 42B, 466
$C_{42}H_{42}N_4O_2 \cdot C_3H_6O$, Sungucine acetone solvate, 46B, 592
$C_{42}H_{52}N_4O_5$, Villalstonine methanol solvate, 30B, 241
$C_{43}H_{48}Br_2N_4O_6$, Dibromovobtusine, 39B, 430
$C_{43}H_{60}N_2O_2$, Staphisine, 46B, 593
$C_{44}H_{38}Cl_4HgN_4O_8$, Perloline mercurichloride hydrate, 31B, 246
$C_{44}H_{48}Br_2O_6$, Pristimerol bis-p-bromobenzoate, 38B, 575
$C_{44}H_{48}N_6O_6$, Ditryptophenaline methanolate, 43B, 766
$C_{44}H_{59}Br_2N_3O_3$, N,O-Di-p-bromobenzoyltetrahydrodeoxyoxolucidine B, 45B, 654
$C_{47}H_{63}IN_4O_{11}$, Leurocristine methiodide dihydrate, 31B, 258
$C_{49}H_{60}N_4O_{11}$, Catharine acetone solvate, 42B, 467

59. MISCELLANEOUS NATURAL PRODUCTS

$C_5H_{14}ClNO$, Choline chloride, 24, 559; 37B, 331
$C_6H_{15}ClN_2O_2$, Carbamoylcholine chloride, 41B, 7
$C_6H_{15}IN_2O_2$, Carbamoylcholine iodide, 41B, 7
$C_7H_{16}BrNO_2$, Acetylcholine bromide, 23, 599; 41B, 8
$C_7H_{16}ClNO_2$, Acetylcholine chloride, 35B, 321
$C_7H_{16}ClNO_6$, Acetylcholine perchlorate, 40B, 28
$C_7H_{16}INO_2$, Acetylcholine iodide, 43B, 37
$C_7H_{16}INO_3$, Methoxycarbonylcholine iodide, 45B, 9
$C_7H_{22}Cl_3N_3$, Spermidine trihydrochloride, 31B, 260
$C_8H_8BrNO_4S$, 7-Deamido-3-bromo-4-hydroxy-3,4-dihydrocephalosporin lactone, 43B, 475
$C_8H_8O_4$, Griffonilide, 42B, 468
$C_8H_{12}ClNO_2$, Dopamine hydrochloride, 33B, 231
$C_8H_{12}ClNO_3$, Noradrenaline hydrochloride, 32B, 263
$C_8H_{12}ClNO_3$, Pyridoxonium chloride, 33B, 232
$C_8H_{12}ClNO_3$, 5-Hydroxydopamine hydrochloride, 38B, 145
$C_8H_{12}ClNO_3$, 6-Hydroxydopamine hydrochloride, 38B, 146
$C_8H_{12}O_7$, 3,4-Dihydroxy-5-(2'-methoxycarbonyl-2'-hydroxyethyl)-tetrahydrofuran-2-one, 45B, 654
$C_8H_{18}BrNO_2$, Acetylhomocholine bromide, 39B, 431
$C_8H_{18}INO_3$, DL-Lactoylcholine iodide, 42B, 18; 43B, 39
$C_8H_{21}Br_2N_5O_5$, Viocidic acid dihydrobromide, 35B, 366
$C_9H_{14}BrNO_2$, Epinine hydrobromide, 42B, 19
$C_9H_{14}ClNO_4$, Adrenalone hydrochloride monohydrate, 37B, 331
$C_9H_{20}CrN_7OS_4$, Choline reineckate, 21, 406
$C_9H_{20}INO_2$, erythro-α,β-Dimethylacetylcholine iodide, 35B, 322
$C_9H_{20}INO_2$, threo-α,β-Dimethylacetylcholine iodide, 35B, 322
$C_9H_{22}I_2N_2O_2$, γ-Aminobutyric acid choline ester di-iodide, 42B, 20
$C_{10}H_8O_4$, 6-Methoxy-2-methyl-3,5-dihydrobenzo[b]furan-4,7-dione, 46B, 593
$C_{10}H_{10}Cl_2O_4$, 2-trans-Allyl-3,5-dichloro-1-hydroxy-4-oxo-(1S,5S)-2-cyclopentene-1-carboxylic acid methyl ester, 34B, 339
$C_{10}H_{12}O_4$, Cantharidin, 43B, 766
$C_{10}H_{12}O_5$, Asperlin, 44B, 542
$C_{10}H_{13}BrO_4$, Jacobine bromodilactone, 28, 560

$C_{10}H_{13}ClO_5$, Chloroasperlin, 44B, 542
$C_{10}H_{14}O_6$, Asperlinol, 44B, 542
$C_{10}H_{16}N_2O_3S$, d-(+)-Biotin, 42B, 468
$C_{10}H_{17}NO_5$, Kainic acid monohydrate, 22, 794
$C_{10}H_{18}N_2O_3$, d,l-Dethiobiotin, 42B, 469
$C_{10}H_{24}NO_5P$, Cyclopentylphosphorylcholine monohydrate, 44B, 542
$C_{10}H_{30}Cl_4N_4$, Spermine tetrahydrochloride, 31B, 261
$C_{11}H_{10}Br_2O_5$, Coarctatin dibromide, 41B, 662
$C_{11}H_{14}O_5$, 3,6-Dimethyl-4,10-dihydroxy-2-oxaspiro[4.5]dec-7-en-1,9-dione, 46B, 594
$C_{11}H_{15}Br_2N_3O_7$, Bromoanhydrotetrodoic lactone hydrobromide, 35B, 367
$C_{11}H_{16}BrCl_2NO$, o-Dichlorophenylcholine ether bromide, 42B, 21
$C_{11}H_{18}BrNO_3$, Mescaline hydrobromide, 39B, 431
$C_{11}H_{18}BrN_3O_8$, Tetrodotoxin hydrobromide, 35B, 368
$C_{11}H_{18}ClNO_3$, Mescaline hydrochloride, 39B, 432
$C_{11}H_{18}O_7$, Pederinolactone, 39B, 432
$C_{12}H_2Cl_6O_2$, 1,2,3,7,8,9-Hexachlorodibenzo-p-dioxin, 34B, 343
$C_{12}H_{20}BrNO \cdot 0.5\ H_2O$, β-Methylphenylcholine ether bromide hydrate, 40B, 34
$C_{12}H_{20}BrNO$, o-Methylphenylcholine ether bromide, 40B, 33
$C_{13}H_{12}O_2$, Goniothalamin, 41B, 662
$C_{13}H_{12}O_5$, 4-Methoxy-6-(2,4-dihydroxy-6-methylphenyl)-2-pyrone, 44B, 543
$C_{13}H_{19}NO_9$, (-)-Adrenaline hydrogen (+)-tartrate, 39B, 433
$C_{13}H_{20}ClN_5O_3$, 2-Chloro-zoanthoxanthin trihydrate, 40B, 535
$C_{13}H_{22}BrNO$, o-Methyl β-methylphenylcholine ether bromide, 40B, 37
$C_{13}H_{26}N_2O_8$, γ-Aminobutyric acid choline ester (DL)-tartrate, 42B, 20
$C_{14}H_6O_8$, Ellagic acid, 33B, 232
$C_{14}H_{12}O_3$, Amyrolin, 35B, 369
$C_{14}H_{14}Cl_2O_4$, Chloromycorrhizin A, 43B, 768
$C_{14}H_{15}BrO_3$, 9β-Bromofraxinellone, 35B, 371
$C_{14}H_{15}ClO_5$, Mikrolin, 42B, 469
$C_{14}H_{18}N_2O_3$, Bohemamine, 46B, 594
$C_{14}H_{20}BrNO_3$, (DL)-trans-2-Cyclobutylamino-5,6-dihydroxy-1,2,3,4-tetrahydro-1-naphthalenol hydrobromide, 46B, 595
$C_{14}H_{20}N_4O_7 \cdot H_2O$, Nitroleonurine monohydrate, 45B, 655
$C_{14}H_{20}O_8$, (-)-1-O-Acetylxylomollin, 44B, 543
$C_{14}H_{21}NO_6$, Acetylcholine β-resorcylate, 41B, 19
$C_{14}H_{30}Cl_2N_2O_{12}$, Succinylcholine perchlorate, 37B, 332
$C_{14}H_{30}I_2N_2O_4$, Succinylcholine iodide, 35B, 329
$C_{14}H_{32}Br_4NO_6U$, Bis(acetylcholine) tetrabromodioxouranate(VI), 40B, 38
$C_{14}H_{34}Cl_2N_2O_6$, Succinylcholine dichloride dihydrate, 42B, 22
$C_{14}H_{59}N_6O_{18}P_3$, Spermidine phosphate trihydrate, 34B, 336
$C_{15}H_{10}O_5$, Genisteine, 41B, 663
$C_{15}H_{12}O_5$, Rubrofusarin, 27, 1005
$C_{15}H_{14}O_6$, Protoaphin-fb quinone A, 43B, 403
$C_{15}H_{15}BrO_5$, (-)-trans-3-Bromokhellactone methyl ether, 34B, 338
$C_{15}H_{15}BrO_6$, α-Bromoisotutinone, 30B, 245
$C_{15}H_{15}BrO_6$, α-Bromopicrotoxinin, 27, 1017
$C_{15}H_{15}BrO_6$, β-Bromopicrotoxinin, 33B, 233
$C_{15}H_{15}O_4$, Germichrysone, 44B, 544
$C_{15}H_{17}BrO_5$, α-Bromoisotutin, 29, 745
$C_{15}H_{18}O_5$, O-Methylasparvenone monoacetate, 41B, 664
$C_{15}H_{19}BrO_3$, 3-Bromoanhydrodehydrodihydropulchellin, 35B, 372
$C_{15}H_{20}N_2O$, Verruculotoxin, 42B, 470
$C_{15}H_{20}O_4$, Phomenone, 41B, 665

$C_{15}H_{22}O_4$, (-)-Avenaciolide, 44B, 544
$C_{15}H_{25}ClN_2O_4$, Anhydro-N-hydroxymethyldeoxyangustifoline perchlorate, 32B, 264
$C_{16}H_{14}Br_2O_6$, D,L-Methyl 5,7-dibromo-2'α,5'α-epoxy-4-methoxy-3-oxo-grisane-6'β-carboxylate, 42B, 471
$C_{16}H_{14}O_5$, 9-Methylrubrofusarin, 43B, 768
$C_{16}H_{14}O_6$, 2-Deacylusnic acid, 40B, 536
$C_{16}H_{16}O_3$, Cunaniol acetate, 45B, 655
$C_{16}H_{16}O_5N_2S$, Phenoxymethyl-Δ^2-desacetoxyl cephalosporin, 35B, 384
$C_{16}H_{20}O_3$, 8-Methoxycarbonyl-6-hydroxy-2,5-dimethyl-2a,3,4,5-tetra-hydroacenaphthene, 46B, 596
$C_{16}H_{22}O_3$, Versiol, 44B, 544
$C_{17}H_9O_6$, Sterigmatin, 41B, 665
$C_{17}H_{10}O_6$, Demethylsterigmatocystin, 45B, 656
$C_{17}H_{12}O_6$, Aflatoxin B_1, 35B, 374
$C_{17}H_{14}O_6$, Aflatoxin B_2, 35B, 374
$C_{17}H_{14}O_7$, Dehydrofomentariol, 42B, 471
$C_{17}H_{16}O_6$, DL-Byak-angelicol, 42B, 247
$C_{17}H_{16}O_7$, (+)-Pisatin monohydrate, 44B, 545
$C_{17}H_{16}O_7$, 2-Hydroxy-1,3,4,7-tetramethoxyxanthone, 34B, 338
$C_{17}H_{18}O_7$, Athrotaxin monohydrate, 38B, 589
$C_{17}H_{19}NO_3$, Piperine, 40B, 537
$C_{17}H_{20}O_6$, PR Toxin, 46B, 596
$C_{17}H_{23}BrO_3$, Laurencin, 34B, 340
$C_{17}H_{25}O_4Br$, O-(Bromoacetyl)tetrahydrodouglanine, 35B, 380
$C_{18}H_8O_5$, Gnidicoumarin, 41B, 666
$C_{18}H_{10}O_6$, 6-Deoxyversicolorin A, 45B, 657
$C_{18}H_{10}O_7$, Versicoloron A, 45B, 657
$C_{18}H_{12}O_6$, Sterigmatocystin, 42B, 471; 44B, 553
$C_{18}H_{12}O_7$, Versicolorin C, 41B, 667
$C_{18}H_{13}Cl_3O_6$, Aflatoxin B_1 - chloroform, 35B, 374
$C_{18}H_{16}O_6$, 2,7-Dihydroxy-5-methyl-[3,4-d](2',3'-dihydro-2',3',3'-tri-methylfurano)naphthalic-1,8-anhydride, 46B, 597
$C_{18}H_{17}BrO_3$, Thujic acid p-bromophenacyl ester, 31B, 262
$C_{18}H_{18}N_2O_7S_2$, Epicorazine A monohydrate, 43B, 769
$C_{18}H_{18}O_6$, Samaderine A, 44B, 546
$C_{18}H_{18}O_7$, Senepoxide, 42B, 472
$C_{18}H_{19}IO_8$, Crotepoxide iodohydrin, 34B, 341
$C_{18}H_{20}N_8O_{16}$, Carbamoylcholine picrate-picric acid complex, 41B, 7
$C_{18}H_{21}NO_4$, Cephalotaxine, 42B, 472
$C_{18}H_{22}O_9$, Olguine, 45B, 657
$C_{18}H_{28}N_2O_4S$, Amphetamine sulfate, 37B, 333
$C_{18}H_{28}O_4$, Diplodiatoxin, 43B, 769
$C_{19}H_{14}O_6$, O-Methylsterigmatocystin, 42B, 471
$C_{19}H_{14}O_7$, 5-Methoxysterigmatocystin, 41B, 668
$C_{19}H_{16}O_5$, 8-(3',4'-Dimethoxyphenyl)-2-methoxynaphtho-1,4-quinone, 45B, 665
$C_{19}H_{18}O_4$, Harringtonolide, 45B, 658
$C_{19}H_{18}O_5$, 7-[3-(4,5-Dihydro-5,5-dimethyl-4-oxo-2-furanyl)-2-butenyl]-oxy-(2H-1-benzopyran-2-one), 45B, 658
$C_{19}H_{18}O_6$, Celebixanthone, 28, 562
$C_{19}H_{18}O_6$, 7-Oxo-deacetamidoisocolchiceine 7-oxo-deacetamidocolchi-ceine, 44B, 157
$C_{19}H_{19}BrO_5$, Bromobruceol, 28, 563
$C_{19}H_{20}ClN_3O_5S_2$, Cephaloridine hydrochloride monohydrate, 35B, 384
$C_{19}H_{20}O_5$, Vafzelin, 46B, 597
$C_{19}H_{20}O_5$, Cudranone, 43B, 770

$C_{19}H_{22}O_5$, Desmethyl-marrubiaketone, 44B, 546
$C_{19}H_{22}O_6$, Strigol, 39B, 434
$C_{19}H_{24}Br_2O_6$, Fukinolidol bisbromoacetate, 35B, 386
$C_{19}H_{24}O_6$, (4R*,8S*,11R*,13S*,14R*)-8,11-Epoxy-14-hydroxy-11-methyl-4-(1-methylvinyl)-6,9-dioxocyclotetradec-1-ene-1,13-carbolactone, 45B, 659
$C_{19}H_{24}O_6$, Eremantholide A, 41B, 670
$C_{19}H_{26}O_6$, (±)-6-epi-Eriolanin, 45B, 659
$C_{20}H_{16}O_6$, Viridin, 38B, 589
$C_{20}H_{18}O_4$, (-)-Phaseollin, 43B, 771; 44B, 547
$C_{20}H_{20}O_5$, Utahin, 42B, 473
$C_{20}H_{21}BrO_6$, Bromomiroestrol, 24, 718
$C_{20}H_{22}Br_2O_5$, Dibromoeriostoic acid, 32B, 265
$C_{20}H_{22}O_6$, 3-Dehydrogibberellin A_3 methyl ester, 46B, 598
$C_{20}H_{22}O_7$, Gnidifolin, 44B, 547
$C_{20}H_{23}ClN_2O_2$, Gelsemine hydrochloride, 23, 727
$C_{20}H_{23}IN_2O_4$, N-p-Iodobenzyl-3-n-octanamidopyridine-2,5,6-trione, 46B, 599
$C_{20}H_{25}N_3O_9S$, Cephaloglycine acetic acid hydrate, 35B, 384
$C_{20}H_{26}Br_2O$, Dibromodehydrobispulegone, 34B, 345
$C_{20}H_{26}Br_2O_3$, Jathrophone dihydrobromide, 37B, 334
$C_{20}H_{26}ClNO_2$, Adiphenine hydrochloride, 39B, 434
$C_{20}H_{26}O_4$, Peunicin, 46B, 599
$C_{20}H_{26}O_5$, Gibberellin A_4 methyl ester, 45B, 660
$C_{20}H_{26}O_{10}$, Anamarine, 45B, 661
$C_{20}H_{27}BrO_6$, Gibberellin bromohydrin, 45B, 661
$C_{20}H_{28}ClNO_2S$, Thiphenamil hydrochloride, 40B, 49
$C_{20}H_{28}O_8$, Soulameanone, 46B, 600
$C_{20}H_{32}O_3$, Asperdiol A, 43B, 771
$C_{20}H_{32}O_4$, Prostaglandin A_1 (monoclinic form), 41B, 672
$C_{20}H_{32}O_4$, Prostaglandin A_1 (orthorhombic form), 41B, 672
$C_{20}H_{32}O_4$, 16-O-Demethyl-19-deoxydideacetyl-3-epifusicoccin aglycone, 41B, 671
$C_{20}H_{32}O_5$, (5Z,11α,13E,15S)-11,15-Dihydroxy-9-oxoprosta-5,13-dien-1-oic acid, 46B, 600
$C_{20}H_{34}O_5$, (5Z,9β,11α,13E,15S)-9,11,15-Trihydroxyprosta-5,13-dien-1-oic acid, 46B, 600
$C_{20}H_{34}O_5$, Prostaglandin E_1, 43B, 772
$C_{21}H_{18}O_6$, Pipoxide, 45B, 662
$C_{21}H_{19}BrN_2O_2$, Dunnione p-bromophenylhydrazone, 46B, 601
$C_{21}H_{26}BrN_3O_4$, 5-Bromo-12S-tetrahydroaustamide, 39B, 436
$C_{21}H_{26}O_5$, Porosin, 44B, 548
$C_{21}H_{28}Br_2O_2 \cdot 0.5\ C_6H_{14}$, Dibromocannabicyclol n-hexane solvate, 40B, 538
$C_{21}H_{28}O_7$, 6-Hydroxypicrasin B, 43B, 772
$C_{21}H_{28}O_8$, Woodhousin, 44B, 548
$C_{21}H_{34}O_3$, Methyl grindelate, 39B, 436
$C_{21}H_{34}O_5$, Cyclo-fusicoccin deacetylaglycone, 45B, 663
$C_{21}H_{34}O_{11} \cdot H_2O$, Patrinoside monohydrate, 45B, 663
$C_{22}H_{19}BrO_6$, Pillaronone monobromoacetate, 35B, 388
$C_{22}H_{21}BrO_3$, p-Bromobenzoylepicatalponol, 46B, 601
$C_{22}H_{22}O_7$, 5'-Demethoxy-β-pectatin A methyl ether, 38B, 590
$C_{22}H_{22}O_8$, Isosteganol, 44B, 548
$C_{22}H_{24}Cl_2N_4O_8$, Aureomycin hydrochloride, 23, 724
$C_{22}H_{25}Br_2NO_8$, 12S,13S-Dibromopseurotin, 42B, 473
$C_{22}H_{25}Cl_4FeO_6$, Atrovenetin orange trimethyl ether ferrichloride, 30B, 242

$C_{22}H_{25}NO_6$, Isocolchicine, 44B, 549
$C_{22}H_{26}ClNO_4$, N-Methylsclerotioramine, 42B, 474
$C_{22}H_{26}O_8$, (-)-Syringaresinol, 42B, 474
$C_{22}H_{27}BrO_4$, Hydroxypelenolide p-bromobenzoate, 35B, 389
$C_{22}H_{28}Br_2O_4$, Nordihydroquaiaretic acid, 34B, 343
$C_{22}H_{28}N_2O_9S_2$, Sirodesmin PL monoacetate, 43B, 773
$C_{22}H_{28}O_6$, rel-(2S,3S,3aS,5S)-3a-Allyl-5-methoxy-3-methyl-2-(3',4',5'-trimethoxyphenyl)-2,3,3a,4,5,6-hexahydro-6-oxobenzofuran, 46B, 602
$C_{22}H_{29}ClO_7$, Dihydroheliangine monochloroacetate, 31B, 264
$C_{22}H_{29}NO_8$, Colchicine dihydrate, 44B, 550
$C_{22}H_{30}ClN_3O_7S_4$, Sporidesmin G etherate, 40B, 539
$C_{22}H_{30}O_4$, Cyclospongiaquinone-2, 44B, 551
$C_{22}H_{30}O_4$, Tridachione rearrangement product, 44B, 550
$C_{22}H_{30}O_4$, Δ^9-Tetrahydrocannabinolic acid B, 41B, 673
$C_{22}H_{30}O_6$, Megaphone, 44B, 551
$C_{22}H_{32}Br_2O_5$, 2-Acetyl-16,17-dibromohydroatractyligenin methyl ester, 44B, 552
$C_{22}H_{32}N_2$, Diisocyanoadociane, 42B, 476
$C_{22}H_{32}O_7$, D-Gibberellin C secodiester photoproduct, 46B, 602
$C_{22}H_{33}BrO_6S$, Fumagillin derivative, 26, 800
$C_{23}H_{22}O_5$, Uvaretin, 42B, 477
$C_{23}H_{23}BrO_7$, Palmarin derivative p-bromophenacyl ester, 34B, 346
$C_{23}H_{24}Br_2N_4O_3S$, N(1')-Carboxy-biotin di(p-bromoanilide), 30B, 242
$C_{23}H_{24}O_8$, β-Peltatin A methyl ether, 41B, 674
$C_{23}H_{26}O_6$, Vismione A, 45B, 664
$C_{23}H_{28}O_8$, Nagilactone A diacetate, 41B, 675
$C_{23}H_{32}O_7$, Gibberellin A_{13} trimethyl ester, 45B, 660
$C_{23}H_{32}O_9$, (+)-Isoolivil acetone hydrate, 43B, 209
$C_{23}H_{34}O_5$, Compactin, 42B, 477
$C_{24}H_{20}O_7$, Glabratephrin, 44B, 553
$C_{24}H_{22}O_7$, Obtusifolin, 37B, 335
$C_{24}H_{23}BrO_7$, Licoricone monobromoacetate, 39B, 438
$C_{24}H_{23}BrO_9$, 9-Bromoisosteganacin, 42B, 478
$C_{24}H_{25}BrO_9$, 2'-Bromopodophyllotoxin - 0.5 ethyl acetate, 39B, 438
$C_{24}H_{26}O_4$, 4a,5,8,8a-Tetrahydro-11,14-dimethoxy-7-methyl-4a-(3-methyl-2-butenyl)-5,8a-o-benzeno-1,4-naphthoquinone, 46B, 603
$C_{24}H_{28}BrN_3O_4$, 5-Bromobrevianamide A acetone solvate, 40B, 538
$C_{24}H_{30}N_2O_{10}S_2$, Sirodesmin A diacetate, 43B, 774
$C_{24}H_{31}BrO_7$, Hydroxypiperenone bromoacetate, 42B, 478
$C_{24}H_{43}NO_5$, Triacetylsphingosine, 34B, 346
$C_{25}H_{15}BrO_7$, Sterigmatocystin p-bromobenzoate, 41B, 676
$C_{25}H_{21}BrO_7$, (-)-7-O-(p-Bromophenacyl)-eucomol, 43B, 774
$C_{25}H_{23}IO_8$, Glauconic acid m-iodobenzoate, 27, 1028
$C_{25}H_{24}BrNO_5$, Cephalotaxine p-bromobenzoate, 40B, 540
$C_{25}H_{26}O_5$, Epishamixanthone, 44B, 553
$C_{25}H_{28}O_5$, Emericellin, 41B, 677
$C_{25}H_{30}O_7$, Phyllanthocin, 43B, 775
$C_{25}H_{34}BrN_3O_5$, Lunarine hydrobromide hydrate, 35B, 390
$C_{25}H_{34}IN_3O_5$, Lunarine hydroiodide hydrate, 35B, 390
$C_{25}H_{39}NO_5$, 7,8-Dihydrobatrachotoxinin A, 38B, 591
$C_{26}H_{19}ClO_{10} \cdot 2\ CH_4O$, Gilmaniellin methanol solvate, 45B, 665
$C_{26}H_{25}BrO_9$, Podolactone A p-bromobenzoate, 41B, 678
$C_{26}H_{28}N_2O_{11}$, 5,12a-Diacetyloxytetracycline, 37B, 336
$C_{26}H_{28}O_{11}$, Leprolomin triacetate, 44B, 554
$C_{26}H_{29}BrO_8$, Bromo-parasiticolide A, 41B, 679
$C_{26}H_{32}O_6$, cis-3-(2',4',5'-Trimethoxyphenyl)-4-[(E)-2''',4''',5'''-

trimethoxystyryl]-cyclohex-1-ene, 45B, 665
$C_{26}H_{32}O_9$, Terretonin, 45B, 666
$C_{26}H_{34}N_8O_{18}$, Succinylcholine dipicrate, 41B, 41
$C_{27}H_{27}BrO_7$, Bisnorquassin m-bromobenzoate, 35B, 391
$C_{27}H_{28}O_6 \cdot 1.5\ H_2O$, Aspulvinone sesquihydrate, 45B, 666
$C_{27}H_{31}BrO_7$, Liatrin diol o-bromobenzoate, 39B, 439
$C_{27}H_{32}O_9$, Verrucarin B, 45B, 667
$C_{27}H_{32}O_9$, Austin, 42B, 479
$C_{27}H_{34}O_{11} \cdot CH_4O$, Undulatone methanol solvate, 45B, 668
$C_{27}H_{37}BrO_8$, Ryanodol p-bromobenzyl ether, 35B, 392
$C_{27}H_{38}O_4 \cdot CH_4O$, Stypoldione methanol solvate, 45B, 668
$C_{28}H_{18}O_5 \cdot CHCl_3$, 1-(p-Hydroxyphenyl)-4-phenyl-7H-benzofur[5,4-c]-[2]benzopyran-2,5-dione methyl ether chloroform solvate, 45B, 669
$C_{28}H_{23}BrO_{13}$, Dothistromin bromoethylether tetraacetate, 38B, 592
$C_{28}H_{24}O_6$, Aniba-dimer-A, 43B, 776
$C_{28}H_{32}Br_2O_{10}$, 4,11-Dibromogomisin D, 45B, 669
$C_{29}H_{20}O_6 \cdot 0.5\ C_6H_6$, 9-Hydroxy-1-(p-hydroxyphenyl)-4-phenyl-6H-benzofura[5,4-c][1]benzopyran-2,5-dione benzene solvate, 45B, 670
$C_{29}H_{21}BrO_{11}$, Monobromoduclauxin, 33B, 235
$C_{29}H_{32}O_7$, Uvafzelin, 46B, 597
$C_{29}H_{33}BrClO_6S$, Ascochlorin p-bromobenzenesulphonate, 37B, 337
$C_{29}H_{38}O_{11} \cdot X\ H_2O$, Baccharin hydrate, 42B, 480
$C_{29}H_{39}NO_{12} \cdot 1.5\ H_2O$, O,O-Dimethylpecoside sesquihydrate, 40B, 541
$C_{29}H_{48}O_9$, Colletotrichin monohydrate methanol solvate, 43B, 777
$C_{30}H_{18}Br_2O_7$, Xylerythrin di(bromoacetate), 31B, 265
$C_{30}H_{24}O_4 \cdot 0.33\ C_3H_6O$, Guayacinin acetone solvate, 42B, 1019
$C_{30}H_{26}O_{14}$, Floccosin, 44B, 555
$C_{30}H_{28}N_6O_6S_4$, Chaetocin, 38B, 593
$C_{30}H_{28}O_{11}$, Xantholaccaic acid B ether dimethyl ester, 44B, 555
$C_{30}H_{30}Br_2N_4O_4$, Byssochlamic acid bis(p-bromophenylhydrazide), 28, 564
$C_{30}H_{34}O_6 \cdot H_2O$, 6-Dehydroxysantoninic acid spiro dimer monohydrate, 46B, 603
$C_{30}H_{36}O_4$, Harunganin, 29, 684
$C_{30}H_{40}O_7$, Anabsin, 45B, 670
$C_{30}H_{41}IO_4$, Triol Q acetonide p-iodobenzoate, 41B, 679
$C_{30}H_{42}O_8$, Stemphone, 41B, 680
$C_{30}H_{44}O_8$, Colletotrichin acetate, 45B, 671
$C_{30}H_{50}BrNO_{11} \cdot X\ C_2H_5OH$, Demycarosyl leucomycin A_3 hydrobromide, 35B, 402
$C_{31}H_{33}N_3O_9$, 3-(3,4-Dimethoxybenzyl)-4-(2,4-dimethoxyphenyl)-6,7,8-trimethoxynaphtha-o-quinone 1-semicarbazone, 46B, 603
$C_{32}H_{32}Br_2O_{13}$, (+)-Dibromodehydrotetrahydrorugulosin, 35B, 398
$C_{32}H_{39}Cl_2IO_{12}$, Phragmalin iodoacetate, 37B, 338
$C_{32}H_{45}BrO_{14}$, 5α-Acetoxy-6β-bromohexahydrophysalin A methanol solvate, 34B, 344; 35B, 400
$C_{33}H_{30}Br_2O_{16}$, 5-Hydroxy-3',6-dibromo-2",3",4',4",6",7-hexaacetyl-vitexin, 41B, 681
$C_{33}H_{30.7}Br_{1.3}O_{16}$, 5-Hydroxy-6-bromo-2",3",4',4",6",7-hexaacetyl-vitexin - 5-hydroxy-3',6-dibromo-2",3",4',4",6",7-hexaacetyl-vitexin mixture, 41B, 681
$C_{33}H_{31}BrO_{16}$, 5-Hydroxy-6-bromo-2",3",4',4",6",7-hexaacetyl-vitexin, 41B, 681
$C_{33}H_{36}N_4O_6 \cdot CHCl_3 \cdot CH_4O$, Bilirubin chloroform methanol solvate, 46B, 604
$C_{33}H_{36}N_4O_6$, Bilirubin, 44B, 556
$C_{33}H_{39}BrO_{15}$, Paeoniflorin bromo derivative, 38B, 594

$C_{33}H_{39}N_3O_7$, Verruculogen benzene solvate, 40B, 542
$C_{34}H_{32}Cl_2O_9$, Di-(p-chlorobenzoyl)-(4'S,5'S)-4',5'-dihydroxyzearalenone 4-methyl ether, 44B, 556
$C_{34}H_{63}BrO_6$, 2-11-Bromoundecanoyl-1,1'-dicaprin, 33B, 236
$C_{35}H_{38}N_4O_6$, Biliverdin dimethyl ester, 42B, 480
$C_{35}H_{42}Cl_6N_4O_6$, Mesobilirubin IXα bis(chloroform), 44B, 556
$C_{36}H_{30}O_7 \cdot C_6H_6$, Uvarinol benzene solvate, 45B, 672
$C_{36}H_{34}O_{14}$, Luteoskyrine acetone solvate, 44B, 557
$C_{36}H_{38}O_{18}$, Secalonic acid A acetic acid solvate, 42B, 481
$C_{36}H_{53}BrO_7$, Platycodigenin bromolactone, 35B, 401
$C_{37}H_{42}O_{10}$, Ohchinolide A, 46B, 605
$C_{38}H_{38}O_8$, Bi-(O-trimethyl-cis-brazilane), 45B, 672
$C_{39}H_{58}BrNO_2$, Bromoindole derivative of 3β-methoxy-21-keto-Δ(1,3)-serratene, 35B, 404
$C_{40}H_{52}BrClO_{14}$, Cleroendrin A p-bromobenzoate chlorohydrin, 39B, 439
$C_{40}H_{52}N_4O_{14}$, Agrobactin-ethyl acetate (1:2), 46B, 605
$C_{41}H_{52}N_4O_6$, Diethoxybilirubin diethyl ester, 45B, 673
$C_{42}H_{47}Br_3O_8$, Tri-p-bromobenzoate of prostaglandin F(2-1) methyl ester, 28, 567
$C_{43}H_{32}O_8$, (DL)-Methyl 4,5-dimethoxy-2-(2,6-dimethoxy-1-oxo-9-phenyl-5-phenalenyl)-1-oxo-8-phenyl-1,2-dihydro-2-acenaphthylenecarboxylate, 46B, 606
$C_{48}H_{40}I_2O_{16}$, Tetra-O-methylergoflavin di-p-iodobenzoate, 31B, 267
$C_{78}H_{72}Br_2O_{12}$, Dibromodeca-O-methylhopeaphenol benzene solvate, 35B, 405

60. MOLECULAR COMPLEXES

CHI_3S_{24}, Iodoform - sulfur, 27, 747
$CH_4N_2O \cdot X\ C_{16}H_{34}$, Urea-cetane, 13, 479
$CH_6ClN_2NaO_2$, Urea - sodium chloride monohydrate, 28, 570
$CH_6N_2O_3$, Urea - hydrogen peroxide, 8, 278
CH_8BrN_3O, Urea - ammonium bromide, 28, 569
$C_2H_3Br_6Hg_3N$, Mercuric bromide - methyl cyanide, 38B, 597
$C_2H_3Cl_5NSb$, Acetonitrile-antimony pentachloride complex, 32B, 266
$C_2H_5NO_3$, Formic acid - formamide, 34B, 349
$C_2H_8Br_2O_2$, Methanol - bromine, 29, 451
$C_2H_8I_2N_4S_2$, Bis(thiourea)iodine(I) iodide, 38B, 597
$C_3H_6Br_2O$, Acetone-bromine (1:1) addition compound, 23, 524
$C_3H_6N_2O_5$, Urea-oxalic acid, 45B, 44
$C_3H_8BaO_8S_5$, Barium pentathionate acetone solvate, 22, 568
$C_3H_9I_2N$, Trimethylamine-iodine (1:1) addition compound, 23, 568
$C_3H_{12}INaO_3$, Methanol-sodium iodide, 30B, 264
$C_4H_6Br_2N_2$, Acetonitrile - bromine, 33B, 238
$C_4H_6N_4O_3S$, Thiourea parabanic acid, 35B, 245
$C_4H_6N_4O_4$, Urea parabanic acid, 35B, 245
$C_4H_8Br_2O_2$, Bromine-1,4-dioxane, 18, 637
$C_4H_8ClLiO_2$, 1,4-Dioxan-lithium chloride complex, 31B, 273
$C_4H_8Cl_2O_2$, 1,4-Dioxane-chlorine, 23, 715
$C_4H_8I_2OSe$, 1,4-Oxaselenane-iodine complex, 31B, 272
$C_4H_8I_2Se$, Tetrahydroselenophene - iodine, 29, 522
$C_4H_8I_6S_2Sb_2$, 1,4-Dithiane-antimony tri-iodide complex, 31B, 273
$C_4H_8N_2O_6$, 1,4-Dioxan-dinitrogen tetroxide complex, 30B, 246
$C_4H_{10}BrN_2NaO_2$, Acetamide-sodium bromide complex, 31B, 271
$C_4H_{10}N_4O_6$, Urea - oxalic acid (2:1), 13, 477; 38B, 598
$C_4H_{10}O_6S$, 1,4-Dioxan - sulfuric acid, 24, 687

$C_4H_{16}Br_4N_2$, Dimethylammonium bromide-bromine (2:1) addition compound, 23, 567
$C_4H_{16}Cl_2I_2N_2$, Dimethylammonium chloride-iodine (2:1) addition compound, 23, 567
$C_4H_{16}N_2O_2$, Hydrazine-bis(ethanol) complex, 32B, 269
$C_4H_{18}ClCsN_8OS_4$, Caesium chloride - thiourea hydrate, 33B, 239
$C_4H_{18}N_8O_4PS_4Tl$, Thallium(I) dihydrogen phosphate - thiourea, 33B, 241
$C_4H_{20}CsFN_8O_2S_4$, Caesium fluoride - thiourea hydrate, 33B, 239
$C_4H_{20}N_2O_4$, Hydrazine-tetra-methanol, 32B, 269
$C_5H_3Cl_2HgNO$, 3,5-Dibromopyridine N-oxide - mercury(II) chloride, 31B, 439
$C_5H_4CuN_3$, Pyridazine copper(I) cyanide (1:1), 38B, 599
C_5H_5ClIN, Pyridine - iodomonochloride, 20, 604; 38B, 600
C_5H_5ClLiN, Pyridine-lithium chloride, 31B, 274
$C_5H_5I_3N_2$, Pyridine - nitrogen triiodide, 40B, 543
$C_5H_5I_4N$, Pyridine - iodine, 26, 778
$C_5H_9I_3O_2$, Iodoform - 1,4-dioxan, 34B, 348
$C_5H_9I_3S_2$, Dithiane - iodoform, 26, 773
$C_5H_{11}BrCl_2O$, Diethyl ether - bromodichloromethane, 29, 450
$C_6H_4Br_4N_2$, Tetrabromoethylene - pyrazine, 33B, 237
$C_6H_4Cl_3HgS_4$, Tetrathiafulvalonium trichloromercurate(II), 46B, 606
$C_6H_4I_4N_2$, Tetraiodoethylene - pyrazine, 33B, 237
$C_6H_6CdCl_2N_2O_2$, p-Nitroaniline - cadmium chloride, 35B, 408
$C_6H_6Cl_2$, Benzene-chlorine (1:1) addition compound, 23, 642
$C_6H_6N_2O_4$, Benzene - dinitrogen tetroxide, 33B, 243
$C_6H_7I_2N$, 4-Picoline - iodine, 26, 763
$C_6H_8Br_2O_4$, Oxalyl bromide-1,4-dioxan complex, 30B, 246
$C_6H_8Cl_2HgO_2$, Cyclohexane-1,4-dione - mercuric chloride, 29, 492
$C_6H_8Cl_2O_4$, Oxalyl chloride-1,4-dioxan complex, 30B, 246
$C_6H_8I_2O_2$, 1,4-Dioxan - diiodoacetylene, 24, 686
$C_6H_8I_2S_2$, Di-iodoacetylene-1,4-dithiane complex, 31B, 270
$C_6H_8I_2Se_2$, Di-iodoacetylene-1,4-diselenane complex, 31B, 270
$C_6H_8I_4Se_2$, 1,4-Diselenane-tetraiodoethylene complex, 30B, 249
$C_6H_8N_4O_5$, Urea - syn-5-nitro-2-furaldehyde oxime (1:1), 38B, 601
$C_6H_{10}ClIN_4$, Pentamethylenetetrazole-iodinemonochloride, 32B, 270
$C_6H_{10}I_6Se_2$, Iodoform - 1,4-diselenane, 28, 571
$C_6H_{12}I_5N_5$, Hexamethylenetetramine - diiodine - nitrogen triiodide, 40B, 544
$C_6H_{14}N_4O_6$, N-Methylurea oxalic acid (monoclinic), 45B, 8
$C_6H_{14}N_4O_6$, N-Methylurea oxalic acid (orthorhombic), 45B, 8
$C_6H_{16}ClN_3O_3S$, L-Cysteine ethyl ester hydrochloride-urea complex, 30B, 208
$C_6H_{20}N_5OPS_{14}$, Tris(dimethylamino)phosphine oxide - bis(heptasulfurimide), 43B, 778
$C_7H_4CuN_3$, 4-Cyanopyridine copper(I) cyanide (1:1), 38B, 602
$C_7H_5Cl_3NO_3$, Pyridine-N-oxide - trichloroacetic acid, 40B, 6
$C_7H_9NO_5$, Allenedicarboxylic acid - acetamide (1:1), 43B, 778
$C_7H_{10}N_2O_3$, Quinol - urea (1:1), 41B, 682
$C_7H_{13}BrN_5S_2$, Bisthiourea pyridinium bromide, 38B, 602
$C_7H_{13}I_2NS$, N-Methylthiocaprolactam - iodine, 40B, 546
$C_7H_{13}I_3N_4$, Hexamethylenetetramine - iodoform, 35B, 202
$C_7H_{15}N_9O_9$, 1,3,5,7-Tetranitro-1,3,5,7-tetraazacyclooctane N,N-dimethylformamide complex (1:1), 41B, 303
$C_7H_{16}N_2O_7$, α-D-Glucose - urea, 37B, 339
$C_7H_{27}Cl_3N_{10}O$, Guanidinium chloride-N,N-dimethylacetamide complex, 30B, 258

$C_7H_{28}Cl_3CoN_{10}S_3$, trans-Dichlorobis(ethylenediamine)cobalt(III) chloride - tris(thiourea), 39B, 440
$C_8H_6K_3MoN_6O_3S_6$, Potassium hexaisothiocyanatomolybdate(III) hydrate-acetic acid, 33B, 422
$C_8H_8I_2O_2$, Cyclohexane-1,4-dione-diiodoacetylene, 30B, 248
$C_8H_8N_4O_2S_2Se$, o-Nitrobenzeneselenenyl thiocyanate - thiourea, 38B, 603
$C_8H_9ClN_2O_2$, Benzenediazonium chloride - acetic acid, 33B, 243
$C_8H_9N_3Ni$, Benzene-ammonia-nickel cyanide complex, 16, 521
$C_8H_{12}N_2O_4S_2$, Cyclo-L-cystine - acetic acid, 40B, 546
$C_8H_{13}N_3O$, 2,6-Lutidine-urea complex, 30B, 259
$C_8H_{14}BrN_2NaO_4$, Bis(diacetamide)-sodium bromide complex, 34B, 351
$C_8H_{14}Br_6N_4$, Bromoform - hexamethylenetetramine, 37B, 339
$C_8H_{14}IKN_2O_4$, Bis(diacetamide)-potassium iodide complex, 34B, 352
$C_8H_{20}N_8S_2$, Hexamethylenetetramine - thiourea (1:2), 44B, 557
$C_9H_8O_7 \cdot 0.22\ C_6H_3N_3O_7$, Trimesic acid monohydrate - picric acid, 43B, 118
$C_9H_{10}Br_4$, Carbon tetrabromide - p-xylene, 27,891
$C_9H_{11}ClN_2O_2S$, S-Methylthiouronium p-chlorobenzoate, 28, 573
$C_9H_{12}O_3$, Quinol-acetone, 23, 650
$C_9H_{16}N_4O_4$, Urea - 5,5-diethylbarbituric acid (1:1), 40B, 547
$C_9H_{18}INaO_3$, Acetone - sodium iodide, 28, 575
$C_9H_{21}IN_3NaO_3$, Dimethylformamide - sodium iodide, 27, 760
$C_{10}H_8Br_2I_2N_2$, 2,2'-Bipyridine - iodine monobromide (1:2), 40B, 548
$C_{10}H_8Cl_2I_2N_2$, 2,2'-Bipyridine - iodine monochloride (1:2), 40B, 548
$C_{10}H_8Cl_6Sb_2$, Naphthalene-antimony trichloride complex, 34B, 357
$C_{10}H_8N_6O_7S$, 2-Thiocytosine picrate, 43B, 520
$C_{10}H_9ClN_2O_2$, 2-Pyridone - 6-chloro-2-hydroxypyridine, 37B, 340
$C_{10}H_{10}Cl_2N_2OSe$, Selenium oxychloride - pyridine, 23, 698
$C_{10}H_{11}N_7O_3$, 9-Ethyladenine - parabanic acid, 42B, 483
$C_{10}H_{12}FeNa_2O_7$, Disodium tetracarbonyliron - dioxane (2:3), 42B, 484
$C_{10}H_{13}BrN_8O_4$, 8-Bromo-9-ethyladenine - cyanuric acid monohydrate, 42B, 485
$C_{10}H_{17}N_3O_4$, Acetamide - 5,5-diethylbarbituric acid (1:1), 40B, 550
$C_{10}H_{26}N_8S_4$, Thiourea - 2,3-dimethylbutadiene (4:1), 42B, 486
$C_{11}H_8N_4O_7$, Pyridine picrate, 42B, 487
$C_{11}H_{10}N_2O_4$, Thymine - p-benzoquinone, 37B, 341
$C_{11}H_{12}N_{10}O$, Purine - urea (2:1), 43B, 523
$C_{11}H_{13}N_5O_5$, trans-5-Amino-3-(2-(5-nitro-2-furyl)vinyl)-1,2,4-oxadiazole - N,N-dimethylformamide, 43B, 779
$C_{11}H_{15}I_2N_5$, 9-Cyclohexyladenine - iodine, 39B, 440
$C_{11}H_{16}N_4O_3$, Imidazole - 5,5-diethylbarbituric acid (1:1), 40B, 551
$C_{11}H_{17}BrN_2O_7$, 5-Bromouridine - dimethylsulfoxide, 32B, 272
$C_{11}H_{19}NO_6$, Glucitol - pyridine, 37B, 342
$C_{11}H_{21}N_7O_{11}$, 1,7-Diacetoxy-2,4,6-trinitro-2,4,6-triazaheptane - N,N-dimethylformamide, 39B, 441
$C_{12}H_2Br_{10}$, Hexabromobenzene - 1,2,4,5-tetrabromobenzene, 29, 644
$C_{12}H_4Cl_4O_2S_4$, Tetrathiafulvalene chloranil, 45B, 674
$C_{12}H_4F_4O_2S_4$, Tetrathiafulvalene fluoranil, 45B, 674
$C_{12}H_8Br_2I_2$, p-Diiodobenzene - p-dibromobenzene, 24, 634
$C_{12}H_8Cl_5NO$, Aniline - pentachlorophenol, 37B, 343
$C_{12}H_8I_2N_2$, Phenazine-iodine complex, 32B, 273
$C_{12}H_9BrO_3$, p-Benzoquinone-p-bromophenol complex, 32B, 275
$C_{12}H_9ClO_3$, p-Benzoquinone-p-chlorophenol complex, 32B, 275
$C_{12}H_9IN_4O_6$, p-Iodoaniline - s-trinitrobenzene, 9, 372
$C_{12}H_{10}Cl_3NO$, Aniline - 2,4,5-trichlorophenol, 37B, 344
$C_{12}H_{10}O_4$, Quinone - resorcinol, 35B, 409

$C_{12}H_{12}N_2O_8$, Furamide - oxalic acid, 39B, 442
$C_{12}H_{12}N_6O_6$, s-Trinitrobenzene - s-triaminobenzene, 35B, 410
$C_{12}H_{14}INO$, Morpholine β-iodophenylacetylene, 29, 744
$C_{12}H_{15}N_6O_7P$, Triethyl phosphate - benzotrifurazan (1:1), 38B, 604
$C_{12}H_{16}BrNO_2$, Piperidine - p-bromobenzoic acid (1:1), 38B, 606
$C_{12}H_{16}ClNO_2$, Piperidine - p-chlorobenzoic acid (1:1), 38B, 606
$C_{12}H_{17}NO_3$, Piperidinium p-hydroxybenzoate, 39B, 443
$C_{12}H_{18}N_4O_2$, Hexamethylenetetramine resorcinol, 45B, 675
$C_{12}H_{18}N_4O_2$, Hexamethylenetetramine - hydroquinone (1:1), 43B, 382; 45B, 675
$C_{12}H_{22}N_6O_{12}$, 1,7-Diacetoxy-2,4,6-trinitro-2,4,6-triazaheptane - 1,4-dioxane, 39B, 442
$C_{12}H_{22}N_8O$, Thymine - N,N-diethylmelamine monohydrate, 39B, 444
$C_{12}H_{24}AgClO_{10}$, Silver perchlorate-1,4-dioxane complex, 20, 620
$C_{12}H_{30}GdN_3O_{18}$, 1,4,7,10,13,16-Hexaoxacyclooctadecane triaquatrinitratogadolinium, 46B, 1146
$C_{12}H_{32}N_2O_{18}U$, Uranyl nitrate tetrahydrate - 18-crown-6 (1:1), 42B, 487
$C_{12}H_{36}Cr_2MgN_8O_{13}$, Magnesium dichromate hexamethylenetetramine hexahydrate, 40B, 554
$C_{12}H_{44}Br_2CaN_8O_{10}$, Hexamethylenetetramine-calcium bromide complex decahydrate, 15, 460; 32B, 276
$C_{13}H_8CuKN_3O_4S$, Bis(pyridine-2-carboxylato)copper(II)-potassium thiocyanate, 35B, 412
$C_{13}H_{11}N_5O_8$, 1-Methylnicotinamide picrate, 40B, 545
$C_{13}H_{15}N_5O_7S_2$, 2-Dimethylamino-4,4-dimethyl-2-thiazolin-5-thione picrate, 42B, 293
$C_{13}H_{16}N_4O_5$, Cytosine - N-benzoylglycine monohydrate, 38B, 606
$C_{13}H_{19}NO_2$, Piperidinium p-toluate, 39B, 444
$C_{14}H_4F_{12}PtS_8$, Tetrathiafulvalinium bis-cis-(1,2-perfluoromethylethylene-1,2-dithiolato)platinum, 42B, 488
$C_{14}H_{10}Cl_6Sb_2$, Antimony trichloride - phenanthrene, 38B, 607
$C_{14}H_{10}F_6$, p-Xylene - hexafluorobenzene, 41B, 683
$C_{14}H_{10}N_4O_6$, Indole - trinitrobenzene, 29, 689
$C_{14}H_{11}F_6N$, Hexafluorobenzene - N,N-dimethylaniline (1:1), 43B, 780
$C_{14}H_{11}N_7S$, 6-Mercapto-3-phenyl-s-triazolo[4,3-b]-s-tetrazine - pyridine, 39B, 445
$C_{14}H_{15}I_3N_2O_2$, Benzamide - hydrogen triiodide, 29, 656
$C_{14}H_{19}N_3O_4$, N-Methyl-2-pyridone - 5,5-diethylbarbituric acid (1:1), 40B, 555
$C_{14}H_{23}N_5O_7S$, Serotonin - creatinine sulfate monohydrate, 30B, 214
$C_{14}H_{26}N_2O_{10}S_2$, 1,4-Dideoxy-1,4-dinitro-neo-inositol - tetrahydrothiophene-1-oxide (1:2), 38B, 608
$C_{14}H_{26}N_4O_{10}$, Caffeine - pyrogallol, 33B, 244
$C_{15}H_9N_3O_7S$, Trinitrobenzene - 3-formylbenzothiophene (1:1), 38B, 609
$C_{15}H_9N_5$, Skatole tetracyanoethylene, 46B, 607
$C_{15}H_{10}N_4O_7$, 8-Hydroxyquinoline - 1,3,5-trinitrobenzene, 37B, 345
$C_{15}H_{11}N_4O_6$, Skatole - trinitrobenzene, 29, 689
$C_{15}H_{12}F_6$, Mesitylene - hexafluorobenzene, 37B, 346
$C_{15}H_{15}ClN_4O_5$, Caffeine - 5-chlorosalicylic acid, 33B, 245
$C_{15}H_{16}N_6$, 9-Ethyladenine - indole (1:1), 42B, 488
$C_{15}H_{18}N_6O_7$, Cytosine - resorcylic acid (2:1) monohydrate, 39B, 446
$C_{15}H_{21}N_7O_3$, 9-Ethyladenine - 5,5-diethylbarbituric acid, 38B, 610
$C_{15}H_{21}N_9O_6$, 2,4,6-Tri(dimethylamino)-1,3,5-triazine-s-trinitrobenzene complex, 31B, 274
$C_{16}H_8N_4$, Tetracyanoethylene-naphthalene complex, 32B, 278
$C_{16}H_{10}Br_6Sb_2$, Bis(antimony tribromide) pyrene, 38B, 611

$C_{16}H_{10}Cl_4N_2O_2$, Chloranil - 1,5-diaminonaphthalene, 43B, 780
$C_{16}H_{10}FeN_4$, Ferrocene-tetracyanoethylene complex, 32B, 279
$C_{16}H_{10}N_6$, p-Phenylenediamine - 1,2,4,5-tetracyanobenzene, 39B, 446
$C_{16}H_{11}BrN_4O_7$, 1-Bromo-2-aminonaphthalene - picric acid, 33B, 245
$C_{16}H_{11}CrN_3O_{10}$, Tricarbonylchromiumanisole-1,3,5-trinitrobenzene complex, 31B, 275
$C_{16}H_{11}N_3O_6$, Azulene-s-trinitrobenzene complex, 30B, 251
$C_{16}H_{12}O_4$, Hydroquinone - 1,4-naphthoquinone (1:1), 43B, 196
$C_{16}H_{14}F_6$, Durene - hexafluorobenzene, 41B, 683
$C_{16}H_{14}N_5O$, Morpholinium - 7,7,8,8-tetracyanoquinodimethane (1:1), 38B, 612
$C_{16}H_{15}N_3O_3$, Indole-3-acetic acid - nicotinamide (1:1), 44B, 558
$C_{16}H_{15}N_5O_7$, Tryptamine picrate, 40B, 557
$C_{16}H_{16}Cl_4N_2O_2$, N,N,N',N'-Tetramethyl-p-diaminobenzene - chloranil, 33B, 247
$C_{16}H_{16}F_6N_2$, N,N,N',N'-Tetramethyl-p-phenylenediamine hexafluorobenzene, 45B, 676
$C_{16}H_{17}N_5O_9$, Serotonin picrate monohydrate, 38B, 614
$C_{16}H_{18}Br_2$, Durene p-dibromobenzene, 44B, 107
$C_{16}H_{18}N_2O_6$, β-Picoline-N-oxide fumaric acid adduct, 41B, 286
$C_{16}H_{18}O_4PdS_4$, Bis(acetylacetonato)palladium(II)-tetrathiafulvalene (1:1), 46B, 607
$C_{16}H_{19}BrN_4O_7S$, 5-Bromocytosine - N-tosyl-L-glutamic acid, 42B, 490
$C_{16}H_{20}N_4O_6$, Cycl[3.2.2]azine - s-trinitrobenzene (1:1), 44B, 559
$C_{16}H_{20}N_6O_5S$, Sulfacetamide - caffeine (1:1), 43B, 138
$C_{16}H_{22}ClI_5N_2$, p-Iodo-N,N-dimethylaniline hydrochloride triiodide, 27, 958
$C_{16}H_{30}ClN_3O_4$, Tetraethylammonium bromide-succinimide addition compound, 22, 793
$C_{16}H_{36}N_6O_4S$, Cyanoguanidin - lupetidin sulfate, 40B, 558
$C_{17}H_{14}N_4O_{10}$, Trinitrobenzene - 1-acetylskatole (1:1), 44B, 559
$C_{17}H_{15}BrN_4O_7 \cdot 0.5\ H_2O$, 5-Bromocytosine - phthaloyl-DL-glutamic acid hemihydrate, 43B, 526
$C_{17}H_{15}Cl_3CuN_2O_2$, N,N'-Ethylenebis(salicylideneiminato)copper(II) - chloroform adduct, 35B, 414
$C_{17}H_{20}N_2O_3$, Nicotinyl salicylate, 37B, 346
$C_{17}H_{20}N_4O_3S$, Sulfanilamide - antipyrine (1:1), 43B, 138
$C_{17}H_{22}BrN_7O_3$, 9-Ethyladenine - 5-isopropyl-5-bromoallylbarbituric acid, 38B, 615
$C_{17}H_{29}CrN_4O_6S_2$, Tris(acetylacetonato)chromium(III) - thiourea (1:2), 39B, 447
$C_{18}Cl_4D_8O_3$, Naphthalene (perdeuterated) tetrachlorophthalic anhydride complex, 41B, 684
$C_{18}H_8Cl_4O_3$, Naphthalene tetrachlorophthalic anhydride complex, 41B, 684
$C_{18}H_8N_4S_4$, Δ(2,2')-Bi-1,3-dithiole - 7,7,8,8-tetracyano-p-quinodimethane, 40B, 559; 42B, 490
$C_{18}H_{10}Cl_4O_2$, Acenaphthene - tetrachloro-p-benzoquinone, 39B, 448
$C_{18}H_{10}N_4PtS_4$, Bis(propene-3-thione-1-thiolato)platinum(II) - 7,7,8,8-tetracyanoquinodimethane, 43B, 781
$C_{18}H_{11}N_3O_6S$, Dibenzothiophene - trinitrobenzene, 43B, 781
$C_{18}H_{12}N_4O_6$, Carbazole - trinitrobenzene, 42B, 491
$C_{18}H_{12}N_4O_6S$, Phenothiazine - 1,3,5-trinitrobenzene, 34B, 351
$C_{18}H_{12}N_6O_6S_2$, Benzotrifuroxan-13,14-dithiatricyclo$[8.2.1.1^{4-7}]$tetradeca-4,6,10,12-tetraene, 30B, 267
$C_{18}H_{14}Br_2O_4$, p-Benzoquinone-di(p-bromophenol) complex, 32B, 280
$C_{18}H_{14}Cl_2O_4$, p-Benzoquinone-di(p-chlorophenol) complex, 32B, 280

$C_{18}H_{14}N_4$, Hexamethylbenzene - tetracyanoethylene (1:1), 44B, 560
$C_{18}H_{14}N_6$, N,N-Dimethyl-p-phenylenediamine - 1,2,4,5-tetracyanobenzene, 39B, 448
$C_{18}H_{14}O_7$, Phloroglucinol - p-benzoquinone, 37B, 347
$C_{18}H_{15}Cl_2O_2PSe$, Triphenylphosphine oxide - seleninyl dichloride, 34B, 354
$C_{18}H_{15}N_5O_6$, Benzidine - s-trinitrobenzene, 40B, 560
$C_{18}H_{16}BrCl_3N_2O_3$, Chloral hydrate - 7-bromo-2,3-dihydro-1-methyl-5-phenyl-1,4-benzodiazepin-2-one, 37B, 348
$C_{18}H_{16}O_4$, Phenol-p-benzoquinone, 17, 720
$C_{18}H_{18}Cl_4O_2$, Chloranil - hexamethylbenzene, 19, 567; 27, 909
$C_{18}H_{18}F_6$, Hexamethylbenzene - hexafluorobenzene (1:1), 38B, 616; 39B, 449
$C_{18}H_{18}N_2O_4$, Antipyrine - salicylic acid, 40B, 561
$C_{18}H_{18}N_4$, Hexamethylbenzene - tetracyanoethylene (1:1), 42B, 491
$C_{18}H_{18}N_6O_{11}$, Deaza-1-isotubercidin picrate, 42B, 492
$C_{18}H_{19}N_5O_{10}$, D,L-Tryptophan picrate methanol, 40B, 557
$C_{18}H_{20}I_2N_2O_2$, N,N'-Dimethyl-4,4'-bipyridylium diiodide - quinol, 42B, 1022
$C_{18}H_{20}N_2O_2$, Tryptamine - phenylacetic acid (1:1), 44B, 560
$C_{18}H_{20}N_2O_6$, Benzamide - succinic acid, 39B, 450
$C_{18}H_{22}I_2N_6O_2S_8$, Merocyanine - iodine, 33B, 248
$C_{18}H_{24}BrN_7O_3$, 8-Bromo-9-ethyladenine - 5-allyl-5-isobutylbarbituric acid (1:1), 42B, 303
$C_{18}H_{24}N_6O_3$, 2-Aminopyridine - 5,5-diethylbarbituric acid (2:1), 40B, 562
$C_{18}H_{25}N_3O_5$, Salicylamide - 5-ethyl-5-isoamylbarbituric acid, 40B, 373
$C_{18}H_{27}ClN_2O_4$, ϵ-Caprolactam - 4-chlororesorcinol (2:1), 40B, 563
$C_{18}H_{28}N_4O_6$, 18-Crown-6 - malononitrile, 43B, 412
$C_{18}H_{42}I_4K_2N_6O_6$, N-Methylacetamide-potassium iodide-potassium triiodide complex, 30B, 257
$C_{19}H_{17}ClN_4O_4$, Lumiflavinium chloride - hydroquinone (1:1), 38B, 617
$C_{19}H_{19}N_5O_7$, 8-Methyl-5,5a,6,7,8,9-hexahydropyrido[2,1-b]quinazolinium picrate, 40B, 563
$C_{19}H_{28}N_2O_3$, Estradiol - urea (1:1), 38B, 618
$C_{19}H_{31}KO_9S$, Potassium p-toluenesulphonate - 1,4,7,10,13,16-hexaoxacyclooctadecane, 37B, 349
$C_{20}H_8F_8$, Naphthalene - octafluoronaphthalene (1:1), 41B, 685
$C_{20}H_9F_{12}NNiOS_4$, Bis(bis(trifluoromethyl)ethylene-1,2-dithiolato)nickel - phenoxazine, 41B, 686
$C_{20}H_9F_{12}NNiS_5$, Bis(bis(trifluoromethyl)ethylene-1,2-dithiolato)nickel - phenothiazine, 41B, 686
$C_{20}H_{10}N_4$, Naphthalene - 1,2,4,5-tetracyanobenzene, 32B, 281
$C_{20}H_{10}N_4O$, α-Naphthol - 1,2,4,5-tetracyanobenzene, 44B, 561
$C_{20}H_{10}N_4O$, β-Naphthol - 1,2,4,5-tetracyanobenzene, 44B, 561
$C_{20}H_{10}O_6$, Naphthalene-pyromellitic dianhydride (orange form), 45B, 676
$C_{20}H_{10}O_6$, Naphthalene-pyromellitic dianhydride (yellow form), 45B, 676
$C_{20}H_{13}N_3O_6$, Anthracene - s-trinitrobenzene, 29, 686
$C_{20}H_{13}N_3O_7$, Anthracene - picric acid, 42B, 492
$C_{20}H_{16}Cl_3O_3P$, Triphenylphosphine oxide - trichloroacetic acid, 42B, 493
$C_{20}H_{18}N_6$, N,N,N',N'-Tetramethyl-p-phenylenediamine-1,2,4,5-tetracyanobenzene complex, 32B, 282
$C_{20}H_{23}BrN_2O_4$, DL-Bromophenirame maleate, 37B, 22

$C_{20}H_{24}N_4O_5$, 1-Methyl-3-carbamidopyridinium N-acetyl-L-tryptophanate monohydrate, 43B, 782
$C_{20}H_{25}ClO_6$, Mesitaldehyde-perchloric acid complex, 30B, 265
$C_{21}H_{10}N_4S_2Se$, 5-Phenyl-[1-thiol]-3-selenol-2-thione - 7,7,8,8-tetracyanoquinodimethane (1:1), 45B, 677
$C_{21}H_{12}Cl_4O_2$, 9-Methylanthracene - tetrachloro-p-benzoquinone, 39B, 450
$C_{21}H_{12}N_4S_4$, Trimethylenetetrathiafulvalene - tetracyanoquinodimethane, 44B, 562
$C_{21}H_{17}N_5O_7$, Pyridinium 1-naphthylamine picrate, 46B, 608
$C_{21}H_{18}Cl_2N_4O_8$, 5-Chlorosalicylic acid - theobromine, 37B, 349
$C_{21}H_{18}N_5O_6$, Benzidine - s-trinitrobenzene (1:1) benzene solvate, 41B, 684
$C_{21}H_{20}N_3O_{10}P$, 4-Carbethoxyanilinium bis-p-nitrophenylphosphate, 38B, 619
$C_{21}H_{23}N_3O_3$, Indole-3-acetic acid - 5-methoxytryptamine, 42B, 494
$C_{21}H_{29}N_5O_4$, Aminopyrine - barbital, 37B, 350
$C_{21}H_{29.82}I_{0.18}O_{4.82}$, 11β,17α,21-Trihydroxy-4-pregnene-3,20-dione - 11β,17α-dihydroxy-21-iodo-4-pregnene-3,20-dione (82:18), 46B, 608
$C_{22}H_{10}Cl_2N_2O_2$, Phenanthrene - 2,3-dichloro-5,6-dicyano-p-benzoquinone, 44B, 562
$C_{22}H_{10}Cl_4O_2$, Pyrene - chloranil, 39B, 451
$C_{22}H_{10}F_4O_2$, Pyrene:fluoranil, 46B, 609
$C_{22}H_{10}N_4$, Pyrene - tetracyanoethylene, 31B, 278; 33B, 250; 41B, 686
$C_{22}H_{10}N_4S_4$, Tetrathiafulvalene - 11,11,12,12-tetracyanonaphtho-2,6-quinodimethane, 41B, 693
$C_{22}H_{10}O_4S_3$, Pyromellitic-bis(thioanhydride) - dibenzothiophene, 44B, 564
$C_{22}H_{11}NO_6S$, Phenothiazine pyromellitic dianhydride complex, 46B, 609
$C_{22}H_{12}BrN_3O_6$, Fluoranthene - picryl bromide (polymorph I), 41B, 687
$C_{22}H_{12}N_4$, 7,7,8,8-Tetracyanoquinodimethane - naphthalene (1:1), 42B, 493
$C_{22}H_{13}N_3O_6$, Acepleiadylene-s-trinitrobenzene complex, 31B, 279
$C_{22}H_{13}N_3O_6$, Pyrene - 1,3,5-trinitrobenzene, 39B, 452
$C_{22}H_{14}O_2$, Pyrene - p-benzoquinone, 41B, 688
$C_{22}H_{16}N_4S_4$, Tetrathiafulvalenium - 2,5-diethyltetracyanoquinodimethane, 42B, 494; 43B, 783
$C_{22}H_{16}N_4S_4$, 4,4',5,5'-Tetramethyl-Δ(2,2')-bis-1,3-dithiolium - 7,7,8,8-tetracyano-p-quinodimethanide, 43B, 784
$C_{22}H_{16}N_4Se_4$, 4,4',5,5'-Tetramethyl-Δ(2,2')-bis-1,3-diselenole - 7,7,8,8-tetracyano-p-quinodimethane, 43B, 785
$C_{22}H_{19}CuN_3O_5$, N,N'-Ethylenebis(salicylideneiminato)copper(II) p-nitrophenol adduct, 35B, 413
$C_{22}H_{20}Cl_4N_2O_2$, N,N,N',N'-Tetramethylbenzidine - chloranil, 39B, 452
$C_{22}H_{20}N_4$, Tetracyanobenzene - hexamethylbenzene, 33B, 250
$C_{22}H_{20}N_6$, 7,7,8,8-Tetracyanoquinodimethan-N,N,N',N'-tetramethyl-p-phenylenediamine complex, 30B, 249
$C_{22}H_{25}N_3O_4$, 5-Methoxyindole-3-acetic acid - 5-methoxytryptamine, 42B, 494
$C_{22}H_{28}N_8O_8S$, Thiamin picrolonate dihydrate, 43B, 529
$C_{22}H_{31}N_9O_9$, Adenine-riboflavine trihydrate, 43B, 786
$C_{22}H_{42}N_7O_7P$, Hexamethylphosphoramide - 5,5-diethylbarbituric acid (1:2), 40B, 564
$C_{23}H_{11}NO_4S_2$, Acridine - pyromellitic dithioanhydride (1:1), 42B, 495
$C_{23}H_{32}Br_2N_4O_{10}$, Riboflavin - quinol bromide, 39B, 453
$C_{24}H_{10}F_{10}$, Biphenyl-perfluorobiphenyl, 45B, 679
$C_{24}H_{12}Cl_4N_2O_3Pd$, Chloranil-bis(8-hydroxyquinolato)palladium(II),

30B, 262
$C_{24}H_{12}CuN_8O_8$, Benzotrifuroxan-bis(8-hydroxyquinolinato)copper(II), 30B, 263
$C_{24}H_{12}N_4$, Anthracene - 1,2,4,5-tetracyanobenzene (1:1) (low temperature form), 46B, 610
$C_{24}H_{12}N_4$, Phenanthrene - 1,2,4,5-tetracyanobenzene, 44B, 563
$C_{24}H_{12}N_4$, Anthracene - 1,2,4,5-tetracyanobenzene (1:1), 38B, 620
$C_{24}H_{12}N_4O_2$, 7,7,8,8-Tetracyanoquinodimethane - dibenzo-p-dioxin, 39B, 454
$C_{24}H_{12}N_6$, 7,7,8,8-Tetracyanoquinodimethane - phenazine, 39B, 453
$C_{24}H_{12}N_6$, 7,7,8,8-Tetracyanoquinodimethane - 1,10-phenanthroline, 39B, 454
$C_{24}H_{12}O_4S_2$, Anthracene - pyromellitic dithioanhydride (1:1), 40B, 565
$C_{24}H_{12}O_6$, Anthracene - pyromellitic dianhydride, 30B, 254; 44B, 563
$C_{24}H_{12}O_6$, Phenanthrene - pyromellitic acid dianhydride, 43B, 786
$C_{24}H_{13}N_5$, Carbazole - 7,7,8,8-tetracyanoquinodimethane, 39B, 456
$C_{24}H_{13}N_5S$, Phenothiazine - 7,7,8,8-tetracyanoquinodimethane (1:1), 40B, 565
$C_{24}H_{14}Cl_4N_2O_4$, 8-Hydroxyquinoline-chloranil complex, 32B, 283
$C_{24}H_{14}N_4$, Acenaphthene - 7,7,8,8-tetracyanoquinodimethane, 39B, 457
$C_{24}H_{14}N_8O_4Pd$, 7,7,8,8-Tetracyanoquinodimethane - bis(1,2-benzoquinonedioximato)palladium(II), 43B, 787
$C_{24}H_{14}O_6$, Pyromellitic dianhydride - trans-stilbene (1:1), 40B, 567
$C_{24}H_{16}Cl_2O_4$, 2-p-Chlorophenylbenzoquinone 2-p-chlorophenylbenzohydroquinone, 45B, 679
$C_{24}H_{16}CuI_2N_4S_8$, Iodobis(1,10-phenanthroline)copper(II) iodide octasulfur, 43B, 788
$C_{24}H_{16}N_4S_4$, Hexamethylenetetrathiafulvene - tetracyanoquinodimethane, 44B, 564
$C_{24}H_{16}N_6 \cdot 1.8\ CH_2Cl_2$, Benzidine - 7,7,8,8-tetracyano-p-quinodimethane (1:1) dichloromethane solvate, 38B, 622
$C_{24}H_{16}N_6$, Benzidine - 7,7,8,8-tetracyano-p-quinodimethane (1:1), 40B, 552
$C_{24}H_{17}AsCl_4O_3$, Triphenylarsine oxide - tetrachlorocatechol, 37B, 351
$C_{24}H_{17}N_3O_4$, N,N'-Dimethylpyromelliticdiimide - carbazole, 44B, 564
$C_{24}H_{18}O_4$, 2-Phenylbenzoquinone 2-phenylbenzohydroquinone, 45B, 679
$C_{24}H_{20}N_4$, Tetracyanoethylene - [3.3]paracyclophane, 37B, 351
$C_{24}H_{20}N_4Se_4$, 2,3,6,7-Tetramethyl-1,4,5,8-tetraselenafulvalenium 2,5-dimethyl-7,7,8,8-tetracyano-p-quinodimethanide, 44B, 565
$C_{24}H_{22}Cl_4O_6$, L-Rotenone carbon tetrachloride complex, 41B, 689
$C_{24}H_{22}N_4$, 3,6-Bis(dicyanomethylene)cyclohexa-1,4-diene - hexamethylbenzene, 35B, 415
$C_{24}H_{28}N_4O_{11}$, 1,2,3,4,4a,6-Hexahydro-10-hydroxy-3,8,9-trimethoxy-5,10b-ethanophenanthridinium picrate, 43B, 757
$C_{24}H_{34}N_8O_8$, 5,5-Diethylbarbituric acid - caffeine (2:1), 40B, 567
$C_{25}H_{13}N_5$, 7,7,8,8-Tetracyanoquinodimethane 7,8-benzoquinoline, 46B, 611
$C_{25}H_{15}N_5S$, N-Methylphenothiazine - 7,7,8,8-tetracyanoquinodimethane, 39B, 457
$C_{25}H_{15}N_6$, N-Methylphenazinium - 7,7,8,8-tetracyanoquinodimethanide, 31B, 136; 41B, 690
$C_{25}H_{15}N_6$, N-Methylphenazinium 7,7,8,8-tetracyanoquinodimethanide (semiconducting form), 42B, 496
$C_{25}H_{22}N_4O_3$, 1,3,7,9-Tetramethyluric acid-pyrene complex, 30B, 253
$C_{25}H_{23}N_3O_7$, 2,4,7-Trinitrofluorenone - hexamethylbenzene, 40B, 568
$C_{25}H_{29}N_4O_{10}P$, Procatine - bis-p-nitrophenyl phosphate, 35B, 416

$C_{25}H_{33}BrO_3$, Testosterone-p-bromophenol complex, 34B, 358
$C_{25.2}H_{15.6}N_6$, Phenazine-doped 5,10-dihydro-5,10-dimethylphenaziniumyl 7,7,8,8-tetracyano-p-quinodimethanide, 46B, 611
$C_{26}H_{12}Br_2Cl_2O_2$, Perylene 2,5-dibromo-3,6-dichloro-p-benzoquinone, 45B, 680
$C_{26}H_{12}Cl_2N_2O_2$, Benzo[c]phenanthrene - 2,3-dichloro-5,6-dicyanobenzoquinone, 43B, 788
$C_{26}H_{12}F_4O_2$, Perylene - fluoranil, 28, 576
$C_{26}H_{12}N_4$, Perylene - tetracyanoethylene, 32B, 284; 35B, 417
$C_{26}H_{12}N_4$, 1,2,4,5-Tetracyanobenzene - pyrene, 39B, 459
$C_{26}H_{12}N_6$, o-Phenanthroline - 7,7,8,8-tetracyanoquinodimethane, 43B, 789
$C_{26}H_{14}N_4$, 7,7,8,8-Tetracyanoquinodimethane - anthracene, 33B, 252
$C_{26}H_{18}N_2O_4$, Pyromellitic N,N'-dimethyldiimide - anthracene, 43B, 790
$C_{26}H_{18}N_6$, 7,7,8,8-Tetracyanoquinodimethane - N,N'-dimethyldihydrophenazine, 39B, 459
$C_{26}H_{18}N_6O_{12}$, Stilbene - 1,3,5-trinitrobenzene (1:2), 44B, 565
$C_{26}H_{20}CrN_4$, 7,7,8,8-Tetracyanoquinodimethane - ditoluenechromium, 41B, 691
$C_{26}H_{20}N_4S_2$, 2,2',6,6'-Tetramethyl-Δ(4,4')-bithiopyran 7,7,8,8-tetracyano-p-quinodimethane, 45B, 681
$C_{26}H_{24}KN_3O_{16}$, 2,2'-Di-o-carboxymethoxyphenoxydiethyl ether potassium picrate, 44B, 312
$C_{26}H_{28}Br_2N_{12}O_3$, Phenobarbital - 8-bromo-9-ethyladenine, 33B, 254
$C_{26}H_{28}N_{10}O_7$, Theophylline - phenobarbital (2:1), 43B, 531
$C_{26}H_{35}NO_5$, Cortisol - pyridine (1:1), 40B, 569
$C_{26}H_{35}N_5O_3$, Deoxycorticosterone - adenine, 41B, 692
$C_{26}H_{37}N_5O_4$, Deoxycorticosterone adenine monohydrate, 41B, 608
$C_{26}H_{39}N_5O_4$, Deoxycorticosterone - adenine monohydrate, 41B, 496
$C_{27}H_{36}O_4$, Progesterone - resorcinol (1:1), 41B, 692
$C_{28}H_{12}F_{12}NiS_4$, Perylene - bis-cis-(1,2-perfluoromethylethylene-1,2-dithiolato)nickel, 34B, 354
$C_{28}H_{12}N_4S_3$, Trithia(5)heterohelicene 7,7,8,8-tetracyano-p-quinodimethane (1:1), 46B, 612
$C_{28}H_{14}N_4$, Pyrene - 7,7,8,8-tetracyanoquinodimethane, 39B, 460
$C_{28}H_{14}N_6O_2Pd$, Palladium(II) 8-hydroxyquinolate-1,2,4,5-tetracyanobenzene complex, 31B, 280
$C_{28}H_{14}O_6$, Benz[a]anthracene - pyromellitic dianhydride (1:1), 42B, 497
$C_{28}H_{16}N_6NiS_4$, Bis(N,N,N',N'-tetramethyl-p-phenylenediamine) bis(ethylenedithiodicyano)nickel, 38B, 623
$C_{28}H_{20}N_4O_4S_2$, 2,3,7,8-Tetramethoxythianthrene - 7,7,8,8-tetracyanoquinodimethane, 43B, 791
$C_{28}H_{22}CuN_8O_{14}$, 1,3,5-Trinitrobenzene - bis(N-methylsalicylaldiminato)copper(II) (2:1), 39B, 461
$C_{28}H_{22}N_8$, Quinolinium 2-dicyanomethylene-1,1,3,3-tetracyanopropanediide, 37B, 352
$C_{28}H_{24}N_4O_2S_4$, 3,4:3',4'-Octamethylene-2,2',5,5'-tetrathiafulvalene 2,5-dimethoxy-7,7,8,8-tetracyano-p-quinodimethane, 46B, 612
$C_{28}H_{28}BrN_5O_3$, 5,5-Diphenylhydantoin 1-(4-bromophenyl)-4-dimethylamino-2,3-dimethyl-3-pyrazolin-5-one, 46B, 613
$C_{28}H_{31}CoN_5O_8$, 1,3,5-Trinitrobenzene - bis-(N-t-butylsalicylideneiminato)cobalt(II), 37B, 353
$C_{28}H_{31}CuN_5O_8$, 1,3,5-Trinitrobenzene - bis-(N-t-butylsalicylideneiminato)copper(II), 37B, 353
$C_{28}H_{31}N_5NiO_8$, 1,3,5-Trinitrobenzene - bis-(N-t-butylsalicylideneiminato)nickel(II), 37B, 353

$C_{29}H_{21}N_5O_7$, 3,5,7-Triphenyl-4H-1,2-diazepine picrate, 38B, 623
$C_{30}H_{14}O_6$, Perylene-pyromellitic dianhydride complex, 30B, 254
$C_{30}H_{16}Cl_4O_5$, 2,3-Dichloro-1,4-naphthoquinone - α-naphthol (2:1), 43B, 199
$C_{30}H_{16}CuN_6O_2$, 7,7,8,8-Tetracyanoquinodimethane-copper(II) 8-hydroxyquinolate complex, 32B, 286
$C_{30}H_{16}CuN_{14}O_{14}$, Picryl azide-bis(8-hydroxyquinolato)copper(II), 30B, 260
$C_{30}H_{16}N_4$, Chrysene - 7,7,8,8-tetracyanoquinodimethane (1:1), 40B, 570
$C_{30}H_{17}BrN_3O_6$, Pyrene - picryl bromide (3:2), 41B, 694
$C_{30}H_{18}N_4$, 7,7,8,8-Tetracyanoquinodimethane - p-terphenyl (1:1), 42B, 497
$C_{30}H_{18}N_4O$, 2,4,6-Triphenylpyrylium 1,1,3,3-tetracyanopropenide, 40B, 570
$C_{30}H_{20}N_9S$, N-Ethyl-2-methylthiazolinium - 7,7,8,8-tetracyanoquinodimethane, 41B, 695
$C_{30}H_{20}N_9S$, N-Methyl-2-ethylthiazoline - 7,7,8,8-tetracyanoquinodimethane (1:2), 43B, 501
$C_{30}H_{20}O_6$, 1,4-Naphthoquinone - 1,4-naphthohydroquinone (2:1), 44B, 566
$C_{30}H_{22}I_3N_3$, Iodoform - tris(quinoline), 27, 986
$C_{30}H_{22}N_6$, Benzidine - 7,7,8,8-tetracyano-p-quinodimethane (1:1) benzene solvate, 40B, 571
$C_{30}H_{33}N_4O_{11}P$, N,N'-Bis-(4-ethoxyphenyl)acetamidinium bis-p-nitrophenylphosphate monohydrate, 37B, 355
$C_{30}H_{34}O_{13}$, Picrotoxin, 42B, 498
$C_{30}H_{43}I_2O_4$, Deoxycholic acid - diiodobenzene, 38B, 626
$C_{31}H_{21}N_9O_7$, Tetramethylammonium p-tricyanovinylphenyldicyanomethide 2,4,7-trinitrofluorenone, 46B, 614
$C_{31}H_{24}N_9O$, N-Methyl-N-ethylmorpholinium bis(7,7,8,8-tetracyanoquinodimethane), 43B, 792
$C_{32}H_{16}Cl_2N_6O_2$, 9,10-Diazaphenanthrene - 2,3-dichloro-5,6-dicyano-1,4-benzoquinone (2:1), 44B, 566
$C_{32}H_{16}N_4$, Perylene - 7,7,8,8-tetracyanoquinodimethane, 39B, 462
$C_{32}H_{26}BF_5$, Phenyltropylium tetrafluoroborate - triphenylmethyl fluoride, 42B, 114
$C_{33}H_{16}N_9$, Quinolinium bis-(7,7,8,8-tetracyanoquinodimethanide), 37B, 356
$C_{33}H_{28}N_4O_6$, Lumiflavin - bis(naphthalene-2,3-diol), 40B, 572
$C_{33}H_{28}N_4O_6$, 10-Propylisoalloxazine - bis(naphthalene-2,3-diol), 40B, 572
$C_{33}H_{34}N_4O_9$, Lumiflavin - bis(naphthalene-2,3-diol) trihydrate, 41B, 695
$C_{34}H_{12}F_4O_2$, Chrysene - fluoranil, 41B, 696
$C_{34}H_{24}N_{10}$, 7,7,8,8-Tetracyanoquinodimethane - N,N,N',N'-tetramethyl-p-phenylenediamine, 33B, 254
$C_{34}H_{26}N_6$, 9-Ethylcarbazole - tetracyanoethylene (2:1), 44B, 567
$C_{35}H_{18}O_{12}$, Pyromellitic dianhydride - trans-4-methylstilbene (2:1), 40B, 573
$C_{35}H_{44}Cl_2N_5OP$, Tri-m-toluidylphosphazenyl oxide bis(m-toluidine hydrochloride), 43B, 792
$C_{36}H_{20}N_{10}$, (1,1'-Ethylene-2,2'-bipyridylium) bis(7,7,8,8-tetracyanoquinodimethane), 38B, 625
$C_{36}H_{22}N_9$, N-(n-Propyl)quinolinium 7,7,8,8-tetracyanoquinodimethane (1:2), 38B, 624
$C_{36}H_{24}N_{11}$, 1-Methyl-3-ethylbenzimidazolium - tetracyanoquinodimeth-

ane (1:2) (acetonitrile solvate), 39B, 462
$C_{36}H_{30}I_6P_2S_2$, Triphenylphosphine sulfide - iodine, 33B, 257
$C_{36}H_{32}HgN_{10}S_3$, Mercury dithizone-pyridine complex, 22, 724
$C_{36}H_{52}Cl_2N_{10}O_8$, 1,1,4,4-Tetraethylpiperazinium dichloride - 4(p-nitroaniline), 35B, 419
$C_{37}H_{18}N_9$, Acridinium bis(7,7,8,8-tetracyanoquinodimethanide), 40B, 574
$C_{37}H_{24}N_8S_{5.2}$, 4,4',5,5'-Tetramethyl-Δ(2,2')-bis-1,3-dithiole - 7,7,8,8-tetracyano-p-quinodimethane, 42B, 484
$C_{37}H_{26}N_{11}$, 1,2-Dimethyl-3-ethylbenzimidazolium - tetracyanoquinodimethane (1:2) (acetonitrile solvate), 39B, 463
$C_{37}H_{29}N_{10}S_2$, 3,3'-Diethylthiazolinocarbocyanine - 7,7,8,8-tetracyanoquinodimethane (1:2), 42B, 498
$C_{38}H_{21}N_{10}$, N-Ethyl-o-phenanthrolinium tetracyanoquinodimethane, 42B, 498
$C_{38}H_{24}CrN_8$, Ditoluenechromium 7,7,8,8-tetracyanoquinodimethane (1:2), 41B, 918
$C_{38}H_{28}N_8S_4$, 4,4',5,5'-Tetraethyltetrathiofulvalene - bis(tetracyanoquinodimethane), 43B, 793
$C_{38}H_{36}N_8O_6$, 1,3,7,9-Tetramethyluric acid - 3,4-benzpyrene, 31B, 111; 32B, 287
$C_{38}H_{40}Cl_4N_4O_2$, N,N,N',N'-Tetramethylbenzidine - chloranil, 37B, 358
$C_{38}H_{49}O_4$, Deoxycholic acid - phenanthrene, 38B, 626
$C_{39}H_{42}ClN_6Na$, Tris-(4,4'-diaminodiphenylmethane) sodium chloride, 38B, 627
$C_{39}H_{68}O_{31}$ · 4.8 H_2O, α-Cyclohexaamylose - 1-propanol 4.8 hydrate, 40B, 575
$C_{39}H_{85}NaO_{42}S$, α-Cyclodextrin - sodium 1-propanesulphonate nonahydrate, 43B, 561
$C_{40}H_{78}O_{35}S$, α-Cyclodextrin - dimethylsulfoxide - methanol - water (1:1:2:2), 44B, 568
$C_{41}H_{23}N_{10}S_2$, 3,3'-Dimethylthiacyanine - 7,7,8,8-tetracyanoquinodimethane (1:2), 40B, 576
$C_{41}H_{30}N_{11}O_6S_2$, 3,3'-Diethylthiazolinocarbocyanine - 7,7,8,8-tetracyanoquinodimethide - 9-dicyanomethylene-2,4,7-trinitrofluorene (1:1:1), 40B, 576
$C_{41}H_{38}O_6P_2$, Dimethylmalonic acid - triphenylphosphine oxide (1:2), 40B, 576
$C_{42}H_{24}Br_6S_{12}Sn$, Dibenzotetrathiofulvalene hexabromostannate (3:1), 46B, 616
$C_{42}H_{32}N_{14}$, Bis(trimethylammonium) tris(7,7,8,8-tetracyanoquinodimethanide), 39B, 463
$C_{42}H_{50}N_6O_4$, Phenylbutazone - piperazine (2:1), 43B, 281
$C_{42}H_{52}N_{20}O_6$, Bis(lumiflavin - 2,6-diamino-9-ethylpurine) - ethanol - water, 43B, 378
$C_{42}H_{71}NO_{36}$, α-Cyclodextrin - p-nitrophenol trihydrate, 43B, 563
$C_{43}H_{25}N_{10}S_2$, 3,3'-Dimethylthiacarbocyanine - 7,7,8,8-tetracyanoquinodimethane (1:2), 40B, 578
$C_{43}H_{26}AsN_8$, Methyltriphenylarsonium bis(tetracyanoquinodimethanide), 37B, 359
$C_{43}H_{26}N_8P$, Methyltriphenylphosphonium bis-7,7,8,8-tetracyanoquinodimethanide, 37B, 359; 39B, 464
$C_{43}H_{72}O_{36}$, α-Cyclodextrin - p-hydroxybenzoic acid trihydrate, 43B, 563
$C_{44}H_{28}N_8P$, Ethyltriphenylphosphonium bis(α,α,α',α'-tetracyanoquinodimethanide), 45B, 682
$C_{44}H_{28}N_{12}Pt$, Bis(2,2'-bipyridyl)platinum(II) bis(7,7,8,8-tetracyano-

quinodimethane), 43B, 794
$C_{44}H_{32}N_{14}O_2$, Morpholinium - 7,7,8,8-tetracyanoquinodimethane (2:3), 38B, 628
$C_{44}H_{38}Br_2N_8O_{10}$, Lumiflavinium bromide hydroquinone (2:3), 38B, 629
$C_{44}H_{64}F_{12}N_2O_{12}P_2$, Bis(18-crown-6)-1,1'-binaphthyl tetramethylenediamine hexafluorophosphate, 43B, 795
$C_{44}H_{66}Cl_4N_4O_4$, 1,6,20,25-Tetraaza[6.1.6.1]paracyclophane tetrahydrochloride-durene tetrahydrate, 46B, 614
$C_{45}H_{29}N_{10}S_2$, 3,3'-Diethylthiacarbocyanin - 7,7,8,8-tetracyanoquinodimethane (1:2), 39B, 465
$C_{46}H_{34}ClN_{10}$, 1-Methyl-3,3-dimethyl-2-((p-N-methyl-N-β-chloroethylamino)styryl)indole bis(7,7,8,8-tetracyanoquinodimethane), 39B, 466
$C_{48}H_{26}N_{14}$, 1,1'-Dimethyl-4,4'-bipyridylium 7,7,8,8-tetracyano-p-quinodimethanide, 45B, 682
$C_{48}H_{28}N_8P$, Tetraphenylphosphonium bis(tetracyanoquinodimethanide), 33B, 254
$C_{48}H_{33}N_6O_{13}$, 4,4'-Dinitrobiphenyl - 4-hydroxybiphenyl, 11, 672
$C_{48}H_{82}N_2O_{42}$, α-Cyclodextrin - m-nitrophenol (1:2) hexahydrate, 44B, 568
$C_{51}H_{30}N_8P$, Methyltriphenylphosphonium - tetracyano-3,7-naphthoquinodimethane (1:2), 41B, 697
$C_{52}H_{32}Cu_2N_8O_4$, N,N'-(1,2-Phenylene)bis(salicylaldiminato)copper(II) 7,7,8,8-tetracyanoquinodimethane (2:1), 46B, 615
$C_{52}H_{34}N_{12}$, N-Ethylphenazinium bis(7,7,8,8-tetracyanoquinodimethanide), 44B, 569
$C_{56}H_{28}N_{12}S_4$, 4,4'-Bithiopyranylidene tetracyanoquinodimethane (2:3 complex), 46B, 615
$C_{58}H_{53}Cl_3F_6NO_8P$, 1,1'-Binaphthylmacrocyclicpolyether (R)-phenylglycine methyl ester hexafluorophosphate chloroform, 43B, 426
$C_{60}H_{34}N_{22}Pd$, Tetrakis(methyl isocyanide)palladium(II) tetrakis-(7,7,8,8-tetracyano-p-quinodimethane) diacetonitrile solvate, 42B, 499
$C_{60}H_{46}N_{18}$, [N,N,N,N',N',N'-Hexamethylhexamethylenediammonium] tetrakis[tetracyanoquinodimethane], 45B, 683
$C_{61}H_{32}N_{18}$, 1,3-Di(N-pyridinium)propane tetrakis(7,7,8,8-tetracyanoquinodimethane), 43B, 796
$C_{62}H_{34}N_{14}$, 1,4-Di(N-quinolinium-methyl)benzene tris(7,7,8,8-tetracyanoquinodimethane), 43B, 797
$C_{62}H_{34}N_{16}$, N-Methylphenazinium - tetracyanoquinodimethane (2:3), 41B, 699
$C_{62}H_{34}N_{18}$, N,N'-Diethyl-4,4'-bipyridylium - 7,7,8,8-tetracyanoquinodimethane (1:4), 41B, 700
$C_{62}H_{34}N_{18}$, 1,2-Di(N-methyl-4-pyridinium)ethane tetrakis(7,7,8,8-tetracyanoquinodimethane, 43B, 798
$C_{62}H_{65}Cl_6N_9Nb_3$, Tris((di-μ-chloro)(hexamethylbenzene)niobium) bis(7,7,8,8-tetracyano-p-quinodimethane) acetonitrile solvate, 43B, 799
$C_{64}H_{36}N_{18}$, 1,2-Di(N-ethyl-4-pyridinium)ethylene - 7,7,8,8-tetracyanoquinodimethane (1:4), 42B, 499
$C_{64}H_{38}N_{18}$, [N,N'-Di(n-propyl)-4,4'-dipyridylium] bis(7,7,8,8-tetracyanoquinodimethanide) bis(7,7,8,8-tetracyanoquinodimethane), 45B, 684
$C_{66}H_{34}N_{18}$, 1,4-Di-(N-pyridinium-methyl)benzene tetrakis(7,7,8,8-tetracyanoquinodimethane), 41B, 701
$C_{72}H_{38}N_{18}$, N,N'-Dibenzyl-4,4'-bipyridylium 7,7,8,8-tetracyanoquinodimethane, 38B, 630
$C_{72}H_{40}N_4$, 7,7,8,8-Tetracyanoquinodimethane - perylene (1:3), 44B,

569
$C_{76}H_{46}N_{22}$, 7,7,8,8-Tetracyanoquinodimethane - 1,2-bis(1-benzyl-4-pyridyl)ethane (5:1), 44B, 570
$C_{77}H_{84}N_{25}O_{28}P_2S \cdot 23\ H_2O$, Proflavine-cytidylyl-(3',5')-guanosine sulfate hydrate, 45B, 685
$C_{86}H_{44}N_{22}$, 1,2-Di(N-benzyl-4-pyridinium)ethylene pentakis(7,7,8,8-tetracyanoquinodimethane), 43B, 799
$C_{92}H_{64}N_{24}$, Bis(1,2-di(N-ethyl-4-pyridinium)ethane) pentakis(7,7,8,8-tetracyanoquinodimethane), 43B, 800
$C_{112}H_{64}Cl_{18}S_{32}Sn_3$, Dibenzotetrathiofulvalene hexachlorostannate (8:3), 46B, 616
$C_{168}H_{84}Cl_4H_{36}$, Anthracene - tetracyanoethylene complex, 35B, 421

61. CLATHRATES

CH_4N_2O, 1,4-Dichlorobutane - urea host structure, 38B, 594
$CH_7Al_2NO_{10}Si_2$, Dickite - formamide, 42B, 500
$C_2H_4O \cdot 7.67\ H_2O$, Ethylene oxide hydrate, 30B, 267; 43B, 802
$C_4H_8O \cdot H_2S \cdot 17\ H_2O$, Tetrahydrofuran hydrogen sulfide hydrate, 30B, 268
$C_4H_{11}N \cdot 8.67\ H_2O$, Diethylamine hydrate, 32B, 288
$C_4H_{11}N \cdot 9.75\ H_2O$, t-Butylamine hydrate, 32B, 289
$C_4H_{20}FNO_4$, Tetramethylammonium fluoride tetrahydrate, 32B, 291
$C_4H_{23}NO_6$, Tetramethylammonium hydroxide pentahydrate, 31B, 282
$C_6H_6O_2 \cdot 0.33\ HCl$, Hydroquinone hydrogen chloride clathrate, 43B, 801
$C_6H_{24}N_4O_6$, Hexamethylenetetramine hexahydrate, 30B, 269
$C_{12}H_{22}CdN_6NiO_4$, Diamminecadmium tetracyanonickelate di-dioxane, 42B, 500
$C_{12}H_{67}FO_{20}S$, Tri-n-butylsulphonium fluoride hydrate, 27, 781
$C_{12}H_{88}N_4O_{26}$, n-Propylamine clathrate hydrate, 39B, 466
$C_{12}H_{118}N_4O_{41}$, Trimethylamine clathrate hydrate, 33B, 260
$C_{13}H_{22}FeN_6N_3$, Ferrocene - thiourea clathrate, 44B, 571
$C_{14}H_{18}CdN_8Ni$, catena-μ-Ethylenediaminecadmium(II) catena-tetra-μ-cyano-nickelate(II) - bis(pyrrole), 41B, 701
$C_{16}H_{18}CdN_6Ni$, Diamminecadmium tetracyanonickelate benzene clathrate, 40B, 1096
$C_{16}H_{18}CuN_6Ni$, Diamminecopper(II) tetracyanonickelate(II) dibenzene clathrate, 39B, 467
$C_{16}H_{18}MnN_6Ni$, Diamminemanganese(II) tetracyanonickelate dibenzene clathrate, 40B, 579
$C_{16}H_{36}FN \cdot 32.8\ H_2O$, Tetra-n-butylammonium fluoride hydrate, 28, 578
$C_{18}H_{18}O_6 \cdot 0.768\ H_2S$, Hydroquinone - hydrogen sulfide clathrate, 42B, 500
$C_{18}H_{18}O_8S$, Quinol - sulfur dioxide, 11, 646
$C_{18}H_{18}O_9S_3$, Phenol-sulfur dioxide clathrate, 22, 697
$C_{18}H_{20}OS \cdot 0.33\ CCl_4$, 4-p-Mercaptophenyl-2,2,4-trimethylchroman carbon tetrachloride solvate, 45B, 687
$C_{18}H_{24}O_3S_3$, Phenol-hydrogen suphide clathrate, 22, 697
$C_{18}H_{30} \cdot 0.39\ C_6H_{12}$, trans-Perhydrotriphenylene cyclohexane solvate, 32B, 292
$C_{19}H_{22}OS \cdot 0.2222\ C_8H_{16}$, 4-p-Hydroxyphenyl-2,2,4,8-tetramethylthiachroman cyclooctane clathrate, 45B, 688
$C_{19}H_{22}O_7$, Quinol - methanol, 11, 649
$C_{20}H_{21}NO_6$, Hydroquinone - acetonitrile (3:1) clathrate, 44B, 571
$C_{20}H_{44}BrP \cdot 32\ H_2O$, Tetraisoamyl phosphonium bromide hydrate, 45B,

688
$C_{20}H_{120}FNO_{38}$, Tetra(iso-amyl)ammonium fluoride hydrate, 26, 613
$C_{23}H_{41}NO_2 \cdot 39.5\ H_2O$, Tetra-n-butylammonium benzoate hydrate, 27, 778
$C_{24}H_{18}N_3O_6P_3$, Tris-(o-phenylenedioxy)cyclotriphosphazene benzene clathrate, 42B, 501
$C_{24}H_{30}N_4O_3$, Hexamethylenetetramine triphenol, 35B, 423
$C_{26}H_{22}N_3O_6P_3$, Tris-(o-phenylenedioxy)cyclotriphosphazene o-xylene clathrate, 42B, 501
$C_{27}H_{34}O_8$, 10,15-Dihydro-2,3,7,8,12,13-hexahydroxy-5H-tribenzo-[a,d,g]cyclononene di-2-propanolate clathrate, 46B, 617
$C_{28}H_{26}N_6Ni_2$, Diamminenickel tetracyanonickelate di-biphenyl clathrate, 40B, 580
$C_{30}H_{250}N_{10}O_{80}$, Isopropylamine octahydrate, 35B, 422
$C_{33}H_{38}O_7$, Cycloveratril-benzene-water inclusion compound, 45B, 689
$C_{34}H_{23}N_3O_6P_3$, Tris(1,8-naphthalenedioxy)cyclotriphosphazene hemi-p-xylene clathrate, 40B, 581
$C_{35}H_{36}N_4O_4S_2$, 5-Methylbenzene-1,3-dicarbaldehyde bis-(p-tolylsulphonylhydrazone) benzene inclusion compound, 46B, 618
$C_{36}H_{60}O_{30} \cdot 0.5\ C_{14}H_{14}KN_3O_3S \cdot 9.75\ H_2O$, Potassium (α-cyclodextrin - methyl orange (2:1)) hydrate, 42B, 501
$C_{36}H_{60}O_{30} \cdot 0.5\ C_{14}H_{14}N_3NaO_3S \cdot 9.75\ H_2O$, Sodium (α-cyclodextrin - methyl orange (2:1)) hydrate, 42B, 501
$C_{36}H_{60}O_{30} \cdot X\ H_2O \cdot Y\ Kr$, α-Cyclodextrin krypton pentahydrate, 42B, 502
$C_{37}H_{74}O_{36}$, α-Cyclodextrin - methanol pentahydrate, 42B, 503
$C_{39}H_{67}NO_{31} \cdot 5\ H_2O$, α-Cyclodextrin N,N-dimethylformamide pentahydrate, 45B, 690
$C_{40}H_{67}NO_{31} \cdot 5\ H_2O$, α-Cyclodextrin 2-pyrrolidone pentahydrate, 45B, 690
$C_{42}H_{30}S_6 \cdot 2\ CCl_4$, Hexakisphenylthiobenzene carbon tetrachloride clathrate, 45B, 691
$C_{42}H_{71}IO_{34}$, α-Cyclodextrin - p-iodophenol trihydrate, 42B, 504
$C_{42}H_{72}INO_{33}$, α-Cyclodextrin - p-iodoaniline trihydrate, 42B, 504
$C_{42}H_{79}N_2O_{38}$, α-Cyclodextrin m-nitroaniline hexahydrate, 46B, 618
$C_{42}H_{85}NaO_{43}S$, α-Cyclodextrin - sodium benzenesulphonate decahydrate, 42B, 505
$C_{46}H_{42}Br_2N_6NiS_2$, Bis(isothiocyanato)tetrakis(4-methylpyridine)nickel(II) bis(2-bromonaphthalene), 46B, 619
$C_{48}H_{36}N_3O_6P_3$, Tris(2,3-naphthalenedioxy)cyclotriphosphazene benzene adduct, 40B, 582
$C_{48}H_{48}N_6NiS_2$, Bis(isothiocyanato)tetrakis(4-methylpyridine)nickel-(II) bis(2-methylnaphthalene), 46B, 619
$C_{58}H_{62}O_2S_6$, Hexakis(benzylthiomethyl)benzene 1,4-dioxan inclusion compound, 46B, 619
$C_{71}H_{70}N_4O_4$, 2-Phenyl-3-p-(2,2,4-trimethylchroman-4-yl)phenylquinazolin-4(3H)-one methylcyclohexane clathrate, 43B, 802
$C_{71}H_{77}NO_{12}$, Bis(tri-o-thymotide) pyridine clathrate, 43B, 804
$C_{86}H_{82}P_6Se_6$, 1,2-Bis(diphenylphosphinoselenoyl)ethane p-xylene inclusion compound, 46B, 620
$C_{115}H_{136}O_{13}$, Dianin's compound - 1-heptanol, 37B, 361

62. BORON COMPOUNDS

CBF_6K, Potassium (trifluoromethyl)trifluoroborate, 43B, 805
CH_3BF_2, Methylboron difluoride (gas-ed), 9, 337

CH_5BF_3N, Methylamine - boron trifluoride, 15, 418
CH_5BN_2, Ammonia-cyanoborane, 44B, 573
CH_7B_5, Monocarbahexaborane (gas-mw), 42B, 1033
CH_8BP, Methylphosphine-borane (gas-mw), 38B, 1069
CH_8B_2, Methyldiborane (gas-mw), 43B, 1485
CH_9B_5, 2-Carbahexaborane(9) (gas-mw), 37B, 695
$CH_9B_{10}Cl_2P$, 9,10-Dichlorophosphacarborane, 41B, 703
$CH_{11}B_5$, 1-Methylpentaborane(9) (gas-mw), 33B, 538
C_2BCsF_8, Cesium bis(trifluoromethyl)difluoroborate, 46B, 621
$C_2H_3BCl_3N$, Acetonitrile-boron trichloride, 40B, 1097
$C_2H_3BF_2$, Vinyldifluoroborane (gas-mw), 40B, 1144; 42B, 1036
$C_2H_3BF_3N$, Acetonitrile-boron trifluoride, 15, 421; 40B, 1097
$C_2H_4B_2Cl_4$, 1,2-Bis(dichloroborane)ethane, 20, 528
$C_2H_4B_{10}Cl_8$, Octachlorocarborane, 29, 529
$C_2H_5B_3$, 1,5-Dicarba-closo-pentaborane(5) (gas-ed), 39B, 888
$C_2H_5B_4Cl$, 2-Chloro-1,6-dicarbahexaborane(6) (gas-mw), 37B, 696
C_2H_6BF, Dimethylboron fluoride (gas-ed), 9, 337
$C_2H_6BF_3O$, Dimethylether - boron trifluoride (gas-ed), 9, 337; 10, 209
C_2H_6BN, Methyl isocyanide-borane (gas-mw), 43B, 1485
$C_2H_6B_2S_3$, Dimethyl-1,2,4-trithia-3,5-diborolane (gas-ed), 39B, 885
$C_2H_6B_4$, closo-2,3-Dicarbahexaborane(6) (gas-mw), 35B, 904
$C_2H_6B_4$, 1,6-Dicarba-closo-hexaborane(6) (gas-ed), 39B, 888
$C_2H_6Cl_2BN$, Dimethylaminodichloroborane (gas-ed), 35B, 890
$C_2H_7BN_4$, Dimethylcyclotetrazenoborane (gas-ed), 40B, 1141
C_2H_8BN, Aziridine borane, 34B, 359
$C_2H_8B_4$, 2,3-Dicarbahexaborane(8), 29, 526
$C_2H_8B_6$, 1,7-Dicarba-closo-octaborane(8) (gas-mw), 42B, 1032
$C_2H_9B_7$, 1,6-Dicarbanonaborane(9) (gas-mw), 42B, 1032
$C_2H_9B_{10}Br_3$, Tribromo-m-carborane, 42B, 507
$C_2H_9B_{10}Br_3$, Tribromo-o-carborane, 31B, 283
$C_2H_9B_{10}Cl_3$, 4,9,10-Trichloro-m-carborane-1,7, 42B, 507
$C_2H_{10}BP$, Dimethylphosphineborane (gas-mw), 40B, 1145
$C_2H_{10}B_2$, 1,1-Dimethyldiborane (gas-ed), 15, 341, 417
$C_2H_{10}B_{10}Br_2$, Dibromo-o-carborane, 31B, 283
$C_2H_{10}B_{10}Br_2$, 1,12-Dibromo-1,2-dicarba-closo-dodecaborane, 41B, 703
$C_2H_{10}B_{10}Br_2$, 9,10-Dibromo-1,7-dicarbadodecaborane(12), 32B, 292
$C_2H_{10}B_{10}I_2$, B,B'-Diiodoneocarborane (gas-ed), 33B, 530
$C_2H_{10}B_{10}I_2$, C,C'-Di-iodine-p-carborane (gas-ed), 42B, 1024
$C_2H_{11}B_2N$, N,N-Dimethylaminodiborane (gas-ed), 15, 340, 417
$C_2H_{11}B_2N$, N,N-Dimethylaminodiborane (gas-mw), 39B, 893
$C_2H_{12}B_{10}$, meta-Carborane (gas-ed), 37B, 689
$C_2H_{12}B_{10}$, ortho-Carborane (gas-ed), 37B, 688
$C_2H_{12}B_{10}$, para-Carborane (gas-ed), 37B, 689
$C_2H_{13}B_5$, 2,3-Dimethylpentaborane, 31B, 287
$C_2H_{14}B_2N_2$, Ethylenediamine-bisborane, 38B, 631
$C_2H_{14}B_4F_2NP$, Dimethylaminodifluorophosphine - tetraborane(8), 34B, 377
$C_2H_{16}B_9N$, Methylcyanoenneaborane, 26, 619
$C_2H_{18}B_{10}$, 1-Ethyldecaborane, 29, 524
$C_3H_6BF_2NO_2$, Difluoroboron N-methylacethydroxamate, 43B, 805
C_3H_8BCl, Dimethyl(chloromethyl)boron, 40B, 583
$C_3H_8B_3MnO_3 \cdot 0.5\ C_7H_8$, Tricarbonyl(octahydrotriborato(1-))manganese (toluene solvate), 44B, 573
C_3H_9B, Trimethylborine (gas-ed), 30B, 400
$C_3H_9BBr_3N$, Trimethylamine - boron tribromide, 37B, 362
$C_3H_9BBr_3P$, Trimethylphosphine boron tribromide, 41B, 727

$C_3H_9BCl_3N$, Trimethylamine - boron trichloride, 34B, 360; 37B, 362
$C_3H_9BCl_3P$, Trimethylphosphine boron trichloride, 41B, 727
$C_3H_9BF_3N$, Trimethylamine - boron trifluoride, 15, 419
$C_3H_9BF_3N$, Trimethylamine-boron trifluoride (gas-mw), 43B, 1485
$C_3H_9BI_3N$, Trimethylamine - boron triiodide, 37B, 362
$C_3H_9BI_3P$, Trimethylphosphine boron triiodide, 41B, 727
$C_3H_9BO_3$, Methyl borate (gas-ed), 42B, 1029
C_3H_9BS, Methylthio-dimethylborane (gas-ed), 39B, 886
$C_3H_9BS_2$, Bis(methylthio)methylborane (gas-ed), 42B, 1024
$C_3H_9BS_3$, Tris(methylthio)borane (gas-ed), 39B, 886
$C_3H_9B_2NS_2$, 3,4,5-Trimethyl-1,2,4,3,5-dithiazadiborolidine, 46B, 621
$C_3H_9B_3Br_6S_3$, Dibromo(methylthio)borane trimer, 42B, 508
$C_3H_9B_3N_3$, β-Trimethyborazole, 31B, 288
$C_3H_9B_4Ga$, 1-Methyl-1-galla-2,4-dicarba-closo-heptaborane(7), 38B, 632
$C_3H_9Cl_6B_3S_3$, Dichloro(methylthio)borane trimer, 42B, 508
$C_3H_{11}B_{10}Br_3$, 1-Methyl-9,10,12-tribromo-m-borane, 40B, 1098
$C_3H_{12}AsB$, Trimethylarsine-borane (gas-mw), 40B, 1144
$C_3H_{12}BN$, Trimethylamine-borane (gas), 39B, 895
$C_3H_{12}BP$, Trimethylphosphine-borane (gas-mw), 38B, 1069
$C_3H_{12}B_3N_3$, N-Trimethylborazole (gas-ed), 19, 545
$C_3H_{12}B_{10}ClI$, B-Chloro-B'-iodo-closocarborane(12), 42B, 508
$C_3H_{14}B_{10}O_2S$, 9-Methylsulphonyl-1,7-dicarba-closo-dodecaborane(12), 46B, 622
$C_3H_{15}B_9$, 11-Methyl-2,7-dicarbo-nido-undecaborane(12), 43B, 806
$C_3H_{16}B_3N$, Trimethylamine triborane, 26, 616
$C_3H_{16}B_5BrSi$, 1-Bromo-μ-trimethylsilylpentaborane(9), 37B, 363
$C_3H_{21}B_9OS$, 4-Methoxy-6-dimethylsulfidododecahydrononaborane, 37B, 363
$C_4H_6B_2$, 2,3,4,5-Tetracarbahexaborane(6) (gas-mw), 40B, 1145
$C_4H_{10}BF_2N$, Diethylaminoborondifluoride, 35B, 424
$C_4H_{12}BLi$, Lithium tetramethylboron, 41B, 704
$C_4H_{12}BN$, (Dimethylamino)dimethylborane, 35B, 425
$C_4H_{12}BNO_2$, Boron nitrogen betaine, 42B, 509
$C_4H_{12}B_2Cl_4N_2$, Dimethylamino-boron dichloride dimer, 28, 580
$C_4H_{12}B_2F_4N_2$, Dimethylaminoboron difluoride, 31B, 289
$C_4H_{12}B_2N_2S$, 2,3,4,5-Tetramethyl-1,3,4,2,5-thiadiazadiborolidine, 46B, 621
$C_4H_{12}B_2O$, Dimethylboric anhydride (gas-ed), 42B, 1029
$C_4H_{12}B_4$, 2,3-Dimethyl-2,3-dicarbahexaborane(8), 29, 526
$C_4H_{12}B_4N_2S_2$, 2,4,6,8-Tetramethyl-3,7-dithia-1,5-diaza-2,4,6,8-tetra-borabicyclo[3.3.0]octane, 41B, 704
$C_4H_{12}B_6$, 1,7-Dimethyl-1,7-dicarbaclovooctaborane(8), 33B, 262
$C_4H_{12}B_{10}Br_4$, 8,9,10,12-Tetrabromo-C,C'-dimethyl-o-carborane, 31B, 283
$C_4H_{13}B_7$, Dimethyl-1,6-dicarbaclovononaborane(9), 33B, 263
$C_4H_{14}B_2$, Tetramethyldiborane (gas-ed), 9, 290; 15, 340, 417; 33B, 531
$C_4H_{14}B_8$, Dimethyl-1,6-dicarba-closo-decaborane(10), 35B, 426
$C_4H_{14}B_9BrO$, 2,3-Dimethyl-4,7-dihydroxy-10-bromo-2,3-dicarba-closo-undecaborane, 46B, 623
$C_4H_{14}B_{10}Br_2$, 1,2-Bis(bromomethyl)carborane, 29, 531
$C_4H_{14}B_{10}Cl_2$, 5,12-Dichloro-1,7-dimethyl-1,7-dicarba-closo-dodecaborane(12), 39B, 468
$C_4H_{15}B_9$, 6,9-Dimethylcarborane, 31B, 291
$C_4H_{15}B_{10}ClHg$, 1-Methylmercury-2-chloromethyl-o-carborane, 44B, 573
$C_4H_{16}AlB_9$, 3-Ethyl-3-alumina-1,2-dicarba-closo-dodecaborane(12),

38B, 633
$C_4H_{16}BClN_2$, Bis(dimethylamino)boronium chloride, 33B, 264
$C_4H_{16}B_2N_2$, Dimethylaminoborine dimer, 27, 771
$C_4H_{17}B_7$, 6,8-Dimethyl-6,8-dicarbanonaborane, 32B, 293
$C_4H_{18}AlB_9$, 7,8-μ-Dimethylalumina-1,2-dicarba-nido-undecaborane(13), 38B, 634
$C_4H_{18}B_2P_2$, Tetramethylbiphosphine-bis(monoborane), 33B, 275
$C_4H_{18}B_{10}N_2$, Bis(methylisocyanato)decaborane, 23, 605; 24, 559
$C_4H_{20}B_2N_8NiO_{12}$, Bis(dihydroxoboroxalenediamide-dioximato)nickel(II) tetrahydrate, 37B, 364
$C_4H_{22}B_{18}CoCs$, Caesium bis(1,2-dicarbollyl)cobaltate, 32B, 294
$C_4H_{22}B_{18}Ni$, 3,3-commo-Bis(undecahydro-1,2-dicarba-3-nickela-closo-dodecaborane), 35B, 429
$C_4H_{22}B_{20}$, Bis(o-dodecacarborane), 30B, 272
$C_4H_{22}B_{20}$, Bis-B-p-carboranyl, 42B, 509
$C_4H_{23}B_8N_2$, 3-Ethylamine-5,6-μ-ethyl-amino-octaborane(12), 28, 581
$C_4H_{24}B_{10}S_2$, Bis(dimethylsulfide) - dodecahydrodecaborane, 27, 772
$C_4H_{25}B_{11}CsN$, Caesium tetramethylammonium tridecahydroundecaborate, 32B, 296
$C_4H_{28}B_{20}S_2$, 10,10'-Bis(dimethyl sulfide)hexadecahydroicosaborane, 43B, 807
$C_5H_5BF_3N$, Pyridine-boron trifluoride, 20, 603
$C_5H_7B_3FeO_3$, Dicarbacyclopentaboranyliron tricarbonyl, 39B, 468
$C_5H_9BCl_2F_6N_2S_2$, Trimethylamine(N-B)-(bis(trifluoromethylthio)amino)-dichloroborane, 42B, 510
$C_5H_{11}BO_3$, meso-2,4-Pentanediol borate, 39B, 469
$C_5H_{11}B_9CsO_3Re$, Caesium π-(1)-2,3-dicarbollylrhenium tricarbonyl, 31B, 294
$C_5H_{16}BNaO_5 \cdot 1.5\ H_2O$, Sodium tetrakis(methoxy)borate methanolate sesquihydrate, 43B, 811
$C_5H_{16}B_{10}S_2$, 9,12-Isopropylidenedithio-1,2-dicarba-closo-dodecaborane(12), 46B, 623
$C_5H_{18}B_9Cs$, Cesium 9,10a,11-trimethyl-7,8-dicarba-nido-undecaborate(-1), 45B, 693; 46B, 624
$C_5H_{21}B_{18}CoS_2$, Bis-dicarbollide cobalt(III), 34B, 360; 37B, 365
$C_5H_{23}AsB_{20}$, 1-Methylarsa-2,3,4,5-bis(o-carboranocyclopentane), 45B, 694
$C_6H_5BCl_2$, Phenylboron dichloride (gas-ed), 15, 417; 19, 545
$C_6H_5BF_2$, Phenylboron difluoride (gas-mw), 40B, 1145
$C_6H_6BBrO_2$, 4-Bromophenyl boric acid, 22, 691
$C_6H_7BN_2OS$, 7-Hydroxy-6-methyl-7,6-borazarothieno[3,2-c]pyridine, 40B, 363
$C_6H_7BO_2$, Phenylboronic acid, 43B, 807
$C_6H_{10}B_2I_2S$, 2,5-Diiodo-3,4-diethyl-1,2,5-thiadiborolene, 42B, 510
$C_6H_{12}BNO_3$, triptych-Boroxazolidine (Triethanolamine borate), 38B, 407; 39B, 470
$C_6H_{13}B_7CoCs$, Caesium 3-η-cyclopentadienyloctahydro-4-carba-3-cobalta-closo-nonaborate, 40B, 583
$C_6H_{14}BN_3$, 1,8,10,9-Triazaboradecalin, 34B, 362
$C_6H_{15}BN_4$, Hexamethylenetetramine-borine, 34B, 162
$C_6H_{18}B_2N_2S_2Si$, 4,5-Bis(dimethylamino)-2,2-dimethyl-1,3-dithia-2-sila-4,5-diboracyclopentane, 46B, 624
$C_6H_{18}B_4Cl_4N_6$, 3,5-Dichloro-1,2,4-trimethyl-1,2,4,3,5-triazadiborolidine dimer, 46B, 621
$C_6H_{18}B_4F_4N_6$, 3,6,6,9-Tetrafluoro-2,4,5,7,8,10-hexamethyl-2,4,8,10-tetraaza-5,7-diazonia-3,9-dibora-1,6-diborato-tricyclo[5.3.0.0$^{1-5}$]-decane, 43B, 808

$C_6H_{20}B_2N_2$, 1,1,4,4-Tetramethyl-1,4-diazonia-2,5-diboratacyclohexane, 41B, 705
$C_6H_{20}B_{10}$, 1-t-Butyl-o-carborane, 45B, 693
$C_6H_{20}B_{20}O_2$, 1,2':1',2-Di-μ-carbonyl-bis(1,2-dicarba-closo-dodecaborane(12)), 35B, 427
$C_6H_{21}BN_6$, Tris(2,2-dimethylhydrazino)borane, 39B, 471
$C_6H_{21}B_3N_6$, B-Tris(dimethylamino)borazine, 37B, 365
$C_6H_{22}B_{10}Si$, 1-Trimethylsilyl-2-methyl-o-carborane, 45B, 693
$C_6H_{24}B_3N_3$, Dimethylaminoborine trimer, 26, 618
$C_6H_{24}B_3P_3$, Dimethylphosphinoborine trimer, 19, 545
$C_6H_{25}B_{20}N_3$, (1)-1',5-Bis(acetonitrile)ollyl di-icosahedralborane acetonitrile solvate, 31B, 292
$C_6H_{25}B_{20}P$, 1-Methylphospha-2,3,5,6-bis(o-carboranocyclohexane), 45B, 694
$C_6H_{28}B_6P_2$, Hexaborane(10)-bis(trimethylphosphine), 41B, 1240; 42B, 511
$C_6H_{34}B_{26}Co_2Cs_2O$, Caesium bis(π-(3)-1,2-dicarbollylcobalt)-π-(3,6)-1,2-dicarbacanastide monohydrate, 40B, 1098
$C_7H_{11}BClN$, 2,6-Lutidine-chloroborane, 40B, 584
$C_7H_{16}B_7Co$, 8-η-Cyclopentadienyl-6,7-dicarba-8-cobalta-nido-nonaborane(11), 40B, 585
$C_7H_{16}B_9Fe$, π-Cyclopentadienyl-π-(1)-2,3-dicarbollyliron(III), 30B, 305
$C_7H_{17}B_{10}Co$, (η^5-Cyclopentadienyl)(7,8,9,10,11,12-η^6-dodecahydro-7,9-dicarba-nido-dodecaborato)cobalt(III), 40B, 585
$C_7H_{20}B_9MnO_4$, 5-Tetrahydrofurano-6-(tricarbonyl)-6-manganadodecahydrononaborane, 40B, 586
C_8H_8BN, Ammonium tetrakis(ethynyl)borate, 42B, 511
$C_8H_9BN_2O$, 4-Methyl-2-phenyl-(3H)-1,3,5,2-oxadiazaboroline, 44B, 574
$C_8H_{10}BKO_{11}$, Potassium boromalate monohydrate, 39B, 471
$C_8H_{12}BKO_8$, Potassium tetraacetatoborate, 41B, 706
$C_8H_{12}B_2O_9$, Tetraacetyl diborate, 38B, 636
$C_8H_{13}BN_4$, Hydrazinium tetraethynylborate - hydrazine, 44B, 574
$C_8H_{18}B_2N_2$, 4,5-Diethyl-3,6-dimethyl-1,2-diaza-3,6-diborinane, 46B, 625
$C_8H_{20}B_2N_2S_2$, Bis(diethylamino)dithiaboretane, 43B, 39
$C_8H_{20}B_2N_4$, B,B'-Bis(1,3-dimethyl-1,3,2-diazaborolidin-2-yl), 42B, 512
$C_8H_{20}B_3CrNO_4$, Tetramethylammonium octahydrotriborotetracarbonylchromium, 35B, 437
$C_8H_{20}B_8$, 2,3,7,8-Tetramethyl-2,3,7,8-tetracarbadodecaborane, 43B, 809
$C_8H_{22}B_8Fe$, Bis(2,3-dimethyl-2,3-dicarbahexaboranyl)dihydridoiron, 45B, 695
$C_8H_{24}B_2N_6$, 3,6-Bis(dimethylamino)-1,2,4,5-tetramethyl-1,2,4,5,3,6-tetraazadiborine, 42B, 512
$C_8H_{24}B_4N_4S_2$, 2,3,5,6-Tetra(dimethylamino)-1,4,2,3,5,6-dithiatetraborinane, 46B, 624
$C_8H_{28}B_3O_2Sc$, Scandium tris(borohydride) tetrahydrofuran, 43B, 908
$C_8H_{28}B_{10}S$, 9-Cyclohexyl-5(7)-(dimethyl sulfide)-nido-decaborane(11), 46B, 625
$C_8H_{28}B_{18}Br_6CoN$, Tetramethylammonium bis-π-(5,9,10-tribromo-(1)-2,3-dicarbollyl)cobaltate(III), 33B, 265
$C_8H_{29}B_{10}N$, Tetramethylammonium C,C'-dimethylundecahydrodicarba-nido-dodecaborate, 39B, 472
$C_8H_{30}B_6N_2$, Tetramethylammonium hexahydrohexaborate, 30B, 271
$C_8H_{30}B_{10}Be$, 2,2'-commo-Bis(2-berylla-nido-hexaborane(11)) m-xylene

solvate, 44B, 573
$C_8H_{30}B_{18}Ni$, (3,4')-Bis(1,2-dimethyl-1,2-dicarbollide)nickel(IV), 35B, 430
$C_8H_{31}B_{10}Cl_3N_2$, Tetramethylammonium 1,6,8-trichloroheptahydro-closo-decaborate(2-), 38B, 635
$C_8H_{32}B_4P_4$, Dimethylphosphinoborine tetramer, 27, 989
$C_8H_{32}B_{18}Cr_2CsO$, Caesium 3,3'-commo-bis(nonahydro-1,2-dimethyl-1,2-dicarba-3-chroma-closo-dodecaborate) hydrate, 37B, 366
$C_8H_{34}B_{12}MgO_2$, Bis((2,3-η)nonahydrohexaborato(1-))bis(tetrahydrofuran)magnesium, 42B, 569
$C_8H_{34}B_{18}NNi$, Tetramethylammonium 3,3'-commo-bis(1,2-dicarba-3-nickela-closo-dodecaborate)(1-), 39B, 473
$C_8H_{38}B_{10}N_2$, Tetramethylammonium tetradecahydrodecaborate, 39B, 473
$C_8H_{44}B_{20}FeN_2S_2$, Bis(tetramethylammonium) di(π-(1)-2-thiollyl)iron(II), 38B, 637
$C_8H_{48}B_{20}N_2Ni$, Tetramethylammonium 7,7-commo-bis(dodecahydro-7-nickela-nido-undecaborate)(2-), 38B, 638
$C_9H_{15}B_4Co$, 1-Cyclopentadienyl-2,3-dimethylcobalta-2,3-dicarbaheptaborane, 43B, 809
$C_9H_{17}BBrF_6N_2P$, (+)-4-Methylpyridinetrimethylaminebromohydroboron hexafluorophosphate, 37B, 367
$C_9H_{18}B_9CoO$, Cyclopentadienyl-cobalt-(8-acetyl-1,2-dicarba-undecaborane), 42B, 513
$C_9H_{18}B_9CoO_2$, Cyclopentadienyl-cobalt-(8-acetoxy-1,2-dicarba-undecaborane), 42B, 513
$C_9H_{18}B_{10}$, 4-(p-Tolyl)-o-carborane, 44B, 1106
$C_9H_{18}B_{10}O$, 4-Benzyloxy-σ-carborane, 44B, 575
$C_9H_{24}B_3N_3$, Tris-1,3,5-(dimethylamino)-1,3,5-triboracyclohexane, 34B, 363
$C_9H_{26}B_9NS$, 9-Triethylamine-6-thiadecaborane(11), 45B, 695
$C_9H_{35}B_5BrP_5$, 4-Bromo-1,1,3,3,5,7,7,9,9-nonamethylbicyclo[4.4.0]pentaborophane, 41B, 733
$C_{10}H_9BF_2O_2$, Benzoylacetonato boron difluoride, 38B, 638
$C_{10}H_{14}BNO_2$, B-Phenyl-diptychboroxazolidine, 41B, 707
$C_{10}H_{15}BFeN_2O$, Heterocyclic boron complex, 39B, 474
$C_{10}H_{15}BN_2$, 3,4-Dihydro-3,4,4-trimethyl-4,3-borazaroisoquinoline, 42B, 299
$C_{10}H_{18}B_2$, Hexamethyl-tetracarbahexaborane(6) (gas-ed), 39B, 892
$C_{10}H_{20}B_{10}$, 1,2-Dimethyl-9-phenyl-o-carborane, 44B, 1106
$C_{10}H_{22}B_2N_2S$, 3,4-Diethyl-2,5-bis(dimethylamino)-1,2,5-thiadiborolene, 43B, 810
$C_{10}H_{22}B_6$, 2,4,6,8,9,10-Hexamethyl-2,4,6,8,9,10-hexabora-adamantane, 43B, 811
$C_{10}H_{24}B_{10}NP$, 7-Trimethylamino-9,10-phenylphosphido-nido-7-carbaundecaborane, 44B, 1107
$C_{10}H_{27}B_9FeN$, Tetramethylammonium 1-η-cyclopentadienyl-1-ferra-2-carbaundecaborate, 44B, 575
$C_{10}H_{30}B_6P_2Pt$, 6,8-Dimethyl-1,1-bis(trimethylphosphine)-6,8-dicarba-1-platina-nonaborane, 42B, 513
$C_{11}H_{18}B_2CrN_2O_3$, 4,5-Diethyl-3,6-dimethyl-1,2-diaza-3,6-diborinane-tricarbonylchromium, 46B, 625
$C_{11}H_{18}B_7CoNi$, 2,3-Di-η-cyclopentadienyl-10-carba-(2,3)(nickelacobalta)decaborane(8)(Ni-Co), 44B, 576
$C_{11}H_{30}B_{18}Cs_2Fe_2O_6$, Dicaesium di-μ-carbonyl-bis(π-(3)-1,2-dicarbollylcarbonyliron) acetone hydrate, 35B, 432
$C_{12}H_9BO_2$, 2-Phenyl-1,3,2-benzodioxaborol, 40B, 587
$C_{12}H_{10}BCl_2N$, Dichloro(diphenylamino)borane, 41B, 708

$C_{12}H_{14}BNOS_2$, Spiro(9H-borepino[2,3-b:7,6-b']dithiophene-9,2'-(1,3,2)oxazaborolidine), 44B, 576
$C_{12}H_{17}B_5Co_2$, 1,7,5,6-Bis(cyclopentadienylcobalta)dicarborane, 42B, 514
$C_{12}H_{17}B_5Co_2$, 1,8,5,6-Bis(cyclopentadienylcobalta)dicarborane, 42B, 514
$C_{12}H_{18}B_2CoF_2N_6O_6$, 1,8-Bis(fluoroboro)-2,7,9,14,15,20-hexaoxa-3,6,10,13,16,19-hexaaza-4,5,11,12,17,18-hexamethylbicyclo[6.6.6]-eicosa-3,5,10,12,16,18-hexaenecobalt(II), 37B, 368
$C_{12}H_{18}B_3CoF_6N_6O_6$, 1,8-Bis(fluoroboro)-2,7,9,14,15,20-hexaoxa-3,6,10,13,16,19-hexaaza-4,5,11,12,17,18-hexamethylbicyclo[6.6.6]-eicosa-3,5,10,12,16,18-hexaenecobalt(III) tetrafluoroborate, 37B, 368
$C_{12}H_{18}B_6Co_2$, 2,6-Di-η-cyclopentadienyloctahydro-1,10-dicarba-2,6-dicobalta-closo-decaborane, 40B, 588
$C_{12}H_{18}B_6Fe_2$, 1,6-Bis(η-cyclopentadienyl)-1,6-diferra-2,3-dicarba-closo-decaborane, 41B, 896
$C_{12}H_{19}B_7CoFe$, 1,8-Di-η-cyclopentadienyl-1-ferra-8-cobalta-2,3-dicarbaundecaborane(9), 44B, 577
$C_{12}H_{20}B_8Co_2$, 2,3-Di-η-cyclopentadienyl-1,7-dicarba-2,3-dicobaltadodecaborane(10), 40B, 588
$C_{12}H_{20}B_9Co$, Cobaltocenium carborane, 40B, 589
$C_{12}H_{22}B_2Hf$, Bis(methylcyclopentadienyl)hafnium bis(tetrahydroborate), 44B, 578
$C_{12}H_{23}B_{10}P$, 5,6-μ-Diphenylphosphino-decaborane(14), 39B, 475
$C_{12}H_{27}BN_2O_3$, 1,7-Diaza-4,10,15-trioxabicyclo[5.5.5]heptadecane-borane, 44B, 577
$C_{12}H_{30}B_3N_3$, Hexaethylborazine, 30B, 270
$C_{12}H_{36}B_3O_3Y$, Tris(tetrahydroborato)tris(tetrahydrofuran)yttrium(III), 44B, 578
$C_{12}H_{38}B_{14}CoN$, Tetraethylammonium 2,2'-commo-bis(nonahydrodicarba-2-cobalta-closo-decaborate), 38B, 639
$C_{12}H_{40}B_{18}N_2Ni$, N,N'-Dimethyltriethylenediammonium bis-(3)-1,2-dicarbollylnickelate, 35B, 431
$C_{12}H_{42}B_{10}N_2$, 6,9-Bis(triethylamino)-nido-decaborane, 44B, 579
$C_{12}H_{50}B_{20}N_2$, Bis(triethylammonium) octadecahydroicosaborate, 37B, 369
$C_{13}H_{12}BBrClN$, Methyl(4-bromophenyl)amino-chlorophenylborane, 43B, 812; 44B, 579
$C_{13}H_{13}BFeN_6O_3$, Acetyl(tri-1-pyrazolylborato)(dicarbonyl)iron, 39B, 476
$C_{13}H_{17}B_3Co_2$, 2-Methyl-1,7-bis(η^5-cyclopentadienyl)-1,7,2,4-dicobaltadicarbaheptaborane(7), 41B, 897
$C_{13}H_{19}B_2F_6N_6P$, Tris(1,2-pyrazolide)di(ethylboronium) hexafluorophosphate, 44B, 580
$C_{13}H_{25}B_7Fe$, (η^5-Cyclopentadienyl)ferra-tetramethyltetracarborane, 43B, 813
$C_{13}H_{25}B_8CoFe$, Iron-cobalt metallocarborane, 42B, 515
$C_{13}H_{35}B_9MnNO_4$, 8-(4-(Triethylammonium)-N-butyloxy)-6-tricarbonyl-6-manganadecaborane, 40B, 590
$C_{14}H_{14}BF_2NO$, B,B-Bis-(p-fluorophenyl)boroxazolidine, 40B, 591
$C_{14}H_{16}BNO$, B,B-Diphenylboroxazolidine (monoclinic), 39B, 476
$C_{14}H_{16}BNO$, B,B-Diphenylboroxazolidine (orthorhombic), 42B, 516
$C_{14}H_{20}B_6Co_2$, Tetracarbon-cobaltacarborane, 44B, 581
$C_{14}H_{20}B_{10}$, 1,7-Diphenyl-m-carborane, 43B, 813
$C_{14}H_{24}B_2Br_2N_4$, 2,6-Dibromo-4,4,8,8-tetraethylpyrazabole, 43B, 814
$C_{14}H_{37}B_{10}N_3Pd$, 1,1-Bis(t-butylisocyanide)-2-(trimethylamine)-2-

carba-1-pallada-closo-decaborane(10), 41B, 830
$C_{14}H_{42}Au_2B_{18}N_2S_4$, Bis(N,N-diethyldithiocarbamato)gold(III) 3,3'-bis(1,2-dicarbollyl)aurate, 43B, 815
$C_{15}H_{16}BBrClN$, Methyl(2-methyl-4-bromophenyl)amino-chloro(2-methylphenyl)borane, 43B, 812; 44B, 579
$C_{15}H_{16}BNO_2$, 4,5-Dimethyl-2,2-diphenyl-1,3-dioxa-4-azonia-2-borata-cyclopent-4-ene, 44B, 581
$C_{15}H_{17}B_3Co_2$, Bis(cyclopentadienylcobalt)metallocarborane, 44B, 582
$C_{15}H_{18}BNO_2$, 4,4-Dimethyl-2,2-diphenyl-1,3-dioxa-4-azonia-2-boranata-cyclopentane, 40B, 591
$C_{15}H_{23}BN_2O_3$, Ethyl 2-boryl-3-(2-(4-methoxybenzylidene)-1-methylhydrazino)butanoate, 43B, 815
$C_{15}H_{28}B_7CoO$, Cobaltocarborane, 44B, 582
$C_{15}H_{32}B_9Co_2N$, Tetramethylammonium 2,3-di-η-cyclopentadienyl-2,3-di-cobalta-1-carbadodecaborate, 43B, 816
$C_{15}H_{42}B_{10}P_2Pt$, (Triethylphosphine)(diethylethylidenephosphine)(2-methyl-1,2-dicarbadodecaboranyl(10))platinum(II), 43B, 817
$C_{16}H_{15}BCl_2Co_3NO_{10}$, Triethylamine-dichloroboron-methoxylidine-trico-balt-nonacarbonyl, 41B, 833
$C_{16}H_{17}BCo_3NO_{10}$, Triethylaminoboroniumdecacarbonyltricobalt, 33B, 266
$C_{16}H_{19}B_2NO_3$, 8,8-Dimethyl-3,5-diphenyl-2,4,6-trioxa-1-azonia-3-bora-5-boranatabicyclo[3.3.0]octane, 42B, 516
$C_{16}H_{20}BNO$, B,B-Bis(p-tolyl)boroxazolidine, 42B, 516
$C_{16}H_{21}B_5Ni_3$, Tris(cyclopentadienylnickel)nidocarborane, 42B, 517
$C_{16}H_{22}B_2N_4S_2$, 1,1,5,5-Tetramethyl-3,7-diphenyl-1λ^6,5λ^6,2,4,6,8,3,7-dithiatetraazadiborocine, 46B, 626
$C_{16}H_{30}B_2$, Bis-9-borabicyclo[3.3.1]nonane, 39B, 477
$C_{16}H_{32}B_4NiS_2$, Bis(2,5-dimethyl-3,4-diethyl-1,2,5-thiaborolene)nic-kel(0), 42B, 701
$C_{16}H_{40}B_4N_{10}$, N,N,N',N'-Tetrakis(1,3-dimethyl-1,3,2-diazaborolid-inyl)hydrazine, 45B, 696
$C_{16}H_{42}B_6P_2Pt$, nido-3,8-Dimethyl-2,2-bis(triethylphosphine)-3,8-dicarba-2-platinanonaborane(6), 43B, 817
$C_{16}H_{43}B_7P_2Pt$, 2,7-Dimethyl-9,9-bis(triethylphosphine)-2,7-dicarba-9-platina-nido-decaborane(7), 41B, 1174
$C_{16}H_{45}B_7NiP_2$, 6,6-Bis(triethylphosphine)-5,9-dimethyl-6,5,9-nickel-adicarba-nido-decaborane(9), 41B, 1175
$C_{16}H_{60}B_{40}CoN$, Tetraethylammonium biscarboranecobalt, 38B, 641
$C_{17}H_{14}BNO_5$, 2-Phenyl-1,3,2-dioxaborolane-4,5-dione 3,4-dihydroiso-quinoline-N-oxide adduct, 42B, 517
$C_{17}H_{18}BNO_2$, L-Prolinatodiphenylboron, 43B, 818
$C_{17}H_{45}B_{16}Co_2N$, Tetraethylammonium cyclopentadienyl-tetracarba-dico-baltadocosaborane, 45B, 696
$C_{17}H_{46}B_{17}CoN_2$, Tetraethylammonium (3,11')-commo-(undecahydro-1,2-dicarba-3-cobalta-closo-dodecaborato)(decahydro-9'-pyridyl-7,8'-dicarba-11'-cobalta-nido-undecaborate), 39B, 478
$C_{18}H_{12}B_3NO_6$, Tri-(1,3,2-benzodioxaborol-2-yl)amine, 35B, 439
$C_{18}H_{15}B$, Triphenylborane, 40B, 592
$C_{18}H_{15}B_3Cl_3N_3$, B-Trichloro-N-triphenylborazine, 43B, 819
$C_{18}H_{20}B_2CoN_{12}$, Bis(hydrotris(1-pyrazolyl)borato)cobalt(II), 35B, 434
$C_{18}H_{24}BMoN_6O_2$, (Diethylbispyrazolylborato)(pyrazolato)(η^3-allyl)(di-carbonyl)molybdenum, 39B, 479
$C_{18}H_{29}B_7Co_2$, Cobaltocenium cobaltacarborane complex, 44B, 583
$C_{18}H_{30}B_8Fe_2$, Bis(η^5-cyclopentadienyl)diferra-tetramethyltetracarbo-rane (isomer I), 43B, 819
$C_{18}H_{30}B_8Fe_2$, Bis(η^5-cyclopentadienyl)diferra-tetramethyltetracarbo-rane (isomer II), 43B, 819

$C_{18}H_{30}B_8Fe_2$, Bis(η^5-cyclopentadienyl)diferra-tetramethyltetracarborane (isomer III), 43B, 819
$C_{18}H_{30}B_8Fe_2$, Bis(η^5-cyclopentadienyl)diferra-tetramethyltetracarborane (isomer IV), 44B, 584
$C_{18}H_{33}B_{10}N$, Tetramethylammonium C,C'-diphenylundecahydrodicarba-nido-dodecaborate(1-), 39B, 479
$C_{18}H_{42}B_2N_2O_6$, 4,7,13,16,21,24-Hexaoxa-1,10-diazabicyclo[8.8.8]hexacosane bisborohydride, 42B, 280
$C_{18}H_{66}B_{20}N_4O$, Tris(triethylammonium) μ-nitrosobis(nonahydrodecaborate), 37B, 370
$C_{19}H_{15}BO_2$, (Salicylaldehydato)diphenylboron, 42B, 584
$C_{19}H_{44}B_5FeNO_3$, Tetra(n-butyl)ammonium ferroborate complex, 42B, 1013
$C_{20}H_{12}B_2OS_4$, Bis-(4-dithieno[3,2:2',3'-f]borepinyl) ether, 40B, 368
$C_{20}H_{18}B_2Cl_2F_6N_2O_3$, 1,5-Dichloro-3,7-bis(trifluoromethyl)-4,8-bis(2',6'-dimethylphenyl)-2,6,9-trioxa-4,8-diaza-1,5-diborabicyclo-[3.3.1]nonadiene, 46B, 626
$C_{20}H_{18}B_2F_{10}N_2$, 1,3-Di-t-butyl-2,4-bis(pentafluorophenyl)-1,3,2,4-diazadiboretidine, 45B, 697
$C_{20}H_{30}B_2N_8$, Bis-(3,5-dimethylpyrazolyl)borane dimer, 40B, 369
$C_{20}H_{36}B_4N_8S_4$, Tetra-B-isothiocyanatotetra-N-t-butylborazocine, 31B, 296
$C_{20}H_{37}B_9P_2Pt$, 1,1-Bis(dimethyphenylphosphine)-2,4-dicarba-1-platina-closo-dodecaborane, 41B, 709
$C_{20}H_{62}B_{18}CuN_2$, Tetraethylammonium bis((3)-1,2-dicarbollyl)cuprate-(II), 32B, 297
$C_{21}H_{26}BNO_4S$, 2,2,3-Triethyl-4-phenyl-5-oxo-5-p-nitrobenzyl-1-oxa-2-bora-5-thia-3-cyclopentene, 44B, 584
$C_{21}H_{36}B_{10}PTl$, Methyltriphenylphosphonium (η^4-decaborato)-dimethylthallate(III), 42B, 585
$C_{22}H_{39}BF_3NO_3$, Cyclohexyl(hydroxy)(trifluoroacetamido(dicyclohexyl)-methyl)-borane methanol solvate, 44B, 585
$C_{22}H_{68}B_{20}N_2O_2Ti$, Bis(tetramethylammonium) 4,4'-commo-bis(decahydro-1,6-dimethyl-1,6-dicarba-4-titana-closo-tridecaborate) bis(acetone), 41B, 850
$C_{23}H_{34}BNO_2$, 2,3,3-Tricyclopentyl-5-phenyl-1-oxa-4-azonia-2-borato-cyclopent-4-en-2-ol, 44B, 585
$C_{23}H_{40}B_{18}CuP$, Triphenylmethylphosphonium bis((3)-1,2-dicarbollyl)-cuprate(III), 33B, 267
$C_{24}H_{12}BClN_6$, 15c-Chloro-triisoindolo[1,2,3-cd:1',2',3'-gh:1",2",3"-kl](2,3a,5,6a,8,9a,9b)-hexaazaboraphenalene, 40B, 681
$C_{24}H_{20}BK$, Potassium tetraphenylborate, 40B, 592
$C_{24}H_{20}BNO$, [(2-Hydroxy-1-naphthyl)methylene-N-methylaminato]diphenylboron, 46B, 627
$C_{24}H_{20}BRb$, Rubidium tetraphenylborate, 26, 735; 27, 884
$C_{24}H_{20}B_2I_4P_2$, 2,2,4,4-Tetraiodo-1,1,3,3-tetraphenylcyclodiborataphosphoniane, 38B, 641
$C_{24}H_{23}B_3O_6$, D-Mannitol tris(benzeneboronic) ester, 43B, 820
$C_{24}H_{24}BN$, Ammonium tetraphenylborate, 46B, 627
$C_{24}H_{24}B_2Cl_2P_2$, 1,4-Dichloro-1,1,3,3-tetraphenyl-catena-di(boraphosphane), 44B, 585
$C_{24}H_{26}B_6MnO_3P$, Triphenylmethylphosphonium 1,1,1-tricarbonyl-4,6-dicarba-1-mangana-closo-nonaborate(1-), 39B, 480
$C_{24}H_{30}BN_3O_2$, Tetramethylammonium (ethylnitrosolato-O)triphenylborate, 46B, 628
$C_{24}H_{55}B_2N$, Tetra-n-butylammonium μ-hydro-1,2-dihydro-1,2:1,2-bis-(tetramethylene)diborate, 41B, 709
$C_{26}H_{38}Cl_2B_2N_2$, 1,3-Di-p-chlorophenyl-2-triethylcarbinyl-4-ethyl-5,5-

diethyl-1,3-diaza-2,4-diborolidine, 35B, 435
$C_{26}H_{44}BLiO_4$, Lithium dimesitylborohydride dimethoxyethane solvate, 40B, 667
$C_{27}H_{33}B$, Trimesitylborane, 39B, 481
$C_{28}H_{24}B_2O$, Bis(4-dibenzoborepinyl) ether, 46B, 628
$C_{28}H_{25}B$, 10-Phenyl-9-mesityl-9,10-dihydro-9-bora-anthracene, 46B, 629
$C_{28}H_{32}BN$, Tetramethylammonium tetraphenylborate, 40B, 593
$C_{31}H_{32}BN$, (Diphenylmethyleneamino)dimesitylborane, 39B, 482
$C_{36}H_{24}B_3N_3$, 1,2:3,4:5,6-Tris-(o,o'-biphenylylene)borazine, 40B, 593
$C_{36}H_{30}B_3N_3$, Hexaphenylborazine, 45B, 698
$C_{36}H_{34}B_2P_2$, Bis(triphenylphosphine)-diborane(4), 39B, 482
$C_{36}H_{36}B_3P_3$, 1,1,3,3,5,5-Hexaphenylcyclotriborataphosphoniane, 39B, 483
$C_{36}H_{38}B_3CuP_2$, Octahydrotriboratobis(triphenylphosphine)copper(I), 40B, 1099
$C_{36}H_{38}B_5CuP_2$, (η^2-Octahydropentaborano)bis(triphenylphosphine)copper(I), 43B, 821
$C_{36}H_{40}B_8P_2PtS$, 9,9-Bis(triphenylphosphine)-6,9-platinathiadecaborane, 43B, 1451
$C_{36}H_{50}BNO$, Tributylammonium tetraphenylborate monohydrate, 43B, 61
$C_{36}H_{86}B_{18}Cl_2Li_2O_8U$, Bis(tetrakis(tetrahydrofuran)lithium) bis(η^5-(3)-1,2-dicarbollyl)dichlorouranium(IV), 43B, 822
$C_{38}H_{42}B_9P_2Rh$, 3-Hydrido-3,3-bis(triphenylphosphine)-3-rhoda-1,2-dicarba-undecaborane, 42B, 641
$C_{38}H_{44}B_8OP_2PtS$, 8-Ethoxy-9,9-bis(triphenylphosphine)-6,9-platinathiadecaborane, 43B, 1451
$C_{38}H_{60}B_{20}CdP_2$, Methyltriphenylphosphonium bis(dodecahydro-nido-decaborato)cadmate, 37B, 371
$C_{38}H_{60}B_{20}HgP_2$, Methyltriphenylphosphonium bis(dodecahydro-nido-decaborato)mercurate, 37B, 371
$C_{38}H_{60}B_{20}P_2Zn$, Methyltriphenylphosphonium bis(dodecahydro-nido-decaborato)zincate, 37B, 371
$C_{38}H_{80}B_{20}O_5Ti_2Zn$, Bis(biscyclopentadienyl(acetone)(tetrahydrofuran)-titanium(III))-bis(dodecahydronidodecaborato)zincate tetrahydrofuran solvate, 42B, 519
$C_{42}H_{73}BO_{16}Rb$, Boromycin (rubidium salt dimethanolate), 37B, 372
$C_{44}H_{45}B_{10}P_2Rh$, Rhodium(I) carborane complex, 40B, 594
$C_{54}H_{57}AuB_9P_3S$, Tris(triphenylphosphine)gold(I)-(dodecahydrido-6-thia-nido-decaborate(1-)), 40B, 594
$C_{62}H_{82}B_4N_{10}Ni_2$, Di-$\mu$-cyanotrihydroborato-bis(2,2',2''-triaminotriethylamine)dinickel(II) tetraphenylborate, 43B, 822
$C_{73}H_{71}B_{10}Cl_3Cu_2P_4$, μ-Decahydrodecaborato-tetrakis(triphenylphosphine)dicopper(I) chloroform solvate, 41B, 1227

63. SILICON COMPOUNDS

CCl_6Si, Trichloromethyltrichlorosilane (gas-ed), 22, 556
CH_3Br_3Si, Methyltribromosilane (gas-ed), 12, 348; 21, 551
CH_3Br_3Si, Methyltribromosilane (gas-mw), 33B, 538
CH_3Cl_3Si, Methyltrichlorosilane (gas-ed), 9, 337
CH_3Cl_3Si, Methyltrichlorosilane (gas-mw), 43B, 1485
CH_3F_3OSi, Methoxytrifluorosilane (gas-ed), 37B, 692
CH_3F_3Si, Methyltrifluorosilane (gas-ed), 22, 553
CH_3F_3Si, Methyltrifluorosilane (gas-mw), 13, 497; 38B, 1072
CH_4Cl_2Si, Methyldichlorosilane (gas-mw), 43B, 1485

CH_5BrSi, Bromomethylsilane (gas-mw), 39B, 893
CH_5ClSi, Chloromethylsilane (gas-mw), 27, 732
CH_5FSi, Methylmonofluorosilane (gas-mw), 22, 610
CH_6OSi, Methyl silyl ether (gas-ed), 35B, 888
CH_6Si, Methylsilane (gas-ed), 18, 681
CH_6Si, Methylsilane (gas-mw), 21, 497
CH_7PSi, Silylmethylphosphine (gas-ed), 38B, 1065
CH_8Si_2, Disilylmethane (gas-ed), 35B, 888
CH_9NSi_2, N-Methyldisilylamine (gas-ed), 40B, 1142
$C_2H_3Cl_3Si$, Vinyltrichlorosilane (gas), 29, 450
$C_2H_4Cl_4Si_2$, 1,1,3,3-Tetrachloro-1,3-disilacyclobutane (gas-ed), 33B, 529; 38B, 1067
C_2H_4Si, Silylacetylene (gas-mw), 28, 664
$C_2H_6Br_2Si$, Dimethyldibromosilane (gas-ed), 12, 348; 21, 551
C_2H_6ClFSi, Dimethylchlorofluorosilane (gas-ed), 18, 680
$C_2H_6Cl_2Si$, Dichlorodimethylsilane (gas-ed), 9, 337
$C_2H_6Cl_3NSi$, Trichlorosilyldimethylamine (gas-ed), 37B, 693
$C_2H_6F_3NSi$, Trifluorosilyldimethylamine (gas-ed), 37B, 693
C_2H_6OSi, Methylsiloxanes (gas-ed), 39B, 891
C_2H_6Si, Vinylsilane (gas-mw), 24, 516; 26, 587
C_2H_7ClSi, Ethylchlorosilane (gas-mw), 40B, 1146
C_2H_8Si, Dimethylsilane (gas-ed), 18, 681
C_2H_9NSi, N-Silyldimethylamine (gas-ed), 35B, 891
C_2H_9PSi, Silyldimethylphosphine (gas-ed), 38B, 1065
$C_3H_3F_5Si$, 2,2-Difluorocyclopropyl-trifluorosilane (gas-ed), 42B, 1031
$C_3H_5F_3Si$, Trifluorosilylcylopropane (gas-ed), 42B, 1030
$C_3H_6Cl_2Si$, 1,1-Dichloro-1-silacyclobutane (gas-ed), 38B, 1067
C_3H_8Si, Silacyclobutane (gas-mw), 37B, 696
C_3H_8Si, 1-Silacyclobutane (gas-ed), 38B, 1067
C_3H_9BrSi, Trimethylbromosilane (gas-ed), 12, 348; 21, 551
C_3H_9ClSi, Chlorotrimethylsilane (gas-ed), 10, 210
C_3H_9ClSi, Trimethylchlorosilane (gas-mw), 17, 679
C_3H_9CsOSi, Caesium trimethylsilanolate, 35B, 440
C_3H_9FSi, Trimethylfluorosilane (gas-ed), 18, 680
C_3H_9KOSi, Potassium trimethylsilanolate, 35B, 440
$C_3H_9N_3O_2S_3Si$, 2-Trimethylsilyl-1,3,5λ^4,2,4,6-trithiatriazine 1,1-dioxide, 45B, 698
$C_3H_9N_3Si$, Trimethylazidosilane (gas-ed), 38B, 1068
C_3H_9ORbSi, Rubidium trimethylsilanolate, 35B, 440
$C_3H_9O_3Si_3$, Hexamethylcyclotrisiloxane, 18, 682
$C_3H_9O_4ReSi$, Trimethylsilyl perrhenate, 34B, 364
$C_3H_{10}Si$, Ethylmethylsilane (gas-mw), 38B, 1069
$C_3H_{10}Si$, Trimethylsilane (gas-ed), 18, 681
$C_4H_4Cl_6OSi_3$, 1,1,3,3,8,8-Hexachloro-2-oxa-1,3,8-trisilabicyclo-[3.3.1]oct-5-ene, 41B, 446
$C_4H_4F_6Si_3$, 2,2,3,3,7,7-Hexafluoro-2,3,7-trisilanorborn-5-ene, 38B, 642
$C_4H_6N_2Si$, Dimethyldicyanosilicon, 38B, 642
C_4H_7NSi, N-Silylpyrrole (gas-ed), 37B, 693
C_4H_9NOSi, Trimethylsilyl isocyanate (gas-ed), 31B, 298
C_4H_9NSSi, Trimethylsilyl isothiocyanate (gas-ed), 31B, 298
C_4H_9NSi, Trimethylsilylcyanide (gas-ed), 40B, 1139
$C_4H_{10}Si$, Tetramethylenesilane (gas-ed), 27, 950
$C_4H_{10}Si$, 1-Silacyclopentane (gas-mw), 42B, 1036
$C_4H_{11}ClSi$, Chloromethyltrimethylsilane (gas-ed), 13, 456
$C_4H_{12}N_4S_2Si_2$, 1,1,5,5-Tetramethyl-1,5-disila-3,7-dithia-2,4,6,8-tet-

raazaoctane, 40B, 362
$C_4H_{12}O_2Si$, Diethylsilanediol, 17, 679
$C_4H_{12}O_3Si$, Methyltrimethoxysilane (gas-ed), 42B, 1030
$C_4H_{12}O_4Si$, Tetramethoxysilane, 2, 803, 818
$C_4H_{12}O_4Si$, Tetramethylsilicate (gas-ed), 13, 497; 21, 551
$C_4H_{12}S_2Si_2$, Tetramethylcyclodisilthiane (gas-ed), 19, 556
$C_4H_{12}Si$, Tetramethylsilane (gas-ed), 4, 277; 37B, 692
$C_4H_{12}Si_4S_6$, Tetra(methylsilicon)hexasulfide, 41B, 710
$C_4H_{13}NSi$, N-Methylaminotrimethylsilane (gas-ed), 27, 770
C_5H_9ClSi, Trimethylsilylchloroacetylene (gas-ed), 40B, 1141
$C_5H_{10}Si$, Trimethylsilylacetylene (gas-ed), 40B, 1141
$C_6H_5Cl_3Si$, Phenyltrichlorosilane (gas-ed), 29, 538
C_6H_7ClSi, Phenylchlorosilane (gas-ed), 30B, 404
$C_6H_8Cl_8Si_6$, Octachlorohexasila-asteran, 39B, 484
C_6H_8OSi, Phenyl silyl ether (gas-ed), 40B, 1143
C_6H_8Si, Phenylsilane (gas-ed), 20, 558
$C_6H_{12}ClNO_3Si$, 1-Chlorosilatrane, 44B, 586
$C_6H_{12}O_2Si$, Diallylsilanediol, 18, 681
$C_6H_{12}Si$, 1-Silabicyclo[2.2.1]heptane (gas-mw), 43B, 1485
$C_6H_{12}Si$, 4-Sila-3,3-spiroheptane (gas-ed), 38B, 1067
$C_6H_{18}F_6N_2P_2Si_2$, N-(Trifluorosilyl)trimethylphosphinimine dimer, 43B, 823
$C_6H_{18}NNaSi_2$, N-Sodiohexamethyldisilazane, 43B, 823
$C_6H_{18}N_2Si_2$, Bis(trimethylsilyl)diimine, 40B, 596
$C_6H_{18}OSi_2$, Hexamethyldisiloxane, 45B, 698
$C_6H_{18}OSi_2$, Hexamethyldisiloxane (gas-ed), 13, 498; 21, 551; 42B, 1030
$C_6H_{18}O_3Si_3$, Hexamethylcyclotrisiloxane (gas-ed), 13, 498
$C_6H_{18}O_6Si_3$, Hexamethylcyclotrisilaperoxane, 46B, 629
$C_6H_{18}S_3Si_3$, Hexamethylcyclotrisilthiane (gas-ed), 19, 557
$C_6H_{18}Si_2$, Hexamethyldisilane (gas-ed), 37B, 692
$C_6H_{21}N_3Si_3$, Hexamethylcyclotrisilazane (gas-ed), 17, 680
$C_6H_{24}N_6NiO_5Si_2 \cdot 8.7\ H_2O$, Trisethylenediaminenickel silicate hydrate, 35B, 441
$C_7H_{14}ClNO_3Si$, 1-(Chloromethyl)-silatrane, 41B, 450
$C_7H_{14}Si$, 1-Methyl-1-silabicyclo[2.2.1]heptane (gas-ed), 42B, 1026
$C_7H_{15}ClN_2O_2Si$, Bis(acetoximato)methylchlorosilane, 45B, 699
$C_7H_{15}NO_3Si$, 1-Methylsilatrane, 44B, 586
$C_8H_6BrF_3O_2Si$, (4-Bromobenzoyloxymethyl)-trifluorosilane, 45B, 700
$C_8H_9O_5ReSi$, Trimethylsilylpentacarbonylrhenium, 42B, 519
$C_8H_{12}O_8Si$, Silicon(IV) acetate, 41B, 711
$C_8H_{14}Si$, Cyclopentadienyltrimethylsilane (gas-ed), 35B, 900
$C_8H_{17}NO_2Si$, Methyl(2,2',3-nitrilodiethoxypropyl)silane, 34B, 366
$C_8H_{17}NO_3Si$, 1-Methoxy-2-carbasilatrane, 43B, 824
$C_8H_{19}NSi$, Trimethyl(pyrrolidinomethyl)silane, 46B, 630
$C_8H_{20}N_2S_2Si_2$, N,N'-Bis(trimethylsilyl)dithiooxamide, 44B, 587
$C_8H_{24}Al_3Br_5O_6Si_4$, Aluminosiloxane, 33B, 268
$C_8H_{24}O_2Si_4$, Bis-tetramethyl-disilylene dioxide, 28, 583
$C_8H_{24}O_4Si_4$, Octamethylcyclotetrasiloxane (gas-ed), 21, 551
$C_8H_{24}O_4Si_4$, Octamethylcyclotetrasiloxane, 19, 549
$C_8H_{24}O_6Si_5$, Octamethylbicyclopentasiloxane, 41B, 712
$C_8H_{24}O_6Si_5$, Octamethylspiro[5.5]pentasiloxane, 11, 620
$C_8H_{24}O_8Si_8$, Octamethylocta(siloxane), 31B, 298
$C_8H_{24}O_{12}Si_8$, Octamethylsilsesquioxane, 24, 563
$C_8H_{25}NO_3Si_4$, 2,2,4,4,6,6-Hexamethyl-5-(hydroxydimethylsilyl)-1,3-dioxa-5-aza-2,4,6-trisilacyclohexane, 42B, 521
$C_8H_{28}N_4Si_4$, Octamethylcyclotetrasilazane (gas-ed), 17, 680

$C_8H_{28}N_4Si_4$, Octamethylcyclotetrasilazane, 28, 584
$C_9H_{18}ClNO_3Si$, 1-Chloromethyl-3,7-dimethylsilatrane, 46B, 631
$C_9H_{21}BrSi_4$, 1-Bromo-3,5,7-trimethyl-1,3,5,7-tetrasila-adamantane, 45B, 700
$C_9H_{21}NSi$, Trimethyl(pyrrolidinoethyl)silane, 46B, 630
$C_9H_{27}N_3SSi_3$, Tris(trimethylsilylimino)sulfur, 45B, 701
$C_{10}H_8F_4N_2Si$, 2,2'-Bipyridyltetrafluorosilicon(IV), 38B, 643
$C_{10}H_{10}F_4N_2Si$, Tetrafluorobis(pyridine)silicon, 34B, 371
$C_{10}H_{18}FeO_4Si_2$, cis-Tetracarbonylbis(trimethylsilyl)iron, 43B, 825
$C_{10}H_{18}O_2Si_2$, p-Bis(dimethylhydroxysilyl)benzene, 32B, 300
$C_{10}H_{24}S_2Si_2$, trans-1,2-Bis(trimethylsilyl)-1,2-bis(methylthio)ethylene, 44B, 588
$C_{10}H_{24}Si_4$, 1,3,5,7-Tetramethyl-tetrasila-adamantane, 38B, 644
$C_{10}H_{28}Hg_2Si_4$, 2,2,4,4,6,6,8,8-Octamethyl-2,4,6,8-tetrasila-1,5-dimercuracyclooctane, 41B, 712; 46B, 632
$C_{10}H_{30}N_2Si_4$, Tetramethyl-N,N'-bistrimethylsilylcyclodisilazane, 27, 942
$C_{10}H_{30}O_7Si_6$, Decamethylbicyclohexasiloxane, 43B, 827
$C_{10}H_{30}O_{15}Si_{10}$, Deca(methylsilasesquioxane), 46B, 631
$C_{10}H_{45}N_5Si_5$, Dimethylsilylamine pentamer, 32B, 301
$C_{11}H_{28}Si_4$, Heptamethyl-tetrasila[2.2.2]barrelane, 42B, 521
$C_{12}H_8Cl_2OSi$, 10,10-Dichloro-10-sila-9-oxaphenanthrene (triclinic), 44B, 588
$C_{12}H_8Cl_2OSi$, 10,10-Dichloro-10-sila-9-oxaphenanthrene (orthorhombic), 41B, 452; 44B, 588
$C_{12}H_8Cl_4Si_2$, 9,9,10,10-Tetrachloro-9,10-disiladihydroanthracene, 46B, 633
$C_{12}H_8Si_2$, 9,10-Disilahydroanthracene, 41B, 714
$C_{12}H_{10}Cl_2Si$, Diphenyldichlorosilane (gas-ed), 20, 558
$C_{12}H_{11}Cl_4N_2Si_2 \cdot 0.5\ CH_3CN$, 1,1,2,2-Tetrachloro-1,2-dimethyldisilane-2,2'-bipyridyl acetonitrile solvate, 42B, 522
$C_{12}H_{12}O_2Si$, Diphenylsilanediol, 43B, 827; 44B, 589
$C_{12}H_{12}O_2Si$, Diphenylsilanediol, 43B, 827
$C_{12}H_{14}FNO_4Si$, p-Fluorophenylsilatranone, 46B, 634
$C_{12}H_{16}N_2O_5Si$, m-Nitrophenyl(2,2',2"-nitrilotriethoxy)silane, 34B, 367
$C_{12}H_{17}NO_3Si$, Phenyl-(2,2',2"-nitrilotriethoxy)silane, 33B, 270
$C_{12}H_{17}NO_3Si$, 1-Phenylsilatrane (β form), 40B, 596
$C_{12}H_{17}NO_3Si$, 1-Phenylsilatrane (γ-form), 41B, 714
$C_{12}H_{19}NO_2Si$, 2,2-Dimethyl-6-phenyl-1,3-dioxa-6-aza-2-silacyclooctane, 46B, 636
$C_{12}H_{20}Cl_3FeNO_4Si$, Tetraethylammonium trichlorosilyl iron tetracarbonyl, 41B, 38
$C_{12}H_{22}Si_2$, p-Bis(trimethylsilyl)benzene, 41B, 715
$C_{12}H_{28}Si_4$, 2,2,4,4,6,6,8,8-Octamethyl-2,4,6,8-tetrasilabicyclo-[3.3.0]oct-1(5)-ene, 34B, 373
$C_{12}H_{32}Si_4$, Octamethyl-tetrasilacyclooctane, 42B, 523
$C_{12}H_{36}HgLi_2Si_4$, Dilithiumtetrakis(trimethylsilyl)mercury(II), 46B, 646
$C_{12}H_{36}N_4Si_4$, Dodecamethylcyclotetrasilazane, 42B, 523
$C_{12}H_{36}N_4Si_4$, Tetrakis(trimethylsilyl)tetrazene, 41B, 716
$C_{12}H_{36}O_4Si_5$, Tetrakis(trimethylsiloxy)silane, 46B, 633
$C_{12}H_{36}O_4Si_5$, Tetra(trimethylsilyl)silicate (gas-ed), 21, 551
$C_{12}H_{36}O_6Si_8$, Tricyclic methyldisilanylenesiloxane, 32B, 302
$C_{12}H_{36}Si_5$, Tetrakis(trimethylsilyl)silane (gas-ed), 35B, 902
$C_{12}H_{36}Si_6$, Dodecamethylcyclohexasilane, 38B, 644
$C_{12}H_{39}N_8Si_6$, 2,2,4,4,6,6,8,8,9,9,11,11-Dodecamethylbicyclo[3.3.3]-

hexasilazane, 45B, 701
$C_{13}H_{12}Si$, 9-Sila-9,10-dihydroanthracene, 42B, 524
$C_{13}H_{14}F_3NO_4Si$, m-Trifluoromethylphenylsilatranone, 46B, 634
$C_{13}H_{21}NO_2Si$, (t-Butyl)dimethylsiloxy-aci-nitro-phenylmethane, 46B, 635
$C_{14}H_{18}Si_2$, 1,1,2,2-Tetramethyl-1,2-disilaacenaphthalene, 42B, 524
$C_{14}H_{18}Si_2$, 1,8-Bis(trimethylsilyl)octatetrayne, 41B, 716
$C_{14}H_{24}O_6Re_2Si_2$, Bis(μ-diethylsilicon)-bis(tricarbonyldihydridorhenium)(Re-Re), 43B, 828
$C_{14}H_{26}O_2Si_4$, 3,7-Dihydro-1,1,3,3,5,5,7,7-octamethyl-1H,5H-benzo[1,2-c:4,5-c']bis(1,2,5)oxadisilole, 39B, 484
$C_{14}H_{27}MnO_5Si_4$, (Tris(trimethylsilyl)silyl)pentacarbonylmanganese, 39B, 485
$C_{14}H_{27}O_5ReSi_4$, (Tris(trimethylsilyl)silyl)pentacarbonylrhenium, 42B, 519
$C_{14}H_{34}KNO_4Si_2$, Bis(dioxane) potassium bis(trimethylsilyl)amide, 40B, 597
$C_{14}H_{36}P_4Si_3$, 2,2,5,5,7,7-Hexamethyl-3,6-di-t-butyl-1,3,4,6-tetraphospha-2,5,7-trisilanobornane, 44B, 589
$C_{14}H_{40}Be_2N_4Si_4$, (Bis(dimethylsilylmethylamido)methyl)beryllium dimer, 45B, 702
$C_{15}H_{22}O_7Re_2Si_2$, Bis(μ-diethylsilicon)-(tricarbonyldihydridorhenium)-(tetracarbonylrhenium)(Re-Re), 43B, 829
$C_{15}H_{23}N_3O_3Si$, Tris(acetone-oximato)phenylsilane, 45B, 703
$C_{15}H_{45}N_3Si_6$, 2,2,4,4,6,6-Hexamethyl-1,3,5-tris(trimethylsilyl)cyclotrisilazane, 35B, 443
$C_{16}H_{12}S_4Si$, Tetra-(2-thienyl)silane, 40B, 598
$C_{16}H_{12}Si$, Diphenyldiethynylsilane, 42B, 525
$C_{16}H_{17}Br_2NSi$, 2,8-Dibromo-5-ethyl-5,10-dihydro-10,10-dimethylphenazasiline, 43B, 830
$C_{16}H_{18}Cl_2Si_2$, Dichloro-9-fluorenyltrimethylsilylsilane, 46B, 635
$C_{16}H_{19}NO_2Si$, (Iminobis(ethyleneoxy))diphenylsilane, 40B, 598
$C_{16}H_{20}Si_2$, 9,9,10,10-Tetramethyl-9,10-disiladihydroanthracene, 40B, 599; 42B, 594
$C_{16}H_{22}N_2Si_2$, Tetramethyl-N,N'-diphenylcyclodisilazane, 42B, 526
$C_{16}H_{22}O_8Si_2W_2$, Bis(diethylsilicon)octacarbonyldihydridoditungsten, 37B, 373
$C_{16}H_{24}O_{12}Si_8$, Vinylsilasesquioxane, 45B, 704
$C_{16}H_{26}Si$, Dimethyldispiro(bicyclo[4.1.0]heptane-7,2'-silacyclopropane-3',7''-bicyclo[4.1.0]heptane), 42B, 526
$C_{16}H_{30}O_6Ru_2Si_4$, Di-μ-(dimethylsilylene)-bis(tricarbonyl(trimethylsilyl)ruthenium(III)), 38B, 645
$C_{16}H_{36}Si_7$, Hexamethyl-heptasila-hexacyclo-heptadecane, 42B, 527
$C_{16}H_{38}N_2O_2Si_2$, 3,6-Bis(di-t-butyl)cyclodisiloxazane, 46B, 636
$C_{16}H_{48}Si_9$, Hexadecamethylbicyclo[3.3.1]nonasilane, 42B, 527
$C_{17}H_{18}Si$, 9-Trimethylsilylphenanthrene, 41B, 1241; 43B, 831
$C_{17}H_{20}Si$, 5,5-Dimethyl-5,6,11,12-tetrahydro-5H-dibenzo[b,f]silocin, 44B, 1106
$C_{17}H_{21}NO_2Si$, 2,2-Diphenyl-6-methyl-1,3-dioxa-6-aza-2-silacyclooctane, 46B, 636
$C_{17}H_{26}O_4Si$, Decahydro-5-methylene-2-oxo-6β-trimethylsiloxy-[3aα,6α,-8aβ]-3a,6-methano-3α-azulenecarboxylic acid methyl ester, 45B, 704
$C_{17}H_{36}Si_8$, 3,7,11,15-Tetramethyl-1,3,5,7,9,11,13,15-octasiladodecascaphan, 40B, 599
$C_{18}H_{15}N_3Si$, Azidotriphenylsilane, 39B, 486
$C_{18}H_{15}Na_3O_6Si_3 \cdot 8\ H_2O$, Trisodium cis-1,3,5-triphenylcyclotrisiloxane-1,3,5-triolate octahydrate, 46B, 637

$C_{18}H_{16}Si$, Triphenylsilane, 45B, 705
$C_{18}H_{22}N_2O_2Si_2$, 2,2,4,4-Tetramethyl-3-benzoyl-6-phenyl-2,4-disila-1,3,5-oxadiazine, 35B, 443
$C_{18}H_{24}F_{18}OSi_2$, Hexa(3,3,3-trifluoropropyl)disiloxane, 42B, 527
$C_{18}H_{26}O_2Si_2$, Bis(dimethylbenzylsilyl) peroxide, 45B, 705
$C_{18}H_{28}O_4Si_4$, 2,6-cis-Diphenylhexamethylcyclotetrasiloxane, 39B, 486
$C_{18}H_{28}O_4Si_4$, 2,6-trans-Diphenylhexamethylcyclotetrasiloxane, 43B, 832
$C_{18}H_{46}I_2N_2Si_2$, N,N'-(4,4,7,7-Tetramethyl-4,7-disiladecamethylene)-bis(trimethylammonium) diiodide, 46B, 638
$C_{18}H_{54}B_2N_4Si_6$, Hexakis(trimethylsilyl)-2,4-diamino-1,3,2,4-diazadiboretin, 34B, 374
$C_{18}H_{54}Li_3N_3Si_6$, N-Lithiohexamethyldisilazane, 44B, 1107
$C_{18}H_{54}Li_6Si_6$, Hexakis(trimethylsilyl-lithium), 46B, 638
$C_{18}H_{54}N_2O_8Si_{10}$, N,N'-Bis(2,2,4,4,6-pentamethylcyclotrisiloxanyl-oxadimethylsilyl)tetramethylcyclodisilazane, 46B, 639
$C_{19}H_{15}BrOSi$, Methyl p-bromophenylsilaoxarophenanthrene, 37B, 374
$C_{19}H_{15}NOSi$, 9-Methyl-9-phenyl-9,10-dihydro-9-sila-3-azaanthrone, 46B, 639
$C_{19}H_{15}NSSi$, Triphenylsilicon isothiocyanate, 41B, 717
$C_{19}H_{17}FSi$, (-)-Naphthyl-2-fluoro-2-sila-1,2,3,4-tetrahydronaphthalene, 39B, 487
$C_{19}H_{21}Br_2N_3O_3Si$, 2,6-Di-p-bromophenyl-1,1-dimethyl-1-sila-2,4,6-triazacyclohexan-3,5-dione acetone, 29, 538
$C_{19}H_{24}N_2Si$, 3-t-Butyl-5-methyl-2,3-diphenyl-1,2-diaza-3-sila-5-cyclopentene, 45B, 705
$C_{19}H_{25}NO_2Si$, (t-Butyl)dimethylsiloxy-aci-nitro-diphenylmethane, 46B, 635
$C_{19}H_{26}ClNSi$, 9-Methyl-9-(3-dimethylaminopropyl)-9,10-dihydrosilaanthracene hydrochloride, 43B, 832
$C_{19}H_{28}Si_3$, 1,1,3-Trimethyl-2-(trimethylsilyl)-3-phenyl-1,3-disila-indan (monoclinic), 46B, 640
$C_{19}H_{28}Si_3$, 1,1,3-Trimethyl-2-(trimethylsilyl)-3-phenyl-1,3-disila-indan (triclinic), 46B, 640
$C_{19}H_{48}Si_7$, Dodecamethyl-heptasila[4.4.4]propellane, 42B, 528
$C_{20}H_{12}Cl_2OSi_2$, Bis-(1,8-naphthylene)dichlorodisiloxane, 41B, 718
$C_{20}H_{12}Cl_2Si_2$, Bis-(1,8-naphthylene)dichlorodisilane, 41B, 718
$C_{20}H_{16}N_4Si$, Bis(2,2'-bipyridyl)silicon, 45B, 706
$C_{20}H_{18}OSi$, Acetyltriphenylsilane, 34B, 373
$C_{20}H_{18}SSi$, Triphenylsilyl-thiirane, 45B, 707
$C_{20}H_{18}Si$, 2,2-Diphenyl-2-silaindane, 39B, 487
$C_{20}H_{22}I_2N_4O_4Si$, Dihydroxybis(bipyridyl)silicon iodide dihydrate, 44B, 589
$C_{20}H_{22}S_4Si_2$, 5,5'-Bis(dimethyl(2-thienyl)silyl)-2,2'-bithienyl, 43B, 833
$C_{20}H_{27}NOSi$, (3-Piperidino-propyl)-diphenyl-silanol, 45B, 707
$C_{20}H_{27}NO_2Si$, 2,2-Diphenyl-6-t-butyl-1,3-dioxa-6-aza-2-silacyclo-octane, 46B, 636
$C_{20}H_{27}NSi$, 6-(1,1-Dimethylethyl)-12,12-dimethyl-5,6-dihydro-7H,12H-dibenzo[c,f](1,5)silazocine, 43B, 834
$C_{20}H_{28}P_2Si$, 1,2-di-t-Butyl-3,3-diphenyl-1,2,3-diphosphasilirane, 46B, 641
$C_{20}H_{28}Si_2$, cis-9,10-Bis(trimethylsilyl)-9,10-dihydroanthracene, 42B, 125
$C_{20}H_{28}Si_2$, trans-9,10-Bis(trimethylsilyl)-9,10-dihydroanthracene, 42B, 125
$C_{20}H_{30}ClNOSi$, 5-Methyl-5-(3-dimethylaminopropyl)-10,11-dihydro-5H-

dibenzo[b,f]silepin hydrochloride hydrate, 42B, 529
$C_{20}H_{34}O_7Si_6$, cis-1,7-Diphenyl-3,3,5,5,9,9,11,11-octamethylbicycloheptasiloxane, 46B, 641
$C_{20}H_{48}Si_4$, 1,2,3,4-Tetra-t-butyl-tetramethylcyclotetrasilane, 41B, 719
$C_{20}H_{60}FN_5Si_9$, 3-[Bis(trimethylsilyl)aminofluoro(methyl)silyl]-2,2,4,4,6,8,8-heptamethyl-5,7-bis(trimethylsilyl)-1,3,5,7-tetraaza-2,4,6,8-tetrasilabicyclo[4.2.0]octane, 46B, 642
$C_{20}H_{60}F_2N_6Si_{10}$, 1-Difluoromethylsilyl-2,2,4,4,6,6,8-heptamethyl-3-trimethylsilyl-7-dimethylsilyl-(11,11,13,13-tetramethyl-12-trimethylsilyl-(10,12-diaza-11,13-disilacyclobutane))-1,3,5,7-tetraaza-2,4,6,8-tetrasilabicyclo[4.2.0]octane, 46B, 642
$C_{21}H_{15}BrOSi$, 3-Bromo-2,2-diphenyl-2-sila-Δ^3-1-tetralone, 38B, 646
$C_{21}H_{18}Si$, 5-Methyl-5-phenyl-5H-dibenzo[b,f]silepin, 43B, 834
$C_{21}H_{20}OSi$, 5-Methoxy-5-phenyl-10,11-dihydro-5H-dibenzo[b,f]silepin, 42B, 529
$C_{21}H_{20}Si$, 2,2-Diphenyl-2-sila-1,3,4-trihydronaphthalene, 38B, 647
$C_{21}H_{20}Si$, 5-Methyl-5-phenyl-10,11-dihydro-5H-dibenzo[b,f]silepin, 43B, 834
$C_{21}H_{24}F_3N_3Si$, 2,4,6-Trifluoro-1,3,5-trimethyl-2,4,6-triphenylcyclotrisilazane, 46B, 643
$C_{21}H_{24}O_3Si_3$, cis-1,2,3-Trimethyl-1,2,3-triphenylcyclotrisiloxane, 40B, 600
$C_{21}H_{24}O_3Si_3$, trans-1,2,3-Trimethyl-1,2,3-triphenylcyclotrisiloxane, 40B, 601
$C_{21}H_{24}S_3Si_3$, trans-1,3,5-Trimethyl-1,3,5-triphenylcyclotrisilthiane, 43B, 835
$C_{21}H_{30}O_2Si$, 17α-Ethynyl-17β-hydroxy-6,6-dimethyl-6-sila-5α-estr-1(10)-en-3-one, 41B, 720
$C_{22}H_{22}OS_2Si$, trans-2-Triphenylsilyl-1,3-dithiane 1-oxide, 44B, 590
$C_{22}H_{23}NO_2Si$, 2,2,6-Triphenyl-1,3-dioxa-6-aza-2-silacyclooctane, 46B, 636
$C_{22}H_{23}NSi$, 5-p-N,N-Dimethylaminophenyl-10,11-dihydro-5H-dibenzo-[b,f]silepin, 41B, 721
$C_{22}H_{25}NO_4Si$, Tetramethylammonium bis(o-phenylenedioxy)phenylsiliconate, 33B, 271
$C_{22}H_{36}Si_4$, 9-Fluorenyltris(trimethylsilyl)silane, 46B, 643
$C_{22}H_{60}N_4Si_6$, 2,4-Bis(t-butyl(trimethylsilyl)amino)-2,4-dimethyl-1,3-bis(trimethylsilyl)-1,3-diaza-2,4-disilacyclobutane, 46B, 644
$C_{22}H_{66}N_6Si_{10}$, 2,2,4,4-Tetramethyl-1,3-bis((2,2,4,4-tetramethyl-3-(trimethylsilyl-1,3-diaza-2,4-disilacyclobutyl))dimethylsilyl)-1,3-diaza-2,4-disilacyclobutane, 46B, 647
$C_{23}H_{24}S_2Si$, 2-Methyl-2-triphenylsilyl-1,3-dithiane, 44B, 590
$C_{24}F_{20}Si$, Tetrakis(pentafluorophenyl)silane, 44B, 591
$C_{24}H_{18}Si$, 9,9-Diphenyl-9-silafluorene, 41B, 722
$C_{24}H_{20}Si$, Tetraphenylsilane (gas-ed), 21, 551
$C_{24}H_{20}Si$, Tetraphenylsilane, 27, 875; 37B, 374; 38B, 648; 40B, 601
$C_{24}H_{22}Si_2$, 1,1,2,2-Tetraphenyldisilane, 46B, 645
$C_{24}H_{44}Si$, Tetracyclohexylsilane, 44B, 591
$C_{24}H_{46}Si_2$, 1,1,2,2-Tetracyclohexyldisilane, 44B, 591
$C_{25}H_{20}N_2Si$, Phenyl(triphenylsilyl)diazomethane, 38B, 85
$C_{25}H_{20}OSi$, 9-Fluorenyldiphenylsilanol, 46B, 643
$C_{25}H_{30}F_3NSi_2$, N-(t-Butylphenylfluorosilyl)-N-(phenyldifluorosilyl)-2,4,6-trimethylaniline, 45B, 708
$C_{25}H_{32}N_2O_4Si$, 5,5'-Dimethyl-10-(4-methylpiperazinyl)-10,11-dihydro-5H-dibenzo[b,f]silepin hydrogen fumarate, 45B, 708
$C_{25}H_{62}Si_7$, 2-(3,4,5,6-Tetrakis(trimethylsilyl)-1-cyclohexen-1-yl)-

heptamethyltrisilane, 40B, 1100
$C_{26}H_{21}Br_2NSi$, 2,8-Dibromo-5-ethyl-9,10-dihydro-10,10-diphenylphenosilazine, 46B, 645
$C_{26}H_{22}Br_2Si$, threo-1,2-Dibromo-1-triphenylsilyl-2-phenylethane, 44B, 592
$C_{26}H_{23}NSi$, 5-Ethyl-5,10-dihydro-10,10-diphenylphenazasiline, 46B, 646
$C_{26}H_{31}FOSi$, Fluoro-(menthoxy)-α-naphthylphenylsilane, 45B, 709
$C_{26}H_{44}Cl_2N_2S_2Si_4$, E-4,4'-Dichlorobis(bis(trimethylsilyl)amidosulphenyl)stilbene, 43B, 836
$C_{26}H_{46}N_4Si_4$, 3,7-Dimesityl-2,2,4,4,6,6,8,8-octamethyl-1,3,5,7-tetraaza-2,4,6,8-tetrasila-bicyclo[3.3.0]octane, 45B, 710
$C_{27}H_{34}O_2Si$, (+)-α-Naphthylphenyl-1-menthoxymethoxysilane, 39B, 488
$C_{28}H_{24}N_2O_6Si$, Pyridinium tris(o-phenylenedioxy)siliconate, 34B, 369
$C_{28}H_{28}Si$, Tetra(m-tolyl)silane, 11, 688
$C_{28}H_{32}O_4Si_4$, r-2,cis-4,trans-6,trans-8-Tetramethyl-2,4,6,8-tetraphenylcyclotetrasiloxane, 44B, 592
$C_{28}H_{32}O_4Si_4$, 1,1,2,2-Tetramethyl-3,3,4,4-tetraphenylcyclotetrasiloxane, 39B, 488
$C_{28}H_{32}O_6Si_5$, trans,trans-d,l-2,4,8,10-Tetramethyl-2,4,8,10-tetraphenylspiro[5.5]pentasiloxane, 44B, 593
$C_{28}H_{32}O_8Si_6$, 1,1,7,7-Tetramethyl-3,5,9,11-tetraphenyltricyclohexasiloxane, 45B, 710
$C_{28}H_{34}N_2O_2Si_4$, 1,3,5',7'-Tetramethyl-1',3',5,7-tetraphenylcyclo-1,3,5,7-tetrasila-2,6-diaza-4,8-dioxane, 40B, 602
$C_{28}H_{34}N_2O_4Si_5$, 2,2,4,4-Tetraphenyl-8,8,10,10-tetramethyl-spiro[5.5]-2,4,6,8,10-pentasila-7,11-diaza-1,3,5,9-tetroxane, 45B, 711
$C_{28}H_{36}N_4Si_4$, 1,3,5',7'-Tetramethyl-1',3',5,7-tetraphenylcyclotetrasilazane, 45B, 711
$C_{30}H_{60}N_4Si_6$, 2,4-Diphenyl-2,4-bis(isopropyl(trimethylsilyl)amino)-1,3-bis(trimethylsilyl)-1,3-diaza-2,4-disilacyclobutane, 46B, 644
$C_{31}H_{25}BrO_2Si$, (+)-α-(1-Naphthylphenylmethylsilyl)benzyl p-bromobenzoate, 38B, 648
$C_{32}H_{20}Mn_2O_8Si_2$, Di-μ-diphenylsilyl-bis(tetracarbonylmanganese), 39B, 489
$C_{32}H_{32}Si_2$, 1,1,4,4-Tetramethyl-2,3,5,6-tetraphenyl-1,4-disilacyclohexadiene, 30B, 273
$C_{32}H_{44}HgLi_2Si_4$, Dilithiumtetrakis(dimethylphenylsilyl)mercury(II), 41B, 712; 46B, 646
$C_{32}H_{44}O_{12}Si_8$, cis-1,1,7,7,9,9,15,15-Octamethyl-3,5,11,13-tetraphenyltricyclodecasiloxane, 44B, 594
$C_{32}H_{45}N_3Si_5$, 1-Methyldiphenylsilyl-3-(1,1,3-trimethyl-3,3-diphenylsilazanyl)-2,2,4,4-tetramethylcyclodisilazane, 33B, 272
$C_{32}H_{68}O_4Si_4$, cis,cis-1,2,11,12-Tetrakis-trimethylsiloxycycloeicosa-1,11-diene, 43B, 837
$C_{34}H_{36}O_2Si_2$, trans-7,7'-Bicyclo-(2,2-diphenyl-1-oxa-2-silacycloheptylidene), 42B, 530
$C_{34}H_{36}O_2Si_2$, 2,2,8,8-Tetraphenyl-1-oxa-2,8-disilaspiro[7.7]tridecan-13-one, 42B, 1023
$C_{34}H_{46}N_4Si_4$, 1,3,3,5,7,7-Hexamethyl-2,4,6,8-tetrabenzyl-1,3,5,7-tetrasila-2,4,6,8-tetraazabicyclo[3.3.0]octane, 45B, 712
$C_{34}H_{50}N_6Si_6$, 2,2,8,8-Tetramethyl-1,3,7,9-tetraphenyl-5,10-bis(trimethylsilyl)dispiro[3.1.3.1]tetrasilazane, 42B, 531
$C_{34}H_{90}N_6Si_{10}$, 2,2,4,4-Tetra-sec-butyl-1,3-bis((2,2,4,4-tetramethyl-3-(trimethylsilyl-1,3-diaza-2,4-disilacyclobutyl))dimethylsilyl)-1,3-diaza-2,4-disilacyclobutane (triclinic form), 46B, 647
$C_{34}H_{90}N_6Si_{10}$, 2,2,4,4-Tetra-sec-butyl-1,3-bis((2,2,4,4-tetramethyl-

3-(trimethylsilyl-1,3-diaza-2,4-disilacyclobutyl))dimethylsilyl)-1,3-diaza-2,4-disilacyclobutane (monoclinic form), 46B, 647
$C_{36}H_{30}CrO_4Si_2$, Bis(triphenylsilyl)chromate, 35B, 445
$C_{36}H_{30}HgSi_2$, Bis(triphenylsilyl)mercury, 46B, 632
$C_{36}H_{30}N_2Si_2$, Hexaphenylcyclodisilazane, 46B, 648
$C_{36}H_{30}OSi_2$, Oxobis(triphenylsilicon(IV)), 44B, 594
$C_{36}H_{30}O_3Si_3$, Hexaphenylcyclotrisiloxane, 38B, 649
$C_{36}H_{46}Si_2$, 1,1,2,2-Tetramesityldisilane, 46B, 645
$C_{37}H_{30}N_2Si_2$, Bis(triphenylsilicon)carbodiimide, 41B, 723
$C_{38}H_{34}N_8O_2Si_3$, Silicon phthalocyanine hexamethyldisiloxane, 41B, 723
$C_{40}H_{36}O_6Si_5$, 1,5-Divinyl-3,3,7,7,10,10-hexaphenylbicyclo[3.3.3]pentasiloxane, 40B, 603
$C_{40}H_{56}Si_6$, 1,1,2,2-Tetraphenyl-3,4-bis[bis(trimethylsilyl)-methylene]-1,2-disilacyclobutane, 45B, 713
$C_{42}H_{45}NSi_3$, Tris(methyldiphenylsilylmethyl)amine, 40B, 603
$C_{44}H_{52}BrN_3O_6Si$, (22R)-3β-Acetoxy-5α,8α-(3,5-dioxo-4-phenyl-1H,2H-1,2,4-triazole-1,2-diyl)-24-trimethylsilylchol-6-en-23-yn-22-yl p-bromobenzoate, 46B, 649
$C_{48}H_{40}O_4Si_4$, Octaphenylcyclotetrasiloxane (monoclinic form), 45B, 713; 46B, 649
$C_{48}H_{40}O_{12}Si_8 \cdot C_3H_6O$, Octa(phenylsilasesquioxane) acetone solvate, 45B, 714
$C_{48}H_{40}O_{12}Si_8$, Octa(phenylsilasesquioxane), 44B, 595
$C_{48}H_{40}Si_4$, Octaphenylcyclotetrasilane, 44B, 596
$C_{52}H_{40}N_4Si$, Tetrakis(diphenylketimine)silicon, 41B, 1242
$C_{54}H_{46}Cl_2N_2Si_4$, 1,3-Bis(diphenylchlorosilyl)-1,3-diaza-2,2,4,4-tetraphenylcyclodisilazane - benzene, 44B, 596
$C_{56}H_{64}O_{12}Si_{10}$, 2,2,4,4,14,14,16,16-Octaphenyl-8,8,10,10,19,19,21,21-octamethyldispiro[5.5.5.5]decasiloxane, 40B, 604
$C_{60}H_{50}Si_5$, Decaphenylcyclopentasilane, 44B, 597
$C_{72}H_{60}O_{18}Si_{12}$, Dodeca(phenylsilasesquioxane), 46B, 649
$C_{72}H_{63}N_3O_9Si_9 \cdot 0.5\ C_6H_6$, 2,2,4,4,10,10,12,12,17,17,19,19-Dodecaphenyl-trispiro[5.1.5.1.5.1]nonasila-7,14,21-triaza-1,3,5,9,11,13,16,18,20-nonaoxane benzene, 46B, 650

64. PHOSPHORUS COMPOUNDS

$CBr_7Cl_2N_3P_3Sb$, Tris(chlorobromophosphazeno)carbenium hexabromoantimony, 46B, 651
$CCl_{15}N_3P_3Sb$, Tris(trichlorophosphazeno)carbenium hexachlorantimony, 46B, 651
$CF_4N_2P_2$, Bis(difluorophosphino)carbodi-imide (gas-ed), 38B, 1064
CH_2Cl_3OP, Chloromethylphosphonic dichloride (gas-ed), 42B, 1029
$CH_2Cl_4O_2P_2$, Methylenebis(phosphonic dichloride), 41B, 724
CH_2F_3P, Trifluoromethylphosphine (gas-mw), 33B, 538
CH_3Cl_2OP, Methyl dichlorophosphite (gas-ed), 37B, 695
CH_3Cl_2OP, Methylphosphonyl dichloride (gas-ed), 37B, 695
CH_3Cl_2OPS, O-Methyldichlorothiophosphate (gas-ed), 42B, 1030
$CH_3Cl_4N_3O_2P_2S$, 1-Methyl-bis(dichlorophosphazo)sulphonimide, 43B, 837
CH_3F_2OP, Methoxydifluorophosphine (gas-mw), 40B, 1144
CH_3F_2P, Methyldifluorophosphine (gas-mw), 40B, 1144
CH_3F_4P, Tetrafluoromethylphosphorane (gas-ed), 30B, 400
CH_5P, Methylphosphine (gas-ed), 24, 542
CH_6NO_3P, Aminomethylphosphonic acid (β form), 41B, 725
$CH_6O_6P_2$, Methylenediphosphonic acid, 39B, 316; 43B, 838
$CH_{12}Na_3O_{11}P$, Trisodium phosphonoformate hexahydrate, 37B, 375

$C_2H_3Cl_2OP$, Vinyl dichlorophosphonate (gas-ed), 42B, 1030
$C_2H_3N_3O_2PS_2Tl$, Thallium(I) P,P-dithiophosphacyanurate, 41B, 726
C_2H_4ClOPS, 3-Oxo-3-chloro-1,3-thiaphosphetane (gas-ed), 37B, 688
$C_2H_4ClO_3P$, Ethylenechlorophosphite (gas-ed), 39B, 892
$C_2H_4ClPS_3$, Ethylenechlorotrithiophosphite (gas-ed), 39B, 892
$C_2H_4ClPS_3$, 2,5-Dithio-1-chloro-1-thiophosphorus(V)-cyclopentane, 37B, 376
$C_2H_5Cl_2P$, Dichloroethylphosphine (gas-ed-ed), 43B, 1484
C_2H_5P, Phosphirane (gas-mw), 40B, 1147
$C_2H_6Cl_2N_2P_2$, N-Methyltrichlorophosphineimine dimer, 31B, 299
$C_2H_6Cl_3N_4O_2PS_2$, 3,5,5-Trichloro-1-dimethylamino-1λ^6,3λ^6,2,4,6,5λ^5-dithiatriazaphosphorine 1,3-dioxide, 46B, 652
$C_2H_6FO_2P$, Methoxymethylphosphoryl fluoride (gas-ed), 39B, 892
$C_2H_6F_2NP$, Dimethylaminodifluorophosphine (gas-ed), 37B, 691
$C_2H_6F_2NP$, Dimethylaminodifluorophosphine (gas-mw), 39B, 893
$C_2H_6F_2NP$, Dimethylaminodifluorophosphine, 34B, 376
$C_2H_6F_3P$, Trifluorodimethylphosphorane (gas-ed), 30B, 400
$C_2H_6F_6N_4P_4$, 2,2,4,4,6,6-Hexafluoro-8,8-dimethylcyclotetraphosphazene, 37B, 376
$C_2H_6KO_2PS_2$, Potassium O,O-dimethylphosphordithioate, 27, 755
$C_2H_6N_3PS$, Dimethylthiophosphinic azide, 43B, 839
$C_2H_6Na_2O_7P_2 \cdot 4\ H_2O$, Disodium dihydrogen 1-hydroxyethylidenediphosphonate tetrahydrate, 45B, 715
$C_2H_6P_2S_4$, Methyl metadithiophosphonate, 27, 941; 29, 539
$C_2H_7N_4O_2PS_2$, Ammonium P,P-dithiophosphacyanurate, 41B, 726
$C_2H_7O_2P$, Dimethylphosphinic acid, 32B, 304
C_2H_7P, Dimethylphosphine (gas-ed), 24, 542
$C_2H_7P_2$, Ethylphosphine (gas-mw), 42B, 1033
$C_2H_8NO_2P$, α-Aminomethylmethylphosphinic acid, 43B, 839
$C_2H_8NO_3P$, 2-Aminoethylphosphonic acid, 31B, 301
$C_2H_8O_6P_2$, Ethane-1,2-diphosphonic acid, 43B, 838
$C_2H_{10}CaO_9P_2$, Calcium dihydrogen ethane-1-hydroxy-1,1-diphosphonate dihydrate, 38B, 650
$C_2H_{10}N_4O_2P_2S_2$, trans-3,6-Dimethoxy-1,2,4,5-tetraaza-3,6-diphosphacyclohexane 3,6-disulfide, 46B, 652
$C_2H_{10}O_8P_2$, Ethane-1-hydroxy-1,1-diphosphonic acid monohydrate, 38B, 651
$C_2H_{13}N_5P_2S_2$, Phosphorus-nitrogen compound, 33B, 273
$C_2H_{16}IN_7P_2$, Phosphorus-nitrogen compound, 33B, 274
C_3F_9P, Tris(trifluoromethyl)phosphine (gas-ed), 18, 678; 42B, 1026
$C_3H_6ClO_2P$, Trimethylenechlorophosphite (gas-ed), 38B, 1068
$C_3H_6ClPS_3$, 2-Chloro-2-thiono-1,3,2-dithiaphosphorinane, 42B, 533
$C_3H_6Cl_5N_2P$, 1,3,3,3,3-Pentachloro-2,4-dimethyl-3-phospha-2,4-diazacyclobutene, 34B, 378
C_3H_6NP, Dimethylcyanophosphine (gas-mw), 40B, 1144
$C_3H_6N_3P$, 1,5-Dimethyl-1H-1,2,4,3-λ^3-triazaphosphole, 45B, 715
$C_3H_6O_3P_2S$, 1-Sulfo-2,6,7-trioxa-1,4-diphosphabicyclo[2.2.2]octane, 41B, 726
C_3H_7ClNOP, N-Methyl-2-chloro-1,3,2-oxaazaphospholane (gas-ed), 38B, 1063
$C_3H_7O_4P$, Propane-1,3-diol cyclic phosphate, 35B, 446
C_3H_7P, Cyclopropylphosphine (gas-mw), 37B, 696
$C_3H_8Cl_4N_3P_3$, 1-Hydrido-1-isopropyltetrachlorocyclotriphosphazene, 46B, 653
$C_3H_8NO_5P$, N-(Phosphonomethyl)glycine, 45B, 716
$C_3H_9Cl_2N_3P_2$, 3,5-Dichloro-1,2,4-trimethyl-1,2,4,3,5-triazadiphospholidine, 42B, 532

$C_3H_9F_2P$, Difluorotrimethylphosphorus (gas-ed), 39B, 890
C_3H_9OP, Trimethylphosphine oxide (gas-ed), 30B, 401
C_3H_9P, Trimethylphosphine (gas-ed), 6, 234; 24, 552
C_3H_9P, Trimethylphosphine (gas-mw), 37B, 697
C_3H_9PSe, Trimethylphosphine selenide, 46B, 653
C_3H_9PSe, Trimethylphosphine selenide (gas-ed), 43B, 1484
$C_3H_{10}NO_3P$, 3-Aminopropylphosphonic acid, 46B, 653
$C_3H_{10}O_6P_2$, Propane-1,3-diphosphonic acid, 43B, 840
$C_3H_{11}N_6O_3PS_2$, Ammonium P,P-dithiophosphacyanurate - urea, 42B, 533
$C_3H_{12}NO_9P_3$, Nitrilomethylenetriphosphoric acid, 32B, 304
$C_4F_{12}P_4$, Trifluoromethylphosphine tetramer, 27, 943
$C_4H_4Cl_5N_2P$, Pyrazine-phosphorus(V) chloride, 46B, 654
C_4H_6ClOP, Divinyl chlorophosphinate (gas-ed), 42B, 1031
C_4H_6ClOP, 1-Oxo-1-chlorophosphacyclopent-3-ene (gas-ed), 35B, 894
$C_4H_6Cl_6F_4N_2P_2$, 1,3-Dimethyl-2,4-bis(trichloromethyl)-2,2,4,4-tetrafluoro-1,3-diaza-2,4-diphosphetidine, 41B, 729
$C_4H_7O_4P$, 1-Oxo-4-methyl-1,2,6,7-phosphatrioxa-bicyclo[2.2.1]heptane, 42B, 535
$C_4H_8ClN_2P$, 2,5-Dimethyl-1,2,3-diazaphospholium chloride, 44B, 597
$C_4H_8Cl_3O_4P$, Trichlorophone, 34B, 378
$C_4H_8N_3Na_2PO_5 \cdot 4.5\ H_2O$, Disodium N-phosphorylcreatine hydrate, 33B, 275
$C_4H_8NaO_6P$, Sodium acetylphosphonate acetic acid solvate, 44B, 598
$C_4H_9O_2P$, Phospholanic acid, 34B, 379
$C_4H_9O_3P$, 4-Methyl-1,3,2-dioxaphosphorinane 2-oxide, 39B, 489
$C_4H_9O_4P$, Tetramethylene phosphoric acid, 41B, 503
$C_4H_9PS_3$, 2-Methyl-2-thiono-1,3,2-dithiaphosphorinane, 42B, 533
$C_4H_{10}ClN_2P$, N,N-Dimethyl-2-chloro-1,3,2-diazaphospholane (gas-ed), 38B, 1064
$C_4H_{10}ClP$, Chlorodiethylphosphine (gas-ed-ed), 43B, 1484
$C_4H_{10}FN_2PS$, 2-Fluoro-1,3-dimethyl-1,3,2-diazaphospholidine 2-sulfide, 44B, 598
$C_4H_{11}N_2O_2P$, 1,6-Dioxa-4,9-diaza-5λ^5-phosphaspiro[4.4]nonane, 42B, 534; 44B, 598
$C_4H_{11}N_2O_4P$, Glycylcilliatine, 45B, 717
$C_4H_{11}P$, Trimethyl(methylene)phosphorane (gas-ed-ed), 43B, 1484
$C_4H_{12}Cl_2N_4P_2$, 3,6-Dichloro-1,2,4,5-tetramethyl-1,2,4,5-tetraaza-3,6-diphosphacyclohexane, 42B, 534
$C_4H_{12}Cl_4N_5P_3$, cis-2,4,6,6-Tetrachloro-2,4-bis(dimethylamino)cyclotri(phosphazene), 46B, 654
$C_4H_{12}Cl_4N_5P_3$, trans-2,4,6,6-Tetrachloro-2,4-bis(dimethylamino)cyclotri(phosphazene), 46B, 654
$C_4H_{12}Cl_6N_6P_4$, 2,4,4-trans-6,8,8-Hexachloro-2,6-bis(dimethylamino)-cyclotetraphosphazatetraene, 39B, 490
$C_4H_{12}F_4N_4P_4$, 2,2,6,6-Tetrafluoro-4,4,8,8-tetramethylcyclotetraphosphazene, 37B, 377
$C_4H_{12}P_2$, Tetramethyldiphosphine (gas-ed), 35B, 895
$C_4H_{12}P_2S_2$, Tetramethyldiphosphine disulfide, 32B, 305; 37B, 378
$C_4H_{14}BP$, Trimethylphosphonium methyleneborontrihydride, 46B, 655
$C_4H_{28}Ca_2O_{22}P_4$, Calcium tetraphosphonate decahydrate, 38B, 687
$C_4H_{28}N_4O_{14}P_4$, Tetraammonium tetraphosphonate dihydrate, 40B, 605
$C_5F_{15}P_5$, Pentakis(trifluoromethyl)cyclopentaphosphine, 24, 552; 26, 620
$C_5H_5F_5NP$, Phosphorus pentafluoride - pyridine, 40B, 606
C_5H_5P, Phosphabenzene (gas-ed), 40B, 1137
C_5H_5P, Phosphabenzene (gas-mw), 38B, 1071
$C_5H_6F_4NP$, 2-Methyl-5-(tetrafluorophosphoranyl)pyrrole, 39B, 491

$C_5H_7F_4O_2P$, (Pentane-2,4-dionato)tetrafluorophosphorane, 44B, 599
$C_5H_8BrO_3P$, 4-Bromomethyl-1,2,6,7-phosphatrioxa-bicyclo[2.2.2]octane, 42B, 535
$C_5H_9Br_2O_3P$, Bis(bromomethyl)acetoxyphosphine oxide, 37B, 379
$C_5H_9Br_2O_3P$, 2α-Bromo-5β-bromomethyl-5α-methyl-2β-oxo-11,3,2-dioxaphosphorinane, 34B, 380
$C_5H_9N_2O_3PS$, 2-Hydroxy-1,3,2-dioxaphospholane 2-sulfide imidazolium salt, 44B, 600
$C_5H_9O_4P$, 1-Oxo-4-methyl-2,6,7-trioxa-1-phosphabicyclo[2.2.2]octane, 33B, 278
$C_5H_9O_4P$, 4,5-Dimethyl-2-methoxy-2-oxo-1,3,2-dioxaphospholene, 32B, 220
$C_5H_{10}ClO_3P$, 5,5-Dimethyl-2-chloro-2-oxo-1,3,2-dioxaphosphorinane, 38B, 652
$C_5H_{10}NPS_3$, 2-Thiono-2-aziridino-1,3,2-dithiaphosphorinane, 43B, 841
$C_5H_{10}N_3O_2PS$, 2-Thia-1,3,5-triazaphosphaadamantane 2,2-dioxide, 42B, 298
$C_5H_{11}O_5P$, 5-Ethoxytrimethylenephosphoric acid, 46B, 655
$C_5H_{12}NO_3PS_2$, O,O-Dimethyl-S-(N-methylcarbamoyl)methylphosphorodithiatedimethoate, 46B, 660
$C_5H_{14}N_3O_3P \cdot 0.5\ CH_6ClN_3$, 2-Diethylphosphorylguanidine - hemi(guanidinium chloride), 45B, 717
C_6H_3P, Triethynylphosphine (gas-irr), 42B, 1037
C_6H_3P, Triethynylphosphine, 37B, 379
$C_6H_4ClO_3P$, Pyrocatechol chlorophosphate (gas-ed), 40B, 1139
$C_6H_5Cl_2P$, Phenyldichlorophosphine (gas-ed), 38B, 1069
$C_6H_5F_2P$, Difluoro(phenyl)phosphine (gas-ed-ed), 43B, 1485
$C_6H_6N_3P$, Phosphinetriyltriacetonitrile, 45B, 718
$C_6H_7O_3P$, Benzenephosphonic acid, 42B, 535
$C_6H_9N_2O_2P$, Phenyl phosphorodiamidate, 39B, 491
$C_6H_9O_3P$, Trivinylphosphite (gas-ed), 30B, 401; 40B, 1143
$C_6H_9O_3PS$, Cyclohexane-1a,3a,5a-thiophosphoric acid ester, 24, 582
$C_6H_9O_6P$, cis-2,4-Dioxo-2-methyl-R-5-acetyl-5-methyl-2,2-dihydro-1,3,2-dioxaphospholane, 40B, 403
$C_6H_{10}O_5P$, 7-Methoxy-3,5,9-trioxa-4-phosphabicyclo[4.3.0]nonan-4-one, 45B, 317
$C_6H_{11}N_2O_3PS$, Imidazolium r-2-hydroxy-c-4-methyl-1,3,2-dioxaphospholan-2-thione, 42B, 536
$C_6H_{11}N_4OPS$, Phenoxythiophosphoryl dihydrazide, 45B, 719
$C_6H_{11}OP$, 1-Phosphabicyclo[2.2.1]heptane 1-oxide, 44B, 600
$C_6H_{11}O_4P$, 3-α-Oxo-3-β-hydrido-7-β-hydroxy-2,4-dioxa-3-phosphabicyclo[3.3.1]nonane, 33B, 279
$C_6H_{12}ClO_2P$, 4,4,5,5-Tetramethyl-2-chloro-1,3,2-dioxaphospholane (gas-ed), 35B, 898
$C_6H_{12}NO_3PS$, 1-Thio-1-phospha-2,8,9-trioxa-5-azabicyclo[3.3.3]undecane, 43B, 842
$C_6H_{12}NO_4P$, 2-Formylamino-2-oxo-5,5-dimethyl-1,3,2-dioxaphosphorinane, 43B, 842
$C_6H_{12}N_3OP$, 1,3,5-Triaza-7-phosphaadamantane-7-oxide, 44B, 601
$C_6H_{12}N_3O_2P$, 2,4,4,5-Tetramethyl(1,3,2λ^5)diazaphospheto[2,1-b]-(1,3,4,2λ^5)oxadiazaphosphol-6(5H)-one, 42B, 536
$C_6H_{12}N_3P$, Tris(ethylenimido)phosphite (gas-ed), 38B, 1068
$C_6H_{12}N_3P$, 1,3,5-Triaza-7-phosphaadamantane, 43B, 843
$C_6H_{12}N_3PS$, Triethylenethiophosphoramide, 40B, 1100
$C_6H_{12}N_3PS$, 1,3,5-Triaza-7-phosphaadamantane-7-sulfide, 44B, 601
$C_6H_{12}N_3PS$, 10-Thio-10-phospha-1,4,7-triazatricyclo-[5.2.1.0$^{4-10}$]decane, 45B, 719

$C_6H_{12}P_2S_2$, 1,5-Diphosphabicyclo[3.3.0]octane-1,5-disulfide (monoclinic), 44B, 601
$C_6H_{12}P_2S_2$, 1,5-Diphosphabicyclo[3.3.0]octane-1,5-disulfide (orthorhombic), 45B, 720
$C_6H_{13}BF_4NO_3P$, 1-Hydro-2,8,9-trioxa-1-phospha-5-aza-tricyclo-[3.3.3.0]undecane tetrafluoroborate, 42B, 537
$C_6H_{13}O_2PS$, 2-Thiono-2,5,5-trimethyl-1,3,2-dioxaphosphorinane, 44B, 602
$C_6H_{13}O_3PSe$, 2-Seleno-2-methoxy-5,5-dimethyl-1,3,2-dioxaphosphorinane, 41B, 729
$C_6H_{13}O_4P$, 5,5-Dimethyl-2-methoxy-2-oxo-1,3,2-dioxaphosphorinane, 46B, 656
$C_6H_{14}KO_4P \cdot H_2O$, Potassium 2-hydroxy-2,3,3-trimethylpropanephosphite monohydrate, 46B, 656
$C_6H_{15}BNO_3P$, 1-Trihydroborano-2,8,9-trioxa-1-phospha-5-azabicyclo-[3.3.3]undecane, 43B, 844
$C_6H_{15}O_3P$, Triethylphosphite (gas-ed), 30B, 401; 40B, 1143
$C_6H_{15}PS$, Triethylphosphorus sulfide, 23, 607
$C_6H_{15}PSe$, Triethylphosphine selenide, 24, 562
$C_6H_{16}BO_3P$, 2-Methoxy-cis-4,6-dimethyl-1,3,2-dioxaphosphorinan-borane, 37B, 380
$C_6H_{18}BNP_2$, [Nitrido-bis(dimethylphosphonium-methylide)]boronate, 45B, 720
$C_6H_{18}Cl_3N_6P_3$, geminal-2,2,4-Trichloro-4,6,6-trisdimethylaminocyclotriphosphazatriene, 38B, 653
$C_6H_{18}Cl_3N_6P_3$, trans-nongeminal-2,4,6-Trichloro-2,4,6-trisdimethylaminocyclotriphosphazatriene, 41B, 730
$C_6H_{18}Cl_3N_6P_3$, 2,4,6-Trichloro-2,4,6-trisdimethylaminocyclotriphosphazatriene, 38B, 655
$C_6H_{18}Cl_8N_6P_4$, 2,2,2,4,6,8,8,8-Octachloro-1,3,5,7,9,10-hexamethyl-1,3,5,7,9,10-hexaza-2,4,6,8-tetraphospha(2,4,6,8-P(V))dispiro-[3.1.3.1]-decane, 32B, 307
$C_6H_{18}I_2N_3P_3$, Hexamethylcyclotriphosphazene - iodine (1:1 adduct), 40B, 607
$C_6H_{18}N_3O_6P_3$, 2,4,6-Trimethoxy-1,3,5-trimethyl-2,4,6-trioxocyclotriphosphazane, 33B, 280
$C_6H_{18}N_3P$, Tris(dimethylamido)phosphite (gas-ed), 38B, 1068
$C_6H_{18}N_3PSe$, Tris(dimethylamino)phosphine selenide, 45B, 735
$C_6H_{18}N_3P_3$, Hexamethylcyclotriphosphazene, 43B, 845
$C_6H_{18}N_6O_2P_2$, Tris(dimethylhydrazino)bis(phosphineoxide), 38B, 655
$C_6H_{18}N_6O_4P_4$, Tetra-P-oxo-closo-tetraphosphorus hexakis(methylimide), 44B, 603
$C_6H_{18}N_6P_2$, Tris(1,2-dimethylhydrazino)diphosphine, 37B, 380
$C_6H_{18}N_6P_4S$, Monothio-closo-tetraphosphorus hexakis(methylimide), 44B, 603
$C_6H_{18}N_6P_4S_4$, Tetra-P-thio-closo-tetraphosphorus hexakis(methylimide), 44B, 603
$C_6H_{18}P_4Si_3$, Tris(dimethylsilyl)tetraphosphorus, 44B, 610
$C_6H_{20}N_5PS_2Si_2$, Cyclo-tri-μ-nitrido-(bis(trimethylsilylamino)phosphorus)disulfur, 43B, 845
$C_7H_7Cl_3NO_3PS$, O,O,-Dimethyl O-(3,5,6-trichloro-2-pyridyl)phosphorothioate, 45B, 721
$C_7H_7Cl_3NO_4P$, Dimethyl 3,5,6-trichloro-2-pyridyl phosphate, 43B, 565
$C_7H_8IO_3P$, 1,3-Dihydro-1-hydroxy-3-methyl-1,2,3-benziodoxaphosphole-3-oxide, 44B, 604
$C_7H_{10}N_5P$, 3,7-Dicyano-3,5,7-triaza-1-phosphabicyclo[3.3.1]nonane, 38B, 656

$C_7H_{13}Cl_2N_2O_3P$, 4-Ketocyclophosphamide, 39B, 492
$C_7H_{13}N_2O_3PS \cdot H_2O$, Imidazolium trans-2-hydroxy-4,5-dimethyl-1,3,2-dioxaphospholane 2-sulfide monohydrate, 46B, 658
$C_7H_{13}N_2O_3PS$, Imidazolium cis-2-hydroxy-4,5-dimethyl-1,3,2-dioxaphospholane 2-sulfide, 46B, 657
$C_7H_{13}N_2O_3PS$, Imidazolium dl-2-hydroxy-4,5-dimethyl-1,3,2-dioxaphospholan-2-thione, 42B, 338
$C_7H_{13}O_3P$, 1-Phospha-2,10-dioxabicyclo[4.4.0]decane 1-oxide, 44B, 604
$C_7H_{15}BF_4N_3P$, 7-Methyl-1,3,5-triaza-7-phosphonioadamantane tetrafluoroborate, 44B, 605
$C_7H_{15}Cl_2N_2O_2P$, 2-(Bis(2-chloroethyl)amino)-2H-1,3,2-oxazaphosphorinane 2-oxide, 43B, 846
$C_7H_{15}Cl_2N_2O_2P$, 3-(2-Chloroethyl)-2-((2-chloroethyl)amino)perhydro-2H-1,3,2-oxazaphosphorine oxide, 43B, 847
$C_7H_{15}Cl_2N_2O_4P$, 4-Hydroperoxycyclophosphamide, 41B, 730; 43B, 848
$C_7H_{15}PS_2$, Triethylphosphine - carbon disulfide, 26, 606; 27, 774
$C_7H_{16}NO_2PS$, 2-Thiono-2-N-dimethylamino-5,5-dimethyl-1,3,2-dioxaphosphorinane, 44B, 602
$C_7H_{16}NO_4PS_2$, O,O-Dimethyl S-(N-2-methoxyethylcarbamoylmethyl) phosphorodithioate, 43B, 848
$C_7H_{17}Cl_2N_2O_3P$, N,N-Bis-2-chloroethyl-N,O-propylene phosphoric ester diamide monohydrate, 38B, 657
$C_7H_{18}P_2$, Hexamethylcarbodiphosphorane (gas-ed-ed), 43B, 1485
$C_7H_{21}IN_6P_4$, Tetraphosphorus hexakis(N-methylimide) methyl iodide, 40B, 608
$C_7H_{27}N_{11}P_4$, 9-Methyl-1,3,3,5,7,7-hexakis(methylamino)bicyclo[3.3.1]-tetra(phosphazene), 45B, 722
$C_8H_4NO_2PS_2$, Dimethylammonium O,O'-diisopropyldithiophosphate, 45B, 722
$C_8H_7F_{12}O_3P$, 2-Hydroxy-1,1,2,2-tetrakis(trifluoromethyl)ethyl(dimethyl phosphinate), 46B, 658
$C_8H_8Cl_3O_4P$, O,O-Dimethyl O-2,4,5-trichlorophenyl phosphate, 44B, 605
$C_8H_9PS_3$, 2,5-Dithio-1-phenyl-1-thiophosphorus(V)-cyclopentane, 37B, 381
$C_8H_{10}NO_5PS$, Methylparathion, 35B, 446
$C_8H_{12}F_{12}N_2P_2$, 1,2,3,4-Tetramethyl-2,2,4,4-tetrakis(trifluoromethyl)-1,3-diaza-2,4-diphosphetidine, 46B, 659
$C_8H_{16}ClOP$, 1-Chloro-2,2,3,4,4-pentamethylphosphetan 1-oxide, 35B, 448
$C_8H_{16}P_2S_2$, Bis(cyclotetramethylene)diphosphine disulfide, 34B, 382
$C_8H_{17}O_3P$, 5-t-Butyl-2-methyl-2-oxo-1,3,2-dioxaphosphorinan, 35B, 449
$C_8H_{17}O_4P$, trans-2-Methoxy-2-oxo-5-t-butyl-1,3,2-dioxaphosphorinane, 44B, 606
$C_8H_{17}PS_4$, 2-t-Butyl-1,3,6,2-trithiaphosphocane, 45B, 723
$C_8H_{18}ClN_2O_5PS$, 3-Chloroethyl-2-(mesyloxyethylamino)tetrahydro-2H-1,3,2-oxazaphosphorine 2-oxide, 45B, 723
$C_8H_{18}Cl_2N_2O_2P_2$, 1,3-Di-t-butyl-2,trans-4-dichloro-2,4-dioxo-cyclodiphosphazene, 40B, 609
$C_8H_{18}Cl_2N_2P_2$, 1,3-Di-t-butyl-2,4-dichlorodiazadiphosphetidine, 41B, 731
$C_8H_{18}NO_2PSe$, cis-2-t-Butylamino-2-seleno-4-methyl-1,3,2-dioxaphosphorinan, 41B, 732
$C_8H_{18}NO_2PSe$, trans-2-t-Butylamino-2-seleno-4-methyl-1,3,2-dioxaphosphorinan, 42B, 538
$C_8H_{20}Cl_4N_5P_3$, 4,4,6,6-Tetrachloro-2,2-di-t-butylaminocyclotriphosphazene, 45B, 724
$C_8H_{20}P_2S_2$, Tetraethyldiphosphine disulfide, 26, 609

$C_8H_{20}P_2S_2Se_2$, Bis(diethylthiophosphoryl) diselenide, 31B, 301
$C_8H_{20}P_2Se_5$, Selenium bis(diethyldiselenophosphinate), 34B, 381
$C_8H_{20}P_6$, 2,3,4,6,7,8-Hexamethyl-2,3,4,6,7,8-hexaphosphabicyclo-[3.3.0]octane, 44B, 607
$C_8H_{22}B_2NO_2P$, 1,5-Bis(borane)-3,3,7,7-tetramethyl-2,8-dioxa-5-aza-1-phosphabicyclo[3.3.0]octane, 46B, 659
$C_8H_{24}Cl_4N_8P_4$, 1,cis-3,cis-5,trans-7-Tetrachloro-1,3,5,7-tetrakis(dimethylamino)cyclotetraphosphazene, 43B, 849
$C_8H_{24}Cl_4N_8P_4$, 2,cis-4,trans-6,trans-8-Tetrachloro-2,4,6,8-tetrakis-(dimethylamino)cyclotetraphosphazene, 38B, 658
$C_8H_{24}Cl_4N_8P_4$, 2,trans-4,cis-6,trans-8-Tetrachloro-2,4,6,8-tetrakis-(dimethylamino)cyclotetraphosphazene, 44B, 607
$C_8H_{24}F_4N_8P_4$, 1,cis-3,trans-5,trans-7-Tetrakis(dimethylamino)-1,3,5,7-tetrafluorotetraphosphonitrile, 39B, 493
$C_8H_{24}F_4N_8P_4$, 1,trans-3,cis-5,trans-7-Tetrakis(dimethylamino)-1,3,5-7-tetrafluorocyclotetraphosphazene, 40B, 610
$C_8H_{24}N_4O_8P_4$, Octamethoxycyclotetraphosphazene, 37B, 382
$C_8H_{24}N_4O_8P_4$, 2,trans-4,cis-6,trans-8-Tetramethoxy-1,3,5,7-tetramethyl-2,4,6,8-tetraoxocyclotetraphosphazane, 43B, 849
$C_8H_{24}N_4P_4$, Octamethylcyclotetraphosphazene, 26, 607
$C_8H_{24}N_4P_4S_3$, Octamethyl-trithiocyclotetra($\lambda^3,\lambda^5,\lambda^5,\lambda^5$)phosphazane, 46B, 659
$C_8H_{25}Cl_3CuN_4P_4$, Trichloro(octamethylcyclotetraphosphonitrilium)copper(II), 35B, 459
$C_8H_{26}Cl_4N_4P_4Pt$, N,N''-Dihydro(octamethylcyclotetraphosphazenium) tetrachloroplatinate(II), 45B, 1155
$C_9H_9F_{15}O_2P_2$, 2,2,2-Trifluoro-4,4,5,5-tetrakis(trifluoromethyl)-1,3,2-dioxaphospholanetrimethylphosphine, 44B, 608
$C_9H_9N_6P$, Tris-(1-pyrazolyl)phosphine, 41B, 732
$C_9H_9O_5P$, Methyl β-D-ribopyranoside phosphite triester, 45B, 724
$C_9H_9O_5PS$, Methyl β-D-ribopyranoside thiophosphate triester, 45B, 724
$C_9H_{11}O_2PS$, 3-Thiono-3-methyl-1,5-dihydro-2,4,3-benzodioxaphosphepin, 44B, 602
$C_9H_{11}O_4P$, O-Methylphenylphosphinylacetic acid, 43B, 850
$C_9H_{11}O_4P$, O-Methylphenylphosphinylacetic acid (optically active form), 44B, 609
$C_9H_{11}O_4P$, 2-Oxo-2-phenoxy-1,3,2-dioxaphosphorinane, 32B, 222
$C_9H_{12}Cl_3N_2O_2PS$, N-Isopropyl-O-methyl-O-(3,5,6-trichloro-2-pyridyl)-phosphoramidothioate, 46B, 660
$C_9H_{12}NO_3P$, 7,8-Dimethyl-5-methoxy-5-oxo-λ^5-5-phospha-6-oxaindolizine, 44B, 609
$C_9H_{13}O_2PS$, Phenyl-α-mercapto-isopropylphosphinic acid, 45B, 725
$C_9H_{13}O_6P$, 3,5,6-Bicyclophosphite-1,2-O-isopropylidene-α-D-glucofuranose, 45B, 725
$C_9H_{14}NO_4P$, Dimethyl N-(p-methoxyphenyl)phosphoramidate, 46B, 661
$C_9H_{16}NO_5P$, 4-Diethylamino-3,5,6-trioxa-4-phosphabicyclo[4.3.0]nonan-4,7-dione, 45B, 726
$C_9H_{17}N_2O_3PS$, Imidazole 2-hydroxy-4,4,5,5-tetramethyl-1,3,2-dioxaphospholane 2-sulfide, 44B, 609
$C_9H_{18}Cl_3N_2O_2P$, 3-(2-Chloroethyl)-2-(bis(2-chloroethyl)amino)perhydro-2H-1,3,2-oxazaphosphorine 2-oxide, 42B, 538; 43B, 851
$C_9H_{18}N_3PS_6$, Tris(N,N-dimethyldithiocarbamato-S)phosphine, 46B, 661
$C_9H_{19}O_4P$, Dimethyl 1-hydroxy-1-cycloheptanephosphonate, 43B, 851; 46B, 477
$C_9H_{20}NPS_3$, 2-Thiono-2-N-(diisopropylamino)-1,3,2-dithiaphosphorinane, 43B, 841
$C_9H_{23}INO_2PS$, (R)-(+)-O-Isopropyl-S-(trimethylammonioethyl)methyl-

phosphonothioate iodide, 43B, 852
$C_9H_{26}N_4P_4$, 2,2,4,4,6,8,8-Heptamethyl-6-methylamino-1,3,5-triaza-2,4,6,8(P(V))-tetraphosphorin, 44B, 610
$C_9H_{27}N_2PSi_3$, (Bis(trimethylsilyl)amino)-(trimethylsilylimino)-phosphane, 43B, 825; 45B, 715
$C_9H_{27}P_7Si_3$, Tris(trimethylsilyl)heptaphosphorus, 44B, 610
$C_9H_{33}N_6PSSi_3$, 1,1',2"-Tris(trimethylsilyl)thiophosphoryltrihydrazide, 46B, 662
$C_{10}H_8F_4NOP$, 2-Methyl-8-oxyquinolinato-tetrafluorophosphorane, 43B, 610
$C_{10}H_{11}Br_2OP$, 2,3-Dibromo-1-phenylphospholane 1-oxide, 45B, 727
$C_{10}H_{11}O_2P$, 4-Hydroxy-1-phenyl-2-phospholene 1-oxide, 46B, 662
$C_{10}H_{12}N_2O_3$, N-(1,2-Dimethyl-ethylenedioxyphosphoryl)imidazole benzene solvate, 44B, 611
$C_{10}H_{13}ClNOPS$, 2-Thio-2-chloro-3,4-dimethyl-5-phenyl-1,3,2-oxazaphospholidine, 43B, 852
$C_{10}H_{13}O_4P$, O-Methyl-α-phenylphosphinylpropionic acid, 43B, 853
$C_{10}H_{14}BrOPS_2$, O-Ethyl S-p-bromophenyl ethylphosphonodithioate, 43B, 857
$C_{10}H_{14}NO_2PS$, N-Methyl-2-phenoxy-1,3,2-oxazaphosphorinan-2-thione, 45B, 727
$C_{10}H_{14}NO_2PS$, 3-Thiono-3-N-dimethylamino-1,5-dihydro-2,4,3-benzodioxaphosphepin, 44B, 602
$C_{10}H_{15}PSe$, P-Methyl-P-phenylpropylphosphine selenide, 43B, 854
$C_{10}H_{19}ClNO_3P$, 2-Piperidino-5-chloromethyl-5-methyl-2-oxo-1,3,2-dioxaphosphorinan, 39B, 494
$C_{10}H_{19}O_2P$, Dipivaloylphosphine, 43B, 854
$C_{10}H_{19}O_7P$, 2r,3t,4t,5c-Tetramethyl-2,5-dimethoxy-3,4-dihydroxytetrahydrofuran-3,4-cyclic hydrogenphosphate, 43B, 566
$C_{10}H_{20}O_4P_2S_2$, 5,5,5',5'-Tetramethyl-2,2'-bi-1,3,2λ^5-dioxaphosphorinane 2,2'-disulfide, 46B, 663
$C_{10}H_{20}O_5P_2S_2$, (a,a)-Bis(4,6-dimethyl-2-thioxo-1,3,2-dioxaphosphorinanyl)oxide, 45B, 728
$C_{10}H_{20}O_5P_2S_2$, (a,e)-Bis(4,6-dimethyl-2-thioxo-1,3,2-dioxaphosphorinanyl)oxide, 45B, 728
$C_{10}H_{20}O_6P_2S$, Bis(5,5'-dimethyl-2-oxo-1,3,2-dioxaphosphorinanyl)sulfide, 42B, 538
$C_{10}H_{20}O_6P_2S$, 2-Oxo-2'-thioxobis(5,5-dimethyl-1,3,2-dioxaphosphorinanyl) oxide, 46B, 663
$C_{10}H_{20}O_7P_2$, Bis(5,5-dimethyl-2-oxo-1,3,2-dioxaphosphorinan-2-yl) oxide, 42B, 338, 539; 44B, 612
$C_{10}H_{20}P_2S_2$, Bis(cyclopentamethylene)disphosphine disulfide, 35B, 451
$C_{10}H_{22}NO_2PSe$, cis-2-t-Butylamino-2-seleno-4,4,6-trimethyl-1,3,2-dioxaphosphorinane, 44B, 612
$C_{10}H_{23}O_2P$, Di-t-pentylphosphinic acid, 44B, 612
$C_{10}H_{24}N_2P_2$, trans-2,4-Di-t-butyl-1,3-dimethyl-1,3,2,4-diazaphosphetidine, 45B, 729
$C_{10}H_{24}N_2P_2S_2$, cis-2,4-Di-t-butyl-1,3-dimethyl-1,3,2,4-diazaphosphetidine-2,4-dithione, 45B, 729
$C_{10}H_{24}N_2P_2S_2$, trans-2,4-Di-t-butyl-1,3-dimethyl-1,3,2,4-diazaphosphetidine-2,4-dithione, 45B, 729
$C_{10}H_{24}N_2P_2Te$, 2-cis-4-Bis(t-butyl)-1,3-dimethyl-2-telluro-1,3,2,4-diazadiphosphetidine, 44B, 613
$C_{10}H_{24}N_2P_2Te_2$, cis-2,4-Di-t-butyl-1,3-dimethyl-1,3,2,4-diazaphosphetidine-2,4-ditellurone, 45B, 729
$C_{10}H_{24}N_2P_2Te_2$, trans-2,4-Di-t-butyl-1,3-dimethyl-1,3,2,4-diazaphosphetidine-2,4-ditellurone, 45B, 729

$C_{10}H_{26}N_4O_4P_2S_2 \cdot C_2H_3N$, 2,4,6,8-Tetraethyl-3,7-dimethyl-1λ[6],5λ[6]-dithia-2,4,6,8-tetraaza-3,7-diphosphorocine 1,1,5,5-tetraoxide acetonitrile solvate, 46B, 664

$C_{10}H_{27}I_2P_3$, Tris(trimethylphosphine)methylide diiodide, 45B, 730

$C_{10}H_{30}Cl_2N_4P_4$, 2-trans-6-Diethyl-2,4,4,6,8,8-hexamethylcyclotetraphosphazene dihydrochloride, 41B, 734

$C_{10}H_{30}Cl_3N_9P_4$, 1,cis-3,cis-5-Trichloro-1,3,5,7,7-pentakis(dimethylamino)cyclotetraphosphazene, 41B, 1243

$C_{10}H_{30}N_5P_5$, Decamethylcyclopentaphosphazene, 43B, 855

$C_{10}H_{34}Cl_4CuN_5OP_5$, 1,5-Dihydrododecamethylcyclopentaphosphonitrilium tetrachlorocuprate(II) monohydrate, 40B, 611

$C_{11}H_6NO_3P$, 2-Oxo-2-methyl-6-aza-6-phenyl-1,3,2-dioxaphosphacynan, 41B, 451

$C_{11}H_{10}N_3P$, 2-(2-Cyanoethyl)-5-phenyl-1,2,3-diazaphosphole, 40B, 612

$C_{11}H_{11}O_2P$, 6-Methyl-4-methylene-2-phenyl-4H-1,3,2-dioxaphosphorin, 46B, 664

$C_{11}H_{11}P$, 1-Benzylphosphole, 39B, 494

$C_{11}H_{13}BrClO_4P$, 2-p-Bromophenoxy-5-chloromethyl-5-methyl-2-oxo-1,3,2-dioxaphosphorinane, 39B, 495

$C_{11}H_{13}Cl_2O_5P$, Diethyl [5,6-dichloro-1,3-benzodioxol-(2)]-phosphonate, 45B, 731

$C_{11}H_{13}OPS$, 1-Phenyl-4-phosphorinanone 1-sulfide, 45B, 731

$C_{11}H_{13}O_2P$, 1-Phenyl-4-phosphorinanone 1-oxide, 45B, 731

$C_{11}H_{14}ClO_4P$, 2-Phenoxy-5-chloromethyl-5-methyl-2-oxo-1,3,2-dioxaphosphorinane, 39B, 495

$C_{11}H_{15}O_2P$, 2-Hydroxy-2-methyl-1-phenylphospholane 1-oxide, 46B, 665

$C_{11}H_{15}O_3P$, 2-Hydroxy-5-hydroxymethyl-1-phenylphospholane 1-oxide, 46B, 665

$C_{11}H_{15}O_3P$, 2-Oxo-2-phenyl-4,4-dimethyl-1,3,2-dioxaphosphorinane, 37B, 383

$C_{11}H_{15}O_3PS$, 2-Phenoxy-2-thiono-5,5-dimethyl-1,3,2-dioxaphosphorinane, 41B, 735

$C_{11}H_{16}IP$, Methyl phenyl phospholanium iodide, 32B, 308

$C_{11}H_{16}NO_2PS$, 4-Methyl-2-(N-methyl-N-phenylamino)-1,3,2λ[5]-dioxaphosphorinane 2-sulfide, 46B, 666

$C_{11}H_{16}NO_3P$, 5,5-Dimethyl-2-oxo-2-aminobenzo-1,3,2-dioxaphosphorinane, 42B, 539

$C_{11}H_{16}NPS$, 1-Methyl-4-phenylperhydro-1,4-azaphosphorine 4-sulfide, 40B, 612

$C_{11}H_{17}Cl_3NO_4P$, (-)-(S)-α-Phenylethylammonium (-)-(R)-2,2,2-trichloro-1-hydroxyethyl-O-methylphosphonate, 44B, 613

$C_{11}H_{17}N_4O_2PS$, 6,8-Dimethyl-7-phenoxy-3-oxa-1,5,6,8-tetraaza-7-phospha-bicyclo[3.3.1]nonane 7-sulfide, 46B, 666

$C_{11}H_{19}OP$, 2,3,5-Trimethyl-3-oxo-3-phosphatricyclo[4.2.1.0$^{2\ 5}$]nonane, 46B, 667

$C_{11}H_{20}ClOP$, 2,3,5-Trimethyl-2-methylchlorophosphorylbicyclo[2.2.1]-heptane, 46B, 667

$C_{11}H_{20}NO_5P$, Phenethylammonium (-)-(1R,2S)-epoxypropylphosphonate monohydrate, 44B, 613

$C_{11}H_{21}O_3P$, 2-Hydroxy-3,5-di-t-butyl-1,2-oxaphosphol-3-ene 2-oxide, 39B, 496

$C_{11}H_{23}O_2PS$, cis-2,5-Di-t-butyl-2-thio-1,3,2-dioxaphosphorinane, 44B, 606

$C_{11}H_{24}MoN_4O_3P_4$, Octamethylcyclotetraphosphazenetricarbonylmolybdenum, 39B, 496

$C_{11}H_{27}N_2PSi$, (t-Butylimino)-(t-butyl(trimethylsilyl)amino)-phosphane, 45B, 715

$C_{11}H_{33}IN_4P_2Si_2$, 2,4-Bis(dimethylamino)-2-methyl-1,3-bis(trimethylsilyl)-1,3,2,4-diazadiphosphetidinium iodide, 44B, 620
$C_{12}H_8ClO_4P$, 2-Chloro-2,2'-spirobis(1,3,2-benzodioxaphosphole), 43B, 855
$C_{12}H_8FO_4P$, 2-Fluoro-2,2'-spirobis-(1,3,2-benzodioxaphosphole), 40B, 613
$C_{12}H_9O_2P$, 5-Hydroxydibenzo-5H-phosphole-5-oxide, 40B, 613
$C_{12}H_9O_4P$, 2-Hydrido-2,2'-spirobi(1,3,2-benzodioxaphosphole), 44B, 614
$C_{12}H_9O_5P$, o-Hydroxyphenyl-o-phenylene phosphate, 44B, 614
$C_{12}H_9PS_3$, Tri(3-thienyl)phosphine, 43B, 856
$C_{12}H_{10}Cl_4N_3P_3$, 2,2-Diphenyl-4,4,6,6-tetrachloro-cyclotriphosphazatriene, 30B, 280
$C_{12}H_{10}Cl_8N_6P_6$, Bi-(2,2,4,4-tetrachloro-6-monophenylcyclotriphosphazatrien-6-yl), 38B, 659
$C_{12}H_{10}F_4N_3P_3$, 1,1-Diphenyltetrafluorocyclotriphosphazene, 34B, 383
$C_{12}H_{10}NO_3P$, 2,3:7,8-Dibenzo-1,4,6-trioxa-9-aza-5λ^5-phosphaspiro-[4.4]nona-2,7-diene, 45B, 732
$C_{12}H_{11}N_2O_2P$, 1,6-Dioxa-4,9-diaza-2,3,7,8-dibenzo-5-λ^5-phosphaspiro-[4.4]nona-2,7-diene, 44B, 598
$C_{12}H_{14}N_4O_2P_2S_2$, cis-Dithio-dihydrazido-dimetaphosphoric acid diphenylester, 42B, 540
$C_{12}H_{14}N_4O_2P_2S_2$, trans-Dithio-dihydrazido-dimetaphosphoric acid diphenylester, 41B, 735
$C_{12}H_{16}ClO_2P$, Phenyl-α-hydroxycyclohexylphosphonic chloride, 45B, 733
$C_{12}H_{17}I_2P$, cis-1-Iodomethyl-3-methyl-1-phenylphospholanium iodide, 42B, 541
$C_{12}H_{17}OP$, (3RS,PSR)-3-Methylphenylphosphinoyl-2-methylbut-1-ene, 40B, 614
$C_{12}H_{17}OP$, 2,2,3-Trimethyl-1-phenylphosphetane 1-oxide, 44B, 615
$C_{12}H_{17}O_6P$, 3,5,6-Bicyclophosphite-1,2-O-cyclohexylidene-α-D-glucofuranose, 45B, 725
$C_{12}H_{18}O_4P_2$, 1,8-Diethoxy-3a,4,7,7a-tetrahydro-4,7-phosphinidenephosphindole-1,8-dioxide, 34B, 385
$C_{12}H_{19}ClNOP$, N,N-Di-isopropyl-P-phenylphosphonamidic chloride, 44B, 615
$C_{12}H_{19}ClNO_3P$, Crufomate, 44B, 616
$C_{12}H_{19}O_2P$, (2RS,PSR)-3-(Methylphenylphosphinoyl)-3-methylbutan-2-ol, 40B, 615
$C_{12}H_{19}O_8P$, 4,6-Di-O-acetyl-2,3-didesoxy-α-D-erythrohex-2-enopyranosyldimethylphosphonate, 45B, 733
$C_{12}H_{22}NO_2PS$, α-Phenylammonium O-ethyl ethylphosphonothioate, 43B, 857
$C_{12}H_{23}O_2P$, Di(cyclohexyl)phosphinic acid, 45B, 734
$C_{12}H_{23}O_4P$, endo-2-Dimethylphosphono-exo-2-hydroxy-(-)-camphane, 45B, 734
$C_{12}H_{24}CoO_4PSi_2$, Bis(trimethylsilyl)dimethylphosphonium tetracarbonylcobalt(-I), 42B, 541
$C_{12}H_{24}N_3O_3P$, Tris(morpholino)phosphine, 44B, 616
$C_{12}H_{24}N_3O_3PSe$, Tris(morpholino)phosphine selenide, 45B, 735
$C_{12}H_{24}N_9P_3 \cdot 3\ CCl_4$, 2,2,4,4,6,6-Hexa(1-aziridinyl)cyclotri(phosphazene) carbon tetrachloride solvate, 46B, 667
$C_{12}H_{26}O_4P_2S_2$, cis-2,8-Dithioxo-2,5,5,8,11,11-hexamethyl-1,3,7,9-tetraoxa-2,8-diphosphacyclododecane, 44B, 617
$C_{12}H_{28}N_4O_2P_2$, Octamethyltetra-amino-2,3-diphosphinyl buta-1,3-diene, 34B, 386
$C_{12}H_{28}O_4P_2S_4$, Bis(O,O'-di-isopropylthiophosphoryl)disulfide, 35B,

452; 42B, 541
$C_{12}H_{30}N_3O_6P_3$, Tris(N,N-diethylmetaphosphoramide) (monoclinic), 40B, 1101
$C_{12}H_{30}N_3O_6P_3$, Tris(N,N-diethylmetaphosphoramide) (orhtorhombic), 37B, 384
$C_{12}H_{30}N_4P_2S_2$, 1,3-Di-t-butyl-2,4-bis(dimethylamino)-2,4-dithio-cyclo-di(phosphazane), 43B, 858
$C_{12}H_{32}Cl_2N_7P_3$, Dichlorotetrakis(isopropylamino)cyclotriphosphazene, 42B, 542
$C_{12}H_{33}Cl_3N_7P_3$, Dichlorotetrakisisopropylaminocyclotriphosphazatriene hydrochloride, 37B, 385
$C_{12}H_{36}Cl_2N_{10}P_4$, 2,trans-6-Dichloro-2,4,4,6,8,8-hexakis(dimethylamino)cyclotetraphosphazatetraene, 40B, 616
$C_{12}H_{36}Cl_{12}Ge_4N_4P_4$, Bis(N-(trichlorogermyl)trimethylphosphinimine) (N-(trichlorogermyl)trimethylphosphinimine dimer), 44B, 618
$C_{12}H_{36}N_3PSi_4$, (Bis(trimethylsilyl)amino)bis(trimethylsilylimino)-phosphorane, 43B, 858
$C_{12}H_{36}N_6O_{12}P_6$, Dodecamethoxycyclohexaphosphazene, 40B, 617
$C_{12}H_{36}N_6P_6$, Dodecamethylcyclohexaphosphazene, 43B, 859
$C_{12}H_{36}N_7O_3P_3 \cdot H_2O$, Dodecamethylbis(imido)triphosphoramide monohydrate, 46B, 668
$C_{12}H_{36}N_9P_3$, Hexakis(dimethylamino)cyclotriphosphazene, 39B, 497
$C_{12}H_{36}P_4Si_6$, Dodecamethylhexasilatetraphosphaadamantane, 44B, 618
$C_{13}H_{10}BrCl_2O_2PS$, O-Methyl-O-(4-bromo-2,5-dichlorophenyl)phenylphosphonothioate, 46B, 668
$C_{13}H_{11}O_2P$, 9,10-Dihydro-9-hydroxy-9-phosphaphenanthrene-9-oxide, 27, 919
$C_{13}H_{11}O_4P$, 2-Oxo-2-phenoxy-4H-1,3,2-benzodioxaphosphorin, 44B, 619
$C_{13}H_{11}O_4P$, 2-Methyl-2,2'-spirobis-(1,3,2-benzodioxaphosphole), 40B, 617
$C_{13}H_{12}ClOP$, Chloromethyldiphenylphosphine oxide, 45B, 736
$C_{13}H_{12}NO_3P$, 2,3:7,8-Dibenzo-9-methyl-1,4,6-trioxa-9-aza-5λ^5-phosphaspiro[4.4]nona-2,7-diene, 45B, 732
$C_{13}H_{13}FNP$, Diphenylfluoro-N-methylphosphine imide, 35B, 454
$C_{13}H_{13}OPS$, Methyl diphenylthiophosphinite, 34B, 388
$C_{13}H_{13}OPSe$, Methyl diphenylselenophosphinite, 34B, 382
$C_{13}H_{13}O_2PS$, 2-Carboxy-3,4-dimethyl-1-phenylphosphole sulfide, 46B, 669
$C_{13}H_{13}P$, 2,6-Dimethyl-4-phenylphosphorin, 35B, 454
$C_{13}H_{26}N_2O_5P_2S_2$, 1,5:2,3-Bis-O-(N-diethylamidothionophosphate)-β-D-ribofuranoside, 46B, 669
$C_{14}H_{10}ClN_2O_4P$, Bis(benzohydroxamato)chlorophosphorus(V), 44B, 619
$C_{14}H_{14}KPS_2$, Potassium O,O-dibenzylphosphorodithioate, 38B, 692
$C_{14}H_{14}NO_3P$, 3-(2,6-Dimethylphenyl)-2,3-dihydro-2-hydroxy-2-oxo-1,3,2-benzoxazaphosph(V)ole, 41B, 739
$C_{14}H_{14}NO_4PS$, O-Ethyl-O-(4-nitrophenyl)benzene phosphonothioate, 46B, 670
$C_{14}H_{15}O_2P$, Methyl 3,4-dimethyl-1-phenylphosphole-2-carboxylate, 46B, 670
$C_{14}H_{15}O_4P$, [1-Hydroxy-1-(2-hydroxyphenyl)ethyl]phenylphosphinic acid, 45B, 754
$C_{14}H_{16}Cl_6N_6P_4$, 2-trans-4,6,6,8,8-Hexachloro-2,4-bis(N-methylanilino)cyclotetraphosphazatetraene, 43B, 861
$C_{14}H_{16}Cl_6N_6P_4$, 2,4,4,trans-6,8,8-Hexachloro-2,6-bis(N-methylanilino)cyclotetraphosphazatetraene, 43B, 860
$C_{14}H_{16}F_4N_2P_2$, Diazadiphosphetidin, 41B, 736
$C_{14}H_{16}NOP$, N,N-Dimethyldiphenylphosphinamide, 42B, 542

$C_{14}H_{16}N_2O_2P_2$, 2,4-Bis(dimethoxy)-1,3-diphenyl-1,3,2,4-diazadiphosphetidine, 44B, 620
$C_{14}H_{16}N_2P_2S_2$, trans-1,3-Dimethyl-2,4-diphenyl-2,4-dithiocyclodiphosphazene, 41B, 737
$C_{14}H_{16}P_2S_2$, 1,2-Dimethyl-1,2-diphenyldiphosphine disulfide, 24, 657
$C_{14}H_{20}ClO_3P$, O-(β-Chloroethyl)phenyl(α-hydroxycyclohexyl)phosphinate, 44B, 621
$C_{14}H_{20}ClO_3P$, O-(β-Chloroethyl)phenyl(α-hydroxycyclohexyl)phosphinate (high melting form), 46B, 671
$C_{14}H_{21}OP$, cis-2,2,3,4,4-Pentamethyl-1-phenylphosphetan 1-oxide, 37B, 386
$C_{14}H_{21}OP$, trans-2,2,3,4,4-Pentamethyl-1-phenylphosphetan 1-oxide, 35B, 456
$C_{14}H_{21}OP$, 2,2,3,3,4-Pentamethyl-1-phenylphosphetane 1-oxide, 44B, 615
$C_{14}H_{21}O_4P \cdot 0.5\ C_6H_{14}$, 4,6-Di-t-butyl-2-hydroxy-2-oxo-benzo[d]-1,3,2-dioxaphosphole n-hexane solvate, 46B, 671
$C_{14}H_{24}BrP_2$, 1-Phenyl-1,2,2,3,4,4-hexamethylphosphetanium bromide, 34B, 387
$C_{14}H_{27}CrIN_4O_5P_4$, Nonamethycyclotetraphosphonitrilium pentacarbonyliodochromate(0), 40B, 618
$C_{14}H_{28}Br_2P_2$, 1,1,4,4-Tetraethyl-2,5-dimethyl-1,4-diphosphoniacyclohexadiene-1,4-dibromide, 35B, 455
$C_{14}H_{28}Cl_4N_4O_6P_2$, 4-Peroxycyclophosphamide, 40B, 619
$C_{14}H_{36}F_6N_6O_7P_2S_2$, Hexakis(dimethylamino)-μ-oxodiphosphonium bis(trifluoromethanesulphonate), 46B, 672
$C_{14}H_{36}N_4O_2P_2Si_2$, (4,5-Dimethyl-1,3-dioxa-2-phosphorole)-2-spiro-2'-(cis-2,4-bis(dimethylamino)-1,3-bis(trimethylsilyl))-1,3,2,4-diazadiphosphetidine, 46B, 673
$C_{14}H_{41}N_{11}P_4$, μ(1-5)-Ethylnitrido-1,3,3,7,7-pentakis(dimethylamino)-5-ethylaminocyclotetraphosphazene, 43B, 861
$C_{14}H_{42}N_7P_7$, Tetradecamethylcycloheptaphosphazene, 43B, 862
$C_{15}H_{11}P$, 2-Phenyl-1-phosphanaphthalene, 40B, 620
$C_{15}H_{15}PO_3$, trans-Methyl meso-hydrobenzoin phosphite, 40B, 620
$C_{15}H_{15}PO_4$, trans-Methyl meso-hydrobenzoin phosphate, 40B, 620
$C_{15}H_{15}PS_3$, Tris(2-methyl-3-thienyl)phosphine, 43B, 862
$C_{15}H_{16}P_2S_2$, 1,2-Diphenyl-1,2-diphospholane-1,2-disulfide, 44B, 1108
$C_{15}H_{17}BrNO_2P$, 4-Methyl-4-p-nitrobenzyl-4-phosphoniotetracyclo-[3.3.0.0$^{2-8}$.0$^{3-6}$]octane bromide, 44B, 621
$C_{15}H_{17}NO_2P_2S_3$, 3-Methyl-2,4-(di-p-methoxyphenyl)-2-trans-4-dithio-1,3,2,4-thiaza-λ^5,λ^5-diphosphetidine, 45B, 736
$C_{15}H_{17}O_2P$, 2-(Diphenylphosphinyl)propan-2-ol, 46B, 673
$C_{15}H_{17}O_3P$, Methyl (1-hydroxy-1-phenylethyl)phenylphosphinate, 46B, 674
$C_{15}H_{19}N_2P$, Phenyl-di-n-propyl-phosphonium-dicyanomethylide, 39B, 497
$C_{15}H_{19}N_6PS$, 9-Methyl-3,7-diphenyl-1,3,4,6,7,9-hexa-aza-5λ^4-phosphabicyclo[3.3.1]nonane-5-thione, 46B, 674
$C_{15}H_{20}NO_2PS$, (S)p-(-)-α-Methylbenzylammonium O-methylphenylphosphinothioate, 46B, 675
$C_{15}H_{23}OP$, trans-4-t-Butyl-1-phenylphosphorinane 1-oxide, 44B, 622
$C_{15}H_{23}OP$, 1,2,3,4-Tetrahydro-1,2,2,3,4,4-hexa-methylphosphinoline 1-oxide, 35B, 457
$C_{15}H_{25}N_2O_2P$, cis-2-Oxo-2-dimethylamino-3-phenyl-5-t-butyl-1,3,2-oxazaphosphorinane, 45B, 737
$C_{15}H_{30}N_3P$, Tris(piperidino)phosphine, 44B, 616
$C_{15}H_{30}N_3PSe$, Tris(piperidino)phosphine selenide, 45B, 735
$C_{16}H_9F_6O_5P$, 2-Phenoxy-2,2-o-phenylenedioxy-4,5-bis(trifluoromethyl)-

2,2-dihydro-1,3,2-dioxaphospholene, 43B, 863
$C_{16}H_{13}F_3NOP$, 2-Methyl-8-(trifluoro-phenyl-phosphoroxy)-quinoline, 40B, 621
$C_{16}H_{13}F_{12}O_4P$, 2,2,3,3-Tetrakis(trifluoromethyl)-5-t-butyl-7,8-benzo-1,4,6,9-tetraoxo-5λ^5-phosphaspiro[4.4]nonene, 44B, 622
$C_{16}H_{15}O_2P$, 2,3-Benzo-5-phenyl-1,4-dioxa-5λ^5-phosphaspiro[4.4]non-7-ene, 45B, 738
$C_{16}H_{17}N_2O_2P$, N-(1-Phenyl-1-phospholanio-4-nitroanilide), 45B, 739
$C_{16}H_{17}O_4P$, 2-t-Butyl-2,2'-spirobi(1,3,2-benzodioxaphosphole), 44B, 623
$C_{16}H_{18}O_2P_2 \cdot H_2O$, cis-1,4-Diphenyl-1,4-diphosphacyclohexane P,P-dioxide monohydrate, 45B, 740
$C_{16}H_{18}O_2P_2$, trans-1,4-Diphenyl-1,4-diphosphacyclohexane P,P-dioxide, 45B, 739
$C_{16}H_{18}O_4P_2S_2$, trans-2,7-Diphenyl-1,3,6,8-tetraoxa-2,7-diphosphacyclodecane-2,7-dithione, 45B, 740
$C_{16}H_{19}O_3P$, Dimethyl 8-endo-phenylbicyclo[5.1.0]octa-2,4-diene-8-phosphonate, 42B, 544
$C_{16}H_{19}O_3P$, Dimethyl 8-exo-phenylbicyclo[5.1.0]octa-2,4-diene-8-phosphonate, 42B, 543
$C_{16}H_{20}N_2P_2S_2$, cis-1,3-Diethyl-2,4-diphenyl-2,4-dithiocyclodiphosphazane, 39B, 499
$C_{16}H_{20}N_2P_2S_2$, trans-1,3-Diethyl-2,4-diphenyl-2,4-dithiocyclodiphosphazane, 39B, 498
$C_{16}H_{25}N_2O_2PS_2$, 3,4-Bis(t-butyl)-2-(p-methoxyphenyl)-1,3,4,2-thiadiazaphospholidin-5-one 2-sulfide, 45B, 741
$C_{16}H_{25}O_2P$, 1,4-Dioxa-5-phenyl-6,6,7,8,8-pentamethyl-5λ^5-phosphaspiro[3.4]octane, 44B, 623
$C_{16}H_{28}BrP$, 1,1,2,3,3,4,4-Heptamethylspiro[4.5]-6,9-diene-1-phosphetanium bromide, 38B, 659
$C_{16}H_{30}N_4OP_4$, 2,2,4,4,6,8,8-Heptamethyl-6-methylamino-7-benzoyl-1,3,5-triaza-2,4,6,8(P(V))-tetraphosphorin, 44B, 610
$C_{16}H_{36}BF_4P$, Tetra(t-butyl)phosphonium tetrafluoroborate, 46B, 675
$C_{16}H_{38}O_4P_2$, Di-t-butylphosphinic acid dimer, 42B, 544
$C_{16}H_{48}N_8O_{16}P_8$, Hexadecamethoxycyclooctaphosphonitrile, 33B, 281
$C_{16}H_{48}N_8P_8$, Hexadecamethylcyclooctaphosphazene, 43B, 864
$C_{16}H_{48}N_{12}P_4$, Octakis(dimethylamino)cyclotetraphosphazatetraene, 27, 989
$C_{16}H_{50}CoCl_4N_8P_8$, Bis(octamethylcyclotetraphosphonitrilium) tetrachlorocobaltate(II), 35B, 462
$C_{16}H_{64}B_{50}P_2 \cdot C_6H_6$, Penta(methyl-o-carboranyl)diphosphine benzene solvate, 45B, 737
$C_{17}H_{14}ClO_3P$, 3-Chloro-4-(diphenylphosphino)-5-methoxy-2(5H)-furanone, 46B, 1255
$C_{17}H_{17}O_2P$, 2,3-Benzo-5-phenyl-7-methyl-1,4-dioxa-5-phospha(P(V))spiro[4.4]non-7-ene, 43B, 864
$C_{17}H_{17}O_2P$, 6-Methyl-2-oxo-1,2-diphenyl-3-oxa-2-phosphabicyclo-[3.1.0]hexane, 44B, 624
$C_{17}H_{17}O_3P$, 3-(Benzyl(phenyl)phosphinyl)-2-butenoic acid, 44B, 625
$C_{17}H_{19}Cl_2O_5P$, Diethyl bis(p-chlorophenoxy)methane-phosphonate, 46B, 675
$C_{17}H_{20}ClN_2OP$, 6-Chloro-5,6,7,12-tetrahydro-2,5,7,10-tetramethyldibenzo[d,g](1,3,2)diazaphosphocine 6-oxide, 38B, 660
$C_{17}H_{20}ClN_2PS$, 6-Chloro-2,5,7,10-tetramethyl-6-thio-6,7-dihydro-5H,12H-dibenzo[d,g]-1,3,2-diazaphospholine, 41B, 738
$C_{17}H_{20}F_6P_2$, 1,2,3,4-Tetrahydro-1,4-dimethyl-1-phenylphosphinolinium hexafluorophosphate, 46B, 676

$C_{17}H_{20}IP$, 3,3-Dimethyl-1,1-diphenylphosphetanium iodide, 45B, 741
$C_{17}H_{21}O_4P$, 7,8-(5-Methoxybenzo)tricyclo[4.3.1.0$^{2-9}$]deca-4,7-diene-9-phosphonic acid, 44B, 625
$C_{17}H_{22}NO_3P$, DL-Diethyl α-anilinobenzylphosphonate, 44B, 626
$C_{17}H_{22}NO_4P$, 3-(2,6-Dimethylphenyl)-2,3-dihydro-2,2,2-trimethoxy-1,3,2-Benzoxazaphosph(V)ole, 41B, 739
$C_{17}H_{29}INO_3P$, O-(2-Trimethylammonio-ethyl)-phenyl-(1-hydroxycyclohexyl)phosphinate iodide, 45B, 742
$C_{17}H_{38}O_2P_2Si_2$, Bis((2,2-dimethyl-1-trimethylsiloxypropylidene)phosphino)methane, 44B, 626
$C_{18}F_{17}P$, Tris(pentafluorophenyl)difluorophosphorane, 41B, 740
$C_{18}H_5Cl_8O_4P \cdot 0.5\ C_6H_5Cl$, 2,3,7,8-Bis(tetrachlorobenzo)-5-phenyl-1,4,6,9-tetraoxa-5λ^5-phosphaspiro[4.4]nona-2,7-diene chlorobenzene solvate, 45B, 743
$C_{18}H_9Cl_4O_4P$, 2,3-Benzo-7,8-tetrachloro-1,4,6,9-tetraoxa-5-phenyl-5λ^5-phosphaspiro[4.4]nona-2,7-diene, 45B, 743
$C_{18}H_9F_{12}O_4P$, 2,2,3,3-Tetrakis(trifluoromethyl)-5-phenyl-7,8-benzo-1,4,6,9-tetraoxa-5λ^5-phosphaspiro[4.4]nonene, 44B, 622
$C_{18}H_{10}F_5OP$, Diphenyl-pentafluorophenyl-phosphine oxide, 45B, 743
$C_{18}H_{12}N_3O_6P_3 \cdot 0.5\ C_6H_5Br$, Tris(o-phenylenedioxy)phosphonitrile trimer bromobenzene inclusion compound, 32B, 309
$C_{18}H_{12}N_3O_6P_3 \cdot 0.5\ C_6H_6$, Tris(o-phenylenedioxy)phosphonitrile trimer benzene inclusion compound, 32B, 309
$C_{18}H_{12}O_2P_2$, 1,6-Diphosphatriptycene dioxide, 41B, 741
$C_{18}H_{12}P_2$, 1,6-Diphosphatriptycene, 41B, 741
$C_{18}H_{13}OP$, 10-Phenylphenoxaphosphine, 42B, 545
$C_{18}H_{13}O_2PS_2$, 2-Phenyl-2,2'-spirobis(1,3,2-benzooxathiaphosphole), 45B, 759
$C_{18}H_{13}O_4P$, 2-Phenyl-2,2'-spirobi(1,3,2-benzodioxaphosphole), 43B, 865; 44B, 626
$C_{18}H_{13}O_5P$, Spirodicatecholphenoxyphosphorane, 42B, 546
$C_{18}H_{13}PS_4$, 2-Phenyl-2,2'-spirobis(1,3,2-benzodithiaphosphole), 45B, 759
$C_{18}H_{14}BrOP$, p-Bromophenyl-diphenyl-phosphine oxide, 37B, 387
$C_{18}H_{14}BrPS$, p-Bromophenyldiphenylphosphine sulfide, 38B, 661
$C_{18}H_{14}ClOP$, p-Chlorophenyl-diphenyl-phosphine oxide, 37B, 387
$C_{18}H_{14}ClPS$, p-Chlorophenyldiphenylphosphine sulfide, 38B, 661
$C_{18}H_{14}NO_5PS$, S-4-Nitrophenyl O,O-diphenyl thiophosphate, 41B, 741
$C_{18}H_{15}Br_3N_3P_3$, cis-1,3,5-Tribromo-1,3,5-triphenylcyclotriphosphazene, 45B, 744
$C_{18}H_{15}ClNOP$, Diphenylphosphinic-4-chloroanilide, 42B, 546
$C_{18}H_{15}ClNOP$, 2-Chloro-4-(diphenylphosphoryl)aniline, 42B, 546
$C_{18}H_{15}Cl_2NOP_2$, N-(Dichlorophosphinoyl)triphenylphosphazene, 45B, 744
$C_{18}H_{15}Cl_5N_4P_4$, Triphenylphosphazenyl-pentachloro-cyclotriphosphazene, 45B, 745
$C_{18}H_{15}Cl_7N_5P_5$, 2,4,4,6,6,8,8-Heptachloro-2-triphenylphosphazenylcyclotetra(phosphazene), 45B, 746
$C_{18}H_{15}OP$, Triphenylphosphine oxide (monoclinic), 39B, 499; 42B, 547
$C_{18}H_{15}OP$, Triphenylphosphine oxide (orthorhombic), 35B, 464
$C_{18}H_{15}O_3P_3S_3$, 2,cis-4,trans-6-Triphenyl-2,4,6-trithioxo-1,3,5,2,4,6-trioxatriphosphorinan, 38B, 661; 39B, 500
$C_{18}H_{15}O_4PS$, 12-Ethoxy-2,3-benzo-6,5-naphtho[b]-(7,12)-thiaphosphorin-7,7,12-trioxide, 41B, 742
$C_{18}H_{15}P$, Triphenylphosphine, 29, 638
$C_{18}H_{15}PS$, Triphenylphosphine sulfide, 44B, 627
$C_{18}H_{15}PSe$, Triphenylphosphine selenide, 45B, 746
$C_{18}H_{16}ClOP$, Triphenylphosphineoxide hydrochloride, 43B, 866

$C_{18}H_{16}FOP$, Triphenylphosphine oxide hydrogen fluoride, 45B, 747
$C_{18}H_{18}ClN_2O_4PS_2$, Bis(N-ethylbenzthiazole(2))-phosphamethinecyanine perchlorate, 31B, 302
$C_{18}H_{18}F_{26}N_2O_4P_2Si_2$, Difluoro-octakis(trifluoromethyl)-bis(trimethylsilyl)dispiro(dioxaphospholan)-diazaphosphetidine-dioxaphospholane, 43B, 866
$C_{18}H_{18}N_{15}P_3$, Hexakis(imidazolyl)cyclotriphosphazene, 46B, 676
$C_{18}H_{21}N_2O_3P$, 3,4-Dimethyl-2,8-diphenyl-1,6-dioxa-4,9-diaza-5-phospha(V)-7-spiro[4.4]nonanone, 45B, 748
$C_{18}H_{21}PS_2$, 6-Phenyl-6-phospha-2,10-dithiabicyclo[9.4.0]pentadeca-11(1),12,14-triene, 46B, 694
$C_{18}H_{22}BrP$, 1-Benzyl-1-phenylphosphorinanium bromide, 46B, 677
$C_{18}H_{22}BrP$, 1,1-Diphenyl-4-methylphosphorinanium bromide, 46B, 677
$C_{18}H_{22}F_6P_2$, 1,2,3,4-Tetrahydro-1-ethyl-4-methyl-1-phenylphosphinolinium hexafluorophosphate, 46B, 676
$C_{18}H_{22}F_6P_2$, 1,2,3,4-Tetrahydro-4,4-dimethyl-1-methyl-1-phenylphosphinolinium hexafluorophosphate, 43B, 867
$C_{18}H_{22}IP$, (1R,3S)-(-)-trans-1-Benzyl-3-methyl-1-phenylphospholanium iodide, 46B, 677
$C_{18}H_{24}NO_3PS \cdot H_2O$, Tetramethylammonium cis-4,5-diphenyl-1,3,2-dioxaphospholane-2-sulfide-2-hydroxylate monohydrate, 45B, 748
$C_{18}H_{25}Cl_2CuNOP$, (2-(Diphenylphosphinoyl)ethyl)diethylammonium dichlorocuprate(I), 40B, 622
$C_{18}H_{26}NOP$, 3,5-Di-(t-butyl)-1-phenylphosphoryl-4-aza-2,5-cyclohexadiene, 38B, 662
$C_{18}H_{28}N_8P_2$, Tris(dimethylhydrazino)bis(phosphine) bis(phenylimide), 41B, 743
$C_{18}H_{33}PS$, Tricyclohexylphosphine sulfide, 43B, 868
$C_{18}H_{36}O_2P_2$, P,P'-Tetra-t-butyloxalic acid diphosphide, 46B, 678
$C_{18}H_{42}N_4O_4P_2Si_2$, (cis-2,4-Bis(dimethylamino)-1,3-bis(trimethylsilyl)-1,3,2,4-diazadiphosphetidine)-2',4'-bisspiro-2,2"-(4,5-dimethyl-1,3-dioxa-2-phosphorole), 46B, 673
$C_{19}H_{13}Cl_2O_3P$, Diphenyl-[5,6-dichloro-1,3-benzodioxol-(2)]-phosphine oxide, 45B, 749
$C_{19}H_{13}P$, Phosphatriptycene, 46B, 703
$C_{19}H_{15}N_2O_7P$, Methyl N-methylphthalimido N-methylphthalimidophosphonate, 42B, 547
$C_{19}H_{15}N_2P$, Triphenylphosphoniocyanoamide, 46B, 678
$C_{19}H_{15}OPS_2$, Diphenyl[1,3-benzodithiolyl-(2)]phosphine oxide, 46B, 678
$C_{19}H_{16}NP$, 5-Methyl-10-phenyl-5,10-dihydrophenophosphazine, 43B, 868
$C_{19}H_{17}F_6NO_4P$, Tetraoxycarbophosphorane benzene solvate, 41B, 744
$C_{19}H_{17}P$, Methylenetriphenylphosphorane, 34B, 389
$C_{19}H_{19}N_2O_4P$, Tetraoxyazaphosphorane caged polycyclic, 39B, 500
$C_{19}H_{19}O_5P$, 2-Oxo-2-(phenanthrenequinoxy)-5,5-dimethyl-1,3,2-dioxaphosphorinane, 46B, 679
$C_{19}H_{20}NOP$, 2-Oxo-2,7-diphenyl-4,5-dimethyl-1-aza-2-phosphacyclohepta-4,6-diene ($P2_1/c$), 41B, 745
$C_{19}H_{20}NOP$, 2-Oxo-2,7-diphenyl-4,5-dimethyl-1-aza-2-phosphacyclohepta-4,6-diene (C2/c), 40B, 622
$C_{19}H_{21}O_2P$, 1-Benzyl-2-phenyl-3-hydroxy-4,5-dimethylphosphol-2-ene-1-oxide, 42B, 547
$C_{19}H_{23}O_3P$, Methyl 2-methyl-3-[benzyl(phenyl)phosphinyl]butyrate, 45B, 750
$C_{19}H_{25}NOSi$, Diphenyl-(2-piperidinoethyl)silanol, 46B, 234
$C_{19}H_{27}ClNOPSi_2$, 4-Chlorophenylimino(trimethylsiloxy)methylphenyltrimethylsilylphosphane, 45B, 750

$C_{19}H_{30}ClN_4O_9P$, 4-Nitrobenzyl-tris(morpholino)phosphonium perchlorate, 46B, 680
$C_{19}H_{57}N_7P_2Si_4$, {Methyl-[tris(dimethylamino)phosphonium]amino}-tris-(trimethylsilyl-amino)-trimethylsilyl-iminophosphoranide, 45B, 751
$C_{20}H_{13}O_4P$, 1-Phenyl-1,1'-spirobi(3H-2,1-benzoxaphosphole)-3,3'-dione, 44B, 627
$C_{20}H_{15}F_{12}O_2P$, 2,2-Diphenyl-2-bis(trifluoromethyl)methoxy-3-methyl-4,4-bis(trifluoromethyl)-2,2-dihydro-1,2,oxaphosphetane, 37B, 388
$C_{20}H_{15}OP$, Triphenylphosphoranylideneketene, 31B, 303
$C_{20}H_{15}PS$, Triphenylphosphoranylidenethioketene, 32B, 310
$C_{20}H_{16}NO_3PS$, 2,3-Dihydro-2-(4-methoxyphenyl)-3-phenyl-4H-1,3,2-benzoxazaphosphorine-4-thione 2-oxide, 44B, 628
$C_{20}H_{17}OP$, 5-Phenyl-10,11-dihydrodibenzo[b,f]phosphepin 5-oxide, 46B, 681
$C_{20}H_{17}OP$, 5,10-Dihydro-10-methyl-5-phenylacridophosphin-10-ol, 43B, 869
$C_{20}H_{17}O_2P$, 10-Methyl-5-phenyldibenzo[b,e]phosphorinan-10-ol 5-oxide, 45B, 751
$C_{20}H_{18}P_2$, 1,4-Diphenyl-1,2,3,4-tetrahydro-1,4-benzodiphosphorin, 38B, 663
$C_{20}H_{20}NP$, p-Dimethylaminophenyl-diphenyl-phosphine, 38B, 664
$C_{20}H_{20}P_2$, meso-o-Phenylenebis(methylphenylphosphine), 46B, 681
$C_{20}H_{21}BF_4P_2$, meso-o-Phenylene(methylphenylphosphine)(methylphenylphosphonium) tetrafluoroborate, 46B, 681
$C_{20}H_{22}BrN_2P$, N',N'-Dimethylhydrazinotriphenylphosphonium bromide, 44B, 628
$C_{20}H_{23}BrNO_2P$, exo-3-p-Nitrobenzyl-endo-3-phenyl-3-phosphoniabicyclo-[3.2.1]octane bromide, 46B, 682
$C_{20}H_{24}N_3OP$, 1',2,3'-Trimethyl-3-phenylspiro[[1,2λ⁵]azaphospholo[2,1-b][1,3,2λ⁵]benzoxazaphosphole-4(1H),2'-[1,3,2λ⁵]diazaphospholidine], 46B, 685
$C_{20}H_{24}N_3OP$, 2-(N',N'-Dimethylethylenediamido-1,5-dihydro-4-methyl-3-phenyl-1,2-azaphospholo)[2,1-b](1,3,2-benzoxaphosphole), 45B, 752
$C_{20}H_{27}N_2O_2P$, Spirophosphorane derivative, 40B, 623
$C_{20}H_{28}N_2O_2P_2$, 2,2-Diethyl-2-phosphabenzoxazole dimer, 40B, 624
$C_{20}H_{29}O_{10}PS$, 2,3:5,6-Di-O-isopropylidene-1-methylphosphono-1-O-p-tolylsulphonyl-D-mannit-δ-phoston, 43B, 870
$C_{20}H_{30}O_4P_2$, trans-1,6-Diphenyl-1λ⁵,6λ⁵-diphosphacyclodecane-1,6-dione dihydrate, 40B, 625
$C_{20}H_{34}P_2Si_4$, 1,4-Diphenyl-2,2,3,3,5,5,6,6-octamethyl-1,4-diphospha-2,3,5,6-tetrasila-cyclohexane, 45B, 752
$C_{20}H_{40}O_5Si$, 7,8,O-Isopropylidene-6,7,8-trihydroxy-2,2,5,7-tetramethyl-3-[(trimethylsilyl)oxy]-4-decanone, 46B, 336
$C_{20}H_{42}N_2P_2S_2$, 1,2-Dicyclohexyl-1,2-bis(diethylamino-1,2-dithioxo-di-λ⁵-phosphane, 46B, 682
$C_{20}H_{44}IP$, Tetraamylphosphonium iodide, 35B, 465
$C_{20}H_{48}N_{12}O_4P_4W$, (Octakisdimethylaminocyclotetraphosphazene)tetracarbonyltungsten, 39B, 501
$C_{20}H_{54}N_4P_2S_2Si_4$, 2r,4c-Bis(bis(trimethylsilyl)amino)-2,4-bis(t-butylimino)-1,3,2λ⁵,4λ⁵-dithiadiphosphetane, 42B, 548
$C_{21}H_{15}N_2P$, (Dicyanomethylene)triphenylphosphorane, 46B, 683
$C_{21}H_{18}NP$, 2,7-Dimethyl-9,10-o-(m-methyl)benzeno-9-phospha-10-azaanthracene, 42B, 549
$C_{21}H_{19}O_2P$, Triphenylphosphinecarbomethoxymethylene, 38B, 664
$C_{21}H_{20}IOP$, cis-5,10-Dihydro-10-hydroxy-5,10-dimethyl-5-phenylacridophosphinium iodide, 43B, 870
$C_{21}H_{21}N_2O_4PS$, Methyl phenyl[syn-*a*-(tosylhydrazono)benzyl]phos-

phinate, 46B, 683
$C_{21}H_{21}OP$, Tri-o-tolyl phosphine oxide, 41B, 746
$C_{21}H_{21}P$, Tri-m-tolylphosphine, 40B, 626; 44B, 629
$C_{21}H_{21}P$, Tri-o-tolyl phosphine, 41B, 746
$C_{21}H_{21}PS$, Tri-m-tolylphosphine sulfide, 44B, 629
$C_{21}H_{21}PS$, Tri-o-tolyl phosphine sulfide, 41B, 746
$C_{21}H_{21}PSe$, Tri-m-tolylphosphine selenide, 44B, 629
$C_{21}H_{21}PSe$, Tri-o-tolyl phosphine selenide, 41B, 746
$C_{21}H_{22}BrN_2PS$, (-)-(S)-6-Benzyl-5,6,7,8-tetrahydro-4-(methylthio)-6-phenylphosphorino[4,3-d]pyrimidinium bromide, 41B, 747
$C_{21}H_{22}BrP$, 1-Ethyl-1,2,3,4-tetrahydro-1-phenylbenzo[h]phosphinolinium bromide, 41B, 748
$C_{21}H_{22}NO_3P$, 4-Acylamino-1,1-dimethoxy-2,6-diphenyl-λ^5-phosphorin, 45B, 753
$C_{21}H_{22}NO_4P$, 5,5-Dimethyl-2,2-diphenyl-3,4-bis(methoxycarbonyl)-1-aza-2-phosphacyclopent-1-ene, 46B, 684
$C_{21}H_{28}N_3O_2P_3$, 2,2,4,6,6-Pentamethyl-5-benzoyl-4-(N-methylbenzamido)-1,3-diaza-2,4,6(P(V))-triphosphorin, 43B, 871
$C_{22}H_{15}ClN_3P$, 1-(Triphenylphosphineimino)-1-chloro-2,2-dicyanoethylene, 45B, 753
$C_{22}H_{15}F_6P$, 2,2,3,3,4,4-Hexafluoro(triphenylphosphoranylidene)cyclobutane, 39B, 502
$C_{22}H_{16}F_6N_2O_4P_2$, 2,2'-Bipyridine-bis(o-phenylenedioxy)phosphonium hexafluorophosphate, 44B, 629
$C_{22}H_{18}NO_4P$, 2',3'-Diphenylspiro[[1,3,2λ^5]dioxaphospholane-2,10'-[1,3,2λ^5]oxazaphospholo[2,3-b][1,3,2λ^5]benzoxazaphosphole], 46B, 685
$C_{22}H_{19}O_3P$, (E)-2-Methyl-4,5,6-triphenyl-1,3,4-dioxaphosphorin-4-one, 44B, 630
$C_{22}H_{19}O_4P$, 3,4;8,9-Dibenzo-5,7-dimethyl-1-phenyl-2,6,10,11-tetraoxa-1-phosphatricyclo[5.3.1.0$^{1-5}$]undecane, 45B, 754
$C_{22}H_{20}F_6OP_2$, 2,3,4,5-Tetrahydro-5-oxo-2,2-diphenyl-1H-2-benzophosphonium hexafluorophosphate, 44B, 630
$C_{22}H_{21}O_4P$, Spirophosphorane derivative, 42B, 549
$C_{22}H_{22}ClN_2O_4P$, Bis(N-ethyl-2-quinolyl)-phosphamethinecyanine perchlorate, 42B, 550
$C_{22}H_{23}O_4P$, 2-(1-Adamantyl)-2,2'-spiro-bis(1,3,2-benzodioxaphosphole), 45B, 754
$C_{22}H_{24}BF_4P$, (Triphenylmethyl)trimethylphosphonium tetrafluoroborate, 46B, 685
$C_{22}H_{25}O_2P$, 7-Benzyl-9-methoxy-8-phenyl-7-phospha-cis-bicyclo[4.3.0]-non-8-ene anti-7-oxide, 45B, 755
$C_{22}H_{25}O_3P$, Methyl 6-(diphenylphosphinyl)-3,4-dimethyl-3-cyclohexene-1-carboxylate, 46B, 686
$C_{22}H_{28}NO_4P$, 2,2,3,3-Tetramethyl-5-dimethylamino-7,8-diphenyl-1,4,6,9-tetraoxa-5-phospha(P(V))spiro[4.4]non-7-ene, 43B, 872
$C_{22}H_{28}O_2P_2$, 1,2-Bis(2',2'-dimethylpropionyl)-1,2-diphenyldiphosphine, 46B, 686
$C_{22}H_{32}ClN_2P$, (p-Chlorobenzylidenemalononitrile)tributylphosphine, 39B, 503
$C_{22}H_{32}N_2O_6P_2$, trans-1,3-Di-p-tolyl-2,4-diethoxy-2,4-bis(ethylenedioxy)-1,3,2λ^5,4λ^5-diazaphosphetidine, 43B, 872
$C_{22}H_{35}INO_5P$, Phenyl-(α-hydroxycyclohexyl)-phosphonic acid N-methyl-N-phenylaminoethyl ester methiodide dihydrate, 43B, 873
$C_{22}H_{36}ClN_4O_6P$, 4-Nitrobenzyl-tris(piperidino)phosphonium perchlorate, 46B, 680
$C_{23}H_{19}P$, Triphenylphosphonium cyclopentadienylide, 39B, 503

$C_{23}H_{21}ClN_5P$, 4-Chloro-4-dimethylamino-6-triphenylphosphazen-1'-yl-1,3,5-triazine, 41B, 749
$C_{23}H_{21}O_4P$, 2-Carboxy-1-methoxycarbonylethyltriphenylphosphorane, 41B, 749
$C_{23}H_{21}P$, 3-t-Butyl-9,10-dihydro-9,10-o-benzeno-9-phosphaanthracene, 46B, 687
$C_{23}H_{23}P$, Triphenyl-3,3-dimethyl-allylidenephosphorane, 39B, 505
$C_{23}H_{25}N_2O_5P$, Methoxy-(erythro-4,5-dicyano-4,5-diphenyl)-(4',4',5',5'-tetramethyl)spirophosphorane, 44B, 631
$C_{23}H_{25}N_2O_5P$, Methoxy-(threo-4,5-dicyano-4,5-diphenyl)-(4',4',5',5'-tetramethyl)spirophosphorane, 44B, 631
$C_{23}H_{29}O_5P$, 2,2,2-Tri-isopropoxy-4,5-(2',2''-biphenyleno)-1,3,2-dioxaphospholene, 32B, 311
$C_{24}F_{20}P_4$, Tetra(pentafluorophenyl)cyclotetraphosphane, 37B, 389
$C_{24}H_{14}F_{11}N_4O_2P$, 1,5-Dimethyl-4-pentafluorophenyl-3,7-bis(3-(trifluoromethyl)phenyl)-1,3,5,7-tetraaza-4λ^5-phosphaspiro[3.3]heptane-2,6-dione, 46B, 687
$C_{24}H_{15}N_6O_3P$, 4-Nitroso-5-Triphenylphosphoranylideneaminobenzo[1,2-c:3,4-c']difurazan, 40B, 1102
$C_{24}H_{15}P$, Tris(phenylethynyl)phosphine, 32B, 314
$C_{24}H_{17}N_6O_5P$, Hydroxytriphenylphosphonium 4-imino-8-aci-nitrobenzo-[1,2-c:4,5-c']difurazan, 40B, 1102
$C_{24}H_{19}BrNP$, p-Bromophenylimino(triphenyl)phosphorane, 37B, 388
$C_{24}H_{19}P_3$, 1,2,3-Triphenyl-1,2,3-triphosphaindane, 31B, 307
$C_{24}H_{20}Br_2IP$, Tetraphenylphosphonium dibromoiodate, 45B, 756
$C_{24}H_{20}Cl_2N_3P_3$, 2,2-Dichloro-4,4,6,6-tetraphenylcyclotriphosphazatriene, 31B, 307
$C_{24}H_{20}Cl_4N_4P_4$, 2,cis-4,cis-6,cis-8-Tetrachloro-2,4,6,8-tetraphenylcyclotetraphosphazatetraene, 38B, 665
$C_{24}H_{20}Cl_4N_4P_4$, 2,cis-4,trans-6,trans-8-Tetrachloro-2,4,6,8-tetraphenylcyclotetraphosphazatetraene, 40B, 627
$C_{24}H_{20}Cl_4N_4P_4$, 2,2,6,6-Tetrachloro-4,4,8,8-tetraphenylcyclotetraphosphazatetraene, 40B, 627
$C_{24}H_{20}Cl_5OPU$, Tetraphenylphosphonium pentachlorooxouranate(VI), 44B, 632
$C_{24}H_{20}IN_6P$, Tetraphenylphosphonium bis(azido)iodine, 46B, 688
$C_{24}H_{20}IP$, Tetraphenylphosphonium iodide, 20, 588
$C_{24}H_{20}N_2P_2S_2$, 1,2,3,4-Tetraphenyl-2,4-dithiocyclodiphosphazane, 39B, 505
$C_{24}H_{20}P_4S$, Tetraphenylcyclotetraphosphine monosulfide, 40B, 628
$C_{24}H_{22}BrP$, (+)-Benzylmethyl-α-naphthylphenylphosphonium bromide, 41B, 750
$C_{24}H_{22}N_4P_2S_2$, trans-2,4-Dithio-2,4-dianilino-1,3-diphenyl-1,3,2,4-diazadiphosphetine, 45B, 756
$C_{24}H_{24}ClO_3P$, α-Triphenylphosphino-acetoacetic ester chloride, 44B, 632
$C_{24}H_{25}F_3NO_5P$, 5,5,7-Trimethyl-2,2-diphenyl-7-(trifluoromethyl)-3,4-bis(methoxycarbonyl)-6-oxa-1-aza-5-phosphabicyclo[3.2.0]hept-3-ene, 46B, 684
$C_{24}H_{25}O_2P$, 2,2-Di(ethoxy)vinylidene-triphenylphosphorane, 40B, 629
$C_{24}H_{25}O_3P$, cis-4,6-Dimethyl-2-oxo-2-triphenylmethyl-1,3,2-dioxaphosphorinan, 38B, 666
$C_{24}H_{27}P$, Tris(2,6-dimethylphenyl)phosphine, 42B, 550
$C_{24}H_{28}NO_6P$, Triethylammonium tris(o-phenylenedioxy)phosphate, 39B, 506
$C_{24}H_{44}P_2$, Tetracyclohexylbiphosphine, 43B, 874
$C_{24}H_{44}P_4$, Tetracyclohexylcyclotetraphosphine, 34B, 392

$C_{24}H_{51}N_3O_6P_2$, t-Butylbis(2-(6-t-butyltetrahydro-4H-1,3-dioxa-6-aza-2-phosphocine-2-oxy)ethyl)amine, 46B, 688
$C_{24}H_{72}Cl_3Cu_2N_{18}P_6$, Chloro(dodeca(dimethylamino)cyclohexaphosphazene-N,N,N,N)-copper(II) dichlorocuprate(I), 37B, 390
$C_{24}H_{72}N_{18}P_6$, Dodecadimethylaminocyclohexaphosphazahexaene (hexameric phosphonitrilic dimethylamide), 33B, 282
$C_{24}H_{74}Cl_4CoN_{18}P_6$, Bis(hexakis(dimethylamino)cyclotriphosphonitrilium) tetrachlorocobaltate(II), 40B, 630
$C_{24}H_{74}Mo_6N_{18}O_{19}P_6$, Bis(hexakisdimethylaminocyclotriphosphonitrilium) hexamolybdate, 39B, 506
$C_{25}H_{19}Br_2OP$, 10-(4-Bromobenzyl)-10-phenylphenoxaphosphonium bromide, 45B, 757
$C_{25}H_{19}Br_2P \cdot 0.5\ H_2O$, 5-(p-Bromobenzyl)-5-phenyldibenzophospholium bromide hemihydrate, 44B, 633
$C_{25}H_{19}P$, Methylbis(2,2'-biphenylylene)phosphorane, 46B, 689
$C_{25}H_{20}Br_5P$, Tetraphenylphosphonium bromide carbon tetrabromide, 42B, 551
$C_{25}H_{20}NOP$, N-(Triphenylphosphoranylidene)benzamide, 46B, 690
$C_{25}H_{22}ClN_2OP$, 10,10'-(5H,5'H)-Spirobiphenophosphazinium chloride, 39B, 507
$C_{25}H_{22}ClP$, Benzyltriphenylphosphonium chloride, 40B, 630
$C_{25}H_{22}Cl_6PU$, Triphenylbenzylphosphonium hexachlorouranate(V), 44B, 634
$C_{25}H_{22}IO_2PS$, Triphenylbenzylphosphonium iodosulfinate, 44B, 634
$C_{25}H_{22}NO_2P$, 5-Methyl-10-phenyl-10-oxo-5,10-dihydrophenophosphazine phenol, 45B, 757
$C_{25}H_{22}NO_2PS$, N-(p-Toluenesulphonyl)iminotriphenylphosphorane, 40B, 631
$C_{25}H_{22}O_2P_2 \cdot 0.5\ C_6H_6$, Dioxotetraphenylmethylenediphosphine benzene, 46B, 690
$C_{25}H_{22}P_2Se_2$, Bis(diphenylphosphinoselenoyl)methane, 42B, 59; 43B, 883
$C_{25}H_{22}P_4$, 1,2,3,4-Tetraphenyl-cyclo-5-carba-1,2,3,4-tetraphosphane, 43B, 874
$C_{25}H_{22}P_4S_2$, 1,4-Dithio-1,2,3,4-tetraphenyl-cyclo-5-carba-1,2,3,4-tetraphosphane, 43B, 874
$C_{25}H_{23}O_2P$, 1,1-Dimethoxy-2,4,6-triphenylphosphorin, 34B, 392
$C_{25}H_{23}P$, 1,1-Dimethyl-2,4,6-triphenylphosphorin, 35B, 465
$C_{25}H_{25}Br_4OP$, 4-endo-Bromomethyl-4,5-dimethyl-1,2,2-triphenyl-3,2-oxaphosphoniabicyclo[3.1.0]hexane tribromide, 46B, 691
$C_{25}H_{29}NP_2$, 2,10-Diphenyl-6-methyl-6-aza-2,10-diphosphabicyclo-[9.4.0]pentadeca-11(1),12,14-triene, 46B, 694
$C_{25}H_{31}O_2P$, 7-Benzyl-4-t-butyl-9-hydroxy-8-phenyl-7-phospha-cis-bicyclo[4.3.0]non-8-ene anti-7-oxide, 46B, 691
$C_{26}H_{20}ClOP$, Benzoyl(triphenylphosphoranylidene)methyl chloride, 30B, 275
$C_{26}H_{20}IOP$, Benzoyl(triphenylphosphoranylidene)methyl iodide, 30B, 275
$C_{26}H_{20}NP$, Triphenyl(phenyliminovinylidene)phosphorane, 43B, 875
$C_{26}H_{20}O_2P_2$, Bis(diphenylphosphonyl)acetylene, 45B, 758
$C_{26}H_{20}O_2P_2$, P,P'-Tetraphenyloxalic acid diphosphide, 46B, 678
$C_{26}H_{20}P_2$, Bis(diphenylphosphino)acetylene, 34B, 390
$C_{26}H_{20}S_2P_2$, Bis(diphenylphosphinothioyl)acetylene, 45B, 758
$C_{26}H_{20}Se_2P_2$, Bis(diphenylphosphinoselenoyl)acetylene, 45B, 758
$C_{26}H_{21}F_3O_3P_2$, (1-Diphenylphosphoryl-1H-trifluoromethyl)-diphenylphosphinate, 43B, 875
$C_{26}H_{23}N_3P_2$, 6-Methyl-2,2,4,4-tetraphenyldiphospha-1,3,5-triazine,

37B, 392
$C_{26}H_{23}O_2PS$, p-Tolyl triphenylphosphoranylidenemethyl sulphone, 30B, 277
$C_{26}H_{24}BrOP$, Benzyl(2-methoxyphenyl)diphenylphosphonium bromide, 43B, 876
$C_{26}H_{24}N_2P_2S$, 1-Methyl-3,3,5,5-tetraphenyl-1-thia-3,5-diphospha(V)-2,6-diazin, 40B, 632
$C_{26}H_{24}P_2$, 1,2-Bis(diphenylphosphino)ethane (form I), 45B, 759
$C_{26}H_{24}P_2$, 1,2-Bis(diphenylphosphino)ethane (form II), 45B, 759
$C_{26}H_{26}F_2N_2P_2$, 1,3-Dimethyl-2,4-difluoro-2,2,4,4-tetraphenyl-1,3-diaza-2,4-diphosphetidine, 43B, 877
$C_{26}H_{26}N_2P_2$, 1,1-Bis(diphenylphosphino)-2,2-dimethylhydrazine, 42B, 48
$C_{26}H_{26}N_3P_3$, 2,2-Dimethyl-4,4,6,6-tetraphenylcyclotri(phosphazene), 46B, 691
$C_{26}H_{30}Br_2NO_2P$, (2-Oxo-4-morpholino)butyltriphenylphosphonium dibromide, 43B, 877
$C_{26}H_{30}ClNO_2P$, Triphenyl-4,4,6-trimethyl-1-oxa-3-azacyclohex-2-enylcarbenyl-phosphonium chloride monohydrate, 38B, 383
$C_{26}H_{30}O_8P_2$, 1-Ethoxy-1,2-diphenyl-3,3,5-tricarbethoxy-1,2-diphosphocyclopenten-5-en-4-one, 39B, 508
$C_{26}H_{43}P$, Phenyldimenthylphosphine, 40B, 632
$C_{27}H_{18}Cl_4NO_2P$, 4',5',6',7'-Tetrachloro-2,8,15-[5,10(1,2)-benzeneophosphazine-10,2'-(1,3,2)-benzodioxaphosphole], 46B, 692
$C_{27}H_{19}O_2PS_2$, 5-Phenyl-2,3-phenanthro-7,8-(2'-methyl-5',6'-benzo)-1,4-dioxa-6,9-dithia-5-phospha(V)spiro[4.4]nona-2,7-diene, 45B, 759
$C_{27}H_{20}HgN_3PS_3$, Tetraphenylphosphonium trithiocyanatomercurate(II), 41B, 751
$C_{27}H_{23}N_2OP$, trans-3,4-Dihydro-2,3,4-triphenyl-5-phenylmethyl-2H-1,2,3-diazaphosphole 3-oxide, 45B, 761
$C_{27}H_{23}N_2P$, cis-3,4-Dihydro-2,3,4-triphenyl-5-phenylmethyl-2H-1,2,3-diazaphosphole, 46B, 692
$C_{27}H_{25}N_2O_3PS$, N,N'-Dibenzyl-N-diphenoxyphosphorylthiourea, 45B, 761
$C_{27}H_{26}N_4P_2$, 6-Dimethylamino-2,2,4,4-tetraphenyldiphospha-1,3,5-triazine, 37B, 393
$C_{27}H_{29}F_6NO_6P$, Triethylammonium diphenoxy-o-phenylenedioxy-1,2-bis(trifluoromethyl)ethenylene-dioxyphosphoride, 45B, 762
$C_{27}H_{29}N_2P$, 1,1-Bis(dimethylamino)-2,4,6-triphenylphosphorin, 38B, 666
$C_{27}H_{29}O_4P$, 1-Methoxycarbonyl-2-t-butoxycarbonylethyltriphenylphosphorane, 41B, 749
$C_{27}H_{34}P$, Trimesitylphosphine, 41B, 752
$C_{28}H_{22}ClO_5P$, 2,4,4,6-Tetraphenyl-4-phosphonia-pyran perchlorate, 40B, 633
$C_{28}H_{23}ClN_4O_{12}P$, Methyltri(p-nitrophenoxy)phosphonium chloride p-nitrophenol hemibenzene, 46B, 693
$C_{28}H_{24}BF_4O_2P$, α-(Triphenylphosphonium)-benzoyl-acetone tetrafluoroborate, 45B, 762
$C_{28}H_{24}ClO_2P$, α-Triphenylphosphonium-benzoylacetone chloride, 44B, 635
$C_{28}H_{24}NO_2PS_2$, 1-Ethylthio-2-(p-nitrophenyl)-2-(triphenylphosphonio)ethenethiolate, 42B, 552
$C_{28}H_{29}BF_4N_3P$, (4-Diethylaminophenyl)diazenyltriphenylphosphonium tetrafluoroborate, 44B, 635
$C_{28}H_{30}Br_2OP_2$, 1,1,4,4-Tetraphenyl-1,4-diphosphoniacyclohexane dibromide hydrate, 38B, 667
$C_{28}H_{30}INP_2$, Bis(diphenylphosphino)ethylamine ethyl iodide, 30B, 278

$C_{28}H_{32}BrP$, (3,7-Dimethyl-2,6-octadienyl)triphenylphosphonium bromide, 39B, 509
$C_{28}H_{32}N_4O_3P_2$, μ-Oxo-bis(phosphenyl-o-ditoluidide), 44B, 636
$C_{28}H_{36}N_8P_4$, 2,cis-4,trans-6,trans-8-Tetrakismethylamino-2,4,6,8-tetraphenylcyclotetraphosphazene, 38B, 668
$C_{30}H_{14}F_{16}N_4O_2P_2$, 3,8-Dimethyl-4,5-bis(pentafluorophenyl)-1,6-bis(m-(trifluoromethyl)phenyl)-1,3,6,8-tetraaza-4λ⁵,5λ³-diphosphaspiro-[3.4]octane-2,7-dione, 45B, 763
$C_{30}H_{15}N_8P$, 2,3,3,4,4,5,5-Heptacyanocyclopentenyliminotriphenylphosphorane, 44B, 637
$C_{30}H_{21}N_4O_5P$, 5-Triphenylphosphoranylidenebenzo[1,2-c:3,4-c']difurazan-4(5H)-one 3,8-dioxide - benzene, 40B, 1101
$C_{30}H_{21}P \cdot 2\ C_6H_6$, Phenylbis(2,2'-biphenylylene)phosphorane-benzene (1:2), 46B, 689
$C_{30}H_{25}BrNO_4P$, N-p-Bromophenyltriphenylphosphine imide-dimethyl acetylenedicarboxylate adduct, 30B, 278
$C_{30}H_{25}NOP_2$, N-(Diphenylphosphinoyl)triphenylphosphazene, 45B, 744
$C_{30}H_{25}O_5P$, Pentaphenoxyphosphorane, 42B, 552
$C_{30}H_{25}P$, Pentaphenylphosphorus, 29, 639
$C_{30}H_{25}P_5$, Phosphobenzene pentamer, 29, 637
$C_{30}H_{26}O_4P_2$, Dimethyl 2,3-bis(diphenylphosphino)fumarate, 41B, 752
$C_{30}H_{30}N_4P_2S_2$, trans-Di-1,2-(6-methyl-2-thio-3-p-tolyl-1,2,3,4-tetrahydro-1,3,2-benzodiazaphosphorine), 43B, 878
$C_{30}H_{31}P_3$, 2,6,10-Triphenyl-2,6,10-triphosphabicyclo[9.4.0]pentadeca-11(1),12,14-triene, 46B, 694
$C_{30}H_{32}Cl_2N_4O_2P_2$, 2-Chloro-6-methyl-3-p-tolyl-1,2,3,4-tetrahydro-1,3,2-benzodiazaphosphorine 2-oxide, 45B, 764
$C_{30}H_{38}IO_2P$, 10-Methoxycarbonyldecyltriphenylphosphonium iodide, 43B, 879
$C_{31}H_{25}ClN_3O_4P$, N²-(p-Chlorophenyl)-N¹-cyano-N²-(1,2-bis(methoxycarbonyl)-2-(triphenylphosphonio)vinyl) hydrazide zwitterion, 38B, 668
$C_{31}H_{27}O_4P$, Tetracyclone - dimethyl phosphonate adduct, 40B, 634
$C_{31}H_{27}O_4P$, 1,2-Bis(methoxycarbonyl)-3-phenylpropen-2-yliden-1-triphenylphosphoran (cis-isomer), 40B, 635
$C_{31}H_{27}O_4P$, 1,2-Bis(methoxycarbonyl)-3-phenylpropen-2-yliden-1-triphenylphosphoran (trans-isomer), 40B, 635
$C_{32}H_{20}F_8P_2$, 1,1'-Bi(3,3,4,4-tetrafluoro-2-diphenylphosphinocyclobutene), 43B, 880
$C_{32}H_{24}NO_2P$, 2,3,8,8-Tetraphenyl-(1,3,2λ⁵)oxazaphospholino[2,3-b]benzo[d](1,3,2λ⁵)oxazaphospholine, 42B, 553
$C_{32}H_{64}N_{12}P_4$, 2,2,4,4,6,6,8,8-Octapyrrolidinylcyclotetra(phosphazene), 45B, 764
$C_{32}H_{96}N_{24}P_8$, Hexadeca(dimethylamino)cyclooctaphosphazene, 42B, 554
$C_{33}H_{33}O_8P$, Trimethyl 3-methoxy-4-oxo-5-(tri-p-tolyl-phosphoranylidene)cyclopentene-1,2,3-carboxylate, 37B, 394
$C_{33}H_{37}NOP_2$, 2,6,10-Triphenyl-6-aza-2,10-diphosphabicyclo[9.4.0]pentadeca-11(1),12,14-triene-acetone, 46B, 694
$C_{34}H_{23}OP$, 11-Oxo-1,11-diphenyl-dibenzo[b,g]indeno[2,3-d]-11,15b-dihydrophosphocin, 43B, 881
$C_{34}H_{23}P$, 1,2-Diphenyl-3-(2-(phenylethynyl)phenyl)-λ³-phosphindole, 43B, 882
$C_{34}H_{23}P \cdot C_3H_6O$, (1-Naphthyl)bis(2,2'-biphenylylene)phosphorane - acetone (1:1), 46B, 696
$C_{34}H_{25}OP$, 1,2,3,4,5-Pentaphenyl-1λ⁵-phosphol-1-one, 42B, 554
$C_{34}H_{25}O_4P$, 2,3,5,7,8-Pentaphenyl-1,4,6,9-tetraoxa-5λ⁵-phosphaspiro-[4.4]nona-2,7-diene, 45B, 738
$C_{34}H_{27}ClNOP$, 1-p-Chlorophenyl-3-p-tolyl-2-triphenylphosphazeno-pro-

penone, 46B, 695
$C_{34}H_{44}N_4P_2$, cis-1,3-Bis(t-butyl)-2,4-bis(methyl(diphenylphosphino)-amino)-cyclodiphosphazane, 45B, 765
$C_{35}H_{26}BrO_3P$, 1,1,5,8-Tetraphenyl-3-p-bromophenyl-1-phospha-2,4,9-trioxabicyclo[4.3.0]nonadiene-5,7, 37B, 395
$C_{35}H_{26}IP$, 1-Methyl-1,2-diphenyl-3-(2-(phenylethynyl)phenyl)-1-phosphindolium iodide, 43B, 882
$C_{35}H_{32}BrO_3P$, Triphenylphosphonia-dibenzoylmethane bromide ethanol, 46B, 695
$C_{36}H_{24}N_3O_6P_3$, Tris(2,2'-dioxybiphenyl)cyclotriphosphazene, 37B, 396
$C_{36}H_{28}NP$, [8-(Dimethylamino)-1-naphthyl]bis(2,2'-biphenylylene)phosphorane, 46B, 696
$C_{36}H_{30}I_6P_2S_2$, Bis(triphenylphosphinesulfide-S-iodine-I)iodine, 44B, 638
$C_{36}H_{30}N_3O_6P_3$, Hexaphenoxycyclotriphosphazene, 37B, 396
$C_{36}H_{30}N_3P_3$, 2,2,4,4,6,6-Hexaphenylcyclotriphosphazatriene, 34B, 393
$C_{36}H_{30}N_6P_2S_4$, Bis(triphenylphosphine-imino)tetrasulfurtetranitride, 45B, 765
$C_{36}H_{30}P_6$, Phosphobenzene hexamer (triclinic), 31B, 309
$C_{36}H_{30}P_6$, Phosphobenzene hexamer (trigonal), 30B, 279
$C_{36}H_{31}AuCl_4O_2P_2$, Bis(triphenylphosphine oxide)hydronium tetrachloroaurate(III), 44B, 639
$C_{36}H_{31}ClO_6P_2$, Triphenylphosphine oxide perchloric acid (2:1), 46B, 697
$C_{36}H_{32}O_4P_2$, Triphenylphosphine oxide hemiperhydrate, 46B, 698
$C_{36}H_{33}BrO_3P_2$, Bis(triphenylphosphine oxide) monohydrate hydrogen bromide, 45B, 766
$C_{36}H_{46}Cl_2O_3P_2$, 1,1,4,4-Tetraphenyl-2,5-di-(t-butyl)-1,4-diphosphoniacyclohexa-2,5-diene dichloride trihydrate, 38B, 669
$C_{37}H_{30}P_2$, Bis(triphenylphosphoranylidene)methane, 38B, 670
$C_{37}H_{31}BrP_2$, ((Triphenylphosphoranylidene)methyl)triphenylphosphonium bromide, 40B, 636; 43B, 883
$C_{37}H_{41}FNO_8P$, 5(S)-(3-Deoxy-3-fluoro-1,2:5,6-di-O-isopropylidene-α-O-glucofuranose-3-yl)-2,4-dioxo-5-hydroxypyrrolidin-3-ylidenetriphenylphosphorane acetone solvate, 44B, 639
$C_{38}H_{36}CdI_4P_2$, Bis(methyltriphenylphosphonium) tetraiodocadmate, 44B, 640
$C_{38}H_{36}Cu_4I_6P_2$, Methyltriphenylphosphonium hexaiodotetracuprate(I), 42B, 555
$C_{39}H_{40}NO_8PS$, Methyl-4,6-O-benzylidene-2,3-dideoxy-2,3-(N-triphenylphosphonioepimino)-α-D-alloside toluene-p-sulphonate hydrate, 44B, 640
$C_{40}H_{30}F_6OP_2$, 4,4-Bis(trifluoromethyl)-2,2,2-triphenyl-3-(triphenylphosphoranylidene)-1,2-oxaphosphetane, 33B, 284
$C_{40}H_{34}Br_2OP_2$, 1,1,2,4,4,5-Hexaphenyl-1,4-diphosphoniacyclohexa-2,5-diene dibromide, 38B, 671
$C_{41}H_{30}FeN_2O_4P_2$, (Bis(triphenylphosphine)imminium) tetracarbonylcyanoiron(0), 40B, 636
$C_{43}H_{37}FeNO_4P_2$, Bis(triphenylphosphine)iminium tetracarbonyl-propylferrate(0), 41B, 753
$C_{48}H_{30}N_6P_2$, 1,2,3,4,5,6-Hexacyanohexa-2,4-diene-1,6-diylidenebis-(triphenylphosphorane), 44B, 637
$C_{48}H_{30}N_8O_4P_2$, 8,8'-Bis(triphenylphosphoranylidene)bi(benzo[1,2-c:4,5-c']difurazan-4-ylidene), 40B, 1102
$C_{48}H_{40}Cl_6Cu_2P_2$, Bis(tetraphenylphosphonium) di-μ-chloro-bis(dichlorocuprate(II)), 40B, 626
$C_{48}H_{40}N_4P_4$, Octaphenylcyclotetraphosphazene, 40B, 637

$C_{48}H_{41}N_6P_5$, Octaphenyl-spiro-bi(cyclotriphosphazene), 45B, 767
$C_{48}H_{47}BF_4O_6P_2$, Triphenyl-(1-ethoxycarbonyl-2-methyl-2-hydroxyvinyl)-phosphonium triphenyl-(1-ethoxycarbonyl-2-methyl-2-olatovinyl)phosphonium tetrafluoroborate, 45B, 767
$C_{50}H_{40}N_2P_2$, Ethylene-1-(diphenylphosphino)-1-(triphenylphosphonium)-2-(diphenylamino)-2-(phenylamide), 38B, 671
$C_{50}H_{40}N_2P_2$, Ethylene-1,1-bis(triphenylphosphonium)-2,2-bis(phenylamide), 37B, 397
$C_{50}H_{46}N_6NiP_6S_2$, Bis-(1-methyl-1-thio-3,3,5,5-tetraphenylcyclotriphosphazene)nickel, 42B, 555
$C_{50}H_{46}N_6P_6$, Bi-(1-methyl-3,3,5,5-tetraphenylcyclotriphosphazene), 42B, 555
$C_{50}H_{146}Cl_{14}Co_4N_{36}P_{12}$, Chloro(dodeca(dimethylamino)cyclohexaphosphazene-N,N,N,N)cobalt(II) di-μ-chloro-bis(dichlorocobaltate(II)) - bischloroform, 39B, 509
$C_{52}H_{40}HgN_4P_2S_4$, Tetraphenylphosphonium tetrathiocyanatomercurate(II), 41B, 754
$C_{52}H_{40}N_2P_2$, (4-(Diphenylamino)-3-(diphenylphosphino)-2-chinolylmethylene)triphenylphosphorane, 46B, 699
$C_{54}H_{39}OP$, 9-Oxo-1,2,3,4,9e-pentaphenyl-9H-tribenzo[b,d,f]phosphepine benzene solvate, 44B, 641
$C_{54}H_{48}Cl_6P_3Pr$, Triphenylphosphonium hexachloropraseodymate(III), 43B, 884
$C_{58}H_{55}O_6P_3$, 4,4-Bis(1,1-dimethoxy-2,6-diphenyl-λ^5-phosphorin-4-ylmethyl)-1-methoxy-1-oxo-2,6-diphenyl-1,4-dihydro-λ^5-phosphorin, 43B, 885
$C_{60}H_{45}N_6O_6P_3 \cdot 0.5\ C_7H_8$, 1,3,5-Trinitro-2,4,6-tris(triphenylphosphoranylideneamino)benzene - toluene, 40B, 1103
$C_{83}H_{60}N_2O_{11}P_4Rh_4$, Bis(triphenylphosphine)iminium hepta-μ-carbonyl-tetracarbonyl-tetrahydro-tetrarhodate, 42B, 652

65. ARSENIC COMPOUNDS

CH_3As, Arsenomethane polymer, 35B, 466; 39B, 510
CH_3AsI_2, Diiodomethylarsine, 28, 585
$C_2H_2AsCl_3$, cis-β-Chlorovinyldichloroarsine (gas-ed), 10, 208
$C_2H_2AsCl_3$, trans-β-Chlorovinyldichloroarsine (gas-ed), 10, 208
$C_2H_4As_4O_4$, 1,3,5,7-Tetraarsa-2,4,6,8-tetraoxaadamantane, 46B, 699
C_2H_6AsBr, Dimethylbromoarsine (gas-ed), 9, 315
C_2H_6AsCl, Dimethylchloroarsine (gas-ed), 9, 315
C_2H_6AsI, Dimethyliodoarsine (gas-ed), 9, 315
$C_2H_7AsO_2$, Cacodylic acid, 30B, 282
C_3AsF_9, Trifluoromethyl arsine (gas-ed), 18, 678
$C_3H_3AsN_2$, Methyldicyanoarsine, 31B, 312
C_3H_6AsN, Cyanodimethylarsine, 28, 586
C_3H_9As, Trimethylarsine (gas-ed), 6, 234
$C_3H_9AsCl_3N$, Arsenic trichloride - trimethylamine, 37B, 397
$C_3H_9AsO_3$, n-Propylarsonic acid, 37B, 398
$C_3H_{27}AsMo_6N_6O_{27} \cdot 6\ H_2O$, Guanidinium hexamolybdomethylarsonate hexahydrate, 45B, 768
$C_4As_4F_{12}$, Trifluoromethylarsenic tetramer, 37B, 396
$C_4H_7AsO_2S_2$, 1,4-Dithiothreitol arsenite complex, 38B, 672
$C_4H_8AsClOS_2$, 5-Chloro-1-oxa-4,6-dithia-5-arsaocane, 41B, 755
$C_4H_8AsClS_3$, 2-Chloro-1,3,6,2-trithia-arsa-ocane, 40B, 639
$C_4H_9AsI_2NS_2$, (Dimethyldithiocarbamato)-iodo-methyl-arsenic(III) iodine, 45B, 769

$C_4H_9AsO_5$, 5-Hydroxy-1,4,6,9-tetraoxa-5-arsaspiro[4.4]nonane, 44B, 642
$C_4H_{12}AsBr$, Tetramethylarsonium bromide, 28, 587
$C_4H_{12}As_2S_2$, Dimethylarsino dimethyldithioarsinate, 29, 539
$C_4H_{13}AsMo_4N_6O_{15} \cdot H_2O$, Guanidinium tetramolybdodimethylarsonate monohydrate, 45B, 768
$C_4H_{14}AsClN_4S_2$, Bis(methyl)-thiourea-S-arsenic(III) chloride - thiourea, 41B, 755
$C_4H_{19}AsMo_4N_6O_{15} \cdot H_2O$, Guanidinium ($\mu_4$-hydroxy)bis($\mu_3$-oxo)tetrakis-($\mu$-oxo)tetrakis(dioxomolybdenum) dimethylarsinate monohydrate, 46B, 700
$C_5H_8AsCl_3N_2S$, Trichloro(1,3-dimethyl-2(3H)-imidazolethione)-arsenic(III), 43B, 261; 44B, 642
$C_5H_9As_3S_3$, 7-Methyl-1,3,5-triarsa-2,4,9-trithiaadamantane, 45B, 770
$C_5H_{10}AsBr_2NS_2$, Dibromo(N,N-diethyldithiocarbamato)arsenic(III), 43B, 886
$C_5H_{13}AsO_3$, Trimethylarsonioacetic acid betaine monohydrate, 43B, 886
$C_5H_{15}As_5$, Arsenomethane (gas-ed), 10, 206
$C_5H_{15}As_5$, Arsenomethane, 21, 511
$C_6H_4As_2Br_2O$, o-Phenylenediarsine oxybromide, 41B, 756
$C_6H_4As_2Cl_2O$, o-Phenylenediarsine oxychloride, 27, 984; 41B, 756
$C_6H_6AsNO_6$, 3-Nitro-4-hydroxyphenylarsonic acid, 43B, 887
$C_6H_7AsO_3$, Phenylarsonic acid, 24, 637
$C_6H_8AsNO_3$, m-Aminophenylarsonic acid, 27, 905
$C_6H_8AsNO_3$, o-Aminophenylarsonic acid, 43B, 888
$C_6H_8AsNO_3$, p-Aminobenzenearsonic acid, 26, 700
$C_6H_{12}As_2Cl_2N_4O_2$, 2,6-Dichloro-tetrahydro-1,3,5,7-tetramethyl-1,3,5,7,2,6-tetraazadiarsocine-4,8(3H,7H)-dione, 46B, 700
$C_6H_{15}AsS$, Triethylarsine sulfide, 45B, 770
$C_6H_{18}As_4N_6$, Hexa(methylamine)tetra-arsenic, 32B, 315
$C_7H_9AsFeO_4$, Trimethylarsenictetracarbonyliron, 37B, 399
$C_8H_{10}As_2I_2$, rac-o-Phenylenebis(iodomethylarsine), 43B, 889
$C_8H_{19}AsO_2$, Di-n-butylarsinic acid, 40B, 1103
$C_9H_{15}AsO_3S_6$, Arsenious xanthate, 24, 620
$C_{10}H_{42}As_2Mo_6N_2Na_2O_{30}$, Bis(tetramethylammonium) disodium hexamolybdobis(methylarsonate) hexahydrate, 42B, 1011
$C_{12}As_2F_8O$, Octafluoro-5,10-epoxy-5,10-dihydroarsanthrene, 46B, 701
$C_{12}H_8AsClO$, 10-Phenoxarsine chloride, 38B, 673
$C_{12}H_8AsCl_6O_2Sb$, 10-Chlorophenoxarsine oxide antimony pentachloride adduct, 38B, 674
$C_{12}H_8As_2O$, 5,10-Epoxy-5,10-dihydroarsanthren, 37B, 178
$C_{12}H_8As_2S$, 5,10-Epithio-5,10-dihydroarsanthren, 35B, 467
$C_{12}H_8As_2Se$, 5,10-Episeleno-5,10-dihydroarsanthren, 37B, 178
$C_{12}H_8As_2Te$, 5,10-Epitelluro-5,10-dihydroarsanthren, 37B, 178
$C_{12}H_9AsBrN$, 10-Bromo-5,10-dihydrophenarsazine, 31B, 312
$C_{12}H_9AsClN$, 10-Chloro-5,10-dihydrophenarsazine, 30B, 283; 31B, 312
$C_{12}H_{10}AsBr$, Bromodiphenylarsine, 27, 867
$C_{12}H_{10}AsCl$, Chlorodiphenylarsine, 27, 867
$C_{12}H_{10}AsI$, Iododiphenylarsine, 28, 588
$C_{12}H_{10}As_2N_4S_2$, 3,7-Diphenyl-3H,7H-1,5,2,4,6,8,3,7-dithia[1,5-S(IV)]-tetraazadiarsocine, 45B, 771
$C_{12}H_{10}As_2S_3$, Diphenyldiarsenic trisulfide, 38B, 674
$C_{12}H_{14}AsCl_4NO_4$, 2,6-Dimethyl-2,2-(tetrachloropyrocatecholato)-perhydro-1,3,6,2-dioxazarsocine, 43B, 890
$C_{12}H_{22}AsNO_2S$, Triethylammonium phenylthioarsenate, 38B, 674
$C_{12}H_{24}AsN_3O_3$, Tris(morpholino)arsine, 46B, 701
$C_{12}H_{36}As_4Si_6$, Dodecamethylhexasilatetraarsaadamatane, 46B, 701

$C_{13}H_{11}AsO_4$, 2-Methyl-2,2'-spiro(1,3,2-benzodioxaarsole), 44B, 642
$C_{13}H_{13}AsN_2O_2$, 2-Methyl-2,2'-spirobi[1,3,2λ^5-benzoxazarsoline], 46B, 702
$C_{13}H_{32}AsN_2O_6P$, Dicyclohexylammonium arsonomethylphosphonate, 43B, 176
$C_{14}H_6As_2F_8$, 5,10-Dihydro-5,10-dimethyloctafluoroarsanthrene, 43B, 891
$C_{14}H_{14}As_2Br_2$, 5,10-Dihydro-5,10-dimethylarsanthren dibromide, 23, 705
$C_{15}H_{30}AsN_3S_6$, Tris(diethyldithiocarbamato)arsenic(III), 41B, 994
$C_{16}H_{24}As_4Cr_2O_8$, Cyclohexa-1,4-di(tetracarbonylchromium)-2,3,5,6-tetra(dimethylarsenic), 40B, 640
$C_{16}H_{25}AsN_2S_4$, Phenylarsine bis(diethyldithiocarbamate), 32B, 316
$C_{17}H_{23}As_4I_3O$, meso-o-Phenylenebis(iodomethylarsine) - 1,3-dimethyl-1,3-dihydro-2,1,3-benzoxadiarsole methiodide, 43B, 889
$C_{18}H_{12}AsKO_6 \cdot 1.5\ H_2O$, Potassium (-)-tris(1,2-benzenediolato)-arsenate(V) sesquihydrate, 38B, 675
$C_{18}H_{13}As$, 9-Phenyl-9-arsafluorene, 28, 589
$C_{18}H_{15}AsF_2$, Triphenylarsenic difluoride, 41B, 757
$C_{18}H_{15}AsN_4S_3$, Triphenylarsine-trisulfur tetranitride, 43B, 891
$C_{18}H_{15}AsS$, Triphenylarsine sulfide, 45B, 772
$C_{18}H_{17}AsO_2$, Triphenylarsine oxide monohydrate, 34B, 394
$C_{18}H_{22}AsNO_4$, 2,6-Dimethyl-4,4-diphenyl-1,4-oxarsenanium nitrate, 42B, 556
$C_{18}H_{22}As_2N_4S_2$, 3,7-Dimesityl-3H-7H-1,5,2,4,6,8,3,7-dithia[1,5-S(IV)]tetraazadiarsocine, 45B, 771
$C_{18}H_{24}AsBrO_2$, 2,6-Dimethyl-4,4-diphenyl-1,4-oxarsenium bromide monohydrate, 41B, 758
$C_{18}H_{29}AsO_4$, 2,2,3,3,7,7,8,8-Octamethyl-5-phenyl-1,4,6,9-tetraoxa-5-arsaspiro[4.4]nonane, 44B, 642
$C_{18}H_{42}As_2O_2Si_2$, 2,4-Di-t-butyl-1,3-dimethyl-2,4-bis(trimethylsiloxy)-1,3-diarsetane, 46B, 702
$C_{19}H_{13}As$, Arsatriptycene, 46B, 703
$C_{20}H_{16}AsCl \cdot H_2O$, 9,10-Dihydro-9-methyl-9,10-o-benzeno-9-arsoniaanthracene chloride monohydrate, 46B, 703
$C_{20}H_{20}As_2Cl_6N_4$, Arsenic trichloride dipyridyl dimer, 38B, 676
$C_{21}H_{21}As$, Tri-p-tolylarsine, 28, 588
$C_{23}H_{17}As$, 2,3,6-Triphenylarsenin, 39B, 510
$C_{24}H_{12}AsN_3O_6$, Tris-(p-nitrophenylethynyl)-arsine, 33B, 285
$C_{24}H_{16}AsF_5O_2$, Triphenylarsine oxide - pentafluorophenol adduct, 42B, 556
$C_{24}H_{16}As_2O_2S$, 10-Phenoxarsine sulfide, 38B, 676
$C_{24}H_{20}AsAuCl_4$, Tetraphenylarsonium tetrachloroaurate(III), 41B, 759
$C_{24}H_{20}AsBr_3$, Tetraphenylarsonium tribromide (polymorph II), 45B, 772
$C_5H_8AsCl_3N_2S$, Trichloro(1,3-dimethyl-2(3H)-imidazolethione)-arsenic(III) (polymorph I), 43B, 261; 44B, 642
$C_{24}H_{20}AsClO_4$, Tetraphenylarsonium perchlorate, 45B, 773
$C_{24}H_{20}AsCl_4Fe$, Tetraphenylarsonium tetrachloroferrate(III), 21, 600; 41B, 760
$C_{24}H_{20}AsCl_4NOs$, Tetraphenylarsonium nitridotetrachloroosmate(VI), 41B, 761
$C_{24}H_{20}AsCl_4NRu$, Tetraphenylarsonium nitridotetrachlororuthenate(VI), 41B, 761
$C_{24}H_{20}AsCl_5SW$, Tetraphenylarsonium pentachlorothiotungstate, 46B, 704
$C_{24}H_{20}AsI$, Tetraphenylarsonium iodide, 8, 312
$C_{24}H_{20}AsI_3$, Tetraphenylarsonium triiodide, 9, 389; 23, 671; 38B, 677

$C_{24}H_{20}AsI_4NOs$, Tetraphenylarsonium nitridotetraiodo-osmate(VI), 41B, 1088; 42B, 557
$C_{24}H_{20}As_2O$, Bis(diphenylarsenic) oxide, 28, 592
$C_{24}H_{20}As_4S_4$, Cyclo-tetrakis(phenylarsene sulfide), 45B, 774
$C_{24}H_{22}AsBr_4MoO_2$, Tetraphenylarsonium oxotetrabromoaquomolybdate, 32B, 317
$C_{24}H_{23}AsCl_4O_2Te$, Tetraphenylarsonium aquotetrachlorohydroxotel-urate(IV), 40B, 640
$C_{24}H_{25}AsCl_2O_2$, Tetraphenylarsonium diaquohydrogen dichloride, 38B, 677
$C_{24}H_{27}As$, Tri-p-xylylarsine, 28, 591
$C_{25}H_{17}As$, 9,10-Dihydro-10-phenyl-9,10-o-benzeno-9-arsaanthracene, 46B, 704
$C_{26}H_{21}As$, trans-10-Benzyl-9-phenyl-9,10-dihydro-9-arsaanthracene, 46B, 704
$C_{28}H_{20}AsCl_3FeHg_2O_4$, Tetraphenylarsonium cis-tetracarbonyl(chloro-mercurio)(dichloromercurio)ferrate, 42B, 557
$C_{30}H_{25}As \cdot 0.5\ C_6H_{12}$, Pentaphenylarsenic hemicyclohexane solvate, 42B, 558
$C_{33}H_{20}AsFN_6$, Tetraphenylarsonium 3-fluoro-1,1,4,5,5-pentacyano-2-azapentadienide, 31B, 317
$C_{36}H_{30}AsN_3P_2$, 1,1,3,3,5,5-Hexaphenyl-1,3-diphospha-5-arsatriazene, 43B, 896
$C_{36}H_{30}As_3N_3$, 2,2,4,4,6,6-Hexaphenylcyclotriarsazine, 39B, 511
$C_{36}H_{30}As_6$, Arsenobenzene, 24, 637; 26, 736
$C_{36}H_{32}As_2Br_4O_2$, Bis(triphenylarsinehydroxy) bromine tribromide, 38B, 679
$C_{36}H_{32}As_2Cl_3IO_2$, μ-Chloro-bis(hydroxytriphenylarsenic) dichloro-iodate, 41B, 762
$C_{36}H_{32}As_2O_5Se$, Triphenylarsine oxide - selenous acid (2:1), 42B, 60, 558
$C_{36}H_{33}As_6N_5 \cdot C_6H_6$, 2,4,6,8,9,11-Hexaphenylbicyclo[3.3.3]-2,4,6,8,-9,11-hexaarsa-1,3,5,7,10-pentaazaundecane benzene solvate, 46B, 705
$C_{37}H_{27}AsO_2$, 1-Triphenylarsonium-3,4-dibenzoylcyclopentadienylide, 42B, 558
$C_{38}H_{20}AsMn_3O_{14}$, Tetraphenylarsonium trimanganesetetradecacarbonyl, 40B, 641
$C_{43}H_{33}AsO$, 1-Acetyl-2,3,4-triphenyl-5-(triphenylarsonio)cyclopenta-dienide, 41B, 763
$C_{43}H_{34}AsClO_5$, Triphenyl-(2-acetyl-3,4,5-triphenylcyclopenta-2,4-dienyl)arsonium perchlorate, 41B, 764
$C_{48}H_{40}As_2Cl_6Cu_2$, Bis(tetraphenylarsonium) di-μ-chloro-bis(dichloro-cuprate(II)), 40B, 642
$C_{48}H_{40}As_2N_4O_{12}Zn$, Tetraphenylarsonium tetranitratozincate, 42B, 559
$C_{48}H_{40}As_4N_4$, Octaphenylcyclotetraarsazene, 40B, 637
$C_{49}H_{22}As_2F_{20}Hg_2$, Bis(pentafluorophenyl)mercury(II) - bis(diphenyl-arsino)methane (2:1), 38B, 679
$C_{53}H_{40}As_2FeN_6O$, Bis(tetraphenylarsonium) nitroprusside, 43B, 896

66. ANTIMONY AND BISMUTH COMPOUNDS

CCl_6N_9Sb, Triazidocarbonium hexachloroantimonate, 35B, 468
CH_3Cl_4OSb, Tetrachloromethoxyantimony(V), 37B, 400
$CH_4F_3N_2SSb$, Thiourea trifluoroantimony, 44B, 643
$C_2F_6Na_2O_4Sb_2$, Sodium bis(trifluoroantimony)oxalate, 44B, 643
$C_2F_{12}O_2Sb_2$, μ-Fluoro-μ-trifluoroacetato-bis[tetrafluoroantimony(V)],

46B, 705
$C_2H_3Cl_6OSb$, Acetyl hexachloroantimonate, 32B, 318
$C_2H_4ClS_2Sb$, 2-Chloro-1,3-dithia-2-stibacyclopentane, 32B, 319
$C_2H_4Cl_6O_4Sb_2$, μ-Acetato-μ-hydroxy-μ-oxo-bis(trichloroantimony(V)), 45B, 774
$C_2H_5Cl_2OSb$, Dichloroethoxyantimony, 46B, 707
$C_2H_5Cl_4OSb$, Ethoxyantimony(V) tetrachloride, 33B, 287
$C_3Cl_9O_3Sb$, Tetrachloroethylenecarbonate - antimony pentachloride, 38B, 680
$C_3Cl_{12}N_3O_3Sb_3$, Antimony tetrachloride cyanurate trimer, 42B, 560
C_3F_9Sb, Trifluoromethyl stibine (gas-ed), 18, 678
$C_3H_3BiO_6$, Bismuth(III) formate, 34B, 395
$C_3H_4Cl_4NO_3Sb$, Tetrachloro-N-methylaminooxalato-O,O'-antimony(V), 45B, 775
$C_3H_6Br_3S_3Sb$, 1,3,5-Trithian antimony(III) bromide, 43B, 426
$C_3H_6Cl_3S_3Sb$, 1,3,5-Trithian antimony(III) chloride, 43B, 426
$C_3H_7Cl_5NOSb$, Dimethylformamide-antimony(V) chloride complex, 31B, 324
C_3H_9Bi, Trimethylbismuth (gas-ed), 39B, 891
$C_3H_9Br_2Sb$, Trimethylantimony dibromide, 6, 200, 219
$C_3H_9Cl_5OPSb$, Trimethylphosphine oxide - antimony pentachloride, 26, 595, 28, 95
$C_3H_9F_2Sb$, Difluorotrimethylantimony, 44B, 645
$C_3H_{12}BiCl_3N_6S_3$, Bismuth trichloride - thiourea, 43B, 898
$C_4F_{12}O_5Sb_2$, μ-Oxo-di-μ-trifluoroacetato-bis[trifluoroantimony(V)], 46B, 705
$C_4H_4Cl_{12}O_2Sb_2$, Succinyl chloride - antimony pentachloride (1:2), 38B, 681
$C_4H_5FNa_3O_{11}Sb$, Sodium fluoro-oxo-dioxalato-antimony oxonium hydrate, 46B, 706
$C_4H_5O_4S_2Sb$, Antimony hydrogen bis(thioglycollate), 33B, 287
$C_4H_7Cl_4N_2O_2Sb$, Tetrachloro-N,N'-dimethyloxamido-antimony(V), 39B, 512
$C_4H_8ClOS_2Sb$, 5-Chloro-1-oxa-4,6-dithia-5-stibocane, 42B, 560
$C_4H_8ClS_3Sb$, 2-Chloro-1,3,6-trithiastibaocane, 40B, 642
$C_4H_8Cl_3S_2Sb$, 1,4-Dithiane - antimony trichloride, 40B, 542
$C_4H_{10}ClO_2Sb$, Chloro-bis(ethoxy)antimony, 46B, 707
$C_4H_{10}Cl_5O_3Sb$, 1,4-Dioxane aquopentachloroantimony, 44B, 644
$C_4H_{12}Cl_4GaSb$, Tetramethylantimony tetrachlorogallate(III), 44B, 645
$C_4H_{12}Cl_6Sb_2$, Di-μ-chloro-bis(dichlorodimethylantimony), 44B, 645
$C_4H_{12}Cl_6Sb_2$, Tetramethylantimony(V) hexachloroantimonate(V), 44B, 645
$C_4H_{12}Cl_8O_8P_2Sb_2$, Tetrachloroantimony(V)-dimethoxyphosphate, 45B, 775
$C_4H_{12}FSb$, Fluorotetramethylantimony, 44B, 645
$C_4H_{20}Bi_3Cl_9N_{12}S_4$, $μ_4$-Chloro(tris(trichloro(thiosemicarbazide)bismuth(III)))(tris(thiosemicarbazide)bismuth(III)) hexachlorobismuthate(III) chloride, 44B, 646
$C_5H_{10}Cl_3S_2Sb$, 1,4-Dithiacycloheptane trichloroantimony(III), 45B, 776
$C_6F_9O_6Sb$, (R)-Tris(trifluoroacetato)antimony(III), 46B, 707
$C_6H_4ClO_2Sb$, o-Phenylenedioxyantimony(III) chloride, 45B, 777
$C_6H_5O_4Sb$, Antimony(III) pyrogallate monohydrate, 42B, 561
$C_6H_6Cl_5Sb$, trans,trans,trans-Tri-2-chlorovinyldichlorostibine, 17, 671
$C_6H_7Cl_3NSb$, Antimony trichloride-aniline, 33B, 288
$C_6H_9O_3S_3Sb$, Antimony(III) tris(monothioacetate), 46B, 708
$C_6H_9O_6Sb$, Antimony(III) triacetate, 46B, 708

$C_6H_{10}Cl_3O_2Sb$, (Acetylacetonato)methyltrichloroantimony(V), 44B, 646
$C_6H_{12}BiCl_3N_4S_2$, catena-μ-Chloro-dichlorobis(ethylenethiourea)bismuth(III), 44B, 646
$C_6H_{18}Cl_2OSb_2$, μ-Oxo-trimethylantimony chloride, 41B, 765
$C_6H_{18}Cl_2O_9Sb_2$, μ-Oxo-trimethylantimony perchlorate, 41B, 765
$C_6H_{18}Cl_8N_2O_6P_2Sb_2$, Tetrachloroantimony(V)-dimethylaminomethoxyphosphate, 45B, 775
$C_6H_{18}N_6OSb_2$, μ-Oxo-trimethylantimony azide, 41B, 765
$C_6H_{18}OPSSb$, Tetramethyl(dimethylthiophosphinato)antimony, 44B, 647
$C_6H_{24}BiBr_6N_3$, Tris(dimethylammonium) hexabromobismuthate(III), 33B, 293
$C_7H_5Cl_6OSb$, Benzoyl chloride - antimony pentachloride, 38B, 682
$C_7H_7O_3Sb$, 2-Methoxy-1,3,2-benzodioxastibole, 43B, 897; 44B, 648
$C_7H_7S_3Sb$, 2-Methylthio-1,3,2-benzodithiastibole, 44B, 648
$C_7H_9FeO_4Sb$, Trimethylantimonytetracarbonyliron, 37B, 399
$C_7H_{13}Br_2O_2Sb$, (Acetylacetonato)dimethyldibromoantimony(V), 40B, 643; 44B, 646
$C_7H_{28}Bi_3Cl_9N_{14}S_7$, Di-μ-chloro-bis(chlorotrithioureabismuth(III)) pentachloromonothioureabismuthate(III), 41B, 767
$C_8H_6Cl_3O_2Sb$, Terephthaldehyde antimony trichloride, 44B, 648
$C_8H_7Cl_6OSb$, Antimony pentachloride p-toluoyl chloride, 38B, 683
$C_8H_7Cl_6OSb$, m-Toluoyl chloride - antimony pentachloride, 38B, 682
$C_8H_7Cl_6OSb$, p-Tolyloxocarbonium hexachloroantimonate(V), 38B, 683
$C_8H_{10}Cl_6Sb_2$, Antimony trichloride - p-xylene (2:1), 42B, 482
$C_8H_{10}K_2O_{15}Sb_2$, Potassium antimony D-tartrate trihydrate, 40B, 644
$C_8H_{10}K_2O_{15}Sb_2$, Potassium di-μ-tartrato-diantimonate(III) trihydrate, 35B, 469
$C_8H_{18}N_2O_{15}Sb_2$, Ammonium antimonyl D-tartrate trihydrate, 32B, 320
$C_9H_{15}O_3S_6Sb$, Antimonious xanthate, 26, 610
$C_9H_{15}Sb$, Trimethyl antimony dipropyne, 44B, 1109
$C_9H_{21}N_2S_4Sb$, Trimethylbis(N,N-dimethyldithiocarbamato)antimony(V), 44B, 649
$C_{10}H_8Cl_3N_2Sb$, 2,2'-Bipyridyltrichloroantimony, 46B, 708
$C_{10}H_{10}Cl_3O_2Sb$, p-Diacetylbenzene antimony trichloride, 44B, 648
$C_{10}H_{13}N_2O_8Sb \cdot 2\ H_2O$, (Hydrogen ethylenediaminetetraacetato)antimony(III) dihydrate, 46B, 709
$C_{10}H_{14}Cl_6O_5Sb_2$, μ-Oxo-bis[trichloro-acetylacetonatonantimony(V)], 45B, 777
$C_{11}H_{23}BiN_2S_4$, Bis(diethyldithiocarbamato)-methyl-bismuth, 45B, 777
$C_{12}H_{10}BrCl_2Sb$, Bromodichlorobisphenylantimony(V), 45B, 778
$C_{12}H_{10}Br_2ClSb$, Dibromochlorobisphenylantimony(V), 45B, 778
$C_{12}H_{10}Br_3Sb$, Tribromobisphenylantimony(V), 45B, 778
$C_{12}H_{10}Cl_6Sb_2$, Biphenyl bis(antimony trichloride), 44B, 650
$C_{12}H_{10}FSb$, Diphenylantimony(III) fluoride, 45B, 771
$C_{12}H_{11}Cl_3NSb$, Diphenylamine trichloroantimony, 46B, 709
$C_{12}H_{11}Cl_6NSb_2$, Diphenylamine bis(antimony trichloride), 44B, 672
$C_{12}H_{12}Cl_3OSb$, Diphenyltrichlorostibine monohydrate, 32B, 321
$C_{12}H_{12}Cl_4NSb$, Diphenylammonium chloride antimony trichloride, 44B, 650
$C_{12}H_{12}Cl_6O_3Sb_2$, Antimony trichloride - 1,3,5-triacetylbenzene (2:1), 44B, 651
$C_{12}H_{16}F_3N_2O_5Sb$, Trifluorobis(4-methoxypyridine N-oxide)antimony(III) hydrate, 41B, 768, 1244
$C_{12}H_{30}O_6P_3S_6Sb$, Tris(O,O-diethylphosphorodithioato)antimony, 46B, 710
$C_{14}H_{13}O_2Sb$, Acetato-diphenylantimony(III), 46B, 711
$C_{14}H_{19}NO_3S_6Sb$, Tris(O-ethylxanthato)antimony(III) 4,4'-bipyridyl

solvate, 44B, 651
$C_{15}H_{24}N_3S_6Sb$, Tris(1-pyrrolidinecarbodithioato)antimony(III), 46B, 711
$C_{15}H_{30}BiN_3S_6$, Tris(diethyldithiocarbamato)bismuth(III), 42B, 561
$C_{15}H_{30}N_3S_6Sb$, Tris(diethyldithiocarbamato)antimony(III), 42B, 561
$C_{16}H_{13}BiN_2O_2S_2$, Bis(1-oxopyridine-2-thiolato)phenylbismuth, 38B, 684
$C_{16}H_{17}BiOS_2$, Isopropyl-xanthogenato-diphenylbismuthine, 46B, 712
$C_{17}H_{17}Cl_2O_2Sb$, (Acetylacetonato)dichlorodiphenylantimony(V), 38B, 685
$C_{17}H_{17}Cl_2O_2Sb$, (Acetylacetonato)dichlorodiphenylantimony(V) (high-melting form), 44B, 646
$C_{18}H_{15}Bi$, Triphenylbismuth, 9, 388; 33B, 295
$C_{18}H_{15}BiCl_2$, Triphenylbismuth dichloride, 33B, 296
$C_{18}H_{15}Cl_2Sb$, Triphenyldichloroantimony, 24, 621; 31B, 326
$C_{18}H_{42}BiO_6P_3S_6$, Bismuth(III) O,O'-diisopropylphosphorodithiote, 40B, 644
$C_{19}H_{18}FSb$, Fluoro(methyl)triphenylantimony, 42B, 561
$C_{20}H_{15}CoNO_3Sb$, Dicarbonylnitrosyltriphenylstibinecobalt(0), 40B, 645
$C_{20}H_{15}N_2O_2Sb$, Bis(isocyanato)triphenylantimony, 41B, 768
$C_{20}H_{19}SSb$, (2,6-Dimethylthiophenolato)diphenylantimony, 44B, 652
$C_{20}H_{21}O_2Sb$, Dimethoxytriphenylantimony, 33B, 290
$C_{21}H_{21}Sb$, Tri-p-tolyl-antimony, 45B, 779
$C_{22}H_{21}O_4Sb$, Triphenylantimony(V) diacetate, 45B, 779
$C_{23}H_{17}ClF_3O_3Sb$, Tris(p-chlorophenyl)-5,5,5-trifluoro-4,4-dihydroxy-2-pentanonatoantimony(V), 43B, 898
$C_{24}H_{20}Cl_6Sb_2$, Di-μ-chloro-bis(dichlorodiphenylantimony), 40B, 646
$C_{24}H_{20}OSb_2$, Oxobis(diphenylantimony(III)), 40B, 647
$C_{24}H_{20}Sb_2$, Tetraphenyldistibine, 46B, 712
$C_{24}H_{21}OSb$, Tetraphenylantimony hydroxide, 34B, 396
$C_{25}H_{20}F_6FeO_2Sb$, (Dicarbonyl(η-cyclopentadienyl)ferrio)triphenylantimony hexafluorophosphate, 44B, 652
$C_{25}H_{21}O_2Sb$, Tetraphenylantimony(V) formate, 45B, 780
$C_{25}H_{23}OSb$, Methoxytetraphenylantimony, 33B, 290
$C_{26}H_{23}N_2O_2Sb$, Acetnitrosolatotetraphenylantimony(V), 40B, 647
$C_{26}H_{33}O_4Sb$, Di(t-butylperoxy)triphenylantimony, 45B, 781
$C_{27}H_{18}N_3S_3Sb$, Tris(8-mercaptoquinolinato)antimony, 39B, 512
$C_{27}H_{33}Bi$, Trimesitylbismuth, 46B, 713
$C_{28}H_{20}Cl_{16}Fe_4O_8Sb_4$, Tetrakis(dicarbonylchloro-cyclopentadienyliron)-tetrakis(trichloroantimony), 42B, 562
$C_{28}H_{20}O_3Sb_2$, 2,3-Bis(diphenylstibino)maleic acid anhydride, 46B, 714
$C_{30}H_{25}Sb$, Pentaphenylantimony, 29, 641; 33B, 292
$C_{30}H_{25}Sb \cdot 0.5\ C_6H_{12}$, Pentaphenylantimony cyclohexane solvate, 40B, 648
$C_{32}H_{30}ClF_3O_3Sb$, (4,4,4-Trifluoro-3,3-dihydroxy-1-phenyl-1-butanonato(2-)O,O',O'')-tris(p-tolyl)antimony(V) 1,2-dichloroethane solvate, 44B, 653
$C_{33}H_{36}BiCl_3N_{12}O_9S_3$, Trichlorotris-(3-sulfanilamido-6-methoxypyridazine)bismuth(III), 38B, 686
$C_{34}H_{20}Cr_2O_{10}Sb_2$, Bis-pentacarbonylchromium-tetraphenyldistibane, 46B, 714
$C_{35}H_{35}Sb$, Penta-p-tolylantimony, 39B, 512
$C_{36}H_{30}Bi_2Cl_2O_9$, μ-Oxo-bis(perchloratotriphenylbismuth(V)), 41B, 769
$C_{36}H_{30}Cl_3O_2P_2Sb$, Trichlorobis(triphenylphosphine oxide)antimony-(III), 44B, 654
$C_{36}H_{30}N_6OSb_2$, μ-Oxy-bis(triphenylazidoantimony), 39B, 512
$C_{36}H_{72}Cd_2I_6N_4S_8Sb_2$, Bis(di-n-butyldithiocarbamato)antimony(III) hexaiododicadmate, 43B, 899

$C_{44}H_{48}O_5Sb_2$, μ-Oxo-bis(t-butylperoxytriphenylantimony), 44B, 654
$C_{46}H_{30}N_6O_3Sb_2 \cdot 0.5\ C_6H_6$, μ-Oxo-bis[1,1,2-tricyanoethenoxo(triphenyl)antimony(III)] benzene solvate, 45B, 781
$C_{48}H_{34}Cl_{12}F_6O_5Sb_2$, μ-Oxo-bis(tris(p-chlorophenyl)(1,1,1-trifluoro-2,4-pentanedionato-O,O')antimony(V)) chloroform solvate, 44B, 655
$C_{48}H_{38}O_4Sb_2 \cdot H_2O$, Bis(triphenylantimony catecholate) hydrate, 46B, 715
$C_{49}H_{40}O_3Sb_2$, μ-Carbonato-bis(tetraphenylantimony), 40B, 648
$C_{55}H_{45}As_3Bi_2I_6O_3$, Triiodobismuth(III)-tri-μ-iodo-tris(triphenylarsineoxide)-bismuth(III), 42B, 562
$C_{72}H_{60}Bi_2I_6O_4P_4$, Di-μ-iodo-bis(diiodobis(triphenylphosphine oxide)-bismuth(III)), 42B, 563

67. GROUPS IA AND IIA COMPOUNDS

CH_3CsO, Caesium methoxide, 35B, 472
CH_3K, Methyl potassium, 35B, 472
CH_3KO, Potassium methoxide, 28, 595; 35B, 472
CH_3Li, Methyllithium, 35B, 471
CH_3LiO, Lithium methoxide, 24, 513
CH_3NaO, Sodium methoxide, 29, 452
CH_3ORb, Rubidium methoxide, 35B, 472
$CH_{10}CaN_4O_{10}$, Calcium nitrate - urea trihydrate, 38B, 687
$C_2Cs_2O_2$, Caesium acetylenediolate, 29, 467
C_2Li_2, Lithium acetylide, 32A, 233
$C_2O_2Rb_2$, Rubidium acetylenediolate, 29, 467
C_2HK, Potassium acetylide, 33B, 296
C_2HNa, Sodium acetylide, 33B, 296; 38B, 54
C_2HRb, Rubidium acetylide, 33B, 296
C_2H_5Li, Ethyl lithium, 28, 593
C_2H_5Na, Ethylsodium, 35B, 473
C_2H_6Be, Dimethylberyllium (gas-ed), 40B, 1140
C_2H_6Be, Dimethylberyllium, 15, 415
$C_2H_6CaN_4O_4$, Calcium hydrazinecarboxylate, 37B, 400
$C_2H_8CaN_4O_5$, Calcium hydrazinecarboxylate monohydrate, 37B, 401
$C_2H_8ILiN_4O_2$, Urea - lithium iodide, 35B, 474
C_3H_3K, Potassium propynylide, 33B, 296
C_3H_3Na, Sodium propynylide, 33B, 296
C_3H_7CsO, Cesium isopropoxide, 45B, 32
C_3H_7KO, Potassium isopropoxide, 45B, 32
C_3H_7ORb, Rubidium isopropoxide, 45B, 32
$C_4BeK_2O_8$, Potassium bis(oxalato)beryllate, 44B, 19
$C_4H_4MgO_5 \cdot 5\ H_2O$, Magnesium malate pentahydrate, 46B, 715
$C_4H_6BeCl_2N_2$, Dichlorobis(acetonitrile)beryllium, 41B, 53
C_4H_9CsO, Caesium t-butylate, 33B, 297
C_4H_9KO, Potassium t-butylate, 33B, 297
C_4H_9NaO, Sodium t-butoxide, 43B, 68
C_4H_9ORb, Rubidium t-butylate, 33B, 297
$C_4H_{10}ClLiO_3$, 1,4-Dioxan lithium chloride monohydrate, 33B, 298
$C_4H_{10}Mg$, Diethylmagnesium, 30B, 284
$C_4H_{12}BeLi_2$, Lithium tetramethylberyllate, 33B, 299
$C_4H_{12}MgO_9 \cdot H_2O$, ((S)-Malato)tetraaquamagnesium(II) hydrate, 45B, 782
$C_4H_{16}BrLiN_4$, Bis(ethylenediamine)lithium bromide, 32B, 267
$C_4H_{16}CaN_2O_{10}$, Calcium nitrate - methanol, 40B, 650
$C_4H_{16}CaN_8O_8S$, Calcium sulfate urea complex, 41B, 60

$C_4H_{16}CaN_{10}O_{10}$, Calcium nitrate tetraurea, 43B, 78
$C_4H_{16}ClLiN_4$, Bis(ethylenediamine)lithium chloride, 31B, 378; 32B, 268
$C_4H_{20}Br_2MgN_8O_6$, Diaquatetrakis(urea)magnesium bromide, 46B, 46
$C_4H_{20}Cl_2MgO_8$, Hexaaquodichloromagnesium-1,4-dioxan (1:1), 42B, 564
C_5H_5BeBr, Cyclopentadienylberyllium bromide (gas-ed), 40B, 1134
C_5H_5BeCl, Cyclopentadienylberyllium chloride (gas-mw-ed), 38B, 1062, 1072
$C_5H_7KO_2 \cdot 0.5\ H_2O$, Potassium acetylacetonate hemihydrate, 41B, 52
$C_5H_7LiO_2$, 2,4-Pentanedionatolithium, 41B, 53
C_5H_9BBe, Cyclopentadienylberyllium borohydride (gas-ed), 38B, 1061
$C_5H_{12}NNaO_2S_2$, Sodium 1-pyrrolidinecarbodithioate dihydrate, 46B, 716
$C_5H_{13}B_5Be$, μ-[η^5-Cyclopentadienylberylla]octahydropentaborate, 45B, 782
$C_6H_4BeK_2O_8 \cdot 0.5\ H_2O$, Potassium bis(malonato)beryllate hemihydrate, 44B, 655
$C_6H_6CaNNaO_6$, Sodium nitrilotriacetatocalcate, 45B, 519
C_6H_8Be, Methyl(cyclopentadienyl)beryllium (gas-ed), 38B, 1062
$C_6H_{12}BaO_5S_2$, Bis(thioacetato)barium(II) trihydrate, 42B, 565
$C_6H_{15}INNaO_3$, Triethanolamine sodium iodide, 40B, 653
$C_6H_{20}Cl_2MgO_{10}$, Myo-inositol magnesium chloride tetrahydrate, 39B, 513
$C_6H_{22}Br_2CaO_{11}$, Myo-inositol calcium bromide pentahydrate, 39B, 514
$C_6H_{24}Br_2CaN_{12}O_6$, Calcium bromide hexurea, 43B, 79
$C_6H_{24}ILiN_6$, Trisethylenediamine lithium iodide, 37B, 402
C_7H_6Be, Cyclopentadienylberyllium acetylide (gas-ed), 40B, 1135
$C_8H_{14}CaN_6O_{14}$, Triaquopurpuratocalcium nitrate dihydrate, 43B, 899
$C_8H_{16}Br_2MgO_2$, Bis(tetrahydrofuran)magnesium bromide, 43B, 408
$C_8H_{16}CaF_4O_8P_2$, Bis(ethylacetato)calcium difluorophosphate, 34B, 397
$C_8H_{18}Be$, Di-t-butylberyllium (gas-ed), 33B, 535
$C_8H_{20}BeCl_2O_2$, Bis(diethyl ether)beryllium chloride, 42B, 568
$C_8H_{20}Br_2MgO_2$, Magnesium bromide diethyletherate, 32B, 322
$C_8H_{20}Cl_2N_{12}O_{16}Sr$, Tetrakis(biuret)strontium perchlorate, 41B, 61
$C_8H_{22}MgN_2$, Dimethyl(N,N,N',N'-tetramethylethylenediamine)magnesium, 46B, 716
$C_8H_{24}Al_2Mg$, Octamethyldialuminummonomagnesium, 34B, 398
$C_8H_{24}BeN_2Si_4$, Bis(di(trimethylsilyl)amino)beryllium (gas-ed), 35B, 901
$C_8H_{32}CaCl_2O_{12}$, Calcium chloride - 1,4,7,10-tetraoxacyclododecane octahydrate, 42B, 569
$C_9H_{18}KNS_2$, Potassium dibutyldithiocarbamate, 42B, 570
$C_9H_{18}NRbS_2$, Rubidium dibutyldithiocarbamate, 42B, 570
$C_9H_{21}INNaO_3$, 2,2',2''-Trimethoxytriethylamine sodium iodide, 40B, 654
$C_{10}H_{10}Be$, Bis(cyclopentadienyl)beryllium, 38B, 777
$C_{10}H_{10}Be$, Dicyclopentadienylberyllium (gas-ed), 29, 524; 33B, 536
$C_{10}H_{10}Ca$, Dicyclopentadienylcalcium, 40B, 745
$C_{10}H_{10}Mg$, Bis(cyclopentadienyl)magnesium, 42B, 571
$C_{10}H_{12}CaN_6O_6$, Calcium hexacyanoisobutene hexahydrate, 32B, 323
$C_{10}H_{12}Ca_2N_2O_8 \cdot 7\ H_2O$, Calcium ethylenediaminetetraacetatocalcate heptahydrate, 45B, 519
$C_{10}H_{12}ClLiN_2O$, Bis(pyridine)lithium chloride hydrate, 32B, 271
$C_{10}H_{14}BeO_4$, Bisacetylacetoneberyllium, 21, 513; 24, 593; 41B, 770
$C_{10}H_{16}CaN_2O_8 \cdot 4\ H_2O$, Calcium di-L-glutamate tetrahydrate, 45B, 520
$C_{10}H_{18}MgO_6$, Diaquobis(acetylacetonato)magnesium(II), 32B, 324
$C_{10}H_{24}MgN_2Na_2O_{14}$, Disodium ethylenediaminetetraacetatoaquo-magnesate(II) pentahydrate, 39B, 514

$C_{10}H_{25}BrMgO_2$, Ethylbis(diethyl ether)magnesium bromide, 29, 532; 33B, 300
$C_{10}H_{26}MgN_2O_{14}$, Hexaaquomagnesium dihydrogenethylenediaminetetraacetate, 39B, 515
$C_{10}H_{28}MgO_2Si_2$, Bis(trimethylsilyl)magnesium 1,2-dimethoxyethane, 43B, 826
$C_{10}H_{30}Mg_2N_2O_{17}$, Magnesium ethylenediaminetetraacetate nonahydrate, 43B, 606
$C_{10}H_{40}Br_2MgN_{20}O_{10}$, Hexakis(urea)magnesium dibromide tetrakis(urea), 45B, 783
$C_{11}H_{21}N_2Na$, Cyclopentadienyl(tetramethylethylenediamine)sodium, 45B, 783
$C_{12}H_4CaN_6O_{14} \cdot 5\ H_2O$, Calcium picrate pentahydrate, 45B, 73
$C_{12}H_8AuKN_4$, Potassium dicyanoaurate(I) 2,2'-bipyridyl, 46B, 228
$C_{12}H_{12}MgN_2O_6$, Magnesium picolinate dihydrate, 39B, 516
$C_{12}H_{15}NaO_4$, (1-Phenylbutane-1,3-dionato)(ethylene glycol)sodium, 37B, 403
$C_{12}H_{16}MgN_2O_8$, Tetraaquo(isonicotinato-N)(isonicotinato-O)magnesium, 40B, 970
$C_{12}H_{18}Al_2Cl_8MgN_6$, Hexakis(methyl cyanide)magnesium tetrachloroaluminate, 38B, 916
$C_{12}H_{20}MgN_2O_{10}$, Magnesium p-nitrosophenolate hexahydrate, 43B, 147
$C_{12}H_{22}MgO_{12}S_2$, Magnesium hexa-aquo benzenesulphonate, 11, 643
$C_{12}H_{30}N_4O_{12}Sr$, Triethanolamine strontium nitrate, 40B, 657
$C_{12}H_{32}Al_2MgO_6$, Bis(dimethyldimethoxyaluminum)magnesium - dioxane, 33B, 301
$C_{12}H_{32}Mg_2N_4$, Bis(2-dimethylaminoethyl(methyl)amino)di(methylmagnesium), 34B, 399
$C_{12}H_{36}Be_3N_6$, Bis(dimethylamino)beryllium trimer, 34B, 401
$C_{13}H_{19}LiN_2$, Benzyllithium triethylenediamine, 35B, 474
$C_{13}H_{24}CsNO_6S$, 1,4,7,10,13,16-Hexaoxacyclooctadecane caesium thiocyanate complex, 40B, 298
$C_{13}H_{24}KNO_6S$, 1,4,7,10,13,16-Hexaoxacyclooctadecane potassium thiocyanate complex, 40B, 299
$C_{13}H_{24}NO_6RbS$, 1,4,7,10,13,16-Hexaoxacyclooctadecane rubidium thiocyanate complex, 40B, 300
$C_{13}H_{26}KN_3O_4S$, 1,7,10,16-Tetraoxa-4,13-diazacyclooctadecane potassium thiocyanate complex, 38B, 378
$C_{13}H_{26}NNaO_7S$, 1,4,7,10,13,16-Hexaoxacyclooctadecane sodium thiocyanate monohydrate, 40B, 300
$C_{13}H_{27}BrMgO_3$, Tris(tetrahydrofuran)bromo(methyl)magnesium, 40B, 1103
$C_{14}H_{11}KN_2O_5$, Potassium o-nitrophenolate-isonitrosoacetophenone, 37B, 404
$C_{14}H_{20}ClNaO_9$, 2,3,5,6,8,9,11,12-Octahydro-1,4,7,10,13-benzopentaoxacyclopentadecin sodium perchlorate, 46B, 1149
$C_{14}H_{22}INaO_6$, Aquo-(2,3-benzo-1,4,7,10,13-pentaoxacyclopentadec-2-ene)-sodium iodide, 38B, 338
$C_{14}H_{22}K_2O_3$, Dipotassium(I) cyclooctatetraenide-1-methoxy-2-(2-methoxyethoxy)ethane, 40B, 658
$C_{14}H_{22}O_3Rb_2$, Dirubidium(I) cyclooctatetraenide-1-methoxy-2-(2-methoxyethoxy)ethane, 41B, 182
$C_{14}H_{25}BrMgO_2$, Phenylbis(diethyl ether)magnesium bromide, 29, 532
$C_{14}H_{28}ILiN_2O_4$, Lithium cryptate, 39B, 517
$C_{14}H_{29}Cl_2Li_2N_5O_{15}$, Sesqui(acetamide)(diacetamide)(perchlorato)-lithium(I) dimer, 42B, 572
$C_{14}H_{30}Be_2N_2$, Methyl-1-propynyl-beryllium-trimethylamine dimer, 37B, 406

$C_{15}H_3F_{18}NaO_6Rb_2$, Dirubidium tris(hexafluoroacetylacetonato)sodate, 39B, 521
$C_{15}H_6CaO_9 \cdot 9\ H_2O$, 4,10-Dioxo-5-methoxy-2,8-dicarboxy-4H,10H-benzodipyran calcium salt nonahydrate, 45B, 335
$C_{15}H_{12}CsNO_5S$, 5-Phenacyl-2-(hydroxymethyl)-4H-pyran-4-one-caesium thiocyanate, 40B, 659
$C_{15}H_{23}LiN_2$, Indenyllithium tetramethylethylenediamine, 41B, 198
$C_{15}H_{45}As_5Cl_2MgO_{13}$, Pentakis(trimethylarsine oxide)magnesium(II) perchlorate, 42B, 884
$C_{15}H_{45}Cl_2MgO_{13}P_5$, Pentakis(trimethylphosphine oxide)magnesium(II) perchlorate, 44B, 656
$C_{15}H_{47}Cl_2MgO_{14}P_5$, Aquopentakis(trimethylphosphine oxide)-magnesium(II) perchlorate, 44B, 657
$C_{16}H_{13}KN_2O_4$, Potassium hydrogen bis(isonitrosoacetophenonate), 37B, 405
$C_{16}H_{16}CaN_{10}O_{16}$, Diaquobis(purpurato)calcium dihydrate, 43B, 512
$C_{16}H_{16}Cl_2MgO_8$, Diaquabis(p-chlorophenoxyacetato)magnesium(II), 46B, 1020
$C_{16}H_{18}MgO_8$, Diaquabis(phenoxyacetato)magnesium(II), 46B, 1020
$C_{16}H_{24}Be_6O_{18}$, Beryllium dioxooctaacetate, 42B, 573
$C_{16}H_{25}ClNNaO_8$, N-Phenyl-4,7,10,13-tetraoxa-1-aza-cyclopentadecane sodium perchlorate, 46B, 717
$C_{16}H_{28}Br_2CaN_4O_8$, Diacetamide-calcium bromide, 35B, 746
$C_{16}H_{28}Cl_2Na_2O_{16}$, Bis(diacetamide)(perchlorato)sodium(I) dimer, 42B, 574
$C_{16}H_{30}CaCl_2N_4O_{15}$, Tetrakis(diacetamide)calcium chlorate monohydrate, 44B, 23
$C_{16}H_{30}Cl_2N_4O_{16}Sr$, Tetrakis(diacetamide)monoaquostrontium(II) perchlorate, 42B, 574
$C_{16}H_{30}N_2O_8S_2Sr$, Heptaethyleneglycol strontium thiocyanate, 45B, 91
$C_{16}H_{32}BeN_2$, Dimethylbis(quinuclidine)beryllium, 37B, 407
$C_{16}H_{32}Br_2MgO_4$, Tetrahydrofurane magnesium bromide, 33B, 301
$C_{16}H_{32}ClNaO_8 \cdot 0.5\ H_2O$, Bis(1,4,7,10-tetraoxacyclododecane)sodium chloride hemihydrate, 40B, 660
$C_{16}H_{32}Cl_2MgN_4O_{18}$, Bis(diacetamide)diaquomagnesium(II) didiacetamide perchlorate, 43B, 900
$C_{16}H_{32}MgN_2$, Dimethylbis(quinuclidine)magnesium, 35B, 709
$C_{16}H_{36}BaN_2O_{10}$, Triethanolamine barium acetate, 40B, 660
$C_{16}H_{36}Br_2MgO_6$, Diaquotetrakis(tetrahydrofuran)magnesium bromide, 43B, 408
$C_{16}H_{38}Br_2Mg_2O_2$, Ethylmagnesium bromide - diisopropyl ether dimer, 40B, 661
$C_{16}H_{40}Br_6Mg_4O_5$, Tetrakis(diethyl ether)hexabromo-oxo-tetramagnesium, 29, 532
$C_{16}H_{44}B_2Be_3O_4$, Tetra-t-butoxytriberyllium bis(tetrahydridoborate), 45B, 36
$C_{16}H_{44}Li_4N_4$, Tetrakis(methyllithium)-bis(tetramethylethylenediamine), 44B, 1110
$C_{16}H_{48}Be_4O_4Si_4$, closo-Tetrakis(methyl(trimethylsilanolato)beryllium), 34B, 368
$C_{16}H_{49}NaO_{17}$, Bis(1,4,7,10-tetraoxacyclododecane)sodium hydroxide octahydrate, 40B, 662
$C_{17}H_{18}CaN_4O_8$, Dinitrato[N,N'-propane-1,3-diylbis(salicyldeneimine)]-calcium(II), 45B, 65
$C_{17}H_{23}CsN_2O_8S$, 4-Nitrobenzo-1,4,7,10,13,16-hexaoxacyclooctadecane-caesium thiocyanate, 43B, 410
$C_{18}H_{13}KN_2O_2$, Potassium quinolin-8-olate quinolin-8-ol (1:1), 45B,

276
$C_{18}H_{14}Mg$, Bis(indenyl)magnesium, 40B, 662
$C_{18}H_{17}Be_3N_3O_{10}$, Tri-μ-hydroxo-tri(pyridine-2-carboxylato)triberyllium monohydrate, 40B, 663
$C_{18}H_{26}MgN_2$, Diphenyl(N,N,N',N'-tetramethylethylenediamine)magnesium, 44B, 657
$C_{18}H_{36}Cs_2N_2S_4$, Caesium di-n-butyldithiocarbamate, 34B, 401
$C_{18}H_{36}IKN_2O_6$, Potassium cryptate, 39B, 518
$C_{18}H_{36}IN_2NaO_6$, Sodium cryptate, 39B, 519
$C_{18}H_{36}N_2Na_2O_6$, Disodium[2.2.2]-cryptate, 40B, 664
$C_{18}H_{42}Br_2CaN_2O_9$, Calcium cryptate, 39B, 519
$C_{18}H_{54}Li_3N_3Si_6$, Cyclo-tris(bis(trimethylsilyl)amidolithium), 34B, 372
$C_{19}H_{11}Li \cdot 0.5\ C_6H_6$, Lithium 7bH-indeno[1,2,3-jk]fluorenide benzene solvate, 45B, 166
$C_{19}H_{25}KN_2$, Fluorenylpotassium tetramethylethylenediamine, 40B, 665
$C_{19}H_{38}CsN_3O_7S$, Caesium cryptate, 39B, 518
$C_{19}H_{38}N_3O_7RbS$, Rubidium cryptate, 39B, 518
$C_{20}H_{16}BeN_2O_2 \cdot 2\ H_2O$, Bis(2-methyl-8-hydroxyquinolinato)beryllium dihydrate, 45B, 784
$C_{20}H_{18}MgN_8O_{10}$, Di(N-methylimidazole)bis(2,4-dinitrophenoxide)-magnesium(II), 45B, 1266
$C_{20}H_{20}Cl_2MgN_4$, Magnesium chloride - pyridine(1:4), 45B, 241
$C_{20}H_{35}BaCl_2N_5O_{18}$, Pentakis(diacetamide)barium(II) perchlorate, 41B, 20
$C_{20}H_{35}CaCl_2N_5O_{18}$, Calcium perchlorate - diacetamide (1:5), 42B, 575
$C_{20}H_{38}BaN_4O_7S_2$, Barium cryptate, 39B, 520
$C_{20}H_{38}MgN_2$, Bis(2,4-dimethyl-2,4-pentadienyl)(N,N,N',N'-tetramethylethylenediamine)magnesium, 46B, 718
$C_{20}H_{40}BaBr_2O_{10} \cdot 2\ H_2O$, Bis(1,4,7,10,13-pentaoxacyclopentadecane) barium bromide dihydrate, 45B, 351
$C_{20}H_{40}BrNaO_8$, Perhydrodibenzo[b,k](1,4,7,10,13,16)hexaoxacyclooctadecin sodium bromide, 39B, 522
$C_{21}H_{24}KN_3O_9S$, 1,11-Bis(2-nitrophenoxy)-3,6,9-trioxaundecane potassium isothiocyanate, 46B, 107
$C_{22}H_{16}MgN_6O_{10}$, Dipyridinebis(2,4-dinitrophenoxide)magnesium(II), 45B, 1266
$C_{22}H_{20}IN_2O_3Rb$, Bis(8-quinolyloxyethyl)ether rubidium iodide, 45B, 284
$C_{22}H_{26}IO_7Rb$, 1,11-Bis(tropolone)-3,6,9-trioxaundecane rubidium iodide, 46B, 107
$C_{22}H_{40}Li_2N_4$, Bis((tetramethylethylenediamine)lithium(I))naphthalenide, 38B, 113
$C_{22}H_{44}BaN_4O_9S_2$, Barium cryptate, 39B, 520
$C_{23}H_{30}KNO_6S$, 2,3-Naphtho-20-crown-6-potassium thiocyanate, 43B, 422
$C_{24}H_{32}BaCl_2N_6O_{10}$, Bis(perchlorato-(1,12-bis(2-acetylpyridine)-1,12-dimethyl-2,5,8,11-tetra-azadodeca-1,11-diene)barium, 46B, 1189
$C_{24}H_{40}Li_2N_4$, Acenaphthylenebis(N,N,N',N'-tetramethylethylenediamine-lithium), 43B, 219
$C_{24}H_{42}Li_2N_4$, Di-μ-phenyl-bis(N,N,N',N'-tetramethylethylenediamine-lithium), 44B, 656
$C_{24}H_{44}K_2O_6$, Bis(potassium-2,2'-dimethoxydiethylether)-1,3,5,7-tetramethylcyclooctatetraene, 40B, 666
$C_{24}H_{72}Cl_2MgN_{12}O_{17}P_6$, Trisoctamethylpyrophosphoramidemagnesium(II) perchlorate, 35B, 742
$C_{25}H_{31}LiN_2$, Triphenylmethyllithium tetramethylethylenediamine, 38B, 114

$C_{25}H_{31}N_2Na$, Sodium triphenylmethanide tetramethylethylenediamine, 45B, 29
$C_{25}H_{32}CsNO_6S$, (7R,9R,18S,20S)-6,7,9,10,17,18,20,21-Octahydro-7,9,18,20-tetramethyldibenzo[b,k](1,4,7,10,13,16)hexaoxacyclooctadecin caesium thiocyanate, 41B, 388
$C_{26}H_{24}KN_3O_{16}$, (2,2'-Di(o-carboxymethoxyphenoxy)diethylether)potassium picrate, 43B, 152
$C_{26}H_{28}IN_2O_5Rb$, Bis[(8-quinolyloxy)ethoxyethyl]ether rubidium iodide, 45B, 287
$C_{26}H_{32}K_2N_2O_2S_2$, Dibenzo-24-crown-8 - potassium isothiocyanate, 39B, 522
$C_{26}H_{42}Li_2N_4$, Bis((tetramethylethylenediamine)-lithium(I))anthracenide, 41B, 195
$C_{26}H_{44}Li_2N_4$, Stilbene bis(lithium tetramethylethylenediamine), 42B, 576
$C_{26}H_{52}Mg_2O_4$, 1,7-Dimagnesiacyclododecane tetrakis(tetrahydrofuran), 43B, 901
$C_{27}H_{20}KN_3O_3$, Potassium quinolin-8-olate quinolin-8-ol (1:2), 45B, 276
$C_{27}H_{35}LiN_2$, Fluorenyllithium bisquinuclidine, 38B, 115
$C_{27}H_{81}As_9Ca_2Cl_4O_{25}$, Tri-μ-(trimethylarsine oxide)-hexakis(trimethylarsine oxide)dicalcium(II) tetraperchlorate, 43B, 895
$C_{28}H_{24}ClNaO_{10}$, Bis(phenacylkojate)sodium chloride, 42B, 577
$C_{28}H_{24}IKO_{10}$, Bis(5-phenacyloxy-2-hydroxymethyl-4H-pyran-4-one)potassium iodide, 40B, 668
$C_{28}H_{32}IN_2O_5Rb$, Bis(2-methyl-8-quinolyloxyethoxyethyl)ether-rubidium iodide, 45B, 289
$C_{28}H_{40}ClNaO_{14}$, Bis(2,3,5,6,8,9,11,12-octahydro-1,4,7,10,13-benzopentaoxacyclopentadecin) sodium perchlorate, 46B, 1149
$C_{28}H_{40}Cl_4Na_2O_{12}U$, Di[(benzo-15-crown-5)sodium] tetrachlorodioxouranate, 45B, 364
$C_{28}H_{40}IKO_{10}$, Benzo-15-crown-5-potassium iodide, 38B, 344
$C_{28}H_{40}IKO_{10}$, Dibenzo-30-crown-10-potassium iodide, 38B, 346
$C_{28}H_{40}MgN_4O_6P_2S_2$, Tetrapyridine-bis(diethylphosphorothioato)magnesium, 39B, 807
$C_{28}H_{58}Cl_6Mg_4O_6$, Grignard reagents, 37B, 408
$C_{29}H_{22}F_{18}MgN_2O_6$, Tris(hexafluoroacetylacetonato)magnesium 1-dimethylammonium-8-dimethylaminonaphthalene, 38B, 831
$C_{29}H_{40}NO_{10}RbS \cdot H_2O$, Dibenzo[b,q][1,4,7,10,13,16,19,22,25,28]decaoxacyclotriacontane rubidium thiocyanate, 45B, 366
$C_{30}H_{20}N_5NaO_3$, o-Nitrophenolatobis(1,10-phenanthroline)sodium, 39B, 523
$C_{30}H_{20}N_5O_3Rb$, o-Nitrophenolatobis(1,10-phenanthroline)rubidium, 39B, 523
$C_{30}H_{22}CaO_4 \cdot 0.5\ C_2H_6O$, Bis-(1,3-diphenyl-1,3-propanedionato)calcium hemiethanolate, 39B, 524
$C_{30}H_{22}O_4Sr \cdot 0.5\ C_3H_6O$, Bis-(1,3-diphenyl-1,3-propanedionato)strontium hemiacetonate, 39B, 525
$C_{30}H_{30}Br_2MgN_6$, Magnesium bromide hexapyridine, 43B, 901
$C_{30}H_{34}I_2MgN_6O_2$, trans-Diaquotetrakis(pyridine)magnesium iodide bis-(pyridine), 45B, 247
$C_{31}H_{36}NO_9S(Na,Rb)$, 2,3,11,12-Dibenzo-1,4,7,10,13,16,-hexaoxocyclooctadeca-2,11-diene(dibenzo-18-crown-6)-rubidium sodium isothiocyanate complex, 35B, 476
$C_{32}H_{40}IN_2O_8Rb$, 1,20-Bis(8-quinolyloxy)-3,6,9,12,15,18-hexaoxaicosane rubidium iodide, 45B, 290
$C_{32}H_{46}N_{20}O_{39}Sr_2$, Di-μ-aquo-tetrakis(purpurato)distrontium trideca-

hydrate, 43B, 512
$C_{32}H_{54}KO_{10}$, Potassium biphenyl bis(2,5,8,11,14-pentaoxapentadecane), 44B, 315
$C_{32}H_{54}O_{10}Rb$, Bis(tetraglyme)rubidium biphenyl, 42B, 67
$C_{32}H_{58}Li_2N_6$, Stilbene bis(lithium pentamethyldiethylenetriamine), 42B, 576
$C_{32}H_{64}I_2N_4Na_2O_{10}$, Sodium (3)-cryptate diiodide, 43B, 466
$C_{32}H_{76}Ba_2Cl_4N_8O_{26}$, Bis($\mu$-(N,N-dimethylacetamide))bis(aquatris(N,N-dimethylacetamide)diperchloratobarium), 46B, 1159
$C_{36}H_{36}IN_4O_3Rb$, Tris[(2-methyl-8-quinolyloxy)ethyl]aminerubidium iodide, 46B, 1194
$C_{36}H_{36}MgN_2O_6$, Bis(dimethylformamido)bis-(1,3-diphenyl-1,3-propanedionato)magnesium, 39B, 525
$C_{36}H_{61}NaO_{11}$, Sodium monensin, 46B, 718
$C_{36}H_{61}NaO_{11} \cdot 2\ H_2O$, Sodium monensin dihydrate, 46B, 718
$C_{36}H_{72}Be_2N_4$, Bis[bis(di-t-butylmethyleneamino)beryllium], 45B, 785
$C_{36}H_{72}Bi_4K_2N_4O_{12}$, 4,7,13,16,21,24-Hexaoxa-1,10-diazabicyclo[8.8.8]-hexacosane-potassium tetrabismuthide(2-), 43B, 467
$C_{36}H_{72}CaCl_2N_4O_8$, N,N,N',N'-Tetrapropyl-3,6-dioxaoctanediamide calcium chloride, 43B, 902
$C_{36}H_{72}N_4Na_2O_{12}Pb_5$, 4,7,13,16,21,24-Hexaoxa-1,10-diazabicyclo[8.8.8]-hexacosane-sodium pentaplumbide(2-), 43B, 467
$C_{36}H_{72}N_4Na_2O_{12}Sn_5$, 4,7,13,16,21,24-Hexaoxa-1,10-diazabicyclo[8.8.8]-hexacosane-sodium pentastannide(2-), 43B, 467
$C_{36}H_{90}Mg_6N_{10}$, μ_6-Nitrido-nonakis(μ-t-butylamino)hexamagnesium, 46B, 719
$C_{38}H_{40}K_2N_4O_{13}S_2$, 1,5-Bis{2-[5-(2-nitrophenoxy)-3-oxapentyloxy]phenoxy}-3-oxapentane potassium, 46B, 110
$C_{38}H_{48}Li_2N_4$, Δ(9,9')-Bifluorenyl bis(lithium tetramethylenediamine), 41B, 208
$C_{38}H_{62}Fe_2Li_4N_6$, Dilithioferrocene-pentamethyldiethylenetriamine, 44B, 658
$C_{38}H_{66}Li_2O_4$, Di-(μ-2,6-di-tert-butyl-4-methylphenoxo)bis(diethylether lithium), 46B, 720
$C_{38}H_{80}K_2N_6O_{12}Te_3$, (4,7,13,16,21,24-Hexaoxa-1,10-diazabicyclo[8.8.8]-hexacosane)potassium tritelluride(2-)-ethylenediamine, 43B, 468
$C_{40}H_{36}Co_3N_4NaO_9$, Bis(N,N'-ethylenebis(salicylideniminato)cobalt)-(tetracarbonylcobalt(I))(tetrahydrofuran)sodium, 44B, 658
$C_{40}H_{38}ClCu_2N_4NaO_8$, Bis-(N,N-ethylenebis(salicylideneiminato)copper-(II))-perchloratosodium p-xylene solvate, 40B, 668
$C_{40}H_{45}NaO_2P_2$, Sodium (diphenylphosphoniumbenzylide)(diphenylphosphonomethylide) tetrahydrofuran diethyl ether, 46B, 698
$C_{40}H_{64}Al_2Na_2O_4$, Tetra(1,4-epoxybutane)disodium(I) tetramethylbis-1,4-dihydro-1,4-naphthylenedialuminate, 35B, 478
$C_{40}H_{67}NaO_{11}$, Sodium nigericine, 46B, 348
$C_{41}H_{64}NNaO_{12}S$, Nonactin sodium thiocyanate complex, 40B, 577
$C_{42}H_{52}CaN_4O_{25}$, Calcium bis(benzo-15-crown-5) bis(3,5-dinitrobenzoate) trihydrate, 43B, 903
$C_{48}H_{40}BaCl_2N_8O_{12}$, Tetraaquobis(1,10-phenanthroline)barium(II) perchlorate bis(1,10-phenanthroline), 43B, 379
$C_{48}H_{40}Cl_2N_8O_{12}Sr$, Tetraaquobis(1,10-phenanthroline)strontium(II) perchlorate bis(1,10-phenanthroline), 43B, 379
$C_{48}H_{60}ClN_6Na$, Tris(p,p'-diamino-2,3-diphenylbutane)sodium chloride, 40B, 669
$C_{48}H_{60}N_7NaO_3$, Tris(p,p'-diamino-2,3-diphenylbutane)sodium nitrate, 40B, 669
$C_{48}H_{62}Li_2Mg_2N_4$, Lithium di-μ-phenyl-bis(diphenylmagnesium)bis-

(N,N,N',N-tetramethylethylenediamine), 44B, 659
$C_{48}H_{78}Li_6$, Cyclohexyllithium hexamer benzene solvate, 40B, 670
$C_{49}H_{60}N_7Na$, Tris(p,p'-diamino-2,3-diphenylbutane)sodium cyanide, 40B, 669
$C_{49}H_{64}CsNO_{12}S$, (7R,9R,18R,20R)-6,7,9,10,17,18,20,21-Octahydro-7,9,18,20-tetramethyldibenzo[b,k](1,4,7,10,13,16)hexaoxacyclooctadecin caesium thiocyanate, 41B, 388
$C_{50}H_{66}K_2N_6O_{15}S_2$, 1,11-Bis(2-acetylaminophenoxy)-3,6,9-trioxaundecane potassium thiocyanate hydrate, 46B, 108
$C_{52}H_{60}BNaO_{10}$, Bis(2,3,5,6,8,9,11,12-octahydro-1,4,7,10,13-benzopentaoxacyclopentadecin) sodium tetraphenylborate, 46B, 1149
$C_{54}H_{108}As_{11}K_3N_6O_{18}$, Tris[(4,7,13,16,21,24-hexaoxa-1,10-diazobicyclo-[8.8.8]hexacosane)potassium] undecaarsenide, 46B, 720
$C_{54}H_{108}N_6Na_3O_{18}Sb_7$, (4,7,13,16,21,24-Hexaoxa-1,10-diazabicyclo-[8.8.8]hexacosane)sodium heptantimonide(3-), 42B, 578
$C_{64}H_{64}BCo_2N_4NaO_6$, Bis-(N,N'-ethylene-bis(salicylideneiminato)cobalt-(II))-bis-tetrahydrofuran-sodium(I) tetraphenylborate, 40B, 671
$C_{66}H_{72}Cl_2MgN_{12}O_{14}$, Magnesium hexa-antipyrine perchlorate, 32B, 324
$C_{68}H_{88}BrN_{10}NaO_{11}$, Sodium (phe(4)-val(6))antamanide bromide ethanol solvate, 40B, 671
$C_{70}H_{87}BrLiN_{13}O_{10}$, Antamanide acetonitrile lithium bromide acetonitrile solvate, 40B, 672
$C_{72}H_{144}N_8Na_4O_{24}Sn_9$, Tetrakis(4,7,13,16,21,24-hexaoxa-1,10-diazabicyclo[8.8.8]hexacosane-sodium) nonastannide(4-), 43B, 468
$C_{81}H_{76}INaO_8P_6$, Tris(bis(diphenylphosphinyl)methane)sodium iodide dihydrate benzene solvate, 42B, 578
$C_{113}H_{236}Ge_{18}K_6N_{17}O_{36}$, Hexakis(4,7,13,16,21,24-hexaoxa-1,10-diazabicyclo[8.8.8]hexacosane-potassium) nonagermanide(2-) nonagermanide(4-) ethylenediamine, 43B, 469

68. GROUP III COMPOUNDS

CH_3AlCl_3K, Potassium methyltrichloroaluminate, 40B, 673
CH_3Cl_2In, Methylindium dichloride, 41B, 771
$C_2H_6AlCsI_2 \cdot C_8H_{10}$, Cesium diiododimethylaluminate p-xylene solvate, 46B, 721
$C_2H_6Al_2Cl_4$, Methylaluminum chloride dimer, 28, 596
C_2H_6BrIn, Dimethylindium bromide, 42B, 579
C_2H_6BrTl, Dimethylthallium bromide, 3, 695
C_2H_6ClIn, Chlorodimethylindium, 40B, 674
C_2H_6ClTl, Dimethylthallium chloride, 40B, 674
$C_2H_6N_3Tl$, Dimethylthallium azide, 41B, 772
$C_2H_{10}AlB$, Dimethylaluminum tetrahydroborate, 45B, 785
$C_2H_{10}BGa$, Dimethylgallium tetrahydroborate, 45B, 785
$C_3H_5AlCl_4O$, Aluminum chloride - propionyl chloride (1:1), 38B, 693
C_3H_6NOTl, Dimethylthallium cyanate (orthorhombic form), 41B, 772
C_3H_6NOTl, Dimethylthallium cyanate (rhombohedral form), 41B, 772
$C_3H_6NO_2STl$, DL-Cysteinatothallium(I), 43B, 587
C_3H_6NSTl, Dimethylthallium thiocyanate (monoclinic form), 41B, 772
C_3H_6NSTl, Dimethylthallium thiocyanate (orthorhombic form), 41B, 772
$C_3H_6NS_2Tl$, Thallium(I) dimethyldithiocarbamate, 39B, 536
C_3H_6NTl, Dimethylthallium cyanide, 41B, 772
C_3H_9Al, Trimethylaluminum (gas-ed), 37B, 686
$C_3H_9AlCl_3N$, Trichlorotrimethylaminealuminum (gas), 39B, 886
$C_3H_9AlCl_3N$, Trichlorotrimethylaminealuminum(III), 34B, 403
$C_3H_9AlCsN_3$, Caesium azidotrimethylaluminate, 41B, 774

$C_3H_9AlN_3Rb$, Rubidium azidotrimethylaluminate, 43B, 904
C_3H_9AlS, Dimethyl(methylthiol)aluminum, 34B, 403
$C_3H_9Cl_3GaP$, Trimethylphosphine-gallium trichloride, 44B, 660
C_3H_9Ga, Trimethylgallium (gas-ed), 40B, 1138
C_3H_9In, Trimethylindium (gas-ed), 8, 286; 40B, 1136
C_3H_9In, Trimethylindium, 22, 571
C_3H_9Tl, Trimethylthallium, 35B, 508
$C_3H_{10}AlK$, Potassium hydridotrimethylaluminate, 40B, 675
$C_3H_{12}AlN$, Trimethylamine-alane (gas-ed), 38B, 1062
$C_3H_{12}GaN$, Trimethylaminegallane, 28, 596
$C_3H_{21}AlB_3N$, Aluminum tetrahydroborate - trimethylamine, 33B, 303
$C_4H_6LiO_7Tl$, Lithium thallium tartrate monohydrate, 44B, 20
$C_4H_6N_3Tl$, Dimethylthallium dicyanamide, 41B, 776
C_4H_9AlKN, Potassium cyanotrimethylaluminate, 39B, 527
$C_4H_9InO_2$, Acetato(dimethyl)indium(III), 39B, 531
$C_4H_9OS_2Tl$, Dimethylthallium methylxanthogenate, 42B, 580
$C_4H_9O_2Tl$, Dimethylthallium acetate, 41B, 775
$C_4H_{10}PSSeTl$, Diethylthioselenophosphinatothallium(I), 39B, 537
$C_4H_{10}PS_2Tl$, Diethyldithiophosphinatothallium(I), 40B, 997
$C_4H_{12}AlRb$, Rubidium tetramethylaluminate, 39B, 526
$C_4H_{12}Al_2Br_4N_2$, Dibromo(dimethylamino)aluminum dimer, 44B, 660
$C_4H_{12}Al_2Cl_2$, Dimethylaluminum chloride dimer (gas-ed), 40B, 1134
$C_4H_{12}Al_2Cl_4N_2$, Dichloro(dimethylamino)alane dimer, 43B, 905
$C_4H_{12}Al_2I_4N_2$, Diiodo(dimethylamino)aluminum dimer, 44B, 660
$C_4H_{12}CsIn$, Caesium tetramethylindate, 39B, 533
$C_4H_{12}GaNO$, Dimethyl-(2-aminoethanolato)-gallium, 45B, 786
$C_4H_{12}InK$, Potassium tetramethylindate, 39B, 533
$C_4H_{12}InLi$, Lithium tetramethylindate, 38B, 693
$C_4H_{12}InNa$, Sodium tetramethylindate, 38B, 693
$C_4H_{12}InRb$, Rubidium tetramethylindate, 39B, 533
$C_4H_{14}AlCl_3N_2$, Aluminum trichloride - dimethylamine (2:1), 43B, 905; 44B, 661
$C_4H_{14}Al_2$, Dimethylaluminum hydride dimer (gas-ed), 38B, 1061
$C_4H_{16}ClN_8O_3S_4Tl$, Thiourea - thallous chlorate, 35B, 791
$C_4H_{16}ClN_8O_4S_4Tl$, Tetrakis(thiourea)thallium(I) perchlorate, 32B, 54
$C_4H_{16}N_9O_3S_4Tl$, Tetrakis(thiourea)thallium(I) nitrate, 32B, 54
$C_4H_{24}Al_2B_4N_2$, Di-μ-ethyleneimino-bis(bis(tetrahydroborato)aluminum), 42B, 580
$C_4H_{24}Cl_7InN_4$, Tetra(methylammonium) hexachloroindium chloride, 38B, 38
$C_5HF_6O_2Tl$, Hexafluoroacetylacetonatothallium(I), 41B, 51
C_5H_5In, Cyclopentadienylindium (gas-ed), 29, 625
C_5H_5In, Cyclopentadienylindium(I), 28, 598
$C_5H_9GaO_4$, Methyl-diacetato-gallium, 44B, 1111
C_5H_9In, Dimethylpropynylindium, 44B, 1111
$C_5H_{10}NS_2Tl$, Thallium(I) diethyldithiocarbamate, 41B, 76
$C_5H_{12}AlN$, Acetonitriletrimethylaluminum, 39B, 528
$C_5H_{15}AlIN$, Iododimethyl(trimethylamine)aluminum, 39B, 526
$C_5H_{15}AlO$, Trimethylaluminum dimethyl ether (gas-ed-ed), 43B, 1484
$C_5H_{15}AsCl_3Ga$, Tetramethylarsonium trichloromethylgallate, 43B, 888
$C_5H_{15}AsCl_3In$, Tetramethylarsonium methyltrichloroindate, 42B, 581
$C_5H_{15}Cl_3InSb$, Tetramethylstibonium methyltrichloroindate, 42B, 581
$C_6H_2N_3O_7Tl$, Thallium(I) picrate (red polymorph), 43B, 128
$C_6H_4N_3Tl$, (Benzotriazolato)thallium(I), 45B, 255
$C_6H_5Cl_2Tl$, Chloro(phenyl)thallium chloride, 17, 700; 20, 212
$C_6H_6AlK_3O_{15}$, Potassium tris(oxalato)aluminate trihydrate, 44B, 864
$C_6H_6Al_2Br_6$, Benzene - aluminum bromide, 26, 653

$C_6H_6D_4N_4Ga_2$, Dideuterio(pyrazol-1-yl)gallane dimer, 39B, 530
$C_6H_6N_3Tl$, Dimethylthallium tricyanomethide, 41B, 776
$C_6H_9O_6Tl$, Thallium(III) triacetate, 44B, 661
$C_6H_{12}Ga_2O_4$, Bis(dimethylgallium)oxalate, 40B, 676
$C_6H_{13}InOS$, Diethylindium thioacetate, 39B, 532
$C_6H_{13}InO_2$, Diethylindium acetate, 39B, 532
$C_6H_{15}AlCl_2N_2$, N-Ethyl-N',N'-dimethylethylenediaminodichloro-aluminum(III), 46B, 721
$C_6H_{15}AlNSTl$, Dimethylthallium isothiocyanatotrimethylaluminate, 40B, 675
$C_6H_{18}AlCl_2N_2PSi_2$, 1,3-Bis(trimethylsilyl)-4,4-dichloro-1,3-diaza-2-phospha-4-alacyclobutane, 43B, 906
$C_6H_{18}AlCl_2N_2PSi_2$, 4,4'-Dichloro-1,3-bis(trimethylsilyl)-1,3-diaza-2-phosphonia-4-aluminatacyclobutane, 45B, 715
$C_6H_{18}AlCl_4N_3Si$, (μ-Dimethylamino)[bis(dimethylamino)chlorosilane]-(trichloroaluminum), 46B, 721
$C_6H_{18}AlN$, Trimethylaluminum-trimethylamine (gas-ed), 38B, 1061
$C_6H_{18}Al_2$, Hexamethyldialuminum (gas-ed), 37B, 686
$C_6H_{18}Al_2$, Trimethylaluminum dimer, 10, 214; 16, 473; 32B, 327
$C_6H_{18}Al_2Br_4O_2Si_2$, Dibromotrimethylsiloxyaluminum, 32B, 326
$C_6H_{18}Al_2KNO_3$, Potassium nitratobis(trimethylaluminate), 44B, 1112
$C_6H_{18}Al_2KN_3$, Potassium μ-azido(bistrimethylaluminate), 40B, 677
$C_6H_{18}Al_2N_2O_2$, {[N-Methyl-N-(nitroso-O')hydroxylamido-O]dimethylaluminum-O'}-trimethylaluminum, 46B, 722
$C_6H_{18}AsBr_2In$, Tetramethylarsonium dibromodimethylindate, 42B, 581
$C_6H_{18}AsCl_2Ga$, Tetramethylarsonium dichlorodimethylgallate, 43B, 888
$C_6H_{18}Ga_3N_3$, Aziridinylgallane trimer, 38B, 694
$C_6H_{19}AlN_2$, Aluminum hydride - N,N,N',N'-tetramethylethylenediamine, 29, 536
$C_6H_{21}AlN_2$, Aluminum hydride - trimethylamine, 28, 598
$C_6H_{24}AlK$, Potassium bis(3,8-cis-cyclooctenyl)aluminate, 44B, 1115
$C_6H_{24}Al_3N_3$, Dimethylaminoalane trimer, 38B, 695
$C_7H_5AlCl_4O$, Benzoyl chloride-aluminum chloride complex, 31B, 326
$C_7H_{11}Al$, Dimethyl(cyclopentadienyl)aluminum (gas-ed), 39B, 887
$C_7H_{11}Ga$, Dimethyl(cyclopentadienyl)gallium, 42B, 581
$C_7H_{13}O_2Tl$, Dimethylthallium acetylacetonate, 41B, 775
$C_7H_{14}NS_2Tl$, Thallium(I) di-n-propyldithiocarbamate, 34B, 404
$C_7H_{18}Al_2KNS$, Potassium thiocyanato-bis(trimethylaluminate), 45B, 786
$C_7H_{21}AlIN$, Tetramethylammonium iodotrimethylaluminate, 46B, 723
$C_7H_{21}AsClGa$, Tetramethylarsonium chlorotrimethylgallate, 43B, 906
$C_7H_{21}AsClIn$, Tetramethylarsonium chlorotrimethylindate, 43B, 906
$C_8H_7AlCl_4O$, m-Toluoyl chloride - aluminum chloride, 38B, 696
$C_8H_7AlCl_4O$, o-Toluoyl chloride - aluminum chloride, 38B, 696
$C_8H_7AlCl_4O$, p-Toluoyl chloride - aluminum chloride, 38B, 696
$C_8H_{12}AlCl_3$, Tetramethylcyclobutadiene aluminum chloride complex, 40B, 677
$C_8H_{12}O_8Tl_2$, Thallium(I) tetraacetatothallate(III), 46B, 1013
$C_8H_{16}ClGaO_4$, Bis(dioxan)gallium(I) chloride, 26, 693
$C_8H_{16}Cl_4Ga_2O_4$, Bis[(1,4-dioxane)dichlorogallium](Ga-Ga), 45B, 787
$C_8H_{20}AlLi$, Lithium aluminum tetraethyl, 29, 535
$C_8H_{24}Al_2N_2$, Cyclodi-μ-dimethylamido-bis(dimethylaluminum), 35B, 479; 38B, 697
$C_8H_{24}Al_4F_4$, Dimethylaluminum fluoride tetramer (gas-ed), 39B, 892
$C_8H_{24}Ga_2N_2O_2$, N,N-Dimethylethanolaminogallane dimer, 41B, 777
$C_8H_{24}In_2N_2$, Dimethylaminodimethylindium dimer, 43B, 907
$C_8H_{28}Ga_4O_4$, Dimethylgallium hydroxide tetramer, 23, 690
$C_9H_8BrCrInO_6$, Bromo(pentacarbonylchromium)tetrahydrofuranindium,

43B, 389
$C_9H_{11}O_2Tl$, Dimethylthallium tropolonate, 41B, 775
$C_9H_{15}AlKNO_3$, Potassium nitratotrimethylaluminate benzene solvate, 44B, 1112
$C_9H_{18}NS_2Tl$, Thallium(I) diisobutyldithiocarbamate, 41B, 78
$C_9H_{18}NS_2Tl$, Thallium(III) dibutyldithiocarbamate, 42B, 765
$C_9H_{20}N_3OS_6Tl$, Tris(N,N-dimethyldithiocarbamato)thallium(III) monohydrate, 41B, 779
$C_9H_{23}GaP_2$, Dimethylgallium-(methanidobis(dimethylphosphonium-methylide)), 43B, 908
$C_9H_{24}AlNO_2$, Tetramethylammonium acetatotrimethylaluminate, 43B, 909
$C_9H_{24}Ga_2N_2O$, (N,N'-Dimethylacetylhydrazino)dimethylgallium trimethylgallium, 46B, 723
$C_9H_{27}Al_3O_3$, Dimethylaluminum methoxide trimer (gas-ed), 39B, 892
$C_9H_{30}Al_3N_3$, cis-Cyclotri-μ-methylamido-tris(dimethylaluminum), 38B, 697
$C_9H_{30}Al_3N_3$, trans-Cyclotri-μ-methylamido-tris(dimethylaluminum), 38B, 697
$C_{10}H_4N_3Tl$, Thallium(I) tricyanovinylcyclopentadienide, 43B, 167
$C_{10}H_{11}GaO_3W$, π-Cyclopentadienyl(dimethylgallium)tricarbonyltungsten, 43B, 909
$C_{10}H_{14}GaNO$, N-Methylsalicylaldiminato-dimethylgallium, 46B, 724
$C_{10}H_{15}Al_3Cl_9N_5 \cdot C_2H_3N$, Pentakis(acetonitrile)chloroaluminum bis-(tetrachloroaluminate) acetonitrile solvate, 45B, 787
$C_{10}H_{16}AlCsI_2$, Cesium diiododimethylaluminate p-xylene solvate, 45B, 788
$C_{10}H_{16}AlKN_2O_{10}$, Potassium ethylenediaminetetraacetatoaluminum(III) dihydrate, 35B, 602
$C_{10}H_{18}Ga_2N_4$, Pyrazolylgallium dimethyl dimer, 41B, 779
$C_{10}H_{22}AlN$, Trimethyl(quinuclidine)aluminum, 37B, 409
$C_{10}H_{24}Al_2N_2$, (Isopropylideneamino)dimethylaluminum dimer, 40B, 678
$C_{10}H_{24}Al_2N_4$, (N,N',N'',N'''-Tetramethyl-oxalamidinato)-bis(dimethylaluminum), 45B, 789
$C_{10}H_{24}Ga_2N_2$, (Isopropylideneamino)dimethylgallane dimer, 43B, 910
$C_{10}H_{24}Ga_2N_2O_4$, N-Methyldiethanolaminogallane dimer, 40B, 679
$C_{10}H_{24}Ga_2N_4$, (N,N',N'',N'''-Tetramethyl-oxalamidinato)-bis(dimethylgallium), 45B, 789
$C_{10}H_{24}In_2N_2$, (Isopropylideneamino)dimethylindium dimer, 45B, 790
$C_{10}H_{24}In_2N_4$, (N,N',N'',N'''-Tetramethyl-oxalamidinato)-bis(dimethylindium), 45B, 789
$C_{10}H_{26}Al_2O_2$, Bis(trimethylaluminum)dioxanate, 32B, 328
$C_{10}H_{27}Ga_2N_3$, (N,N',N"-Trimethylacetimidohydrazino)dimethylgallium trimethylgallium, 46B, 723
$C_{10}H_{30}Al_8Cl_4N_6$, Aluminum-nitrogen cage compound, 35B, 480
$C_{11}H_{15}O_2Tl$, Diethyl(salicylaldehydato)thallium(III), 32B, 329
$C_{11}H_{16}NO_2Tl$, (L-Phenylalaninato)dimethylthallium(III), 44B, 662
$C_{11}H_{19}O_4Tl$, Cyclopropyl-bis(isobutyrate)thallium(III), 45B, 790
$C_{11}H_{21}Cl_3N_2S_4Tl_2$, Thallium(I) diethyldithiocarbamate chloroform solvate, 44B, 662
$C_{11}H_{21}N_8O_2S_4Tl$, Thallium(I) benzoate-tetra(thiourea) complex, 34B, 356
$C_{12}HF_{10}OTl$, Hydroxobis(pentafluorophenyl)thallium(III), 35B, 510
$C_{12}H_8Cl_3N_2Tl$, Trichloro(1,10-phenanthroline)thallium(III), 38B, 698
$C_{12}H_8Cl_3N_4O_2Tl$, Trichlorobis(4-pyridinecarbonitrile 1-oxide-O)thallium, 46B, 725
$C_{12}H_{22}Ga_2N_4$, 3-Methylpyrazolylgallium dimethyl dimer, 41B, 779
$C_{12}H_{24}Al_2FK$, Potassium bis(trimethylaluminum)fluoride benzene sol-

vate, 40B, 679
$C_{12}H_{30}AlN$, Tributylaminalane, 41B, 780
$C_{12}H_{30}Al_2FK$, Triethylaluminum - potassium fluoride, 26, 611; 28, 600
$C_{12}H_{30}Al_2N_4$, Dimethylaluminum-N,N'-dimethylacetamidine, 44B, 1113
$C_{12}H_{30}Al_3N_3$, Ethyleniminodimethylaluminum trimer, 35B, 481
$C_{12}H_{30}Cl_3InO_9$, Indium trichloride trihydrate trisdioxane, 41B, 366
$C_{12}H_{30}Ga_2N_4$, Dimethylgallium-N,N'-dimethylacetamidine, 44B, 1113
$C_{12}H_{32}AlCl_3O_4$, Aluminum chloride - iso-propanol, 43B, 67
$C_{12}H_{32}Al_2O_{28}$, Aluminum benzenehexacarboxylate hydrate, 39B, 528
$C_{12}H_{32}Al_4N_4$, N-Isopropyliminoalane tetramer, 43B, 911
$C_{12}H_{32}Ga_2N_2O_2$, N,N-Dimethylethanolaminogallium dimethyl dimer, 41B, 777
$C_{12}H_{32.25}Cl_{0.75}Al_3N_4$, Bis($\mu$-N,N'-diethylethylenediamido)bis(dihydridoaluminum)hydridoaluminum (partially-chlorinated), 43B, 48
$C_{12}H_{33}Al_3N_4$, Bis(μ-N,N'-diethylethylenediamido)bis(dihydridoaluminum)hydridoaluminum, 43B, 48
$C_{12}H_{36}Al_3N_3O_3$, N-Dimethylhydroxylaminodimethylaluminum, 44B, 1114
$C_{13}H_{19}N_2O_3Tl$, (Di-μ-DL-tryptophanato)bis(dimethylthallium(III)) dihydrate, 44B, 663
$C_{14}H_{14}ClGaN_2$, Chloro(dimethyl)-1,10-phenanthrolinegallium(III), 42B, 582
$C_{14}H_{14}ClN_2O_4Tl$, 1,10-Phenanthroline-dimethylthallium(III) perchlorate, 38B, 699
$C_{14}H_{22}AlY$, Bis(cyclopentadienyl)yttrium-di-μ-methyl-(dimethylaluminum), 44B, 663
$C_{14}H_{26}Ga_2N_4$, Di-μ-3,5-dimethylpyrazolyl-bis(dimethylgallium), 41B, 781
$C_{14}H_{28}N_2S_4Tl_2$, Thallium(I) diisopropyldithiocarbamate dimer, 38B, 863
$C_{14}H_{30}Ga_2N_6NiO_2$, mer-Bis(dimethyl(ethanolamino)(1-pyrazolyl)gallato-(N^2,O,N))nickel(II), 44B, 664
$C_{14}H_{30}Ga_2N_6NiO_2$, sym-fac-Bis(dimethyl(ethanolamino)(1-pyrazolyl)gallato(N^2,O,N))nickel(II), 44B, 664
$C_{14}H_{42}Al_7N_7$, Hepta-μ_3-methylimido-heptakis(methylaluminum), 42B, 583
$C_{15}H_{15}Cl_2InN_2O_2$, Dichloro(acetylacetonato)-2,2'-bipyridylindium(III), 40B, 976
$C_{15}H_{15}In$, Tris(cyclopentadienyl)indium(III), 38B, 700
$C_{15}H_{17}GaMoN_6O_2$, Dicarbonyl(methyltris(1-pyrazolyl)gallato)(η^3-allyl)molybdenum, 44B, 665
$C_{15}H_{21}AlO_6$, Tris(acetylacetonato)aluminum, 24, 598; 39B, 530
$C_{15}H_{21}AlO_6$, Tris(2,4-pentanedionato)aluminum (γ-form), 41B, 782
$C_{15}H_{21}GaO_6$, Tris(acetylacetonato)gallium(III), 40B, 821
$C_{15}H_{21}InO_6$, Tris(acetylacetonato)indium(III) (monoclinic), 45B, 1055
$C_{15}H_{21}InO_6$, Tris(2,4-pentanedionato)indium(III) (orthorhombic), 46B, 974
$C_{15}H_{30}GaN_3S_6$, Tris(diethyldithiocarbamato)-gallium(III), 42B, 584
$C_{15}H_{30}InN_3S_6$, Tris(diethyldithiocarbamato)-indium(III), 42B, 584
$C_{15}H_{30}N_3S_6Tl$, Tris(N,N-diethyldithiocarbamato)thallium(III), 44B, 665
$C_{15}H_{46}Al_5N_5$, Bis(isopropylamido-hydridoaluminum)-tris(isopropylaminodihydridoaluminum), 41B, 783
$C_{16}H_{17}InN_2O_6$, Tri(acetato)-(2,2'-bipyridyl)indium(III), 45B, 1117
$C_{16}H_{20}Br_2Co_2InNO_8$, Tetraethylammonium dibromobis(tetracarbonylcobalt)indate(III), 37B, 409
$C_{16}H_{22}Al_2$, Di-μ-phenyl-bis(dimethylaluminum), 38B, 700
$C_{16}H_{22}In_2N_4O_2 \cdot 0.5\ C_6H_6$, Di-$\mu$-(pyridine-2-carbaldehyde-oximato)-bis(dimethylindium) benzene solvate, 46B, 725

$C_{16}H_{24}CuGa_2N_8$, Bis(dimethylbis-(1-pyrazolyl)-gallato)copper(II), 41B, 1075
$C_{16}H_{24}Ga_2N_8Ni$, Bis(dimethylbis(pyrazol-1-yl)gallato)nickel(II), 41B, 1076
$C_{16}H_{40}Al_4N_4$, N-Isopropyliminomethylalane tetramer, 43B, 911
$C_{16}H_{44}Cl_2Si_4Tl_2$, Di-μ-chlorotetrakis(trimethylsilylmethyl)-dithallium(III), 46B, 726
$C_{16}H_{48}AlO_4SbSi_4$, Tetramethylstibonium tetrakis(trimethylsiloxy)aluminate, 28, 601
$C_{17}H_{16}GaNOS$, 2-(N-Phenylaminomethylene)-3(2H)-benzo[b]furanthionato-dimethylgallium, 46B, 724
$C_{17}H_{20}NS_2Tl$, Diethyldithiocarbamatodiphenylthallium(III), 46B, 726
$C_{17}H_{25}Al_2N$, μ-Diphenylamino-μ-methyl-tetramethyldialuminum, 34B, 406
$C_{18}H_{15}Ga$, Triphenylgallium, 35B, 482
$C_{18}H_{15}In$, Triphenylindium, 35B, 482
$C_{18}H_{17}InN_2O_6$, Tri(acetato)-(1,10-phenanthroline)indium(III), 45B, 1117
$C_{18}H_{22}Ga_2N_4$, Indazolylgallium dimethyl dimer, 41B, 779
$C_{18}H_{22}Ga_2O_4$, Salicylaldehydatogallium dimethyl dimer, 42B, 584
$C_{18}H_{24}ClInN_2$, Bis[2-[(dimethylamino)methyl]phenyl]chloroindium(III), 46B, 727
$C_{18}H_{25}AlN_2O_2$, Dimethylaluminum N-phenylbenzimidate - trimethylamine oxide, 38B, 701
$C_{18}H_{30}Al_2$, Tricyclopropylaluminum dimer, 46B, 728
$C_{18}H_{30}InN_3S_6$, Tris(pentamethylenedithiocarbamato)indium(III), 39B, 534
$C_{18}H_{42}Al_6Cl_6N_6$, Hexa(N-isopropyliminochloroalane), 43B, 911
$C_{18}H_{48}Al_6N_6$, N-Isopropyliminoalane hexamer, 40B, 680
$C_{18}H_{48}Al_6N_6$, N-Propyliminoalane hexamer, 43B, 912
$C_{18}H_{51}Al_7N_6$, Hexakis(isopropylamido-hydrido-aluminum)trihydridoaluminum, 41B, 784
$C_{18}H_{54}AlN_3Si_6$, Tris(bis(trimethylsilyl)amino)aluminum, 34B, 370
$C_{18}H_{54}N_3Si_6Tl$, Tris(bis(trimethylsilyl)amine)thallium, 44B, 1115
$C_{18}H_{56}Al_8N_8$, Bis-μ-methylamido-hexa-μ_3-methylimido-bis(dimethylaluminum)-hexakis(methylaluminum), 45B, 791
$C_{18}H_{56}Ga_8N_8$, Bis-μ-methylamido-hexa-μ_3-methylimido-bis(dimethylgallium)-hexakis(methylgallium), 45B, 791
$C_{19}H_{13}N_2O_3Tl$, Salicylato-(1,10-phenthroline)thallium(I), 38B, 836
$C_{19}H_{36}Ga_2N_6O$, μ-Hydroxy-μ-3,5-dimethylpyrazolyl-bis(dimethylgallium) 3,5-dimethylpyrazole solvate, 41B, 781
$C_{19}H_{39}N_2O_9Tl$, Thallium(I) cryptate, 39B, 521
$C_{19}H_{54}Al_6LiN_5O$, Lithiumpentakis(isopropylamido)-tris(hydroaluminum)tris(dihydridoaluminum), 41B, 786
$C_{20}H_{16}ClGaN_2O_2$, Bis(2-methyl-8-quinolinolato)chlorogallium(III), 37B, 604
$C_{20}H_{16}Cl_6Ga_2N_4$, cis-Dichlorobis-(2,2'-bipyridyl)gallium(III) tetrachlorogallate(III), 38B, 704
$C_{20}H_{22}Al_2O_6W_2$, Di-μ-(tricarbonyl-η^5-cyclopentadienyltungsten-O,O')-bis(dimethylaluminum), 41B, 904
$C_{20}H_{26}Ga_2N_2O_2$, N,N-Ethylenebis(salicylideneiminato)bis(dimethylgallium), 43B, 913
$C_{20}H_{29}Al_2NO_2$, Dimethylaluminum N-phenylbenzimidate-O-(trimethylaluminum) - acetaldehyde, 38B, 701
$C_{20}H_{33}Al_2ClZr$, μ-Chloro-1-[dicyclopentadienylzirconio(IV)]-2,2-bis(diethylalumino)ethane, 46B, 728
$C_{20}H_{47}Al_3MgN_4O$, Tetrahydrofuranmagnesium-t-butylimino-tris(t-butyliminoalane), 43B, 914

$C_{21}H_{15}AlO_6$, Tris(tropolonato)aluminum(III), 38B, 704

$C_{21}H_{29}B_9PTl \cdot 0.5\ C_4H_8O$, Methyltriphenylphosphonium undecahydro-1,2-dicarba-3-thalla-closo-dodecaborate tetrahydrofuran, 44B, 665

$C_{21}H_{31}AlZr$, μ-Hydrido-(triethylalumino)-[tri(cyclopentadienyl)-zirconium(IV)], 46B, 729

$C_{21}H_{34}AlClTi$, μ-Chloro-μ-methylenebis(η^5-cyclopentadienyl)bis(neo-pentyl)aluminumtitanium, 46B, 729

$C_{21}H_{57}InSi_6 \cdot C_4H_{10}O$, Tris[bis(trimethylsilyl)methyl]indium(III) di-ethyl ether, 46B, 730

$C_{22}H_{19}AlCl_3N_5$, cis-Dichlorobis(2,2'-bipyridine)aluminum chloride acetonitrile, 43B, 915

$C_{22}H_2ON_3S_2Tl$, (2,6-Bis(1-methyl-2-(2-thiolophenyl)-2-azaethene)pyr-idine)methylthallium(III), 43B, 1400

$C_{22}H_{30}Al_2$, 1-Ethyl-3-methyl-1-alumina-indan dimer, 45B, 791

$C_{22}H_{36}Al_2Cl_2$, Di-μ-chloro-bis(η^3-pentamethylcyclopentadienyl)methyl-aluminum)), 45B, 792

$C_{22}H_{42}Al_3KSe$, Potassium methyl-tris(trimethylalumino)selenide di-benzene solvate, 42B, 585

$C_{24}H_4Br_2F_{16}Tl_2$, Bis(bromobis(2,3,5,6-tetrafluorophenyl)-thallium(III)), 45B, 792

$C_{24}H_{12}F_9InO_6S_3$, Tris(1-(2-thienyl)-4,4,4-trifluoro-1,3-butanedione)-indium(III), 42B, 738

$C_{24}H_{15}GaO_9W_3$, Gallio-tris(tricarbonyl-η^5-cyclopentadienyltungsten), 41B, 853

$C_{24}H_{15}Mo_3O_9Tl$, Tris(cyclopentadienyltricarbonylmolybdenum)-thallium(III), 39B, 538

$C_{24}H_{18}AlMn_3O_{18}$, Tris(cis-diacetyltetracarbonylmanganate)aluminum, 41B, 854

$C_{24}H_{20}InNa$, Sodium tetraphenylindate, 39B, 535

$C_{24}H_{21}InS_6$, Tris(dithiophenylacetato)indium(III), 35B, 816

$C_{24}H_{40}CuGa_2N_8$, Bis(dimethylbis-(3,5-dimethyl-1-pyrazolyl)-gallato)-copper(II), 41B, 1075

$C_{24}H_{60}Al_6N_6$, Hexa(N-isopropyliminomethylalane) (partially-hydro-genated), 43B, 911

$C_{24}H_{64}Al_8N_8$, N-Propyliminoalane octamer, 43B, 912

$C_{24}H_{72}Ga_8N_{12}O_2$, Tetramethyl-tetra(N,N-dimethylaminoethyl)-dioxonia-octaazonia-octagallanata-nonocyclo-docosane, 41B, 787

$C_{25}H_{35}Al_3Mo_2$, Bis(hydridobiscyclopentadienylmolybdenum)-pentamethyl-trialuminum, 40B, 682, 769

$C_{25}H_{38}Al_2Zr$, Di-(η^5-cyclopentadienyl)-2,2-bis(diethylalumino)ethyl-zirconiumcyclopentadienide, 46B, 730

$C_{26}H_{34}Al_4Mo_2$, Bis(biscyclopentadienylmolybdenumtrimethyldialuminum), 40B, 769

$C_{26}H_{40}Al_2Fe_2O_4$, Bis(cyclopentadienyldicarbonyliron)-bis(triethyl-aluminum), 39B, 530

$C_{28}H_{40}Al_2Ti_2$, Dicyclopentadienyltitanium diethylaluminum dimer, 32B, 362

$C_{30}H_{32}Al_2N_2O_2$, Dimethylaluminum N-phenylbenzimidate dimer, 37B, 410

$C_{32}H_{60}Ga_4N_8$, Tetrakis-μ-(2-methylimidazolyl)-tetrakis(diethylgal-lium), 41B, 788

$C_{32}H_{71}Al_3CaN_4O_4$, Tris(tetrahydrofuran)calcium-t-butylimino-tris(t-butyliminoalane) tetrahydrofuran, 43B, 914

$C_{32}H_{77}Al_5O_9$, μ_5-Oxo-octakis(μ-isobutoxy)pentahydridopentaaluminum, 46B, 731

$C_{32}H_{100}Al_4Cl_4O_{32}P_4$, Aluminum phosphate chloride ethanol solvate tet-ramer, 41B, 789

$C_{34}H_{40}Al_2N_2O_4$, Bis(dimethylaluminum N-phenylbenzimidate -

acetaldehyde), 38B, 701
$C_{36}H_{30}Al_2$, Di-μ-phenyl-bis(diphenylaluminum), 38B, 705
$C_{36}H_{60}InN_9S_6$, Tris(tetraethylammonium) tris-(1,2-dicyanoethylene-1,2-dithiolato)indate(III), 37B, 640
$C_{36}H_{84}Al_4O_{12}$, Aluminum isopropoxide tetramer, 45B, 793
$C_{38}H_{30}O_4Tl_2$, Di-μ-2-hydroxycyclohepta-2,4,6-trien-1-onato-bis[diphenylthallium(III)], 46B, 726
$C_{38}H_{35}AlO$, Pentaphenyl(diethyl ether)aluminacyclopentadiene, 43B, 916
$C_{40}H_{32}Al_2N_4O_5$, Bis(2-methyl-8-quinolinolato)aluminum(III)-μ-oxo-bis(2-methyl-8-quinolinolato)aluminum(III), 35B, 732
$C_{42}H_{30}AlO_6P_3$, Aluminum-tris(dibenzoylphosphide), 43B, 916
$C_{46}H_{47}AlNiO$, Pentaphenyl(diethyl ether)aluminacyclopentadiene-1,5-cyclooctadienenickel, 43B, 916
$C_{48}H_{40}Al_4N_4$, Diphenylaluminum nitride tetramer, 28, 599; 38B, 706
$C_{48}H_{60}Al_6N_6 \cdot 0.33\ C_6H_{14}$, Hexa[(1-phenylethylimino)-hydrido-aluminum] n-hexane solvate, 45B, 794
$C_{48}H_{64}Al_2Na_2O_4$, Bis(tetrahydrofuran)sodium (9,10-dihydro-9,10-anthrylene)dimethylaluminate, 38B, 707
$C_{50}H_{22}F_{20}N_6Tl_2$, μ-(2,2'-Dipyridylamido)-μ-(2'',2'''-dipyridylamido)-bis(bis(pentafluorophenyl)thallium(III)) benzene, 43B, 917
$C_{60}H_{34}Cl_2F_{16}O_2P_2Tl_2$, di-μ-Chlorobis[bis(2,3,5,6-tetrafluorophenyl)-(triphenylphosphine oxide)thallium(III)], 46B, 731
$C_{62}H_{50}Al_2Br_2N_2$, Diphenylaluminum-phenyl-p-bromophenylketimine dimer benzene solvate, 34B, 406
$C_{114}H_{90}Ag_3AlO_6P_6S_6$, Tris(bis(triphenylphosphine)silver(I))tris(dithiooxalato)aluminum(III), 40B, 870

69. GERMANIUM, TIN, AND LEAD COMPOUNDS

CCl_6Ge, Trichloromethyltrichlorogermane (gas-ed), 40B, 1139
CH_2Cl_4Sn, Chloromethyltrichlorostannane (gas-ed), 38B, 1068
CH_3Br_3Sn, Methyltin tribromide (gas-ed), 9, 315
CH_3Cl_3Ge, Trichloro(methyl)germane (gas-ed-ed), 43B, 1484
CH_3Cl_3Sn, Methyltin trichloride (gas-ed), 9, 315
CH_3I_3Sn, Methyltin triiodide (gas-ed), 9, 315
CH_5ClGe, (Chloromethyl)germane (gas-mw), 42B, 1032
CH_5FGe, Methyl monofluorogermane (gas-mw), 42B, 1033
CH_6Ge, Methylgermane (gas-mw), 22, 554
C_2O_4Sn, (μ-Oxalato)tin, 45B, 794
$C_2H_2O_4Sn$, Tin(II) formate, 44B, 666
$C_2H_4Cl_4Sn$, Bis(chloromethyl)dichlorostannane, 38B, 707
$C_2H_4PbO_4$, Lead formate, 15, 383
$C_2H_6Br_2Sn$, Dimethyltin dibromide (gas-ed), 9, 315
$C_2H_6Cl_2Ge$, Dichloro(dimethyl)germane (gas-ed-ed), 43B, 1484
$C_2H_6Cl_2Sn$, Dimethyltin dichloride (gas-ed), 9, 315; 37B, 688
$C_2H_6Cl_2Sn$, Dimethyltin dichloride, 35B, 485
$C_2H_6F_2O_6S_2Sn$, Dimethyltin bisfluorosulfate, 37B, 411
$C_2H_6F_2Sn$, Dimethyltin difluoride, 31B, 328
C_2H_6Ge, Vinylgermane (gas-mw), 40B, 1144
$C_2H_6I_2Sn$, Dimethyltin diiodide (gas-ed), 9, 315
$C_2H_6N_2O_6Sn$, Dimethyldinitratotin(IV), 39B, 540
$C_2H_7Br_3O_2Sn$, Tribromostannic acid methyl ester, 28, 602
$C_2H_7Cl_3O_2Sn$, Trichlorostannic acid methyl ester, 28, 602
$C_2H_8Cl_2N_4PbS_2$, Bis(thiourea)lead(II) dichloride, 23, 586
$C_2H_8Cl_2O_2Sn$, Dichlorohydroxy(ethyl)tin(IV) monohydrate, 42B, 586

$C_2H_8N_4O_4S_3Sn$, Sulfatobis(thiourea)tin(II), 39B, 540
$C_3H_3KO_6Sn$, Potassium triformatostannate(II), 40B, 1104
C_3H_9BrSn, Trimethyltin bromide (gas-ed), 9, 315
C_3H_9ClSn, Trimethyltin chloride, 45B, 795
C_3H_9ClSn, Trimethyltin chloride (gas-ed), 9, 315
$C_3H_9Cl_4GeN$, Germanium tetrachloride - trimethylamine, 38B, 708
C_3H_9FSn, Trimethyltin fluoride, 29, 625
C_3H_9ISn, Trimethyltin iodide (gas-ed), 9, 315
$C_3H_9N_3O_2S_3Sn$, 2-Trimethyltin-1,3,5λ^4,2,4,6-trithiazine 1,1-dioxide, 45B, 795
$C_3H_9N_3Pb$, Trimethyllead azide, 46B, 732
$C_3H_9N_3Sn$, Trimethyltin azide, 46B, 733
$C_3H_{10}Ge$, Trimethylgermane (gas-mw), 39B, 895
$C_3H_{11}NO_4Sn$, Trimethyltin nitrate monohydrate, 38B, 708
$C_4Na_2O_8Sn$, Disodium bisoxalatostannate(II), 42B, 586
$C_4O_8K_2Sn \cdot H_2O$, Potassium bisoxalatostannate(II) monohydrate, 45B, 794
$C_4H_4Cl_4Ge_2$, 1,1,4,4-Tetrachloro-1,4-digermaniacyclohexa-2,5-diene, 33B, 304
$C_4H_4Ge_2I_4$, 1,1,4,4-Tetraiodo-1,4-digermacyclohexa-2,5-diene, 32B, 330
$C_4H_4O_5Sn$, Tin(II) maleate monohydrate, 43B, 918
$C_4H_6Cl_4N_2Sn$, Bisacetonitriletetrachlorotin(IV), 40B, 1104
$C_4H_6GeN_2$, Dimethyldicyanogermanium, 38B, 709
$C_4H_6N_2S_2Sn$, Dimethyltin diisothiocyanate, 35B, 486
$C_4H_6N_2Sn$, Dimethyldicyanotin, 38B, 710
$C_4H_6O_2S_4Sn$, Bis-(O-methyl-dithiocarbonate)tin(II), 42B, 587
$C_4H_7Cl_3O_2Sn$, Trichloro-β-carbomethoxyethyltin, 45B, 796
$C_4H_8Br_2O_2Sn$, Dibromo(1,4-dioxan)tin(II), 43B, 918
$C_4H_8Cl_2GeOS_2$, 5,5-Dichloro-1-oxa-4,6-dithia-5-germocane, 42B, 587
$C_4H_8Cl_2GeO_2$, 1,4-Dioxane-germanium dichloride, 35B, 407
$C_4H_8Cl_2GeS_3$, 2,2-Dichloro-1,3,6,2-trithiagermocane, 41B, 790
$C_4H_8Cl_2OS_2Sn$, 5,5-Dichloro-1-oxa-4,6-dithia-5-stannaocane, 41B, 791
$C_4H_8Cl_2O_2Sn$, Dichloro(1,4-dioxan)tin(II), 42B, 588
$C_4H_8Cl_2S_3Sn$, 2,2-Dichloro-1,3,6,2-trithiastannaocane, 41B, 447
C_4H_9GeN, Trimethylcyanogermane (gas-mw), 40B, 1144
C_4H_9GeN, Trimethylcyanogermane, 31B, 329
C_4H_9NSSn, Trimethytin isothiocyanate, 35B, 487
C_4H_9NSn, Trimethyltin cyanide, 31B, 330
$C_4H_{10}Br_2Sn$, Diethyltin dibromide, 43B, 919
$C_4H_{10}Br_3NOSn$, Tribromomethyl(dimethylformamide)tin(IV), 44B, 668
$C_4H_{10}Cl_2Sn$, Diethyltin dichloride, 43B, 919
$C_4H_{10}I_2Sn$, Diethyltin diiodide, 43B, 919
$C_4H_{12}Br_2O_2PbS_2$, Bis(dimethylsulfoxide)lead(II) bromide, 43B, 919
$C_4H_{12}Ge$, Tetramethylgermane (gas-ed), 4, 277
$C_4H_{12}Ge_4S_6$, Tetra(methylgermanium) hexasulfide, 39B, 538
$C_4H_{12}N_2O_2Sn$, N-(Trimethylstannyl)-N-nitromethylamine, 40B, 683
$C_4H_{12}N_4O_5PbS_2$, Bis(thiourea)lead(II) formate monohydrate, 39B, 552
$C_4H_{12}OSn$, Trimethyltin methoxide, 40B, 683
$C_4H_{12}O_2SSn$, Trimethyltin methylsulfinate, 42B, 588; 43B, 920
$C_4H_{12}O_2SeSn$, Methaneseleninatotrimethyltin, 44B, 667
$C_4H_{12}Pb$, Tetramethyllead (gas-ed), 4, 277; 22, 580; 37B, 688
$C_4H_{12}S_6Sn_4$, Tetra(methyltin)hexasulfide, 34B, 408; 38B, 710
$C_4H_{12}Sn$, Tetramethyltin (gas-ed), 4, 277; 39B, 888
$C_4H_{14}N_2O_8Sn_2$, Di-μ-hydroxo-bis(dimethylnitratotin), 40B, 684
C_5H_5ClSn, Cyclopentadienyltin(II) chloride, 41B, 791
$C_5H_6Cl_4N_2Sn$, Tin(IV) chloride glutaronitrile, 33B, 307

C_5H_8Ge, 5-Germylcyclopenta-1,3-diene, 46B, 733
C_5H_9ClGe, Trimethylgermylchloroacetylene (gas-mw-ed), 40B, 1141
$C_5H_9F_3O_2Sn$, Trimethyltin(IV) trifluoroacetate, 39B, 541
$C_5H_9N_3Sn$, Trimethyltin(IV) dicyanamide, 37B, 412
$C_5H_{10}N_2O_4PbS$, Thiourealead(II) acetate, 23, 585; 24, 532
$C_5H_{12}ClNS_2Sn$, Dimethyltin chloride N,N-dimethyldithiocarbamate, 35B, 488
$C_5H_{12}O_2Pb$, Trimethyllead acetate, 41B, 792
$C_5H_{12}O_2Sn$, Trimethyltin(IV) acetate, 39B, 541
$C_5H_{13}NO_2Sn$, Glycinato(trimethyl)tin(IV), 46B, 734
$C_6GeK_2O_{12} \cdot H_2O$, Potassium trioxalatogermanate(IV) monohydrate, 46B, 734
$C_6H_6Cl_3KO_6Sn$, Potassium tris(monochloroacetato)stannate(II), 45B, 797
$C_6H_6N_6Sn$, Dimethyltin(IV) bisdicyanamide, 37B, 412
$C_6H_{10}Ge_2O_7$, Carboxyethylgermanium sesquioxide, 42B, 589
$C_6H_{10}O_2PbS_4$, Lead ethylxanthate, 31B, 331
$C_6H_{12}Cl_2N_2O_2Sn$, Dichloro-bis(β-amidoethyl)tin, 45B, 796
$C_6H_{12}F_3N_2O_4S_3Sn$, N,N'-Bis(trifluoromethanesulphonyl)-N-(trimethylstannyl)methanesulfinamidine, 44B, 667
$C_6H_{12}N_2O_2PbS_4$, Di-isothiocyanatolead(II)-bis(dimethylsulfoxide), 42B, 590
$C_6H_{12}N_2PbS_4$, Bis(N,N-dimethyldithiocarbamato)lead(II), 46B, 987
$C_6H_{12}O_2SSn$, Trimethyltin prop-2-yne-sulfinate, 39B, 542
$C_6H_{14}N_2O_4Sn$, Bis(N-acetylhydroxylamino)dimethyltin(IV), 42B, 590
$C_6H_{15}NS_2Sn$, N,N-Dimethyldithiocarbamatotrimethylstannane, 35B, 793
$C_6H_{16}N_2O_5Sn$, Bis(N-acetylhydroxylamino)dimethyltin(IV) monohydrate, 42B, 590
$C_6H_{18}Br_2O_2S_2Sn$, Dibromodimethylbis(dimethylsulfoxide)tin, 44B, 668
$C_6H_{18}Cl_2O_2S_2Sn$, cis-Dichloro-cis-bis(dimethylsulfoxide)-trans-dimethyltin(IV), 35B, 735; 44B, 688
$C_6H_{18}N_4O_2S_4Sn_2 \cdot 0.5\ C_6H_6$, 1,3-Bis(trimethyltin)-1,3,5,7-tetraaza-2,4,6,8-tetrathiocin-2,2-dioxide benzene solvate, 45B, 797
$C_6H_{18}Pb_2$, Hexamethyldilead (gas-ed), 8, 286
$C_6H_{18}S_3Sn_3$, Hexamethylcyclotristannathiane (monoclinic), 43B, 921
$C_6H_{18}S_3Sn_3$, Hexamethylcyclotristannathiane (tetragonal), 41B, 793
$C_6H_{18}Se_2Sn_3$, 2,2,4,4,5,5-Hexamethyl-1,3-diselena-2,4,5-tristannolane, 46B, 735
$C_6H_{18}Se_3Sn_3$, Hexamethylcyclotristannaselenane, 44B, 1116
$C_6H_{24}Cl_2N_{12}O_8PbS_6$, Hexakis(thiourea)lead(II) perchlorate (tetragonal), 38B, 712
$C_6H_{24}Cl_2N_{12}O_8PbS_6$, Hexakis(thiourea)lead(II) perchlorate (triclinic), 32B, 54
$C_6H_{24}N_{14}O_6PbS_6$, Hexakis(thiourea)lead(II) nitrate, 32B, 54
$C_6H_{28}Cl_2N_{12}O_{10}PbS_6$, Hexakis(thiourea)lead(II) perchlorate dihydrate, 32B, 54
$C_6H_{34}O_{16}P_2Sn_3$, Tris(dimethyltin(IV)) bis(orthophosphate) octahydrate, 43B, 921
$C_7H_6N_2O_5Pb$, Nitrato(p-aminobenzoato)lead(II), 46B, 735
$C_7H_6N_3O_3SSn$, (2-Aminobenzothiazolato)nitratotin(II) polymer, 41B, 794
$C_7H_{14}N_2O_2Pb$, Trimethylleaddiazoethyl acetate, 46B, 736
$C_7H_{16}ClNO_2SSn$, Chloro(ethyl L-cysteinato-N,S)dimethyltin(IV), 45B, 798, 799
$C_7H_{17}Cl_3N_2O_2Sn$, Trichloromethylbis(dimethylformamide)tin(IV), 44B, 668
$C_7H_{18}N_2Sn_2$, Bis(trimethylstannyl)carbodiimide, 37B, 413

$C_7H_{19}NO_2Sn_2$, Trimethylstannyl trimethylstannylcarbamoate, 46B, 737
$C_7H_{19}NO_2Sn_2$, Trimethyltin isocyanate hydroxide, 38B, 713
$C_8H_9Cl_3O_2Sn$, (Trichloroacetato)trivinyltin, 45B, 799
$C_8H_9MnO_5Sn$, Trimethyltin-pentacarbonylmanganese, 33B, 308
$C_8H_{10}N_2O_4Sn$, Bis(N-methyl-N-acetylhydroxylamino)dimethyltin(IV), 41B, 795
$C_8H_{12}F_6O_4Sn_2$, Bis(μ-trifluoroacetato-O,O')-bis[dimethyltin(IV)], 45B, 800
$C_8H_{12}O_8Sn$, Tin tetraacetate, 45B, 800
$C_8H_{14}ClNSn$, Chloro(trimethyl)pyridinetin, 28, 603
$C_8H_{14}Cl_2O_4Sn$, Dichloro-bis(β-carbomethoxyethyl)tin, 45B, 796
$C_8H_{14}Ge$, Cyclopentadienyltrimethylgermane (gas-ed), 35B, 900
$C_8H_{16}Cl_2O_4Sn_2$, Bis(μ-chloroacetato-O,O')-bis(dimethyltin(IV)), 44B, 668
$C_8H_{16}Ge_2$, 1,1,4,4-Tetramethyl-1,4-digermacyclohexa-2,5-diene, 32B, 330
$C_8H_{17}GeNO_3$, 1-Ethylgermatrane, 35B, 484
$C_8H_{18}Ge_2O$, Bis(trimethylgermyl)ketene (gas-ed), 39B, 891
$C_8H_{18}N_2S_4Sn$, Dimethyltin bis(N,N-dimethyldithiocarbamate), 38B, 714
$C_8H_{20}Br_4O_6Sn$, cis-Diaquotetrabromotin(IV)-1,4-dioxan (1:2), 42B, 564
$C_8H_{20}Cl_2N_2O_2Sn$, Dichlorodimethylbis(dimethylformamide)tin, 44B, 668
$C_8H_{20}Cl_4Ge_2Si_2$, 2,2,4,4-Tetrachloro-1,3-bis(trimethylsilyl)-2,4-digermacyclobutane, 42B, 520
$C_8H_{20}O_4P_2PbS_4$, Lead(II) bis-(O,O'-diethyldithiophosphate), 38B, 714
$C_8H_{22}Cl_6O_4Sn_2$, Trichloroethoxytin(IV) ethanolate dimer, 40B, 682
$C_8H_{24}Cl_4O_2Sn_4$, Di-μ_3-oxo-bis-(μ-dichloro)-bis[μ-dimethyltin(IV)]-bis[chlorodimethyltin(IV)], 46B, 737
$C_8H_{24}N_4Sn$, Tetra(dimethylamino)tin (gas-ed), 35B, 901
$C_9H_{12}Cl_2O_2Sn$, Salicylaldehyde-dichlorodimethyltin(IV), 41B, 795
$C_9H_{15}NO_3Sn$, Aquo-trimethyl-(2-pyridyl-carboxylato)tin(IV), 45B, 801
$C_9H_{16}O_4SSn$, Trimethyltin-benzenesulphonate monohydrate, 42B, 591
$C_9H_{19}NOSn$, catena-(Cyclohexanone-oximato)trimethyltin, 41B, 796
$C_9H_{20}O_4Sn_2$, Malonato-bis(trimethyltin), 44B, 1117
$C_9H_{27}ClN_3OPSn$, (Hexamethyltriamidophosphato)chlorotrimethyltin(IV), 43B, 922
$C_9H_{28}CrO_5Sn_3$, Tris(trimethyltin) chromate hydroxide, 40B, 684
$C_{10}H_8Cl_2N_2S_2Sn$, Dichlorobis(2-pyridinethiolato)tin(IV), 44B, 669
$C_{10}H_{10}Cl_4GeN_2$, trans-Dipyridinetetrachlorogermanium, 24, 619
$C_{10}H_{10}Pb$, Dicyclopentadienyllead, 31B, 332
$C_{10}H_{14}Cl_2O_4Sn$, Dichlorobis(2,4-pentanedionato)tin(IV), 44B, 670
$C_{10}H_{14}N_2O_8Sn$, Dihydrogen ethylenediaminetetraacetatostannate(II), 39B, 542
$C_{10}H_{14}N_2O_9Sn$, Stannic ethylenediaminetetraacetate monohydrate, 37B, 414
$C_{10}H_{15}BF_4Sn$, (Pentamethylcyclopentadienyl)tin tetrafluoroborate, 46B, 748
$C_{10}H_{16}N_2O_{10}Sn_2$, Distannous ethylenediaminetetraacetate dihydrate, 37B, 415
$C_{10}H_{18}O_2PbS_4$, Lead n-butylxanthate, 33B, 319
$C_{10}H_{20}N_2PbS_4$, Lead(II) diethyldithiocarbamate, 38B, 715
$C_{10}H_{20}N_2S_4Sn$, Bis(N,N-diethyldithiocarbamato)tin(II), 39B, 543; 42B, 766
$C_{10}H_{22}ClNO_4Sn$, 4-Trimethyltin-quinuclidinium perchlorate, 46B, 738
$C_{10}H_{24}Cl_2N_2OSiSn_3$, (1,5-Di-t-butyl-6,6-dimethyl-3-oxa-1,5-diaza-6-sila-2,4-distannatricyclo[2.2.0.0$^{2-5}$]hexane)dichlorotin, 44B, 670
$C_{10}H_{24}N_2SiSn$, 1,3-Di-t-butyl-2,2-dimethyl-1,3,2,4-diazasilastannetidine, 44B, 671

$C_{10}H_{24}N_6O_6Pb$, (1,4,8,11-Tetra-azacyclotetradecane)dinitratolead(II), 45B, 801, 1243
$C_{11}H_6GeMn_2O_9$, μ-Carbonyl-μ-(dimethylgermylene)-bis(tetracarbonylmanganese)(Mn-Mn), 41B, 797
$C_{11}H_{11}F_4GeN_3O_2$, 2,2'-Bipyridyltetrafluorogermanium(IV) nitromethane solvate, 38B, 716
$C_{11}H_{11}F_4N_3O_2Sn$, 2,2'-Bipyridyltetrafluorotin(IV) nitromethane solvate, 38B, 717
$C_{11}H_{13}Br_3N_2Sn$, Tribromomethylbis(pyridine)tin, 44B, 671
$C_{11}H_{13}NO_2Sn$, N-Trimethyltin(IV)-phthalimide, 44B, 672
$C_{11}H_{14}Cl_3NSn$, Quinolinium trichlorodimethylstannate, 41B, 797
$C_{11}H_{22}Sn_2$, 1,1-Bis(trimethylstannyl)-cyclopentadiene, 41B, 798
$C_{11}H_{22}Sn_2$, 1,1'-Bis(trimethylstannyl)cyclopentadiene (gas-ed), 37B, 695
$C_{12}H_8Cl_2GeO$, 10,10-Dichloro-10-germa-9-oxa-9,10-dihydroanthracene, 40B, 685
$C_{12}H_8N_2O_4Pb$, Bis(nicotinato-N,O)-lead(II), 41B, 799
$C_{12}H_{10}Cl_2Sn$, Dichloro(diphenyl)tin, 37B, 416
$C_{12}H_{10}N_2O_5Pb$, Bis(N,O-isonicotinato)monoaquolead(II), 38B, 718
$C_{12}H_{12}Al_2Cl_8Pb$, π-Benzene-lead bis(tetrachloroaluminate) benzene solvate, 40B, 686
$C_{12}H_{12}Al_2Cl_8Sn$, π-Benzene-bis(tetrachloroaluminate)-tin(II) benzene solvate, 41B, 800
$C_{12}H_{12}Al_2Cl_{10}Sn_2$, Di-μ-chlorobis[$\eta^6$-benzenetin(II)] tetrachloroaluminate, 45B, 805
$C_{12}H_{12}Cl_6O_{12}Sn_2Sr$, Strontium bis[tris(monochloroacetato)-stannate(III)], 46B, 738
$C_{12}H_{12}Fe_2O_8Sn_2$, Di-μ-dimethylstannylene-bis(tetracarbonyliron), 38B, 718
$C_{12}H_{16}Br_2N_2Sn$, Dibromodimethylbis(pyridine)tin, 44B, 671
$C_{12}H_{16}Cl_2N_2O_2Sn$, Dichlorodimethylbis(pyridine-N-oxide)tin(IV), 34B, 410
$C_{12}H_{16}Cl_2N_2Sn$, Dichlorodimethylbis(pyridine)tin, 44B, 671
$C_{12}H_{18}CaO_{12}Sn_2$, Calcium bis(triacetatostannate(II)), 43B, 923
$C_{12}H_{18}Cl_2O_6Sn$, Dichlorobis(ethyl 3-oxobutanoato)tin(IV), 45B, 802
$C_{12}H_{18}Cl_4N_2Sn$, Pyridinium tetrachlorodimethylstannate(IV), 42B, 592
$C_{12}H_{18}Fe_2Ge_3O_6$, Tri-μ-dimethylgermaniumbis(tricarbonyliron), 34B, 408
$C_{12}H_{18}N_2O_2Sn$, N,N'-Ethylenebis(acetylideneimine)tin(II), 42B, 592
$C_{12}H_{20}O_4S_8Sn$, Tetrakis(O-ethylxanthato)tin(IV), 44B, 673
$C_{12}H_{20}O_4Sn$, Bis(2,4-pentanedionato)dimethyltin(IV), 39B, 544
$C_{12}H_{21}AsN_2O_4Sn$, N-(Trimethyltin)dimethylarsinobis(carbomethoxy)pyrazole, 46B, 739
$C_{12}H_{24}GeO$, 6,6-Dimethyl-6-germacycloundecanone, 44B, 674
$C_{12}H_{24}N_4O_2S_4Sn_4$, Tetramethyl-1,3-diisothiocyanatodistannoxane, 37B, 416
$C_{12}H_{24}N_4S_8Sn$, Tin(IV) tetrakis-(N,N-dimethyldithiocarbamate), 42B, 767
$C_{12}H_{26}N_2O_6Sn$, Bis(triethanolamine)tin(IV) (α form), 38B, 983
$C_{12}H_{26}N_2O_6Sn$, Bis(triethanolamino)tin(IV) (β form), 43B, 924
$C_{12}H_{26}N_2S_4Sn$, Bis(N,N-diethyldithiocarbamato)dimethyl-tin(IV) (triclinic), 45B, 803
$C_{12}H_{26}N_2S_4Sn$, Bis(N,N-diethyldithiocarbamato)dimethyl-tin(IV) (monoclinic), 45B, 803
$C_{12}H_{28}O_4P_2PbS_4$, Lead(II) O,O'-diisopropylphosphorodithioate, 34B, 412; 38B, 719
$C_{12}H_{30}Cl_4P_2Sn$, trans-Tetrachlorobis(triethylphosphine)tin(IV), 39B,

545
$C_{12}H_{30}Cl_6O_{10}Sn_2$, Di-μ-hydroxo-bis[aquatrichlorotin(IV)] - 1,4-dioxan, 37B, 417; 46B, 740
$C_{12}H_{33}ISi_3Sn$, Iodotris(trimethylsilylmethyl)tin(IV), 44B, 674
$C_{12}H_{36}Br_4N_6O_2P_2Sn$, Bis(hexamethyltriamidophosphato)tetrabromotin(IV), 43B, 925
$C_{12}H_{36}Cl_4N_6O_2P_2Sn$, Bis(hexamethyltriamidophosphato)tetrachlorotin(IV), 43B, 925
$C_{12}H_{36}Ge_6$, Dodecamethylcyclohexagermane, 41B, 800
$C_{12}H_{36}Ge_6P_4$, Hexa(dimethylgerma)tetraphosphide, 41B, 801
$C_{13}H_{19}NO_2PbS_4$, Lead isopropylxanthate-pyridine complex, 34B, 413
$C_{13}H_{39}Br_3N_6O_2P_2Sn$, Bis(hexamethyltriamidophosphato)tribromomethyltin(IV), 43B, 925
$C_{13}H_{39}Cl_3N_6O_2P_2Sn$, Bis(hexamethyltriamidophosphato)trichloromethyltin(IV), 43B, 925
$C_{14}H_8K_6O_{32}Sn_2$, Potassium μ-oxalato-di(trisoxalatostannate(IV)) tetrahydrate, 42B, 593
$C_{14}H_{10}Cl_2Fe_2GeO_4$, Bis(dicarbonyl-π-cyclopentadienyliron)dichlorogermane, 32B, 355
$C_{14}H_{10}Cl_2Fe_2O_4Sn$, Dichlorobis(dicarbonyl-π-cyclopentadienyliron)-tin(IV), 32B, 356
$C_{14}H_{10}Fe_2N_2O_8Sn$, Bis(dicarbonyl-π-cyclopentadienyliron)tin dinitrite, 33B, 310
$C_{14}H_{11}Cl_3MoN_2O_3Sn$, μ-Chloro-(dichloromethyltin)-(2,2'-bipyridyltricarbonylmolybdenum), 34B, 409
$C_{14}H_{11}Cl_4NO_2Sn$, Tetrachloro(benzoylbenzamido-O,O')tin (monoclinic form 1), 46B, 741
$C_{14}H_{11}Cl_4NO_2Sn$, Tetrachloro(benzoylbenzamido-O,O')tin (monoclinic form 2), 46B, 741
$C_{14}H_{11}GeMoNO_5$, μ-Cyano-dimethylphenylgermanium-pentacarbonylmolybdenum(0), 43B, 925
$C_{14}H_{12}N_2O_4Pb$, Bis(p-aminobenzoato)lead(II), 46B, 741
$C_{14}H_{14}Cl_4N_4O_3Sn \cdot 2\ H_2O$, Tetrachloro(N-pyridoxylidene-N'-picolinoylhydrazine)tin(IV) dihydrate, 46B, 742
$C_{14}H_{14}GeS$, 10,10-Dimethyl-10-germa-9-thio-9,10-dihydroanthracene, 45B, 804
$C_{14}H_{14}N_2O_{12}Pb \cdot 2\ C_7H_5NO_4 \cdot 2\ H_2O$, Bis(diaquo)(pyridine-2,6-dicarboxylato)lead(II) pyridine-2,6-dicarboxylic acid solvate hydrate, 45B, 1251
$C_{14}H_{18}Cl_2N_2Sn$, 2,2'-Bipyridyl-dichloro-diethyltin, 46B, 743
$C_{14}H_{18}O_8Ru_2Sn_2$, Bis(tetracarbonyl(trimethylstannio)ruthenium)(Ru-Ru), 41B, 801
$C_{14}H_{25}BrN_2Sn$, (2,6-Bis((dimethylamino)methyl)phenyl)dimethyltin bromide, 44B, 1117
$C_{14}H_{26}N_4O_4PbS_2$, 1,7,10,16-Tetraoxa-4,13-diazacyclooctadecanelead(II) thiocyanate, 39B, 553
$C_{14}H_{27}NO_{10}Sn$, Tetramethylammonium pentaacetatostannate(IV), 42B, 593
$C_{14}H_{28}N_2PbS_4$, Bis(N,N-diisopropyldithiocarbamato)lead(II), 46B, 743
$C_{14}H_{42}B_2N_4Si_4Sn_2$, 1,3-Bis(trimethylsilyl)-4-methyl-diazastannaboretidine dimer, 45B, 703
$C_{14}H_{42}Br_2N_6O_2P_2Sn$, Bis(hexamethyltriamidophosphato)dibromodimethyltin(IV), 43B, 925
$C_{14}H_{42}Cl_2N_6O_2P_2Sn$, Bis(hexamethyltriamidophosphato)dichlorodimethyltin(IV), 43B, 925
$C_{15}H_{12}Cl_2MoO_2Sn$, ((π-Cycloheptatrienyl)dicarbonylmolybdenum)phenyldichlorotin, 39B, 549
$C_{15}H_{13}N_{13}O_{14}Pb_2$, 2,4,6-Tris(2-pyrimidyl)-1,3,5-triazinedilead(II)

nitrate dihydrate, 43B, 524
$C_{15}H_{15}NO_2Sn$, N-(2-Hydroxyphenyl)salicylaldiminedimethyltin(IV), 41B, 802
$C_{15}H_{15}NO_2Sn$, 2-Hydroxy-N-(2-hydroxybenzylidene)aniline-dimethyltin, 40B, 977
$C_{15}H_{18}GeO_9Ru_3$, Tri-μ-(dimethylgermanio)-tris(tricarbonylruthenium), 37B, 422
$C_{15}H_{21}ClGeO_{10}$, Tris(acetylacetonato)germanium(IV) perchlorate, 46B, 744
$C_{16}H_{10}Cl_2Cr_2O_6Sn$, μ-(Dichlorostannio)-bis(tricarbonyl-π-cyclopentadienylchromium), 41B, 900
$C_{16}H_{10}N_{10}O_{13}Pb$, Aquobis(purpurato)lead(II), 43B, 511
$C_{16}H_{12}GeS_4$, Tetra(2-thienyl)germanium, 43B, 926
$C_{16}H_{12}PbS_4$, Tetra(2-thienyl)lead, 43B, 926
$C_{16}H_{12}S_4Sn$, Tetra(2-thienyl)tin, 43B, 926
$C_{16}H_{16}Fe_2O_4Pb$, Bis(dicarbonyl-π-cyclopentadienyliron)dimethyllead, 33B, 310
$C_{16}H_{16}Fe_2O_4Sn$, Bis(dicarbonyl-π-cyclopentadienyliron)dimethyltin, 33B, 310
$C_{16}H_{16}N_2O_3Sn$, Diphenyltin glycylglycinate, 43B, 926
$C_{16}H_{16}N_{10}O_{16}Pb$, Aquobis(purpurato)lead(II) trihydrate, 43B, 511
$C_{16}H_{18}Ge_2O_{10}Re_2$, Dimethylgermyl-acetyl(tetracarbonyl)rhenium dimer, 40B, 686
$C_{16}H_{19}Cl_3N_2Sn$, Butyltrichloro(N'-phenylpyridine-2-carbaldimine-N,N')tin(IV), 45B, 804
$C_{16}H_{19}GeNO_3$, 1-(α-Naphthyl)germatrane, 39B, 539
$C_{16}H_{19}NO_2Sn$, Trimethyl(N-phenyl-N-benzoyl-hydroxylamine)tin, 46B, 744
$C_{16}H_{20}Al_2Cl_{10}Sn_2$, Di-μ-chlorobis[$\eta^6$-p-dimethylbenzenetin(II)] tetrachloroaluminate, 45B, 805
$C_{16}H_{20}GeO_{10}$, Mandelic acid-germanium dioxide complex dihydrate, 32B, 331
$C_{16}H_{20}Ge_2$, 9,9,10,10-Tetramethyl-9,10-digermadihydroanthracene, 42B, 594
$C_{16}H_{20}N_{14}O_{14}PbS_4$, Tetrakis(thiourea)lead(II) picrate, 38B, 721
$C_{16}H_{22}Cl_2O_2S_2Sn$, cis-Dichloro-cis-bis(dimethylsulfoxide)-trans-diphenyltin, 40B, 687
$C_{16}H_{22}Co_2O_2Sn_2$, Di-μ-dimethylstannylene-bis(carbonyl-π-cyclopentadienylcobalt), 39B, 546
$C_{16}H_{22}Sn_2$, 1,2-Phenylene-diethynyl-bis(trimethyltin), 46B, 745
$C_{16}H_{24}Cl_3IrSn$, Bis(cyclo-octa-1,5-diene)iridium trichlorotin, 32B, 376
$C_{16}H_{24}F_{12}O_{10}Sn_4$, Di-$\mu_3$-oxo-bis(μ-trifluoroacetato-O,O')-bis(trifluoroacetato)tetrakis(dimethyltin(IV)), 44B, 674
$C_{16}H_{24}Ge_2$, 1,8-Bis(trimethylgermyl)naphthalene, 46B, 745
$C_{16}H_{24}Sn_2$, 1,8-Bis(trimethylstannyl)naphthalene, 46B, 745
$C_{16}H_{25}ClN_2S_4Sn$, Phenylchlorobis(diethyldithiocarbamato)tin(IV), 42B, 595
$C_{16}H_{26}Br_2O_8Sn$, Bis(1,2-diethoxycarbonyl-ethyl)tin dibromide (form 1), 33B, 313
$C_{16}H_{26}Br_2O_8Sn$, Bis(1,2-diethoxycarbonyl-ethyl)tin dibromide (form 2), 40B, 1105
$C_{16}H_{30}O_6Ru_2Sn_4$, Di-μ-dimethylstannylenebis(tricarbonyl(trimethylstannyl)ruthenium), 34B, 411
$C_{16}H_{33}N_3S_6Sn$, Tris(N,N-diethyldithiocarbamato)methyltin(IV), 44B, 854
$C_{16}H_{38}N_4Sn_3$, Tetrakis(t-butylimino)dihydridotritin, 46B, 746

$C_{16}H_{40}Cl_4O_2Sn_4$, Di-μ_3-oxo-bis-(μ-dichloro)-bis[μ-diethyltin(IV)]-bis[chlorodiethyltin(IV)], 46B, 737
$C_{17}H_{14}Co_2O_{11}Sn$, Bis(2,4-pentanedionato)(heptacarbonyldicobalt)-tin(IV), 39B, 546
$C_{17}H_{18}FeO_2Sn$, Poly(trivinylferrocenoatotin(IV)), 46B, 747
$C_{17}H_{21}N_5O_3PbS_2$, Isothiocyanatothiocyanato{3,15,21-triaza-6,9,12-trioxabicyclo[15.3.1]heneicosa-1(21),2, 5,17,19-pentaene}lead(II), 45B, 806
$C_{17}H_{22}BrNSn$, (S)-(2-(1-(S)-(Dimethylamino)ethyl)phenyl)methylphenyl-tin bromide, 44B, 1118
$C_{18}H_{12}N_2PbS_2$, Lead 8-mercaptoquinolate, 37B, 631
$C_{18}H_{14}F_6N_2O_4Sn$, (2,2'-Bipyridyl)bis(trifluoroacetato)divinyltin, 41B, 803
$C_{18}H_{15}BrGe$, Triphenylgermanium bromide, 45B, 806
$C_{18}H_{15}BrPb$, Triphenyllead bromide, 43B, 927
$C_{18}H_{15}BrSn$, Triphenyltin bromide, 45B, 807
$C_{18}H_{15}ClPb$, Triphenyllead chloride, 43B, 927
$C_{18}H_{15}ClSn$, Triphenyltin chloride, 35B, 489
$C_{18}H_{15}Cl_2GeP$, Dichloro(triphenylphosphine)germanium(II), 42B, 595
$C_{18}H_{15}NO_3Sn_2$, Nitratotriphenylstannyltin(II), 41B, 804
$C_{18}H_{15}NO_3Sn_2$, Nitratotriphenylstannyltin(II), 43B, 928
$C_{18}H_{16}Cl_2Co_2Sn$, Dichlorobis(norbornadienedicarbonylcobalt)tin(IV), 37B, 418
$C_{18}H_{16}OPb$, Hydroxotriphenyllead(IV), 44B, 675
$C_{18}H_{16}OSn$, Hydroxotriphenyltin(IV), 44B, 675
$C_{18}H_{20}N_2O_2Sn$, (N,N'-Ethylenebis(salicylideneiminato))dimethyl-tin(IV), 38B, 835
$C_{18}H_{22}Fe_2Ge_2O_5$, Bis((η^5-cyclopentadienyldicarbonyliron)dimethyl-germyl)oxide, 40B, 688
$C_{18}H_{23}CrNO_5Sn$, (Di-t-butylstannylene)pyridinopentacarbonylchromium, 39B, 547
$C_{18}H_{24}N_4O_4S_2Sn$, Bis(2-thio-5-nitropyridine)-S-di-n-butyl-stannane(IV), 45B, 807
$C_{18}H_{28}N_2O_9S_3Sn$, Tris(dimethylsulfoxide)nitratodiphenyltin(IV) nitrate), 40B, 689; 42B, 595
$C_{18}H_{30}Cl_4O_2Sn_2$, Tetrachloro-1,4-bis(triethylstannyloxy)benzene, 26, 721
$C_{18}H_{33}ClSn$, Chloro-tricyclohexyl-tin(IV), 45B, 808
$C_{18}H_{54}Br_2N_6O_2P_2Sn_2$, Bis(hexamethyltriamidophosphato)trimethyltin(IV) dibromotrimethylstannate(IV), 43B, 922
$C_{19}H_{15}Cl_2NSSn$, Dichloro(benzothiazole)diphenyltin(IV), 44B, 1119
$C_{19}H_{15}GeNO$, Triphenylgermanium isocyanate, 42B, 596
$C_{19}H_{15}NOPb$, Triphenylisocyanatolead(IV), 44B, 676
$C_{19}H_{15}NOSn$, Triphenyltin isocyanate, 43B, 929
$C_{19}H_{15}NSSn$, Triphenyltin isothiocyanate, 40B, 690
$C_{19}H_{17}ISn$, Triphenyliodomethyltin(IV), 44B, 1119
$C_{19}H_{17}N_5S_2Sn$, Dimethyldiisothiocyanato(terpyridyl)tin(IV), 39B, 548
$C_{19}H_{23}Cl_4N_3Sn_2$, 2,2',2"-Terpyridyl(dimethyl)chlorotin(IV) dimethyl-(trichloro)stannate(IV), 33B, 314
$C_{19}H_{27}N_7PbS_2$, {2,15-Dimethyl-3,7,10,14,20-pentaazabicyclo[14.3.1]-eicosa-1(20),2,14,16,18-pentaene}(thiocyanato)lead(II) thiocyanate, 45B, 808
$C_{19}H_{39}N_3S_6Sn$, Tris(N,N-diethyldithiocarbamato)-n-butyl-tin(IV), 45B, 809
$C_{20}H_{12}Fe_4O_{16}Sn_3$, Tetramethyltritintetra(iron tetracarbonyl), 32B, 332
$C_{20}H_{18}N_2O_2Sn$, Dimethyltin bis(8-hydroxyquinolinate), 32B, 333

$C_{20}H_{18}OGe$, Acetyltriphenylgermane, 33B, 306
$C_{20}H_{18}O_4Sn$, Bis(1-phenylbutane-1,3-dionato)tin(II), 41B, 804
$C_{20}H_{20}Sn$, Tetracyclopentadienyltin, 41B, 805
$C_{20}H_{28}ClGeNO_4$, Tetraethylammonium 2-chloro-2,2'-spirobis(1,3,2-benzodioxagermole), 46B, 747
$C_{20}H_{30}O_4P_2S_4Sn$, Bis(O,O-diethyl dithiophosphato)diphenyltin, 44B, 676
$C_{20}H_{30}Sn$, Bis(pentamethylcyclopentadienyl)tin, 46B, 748
$C_{20}H_{33}F_3O_2Sn$, Tricyclohexyltin trifluoroacetate, 46B, 748
$C_{20}H_{36}N_4O_6PbS_2$, (4,7,13,16,21,24-Hexaoxa-1,10-diazabicyclo[8.8.8]-hexacosane)lead(II) thiocyanate, 40B, 691
$C_{20}H_{36}O_2Sn$, Tricyclohexyltin acetate, 33B, 318
$C_{20}H_{40}N_4S_8Sn$, Tetrakis-(N,N-diethyldithiocarbamato)tin(IV), 37B, 419
$C_{20}H_{48}N_4Si_2Sn$, 1,3,5,7-Tetrakis(t-butyl)-2,2,6,6-tetramethyl-1,3,5,7-tetraaza-2,6-disila-4-stannaspiro[3.3]heptane, 44B, 677
$C_{21}H_{10}CoMnO_9Sn$, Tetracarbonylcobalt-diphenyltin-manganesepentacarbonyl, 33B, 316
$C_{21}H_{15}CoGeO_4$, (+)-Tetracarbonyl(methyl-1-naphthylphenylgermyl)cobalt, 43B, 930
$C_{21}H_{16}O_7Sn \cdot 2\ H_2O \cdot 0.5\ CH_3OH$, Tris(tropolonato)monohydroxotin(IV), 35B, 751
$C_{21}H_{22}BrNSn$, C,N-(2-((Dimethylamino)methyl)phenyl)diphenyltin bromide, 42B, 596
$C_{21}H_{22}Ge$, Tri-o-tolylgermane, 41B, 806
$C_{21}H_{24}BrNOSn$, Bromo(8-dimethylaminomethyl-5-methoxynaphthyl-C[1],N)-methylphenyltin(IV), 46B, 749
$C_{21}H_{27}ClMoO_2Sn$, ((π-Cycloheptatrienyl)dicarbonylmolybdenum)diphenylchlorotin, 39B, 549
$C_{21}H_{57}ClSi_6Sn$, Tris(bis(trimethylsilyl)methyl)chlorotin(IV), 43B, 931
$C_{22}H_{16}Cl_4O_6Sn$, Tris(tropolonato)monochlorotin(IV), 35B, 751
$C_{22}H_{18}Cl_2N_2Sn$, 2,2'-Bipyridyldichlorodiphenyltin, 40B, 691
$C_{22}H_{24}Cl_2N_8O_{16}Pb$, Bis(10-methylisoalloxazine)lead(II) perchlorate tetrahydrate, 41B, 807
$C_{22}H_{24}Sn$, (+)-2-Triphenylstannylbutane, 43B, 931
$C_{22}H_{25}O_2PS_2Sn$, (O,O'-Diethyl dithiophosphato)triphenyltin(IV), 45B, 810
$C_{22}H_{30}N_2S_4Sn$, Bis(N,N-diethyldithiocarbamato)diphenylstannane, 40B, 852
$C_{22}H_{31}NO_6Pb$, Tetramethylammonium triacetatodiphenylplumbate(IV), 38B, 721
$C_{22}H_{70}N_{32}O_{12}Pb_3S_{16}$, Tetrakis(thiourea)lead(II) formate, 38B, 711
$C_{23}H_{15}MnO_5Sn$, Triphenyltin pentacarbonyl manganese, 32B, 335
$C_{23}H_{19}NPbS$, Triphenyl(pyridine-4-thiolato)lead, 45B, 810
$C_{23}H_{19}NSSn$, (4-Thiopyridone)triphenyltin, 39B, 549
$C_{23}H_{20}N_2O_4Sn$, Nitratotriphenyl(pyridine N-oxide)tin(IV) (monoclinic), 43B, 932
$C_{23}H_{20}N_2O_4Sn$, Nitratotriphenyl(pyridine N-oxide)tin(IV) (triclinic), 43B, 933
$C_{23}H_{22}CrGeO_5S_2$, Pentacarbonyl(bis(mesitylthio)germylene)chromium(0), 44B, 678
$C_{23}H_{24}O_2Sn$, Tribenzyltin acetate, 33B, 317
$C_{24}F_{20}Ge$, Tetrakis(pentafluorophenyl)germanium(IV), 39B, 539; 40B, 692
$C_{24}F_{20}Sn$, Tetrakis(pentafluorophenyl)tin(IV), 40B, 692
$C_{24}H_{17}Br_2FSSn$, Triphenyltin 2,6-dibromo-4-fluorothiophenolate, 40B, 693

$C_{24}H_{18}FNO_2Pb$, Triphenyllead 2-fluoro-4-nitrosophenolate, 43B, 933
$C_{24}H_{18}Ge$, 9,9-Diphenyl-9-germafluorene, 41B, 722
$C_{24}H_{19}BrPbS$, Triphenyllead 2-bromothiophenolate, 45B, 811
$C_{24}H_{20}Fe_2O_4Sn$, Bis(dicarbonyl-π-cyclopentadienyliron)dicyclopentadienyltin, 34B, 414
$C_{24}H_{20}Ge$, Tetraphenylgermanium, 37B, 423; 38B, 722
$C_{24}H_{20}O_4P_2Pb$, Lead(II) bis(diphenylphosphinate), 42B , 597
$C_{24}H_{20}Sn$, Tetraphenyltin, 27, 875; 35B, 490
$C_{24}H_{22}N_2O_4S_4Sn$, Bis(O-ethylxanthato)bis(quinolin-8-olato)tin(IV), 44B, 678
$C_{24}H_{28}ClOPSn$, Chlorotrimethyl(triphenylphosphoranylideneacetone)-tin(IV), 41B, 808
$C_{24}H_{36}Sn_2$, 7,7,14,14-Tetramethyldinaphtho[1,8-bc:1',8-fg](1,5)distannocin, 43B, 934
$C_{24}H_{38}O_4P_2S_4Sn$, Bis(O,O'-diisopropyl dithiophosphato)diphenyl-tin(IV), 46B, 750
$C_{25}H_{19}NOSSn$, Diphenyltin - schiff base complex, 40B, 989
$C_{25}H_{19}NO_2Sn$, (2-(((2-Hydroxyphenyl)imino)methyl)phenolato(2-)-N,O,O')diphenyltin, 42B, 598
$C_{25}H_{20}Cl_8MnO_2PSn_2$, Dicarbonylcyclopentadienyl(trichlorostannyl)(triphenylphosphine)manganese pentachlorostannate, 43B, 934
$C_{25}H_{22}SSn$, Triphenyltin 2-methylthiophenolate, 40B, 693
$C_{25}H_{22}Sn$, Triphenyl-7-cyclohepta-1,3,5-trienyltin, 38B, 720
$C_{25}H_{23}AsCl_4Sn$, Tetraphenylarsonium methyltetrachlorostannate, 42B, 598
$C_{26}H_{20}Cl_2N_2O_4Sn$, Bis(N-benzoyl-N-phenylhydroxylaminato)dichloro-tin(IV), 42B, 599
$C_{26}H_{22}Ge$, 5,5-Diphenyl-10,11-dihydro-5H-dibenzo[b,f]germepin, 38B, 723
$C_{26}H_{24}PbS$, Triphenyllead 2,6-dimethylthiophenolate, 45B, 811
$C_{27}H_{21}NPbS$, (8-Mercaptoquinolinato-S)triphenyllead(IV), 46B, 750
$C_{27}H_{21}NSSn$, (8-Mercaptoquinolinato-S)triphenyltin(IV), 46B, 750
$C_{27}H_{26}SSn$, Triphenyltin 2,4,6-trimethylthiophenolate, 39B, 550
$C_{28}H_{20}Fe_2O_8Sn_2$, Di-μ-bis(cyclopentadienyl)stannyl-bis(tetracarbonyl-iron), 41B, 808
$C_{28}H_{24}Ge_2$, 1,1,4,4-Tetraphenyl-1,4-digermanacyclohexa-2,5-diene, 31B, 333; 32B, 336
$C_{28}H_{28}GeS$, Triphenylgermanium p-t-butylphenylmercaptide, 40B, 694
$C_{28}H_{28}I_2Sn_2$, 1,4-Bis(iododiphenyltin)butane, 34B, 414
$C_{28}H_{28}SSn$, Triphenyltin p-t-butylphenylsulfide, 39B, 550
$C_{28}H_{28}Sn$, Tetra(3-methylphenyl)tin, 43B, 935
$C_{28}H_{28}Sn$, Tetra-p-tolyltin, 41B, 809
$C_{29}H_{33}N_5O_4Sn$, 2,6-Diacetylpyridine-bis(salicyloylhydrazonato)dipropyltin(IV), 46B, 751
$C_{30}H_{22}O_4Sn$, Bis(1,3-diphenylpropane-1,3-dionato)tin(II), 43B, 936
$C_{30}H_{25}AsN_2O_7Sn$, Dinitratodiphenyl(triphenylarsine oxide)tin(IV), 44B, 679
$C_{30}H_{25}N_2O_7PSn$, Dinitratodiphenyl(triphenylphosphine oxide)tin(IV), 44B, 679
$C_{30}H_{26}Br_2Sn$, (4-Bromo-1,2,3,4-tetraphenyl-cis,cis-1,3-butadienyl)dimethyltin bromide, 35B, 491
$C_{30}H_{26}Co_2Sn$, Diphenylbis(norbornadienedicarbonylcobalt)tin(IV), 37B, 418
$C_{30}H_{46}GeO_2$, Bis(2,6-di-tert-butyl-4-methylphenoxo)germanium(II), 46B, 752
$C_{30}H_{46}O_2Sn$, Bis(2,6-di-tert-butyl-4-methylphenoxo)tin(II), 46B, 752
$C_{31}H_{25}NO_2Sn$, N-Benzoyl-N-phenyl-O-(triphenylstannyl)hydroxylamine,

40B, 695
$C_{33}H_{26}O_2Sn$, (1,3-Diphenylpropane-1,3-dionato)triphenyltin(IV), 41B, 810
$C_{36}Bi_2F_{30}Ge_3$, Tris(μ-bis(pentafluorophenyl)germyl)dibismuth, 46B, 752
$C_{36}H_{30}AsNO_4Sn$, Nitratotriphenyl(triphenylarsine oxide)tin(IV), 43B, 937
$C_{36}H_{30}GeO_2Si$, Triphenyl(triphenylgermanylperoxy)silane, 45B, 811
$C_{36}H_{30}Ge_2$, Hexaphenyldigermanium, 46B, 753
$C_{36}H_{30}Ge_2 \cdot 2\ C_6H_6$, Hexaphenyldigermanium benzene solvate, 46B, 754
$C_{36}H_{30}Ge_2O$, Oxobis(triphenylgermanium(IV)), 38B, 723; 44B, 680
$C_{36}H_{30}Ge_2S$, Hexaphenyldigermanium sulfide (monoclinic), 45B, 812
$C_{36}H_{30}Ge_2S$, Hexaphenyldigermanium sulfide (orthorhombic), 45B, 812
$C_{36}H_{30}NO_4PSn$, Nitratophenyl(triphenylphosphine oxide)tin(IV), 42B, 600
$C_{36}H_{30}OPbSi$, Triphenylsiloxytriphenyllead(IV), 42B, 531
$C_{36}H_{30}OSn_2$, Oxobis(triphenyltin(IV)), 44B, 681
$C_{36}H_{30}Pb_2$, Hexaphenyldiplumbane, 38B, 723; 42B, 600
$C_{36}H_{30}SSn_2$, μ-Thio-bis(triphenyltin(IV)), 43B, 938
$C_{36}H_{30}SeSn_2$, Bis(triphenyltin)selenide, 45B, 812
$C_{36}H_{30}Sn_2$, Hexaphenyldistannane, 39B, 550
$C_{36}H_{31}BrSn$, (4-Bromo-1,2,3,4-tetraphenyl-cis,cis-1,3-butadienyl)dimethylphenyltin, 37B, 419
$C_{36}H_{31}ClSn$, (4-Chloro-1,2,3,4-tetraphenyl-cis,cis-1,3-butadienyl)dimethylphenyltin, 37B, 419
$C_{37}H_{49}Cl_2PSn$, Triphenyl(benzyl)phosphonium tributyldichlorostannate, 44B, 1120
$C_{38}H_{32}Fe_2O_{10}S_2Sn_2$, Di((benzenesulfinato)-μ-hydroxo-phenylstannio)-tetracarbonyl-di-π-cyclopentadienyl-di-iron, 37B, 420
$C_{38}H_{34}F_6FeOP_2Sn$, (1,2-Bis(diphenylphosphino)hexafluorocyclopentene)-carbonyl-π-cyclopentadienyl(trimethylstannyl)iron, 37B, 421
$C_{38}H_{48}N_2O_{12}S_2Sn_2$, μ-Oxalato-bis((di-n-propyl sulfoxide)nitratodiphenyltin(IV)), 39B, 551
$C_{40}H_{30}FeO_4Sn_2$, Tetracarbonylbis(triphenylstannyl)iron(II), 43B, 938
$C_{40}H_{30}MnO_4PSn$, Triphenyltin-tetracarbonyltriphenylphosphinemanganese, 32B, 481
$C_{40}H_{30}O_4OsSn_2$, trans-Bis(triphenyltin)tetracarbonylosmium, 40B, 696
$C_{40}H_{30}Pb_2 \cdot CH_2Cl_2$, μ-Butadiynylenebis(triphenyllead(IV)) dichloromethane solvate, 46B, 754
$C_{40}H_{30}Sn_2 \cdot CHCl_3$, μ-Butadiynylenebis(triphenyltin(IV)) chloroform solvate, 46B, 754
$C_{40}H_{72}Cl_{12}O_{10}Sn_4$, Tetrabutyl-1,3-trichloroacetoxydistannoxane dimer, 43B, 939
$C_{40}H_{78}Br_6O_8Sn_2$, Di-μ-hydroxo-bis[aquatribromotin(IV)]-1,8-epoxy-p-menthane, 46B, 740
$C_{40}H_{78}Cl_6O_8Sn_2$, Di-μ-hydroxo-bis[aquatrichlorotin(IV)]-1,8-epoxy-p-menthane, 46B, 740
$C_{40}H_{96}N_8O_4Si_4Sn_6$, Hexakis(μ-t-butylamido)-bis(μ₃-t-butylamido)-tetra(μ-dimethylsilyl)-tetra(μ₃-oxo)-hexatin, 45B, 813
$C_{41}H_{33}NOSn$, 4-(N-Methylanilino)-1-phenyl-3-(triphenyltin)-2-naphthol, 46B, 754
$C_{42}H_{26}Fe_2GeO_8$, μ-(1-(1,2,3,4-Tetraphenylbutadienyl)phenylgermylene)octacarbonyldiiron(Fe-Fe), 46B, 755
$C_{42}H_{42}Ge_2O$, Oxo-bis[tribenzylgermanium(IV)], 45B, 814
$C_{42}H_{42}OSn_2$, μ-Oxo-bis[tribenzyltin(IV)], 45B, 814
$C_{44}H_{37}AsCl_2OSn$, Triphenyl(phenacyl)arsonium triphenyldichlorostannate, 44B, 1120

$C_{44}H_{68}O_{12}Sn_2$, Di-μ-(4,6-di-O-benzylidene-α-D-glucopyranoside-2,3-D-O-yl)-bis(dibutyl-tin(IV)), 45B, 815
$C_{48}H_{40}Ge_4$, Octaphenylcyclotetragermane, 46B, 756
$C_{48}H_{40}Ge_4S$, Octaphenyl-thia-tetragermacyclopentane, 46B, 756
$C_{49}H_{40}Br_4Pb_4$, Tetrakis(diphenylbromoplumbyl)methane, 35B, 492
$C_{50}H_{35}MnO_4Sn$, Triphenylstannoxytetraphenylcyclopentadienyl tricarbonylmanganese, 32B, 367
$C_{52}H_{40}GeN_4$, Tetrakis(diphenylketimine)germanium, 41B, 1242
$C_{52}H_{40}N_4Sn$, Tetrakis(diphenylketimine)tin, 41B, 1242
$C_{54}H_{45}NO_3Sn_4$, Nitratotris(triphenylstannyl)tin(IV), 41B, 811 43B, 928
$C_{54}H_{54}CuKO_8P_2S_6Sn$, Potassium bis(tri-p-tolylphosphine)copper(I) tris(dithiooxalato-O,O')stannate(IV) bis(acetone), 43B, 940
$C_{62}H_{54}Cl_2O_2P_2Sn_2$, μ-Bis(diphenylphosphinyl)ethanebis(chlorotriphenyltin), 46B, 757
$C_{62}H_{54}N_2O_8P_2Sn_2$, μ-Bis(diphenylphosphinyl)ethane-bis(nitratotriphenyltin), 45B, 815
$C_{64}H_{48}N_8O_{36}Sn_4$, Octakis-μ-(o-nitrobenzoato)-di-μ_3-oxo-bis(tetrahydrofuran)ditin(II)-ditin(IV), 42B, 601
$C_{66}H_{72}Cl_2N_{12}O_{14}Pb$, Lead hexa-antipyrine perchlorate, 31B, 334
$C_{72}H_{30}Bi_2F_{30}Ge_3P_2Pt \cdot C_6H_6$, Bis(μ-bis(pentafluorophenyl)germyl)-(μ-(bis(pentafluorophenyl)germyl-bis(triphenylphosphine)platinum)dibismuth benzene solvate, 46B, 752
$C_{72}H_{60}As_2N_2O_6Sn_4$, Di-μ-nitrato-bis((triphenylarsine)(triphenylstannio)tin(II)), 43B, 941
$C_{72}H_{60}Ge_6 \cdot 7\ C_6H_6$, Dodecaphenylcyclohexagermanium heptabenzene solvate, 46B, 758
$C_{82}H_{70}Ge_4Hg_3Ni_2 \cdot C_7H_8$, Bis(($\eta^5$-cyclopentadienyl)-(triphenylgermyl)-(triphenylgermylmercury)-nickel)mercury toluene solvate, 46B, 758
$C_{88}H_{80}Sn_6$, Dodecaphenylcyclohexastannane m-xylene solvate, 28, 604

70. TELLURIUM COMPOUNDS

$C_2H_6Cl_2Te$, Dimethyltellurium dichloride, 22, 563
$C_2H_6I_2Te$, α-Di-iodo(dimethyl)tellurium, 38B, 724
$C_2H_6I_4Te$, Dimethyltellurium tetraiodide, 45B, 816
$C_2H_6O_4S_4Te$, Tellurium dimethanethiosulphonate, 18, 678
$C_2H_8Br_2N_4S_2Te$, Bis(thiourea)tellurium(II) bromide, 31B, 335
$C_2H_8Cl_2N_4S_2Te$, Bis(thiourea)tellurium(II) chloride, 31B, 335
$C_4H_6O_2S_4Te$, Tellurium di(methylxanthate), 41B, 811
$C_4H_8Br_2STe$, 1-Thia-4-telluracyclohexane 4,4-dibromide, 38B, 724
$C_4H_8I_2OTe$, 1-Oxa-4-telluracyclohexane 4,4-diiodide, 39B, 553
$C_4H_{12}I_4Te_2$, β-Dimethyltellurium diiodide, 32B, 337
$C_4H_{12}O_4P_2S_4Te$, Tellurium(II) dimethyldithiophosphate, 31B, 336
$C_4H_{14}N_4O_4S_6Te$, Tellurium dimethanethiosulphonate thiourea complex, 30B, 284
$C_4H_{16}Cl_2N_8S_4Te$, Tetrathioureatellurium(II) dichloride, 30B, 285
$C_4H_{16}Cl_2N_8Se_4Te$, Tetrakis(selenourea)tellurium(II) dichloride, 37B, 423
$C_4H_{20}Cl_2N_8O_2S_4Te$, Tetrathioureatellurium(II) dichloride dihydrate, 30B, 285
$C_5H_4O_2Te$, α-Tellurophene-carboxylic acid, 38B, 725
$C_5H_6Cl_2O_2Te$, 1,1-Dichloro-1-telluracyclohexane-3,5-dione, 42B, 601
$C_5H_6O_2Te$, Acetylacetonetellurium(II), 42B, 601; 43B, 942
$C_6H_5Br_{1.3}Cl_{1.7}Te$, Phenyl tellurium trihalide, 46B, 759
$C_6H_{10}Br_6Te_2$, Di-μ-bromo-μ-1,2-cyclohexylenetetrabromoditellurium,

38B, 725
$C_6H_{10}O_2S_4Te$, Tellurium di(ethylxanthate), 32B, 338
$C_6H_{12}Br_2N_4S_2Te$, Tellurium dibromide ethylenethiourea complex, 30B, 287
$C_6H_{12}I_2N_4S_2Te$, Tellurium diiodide ethylenethiourea complex, 30B, 287
$C_6H_{28}F_8N_{12}S_6Te_2$, Trithioureatellurium(II) hydrogen difluoride, 30B, 286
C_7H_5BrOTe, o-Formylphenyltellurenyl bromide, 40B, 696
$C_7H_6BrNOTe \cdot C_2H_6OS$, 2-(Bromotelluro)benzamide dimethylsulfoxide solvate, 45B, 817
C_7H_7NTe, 1,2-Benzisotellurazole, 44B, 355
$C_7H_9BrN_2STe$, Benzenetellurenyl bromide thiourea complex, 31B, 337
$C_7H_9ClN_2STe$, Benzenetellurenyl chloride thiourea complex, 31B, 337
$C_7H_{10}O_2Te$, (Heptane-3,5-dionato(2-))tellurium(II), 43B, 943
$C_7H_{10}O_2Te$, 1,3-Dimethylacetylacetone tellurium(II), 43B, 942
$C_7H_{10}O_2Te$, 3,3-Dimethylpentane-2,4-dionato(2-)-C(1),C(5)-tellurium(II), 43B, 944
$C_8H_8ClNOTe$, 2-(Chlorotelluro)-N-methylbenzamide, 45B, 817
$C_8H_8I_2Te$, 1,1-Di-iodo-3,4-benzo-1-tellura-cyclopentane (α-form), 45B, 817
$C_8H_9Cl_3OTe$, (4-Ethoxyphenyl)tellurium(IV) trichloride, 46B, 760
$C_8H_{12}N_6S_2Se_2Te$, trans-Diselenocyanato-bis(ethylenethiourea)-tellurium(II), 37B, 424
$C_8H_{12}N_6S_4Te$, trans-Dithiocyanato-bis(ethylenethiourea)tellurium(II), 37B, 424
$C_8H_{12}O_2Te$, 4-Methylheptane-3,5-dionato(2-)-C(2),C(6)-tellurium(II), 43B, 944
$C_8H_{13}ClN_4S_2Te$, Phenylbis(thiourea)tellurium(II) chloride, 31B, 340; 44B, 681
$C_8H_{13}ClN_4Se_2Te$, Phenylbis(selenourea)tellurium(II) chloride, 44B, 681
$C_8H_{16}Br_2N_4S_2Te$, cis-Dibromo-bis(trimethylenethiourea)-tellurium(II), 41B, 812
$C_8H_{16}Cl_2N_4S_2Te$, cis-Dichloro-bis(trimethylenethiourea)-tellurium(II), 41B, 812
$C_8H_{18}N_4O_4S_6Te$, trans-Dimethanethiosulphonatobis(ethylenethiourea)tellurium(II), 39B, 554
$C_8H_{20}P_2S_2Se_2Te$, Tellurium bis(diethylthioselenophosphinate), 34B, 380
$C_9H_8Cl_3NO_2Te$, 2,6-Diacetylpyridine(C,N,O)-tellurium(IV) trichloride, 46B, 761
$C_9H_{11}BrN_2STe$, Bromo(ethylenethiourea)phenyltellurium(II) (polymorph I), 41B, 813
$C_9H_{11}BrN_2STe$, Bromo(ethylenethiourea)phenyltellurium(II) (polymorph II), 41B, 813
$C_9H_{11}BrN_2SeTe$, Bromo(ethyleneselenourea)phenyltellurium(II), 41B, 814
$C_9H_{11}ClN_2STe$, Chloro(ethylenethiourea)phenyltellurium(II), 41B, 814
$C_9H_{11}IN_2STe$, Iodo(ethylenethiourea)phenyltellurium(II), 41B, 814
$C_9H_{11}IN_2SeTe$, Iodo(ethyleneselenourea)phenyltellurium(II), 41B, 814
$C_9H_{17}Br_3OTe$, cis-2-Ethoxycycloheptyltribromotellurium(IV), 46B, 761
$C_{10}H_{16}N_2O_2S_4Te$, Tellurium di(morpholyldithiocarbamate), 35B, 498
$C_{10}H_{16}N_6S_2Se_2Te$, trans-Diselenocyanatobis(trimethylenethiourea)tellurium(II), 37B, 424
$C_{10}H_{17}Cl_3OTe$, Trichloro-(8-ethoxy-4-cyclooctenyl)tellurium, 45B, 818
$C_{10}H_{20}N_2S_4Te$, Bis-(N,N-diethyldithiocarbamato)tellurium(II), 37B, 425

$C_{10}H_{24}Br_4N_4S_2Te$, trans-Tetrabromobis(tetramethylthiourea)tellurium(IV), 34B, 416
$C_{10}H_{24}Cl_4N_4S_2Te$, trans-Tetrachloro-bis(tetramethylthiourea)-tellurium(IV) (monoclinic form, B), 41B, 815
$C_{10}H_{24}Cl_4N_4S_2Te$, trans-Tetrachloro-bis(tetramethylthiourea)-tellurium(IV), 34B, 416
$C_{11}H_{15}BrTe$, 1-Phenyl-1-telluroniacyclohexane bromide, 45B, 818
$C_{12}H_8Cl_2I_2Te$, Di-p-chlorodiphenyltellurium diiodide, 27, 864
$C_{12}H_8Cl_2OTe$, Phenoxatellurin 10,10-dichloride, 46B, 761
$C_{12}H_8Cl_2Te_2$, 4,4'-Dichlorodiphenyl ditelluride, 21, 595
$C_{12}H_8I_2OTe$, Phenoxatellurin 10,10-diiodide, 39B, 555
$C_{12}H_8I_2Te$, Dibenzotellurophene diiodide, 41B, 816
$C_{12}H_8N_2O_7Te$, Phenoxatellurine dinitrate, 39B, 556
$C_{12}H_8OTe$, Phenoxatellurine, 39B, 556
$C_{12}H_8O_2S_2Te$, Bis[monothiopyrocatecholato(2-)]tellurium(IV), 46B, 762
$C_{12}H_8O_4Te$, Tellurium(IV) catecholate, 32B, 339
$C_{12}H_8Te$, Dibenzotellurophene, 41B, 816
$C_{12}H_9Br_3Te$, 2-Biphenylyltellurium tribromide, 43B, 945
$C_{12}H_9ClN_2Te$, (2-Phenylazophenyl-C,N')-tellurium(II) chloride, 45B, 819
$C_{12}H_9I_3Te$, 2-Biphenylyltellurium triiodide (α-form), 42B, 604
$C_{12}H_9I_3Te$, 2-Biphenylyltellurium triiodide (β-form), 43B, 946
$C_{12}H_{10}Br_2Te$, Diphenyltellurium dibromide, 22, 704
$C_{12}H_{10}F_2Te$, Diphenyltellurium(IV) difluoride, 46B, 762
$C_{12}H_{10}O_4S_4Te$, Tellurium dibenzenethiosulphonate, 20, 583; 37B, 426
$C_{12}H_{10}Te_2$, Diphenyl ditelluride, 38B, 726
$C_{12}H_{17}N_3S_2Te$, Tetramethylammonium phenyldithiocyanatotellurate(II), 41B, 817
$C_{12}H_{17}N_3Se_2Te$, Tetramethylammonium phenyl-diselenocyanatotellurate(II), 41B, 817
$C_{12}H_{20}Cl_4Te$, Di-cis-2-chlorocyclohexyldichlorotellurium(IV), 46B, 763
$C_{12}H_{24}BrN_3O_3S_6Te$, Bromotris[N-(2-hydroxyethyl)-N-methyldithiocarbamato]tellurium(IV), 45B, 821
$C_{12}H_{24}Br_4N_8S_4Te_2$, Di-$\mu$-bromo-bis(bis(ethylenethiourea)tellurium(II)) dibromide (form I), 42B, 603
$C_{12}H_{24}Br_4N_8S_4Te_2$, Di-$\mu$-bromo-bis(bis(ethylenethiourea)tellurium(II)) dibromide (form II), 42B, 603
$C_{12}H_{24}Cl_6N_8S_4Te_2$, Tetrakis(ethylenethiourea)tellurium(II) hexachlorotellurate(IV), 44B, 682
$C_{12}H_{24}N_3O_3PTe$, Tris(morpholino)phosphine telluride, 46B, 763
$C_{12}H_{24}N_6S_2Se_2Te$, trans-Diselenocyanatobis(tetramethylthiourea)tellurium(II), 37B, 424
$C_{12}H_{28}Cl_2N_8O_2S_4Te$, Tetrakis(ethylenethiourea)tellurium(II) chloride dihydrate, 43B, 947
$C_{12}H_{32}Br_2N_8S_4Te$, trans-Dithioureabis(tetramethylthiourea)tellurium(II) bromide, 37B, 426
$C_{12}H_{32}Cl_2N_8S_4Te$, Tetrakis(dimethylthiourea)tellurium(II) dichloride, 46B, 764
$C_{12}H_{32}Cl_2N_8S_4Te$, trans-Dithioureabis(tetramethylthiourea)tellurium(II) chloride, 37B, 426
$C_{13}H_{17}Cl_3Te$, p-Tolyl-2-chlorocyclohexyldichlorotellurium(IV), 46B, 764
$C_{14}H_{10}Cl_4N_2S_4Te \cdot 2\ C_4H_8O_2$, trans-Bis(benzothiazolethione-S)tetrachlorotellurium(IV) dioxane solvate, 45B, 819
$C_{14}H_{12}Cl_4N_4S_2Te \cdot 3\ C_4H_8O_2$, cis-Bis(benzimidazole-thione-S)tetrachlorotellurium(IV) dioxane solvate, 45B, 820

$C_{14}H_{14}I_6O_2Te_2$, (4-Methoxyphenyl)tellurium(IV) triiodide dimer, 46B, 760
$C_{14}H_{14}O_4S_4Te$, Tellurium di-p-tolylthiosulphonate, 19, 574
$C_{14}H_{14}Te_2$, p,p'-Ditolyl ditelluride, 45B, 820
$C_{14}H_{16}OTe$, 9-Telluro-1-formyl-3,4,5,6,7,8-hexahydro-2H-anthracene, 43B, 947
$C_{15}H_{13}NO_5Te$, 10-Acetonylphenoxatellurine nitrate, 39B, 556
$C_{15}H_{30}ClN_3S_6Te \cdot C_4H_8O_2$, Chlorotris(N,N-diethyldithiocarbamato-S,S')tellurium(IV) dioxane solvate, 46B, 765
$C_{16}H_8F_6O_5Te$, Phenoxatellurine bis(trifluoroacetate), 39B, 556
$C_{16}H_{18}Br_6O_2Te_2$, (4-Ethoxyphenyl)tellurium(IV) tribromide dimer, 46B, 760
$C_{16}H_{32}N_4O_4S_8Te$, Tetrakis[N-(2-hydroxyethyl)-N-methyldithiocarbamato]tellurium(IV), 45B, 821
$C_{17}H_{35}NO_3S_6Te$, Tetraethylammonium tris(O-ethylxanthato)tellurate(II), 42B, 604
$C_{18}H_{22}N_4O_4S_6Te$, trans-Dibenzenethiosulphonatobis(ethylenethiourea)-tellurium(II), 37B, 427
$C_{18}H_{36}Cl_4N_{12}O_{16}S_6Te_2$, Hexakis(ethylenethiourea-S)ditellurium(II) perchlorate, 46B, 765
$C_{19}H_{15}NSTe$, Triphenyltelluronium thiocyanate, 43B, 948
$C_{20}H_{26}N_4O_4S_6Te$, trans-Dibenzenethiosulphonato-bis(trimethylenethiourea)tellurium(II), 40B, 1105
$C_{20}H_{40}N_4S_8Te$, Tetrakis(diethyldithiocarbamato)tellurium(IV), 39B, 557
$C_{21}H_{35}N_3S_6Te$, Tris(diethyldithiocarbamato)phenyltellurium(IV), 38B, 726
$C_{22}H_{24}O_4Te$, Di(p-anisyl)telluronium 4,4-dimethyl-2,6-dioxocyclohexanilide, 44B, 682
$C_{22}H_{34}N_4O_4S_6Te$, Tellurium di-benzenethiosulphonate di-tetramethylthiourea, 40B, 697
$C_{24}H_{16}N_2O_9Te_2$, Bisphenoxatelluronium dinitrate, 39B, 556
$C_{24}H_{48}Cl_4N_{12}O_{16}S_6Te_2$, Hexakis(trimethylenethiourea-S)ditellurium(II) perchlorate (monoclinic), 46B, 765
$C_{24}H_{48}Cl_4N_{12}O_{16}S_6Te_2$, Hexakis(trimethylenethiourea-S)ditellurium(II) perchlorate (triclinic), 46B, 765
$C_{26}H_{20}N_2OS_2Te_2$, Bis(isothiocyanatodiphenyltellurium(IV)) oxide, 43B, 948
$C_{26}H_{60}N_2PdS_2Si_4Te_2$, trans-Bis(thiocyanato)bis[di(dimethylsilyl-3-propyl)tellurium]palladium(II), 45B, 822
$C_{27}H_{29}BTe$, Trimethyltelluronium tetraphenylborate, 45B, 822
$C_{36}H_{30}Cl_2Te_2$, Triphenyltelluronium chloride, 46B, 766
$C_{38}H_{50}N_4O_4S_8Te$, Tetrakis-(4-morpholinecarbodithioato)tellurium(IV) tribenzene solvate, 41B, 429
$C_{42}H_{35}HgPTe_3$, Tetraphenylphosphonium tris(tellurophenolato)-mercurate(II), 43B, 949
$C_{78}H_{62}Cl_6N_4O_4Te_4$, Triphenyltelluronium cyanate tetramer chloroform solvate, 42B, 605

71. TRANSITION METAL-CARBON COMPOUNDS

$CBrCl_3Hg$, Trichloromethylmercury(II) bromide, 33B, 320
CF_3HgN_3, Trifluoromethyl(azido)mercury, 44B, 1121
CH_3BrHg, Methyl mercury(II) bromide (gas-mw), 18, 677
CH_3ClHg, Methyl mercury(II) chloride (gas-mw), 18, 677
CH_3ClHg, Methyl mercury(II) chloride, 13, 449

CH_3HgN_3, Methylmercury azide, 40B, 698
C_2F_3HgNO, Trifluoromethyl(isocyanato)mercury, 44B, 1121
C_2F_6Hg, Bis(trifluoromethyl)mercury(II), 43B, 950
$C_2H_2BrClHg$, β-Chlorovinylmercury(II) bromide, 11, 609
$C_2H_2Cl_2Hg$, cis-2-Chlorovinylmercury chloride (gas-mw-ed), 38B, 1068
$C_2H_2Cl_2Hg$, trans-β-Chlorovinyl mercury chloride, 31B, 341
C_2H_3ClHgO, (Chloromercurio)acetaldehyde, 45B, 823
$C_2H_3ClHgO_2$, Methoxycarbonylmercury(II) chloride, 27, 816
C_2H_3CuIN, Methyl isocyanide - copper(I) iodide, 24, 547
C_2H_3HgN, Methylmercury(II) cyanide, 33B, 320
C_2H_5BrHg, Ethyl mercury(II) bromide, 13, 449
C_2H_5ClHg, Ethyl mercury(II) chloride, 13, 449
C_2H_5IZn, Ethylzinc iodide, 39B, 558
C_2H_6Hg, Dimethylmercury (gas-ed), 5, 151; 39B, 888
C_2H_7AuO, Dimethylgold(III) hydroxide, 33B, 321
$C_2H_7HgN_2Se$, Methyl-(selenourea)-mercury(II) nitrate, 45B, 1283
$C_3H_3AuN_2$, Cyano(methylisocyanide)gold(I), 42B, 605
C_3H_5BrHgO, 2-Oxopropylbromomercury, 44B, 682
C_3H_7ClHg, Propyl mercury(II) chloride, 13, 449
$C_3H_8AuF_3O_4S$, Dimethylgold trifluoromethanesulphonate monohydrate, 43B, 951
C_3H_9ClPt, Trimethylplatinum chloride, 11, 607
C_3H_9IPt, Trimethylplatinum(IV) iodide, 33B, 322
$C_3H_{10}OPt$, Trimethylplatinum(IV) hydroxide, 33B, 323
C_4H_9ClHg, Butyl mercury(II) chloride, 13, 449
$C_4H_{12}Li_2Zn$, Lithium tetramethylzincate, 33B, 299
$C_4H_{12}Pt$, Tetramethylplatinum, 11, 608
$C_5H_2Hg_4N_4O$, Tetrakis(cyanomercuri)methane hydrate, 44B, 1122
$C_5H_6Cl_2Hg_2O_2$, 3,3-Bis(chloromercury)-2,4-pentadione, 41B, 817
$C_5H_6HgN_2S$, Methyl 2-mercaptopyrimidinatomercury(II), 44B, 683
$C_5H_7HgN_3OS \cdot H_2O$, (4-Amino-2-mercapto-6-pyrimidionato)methylmercury-(II) monohydrate, 46B, 767
$C_6H_3AlBr_3MnO_5$, Tetracarbonylmanganese[(μ-acetyl)(μ-bromo)dibromoaluminum], 46B, 767
$C_6H_3MnO_5$, Methylmanganese pentacarbonyl (gas-ed), 35B, 897
$C_6H_3O_5Re$, Methylrhenium pentacarbonyl (gas-ed), 42B, 1029
C_6H_5BrHg, Phenyl mercuric bromide (gas-ed), 33B, 532
$C_6H_8Cl_4K_2O_8Pd$, Potassium dichloro(chloranylic acid)palladate(II) tetrahydrate, 40B, 698
$C_6H_8HgN_2O_3$, Methyl(pyridine)mercury(II) nitrate, 44B, 683
$C_6H_9HgN_3S$, (4-Amino-5-methyl-2-pyrimidinethiolato)methylmercury(II), 46B, 767
$C_6H_{14}ClHgN$, N-(2-(Chloromercuri)ethyl)diethylamine, 39B, 559
$C_6H_{14}Cl_2N_2O_2Pd$, cis-Dichlorobis(methylamino(methoxy)carbene)palladium(II), 41B, 818
$C_6H_{17}AuOS$, Trimethylgold(III)-methylene-dimethylsulfoxoniumylide, 43B, 954
$C_7HF_2MnO_5$, cis-1,2-Difluorovinylpentacarbonylmanganese, 32B, 341
C_7H_5HgN, Phenylmercury(II) cyanide, 42B, 606
$C_7H_6BrMnN_2O_3$, Bromotricarbonylbis(methyl isocyanide)manganese, 38B, 727
$C_7H_6Cl_2FeO_4Si$, 2,2,2,2-Tetracarbonyl-1,1-dichloro-1-sila-2-ferracyclopentane, 45B, 823
$C_7H_6FeO_3$, Tris(methylene)methane-tricarbonyliron (gas-ed), 41A, 1140
C_7H_7ClHg, Benzyl mercury chloride, 45B, 824
$C_7H_7MnN_2O_3$, (η^5-Cyclopentadienyl)-carbonyl-nitrosyl-carbamoyl-manganese, 45B, 824

$C_7H_9MnN_2O_5$, (N-Methylcarboxamido)(methylamine)tetracarbonylmanganese, 34A, 164
$C_7H_{10}ClHg_2N_5O_4 \cdot H_2O$, μ-Adeninatobis(methylmercury(II)) perchlorate monohydrate, 46B, 1049
$C_7H_{10}HgN_6O_3$, Methyl(9-methyladenine)mercury(II) nitrate, 46B, 1050
$C_7H_{10}HgN_6O_4$, (9-Methylguanine)methylmercury(II) nitrate, 46B, 1050
$C_7H_{10}Hg_2N_6O_3 \cdot 2\ H_2O$, Adeninyl-bis(methylmercury) nitrate dihydrate, 45B, 1152
$C_7H_{12}ClO_3Pt$, Acetylacetonatochloro(1,2-dihapto(vinyl alcohol))platinum(II), 39B, 560
$C_7H_{13}ClHgO$, α-2-Methoxycyclohexyl mercury(II) chloride, 15, 454
$C_7H_{13}ClHgO$, β-2-Methoxycyclohexyl mercury(II) chloride, 15, 455
$C_7H_{14}AuN$, Dipropylcyanogold, 7, 216, 232
$C_8H_2Br_2Fe_2O_6$, 1-H1:1,2-H2-trans-2-Bromovinyl-μ-bromo-bis(tricarbonyliron), 38B, 728
$C_8H_2F_8FeO_4$, Bis(1,1,2,2-tetrafluoroethyl)iron(II) tetracarbonyl, 32B, 341
$C_8H_2FeO_6$, Tetracarbonyl-ferra-3-cyclopentene-2,5-dione, 43B, 954
$C_8H_3MnO_7$, Pyruvoylpentacarbonylmanganese(I), 42B, 606
$C_8H_4CdK_2$, Potassium tetraethinylcadmate(II), 33B, 324
$C_8H_4K_2Zn$, Potassium tetraethinylzincate(II), 33B, 324
$C_8H_6HgS_2$, Di(2-thienyl)mercury, 45B, 825
$C_8H_7NO_3Ru$, Dicarbonyl-carbamoyl-(η^5-cyclopentadienyl)ruthenium, 45B, 825
$C_8H_7O_6Re$, Diacetyltetracarbonylrhenium, 42B, 607
$C_8H_8BrO_2Re$, π-Cyclopentadienyl-σ-methyl-bromodicarbonylrhenium, 39B, 561
$C_8H_8HgO_2$, Phenylmercury(II) acetate, 38B, 728
$C_8H_{10}HgN_4O_4$, Bis(ethyl diazoacetate)mercury(II), 43B, 1224
$C_8H_{12}ClHg_2N_5O_4$, 9-Methyladeninyl-bis(methylmercury) perchlorate, 45B, 1152
$C_8H_{12}ClI_3MoN_4O$, trans-Oxochlorotetrakis(methylisocyanide)molybdenum(IV) triiodide, 42B, 607
$C_8H_{12}FeN_2O_2S_3$, Dicarbonyl(dimethylthiocarboxamido)dimethyldithiocarbamatoiron, 44B, 1123
$C_8H_{13}Cl_2Hg_3N_5O_8$, μ-Adeninato-3,7,9-tris(methylmercury(II)) perchlorate, 46B, 1052
$C_8H_{13}Hg_3N_7O_6$, μ_3-(Adeninato-N^3,N^7,N^9)tris(methylmercury(II)) nitrate, 46B, 1036
$C_8H_{14}HgN_2O_2$, t-Butyl(ethyl diazoacetate)mercury(II), 43B, 1282
$C_8H_{14}Pt$, Cyclopentadienyl(trimethyl)platinum(IV), 34B, 418; 37B, 428
$C_8H_{16}Hg_2S_2$, trans-1,2-Dimercaptocyclohexane-bis[methylmercury(II)], 46B, 1207
$C_8H_{18}Cl_2O_5Re_2S$, Di-μ-acetato-O,O'-bis[chloro(methyl)rhenium(III)]-dimethylsulfoxide, 45B, 828
$C_8H_{20}D_6F_{12}N_6O_2P_2RuS$, trans-Tetraamminecarbonyl-2-(4,5-dimethylimidazolium)ruthenium(II) hexafluorophosphate (deuterated dimethyl sulfoxide solvate), 40B, 700
$C_8H_{22}AuNP_2$, Dimethylgold nitridobis(dimethylphosphoniummethylide), 43B, 955
$C_8H_{23}Cl_3N_7O_3Ru$, trans-Chloro-8-caffeinechlorotriammineruthenium(III) chloride monohydrate, 41B, 819
$C_8H_{24}Br_2Pt_2Se_2$, Di-μ-bromo[μ-(dimethyldiselenide-Se:Se')]hexamethyldiplatinum, 46B, 1209
$C_8H_{24}O_4Zn_4$, Tetrameric methylzinc methoxide, 46B, 1144
$C_9H_2Fe_2O_7S$, Tricarbonyl(1,2:4-5-η-(1,1,1-tricarbonyl-1-ferra-2-thia-4-cyclopenten-3-one))iron, 43B, 955

$C_9H_6Cl_2HgO_2$, 4-Chloro-3-chloromercuri-3,4-dihydrocoumarin, 17, 707
C_9H_7ClHgO, cis-β-Benzoylvinylmercury(II) chloride, 37B, 429
$C_9H_8FeN_2O_4$, Tetracarbonyl(1,3-dimethylimidazolinylidene)iron(0), 38B, 729
$C_9H_8FeO_4$, (4,4-Dimethylbuta-1,3-dienone)tricarbonyl-iron, 44B, 1123
$C_9H_8FeO_4$, π-Cyclopentadienyl(dicarbonyl)carboxymethyliron, 34B, 418
$C_9H_9BClO_6Re$, fac-(Triacetyltricarbonylrhenato)boronchloride, 46B, 769
$C_9H_{11}CrNO_6$, Pentacarbonyl(dimethylamino(ethoxy)carbene)chromium(0), 38B, 730
$C_9H_{12}Hg_4O_8 \cdot 2\ H_2O$, Tetrakis(acetoxymercury)methane dihydrate, 45B, 826
$C_9H_{13}ClHgO_2$, 2-exo-(Chloromercurio)-3-exo-acetoxybicyclo[2.2.1]heptane, 44B, 684
$C_9H_{15}BrHgO_2$, 1-Bromomercurimethyl-2,3-dioxadecalin, 46B, 769
$C_9H_{18}ClN_3RuS_5 \cdot C_3H_6O$, Chlorobis(N,N-dimethylthiocarbamato)(N,N-dimethylthiocarboxamido)ruthenium(IV) acetone solvate, 45B, 1084
$C_9H_{18}O_3Pt$, Ethyl(trimethylplatini)acetoacetate, 24, 628
$C_9H_{19}CoN_4O_5$, Aquo-methyl-bis(dimethylglyoximato)-cobalt(III), 40B, 701; 41B, 820
$C_9H_{22}Br_3NOPt$, Tetraethylammonium tribromodihydrocarbonylplatinum-(IV), 40B, 701
$C_9H_{23}AuP_2$, Dimethylgold methanidobis(dimethylphosphoniummethylide), 43B, 955
$C_{10}Cl_8Ni_4O_4$, Di-μ-chloro-di-μ-trichloropropenyl-tetra(carbonylnickel)(2Ni-Ni), 43B, 957; 44B, 684
$C_{10}H_2O_{10}Os_3$, Dihydridodecacarbonyltriosmium, 43B, 1355
$C_{10}H_4Fe_2O_6$, Tricarbonylferracyclopentadiene-tricarbonyliron, 42B, 697
$C_{10}H_6F_4Fe_2O_6S_2$, Di-μ-methylthio-μ-(tetrafluoroethane-1,1-diyl)-bis(tricarbonyliron), 45B, 827
$C_{10}H_6F_4Fe_2O_6S_2$, Di-μ-methylthio-μ-(tetrafluoroethane-1,2-diyl)-bis(tricarbonyliron), 45B, 827
$C_{10}H_6F_6FeN_2O$, Carbonyl(η-cyclopentadienyl)-1-((1-iminotrifluoroethyl)imino)-trifluoroethyl-N(ω)-iron, 41B, 821
$C_{10}H_7CrIO_4 \cdot CH_2Cl_2$, trans-Tetracarbonyl[(1-cyclopentenyl)carbyne]-iodochromium dichloromethane solvate, 45B, 827
$C_{10}H_7HgI$, α-Naphthylmercury(II) iodide, 28, 662
$C_{10}H_8MoO_5$, π-Cyclopentadienyl(tricarbonyl)carboxymethylmolybdenum, 34B, 418
$C_{10}H_{10}ClCrNO_5$, Pentacarbonyl(diethylaminochlorocarbene)chromium, 43B, 956
$C_{10}H_{10}FeO_5$, Ferrelactone ring complex, 42B, 608
$C_{10}H_{11}MnO_2$, Dimethylcarbene-dicarbonyl(η^5-cyclopentadienyl)manganese, 43B, 957
$C_{10}H_{11}MoNO_3$, Dicarbonyl(η-cyclopentadienyl)(3-aminopropionyl)molybdenum(II), 41B, 822
$C_{10}H_{12}FeN_6$, β-Tetramethyl ferrocyanide, 21, 546
$C_{10}H_{12}HgN_4O_4 \cdot H_2O$, (Creatinine)phenylmercury(II) nitrate monohydrate, 45B, 828
$C_{10}H_{14}N_2O_3W$, (N-Methylcarbamoyl)(η^5-cyclopentadienyl)(methylamine)-dicarbonyltungsten, 44B, 685
$C_{10}H_{15}ClCoN_5O_4$, Pentakis(methylisonitrile)cobalt(I) perchlorate, 30B, 295
$C_{10}H_{18}N_2NiO_2$, Dioxygen-bis-(t-butylisocyanide)nickel, 41B, 822
$C_{10}H_{18}O_8Re_2$, Di-μ-acetato-O,O'-bis[(acetato-O,O')methyl-rhenium(III)], 45B, 828

$C_{10}H_{20}B_{10}Hg$, 1-Methylmercury-2-benzyl-1,2-dicarbadodecaborane, 46B, 770
$C_{10}H_{20}Cl_2O_2Pt$, Dichloro-bis[methyl(isopropoxy)carbene]-platinum, 45B, 856
$C_{10}H_{20}HgN_2O_2$, Bis(diethylcarbamoyl)mercury, 41B, 823
$C_{10}H_{21}ClN_2O_2Pt$, Chloro-(2-methyl-2-nitrosopropane)(2-(N-oxo-t-butylimino)ethyl)platinum, 46B, 771
$C_{10}H_{26}Cl_2N_2Pt$, 1-(trans-(Dichloro(diethylamine)platinio))-2-diethylammoniumethane, 38B, 731
$C_{10}H_{29}ClNiP_2Si$, Chlorobis(trimethylphosphine)trimethylsilylnickel, 46B, 771
$C_{10}H_{30}Mo_2N_4$, Dimethyltetrakis(dimethylamino)dimolybdenum, 44B, 685
$C_{11}H_2Mn_2N_2$, Nonacarbonyl-[μ-[methylenehydrazinecarboxyaldehydato(2-)-C^1,N:N]]dimanganese, 45B, 835
$C_{11}H_3Co_3O_9$, Ethylidyne tri(cobalt tricarbonyl), 32B, 341
$C_{11}H_4O_{10}Os_3$, Di-μ-hydrido-μ-methylene-decacarbonyltriosmium, 45B, 829
$C_{11}H_5Fe_3NO_9$, triangulo-Nonacarbonyl-[μ_3-[(N,1-η)-ethaniminato-N:N]-μ-hydrido-triiron, 45B, 830
$C_{11}H_5Fe_3NO_9$, triangulo-Nonacarbonyl-[μ_3-[(N,1-η)-1-iminoethyl-C:N]-μ-hydrido-triiron, 45B, 830
$C_{11}H_6Fe_2O_6S$, Tricarbonyl(tricarbonyl(methylferrathiacyclohexadiene))iron, 43B, 958
$C_{11}H_6O_9Ru_3$, Nonacarbonyl-μ_3-ethylidyne-tri-μ-hydrido-triruthenium, 41B, 824
$C_{11}H_8F_6FeO_3$, Tricarbonyl(isoprene)iron-hexafluoropropene, 41B, 825
$C_{11}H_{10}S_2V$, (Carbon disulfide-C,S)bis(η-cyclopentadienyl)vanadium, 45B, 830
$C_{11}H_{11}HgN_3O_3$, (2,2'-Bipyridyl)methylmercury(II) nitrate, 42B, 609
$C_{11}H_{12}FeO_5$, (1,4,5,6-η-Butadienemethyl acrylate)tricarbonyliron, 42B, 657
$C_{11}H_{13}FeNO_2$, 1,3-Cyclohexadiene-dicarbonyl-ethylisonitrile-iron(0), 44B, 1124
$C_{11}H_{13}MoNO$, Bis(cyclopentadienyl)nitrosylmethylmolybdenum, 38B, 732
$C_{11}H_{13}NbS_2$, Bis(η^5-cyclopentadienyl)(η^2-disulfuro)methylniobium, 46B, 774
$C_{11}H_{14}AuN$, Phenylethynyl(isopropylamine)gold(I), 32B, 343
$C_{11}H_{14}CrNO_5$, Diethylamino(methyl)carbene-pentacarbonylchromium, 34B, 423
$C_{11}H_{21}As_2ClF_6Pt$, trans-Chloromethylbis(trimethylarsine)platinum(II) hexafluorobut-2-yne, 38B, 732
$C_{11}H_{23}ClN_2Zn$, Chloro(N,N,N',N'-tetramethylethylenediamine)penta-2,4-dienyl-zinc(II), 46B, 772
$C_{11}H_{25}ClN_2OPd$, Chloro(3-diethylaminopropionyl)(diethylamine)palladium(II), 43B, 958
$C_{11}H_{27}ClIrO_4P_3 \cdot 0.5\ C_6H_6$, ((Carbonic formic monoanhydridato)(2-))-chlorotris(trimethylphosphine)iridium hemibenzene solvate, 42B, 610
$C_{11}H_{29}ClNiOP_2Si$, Chlorobis(trimethylphosphine)trimethylsilylacetylnickel, 46B, 771
$C_{12}Co_2F_6O_6$, Hexacarbonyl-3,4,5,5,6,6-hexafluorocyclohexa-1-yne-3-enedicobalt, 33B, 325
$C_{12}Co_2F_{12}O_4S$, μ-[1,6-Bis(trifluoro)-3,4-bis(trifluoromethyl)hexene-2-thionato]tetracarbonyldicobalt, 45B, 831
$C_{12}H_2F_8Hg$, Bis(2,3,4,5-tetrafluorophenyl)mercury, 46B, 772
$C_{12}H_3Co_3O_{11}$, μ_3-(Acetoxymethylidyne)-cyclo-tris(tricarbonylcobalt)-(3Co-Co), 42B, 610
$C_{12}H_4BrCrF_3O_4$, trans-Bromotetracarbonyl(4-(trifluoromethyl)phenyl-

carbyne)chromium, 43B, 959
$C_{12}H_4ClO_6Re$, p-Chlorobenzoylpentacarbonylrhenium, 34B, 420
$C_{12}H_4Cl_6K_2O_{10}Pd_2$, Potassium di-μ-chloro-bis(chloranylic acid)dipalladate(II) dihydrate, 40B, 698
$C_{12}H_4O_{10}Os_3$, Decacarbonyl-μ-hydrido-μ-vinyl-triangulo-triosmium, 44B, 686
$C_{12}H_4O_{10}Os_3$, Nonacarbonyl-μ-hydrido-μ_3-[1-3-η-(1-hydroxyprop-1-enyl-3-ylidene)]-triangulo-triosmium, 46B, 773
$C_{12}H_4O_{11}Ru_3$, μ-Methoxycarbyne-μ-hydrido-μ-tetracarbonyl-ruthenium-bis(tricarbonylruthenium), 45B, 832
$C_{12}H_6FeO_6$, Tetracarbonyl-(2-methyl-3-prop-1-ynylmaleoyl)iron(0), 42B, 611
$C_{12}H_6Fe_2O_6$, Methylenecyclopentadienehexacarbonyldiiron, 30B, 303
$C_{12}H_6Fe_3O_{12}$, Hydrido-(O-methyl-carbonyl)decacarbonyltriiron hydrate, 41B, 825
$C_{12}H_6MnO_{10}Re$, cis-(Pentacarbonylmangano)(methylmethoxycarbeno)tetracarbonylrhenium, 41B, 826
$C_{12}H_7MnO_5$, Tetracarbonyl-2-acetylphenylmanganese, 41B, 827
$C_{12}H_8BrClHgO$, (2-Chloro-4-bromophenolate)phenylmercury, 39B, 561
$C_{12}H_{10}F_2FeO$, 1-(1,1-Difluoro-2-oxo-2-(cyclopentadienyliron)ethyl)-cyclopenta-2,4-diene, 42B, 627
$C_{12}H_{10}FeO_6$, Tetracarbonyl(2,3-diethyl-1,4-dioxobut-2-ene-1,4-diyl)-iron, 45B, 837
$C_{12}H_{10}Hg$, Diphenylmercury (gas-ed), 33B, 537
$C_{12}H_{10}Hg$, Diphenylmercury, 29, 628; 43B, 960
$C_{12}H_{10}HgN_2O_3S$, 1-Methyl-4-thiouracilyl-p-mercuribenzoic acid, 41B, 828
$C_{12}H_{10}NO_5Re$, cis-Acetyl(aniline)tetracarbonylrhenium, 43B, 960
$C_{12}H_{12}CrO_6S_2$, cis-(1,3-Dithian-2-ylidene(hydroxy)methyl(ethoxy)-carbene-C,S)-tetracarbonylchromium(0), 43B, 961
$C_{12}H_{12}FeO_8$, Dimethyl trans-1,1,1,1-tetracarbonylferracyclopentane-2,5-dicarboxylate, 42B, 612
$C_{12}H_{12}MoN_2O_5$, π-Cyclopentadienyl-dicarbonyl(4-carboethoxy-5-hydroxy-1,2,3-molybdeno-diazacyclopenta-3,5-diene), 34B, 426
$C_{12}H_{12}MoN_8$, Tetracyanotetrakis(methyl isocyanide)molybdenum(IV), 38B, 733
$C_{12}H_{13}BF_4N_6Pt$, Methyl(hydrotris(1-pyrazolyl)borato)tetrafluoroethyleneplatinum, 44B, 687
$C_{12}H_{13}Br_3F_8N_2O_2P_4W_2$, μ-Bromo-μ-p-tolylmethylidene-μ-bis(methylamino-bis(difluorophosphine))-bis(bromocarbonyltungsten), 46B, 1248
$C_{12}H_{13}ClOTi$, Acetylchlorobis(η^5-cyclopentadienyl)titanium(IV), 43B, 962
$C_{12}H_{13}I_3S_2V$, Bis(η-cyclopentadienyl)(dithiomethoxycarbonyl-C,S)vanadium(IV) triiodide, 45B, 830
$C_{12}H_{13}NbS_2$, Bis(η^5-cyclopentadienyl)(η^2-carbon disulfide)methylniobium, 46B, 774
$C_{12}H_{14}CoF_8O_5P$, Dicarbonyl[4,5-η-1,2-difluoro-1,2-bis(trifluoromethyl)pent-4-enyl]-(trimethyl phosphite)cobalt, 45B, 833
$C_{12}H_{14}FeO_5$, (3,3,6-Trimethyl-1-oxo-2-oxaheptyl-[4-6-eta]-enyl)tricarbonyliron (form I), 45B, 833
$C_{12}H_{14}FeO_5$, (3,3,6-Trimethyl-1-oxo-2-oxaheptyl-[4-6-eta]-enyl)tricarbonyliron (form II), 45B, 833
$C_{12}H_{15}BF_2FeO_3$, 3-Carbonyl-3-(η^5-cyclopentadienyl)-6,6-difluoro-2-isopropyl-4-methyl-1,5-dioxa-3-ferra-6-borinane, 46B, 774
$C_{12}H_{15}FeO_3P$, η^1-Trimethylphosphoranylidene-acetyl-dicarbonyl-(η^5-cyclopentadienyl)iron, 46B, 874
$C_{12}H_{17}AsMoO_2$, Dicarbonyl-η^5-cyclopentadienyl((β-dimethylarsino)pro-

pyl)molybdenum, 42B, 612
$C_{12}H_{17}IMoN_2O$, (η^5-Cyclopentadienyl)carbonyliodo(iminodimethylaminocarbene)molybdenum, 42B, 613; 43B, 962
$C_{12}H_{18}Cl_8Fe_3N_6$, Hexakis(methylisonitrile)iron(II) tetrachloroferrate(III), 39B, 561
$C_{12}H_{18}F_{12}N_6P_2Pd_2 \cdot 0.5\ C_3H_6O$, Hexakis(methyl isocyanide)dipalladium(I) bis(hexafluorophosphate) hemiacetone, 42B, 614
$C_{12}H_{20}ClF_6FeN_3O_2PS_3$, Dicarbonyl(dimethyldithiothiocarbamato)(thiobis(dimethylaminocarbene))iron hexafluorophosphate 1,2-dichloroethane solvate, 44B, 1125
$C_{12}H_{24}Cl_2FeN_6O_3$, Hexa(methylisocyano)iron(II) chloride trihydrate, 10, 214
$C_{12}H_{24}Cl_2N_2O_2Pd_2$, Di-$\mu$-chloro-bis[(3-hydroxyimino-2,2-dimethylbutyl-C[1]N)palladium(II)], 46B, 775
$C_{12}H_{24}F_{12}N_4P_2PtS_2$, trans-Bis(methyl isocyanide)bis(methylamino(thioethoxy)carbene)platinum(II) hexafluorophosphate, 39B, 562
$C_{12}H_{24}Ni_2$, Di-μ-methylbis(1,3-dimethyl-η^3-allylnickel), 44B, 687
$C_{12}H_{24}O_2Pt$, Trimethyl-4,6-dioxononylplatinum, 24, 627
$C_{12}H_{25}F_3N_2O_4PdS$, (Trifluoromethylsulphonato)(3-diethylaminopropionyl)(diethylamine)palladium(II), 45B, 834
$C_{12}H_{28}Au_2Cl_2P_2$, Dichloro-bis(μ-diethylphosphonium-bis(methylido))digold(II), 42B, 614
$C_{12}H_{28}Au_2P_2$, Bis-μ-(diethylphosphonium-bis(methylido))-digold(I), 43B, 963
$C_{12}H_{28}N_8O_{10}Pd_2$, Bis(ethylenediamine(barbiturato)palladium(II)) tetrahydrate, 44B, 687
$C_{12}H_{32}F_{12}N_8P_2Pt$, Tetrakis(bis(methylamino)carbene)platinum(II) hexafluorophosphate, 43B, 964
$C_{12}H_{32}N_2NiP_4$, Bis(nitridobis(dimethylphosphonium-methylide))nickel, 43B, 964
$C_{12}H_{34}CoP_3$, Dimethylbis(trimethylphosphine)cobalt(III)-(dimethylphosphonium-bis(methylide)), 40B, 1036
$C_{12}H_{36}B_2NiP_4$, Bis(boranato-bis(dimethylphosphonium-methylide))nickel, 45B, 834
$C_{12}H_{36}N_{12}Pt_4$, Trimethylplatinumazide tetramer, 40B, 701
$C_{13}H_2Fe_4O_{12}$, μ-Hydrido-(μ_4-η^2-methylidyne)-dodecacabronyltetrairon-(5Fe-Fe), 46B, 775
$C_{13}H_6O_{10}Os_3$, Nonacarbonyl-μ-hydrido-μ_3-[1-3-η-(1-methoxyprop-1-enyl-3-ylidene)]-triangulo-triosmium, 46B, 773
$C_{13}H_7NO_{10}Ru_3$, μ-(N,N-Dimethylcyano)-μ-hydrido-decarbonyltriruthenium, 41B, 828; 42B, 615
$C_{13}H_7NO_{11}Os_4$, Tetra(μ-hydrido)(methyl isocyanide)undecacarbonyltetraosmium, 46B, 776
$C_{13}H_7NO_{11}Ru_3$, (η^2-Dimethylformamide)(η^2-hydrido)decacarbonyltriruthenium, 46B, 776
$C_{13}H_8CrO_5S$, Pentacarbonyl(methyl(phenylthio)carbene)chromium, 38B, 734
$C_{13}H_8CrO_6$, Acetophenonepentacarbonylchromium, 30B, 294
$C_{13}H_8Fe_2O_7Se$, μ-{1-2-η-[Hydroseleno-1-cyclohexene-1-carbaldehydato(2-)]-μ-Se}bis(tricarbonyliron)(Fe-Fe), 46B, 777
$C_{13}H_8O_8W_2$, μ-(3-Methyl-but-2-enylidene)-bis(tetracarbonyltungsten), 46B, 778
$C_{13}H_9FeN_3O_6$, 4-Methyl-3,5,11-trioxo-2,4,6-triazatricyclo-[5.3.2.0$^{2-6}$]dodeca-(8-10-η)-dienyl-12-yl-tricarbonyliron, 44B, 1125
$C_{13}H_{10}F_3HgNO_2$, Phenyl(pyridine)mercury(II) trifluoroacetate, 46B, 778
$C_{13}H_{10}Fe_2O_8$, μ-[1-σ:1-2-η-2-Carboxylato-1-ethylbut-1-enyl-O(2Fe)]-

bis(tricarbonyliron)(Fe-Fe), 45B, 836
$C_{13}H_{11}Cl_7HgO$, exo-6-Chloromercurio-6,7-dihydro-exo-7-methoxyaldrin, 44B, 688
$C_{13}H_{11}HgNO$, Benzamido(phenyl)mercury(II), 46B, 1069
$C_{13}H_{11}I_2IrN_2O_3$, Di-iodocarbomethoxycarbonyl(2,2'-bipyridyl)iridium, 34B, 430
$C_{13}H_{12}MnNO_4$, Tetracarbonyl-2-(dimethylaminomethyl)phenylmanganese, 39B, 563
$C_{13}H_{12}O_2Rh_2$, μ-Methylene-bis(carbonyl-η^5-cyclopentadienylrhodium)-(Rh-Rh), 43B, 965
$C_{13}H_{13}BrO_2Rh_2$, Bromo(methyl)bis[μ-carbonyl(η^5-cyclopentadienyl)rhodium](Rh-Rh), 46B, 779
$C_{13}H_{14}F_6MoN_2O_5P$, Nitrogeno-molybdenum chelate, 38B, 734
$C_{13}H_{14}HgN_2O_3$, (2-Benzylpyridine)methylmercury(II) nitrate, 46B, 1069
$C_{13}H_{15}ClHgO_2$, 1,2,3,4,4a,9b-Hexahydro-4-chloromercury-8-methoxydibenzofuran, 44B, 689
$C_{13}H_{15}ClHgO_2$, 2,4-Propano-3-chloromercury-6-methoxychroman, 44B, 689
$C_{13}H_{15}HgN_3O_3$, (3,3'-Dimethyl-2,2'-bipyridyl)methylmercury(II) nitrate, 44B, 924
$C_{13}H_{16}Cl_2N_2Pt$, Dichloro(pyridinium propylide)pyridineplatinum(II), 37B, 430
$C_{13}H_{16}Cl_2N_2Pt$, 1,6-Dichloro-2,3-trimethylene-4,5-bis(pyridine)platinum(IV), 37B, 429
$C_{13}H_{17}Cl_2N_2Ta$, 2,2'-Bipyridyldichloro(trimethyl)tantalum(V), 39B, 564
$C_{13}H_{20}Cl_2NPPt$, cis-Dichloro(diethyl(phenyl)phosphine)(ethyl isocyanide)platinum(II), 38B, 735
$C_{13}H_{22}HgO_4$, Dipivaloylmethanemercury acetate, 39B, 565
$C_{13}H_{24}B_7Co$, 3,7,9,13-Tetramethyl-6-(η^5-cyclopentadienyl)-3,7,9,13-tetracarba-6-cobaltadodecaborane, 46B, 822
$C_{13}H_{24}N_2O_2Pd$, (3,3-Dimethyl-2-N,N-dimethylhydrazonobutyl-C^1,N)pentane-2,4-dionatopalladium(II), 46B, 775
$C_{13}H_{26}Cl_4O_3Rh_2$, Di-μ-chloro-bis(chloro-(2-(hydroxymethyl)pent-4-enyl)rhodium(III))-methanol, 40B, 702
$C_{13}H_{26}NO_2P_2Re$, Methyl(η^1-cyclopentadienyl)carbonylbis(trimethylphosphine)nitrosylrhenium(I), 46B, 779
$C_{13}H_{28}ClN_3OPd$, Chloro(3-diethylaminopropionyl)((diethylamino)(methylamino)carbene)palladium(II), 44B, 689
$C_{14}F_4Fe_2O_8$, μ-(o-Tetrafluorophenylene)-diiron octacarbonyl, 39B, 565
$C_{14}H_5ClCrO_7W$, trans-Tetracarbonylchloro((tricarbonylchromium)-η^6-phenylcarbene)tungsten, 46B, 780
$C_{14}H_6Fe_2O_6S_2$, 2,2'-Bithienyl-bis(tricarbonyliron), 43B, 966
$C_{14}H_6O_6Os_2$, η-(1,1,1-Tricarbonyl-1-osmainden-1-yl)tricarbonylosmium(Os-Os), 42B, 616
$C_{14}H_8O_9W_2$, Tetracarbonyltungsten(μ-1-methyl-2-but-2-ene-di-ylidene)-pentacarbonyltungsten, 46B, 780
$C_{14}H_{10}Cl_2N_2Pt$, cis-Dichlorobis(phenyl isocyanide)platinum(II), 38B, 736
$C_{14}H_{10}FeO_6$, Tetracarbonyl-3,4-dicyclopropyl-1-ferracyclopent-3-ene-2,5-dione, 43B, 967
$C_{14}H_{10}Fe_2O_7$, Hexacarbonyl(2-4-η:5-6-η-(7-oxo-3,5-octadiene-2,2-diyl))diiron, 41B, 829
$C_{14}H_{10}IMoNO_2$, trans-Dicarbonyl(η-cyclopentadienyl)iodo(phenylisocyanide)molybdenum, 45B, 836
$C_{14}H_{11}NO_9Os_3$, Bis-(μ-hydrido)nonacarbonyl(t-butyl isocyanide)triosmium, 45B, 843
$C_{14}H_{12}Fe_2N_2NiO_{10}$, Nickel-bis[tetracarbonyl-μ-(dimethylcarbamoyl-O)-

iron], 45B, 837
$C_{14}H_{12}Fe_2O_8$, μ-[1-4-η-2,3-Diethyl-1,4-dihydroxybutadiene-1,4-diyl]C^{1}C^4(Fe^1):η^4(Fe^2)-bis(tri-carbonyliron), 45B, 837
$C_{14}H_{13}O_2Re$, Benzyldicarbonyl(cyclopentadienyl)hydride-rhenium, 44B, 690
$C_{14}H_{14}Hg$, Di(o-tolyl)mercury, 46B, 781
$C_{14}H_{14}Hg$, Dibenzylmercury, 45B, 838
$C_{14}H_{14}Hg$, Di-p-tolylmercury, 35B, 511
$C_{14}H_{14}HgN_4S$, Methylmercury(II) dithizonate, 46B, 997
$C_{14}H_{14}HgS$, Phenylmercury 2,6-dimethylthiophenolate, 40B, 84
$C_{14}H_{15}CoO_6$, Carbonyl-cyclopentadienyl-(1-oxo-2-ethoxycarbonyl-3-ethoxy-prop-2-ene-1-yl-3-olato)cobalt, 45B, 838
$C_{14}H_{15}NbS_2$, Bis(cyclopentadienyl)(σ-allyl)(carbon disulfide)niobium, 43B, 967
$C_{14}H_{16}Cl_{12}FeN_6$, cis-Dicyanotetra(methyl isocyanide)iron(II)-chloroform, 33B, 326
$C_{14}H_{16}N_2Ni$, 2,2'-Bipyridyl-nickelacyclopentane, 45B, 839
$C_{14}H_{16}N_2Ni_2$, Di-(μ-methylisonitrile)di-(η^5-cyclopentadienyl)dinickel, 38B, 736
$C_{14}H_{17}Cl_7N_2Pt$, Tetrachloro(pyridinium propylide)pyridineplatinum-(IV), 37B, 430
$C_{14}H_{19}B_3Co_2$, Bis(η^5-cyclopentadienyl)-(μ-(η^3,η^4-trihydro-C,C'-dimethyldicarbapentaborato))dicobalt(Co-Co), 46B, 782
$C_{14}H_{20}As_2ClF_{12}ORh$, Chloroaquobis(trimethylarsine)tetrakis(trifluoromethyl)rhodiacyclopentadiene, 39B, 566
$C_{14}H_{20}ClNOPt$, Chloro-(2-methoxycyclo-octa-1,5-dienyl)pyridineplatinum, 38B, 737
$C_{14}H_{20}Hg_2I_4S_2$, trans-Bis(3-dimethylsulphoniocyclopentadienylide)-di-μ-iodo-diiododimercury(II), 46B, 782
$C_{14}H_{20}MoN_4O_4$, cis-Tetracarbonylbis(1,3-dimethylimidazolidin-2-ylidene)molybdenum(0), 43B, 968
$C_{14}H_{21}B_2F_8MoN_7$, Hepta(methyl isocyano)molybdenum(II) bis(tetrafluoroborate), 45B, 839
$C_{14}H_{22}AlYb$, Di-μ-methyl-bis(η^5-cyclopentadienyl)ytterbium(III)-dimethylaluminum(III), 45B, 840
$C_{14}H_{23}Cl_2Ta$, (η^5-Pentamethylcyclopentadienyl)(butane-1,4-diyl)dichlorotantalum, 45B, 841; 46B, 783
$C_{14}H_{23}CoN_6O_4$, Bis(dimethylglyoximato)(N-iminopyridine)methylcobalt-(III), 46B, 1073
$C_{14}H_{23}MnO_5P_2Pt$, μ-2-σ:2-3-η-(4,5-Dihydro-2-furyl)-[bis(trimethylphosphine)platinum][tetracarbonylmanganese](Mn-Pt), 46B, 1252
$C_{14}H_{24}Cl_2N_2OPt$, Dichloro-(2,4,6-trimethylpyridine)(2-(N-hydroxy-t-butylimino)ethyl)platinum, 46B, 771
$C_{14}H_{24}F_{12}FeN_8P_2$, Tetrakis(methyl isocyanido)-bis(methyl isocyanido)-acetamide-iron(II) hexafluorophosphate, 44B, 690
$C_{14}H_{30}Cl_2N_2O_2Pt$, Diisopropylammonium cis-dichloro(N,N-diisopropylcarbamoyl)carbonylplatinate(II), 45B, 841
$C_{14}H_{30}P_2Pt$, Bis(trimethylphosphine)-(1,4-trans-divinylbutane-1,4-diyl)platinum, 44B, 694
$C_{14}H_{34}O_3P_2Pd$, trans-Bis(triethylphosphine)(methyl)hydrogencarbonato-palladium(II), 43B, 1425
$C_{15}H_8F_6Ni_3O_3$, Tricarbonyl-μ$_3$-(η-cyclooctatetraene)-μ$_3$-(η^2-hexafluorobut-2-yne)-triangulo-trinickel, 45B, 845
$C_{15}H_9BrCrFeO_4$, trans-Bromotetracarbonyl(ferrocenylcarbyne)chromium, 44B, 691
$C_{15}H_9Ir_4O_{11}$, (t-Butyl isocyanide)(undecacarbonyl)tetrairidium, 45B, 842

$C_{15}H_9MnMoO_5$, μ-(η⁵:η¹-Cyclopentadienyl)[(η⁵-cyclopentadienyl)carbonylmolybdenum][tetracarbonyl-manganese], 45B, 842
$C_{15}H_{10}FeO_2$, 1-(Dicarbonyl-π-cyclopentadienyl-ferrio)-2-(phenyl)-ethyne, 40B, 703
$C_{15}H_{10}O_9Ru_3$, μ_3-η-(1-Methyl-3-ethyl-3-allenyl)-μ-hydrido-triangulo-tris(tricarbonylruthenium)(3Ru-Ru), 42B, 616
$C_{15}H_{11}HgNO$, Phenyl(quinolin-8-olato)mercury(II), 44B, 1010
$C_{15}H_{11}MnO_2$, (η^5-Cyclopentadienyl)-(η^1-phenylvinylidene)manganese dicarbonyl, 43B, 969
$C_{15}H_{11}NO_{10}Os_3$, μ-Hydrido(hydrido)decacarbonyl(t-butyl isocyanide)-triosmium, 45B, 843
$C_{15}H_{11}NO_{10}Os_3$, μ-Hydrido-μ-η'-(N-hydro-t-butyl isonitrile)decacarbonyl-triangulo-triosmium, 45B, 843
$C_{15}H_{12}Fe_2O_7$, 2,6-Di(tricarbonyliron)-3,5-dimethylhepta-2,5-diene-4-one, 34B, 422
$C_{15}H_{12}O_2W$, Dicarbonyl(π-cyclopentadienyl)(p-tolylcarbyne)tungsten, 43B, 970
$C_{15}H_{14}BF_4Fe_2NO_3$, cis-μ-Carbonyl-μ-methyliminiomethylene-bis[carbonyl(η-cyclopentadienyl)iron], 45B, 844
$C_{15}H_{14}Cl_3HgNO_3$, 2-Nitroso-4-methylphenol-2-mercuri-4-methylphenol - chloroform, 40B, 1106
$C_{15}H_{14}NO_5Re$, cis-Acetyltetracarbonyl(N-phenylmethyliminium-C)rhenium, 44B, 691
$C_{15}H_{14}O_7SW$, Dicarbonyl(cyclopentadienyl)(1,2-bis(methoxycarbonyl)-3-oxo-4-thiapent-1-enyl)tungsten, 45B, 845
$C_{15}H_{17}CrNO_6$, Pentacarbonyl-(cyclohexylamino-(1-methoxy-vinyl)-carbene)chromium(0), 35B, 501
$C_{15}H_{20}ClTa$, Bis(η^5-cyclopentadienyl)chloro(neopentylidene)tantalum, 44B, 692
$C_{15}H_{22}CoN_5O_4$, trans-Bis(dimethylglyoximato)(vinyl)(pyridine)cobalt-(III), 46B, 1076
$C_{15}H_{23}ClMoN_2O_6P_2$, (Dicyanovinylidene)(π-cyclopentadienyl)-trans-bis(trimethyl phosphite)chloromolybdenum(II), 40B, 753
$C_{15}H_{24}ClNO_2Pt$, Allyl ether platinum(II) chloride (α-methylbenzyl)amine complex, 38B, 738
$C_{15}H_{24}CoN_5O_4$, Ethyl-bis(dimethylglyoximato)pyridine-cobalt(III) (form I), 46B, 783
$C_{15}H_{24}CoN_5O_4$, Ethyl-bis(dimethylglyoximato)pyridine-cobalt(III) (form II), 46B, 783
$C_{15}H_{24}F_{12}N_6OP_2Pd_2$, Hexakis(methyl isocyanide)dipalladium(I) hexafluorophosphate acetone solvate, 41B, 831
$C_{15}H_{26}Cl_2NOPPt$, cis-(Anilino(ethoxy)carbene)dichloro(triethylphosphine)platinum(II), 42B, 617
$C_{15}H_{27}Cl_3N_3V$, Trichlorotris(tert-butyl isocyanide)vanadium(III), 46B, 784
$C_{15}H_{35}BrP_2Pt$, trans-η^1-Allyl(bromo)bis(triethylphosphine)platinum-(II), 43B, 970
$C_{15}H_{37}Cl_3NP_2Rh$, Trichloro(dimethylaminomethylene)bis(triethylphosphine)rhodium(III), 40B, 704
$C_{15}H_{42}ClMoPSi_3$, Chloro(trimethylphosphine)tris(trimethylsilylmethyl)molybdenum(IV), 46B, 785
$C_{16}F_{18}Ni_4O_4$, Tris-μ_3-(η^2-hexafluorobut-2-yne)tetrakis(carbonylnickel), 45B, 845
$C_{16}H_7NO_9Os_3$, μ-Hydrido-(μ_3-η^2-N-phenyl-formimidoyl)-tris(tricarbonylosmium), 44B, 693; 45B, 846
$C_{16}H_8F_{12}FeO_4$, 1,1,1,1-Tetracarbonyl-2,3,3-trifluoro-3a,4,5,7a-tetrahydro-5-(1,1,2,3,3,3-hexafluoropropyl)-2-trifluoromethyl-1-ferrain-

dane, 43B, 971
$C_{16}H_8FeO_5$, Naphtho[b-3,4]ferracyclopentenone tetracarbonyl, 41B, 832
$C_{16}H_{10}CrO_6$, Pentacarbonyl(ethoxy(phenylethynyl)carbene)chromium(0), 41B, 833
$C_{16}H_{10}Fe_2O_6$, (μ-Bicyclo[3.3.2]deca-3,7,9-triene)-hexacarbonyldiiron-(Fe-Fe), 44B, 1126
$C_{16}H_{12}CuO_{12}Re_2$, Bis(3,3,3,3-tetracarbonyl-3-rheniaacetylacetonato)-copper(II), 46B, 974
$C_{16}H_{12}Mn_2O_4$, μ-(Vinylidene)-bis(dicarbonylcyclopentadienyl manganese), 45B, 847
$C_{16}H_{12}O_4W_2$, μ-(η^2-Acetylene)bis((η^5-cyclopentadienyl)dicarbonyltungsten)(W-W), 44B, 1127
$C_{16}H_{13}Co_3O_4$, Bis(η-cyclopentadienyl)(methylmethinyl)tetracarbonyl-tricobalt, 44B, 693
$C_{16}H_{13}NO_{10}Os_3$, μ-Hydrido-μ-(2-diethylimmonium-ethylido)-μ-(tetracarbonylosmium)bis(tricarbonylosmium), 45B, 847
$C_{16}H_{14}CrO_3$, π-Benzenedicarbonyl(methoxyphenylcarbene)chromium, 46B, 786
$C_{16}H_{15}BBr_2Co_3NO_{10}$, Methylidyne-tricobalt nonacarbonyl cluster derivative, 42B, 617
$C_{16}H_{16}HgO_2$, Di-(ω,ω'-ethylenedioxy)-o-tolylmercury, 33B, 328
$C_{16}H_{19}ClHgO_6$, trans-3-endo-Chloromercur-4-acetoxy-9,10-cis-endo-dimethoxycarbonyltricyclo[4.2.2.0^{2-5}]dec-7-ene, 46B, 786
$C_{16}H_{19}Cl_2HgN$, (1,1-Dichloro-1-(2-methyl-5-t-butyl-2H-2-pyrryl))phenylmercury, 46B, 787
$C_{16}H_{20}Fe_2N_2O_6$, Bis(μ-diethylimmoniocarbene)-hexacarbonyldiiron(-I), 43B, 972
$C_{16}H_{20}MnO_8PS_2$, Carbonyl-η^5-cyclopentadienyltrimethoxyphosphine-(4,5-bis(methoxycarbonyl)-1,3-dithiol-2-ylidene)manganese(I), 46B, 881
$C_{16}H_{22}Cl_2Pd_2$, Di-μ-chloro-bis(1,5-cyclooctadienylpalladium), 42B, 618
$C_{16}H_{24}Pt$, (η-Cycloocta-1,5-diene)(1,4-trans-divinylbutane-1,4-diyl)-platinum, 44B, 694
$C_{16}H_{25}ClPdS_2$, [2,6-Bis(t-butylthiomethyl)phenyl-C^1,S,S']chloropalladium(II), 46B, 787
$C_{16}H_{26}CoN_5O_4$, trans-Bis(dimethylglyoximato)(isopropyl)(pyridine)cobalt(III), 45B, 848
$C_{16}H_{30}Cl_2PdS_2$, Chloro(2,5-dithiahexane)-(1-(1,4-di-t-butyl-4-chloro)butadienyl)palladium, 41B, 834
$C_{16}H_{32}Cl_2N_2O_2Pd_2$, Di-μ-chloro-bis(3,3,N,N-tetramethyl-1-oxo-5-amino-pent-2-ylpalladium(II)), 43B, 972
$C_{16}H_{32}Li_4O_4W_2 \cdot 4.8\ CH_3 \cdot 3.2\ Cl$, Tetralithium trichloropentamethyl-ditungsten - tetrakis(tetrahydrofuran), 44B, 710
$C_{16}H_{38}Cl_6P_2Ta_2$, Di-μ-chloro-bis(dichloro-trimethylphosphine-neopentylidene-tantalum), 45B, 848
$C_{16}H_{40}Cr_2P_4$, Tetrakis(dimethylphosphoniumdimethylido)dichromium, 45B, 849
$C_{16}H_{40}Mo_2P_4$, Tetrakis(dimethylphosphoniumdimethylido)dimolybenum, 45B, 849
$C_{16}H_{40}Ni_2P_4$, Dinickel-tetrakis(dimethylphosphonium-bis(methylide)), 40B, 1036
$C_{16}H_{44}Cu_4Si_4$, Tetrakis((trimethylsilylmethyl)copper(I)), 43B, 973
$C_{16}H_{44}Li_2O_2Re_2$, Dilithium octamethyldirhenate(III) - bis(diethyl-ether), 42B, 619
$C_{16}H_{48}Hg_4O_4Si_4$, Methyl(trimethylsiloxo)mercury tetramer, 34B, 368
$C_{17}H_5Co_3O_{10}$, Benzoylmethylidynetricobaltnonacarbonyl, 45B, 850
$C_{17}H_7O_{17}Os_5P$, Carbidotetradecacarbonylhydrido(dimethyl phosphonato)-

pentaosmium, 44B, 694
$C_{17}H_8Mn_2O_{10}$, Nonacarbonyl(phenyl(methoxy)carbene)dimanganese(0), 38B, 739
$C_{17}H_{10}BrMnN_2O_3$, Bromotriscarbonylbis(phenylisocyanide)manganese(I), 40B, 704
$C_{17}H_{10}Fe_2N_2O_3$, μ-Carbonyl-μ-dicyanovinylidene-bis(carbonylcyclopentadienyliron(0)), 40B, 705
$C_{17}H_{10}Fe_2O_6$, (5-Cyclopentadienyl-π-(6-keto)irondicarbonyl)cyclohexadienylirontricarbonyl, 41B, 835
$C_{17}H_{10}MnNO_4$, Tetracarbonyl-2-(N-phenylformimidoyl)phenylmanganese, 39B, 567
$C_{17}H_{10}O_{18}Os_5P_2$, Carbidotridecacarbonylhydrido(trimethyl diphosphito)pentaosmium, 44B, 695
$C_{17}H_{11}Fe_2N_3O_7$, (2-Amino-3-cyano-4-(1',2'-dimethylvinyl)furan-2',5-diyl)tricarbonyliron-μ-carbonyl-dicarbonyliron acetonitrile solvate, 44B, 1127
$C_{17}H_{11}NO_9Os_3$, Bis(μ-hydrido)-(μ-o-methylbenzylamido)nonacarbonyltriosmium(3Os-Os), 46B, 1084
$C_{17}H_{12}FeN_4O_2$, Dicarbonyl-η⁵-cyclopentadienyl(1-η¹-3,3,4,4-tetracyano-1-methylcyclopentyl)iron(II), 39B, 568
$C_{17}H_{12}Fe_3O_8$, Octacarbonyl-μ₃-(1,3,6-trimethylhexa-1,3,5-triene-1,5-diyl)-triangulo-tri-iron, 42B, 666
$C_{17}H_{12}O_9Os_3$, μ-Hydrido-(1-oxa-2-ethylidene-5-ethylcyclopentadienyl)-octacarbonyltriosmium, 44B, 695
$C_{17}H_{12}O_{11}Os_4$, 1,1,1,2,2,3,3,3,4,4,4-Undecacarbonyl-1,2-μ-(1'-σ,1'-2'-η-cyclohexenyl)tri-μ-hydrido-tetrahedro-tetraosmium, 45B, 850
$C_{17}H_{13}Co_3O_7$, (2,3:5-6-η-Norbornadiene)propylidyneheptacarbonyltricobalt, 44B, 696
$C_{17}H_{13}Fe_2NO_6$, 3-Dimethylamino-1-phenyl-allyl-(hexacarbonyl-diiron), 43B, 975
$C_{17}H_{13}O_8Os_3PS_2$, μ₃-Methylenethiolato-μ₃-sulfido-(octacarbonyl(dimethylphenylphosphine)triosmium), 45B, 854
$C_{17}H_{14}Fe_2NiO_6$, (η⁵-Cyclopentadienylnickel)-(η-t-butylethynyl)hexacarbonyldiiron, 46B, 845
$C_{17}H_{14}O_2Ti$, Bis(π-cyclopentadienyl)-2-carboxyphenyltitanium, 37B, 431
$C_{17}H_{15}As_2Co_3F_4O_7$, Ethylidyneheptacarbonyl-μ-(1,2-bis(dimethylarsino)tetrafluorocyclobutene)-triangulo-tricobalt, 38B, 740
$C_{17}H_{15}FeNbO_2$, Niobocenehydrido-(σ,η⁵-cyclopentadienylidene)-dicarbonyliron, 46B, 788
$C_{17}H_{16}Fe_2O_4$, (Di-η⁵-cyclopentadienyldicarbonyliron)propane, 42B, 619
$C_{17}H_{16}I_3N_2ORh$, Carbonyltri-iodo-(α-(N-methyl-α-methyliminobenzylamino)benzylidene-N,C)rhodium, 40B, 706
$C_{17}H_{20}MnNO_6$, Tetramethylammonium cis-acetylbenzoyltetracarbonylmanganate(I), 42B, 620
$C_{17}H_{21}F_6O_3Rh \cdot 0.5\ H_2O$, (Acetylacetonato)aquo(7-8-η-7,8-bis(trifluoromethyl)bicyclo[4.2.2]dec-7-ene-2,5-diyl)rhodium(III) hemihydrate, 42B, 620
$C_{17}H_{27}Cl_2Ta$, (η⁵-Pentamethylcyclopentadienyl)(cyclopentane-1,2-diylbis(methylene))dichlorotantalum, 45B, 841; 46B, 783
$C_{17}H_{27}IMnO_5PPt$, defg-Tetracarbonyl-c-μ-iodo-a-(methyldi-t-butylphosphine)-b-(2-oxacyclopentylidene)platinummanganese(Pt-Mn), 46B, 788
$C_{17}H_{30}Co_3O_{15}P_3$, Hexacarbonyl-μ₃-ethylidene-tris(trimethyl phosphine)-triangulo-tricobalt, 45B, 851
$C_{17}H_{50}Co_2P_6$, μ-(Dimethylphosphido)-μ-(dimethylphosphinomethyl)-bis[di(trimethylphosphine)cobalt], 45B, 851
$C_{18}Co_6O_{16}S_3$, (Nonacarbonyltricobalt)-μ-(dithiocarboxylatocarbido)-

(μ_3-thioheptacarbonyltricobalt), 46B, 789
$C_{18}HCo_5O_{15}$, (Methinyltricobalt enneacarbonyl)-acetylene-(dicobalt hexacarbonyl), 35B, 502
$C_{18}H_4F_{10}O_4Os$, Bis(pentafluorobenzyl)tetracarbonylosmium, 46B, 790
$C_{18}H_9NO_{10}Os_3$, Hydrido(μ-η^2-N-methylphenylimino)decacarbonyltriosmium, 44B, 696
$C_{18}H_{10}F_{12}Mo$, Bis(cyclopentadienyl)bis(hexafluorobut-2-yne)molybdenum, 43B, 977
$C_{18}H_{10}O_5W$, Diphenylcarbene(pentacarbonyl)tungsten(0), 43B, 978
$C_{18}H_{12}Cl_2HgN_2$, Bis(3-chloropropynyl)mercury-1,10-phenanthroline, 45B, 852
$C_{18}H_{12}Fe_4N_2O_{12}S_2$, ($\mu$-(N,N-Dimethylimmoniocarbene)bis(tricarbonyliron))-μ_4-thio-(μ-N,N-dimethylthiocarboxamido)bis(tricarbonyl)), 44B, 1128
$C_{18}H_{12}Hg_3$, Tribenzo[b,e,h][1,4,7]trimercuronin (monoclinic), 46B, 790
$C_{18}H_{12}Hg_3$, Tribenzo[b,e,h][1,4,7]trimercuronin (orthorhombic), 44B, 697
$C_{18}H_{12}O_8Ru_3$, (2-(Isopropenyl)-5-methyl-hexa-1,3,5-triene-1,3-diyl)-octacarbonyl-triruthenium, 45B, 852
$C_{18}H_{12}O_{10}Os_3$, η^2,η^2,η^1-(3-Ethylidene-4-ethyl-1-oxocyclopent-4-en-2-yl)-μ-hydrido-nonacarbonyl-triosmium, 45B, 853
$C_{18}H_{13}IN_2PtS_2$, Iodo(2-(2'-thienyl)pyridine)dehydro(2-(2'-thienyl)-pyridine)platinum(II), 41B, 836
$C_{18}H_{13}O_9Os_3PS_2$, μ-Methylenethiolato-μ_3-sulfido-(nonacarbonyl(dimethylphenylphosphine)triosmium), 45B, 854
$C_{18}H_{14}BrHgNO_2S$, N-(Phenylmercury)-2'-bromobenzenesulphonanilide, 42B, 621
$C_{18}H_{14}ClHgNO_2S$, N-(Phenylmercury)-2'-chlorobenzenesulphonanilide, 42B, 621
$C_{18}H_{14}FHgNO_2S$, N-(Phenylmercury)-2'-fluorobenzenesulphonanilide, 42B, 622
$C_{18}H_{14}F_{12}O_4Pd$, cis-Bis-(1,2-bis(trifluoromethyl)-3-acetyl-4-oxopent-1-enyl-O,C[1])palladium(II), 41B, 837, 1244
$C_{18}H_{14}Fe_2O_4$, trans-1,4-Bis(π-cyclopentadienyldicarbonyliron)buta-1,3-diene, 34B, 425
$C_{18}H_{14}Fe_2O_{10}$, μ-(2,6-Dimethoxyphenyl(ethoxy)carbene)heptacarbonyldiiron, 38B, 741
$C_{18}H_{14}Ir_2O_2$, Bis(η^5-cyclopentadienyl)bis(carbonyl)-μ-(o-phenylene)-diiridium(Ir-Ir), 43B, 979
$C_{18}H_{14}O_6RuW$, Pentacarbonyl(ethoxy(ruthenocenyl)carbene)tungsten, 46B, 790
$C_{18}H_{16}F_6O_2PRh_3$, Bis(μ-carbonyl)tris(η^5-cyclopentadienyl)(μ_3-methylidene)trirhodium(3Rh-Rh) hexafluorophosphate, 46B, 791
$C_{18}H_{18}F_{18}NiO_6P_2$, Bis(trimethylphosphite)nickelahexakis(trifluoromethyl)cyclohepta-cis-trans-cis-triene, 40B, 706
$C_{18}H_{18}Fe_2O_4$, (Di-η^5-cyclopentadienyldicarbonyliron)butane, 42B, 619
$C_{18}H_{19}F_6NO_2Pd$, ab(1,2-Bis(trifluoromethyl)-3-acetyl-4-oxopent-1-enyl-O,C(1))-cd-2-((dimethylamino)methyl)phenyl-C(1),N)palladium-(II), 41B, 837, 1244
$C_{18}H_{19}Fe_2NO_3$, cis-(Isobutylisocyanide)di-μ-carbonyl-carbonylbis(π-cyclopentadienyl)diiron, 41B, 838
$C_{18}H_{19}MoNO_3$, η-Cyclopentadienyldicarbonyl(2-phenyl-3-methyl-3-methylaminopropionyl-C,N)molybdenum, 43B, 980
$C_{18}H_{19}Ti$, (2,6-Dimethylphenyl)dicyclopentadienyltitanium(III), 42B, 623
$C_{18}H_{20}Cl_2N_2O_4Pd$, trans-Chloro-η^1-(3-chloro-cis-1,2-bis(methoxycar-

bonyl)but-3-enyl)bis(pyridine)palladium(II), 42B, 624
$C_{18}H_{20}Fe_2INO_3$, μ-Carbonyl-μ-dimethyliminiomethylene-bis[carbonyl(η-methylcyclopentadienyl)iron] iodide, 46B, 792
$C_{18}H_{20}O_2Ru$, Dicarbonylbis(dimethylfulvene)ruthenium, 42B, 624
$C_{18}H_{21}Cr_2NO_{10}$, Tetraethylammonium μ-hydridobis(pentacarbonylchromium), 31B, 31; 35B, 506
$C_{18}H_{21}NO_{10}W_2$, Tetraethylammonium μ-hydridobis(pentacarbonyltungsten), 41B, 839
$C_{18}H_{22}HgN_2$, Di-(ω,ω'-(N,N-dimethyl)ethylenediamino)-o-tolylmercury, 33B, 328
$C_{18}H_{22}MnNO_3$, Tetramethylammonium cyclopentadienyl-phenyloxycarbene-dicarbonyl-manganese, 37B, 432
$C_{18}H_{22}MoNO_5P$, ($η^1$-N-Phenyliminoethyl)($η^5$-cyclopentadienyl)(trimethylphosphite)dicarbonylmolybdenum, 44B, 745
$C_{18}H_{23}Cl_2OPPt$, cis-[Benzyl(ethoxy)carbene]dichloro(dimethylphenylphosphine)platinum(II), 45B, 854
$C_{18}H_{24}N_2O_2Pt$, Trimethyl(acetylacetonyl)-2,2'-bipyridylplatinum(V), 27, 886
$C_{18}H_{25}CrNO_5S_2$, fac-((1,3-Dithian-2-ylidene(ethoxy)methyl(ethoxy)-carbene-C,S)-t-butyl isocyanide)tricarbonylchromium(0), 43B, 961
$C_{18}H_{25}Ta$, ($η^2$-Benzyne)($η^5$-pentamethylcyclopentadienyl)dimethyltantalum, 45B, 855
$C_{18}H_{30}CdN_2Si_2 \cdot 0.5\ C_{10}H_8N_2$, Bipyridylbis(trimethylsilylmethyl)cadmium(II) 2,2'-bipyridyl solvate, 46B, 792
$C_{18}H_{30}ClF_5P_2Pt$, cis-Chloro-perfluorophenyl-bis(triethylphosphine)-platinum(II), 44B, 1061
$C_{18}H_{30}I_3MnN_6$, Hexakis(ethylisocyanide)manganese(I) tri-iodide, 45B, 855
$C_{18}H_{36}Cl_2NO_2Rh$, Tetra-n-butylammonium cis-dichlorodicarbonyl-rhodate(I), 37B, 433
$C_{18}H_{36}Cl_4O_4Pt_2$, Di-μ-chloro-bis[chloro{neopentyl(isopropoxy)-carbene}-platinum], 45B, 856
$C_{18}H_{39}Cl_5N_2Re$, Tetrabutylammonium pentachloro(methyl isocyano)-rhenate(IV), 45B, 856
$C_{18}H_{42}CrLi_3O_6$, Lithium hexamethylchromium(III) dioxane solvate, 40B, 665
$C_{18}H_{46}Mo_2O_4P_2Si_2$, Di-μ-acetato-bis[(trimethylphosphine)trimethylsilyl)molybdenum(II)], 45B, 857
$C_{18}H_{46}N_4W_2$, Dimethyltetrakis(diethylamido)ditungsten, 42B, 625
$C_{19}H_8CoF_{10}$, $η^6$-Toluene-bis(perfluorophenyl)cobalt, 42B, 693; 46B, 793
$C_{19}H_8F_{10}Ni$, $η^6$-Toluene-bis(perfluorophenyl)nickel, 46B, 793
$C_{19}H_8O_9Ru_3$, (μ-Hydrido)-[(α-methylenebenzyl)acetylido]tris(tricarbonylruthenium), 46B, 793
$C_{19}H_{10}MnO_7Re$, (Dicarbonylcyclopentadienylmanganese)-μ-phenylketenyl-(tetracarbonylrhenium), 46B, 794
$C_{19}H_{12}Co_6NO_{14}$, Tetramethylammonium hexa-cobalt-tetradecacarbonyl-carbide, 46B, 795
$C_{19}H_{12}Fe_4O_{11}$, Bis($μ_4$-but-1-yne)undecacarbonyl-quadro-tetrairon, 44B, 697
$C_{19}H_{13}HgN_3$, Phenylmercury cyanide phenanthroline, 44B, 698
$C_{19}H_{14}NiO_8Ru_3$, $η^5$-Cyclopentadienylnickelbis(μ-carbonyl)-$μ_4$-(σ,σ,$η^3$,$η^3$-1-methyl-3-ethylallyl)hexacarbonyltriruthenium, 46B, 885
$C_{19}H_{16}HgN_4S$, Phenylmercury(II) dithizonate, 46B, 997
$C_{19}H_{16}MnO_5P$, Tetracarbonyl-(3-diphenylphosphineoxy)propyl)manganese, 46B, 1260
$C_{19}H_{16}NO_{12}Os_3P$, μ-Hydrido-μ-(n-phenylformimidoyl)triphenylphosphite-

nonacarbonyl-triosmium, 45B, 846
$C_{19}H_{18}AuP$, Methyl(triphenylphosphine)gold, 43B, 981
$C_{19}H_{18}N_2Ni$, π-Cyclopentadienylazotoluenenickel, 38B, 742
$C_{19}H_{20}FeO_{11}$, Cyclobutadieneirontricarbonyl dimethyl maleate (1:2) photoadduct, 41B, 840
$C_{19}H_{20}Fe_2O_5$, μ-Carbonyl-μ-t-butoxycarbonylmethylene-bis[carbonyl(η⁵-cyclopentadienyl)iron], 46B, 796
$C_{19}H_{20}O_7Ru_3$, Heptacarbonylhydrido(t-butylethynyl)(2,4-hexadiene)-triangulo-triruthenium, 45B, 857
$C_{19}H_{20}O_8Pd$, Tetrakis(methoxycarbonyl)palladiacyclopentadiene norbornadiene complex, 42B, 626
$C_{19}H_{22}CrN_2O_4$, cis-Tetracarbonyl[(3-aza-1-methyl-2-phenyl-2-butenyl)diethylaminocarbene]chromium, 45B, 858
$C_{19}H_{22}NTi$, 2-(Dimethylaminoethyl)phenyldicyclopentadienyltitanium-(III), 44B, 1129
$C_{19}H_{22}U$, Tris(η^5-cyclopentadienyl)-η^1-2-methylallyluranium(IV), 41B, 903
$C_{19}H_{23}BrCrN_2O_2$, Bromobis(t-butylisonitrile)dicarbonyl(phenylcarbyne)chromium, 43B, 981
$C_{19}H_{24}IrNO_2$, Acetylacetonato(η^2-allene)pyridine-3,4-dimethylene-iridocyclopentane, 43B, 976
$C_{19}H_{24}U$, Tricyclopentadienyl-n-butyl-uranium(IV), 41B, 851
$C_{19}H_{26}O_6P_2PtW$, μ-Methoxy(phenyl)carbene-[bis(trimethylphosphine)-platinum][pentacarbonyltungsten](Pt-W), 46B, 796
$C_{19}H_{27}MnO_3$, (-)-π-Cyclopentadienyldicarbonylmenthylmethoxycarbene-manganese, 44B, 1129
$C_{19}H_{29}As_2IPt$, Iodo-(1,2-bis(phenylmethylarsino)ethane)-trimethyl-platinum(IV), 39B, 569
$C_{19}H_{29}CoN_6O_4$, [(R)-1-Cyanoethyl][(S)-α-methylbenzylamine]bis(dimethylglyoximato)cobalt(III), 46B, 797
$C_{19}H_{29}CoN_6O_4$, [(S)-1-Cyanoethyl][(S)-α-methylbenzylamine]bis(dimethylglyoximato)cobalt(III), 46B, 798
$C_{19}H_{30}CrO_7Si_4$, Pentacarbonyl(2-furyl(tris(trimethylsilyl)siloxy)-carbene)chromium(0), 45B, 859
$C_{19}H_{34}ClF_6OP_3Pt$, trans-Carbonyl(p-chlorophenyl)bis(triethylphosphine)platinum(II) hexafluorophosphate, 40B, 707
$C_{19}H_{36}BrIrOP_2$, Bromocarbonylhydridophenylbis(triethylphosphine)iridium(III), 42B, 626
$C_{19}H_{37}ClP_2Pt$, cis-Chloro-p-tolyl-bis(triethylphosphine)platinum(II), 44B, 1061
$C_{19}H_{38}IrO_{10}P_3$, 5,7-Dimethyl-2,2,2-tris(trimethylphosphito)-2H-2-irida-indan-1-one, 46B, 798
$C_{19}H_{39}Cl_2N_2PPd$, cis-Dichloro-(tri-n-butylphosphine)-(bis(dimethylamino)-cyclopropenylidene)-palladium(II), 45B, 859
$C_{20}Co_6O_{18}$, Bis(methinyltricobalt enneacarbonyl), 38B, 742
$C_{20}H_6Cu_2N_2O_{15}Rh_6 \cdot 0.5\ CH_4O$, Di-$\mu_3$-acetonitrilecuprio-carbido-ennea-μ-carbonyl-hexacarbonyl-polyhedro-hexarhodium methanol solvate, 46B, 799
$C_{20}H_{10}ClCo_3HfO_{10}$, μ_3-((Chlorobis(η-cyclopentadienyl)hafniumoxy)methylidyne)cyclotris(tricarbonylcobalt), 44B, 699
$C_{20}H_{10}ClCo_3O_{10}Zr$, μ_3-((Chlorobis(η-cyclopentadienyl)zirconiumoxy)-methylidyne)cyclotris(tricarbonylcobalt), 44B, 699
$C_{20}H_{10}CrO_5$, Pentacarbonyl(2,3-diphenylcyclo-propenylidene)chromium(0), 34B, 426
$C_{20}H_{10}F_{12}Fe_2O_2$, 1-Carbonyl-1-cyclopentadienyl-1-(cyclopentadienyl-iron)-2,3,5,6-tetra(trifluoromethyl)-1-ferra-cyclohexa-2,5-diene-4-one, 42B, 627

$C_{20}H_{10}Fe_2O_8$, Bis(μ-phenyloxycarbenetricarbonyliron), 34B, 427
$C_{20}H_{10}Fe_3O_7$, Phenylethynyl(cyclopentadienyl)heptacarbonyltriiron, 41B, 840
$C_{20}H_{12}FeO_4$, Dibenzosemibullvaleneiron tetracarbonyl, 40B, 708
$C_{20}H_{15}NO_8Os_3$, Bis(μ-hydrido)(μ_3-η^4-1,3-dimethyl-1-(phenylimino)-butane)octacarbonyltriosmium, 46B, 1102
$C_{20}H_{15}NiO_9Ru_3$, (η^5-Cyclopentadienylnickel)(μ_4-3,3-dimethylbut-1-enyl)nonacarbonyltriruthenium, 46B, 799
$C_{20}H_{15}O_{10}Os_3P$, μ-Hydrido-2-(dimethylphenylphosphino)ethylidene-decacarbonyltriosmium, 42B, 628; 43B, 983
$C_{20}H_{16}Cl_6HgN_2$, Bis(trichlorovinyl)mercury 3,4,7,8-tetramethyl-1,10-phenanthroline, 46B, 800
$C_{20}H_{16}Fe_3O_6S_3$, μ-(Thioferrocenylmethylmethane-thiomethylene-C^1,S^2)-1,1,1,2,2,2-hexacarbonyl-μ-methylthioiron(Fe-Fe), 46B, 801
$C_{20}H_{16}Fe_3O_8$, Octacarbonyl-(η-1,3-dimethyl-2-vinylcyclopentadienyl)-μ_3-propylidyne-triangulo-triiron, 43B, 983
$C_{20}H_{17}FeNO$, Carbonyl-cyclopentadienyl-[5-methyl-2-(phenylcarbenylamino)phenyl]iron, 45B, 899
$C_{20}H_{18}Mo_2$, Di-μ-(σ:η-cyclopentadienyl)bis(η-cyclopentadienyl)dimolybdenum(Mo-Mo), 45B, 860
$C_{20}H_{20}ClPPdS$, Chlorothiomethoxymethyltriphenylphosphinepalladium-(II), 43B, 984
$C_{20}H_{20}Hf$, Bis(η^5-indenyl)dimethylhafnium, 41B, 841
$C_{20}H_{20}Ti$, Bis(η^5-indenyl)dimethyltitanium, 41B, 841
$C_{20}H_{20}W_2$, trans-Di-μ-(σ:η^5-cyclopentadienyl)-bis(η^5-cyclopentadienyl)dihydridoditungsten, 45B, 861
$C_{20}H_{20}Zr$, Bis(η^5-indenyl)dimethylzirconium, 41B, 841
$C_{20}H_{24}Cl_4N_8Rh_2 \cdot 8\ H_2O$, Dichlorotetrakis(1,3-diisocyanopropane)dirhodium dichloride octahydrate, 45B, 862
$C_{20}H_{24}N_2Ni$, α,α'-Bipyridyl-5-nickela-3,3,7,7-tetramethyl-trans-tricyclo[4.1.0.0$^{2-4}$]heptane, 43B, 985
$C_{20}H_{26}Cl_2P_2Pt$, trans-Bis(α-chlorovinyl)bis(dimethylphenylphosphine)-platinum(II), 42B, 628
$C_{20}H_{26}F_6N_2O_4Pt_2$, Bis($\mu$-trifluoroacetato-O,O')-bis(dimethyl(4-methylpyridine)platinum)(Pt-Pt), 44B, 700
$C_{20}H_{27}BN_4Pt$, (Diethylbis(1-pyrazolyl)borato)methyl(1-phenylpropyne)-platinum(II), 41B, 842
$C_{20}H_{28}Br_2Cl_2O_2P_2Rh_2$, Bromocarbonylchloromethyl(dimethylphenylphosphine)rhodium dimer, 44B, 1130
$C_{20}H_{29}F_{12}N_3O_2Pd$, 4-((Diethylamino)(t-butylamino)-methylene)-4-(t-butyl isocyanide)-2,2,5,5-tetrakis(trifluoromethyl)-1,3,4-dioxopalladolan, 40B, 709
$C_{20}H_{30}Cl_2Pd_2$, Di-μ-chloro-bis[(2'-3'-η-exo-3-allylnorborn-2-yl)palladium], 45B, 862
$C_{20}H_{30}Cl_4Co_2N_{10}O_{16}$, Deca(methylisonitrile)dicobalt(II) perchlorate, 29, 563
$C_{20}H_{30}I_2P_2Pt$, Diiodo(butane-1,4-diyl)bis(dimethylphenylphosphine)-platinum(IV), 42B, 629
$C_{20}H_{31}F_6OP_3Pt$, trans-((Methyl)(methylmethoxycarbene)bis(dimethylphenylphosphine)platinum(II)) hexafluorophosphate, 39B, 569
$C_{20}H_{32}CoN_4O_9$, Dioxo-hydroximino-oxanonyl-oximato-2-(dioxo-hydroximino-oxanonyl-oximato)propyl-methoxidocobalt(III), 44B, 700
$C_{20}H_{32}CoN_5O_6$, (R)-1-(Methoxycarbonyl)ethyl(R)(+)-α-methylbenzylaminebis(dimethylglyoximato)cobalt(III), 43B, 986
$C_{20}H_{33}ClF_6P_2PdSi$, Chloro[dimethylphenylphosphonium(trimethylsilyl)-methylide](1,5-cyclooctadiene)-palladium(II) hexafluorophosphate, 45B, 863

$C_{20}H_{33}ClP_2PtSi$, trans-Chlorobis(dimethylphenylphosphine)(trimethylsilyl)platinum(II), 40B, 709
$C_{20}H_{37}ClINP_2Pt$, trans-Iodo(p-chlorophenyl-methylisocyanide)bis(triethylphosphine)platinum(II), 40B, 710
$C_{20}H_{37}N_2OP_2RhS_2$, Butanoylbis(triethylphosphine)(maleonitriledithiolato)rhodium(III), 45B, 864
$C_{20}H_{38}Cl_4N_2O_4P_2Pt$, ab-Dichloro-df-(methylamino)(4-chlorophenylamino)carbene-C,C(2)-ce-bis(triethylphosphine)platinum(IV) perchlorate, 41B, 843
$C_{20}H_{39}F_{12}OP_3Pt$, Bis(triethylphosphine)-{1,1,1-trifluoro-5-methoxy-3-(trifluoromethyl)-4-methyl-pent-3-ene-2-yl}platinum hexafluorophosphate, 45B, 864
$C_{20}H_{40}Cl_2Mo_2O_6P_2Si_2$, Di-μ-chloro-bis[bis(carbonyl)trimethylphosphine(1-2-η-trimethylsilylmethylcarbonyl)molybdenum(II)], 46B, 785
$C_{20}H_{41}F_6O_4P_3Pt$, [η³-3-Methylallyl-1,2-bis(methylcarboxylate)]bis-(triethylphosphine)platinum hexafluorophosphate, 46B, 828
$C_{20}H_{44}Cl_4Li_4O_4W_2$, Lithium tetrachlorotetramethylditungsten tetrahydrofuran solvate, 43B, 987
$C_{20}H_{45}Cl_4OPTa_2$, μ-Oxo-μ-(trimethylphosphine-methyl)-bis(η⁵-1-ethyl-tetramethyl-cyclopentadienyl)-hydrido-bis(dichloro-tantalum), 46B, 801
$C_{20}H_{48}Mo_2N_4$, Bis(μ-N-tert-butylimido)bis[(N-tert-butylimido)dimethylmolybdenum], 46B, 1101
$C_{20}H_{54}HgSi_6$, Bis(tris(trimethylsilyl)methyl)mercury(II), 43B, 987
$C_{20}H_{58}B_2F_8P_6Ru_2$, Hexakis(trimethylphosphine)bis(μ-methylene)-diruthenium(III)(Ru-Ru) bistetrafluoroborate, 45B, 866
$C_{21}Co_6O_{19}$, Bis(tricobalt enneacarbonyl)acetone, 34B, 430
$C_{21}H_{12}F_{10}Ni$, (η⁶-Mesitylene)bis(pentafluorophenyl)nickel(II), 46B, 802
$C_{21}H_{13}NO_8Os_3$, μ-η²-Benzyne-μ-η²-formimidoyl-octacarbonyltri(hydrido-osmium), 45B, 865
$C_{21}H_{14}Co_2O_4$, μ-(1,3-Dioxoindane-2-ylidene)bis[carbonyl-(η⁵-cyclopentadienyl)cobalt], 46B, 802
$C_{21}H_{14}Co_2O_7$, μ-Carbonyl-μ-(1-3-η:1-σ,4-2'-η-{1,3-dimethyl-4-[5-oxo-4-phenyl-2(5H)-furan-2-ylidene]-but-1-ene-1,3-diyl})-bis(dicarbonylcobalt)(Co-Co), 46B, 803
$C_{21}H_{14}O_9Ru_4$, (η⁶-Benzene)nonacarbonyl-1-cyclohexen-1,2-ylenetetraruthenium, 45B, 865
$C_{21}H_{15}Co_3CrO_5S \cdot 0.5\ C_4H_8O$, Tris(η⁵-cyclopentadienyl)-μ₃-((pentacarbonylchromiumthio)methylidine)-μ₃-sulfido-triangulo-tricobalt tetrahydrofuran solvate, 46B, 1218
$C_{21}H_{15}MnO_3$, (π-Cyclopentadienyl)(benzoylphenylcarbene)dicarbonylmanganese, 41B, 843
$C_{21}H_{17}NO_9Os_3$, Bis(μ-hydrido)(μ₃-η²-1-isopropyl-2-methyl-2-(phenylamino)ethylene)nonacarbonyltriosmium(3Os-Os), 46B, 1102
$C_{21}H_{19}F_{12}IrO_2$, acb(1,2-Bis(trifluoromethyl)-3-acetyl-4-oxopent-1-enyl-O',O,C(1))-fde(1,4-5,β-η-(bis(trifluoromethyl)ethylene)oct-4-enyl)iridium(III), 41B, 844
$C_{21}H_{20}Fe_2O_5$, Bis(dimethylfulvene)pentacarbonyldiiron, 41B, 845
$C_{21}H_{20}HgN_2$, N-Phenylmercury-N,N'-di-p-tolylformamidine, 43B, 989
$C_{21}H_{20}O_5PRe$, 3,3,3,3-Tetracarbonyl-5,5-dimethyl-2,2-diphenyl-1-oxa-2-phospha-3-rhenacyclohexane, 46B, 804
$C_{21}H_{21}ClN_2O_3Pd$, cis-Chloro(benzylacetoacetate)dipyridinepalladium-(II), 41B, 847
$C_{21}H_{24}Co_2O_5$, Dicarbonyl(bis(hexamethylene)(tricarbonylcobalta)cyclopentadiene)cobalt(Co-Co), 44B, 701
$C_{21}H_{28}F_6P_2W$, Bis-(η-cyclopentadienyl)-σ-((2-dimethylphenylphos-

phoniumethyl)methyltungsten hexafluorophosphate, 41A, 848
$C_{21}H_{29}Cl_2N_2PPt$, cis-Dichloro(1,3-diphenylimidazolidin-2-ylidene)-(triethylphosphine)platinum(II), 40B, 711
$C_{21}H_{29}Cl_2N_2PPt$, trans-Dichloro(1,3-diphenylimidazolidin-2-ylidene)-(triethylphosphine)platinum(II), 40B, 711
$C_{21}H_{29}NO_6W$, Tetraethylammonium·(α-methoxybenzyl)pentacarbonyltungstate, 44B, 702
$C_{21}H_{31}F_6OPPt$, trans-(Methyl(2-oxacyclopentylidene)bis(dimethylphenylphosphine)platinum(II)) hexafluorophosphate, 40B, 712
$C_{21}H_{34}F_6NP_3Pt$, trans-(Methyl(methyl-N,N-dimethylaminocarbene)bis(dimethylphenylphosphine)platinum(II)) hexafluorophosphate, 40B, 713
$C_{21}H_{39}PPd$, (η^1-Cyclopentadienyl)(η^3-2-t-butylallyl)triisopropylphosphinopalladium(II), 46B, 1263
$C_{21}H_{45}N_3W$, 2-t-Butyl-1-t-butylimino-1,3,3-trimethyl-1-(N-(3-methylbut-2-en-2-yl)-t-butylamino)-2-aza-1-tungstacyclopropane, 46B, 804
$C_{21}H_{57}CrSi_6$, Tris(bis(trimethylsilyl)methyl)chromium(III), 44B, 702
$C_{21}H_{60}P_6Ru_2$, Hexakis(trimethylphosphine)tris(μ-methylene)-diruthenium(III), 45B, 866
$C_{21}H_{61}BF_4P_6Ru_2$, Hexakis(trimethylphosphine)bis(μ-methylene))-μ-methyl-diruthenium(III)(Ru-Ru) tetrafluoroborate, 45B, 866
$C_{22}Co_6O_{18}$, Bis(methinyltricobalt enneacarbonyl)acetylene, 35B, 504
$C_{22}H_8Fe_4N_2O_{12}S_2$, Iron carbonyl 2-mercapto-pyridine complex, 41B, 849
$C_{22}H_{16}Mn_2O_4$, Phenylvinylidenebis(cyclopentadienylmanganesedicarbonyl)(Mn-Mn), 43B, 989
$C_{22}H_{17}Co_3O_6$, Phenylmethinyltricobalt hexacarbonyl - mesitylene, 38B, 744
$C_{22}H_{17}F_6PPt$, Bis(trifluoromethyl)[(2-vinylphenyl)diphenylphosphine]-platinum(II), 45B, 867
$C_{22}H_{17}Fe_2NO_{10}$, 2,2,2,3,3,3-Hexacarbonyl-4,5-bis(methoxycarbonyl)-1-(2,2-dimethyl-1-phenylvinyl)-1,2,3-azadiferracyclopent-4-ene, 44B, 703
$C_{22}H_{20}Fe_2Ni_2O_6$, ($\mu_4,\sigma,\sigma,\eta^2,\eta^2$-Hex-3-ene-3,4-diyl)-bis(η^5-cyclopentadienylnickel)-bis(tricarbonyliron), 46B, 882
$C_{22}H_{23}ClN_2O_5Pd$, Aquo(benzo[h]quinoline)(2-(dimethylaminomethyl)phenyl-N)palladium(II) perchlorate, 44B, 703
$C_{22}H_{23}PPt$, Dimethyl[(2-vinylphenyl)diphenylphosphine]platinum(II), 45B, 867
$C_{22}H_{26}Y_2$, Bis[bis(η^5-cyclopentadienyl)methylyttrium], 45B, 868
$C_{22}H_{26}Yb_2$, Bis[bis(η^5-cyclopentadienyl)methylytterbium], 45B, 868
$C_{22}H_{27}F_6Ir_3O_6S_3$, Tris(μ-t-butylthiolato)(μ-bis(trifluoromethyl)acetylene)hexacarbonyltriiridium(Ir-Ir), 46B, 1225
$C_{22}H_{28}CoN_5O_5 \cdot C_6H_6$, [Benzoyl(1-pyridinio)methanide]bis(dimethylglyoximato)methylcobalt benzene solvate, 46B, 805
$C_{22}H_{28}N_4Ni$, Bis(t-butyl isocyanide)(azobenzene)nickel(0), 38B, 744
$C_{22}H_{32}Cl_2O_2P_2Pt_2$, trans-Di-μ-chloro-bis(propionyl-dimethylphenylphosphine-platinum(II)), 45B, 868
$C_{22}H_{32}Co_2O$, (μ-Carbonyl)(μ-methylene)-bis(η^5-pentamethylcyclopentadienylcobalt)(Co-Co), 46B, 805
$C_{22}H_{33}ClP_2Pt$, trans-Chlorobis(diethylphenylphosphine)(vinyl)platinum(II), 43B, 990
$C_{22}H_{35}ClF_6NOP_3Pt$, trans-Chloro(3-hydroxypropyl-N,N-dimethylaminocarbene)bis(dimethylphenylphosphine)platinum(II) hexafluorophosphate, 44B, 704
$C_{22}H_{38}HgO_4$, Bis(dipivaloylmethyl)mercury, 38B, 829
$C_{22}H_{49}I_3S_2Zn$, Methyl-di(neopentyl)sulphonium tri-iododi(neopentyl)-sulphonium-methylzincate, 34B, 47
$C_{22}H_{56}Cl_4N_6O_8Pt_2$, Dichloro(3,3,6,6-tetramethyl-3,6-diazaoctane-1,8-

diyl)bis(N,N,N',N'-tetramethyl-1,2-diaminoethane)diplatinum(II) diperchlorate, 45B, 869
$C_{22}H_{62}Cr_2P_2Si_4$, Di-μ-trimethylsilylmethyl-bis((trimethylphosphine)-(trimethylsilylmethyl)chromium(II))(4Cr-Cr), 44B, 705
$C_{23}H_{14}O_5Ru_2$, Diphenylfulvenepentacarbonyldiruthenium, 41B, 851
$C_{23}H_{16}FeO_3$, η^5-Cyclopentadienyl-(2,4-diphenylcyclobut-1-ene-3-one-1-yl)dicarbonyliron, 46B, 895
$C_{23}H_{18}CoN_3O_4Pd$, (Tetracarbonylcobalto)pyridine(N-(phenylimino)-α-methylbenzylidenimino-2-C,N)palladium(II), 43B, 991
$C_{23}H_{22}ClPZr$, Chlorobis(η^5-cyclopentadienyl)(diphenylphosphinomethyl)zirconium(IV), 46B, 805
$C_{23}H_{23}Fe_4NO_{14}$, Tetraethylammonium μ_4-carbomethoxymethylidene-tetrakis(tricarbonyliron), 45B, 869
$C_{23}H_{24}Fe_3O_7$, Heptacarbonyl-μ_3-(pent-1-en-1-yl-3-ylidene)-(η-1,2,3-triethylcyclopentadienyl)-triangulo-triiron(3Fe-Fe), 44B, 705
$C_{23}H_{24}HgN_2O_3S$, O-Methylbenzoin methylmercurio(p-tolyl-sulphonyl)hydrazone, 44B, 706
$C_{23}H_{24}U$, Tricyclopentadienyl-p-methylbenzyl-uranium(IV), 41B, 851
$C_{23}H_{26}N_4Ni$, (Diazofluorene)bis(t-butyl isocyanide)nickel(0), 43B, 992
$C_{23}H_{34}CoN_5O_4$, Methyl(R(+)-α-methylbenzylamine)bis(dimethylglyoximato)cobalt(III) benzene solvate, 43B, 993
$C_{23}H_{35}ClO_{14}Pd \cdot 0.66\ CHCl_3$, ((Chloro(methoxycarbonyl)(1,2,3,4,5-pentakismethoxycarbonylcyclopenta-2,4-dienyl)-2-MeOCO)methyl)pentane-2,4-dionatopalladium(II) chloroform solvate, 41B, 852
$C_{23}H_{35}NP_2PdS$, o-Ethynylphenylethynyl-isothiocyanato-trans-bis(triethylphosphine)palladium(II), 43B, 993
$C_{23}H_{37}NiO_3P$, Tricyclohexylphosphineethylidenenickel tricarbonyl, 38B, 746
$C_{23}H_{38}BN_3Ni$, (2,2'-Bipyridyl)-(ethyl)-(neopentyliminotriethylborato)nickel, 45B, 870
$C_{23}H_{38}ClP_2Ta$, Chloro(η^5-pentamethylcyclopentadienyl)bis(trimethylphosphine)benzylidynetantalum, 44B, 707; 45B, 871
$C_{23}H_{43}Cl_2N_4Ta$, Dichlorobis-(N,N-diisopropylacetamidinato)methyltantalum(V) benzene solvate, 41B, 836
$C_{24}H_{10}F_{10}Ni_2$, μ-(Perfluorodiphenylethyne)-bis(η^5-cyclopentadienylnickel), 46B, 806
$C_{24}H_{10}O_{10}Os_3$, μ_3-(η-Diphenylacetylene)-decacarbonyltriosmium, 41B, 853
$C_{24}H_{15}AuF_5P$, (Pentafluorophenyl)(triphenylphosphine)gold(I), 38B, 747
$C_{24}H_{15}BrMnO_4P$, 1-(Triphenylphosphonium)-2-(bromotetracarbonylmanganese(II))-acetylene, 38B, 747
$C_{24}H_{16}F_{18}Ir_2O_2$, Dicarbonyl(hexafluorobut-2-ene-2-yl)(hexafluoro-3,4-di(trifluoromethyl)hexa-2,4-diene-2,4-diyl)(pentamethylcyclopentadienyliridium)iridium (form I), 44B, 707
$C_{24}H_{16}FeO_6$, Tricarbonyl(5-acetyl-2,3-diphenyl-hepta-2,4-diene-1,6-dione-1,4-diyl)iron, 44B, 1131
$C_{24}H_{18}Cl_2N_4Pt_2$, trans-μ-Dichloro-bis(phenylazophenyl-2C,N')diplatinum(II), 42B, 630
$C_{24}H_{21}MnNO_3PS$, Dimethylthiocarboxamido(triphenylphosphine)tricarbonylmanganese, 42B, 630
$C_{24}H_{22}O_2Pd$, Pentane-2,4-dionato(α,1,2-η-triphenylmethyl)palladium, 45B, 871
$C_{24}H_{22}O_2Pt$, Pentane-2,4-dionato(α,1,2-η-triphenylmethyl)platinum, 45B, 871
$C_{24}H_{22}O_8Ru_3$, (2-Methyl-5-neopentyl-6-isopropenyl-(1-7-η)-hepta-

trienyl)-octacarbonyl-triruthenium, 45B, 872
$C_{24}H_{23}Ta$, Bis(η^5-cyclopentadienyl)(benzyl)(benzylidenyl)tantalum, 44B, 708
$C_{24}H_{24}F_{12}Pt_2$, Bis(cycloocta-1,5-diene)bis(hexafluorobut-2-yne)diplatinum, 43B, 1008
$C_{24}H_{26}Mo_2O_4S$, Bis(η^5-cyclopentadienyl)(thiocamphor)tetracarbonyldimolybdenum, 45B, 976
$C_{24}H_{28}NPPtSe_2$, Methyl(diethyldiselenocarbamato)(triphenylphosphine)platinum(II), 45B, 872
$C_{24}H_{28}N_2Ni$, Bis(t-butyl isocyanide)(diphenylacetylene)nickel(0), 38B, 748
$C_{24}H_{28}N_4P_2Pt$, trans-Bis(trimethylphosphine)propynyl-1-(4'-dicyanomethylene-cyclohexa-2',5'-dien-1-yliden)-3,3-dicyano-2-methyl-prop-2-en-1-ylplatinum, 41B, 855
$C_{24}H_{30}ClIrN_2O_3P_2 \cdot C_3H_6O$, (2-aci-Nitrato-propan-1-one-1-yl)-bis(dimethylphenylphosphine)pyridine-chloroiridium acetone solvate, 45B, 873
$C_{24}H_{30}F_{18}P_2Pt$, Bis(triethylphosphine)-(hexakis(trifluoromethyl)benzene)-platinum, 40B, 713
$C_{24}H_{30}Fe_2O_6$, 2,4-Di-t-butyl-8,8-dimethylnona-1,5,6-trien-3-one-1,4,7-triyl-pentacarbonyldiiron(Fe-Fe), 41B, 855
$C_{24}H_{30}SiW_2$, cis-Di-μ-(σ:η^5-cyclopentadienyl)-bis(η^5-cyclopentadienyl)(hydrido)trimethylsilyl-methylditungsten, 45B, 861
$C_{24}H_{30}SiW_2$, trans-Di-μ-(σ:η^5-cyclopentadienyl)-bis(η^5-cyclopentadienyl)(hydrido)trimethylsilyl-methylditungsten, 45B, 861
$C_{24}H_{32}Cr_2O_6$, Bis(o-t-butoxyphenyl)diacetatodichromium(Cr-Cr), 44B, 708
$C_{24}H_{34}MoP_2$, (η^6-Benzene)bis(dimethylphenylphosphine)dimethylmolybdenum, 45B, 873
$C_{24}H_{34}O_2Pd$, dihapto-Cyclopentadiene-acetylacetone-palladium complex, 39B, 570
$C_{24}H_{36}O_2P_2Pt$, trans-Bis(dimethylphenylphosphine)bis-(3-Z-methoxy-1-propenyl)platinum(II), 42B, 631
$C_{24}H_{43}NiO_2P$, (Tricyclohexylphosphine)methylnickel(II) 2,4-pentanedionate, 38B, 749
$C_{24}H_{48}O_4Pt_2$, Trimethyl(nonane-4,6-dionato)platinum(IV) dimer, 24, 627; 34B, 432
$C_{24}H_{56}Cr_2Li_4O_4$, Tetralithium octamethyldichromium tetrakis(tetrahydrofuran), 35B, 507
$C_{24}H_{56}Li_4Mo_2O_4$, Lithium octamethyldimolybdate tetrahydrofuran solvate, 40B, 714
$C_{24}H_{60}Cl_4Pt_4$, Tetra-μ_3-chlorotetrakis(triethylplatinum(IV)), 37B, 433
$C_{24}H_{62}Re_2Si_6$, Bis(trimethylsilylmethylidyne)tetrakis(trimethylsilylmethyl)dirhenium(Re-Re), 46B, 806
$C_{24}H_{62}Si_6W_2$, Bis(μ-trimethylsilylmethylidyne)-tetrakis(trimethylsilylmethyl)ditungsten(W-W), 44B, 709
$C_{24}H_{64}Li_4O_4W_2$, Tetralithium octamethylditungstate - tetrakis(diethyl ether), 44B, 710
$C_{24}H_{64}P_8Ru_2$, Bis(tetramethyldiphosphinoethane)ruthenium dimer, 42B, 631
$C_{24}H_{66}Cl_3N_2O_2Re_3Si_6$, Tri-$\mu$-chloro-(N-nitroso-N-trimethylsilylmethylhydroxylaminato)-pentakis(trimethylsilylmethyl)-triangulo-trirhenium(III), 46B, 818
$C_{24}H_{66}Cl_3Re_3Si_6$, Tri-μ-chloro-hexakis(trimethylsilylmethyl)-triangulo-trirhenium(III), 44B, 710
$C_{24}H_{66}Si_6W_2$, Hexakis(trimethylsilylmethyl)ditungsten, 42B, 632

$C_{25}H_5Co_7O_{24}Ti$, (η-Cyclopentadienyl)bis(μ_3-oxymethylidyne-cyclo-tris(tricarbonylcobalt))(tetracarbonylcobalt)titanium, 44B, 718
$C_{25}H_{16}FeO_4$, Tricarbonyl(1,2,3-triphenylallylcarbonyl)iron, 44B, 711
$C_{25}H_{18}Fe_2O_8$, Tricarbonyl-π-(1,1,1-tricarbonyl-2,3-dimethoxy-5-(diphenylmethyl)ferracyclopentadiene)iron, 38B, 749
$C_{25}H_{20}MnO_4P$, Tetracarbonyl-2-(bis-p-tolylphosphino)-5-methylphenylmanganese, 43B, 997
$C_{25}H_{20}OW$, Oxo-π-cyclopentadienyl-π-diphenylacetylene-σ-phenyltungsten, 39B, 570
$C_{25}H_{21}ClNOPPt$, Chloro(2-(diphenylphosphinato)phenoxymethyl)pyridineplatinum(II), 43B, 997
$C_{25}H_{21}CrO_5P$, (Methoxy(methyl)carbene)triphenylphosphinetetracarbonylchromium, 34B, 433
$C_{25}H_{22}ClN_2OPPd \cdot CH_2Cl_2$, Chlorotriphenylphosphine(O-N-methyl-N-nitrosylaniline)palladium methylene chloride, 46B, 807
$C_{25}H_{22}CoN_5O_2$, Malononitrilato(propane-1,2-bis(salicylideneiminato))-pyridinecobalt(III), 41B, 856
$C_{25}H_{22}NO_2PPd \cdot 0.6\ CH_2Cl_2$, Carboxymethyl(triphenylphosphine)pyridinepalladium(II) dichloromethane solvate, 45B, 1275
$C_{25}H_{22}Ti$, 1,1-Bis(η^5-cyclopentadienyl)-2,3-diphenyl-1-titanacyclobut-2-ene, 46B, 808
$C_{25}H_{23}FeNO_4$, (2-Cyclohexyl-3,5-diphenyl-1-oxo-2-azapenta-(3-5-η)-enyl-1-yl)tricarbonyliron, 44B, 1131
$C_{25}H_{24}BF_4MnNO_3PS$, (Dimethylamino(thiomethoxy)carbene)(triphenylphosphine)tricarbonylmanganese tetrafluoroborate, 42B, 630
$C_{25}H_{24}Cl_2N_2Pt \cdot 0.5\ C_2H_6O$, Dichlorobis(1,2-diphenylcyclopropane)bis-(pyridine)platinum ethanolate, 39B, 571
$C_{25}H_{24}Fe_6N_2O_{16}$, Tetramethylammonium carbidohexadecacarbonylhexaferrate(2-), 40B, 715
$C_{25}H_{24}N_2O_{16}Ru_6$, Tetramethylammonium η^6-carbidohexadecacarbonylhexaruthenate, 46B, 808
$C_{25}H_{25}IrN_4O$, Fumaronitrile(8-methoxycyclo-oct-4-enyl)(1,10-phananthroline)iridium(I), 42B, 633
$C_{25}H_{25}O_4PPd$, (Acetylacetonato)(carboxymethyl)(triphenylphosphine)-palladium(II), 42B, 633
$C_{25}H_{27}BF_4IrN_2OPS_2$, Carbonylbis(N,N-dimethylthiocarboxamido)triphenylphosphineiridium tetrafluoroborate, 44B, 1132
$C_{25}H_{27}F_6IrN_2OP_2S_3$, Carbonyl(dimethyldithiocarbamato)(dimethylthiocarboxamido)triphenylphosphineiridium(III) hexafluorophosphate, 45B, 1091
$C_{25}H_{27}NiO_2P$, (2,4-Pentanedionato)(triphenylphosphine)ethylnickel-(II), 40B, 716
$C_{25}H_{28}Cl_4N_2PRhS_3$, Chloro-(N,N-dimethylthiocarbamato)-(N,N-dimethylthiocarboxamido)-triphenylphosphinerhodium(III)-chloroform, 41B, 857
$C_{25}H_{32}B_7PRu$, hyper-closo-2,3-Dimethyl-6-(η^3-o-allylphenyl(diphenyl)-phosphine)-6-ruthena-2,3-dicarbadecaborane, 46B, 809
$C_{25}H_{33}Co_3Si_2$, μ_3-(Trimethylsilylmethylidene)-μ_3-(trimethylsilylpropynylidene)-tris(cyclopentadienylcobalt), 45B, 875
$C_{25}H_{36}MoP_2$, (η^6-Toluene)bis(dimethylphenylphosphine)dimethylmolybdenum, 45B, 873
$C_{25}H_{45}CoN_5O_4P$, Bis(dimethylglyoximato)(tri-n-butylphosphine)(4-pyridyl)cobalt(I), 39B, 572
$C_{25}H_{45}FeN_5$, Pentakis(t-butyl isocyanide)iron, 45B, 898
$C_{26}H_{10}O_{12}Ru_4$, μ-Diphenylacetylene-closo-tetrakis(tricarbonylruthenium)(4Ru-Ru), 43B, 999
$C_{26}H_{15}Fe_2O_7P$, Hexacarbonyl(diphenylphenylenephosphino-aldehydo-

carbido)diiron, 43B, 1000; 44B, 712
$C_{26}H_{18}Cl_2N_4O_2Rh_2$, Di-μ-chloro-dicarbonylrhodium(I)-bis(phenylazophenyl-2C,N')rhodium(III), 38B, 750
$C_{26}H_{18}Mn_2O_6Ti$, Bis(1,5-η-cyclopentadienyl)bis[tricarbonyl(2,5-η-cyclopentadienylidene)manganese-C^1]-titanium, 46B, 810
$C_{26}H_{18}N_2O_{16}Os_6$, Bis(t-butyl isocyanide)hexadecacarbonylhexaosmium, 44B, 712
$C_{26}H_{20}AsAuN_2O_2$, Tetraphenylarsonium difulminatogold, 46B, 811
$C_{26}H_{20}CrO_6Si$, Pentacarbonyl(ethoxy(triphenylsilyl)carbene)chromium, 43B, 1001
$C_{26}H_{20}Fe_2N_2O_2$, trans-anti-Bis(η^5-cyclopentadienyl)dicarbonylbis(μ-phenylisonitrile)-diiron(Fe-Fe), 43B, 1001
$C_{26}H_{20}MoO_6Si$, Pentacarbonyl(ethoxy(triphenylsilyl)carbene)molybdenum, 43B, 1001
$C_{26}H_{21}N_4O_2Rh$, Acetatobis(phenylazophenyl-2C,N')rhodium(III), 38B, 751
$C_{26}H_{21}O_2ReSi \cdot CH_2Cl_2$, Dicarbonyl(cyclopentadienyl)(triphenylsilylcarbene)rhenium dichloromethane solvate, 46B, 811
$C_{26}H_{22}Mn_2O_4$, μ-(Cyclobuta[1,2-a:3,4-a']dicyclopentene)-bis[dicarbonyl-(η^5-methylcyclopentadienyl)-manganese], 46B, 812
$C_{26}H_{24}AuO_2P$, (2,6-Dimethoxyphenyl)(triphenylphosphine)gold(I), 46B, 813
$C_{26}H_{25}CoGeO_4$, Tricarbonyl-(triphenylgermyl)-(ethyl(ethoxy)carbene)-cobalt, 45B, 875
$C_{26}H_{26}I_2IrPS$, η^5-Cyclopentadienyliodomethyl(methylthio)carbene(triphenylphosphine)iridium(III), 46B, 903
$C_{26}H_{26}OPdS$, 3-Thio-1,3-diphenylprop-2-en-1-one-norbornenyl-2-methylallylpalladium(II), 39B, 573
$C_{26}H_{27}CoN_{10}O_4$, trans-3,3,4,4-Tetracyano-2-phenylcyclopentylbis(dimethylglyoximato)imidazolecobalt(III), 44B, 713
$C_{26}H_{29}F_5Fe_2O_6P$, (3,4,4,5,5-Pentafluoro-1-dicyclohexylphosphino-cyclopent(1-3-7)enyl-2-yl)hexacarbonyldiiron hexane solvate, 44B, 1133
$C_{26}H_{30}O_8Ru_3$, Bis(μ-carbonyl)-μ_3-(σ^2,η^4,η^4-1,3,6-tri-t-butylhexa-1,3,5-triene-1,4-diyl)hexacarbonyltriruthenium, 46B, 810
$C_{26}H_{31}IrO_2$, [2,2'-Bis(bicyclo[2.2.1]hept-5-ene)-3,3'-diyl](η^4-bicyclo[2.2.1]hepta-2,5-diene)acetylacetonatoiridium, 46B, 978
$C_{26}H_{33}Co_3O_4$, Bis(η^5-pentamethylcyclopentadienyl)tetracarbonyl(μ_3-ethylidene)-triangulo-tricobalt, 45B, 876
$C_{26}H_{34}F_6Ni_2O_4$, Di-μ-trifluoroacetatobis((2-methylallyl-3-norbornyl)-nickel(II)), 41B, 858
$C_{26}H_{38}P_2Ta$, Mesitylbis(neopentylidene)bis(trimethylphosphine)-tantalum(V), 45B, 877
$C_{26}H_{39}FeN_3O$, Cyclopentadienylironcarbonyl cyclic carbene complex, 42B, 634
$C_{26}H_{40}Cl_4O_2Rh_4$, cis-Tetraethylbutadienedichlorocarbonyldirhodium dimer, 40B, 1106
$C_{26}H_{42}ClNO_3P_2Pd$, trans-(1-p-Tolyl-3-oxo-2-(E)-carbomethoxymethylidene-2,3-dihydropyrrole)chlorobis(triethylphosphine)palladium(II), 44B, 713
$C_{26}H_{42}O_4P_2Pd$, trans-Bis(triethylphosphine)(1,2-bis(methoxycarbonyl)-vinyl)(phenylethynyl)palladium(II), 43B, 1002
$C_{26}H_{43}ClN_2P_2Pd$, trans-Chloro(N-(phenylamino)-α-methylbenzylidenimino-2-C,N)bis(triethylphosphine)palladium(II), 42B, 635
$C_{26}H_{46}P_2W$, [1,2-Bis(dimethylphosphino)ethane](neopentylidyne)(neopentylidene)(neopentyl)-tungsten(VI), 45B, 877
$C_{26}H_{56}Cl_4N_2O_2Pt_2$, Bis(tetrapropylammonium) bis(carbonyl)tetrachlorodiplatinate(I), 41B, 859

$C_{27}O_{25}Rh_{12}$, μ_3-Carbonyl-deca-μ-carbonyl-dicarbidotetradecacarbonyl-polyhedra-dodecarhodium, 44B, 714
$C_{27}H_{13}MnO_9PRe$, Tetracarbonyl-(μ-(carbonyl(6-(diphenylphosphino)-o-phenylene)))(tetracarbonylmanganese)rhenium, 43B, 1024
$C_{27}H_{16}Fe_2O_6$, μ-(1,2,3-Triphenylallyl)bis(tricarbonyliron), 44B, 711
$C_{27}H_{22}ClF_3P_2Pt$, Chloro(trifluoromethyl)[1,2-cis-bis(diphenylphosphino)ethene]platinum(II), 45B, 878
$C_{27}H_{24}O_6W$, Pentacarbonyl[3-(1,1-diphenylvinyl)cyclopentyl]ethoxymethylenetungsten(0), 46B, 813
$C_{27}H_{25}CrNO_4Sn$, trans-Triphenylstannyl-tetracarbonyl-diethylaminocarbynechromium, 46B, 814
$C_{27}H_{26}ClF_3P_2Pt$, trans-Bis(diphenylmethylphosphine)chloro(trifluoromethyl)platinum(II), 45B, 880
$C_{27}H_{27}Cl_3CoN_5O_4$, (1-Chloro-2,2-bis(p-chlorophenyl)vinyl)bis(dimethylglyoximato)pyridinecobalt(III), 41B, 860
$C_{27}H_{27}I_2N_3O_3W$, Diiodo-dicarbonyltris(t-butyl isocyanide)tungsten-(II), 45B, 878
$C_{27}H_{28}BrNPd$, Bromo{3-4-η-4-[(2-dimethylaminomethyl)phenyl]-2,4-dimethyl-1,3-diphenyl-buta-1,3-dienyl-C'N}palladium(II), 45B, 879
$C_{27}H_{28}ClNPd$, Chloro{3-4-η-4-[(2-dimethylaminomethyl)phenyl]-1,4-dimethyl-2,3-diphenyl-buta-1,3-dienyl-C'N}palladium(II), 45B, 879
$C_{27}H_{28}CoP$, 1-(η^5-Cyclopentadienyl)-1-triphenylphosphine-1-cobaltolane, 46B, 820
$C_{27}H_{28}SiTi$, 1,1-Bis(η^5-cyclopentadienyl)-2-trimethylsilyl-3-phenylbenzotitanole, 44B, 719
$C_{27}H_{29}ClP_2Pt$, trans-Bis(diphenylmethylphosphine)chloro(methyl)platinum(II), 45B, 880
$C_{27}H_{30}BCoF_7N_3O_6P_2$, Tris(p-fluorophenyl isocyanide)bis(trimethylphosphite)cobalt(I) tetrafluoroborate, 44B, 715
$C_{27}H_{32}CoN_4O_4P$, trans-Bis(dimethylglyoximato)methyl(triphenylphosphine)cobalt(III), 45B, 881
$C_{27}H_{33}F_6N_2O_2Rh$, 1-((Di(pivaloyl)methanato)bis(pyridine)rhoda)-2,3-bis(trifluoromethyl)cyclopent-2-ene, 46B, 814
$C_{27}H_{34}Cl_2NPPt$, Dichloro(1-(N,N-diethylammoniomethyl)-2-methylprop-1-enyl)(triphenylphosphine)platinum(II), 44B, 715
$C_{27}H_{36}Fe_2O_7Si_4$, Bis[1,4-bis(trimethylsilyl)buta-1,3-diyne]heptacarbonyldiiron, 45B, 881
$C_{27}H_{37}ClNPPd$, ((S)-Isopropyl-t-butylphenylphosphine)-((R)-N,N-dimethyl-α-(2-naphthyl)ethylamine-3C,N)chloropalladium(II), 43B, 1003
$C_{27}H_{43}IO_2P_2Rh_2S_2$, Bis[($\mu$-t-butylthiolato)(dimethylphenylphosphino)]-acetylcarbonyliododirhodium, 46B, 1228
$C_{28}H_{15}F_9MnO_4P$, Tris(trifluoromethyl)allenyltetracarbonyltriphenylphosphinemanganese, 44B, 716
$C_{28}H_{18}HgN_2$, 1,10-Phenanthroline-bis(phenylethynyl)mercury, 44B, 716
$C_{28}H_{20}FeO_2$, Dicarbonyl(η^5-cyclopentadienyl)(1,2,3-triphenylcyclopropenyl)iron, 45B, 882
$C_{28}H_{22}F_{12}OTi_2$, (μ-Oxo)bis[(1,1,1,4,4,4-hexafluorobut-2-en-3-yl)titanocene], 46B, 816
$C_{28}H_{25}CrNO_5Sn \cdot 0.5\ CH_2Cl_2$, Pentacarbonyl[diethylamino(triphenyltin)carbene]chromium(0) dichloromethane solvate, 45B, 882
$C_{28}H_{26}ClF_5P_2Pt$, trans-Bis(diphenylmethylphosphine)chloro(pentafluoroethyl)platinum(II), 45B, 880
$C_{28}H_{32}W$, Bis-(η-cyclopentadienyl)bis-(3,5-dimethylbenzyl)tungsten, 41B, 861
$C_{28}H_{33}ClCoN_5O_5$, (Benzoyl(1-pyridino)methanide)chlorobis(dimethylglyoximato)cobalt(III) - toluene, 44B, 716
$C_{28}H_{35}ClP_2Pt$, trans-Chlorobis(diethylphenylphosphine)(phenyl-

ethynyl)platinum(II), 44B, 717
$C_{28}H_{36}BFeNO_2P_2$, Dicarbonyl(methyl)bis(trimethylphosphine)(triphenylboratonitrilomethyl)iron(II), 46B, 1271
$C_{28}H_{36}Co_3O_8P$, Octacarbonyl(tricyclohexylphosphine)methylmethinyltricobalt, 43B, 1003
$C_{28}H_{38}CrIN_4Si_2$, Bis((trimethylsilyl)methyl)bis-(2,2'-bipyridyl)-chromium(III) iodide, 39B, 574
$C_{28}H_{45}NO_3P_2Pt$, trans-Nitrato(2-(di-t-butylphosphino)phenyl)di-t-butylphenylphosphineplatinum(II), 43B, 1004
$C_{28}H_{56}Cl_2N_8Ru \cdot 0.5\ C_8H_{18}O$, trans-Dichlorotetrakis(1,3-diethylimidazolidin-2-ylidene)ruthenium(II) di-n-butyl ether solvate, 44B, 717
$C_{28}H_{80}B_2F_8P_8Ru_3$, Bis[bis(μ-methylene)-tetrakis(trimethylphosphine)-ruthenium(III)]ruthenium(IV)(Ru-Ru-Ru) bis(tetrafluoroborate), 46B, 815
$C_{29}H_{14}O_{14}Ru_6$, μ_3-(1-6-η-Bitropyl)-carbido-μ-carbonyl-tridecacarbonyl-octahedro-hexaruthenium, 46B, 816
$C_{29}H_{24}MnO_3PS$, Tricarbonyl((methylsulfidomethyl)phenyl-2-C,S)triphenylphosphinemanganese, 41B, 862
$C_{29}H_{29}CoF_6N_5P \cdot C_2H_3N$, [(7-(cis-β-Vinylide)-7,16-dihydro-6,18,15,17-tetramethyldibenzo[b,i]-[1,4,8,11]-tetraazacyclotetradecinato)pyridinecobalt(III)] hexafluorophosphate - acetonitrile, 45B, 883
$C_{29}H_{31}P_3Pt$, [Bis(diphenylphosphino)methanide][dimethylphosphonium bis(methylide)]platinum(II), 46B, 1274
$C_{29}H_{35}BCl_3F_4IrP_2$, Carbonylchlorobis(trimethylphosphine)(1,2,3-triphenylpropenylium-1,3-diyl)iridium(1+) tetrafluoroborate(1-) - dichloromethane, 38B, 752
$C_{29}H_{36}CoN_4O_4P$, trans-Bis(dimethylglyoximato)(isopropyl)(triphenylphosphine)cobalt(III), 46B, 817
$C_{29}H_{47}ClN_2P_2Ru$, [1,3-Bis(4-tolyl)imidazolidin-2-ylidene-C^2,$C^{2'}$]chlorobis(triethylphosphine)-ruthenium(II), 45B, 883
$C_{29}H_{57}P_2Re_3$, Hexamethylbis(diethylphenylphosphine)tri-μ-methyl-triangulo-trirhenium(III), 46B, 818
$C_{30}Co_8O_{24}$, Hexacarbonoctacobalt tetracosacarbonyl, 37B, 434
$C_{30}H_{10}Co_6HfO_{20}$, Bis(η-cyclopentadienyl)bis((μ_3-oxymethylidyne)cyclo-(tris(tricarbonylcobalt)))hafnium, 44B, 699
$C_{30}H_{10}Co_6O_{20}Zr$, Bis(η-cyclopentadienyl)bis((μ_3-oxymethylidyne)cyclo-(tris(tricarbonylcobalt)))zirconium, 44B, 699
$C_{30}H_{10}O_{16}Os_6$, μ_4-Benzylidyne-μ_3-benzylidyne-hexadecacarbonylhexaosmium, 44B, 719
$C_{30}H_{14}F_{10}Ti \cdot 0.5\ C_6H_{14}$, 1,1-Bis($\eta^5$-cyclopentadienyl)-2,3-bis(pentafluorophenyl)benzotitanole - 0.5 hexane, 44B, 719
$C_{30}H_{18}Br_2Hg_2O_{18}Ru_6$, Bis((μ-bromo)-((μ-1-t-butylacetylene)-nonacarbonyl-triruthenium)-mercury), 46B, 818
$C_{30}H_{20}Cl_2Cu_2Fe_2O_4$, Di-μ-chloro-bis-((1-(dicarbonyl-π-cyclopentadienylferrio)-2-phenylethyne)copper(I)), 40B, 716
$C_{30}H_{22}O_4Re_2$, μ-2,3-Diphenylbutadienylidene-bis(dicarbonylcyclopentadienylrhenium), 43B, 1004
$C_{30}H_{24}Cl_2HgN_4O_8$, Mercury-carbene complex, 37B, 436
$C_{30}H_{25}FeOP$, π-Cyclopentadienyltriphenylphosphine-σ-phenyl-carbonyl-iron, 34B, 434
$C_{30}H_{28}Cl_6O_{18}Ru_6$, Tetra-μ-chloro-tetra-μ-ethyloxycarbene-di-μ_3-hydroxododecacarbonyl-dichloro-hexaruthenium(II) benzene solvate, 44B, 720
$C_{30}H_{28}Fe_2Ti$, Diferrocenyltitanocene, 45B, 884
$C_{30}H_{28}Zr_2$, μ-(2-η^1:1-2-η^2-Naphthyl)-hydrido-bis[bis(η^5-cyclopentadienyl)zirconium](Zr-Zr), 45B, 885
$C_{30}H_{34}As_2F_6NPPd$, [S-Dimethyl-(α-methylbenzyl)aminato-C^2,N][SS-ortho-

phenylenebis(methylphenylarsine)]palladium(II) hexafluorophosphate, 46B, 1275

$C_{30}H_{36}Au_3N_3O_3$, Tris-μ-[(ethoxy)(N-p-tolylimino)methyl-N,C]trigold(I), 45B, 885

$C_{30}H_{42}AlCoN_4$, cis-Bis(2,2'-bipyridine)dimethylcobalt(III) tetraethylaluminate, 45B, 886

$C_{30}H_{44}B_2F_8O_4Rh_2$, Bis[(pentamethylcyclopentadienyl)(2,4-pentanedionato)rhodium] tetrafluoroborate, 45B, 983

$C_{30}H_{46}GeOP_2Pt$, cis-(Hydroxydiphenylgermyl)phenylbis(triethylphosphine)platinum(II), 37B, 436

$C_{30}H_{47}ClN_2OP_2Ru$, [1,3-Bis(4-tolyl)imidazolidin-2-ylidene-C^2,$C^{2'}$]carbonylchlorobis(triethylphosphine)-ruthenium(II), 45B, 883

$C_{30}H_{49}BrClN_3Pd$, Tetra-n-butylammonium chlorobromo(N-(phenylamino)-α-methylbenzylidenimino-2-C,N)palladate(II), 42B, 635

$C_{30}H_{54}Br_2MoN_6$, Bromohexakis(t-butyl isocyanide)molybdenum(II) bromide, 44B, 721

$C_{30}H_{54}I_2MoN_6$, Iodohexakis(t-butyl isocyanide)molybdenum(II) iodide, 38B, 752

$C_{30}H_{56}I_2MoN_6$, Iodo-tetrakis(t-butylisocyanide)-(di(t-butylisoamino)-acetylene)molybdenum iodide, 43B, 1007

$C_{30}H_{62}N_6Si_6Ti$, Dibenzyl-bis(octamethyl-2,4,6-trisila-1,3,5-triazacyclohex-1-yl)titanium, 44B, 1134

$C_{31}H_{25}FeO_2P$, π-Cyclopentadienyltriphenylphosphine-σ-benzoyl-carbonyliron, 34B, 435

$C_{31}H_{25}OPSW$, Carbonyl-(phenylthiocarbyne)-(η^5-cyclopentadienyl)-triphenylphosphinetungsten, 45B, 887

$C_{31}H_{34}O_2P_2PtW$, de-Dicarbonyl-f-1-5-η-cyclopentadienyl-ab-bis(dimethylphenylphosphine)-c-μ-p-methylbenzylidyne-platinumtungsten(Pt-W), 46B, 819

$C_{31}H_{35}Ta$, (η^5-Pentamethylcyclopentadienyl)(benzylidene)dibenzyltantalum(V), 46B, 820

$C_{31}H_{40}O_7Ru_3$, (2,4-Di-t-butyl-buta-(1,3-η)-enyl-3-ene-4-yl)-(2,5-di-t-butyl-3-oxahexa-(4,6-η)-enyl-1-ene-1,4,6-triyl)-tris-triangulo(dicarbonylruthenium), 44B, 721

$C_{31}H_{41}OWZr$, Bis(cyclopentadienyl)tungsten-carbene-oxy-bis(pentamethylcyclopentadienyl)hydrido-zirconium, 45B, 888

$C_{31}H_{44}N_2O_{15}Re_4$, Tetraethylammonium tetrahydridopentadecacarbonyltetrarhenate(-2) ($P2_1/c$), 42B, 636

$C_{31}H_{44}N_2O_{15}Re_4$, Tetraethylammonium tetrahydridopentadecacarbonyltetrarhenate(-2) ($P2_1/n$), 42B, 636

$C_{32}H_{22}F_9PRu$, (η^5-Cyclopentadienyl)(triphenylphosphine)(1,1,1,8,8,8-hexafluoro-4-trifluoromethylocta-2,4,5,6-tetraen-2-yl)ruthenium, 43B, 1007

$C_{32}H_{23}MoN_4O_2P$, Dicarbonyl(η^5-cyclopentadienyl)triphenylphosphine-(1-η-1,1,2,2-tetracyanopropyl)molybdenum, 41B, 862

$C_{32}H_{24}F_{24}Pt_3$, Bis(cycloocta-1,5-diene)tetrakis(hexafluorobut-2-yne)triplatinum, 43B, 1008

$C_{32}H_{26}BrF_5NiP_2$, trans-Bromobis(methyldiphenylphosphine)(σ-pentafluorophenyl)nickel(II), 39B, 574

$C_{32}H_{26}CrIN_4$, cis-Diphenylbis(2,2'-bipyridyl)chromium(III) iodide, 39B, 575

$C_{32}H_{26}Hg_2N_2O_2$, Di-μ-(2-methylquinolin-8-olato)bis(phenylmercury), 44B, 1010

$C_{32}H_{26}IrMnO_5P$, μ-(Acetyl-O:C)-μ-(benzoyl-O:C)-μ-(diphenylphosphido-P)-tricarbonyl-((η^5-cyclopentadienyl)iridium(III))manganese hemibenzene, 41B, 863

$C_{32}H_{26}N_2O_4Pd_2S_2$, Di-μ-acetatobis[2-p-tolylbenzthiazolato(N,C)pallad-

ium(II)], 46B, 1027
$C_{32}H_{26}N_2O_6Pd_2$, Di-μ-acetatobis[2-p-tolylbenzoxazolato(N,C)palladium-(II)], 46B, 1027
$C_{32}H_{27}F_6O_5PRu$, Triphenylphosphine-(1-di(trifluoromethyl)-hydroxy-methyl-cyclopentadienyl)-(1,2-di(carboxymethyl)ethylene-1-yl)ruthenium(0), 41B, 864
$C_{32}H_{28}Br_3N_4Re$, Tribromo-tetra(p-tolyl isocyanide)rhenium, 45B, 888
$C_{32}H_{28}CoI_2N_4$, Diiodotetrakis(p-tolylisocyano)cobalt(II), 40B, 1107
$C_{32}H_{28}Fe_2NO_6P$, μ-Diphenylphosphido-μ-(iminium-ion)-bis(tricarbonyl-iron), 44B, 722
$C_{32}H_{36}Cr_2O_4$, Tetrakis(2-methoxy-5-methyl)dichromium, 44B, 723
$C_{32}H_{36}Cr_2O_8$, Tetrakis(2,6-dimethoxyphenyl)dichromium(Cr-Cr), 43B, 1009; 44B, 723
$C_{32}H_{36}Mo_2O_8$, Tetrakis(2,6-dimethoxyphenyl)dimolybdenum(Mo-Mo), 43B, 1009; 44B, 723
$C_{32}H_{37}Cl_3IrO_2P$, Carbonyl[carbonyl(5,5'.6',5"-ternorborn-2-ene)-6,6"-diyl]chloro(dimethylphenylphosphine)iridium-dichloromethane, 46B, 978
$C_{32}H_{38}IrP$, 1-(η^5-Pentamethylcyclopentadienyl)-1-triphenylphosphine-1-iridolane, 46B, 820
$C_{32}H_{38}Mo_2N_2O_7Pt \cdot 0.5\ C_6H_{12}$, cis(Cyclohexylisocyanido)[ethoxy(cyclohexylamino)carbene]-bis(η^5-cyclopentadienyl-tricarbonylmolybdato)-platinum(II) hemicyclohexane solvate, 45B, 889
$C_{32}H_{38}PRh$, 1-(η^5-Pentamethylcyclopentadienyl)-1-triphenylphosphine-1-rhodolane, 46B, 820
$C_{32}H_{44}F_6IrP_5$, Tris(dimethylphenylphosphine)-(o-dimethylphosphino-phenyl)hydridoiridium hexafluorophosphate, 46B, 1276
$C_{32}H_{50}IrNO_4P_2$, (2-Di-t-butylphosphino-3-methoxyphenoxo-O,P)(2-((2-hydroxy-6-methoxyphenyl)-t-butylphosphino)-2-methylpropanato(2-)-C^1,O,P^2)(methylisocyanide)iridium(III), 44B, 724
$C_{32}H_{50}O_7P_2Pt_2W$, μ-Methoxyphenylmethylene-μ-tetracarbonyltungstenio-bis[carbonyl(methyldi-t-butyl-phosphine)platinum], 46B, 821
$C_{32}H_{72}Cr_2Li_4O_4$, Tetralithium tetrakis(tetramethylene)dichromate tetrakis(diethylether), 37B, 437
$C_{33}H_3Co_8O_{24}$, 1,6-Bis(tricobalt nonacarbonyl)-2,3-(dicobalt hexacarbonyl)hexa-4-yne - benzene, 35B, 505
$C_{33}H_{26}BF_4FeO_2P$, (1-Triphenylphosphonium-2-phenyl-vinyl)-(η^5-cyclopentadienyl)dicarbonyliron tetrafluoroborate, 45B, 890
$C_{33}H_{27}F_6O_4PRu$, η^5-Cyclopentadienyl(1,3,4-η-(1,2-dimethoxycarbonyl-5,5,5-trifluoro-3-trifluoromethylpent-1,3-dienyl))triphenylphosphineruthenium, 42B, 637
$C_{33}H_{29}FeO_2P \cdot C_6H_6$, Acetyl-carbonyl-($\eta^5$-1-methyl-3-phenyl-cyclopentadienyl)-(triphenylphosphine)iron benzene solvate, 45B, 887
$C_{33}H_{35}ClIrN_2O_4P$, Chlorobis(γ-picoline)tris(o-tolyl)phosphite-P,C,C-iridium monohydrate, 43B, 1010
$C_{33}H_{38}BN_3NiS_3$, (Dimethylthiocarboxamido)(dimethylamino(dimethyldithiocarbamato)carbene)nickel(II) tetraphenylborate, 43B, 1011
$C_{33}H_{47}BN_6Pt$, Chugaev's red salt, 39B, 576
$C_{34}H_{14}N_2O_{18}Os_6$, Bis(p-tolyl isocyanide)octadecacarbonylhexaosmium, 44B, 725
$C_{34}H_{20}Fe_2O_6$, Tetraphenylferrole-iron tricarbonyl, 41B, 865; 42B, 638
$C_{34}H_{24}O_4PRe$, (2-Benzoylphenyl)tricarbonyl(triphenylphosphine)-rhenium, 46B, 821
$C_{34}H_{30}GeMoO_3$, η^5-Cyclopentadienyl(triphenylgermyl)dicarbonyl(phenyl-(ethoxy)carbene)molybdenum(II), 43B, 1012
$C_{34}H_{31}Cl_2OP_2Rh$, Benzoyl-dichloro-(1,3-bis(diphenylphosphino)-propane)rhodium, 45B, 890; 46B, 823

$C_{34}H_{31}Cl_2OP_2Rh$, Benzoyldichloro(1,3-diphenylphosphinopropane)rhodium(III), 45B, 890
$C_{34}H_{32}CrIN_4O_3$, cis-Bis-(2-methoxyphenyl)bis-(2,2'-bipyridyl)-chromium(III) iodide monohydrate, 38B, 753
$C_{34}H_{34}N_2Ti_2$, μ-Dinitrogen-bis(p-tolyl-dicyclopentadienyl-titanium-(III)), 45B, 891
$C_{34}H_{40}O_{12}Re_2$, Di-μ-p-tolylmethoxymethylidene-octacarbonyldirhenium bis(diethyl ether), 43B, 1482
$C_{34}H_{41}IrO_6P_2$, (Cyclo-octa-1,5-diene)-(1,1,1-trimethoxyphosphinoethane)-(3-methyl-2-(bis-(2-methylphenoxy)phosphineoxy)phenyl)iridium, 44B, 1134
$C_{34}H_{44}B_8NiP_2$, 3,7,9,13-Tetramethyl-6-[bis(diphenylphosphino)ethane]-3,7,9,13-tetracarba-6-nickelatridecaborane, 46B, 822
$C_{34}H_{68}FeN_6Ti_2$, Bis(tris(diethylamido)titano)ferrocene, 41B, 866
$C_{35}H_{18}F_9N_2PRu$, (Diphenyl(2-(η^5-cyclopentadienyl)phenyl)phosphine)-(nonafluoro(phenylazo)phenyl-C(2),N')ruthenium, 42B, 638
$C_{35}H_{31}Fe_2NO_6P$, μ-Diphenylphosphido-μ-(iminium-ion)-bis(tricarbonyl-iron) benzene solvate, 44B, 722
$C_{35}H_{31}O_8PPd$, (Triphenylphosphonium cyclopentadienyide)tetrakis(methoxycarbonyl)palladiacyclopentadiene, 43B, 1012
$C_{35}H_{34}FeP_2$, (η^5-Cyclopentadienyl)(2-methyl-4,5-bis(diphenylphosphino)-2-penten-3-yl)iron(II), 45B, 893
$C_{35}H_{39}FeO_3P$, (-)-(Menthyl ester)(η^5-cyclopentadienyl)(triphenylphosphine)carbonyliron, 44B, 725
$C_{35}H_{41}FeO_2P$, (+)-(η^5-Cyclopentadienyl)((-)-menthoxymethyl)(triphenylphosphine)carbonyl-iron, 44B, 726
$C_{35}H_{46}N_4OPt_2$, μ-2-Oxo-1,3-diphenylpropanediylidene-bis(bis(t-butyl isocyanide)platinum), 44B, 727
$C_{35}H_{63}F_{12}MoN_7P_2$, Heptakis(t-butyl isocyanide)molybdenum(II) bis(hexafluorophosphate), 41B, 866
$C_{35}H_{63}N_7O_{19}W_7$, Heptakis(tert-butyl isocyanide)tungsten(II) hexatungstate, 46B, 823
$C_{36}H_{26}Cl_4CoN_5O_4$, Pentakis(phenylisocyanide)cobalt(I) perchlorate chloroform, 41B, 867
$C_{36}H_{27}Cl_3CoN_5O_8$, Pentakis(phenylisocyanide)cobalt(II) perchlorate dichloroethane solvate, 41B, 868
$C_{36}H_{32}Hf$, Bis(η^5-cyclopentadienyl)bis(benzhydryl)hafnium(IV), 43B, 1014
$C_{36}H_{32}N_2O_{15}Rh_6$, Bis(benzyltrimethylammonium) carbidopentadecacarbonylhexarhodate, 39B, 578
$C_{36}H_{32}Zr$, Bis(η^5-cyclopentadienyl)bis(benzhydryl)zirconium(IV), 43B, 1013
$C_{36}H_{33}Cr_2O_9$, Tetrakis(2,4,6-trimethoxyphenyl)dichromium, 43B, 1014
$C_{36}H_{33}Ir_7O_{12}$, Dodecacarbonyl(cycloocta-1,5-diene)(dehydro-cycloocta-1,5-diene)(cycloocta-1-en-5-yne)heptairidium, 45B, 893
$C_{36}H_{35}I_2N_2O_2PRu$, Carbonyldiiodo(triphenylphosphine)(p-tolylisonitrile)(N,4-dimethylanilinocarbene)ruthenium, 43B, 1015
$C_{36}H_{36}Cl_2F_{12}Ir_2$, μ,μ'-Dichlorobis(hexafluorobut-2-enyl)-bis(cycloocta-1,5-dienyl)diiridium(III) bis(benzene), 43B, 1013
$C_{36}H_{38}F_{18}O_6Rh_2$, μ-(Hexafluorobut-2-ene-2,3-diyl)-bis(carbonyl-(4,4,4-trifluoro-1,1-dipivaloyl-2-trifluoromethylbut-2-ene-3-yl)-rhodium), 44B, 1135
$C_{36}H_{41}FeO_3P$, (+)-(η^5-Cyclopentadienyl)((-)-menthylacetyl)(triphenylphosphine)carbonyl-iron, 44B, 726
$C_{36}H_{44}Cr_2O_{12}$, Tetrakis(2,4,6-trimethoxyphenyl)dichromium(Cr-Cr), 44B, 723
$C_{36}H_{44}V$, Tetramesitylvanadium, 44B, 1136

$C_{36}H_{48}O_4Ti_2$, Di-μ-ethoxy-bis(dibenzyl(ethoxy)titanium(IV)), 41B, 1245

$C_{36}H_{58}Hg$, Bis(2,4,6-tri-tert-butylphenyl)mercury(II), 46B, 824

$C_{36}H_{64}N_2P_4Pt_2S_2$, p-Phenylene-bis(ethynylisothiocyanato-trans-bis(triethylphosphine)platinum(II)), 42B, 636

$C_{37}H_{20}F_6Fe_3O_7P_2$, 3-Diphenylphosphino-1,1,1,6,6,6-hexafluorohexa-2,4-diene-2,4,5-triyl-(iron tricarbonyl)(iron dicarbonyl)-μ-diphenylphosphinodicarbonyl-iron, 42B, 639

$C_{37}H_{25}Fe_2O_5PS$, Dicarbonyl(triphenylphosphine)iron-μ-(1,2-diphenylethylene-1-thiolato-2-yl)-tricarbonyliron, 44B, 727

$C_{37}H_{30}ClOP_2Rh$, trans-Carbonyl-chloro-bis(triphenylphosphine)rhodium(I), 45B, 894

$C_{37}H_{30}ORh_2$, μ-Carbonyl-di-μ-diphenylmethylene-bis(π-cyclopentadienylrhodium), 43B, 1016

$C_{37}H_{31}F_3P_2Pt$, trans-Hydridotrifluoromethylbis(triphenylphosphine)platinum(II), 45B, 894

$C_{37}H_{33}IO_2P_2PtS$, Iodo(sulfur dioxide)methylbis(triphenylphosphine)platinum, 39B, 576

$C_{37}H_{37}Cl_2P_2Rh$, (1,2,3-Triphenylpropylium-1,3-diyl)dichlorodi(dimethylphenylphosphine)rhodium, 45B, 895

$C_{37}H_{62}N_6Pt_3$, Tris-μ-(t-butyl isocyanide)-tris(t-butyl isocyanide)-triangulo-triplatinum toluene, 43B, 1016

$C_{38}H_{23}Fe_2O_7$, Hexacarbonyl(μ-(1,2,5-η:1,4,5-η)-3-oxo-1,2,4,5-tetraphenyl-1,4-pentadiene-1,5-diyl)diiron(Fe-Fe) benzene solvate, 42B, 640

$C_{38}H_{26}Cl_5F_5NiP_2$, trans-Bis(methyldiphenylphosphine)-(σ-pentachlorophenyl)-(σ-pentafluorophenyl)nickel(II), 37B, 438

$C_{38}H_{26}F_{10}NiP_2$, trans-Bis(methyldiphenylphosphine)bis(σ-pentafluorophenyl)nickel(II), 38B, 754

$C_{38}H_{30}Fe_2N_4$, trans-anti-Bis[η-cyclopentadienyl(μ-phenyl isocyanide)-(phenyl isocyanide)iron](Fe-Fe), 46B, 824

$C_{38}H_{30}Hf$, 1,1-Bis(η^5-cyclopentadienyl)-2,3,4,5-tetraphenylhafnole, 42B, 640

$C_{38}H_{30}O_2Ti$, Bis(diphenylketene)bis(η^5-cyclopentadienyl)titanium, 44B, 736

$C_{38}H_{30}Ti$, 1,1-Bis(η^5-cyclopentadienyl)-2,3,4,5-tetraphenyltitanole, 42B, 640

$C_{38}H_{31}Cl_2F_2IrOP_2$, Dichloro(difluoromethyl)carbonylbis(triphenylphosphine)iridium(III), 40B, 717

$C_{38}H_{31}IN_4O_5PRh$, Iodo(fumaronitrile)(triphenylphosphite)bis(p-methoxyphenylisocyanide)rhodium(I), 41B, 868

$C_{38}H_{32}ClNP_2Pt$, trans-Chlorobis(triphenylphosphine)-cyanomethyl-platinum(II), 45B, 896

$C_{38}H_{33}ClO_2P_2Pd$, trans-Chloro(methoxycarbonyl)bis(triphenylphosphine)palladium, 44B, 728

$C_{38}H_{33}NP_2Pt$, trans-Cyanomethyl(hydrido)bis(triphenylphosphine)platinum(II), 45B, 896

$C_{38}H_{33}N_2P_3Pt_2S$, Phenyl(triphenylphosphine)platinum-μ-diphenylphosphido-μ-(2-thionitrosyl-phenylamido)-triphenylphosphineplatinum, 44B, 1165

$C_{38}H_{34}FeOP_2$, Benzoyl(1,2-bis(diphenylphosphino)ethane)(η^5-cyclopentadienyl)iron, 43B, 1017

$C_{38}H_{35}ClP_2PdS \cdot CH_2Cl_2$, Chlorothiomethoxymethylbis(triphenylphosphine)palladium(II) methylene chloride, 45B, 897

$C_{38}H_{38}Fe_2O_6P_2$, μ-(Phenyl(dicyclohexylphosphonio)ethenidyl)-μ-(diphenylphosphido)hexacarbonyldiiron, 44B, 1136

$C_{38}H_{42}HgN_4O_2$, Bis((acetyl(2,6-dimethylphenyl)amino)((2,6-dimethyl-

phenyl)imino)methyl-C-O)-(T-4)-mercury, 42B, 642
$C_{38}H_{42}N_2O_{16}Pd_2$, Tetrakis(methoxycarbonyl)palladiacyclopentadiene-(2,6-lutidine) dimer, 44B, 728
$C_{38}H_{51}N_4PRu$, Tetrakis(t-butyl isocyanide)(triphenylphosphine)ruthenium, 45B, 898
$C_{38}H_{52}CoP_2$, trans-Bis(diethylphenylphosphine)dimesitylcobalt(II), 28, 660; 45B, 898
$C_{39}H_{30}Cl_2F_4OP_2Pt$, cis-Chloro-(3-chloro-1,1,3,3-tetrafluoropropan-2-one)bis(triphenylphosphine)platinum(II), 41B, 869
$C_{39}H_{30}F_6NiOP_2$, Bis(triphenylphosphine)hexafluoroacetonenickel(0), 38B, 755
$C_{39}H_{31}Co_2N_2O_5P$, (Dicarbonyl-dimethylphenylphosphinecobalt)-μ-carbonyl-μ-[N,N'-diphenyl-N-(phenyl-carbenyl)benzadmidine]-(dicarbonylcobalt), 45B, 899
$C_{39}H_{31}Co_2N_2O_5P$, Pentacarbonyl(dimethylphenylphosphine)(tetraphenylbisimidoyl)dicobalt, 43B, 1018
$C_{39}H_{31}F_7P_2Pt$, cis-Fluoro-(1,1,1,3,3,3-hexafluoroisopropyl)bis(triphenylphosphine)platinum, 39B, 578
$C_{39}H_{32}Co_8N_2O_{18}$, Bis(benzyltrimethylammonium) deca-μ-carbonylcarbidoocta(carbonyl cobalt), 44B, 729
$C_{39}H_{35}ClP_2Pt$, trans-(η^1-Allyl)chlorobis(triphenylphosphine)platinum-(II), 43B, 1018
$C_{39}H_{36}CoO_6P \cdot 0.25\ CH_2Cl_2$, Cyclopentadienyl-[(5-6-η)-1,2,3-tri(methoxycarbonyl)-6-cyano-4-phenyl-hexa-3,5-diene]-triphenylphosphine-cobalt methylene dichloride solvate, 45B, 900
$C_{39}H_{39}Fe_2NaO_8P_2$, Sodium tricarbonyliron-di-μ-(diphenylphosphido)-(acetyldicarbonyliron) tetrahydrofuran, 45B, 904
$C_{39}H_{55}FeN_5 \cdot C_7H_8$, 1-4-η-(2,3-Diphenyl-N,N'-di-t-butylbuta-1,3-diene-1,4-diimine)-tris(t-butyl isocyanide)iron toluene solvate, 46B, 825
$C_{40}H_{30}FeHNO_4P_2$, Bis(triphenylphosphine)iminium hydridotetracarbonyliron, 39B, 579
$C_{40}H_{33}ClP_2Pd$, (1,3-Bis((o-diphenylphosphino)phenyl)-2-buten-1-yl)-chloropalladium(II), 42B, 642
$C_{40}H_{33}O_6P_2RuS_2 \cdot C_6H_{12}$, ($\eta^2$-Dithiomethyl ester)dicarbonylbis(triphenylphosphine)ruthenium(II) perchlorate cyclohexane solvate, 46B, 828
$C_{40}H_{34}HgN_4$, Diphenylbis-(2,9-dimethyl-1,10-phenanthroline)mercury-(II), 38B, 756
$C_{40}H_{34}O_3P_2Pd$, Oxydi(carboxymethyl)bis(triphenylphosphine)palladium-(II), 44B, 730
$C_{40}H_{36}O_4P_2Pd$, trans-Acetato(methoxycarbonyl)bis(triphenylphosphine)-palladium(II), 45B, 901
$C_{40}H_{37}ClOP_2Pd$, trans-Chloro(butanoyl)bis(triphenylphosphine)palladium(II), 45B, 901
$C_{40}H_{38}P_2Pt$, Bis(triphenylphosphine)tetramethyleneplatinum(II), 39B, 579
$C_{40}H_{38}Ti_2$, μ-(1-3-η:2-4η-trans,trans-1,4-Diphenylbutadiene)bis(bis-(η^5-methylcyclopentadienyl)titanium), 42B, 643
$C_{40}H_{48}AsAuN_{16}$, Tetraphenylarsonium tetrakis(1-isopropyltetrazol-5-ato)aurate(III), 38B, 757
$C_{40}H_{51}Cl_{17}Co_4N_{16}Rh_4 \cdot 6\ H_2O$, Octakis(1,3-di-isocyanopropane)-chlorotetrarhodium tetrakis(tetrachloro-cobalt) trihydrogen hexahydrate, 46B, 826
$C_{40}H_{52}Cr$, Tetrakis(2-methyl-2-phenyl-propyl)chromium, 39B, 580
$C_{40}H_{52}O_{10}V_2$, Tetrakis(2,6-dimethoxyphenyl)divanadium bis(tetrahydrofuran) solvate, 43B, 1019

$C_{40}H_{66}F_{12}N_8P_2Rh_2 \cdot 2\ C_2H_3N$, Tetrakis($\mu$-2,5-dimethyl-2,5-di-isocyanohexane)dirhodium bis(hexafluorophosphate) acetonitrile solvate, 46B, 833

$C_{40}H_{72}Co_2N_8$, Octakis(t-butyl isocyanide)dicobalt, 46B, 826

$C_{41}H_{31}N_3OP_2Pt$, 3,3,4-Tricyano-2,2-bis(triphenylphosphine)-1-oxa-2-platinacyclobutane, 44B, 730

$C_{41}H_{35}ClP_2Pt \cdot 0.67\ CHCl_3$, trans-Chloro(isopropenylacetylido)bis-(triphenylphosphine)platinum(II) chloroform (3:2), 43B, 1020

$C_{41}H_{38}Cl_2OP_2Pd$, ((Benzoylmethylene)diphenyl-2-(diphenylphosphino)-ethylphosphorane)dichloropalladium(II) toluene solvate, 43B, 1021

$C_{41}H_{39}ClN_2O_4P_2Pt$, (1,2-Bis(diphenylphosphino)ethane)-bis(N-4-tolylamino)carbene-2'-ylplatinum perchlorate, 44B, 731

$C_{41}H_{42}ClCoN_6O_{14}P_2$, Tris(4-nitrophenyl isocyanide)bis(diethyl phenylphosphonite)cobalt(I) perchlorate, 42B, 644

$C_{41}H_{69}N_7Ni_4$, Hepta(t-butylisocyanide)tetranickel benzene solvate, 41B, 871

$C_{42}H_{30}As_2N_4OPt$, 1,1-Bis(triphenylarsine)-3,3,4,4-tetracyano-1-platina-2-oxacyclobutane, 40B, 718

$C_{42}H_{30}CrN_6$, Hexakis(phenyl isocyanate)chromium(0), 44B, 731

$C_{42}H_{30}F_{12}N_2P_2Pt$, Hexafluoroacetoneazinebis(triphenylphosphine)platinum, 43B, 1021

$C_{42}H_{30}I_3MnN_6$, Hexakis(phenyl isocyanide)manganese(I) triiodide, 46B, 827

$C_{42}H_{36}O_4P_2Pt$, o-(Diphenylphosphino)phenyl-cis-1,2-dicarbomethoxyethenyltriphenylphosphineplatinum, 44B, 1137

$C_{42}H_{38}O_2P_2Pt$, 1-Oxo-2-oxa-3,3-dimethylpent-4-ene-1,4-diylbis(triphenylphosphine)platinum(II), 42B, 645

$C_{42}H_{40}O_4P_2Pt$, trans-Bis(ethoxycarbonyl)bis(triphenylphosphine)platinum, 40B, 719

$C_{42}H_{40}P_2Pt$, Bis(dibenzyl-cis-benzyl-2-yl-phosphine)platinum(II), 46B, 1287

$C_{42}H_{42}OTi_2$, μ-Oxo-bis(tribenzyltitanium(IV)), 40B, 719

$C_{43}H_{30}F_5IrOP_2$, trans-Pentafluorophenylcarbonylbis(triphenylphosphine)iridium(I), 41B, 871

$C_{43}H_{32}N_4P_2Pt$, 1,1,3,3-Tetracyanopropylidene-bis(triphenylphosphine)-platinum(II), 40B, 720

$C_{43}H_{35}BF_4IrN_3O_3P_2$, (2-Nitrophenylhydrazido)carbonylbis(triphenylphosphine)iridium tetraphenylborate, 43B, 1022

$C_{43}H_{35}BF_5IrN_2O_2P_2$, Fluorocarbonyl(4-fluorodiphenyldiazene-2-yl)-bis(triphenylphosphine)iridium hydroxotrifluoroborate, 43B, 1022

$C_{43}H_{37}ClN_2OP_2Pd \cdot CH_2Cl_2$, Chlorobis(triphenylphosphine)(O-N-methyl-N-nitrosylaniline)palladium methylene chloride solvate, 46B, 807

$C_{43}H_{37}O_2OsP_2S_2$, Hydrido(dithiomethyl ester)bis(carbonyl)bis(triphenylphosphine)osmium hemi(benzene), 43B, 1023

$C_{43}H_{43}O_3PPd$, Acetylacetonato(1,cis-3,trans-tetraphenyl-4-ethoxybutadien-1-yl)(dimethylphenylphosphine)palladium(II), 43B, 1024

$C_{44}H_{28}Mn_2O_8P_2$, Tetracarbonyl-(μ-(carbonyl(6-(diphenylphosphino)-o-phenylene)))(tricarbonyltriphenylphosphinemanganese)manganese, 43B, 1025

$C_{44}H_{28}Mn_2O_8P_2$, Tricarbonyltriphenylphosphine-(μ-(carbonyl(6-(diphenylphosphino)-o-phenylene)))(tetracarbonylmanganese)manganese, 43B, 1024

$C_{44}H_{36}Cl_5IrN_2OP_2$, Dichloro-(4-methoxyphenyldiimide-C(2),N^1)bis(triphenylphosphine)iridium chloroform solvate, 40B, 722

$C_{44}H_{36}O_6P_2Pt_2$, μ-(Dimethyl acetylenedicarboxylate)bis[carbonyl(triphenylphosphine)platinum(I)](Pt-Pt), 46B, 1291

$C_{44}H_{37}BrP_2Pt$, trans-Bromo-(trans-styryl)bis(triphenylphosphine)plat-

inum(II), 40B, 721
$C_{44}H_{39}Cl_3IIrN_2OP_2$, Hydridoiodobis(triphenylphosphine)-p-methoxyphenyldiazene-C(2),N¹-iridium chloroform solvate, 44B, 1138
$C_{44}H_{40}F_6N_3P_2Rh$, Bis(triphenylphosphine)hexafluoroacetonimido-(N,N'-dimethylimidazolidin-2-ylideno)rhodium, 40B, 722
$C_{44}H_{41}BF_4O_4P_2Pt$, (2,3-Bis(methoxycarbonyl)-1-methylcyclopropyl)-bis(triphenylphosphine)platinum(II) tetrafluoroborate, 46B, 1164
$C_{44}H_{42}HgN_4$, Diphenylbis-(2,4,7,9-tetramethyl-1,10-phenanthroline)-mercury(II), 38B, 756
$C_{44}H_{42}O_2P_2Pt$, (1,3-Diacetylbutane-1,4-diyl)bis(triphenylphosphine)-platinum, 45B, 902
$C_{44}H_{52}CoI_2N_4$, Di-iodo-tetrakis(2,6-diethylphenylisocyanide)cobalt, 41B, 860
$C_{44}H_{58}BCuN_4$, (N,N,N',N'-Tetramethylethylenediamine)bis(cyclohexyl isocyanide)copper(I) tetraphenyl-borate, 45B, 892
$C_{44}H_{61}BO_4P_2Pt$, [η^2-3-Methylallyl-1,2-bis(methylcarboxylate)(C,O)]-bis(triethylphosphine)platinum tetraphenylborate, 46B, 828
$C_{44}H_{64}NiP_2$, trans-Bis(o-ethynylphenylethynyl)-bis(tris(n-butyl)phosphine)nickel(II), 45B, 902
$C_{44}H_{80}Na_2Nb_2O_6Si_4$, Disodium bis($\eta^5$-cyclopentadienyl-(1,1,3,3-tetramethyldisiloxane)-η^5-cyclopentadienylhydridoniobium) tetrakis(diethylether), 46B, 911
$C_{45}H_{31}IrN_4OP_2$, (α-Dicyanovinyl)(carbonyl)(π-dicyanoacetylene)bis-(triphenylphosphine)iridium(I), 39B, 581
$C_{45}H_{32}Cl_3F_{12}RhSb_2$, Chlorobis(triphenylstibine)tetrakis(trifluoromethyl)-rhodiacyclopentadiene - dichloromethane solvate, 35B, 512
$C_{45}H_{35}BrCrO_8P_2$, Bromodicarbonyl(phenylcarbyne)bis(triphenylphosphite)chromium, 43B, 981
$C_{45}H_{37}ClN_2O_6OsP_2$, Carbonylnitrosyl(p-tolyl isocyanide)-bis(triphenylphosphine)osmium(0) perchlorate, 42B, 645
$C_{45}H_{42}P_2Pd_2$, μ-(Cyclopentadienyl)-μ-(2-methylallyl)-bis(triphenylphosphine)-dipalladium(Pd-Pd), 43B, 1026
$C_{45}H_{43}ClCuN_3O_4P_2$, (tert-Butylisocyanide)[N,N'-bis(o-(diphenylphosphino)benzylidene]ethylenediamine-P,P',N]copper(I) perchlorate, 46B, 972
$C_{45}H_{47}NPdS_2$, (1:3,4-η^3-1,2,3,4-Tetra(p-tolyl)-4-phenyl-1,3-butadienyl)(N,N-diisopropyldithiocarbamato)palladium, 44B, 735
$C_{45}H_{54}B_{10}P_2Pt$, cis-1-((Tribenzylphosphine)-(dibenzylphosphino-phenyl-methyl)platinum)-2-methyl-1,2-dicarbadodecaborane, 45B, 902
$C_{46}H_{30}Cr_2INO_{10}P_2$, Bis(triphenylphosphine)iminium μ-iodo-bis(pentacarbonylchromium), 35B, 506
$C_{46}H_{31}NO_{10}P_2W_2$, Bis(triphenylphosphino)imminium μ-hydridobis(pentacarbonyltungsten), 41B, 839
$C_{46}H_{37}FeNO_8P_2$, Bis(triphenylphosphine)iminium η^3-(trans-2,3-bis(methoxycarbonyl)acryloyl)tricarbonylferrate, 44B, 1139
$C_{46}H_{38}Hg_2I_4P_2$, trans-Di-μ-diiodobis(triphenylphosphoniumcyclopentadienylide)dimercury(II), 44B, 732
$C_{46}H_{38}N_4OP_2Pt$, 2,2,5,5-Tetracyano-3-ethoxy-1,1-bis(triphenylphosphine)-1-platina-cyclopentane, 42B, 646
$C_{46}H_{40}P_2Pt$, trans-Bis(isopropenylacetylide)-bis(triphenylphosphine)-platinum(II), 42B, 987
$C_{46}H_{42}BF_8IrN_2O_3P_2$, (2-Trifluoromethylphenylhydrazido)fluorocarbonyl-bis(triphenylphosphine)iridium tetrafluoroborate bis(methanol), 43B, 1022
$C_{46}H_{42}Cl_3NO_5P_2Pt$, Bis(triphenylphosphine)iminium carbonyldichloro-(trans-1-ethoxycarbonyl-2-isopropoxycarbonyl-2-chloroethenyl)platinum(II), 44B, 1139

$C_{46}H_{42}Co_2O_4$, μ-(1,6-Bis(ethoxycarbonyl)-2,3,4,5-tetraphenylhexa-2,4-diene-1,1-diyl)-bis(η[5]-cyclopentadienylcobalt)(Co-Co), 44B, 1140
$C_{46}H_{42}P_2Pt$, trans-(Isopropenylethynyl)(1-isopropenylvinyl)bis(triphenylphosphine)platinum(II), 43B, 1027
$C_{46}H_{45}ClOP_2Pt$, trans-Chloro(1-ethinoxycyclohexylethynyl)bis(triphenylphosphine)platinum(II), 45B, 903
$C_{46}H_{59}Cl_2FeN_2O_{10}P_3$, Chlorotris(diethyl phenyl phosphonite)bis(4-tolyl isocyanide)iron(II) perchlorate, 42B, 647
$C_{46}H_{63}CrNa_2O_4$, Sodium pentaphenylchromate tris(diethyl ether) mono-(tetrahydrofuran), 38B, 758
$C_{46}H_{64}N_4Pt_2Si_2$, Bis-μ-(N-t-butylformimidoyl-CN)bis[(methyldiphenylsilyl)(t-butyl isocyanide)platinum], 45B, 904
$C_{46}H_{69}BFe_2P_4O_6S_2$, (μ-1,2-Bis(carbomethoxyethylene))bis[(μ-methylthio)(trimethylphosphine)carbonyliron] tetraphenylborate, 46B, 1235
$C_{47}H_{40}ClNO_5P_2RuS_2 \cdot 0.5\ H_2O \cdot 0.5\ CHCl_3$, (η[2]-Dithiomethyl ester)carbonyltolylisocyanide-bis(triphenylphosphine)ruthenium(II) perchlorate hemihydrate hemichloroform solvate, 46B, 828
$C_{47}H_{41}NO_3P_2Ru$, Acetatocarbonyl(N-p-tolylformimidoyl)bis(triphenylphosphine)ruthenium(II), 41B, 872
$C_{47}H_{43}ClP_2Pt$, trans-Chloro(1-isopropenylvinyl)bis(triphenylphosphine)platinum(II) - benzene, 43B, 1027
$C_{48}H_{38}Hg_2N_2O_{12}P_4 \cdot 2\ C_2H_4O_2$, 7,15-Bis(diphenoxyphosphinyl)-6,7,14,15-tetrahydro-6,14-diphenoxydibenzo[e.k](1,7,3,9,2,8,4,10)-dioxadiazadiphosphadimercuracyclododecan-6,14-dioxide acetic acid solvate, 46B, 1135
$C_{48}H_{39}O_4P_2ReSi$, fac-(1,2-Bis(diphenylphosphino)ethane)(triphenylsilylcarbonyl)tricarbonylrhenium, 43B, 1028
$C_{48}H_{46}P_2U_2 \cdot 0.5\ C_4H_{10}O$, Tetrakis(η[5]-cyclopentadienyl)bis[μ-(methylene(diphenylphosphino)methylidyne)]-diuranium diethyl ether solvate, 46B, 829
$C_{48}H_{46}P_2U_2 \cdot C_5H_{12}$, Tetrakis(η[5]-cyclopentadienyl)bis[μ-(methylene(diphenylphosphino)methylidyne)]-diuranium pentane solvate, 46B, 829
$C_{48}H_{49}Fe_2LiO_9P_2$, Lithium tricarbonyliron-di-μ-(diphenylphosphido)-(benzoyldicarbonyliron) tetrahydrofuran, 45B, 904
$C_{48}H_{72}N_2O_{16}Re_4$, Tetra-n-butylammonium hexadecacarbonyltetrarhenium, 33B, 329
$C_{48}H_{76}Br_2Cr_2Li_6O_{10}$, Hexalithium tetrakis(o-oxyphenyl)dibromodichromate hexakis(diethyl ether), 44B, 732
$C_{49}H_{36}MnNO_7P_2$, μ-Nitrido-bis(triphenylphosphorus) 3-phenyl-1,1,1,1-tetracarbonyl-1-mangana-3-oxacyclopenta-2,5-dionate, 44B, 733
$C_{49}H_{36}N_4P_2Pt$, 2,2,4,4-Tetracyano-3-phenyl-1,1-di(triphenylphosphine)-1-platinacyclobutane, 44B, 733
$C_{49}H_{40}AsIN_2OPRhS_2$, Tetraphenylarsonium iodopropanoylmaleonitriledithiolato(triphenylphosphine)rhodate(III), 43B, 1028
$C_{49}H_{42}Fe_2O_5P_2$, (Phenylacetylido)(diphenylphosphido)(triphenylphosphine)pentacarbonyldiiron cyclohexane, 43B, 1029
$C_{50}H_{54}Cu_6N_4$, Tetrakis-μ_3-2-dimethylaminophenyl-bis-μ_2-4-tolylethynyl-octahedro-hexacopper(I), 44B, 734
$C_{51}H_{20}F_{20}PRh$, 1-(η[5]-Cyclopentadienyl)-1-triphenylphosphine-2,3,4,5-tetrakis(pentafluorophenyl)rhodole, 42B, 648
$C_{51}H_{35}Fe_2O_5P$, Dicarbonyl(triphenylphosphine)-2,3,4,5-tetraphenylferracyclopentadiene-tricarbonyl, 42B, 712
$C_{51}H_{41}AsPO_4V$, (η[2]-(C,O)-Triphenylcyclopropylcarbonyl)tricarbonyl(1-diphenylphosphino-2-diphenylarsinoethane)vanadium, 45B, 905
$C_{51}H_{41}N_3O_2P_2Pt \cdot 0.78\ CHCl_3$, 2-Carboethoxy-2,4,4-tricyano-trans-3-phenyl-1,1-di(triphenylphosphine)-1-platinacyclobutane chloroform

solvate, 44B, 733
$C_{51}H_{51}ClN_3O_7P_2RhS_3$, Chlorotris(ethoxycarbonyl isothiocyanate)bis-(triphenylphosphine)rhodium(III) acetone solvate, 42B, 647; 43B, 1030
$C_{51}H_{51}O_2PPd$, (1-η^1-1,2,3,4-Tetra(p-tolyl)-4-phenyl-1,3-butadienyl)-(acetylacetonato)(dimethylphenylphosphine)palladium, 44B, 735
$C_{51}H_{55}O_6P_2Pd_2$, μ-(Cyclopentadienyl)-μ-(2-methylallyl)-bis(tri-o-tolylphosphite-P)-dipalladium(Pd-Pd), 43B, 1026
$C_{52}H_{40}P_2Pt$, cis-Bis(phenylacetylido)bis(triphenylphosphine)platinum-(II), 43B, 1030
$C_{52}H_{42}P_2Pt \cdot 0.5\ CHCl_3$, trans-(Phenylacetylido)(styryl-C(2))-bis(triphenylphosphine)platinum(II) chloroform (2:1), 43B, 1031
$C_{52}H_{44}F_{24}O_8Rh_4$, 1,6-8-$\eta^4$-5-Allylcyclopent-2-enyl(hexafluoroacetyl-acetonato)rhodium(II) tetramer, 45B, 906
$C_{52}H_{46}ClCuOP_2Ru$, Bis(triphenylphosphine)cyclopentadienylruthenium-phenylacetylide-chlorocopper(I) acetone solvate, 42B, 648
$C_{52}H_{54}O_2P_2Pt$, Bis(1-ethynylcyclohexanol)bis(triphenylphosphine)plat-inum, 39B, 582
$C_{52}H_{56}OP_2U_2$, Biscyclopentadienyluranium(IV) phosphoylide dimer, 44B, 735
$C_{52}H_{64}Ag_4Fe_4N_4$, 2-Silver(dimethylaminomethyl)ferrocene tetramer, 44B, 1141
$C_{53}H_{47}ClN_2O_2P_2Pd$, trans-Chloro-(1,4,-bis(p-methoxyphenyl)-3-methyl-1,4-diazabutadiene-2-yl)-bis(triphenylphosphine)palladium(II), 46B, 830
$C_{53}H_{47}ClP_2Pt$, trans-Chloro(isopropenylacetylido)bis(triphenylphos-phine)platinum(II) - benzene (1:2), 43B, 1033
$C_{53}H_{47}Cl_3CuN_2O_2P_2Pd$, 2-(Chlorobis(triphenylphosphine)palladium)-3-methyl-1,4-bis(methoxyphenyl)-1,4-diaza-butadien-1,4-diyl-dichloro-copper, 46B, 831
$C_{54}H_{43}ClIrO_9P_3$, Chlorobis-(2-(diphenoxyphosphino-oxy)phenyl)(tri-phenyl phosphite)iridium(III), 38B, 759
$C_{54}H_{45}BrIrP_3$, Bromo[2-(diphenylphosphino)phenyl-C,P]hydridobis(tri-phenylphosphine)iridium(III), 46B, 832
$C_{54}H_{51}ClIrNO_4P_4$, (Methyl isocyanide)bis(1,2-bis(diphenylphosphino)-ethane)iridium(I) perchlorate, 42B, 649
$C_{54}H_{82}P_2Pt_2Si$, (μ-Dimethylsilanediyl)(σ-phenylethynyl)[μ-(1-σ:1-2-η-phenylethynyl)]-bis(tricyclo-hexylphosphine)diplatinum(Pt-Pt), 45B, 907
$C_{55}H_{46}MoN_6O_2P_2S_4$, Tetraphenylphosphonium nitrosyl(N,N)dimethylcarb-amido)tetrakis(thiocyanato-N)-molybdate(VI), 45B, 907
$C_{56}H_{32}Cl_2F_8N_8Rh_2 \cdot 2\ H_2O$, Tetrakis(p-fluorophenyl isocyanide)rhodi-um(I) chloride dimer dihydrate, 45B, 908
$C_{56}H_{32}Cl_2N_{16}O_{16}Rh_2$, Tetrakis(p-nitrophenyl isocyanide)rhodium(I) chloride dimer, 45B, 908
$C_{56}H_{40}Co_2I_4N_8$, μ-Iodo-bis(iodo-tetrakis(phenylisocyanide)cobalt) io-dide, 41B, 860
$C_{56}H_{42}Au_2O_2$, 1-Hydroxy-2,3,4,5-tetraphenylauracyclopentadiene dimer, 41B, 873
$C_{56}H_{50}ClN_2O_3P_2RhS_2$, Chloro(di(benzoylisocyanato))bis(triphenylphos-phine)rhodium(III) diethyl ether solvate, 41B, 873; 42B, 649
$C_{56}H_{53}F_{12}N_3P_6Pd_2 \cdot C_3H_6O$, μ-Methylisocyano-di-μ-bis(diphenylphos-phino)methane-bis(methylisocyanopalladium) hexafluorophosphate ace-tone, 43B, 1033
$C_{56}H_{56}O_4Ti_2$, Bis-μ-(η^2-diphenylketene-C,O)-bis(bis(η^5-cyclopentadi-enyl)titanium) - bis(tetrahydrofuran), 44B, 736
$C_{57}H_{50}Cl_2O_5P_4Rh_2$, ($\mu$-Carbonyl)-($\mu$-dimethylacetylenedicarboxylate)-

bis(μ-bis(diphenylphosphino)-methane)-dichlororhodium, 46B, 832
$C_{57}H_{58}Cl_2Pd_4S_8$, Bis(thiophenoxymethyl)palladium(II) tetramer dichloromethane solvate, 44B, 1156
$C_{57}H_{87}ClN_3Rh$, Chlorotris(2,4,6-tri-t-butylphenyl isocyanide)rhodium(I), 45B, 909
$C_{58}H_{40}Hg_3N_2$, Bis(1,3-tetraphenylbutadienecyanomercury(II))mercury(II), 41B, 874
$C_{59}H_{38}CoF_{20}P$, 1-(π-Cyclopentadienyl)-1-triphenylphosphine-2,3,4,5-tetrakis(pentafluorophenyl)cobaltole octane solvate, 42B, 650
$C_{60}H_{40}O_{12}P_2Pt_6$, Bis(tetraphenylphosphonium) bis(triplatinumhexacarbonyl), 40B, 638
$C_{60}H_{50}P_2PbPt$, Bis(triphenylphosphine)(σ-phenyl)(triphenyllead)platinum, 39B, 582
$C_{60}H_{51}Cl_5Fe_2N_3P_2$, Chlorobis(triphenylphosphine)tris(4-tolyl isocyanide)iron(II) tetrachloroferrate(III), 43B, 1034
$C_{62}H_{45}Cl_2Mn_2O_7P_3$, Tricarbonyltriphenylphosphine-(μ-(carbonyl(6-(diphenylphosphino)-o-phenylene)))(tricarbonyltriphenylphosphinemanganese)manganese dichloromethane, 43B, 1025
$C_{62}H_{59}O_2P_3Ru$, Hydrido(2-n-butoxycarbonylpropenyl-C',O)tris(triphenylphosphine)ruthenium(II), 42B, 650
$C_{64}H_{31}NO_{29}P_2Os_9$, Bis(triphenylphosphine)iminium μ-hydrido-decacarbonyltriosmium-μ-carboxylato-heptadecacarbonylhexaosmate, 44B, 737
$C_{64}H_{70}N_6O_{12}Zn_4$, Hexakis(N-phenylmethylcarbamato)diethyltetrazinc(II) - bis(benzene), 44B, 737
$C_{65}H_{57}Cl_6Mn_2NO_4P_4$, ($\mu$-p-Isonitriletoluene)di(bis(diphenylphosphino)-methane)tetracarbonyldimanganese(Mn-Mn) dichloromethane solvate, 44B, 1141
$C_{66}H_{40}O_{18}P_2Pt_9$, Bis(tetraphenylphosphonium) tris(triplatinumhexacarbonyl), 40B, 638
$C_{67}H_{53}Fe_2N_2O_6P_4$, Bis(triphenylphosphine)immonium tricarbonyliron-di-μ-(diphenylphosphido)-(acetyldicarbonyliron), 45B, 904
$C_{68}H_{56}B_2F_8N_2P_4Pt_2$, o-Cyanobenzylbis(diphenylphosphino)ethyleneplatinum(II) tetrafluoroborate dimer, 43B, 1035
$C_{68}H_{64}B_2N_8Rh_2 \cdot C_2H_3N$, Tetrakis($\mu$-1,3-di-isocyanopropane)dirhodium bis(tetraphenylborate) acetonitrile solvate, 46B, 833
$C_{68}H_{64}O_{16}P_4W_4$, Tetrakis-(μ_3-hydroxotricarbonylhydridotungsten) - diphenylethylphosphine oxide, 38B, 760
$C_{68}H_{65}BNNiOP_3 \cdot 2.5\ C_4H_8O$, Acetyl(tris(2-diphenylphosphinoethyl)amine)nickel(II) tetraphenylborate - 2.5 tetrahydrofuran, 44B, 738
$C_{70}H_{71}BNNiOP_3$, Methyl(tris(2-diphenylphosphinoethyl)amine)nickel(II) tetraphenylborate acetone, 43B, 1036
$C_{71}H_{57}P_2Rh$, 5,5,7,8,9,10-Hexaphenyl-6-(triphenylphosphine)-5,6-dihydro-5-phospha-6-rhodabenzocyclooctene (toluene solvate), 37B, 438
$C_{71}H_{78}N_8O_2W_2$, Di-μ-(N,N'-di-3,5-xylylformamidino)-di-μ-carbonyl-(N,N'-di-3,5-xylylformamidino)-(N^1-methylene-N^1-3,5-xylyl-N^2-3,5-xylylformamidino-N^2,C)ditungsten, 44B, 739
$C_{72}H_{67}As_3BNNi$, σ-Phenyl(tris(2-diphenylarsinoethyl)amine)nickel(II) tetraphenylborate, 42B, 651
$C_{73}H_{60}BClF_4P_4Pt_2S_2 \cdot 0.2\ CH_2Cl_2$, Bis(triphenylphosphine)((trans-bis(triphenylphosphine)chloroplatino)dithiocarboxylato)platinum tetrafluoroborate dichloromethane, 41B, 875
$C_{78}H_{40}As_2O_{30}Pt_{15}$, Bis(tetraphenylarsonium) pentakis(triplatinumhexacarbonyl), 40B, 638
$C_{83}H_{62}Cl_2Cr_2N_2O_{10}P_4$, Bis(triphenylphosphine)iminium decacarbonyldichromium - dichloromethane, 35B, 506
$C_{83}H_{62}Cl_2Mo_2N_2O_{10}P_4$, Bis(triphenylphosphine)iminium decacarbonyldimolybdenum - dichloromethane, 35B, 506

$C_{94}H_{45}Ag_2F_{25}P_3Rh$, Tris(triphenylphosphine)pentakis(pentafluorophenylethynyl)rhodiumdisilver, 41B, 876
$C_{96}H_{84}Br_3I_2N_{12}Rh_3$, Tetrakis(benzyl isocyanide)-bis(iodo-tetrakis-(benzyl-isocyanide)-rhodium)rhodium bromide, 45B, 909
$C_{100}H_{70}Cu_4Ir_2P_2$, Bis(triphenylphosphine)octakis(phenylethynyl)tetracopperdiiridium(4Cu-Cu)(8Cu-Ir), 40B, 1071
$C_{100}H_{150}Li_{12}N_4Na_6Ni_4O_{14}$, Dinitrogen nickel complex, 42B, 652
$C_{104}H_{80}B_2N_8Rh_2$, Octakis(phenylisocyanide)dirhodium(I)(Rh-Rh) tetraphenylborate, 44B, 739
$C_{151}H_{159}B_2N_2Ni_2O_6P_6$, Acetyl(tris(2-diphenylphosphinoethyl)amine)nickel - carbonyl(tris(2-diphenylphosphinoethyl)amine)nickel tetraphenylborate - 4 tetrahydrofuran, 44B, 738

72. METAL π-COMPLEXES (OPEN-CHAIN)

$C_2H_4Cl_3KPt$, Potassium trichloro(ethylene)platinate(II), 43B, 1037
$C_2H_6Br_3KOPt$, Potassium tribromo(ethylene)platinate(II) monohydrate, 30B, 292
$C_2H_6Cl_3KOPt$, Potassium ethylenetrichloroplatinate(II) monohydrate (Zeise's salt - $P2_1$), 18, 633; 21, 503; 30B, 292; 34B, 437; 37B, 439
$C_2H_6Cl_3KOPt$, Potassium trichloro(ethylene)platinate(II) monohydrate (Zeise's salt - $P2_1/c$), 41B, 877
$C_2H_7Br_2NPt$, cis-Amminedibromoethyleneplatinum(II), 29, 586
$C_3H_8Cl_3NPt$, Trichloro(π-allylammonium)platinum(II), 43B, 1037
$C_4H_8Cl_4Pd_2$, trans-Di-μ-chlorobis(π-ethylene)dichlorodipalladium(II), 19, 538
$C_4H_{10}Cl_3NPt$, Trichloro(cis-but-2-enylammonium)platinum(II), 35B, 513
$C_4H_{10}Cl_3NPt$, Trichloro(π-but-3-enylammonium)platinum(II), 43B, 1037
$C_4H_{10}Cl_3NPt$, Trichloro(trans-but-2-enylammonium)platinum(II), 35B, 513
$C_4H_{11}Cl_2NPt$, trans-Dichloro(dimethylamino)ethyleneplatinum, 24, 623
$C_4H_{12}Cl_4N_2Pt \cdot 0.5\ H_2O$, Trichloro(π-trans-but-2-en-1,4-diammonium)-platinum(II) chloride, 37B, 440
$C_4H_{12}Cl_4N_2Pt$, Trichloro(π-cis-but-2-en-1,4-diammonium)platinum(II) chloride, 43B, 1038
$C_4H_{19}N_5O_{10}RuS_2 \cdot 2\ H_2O$, Penta-ammine-ruthenium(II) fumaric acid dithionate dihydrate, 45B, 910
$C_5H_4FeO_3$, Ethyleneiron tricarbonyl (gas-ed), 35B, 895
$C_5H_9Cl_2NOPdS$, Dichloro(O-methyl N-allylthiocarbamato)palladium(II), 37B, 441
$C_5H_{12}Cl_3NPt$, Trichloro(pent-4-enylammonium)platinum(II) (orange form), 38B, 761
$C_5H_{12}Cl_3NPt$, Trichloro(pent-4-enylammonium)platinum(II) (yellow form), 38B, 761
$C_5H_{12}Cl_3NPt$, Trichloro(π-cis-pent-2-enylammonium)platinum(II), 43B, 1038
$C_5H_{12}Cl_3NPt$, Trichloro(π-trans-pent-2-enylammonium)platinum(II), 39B, 583
$C_5H_{12}Cl_3NPt$, Trichloro-(π-cis-pent-3-enylammonium)platinum(II), 42B, 653
$C_5H_{13}ClN_4NiS_2$, π-Allyldi(thiourea)nickel(II) chloride, 35B, 515
$C_6F_4FeO_4$, Tetrafluoroethylenetetracarbonyliron (gas-ed), 39B, 887
$C_6H_5BrFeO_3$, (η^3-Allyl)bromotricarbonyliron, 46B, 851
$C_6H_5CoO_3$, π-Allyltricarbonylcobalt (gas-ed), 38B, 1061
$C_6H_5FeIO_3$, π-Allyltricarbonyliron iodide, 31B, 343; 33B, 330

$C_6H_{10}Cl_2Pd_2$, Di-μ-chlorobis(allylpalladium), 27, 820; 30B, 299
$C_6H_{11}NO_2Pd$, (π-2-Methylallyl)(glycinato)palladium(II), 39B, 584
$C_6H_{14}Cl_3NPt$, Trichloro(η-N-methyl-4-pentenylammonium)platinum(II), 44B, 740
$C_6H_{14}Cl_3NPt$, Trichloro(π-hex-5-enylammonium)platinum(II), 43B, 1037
$C_7H_3FeNO_4$, Tetracarbonyl(acrylonitrile)iron, 26, 684; 27, 824
$C_7H_6BF_3FeO_5S$, Butadieneiron tricarbonyl - sulfur dioxide - boron trifluoride adduct, 35B, 516
$C_7H_6FeO_3$, Butadieneiron tricarbonyl (gas-ed), 35B, 895
$C_7H_6FeO_3$, Butadienetricarbonyliron, 28, 605
$C_7H_9O_2RhS$, η⁵-Cyclopentadienyl(sulfur dioxide)ethenylrhodium(I), 42B, 654
$C_7H_{15}ClNORh$, Chloro(diethylamine)ethylenecarbonylrhodium, 42B, 654
$C_8H_4FeO_8$, (-)-Tetracarbonyl(fumaric acid)iron, 33B, 407
$C_8H_4FeO_8$, DL-Tetracarbonyl(fumaric acid)iron, 32B, 343
$C_8H_8Cl_2N_2Pt$, 1-(4-Cyanopyridine)-3-ethylene-2,4-dichloroplatinum-(II), 41B, 877
$C_8H_{10}IMoNO$, η³-Allyl-(η⁵-cyclopentadienyl)iodonitrosylmolybdenum, 46B, 835
$C_8H_{10}INOW$, (η⁵-Cyclopentadienyl)(η³-allyl)(nitrosyl)iodotungsten, 45B, 911
$C_8H_{10}Pd$, π-Allyl-π-cyclopentadienylpalladium, 31B, 343; 33B, 331
$C_8H_{11}Cl_2NPt$, trans-Dichloro(ethylene)(4-methylpyridine)platinum(II), 41B, 877; 44B, 740
$C_8H_{12}BrNNiO_2$, Acetonitrilebromo(1-3-η-(carbomethoxy-2-methylallyl))-nickel, 40B, 723
$C_8H_{12}ClIr$, Chlorobis(butadiene)iridium(I), 39B, 584
$C_8H_{12}ClRh$, Bis(butadiene)rhodium(I) chloride, 34B, 441
$C_8H_{14}Cl_2OPt$, cis-Dichloro(2,2'-oxydi-3-butene)platinum(II), 39B, 584
$C_8H_{14}Cl_2Pd_2$, π-2-Methylallylpalladium chloride dimer, 33B, 333
$C_8H_{14}Ni$, Bis(methallyl)nickel, 30B, 299
$C_8H_{15}ClPdS$, Chloro[1,3-η-syn-1-(1,1'-dimethyl-2'-methylthioethyl)-allyl]palladium(II), 46B, 834
$C_8H_{16}Cl_2PdS$, Dichloro[2,2-dimethylpent-(E)-3-enyl methyl sulfide]-palladium(II), 46B, 834
$C_8H_{18}Cl_2N_4Pt$, (Butane-2,3-dionebis(methylhydrazone))dichloro(η-ethylene)platinum(II), 42B, 814
$C_8H_{20}Cl_2N_2O_4Pt$, Chloro(η-ethylene)(N,N,N',N'-tetramethylethylenediamine)platinum(II) perchlorate, 45B, 911
$C_9H_8FeO_5$, Sorbic acid - iron tricarbonyl, 35B, 518
$C_9H_9F_4Rh$, π-Cyclopentadienylethylenetetrafluoroethylenerhodium, 38B, 730
$C_9H_9F_6FePO_4$, Tricarbonyl(1-3-η-hexen-5-one)iron hexafluorophosphate, 38B, 762
$C_9H_{10}MoO_2S$, Cyclopentadienyldicarbonylmolybdenum methylenemethylsulfide, 34B, 438
$C_9H_{12}FeO$, Bis(butadiene)monocarbonyl iron, 38B, 763
$C_9H_{12}MnO$, Bis(1,3-butadiene)-monocarbonylmanganese, 42B, 655
$C_9H_{13}Cl_2NPt$, trans-Dichloro(η-ethylene)(2,6-dimethylpyridine)platinum(II), 45B, 912
$C_9H_{19}Cl_2NPt$, cis-Dichloro(η-ethylene)(2,6-dimethylpiperidine)platinum(II), 46B, 834
$C_9H_{19}NiP$, Bis(η³-allyl)trimethylphosphinenickel, 46B, 834
$C_{10}H_8FeO_5$, Tricarbonyl-η⁴-(3-methylene-endo-4-vinyldihydrofuran-2(3H)-one)iron(0), 42B, 655
$C_{10}H_{10}MoO_2$, η³-Allyl-dicarbonyl(η⁵-cyclopentadienyl)molybdenum, 46B, 835

$C_{10}H_{12}FeO_4$, trans,trans-3,5-Heptadien-2-ol-iron tricarbonyl, 42B, 656
$C_{10}H_{14}Cl_2FeGe$, π-Butadiene-π-cyclopentadienyldichloromethylgermyl-iron, 37B, 442
$C_{10}H_{14}Ni_2$, Bis(pentadienyl)dinickel, 34B, 438
$C_{10}H_{15}Cl_2NPt$, trans-Dichloro(ethylene)(2,4,6-trimethylpyridine)platinum(II), 44B, 740
$C_{10}H_{18}BF_4N_2Rh$, Tris(ethylene)bis(acetonitrile)rhodium(I) tetrafluoroborate, 40B, 724
$C_{10}H_{18}Cl_3KO_2Pt$, Potassium 3,6-dimethyloct-4-yne-3,6-dioltrichloroplatinate(II), 42B, 657
$C_{10}H_{22}Cl_2N_2PtS$, (Ethylene)platinum dichloride - di-t-butylsulfurdiimine, 39B, 585
$C_{11}H_8FeO_6$, Tetracarbonyl-η^2-(3-methylene-exo-4-vinyldihydrofuran-2(3H)-one)iron(0), 42B, 655
$C_{11}H_{11}MnO_3$, (π-Cyclopentadienyl)dicarbonyl(methylvinylketone)manganese(I), 41B, 878
$C_{11}H_{13}F_6FeO_4P$, Tricarbonyl[2-4-η-(5-methylhepten-6-one)]iron hexafluorophosphate, 45B, 912
$C_{11}H_{14}AgP$, Phenylethynyl(trimethylphosphine)silver(I), 31B, 344
$C_{11}H_{14}CuP$, Phenylethynyl(trimethylphosphine)copper(I), 31B, 345
$C_{11}H_{15}Cl_2OTa$, Dichloro(η^5-cyclopentadienyl)[(2-oxo-4-methyl-pent-3-ene)]tantalum, 46B, 868
$C_{11}H_{15}F_3MoO_6$, π-Allyldicarbonylmolybdenum trifluoroacetate 1,2-dimethoxyethane, 40B, 725
$C_{11}H_{15}F_3O_6W$, π-Allyldicarbonyltungsten trifluoroacetate 1,2-dimethoxyethane, 40B, 725
$C_{11}H_{16}ClNPd$, Chloro-(1-3-η-1,2-dimethylallyl)(3-picoline)palladium, 43B, 1039
$C_{11}H_{17}Cl_2OPPt$, Dichloro(dimethylphenylphosphine)(prop-2-en-1-ol)-platinum(II), 46B, 836
$C_{11}H_{23}ClMoO_8P_2$, η^3-Allylbis(trimethyl phosphite)chlorodicarbonylmolybdenum, 45B, 913; 46B, 836
$C_{11}H_{23}MnO_8P_2$, 1-3-η-Allyldicarbonylbis(trimethyl phosphite)manganese(I), 46B, 837
$C_{12}H_6Fe_2O_6$, 1,5-Di(iron tetracarbonyl)-3-methylene-penta-1,4-diene, 30B, 296
$C_{12}H_6MnO_8$, μ-Butadiene-dimanganese-octacarbonyl, 38B, 764
$C_{12}H_8FeO_4$, Tricarbonyl(cinnamaldehyde)iron, 38B, 763
$C_{12}H_8Fe_2O_8$, But-2-ynedihydridooctacarbonyldiiron, 26, 682
$C_{12}H_8O_9Os_3S$, 1,1,1,2,2,2,3,3,3-Nonacarbonyl-2-ethylene-1,3-μ-hydrido-1,3-μ-(methylthiolato)-triangulo-triosmium, 46B, 838
$C_{12}H_9ClFeO_4$, (exo-2-Chloro-5,6-dimethylene-syn-7-norbornanone)-endo-tricarbonyliron, 45B, 913
$C_{12}H_9FeN_3O_3$, 1-N-Allylbenzotriazoletricarbonyliron, 43B, 1040
$C_{12}H_{10}FeO_4$, (6,7-Methylene-exo-3-oxatricyclo[3.2.1.0^{2-4}]octane)-exo-tricarbonyliron, 45B, 913
$C_{12}H_{10}FeO_8$, η^2-(cis-2,3-Dicarbomethoxymethylenecyclopropane)iron tetracarbonyl, 40B, 726
$C_{12}H_{10}Fe_2O_6$, Bis(tricarbonyl(η^3-allyl)iron)(Fe-Fe), 44B, 740
$C_{12}H_{11}F_6O_2Rh$, (1-3-η:5-7-η-Heptadienediyl)rhodium(I) hexafluoroacetylacetonate, 43B, 1041
$C_{12}H_{12}Co_2O_4$, Butadiene cobalt dicarbonyl, 31B, 346
$C_{12}H_{12}Ni_2$, μ-Acetylene-bis(cyclopentadienylnickel), 42B, 658
$C_{12}H_{12}O_8W$, trans-Bis(η^2-methyl acrylate)tetracarbonyltungsten, 46B, 838
$C_{12}H_{16}Cl_4Pd_2$, Di-μ-chloro-di-π-(β-(3-chloro-1-propen-2-yl)-allyl)di-

palladium(II), 37B, 442
$C_{12}H_{17}Cl_2O_2PPt$, cis-Dichloro(dimethylphenylphosphine)(vinyl acetate)platinum(II), 45B, 914
$C_{12}H_{18}Ag_2N_2O_6$, 1,2-η^2:7,8,11,12-η^4-3-Isopropenyl-4-methyl-4-vinylcyclohexenedisilver nitrate, 35B, 520
$C_{12}H_{18}Br_2Ni_2O_4$, 2-Carboxyethyl-π-allylnickel bromide dimer, 32B, 345
$C_{12}H_{18}Cl_2Pd_2O_6$, Di-μ-chloro-bis(π-1-ethoxycarbonyl-2-hydroxyallyl)dipalladium(II), 35B, 521
$C_{12}H_{18}Cl_2Ru$, Dichloro(dodeca-2,6,10-triene-1,12-diyl)ruthenium(IV), 33B, 335
$C_{12}H_{19}Cl_2NPt$, (+)-cis-Dichloro((S)-1-butene)((S)-α-methylbenzylamine)platinum(II), 37B, 443
$C_{12}H_{19}Cl_2NPt$, (-)-cis-Dichloro(trans-2-butene)((S)-α-phenylethylamine)platinum(II), 34B, 439
$C_{12}H_{20}Br_2Rh_2$, Di-μ-bromo-tetrakis(η^3-2-propenyl)dirhodium, 42B, 659
$C_{12}H_{20}Cl_4Pt_4$, Bis-(di-μ-allyl-(di-μ-chloro-diplatinum)), 38B, 764
$C_{12}H_{20}Co_2O_4P_2$, (Acetylene)bis(trimethylphosphine)(tetracarbonyl)dicobalt(0), 44B, 741
$C_{12}H_{20}Cr_2$, Tetraallyldichromium, 34B, 441
$C_{12}H_{20}Mo_2$, Tetraallyldimolybdenum(II), 37B, 444
$C_{12}H_{20}Re_2$, Tetraallyldirhenium, 44B, 741
$C_{12}H_{21}Cl_2NOPd_2$, μ-Chloro-chloro-di-π-allyl(cyclohexanone oxime)dipalladium, 37B, 444
$C_{12}H_{22}Cl_3KO_2Pt$, Potassium trichloro(3,6-diethyloct-4-yne-3,6-diol)platinate(II), 39B, 586
$C_{12}H_{26}N_6Pd_2$, Bis(dimethyltriazenido-π-methallylpalladium), 41B, 879
$C_{12}H_{28}Cl_3NOPt$, Tetraethylammonium trichloro(ethyl vinyl ether)platinum(II), 44B, 742
$C_{13}H_3Co_2F_9O_4$, μ-(1-3,6-η:1,4-6-η-1,3,6-Tris(trifluoromethyl)hexa-1,3,5-trien-1,6-diyl)-bis(dicarbonylcobalt), 38B, 766
$C_{13}H_5ClF_{12}W$, Chloro(cyclopentadienyl)bis(di(trifluoromethyl)acetylene)tungsten, 42B, 659
$C_{13}H_7Fe_3NO_9$, Triangulo-[μ_3-[(N(3)1-η:N,1-η)-butyronitrile-N]]nonacarbonyltriiron, 45B, 915
$C_{13}H_8O_6Cr$, Phenylmethoxycarbenepentacarbonylchromium, 33B, 336
$C_{13}H_{10}FeO_3$, Phenyltrimethylenemethane-iron tricarbonyl, 34B, 445
$C_{13}H_{10}FeO_4$, η^4-(2,3,5,6-Tetrakis(methylene)-7-oxabicyclo[2.2.1]heptane)-endo-tricarbonyliron, 46B, 839
$C_{13}H_{11}Cl_3N_2O_2Pt$, Dichloro-(p-nitrosostyrene)(4-chloropyridine)platinum(II), 42B, 664
$C_{13}H_{14}FeO$, Butadiene(cyclooctatetraene)iron monocarbonyl, 38B, 766
$C_{13}H_{14}FeO_5$, (exo-2-Methoxy-5,6-dimethylene-syn-7-norbornanol)-endo-tricarbonyliron, 45B, 913
$C_{13}H_{17}BrNNiO_2$, Bromo(lutidine)-η^3-(3-carbomethoxy-2-methylallyl)nickel(II), 42B, 660
$C_{13}H_{19}O_2Rh$, Acetylacetonatobis(η^2-methylenecyclopropane)rhodium(I), 41B, 880
$C_{13}H_{20}Cl_2OPtS$, cis-Dichloro((S)-methyl p-tolyl sulfoxide)(3-methyl-1-butene)platinum(II), 43B, 1041
$C_{13}H_{21}O_3Rh$, (But-2-enyl 1-methylallyl ether)acetylacetonatorhodium, 44B, 742
$C_{13}H_{22}N_2OS_4W$, Bis(N,N-diethyldithiocarbamato)(acetylene)carbonyltungsten, 44B, 743
$C_{13}H_{24}GaMoN_3O_3S$, [Dimethyl(N,N-dimethylethanolamino)(1-pyrazolyl)gallato](η^2-thiomethoxymethyl)dicarbonylmolybdenum, 46B, 840
$C_{14}H_6O_{10}Os_3$, (s-cis-1,3-Butadiene)triosmium decacarbonyl, 44B, 744
$C_{14}H_6O_{10}Os_3$, (s-trans-1,3-Butadiene)triosmium decacarbonyl, 44B, 744

$C_{14}H_8O_{10}Os_3$, μ-But-1-enyl-μ-hydrido-decacarbonyl-triangulo-triosmium, 42B, 660
$C_{14}H_{10}Fe_2O_6$, Iron carbonyl complex, 33B, 337
$C_{14}H_{10}Fe_2O_7$, Hexacarbonyl(μ-((1,2,3-η:3,4,5-η)-2-ethyl-4-methyl-1-oxo-2,3-pentadiene-1,5-diyl))diiron, 42B, 662
$C_{14}H_{12}Co_2O_6$, Bis(η³-2-allyl)-1,2-ethanobis(tricarbonylcobalt), 44B, 744
$C_{14}H_{13}BF_6N_6Pt$, Methyl(hydrotris(1-pyrazolyl)borato)hexafluorobut-2-yneplatinum(II), 40B, 726
$C_{14}H_{15}Cl_2NPt$, Dichloro-(styrene)(4-methylpyridine)platinum(II), 42B, 664
$C_{14}H_{16}F_8FeN_2O_2P_4$, Carbonyl-(1-difluorophosphino(methyl)amino)-2-methyl-4-phenyl-1-(difluorophosphonyl)-butadiene-methylamino-bis(difluorophosphine)iron, 45B, 915
$C_{14}H_{16}Ni_2$, Di-π-allyl(dihydropentalenylene)dinickel, 39B, 586
$C_{14}H_{16}Zr$, Bis(η⁵-cyclopentadienyl)(η⁴-s-trans-buta-1,3-diene)-zirconium(II), 46B, 841
$C_{14}H_{19}MoNO_3$, Carbonyl-(η⁵-cyclopentadienyl)-(2,2-dimethyl-hex-4-ene-aldehyde)nitrosylmolybdenum, 45B, 916
$C_{14}H_{19}Nb$, Bis(cyclopentadienyl)ethyl(ethylene)niobium, 40B, 750
$C_{14}H_{21}Cl_2NPt$, cis-Dichloro(1,5-hexadiene)platinum(II) - (S)-α-methylbenzylamine complexes, 37B, 476
$C_{14}H_{22}Cl_2O_4Pd_2$, Di-μ-chloro-bis(1-acetoxy-3-methylbut-(2-4-η)-enyl-palladium), 44B, 745
$C_{14}H_{26}Cl_2Pd_2$, π-1,1,3,3-Tetramethylallylpalladium chloride dimer, 33B, 334
$C_{14}H_{32}O_6P_2Ru$, Bis-(π-2-methylallyl)bis(trimethylphosphite)ruthenium, 39B, 587
$C_{15}H_5F_{11}MoOS$, Cyclopentadienyl(hexafluorobut-2-yne)oxo(pentafluorophenylthio)molybdenum(IV), 42B, 661
$C_{15}H_{10}F_6ORh_2$, μ-Carbonyl-bis(η-cyclopentadienyl)-μ-η-(hexafluoro-2-butyne)-dirhodium (Rh-Rh), 45B, 917
$C_{15}H_{10}O_9Ru_3$, Nonacarbonyl-μ-(t-butylacetylene)-triangulo-triruthenium, 39B, 588; 43B, 968
$C_{15}H_{12}FeO_3$, (5,6,7,8-Tetrakis(methylene)bicyclo[2.2.2]oct-2-ene)tricarbonyliron, 45B, 917
$C_{15}H_{13}FeNO_6$, Tricarbonyl-(methyl α-acetamidocinnamate)iron, 46B, 841
$C_{15}H_{13}MoNO_2$, (η²-N-Phenyliminoethyl)(η⁵-cyclopentadienyl)dicarbonyl-molybdenum, 44B, 745
$C_{15}H_{14}FeO_3$, Tricarbonyl[2,3,5,6-tetrakis(methylene)bicyclo[2.2.2]-octane]iron, 45B, 918
$C_{15}H_{17}BMoN_6O_2$, Dicarbonyl(hydrotris(pyrazol-1-yl)borato-N²,N²',N²")-π-(2-methylallyl)molybdenum, 39B, 588
$C_{15}H_{18}ORh$, (+)-(4S-Carvone)-cyclopentadienylrhodium(I), 44B, 1142
$C_{15}H_{19}Ti$, 1,2-Dimethylallyl-dicyclopentadienyltitanium(III), 32B, 346
$C_{15}H_{20}FeO_3$, Tricarbonyl(η⁴-1,5-dimethylene-2,6-dimethylcyclooctane)-iron, 46B, 842
$C_{15}H_{21}BMoN_4O_2$, π-Allyl-dihydrobis-(3,5-dimethyl-1-pyrazolyl)-boratodicarbonylmolybdenum, 37B, 445
$C_{15}H_{21}MoNO_3$, Carbonyl-(η⁵-cyclopentadienyl)-(2,2,3-trimethyl-hex-4-ene-aldehyde)nitrosylmolybdenum, 45B, 916
$C_{15}H_{26}GaMoN_3O_3$, (η³-2-Methylallyl)[dimethyl(ethanolamino)-(3,5-dimethyl-1-pyrazolyl)gallato]-dicarbonyl-molybdenum, 45B, 919
$C_{15}H_{28}GaMoN_3O_3S$, [Dimethyl(N,N-dimethylethanolamino)(3,5-dimethyl-1-pyrazolyl)gallato](η²-thiomethoxymethyl)dicarbonylmolybdenum, 46B, 840

$C_{15}H_{32}ClPPt$, Chloro(η^3-allyl)(tri-t-butylphosphine)platinum(II), 45B, 920
$C_{16}F_{18}Ni_4O_4$, Tris(hexafluorobut-2-yne)tetracarbonyltetranickel, 41B, 880
$C_{16}H_5F_{11}MoOS$, Carbonylcyclopentadienylhexafluorobut-2-yne(pentafluorophenylthio)molybdenum(II), 42B, 661
$C_{16}H_7CoFeO_4$, Di-μ-carbonyl-carbonyl(carbonyl-(2,3-dimethylbuta-1,3-diene)cobalt)(π-methylcyclopentadienyl)iron, 40B, 753
$C_{16}H_{10}Co_4O_{10}$, Tetracobaltdecacarbonyl(diethylacetylene), 27, 826
$C_{16}H_{10}F_6O_2Rh_2$, Dicarbonylbis(η-cyclopentadienyl)-μ-(2,3-η-1,1,1,4,4,4-hexafluoro-2-butenyl)-dirhodium, 43B, 972
$C_{16}H_{10}Fe_2O_7$, (2-Methyl-3,5,6-tri(methylene)-cyclohexane-(2-4-η)-one)-bis(tricarbonyliron), 46B, 839
$C_{16}H_{12}Mo_2O_4$, μ-Acetylene-bis(η^5-cyclopentadienyl)tetracarbonyldimolybdenum, 44B, 746
$C_{16}H_{13}ClF_5NOPt$, Chloropentafluorophenolato-o-vinyl-N,N-dimethylanilineplatinum(II), 44B, 1143
$C_{16}H_{13}ClF_5NPtS$, Chloropentafluorothiophenolato-o-vinyl-N,N-dimethylanilineplatinum(II), 44B, 1143
$C_{16}H_{13}MoN_3O_2S$, (Isothiocyanato)dicarbonyl-2,2'-bipyridyl-π-allylmolybdenum, 34B, 443
$C_{16}H_{14}Fe_2O_6$, Hexacarbonyldeca-1,3,7,9-tetraenediiron (form I), 46B, 842
$C_{16}H_{14}Fe_2O_6$, Hexacarbonyldeca-1,3,7,9-tetraenediiron (form II), 46B, 842
$C_{16}H_{14}Fe_2O_7$, Hexacarbonyl(μ-((1,2,3-η:3,4,5-η)-2-(1,1-dimethylethyl)-4-methyl-1-oxo-2,3-pentadiene-1,5-diyl))diiron, 42B, 662
$C_{16}H_{14}Mn_2O_3$, μ-Allene-μ-carbonyl-bis(carbonyl-η^5-cyclopentadienylmanganese), 46B, 843
$C_{16}H_{16}Cl_4Pd_2$, Styrenepalladium chloride, 19, 542; 26, 689
$C_{16}H_{16}FeN_2O_5$, (2-((m-Nitrophenyl)amino)-trans-trans-3,5-heptadiene)-iron tricarbonyl, 40B, 727
$C_{16}H_{16}O_4V$, (η^2-Dimethyl acetylenedicarboxylato-C,C)bis(η-cyclopentadienyl)vanadium, 45B, 922; 46B, 844
$C_{16}H_{17}FeNO_3$, (2-((Phenyl)amino)-cis-trans-3,5-heptadiene)iron tricarbonyl, 40B, 727
$C_{16}H_{18}As_3F_5Mn_2O_6$, μ-(Dimethylarsino)-μ-(1-3-η-(2,3-bis(dimethylarsino)-1,1-difluoro-3-trifluoromethylallyl)-(As,As',C,C',C"))bis-(tricarbonylmanganese(I)), 41B, 882
$C_{16}H_{18}Cl_2OPtS$, cis-Dichloro((S)-methyl p-tolyl sulfoxide)((R)-styrene)platinum(II), 42B, 662
$C_{16}H_{18}Co_2O_6$, (Di-t-butylacetylene)hexacarbonyldicobalt, 42B, 663
$C_{16}H_{18}Fe_2O_6$, (Di-t-butylacetylene)hexacarbonyldiiron, 42B, 663
$C_{16}H_{18}N_6Ni$, Bis(t-butyl isocyanide)(tetracyanoethylene)nickel(0), 35B, 523
$C_{16}H_{18}Ni_2$, Bis(cyclopentadienyl)-2,2'-bi-π-allyl-bis(nickel), 38B, 767
$C_{16}H_{19}FeNO_6$, η^4-(1-Cyclohexylimino-2-methoxy-3-methoxycarbonylbutene)tricarbonyliron, 46B, 844
$C_{16}H_{20}Cl_2N_2Pt$, Dichloro-(p-dimethylaminostyrene)(4-methylpyridine)-platinum(II), 42B, 664
$C_{16}H_{20}Co_2O_4$, trans-Di-μ-carbonyl-bis-(π-2,3-cis-dimethylbutadiene carbonylcobalt), 35B, 525
$C_{16}H_{20}Zr$, Bis(η^5-cyclopentadienyl)(η^4-s-cis-2,3-dimethylbuta-1,3-diene)zirconium(II), 46B, 841
$C_{16}H_{24}Cl_2Cu_2N_4$, (1-Allyl-3,5-dimethylpyrazole)copper(I) chloride, 42B, 665

$C_{16}H_{24}O_4Pt_2$, Bis(acetylacetonato-μ-allyl-platinum), 38B, 764
$C_{16}H_{27}Cl_2NOPt$, cis-Dichloro((R)-α-methylbenzylamine)-((S)-1,2,2-trimethylpropyl(R)-vinyl ether)platinum(II), 42B, 665
$C_{17}H_{11}MnO_5$, Tricarbonyl-1-syn-(1',2'-dihydro-2'-oxo-1'-oxaazulen-3'-yl)η^5-pentadienylmanganese, 40B, 728
$C_{17}H_{12}Br_2FeO_3$, 1-(p-Bromobenzyl)-2-bromo-5-methylenecyclohexa-2,5-diene-4-yl-η^4-4,5,6,7-iron tricarbonyl, 43B, 974
$C_{17}H_{14}Fe_2O_5$, Pentacarbonyl-(6-cyclopropyl-6-(trans-1-propenyl)fulvene)diiron, 44B, 747
$C_{17}H_{14}Mo_2O_4$, μ-Allene-bis(cyclopentadienyl)tetracarbonyldimolybdenum, 43B, 1042; 44B, 748
$C_{17}H_{16}Mo_2N_2O_4$, μ-Dimethylaminocyanamide-bis(cyclopentadienyl)tetracarbonyldimolybdenum, 44B, 749
$C_{17}H_{17}F_6IrO_2$, Hexafluoroacetylacetonato(2,3,6,7-tetramethylene-octane-1,8-diyl)iridium(III), 43B, 976
$C_{17}H_{21}ClN_2O_4Pd$, (η-Allyl)(8-isopropylquinoline-2-carboxaldehyde-N-methylimine-N,N')palladium(II) perchlorate, 45B, 920
$C_{17}H_{23}As_2O_3V$, Tricarbonyl-(η^3-3-methylallyl)-[o-phenylene-bis(dimethylarsine)]vanadium, 45B, 921
$C_{17}H_{23}NOPd$, (π-Methylallyl)(2-(R,S)-α-phenylethylimino-3-penten-4-olato)palladium(II), 39B, 589
$C_{17}H_{25}GaMoN_4O_3$, (Hydroxybis(2,5-dimethylpyrazolyl)gallium)-dicarbonyl(2-methylallyl)molybdenum, 45B, 921
$C_{17}H_{27}Cl_2NPt$, trans-Dichloro-(π-di-t-butylacetylene)-p-toluidine-platinum(II), 35B, 526
$C_{18}H_{13}FeNO_3$, Tricarbonyl(N-cinnamylideneaniline)iron, 38B, 768
$C_{18}H_{15}Fe_2NO_8$, (Ethyl N-α-methylbenzyliminoacetato)-bis(tricarbonyl-iron), 45B, 1258
$C_{18}H_{16}Mn_2O_4$, μ-Butadiene-bis(cyclopentadienylmanganesedicarbonyl), 32B, 347
$C_{18}H_{18}Cl_2Pd_2$, Bis(2-phenyl-π-allylpalladium chloride), 34B, 444
$C_{18}H_{18}Cl_2Pt$, Dichloro(2,5-diphenyl-1,5-hexadiene)platinum(II), 46B, 846
$C_{18}H_{22}Cl_2Mo_2$, μ-Dichloro-bis-((π-benzene-π-allyl)-molybdenum(II)), 40B, 729
$C_{18}H_{22}O_4V$, (η^2-Diethyl fumarato-C,C)bis(η-cyclopentadienyl)vanadium, 45B, 922
$C_{18}H_{27}NNiO_8$, Bis(ethyl fumarate)(acetonitrile)nickel(0), 42B, 667
$C_{18}H_{27}N_5Ni$, Bis(t-butyl isocyanide)(N-t-butyldicyanoketenimine)nickel(0), 39B, 590
$C_{18}H_{28}IMoNO$, (η^3-Allyl)(η^5-neomenthylcyclopentadienyl)iodonitrosylmolybdenum, 46B, 847
$C_{18}H_{29}Ir$, (η^4-Cycloocta-1,5-diene)(η^3-1,3-dimethylallyl)(η^2-penta-1,3-diene)iridium(I), 46B, 846
$C_{18}H_{30}Ag_2Cl_2O_8$, 1,5-Hexadiene-silver perchlorate, 33B, 376
$C_{18}H_{30}Ni_6S_3$, Trithio-hexa(η^3-allyl-nickel), 45B, 923
$C_{18}H_{31}NiP$, (π-Pentenyl)(diisopropylphenylphosphine)methylnickel(II), 40B, 730
$C_{18}H_{34}Cl_6O_2Ta_2$, Di(μ-chloro)(μ-di-t-butylacetylene)bis[dichloro(tetrahydrofuran)tantalum](Ta-Ta), 46B, 847
$C_{19}H_{14}FeO_3$, 1,4-Diphenylbutadienetricarbonyliron, 38B, 769
$C_{19}H_{15}Fe_2NO_5$, μ-(6',1,2-η:1',5'-η-(6-Dimethylamino-1-(fulvenyl-6')-fulvene))-pentacarbonyldiiron(Fe-Fe), 39B, 591
$C_{19}H_{15}Fe_2NO_{10}S$, Tricarbonyl iron N-p-tolylsulphonyl-2,4-dimethoxy-4-imino-1-but-2-enyl iron tricarbonyl, 34B, 451
$C_{19}H_{15}MoN_3O_2S$, (Isothiocyanato)dicarbonyl-1,10-phenanthroline(2-methylallyl)molybdenum, 35B, 527

$C_{19}H_{16}F_6ORh_2$, Bis(η-cyclopentadienyl)-μ-((1,2,5-η:1,4,5-η)-1,2-dimethyl-3-oxo-4,5-bis(trifluoromethyl)-1,4-pentadiene-1,5-diyl)-dirhodium(Rh-Rh), 43B, 980
$C_{19}H_{18}OPdS$, Monothiodibenzoylmethanato-π-methylallylpalladium(II), 34B, 444
$C_{19}H_{26}Cl_2OP_2Ru$, Bis(dimethylphenylphosphine)(ethylene)(carbonyl)dichlororuthenium(II), 44B, 749
$C_{19}H_{28}F_6MoNO_2P$, (η^3-Allyl)(η^5-neomenthylcyclopentadienyl)carbonylnitrosylmolybdenum hexafluorophosphate, 46B, 847
$C_{19}H_{29}BrN_2O_2W$, (π-Allyl)bromodicarbonyl(N,N'-dicyclohexylethylenediimine)tungsten(II), 43B, 1044
$C_{19}H_{30}Co_2O_{11}P_2$, Bis($\eta^3$-2-allyl)-1,3-propanonobis((trimethylphosphite)dicarbonylcobalt), 44B, 744
$C_{19}H_{31}O_2Rh$, Acetylacetonatobis(tetramethylallene)rhodium(I), 37B, 446
$C_{20}H_{10}Fe_2O_6$, Di-iron hexacarbonyl diphenylacetylene complex, 32B, 348
$C_{20}H_{12}FeO_6$, Tetracarbonyl(trans-1,2-dibenzoylethylene)iron, 38B, 769
$C_{20}H_{14}FeO_5$, 1,2,3,4-η^4-(1-Phenyl-1-benzoxy-1,3-butadiene)tricarbonyliron, 44B, 750
$C_{20}H_{17}AsCl_2Pt$, Dichloro(1-2-η-o-diphenylarsinophenylethylene-As)-platinum(II), 46B, 848
$C_{20}H_{18}BF_4MoN_3O_2$, Dicarbonyl-2,2'-bipyridinepyridine-π-allylmolybdenum tetrafluoroborate, 38B, 770
$C_{20}H_{20}Mo_2O_4$, μ-Diethylacetylene-bis(η^5-cyclopentadienyl)tetracarbonyldimolybdenum, 44B, 746
$C_{20}H_{22}Fe_2O_8$, μ-(1,3,11,13-Tetradecatetraene-5,10-diol)-bis(tricarbonyliron), 44B, 750
$C_{20}H_{24}O_{10}Pd_2S_2$, Diacetato-bis($\mu_2$-bisulfito)-bis($\eta^2$-styrene)-dipalladium, 46B, 849
$C_{20}H_{27}ClF_6IrOP_3$, (η^3-Allyl)carbonylchlorobis(dimethylphenylphosphine)iridium(III) hexafluorophosphate, 43B, 1045
$C_{20}H_{30}CoN_2O_8$, Bis(ethylfumarate)bis(acetonitrile)cobalt(0), 43B, 1046
$C_{20}H_{31}ClMoN_2O_2$, Chlorodicarbonyl(N,N'-dicyclohexylethylenediimine)-(π-2-methylallyl)molybdenum(II), 42B, 667
$C_{20}H_{32}Cl_4Ru_2$, Dichloro(2,7-dimethyl-octa-2,6-diene-1,8-diyl)ruthenium(IV) dimer, 37B, 447
$C_{21}H_{20}Cl_3PPdSn \cdot 0.4\ C_3H_6O$, π-Allyl(triphenylphosphine)palladiumtrichlorotin acetonate, 40B, 1108
$C_{21}H_{23}ClNNiP$, Chloro(triphenylphosphine)(η^2-dimethylmethyleniminium)nickel(0), 42B, 668
$C_{21}H_{23}Cl_2N_3O_4Pt$, Bipyridine-chloro-(o-isopropenyl-N,N-dimethylaniline)platinum(II) perchlorate, 45B, 923
$C_{21}H_{24}Cl_3Mo_3N_3O_6 \cdot C_6H_6$, ($\eta^3$-Allyl)dicarbonyltris(methyl cyanide)molybdenum(II) bis(η^3-allyl)tetracarbonyl-μ-trichlorodimolybdate(II) benzene solvate, 45B, 924
$C_{21}H_{25}BMoN_4O_2$, (Diethyl-1-pyrazolylborato)-(η^3-2-phenylallyl)(dicarbonyl)molybdenum, 40B, 731
$C_{21}H_{31}F_6P_3Pt$, trans-(Methyl)(but-2-yne)-bis(dimethylphenylphosphine)platinum(II) hexafluorophosphate, 39B, 592
$C_{21}H_{47}IrP_2$, π-Allylbis(triisopropylphosphine)iridium(I), 40B, 732
$C_{22}H_{10}Fe_2O_8$, Octacarbonyldiphenylvinylidenedi-iron, 33B, 339
$C_{22}H_{15}CoNiO_3$, μ-(Diphenylacetylene)-(tricarbonylcobaltio)(η^5-cyclopentadienyl)nickel(Ni-Co), 46B, 850
$C_{22}H_{20}Mn_2O_4$, μ-(2-3-η:4-5-η)-Hexa-2,4-diyne-bis[dicarbonyl(η-methylcyclopentadienyl)manganese(I)], 45B, 924

$C_{22}H_{31}F_6O_2Rh$, (Di(pivaloyl)methanato)(η-1,2,4,5-(1,2-bis(trifluoromethyl)cyclopent-2-ene, 46B, 814
$C_{22}H_{32}Cl_2N_2Pt$, Dichloro[(R,R)-N,N'-dimethyl-N,N'-bis(α-methylbenzyl)-1,2-ethanediamine](ethene)-platinum(II), 46B, 850
$C_{22}H_{36}Ir_2N_4O_8P_2S_2$, Bis(μ-t-butylthiolato)(bis(trimethyl phosphite)-carbonyliridium)((η^2-tetracyanoethylene)carbonyliridium), 46B, 851
$C_{22}H_{46}ClP_2Rh$, Chloro(hex-3-ene-1,6-diylbis(di-t-butylphosphine))-rhodium(I), 44B, 1067
$C_{22}H_{49}IrP_2$, Butadiene(hydrido)bis(triisopropylphosphine)iridium(I), 41B, 882
$C_{23}H_{10}Fe_3O_9$, Diphenylacetylene-tri(iron tricarbonyl) complex, 31B, 347
$C_{23}H_{14}Fe_2O_5$, Diphenylfulvenepentacarbonyldiiron, 41B, 845
$C_{23}H_{15}MnO_4$, Dicarbonyl(η^5-methylcyclopentadienyl)(η^2-anthronylketone)manganese, 45B, 925
$C_{23}H_{19}FeO_2PS$, Dicarbonyltriphenylphosphine-(π-thioacrolein)iron(0), 39B, 592
$C_{23}H_{20}CoO_2P$, (η^3-Allyl)dicarbonyl(triphenylphosphine)cobalt, 45B, 925
$C_{23}H_{20}FeIO_2P$, π-Allyltriphenylphosphinedicarbonyliron iodide, 33B, 340
$C_{23}H_{20}MoN_2O_3$, (η-Allyl)dicarbonyl(N-phenylsalicylideneiminato)pyridinemolybdenum(II), 45B, 926
$C_{23}H_{26}OPdS$, Monothiodibenzoylmethanato-π-syn-1-t-butyl-2-methallyl-palladium(II), 38B, 745
$C_{23}H_{28}FeO_4$, (Vitamin-A aldehyde)tricarbonyliron, 35B, 529
$C_{23}H_{34}Cl_2N_2Pt$, Dichloro[(R,R)-N,N'-dimethyl-N,N'-bis(α-methylbenzyl)-1,2-ethanediamine](propene)-platinum(II), 46B, 850
$C_{24}H_{10}F_{10}V$, Bis(η^5-cyclopentadienyl)bis(pentafluorophenyl)acetylene-vanadium, 46B, 896
$C_{24}H_{10}O_{10}Os_3$, μ_3-(η-Diphenylacetylene)-decacarbonyltriosmium, 43B, 994
$C_{24}H_{20}AuFeO_3P$, (η^3-Allyl)tricarbonyl(triphenylphosphine-gold)iron (Au-Fe), 46B, 851
$C_{24}H_{20}Cl_2Rh_2$, Di-μ-chloro-tetrakis-(4-methylpenta-1,3-diene)dirhodium(I), 39B, 594
$C_{24}H_{20}Ni_2$, (Diphenylacetylene)bis(cyclopentadienylnickel), 33B, 341
$C_{24}H_{21}BrPRh$, Bromo(tris(2-vinylphenyl)phosphine)rhodium(I), 39B, 593
$C_{24}H_{21}Cl_4N_2Ta$, Pyridinium tetrachloro(pyridine)(tolane)tantalate, 46B, 852
$C_{24}H_{23}BMoN_4O_2$, (Diphenyldipyrazolylborato)(2-methylallyl)dicarbonyl-molybdenum, 41B, 883
$C_{24}H_{26}ClPPt$, Chloro(η^3-allyl)(tri-p-tolylphosphine)platinum(II), 45B, 926
$C_{24}H_{26}MoN_2O_3S_4 \cdot C_6H_6$, Bis(N,N-dimethyldithiocarbamato)(ditoluoyl-acetylene)oxomolybdenum(IV) benzene solvate, 46B, 999
$C_{24}H_{44}O_4Pt$, Bis-(3,6-diethyl-4-octyne-3,6-diol)platinum(0), 41B, 884
$C_{24}H_{48}O_4U_2$, Di-μ-isopropoxo-bis[di(η-allyl)isopropoxouranium(IV)], 45B, 927
$C_{24}H_{54}Li_2N_6Ni$, Dilithium tris(N,N,N',N'-tetramethyl-but-2-ene-1,4-diamine)nickel(II), 45B, 928
$C_{25}H_{17}MoO_4P$, Tetracarbonyl(diphenyl-2-(prop-cis-1-enyl)phenylphosphine)molybdenum(0), 34B, 451
$C_{25}H_{18}Fe_2O_8$, Tricarbonyl-π-(1,1,1-tricarbonyl-2-methyl-3-diphenyl-methylene-6-methoxyferra-2-oxacyclohexenyl)iron(Fe-Fe), 39B, 594
$C_{25}H_{20}F_3Ni_2OP$, Bis(η^5-cyclopentadienyl)-μ-(trifluoromethyl(oxodiphenylphosphino)acetylene)-dinickel(0), 43B, 1046

$C_{25}H_{20}MnO_4P$, 2-(1',2'-Dimethyloxopropenyl)phenyldiphenylphosphinomanganese tricarbonyl, 39B, 595
$C_{25}H_{20}OTi$, Carbonyl-bis(η^5-cyclopentadienyl)(η^2-diphenylacetylene)-titanium, 44B, 750
$C_{25}H_{20}O_4PV$, η-Allyltetracarbonyltriphenylphosphinevanadium, 42B, 669
$C_{25}H_{21}AsF_6O_3Pt$, [1-(o-Diphenylarsinophenyl)-2-methoxyethyl-As,C[1]]-(1,1,1,5,5,5-hexafluoropentane-2,4-dionato)platinum(II), 46B, 848
$C_{25}H_{21}CrO_3P$, Trimethylenemethanetricarbonyltriphenylphosphine-chromium, 40B, 732
$C_{25}H_{21}FeO_5P$, (η-Methyl acrylate)(triphenylphosphine)tricarbonyliron, 42B, 669
$C_{25}H_{24}NTi$, Bis(η^5-cyclopentadienyl)-[2,6-dimethylphenylimino(phenyl)methyl]titanium, 45B, 874
$C_{25}H_{28}ClP_2Rh$, Chloro((3-(diphenylphosphino)propyl)(3-butenyl)phenylphosphine)rhodium(I), 44B, 751
$C_{25}H_{40}NiOP_2$, (Benzophenone)bis(triethylphosphine)nickel, 45B, 929
$C_{26}H_{26}CoO_2P$, Dicarbonyl[1-η^3-(2,3-dimethylbut-1-enyl)]triphenylphosphinecobalt, 46B, 852
$C_{26}H_{27}FeMnO_4P_2S_2$, (μ-(C,S-$\eta^2$)-Carbon disulfide-S')dicarbonyl(η^5-cyclopentadienyl)(dicarbonylbis-(dimethylphenylphosphine)iron)manganese, 46B, 853
$C_{26}H_{40}Ni_2O_4$, Di-μ-acetato-bis((2-methylallyl-3-norbornyl)nickel-(II)), 39B, 597
$C_{26}H_{40}O_4Pd_2$, Di-μ-acetato-bis((2-methylallyl-3-norbornyl)palladium-(II)), 39B, 596
$C_{26}H_{42}Cl_2O_4Pd_2$, Di-μ-chloro-di-(1,3-η-allyl-di-t-butylketone)dipalladium(II), 42B, 670
$C_{26}H_{54}Br_2Ni_2P_2$, μ-(α,ω-Octadi-π-enyl)bisbromotriisopropylphosphinenickel(II), 38B, 771
$C_{27}H_{19}Fe_2O_5P$, μ-Butatriene-(triphenylphosphine)pentacarbonyldiiron, 42B, 670
$C_{27}H_{21}FeO_3$, 1,4-Diphenylbutadienetricarbonyliron 1,4-diphenylbutadiene, 39B, 597
$C_{27}H_{22}FeN_2O_5$, (Diethyl muconate)(2,2'-bipyridyl)carbonyliron, 43B, 1047
$C_{27}H_{29}ORh$, (Dibenzylideneacetone)(pentamethylcyclopentadienyl)rhodium(I), 40B, 770
$C_{27}H_{53}NiP$, (π-Pentenyl)(dimenthylmethylphosphine)methylnickel(II), 40B, 733
$C_{28}H_{20}Mo_2O_4$, μ-Diphenylacetylene-bis(η^5-cyclopentadienyl)tetracarbonyldimolybdenum, 44B, 746
$C_{28}H_{20}Pt$, Bis(diphenylacetylene)platinum, 46B, 853
$C_{28}H_{28}Cl_3O_2Pt$, Tetraphenylphosphonium trichloro-(cis-but-2-en-1,4-diol)platinate(II), 37B, 448
$C_{28}H_{40}F_6P_4Ru \cdot 0.5\ CH_2Cl_2$, (η-Buta-1,3-diene)tris(dimethylphenylphosphine)hydridoruthenium(II) hexafluorophosphate dichloromethane solvate, 44B, 752
$C_{28}H_{52}Ni_2P_2$, Bis-μ-dicyclohexylphosphido-bis(π-ethylenenickel), 39B, 597
$C_{29}H_{23}FeO_3P$, Dicarbonyl(η^4-cinnamaldehyde)(triphenylphosphine)iron-(0), 46B, 854
$C_{29}H_{29}Cl_3NPPt$, cis-Dichloro-η^2-[(Z)-2-chloro-N-(3-methylbut-1-enyl)benzenamine](triphenylphosphine) platinum(II), 46B, 855
$C_{30}H_{15}F_{10}ORh_3$, Carbonyl(bis(pentafluorophenyl)acetylene)tris(η^5-cyclopentadienylrhodium), 43B, 1006
$C_{30}H_{18}Fe_2O_6$, Irontricarbonyl - phenylacetylene, 27, 877
$C_{30}H_{20}F_{11}FeP$, 2,4-Bis(trifluoromethyl)-perfluoropenta-2,4-dienyl-

cyclopentadienyltriphenylphosphineiron, 43B, 1048
$C_{30}H_{25}ORh_3 \cdot 0.5\ C_6H_6$, Carbonyl(diphenylacetylene)tris(η^5-cyclopentadienylrhodium) hemibenzene, 43B, 1006
$C_{30}H_{27}MoO_2P$, Carbonyl-cyclopentadienyl-(1,5a-η^3-1-methylene-cyclopentan-2-onato)triphenylphosphinemolybdenum(II), 45B, 929
$C_{30}H_{27}N_2OPd$, 2,2'-Bipyridyldibenzylideneacetonepalladium(0) hemi-(benzene), 43B, 1049
$C_{30}H_{30}ClIrP_2$, Chloro-(1,6-bis(diphenylphosphino)-trans-hex-3-ene)iridium(I), 45B, 930
$C_{30}H_{30}Cl_3IrP_2$, Trichloro-1,6-bis(diphenylphosphino)-trans-hex-3-ene-iridium(III), 45B, 931
$C_{30}H_{31}BrNiP_2$, π-Methallyl(bis-1,2-(diphenylphosphino)ethane)-nickel bromide, 35B, 532
$C_{30}H_{32}ClIrP_2 \cdot 0.5\ C_6H_6$, Dihydrido-chloro(1,6-bis(diphenylphosphino)-trans-hex-3-ene)iridium(III) benzene solvate, 45B, 930
$C_{30}H_{34}Ni_2$, (μ_2(η^2)-Diphenylacetylene)bis(1,5-cyclooctadienenickel)-(Ni-Ni), 42B, 671
$C_{30}H_{34}PRh$, (η^5-Pentamethylcyclopentadienyl)(triphenylphosphine)ethylenerhodium(I), 44B, 752
$C_{31}H_{25}MoO_4P$, (Allylidenetriphenylphosphorane)tetracarbonylmolybdenum - benzene, 39B, 598
$C_{31}H_{29}ClMoO_2P_2$, η-Allyl-(dicarbonyl)-chloro-(1,2-bis(diphenylphosphino)ethane)molybdenum, 43B, 1049; 45B, 931
$C_{31}H_{40}Cl_2NO_2PW$, Tetraethylammonium η-allyldicarbonyldichlorotriphenylphosphinetungsten, 46B, 855
$C_{32}H_{20}FeO_4$, (Tetraphenylbutatriene)tetracarbonyliron, 37B, 449
$C_{32}H_{29}O_3P_2V$, η^3-Allyltricarbonylbis(diphenylphosphino)ethane-vanadium(0), 43B, 1009
$C_{32}H_{32}NbO_2 \cdot 0.5\ C_6H_6$, Dicyclopentadienyl-(diphenylacetylene)-pivaloyl-niobium benzene solvate, 45B, 918
$C_{32}H_{34}Cl_3O_2PPt$, Tetraphenylphosphonium trichloro-(2,5-dimethylhex-3-yne-2,5-diol)platinate(II), 39B, 599
$C_{32}H_{60}NiP_2$, Tetramethylethylene-1,2-bis(dicyclohexylphosphino)ethanenickel, 40B, 735
$C_{34}H_{16}Fe_2O_6$, (Bis(biphenylylidene)butatriene)hexacarbonyldi-iron, 38B, 773
$C_{34}H_{25}FeNiO_3P$, Diphenylacetylene-tricarbonyliron-diphenylphosphido-π-cyclopentadienylnickel complex, 39B, 600
$C_{34}H_{25}NbO$, π-Cyclopentadienyldi(tolane)carbonylniobium, 34B, 447
$C_{34}H_{38}N_6Pd_2$, Di-μ-(1,3-di-p-tolyltriazenido)di-(1,3-η-allyl)dipalladium(II), 42B, 672
$C_{34}H_{45}BCuN_3$, (Diethylenetriamine)(1-hexene)copper(I) tetraphenylborate, 45B, 892
$C_{35}H_{33}NNbP$, (π-η^5-Cyclopentadienyl)(π-N,2,3-η^3-π-4,5,6-η^3-3,4,5,6-tetraphenyl-3,5-hexadiene-2-oneimine)(phosphine)-niobium, 39B, 600
$C_{36}H_{20}Fe_3O_8$, Iron carbonyl-diphenylacetylene complex, 30B, 297
$C_{36}H_{20}O_8Os_3$, Octacarbonyl-μ-(1,2,3,4-tetraphenylbut-2-ene-1,1,4,4-tetrayl)-triangulo-triosmium, 38B, 774
$C_{37}H_{20}O_9Os_3$, Nonacarbonyl-μ-(1,2,3,4-tetraphenyl-butadiene-1,4-diyl)-triangulo-triosmium (monoclinic form), 40B, 736
$C_{37}H_{20}O_9Os_3$, Nonacarbonyl-μ-(1,2,3,4-tetraphenyl-butadiene-1,4-diyl)-triangulo-triosmium (orthorhombic form), 40B, 736
$C_{38}H_{30}As_2F_4Pt$, Tetrafluoroethylenebis(triphenylarsine)platinum(0), 41B, 884
$C_{38}H_{30}ClF_3P_2Pt$, Bis(triphenylphosphine)(chlorotrifluoroethylene)-platinum(0), 37B, 450
$C_{38}H_{30}ClP_2Rh \cdot CH_2Cl_2$, Chloro-(2,2'-bis(o-diphenylphosphino)-trans-

stilbene)rhodium(I) dichloromethane, 46B, 856
$C_{38}H_{30}Cl_2F_2P_2Pt$, Bis(triphenylphosphine)(dichlorodifluoroethylene)-platinum(0), 37B, 450
$C_{38}H_{30}Cl_3IrP_2$, Trichloro-(2,2'-bis(o-diphenylphosphino)-trans-stilbene)iridium(III), 46B, 856
$C_{38}H_{30}Cl_3P_2Rh$, Trichloro-(2,2'-bis(o-diphenylphosphino)-trans-stilbene)rhodium(III), 46B, 856
$C_{38}H_{30}Cl_4P_2Pt$, Bis(triphenylphosphine)(tetrachloroethylene)platinum-(0), 37B, 450
$C_{38}H_{34}ClIrP_2$, trans-Chloro(ethylene)bis(triphenylphosphine)iridium-(I), 41B, 885
$C_{38}H_{34}NiP_2$, Bistriphenylphosphine-ethylenenickel(0), 33B, 343; 37B, 449
$C_{38}H_{34}P_2Pt$, Bis(triphenylphosphine)-(ethylene)platinum, 38B, 754
$C_{38}H_{37}BF_4NO_3P_2Rh$, 1,2-Bis(diphenylphosphino)ethane[methyl (Z)-α-acetamidocinnamate]rhodium tetrafluoroborate, 45B, 932
$C_{39}H_{32}O_3OsP_2 \cdot H_2O$, Dicarbonyl-(formaldehyde)-bis(triphenylphosphine)osmium monohydrate, 45B, 933
$C_{39}H_{34}IP_2Rh$, Iodobis(triphenylphosphine)allene-rhodium, 40B, 737
$C_{39}H_{34}P_2Pd$, Bis(triphenylphosphine)allenepalladium, 40B, 737
$C_{39}H_{35}NOP_2Ru$, Allylnitrosylbis(triphenylphosphine)ruthenium, 43B, 1050; 44B, 753
$C_{39}H_{72}F_6P_3Pt \cdot C_6H_4Cl_2$, trans-Hydrido(1,1-dimethylallene)bis(tricyclohexylphosphine)platinum(II) hexafluorophosphate o-dichlorobenzene solvate, 45B, 933
$C_{40}H_{30}Cl_2N_2P_2Pt$, Bis(triphenylphosphine)(1,1-dichloro-2,2-dicyano-ethylene)platinum(0), 37B, 451
$C_{40}H_{30}F_6P_2Pt$, Bis(triphenylphosphine)hexafluorobut-2-yneplatinum(0), 40B, 738
$C_{40}H_{30}F_8P_2Pt$, Octafluoro-trans-but-2-enebis(triphenylphosphine)platinum, 41B, 870
$C_{40}H_{30}Nb_2O_2$, Bis(π-cyclopentadienyltolanecarbonylniobium), 34B, 448
$C_{40}H_{37}As_2Cl_2Rh$, Dichloro-π-methylallyl-bis-(triphenylarsine)rhodium-(III), 35B, 534
$C_{40}H_{66}F_6P_2Pt$, Bis(tricyclohexylphosphine)(hexafluorobut-2-yne)platinum(0), 43B, 1050
$C_{41}H_{38}P_2Pt$, Bis(triphenylphosphine)-1,1-dimethylalleneplatinum, 41B, 886
$C_{42}H_{32}O_3P_2W \cdot 0.5\ CH_2Cl_2$, ((E)-1,3-Bis(2-(diphenylphosphino)phenyl)-propene)tricarbonyltungsten(0) hemi(methylene chloride) solvate, 46B, 1286
$C_{42}H_{34}FeO_2P_2$, Dicarbonylbis((2-vinylphenyl)diphenylphosphine)iron-(0), 39B, 601
$C_{42}H_{36}O_4P_2Pd$, (Dimethyl acetylenedicarboxylate)bis(triphenylphosphine)palladium, 40B, 739
$C_{42}H_{39}CoP_3S_2$, [1,1,1-Tris(diphenylphosphinomethyl)ethane](η^2-carbon disulfide)cobalt, 46B, 857
$C_{42}H_{40}NiO_2P_2$, (Ethyl methacrylate)bis(triphenylphosphine)nickel(0), 46B, 858
$C_{42}H_{40}P_2Pd_2$, Bis(η^3-allyltriphenylphosphinopalladium), 46B, 858
$C_{43}H_{30}As_2ClIrN_4O$, Carbonylchloro(tetracyanoethylene)bis(triphenylarsine)iridium, 38B, 757
$C_{43}H_{30}CrO$, Tris(η^2-diphenylacetylene)carbonylchromium, 46B, 859
$C_{43}H_{32}Co_2O_4P_2$, μ-Bis(diphenylphosphino)methane-μ-diphenylethyne-tetracarbonyldicobalt, 43B, 1051
$C_{44}H_{42}OP_2Pt$, (1-Ethynylcyclohexanol)bis(triphenylphosphine)platinum-(0), 42B, 672

$C_{44}H_{46}ClP_2Rh$, Chlorobis(3-butenyldiphenylphosphino)rhodium(I) bis-(benzene) solvate, 41B, 887
$C_{44}H_{46}NiO_6P_2$, (Ethylene)bis(tri-o-tolyl phosphite)nickel(0), 39B, 602
$C_{45}H_{38}P_2Pt$, Bis(triphenylphosphine)(1-phenylpropyne)platinum(0), 41B, 888
$C_{45}H_{40}ClIrP_2$, (η^3-1-Phenyl-allyl)-chloro-hydrido-bis(triphenylphosphine)iridium, 44B, 753; 45B, 934
$C_{45}H_{41}IP_2Pd_2$, μ-Allyl-μ-iodo-bis(triphenylphosphinepalladium) benzene solvate, 38B, 762
$C_{45}H_{45}NNiO_6P_2$, (Acrylonitrile)bis(tri-o-tolyl phosphite)nickel(0), 39B, 602
$C_{45}H_{51}NOP_2PtS$, Bis(triphenylphosphine)-2,4,6-trimesityl-sulfinylanilineplatinum(0), 44B, 1143
$C_{46}H_{30}Fe_2O_6P_2$, Bis(diphenylphosphino(phenyl)acetylene)hexacarbonyldiiron(0), 43B, 1052
$C_{46}H_{79}F_6P_3Pt$, (η^3-Allyl)bis(tricyclohexylphosphine)platinum hexafluorophosphate - toluene, 44B, 754
$C_{47}H_{39}CoCrO_5P_3S_2 \cdot 0.25\ CH_2Cl_2$, [1,1,1-Tris(diphenylphosphinomethyl)ethane]cobalt(μ-carbon disulfide)pentacarbonylchromium methylene chloride, 46B, 857
$C_{47}H_{40}NiO_3P_2 \cdot C_6H_6$, ($\eta^2$-Methyl 3-benzoylacrylate)bis(triphenylphosphine)nickel(0) benzene solvate, 46B, 860
$C_{48}H_{35}NbO$, π-Cyclopentadienyltolanecarbonyl-π-tetraphenylcyclobutadieneniobium, 34B, 450
$C_{49}H_{30}O_7Os_3$, Heptacarbonyl-μ_3-diphenylacetylene-μ-(1,2,3,4-tetraphenylbutadiene-1,4-diyl)-triangulo-triosmium, 39B, 603
$C_{49}H_{31}IrN_8P_2 \cdot 0.5\ C_6H_6$, (Cyano(dicyanomethyl)keteniminato)carbonyl-(tetracyanoethylene)bis(triphenylphosphine)iridium, 37B, 451
$C_{49}H_{48}P_2Ru$, Bis(triphenylphosphine)bis(π-allyl)ruthenium - toluene solvate, 38B, 774
$C_{50}H_{40}N_2O_4P_2Pt$, 4,4'-Dinitro-trans-stilbenebis(triphenylphosphine)-platinum, 40B, 739
$C_{52}H_{44}Cl_2O_3Pd_2$, Tris(dibenzylideneacetone)dipalladium(0) methylene chloride solvate, 40B, 740
$C_{52}H_{80}P_4Pt_3$, cd-Bis(μ-diphenylacetylene)-abef-tetrakis(triethylphosphine)triplatinum, 46B, 1299
$C_{54}H_{50}F_{12}OP_6Rh_2$, μ-Diphenylacetylene-bis(triphenylphosphine)tetrakis(trifluorophosphine)dirhodium diethylether solvate, 42B, 673
$C_{56}H_{54}NiP_2 \cdot 0.5\ C_4H_8O$, Bis(tri-p-tolylphosphine)(trans-stilbene)-nickel(0) hemitetrahydrofuranate, 40B, 741
$C_{56}H_{56}Cl_6P_4Pt_3$, Bis(1,2-bis(diphenylphosphino)ethane)platinum(II) trichloro(ethylene)platinate(II), 43B, 1053
$C_{56}H_{56}O_4Ti_2$, Di-μ-diphenylketene-bis(dicyclopentadienyltitanium) - tetrahydrofuran, 44B, 754
$C_{57}H_{48}O_3Pd$, Tris(dibenzylideneacetone)palladium(0) benzene solvate, 39B, 604
$C_{58}H_{45}Ge_2NbO$, π-Cyclopentadienylbis(π-triphenylgermanylphenylacetylene)carbonylniobium, 40B, 741
$C_{58}H_{114}Mo_2N_2O_4P_4$, Di-$\mu$-acrylonitrile-bis(dicarbonyl-di(tributylphosphine)-molybdenum), 45B, 934
$C_{59}H_{49}Co_2O_4P_3$, Tricarbonylcobalt-μ-(diphenylacetylene)-(1,1,1-tris(diphenylphosphinomethyl)ethane)carbonylcobalt, 45B, 935
$C_{60}H_{62}O_9Pd_3$, Palladiumbis(μ-1,3-bis(p-methoxyphenyl)phenylpropenium-acetylacetonatopalladium) diethyl ether solvate, 44B, 755
$C_{62}H_{64}Br_2Cl_6Ni_2P_4$, μ-(α,ω-Octadi-π-enyl)bisbromodiphenylphosphinoethanenickel(II) chloroform solvate, 38B, 771

$C_{66}H_{70}ClO_2P_2Rh \cdot 1.5\ C_3H_6O$, Chloro-(bis(3,5-di-t-butyl-4-oxo-cyclohexadiene-1-ylidene)ethylene)-bis(triphenylphosphine)rhodium acetone solvate, 45B, 936
$C_{68}H_{58}As_4Cl_2Co_2O_2$, Di-$\mu$-bis(diphenylarsino)methane-μ-diphenylethyne-dicarbonyldicobalt 1,2-dichloroethane, 43B, 1051
$C_{73}H_{70}Li_3N_5Ni_2O_2$, Lithium ($\mu$-benzophenonimino)-tetrakis(benzophenonimino)dinickel bis(diethylether), 46B, 1201
$C_{74}H_{50}Mo_2O_4 \cdot 0.8\ CHCl_3$, μ-Diphenylacetylene-(1-(carbonyl(η^4-tetraphenylcyclobutadiene)molybdenum(0))-2-(dicarbonyl-μ-(η^4-tetraphenylcyclopentadienone)-molybdenum(0))) chloroform, 43B, 1053
$C_{78}H_{66}N_6Ni_3$, Bis(benzophenone-imine)nickel(0) trimer, 45B, 937
$C_{128}H_{122}N_4Ni_4P_4$, Tetrakis((benzonitrile)(triphenylphosphorane)nickel(0)) toluene n-hexane cycloocta-1,5-diene solvate, 44B, 1144
$C_{131}H_{98}B_2Co_2P_6S_2 \cdot 2\ C_3H_6O$, ($\mu$-Carbon disulfide)bis[(1,1,1-tris(diphenylphosphinomethyl)ethane)cobalt] bis(tetraphenylborate) acetone, 46B, 857

73. METAL π-COMPLEXES (CYCLOPENTADIENE)

$C_5H_5Br_3Ti$, Cyclopentadienyltitanium tribromide (gas-ed), 40A, 1141
$C_5H_5ClCrN_2O_2$, Chloro(η^5-cyclopentadienyl)dinitrosylchromium, 31B, 349; 46B, 860
$C_5H_5ClN_2O_2W$, Chloro(η^5-cyclopentadienyl)dinitrosyltungsten, 46B, 860
$C_5H_5Cl_3Ti$, Cyclopentadienyltitanium(IV) trichloride, 28, 609
$C_5H_{14}B_9CoSe_2$, 7-Cyclopentadienyl-nido-7-cobalta-8,12-diselenaborane(12), 46B, 861
$C_5H_{18}B_9Co$, 5-(η^5-Cyclopentadienyl)-5-cobalta-nido-decaborane(14), 43B, 1053
$C_6H_5CrN_3O_3$, Isocyanatocyclopentadienyldinitrosylchromium, 35B, 537
$C_6H_6Cl_6FeOSi_2$, Hydridobis(trichlorosilyl)carbonyl-π-cyclopentadienyliron, 35B, 535
$C_7H_4NO_4Rh$, (η^5-Nitrocyclopentadienyl)dicarbonylrhodium, 46B, 861
$C_7H_5Br_3FeO_2Sn$, Dicarbonylcyclopentadienyl(tribromostannyl)iron, 35B, 539
$C_7H_5Cl_2CoHgO_2$, Dicarbonyl(π-cyclopentadienyl)cobalt - mercury(II) chloride, 38B, 775
$C_7H_5Cl_3FeO_2Sn$, Dicarbonylcyclopentadienyl(trichlorostannyl)iron, 35B, 538
$C_7H_5Cl_6CoHg_3O_2$, Dicarbonyl(π-cyclopentadienyl)cobalt - tris(mercury(II) chloride), 38B, 776
$C_7H_5CrNO_2S$, Dicarbonyl(η^5-cyclopentadienyl)(thionitrosyl)chromium, 45B, 937
$C_7H_5CrNO_3$, Dicarbonylcyclopentadienylnitrosylchromium, 45B, 938
$C_7H_5MnO_4S$, Cyclopentadienyldicarbonyl(sulfur dioxide)manganese, 39B, 605
$C_7H_7IMnNOS$, Iodo(η^5-methylcyclopentadienyl)nitrosyl(thiocarbonyl)-manganese(II), 46B, 862
$C_7H_{11}ClF_8MoN_2P_4$, Chloro-η^5-cyclopentadienylbis(methylamino-bis(difluorophosphine))molybdenum, 43B, 1054
$C_8H_4FeO_4$, Tricarbonyl(η^4-cyclopentadienone)iron, 43B, 1055
$C_8H_5ClHgMoO_3$, Tricarbonyl-cyclopentadienyl(chloro-mercury)molybdenum, 44B, 1144
$C_8H_5ClMoO_3$, π-Cyclopentadienyl molybdenum tricarbonyl chloride, 33B, 343
$C_8H_5F_5IORh$, Iodocarbonyl-π-cyclopentadienyl-pentafluoroethylrhodium, 30B, 313

$C_8H_5F_6FeO_3P$, π-Cyclopentadienyliron(II) tricarbonyl hexafluorophosphate, 39B, 605
$C_8H_5FeNO_2S$, Dicarbonyl-cyclopentadienyl-isothiocyanato-iron, 46B, 863
$C_8H_5MnO_3$, Cyclopentadienyltricarbonylmanganese, 28, 608
$C_8H_8N_2O_2W$, (η^5-Cyclopentadienyl)dicarbonyl(methyldiazo)tungsten, 45B, 939
$C_8H_{12}F_4FeOSi_2$, trans-Hydridobis(difluoromethylsilyl)-(η-cyclopentadienyl)monocarbonyliron, 43B, 1055
$C_8H_{14}CoPS_5$, η^5-Cyclopentadienyl-trimethylphosphine-cyclopentasulfur-cobalt, 46B, 863
$C_9H_5CoF_6S_2$, π-Cyclopentadienyl-cis-1,2-bis(trifluoromethyl)ethanedithione-cobalt, 31B, 348
$C_9H_5CoN_2S_2$, π-Cyclopentadienyl(1,2-dicyanoethene-1,2-dithiolato)cobalt, 33B, 344
$C_9H_5F_6FeO_2P$, Dicarbonyl(π-cyclopentadienyl)(bis(trifluoromethyl)-phosphino)iron, 41B, 888
$C_9H_5F_6FeO_3P$, Dicarbonyl(π-cyclopentadienyl)(bis(trifluoromethyl)oxo-phosphino)iron, 41B, 888
$C_9H_5F_6O_4V$, π-Cyclopentadienylbis(trifluoroacetato)vanadium, 35B, 540
$C_9H_8BF_4FeNO_2$, Acetonitrile-dicarbonyl-cyclopentadienyliron tetrafluoroborate, 45B, 939
C_9H_9Co, (π-Cyclopentadienyl)(π-cyclobutadiene)cobalt, 42B, 673
$C_9H_{10}FeO_2S$, Dicarbonyl(η-cyclopentadienyl)(ethylthio)iron, 44B, 756
$C_9H_{14}CoPS_2$, Carbondisulfide(η^5-cyclopentadienyl)trimethylphosphine-cobalt, 44B, 1145
$C_9H_{15}B_9F_3FeO_2$, 8-Trifluoroacetoxy-π-cyclopentadienyl-(3)-1,2-dicarbollyliron(III), 45B, 940
$C_{10}H_7MnO_4$, Acetylcyclopentadienyl-manganese tricarbonyl, 41B, 889
$C_{10}H_7O_4Re$, Acetylcyclopentadienyl-rhenium tricarbonyl, 41B, 889
$C_{10}H_8Cl_2FeO_4S_2$, Ferrocenyldisulphonyl chloride, 24, 599; 29, 552
$C_{10}H_8FeS_3$, 1,2,3-Trithia-[3]-ferrocenophane, 38B, 777
$C_{10}H_9FeNO_2S$, [2]Ferrocenophanethiazine 1,1-dioxide, 41B, 890
$C_{10}H_{10}BBr_2F_4Re$, Bis-π-cyclopentadienyldibromorhenium(V) tetrafluoroborate, 40B, 742
$C_{10}H_{10}BCl_2F_4Mo$, Bis-π-cyclopentadienyldichloromolybdenum(V) tetrafluoroborate, 40B, 742
$C_{10}H_{10}BiCl_4Fe$, Ferricenium tetrachlorobismuthate, 43B, 1056
$C_{10}H_{10}Br_4MoSn$, Bromobis(π-cyclopentadienyl)(tribromostannyl)molybdenum(IV), 38B, 778
$C_{10}H_{10}ClNbO_2$, Bis(η^5-cyclopentadienyl)peroxochloroniobium(V), 46B, 864
$C_{10}H_{10}ClV$, Bis(cyclopentadienyl)vanadium monochloride, 43B, 1057
$C_{10}H_{10}Cl_2Mo$, Bis-π-cyclopentadienyldichloromolybdenum(IV), 40B, 742
$C_{10}H_{10}Cl_2Nb$, Bis-π-cyclopentadienyldichloroniobium(IV), 40B, 742
$C_{10}H_{10}Cl_2Ti$, Bis(cyclopentadienyl)titanium dichloride (gas-ed), 43B, 1485
$C_{10}H_{10}Cl_2Ti$, Bis(π-cyclopentadienyl)titanium dichloride, 38B, 779; 41B, 891
$C_{10}H_{10}Cl_2Zr$, Bis(cyclopentadienyl)zirconium dichloride (gas-ed), 35B, 901; 43B, 1485
$C_{10}H_{10}Cl_2Zr$, Bis-π-cyclopentadienyldichlorozirconium(IV), 40B, 742
$C_{10}H_{10}Cl_4Fe_2$, Ferricinium tetrachloroferrate(III), 44B, 1146
$C_{10}H_{10}Cl_4OTi_2$, μ-Oxo-bis(cyclopentadienyltitanium(IV) dichloride), 28, 610; 43B, 1057
$C_{10}H_{10}Co$, Dicyclopentadienylcobalt (gas-ed), 42B, 1030
$C_{10}H_{10}Co$, Dicyclopentadienylcobalt, 41B, 891

$C_{10}H_{10}Co_2N_2O_2$, Di-μ-nitrosyl-bis(cyclopentadienylcobalt), 43B, 1062
$C_{10}H_{10}CrO_3S$, Tricarbonyl(dimethysulphonium cyclopentadienylide)-chromium, 43B, 1058
$C_{10}H_{10}F_2Zr$, Difluorobis-(π-cyclopentadienyl)zirconium(IV), 37B, 452
$C_{10}H_{10}Fe$, Ferrocene (monoclinic), 20, 550; 45B, 941
$C_{10}H_{10}Fe$, Ferrocene (triclinic), 45B, 942
$C_{10}H_{10}Fe$, Ferrocene (gas-ed), 19, 591; 26, 650; 33B, 536
$C_{10}H_{10}FeI_3$, Ferricinium triiodide, 33B, 345
$C_{10}H_{10}I_2Mo_2O_3$, μ-Oxo-bis[(η^5-cyclopentadienyl)iodooxomolybdenum(V)], 46B, 1145
$C_{10}H_{10}I_2Zr$, Diiodobis-(π-cyclopentadienyl)zirconium(IV), 37B, 452
$C_{10}H_{10}I_4Ru$, Iodobis(cyclopentadienyl)ruthenium triiodide, 40B, 746
$C_{10}H_{10}Mn$, Dicyclopentadienylmanganese, 44B, 756
$C_{10}H_{10}MoO_3$, Tricarbonyl-π-cyclopentadienylethylmolybdenum, 28, 612
$C_{10}H_{10}MoS_4$, Bis-(π-cyclopentadienyl)molybdenum tetrasulfide, 41B, 892
$C_{10}H_{10}Mo_2O_2S_2$, Bis(cyclopentadienyl oxomolybdenum sulfide), 32B, 350
$C_{10}H_{10}Mo_2O_4$, Di-μ-oxo-bis(oxo(η^5-cyclopentadienyl)molybdenum(V)), 44B, 757
$C_{10}H_{10}N_6Ti$, Diazidobis(η-cyclopentadienyl)titanium, 43B, 1059
$C_{10}H_{10}Ni$, Nickelocene, 46B, 864
$C_{10}H_{10}Ni$, Nickelocene (gas-ed), 19, 591; 35B, 902; 40B, 1142
$C_{10}H_{10}Os$, Osmocene, 23, 692
$C_{10}H_{10}Ru$, Ruthenocene, 23, 692; 46B, 865
$C_{10}H_{10}Ru$, Ruthenocene (gas-ed), 33B, 536
$C_{10}H_{10}S_4W$, Bis(cyclopentadienyl)tungsten tetrasulfide, 38B, 1011
$C_{10}H_{10}S_5Ti$, Di-π-cyclopentadienyltitanium pentasulfide, 37B, 453; 42B, 674
$C_{10}H_{10}S_5V \cdot 0.5\ H_2O$, Dicyclopentadienylvanadium pentasulfide hemihydrate, 42B, 674
$C_{10}H_{10}V$, Bis(η^5-cyclopentadienyl)vanadium, 45B, 943
$C_{10}H_{11}Cl_5FeO_2Si_3$, Trisilapentachlorocyclohexyldicarbonylcyclopentadienyliron, 46B, 865
$C_{10}H_{11}MoNO_3$, Dicarbonyl(η-cyclopentadienyl)(2-propanoneoximato-O,N)-molybdenum, 43B, 1059
$C_{10}H_{12}Cr_2N_4O_3$, trans-μ-Amido-μ-nitrosyl-bis(π-cyclopentadienyl-nitrosylchromium), 35B, 543
$C_{10}H_{12}Mo$, Dihydridodi(π-cyclopentadienyl)molybdenum, 30B, 311; 31B, 352; 43B, 1060
$C_{10}H_{12}MoP_2$, Bis(η^5-cyclopentadienyl)(diphosphino)molybdenum, 43B, 1061
$C_{10}H_{13}Nb$, Trihydridobis(cyclopentadienyl)niobium, 43B, 1061
$C_{10}H_{13}Ta$, Trihydridobis(cyclopentadienyl)tantalum, 43B, 1061
$C_{10}H_{14}BNb$, Bis-(π-cyclopentadienyl)niobium tetrahydroborate, 40B, 747
$C_{10}H_{14}BTi$, Tetrahydroboratobis(cyclopentadienyl)titanium(III), 39B, 606
$C_{10}H_{14}Cl_2O_{10}Ti \cdot 3\ C_4H_8O$, Diaquabis($\eta^5$-cyclopentadienyl)titanium(IV) perchlorate tetrahydrofuran, 46B, 866
$C_{10}H_{14}PRhS_4$, (2,4-Dithio-1,3-dithia-1,4-diyl)-η^5-cyclopentadienyl-(trimethylphosphine)rhodium, 46B, 866
$C_{10}H_{15}N_2O_6Rh$, (Nitrato-O)(nitrato-O,O')(η-pentamethylcyclopentadienyl)rhodium(III), 46B, 905
$C_{10}H_{16}B_4Co_2$, 1,2-Bis(η^5-cyclopentadienyl)-1,2-dicobaltahexaborane, 45B, 944
$C_{10}H_{18}B_2Cl_2Ti_2$, Di-μ-chloro-bis(cyclopentadienyl(tetrahydroborato)-titanium), 42B, 675

$C_{10}H_{28}B_{10}CoN$, Tetramethylammonium 2-(η-cyclopentadienyl)-1-carba-2-cobalta-closo-undecahydrododecaborate(1-), 45B, 944
$C_{11}H_5CoFeO_6$, Di-μ-carbonyl-(tricarbonylcobaltio)-carbonyl(π-cyclopentadienyl)iron, 41B, 893
$C_{11}H_5F_7MoO_3$, Tricarbonyl-π-cyclopentadienylheptafluoropropylmolybdenum, 32B, 351
$C_{11}H_5MnNiO_6$, (Pentacarbonylmanganese)carbonylcyclopentadienylnickel, 46B, 870
$C_{11}H_9O_3Re$, Tricarbonyltrimethylene-1,2-cyclopentadienylrhenium, 33B, 347
$C_{11}H_{10}Cl_2Ti$, (1,1'-Methylenedicyclopentadienyl)dichlorotitanium, 45B, 949
$C_{11}H_{10}CoF_6O_2P$, Carboxycobaltocenium hexafluorophosphate, 44B, 1146
$C_{11}H_{10}Co_2NO_2$, μ-Carbonyl-μ-nitrosyl-bis(cyclopentadienylcobalt), 43B, 1062
$C_{11}H_{10}FeO_4S$, 3-Methyl-4-(π-cyclopentadienyldicarbonyliron)-3,4-dehydro-1,2-oxothiolane 2-oxide, 37B, 454
$C_{11}H_{11}Co_2O_4P$, Carbonylcyclopentadienylcobalt-μ-(dimethylphosphido)-tricarbonylcobalt, 44B, 757
$C_{11}H_{11}CrNO_3$, Tricarbonyl(6-dimethylaminofulvene)chromium, 44B, 1147
$C_{11}H_{11}NbO$, Di(π-cyclopentadienyl)niobium carbonyl hydride, 38B, 780
$C_{11}H_{11}NbOS$, Di-π-cyclopentadienyl-hydrosulfide-carbonylniobium, 39B, 607
$C_{11}H_{13}F_6N_2O_3PW$, (Acetone hydrazone)tricarbonyl(η-cyclopentadienyl)-tungsten hexafluorophosphate, 44B, 758
$C_{11}H_{13}FeP$, 3,4-Dimethylphospholyl-cyclopentadienyliron, 43B, 1063
$C_{11}H_{13}O_3ReSi$, Tricarbonyl((trimethylsilyl)-π-cyclopentadienyl)-rhenium, 38B, 781
$C_{11}H_{15}B_2Co$, $η^5$-Cyclopentadienyl($η^4$-1,4-dimethyl-1,4-dibora-2,5-cyclohexadiene)cobalt, 46B, 867
$C_{11}H_{16}F_6MoNOP$, Bis-π-cyclopentadienylhydroxymethylaminomolybdenum-(IV) hexafluorophosphate, 40B, 742
$C_{11}H_{19}B_2ONbZn \cdot 0.5\ C_6H_6$, [Carbonylbis($η^5$-cyclopentadienyl)-niobium]-μ-hydrido-[bis(tetrahydridoborato)zinc] benzene solvate, 45B, 945
$C_{11}H_{20}CuP$, π-Cyclopentadienyl(triethylphosphine)copper(I), 35B, 541
$C_{12}F_{12}FeO_4$, Tricarbonyl tetrakis(trifluoromethyl) cyclopentadienone iron, 31B, 352
$C_{12}H_6BrMnO_3$, Tricarbonyl($η^5$-1-bromoindenyl)manganese, 45B, 946
$C_{12}H_{10}FeO_2$, π-Cyclopentadienyliron dicarbonyl 2,4-cyclopentadiene, 31B, 353
$C_{12}H_{10}FeO_4$, 1,1'-Ferrocenedicarboxylic acid (triclinic), 45B, 946
$C_{12}H_{10}FeO_4$, 1,1'-Ferrocenedicarboxylic acid (monoclinic), 40B, 1108
$C_{12}H_{10}HfO_2$, Dicarbonyl-bis(cyclopentadienyl)hafnium, 45B, 947
$C_{12}H_{10}Mn_2N_2O_4$, trans-Di(π-cyclopentadienyl)(dicarbonyl)(dinitrosyl)-dimanganese, 39B, 607
$C_{12}H_{10}N_2O_2Ti$, Bis($η^5$-cyclopentadienyl)diisocyanatotitanium(IV), 45B, 948
$C_{12}H_{10}N_2O_2Zr$, Bis($η^5$-cyclopentadienyl)diisocyanatozirconium(IV), 45B, 948
$C_{12}H_{10}N_2S_2Ti$, Bis-(π-cyclopentadienyl)diisothiocyanatotitanium(IV), 42B, 675
$C_{12}H_{10}Ni_2O_2$, Di-μ-carbonylbis($η^5$-cyclopentadienylnickel)(Ni-Ni), 46B, 869, 870
$C_{12}H_{10}O_2Ti$, Dicarbonylbis(cyclopentadienyl)titanium(II), 41B, 893; 43B, 1064
$C_{12}H_{10}O_2W$, Dicarbonyl-$η^5$-cyclopentadienyl-$η^3$-cyclopentadienyltung-

sten, 44B, 1148

$C_{12}H_{10}O_2Zr$, Bis(η^5-cyclopentadienyl)dicarbonylzirconium(II), 46B, 868

$C_{12}H_{11}Fe_2O_5P$, Cyclopentadienyl(carbonyl)iron-(μ-dimethylphosphide-μ-carbonyl)tricarbonyliron, 39B, 608

$C_{12}H_{12}Cl_2Ti$, (1,1'-Ethylenedicyclopentadienyl)dichlorotitanium, 45B, 949

$C_{12}H_{12}MnNO_4$, (R)-α-(N-Acetylamino)ethylcymantrene, 46B, 871

$C_{12}H_{12}MnNO_4$, (S)-1-(N,N-Dimethylaminomethyl)-2-formylcymantrene, 46B, 870

$C_{12}H_{12}S_2Ti$, Ethylene-1,2-dithiolato-di-π-cyclopentadienyl-titanium-(IV), 39B, 608

$C_{12}H_{13}CoN_2O_2$, Bicyclo[2.2.1]hept-2-en-5,6-dinitrosocobalt-η^5-cyclopentadienide, 43B, 1064

$C_{12}H_{13}CoO_2$, (η-Cyclopentadienyl)(4-5-η-2-methoxycarbonylcyclopent-4-en-1,3-ylene)cobalt, 45B, 832

$C_{12}H_{13}FeNO$, N-Formylaminomethylferrocene, 37B, 455

$C_{12}H_{14}Cl_2Ti$, Bis(methylcyclopentadienyl)titanium dichloride, 41B, 894

$C_{12}H_{14}Cl_2V$, Bis(methylcyclopentadienyl)vanadium dichloride, 41B, 894

$C_{12}H_{14}I_3Fe$, 1,1'-Dimethylferricinium tri-iodide, 37B, 455

$C_{12}H_{14}Mo_2S_4$, Bis(η-methylcyclopentadienyl)molybdenum(IV)-di-μ-sulfido-disulfidomolybdenum(VI), 44B, 758

$C_{12}H_{15}ClHg_2MoS$, (μ-Ethylthiolato)bis(η^5-cyclopentadienyl)molybdenum-dimercury chloride, 46B, 871

$C_{12}H_{15}ClMo$, Bis-π-cyclopentadienylethylchloromolybdenum(IV), 40B, 742

$C_{12}H_{15}ClOTi$, Chloro-bis(cyclopentadienyl)-ethoxo-titanium, 46B, 872

$C_{12}H_{15}CoO_2$, Dicarbonyl(pentamethylcyclopentadienyl)cobalt(I), 46B, 873

$C_{12}H_{15}CrNO_3$, Dicarbonylnitrosyl(η^5-pentamethylcyclopentadienyl)-chromium, 46B, 873

$C_{12}H_{15}MoNO_3$, Dicarbonylnitrosyl(η^5-pentamethylcyclopentadienyl)molybdenum, 46B, 873

$C_{12}H_{15}NO_3W$, Dicarbonylnitrosyl(η^5-pentamethylcyclopentadienyl)tungsten, 46B, 873

$C_{12}H_{15}Ta$, Bis(cyclopentadienyl)methylmethylenetantalum, 41B, 895

$C_{12}H_{16}Cr_2N_2O_4$, cis-Di-μ-methoxo-bis(η^5-cyclopentadienylnitrosylchromium), 45B, 949

$C_{12}H_{16}Hf$, Dimethylhafnocene, 42B, 676

$C_{12}H_{16}IMoNS$, Bis-π-cyclopentadienyl(2-aminoethanethiolato)molybdenum iodide, 34B, 458

$C_{12}H_{16}I_2Mo_2N_4O_2$, μ-(Dimethylhydrazido)-bis(η-cyclopentadienyliodonitrosylmolybdenum), 46B, 875

$C_{12}H_{17}Cl_3Mo_2OS_2 \cdot 0.5\ C_7H_8$, Trichlorobis($\eta^5$-cyclopentadienyl)-μ-hydroxo-di-μ-methanethiolatodimolybdenum hemitoluene solvate, 45B, 950

$C_{12}H_{17}CrNO_3$, Tetramethylammonium tricarbonylcyclopentadienylchromium, 44B, 759

$C_{12}H_{18}Cl_6Nb_2O_3$, μ-Oxo-bis[aquatrichloro(η^5-methylcyclopentadienyl)-niobium(V)], 45B, 983

$C_{12}H_{20}CrO_8P_2$, (η^5-Cyclopentadienyl)dicarbonyl(trimethyl phosphite)-(dimethyl phosphonato)chromium, 45B, 971

$C_{13}H_5F_5FeO_4S$, Dicarbonyl(π-cyclopentadienyl)(pentafluorophenylsulphonyl-S)iron, 40B, 748

$C_{13}H_5MnMoO_8$, π-Cyclopentadienyltricarbonylmolybdenum-manganesepentacarbonyl, 33B, 347

$C_{13}H_{10}Cl_2FeO_2Sn$, Dicarbonyl-π-cyclopentadienyl(dichloro(phenyl)-

stannyl)iron, 35B, 545
$C_{13}H_{10}FeNi_2O_3S$, μ_3-Sulfidobis(η^5-cyclopentadienyl)dinickeltricarbonyliron, 46B, 875
$C_{13}H_{10}Fe_2O_5S$, cis-μ-Carbonyl-μ-(sulfur dioxide)-bis-(π-cyclopentadienylcarbonyliron), 39B, 609
$C_{13}H_{10}O_3Rh_2$, Dicyclopentadienyltricarbonyldirhodium, 32B, 352
$C_{13}H_{11}FeN$, trans-β-Ferrocenylacrylonitrile, 35B, 544
$C_{13}H_{12}FeO$, α-Keto-1,1'-trimethyleneferrocene, 30B, 307
$C_{13}H_{14}Cl_2Hf$, Dichloro-(1,1'-trimethylene-π-cyclopentadienyl)hafnium, 40B, 748
$C_{13}H_{14}Cl_2Ti$, (1,1'-Trimethylenedicyclopentadienyl)titanium dichloride, 37B, 455; 39B, 610
$C_{13}H_{14}Cl_2Zr$, 1,1'-Trimethylenecyclopentadienyldichlorozirconium, 40B, 749
$C_{13}H_{14}Fe_2OS_4$, Bis(methylcyclopentadienyliron)monocarbonyl tetrasulfide, 43B, 1065
$C_{13}H_{15}Mo_2NO_5$, μ-Ethoxycarbonylimido-μ-oxo-bis[(η-cyclopentadienyl)-oxomolybdenum], 45B, 959
$C_{13}H_{18}OPd$, Cyclopentadienyl(2-hydroxycyclooct-5-enyl)palladium(II), 39B, 610
$C_{13}H_{19}Br_2MoNOZn$, Dihydridobis(cyclopentadienyl)molybdenum-dibromodimethylformamidezinc(Mo-Zn), 43B, 1066
$C_{13}H_{19}ClSiTi$, Dicyclopentadienyl(trimethylsilyl)titanium chloride, 46B, 634
$C_{13}H_{19}Cl_3Mo_2S_3$, Trichlorobis(η^5-cyclopentadienyl)tri-μ-methanethiolatodimolybdenum, 45B, 950
$C_{13}H_{19}NSiV$, Trimethylsilylnitrenebis(η^5-cyclopentadienyl)vanadium, 46B, 876
$C_{13}H_{19}Re$, Dimethyl-π-cyclopentadienyl(methylcyclopentadiene)rhenium, 32B, 354
$C_{14}H_5CoF_{12}O$, Tetrakis(trifluoromethyl)cyclopentadienone-π-cyclopentadienylcobalt, 29, 556
$C_{14}H_9MnNO_5Re$, (π-Pyrrolyltricarbonylmanganese)cyclopentadienyldicarbonylrhenium, 44B, 1148
$C_{14}H_{10}BF_4Fe_2IO_4$, μ-Iodobis(π-cyclopentadienyldicarbonyliron) fluoroborate, 39B, 611
$C_{14}H_{10}Cl_9Fe_2O_4Sb_3$, μ-Dichloroantimony(V)-bis((π-cyclopentadienyl)dicarbonyliron(0)) heptachlorodiantimonate, 39B, 612
$C_{14}H_{10}CrNiO_4$, (Tricarbonylcyclopentadienylchromium)carbonylcyclopentadienylnickel, 46B, 870
$C_{14}H_{10}Cr_2O_4$, Bis(cyclopentadienyldicarbonylchromium), 44B, 1149
$C_{14}H_{10}Cr_2O_4S$, Bis(η^5-cyclopentadienyl)dicarbonylchromium sulfide, 45B, 951
$C_{14}H_{10}FeN_2$, 2,2-Dicyanovinylferrocene,` 38B, 781
$C_{14}H_{10}Fe_2O_3S$, μ-Carbonyl-μ-thiocarbonyl-bis(carbonylcyclopentadienyliron), 45B, 951
$C_{14}H_{10}Fe_2O_4$, cis-Di-μ-carbonyl-dicarbonyldi-π-cyclopentadienyldiiron(Fe-Fe), 35B, 546
$C_{14}H_{10}Fe_2O_4$, trans-Di-μ-carbonyl-dicarbonyldi-π-cyclopentadienyldiiron(Fe-Fe), 22, 680; 35B, 546; 44B, 759
$C_{14}H_{10}Fe_2O_6S$, μ-(Sulfur dioxide)-bis(π-cyclopentadienyldicarbonyliron), 39B, 613
$C_{14}H_{10}Mn_2O_4S_2$, (μ-Disulfur)-bis(η^5-cyclopentadienyldicarbonylmanganese), 46B, 876
$C_{14}H_{10}Mo_2O_4$, Dicyclopentadienyltetracarbonyldimolybdenum(Mo-Mo), 41B, 898; 44B, 760
$C_{14}H_{10}N_2S_2Ti$, Bis-(π-cyclopentadienyl)maleonitriledithiolatotitani-

um(IV), 42B, 675
$C_{14}H_{10}O_3W$, π-Cyclopentadienyltricarbonyl-σ-phenyltungsten, 33B, 349
$C_{14}H_{10}O_4Ru_2$, Bis(cyclopentadienyldicarbonylruthenium), 32B, 357
$C_{14}H_{11}AsCr_2O_7$, Dicarbonylcyclopentadienylchromium-μ-diphenylarsido-pentacarbonylchromium, 44B, 761
$C_{14}H_{12}Cr_2N_4O_2$, Di-μ-dimethylamido-bis(cyclopentadienylnitrosyl-chromium), 35B, 548
$C_{14}H_{13}MnO_2$, Dicarbonyl-π-cyclopentadienylbicyclo[2.2.1]hepta-2π,5-dienemanganese(I), 33B, 350
$C_{14}H_{14}FeO$, α-Keto-1,5-tetramethyleneferrocene, 32B, 358
$C_{14}H_{14}FeO$, 1,4-(1,1'-Ferrocenediyl)butan-1-one, 45B, 952
$C_{14}H_{14}FeO_2$, Diacetylferrocene, 35B, 550
$C_{14}H_{14}Ni_2O_2$, Di-μ-carbonylbis(η⁵-methylcyclopentadienylnickel)(Ni-Ni), 46B, 869
$C_{14}H_{14}O_2Ru$, Diacetylruthenocene, 28, 612
$C_{14}H_{15}F_6FeO_3P$, Dicarbonyl-cyclopentadienyl(3-methylcyclohexenone)-iron hexafluorophosphate, 46B, 877
$C_{14}H_{15}F_6FeP$, 1,1'-Trimethylenebenzenecyclopentadienyliron hexafluorophosphate, 43B, 1067
$C_{14}H_{16}FeS_3$, (R)-Ferrocenylmethylmethane-S-methyl-trithiocarbonate, 46B, 877
$C_{14}H_{17}CoO$, Cyclopentadienyl(tetramethylcyclopentadienone)cobalt(II), 26, 652; 27, 862
$C_{14}H_{20}AlCl_2Ti$, Diethylaluminum-biscyclopentadienyl titanium dichloride, 22, 592
$C_{14}H_{20}Mo_2N_2O_2S_2$, trans-Di-μ-thioethyl-bis(nitrosyl-η⁵-cyclopentadienylmolybdenum), 45B, 952
$C_{14}H_{20}Mo_2S_4$, Bis(μ-methylthio)bis(μ-thio)bis(η⁵-methylcyclopentadienyl)dimolybdenum(Mo-Mo), 46B, 878
$C_{14}H_{21}Al_2ClFe$, 1,1-[μ-Chlorobis(dimethylaluminum)]ferrocene, 45B, 953
$C_{14}H_{22}Cl_4OTi_2$, (μ-Pinacolato)-bis(cyclopentadienyl-dichloro-titanium), 46B, 872
$C_{14}H_{22}F_{12}MoN_2P_2$, Amminebis-(η-cyclopentadienyl)-N-(C-methyl-C-ethyl-ketimino)molybdenum bis(hexafluorophosphate), 41B, 898
$C_{14}H_{23}Cl_3O_2U$, Trichloro-methylcyclopentadienyl-bis(tetrahydrofuran)-uranium(IV), 45B, 954
$C_{15}H_9CrNO_3$, Dicarbonylfluorenylnitrosylchromium, 45B, 938
$C_{15}H_{10}CoNbO_5$, μ-Carbonyl[bis(η⁵-cyclopentadienyl)carbonylniobium]-[tricarbonylcobalt](Nb-Co), 45B, 954
$C_{15}H_{10}O_5V_2$, Dicyclopentadienylpentacarbonyldivanadium, 44B, 1150; 46B, 878
$C_{15}H_{11}BF_7MnN_2O_2$, (η⁵-Methylcyclopentadienyl)(α,α,α-trifluorotolyl)-diazenido-dicarbonylmanganese tetrafluoroborate, 46B, 1075
$C_{15}H_{11}FeNbO_5$, Bis(η⁵-cyclopentadienyl)carbonylniobium-(μ-hydrido)-tetracarbonyliron, 45B, 955
$C_{15}H_{13}Fe_2O_4P$, Tetracarbonyl(3,4-dimethylphosphaferrocene-P)iron, 46B, 879
$C_{15}H_{14}MoO_2$, p-Methyl-π-benzyl-π-cyclopentadienyldicarbonylmolybdenum, 33B, 350
$C_{15}H_{15}ClU$, π-Tri-cyclopentadienyluranium chloride, 30B, 315
$C_{15}H_{15}Co_3S_2$, Tris(η-cyclopentadienyl)-di-μ₃-thio-tricobalt, 45B, 955
$C_{15}H_{15}Co_3S_2$, Tris(η-cyclopentadienyl)-di-μ₃-thio-tricobalt (low temperature form), 45B, 955
$C_{15}H_{15}FU$, Tris(cyclopentadienyl)fluorouranium(IV), 41B, 899
$C_{15}H_{15}F_6Fe_2O_3SSb$, μ-Ethylthio-μ'-carbonyl-bis(η-cyclopentadienylcarbonyliron)(Fe-Fe) hexafluoroantimonate, 43B, 1067

$C_{15}H_{15}FeNO_4$, 2-Nitro-3-ferrocenylacrylic acid ethyl ester, 46B, 880
$C_{15}H_{15}Mn_3N_4O_4$, Tris(cyclopentadienylmanganese)tetranitrosyl, 40B, 894
$C_{15}H_{15}MoNO$, Tris(cyclopentadienyl)nitrosylmolybdenum, 34B, 453
$C_{15}H_{15}Sc$, Tris(cyclopentadienyl)scandium(III), 39B, 613
$C_{15}H_{15}Sm$, Tris(cyclopentadienyl)samarium(III), 34B, 459
$C_{15}H_{15}Ti$, Tris(cyclopentadienyl)titanium, 40B, 751
$C_{15}H_{16}Fe$, 1,1-(1'',3''-Cyclopentylene)ferrocene, 41B, 899
$C_{15}H_{16}FeO$, (-)-2,3-Ferroceno-5-exo-methylcyclohex-2-en-1-one, 39B, 615
$C_{15}H_{16}Fe_2GeO_3$, μ-Dimethylgermyl-μ-carbonyl-dicyclopentadienyldicarbonyldiiron, 40B, 751
$C_{15}H_{19}B_4Co_3$, 1,2,3-Tris(η^5-cyclopentadienyl)tricobaltaheptaborane, 43B, 1070
$C_{15}H_{19}ClO_4Zr$, π-Cyclopentadienylbis(acetylacetonato)chlorozirconium(IV), 34B, 464
$C_{15}H_{20}B_3Co_3$, 1,2,3-Tris(η^5-cyclopentadienyl)tricobaltahexaborane, 43B, 1070
$C_{15}H_{20}Cl_2Ti$, Pentamethylcyclopentadienyl(cyclopentadienyl)titanium dichloride, 39B, 615
$C_{15}H_{20}CoNiO_4P$, Di-μ-carbonyl-dicarbonyl(triethylphosphine)cobalt(π-cyclopentadienylnickel), 40B, 752
$C_{16}H_5Co_3MoO_{11}$, (η^5-Cyclopentadienyl)dicarbonylmolybdenum-nonacarbonyltricobalt, 43B, 1070
$C_{16}H_5Fe_3O_{11}Rh$, Di-μ-carbonyl-nonacarbonyl-(π-cyclopentadienylrhodio)tri-iron(3Rh-Fe)(3Fe-Fe), 37B, 457
$C_{16}H_8Cl_2Mn_2O_6Sn$, Dichloro-bis(tricarbonylmanganesecyclopentadienyl-1-yl)tin(IV), 44B, 761
$C_{16}H_8O_{11}Os_3W$, Tri-μ-hydrido-undecacarbonylcyclopentadienyltriosmiumtungsten, 45B, 956
$C_{16}H_{10}BrN_4V$, Bromobis(π-cyclopentadienyl)(tetracyanoethylene-N)vanadium, 40B, 1109
$C_{16}H_{10}Cl_4Co_2FeGe_2O_6$, μ-(Tetracarbonyliron)-μ-(di-μ-carbonyl-di(cyclopentadienyl-cobalt))-di(dichlorogermanium), 38B, 783
$C_{16}H_{10}Cr_2O_6$, Dicyclopentadienylhexacarbonyldichromium, 40B, 754
$C_{16}H_{10}HgMo_2O_6$, Bis(tricarbonyl(η-cyclopentadienyl)molybdato(0))mercury(II), 43B, 1071
$C_{16}H_{10}Mo_2O_6$, Bis(η^5-cyclopentadienyl)hexacarbonyldimolybdenum, 21, 572, 40B, 755
$C_{16}H_{10}Mo_2O_6Zn$, Bis(tricarbonyl-cyclopentadienyl-molybdenum)zinc, 40B, 765
$C_{16}H_{10}O_6W_2$, Bis(η^5-cyclopentadienyl)hexacarbonylditungsten, 40B, 755
$C_{16}H_{11}F_9FeO$, [1-5-η-exo-1-Acetyl-2,4,6-tris(trifluoromethyl)cyclohexadienyl](η-cyclopentadienyl)iron, 46B, 880
$C_{16}H_{12}FeO_4$, (3,4-Bis(1-propynyl)-2,5-dimethylcyclopentadienone)tricarbonyliron(0), 44B, 762
$C_{16}H_{13}FeP$, (1,1'-Ferrocenediyl)phenylphosphine, 46B, 892
$C_{16}H_{14}CrO_3$, Tricarbonyl(4,6,8-trimethylazulene)chromium, 45B, 957
$C_{16}H_{14}Fe_2O_4Si$, Di-μ-carbonyl-cis-μ-(1-5-η:1'-5'-η-dicyclopentadienyldimethylsilane)bis(carbonyliron)(Fe-Fe), 39B, 616; 43B, 1072
$C_{16}H_{14}Mn_2N_2O_4$, μ-Dinitrogen-bis[dicarbonyl-η^5-methylcyclopentadienylmanganese], 45B, 957
$C_{16}H_{14}MoS_2$, Benzene-1,2-dithiolatodi(π-cyclopentadienyl)molybdenum(IV), 35B, 553
$C_{16}H_{14}S_2Ti$, Benzene-1,2-dithiolato-di(π-cyclopentadienyl)titanium(IV), 38B, 784
$C_{16}H_{14}S_2W$, Benzene-1,2-dithiolene-di-(π-cyclopentadienyl)tungsten-

(VI), 39B, 617
$C_{16}H_{15}BF_4Fe_2O_4S$, μ-Ethylthio-bis(dicarbonyl(η-cyclopentadienyl)iron) tetrafluoroborate, 44B, 756
$C_{16}H_{15}Co$, π-Cyclopentadienyl-1-phenylcyclopentadienecobalt, 29, 558
$C_{16}H_{15}Co_3S_2$, Tris(η^5-cyclopentadienyl)-μ_3-sulfido-μ_3-thiocarbonyl-triangulo-tricobalt, 46B, 1218
$C_{16}H_{16}Co_3O_2$, μ_3-Oxo-μ_3-carbonyltricyclopentadienyltricobalt, 34B, 466
$C_{16}H_{16}FeIN$, Ferrocenylmethylpyridinium iodide, 43B, 1073
$C_{16}H_{16}Fe_2N_2O_2$, cis-anti-Bis(η^5-cyclopentadienyl)dicarbonylbis(μ-methylisocyanide)-diiron, 40B, 756
$C_{16}H_{17}BF_4N_2W$, Bis(η^5-cyclopentadienyl)(phenylhydrazido-N,N')tungsten tetrafluoroborate, 46B, 1079
$C_{16}H_{17}Mo_2O_4P$, μ-Hydrido-μ-dimethylphosphido-bis(η^5-cyclopentadienyl-dicarbonylmolybdenum), 30B, 312; 40B, 756; 44B, 762
$C_{16}H_{18}AsHgIMoO_2$, trans-Dicarbonyl(dimethylphenylarsine)(iodo-mercurio(0))(η-methylcyclopentadienyl)molybdenum(II), 43B, 1071
$C_{16}H_{19}O_2Rh \cdot 2\ C_6H_6O_2$, Catecholato($\eta^5$-pentamethylcyclopentadienyl)-rhodium(III) catechol solvate, 45B, 958
$C_{16}H_{20}Fe$, 1,1'-Tetramethylethyleneferrocene, 30B, 306
$C_{16}H_{20}Mo_2N_2O_6$, Bis(μ-ethoxycarbonylimido)bis[(η-cyclopentadienyl)-oxomolybdenum], 45B, 959
$C_{16}H_{21}BrMoO_5Zn$, η-Cyclopentadienyltricarbonyl(bromozinc)molybdenum bis(tetrahydrofuran), 43B, 1073
$C_{16}H_{22}BF_4Mo_2S_4$, Di-μ-(propane-1,2-dithiolato)bis(cyclopentadienylmo-lybdenum) tetrafluoroborate, 45B, 959
$C_{16}H_{23}MoNO_2$, Dicarbonyl(π-cyclopentadienyl)(di-(t-butyl)methyleneam-ino)molybdenum(II), 39B, 618
$C_{16}H_{24}Mo_2N_2O_2S_2$, cis-Di-μ-thioisopropyl-bis(nitrosyl-η^5-cyclopenta-dienylmolybdenum), 45B, 952
$C_{16}H_{25}BCr_2N_4O_3$, (μ-Nitrosyl)(μ-diethylethylaminoborane-N)bis(η^5-cyc-lopentadienyl)dinitrosyldichromium(Cr-Cr), 46B, 1081
$C_{17}H_6O_{12}Os_3W$, μ-Hydrido-dodecacarbonyltriosmium(η^5-cyclopentadien-yl)tungsten, 45B, 960
$C_{17}H_{10}Fe_2Ni_2O_7$, (μ_3-Carbonyl)-bis(η^5-cyclopentadienyl)dinickel-hexa-carbonyldiiron, 46B, 882
$C_{17}H_{13}Mo_2NO_5$, trans-Di(π-cyclopentadienyl)(methylisocyanide)(penta-carbonyl)dimolybdenum, 39B, 618
$C_{17}H_{15}CrIN_2$, η^5-Cyclopentadienyl(diphenylamido)iodonitrosylchromium, 45B, 960
$C_{17}H_{16}Cl_4Fe_2O_4Si_3$, Trisilatetrachlorocyclohexyl-bis(dicarbonylcyclo-pentadienyliron), 46B, 865
$C_{17}H_{16}FeMnNO_4$, N-Ferrocenylmethyl-N-methylaminomethylene(tetracar-bonyl)manganese, 43B, 976
$C_{17}H_{16}U$, Tricyclopentadienylethynyluranium(IV), 42B, 677
$C_{17}H_{17}F_3MoO_3$, Carbonyl(η-cyclopentadienyl)(3-4:5-6-η-2,3,4,5-tetra-methyl-6-oxo-1-trifluoromethylhexa-1,3,5-trienyloxo)molybdenum, 44B, 764
$C_{17}H_{20}FeO_2$, (Z)-2-Ferrocenyl-2-methylcyclopropanecarboxylic acid ethyl ester, 45B, 961
$C_{17}H_{24}F_{12}NP_2Rh$, η^5-Pentamethylcyclopentadienyl-(6-methylimino-1-5-η-cyclohexadienyl)rhodium hexafluorophosphate, 46B, 882
$C_{18}H_{10}ClCoFe_2O_8Sn$, Chloro-bis($\eta^5$-cyclopentadienyldicarbonyliron)-(tetracarbonylcobalt)tin(IV), 46B, 883
$C_{18}H_{10}Fe_2O_8Rh_2$, Tri-μ-carbonyl-pentacarbonylbis-(π-cyclopentadienyl-rhodio)di-iron(Rh-Rh)(Fe-Fe)(4Rh-Fe), 37B, 458
$C_{18}H_{10}O_8Ru_3$, Tetracarbonylbis[dicarbonyl-(η-cyclopentadienyl)-

ruthenio]ruthenium, 45B, 961
$C_{18}H_{15}MoNbO_3$, Tricarbonyl-cyclopentadienylmolybdenum-(dicyclopentadienylniobium), 45B, 962
$C_{18}H_{15}MoO_3Re$, Bis(cyclopentadienyl)molybdenumdi(μ-carbonyl)cyclopentadienylcarbonylrhenium(III), 45B, 963
$C_{18}H_{15}O_3Rh_3$, Tris(π-cyclopentadienylcarbonylrhodium), 32B, 360; 34B, 461
$C_{18}H_{16}FeO$, Phenylacetyl-ferrocene, 41B, 902
$C_{18}H_{18}N_2SU$, Tris(cyclopentadienyl)isothiocyanatoacetonitrileuranium-(IV), 44B, 763
$C_{18}H_{18}N_2Ti$, Bis(η^5-cyclopentadienyl)dipyrrolyltitanium(IV), 46B, 1087
$C_{18}H_{18}N_2Zr$, Bis(η^5-cyclopentadienyl)dipyrrolylzirconium(IV), 46B, 1087
$C_{18}H_{19}ClMn_2O_8S$, μ-Ethylthio-bis(dicarbonyl(η-methylcyclopentadienyl)manganese)(Mn-Mn) perchlorate, 43B, 1075
$C_{18}H_{19}Fe_2NO_3$, Bis(η^5-cyclopentadienyl)tricarbonyl(t-butyl isocyanide)diiron, 40B, 757
$C_{18}H_{20}CrO_2$, Dicarbonyl(1-(1-((2,3,4-η)-5-isopropylidene-3-cyclopentene-1,2-diyl)-1-methylethyl)-η^5-cyclopentadienyl)chromium(0), 46B, 883
$C_{18}H_{20}Fe$, 2,3-exo-Ferroco-4,4-dimethylbicyclo[3.2.1]octa-2,6-diene, 40B, 758
$C_{18}H_{21}Nd$, Tris(methylcyclopentadienyl)neodymium, 40B, 759
$C_{18}H_{22}Mn_2O_4P_2S$, μ-Thio-bis(dimethylphosphinodicarbonyl-(η^5-cyclopentadienyl)manganese), 44B, 763
$C_{18}H_{22}Mn_2O_4P_2S_2$, μ-Dithio-bis(dimethylphosphinedicarbonyl-(η^5-cyclopentadienyl)manganese), 44B, 763
$C_{18}H_{23}MoNO_3$, Carbonyl(η-cyclopentadienyl)(3-5-η-2,3-dihydro-3,4,5-trimethyl-2-oxo-3-furyl)(t-butylisocyanide)molybdenum, 44B, 764
$C_{18}H_{26}Cl_4Mo_2S_2 \cdot 0.5\ CH_2Cl_2$, μ-Disulfidobis[η-(n-butylcyclopentadienyl)dichloromolybdenum] dichloromethane solvate, 45B, 963
$C_{18}H_{26}F_6MoPRhS_2$, Bis-π-cyclopentadienylmolybdenum-(μ-bismethanethiolato)-bis-π-allylrhodium hexafluorophosphate, 40B, 759
$C_{18}H_{26}Fe$, sym-Octamethylferrocene, 44B, 765
$C_{18}H_{27}BrMgMoO_2$, [Bromobis(tetrahydrofuran)magnesio]-bis(η-cyclopentadienyl)hydridomolybdenum, 45B, 964
$C_{18}H_{28}B_6Co_2$, Bis(η^5-cyclopentadienylcobalt)tetra(methylcarbo)hexaborane, 45B, 964
$C_{18}H_{28}B_8Fe$, 4-Ferrocenyl-2,3,7,8-tetramethyl-2,3,7,8-tetracarbadodecaborane, 46B, 884
$C_{18}H_{28}Cl_2FeMoS_2$, Di-μ-thio-n-butyl(bis-π-cyclopentadienylmolybdenum)dichloroiron, 38B, 1017
$C_{18}H_{28}Cr_2O_2$, Di(cyclopentadienyl)-di-(t-butoxydichromium), 45B, 965
$C_{18}H_{28}Cr_2S_3$, μ-Sulfido-di-μ-(t-butylmercapto)-bis(cyclopentadienylchromium), 45B, 966
$C_{18}H_{29}B_8Co$, Cobaltocenyl-tetra(methylcarba)-dodecaborane, 45B, 966
$C_{18}H_{29}ClNP_2Ta$, Bis(η^5-cyclopentadienyl)(bis(1,2-dimethylphosphino)-ethane)tantalum chloride - acetonitrile, 44B, 765
$C_{18}H_{30}ClMoP$, η^4-(1-endo-Ethyl)cyclopentadiene-η^5-cyclopentadienyl-(triethylphosphine)chloromolybdenum(II), 43B, 1076
$C_{18}H_{38}FeO_2Si_6$, Nonamethyl-cyclopentasilane-dimethylsilyl-η^5-cyclopentadienyl-dicarbonyl-iron, 46B, 885
$C_{19}H_{16}FeO_2$, 1'-Acetyl-1-benzoylferrocene, 34B, 454; 37B, 459
$C_{19}H_{16}Mn_2O_4$, μ-Cyclopentadiene-bis(η^5-cyclopentadienyldicarbonylmanganese), 42B, 677
$C_{19}H_{16}Mo_2O_4$, [Hydrido(carbonyl)bis(η-cyclopentadienyl)molybdenum-

(IV)] [tricarbonyl-η-cyclopentadienylmolybdate(0)], 45B, 967
$C_{19}H_{17}BMoN_8O_2$, Dicarbonyl-π-cyclopentadienyl(tetrakis(pyrazol-1-yl)borato)molybdenum, 39B, 619
$C_{19}H_{22}Fe$, Bis(μ-trimethylene)-1,2-trimethylene-ferrocene, 46B, 886
$C_{19}H_{22}Fe$, 1,1',2,2',4,4'-Tris(trimethylene)ferrocene, 44B, 1150
$C_{19}H_{23}GdO$, Tris(cyclopentadienyl)tetrahydrofurangadolinium, 46B, 887
$C_{19}H_{24}NNi_3$, Tris(π-cyclopentadienylnickel)-t-butylammonium, 40B, 760
$C_{19}H_{24}U$, Tricyclopentadienyl(n-butyl)uranium, 42B, 682
$C_{19}H_{25}BF_6N_6PRh$, (Hydrotris(1-pyrazolyl)borato)(η^5-pentamethylcyclopentadienyl)rhodium(II) hexafluorophosphate, 41B, 902
$C_{19}H_{25}FeNO_2Pd$, (-)-RS-Cyclo-1-(1'-dimethylaminoethyl)ferrocene-2-(acetylacetonato)palladium, 45B, 968
$C_{19}H_{28}Co_2N_2O$, Bis-N,N'-(π-cyclopentadienylcobalt)-di-t-butylurea, 34B, 455
$C_{19}H_{29}F_6O_3P_2Rh$, (1-5-η-6-Dimethoxyphosphorylcyclohexadienyl)(1-5-η-ethyltetramethylcyclopentadienyl)-rhodium(III) hexafluorophosphate, 46B, 887
$C_{19}H_{32}IMoO_2P$, Dicarbonyl-π-cyclopentadienyliodotri(n-butyl)phosphine-molybdenum(II), 37B, 459
$C_{20}H_{10}ClCo_3O_{10}Ti$, μ_3-((Chloro-bis(η-cyclopentadienyl)titanoxy)methylidyne)cyclo-tris(tricarbonylcobalt)(3Co-Co), 42B, 677
$C_{20}H_{14}O_2W$, (η^3-Indenyl)(η^5-indenyl)dicarbonyltungsten, 44B, 1151
$C_{20}H_{15}As_4BF_4Fe_3O_{10}$, Tris($\eta^5$-cyclopentadienyl)(pentaoxatetraarsenic)-pentacarbonyltriiron tetrafluoroborate, 43B, 1077
$C_{20}H_{16}Cl_2Fe_2$, Bis(1-(2'-chloroferrocenyl)), 30B, 310; 31B, 354
$C_{20}H_{16}Fe_2$, Bis(fulvalene)diiron, 34B, 456
$C_{20}H_{18}ClNb_2$, (μ-Chloro)(η^5,η^5-1,1'-bicyclopentadienyl)bis(η^5-cyclopentadienyl)diniobium, 46B, 888
$C_{20}H_{18}Cl_2Ti_2$, Di-μ-chloro-μ-fulvalene-bis(cyclopentadienyltitanium), 43B, 1078
$C_{20}H_{18}FeO$, α-Oxo-γ-phenyl-1,1'-trimethylene-2'-methylferrocene, 39B, 620
$C_{20}H_{18}Fe_2$, Biferrocenyl, 29, 549
$C_{20}H_{18}Fe_2I_5Se \cdot 0.5\ CH_2Cl_2$, Diferrocenylselenium iodide triiodide hemi(methylene chloride), 46B, 889
$C_{20}H_{20}B_2Cl_2F_8Nb_2O$, μ-Oxo-bis(bis-π-cyclopentadienylchloroniobium(V)) tetrafluoroborate, 40B, 742
$C_{20}H_{20}BrFe_4S_4$, Tetra-μ_3-sulfido-tetrakis(η^5-cyclopentadienyliron) bromide, 43B, 1078
$C_{20}H_{20}Cl_2OTi_2$, Bis(biscyclopentadienylchlorotitanium)oxide, 46B, 889
$C_{20}H_{20}Cl_2OZr_2$, Bis(chlorobiscyclopentadienylzirconium)oxide, 40B, 761
$C_{20}H_{20}Cl_2Sc_2$, Di-μ-chloro-bis(di-η-cyclopentadienylscandium(III)), 39B, 620
$C_{20}H_{20}Cl_2Ti_2$, Di-μ-chloro-bis(bis(η^5-cyclopentadienyl)titanium-(III)), 43B, 1079
$C_{20}H_{20}Cl_4O_4Ti_4$, Cyclotetra(μ-oxo-chloro-π-cyclopentadienyltitanium-(IV)), 35B, 554
$C_{20}H_{20}Co_4S_6 \cdot 0.5\ CHCl_3$, Tetra(cyclopentadienylcobalt)hexasulfide - chloroform, 34B, 457
$C_{20}H_{20}F_{12}Fe_4P_2S_4$, Tetra-$\mu_3$-sulfido-tetrakis($\eta^5$-cyclopentadienyliron) bis(hexafluorophosphate), 43B, 1080
$C_{20}H_{20}F_{12}Mo_2O_4P_3$, Bis(bis-η-cyclopentadienylmolybdenumdi-μ-oxo)phosphorus hexafluorophosphate, 43B, 1081
$C_{20}H_{20}FeO$, trans-1-Ferrocenyl-2-methoxyphenyl-cyclopropane, 45B, 968
$C_{20}H_{20}Fe_4S_4$, Cyclopentadienyliron sulfide, 31B, 356
$C_{20}H_{20}Hf$, Tetracyclopentadienylhafnium, 38B, 784

$C_{20}H_{20}Nb_2$, μ-(η⁵(η¹)-Cyclopentadienyl)-(η⁵-cyclopentadienyl)hydridoniobium dimer, 39B, 621
$C_{20}H_{20}Ti$, Tetra(cyclopentadienyl)titanium, 37B, 460
$C_{20}H_{20}U$, Tetrakis(cyclopentadienyl)uranium(IV), 40B, 761
$C_{20}H_{20}Zr$, Tetrakis(cyclopentadienyl)zirconium(IV) (monoclinic), 44B, 766
$C_{20}H_{20}Zr$, Tetrakis(cyclopentadienyl)zirconium(IV) (orthorhombic), 35B, 556
$C_{20}H_{21}Rh_3$, Tetracyclopentadienylhydridotrirhodium, 33B, 351
$C_{20}H_{22}Cl_2OTi$, η^5-Cyclopentadienyl-(η^5-1-methyl-3-isopropylcyclopentadienyl)(2-chlorophenoxy)chlorotitanium, 43B, 1085
$C_{20}H_{22}Fe$, syn-Bis(cyclopentyl-1",3"-ene)-(1,1'),(3,3')ferrocenophane, 42B, 703
$C_{20}H_{22}Fe_2N_2O_4$, Di-μ-carbonyl-μ-(1,2-di-π-cyclopentadienyl-N,N,N',N'-tetramethylethylenediamine)-bis-(carbonyliron), 35B, 557
$C_{20}H_{22}Ni$, Bis(η^5-isodicyclopentadienyl)nickel(II), 42B, 678
$C_{20}H_{23}ClO_4W_2 \cdot C_3H_6O$, Tungstenocene hydride dimer cation perchlorate acetone, 46B, 890
$C_{20}H_{23}N_2Ti_2$, μ-Diimidohydrido-bis(dicyclopentadienyltitanium)(Ti-Ti), 44B, 766
$C_{20}H_{23}Ni_4$, Tris(μ_3-hydrido)-tetra(cyclopentadienyl-nickel), 40B, 762; 45B, 969
$C_{20}H_{24}B_4Co_4$, Tetrakis(η^5-cyclopentadienyl)tetracobaltaoctaborane, 45B, 969
$C_{20}H_{24}B_4Ni_4$, (Tetra-cyclopentadienyl-nickela)octaborane, 45B, 970
$C_{20}H_{24}Co_4$, Tetrakis(1-5-η-cyclopentadienyl)tetrahydridotetracobalt, 41B, 905
$C_{20}H_{24}Si_2Ti_2$, Tetrakis(π-cyclopentadienyl)-di-μ-silyleno-dititanium, 39B, 622
$C_{20}H_{28}ClN_3S_6Zr$, η^5-Cyclopentadienyltris(N,N-dimethyldithiocarbamato)zirconium(IV) - chlorobenzene, 42B, 681
$C_{20}H_{28}Cl_2O_{13}Ti_2$, μ-Oxo-bis(aquodicyclopentadienyltitanium) perchlorate dihydrate, 44B, 1151
$C_{20}H_{28}Cl_2Ti$, Dichloro-cyclopentadienyl(menthylcyclopentadienyl)titanium(IV), 44B, 1152
$C_{20}H_{28}CoF_6O_2P$, 4-Hydroxy-1-oxo-2,3,5,6-tetramethylcyclohexadienyl-pentamethylcyclopentadienyl-cobalt(III) hexafluorophosphate, 46B, 890
$C_{20}H_{28}Cr_2O_{10}P_2$, Bis[($\eta^5$-cyclopentadienyl)dicarbonyl(trimethyl phosphite)chromium](Cr-Cr), 45B, 971
$C_{20}H_{28}FeSi_4$, 1,1'-Bis(pentamethyldisilanyl)ferrocene, 33B, 353
$C_{20}H_{30}Br_4Ir_2$, Di-μ-bromobis[bromo(η^5-pentamethylcyclopentadienyl)iridium], 45B, 971
$C_{20}H_{30}Br_4Rh_2$, Di-μ-bromo-bis(bromo(η^5-pentamethylcyclopentadienyl)-rhodium), 44B, 767
$C_{20}H_{30}Br_{0.76}Cl_{3.24}Rh_2$, Di-μ-halo-bis(halo($\eta^5$-pentamethylcyclopentadienyl)rhodium), 44B, 767
$C_{20}H_{30}Cl_2Ti$, Bis(pentamethylcyclopentadienyl)dichlorotitanium, 41B, 905
$C_{20}H_{30}Cl_4Ir_2$, Di-μ-chloro-dichlorobis(pentamethylcyclopentadienyl)-diiridium(III), 43B, 1082
$C_{20}H_{30}Cl_4Rh_2$, Di-μ-chloro-dichlorobis(pentamethylcyclopentadienyl)-dirhodium, 43B, 1081
$C_{20}H_{30}Co_2O_2$, Bis(pentamethylcyclopentadienyl)dicarbonyldicobalt, 45B, 972
$C_{20}H_{30}Fe$, Bis(pentamethylcyclopentadienyl)iron(II), 44B, 765; 45B, 973

$C_{20}H_{30}I_4Ir_2$, Di-μ-iodobis[iodo(η[5]-pentamethylcyclopentadienyl)iridium], 45B, 971
$C_{20}H_{30}I_4Rh_2 \cdot 2\ C_7H_8$, Bis-μ-iodo-bis[(η[5]-pentamethylcyclopentadienyl)iodorhodium] toluene solvate, 45B, 973
$C_{20}H_{30}Mn$, Bis(pentamethylcyclopentadienyl)manganese(II), 45B, 973
$C_{20}H_{31}Cl_3Ir_2$, μ-Chloro-μ-hydrido-dichlorobis(pentamethylcyclopentadienyl)diiridium(III), 43B, 1082
$C_{20}H_{31}Cl_3Rh_2$, μ-Chloro-μ-hydridobis(chloro(pentamethylcyclopentadienyl)rhodium(III)), 39B, 622
$C_{21}H_{14}CrO_3$, (η[6]-Diphenylfulvene)tricarbonylchromium, 43B, 1053
$C_{21}H_{15}BF_4Mo_3O_7$, μ₃-Oxo-tris[dicarbonyl(η[5]-cyclopentadienyl)molybdenum] tetrafluoroborate, 46B, 1153
$C_{21}H_{15}ClMn_3O_6Sb$, Chloro-tris(cyclopentadienyl-dicarbonyl-manganese)-antimony, 46B, 891
$C_{21}H_{15}Cl_4Mn_3O_6Sb_2$, Dicarbonyl-cyclopentadienyl-bis(dicarbonyl-cyclopentadienyl-manganese)-bis(μ-dichloro-antimony)manganese, 46B, 891
$C_{21}H_{18}FeORu$, Ferrocenyl ruthenocenyl ketone, 29, 551
$C_{21}H_{18}Fe_2O$, Diferrocenyl ketone, 31B, 359
$C_{21}H_{18}MoN_2O_2S$, Dicarbonylcyclopentadienyl-(N-(S)-α-phenylethyl-pyridine-2-thiocarbamido)molybdenum, 45B, 974
$C_{21}H_{19}BF_4Fe_2$, α,α-Diferrocenylmethylium tetrafluoroborate, 44B, 767
$C_{21}H_{20}OS_2V_2 \cdot C_6H_6$, (μ-Dithiocarbonato-O,S,S')bis(vanadocene) benzene, 46B, 891
$C_{21}H_{21}F_3FeO$, Tris(μ-trimethylene)-trifluoroacetyl-ferrocene, 46B, 886
$C_{21}H_{24}FeO_4S$, Iron allyl complex, 37B, 461
$C_{22}H_{10}Cr_3F_{18}O_{12}$, Bis(cyclopentadienyl)hexakis(trifluoroacetato)trichromium, 44B, 768
$C_{22}H_{18}Fe$, 2-Biphenylylferrocene, 32B, 361
$C_{22}H_{18}Fe$, 4-Biphenylylferrocene, 35B, 558
$C_{22}H_{18}FeGe$, (1,1'-Ferrocenediyl)diphenylgermane, 46B, 892
$C_{22}H_{18}FeSi$, (1,1'-Ferrocenediyl)diphenylsilane, 42B, 530
$C_{22}H_{20}Cr_2N_2O_2S_2$, trans-Di-μ-phenylthio-dinitrosylbis-(π-cyclopentadienyl)dichromium(I), 33B, 354
$C_{22}H_{20}FeSi$, Ferrocenyldiphenylsilane, 43B, 1084
$C_{22}H_{20}N_2Ti$, η[2]-cis-Azobenzenebis(η[5]-cyclopentadienyl)titanium, 46B, 1105
$C_{22}H_{20}O_2W$, Dicarbonyl(η-cyclopentadienyl)[1-3-η-(4-cyclopenta-1',3'-dienyl)-5-(cyclopenta-1",4"-dienyl)cyclopentenyl]tungsten, 46B, 893
$C_{22}H_{20}S_2Ti$, Dicyclopentadienyl(diphenylmercapto)titanium(IV), 42B, 679
$C_{22}H_{20}S_2V$, Dicyclopentadienyl(diphenylmercapto)vanadium(IV), 42B, 679
$C_{22}H_{22}As_2Mn_2O_4$, meso-μ-(1,2-Diphenyldiarsane)-bis(dicarbonyl(cyclopentadienyl)manganese), 42B, 680
$C_{22}H_{22}Fe_2$, 1,2-Bis(ferrocenyl)ethane, 40B, 763
$C_{22}H_{24}FeO$, 2,1'-Trimethylene-1-(α-phenyl-α-hydroxypropyl)ferrocene, 39B, 623
$C_{22}H_{26}Fe$, 1,1',2,2',3,4,4',5'-Tetrakis(trimethylene)ferrocene, 44B, 1153
$C_{22}H_{26}Hf_2O$, μ-Oxo-bis(methylhafnocene), 42B, 676
$C_{22}H_{26}NPr$, Tris(η[5]-cyclopentadienyl)(isocyanocyclohexane)praseodymium, 42B, 681
$C_{22}H_{27}As_2CoFe_2O_3$, Di-μ-carbonyl-μ-[carbonyl(cyclopentadienyl)-bis-(dimethylarsenido)iron]cyclopentadienyl(cyclopentadienylcobalt)-iron, 46B, 1266
$C_{22}H_{27}ClOTi$, Chloro-(η[5]-cyclopentadienyl)-(η[5]-1-methyl-3-isopropyl-

cyclopentadienyl)-(2,6-dimethylphenoxy)titanium, 43B, 1085
$C_{22}H_{27}FeNO_2$, (S,R,S)-2-(p-Methoxyphenyl)hydroxymethyl-N,N-dimethyl-1-ferrocenylethylamine, 39B, 624
$C_{22}H_{34}Cl_7Co_2Fe$, Tri-μ-chloro-bis(η^5-tetramethylethylcyclopentadienyl)dicobalt(II) tetrachloroferrate, 44B, 772
$C_{22}H_{34}I_4Rh_2$, Di-μ-iodo-diiodobis(η^5-ethyltetramethylcyclopentadienyl)dirhodium(III), 46B, 894
$C_{22}H_{46}Cr_2N_4Si_4$, Di-μ-(trimethylsilylimido)bis(cyclopentadienyltrimethylsilylimidochromium), 44B, 768
$C_{22}H_{46}Mn_2N_4Si_4$, Di-μ-bis(trimethylsilyl)isodiazenedi(cyclopentadienylmanganese), 44B, 769
$C_{23}H_{11}F_{12}MoN_3O_2$, Bis-(3,5-di(trifluoromethyl)phenyl)triazenido(dicarbonyl)(η-cyclopentadienyl)molybdenum(II), 41B, 906
$C_{23}H_{17}ClN_2O_2Ti$, Chloro-π-cyclopentadienylbis-8-quinolatotitanium-(IV), 35B, 559
$C_{23}H_{18}CoFeO_5P$, Di-μ-carbonyl-carbonyl(dicarbonyl(methyldiphenylphosphine)cobaltio)-π-cyclopentadienyliron, 40B, 763
$C_{23}H_{19}BF_4Fe$, Ferrocenyldiphenylcarbenium tetrafluoroborate, 45B, 975
$C_{23}H_{20}Cl_4MoNOPSn$, Chloro-η^5-cyclopentadienyl-nitroso-triphenylphosphino-molybdenum-trichlorotin, 46B, 898
$C_{23}H_{20}CuP$, Triphenylphosphine(pentahaptocyclopentadienyl)copper(I), 35B, 561
$C_{23}H_{24}U$, Tricyclopentadienyl(p-methylbenzyl)uranium, 42B, 682
$C_{23}H_{30}Mo_2N_2O_4$, Tetraethylammonium μ-cyano-bis(cyclopentadienyldicarbonylmolybdate)(Mo-Mo), 46B, 895
$C_{24}H_{12}BiMn_3O_9$, Tris(tricarbonylmanganese-σ^1,η^5-cyclopentadienyl-C)-bismuth, 46B, 896
$C_{24}H_{12}ClMn_3O_9Sn$, Chloro-tris[tricarbonyl-(σ,η^5-cyclopentadienyl)manganese]tin, 46B, 897
$C_{24}H_{18}FeO_2$, Dibenzoylferrocene, 20, 551
$C_{24}H_{18}N_2O_8Ti$, Dicyclopentadienyldi(p-nitrobenzoyl)titanium, 42B, 683
$C_{24}H_{20}F_3NiP$, π-Cyclopentadienyl-σ-trifluoromethyl(triphenylphosphine)-nickel, 35B, 530
$C_{24}H_{20}F_6Fe_4O_4P$, Tetrakis(cyclopentadienyliron carbonyl) hexafluorophosphate, 38B, 785
$C_{24}H_{20}Fe$, (Z)-(1,2-Diphenylethenyl)ferrocene, 45B, 975
$C_{24}H_{20}Fe_4O_4$, Cyclopentadienyliron carbonyl tetramer, 38B, 784
$C_{24}H_{20}IrOP$, η^5-Cyclopentadienyl(triphenylphosphine)carbonyliridium, 40B, 764
$C_{24}H_{22}Cl_2Zr$, Dichloro-bis(η^5-benzylcyclopentadienyl)zirconium, 44B, 1153
$C_{24}H_{22}Cl_8MoNO_3PSn_2$, Carbonyl-$\eta^5$-cyclopentadienyl-nitroso-triphenylphosphino-molybdenum-trichlorotin aquapentachlorotin, 46B, 898
$C_{24}H_{22}Fe_2O_4$, Bis-(1-(1'-carbomethoxyferrocenyl)), 33B, 355
$C_{24}H_{22}MnO_4P$, Cyclopentadienyl-dicarbonyl-[(1-methyl-3-oxo-1-butenyloxy)diphenylphosphane]manganese, 46B, 898
$C_{24}H_{22}Nb_2O_{10}$, Bis-(π-cyclopentadienyldicarbomethoxyacetylenecarbonylniobium), 35B, 562
$C_{24}H_{23}F_{15}MoN_2$, (η^5-Cyclopentadienyl)(η^4-tetrakis(trifluoromethyl)-t-butyliminocyclopentadiene)(t-butylisocyano)trifluoromethylmolybdenum, 43B, 995
$C_{24}H_{24}Fe_2$, 1,12-Dimethyl(1,1)ferrocenophane, 39B, 624
$C_{24}H_{26}Fe_2$, Bis(1'-(1-ethylferrocenyl)), 30B, 308; 31B, 361
$C_{24}H_{28}Br_2Ti_2$, Di-μ-bromo-bis(bis(η^5-methylcyclopentadienyl)titanium-(III)), 43B, 1079
$C_{24}H_{28}Cl_2Ti_2$, Di-μ-chloro-bis(bis(η^5-methylcyclopentadienyl)titanium(III)), 43B, 1079

$C_{24}H_{28}Cl_2Yb_2$, Di-μ-chloro-bis(bis(methylcyclopentadienyl)ytterbium), 41B, 907
$C_{24}H_{28}Cl_4O_4Ti_4$, cyclo-Tetrakis(μ-oxo-chloro(methylcyclopentadienyl)-titanium(IV)), 46B, 899
$C_{24}H_{28}Mo_4O_8$, Di-μ-oxo-di-$μ_3$-oxo-bis(bis(η-methylcyclopentadienyl)molybdenum(IV)-dioxomolybdenum(VI)), 44B, 758
$C_{24}H_{28}O_3Ti_2$, μ-($η^5$:$η^5$-Fulvalene)-di-μ-hydroxyl-bis(cyclopentadienyl-titanium) - tetrahydrofuran, 42B, 683
$C_{24}H_{30}Cl_2Mo_2O_8Zn_2$, Di-μ-chloro-bis((tricarbonyl-cyclopentadienyl-molybdenum)-diethylether zinc), 40B, 765
$C_{24}H_{30}Cr_2O_4$, Dicarbonylpentamethylcyclopentadienylchromium dimer, 40B, 766
$C_{24}H_{30}Fe_2O_4$, Bis(μ-carbonyl)bis[($η^5$-pentamethylcyclopentadienyl)carbonyliron](Fe-Fe), 46B, 899
$C_{24}H_{32}B_2F_8Mo_2NiS_4$, Bis-(bis-(π-cyclopentadienyl)molybdenum(IV)-bis-μ-methanethiolato)nickel(II) tetrafluoroborate, 40B, 767
$C_{24}H_{32}F_{12}P_2PtS_4Ta_2$, Bis[bis($η^5$-cyclopentadienyl)tantalum(V)bis(μ-methanethiolato)]platinum(0) hexafluorophosphate, 45B, 976
$C_{24}H_{36}B_2F_8Nb_2NiO_2S_4$, Bis-(bis-(π-cyclopentadienyl)niobium(V)-bis-μ-methanethiolato)nickel(0) tetrafluoroborate dihydrate, 40B, 767
$C_{24}H_{38}OYb \cdot 0.5\ C_7H_8$, Bis($η^5$-pentamethylcyclopentadienyl)(tetrahydrofuran)ytterbium(II) hemitoluene, 46B, 900
$C_{24}H_{40}Fe_2O_4Si_6$, (Octamethyl-($η^5$-cyclopentadienyl-dicarbonyl-iron)-cyclopentasilane)-dimethylsilyl-$η^5$-cyclopentadienyl-dicarbonyl-iron, 46B, 885
$C_{24}H_{41}MoNO_3$, Tetra-n-butylammonium η-cyclopentadienyltricarbonylmolybdate, 43B, 1073
$C_{25}H_{19}BF_4Fe$, Ferrocenyldiphenylcyclopropenium tetrafluoroborate, 40B, 768
$C_{25}H_{20}BrMoO_2P \cdot 0.25\ CH_2Cl_2$, cis-Bromodicarbonyl($η^5$-cyclopentadienyl)(triphenylphosphine)molybdenum dichloromethane solvate, 45B, 977
$C_{25}H_{20}IMoPO_2$, π-Cyclopentadienyl-trans-dicarbonyliodo(triphenylphosphine)molybdenum, 37B, 461
$C_{25}H_{20}MnO_2P$, Dicarbonyl-π-cyclopentadienyltriphenylphosphinemanganese, 39B, 625
$C_{25}H_{21}O_2ReSi$, cis-Hydridotriphenylsilyl(η-cyclopentadienyl)dicarbonylrhenium, 43B, 1086
$C_{25}H_{23}MnNO_2P$, Carbonylnitrosyl(triphenylphosphine)(5-exo-methylcyclopentadiene)manganese, 42B, 684
$C_{25}H_{24}Co_2$, 2,4-Bis(π-cyclopentadienylcobalt-cyclopentadien)yl-cyclopentadiene, 26, 653; 30B, 302
$C_{25}H_{25}ClOTi$, Chloro-(o-cresolato)-(1-α-methylbenzyltitanocene), 41B, 908
$C_{25}H_{30}Fe$, 1,1',2,2',3,3',4,5,4',5'-Pentakis(trimethylene)ferrocene, 44B, 1154
$C_{25}H_{30}O_8Ru_3Si_3$, Octacarbonyl-1,3,5-tris(trimethylsilyl)pentalene-triruthenium (form I), 45B, 978
$C_{25}H_{30}O_8Ru_3Si_3$, Octacarbonyl-1,3,5-tris(trimethylsilyl)pentalene-triruthenium (form II), 45B, 978
$C_{25}H_{32}Fe$, [4](1,1')[4](3,3')[4](5,5')[3](4,4')Ferrocenophane, 45B, 978
$C_{25}H_{33}OTi$, Bis(cyclopentadienyl)(2,6-di-t-butyl-4-methylphenoxy)titanium(III), 46B, 901
$C_{26}H_{18}FeO_4$, 1,1'-Bis(phenylglyoxyloyl)-ferrocene, 45B, 979
$C_{26}H_{19}CrO_3P$, Tricarbonyl(triphenylphosphonium cyclopentadienylide)chromium, 46B, 902
$C_{26}H_{20}AuO_3PW$, Tricarbonyl-π-cyclopentadienyltungstio(triphenylphos-

phine)gold, 34B, 463
$C_{26}H_{20}Ni$, π-Cyclopentadienyl-π-triphenylcyclopropenylnickel, 37B, 462
$C_{26}H_{22}Co_2MnO_4P$, Dicarbonyl-cyclopentadienyl-manganese-μ_3-(benzylphosphido)-bis(carbonyl-cyclopentadienyl-cobalt), 45B, 980
$C_{26}H_{25}F_{12}N_3P_2V_2$, Bis(fulvalene)bis(acetonitrile)divanadium(III)(V-V) bis(hexafluorophosphate) acetonitrile, 45B, 980
$C_{26}H_{25}N_2Ti$, 4-Methylene-3,7,8-trimethyl-1,10-phenanthrolinebis(η^5-cyclopentadienyl)titanium, 46B, 902
$C_{26}H_{26}N_4Ti_2$, μ-Pyrazole-bis(η^5-cyclopentadienyl)titanium(III), 44B, 770
$C_{26}H_{28}FeNP$, [1-(2-Diphenylphosphinoferrocenyl)ethyl]dimethylamine, 46B, 903
$C_{26}H_{28}Fe_2$, α,α,γ-Trimethyl-γ-ferrocenyl-1,2-trimethyleneferrocene, 40B, 770
$C_{26}H_{30}Co_2FeO_6$, (μ_3-Carbonyl)bis(μ-carbonyl)[tricarbonylironbis((η^5-pentamethylcyclopentadienyl)cobalt)](2Fe-Co,Co-Co), 46B, 907
$C_{26}H_{34}O_4Ru_2$, Bis(ethyltetramethylcyclopentadienyldicarbonylruthenium(I)), 44B, 1155
$C_{26}H_{38}Cl_4Fe_2N_2OZn$, Ferrocenylmethyl(dimethyl)ammonium tetrachlorozincate monohydrate, 37B, 463
$C_{26}H_{40}Al_2Fe_2O_4$, Di-μ-triethylaluminumcarbonylbis(cyclopentadienylcarbonyliron), 34B, 459
$C_{26}H_{42}Fe$, Tetra-t-butylferrocene, 38B, 786
$C_{27}H_{17}CoF_3NiO_4P$, Di-μ-carbonyl-dicarbonyl(π-cyclopentadienylnickelio)(tris-p-fluorophenylphosphine)cobalt, 41B, 908
$C_{27}H_{23}AsBCo_4F_4O_4$, Tetrakis($\eta^5$-cyclopentadienylcarbonylcobalt)arsonium tetrafluoroborate hemi(benzene), 43B, 1088
$C_{27}H_{23}MoO_3P$, trans-Dicarbonyl-π-cyclopentadienyl(triphenyl-phosphine)molybdenum acetyl, 33B, 356
$C_{27}H_{27}MoN_2O_2P$, (-)-Carbonyl(η-cyclopentadienyl)(((S)-methyl(1-phenylethyl)amino)diphenylphosphine)nitrosylmolybdenum, 44B, 770
$C_{27}H_{33}Co$, (π-Cyclopentadienyl)(trans-diphenyldi(trimethylsilyl)cyclobutadiene)cobalt, 39B, 625
$C_{28}H_{20}Bi_2Cl_2Mn_4O_8$, μ-Dichloro-bis[bis(dicarbonylcyclopentadienylmanganese)bismuth], 46B, 904
$C_{28}H_{20}Cl_{10}Fe_4O_8Sb_2$, Bis-μ-(dicarbonyl(π-cyclopentadienyliron)ferriochloro)-bis-((dicarbonyl(π-cyclopentadienyliron)ferriochloro)trichloroantimony), 40B, 771
$C_{28}H_{21}FeNiO_3P$, (π-Cyclopentadienylnickel)(ethynyl)(triphenylphosphine)iron tricarbonyl, 41B, 909
$C_{28}H_{25}ClSiZr$, Chlorobis(π-cyclopentadienyl)(triphenylsilyl)-zirconium(IV), 37B, 464
$C_{28}H_{25}Co_2I_3P$, Cobalticenium (triphenylphosphine)triiodocobaltate(II), 38B, 786
$C_{28}H_{28}As_3CoCr_2FeO_{12}$, Tricarbonyl-(dicarbonyl(tricarbonyl(cyclopentadienyl)chromium)cyclopentadienyl-μ-(dimethylarsenido)chromium)-bis-μ-(dimethylarsenido)-(tetracarbonyliron)cobalt, 46B, 905
$C_{28}H_{28}CoNiO_4P$, Di-μ-carbonyl-dicarbonyl(cyclohexyl(diphenyl)phosphine)(π-methylcyclopentadienylnickelio)cobalt, 41B, 910
$C_{28}H_{28}N_2S_2Ti_2$, (2,4-Dithiopyrimidinato)bis[bis(η^5-methylcyclopentadienyl)titanium(III)], 45B, 981
$C_{28}H_{29}FeO_3PS$, (-)-(η^5-Cyclopentadienyl)(isobutylsulfinato)(triphenylphosphine)carbonyl-iron, 44B, 771
$C_{28}H_{30}N_2O_6PRh$, Di(nitrato-O)(η-pentamethylcyclopentadienyl)triphenylphosphinerhodium(III), 46B, 905
$C_{28}H_{31}ClOTi$, 2,6-Dimethylphenoxy(3-methyl-1-(α,α-dimethylbenzyl)cyc-

lopentadienyl)cyclopentadienyltitanium chloride, 40B, 772

$C_{28}H_{31}ClO_2Ti$, η^5-Cyclopentadienyl-(η^5-1-methyl-3-isopropylcyclopentadienyl)(2-chlorophenoxy)(2,6-dimethylphenoxy)titanium, 43B, 1085

$C_{28}H_{35}O_2Ti_2$, μ-(η^1:η^5-Cyclopentadienyl)tris(η^5-cyclopentadienyl)-(tetrahydrofuran)dititanium(Ti-Ti) - tetrahydrofuran, 42B, 684

$C_{28}H_{36}Cl_4MnO_2Ti_2$, Bis(di-$\mu$-chloro-bis(cyclopentadienyl)titanium-(III))manganese(II)-bis(tetrahydrofuran), 44B, 771

$C_{29}H_{19}Co_2O_6P$, Dicarbonyl(η-triphenylphosphoniumcyclopentadienylide)-cobalt(I) tetracarbonyl-cobaltate(-1), 45B, 981

$C_{29}H_{20}F_5NiP$, π-Cyclopentadienyl-σ-pentafluorophenyl(triphenylphosphine)nickel, 33B, 357

$C_{29}H_{21}CoS_2$, (η^5-Cyclopentadienyl)(η^4-trans-diphenyldi-2-thienylcyclobutadienyl)cobalt, 43B, 1088

$C_{29}H_{25}NiP$, π-Cyclopentadienyl(triphenylphosphine)-phenylnickel, 34B, 463

$C_{29}H_{37}Co_3O_3$, (μ_3-Carbonyl)bis(μ-carbonyl)[(η^5-methylcyclopentadienyl)cobaltbis((η^5-pentamethylcyclopentadienyl)cobalt)](3Co-Co), 46B, 907

$C_{30}H_{22}CrFeO_5$, Tricarbonyl[(5-ferrocenyl-2-methoxy-3-η^6-phenyl-4-phenyl)furan]chromium(0), 45B, 982

$C_{30}H_{22}O_{12}V_2$, Bis(π-cyclopentadienylvanadium-bis-α-furancarboxylate), 38B, 787

$C_{30}H_{25}FeOP$, π-Cyclopentadienyltriphenylphosphine-monocarbonyl-σ-phenyl iron, 31B, 361

$C_{30}H_{28}Th_2$, Di-μ-(η^5:η^1-cyclopentadienyl)-bis(dicyclopentadienylthorium(IV)), 40B, 773

$C_{30}H_{30}F_6Ni_6P$, Hexakis[(η^5-cyclopentadienyl)nickel](12Ni-Ni) hexafluorophosphate, 46B, 906

$C_{30}H_{30}Ni_6$, Hexakis[(η^5-cyclopentadienyl)nickel](12Ni-Ni), 46B, 906

$C_{30}H_{30}O_3Zr \cdot C_7H_8$, Tri-$\mu$-oxo-tris(dicyclopentadienylzirconium) toluene solvate, 45B, 991

$C_{30}H_{30}O_8Ti_6$, Octa-μ_3-oxo-hexakis(cyclopentadienyltitanium), 43B, 1090

$C_{30}H_{37}Co_2MnO_4$, (μ_3-Carbonyl)tris(μ-carbonyl)[(η^5-methylcyclopentadienyl)manganesebis((η^5-pentamethylcyclopentadienyl)cobalt)](2Mn-Co,Co-Co), 46B, 907

$C_{30}H_{39}AsMo_3O_7P_2$, μ-Dimethylarsonio-bis(dicarbonyl-cyclopentadienyl-trimethylphosphine)molybdenum tricarbonyl(cyclopentadienyl)molybdate, 46B, 909

$C_{30}H_{50}Cl_6Cr_2Li_2O_6$, Bis(lithiumcyclopentadienylchromiumtrichloride-ditetrahydrofuran)dioxane, 38B, 787

$C_{31}H_{25}Fe_2O_6P$, cis-Bis(η^5-cyclopentadienyl)tricarbonyl(triphenylphosphito)diiron, 40B, 774

$C_{31}H_{26}FeIOP$, Carbonyl-(η^5-1-methyl-3-phenyl-cyclopentadienyl)-iodo-(triphenylphosphine)iron, 45B, 887

$C_{31}H_{29}Cl_3NbP_2 \cdot 2\ C_7H_8$, [1,2-Bis(diphenylphosphino)ethane]trichloro-(η^5-cyclopentadienyl)niobium(IV) bistoluene solvate, 45B, 983

$C_{32}H_{20}AgBF_4I_4O_{12}W_4$, Tetrakis(cyclopentadienyliodotungsten-tricarbonyl)silver tetrafluoroborate, 43B, 1090

$C_{32}H_{20}Hg_4Mo_8O_{12}$, ($\eta^5$-Cyclopentadienyl-tricarbonyl-molybdenum)-mercury-molybdenum tetramer, 45B, 984

$C_{32}H_{25}O_4PV_2$, (Cyclopentadienyltriphenylphosphinevanadium)bis(μ-carbonyl)(cyclopentadienyldicarbonylvanadium), 46B, 878

$C_{32}H_{26}CeN$, Tris(indenyl)pyridinecerium, 45B, 985

$C_{32}H_{27}FeO_2P$, (1-4-η-5-exo-Benzylcyclopenta-1,3-diene)dicarbonyl(triphenylphosphine)iron(0), 45B, 985

$C_{32}H_{29}ClMoP_2O$, π-Cyclopentadienyl-cis-carbonylchloro-(1,2-bis(di-

phenylphosphino)ethane)molybdenum, 37B, 461
$C_{32}H_{30}F_{18}N_{12}Rh_2$, Azidobis($\eta^5$-pentamethylcyclopentadienyl)tris(1,2-bis(trifluoromethyl)triazolato-cyclopentane)dirhodium, 45B, 986
$C_{32}H_{30}OS_2Zr_2$, μ-Oxo-bis((di-η^5-cyclopentadienyl)-benzenethiolato-zirconium), 45B, 986
$C_{32}H_{32}Cl_4Ti_2Zn$, Bis(di-μ-chloro-biscyclopentadienyltitanium)zinc benzene solvate, 39B, 626; 42B, 685
$C_{33}H_{20}FeN_5O_2P$, Dicarbonyl(η^5-cyclopentadienyl)(triphenylphosphine)-iron(II) 1,1,2,3,3-pentacycano-propenide, 45B, 985
$C_{33}H_{24}FeO_9$, Dicarbonyl-1-carbomethoxy-2-phenyl-2-(π-2',4'-dicarbomethoxy-3',5'-diphenyl-1'-cyclopentadienyloxy)-σ-vinyl iron, 31B, 362
$C_{33}H_{25}Co$, (η^5-Cyclopentadienyl)(η^4-1,2,3,4-tetraphenylcyclobutadiene)cobalt, 45B, 988
$C_{33}H_{28}B_3Co$, 1-(η^5-Cyclopentadienyl)-4,5,7,8-tetraphenyl-4,5,7,8-tetracarba-1-cobaltaoctaborane, 46B, 908
$C_{33}H_{29}O_2P_2V$, cis-Dicarbonylcyclopentadienyl-(1,2-bis(diphenylphosphino)ethane)vanadium, 44B, 772
$C_{33}H_{30}Fe_3$, (1.1.1)Ferrocenophane, 35B, 563
$C_{33}H_{30}N_2OTi_2 \cdot 0.5\ C_7H_8$, [$\mu$-Diphenylureylene]-bis[bis(η^5-cyclopentadienyl)titanium] toluene solvate, 45B, 994
$C_{33}H_{40}AlFeNO_2$, Tetraethylammonium triphenyl[(η^5-cyclopentadienyl)dicarbonyliron]aluminate, 45B, 987
$C_{33}H_{40}FeN_2O_5$, Quinidine(-)-1,1'-dimethylferrocene-3-carboxylate monohydrate, 32B, 363
$C_{34}H_{26}Cl_2Ti$, Bis(1,3-diphenylcyclopentadienyl)titanium dichloride, 39B, 626
$C_{34}H_{30}Co_2P_2$, Diphenylphosphino-cyclopentadienyl-cobalt dimer, 32B, 364
$C_{34}H_{30}Ni_2P_2$, Diphenylphosphino-cyclopentadienyl-nickel dimer, 32B, 364
$C_{34}H_{34}N_2Yb_2$, μ-Pyrazine-bis(tris(cyclopentadienyl)ytterbium(III)), 43B, 1091
$C_{34}H_{46}Cl_6O_4Ti_2Zn_2$, Bis(di(cyclopentadienyl)-dimethoxyethane-titanium) hexachlorodizincate(II) benzene solvate, 42B, 685
$C_{35}H_{34}Fe_3O$, (Triferrocenyl)(tetrahydrofuran)methane, 35B, 564
$C_{36}H_{30}BO_2V$, Bis(η^5-cyclopentadienyl)dicarbonylvanadium(III) tetraphenylborate, 46B, 868
$C_{36}H_{33}ClU$, Tris(benzylcyclopentadienide)chlorouranium(IV), 39B, 626
$C_{36}H_{33}CoSi$, (η^5-Trimethylsilylcyclopentadienyl)(η^4-tetraphenylcyclobutadiene)cobalt, 45B, 988
$C_{36}H_{37}FeIO_2P_2$, Cyclopentadienyl-iodo-(O-isopropylidene-2,3-dihydroxy-1,4-bis(diphenylphosphino)butane)iron, 46B, 910
$C_{36}H_{40}Cl_2CoP$, Dichloro(triphenylphosphine)(η^5-tetramethylethylcyclopentadienyl)cobalt toluene, 44B, 772
$C_{37}H_{25}MoN_4PS_4$, Tetraphenylphosphonium π-cyclopentadienylbis-(1,2-dicyanoethylene-1,2-dithiolato)molybdenum, 35B, 566
$C_{38}H_{30}FeOSn$, π-Cyclopentadienyl-π-diphenylacetylene-monocarbonyltriphenylstannyliron, 38B, 789
$C_{38}H_{42}Fe_2O$, (μ-Oxo)-bis(tris(μ-trimethylene)-ferrocene), 46B, 886
$C_{39}H_{35}BFe_2O_3S$, μ-Ethylthio-μ'-carbonyl-bis(η-cyclopentadienylcarbonyliron)(Fe-Fe) tetraphenylborate, 44B, 773
$C_{39}H_{37}Co$, (η^5-Cyclopentadienyl)(η^4-1,3-dimesityl-2,4-diphenylcyclobutadiene)cobalt, 45B, 988
$C_{40}H_{28}Fe$, 1,1'-(Tetraphenyl-o-phenylene)ferrocene, 43B, 1092
$C_{40}H_{36}AsMo_2O_4P$, Tetraphenylarsonium μ-dimethylphosphidobis[η^5-cyclopentadienyldicarbonylmolybdate](Mo-Mo), 46B, 910

$C_{40}H_{36}BFe_4$, Ferricenyl(III)tris(ferrocenyl(II))borate, 45B, 989
$C_{40}H_{37}CoO$, (η^5-Cyclopentadienyl)(η^4-2,4-dimesityl-3,5-diphenylcyclopentadienone)cobalt, 46B, 911
$C_{40}H_{44}BFe_2NO_4 \cdot 0.5\ C_4H_8O$, Tetraethylammonium {di-μ-carbonyl[η^5-cyclopentadienyl(carbonyl)iron]-[(η^5-triphenyl-boronylcyclopentadienyl)(carbonyl)iron](Fe-Fe)} tetrahydrofuran solvate, 45B, 989
$C_{40}H_{44}Li_4Mo_4$, Cyclotetra-μ-lithio-tetra(hydrido(bis-(π-cyclopentadienyl))molybdenum), 40B, 775
$C_{40}H_{44}Li_4W_4$, Cyclotetra-μ-lithio-tetra(hydrido(bis-(π-cyclopentadienyl))tungsten), 40B, 775
$C_{40}H_{60}N_2Ti_2$, μ-Dinitrogen-bis(bis(pentamethylcyclopentadienyl)titanium(II)), 42B, 686
$C_{40}H_{60}N_6Zr_2$, μ-Dinitrogen-bis(bis(pentamethylcyclopentadienyl)dinitrogenzirconium(II)), 42B, 687
$C_{40}H_{64}B_2F_8Rh_4$, {Tetrakis[hydrido(η^5-pentamethylcycloentadienyl)rhodium]}(+2) tetrafluoroborate, 45B, 990
$C_{41}H_{29}Co$, (η^5-Cyclopentadienyl)-(η^4-1,3-di-(1-naphthyl)-2,4-diphenylcyclopbutadiene)cobalt, 45B, 990
$C_{41}H_{35}FeIO_6P_2$, π-Cyclopentadienyl-bis(triphenylphosphite)iron iodide, 33B, 358
$C_{42}H_{37}NbOP_2$, (π-Cyclopentadienyl)bis(triphenylphosphine)niobium-carbonyl dihydride, 40B, 775
$C_{42}H_{40}O_6Ti_4$, Di-μ-(dicyclopentadienyltitanium)-bis(carbonato-dicyclopentadienyltitanium), 45B, 991
$C_{43}H_{32}Cl_8Fe_7Sb_2$, Hexacarbonyl(μ_3-chloroantimonio)tri-π-cyclopentadienyltriiron tetrachloroferrate - dichloromethane, 37B, 465
$C_{43}H_{35}MoNO_2P_2 \cdot 0.5\ CH_2Cl_2$, Isocyanatocarbonyl-π-cyclopentadienyl-bis(triphenylphosphine)molybdenum, 37B, 466
$C_{44}H_{39}CoFe_2$, (η^5-Cyclopentadienyl)(η^4-1,3-diferrocenyl-2,4-diphenylcyclobutadiene)cobalt hexane solvate, 44B, 1155
$C_{44}H_{74}Br_4Mg_4Mo_2O_3$, Bis-μ-(bis-(η-cyclopentadienyl)hydridomolybdenum)-bis-(di-μ-bromo-(cyclohexylmagnesium)((diethyl ether)magnesium)), 41B, 911
$C_{46}H_{38}Hg_2I_4P_2$, Di-μ-iodo-bis(iodo(3-triphenylphosphoniumcyclopentadienylide)mercury), 42B, 688
$C_{46}H_{39}Au_2BF_4FeP_2$, Ferrocenyl-(triphenylphosphine)gold complex, 40B, 776
$C_{46}H_{46}F_6P_3Rh$, (η^5-Pentamethylcyclopentadienyl)-bis(triphenylphosphine)-hydridorhodium(III) hexafluorophosphate, 45B, 992
$C_{48}H_{40}Co_2NO_2P_2$, Bis(triphenylphosphine)iminium bis(η^5-cyclopentadienyl)di-μ-carbonyl-dicobaltate, 42B, 688; 43B, 1094
$C_{50}H_{36}Co_2S_2$, Bis((η^5-cyclopentadienyl)(η^4-trans-diphenyl-2-thienylcyclobutadienyl)cobalt), 43B, 1088
$C_{50}H_{40}NO_4P_2Rh_3$, μ-Nitrido-bis(triphenylphosphorus) bis(η^5-cyclopentadienyl)tetracarbonyltrirhodate(Rh-Rh), 44B, 773
$C_{50}H_{48}N_4Ti_2$, N,N',N'',N'''-Tetra-p-tolyloxalylamidine-bis(di(cyclopentadienyl)-titanium), 45B, 992
$C_{52}H_{68}Cu_4Fe_4N_4$, 2-Copper-1-(dimethylaminomethyl)ferrocene tetramer, 43B, 1032
$C_{53}H_{52}F_6Fe_2INOP_4$, Carbonyl-cyclopentadienyl-iodo-iron-μ-(tris(2-diphenylphosphinoethyl)-amine)-cyclopentadienyl-iron hexafluorophosphate, 45B, 993
$C_{55}H_{47}BFeOP_2$, η^5-Cyclopentadienylcarbonylbis(diphenylphosphino)methaneiron tetraphenylborate, 45B, 994
$C_{56}H_{50}N_4O_2Ti_3$, 1,2:2,3-Bis[μ-diphenylureylene]-tris[bis(η^5-cyclopentadienyl)titanium], 45B, 994
$C_{57}H_{52}BFeNP_2$, (η^5-Cyclopentadienyl)(bis(diphenylphosphino)ethane)-

(acetonitrile)iron tetraphenylborate, 44B, 774
$C_{64}H_{44}Ag_2F_{12}Fe_2O_{10}P_2 \cdot X\ CH_2Cl_2$, μ-1-5-η:1'-5'-η-[Bis(μ-tetraphenylcyclopentadienediyl-oxo-O)-di(aquasilver)]-bis(tricarbonyliron) bis(hexafluorophosphate) dichloromethane solvate, 46B, 912
$C_{64}H_{68}Fe_2N_8$, Decamethylferrocenium 7,7,8,8-tetracyano-p-quinodimethanide (1:1) dimer, 45B, 685
$C_{82}H_{70}Cd_3Ge_4Ni_2 \cdot C_7H_8$, Bis($\eta^5$-cyclopentadienyl-(triphenylgermyl)-(triphenylgermylcadmium)nickel)cadmium toluene solvate, 46B, 913

74. METAL π-COMPLEXES (ARENE)

$C_6H_6AgAlCl_4$, Silver(I) aluminum chloride-benzene complex, 31B, 364
$C_6H_6AgClO_4$, Silver perchlorate-benzene complex, 13, 530; 22, 689
$C_6H_6AlCl_4Cu$, Benzene-copper(I) aluminum tetrachloride complex, 31B, 365
$C_6H_6Al_3Cl_{12}U$, π-Benzene-uranium(III)-aluminum chloride, 37B, 467
$C_6H_{12}ClF_6N_2PRu \cdot 0.33\ NH_4PF_6$, Diammine($\eta^6$-benzene)chlororuthenium-(II) hexafluorophosphate ammonium hexafluorophosphate, 44B, 774
$C_7H_4FeO_3$, Cyclobutadienetricarbonyliron (gas-ed), 35B, 895; 40B, 1143
$C_7H_6Cl_6Ge_2ORu$, (π-Benzene)carbonylbis(trichlorogermyl)ruthenium, 41B, 911
$C_9H_6Cl_6CrHg_3O_3$, Benzenechromium tricarbonyl tris(mercuric chloride), 43B, 1094
$C_9H_6CrO_3$, Benzenetricarbonylchromium, 26, 684; 30B, 316; 39B, 627
$C_9H_7BrMoO_2$, π-Cycloheptatrienylbromodicarbonylmolybdenum, 42B, 695
$C_9H_7ClMoO_2$, π-Cycloheptatrienylchlorodicarbonylmolybdenum, 42B, 695
$C_9H_7Cl_3MoO_2Sn$, π-Cycloheptatrienyldicarbonyl(trichlorostannyl)molybdenum, 42B, 696
$C_{10}H_7BF_4MoO_3$, π-Cycloheptatrienyliumtricarbonylmolybdenum(0) tetrafluoroborate, 39B, 628
$C_{10}H_8CrO_3$, π-(Tricarbonylchromium)toluene, 43B, 1095
$C_{10}H_9CrNO_3$, Tricarbonylchromium-o-toluidine, 32B, 368
$C_{10}H_{16}Ag_4Cl_4O_{20}$, Naphthalene-tetrakis(silver perchlorate) tetrahydrate, 40B, 549
$C_{11}H_8CrO_4S$, Methylbenzoatechromium(0) dicarbonyl thiocarbonyl, 41B, 912
$C_{11}H_8CrO_4Se$, (1,6-η-(Methyl benzoate))dicarbonylselenocarbonylchromium, 44B, 775
$C_{11}H_8CrO_5$, Tricarbonylchromium-methylbenzoate, 32B, 369; 42B, 689
$C_{11}H_8CrO_5$, o-Hydroxyacetylbenchrotrene, 39B, 628
$C_{12}H_7FeO_5Rh$, μ-(1-3-η:4-7-η-Cycloheptatrienyl)-tricarbonylirondicarbonylrhodium(Fe-Rh), 42B, 698
$C_{12}H_7IMoO_3$, Iodo-π-indenyltricarbonylmolybdenum(II), 38B, 789
$C_{12}H_8AgClO_4$, Acenaphthylenesilver perchlorate, 38B, 790
$C_{12}H_8F_4V$, Bis(1,4-difluorobenzene)vanadium(0), 42B, 689
$C_{12}H_{10}AgClO_4$, Acenaphthenesilver perchlorate, 38B, 791
$C_{12}H_{10}CrO_5$, o-Methoxyacetylbenchrotrene, 39B, 628
$C_{12}H_{12}Al_2Cl_8Pd_2$, Benzene sandwich complex, 35B, 567
$C_{12}H_{12}Al_4Cl_{14}Pd_2$, Benzene sandwich complex, 35B, 567
$C_{12}H_{12}Cr$, Dibenzenechromium, 20, 557; 24, 595; 28, 613
$C_{12}H_{12}CrI$, Bis(benzene)chromium(I) iodide, 40B, 777
$C_{12}H_{12}CrO_5$, (+)-o-Methoxy-(1'-hydroxyethyl)benzene chromium tricarbonyl, 38B, 791
$C_{12}H_{12}CrO_6$, (η^6-1,2,3-Trimethoxybenzene)tricarbonylchromium, 46B, 914

$C_{12}H_{12}MoO_3$, 1,3,5-Trimethylbenzenetricarbonylmolybdenum, 43B, 1096
$C_{12}H_{12}Ti$, Cyclopentadienylcycloheptatrienyltitanium, 39B, 630
$C_{12}H_{12}V$, π-Cyclopentadienyl-π-cycloheptatrienylvanadium, 28, 614
$C_{12}H_{14}CrO_3Si$, Tricarbonyl[trimethyl(η^6-phenyl)silane]chromium, 45B, 995
$C_{12}H_{15}CrF_6O_3P$, 1,3,5-Tris(2-difluorophosphito-ethyl)benzenechromium, 45B, 995
$C_{12}H_{16}BF_4MoO_3$, Acetylacetonatoaquo(η^7-cycloheptatrienyl)molybdenum tetrafluoroborate, 43B, 1096
$C_{12}H_{16}B_2Co$, Bis(1-methylborinato)cobalt, 38B, 792
$C_{12}H_{16}B_2CoO_2$, Bis(1-methoxyborinato)cobalt, 38B, 792
$C_{12}H_{18}Al_2Cl_8Ti \cdot C_6H_6$, ($\eta^6$-Hexamethylbenzene)-titanium-bis(di-μ-chloro-dichloro-aluminum) benzene solvate, 45B, 996
$C_{13}H_8CrO_3$, Naphthalene chromium tricarbonyl, 32B, 370
$C_{13}H_9CrNO_3$, 1-Aminonaphthalenetricarbonylchromium, 33B, 361
$C_{13}H_{12}CrO_4$, endo-2-Methyl-1-indanoltricarbonylchromium, 44B, 1156
$C_{13}H_{12}CrO_4$, exo-2-Methyl-1-indanoltricarbonylchromium, 44B, 1156
$C_{13}H_{14}CrO_5$, (1,2-Bis(1-hydroxyethyl)benzene)tricarbonylchromium, 41B, 913
$C_{13}H_{14}MoNO_2S$, Acetylacetonato(η^7-cycloheptatrienyl)isothiocyanatomolybdenum, 44B, 776
$C_{13}H_{15}CrNO_3$, (η^6-N,N-Diethylaniline)tricarbonylchromium, 46B, 914
$C_{14}H_8Fe_2O_5$, Pentacarbonyl-7H-indenediiron, 43B, 1066
$C_{14}H_{12}Ag_4Cl_4O_{17}$, Anthracenetetrakis(silver perchlorate) monohydrate, 40B, 777
$C_{14}H_{12}CrO_3S$, (1-3,3a,8a-η)(5,7-Dimethyl-4H-cyclohepta[c]thiophene)tricarbonylchromium, 39B, 630
$C_{14}H_{14}AgClO_4$, 1,2-Diphenylethanesilver(I) perchlorate, 41B, 913
$C_{14}H_{14}BCl_3F_4Mo_2$, Tri-μ-chloro-bis(η-cycloheptatrienyl)dimolybdenum tetrafluoroborate, 43B, 1097
$C_{14}H_{14}CrO_5$, 4-t-Butyl-π-(tricarbonylchromium)benzoic acid, 40B, 778
$C_{14}H_{16}CrI$, Bis(toluene)chromium iodide, 26, 679
$C_{14}H_{16}CrO_4$, Methyl-ethyl-o-methylbenchrotrenylmethanol, 39B, 631
$C_{14}H_{16}Ru$, (1-5-η-Cycloheptadienyl)(1-5-η-cycloheptatrienyl)ruthenium(II), 42B, 699
$C_{14}H_{18}CrN_2$, Bis(2,6-dimethylpyridine)chromium (Form A), 42B, 690
$C_{14}H_{18}CrN_2$, Bis(2,6-dimethylpyridine)chromium (Form B), 42B, 690
$C_{14}H_{18}Ru$, (Benzene)(1,5-cyclooctadiene)ruthenium(0), 42B, 691
$C_{15}H_6Co_4O_9$, Benzeneenneacarbonyltetracobalt, 40B, 779
$C_{15}H_{12}F_6FeO_3$, Tricarbonyl(η^4-tetramethylbis(trifluoromethyl)benzene)iron, 43B, 1098
$C_{15}H_{14}CrO_3S$, (3a,4-8,8a-η)-(4,5,7-Trimethyl-4H-cyclohepta[b]thiophene)tricarbonylchromium, 39B, 631
$C_{15}H_{14}CrO_3S$, (3a,4-8,8a-η)-(5,7,8-Trimethyl-8H-cyclohepta[b]thiophene)tricarbonylchromium, 39B, 631
$C_{15}H_{17}IrO_2$, π-Cyclopentadienyl-π-duroquinoneiridium, 37B, 456
$C_{15}H_{17}O_2Rh$, π-Cyclopentadienyl-π-duroquinonerhodium, 35B, 573
$C_{15}H_{18}CrO_3$, Hexamethylbenzenechromium tricarbonyl, 30B, 318
$C_{15}H_{18}MoO_3$, Hexamethylbenzenetricarbonylmolybdenum, 43B, 1096
$C_{15}H_{19}O_4Rh$, Acetylacetonato-π-duroquinonerhodium, 35B, 574
$C_{15}H_{21}CoO_4$, Cyclopentadienyl(duroquinone)cobalt dihydrate, 38B, 782
$C_{15}H_{30}B_3CrN_3O_3$, Tricarbonyl(hexaethylborazine)chromium(0), 38B, 793
$C_{16}H_8Mn_2O_6$, trans-Azulenedimanganese hexacarbonyl, 33B, 364
$C_{16}H_{12}Fe_2O_6$, 3,α-Dimethylstyrenebis(tricarbonyliron), 43B, 829
$C_{16}H_{18}CrI$, 1,1'-Tetramethylene-dibenzenechromium(I) iodide, 44B, 776
$C_{16}H_{20}AgClO_4$, Bis(o-xylene)silver(I) perchlorate, 41B, 914
$C_{16}H_{22}CrIO$, Bis(ethylbenzene)chromium iodide monohydrate, 43B, 1098

$C_{17}H_5F_{18}Rh$, π-Cyclopentadienylhexakis(trifluoromethyl)benzene rhodium, 31B, 366
$C_{17}H_6Fe_3O_9$, 1,1,1-Tricarbonylferraindene-bis(tricarbonyliron), 43B, 829
$C_{17}H_8Fe_2O_5$, Acenaphthylenediiron pentacarbonyl, 35B, 572
$C_{17}H_8O_7Ru_3$, Azulenetriruthenium heptacarbonyl, 39B, 632
$C_{17}H_{10}Co_4O_9$, Xyleneenneacarbonyltetracobalt, 40B, 779
$C_{17}H_{10}CrO_3$, Anthracenechromium tricarbonyl, 33B, 365
$C_{17}H_{10}CrO_3$, Tricarbonylphenanthrenechromium(0), 29, 542; 33B, 366; 39B, 633
$C_{17}H_{12}CrO_3$, Tricarbonyl-1,4-dihydrophenanthrenechromium(0), 39B, 633
$C_{17}H_{12}CrO_3$, 9,10-Dihydrophenanthrenechromium tricarbonyl, 33B, 366
$C_{17}H_{16}CrO_5$, Tricarbonyl(1-4:9-10-η-(2-ethyl-4-methoxy-3-methyl-1-naphthol))chromium(0), 42B, 692
$C_{18}H_{10}Cr_2O_6$, Bis(tricarbonylchromium)biphenyl, 26, 650
$C_{18}H_{14}Fe$, Diindenyliron, 22, 742
$C_{18}H_{14}Fe$, η^5-Cyclopentadienyl-η^6-fluorenyliron, 43B, 1075
$C_{18}H_{14}FeO_2$, Bicyclooctatetraenyldicarbonyliron, 42B, 705
$C_{18}H_{14}Ru$, Bis-indenylruthenium, 32B, 359
$C_{18}H_{16}Ag_2Cl_2O_8$, Indene silver perchlorate dimer, 38B, 794
$C_{18}H_{18}BF_6N_8PRu$, (Tetrakis(1-pyrazolyl)borato)(η^6-benzene)ruthenium-(II) hexafluorophosphate, 41B, 902
$C_{18}H_{20}CrF_6P$, (η^{12}-[3.3]Paracyclophane)chromium(I) hexafluorophosphate, 46B, 914
$C_{18}H_{20}CrI_3$, (η^{12}-[3.3]Paracyclophane)chromium(I) triiodide, 46B, 914
$C_{18}H_{24}BBrF_4Mo_2O_3$, μ-Bromo-di-μ-hydroxo-bis((η^7-cycloheptatrienyl)molybdenum(II)) tetrafluoroborate tetrahydrofuran, 43B, 1099
$C_{18}H_{24}CrO_3$, (1'-t-Butyl-2',2'-dimethylpropyl)-π-tricarbonylchromiumbenzene, 42B, 693
$C_{18}H_{24}NiO_2$, 1,5-Cyclooctadiene-duroquinone-nickel, 30B, 323
$C_{19}H_{16}CrO_3$, Tricarbonyl(3-8-η-[2.2]paracyclophane)chromium, 44B, 777
$C_{19}H_{18}CrO_4$, Ethyl-phenyl-o-methylbenchrotrenylmethanol, 39B, 634
$C_{19}H_{19}Mn$, (η-Methylcyclopentadienyl)(η-7-exo-phenylcyclohepta-1,3,5-triene)manganese, 44B, 777
$C_{19}H_{19}O_2Rh$, π-Indenyl-π-duroquinone-rhodium, 37B, 468
$C_{19}H_{25}O_2Rh$, π-Cyclopentadienyl-pi(2,6-di-t-butyl-1,4-benzoquinone)-rhodium, 34B, 467
$C_{20}H_{16}Fe$, 4-endo,6'-endo-Bis(π-azulene)iron, 34B, 473
$C_{20}H_{18}Fe_2O_5$, Guaiazulenepentacarbonyldiiron (isomer I), 43B, 1100
$C_{20}H_{18}Fe_2O_5$, Guaiazulenepentacarbonyldiiron (isomer II), 43B, 1100
$C_{20}H_{20}F_{12}Mo_2OP_2 \cdot 0.5\ H_2O$, μ-($\eta^5$:$\eta^5$-Fulvalene)-μ-hydrido-μ-hydroxyl-bis(η^5-cyclopentadienylmolybdenum) hexafluorophosphate hemihydrate, 43B, 1080
$C_{20}H_{24}NiO_4$, Bis(π-duroquinone)nickel, 39B, 635
$C_{20}H_{26}CrO_5Si_2$, Tricarbonyl-[4-methoxy-4-η^6-phenyl-2,3-bis(trimethylsilyl)-1,3-butadien-1-one]-chromium, 45B, 997
$C_{20}H_{26}Ru$, (η^4-Cyclooctatetraene)(η^6-hexamethylbenzene)ruthenium(0), 46B, 915
$C_{21}H_{15}AsCrO_3$, (η^6-Diphenylarsinophenyl)tricarbonylchromium(0), 46B, 916
$C_{21}H_{18}Mo_2O_6$, Guaiazulenedimolybdenum hexacarbonyl, 33B, 379
$C_{21}H_{18}O_6W_2$, Guaiazulenehexacarbonylditungsten, 42B, 694
$C_{21}H_{30}CrO_3$, Tricarbonyl(hexaethylbenzene)chromium(0), 46B, 916
$C_{22}H_{14}Mn_2O_6$, (3,5-Dimethylaceheptylene)hexacarbonyldimanganese, 44B, 778
$C_{22}H_{14}O_6Ru_3$, Bicyclooctatetraenylhexacarbonyl-triangulo-triruthenium, 42B, 705

$C_{22}H_{14}O_9Ru_4$, 4,6,8-Trimethylazuleneenneacarbonyltetraruthenium, 34B, 475
$C_{22}H_{22}F_6FeP$, Bis(1,3-dimethylindenyl)iron(III) hexafluorophosphate, 41B, 915
$C_{22}H_{25}CrO_2P$, Dicarbonyl[1-4a(10a)-η-phenanthrene](triethylphosphine)chromium(0), 46B, 917
$C_{22}H_{40}ClP_4Ta$, η^4-Naphthalene-bis[1,2-bis(dimethylphosphino)ethane]-chlorotantalum, 43B, 1101; 45B, 998
$C_{23}H_{13}CrN_3O_9$, Tricarbonylphenanthrenechromium 1,3,5-trinitrobenzene, 45B, 678
$C_{23}H_{21}CrO_5P$, (3,5-Dimethylbenzyl)diphenylphosphitodicarbonyl-chromium, 45B, 998
$C_{23}H_{24}CrO_3$, 1,3,3,5-Tetramethyl-6-(1',2'-naphtho)-bicyclo[3.2.1]-octenechromium(0) tricarbonyl, 40B, 780
$C_{24}H_{24}Cl_8Hg_4Mo_2O_6$, Tricarbonylmesitylenemolybdenum - mercury(II) chloride (1:2) dimer, 42B, 695
$C_{24}H_{24}Cr_2$, Tris(cyclooctatetraene)dichromium, 42B, 706
$C_{24}H_{26}CoF_6P$, Bis(η^6-hexamethylbenzene)cobalt hexafluorophosphate, 46B, 918
$C_{24}H_{32}AgClO_4$, Bis(cyclohexylbenzene)silver(I) perchlorate, 37B, 468
$C_{24}H_{36}Ru$, Bis(hexamethylbenzene)ruthenium(0), 38B, 795
$C_{24}H_{38}F_6PRh$, (η^6-Hexamethylbenzene)-(η^4-hexamethylcyclohexadiene)-rhodium hexafluorophosphate, 46B, 918
$C_{25}H_{33}Fe_2O_4P$, Guaiazulenetetracarbonyl(triethylphosphine)diiron, 43B, 1100
$C_{26}H_{16}Mn_2O_6$, Azulenetricarbonylmanganese dimer, 43B, 1087
$C_{27}H_{21}ClU$, Triindenyluranium chloride, 37B, 470
$C_{27}H_{21}Sm$, Triindenylsamarium, 39B, 636
$C_{27}H_{28}CrO_6$, Tricarbonyl(5,10-η-(4-(2-methoxy-1-propenyl)-2-phenyl-1-naphthol))chromium ether, 44B, 778
$C_{28}H_{22}Mo_2O_6$, Azulenetricarbonylmethylmolybdenum dimer, 33B, 382
$C_{28}H_{23}CrO_4P$, (π-Methylbenzoate)(triphenylphosphine)chromium dicarbonyl, 41B, 916
$C_{29}H_{25}BRu$, π-Cyclopentadienyl-η-(tetraphenylborato)ruthenium(II), 40B, 780
$C_{30}H_{23}Cr_2O_5P$, Dicarbonyl[1-4a(8a)-η-naphthalene](triphenyl phosphite)chromium(0), 46B, 917
$C_{30}H_{25}CrO_2PS$, η^6-Tetralone-carbonyl-thiocarbonyl-(triphenylphosphine)chromium, 46B, 918
$C_{30}H_{38}BO_6P_2Rh$, Bis(trimethylphosphite)tetraphenylboronrhodium(I), 40B, 781
$C_{30}H_{56}Mo_2N_2P_4$, μ-Dinitrogen-bis((π-mesitylene)(1,2-bis(dimethylphosphino)ethane)molybdenum), 40B, 781
$C_{31}H_{27}BMo$, Cycloheptatrienyl(tetraphenylborato)molybdenum, 44B, 779
$C_{32}H_{20}Cl_2Fe_4O_{10}$, 4,4'-Diazulenedecacarbonyltetrairon - dichloroethane, 34B, 477
$C_{32}H_{44}MoP_4$, Tetrakis(dimethylphenylphosphine)molybdenum(0), 41B, 917
$C_{35}H_{40}BMoNO_3$, Tetraethylammonium tricarbonyl(tetraphenylborato)molybdate, 44B, 779
$C_{38}H_{45}CrO_2P$, Dicarbonyl(hexaethylbenzene)(triphenylphosphine)-chromium(0), 46B, 916
$C_{42}H_{34}O_2P_2Pt$, Bis(triphenylphosphine)-(2-3-η-(1,4-benzoquinone))-platinum(0), 43B, 1102
$C_{43}H_{38}ClO_{10}P_2Rh$, η^6-Toluenebis(triphenyl phosphite)rhodium perchlorate, 46B, 919
$C_{50}H_{44}BP_2Rh$, (η^6-Tetraphenylborato)[1,2-bis(diphenylphosphino)ethane]rhodium(I), 46B, 919

$C_{57}H_{84}NiP_2$, Bistricyclohexylphosphine(1,2-η^2-anthracene)nickel(0) - toluene, 43B, 1102

75. METAL π-COMPLEXES (MISCELLANEOUS RING SYSTEMS)

$C_7H_4CrO_3S$, Thiophenechromium tricarbonyl, 30B, 318
$C_7H_6FeO_4S$, Tricarbonyl(2,5-dihydrothiophene 1-oxido)iron, 43B, 1103
$C_7H_8Ag_2N_2O_6$, Norbornadiene-silver nitrate complex, 31B, 367
$C_7H_8Cl_2Pd$, Norbornadienedichloropalladium(II), 26, 687; 30B, 324
$C_8H_7CrNO_3$, Tricarbonyl(N-methylpyrrole)chromium(0), 38B, 796
$C_8H_8AgNO_3$, Cyclooctatetraenesilver nitrate, 23, 639; 24, 586
C_8H_8ClCu, Cyclooctatetraenecopper(I) chloride, 29, 587
$C_8H_8Cl_2Pd$, Dichloro(1-2:5-6-η-cyclooctatetraene)palladium(II), 44B, 780
$C_8H_{10}AgNO_3$, exo-Tricyclo[3.2.1.0$^{2-4}$]oct-6-ene-silver nitrate, 37B, 471
$C_8H_{12}Cl_2Pd$, 1,5-Cyclooctadienedichloropalladium(II), 43B, 1103
$C_8H_{12}Cl_2Pt$, Dichloro(5-methylenecycloheptene)platinum(II), 46B, 920
$C_9F_8FeO_3$, Octafluorocyclohexa-1,3-dieneiron tricarbonyl, 32B, 371
$C_9H_7FeNO_3$, Azepine iron tricarbonyl, 38B, 797
$C_9H_7MnO_3$, π-Cyclohexadienylmanganese tricarbonyl, 34A, 162
$C_9H_8FeN_2O_5$, (1,2-Dimethyl-1,2-dihydropyridazine-3,6-dione)iron tricarbonyl, 40B, 782
$C_9H_8FeO_5S$, Tricarbonyl(η^4-3,4-dimethylthiophene-1,1-dioxide)iron, 43B, 1104
$C_9H_{10}Cl_4Cu_3NO_3 \cdot H_2O$, Ammonium ($\mu_3$-chloro)tris($\mu$-chloro)tricuprate-(I)-sesqui(p-benzoquinone) monohydrate, 46B, 920
$C_9H_{12}Ag_3N_3O_9$, cis,cis,cis-1,4,7-Cyclononatriene silver nitrate complex, 32B, 373
$C_{10}F_{10}FeO_3$, Tricarbonyl(decafluorocyclohepta-1,3-dienyl)iron, 40B, 783
$C_{10}H_4Co_2O_6$, (π-Cyclobutadiene)dicobalt hexacarbonyl, 43B, 1105
$C_{10}H_6FeO_4$, Troponetricarbonyliron, 29, 552
$C_{10}H_7F_3FeN_2O_5$, Tricarbonyl(η^4-1(1H),2(2H)-diazepinium)iron trifluoroacetate, 42B, 697
$C_{10}H_7FeNO_3$, (5-exo-Cyanocyclohexa-1,3-diene)tricarbonyliron, 44B, 1157
$C_{10}H_8MoO_3$, Cycloheptatrienetricarbonylmolybdenum, 24, 585
$C_{10}H_{11}CrNO_3$, Tricarbonyl(1,4-dimethyl-1,2-dihydropyridine)-chromium(0), 38B, 798
$C_{10}H_{12}AgBF_4O$, Bullvalene - silver tetrafluoroborate monohydrate, 33B, 368
$C_{10}H_{12}Cl_2Pt$, (+)-Dichloro(endo-dicyclopentadiene)platinum(II), 39B, 637
$C_{10}H_{14}Cl_2O_3Pt$, Dichloro(1,3,5,7-tetramethyl-4,8,9-trioxa-bicyclo-[3.3.1]nona-2,6-diene)-platinum(II), 34B, 468
$C_{10}H_{16}Cl_2Pt$, Dipentene platinum(II) chloride, 30B, 327
$C_{10}H_{16}Cl_4Pt_2$, Di-μ-chloro-dichlorobis(cyclopentene)diplatinum(II), 40B, 784
$C_{11}H_8Br_2FeO_3$, Tricarbonyl-(8,8-dibromobicyclo[5.1.0]octa-2,4-diene)-iron, 41B, 919
$C_{11}H_8FeO_3$, Cyclooctatetraenetricarbonyliron, 26, 681; 27, 855
$C_{11}H_8FeO_3$, Dicarbonyl-3-(π-(2-cyclohexadienyl))-σ-propenoyliron, 37B, 472
$C_{11}H_8FeO_3$, 1,6,7,8-tetrahapto-Heptafulveneiron tricarbonyl, 39B, 637
$C_{11}H_8MoO_3$, Cyclo-octatetraenemolybdenum tricarbonyl, 31B, 368

$C_{11}H_8O_3Ru$, (Cyclooctatetraene)tricarbonylruthenium, 40B, 1110
$C_{11}H_9FeNO_4$, (3-Acetyl-1H-azepine)tricarbonyliron(0), 37B, 473
$C_{11}H_9FeNO_5$, Tricarbonyliron complex of N-methoxycarbonylazepine, 35B, 569
$C_{11}H_{10}CrO_3$, Tricarbonylcyclo-octa-1,3,5-trienylchromium, 27, 851
$C_{11}H_{10}FeN_2O_5$, 1-Carbethoxy-(1H)-1,2-diazepinetricarbonyliron, 39B, 638
$C_{11}H_{13}CrNO_3$, 3-Ethyl-1,2-dihydro-1-methylpyridine(tricarbonyl)-chromium, 39B, 639
$C_{11}H_{13}CrNO_3$, 5-Ethyl-1,2-dihydro-1-methylpyridine(tricarbonyl)-chromium, 39B, 639
$C_{11}H_{15}Cl_2NORu$, Acetonitrilecarbonyldichloro(cycloocta-1,5-diene)-ruthenium(II), 43B, 1106
$C_{12}H_6FeO_4$, (1,7,8,9-η-Bicyclo[5.2.0]nona-1(7),2,5,8-tetraen-4-one)tricarbonyliron(0), 44B, 780
$C_{12}H_6FeO_6S$, Benzo[b]thiophen-1,1-dioxide-tetracarbonyliron, 40B, 785
$C_{12}H_8FeO_4$, Barbaralonetricarbonyliron, 40B, 786
$C_{12}H_8FeO_4$, [(2,3,6,7-η)-Bicyclo[3.2.2]nona-2,6,8-trien-4-one]tricar-bonyliron, 46B, 921
$C_{12}H_8MnNO_3$, Tricarbonyl(2-methylindolyl)manganese, 43B, 1106
$C_{12}H_{10}MoO_3$, 5,6-Dimethylenebicyclo[2.2.1]hept-2-enetricarbonylmolyb-denum, 42B, 699
$C_{12}H_{11}NO_3Ru$, 1,2,3,6-η^4-(5-Cyanocyclooctadienyl)tricarbonyl ruthen-ium, 38B, 799
$C_{12}H_{12}FeO_5$, Methyl 8-tricarbonylferra-2,3-η-bicyclo[4.2.0]oct-2-ene-7-carboxylate, 43B, 1107
$C_{12}H_{12}Mn_2N_2O_6$, (η^4-σ-1,1,1-Tricarbonyl-2,3,4,5-tetramethyl-2,5-diaza-1-manganole)tricarbonylmanganese(Mn-Mn), 46B, 921
$C_{12}H_{16}As_2Fe$, 2,2',5,5'-Tetramethyl-1,1'-diarsaferrocene, 46B, 922
$C_{12}H_{16}Cl_2OZr$, π-Cyclooctatetraenyl(tetrahydrofuran)dichlorozir-conium, 41B, 919
$C_{12}H_{16}FeP_2$, 3,3',4,4'-Tetramethyl-1,1'-diphosphaferrocene, 46B, 922
$C_{12}H_{18}Ni$, trans,trans,trans-1,5,9-Cyclododecatrienenickel, 38B, 800
$C_{12}H_{20}AgNO_4$, Pregeijerene silver nitrate, 35B, 570
$C_{13}H_6CoF_{12}O_2P$, π-Cyclopentadienyl(1-hydroxy-2,3,4,5-tetrakis-(trifluoromethyl)phosphole-1-oxide)cobalt, 41B, 920
$C_{13}H_8Co_2O_6$, Di-μ-carbonyl-tetracarbonyl-(π-norbornadiene)dicobalt, 38B, 801
$C_{13}H_8Fe_2O_5$, Cyclo-octatetraenedi-iron pentacarbonyl, 31B, 369
$C_{13}H_8Fe_2O_6$, Cycloheptatrienediiron hexacarbonyl, 37B, 474
$C_{13}H_{11}F_6FeO_4P$, (2-4:6-7-η-(8-Acetylbicyclo[3.2.1]octadienylium))tri-carbonyliron hexafluorophosphate, 44B, 781
$C_{13}H_{11}FeNO_6$, (3-Formyl-N-ethoxycarbonylazepine)tricarbonyliron(0), 40B, 787
$C_{13}H_{12}Cl_2FeO_4$, Tetracarbonyl(η^2-(9,9-dichlorobicyclo[6.1.0]non-3-ene))iron(0), 44B, 781
$C_{13}H_{12}FeO_3$, Tricyclo[6.2.0.0$^{2-7}$]deca-3,5-dienetricarbonyliron, 40B, 788
$C_{13}H_{13}Ti$, Cyclopentadienylcyclooctatetraenetitanium, 35B, 575
$C_{13}H_{14}CrO_6$, Tetracarbonyl-(7,7-dimethoxynorborn-2-ene)chromium(0), 40B, 788
$C_{13}H_{15}N_2RhS_2$, 1,5-Cyclooctadienyl-methylmaleonitriledithiolatorhodi-um(I), 40B, 1006
$C_{13}H_{16}FeO$, Bis(1,3-cyclohexadiene)monocarbonyliron, 37B, 475
$C_{13}H_{17}Cl_2O_2Rh$, (1,6-Dichloro-1,5-cyclooctadiene)(2,4-pentanedion-ato)rhodium(I), 45B, 999
$C_{13}H_{18}O_2Pd$, Acetylacetonato(cyclo-octa-2,4-dienyl)palladium, 31B,

370
$C_{13}H_{19}O_2Rh$, 1,5-Cyclooctadieneacetylacetonatorhodium(I), 41B, 921
$C_{13}H_{22}B_2FeN_2O_3S$, 3,4-Diethyl-2,5-bis(dimethylamino)-1,2,5-thiadiborolene-tricarbonyliron, 43B, 810
$C_{14}H_8FeO_5$, Tetracarbonyl(2,3-η-1,4-dihydro-1,4-epoxynaphthalene)-iron(0), 43B, 1107
$C_{14}H_8Fe_2O_6$, Cyclooctatetraenebis(tricarbonyliron), 26, 681; 27, 854
$C_{14}H_8Mn_2O_6$, Cyclooctatetraenehexacarbonyldimanganese, 41B, 922
$C_{14}H_8O_6Ru_2$, Cyclooctatetraenediruthenium hexacarbonyl, 33B, 369
$C_{14}H_{10}BMnO_3$, Tricarbonyl(1-phenylborinato)manganese, 40B, 790
$C_{14}H_{10}CrO_3$, Tricarbonyl-(1,6-methanocyclodecapentaene)-chromium, 34B, 472
$C_{14}H_{10}CrO_5$, 3a-4:8-8a-η-(5,7-Dimethylcyclohepta[b]furan-6-one)-chromium tricarbonyl, 44B, 782
$C_{14}H_{10}Fe_2O_6$, (1,3,5-Cyclo-octatriene) di-iron hexacarbonyl, 34B, 475
$C_{14}H_{12}CrO_4$, (5,7-Dimethyl-4H-cyclohepta[b]furan)tricarbonylchromium, 43B, 1108
$C_{14}H_{12}F_{12}O_2Pt$, Bis(hexafluoroacetone)(cycloocta-1,5-diene)platinum, 43B, 982
$C_{14}H_{12}FeO_3S$, (5-8-η)-5,7-Dimethyl-4H-cyclohepta[b]thiophenetricarbonyliron, 44B, 1158
$C_{14}H_{12}FeO_4$, (5-8-η)-5,7-Dimethyl-4H-cyclohepta[b]furantricarbonyl-iron, 44B, 1158
$C_{14}H_{12}Fe_2O_6Se$, Hexacarbonyl[μ-[(1,2-η:2-η)-1-Cyclooctene-1-selenolato(2-)-Se:Se]]diiron(Fe-Fe), 46B, 923
$C_{14}H_{14}FeO_3$, Tricyclo[6.3.0.0$^{2-7}$]undeca-3,5-dienetricarbonyliron, 39B, 639
$C_{14}H_{16}AgNO_3$, Norbornadiene dimer - silver nitrate, 42B, 699
$C_{14}H_{16}MoO_3$, (1-6-η-7-t-Butyl-cycloheptatriene)(tricarbonyl)molybdenum(0), 41B, 922
$C_{14}H_{19}Ir$, (η^4-1,3-Cycloheptadienyl)-(η^5-cycloheptadienyl)iridium(I), 46B, 923
$C_{14}H_{24}Cl_4Pt_2$, Di-μ-chloro-dichlorobis(cycloheptene)diplatinum(II), 40B, 784
$C_{14}H_{37}F_6N_6PRu$, (η-Cycloocta-1,5-diene)tris(N,N-dimethylhydrazine)-hydridoruthenium(II) hexafluorophosphate, 44B, 782
$C_{15}H_7CoFeO_6$, Di-μ-carbonyl-carbonyl(tricarbonylcobalt)-π-indenyl-iron, 40B, 789
$C_{15}H_8F_6Ni_3O_3$, Hexafluorobut-2-yne-cyclooctatetraene-tricarbonyltrinickel, 41B, 880
$C_{15}H_8Fe_2O_5$, Azulene di-iron pentacarbonyl, 32B, 376
$C_{15}H_8Fe_2O_8$, 7-Oxabicyclo[2.2.1]hept-2-enediironheptacarbonyl, 44B, 783
$C_{15}H_{10}CrO_3$, Tricarbonylelassovalenechromium, 43B, 1109
$C_{15}H_{10}Fe_2O_6$, μ-(1,2,6-η:3-5-η-Bicyclo[6.1.0]nona-1,3,5-triene)hexacarbonyldiiron, 41B, 830
$C_{15}H_{14}AsCoFeO_6$, Tetracarbonyliron-μ-(dimethylarsenido)-(dicarbonyl-norbornadiene-cobalt), 45B, 1000
$C_{15}H_{15}BFeO_3Si$, (1,1-Dimethyl-4-phenyl-1-sila-4-bora-2,5-cyclohexadiene)tricarbonyliron, 43B, 1109
$C_{15}H_{16}CrO_3$, (1,3,5,7-Tetramethylcyclooctatetraene)chromium tricarbonyl, 33B, 371
$C_{15}H_{16}FeO_3$, Tricyclo[6.4.0.0$^{2-7}$]dodeca-3,5-dienetricarbonyliron, 41B, 923
$C_{15}H_{19}ClO_2PdS$, 1,5-Cyclooctadiene(phenylsulphonylmethanato)chloro-palladium(II), 43B, 1103
$C_{15}H_{19}O_2Rh$, Cycloocta-1,5-diene(η-methoxycarbonylcyclopentadienyl)-

rhodium(I), 43B, 1069
$C_{15}H_{20}Ag_2N_2O_8$, Costunolide bis(silver nitrate), 44B, 784
$C_{15}H_{24}AgNO_3$, Germacratriene - silver nitrate, 37B, 477
$C_{15}H_{24}AgNO_3$, β-Gorgonene-silver nitrate, 33B, 372
$C_{16}H_8FeN_4O_3$, Tricarbonyl(7,7,8,8-tetracyano-π,σ-bicyclo[4.2.1]non-3-en-2,9-ylene)iron, 37B, 478
$C_{16}H_8FeO_4$, Acenaphthyleneiron tetracarbonyl, 41B, 924
$C_{16}H_{10}Fe_2O_6$, Bullvalene-hexacarbonyldiiron, 38B, 802
$C_{16}H_{12}CrO_3$, Tricarbonyl-exo-7-phenylcyclohepta-1,3,5-trienechromium, 33B, 373
$C_{16}H_{12}Fe$, (4,5,6,7,8,4',5',6',7',8'-η^{10}-1,1'-Dihydro-1,1'-bipentalenyl)iron, 39B, 640
$C_{16}H_{12}FeO_3$, Tricarbonyl(exo-7-phenylcyclohepta-1,3,5-triene)iron, 43B, 1110
$C_{16}H_{12}Fe_2O_6$, Bicyclo[6.2.0]deca-2,4,6-triene iron carbonyl complex, 39B, 641
$C_{16}H_{12}Fe_2O_6$, cis-(1,2,6-η^3-3,4,5-η^3-Bicyclo[6.2.0]deca-1,3,5-triene)hexacarbonyldiiron(Fe-Fe), 39B, 642
$C_{16}H_{12}Fe_2O_8$, 1,5-Cyclooctadienebis(iron tetracarbonyl), 35B, 575
$C_{16}H_{12}MnO_4P$, 2-Benzoyl-3,4-dimethylphosphacymantrene, 44B, 784
$C_{16}H_{13}CoFeO_4$, Di-μ-carbonyl-(carbonyl-π-norbornadienecobaltio)carbonyl-π-cyclopentadienyliron, 41B, 901
$C_{16}H_{14}CrO_5$, π-(exo-2-Acetoxybenzonorbornenyl)-exo-tricarbonylchromium, 42B, 691
$C_{16}H_{16}AgNO_3$, Cyclooctatetraene dimer-silver nitrate, 23, 640
$C_{16}H_{16}B_2Fe_2O_4$, Di-μ-carbonyl-dicarbonylbis(1-methylborinato)diiron, 40B, 790
$C_{16}H_{16}Cl_4Pd_2$, Di-μ-chloro-bis(4,6-η-(1-chlorocyclooctatrienyl))dipalladium(II), 44B, 785
$C_{16}H_{16}Co_2O_4$, cis-Di-μ-carbonyl-bis(carbonyl-(π-cyclohexa-1,3-diene)-cobalt), 38B, 802
$C_{16}H_{16}Fe$, Bis(cyclooctatetraene)iron, 38B, 803
$C_{16}H_{16}NiO_4$, (2,3-Bis(methoxycarbonyl)-2pi,5-norbornadien-7-yl)-(π-cyclopentadienyl)nickel(II), 28, 606
$C_{16}H_{16}Ni_2$, Di-η^3,$\eta^{3'}$-cyclooctatetraenedinickel, 42B, 700
$C_{16}H_{16}Th$, Bis(cyclooctatetraenyl)thorium(IV), 38B, 803
$C_{16}H_{16}Ti$, Bis(cyclooctatetraene)titanium, 34B, 469
$C_{16}H_{16}U$, Bis(cyclooctatetraenyl)uranium(IV), 34B, 470; 38B, 803
$C_{16}H_{18}CrO_4$, Tetracarbonyl(hexamethylbicyclo[2.2.0]hexa-2,5-diene)-chromium, 37B, 479
$C_{16}H_{18}Ni$, Bis(η^3-cyclooctatrienyl)nickel, 46B, 924
$C_{16}H_{19}BFeINO_2$, (1-t-Butyl-3-methyl-2-phenyl-η^5-1,2-azaborolinyl)dicarbonyliodoiron, 46B, 935
$C_{16}H_{19}O_6Rh$, Pentane-2,4-dionato-(2,3,4,6-η^4-2,3-dicarbomethoxobicyclo[2.2.1]heptadiene)-rhodium(I), 41B, 925
$C_{16}H_{20}Fe$, (1-6η-(1,3,5-Cyclooctatriene))(2-5η-bicyclo[4.2.0]octa-2,4-diene)iron(0), 40B, 791
$C_{16}H_{21}F_6FeO_3P$, Heptamethylcyclohexadienyltricarbonyliron hexafluorophosphate, 41B, 925
$C_{16}H_{23}Rh$, (1,2:5,6-η-1,5-Cyclooctadiene)(1,3-η-1,4-cyclooctadienyl)-rhodium(I), 46B, 925
$C_{16}H_{24}AgBF_4$, Bis(1,5-cyclooctadiene)silver(I) tetrafluoroborate, 45B, 1000
$C_{16}H_{24}B_4F_4Ni_{0.8}$, Bis(1,4-difluoro-2,3,5,6-tetramethyl-1,4-diboracyclohexa-2,5-diene)nickel(0), 41B, 926
$C_{16}H_{24}Br_4Ni_2$, Di-μ-bromo-bis(bromo-(tetramethylcyclobutadiene)nickel(II)), 45B, 1001

$C_{16}H_{24}Cl_2Cu_2$, Di-μ-chloro-bis(cycloocta-1,5-dienecopper(I)), 28, 614
$C_{16}H_{24}Cl_2O_2Th$, (η^8-Cyclooctatetraene)thorium(IV) dichloride bis(tetrahydrofuran) (α form), 46B, 925
$C_{16}H_{24}Cl_2O_2Th$, (η^8-Cyclooctatetraene)thorium(IV) dichloride bis(tetrahydrofuran) (β form), 46B, 925
$C_{16}H_{24}Cl_2Rh_2$, μ-Dichlorobis(cyclo-octa-1,5-dienylrhodium(I)), 27, 853
$C_{16}H_{24}Ni$, Bis(cyclo-octa-1,5-diene)nickel(0), 30B, 321
$C_{16}H_{25}Co$, π-Cyclooctenyl-π-cycloocta-1,5-dienecobalt, 37B, 479
$C_{16}H_{26}Cl_2Pd_2$, Bis(η^3-2-methylene-6-methylcyclohexyl)-di-μ-chloro-dipalladium, 42B, 701
$C_{16}H_{26}O_2RuSi_2$, Dicarbonyl(trimethylsilyl)(1-3,6-7,η-8-endo-trimethylsilylcyclooctatrienyl)ruthenium, 44B, 785
$C_{16}H_{30}Cl_2N_2O_2Pd \cdot H_2O$, Dichloro[$\eta^4$-1,3-bis(diethylamino)-2,4-bis-(methoxymethyl)cyclobutadiene]palladium monohydrate, 45B, 1001
$C_{17}F_{17}MnO_3S$, Tricarbonyl-η^4-(1-pentafluorophenyl-2,3,4,5-tetrakis-(trifluoromethyl)thiophen)manganese, 41B, 920
$C_{17}H_8IMn_3N_2O_9$, Bis(σ-pyrrolyltricarbonylmanganese)tricarbonyliodomanganese, 46B, 926
$C_{17}H_8Mn_3O_{11}P$, (3,4-Dimethylphospholyl)undecacarbonyltrimanganese, 43B, 1111
$C_{17}H_{11}IO_4Ru_2$, μ-Iodo-μ-3-phenylcycloheptatrienyl-bis(dicarbonylruthenium), 43B, 973
$C_{17}H_{12}F_6FeO_5$, Tricarbonyl(tetramethylbis(trifluoromethyl)bicyclo-[3.3.0]octadiendione)iron, 43B, 1098
$C_{17}H_{16}Fe_2O_5$, (1,3,5-Trimethyl-7-methylene-1,3,5-cyclo-octatriene)diiron pentacarbonyl, 33B, 375
$C_{17}H_{16}Fe_2O_5$, (1,3,5,7-Tetramethylcyclooctatetraene)diiron pentacarbonyl, 33B, 374
$C_{17}H_{17}F_9OPt$, (Cyclo-octa-1,5-diene)[2-methyl-2,4,6-tris(trifluoromethyl)pyran]platinum, 46B, 926
$C_{17}H_{17}Mn_2O_7P$, (1-t-Butyl-3,4-dimethylphosphole)heptacarbonyldimanganese, 43B, 1111
$C_{17}H_{20}FeO_5$, 3-Methoxy-1-methyl-5-(2-oxocyclohexyl)cyclohexadieneiron tricarbonyl, 40B, 792
$C_{17}H_{24}O_2Pd$, Acetylacetonato(π-pentamethylbicyclo[2.2.0]hexa-2,5-dienylmethyl)palladium(II), 35B, 576
$C_{17}H_{28}B_2Ni$, Cyclopentadienyl(1,3,4,5-tetraethyl-2-methyl-1,3-diborolenyl)nickel, 44B, 786
$C_{17}H_{30}Cl_2N_2Ru$, Dichlorodipiperidine(π-norbornadienyl)ruthenium, 42B, 702
$C_{17}H_{40}BF_4FeO_9P_3$, Tris(trimethyl phosphite)(η^3-cyclooctenyl)iron tetrafluoroborate, 46B, 927
$C_{17}H_{40}FeO_9P_3$, η^3-Cyclooctenyltris(trimethyl phosphite)iron(I), 45B, 1002
$C_{18}H_{10}FeO_7$, Tricarbonyl[methyl 1-(η-2-5,4-methoxy-1-methylcyclohexa-2,4-dienyl)-2-oxocyclopentanecarboxylate]iron, 46B, 928
$C_{18}H_{10}Fe_2O_6$, Bis(tricarbonyliron)heptalene, 45B, 1003
$C_{18}H_{12}Fe_2O_7$, 12-Oxa[4.4.3]propella-2,4,7,9-tetraene bis(irontricarbonyl) (triclinic), 38B, 805
$C_{18}H_{12}Fe_2O_7$, 12-Oxa[4.4.3]propella-2,4,7,9-tetraene bis(irontricarbonyl) (monoclinic), 38B, 804
$C_{18}H_{12}Mo_2O_6$, μ-(π,π-1,6-Dihydroheptalene)-bis(tricarbonylmolybdenum), 34B, 471
$C_{18}H_{16}Co_2O_4$, Bis(norbornadiene)dicobalt tetracarbonyl, 44B, 786
$C_{18}H_{16}CrO_3$, Tricarbonyl(1,2,3,4,5,6-η^6-3-ethyl-endo-7-phenyl-1,3,5-cycloheptatriene)chromium, 45B, 1003

$C_{18}H_{16}Fe_2O_6$, (1,7-Cyclododecadiyne)hexacarbonyldiiron, 39B, 643
$C_{18}H_{18}F_3O_2Rh$, Benzoyl-1,1,1-trifluoroacetonatocycloocta-1,5-diene-rhodium(I), 46B, 928
$C_{18}H_{18}Fe_2O_7$, μ-Cycloundeca-allyl-heptacarbonyldi-iron, 35B, 578
$C_{18}H_{19}Cl_2O_2Rh$, 1-Phenyl-1,3-butanedionato(1,6-dichloro-1,5-cyclo-octadiene)rhodium(I), 44B, 787
$C_{18}H_{22}Ni$, Cyclopentadienyl(1-(cyclopentadienyl)-1,2,3,4-tetramethyl-π-cyclobutenyl)nickel, 30B, 322
$C_{18}H_{22}NiO_2$, (η⁵-Cyclopentadienyl)-1,3',4'η³-(2-methyl-2-(6',6'-dimethylbicyclo[3.2.0]hept-3'-en-7'-on-2'-yl)propionyl)nickel(II), 40B, 793
$C_{18}H_{22}O_4Rh_2$, Bis(μ-acetato)bis(norbornadiene)dirhodium(I), 45B, 1004
$C_{18}H_{24}B_4F_4NiO_2$, Dicarbonyl(1,4-difluoro-2,3,5,6-tetramethyl-1,4-diboracyclohexa-1,5-diene)nickel(0) solvate, 41B, 927
$C_{18}H_{24}Ge_2O_4Ru_2$, μ-Pentalene-bis(dicarbonyl(trimethylgermyl)ruthenium), 44B, 787
$C_{18}H_{24}O_5Ru_2Si_2$, μ-Trimethylsilylcycloheptatrienyl-pentacarbonyltrimethylsilyldiruthenium(Ru-Ru), 41B, 928
$C_{18}H_{26}O_4Ru_2Si_2$, Tetracarbonyl(trimethylsilyl)-μ-[1-4-η:1-σ,5-8-η-(8-trimethylsilylocta-1,3,5,7-tetraenyl)]diruthenium(Ru-Ru), 46B, 929
$C_{18}H_{34}Cl_2N_2Ru_2$, Hydrido(η-cycloocta-1,5-diene)ruthenium-μ-chloro-μ-hydrido-μ-(N,N-dimethylhydrazine)-chloro(η-cycloocta-1,5-diene)-ruthenium, 44B, 788
$C_{18}H_{35}ClN_2Ru$, Chloro(cycloocta-1,5-diene)hydridobis(piperidine)ruthenium, 44B, 1158
$C_{19}H_{16}FeO_3$, π-Tetracyclo[8.6.0.0²⁻⁹.0³⁻⁸]hexadecapenta-4,6,11,13,15-enetricarbonyliron, 33B, 377
$C_{19}H_{16}O_3Ru$, Tricarbonyl(cyclooctatetraene dimer)ruthenium, 45B, 1005
$C_{19}H_{22}Cl_2N_2Ru$, Dichlorobis(phenylamine)(bicyclo[2.2.1]hepta-2,5-diene)ruthenium, 40B, 794
$C_{19}H_{23}Mo_2O_5$, η³-Cycloheptatrienyldicarbonylmolybdenum-tris(μ-methoxy)-η⁷-cycloheptatrienylmolybdenum, 45B, 1005
$C_{19}H_{24}F_6OPt_2$, μ-Hexafluoroacetone-bis((cycloocta-1,5-diene)platinum)(Pt-Pt), 43B, 982
$C_{19}H_{24}FeNiO_3$, Iron nickel cyclobutadiene complex, 35B, 585
$C_{19}H_{24}O_5Ru_2Si_2$, Pentacarbonyl-μ-{2'-3'-η,7'-8'-η:4'-6'-η-[4-(cycloocta-2',5',7'-triene-1',4'-diyl)-1,1,4,4-tetramethyl-1,4-disilapentyl]}diruthenium(Ru-Ru), 46B, 929
$C_{19}H_{26}CrN_2O_3$, (2SR,2'RS)-Tricarbonyl[5-ethyl-2-(5'-ethyl-1',2',3',4'-tetrahydro-1'-methyl-2'-pyridyl)-1,6-dihydro-1-methylpyridine]chromium, 46B, 930
$C_{19}H_{26}CrN_2O_3$, (2SR,2'SR)-Tricarbonyl[5-ethyl-2-(5'-ethyl-1',2',3',4'-tetrahydro-1'-methyl-2'-pyridyl)-1,6-dihydro-1-methylpyridine]chromium, 46B, 930
$C_{19}H_{26}CrN_2O_3$, Tricarbonyl[5-ethyl-2-(5'-ethyl-1',2',3',6'-tetrahydro-1'-methyl-2'-pyridyl)-1,6-dihydro-1-methylpyridine]chromium, 46B, 929
$C_{20}H_{14}Cr_2O_6$, Hexacarbonyl-trans-6a,12a-dihydro-octalenechromium(0), 39B, 643
$C_{20}H_{16}O_4Ru_3$, Bis(cyclooctatetraene)triruthenium tetracarbonyl, 33B, 378
$C_{20}H_{20}U$, Bis[η⁸-(cyclobutenocyclotetraoctatetraene)]uranium(IV), 45B, 1006
$C_{20}H_{24}OZr$, Bis(cyclooctatetraene)(tetrahydrofuran)zirconium, 38B, 806
$C_{20}H_{32}Br_2NiS_2$, Dibromo(1-9:2-8-η-3,3,7,7,10,10,14,14-octamethyl-5,12-dithiatricyclo[7.5.0.0²⁻⁸]tetradeca-1(9),2(8)-diene)nickel,

42B, 703
$C_{20}H_{36}AgNO_3$, Bis(trans-cyclodecene)silver nitrate, 32B, 378
$C_{20}H_{36}AgNO_3$, cis-Cyclodecene - silver nitrate, 37B, 480
$C_{20}H_{42}Cl_2N_2Ru$, Dichlorodihexylamine(π-cyclooctadienyl)ruthenium, 42B, 702
$C_{20}H_{44}P_4Zr$, Hydrido(η^5-cyclooctadienyl)bis[1,2-bis(dimethylphosphino)ethane]zirconium, 46B, 931
$C_{21}H_{16}CrO_4S$, (1-Methyl-3,5-diphenylthiabenzene-1-oxide)tricarbonylchromium, 44B, 788
$C_{21}H_{16}CrO_5$, Chromium tricarbonyl-1-ethoxy-3-oxo-3a-phenyl-3,3a-dihydroazulene, 31B, 372
$C_{21}H_{16}O_5Ru$, Pentacarbonyl(cyclooctatetraene dimer)diruthenium(Ru-Ru), 45B, 1005
$C_{21}H_{16}O_9Ru_3$, Nonacarbonyl(cyclododecatrienetriyl)hydrido-triangulo-triruthenium, 37B, 480
$C_{21}H_{17}F_{12}O_5PRu$, Dicarbonyl(4-methyl-2,6,7-trioxa-1-phosphabicyclo[2.2.2]octane)(cyclohexa-1,5-diene-hexafluorobut-2-yne)ruthenium, 43B, 988
$C_{21}H_{18}F_6NO_6Rh$, (1,1,1,5,5,5-Hexafluoropentane-2,4-dionato)pyridine-(2,3,5,6-η^4-2,3-dicarbomethoxybicyclo[2.2.1]heptadiene)rhodium(I), 43B, 1111
$C_{21}H_{20}FeO_4S$, Tricarbonyl(3-4:6-7-η-(2-isopropylthio-8-benzolbicyclo[3.2.1]octadiene))iron, 44B, 789
$C_{21}H_{24}O_7W_2$, η^3-Cycloheptatrienyldicarbonyltungsten-tris(μ-methoxy)-η^4-cycloheptatrienyldicarbonyl tungsten, 45B, 1005
$C_{21}H_{27}Fe_2O_5P$, η^6-(Bicyclo[6.2.0]dodeca-1,3,5-triene)pentacarbonyltriethylphosphinediiron(Fe-Fe), 41B, 847
$C_{21}H_{28}CrN_2O_4$, Tricarbonyl-(2,3-bis(diethylamino)-1-methoxyindene)chromium(0), 45B, 148
$C_{21}H_{28}FeO_8$, Tricarbonyl-[methyl 3-hydroxymethyl-1-(2-5-η-4-methoxy-1-methylcyclohexa-2,4-dienyl)-3-methyl-2-oxocyclohexane-1-carboxylate]iron, 46B, 931
$C_{21}H_{30}Pt$, Tris(bicyclo[2.2.1]heptene)platinum, 43B, 1112
$C_{21}H_{32}BrNiP$, ((S)-t-Butylmethylphenylphosphine)((+)-(1R,5R)-3-2-10-η-pinenyl)nickel bromide, 42B, 704
$C_{22}H_{14}Fe_2O_6$, Heptafulveneiron tricarbonyl dimer, 40B, 795
$C_{22}H_{15}F_{12}Rh$, η^5-Indenyl{1-2:3-4-η^4-[6-endo-propen-2-yl-1,2,3,4-tetrakis-(trifluoromethyl)cyclohexa-1,3-diene]}rhodium, 46B, 932
$C_{22}H_{16}O_{10}Ru_4$, Decacarbonyl(cyclododecatrienyl)-tetrahedro-tetraruthenium, 38B, 807
$C_{22}H_{17}F_{12}FeO_5P$, (4-Methyl-2,6,7-trioxa-1-phosphabicyclo[2.2.2]-octane)-(1,2,3,8-tetra(trifluoromethyl)-tricyclo[6.3.0.0^{4-11}]-undeca-2,5-diene)dicarbonyliron, 45B, 1008
$C_{22}H_{18}Fe_2O_6$, 3,3'-Bis(bicyclo[4.2.0]octa-2,4-dienetricarbonyliron), 40B, 795
$C_{22}H_{28}N_4Rh_2$, (2,2'-Biimidazolyl)bis((1,5-cyclooctadiene)rhodium(I)), 42B, 706
$C_{22}H_{30}CeKO_3$, Bis(cyclooctatetraenyl)cerium(III) - potassium diglyme, 38B, 808
$C_{22}H_{30}Cl_4Ni_2$, 1,2,3,4-Tetramethylcyclobutadienenickel(II) chloride - benzene, 27, 847
$C_{22}H_{34}Mo_2O_3S_3$, Tricarbonylmolybdenum-tris-(μ-n-butylthio)-η^7-cycloheptatrienylmolybdenum, 45B, 1008
$C_{22}H_{34}Mo_2O_3S_3$, Tricarbonylmolybdenum-tris-(μ-t-butylthio)-η^7-cycloheptatrienylmolybdenum, 45B, 1008
$C_{23}H_{14}FeO_5$, Tricarbonyl(1,2,2a,12a-η-5,10-dimethyldibenzo[a,c]cyclobuta[f]cycloocten-3,12-dione)iron, 43B, 1113

$C_{23}H_{21}Cl_2O_2Rh$, 1,3-Diphenyl-1,3-propanedione-(1,6-dichloro-1,5-cyclooctadiene)rhodium(I), 40B, 796
$C_{23}H_{30}Cl_3IrP_2Sn$, (Norbornadiene)bis(dimethylphenylphosphine)(trichlorostannato)iridium(I), 40B, 797
$C_{23}H_{30}FeO_3$, Tricarbonyl-(1,2-di-t-butyl-3,4,5,6-tetramethylbenzobutadiene)iron, 45B, 1009
$C_{24}H_{14}Fe_3O_8$, (3,5-Dimethylaceheptylene)octacarbonyltriiron(Fe-Fe), 44B, 789
$C_{24}H_{15}CoO_3$, (η^3-Triphenylcyclopropenyl)tricarbonylcobalt, 45B, 1010
$C_{24}H_{21}CrF_2O_3P$, Tricarbonyl(4-t-butyl-1,1-difluoro-2,6-diphenyl-λ^5-phosphorin)chromium, 45B, 1011
$C_{24}H_{22}Fe_2O_4$, trans-Di-μ-carbonyl-bis(2,3,4,5,6-pentahapto-tricyclo-[6.2.0.0$^{2-6}$]deca-2,4-dien-6-yl)dicarbonyldiiron(Fe-Fe), 39B, 644
$C_{24}H_{24}Mo_2$, Tris(cyclooctatetraene)dimolybdenum, 44B, 790
$C_{24}H_{24}Ti_2$, Tris(cyclooctatetraene)dititanium, 33B, 380
$C_{24}H_{24}W_2$, Tris(cyclooctatetraene)ditungsten, 43B, 1113; 44B, 790
$C_{24}H_{28}MoO_2$, Bis(tricyclo[6.3.0.0$^{2-7}$]undeca-3,5-diene)dicarbonylmolybdenum, 40B, 798
$C_{24}H_{28}O_6Os_2$, Di-μ-cyclononaallyl-hexacarbonyldiosmium, 43B, 996
$C_{24}H_{30}B_2NiSi_2$, (1,1-Dimethyl-4-phenyl-1-sila-4-bora-2,5-cyclohexadiene)nickel, 46B, 932
$C_{24}H_{30}Rh_2$, (1,5-Cyclooctadiene)-μ-(3,5-η-1-2:6-7-η-cyclooctatrienyl)(3-5-η-cyclooctatrienyl)dirhodium, 42B, 707
$C_{24}H_{32}Cl_2O_2Ti_2$, Cyclooctatetraenetetrahydrofurantitanium chloride, 41B, 928
$C_{24}H_{32}Fe_2O_4S_2$, Bis(3,3,6,6-Tetramethyl-1-thiacyclohept-4-yne)bis(dicarbonyliron)(Fe-Fe), 40B, 798
$C_{24}H_{32}U$, Bis(1,3,5,7-tetramethylcyclooctatetraenyl)uranium(IV), 39B, 645
$C_{24}H_{34}Cl_2Pt_2$, Dichloro-di-μ-(pentamethylbicyclo[2.2.0]hexa-2,5-dienylmethyl)-diplatinum(II), 35B, 579
$C_{24}H_{40}Br_4Ni_2$, Di-μ-bromo-dibromobis(tetraethylcyclobutadiene)dinickel, 43B, 1114
$C_{24}H_{42}Cl_2Cu_2$, Di-μ-chlorotris(trans-cyclooctene)dicopper(II), 35B, 581
$C_{24}H_{50}Li_2N_4Ni$, (Dilithium trans-1,5,9-cyclododecatriene)nickel bis(N,N,N',N'-tetramethylethylene-diamine), 45B, 928
$C_{25}H_{16}Br_2FeO_3$, 1-Bromo-2-(bromomethyl)naphthalene enneacarbonyldiiron reaction product, 40B, 799
$C_{25}H_{18}FeO_3$, Iron π-complex, 35B, 582
$C_{25}H_{34}F_9P_2PtSb$, Tetramethylcyclobutadienetrifluoromethylbis(dimethylphenylphosphine)platinum(II) hexafluoroantimonate, 39B, 646
$C_{25}H_{37}CrLiO_9S_2$, Lithium tricarbonyl-(1-5-η^5)-6-(1,3-dithian-2-yl)-cyclohexadienyl-chromium(0) dioxane solvate, 45B, 1010
$C_{25}H_{37}IrP_2$, (Cycloocta-1,5-diene)bis(dimethylphenylphosphino)methyl-iridium(I), 38B, 808
$C_{26}H_{14}Fe_2O_8$, Diphenylfulveneoctacarbonyldiiron, 42B, 707
$C_{26}H_{17}CrO_3P$, 2,4,6-Triphenylphosphorinchromium(0) tricarbonyl, 38B, 809
$C_{26}H_{18}Cl_2Zr$, (η^5-Fluorenyl)(η^3-fluorenyl)dichlorozirconium(IV), 42B, 708
$C_{26}H_{28}ClPPd$, Chloro-(η^3-1-methylene-3-methylcyclohexyl)(triphenylphosphine)palladium(II), 43B, 1101
$C_{26}H_{33}Mo_2O_5P \cdot CH_2Cl_2$, Guaiazulene(triethylphosphine)pentacarbonyldimolybdenum (isomer II), 42B, 708
$C_{26}H_{33}Mo_2O_5P$, Guaiazulene(triethylphosphine)pentacarbonyldimolybdenum (isomer I), 42B, 708

$C_{26}H_{40}Fe$, Bis(6-t-butyl-1,3,5-trimethylcyclohexadienyl)iron(II), 38B, 810
$C_{27}H_{21}U$, Triindenyluranium, 46B, 933
$C_{27}H_{22}BF_4FeO_3P$, (5-exo-Triphenylphosphinocyclohexa-1,3-diene)tricarbonyliron tetrafluoroborate, 42B, 709
$C_{28}H_{18}FeO_4$, 2,4,6-Triphenyltroponeirontricarbonyl, 27,876
$C_{28}H_{23}CrO_3P$, π-Norbornadienetricarbonyltriphenylphosphinechromium(0), 41B, 929
$C_{28}H_{23}CrO_5P$, Tricarbonyl(1,1-dimethoxy-2,4,6-triphenyl-λ^5-phosphorin)chromium, 45B, 1011
$C_{28}H_{32}Cl_4Cu_4$, Norbornadienecopper(I) chloride, 29, 587
$C_{28}H_{34}Co_2FeO_4$, (μ_3-Carbonyl)tris(μ-carbonyl)[(η^4-cyclobutadiene)ironbis((η^5-pentamethylcyclopentadienyl)cobalt)](2Fe-Co,Co-Co), 46B, 907
$C_{28}H_{52}AgNO_3$, 1,1,4,4-Tetramethyl-cis-cyclodec-7-ene - silver nitrate, 37B, 480
$C_{29}H_{29}IMoNOP$, (η^7-Cycloheptatrienyl)-carbonyl-iodo-(α-methylbenzyl-N-methylamino(diphenyl)-phosphine)-molybdenum, 46B, 934
$C_{29}H_{34}Ir_4O_5$, (Cycloocta-1-ene-5-yne)bis(cycloocta-1,5-diene)pentacarbonyltetrairidium, 44B, 791
$C_{31}H_{19}Mn_3N_2O_{11}$, fac-Tricarbonyl-(diphenylacetato)-bis(tricarbonyl-(η^5-pyrrolyl)manganese)manganese, 45B, 1011
$C_{31}H_{20}FeO_3$, (Tetraphenylcyclobutadiene)tricarbonyliron, 24, 601; 30B, 320
$C_{31}H_{28}FeOP_2$, (Cyclobutadiene)(1,2-bis(diphenylphosphino)ethane-P,P')iron carbonyl, 46B, 934
$C_{31}H_{30}Br_3O_2PU$, (η^5-Indenyl)-tetrahydrofuran-(triphenylphosphine oxide)uranium tribromide, 46B, 935
$C_{32}H_{32}Cl_4Ti_4$, Cyclooctatetraenetitanium chloride, 41B, 928
$C_{32}H_{37}As_2Nb$, (η^4-Cyclooctatetraenyl)(2,3,4,5,6-η-endo-8-phenylbicyclo[5.1.0]octadienyl)(o-phenylenebis(dimethylarsine))niobium, 42B, 710
$C_{32}H_{38}B_2Fe_2N_2O_4$, cis-Di-$\mu$-carbonylbis[(1-t-butyl-3-methyl-2-phenyl-η^5-1,2-azaborolinyl)carbonyliron)], 46B, 935
$C_{32}H_{40}Nd_2O_2$, Cyclooctatetraenylbis(tetrahydrofuran)neodymium(III) bis(cyclooctatetraenyl)neodymate(III), 42B, 710; 44B, 791
$C_{32}H_{48}Ce_2Cl_2O_4$, Di-μ-chloro-di(π-cyclooctatetraene-bis(tetrahydrofuran)cerium), 38B, 811
$C_{33}H_{23}CoI_2$, (π-1,2-Diiodocyclopentadienyl)tetraphenylcyclobutadienecobalt, 40B, 799
$C_{33}H_{23}Rh$, (η^5-Cyclopentadienyl)-(η^4-1,2-diphenylcyclobuta[l]phenanthrene)rhodium, 46B, 909
$C_{33}H_{24}CoI$, (π-Iodocyclopentadienyl)tetraphenylcyclobutadienecobalt, 40B, 799
$C_{33}H_{25}Rh$, π-Cyclopentadienyl-π-tetraphenylcyclobutadienerhodium(I), 39B, 647
$C_{33}H_{30}Cl_3OPRu$, Benzyltriphenylphosphonium bicyclo[2.2.1]hepta-2,5-dienecarbonyltrichlororuthenate, 44B, 792
$C_{33}H_{30}FeO_3Si$, (1-Trimethylsilyl-7-triphenylmethylbicyclo[4.2.0]octa-2,4,7-triene)tricarbonyliron, 42B, 711
$C_{33}H_{36}F_6FeNP_2Rh$, [(R,S:S,R)-α-(2-Diphenylphosphinoferrocenyl)ethyldimethylamine](norbornadiene)rhodium hexafluorophosphate, 46B, 936
$C_{34}H_{24}CoN$, (π-Cyanocyclopentadienyl)tetraphenylcyclobutadienecobalt, 40B, 799
$C_{34}H_{27}AsFeO_3$, Tetraphenylarsonium cycloheptatrienyltricarbonyliron(0), 44B, 792
$C_{34}H_{28}CoPO_2$, (η^5-Cyclopentadienyl)(η^4-exo-1-methoxy-2,3,4,5-tetra-

phenylphosphole-1-oxide)cobalt, 46B, 908

$C_{35}H_{25}O_2V$, π-Cyclopentadienyl-π-tetraphenylcyclobutadienedicarbonylvanadium, 34B, 479

$C_{35}H_{39}IrP_2$, (Cycloocta-1,5-diene)(1,2-bis(diphenylphosphino)ethane)methyliridium(I), 39B, 648

$C_{35}H_{43}BCuN_3$, Norbornene(diethylenetriamine)copper(I) tetraphenylborate, 44B, 793

$C_{36}H_{30}ClN_3Ni$, π-Triphenylcyclopropenylchlorodipyridinenickel(0) - pyridine, 37B, 482

$C_{36}H_{33}Ir_7O_{12}$, Iridium heptanuclear cluster, 44B, 793

$C_{36}H_{41}IrP_2$, (Cycloocta-1,5-diene)(1,3-bis(diphenylphosphino)propane)methyliridium(I), 39B, 648

$C_{37}H_{37}F_6MoO_2P_3S$, ((η-Cyclohexadiene)(η-cyclopentadiene)(1,2-bis(diphenylphosphino)ethane)molybdenum(V)) hexafluorophosphate sulfur dioxide, 43B, 1092

$C_{40}H_{36}P_2Pt$, 3-Methylcyclopropenebis(triphenylphosphine)platinum(0), 41B, 930

$C_{40}H_{48}ClO_5P_2Rh$, (1,5-Cyclooctadiene)-(2S,3S)-2,3-bis(diphenylphosphino)butanerhodium(I) perchlorate tetrahydrofuran solvate, 43B, 1114

$C_{41}H_{38}P_2Pt$, 1,2-Dimethylcyclopropenebis(triphenylphosphine)platinum(0), 41B, 930

$C_{42}H_{38}P_2Pt$, Cyclohexynebis(triphenylphosphine)platinum(0), 41B, 932

$C_{42}H_{38}P_2Pt$, Δ(1,4)-Bicyclo[2.2.0]hexenebis(triphenylphosphine)platinum, 41B, 931

$C_{43}H_{40}P_2Pt$, Cycloheptynebis(triphenylphosphine)platinum(0), 41B, 932

$C_{44}H_{32}Fe_2O_4$, (10,10'-Biazulenyl)diiron tetracarbonyl, 42B, 687

$C_{44}H_{42}Cl_2O_6P_2Rh_2$, Di-μ-chloro-bis(triphenyl phosphite)(cyclo-octa-1,5-diene)dirhodium(I), 35B, 583

$C_{44}H_{42}P_2Pt$, (1,2-η-Cyclooctyne)bis(triphenylphosphine)platinum(0), 46B, 937

$C_{44}H_{42}P_2Pt \cdot 0.5\ C_6H_6$, Cyclooctynebis(triphenylphosphine)platinum(0) benzene solvate, 45B, 1012

$C_{47}H_{38}O_2P_2Pt$, (1-Methyl-2-phenylcyclobutenedione)bis(triphenylphosphine)platinum(0), 42B, 712

$C_{48}H_{28}Fe_2O_4$, Iron carbonyl complex, 35B, 584

$C_{48}H_{30}F_{12}P_2Pt \cdot 0.5\ C_5H_{12}$, Bis(triphenylphosphine)(perfluoro-1,2-3,4-5,6-triethanobenzene)platinum(0) pentane solvate, 44B, 794

$C_{48}H_{44}AsNb$, Tetraphenylarsonium tris(cyclooctatetraene)niobium, 41B, 933

$C_{48}H_{47}ClCoO_4P_3 \cdot 0.5\ CH_2Cl_2$, 1,1,1-Tris(diphenylphosphinomethyl)ethane-(η^4-1,3,5-cycloheptatriene)cobalt perchlorate methylene chloride solvate, 46B, 937

$C_{48}H_{52}Cl_4Mo_6O_5$, Tetrakis(μ_3-hydroxycycloheptatrienylmolybdenum) di-μ-chloro-μ-hydroxy-bis(cycloheptatrienylmolybdenum) chloride chlorobenzene solvate, 44B, 795

$C_{51}H_{53}OP_2Rh \cdot 2\ C_6H_6$, Bis(triphenylphosphine)(η^5-2,6-di-t-butyl-4-methylcyclohexadienonyl)rhodium(I) benzene solvate, 46B, 901

$C_{52}H_{72}Cl_2Hg_2Ir_2N_{12}$, Di-μ-chloro-bis(mercury-μ-(p-tolyl(ethyl)triazenido)-(p-tolyl(ethyl)triazenidocyclooctadiene-iridium)), 46B, 938

$C_{58}H_{40}MoO_2$, Dicarbonylbis(tetraphenylcyclobutadiene)molybdenum, 44B, 1159

$C_{60}H_{40}Br_2Mo_2O_4$, Di-μ-bromo-bis(π-tetraphenylcyclobutadiene)tetracarbonyldimolybdenum(I), 34B, 479; 39B, 649

$C_{60}H_{50}Cl_2O_2Pd_2$, endo-1-Ethoxy-1,2,3,4-tetraphenylcyclobutenyl palladium chloride dimer, 30B, 326

$C_{60}H_{50}Cl_2O_2Pd_2$, exo-1-Ethoxy-1,2,3,4-tetraphenylcyclobutenyl palladium chloride dimer, 30B, 326
$C_{64}H_{48}U$, Octaphenyluranocene, 42B, 713

76. METAL COMPLEXES (ETHYLENEDIAMINE)

$C_2H_8AgClN_2O_4$, Ethylenediaminesilver(I) perchlorate, 44B, 795
$C_2H_8Br_3N_2Pt$, Ethylenediaminetribromoplatinum, 26, 689
$C_2H_8Cl_2HgN_2$, Mercury(II) ethylenediamine dichloride, 23, 571
$C_2H_8Cl_2N_2Pd$, cis-Dichloro(ethylenediamine)palladium(II), 41B, 933
$C_2H_8Cl_2N_2Pt$, cis-Dichloro(ethylenediamine)platinum(II), 41B, 933
$C_2H_8Cl_2N_4O_4Pt$, trans-Dichlorodinitroethylenediamineplatinum(IV), 40B, 801
$C_2H_{11}ClCoN_5O_4$, Chlorodinitro(ethylenediamine)amminecobalt(III), 31B, 380
$C_2H_{11}CoN_6O_6$, mer-Trinitro(ammine)ethylenediaminecobalt(III), 35B, 586
$C_2H_{12}Cl_3CrN_2O_2$, trans-Dichlorodiaquoethylenediaminechromium(III) chloride, 34B, 480
$C_2H_{12}CrN_2O_6$, Diperoxoaquoethylenediaminechromium(IV) monohydrate, 30B, 328
$C_2H_{12}CuN_2O_6S$, Diaquoethylenediaminesulfatocopper(II), 44B, 1160
$C_2H_{14}CoN_7O_7$, cis-Dinitro-cis-diammine(ethylenediamine)cobalt(III) nitrate (triclinic), 41B, 934
$C_2H_{14}CoN_7O_7$, cis-Dinitro-cis-diammine(ethylenediamine)cobalt(III) nitrate (orthorhombic), 44B, 796
$C_2H_{16}N_4NiO_{10}$, Tetra-aqua(ethylenediamine)nickel(II) nitrate, 46B, 939
$C_4H_8CuN_4S_2$, Ethylenediaminedithiocyanatocopper(II), 37B, 482
$C_4H_8HgN_4S_2$, Dithiocyanato(ethylenediamine)mercury(II), 42B, 716
$C_4H_{12}AuCl_2N_3O_4$, Chloro-bis(2-aminomethyl)amido-gold(III) perchlorate, 46B, 940
$C_4H_{12}Br_2CuN_2O$, Bromo(N-(2-hydroxyethyl)ethylenediamine)copper(II) bromide, 46B, 939
$C_4H_{12}Br_2Cu_2N_2$, Dibromo(2-methyl-1,2-diaminopropane)copper(II), 43B, 1118
$C_4H_{12}Cl_2N_2Pt$, cis-Dichloro(1-methylamino-2(S)-aminopropane)platinum-(II), 42B, 714
$C_4H_{13}AuCl_3N_3O_4$, Chloro-bis(2-aminomethyl)amine-gold(III) chloride perchlorate, 46B, 940
$C_4H_{13}Br_2N_3Pt$, Bromo(diethylenetriamine)platinum(II) bromide, 41B, 935
$C_4H_{13}CdCl_2N_3$, Bis(2-aminoethyl)aminedichlorocadmium, 46B, 941
$C_4H_{13}Cl_4N_3Pt \cdot H_2O$, fac-Trichloro(diethylenetriamine-N^1,N^2,N^3)platinum(IV) chloride monohydrate, 46B, 1043
$C_4H_{13}CoN_{12}$, mer-Triazidodiethylenetriaminecobalt(III), 40B, 801
$C_4H_{13}N_5O_5Pd$, Nitro-(diethylenetriamine)palladium(II) nitrate, 42B, 714
$C_4H_{14}Cl_2N_5O_3Rh$, cis-Dichlorobis(ethylenediamine)rhodium(III) nitrate, 45B, 1013
$C_4H_{14}Cl_2N_5O_3Rh$, trans-Dichlorobis(ethylenediamine)rhodium(III) nitrate, 45B, 1013
$C_4H_{16}Ag_2Br_4N_4Ni$, Bis(ethylenediamine)nickel dibromoargentate(I), 40B, 1110
$C_4H_{16}Ag_2I_4N_4Ni$, Bis(ethylenediamine)nickel diiodoargentate(I), 40B, 1110

$C_4H_{16}BF_4N_5NiO_2$, Nitritobis(ethylenediamine)nickel tetrafluoroborate, 30B, 331
$C_4H_{16}B_2CuF_8N_4$, Bis(ethylenediamine)copper(II) fluoroborate, 33B, 383
$C_4H_{16}ClCoN_4Na_2O_{10}S_2 \cdot 3\ H_2O$, Disodium(I) cis-bis(ethylenediamine)-disulfitocobaltate(III) perchlorate trihydrate, 45B, 1013
$C_4H_{16}ClCoN_4O_3S \cdot H_2O$, trans-Chlorobis(ethylenediamine)sulfitocobalt-(III) monohydrate, 46B, 941
$C_4H_{16}ClN_4O_2Re$, trans-Dioxobis(ethylenediamine)rhenium(V) chloride, 38B, 812; 40B, 802; 44B, 796
$C_4H_{16}ClN_5NiO_2$, Nitrobisethylenediaminenickel chloride, 31B, 381
$C_4H_{16}ClN_5NiO_6$, Nitritobis(ethylenediamine)nickel(II) perchlorate, 27, 831
$C_4H_{16}Cl_2CoN_5O_3$, trans-Dichloro-bis(ethylenediamine)cobalt(III) nitrate, 28, 615
$C_4H_{16}Cl_2CoN_5O_5$, trans-Chloronitrosylbis(ethylenediamine)cobalt(III) perchlorate, 35B, 588
$C_4H_{16}Cl_2CoN_5O_9$, trans-Perchloratonitrosylbis(ethylenediamine)cobalt perchlorate, 42B, 715
$C_4H_{16}Cl_2CuN_4O_8$, Bis(ethylenediamine)copper(II) perchlorate, 32B, 379
$C_4H_{16}Cl_2HgN_4O_8$, Bis(ethylenediamine)mercury(II) diperchlorate, 42B, 716
$C_4H_{16}Cl_2N_4NiO_4$, Chlorobis(ethylenediamine)nickel perchlorate, 37B, 483
$C_4H_{16}Cl_2N_4Pd$, Bis(ethylenediamine)palladium(II) chloride, 31B, 375
$C_4H_{16}Cl_3CoN_4$, trans-Dichlorobis(ethylenediamine)cobalt(III) chloride, 23, 572
$C_4H_{16}Cl_3IrN_4$, Dichlorobis(ethylenediamine)iridium(III) chloride, 41B, 935
$C_4H_{16}Cl_3N_5O_2Pt \cdot 3\ H_2O$, trans-Nitrochloroamminediethylenetriamine-platinum(IV) chloride trihydrate, 45B, 1014
$C_4H_{16}Cl_4N_4O_3Re_2$, Oxorhenium(V) ethylenediamine complex, 38B, 812
$C_4H_{16}Cl_4N_4Pt$, cis-Dichlorobis(ethylenediamine)platinum(IV) chloride, 35B, 587
$C_4H_{16}Cl_6CoN_4Tl$, trans-Bis(ethylenediamine)dichlorocobalt(III) tetrachlorothallate, 38B, 813
$C_4H_{16}CoN_4NaO_6S_2 \cdot 3\ H_2O$, Sodium trans-bis(ethylenediamine)disulfito-cobalt(III) trihydrate, 46B, 942
$C_4H_{16}CoN_4O_4P \cdot 2.5\ H_2O$, Phosphatobis(ethylenediamine)cobalt(III) hydrate, 43B, 1115
$C_4H_{16}CoN_7O_7$, trans-Bis(ethylenediamine)dinitrocobalt(III) nitrate, 42B, 717
$C_4H_{16}CoN_{11}O_3$, cis-Diazidobisethylenediaminecobalt(III) nitrate, 33B, 384
$C_4H_{16}CuN_4O_3SSe$, Bis(ethylenediamine)copper(II) selenosulfate, 37B, 483
$C_4H_{16}CuN_4O_3S_2$, Bis(ethylenediamine)copper(II) thiosulfate, 37B, 483
$C_4H_{16}CuN_6O_6$, Bisethylenediaminecopper(II) nitrate, 29, 593
$C_4H_{16}Cu_3I_4N_4$, Bis(ethylenediamine)copper(II) diiodocopper(I), 46B, 943
$C_4H_{16}F_6N_9PRu$, Azidodinitrogenbis(ethylenediamine)ruthenium(II) hexafluorophosphate, 35B, 592
$C_4H_{16}Hg_3I_6N_4$, Bis(ethylenediamine)triiododimercury(II) triiodo-mercurate(II), 43B, 1116
$C_4H_{16}N_4O_6Re_2$, trans-Dioxobis(ethylenediamine)rhenium(V) perrhenate, 43B, 1116
$C_4H_{16}N_5O_5Re$, trans-Dioxobis(ethylenediamine)rhenium(V) nitrate, 41B, 937

$C_4H_{16}N_6NiO_4$, Dinitrobis(ethylenediamine)nickel(II), 34B, 481
$C_4H_{16}N_6NiO_5$, Bisethylenediaminenitritonickel nitrate, 40B, 803
$C_4H_{16}N_6O_4Zn$, Bisethylenediaminenitritozinc nitrite, 40B, 803
$C_4H_{18}B_2F_8N_4NiO$, Aquobis(ethylenediamine)(tetrafluoroborato)nickel-(II) tetrafluoroborate, 38B, 813
$C_4H_{18}Br_2CuN_4O$, Bromobis(ethylenediamine)aquocopper(II) bromide, 17, 670
$C_4H_{18}ClCoN_4O_6Se$, Bis(ethylenediamine)(2-selenoacetato-O,Se)cobalt-(III) perchlorate, 45B, 1015
$C_4H_{18}Cl_2CuN_4O$, Chloroaquobis(ethylenediamine)copper(II) chloride, 17, 688; 32B, 379
$C_4H_{18}Cl_3CoN_4O$, cis-(+)$_{589}$-Dichlorobis(ethylenediamine)cobalt(III) chloride monohydrate, 35B, 591
$C_4H_{18}Cl_3CoN_4O$, cis-Dichloro-diethylenediamine cobalt(III) chloride monohydrate, 30B, 331
$C_4H_{18}Cl_6CuN_4OPt$, trans-Dichlorobis(ethylenediamine)platinum(IV) tetrachlorocuprate(II) monohydrate, 41B, 936
$C_4H_{18}I_2N_4NiO$, cis-Iodoaquobis(ethylenediamine)nickel(II) iodide, 38B, 814
$C_4H_{18}N_4O_{10}OsS_2$, trans-Dioxobis(ethylenediamine)osmium(VI) bis(hydrogen sulfate), 43B, 1117
$C_4H_{18}N_6NiO_5$, Bisethylenediaminedinitronickel monohydrate, 40B, 803
$C_4H_{19}ClCoN_5O_7S$, trans-Amminebis(ethylenediamine)sulfitocobalt(III) perchlorate, 46B, 943
$C_4H_{19}Cl_3CoN_5O_4$, Diamminechlorodiethylenetriaminecobalt(III) chloride perchlorate, 44B, 796
$C_4H_{20}ClCoN_4O_9S$, trans-Aquobis(ethylenediamine)sulfitocobalt(III) perchlorate monohydrate, 41B, 938
$C_4H_{20}Cl_2N_4NiO_{10}$, Diaquobisethylenediaminenickel perchlorate, 40B, 803
$C_4H_{21}Br_4CoN_4O_2$, trans-Dibromobis(ethylenediamine)cobalt(III) bromide hydrobromide dihydrate, 23, 572
$C_4H_{21}Cl_4CoN_4O_2$, trans-Dichlorobis(ethylenediamine)cobalt(III) hydrochloride dihydrate, 16, 457; 42B, 717
$C_4H_{21}Cl_4CrN_4O_2$, trans-Dichlorobis(ethylenediamine)chromium(III) chloride hydrochloride dihydrate, 24, 596
$C_4H_{22}ClCoN_6O_6S_2$, Tris(ammine)diethylenetriaminecobalt(III) chloride dithionate, 44B, 797
$C_4H_{46}Cl_6N_{14}O_2Ru_3$, Di-μ-oxo-bis(pentaammineruthenium)bis-(ethylenediamine)ruthenium hexachloride, 37B, 484
$C_5H_{14}Cl_2N_2Pd \cdot 0.5\ H_2O$, Dichloro(N,N-dimethyl-1,2-propanediamine)-palladium(II) hemihydrate, 46B, 944
$C_5H_{16}CdClN_5S$, Bisethylenediamineisothiocyanatocadmium(II) chloride, 40B, 807
$C_5H_{16}ClCoN_6O_6S$, trans-Bis(ethylenediamine)(isothiocyanato)-nitrocobalt(III) perchlorate, 45B, 1015
$C_5H_{16}ClCoN_6O_6S$, trans-Bis(ethylenediamine)(isothiocyanato)-nitritocobalt(III) perchlorate, 45B, 1015
$C_5H_{16}ClCuN_5O_4S$, Bis(ethylenediamine)copper(II) thiocyanate perchlorate, 39B, 650
$C_5H_{16}ClN_5NiS$, Chloroisothiocyanodiethylenediaminenickel, 33B, 385
$C_5H_{16}CoIN_4O_3 \cdot 2\ H_2O$, Carbonatobis(ethylenediamine)cobalt(III) iodide dihydrate, 46B, 944
$C_5H_{16}CoIN_6O_2S$, trans-Bis(ethylenediamine)(isothiocyanato)-nitritocobalt(III) iodide, 45B, 1015
$C_5H_{16}CoIN_6O_2S$, trans-Bis(ethylenediamine)(isothiocyanato)-nitrocobalt(III) iodide, 45B, 1015

$C_5H_{16}IN_5NiS$, Bis(ethylenediamine)thiocyanatonickel iodide, 34B, 481
$C_5H_{18}N_6NiO_4S$, Bisethylenediamineaquoisothiocyanatonickel nitrate, 40B, 957
$C_5H_{19}Cl_2CoN_4O_4S_2$, trans-Dichlorobis(ethylenediamine)cobalt(III) S-hydroxymethyl thiosulfate, 46B, 945
$C_5H_{20}Cl_5N_4PtSb$, $(+)_{350}$-Diammine-N,N,N'-trimethylethylenediamineplatinum(II) pentachloroantimonate, 45B, 1049
$C_5H_{20}CoN_5O_5S_2$, trans-Sulfiteisothiocyanate-bis(ethylenediamine)cobalt(III) dihydrate, 34B, 483
$C_6H_{12}CrIK_2N_2O_{10}$, Potassium ethylenediaminebis(oxalato)chromate(III) - potassium iodide dihydrate, 42B, 896
$C_6H_{13}CuN_5S_2$, (Di-(2-aminoethyl)amine)di-isothiocyanatocopper(II), 40B, 804
$C_6H_{14}CuN_2O_5$, Malonato(N-methylethylenediamine)copper(II) monohydrate, 40B, 804
$C_6H_{14}CuN_2O_5$, Oxalato(N,N'-dimethylethylenediamine)copper(II) monohydrate, 40B, 805
$C_6H_{15}Cl_2CuN_3$, Dichloro-(1,4,7-triazacyclononane)-copper(II), 46B, 946
$C_6H_{15}CuBr_2N_3$, Dibromo(1,4,7-triazacyclononane)copper(II), 45B, 1016
$C_6H_{16}Br_2CdN_2$, Dibromo-(N,N,N',N'-tetramethylethylenediamine)cadmium-(II), 40B, 806
$C_6H_{16}Br_2CuN_2$, Dibromo(N,N,N',N'-tetramethylethylenediamine)copper-(II), 40B, 806
$C_6H_{16}CdN_6S_2$, Bisethylenediaminebisisothiocyanatocadmium(II), 40B, 807
$C_6H_{16}ClCoN_4O_3S \cdot H_2O$, Bis(ethylenediamine)(thiooxalato-O,S)cobalt chloride hydrate, 46B, 946
$C_6H_{16}ClCoN_6$, trans-Dicyanobis(ethylenediamine)cobalt(III) chloride, 35B, 592
$C_6H_{16}Cl_2N_2Zn$, Dichloro-(N,N,N',N'-tetramethylethylenediamine)-zinc(II), 39B, 652
$C_6H_{16}Cl_2N_4O_4Pt$, Pentahydrodioxonium chloro(uracilato-N^1)(ethylenediamine)platinum(II) chloride, 46B, 947
$C_6H_{16}CoN_5O_7$, Ammine(diethylenetriamine)oxalatocobalt(III) nitrate, 41B, 938
$C_6H_{16}CoN_7O_2S_2$, Bis(ethylenediamine)isothiocyanatonitrocobalt(III) thiocyanate, 42B, 810
$C_6H_{16}CuN_4O_4$, Dinitrito(N,N,N',N'-tetramethylethylenediamine)copper-(II), 40B, 807
$C_6H_{16}CuN_4O_4$, Dinitrito(N,N-diethylethylenediamine)copper(II) (triclinic), 44B, 798
$C_6H_{16}CuN_4O_4$, Dinitrito(N,N'-diethylethylenediamine)copper(II) (orthorhombic), 43B, 1118
$C_6H_{16}CuN_4O_6$, Dinitrato(N,N,N',N'-tetramethylethylenediamine)copper-(II), 43B, 1119
$C_6H_{16}CuN_6S_2$, Bisethylenediaminecopper(II) thiocyanate, 26, 670; 29, 595
$C_6H_{16}N_4NiO_6$, Dinitrato(N,N,N',N'-tetramethylethylenediamine)nickel-(II), 40B, 808
$C_6H_{16}N_6NiS_2$, trans-Bis(ethylenediamine)bis(isothiocyanato)nickel-(II), 26, 670; 28, 616
$C_6H_{18}BrCoN_4O_5$, (-)-Oxalatobis(ethylenediamine)cobalt(III) bromide monohydrate, 39B, 651
$C_6H_{18}ClCoN_6O$, $(+)_{589}$-Dicyanobis(ethylenediamine)cobalt(III) chloride monohydrate, 37B, 484
$C_6H_{18}ClCrN_4O_6S$, Mercaptoacetatobis(ethylenediamine)chromium(III)

perchlorate, 39B, 652
$C_6H_{18}F_{18}KN_4P_3Pd$, Potassium triethylenetetraminepalladium(II) tris(hexafluorophosphate), 43B, 1119
$C_6H_{19}AgN_{10}O_9S_2$, Ethylenebis(biguanide)silver(III) sulfate hydrogen sulfate monohydrate, 41B, 939
$C_6H_{20}Br_3N_4Pt$, Bis(1,2-diaminopropane)platinum(II,IV) tribromide, 46B, 947
$C_6H_{20}Br_6Cu_3N_4Pt$, Bis(1,2-diaminopropane)platinum(II) dibromobis(1,2-diaminopropane)platinum(IV) pentabromotricopper, 44B, 798
$C_6H_{20}ClCoN_4O_3S$, Mercaptoacetatobis(ethylenediamine)cobalt(III) chloride monohydrate, 39B, 652
$C_6H_{20}ClCoN_6O_4 \cdot H_2O$, cis-Dinitrobis(1,3-diaminopropane)cobalt(III) chloride monohydrate, 45B, 1151
$C_6H_{20}ClCrN_4O_4S_2$, Bis(2-mercaptoethylamine)ethylenediamine-chromium(III) perchlorate, 42B, 718
$C_6H_{20}Cl_2CuN_4O_8$, Bis(N-methylethylenediamine)copper(II) perchlorate, 35B, 677
$C_6H_{20}Cl_2N_4Pt \cdot 2\ H_2O$, Bis[(R)-1,2-propanediamine]platinum(II) chloride dihydrate (monoclinic), 46B, 948
$C_6H_{20}Cl_2N_4Pt \cdot 2\ H_2O$, Bis[(R)-1,2-propanediamine]platinum(II) chloride dihydrate (triclinic), 46B, 948
$C_6H_{20}Cl_3CoN_4O_9$, DL-cis-β-(Chloroaquotriethylenetetramine)cobalt(III) perchlorate, 34B, 482
$C_6H_{20}CoN_7O_6$, Dinitrobis(trimethylenediamine)cobalt(III) nitrite, 43B, 1120
$C_6H_{20}CuN_6O_6$, Bis(N-methylethylenediamine)copper(II) nitrate, 40B, 1110
$C_6H_{21}Cl_2CuN_5O_8$, Ammine[tris(2-aminoethyl)amine]copper(II) perchlorate, 46B, 948
$C_6H_{21}Cl_3CoN_5 \cdot 0.5\ H_2O$, ω-Chloro(diethylenetriamine)(ethylenediamine)cobalt(III) dichloride hemihydrate, 41B, 940
$C_6H_{21}Cl_5CoN_5Zn$, κ-Chloro(ethylenediamine)(diethylenetriamine)cobalt-(III) tetrachlorozincate(II), 37B, 485
$C_6H_{21}Cl_5CoN_5Zn$, π(racemic)-Chloro(ethylenediamine)(diethylenetriamine)cobalt(III) tetrachlorozincate(II), 37B, 485
$C_6H_{22}ClCoN_6O_9S$, (2-Sulfinatoethylamine-N,S)bis(ethylenediamine)cobalt(III) perchlorate nitrate, 42B, 718
$C_6H_{22}CoN_7O_6Se$, (2-Selenolatoethylamine-N,Se)bis(ethylenediamine)cobalt(III) nitrate, 42B, 719
$C_6H_{22}CuN_2O_7S$, Diaquosulfato(N,N,N',N'-tetramethylethylenediamine)-copper(II) hydrate, 42B, 720
$C_6H_{24}Br_3CrN_6 \cdot 0.6\ H_2O$, Tris(ethylenediamine)chromium(III) bromide hydrate, 45B, 1016
$C_6H_{24}CdN_6O_3S_2$, Trisethylenediaminecadmium thiosulfate, 40B, 1111
$C_6H_{24}Cl_2CuN_6 \cdot 0.75\ C_2H_8N_2$, Tris(1,2-diaminoethane)copper(II) chloride 1,2-diaminoethane solvate, 45B, 1017
$C_6H_{24}Cl_3CoN_6 \cdot 2.8\ H_2O$, DL-Tris(ethylenediamine)cobalt(III) chloride hydrate, 41B, 940
$C_6H_{24}Cl_3CrN_6 \cdot 3\ H_2O$, DL-Tris(ethylenediamine)chromium(III) chloride hydrate, 41B, 940
$C_6H_{24}Cl_5CoN_6Sn$, Tris(ethylenediamine)cobalt(III) trichloro-stannate(II) dichloride, 42B, 720
$C_6H_{24}CoI_3N_6 \cdot H_2O$, Tris(ethylenediamine)cobalt(III) iodide monohydrate, 46B, 948
$C_6H_{24}CoN_9O_9$, (+)-Tris(ethylenediamine)cobalt(III) nitrate, 38B, 815
$C_6H_{24}Co_2N_{10}O_{13}$, Dinitrobis(ethylenediamine)cobalt(III) dinitro-oxalatodiamminecobaltate(III) monohydrate, 42B, 720

$C_6H_{24}CuN_6O_4S$, Tris(1,2-diaminoethane)copper(II) sulfate (triclinic-low temperature), 45B, 1017
$C_6H_{24}CuN_6O_4S$, Tris(1,2-diaminoethane)copper(II) sulfate (trigonal), 26, 605; 35B, 594
$C_6H_{24}N_6NiO_4S$, Tris(ethylenediamine)nickel(II) sulfate, 35B, 595
$C_6H_{24}N_6O_6Pd_2S_4$, Bisethylenediaminepalladium(II) cis-dithiosulfate-ethylenediaminepalladite, 35B, 598
$C_6H_{24}N_8NiO_6$, (-)-Tris(ethylenediamine)nickel(II) dinitrate, 24, 605; 46B, 949
$C_6H_{24}N_8NiO_6$, Trisethylenediaminenickel nitrate, 24, 605
$C_6H_{25}Cl_4CoN_2O_2$, trans-Dichlorobis(l-propylenediamine)cobalt(III) chloride hydrochloride dihydrate, 27, 833
$C_6H_{26}Au_2N_6O_{12}S_4$, Ethylenediammonium bis-cis-(ethylenediamine-disulfitoaurate(III)), 41B, 941
$C_6H_{26}Br_3CoN_6O$, Tris(ethylenediamine)cobalt(III) bromide monohydrate, 27, 830
$C_6H_{26}Cl_2N_6NiO_9$, Tris(ethylenediamine)nickel(II) diperchlorate monohydrate, 44B, 799
$C_6H_{26}Cl_3CoN_6O$, (+)-D-Tris(ethylenediamine)cobalt(III) chloride monohydrate, 34B, 485
$C_6H_{26}CrI_3N_6O$, DL-Tris(ethylenediamine)chromium(III) iodide monohydrate, 44B, 799
$C_6H_{28}Cl_3CrN_6O_2$, (+)-D-Tris(ethylenediamine)chromium(III) chloride dihydrate, 43B, 1121; 44B, 799
$C_6H_{30}Cl_3CoN_6O_3$, D-Tris-(ethylenediamine)cobalt(III) chloride trihydrate, 20, 503
$C_6H_{30}Cl_3N_6O_3Rh$, Tris(ethylenediamine)rhodium(III) chloride trihydrate, 42B, 721
$C_7H_{10}CdN_6Ni \cdot 1.5\ C_4H_5N$, (dl-1,2-Diaminopropane)cadmium(II) tetracyanonickelate(II) pyrrole, 46B, 1048
$C_7H_{10}CdN_6Ni \cdot 1.5\ C_4H_5N$, (l-1,2-Diaminopropane)cadmium(II) tetracyanonickelate(II) pyrrole, 46B, 1048
$C_7H_{12}BrN_2O_3Re$, Bromotricarbonyl(N,N'-dimethylethane-1,2-diamine)-rhenium(I), 46B, 949
$C_7H_{13}ClN_4O_2Pt$, Chloro(thyminato-N^1)(ethylenediamine)platinum(II), 46B, 947
$C_7H_{13}CrN_3O_3$, cis-Diethylenetriaminechromium tricarbonyl, 31B, 376
$C_7H_{13}MoN_3O_3$, cis-(Diethylenetriamine)molybdenum tricarbonyl, 30B, 355
$C_7H_{18}CdI_2N_2$, 1,3-Bis(methylamino)-2,2-dimethylpropanecadmium iodide, 43B, 1122
$C_7H_{18}ClCuN_5O_4S$, ((3,6-Diazaoctane)-1,8-diamine)isothiocyanatocopper-(II) perchlorate, 45B, 1018
$C_7H_{18}Cl_2CoN_2$; Dichloro(R-N,N,N',N'-tetramethyl-1,2-propylenediamine)cobalt(II), 46B, 950
$C_7H_{18}Cl_2N_2Pd$, Dichloro(N,N-diethyl-1,2-propanediamine)palladium(II), 46B, 944
$C_7H_{18}I_2N_2Zn$, 1,3-Bis(methylamino)-2,2-dimethylpropanezinc iodide, 43B, 1122
$C_7H_{20}Cl_2CuN_4O_8$, [N,N'-Bis(2'-aminoethyl)propylene-1,3-diamine]copper(II) perchlorate, 46B, 950
$C_7H_{20}Cl_4CoN_5O_2SZn$, (Ethylenediamine-cysteinesulfenamide)(ethylenediamine)cobalt(III) tetrachlorozincate, 43B, 1123
$C_7H_{23}Cl_5CoN_5Zn$, B-Chloro(RS-1,2-diaminopropane)(diethylenetriamine)-cobalt(III) tetrachlorozincate(II), 44B, 799
$C_7H_{24}ClCoN_6O_9S$, trans-Bis(ethylenediamine)imidazolesulfitocobalt-(III) perchlorate dihydrate, 44B, 800

$C_7H_{25}Cl_5CoN_5OZn$, cis-(1-Aminopropane-2-ol-N)-chlorobis(1,2-diaminoethane)cobalt(III) tetrachlorozincate, 46B, 951
$C_7H_{26}Br_3CoN_6$, Bis(ethylenediamine)trimethylenediaminecobalt(III) bromide, 38B, 816
$C_7H_{26}Cl_2CoN_5O_{10}$, N-Methylaminoethanolatobis(ethylenediamine)cobalt-(III) perchlorate monohydrate, 42B, 897
$C_7H_{28}Br_3CrN_6O$, Bis(ethylenediamine)(1,3-propanediamine)chromium(III) bromide monohydrate, 44B, 801
$C_8H_{16}Br_2N_4O_4Pt \cdot 2\ H_2O$, Dibromobis(1,2-diaminoethane)platinum(IV) 3,4-dihydroxy-3-cyclobutene-1,2-dionate dihydrate, 45B, 1018
$C_8H_{16}CoN_2NaO_{10}$, Sodium ethylenediaminebis(malonato)cobaltate(III) dihydrate, 42B, 899
$C_8H_{16}CuHgN_8S_4$, Mercury tetrathiocyanate-copper diethylenediamine, 17, 673
$C_8H_{16}CuN_4O_8Pt$, Bis(ethylenediamine)copper(II) bis(oxalato)platinate-(II), 42B, 899
$C_8H_{16}HgN_4S_2$, (N,N,N',N'-Tetramethylethylenediamine)-bis(thiocyanato)mercury(II), 44B, 801
$C_8H_{16}N_4NiS_2$, catena-Di-μ-(thiocyanato-S,N)-(N,N,N',N'-tetramethylethylenediamine)nickel(II), 43B, 1123
$C_8H_{16}N_4NiSe_2$, catena-Di-μ-(selenocyanato-Se,N)-(N,N,N',N'-tetramethylethylenediamine)nickel(II), 43B, 1124
$C_8H_{16}N_8NiPd$, Bis(ethylenediamine)nickel(II) tetracyanopalladate(II), 37B, 486
$C_8H_{17}N_5CdS_2$, Di-(3-aminopropyl)aminediisothiocyanatocadmium(II), 43B, 1124
$C_8H_{18}CuN_6S_2$, β,β',β''-Triaminotriethylamineisothiocyanatocopper(II) thiocyanate, 32B, 381
$C_8H_{18}Cu_3N_8O$, Aquobis(ethylenediamine)copper(II) di-(catena-di-μ-cyanocuprate(I)), 38B, 816
$C_8H_{18}Cu_3N_8OSe_2$, Aquabis(ethylenediamine)copper(II) dicyanodi(selenocyanato)cuprate(I), 45B, 1019
$C_8H_{18}N_6NiS_2$, 2,2',2"-Triaminotriethylamine nickel(II) di-thiocyanate, 23, 574; 35B, 682
$C_8H_{18}N_6S_2Zn$, β,β',β''-Triaminotriethylamineisothiocyanatozinc(II) thiocyanate, 34B, 486
$C_8H_{20}CuN_6S_2$, Bis(N-methylethylenediamine)copper(II) thiocyanate, 38B, 903
$C_8H_{20}N_4O_6Pt$, Bis(ethylenediamine)platinum(II) (R)-tartrate, 42B, 721
$C_8H_{21}Cl_2CoN_6O_{10}S$, Δ-[Bis(ethylenediamine)-(R)-2-aminothiazoline-4-carboxylatocobalt(III)] bis(perchlorate), 45B, 1020
$C_8H_{22}Cl_2CoN_5O_2S$, Bis(ethylenediamine)[(3R,4R)-thiazolidine-4-carboxylato-N,O]cobalt(III) dichloride, 46B, 952
$C_8H_{22}Cl_2N_4Pt \cdot 2\ H_2O$, $(+)_{240}$-[(2S,7S)-2,7-Dimethyl-3,6-diazaoctane-1,8-diamine]platinum(II) chloride dihydrate, 45B, 1154
$C_8h_{22}CoN_7OS_3$, (2-Sulfenatoethylamine-N,S)bis(1,2-diaminoethane)cobalt(III) thiocyanate, 45B, 1020
$C_8H_{22}CoN_7S_3$, β-Mercaptoethylaminebis(ethylenediamine)cobalt(III) dithiocyanate, 39B, 652
$C_8H_{22}MoN_8O_4Pt$, Bis(ethylenediamine)platinum(II) oxo(aquo)tetracyanomolybdate(IV) dihydrate, 41B, 942
$C_8H_{23}Br_2CuN_5$, N,N-Bis(2-aminoethyl)diethylenetriaminecopper(II) bromide, 42B, 722
$C_8H_{23}Cl_2CuN_4O_{12} \cdot H_2O$, trans-Bis(ethylenediamine)hydrogendiacetatocobalt(III) diperchlorate monohydrate, 45B, 1021
$C_8H_{24}Br_2N_4Ni$, Bis(N,N'-Dimethylethylenediamine)nickel(II) bromide, 34B, 489

$C_8H_{24}Br_4Cu_2N_4$, Bis(dibromo(N,N-dimethylethylenediamine)copper(II)), 42B, 723
$C_8H_{24}ClCoN_6O_4$, $(-)_{589}$-cis-Dinitrobis[(2S)-2-amino-4-azapentane]cobalt(III) chloride, 45B, 1021
$C_8H_{24}Cl_2CoN_5O_3$, racemic-β(2)-(RS,SR)-Glycinato(1,4,7,10-tetraazadecane)cobalt(III) dichloride hydrate, 41B, 943
$C_8H_{24}Cl_2CuN_4O_2$, Bis(N-(2-hydroxyethyl)ethylenediamine)copper(II) chloride, 38B, 817
$C_8H_{24}Cl_2CuN_4O_8$, Bis(N-ethylethylenediamine)copper(II) perchlorate, 45B, 1034
$C_8H_{24}Cl_4Cu_2N_4$, Bis(dichloro(N,N-diethylethylenediamine)copper(II)), 42B, 723
$C_8H_{24}CoCl_2N_5O_{10}$, N-(Methylalaninato)bis(ethylenediamine)cobalt(III) perchlorate, 43B, 1126
$C_8H_{24}CuN_4O_6$, Bis(N-methylethylenediamine)copper(II) oxalate dihydrate, 40B, 808
$C_8H_{24}CuN_6O_6$, Nitratobis(N-methyl-1,3-propanediamine)copper(II) nitrate, 44B, 802
$C_8H_{24}CuN_6O_6$, Bis(N,N'-dimethylethylenediamine)copper(II) nitrate, 34B, 488
$C_8H_{25}Cl_5CoN_5Zn$, β-Chloro(ethylenediamine)(dipropylenetriamine)cobalt(III) tetrachlorozincate(II), 35B, 599
$C_8H_{26}Br_3CoN_6 \cdot 1.6\ H_2O$, $(+)_{589}$-mer-Bis(diethylenetriamine)cobalt(III) bromide 1.6-hydrate, 45B, 1022
$C_8H_{26}Br_6N_6Pt_3$, Bis[bromo(diethylenetriamine)platinum(II)] tetrabromoplatinate(II), 45B, 1023
$C_8H_{26}Cl_3CoN_6 \cdot 2\ H_2O$, (2,2',2"-Triaminotriethylamine)-(ethylenediamine)cobalt(III) chloride dihydrate, 45B, 1023
$C_8H_{26}CuN_8O_6$, Bis(diethylenetriamine)copper(II) nitrate, 34B, 486
$C_8H_{27}ClCoI_2N_5O$, α-Chloro(ethylenediamine)(dipropylenetriamine)cobalt(III) iodide monohydrate, 37B, 539
$C_8H_{27}Cl_3CoN_5O_{12}S_2 \cdot H_2O$, [(S[1]-Ethyl-S[2]-(2-aminoethyl)disulfide)-N,S[2]]bis(ethylenediamine)cobalt(III), 46B, 952
$C_8H_{28}Br_2N_6OZn$, Bis(diethylenetriamine)zinc(II) dibromide monohydrate, 40B, 809
$C_8H_{28}CoI_2N_5O_5$, (Triethylenetetramine)(glycinato)cobalt(III) iodide trihydrate, 43B, 1126
$C_8H_{28}CoN_9O_{10}$, mer-Bis(diethylenetriamine)cobalt(III) nitrate monohydrate, 42B, 723
$C_8H_{30}CrI_3N_6O$, Ethylenediamine-bis(1,3-propanediamine)chromium(III) iodide monohydrate, 44B, 801
$C_8H_{32}Br_2Cl_4N_8O_{16}Pt_2$, Tetrakis(1,2-diaminoethane)dibromodiplatinum-(II,IV) tetraperchlorate, 46B, 953
$C_8H_{32}Cl_4Co_2N_8O_6S_2Se_2 \cdot H_2O$, trans-Dichlorobis(ethylenediamine)cobalt(III) diselenotetrathionate monohydrate (form I), 46B, 954
$C_8H_{32}Cl_4Co_2N_8O_6S_2Se_2 \cdot H_2O$, trans-Dichlorobis(ethylenediamine)cobalt(III) diselenotetrathionate monohydrate (form II), 46B, 954
$C_8H_{32}Cl_4Co_2N_8O_6S_3Se$, trans-Dichlorobis(ethylenediamine)cobalt(III) selenotetrathionate, 46B, 953
$C_8H_{32}Cl_4I_2N_8O_{16}Pt_2$, Bis(1,2-diaminoethane)platinum(II)-bis(1,2-diaminoethane)diiodoplatinum(IV) tetraperchlorate, 45B, 1024
$C_8H_{32}Cl_4N_8Ni_2$, (RS)-Di-μ-chloro-tetrakis(ethane-1,2-diamine)nickel-(II) dichloride, 44B, 802; 45B, 1025
$C_8H_{32}Cl_4N_8Ni_2$, Di-μ-chloro-bis(bisethylenediaminenickel) dichloride, 45B, 1025
$C_8H_{32}Cl_4N_8Ni_2O_8$, (RS)-Di-μ-chloro-tetrakis(ethane-1,2-diamine)nickel(II) diperchlorate, 44B, 802

$C_8H_{32}CuN_4O_8S$, Bis(N,N'-dimethylethylenediamine)copper(II) sulfate tetrahydrate, 40B, 810
$C_8H_{33}Cl_3Co_2N_8O_{15}$ · H_2O, μ-Peroxo-μ-hydroxo-bis[bis(ethylenediamine)-cobalt(III)] perchlorate monohydrate, 46B, 953
$C_8H_{33}Cr_2N_8O_{14}S_4$ · H_2O, Δ,Λ-μ-Hydroxo-μ-sulfatobis[bis(ethylenediamine)chromium(III)] dithionate hydrate, 45B, 1026
$C_8H_{34}Br_3Co_2N_9O_4S$, DL-μ-Amido-μ-sulfato-bis(bis(ethylenediamine)cobalt(III)) tribromide, 37B, 486
$C_8H_{34}Br_4Cr_2N_8O_2$ · 2 H_2O, Di-μ-hydroxo-bis[bis(1,2-diaminoethane)-chromium(III)] bromide dihydrate, 45B, 1026
$C_8H_{34}Cl_4Co_2N_8O_7S_6$, trans-Dichlorobis(ethylenediamine)cobalt(III) hexathionate monohydrate, 30B, 330; 39B, 654
$C_8H_{34}Cl_4Cr_2N_8O_2$ · 2 H_2O, Di-μ-hydroxo-bis[bis(1,2-diaminoethane)-chromium(III)] chloride dihydrate, 45B, 1026
$C_8H_{34}Cr_2N_8O_{14}S_4$, Di-μ-hydroxo-bis(bis(ethylenediamine)chromium(III)) dithionate, 43B, 1127
$C_8H_{35}Cl_3Cr_2N_8O_{15}$ · H_2O, Δ,Λ-μ-Hydroxo-bis[bis(ethylenediamine)hydroxochromium(III)] triperchlorate monohydrate, 45B, 1027
$C_8H_{35}Co_2N_{12}O_{16}$, μ-Hydroxo-μ-superoxo-bis(bis(ethylenediamine)cobalt-(III)) tetranitrate monohydrate, 42B, 901
$C_8H_{36}Co_2N_{13}O_{15}$, μ-Amido-μ-superoxo-bis(bis(ethylenediamine)cobalt-(III)) tetranitrate hydrate, 38B, 817
$C_8H_{37}Co_2N_9O_{14}S_2$, μ-Hydroxo-μ-peroxo-bis(bis(ethylenediamine)cobalt-(III)) nitrate dithionate dihydrate, 42B, 901
$C_8H_{37}Co_2N_{13}O_{14}$, μ-Amido-μ-hydroxo-bis(bis(ethylenediamine)cobalt-(III)) tetranitrate hydrate, 37B, 487
$C_8H_{38}Cl_4Cr_2N_8O_{12}$, meso-Di-μ-hydroxo-bis(bis(ethylenediamine)-chromium(III)) diperchlorate dichloride dihydrate, 42B, 724
$C_8H_{40}Co_2N_{12}O_{16}$, μ-Peroxo-bis(nitrobis(ethylenediamine)cobalt(III)) dinitrate tetrahydrate, 39B, 655
$C_9H_{16}BrN_2O_3Re$, Bromotricarbonyl(N,N,N',N'-tetramethylethane-1,2-diamine)rhenium(I), 45B, 1028
$C_9H_{20}CoN_7OS_3$, 2,2',2"-Triaminotriethylaminebis(isothiocyanato)cobalt(III) thiocyanate monohydrate, 41B, 944
$C_9H_{21}BrCoN_7O_6$, Bromobis(ethylenediamine)pyridinecobalt(III) dinitrate, 42B, 725
$C_9H_{22}BrClCoN_9O$, cis-(Adeninato-chloro-bis(ethylenediamine)cobalt-(III)) bromide monohydrate, 40B, 810
$C_9H_{22}ClCoN_4O_5$, Ethylenediamine-N,N'-diacetato-((R)-1,2-diaminopropane)cobalt(III) chloride monohydrate, 41B, 946
$C_9H_{22}ClCoN_4O_7$, cis-β-Carbonato-(3S,8S-dimethyltriethylenetetramine)-cobalt(III) perchlorate, 41B, 946
$C_9H_{22}CoIN_4O_5$, (Bis(ethylenediamine)acetylpyruvato)cobalt(III) iodide monohydrate, 42B, 904
$C_9H_{22}CuN_2O_6$, Malonato(aquo)(N,N'-diethylethylenediamine)copper(II) monohydrate, 43B, 1127
$C_9H_{23}ClCoN_5O_8$, (+)-4,5-(L-Glutamatobis(ethylenediamine)cobalt(III)) perchlorate, 35B, 600
$C_9H_{24}BrCuN_5O_5$, Aquo(diethylenetriamine)thyminatocopper(II) bromide dihydrate, 41B, 945
$C_9H_{24}CoN_9S_3$, DL-Tris(thiocyanato)tris(ethylenediamine)cobalt(III), 42B, 725
$C_9H_{24}CrN_9S_3$ · 0.75 H_2O, DL-Tris(ethylenediamine)chromium(III) thiocyanate hydrate, 43B, 1129
$C_9H_{24}CrN_9S_3$, (+)-Tris(ethylenediamine)chromium(III) thiocyanate, 44B, 803
$C_9H_{24}CrN_9S_3$, (-)-Tris(ethylenediamine)chromium(III) thiocyanate,

43B, 1128
$C_9H_{26}CuN_4O_6$, Bis(N-methylethylenediamine)copper(II) malonate dihydrate, 40B, 804
$C_9H_{30}Br_3CoN_6$, Tris-(R-propylenediamine)cobalt(III) bromide, 31B, 378; 41B, 947
$C_9H_{32}Cl_3CoN_6O$, (-)-Tris-(1,3-diaminopropane)cobalt(III) chloride monohydrate, 39B, 656
$C_{10}H_{12}KMnN_2O_8 \cdot 2\ H_2O$, Potassium ethylenediaminetetraacetatomanganese(III) dihydrate, 45B, 1028
$C_{10}H_{16}Cl_2CuN_2$, Dichloro(N,N,N',N'-tetramethyl-o-phenylenediamine)-copper(II), 46B, 956
$C_{10}H_{16}CrN_2O_4$, Tetracarbonyl[N,N,N',N'-tetramethylethylenediamine]-chromium(0), 46B, 955
$C_{10}H_{16}CsN_2O_{10}Yb \cdot 3\ H_2O$, Cesium diaqua(ethylenediaminetetraacetato)-ytterbate(III) trihydrate, 45B, 1029
$C_{10}H_{16}CuN_2O_9$, Aquo(dihydrogenethylenediaminetetraacetato)copper(II), 34B, 492
$C_{10}H_{16}FeN_2O_{10}Rb$, Rubidium ethylenediaminetetraacetatoaquoferrate, 26, 680
$C_{10}H_{16}KMnN_2O_{10}$, Potassium ethylenediaminetetraacetatomanganate(III) dihydrate, 44B, 803
$C_{10}H_{18}Cl_6CuN_2O_5$, Bis(trichloroacetato)aquo(N,N,N',N'-tetramethylethylenediamine)copper(II), 44B, 804
$C_{10}H_{18}DyNaN_2O_{11} \cdot 5\ H_2O$, Sodium triaqua(ethylenediaminetetraacetato)dysprosate(III) pentahydrate, 45B, 1029
$C_{10}H_{20}Cl_2CuN_2O_4$, Bis(chloroacetato)-(N,N,N',N'-tetramethylethylenediamine)copper(II), 44B, 804
$C_{10}H_{21}CoN_4O_{10} \cdot H_2O$, $(-)_{589}$-trans(O)-Ethylenediaminebis(glycinato)-cobalt(III) hydrogen-D-tartrate monohydrate, 45B, 1030
$C_{10}H_{21}CoN_4O_{10} \cdot 2\ H_2O$, Oxalatobis(ethylenediamine)cobalt(III) hydrogen-d-tartrate dihydrate, 46B, 958
$C_{10}H_{21}CoN_4O_{10} \cdot 3\ H_2O$, $(+)_{589}$-trans(O)-Ethylenediaminebis(glycinato)cobalt(III) hydrogen-D-tartrate trihydrate, 45B, 1030
$C_{10}H_{22}Cl_2CoN_5O_4S$, (2-Mercaptoaniline-N,S)bis(ethylenediamine)cobalt-(III) chloride perchlorate, 46B, 956
$C_{10}H_{22}Cr_2N_4O_{10} \cdot 6\ H_2O$, Bis($\mu$-hydroxo)-bis[(ethylenediamine)malonatochromium(III)] hexahydrate, 46B, 957
$C_{10}H_{23}CoN_4O_{11}$, (+)-Oxalatobis(ethylenediamine)cobalt(III) hydrogen-d-tartrate monohydrate, 44B, 805
$C_{10}H_{24}Cl_2CuN_2$, Dichloro(N,N,N',N'-tetraethylethylenediamine)copper-(II), 41B, 948
$C_{10}H_{24}Cl_2N_6O_8Pt$, trans-Bis(2,4-pentanediiminato)diammineplatinum perchlorate, 44B, 806
$C_{10}H_{24}Co_2N_{11}O_3S_4$, Tris(ethylenediamine)cobalt(III) tetrakis(isothiocyanato)cobaltate(II) nitrate, 43B, 1132
$C_{10}H_{24}CuN_2O_6$, Malonato(N[1]-isopropyl-2-methylpropane-1,2-diamine)copper(II) dihydrate, 43B, 1131
$C_{10}H_{24}CuN_4O_6$, Dinitrato(N,N,N',N'-tetraethylethylenediamene)copper-(II), 46B, 957
$C_{10}H_{24}CuN_6Se_2$, Bis(N,N'-dimethylethylenediamine)copper(II) selenocyanate, 39B, 656
$C_{10}H_{24}MgN_2O_{14}Zn$, Magnesium zinc ethylenediaminetetraacetate hexahydrate, 38B, 818
$C_{10}H_{26}Cl_2Cu_2N_6O_{12}$, μ-Oxalato-bis(di-(2-aminoethyl)aminecopper(II)) diperchlorate, 39B, 657
$C_{10}H_{26}Cl_2Cu_2N_8O_8$, Di-$\mu$-thiocyanato-bis(di-(2-aminoethyl)amine)dicopper(II) diperchlorate, 40B, 811

$C_{10}H_{26}Cl_2N_6O_8Pd$, Bis(cis-3,5-diaminopiperidine)palladium(II) perchlorate, 45B, 1160
$C_{10}H_{26}CoI_2N_5O_2$, (Pipecolato)bis(ethylenediamine)cobalt(III) iodide, 43B, 1133
$C_{10}H_{26}Co_2N_{10}O_5$, cis-Diaquobis(ethylenediamine)cobalt(III) hexacyanocobaltate(III) trihydrate, 42B, 725
$C_{10}H_{27}CrMoN_{10}O_3$, Tris(ethylenediamine)chromium(III) oxo(hydroxo)tetracyanomolybdate(IV) monohydrate, 41B, 942
$C_{10}H_{27}N_9O_9Pd \cdot H_2O$, (cis-3,5-Diaminopiperidine)-(cis-3,5-diaminopiperidinium)palladium(II) nitrate monohydrate, 45B, 1160
$C_{10}H_{28}ClCoN_6O_6 \cdot 5\ H_2O$, (+)-Tris(ethylenediamine)cobalt(III) chloride (+)-tartrate pentahydrate, 45B, 1030
$C_{10}H_{28}Cl_2N_4Pt \cdot H_2O$, Bis((S,S)-2,4-pentanediamine)platinum(II) chloride monohydrate, 45B, 1159
$C_{10}H_{28}Cl_4N_6O_{16}Pd \cdot 2\ H_2O$, Bis(cis-3,5-diaminopiperidinium)palladium(II) perchlorate dihydrate, 45B, 1160
$C_{10}H_{28}CuN_4O_8$, Bis(N-methylethylenediamine)copper(II) d-tartrate dihydrate, 40B, 812
$C_{10}H_{30}Cd_2Cl_4N_6$, Bis(μ-chloro)-bis(2-aminoethyl-3-aminopropylamine-chlorocadmium), 46B, 941
$C_{10}H_{32}Cl_2N_8Ni_2O_{12}$, μ-Oxalato-bis(bisethylenediaminenickel) perchlorate, 40B, 968
$C_{10}H_{32}N_{10}Ni_2O_{10}$, μ-Oxalato-bis(bisethylenediaminenickel) nitrate, 39B, 657; 40B, 968
$C_{10}H_{34}Cl_3CoN_6 \cdot H_2O$, $(+)_{470}$-cis-Diamminebis(R,R-2,4-diaminopentane)-cobalt(III) trichloride monohydrate, 45B, 1160
$C_{10}H_{34}N_6NiO_6$, Tris(ethylenediamine)nickel(II) acetate dihydrate, 42B, 726
$C_{11}H_{16}ClN_2O_3Re$, Tricarbonylchloro(N,N'-diisopropylethylenediimine)-rhenium(I), 43B, 1133
$C_{11}H_{21}Cl_2N_5O_5Pt$, Chloro-ethylenediamine-(1-(β-D-arabinofuranosyl)-cytosine)platinum(II) chloride, 44B, 806
$C_{11}H_{22}CuN_2O_4$, Glutarato(N,N,N',N'-tetramethylethylenediamine)copper(II), 43B, 1134
$C_{11}H_{23}CdN_5S_2$, (Bis(2-dimethylaminoethyl)methylamine)di-isothiocyanatocadmium(II), 42B, 726
$C_{11}H_{23}Cl_2CoN_8O_6$, cis-(Theophyllinatochlorobis(ethylenediamine)cobalt(III) perchlorate, 41B, 949
$C_{11}H_{23}Cl_3CoN_5O_4$, (1,8-Diamino-3,6-diazaoctane)cobalt(III) chloride perchlorate, 45B, 1031
$C_{11}H_{24}ClCoN_4O_8$, (-)-β-Oxalato((4R,6R)-dimethyl-3,7-diazanonane-1,9-diamine)cobalt(III) perchlorate, 44B, 806
$C_{11}H_{24}CrN_{11}Ni \cdot 1.5\ H_2O$, Tris(ethylenediamine)chromium(III) pentacyanonickelate(II) sesquihydrate, 33B, 386
$C_{11}H_{24}N_5NdO_{11}$, Guanidinium neodymium ethylenediaminetetraacetate trihydrate, 38B, 818
$C_{11}H_{27}Cl_2CoN_8O_4$, trans-(Theophyllinatochlorobis(ethylenediamine)cobalt(III)) chloride dihydrate, 41B, 948
$C_{11}H_{28}CoCrN_{12}O_3$, Tris(ethylenediamine)cobalt pentacyanonitrosylchromate dihydrate, 35B, 602
$C_{11}H_{29}B_2CuN_5$, Bis(cyanotrihydroborato)-1,1,4,7,7-pentamethyldiethylenetriaminecopper(II), 40B, 812
$C_{11}H_{30}Co_2N_8O_{11}$, (-)-Oxalatobis(ethylenediamine)cobalt(III) (+)-dicyanomalonatodiamminecobaltate(III) trihydrate, 43B, 1135
$C_{11}H_{36}Co_2N_{12}O_3S_3$, μ-Amido-μ-peroxo-bis(bis(ethylenediamine)cobalt-(III)) trithiocyanate monohydrate, 38B, 819
$C_{12}H_{17}CuN_3O_3$, N'-(1-Methyl-3-oxo-butyliden)-N'-(1-methyl-2-isonitro-

so-3-oxo-butyliden)-ethylenediaminecopper(II), 38B, 820
$C_{12}H_{17}KMo_2N_2O_{12}$, Potassium di-μ-hydroxo-μ-acetato-μ-ethylenediamine-tetraacetato-bis(molybdenum(III)), 41B, 950
$C_{12}H_{18}CuN_2S_2$, N,N'-Ethylenebis(monothioacetylacetoneiminato)copper-(II), 46B, 958
$C_{12}H_{22}Cr_2N_4O_{10} \cdot 4\ H_2O$, α-cis-Di-μ-hydroxobis(ethylenediamine-N,N'-diacetatochromium(III)) tetrahydrate, 46B, 963
$C_{12}H_{23}Cl_{1.5}I_{0.5}N_7O_6Pt$, (Ethylenediamine)bis(guanosine)platinum chloride iodide dihydrate, 41B, 950
$C_{12}H_{24}CuN_2O_4$, Adipato(N,N-diethylethylenediamine)copper(II), 43B, 1138
$C_{12}H_{24}CuN_2O_5 \cdot H_2O$, Aqua(cyclobutane-1,1-dicarboxylato)(N,N,N',N'-tetramethylethylenediamine)copper(II) monohydrate, 45B, 1032
$C_{12}H_{24}N_4O_{13}PtSb_2$, (+)-Diammine((S)-1-N-methyl-1,2-propanediamine)-platinum(II) di-μ-(+)-tartrato(4-)-bis(antimonate(III)) monohyd-rate, 43B, 1137
$C_{12}H_{25}Cl_4NiO_6$, 2,2'-(Ethylenediamino)bis(2-methyl-3-butanone oxim-ato)nickel(II) perchlorate, 44B, 807
$C_{12}H_{26}Cd_2N_{10}S_4$, catena-Bis(μ-bis(2-aminoethyl)amine)-bis(μ-thio-cyanato)-bis(isothiocyanato)dicadmium(II), 43B, 1139
$C_{12}H_{26}Cl_2N_6NiO_{10}$, Bis[2-(2-aminoethyl)imino-3-butanone oxime]nickel-(II) diperchlorate, 45B, 1033
$C_{12}H_{26}CoF_{12}N_4OP_2$, 12,14-Dimethyl-1,4,8,11-tetraazacyclotetradeca-11,13-dieneaquocobalt(II) hexafluorophosphate, 42B, 727
$C_{12}H_{26}CuN_2O_6$, Bis(L-lactato)(N,N,N',N'-tetramethylethylenediamine)-copper(II), 43B, 1140
$C_{12}H_{26}N_2NiO_6$, Bis(L-lactato)(N,N,N',N'-tetramethylethylenediamine)-nickel(II), 43B, 1141
$C_{12}H_{27}N_5NiO_6$, 2,2'-(1,2-Diaminoethane)bis(2-methyl-3-butanoneoxim-ato)nickel(II) nitrate monohydrate, 41B, 951
$C_{12}H_{28}CoFeN_{12}O_2$, Tris(ethylenediamine)cobalt(III) hexacyano-ferrate(III) dihydrate, 38B, 820
$C_{12}H_{28}Co_2N_8O_{12}$, (-)-Dinitrobis(ethylenediamine)cobalt(III) (+)-bis-(malonato)ethylenediaminecobaltate(III), 38B, 821
$C_{12}H_{28}Cr_2N_6O_{14}$, Oxalatobis(ethylenediamine)chromium(III) bis-(oxalato)ethylenediaminechromate(III) dihydrate, 35B, 604
$C_{12}H_{28}CuN_4O_6$, Bis(N-(2-hydroxyethyl)ethylenediamine)copper(II) suc-cinate, 44B, 807
$C_{12}H_{29}Cl_2CoN_3$, 1,1,7,7-Tetraethyldiethylenetriamine cobalt(II) chloride, 32B, 381
$C_{12}H_{29}I_2N_3Pt$, Iodo-N,N,N',N'-tetraethyldiethylenetriamineplatinum-(II) iodide, 46B, 959
$C_{12}H_{29}N_7O_3Pd \cdot 0.5\ H_2O$, Azido-1,1,7,7-tetraethyldiethylenetriamine-palladium(II) nitrate hemihydrate, 41B, 952; 42B, 714
$C_{12}H_{30}Br_2CoN_4$, Tris(2-dimethylaminoethyl)aminecobalt(II) bromide, 32B, 383
$C_{12}H_{30}N_8O_{12}PtS_4 \cdot 2.5\ H_2O$, (1,3,6,8,10,13,16,19-Octaazabicyclo-[6.6.6]eicosane)platinum(IV) dithionate hydrate, 45B, 1033
$C_{12}H_{31}Cl_2F_6N_4PPd$, Chloro(N,N-bis(2-dimethylaminoethyl)-2-dimethyl-ammonioethylamine)palladium(II) chloride hexafluorophosphate, 44B, 808
$C_{12}H_{31}N_5O_6Pd$, Nitro-(tetraethyl-diethylenetriamine)palladium(II) nitrate monohydrate, 42B, 714
$C_{12}H_{32}BF_4N_5NiO_2$, Bis(N,N'-diethylethylenediamine)(nitrito-O,O')nic-kel(II) tetrafluoroborate, 43B, 1141
$C_{12}H_{32}Cl_2CuN_4O_8$, Bis(N,N-diethylethylenediamine)copper(II) perchlor-ate (triclinic red form), 42B, 728; 45B, 1034

$C_{12}H_{32}Cl_2CuN_4O_8$, Bis(N,N-diethylethylenediamine)copper(II) perchlorate (monoclinic violet form), 45B, 1034
$C_{12}H_{32}Cl_2CuN_4O_8$, Bis(tetramethylethylenediamine)copper(II) perchlorate, 44B, 809
$C_{12}H_{32}Cl_2N_4NiO_2$, trans-Diaqua-bis(1,2-cyclohexanediamine)nickel(II) chloride, 46B, 959
$C_{12}H_{32}Cl_4Cu_2N_4$, Bis(dichloro(N,N,N',N'-tetramethylethylenediamine)-copper(II)), 41B, 953
$C_{12}H_{32}Co_2N_8O_{12}S_4 \cdot 2\ H_2O$, Bis[(thiooxalato-O,S)bis(ethylenediamine)-cobalt(III)] dithionate dihydrate, 46B, 961
$C_{12}H_{32}CuN_4O_6$, Diaquo-bis(N-methylethylenediamine)copper(II) adipate, 45B, 1035
$C_{12}H_{32}N_6NiO_4$, Bis(N,N'-diethylethylenediamine)(nitrito-O,O')nickel-(II) nitrite, 45B, 1035
$C_{12}H_{32}N_{10}O_4S_2Zn_2$, μ-Oxalato-bis(bisethylenediaminezinc) thiocyanate, 40B, 969
$C_{12}H_{34}Cd_2Cl_4N_6$, Bis(μ-chloro)-bis[bis(3-aminopropyl)aminechlorocadmium], 46B, 941
$C_{12}H_{34}Cl_2Cu_2N_4O_{10}$, Di-μ-hydroxo-bis(N,N,N',N'-tetramethylethylenediamine)dicopper(II) perchlorate, 43B, 1142
$C_{12}H_{34}Cu_2N_6O_8$, Di-μ-hydroxo-bis(N,N,N',N'-tetramethylethylenediamine)dicopper(II) nitrate, 43B, 1143
$C_{12}H_{36}AgCl_3Co_2N_8O_{16}$, Bis[2-thioacetato(O,S)bis(ethylenediamine)cobalt(III)-(S)]silver(I) perchlorate, 46B, 960
$C_{12}H_{36}Cl_2N_4NiO_2$, Bis(tetramethylethylenediamine)nickel(II) dichloride dihydrate, 43B, 1144
$C_{12}H_{36}Cl_4Co_2N_6O_{20} \cdot 4\ H_2O$, trans-Diaqua-di-μ-hydroxo-bis[(1,4,7-triazacyclononane)cobalt(III)] tetraperchlorate tetrahydrate, 45B, 1036
$C_{12}H_{36}Cl_4N_6O_{20}Rh_2 \cdot 4\ H_2O$, trans-Diaqua-di-μ-hydroxy-bis[(1,4,7-triazacyclononane)rhodium(III)] perchlorate tetrahydrate, 46B, 960
$C_{12}H_{36}CoCrN_{12}O_6$, Tris(ethylenediamine)chromium(III) hexacyanocobaltate(III) hexahydrate, 33B, 387
$C_{12}H_{36}CuN_6O_8$, Bis(N,N'-Diethylethylenediamine)diaquocopper(II) nitrate, 34B, 494
$C_{12}H_{38}Cl_3CoN_6O_3$, cis-(1,4,8,11-Tetraazacyclotetradecane)ethylenediaminecobalt(III) chloride trihydrate, 42B, 728
$C_{12}H_{40}B_4Br_2F_{16}N_8Pt_2$, Bis(trimethylenediamine)platinum(II) dibromo-(trimethylenediamine)platinum(IV) tetrafluoroborate, 44B, 809
$C_{12}H_{40}B_4Cl_2F_{16}N_8Pt_2$, Bis(trimethylenediamine)platinum(II) dichloro-(trimethylenediamine)platinum(IV) tetrafluoroborate, 44B, 809
$C_{12}H_{40}Br_2Cl_4N_8O_{16}Pt_2$, Bis(trimethylenediamine)platinum(II) dibromo-(trimethylenediamine)platinum(IV) perchlorate, 44B, 809
$C_{12}H_{40}Cl_4I_2N_8O_{16}Pt_2$, Bis(1,2-diaminopropane)platinum(II) diiodo-bis(1,2-diaminopropane)platinum(IV) perchlorate, 44B, 810
$C_{12}H_{40}I_6N_8Pt_2$, Bis(1,2-diaminopropane)platinum(II)bis(1,2-diaminopropane)diiodoplatinum(IV) tetraiodide, 46B, 953
$C_{12}H_{42}Cl_4Co_2N_{10}O_{18}$, μ-Peroxo-bis((ethylenediamine)(diethylenetriamine)cobalt(III)) perchlorate, 39B, 563
$C_{12}H_{48}Cl_6CoCrN_{12} \cdot 6.1\ H_2O$, (+)-Tris(ethylenediamine)cobalt(III) (-)-tris(ethylenediamine)chromium(III) hexachloride hydrate, 44B, 810
$C_{12}H_{52}CdCl_8Co_2N_{12}O_2$, Tris(ethylenediamine)cobalt(III) hexachlorocadmate(II) dichloride dihydrate, 38B, 821
$C_{12}H_{52}Cl_{10}Co_2Cu_2N_{12}O_2$, Tris(ethylenediamine)cobalt(III) di-μ-chloro-bis(trichlorocuprate(II)) dichloride dihydrate, 37B, 488
$C_{12}H_{54}Cl_{10}Co_2N_{12}O_3Pb_2$, Bis(tris(ethylenediamine)cobalt(III)) nonachlorodiplumbate(II) chloride trihydrate, 44B, 811

$C_{12}H_{58}Br_6Cr_4N_{12}O_8$, μ-(Di-μ-hydroxo-bis(mono(ethylenediamine)-cis-dihydroxochromium(III)))-bis(bis(ethylenediamine)chromium(III)) hexabromide dihydrate, 44B, 811
$C_{12}H_{60}Cl_6CoCrN_{12}O_6$, (+)-Tris(ethylenediamine)cobalt(III) (-)-tris-(ethylenediamine)chromium(III) chloride hydrate, 42B, 729
$C_{12}H_{60}Cl_6CrN_{12}O_6Rh$, (+)-Tris(ethylenediamine)chromium(III) (+)-tris-(ethylenediamine)rhodium(III) chloride hydrate, 42B, 729
$C_{12}H_{60}Cl_7Co_2N_{12}NaO_6$, Sodium D-tris(ethylenediamine)cobalt(III) chloride hexahydrate, 19, 539; 21, 547
$C_{12}H_{60}Cl_7Co_2N_{12}NaO_6$, Sodium L-tris-(ethylenediamine)cobalt(III) chloride hexahydrate, 19, 539
$C_{12}H_{69}Co_2N_{12}O_{21}P_3$, Tris(ethylenediamine)cobalt(III) monohydrogen phosphate nonahydrate, 37B, 489
$C_{12}H_{70}Co_4N_{12}O_{32}S_6$, racemic-Tris(di-μ-hydroxo-bis(ethylenediamine)cobalt(III))cobalt(III) trisdithionate octahydrate, 37B, 490
$C_{13}H_{24}Cl_2CoN_5O_{10}$, Bis(ethylenediamine)(1-phenyl-1,2-propanedione 2-oximato)cobalt(III) diperchlorate, 45B, 1036
$C_{13}H_{26}N_4O_{13}PtSb_2$, Diammine-N,N,N'-trimethylethylenediamineplatinum-(II) di-μ-tartrato(4-)-bis(antimonate(III)) monohydrate, 42B, 730
$C_{13}H_{28}ClCoN_4O_8$, (+)-β-Oxalato-((6R,8S)-dimethyl-2,5,9,12-tetra-azatridecane)cobalt(III) perchlorate, 43B, 1144
$C_{13}H_{31}CoI_2N_4O_3$, uns-cis-2,4-Pentanedionato(4,7-diaza-1,10-decanediamine)cobalt(III) iodide monohydrate, 42B, 730
$C_{13}H_{33}CoFeN_{11}O_4S$, ((Methyl 2-aminoethyl thioether)-N,S)bis(ethylenediamine)cobalt(III) hexacyanoferrate(III) tetrahydrate, 44B, 812
$C_{13}H_{42}Cu_3I_4N_8O_5$, Bis(1,3-diamino-2-propanol)bis(1,3-diamino-2-propanolato)tricopper tetraiodide - methanol (1:1), 43B, 1362
$C_{14}H_{20}Cl_2CuN_4O_8S_2$, Bis(N-(2-aminoethyl)thiophen-2-aldimine)diperchloratocopper(II), 43B, 1145
$C_{14}H_{20}Cl_4Cu_2N_4S_2$, Di-μ-chloro-bis((N-2-(aminoethyl)thiophen-2-aldimine)chlorocopper(II)), 43B, 1145
$C_{14}H_{20}CuN_6O_7$, (N-Salicylidene-N'-methylethylenediamine)(cytosine)-copper(II) nitrate monohydrate, 41B, 953
$C_{14}H_{22}Cl_2N_4O_6U$, Dioxo(nitrato)(2-(2-methyl-2,5,8-triazanon-8-enyl)phenoxy)uranium(VI) methylene chloride solvate, 41B, 955
$C_{14}H_{23}CuN_3O_{11}$, Trihydrogen diethylenetriaminepentaacetatocuprate(II) monohydrate, 41B, 954
$C_{14}H_{23}CuN_{13}O$, fac-Bis(adeninato)(diethylenetriamine)copper(II) monohydrate, 44B, 812
$C_{14}H_{24}CuN_4O_6$, Bis(2-hydroxy-1,3-propanediamine)copper(II) terephthalate, 41B, 1020
$C_{14}H_{25}CuN_7O_8$, Bis(2-aminoethyl)amine(2,2'-bipyridyl)copper(II) nitrate dihydrate, 45B, 1037
$C_{14}H_{26}Cl_2N_8O_{13}Pd$, (Diethylenetriamine)(guanosine)palladium(II) bis-(perchlorate), 46B, 961
$C_{14}H_{26}Cl_2N_8O_{13}Pt$, (Diethylenetriamine)-(guanosine)-platinum(II) perchlorate, 45B, 1038
$C_{14}H_{26}Cu_2FeN_{12} \cdot 6\ H_2O$, Bis(diethylenetriamine-copper)-μ-cyano)-tetracyanoiron hexahydrate, 46B, 964
$C_{14}H_{27}N_9O_{12}Pt$, (Diethylenetriamine)(inosine)platinum(II) dinitrate monohydrate, 44B, 813
$C_{14}H_{28}Cl_2CuN_2O_4$, Bis(chloroacetato)(N,N,N',N'-tetraethylethylenediamine)copper(II), 46B, 962
$C_{14}H_{30}N_6PdS_2$, Isothiocyanato[tris(2-dimethylaminoethyl)amine]palladium(II) thiocyanate, 45B, 1038
$C_{14}H_{31}ClCuN_2O_8 \cdot H_2O$, (N,N,N',N'-Tetra(2-hydroxypropyl)ethylenediamine)copper(II) perchlorate monohydrate, 45B, 1039

$C_{14}H_{34}Cl_2N_6O_{12}Zn_2$, μ-Oxalato-bis(di-(3-aminopropyl)aminezinc(II)) diperchlorate, 39B, 657
$C_{14}H_{36}CuN_6O_4$, Bis(N-isopropyl-2-methyl-1,2-propanediamine)nitrocopper(II) nitrite, 45B, 1040
$C_{14}H_{39}CoN_6O_{15}$, Hydrogen tris(ethylenediamine)cobalt(III) d-tartrate trihydrate, 43B, 1146
$C_{15}H_{25}CoN_6O_2S_3$, ((Benzyl 2-carboxymethyl thioether)-O,S)bis(ethylenediamine)cobalt(III) thiocyanate, 44B, 812
$C_{15}H_{30}N_4O_{14}PtSb_2$, (R)-Propylenediamine-N,N'-dimethylethylenediamine-platinum(II) di-μ-(+)-tartrato(4-)-bis(antimonate(III)) dihydrate, 42B, 732
$C_{15}H_{34}Cl_3CoN_8O_{12} \cdot H_2O$, 2,3-Bis[(2-aminoethyl-methyl-aminoethylimino]-1-methylpiperazinecobalt(III) perchlorate monohydrate, 46B, 962
$C_{15}H_{34}Co_2N_{12}O_2$, facial-Tris-(R-propylenediamine)cobalt(III) hexacyanocobaltate(III) dihydrate, 40B, 813
$C_{15}H_{46}Cl_3CoN_6O_2$, (-)-Tris-(R,R-2,4-diaminopentane)cobalt(III) chloride dihydrate, 39B, 658
$C_{16}H_{21}CuN_7O_6$, Bis(2-aminoethyl)amine(1,10-phenanthroline)copper(II) nitrate, 45B, 1037
$C_{16}H_{26}Cl_2F_6FeN_4O_2P$, N,N'-Bis(2-(3'-chloroacetylacetiminato))ethylenediamineiron(III) hexafluorophosphate, 44B, 813
$C_{16}H_{26}CuN_8O_7$, (N-Salicylidene-N'-methylethylenediamine)(aquo)(9-methyladenine)copper(II) nitrate dihydrate, 41B, 955
$C_{16}H_{28}F_6FeN_4O_2P$, N,N'-Bis(2-acetylacetiminato)ethylenediamineiron-(III) hexafluorophosphate, 44B, 813
$C_{16}H_{29}Cl_2N_5NiO_{10}$, [11-(6-Acetyl-2-pyridyl)-3,7,10-triazadodec-10-enylamine-N,N',N'',N''',N'''']aqua-nickel(II) perchlorate, 45B, 1041
$C_{16}H_{30}Cl_3CoN_6O_9$, (+)-(3,3'-Dimethyl-2,2'-bipyridine)bis(ethylenediamine)cobalt(III) chloride diperchlorate monohydrate, 44B, 814
$C_{16}H_{32}Cl_6N_4NiO_4$, Bis(tetramethylethylenediamine)nickel(II) trichloroacetate, 44B, 815
$C_{16}H_{34}Co_2N_{12}O_3$, Pentaethylenehexaminecobalt(III) hexacyanocobaltate(III) trihydrate, 41B, 957
$C_{16}H_{34}CuN_2O_6 \cdot 0.5\ H_2O$, Bis(L-lactato)(N,N,N',N'-tetraethylethylenediamine)copper(II) hemihydrate, 43B, 1147
$C_{16}H_{36}AgN_6O_6$, meso-5,5,7,12,12,14-Hexamethyl-1,4,8,11-tetraazacyclotetradecanesilver(II) nitrate, 44B, 815
$C_{16}H_{36}Cl_2CuN_4O_8$, (C-meso-5,5,7,12,12,14-Hexamethyl-1,4,8,11-tetraaza-cyclotetradecane)copper(II) perchlorate (red form), 45B, 1041
$C_{16}H_{36}Cl_2CuN_4O_8$, (C-meso-5,5,7,12,12,14-Hexamethyl-1,4,8,11-tetraaza-cyclotetradecane)copper(II) perchlorate (blue form), 45B, 1041
$C_{16}H_{36}Cl_2CuN_4O_8$, Bis(perchlorato)(γ-C-meso-5,5,7,12,14,14-hexamethyl-1,4,8,11-tetraazacyclotetradecane)copper(II), 44B, 815
$C_{16}H_{38}CoKN_4O_{12}$, Potassium N,N'-ethylenebis(acetylacetoniminato)-trans-diglycinatocobaltate(III) hexahydrate, 41B, 956
$C_{16}H_{38}Cu_3N_{12}O_4S_4$, Di-μ-(1,3-diamino-2-propanolato)bis(1,3-diamino-2-propanol)tricopper(II) tetrakis(thiocyanate), 43B, 1394
$C_{16}H_{45}Cl_3Co_2N_8O_{15} \cdot 2\ H_2O$, μ-Peroxo-μ-hydroxy-bis(4,7-dimethyl-1,4,7,10-tetraazadecanecobalt(III)) perchlorate dihydrate, 45B, 1042
$C_{16}H_{46}Fe_2I_4N_{10}O$, μ-Oxo-bis(tetraethylenepentamineiron(III)) iodide, 41B, 957
$C_{16}H_{58}Co_4N_{18}O_{14}S_4$, Tris(di-μ-hydroxo-bis(ethylenediaminecobalt-(II)))cobalt(II) tetrakisthiocyanate bisnitrate dihydrate, 44B, 816
$C_{16}H_{140}Cu_4N_{16}O_{58}Si_8$, Bis(ethylenediamine)copper silicate complex,

38B, 822
$C_{17}H_{18}N_4Ni$, 1,2-Bis(2-iminobenzylideneimino)propanenickel(II), 46B, 963
$C_{17}H_{22}CuN_6O_4$, (N-Salicylidene-N-methylethylenediamine)-(theophyllinato)copper(II) monohydrate, 41B, 958
$C_{17}H_{23}CoN_6O_2$, trans-Azidobis(acetylacetonato)ethylenediiminepyridinecobalt(III), 44B, 816
$C_{17}H_{23}FeN_7S_2$, Di-isothiocyanato-(N,N'-(3,6-diazaocta-1,8-diyl)-2,6-di(iminoacetyl)-pyridineiron, 42B, 732
$C_{17}H_{24}BrMnN_2O_3$, Bromotricarbonyl(N,N'-dicyclohexylethylenediimine)-manganese(I), 43B, 1149
$C_{17}H_{26}N_2O_7U$, (N,N'-Bis(hept-5-ene-2,4-dionato)ethylenediamine)dioxomethanoluranium(VI), 42B, 733
$C_{17}H_{27}Br_2CdN_5 \cdot 0.5\ H_2O$, Bromo(2,15-dimethyl-3,7,10,14,20-pentaazabicyclo[14.3.1]eicosa-1(20),2,14,16,18-pentaene)cadmium bromide hemihydrate, 45B, 1042
$C_{17}H_{34}Cl_2N_6NiO_8$, (2,12-Dimethyl-3,7,11,17-tetraazabicyclo[11.3.1]-heptadeca-1(17),13,15-triene)(ethylenediamine)nickel(II) diperchlorate, 46B, 964
$C_{17}H_{54}BrCo_4N_{17}O_9S_5$, Tris(di-μ-hydroxo-bis(ethylenediamine)cobalt-(III))cobalt(III) bromide pentathiocyanate trihydrate (racemate), 41B, 959
$C_{18}H_{13}CoCuN_2O_6 \cdot 6\ H_2O$, [μ-3,3'-(Ethane-1,2-diyl-bis(nitrilomethylidyne))-bis(2-hydroxybenzoato)]cobaltcopper hexahydrate, 46B, 965
$C_{18}H_{20}CdN_6Ni$, catena-μ-Ethylenediaminecadmium(II)-tetracyanonickelate(II) dibenzene clathrate, 38B, 824
$C_{18}H_{22}CuN_6O_3 \cdot 3.5\ H_2O$, (N-Salicylidene-N',N'-dimethylethylenediamine)(theophyllinato)copper(II) - 3.5 water, 44B, 817
$C_{18}H_{25}FeN_7S_2$, Di-isothiocyanato-(N,N'-(3,7-diazanonan-1,9-diyl)-2,6-di(iminoacetyl)-pyridineiron, 42B, 732
$C_{18}H_{29}Br_2CoN_6O_6 \cdot 2\ H_2O$, [3-(6-Methyl-2,4-dinitrophenyl)pentane-2,4-dionato][tris(2-aminoethyl)amine]cobalt(III) dibromide dihydrate, 46B, 965
$C_{18}H_{31}CuN_{11}O_6$, Bis(theophyllinato)(diethylenetriamine)copper(II) dihydrate, 42B, 734
$C_{18}H_{36}Cu_2N_4O_8 \cdot H_2O$, Bis(N'-isopropyl-2-methyl-1,2-propanediamine)-μ-oxalato-oxalatodicopper(II) monohydrate, 46B, 966
$C_{18}H_{42}CoCrN_6O_{15}$, Tris(1,2-diaminopropane)cobalt(III) tris(malonato)-chromate(III) trihydrate, 42B, 1023
$C_{18}H_{44}CoCl_3N_6O$, (+)-Tris((-)-trans-1,2-diaminocyclohexane)cobalt-(III) chloride monohydrate, 38B, 824
$C_{18}H_{48}CoCrN_{12}$, Tris(ethylenediamine)cobalt(III) tris(ethylenediamine)chromium(III) hexakis-(thiocyanate), 46B, 966
$C_{18}H_{48}CoCrN_{18}S_6 \cdot X\ H_2O$, (+)-Tris(ethylenediamine)cobalt(III) (-)-tris(ethylenediamine)chromium(III) thiocyanate hydrate, 43B, 1149
$C_{18}H_{48}N_9O_{12}Rh$, (+)-Tris(trans-1(R),2(R)-diaminocyclohexane)rhodium-(III) nitrate trihydrate, 43B, 1150
$C_{18}H_{50}Cr_3I_4N_9O_5 \cdot 5\ H_2O$, μ-[cis-Dihydroxo(O,O')-hydroxo(1,4,7-triazacyclononane)chromium(III)]-di-μ-hydroxo-bis[(1,4,7-triazacyclononane)chromium(III)] tetraiodide pentahydrate, 45B, 1044
$C_{18}H_{54}Cl_4Co_2N_{12}O_{18} \cdot 2\ H_2O$, μ-Peroxo-μ-(2,2',2"-triaminotriethylamine)-bis(2,2',2"-triaminotriethylaminecobalt(II)) perchlorate dihydrate, 45B, 1044
$C_{19}H_{26}N_4O_5PRe$, Tricarbonylbis(ethlyenediamine)rhenium(I) diphenyl phosphinate, 46B, 967
$C_{19}H_{45}Cl_3Cu_2N_8O_{12}$, μ-Imidazolato-bis(perchlorato-1,1,7,7-tetramethyldiethylenetriaminecopper(II)) perchlorate, 44B, 818

$C_{20}H_{16}N_4Ni$, Bis(N,N'-bis(8-quinolyl)ethylenediaminato)nickel(II), 43B, 1151
$C_{20}H_{24}Co_2N_{12} \cdot 2\ H_2O$, Ethylenediaminebis[2-(aminomethyl)pyridine]cobalt(III) hexacyanocobalt(III) dihydrate, 45B, 1045
$C_{20}H_{26}CuN_2O_6$, Bis(salicylato)(N,N,N',N'-tetramethylethylenediamine)-copper(II), 44B, 818
$C_{20}H_{26}FeN_5O_6$, N,N'-Bis(2-salicylaldiminato)ethylenediamineiron(III) nitrate hydrate, 44B, 813
$C_{20}H_{28}ClFeN_4O_4$, N,N'-Bis(2-salicylaldiminato)ethylenediamineiron-(III) chloride dihydrate, 44B, 813, 819
$C_{20}H_{28}Cl_2CuN_4O_4$, Bis(1,3-propanediamine)copper(II) p-chlorobenzoate, 40B, 814
$C_{20}H_{28}N_4NiO_4$, Bis((aci-nitromethyl)benzenato)(N,N,N',N'-tetramethyl-1,2-diaminoethane)nickel(II), 41B, 959
$C_{20}H_{30}CuI_2N_4O_5$, Aquobis(1,3-propanediamine)copper(II) di-m-iodobenzoate, 43B, 1320
$C_{20}H_{30}N_4NiO_4$, Bis(1,3-propanediamine)nickel(II) benzoate, 40B, 815
$C_{20}H_{32}CoIN_4O_4 \cdot H_2O$, (1,2-Diaminoethane)bis(1S,2S-(+)-1-phenyl-2-amino-1,3-dihydroxypropane)cobalt(III) iodide hydrate, 45B, 1045
$C_{20}H_{32}CuN_4O_{10}$, Bis(2-methoxy-4-nitrophenolato)-N,N,N',N'-tetramethylethylenediaminecopper(II) dihydrate, 42B, 915
$C_{20}H_{36}Co_2Li_2N_4O_{22}$, Dilithium bis(ethylenediamine-N,N'-disuccinatocobaltate(III)) hexahydrate, 43B, 1153
$C_{20}H_{38}Cl_8N_4Ni_2O_9$, μ-Aquo-bis(μ-dichloracetato-O,O')bis(dichloracetato)bis(N,N,N',N'-tetramethylethylenediamine)dinickel(II), 44B, 819
$C_{20}H_{46}N_4Ni_2O_9$, μ-Aquo-bis(μ-acetato-O,O')-bis(acetato)-bis-(N,N,N',N'-tetramethylethylenediamine)dinickel, 43B, 1154
$C_{20}H_{50}Cl_2Cu_2N_4O_{10}$, Di-μ-hydroxo-bis(N,N,N',N'-tetraethylethylenediamine)dicopper(II) perchlorate (α form), 40B, 815
$C_{20}H_{50}Cl_2Cu_2N_4O_{10}$, Di-μ-hydroxo-bis(N,N,N',N'-tetraethylethylenediamine)dicopper(II) perchlorate (β form), 43B, 1143
$C_{20}H_{54}Co_2N_{12}O_{14}S_2 \cdot 4\ H_2O$, μ-Dioxo-bis(1,5,8,11,15-pentaazapentadecanecobalt) dithionate dinitrate tetrahydrate, 46B, 968
$C_{21}H_{48}Cl_2Cu_2N_4O_3$, μ-Carbonato-dichlorobis(N,N,N',N'-tetraethylethylenediamine)dicopper(II), 46B, 968
$C_{22}H_{29}CoN_4PS_2$, Bis(isothiocyanato)(N-(2-diphenylphosphino-ethyl)-N',N'-diethylethylenediamine)cobalt(II), 39B, 659; 44B, 820
$C_{22}H_{34}CuN_4O_4$, Bis(1,3-propanediamine)copper(II) m-methylbenzoate, 43B, 1244
$C_{22}H_{44}N_{10}O_{18}P_2Pt_2$, Di-μ-(5'-cytidine monophosphate)-bis(ethylenediamineplatinum) dihydrate, 43B, 1325
$C_{23}H_{29}N_5O_7Zn$, (Ethylenediamine)zinc(II) benzohydroxamate benzohydroxamic acid monohydrate, 40B, 816
$C_{23}H_{31}CoN_4PS_2$, Bis(isothiocyanato)-N-((diphenylphosphino)ethyl)-N-methyl-N'-diethylethylenediaminecobalt(II), 39B, 659
$C_{24}H_{32}Cl_2N_6NiO_{10}$, N,N,N'-Tris(2-(2'-pyridyl)ethyl)ethane-1,2-diaminenickel(II) perchlorate nitromethane, 41B, 960
$C_{24}H_{32}CuN_6O_6$, Dinitrato-(5,12-dimethyl-7,14-diphenyl-1,4,8,11-tetraazatetradeca-4,11-diene)copper(II), 44B, 820
$C_{24}H_{40}BrCl_2CoN_6O_8 \cdot H_2O$, $(-)_{589}$-Bis[(1R,2R)-1,2-cyclohexanediamine]-(3,3'-dimethyl-2,2'-bipyridine)cobalt(III) bromide diperchlorate monohydrate, 45B, 1046
$C_{24}H_{48}Cu_2N_{14}O_{12}$, μ-Benzidine-bis(2,2',2"-triaminotriethylaminecopper(II)) nitrate, 44B, 821; 45B, 1047
$C_{24}H_{49}Br_2ClCoN_5O_9S_2$, (+)-cis-Chloroamminebis(ethylenediamine)cobalt-(III) d-3-bromocamphor-9-sulphonate monohydrate, 44B, 821
$C_{24}H_{50}Cl_2Cu_2N_8O_{12}$, [2,2'-(1,2-Diaminoethane)bis(2-methyl-3-butanone-

oximato)]copper(II) perchlorate dimer, 45B, 1047
$C_{24}H_{50}Cl_4N_4Ni_2O_9$, μ-Aquo-di-μ-(2-chloropropionato)-bis(2-chloropropionato)-bis(N,N,N',N'-tetramethylethylenediamine)dinickel(II), 43B, 1154
$C_{24}H_{50}Cl_4N_4Ni_2O_9$, μ-Aquo-di-μ-(3-chloropropionato)-bis(3-chloropropionato)-bis(N,N,N',N'-tetramethylethylenediamine)dinickel(II), 43B, 1155
$C_{24}H_{54}N_4Ni_2O_9$, μ-Aquo-bis(μ-propionato-O,O')-bispropionatobis-(N,N,N',N'-tetramethylethylenediamine)dinickel(II), 43B, 1156
$C_{24}H_{56}Cl_6N_8Pt_2$, trans-Dichlorobis(1,2-cyclohexanediamine)platinum-(IV) bis(1,2-cyclohexanediamine)platinum(II) tetrachloride, 43B, 1156
$C_{24}H_{60}N_{12}U_3$, Tris(bis(N,N'-dimethylethylenediamido)uranium(IV)), 43B, 1157
$C_{24}H_{96}Cl_{22}Cu_4N_{24}Pt_6$, Tris[bis(1,2-diaminoethane)platinum(II)dichlorobis(1,2-diaminoethane)platinum(IV)] tetrakis[tetrachlorocuprate-(I)], 45B, 1048
$C_{26}H_{16}Cu_2F_{12}N_2O_6$, Bis(1,1,1,5,5,5-hexafluoropentane-2,4-dionato)copper(II)-N,N'-ethylenebis(salicylodeneaminato)copper(II), 46B, 969
$C_{26}H_{48}Br_2N_4O_8PtS_2$, $(+)_{350}$-Ethylenediamine-N,N'-dimethylethylenediamineplatinum(II) (+)-α-bromocamphor-π-sulphonate (1:2), 45B, 1049
$C_{27}H_{28}BCuN_2O$, Carbonyl-ethylenediamine-copper(I) tetraphenylborate, 46B, 969
$C_{28}H_{28}CeF_{24}N_4O_4$, Bis[1,1,1,12,12,12-hexafluoro-2,11-bis(trifluoromethyl)-4,9-dimethyl-2,11-diolato-5,8-diazadodeca-4,8-diene(2-)]-cerium(IV), 46B, 970
$C_{28}H_{36}Cu_2N_4O_4 \cdot 2\ C_6H_6O$, Bis((μ-phenoxo)ethylenediaminephenoxycopper(II)) phenol solvate, 46B, 971
$C_{28}H_{60}Cu_2N_{12}Ni_2O_{18} \cdot 2\ H_2O$, Bis(ethylenediamine)diaquocopper(II) bis[(ethylenediaminetetraacetato)nickel(II)-O] bis(ethylenediamine)cuprate(II) dihydrate, 46B, 971
$C_{28}H_{66}CuN_4O_8S_2$, Bis(ethylenediamine)bis(dodecylsulfato)copper(II), 43B, 1157
$C_{29}H_{33}BCuN_3O$, (Diethylenetriamine)carbonylcopper(I) tetraphenylborate, 44B, 822
$C_{30}H_{36}Mo_6N_6Na_4O_{32} \cdot 14\ H_2O$, Sodium tri-μ-(ethylenediaminetetraacetato)-bis(tetraoxotrimolybdenum) hydrate, 45B, 1050
$C_{30}H_{60}Cu_2N_4O_8$, Di-μ-glutarato-bis[(N,N,N',N'-tetraethylethylenediamine)copper(II)], 45B, 1050
$C_{31}H_{29}N_3NiO_4Zn$, (N,N'-Bis(1-phenylhexa-1,3-dione-5-yl)ethylenediaminenickel)pyridinezinc, 41B, 961
$C_{32}H_{22}Co_2I_4N_{10}O_2$, μ-Peroxo-bis[(1,9-bis(2-pyridyl)-2,5,8-triazanonane)cobalt(III)] tetraiodide, 45B, 1051
$C_{32}H_{80}N_{16}U_4$, Tetrakis(bis(N,N'-dimethylethylenediamido)uranium(IV)), 43B, 1159
$C_{34}H_{54}Br_8Hg_4N_{10}$, [Bromo(2,15-dimethyl-3,7,10,14,20-pentaazabicyclo-[14.3.1]eicosa-1(20),2,14,16,18-pentaene)mercury] [μ-dibromobis(dibromomercurate)], 45B, 1042
$C_{40}H_{34}ClCuN_2O_4P_2 \cdot CH_2Cl_2$, [N,N'-Bis[o-(diphenylphosphino)benzylidene]ethylenediamine]copper(I) perchlorate dichloromethane, 46B, 972
$C_{42}H_{50}CoN_9O_{10}$, (+)-Tris((-)-1,2-diphenylethylenediamine)cobalt(III) nitrate monohydrate, 43B, 1160
$C_{54}H_{80}Cl_4Cu_4N_{12}O_{22}$, Bis(perchlorato)bis[1-(N,N-dimethyl-2-aminoethyl)-1-phenyl-2-oximatopropane]dicopper bis(methanol)bis[1-(N,N-dimethyl-2-aminoethyl)-1-phenyl-2-oximatopropane]dicopper bis(perchlorate), 45B, 1051

$C_{56}H_{64}B_2Cu_2N_6O_2$, μ-Ethylenediamine-bis(carbonyl-ethylenediamine-copper(I)) bis(tetraphenylborate), 46B, 969
$C_{60}H_{76}B_2Cl_2Cu_2N_8$, Bis(chloro(2,2',2"-triaminotriethylamine)copper-(II)) tetraphenylborate, 41B, 962
$C_{60}H_{76}B_2N_{14}Ni_2$, Di-μ-azido-bis(2,2',2"-triaminotriethylamine)dinickel(II) tetraphenylborate, 41B, 962
$C_{60}H_{82}B_2N_6NiO_3S_3$, Tris(ethylenediamine)nickel(II) tetraphenylborate - tris(dimethyl sulfoxide), 44B, 823
$C_{62}H_{76}B_2Cu_2N_{10}$, Bis(cyano-(2,2',2"-triaminotriethylamine)copper(II)) bis(tetraphenylborate), 40B, 817
$C_{62}H_{76}B_2Cu_2N_{10}O_2$, Bis(cyanato(2,2',2"-triaminotriethylamine)copper-(II)) tetraphenylborate, 41B, 962
$C_{62}H_{76}B_2Cu_2N_{10}S_2$, Bis(thiocyanato(2,2',2"-triaminotriethylamine)copper(II)) tetraphenylborate, 41B, 962
$C_{62}H_{76}B_2Mn_2N_{10}O_2$, Bis(isocyanato(2,2',2"-triaminotriethylamine)manganese(II)) tetraphenylborate, 44B, 823
$C_{62}H_{76}B_2Mn_2N_{10}S_2$, Bis(isothiocyanato(2,2',2"-triaminotriethylamine)-manganese(II)) tetraphenylborate, 44B, 823
$C_{62}H_{76}B_2N_{10}Ni_2O_2$, Di-μ-cyanato-bis(2,2',2"-triaminotriethylamine)dinickel(II) tetraphenylborate, 40B, 818
$C_{66}H_{66}B_2Cu_2N_{12}$, Di-μ(1,3)-azido-bis((1,1,4,7,7-pentamethyldiethylenetriamine)copper(II)) tetraphenylborate, 44B, 824
$C_{66}H_{70}B_2Cu_2N_6O_4$, μ-(5,8-Dihydroxy-1,4-naphthoquinonato(2-))-bis(diethylenetriaminecopper(II)) tetraphenylborate, 44B, 824
$C_{72}H_{90}B_2Cu_2N_{10}$, μ-Biimidazolato-bis(1,1,4,7,7-pentamethyldiethylenetriamine)dicopper(II) tetraphenylborate, 45B, 1052

77. METAL COMPLEXES (ACETYLACETONE)

$C_7H_7O_4Rh$, Acetylacetonatodicarbonylrhodium(I), 40B, 819
$C_9H_9CrO_6$, Tris-(1,3-propanedionato)chromium(III), 41B, 1113
$C_{10}H_2Cl_4F_{12}Mo_2O_6$, Di-μ-chloro-bis(dichloro(1,1,1,5,5,5-hexafluoropentane-2,4-dionato)oxomolybdenum(V)), 44B, 825
$C_{10}H_2CuF_{12}O_4$, Bis(1,1,1,5,5,5-hexafluoroacetylacetonato)copper(II) (gas-ed), 42B, 1029
$C_{10}H_5F_6NO_6U$, Dioxo(ammine)bis(1,1,1,5,5,5-hexafluoropentane-2,4-dionato)uranium(VI), 45B, 1053
$C_{10}H_{14}CdO_4$, catena-Di-μ-acetylacetonato-cadmium(II), 41B, 963
$C_{10}H_{14}ClKO_4Pt$, Potassium bis(acetylacetonato)chloroplatinate(II), 27, 822; 34B, 495
$C_{10}H_{14}Cl_2O_4Re$, trans-Dichlorobis(pentane-2,4-dionato)rhenium(IV), 39B, 659
$C_{10}H_{14}Cl_6O_4Ti_2$, Di-μ-chloro-tetrachlorobis(2,4-pentanedionato)dititanium(IV), 43B, 1162
$C_{10}H_{14}CrO_4$, Bis(2,4-pentanedionato)chromium(II), 43B, 1162
$C_{10}H_{14}CuO_4$, Copper acetylacetonate, 15, 423; 34B, 497
$C_{10}H_{14}I_2O_4Pt$, trans-Bis(acetylacetonato)diiodoplatinum(IV), 39B, 660
$C_{10}H_{14}MnN_3O_4$, Azidobis(acetylacetonato)manganese(III), 41B, 964
$C_{10}H_{14}MoO_6$, Bis(acetylacetonato)dioxomolybdenum, 39B, 661; 41B, 964
$C_{10}H_{14}N_2O_{10}Zr$, Bis(acetylacetonato)dinitratozirconium(IV), 41B, 965
$C_{10}H_{14}NiO_4$, Bis(acetylacetonato)nickel(II) (gas-ed), 21, 514
$C_{10}H_{14}O_4Pd$, Palladium acetylacetonate, 35B, 605
$C_{10}H_{14}O_4Pt$, Bis(acetylacetonato)platinum(II), 46B, 973
$C_{10}H_{14}O_5V$, Vanadyl bisacetylacetonate, 26, 686; 40B, 1111
$C_{10}H_{16}O_5Zn$, Bis(acetylacetonato)aquozinc(II), 24, 616; 28, 618
$C_{10}H_{18}CoO_6$, Cobalt bis(acetylacetonate) dihydrate, 23, 527

$C_{10}H_{18}MnO_6$, Diaquobisacetylacetonatomanganese(II), 33B, 388; 35B, 606
$C_{10}H_{18}NiO_6$, Diaquobisacetylacetonatonickel(II), 29, 567
$C_{10}H_{20}Cl_2NiO_{14}$, trans-Diaquobis(2,4-pentanedione)nickel(II) perchlorate, 37B, 490; 43B, 1162
$C_{10}H_{20}NiO_7$, Diaquobis(acetylacetonato)nickel(II) monohydrate, 39B, 661
$C_{11}H_{14}MnNO_4S$, catena-Thiocyanatobis(acetylacetonato)manganese(III), 45B, 1053
$C_{12}H_6F_3O_4Rh$, Benzoyl-1,1,1-trifluoroacetonatodicarbonylrhodium(I), 43B, 1163
$C_{12}H_{12}CuF_6N_2O_2$, N,N'-Ethylenebis(1,1,1-trifluoro-2,4-pentanedioneiminato)copper(II), 45B, 1054
$C_{12}H_{18}CuN_2O_2 \cdot 0.5\ H_2O$, N,N-Ethylenebis(acetylacetoneiminato)copper-(II) hemihydrate, 33B, 389
$C_{12}H_{18}CuO_4$, Bis(3-methylpentane-2,4-dionato)copper(II), 32B, 384
$C_{12}H_{18}CuO_6$, Copper(II) ethylacetoacetate, 30B, 336
$C_{12}H_{18}N_2O_3V$, N,N-Ethylenebis(acetylacetoneiminato)oxovanadium(IV), 35B, 610
$C_{12}H_{20}CuN_2O_3$, Aquo-N,N'-ethylenebis(acetylacetoneiminato)copper(II), 34B, 500
$C_{13}H_{11}F_{12}O_{10}PU$, Bis(1,1,1,5,5,5-hexafluoropentane-2,4-dionato)dioxo-(trimethyl phosphate)uranium(VI) (α-form), 43B, 1164
$C_{13}H_{11}F_{12}O_{10}PU$, Bis(1,1,1,5,5,5-hexafluoropentane-2,4-dionato)dioxo-(trimethyl phosphate)uranium(VI) (β-form), 43B, 1164
$C_{13}H_{24}ClCuN_3O_6$, N,N'-Ethylenebis(acetylacetoneiminato)copper(II) methylammonium perchlorate, 35B, 615
$C_{14}H_6CuF_{12}N_2O_4$, Bis(hexafluoroacetylacetonato)(pyrazine)copper(II), 40B, 819
$C_{14}H_8F_{12}O_7U$, Bis(1,1,1,5,5,5-hexafluoro-2,4-pentanedionato)-dioxouranium tetrahydrofuran, 46B, 973
$C_{14}H_{14}N_4NiO_4$, Bis[2-(1-iminoethyl)-1-cyano-1,3-butanedionato]nickel-(II), 45B, 1054
$C_{14}H_{16}F_6N_2NiO_2$, N,N'-Ethylenebis(1,1,1-trifluoro-2,4-hexanedioneiminato)nickel(II), 45B, 1054
$C_{14}H_{19}CoN_4O_6$, Bis(acetylacetonato)(nitro)(2-aminopyrimidine)cobalt-(III), 44B, 825
$C_{14}H_{26}NiO_6$, Bis(acetylacetonato)diethanolnickel(II), 39B, 661
$C_{15}H_{17}MoNO_4$, (η-Allyl)dicarbonyl(acetylacetonato)pyridinemolybdenum-(II), 44B, 826
$C_{15}H_{19}ClO_4Zr$, Chlorobis(acetylacetonato)cyclopentadienylzirconium, 40B, 1111
$C_{15}H_{20}CuN_2O_4$, Bis-(2,4-pentanedionato)-4-aminopyridinecopper(II), 37B, 491
$C_{15}H_{21}ClO_6Zr$, Chlorotris(acetylacetonato)zirconium(IV), 37B, 491
$C_{15}H_{21}ClO_{10}Ti$, Tris(acetylacetonato)titanium(IV) perchlorate, 44B, 827
$C_{15}H_{21}CoO_6$, Tris(acetylacetonato)cobalt(III), 22, 594; 24, 598; 39B, 530; 40B, 820
$C_{15}H_{21}CrO_6$, Δ(-)-Tris(pentane-2,4-dionato)chromium(III), 45B, 1055
$C_{15}H_{21}CrO_6$, Tris(acetylacetonato)chromium(III), 24, 597; 30B, 332
$C_{15}H_{21}FeO_6$, Tris(acetylacetonato)iron(III), 20, 474; 32B, 385
$C_{15}H_{21}MnO_6$, Tris(acetylacetonato)manganese(III) (β form), 29, 543; 39B, 530; 40B, 821
$C_{15}H_{21}MnO_6$, Tris(acetylacetonato)manganese(III) (γ form), 45B, 1056
$C_{15}H_{21}MoO_6$, Tris(acetylacetonato)molybdenum(III), 45B, 1056
$C_{15}H_{21}NO_9Zr$, Nitratotris(acetylacetonato)zirconium(IV), 42B, 735

$C_{15}H_{21}O_6Rh$, Tris-(2,4-pentanedionato)rhodium(III), 39B, 662

$C_{15}H_{21}O_6Ru$, Tris-acetylacetonatoruthenium(III), 39B, 662

$C_{15}H_{21}O_6Sc$, Tris(acetylacetonato)scandium(III), 39B, 662

$C_{15}H_{21}O_6V$, Tris(acetylacetonato)vanadium(III), 34B, 507; 35B, 606

$C_{15}H_{22}O_8U$, Acetylacetonebis(acetylacetonato)dioxouranium(VI), 45B, 1057

$C_{15}H_{23}AgClFeO_{11}$, Tris(acetylacetonato)iron(III) silver perchlorate monohydrate, 40B, 556

$C_{15}H_{23}Ag_3N_2NiO_{13}$, Trisilver dinitrate tris(acetylacetonato)nickelate(II) monohydrate, 31B, 381

$C_{15}H_{23}CdKO_7$, Potassium tris(acetylacetonato)cadmate(II) monohydrate, 41B, 966

$C_{15}H_{23}NO_7U$, Dioxobis(2,4-pentanedionato)mono(2-aminopentan-4-one)uranium(VI), 43B, 1165

$C_{15}H_{23}O_7Yb \cdot 0.5\ C_6H_6$, Tris(acetylacetonato)aquoytterbium(III) benzene solvate, 34B, 510

$C_{15}H_{23}O_7Yb$, Tris(acetylacetonato)aquoytterbium(III), 34B, 508

$C_{15}H_{25}LaO_8$, Diaquotris(acetylacetonato)lanthanum(III), 33B, 390

$C_{15}H_{25}NdO_8$, Tris(acetylacetonato)neodymium dihydrate, 37B, 492

$C_{15}H_{27}EuO_9$, Diaquotris(acetylacetonato)europium(III) monohydrate, 34B, 497

$C_{15}H_{27}O_9Y$, Tris(acetylacetonato)diaquoyttrium monohydrate, 32B, 386

$C_{15}H_{29}HoO_{10}$, Tris(acetylacetonato)holmium tetrahydrate, 37B, 493

$C_{16}H_8CuF_6O_4S_2$, Bis(1-(2-thienyl)-4,4,4-trifluoro-1,3-butanedionato)-copper(II), 43B, 1166

$C_{16}H_8F_6O_4PdS_2$, Bis(4,4,4-trifluoro-1-(2-thienyl)-1,3-butanedionato)-palladium(II), 42B, 735

$C_{16}H_{12}Cl_3F_3NNbO_3$, Tetraethylammonium trichlorooxo(1,1,1-trifluoro-4-thenoyl-2,4-butanedionato)niobate(V), 45B, 1057

$C_{16}H_{14}CuF_{12}N_2O_4$, Bis-(1,1,1,5,5,5-hexafluoropentane-2,4-dionato)copper(II)-1,4-diazabicyclo[2.2.2]octane, 38B, 825

$C_{16}H_{25}NO_7U$, Dioxobis-(2,4-pentanedionato)mono-(2-N-methylaminopentan-4-one)uranium(VI), 42B, 736

$C_{17}H_{15}N_2O_4V$, cis-Dioxo(acetylacetonato)(1,10-phenanthroline)-vanadium(V), 41B, 967

$C_{17}H_{27}NO_7U$, Dioxobis(2,4-pentanedionato)mono(2-N,N-dimethylaminopentan-4-one)uranium(VI), 43B, 1166

$C_{17}H_{29}CoN_4O_6Se_2$, Tris(acetylacetonato)cobalt(III) selenourea, 46B, 975

$C_{18}H_{29}NO_7U$, Dioxobis(2,4-pentanedionato)mono(2-N-isopropylaminopentan-4-one)uranium(VI), 43B, 1167

$C_{18}H_{36}N_2O_4Pd$, (Acetylacetonato)bis(diethylamine)palladium(II) acetylacetonate, 46B, 975

$C_{18}H_{40}N_2O_4Pt_2$, µ-Ethylenediamine-bis(trimethyl(acetylacetonato)platinum(IV)), 30B, 339

$C_{19}H_{16}CoF_6O_6S_2$, cis-Bis(1-(2-thienyl)-4,4,4-trifluoro-1,3-butanedionato)bis(methanol)cobalt(II), 44B, 827

$C_{19}H_{16}F_6O_6S_2Zn$, cis-Bis(1-(2-thienyl)-4,4,4-trifluoro-1,3-butanedionato)bis(methanol)zinc(II), 44B, 827

$C_{19}H_{20}CuF_{12}NO_6$, Bis(hexafluoroacetylacetonato)(4-hydroxy-2,2,6,6-tetramethylpiperidinyl-N-oxy)copper(II), 46B, 976

$C_{19}H_{21}CuNO_4$, Bis(acetylacetonato)quinolinecopper(II), 34B, 498

$C_{19}H_{33}AsCoN_4O_{15}$, (-)-Acetylacetonatobis(trimethylenediamine)cobalt-(III) arsenic(III) (+)-tartrate monohydrate, 39B, 663

$C_{20}H_4AmCsF_{24}O_8$, Caesium tetrakis(hexafluoroacetylacetonato)americate, 34B, 500

$C_{20}H_4CsEuF_{24}O_8$, Caesium tetrakis(hexafluoroacetylacetonato)europate,

34B, 500
$C_{20}H_4CsF_{24}O_8Y$, Caesium tetrakis(hexafluoroacetylacetonato)-yttrate(III), 31B, 383; 33B, 391
$C_{20}H_8F_{18}O_6Zr$, Tris(hexafluoroacetylacetonato)-π-cyclopentadienylzirconium, 40B, 1112
$C_{20}H_{10}CuF_{12}N_2O_4$, Bis(hexafluoroacetylacetonato)-2,2'-bipyridinecopper(II), 34B, 503
$C_{20}H_{12}F_{12}N_2O_4Zn$, cis-Bis(1,1,1,5,5,5-hexafluoro-2,4-pentanedionato)bis(pyridine)zinc(II), 39B, 664
$C_{20}H_{16}CuO_4$, Copper(II) benzoylacetonate, 31B, 383
$C_{20}H_{16}F_{12}O_8Th$, Tetrakis(trifluoroacetylacetonato)thorium(IV), 38B, 826
$C_{20}H_{18}F_{12}O_9Th$, Aquotetrakis(trifluoroacetylacetonato)thorium(IV), 44B, 827
$C_{20}H_{18}O_4Pd$, Bis(1-phenyl-1,3-butanedionato)palladium(II), 32B, 387
$C_{20}H_{18}O_5V$, Vanadyl bisbenzoylacetonate, 30B, 340; 40B,1111
$C_{20}H_{24}CoN_2O_4$, trans-Bis(2,4-pentanedionato)dipyridinecobalt(II), 33B, 392
$C_{20}H_{24}N_2NiO_4$, trans-Bis(2,4-pentanedionato)dipyridinenickel(II), 33B, 393
$C_{20}H_{24}N_2NiO_6$, Bis(2,4-pentanedionato)bis(pyridine N-oxide)nickel-(II), 33B, 394
$C_{20}H_{27}CoN_6O_9 \cdot 3.5\ H_2O$, Bis(acetylacetonato)(nitro)(deoxyadenosine)-cobalt(III) hydrate, 43B, 1168
$C_{20}H_{28}CeO_8$, Tetrakis(acetylacetonato)cerium(IV) (α form), 28, 622; 40B, 822
$C_{20}H_{28}CeO_8$, Tetrakis(acetylacetonato)cerium(IV) (β form), 34B, 503
$C_{20}H_{28}Fe_2O_8$, Bis(acetylacetonato)iron(II) dimer, 43B, 1169
$C_{20}H_{28}NpO_8$, β-Tetrakis(acetylacetonato)neptunium(IV), 38B, 826
$C_{20}H_{28}O_8Th$, Tetrakis(acetylacetonato)thorium(IV) (α-form), 22, 594; 42B, 736
$C_{20}H_{28}O_8U$, β-Tetrakis(acetylacetonato)uranium(IV), 35B, 608
$C_{20}H_{28}O_8Zr$, Tetrakis(acetylacetonato)zirconium(IV), 28, 619
$C_{20}H_{28}O_{10}Ti_2$, Di-μ-oxo-bis(diacetylacetonatotitanium(IV)), 38B, 827
$C_{20}H_{30}NO_7Yb$, Acetylacetoniminotris(acetylacetonato)ytterbium(III), 35B, 612
$C_{20}H_{32}Co_2O_{10}$, Dimeric bis(acetylacetonato)aquocobalt(II), 31B, 384
$C_{21}H_{21}CoN_2O_4 \cdot 0.7\ H_2O$, (N,N'-ethylenebis-(salicylideneiminato))-(acetylacetonato)cobalt(III) - 0.7 water, 38B, 828
$C_{21}H_{21}N_2NbO_4S_2$, Diisothiocyanatodiethoxy-(1,3-diphenylpropane-1,3-dionato)niobium(V), 42B, 737
$C_{21}H_{25}N_2ORh$, Acetylacetonato-2,3-dimethylenebutadienebis(pyridine)-rhodium, 40B, 822
$C_{21}H_{29}Cl_5O_9Ti_2$, μ-Oxo-bis(chlorobis(2,4-pentanedionato)titanium-(IV))-chloroform, 32B, 388
$C_{21}H_{36}Mo_3O_{15}$, Tris(acetylacetonato)oxomolybdenum triethanolate, 32B, 390
$C_{21}H_{43}ClN_4NiO_6$, Acetylacetonato-C-meso-(5,5,7,12,12,14-hexamethyl-1,4,8,11-tetra-azacyclotetradecane)nickel(II) perchlorate, 39B, 666
$C_{22}H_{16}CuF_6N_2O_4$, (Acetylacetonato)-(hexafluoroacetylacetonato)-(o-phenanthroline)copper(II), 46B, 976
$C_{22}H_{18}CuF_6N_2O_5 \cdot H_2O$, Aqua-(acetylacetonato)-(o-phenanthroline)copper hexafluoroacetylacetone hydrate, 46B, 976
$C_{22}H_{21}CuNO_4$, 4-Methylpyridine bis(o-hydroxyacetophenonato)copper-(II), 34B, 512
$C_{22}H_{22}CuO_4$, Bis(3-phenyl-2,4-pentanedionato)copper(II), 30B, 337
$C_{22}H_{22}MnN_2O_4$, 1,10-Phenanthrolinebis(acetylacetonato)manganese(II),

43B, 1170
$C_{22}H_{24}N_2NiO_2S_2$, Bis(monothioacetoacet-p-toluidide)nickel, 41B, 967
$C_{22}H_{28}N_2NiO_4$, trans-Bis(2,4-pentanedionato)bis(4-methylpyridine)nickel(II), 41B, 968
$C_{22}H_{30}Br_4Cr_2O_{10}$, Di-$\mu$-methoxy-bis(bis(3-bromo-2,4-pentanedionato)-chromium(III)), 43B, 1170
$C_{22}H_{30}Cl_4Cr_2O_{10}$, Di-$\mu$-methoxo-bis(bis(3-chloro-2,4-pentanedionato)-chromium(III)), 42B, 737
$C_{22}H_{38}NiO_4$, Bis(dipivaloylmethanido)nickel(II), 31B, 386
$C_{23}H_{15}CuF_{12}N_2O_4$, cis-Bis(1,1,1,5,5,5-hexafluoro-2,4-pentanedionato)-bis(pyridine)copper(II) hemibenzate, 39B, 664
$C_{23}H_{22}Cl_2O_3PRe$, Oxodichloropentane-2,4-dionatotriphenylphosphine-rhenium(V), 41B, 969
$C_{23}H_{31}O_8Th$, Tetrakis(acetylacetonato)thorium(IV) hemibenzene solvate, 44B, 828
$C_{23}H_{36}CoN_7O_8$, Bis(acetylacetonato)nitro(triacanthine)cobalt(III) hydrate - dimethylformamide, 44B, 829
$C_{24}H_8Cu_2F_{24}N_2O_8$, Tetrakis(hexafluoroacetylacetonato)-μ-pyrazine-dicopper(II), 40B, 819
$C_{24}H_{12}F_9FeO_6S_3$, Tris(1-(2-thienyl)-4,4,4-trifluoro-1,3-butanedione)-iron(III), 42B, 738
$C_{24}H_{16}CdF_9NO_9$, Ammonium tris(4,4,4-trifluoro-1-(2-furyl)-1,3-butanedionato)cadmate, 46B, 977
$C_{24}H_{16}EuF_9O_8S_3$, Tris(4,4,4-trifluoro-1-(2-thienyl)-1,3-butanedione)-europium dihydrate, 42B, 739
$C_{24}H_{22}O_3PRh$, Acetylacetonatocarbonyl(triphenylphosphine)rhodium(I), 44B, 829
$C_{24}H_{34}Br_4Cr_2O_{10}$, Di-$\mu$-ethoxy-bis(bis(3-bromo-2,4-pentanedionato)-chromium(III)), 43B, 1170
$C_{25}H_{32}NNbO_5S$, fac-(1,3-Diphenylpropane-1,3-dionato)tris(isopropoxo)-isothiocyanatoniobium(V), 42B, 739
$C_{26}H_{16}CoCuF_{12}N_2O_6$, N,N'-Ethylenebis(salicylideniminato)copper(II)-bis(hexafluoroacetylacetonato)cobalt(II), 39B, 666
$C_{26}H_{16}Cu_2F_{12}N_2O_6$, N,N'-Ethylenebis(salicylideniminato)copper(II)-bis(hexafluoroacetylacetonato)copper(II), 45B, 1058
$C_{26}H_{18}CoF_6N_2O_4S_2$, Bis(thenoyltrifluoroacetonato)bis(pyridine)cobalt-(II), 44B, 830
$C_{26}H_{18}Cu_2F_{12}N_2O_6$, Bis[1,1,1,5,5,5-hexafluoro-2,4-pentanedionato-μ-(N-methyl-2-hydroxybenzylideneiminato)-μ-O]dicopper(II), 45B, 1059
$C_{26}H_{18}F_6N_2NiO_4S_2$, Bis(thenoyltrifluoroacetonato)bis(pyridine)nickel-(II), 44B, 830
$C_{26}H_{18}F_6N_2O_4S_2Zn$, Bis(thenoyltrifluoroacetonato)bis(pyridine)-zinc(II), 44B, 830
$C_{27}H_{19}F_3O_3PRhS$, Thenoyltrifluoroacetonatocarbonyltriphenylphosphine-rhodium(I), 44B, 831
$C_{27}H_{29}EuN_2O_6$, Tris(acetylacetonato)(1,10-phenanthroline)-europium(III), 38B, 830
$C_{27}H_{38}Co_2N_8O_{14}$, [Bis(acetylacetonato)(nitro)(1,9-dimethyladeninium)-cobalt(III)] [trans-bis(acetyl-acetonato)(dinitro)cobaltate(III)], 45B, 1059
$C_{28}H_{22}CoF_6N_2O_4S_2$, cis-Bis(1-(2-thienyl)-4,4,4-trifluoro-1,3-butanedionato)bis(4-methylpyridine)cobalt(II), 44B, 832
$C_{28}H_{22}CuF_6N_2O_4S_2$, cis-Bis(1-(2-thienyl)-4,4,4-trifluoro-1,3-butanedionato)bis(4-methylpyridine)copper(II), 44B, 832
$C_{28}H_{22}F_6N_2O_4S_2Zn$, cis-Bis(1-(2-thienyl)-4,4,4-trifluoro-1,3-butanedionato)bis(4-methylpyridine)zinc(II), 44B, 832
$C_{28}H_{27}Cl_2O_7PU$, Bis(3-chloropentane-2,4-dionato)dioxo(triphenylphos-

phine oxide)uranium(VI), 45B, 1060

$C_{28}H_{29}ClO_4PTc$, trans-Chlorobis(pentane-2,4-dionato)(triphenylphosphine)technetium(III) (*α* form), 43B, 1171

$C_{28}H_{29}ClO_4PTc$, trans-Chlorobis(2,4-pentanedionato)(triphenylphosphine)technetium(III) (*β* form), 44B, 832

$C_{28}H_{29}O_4PPd \cdot 0.5\ C_6H_6$, Acetylacetonatoacetylacetonyltriphenylphosphinepalladium(II) benzene solvate, 40B, 824

$C_{28}H_{29}O_7PU \cdot C_6H_6$, Dioxobis(pentane-2,4-dionato)(triphenylphosphine oxide)uranium(VI) benzene solvate, 45B, 1060

$C_{28}H_{44}O_{14}Ti_2$, Di-*μ*-oxo-bis(diacetylacetonatotitanium(IV)) dioxane solvate, 38B, 827

$C_{28}H_{46}Co_4O_{16}$, Di-*μ*-acetato-tetrakis(*μ*$_3$-methoxo-2,4-pentanedionatocobalt(II,III)), 39B, 667

$C_{28}H_{56}Co_4O_{16}$, Tetrakis(*μ*$_3$-methoxo-2,4-pentanedionatomethanolcobalt-(II)), 37B, 566

$C_{29}H_{22}CuF_{18}N_2O_6$, Tris(hexafluoroacetylacetonato)copper(II) 1-dimethylammonium-8-dimethylaminonaphthalene, 38B, 830

$C_{29}H_{32}N_2O_6V$, ((5,5'-(1,2-Ethanediyldinitrilo)bis(1-phenyl-1,3-hexanedionato))(2-)-O,O',O'',O''')oxovanadium - acetone, 42B, 740

$C_{30}H_6F_{36}O_{18}U_3$, Tri-*μ*-oxo-tris(bis(1,1,1,5,5,5-hexafluoropentane-2,4-dionato)oxouranium(VI)), 44B, 833

$C_{30}H_{22}CuO_4$, Copper(II) 1,3-diphenyl-1,3-propanedionate, 31B, 387

$C_{30}H_{24}CoCuF_{12}N_2O_6$, Bis(hexafluoroacetylacetonato){N,N'-ethylenebis-[2-hydroxypropiophenoneiminato-N,O(2-)]-copper(II)}cobalt(II), 45B, 1060

$C_{30}H_{24}CoF_{12}N_2NiO_6$, Bis(hexafluoroacetylacetonato){N,N'-ethylenebis-[2-hydroxypropiophenoneiminato-N,O(2-)]-nickel(II)}cobalt(II), 45B, 1060

$C_{30}H_{24}CuF_{12}MnN_2O_6$, Bis(hexafluoroacetylacetonato){N,N'-ethylenebis-[2-hydroxypropiophenoneiminato-N,O(2-)]-copper(II)}manganese(II), 45B, 1060

$C_{30}H_{24}CuF_{12}N_2NiO_6$, Bis(hexafluoroacetylacetonato){N,N'-ethylenebis-[2-hydroxypropiophenoneiminato-N,O(2-)]-nickel(II)}copper(II), 45B, 1060

$C_{30}H_{24}Cu_2F_{12}N_2O_6$, Bis(hexafluoroacetylacetonato){N,N'-ethylenebis[2-hydroxypropiophenoneiminato-N,O(2-)]-copper(II)}copper(II), 45B, 1060

$C_{30}H_{24}F_{12}MnN_2NiO_6$, Bis(hexafluoroacetylacetonato){N,N'-ethylenebis-[2-hydroxypropiophenoneiminato-N,O(2-)]-nickel(II)}manganese(II), 45B, 1060

$C_{30}H_{32}CoN_2O_4$, Bis(acetylacetonato)bis(6-methylquinoline)cobalt(III), 44B, 833

$C_{30}H_{42}Ni_3O_{12}$, Tris(bisacetylacetonenickel), 26, 673

$C_{30}H_{42}O_{12}Zn_3$, Trimeric bis(acetylacetonato)zinc(II), 33B, 395

$C_{30}H_{44}Co_3O_{13}$, Hexa(acetylacetonato)aquotricobalt(II), 33B, 396

$C_{32}H_{16}F_{12}O_8S_4Th$, Tetrakis(thienoyltrifluoroacetonato)thorium(IV), 45B, 1061

$C_{32}H_{22}F_{12}NO_9PrS_4$, Ammonium tetrakis(4,4,4-trifluoro-1-(2-thienyl)-1,3-butanedione)praseodymate(III) monohydrate, 32B, 391

$C_{32}H_{38}O_8Zn_3$, Bis(phenyl(acetylacetonato)zinc(II))-bis(acetylacetonato)zinc(II), 39B, 668

$C_{32}H_{54}Co_2N_2O_8$, Bis(2,4-pentanedionato)cyclohexylaminecobalt(II) dimer, 37B, 493

$C_{33}H_{57}LuO_6$, Tris-(2,2,6,6-tetramethylheptane-3,5-dionato)-lutetium(III), 42B, 740

$C_{34}H_{34}AsCl_2O_4Re$, Tetraphenylarsonium trans-dichlorobis(2,4-pentanedionato)rhenate(III), 45B, 1062

$C_{34}H_{48}O_6Ti$, Bis(2,4-pentanedionato)bis(2,6-diisopropylphenoxo)titanium(IV), 39B, 669
$C_{36}H_{52}Cu_2O_6$, Di-μ-(benzyloxo)-bis[(2,2,6,6-tetramethylheptane-3,5-dionato)copper(II)], 45B, 1063
$C_{37}H_{22}F_{12}NNdO_8S_4$, Pyridinium tetrakis(thenoyltrifluoroacetonato)-neodymate(III), 43B, 1172
$C_{38}H_{38}MnO_6$, Bis(1,3-diphenyl-1,3-propanedionato)bis(tetrahydrofuran)manganese(II), 45B, 1063
$C_{38}H_{42}N_2O_9U$, Tetraethylammonium bis(1,3-diphenylpropane-1,3-dionato)nitratodioxouranate(VI), 44B, 834
$C_{38}H_{67}EuO_7S$, 3,3-Dimethylthietane-1-oxide-tris(dipivalomethano)-europium(III), 39B, 671
$C_{39}H_{64}LuNO_6$, 3-Methylpyridine-tris(2,2,6,6-tetramethyl-3,5-heptanedionato)lutetium(III), 39B, 672
$C_{39}H_{71}EuN_2O_8$, Tris(2,2,6,6-tetramethyl-3,5-heptanedionato)bis(dimethylformamide)europium(III), 46B, 978
$C_{40}H_{24}F_{12}O_8U$, Tetrakis(4,4,4-trifluoro-1-phenyl-1,3-butanedionato)-uranium, 46B, 979
$C_{40}H_{41}EuO_8 \cdot X\ C_4H_{11}N$, Diethylamine - hydrogen tetrakis(benzoylacetonato)europate, 38B, 832
$C_{40}H_{56}Co_4O_{16}$, Cobalt(II) acetylacetonate, 30B, 334
$C_{40}H_{59}F_6O_7PS_2U$, (Tri-n-octylphosphine oxide)bis(thenoyltrifluoroacetonato)dioxouranium(VI), 43B, 1172
$C_{41}H_{37}Cl_2O_2P_2Re$, cis-Dichloro-pentane-2,4-dionato-trans-bis(triphenylphosphine)rhenium(III), 40B, 824
$C_{43}H_{67}EuN_2O_6$, Tris-(2,2,6,6-tetramethylheptane-3,5-dionato)-dipyridine-europium(III), 38B, 833
$C_{44}H_{48}Cr_2O_{12}P_2$, Di-μ-diphenylphosphinatoacetylacetonato-chromium-(III) (phase I), 30B, 333
$C_{44}H_{48}Cr_2O_{12}P_2$, Di-μ-diphenylphosphinatoacetylacetonatochromium(III) (phase II), 45B, 1064
$C_{44}H_{50}O_6P_2Rh_2$, ((p-Phenylenedimethylene)bis(diphenylphosphine))bis-((acetylacetonato)carbonylrhodium(I)), 43B, 1173
$C_{44}H_{76}NbO_8$, Tetrakis(2,2,6,6-tetramethyl-3,5-heptanedionato)-niobium(IV), 41B, 969
$C_{45}H_{33}FeO_6$, Tris(1,3-diphenyl-1,3-propanedionato)iron(III), 46B, 979
$C_{45}H_{52}EuNO_8$, Piperidinium tetrakis(benzoylacetonato)europate, 38B, 832
$C_{48}H_{56}Cu_4O_{12}$, Bis-(di-μ-(phenylmethoxo)-bis(pentane-2,4-dionato)di-copper(II)), 39B, 673
$C_{50}H_{38}Cu_2F_{12}N_6O_8S_2$, Bis(thenoyltrifluoroacetonato)bis(pyridine)copper(II) bis(trifluoroacetato)tetrakis(pyridine)copper(II), 44B, 830
$C_{51}H_{72}O_7PPr$, Tris(dipivaloylmethanato)(triphenylphosphine oxide)praseodymium, 45B, 1064
$C_{52}H_{52}O_2P_2Rh$, Bis(acetylacetonato)-2,3-dimethylenebutadienebis(triphenylphosphine)dirhodium, 40B, 822
$C_{56}H_{67}F_{12}O_9PS_4Th$, Tetrakis(1,1,1-trifluoro-3-2'-thenoylacetonato)-(tri-n-octylphosphineoxide)thorium(IV), 41B, 970
$C_{57}H_{54}F_{30}Ni_6O_{24}$, Di-μ-aquo-di-$\mu_3$-hydroxodeca(1,1,1-trifluoropentan-2,4-dionato)hexanickel(II) - toluene, 34B, 511
$C_{60}H_{42}F_9NdO_8P_2S_3$, Tris(thenoyltrifluoroacetonato)bis(triphenylphosphineoxide)neodymium(III), 41B, 971
$C_{60}H_{44}O_8Zr$, Tetrakis(1,3-diphenyl-1,3-propanedionato)zirconium(IV), 45B, 1065
$C_{74}H_{64}Co_2N_8O_6$, Bis(1,5-diphenyl-1,3,5-pentanetrionato)tetrapyridine-dicobalt(II) - tetrapyridine, 39B, 674
$C_{110}H_{202}Er_8O_{33}$, Erbium tetramethylheptanedionate, 38B, 835

78. METAL COMPLEXES (SALICYLIC DERIVATIVES)

$C_9H_9CuNO_4 \cdot 0.5\ H_2O$, N-Salicylideneglycinatoaquocopper(II) hemihydrate, 32B, 391

$C_9H_9NO_2Zn$, (N-Salicylidene ethanolaminato)zinc(II) complex, 45B, 1065

$C_9H_{13}CuN_2O_4 \cdot 0.5\ SO_4$, N-(Carbamoylmethyl)salicylideneiminatoaquocopper(II) sulfate monohydrate, 39B, 675

$C_9H_{13}N_4O_3SV \cdot H_2O$, Ammonium (salicylaldehyde S-methylthiosemicarbazonato)dioxovanadate(V) monohydrate, 46B, 980

$C_9H_{17}CuNO_8$, N-Salicylideneglycinatoaquocopper(II) tetrahydrate, 34B, 513

$C_{10}H_{11}ClCuNO_2$, Chloro-N-(2-hydroxypropyl)-salicylaldiminocopper(II), 35B, 617

$C_{10}H_{11}CuN_3O_3$, Salicylideneglycinato(thiourea)copper(II), 43B, 1174

$C_{11}H_{13}CuNO_4$, N-Salicylidene-α-aminoisobutyratoaquocopper(II), 37B, 494

$C_{14}H_{10}Cl_2CuN_2O_4$, Bis-(5-chlorosalicylaldoximato)copper(II), 29, 598

$C_{14}H_{10}Cl_2N_2NiO_4$, Bis(5-chlorosalicylaldoximato)nickel(II), 27, 871

$C_{14}H_{10}Cl_2N_2O_4Pd$, Bis(5-chlorosalicylaldoximato)palladium(II), 27, 871

$C_{14}H_{10}CuN_4O_2S$, Isothiocyanato-(N'-pyridyl-methylene-N"-salicyloylhydrazinato-N,N',O)copper(II), 41B, 972

$C_{14}H_{10}CuO_4$, Bis(salicylaldehydato)copper(II) (form 1), 29, 599

$C_{14}H_{10}CuO_4$, Bis(salicylaldehydato)copper(II) (form 2), 30B, 369

$C_{14}H_{12}CuN_2O_2$, Bis(salicylaldiminato)copper(II), 23, 655; 31B, 389

$C_{14}H_{12}CuN_2O_4$, Bis(salicylaldoximato)copper(II), 29, 596

$C_{14}H_{12}N_2NiO_2$, Bis(salicylaldiminato)nickel(II), 23, 655

$C_{14}H_{12}N_2NiO_4$, Bis(salicylaldoximato)nickel(II), 20, 572; 32B, 392

$C_{14}H_{12}N_2O_4Pd$, Bis(salicylaldoximato)palladium(II), 35B, 619

$C_{14}H_{13}HgNO$, Phenylmercury salicylalmethyliminate, 40B, 825

$C_{14}H_{14}CuO_8$, Diaquobis(salicylato)copper(II), 42B, 741

$C_{14}H_{14}NiO_6$, Diaquobis(salicylaldehydato)nickel, 26, 674

$C_{14}H_{14}O_8Zn$, Zinc salicylate dihydrate, 22, 711

$C_{14}H_{18}CoO_{10}$, Tetraaquobis(salicylato)cobalt(II), 44B, 834

$C_{14}H_{18}CuO_{10}$, Copper(II) salicylate tetrahydrate, 24, 609

$C_{14}H_{22}Cl_2N_4O_6U$, (N-(2'-Dimethylaminoethyl)-2-aminoethyl)salicylaldiminato-uranyl nitrate methylene dichloride solvate, 42B, 731

$C_{14}H_{22}N_4NiO_4 \cdot 0.5\ H_2O$, Salicylidene-γ-iminopropyl-methyl-γ-aminopropylaminonickel nitrate hemihydrate, 40B, 975

$C_{15}H_{14}ClCuN_3O_4 \cdot H_2O$, Chloro(N-pyridoxylidene-N'-salicyloylhydrazinato)copper(II) monohydrate, 45B, 1066

$C_{15}H_{20}BrClCuN_2O_2$, 2,2,5,5-Tetramethylpyrrolidin-1-oxyl-3-N-(2-hydroxy-5-bromosalicylaldiminato)chlorocopper(II), 43B, 1175

$C_{16}H_{12}FeN_3O_3$, N,N'-Ethylenebis(salicylideneiminato)nitrosyliron, 45B, 1067

$C_{16}H_{13}CuNO_4$, catena-μ-(N-Salicylidene-L-tyrosinato-O,O')copper(II), 44B, 835

$C_{16}H_{14}ClFeN_2O_2$, Chloro-N,N'-bis(salicylidene)ethylenediamineiron(III), 32B, 394

$C_{16}H_{14}CoN_2O_2$, Bis(salicylaldehyde)ethylenediiminecobalt(II), 37B, 494

$C_{16}H_{14}CoN_2S_2$, N,N'-Ethylenebis(thiosalicylideneiminato)cobalt(II), 42B, 742

$C_{16}H_{14}CoN_3O_3$, N,N'-Ethylene-bis(salicylideneiminato)nitrosylcobalt(II), 44B, 835

$C_{16}H_{14}CuN_2O_2$, N,N'-Disalicylideneethylenediaminecopper, 24, 611

$C_{16}H_{14}N_2NiO_2$, N,N'-Bis(salicylal)ethylenediiminenickel, 35B, 620
$C_{16}H_{16}Cl_2Cu_2N_2O_2$, Dichlorobis(N-methylsalicylaldimino)dicopper(II), 40B, 827
$C_{16}H_{16}Cl_2FeN_7O_2S_2$, Ammonium bis(5-chlorosalicylaldehydethiosemicarbazonato)iron(III), 44B, 836
$C_{16}H_{16}Cl_2MoN_2O_2$, trans-Dichlorobis-(N-methylsalicylaldiminato)molybdenum(IV), 40B, 826
$C_{16}H_{16}CuN_2O_2$, Bis(N-methylsalicylaldiminato)copper(II) (α-form), 26, 667
$C_{16}H_{16}CuN_2O_2$, Bis(N-methylsalicylaldiminato)copper(II) (γ-form), 33B, 397
$C_{16}H_{16}MoN_2O_4$, Bis-(N-methylsalicylideneiminato)dioxomolybdenum(VI), 41B, 972
$C_{16}H_{16}N_2NiO_2$, Bis(N-methylsalicylaldiminato)nickel(II) (monoclinic), 23, 656
$C_{16}H_{16}N_2NiO_2$, Bis(N-methylsalicylaldiminato)nickel(II) (orthorhombic), 32B, 395
$C_{16}H_{16}N_2O_2Zn$, Zinc N-methylsalicylaldiminate, 31B, 389
$C_{16}H_{16}N_2O_3Zn$, N,N'-Disalicylideneethylenediaminezinc(II) monohydrate, 24, 618; 31B, 392
$C_{16}H_{18}ClCrN_2O_4$, N,N'-Ethylenebis(salicylideneiminato)diaquochromium(III) chloride, 35B, 621
$C_{16}H_{18}CoO_6$, Dimethanolbis(salicylaldehydato)cobalt(II), 40B, 827
$C_{17}H_{15}Cl_3CoN_2O_2$, Bis(salicylaldehyde)ethylenedi-imine cobalt(II) monochloroformate, 34B, 514
$C_{17}H_{16}MoN_2O_4$, Dioxo(bis(salicylaldehyde)trimethylenediiminato)molybdenum(VI), 40B, 980
$C_{17}H_{17}ClFeN_3O_4$, N,N'-Bis(salicylideneiminato)iron(III) chloride nitromethane solvate, 32B, 396
$C_{17}H_{18}CuN_2O_3$, N,N'-Disalicylidenepropane-1,2-diaminecopper monohydrate, 24, 612
$C_{17}H_{18}CuN_4O_2S$, N,N'-Ethylenebis(salicylaldiminato)copper(II) - thiourea, 42B, 742
$C_{17}H_{18}N_2O_5U$, (N,N'-Ethylenebis(salicylideneiminato))(methanol)dioxouranium, 39B, 676
$C_{17}H_{20}N_2O_2Pt$, 2-Hydroxy-N-salicylideneaniline(diethylamine)platinum-(II), 46B, 980
$C_{18}H_{16}CuMgN_2O_8 \cdot H_2O$, (N,N'-Ethylene-bis(3-formyl-salicylideneiminato))copper(II)-diaquo-magnesium monohydrate, 45B, 1068
$C_{18}H_{17}MnN_2O_4$, catena-μ-Acetato-(N,N'-ethylenebis(salicylaldiminato))manganese(III), 39B, 676
$C_{18}H_{18}Cl_2Cu_2N_2O_2$, Dichloro(N,N'-ethylenebis(2-hydroxyacetophenimino)copper(II))copper(II), 40B, 828
$C_{18}H_{18}CoCuN_2O_9$, (N,N'-Bis(3-carboxysalicylidene)ethylenediamino(4-))copper(II)diaquocobalt(II) hydrate, 44B, 836
$C_{18}H_{18}CoN_2O_2$, (+)-N,N'-Butylenebis(salicylideneiminato)cobalt(II), 39B, 677
$C_{18}H_{18}CoN_2O_2$, meso-N,N'-Butylenebis(salicylideneiminato)cobalt(II), 39B, 677
$C_{18}H_{18}N_2O_5U$, N,N'-Bissalicylidene-1,5-diamino-3-oxapentanedioxouranium(VI) (α form), 43B, 1175
$C_{18}H_{18}N_2O_5U$, N,N'-Bissalicylidene-1,5-diamino-3-oxapentanedioxouranium(VI) (β form), 43B, 1175
$C_{18}H_{18}N_2O_5U \cdot CHCl_3$, N,N'-Bis-salicylidene-1,5-diamino-3-oxapentanedioxouranium(VI) chloroform solvate, 45B, 1068
$C_{18}H_{19}N_3O_4U$, [N,N'-(3-Aza-1,5-pentanediyl)bis(salicylideneiminato)]-dioxouranium(VI), 39B, 678; 45B, 1043

$C_{18}H_{20}Br_2Cu_2N_2O_2$, Dibromobis(N-ethylsalicylaldimino)dicopper(II), 40B, 828
$C_{18}H_{20}Cl_2Cu_2N_2O_2$, Dichlorobis(N-ethylsalicylaldimino)dicopper(II), 40B, 827
$C_{18}H_{20}CoN_2NaO_9$, Sodium bis(N-salicylidene(glycinato))cobaltate(III) trihydrate, 41B, 1079
$C_{18}H_{20}CoN_2O_5$, Aquo-N,N'-ethylenebis-(3-methoxysalicylideneiminato)-cobalt(II), 40B, 829
$C_{18}H_{20}CuN_2O_2$, Bis(N-ethylsalicylaldiminato)copper(II) (monoclinic form), 32B, 397
$C_{18}H_{20}CuN_2O_2$, Bis(N-ethylsalicylaldiminato)copper(II) (orthorhombic form), 40B, 1112
$C_{18}H_{20}CuN_2O_4$, Bis(N-2-hydroxyethylsalicylaldiminato)copper(II), 31B, 393; 35B, 622
$C_{18}H_{20}Cu_2N_4O_8$, Bis(nitrato(N-ethylsalicylaldiminato)copper(II)), 42B, 743
$C_{18}H_{20}N_2O_2Pd$, Bis-N-ethylsalicylaldiminepalladium(II), 29, 582
$C_{18}H_{22}ClFeN_4S_2$, Bis(2-aminoethylthiosalicylideneiminato)iron(III) chloride, 41B, 973
$C_{18}H_{22}Co_{1.5}N_2O_{10}$, Hexaaquocobalt(II) bis-(N-salicylideneglycinato)-cobaltate(III) hydrate, 42B, 912
$C_{18}H_{22}CrIN_4O_2$, Bis-(2-aminoethylsalicylideneiminato)chromium(III) iodide, 37B, 495
$C_{18}H_{24}ClFeN_4O_3$, Bis(N-(2-aminoethyl)salicylaldiminato)iron(III) chloride monohydrate, 44B, 837
$C_{18}H_{24}CoIN_4O_3$, Bis(N-(2-aminoethyl)salicylaldiminato)cobalt(III) iodide monohydrate, 43B, 1176
$C_{19}H_{16}Cu_2N_2O_7$, [N,N'-Bis(2-oxido-3-carboxylatobenzylidene)-1,2-diaminoethane]-methanol-dicopper(II), 45B, 1069
$C_{19}H_{19}CoF_2N_2O_4 \cdot 0.39\ C_3H_6O \cdot 0.32\ H_2O$, Aquo-acetonyl-N,N'-ethylenebis(3-fluorosalicylideniminato)cobalt(III) - 0.32 water - 0.39 acetone, 44B, 837
$C_{20}H_{12}N_2O_2Pd$, N,N'-o-Phenylenebis(salicylaldiminato)palladium(II), 42B, 744
$C_{20}H_{14}CoN_2O_2$, N,N'-(o-Phenylene)bis(salicylideneamine)cobalt(II) (monoclinic), 42B, 745
$C_{20}H_{14}CoN_2O_2$, N,N'-(o-Phenylene)bis(salicylideneamine)cobalt(II) (orthorhombic), 42B, 745
$C_{20}H_{18}CuN_4O_2S$, N,N'-o-Phenylenebis(salicylaldiminato)copper(II) - thiourea, 42B, 742
$C_{20}H_{20}Cl_2Cu_2N_2O_4$, Bis(N-(3'-hydroxy-n-propyl)-2-hydroxy-5-chlorobenzylideniminato-μ-O')copper(II)), 42B, 745
$C_{20}H_{20}CoN_2O_2$, meso-N,N'-Cyclohexylenebis(salicylideneiminato)cobalt-(II), 40B, 830
$C_{20}H_{20}Cu_2N_4O_8$, Bis(N-(3'-hydroxy-n-propyl)-2-hydroxy-3-nitrobenzylideniminato-μ-O')copper(II)), 42B, 745
$C_{20}H_{20}Cu_2N_4O_8$, Bis(2-hydroxy-N-3-hydroxypropyl-5-nitrobenzylideneaminato-(μ-O)-copper(II)), 42B, 746
$C_{20}H_{20}N_2O_2Pd$, Bis(N-allylsalicylaldiminato)palladium(II), 44B, 838
$C_{20}H_{22}Cl_2N_2O_3Ti$, Dichloro-N,N'-ethylenebis(salicylideneiminato)titanium(IV) -tetrahydrofuran, 38B, 837
$C_{20}H_{22}CoN_2NaO_3$, Sodium tetrahydrofuran [N,N'-ethylenebis(salicylideniminato)]cobaltate, 45B, 1007
$C_{20}H_{22}Cu_2N_2O_4$, (3-(Salicylideneamino-1-propanolato(2-))copper(II) dimer, 40B, 985
$C_{20}H_{22}O_4Pt_2$, Trimethyl(salicylaldehydato)platinum(IV), 32B, 399
$C_{20}H_{23}ClFeN_3O_2$, Bis-(γ-salicylideneiminopropyl)amineiron(III) chlor-

ide, 45B, 1069
$C_{20}H_{23}N_3NiO_2$, Bis(salicylidene-γ-iminopropyl)aminenickel(II), 37B, 496
$C_{20}H_{24}ClFeN_2O_2$, Chlorobis-(N-n-propylsalicylaldiminato)iron(III), 38B, 838
$C_{20}H_{24}Cl_2Cu_2N_2O_2$, Bis(chloro(N-isopropyl-2-hydroxybenzylidene)-aminato-μ-O-copper(II)), 42B, 747
$C_{20}H_{24}CuN_2O_2$, Bis(N-iso-propylsalicylaldiminato)copper(II), 31B, 393; 42B, 747
$C_{20}H_{24}CuN_2O_2$, Bis(N-n-propylsalicylaldiminato)copper(II), 34B, 517
$C_{20}H_{24}N_2NiO_2$, Bis-(N-isopropylsalicylaldiminato)nickel(II), 29, 568
$C_{20}H_{30}N_4NiO_5$, N,N'-Bis(2-salicylideneaminoethyl)(ethylenediamine)-nickel(II) trihydrate, 43B, 1152
$C_{20}H_{36}N_4NiO_8$, Bis-salicylidene-triethylenetetramine-nickel hexahydrate, 38B, 839
$C_{21}H_{15}Cl_3CuN_2O_2$, Di-μ-N,N'-m-phenylene-tetrakis(salicylideneiminato)dicopper(II) chloroform solvate, 35B, 624
$C_{21}H_{17}AmO_{10}$, Aquotris(salicylato)americium(III), 43B, 1178
$C_{21}H_{17}O_{10}Sm$, Aquotris(salicylato)samarium(III), 43B, 1178
$C_{21}H_{18}CuN_6O_4S$, (o-Phenanthroline)(salicylaldehyde S-methylthiosemicarbazone)copper(II) nitrate, 45B, 1269
$C_{21}H_{19}CoN_3O_2$, Monopyridine-N,N'-ethylenebis(salicylideneiminato)cobalt(II), 35B, 625
$C_{21}H_{25}N_3NiO_2$, Bis(salicylidene-γ-iminopropyl)methylaminenickel(II), 37B, 497
$C_{22}H_{20}CuN_2NiO_4$, Aquo(3,8-dimethyl-1,10-di(o-phenato)-4,7-diazadeca-2,8-diene-1,10-dionato)copper(II)-nickel(II), 45B, 1270
$C_{22}H_{22}N_2OPd$, (N,N-Dimethylbenzylamine-2C,N)(N-phenysalicylaldiminato)palladium(II), 40B, 830
$C_{22}H_{26}CoIN_2O_2S_2$, 1,10-Bis(salicylideneamino)-4,7-dithiadecanecobalt-(III) iodide, 42B, 748
$C_{22}H_{26}CuN_2O_2$, 2,5-Bis(salicylaldiminato)2,5-dimethylhexanecopper-(II), 45B, 1070
$C_{22}H_{26}CuN_6O_4$, (N-3,4-Benzosalicylidene-N,N'-dimethylethylenediamine)(theophyllinato)copper(II) monohydrate, 42B, 748
$C_{22}H_{26}Cu_2N_2O_4$, Bis(2-hydroxy-N-3-hydroxypropyl-α-methylbenzylideneaminato-(μ-O)-copper(II)), 42B, 746
$C_{22}H_{26}N_2O_2Pd$, Bis(N-isopropyl-3-methylsalicylaldiminato)palladium, 32B, 400
$C_{22}H_{28}Cl_2N_2O_2V \cdot C_6H_6$, Bis[N-n-butyl(salicylideniminato)]dichlorovanadium(IV) benzene solvate, 45B, 1070
$C_{22}H_{28}CoN_2O_4$, Bis(N-(3-methoxysalicylidene)isopropylaminato)cobalt-(II), 42B, 749
$C_{22}H_{28}CoN_2O_7 \cdot C_4H_{10}O_2$, Dioxygenaquo[N,N'-(1,1,2,2-tetramethylethylene)bis(3-methoxysalicylideniminato)]-cobalt(II) dimethoxyethane solvate, 45B, 1071
$C_{22}H_{28}CuN_2O_2$, Bis(N-n-butylsalicylaldiminato)copper(II), 34B, 518
$C_{22}H_{28}CuN_2O_2$, Bis(N-t-butylsalicylaldiminato)copper(II), 31B, 395
$C_{22}H_{28}CuN_2O_4$, Bis(N-isopropyl-3-methoxysalicylideneaminato)copper-(II) (low temperature form), 46B, 981
$C_{22}H_{28}N_2NiO_2$, Bis(N-isopropyl-3-methylsalicylaldiminato)nickel(II), 31B, 397
$C_{22}H_{28}N_2NiO_4$, Bis(N-(3-methoxysalicylidene)isopropylaminato)nickel-(II), 42B, 749
$C_{22}H_{28}N_2O_2Pd$, Bis(N-t-butylsalicylaldiminato)palladium(II), 33B, 399
$C_{22}H_{28}N_2O_2Pd$, Bis-N-n-butylsalicylaldiminepalladium(II), 29, 582
$C_{22}H_{30}CuN_4O_2$, Bis(N-β-dimethylaminoethylsalicylaldiminato)copper-

(II), 38B, 841
$C_{22}H_{30}N_4O_4U$, Bis-(N-ethylenedimethylaminesalicylaldiminato)dioxouranium(VI), 40B, 985
$C_{23}H_{21}FeN_3O_2$, N,N'-(2-(2'-Pyridyl)ethyl)ethylenebis(salicylideneiminato)iron(II), 44B, 839
$C_{23}H_{22}CoN_3O_2$, (N,N'-Ethylenebis(salicylideneiminato))pyridine(vinyl)cobalt(III), 38B, 842
$C_{23}H_{23}ClN_3O_2Rh$, Chloropyridine-N,N'-tetramethylene-bis(salicylaldiminato)rhodium(III), 42B, 750
$C_{23}H_{23}CoN_3O_2$, (N,N'-Butylenebis(salicylideneiminato))pyridinecobalt-(II), 40B, 831
$C_{24}H_{20}CuN_2O_2$, Bis(N-methyl-2-hydroxy-1-naphthaldiminato)copper(II) (β form), 41B, 974
$C_{24}H_{20}CuN_2O_2$, Bis(N-methyl-2-hydroxy-1-naphthaldiminato)copper(II) (γ form), 43B, 1178
$C_{24}H_{20}CuN_2O_2$, Bis-(2-hydroxy-N-methyl-1-naphthylmethyleneiminato)-copper(II) (δ form), 39B, 804
$C_{24}H_{28}N_2O_2Pd_2$, μ-N,N'-Ethylenebis(salicylaldiminato)bis-(2-methylallylpalladium(II)), 40B, 832
$C_{24}H_{30}Br_3Co_2N_4O_4$, (Bis(propane-1,3-diyl)-di(5-methylphalaldimin-2-olato))aquobromocobalt(II)-aquobromocobalt(III) bromide (monoclinic), 41B, 974
$C_{24}H_{30}Br_3Co_2N_4O_4$, (Bis(propane-1,3-diyl)-di(5-methylphalaldimin-2-olato))aquobromocobalt(II)-aquobromocobalt(III) bromide (orthorhombic), 41B, 974
$C_{24}H_{30}Cl_2N_2O_4U$, Dichloro(N,N'-ethylenebissalicylideneiminato)uranium(IV) tetrahydrofuran solvate, 42B, 750
$C_{24}H_{30}CoN_2O_2$, [N,N'-Ethylenebis(3-tert-butylsalicylideniminato)]cobalt(II), 46B, 981
$C_{24}H_{30}Cu_2N_2O_8$, μ-(N-Salicylidene-L-valinato-O)-N-salicylidene-L-valinatodiaquadicopper(II), 45B, 1066
$C_{24}H_{31}BrCoN_5O_3$, (N,N'-Bis(salicylidene)dipropylenetriamine)(1-methylimidazole)cobalt(III) bromide monohydrate, 40B, 832
$C_{24}H_{32}N_2NiO_2$, Bis(N-isopropyl-3-ethylsalicylaldiminato)nickel(II), 32B, 403
$C_{24}H_{32}N_2O_2Pd$, Bis(N-isopropyl-3-ethylsalicylaldiminato)palladium-(II), 32B, 404
$C_{24}H_{34}N_4NiO_2$, Bis(N-γ-dimethylaminopropylsalicylaldiminato)nickel-(II), 32B, 406
$C_{25}H_{24}CoN_4O_4$, Dioxygen(N,N'-(2-(2'-pyridyl)ethyl)ethylenebis(salicylideneiminato))cobalt acetonitrile solvate, 44B, 839
$C_{25}H_{27}ClN_3O_3Ti$, N,N'-Ethylenebis(salicylideneiminato)chloropyridine-titanium(III) tetrahydrofuran solvate, 44B, 840
$C_{25}H_{27}CoN_3O_3$, N,N'-(2-(2'-Pyridyl)ethyl)ethylenebis(salicylideneiminato)cobalt(II) ethanol solvate, 44B, 839
$C_{25}H_{30}Br_2Co_2N_4O_3$, (Bis(propane-1,3-diyl)-di(5-methylphalaldimin-2-olato))bis(bromocobalt(II)) methanol solvate, 41B, 976
$C_{26}H_{18}ClFeN_2O_2S_2$, Chloro(bis(salicylideniminephenyl)disulfido)iron-(III), 40B, 1018
$C_{26}H_{18}Cl_2N_2O_3V$, Bis(N-(4-chlorophenyl)salicylideneiminato)oxovanadium(IV), 43B, 1179
$C_{26}H_{18}CuN_2O_2$, N,N'-(2,2'-Biphenyl)-bis(salicylaldiminato)copper(II), 31B, 398
$C_{26}H_{18}CuN_4O_6 \cdot$ 1-2 $C_4H_8O_2$, Bis(N-p-nitrophenylsalicylideneiminato)-copper(II) 1,4-dioxane, 44B, 841
$C_{26}H_{20}CuN_2O_2$, Bis-(N-phenylsalicylaldiminato)copper(II), 29, 597
$C_{26}H_{20}N_6NiO_4$, Bis(N-picolinylidene-N'-salicyloylhydrazinato)nickel-

(II), 41B, 977
$C_{26}H_{22}N_3NiOPS$, Salicylaldehydethiosemicarbazone-triphenylphosphine-nickel, 40B, 990
$C_{26}H_{22}N_3NiOPSe$, Salicylaldehydeselenosemicarbazonato-triphenylphosphinenickel, 40B, 990
$C_{26}H_{23}ClCuN_6O_8 \cdot C_2H_6O$, (N-Picolinylidene-N'-salicyloylhydrazinato)-(N-picolinylidene-N'-salicyloylhydrazine)-copper(II) perchlorate ethanol solvate, 45B, 1071
$C_{26}H_{23}CoN_2O_4 \cdot 1.5\ H_2O$, Benzoylacetonato-(N,N'-ethylenebis-(salicylideneiminato))cobalt(III) hydrate, 38B, 843
$C_{26}H_{26}CoIN_2O_2S_2 \cdot 0.5\ CH_3OH$, 1,10-Bis(salicylideneamino)-4,7-dithiadecanecobalt(III) iodide methanol solvate, 42B, 748
$C_{26}H_{32}CoN_2NaO_9 \cdot 1.25\ H_2O$, Sodium bis(N-3-methylsalicylidene-threoninato)cobaltate(III) ethanolate hydrate, 43B, 1180
$C_{26}H_{32}CuN_2O_2$, Bis(N-cyclohexylsalicylaldiminato)copper(II) (brown form), 43B, 1180
$C_{26}H_{36}Cl_2N_4NiO_2$, N-β-Diethylaminoethyl-5-chlorosalicylaldiminatonickel, 31B, 400
$C_{27}H_{24}Cl_3CrN_4O_3 \cdot 3\ H_2O$, (Tris(2-(5'-chlorosalicylaldiminato)ethyl)-amine)chromium(III) trihydrate, 46B, 982
$C_{27}H_{24}Cl_3MnN_4O_3 \cdot 3\ H_2O$, (Tris(2-(5'-chlorosalicylaldiminato)ethyl)-amine)manganese(III) trihydrate, 46B, 982
$C_{27}H_{24}CoN_3O_3$, (Cyclohexane-cis,cis-1,3,5-tris(salicylaldiminato))cobalt(III), 41B, 978
$C_{27}H_{30}Cl_2Fe_2N_2O_4$, Bis(chloro-N-(3-propanolato)salicylaldiminatoiron-(III))-toluene, 40B, 833
$C_{27}H_{30}Cl_3FeN_4O_6$, 2,2,2-Tri(5-chlorosalicylaldiminato)ethylamineiron-(III) trihydrate, 41B, 977; 42B, 751
$C_{28}H_{20}O_8Th$, Tetrakis(salicylaldehydato)thorium(IV), 43B, 1181
$C_{28}H_{22}CoN_2O_2$, Δ-2,2'-Bis(salicylidenaminato)-(+)-(R)-6,6'-dimethyl-biphenylcobalt(II), 34B, 516
$C_{28}H_{24}CoN_2O_4$, Bis(salicylal-N-m-anisidinate)cobalt(II), 39B, 679
$C_{28}H_{24}CoN_2O_4$, Bis(salicylal-N-o-anisidinate)cobalt(II), 39B, 678
$C_{28}H_{24}CuN_2O_2$, Bis-(N-p-tolylsalicylaldiminato)copper(II), 42B, 752
$C_{28}H_{24}N_2NiO_2S_2$, Bis(salicylal-N-o-thioanisidinate)nickel(II), 43B, 1181
$C_{28}H_{24}N_2O_4Zn$, Zinc salicylal-o-anisidinate, 35B, 627
$C_{28}H_{26}Cu_2N_2O_4$, Bis(N-(3'-hydroxy-n-propyl)-2-hydroxy-5,6-benzobenzylideniminato-μ-O')copper(II)), 42B, 745
$C_{28}H_{28}Cd_2O_{16}$, Tetrakis(salicylate)tetraaquodicadmium(II), 38B, 842
$C_{28}H_{28}CuN_2O_2$, Bis(N-isopropyl-2-oxyl-1-naphthylideneaminato)copper-(II), 45B, 1072
$C_{28}H_{30}N_6O_{16}U_2$, Di-μ-salicylicacidatobis(nitratodioxouranium(VI)) bis(4-N,N'-dimethylaminopyridine) solvate, 42B, 752
$C_{28}H_{36}CuN_2O_4$, Bis(N-cyclohexyl-3-methoxysalicylideneiminato)copper-(II), 44B, 842
$C_{28}H_{38}CoN_2O_2$, N,N'-(1,1,2,2-Tetramethyl)ethylenebis(3-t-butylsalicylideniminato)cobalt(II), 42B, 753
$C_{28}H_{40}CuN_2O_2$, Bis-(N-n-hexyl-7-methylsalicylaldiminato)copper(II), 39B, 679
$C_{29}H_{35}CoN_3O_4$, (Dioxygen)-[N,N'-ethylenebis(3-tert-butylsalicylideniminato)](pyridine)cobalt, 46B, 981
$C_{30}H_{24}CuN_2O_4$, Bis(N-phenyl-5-methyl-3-formylsalicylideneiminato)copper(II), 42B, 753
$C_{30}H_{28}ClFeN_2O_2$, Chloro-bis(N-(2-phenylethyl)salicylaldiminato)iron-(III), 40B, 834
$C_{30}H_{28}CuN_2O_2$, Bis(N-α-phenylethylsalicylaldiminato)copper, 38B, 844

$C_{30}H_{30}CuN_4O_2$, Bis(N-p-dimethylaminophenylsalicylidene-iminato)copper(II), 43B, 1182
$C_{30}H_{30}N_6Ni_2O_{14}$, Bis-[μ-2-(p-nitrophenyliminomethyl)phenolato-N,O]-bis(ethanol)bis(nitrato-O,O')-dinickel(II), 45B, 1073
$C_{30}H_{35}Cl_4Fe_2N_4O_5$, μ-Hydroxo-μ-(2-(o-phenolato)-1,2-bis((2-salicylaldiminato)ethylimidazolidine))-bis(chloroiron(III)) (methanol - 1,2-dichloroethane) solvate, 43B, 1158
$C_{30}H_{38}CoF_2N_4O_6$, N,N'-(1,1,2,2-Tetramethyl)ethylenebis(3-fluorosalicylideniminato)(1-methylimidazole)superoxocobalt(III) diacetone solvate, 42B, 754
$C_{31}H_{24}NO_2PPt$, 2-Hydroxy-N-salicylideneaniline(triphenylphosphine)-platinum(II), 46B, 980
$C_{31}H_{25}CuN_3O_2$, Bis-(N-phenylsalicylaldiminato)copper(II) pyridine solvate, 33B, 401
$C_{31}H_{36}Cl_2Fe_2N_4O_5$, μ-Hydroxo-μ-(2-(o-phenolato)-1,2-bis(2-salicylaldiminato)ethylimidazolidine)-bis(chloroiron(III)) tetrahydrofuran solvate, 43B, 1158
$C_{32}H_{28}Cl_2Co_2N_4O_4 \cdot 4\ CHCl_3$, Bis[μ-bis(salicylaldehyde)ethylenediimine]dicobalt(III) dichloride chloroform solvate, 45B, 1074
$C_{32}H_{28}Cl_4Cu_4N_4O_4$, Di-μ-chloro-bis(chloro(N,N'-ethylenebis(salicylideneiminato)copper(II))copper(II)), 40B, 835
$C_{32}H_{28}Co_2N_4O_4$, N,N'-Ethylenebis(salicylaldehydeiminato)cobalt(II), 34B, 520
$C_{32}H_{28}Co_2N_4S_4$, N,N'-Ethylenebis(thiosalicylaldiminato)cobalt(II), 41B, 979
$C_{32}H_{28}Fe_2N_4O_5$, μ-Oxo-bis-(N,N'-ethylenebis(salicylaldiminato)iron-(III)), 39B, 680
$C_{32}H_{34}Cl_2Cu_3N_4O_{15}$, Bis(N,N'-ethylene-bis(salicylaldiminato)copper-(II))-diaquo-copper(II) perchlorate monohydrate, 40B, 835
$C_{32}H_{36}Mn_2N_6O_8$, Di-μ-methoxo-bis(salicylaldehyde anthraniloylhydrazonato(2-))dimanganese(III) - bismethanol, 39B, 680
$C_{32}H_{44}N_4O_4Zn$, Bis(salicylidene-1-oxyl-4-amino-2,2,6,6-tetramethylpiperidonato)zinc(II), 44B, 842
$C_{32}H_{46}CuN_4O_2$, Bis(N-(2,2,6,6-tetramethylpiperidin-4-yl)-salicylaldiminato)copper(II), 41B, 980
$C_{32}H_{52}N_4O_{10}Pd_2S_4$, (N(*a*)-Salicylidene-D-ornithinato)(N(*a*)-salicylidene-L-ornithinato)dipalladium(II)-dimethylsulfoxide, 42B, 754
$C_{33}H_{30}Cl_2Fe_2N_4O_5$, μ-Oxo-bis(N,N'-ethylenebis(salicylideneiminato)-iron(III)) - dichloromethane, 37B, 497
$C_{33}H_{31}Cl_2CuN_3O_2$, (3,3'-Dichloro-2,2'-(1,12-diphenyl-2,6,11-triazadodeca-1,11-diene-1,12-diyl)bis(phenolato))copper(II), 45B, 1074
$C_{33}H_{31}Cl_2N_3NiO_2$, (3,3'-Dichloro-2,2'-(1,12-diphenyl-2,6,11-triazadodeca-1,11-diene-1,12-diyl)bis(phenolato))nickel(II), 45B, 1074
$C_{34}H_{24}CuN_2O_2$, Bis(N-phenyl-2-hydroxy-1-naphthylideneiminato)copper-(II), 44B, 843
$C_{34}H_{34}Br_2Cl_2Cu_2N_2O_2$, Bis(bromo-(N-n-butyl-5-chloro-*a*-phenyl-2-hydroxybenzylidene)aminato-μ-O-copper(II)), 40B, 837
$C_{34}H_{34}Cu_2N_2O_6$, Di-μ-propionato-O,O'-bis(N-p-tolylsalicylideneaminatocopper(II)), 43B, 1184
$C_{34}H_{40}CoN_4O_5$, Dioxygen(N,N'-(1,1,2,2-tetramethylethylene)bis(salicylideniminato))(1-benzylimidazole)cobalt(II) - tetrahydrofuran, 42B, 868
$C_{35}H_{25}CoN_2O_4$, Dibenzoylmethanato-O,O'-(N,N'-o-phenylenebis(salicylideneiminato))cobalt(III), 42B, 924
$C_{35}H_{37}CuN_3O_2 \cdot 2\ H_2O$, N,N'-Bis[(2-hydroxy-5-methylphenyl)(4-methylphenyl)methylene]-3-azahexane-1,6-diaminecopper(II) dihydrate, 45B, 1075

$C_{35}H_{39}N_3NiO_3$, N,N'-Bis[(2-hydroxy-5-methylphenyl)(4-methylphenyl)-methylene]-3-azahexane-1,6-diamineaquanickel(II), 45B, 1075
$C_{36}H_{30}N_4O_4Pd_2$, μ-N,N'-o-Phenylenebis(salicylaldiminato)-bis((2-C,N-acetophenone oxime)palladium(II)), 42B, 755
$C_{36}H_{38}Cl_2Cu_3N_4O_{13}$, Bis(N,N'-ethylene-bis(o-hydroxyacetopheniminato)-copper(II)-aquo-copper(II) perchlorate, 40B, 835
$C_{36}H_{38}Co_2N_4O_4$, N,N'-Ethylenebis(salicylideneiminato)ethylcobalt-(III), 37B, 498
$C_{37}H_{54}CuN_4O_7$, Bis(2-hydroxy-4-methoxybenzylidene-1-oxyl-4-amino-2,2,6,6-tetramethylpiperidinato)copper(II) acetone solvate, 44B, 844
$C_{38}H_{30}N_2NiO_4$, 2,2'-Tetramethylenedioxydi(8-N-salicylideneiminonaphthalene)nickel(II), 45B, 1076
$C_{38}H_{34}CuN_4O_2$, Bis(N-p-dimethylaminophenyl-2-hydroxy-1-naphthaldiminato)copper(II), 43B, 1185
$C_{38}H_{42}Co_2N_6O_8$, N,N'-Ethylenebis(salicylideneiminato)cobalt(II) - oxygen - dimethylformamide adduct, 35B, 781
$C_{38}H_{48}CoN_4O_4 \cdot 1.5\ C_3H_6O$, Dioxygen-N,N'-(1,1,2,2-tetramethyl)ethylenebis(3-t-butylsalicylideniminato)(1-benzylimidazole)cobalt(II) acetone solvate, 42B, 753
$C_{40}H_{28}N_4O_4Th$, Bis(N,N'-o-phenylenebis(salicylaldiminato))-thorium(IV), 44B, 844
$C_{40}H_{28}N_4O_4Zr \cdot 2.5\ C_6H_6$, Bis(N,N'-disalicylidene-1,2-phenylenediamino)zirconium(IV) benzene solvate, 45B, 1077
$C_{40}H_{56}Ni_4O_{16}$, Tetra-μ_3-methoxy-tetrakis(salicylaldehydato(ethanol)-nickel(II)), 34B, 517
$C_{42}H_{36}MnN_3O_3$, mer-Tris-(N-benzylsalicylaldiminato)manganese(III), 40B, 838
$C_{42}H_{38}Fe_2N_6O_5$, μ-Oxo-bis(N,N'-ethylenebis(salicylideneiminato)iron-(III))-bispyridine, 40B, 1113
$C_{42}H_{50}Co_2N_6O_6 \cdot 0.67\ C_3H_6O \cdot 0.33\ C_5H_{11}N$, μ-Dioxygen-bis(N,N'-ethylenebis(salicylideniminato)piperidinecobalt) acetonate piperidinate, 42B, 755
$C_{44}H_{52}Co_2Li_2N_4O_7$, Lithium tetrahydrofuran [N,N'-ethylenebis(salicylideniminato)]cobaltate, 45B, 1007
$C_{47}H_{54}Co_2N_6O_6$, μ-Peroxo-bis-(3,3'-diimino-di-n-propylamine-bis-salicylaldehydecobalt(III)) - toluene, 37B, 499
$C_{47}H_{54}Fe_2N_4O_8$, Di-iron dihydrogen tetra(N-(3-propanolato)-salicylaldiminate) - toluene, 40B, 838
$C_{48}H_{44}Cu_4N_{12}O_{20}$, Tetrakis((aquo)-(N-2-pyridylsalicylaldiminato)copper(II)) tetranitrate, 38B, 845
$C_{52}H_{36}Cl_4Fe_2N_4O_5$, μ-Oxo-bis(bis-(N-p-chlorophenylsalicylaldiminato)-iron(III)), 39B, 681
$C_{52}H_{36}CuN_{10}O_2$, Bis(7,7',8,8'-tetracyanoquinodimethane) bis(N-isopropyl-2-oxy-1-naphthylideneaminato)-copper(II), 45B, 1072
$C_{52}H_{64}Cu_2N_4O_4$, Bis(N-cyclohexylsalicylideneaminato)copper(II) dimer, 45B, 1077
$C_{53}H_{46}Cl_6N_2O_4P_2Re_2$, Dioxotetrachloro(N,N'-ethylenebis(salicylideneiminato)bis(triphenylphosphine))dirhenium(V) dichloromethane solvate, 44B, 1160
$C_{56}H_{46}N_8Ni_2O_{17} \cdot 2\ H_2O$, μ-Aquo-bis(bis[N-(4-methoxyphenyl)-5-nitrosalicylideneiminato]nickel(II)) dihydrate, 45B, 1078
$C_{56}H_{48}BN_4NaO_6V_2$, Sodium N,N'-ethylenebis(salicylideniminato)-oxovanadium(IV) tetraphenylborate, 46B, 983
$C_{64}H_{60}BN_8NaNi_2O_4$, Bis(N,N'-ethylenebis(salicylideneiminato)nickel-(II)) sodium tetraphenylborate diacetonitrile acetonitrile solvate, 42B, 756

$C_{80}H_{82}BN_7Ni_3O_8$, Tris(N,N'-ethylenebis(salicylideneiminato)nickel-(II)) ammonium tetraphenylborate tetrahydrofuran solvate, 42B, 756

79. METAL COMPLEXES (THIOUREA)

$CH_4Cl_3Hg_{1.5}N_2S$, Dichloromercury-thiourea, 39B, 681
$CH_5Cl_2HgN_3S$, Thiosemicarbazidedichloromercury(II), 43B, 1381
$CH_6Cl_3N_2O_2ReS$, trans-(Oxoaquo)trichlorothiourearhenium(V), 43B, 1187
$CH_7Cl_3CuN_4S$, Dichloro(thiocarbonohydrazidium-N,S)copper(II) chloride, 41B, 981
$CH_8CdN_2O_6S_2$, Thioureacadmium sulfate dihydrate, 32A, 346; 32B, 407
$CH_9Cl_3CuN_4OS$, Trichloro(1H(+)-thiocarbonohydrazidium-N,S)copper(II) monohydrate, 44B, 845
$C_2H_8AgClN_4S_2$, Bis(thiourea)silver(I) chloride, 33B, 469
$C_2H_8CdCl_2N_4S_2$, Dichlorobis(thiourea)cadmium(II), 21, 529
$C_2H_8ClCuN_4S_2$, Bis(thiourea)copper(I) chloride, 35B, 784
$C_2H_8Cl_2CoN_4S_2$, Dichlorobis(thiourea)cobalt(II), 34B, 521; 38B, 846
$C_2H_8Cl_2HgN_4S_2$, Chlorobis(thiourea)mercury(II) chloride, 39B, 682
$C_2H_8Cl_2N_4S_2Zn$, Dichlorobis(thiourea)zinc(II), 22, 605
$C_2H_8HgI_2N_4S_2$, Bis(thiourea)mercury(II) iodide, 33B, 467
$C_2H_{10}Br_2HgN_6S_2$, Dibromobis(thiosemicarbazide)mercury(II), 44B, 846
$C_2H_{10}CdN_6O_4S_3$, Bis(thiosemicarbazide)cadmium(II) sulfate, 41B, 981
$C_2H_{10}Cl_2HgN_6S_2$, Dichlorobis(thiosemicarbazide)mercury(II), 43B, 1188
$C_2H_{10}Cl_3N_4O_2ReS_2$, trans-Oxoaquo-cis-dichloro-cis-bis(thiourea)-rhenium(V) chloride, 42B, 757
$C_2H_{12}AgN_9O_3S_2$, Bis(thiocarbonohydrazide-N,S)silver(I) nitrate, 41B, 982
$C_2H_{12}Cl_2CuN_8O_8S_2$, Diperchloratobis(thiocarbonohydrazide-N,S)copper-(II), 41B, 983
$C_2H_{20}CuN_8O_8S_3$, Sulfatobis(thiocarbonohydrazide-N,S)copper(II) tetrahydrate, 41B, 983
$C_3H_{12}AgClN_6O_4S_3$, Tris(thiourea)silver(I) perchlorate, 39B, 683
$C_3H_{12}CdN_6O_4S_4$, Tris(thiourea)cadmium sulfate, 35B, 627; 40B, 1114
$C_3H_{12}ClCuN_6S_3$, Tris(thiourea)copper(I) chloride, 23, 587; 29, 593
$C_3H_{12}Cl_2HgN_6S_3$, Tristhioureamercury(II) chloride, 33B, 477
$C_3H_{12}N_6O_4S_4Zn$, Tris(thiourea)zinc(II) sulfate, 33B, 479
$C_3H_{17}Cl_2N_9NiOS_3$, Tris(thiosemicarbazide)nickel(II) dichloride monohydrate, 40B, 996
$C_4H_8FeN_6S_4$, Bis(isothiocyanato)bis(thiourea)iron(II), 46B, 983
$C_4H_8HgN_6S_2$, Bis(thiourea)dicyanomercury(II), 37B, 615
$C_4H_8HgN_6S_4$, Mercury(II) thiocyanate-thiourea complex, 28, 113; 31B, 401
$C_4H_8N_6NiS_4$, Bis(thiourea)nickel(II) thiocyanate, 21, 532; 31B, 402
$C_4H_{10}CdN_4O_4S_2$, Bis(thiourea)cadmium formate, 27, 835; 30B, 344
$C_4H_{12}MoN_4O_4S$, Bis[N-methylhydroxylamido-(O,N)][N-methyl-N-hydroxylthiourea-(O,N)]oxomolybdenum(VI), 46B, 984
$C_4H_{16}Cl_2CoN_8S_4$, trans-Dichlorotetrakis(thiourea)cobalt(II), 40B, 1114
$C_4H_{16}Cl_2FeN_8S_4$, Dichlorotetrakis(thiourea)iron(II), 45B, 1078
$C_4H_{16}Cl_2HgN_8S_4$, Tetrakis(thiourea)mercury(II) chloride (*α* form), 38B, 850; 39B, 683
$C_4H_{16}Cl_2HgN_8S_4$, Tetrakis(thiourea)mercury(II) chloride (*β* form), 39B, 684
$C_4H_{16}Cl_2N_8NiS_4$, Dichlorotetrakis(thiourea)nickel(II), 20, 526; 28, 623
$C_4H_{16}Cl_2N_8PdS_4$, Tetrakis(thiourea)palladium(II) chloride

(monoclinic), 24, 535
$C_4H_{16}Cl_2N_8PdS_4$, Tetrakis(thiourea)palladium(II) chloride (orthorhombic), 35B, 628
$C_4H_{16}Cl_2N_8PtS_4$, Tetrakis(thiourea)platinum(II) chloride, 39B, 684
$C_4H_{16}Cl_2N_8PtSe_4$, Tetrakis(selenourea)platinum(II) chloride, 43B, 1190
$C_4H_{16}Cl_4CuN_{10}O_2$, Bis(2-chlorobiguanide)copper(II) dichloride dihydrate, 41B, 984
$C_4H_{16}CoN_{10}O_6S_4 \cdot 0.5\ H_2O$, Tetrakis(thiourea)cobalt(II) nitrate hemihydrate (monoclinic form), 41B, 985
$C_4H_{16}N_{10}O_6S_4Zn$, Tetrakis(thiourea)zinc(II) nitrate, 44B, 846
$C_4H_{17}ClHgN_8S_5$, Tetrakis(thiourea)mercury(II) chloride hydrogen sulfide, 42B, 757
$C_4H_{18}CoN_{10}O_7S_4$, Tetrakis(thiourea)cobalt(II) nitrate monohydrate, 42B, 930
$C_4H_{18}N_8NiO_4S_6$, Tetrakis(thiourea)thiosulfatonickel monohydrate, 34A, 315
$C_4H_{20}CuN_8O_8S_2$, Diaquobis(thiocarbonohydrazide-N,S)copper(II) oxalate dihydrate, 41B, 986
$C_5H_{13}MoN_5O_4S_2 \cdot H_2O$, [N-Methylhydroxylamido-(O,N)][N-methyl-N-hydroxylthiourea-O,N)][N-methyl-N-hydroxylthiourea-(O,S)]oxomolybdenum-(VI) monohydrate, 46B, 984
$C_5H_{20}N_{12}O_{13}U$, Pentakis(urea)dioxouranium(VI) nitrate, 45B, 1079
$C_5H_{24}Cu_2N_{10}O_6S_6$, Pentakis(thiourea)dicopper(I) sulfate dihydrate, 40B, 1115
$C_5H_{26}Cu_2N_{10}O_7S_6$, Penta(thiourea)copper(I) sulfate trihydrate, 42B, 758
$C_6H_{12}N_4O_3S_4Zn$, Bis(ethylenethiourea)zinc(II) thiosulfate, 40B, 839
$C_6H_{14}AuClN_4OS_2$, Bis(ethylenethiourea)gold(I) chloride hydrate, 42B, 759
$C_6H_{14}N_4O_4S_2Zn$, Bis(thiourea)zinc acetate, 32B, 408
$C_6H_{16}Ag_2N_{10}S_6$, Thiocyanatobis(thiourea)silver(I) dimer, 42B, 759
$C_6H_{18}AgClN_6S_3$, Chlorotris(monomethylthiourea)silver(I), 38B, 1005
$C_6H_{21}CoN_{10}O_{10}S_2$, Dimethylglyoximato-bis(thiosemicarbazide)-cobalt-(III) nitrate dihydrate, 43B, 1192
$C_6H_{24}B_2Cu_2F_8N_{12}S_6$, Tris(thiourea)copper(I) tetrafluoroborate dimer, 40B, 840
$C_6H_{24}Br_2N_{12}NiS_6$, Hexakis(thiourea)nickel(II) bromide, 34B, 523
$C_8H_{12}CdN_6S_4$, Bis(ethylenethiourea)cadmium thiocyanate, 24, 536
$C_8H_{16}CdN_{12}S_8Zn$, Tetrakisthioureacadmium tetraisothiocyanatozincate, 32A, 240; 32B, 409
$C_8H_{16}CoHgN_{12}S_8$, Tetrathioureamercury(II) tetraisothiocyanatocobaltate(II), 30A, 299; 30B, 342
$C_8H_{16}N_6O_2S_2Zn$, Aqua[3-ethoxy-2-oxobutyraldehyde bis(thiosemicarbazonato)]zinc(II), 46B, 987
$C_8H_{20}Cl_2CoN_6OS_2$, Chlorobis(acetonethiosemicarbazone)cobalt(II) chloride monohydrate, 44B, 846
$C_8H_{20}Cl_2FeN_6OS_2$, Chlorobis(acetonethiosemicarbazone)iron(II) chloride hydrate, 44B, 847
$C_8H_{32}Cu_2F_6N_{16}S_8Si$, Tetrakis(thiourea)copper(I) hexafluorosilicate, 40B, 841; 45B, 1079
$C_9H_{24}ClCuN_6S_3$, Chlorotris(N,N'-dimethylthiourea)copper(I), 37B, 620
$C_{10}H_{18}CoN_4O_4S_2$, Bisacetatobis(ethylenethiourea)cobalt(II), 35B, 632
$C_{10}H_{20}ClCuN_4S_2$, Chlorobis(N-ethyl-1,3-imidazolidine-2-thione)copper-(I), 45B, 1079
$C_{10}H_{20}N_6NiS_4 \cdot 2\ C_3H_7NO$, Bis(1-isopropyl-2,4-dithiobiuretato)nickel-(II) dimethyl formamide solvate, 45B, 1080

$C_{10}H_{24}Cl_2CoN_4S_2$, Dichlorobis-(N,N'-diethylthiourea)cobalt(II), 39B, 685
$C_{10}H_{24}Cl_2CuN_4S_2$, Dichlorobis(tetramethylthiourea)copper(II), 44B, 847
$C_{10}H_{24}Cl_2N_4S_2Zn$, Dichlorobis(N,N'-diethylthiourea)zinc(II), 37B, 622
$C_{10}H_{24}CoN_6O_6S_2$, Dinitratobis(tetramethylthiourea)cobalt(II), 44B, 847
$C_{11}H_{17}CuN_6O_4S_3$, Tris(thiourea)copper(I) hydrogen o-phthalate, 43B, 1194
$C_{12}H_{24}Cl_2N_8NiS_4$, trans-Dichlorotetrakis(ethylenethiourea)nickel(II), 32B, 410
$C_{12}H_{32}Cl_2N_8PtS_4$, Tetrakis(N,N'-dimethylthiourea)platinum(II) chloride, 43B, 1195
$C_{12}H_{32}I_2N_8PtS_4$, Tetrakis(N-ethylthiourea)platinum(II) iodide, 42B, 761
$C_{12}H_{36}Br_2N_8NiO_2S_4$, Tetrakis(N,N'-dimethylthiourea)nickel(II) bromide dihydrate, 42B, 761
$C_{14}H_{14}N_8NiS_2$, Bis(2-formylpyridine thiosemicarbazonato)nickel(II), 46B, 984
$C_{14}H_{16}CuN_6O_4S_2$, Bis(picolinato)bis(thiourea)copper(II), 38B, 851
$C_{16}H_{14}CuN_6S_2$, Benzil(bisthiosemicarbazonato)copper(II), 45B, 1081
$C_{16}H_{24}N_{10}O_6S_4Zn \cdot H_2O$, Tetrakis[1-methyl-2(3H)-imidazolinethione]-zinc(II) nitrate monohydrate, 45B, 1081
$C_{16}H_{32}Cl_2N_8NiS_4$, Dichlorotetrakis(trimethylenethiourea)nickel(II), 33B, 492
$C_{17}H_{22}Cl_2CuN_6O_8S$, N,N'-Tetramethylenebis-(2-pyridinaldimine)thioureacopper(II) perchlorate, 39B, 685
$C_{18}H_{22}CoI_2N_6S_2$, Diiodobis(acetophenonethiosemicarbazone)cobalt(II), 41B, 1158
$C_{18}H_{48}B_2Cu_2F_8N_{12}S_6$, Tris(S-dimethylthiourea)copper(I) tetrafluoroborate dimer, 40B, 840
$C_{20}H_{24}B_2Br_2F_8Hg_2N_8$, Di-μ-bromo-bis[bis(tetramethylthiourea-S)mercury(II)] tetrafluoroborate, 45B, 1082
$C_{20}H_{24}CuN_6O_4S_2$, Bis(picolinato)bis(allylthiourea)copper(II), 38B, 851
$C_{20}H_{26}Br_2CoN_6S_2$, Bromo-bis(acetone 4-phenylthiosemicarbazone)cobalt-(II) bromide (green form), 44B, 848
$C_{20}H_{48}Cl_2CoN_8O_8S_4$, Tetrakis(N,N'-diethylthiourea)cobalt(II) diperchlorate, 45B, 1083
$C_{20}H_{50}Cl_2N_8OPtS_4$, Tetrakis(N,N'-diethylthiourea)platinum(II) chloride monohydrate, 41B, 987
$C_{21}H_{20}Cl_2CuN_6O_8S$, Bis-(2,2'-bipyridyl)thioureacopper(II) perchlorate, 39B, 685
$C_{22}H_{26}N_4NiO_2S_2$, Bis(N-phenyl-N'-morpholino-thiourea-N,S)nickel, 46B, 985
$C_{22}H_{48}N_{10}NiS_6$, trans-Dithiocyanatotetrakis(N,N'-diethylthiourea)nickel, 34B, 522
$C_{24}H_{30}CuN_4O_2S_2$, Bis(1,1-diethyl-3-benzoylthiourea)copper(II), 46B, 986
$C_{24}H_{32}N_4O_2PdS_2$, Bis(1,1-diethyl-3-benzoyl-thioureato)palladium(II), 43B, 1196
$C_{25}H_{22}Cl_2CuN_6O_9S$, Bis(1,10-phenthroline)thioureacopper(II) perchlorate monohydrate, 41B, 988
$C_{26}H_{24}Br_2CoN_4S_2$, Dibromobis(N,N'-diphenylthiourea)cobalt(II), 39B, 687
$C_{26}H_{24}Br_2N_4NiS_2$, Dibromobis(N,N'-diphenylthiourea)nickel(II), 39B, 688

$C_{26}H_{24}CuIN_8S_2$, o-Phenanthrolinebis(thiourea)copper(I) iodide o-phenanthroline, 43B, 1197
$C_{28}H_{48}I_2N_2NiS$, Tetrakis(N,N'-diallylthiourea)nickel(II) iodide, 37B, 638
$C_{30}H_{32}ClN_3PRhS_2 \cdot 0.8\ CHCl_3$, Chloro-(N,N-dimethylthiocarboxamido)-(N,N-dimethyl-N'-phenylthioureido)triphenylphosphinerhodium(III) chloroform solvate, 42B, 761
$C_{42}H_{96}Cl_2N_{12}NiO_8S_6$, Hexakis(diisopropylthiourea)nickel(II) perchlorate, 40B, 1023
$C_{60}H_{60}Cr_2N_8O_6 \cdot C_6H_{12}$, Tetrakis($\mu$-(N,N')-diphenylurea)-bis(tetrahydrofuran)-dichromium cyclohexane solvate, 46B, 1197

80. METAL COMPLEXES (THIOCARBAMATE OR XANTHATE)

$C_2H_4N_2NiS_4$, Nickel(II) dithiocarbamate, 32B, 411
$C_2H_{12}Cl_2CdN_8O_8S_2$, Diperchloratobis(thiocarbonohydrazide-N,S)cadmium-(II), 42B, 762
$C_3H_6CoN_3S_6$, Tris(dithiocarbamato)cobalt(III), 41B, 988
$C_3H_{15}N_{11}NiO_6S_3$, Tris(thiosemicarbazide)nickel(II) dinitrate, 42B, 763
$C_3H_{17}N_{11}NiO_7S_3$, Tris(thiosemicarbazide)nickel(II) dinitrate monohydrate, 42B, 763
$C_3H_{18}Cl_2CdN_{12}O_8S_3$, Tris(thiocarbonohydrazide-N,S)cadmium(II) perchlorate, 42B, 763
$C_4H_8N_2NiS_4$, Bis(N-methyldithiocarbamato)nickel(II), 38B, 852
$C_4H_8N_6PdS_4$, Bis(dithiobiureto)palladium, 42B, 764
$C_4H_8N_6PtS_4$, Bis(dithiobiureto)platinum, 42B, 764
$C_6H_4F_6O_2PdS_4$, Bis(trifluoroethylxanthato)palladium(II), 45B, 1083
$C_6H_{10}CdO_2S_4$, Cadmium ethylxanthate, 38B, 853
$C_6H_{10}HgO_2S_4$, Mercury(II) ethylxanthate, 46B, 986
$C_6H_{10}HgO_2S_4$, Mercury ethylxanthate (chiral form), 43B, 1198
$C_6H_{10}NiO_2S_4$, Nickel xanthate, 28, 627
$C_6H_{10}O_2S_4Zn$, Zinc ethylxanthate, 31B, 404
$C_6H_{12}CoN_3OS_4$, Bis(dimethyldithiocarbamato)nitrosylcobalt, 27, 846; 38B, 854
$C_6H_{12}CuN_2S_4$, Bis-(N,N-dimethyldithiocarbamato)copper(II), 40B, 842
$C_6H_{12}FeN_3OS_4$, Bis-(N,N-dimethyldithiocarbamato)nitrosyliron, 35B, 792
$C_6H_{12}HgI_2N_2S_3$, Diiodo(N,N,N',N'-tetramethylthiuram monosulfide)mercury(II), 43B, 9
$C_6H_{12}N_2NiS_4$, Bis(dimethyldithiocarbamato)nickel(II), 46B, 987
$C_6H_{12}N_2S_4Zn$, Zinc dimethyldithiocarbamate, 31B, 405
$C_6H_{13}HgNS_2$, Methyl N,N-diethyldithiocarbamatomercury(II), 42B, 764
$C_6H_{14}Cl_2HgN_2O_2S_2$, Dichlorobis(O-ethylthiocarbamato)mercury(II), 40B, 842; 41B, 989
$C_6H_{14}N_4NiS_4$, (S,S-2-Ethyldithiocarbazato)(S,N-2,3-dimethyldithiocarbazato)nickel(II), 41B, 990
$C_7H_{14}AgNOS$, Silver(I) dipropylmonothiocarbamate, 37B, 616
$C_7H_{21}AuB_9NS_2$, 3-Diethyldithiocarbamato-1,2-dicarba-3-auradodecaborane, 44B, 848
$C_8H_{14}CdO_2S_4$, Cadmium isopropylxanthate, 39B, 688
$C_8H_{14}MoOS_6$, Oxobis(isopropylthioxanthato)molybdenum(IV), 44B, 849
$C_8H_{16}N_2NiS_4$, cis-Bis-(N-isopropyldithiocarbamato)nickel(II), 40B, 843
$C_8H_{18}MoN_4OS_4$, Bis(dimethyldithiocarbamato)[N,N-dimethylhydrazido(2-)-N']oxomolybdenum(VI), 45B, 1084

$C_8H_{20}Cl_2CoN_6OS_2$, Dichlorobis(acetonethiosemicarbazone)cobalt(II) monohydrate, 42B, 764

$C_9H_{12}N_2O_3S_4W$, Tricarbonylbis(N,N-dimethyldithiocarbamato)tungsten-(II), 46B, 988

$C_9H_{15}CoO_3S_6$, Cobalt(III) tris(O-ethylxanthate), 34B, 524

$C_9H_{15}CoS_9$, Tris(ethylthioxanthato)cobalt(III), 38B, 854

$C_9H_{15}CrO_3S_6$, Tris-(O-ethylxanthato)chromium(III), 38B, 855

$C_9H_{18}Br_2CuNS_2$, N,N-Dibromo-di-n-butyldithiocarbamatocopper(III), 33B, 485

$C_9H_{18}ClN_3S_6Ti$, Chlorotris(N,N-dimethyldithiocarbamato)titanium(IV), 40B, 844

$C_9H_{18}CoN_3S_6$, Tris(N,N-dimethyldithiocarbamato)cobalt(III), 46B, 988

$C_9H_{18}FeN_3O_3S_3$, Tris(N,N-dimethylthiocarbamato)iron(III), 43B, 1198

$C_9H_{18}FeN_3S_6$, Tris(N,N-dimethyldithiocarbamato)iron(III), 43B, 1200

$C_9H_{18}I_3N_3RuS_6$, Iodotris(N,N-dimethyldithiocarbamato)ruthenium(IV)-iodine, 43B, 1199

$C_9H_{18}MoN_4S_7$, Tris(N,N-dimethyldithiocarbamato)thionitrosylmolybdenum, 45B, 1089

$C_{10}H_{16}ClCuN_2O_4S_4$, Bis(N-pyrrolidyldithiocarbamato)copper(III) perchlorate, 45B, 1085

$C_{10}H_{16}CuN_2S_4$, Bis(pyrrolidonecarbodithioato)copper(II), 39B, 814

$C_{10}H_{16}FeI_2N_2S_4$, Iodobis(pyrrolidinyldithiocarbamato)iron(III) iodine, 46B, 989

$C_{10}H_{16}N_2NiO_2S_4$, Bis(morpholine-N-carbodithioato)nickel(II), 45B, 1085

$C_{10}H_{16}N_2NiS_4$, Bis(pyrrolidonecarbodithioato)nickel(II), 39B, 814

$C_{10}H_{18}CdO_2S_4$, Cadmium n-butylxanthate, 30B, 371

$C_{10}H_{20}As_2I_6N_2NiS_4$, Bis(diethyldithiocarbamato)bis(triiodoarsenido)-nickel(II), 46B, 989

$C_{10}H_{20}Br_2Hg_2N_2S_4$, Dibromo(N,N-diethyldithiocarbamato)dimercury polymer, 44B, 850

$C_{10}H_{20}Br_2MoN_2OS_4$, Oxodibromobis(diethyldithiocarbamato)molybdenum-(VI), 42B, 765

$C_{10}H_{20}CdN_2S_4$, Cadmium(II) N,N-diethyldithiocarbamate, 33B, 486

$C_{10}H_{20}ClFeN_2S_4$, Monochlorobis(diethyldithiocarbamato)iron(III), 35B, 798

$C_{10}H_{20}Cl_2MoN_2OS_4$, Oxodichlorobis(diethyldithiocarbamato)molybdenum-(VI), 42B, 765

$C_{10}H_{20}CuN_2S_2Se_2$, Bis-(N,N-diethylthioselenocarbamato)copper(II), 41B, 990

$C_{10}H_{20}CuN_2S_4$, Copper(II) diethyldithiocarbamate, 30B, 373; 31B, 404

$C_{10}H_{20}CuN_2Se_4$, Bis(N,N-diethyldiselenocarbamato)copper(II), 37B, 621

$C_{10}H_{20}FeIN_2S_4$, Iodobis(N,N-diethyldithiocarbamato)iron(III), 38B, 856

$C_{10}H_{20}FeN_4O_3S_4$, cis-Bis(diethyldithiocarbamato)(nitro)nitrosyliron, 43B, 1200

$C_{10}H_{20}HgI_2N_2S_4$, Diiodotetraethylthiuramdisulfidemercury(II), 43B, 1390

$C_{10}H_{20}HgN_2S_4$, Bis(N,N-diethyldithiocarbamato)mercury(II), 39B, 816, 827

$C_{10}H_{20}Hg_2I_2N_2S_4$, Bis(iodo-N,N-diethyldithiocarbamatomercury(II)), 43B, 1201

$C_{10}H_{20}MnN_2S_4$, Bis-(N,N-diethyldithiocarbamato)manganese(II), 41B, 991

$C_{10}H_{20}MoN_2O_2S_4$, Dioxo(N,N-diethyldithiocarbamato)molybdenum(VI), 34B, 524; 38B, 858; 46B, 990

$C_{10}H_{20}Mo_2N_2OS_7$, Di-μ-sulfido-[oxo(N,N-diethyldithiocarbamato)molyb-

denum(V)][thio(N,N-diethyldithiocarbamato)molybdenum(V)], 45B, 1086
$C_{10}H_{20}Mo_2N_2O_4S_4$, Di-μ-oxo-bis(oxodiethyldithiocarbamatomolybdenum-(V)), 41B, 991
$C_{10}H_{20}Mo_2N_2S_8$, Di-μ-sulfido-bis(N,N-diethyldithiocarbamato-thiomolybdenum), 44B, 850; 45B, 1086
$C_{10}H_{20}N_2NiS_4$, Bis(N,N-diethyldithiocarbamato)nickel, 24, 604; 26, 672; 30B, 378
$C_{10}H_{20}N_2NiSe_4$, Bis(N,N-diethyldiselenocarbamato)nickel(II), 37B, 621
$C_{10}H_{20}N_2OS_4V$, Bis(diethyldithiocarbamato)oxovanadium(IV), 42B, 766
$C_{10}H_{20}N_2PdS_4$, Bis(N,N-diethyldithiocarbamato)palladium(II), 37B, 620
$C_{10}H_{20}N_2S_4Zn$, Zinc diethyldithiocarbamate, 30B, 381
$C_{10}H_{20}N_2Se_4Zn$, Bis(N,N-diethylselenocarbamato)zinc(II), 37B, 621
$C_{10}H_{20}N_3ReS_4$, Nitridobis(N,N-diethyldithiocarbamato)rhenium(V), 38B, 857
$C_{10}H_{22}Br_2N_2O_2PdS_2$, trans-Dibromo-bis(O-ethyl-N,N-dimethylthiocarbamato)palladium(II), 46B, 990
$C_{10}H_{22}Cl_2N_2O_2PdS_2$, trans-Dichloro-bis(O-ethyl-N,N-dimethylthiocarbamate)palladium(II), 46B, 991
$C_{11}H_{15}NO_2S_4Zn$, Bis(O-ethylxanthato)pyridinezinc(II), 42B, 767
$C_{11}H_{20}FeN_3S_5$, Bis(N,N-diethyldithiocarbamato)(isothiocyanato)iron-(III), 46B, 991
$C_{12}H_{20}Br_2Mo_2O_4S_8$, Tetrakis(ethylxanthato)dimolybdenum bromine adduct, 44B, 851
$C_{12}H_{20}Br_4Cu_5N_2S_4$, Bis(piperidyldithiocarbamato)copper(II) tetrakis-(copper(I) bromide), 40B, 845
$C_{12}H_{20}Br_6Cu_7N_2S_4$, Bis(piperidyldithiocarbamato)copper(II) hexakis-(copper(I) bromide), 40B, 845
$C_{12}H_{20}CuN_2S_4$, Bis(cyclopentamethylenedithiocarbamato)copper(II), 43B, 1202
$C_{12}H_{20}I_2Mo_2O_4S_8$, Tetrakis(ethylxanthato)dimolybdenum iodine adduct, 44B, 851
$C_{12}H_{20}Mo_2O_7S_8$, μ-Oxo-bis(bis(ethyl xanthate)oxomolybdenum(V)), 29, 622
$C_{12}H_{20}N_2NiS_4$, Bis(cyclopentamethylenedithiocarbamato)nickel(II), 44B, 851
$C_{12}H_{22}Cl_4Hg_2N_2S_4$, Di-μ-chloro-dichlorobis(methyl pyrrolidine-1-carbodithioate)dimercury(II), 40B, 846
$C_{12}H_{24}Cl_6N_4S_8Ta_2 \cdot 0.5\ CH_2Cl_2$, Tetrakis(N,N-dimethyldithiocarbamato)tantalum(V) hexachlorotantalate(V) - hemidichloromethane, 42B, 768
$C_{12}H_{24}Mo_2N_2O_3S_6$, μ-(2-Mercaptoethanolato-S,O)-μ-sulfido-bis(oxo(N,N-diethyldithiocarbamato)molybdenum(V)), 44B, 852
$C_{13}H_{18}AuN_3S_4$, N,N-Di-n-butyldithiocarbamato-1,2-dicyanoethene-1,2-dithiolato-gold(III), 39B, 820
$C_{13}H_{26}Cl_3N_4S_8Ta$, Tetrakis(N,N-dimethyldithiocarbamato)tantalum(V) chloride - dichloromethane, 42B, 768
$C_{14}H_7Fe_2NO_7S$, μ-(N-p-Tolymonothiocarbamato)-bis(tricarbonyliron), 45B, 1168
$C_{14}H_{20}ClCoN_8O_3S_2$, Bis(pyridinealdehydethiosemicarbazonato)cobalt-(III) chloride trihydrate, 40B, 847
$C_{14}H_{20}FeN_2O_2S_4$, Bis-(cyclopentamethylenedithiocarbamato)iron(II) dicarbonyl, 38B, 858
$C_{14}H_{20}N_3S_4Zn$, Bis(N,N-dimethyldithiocarbamato)pyridinezinc benzene solvate, 32B, 412
$C_{14}H_{24}CdN_2S_4$, Bis(hexamethylenedithiocarbamato)cadmium, 38B, 859
$C_{14}H_{24}CuN_2S_4$, Copper(II) hexamethylenedithiocarbamate, 33B, 488
$C_{14}H_{24}N_2NiS_4$, Bis(hexamethylenedithiocarbamato)nickel, 38B, 860

$C_{14}H_{28}Au_2N_2S_4$, Gold(I) dipropyldithiocarbamate dimer, 38B, 860
$C_{14}H_{28}ClFeN_2S_4 \cdot 0.87\ CHCl_3$, Chlorobis(N,N-di-isopropyldithiocarbamato)iron(III) chloroform solvate, 45B, 1086
$C_{14}H_{28}CuN_2S_4$, Bis(N,N-diisopropyldithiocarbamato)copper(II), 46B, 991
$C_{14}H_{28}CuN_2S_4$, Bis(N,N-di-n-propyldithiocarbamato)copper(II), 27, 844; 38B, 861
$C_{14}H_{28}FeN_3OS_4$, Nitrosylbis(N,N-diisopropyldithiocarbamato)iron, 46B, 992
$C_{14}H_{28}HgN_2S_4$, Bis(N,N-diisopropyldithiocarbamato)mercury(II) (α-form), 45B, 1087
$C_{14}H_{28}MoN_2OS_4$, Oxo-bis-di-n-propyldithiocarbamatomolybdenum(IV), 40B, 847
$C_{14}H_{28}MoN_2OS_6$, Bis(N,N-dipropylcarbamato)oxo(disulfur)molybdenum-(IV), 41B, 992
$C_{14}H_{28}MoN_2O_2S_4$, Di(oxo-di-n-propyldithiocarbamato)molybdenum(VI), 40B, 847
$C_{14}H_{28}N_2NiS_4$, Bis(N,N-di-isopropyldithiocarbamato)nickel(II), 38B, 862
$C_{14}H_{28}N_2NiS_4$, Bis(N,N-di-n-propyldithiocarbamato)nickel, 8, 326; 32B, 414
$C_{14}H_{32}N_2O_4S_6W_2$, Di-μ-sulfido-bis((diethyldithiocarbamato)dimethoxy-tungsten(V))(W-W), 44B, 857
$C_{15}H_{12}FeN_3S_6 \cdot 0.5\ CH_2Cl_2$, Tris(pyrrole-N-carbodithioato)iron(III) dichloromethane solvate, 45B, 1088
$C_{15}H_{22}MoN_6O_2S_6 \cdot 0.25\ CH_2Cl_2 \cdot 0.25\ H_2O$, Tris(dimethyldithiocarbamato)(m-nitrophenyldiazenato)molybdenum methylene chloride solvate hydrate, 46B, 992
$C_{15}H_{23}MoN_5S_6 \cdot CH_2Cl_2$, Tris(dimethyldithiocarbamato)(benzene-diazenato)molybdenum methylene chloride solvate, 46B, 992
$C_{15}H_{24}CrN_3S_6 \cdot 0.5\ C_6H_6$, Tris(1-pyrrolidinecarbodithioato)-chromium(III) - hemibenzene, 42B, 769
$C_{15}H_{24}FeN_3S_6 \cdot 0.5\ C_6H_6$, Tris(1-pyrrolidinecarbodithioato)iron(III) - hemibenzene, 42B, 769
$C_{15}H_{24}FeN_3S_6$, Tris(1-pyrrolidenecarbodithioato)iron(III) (form 1), 38B, 863
$C_{15}H_{24}FeN_3S_6$, Tris(1-pyrrolidinecarbodithioato)iron(III) (form 2), 44B, 852
$C_{15}H_{24}IrN_3S_6 \cdot 0.5\ C_6H_6$, Tris(1-pyrrolidinecarbodithioato)iridium-(III) - hemibenzene, 42B, 769
$C_{15}H_{26}FeN_3O_4S_6$, Tris(1-morpholinecarbodithiolato-S,S')iron(III) monohydrate, 42B, 771
$C_{15}H_{27}FeS_9$, Tris(t-butylthioxanthato)iron(III), 38B, 864
$C_{15}H_{30}AuN_3S_6$, Tris(N,N-diethyldithiocarbamato)gold(III); 39B, 823
$C_{15}H_{30}Br_2Cu_2N_3S_6$, Dibromo-tris(N,N-diethyldithiocarbamato)dicopper, 42B, 769
$C_{15}H_{30}ClN_3O_3Ti$, Chlorotris(N,N-diethylmonothiocarbamato)titanium-(IV), 45B, 1088
$C_{15}H_{30}ClN_3RuS_6$, Chlorotris(N,N-diethyldithiocarbamato)ruthenium(IV), 42B, 770
$C_{15}H_{30}CoNO_3S_6$, Tetraethylammonium (dithiocarbonato-S,S')bis(ethyl-xanthato)cobaltate(III), 44B, 853
$C_{15}H_{30}CoN_3S_6$, Tris(diethyldithiocarbamato)cobalt(III), 33B, 491; 34B, 527
$C_{15}H_{30}CrN_3S_6$, Tris(N,N-diethyldithiocarbamato)chromium(III), 43B, 1202
$C_{15}H_{30}CrN_4OS_6$, Tris(diethyldithiocarbamato-S,S')nitrosylchromium,

46B, 994
$C_{15}H_{30}F_6N_5Os_2PS_{12} \cdot C_7H_8$, Bis(μ-N,N-dimethyltrithiocarbamato)-tris(N,N-dimethyldithiocarbamato)diosmium(Os-Os) hexafluorophosphate toluene solvate, 46B, 998
$C_{15}H_{30}FeI_5N_3S_6$, Tris(N,N-diethyldithiocarbamato)iron(IV) pentaiodide, 46B, 993
$C_{15}H_{30}FeN_3O_6S_6$, Tris(N,N-bis(2-hydroxyethyl)dithiocarbamato)iron-(III), 45B, 1089
$C_{15}H_{30}FeN_3S_6$, Tris(N,N-diethyldithiocarbamato)iron(III), 39B, 688
$C_{15}H_{30}IrN_3S_6$, Tris(diethyldithiocarbamato)iridium(III), 42B, 771
$C_{15}H_{30}MnN_3S_6$, Tris-(N,N-diethyldithiocarbamato)manganese(III), 38B, 865
$C_{15}H_{30}MoN_4S_6$, Tris(N,N-diethyldithiocarbamato)nitridomolybdenum, 45B, 1089
$C_{15}H_{30}N_3NbOS_6$, Tris(N,N-diethyldithiocarbamato)oxo-niobium(V), 39B, 689
$C_{15}H_{30}N_3OS_6V$, Tris(N,N-diethyldithiocarbamato)oxo-vanadium(V), 39B, 689
$C_{15}H_{30}N_3RhS_6$, Tris(diethyldithiocarbamato)rhodium(III), 41B, 994
$C_{15}H_{30}N_3RuS_6$, Tris(N,N-diethyldithiocarbamato)ruthenium(III), 40B, 849
$C_{15}H_{30}N_3S_7Ta$, Sulfidotris(N,N-diethyldithiocarbamato)tantalum(V), 44B, 854
$C_{15}H_{30}Pd_3S_{12}$, Tris(ethyl trithiocarbonato)tri-μ-(ethylthio)-tripalladium(II), 39B, 694
$C_{16}H_{16}CuN_2S_4$, Bis-(N-methyl-N-phenyldithiocarbamato)copper(II), 38B, 865
$C_{16}H_{16}N_2NiS_4$, Bis-(N-methyl-N-phenyldithiocarbamato)nickel(II), 38B, 865
$C_{16}H_{24}N_6NiS_4$, Tetramethylammonium bis(maleonitriledithiolato)nickel-(II), 30B, 377
$C_{16}H_{25}Cl_3FeN_3O_3S_6$, Tris(1-morpholinecarbodithiolato-S,S')iron(III) chloroform solvate, 42B, 771
$C_{16}H_{25}Cl_3MnN_3O_3S_6$, Tris(1-morpholinecarbodithiolato-S,S')manganese-(III) chloroform solvate, 42B, 771
$C_{16}H_{25}Cl_4MoN_3S_4 \cdot CHCl_3$, Dichlorobis(diethyldithiocarbamato)(phenylnitreno)molybdenum chloroform, 46B, 994
$C_{16}H_{26}Cl_2CoN_3O_3S_6$, Tris(4-morpholinecarbodithioato-S,S')cobalt(III) dichloromethane solvate, 41B, 992
$C_{16}H_{26}Cl_2CrN_3O_3S_6$, Tris(morpholinocarbodithioato)chromium(III) dichloromethane solvate, 41B, 993
$C_{16}H_{26}Cl_2FeN_3O_3S_6$, Tris(4-morpholinecarbodithioato-S,S')iron(III) dichloromethane solvate, 41B, 992
$C_{16}H_{26}Cl_2MnN_3O_3S_6$, Tris(morpholinocarbodithioato)manganese(III) dichloromethane solvate, 41B, 993
$C_{16}H_{26}Cl_2N_3O_3RhS_6$, Tris(morpholinocarbodithioato)rhodium(III) dichloromethane solvate, 41B, 993
$C_{16}H_{28}MoN_2O_2S_4 \cdot CH_2Cl_2$, Bis(N,N-diisopropyldithiocarbamato)dicarbonylmolybdenum(II) dichloromethane, 46B, 995
$C_{16}H_{30}Co_2S_{14}$, Tetrakis(ethylthioxanthato)-μ-bis(ethylthio)-dicobalt-(III), 38B, 866
$C_{16}H_{30}Fe_2S_{14}$, Bis(ethyl thioxanthato)-μ-bis(ethyl thioxanthato)-μ'-bis(ethylthio)-diiron(III), 34B, 597; 35B, 802
$C_{16}H_{30}N_3OReS_6$, Carbonyltris(N,N-diethyldithiocarbamato)rhenium(III), 40B, 849
$C_{16}H_{37}N_3O_5S_2U$, Diethylammonim ethoxybis(diethylmonothiocarbamato)dioxouranate(VI), 44B, 855

$C_{17}H_{20}MoN_3S_4$, Bis(2-aminobenzenethiolato)(diethyldithiocarbamato)molybdenum(V), 44B, 855
$C_{17}H_{38}N_4S_6Zn$, Tetraethylammonium (dimethyldithiocarbamato-S,S')bis-(dimethyldithiocarbamato-S)zincate, 46B, 996
$C_{18}H_{20}F_{12}N_2Ni_2S_8$, Nickel thiete diethyldithiocarbamate dimer, 38B, 867
$C_{18}H_{20}N_2O_2PdS_2$, Bis(O-ethyl-N-phenylthiocarbamato)palladium(II), 37B, 632
$C_{18}H_{22}Br_2CoN_2O_2S_2$, Dibromo-bis(O-ethyl-N-phenylthiocarbamate)cobalt-(II), 37B, 632
$C_{18}H_{28}CdN_2O_2S_4$, Bis(O-ethylxanthato)-1,10-phenanthrolinecadmium(II), 42B, 772
$C_{18}H_{34}CoN_4O_2S_2$, Bis(cyclotetramethylenethiocarbamato)bis(pyrrolidine)cobalt(II), 40B, 850
$C_{18}H_{36}AgAuBr_2N_2S_4$, Bis(N,N-di-n-butyldithiocarbamato)gold(III) dibromoargentate(I), 37B, 633
$C_{18}H_{36}Au_2Br_2N_2S_4$, Bis(N,N-di-n-butyldithiocarbamato)gold(III) dibromoaurate(I), 33B, 493
$C_{18}H_{36}CuI_3N_2S_4$, Bis(N,N-di-n-butyldithiocarbamato)copper(III) triiodide, 38B, 1018
$C_{18}H_{36}I_2N_2PtS_4$, cis-Bis(N,N-di-n-butyldithiocarbamato)diiodoplatinum(IV), 39B, 824
$C_{18}H_{36}Mo_2N_2O_2S_6$, Di-$\mu$-sulfido-bis(oxo-(di-n-butyldithiocarbamato)molybdenum(V)), 42B, 772
$C_{18}H_{36}Mo_2N_2S_8$, Di-μ-sulfido-bis(di-n-butyldithiocarbamato-thiomolybdenum(V)), 41B, 995
$C_{18}H_{36}N_2NiS_4$, Bis(N,N-diisobutyldithiocarbamato)nickel(II), 42B, 773
$C_{18}H_{36}N_2NiSe_4$, Bis(N,N-di-n-butyldiselenocarbamato)nickel(II), 40B, 1011
$C_{18}H_{36}Pd_2S_8$, Bis(t-butyl trithiocarbonato)di-μ-(t-butylthio)-dipalladium(II), 39B, 694
$C_{19}H_{31}Cl_4MnN_3O_4S_6$, Tris(pentamethylenedithiocarbamato)manganese(IV) perchlorate chloroform solvate, 40B, 851
$C_{19}H_{39}NNiO_3S_6$, Tetramethylammonium tris(O-isobutyldithiocarbonato)-nickelate(II), 44B, 856
$C_{20}H_{29}N_3S_6Ti$, η^5-Cyclopentadienyltris(N,N-diethyldithiocarbamato)titanium(IV) - benzene, 44B, 856
$C_{20}H_{32}N_4S_8Zn_2$, Bis(pyrrolidine-carbodithioato)zinc(II) dimer, 45B, 1090
$C_{20}H_{36}Mo_2O_6S_8$, Tetrakis(O-ethyl-dithiocarbonato)dimolybdenum-bis-(tetrahydrofuranate), 39B, 690
$C_{20}H_{40}BrN_4S_8W$, Tetrakis(N,N-diethyldithiocarbamato)tungsten(V) bromide, 39B, 826
$C_{20}H_{40}ClMoN_4S_8$, Tetrakis(diethyldithiocarbamato)molybdenum(V) chloride, 44B, 861
$C_{20}H_{40}Cl_2Hg_3N_4S_8$, catena-{Di-$\mu$-chloro-tetrakis[$\mu$-(N,N-diethyldithiocarbamato-S,S')]-trimercury(II)}, 46B, 997
$C_{20}H_{40}Cu_4N_4S_8$, Copper(I) diethyldithiocarbamate, 28, 625
$C_{20}H_{40}Fe_2N_4S_8$, Bis(diethyldithiocarbamato)iron(II), 41B, 995
$C_{20}H_{40}Hg_2N_4S_8$, Tetrakis(N,N-diethyldithiocarbamato)dimercury(II), 39B, 827
$C_{20}H_{40}LaN_4NaS_8$, Sodium tetrakis(diethyldithiocarbamato)lanthanate, 43B, 1203
$C_{20}H_{40}MoN_4S_8$, Tetrakis(N,N-diethyldithiocarbamato)molybdenum(IV), 40B, 851
$C_{20}H_{40}Mo_2N_4O_3S_8$, μ-Oxo-bis[bis(diethyldithiocarbamato)oxomolybdenum-(V)], 45B, 1095

$C_{20}H_{40}N_4O_3Re_2S_8$, μ-Oxo-bis(oxobis-(N,N-diethyldithiocarbamato)-rhenium(V)), 38B, 868
$C_{20}H_{40}N_4O_4S_4Ti$, Tetrakis(N,N-diethylmonothiocarbamato)titanium(IV), 42B, 773; 44B, 857
$C_{20}H_{40}N_4O_4S_4Zr$, Tetrakis(N,N-diethylmonothiocarbamato)zirconium(IV), 44B, 857
$C_{20}H_{40}N_4Os_2S_{14}$, (μ-Pentasulfur)(μ-N,N-diethyltrithiocarbamato)-tris(N,N-diethyldithiocarbamato)diosmium(Os-Os), 46B, 998
$C_{20}H_{40}N_4S_8Th$, Thorium(IV) tetrakis-(N,N-diethyldithiocarbamate), 35B, 813
$C_{20}H_{40}N_4S_8Ti$, Tetrakis(N,N-diethyldithiocarbamato)titanium(IV), 38B, 868
$C_{20}H_{40}N_4S_{10}W_2$, Di-μ-sulfido-di-μ-diethyldithiocarbamato-bis(diethyldithiocarbamatotungsten(IV))(W-W), 44B, 857
$C_{21}H_{29}FeN_4O_5S_6$, Tris(4-morpholinecarbodithioato-S,S')iron(III) nitrobenzene solvate, 43B, 1204
$C_{22}H_{21}O_3PS_4U$, Bis(dithioacetato)dioxo(triphenylphosphine oxide)uranium(VI), 38B, 869
$C_{22}H_{30}MoN_4S_4$, cis-Bis(phenylnitrene)-bis(diethyldithiocarbamato)molybdenum, 45B, 1090
$C_{22}H_{30}Mo_2N_2O_3S_6 \cdot 0.25\ C_3H_6O$, μ-Oxo-di-μ-thiophenolato-bis(N,N-diethyldithiocarbamato-oxo-molybdenum acetone solvate, 43B, 1205
$C_{22}H_{40}BF_4N_4O_2Ru_2S_8$, μ-Bis(N,N-diethyldithiocarbamato)-bis(N,N-diethyldithiocarbamato)dicarbonyldiruthenium(II,III) tetrafluoroborate, 46B, 999
$C_{22}H_{40}N_4O_2Ru_2S_8$, Di-(μ-diethyldithiocarbamato-carbonyl-diethyldithiocarbamatoruthenium(II)), 41B, 996
$C_{22}H_{42}N_4O_2S_2Zn$, Bis(cyclopentamethylenethiocarbamato)bis(piperidine)zinc(II), 39B, 691
$C_{22}H_{45}Co_2N_3S_{11}$, Bis(ethylmercaptide)(ethylthioxanthato)tris(diethyldithiocarbamato)dicobalt(III), 41B, 997
$C_{22}H_{46}N_6Os_2S_{10}$, Tetrakis(N,N-dimethyldithiocarbamato)-μ-nitrido-μ-(N,N-dimethyldithiocarbamato)-diosmium(IV) n-heptane, 43B, 1205
$C_{22}H_{49}N_3O_5S_2U$, Di-n-propylammonium ethoxybis(di-n-propylmonothiocarbamato)dioxouranate(VI), 45B, 1091
$C_{23}H_{25}AuNPS_2$, Triphenylphosphine-(N,N-diethyldithiocarbamato)gold-(I), 38B, 869
$C_{23}H_{40}Cl_2N_4O_3Ru_3S_8$, Tetrakis(diethyldithiocarbamato)-dichloro-tricarbonyl-triruthenium(II), 41B, 997
$C_{24}H_{21}MnNO_3PS_2$, Dimethyldithiocarbamato(triphenylphosphine)tricarbonylmanganese, 42B, 774
$C_{24}H_{24}FeN_3S_6$, Tris(N-methyl-N-phenyldithiocarbamato)iron(III), 38B, 864
$C_{24}H_{28}Cl_2FeI_2N_4S_4Zn$, Di-μ-diethyldithiocarbamato-bis(di-p-chlorophenylisocyano)iron-diiodozinc, 43B, 996
$C_{25}H_{29}Cl_5Mo_2N_6OS_6$, Bis(p-chlorothiobenzoyldiazenido)bis(diethyldithiocarbamato)oxodimolybdenum - chloroform, 44B, 858
$C_{25}H_{32}Cl_2Mo_2N_6O_3S_4$, Bis(benzoyldiazenido)bis(diethyldithiocarbamato)oxodimolybdenum - dichloromethane, 44B, 858
$C_{25}H_{45}NNiO_3S_6 \cdot C_3H_6O$, Tetramethylammonium tris(O-cyclohexyldithiocarbonato)nickelate(II) acetone solvate, 45B, 1092
$C_{25}H_{50}BCo_2F_4N_5S_{10}$, Pentakis(diethyldithiocarbamato)dicobalt(III) tetrafluoroborate, 41B, 998
$C_{26}H_{25}NiOPS_2$, (η^5-Cyclopentadienyl)(triphenylphosphine)(η^1-ethylxanthato)nickel(II), 46B, 1000
$C_{26}H_{36}Au_2N_6S_8$, Bis(N,N-di-n-butyldithiocarbamato)gold(III) bis(1,2-dicyanoethane-1,2-dithiolato)aurate(III), 37B, 636

$C_{26}H_{36}N_6Ni_2S_8$, N,N-Di-n-butyldithiocarbamato-1,2-dicyanoethene-1,2-dithiolato-nickel(III) dimer, 40B, 853

$C_{27}H_{36}CoN_3O_3S_6$, Tris(4-morpholinecarbodithioato-S,S')cobalt(III) benzene solvate, 42B, 774

$C_{27}H_{36}CrN_3O_3S_6$, Tris(4-morpholinecarbodithioato-S,S')chromium(III) benzene solvate, 42B, 774

$C_{27}H_{36}FeN_3O_3S_6$, Tris(4-morpholinecarbodithioato-S,S')iron(III) benzene solvate, 42B, 774

$C_{27}H_{36}IrN_3O_3S_6$, Tris(4-morpholinecarbodithioato-S,S')iridium(III) benzene solvate, 42B, 774

$C_{27}H_{36}N_3O_3RhS_6$, Tris(4-morpholinecarbodithioato-S,S')rhodium(III) benzene solvate, 42B, 774

$C_{27}H_{54}BrN_3NiS_6$, Tris(N,N-di-n-butyldithiocarbamato)nickel(IV) bromide, 39B, 692

$C_{27}H_{54}BrN_3NiSe_6$, Tris(N,N-di-n-butyldiselenocarbamato)nickel(IV) bromide, 37B, 637

$C_{27}H_{54}MoN_4OS_6$, Nitrosyltris(N,N-di-n-butyldithiocarbamato)molybdenum(II), 39B, 830

$C_{28}H_{20}S_8V$, Tetrakis(dithiobenzoato)vanadium(IV), 40B, 854

$C_{28}H_{35}AsN_2O_3S_4U$, Bis(diethyldithiocarbamate)dioxo(triphenylarsine oxide)uranium(VI), 35B, 819

$C_{28}H_{35}N_2O_3PS_4U$, Bis(diethyldithiocarbamate)dioxo(triphenylphosphine oxide)uranium(VI), 35B, 819

$C_{28}H_{48}N_4S_8Zn_2$, Zinc hexamethylenedithiocarbamate dimer, 38B, 870

$C_{28}H_{53}N_2PPtS_4 \cdot C_6H_{12}$, Bis(N-t-butyldithiocarbamato)(tricyclohexylphosphine)platinum(II) cyclohexane solvate, 45B, 1093

$C_{28}H_{56}BF_4N_5ORu_2S_{10}$, Bis(N,N-diethyldithiocarbamato)-μ-tris(N,N-diethyldithiocarbamato)diruthenium(III) tetrafluoroborate acetone solvate, 42B, 775

$C_{28}H_{56}Mo_2N_4O_3S_8$, μ-Oxo-bis-(di(oxo-di-n-propyldithiocarbamato)molybdenum(V)), 40B, 847

$C_{28}H_{56}Mo_2N_4S_8$, Bis(μ-sulfido-thiocarboxamido(di-n-propyldithiocarbamato)molybdenum), 39B, 693

$C_{28}H_{56}N_4S_8Zn_2$, Bis[μ-(N,N-diisopropyldithiocarbamato)-μ-S,S']-bis(N,N-diisopropyldithiocarbamato)dizinc(II), 45B, 1093

$C_{28}H_{60}N_5NpS_8$, Tetraethylammonium neptunium(III) tetrakis-(N,N-diethyldithiocarbamate), 35B, 818

$C_{29}H_{46}MoN_6S_6$, Tetrabutylammonium N,N-diethyldithiocarbamatobis-(maleonitriledithiolato)molybdate(IV), 42B, 776

$C_{29}H_{52}N_4S_6Zn \cdot C_2H_6O$, Tetra-n-butylammonium (benzothiazole-2-thiolato-N)bis(dimethyldithiocarbamato-S,S')zincate ethanol solvate, 46B, 996

$C_{29}H_{54}CuN_4S_4$, Tetra-n-butylammonium (N,N-di-n-butyldithiocarbamato)-(1,2-dicyanoethene-1,2-dithiolato)cuprate(II), 39B, 831

$C_{30}H_{28}Ni_2S_8$, Bis(benzyl trithiocarbonato)di-μ-(benzylthio)-dinickel-(II), 39B, 694

$C_{30}H_{32}MoN_6OS_4 \cdot C_3H_6O$, Bis[N,N-diphenylhydrazido(2-)]bis(N,N-dimethyldithiocarbamato)molybdenum(VI) acetone solvate, 45B, 1094

$C_{30}H_{60}Ag_6N_6S_{12}$, Silver(I) N,N-diethyldithiocarbamate hexamer, 42B, 776

$C_{30}H_{60}F_6Mo_4N_6O_6S_{12}$, Bis(tris(N,N-diethyldithiocarbamato)oxomolybdenum(VI)) di-μ-fluoro-bis(difluorodioxomolybdate(VI)), 41B, 999; 42B, 777

$C_{30}H_{60}F_{12}N_6Os_2P_2S_{12} \cdot CH_2Cl_2$, Tris(N,N-diethyldithiocarbamato)-osmium(IV) hexafluorophosphate dichloromethane, 46B, 1000

$C_{30}H_{60}Mo_3N_6O_{4.5}S_{12} \cdot 0.5\ H_5O_2$, Diaquahydrogen μ-oxo-[bis(diethyldithiocarbamato)oxomolybdenum(V)][bis(diethyldithiocarbamato)oxomo-

lybdenum(IV)] bis{μ-oxo-bis[bis(diethyldithiocarbamato)oxomolybdenum(V)]}, 45B, 1095
$C_{30}H_{63}N_5O_2S_8Zn_2$, Tetra-n-butylammonium μ-acetato-O,O'-bis[bis(dimethyldithiocarbamato)zincate], 46B, 1001
$C_{31}H_{63}NNiO_3S_6$, Tetra-n-butylammonium tris(n-butylxanthogenato)nickelate(II), 43B, 1206
$C_{32}H_{28}S_8V$, Tetrakis(phenyldithioacetato)vanadium(IV), 40B, 854
$C_{32}H_{48}Au_2Br_4N_2S_4$, Dibromo-(N,N-di-n-butyldithiocarbamato)gold(III) - trans-stilbene (2:1), 45B, 1095
$C_{32}H_{56}O_8S_{16}Zn_4$, Zinc isopropylxanthate tetramer, 38B, 871
$C_{33}H_{50}N_4S_6Zn$, Tetra-n-butylammonium bis(benzothiazole-2-thiolato-N)-(dimethyldithiocarbamato-S,S')zincate, 46B, 996
$C_{33}H_{60}FeN_3S_6$, Tris(N,N-dibutyldithiocarbamato)iron(II) benzene solvate, 42B, 778
$C_{34}H_{32}N_2O_2PPdS_2$, Bis-(O-methyl phenylthiocarbamato)(triphenylphosphine)palladium(II), 37B, 639
$C_{34}H_{50}Au_2Br_4N_2S_4$, Dibromo-(N,N-di-n-butyldithiocarbamato)gold(III) - trans,trans-1,4-diphenyl-1,3-butadiene (2:1), 45B, 1095
$C_{34}H_{58}N_2P_2PtS_4$, Bis(N,N-diisobutyldithiocarbamato)bis(dimethylphenylphosphine)platinum(II), 44B, 860
$C_{35}H_{53}Cl_{15}N_6O_6Ru_2S_{12}$, Bis(tris(morpholyldithiocarbamato)ruthenium-(III)) penta-chloroform, 41B, 999
$C_{40}H_{35}NO_2P_2PdS_2$, Bis(triphenylphosphine)(N-ethoxycarbonyldithiocarbimato)palladium(II), 43B, 1207
$C_{40}H_{80}Mo_8N_8O_{19}S_{16}$, Bis(tetrakis(diethyldithiocarbamato)molybdenum-(V)) hexamolybdate, 44B, 861
$C_{42}H_{84}Ag_6N_6S_{12}$, Silver di-n-propyldithiocarbamate, 34B, 525
$C_{42}H_{84}Cu_6N_6O_6S_6$, Copper(I) dipropylthiocarbamate hexamer, 35B, 828
$C_{45}H_{42}FeN_3S_6$, Tris(N,N-dibenzyldithiocarbamato)iron(III), 45B, 1097
$C_{49}H_{70}BN_5Os_2S_{12}$, Tris(N,N-diethyldithiocarbamato)bis[μ-(N,N-diethyltrithiocarbamato)]-diosmium(III) tetraphenylborate, 45B, 1097
$C_{50}H_{60}BMoN_5S_6$, N-Ethyl-N-phenylhydrazido(2-)-N'-tris(N,N-pentamethylenedithiocarbamato)molybdenum(VI) tetraphenylborate, 41B, 1001
$C_{50}H_{85}ClN_5Ru_2S_{10}$, Pentakis(diisopropyldithiocarbamato)diruthenium-(III) chloride benzene solvate, 41B, 1000
$C_{52}H_{40}As_2N_4NiS_4$, Tetraphenylarsonium bis(N-cyanodithiocarbimato)nickelate(II), 33B, 498
$C_{54}H_{108}Br_6Cd_2Cu_3N_6S_{12}$, Bis(bis(N,N-di-n-butyldithiocarbamato)copper-(III))bis(N,N-di-n-butyldithiocarbamato)copper(II) tetrabromo-di-μ-bromo-cadmate(II), 39B, 838
$C_{54}H_{108}Br_6Cu_3Hg_2N_6S_{12}$, Bis[bis(N,N-di-n-butyldithiocarbamato)copper-(III)]bis(N,N-di-n-butyldithiocarbamato)copper(II) tetrabromo-di-μ-bromo-dimercurate(II), 45B, 1098
$C_{72}H_{142}Cl_{12}N_{10}Ru_6S_{20}$, Bis(pentakis(diisopropyldithiocarbamato)diruthenium(III)) hexachlorodiruthenate(II) chloroform solvate, 41B, 1000

81. METAL COMPLEXES (CARBOXYLIC ACID)

CH_4O_6U, Aqua-formato-hydroxy-dioxouranium(VI), 45B, 1099
$CH_{17}CoN_8O_8$, Pentamminecarbamatocobalt(III) nitrate, 45B, 1099
$C_2Ag_2O_4$, Silver oxalate, 9, 319
$C_2CuD_2O_4 \cdot 4\ D_2O$, Copper formate tetradeuterohydrate (low temperature phase), 45B, 1099
$C_2CuD_2O_4 \cdot 4\ D_2O$, Copper formate tetradeuterohydrate, 45B, 1099
$C_2CuO_4 \cdot n\ H_2O$, Copper(II) oxalate hydrates, 33B, 402

$C_2K_2MoO_9$, Potassium oxodiperoxooxalatomolybdate(VI), 35B, 635
$C_2H_2Cd_2O_4$, Cadmium formate, 46B, 1001
$C_2H_2CuO_4$, Copper(II) formate, 26, 547
$C_2H_2K_3O_{10}V$, Potassium bisperoxo(oxalato)oxovanadate(V) monohydrate, 41B, 1003
$C_2H_3As_3Cu_2O_8$, Copper arsenate acetate, 43B, 1208
$C_2H_3CuNO_4$, catena-μ-Oxalato-amminecopper(II), 38B, 965
$C_2H_4CoO_6$, Cobalt(II) oxalate dihydrate, 39B, 696
$C_2H_4FeO_3S$, Iron(II) thioglycolate hydrate, 38B, 872
$C_2H_4FeO_6$, Iron(II) oxalate dihydrate, 21, 505; 23, 537
$C_2H_4MnO_6$, Manganese(II) oxalate dihydrate, 39B, 696
$C_2H_4NiO_6$, Nickel oxalate dihydrate, 39B, 696
$C_2H_4O_6Zn$, Zinc oxalate dihydrate, 39B, 696
$C_2H_4O_7U$, Uranyl diformate monohydrate, 43B, 1209
$C_2H_6CdO_6$, Cadmium(II) formate dihydrate, 40B, 855
$C_2H_6CoO_6$, Cobalt formate dihydrate, 32B, 414
$C_2H_6CuN_2O_4$, Oxalato-diamminecopper(II) (α-form), 43B, 1209
$C_2H_6CuO_6$, Copper(II) formate dihydrate, 30B, 346; 33B, 402
$C_2H_6Cu_{0.5}O_6Zn_{0.5}$, Copper zinc formate dihydrate, 43B, 1214
$C_2H_6Fe_2O_6$, Iron(II) formate dihydrate, 46B, 1002
$C_2H_6MnO_6$, Manganese(II) formate dihydrate, 29, 455; 33B, 402
$C_2H_6NiO_6$, Nickel formate dihydrate, 28, 628
$C_2H_6O_6Zn$, Zinc formate dihydrate, 43B, 1207
$C_2H_6O_9U$, Uranyl oxalate trihydrate, 38B, 873
$C_2H_8FN_2O_6V$, Ammonium dioxofluorooxalatovanadate, 42B, 778
$C_2H_8MoNNaO_9$, Sodium ammonium oxalatotrioxomolybdate(VI) dihydrate, 28, 629
$C_2H_9BrCuN_2O_2$, catena-μ-Acetato-bromodiamminecopper(II), 38B, 966
$C_2H_{10}Cl_6N_2O_4Re_2$, Ammonium hexachlorodirhenate diformate, 40B, 855
$C_2H_{10}CuN_2O_6$, β-Oxalatodiamminecopper(II) dihydrate, 38B, 967
$C_2H_{10}CuO_8$, Copper(II) formate tetrahydrate, 18, 639; 40B, 1115; 41B, 1002
$C_2H_{14}CoN_8O_4$, Bis(hydrazine)bis(hydrazinecarboxylato)cobalt(II), 41B, 1004
$C_2H_{18}Cl_2CoN_5O_6$, Acetatopentaamminecobalt(III) chloride perchlorate, 30B, 344
$C_2H_{29}Br_3Co_2N_6O_7$, Hexaammine-μ-acetato-di-μ-hydroxo-dicobalt tribromide trihydrate, 43B, 1210
$C_3H_3Cl_3O_6Re_2$, Trichlorotris(μ-formato)dirhenium, 46B, 1003
$C_3H_3GdO_6$, Gadolinium formate, 9, 313
$C_3H_3O_6Sc$, Scandium formate, 33B, 402
$C_3H_4CdO_5$, Cadmium(II) malonate monohydrate, 40B, 856
$C_3H_5NaO_9U$, Sodium uranyl triformate monohydrate, 43B, 1211
$C_3H_6MnO_6$, Diaquo-malonato-manganese(II), 45B, 1100
$C_3H_7O_7Sc$, Scandium hydroxide malonate dihydrate, 39B, 696
$C_3H_7O_8Y$, Yttrium formate dihydrate, 43B, 4
$C_4Ag_2F_6O_4$, Silver trifluoroacetate dimer, 38B, 24
$C_4F_6Hg_2O_4$, Mercury(I) trifluoroacetate, 40B, 857
$C_4O_8PtRb_{1.67} \cdot 1.5\ H_2O$, Rubidium bis(oxalato)platinate(II) bis(oxalato)platinate(III) hydrate, 45B, 1103
$C_4H_4Cl_2KO_8Ru_2$, Potassium dichloro(tetraformato)diruthenate(Ru-Ru)(-1), 45B, 1106
$C_4H_4Cs_2CuO_{10}$, Caesium copper oxalate dihydrate, 40B, 22
$C_4H_4CuK_2O_{10}$, Potassium bisoxalatocuprate(II) dihydrate, 27, 817; 40B, 22
$C_4H_4CuNa_2O_{10}$, Disodium catena-bis(μ-oxalato)-cuprate(II) dihydrate, 46B, 1003

$C_4H_4CuO_5$, Copper(II) maleate hydrate, 37B, 500
$C_4H_4CuO_5 \cdot 0.5\ H_2O$, Copper(II) oxydiacetate hemihydrate, 41B, 1005
$C_4H_4K_2MnO_{10}$, Potassium manganese oxalate dihydrate, 40B, 23
$C_4H_4K_2Mo_2O_{15}$, Potassium μ-oxo-bis(aquodioxooxalatomolybdate(VI)), 29, 621
$C_4H_4K_2O_{10}Pt$, Potassium dioxalatoplatinate(II) dihydrate, 29, 586
$C_4H_4Mo_2O_8$, Dimolybdenum tetraformate, 42B, 778
$C_4H_4O_7U$, Oxydiacetatodioxouranium(VI), 40B, 857
$C_4H_4O_{10}SrU \cdot (1+x)\ H_2O$, Strontium uranyl tetraformate hydrate, 44B, 862
$C_4H_6CdO_5S$, Cadmium(II) thiodiacetate monohydrate, 41B, 1006
$C_4H_6CdO_6$, Diaquo-maleato-cadmium(II) (Cc), 45B, 1101
$C_4H_6CdO_6$, Diaquo-maleato-cadmium(II) ($P2_1/c$), 40B, 858
$C_4H_6CoO_4 \cdot 0.4\ H_2O$, Cobalt(II) acetate, 40B, 859
$C_4H_6CuO_6$, Bis(glycollato)copper(II), 33B, 403
$C_4H_6Cu_2O_4$, Copper(I) acetate, 40B, 859
$C_4H_6HgO_4$, Mercury(II) acetate, 39B, 697
$C_4H_6K_3O_{13}V$, Tripotassium bis(oxalato)dioxovanadate(V) trihydrate, 40B, 860
$C_4H_6NO_9Y$, Ammonium yttrium(III) oxalate hydrate, 32B, 415
$C_4H_6O_4Zn$, Zinc(II) acetate, 45B, 1101
$C_4H_6O_7U$, Aquo-dioxo(succinato)uranium(VI), 45B, 1102
$C_4H_6O_8U$, Bis(hydroxyacetato)dioxouranium(VI), 46B, 1004
$C_4H_7ErO_{11}$, Erbium oxalate trihydrate, 35B, 636
$C_4H_7O_{11}Y$, Yttrium oxalate trihydrate, 37B, 504; 41B, 1004
$C_4H_8CsNbO_{13}$, Caesium oxobisoxalatobisaquoniobate(V) dihydrate, 39B, 698
$C_4H_8Cs_2Mo_2O_{14}S_2$, Caesium di-μ-sulfido-bis(aquooxalatooxo-molybdate(V)) dihydrate, 44B, 863
$C_4H_8F_4O_8Zn$, Tetraaqua(2,2,3,3-tetrafluorosuccinato)zinc(II), 46B, 1005
$C_4H_8K_2O_{12}Pd$, Potassium bisoxalatopalladate(II) tetrahydrate, 31B, 406
$C_4H_8MnO_7 \cdot H_2O$, Diaquo(1-malato)manganese(II) hydrate, 46B, 1005
$C_4H_8N_2O_8Zn$, Bis(amidoxalato-O,O)-zinc dihydrate, 37B, 504
$C_4H_8N_2O_{10}U$, Ammonium uranyl dioxalate, 39B, 698
$C_4H_9ErO_8$, Hydroxyacetatooxyacetatoaquoerbium(III) monohydrate, 34B, 527
$C_4H_{10}BaMo_2O_{17}$, Barium bis-dioxomolybdenum(V) oxalate pentahydrate, 30B, 346
$C_4H_{10}CdO_6$, Cadmium acetate dihydrate, 38B, 874
$C_4H_{10}CdO_8$, Cadmium oxydiacetate trihydrate (monoclinic), 43B, 1212
$C_4H_{10}CdO_8$, Cadmium oxydiacetate trihydrate (orthorhombic), 43B, 1213
$C_4H_{10}CoO_8$, Cobalt(II) (-)-malate trihydrate, 38B, 875
$C_4H_{10}CrKO_{13}$, Potassium trans-dioxalatodiaquochromate(III) trihydrate, 15, 424
$C_4H_{10}CsO_{13}Ti$, Caesium triaquobis(oxalato)titanate(III) dihydrate, 43B, 1213
$C_4H_{10}CuO_9$, Copper(II) D-tartrate trihydrate, 37B, 501
$C_4H_{10}CuO_9$, Copper(II) meso-tartrate trihydrate, 37B, 501
$C_4H_{10}MnO_8$, Diaquabis(glycollato)manganese(II), 46B, 1140
$C_4H_{10}MnO_8$, Malatodiaquomanganese(II) hydrate, 42B, 779
$C_4H_{10}N_2O_{10}Ti$, Ammonium titanyl oxalate monohydrate, 40B, 861
$C_4H_{10}O_6Zn$, Zinc acetate dihydrate, 17, 623
$C_4H_{10}O_8U$, Uranyl acetate dihydrate, 42B, 779
$C_4H_{10}O_8Zn$, S-Malatodiaquozinc(II) hydrate, 42B, 780
$C_4H_{10}O_{11}Th$, Thorium formate trihydrate, 31B, 408

$C_4H_{10.6}Mg_{0.82}PtO_{13.3}$, Magnesium dioxalatoplatinate, 33B, 406
$C_4H_{12}ClNdO_8$, Neodymium thiodiacetate chloride tetrahydrate, 39B, 700
$C_4H_{12}Cl_4O_6Re_2$, Diacetatodiaquotetrachlorodirhenium(II), 35B, 637
$C_4H_{12}CoN_2O_{10}$, Bis(amidoxalato-O,O)-cobalt(II) tetrahydrate, 38B, 876
$C_4H_{12}Co_{0.83}O_{14}Pt$, Cobalt bis(oxalato)platinate hexahydrate, 44B, 864
$C_4H_{12}CuN_2O_4$, Diamminecopper(II) acetate, 28, 631
$C_4H_{12}CuN_2O_{10}$, Ammonium bisoxalatocuprate(II) dihydrate, 27, 817
$C_4H_{12}FNa_3O_{15}V$, Sodium oxofluorobis(oxalato)vanadate hexahydrate, 42B, 781
$C_4H_{12}Mo_2N_2O_{11}$, Ammonium dimolybdomalate hydrate, 31B, 407
$C_4H_{12}N_2O_{10}U$, Diammonium uranyl tetraformate, 44B, 863
$C_4H_{12}N_2O_{11}V$, Diammonium bis(oxalato)monaquooxovanadate(IV) monohydrate, 42B, 781
$C_4H_{12}O_8SZn$, Zinc(II) thiodiglycolate tetrahydrate, 41B, 1007
$C_4H_{14}CoO_8$, Cobalt acetate tetrahydrate, 17, 627
$C_4H_{14}CuN_4O_8$, Copper(II) formate - urea - water (1:2:2), 42B, 781
$C_4H_{14}MnO_8$, Manganese acetate tetrahydrate, 40B, 862; 43B, 1215
$C_4H_{14}NNbO_{14}$, Ammonium oxobisoxalatobisaquoniobate(V) trihydrate, 43B, 1211
$C_4H_{14}N_3NbO_{13}$, Ammonium diperoxodioxalatoniobate monohydrate, 37B, 507
$C_4H_{14}N_4O_4Zn$, Zinc dihydrazine diacetate, 30B, 365
$C_4H_{14}NiO_8$, Nickel acetate tetrahydrate, 17, 627; 37B, 506; 41B, 1019; 46B, 1006
$C_4H_{16}N_3O_{12}V$, Triammonium bis(oxalato)dioxovanadate(V) dihydrate, 37B, 505
$C_4H_{18}ClCoN_4O_8$, trans-Diacetatotetraamminecobalt(III) perchlorate, 40B, 956
$C_5H_{10}CuN_2O_5$, Oxalato(1,3-diamino-2-propanol)copper(II), 43B, 1215
$C_5H_{10}HgO_2S$, n-Propylmercaptomercury(II) acetate, 41B, 1007
$C_5H_{11}CuNO_6$, Dimethylammonium copper(II) formate, 39B, 701
$C_5H_{24}Cl_4CoN_8O_{20}$, μ-Pyrimidine-5-carboxylato(O,O')-di-μ-hydroxo-bis(triamminecobalt(III)) perchlorate, 43B, 1216
$C_6K_3MnO_{12} \cdot 3\ H_2O$, Potassium tris(oxalato)manganate(III) trihydrate, 46B, 1006
$C_6K_3MoNO_{13} \cdot 4\ H_2O$, Tripotassium nitrosotris(oxalato)molybdenum tetrahydrate, 46B, 1007
$C_6Nd_2O_{12} \cdot 10.5\ H_2O$, Neodymium(III) oxalate 10.5-hydrate, 35B, 639
$C_6H_4IrK_3O_{14}$, Potassium trisoxalatoiridate(III) dihydrate, 22, 582
$C_6H_4K_3O_{14}Rh$, Potassium trisoxalatorhodate(III) dihydrate, 22, 582
$C_6H_6Co_6O_{18}$, Hexa-μ-oxo-hexacobalt hexaformate, 42B, 782
$C_6H_6CrK_3O_{15}$, Potassium tris(oxalato)chromate(III) trihydrate, 16, 469; 44B, 864
$C_6H_6CrO_{15}Rb_3$, Rubidium trioxalatochromate(III) trihydrate, 16, 470
$C_6H_6Cs_2Mn_2O_{15}$, Dicaesium dimanganese trioxalate trihydrate, 38B, 33
$C_6H_6F_9LaO_9$, Diaquo-tris(trifluoroacetato)lanthanum(III) monohydrate, 43B, 1217
$C_6H_6FeK_3O_{15}$, Potassium trisoxalatoferrate(III) trihydrate, 22, 584
$C_6H_8Cu_2O_9$, Copper citrate dihydrate, 42B, 782
$C_6H_8K_2O_{20}U_2$, Potassium diuranyl trisoxalate tetrahydrate, 41B, 1008
$C_6H_8K_4N_2O_{12}PtS_2$, Potassium bis(oxalato)bis(thiocyanato)platinate(II) tetrahydrate, 41B, 1009
$C_6H_8N_2O_{16}U_2$, Ammonium diuranyl trioxalate, 39B, 699
$C_6H_9AgO_8U \cdot x\ H_2O$, Silver uranyl acetate, 4, 264, 287
$C_6H_9AmNaO_8$, Sodium americyl acetate, 13, 496; 18, 641
$C_6H_9EuO_9$, Tris(hydroxyacetato)europium(III), 37B, 508
$C_6H_9GdO_9 \cdot 0.5\ H_2O$, Tris(hydroxyacetato)gadolinium(III) hemihydrate,

34B, 528
$C_6H_9GdO_9$, Tris(hydroxyacetato)gadolinium(III), 38B, 877
$C_6H_9LaO_9$, Tris(hydroxyacetato)lanthanum(III), 38B, 877
$C_6H_9NaO_8U$, Sodium uranyl acetate, 3, 668, 717; 23, 532
$C_6H_{10}BaO_{13}U$, Barium dioxobis(malonato)uranium trihydrate, 46B, 1007
$C_6H_{10}Cr_3O_{14}$, Hexakis(μ-formato)bis(aquo)trichromium(Cr-Cr), 44B, 865
$C_6H_{10}CuO_4$, Copper propionate, 35B, 641
$C_6H_{10}CuO_6 \cdot 1.5\ H_2O$, Aquobis(DL-lactato)copper(II) hemihydrate, 33B, 403
$C_6H_{10}O_4Zn$, Zinc(II) propionate, 43B, 1218
$C_6H_{10}O_{13}SrU$, Strontium dioxobis(malonato)uranium trihydrate, 46B, 1007
$C_6H_{12}CdCl_2N_4O_4 \cdot 0.5\ CH_4O$, Dichlorobis(diamidomalonato)cadmium methanol solvate, 46B, 1008
$C_6H_{12}CuN_2O_4$, Malonato(1,3-diaminopropane)copper(II), 43B, 1219
$C_6H_{12}Cu_2N_4O_{10}$, Copper(II) formate monourea, 35B, 633
$C_6H_{12}HgO_2S$, n-Butylmercaptomercury(II) acetate, 41B, 1007
$C_6H_{12}KMnO_{12}$, Potassium trans-diaquobis(malonato)manganate(III) dihydrate, 43B, 1219
$C_6H_{12}MnN_6O_{10}$, Dinitratobis(diamidomalonato)manganese, 46B, 1008
$C_6H_{12}NNdO_9$, Neodymium nitrilotriacetate trihydrate, 31B, 409
$C_6H_{12}N_2O_4Pt$, cis-Diammine(1,1-cyclobutanedicarboxylato)platinum(II), 46B, 1009
$C_6H_{12}N_2O_{10}U \cdot H_2O$, Ammonium dioxobis(malonato)uranate(VI) monohydrate, 45B, 1104
$C_6H_{12}NiO_6S_2$, Nickel ethylenedithiodiacetate dihydrate, 38B, 878
$C_6H_{12}O_{18}Sc_2$, Tetraaquo-trisoxalato-discandium(III) dihydrate, 38B, 879
$C_6H_{12}O_{18}Yb_2$, Ytterbium oxalate hexahydrate, 39B, 701
$C_6H_{13}BaNpO_{10}$, Barium neptunyl(V) triacetate dihydrate, 43B, 1220
$C_6H_{13}ErO_{11}$, Tris(hydroxyacetato)erbium(III) dihydrate, 37B, 508
$C_6H_{13}GdO_8 \cdot 2\ H_2O$, Tris($\mu_2$-acetato)diaquagadolinium(III) dihydrate, 46B, 1009
$C_6H_{14}Ba_2CuO_{16}$, Dibarium cupric formate tetrahydrate, 22, 587
$C_6H_{14}CuO_8$, Diaquobis(methoxyacetato)copper(II), 33B, 403
$C_6H_{14}MnO_6$, Manganese(II) propionate dihydrate, 43B, 1221
$C_6H_{14}N_3NbO_{13}$, Ammonium oxotrioxalatoniobate monohydrate, 37B, 509
$C_6H_{14}NiO_8$, Diaquobis(methoxyacetato)nickel(II), 37B, 510
$C_6H_{15}Cl_5Mo_3O_{14}$, Tri(aquo)trichlorotris(acetato)oxotrimolybdenum chloride perchlorate, 45B, 1105
$C_6H_{15}CuN_3O_4$, μ-Formato-diethylenetriaminecopper(II) formate, 37B, 510
$C_6H_{16}CrN_3O_{14}$, Ammonium trioxalatochromate(III) dihydrate, 16, 466
$C_6H_{16}N_4O_{14}U$, Ammonium uranyl trioxalate, 39B, 700
$C_6H_{16}O_9Zn$, Zinc lactate trihydrate, 41B, 1010
$C_6H_{17}HoO_{10}$, Holmium triacetate tetrahydrate, 45B, 1105
$C_6H_{20}CoN_5O_8S_2$, (2-Amino-4-imino-2-methylpentanoato)triamminecobalt(III) dithionate, 41B, 1011
$C_6H_{21}CuN_3O_8$, Oxalatodiethylenetriaminecopper(II) tetrahydrate, 40B, 1115
$C_6H_{21}Nd_2O_{23}$, Neodymium oxalate hydrate, 39B, 703
$C_7H_5NO_7U$, Pyridine-2,6-dicarboxylatodioxouranium(VI) monohydrate, 41B, 1012
$C_7H_6AgNO_2$, Silver(I) p-aminobenzoate, 46B, 1010
$C_7H_{10}FNO_7V$, Pyridinium oxofluoro-oxalatodiaquovanadate(IV), 41B, 1013
$C_7H_{15}Cs_2Mo_3O_{25}$, Caesium μ_3-oxo-μ-oxo-tris(aquooxalatomolybdate(IV))

tetrahydrate hemi(oxalic acid), 44B, 867
$C_7H_{17}LaN_2O_8S$, Tris(acetato)diaquolanthanum(III) thiourea, 46B, 1011
$C_7H_{18}CuN_2O_{12}$, 1,3-Diammonium-2-propanol bis(oxalato)cuprate(II) trihydrate, 43B, 1222
$C_7H_{24}Cl_2CoN_5O_{13}$, (RS,RS)-Tetraammine(4,5-dihydroxy-4,5-dimethyl-1-pyrroline-2-carboxylato)cobalt(III) perchlorate monohydrate, 41B, 1110
$C_8Ag_2F_{14}O_4$, Silver perfluorobutyrate, 20, 477
$C_8F_{12}Mo_2O_8$, Molybdenum(II) trifluoroacetate dimer, 38B, 880
$C_8H_4Hg_2O_4$, Mercury(I) o-phthalate, 32B, 417
$C_8H_6CuO_5$, Copper(II) phthalate monohydrate, 37B, 513
$C_8H_6HfNa_4O_{19}$, Sodium tetrakisoxalatohafnate trihydrate, 26, 691
$C_8H_6K_4O_{21}Re_2$, Tetrapotassium μ,μ'-dioxo-tetraoxalato-dirhenate(IV) trihydrate, 41B, 1014
$C_8H_6Na_4O_{19}Zr$, Sodium tetrakisoxalatozirconate(IV) trihydrate, 26, 691; 28, 632
$C_8H_8CuO_6$, o-Phthalatocopper(II) dihydrate, 44B, 867
$C_8H_8K_4O_{20}Th$, Tetrapotassium tetraoxalatothorate(IV) tetrahydrate, 41B, 1014
$C_8H_{10}HfK_4O_{21}$, Potassium tetrakis(oxalato)hafnate pentahydrate, 43B, 1223
$C_8H_{10}K_4O_{21}Zr$, Tetrapotassium tetrakis(oxalato)zirconate(IV) pentahydrate, 44B, 868
$C_8H_{11}CuNO_4$, Acetato-(2-pyridylmethanolato)copper(II) monohydrate, 41B, 1016
$C_8H_{12}ClO_8Ru_2 \cdot 2\ H_2O$, catena(μ-Chloro)(tetraacetato)diruthenium(Ru-Ru) dihydrate, 45B, 1106
$C_8H_{12}ClO_8Ru_2$, Tetrakis(μ-acetato)diruthenium chloride (C2/c), 46B, 1013
$C_8H_{12}ClO_8Ru_2$, Tetrakis(μ-acetato)diruthenium chloride (I2/m), 46B, 1013
$C_8H_{12}Cl_2CsO_8Ru_2$, Cesium dichloro(tetraacetato)diruthenate(Ru-Ru)(-1), 45B, 1106
$C_8H_{12}Cl_2O_8Re_2$, Tetrakis(μ_2-acetato)dichlorodirhenium, 46B, 1012
$C_8H_{12}CrMoO_8$, Chromium(II) molybdenum(II) tetraacetate, 42B, 783
$C_8H_{12}Cu_2O_{10}$, Copper(II) formate dioxan complex, 31B, 406
$C_8H_{12}Mo_2O_8$, Dimolybdenum tetraacetate, 40B, 864
$C_8H_{12}O_8U$, Uranium(IV) acetate, 29, 458
$C_8H_{14}CuO_{12}$, Bis(hydrogen malato)copper(II) dihydrate, 44B, 868
$C_8H_{14}CuO_{12}$, Copper(II) hydrogen maleate tetrahydrate, 37B, 500
$C_8H_{14}MnO_{12}$, Tetraaquobis(hydrogenmaleato)manganese(II), 44B, 869
$C_8H_{14}NiO_{12} \cdot 2\ H_2O$, Diaqua-bis(hydrogen malate)nickel(II) dihydrate, 46B, 1013
$C_8H_{16}BF_4O_{10}Ru_2$, Diaquo(tetraacetato)diruthenium(Ru-Ru)(+1) tetrafluoroborate, 45B, 1106
$C_8H_{16}Cr_2O_{10}$, Tetrakis(μ-acetato)bis(aquochromium), 17, 624; 37B, 514; 46B, 1014
$C_8H_{16}Cr_2O_{12}$, Chromium(II) succinate dihydrate, 39B, 703
$C_8H_{16}Cu_2O_{10}$, Copper(II) acetate dihydrate, 17, 625; 39B, 704
$C_8H_{16}Cu_2O_{12}$, Copper(II) succinate dihydrate, 31B, 408
$C_8H_{16}K_4O_{32}V_4$, Potassium di-μ-aquo-tetrakis(oxalato)tetra-μ-oxo-tetraoxotetravanadate hexahydrate, 42B, 784
$C_8H_{16}N_2O_4Pt$, (Malonato)(2,2-dimethyl-1,3-diaminopropane)platinum-(II), 46B, 1015
$C_8H_{16}O_{10}Rh_2$, Dirhodium tetra-acetate dihydrate, 37B, 514
$C_8H_{16}O_{14}U$, Diaquatetra(glycolato-O,O')uranium(IV), 46B, 1015
$C_8H_{18}Cl_2N_2O_2Zn$, Dichlorobis-(N,N'-dimethylacetamide)zinc(II), 40B,

865
$C_8H_{18}Cr_2O_8$, Tetraacetatodichromium, 43B, 1224
$C_8H_{18}CuO_8$, Diaquobis(ethoxyacetato)copper(II), 37B, 512
$C_8H_{18}CuO_8$, Diaquobis(2-hydroxy-2-methylpropionato)copper(II), 33B, 403
$C_8H_{22}Cd_2O_{17}$, Cadmium oxydiacetate hydrate, 43B, 1225
$C_8H_{24}Cu_2Na_4O_{22}$, Sodium dicopper(II)-DL-tartrate decahydrate, 38B, 881
$C_8H_{24}Mo_4N_4O_{22}$, Ammonium μ-oxo-bis((μ_3-malato)-di-μ-oxo-bis(dioxomolybdate(VI))) monohydrate, 43B, 1226
$C_8H_{24}N_4O_{16}V_2$, Ammonium vanadyl (+)-tartrate monohydrate, 32B, 416
$C_8H_{28}Na_4O_{26}V_2$, Tetrasodium divanadyl(IV) D-tartrate L-tartrate dodecahydrate, 33B, 409
$C_8H_{30}F_3N_{15}O_9Th$, Guanidinium tricarbonatotrifluorothorate(IV), 45B, 1107
$C_8H_{32}FeNa_5O_{26}$, Sodium iron(III) tartrate tetradecahydrate, 41B, 1015
$C_8H_{38}Mo_4N_4O_{27}$, Ammonium dimolybdomalate, 33B, 410
$C_9H_{10}K_3MnO_{14}$, Potassium tris(malonato)manganate(III) dihydrate, 43B, 1219
$C_9H_{11}CuNO_4$, Pyridinecopper(II) acetate, 26, 776
$C_9H_{11}HgNO_4$, Pyridinemercury(II) acetate, 42B, 901
$C_9H_{18}Nd_2O_{18}$, Neodymium malonate hexahydrate, 39B, 705
$C_9H_{20}CeN_3O_9$, Guanidinium tetraacetatocerate monohydrate, 37B, 515
$C_9H_{22}Eu_2O_{20}$, Europium(III) malonate octahydrate, 39B, 706
$C_9H_{22}Nd_2O_{20}$, Neodymium malonate octahydrate, 39B, 705
$C_{10}H_6O_{10}Os_2$, Bis-(μ-acetato)hexacarbonyldiosmium, 37B, 516
$C_{10}H_{12}N_4O_4Zn \cdot 0.5\ H_2O$, Bis(2-methylimidazole)-catena-μ-oxalatozinc(II) hemihydrate, 45B, 1107
$C_{10}H_{13}LiO_{10}U \cdot 4\ H_2O$, Lithium glutarato(hydrogenglutarato)dioxouranate(VI) tetrahydrate, 45B, 1108
$C_{10}H_{14}CuN_4O_4$, Bis(imidazole)copper(II) diacetate, 43B, 1226
$C_{10}H_{16}Cl_4Cu_2N_4O_{10}$, Copper chloroacetate - urea, 40B, 866
$C_{10}H_{18}Cl_4O_4Re_2$, Tetrachlorobis-μ-(2,2-dimethylpropionato)dirhenium, 45B, 1109
$C_{10}H_{18}CrKO_8$, Potassium bis(2-hydroxy-2-methylbutyrato)oxochromate(V) monohydrate, 44B, 869
$C_{10}H_{20}Cl_4N_2O_6Re_2$, Di-$\mu$-acetato-bis(dichloro-dimethylformamide-rhenium(III)), 45B, 1109
$C_{10}H_{20}CuO_{10}$, Bis(methanol)tetraacetatodicopper(II), 45B, 1112
$C_{10}H_{20}K_6O_{34}U_2$, Potassium uranyl oxalate hydrate, 42B, 784
$C_{10}H_{24}CuN_2O_{10}$, Bis(2-hydroxyethyliminopyruvato)copper(II) tetrahydrate, 34B, 529
$C_{10}H_{26}Cl_4CuN_4O_6$, Bis(1,3-diamino-2-propanol)(chloroacetic acid)copper(II) chloroacetic acid dichloride, 43B, 1228
$C_{12}H_{11}AgN_2O_5$, Silver(I) pyridine-2-carboxylate hydrate, 43B, 1229
$C_{12}H_{11}CuN_3O_5 \cdot H_2O$, Aqua(2,2'-bipyridine)[nitroacetato(2-)]copper(II) hydrate, 46B, 1063
$C_{12}H_{12}CuN_2O_4 \cdot H_2O$, catena-$\mu$-Formato-(formatobis(pyridine)copper(II)) monohydrate, 45B, 1110, 1111
$C_{12}H_{12}EuN_3O_{11}$, Bis(isonicotinato)tetraaquoeuropium(III) nitrate, 40B, 866
$C_{12}H_{12}F_{12}O_{10}Rh_2S_2$, Bis(dimethyl sulfoxide)tetrakis(trifluoroacetato)dirhodium(II), 46B, 1022
$C_{12}H_{12}F_{18}O_{18}Pr_2$, Tetrakis-$\mu$-trifluoroacetato-bis(triaquo(trifluoroacetato)praseodymium(III)), 44B, 870
$C_{12}H_{16}Cr_2N_2O_8$, Tetrakis(μ-acetato)(pyrazine)dichromium(Cr-Cr), 46B, 1022

$C_{12}H_{16}CuN_2O_6$, Diaqua-bis(formato)-bis(pyridine)copper, 45B, 1111
$C_{12}H_{16}Cu_2N_2O_8$, Pyrazine copper acetate, 41B, 1017
$C_{12}H_{17}CuNO_4$, Tetrakis-μ-propionato-bis(γ-picoline)copper(II), 44B, 871
$C_{12}H_{18}O_4S_2Zn$, Bis(1-carbethoxypropan-2-thionato)zinc, 41B, 1018
$C_{12}H_{18}O_{13}Zn_4$, Zinc oxyacetate, 18, 640
$C_{12}H_{19}Mo_2NaO_{12}$, Tetrakis(μ-acetato)dimolybdenum(II) sodium acetate acetic acid, 46B, 1016
$C_{12}H_{20}ClO_8Ru_2$, catena(μ-Chloro)(tetrapropionato)diruthenium(Ru-Ru), 45B, 1106
$C_{12}H_{20}Cr_2Na_4O_{23}$, Sodium di-μ-hydroxo-bis(bis(malonato)chromate(III)) pentahydrate, 43B, 1232
$C_{12}H_{20}Cr_2O_{12}$, Tetra-μ-acetato-dichromium - bis(acetic acid), 44B, 871
$C_{12}H_{20}F_6Nd_2O_{16}$, Tetra-μ-fluoroacetato-bis(diaquofluoroacetato-neodymium(III)), 43B, 1217
$C_{12}H_{20}UO_{12}$, Bis(acetic acid)tetraacetatodicopper(II), 45B, 1112
$C_{12}H_{21}Cl_2O_{10}Re_3$, Tris-μ-i-butyrato-bis(chlororhenium(III)) perrhenate, 40B, 1116
$C_{12}H_{22}CuO_{10}S_2$, Diaquabis[3,3'-(thiopropionatopropionic acid)]copper-(II), 46B, 1017
$C_{12}H_{24}B_2F_8NiO_{12}$, Hexakis(acetic acid)nickel(II) tetrafluoroborate, 41B, 1019
$C_{12}H_{24}CeCl_2Na_5O_{29}$, Sodium tris(oxydiacetato)cerate(III) perchlorate hydrate, 43B, 1230
$C_{12}H_{24}Cl_2GdNa_5O_{29}$, Trisodium tris(oxydiacetato)gadolinate di(sodium perchlorate) hydrate, 33B, 412
$C_{12}H_{24}Cl_2Na_5NdO_{29}$, Trisodium tris(oxydiacetato)neodymate di(sodium perchlorate) hexahydrate, 33B, 412; 35B, 642
$C_{12}H_{24}Cl_2Na_5O_{29}Yb$, Trisodium tris(oxydiacetato)ytterbate di(sodium perchlorate) hexahydrate, 33B, 412; 35B, 642
$C_{12}H_{24}Cl_6Cr_3NO_{22}$, Triaquooxohexakis(chloroacetato)trichromium(III) nitrate trihydrate, 43B, 1231
$C_{12}H_{24}Cl_7O_{23}V_3$, Triaquooxohexakis(chloroacetato)trivanadium(III) perchlorate trihydrate, 43B, 1231
$C_{12}H_{24}Er_2O_{21}$, Erbium tris(oxydiacetato)erbate hexahydrate, 43B, 1230
$C_{12}H_{24}N_2O_{10}V_2 \cdot 2\ H_2O$, N,N,N',N'-Tetramethylethylenediammonium aqua-bis(malonato)oxovanadate(IV) dihydrate, 46B, 1147
$C_{12}H_{24}O_{10}Rh_2S_2$, Tetrakis(μ-acetato)bis(dimethylsulfoxide)dirhodium-(II)(Rh-Rh), 46B, 1023
$C_{12}H_{25}CoIN_6O_2S_2$, Acetato-bis(acetone S-methylthiosemicarbazone)co-balt(II) iodide, 44B, 872
$C_{12}H_{26}ClO_{21}Rh_3$, trans-μ_3-Oxo-tris(bis(acetato)aquarhodium(III)) perchlorate monohydrate, 43B, 1234
$C_{12}H_{26}MnO_{16}$, Diaquabis(D-gluconato)manganese(II), 45B, 1112
$C_{12}H_{28}ClCoN_4O_8$, cis-anti-(Oxalatobis(2R,4S-diaminopentane)cobalt-(III) perchlorate, 44B, 872
$C_{12}H_{28}ClO_{22}Rh_3$, trans-μ_3-Oxo-tris(bis(acetato)aquarhodium(III)) perchlorate dihydrate, 43B, 1234
$C_{12}H_{30}CeNa_3O_{24}$, Trisodium tris(oxydiacetato)cerate(III) nonahydrate, 42B, 785
$C_{12}H_{30}ClCoN_4O_9$, (+)-Oxalatobis-(R,S-2,4-diaminopentane)cobalt(III) perchlorate monohydrate, 40B, 891
$C_{12}H_{30}Fe_3O_{24}$, Hexaaquoiron(II) bis(aquocitrato(3-)ferrate(II)) dihydrate, 43B, 1235
$C_{12}H_{30}Mn_3O_{24}$, Manganese(II) citrate decahydrate, 39B, 707
$C_{12}H_{30}Nd_2O_{24}$, Hexaaquo-fumarato-dimaleato-dineodymium(III) hexahyd-

rate, 41B, 1018
$C_{12}H_{31}Cl_2N_3O_{19}Pr_2$, Tetraaquobis(hydrogen-iminodiacetato)iminodiacetatodipraseodymium(III) dichloride trihydrate, 40B, 867
$C_{12}H_{34}Er_2O_{20}$, Erbium acetate tetrahydrate dimer, 38B, 881
$C_{12}H_{36}ClCr_3O_{22}$, trans-μ_3-Oxo-tris(bisacetatoaquochromium(III)) chloride hexahydrate, 35B, 738
$C_{12}H_{44}Cr_6O_{40}$, Tris(tetra-μ-formato-diaquodichromium(II)) decahydrate, 44B, 873
$C_{13}H_{19}CuNO_4$, Bis(propionato)-p-toluidinecopper(II), 39B, 707
$C_{14}H_8Cl_2O_4Zn$, Bis(2-chlorobenzoato)zinc(II), 42B, 786
$C_{14}H_{10}F_6HgN_2O_4$, Bis(pyridine)mercury(II) bis(trifluoroacetate), 44B, 873
$C_{14}H_{10}I_4O_4Re_2$, Tetraiododibenzoatodirhenium, 34B, 530
$C_{14}H_{12}Cr_2N_2O_{12}$, Di-$\mu$-hydroxo-bis(pyridine-2,6-dicarboxylatoaquochromium(III)), 43B, 1237
$C_{14}H_{12}Fe_2N_2O_{12}$, Di-$\mu$-hydroxo-bis(2,6-pyridinedicarboxylatoaquoiron(III)), 42B, 786
$C_{14}H_{12}N_2O_4Zn \cdot 1.5\ H_2O$, Bis(p-aminobenzoato)zinc(II) sesquihydrate, 46B, 1017
$C_{14}H_{14}CdN_2O_5 \cdot 2\ H_2O$, Aquo-bis(p-aminobenzoato)cadmium(II) dihydrate, 45B, 1114
$C_{14}H_{14}Cr_2N_2O_8$, Tetrakis(μ-formato)bis(pyridinechromium)(Cr-Cr), 44B, 865
$C_{14}H_{14}Cu_2N_2O_8$, Bis(formato)-pyridinecopper dimer, 45B, 1111
$C_{14}H_{14}O_4S_2Zn$, Zinc monothiobenzoate dihydrate, 39B, 708
$C_{14}H_{16}Cl_2NiO_8$, Tetraaquobis(m-chlorobenzoato)nickel(II), 44B, 874
$C_{14}H_{16}CuN_2O_6$, Bis-(2-pyridineacetato)copper(II) dihydrate, 41B, 1019
$C_{14}H_{16}CuO_7$, Copper(II) benzoate trihydrate, 28, 634
$C_{14}H_{16}K_2Mn_2O_{18}$, Potassium tetrakis(malonato)-bis(methanol)-di-manganese(III), 46B, 1018
$C_{14}H_{16}N_2NiO_6$, Bis(p-aminobenzoato)diaquonickel(II), 44B, 874
$C_{14}H_{18}CuN_2O_5$, Bis(β-picoline)bis(formato)aquocopper(II), 43B, 1237
$C_{14}H_{20}CoN_2O_8$, Tetraaqua-bis(p-aminobenzoato)cobalt(III), 45B, 1115
$C_{14}H_{20}Fe_2N_2O_{18}$, Di-$\mu$-hydroxo-bis(4-hydroxo-2,6-pyridinedicarboxylatoaquoiron(III)) tetrahydrate, 42B, 786
$C_{14}H_{20}MnN_2O_8$, Bis(p-aminobenzoato)tetraaquomanganese(II), 44B, 875
$C_{14}H_{24}F_6O_{23}S_2W_3$, Hexa-$\mu$-acetato-di-$\mu_3$-oxo-tris[aquatungsten(IV)] trifluoromethylsulphonate, 44B, 876; 46B, 1018
$C_{14}H_{28}Cr_2FeNO_{21}$, Hexakis(acetato)triaquooxodichromiumiron nitrate acetic acid, 43B, 1239
$C_{14}H_{37}F_3MoO_{14}P_4$, Hydrido(trifluoroacetato)tetrakis(trimethyl phosphite)molybdenum(II), 46B, 1019
$C_{15}H_{13}EuN_2O_7$, Aquo-tris(formato)-(1,10-phenanthroline)europium(III), 45B, 1116
$C_{15}H_{19}NbO_2$, Dicyclopentadienyl-pivaloyl-niobium, 45B, 918
$C_{15}H_{27}Cl_3O_6Re_2$, Trichlorotri-μ-(2,2-dimethylpropionato)dirhenium, 45B, 1109
$C_{16}H_{10}Cl_4CuO_6 \cdot 2.5\ C_4H_8O_2$, Bis(2,4-dichlorophenoxyacetato)copper(II) dioxane solvate, 45B, 1116
$C_{16}H_{12}CuK_2O_{10}$, Dipotassium catena-di-μ-(o-phthalato)-cuprate(II) dihydrate, 44B, 878
$C_{16}H_{12}CuMgO_{10}$, Magnesium di-o-phthalatocuprate(II) dihydrate, 44B, 877
$C_{16}H_{12}CuNa_2O_{10}$, Disodium di-o-phthalatocuprate(II) dihydrate, 44B, 878
$C_{16}H_{12}CuO_{10}Rb_2$, Dirubidium catena-di-μ-(o-phthalato)-cuprate(II) dihydrate, 43B, 1239

$C_{16}H_{14}CuO_6$, Copper(II) phenoxyacetate, 41B, 1021
$C_{16}H_{14}CuO_{10}$, Bis(hydrogen o-phthalato)diaquocopper(II), 34B, 535; 46B, 1019
$C_{16}H_{14}CuO_{11}Sr$, Strontium di-o-phthalatocuprate(II) trihydrate, 44B, 877
$C_{16}H_{14}MnO_{10}$, Manganese dihydrogen diphthalate dihydrate, 44B, 879
$C_{16}H_{16}BaCuO_{12}$, Barium diaquodi(o-phthalato)cuprate(II) dihydrate, 44B, 879
$C_{16}H_{16}Cl_2CoO_8$, Diaquabis(p-chlorophenoxyacetato)cobalt(II), 46B, 1020
$C_{16}H_{16}Cl_2MnO_8$, Diaquabis(p-chlorophenoxyacetato)manganese(II), 46B, 1020
$C_{16}H_{16}CuLi_2O_{12}$, Dilithium catena-di-μ-(o-phthalato)-cuprate(II) tetrahydrate, 43B, 1239
$C_{16}H_{18}CoO_8$, Diaquabis(phenoxyacetato)cobalt(II), 46B, 1020
$C_{16}H_{18}CuO_8$, Diaquobis(phenoxyacetato)copper(II), 33B, 403
$C_{16}H_{18}MnO_8$, Diaquabis(phenoxyacetato)manganese(II), 46B, 1020
$C_{16}H_{18}O_{45}K_8Ti_4$, Potassium tetra-$\mu$-oxo-tetrakis(bisoxalatotitanate) nonahydrate, 44B, 880
$C_{16}H_{20}Cr_2F_{12}O_{10}$, Tetrakis($\mu$-trifluoroacetato)bis(diethylether-chromium)(Cr-Cr), 44B, 865
$C_{16}H_{20}CuN_2O_5S_2$, Aquodipyridinecopper(II) dithiodipropionate, 42B, 788
$C_{16}H_{20}CuO_9$, Bis(phenoxyacetato)triaquocopper(II), 37B, 516
$C_{16}H_{22}CuN_2O_5 \cdot H_2O$, Aquo-bis(methylamine)copper(II) dibenzoate monohydrate, 45B, 1118
$C_{16}H_{22}NiO_{14}$, Nickel dihydrogen diphthalate hexahydrate, 42B, 788
$C_{16}H_{24}O_{16}Pt_4$, Cyclo-tetrakis(di-μ-acetato-platinum(II)) (monoclinic form), 44B, 881
$C_{16}H_{24}O_{16}Pt_4$, Cyclo-tetrakis(di-μ-acetato-platinum(II)) (tetragonal form), 44B, 880
$C_{16}H_{25}Mn_3O_{17}$, Manganese(III) acetate, 34B, 531
$C_{16}H_{26}N_2O_{18}Pt_4$, Cyclo-bis-($\mu$-acetato-$\mu$-nitrosyl)-bis(di-$\mu$-acetato-diplatinum(II)), 39B, 709
$C_{16}H_{28}ClO_8Ru_2$, Tetra-n-butyratodiruthenium chloride, 34B, 536
$C_{16}H_{28}CuN_4O_8 \cdot 2\ H_2O$, Diaquabis(2,2,5,5-tetramethyl-3-imidazoline-1-oxo-4-carboxylato)copper(II) dihydrate, 46B, 1021
$C_{16}H_{28}Cu_2O_8$, Copper butyrate dimer, 38B, 882
$C_{16}H_{28}Cu_2O_{10}$, Tetrakis-μ-propionato-dicopper(II)-μ-dioxane, 42B, 789
$C_{16}H_{28}N_4O_8Zn \cdot 2\ H_2O$, Diaquabis(2,2,5,5-tetramethyl-3-imidazoline-1-oxo-4-carboxylato)zinc(II) dihydrate, 46B, 1021
$C_{16}H_{28}O_8Rh_2S_2$, Tetrakis(μ-acetato)bis(tetrahydrothiophene)dirhodium-(II)(Rh-Rh), 46B, 1023
$C_{16}H_{28}O_{16}Re_4$, Tetra-μ-n-butyratodirhenium(III) diperrhenate, 35B, 643
$C_{16}H_{30}CuO_4$, Copper(II) octanoate, 40B, 940
$C_{16}H_{32}O_{10}Rh_2S_2$, Bis(dimethyl sulfoxide)tetrakis(propionato)dirhodium(II), 46B, 1022
$C_{16}H_{33}HgO_4P$, Diacetato(tris(t-butyl)phosphine)mercury(II), 44B, 1060
$C_{16}H_{34}N_2O_8Rh_2$, Bis(diethylamine)tetra-μ-acetatodirhodium(II) (Rh-Rh), 45B, 1119
$C_{16}H_{34}N_8O_{16}Pr_2 \cdot 2\ CH_4N_2O \cdot 2\ H_2O$, Bis($\mu$-acetato)-bis(di(urea)-di-(acetato)-praseodymium(III)) urea dihydrate, 45B, 1217
$C_{18}H_8Cl_{14}Cu_2N_2O_8$, Bis(2-chloropyridine)tetrakis(μ-trichloroacetato-(O,O'))dicopper(II), 44B, 881
$C_{18}H_{14}CuO_8$, Copper(II) aspirinate, 39B, 709
$C_{18}H_{16}ErN_3O_8$, Erbium isonicotinate dihydrate, 39B, 710

$C_{18}H_{16}LaN_3O_8$, Diaquotris(isonicotinato)lanthanum(III), 38B, 884
$C_{18}H_{16}N_3O_8Pr$, Praseodymium nicotinate dihydrate, 38B, 883
$C_{18}H_{20}O_7Zn$, Zinc o-ethoxybenzoate monohydrate, 40B, 867
$C_{18}H_{22}Cr_2N_2O_8$, Tetrakis(μ-acetato)bis(pyridine)dichromium(Cr-Cr), 46B, 1022
$C_{18}H_{22}CuO_{10}$, Copper(II) p-methoxyphenoxyacetate dihydrate, 41B, 1022
$C_{18}H_{22}Cu_2N_2O_8$, Bis(pyridine)tetraacetatodicopper(II), 26, 669; 29, 606
$C_{18}H_{22}K_2Mn_7O_{40}$, Potassium manganese formate hydrate polymer, 43B, 1242
$C_{18}H_{22}N_2O_8Rh_2$, Tetra-μ-acetato-bis(pyridinerhodium(II)), 44B, 882
$C_{18}H_{22}Nb_3O_{10}$, Hydrido-μ_3-oxo-tri-μ-hydroxo-tri-μ-formato-tri-triangulo-cyclopentadienylniobium(IV), 39B, 710
$C_{18}H_{24}CuI_2N_2O_{11}$, Copper(II) m-iodohippurate pentahydrate, 41B, 1121
$C_{18}H_{26}CuO_{12}$, Copper(II) p-methoxyphenoxyacetate tetrahydrate, 41B, 1022
$C_{18}H_{26}Fe_2N_4O_{14}$, Di-μ-hydroxo-bis(4-dimethylamino-2,6-pyridinedicarboxylatoaquoiron(III)) dihydrate, 44B, 882
$C_{18}H_{30}N_3O_9Pd_3$, Tris(μ-acetato-μ-acetoximatopalladium(II)) - 0.5 benzene, 37B, 517
$C_{18}H_{33}CsO_{23}W_3$, Caesium di-μ_3-oxo-hexakis(μ-acetato)triacetato-tritungstate(IV) trihydrate, 44B, 876
$C_{18}H_{34}Cr_2N_2O_8$, Tetra-μ-acetato-dichromium - bis(piperidine), 44B, 871
$C_{18}H_{36}B_2F_8O_{17}W_3 \cdot 5.5\ H_2O$, Di-$\mu_3$-oxo-hexakis(μ-n-propionato)triaquo-tritungsten(IV)(W-W) bis(tetrafluoroborate) - 5.5 water, 44B, 876
$C_{19}H_{15}F_6HgN_3O_4$, Tris(pyridine)mercury(II) bis(trifluoroacetate), 44B, 883
$C_{19}H_{21}CuN_3O_4$, Bispyridinebisacetatocopper(II) pyridine, 32B, 419
$C_{19}H_{45}O_3P_2Rh$, Bis(triisopropylphosphine)dihydridobicarbonatorhodium-(III), 45B, 1119
$C_{20}H_{20}La_2NaO_{14} \cdot 4\ H_2O$, Sodium diaquobis(benzene-1,2-dioxydiacetato)lanthanate(III) tetrahydrate, 45B, 1120
$C_{20}H_{23}Mo_2N_7O_8S_4$, Tripyridinium di-μ-oxo-μ-formato-bis(oxobis(isothiocyanato)molybdate(V)) dihydrate, 43B, 1227
$C_{20}H_{25}Cl_2O_4P_2Pd_2 \cdot 0.5\ CHCl_3$, Bis(μ-acetato)-dichlorobis(dimethylphenylphosphine)dipalladium(II) chloroform solvate, 45B, 1120
$C_{20}H_{26}Cu_2N_2O_8$, Tetrakis-μ-acetato-bis(2-methylpyridinecopper(II), 44B, 884
$C_{20}H_{28}N_6NiO_8$, Bis[(m-nitrobenzoato)(1,3-propanediamine)]nickel(II), 46B, 1023
$C_{20}H_{36}Br_2O_8Re_2$, Tetrakis(2,2-dimethylpropionato)dibromodirhenium-(III), 45B, 1122
$C_{20}H_{36}Cl_2O_8Re_2$, Tetrakis(2,2-dimethylpropionato)dichlorodirhenium-(III), 45B, 1122
$C_{20}H_{36}Cr_2O_8$, Tetrakis(μ-pivalato)dichromium(Cr-Cr), 44B, 865
$C_{20}H_{36}MoO_8W$, Tetrakis(t-butylcarboxylato)molybdenumtungsten(Mo-W), 44B, 884
$C_{20}H_{36}Mo_2O_8$, Dimolybdenum tetrapivalate, 44B, 886
$C_{20}H_{38}CuO_4$, Copper(II) decanoate, 40B, 868
$C_{20}H_{40}O_{10}Rh_2$, Tetrakis(μ-pivalato)diaquodirhodium(II)(Rh-Rh), 46B, 1023
$C_{21}H_{17}Cl_3O_9Zn_2$, μ_3-Hydroxo-tri-μ-(2-chlorobenzoato)-dizinc(II) dihydrate, 42B, 789
$C_{21}H_{21}Mo_2N_7O_6S_4$, Tripyridinium di-μ-oxo-μ-acetato-bis(oxobis(isothiocyanato)molybdate(V)), 43B, 1227
$C_{21}H_{37}Na_3O_{26}Yb$, Trisodium tris(pyridine-2,6-dicarboxylato)ytter-

bate(III)-14-hydrate, 35B, 644
$C_{22}H_{19}BCuF_4N_4O_2$, Acetatobis(2,2'-bipyridyl)copper(II) tetrafluoroborate, 46B, 1105
$C_{22}H_{19}ClCuN_4O_6 \cdot H_2O$, Acetatobis(2,2'-bipyridyl)copper(II) perchlorate monohydrate, 46B, 1105
$C_{22}H_{26}Cl_4Co_2N_2O_8$, Dichloro(diethyl pyridine-2,6-dicarboxylato)cobalt(II) dimer, 43B, 1244
$C_{22}H_{26}N_4O_8W_2$, Dimethyltetrakis(N,N-diethylcarbamato)ditungsten, 43B, 999
$C_{22}H_{30}Cu_2N_2O_8$, Bis(propionato)pyridinecopper(II) dimer, 42B, 917
$C_{22}H_{34}N_4NiO_4$, Bis[(m-methylbenzoato)(1,3-propanediamine)]nickel(II), 46B, 1024
$C_{22}H_{34}N_4NiO_4$, Bis[(p-methylbenzoato)(1,3-propanediamine)]nickel(II), 46B, 1024
$C_{22}H_{39}IMoNO_8W$, Tetra-μ-pivalato-(iodotungsten)molybdenum acetonitrile solvate, 41B, 1023
$C_{24}H_{20}CuF_6N_4O_4$, Tetrakis(pyridine)bis(trifluoroacetato)copper(II), 44B, 830; 45B, 1122
$C_{24}H_{23}CrHoN_9O_{10}S_6$, Diaquotri(nicotinic acid)holmium(III) hexa(isothiocyanato)chromate(III) dihydrate, 38B, 884
$C_{24}H_{34}Cu_2N_2O_8$, Tetrakis-μ-propionato-bis(β-picoline)dicopper(II), 42B, 790
$C_{24}H_{36}Cl_{12}O_{16}Pd_4$, Palladium(II) tert-butyl peroxide trichloroacetate, 46B, 1025
$C_{24}H_{38}CdN_4O_8$, Diacetatodiaquabis(diethylnicotinamide)cadmium, 46B, 1025
$C_{24}H_{48}Cl_4Cu_4N_4O_{12}$, Tetrakis[chloroacetato-μ-(2-dimethyl-aminoethanolato)copper(II)], 45B, 1123
$C_{24}H_{56}Cl_6Mn_3N_8O_{40} \cdot 2\ H_2O$, catena-Octakis-μ-(β-alanine)-trimanganese(II) hexaperchlorate dihydrate, 46B, 1026
$C_{25}H_{44}N_6O_{12}W_2$, Hexakis(N,N-dimethylcarbamato)ditungsten xylene, 43B, 999
$C_{26}H_{14}Cu_2F_{12}N_2O_8$, Tetra-μ-trifluoroacetato-bis(quinolinecopper(II)), 41B, 1024
$C_{26}H_{22}Cl_4Cu_2N_2O_8$, Tetrakis(μ-chloroacetato)bis(quinolinecopper(II)), 42B, 790
$C_{26}H_{22}Cu_2F_4N_2O_8$, Tetrakis(μ-monofluoroacetato)-bis(quinolinecopper-(II)), 44B, 885
$C_{26}H_{24}N_2NiO_4$, Bisbenzoatobis(2-methylpyridine)nickel(II), 43B, 1245
$C_{26}H_{26}CuN_2O_5$, Aquobis(benzoato)bis(γ-picoline)copper(II), 44B, 885; 45B, 1123
$C_{26}H_{26}Cu_2N_2O_8$, Tetrakis(μ-acetato)bis(quinolinecopper(II)), 33B, 412; 42B, 790
$C_{26}H_{30}Mo_2N_2O_6$, Diacetatobis(4-phenylimino-2-pentanonato)dimolybdenum, 45B, 1124
$C_{28}H_{20}Br_4Cu_2O_{10}$, Copper(II) o-bromobenzoate monohydrate, 38B, 886
$C_{28}H_{20}Cu_4O_8$, Tetrakis(copper(I) benzoate), 43B, 1246
$C_{28}H_{20}Mo_2O_8$, Dimolybdenum tetrabenzoate, 44B, 886
$C_{28}H_{24}Cl_4N_2O_6Re_2$, Tetrachlorobis($μ_2$-formato)bis(N,N-diphenylformamide-O)dirhenium, 46B, 1026
$C_{28}H_{24}Mo_2N_8O_{11}S_4$, Tetrapyridinium μ-oxo-bis(oxo(oxalato)bis(isothiocyanato)molybdate(V)), 43B, 1227
$C_{28}H_{24}O_4P_2Pd$, Bis(diphenylphosphinoacetato)palladium(II), 46B, 1027
$C_{28}H_{26}Mo_2N_{10}O_7S_6$, Tetrapyridinium μ-oxo-bis(oxo(formato)tris(isothiocyanato)molybdate(V)), 43B, 1227
$C_{28}H_{28}Cl_6N_4NiO_4$, Bis(trichloroacetato)tetrakis(aniline)nickel, 44B, 886

$C_{28}H_{52}Cr_2O_{14}$, Tetra-(μ-pivalato)-bis(tetrahydrofuranchromium), 45B, 965
$C_{30}H_{22}Cl_8O_8Re_2$, Chlororhenium(III)-μ-tetrabenzoato-chlororhenium-(III)-chloroform, 33B, 414
$C_{30}H_{40}N_2O_{22}Zn_7$, Deca-μ-acetato-dioxobis(pyridine)heptazinc(II), 45B, 1124
$C_{32}H_{24}N_2NiO_4$, Bisbenzoatobis(quinoline)nickel(II), 43B, 1245
$C_{32}H_{28}Cu_2O_{10} \cdot 2\ C_4H_8O_2$, Tetrakis(μ-benzoato)-dioxane-dicopper dioxane solvate, 45B, 1125
$C_{32}H_{36}Cl_6N_4NiO_4$, Bis(trichloroacetato)tetrakis(p-toluidine)nickel-(II), 44B, 887
$C_{32}H_{36}Cl_6N_4NiO_8$, Bis(trichloroacetato)tetrakis(p-anisidine)nickel-(II), 44B, 887
$C_{32}H_{38}Ag_2Cu_4O_{28}$, Silver(I) di-$\mu_3$-hydroxo-tetrakis-μ-phthalato-tetracopper(II) decahydrate, 45B, 1126
$C_{32}H_{56}Cl_{12}Cu_4N_4O_{12}$, 2-Diethylaminoethanolato(trichloroacetato)copper(II) tetramer, 45B, 1126
$C_{32}H_{56}Mn_{12}O_{48} \cdot 2\ C_2H_4O_2 \cdot 4\ H_2O$, Tetraaqua-dodecaoxo-hexadecaacetato-dodecamanganese acetic acid tetrahydrate, 46B, 1028
$C_{33}H_{66}ClFe_3O_{16}$, μ_3-Oxo-hexakis(μ-trimethylacetato)-trismethanoltriiron(III) chloride, 41B, 1025
$C_{34}H_{24}CuINO_6$, Bis(o-benzoylbenzoato)copper(II) - p-iodoaniline, 45B, 1127
$C_{34}H_{30}O_8Ti_2$, Tetra-μ-benzoato-di(η^5-cyclopentadienyltitanium), 42B, 791
$C_{34}H_{36}Cu_2N_4O_{10}$, Copper(II) tetrakis(phenylacetate) bisurea solvate, 43B, 1247
$C_{34}H_{47}F_6O_2P_5Ru$, Acetatotetrakis(dimethylphenylphosphine)ruthenium-(II) hexafluorophosphate, 42B, 792
$C_{36}H_{32}La_2N_6O_{16}$, Lanthanum nicotinate dihydrate dimer, 38B, 885
$C_{36}H_{32}N_6O_{16}Sm_2$, Samarium nicotinate dihydrate dimer, 38B, 885
$C_{36}H_{68}O_8P_2Ru_2$, Di-μ-(n-butyrato)-bis(dicarbonyl(tri-t-butylphosphine)ruthenium(I))(Ru-Ru), 43B, 1248
$C_{36}H_{76}Cu_6N_4O_{18} \cdot H_2O$, Bis[μ-acetato-(O,O')-diacetato-μ_3-(2-diethylaminoethanolato-N,μ_3-O)-μ-(2-diethylaminoethanolato-N,μ-O)-μ_3-hydroxo-tricopper(II)] hydrate B , 46B, 1028
$C_{38}H_{30}Cu_2N_2O_8$, Tetrakis(μ_2-benzoato)bis(pyridine)dicopper(II), 46B, 1029
$C_{38}H_{33}CuO_2P_2$, Acetatobis(triphenylphosphine)copper(I), 41B, 1025
$C_{38}H_{64}Br_2Mo_2O_4P_2$, Bis-μ-(benzoato)-1,2-dibromo-1,2-bis(tri-n-butylphosphine)dimolybdenum(II), 42B, 792
$C_{38}H_{64}Cl_4N_4O_{12}$, Tetrakis(chloroacetato)-tetra-μ-(2-diethylaminoethanolato)-tetracopper(II), 45B, 1128
$C_{39}H_{35}Cl_5O_3P_2Re_2$, Oxopentachloropropionatobis(triphenylphosphine)-dirhenium(IV), 33B, 416
$C_{40}H_{33}MnO_4P_2$, Acetatobis(triphenylphosphine)dicarbonylmanganese(I), 39B, 711
$C_{40}H_{33}O_4P_2Re$, Acetatobis(triphenylphosphine)dicarbonylrhenium(I), 42B, 793
$C_{40}H_{48}Mo_2O_{14}$, Tetrakis(benzoato)dimolybdenum(II) - bis(diglyme), 42B, 794
$C_{40}H_{54}N_2Ni_2O_8$, Tetrakis(μ-trimethylacetato)bis(quinaldine)dinickel, 46B, 1029
$C_{40}H_{76}Cu_2N_4O_{16}$, Tetrakis(μ-2,2,5,5-tetramethylpyrroline-oxy-3-carboxylato)-bis(ethanolcopper(II)), 44B, 888
$C_{40}H_{76}HgO_6P_2 \cdot 2\ H_2O$, Diacetatobis(tricyclohexylphosphine)mercury-(II) dihydrate, 45B, 1359

$C_{40}H_{84}N_4O_{14}V_2 \cdot 8\ H_2O$, Tetraethylammonium vanadyl(IV) tartrate octahydrate, 46B, 1031

$C_{42}H_{32}Cr_2O_{12}$, Tetrakis(μ-benzoato)bis(benzoic acid chromium)(Cr-Cr), 44B, 865

$C_{42}H_{40}Cl_3O_5P_2Re_2$, μ-Oxo-μ-chloro-di-μ-propionato-bis(chlorotriphenylphosphinerhenium), 34B, 532

$C_{42}H_{40}N_{10}Ni_2O_8$, μ-Oxalato-bis(nitrito-mer-tris(pyridine)nickel(II)) - pyridine (1:2), 43B, 1248

$C_{42}H_{42}F_6O_3P_4Ru$, Acetatocarbonyltris(methyldiphenylphosphine)ruthenium(II) hexafluorophosphate, 44B, 889

$C_{42}H_{46}O_{14}U$, Bis(anthracene-9-carboxylate)diaquodioxouranium(VI) dioxane solvate, 44B, 889

$C_{42}H_{56}NNaO_9V$, Sodium tetraethylammonium bis(benzilato)oxovanadium-(IV) - 2-propanol, 34B, 532

$C_{44}H_{35}ClO_3P_2Ru$, Benzoatochlorocarbonylbis(triphenylphosphine)ruthenium(II), 44B, 890

$C_{44}H_{42}O_{14}P_2U_2$, trans-Di-μ-acetato-bis(dioxotriphenylphosphineoxide-acetato)diuranium(VI), 34B, 533

$C_{44}H_{56}Mo_2N_4O_4$, Tetrakis(N-phenylpivalamido)dimolybdenum(II)(Mo-Mo), 46B, 1031

$C_{44}H_{56}Mo_2N_4O_6$, Tetrakis[N-(2,6-dimethylphenyl)formamido]bis(tetrahydrofuran)dimolybdenum(II)(Mo-Mo), 46B, 1031

$C_{44}H_{78}Hg_2O_8P_2$, Bis(μ-acetato-O,O)bis[acetato(tricyclohexylphosphine)mercury(II)], 45B, 1128

$C_{45}H_{84}O_{21}W_3$, Di-μ_3-oxo-hexakis(μ-pivalato)dipivalatoaquotritungsten-(IV)-pivalic acid, 44B, 876

$C_{46}H_{34}Co_2N_2O_8$, Tetra-μ-benzoato-bisquinolinedicobalt(II), 43B, 1249

$C_{46}H_{34}Cu_2N_2O_8$, Tetra-μ-benzoato-bis(quinolinecopper(II)), 45B, 1129

$C_{46}H_{54}Cu_2N_2O_8$, Tetrakis(μ-trimethylacetato)bis(acridine)dicopper, 46B, 1029

$C_{46}H_{62}O_4P_2Pd_2$, sym-trans-Di-μ-acetatobis(o-(t-butyl-o-tolylphosphino)benzyl)dipalladium(II) cyclohexane solvate, 40B, 868

$C_{48}H_{30}Cu_4F_{18}N_4O_{14}$, Bis(hydroxy-tris(trifluoroacetato)di(quinoline-copper(II))), 40B, 994

$C_{48}H_{38}Co_2N_2O_8$, Tetra-μ-benzoato-bis(4-methylquinoline)dicobalt(II), 43B, 1250

$C_{50}H_{54}Hg_2O_8P_2$, Bis(μ-acetato-O,O)bis[acetato(tri-o-tolylphosphine)-mercury(II)], 45B, 1128

$C_{52}H_{46}As_2Cl_4Mo_2O_4 \cdot 2\ CH_4O$, Tetraphenylarsonium trans-bis(μ-acetato)tetrachlorodimolybdate(II) methanol solvate, 45B, 1130

$C_{53}H_{55}AgIrN_3O_5P_2$, Isobutyrato-μ-N,N'-methyl-p-tolyltriazenido-silver(I)-carbonylbis(triphenylphosphine)iridium(I) - isobutyric acid, 42B, 794

$C_{55}H_{47}O_2P_3Ru$, Hydrido-formato-tris(triphenylphosphine)ruthenium(II), 39B, 712

$C_{56}H_{40}As_2CoF_{12}O_8$, Tetraphenylarsonium tetrakis(trifluoroacetato)cobaltate(II), 31B, 410

$C_{56}H_{49}O_2P_3Ru$, Acetatohydridotris(triphenylphosphine)ruthenium(II), 40B, 869

$C_{61}H_{50}O_2P_3Rh \cdot 0.5\ C_6H_6$, Benzoatotris(triphenylphosphine)rhodium(I) benzene solvate, 40B, 870

$C_{66}H_{63}O_{13}P_3Ru_3$, Ruthenium acetate tris(triphenylphosphine) adduct, 38B, 887

$C_{68}H_{56}Cr_2O_{12}$, Tetrakis(μ-anthracene-9-carboxylato)bis(1,2-dimethoxyethanechromium)(Cr-Cr), 44B, 865

$C_{76}H_{222}N_{10}Ni_8O_{82}$, Bis(pentakis(tetramethylammonium) aquohydroxotris-(citrato(4-))tetranickelate(II) octadecahydrate), 43B, 1235

$C_{78}H_{106}B_2Cu_2N_6O_4$, μ-Oxalato-bis(1,1,4,7,7-pentaethyldiethylenetriaminecopper(II)) bis(tetraphenylborate), 43B, 1161
$C_{80}H_{72}Ag_4O_8P_4$, Tetrakis(acetatotriphenylphosphinesilver(I)), 43B, 1251
$C_{96}H_{72}Ag_4O_8P_4 \cdot 2\ C_6H_6$, Bis(1,8-naphthalenedicarboxylato)tetrakis-[triphenylphosphinesilver(I)] benzene solvate, 45B, 1130
$C_{114}H_{90}Ag_3FeO_6P_6S_6$, Tris(bis(triphenylphosphine)silver(I))tris(dithiooxalato)iron(III), 40B, 870

82. METAL COMPLEXES (AMINO-ACID)

$C_2H_4AgNO_2$, Silver aminoacetate, 8, 281
$C_2H_7ClN_2O_2Pt$, cis-Amminechloro(glycinato)platinum(II), 46B, 1032
$C_2H_8Cl_2N_2O_2Pt$, cis-Amminedichloro(glycine)platinum(II), 46B, 1032
$C_2H_9Br_2MnNO_4$, Glycinemanganese(II) bromide dihydrate, 43B, 1251
$C_2H_9Cl_2MnNO_4$, catena-Diaquadichloro-μ-glycine-manganese(II), 46B, 1003
$C_2H_9CoKN_5O_9$, Potassium trinitroglycinatoamminecobaltate(III) monohydrate, 42B, 795
$C_2H_{15}FeNO_{11}S$, Glycinatoiron(II) sulfate pentahydrate, 24, 601
$C_3H_7Cl_2HgNO_2S$, Dichloro-(L-cysteinato)mercury(II), 43B, 1252
$C_4H_5NO_6U$, (Iminodiacetato)dioxouranium(VI), 45B, 1131
$C_4H_8BaCl_4N_2O_4Pd_2 \cdot 2\ H_2O$, Barium bis(dichloroglycinatopalladium(II)) dihydrate, 46B, 1033
$C_4H_8N_2O_4Pd \cdot 3\ H_2O$, cis-Bis(glycinato)palladium(II) trihydrate, 45B, 1132
$C_4H_8N_2O_4Pt$, α-cis-Diglycinatoplatinum(II), 41B, 1027
$C_4H_8N_2O_4Pt$, trans-Bis(glycinato)platinum(II), 34B, 538
$C_4H_9Cl_2NO_3PdS \cdot H_2O$, Dichlorobis((S-methyl-L-cysteine)sulfoxide)palladium(II) monohydrate, 46B, 1033
$C_4H_9Cl_2NO_3PtS \cdot H_2O$, Dichloro[(2S,SR)-S-methylcysteine S-oxide]platinum(II) hydrate, 45B, 1132
$C_4H_9CuNO_6$, Diaquo-iminodiacetatocopper, 45B, 1133
$C_4H_9NO_7Zn$, Zinc aspartate trihydrate, 21, 537
$C_4H_{10}Ag_2N_4O_{10}$, Glycine silver(I) nitrate dimer, 38B, 888
$C_4H_{10}Br_2N_2O_4Pt \cdot H_2O$, cis-Dibromobis(glycine)platinum(II) monohydrate, 45B, 1134
$C_4H_{10}CdN_2O_5$, Cadmium glycinate monohydrate, 23, 592
$C_4H_{10}Cl_2MnN_2O_4$, Bisglycinemanganese(II) dichloride, 41B, 527
$C_4H_{10}Cl_2N_2O_4Pt \cdot H_2O$, cis-Dichlorobis(glycine)platinum(II) monohydrate, 45B, 1134
$C_4H_{10}CuN_2O_5$, Bisglycinatocopper(II) monohydrate, 26, 661; 29, 605
$C_4H_{10}CuN_2O_5$, Glycylglycinatocopper(II) dihydrate, 41B, 1026
$C_4H_{11}Cl_2NO_3PdS$, Dichloro-(S-methyl-L-cysteine)palladium(II) monohydrate, 39B, 712
$C_4H_{11}HgNO_3S$, L-Cysteinato(methyl)mercury(II) monohydrate, 41B, 1028
$C_4H_{11}NO_7Zn$, Zinc(II) (+)-aspartate trihydrate, 39B, 714
$C_4H_{12}CuN_2O_6$, Glycylglycinatocopper(II) trihydrate, 26, 657
$C_4H_{12}N_2NiO_6$, Diaquobisglycinatonickel(II), 10, 216; 33B, 416
$C_4H_{13}CoN_4O_7$, Nitrodiglycinatoamminecobalt(III) monohydrate, 42B, 796
$C_4H_{14}Br_2MnN_2O_6$, Diaquo-bis(glycine)manganese(II) bromide, 44B, 890
$C_5H_8CuN_4O_4$, Diammine(orotato)copper(II), 46B, 1006
$C_5H_{10}CuN_2O_4$, Glycyl-L-alaninatocopper(II) hydrate, 43B, 1253
$C_5H_{11}CdNO_6$, Aquo(L-glutamato)cadmium(II) monohydrate, 40B, 872; 43B, 1254
$C_5H_{11}Cl_2NO_2PdS \cdot H_2O$, Dichloro(S-methyl-L-cysteine methyl ester)pal-

ladium(II) monohydrate, 46B, 1034
$C_5H_{11}Cl_2NO_2PdS$, Dichloro-DL-methioninepalladium(II), 35B, 646
$C_5H_{11}CuNO_6$, Copper glutamate dihydrate, 31B, 411
$C_5H_{11}NO_6Zn$, Zinc glutamate dihydrate, 31B, 412
$C_5H_{12}Cl_2N_2O_2Pt$, trans-Aminodichloro(L-proline)platinum, 46B, 1034
$C_5H_{13}Cl_2NO_4PtS$, Dichloro((2S,SR)-methionine sulfoxide)platinum(II) hydrate, 43B, 1255
$C_5H_{16}BrCdNO_5S$, Aquo(bromo)-DL-penicillaminatocadmium(II) dihydrate, 43B, 1255
$C_6H_8AgNO_6$, Silver di(carboxymethyl)glycinate, 29, 483
$C_6H_9BrKNO_5Pd \cdot H_2O$, Potassium bromo-β-hydroxyethyliminodiacetato-palladate(II) monohydrate, 45B, 1135
$C_6H_{10}ClCuN_3O_4 \cdot 1.5\ H_2O$, Glycylglycylglycinocopper(II) chloride sesquihydrate, 29, 602
$C_6H_{10}Cl_3HgN_3O_2$, Histidine hydrochloride - mercury(II) chloride, 35B, 647
$C_6H_{12}AgCoN_4O_8$, Silver (+)-cis-dinitrobis-(D-alaninato)cobaltate-(III), 40B, 863
$C_6H_{12}CoKNO_9$, Potassium aquo(nitrilotriacetato)cobaltate(II) dihydrate, 41B, 1029
$C_6H_{12}CoKN_4O_8$, trans,trans,trans-Potassium dinitrobis(β-alaninato)cobaltate(III), 39B, 715
$C_6H_{12}CuKNO_9$, Potassium nitrilotriacetatocuprate(II) trihydrate, 41B, 1028
$C_6H_{12}CuN_2O_4$, cis-Bis(D-alaninato)copper(II), 34B, 537
$C_6H_{12}CuN_2O_4$, trans-Bis-(L-α-alaninato)copper(II), 31B, 413
$C_6H_{12}CuN_2O_6$, Bis(L-serinato)copper(II), 34B, 541
$C_6H_{12}N_2O_6Pd$, Bis(L-serinato)palladium(II), 45B, 1135
$C_6H_{12}N_2O_6Zn$, Bis(L-serinato)zinc, 35B, 651
$C_6H_{13}ClHgN_2O_4S_2 \cdot 0.5\ H_2O$, Hydrogen bis(L-cysteinato)mercury(II) chloride hemihydrate, 43B, 1252
$C_6H_{13}Cl_2NO_2PdS$, Dichloro-DL-methioninepalladium(II), 45B, 1136
$C_6H_{13}CoNO_7S$, Triaquo(2-methyl-2,4-dicarboxy-thiazolidinato)cobalt-(II), 45B, 1104
$C_6H_{13}CuNO_7$, Pyruvidine-β-alaninatoaquocopper(II) dihydrate, 33B, 418
$C_6H_{13}HgNO_2S$, (DL-Methioninato)methylmercury(II), 43B, 952
$C_6H_{14}BrCoN_4O_4$, Bis(diaminopropionato)cobalt(III) bromide, 34B, 539
$C_6H_{14}CrN_2NaO_6S_2$, Sodium bis(L-cysteinato)chromate(III) dihydrate, 43B, 1256
$C_6H_{14}CrN_3O_7$, Tris(glycinato)chromium(III) monohydrate, 37B, 518
$C_6H_{14}Mo_2N_2Na_2O_8S_4$, Sodium di-μ-sulfido-bis-((L-cysteinato)oxo-molybdate(V)) dihydrate, 39B, 714
$C_6H_{15}HgNO_3S$, Methyl-DL-penicillaminatomercury(II) monohydrate, 43B, 952
$C_6H_{16}CoN_3O_6$, Bis(glycinato)aminoethanolatocobalt(III) hydrate, 44B, 891
$C_6H_{16}CoN_5O_6$, trans-Dinitro(β-alaninato)(1,3-diaminopropane)cobalt-(III), 43B, 1257
$C_6H_{16}N_2NiO_6$, Bis(sarcosinato)nickel(II) dihydrate, 39B, 715
$C_6H_{16}N_2NiO_6$, Nickel β-alanine dihydrate, 29, 481
$C_6H_{16}N_2NiO_8$, Diaquobis(L-serinato)nickel(II), 34B, 542
$C_6H_{18}Br_2MnN_2O_6 \cdot 2\ H_2O$, Bis(DL-α-alanine)diaquomanganese(II) dibromide dihydrate, 46B, 1035
$C_6H_{18}N_3O_8Zn \cdot 0.5\ SO_4$, Diaquo zinc glycylgylcylglycinato hemisulfate dihydrate, 35B, 648
$C_6H_{20}Cl_2CuN_4O_{14}$, Bis(2,3-diaminopropionato-perchlorato)copper(II) dihydrate, 42B, 796

$C_6H_{20}CuN_2O_8$, Bis-(β-alaninato)copper(II) tetrahydrate, 42B, 797
$C_6H_{20}Mo_2N_2Na_2O_{13}S_2$, Disodium di-μ-oxo-bis(oxocysteinatomolybdenum-(V)) pentahydrate, 34B, 540
$C_6H_{22}N_4O_{14}Zn$, Tetraaquobis(β-alanine)zinc(II) nitrate, 44B, 891
$C_6H_{24}CuN_2O_{10}$, Bis(β-alanine)copper hexahydrate, 26 , 662
$C_7H_{12}CuN_3O_3S$, (L-Methionylglycinato)copper(II), 45B, 1137
$C_7H_{12}N_4NiO_4S \cdot H_2O$, Diaquahistidinatoisothiocyanatonickel(II) monohydrate, 46B, 1035
$C_7H_{13}CuN_2O_3S$, Glycyl-L-methioninatocopper(II), 42B, 797
$C_7H_{18}Hg_2NO_2S$, μ-DL-Penicillaminato-bis(methylmercury(II)), 43B, 952
$C_7H_{21}ClCoN_5O_6S$, Λ-(R)-Cysteinato(N,S)-bis(ethylenediamine)cobalt-(III) perchlorate, 44B, 891
$C_7H_{23}ClCoN_5O_7S$, Δ-(R)-Cysteinato(N,S)-bis(ethylenediamine)cobalt-(III) perchlorate monohydrate, 44B, 891
$C_8H_{10}CoKN_2O_8 \cdot 2.5\ H_2O$, Potassium cis-bis(iminodiacetato)cobaltate(III) hydrate, 39B, 716
$C_8H_{10}CrKN_2O_8 \cdot 3\ H_2O$, Potassium cis-bis(iminodiacetato)chromate(III) trihydrate, 46B, 1036
$C_8H_{10}CuN_4O_3 \cdot 1.5\ H_2O$, (Glycyl-L-histidinato)copper(II) sesquihydrate, 32B, 420
$C_8H_{10}Mo_2N_6O_4S_4 \cdot H_2O$, cis-Bis(glycine)tetrakis(thiocyanato)dimolybdenum(II) hydrate, 45B, 1137
$C_8H_{12}N_2O_{10}U$, Bis(iminodiacetato)dioxouranium(VI), 40B, 864
$C_8H_{14}CoN_3O_8$, L-Asparaginato-D-aspartatocobalt(III) monohydrate, 44B, 892
$C_8H_{14}CuN_4O_6$, Bis-(L-asparaginato)copper(II), 41B, 1031
$C_8H_{14}N_4O_6Zn$, Bis(L-asparaginato)zinc(II), 43B, 1258
$C_8H_{15}CuNO_8$, Diaquo-(N-(o-hydroxyphenyl)iminodiacetato)copper(II) monohydrate, 43B, 1257
$C_8H_{15}CuN_5O_6$, (Glycylglcinato)(cytosine)copper(II) dihydrate, 41B, 1030
$C_8H_{16}CdN_2O_4S_2$, Bis(S-methyl-L-cysteinato)cadmium(II), 43B, 1259
$C_8H_{16}CuN_2O_4$, Bis(D,L-2-aminobutyrato)copper(II), 10, 217; 45B, 1142
$C_8H_{16}CuN_2O_4$, Copper(II) di-γ-aminobutyrate, 39B, 716
$C_8H_{16}MoN_2O_6S_4$, Di-μ-sulfido-bis(oxo-(L-cysteinato methyl ester-N,S)-molybdenum(V)), 37B, 518
$C_8H_{16}N_2O_4S_2Zn$, Bis(S-methyl-L-cysteinato)zinc(II), 43B, 1259
$C_8H_{18}BrCoN_4O_4 \cdot 1.7\ H_2O$, Bis((S)-2,4-diaminobutyrato-O,N,N')cobalt-(III) bromide hydrate, 44B, 893
$C_8H_{18}ClCoN_2O_9S_2$, Bis(S-methyl-L-cysteinato)cobalt(III) perchlorate monohydrate, 42B, 798
$C_8H_{18}Cl_2N_6NiO_{12}$, Bis(iminodiacetamide)nickel(II) perchlorate, 42B, 799
$C_8H_{18}Cr_2N_4O_{10}$, Di-μ-hydroxo-tetraglycinatodichromium(III), 39B, 717
$C_8H_{18}Cs_2N_2NiO_{12}$, Caesium bis(iminodiacetato)nickelate(II) tetrahydrate, 43B, 1259
$C_8H_{18}CuN_2O_7$, Bis-(L-threonine)copper(II) monohydrate, 41B, 1032
$C_8H_{18}Li_2N_2NiO_{12}$, Lithium bis(iminodiacetato)nickelate(II) tetrahydrate, 43B, 1259
$C_8H_{20}Cl_4Mo_2N_4O_8 \cdot 2.67\ H_2O$, Tetrakis(glycine)tetrachlorodimolybdenum(II) 2.67-hydrate, 45B, 1138
$C_8H_{20}Cl_4Mo_2N_4O_8 \cdot 3\ H_2O$, Tetrakis(glycine)tetrachlorodimolybdenum-(II) trihydrate, 45B, 1138
$C_8H_{20}CuN_2O_6$, Bis(β-aminobutyrato)copper(II) dihydrate, 26, 663
$C_8H_{20}CuN_2O_6$, Copper(II) di-γ-aminobutyrate dihydrate, 39B, 716
$C_8H_{20}CuN_2O_{10}$, Diaquobis(N-acetylglycinato)copper(II) dihydrate, 42B. 799; 44B, 893

$C_8H_{20}N_2O_8Zn$, Diaquo-cis-bis(L-threoninato)zinc(II), 43B, 1260
$C_8H_{22}Cl_2CoN_5O_6$, Glycinato(tris-(2-aminoethyl)amine)cobalt(III) chloride perchlorate, 42B, 900
$C_8H_{22}Cl_2CoN_5O_{10}$, Glycinato(tris-(2-aminoethyl)amine)cobalt(III) diperchlorate, 42B, 900
$C_8H_{22}CoI_2N_5O_2 \cdot 0.5\ H_2O$, (Glycinatotriethylenetetramine)cobalt(III) iodide hemihydrate, 40B, 872
$C_8H_{22}CuN_2O_7$, Aquobis(N,N-dimethylglycinato)copper(II) dihydrate, 39B, 718
$C_8H_{23}CoN_6O_3 \cdot 1.6\ I \cdot 0.4\ NO_3$, Λ-Asparaginatobis(ethylenediamine)cobalt(III) iodide-nitrate, 42B, 800
$C_8H_{24}CoI_2N_5O_2$, Δ-R-(N-Methylalaninatobis-(1,2-diaminoethane)cobalt-(III)) iodide, 41B, 1033
$C_8H_{24}CuK_2N_4O_{12}$, Dipotassium bis(glycylglycinato)cuprate(II) hexahydrate, 33B, 418
$C_8H_{24}N_2NiO_8$, Bis(2-amino-2-methylpropionato)nickel(II) tetrahydrate, 27, 839
$C_8H_{26}N_4Na_2NiO_{13}$, Disodium triglycylglycinatonickelate(II) octahydrate, 44B, 894
$C_8H_{28}Mo_2N_4O_{20}S_2$, Tetrakis(glycine)dimolybdenum disulfate tetrahydrate, 42B, 801
$C_8H_{28}N_4Na_2NiO_{14}$, Disodium bis(glycylglycinato)nickelate(II) octahydrate, 44B, 894
$C_8H_{30}CuN_4Na_2O_{15}$, Disodium glycylglycylglycylglycinatocuprate(II) decahydrate, 30B, 349
$C_8H_{30}N_4Na_2NiO_{15}$, Disodium bis(glycylglycinato)nickelate(II) nonahydrate, 44B, 894
$C_8H_{30}N_4Na_2NiO_{15}$, Disodium triglycylglycinatonickelate(II) decahydrate, 44B, 894
$C_9H_{15}CuN_5O_4$, (Decarboxylatoglycylglycyl-L-histidine)copper(II) hydrate, 42B, 815
$C_9H_{15}CuN_5O_4$, Didehydroglycylglycylhistaminatocopper(II) dihydrate, 44B, 895
$C_9H_{16}CuN_4O_5$, Monoaquo(β-alanyl-L-histidinato)copper(II) monohydrate, 32B, 421
$C_9H_{20}CoN_3O_7$, (-)-mer-Tris(L-alaninato)cobalt(III) monohydrate, 44B, 895
$C_9H_{26}CoN_3O_{10}$, Tris(β-alaninato)cobalt(III) tetrahydrate, 44B, 896
$C_{10}H_{12}N_2O_6V \cdot 2\ H_2O$, Aquaoxo[N-(2-pyridylmethyl)iminodiacetato]-vanadium(IV) dihydrate, 45B, 1139
$C_{10}H_{13}CuN_5Na_2O_6 \cdot 4.5\ H_2O$, Disodium tetraglycylglycinatocuprate(II) 4.5-hydrate, 35B, 653
$C_{10}H_{15}CuN_5O_5$, Glycyl-L-histidyl-glycinatocopper(II) hydrate, 43B, 1260
$C_{10}H_{15}CuN_5O_5$, L-Asparaginato-L-histidinatocopper(II), 45B, 1139
$C_{10}H_{15}HgNO_4$, Methyl(L-tyrosinato)mercury(II) monohydrate, 44B, 897
$C_{10}H_{16}N_2O_4Pd$, Bis-(L-prolinato)palladium(II), 37B, 520
$C_{10}H_{16}N_6NiO_4$, Bis(glycinato)bis(imidazole)nickel(II), 38B, 890
$C_{10}H_{17}CuN_5O_6 \cdot 3\ H_2O$, L-Asparaginato-L-histidinatoaquocopper(II) trihydrate, 45B, 1139
$C_{10}H_{17}NO_6Zn$, Zinc kainate dihydrate, 21, 613
$C_{10}H_{18}N_2O_6Pt_2S_2$, Bis(μ-(2-amino-4-(methylsulfinyl)butanato(2-)))-diplatinum(II), 44B, 897
$C_{10}H_{20}CuN_2O_4S_2$, trans-Bis(L-methionato)copper(II), 44B, 898
$C_{10}H_{20}CuN_2O_6$, Copper proline dihydrate, 16, 447; 39B, 719
$C_{10}H_{20}Mo_2N_2O_8S_2$, Di-μ-oxo-bis(oxo(L-cysteinato ethyl ester-N,S)-molybdenum(V)), 37B, 519

$C_{10}H_{20}N_2O_4S_2Zn$, Bis(L-methionato)zinc(II), 43B, 1261
$C_{10}H_{22}Br_2MnN_2O_6$, Bis(DL-proline)manganese(II) dibromide dihydrate, 43B, 1262
$C_{10}H_{22}CoKN_2O_6S_2$, Potassium (D-penicillaminato)(L-penicillaminato)cobaltate(III) dihydrate, 43B, 1262
$C_{10}H_{22}N_2O_5Pd$, cis-Bis(L-valinato)palladium(II) monohydrate, 39B, 719
$C_{10}H_{22}N_4NiO_4$, Bis(L-prolinamidato)nickel dihydrate, 38B, 891
$C_{10}H_{23}CuN_7O_8$, (Glycylglycinato)(aquo)(9-methyladenine)copper(II) tetrahydrate, 42B, 905
$C_{10}H_{24}CoN_5O_8$, Bis(L-ornithinato)cobalt(III) nitrate monohydrate, 42B, 801
$C_{10}H_{28}Cl_2CuN_4O_6$, Dichlorobis(δ-N-hydro-L-ornithinato)copper(II) dihydrate, 35B, 656; 43B, 1263
$C_{11}H_{14}CuN_6O_4 \cdot 4\ H_2O$, (Glycylglycinato)(7,9-dimethylhypoxanthine)-copper(II) tetrahydrate, 45B, 1140
$C_{11}H_{15}HgNO_2$, [L-(2-Amino-4-phenylbutanoato)]methylmercury(II), 44B, 897
$C_{11}H_{18}CuN_6O_4$, Glycylglycyl-L-histidine-N-methylamide-copper(II) monohydrate, 42B, 802
$C_{11}H_{19}CoN_4O_5S$, L-Histidinyl-D-penicillaminatocobalt(III) monohydrate, 43B, 1264
$C_{11}H_{19}CrN_4O_5S$, L-Histidinato-D-penicillaminatochromium(III) monohydrate, 43B, 1264
$C_{11}H_{20}CuN_2O_8$, Diaquo(glycyl-L-tyrosinato)copper(II) dihydrate, 43B, 1265
$C_{11}H_{26}Cl_4CoN_5O_2Zn$, D-β2-(SSS)-(Triethylenetetramine-(S)-prolinato)-cobalt(III) tetrachlorozincate, 35B, 660
$C_{11}H_{30}CoI_2N_5O_4$, L(-)$_{589}$-β_2-(RRS)-(Triethylenetetramine-(S)-prolinato)cobalt(III) diiodide dihydrate, 35B, 658
$C_{12}H_{16}BrCoN_6O_4$, D-Histidinato-L-histidinatocobalt(III) bromide, 43B, 1266
$C_{12}H_{16}ClCoN_6O_8 \cdot 2\ H_2O$, trans-Ammine-bis(L-histidinato)cobalt(III) perchlorate dihydrate, 45B, 1141
$C_{12}H_{16}ClCuN_3O_4$, 2,2'-Bipyridylglycinatochlorocopper(II) dihydrate, 41B, 1035
$C_{12}H_{16}ClMoNO_3$, Bis-π-cyclopentadienylglycinatomolybdenum(IV) chloride monohydrate, 38B, 896
$C_{12}H_{16}Mo_2N_6O_6S_2 \cdot 1.5\ H_2O$, Di-μ-sulfido-bis((L-histidinato)oxomolybdenum(V)) sesquihydrate, 41B, 1035
$C_{12}H_{18}CoN_6O_5$, Bis-(L-histidino)cobalt(II) monohydrate, 33B, 421
$C_{12}H_{18}N_6NiO_5$, Bis(L-histidinato)nickel(II) monohydrate (monoclinic), 44B, 899
$C_{12}H_{18}N_6NiO_5$, Bis(L-histidinato)nickel(II) monohydrate (orthorhombic), 32B, 424
$C_{12}H_{20}CdN_6O_6$, Bis-(L-histidinato)cadmium dihydrate, 32B, 422; 42B, 803
$C_{12}H_{20}CoN_6O_6$, D-Histidino-L-histidinocobalt(II) dihydrate, 35B, 664
$C_{12}H_{20}Cu_2N_6Na_2O_{10}$, Sodium glycylglycylglycinatocuprate(II) monohydrate, 30B, 348
$C_{12}H_{20}N_6O_6Zn$, Bis(L-histidino)zinc(II) dihydrate, 28, 636; 38B, 892
$C_{12}H_{21}CoN_6O_9 \cdot 3\ H_2O$, fac-Δ-Tris(L-asparaginato)cobalt(II) trihydrate, 45B, 1141
$C_{12}H_{22}CuN_8O_{12}$, Bis(L-histidine)copper(II) nitrate dihydrate, 34B, 544
$C_{12}H_{24}CuN_2O_4$, Bis(L-leucinato)copper(II), 45B, 1142
$C_{12}H_{24}CuN_2O_8$, trans-Bis(N-ethylidene-DL-threoninato)copper(II) dihydrate, 26, 660

$C_{12}H_{26}CoCs_2N_2O_{14}$, Caesium bis(β-hydroxyethyliminodiacetato)cobaltate(II) tetrahydrate, 44B, 899
$C_{12}H_{26}CuN_2O_5$, Aquobis(L-leucinato)copper(II), 34B, 545
$C_{12}H_{26}LiN_2O_{14}Yb$, Lithium bis(β-hydroxyethyliminodiacetato)-ytterbate(III) tetrahydrate, 44B, 899
$C_{12}H_{26}N_2NaNiO_{14}$, Sodium bis(β-hydroxyethyliminodiacetato)nickelate-(II) tetrahydrate, 44B, 900
$C_{12}H_{26}N_6O_9Zn$, Bis(histidino)zinc(II) pentahydrate, 28, 636
$C_{12}H_{28}CuN_6O_{10}$, L-Histidinato-D-histidinatodiaquocopper(II) tetrahydrate, 44B, 900
$C_{12}H_{28}K_4N_2NiO_{20}$, Potassium bis(nitrilotriacetato)nickelate(II) octahydrate, 41B, 1036
$C_{12}H_{29}ClCoN_5O_9$, 2,9-Diamino-4,7-diazadecanecobalt(III) aminomethylmalonate perchlorate monohydrate, 40B, 873
$C_{12}H_{32}Cl_4N_4O_{26}Rh_2 \cdot 2\ H_2O$, Diaquo-tetrakis-μ-β-alaninatoniumdirhodium(II) perchlorate dihydrate, 45B, 1143
$C_{12}H_{32}CoN_{11}O_{13}$, (+)-Dinitrobis-(L-argininato)cobalt(III) nitrate dihydrate, 38B, 892
$C_{12}H_{36}Cl_7Fe_3N_6O_{44}$, Tri-$\mu_3$-oxo-triaquohexakis(glycine)triiron(III) perchlorate, 43B, 1267
$C_{13}H_{19}CuN_3O_6$, Aquo(glycyl)-L-tryptophanatocopper(II) dihydrate, 41B, 1037
$C_{13}H_{19}CuN_5O_8$, Cytidine(glycylglycinato)copper(II), 41B, 1038
$C_{13}H_{19}CuN_5O_{10}$, (Glycylglycinato)(cytidine)copper(II) dihydrate, 43B, 1267
$C_{14}H_6CrN_2O_8Rb$, Rubidium bis(pyridine-2,6-dicarboxylato)chromate(III), 45B, 1113
$C_{14}H_8Cr_2N_2O_{14} \cdot 4\ H_2O$, Di-μ-hydroxy-bis[(4-hydroxy-2,6-dicarboxylatopyridine)aquachromium(III)] tetrahydrate, 45B, 1113
$C_{14}H_{10}Cl_2Cr_2N_2O_{12} \cdot 2\ H_2O$, Di-μ-hydroxy-bis[(4-chloro-2,6-dicarboxylatopyridine)aquachromium(III)] dihydrate, 45B, 1113
$C_{14}H_{12}CuN_2O_4$, Bis(o-aminobenzoato)copper(II), 42B, 804
$C_{14}H_{20}ClMoNO_3$, Bis-π-cyclopentadienylsarcosinatomolybdenum(IV) chloride methanolate, 38B, 896
$C_{14}H_{22}Cd_2Cl_2N_8O_6$, Chloroglycylglycinato-(imidazole)-cadmium, 46B, 1037
$C_{14}H_{22}CrN_2NaO_8 \cdot 2\ H_2O$, Sodium trans-bis(N-isopropyliminodiacetato)-chromate(III) dihydrate, 46B, 1037
$C_{14}H_{28}CuN_2O_4$, Bis(DL-N,N-diethyl-α-alaninato)copper(II), 34B, 543
$C_{14}H_{28}CuN_2O_8$, (L-Valine-L-tyrosine)copper(II) tetrahydrate, 42B, 805
$C_{14}H_{42}Cl_4Co_2N_{10}O_{20}S_2 \cdot 6\ H_2O$, Δ-(R)-μ-Cystinato-tetrakis(ethylenediamine)dicobalt(III) tetraperchlorate hexahydrate, 45B, 1040
$C_{15}H_{18}F_6MoNO_2P$, Bis-η-cyclopentadienyl-L-prolinatomolybdenum hexafluorophosphate, 43B, 1068
$C_{16}H_{16}Hg_2N_4O_4$, Creatininatobis(phenylmercury(II)) nitrate, 44B, 900
$C_{16}H_{18}CoN_5O_7 \cdot 0.5\ H_2O$, Bis[D-β-(2-pyridyl)-α-alaninato]cobalt(III) nitrate hemihydrate, 45B, 1143
$C_{16}H_{18}N_4NiO_4 \cdot 2\ H_2O$, Bis[D-β-(2-pyridyl)-α-alaninato]nickel(II) dihydrate, 46B, 1038
$C_{16}H_{19}AgN_4O_8S_2$, Silver o-aminobenzenesulphonyl-glycinate, 44B, 901
$C_{16}H_{20}CuN_4O_6$, Glycylglycinato(1,10-phenanthroline)copper(II) trihydrate, 42B, 805
$C_{16}H_{21}CoN_4O_7$, (α-N-(o-Hydroxybenzyl)-L-histidinato)(L-alaninato)cobalt(III) dihydrate, 44B, 901
$C_{16}H_{22}F_6MoNO_2P$, Bis-η-cyclopentadienyl-L-leucinatomolybdenum hexafluorophosphate, 43B, 1068
$C_{16}H_{26}Mo_2N_6O_4S_4 \cdot 4.5\ H_2O$, cis-Bis(L-isoleucine)tetrakis(thiocyan-

ato)dimolybdenum(II) hydrate, 45B, 1137
$C_{16}H_{28}CuN_2O_{14}$, Bis(O-(β-D-xylopyranosyl)-L-serinato)copper(II), 41B, 538
$C_{16}H_{28}Cu_2N_4O_{14}$, Tetra-μ-N-acetylglycinatodiaquodicopper(II), 45B, 1144
$C_{16}H_{32}Cl_4Mo_2N_8O_{12} \cdot 6\ H_2O$, Tetrakis(glycylglycine)-dimolybdenum(II) tetrachloride hexahydrate, 46B, 1038
$C_{16}H_{32}Co_2N_6O_{13}$, (+)-cis(O)-Triammine(sarcosinate-N-propionato)cobalt(III) (+)-(ethylenediaminetetraacetato)cobaltate(III) monohydrate, 44B, 902
$C_{16}H_{34}Cl_2CoN_5O_{10} \cdot 2\ H_2O$, β_2-((R)-Alaninato)(1,7-bis(2(S)-pyrrolidyl)-2,6-diazaheptane) bis(perchlorate) dihydrate, 46B, 1039
$C_{16}H_{40}CaCo_2N_4O_{26}$, Calcium cis(N)-trans($O^5$)-bis-(L-aspartato)cobaltate(III) cis(N)-trans(O^6)-bis-(L-aspartato)cobaltate(III) decahydrate, 39B, 720
$C_{17}H_{32}CuN_2O_8$, Aquo-L-leucyl-L-tyrosinecopper(II) dihydrate ethanol solvate, 41B, 1038
$C_{18}H_{10}N_2O_6Pd \cdot 0.5\ H_2O$, cis-Bis(L-tyrosinato)palladium(II) hemihydrate, 39B, 721
$C_{18}H_{20}CuN_2O_4$, Bis-(L-phenylalalaninato)copper(II), 37B, 521
$C_{18}H_{20}CuN_2O_6$, Bis-(L-tyrosinato)copper(II), 38B, 893
$C_{18}H_{20}N_2O_6Pd$, Bis(L-tyrosinato)palladium(II), 45B, 1144
$C_{18}H_{22}CoN_2O_9 \cdot 2\ H_2O$, Triaquo-bis(N-benzoylglycinato)cobalt(II) dihydrate, 45B, 1145
$C_{18}H_{22}N_2NiO_9 \cdot 2\ H_2O$, Triaquo-bis(N-benzoylglycinato)nickel(II) dihydrate, 45B, 1145
$C_{18}H_{26}CoN_2O_{11}$, Cobalt(II) dihippurate pentahydrate, 43B, 1268
$C_{18}H_{26}N_2NiO_9$, Diaquobis(L-tyrosinato)nickel(II) monohydrate, 44B, 902
$C_{18}H_{28}CuN_4O_8$, (2,9-Dimethyl-1,10-phenanthroline)(glycylglycinato)-copper(II) pentahydrate, 44B, 903
$C_{19}H_{20}CoN_3O_6 \cdot 3\ H_2O$, [N-(2-Pyridylmethyl)-L-aspartato](L-phenylalaninato)cobalt(III) trihydrate, 46B, 1039
$C_{20}H_{20}CuN_4O_6$, Bis(N-acetylglycinato)-1,10-phenanthrolinecopper(II), 43B, 1269
$C_{20}H_{22}CuN_4O_{12}$, Bis(p-nitrophenaceturo)diaquocopper(II), 43B, 1243
$C_{20}H_{24}Cu_2N_6Na_4O_{12}S_2 \cdot 6\ H_2O$, Potassium glutathionato(8-)biscuprate-(II) hexahydrate, 46B, 1040
$C_{20}H_{25}Cl_3CoNO_6$, L-Phenylalaninatobis(acetylacetonato)cobalt(III) - chloroform, 43B, 1168
$C_{20}H_{28}CuN_2O_2 \cdot 0.7\ C_6H_6$, Bis-l-ephedrinecopper(II) benzene, 29, 609
$C_{20}H_{32}N_2NiO_{10}$, Diaquobis(L-tyrosinato)nickel(II) bis(methanol), 43B, 1270
$C_{20}H_{54}Cu_2N_4O_{17}S_4$, Bis(aquocopper(II)-D-penicillamine disulfide) heptahydrate, 40B, 874
$C_{24}H_{28}Cl_2Cu_2N_{12}O_{13} \cdot 3.5\ H_2O$, Aquobis(cyclo-L-histidyl-L-histidylato)dicopper(II) perchlorate hydrate, 43B, 1270
$C_{24}H_{28}CuN_2O_4$, (N-Benzyl-D-prolinato)(N-benzyl-L-prolinato)copper-(II), 38B, 894
$C_{24}H_{28}CuN_2O_4$, Bis-(N-benzyl-L-prolinato)copper(II), 39B, 722
$C_{24}H_{38}CuN_2O_6$, (N-Benzyl-D-valinato)(N-benzyl-L-valinato)copper(II) dihydrate (red isomer), 41B, 1039
$C_{24}H_{38}CuN_2O_6$, (N-Benzyl-D-valinato)(N-benzyl-L-valinato)copper(II) dihydrate (blue isomer), 41B, 1039
$C_{24}H_{38}CuN_2O_7$, Bis(N-benzyl-L-valine)copper(II) trihydrate, 42B, 806
$C_{26}H_{31}ClMo_2N_2O_4S_2$, Hydrogendi(bis-π-cyclopentadienyl)-L-cysteinatomolybdenum(IV) chloride, 38B, 895

$C_{26}H_{31}F_6Mo_2N_2O_4PS_2$, Hydrogendi(bis-π-cyclopentadienyl)-L-cysteinato-molybdenum(IV) hexafluorophosphate, 38B, 896
$C_{26}H_{42}MnN_4O_{12}$, Bis(pyridoxylidenevaline)manganese(II) tetrahydrate, 28, 639
$C_{32}H_{30}Cu_2N_2O_8$, Di-μ-(N-salicylidene-L-phenylalaninato)bis(aquocopper(II)), 44B, 903
$C_{32}H_{70}Ca_2Co_4N_8O_{47}$, Bis(calcium cis(N)-trans(O(6))-bis(L-aspartato)-cobaltate(III)) pentadecahydrate, 41B, 1040
$C_{36}H_{40}CuN_6O_8$, Diaquabis(N-acetyl-DL-tryptophanato)bis(pyridine)copper(II), 46B, 1040
$C_{38}H_{62}Cl_2Mo_2N_4O_{14}S_2 \cdot 2\ H_2O$, Tetrakis(L-leucine)-dimolybdenum(II) dichloride bis(p-toluenesulphonate) dihydrate, 46B, 1040
$C_{38}H_{72}Cu_2N_6O_{19}$, Glycyl-L-leucyl-L-tyrosinecopper(II) dimer, 37B, 521
$C_{41}H_{66}FeN_9O_{24}$, Ferrichrome-A tetrahydrate, 31B, 414
$C_{60}H_{108}ClCu_{14}N_{12}O_{24}S_{12}Tl_5 \cdot 55\ H_2O$, Pentathallium(I) μ_8-chloro-dodeca(D-penicillaminato)octacuprate(I)hexacuprate(II) hydrate, 43B, 1271

83. METAL COMPLEXES (NITROGEN LIGAND)

$CHCl_5NNb$, Pentachloroformonitrileniobium(V), 41B, 1041
$CH_5Cl_2CuN_3O$, catena-Di-μ-chloro-semicarbazidecopper(II), 37B, 523
$C_2H_2Cl_4N_2Ti$, cis-Tetrachlorodiformonitriletitanium(IV), 37B, 523
$C_2H_2Cl_4N_2V$, Tetrachloro-cis-diformonitrilevanadium(IV), 39B, 722
C_2H_3BrCuN, Acetonitrile copper(I) bromide, 37B, 524
C_2H_3ClCuN, Copper(I) chloride-acetonitrile complex, 34B, 546
$C_2H_3Cl_2CuN_3$, Copper(II) chloride - 1,2,4-triazole, 27, 859
$C_2H_3Cl_3NOV$, Acetonitriletrichlorooxovanadium(V), 41B, 1042
$C_2H_3Cl_4N_2ORe$, Nitroso-acetonitrile-tetrachlororhenium, 41B, 1043
$C_2H_4CdCl_2N_4$, Di-μ-chloro(dicyandiamide)cadmium(II), 37B, 525
$C_2H_6Cl_2Cu_2N_2$, Azomethane copper(I) chloride, 24, 607
$C_2H_6N_4NiOS_4$, (N-[Methoxylmethyl]disulfur-dinitrogen)-(N-hydrogen-disulfur-dinitrogen)nickel, 45B, 1145
$C_2H_{19}Cl_2CoN_6O_9$, Acetamidopentaamminecobalt(III) perchlorate, 39B, 722
$C_3H_3Cl_2Cu_2N$, Acrylonitrile dicopper(I) dichloride, 43B, 1272
$C_3H_4CdCl_2N_2$, Di-μ-chloro-imidazolocadmium(II) polymer, 42B, 807
$C_3H_6N_4NiO_2S_4$, 5-Hydroxy-6-methoxy-1,3,8,10-tetrathia-2,4,7,9-tetra-azadecane-4,7-diato-S(1),S(10)-nickel(II), 32B, 426
$C_3H_{10}CdCl_2N_2$, Dicatena-di-μ-chloromonochloromono(1,3-diamino-propane)cadmium(II), 37B, 525
C_4AgN_3, Silver(I) tricyanomethanide, 31B, 41
$C_4H_4Ag_2N_4O_6$, Succinonitrile-silver nitrate complex, 31B, 418
$C_4H_4Hg_2N_4O_6$, 1,4-Diazinedimercury(I) dinitrate, 40B, 876
C_4H_5BrCuN, (Methylacrylonitrile)copper(I) bromide, 44B, 904
$C_4H_5Cl_3N_6Zn$, Trichloro(8-azaadeninium)zinc(II), 43B, 1273
$C_4H_6Cl_2N_2Zn$, Dichlorobis(acetonitrile)zinc(II), 27, 829
$C_4H_6Cl_3N_2NbO$, Diacetonitriletrichlorooxoniobium(V), 41B, 1044
$C_4H_6Cl_4Cu_2N_2$, Di-μ-chloro-bis(chloro(methyl cyanide)copper), 29, 590
$C_4H_6Cl_4Cu_2N_4S_2$, Bis[(methylcyano)-(μ-chloro)-(μ-1,3-diaza-2,4-dithietidine)chlorocopper], 46B, 1042
$C_4H_6Cl_6Cu_3N_2$, Hexachlorobis(methyl cyanide)tricopper, 29, 590
$C_4H_6IN_4O_4Pd$, Bis(glyoximato)palladium(II) iodine, 42B, 807
$C_4H_6N_4NiO_4$, Bis(glyoximato)nickel(II), 23, 600; 32B, 428
$C_4H_6N_4O_4Pd$, Bis(glyoximato)palladium(II), 38B, 898
$C_4H_6N_4O_4Pt$, Bis(glyoximato)platinum(II), 34B, 550

$C_4H_8CdI_2N_8$, Diodobis(dicyandiamide)cadmium(II), 40B, 875
$C_4H_8Cl_2CuN_2O_2$, Copper dimethylglyoxime dichloride, 40B, 1117
$C_4H_8Cl_2N_8Zn$, Dichlorobis(1-methyltetrazole)zinc(II), 37B, 526
$C_4H_8HgN_2O_2$, Mercury(II) acetamide, 34B, 548
C_4H_9AgINO, Morpholine silver iodide, 42B, 269
$C_4H_9CoN_5OS_4$, Bis(S-aminodithionitrito)cobalt complex, 40B, 876
$C_4H_9N_5NiOS_4$, Bis(S-aminodithionitrito)nickel(II) complex, 38B, 898
$C_4H_{10}Cl_2N_2Pt$, cis-Dichlorobis(ethyleneimine)platinum(II), 41B, 1045
$C_4H_{10}Cl_2N_2Pt$, trans-Dichlorobis(ethyleneimine)platinum(II), 41B, 1045
$C_4H_{10}Cl_3NTi$, Diethylaminotitanium trichloride, 37B, 527
$C_4H_{10}CrN_2O_8U$, Diacetamido-chromato-uranyl, 43B, 1273
$C_4H_{10}N_8NiO_4 \cdot 2\ H_2O$, Bis(oxamide oximato)nickel(II) dihydrate, 45B, 1146
$C_4H_{10}N_8O_4Pt \cdot 0.33\ C_2H_3NaO_2 \cdot 0.33\ NaCl \cdot 2\ H_2O$, Bis(oxamide oximato)platinum(II) sodium acetate sodium chloride water, 46B, 1042
$C_4H_{12}Cl_2N_2Pd$, cis-Dichloro(meso-2,3-diaminobutane)palladium(II), 37B, 528
$C_4H_{12}Cl_2N_8O_4Pt$, Bis(oxamide oximato)platinum(II) hydrogen chloride (1:2) (form I), 45B, 1147
$C_4H_{12}Cl_2N_8O_4Pt$, Bis(oxamide oximato)platinum(II) hydrogen chloride (1:2) (form II), 46B, 1043
$C_4H_{12}CoN_4OS_4Si_2$, Bis(S-aminodithionitrito)cobalt silicon complex, 40B, 880
$C_4H_{13}Cl_3CrN_3$, Trichloro-1,5-diamino-3-azapentanechromium(III), 35B, 665
$C_4H_{14}ClN_9O_4Pt$, Bis(oxamide oximato)platinum(II) ammonium chloride, 45B, 1147
$C_4H_{14}CuK_2N_6O_8$, Potassium bis(biureto)cuprate(II) tetrahydrate, 26, 658
$C_4H_{20}Br_2CuN_4$, Tetrakis(methylamine)dibromocopper(II), 18, 661
$C_4H_{20}Br_3CoN_8 \cdot 0.67\ H_2O$, (1,2,3-Triamino-1,3-diimino-2-methylpropane)triamminecobalt(III) tribromide hydrate, 45B, 1148
$C_4H_{20}Cl_2N_4Pt$, trans-Bis(dimethylamine)diammineplatinum(II) chloride, 35B, 666
$C_4H_{20}Cl_2N_6OPt$, Bis(acetamidine)diammineplatinum(II) chloride monohydrate, 27, 834
$C_4H_{20}Cl_4N_4Pt_2$, Tetra(methylamine)platinum(II) tetrachloroplatinate(II), 37B, 529
$C_4H_{42}(Br,Cl)_5N_{12}O_4Ru_2$, μ-Pyrazine-bis(pentaammineruthenium) penta(bromide-chloride) tetrahydrate, 43B, 1274
C_5H_5ClCuN, Pyridinecopper(I) chloride, 43B, 1274
C_5H_5CuIN, Cuprous iodide pyridine polymer, 46B, 1044
$C_5H_5NO_4S_2Zn$, Zinc dithionite pyridine, 44B, 904
$C_5H_6Cl_3N_5Zn$, Trichloroadeniniumzinc(II), 39B, 723
$C_5H_7Cl_2Cu_2NO$, catena-3,5-Dimethylisoxazole-μ-dichloro-dicopper(I), 43B, 1275
$C_5H_8AgCl_2N_5O_9$, Di-μ-adeninium-disilver(I) perchlorate monohydrate, 43B, 1290
$C_5H_8ClCuN_2$, Chloro-(2,3-diazabicyclo[2.2.1]hept-2-ene)copper(I), 40B, 877
$C_5H_8Cl_3CuN_5O_2$, Guanine copper(II) chloride monohydrate, 37B, 530
$C_5H_9Cl_2CuN_3$, Dichloro-histamino-copper(II), 46B, 1044
$C_5H_{11}AgIN$, Piperidine - silver iodide, 37B, 531
$C_5H_{11}AuClN$, Chloro(piperidine)gold(I), 43B, 1276
$C_5H_{19}Br_2CoN_6O_3$, trans-Tetraamminenitropyridinecobalt(III) dibromide monohydrate, 43B, 1277

$C_5H_{21}F_{12}N_8OP_2Ru$, (1-Methylcytosinato)pentaammineruthenium(III) hexafluorophosphate, 45B, 1149
$C_5H_{22}CuN_{10}O_7$, Bis(biguanine)copper(II) carbonate tetrahydrate, 44B, 907
$C_5H_{23}Cl_3N_5O_8Re$, trans-Chloro(methylimido)tetrakis(methylamine)-rhenium(V) perchlorate, 40B, 878
$C_5H_{25}ClCoN_7O_6$, Chloropentakis(methylamine)cobalt(III) dinitrate, 44B, 905
$C_5H_{25}Cl_3CrN_5$, Chloropentakis(methylamine)chromium(III) dichloride, 44B, 905
$C_6H_4Cl_5IrK_2N_2 \cdot 0.5\ H_2O$, Potassium pentachloro(pyrazine)iridate(III) hemihydrate, 39B, 725
$C_6H_6Br_8Fe_3N_6$, Hexakis(formonitrile)iron(II) tetrabromoferrate(III), 39B, 725
$C_6H_6Cl_2CuN_2S_2$, Dichlorobis(thiazole)copper(II), 44B, 905
$C_6H_6Cl_2N_2O_2Pd$, trans-Dichlorobis(oxazole)palladium(II), 45B, 1149
$C_6H_6CoN_4$, Diimidazolylcobalt, 41B, 1047
$C_6H_6CoN_8S_2$, poly-Bis(thiocyanato-N)bis-μ-(1,2,4-triazole-N^2,N^4)co-balt(II), 45B, 1150
$C_6H_6CuN_8S_2$, poly-Bis(thiocyanato-N)bis-μ-(1,2,4-triazole-N^2,N^4)cop-per(II), 45B, 1150
$C_6H_6N_8S_2Zn$, poly-Bis(thiocyanato-N)bis-μ-(1,2,4-triazole-N^2,N^4)zinc-(II), 45B, 1150
$C_6H_7Cl_2N_5Zn$, catena-Dichloro-μ-(9-methyladenine)-zinc(II), 42B, 808
$C_6H_8AgClN_4O_4$, Silver(I) imidazole perchlorate, 46B, 1045
$C_6H_8AgN_5O_3$, Diimidazolesilver nitrate, 37B, 532
$C_6H_8Br_2CuN_2$, Dibromo(2-(2-aminomethyl)pyridine)copper(II), 43B, 1118
$C_6H_8Br_2N_2Ni$, Dibromo-2,5-dimethylpyrazinenickel(II), 29, 570
$C_6H_8CdCl_2N_4$, catena-μ-Dichlorobisimidazolecadmium(II), 39B, 725
$C_6H_8Cl_2CoN_4$, Diimidazolecobalt(II) dichloride, 38B, 899
$C_6H_8Cl_2CuN_4$, Diimidazolecopper(II) dichloride, 38B, 899
$C_6H_8Cl_2MnN_4$, catena-μ-Dichlorobis(pyrazole)manganese(II), 40B, 878
$C_6H_8Cl_2N_4Zn$, Zinc chloride-bis(imidazole) complex, 31B, 421
$C_6H_8Cl_3N_5Pt$, Trichloro(9-methyladeninium)platinum(II), 42B, 809
$C_6H_8CuN_6O_6$, Diaquobis(6-azauracilato)copper(II), 43B, 1278
$C_6H_9AgN_6O_4$, catena-(μ-9-Methyladenine)silver(I) nitrate hydrate, 43B, 1279
$C_6H_9CoK_3N_9O_6 \cdot 6.38\ H_2O$, Potassium tris(biuretato)cobaltate(III) hydrate, 39B, 726
$C_6H_{10}CuN_8O_6$, Bis(isocyanurato)diamminecopper(II), 39B, 726
$C_6H_{11}AgClNO_3 \cdot 0.5\ H_2O$, Silver N-cyclohexylamidoperchlorate hemihyd-rate, 42B, 809
$C_6H_{12}CuN_8S_2$, 1-(2-Aminoethyl)biguanide-isothiocyanato-copper(II) thiocyanate, 37B, 533
$C_6H_{14}Cl_2N_2O_2Pd$, Dichlorobis(acetoxime)palladium, 40B, 879
$C_6H_{14}Cl_2N_2Pt$, trans-Dichloro-ethylene-acetaldehydedimethylhydrazone-platinum(II), 42B, 827
$C_6H_{14}CuN_4O_6$, (1,5-Diazacyclooctane)dinitratocopper(II), 46B, 1046
$C_6H_{16}Cl_2CuN_8O_8$, Bis(malondiamidine)copper(II) diperchlorate, 44B, 906
$C_6H_{16}Cl_2N_8Ni$, Bis(malondiamidine)nickel(II) dichloride, 44B, 906
$C_6H_{16}Cl_4CuN_{12}O$, Tetrachlorobis-2-((5-amino-4-carboxamidinium)-(1,2,3)triazole)copper(II) monohydrate, 41B, 1047
$C_6H_{16}CoN_{12}O_6$, Bis(diaminoglyoximato)cobalt(II) diaminoglyoxime, 43B, 1277
$C_6H_{16}N_{12}NiO_6$, Bis(oxamide oximato)nickel(II) oxamide oxime, 46B, 1046

$C_6H_{17}Cl_3CoN_3$, mer-Trichloro-(N-(3-aminopropyl)-1,3-diaminopropane)-cobalt(III), 46B, 1047
$C_6H_{17}CuN_5O_9S$, Tetraaquo-(9-methyladenine)copper(II) sulfate monohydrate, 40B, 960
$C_6H_{17}N_5NiS_4Si_2$, Bis(S-aminodithionitrito)nickel silicon complex, 40B, 880
$C_6H_{18}Br_3N_2Ti$, Tribromobis(trimethylamine)titanium, 37B, 534
$C_6H_{18}Cl_2CuN_{10}O$, Ethylenebidiguanide copper(II) chloride monohydrate, 35B, 670
$C_6H_{18}Cl_2N_4NiO_8$, Triethylenetetraminenickel(II) perchlorate, 38B, 899
$C_6H_{18}Cl_2N_{10}NiO$, Ethylenebis(biguanidine)nickel(II) dichloride monohydrate, 35B, 667; 37B, 534; 40B, 1117
$C_6H_{18}Cl_2N_{12}NiO_6 \cdot 0.5\ H_2O$, Tris(oxamide oxime)nickel(II) dichloride hemihydrate, 46B, 1047
$C_6H_{18}Cl_3CoN_{12}O_6$, Tris(oxamide oxime)cobalt(III) trichloride, 44B, 907
$C_6H_{18}Cl_3CrN_2$, Trichlorobis(trimethylamine)chromium, 37B, 534
$C_6H_{18}Cl_3N_2V$, Trichlorobis(trimethylamine)vanadium(III), 34B, 549
$C_6H_{18}Cl_4N_2Si_2Ti_2$, catena-Di-μ-chloro-bis-μ-(trimethylsilylamino)-di-(chlorotitanium(IV)), 42B, 810
$C_6H_{18}I_2N_4Zn$, Iodo(triethylenetetramine)zinc(II) iodide, 39B, 727
$C_6H_{20}ClCoN_6O_4$, $(+)_{589}$-cis-Dinitrobis($(-)_{589}$-1,2-propylenediamine)cobalt(III) chloride, 35B, 672
$C_6H_{20}CoN_7O_9$, trans-Dinitratobis(trimethylenediamine)cobalt(III) nitrate, 35B, 678
$C_6H_{20}CrN_{15}O$, Tris(biguanidato)chromium(III) monohydrate, 42B, 811
$C_6H_{20}Cr_2CuN_2O_7$, Bis(1,2-propanediamine)copper(II) dichromate(VI), 35B, 675
$C_6H_{20}CuCl_2N_4$, Bis(2-hydroxy-1,3-propanediamine)copper(II) chloride, 40B, 880
$C_6H_{20}CuN_6O_4$, Bis(1,3-diaminopropane)copper(II) nitrite, 35B, 676
$C_6H_{20}CuN_6O_6$, Bis(1,2-propanediamine)copper(II) nitrate, 35B, 674
$C_6H_{20}CuN_6O_6$, Bis(1,3-diaminopropane)copper(II) nitrate, 34B, 555
$C_6H_{20}K_2N_8NiO_{12}$, Dipotassium bis(trimethylenedinitramine)nickelate(II) tetrahydrate, 30B, 353
$C_6H_{21}ClCoN_7O_6$, racemic-α-(Amminechlorotriethylenetetramine)cobalt-(III) nitrate, 35B, 679
$C_6H_{22}CoN_{15}O_2$, Tris(biguanide)cobalt(III) dihydrate, 44B, 907
$C_6H_{22}CuN_4O_5S$, Bis-(1,3-propanediamine)copper(II) sulfate monohydrate, 35B, 681
$C_6H_{22}CuN_4O_5Se$, Bis-(1,3-propanediamine)copper(II) selenate monohydrate, 35B, 681
$C_6H_{23}Cl_3CoN_{15}O$, (+)-Tris(biguanide)cobalt(III) trichloride monohydrate, 40B, 881
$C_6H_{24}Cl_2N_4NiO_{10}$, Bis(1,3-diaminopropane)nickel perchlorate dihydrate, 40B, 1118
$C_7H_3CrNO_5S$, Pentacarbonyl(methyl thiocyanate)chromium, 43B, 1280; 44B, 908
C_7H_5BrCuN, Benzonitrilecopper(I) bromide, 42B, 811
C_7H_5ClCuN, Benzonitrilecopper(I) chloride, 42B, 811
$C_7H_5ClIrNO_2$, Chlorodicarbonylpyridyliridium(I), 40B, 881
$C_7H_5Cl_3NOV$, (Benzonitrile)trichlorooxovanadium(V), 46B, 1048
$C_7H_9Br_2HgN$, Dibromo(2,4-dimethylpyridine)mercury(II), 46B, 1048
$C_7H_9Cl_3KNPt$, Potassium trichloro(2,6-lutidine)platinate(II), 42B, 812
C_7H_9CuIN, 2,6-Lutidinecopper(I) iodide, 43B, 1274
$C_7H_{10}BF_4N_2O_2Rh$, Dicarbonyl(2,4-pentanediimine)rhodium(I) tetra-

fluoroborate, 45B, 1151
$C_7H_{10}Br_2CuN_2$, Dibromo(2-(2-aminoethyl)pyridine)copper(II), 38B, 900
$C_7H_{10}Cl_2CuN_2$, Dichloro(2-(2-aminoethyl)pyridine)copper(II), 39B, 727
$C_7H_{17}Cl_3N_2NiO$, N-Methyl-1,4-diazabicyclo[2.2.2]octonium-trichloro-aquonickelate, 40B, 1118
$C_7H_{19}N_9NiO_6$, Bis(oxamide oximato)nickel(II) dimethylformamide hydrate, 44B, 908
$C_7H_{20}ClCoN_6O_4$, (+)-cis-β-Dinitro-(R-5-methyltriethylenetetramine)cobalt(III) chloride, 39B, 728
$C_7H_{20}ClCrF_2N_4O_4$, trans-Difluoro(1,4,8,11-tetraazaundecane)chromium-(III) perchlorate, 44B, 909
$C_7H_{20}ClCuN_5O_4S$, Bis-(1,3-diaminopropane)isothiocyanatocopper(II) perchlorate, 40B, 882
$C_7H_{20}N_4NiO_5$, (N,N'-Di(2-aminoethyl)malondiamidato)nickel(II) trihydrate, 38B, 901
$C_8H_4FeN_2O_4$, Tetracarbonyl(1,2-diazine)iron, 43B, 1280; 44B, 909
$C_8H_4FeN_2O_4$, Pyrazinetetracarbonyliron, 40B, 885
$C_8H_6ClCuN_4S_2 \cdot H_2O$, Chloro(bis(2-pyrimidyl)disulfide)copper(I) monohydrate, 45B, 1152
$C_8H_6Fe_2N_2O_6$, μ-Methazo-bis(tricarbonyliron), 40B, 1119
$C_8H_7Cl_3NOV$, Trichlorooxo(phenylacetonitrile)vanadium(V), 46B, 1048
$C_8H_8ClCuN_4O_4$, Bis(succinonitrile)copper(I) perchlorate, 34B, 556
$C_8H_8Cl_2N_4Pd$, Bis(diaminomaleonitrile)dichloropalladium(II), 37B, 535
$C_8H_8CuN_5O_3$, Bis(succinonitrilo)copper(I) nitrate, 23, 564
$C_8H_{10}Cl_2CuN_6O_2$, Cytosine copper(II) chloride, 37B, 530
$C_8H_{10}Cl_8N_2O_4Ti_2$, Di-μ-chloro-hexachloro-bis(ethyl-cyanoformate-N)-dititanium(IV), 40B, 882
$C_8H_{11}Cl_2HgN$, (Trimethylpyridine)mercury(II) chloride, 32B, 429
$C_8H_{11}CuIN$, Collidinecopper(I) iodide, 43B, 1274
$C_8H_{12}B_2F_4N_4NiO_4$, Bis(difluoroborondimethylglyoximato)nickel(II), 43B, 1281
$C_8H_{12}Br_2CuN_4$, catena-Di-μ-bromobis(N-methylimidazole)copper(II), 44B, 910
$C_8H_{12}CdN_4O_7$, Triaquabis(uracilato)cadmium(II), 46B, 1051
$C_8H_{12}CdN_{10}O_6$, Bis(8-azahypoxanthinato)tetraaquocadmium(II), 42B, 813
$C_8H_{12}ClCuN_4O_4$, Tetraacetonitrilecopper(I) perchlorate, 41B, 1048
$C_8H_{12}Cl_2CuN_2$, Dichloro(2-(2-methylaminoethyl)pyridine)copper(II), 41B, 1049
$C_8H_{12}Cl_2CuN_4$, trans-Dichlorobis(N-methylimidazole)copper(II), 45B, 1153
$C_8H_{12}Cl_2CuN_4O_3$, Aquodichlorocaffeinecopper(II), 42B, 898
$C_8H_{12}Cl_2N_4Pt$, β-cis-Dichlorobis(N-methylimidazole)platinum(II), 44B, 910
$C_8H_{12}I_2N_2Ni$, Diiodo(2-(β-methylaminoethyl)pyridine)nickel(II), 39B, 729
$C_8H_{12}N_4Na_2NiO_6 \cdot H_2O$, Disodium bis(N-(hydroxyethyl)oxamide)nickel-(II) monohydrate, 46B, 1051
$C_8H_{14}Cl_2CuN_{10}O_{10}$, Bis-(4-aminoimidazole-5-carboxamidoxime)copper(II) perchlorate, 38B, 902
$C_8H_{14}Cl_2N_4O_4RbRh \cdot H_2O$, Rubidium dichloro-bis(dimethylglyoximato)-rhodate(III) monohydrate, 45B, 1239
$C_8H_{14}CuN_4O_4$, Copper dimethylglyoxime, 23, 600; 35B, 760
$C_8H_{14}N_4NiO_4$, Nickel dimethylglyoxime, 17, 760; 23, 602
$C_8H_{14}N_4O_4Pd$, Bis(dimethylglyoximato)palladium(II), 23, 602; 45B, 1154
$C_8H_{14}N_4O_4Pt$, Bis(dimethylglyoximato)platinum(II), 23, 602; 45B, 1154
$C_8H_{16}Br_4Cu_2N_4O_4$, Bis(μ-bromo)bis(bromodimethylglyoximatocopper),

44B, 910; 46B, 55
$C_8H_{16}Br_4Cu_2N_4O_4$, Dibromo(2,3-butanedione dioximato)copper(II) dimer, 44B, 910
$C_8H_{16}Cl_4Cu_2N_4O_4$, Bis(μ-chloro)bis(chlorodimethylglyoximatocopper), 46B, 55
$C_8H_{16}N_6NiO_6$, Diamminediaquabis(uracilato)nickel(II), 46B, 1052
$C_8H_{17}ClCoN_5O_4$, Chloroamminebis(dimethylglyoxime)cobalt(III), 40B, 1120
$C_8H_{17}CuN_5S_2$, Di-(3-aminopropyl)aminebis(isothiocyanato)copper(II), 40B, 883
$C_8H_{18}ClCoN_6O_4$, trans-Dicyanotriethylenetetraminecobalt(III) perchlorate, 39B, 730
$C_8H_{18}Cl_2N_2O_2Pt$, Bis(trimethylnitrosomethane)dichloroplatinum, 44B, 1161
$C_8H_{18}Cl_2N_2PdS_2$, Dichlorobis(thiomorpholine)palladium(II), 44B, 911
$C_8H_{18}CoN_6NaO_{10}$, Sodium trans-dinitro(dimethylglyoximato)(dimethylglyoxime)cobaltate(III) dihydrate, 44B, 912
$C_8H_{18}CuN_6S_2$, Thiocyanatotriethylenetetraminecopper(II) thiocyanate, 40B, 1119
$C_8H_{18}Fe_4N_6O_4S_2$, Di-(μ_3-t-butylamido)-di-(μ_3-sulfido)-tetra(nitrosyliron), 40B, 999
$C_8H_{18}N_6NiS_2$, cis-α-Triethylenetetraminediisothiocyanatonickel(II), 35B, 683
$C_8H_{19}ClCoN_5O_5$, trans-Chlorobis(dimethylglyoximato)(ammine)cobalt-(III) monohydrate, 40B, 917
$C_8H_{20}BrCoN_6O_4$, trans-Bis(dimethylglyoximato)diamminecobalt(III) bromide, 46B, 1053
$C_8H_{20}Cl_2N_4NiO_4$, Dichlorotetrakis(acetaldoxime)nickel(II), 37B, 535
$C_8H_{20}Cl_2N_6NiOS_2$, Chlorobis(acetone thiosemicarbazone)nickel(II) chloride monohydrate, 39B, 730
$C_8H_{20}CoN_7O_7$, trans-Diamminebis[dimethylglyoximato(1-)]cobalt(III) nitrate, 26, 676; 46B, 1053
$C_8H_{20}CuN_6S_2$, Bis(1,3-diaminopropane)copper(II) thiocyanate, 37B, 617
$C_8H_{20}IN_6O_4Rh$, Diamminebis(dimethylglyoximato)rhodium(III) iodide, 44B, 912
$C_8H_{20}I_3N_4Rh$, trans-Diiodotetra(ethyleneimine)rhodium(III) iodide, 34B, 550
$C_8H_{20}N_2NiO_{10}$, Tetraaquobis(succinimidato)nickel(II) dihydrate, 42B, 815
$C_8H_{20}N_6NiS_2$, Bispropylenediaminebisisothiocyanatonickel, 40B, 958
$C_8H_{20}N_8NiO_7S_2$, Nitratobis(acetone thiosemicarbazone)nickel(II) nitrate monohydrate, 39B, 730
$C_8H_{22}BrCoN_6O_4$, (+)-Δ-β-cis-Dinitro(5-methyl-1,4,8,11-tetraazaundecane)cobalt(III) bromide, 39B, 731
$C_8H_{22}BrCoN_6O_4$, (+)-trans-Dinitro(1,10-diamino-4,7-diazadecane)cobalt(III) bromide, 38B, 903
$C_8H_{22}ClCoN_6O_5$, Dinitro(1,4,7,10-tetraazacyclododecane)cobalt(III) chloride monohydrate, 40B, 884
$C_8H_{22}ClCoN_6O_8$, (-)-cis-α-Dinitro-(L-3,8-dimethyltriethylenetetramine)cobalt(III) perchlorate, 38B, 904
$C_8H_{22}ClCoN_6O_8$, (-)-trans-Dinitro-(L-3,8-dimethyltriethylenetetramine)cobalt(III) perchlorate, 38B, 905
$C_8H_{22}ClCoN_6O_8$, $(-)_{546}$-cis-β-Dinitro-(L-3,8-dimethyltriethylenetetramine)cobalt(III) perchlorate, 35B, 684
$C_8H_{22}Cl_2CoN_5O_3$, (-)-trans-Dichloro(1,10-diamino-4,7-diazadecane)cobalt(III) nitrate, 39B, 732
$C_8H_{22}Cl_2CuN_{10}O_4$, Bis(2,2'-iminobis(acetamidoxime))copper(II) chlor-

ide, 35B, 687
$C_8H_{22}Cl_2N_6O_2Pt$, trans-Diamminebis(N-methylimidazole)platinum(II) chloride dihydrate, 38B, 906
$C_8H_{22}N_{16}O_{12}Pd_2S \cdot H_2O$, Bis[(oxamide oxime)(oximide oximato)palladium(II) sulfate hydrate, 46B, 1054
$C_8H_{23}Cl_3CoN_5O_4$, α,α-Chlorotetraethylenepentaminecobalt(III) chloride perchlorate, 35B, 686
$C_8H_{23}Cl_3CoN_5O_8$, α,β-R-Chloro(tetraethylenepentamine)cobalt(III) perchlorate, 38B, 907
$C_8H_{23}Cl_3CoN_5O_8$, α,β-S-Chloro(tetraethylenepentamine)cobalt(III) perchlorate, 38B, 907
$C_8H_{23}Cl_5CoN_5Zn$, α,β-S-Monochlorotetraethylenepentaminecobalt(III) tetrachlorozincate(II), 37B, 537
$C_8H_{24}Cl_2Mo_2N_4$, Dichlorotetrakis(dimethylamido)dimolybdenum, 43B, 1284
$C_8H_{24}Cl_2N_4NiO_6$, Tetrakis(acetamide)bisaquonickel(II) dichloride, 37B, 535
$C_8H_{24}Cl_2N_4P_4Pt \cdot C_2H_3N$, cis-Dichloro(octamethylcyclotetraphosphazene-N,N'')platinum(II) acetonitrile solvate, 45B, 1155
$C_8H_{24}Cl_2N_4W_2$, Dichlorotetrakis(dimethylamido)ditungsten, 43B, 1284
$C_8H_{24}MoN_4$, Tetrakis(dimethylamido)molybdenum(IV), 44B, 913
$C_8H_{25}CoN_{10}O_7$, (4-(2-Aminoethyl)-1,4,7,10-tetraazadecane)azidocobalt(III) nitrate hydrate, 37B, 538
$C_8H_{26}Br_3CoN_6$, s-facial-Bis(diethylenetriamine)cobalt(III) bromide, 38B, 908
$C_8H_{26}Cl_2N_{10}NiO_6$, Bis(2,2'-iminobis(acetamidoxime))nickel(II) chloride dihydrate, 35B, 687
$C_8H_{26}N_8O_6Zn$, Bis(diethylenetriamine)zinc(II) nitrate, 38B, 909
$C_8H_{28}Br_2CuN_6O$, Bis(diethylenetriamine)copper(II) bromide monohydrate, 34B, 489
$C_8H_{28}Br_4N_4Pt_2$, Tetra(ethylamine)platinum(II) tetrabromoplatinate(II), 37B, 529
$C_8H_{28}Cl_2N_2NiO$, Bis(di-(2-aminoethyl)amine)nickel(II) chloride monohydrate, 35B, 702
$C_8H_{28}Cl_4N_4Pt_2$, Tetra(ethylamine)platinum(II) tetrachloroplatinate(II), 37B, 529
$C_8H_{32}Cl_2N_{12}P_4Pt$, cis-Dichloro(octa(methylamino)cyclotetraphosphazene-N,N")platinum(II), 43B, 1284
$C_9H_5FeNO_4$, Pyridinetetracarbonyliron, 40B, 885
$C_9H_6Fe_2N_2O_7$, μ-Dimethylureylene-bis(tricarbonyliron), 33B, 423
$C_9H_9BF_4N_3O_3Re$, Tris(acetonitrile)tricarbonylrhenium(I) tetrafluoroborate, 43B, 1285
$C_9H_{11}ClCuN_6$, catena-μ-Imidazolato-chloro-diimidazolocopper(II), 38B, 910
$C_9H_{11}N_7O_3Zn$, catena-μ-Imidazolato-bis(imidazole)zinc nitrate, 44B, 913
$C_9H_{14}Br_2N_2Ni$, Dibromo(2-(β-dimethylaminoethyl)pyridine)nickel(II), 39B, 729
$C_9H_{14}CdN_6O_5S$, cis-catena-μ-Sulfato-aquotris(imidazole)cadmium(II), 42B, 902
$C_9H_{14}Cl_2CuN_2$, Dichloro(2-(2-dimethylaminoethyl)pyridine)copper(II), 44B, 914
$C_9H_{14}CuN_6O_5S$, Aquotrisimidazolecopper(II) sulfate, 40B, 965
$C_9H_{20}CoN_7O_4S \cdot H_2O$, trans-Diamminebis[dimethylglyoximato(1-)]cobalt(III) thiocyanate hydrate, 46B, 1053
$C_9H_{22}ClCoN_4O_8$, Carbonato(1,4,7,10-tetraazacyclododecane)cobalt(III) perchlorate monohydrate, 42B, 903

$C_9H_{23}Cl_2CoN_3$, Dichlorobis(2-dimethylaminoethyl)methylaminecobalt-(II), 40B, 1120
$C_9H_{24}Br_2CoN_4 \cdot 0.5\ C_2H_5OH$, Bromotris(3-aminopropyl)aminecobalt(II) bromide hemiethanolate, 37B, 540
$C_9H_{32}Br_3CoN_6O$, Tris(trimethylenediamine)cobalt(III) bromide monohydrate, 34B, 491
$C_{10}H_8Br_2MoN_2O_2$, Dioxodibromo-2,2'-bipyridylmolybdenum(VI), 34B, 551
$C_{10}H_8Cl_2CoN_2S_2$, Dichloro(bis(2-pyridyl)disulfide)cobalt(II), 42B, 816
$C_{10}H_8Cl_2HgN_2S_2$, Dichloro(bis(2-pyridyl)disulfide)mercury(II), 42B, 817
$C_{10}H_8Cl_2N_2Pt$, 2,2'-Bipyridyldichloroplatinum(II) (red form), 40B, 886
$C_{10}H_8Cl_4Hg_2N_2O_8$, Bis((3-chloropyridine)mercury(I)) diperchlorate, 39B, 723
$C_{10}H_8Cl_4N_3V$, (2,2'-Bipyridyl)-trichloro-chloroimidovanadium, 45B, 1155
$C_{10}H_8CuN_4O_4$, Bisnitrito-2,2'-bipyridylcopper(II), 34B, 552
$C_{10}H_8CuN_8$, Diazido-2,2'-bipyridinecopper(II), 40B, 886
$C_{10}H_8FN_2O_2V$, 2,2'-Bipyridylfluorodioxovanadium(V), 43B, 1286
$C_{10}H_8HgN_4O_6$, 2,2'-Bipyridylmercury(II) nitrate, 44B, 914
$C_{10}H_8MoN_{12}$, 2,2'-Bipyridyltriazidonitridomolybdenum, 46B, 1054
$C_{10}H_{10}BCuN_6O$, (Hydrotris(1-pyrazolyl)borato)copper(I) carbonyl, 41B, 1049
$C_{10}H_{10}Br_2CuN_2$, Dibromobis(pyridine)copper(II), 24, 608; 41B, 1050
$C_{10}H_{10}Br_4Cu_2N_2$, Di-μ-bromo-bis[bromo(pyridine)copper(II)], 46B, 1055
$C_{10}H_{10}CdCl_2N_2$, Bis(pyridine)cadmium chloride, 24, 618; 34B, 553
$C_{10}H_{10}CdN_8$, Diazidodipyridinecadmium, 35B, 691
$C_{10}H_{10}Cl_2CoN_2$, Dichlorobis(pyridine)cobalt(II) (α form), 21, 616; 41B, 1056
$C_{10}H_{10}Cl_2CoN_2$, Dichlorobis(pyridine)cobalt(II) (γ form), 41B, 1056
$C_{10}H_{10}Cl_2CuN_2$, Dichlorobis(pyridine)copper(II), 20, 605; 21, 616; 41B, 1050
$C_{10}H_{10}Cl_2N_2Pt$, cis-Dichlorobis(pyridine)platinum(II), 41B, 1051
$C_{10}H_{10}Cl_2N_2Pt$, trans-Dichlorobis(pyridine)platinum(II), 41B, 1051
$C_{10}H_{10}Cl_2N_2Zn$, Dichlorobis(pyridine)zinc(II), 24, 614; 31B, 424; 42B, 818
$C_{10}H_{10}Cl_2N_6O_2Zn$, Bis(pyrazinamide)zinc(II) chloride, 38B, 910
$C_{10}H_{10}Cl_4N_2W$, trans-Tetrachlorobis(pyridine)tungsten, 45B, 1156
$C_{10}H_{10}CrN_2O_5$, Pentacarbonyldiethylaminonitrilechromium(0), 44B, 915
$C_{10}H_{10}CuN_4$, Bis-(H-pyrrole-2-aldimine)copper(II), 37B, 540
$C_{10}H_{10}CuN_8$, Bis(pyridine)copper(II) azide, 34B, 554
$C_{10}H_{10}I_2N_2Pt$, trans-Diiodobis(pyridine)platinum(II), 44B, 915
$C_{10}H_{10}I_2N_2Zn$, Diiodobis(pyridine)zinc, 43B, 1286
$C_{10}H_{10}N_8Zn$, Diazidodipyridinezinc, 35B, 693
$C_{10}H_{12}Br_4CuN_{10}$, Dibromodiadeniniumcopper(II) dibromide, 39B, 733
$C_{10}H_{12}Cl_4CuN_{10}$, Dichlorobis(adeninium)copper(II) chloride, 43B, 1287
$C_{10}H_{12}CuN_2O_8S_2$, Diaquo-bis(pyridine-3-sulphonato)copper(II), 46B, 1057
$C_{10}H_{12}CuN_5O_3$, Bis(glutaronitrilo)copper(I) nitrate, 23, 565
$C_{10}H_{12}Hg_2N_2O_8S_2$, Diaquabis(pyridine-3-sulphonate)dimercury(I), 46B, 1055
$C_{10}H_{12}MnN_2O_4P_2$, catena-Bis(phosphinato)(2,2'-bipyridyl)manganese-(II), 44B, 915
$C_{10}H_{12}N_2NiO_6S$, Diaquo-2,2'-bipyridylnickel sulfate, 40B, 965
$C_{10}H_{13}Cl_2HgN_5O_5$, catena-(μ-Chloro)-chloro(guanosine-N^7)mercury(II), 44B, 916

$C_{10}H_{14}Cl_2CuN_2S_2$, trans-Dichlorobis(2,4-dimethylthiazole)copper(II), 45B, 1156
$C_{10}H_{14}Cl_2CuN_8O_5$, Bis-(6-hydroxypurine)copper(II) chloride trihydrate, 35B, 694
$C_{10}H_{14}Cl_2HgN_2$, Dichloronicotinemercury(II), 46B, 1056
$C_{10}H_{14}Cl_2N_6O_2Pd$, Dichlorobis(1-methylcytosine)palladium(II), 43B, 1288
$C_{10}H_{14}Co_2N_8O_4$, Di-μ-(3,5-dimethylpyrazolyl)bis(dinitrosylcobalt), 45B, 1157
$C_{10}H_{14}CuN_4O_9$, Nitratotriaquo(2,2'-bipyridine)copper(II) nitrate, 43B, 1290
$C_{10}H_{14}Fe_2N_8O_4$, Di-μ-(3,5-dimethylpyrazolyl)bis(dinitrosyliron), 45B, 1157
$C_{10}H_{14}HgN_4O_8$, Bispyridinemercury(II) nitrate dihydrate, 42B, 818
$C_{10}H_{14}N_6Ni_2O_2$, Bis(μ-3,5-dimethylpyrazolyl(N,N'))bis(nitrosylnickel-(I)), 45B, 1158
$C_{10}H_{14}N_8Ni$, Dihydrooctaaza[14]annulenenickel, 42B, 851
$C_{10}H_{15}Br_2N_3Zn$, Dibromo-1-(2-pyridyl)-2,5-diaza-5-methyl-hexa-1-ene-zinc(II), 35B, 696
$C_{10}H_{16}CoN_6O_6S_2$, Dinitratobis(2-methylthio-3-methylimidazole)cobalt-(II), 44B, 916
$C_{10}H_{16}CuN_6O_{10}$, Bis(2-pyrimidylcarbonyl)aminatotriaquocopper(II) nitrate dihydrate, 43B, 1288
$C_{10}H_{16}CuN_{10}O_4$, Bis(6-aminopurine)copper(II) tetrahydrate, 34B, 559
$C_{10}H_{16}Cu_{0.5}N_2O_{10}S_2Zn_{0.5}$, Tetraaquo-bis(pyridine-3-sulphonato)copper-(II)zinc(II), 46B, 1057
$C_{10}H_{16}N_2O_{10}S_2Zn$, Tetraaquo-bis(pyridine-3-sulphonato)zinc(II), 46B, 1057
$C_{10}H_{16}N_6NiO_6S_2$, Dinitratobis(2-methylthio-3-methylimidazole)nickel-(II), 44B, 917
$C_{10}H_{16}N_6O_6S_2Zn$, Dinitratobis(2-methylthio-3-methylimidazole)zinc(II), 44B, 916
$C_{10}H_{18}ClN_7O_7Pt \cdot 3\ H_2O$, cis-Bisammine(thiaminato)-1-methylcytosinatoplatinum(II) perchlorate trihydrate, 45B, 1158
$C_{10}H_{18}Cl_2N_6Pd$, Bis(histamino)palladium dichloride, 42B, 819
$C_{10}H_{18}D_5ClN_4NiO_2 \cdot D_2O$, Bis(3-amino-3-methyl-2-butanone oximato)nickel(II) chloride monohydrate, 46B, 1058
$C_{10}H_{18}N_4NiO_2$, Dicarbonyldiacetylbis(dimethylhydrazone)nickel(0), 38B, 911
$C_{10}H_{18}N_4NiO_4$, Bis(methylethylglyoximato)nickel, 24, 603; 38B, 911
$C_{10}H_{18}N_8Ni$, (6,7,13,14-Tetramethyl-1,2,4,5,8,9,11,12-octaazacyclotetradeca-1,3,8,10-tetraenato(2-))nickel(II), 44B, 918
$C_{10}H_{19}Br_2CoN_4O_4$, (O-Methyl-diacetyldioxime)(O-methyl-diacetyldioximato)dibromocobalt(III), 41B, 1053
$C_{10}H_{20}Cl_2CoN_2O_2$, Dichloro(ethylenedimorpholine)cobalt(II), 40B, 1121
$C_{10}H_{20}Cl_2N_6NiO_9$, Aquo-cis-bis(histamino)perchloratonickel(II) perchlorate, 35B, 655
$C_{10}H_{20}Cl_8Cu_3N_{10}O_4$, Octachlorobis(adeninium)tricopper(II) tetrahydrate, 38B, 912
$C_{10}H_{20}N_{12}Ni$, (6,14-Dihydrazyl-6,7,13,14-tetramethyl-1,2,4,5,8,9,11,-12-octaazacyclotetradeca-1,3,7,9,12-pentaenato(2-))nickel(II), 44B, 918
$C_{10}H_{23}Br_4N_3ORe$, Tetraethylammonium acetonitriletetrabromonitrosylrhenate(II), 41B, 1117
$C_{10}H_{23}ClN_4O_2Pt \cdot 3.5\ H_2O$, Bis(2-amino-2-methyl-3-butanone oximato)-platium(II) chloride 3.5 hydrate, 43B, 1291
$C_{10}H_{23}Cl_2CoN_4O_2$, Dichlorobis(2-amino-2-methyl-3-butanone oximato)co-

balt(III), 44B, 918
$C_{10}H_{24}Cl_2CoN_4O_8$, (1,4,8,11-Tetraazacyclotetradecane)cobalt(II) diperchlorate, 43B, 1292
$C_{10}H_{24}Cl_2CuN_4O_8$, Di(perchlorato)(1,4,8,11-tetraazacyclotetradecane)-copper(II), 41B, 1053
$C_{10}H_{24}Cl_2NPPt$, trans-Dichloro(pyrrolidine)(triethylphosphine)platinum(II), 46B, 1058
$C_{10}H_{24}Cl_2N_4Ni$, Chloro(N,N'-di-(3-aminopropyl)-piperazine)nickel(II) chloride, 40B, 887
$C_{10}H_{24}Cl_2N_8O_6Zn$, Diaquobis(2-hydrazino-4-hydroxy-6-methylpyrimidine)zinc chloride dihydrate, 43B, 1293
$C_{10}H_{24}CuN_6O_6 \cdot 0.5\ H_2O$, Bis(1,4-diazacycloheptane)copper(II) nitrate hemihydrate, 34B, 562
$C_{10}H_{24}CuN_6S_2$, Bis(N,N'-dimethylethylenediamine)copper(II) thiocyanate, 35B, 678
$C_{10}H_{25}ClN_4NiO_3$, Bis(2-amino-2-methyl-3-butanone oximato)nickel(II) chloride monohydrate, 33B, 424; 37B, 540
$C_{10}H_{25}ClN_4O_3Pt$, Bis(2-amino-2-methyl-3-butanoneoximato)platinum(II) chloride monohydrate, 40B, 1121
$C_{10}H_{25}CoN_4O_{15}P$, (Inosine 5'-phosphate)cobalt hydrate, 41B, 1052
$C_{10}H_{25}N_4NiO_{15}P$, (Inosine 5'-phosphate)nickel hydrate, 41B, 1052
$C_{10}H_{26}Cl_2CuN_4O_8$, Sperminecopper(II) perchlorate, 41B, 1054
$C_{10}H_{26}Cl_2N_2O_{13}Pd$, Dichloro(tetrahydrogenethylenediaminetetra-acetato)palladium(II) pentahydrate, 35B, 697
$C_{10}H_{28}CdN_5O_{16}P$, Pentaaquo-guanosine-5'-phosphate-cadmium(II) trihydrate, 42B, 905
$C_{10}H_{28}N_4Zn_2$, Bis(μ-(2-dimethylamino-N-methylethylamido)-N',μ-N)bis-(hydridozinc), 46B, 1059
$C_{10}H_{31}Cl_4CoN_6 \cdot 2\ H_2O$, (1,1,1-Tris(aminomethyl)ethane-N,N')-(1,1,1-tris(aminomethyl)ethane-N,N',N")chlorocobalt(III) trichloride dihydrate, 46B, 1060
$C_{11}H_4Fe_2N_2O_7$, (1,2-Diazine)heptacarbonyldiiron(Fe-Fe), 43B, 1294
$C_{11}H_6Fe_3N_2O_9$, Di-μ_3-methylimido-tris(tricarbonyliron), 34B, 564
$C_{11}H_8Fe_2N_2O_6$, μ-2,3-(2,3-Diazabicyclo[2.2.1]heptane)diyl-bis(tricarbonyliron), 38B, 913
$C_{11}H_{10}BMoN_7O_3$, Dicarbonylnitrosyl(hydrotris(pyrazol-1-yl)borato)molybdenum, 42B, 819
$C_{11}H_{10}Cl_2CoN_4$, Dichloro(pyridine-2-aldehyde 2-pyridylhydrazone)cobalt(II), 31B, 425
$C_{11}H_{13}BN_6OPt$, Methyl(hydrotris(pyrazol-1-yl)borato)carbonylplatinum-(II), 42B, 820
$C_{11}H_{17}Cl_5Hg_2N_2$, (N-Benzylpiperazinium)pentachlorodimercury(II), 46B, 1060
$C_{11}H_{18}BCuF_2IN_4O_2$, Iodo[difluoro(3,3'-(trimethylenedinitrilo)bis(2-butanone oximato))borato]copper(II), 45B, 1161
$C_{11}H_{18}BCuF_4N_4O_2$, (1,1-Difluoro-4,5,11,12-tetramethyl-1-bora-3,6,10,13-tetraaza-2,14-dioxacyclotetradeca-3,5,10,12-tetraenato)-copper(I), 44B, 919
$C_{11}H_{18}CuN_6O_4$, Bis(dimethylglyoximato)(imidazole)copper(II), 46B, 1060
$C_{11}H_{19}BrN_4NiO_2$, Bromo-2,2'-(1,3-diiminopropane)bis(3-butanone oximato)nickel(II), 45B, 1161
$C_{11}H_{19}ClN_4NiO_6$, 2,2'-(1,3-Diiminopropane)bis(3-butanone oximato)nickel(II) perchlorate, 45B, 1031
$C_{11}H_{19}Cl_2CoN_4O_2$, Dichloro-2,2'-(1,3-diiminopropane)bis(3-butanone oximato)cobalt(III), 45B, 1162
$C_{11}H_{19}Cl_2N_3O_6PtS$, trans-Dichloro(dimethylsulfoxide)(cytidine)platin-

um(II), 44B, 919
$C_{11}H_{21}ClN_4NiO_4$, 11,13-Dimethyl-1,4,7,10-tetraazacyclotrideca-10,12-dienatonickel(II) perchlorate, 38B, 913
$C_{11}H_{21}CuNO_8$, DL-(Ethylvalinate-N,N'-diacetato)diaquocopper(II), 41B, 1034
$C_{11}H_{23}ClCoN_4O_7P$, trans-Chlorobis(dimethylglyoximato)(trimethyl phosphite)cobalt(III), 46B, 1099
$C_{11}H_{26}Cl_2N_4NiO_8$, (3,3-Dimethyl-1,5,8,11-tetraazacyclotridecane)nickel(II) diperchlorate, 38B, 914
$C_{12}H_6CoHg_2N_6S_6$, Hexathiocyanato-cobalt(II)-dimercury(II) benzene, 29, 629
$C_{12}H_8Br_3N_2O_2Re$, Tribromodicarbonyl(2,2'-bipyridyl)rhenium(III), 44B, 920
$C_{12}H_8Cl_2Hg_2N_4O_8$, 4-Cyanopyridinemercury(I) perchlorate, 38B, 917
$C_{12}H_8Cl_2N_2Zn$, Dichloro(1,10-phenanthroline)zinc, 31B, 427
$C_{12}H_8Cl_2N_4Zn$, Dichlorobis(4-cyanopyridine)zinc(II), 43B, 1301
$C_{12}H_8Cl_3FeN_4$, Trichlorobis(4-cyanopyridine-N[1])iron(III), 46B, 1062
$C_{12}H_8CrN_2O_5$, Oxodiperoxo-1,10-phenanthrolinechromium(VI), 30B, 350
$C_{12}H_8CuN_4O_6$, Dinitrato(1,10-phenanthroline)copper(II), 41B, 1055
$C_{12}H_8Hg_2N_4O_6$, 1,10-Phenanthrolinemercury(I) nitrate, 32B, 430
$C_{12}H_8I_6N_2Pt$, Tetraiodo(1,10-phenanthroline)platinum(IV) iodine, 43B, 1295
$C_{12}H_8MoN_2O_6S_2$, Dicarbonyl(2,2'-bipyridyl)bis(η^2-sulfur dioxide)molybdenum(0), 46B, 1075
$C_{12}H_{10}AgN_3O_3$, Nitrato-2-pyridyleneanilinesilver, 46B, 1062
$C_{12}H_{10}Cl_2CuN_6O_8$, Dihistaminocopper(II) perchlorate, 35B, 662
$C_{12}H_{10}Cl_2N_2O_2Pd$, Dichlorobis(nitrosobenzene)palladium(II), 39B, 734
$C_{12}H_{10}Cl_2N_4Zn$, β-Dichloro-(2-pyridinaldazine)zinc(II), 41B, 1057
$C_{12}H_{10}CuN_3$, Diazoaminobenzenecopper(I), 26, 655
$C_{12}H_{10}I_{0.018}N_4NiO_4$, Bis(o-benzoquinone dioximato)nickel(II) iodine, 45B, 1162
$C_{12}H_{10}I_{0.48}N_4O_4Pd \cdot 0.91\ C_6H_4Cl_2$, Bis(o-benzoquinone dioximato)palladium(II) iodine o-dichlorobenzene solvate, 45B, 1162
$C_{12}H_{10}I_{0.5}N_4NiO_4$, Bis-(1,2-benzoquinonedioximato)nickel(II) iodine, 41B, 1059
$C_{12}H_{10}I_2N_4O_4Pt$, Diiodo-bis(1,2-benzoquinonedioximato)platinum (form I), 42B, 821
$C_{12}H_{10}I_2N_4O_4Pt$, Diiodo-bis(1,2-benzoquinonedioximato)platinum (form II), 42B, 821
$C_{12}H_{10}N_4NiO_4$, Bis-(1,2-benzoquinonedioximato)nickel(II), 41B, 1060
$C_{12}H_{10}N_4O_4Pd$, α-Bis-(1,2-benzoquinonedioximato)palladium(II), 41B, 1060
$C_{12}H_{10}N_4O_4Pd$, β-Bis(1,2-benzoquinone dioximato)palladium(II), 45B, 1163
$C_{12}H_{10}N_4O_4Pt$, Bis(1,2-benzoquinone dioximato)platinum(II), 45B, 1163
$C_{12}H_{10}N_4O_4Pt \cdot 0.5\ AgClO_4$, Bis-(1,2-benzoquinonedioximato)platinum-(II) hemi(silver perchlorate), 42B, 822
$C_{12}H_{10}N_4PtS_2$, trans-Dithiocyanatobis(pyridine)platinum(II), 41B, 1061
$C_{12}H_{12}AgClN_8O_6 \cdot H_2O$, Bis(9-methylhypoxanthine)silver(I) perchlorate monohydrate, 46B, 1063
$C_{12}H_{12}Cl_3FeN_8 \cdot H_2O$, cis-Dichlorobis(2,2'-biimidazole)iron(III) chloride monohydrate, 46B, 1064
$C_{12}H_{12}CuN_4O_8$, Diaquonitrato(1,10-phenanthroline)copper(II) nitrate, 41B, 1055
$C_{12}H_{12}N_4Ni$, Bis(o-phenylenediamino)nickel, 33B, 425
$C_{12}H_{12}N_8Zn_2$, Bis(diimidazolyl zinc), 46B, 1113

$C_{12}H_{13}Cl_3N_2OZn$, Trichloro(N-(4'-pyridyl)-4-ethoxypyridinium)zinc(II), 44B, 920
$C_{12}H_{14}Br_2CuN_2$, Dibromobis(2-methylpyridine)copper(II), 38B, 915
$C_{12}H_{14}Br_2N_2Zn$, Bis-(γ-picoline)zinc(II) dibromide, 38B, 915
$C_{12}H_{14}Cl_2CoN_2$, Dichlorobis-(4-methylpyridine)cobalt(II), 41B, 1061
$C_{12}H_{14}Cl_2CuN_2$, Dichlorobis(2-methylpyridine)copper(II), 34B, 566
$C_{12}H_{14}Cl_2N_2Zn$, Dichlorobis(4-methylpyridine)zinc(II), 37B, 541
$C_{12}H_{14}Cl_3CoN_2O_3$, cis-Dichloro-bis-phenanthrolinecobalt(III) chloride trihydrate, 31B, 427
$C_{12}H_{14}Cu_4I_4N_2$, Bis(μ-iodo)-bis(μ₃-iodo)-bis(2-methylpyridine)tetracopper(I), 46B, 1064
$C_{12}H_{14}N_4NiO_4$, Bis(pyridine-2-carboxamido)nickel(II) dihydrate, 33B, 427
$C_{12}H_{14}N_4O_4Pt$, Bis(pyridine-2-carboxaldoximinato)platinum(II) dihydrate, 42B, 824
$C_{12}H_{16}AgClN_4O_4$, Adiponitrile-silver perchlorate, 40B, 1122
$C_{12}H_{16}B_2CoN_8$, Bis(dihydrobis(1-pyrazolyl)borato)cobalt(II), 39B, 735
$C_{12}H_{16}B_2CrN_{12}$, Bis(dihydrobis(1-pyrazolyl)borato)chromium(II), 44B, 941
$C_{12}H_{16}B_2N_8Ni$, Bis(dihydrobis-(1-pyrazolyl)borato)nickel(II), 42B, 824
$C_{12}H_{16}Br_2N_8Ni$, Dibromotetra(pyrazole)nickel(II), 34B, 568
$C_{12}H_{16}Cl_2CuN_8$, Dichlorotetrapyrazolecopper(II), 41B, 1062
$C_{12}H_{16}Cl_2N_4Pt \cdot H_2O$, Bis[2-(aminomethyl)pyridine]platinum(II) dichloride monohydrate, 45B, 1164
$C_{12}H_{16}Cl_2N_8Ni$, Tetrakispyrazolenickel chloride, 32B, 432
$C_{12}H_{16}Cl_2N_8O_8Zn$, Tetraimidazolezinc(II) perchlorate, 41B, 1063
$C_{12}H_{16}CuN_5O_3$, Bis(adiponitrilo)copper(I) nitrate, 23, 566
$C_{12}H_{16}CuN_6O_2$, α-Bis(1-carbamoyl-3,5-dimethylpyrazolato)copper(II), 45B, 1165
$C_{12}H_{16}CuN_{10}O_6$, Tetrakis(imidazole)dinitratocopper(II), 42B, 825
$C_{12}H_{16}N_2O_8Zn$, Zinc picolinate tetrahydrate, 40B, 1126
$C_{12}H_{16}N_{10}NiO_8$, Diaquo-bis(2,2'-biimidazole)nickel(II) nitrate, 34B, 569
$C_{12}H_{17}Cl_3N_{10}OZn$, 9-Methyladeninium trichloro(9-methyladenine)-zincate(II) monohydrate, 42B, 825
$C_{12}H_{18}BCuF_2N_4O_3$, Carbonyl(difluoro-3,3'-(trimethylenedinitrilo)bis-(2-butanone oximato)borate)copper(I), 43B, 1296
$C_{12}H_{18}BCuF_2N_5O_2 \cdot CH_4O$, Cyano[difluoro-3,3'-(trimethylenedinitrilo)-bis(2-butanoneoximato)borato]copper(III) methanol, 46B, 1065
$C_{12}H_{18}BCuF_2N_5O_3$, (Cyanato-N)(difluoro-3,3'-(trimethylenedinitrilo)-bis(2-butanone oximato)borato)copper(II), 44B, 921
$C_{12}H_{18}Br_2CoN_2O_2S_2$, Dibromobis-(5-(2-hydroxyethyl)-4-methylthiazole)-cobalt(II), 40B, 888
$C_{12}H_{18}Cl_2MnN_6$, Dichlorotris-(2-methylimidazole)manganese(II), 42B, 826
$C_{12}H_{18}Cl_2N_2Pt$, trans-Dichloro-ethylene-acetonemethylphenylhydrazone-platinum(II), 42B, 827
$C_{12}H_{18}Cl_4N_6NiZn$, Hexaacetonitrilenickel(II) tetrachlorozincate, 42B, 827
$C_{12}H_{18}Cl_8Fe_3N_6$, Hexakis(acetonitrile)iron(II) bis(tetrachloroferrate), 38B, 916, 917; 46B, 1066
$C_{12}H_{18}Cl_8Ga_2N_6Ni$, Hexakis(methyl cyanide)nickel tetrachlorogallate, 38B, 916
$C_{12}H_{18}CuN_{12}O_{10}$, Dinitratodiaquobis(9-methylguanine)copper(II), 43B, 1296
$C_{12}H_{18}N_2O_4SW$, Di-t-butylsulfurdiiminetetracarbonyltungsten(0), 41B,

1063
$C_{12}H_{19}Br_3CuN_2O_2S_2$, 5-(2-Hydroxyethyl)-4-methylthiazolium tribromo-5-(2-hydroxyethyl)-4-methyl-3-thiazolocuprate(II), 40B, 889
$C_{12}H_{20}Cl_2N_4NiO_4$, cis-Dichloro-bis(1,2-cyclohexadione dioxime)nickel-(II), 45B, 1165
$C_{12}H_{20}CuN_8O_{10}S$, Tetraaquabis(9-methylhypoxanthine)copper(II) sulfate, 43B, 1297
$C_{12}H_{20}MoNO_7P$, cis-Piperidine(trimethyl phosphite)tetracarbonylmolybdenum, 43B, 1298
$C_{12}H_{21}CoN_6O_5$, trans-Bis(dimethylglyoximato)hydroxo(3-N-methylimidazole)cobalt(III), 42B, 833
$C_{12}H_{22}Cl_2CuN_8O_7$, Dichlorobis-(9-methyl-6-oxypurine)diaquocopper(II) trihydrate, 40B, 970
$C_{12}H_{22}Cl_2N_2O_2Pd$, Palladium(II) chloride bis(cyclohexanone oxime), 32B, 432
$C_{12}H_{22}Cl_3CoN_4O_5$, Dichloro(5,7-dimethyl-1,4,8,11-tetraazacyclotetradeca-4,7-diene-6-one)cobalt(III) perchlorate, 43B, 1299
$C_{12}H_{22}FeN_8O$, Carbonyl-methyl-(6,7,13,14-tetramethyl-1,2,4,5,8,9,11,-12-octa-azacyclotetradeca-1(14),5,7,12-tetraenyl)iron(II), 40B, 890
$C_{12}H_{24}Cl_2N_4Pd$, Dichlorobis(1,3-dimethyl-1,4,5,6-tetrahydropyridazine)palladium(II), 46B, 1066
$C_{12}H_{24}Cl_2N_6NiO_{10}$, Bis[2-(2-aminoethyl)imino-3-butanone oximato]nickel(IV) diperchlorate, 46B, 1067
$C_{12}H_{24}Cl_2N_8O_{10}Ru$, cis-Tetraamminebis(isonicotinamide)ruthenium(II) perchlorate, 45B, 1166
$C_{12}H_{24}Cl_3N_8O_{14}Ru \cdot H_2O$, cis-Tetraamminebis(isonicotinamide)ruthenium(III) perchlorate monohydrate, 45B, 1166
$C_{12}H_{26}Br_2N_2Pt$, trans-Dibromobis(cyclohexylamine)platinum(II), 46B, 1067
$C_{12}H_{26}Cl_2CoN_4O_{11}$, Diaquo(5,7-dimethyl-1,4,8,11-tetraazacyclotetradeca-4,7-diene-6-one)cobalt(II) perchlorate, 43B, 1299
$C_{12}H_{26}Cl_2CuN_{10}O_6$, Tetraaquo-bis-(9-methyladenine)copper(II) dichloride dihydrate, 41B, 1064
$C_{12}H_{26}Cl_2N_2Pt$, cis-Dichlorobis(cyclohexylamine)platinum(II), 43B, 1299
$C_{12}H_{26}Cl_2N_2Pt$, trans-Dichlorobis(cyclohexylamine)platinum(II), 44B, 922
$C_{12}H_{26}Cl_3CoN_8O$, (1,3,6,8,10,13,16,19-Octaazabicyclo[6.6.6]eicosane)-cobalt(III) chloride monohydrate, 43B, 1137
$C_{12}H_{26}CuN_{10}O_{12}S$, Di-9-methylguaninetriaquocopper(II) sulfate trihydrate, 42B, 827
$C_{12}H_{28}Br_2N_4Ni$, Bis(2-imino-4-amino-4-methylpentane)nickel(II) bromide, 38B, 918
$C_{12}H_{28}Cl_2NPPt$, trans-Dichloro(trans-2,5-dimethylpyrrolidine)(triethylphosphine)platinum(II), 46B, 1058
$C_{12}H_{28}Cl_2N_4NiO_8$, Bis(2,4,4-trimethyl-1,5-diazapent-1-ene)nickel(II) perchlorate, 41B, 1065
$C_{12}H_{28}Cl_2N_4NiO_8$, 5,12-Dimethyl-1,4,8,11-tetraazacyclotetradecanenickel(II) diperchlorate, 43B, 1300
$C_{12}H_{28}Cl_2N_4NiO_{10}$, Bis(1,5-diazacyclooctane)nickel(II) perchlorate dihydrate, 38B, 918
$C_{12}H_{28}Cl_2N_4Pd$, Bis(cis-1,3-diaminocyclohexane)palladium(II) chloride, 44B, 922
$C_{12}H_{28}Cl_2N_4Pd$, Dichlorobis(azoisopropane)palladium(II), 44B, 947
$C_{12}H_{28}CoN_{13}$, trans-Diazido(C-meso-5,12-dimethyl-1,4,8,11-tetraazacyclotetradecane)cobalt(III) azide, 44B, 922
$C_{12}H_{28}CuN_6O_6$, Bis(2-imino-4-amino-4-methylpentane)copper(II) ni-

trate, 29, 601
$C_{12}H_{28}N_6NiO_6$, Bis(2-imino-4-amino-4-methylpentane)nickel(II) nitrate, 34B, 571
$C_{12}H_{29}BrCuN_6$, Bromoazido-1,1,7,7-tetraethyldiethylenetriaminecopper-(II), 38B, 919
$C_{12}H_{29}N_{12}Rh \cdot 0.81\ C_2H_6O$, Triazido-1,1,7,7-tetraethyldiethylenetriaminerhodium(III) ethanol solvate, 40B, 890
$C_{12}H_{30}Br_2CuN_4$, Tris-(2-dimethylaminoethyl)aminecopper(II) bromide, 33B, 427
$C_{12}H_{30}Br_2FeN_4$, Tris-(2-dimethylaminoethyl)amineiron(II) bromide, 33B, 427
$C_{12}H_{30}Br_2MnN_4$, Tris-(2-dimethylaminoethyl)aminemanganese(II) bromide, 33B, 427
$C_{12}H_{30}Br_2N_4Ni$, Tris-(2-dimethylaminoethyl)aminenickel(II) bromide, 33B, 427
$C_{12}H_{30}Br_2N_4Zn$, Tris-(2-dimethylaminoethyl)aminezinc(II) bromide, 33B, 427
$C_{12}H_{30}Cl_2N_4NiO_8$, 2,10-Diamino-5,5,7-trimethyl-4,8-diaza-undecanenickel(II) perchlorate, 42B, 828
$C_{12}H_{30}CuCl_2N_6O_8$, Bis(cis,cis-1,3,5-triaminocyclohexane)copper(II) perchlorate, 45B, 1167
$C_{12}H_{30}CuN_8O_6$, Bis(cis,cis-1,3,5-triaminocyclohexane)copper(II) nitrate, 45B, 1167
$C_{12}H_{30}Mo_2N_6$, Tris(N,N'-dimethylethylenediamido)dimolybdenum, 46B, 1068
$C_{12}H_{30}N_8NiO_6$, Bis(cis,cis-1,3,5-triaminocyclohexane)nickel(II) nitrate, 45B, 1167
$C_{12}H_{32}ClN_7NiO_4$, Bis(1,4,7-triazacyclononane)nickel(II) nitrate chloride monohydrate, 44B, 923
$C_{12}H_{34}Cl_2N_6NiO_8$, Bis(di-(3-aminopropyl)amine)nickel(II) perchlorate, 35B, 702
$C_{12}H_{36}Br_3CoN_6$, Tris-(1,4-diaminobutane)cobalt(III) bromide, 41B, 1067
$C_{12}H_{36}Mo_2N_6$, Hexakis(dimethylamido)dimolybdenum, 42B, 829
$C_{12}H_{36}N_4Si_4Ti$, Bis-(1,2-bis(methylimino)tetramethyldisilane)titanium, 41B, 1067
$C_{12}H_{36}N_6W_2$, Hexakis(dimethylamido)ditungsten(III), 42B, 871
$C_{12}H_{41}Cl_3Co_2N_8O_{18}$, Bis(N,N-bis(2-aminoethyl)-1,2-ethanediamine-N,N',N'',N''')-μ-hydroxy(μ-(peroxy)-O:O))dicobalt(III) perchlorate trihydrate, 42B, 908
$C_{13}F_{12}Mn_2N_2O_7$, μ-Carbonyl-bis-(μ-hexafluoroisopropylidenimino)-hexacarbonyldimanganese, 41B, 1069
$C_{13}H_3Mn_3N_2$, μ-Methanediazo-(N,N,N')-tris(tetracarbonylmanganese), 45B, 835
$C_{13}H_8Cl_2HgMoN_2O_3$, (2,2'-Bipyridyl)tricarbonylchloro(chloromercurio)-molybdenum(II), 40B, 892
$C_{13}H_8F_2N_2O_5PRe$, Triscarbonyl-2,2'-bipyridyl-difluorophosphato-rhenium, 46B, 1068
$C_{13}H_{10}CoN_8O_4$, trans-Azidobis(dimethylglyoximato)pyridinecobalt(III), 44B, 925
$C_{13}H_{17}Br_2CuN_3O_2$, Dibromo(2-picolyl-(2-pyridylmethanol)methylamine)-copper(II) monohydrate, 44B, 924
$C_{13}H_{21}N_5NiO_6$, (2,2,3,9,10,10-Hexamethyl-5,7-dioxa-6-hydra-1,4,8,11-tetraaza-13-nitrocyclotetradeca-3,8,11,13-tetraene)nickel(II), 46B, 1070
$C_{13}H_{23}CoN_6O_4$, trans-Bis(dimethylglyoximato)methyl(3-N-methylimidazole)cobalt(III), 42B, 833

$C_{13}H_{24}Cr_2I_2N_6O_5 \cdot H_2O$, μ-Carbonato-bis(μ-hydroxo)-bis[(1,4,7-triazacyclononane)chromium(III)] diiodide hydrate, 46B, 1070
$C_{13}H_{25}Cl_4N_3ORe$, Tetraethylammonium tetrachloronitrosyl(pyridine)-rhenate(-1), 44B, 925
$C_{13}H_{25}CoN_4O_2$, Dimethyl-(3,3'-(trimethylenedinitrilo)-bis(butan-2-one oximato))cobalt(III), 40B, 892
$C_{13}H_{27}ClN_4NiO_2 \cdot H_2O$, 2,2'-(1,3-Diaminopropane)bis(2-methyl-3-butanone oxime)nickel(II) chloride hydrate, 45B, 1167
$C_{13}H_{27}CoN_6O_6$, trans-Dinitro-2,2'-(1,3-diaminopropane)bis(2-methyl-3-butanone)dioximatocobalt(III), 39B, 736
$C_{13}H_{27}CuN_4O_6Re$, Perrhenato-2,2'-(1,3-diaminopropane)bis(2-methyl-3-butanoneoximato)copper(II), 41B, 1068
$C_{13}H_{27}N_5O_5Pd$, [N,N'-(1,3-Propandiyl)bis(2-amino-2-methyl-3-butanone oximato)]palladium(II) nitrate, 45B, 1168
$C_{13}H_{28}Cl_3CoN_4O_4$, trans-Dichloro(1,7-bis(2(S)-pyrrolidyl)-2,6-diazaheptane)cobalt(III) perchlorate, 44B, 926
$C_{14}H_4N_2O_{10}Ru_3$, (1,2-Diazine)decacarbonyl-triangulo-triruthenium, 42B, 830; 43B, 1301
$C_{14}H_9Br_2MnN_2O$, Aquodibromo(2-(2-pyridyl)quinolinyl)manganese(II), 42B, 831
$C_{14}H_9CuN_5S_2$, Bis(isothiocyanato)(2-(2-pyridyl)benzimidazole)copper-(II), 42B, 843
$C_{14}H_{10}AuBr_3N_2$, Tribromo(2-(2'-pyridyl)quinoline)gold(III), 44B, 927
$C_{14}H_{10}AuCl_3N_2$, Trichloro(2-(2'-pyridyl)quinoline)gold(III), 44B, 927
$C_{14}H_{10}CdN_6Ni$, Bispyridinecadmium tetracyanonickelate, 42B, 831
$C_{14}H_{10}Cl_3N_2OV$, Bis(benzonitrile)trichlorooxovanadium(V), 46B, 1071
$C_{14}H_{10}CoHgN_6S_4$, Bispyridinecobalt(II) mercury(II) tetrathiocyanate, 42B, 832
$C_{14}H_{11}FeNO_4$, (N-Methyl-cinnamaldehyde-imine)-tetracarbonyl-iron, 45B, 1169
$C_{14}H_{12}AuBr_3N_2$, Tribromo(2,9-dimethyl-1,10-phenanthroline)gold(III), 41B, 1070
$C_{14}H_{12}AuCl_3N_2$, Trichloro(2,9-dimethyl-1,10-phenanthroline)gold(III), 41B, 1070
$C_{14}H_{12}Cl_2N_2Zn$, Dichloro(2,9-dimethyl-1,10-phenanthroline)zinc, 34B, 557
$C_{14}H_{12}HgN_2O_2$, Bis(benzamido)mercury(II), 46B, 1071
$C_{14}H_{12}I_2N_2Ni$, (2,9-Dimethyl-1,10-phenanthroline)diiodonickel(II), 45B, 1170
$C_{14}H_{12}N_4O_4Pd$, Dinitro(2,9-dimethyl-1,10-phenanthroline)palladium-(II), 37B, 542
$C_{14}H_{14}AuCl_3N_2$, (4,4'-Azotoluene)trichlorogold(III), 44B, 927
$C_{14}H_{14}Cl_2CuN_2O$, Dichloroaquo-(2,9-dimethyl-1,10-phenanthroliine)copper(II), 40B, 1126
$C_{14}H_{14}Cl_2N_2O_2Zn$, Dichlorobis(4-acetylpyridine)zinc(II), 43B, 1301
$C_{14}H_{14}Cl_2N_2Zn$, Dichlorobis(4-vinylpyridine)zinc(II), 43B, 1301
$C_{14}H_{14}N_4O_2Pt$, Bis(pyridine-2-methylcarboxaldoximinato)platinum(II), 42B, 832
$C_{14}H_{14}N_6Ni$, Butan-2,3-dione-bis(2'-pyridylhydrazonato)nickel, 34B, 558
$C_{14}H_{16}BCuN_2$, (2,9-Dimethyl-1,10-phenanthroline)tetrahydroboratocopper(I), 46B, 1072
$C_{14}H_{17}Br_2N_3Ni$, Dibromo-(6,6'-dimethyldi-(2-pyridylmethyl)amine-NNN)-nickel(II), 35B, 702
$C_{14}H_{17}CrNO_6S_2$, Tetracarbonyl-cis-[ethyl-α-ethoxy-α-(1,3-dithian-2-ylidene)acetimidate]chromium(0), 46B, 955
$C_{14}H_{18}Br_2CuN_2$, Dibromobis-(2,3-dimethylpyridine)copper(II), 37B, 542

$C_{14}H_{18}Br_2CuN_2$, catena-Dibromobis(3,5-dimethylpyridine)copper(II), 45B, 1170
$C_{14}H_{18}Br_2CuN_2$, trans-Dibromobis(2,6-dimethylpyridine)copper(II), 43B, 1302
$C_{14}H_{18}Cl_2CoN_2$, Dichlorobis(p-toluidine)cobalt(II), 20, 364, 606; 21, 598
$C_{14}H_{18}Cl_2CuN_2$, Dichlorobis-(2,3-dimethylpyridine)copper(II), 37B, 542
$C_{14}H_{18}Cl_2CuN_2$, Dichlorobis-(4-ethylpyridine)copper(II), 37B, 543
$C_{14}H_{18}Cl_2CuN_2$, trans-Dichlorobis(2,6-dimethylpyridine)copper(II), 43B, 1302
$C_{14}H_{18}Cl_2CuN_4 \cdot x\ H_2O$, Chloro(1,6-bis-(2'-pyridyl)-2,5-diazahexane)-copper(II) chloride solvate, 39B, 737
$C_{14}H_{20}Br_2CuN_4$, Bromobis(2-(2-aminoethyl)pyridine)copper(II) bromide, 38B, 919
$C_{14}H_{20}Cl_2CuN_4O_8$, Diperchloratobis(2-(2-aminoethyl)pyridine)copper-(II), 40B, 973
$C_{14}H_{20}Cl_2N_8O_4PdS_2$, trans-Dichlorobis(sulfaguanidine)palladium(II), 42B, 812
$C_{14}H_{20}CrN_4O_4$, Tetracarbonyl(N,N',N'',N'''-tetramethylbi(imidazolidin-2-ylidene)-N,N'')chromium(0), 43B, 1302
$C_{14}H_{20}CuI_2N_4$, Bis(2-(2-aminoethyl)pyridine)copper(II) iodide, 39B, 738
$C_{14}H_{20}CuN_{10}O_{12}$, Nitrato-diaquo-bis(1,3-dimethyl-2,6-dioxopurine)copper(II) nitrate, 42B, 909
$C_{14}H_{20}N_2NiO_4P_2S_4$, (2,2'-Bipyridyl)bis(O,O'-dimethyl dithiophosphato)nickel(II), 43B, 1303
$C_{14}H_{21}CoIN_5O_4$, (Aniline)bis(dimethylglyoximato)iodocobalt, 44B, 928
$C_{14}H_{22}CoN_5O_4$, trans-Bis(dimethylglyoximato)methylpyridinecobalt-(III), 40B, 974; 42B, 833
$C_{14}H_{22}N_8O_6Zn$, Bis(2-aminoethyl)amine(di-2-pyridylamine)zinc(II) nitrate, 46B, 1072
$C_{14}H_{24}ClCoN_6O_7S$, Chloro(dimethylglyoximato)(dimethylglyoxime)-(sulfanilamide)cobalt(III) monohydrate, 42B, 834
$C_{14}H_{24}Cl_2CoN_5O_6$, Chloro(dimethylglyoximato)(dimethylglyoxime)(4-chloroaniline)cobalt(III) dihydrate, 42B, 834
$C_{14}H_{25}BF_2IN_4O_2Rh$, trans-Methyliodo(difluoro(3,3'-(trimethylenedinitrilo)bis(2-pentanoneoximato))borate)rhodium(III), 42B, 835
$C_{14}H_{25}N_5NiS_2$, Diisothiocyanato(2,4,4-trimethyl-1,5,9-triazacyclododec-1-ene)nickel(II), 44B, 928
$C_{14}H_{28}Cl_2CoN_4O_{10}$, Diaquo(2,3,9,10-tetramethyl-1,4,8,11-tetraazacyclotetradeca-1,3,8,10-tetraene)cobalt(II) bis(perchlorate), 43B, 1292
$C_{14}H_{28}Cl_2N_4NiO_8$, N-meso-C-meso(5,6,12,13-Tetramethyl-1,4,8,11-tetraaza-4,11-cyclotetradecadiene)-nickel(II) diperchlorate (isomer b), 45B, 1170
$C_{14}H_{28}Cl_2N_4NiO_8$, N-rac-C-rac(5,6,12,13-Tetramethyl-1,4,8,11-tetraaza-4,11-cyclotetradecadiene)-nickel(II) diperchlorate (isomer c), 45B, 1171
$C_{14}H_{28}Cl_2N_4NiO_8$, racemic-5,7,12,14-Tetramethyl-1,4,8,11-tetraazacyclotetradeca-4,11-dienenickel(II) diperchlorate, 40B, 893
$C_{14}H_{28}Cl_2N_4NiO_8$, 3,5,10,12-Tetramethyl-1,4,8,11-tetraazacyclotetradeca-4,11-dienenickel(II) perchlorate, 43B, 1305
$C_{14}H_{30}Cl_4CuN_4O_4$, Bis(N-methyl-1,4-diazabicyclo[2.2.2]octonium)trichlorocopper(II) perchlorate, 43B, 1306
$C_{14}H_{30}Cl_4N_4NiO_4$, Bis(N-methyl-1,4-diazabicyclo[2.2.2]octonium)trichloronickel(II) perchlorate, 43B, 1306

$C_{14}H_{30}Cl_6N_6Pt_3$, Tetraammineplatinum(II) bis[trichloro(2,6-dimethylpyridine)platinate(II)], 46B, 1073

$C_{14}H_{30}Co_2N_{12}O_2$, (-)-u-facial-Bis(diethylenetriamine)cobalt(III) hexacyanocobaltate(III) dihydrate, 39B, 738

$C_{14}H_{32}B_2F_8FeN_5O$, Nitrosyl-(1,4,8,11-tetramethyl-1,4,8,11-tetraazacyclotetradecane)iron tetrafluoroborate, 45B, 1172

$C_{14}H_{32}ClN_7NiO_4$, Azido-(1,4,8,11-tetramethyl-1,4,8,11-tetraazacyclotetradecane)nickel(II) perchlorate, 41B, 1071

$C_{14}H_{32}Cl_2N_4NiO_8$, (5,6,12,13-Tetramethyl-1,4,8,11-tetraazacyclotetradecane)nickel(II) perchlorate, 44B, 929

$C_{14}H_{32}Cl_2N_4O_2Pd$, Bis(1-t-butyl-2,3-dimethylisourea)dichloropalladium(II), 45B, 1172

$C_{14}H_{32}Cl_2N_4O_4Zn$, Chloro(1,4,8,11-tetramethyl-1,4,8,11-tetraazacyclotetradecane)zinc(II) perchlorate, 44B, 930

$C_{14}H_{32}N_6NiOS_2$, Isothiocyanato(tris(2-dimethylaminoethyl)amine)nickel(II) thiocyanate monohydrate, 39B, 739

$C_{14}H_{33}Cl_2FeN_5O_{10} \cdot C_2H_3N$, Hydroxyl-nitrosyl-(1,4,8,11-tetramethyl-1,4,8,11-tetraazacyclotetradecane)iron perchlorate acetonitrile solvate, 45B, 1172

$C_{14}H_{34}Br_3CoN_6O_5$, Triethylenetetramine-bis(dimethylglyoxalylidenemonoxime)cobalt(III) tribromide trihydrate, 43B, 1304

$C_{14}H_{34}ClCoN_6O_6 \cdot 5.4\ H_2O$, Bis(1,1,1-tris(aminoethyl)ethane)cobalt(III) chloride tartrate hydrate, 43B, 1307

$C_{14}H_{34}Cl_2Cu_2N_8O_8S_2$, Di-$\mu$-thiocyanato-bisdi(3-aminopropyl)aminedicopper(II) perchlorate, 40B, 976

$C_{14}H_{34}Cl_2N_6NiO_8$, Bis(1,4,7-triazacyclodecane)nickel(II) diperchlorate, 46B, 1074

$C_{14}H_{34}CrN_{11}NiO_2$, Tris(1,3-propanediamino)chromium(III) pentacyanonickelate(II) dihydrate, 40B, 893

$C_{14}H_{35}Br_2CoN_6O_6$, Triethylenetetramine-N,N-bis(dimethylglyoxylidenemonoxime)cobalt(III) dibromide tetrahydrate, 43B, 1304

$C_{14}H_{44}CoI_3N_6O_5$, Bis((R)-2-methyl-1,4,7-triazacyclononane)cobalt(III) iodide pentahydrate, 43B, 1146

$C_{15}H_8MoN_2O_5S$, Tricarbonyl(1,10-phenanthroline)(η^2-sulfur dioxide)molybdenum(0), 46B, 1075

$C_{15}H_{11}Cl_2CoN_3$, Dichloro-2,2',2"-terpyridylcobalt(II), 35B, 704

$C_{15}H_{11}Cl_2N_3Zn$, Terpyridylzinc chloride, 20, 611; 31B, 431

$C_{15}H_{15}Cl_2CoN_3OS$, Dichloro(N-(methoxy(6-methyl-2-pyridyl)methyl)benzothiazolin-2-ylideneimine)cobalt(II), 38B, 920

$C_{15}H_{15}Cl_2N_3O_2Pd$, Chloro(2,2',2"-terpyridine)palladium(II) chloride dihydrate, 39B, 740

$C_{15}H_{15}Cl_3MoN_3$, mer-Trispyridinetrichloromolybdenum, 40B, 894

$C_{15}H_{16}Br_{1.5}Cl_{1.5}N_3Zn$, Pyridine-pyridinium dibromochloropyridinezincate bromodichloropyridinezincate, 44B, 930

$C_{15}H_{18}BrClCuN_4O_5$, Five-covalent copper(II) complex, 35B, 705

$C_{15}H_{18}N_4Ni$, DL-N,N'-Bispyrrol-2-ylmethylene-2,4-diaminopentanenickel(II), 42B, 835

$C_{15}H_{19}Cl_2N_3Pd$, Chloro(methyldi-((6-methyl-2-pyridyl)methyl)amine)-palladium(II) chloride, 38B, 921

$C_{15}H_{19}Cl_3N_3Ti$, Trichloro-(N,N-bis-(6-methyl-2-pyridylmethyl)methylamine-N,N,N)titanium(III), 38B, 921

$C_{15}H_{20}Cl_2CuN_4O_8$, 1,7-Bis(2-pyridyl)-2,6-diazaheptanecopper(II) diperchlorate, 46B, 1076

$C_{15}H_{20}Cl_2N_4O_8Pd$, 1,7-Bis(2-pyridyl)-2,6-diazaheptanepalladium(II) diperchlorate, 46B, 1076

$C_{15}H_{21}N_8NaNi_2O_2 \cdot 2\ C_4H_8O$, Sodium tri-$\mu$-(3,5-dimethylpyrazolyl)-bis-(nitrosylnickel(I)) tetrahydrofuran solvate, 45B, 1182

$C_{15}H_{22}BN_7Pt$, Methyl(hydrotris(1-pyrazolyl)borato)(t-butylisocyano)-platinum(II), 42B, 836
$C_{15}H_{24}Br_2N_4NiO$, Bromo(2,12-dimethyl-3,7,11,17,-tetraazabicyclo-[11.3.1]heptadeca-1(17),2,11,13,15-pentaene)nickel(II) bromide monohydrate, 33B, 428
$C_{15}H_{24}Cl_2CuN_6$, Dichlorotris-(1,2-dimethylimidazole)copper(II), 37B, 543
$C_{15}H_{26}Cl_2CoN_2$, Dichloro[(-)-α-isospartein]cobalt(II), 45B, 1173
$C_{15}H_{26}Cl_2CoN_2$, Dichloro((-)-spartein)cobalt(II), 43B, 1308
$C_{15}H_{26}Cl_2CuN_2$, (-)-β-Isosparteinecopper(II) chloride, 41B, 1072
$C_{15}H_{26}Cl_2N_4NiO_8$, (2,12-Dimethyl-3,7,11,17-tetraazabicyclo[11.3.1]-heptadeca-1(17),13,15-triene)nickel perchlorate, 34B, 560
$C_{15}H_{26}Cl_2N_4NiO_8 \cdot H_2O$, {2,12-Dimethyl-3,7,11,17-tetraazabicyclo-[11.3.1]heptadeca-1(17),13,15-triene}nickel(II) monohydrate diperchlorate (β form), 46B, 1077
$C_{15}H_{26}CoN_7OS_3$, trans-Di-isothiocyanato-N-rac-(5,12-dimethyl-1,4,8,11-tetraazacyclotetradeca-4,11-diene)cobalt(III) thiocyanate monohydrate, 42B, 836
$C_{15}H_{26}N_6NiO_4 \cdot 0.5\ H_2O$, {2,12-Dimethyl-3,7,11,17-tetraazabicyclo-[11.3.1]heptadeca-1(17),13,15-triene}nitritonitronickel(II) hemihydrate, 46B, 1077
$C_{15}H_{27}Cl_2FeN_5O_6$, Diaquo-2,13-dimethyl-3,6,9,12,18-pentaazabicyclo-[12.3.1]octadeca-1(18),2,12,14,16-pentaeneiron(II) chloride perchlorate, 42B, 910
$C_{15}H_{27}Cl_2N_9NiO_8$, Trihistaminenickel(II) perchlorate, 38B, 922
$C_{15}H_{28}ClCoN_4O_8$, Λ-cis-β-Oxalato(1,7-bis(2(S)-pyrrolidyl)-2,6-diazaheptane)cobalt(III) perchlorate, 44B, 926
$C_{15}H_{44}Cl_3CoN_6O$, Tris-(2,4-diaminopentane)cobalt(III) chloride monohydrate, 38B, 922
$C_{15}H_{44}Cl_3CoN_6O_4$, $(-)_{589}$-Tris-(+)-trans-1,2-diaminocyclopentane)cobalt(III) chloride tetrahydrate, 37B, 544
$C_{16}F_{24}HgN_2S_8$, Bis[2,3,4,5-tetrakis(trifluoromethylthio)pyrrolyl]mercury, 45B, 1173
$C_{16}H_7Fe_3NO_9S$, μ_3-Thio-μ_3-(p-tolylimino)-tris(tricarbonyliron), 45B, 1168
$C_{16}H_{10}CrMoN_2O_{10}$, ($\eta^2$-Pentamethylenediazirine)(pentacarbonylchromium)(pentacarbonylmolybdenum), 46B, 1078
$C_{16}H_{12}Cl_2CuN_4$, Dichloro-bis(1,8-naphthyridine)copper(II), 45B, 1174
$C_{16}H_{12}Cl_2Hg_2N_4O_8$, Bis-((1,8-naphthyridine)mercury(I)) diperchlorate, 41B, 1072
$C_{16}H_{12}MoN_4O_4$, (E)-5-Methylpyridine-2-carboxaldehyde-2'-pyridylhydrazone-tetracarbonyl-molybdenum(0), 40B, 895
$C_{16}H_{12}N_4Zn$, Dicyano(2,9-dimethyl-1,10-phenanthroline)zinc, 43B, 1308
$C_{16}H_{13}CuN_5O_4$, Cyano(2,2',2"-terpyridine)copper(II) nitrate monohydrate, 42B, 837
$C_{16}H_{14}Cl_2CoN_4$, (N,N'-Ethylenebis-(2-amino-5-chlorobenzylideneiminato))cobalt(II), 38B, 923
$C_{16}H_{16}N_4PtS_4$, Bis(methyl N'-phenyldiazenecarbodithioato-N',S)platinum, 46B, 88
$C_{16}H_{18}Cl_2N_2O_4Pt$, trans-Dichlorobis(γ-ethoxycarbonylpyridine)platinum(II), 46B, 1079
$C_{16}H_{18}Cl_2N_2Pd$, Dichloro-(N-methylbenzaldimine)-palladium(II), 45B, 1174
$C_{16}H_{18}CuLi_2N_4O_9$, Lithium tetrakis(succinimidato)cuprate(II) monohydrate, 40B, 897
$C_{16}H_{18}N_4S_2Zn$, Diisothiocyanatobis(p-toluidine)zinc(II), 42B, 838
$C_{16}H_{20}Cl_2CoN_4O_2$, Dichlorobis(4-nitroso-N,N-dimethylaniline)cobalt-

(II), 45B, 1175

$C_{16}H_{20}CoN_3O_8$, Carbonatohydroxo-(2,2':6',2"-terpyridyl)cobalt(III) tetrahydrate, 44B, 930

$C_{16}H_{20}Cs_2CuN_4O_{10}$, Caesium tetrakis(succinimidato)copper(II) dihydrate, 38B, 923

$C_{16}H_{20}CuN_6S_2$, Bis-(2-(2-aminoethyl)pyridine)-di-isothiocyanatocopper(II), 41B, 1073

$C_{16}H_{22}Cl_2CuN_4O_8$, 1,8-Bis-(2'-pyridyl)-3,6-diazaoctanecopper(II) perchlorate, 40B, 896

$C_{16}H_{22}Cl_2N_4NiO_4S_2$, Chloro(bis(2-((2-pyridylmethyl)amino)ethyl) disulfide)nickel(II) perchlorate, 38B, 924

$C_{16}H_{23}BClCuF_2N_5O_6$, Difluoro-3,3'-(trimethylenedinitrilo)bis(2-butanone oximato)borato](pyridine)copper(II) perchlorate, 46B, 1079

$C_{16}H_{24}BCrF_6N_8$, Difluoro-tetrakis(5-methylpyrazole)chromium(III) tetrafluoroborate, 45B, 1175

$C_{16}H_{24}Br_2MnN_8$, Dibromotetrakis(5-methylpyrazole)manganese(II), 37B, 545

$C_{16}H_{24}Br_4Cu_2N_4$, Bis(dibromo(2-(2-methylaminoethyl)pyridine)copper-(II)), 42B, 838

$C_{16}H_{24}Cl_6N_8Pt_2$, Tetra(N-methylimidazole)platinum(II) hexchloroplatinate(IV), 46B, 1080

$C_{16}H_{24}N_4Pd_2$, Bis((1,2-η^2-3,5-dimethylpyrazolide)-π-allylpalladium-(II)), 43B, 1042

$C_{16}H_{25}CuN_7O_6$, Bis(3-aminopropyl)amine(2,2'-bipyridyl)copper(II) nitrate, 44B, 931

$C_{16}H_{26}Cl_2N_4O_2Pt$, (1,3-Diaminocyclohexane)(2,2'-bipyridine)platinum-(II) chloride dihydrate, 43B, 1309

$C_{16}H_{26}CuN_8O_6$, Bis(3-aminopropyl)amine(di-2-pyridylamine)copper(II) nitrate, 44B, 931

$C_{16}H_{28}ClCuN_5O_5$, Chloro-(2,7,12-trimethyl-3,7,11,17-tetraazabicyclo-[11.3.1]heptadeca-1(17),2,11,13,15-pentaene)copper(II) nitrate dihydrate, 41B, 1074

$C_{16}H_{28}Cl_2CoN_4$, Dichloro-bis(3,5-diethyl-4-methyl-1H-pyrazole)cobalt-(II), 45B, 1176

$C_{16}H_{28}Cl_2N_4NiO_8$, 5,7,7,12,12,14-Hexamethyl-1,4,8,11-tetra-azacyclotetradeca-4,8,10,14-tetraenenickel(II) perchlorate, 38B, 925

$C_{16}H_{28}Cl_2N_8NiO_{10}$, Diaquobis(3-acetyl-5,6-dimethylpyridazine hydrazone)nickel(II) perchlorate, 43B, 1309

$C_{16}H_{28}CuK_2N_4O_{14}$, Potassium tetrakis(succinimidato)cuprate(II) hexahydrate, 40B, 897

$C_{16}H_{28}CuN_6O_6$, Bis(1H-3,5-diethyl-4-methylpyrazole-N^2)dinitratocopper(II), 46B, 1081

$C_{16}H_{29}Cl_2N_5O_4Pt$, Aquo-(1,3-diamido-5-aminocyclohexane)-(2,2'-bipyridine)platinum(IV) chloride trihydrate, 43B, 1309

$C_{16}H_{30}FeN_8O_6$, Bis(dimethylglyoximato)diimidazoleiron(II)-dimethanol, 38B, 926

$C_{16}H_{32}BrN_2Ni$, Bromo-(3,8-diisopropyl-2,9-dimethyl-4,7-diaza-4,6-decadienyl-N,N')nickel(II), 42B, 839

$C_{16}H_{32}ClFeIN_4$, Chloro(5,7,7,12,14,14-hexamethyl-1,4,8,11-tetraazacyclotetradeca-4,11-diene)iron(II) iodide, 39B, 740

$C_{16}H_{32}Cl_2N_4NiO_8$, racemic-5,7,7,12,12,14-Hexamethyl-1,4,8,11-tetraazacyclotetradeca-4,14-dienenickel(II) perchlorate (Pbcn), 40B, 1122

$C_{16}H_{32}Cl_2N_4NiO_8$, racemic-5,7,7,12,14,14-Hexamethyl-1,4,8,11-tetraazacyclotetradeca-4,11-dienenickel(II) perchlorate (Pbca), 38B, 926

$C_{16}H_{32}Cl_4N_{12}Ni_2O_6$, Di-$\mu$-(1,4-dihydrazinophthalazine)bis(diaquonickel) tetrachloride dihydrate, 34B, 561

$C_{16}H_{32}Co_2N_{12}O_2$, (+)D-(N,N,N',N'-Tetrakis(2'-aminoethyl)-1,2-diaminoethane)cobalt(III) hexacyanocobaltate(III) dihydrate, 35B, 707
$C_{16}H_{32}CrF_6N_6O_3P$, Nitro(5,7,7,12,14,14-hexamethyl-1,4,8,11-tetraazacyclotetradeca-4,11-diene)nitrosylchromium(I) hexafluorophosphate, 43B, 1310
$C_{16}H_{34}Cl_3N_4Ta$, Trichlorobis(N,N'-di-isopropylacetamidinato)-tantalum(V) (monoclinic), 40B, 897
$C_{16}H_{34}Cl_3N_4Ta$, Trichlorobis(N,N'-diisopropylacetamidinato)-tantalum(V) (orthorhombic), 41B, 1077
$C_{16}H_{36}B_2CoF_8N_4O_2$, Diaquo(5,7,7,12,14,14-hexamethyl-1,4,8,11-tetraazacyclotetradeca-4,11-diene)cobalt(II) bis(tetrafluoroborate), 43B, 1292
$C_{16}H_{36}Cl_2N_4NiO_8$, α-C-rac-5,5,7,12,12,14-Hexamethyl-1,4,8,11-tetraazacyclotetradecanenickel(II) bis-perchlorate, 39B, 742
$C_{16}H_{36}Cl_2N_4NiO_8$, γ-C-rac-5,5,7,12,12,14-Hexamethyl-1,4,8,11-tetraazacyclotetradecanenickel(II) bis-perchlorate, 39B, 742
$C_{16}H_{36}Cl_2N_4NiO_8$, 5,12-Diethyl-7,14-dimethyl-1,4,8,11-tetraazacyclotetradecanenickel(II) diperchlorate, 43B, 1311
$C_{16}H_{36}Co_4N_8O_4$, Tetra(μ_3-t-butylamido)tetra(nitrosylcobalt), 40B, 898
$C_{16}H_{38}Br_3CoN_4O$, cis-(SSSR)-{Bromo[dibromoaqua(2-)][(2R,5R,8R,11R)-2,5,8,11-tetraethyl-1,4,7,10-tetraazacyclododecane]}cobalt(III), 46B, 1082
$C_{16}H_{38}Cl_4N_4NiOZn$, β-C-rac-5,5,7,12,12,14-Hexamethyl-1,4,8,11-tetraazacyclotetradecanenickel(II) tetrachlorozincate hydrate, 39B, 742
$C_{16}H_{40}Br_2N_4W_2$, Dibromotetrakis(diethylamido)ditungsten, 43B, 1312
$C_{16}H_{40}Cl_2N_4W_2$, Dichlorotetrakis(diethylamido)ditungsten(III), 42B, 839
$C_{16}H_{40}I_2N_4W_2$, Diiodotetrakis(diethylamido)ditungsten, 43B, 1312
$C_{16}H_{42}Cl_2N_{10}O_{12}Rh_2$, Bis(dimethylglyoximato)diamminerhodium(III) dichlorobis(dimethylglyoximato)rhodate(III) tetrahydrate, 42B, 840
$C_{16}H_{42}Co_2Cl_2N_{10}O_{12}$, Diamminebis(dimethylglyoximato)cobalt(III) dichlorobis(dimethylglyoximato)cobaltate(III) tetrahydrate, 44B, 932
$C_{16}H_{42}N_8Zr_2$, Di-μ-(N-t-butylimido)-bis(bis(dimethylamido)zirconium), 45B, 1177
$C_{16}H_{44}CuN_{10}O_{12} \cdot 0.75\ NH_4ClO_4$, Diammonium aqua(hexamethyldioxaoctaazatricycloeicosatetronato)copper(2+) pentahydrate ammonium perchlorate, 42B, 840
$C_{16}H_{48}F_8N_8Ti_4$, Bis-dimethylaminotitanium difluoride tetramer, 40B, 899
$C_{16}H_{52}Br_2N_8NiO_8P_4S_4$, Dibromo-tetrakis(diethoxy-hydrazido-thio-phosphorus)nickel(II), 45B, 1177
$C_{16}H_{56}Br_6N_8Pt_2$, Tetrakis(ethylamine)platinum(II) dibromotetrakis-(ethylamine)platinum(IV) tetrabromide, 31B, 423; 46B, 1082
$C_{16}H_{64}Br_6N_8O_4Pt_2$, Tetrakis(ethylamine)platinum(II) dibromotetrakis-(ethylamine)platinum(IV) tetrabromide tetrahydrate, 42B, 841
$C_{16}H_{64}Cl_6N_8O_4Pt_2$, Wolffram's red salt, 26, 611
$C_{17}H_9NO_{12}Os_3S$, (μ-Hydrido)-(μ-p-toluenesulphonylamide)decacarbonyltriosmium, 46B, 1083
$C_{17}H_{11}CoN_5O_2$, Bis(isocyanato)(2,2':6',2"-terpyridyl)cobalt(II), 46B, 1083
$C_{17}H_{14}FeN_2O_8$, 3,3-Bis(methoxycarbonyl)-4-phenyl-1-pyrazoline-N-(tetracarbonyliron), 39B, 743
$C_{17}H_{15}BMoN_8O_2$, Hydrotris(1-pyrazolyl)boratobenzenediazodicarbonylmolybdenum, 37B, 545
$C_{17}H_{16}Cl_2CoN_2O$, Dichloro-(trans-2-(2-quinolyl)methylenequinuclidin-3-one)cobalt(II), 41B, 1077
$C_{17}H_{16}N_4O_4PtS$, 2-Hydroxyethanethiolato(2,2',2''-terpyridine)platin-

um(II) nitrate, 42B, 842
$C_{17}H_{18}CuN_4$, (1,2-Bis-((2-aminobenzylidene)amino)propanato(2-))copper(II), 39B, 743
$C_{17}H_{18}N_4Ni$, 1,3-Bis(2-iminobenzylideneimino)propanenickel(II), 40B, 899
$C_{17}H_{20}CuN_6S_2$, Isothiocyanato-(1,7-bis-(2-pyridyl)-2,6-diazaheptane)-copper(II) thiocyanate, 38B, 927
$C_{17}H_{23}ClFeN_7O_4S_2$, (2,13-Dimethyl-3,6,9,13,18-penta-azabicyclo-[12.3.1]octadeca-1(18),2,12,14,16-pentaene)di-isothiocyanatoiron-(III) perchlorate, 41B, 1078
$C_{17}H_{27}F_{12}FeN_5OP_2$, Acetonitrile(carbonyl)(2,3,9,10-tetramethyl-1,4,8,11-tetraaza-1,3,8,10-cyclotetradecatetraene)iron(II) hexafluorophosphate, 45B, 1178
$C_{17}H_{35}CoN_5S_3 \cdot 0.5\ C_4H_{10}O$, Diisothiocyanato-(N,N-bis-(2-diethylaminoethyl)-2-methylthioethylamine-N,N,N)cobalt(II), 37B, 547
$C_{17}H_{36}Co_2N_{12}O_3$, (-)-(L-N,N,N',N'-Tetrakis-(2'-aminoethyl)-1,2-diaminopropane)cobalt(III) hexacyanocobaltate(III) trihydrate, 40B, 900
$C_{18}H_8Fe_2N_2O_6$, Benzo[c]cinnolinebis(tricarbonyliron), 35B, 699
$C_{18}H_{12}B_2F_5N_6NiO_3P$, (Fluoroborotris(2-aldoximo-6-pyridyl)phosphine)-nickel(II) tetrafluoroborate, 38B, 928
$C_{18}H_{12}B_2F_5N_6O_3PZn$, (Fluoroborotris(2-aldoximato-6-pyridyl)phosphine)zinc(II) tetrafluoroborate, 39B, 745
$C_{18}H_{12}Br_2N_2Ni$, (2,2'-Biquinolyl)dibromonickel(II), 43B, 1334
$C_{18}H_{13}MoN_3O_3$, (2,2'-Bipyridine)pyridine-tricarbonylmolybdenum(III), 37B, 548
$C_{18}H_{14}Br_2N_2PdS_2$, trans-Dibromo(2-(2'-thienyl)pyridine)palladium(II), 44B, 932
$C_{18}H_{14}CuN_4O_3$, Aqua(N,N'-bis(2'-pyridinecarboxamido)-1,2-benzene)copper(II), 46B, 1085
$C_{18}H_{14}Hg_2N_4O_6$, Bis(quinoline)dimercury(I) dinitrate, 41B, 1080
$C_{18}H_{14}N_4Ni$, Dibenzo[b,i](1,4,8,11)tetraaza[14]annulenenickel(II), 43B, 1313
$C_{18}H_{16}Cl_2CuN_8O_8$, Bis(2-pyridyl-2-pyrimidylamine)copper(II) perchlorate, 46B, 1086
$C_{18}H_{16}Cl_4Co_2N_6O_2$, Tetrachloroaquo(2,4,6-tris(2'-pyridyl)-1,3,5-triazine)dicobalt(II) monohydrate, 44B, 932
$C_{18}H_{18}Cl_2N_2Pd$, Dichlorobis(2-p-tolyl-1-aziridine)palladium(II), 44B, 933
$C_{18}H_{18}CuN_6S_2$, Isothiocyanato(N,N'-bis(2-pyridylmethylene)tetramethylenediamine)copper(II) thiocyanate, 42B, 843
$C_{18}H_{18}N_4Ni$, (7,8,15,16,17,18-Hexahydrodibenzo[e,m](1,4,8,11)tetra-azacyclotetradecinato(2-))nickel(II), 40B, 908
$C_{18}H_{18}N_6NiO_4$, Bis(2,6-diacetylpyridine dioximato)nickel(IV), 39B, 746
$C_{18}H_{20}B_2Cl_2Cu_2N_{12}$, Di-μ-chlorobis[(hydrotris(1-pyrazolyl)borato)copper(II)], 45B, 1178
$C_{18}H_{20}B_2CuN_{12}$, Bis[hydrotris(pyrazol-1-yl)borato]copper(II), 45B, 1179
$C_{18}H_{20}B_2Cu_2N_{12}$, Bis((hydridotris(1-pyrazolyl)borato)copper(I)), 40B, 901; 42B, 843
$C_{18}H_{20}B_2FeN_{12}$, Bis[hydrotris(1-pyrazolyl)borato]iron(II), 46B, 1088
$C_{18}H_{20}B_2N_{12}Ni$, Bis[hydrotris(1-pyrazolyl)borato]nickel(II), 45B, 1179
$C_{18}H_{20}Cl_2CuN_6O$, Dichlorobis(1-(2-pyridylmethyl)imidazole)copper(II) monohydrate, 44B, 934
$C_{18}H_{20}Fe_2N_2O_6$, Di-μ-cyclohexylideneiminato-bis(tricarbonyliron), 43B, 1313; 44B, 934

$C_{18}H_{21}Br_3MoN_3$, Tribromotris(4-methylpyridine)molybdenum, 44B, 935
$C_{18}H_{21}Br_3MoN_3 \cdot 0.5\ C_6H_7N$, Tribromotris(4-methylpyridine)molybdenum-(III) 4-methylpyridine, 46B, 1088
$C_{18}H_{21}Br_3N_6NiO_4$, Triaquo(2,6-bis(2'-pyridyl)-4-(2'-pyridinio)-1,3,5-triazine)nickel(II) bromide monohydrate, 34B, 565; 43B, 1314
$C_{18}H_{21}Cl_3MoN_3 \cdot 0.5\ C_6H_7N$, Trichlorotris(4-methylpyridine)molybdenum(III) 4-methylpyridine, 46B, 1089
$C_{18}H_{22}Cl_2CuN_8O_{13}$, Tetrakis(2-pyrimidinone)copper(II) perchlorate - ethanol, 43B, 1314
$C_{18}H_{24}B_2F_8N_6Ni$, Tris(2-aminomethylpyridine)nickel(II) tetrafluoroborate, 46B, 1089
$C_{18}H_{24}B_2F_8N_{12}Ni$, Hexakis(imidazole)nickel(II) bis(tetrafluoroborate), 41B, 1081
$C_{18}H_{24}CdN_{14}O_6$, Hexakis(imidazole)cadmium(II) nitrate, 37B, 548
$C_{18}H_{24}Cl_2FeN_6 \cdot CH_4O$, Tris(2-picolylamine)iron(II) dichloride methanolate, 45B, 1179
$C_{18}H_{24}Cl_2FeN_6 \cdot C_2H_6O$, Tris(2-picolylamine)iron(II) dichloride ethanolate, 45B, 1179, 1180; 46B, 1090
$C_{18}H_{24}Cl_2FeN_6 \cdot 2\ H_2O$, Tris(2-picolylamine)iron(II) dichloride dihydrate, 46B, 1090
$C_{18}H_{24}Cl_2N_{12}NiO_8$, Hexakis(imidazole)nickel(II) diperchlorate, 45B, 1181
$C_{18}H_{24}CoN_{14}O_6$, Hexakis(imidazole)cobalt(II) nitrate, 38B, 929
$C_{18}H_{24}CuN_{14}O_6$, Hexakis(imidazole)copper(II) nitrate, 41B, 1081
$C_{18}H_{24}FeI_2N_6$, Tris(2-picolylamine)iron(II) diiodide, 46B, 1090
$C_{18}H_{24}N_{14}NiO_6$, Hexakis(imidazole)nickel(II) nitrate, 34B, 572
$C_{18}H_{24}N_{14}NiO_6$, Hexapyrazolenickel(II) nitrate, 35B, 710
$C_{18}H_{25}ClFeN_7O_4S_2$, (2,14-Dimethyl-3,6,10,13,19-penta-azabicyclo-[13.3.1]nonadeca-1(19),2,13,15,17-pentaene)di-isothiocyanatoiron-(III) perchlorate, 41B, 1078
$C_{18}H_{26}CoN_4$, Bis(N-t-butylpyrrole-2-carboxaldimino)cobalt(II), 38B, 930
$C_{18}H_{26}CoN_6O_4$, trans-Bis(dimethylglyoximato)bis(pyridine)cobalt(II), 44B, 935
$C_{18}H_{26}CuN_4$, Bis(N-t-butylpyrrole-2-carbaldimino)copper(II), 38B, 930
$C_{18}H_{26}N_4Ni$, Bis(N-t-butylpyrrole-2-carbaldimino)nickel(II), 38B, 931
$C_{18}H_{27}Cl_5Co_2N_5 \cdot 0.5\ H_2O$, Chloro(1,11-bis(2-pyridyl)-2,6,10-triazaundecane)cobalt(III) tetrachlorocobaltate(II) hemihydrate, 46B, 1091
$C_{18}H_{28}Cl_2FeN_6O_2$, Tris(α-picoline)iron(II) dichloride dihydrate, 44B, 935
$C_{18}H_{28}CoI_2N_{10}O$, Bis(2,6-diacetylpyridine dihydrazone)cobalt(II) diiodide monohydrate, 44B, 936
$C_{18}H_{28}FeN_8O_2$, N,N'-Dimethyl-1,4-diazabicyclo[2.2.2]octonium (o-benzoquinone diimine)tetracyanoferrate(II) dihydrate, 39B, 747
$C_{18}H_{28}Mo_2N_2S_2$, Di-μ-sulfido-bis(cyclopentadienyl-t-butylimidomolybdenum(V)), 40B, 902
$C_{18}H_{30}Cl_3CoN_{10}O_{14}$, Bis(acetonitrile)(octaaza-macrocycle)cobalt(III) perchlorate, 44B, 937
$C_{18}H_{30}F_{12}FeN_6P_2$, Bis(acetonitrile)(2,3,9,10-tetramethyl-1,4,8,11-tetraazacyclotetradeca-1,3,8,10-tetraene)iron(II) hexafluorophosphate, 45B, 1181
$C_{18}H_{30}F_{12}FeN_6P_2$, cis-(3,11-Bis(1-iminoethyl)-2,12-dimethyl-1,5,9,13-tetraazacyclohexadeca-1,4,8,12-tetraeneiron(II)) bis(hexafluorophosphate), 41B, 1082; 43B, 1315
$C_{18}H_{30}FeN_8O_6$, Di(cyclohexane-1,2-dioximato(1-))di-imidazoleiron(II), 29, 554

$C_{18}H_{32}ClCoN_6O_4$, Dicyano-5,7,7,12,12,14-hexamethyl-1,4,8,11-tetraazacyclotetradeca-4,14-dienecobalt(III) perchlorate, 35B, 712

$C_{18}H_{32}Cl_2N_{12}NiO_4$, Hexakis(imidazole)nickel(II) chloride tetrahydrate, 42B, 845

$C_{18}H_{32}Cl_2N_{12}O_4Zn$, Hexaimidazolezinc(II) dichloride tetrahydrate, 32B, 437

$C_{18}H_{33}CdN_{13}O_8$, Hexakis(imidazole)cadmium(II) hydroxide nitrate tetrahydrate, 37B, 548

$C_{18}H_{34}IN_7Ni_2O_2$, Tetraethylammonium μ-iodo-di-μ-(3,5-dimethylpyrazolyl)-bis(nitrosylnickel), 45B, 1182

$C_{18}H_{34}N_6NiOS_2$, Hexamethyl-1,4,8,11-tetraazacyclotetradeca-4,11-dienenickel(II) thiocyanate, 38B, 933

$C_{18}H_{35}Cl_2FN_2O_4P_2Pt$, trans-Chlorobis(triethylphosphine)(p-fluorophenyldiazine)platinum(II) perchlorate, 40B, 902

$C_{18}H_{35}Cl_4N_5O_2Pt_2S_2$, μ-(9-Methyladenine-N^1,N^7)-bis(isopropylsulfoxide-S-trans-dichloroplatinum(II)), 42B, 845

$C_{18}H_{36}Cl_2N_4NiO_8$, (3,5,7,7,10,12,14,14-Octamethyl-1,4,8,11-tetra-azacyclotetradeca-4,11-diene)nickel(II) perchlorate, 38B, 934

$C_{18}H_{36}Cl_2N_4Pd$, Dichlorobis(di-t-butylcarbodiimide)palladium(II), 44B, 937

$C_{18}H_{40}Cl_4N_8Pd_2$, Di-μ-[glutaraldehydebis(dimethylhydrazone)]-bis[dichloropalladium(II)], 45B, 1182

$C_{18}H_{48}N_9O_{12}Rh$, (+)-Tris-((-)-trans-1,2-diaminocyclohexane)rhodium(III) nitrate trihydrate, 40B, 903

$C_{18}H_{52}Cl_3CoN_6O_5$, $(-)_{589}$-Tris(+trans-1,2-diaminocyclohexane)cobalt(III) chloride pentahydrate, 35B, 713

$C_{18}H_{52}Cl_3CoN_6O_5$, Bis((S,S)-trans-1,2-diaminocyclohexane)-(R,R)-trans-1,2-diaminocyclohexanecobalt(III) chloride pentahydrate, 43B, 1151

$C_{18}H_{53}N_3NdSi_6$, Tris(bis(trimethylsilyl)amido)neodymium(III), 44B, 938

$C_{18}H_{54}ClHfN_3Si_6$, Chlorotris[bis(trimethylsilyl)amido]hafnium, 46B, 1092

$C_{18}H_{54}ClN_3Si_6Ti$, Chlorotris[bis(trimethylsilyl)amido]titanium, 46B, 1092

$C_{18}H_{54}ClN_3Si_6Zr$, Chlorotris[bis(trimethylsilyl)amido]zirconium, 46B, 1092

$C_{18}H_{54}FeN_3Si_6$, Tris(hexamethyldisilylamido)iron(III), 38B, 935

$C_{18}H_{58}BN_3Si_6Th$, (Tetrahydroborato)tris((hexamethyldisilyl)amido)-thorium(IV), 45B, 1183

$C_{19}H_{10}Fe_2N_2O_7$, μ-Diphenylureylene-bis(tricarbonyl iron), 32B, 437

$C_{19}H_{14}B_2Cl_2F_5FeN_6O_3P$, (Fluoroborotris(2-aldoximo-6-pyridyl)phosphine)iron(II) tetrafluoroborate dichloromethane solvate, 38B, 935

$C_{19}H_{15}CdCl_2N_5O$, Dichloro-2-(2'-pyridyl)-3-(N-2-picolylimino)-4-oxo-1,2,3,4-tetrahydroquinazolinecadmium, 43B, 1316

$C_{19}H_{15}Cl_2CuN_5O$, Dichloro-2-(2'-pyridyl)-3-(N-2-picolylimino)-4-oxo-1,2,3,4-tetrahydroquinazolinecopper(II), 40B,904

$C_{19}H_{15}Cl_2MnN_5O$, Dichloro-2-(2'-pyridyl)-3-(N-2-picolylimino)-4-oxo-1,2,3,4-tetrahydroquinazolinemanganese(II), 40B, 905

$C_{19}H_{17}BMoN_8O_2$, Tetrakis(1-pyrazolyl)boratecyclopentadienyldicarbonylmolybdenum, 38B, 936

$C_{19}H_{21}ClCu_2N_8O_{11} \cdot H_2O$, Dinitrato(perchlorato)[4,5-bis(2-(2-pyridyl)ethylimino)methyl)imidazolato](aqua)dicopper(II) monohydrate, 46B, 1093

$C_{19}H_{21}N_9O_8Zn$, (8,15-Dihydro-2,6-dimethyltripyridoheptaazapentadecin)diaquozinc nitrate, 43B, 1316

$C_{19}H_{23}BMoN_4O_2$, (Dihydrobis-(3,5-dimethyl-1-pyrazolyl)borate)(η^3-cycloheptatrienyl)dicarbonylmolybdenum, 38B, 937

$C_{19}H_{23}Cl_2CoN_7O_2$, Diaquo(2,6-diacetylpyridinebis(2'-pyridylhydrazone))cobalt(II) dichloride, 42B, 847

$C_{19}H_{23}Cl_2N_7O_2Zn$, Diaquo(2,6-diacetylpyridinebis(2'-pyridylhydrazone))zinc(II) dichloride, 42B, 847

$C_{19}H_{25}Mn_2N_{19}S_4$, Tris-$\mu$-(4-methyl-1,2,3-triazole-N[1],N[2])-bis[(4-methyl-1,2,4-triazole-N[1])bis(thiocyanato-N)manganese(II)], 45B, 1183

$C_{19}H_{27}MnN_7S_2$, (2,15-Dimethyl-3,7,10,14,20-pentaazabicyclo[14.3.1]-eicosa-1(20),2,14,16,18-pentaene)bis(isothiocyanato)manganese(II), 43B, 1317

$C_{19}H_{28}Cl_2FeN_6O$, Tris(α-picoline)iron(II) dichloride methanolate, 44B, 935

$C_{19}H_{28}Cl_2N,PPd \cdot 0.8\ CH_2Cl_2$, Dichloro[(8-methyl-2-quinolylmethyl)di-t-butylphosphine-NP]palladium(II) dichloromethane solvate, 46B, 1093

$C_{19}H_{34}CoN_{12}O_8$, Hexaimidazolecobalt(II) carbonate pentahydrate, 37B, 549

$C_{20}H_8CdMn_2N_2O_{10}$, μ-(2,2'-Bipyridylcadmium)-bis(pentacarbonylmanganese) (2Cd-Mn), 40B, 906

$C_{20}H_8N_8O_8Rh_4$, Bis(2,2'-biimidazolyl)octacarbonyltetrarhodium(I), 42B, 848

$C_{20}H_{10}Mn_2N_4O_8$, Di-μ-phenyldiazo-bis(tetracarbonylmanganese), 41B, 1083

$C_{20}H_{12}CuN_2S_2$, cis-(Dichloro-(2,2'-o-phenylene-bis(benzothiazole)))-copper(II), 45B, 1184

$C_{20}H_{13}N_2O_{10}Re_3$, Tri-$\mu$-hydridodecacarbonyldipyridinetrirhenium, 46B, 1094

$C_{20}H_{14}Cl_2CuN_6S_2 \cdot 2\ H_2O$, Chloro-di(2-(4'-thiazolyl)benzimidazole)-copper(II) chloride dihydrate, 45B, 1185

$C_{20}H_{15}B_2CoF_5N_7O_3P$, (Fluoro-(6,6',6''-phosphinidynetris-(pyridine-2-carbaldehyde oximato)(3-))-borato(1-))cobalt(1+) tetrafluoroborate - acetonitrile, 39B, 744

$C_{20}H_{16}BBrCuF_4N_4$, Bis(2,2'-bipyridyl)monobromocopper(II) tetrafluoroborate, 46B, 1095

$C_{20}H_{16}Br_4Hg_2N_4$, Bis-μ-bromo-bis(bromo-2,2'-bipyridine-mercury(II)), 40B, 907

$C_{20}H_{16}ClCuIN_4O_4$, Bis(2,2'-bipyridyl)monoiodocopper(II) perchlorate, 46B, 1095

$C_{20}H_{16}ClCuN_4O_4S_4 \cdot x\ H_2O$, Bis(bis(2-pyridyl)disulfide)copper(I) perchlorate hydrate, 42B, 848

$C_{20}H_{16}Cl_2CoN_6OS_2$, Dichlorodithiabendazolecobalt(II) hydrate, 39B, 748

$C_{20}H_{16}Cl_4Cu_3N_4$, Bis-(2,2'-bipyridyl)copper(II) bis(dichlorocuprate-(I)), 40B, 907

$C_{20}H_{16}CuI_2N_4$, Iodobis-(2,2'-bipyridyl)copper(II) iodide, 26, 657; 28, 643

$C_{20}H_{16}CuN_4O_6S_4$, Bis(2,2'-bipyridyl)-catena-μ-tetrathionatocopper-(II), 44B, 938

$C_{20}H_{16}CuN_4O_8S_2 \cdot H_2O$, Bis(2,2'-bipyridyl)peroxodisulfatocopper(II) monohydrate, 46B, 1096

$C_{20}H_{16}CuN_6O_5$, Nitritobis(2,2'-bipyridyl)copper(II) nitrate, 34B, 569

$C_{20}H_{16}CuN_6O_6 \cdot H_2O$, Bis(2,2'-bipyridyl)nitratocopper(II) nitrate monohydrate, 46B, 1096

$C_{20}H_{16}N_6Pd$, Bis(2,2'-dipyridyliminato)palladium(II), 30B, 354

$C_{20}H_{18}AgN_6O_7$, Bis(2,2'-bipyridine)silver(II) nitrate monohydrate, 39B, 748

$C_{20}H_{18}Cl_2CuN_6O_8$, Bis-(2,2-bipyridylamine)copper(II) perchlorate,

37B, 550
$C_{20}H_{18}Cl_2Cu_2N_4O_{10}$, Di-$\mu$-hydroxo-bis(bipyridylcopper(II)) diperchlorate, 42B, 913
$C_{20}H_{18}CoN_5O_8S$, Aquo-(2,2':6',2'':6'',2'''-quaterpyridyl)sulfitocobalt(III) nitrate monohydrate, 41B, 1084
$C_{20}H_{18}N_4Ni$, 2,11-Dimethyldibenzo(6,7,13,14)-1,5,8,12-tetraazacyclotetradeca-2,4,6,8,10,13-hexaene-nickel(II), 38B, 938
$C_{20}H_{18}N_6Ni$, (7,8,15,17,18,20-Hexahydrodibenzo[e,m]pyrazino[2,3-b]-(1,4,8,11)-tetraazacyclotetradecinato(2-))nickel(II), 40B, 908
$C_{20}H_{18}N_6O_7Pd$, Bis-(2,2'-bipyridyl)palladium dinitrate monohydrate, 38B, 938
$C_{20}H_{19}BMoN_6O_2$, (Phenyltrispyrazolylborato)(allyl)dicarbonylmolybdenum, 43B, 1044
$C_{20}H_{19}B_2CuF_8N_5$, Amminebis-(2,2'-bipyridyl)copper(II) tetrafluoroborate, 38B, 938
$C_{20}H_{19}Cl_2N_4O_{12}Ru$, trans-Hydroxybis(2,2'-bipyridyl)aquoruthenium(III) perchlorate, 46B, 1097
$C_{20}H_{19}Cl_3Cu_2N_6$, catena-μ-Chloro-dichloro-μ-(N,N'-bis(2-(2-pyridyl)-ethyl)-2,3-pyrazinedicarboxamidato-N,N(N),N^1:N',N(N'),N(4))-dicopper(II), 40B, 909
$C_{20}H_{20}Br_2N_4Ni$, Nickel(II) tetrapyridine dibromide, 22, 782
$C_{20}H_{20}Br_6Cu_4N_4O$, μ_4-Oxo-hexa-μ-bromotetrakis(pyridine copper(II)), 39B, 749
$C_{20}H_{20}Cl_2CoN_4$, trans-Tetrakis(pyridine)dichlorocobalt(II), 44B, 939
$C_{20}H_{20}Cl_2CuN_4$, trans-Bis(2,7-dimethyl-1,8-naphthyridine)dichlorocopper(II), 43B, 1318
$C_{20}H_{20}Cl_2FeN_4$, trans-Tetrakis(pyridine)dichloroiron(II), 44B, 939
$C_{20}H_{20}Cl_2N_4Ni$, trans-Tetrakis(pyridine)dichloronickel(II), 18, 749; 44B, 939
$C_{20}H_{20}Cl_2N_4V$, Tetrapyridinedichlorovanadium(II), 39B, 749
$C_{20}H_{20}Cl_4N_4O_3Re_2$, μ-Oxo-bis(cis-dichloro-cis-di(pyridine)-trans-oxorhenium(V)), 44B, 940
$C_{20}H_{20}Cl_6N_4W_2 \cdot x\ C_3H_6O$, trans-Dichloro-cis-dipyridinetungsten(III)-di-μ-chloro-cis-dichloro-trans-dipyridinetungsten(III) - acetone, 37B, 550
$C_{20}H_{20}CrN_7O_5$, Dinitritonitrosyltripyridinechromium monopyridine solvate, 43B, 1318
$C_{20}H_{20}Cu_4I_4N_4$, Copper(I) iodide - pyridine tetramer, 42B, 849
$C_{20}H_{20}I_2N_4Ni$, Diiodotetrakis(pyridine)nickel(II), 39B, 750
$C_{20}H_{20}Mo_2N_4O_8P_2$, Di-$\mu$-oxo-bis(2,2'-bipyridyloxohypophosphitomolybdenum(V)), 42B, 914
$C_{20}H_{20}N_6NiO_6$, Bis(8-amino-2-methylquinoline)nitratonickel(II) nitrate, 41B, 1084
$C_{20}H_{21}Br_4N_4W$, Pyridinium tetrabromobis(pyridine)tungstate(III), 45B, 1185
$C_{20}H_{21}CeN_8O_{14}$, 4,4'-Bipyridylium tetranitratodiaquo-4,4'-bipyridylcerium(III), 44B, 940
$C_{20}H_{21}Cl_2N_6O_6Rh$, trans-Dichlorotetrakis(pyridine)rhodium(III) hydrogendinitrate, 45B, 1186
$C_{20}H_{22}Cl_2FeN_4O$, trans-Tetrakis(pyridine)dichloroiron(II) monohydrate, 44B, 939
$C_{20}H_{24}Br_2N_2Ni$, Dibromo(cis-endo-N,N'-di(4-methylbenzylidene)-meso-2,3-butanediamine)nickel(II), 37B, 551
$C_{20}H_{24}ClN_4O_4Re$, Dioxotetrakis(pyridine)rhenium(V) chloride dihydrate, 44B, 796
$C_{20}H_{24}Cl_2CoN_8$, Dichlorotetrakis(1-vinylimidazole)cobalt(II), 44B, 941

$C_{20}H_{24}Cl_2CuN_8$, Dichlorotetrakis(1-vinylimidazole)copper(II), 44B, 941

$C_{20}H_{24}Cl_2N_8O_8S_4Zn \cdot 2\ C_3H_6O$, Tetrakis(1-methylpyrimidine-2-thione)-zinc(II) perchlorate bisacetone solvate, 45B, 1186

$C_{20}H_{24}CuN_4O_5$, Aquobis(γ-picoline)bis(succinimidato)copper(II), 41B, 1085

$C_{20}H_{24}Ga_2N_{12}Ni$, Bis(methyl-tris(1-pyrazolyl)-gallato)nickel(II), 45B, 1187

$C_{20}H_{25.36}N_{9.12}Na_{2.88}O_{16}P_{1.12}Pt_{0.56} \cdot 16\ H_2O$, Sodium diammine-platinum-(inosine-5'-phosphate) inosine-5'-phosphate hydrate, 42B, 850

$C_{20}H_{26}CoKN_6O_6S_2$, Potassium bis(3-phenylbiuretato)cobaltate(III) dimethylsulfoxide solvate, 40B, 910

$C_{20}H_{27}Cl_2CuN_5O_8$, [2,12-Di(2-pyridyl)-3,7,11-triazatrideca-2,11-diene-N1',N1",N3,N7,N11]copper(II) perchlorate, 45B, 1187

$C_{20}H_{27}Cl_2MnN_6O_9$, [2,12-Di(2-pyridyl)-3,7,11-triazatrideca-2,11-diene]nitrosylmanganese(II) diperchlorate, 45B, 1188

$C_{20}H_{27.44}N_{9.72}Na_2O_{16}P_2Pt_{0.86} \cdot 16\ H_2O$, Sodium diammine-bis(inosine-5'-monophosphate)platinum(II) inosine-5'-monophosphate hydrate, 45B, 1189

$C_{20}H_{28}ClCoN_6O_4$, Bisanilinebis(dimethylglyoximato)cobalt(III) chloride, 40B, 910

$C_{20}H_{28}Cl_2CuN_4O_6$, Chlorobis(2,2'-bipyridyl)copper(II) chloride hexahydrate, 39B, 750

$C_{20}H_{28}Cl_2N_4O_2Zn$, Dichlorobis(N,N-diethylnicotinamide)zinc(II), 42B, 851; 45B, 1189

$C_{20}H_{28}Cl_2N_4Pd$, trans-Bis(acetone methylphenylhydrazone)dichloropalladium(II), 43B, 1320

$C_{20}H_{28}Cl_4Cu_2N_{20}O_{20}$, Tetra-μ-adenine-diaquodicopper(II) perchlorate dihydrate, 39B, 751

$C_{20}H_{28}Cl_4N_4Pd_2 \cdot CHCl_3$, Bis(acetone methylphenylhydrazone)tetrachlorodipalladium(II) chloroform solvate, 45B, 1190

$C_{20}H_{28}Cl_4N_8Ni_2O_6$, μ-Chloro-tetraaquo(1,4-dihydrazinophthalazine bis-(2'-pyridinecarboxaldimine))dinickel(II) chloride dihydrate, 43B, 345

$C_{20}H_{28}Cu_2N_4O_{11}S$, Aquobis(2,2'-bipyridine)di-μ-hydroxo-sulfatodicopper(II) tetrahydrate, 41B, 1086

$C_{20}H_{28}I_2N_4O_2Zn$, Diiodo-bis(N,N-diethylnicotinamide)zinc, 45B, 1190

$C_{20}H_{28}N_{10}Ni_3O_2$, Bis[bis(3,5-dimethylpyrazolyl)nitrosylniccollate(I)-(N,N')]-nickel(II), 45B, 1158

$C_{20}H_{28}N_{16}Ni_2$, Dihydrooctaaza[14]annulenenickel dimer, 42B, 851

$C_{20}H_{30}B_3KN_{12}OV$, Potassium tris(dihydrobis(1-pyrazolyl)borato)-vanadate(II) monoethanolate, 44B, 941

$C_{20}H_{32}B_2N_8Ni$, Bis(diethylbis-(1-pyrazolyl)borato)nickel(II), 40B, 911

$C_{20}H_{32}Cl_2N_2O_4Rh_2$, cis-Dicarbonyl-glyoxal-bis(2,4-dimethylpentyl-3-imine)rhodium(I) cis-dicarbonyl-dichlororhodate(I), 46B, 1097

$C_{20}H_{32}Cl_2N_{12}O_{14}Pt \cdot 7\ H_2O$, cis-Diamminebis(guanosine)platinum(II) chloride perchlorate heptahydrate, 46B, 1098

$C_{20}H_{32}Cl_4Cu_2N_{20}O_6$, Dichlorotetra-μ-adenine-dicopper(II) chloride hexahydrate, 37B, 552

$C_{20}H_{32}N_4O_4Ru_2$, Tetracarbonylbis[glyoxal bis(isopropylimine)]diruthenium, 46B, 1098

$C_{20}H_{36}BClF_4N_2P_2Pt$, cis-Chlorobis(triethylphosphine)-1,8-naphthyridineplatinum(II) tetrafluoroborate, 44B, 942

$C_{20}H_{36}BClF_4N_2P_2Pt$, cis-Chlorobis(triethylphosphine)phthalazineplatinum(II) tetrafluoroborate, 44B, 942

$C_{20}H_{40}Cl_2CuN_4O_8$, [5R(S),14R(S)-2,4,4,11,13,13-Hexamethyl-1,5,10,14-

tetraazacyclooctadeca-1,10-diene]copper(II) perchlorate, 45B, 1191

$C_{20}H_{40}Cl_2CuN_4O_8$, cis-[(5S(R),10R(S))-2,4,4,11,11,13-Hexamethyl-1,5,10,14-tetraazacyclooctadeca-1,13-diene]copper(II) perchlorate, 46B, 1099

$C_{20}H_{40}Cl_2CuN_4O_8$, trans-[(5R(S),14R(S))-2,4,4,11,13,13-Hexamethyl-1,5,10,14-tetraazacyclooctadeca-1,10-diene]copper(II) perchlorate, 46B, 1099

$C_{20}H_{41}ClCoN_4O_4P$, trans-Chlorobis(dimethylglyoximato)tributylphosphinecobalt(III), 46B, 1099

$C_{20}H_{42}CoN_{20}O_{18}S_2$, Bis(adeninium) trans-bis(adenine)tetraaquocobalt-(II) bis(sulfate) hexahydrate, 39B, 752

$C_{20}H_{44}Cl_6Cu_4N_4O$, Hexakis(μ-chloro)(μ$_4$-oxo)tetrakis(piperidine-copper), 46B, 1100

$C_{20}H_{44}Cu_4I_4N_4$, Copper(I) iodide - piperidine tetramer, 44B, 943

$C_{20}H_{48}Cl_6Hg_3N_8$, Bis[chloro(1,4,8,11-tetra-azacyclotetradecane)mercury] tetrachloromercurate, 45B, 1191

$C_{20}H_{48}N_4Si_2Ti$, 1,3,5,7-Tetra-t-butyl-2,2,6,6-tetramethyl-1,3,5,7-tetraaza-2,6-disila-4-titaniaspiro[3.3]heptane, 46B, 1101

$C_{20}H_{48}N_4Si_2Zr$, 1,3,5,7-Tetra-t-butyl-2,2,6,6-tetramethyl-1,3,5,7-tetraaza-2,6-disila-4-zirconiaspiro[3.3]heptane, 46B, 1101

$C_{21}H_{15}MnNO_4P$, Nitrosyltricarbonyltriphenylphosphinemanganese, 33B, 429

$C_{21}H_{17}N_3S_2Zn$, Bis(2-thiobenzaldimino)-2,6-diacetylpyridinezinc(II), 39B, 752

$C_{21}H_{19}Cl_2Cu_2N_7O_3$, Bis(dichloro-(2,6-diacetylpyridinebis(picolinoylhydrazonato)))dicopper(II) dihydrate, 40B, 913

$C_{21}H_{19}CuN_5O_6$, Dinitrato(2,6-bis(1-(phenylimino)ethyl)pyridine)copper(II), 42B, 915

$C_{21}H_{19}F_6MoN_2O_2P$, (+)-(η^5-Cyclopentadienyl)(N-((S)-α-phenylethyl)pyridine-2-carbimine)dicarbonylmolybdenum hexafluorophosphate, 44B, 943

$C_{21}H_{19}N_5NiO_6$, Dinitrato(2,6-diacetylpyridine-bis(anil))nickel(II), 41B, 1087

$C_{21}H_{20}ClIrNOP$, Chlorocarbonyl(o-(diphenylphosphino)-N,N-dimethylaniline)iridium(I), 46B, 1103

$C_{21}H_{24}FeN_7$, Tris(1-(2-azoyl)-2-azabut-4-yl)amineiron(III), 44B, 944

$C_{21}H_{28}I_2N_2O_2Zn$, (3,4.9,10-Dibenzo-1,12-dimethyl-5,8-dioxa-1,12-diazacyclopentadecane-3,9-diene-N',N')diiodozinc(II), 46B, 1103

$C_{21}H_{34}Br_2N_4OZn \cdot C_3H_7NO$, [Bis(4,5-diisopropyl-1-methyl-2-imidazolyl) ketone]dibromozinc(II) dimethylformamide solvate, 46B, 1104

$C_{21}H_{34}Cl_3CoN_6O_{14}$, (+)-Tris-((-)-1-(2-pyridyl)ethylamine)cobalt(III) perchlorate dihydrate, 43B, 1323

$C_{21}H_{34}CoI_2N_5O_2$, 2,12-Di-2-pyridyl-3,7,11-triazatrideca-2,11-diene-(methyl)cobalt(III) diiodide dihydrate, 43B, 1323

$C_{21}H_{52}Co_2N_{12}O_5$, (+)-fac-Tris(meso-2,4-pentanediamine)cobalt(III) hexacyanocobaltate(III) pentahydrate, 44B, 944

$C_{22}H_8CdMn_2N_2O_{10}$, μ-(1,10-Phenanthrolinecadmium)-bis(pentacarbonylmanganese) (2Cd-Mn), 40B, 906

$C_{22}H_{14}Cl_2N_{10}Pt_2$, Bis(diiminosuccinonitrilo)platinum(II) trans-dichlorobis(benzonitrile)platinum(II), 41B, 1088

$C_{22}H_{16}Cl_3Cu_2N_4$, Trichlorobis(4-methyl-1,8-naphthyridine)dicopper, 42B, 853

$C_{22}H_{16}MnN_6O_2$, Diisocyanatobis(2,2'-bipyridyl)manganese(II), 41B, 1091

$C_{22}H_{20}CoI_4N_6S_2$, Tetrapyridine-di(diiodothiocyanato)cobalt(II), 43B, 1324

$C_{22}H_{20}CoN_6S_2$, trans-Bis(isothiocyanato)tetrakis(pyridine)cobalt(II),

46B, 1105
$C_{22}H_{20}FeN_6S_2$, Iron(II) thiocyanate tetra(pyridine), 32B, 440
$C_{22}H_{20}N_6NiS_2$, Nickel(II) tetrapyridine isothiocyanate, 22, 784
$C_{22}H_{22}FeN_4$, 7,16-Dihydro-6,8,15,17-tetramethyldibenzo[b,i]-(1,4,8,11)tetraazacyclotetradecinatoiron(II), 42B, 853
$C_{22}H_{22}N_4Pd$, 5,7,12,14-Tetramethyldibenzo[b,i][1,4,8,11]tetraaza[14]-annulenepalladium, 45B, 1192
$C_{22}H_{24}Cl_2CuN_4$, Dichlorobis(1-phenyl-3,5-dimethylpyrazole)copper(II), 46B, 1106
$C_{22}H_{24}Cl_2FeN_8O_8$, Bis(2,6-diacetylpyridine-hydrazinedi-imine)-bis(acetonitrile)iron(II) diperchlorate, 40B, 913
$C_{22}H_{26}Cl_2CoN_6$, Dichlorobis(1-(2-pyridylmethyl)-4,5-dimethylimidazole)cobalt, 43B, 1328
$C_{22}H_{26}HgI_2N_2O_3$, Diiodo(2,3:12,13-dibenzo-1,14,17-trioxa-5,10-diaza-2,3:4,5:10,11:12,13-cycloenneadecatetraene)mercury(II), 46B, 1107
$C_{22}H_{30}CoN_5O_6$, (Cumyl peroxide)(pyridine)cobaloxime, 42B, 916
$C_{22}H_{32}Cl_4N_8Ni_2$, Di-μ-chloro-dichlorobis(bis(3,5-dimethylpyrazolyl)-methane)dinickel(II), 46B, 1107
$C_{22}H_{32}CoN_{12}O_5$, Hexaimidazolecobalt(II) acetate monohydrate, 40B, 914
$C_{22}H_{32}N_6O_4S_2Zn$, Diaquobis-(N,N-diethylnicotinamide)diisothiocyanatozinc, 39B, 753
$C_{22}H_{34}Mo_2N_4O_6$, Hexacarbonylbis[μ-(1-(isopropylamino)-2-(isopropylimino)ethane-N,N')]dimolybdenum, 45B, 1192
$C_{22}H_{42}Cl_4N_8O_{12}Yb_2$, Di-μ-hydroxo-bis(triaquo(pyridine-2-carboxaldehyde-2'-pyridylhydrazone)ytterbium(III)) tetrachloride tetrahydrate, 42B, 854
$C_{23}H_{21}BClF_4MnN_7$, (2,6-Diacetylpyridine)-(2,9-bis(1-methylhydrazone)-1,10-phenanthroline)chloromanganese tetrafluoroborate, 46B, 1108
$C_{23}H_{22}FeN_4O \cdot 0.5\ C_7H_8$, (7,16-Dihydro-6,8,15,17-tetramethyldibenzo-[b,i](1,4,8,11)tetraazacyclotetradecinato)carbonyliron(II) hemi(toluene), 42B, 855
$C_{23}H_{22}MnN_5S$, (7,16-Dihydro-6,8,15,17-tetramethyldibenzo[b,i]-[1,4,8,11]tetraazacyclotetradecinato)-(isothiocyanate)manganese-(III), 45B, 1193
$C_{23}H_{23}CdI_2N_3S_2$, Diiodo(S,S'-dimethyl-2,6-bis(1-(2-thiophenylimino)-ethyl)pyridine)cadmium(II), 44B, 945
$C_{23}H_{23}Cl_3CoIN_4$, Iodo(7,16-dihydro-6,8,15,17-tetramethyldibenzo[b,i]-(1,4,8,11)tetraazacyclotetradecinato)cobalt(III) - chloroform, 42B, 856
$C_{23}H_{24}Cl_2CoN_6O_8$, 1,1,1-Tris(pyridine-2-aldiminomethyl)ethanecobalt-(II) perchlorate, 43B, 1325
$C_{23}H_{24}Cl_2MnN_6O_8$, 1,1,1-Tris(pyridine-2-aldiminomethyl)ethanemanganese(II) perchlorate, 43B, 1325
$C_{23}H_{24}Cl_2N_6NiO_8$, 1,1,1-Tris(pyridine-2-aldiminomethyl)ethanenickel-(II) perchlorate, 43B, 1325
$C_{23}H_{25}BCoF_4N_7O_2$, Diaqua(7,15-dihydro-7,9,13,15-tetramethylpyrido-[2',1',6':12,13,14][1,2,4,7,9,10,13]-heptaazacyclopentadeca-[3,4,5,6,7,8-aklmn][1,10]phenanthroline)cobalt(II) bis(tetrafluoroborate), 46B, 1109
$C_{23}H_{25}B_2F_8FeN_7O_2$, Diaqua(2,6-diacetylpyridine)-(2,9-bis(1-methylhydrazone)-1,10-phenanthroline)iron bis(tetrafluoroborate), 46B, 1108
$C_{23}H_{25}Cl_2FeN_6O_8$, 1,1,1-Tris(pyridine-2-aldiminomethyl)ethaneiron(II) perchlorate, 38B, 939
$C_{23}H_{25}Cl_2N_6O_8Zn$, 1,1,1-Tris(pyridine-2-aldiminomethyl)ethanezinc(II) perchlorate, 38B, 939
$C_{23}H_{26}FeN_6O$, (7,16-Dihydro-6,8,15,17-tetramethyldibenzo[b,i]-(1,4,8,11)tetraazacyclotetradecinato)(hydrazine)carbonyliron(II),

42B, 855

$C_{23}H_{31}Cl_2CuN_5O_8$, Dibenzopentaazaeicosa(9,20)dienecopper(II) perchlorate, 43B, 1326

$C_{24}H_{16}Cl_2N_4O_8Pd$, Bis(1,10-phenanthroline)palladium(II) diperchlorate, 46B, 1109

$C_{24}H_{16}Cl_2N_4Pt \cdot 3\ H_2O$, Bis(1,10-phenanthroline)platinum(II) dichloride trihydrate, 46B, 1110

$C_{24}H_{16}Cl_6Fe_2N_4$, Dichlorobis(1,10-phenanthroline)iron(III) tetrachloroferrate(III), 43B, 1327

$C_{24}H_{16}HgN_6O_6$, Bis(1,10-phenanthroline)dinitratomercury(II), 44B, 945

$C_{24}H_{16}I_{10}N_4Pt_2$, Bis(tetraiodo(1,10-phenanthroline)platinum(IV)) iodine, 43B, 1295

$C_{24}H_{16}N_8NbS_4$, Tetraisothiocyanatobis(2,2'-bipyridine)niobium(IV), 42B, 857

$C_{24}H_{16}N_8S_4Zr$, Tetraisothiocyanatobis(2,2'-bipyridine)zirconium(IV), 42B, 857

$C_{24}H_{18}B_2CuF_8N_4O$, Aquobis(1,10-phenanthroline)copper(II) tetrafluoroborate, 44B, 946

$C_{24}H_{18}Cl_2HgN_6O_8$, Tris(1,8-naphthyridine)(perchlorato)mercury(II) perchlorate, 40B, 915

$C_{24}H_{18}CuN_6O_7$, Aquobis(1,10-phenanthroline)copper(II) nitrate, 41B, 1058

$C_{24}H_{18}I_2MoN_6O_2 \cdot H_2O \cdot C_{12}H_8N_2$, (Hydroxylamido)nitrosylbis(1,10-phenanthroline)molybdenum diiodide monohydrate 1,10-phenanthroline solvate, 46B, 1186

$C_{24}H_{19}CuN_6O_{6.5}S_3 \cdot 0.5\ H_2O$, Nitrato-tris(2-benzothiazolylmethyl)aminecopper nitrate hemihydrate, 46B, 1111

$C_{24}H_{20}Br_2CoN_4O_4$, Dibromo(N-tetrakis-4-pyridinecarbaldehyde)cobalt(II), 43B, 1327

$C_{24}H_{20}Cl_2N_4Pd$, Dichlorobis(azobenzene)palladium(II), 41B, 1089

$C_{24}H_{21}BMoN_6O_2$, (Phenyltrispyrazolylborato)(cycloheptatriene)dicarbonylmolybdenum, 43B, 1044

$C_{24}H_{22}CoI_6N_6S_2$, trans-Bis(isothiocyanato)tetrakis(pyridine)cobalt(II) diiodoform, 46B, 1111

$C_{24}H_{24}BCuF_4N_4$, Bis(6,6'-dimethyl-2,2'-bipyridyl)copper(I) tetrafluoroborate, 46B, 1112

$C_{24}H_{24}CoN_6S_2$, Bis(3,5-dimethyl-1-phenylpyrazole)bis(isothiocyanato)cobalt(II), 46B, 1112

$C_{24}H_{24}CuN_6$, Bis(3,3'-dimethyl-2,2'-dipyridylamine)copper(II), 45B, 1193

$C_{24}H_{25}ClFeN_5$, Chloro(7,16-dihydro-6,8,15,17-tetramethyldibenzo[b,i]-(1,4,8,11)tetraazacyclotetradecinato)iron(III) - acetonitrile, 42B, 856

$C_{24}H_{26}Fe_3N_{16}$, Bis(imidazole)tris(di(imidazolo)iron), 44B, 947

$C_{24}H_{26}Mn_3N_{16}$, Tetraimidazolyl trimanganese, 46B, 1113

$C_{24}H_{27}Cl_2N_6NiO_8$, cis,cis-1,3,5-Tris(pyridine-2-aldimino)cyclohexanenickel(II) perchlorate, 38B, 939

$C_{24}H_{28}Br_4Cu_4N_4$, Tetra-μ_3-bromo-tetrakis(2-picoline)-tetrahedro-tetracopper(I), 46B, 1114

$C_{24}H_{28}Cl_2CuN_{12}O_{12} \cdot 4\ H_2O$, Bis(cyclo-L-histidyl-L-histidyl)copper(II) perchlorate tetrahydrate, 45B, 1194

$C_{24}H_{28}Cl_2N_4NiO_8$, Tetrakis(4-methylpyridine)nickel(II) perchlorate, 39B, 754

$C_{24}H_{28}CoF_{12}N_4P_2$, Tetrakis(4-methylpyridine)cobalt(II) hexafluorophosphate, 45B, 1195

$C_{24}H_{28}F_{12}N_4NiP_2$, Tetrakis(4-methylpyridine)nickel(II) hexafluorophosphate, 46B, 1115

$C_{24}H_{29}Br_4N_4W$, 4-Methylpyridinium 4-methylpyridine tetrabromobis(4-methylpyridine)tungsten(III), 46B, 1115
$C_{24}H_{31}N_2O_2RhS$, (S-t-Butyl-N,N'-dimesitylsulfurdiimine)dicarbonylrhodium, 42B, 858
$C_{24}H_{36}B_2CdF_8N_{12}$, Hexakis-(2-methylimidazole)cadmium(II) tetrafluoroborate, 39B, 754
$C_{24}H_{36}Cl_2N_4Pd$, trans-Dichlorobis(isopropylmethylketone methylphenylhydrazone)palladium(II), 45B, 1195
$C_{24}H_{36}Cl_3CoN_4O_4$, $(+)_{589}$-trans-($(-)_{589}$-N,N'-Bis((S)-2-amino-3-phenylpropyl)-trans-(R)-1,2-cyclohexanediamine-δ λ δ)dichlorocobalt(III) perchlorate, 37B, 554
$C_{24}H_{38}BClF_4N_2P_2Pt$, cis-Chlorobis(triethylphosphine)-1,10-phenanthrolineplatinum tetrafluoroborate, 40B, 915
$C_{24}H_{39}Cl_3CoN_5O_4$, Chloro-pyridino-cis-1,12-bis-methylglyoximato-dodecane-cobalt(III), 37B, 554
$C_{24}H_{43}Cl_3N_{20}O_{16}Pt_2$, (9-Ethylguanine)(1-methylcytosine)diammineplatinum(II) (9-ethylguaninato)(1-methylcytosine)diammineplatinum(II) triperchlorate, 46B, 1116
$C_{24}H_{44}Cl_2N_4Pd$, Dichlorobis(azocyclohexane)palladium(II), 44B, 947
$C_{24}H_{50}Cl_4N_4Pd_2 \cdot C_6H_6$, Di-μ-(N,N,N',N'-tetramethylpentane-1,5-diamine-N,N')bis[dichloropalladium(II)] benzene solvate, 45B, 1196
$C_{24}H_{56}CuN_{10}O_{16}$, Diammonium aqua(tetramethyldiphenyloxaoctaazabicycloheptadecatetronato)copper(2+) decahydrate, 42B, 840
$C_{24}H_{62}Cl_2N_8O_2P_2Pt$, cis-Dichlorobis(cyclohexylamine-N)platinum(II) bis(hexamethylphosphoramide), 46B, 1116
$C_{24}H_{78}N_{16}Ni_4O_{23}$, Tetra-$\mu_3$-hydroxo-tetrakis-(1,3,5-triaminocyclohexane)-tetranickel tetranitrate heptahydrate, 43B, 1327
$C_{25}H_{11}CdMn_2N_3O_{10}$, μ-(2,2':6',2"-Terpyridylcadmium)-bis(pentacarbonylmanganese)(2Cd-Mn), 39B, 755
$C_{25}H_{16}BrCoN_4O_3 \cdot 4\ H_2O$, Bis(phenanthroline)carbonatocobalt(III) bromide tetrahydrate, 46B, 1117
$C_{25}H_{16}ClCoN_4O_3 \cdot 3\ H_2O$, Carbonatobis(1,10-phenanthroline)cobalt(III) chloride trihydrate, 46B, 1117
$C_{25}H_{16}N_6O_3Pt \cdot H_2O$, Cyanobis(1,10-phenanthroline)platinum(II) nitrate hydrate, 46B, 1118
$C_{25}H_{18}CuN_6O_4$, Cyanobis(1,10-phenanthroline)copper(II) nitrate monohydrate, 41B, 1090
$C_{25}H_{25}I_2N_3S_2Zn$, Diiodo(6,7,8,9-tetrahydro-16,22-dimethyl-5,10-dithia-15,23,24-triaza-17,21-methenodibenzo[a,i]cyclononadecene-N,N',N")zinc(II), 44B, 948
$C_{25}H_{40}CoN_5O_4$, Cobaloxime (intramolecularly-alkylated), 43B, 998
$C_{25}H_{40}N_{12}O_{16}P_2Pt \cdot 11\ H_2O$, Bis(guanosine 5'-monophosphate methyl ester)trimethylenediamineplatinum(II) undecahydrate, 46B, 1118
$C_{25}H_{43}Br_3N_2Ni$, Tetra-n-butylammonium tribromo(quinoline)nickelate-(II), 33B, 430
$C_{26}H_{16}CuN_2O_4$, Copper(II) 2-(o-hydroxyphenyl)benzoxazole, 34B, 572
$C_{26}H_{16}HgN_6S_2$, cis-Dithiocyanatobis(1,10-phenanthroline)mercury(II), 40B, 1017
$C_{26}H_{16}N_6NiO_2$, Diisocyanatobis(1,10-phenanthroline)nickel, 41B, 1091
$C_{26}H_{18}Cl_2Hg_2N_2O_8$, Bis(acridine)dimercury(I) perchlorate, 42B, 860
$C_{26}H_{18}CuN_6O_3$, Aquobis(2-(2-pyridyl)quinazolin-4(3H)-onato)copper-(II), 42B, 861
$C_{26}H_{20}AsCl_{10}NW$, Tetraphenylarsonium pentachloro-N-pentachloroethylnitridotungsten, 45B, 1197
$C_{26}H_{20}I_2N_2Ni$, (2,9-Dimethyl-4,7-diphenyl-1,10-phenanthroline)diiodonickel(II), 43B, 1334
$C_{26}H_{22}Cl_2CoN_6$, Dichlorobis(1-(2-pyridylmethyl)benzimidazole)cobalt,

43B, 1328
$C_{26}H_{22}N_2S_2Zn$, (1,10-Phenanthroline)bis(4-toluenethiolato)zinc(II), 46B, 1119
$C_{26}H_{22}N_4O_4Rh_2$, (7,16-Dihydro-6,8,15,17-tetramethyldibenzo[b,i]-(1,4,8,11)tetraazacyclotetradecinato)tetracarbonyldirhodium(I), 44B, 949
$C_{26}H_{24}ClCuN_8O_4$, Bis(1,5-diphenylformazan)copper(I) perchlorate, 42B, 861
$C_{26}H_{28}N_6NiS_2$, Tetrakis(4-methylpyridine)diisothiocyanatonickel(II) (α form), 43B, 1329
$C_{26}H_{28}N_6NiS_2$, Bis(isothiocyanato)tetrakis(4-methylpyridine)nickel-(II) (β form), 38B, 940
$C_{26}H_{29}ClCoN_4O_4P$, trans-Chlorobis(dimethylglyoximato)(triphenylphosphine)cobalt(III), 40B, 917
$C_{26}H_{29}ClCoN_4O_4Sb$, Chlorobis(dimethyglyoximato)triphenylstibinecobalt(III), 39B, 755
$C_{26}H_{29}CoN_5O_6P$, trans-Nitrobis(dimethylglyoximato)(triphenylphosphine)cobalt(III), 44B, 912
$C_{26}H_{30}N_4Ni$, Bis(3,3',5,5'-tetramethyldipyrromethenato)nickel(II), 35B, 715
$C_{26}H_{32}B_4F_8N_{10}Ni_2O_8 \cdot C_{10}H_8N_2$, ($\mu$-4,4'-Bipyridyl)bis(bis(difluoroborondimethylglyoximato)nickel(II)) 4,4'-bipyridine solvate, 46B, 1119
$C_{26}H_{32}CuN_6O_6$, Bis-(5,5'-diethylbarbiturato)bispyridinecopper, 39B, 756
$C_{26}H_{36}F_{12}N_6NiP_2$, 2,11,20,26-Tetramethyl-3,10,14,18,21,25-hexaazatricyclo[10.7.7.2$^{5-8}$]octacosa-1,5,7,11,13,18,20,25,27-nonaene(N4)nickel(II) hexafluorophosphate, 45B, 1198
$C_{26}H_{37}ClCoN_7O_{11}S \cdot H_2O$, Aniline-bis(dimethylglyoximato)sulphonatocobalt(III) bis(anilinium) monoperchlorate monohydrate, 45B, 1198
$C_{26}H_{47}ClCoN_4O_4P \cdot C_7H_8$, trans-Bis(dimethylglyoximato)chloro(tricyclohexylphosphine)cobalt(III) toluene solvate, 45B, 881
$C_{27}H_{45}Cl_3CuN_7O_4$, Tetra-n-butylammonium o-phenylenebis(biuretato)cuprate(III) - chloroform, 43B, 1329
$C_{28}H_{16}B_2F_8N_4Pd$, Tetrabenzo[b,f,j,n][1,5,9,13]tetraazacyclohexadecinepalladate(II) bis(tetrafluoroborate), 45B, 1199
$C_{28}H_{16}Fe_2N_4O_8Zn_2$, (2,2'-Bipyridyl)zinc-tetracarbonyliron(Fe-Zn), 46B, 1120
$C_{28}H_{20}B_2F_8N_4Ni$, Tetrabenzo[b,f,j,n](1,5,9,13)tetraazacyclohexadecinenickel tetrafluoroborate, 40B, 1123
$C_{28}H_{20}MnN_{10}O_6$, Bis-(3,4-di-2-pyridylpyridazine)dinitratomanganese-(II), 41B, 1091
$C_{28}H_{20}N_6Ni$, Bis-(3-phenyl-5-pyridyl(2)-pyrazolato)-nickel(II), 37B, 556
$C_{28}H_{22}ClN_4O_8Pt$, Bis(diphenylglyoximato)platinum(III) perchlorate, 44B, 949
$C_{28}H_{22}Cl_3CoN_4 \cdot 0.5\ CHCl_3 \cdot 0.5\ CH_2Cl_2$, (N,N'-Bis($\alpha$-(2-amino-5-chlorophenyl)benzylidene)-ethane-1,2-diaminato(2)-N,N',N'',N''')chlorocobalt(III) chloroform dichloromethane, 43B, 1330
$C_{28}H_{22}IN_4NiO_4$, Bis(diphenylglyoximato)nickel(III) iodide, 41B, 1092; 45B, 1200
$C_{28}H_{22}I_2N_4NiO$, Tetrabenzo[b,f,j,n](1,5,9,13)tetraazacyclohexadecinenickel iodide monohydrate, 40B, 1123
$C_{28}H_{23}AuF_6N_2P_2$, 2,2'-Bipyridyltriphenylphosphinegold(I) hexafluorophosphate, 42B, 863
$C_{28}H_{24}Br_4N_4Ni_2$, Bis(2,9-dimethyl-1,10-phenanthroline)di-μ-bromo-bis-(bromonickel(II)), 43B, 1334

$C_{28}H_{24}ClCuN_4O_4$, Bis(2,9-dimethyl-1,10-phenanthroline)copper(I) perchlorate, 45B, 1200
$C_{28}H_{24}Cl_4N_4Ni_2$, Bis(2,9-dimethyl-1,10-phenanthroline)di-μ-chloro-bis(chloronickel(II)), 43B, 1334
$C_{28}H_{24}CuN_5O_3$, Bis(2,9-dimethyl-1,10-phenanthroline)copper(I) nitrate, 45B, 1200
$C_{28}H_{26}Cl_4N_4Re_2$, Tetrachloro-bis(N,N'-diphenylacetamidinato)dirhenium(Re-Re), 46B, 1121
$C_{28}H_{27}FeN_5O$, (7,16-Dihydro-6,8,15,17-tetramethyldibenzo[b,i]-(1,4,8,11)tetraazacyclotetradecinato)(pyridine)carbonyliron(II), 42B, 855
$C_{28}H_{28}Mo_2N_8O_4$, [N-(2-Pyridyl)acetamido-N,N']dimolybdenum, 45B, 1201
$C_{28}H_{36}Cl_2N_4NiO_8$, Tetrakis(3,4-dimethylpyridine)nickel(II) perchlorate, 33B, 449
$C_{28}H_{36}N_6O_6Zn$, Bis-(5,5'-diethylbarbiturato)bis(picoline)zinc(II), 40B, 918
$C_{28}H_{37}MnN_5$, (7,16-Dihydro-6,8,15,17-tetramethyldibenzo[b,i]-(1,4,8,11)tetraazacyclotetradecinato)triethylaminemanganese(II), 42B, 856
$C_{28}H_{38}B_4F_8N_{10}Ni_2O_8$, Anilinebis(difluoroborondimethylglyoximato)nickel(II) dimer, 44B, 950
$C_{28}H_{40}CuN_6O_8$, Bis-(5,5'-diethylbarbiturato)bis(picoline)copper(II) dihydrate, 39B, 756; 40B, 919
$C_{28}H_{42}AsN_5NiS_2$, Diisothiocyanato-(N,N-bis-(2-diethylaminoethyl)-2-diphenylarsino-ethylamine-N,N,N)-nickel(II), 37B, 555
$C_{28}H_{44}Cl_6Cu_4N_4O_5$ · Solvent, μ_4-Oxo-hexa-μ-chloro-tetrakis((3-quinuclidinone)copper(II)) solvent, 43B, 1331
$C_{28}H_{50}Cl_3N_4Ta$, Trichlorobis(N,N'-dicyclohexylacetamidinato)-tantalum(V), 41B, 1132
$C_{28}H_{58}Cl_4N_4Pd_2$ · C_6H_6, Di-μ-(N,N,N',N'-tetramethylheptane-1,7-diamine-N,N')bis[dichloropalladium(II)] benzene solvate, 45B, 1196
$C_{28}H_{64}IN_{17}Ni_2$, μ-Azido-bis-((N-tetramethylcyclam)azidonickel(II)) iodide, 40B, 920
$C_{29}H_{53}Cl_2N_4Ta$, Dichlorobis-(N,N-dicyclohexylacetamidinato)methyl-tantalum(V), 40B, 920
$C_{30}H_{22}Cl_6N_6Pd_3$, Chloro(2,2',2"-terpyridine)palladium(II) tetrachloropalladate(II), 39B, 757
$C_{30}H_{22}CuN_8O_6$, Bis(terpyridine)copper(II) nitrate, 44B, 951
$C_{30}H_{23}CoN_8O_2S_2$ · 2 H_2O, Dimethylglyoximatobis(o-phenanthroline)cobalt(III) dithiocyanate dihydrate, 45B, 1201
$C_{30}H_{24}Cl_2CoN_8O_8$, Bis(tris(2-pyridyl)amine)cobalt(II) perchlorate, 44B, 951
$C_{30}H_{24}Cl_2CuN_6O_8$, Tris-(2,2'-bipyridyl)copper(II) perchlorate, 38B, 941
$C_{30}H_{24}Cl_2FeN_8O_8$, Bis(tri-2-pyridylamine)iron(II) diperchlorate, 44B, 951
$C_{30}H_{24}N_6NiO_4S$ · 7.5 H_2O, Tris-(2,2'-bipyridyl)nickel(II) sulfate hydrate, 42B, 865
$C_{30}H_{26}Cl_{10}N_4Ni_2$, Di-$\mu$-chloro-sym-trans-dichloro-bis(2,9-dimethyl-1,10-phenanthroline)dinickel bischloroform, 40B, 1124
$C_{30}H_{28}Br_2CoN_6O_3$, Bis-(2,2':6',2"-terpidyl)cobalt(II) bromide trihydrate, 40B, 921
$C_{30}H_{28}CoN_6S_2$, Bis(isothiocyanato)tetrakis(4-vinylpyridine)cobalt-(II), 38B, 942
$C_{30}H_{28}CuN_6O_4$, Bis(5,5-diphenylhydantoinato)diamminecopper(II), 46B, 1121
$C_{30}H_{30}B_2F_8N_6Ru$, Hexakis(pyridine)ruthenium(II) tetrafluoroborate,

45B, 1202
$C_{30}H_{30}Fe_2N_2O_{10}$, 2,7-Diphenyl-4,5-bis(ethoxycarbonyl)-3,6-diazaocta-3,6-diyl-bis(tricarbonyliron), 45B, 1258
$C_{30}H_{30}N_6O_3Zn$, Dicyano(2,9-dimethyl-1,10-phenanthroline)zinc 2,9-dimethyl-1,10-phenanthroline trihydrate, 44B, 952
$C_{30}H_{30}N_8Ni$, Di-(3-methyl-1-phenyl-5-p-tolylformazyl)nickel(II), 32B, 443
$C_{30}H_{32}Br_2CuN_6O_8$, Bis-(5-allyl-5-(2-bromoallyl)barbiturato)-bispyridinecopper(II) dihydrate, 40B, 922
$C_{30}H_{32}N_6NiO_6$, Bis(5,5-diphenylhydantoinato)diaquadiaminenickel(II), 46B, 1122
$C_{30}H_{36}B_4F_8N_{12}Ni_2$, Bis(difluoroboron-dimethylglyoximato)nickel dimer bis(benzimidazole) adduct, 46B, 1122
$C_{30}H_{36}F_6N_8Ni_2O_6S_2$, Bis((5,7,12,14-tetramethyl-1,4,8,11-tetraazacyclotetradecahexaenato)nickel)(Ni-Ni) trifluoromethanesulphonate, 42B, 865
$C_{30}H_{38}BClN_4Zn$, Tris(2-aminoethyl)aminochlorozinc(II) tetraphenylborate, 37B, 557
$C_{30}H_{42}Cl_2CoN_6O_{12} \cdot H_2O$, Bis(2,6-diacetyliminopyridine)-N,N'-bis(μ-3,6-dioxaoctane)cobalt(II) diperchlorate monohydrate, 46B, 1123
$C_{30}H_{42}Cl_2FeN_6O_{12} \cdot H_2O$, Bis(2,6-diacetyliminopyridine)-N,N'-bis(μ-3,6-dioxaoctane)iron(II) diperchlorate monohydrate, 46B, 1123
$C_{30}H_{44}B_2Cu_2N_{12}$, Bis((hydrotris(3,5-dimethyl-1-pyrazolyl)borato)copper(I)), 42B, 843
$C_{30}H_{44}B_2FeN_{12}$, Bis[hydrotris(3,5-dimethyl-1-pyrazolyl)borato]iron(II), 46B, 1088
$C_{30}H_{48}B_2Co_2F_{10}N_{12}$, Di-$\mu$-fluoro-hexakis(3,5-dimethylpyrazole)dicobalt(II) bis(tetrafluoroborate), 42B, 866; 44B, 952
$C_{30}H_{59}Cl_2Cu_2N_7O_{12}$, μ-Carbonato-bis(2,4,4,9-tetramethyl-1,5,9-triazacyclododec-1-ene)dicopper(II) perchlorate dimethylformamide, 46B, 1123
$C_{30}H_{62}Cu_3N_{15}O_{37}P_3$, Octaaquotris(guanosine-5'-monophosphatecopper(II)) pentahydrate, 42B, 921
$C_{31}H_{27}BMoN_9O_2P$, Tetrakis(1-pyrazolyl)borato(triphenylphosphine)(carbonyl)(nitrosyl)molybdenum, 43B, 1331
$C_{32}H_{24}CdCl_2N_8O_8$, Tetrakis(1,8-naphthyridine)cadmium(II) bis(perchlorate), 40B, 915
$C_{32}H_{24}Cl_2FeN_8O_8$, Tetrakis(1,8-naphthyridine)iron(II) perchlorate, 38B, 942
$C_{32}H_{26}Cl_2CuO_2 \cdot 0.33\ CHCl_3 \cdot 2\ H_2O$, Dichlorobis(7-chloro-1,3-dihydro-1-methyl-5-phenyl-3H-1,4-benzodiazepin-2-one)copper(II) dihydrate chloroform, 46B, 1124
$C_{32}H_{26}CoN_8O_2S_2$, Bis(terpyridyl)cobalt(II) thiocyanate dihydrate (Form I), 42B, 866
$C_{32}H_{26}CoN_8O_2S_2$, Bis(terpyridyl)cobalt(II) thiocyanate dihydrate (Form II), 42B, 866
$C_{32}H_{32}N_{12}Ni_2 \cdot 1.91\ C_6H_6$, Bis-$\mu$-(cyclohexane-1,2-bis-2'-pyridylhydrazonato)dinickel(II) benzene solvate, 46B, 1124
$C_{32}H_{34}Cu_2N_{12}O_4$, [(Phenylazo)acetaldoximato-N,N'][(phenylazo)-acetaldoxime-N,N']copper(I), 46B, 1125
$C_{32}H_{36}Fe_3N_{12}O_8$, Hexakis(1-methylimidazole)iron(II) octacarbonyldiiron(I), 46B, 1126
$C_{32}H_{36}N_8Ni$, Bis[1,4-di(2,6-dimethylphenyl)tetraazadiene]nickel(0), 46B, 1127
$C_{32}H_{42}CoN_8O_4$, Bis(viridaminato)cobalt(II), 43B, 1332
$C_{32}H_{42}CuN_8O_4 \cdot 0.67\ H_2O$, Bis(viridaminato)copper(II) hydrate, 43B, 1332

$C_{32}H_{44}Cr_2N_8O_2$, Tetrakis(2-amino-6-methylpyridinato)dichromium(Cr-Cr) - bis(tetrahydrofuran), 44B, 953
$C_{32}H_{44}Mo_2N_8O_2$, Tetrakis(2-amino-6-methylpyridinato)dimolybdenum(Mo-Mo) - bis(tetrahydrofuran), 44B, 953
$C_{32}H_{44}N_8O_2W_2$, Tetrakis(2-amino-6-methylpyridinato)ditungsten(W-W) - bis(tetrahydrofuran), 44B, 953
$C_{32}H_{44}N_{10}W_2$, Bis(1,3-diphenyltriazenido)tetrakis(dimethylamido)ditungsten, 45B, 1202
$C_{32}H_{56}Br_2N_8Ni$, Tetrakis(1H-3,5-diethyl-4-methylpyrazole-N²)nickel(II) dibromide, 46B, 1127
$C_{32}H_{64}Br_2N_4Ni_2$, Di-μ-bromobis(N,N'-bis(2,4-dimethyl-3-pentyl)ethanediimine)dinickel(I), 44B, 954
$C_{32}H_{80}N_8U_2$, Tetrakis(diethylamido)uranium(IV) dimer, 42B, 868
$C_{33}H_{31}ClN_4O_8Rh_2$, (7,16-Dihydro-6,8,15,17-tetramethyldibenzo[b,i]-(1,4,8,11)tetraazacyclotetradecinato)tetracarbonyldirhodium(I) perchlorate toluene solvate, 44B, 949
$C_{33}H_{36}AgN_7O_3$, Tris(1-phenyl-3,5-dimethylpyrazole)silver(I) nitrate, 45B, 1203
$C_{33}H_{40}N_{10}S_5U$, Tetraethylammonium pentakis(isothiocyanato)bis(2,2'-bipyridine)uranate(IV), 46B, 1129
$C_{33}H_{64}Cl_3Cu_2N_9O_{12}$, μ-Cyano-bis(5,7,7,12,14,14-hexamethyl-1,4,8,11-tetraazacyclotetradeca-4,11-diene)dicopper(II) perchlorate, 40B, 923
$C_{34}H_{18}CoN_8O_2$, Bis(cyanobis(isoindolinone)methine)cobalt(II), 43B, 1333
$C_{34}H_{30}N_4NiO_2$, (4,12-Bis(acetonyl)-tetrabenzo[b,f,j,n][1,5,9,13]tetraazacyclohexadecine)nickel(II), 45B, 1204
$C_{34}H_{32}Cl_4N_4Ni_2O_2$, trans-2-(2'-Quinolyl)methylene-3-quinuclidinonedichloronickel(II) dimer, 40B, 923
$C_{34}H_{34}CuN_4$, Bis(4-phenylamino-2-phenyliminopent-3-enato-N,N')copper(II), 45B, 1204
$C_{34}H_{34}N_4Ni$, Bis[{N,N'-(1,3-dimethylpropanediylidene)dianilinato)(1-)]nickel(II), 45B, 1205
$C_{34}H_{48}Cl_2N_4Pd$, trans-Dichloro-bis(2-phenyl-3,5-dipropyl-4-ethylpyrazole)palladium(II), 46B, 1129
$C_{36}H_{24}Cl_2CuN_6O_8$, Tris-(1,10-phenanthroline)copper(II) perchlorate, 39B, 757
$C_{36}H_{24}Cl_4Mn_2N_4$, Bis(2,2'-biquinolyl)-di-μ-chloro-dimanganese(II), 42B, 869
$C_{36}H_{24}Cl_4N_4Ni_2$, Bis(2,2'-biquinolyl)di-μ-chloro-bis(chloronickel-(II)), 43B, 1334
$C_{36}H_{26}Cl_3FeN_6O_{13}$, Tris(o-phenanthroline)iron(III) perchlorate monohydrate, 41B, 1093
$C_{36}H_{28}FeI_2N_6O_2$, Tris(1,10-phenanthroline)iron(II)·iodide dihydrate, 44B, 954
$C_{36}H_{30}Cl_8N_2Nb_2P_2 \cdot C_2H_4Cl_2$, Bis(μ-chloro)-bis(trichloro-triphenylphosphinimino-niobium) 1,2-dichloroethane solvate, 45B, 1205
$C_{36}H_{30}CoN_9$, Tris(diphenyltriazine)cobalt(III), 39B, 758
$C_{36}H_{30}CrN_9 \cdot C_7H_8$, Tris(1,3-diphenyltriazino)chromium(III) toluene solvate, 45B, 1208
$C_{36}H_{32}B_2N_8Ni$, Bis(diphenyldipyrazolylborato)nickel(II), 42B, 870
$C_{36}H_{36}Cl_3N_6O_{12}Rh \cdot H_2O$, $(-)_{589}$-Tris(3,3'-dimethyl-2,2'-bipyridine)-rhodium(III) triperchlorate monohydrate, 45B, 1206
$C_{36}H_{41}Cl_3N_4O_8Pd$, Bis-(3,4'-bis(ethoxycarbonyl)-5-chloro-3',4,5'-trimethyldipyrromethene)chloropalladium(II), 38B, 943
$C_{36}H_{44}Cl_2N_8Re_2 \cdot CCl_4$, Dichloro-tetrakis(N,N'-dimethylbenzamidinato)dirhenium(Re-Re) carbon tetrachloride solvate, 46B, 1121

$C_{36}H_{44}Cr_2N_8$, Tetrakis(N,N'-dimethylbenzamidinato)dichromium, 45B, 1207
$C_{36}H_{48}Cl_2N_{12}Ni$, Tetrakis(o-phenylenediamine)nickel(II) chloride - bis(o-phenylenediamine), 40B, 925
$C_{36}H_{60}Cl_{12}CuN_6Sb_2$, catena-Tri-$\mu_2$-(1,12-dodecanedinitrile)-copper(II) hexachloroantimonate(V), 42B, 870
$C_{36}H_{69}LaN_3OPSi_6$, Tris(bis(trimethylsilyl)amido)(triphenylphosphine oxide)lanthanum(III), 43B, 1335
$C_{36}H_{108}N_{18}W_5$, Hexakis(dimethylamido)tungsten bis(hexakis(dimethylamido)ditungsten), 41B, 1093; 42B, 871
$C_{37}H_{30}BClF_4IrNO_2P_2$, Chlorocarbonylnitrosylbis(triphenylphosphine)iridium tetrafluoroborate, 33B, 430
$C_{37}H_{50}N_4OS_6Zn$, Tetra-n-butylammonium aqua(benzothiazole-2-thiolato-N)bis(benzothiazole-2-thiolato-S")zincate, 46B, 996
$C_{38}H_{24}F_6HgN_6O_6S_2 \cdot C_2H_6O$, Tris(1,10-phenanthroline)mercury(II) trifluoromethanesulphonate ethanol solvate, 45B, 1207
$C_{38}H_{30}Cl_4N_4Re_2$, Bis(N,N'-diphenylbenzamidinato)bis(dichlororhenium(III)), 41B, 1094
$C_{38}H_{32}N_6NiO_2S_2$, Diaquobis(quinoline)bisisothiocyanatonickel(II) diquinoline, 44B, 955
$C_{38}H_{38}N_{20}O_{24}U_3$, Di[nitrato-bis(dioxo-2,6-diacetylpyridine-bis(2'-pyridylhydrazone))uranium] dioxouranium tetranitrate, 46B, 1130
$C_{38}H_{46}CuN_4O_8$, Bis(dipyrromethene)copper(II), 40B, 1124
$C_{38}H_{46}N_4O_8Pd$, Bis-(4,4'-dicarboxyethyl-3,3',5,5'-tetramethyldipyrromethenato)-palladium(II), 37B, 557
$C_{38}H_{48}Fe_3N_{12}O_8$, Hexakis(1-ethylimidazole)iron(II) bis(tetracarbonyliron(I)), 46B, 1126
$C_{39}H_{27}Fe_2N_2O_5P$, 3,6-Diphenylpyridazinotriphenylphosphinepentacarbonyldiiron, 37B, 558
$C_{40}H_{20}N_{10}O_2U$, Dioxocyclopentakis(2-iminoisoindoline)uranium(VI), 41B, 1095
$C_{40}H_{32}Cl_2Cu_2N_8O_6S_5 \cdot 6\ H_2O$, Bis[bis(2,2'-bipyridyl)monochlorocopper-(II)] disulphonatotrisulfane hexahydrate, 45B, 1265
$C_{40}H_{32}N_8U$, Tetrakis-(2,2'-bipyridyl)uranium(0), 41B, 1096
$C_{40}H_{33}Cl_4FeN_8O_{16}$, 2,2'-Bipyridylium tris(2,2'-bipyridyl)iron(III) tetraperchlorate, 44B, 955
$C_{40}H_{36}ClCu_2I_3N_{12}O_4$, Di(iodobis(2,2'-bipyridylamine)copper(II)) iodide perchlorate, 39B, 759
$C_{40}H_{36}Cl_2N_{10}O_{15}Ru_2$, μ-Oxo-bis(nitrobis(2,2'-bipyridine)ruthenium-(III)) perchlorate dihydrate, 41B, 1097
$C_{40}H_{36}FeI_5N_6O_4$, Bis(diphenylglyoximato)-bis(β-picoline)iron(III) iodide bis(iodine), 43B, 1336
$C_{40}H_{40}Cl_6N_{14}O_2Zn_2$, (2,6-Diacetylpyridinebis(2'-pyridylhydrazono))-zinc(II) hydrate chloroform dimer, 42B, 847
$C_{40}H_{42}N_{10}Ni_2S_4$, Di-$\mu$-isothiocyanato-bis(isothiocyanatotris(4-methylpyridine)nickel(II)), 44B, 956
$C_{40}H_{56}Cl_2MnN_8O_4$, Tetrakis(N,N'-diethylnicotinamide)manganese(II) chloride, 43B, 1337
$C_{40}H_{80}N_{12}S_8U$, Tetraethylammonium octathiocyanato-N-uranate(IV), 37B, 559
$C_{41}H_{30}Cl_5IrN_2P_2 \cdot C_7H_8$, Bis(triphenylphosphine)-(tetrachlorodiazocyclopentadiene)-chloro-iridium(I) toluene solvate, 46B, 1132
$C_{41}H_{37}BFeN_2O_3$, Benzimidazoledicarbonyl-π-cyclopentadienyliron tetraphenylborate acetone, 43B, 1093
$C_{42}H_{24}Cd_3Fe_3N_6O_{12} \cdot 0.75\ C_6H_3Cl_3$, Tris((2,2'-bipyridyl)cadmiumtetracarbonyliron) 1,2,4-trichlorobenzene solvate, 43B, 1338
$C_{42}H_{30}Cl_4N_4Ni$, Bis{N-[3-(4-chlorophenylimino)-2-phenyl-1-propenyl]-

4-chloroanilinato-N,N'}nickel(II), 45B, 1207
$C_{42}H_{30}Co_2N_{14}O_2S_4$, Di-μ-thiocyanato(S,N)-diisothiocyanatobis-(2-(2'-pyridyl)-3-(N-2-picolylimino)-4-oxo-1,2,3,4-tetrahydroquinazoline)dicobalt(II), 40B, 912
$C_{42}H_{36}CoI_3N_6O_3$, $(+)_{546}$-Bis(tribenzo[b,f,j](1,5,9)triazacycloduodecine)cobalt(III) iodide trihydrate, 35B, 716
$C_{42}H_{38}BCoN_6 \cdot 5\ H_2O$, Bis(1,2-benzoquinonediimide)-1,2-benzosemiquinonediimidecobalt(II) tetraphenylborate pentahydrate, 46B, 1132
$C_{42}H_{39}F_{12}IrN_4OP_4$, Tris(acetonitrile)nitrosylbis(triphenylphosphine)-iridium(III) dihexafluorophosphate, 46B, 1133
$C_{42}H_{39}F_{12}N_4OP_4Rh$, Tris(acetonitrile)nitrosylbis(triphenylphosphine)-rhodium(III) bis(hexafluorophosphate), 43B, 1339
$C_{42}H_{54}Cl_{10}Co_2N_{12}O_6Zn_3$, Tris-μ-(2,5-di(2-pyridyl)-3,4-diazahexa-2,4-diene)-dicobalt(II) di(aquotrichlorozincate(II)) tetrachlorozincate(II) tetrahydrate, 40B, 925
$C_{43}H_{30}Cl_4N_2O_2P_2Ru \cdot CH_2Cl_2$, Bis(triphenylphosphine)(η^2-tetrachlorodiazocyclopentadiene)dicarbonylruthenium dichloromethane solvate, 46B, 1134
$C_{43}H_{30}Fe_5N_6O_{13}$, Hexapyridineiron(II) tetrairon-tridecacarbonyl, 31B, 433
$C_{43}H_{34}IrN_3OP_2$, Bis(triphenylphosphine)carbonyl(benzotriazenido)iridium(I), 44B, 956
$C_{43}H_{38}CoN_9$, Tris(1,3-diphenyltriazenato)cobalt(III) toluene solvate, 40B, 926
$C_{44}H_{35}Co_2N_{19}S_4 \cdot 2.7\ H_2O$, Tris-μ-(4-phenyl-1,2,4-triazole-N^1,N^2)-bis[bis(isothiocyanato)(4-phenyl-1,2,4-triazole-N^1)cobalt(II)] hydrate, 46B, 1041
$C_{44}H_{42}N_8Pd$, Bis(1,3,5-tri-p-tolylformazanyl)palladium(II), 46B, 1134
$C_{44}H_{44}FeN_6O_{20}Sb_2$, (-)-Tris(1,10-phenanthroline)iron(II) bis(antimony(III) (+)-tartrate) octahydrate, 39B, 760
$C_{46}H_{24}Mn_2N_6NiO_{10}$, Tris(1,10-phenanthroline)nickel(II) pentacarbonylmanganate(-I), 38B, 945
$C_{46}H_{40}B_2F_4N_6NiO_6$, 1,10-Phenanthrolinebis(difluoroborondiphenylglyoximato)nickel(II) bis(acetone), 43B, 1339
$C_{48}H_{36}Cl_3N_{12}O_{12}Pr$, Hexakis(1,8-naphthyridine)praseodymium(III) perchlorate, 43B, 1340
$C_{48}H_{40}As_2N_{18}Pd_2$, Bis(tetraphenylarsonium) hexaazidodipalladate(II), 38B, 945
$C_{48}H_{40}Cr_2N_{12}$, Tetrakis(1,3-diphenyltriazino)dichromium(II), 45B, 1208
$C_{48}H_{40}Cu_2N_{12}$, Copper 1,3-diphenyltriazine, 41B, 1099
$C_{48}H_{40}Mo_2N_{12} \cdot 0.5\ C_7H_8$, Tetrakis(1,3-diphenyltriazino)dimolybdenum-(II) toluene solvate, 45B, 1208
$C_{48}H_{40}N_4U$, Tetrakis(diphenylamido)uranium(IV), 43B, 1340
$C_{48}H_{40}N_{12}Ni_2$, Nickel 1,3-diphenyltriazine, 41B, 1099
$C_{48}H_{40}N_{12}Pd_2$, Palladium 1,3-diphenyltriazene, 41B, 1099
$C_{48}H_{42}Cr_2I_4N_8O_6$, Di-μ-hydroxo-bis(bis(1,10-phenanthroline)chromium-(III)) iodide tetrahydrate, 41B, 1098
$C_{48}H_{44}Cl_2Hg_2N_4O_8$, Tetrakis(4-benzylpyridine)dimercury(I) perchlorate, 43B, 1341
$C_{49}H_{46}Cl_2IrN_3O_5P_2$, Carbonyldichloro(2-nitrophenyldiazenato)bis(triphenylphosphine)iridium(III) bis(acetone), 43B, 1342
$C_{50}H_{43}As_2Cl_8Mo_2NO_2$, Bis(tetraphenylarsonium) bis(tetrachlorooxomolybdenum)methylnitrile, 46B, 1135
$C_{50}H_{45}N_3P_2Pt$, trans-Hydridobis(triphenylphosphine)(1,3-di-p-tolyltriazenido)platinum(II), 42B, 873
$C_{50}H_{54}FeN_8$, (6,7,13,14-Bis(9',10'-dihydroanthr-9',10'-yl)-3,10-di-n-

octyl-1,2,4,5,8,9,11,12-octaazacyclotetradeca-9(14),3,5,7,10,12-hexaen-2,9-diyl)iron(II), 41B, 1100
$C_{51}H_{41}IrN_3OP_2$, trans-Carbonyl(1,3-di-p-tolyltriazenido)bis(triphenylphosphine)iridium(I), 46B, 1136
$C_{51}H_{45}Cl_4N_3P_2Pt$, cis-Chlorobis(triphenylphosphine)-(1,3-di-p-tolyltriazenido)platinum(II) chloroform solvate, 42B, 874
$C_{51}H_{45}N_3OP_2Ru$, Hydrido(1,3-di-p-tolyltriazenido)bis(triphenylphosphine)ruthenium(II) carbonyl, 42B, 874
$C_{51}H_{49}BF_4N_2P_2Pt$, trans-Hydrido(acetone phenylhydrazone)bis(triphenylphosphine)platinum(II) tetrafluoroborate benzene, 43B, 1343
$C_{51}H_{62}AsBN_4NiS$, Isothiocyanato-(N,N-bis-(2-(diethylamino)ethyl)-2-(diphenylarsino)ethylamine-N,N,N)-nickel(II) tetraphenylborate, 38B, 946
$C_{52}H_{46}N_2OP_2Ru$, trans-Bis(triphenylphosphine)carbonyl(N,N'-di-p-tolylformamidinato)hydridoruthenium(II), 43B, 1343
$C_{52}H_{52}N_2O_2P_2PtS$, Bis(triphenylphosphine)-N-(3,5-xylyl)-N-(2-amino-3,5-xylyl)aminesulfide-N',S-platinum dihydrate, 44B, 957
$C_{54}H_{54}ClIrP_2$, Hydridochloro(phenylazophenyl-2C,N')bis(triphenylphosphine)iridium(III) n-hexane solvate, 40B, 927
$C_{55}H_{44}BF_6IrN_4OP_2$, Carbonylbis(triphenylphosphine)(1,4-p-fluorophenyltetrazene)iridium tetrafluoroborate benzene solvate, 38B, 948
$C_{55}H_{54}Br_2N_2P_4W \cdot 0.5\ CH_4O$, Bis[1,2-bis(diphenylphosphino)ethane]bromo-1-diazo-1-methylethanetungsten bromide hemimethanol solvate, 46B, 1137
$C_{56}H_{44}BBr_2N_8Ni_2$, Bis(bis(naphthyridino)bromonickel) tetraphenylborate, 40B, 928
$C_{56}H_{56}BrF_6N_2OP_5W \cdot 0.5\ C_2H_6O$, Bis[1,2-bis(diphenylphosphino)ethane]-bromo-4-diazobutanoltungsten hexafluorophosphate hemiethanol solvate, 46B, 1137
$C_{58}H_{103}N_{13}Re_2S_{10}$, Tris(tetrabutylammonium) di-μ-thiocyanatobis(tetrathiocyanatorhenate)(Re-Re), 45B, 1209
$C_{59}H_{53}ClMoN_2OP_4$, Benzoyldiazenido(chloro)bis(diphenyl(diphosphino)-ethane)molybdenum(III), 44B, 1165
$C_{60}H_{48}Mo_2N_3O_6P_3 \cdot 0.5\ C_4H_8O$, Tris($\mu$-triphenylphosphinimino)(hexacarbonyl)dimolybdenum tetrahydrofuran solvate, 40B, 929
$C_{60}H_{102}La_2N_4O_4P_2Si_8$, Tetrakis(bis(trimethylsilyl)amido)peroxobis-(triphenylphosphine oxide)dilanthanum(III), 43B, 1335
$C_{61}H_{50}NNiP_3$, (Benzonitrile)tris(triphenylphosphorane)nickel(0), 42B, 874
$C_{61}H_{64}BF_5N_2O_2P_4W$, (2-Diazo-4-oxopentanato(2-))fluorobis(1,2-bis(diphenylphosphino)ethane)tungsten tetrafluoroborate - tetrahydrofuran, 44B, 957
$C_{64}H_{62}BClMoN_4P_2 \cdot 0.5\ CH_2Cl_2$, Chloro-bis(N,N-dimethylhydrazido)bis-(triphenylphosphine)molybdenum tetraphenylborate methylene chloride solvate, 46B, 1138
$C_{64}H_{76}Cl_2N_{12}Ni_2O_{18}$, Tris(2,2'-bipyridyl)nickel(II) chloride tartrate hydrate, 42B, 875
$C_{65}H_{67}B_2Cu_2N_9O_2$, ($\mu$-Histamine)-bis[histamine(carbonyl)copper(I)] bis(tetraphenylborate), 46B, 1138
$C_{66}H_{56}N_6P_2Pt$, cis-Bis(1,3-diphenyltriazenido)bis(triphenylphosphine)platinum(II) benzene solvate, 42B, 874, 998
$C_{68}H_{72}Cl_4N_{16}Ni_2O_4$, μ-Chloro-bis(chlorotetra(benzimidazole)nickel-(II)) chloride-4-acetone, 33B, 435
$C_{72}H_{56}Cl_{16}CoN_{10}O_{10}$, trans-Tetrakis(miconazole)cobalt(II) nitrate, 44B, 958
$C_{72}H_{60}Cl_{18}CuN_8O_6$, trans-Dichlorotetra(miconazole)copper(II) dihydrate, 44B, 958

$C_{72}H_{120}N_6Na_2O_{12}Zr$, Bis[hexakis(tetrahydrofuran)sodium] hexapyrrolylzirconium(IV), 46B, 1087
$C_{74}H_{79}B_{10}N_2P_3RuS$, Dihydrido(dimethylsulfidodiazodecaborane)tris(triphenylphosphine)ruthenium tris(benzene), 43B, 1344
$C_{76}H_{60}Mo_2N_8$, Tetrakis(N,N'-diphenylbenzamidinato)dimolybdenum(II), 41B, 1101
$C_{78}H_{84}Ag_2B_2N_8$, (Cyclo-bis(dipropylamine)-3,3'-bis(2,6-diacetyliminopyridine))-di-silver(I) tetraphenylborate, 45B, 1210
$C_{80}H_{80}Li_2N_6O_4U_2$, Bis(μ-oxo-tris(diphenylamido)uranium(IV)lithium diethyl etherate), 43B, 1340
$C_{84}H_{77}BNNiP_3Sn$, Triphenylstannyl(tris(2-diphenylphosphinoethyl)amine)nickel(II) tetraphenylborate, 44B, 1121
$C_{90}H_{96}N_{24}Ni_3O_2P_2$, Hexakis(benzotriazolyl)-hexakis(allylamine)-trinickel(II) bis(triphenylphosphine oxide), 42B, 876
$C_{92}H_{82}Cl_{10}N_6Nb_2P_4$, Bis(bis(triphenylphosphine)iminium) μ-(buta-2,3-diimino-N,N')-bis(acetonitriletetrachloroniobate) bis(chlorobenzene), 41B, 1102

84. METAL COMPLEXES (OXYGEN LIGAND)

CH_2AgBrO_3S, Silver(I) bromomethanesulphonate, 44B, 959
CH_3AgO_3S, Silver(I) methanesulphonate, 43B, 1345
$CH_3CuN_3O_8$, Copper(II) nitrate-nitromethane complex, 31B, 435
$CH_3CuO_3P \cdot H_2O$, Copper(II) methylphosphonate monohydrate, 45B, 1211
$CH_{34}Cl_5Co_2N_{11}O_2$, μ-Formamido-bis(pentaamminecobalt(III)) pentachloride monohydrate, 41B, 1102
$C_2H_6CdCl_2N_2O$, Methylurea cadmium chloride complex, 22, 640
$C_2H_6CdCl_2N_2O_2$, Cadmium chloride-formamide complex, 32B, 445
$C_2H_6Cl_2CuN_2O$, Dimethylnitrosamine copper(II) chloride, 34B, 573
$C_2H_6Cl_2CuOS$, Dichloro(dimethylsulfoxide)copper(II), 45B, 1211
$C_2H_6Cl_2CuO_2 \cdot 0.5\ H_2O$, Dichloro-(1,2-ethanediol)copper(II) hemihydrate, 42B, 877
$C_2H_6Cl_2CuO_2$, Dichloro-(1,2-ethanediol)copper(II), 42B, 877
$C_2H_6Cl_2FeN_2O_2$, catena-μ-Dichlorobis(formamido-O)iron, 37B, 559
$C_2H_6Cl_8N_2O_4Ti_2$, Di-μ-chloro-bis(trichloronitromethanetitanium), 44B, 959
$C_2H_6F_2O_3SU$, catena-Di-μ-fluoro-(dimethylsulfoxide)dioxouranium(VI), 41B, 1103
$C_2H_8CdCl_2N_4O_2$, Dichlorobis(urea)cadmium(II), 21, 530
$C_2H_8Cl_2HgO_2$, Bis(methanol)mercury(II) chloride, 30B, 368
$C_2H_8O_9S_2U$, Uranyl methanesulphonate monohydrate, 44B, 960
$C_2H_{10}CdO_8S_2$, Diaquobis(methanesulphonato)cadmium(II), 44B, 960
$C_2H_{10}CuN_2O_6P_2$, Bis(aminomethyl phosphonate)copper(II), 46B, 1139
$C_2H_{10}N_2O_6U$, Bis(hydroxylamido)(1,2-ethanediol-O,O)dioxouranium(VI), 44B, 961
$C_2H_{12}CuO_{10}S_2$, Copper(II) 1,2-ethanedisulphonate tetrahydrate, 43B, 1345
$C_2H_{12}N_6O_8S_2Zn$, Zinc guanidinium sulfate, 35B, 733
$C_2H_{14}CoN_2O_8P_2 \cdot H_2O$, Diaquabis(aminomethyl phosphonate)cobalt(II) monohydrate, 46B, 1139
$C_2H_{14}CuO_{10}S_2$, Tetraaquobis(methanesulphonato)copper(II), 41B, 1104; 43B, 1346
$C_2H_{16}N_2O_{10}P_2Zn$, Bis(aminomethylphosphonato)zinc tetrahydrate, 44B, 961
$C_2H_{18}Ce_2N_4O_{19}S_3$, Cerium sulfate 2-urea 5-water, 44B, 962
$C_2H_{22}B_4OU$, Uranium(IV) borohydride - dimethyl ether, 44B, 963

$C_2H_{32}Mo_5N_4O_{26}P_2$, Ammonium pentamolybdobis(methylphosphonate) pentahydrate, 42B, 878
$C_3H_6CoO_3$, Cobalt(II) monoglycerolate, 37B, 560
$C_3H_9Cl_2N_4O_4Ta$, Methyldichlorobis(N-methyl-N-nitrosohydroxylaminato)tantalum(V), 40B, 931
$C_3H_{12}CuN_6O_7S$, Tris(urea)copper(II) sulfate, 43B, 1188
$C_3H_{12}N_6O_9SU$, Tris(urea)dioxouranium(VI) sulfate, 46B, 1140
$C_4H_4NiO_6$, Nickel squarate dihydrate, 40B, 931
$C_4H_6CuN_4O_6$, Copper(II) nitrate-methyl cyanide, 33B, 453
$C_4H_8Br_2CdO_2$, Dibromo(1,4-dioxan)cadmium, 43B, 1353
$C_4H_8Br_2HgO$, Tetrahydrofuran mercury(II) bromide, 37B, 561
$C_4H_8CdN_6O_2S_2$, catena-Di-μ-isothiocyanato-di(urea)cadmium, 41B, 1045
$C_4H_8Cl_2CuOS$, Dichloro(tetramethylenesulfoxide)copper(II), 45B, 1211
$C_4H_8N_2NiO_2S_2$, trans-Bis(thioacetohydroxamato)nickel(II), 34B, 574
$C_4H_8O_5Os$, Oxobis(ethane-1,2-diolato)osmium(VI), 41B, 1105
$C_4H_{10}CdCl_2N_6O_4$, Bis(biuret)cadmium chloride, 24, 539
$C_4H_{10}CdCl_4N_2O_2$, Cadmium(II) diacetamido dichloride, 21, 541
$C_4H_{10}Cl_2CuN_6O_4$, Bis(biuret)copper(II) chloride, 31B, 437
$C_4H_{10}Cl_2HgN_6O_4$, Bis(biuret)mercury(II) dichloride, 43B, 1189
$C_4H_{10}Cl_2N_6O_4Zn$, Bis(biuret)dichlorozinc(II), 28, 645
$C_4H_{10}Cl_2O_2Ti$, Diethoxytitanium dichloride, 33B, 454
$C_4H_{10}MoO_6$, cis-Dioxobis-(2-hydroxyethyl-1-oxo)molybdenum(VI), 41B, 1105
$C_4H_{12}AgClO_6S_2$, Dimethylsulfoxidesilver(I) perchlorate, 42B, 878
$C_4H_{12}Br_2CuO_2S_2$, Copper(II) bromide bis(dimethylsulfoxide), 43B, 1347
$C_4H_{12}Cl_2CuO_2S_2$, Copper(II) chloride bis(dimethylsulfoxide), 35B, 734
$C_4H_{12}Cl_2MnO_4$, cis-Dichlorobis(1,2-ethanediol)manganese(II), 39B, 761
$C_4H_{12}Cl_2O_4W$, cis-Dichlorotetramethoxytungsten(VI), 41B, 1106
$C_4H_{12}Cl_6Hg_3O_2S_2$, Dimethyl sulfoxide - mercury(II) chloride (2:3), 40B, 932
$C_4H_{12}O_4Ti$, Titanium tetramethoxide, 33B, 455
$C_4H_{16}Cl_2CoO_4$, Dichlorotetramethanolcobalt(II), 31B, 438; 32B, 445; 33B, 239
$C_4H_{16}CuO_{10}S_2$, Copper(II) 1,4-butanedisulphonate tetrahydrate, 43B, 1347
$C_4H_{16}F_4N_8O_8U_2$, Tetra(urea)tetrafluorodiuranyl, 43B, 1191
$C_4H_{16}N_{11}O_{13}Sc$, Tetrakis(urea)bis(nitrato)scandium nitrate, 44B, 963
$C_4H_{18}CuO_{12}S_2$, Copper ethylsulfate tetrahydrate, 34A, 306
$C_4H_{20}Cl_2MnN_2O_6P_2$, Bis(α-aminomethylmethylphosphinic acid)manganese(II) dichloride dihydrate, 43B, 1348
$C_4H_{24}Br_2MnN_2O_8P_2$, Diaquobis(α-aminomethylmethylphosphinic acid)manganese(II) dibromide dihydrate, 43B, 1348
$C_4H_{26}B_4OU$, Uranium(IV) borohydride - diethyl ether, 44B, 963
$C_5H_3Br_2Cl_2HgNO$, 3,5-Dibromopyridine oxide-mercury(II) chloride complex, 34B, 576
$C_5H_5CdI_2NO$, (Pyridine-N-oxide)cadmium diiodide, 40B, 934
$C_5H_5Cl_2HgNO$, (Pyridine-N-oxide)mercury dichloride, 40B, 934
$C_5H_6CuO_8$, Copper croconate, 29, 669
$C_5H_6MnO_8$, Manganese(II) croconate trihydrate, 31B, 440
$C_5H_6O_8Zn$, Zinc croconate, 29, 669
$C_5H_{11}MoNO_7$, Oxo-diperoxo-((S)-N,N-dimethyllactamido)molybdenum(VI), 46B, 1141
$C_5H_{15}Cl_2CoO_{2.5}$, Cobalt chloride methanol, 43B, 1351
$C_6H_{12}CuN_2O_4$, Bis(propane-2-nitronato)copper(II), 39B, 761
$C_6H_{12}CuO_{10}S$, Tris(1,2-ethanediol)copper(II) sulfate, 38B, 949
$C_6H_{12}O_6W$, Tris(ethylene-1,2-dioxo)tungsten(VI), 40B, 932
$C_6H_{14}BrCuNO$, (2-Diethylaminoethanolato)copper(II) bromide, 37B, 561

$C_6H_{14}ClNOZn$, Chloro(2-diethylaminoethanolato)zinc(II), 41B, 1108
$C_6H_{14}Cl_2MoN_2O_4$, Dioxodichlorobis(N,N-dimethylformamido)molybdenum-(VI), 33B, 456
$C_6H_{14}Cl_3O_3Ti$, Trichloro(2,5,8-trioxanonane)titanim(III), 44B, 964
$C_6H_{14}N_4NiO_2S_2$, Bisethanolaminebisisothiocyanatonickel, 40B, 958
$C_6H_{14}N_4O_{12}U$, Bis(ethyl carbamate)dinitratodioxouranium(VI), 39B, 762
$C_6H_{15}Mo_2NaO_{13}$, Sodium D-mannitolatopentaoxodimolybdate dihydrate, 43B, 1351
$C_6H_{16}Cl_{10}Cu_5O_2$, Decachlorobis(propyl alcohol)pentacopper, 29, 590
$C_6H_{16}N_{10}NiO_4S_2$, Bis(isothiocyanato)tetrakis(urea)nickel(II), 46B, 1142
$C_6H_{17}Mo_2NO_{12}$, Ammonium μ-oxo-μ-mannitolate-tetraoxodimolybdate(VI) monohydrate, 41B, 1108
$C_6H_{18}Cl_4Mn_2O_3$, Ethanol - manganese chloride, 38B, 950
$C_6H_{18}CoN_2O_8P_2$, Bis(trimethylphosphine oxide)cobalt(II) dinitrate, 28, 647
$C_6H_{18}CoO_{10}S$, Tris-(1,2-ethanediol)cobalt(II) sulfate, 42B, 879
$C_6H_{18}ErN_3O_{12}S_3$, Dimethyl sulfoxide - erbium nitrate, 38B, 950
$C_6H_{18}LuN_3O_{12}S_3$, Tris(nitrato)tris(dimethylsulfoxide)lutetium(III), 39B, 763
$C_6H_{18}N_3O_{12}S_3Yb$, Tris(dimethylsulfoxide)trinitratoytterbium, 39B, 763; 41B, 1109
$C_6H_{18}N_4O_4W$, Tetramethylbis(N-methyl-N-nitrosohydroxylaminato)tungsten(VI), 39B, 764
$C_6H_{18}O_8V_2$, Methyl vanadate, 31B, 440
$C_6H_{18}O_9Re_2$, Di-μ-methoxo-tetramethoxo-μ-oxo-dioxorhenium(VI)(Re-Re), 46B, 1143
$C_6H_{18}O_{10}SZn$, Tris-(1,2-ethanediol)zinc(II) sulfate, 42B, 879
$C_6H_{20}MoN_3O_7P$, Oxodiperoxohexamethylphosphoramido-aquo-molybdate(VI), 38B, 951
$C_6H_{24}Cl_3CrN_{12}O_6 \cdot 3\ H_2O$, Hexakis(urea)chromium(III) chloride trihydrate, 46B, 1143
$C_6H_{24}Cl_3N_{12}O_{18}Ti$, Hexaureatitanium perchlorate, 38B, 952
$C_6H_{24}I_3N_{12}O_6Ti$, Hexakis(urea)titanium(III) iodide, 35B, 630
$C_6H_{24}I_3N_{12}O_6V$, Hexaureavanadium triiodide, 38B, 953
$C_6H_{24}N_{14}O_{12}Zn$, Hexaureazinc nitrate, 38B, 954
$C_6H_{26}CoN_{12}O_{11}S$, Hexakis(urea)cobalt(III) sulfate monohydrate, 43B, 1193
$C_6H_{33}ErO_{21}S_3$, Nona-aquo erbium ethyl sulfate, 23, 609
$C_6H_{33}HoO_{21}S_3$, Holmium ethylsulfate enneahydrate, 40B, 78
$C_6H_{33}O_{21}PrS_3$, Nona-aquo praseodymium ethyl sulfate, 23, 609
$C_6H_{33}O_{21}S_3Y$, Nona-aquo yttrium ethyl sulfate, 23, 609
$C_7H_9NO_{10}U$, Pyridine-2,6-dicarboxylato-N-oxidedioxouranium(VI) trihydrate, 43B, 1238
$C_8H_{11}CdCl_2NO_3$, Dichloro(pyridoxine)cadmium(II), 44B, 964
$C_8H_{16}Br_2HgO_4$, 1,4-Dioxane mercury(II) bromide, 37B, 562
$C_8H_{16}Br_2NiO_4$, Dibromobis(1,4-dioxan)nickel(II), 43B, 1353
$C_8H_{16}Cl_2CuO_4$, 1,4,7,10-Tetraoxacyclododecanecopper(II) chloride, 41B, 1111
$C_8H_{16}Cl_4MnO_8P_2$, Manganese(II) dichlorophosphate di-ethylacetate, 28, 649
$C_8H_{16}Cl_6Cu_3O_4$, Copper(II) chloride - 1,4-dioxan (3:2), 43B, 1353
$C_8H_{16}Cl_8O_4Ti_2$, Titanium(IV) chloride-ethyl acetate complex, 31B, 442
$C_8H_{16}N_2O_{10}U$, Uranyl nitrate - bis(tetrahydrofuran), 43B, 1283
$C_8H_{18}Cl_2CoN_2O_2$, Dichlorobis(N,N-dimethylacetamide)cobalt(II), 39B, 765
$C_8H_{18}CuO_{10}$, Diaquobis(2,3-dihydroxy-2-methylpropanoato)copper(II),

42B, 879
$C_8H_{18}MoO_6 \cdot 2\ C_4H_{10}O_2$, cis-Bis(butane-2,3-diolato)dioxomolybdenum-(VI) butane-2,3-diol solvate, 45B, 1224
$C_8H_{20}Br_6N_2O_4Ru_2S_2$, Di-μ-bromo-bis(dibromo(diethyl sulfoxide)nitrosylruthenium(II)), 42B, 880
$C_8H_{20}N_2O_{14}U$, 1,4,7,10-Tetraoxacyclododecane-uranyl nitrate dihydrate, 43B, 1354
$C_8H_{24}Cl_2O_4RuS_4$, Dichlorotetrakis(dimethylsulfoxide)ruthenium(II), 41B, 1112
$C_8H_{24}Cl_4Co_2O_8$, Di-μ-chloro-bis(di(1,2-ethanediol)cobalt(II)) dichloride, 41B, 1112
$C_8H_{24}Cl_4Ni_2O_8$, Di-μ-chloro-bis(di(1,2-ethanediol)nickel(II)) dichloride, 41B, 1112
$C_8H_{24}Cl_6Fe_2O_4S_4$, trans-Dichlorotetrakis(dimethylsulfoxide)iron(III) tetrachloroferrate(III), 32B, 446
$C_8H_{24}LaN_3O_{13}S_4$, Tris(nitrato)tetrakis(dimethylsulfoxide)lanthanum, 39B, 763
$C_8H_{24}Mo_4O_{12}P_4S_4 \cdot CCl_4 \cdot 2\ CHCl_3$, Tetra-$\mu_3$-oxotetrakis-μ-(dimethylthiophosphinato)tetra(oxomolybdenum) chloroform carbon tetrachloride solvate, 46B, 1144
$C_8H_{24}Mo_4O_{12}P_4S_4 \cdot 2\ CCl_4 \cdot 2\ CHCl_3$, Tetra-$\mu_3$-oxotetrakis-μ-(dimethylthiophosphinato)tetra(oxomolybdenum) chloroform carbon tetrachloride solvate, 46B, 1144
$C_8H_{24}N_3NdO_{13}S_4$, Dimethyl sulfoxide - neodymium nitrate, 38B, 955
$C_8H_{32}B_4O_2U$, Uranium(IV) borohydride - bis(tetrahydrofuran), 44B, 965
$C_8H_{36}Mo_5N_3NaO_{26}P_2$, Sodium tetramethylammonium pentamolybdobis(ethylammoniumphosphonate) pentahydrate, 42B, 878
$C_9H_8CrO_6$, Pentacarbonyltetrahydrofuranchromium(0), 44B, 1161
$C_9H_{13}CuN_2O_6P$, Aquo-(N-pyridoxylidene-aminomethylphosphonato)copper-(II), 44B, 965
$C_9H_{16}Cl_2HgO$, Cyclononanone mercury(II) chloride, 37B, 562
$C_{10}H_{10}Br_4Cu_2N_2O_2$, Di-μ-(pyridine-1-oxide)-bis(dibromocopper(II)), 38B, 956
$C_{10}H_{10}Cl_4Cu_2N_2O_2$, Di-μ-(pyridine-N-oxide)-bis(dichlorocopper(II)), 30B, 358; 32B, 446; 34B, 575
$C_{10}H_{10}I_2N_2O_2Zn$, Bis(pyridine-N-oxide)zinc diiodide, 40B, 934
$C_{10}H_{12}Cl_2CuN_4O_4$, Bis(2-methylpyrazine di-N-oxide)copper(II) chloride, 41B, 1114
$C_{10}H_{13}Mo_2N_2O_{14}$, Dipyridinium di-μ-hydroperoxo-tetraperoxo-dioxo-dimolybdate(VI), 38B, 957
$C_{10}H_{14}Cl_4Cu_2N_2O_4$, Di(μ-pyridine N-oxide)bis(dichloroaquacopper(II)), 42B, 881; 46B, 1145
$C_{10}H_{14}CoHgN_6O_2S_4$, Mercury(II)-tetrathiocyanatobis(dimethylformamide)cobalt(II), 46B, 1056
$C_{10}H_{15}Mo_2N_2O_{13}$, Dipyridinium μ-oxo-tetraperoxo-dioxo-diaquo-dimolybdate(VI), 38B, 957
$C_{10}H_{22}Cl_2HgO_5$, Tetraethyleneglycol dimethyl ether - mercury(II) chloride (1:1), 39B, 765
$C_{10}H_{24}Cl_2N_4O_3V$, Bis(tetramethylurea)dichloro-oxovanadium(IV), 35B, 737
$C_{10}H_{24}Mo_2O_{10}$, Bis-μ-(2,2-dimethylpropane-1,3-diolato)-bis(aquodioxomolybdenum(VI)), 41B, 1116
$C_{10}H_{26}Br_4N_2O_2Re$, Tetraethylammonium tetrabromo(ethanol)nitrosylrhenate(II), 41B, 1117
$C_{10}H_{30}Cl_2O_{14}S_5V$, Oxo-pentakis(dimethylsulfoxide)vanadium perchlorate, 45B, 1213
$C_{10}H_{30}Cl_7Fe_3O_6S_5$, Chloro-pentakis(dimethylsulfoxide)iron(III) μ-oxo-

bis(trichloroferrate(III)), 44B, 966
$C_{10}H_{30}Cl_9Fe_3O_5S_5$, Chloro-pentakis(dimethylsulfoxide)iron(III) bis-(tetrachloroferrate(III)), 44B, 966
$C_{10}H_{30}Nb_2O_{10}$, Niobium pentamethoxide dimer, 42B, 881
$C_{11}H_4O_{11}Os_3$, (μ-Hydrido)(μ-methoxy)decarbonyltriosmium(3Os-Os), 46B, 1146
$C_{11}H_{23}MoN_4O_6P$, Oxodiperoxohexamethylphosphoramido-pyridino-molybdate(VI), 38B, 951
$C_{12}H_6O_{12}Os_3$, Dimethoxodecacarbonyltriosmium, 43B, 1355
$C_{12}H_{10}Cl_2HgOS$, Diphenyl sulfoxide - mercury(II) chloride, 39B, 767
$C_{12}H_{10}Cl_4N_2O_2W$, Azoxybenzene-oxotetrachlorotungsten(VI), 41B, 1118
$C_{12}H_{10}CuN_4O_4$, Copper(II) nitrosophenylhydroxylamine, 28, 648
$C_{12}H_{14}Cl_2CuN_2O_2$, Dichlorobis(4-methylpyridine 1-oxide)copper(II) (green modification), 37B, 563
$C_{12}H_{16}Cl_2CuN_4O_4$, Dichlorobis(1,3-dimethyluracil)copper(II), 44B, 967
$C_{12}H_{18}Cl_6Cu_3NO_3$, Polybis(μ-(2-picoline N-oxide)-chlorocopper(II)-di-μ-chloro-)diaquocopper(II), 33B, 458
$C_{12}H_{20}Mo_2N_2O_{11}$, Diammonium μ-oxo-di-μ-catecholato-bis(dioxo-molybdate) dihydrate, 43B, 1359
$C_{12}H_{22}O_{12}S_2Zn$, Zinc hexa-aquo benzenesulphonate, 11, 641
$C_{12}H_{24}Cl_3O_3Sc$, Trichlorotris(tetrahydrofuran)scandium(III), 40B, 936
$C_{12}H_{24}LaN_3O_{15}$, 1,4,7,10,13,16-Hexaoxacyclooctadecane-trinitratolan-thanum, 46B, 1146
$C_{12}H_{24}N_3NdO_{15}$, 1,4,7,10,13,16-Hexaoxacyclooctadecanetrinitratoneody-mium, 46B, 1148
$C_{12}H_{24}O_5Os$, Oxo-bis(2,3-dimethylbutane-2,3-diolato)osmium(VI), 45B, 1214
$C_{12}H_{24}O_8Os_2$, Di-μ-oxo-bis(oxo(tetramethylethane-1,2-diolato)-osmium(VI)), 41B, 1118
$C_{12}H_{26}Cl_2HgO_5$, Tetraethyleneglycol diethyl ether - mercury(II) chloride (1:1), 39B, 766
$C_{12}H_{28}Cl_4CuN_4O_4$, Bis(μ_2-chloro)-dichloro-tetrakis(N,N'-dimethyl-formamide)dicopper(II), 46B, 1066
$C_{12}H_{28}Cl_6Fe_2N_4O_4$, cis-Dichlorotetrakis(dimethylformamide)iron(III) tetrachloroferrate(III), 44B, 968
$C_{12}H_{30}As_2Cl_4O_2U$, Tetrachlorobis(triethylarsine oxide)uranium(IV), 42B, 882
$C_{12}H_{30}N_2O_{16}P_2U$, Triethylphosphate - uranyl nitrate, 24, 630
$C_{12}H_{32}Cl_3N_3NbO_3P$, Bis(isopropyloxo)trichloroniobium(V) - hexamethyl-phosphortriamide adduct, 44B, 968
$C_{12}H_{32}Cl_3O_4Ti$, Dichlorotetrakis(propanol-2)titanium(III) chloride, 40B, 936
$C_{12}H_{36}Br_4Co_2O_6$, Hexakis(ethanolato)cobalt(II) tetrabromocobaltate-(II), 45B, 1214
$C_{12}H_{36}Br_4N_6O_2P_2U$, Tetrabromobis(hexamethylphosphoramide)uranium(IV), 44B, 970
$C_{12}H_{36}CdCl_2O_{14}S_6$, Hexakis(dimethylsulfoxide)cadmium(II) perchlorate, 44B, 969
$C_{12}H_{36}Cl_2HgO_{14}S_6$, Hexakis(dimethylsulfoxide)mercury(II) perchlorate, 44B, 969
$C_{12}H_{36}Cl_2MoN_6O_4P_2$, Dioxodichlorobis(hexamethylphosphoramide)molyb-denum(VI) (monoclinic), 43B, 1360
$C_{12}H_{36}Cl_2MoN_6O_4P_2$, Dioxodichlorobis(hexamethylphosphoramide)molyb-denum(VI) (tetragonal), 43B, 1360
$C_{12}H_{36}Cl_2N_6O_3V$, Bis(hexamethylphosphoramide)dichlorooxovanadium(IV), 41B, 1119
$C_{12}H_{36}Cl_2N_6O_4P_2U$, Dichlorobis(hexamethylphosphoramide)dioxouranium-

(VI), 43B, 1360

$C_{12}H_{36}Cl_3MoN_6O_3P_2$, Oxotrichlorobis(hexamethylphosphorotriamide)molybdenum(V), 43B, 1361

$C_{12}H_{36}Cl_3N_6NbO_3P_2$, Trichlorooxobis(hexamethylphosphorotriamide)-niobium(V), 43B, 1361

$C_{12}H_{36}Cl_4N_6O_2P_2U$, Tetrachlorobis(hexamethylphosphoramide)uranium-(IV), 44B, 970

$C_{12}H_{36}CoN_{12}O_9S_2$, Hexakis(N-methylurea)cobalt(II) thiosulfate, 46B, 1148

$C_{12}H_{36}CoN_{12}O_{10}S$, Hexakis(N-methylurea)cobalt(II) sulfate, 46B, 1148

$C_{12}H_{60}B_8O_2U_2$, Uranium(IV) borohydride - n-propyl ether, 44B, 971

$C_{13}H_{10}FMn_3O_{11}$, 1,1,1,2,2,2,3,3,3-Nonacarbonyl-di-μ_3-ethoxy-1,2-μ-fluoro-trimanganese, 45B, 1215

$C_{13}H_{10}IMn_3O_{11}$, 1,1,1,2,2,2,3,3,3-Nonacarbonyl-di-μ_3-ethoxy-1,2-μ-iodo-trimanganese, 45B, 1215

$C_{13}H_{14}Cl_4Hg_2N_2O_2$, (2,2'-Trimethylenedipyridine-1,1'-dioxide)bis(dichloromercury(II)), 44B, 971

$C_{14}H_8Cl_2MoO_4$, cis-Dioxodichloro(9,10-phenanthrenequinone)molybdenum-(VI), 43B, 1363

$C_{14}H_{10}CuO_4$, Bis(tropolonato)copper(II), 15, 473; 30B, 370; 44B, 972

$C_{14}H_{10}HgO_4$, Mercury(II) tropolonate, 42B, 883

$C_{14}H_{16}Br_2O_8Re_2$, Di-μ-bromo-bis(tricarbonyl-tetrahydrofuranrhenium), 44B, 1162

$C_{14}H_{16}Cl_2Mn_2O_8$, μ-Dichloro-bis(tetrahydrofuran)hexacarbonyldimanganese, 46B, 1149

$C_{14}H_{16}N_2O_{18}K_2Ti_2$, Chelated dinuclear peroxytitanium(IV), 35B, 740

$C_{14}H_{18}Cl_2CuN_2O_2$, Dichlorobis(2,6-lutidine-N-oxide)copper(II), 34B, 578

$C_{14}H_{18}Cl_2N_2O_2Zn$, Dichlorobis(2,6-lutidine N-oxide)zinc(II), 33B, 458

$C_{14}H_{22}Cl_4Cu_2N_2O_4S_2$, Di-$\mu$-(pyridine oxide)-bis(dichloro(dimethyl sulfoxide)copper(II)), 41B, 1120

$C_{14}H_{24}Mo_3N_4O_{13}$, Tetramethylammonium ($\mu_3$-methoxy)-tris($\mu$-methoxy)-tris(dicarbonyl-nitroso-molybdenum), 46B, 1165

$C_{14}H_{26}O_{12}S_2Zn$, Zinc hexa-aquo p-toluenesulphonate, 21, 585

$C_{14}H_{27}F_4N_2O_3Ta$, Tetraethylammonium tetrafluoro(2-methylpyridine N-oxide)peroxotantalate(V), 43B, 1364

$C_{15}H_{28}Cl_6Cu_4N_3O_5 \cdot C_5H_9NO$, μ_4-Oxa-hexa-μ-chloro-(aquo-copper(II))-tris(N-methyl-2-pyrrolidinone-copper(II)) N-methyl-2-pyrrolidinone solvate, 45B, 1216

$C_{15}H_{30}Cl_2CuN_{10}O_{13}$, Pentakis(2-imidazolidinone)copper(II) perchlorate, 40B, 937

$C_{15}H_{30}N_5NbO_{10}$, Pentakis(N,N-dimethylcarbamato)niobium(V), 41B, 1121

$C_{15}H_{36}EuN_9O_{12}$, Tris(tetramethylurea)trinitratoeuropium(III), 42B, 883

$C_{15}H_{45}As_5Cl_2NiO_{13}$, Pentakis(trimethylarsine oxide)nickel(II) perchlorate, 42B, 884

$C_{16}F_{36}MoO_5$, Tetrakis(t-perfluorobutoxy)oxomolybdenum(VI), 46B, 1150

$C_{16}H_{10}Cu_2F_{12}O_8$, Bis(1,1,1,7,7,7-hexafluoroheptane-2,4,6-trionato)-bis(methanol)dicopper(II), 46B, 1151

$C_{16}H_{10}Hg_3N_8O_2S_6$, Mercury(II) thiocyanate - pyridine-N-oxide (3:2), 42B, 489

$C_{16}H_{12}CuN_2O_6$, Bis(ω-nitroacetophenonato)copper(II), 38B, 958

$C_{16}H_{14}CdMn_2O_{13}$, μ-(Di-(2-methoxyethyl)ethercadmio)-bis(pentacarbonylmanganese)(2Cd-Mn), 40B, 938

$C_{16}H_{16}O_7U$, Bis(tropolonato)dioxo(ethanol)uranium(VI), 39B, 768

$C_{16}H_{21}Cl_3CuNOP$, Trichloro((2-(diphenylphosphinyl)ethyl)dimethylammonium)copper(II), 40B, 939

$C_{16}H_{30}N_6O_6PtS_{11}V$, Pentakis(dimethylsulfoxide)oxovanadium(IV) hexakis(thiocyanato)platinate(IV), 44B, 972
$C_{16}H_{32}Br_6MnO_8$, Bis(1,4,7,10-tetraoxacyclododecane)manganese(II) tribromide, 46B, 1152
$C_{16}H_{32}Cl_4Cu_2O_4S_4$, Di-μ-chloro-bis[chlorobis(tetramethylene sulfoxide)copper(II)], 45B, 1216
$C_{16}H_{34}Cl_4Hg_2O_7$, Hexaethyleneglycol diethyl ether - mercury(II) chloride (1:2), 39B, 766
$C_{16}H_{36}Cl_2CuN_4O_{12}$, Tetrakis(N,N-dimethylacetamide)copper(II) diperchlorate, 46B, 1152
$C_{16}H_{36}Cl_3FeN_8O_6 \cdot 2\ H_2O$, Diaquatetrakis(perhydropyrimidin-2-one)iron trichloride dihydrate, 46B, 1153
$C_{16}H_{36}CuO_4P_2$, Bis(μ-dibutylphosphinato)-copper(II), 43B, 1366
$C_{16}H_{36}Cu_4O_4$, Copper(I) t-butoxide, 42B, 885
$C_{16}H_{36}O_4Ti_2$, Bis-(2-methylpentane-2,4-dioxydimethyltitanium), 35B, 747
$C_{16}H_{40}Cl_4N_2OTi$, Di-tetraethylammonium tetrachlorooxotitanate(IV), 33B, 460
$C_{16}H_{40}Cl_4N_2OV$, Di-tetraethylammonium tetrachlorooxovanadate(IV), 33B, 460
$C_{16}H_{44}As_2Mo_6N_{12}O_{29}$, Guanidinium hexamolybdobis(phenylarsonate) tetrahydrate, 44B, 973
$C_{16}H_{48}Br_4Ni_2O_8$, Di-μ-bromo-bis(tetraethanolnickel(II)) bromide, 44B, 973
$C_{16}H_{48}Cl_2CuN_8O_{14}P_4$, Bis(perchlorato)bis(octamethylpyrophosphoramide)copper(II), 35B, 744
$C_{16}H_{48}Cl_4Hg_2O_{24}S_8$, Octakis(dimethylsulfoxide)dimercury(II) perchlorate, 44B, 974
$C_{16}H_{64}B_{20}Cd_2O_4$, Bis(bis(diethyl ether)-μ-(dodecahydro-nido-decaborato)-cadmium), 38B, 959
$C_{17}H_{29}NO_9Re_2$, Tetraethylammonium tri(μ-methoxy)hexacarbonyldirhenate(I), 45B, 1217
$C_{18}H_{12}CrK_3O_6 \cdot 1.5\ H_2O$, Potassium tris(catecholato)chromate(III) sesquihydrate, 42B, 885
$C_{18}H_{12}FeK_3O_6 \cdot 1.5\ H_2O$, Potassium tris(catecholato)ferrate(III) sesquihydrate, 42B, 885
$C_{18}H_{15}Cl_5OPU$, Pentachloro(triphenylphosphine oxide)uranium(V), 42B, 886
$C_{18}H_{15}FeN_6O_6$, Iron cupferron, 30B, 358
$C_{18}H_{18}NiO_4$, Bis-(2-hydroxy-5-methylacetophenato)nickel(II), 40B, 941
$C_{18}H_{24}BaMo_2O_{16}$, Barium μ-oxo-di-μ-catecholato-bis(dioxo-molybdate) pyrocatechol pentahydrate, 43B, 1359
$C_{18}H_{28}O_{12}P_2Re_2 \cdot 2\ C_4H_8O$, Bis[tricarbonyl(μ-dimethylphosphinato-O,O')-tetrahydrofuranrhenium(I)] tetrahydrofuran solvate, 45B, 1218
$C_{18}H_{36}CdCl_2N_{12}O_{14}$, Hexakis(2-imidazolidinone)cadmium(II) perchlorate, 38B, 960
$C_{18}H_{36}N_6NiO_4S_2$, Tetra(N,N-dimethylacetamide)isothiocyanatonickel, 41B, 1045
$C_{18}H_{42}Cl_3FeN_6O_{18}$, Hexa(N,N-dimethylformamido)iron(III) perchlorate, 45B, 1219
$C_{18}H_{42}Cl_4Mo_4O_{12}$, Tetrachloro-di-μ₃-oxo-tetra-μ-propoxo-tetraoxodipropoxotetramolybdenum(2Mo-Mo), 39B, 769
$C_{18}H_{42}Mo_2N_2O_8$, Bis(nitrosyl)hexakis(isopropoxy)dimolybdenum, 44B, 974
$C_{18}H_{50}Cl_3ErN_{12}O_{19}$, Aquohexakis(N,N'-dimethylurea)erbium(III) perchlorate, 43B, 1196
$C_{18}H_{54}Cl_3N_9O_3P_3Pr$, Trichlorotris(hexamethylphosphoramide)-

praseodymium(III), 39B, 769
$C_{18}H_{54}Cl_4O_6P_6U$, Chlorohexakis(trimethylphosphine oxide)uranium(IV) trichloride, 42B, 887
$C_{19}H_{31}Cl_2CrO_3$, Dichlorotris(tetrahydrofuran)-p-tolylchromium(III), 31B, 444; 32B, 448
$C_{20}H_{20}B_2CuF_8N_4O_4$, Tetra(pyridine oxide)copper(II) fluoroborate, 34B, 576
$C_{20}H_{20}Br_4Cu_2N_4O_4$, Dibromobis(pyridine N-oxide)copper(II) dimer, 38B, 960
$C_{20}H_{20}Cl_2CuN_4O_{12}$, Tetra(pyridine oxide)copper(II) perchlorate, 34B, 580
$C_{20}H_{20}Cl_2Hg_2N_4O_{12}$, Pyridine 1-oxide mercury(I) perchlorate dimer, 39B, 770
$C_{20}H_{20}Cl_4Cu_2N_4O_4$, Dichlorobis(pyridine-1-oxide)copper(II) dimer, 40B, 941
$C_{20}H_{20}Cu_2N_8O_{16}$, Bis(pyridine-N-oxide)copper(II) nitrate, 34B, 580
$C_{20}H_{24}Cl_3O_{18}Sm$, Tris-perchlorato-5,6,14,15-tetrahydrodibenzo[b,k]-[1,4,7,10,13,16-hexaoxacyclooctadecin]samarium(III), 45B, 1219
$C_{20}H_{24}Cl_8O_6Ti_2$, Di-μ-chloro-bis(ethylanisatotrichlorotitanium(IV)), 43B, 1356
$C_{20}H_{28}CuN_2O_7$, Copper bis(tenuazonate) monohydrate, 43B, 1369
$C_{20}H_{28}O_4P_2Zn$, Zinc n-butylphenylphosphinate, 34B, 582
$C_{20}H_{33}ClO_3PRh$, cis-Chlorodicarbonyl(tricyclohexylphosphine oxide)-rhodium(I), 40B, 942
$C_{20}H_{34}Cl_4N_4O_2P_2U$, Tetrachlorobis(N,N,N',N'-tetramethylphenylphosphoramide)uranium(IV), 43B, 1370
$C_{20}H_{44}Cd_4Cl_8O_{10}$, Tetraethylene glycol dimethyl ether - cadmium chloride (1:2), 42B, 887
$C_{20}H_{54}O_8Zn_7$, Zinc dialkoxide complex, 42B, 888
$C_{21}H_{15}CoO_6$, Tris(tropolonato)cobalt(III), 45B, 1220
$C_{21}H_{15}MnO_6 \cdot 0.25\ C_7H_8$, Tris(tropolonato)manganese(III) toluene solvate, 40B, 943
$C_{21}H_{15}O_6Sc$, Tris(tropolonato)scandium(III), 40B, 944
$C_{21}H_{18}CrN_3O_6 \cdot 2\ C_3H_8O$, trans-Tris(benzohydroxamato)chromium(III) propanol solvate, 45B, 1220
$C_{21}H_{24}FeN_3O_9$, Iron(III) benzhydroxamate trihydrate, 34B, 583
$C_{21}H_{26}ClNO_6U$, Chlorodioxo-N-phenylbenzohydroxamatobis(tetrahydrofuran)uranium(VI), 45B, 1220
$C_{22}H_3Cl_{12}CrO_6S_2$, Tris(tetrachloro-1,2-benzoquinone)chromium(0) - carbon disulfide - hemibenzene, 42B, 889
$C_{22}H_{18}Cl_4Cu_2N_2O_2$, Di-$\mu$-4-phenylpyridine-1-oxide-bis(dichlorocopper-(II)), 38B, 958
$C_{22}H_{22}Cl_4Cu_2N_2O_4$, Di-$\mu$-4-phenylpyridine-1-oxide-bis(aquodichlorocopper(II)), 38B, 958
$C_{22}H_{24}Cl_2N_4O_2Zn$, Dichlorobis(antipyrine)zinc, 38B, 962
$C_{22}H_{24}CoN_6O_8$, Dinitratobis(antipyrine)cobalt(II), 40B, 945
$C_{22}H_{24}CuN_6O_8$, Dinitratobis(antipyrine)copper(II) (form 1), 40B, 945
$C_{22}H_{24}CuN_6O_8$, Dinitratobis(antipyrine)copper(II) (form 2), 40B, 945
$C_{22}H_{24}N_6O_8Zn$, Dinitratobis(antipyrine)zinc(II), 40B, 946
$C_{22}H_{24}O_5Zn$, Bis(benzoylacetonato)zinc monoethanolate, 34B, 504
$C_{22}H_{26}Mn_3O_{11}P$, 1,1,2,2,2,3,3,3-Octacarbonyl-dimethylphenylphosphine-1,2-μ-ethoxy-di-μ_3-ethoxy-trimanganese, 45B, 1215
$C_{22}H_{28}Br_4HgO_5$, Dibromo-1,15-bis(2-bromophenyl)-2,5,8,11,14-pentaoxapentadecanemercury, 46B, 1154
$C_{22}H_{38}O_4Zn$, Bis(dipivaloylmethanido)zinc, 29, 616
$C_{23}H_{37}CrN_3Na_3O_{15}$, Sodium trans-tris(benzohydroximato)chromium(III) octahydrate ethanol, 46B, 1154

$C_{24}H_{15}Br_4IrNO_3P \cdot CH_2Cl_2$, (Tetrabromocatecholato)(triphenylphosphine)nitrosyliridium(I) dichloromethane solvate, 46B, 1155

$C_{24}H_{16}CeNa_4O_8 \cdot 21\ H_2O$, Sodium tetrakis(catecholato)cerate(IV) hydrate, 45B, 1221

$C_{24}H_{16}HfNa_4O_8 \cdot 21\ H_2O$, Sodium tetrakis(catecholato)hafnate(IV) hydrate, 45B, 1221

$C_{24}H_{20}Cl_2O_4W$, trans-Dichlorotetraphenoxytungsten(VI), 41B, 1126

$C_{24}H_{20}Cl_4O_4Ti_2$, Dichlorodiphenoxytitanium(IV), 31B, 445

$C_{24}H_{20}N_8O_8Zr$, Zirconium cupferrate, 35B, 748

$C_{24}H_{28}Cl_4Cu_2N_4O_4$, Dichlorobis(4-methylpyridine 1-oxide)copper(II) (yellow modification), 37B, 564

$C_{24}H_{28}Cu_2N_8O_{20}$, Di-μ-(4-methoxypyridine-1-oxide)-bis(dinitratocopper(II)), 41B, 1127

$C_{24}H_{46}Cl_2N_4O_{15}$, Aquatetrakis(N-formylpiperidine)dioxouranium(VI) bis(perchlorate), 46B, 1156

$C_{24}H_{48}Cl_2MnO_{14}S_{12}$, Hexakis(1,4-dithiane monosulfoxide)manganese(II) perchlorate, 39B, 771

$C_{24}H_{48}N_{12}O_4S_4Th$, Tetrakis(isothiocyanato)tetrakis(tetramethylurea)-thorium(IV), 46B, 1156

$C_{24}H_{50}Mo_2O_{11}$, μ-Oxo-bis{[2,3-dimethyl-2,3-butanediolato(1-)][2,3-dimethyl-2,3-butanediolato(2-)]oxomolybdenum(VI)}, 45B, 1221

$C_{24}H_{50}N_8O_{50}Zn_{13}$, 2,4-Dinitrophenolate zinc hydroxide, 26, 738

$C_{24}H_{56}Mo_2O_8$, Octaisopropoxydimolybdenum(IV), 44B, 975

$C_{24}H_{58}Na_4O_{29}Th$, Sodium tetrakis(catecholato)thorate(IV) - 21-water, 44B, 975

$C_{24}H_{58}Na_4O_{29}U$, Sodium tetrakis(catecholato)uranate(IV) - 21-water, 44B, 975

$C_{24}H_{60}Cl_6Cu_4O_5P_4$, $μ_4$-Oxo-hexa-$μ_2$-chloro-tetrakis(triethylphosphine oxide)tetracopper(II), 41B, 1129

$C_{24}H_{72}Cl_2CoN_{12}O_{17}P_6$, Trisoctamethylpyrophosphoramidecobalt(II) perchlorate, 35B, 742

$C_{24}H_{72}Cl_2CuN_{12}O_{17}P_6$, Trisoctamethylpyrophosphoramidecopper(II) perchlorate, 35B, 742

$C_{24}H_{72}Cl_2N_{12}O_{14}P_4U$, Tetrakis(hexamethylphosphoramide)dioxouranium-(VI) perchlorate, 42B, 889

$C_{24}H_{72}N_{12}O_{44}P_8Th_3$, Bis(trinitratotetrakis(trimethylphosphine oxide)-thorium(IV)) hexanitratothorate(IV), 44B, 976

$C_{25}H_{53}Cl_9NO_{18}U_3$, Trichloro(1,4,7,10,13,16-hexaoxacyclooctadecane)uranium aquotrichlorohydroxydioxouranate nitromethane solvate, 44B, 976

$C_{25}H_{54}Mo_2O_7$, (μ-Carbonyl)-bis-(μ-t-butoxy)-bis(di-t-butoxy-molybdenum), 44B, 977; 45B, 1222

$C_{26}H_{20}Cl_2O_{10}Re_2$, Di-μ-chloro-bis(tricarbonyl(1-phenyl-1-hydroxobut-1-en-3-one)rhenium(I)), 39B, 771

$C_{26}H_{20}O_3PRh$, Tropolonatocarbonyltriphenylphosphinerhodium(I), 46B, 1157

$C_{26}H_{24}Cl_2CoO_2P_2$, catena-Dichloro-μ-[1,2-bis(diphenylphosphine oxide)ethane]cobalt(II), 45B, 1222

$C_{26}H_{24}Cl_2CuO_2P_2$, catena-μ-Bis(1,2-diphenylphosphinyl)ethane-dichlorocopper(II), 37B, 565

$C_{26}H_{32}CuN_2O_3S_2U$, (3,8-Dimethyl-1,10-di(o-phenolato)-4,7-diazadeca-2,8-diene-1,10-dionato)bis(dimethylsulfoxide)dioxouranium(VI)copper(II), 44B, 978

$C_{26}H_{52}CrIN_3Na_5O_{20}$, Sodium cis-tris(benzohydroximato)chromium(III) sodium iodide sodium hydroxide nonahydrate methanol ethanol, 46B, 1154

$C_{26}H_{54}Mo_2O_{10}$, Bis(t-butylcarbonato)tetrakis(t-butyloxy)dimolybdenum-

(Mo-Mo), 44B, 978
$C_{27}H_{42}FeN_9O_{12}$ · X CH_3OH · H_2O, Ferrichrome solvate, 46B, 1158
$C_{27}H_{45}FeN_6O_9$ · 1.5 H_2O, Ferrioxamine E, 42B, 889
$C_{27}H_{78}Cl_2CuN_{12}O_{14}P_6$, Tris(octamethylmethylenediphosphonic diamide)-copper(II) perchlorate, 38B, 962
$C_{28}H_{20}O_8Zr$ · 2.25 $CHCl_3$, Tetrakis(tropolonato)zirconium(IV) chloroform solvate, 44B, 979
$C_{28}H_{22}O_9Th$, Tetrakis(tropolonato)aquothorium(IV), 41B, 1131
$C_{28}H_{24}Ni_2O_{10}$, Di-μ-tropolonato-bis(aquo(tropolonato)nickel(II)), 39B, 772
$C_{28}H_{32}O_{16}V_3$, Tetrakis(tetrahydrofuran)vanadium(II)-bis(hexacarbonylvanadate(-I)), 42B, 890
$C_{28}H_{84}Cl_{10}Mn_5O_{14}$, Manganese(II) chloride ethanol, 39B, 772
$C_{29}H_{18}Cl_2Mo_2O_9$, Bis(9,10-phenanthrenequinone)pentaoxodimolybdenum methylene chloride, 41B, 1133
$C_{30}H_{22}CuO_4$, Copper dibenzoylmethanate, 34B, 505
$C_{30}H_{22}MoO_6$, Dioxobis-(1,3-diphenylpropanedionato)molybdenum(VI), 40B, 947
$C_{30}H_{22}O_4Pd$, Palladium dibenzoylmethanate, 33B, 462
$C_{30}H_{24}Cl_4Cu_2N_{12}O_6$, catena-μ-(Azobispyridine-4,4'-di-N-oxide)-bis(dichloro(azobispyridine-4,4'-di-N-oxide)copper(II)), 41B, 1133
$C_{30}H_{24}Cr_4N_3O_{18}$, Tris(pentacarbonylcyanochromato(0))-tris(tetrahydrofuran)-chromium(III), 43B, 1005
$C_{30}H_{29}O_7Y$, Tris(1-phenyl-1,3-butanedionato)aquoyttrium(III), 33B, 463
$C_{30}H_{30}B_2CuF_8N_6O_6$, Hexakis(pyridine N-oxide)copper(II) tetrafluoroborate, 43B, 1376
$C_{30}H_{30}B_2F_8N_6NiO_6$, Hexakis(pyridine-N-oxide)nickel(II) bis(tetrafluoroborate), 40B, 948
$C_{30}H_{30}B_2F_8N_6O_6Zn$, Hexakis(pyridine N-oxide)zinc(II) tetrafluoroborate, 43B, 1376
$C_{30}H_{30}Cl_2CoN_6O_{14}$, Hexakis(pyridine 1-oxide)cobalt(II) perchlorate, 41B, 1134; 44B, 980
$C_{30}H_{30}Cl_2CuN_6O_{14}$, Hexakis(pyridine 1-oxide)copper(II) perchlorate, 43B, 1376; 44B, 980
$C_{30}H_{30}Cl_2CuN_6O_{14}$, Hexakis(2-pyridone)copper(II) perchlorate, 41B, 1135
$C_{30}H_{30}Cl_2FeN_6O_{14}$, Hexakis(pyridine 1-oxide)iron(II) perchlorate, 44B, 980
$C_{30}H_{30}Cl_2HgN_6O_{14}$, Hexakis(pyridine 1-oxide)mercury(II) bisperchlorate, 39B, 773
$C_{30}H_{30}Cl_2N_6O_{14}Zn$, Hexakis(pyridine N-oxide)zinc(II) perchlorate, 43B, 1376
$C_{30}H_{32}CuN_2O_6$, Bis(N-phenyl-2-carbamoyl-5,5'-dimethylcyclohexane-1,3-dionato)copper(II), 37B, 567
$C_{30}H_{32}F_{21}LuO_7$, Tris(1,1,1,2,2,3,3-heptafluoro-7,7-dimethyl-4,6-octanedionato)aquolutetium(III), 37B, 568
$C_{30}H_{32}F_{21}O_7Pr$, Tris(1,1,1,2,2,3,3-heptafluoro-7,7-dimethyl-4,6-octanedionato)praseodymium monohydrate, 37B, 567
$C_{30}H_{35}Cl_2CoN_5O_{13}$, Pentakis(2-picoline-N-oxide)cobalt(II) perchlorate (form 1), 35B, 749
$C_{30}H_{35}Cl_2CoN_5O_{13}$, Pentakis(2-picoline-N-oxide)cobalt(II) perchlorate (form 2), 41B, 1136
$C_{30}H_{38}N_5O_{17}P_3Th$, Tetraphenylphosphonium pentanitratobis(trimethylphosphine oxide)thorium(IV), 44B, 976
$C_{30}H_{38}O_4P_2Re$, trans-Tetraphenoxobis(trimethylphosphine)rhenium(IV), 46B, 1143

$C_{30}H_{66}Mo_2O_6$, Hexakis(2,2-dimethylpropoxy)dimolybdenum, 43B, 1377
$C_{30}H_{70}Cl_2Li_4O_{12}Re_2$, Bis[lithium pentaisopropoxo-oxorhenate(VI) lithium chloride tetrahydrofuran], 46B, 1143
$C_{30}H_{72}Cl_2FeN_{12}O_{14}$, Hexakis(diethylurea)iron(II) perchlorate, 46B, 1158
$C_{30}H_{72}Cl_3ErN_{12}O_{18}$, Tetramethylureaerbium perchlorate, 42B, 891
$C_{31}H_{27}HfNO_9$, Tetrakis(tropolonato)hafnium(IV) N,N-dimethylformamide, 43B, 1377
$C_{31}H_{27}NO_9Th$, Tetrakis(tropolonato)-N,N'-dimethylformamidethorium(IV), 35B, 751
$C_{32}H_{20}O_{12}P_2Re_2$, Di-μ-diphenylphosphinato-bis(tetracarbonylrhenium), 44B, 638
$C_{32}H_{80}O_{16}Ti_4$, Titanium(IV) ethoxide tetramer, 28, 651
$C_{33}H_{36}N_9NdO_{12}$, Tris(antipyrine)trinitratoneodymium, 42B, 923
$C_{33}H_{57}ErO_6$, Tris-(2,2,6,6-tetramethylheptane-3,5-dionato)erbium(III), 37B, 569
$C_{33}H_{57}O_6Pr$, 2,2,6,6-Tetramethyl-3,5-heptanedionatopraseodymium, 35B, 752
$C_{33}H_{59}DyO_7$, Tris(2,2,6,6-tetramethyl-3,5-heptanedionato)aquodysprosium(III), 37B, 569
$C_{34}H_{28}Cl_2O_4U$, trans-Dichlorobis(trans,trans-dibenzylideneacetone)dioxouranium(VI), 45B, 1223
$C_{34}H_{28}Cl_2O_4U \cdot 2\ C_2H_4O_2$, trans-Dichlorobis(trans,trans-dibenzylideneacetone)dioxouranium(VI) acetic acid solvate, 45B, 1223
$C_{35}H_{50}F_6NO_2P_5Ru$, Carbonyl(dimethylcarbamato)tetra(dimethylphenylphosphine)ruthenium(II) hexafluorophosphate, 44B, 861
$C_{36}H_{22}O_{10}Re_2$, Bis(μ-O-1,3-diphenylpropane-1,3-dionatotricarbonylrhenium(I)), 39B, 774
$C_{36}H_{26}O_6P_2W$, ((Triphenylphosphinemethylide)diphenylphosphine oxide)pentacarbonyltungsten(0), 39B, 775
$C_{36}H_{30}As_2Cl_2CoO_2$, Bis(triphenylarsine oxide)dichlorocobalt(II), 42B, 892
$C_{36}H_{30}As_2Cl_2HgO_2$, Di(triphenylarsenic oxide)mercury(II) chloride, 28, 652
$C_{36}H_{30}As_2Cl_4Hg_2O_2$, Triphenylarsine oxide mercury chloride, 29, 632
$C_{36}H_{30}Br_2CuO_2P_2$, Dibromobis(triphenylphosphine oxide)copper(II), 42B, 892
$C_{36}H_{30}Br_2MoO_4P_2$, cis-Dibromodioxobis(triphenylphosphine oxide)molybdenum(VI), 45B, 1224
$C_{36}H_{30}CeN_4O_{14}P_2$, Tetranitratobis(triphenylphosphine oxide)cerium(IV), 37B, 570
$C_{36}H_{30}Cl_2CoO_2P_2$, Dichlorobis(triphenylphosphine oxide)cobalt(II), 42B, 893
$C_{36}H_{30}Cl_2CuO_2P_2$, Dichlorobis(triphenylphosphine oxide)copper(II), 37B, 570
$C_{36}H_{30}Cl_2MoO_4P_2$, cis-Dichlorodioxobis(triphenylphosphine oxide)molybdenum(VI), 45B, 1224
$C_{36}H_{30}Cl_2O_2P_2Zn$, Dichlorobis(triphenylphosphine oxide)zinc(II), 46B, 1159
$C_{36}H_{30}Cl_2O_3P_2V$, Dichlorooxobis(triphenylphosphine oxide)vanadium(IV), 46B, 1159
$C_{36}H_{30}Cl_2O_4P_2U$, Dichlorodioxobis(triphenylphosphine oxide)uranium(VI), 44B, 982
$C_{36}H_{30}Cl_2O_4P_2W$, Dichlorodioxobis(triphenylphosphine oxide)tungsten(VI), 44B, 983
$C_{36}H_{30}Cl_3MoO_3P_2$, Trichloro-oxo-bis(triphenylphosphine oxide)molybdenum(V), 44B, 981

$C_{36}H_{30}Cl_3O_3P_2W$, Trichloro-oxo-bis(triphenylphosphine oxide)tungsten-(V), 46B, 1160
$C_{36}H_{30}Cl_4Mo_2O_6$, Di-μ-phenoxo-bis(dichlorodiphenoxomolybdenum(V)), 43B, 1379
$C_{36}H_{30}Cl_4O_2P_2U$, Tetrachlorobis(triphenylphosphine oxide)uranium(IV), 41B, 1138
$C_{36}H_{30}N_4O_{14}P_2Th$, Tetranitratobis(triphenylphosphine oxide)-thorium(IV), 39B, 776
$C_{36}H_{48}Cu_2N_4O_{10}$, Tetraaquobis(μ-hippurato-O)-bis(hippurato-O)dicop-per(II) tetrahydrate, 39B, 776
$C_{36}H_{65}AgO_{13}$, Monensin silver complex, 35B, 754
$C_{36}H_{76}CrO_4$, Tetrakis[bis(t-butyl)methoxo]chromium(IV), 46B, 1161
$C_{36}H_{108}O_{44}Zr_{13}$, Zirconium oxide methoxide, 43B, 1379
$C_{38}H_{36}As_2Cl_2CoO_3$, Bis(triphenylarsine oxide)dichlorocobalt(II) - ethanol, 43B, 1380
$C_{38}H_{36}Cl_2CoN_2O_2P_2$, Bis-(N-benzyldiphenylphosphinic amide)dichloroco-balt(II), 39B, 778
$C_{38}H_{67}EuO_7S$, (3,3-Dimethylthietane 1-oxide)-tris(dipivalomethanato)-europium(III), 38B, 963
$C_{39}H_{57}FeN_6O_{15}$ · 7 C_6H_6, Ferric N,N',N"-triacetylfusarinine, 46B, 1161
$C_{39}H_{90}Cl_2CuO_{26}P_6$, Tris(tetraisopropylmethylenediphosphonate)copper-(II) perchlorate, 39B, 778
$C_{40}H_{32}Cl_3LaN_8O_{20}$, Tetrakis(2,2'-bipyridine dioxide)lanthanum(III) perchlorate, 44B, 983
$C_{40}H_{40}Cl_3LaN_8O_{20}$, Octakis(pyridine-N-oxide)lanthanum perchlorate (form i), 45B, 1225
$C_{40}H_{40}Cl_3N_8NdO_{20}$, Octakis(pyridine-N-oxide)neodymium perchlorate (form ii), 45B, 1225
$C_{40}H_{46}O_9Th$, Tetrakis(γ-isopropyltropolonato)aquothorium(IV), 41B, 1139
$C_{40}H_{67}AgO_{11}$, Silver polyetherin A, 35B, 755
$C_{40}H_{80}Cl_4Co_2O_{26}$, μ-Bis(2(1H)-tetrahydropyrimidinone)-octakis(2(1H)-tetrahydropyrimidinone)dicobalt(II) perchlorate, 38B, 964
$C_{40}H_{84}CrLiO_5$, Lithium tetrakis[bis(t-butyl)methoxo]chromate(III) tetrahydrofuran, 46B, 1161
$C_{41}H_{24}CeF_{12}NO_8S_4$, Isoquinolinium tetrakis(4,4,4-trifluoro-1-(2-thienyl)butane-1,3-dionato)cerate(III), 40B, 949
$C_{42}H_{24}MoO_6$, Tris(9,10-phenanthrenequinone)molybdenum, 41B, 1139
$C_{42}H_{42}Cl_2CoO_2P_2$, Dichloro-bis(tribenzylphosphine oxide)cobalt(II), 45B, 1226
$C_{42}H_{60}CrO_6$, Tris(3,5-di-t-butyl-o-benzoquinone)chromium(0), 45B, 1226
$C_{42}H_{84}Co_3O_{24}P_6$, Cobalt(II) diethoxyphosphonylacetylmethane, 33B, 464
$C_{42}H_{100}O_{14}W_4$, Tetradeca(isopropoxy)-dihydro-tetra(tungsten(IV)), 45B, 1227
$C_{42}H_{126}Cu_{18}O_{16}Si_{14}$, Dioxotetradecakis(trimethylsiloxo)octadecacop-per(I), 44B, 984
$C_{43}H_{35}NO_2P_2Pt$, [Benzoylhydroxylamido(2-)-O,O']bis(triphenylphos-phine)platinum, 46B, 1163
$C_{44}H_{84}Br_4Mn_2N_4O_8$, Bis(N,N,N',N'-tetrapropyl-1,2-cyclohexylenedioxy-diacetamide)manganese(II) tetrabromomanganate(II), 43B, 902
$C_{45}H_{35}HoO_7$, Tris(diphenylpropanedionato)aquoholmium, 40B, 1125
$C_{45}H_{48}EuNO_8$, Piperidinium tetra(benzoylacetonato)europium(III), 34B, 508
$C_{45}H_{96}FeLiO_5$, Lithium tetrakis[bis(t-butyl)methoxo]ferrate(III) bis(t-butyl)methanol, 46B, 1161

$C_{48}H_{40}As_2MnN_4O_{12}$, Tetraphenylarsonium tetranitratomanganate(II), 35B, 756
$C_{48}H_{40}Cl_4O_4S_4Th$, Tetrachlorotetrakis(diphenyl sulfoxide)thorium(IV), 46B, 1164
$C_{48}H_{52}Co_3O_{17}$, Cobalt(II) 4-methyltropolonate hydrate, 41B, 1140
$C_{48}H_{144}N_{36}O_{44}P_4Th_3$, Bis[trinitrato-tetrakis{tris(dimethylamido)phosphine-oxide}thorium] hexanitratothorate, 45B, 1228
$C_{51}H_{119}ClNd_6O_{17}$, Hexaisopropoxynona-$\mu$-isopropoxy-di-$\mu_3$-isopropoxy-$\mu_6$-chloro-hexaneodymium, 44B, 984
$C_{52}H_{40}HfN_4O_8$, Tetrakis(N-benzoyl-N-phenylhydroxylaminato)hafnium(IV), 44B, 985
$C_{54}H_{18}Cl_{24}Mo_2O_{12}$, Hexakis(tetrachloro-1,2-benzoquinone)dimolybdenum benzene solvate, 41B, 1141
$C_{54}H_{45}ClHgO_7P_3$, Hexakis(triphenylphosphine oxide)dimercury bisperchlorate, 39B, 779
$C_{56}H_{32}FeO_8$, Tris(9,10-phenanthrenesemiquinone)iron(III) - 9,10-phenanthrenequinone, 44B, 986
$C_{56}H_{42}O_{16}Sc_2$, Tetrakis(tropolonato)scandium(III) acid dimer, 40B, 949
$C_{56}H_{44}Co_4O_{18}$, Tetrakis(bistropolonatocobalt(II)) dihydrate, 41B, 1141
$C_{56}H_{64}Cl_3LaO_{28}$, Octa(2,6-dimethyl-4-pyrone)lanthanum(III) perchlorate, 40B, 950
$C_{56}H_{72}O_{10}Zn_4$, Bis(2,2'-dimethyl-3,5-hexanedionato)zinc phenylzinc phenoxide dimer, 40B, 951
$C_{56}H_{80}Mo_2O_{10} \cdot C_7H_8$, Bis[bis(3,5-di-t-butylcatecholato)oxomolybdenum(VI)] toluene solvate, 45B, 1229
$C_{58}H_{46}Cl_3NNb_2O_{17}$, Oxonium bis(tetrakistropolonato)niobium(V) trichloride acetonitrile, 41B, 1142
$C_{60}H_{44}O_8Ce$, Tetrakis(dibenzoylmethane)cerium, 24, 621
$C_{60}H_{52}O_{10}Ti_2$, sym-trans-Di-μ-phenoxy-hexaphenoxydiphenolatodititanium(IV), 38B, 964
$C_{60}H_{120}N_{12}O_{26}U_4$, Di-$\mu_3$-oxo-dodeca(diethylcarbamato)tetrauranium(IV), 44B, 986
$C_{66}H_{72}CaCl_2N_2O_{14}$, Calcium hexa-antipyrine perchlorate, 33B, 465
$C_{66}H_{72}I_3N_{12}O_6Y$, Hexakis(antipyrine)yttrium tri-iodide, 40B, 952
$C_{72}H_{60}Cl_6Cu_4O_5P_4$, μ_4-Oxo-hexa-μ-chloro-tetrakis((triphenylphosphine oxide)copper(II)), 31B, 446; 32B, 449
$C_{81}H_{105}F_{18}N_3O_{15}Pr_2$, Tris-$\mu$-(N,N'-dimethylformamide)-bis(tris(trifluoroacetyl-D-camphor)praseodymium), 41B, 1143
$C_{112}H_{160}Co_4O_{16} \cdot 2\ C_6H_6$, Octakis(3,5-di-t-butyl-1,2-semiquinone)tetracobalt(II) benzene solvate, 45B, 1230
$C_{126}H_{108}Mo_6N_9NaO_{26}P_6$, Tris(triphenylphosphine-iminium) sodium (μ_3-oxo)-tris(μ-methoxy)-tris(dicarbonyl-nitroso-molybdenum), 46B, 1165

184. METAL COMPLEXES (NITROGEN AND OXYGEN LIGANDS)

$C_2Cl_{12}MoNOP$, Tetrachloro(trichlorophosphine oxide)-N-pentachloroethylnitridomolybdenum, 45B, 1231
$C_2Cl_{12}NOPRe$, Tetrachloro(trichlorophosphine oxide)-N-pentachloroethylnitridorhenium, 45B, 1231
$C_2H_3N_4O_{10}V$, Vanadium(V) oxide trinitrate-acetonitrile, 37B, 571
$C_2H_6CdN_4O_4$, Bis(hydrazinecarboxylato-N',O)cadmium, 33B, 436
$C_2H_6N_4O_4Zn$, Bis(hydrazinecarboxylato-N',O)-zinc, 37B, 571
$C_2H_8CdN_4O_5$, Bis(hydrazinecarboxylato-N',O)cadmium monohydrate, 34B, 547

$C_2H_8CuN_2O_6S_2$, Bis(aminomethanesulphonato)copper(II), 39B, 780
$C_2H_{10}Cl_2CuN_6O_2$, Copper(II) chloride bis(semicarbazide) complex, 30B, 360
$C_2H_{10}Cl_2N_6O_2Zn$, Zinc chloride bis(semicarbazide) complex, 30B, 360
$C_2H_{10}MnN_4O_6$, Bis(hydrazinecarboxylato-N',O)manganese(II) dihydrate, 32B, 425
$C_2H_{10}N_4NiO_6$, Bis(hydrazinecarboxylato-N',O)-nickel dihydrate, 35B, 720
$C_2H_{14}N_8O_4Zn$, Bis(hydrazine)bis(hydrazinecarboxylato-N',O)zinc, 30A, 409; 30B, 367
C_3AgN_3O, Silver dicyanonitrosomethanide, 40B, 56
$C_3AgN_3O_2$, Silver dicyanonitromethanide, 40B, 57
$C_3H_{16}N_8NiO_7$, Hydrazinium tris(hydrazinecarboxylato-N,O)-nickelate(II) monohydrate, 31B, 417; 32A, 151; 32B, 427
$C_4H_4AgN_3O_3$, Silver nitrate-pyrazine complex, 31B, 417
$C_4H_4CuN_2O_6$, Copper(II) nitrate - pyrazine, 35B, 721
$C_4H_8N_4NiO_4$, Ethylnitrosolato-O,N-(hydroxyacetamidoximato)-O,N-nickel(II), 42B, 894
$C_4H_{10}N_4NiO_2$, Tetramethylenedinitraminenickel(II) dihydrate, 31B, 419
$C_4H_{11}ClNNdO_7$, Triaquoiminodiacetatoneodymium(III) chloride, 37B, 573
$C_4H_{12}Cl_2CuN_8O_2$, Dichlorodiaquobis(dicyandiamide)copper(II), 37B, 572
$C_4H_{13}MoN_3O_3$, Trioxo(diethylenetriamine)molybdenum(VI), 29, 620
$C_4H_{16}CoN_{10}O_{10}$, Tetrakis(urea)cobalt(II) nitrate, 40B, 955
$C_4H_{16}Cu_2K_2N_6O_{10}$, Dipotassium di-$\mu$-hydroxo-bis(biuretato)dicuprate-(II) tetrahydrate, 40B, 955
$C_4H_{20}CuN_2O_6$, Dihydroxobis(ethanolamine)copper(II) dihydrate, 41B, 1046
$C_4H_{23}N_{10}O_{11}Sc$, Hydrazinium tetrahydrazincarboxylatoscandate(III) trihydrate, 40B, 957
$C_4H_{24}Cu_2N_4O_7S$, Di-μ-hydroxobis(dimethylaminecopper(II)) sulfate monohydrate, 31B, 421
$C_5H_5CrNO_5$, Oxodiperoxopyridinechromium(VI), 29, 541
$C_5H_7AgN_4O_4$, Nitrato(1-methylcytosine)silver(I), 43B, 1349; 45B, 1232
$C_5H_{10}CuN_2OS$, Thiocyanato(2-dimethylaminoethanolato)copper(II), 42B, 895
$C_5H_{10}CuN_2O_2$, Cyanato(2-dimethylaminoethanolato)copper(II), 43B, 1350
$C_5H_{10}N_2NiO_8 \cdot H_2O$, Tetraaquaorotatonickel(II) monohydrate, 46B, 1166
$C_5H_{11}CuNO_7$, Pyruvylideneglycinatoaquocopper(II) dihydrate, 37B, 574
$C_6H_4CdN_2O_4$, Cadmium(II) cyanoacetate, 40B, 958
$C_6H_6CoN_2O_6$, Diaquopyrazine-2,3-dicarboxylatocobalt(II), 39B, 781
$C_6H_7AgN_2O_2$, (1-Methylthymine)silver(I), 45B, 1232
$C_6H_8AgClN_4O_6S_2$, Bis(2-imino-4-oxo-1,3-thiazolidine)silver(I) perchlorate, 42B, 895
$C_6H_8CuNNaO_7$, Sodium nitrilotriacetatocuprate(II) monohydrate, 39B, 781
$C_6H_8K_3MoNO_{10}$, Potassium cis-trioxo(nitrilotriacetato)molybdate(VI) monohydrate, 42B, 896
$C_6H_8N_2O_7Zn$, Zinc pyrazine-2,3-dicarboxylate trihydrate, 40B, 959
$C_6H_9CsFeN_6O_6 \cdot H_2O$, Cesium tris(ethylnitrosolato)ferrate(II) monohydrate, 45B, 1233
$C_6H_{11}MnN_5O_8$, Aquo-(isonicotinic acid hydrazide)manganese(II) nitrate, 45B, 1234
$C_6H_{12}Cl_2HgN_4O_2$, Bis(2-imidazolidinone)mercury(II) chloride, 38B, 967
$C_6H_{12}CuLiNO_9$, Lithium nitrilotriacetatocuprate trihydrate, 40B, 959
$C_6H_{12}MnN_2O_2$, Bis(ethanol)bis(thiocyanato)manganese(II), 45B, 1234
$C_6H_{12}NO_9Pr$, Nitrilotriacetatodiaquopraseodymium(III) monohydrate, 38B, 968

$C_6H_{14}ClNO_3Zn$, Chlorotriethanolaminezinc, 38B, 970
$C_6H_{14}CuN_2O_6 \cdot 2\ H_2O$, Aqua(N,N-dimethylethylenediamine N-oxide)-oxalatocopper(II) dihydrate, 46B, 1166
$C_6H_{14}DyNO_{10}$, Nitrilotriacetatodiaquodysprosium(III) dihydrate, 38B, 969
$C_6H_{16}CuN_4O_9S$, catena-Tetraaqua-μ-(9-methylpurine)-copper(II) sulfate monohydrate, 45B, 1235
$C_6H_{17}Cl_4NO_4Zn_2$, μ-Chloro-trichlorozinc-aquotriethanolaminezinc, 38B, 971
$C_6H_{18}Cl_2N_2OV$, Bis(trimethylamine)oxovanadium(IV) dichloride, 33B, 437
$C_6H_{20}Cl_2CuN_4O_8$, Bis(1,2-propanediamine)copper(II) perchlorate, 38B, 971
$C_6H_{42}Co_3CrN_{30}O_{30} \cdot 3\ H_2O$, Hexakis(urea)chromium(III) diamminetetra-(nitro)cobaltate(III) trihydrate, 45B, 1236
$C_7H_5CsNO_7V$, Caesium cis-dioxodipicolinatovanadate(V) monohydrate, 44B, 987
$C_7H_5MoNO_8$, Oxoperoxo(pyridine-2,6-dicarboxylato)aquomolybdenum(VI), 44B, 996
$C_7H_7CuNO_4$, (Pyridine-2,6-dicarboxylato)diaquocopper(II), 37B, 574
$C_7H_7FMoN_2O_7$, Ammonium fluoro-oxoperoxo(pyridine-2,6-dicarboxylato)-molybdate(VI), 43B, 1221
$C_7H_7F_2K_2NO_8Ti$, Potassium difluoroperoxotitanium dipicolinate dihydrate, 38B, 972
$C_7H_8Br_2HgN_2O$, Dibromo[N-(2-pyridyl)acetamide]mercury(II), 46B, 1166
$C_7H_{11}CuNO_2$, Oxygen - bridged copper(II) tetramer, 35B, 723
$C_7H_{11}NNiO_2$, Oxygen - bridged nickel(II) dimer, 35B, 722
$C_7H_{11}NO_9V$, Vanadyl(IV) pyridine-2,6-dicarboxylate tetrahydrate, 33B, 437
$C_7H_{11}NO_{10}Ti$, Diaquoperoxotitanium(IV) dipicolinate dihydrate (triclinic), 38B, 972
$C_7H_{11}NO_{10}Ti$, Diaquoperoxotitanium(IV) dipicolinate dihydrate (orthorhombic), 40B, 961
$C_7H_{11}N_2O_9V$, Ammonium oxoperoxo(pyridine-2,6-dicarboxylato)-vanadate(V) hydrate, 39B, 782
$C_7H_{12}Cl_2CuN_4O_4$, Diaquadichloro(theophylline)copper(II), 45B, 1236
$C_7H_{12}CoN_4O_5$, Carbonatodiaquobisimidazolecobalt(II), 35B, 759
$C_7H_{14}Cl_2CuN_6O_2 \cdot CH_4O$, Dichloro-(5-(3,3-dimethyl-1-triazenyl)-imidazole-4-carboxamido)methanolcopper(II) methanol solvate, 45B, 1237
$C_7H_{14}CuN_2OS$, (2-Diethylaminoethanolato)copper(II) thiocyanate, 40B, 962
$C_7H_{20}ClCoN_4O_7$, cis-Carbonatobis(trimethylenediamine)cobalt(III) perchlorate, 37B, 575
$C_8H_8Cl_2CuN_4O_8$, Bis(perchlorato)-bis(pyrazine)copper(II), 45B, 1238
$C_8H_{10}CuN_2O_4$, Diammine-(o-phthalato)copper(II), 35B, 725
$C_8H_{11}CoKN_3O_8 \cdot 1.5\ H_2O$, Potassium nitro(ethylenediaminetriacetato)-cobaltate(III) sesquihydrate, 39B, 783
$C_8H_{12}CoN_6O_{13}$, Diaquonitratopurpuratocobalt(II) dihydrate, 43B, 1352
$C_8H_{12}CuN_6O_{12}$, Copper(II) violurate tetrahydrate, 38B, 973
$C_8H_{12}MoN_4O_{10}$, Ammonium cis-dioxo(uramil-N,N-diacetato)molybdate(VI) monohydrate, 42B, 898
$C_8H_{12}N_6O_{13}Zn$, Diaquonitratopurpuratozinc(II) dihydrate, 43B, 1352
$C_8H_{13}CuNO_2$, Oxygen - bridged copper(II) dimer, 35B, 723
$C_8H_{14}CuN_2O_6S_2$, trans-Diaqua(L-1,3-thiazolidine-4-carboxylato-O,N)-(L-Δ^2-1,3-thiazoline-4-carboxylato-O,N)copper(II), 46B, 1167
$C_8H_{16}CuN_2O_7$, Aqua[(R,S)-N,N'-ethylenebis(serinato)]copper(II), 46B, 1167

$C_8H_{16}CuN_6O_{11}$, Triaquonitratocaffeinecopper(II) nitrate, 38B, 973
$C_8H_{18}Cl_2CuN_6O_{12}$, Bis(iminodiacetamide)copper(II) perchlorate, 40B, 963
$C_8H_{18}N_2O_2Os$, Bis(N-t-butylimido)dioxoosmium, 45B, 1241
$C_8H_{20}ClCuN_3O_6$, catena-μ-Acetato-(di-(3-aminopropyl)amine)copper(II) perchlorate, 38B, 974
$C_8H_{20}CuN_6O_6$, Nitrato(1,4,7,10-tetraazacyclododecane)copper(II) nitrate, 45B, 1239
$C_8H_{20}MoN_2O_4$, Dioxo-bis(N,N-diethylhydroxylaminato)molybdenum(VI), 46B, 1168
$C_8H_{22}ClCoN_4O_6$, trans-Bis(ethylenediaminomonoacetato)cobalt(III) chloride dihydrate, 40B, 964
$C_8H_{22}CoI_2N_5O_2 \cdot 0.84\ H_2O$, Ammineglycinato(1,4,7-triazacyclononane)-cobalt(III) diiodide hydrate, 46B, 1169
$C_8H_{24}Cl_2CuN_4O_8$, Bis(1,3-diaminobutane)copper(II) perchlorate, 38B, 975
$C_8H_{24}Cl_2CuN_4O_{10}$, Bis(N-(2-hydroxyethyl)ethylenediamine)copper(II) perchlorate, 39B, 784
$C_8H_{26}Cu_2N_6O_{12}$, Bis(cis-bis(2-aminoethanolato)copper(II)) dinitrate, 46B, 1169
$C_9H_{10}BCl_2N_6OTc$, Dichloro[hydrotris(1-pyrazolyl)borato]oxo-technetium(V), 45B, 1240
$C_9H_{14}CdN_3O_9P \cdot 1.5\ H_2O$, Aquo(cytidine-5'-phosphato)cadmium(II) sesquihydrate, 44B, 987
$C_9H_{14}CoN_3O_9P$, Aquo(cytidine-5'-phosphato)cobalt(II), 44B, 988
$C_9H_{14}CuN_4O_6$, Dinitrato(2-(2-dimethylaminoethyl)pyridine)copper(II), 39B, 785
$C_9H_{16}CdN_3O_{10}P$, Aquo(cytidine-5'-phosphato)cadmium(II) monohydrate, 44B, 988
$C_9H_{16}CsN_9O_{11}$, Caesium tris(isonitrosomalonamido)ferrate(II) dihydrate, 43B, 1355
$C_9H_{16}FeKN_9O_{11}$, Potassium tris(isonitrosomalonamido)ferrate(II) dihydrate, 43B, 1355
$C_9H_{24}Cl_2N_6NiO_6$, Tris(N,O-carbazic ethyl ester)nickel(II) chloride, 42B, 904
$C_9H_{27}Cu_3N_9O_{12}$, Tri-μ-(1,3-diamino-2-propanolato)-tricopper(II) nitrate, 43B, 1143
$C_{10}H_8AgN_4O_6$, Dinitrato-2,2'-bipyridylsilver(II), 38B, 976
$C_{10}H_8CrN_2O_5$, Oxidodiperoxido-2,2'-dipyridylchromium(VI), 33B, 439
$C_{10}H_8CuN_2O_4S_2$, Bis(2-pyridinesulfinato)copper(II), 43B, 1356
$C_{10}H_8MnN_5O_9$, Trinitrato-2,2'-bipyridylmanganese(III), 38B, 977
$C_{10}H_9AgN_4O_2S$, 2-Sulfanilamidopyrimidinesilver(I), 42B, 817
$C_{10}H_{10}Cl_2CuN_6O_{10}$, Bis(pyrazine-2-carboxamide)copper(II) perchlorate, 39B, 786
$C_{10}H_{10}N_4O_6Zn$, Dinitratobis(pyridine)zinc(II), 37B, 576
$C_{10}H_{11}CuN_4O_8P \cdot H_2O$, (Inosine 5'-monophosphate)copper(II) monohydrate polymer, 45B, 1212
$C_{10}H_{12}AgN_3O_4$, Bispyridinesilver(I) nitrate monohydrate, 35B, 762
$C_{10}H_{12}BaN_2O_{12}W_2 \cdot 3.5\ H_2O$, Barium μ-N,N'-ethylenediaminetetraacetato-di-μ-oxo-dioxoditungstate(V) hydrate, 43B, 1130; 44B, 988
$C_{10}H_{12}Cl_2CuN_4O_8$, trans-Dichlorodiaquocopper(II) - 4-nitropyridine N-oxide, 37B, 576
$C_{10}H_{12}N_4NiO_6$, Diaquobis(imidazole)-catena-μ-(squarato-1,3)nickel(II) (polymer), 45B, 1240
$C_{10}H_{13}NNiO_9S \cdot 1.5\ H_2O$, Tetraaquo(1,2-naphthoquinone-1-oximato-6-sulphonate)nickel(II) sesquihydrate, 43B, 1357
$C_{10}H_{14}CdCl_2N_6O_2$, Dichloro-bis(1-methylcytosine)-cadmium(II), 45B,

1241
$C_{10}H_{14}Cl_4Hg_2N_6O_2$, Di-μ-chloro-bis(chloro(1-methylcytosine-O,N[3])mercury(II)), 43B, 1289
$C_{10}H_{14}CoN_4O_4$, Bis(imidazole)bis(acetato)cobalt, 40B, 966
$C_{10}H_{14}CuN_2O_6S$, Dipyridinecopper(II) sulfate dihydrate, 29, 608
$C_{10}H_{14}N_4NiO_8$, Diaquodinitratobispyridinenickel(II), 38B, 978
$C_{10}H_{15}CuNO_{10}S$, Diaquo-(1,2-naphthoquinone-2-oximato-4-sulphonate)-copper(II) trihydrate, 41B, 1113
$C_{10}H_{15}NO_3Os$, N-1-Adamantylimidotrioxoosmium, 45B, 1241
$C_{10}H_{15}N_2O_9Rh$, Aquo(ethylenediaminetriacetatoacetic acid)rhodium-(III), 39B, 786
$C_{10}H_{16}Cl_2N_2OPt$, trans-Dichloro(dimethylformamide)(2,6-lutidine)platinum(II), 46B, 1057
$C_{10}H_{16}CoN_2O_{10}Rb$, Rubidium ethylenediaminetetraacetatocobaltate(III) dihydrate, 23, 577
$C_{10}H_{16}Cs_2Mo_2N_2O_{12}S_2$, Dicaesium μ-(ethylenediaminetetraacetato)-di-μ-sulfido-bis(oxomolybdate(V)) dihydrate, 39B, 787
$C_{10}H_{16}FeN_2O_{10}Rb$, Rubidium aquoethylenediaminetetraacetatoferrate monohydrate, 29, 544
$C_{10}H_{16}N_2NiO_9$, Dihydrogen ethylenediaminetetraacetatoaquonickel(II), 23, 576
$C_{10}H_{16}N_4NiO_6$, Dinitrato-(N,N,N',N'-tetramethyl-o-phenylenediamine)-nickel(II), 35B, 762
$C_{10}H_{18}Cl_2CuN_8O_3 \cdot 2\ H_2O$, Aquo-bis(2-hydrazino-4-hydroxy-6-methyl-pyrimidine)copper(II) chloride dihydrate, 45B, 1242
$C_{10}H_{18}CoLiN_2O_{11}$, Lithium ((S,S)-N,N'-ethylenediaminesuccinato)cobalt(III) trihydrate, 44B, 989
$C_{10}H_{18}CoLiN_2O_{11}$, Lithium [(R,S)-N,N'-ethylenediaminedisuccinato]cobaltate(III) trihydrate, 46B, 1170
$C_{10}H_{18}CuK_2N_2O_{11}$, Potassium ethylenediaminetetraacetatocuprate trihydrate, 39B, 788
$C_{10}H_{18}FeLiN_2O_{11}$, Lithium aquoethylenediaminetetraacetatoferrate dihydrate, 29, 544
$C_{10}H_{20}CoN_3O_{10}$, Ammonium ethylenediaminetetraacetatocobaltate(III) dihydrate, 23, 577
$C_{10}H_{20}CuN_2O_4 \cdot 4\ H_2O$, Succinato(N,N-diethylethylenediamine)copper-(II) tetrahydrate, 45B, 1243
$C_{10}H_{20}Cu_2N_2O_{12}$, Copper ethylenediaminetetraacetate tetrahydrate, 39B, 788
$C_{10}H_{20}FeN_2NaO_{11}$, Sodium aquo(ethylenediaminetetraacetato)-ferrate(III) dihydrate, 41B, 1033
$C_{10}H_{20}N_2Na_3O_{14}V$, Trisodium (ethylenediaminetetraacetato)dioxo-vanadate(V) tetrahydrate, 37B, 577
$C_{10}H_{20}N_2O_{12}Zr$, Zirconium ethylenediaminetetraacetate tetrahydrate, 40B, 967
$C_{10}H_{22}CaCoN_2O_{13}$, Calcium cobalt(II) ethylenediaminetetraacetate pentahydrate, 43B, 1130
$C_{10}H_{22}CuN_2O_9S_2$, Copper yunainate trihydrate, 35B, 763
$C_{10}H_{22}Li_2MnN_2O_{13}$, Lithium ethylenediaminetetraacetatomanganate(II) pentahydrate, 40B, 967
$C_{10}H_{22}N_2NiO_7$, 1,5-Diazacyclooctane-N,N'-diacetatoaquonickel(II) dihydrate, 37B, 579
$C_{10}H_{24}Co_2N_2O_{14}$, Tetraaquocobalt(II) (ethylenediaminetetraacetato(4-))cobaltate(II) dihydrate, 44B, 1008
$C_{10}H_{24}N_2O_{14}Zn_2$, Zinc ethylenediaminetetraacetate hexahydrate, 39B, 789
$C_{10}H_{24}N_3O_{13}V$, Ammonium (dihydrogenethylenediaminetetraacetato)dioxo-

vanadate(V) trihydrate, 37B, 578
$C_{10}H_{24}N_4O_4Ti$, Bis(dimethylamido)bis(N,N-dimethylcarbamato)titanium-(IV), 43B, 1201
$C_{10}H_{26}Cu_2I_2N_4O_2 \cdot 2\ H_2O$, catena-μ-Iodo-bis-{μ-{[2-(3-aminopropyl)amino]-ethanolato-N,N',μ-O}copper(II)} iodide dihydrate, 46B, 1170
$C_{10}H_{28}Mo_2N_2Na_4O_{22}$, Sodium hexaoxo-μ-ethylenediaminetetraacetato-dimolybdate octahydrate, 34A, 279
$C_{11}H_{10}AgN_3O_2S$, Silver N-(3-pyridyl)sulfanilamide (α polymorph), 42B, 906
$C_{11}H_{10}AgN_3O_2S$, Silver N-(3-pyridyl)sulfanilamide (β polymorph), 42B, 906
$C_{11}H_{11}CuNO_2$, Acetylacetone-mono-(o-hydroxyanil)copper(II), 30B, 361
$C_{11}H_{14}CoKN_2O_9$, Potassium (2-hydroxy-1,3-propanediamine-N,N,N',N'-tetraacetato)cobaltate(III), 44B, 990
$C_{11}H_{16}N_4NiO_4$, (4-Methyliminopentane-2,3-dione 3-oximato)(4-iminopentane-2,3-dione 3-oximato)nickel(II), 39B, 789
$C_{11}H_{18}CoKN_2O_{10}$, Potassium (-)-trimethylenediaminetetraacetatocobaltate(III) dihydrate, 38B, 978
$C_{11}H_{18}CuN_4O_6 \cdot 2\ H_2O$, N,N'-Methylethylenebis(N-methylcarbamylglycinato)copper(II) dihydrate, 45B, 1244
$C_{11}H_{19}ClCuN_4O_6 \cdot 0.5\ CH_4O$, (3,9-Dimethyl-4,8-diaza-3,8-undecadiene-2,10-dione dioxime)copper(II) perchlorate hemimethanolate, 43B, 1294
$C_{11}H_{19}Cl_3FeN_7O_4$, Dichloro(2,6-diacetylpyridine bis(semicarbazone))-iron(III) chloride dihydrate, 42B, 821
$C_{11}H_{20}BClCuF_2N_4O_7$, Aqua(difluoro-3,3'-(trimethylenedinitrilo)bis(2-butanone oximato)borato)copper(II) perchlorate, 45B, 1244
$C_{11}H_{20}F_{12}N_8O_2P_2Ru \cdot 2\ H_2O$, 10-Methylisoalloxazinotetrakisammineruthenium bis(hexafluorophosphate) dihydrate, 45B, 1245; 46B, 1171
$C_{11}H_{21}ClCuN_4O_7 \cdot H_2O$, Aqua(3,3'-trimethylenedinitrilo)bis(2-butanone oximato)copper(II) perchlorate monohydrate, 45B, 1244
$C_{11}H_{21}Cl_2CoN_7O_5$, Chloroaquo(2,6-diacetylpyridine bis(semicarbazone))cobalt(II) chloride dihydrate, 44B, 990
$C_{11}H_{21}Cl_2FeN_7O_5$, Chloroaquo(2,6-diacetylpyridine bis(semicarbazone))iron(II) chloride dihydrate, 44B, 990
$C_{11}H_{21}Cl_2MnN_7O_5$, Chloroaquo(2,6-diacetylpyridine bis(semicarbazone))manganese(II) chloride dihydrate, 44B, 990
$C_{11}H_{21}Cl_2N_7O_5Zn$, Chloroaquo(2,6-diacetylpyridine bis(semicarbazone))zinc(II) chloride dihydrate, 44B, 990
$C_{11}H_{22}Cl_2CuN_4O_8$, 1,4,8,12-Tetraazacyclopentadecanecopper(II) perchlorate, 45B, 1246
$C_{11}H_{22}CrN_9O_{12}$, Diaquo(2,6-diacetylpyridine bis(semicarbazone))-chromium(III) hydroxide dinitrate hydrate, 42B, 821
$C_{11}H_{23}FeGaN_5O_3$, (Dimethyl(N,N-dimethylethanolamino)(3,5-dimethylpyrazolyl)gallato)dinitrosyliron, 45B, 1246
$C_{11}H_{23}GaN_4NiO_2$, (Dimethyl(N,N-dimethylethanolamino)(3,5-dimethylpyrazolyl)gallato)nitrosylnickel, 45B, 1247
$C_{12}H_5CuN_4O_2$, Dicyanatobis(pyridinato)copper(II), 46B, 1061
$C_{12}H_8CdN_2O_4$, Cadmium pyridine-2-carboxylate, 40B, 969
$C_{12}H_8Cl_2MoN_2O_2$, Dichlorodioxo(o-phenanthroline)molybdenum(VI), 45B, 1247
$C_{12}H_{10}ClFeN_2O_5$, Aquochlorobis(α-picolinato)iron(III), 42B, 802
$C_{12}H_{10}FeN_{10}O_{12} \cdot 4.5\ H_2O$, Ammonium tris(violurato)ferrate(II) hydrate, 42B, 907
$C_{12}H_{12}Cl_2CuN_8O_{12}$, Bis(pyrazine-2,3-dicarboxamide)copper(II) perchlorate, 43B, 1358
$C_{12}H_{12}CuN_2O_6$, Bis(pyridine-2-carboxylato)copper(II) dihydrate, 35B,

765; 39B, 790
$C_{12}H_{12}CuN_6O_6S$, Dinitrato-(S-methylthiosemicarbazone-8-quinolinealdehyde)copper(II), 45B, 1248
$C_{12}H_{12}F_6N_2NiO_2$, N,N'-Ethylenebis(1,1,1-trifluoroacetylacetoneiminato)nickel(II), 42B, 822
$C_{12}H_{12}Fe_2N_2O_7$, μ-Isopropylidenamido-μ-2-propanoximato(O,N)-bis(tricarbonyliron), 42B, 823
$C_{12}H_{12}MnN_2O_4P_2$, 1,10-Phenanthrolinebis(phosphinato)manganese(II), 44B, 991
$C_{12}H_{13}Cl_2N_2NbO_2$, Dichloro(oxo)-2,2'-bipyridylethoxyniobium(V), 35B, 766
$C_{12}H_{14}CuN_4O_6$, Dinitratobis(α-picoline)copper(II), 38B, 980
$C_{12}H_{14}Fe_2N_2O_7$, μ-(Acetone oximato(1-)-N,O)-μ-isopropylamido-bis(tricarbonyliron)(Fe-Fe), 44B, 966
$C_{12}H_{14}KN_2NbO_9$, Potassium triperoxo-(o-phenanthroline)niobate trihydrate, 37B, 580
$C_{12}H_{14}K_2N_2O_{13}Zr$, Potassium bis(nitriloacetato)zirconate(IV) monohydrate, 33B, 440
$C_{12}H_{15}Cl_2N_2O_2Re$, 1-Oxo-6-ethoxo-2,4-dichloro-3,5-pyridinerhenium(V), 43B, 1358
$C_{12}H_{15}N_9O_{16}Ru$, Tris(dihydrogenoviolurato)ruthenium tetrahydrate, 44B, 991
$C_{12}H_{16}CdN_2O_8$, Bis(N,isonicotinato)tetraaquocadmium(II), 37B, 583
$C_{12}H_{16}ClMnN_2O_2$, Chloro-N,N'-ethylenebis(acetylacetoniminato)manganese(III), 41B, 571
$C_{12}H_{16}Cl_2CuN_4O_4$, trans-Diaquobis(pyridine-2-carboxamide)copper(II) chloride, 37B, 581
$C_{12}H_{16}Cl_2CuN_8O_8$, Diperchlorato-tetraimidazolo-copper(II), 39B, 791
$C_{12}H_{16}Cl_2N_4NiO_4$, trans-Diaquo-bis(pyridine-2-carboxamide)nickel(II) chloride, 32B, 431
$C_{12}H_{16}CoN_2O_8$, Diaquobis(picolinato)cobalt(II) dihydrate, 38B, 980
$C_{12}H_{16}CuN_2O_8$, Bis(dihydro-1H,3H,5H-oxazolo[3,4-c]oxazole-7a-carboxylato)copper(II), 43B, 1231
$C_{12}H_{16}CuN_8O_4S$, Tetrakis(imidazole)copper(II) sulfate, 38B, 981
$C_{12}H_{16}KN_2NbO_{11}$, Potassium triperoxo-(o-phenanthroline)niobate trihydrate perhydrate, 37B, 580
$C_{12}H_{16}N_2NiO_8$, trans-Diaquo-bis(picolinato)nickel(II) dihydrate, 35B, 768; 38B, 982
$C_{12}H_{16}N_2O_8Zn$, Bis(N-isonicotinato)tetraaquozinc(II), 37B, 583
$C_{12}H_{16}N_2O_8Zn$, Bis(N-nicotinato)tetraaquozinc(II), 37B, 582
$C_{12}H_{16}N_2O_8Zn$, trans-Diaquo-bis(picolinato)zinc(II) dihydrate, 35B, 768
$C_{12}H_{18}CoN_3O_3$, N,N'-Ethylenebis(acetylacetoneiminato)nitrosylcobalt-(III), 38B, 982
$C_{12}H_{18}N_2NiO_2$, N,N'-Ethylenebis(acetylacetoneiminato)nickel(II) (cobalt(II)-doped), 42B, 907
$C_{12}H_{20}Cl_4Cu_2N_{12}O_2$, Di-μ-chlorobis(chloro-(5-(3,3-dimethyl-1-triazenyl)imidazole-4-carboxamido)copper(II)), 45B, 1237
$C_{12}H_{20}CuN_2O_2$, N,N-Ethylenebis(acetylacetoneiminato)copper(II), 28, 640
$C_{12}H_{20}CuN_2O_4$, Bis-1-aminocyclopentanecarboxylatocopper(II), 28, 642
$C_{12}H_{20}CuN_4O_6 \cdot 2\ H_2O$, N,N'-Ethylethylenebis(N-methylcarbamylglycinato)copper(II) dihydrate, 45B, 1244
$C_{12}H_{20}Cu_2N_2O_3$, N,N'-Ethylenebis(acetylacetoneiminato)copper(II) monohydrate, 31B, 428
$C_{12}H_{24}CuN_2O_8$, Bis(2,5-dimethyl-4-carboxylate-oxazolidine)copper(II) dihydrate, 37B, 583

$C_{12}H_{26}Br_2CuN_2O_4$, Dibromo(1,7,10,16-tetraoxa-4,13-diaza-cyclooctadecane)copper(II), 39B, 791
$C_{12}H_{26}CdI_2N_2O_4$, 1,4,10,13-Tetraoxa-7,16-diazacyclooctadecanecadmium-(II) diiodide, 45B, 1248
$C_{12}H_{26}Cl_2CuN_2O_4$, Dichloro(1,7,10,16-tetraoxa-4,13-diaza-cyclooctadecane)copper(II), 39B, 792
$C_{12}H_{26}Cl_2Cu_2N_4O_{14}$, Bis(perchlorato(2-(2-hydroxyethyl)imino-3-oximobutanato)aquocopper(II)), 40B, 971
$C_{12}H_{26}CrLiN_2O_{13}$, Lithium ethylenediamine-N,N'-diacetato-N,N'-di-3-propionatochromate(III) pentahydrate, 43B, 1139
$C_{12}H_{26}HgI_2N_2O_4$, 1,4,10,13-Tetraoxa-7,16-diazacyclooctadecanemercury-(II) diiodide, 45B, 1248
$C_{12}H_{28}Br_2Cu_2N_2O_2$, Bromo(2-diethylaminoethanolato)copper(II), 39B, 805
$C_{12}H_{28}CuN_2O_6$, Bis-(6-aminohexanoato)copper(II) dihydrate, 39B, 792
$C_{12}H_{28}CuN_2O_8$, Bisacetatobismorpholinecopper(II) dihydrate, 39B, 793
$C_{12}H_{28}N_{10}O_{11}Pt_2$, Bis($\mu$-1-methylthyminato-$N^3$,$O^4$)bis(cis-diammineplatinum(II)) dinitrate hydrate, 44B, 992
$C_{12}H_{28}N_{16}Ni_2O_{12}$, Di-$\mu$-(malonamide-dioximato)-bis[aquo-(2-ketomalonamide-dioximato)nickel(II)], 45B, 1249
$C_{12}H_{30}N_4NiO_{12}$, Bis(triethanolamine)nickel(II) dinitrate, 38B, 983
$C_{12}H_{34}Br_2Cu_2N_4O_2$, Di-$\mu$-hydroxo-bis-(N,N,N',N'-tetramethylethylenediamine)dicopper(II) bromide, 35B, 700
$C_{12}H_{34}N_2Na_4O_{28}Ti_2$, Sodium μ-oxo-bis(nitrilotriacetatoperoxotitanate(IV)) undecahydrate, 41B, 1066
$C_{12}H_{34}N_{24}Ni_3O_{26}$, Bis-($\mu$-(tri-1,2,4-triazolo-$N^1$,$N^2$)-triaquonickel)-nickel hexanitrate dihydrate, 37B, 584
$C_{12}H_{38}Cu_3N_{12}O_{12}$, Di-$\mu$-(1,3-diamino-2-propanolato)bis(1,3-diamino-2-propanol)tricopper(II) nitrite, 45B, 1250
$C_{12}H_{39}Cl_3Co_2N_6O_{18} \cdot 0.5\ H_2O$, Tris(2-aminoethanol)tris(2-aminoethoxide)dicobalt(III) triperchlorate hemihydrate, 45B, 1250
$C_{12}H_{40}Cl_2N_6Ni_2O_{14}$, Tetrakis(2-aminoethanol)bis(2-aminoethoxide)dinickel(II) diperchlorate, 45B, 1250
$C_{13}H_{11}Cl_3MoN_3O_3 \cdot C_2H_3N$, Trichloro-oxo-(N-pyridylmethylene-N-salicyloylhydrazine-N,O)molybdenum(V) acetonitrile solvate, 46B, 1172
$C_{13}H_{11}NO_5SW$, Tetracarbonyl(S,S-dimethylsulphonium 2-picolinylmethylide)tungsten(0), 46B, 1171
$C_{13}H_{12}ClCuN_3O_7S$, 1-(2-Thiazolylazo)-2-naphtholato-diaquocopper(II) perchlorate, 42B, 829
$C_{13}H_{20}CuN_2O_2$, 4,4'-(R-Propylene-diiminato)-di-(3-penten-2-one)copper(II), 42B, 908
$C_{13}H_{22}Br_2N_2NiO$, Dibromo-(1-(o-methoxyphenyl)-2,6-diazaoctane)nickel-(II), 33B, 441
$C_{13}H_{24}CoN_2NaO_{10}$, cis-$\alpha$-Sodium carbonato[(2S,2'S)-1,1'-ethylenedi-2-pyrrolidinecarboxylato(2-)]-cobaltate(III) trihydrate, 46B, 1172
$C_{13}H_{24}CrIN_2O_4 \cdot H_2O$, $(-)_{589}$-Bis(2,4-pentanedionato)(1,3-propanediamine)chromium(III) iodide monohydrate, 45B, 1251
$C_{13}H_{29}CoN_4O_4$, N,N'-Di(2-hydroxyethyl)-2,4-pentanediimine-bis-(2-aminoethanol)cobalt(III), 40B, 972
$C_{13}H_{30}N_5O_6Ta$, Bis(dimethylamido)tris(N,N-dimethylcarbamato)-tantalum(V), 44B, 992
$C_{13}H_{30}N_5O_{13}Yb$, Guanidinium bis-β-hydroxyethyliminodiacetato-ytterbate(III) trihydrate, 43B, 1267
$C_{14}H_{10}AgN_2O_9$, Silver(II) bis(pyridine-2,6-dicarboxylate) monohydrate, 35B, 770
$C_{14}H_{10}CuN_6O_3$, Bis(pyridine-2-(N-cyanocarboxamidato))aquocopper(II), 37B, 585

$C_{14}H_{11}ClN_2OPt \cdot CHCl_3$, cis-[N-(o-Aminobenzylidene)anthranilaldehydato-O,N,N']chloroplatinum(II) chloroform solvate, 46B, 1173
$C_{14}H_{12}AgN_2O_{10}$, Bis(pyridine-2,3-dicarboxylato)silver(II) dihydrate, 37B, 586
$C_{14}H_{12}N_2O_5U$, Aquo(bis(2-hydroxyphenylimino)ethanato-O,O',N,N')dioxouranium, 41B, 1120
$C_{14}H_{13}CuN_3O_4 \cdot 6\ H_2O$, (2,2'-Bipyridyl)(iminodiacetato)copper(II) hexahydrate, 46B, 1173
$C_{14}H_{13}HgNO_3$, 2-Mercuri-4-methylphenol - 2-nitroso-4-methylphenol, 35B, 771
$C_{14}H_{13}N_5NiO_2$, Benzamidoximato-N'-oxy-N²-benzoyliminodiazenatonickel-(II), 46B, 1174
$C_{14}H_{14}Cl_2CoN_6O_2$, Bis-(1-methyl-3-o-chlorophenyltriazine 1-oxide)cobalt(II), 37B, 587
$C_{14}H_{14}Cl_2N_6NiO_2$, Bis-(1-methyl-3-o-chlorophenyltriazene-1-oxide)nickel(II), 42B, 909
$C_{14}H_{14}CuN_2O_{11}$, ((Pyridine-2,6-dicarboxylato)(pyridine-2,6-dicarboxylic acid))copper(II) hydrate, 38B, 985; 39B, 793
$C_{14}H_{14}GdN_7O_9$, Trinitrato-1,2-bis(pyridine-2-aldimino)ethanegadolinium(III), 39B, 794
$C_{14}H_{14}N_2NiO_{11}$, Bis(hydrogen-pyridine-2,6-dicarboxylato)nickel(II) trihydrate, 38B, 984
$C_{14}H_{16}Cl_2CuN_4O_{10}$, Bis(pyridine-2-acetamide)copper(II) perchlorate, 37B, 590
$C_{14}H_{16}Cl_5Cu_4N_4O_2$, Bis(N-benzoylhydrazine)copper(II) pentachlorotricuprate(I), 37B, 588
$C_{14}H_{16}CuN_2O_6$, Copper 2-pyridylacetate dihydrate, 38B, 986
$C_{14}H_{16}N_2O_6Zn$, Zinc 2-pyridylacetate dihydrate, 38B, 986
$C_{14}H_{20}Cl_2CuN_4O_{10}$, Bis(2-dimethylaminopyridine 1-oxide)copper(II) diperchlorate, 46B, 1175
$C_{14}H_{20}Cl_2Cu_2N_2O_2$, Di-$\mu$-methoxo-bis(chloro-2-methylpyridine-copper-(II)), 37B, 591
$C_{14}H_{20}Cl_2Fe_2N_2O_{17}$, μ-Oxo-bis(4-chloro-2,6-pyridinedicarboxylatodiaquoiron(III)) tetrahydrate, 44B, 993
$C_{14}H_{20}CuN_6O_6$, Dinitratobis(methyl-α-picolylamine)copper(II), 37B, 591
$C_{14}H_{20}MnN_2O_9$, Potassium trans-1,2-diaminocyclohexane-N,N'-tetraacetatomanganate(III) monohydrate, 39B, 795
$C_{14}H_{20}N_2NiO_6$, Bis((acetato)aquo(pyridine))nickel(II), 38B, 987
$C_{14}H_{28}N_6NiO_4Se_2$, Tetrakis(dimethylformamido)nickel di-isoselenocyanate, 32B, 433
$C_{14}H_{30}BrCoN_4O_4 \cdot 3\ H_2O$, $(+)_{546}$-β-(Oxalato)-[(2S,4S,9S,11S)-4,9-dimethyl-5,8-diazadodecane-2,11-diamine]cobalt(III) bromide trihydrate, 45B, 1115
$C_{14}H_{32}Cl_2Cu_2N_2O_2$, Di-$\mu$-(1-diethylamino-2-propanolato)-dichlorodicopper(II), 43B, 1143
$C_{14}H_{34}Cl_2Cu_2N_4O_{10} \cdot 2\ CH_4O$, Di-$\mu$-(N-2-ethanolato-N',N'-dimethylpropane-1,3-diamine)-di-copper perchlorate methanol solvate, 45B, 1252
$C_{14}H_{36}Cl_2Cu_2N_2O_4S_2$, Diaquabis(N-(2-ethylthioethyl)-3-aminopropanolato-S,N,O)dichlorodicopper(II), 46B, 1175
$C_{14}H_{36}MoN_8O_4P_2S_2$, Bis(hexamethylphosphoramide)bis(isothiocyanato)dioxomolybdenum(VI), 45B, 1253
$C_{15}H_{10}Fe_4N_2O_{12}$, Tri-$\mu$-carbonyl-($\mu_4$-ethylamino)-($\mu_4$-N-oxyethylamino)-tetrakis(dicarbonyl-iron), 45B, 1253
$C_{15}H_{15}CdN_5O_6$, Dinitratotrispyridinecadmium(II), 38B, 988
$C_{15}H_{15}Cl_3CrN_3O$, mer-Trichloro(dimethylformamide)(1,10-phenanthro-

line)chromium(III), 43B, 1364
$C_{15}H_{15}CoN_5O_6$, Dinitratotrispyridinecobalt(II), 38B, 988
$C_{15}H_{15}CuN_5O_6$, Dinitratotrispyridinecopper(II), 38B, 988
$C_{15}H_{15}N_5O_6Zn$, Dinitratotrispyridinezinc(II), 38B, 988
$C_{15}H_{18}CoN_3O_9$, fac-Tris(pentane-2,3,4-trione 3-oximato)cobalt(III), 44B, 993
$C_{15}H_{27}Cl_3N_3O_8Pr$, Chloropentaaquoterpyridylpraseodymium(III) dichloride trihydrate, 37B, 593
$C_{15}H_{36}Cl_2Cu_2N_4O_3$, μ-Carbonato-dichlorobis(N,N,N',N'-tetramethyl-1,3-propanediamine)dicopper(II), 45B, 1254
$C_{15}H_{36}N_6O_6W$, Tris(dimethylaminato)tris(N,N-dimethylcarbamato)tungsten(VI), 40B, 978
$C_{15}H_{56}Cl_4N_{10}Ni_2O_3$, Tris(1,3-diaminopropane)nickel(II) diaquobis(1,3-diaminopropane)nickel(II) tetrachloride monohydrate, 37B, 593
$C_{16}H_{16}CdN_{10}O_{16}$, Diaquobis(purpurato)cadmium dihydrate, 43B, 512
$C_{16}H_{16}Cl_4CuN_2O_4$, Copper(II) dichloroacetate - α-picoline, 37B, 594
$C_{16}H_{16}CuN_2O_2$, Bis(o-hydroxyacetophenone imine)copper(II), 37B, 595
$C_{16}H_{16}CuN_6O_8 \cdot H_2O$, Nitratobis(benzimidazol-2-ylmethanol-N',O)copper(II) nitrate monohydrate, 45B, 1254
$C_{16}H_{16}N_2NiO_6$, Nickel(II) dehydroacetic acid monoimide complex, 32B, 434
$C_{16}H_{16}N_{10}O_{16}Zn$, Bis(purpurato)zinc(II) tetrahydrate, 43B, 1365
$C_{16}H_{18}Cl_2CuN_2O_4$, trans-Bis((chloroacetato)-(α-picoline))copper(II), 37B, 596
$C_{16}H_{18}Cl_2N_6NiO_2$, Bis[1-methyl-3-(2-chloro-6-methylphenyl)triazine 1-oxidato]nickel(II), 45B, 1255
$C_{16}H_{18}Cl_2N_6NiO_2 \cdot C_6H_6$, Bis[1-methyl-3-(2-chloro-6-methylphenyl)-triazine 1-oxidato]nickel(II) benzene solvate, 45B, 1255
$C_{16}H_{18}CuN_{10}O_{17}$, Bis(purpurato)copper(II) pentahydrate, 43B, 1364
$C_{16}H_{20}Cu_2N_4O_2$, Schiff base copper(II) complex, 38B, 989
$C_{16}H_{22}CuN_4O_4 \cdot 2\ H_2O$, Bis(pyridoxaminato)copper(II) dihydrate, 46B, 1176
$C_{16}H_{22}F_6N_4NiO_2$, Bis(trifluoroacetylacetone)triethylenetetraminenickel(II), 40B, 978
$C_{16}H_{24}CoN_{10}O_6 \cdot 0.5\ C_2H_6O$, Nitratotetrakis(2-methylimidazole)cobalt-(II) nitrate - ethanol, 38B, 990
$C_{16}H_{24}CuN_6O_{10} \cdot H_2O$, Bis(pyridoxamine)copper(II) dinitrate monohydrate, 46B, 1176
$C_{16}H_{26}Cl_2Cu_2N_8O_2 \cdot 2\ H_2O$, Di-μ-hydroxybis[di(2-methylimidazole)copper(II)] diperchlorate dihydrate, 45B, 1256
$C_{16}H_{26}CrKN_2O_8 \cdot 4\ H_2O$, Potassium trans-bis(tert-butyliminodiacetato)chromate(III) tetrahydrate, 46B, 1176
$C_{16}H_{26}FeN_{10}O_{21}$, Triaquopurpuratoiron(II) purpurate hexahydrate, 43B, 1366
$C_{16}H_{28}CuN_2O_{10}$, Bis(methoxyacetato)bis(pyridine)copper(II) tetrahydrate, 37B, 596
$C_{16}H_{28}MnN_{10}O_{22}$, Tetraaquopurpuratomanganese(II) purpurate hexahydrate, 43B, 1366
$C_{16}H_{28}N_6O_{12}Zn$, Diaquabis(pyridoxamine-O,N)zinc dinitrate, 46B, 1177
$C_{16}H_{30}Cl_2CuN_{10}O_2$, Dichloro-bis(2,4-bis(ethylamino)-6-methoxy-1,3,5-triazine)copper(II), 46B, 1179
$C_{16}H_{31}CoN_5NaO_{10}S$, Sodium trans-sulfitobis(dimethylglyoximato)-(aniline)cobaltate(III) dihydrate, 44B, 994
$C_{16}H_{34}Cl_2CuN_4O_{10}$, (4,4,9,9-Tetramethyl-5,8-diazadodecane-2,11-dione bis(O-methyloxime))copper(II) perchlorate, 44B, 994
$C_{16}H_{36}N_4O_4P_4S_4U$, Tetraisothiocyanatotetrakis(trimethylphosphine oxide)uranium(IV), 45B, 1256

$C_{16}H_{40}Br_4Cu_4N_4O_4$, (2-Amino-2-methyl-1-propanolato)bromocopper(II) tetramer, 46B, 1178
$C_{16}H_{40}Cl_4Cu_4N_4O_4$, (2-Amino-2-methyl-1-propanolato)chlorocopper(II) tetramer, 46B, 1178
$C_{16}H_{40}N_4O_7V_2$, μ-Oxo-bis(bis(N,N-diethylhydroxylaminato)oxo-vanadium(V)), 46B, 1168
$C_{16}H_{42}Co_3N_6O_{10}$, Di(tris(2-aminoethoxido)cobalt(III))cobalt(II) acetate, 34B, 556
$C_{16}H_{46}Cu_2N_6O_{12}$, Bis(aquabis(2-amino-2-methylpropanolato)copper(II)) dinitrate, 46B, 1169
$C_{16}H_{50}MoN_2O_4Si_4$, Tetrakis(trimethylsiloxy)bis(dimethylamine)molybdenum(IV), 44B, 995
$C_{17}H_{16}ClN_3O_3PdS$, Chloro-1-(2-thiazolylazo)-2-naphtholatopalladium-(II) dioxane solvate, 40B, 979
$C_{17}H_{20}AgClN_4O_{10}$ · 0.45 CO · 0.5 @H_2O, Riboflavin silver perchlorate hemihydrate, 39B, 804
$C_{17}H_{21}MnN_5O_2S_2$, (2,13-Dimethyl-6,9-dioxa-3,12,18-triazabicyclo-[12.3.1]octadeca-1(18),2,12,14,16-pentaene)diisothiocyanatomanganese(II), 43B, 1367
$C_{17}H_{32}N_2O_4W$, Tri-t-butoxy(nitrosyl)(pyridine)tungsten, 45B, 1257
$C_{18}H_{10}F_{12}Mo_2N_2O_8$, Tetrakis-μ-trifluoroacetato-bis(pyridinemolybdenum(II)), 38B, 991
$C_{18}H_{10}N_2O_6Rh_2$, μ-N,N'-Dibenzoylhydrazido-bis(dicarbonylrhodium), 44B, 995
$C_{18}H_{12}Cl_2N_2O_2Ti$, Dichlorobis-8-quinolinolatotitanium(IV), 33B, 445
$C_{18}H_{12}CuN_2O_2$, Copper(II) 8-hydroxyquinolinate (form A), 32B, 435
$C_{18}H_{12}CuN_2O_2$, Copper(II) 8-hydroxyquinolinate (form B), 29, 612
$C_{18}H_{12}MoN_2O_4$, Dioxo-bis-(8-hydroxyquinolinato)molybdenum(VI), 37B, 598
$C_{18}H_{12}N_2O_2Pd$, 8-Hydroxyquinolinatopalladium(II), 31B, 432
$C_{18}H_{14}Cl_3MoN_2O_3$, 8-Hydroxyquinolinium trichloro(8-hydroxyquinolinato)oxomolybdate(V), 45B, 1258
$C_{18}H_{14}Cl_6CuN_4O_3$, Bis-(2,4,6-trichlorophenolato)diimidazolecopper(II) monohydrate, 42B, 910
$C_{18}H_{14}Cu_2F_{12}O_9$, Bis(2,5-bis(trifluoroacetyl)cyclopentanato)diaquodicopper(II) monohydrate, 42B, 928
$C_{18}H_{14}MnN_3O_7$, Tris(pyridine-2-carboxylato)manganese(III) monohydrate, 44B, 996
$C_{18}H_{16}CdN_4O_7$, Aquodinitratobis(quinoline)cadmium(II), 39B, 796
$C_{18}H_{16}CoN_2O_2$, 1,6-Bis-(2-hydroxyphenyl)-2,5-diaza-3,4-dimethyl-1,3,5-hextrienatocobalt(II), 42B, 911
$C_{18}H_{16}Cu_3N_6O_8S$ · 16.3 H_2O, $μ_3$-Hydroxo-tri-μ-(pyridine-2-carbaldehyde oximato)-$μ_3$-sulfato-tri-copper(II) hydrate, 38B, 991
$C_{18}H_{16}N_2O_4Zn$, Zinc 8-hydroxyquinolate dihydrate, 18, 761; 29, 614
$C_{18}H_{17}MoN_3O_{12}$, Hydrogen oxodiperoxo(pyridine-2-carboxylato)molybdate(VI) - bis(pyridine-2-carboxylic acid) monohydrate, 44B, 996
$C_{18}H_{18}Cl_2FeNO_3$, Dichloro(1,3-dibenzoyl-2-azapropenato)(ethanol)iron-(III), 44B, 997
$C_{18}H_{18}I_2N_2NiO_2$, 1,8-Diaza-3,4,10,11-dibenzo-5,12-dioxo-1,3,8,10-cyclotetradecatetraenenickel(II) diiodide, 37B, 599
$C_{18}H_{18}N_2O_5U$, (N,N'-Bis(salicylidene)-1,5-diamino-3-oxapentane)dioxouranium(VI) (α-form), 44B, 998
$C_{18}H_{18}N_2O_5U$, (N,N'-Bis(salicylidene)-1,5-diamino-3-oxapentane)dioxouranium(VI) (β-form), 44B, 998
$C_{18}H_{20}B_2Cu_2F_8N_6O_4$ · 2 H_2O, Bis(2,6-diacetylpyridine dioxime(1-))dicopper(II) tetrafluoroborate dihydrate, 46B, 1178
$C_{18}H_{20}CuN_2O_2$, Bis(N-methyl-o-hydroxyacetophenoneimine)copper(II),

39B, 797
$C_{18}H_{24}Cu_2N_4O_4$, N,N'-Bis(2-hydroxyethyl)-2,4-pentanediiminecopper(II) dimer, 42B, 844
$C_{18}H_{26}BCl_4MoN_7O_2$, Chloronitrosylisopropoxy(tris(4-chloro-3,5-dimethylpyrazolyl)borato)molybdenum, 42B, 912
$C_{18}H_{26}CuF_{12}N_4O_4$, Bis-(1,1,1,5,5,5-hexafluoropentane-2,4-dionato)bis-(N,N-dimethylethylenediamine)copper(II), 37B, 600
$C_{18}H_{26}CuN_2O_5 \cdot 4\ H_2O$, Aquabis[L-(+)-threo-2-amino-1-phenyl-1,3-propanediolato-N,O']copper(II) tetrahydrate, 45B, 1259
$C_{18}H_{26}CuN_8O_6$, Bis(methoxyacetato)tetrakis(imidazole)copper(II), 37B, 601
$C_{18}H_{27}Cl_4Mn_2N_9O_6 \cdot 0.5\ C_6H_7N_3O$, Triaqua-chlorotris(iso-nicotinehydrazido)dimanganese trichloride iso-nicotine hydrazide solvate, 46B, 1179
$C_{18}H_{27}N_{12}O_{15}Sm$, Triaquotris(isonicotinic acid hydrazide)samarium trinitrate, 45B, 1259
$C_{18}H_{28}Cu_2N_4O_8$, Bis(N-(picolinoyl)-3-amino-1-propoxidoaquocopper(II)) dihydrate, 40B, 980
$C_{18}H_{30}Cl_2Cu_2N_4O_{10}$, Di-$\mu$-hydroxo-bis(2-(2-ethylaminoethyl)pyridine)-dicopper(II) perchlorate, 38B, 993
$C_{18}H_{30}Cl_2Cu_2N_4O_{10}$, α-Di-μ-hydroxo-bis(2-(2-dimethylaminoethyl)pyridine)dicopper(II) perchlorate, 40B, 953
$C_{18}H_{30}Cl_2Cu_2N_4O_{10}$, β-Di-μ-hydroxo-bis(2-(2-dimethylaminoethyl)pyridine)dicopper(II) perchlorate, 40B, 954
$C_{18}H_{30}CuN_2O_5$, Aquobis(norephedrinato)copper(II) dihydrate, 44B, 998
$C_{18}H_{30}CuN_2O_7$, Diacetatocopper(II)-bis(p-toluidine) trihydrate, 37B, 600
$C_{18}H_{30}N_2NiO_2$, N,N'-Ethylenebis(5,5-dimethyl-4-oxohexan-2-iminato)-nickel(II), 43B, 1367
$C_{18}H_{32}CoF_9N_4O_6$, Bis(tetramethylethylenediamine)bis(trifluoroacetato)cobalt(III) trifluoroacetate, 45B, 1260
$C_{18}H_{34}Cl_2CuN_{10}O_2$, Dichloro-bis(2-ethylamino-4-isopropylamino-6-methoxy-1,3,5-triazine)copper(II), 46B, 1179
$C_{18}H_{34}Cu_2N_8O_4S_2$, μ-N,N'-Bis-(2-dimethylaminoethyl)oxamidato-diisothiocyanato-bis(dimethylformamidecopper(II)), 40B, 981
$C_{18}H_{36}Cl_2EuN_2O_{18} \cdot C_2H_3N$, 4,7,13,16,21,24-Hexaoxa-1,10-diazabicyclo-[8.8.8]hexacosane-perchlorato-europium(III) perchlorate acetonitrile solvate, 45B, 1260
$C_{18}H_{36}Cu_2N_4O_2S_2$, Thiocyanato(2-dipropylaminoethanolato)copper(II), 43B, 1368
$C_{18}H_{36}Cu_2N_4O_4$, Isocyanato(2-dipropylaminoethanolato)copper(II) dimer, 44B, 999
$C_{18}H_{38}N_8O_{25}Sm_2$, Nitrato[4,7,13,16,21,24-hexaoxa-1,10-diazabicyclo-[8.8.8]hexacosane]samarium aquapentanitratosamarate, 45B, 1261
$C_{18}H_{39}ClN_4NiO$, Acetato-C-rac-(5,7,7,12,14,14-hexamethyl-1,4,8,11-tetra-azacyclotetradecane)nickel(II) perchlorate, 35B, 772
$C_{18}H_{40}Cr_2N_4O_{20}$, Diaquo-$\mu$-triethylenetetraminehexaacetato-dichromium(III) hexahydrate, 40B, 982
$C_{18}H_{44}Cl_2Cu_2N_4O_{11}$, Bis($\mu$-N-(2-diethylaminoethyl)-3-aminopropanolato-O)-aquoperchloratodicopper(II) perchlorate, 40B, 983
$C_{18}H_{44}Cl_2Cu_3N_4O_{12}$, Trinuclear copper complex, 42B, 846
$C_{18}H_{44}N_4Na_2O_{24}V_2$, Sodium μ-triethylenetetraaminehexaacetatodi-(oxovanadate(IV)) decahydrate, 42B, 913
$C_{18}H_{50}Cl_2N_6Ni_2O_7$, N-(6-Amino-4-aza-3,3,6-trimethyl-2-heptylidenyl)-hydroxanatonickel(II) chloride pentahydrate, 40B, 983
$C_{19}H_{15}NO_6U$, Dioxo(pyridine)bis(tropolonato)uranium(VI), 40B, 984
$C_{19}H_{17}ClCuN_4O_7S$, Aquo-2-benzothiazolin-2-ylideneaminomethylpyridine-

(picolinato)copper(II) perchlorate, 38B, 994
$C_{19}H_{17}CuN_5O$, Ammine[1-(2-hydroxyphenyl)-3,5-diphenylformazanato]copper(II), 45B, 1261
$C_{19}H_{17}N_5NiO$, Ammine[1-(2-hydroxyphenyl)-3,5-diphenylformazanato]nickel(II), 45B, 1261
$C_{19}H_{20}CuN_2O_2$, Bis-(2-hydroxyacetophenone)trimethylenediiminocopper-(II), 35B, 727
$C_{19}H_{21}N_9O_8Zn$, {2,6-Dihydroxy-8,15-dimethyltripyrido[c,d:i,j:l,m]-[1,4,7,8,10,13,15]heptaaza[1,2,6,7,8,15]hexahydrocyclopentadecine-N^1,N(2b),N^7,N(8b),N(12b)}nitratozinc(II) nitrate, 45B, 1262
$C_{19}H_{24}Cl_2CuN_2O_6$, Chloro(3,4,9,10-dibenzo-1,12-diaza-5,8-dioxacyclopentadecane)copper(II) perchlorate, 46B, 1180
$C_{19}H_{24}Cl_2N_2NiO_2$, Dichloro(6,7,8,9,10,11,17,18-octahydro-5h-dibenzo-[e,n][1,4]dioxa[8,12]diazacyclopentadecine)nickel(II), 45B, 1263
$C_{20}H_{12}Cl_2Fe_2N_2O_8$, Bis(tricarbonyl-(3-chloro-2-methylnitrosobenzene)-iron), 37B, 603
$C_{20}H_{12}Cl_4Cr_2N_4O_4$, Tetrakis(μ-6-chloro-2-pyridinolato)dichromium(Cr-Cr), 46B, 1181
$C_{20}H_{12}Cl_4Mo_2N_4O_4$, Tetrakis(μ-6-chloro-2-pyridinolato)dimolybdenum-(Mo-Mo), 46B, 1181
$C_{20}H_{12}Cl_4N_4O_4W_2$, Tetrakis(μ-6-chloro-2-pyridinolato)ditungsten(W-W), 46B, 1181
$C_{20}H_{14}CuN_2O_5$, Copper(II) di-2-nitroso-1-naphtholate monohydrate, 41B, 1122
$C_{20}H_{16}ClN_4O_4Tc_2$, Tetra(2-oxopyridinato)chloroditechnitate, 46B, 1181
$C_{20}H_{16}Cl_2CuN_4O_8$, Perchlorato-bis(2,2'-bipyridine)copper(II) perchlorate, 37B, 603
$C_{20}H_{16}Cl_2N_4O_4Os_2 \cdot C_4H_{10}O$, Dichlorotetrakis(2-hydroxypyridinato)-diosmium(III)(Os-Os) monoetherate, 46B, 1182
$C_{20}H_{16}Cl_2N_4O_4Os_2 \cdot 2\ C_2H_3N$, Dichlorotetrakis(2-hydroxypyridinato)-diosmium(III)(Os-Os) bis(acetonitrile) solvate, 46B, 1182
$C_{20}H_{16}Cl_2N_4O_4Re_2$, Dichlorotetrakis(o-oxypyridine)dirhenium(III), 45B, 1263
$C_{20}H_{16}Cl_6CoN_4O_2$, Bis(N-methylimidazole)bis(2,4,6-trichlorophenolato)cobalt(II), 46B, 1095
$C_{20}H_{16}HgN_6O_6 \cdot 2\ H_2O$, Bis(2,2'-bipyridyl)mercury(II) nitrate dihydrate, 45B, 1264
$C_{20}H_{16}LaN_7O_9$, Trinitratobis(bipyridyl)lanthanum(III), 38B, 995
$C_{20}H_{16}N_2O_3V$, Bis(2-methyl-8-quinolinolato)oxovanadium(IV), 37B, 604
$C_{20}H_{18}CuN_4O_7S_2$, Monoaquabis(2,2'-bipyridyl)copper(II) dithionate, 45B, 1264
$C_{20}H_{18}CuN_4O_7S_5$, Monoaquabis(2,2'-bipyridyl)copper(II) disulphonato-trisulfane, 45B, 1265
$C_{20}H_{18}CuN_6O_8$, Bisbipyridyl-μ-dihydroxo-dicopper(II) nitrate, 35B, 728
$C_{20}H_{20}CoN_6O_6$, cis-Bis(1-methylimidazole)bis(o-nitrophenolato)cobalt-(II), 45B, 1266
$C_{20}H_{20}CuN_2O_2$, Bis-(3-amino-1-phenyl-2-buten-1-ono)copper(II), 33B, 446
$C_{20}H_{22}Cl_4Cu_2N_2O_8$, Mono-α-picolinecopper(II) chloroacetate, 35B, 774
$C_{20}H_{22}I_2N_2NiO_4$, Nickel(II) iodide - tetradentate ligand complex, 42B, 850
$C_{20}H_{22}N_4O_{14}S_2Zn$, Diammonium diaquobis(1,2-naphthoquinone-2-oximato-5-sulphonate)zincate(2-), 43B, 1369
$C_{20}H_{24}ClN_4O_4Re$, trans-Dioxotetrapyridinerhenium(V) chloride dihydrate, 37B, 606
$C_{20}H_{25}CoN_6O_{11}$, Nitratobis-(2,2'-bipyridyl)cobalt(III) hydroxide ni-

trate tetrahydrate, 37B, 607

$C_{20}H_{26}K_4Mo_4N_4O_{22}$ · 16.4 H_2O, Tetrapotassium di-μ-oxo-bis(μ-hydroxo-μ-oxo-μ-N,N'-ethylenediaminetetraacetato-bis(molybdate(III,IV))) hydrate, 42B, 914

$C_{20}H_{28}Cl_2CuN_4O_2$, Bis(N,N'-diethylnicotinamide)copper(II) chloride, 43B, 1319

$C_{20}H_{28}CuN_6O_8$, Bis(1,3-propanediamine)-bis(m-nitrobenzoato)copper(II), 45B, 1121

$C_{20}H_{28}MoN_2O_4$, trans-Bis(t-butoxy)-cis-dicarbonyl-cis-bis(pyridine)-molybdenum, 45B, 1267

$C_{20}H_{29}Cl_2N_5NiO_9$, 2-Acetylpyridine(9-(2-pyridyl)-4,8-diazadec-8-en-1-amine)nickel(II) diperchlorate, 41B, 1123

$C_{20}H_{30}CoN_5O_6$, 1-(p-Methylphenyl)ethyldioxybis(dimethylglyoximato)-pyridine-cobalt(III), 39B, 798

$C_{20}H_{30}CuN_4O_4$, Bis(α-hydroxy-α-phenylbutyramidine)copper(II) dihydrate, 32B, 438

$C_{20}H_{30}CuN_4O_4$, Bis(1,3-propanediamine)copper(II) benzoate, 38B, 996

$C_{20}H_{30}N_4NiO_4$, Bis(camphorquinone dioximato)nickel(II), 44B, 1000

$C_{20}H_{34}Ce_2N_4O_{18}$ · 6 H_2O, Hydroxyethylethylenediaminetriacetatodiaquacerium(III) trihydrate dimer, 45B, 1121

$C_{20}H_{38}Cl_6N_2O_4Ta_2$, μ-(1,2-Dimethyl-1,2-diimidoethenato(2-))-bis(trichlorobis(tetrahydrofuran)tantalum), 44B, 1000

$C_{20}H_{40}N_{17}O_{19}Pt_4$ · H_2O, cis-Diammineplatinum α-pyridone blue nitrate hydrate, 45B, 1268

$C_{20}H_{42}Cl_2Cu_2N_4O_{12}$, Di-μ-(4,6,6-trimethyl-3,7-diazanon-3-ene-1,9-diolato(1-)-O,N,N',μ-O')-dicopper(II) diperchlorate, 43B, 1371

$C_{20}H_{42}N_{17}O_{20}Pt_4$, Tetrakis(cis-diammineplatinum)-tetrakis(α-pyridonyl) pentanitrate monohydrate, 43B, 1371

$C_{20}H_{44}Br_2Cu_2N_2O_2$, Bromo(2-dibutylaminoethanolato)copper(II), 41B, 1107, 1123

$C_{20}H_{44}Co_3N_4O_{22}$ · 8 H_2O, Hexaaquocobalt(II) bis((R,S)-ethylenediamine-N,N'-disuccinato)dicobaltate(III) octahydrate, 46B, 1183

$C_{20}H_{46}Mn_3N_4O_{26}$, Bis(ethylenediaminetetraacetato)trimanganese decahydrate, 29, 544

$C_{20}H_{50}La_2N_4O_{24}$, ((Hydroxyethyl)ethylenediaminetriacetato)diaquolanthanum(III) trihydrate, 44B, 1001

$C_{20}H_{52}Cu_4N_8O_{12}S_2$ · 8 H_2O, {2-[(3-Aminopropyl)amino]ethanolato}copper(II) tetramer sulfate octahydrate, 45B, 1268

$C_{20}H_{56}Cu_4N_{12}O_{18}$, Tetrakis-$\mu_3$-(2-((3-aminopropyl)amino)ethanolato)-tetracopper(II) tetranitrate dihydrate, 43B, 1372

$C_{20}H_{58}Br_4Cu_4N_8O_7$, Tetrakis(2-((3-aminopropyl)amino)ethanolato)copper(II) bromide trihydrate, 44B, 1001

$C_{20}H_{60}Cl_4Cu_4N_8O_8$, (2-((3-Aminopropyl)amino)ethanolato)-copper(II) chloride monohydrate tetramer, 41B, 1124

$C_{21}H_{19}Cu_2N_7O_7$, Bis(2,2'-bipyridine)-μ-methoxo-μ-nitrito-dinitritodicopper(II), 43B, 1321

$C_{21}H_{19}N_2O_4V$, Oxoisopropoxobis(8-hydroxyquinolinato)vanadium(V), 39B, 799

$C_{21}H_{19}N_5NiO_8$, Tribenzo[b,f,j](1,5,9)triazacycloduodecine nickel(II) nitrate dihydrate, 30B, 363

$C_{21}H_{20}CoN_4O_2S_2$, cis-Dithiocyanato(5,6:14,15-dibenzo-1,4-dioxa-8,12-diazacyclopentadeca-7,12-diene)cobalt(II), 43B, 1177

$C_{21}H_{22}ClCoN_4O_8$, Carbonatotetrakis(pyridine)cobalt(III) perchlorate monohydrate, 39B, 799

$C_{21}H_{24}CuN_6O_4$, (N-3,4-Benzosalicylidene-N'-methylethylenediamine)-(theophyllinato)(aquo)copper(II), 42B, 852

$C_{21}H_{24}N_4NiO_2S_2$, Bis(isothiocyanato)(2,3:11,12)-dibenzo-1,13-dioxa-

5,9-diaza-2,11-cyclopentadecadienenickel(II), 46B, 1183
$C_{21}H_{29}ClN_3Na_4O_{26}Yb$, Trisodium tris(pyridine-2,6-dicarboxylato)-ytterbate(III) mono(sodium perchlorate) decahydrate, 38B, 996
$C_{21}H_{29}Cl_2MnN_7O_7$, 2,6-Diacetylpyridinebis(picolinoyl hydrazone)manganese chloride pentahydrate, 43B, 1322
$C_{21}H_{35}N_3Na_3O_{25}Yb$, Trisodium tris(pyridine-2,6-dicarboxylato)-ytterbate(III) 13-hydrate, 38B, 997
$C_{21}H_{39}N_3Na_3NdO_{27}$, Trisodium tris(pyridine-2,6-dicarboxylato)-neodymate(III) 15-hydrate, 38B, 998
$C_{22}H_{14}Cl_6CuN_2O_2$, Bis(pyridine)bis(2,4,6-trichlorophenolato)copper-(II), 46B, 1104
$C_{22}H_{19}CuN_3O_6$, Bis(ω-nitroacetophenonato)-(2-methylpyridine)copper-(II), 38B, 998
$C_{22}H_{22}CoN_3O_3$, N,N'-Ethylenebis(benzoylacetoneiminato)nitrosylcobalt-(III), 38B, 982
$C_{22}H_{22}Cu_2N_2O_4$, Tricoordinated copper(II) complex, 26, 666
$C_{22}H_{22}Cu_2N_2O_5$, (3,8-Dimethyl-1,10-di(o-phenolato)-4,7-diazadeca-2,8-diene-1,10-dionato)aquocopper(II), 44B, 1002
$C_{22}H_{24}AgN_9O_{10}$, Bis-(10-methylisoalloxazine)silver nitrite tetrahydrate, 39B, 802
$C_{22}H_{24}AgN_9O_{10.45}$, Bis-(10-methylisoalloxazine)silver nitrite nitrate tetrahydrate, 39B, 802
$C_{22}H_{24}Cl_2CuN_8O_{16}$, Bis-(10-methylisoalloxazine)copper(II) perchlorate tetrahydrate, 39B, 803
$C_{22}H_{24}CuN_6O_6$, Bis(3,5-dimethyl-1-phenylpyrazole)dinitratocopper(II), 45B, 1270
$C_{22}H_{26}B_2Mo_2N_{12}O_4$, Diacetatobis(trispyrazolylborato)dimolybdenum, 42B, 858
$C_{22}H_{26}FMo_2N_3O_{14}$, Tetraethylammonium μ-fluorobis[oxoperoxo(pyridine-2,6-dicarboxylato)molybdate(VI)], 46B, 1184
$C_{22}H_{26}FeN_4O_6$, Bis(3-methylpyridine)bis(pentane-2,3,4-trione 3-oximato)iron(II), 44B, 1002
$C_{22}H_{28}MnN_6O_2S_2$, Di-μ-(N,N-diethylnicotinamide-O,N)diisothiocyanato-manganese(II), 39B, 800
$C_{22}H_{29}CoN_4OPS_2$, ((2-((2-(Diethylamino)ethyl)amino)ethyl)diphenyl-phosphine oxide)di-isothiocyanatocobalt(II), 38B, 999
$C_{22}H_{30}CuN_2O_6$, Bis(4-formyl-2-methoxyphenolato)(N,N,N',N'-tetramethylethylenediamine)copper(II) (red isomer), 44B, 1003
$C_{22}H_{32}CuN_2O_6$, Bis(2-amino-2-methyl-1-propanol)copper(II) dibenzoate, 44B, 1003
$C_{22}H_{34}CuN_2O_4$, N,N'-Ethylenebis(l-ephedrine)copper(II) dihydrate, 33B, 447
$C_{22}H_{38}CuN_2O_{10}$, Bis(4-formyl-2-methoxyphenolato)(N,N,N',N'-tetramethylethylenediamine)copper(II) tetrahydrate, 44B, 1004
$C_{22}H_{38}CuN_8O_{12}$, Dinitratobis(2-(2-diethylaminoethyl)pyridine)copper-(II) nitrate, 39B, 801
$C_{22}H_{38}Cu_3N_4O_4$, Bis(N,N-bis(3-propanolato)acetylacetonatodiimine)tri-copper, 41B, 1125
$C_{22}H_{52}Fe_2N_6O_{21}$, Ethylenediammonium μ-oxo-bis(N-hydroxyethylethylenediaminetriacetatoiron(III)) hexahydrate, 32B, 441
$C_{22}H_{56}Mo_2N_4O_8$, Bis(μ-isopropoxy)bis(bis(isopropoxy)dimethylaminonit-rosylmolybdenum), 46B, 1184
$C_{22}H_{68}Mo_2N_2O_6Si_6$, Bis(dimethylamine) dimolybdenum hexamethylsiloxide, 44B, 1004
$C_{23}H_{18}AgN_3O_2$, Bis(8-hydroxyquinoline)silver(I) (pyridine solvate), 33B, 443
$C_{23}H_{18}ClCuN_8O_{10}$, Bis(10-methylisoalloxazine)copper(I) perchlorate

formic acid, 41B, 1125
$C_{23}H_{21}CuN_3O_6$, Bis(2-nitroacetophenonato)-(2,6-dimethylpyridine)copper(II), 40B, 986
$C_{23}H_{21}LaN_8O_{11}$, Tris-nitrato[2,6-diacetylpyridinebis(benzoic acid hydrazone)]lanthanum(III), 45B, 1271
$C_{23}H_{23}CoN_7O_9$, Aqua(nitrato)-2,6-diacetylpyridinebis(benzoic acid hydrazone)cobalt(II) nitrate, 45B, 1271
$C_{23}H_{25}N_7NiO_{10} \cdot 2\ H_2O$, Diaqua-2,6-diacetylpyridinebis(benzoic acid hydrazone)nickel(II) nitrate dihydrate, 45B, 1271
$C_{23}H_{30}FeN_5O_8$, Bis(methanol)bis[2-(5-methylpyrazol-3-yl)phenolato]-iron(III) nitrate methanol, 46B, 1185
$C_{23}H_{43}Ga_2N_7NiO$, (Dimethylbis(3,5-dimethyl-1-pyrazolyl)gallato)(dimethyl(3,5-dimethyl-1-pyrazolyl)dimethylethanolamino)gallato)nickel(II), 46B, 1185
$C_{24}H_{12}BaN_{18}O_{24}Ru_2 \cdot 9\ H_2O$, Barium tris(dihydrogenviolurato)ruthenate nonahydrate, 46B, 1186
$C_{24}H_{12}F_{24}N_8O_4Th$, Tetrakis(hexafluoroacetonylpyrazolide)thorium(IV), 42B, 857
$C_{24}H_{12}F_{24}N_8O_4U$, Tetrakis(hexafluoroacetonylpyrazolide)uranium(IV), 42B, 857
$C_{24}H_{18}CuN_2O_2$, N,N'-Ethylenebis(1-iminomethyl-2-naphtholato)copper-(II), 46B, 1110
$C_{24}H_{18}N_2NiO_2$, N,N'-Ethylenebis(1-iminomethyl-2-naphtholato)nickel-(II), 46B, 1110
$C_{24}H_{20}N_6O_2Pd$, Bis(3-hydroxyl-1,3-diphenyltriazabenzene)palladium, 28,
$C_{24}H_{22}ClCoN_5O_4 \cdot H_2O$, Aquachlorbis[N-(2-pyridylmethylene)aniline]cobalt(II) nitrate monohydrate, 45B, 1272
$C_{24}H_{22}CuN_4O_8$, Bis-(2-methoxy-4-nitrophenolato)bis(pyridine)copper-(II), 40B, 987
$C_{24}H_{22}N_4NiO_8$, Bis(2-methoxy-4-nitrophenolato)bis(pyridine)nickel-(II), 43B, 1372
$C_{24}H_{24}Cs_4N_4O_{28}Ti_4 \cdot 6\ H_2O$, Cesium tetra-$\mu$-oxo-tetrakis[(nitrilotriacetato)titanate(IV)] hexahydrate, 46B, 1187
$C_{24}H_{24}N_4O_4Rh_2$, Tetrakis[μ-(6-methyl-2-pyridinolato)]dirhodium, 46B, 1187
$C_{24}H_{24}N_4O_4Ru_2 \cdot CH_2Cl_2$, Tetrakis[$\mu$-(6-methyl-2-pyridinolato)]diruthenium dichloromethane solvate, 46B, 1188
$C_{24}H_{26}ClCu_2N_4O_6 \cdot 0.5\ CH_4O$, Copper(II)-copper(I)-pseudo-porphyrin-macrocyclic ligand complex, 46B, 1188
$C_{24}H_{26}Cu_2N_2O_6$, Bis(heptanetrionato(2-))bis(pyridine)dicopper(II), 40B, 987
$C_{24}H_{26}N_2NiO_7SU$, Nickel(II)(3,8-dimethyl-4,7-diazadeca-2,8-diene-1,10-dione-1,10-di(o-phenolato))uranyl dimethylsulfoxide, 44B, 1005
$C_{24}H_{28}Cr_2N_8O_4 \cdot C_6H_{14}O_3$, Tetrakis(2,4-dimethyl-6-hydroxypyrimidino)-dichromium(II) 2,2'-dimethoxydiethylether solvate (form A), 45B, 1273
$C_{24}H_{28}Cr_2N_8O_4 \cdot 0.5\ C_6H_{14}O_3$, Tetrakis(2,4-dimethyl-6-hydroxypyrimidino)dichromium(II) 2,2'-dimethyoxydiethylether solvate (form B), 45B, 1273
$C_{24}H_{28}CuN_8O_{14}P_2S \cdot H_2O$, (Thiamine pyrophosphate)(1,10-phenanthroline)aquacopper(II) dinitrate water, 46B, 1114
$C_{24}H_{28}Mo_2N_8O_4 \cdot C_6H_{14}O_3$, Tetrakis(2,4-dimethyl-6-hydroxypyrimidino)-dimolybdenum(II) 2,2'-dimethoxydiethylether solvate (form A), 45B, 1273
$C_{24}H_{28}N_8O_2S_4Zn_2$, Di-$\mu$-(N,N-diethylnicotinamide-O,N)-tetraisothiocyanatodizinc, 39B, 805

$C_{24}H_{28}N_8O_4W_2 \cdot C_6H_{14}O_3$, Tetrakis(2,4-dimethyl-6-hydroxypyrimidino)-ditungsten(II) 2,2'-dimethoxydiethylether solvate (form A), 45B, 1273
$C_{24}H_{28}N_8O_4W_2 \cdot 0.5\ C_6H_{14}O_3$, Tetrakis(2,4-dimethyl-6-hydroxypyrimidino)ditungsten(II) 2,2'-dimethoxydiethylether solvate (form B), 45B, 1273
$C_{24}H_{30}Cl_6Cu_4N_4O_2$, μ_4-Oxo-hexa-μ-chloro-tetrakis((2-methylpyridine)-copper(II)) hydrate, 35B, 775
$C_{24}H_{30}N_2O_2Pt_2$, Trimethyl-(8-quinolinolato)platinum(IV), 30B, 364
$C_{24}H_{32}CuN_2O_2$, o-Hydroxyacetophenone-isobutyliminecopper(II), 35B, 730
$C_{24}H_{36}ClN_4NaNi_2O_8$, N,N'-Ethylenebis(acetylacetoneiminato)nickel(II) sodium perchlorate, 43B, 1373
$C_{24}H_{38}Cl_2Cu_2N_4O_8$, (Tetra-Schiff base)dicopper(II) hexahydrate, 42B, 918
$C_{24}H_{40}Ag_2N_{18}O_6$, Bis(nitratobis(pentamethylenetetrazole)silver(I)), 38B, 999
$C_{24}H_{44}AgN_{13}O_{11}Pt_2 \cdot 5\ H_2O$, Bis(bis($\mu$-1-methylthyminato)-cis-diammineplatinum(II))silver nitrate pentahydrate, 46B, 1190
$C_{24}H_{44}Cl_8Cu_4N_4O_{12}$, Tetrakis[dichloroacetato-μ-(2-dimethylaminoethanolato)copper(II), 46B, 1190
$C_{24}H_{46}Co_2N_4O_{14}P_2$, Nitrato($\gamma$-piperidino β-ketophosphonate)cobalt(II) dimer, 43B, 1374
$C_{24}H_{48}Cl_4Cu_4N_{12}O_{20}$, Diperchloratotetrakis(2-(2-aminoethyl)imino-3-butanone oxime)tetracopper(II) diperchlorate, 43B, 1374
$C_{24}H_{54}Cl_2Cu_2N_4O_{10}$, Tetrakis(cyclohexylamine)di-μ-hydroxo-dicopper-(II) perchlorate, 45B, 1274
$C_{24}H_{55}CuN_{11}Na_2O_{26}P_2$, Sodium (bis(inosine 5'-monophosphate)(diethylenetriamine)copper(II)) decahydrate, 44B, 1006
$C_{24}H_{56}Br_4Cu_4N_4O_4 \cdot 4\ CCl_4$, Tetrakis[bromo(2-diethylaminoethanolato)-copper(II)] tetrachloromethane solvate, 46B, 1191
$C_{24}H_{56}Cl_4Cu_4N_4O_4$, Chloro(2-diethylaminoethanolato)copper(II), 39B, 805; 41B, 1107
$C_{24}H_{56}Cl_4N_4O_4Zn_4$, Chloro-(2-diethylaminoethanolato)zinc(II), 41B, 1108
$C_{24}H_{58}Cl_2Co_4N_6O_{20}$, Bis(diethylamine-2,2'-diol)tetrakis(diethylamine-2,2'-diolato)tetracobalt(II,III) perchlorate, 45B, 1274
$C_{24}H_{65.5}Cl_{1.5}Cu_3N_6O_{17}$, μ-Hyroxo(oxo)-tri-μ-(2-propylamino-2-methyl-3-butanoneoximato)-triaquo-tricopper(II) perchlorate tetrahydrate, 40B, 988
$C_{25}H_{23}N_5O_6U$, (2,6-Diacetylpyridine-bis(4-methoxybenzoylhydrazone))-dioxouranium, 46B, 1191
$C_{25}H_{25}Cu_2N_9O_{12}$, Bis(dinitratobis(pyridine)copper(II)) - pyridine, 37B, 605
$C_{25}H_{26}Cl_2Cr_2N_4O_4$, Tetrakis(6-methylpyridine-2-olato)dichromium dichloromethane, 44B, 1006
$C_{25}H_{26}Cl_2MoN_4O_4W$, Tetrakis(6-methylpyridine-2-olato)molybdenumtungsten dichloromethane, 44B, 1007
$C_{25}H_{26}Cl_2Mo_2N_4O_4$, Tetrakis(6-methylpyridine-2-olato)dimolybdenum dichloromethane, 44B, 1006
$C_{25}H_{26}Cl_2N_4O_4W_2$, Tetrakis(6-methylpyridine-2-olato)ditungsten dichloromethane, 44B, 1006
$C_{25}H_{38}B_2Mo_2N_8O_4S_2$, Diacetatobis(diethyldipyrazolylborato)dimolybdenum carbon disulfide solvate, 42B, 858
$C_{26}H_{16}ClCoN_6O_6S_2$, Bis(1-(2-thiazolylazo)-2-naphtholato)cobalt(III) perchlorate, 42B, 859
$C_{26}H_{16}Cu_2F_{24}N_2O_8$, Bis[bis(1,1,1,5,5,5-hexafluoro-2,4-pentanedion-

ato)copper(II)]1,4-diazabicyclo[2.2.2]-octane, 46B, 1157
$C_{26}H_{16}FeN_6O_2S_2 \cdot 1.5\ CHCl_3$, mer-Bis(1-(2-thiazolylazo)-2-naphtholato)iron(II) - chloroform (2:3), 42B, 860
$C_{26}H_{16}N_6NiO_2S_2$, Bis-(1-(2-thiazolylazo)-2-naphtholato)nickel(II), 40B, 989
$C_{26}H_{20}CuN_2O_2$, Bis-(2-picolyl-phenylketonato)copper(II), 42B, 918
$C_{26}H_{22}CuN_2O_2$, 2,2'-Biphenylbis-(2-iminoethylenephenolato)copper(II), 28, 645
$C_{26}H_{23}AsBr_4NORe$, Tetraphenylarsonium oxo-tetrabromo-acetonitrile-rhenate(V), 31B, 415
$C_{26}H_{24}CuN_2O_6$, Copper(II) di-1-nitroso-2-naphtholate acetone (1:2), 41B, 1129
$C_{26}H_{24}N_2NiO_5V$, (μ-((5,5'-(1,2-Ethanediyldinitrilo)bis(1-phenyl-1,3-hexanedionato))(4-)))(nickel)oxovanadium, 42B, 927
$C_{26}H_{26}CuN_2O_7$, Bis(phenoxyacetato)aquobis(pyridine)copper(II), 37B, 596
$C_{26}H_{26}CuN_2O_7$, Bis(4-formyl-2-methoxyphenolato)bis(pyridine)copper-(II) monohydrate, 39B, 806
$C_{26}H_{26}CuN_4O_8$, Bis(2-methoxy-4-nitrophenolato)bis(γ-picoline)copper-(II), 46B, 1104
$C_{26}H_{26}Cu_2N_6O$, μ-(N,N'-Bis(2-(2-pyridyl)ethyl)-2-hydroxy-5-methyl-isophthalaldiminato)-μ-pyrazolato-dicopper(I), 45B, 1275
$C_{26}H_{32}Hg_5N_{10}O_4$, Tetrahydrofuran mercury(II) cyanide, 37B, 609
$C_{26}H_{32}N_8NiO_{13}$, Aquo-(o-((2-pyridylmethylene)amino)benzamide)-(2-(2-pyridyl)-1,2,3,4-tetrahydroquinazolin-4-one)nickel(II) dinitrate tetrahydrate, 40B, 991
$C_{26}H_{34}Br_5Co_2N_4O_4$, Dibromo($\mu$-(11,23-dimethyl-3,7,15,19-tetraazatricyclo[19.3.1.1$^{9-13}$]hexacosadecaene-25,26-diolato(2-)))bis(methanol)dicobalt(1+) tribromide, 42B, 919
$C_{26}H_{34}Cl_2Cu_2N_8O_{10}$, Bis(2-(2-($\alpha$-pyridyl)ethyl)imino-3-butanone oxime)bis(acetonitrile)dicopper(II) diperchlorate, 43B, 1374
$C_{26}H_{34}N_4NiO_8 \cdot 7.4\ H_2O$, Bis(pyridoxylidene-DL-valinato)nickel(II) hydrate, 40B, 992
$C_{26}H_{34}N_4O_8Zn \cdot 8.1\ H_2O$, Bis(pyridoxylidene-L-valinato)zinc(II) hydrate, 40B, 992
$C_{26}H_{37}CuN_3O_6$, Bis(3-nitrocamphorato)(2-methylpyridine)copper(II), 43B, 1375
$C_{26}H_{56}Cu_4N_8O_{12} \cdot 5\ H_2O$, Tetrakis[{2-[(3-aminopropyl)amino]ethanolato}copper(II)] dimalonate pentahydrate, 45B, 1276
$C_{27}H_{18}MnN_3O_3 \cdot 0.5\ C_6H_{14}O$, Tris(8-quinolinolato)manganese(III) hexanol, 41B, 1130
$C_{27}H_{48}Cl_3N_4OTa$, Trichloro(N,N'-dicyclohexylacetamidinato)(N,N'-dicyclohexylureato)tantalum(V), 41B, 1132
$C_{28}H_{18}Cl_{6.7}CoI_{0.3}N_6O_2S_2$, 1-(2-Thiazolylazo)-2-naphthol-cobalt(III) halide chloroform solvate, 42B, 920
$C_{28}H_{20}Cl_2Cu_2F_{12}N_2O_6$, Bis(1,1,1,5,5,5-hexafluoro-2,4-pentanedionato-(N-ethyl-5-chloro-2-hydroxybenzylideniminato-μ-O)copper(II)), 42B, 863
$C_{28}H_{20}Cl_3N_3O_5U$, Dioxodi-8-quinolinato-8-quinolinoluranium(VI) chloroform solvate, 32B, 442
$C_{28}H_{21}NO_2PRh$, 8-Hydroxyquinolinatocarbonyltriphenylphosphinerhodium-(I), 37B, 609
$C_{28}H_{22}F_6N_2NiO_4S_2$, cis-Bis(1-(2-thienyl)-4,4,4-trifluoro-1,3-butanedionato)-bis(4-methylpyridine)nickel(II), 42B, 920
$C_{28}H_{22}MnN_3O_4$, Tris(8-quinolinolato)manganese(III) methanol, 41B, 1130
$C_{28}H_{22}N_4NiO_2$, Bis(benzoylbenzamidine)nickel(II), 44B, 1007

$C_{28}H_{26}CuN_4O_6$, Bis(ω-nitroacetophenonato)bis(4-methylpyridine)copper-(II), 38B, 1000
$C_{28}H_{28}Cu_2N_4O_8$, Bis(nitrato(N-n-propyl-2-hydroxybenzylideniminato-μ-O)copper(II)), 42B, 863
$C_{28}H_{31}Cl_2MoNO_3P_2$, Dichlorobis(ethyldiphenylphosphine oxide)(imino)-oxomolybdenum, 45B, 1277
$C_{28}H_{36}Cl_2N_4NiO_8$, Bisperchloratotetrakis-(3,5-dimethylpyridine)nickel(II), 33B, 447
$C_{28}H_{42}N_8NiO_6$, Bis-(5-ethyl-5-isoamylbarbiturato)bis(imidazole)nickel(II), 40B, 993
$C_{28}H_{44}Co_3N_4Na_2O_{26}$, Disodium hexaaquocobalt(II) (o-phenylenediamine-tetraacetato(4-))cobaltate(II), 44B, 1008
$C_{28}H_{52}N_2O_6W_2$, Hexaisopropoxybis(pyridine)ditungsten, 45B, 1277
$C_{28}H_{56}CaFe_2N_4O_{16}$, Calcium trans-1,2-diaminocyclohexane-N,N'-tetra-acetatoaquoferrate(III) octahydrate, 31B, 429
$C_{28}H_{56}Cu_4N_8O_8$, Tetrakis((2-diethylaminoethanolato)isocyanatocopper-(II)), 44B, 980
$C_{28}H_{58}Br_2Cu_2N_8O_4$, Bis(4,4,9,9-tetramethyl-5,8-diazadodecane-2,11-dione 2-oxime μ-11-oximato)copper(II) bromide, 38B, 1001
$C_{28}H_{60}Cr_2N_6O_8$, Bis(diethylamido)tetrakis(diethylcarbamato)-dichromium(III), 44B, 859
$C_{28}H_{62}Cr_2N_6O_8$, Bis(diethylamine)tetrakis(diethylcarbamato)-dichromium(II)(Cr-Cr), 44B, 859
$C_{28}H_{64}Cl_2Cu_2N_6O_{10}$, Bis(N,N-bis(2-(diethylamino)ethyl)-2-hydroxyethylamino-O)dicopper(II) diperchlorate, 46B, 1192
$C_{28}H_{64}Cl_2N_6Ni_2O_{10}$, N,N-Bis(2-diethylaminoethyl)-2-oxyethylaminonickel(II) perchlorate, 35B, 779
$C_{30}H_{18}N_4O_8Re_2$, Bis(o-phenylazophenolato(tricarbonyl)rhenium(I), 46B, 1192
$C_{30}H_{19}Cl_3NO_4PW$, Tricarbonylchloro(5,7-dichloro-8-quinolinato-N,O(1-))(triphenylphosphine)tungsten(II), 46B, 1193
$C_{30}H_{20}Cl_2Cu_2N_6O_2$, 1-(2-Pyridylazo)-2-naphtholatochlorocopper(II), 44B, 1009
$C_{30}H_{26}Cu_2F_{12}N_2O_6$, Bis(1,1,1,5,5,5-hexafluoro-2,4-pentanedionato(N-(2-methylethyl)-2-hydroxybenzylideniminato-μ-O)copper(II)), 42B, 863
$C_{30}H_{28}CuN_2O_2$, Bis(N-benzyl-o-hydroxyacetophenoneimine)copper(II), 39B, 797
$C_{30}H_{38}Cl_4Cu_3N_{20}O_{16}$, catena-Di-μ-perchlorato(di-μ-imidazolato-diperchlorato-octaimidazolo-tricopper(II)), 38B, 1002
$C_{30}H_{48}Cu_2Ga_4N_{12}O_2$, Copper(II) 3-methylpyrazolylgallate dimer, 42B, 920
$C_{30}H_{60}Cu_3N_{15}O_{36}P_3$, Octaaquo-tris(guanosine-5'-phosphato)tricopper-(II) tetrahydrate, 44B, 981
$C_{30}H_{61}Cl_2Cu_2N_7O_{12}$, μ-Carbonato-bis(2,4,4,7-tetramethyl-1,5,9-triaza-cyclododec-1-ene)dicopper(II) perchlorate - dimethylformamide, 44B, 1009
$C_{31}H_{35}ClCoF_6NOP_3$, Chloro(N,N-bis(2-diphenylphosphinoethyl)-2-methoxyethylamine)cobalt(II) hexafluorophosphate, 38B, 1002
$C_{32}H_{22}CuN_4O_2$, Bis(benzene azo-β-naphthol)copper(II), 26, 663
$C_{32}H_{30}Cu_2F_{12}N_2O_6$, Bis(1,1,1,5,5,5-hexafluoro-2,4-pentanedionato(N-(2,2-dimethylethyl)-2-hydroxybenzylideniminato-μ-O)copper(II)), 42B, 863
$C_{32}H_{31}Cl_2N_3O_3Zn$, N,N-Bis(2-hydroxy-5-chlorophenyl)phenylmethylene-4-azaheptane-1,7-diaminezinc(II) monohydrate, 42B, 922
$C_{32}H_{32}Cr_2N_4O_4$, Tetraacetanilidodichromium, 45B, 1278
$C_{32}H_{32}CuN_2O_2$, Bis(N-2-phenylethyl-o-hydroxyacetophenoneimine)copper-

(II), 39B, 797

$C_{32}H_{32}Mo_2N_4O_4 \cdot 2\ C_4H_8O$, Tetraacetanilidodimolybdenum tetrahydrofuran solvate, 45B, 1278

$C_{32}H_{34}Cl_2N_4Ni_2O_{10}$, Bis-μ-(5-chloro-2-hydroxy-N-methyl-α-phenylbenzylideneiminato-N,O)-bis(ethanol(nitrato-O,O')nickel(II)), 42B, 891

$C_{32}H_{34}Cl_4N_4NiO_4 \cdot 0.7\ C_2H_6O \cdot 1.3\ H_2O$, Bis(meso-stilbenediamine)nickel(II) dichloroacetate ethanolate hydrate, 29, 571

$C_{32}H_{42}Cl_4N_4NiO_8$, Bis(meso-stilbenediamine)nickel(II) dichloroacetate tetrahydrate, 29, 571

$C_{32}H_{44}CuN_4O_6$, Bis(2,2,5,5-tetramethyl-3-aminopyrrolidine-1-iminoxy-1-σ-vanillato)copper(II), 44B, 1011

$C_{32}H_{48}CuN_2O_2$, Bis(N-octyl-o-hydroxyacetophenoneimine)copper(II), 39B, 797

$C_{32}H_{60}N_4Ni_4O_{12}$, Tetrakis(4-methyl-5-azahept-2,4-diene-2,7-diolato-methanolatonickel(II)), 44B, 1011

$C_{32}H_{66}MoN_2O_4Si_4$, Bis(1-adamantylamido)tetrakis[(trimethylsilyl)oxo]-molybdenum, 46B, 1128

$C_{32}H_{72}Cl_4Cu_4N_4O_4$, N,N-Di-(n-propyl)-2-aminoethanolatocopper(II) chloride tetramer, 42B, 867

$C_{33}H_{30}CoN_3O_6$, Tris(1-phenyl-1,2-propanedione-2-oximato)cobalt(III) benzene solvate, 44B, 1012

$C_{34}H_{20}F_9N_2NdO_6S_3$, Tris(thenoyltrifluoroacetonato)-2,2'-dipyridyl-neodymium(III), 42B, 923

$C_{34}H_{22}Cu_2N_6O_8 \cdot 4\ H_2O$, Bis(2,2'-bipyridyl)-bis(pyridine-2,6-dicarboxylato)dicopper(II) tetrahydrate, 46B, 1173

$C_{34}H_{26}N_4NiO_2$, Bis-(1-m-tolylazo-2-naphtholato)nickel(II), 33B, 450

$C_{34}H_{30}F_6N_4Ni_2O_6$, Bis(1,1,1-trifluoroheptane-2,4,6-trionato)tetrakis-(pyridine)dinickel(II), 46B, 1151

$C_{34}H_{34}Cl_2CuN_2O_2$, Bis(N-isobutyl-(5-chloro-α-phenyl-2-hydroxybenzylidene)aminato)copper(II) (green isomer), 43B, 1183

$C_{34}H_{34}Cl_2CuN_2O_2$, Bis(N-n-butyl-(5-chloro-α-phenyl-2-hydroxybenzylidene)aminato)copper(II), 43B, 1183

$C_{34}H_{34}Cl_2Cu_2N_2O_2$, Bis(N-isobutyl-(5-chloro-α-phenyl-2-hydroxybenzylidene)aminato)copper(II) (red isomer), 43B, 1183

$C_{34}H_{35}CuN_3O_2$, N,N'-Bis((2-hydroxy-5-methylphenyl)phenylmethylene)-4-azaheptane-1,7-diaminato(2-)-copper(II), 41B, 1137

$C_{34}H_{35}N_3NiO_2$, N,N'-Bis((2-hydroxy-5-methylphenyl)phenylmethylene)-4-azaheptane-1,7-diaminato(2-)-nickel(II), 41B, 1137

$C_{34}H_{37}N_3O_3Zn$, N,N-Bis(2-hydroxy-5-methylphenyl)phenylmethylene-4-azaheptane-1,7-diaminezinc(II) monohydrate, 42B, 922

$C_{34}H_{48}N_6Ni_2O_4P_2S_4$, Bis(4-hydroxy-4-methyl-2-pentanone)nickel(II) tetrakisthiocyanato-bis(tris(2-cyano-ethyl)phosphine)nickelate(II), 45B, 1278

$C_{35}H_{27}N_3O_4PV$, Tetraphenylphosphonium dioxo[4-(2-pyridylazo)resorcinolato]vanadate(V), 45B, 1279

$C_{36}H_{22}Cu_2F_{12}N_2O_6$, Bis(1,1,1,5,5,5-hexafluoro-2,4-pentanedionato(N-phenyl-2-hydroxybenzylideniminato-μ-O)copper(II)), 42B, 863

$C_{36}H_{24}N_2O_4Os$, Osmium tetroxide - 9-methylbenzanthracene - bis(pyridine) adduct (toluene solvate), 43B, 1378

$C_{36}H_{26}F_9N_2NdO_7S_3$, Tris(thenoyltrifluoroacetonato)-1,2-di(4-pyridyl)-ethaneaquoneodymium(III), 42B, 924

$C_{36}H_{30}Cu_2N_8O_{11}$, μ-Acetato-(O,O')-μ-hydroxo-μ-phenyldinitromethanato-(O,O')-phenyldinitromethanato-bis(2,2'-bipyridinecopper(II)), 42B, 925

$C_{36}H_{32}Cu_2F_6N_2O_6S_2$, Bis(4,4,4-trifluoro-1-(2-thienyl)-1,3-butanedionato(N-(2-methylethyl)-2-hydroxybenzylideniminato-μ-O)copper(II)), 42B, 863

$C_{36}H_{34}N_{10}O_2W_2 \cdot 2\ C_4H_8O$, Bis(2,4-dimethyl-6-oxopyrimidine)bis(1,3-diphenyltriazino)ditungsten(II)(W-W) bis(tetrahydrofuran) solvate, 46B, 1194
$C_{36}H_{44}F_{24}N_4Ni_2O_4$, Iminoalkoxy nickel(II) complex, 44B, 1013
$C_{36}H_{54}Co_2I_4N_{10}O_2 \cdot 3\ H_2O$, μ-Peroxo-bis{[1,11-bis(2-pyridyl)-2,6,10-triazaundecane]cobalt(III)} tetraiodide trihydrate, 45B, 1279
$C_{36}H_{56}CuN_2O_2$, Bis(N-n-decyl-o-hydroxyacetophenoniminato)copper(II), 42B, 926
$C_{36}H_{72}Cu_4N_8O_8$, Isocyanato(2-dipropylaminoethanolato)copper(II) tetramer, 44B, 999
$C_{38}H_{16}CuN_{10}O_2$, Bis-8-hydroxyquinolinatocopper(II)-1,2,4,5-tetracyanobenzene, 33B, 451
$C_{38}H_{23}Cl_4N_2O_4PW \cdot CH_2Cl_2$, Dicarbonylbis(5,7-dichloro-8-quinolinato-N,O(1-))(triphenylphosphine)tungsten(II) dichloromethane solvate, 46B, 1193
$C_{38}H_{42}Cl_2Fe_2N_6O_{22} \cdot 2.6\ H_2O$, Bis{{α,3-dihydroxy-β-[[[3-hydroxy-5-(hydroxymethyl)-2-methyl-4-pyridyl]methylene]amino]-5-(hydroxymethyl)-α,2-dimethyl-4-pyridinepropanoato(2-)}iron(III)} perchlorate hydrate, 46B, 1195
$C_{38}H_{48}B_2Cu_4F_{14}N_6O_3Si \cdot 0.84\ H_2O$, Aqua-(2,6-bis(N-(2-(4-imidazolyl)-ethyl)iminomethyl)-4-methylphenol)-(μ-2-hydroxo)dicopper(II) dimer tetrafluoroborate silicon hexafluoride hydrate, 46B, 1196
$C_{38}H_{48}CoN_2O_4 \cdot 0.5\ C_7H_8$, (3,5-Di-tert-butylcatecholato)(3,5-di-tert-butylsemiquinone)(bipyridyl)cobalt(III) toluene solvate, 46B, 1160
$C_{38}H_{54}Cu_2N_{10}O_{25}P_2$, (Uridine 5'-monophosphate)-2,2'-dipyridylamine-copper(II) pentahydrate, 44B, 1013
$C_{40}H_{36}CuN_6O_4S_2$, Bis(2-tosylamino-4'-p-tolyl-benzene-1-azoato)copper-(II), 46B, 1131
$C_{40}H_{40}N_8O_2W_2 \cdot 2\ C_4H_8O$, Bis(N,N'-diphenylacetamidinato)bis(2,4-dimethyl-6-hydroxypyrimidinato)ditungsten tetrahydrofuran solvate, 45B, 1280
$C_{40}H_{48}Cr_2N_4O_4 \cdot 2\ CH_2Cl_2$, Tetrakis(μ-N-2,6-xylyl-acetamidato)dichromium bis(methylene chloride) solvate, 46B, 1196
$C_{40}H_{48}Mo_2N_4O_4 \cdot 2\ CH_2Cl_2$, Tetrakis(μ-N-2,6-xylyl-acetamidato)dimolybdenum bis(methylene chloride) solvate, 46B, 1196
$C_{40}H_{50}CuN_4O_4$, Bis(N-(2,2,6,6-tetramethyl-piperidine-1-iminoxyl-4-yl)-naphthalene-1-aldimine-2-hydroxylato)copper(II), 41B, 1097
$C_{40}H_{56}Cu_4N_4O_8$, (7-Hydroxy-4-methyl-5-azahept-4-en-2-one)copper(II) tetramer dibenzene solvate (β-form), 42B, 926
$C_{40}H_{70}EuNO_6$, Tris-(2,2,6,6-tetramethylheptan-3,5-dionato)quinuclidineeuropium(III), 40B, 993
$C_{40}H_{88}Br_4Cu_4N_4O_4$, Bromo-(2-dibutylaminoethanolato)copper(II) (γ form), 41B, 1115
$C_{40}H_{88}Cl_4Cu_4N_4O_4$, Chloro-(2-dibutylaminoethanolato)copper(II), 41B, 1115
$C_{41}H_{32}Cu_2N_{10}O_3$, μ-Carbonatobis(1,3-bis(2-(4-methylpyridyl)imino)-isoindolinato)copper(II), 45B, 1281
$C_{41}H_{33}Cl_3Fe_2N_4O_5$, μ-Oxo-bis(bis-(2-methyl-8-hydroxyquinolinato)iron-(III)) - chloroform, 39B, 807
$C_{41}H_{36}N_2O_5P_2S_2U$, Acetone-bis(isothiocyanato)-dioxo-bis(triphenylphosphine oxide)uranium(VI), 45B, 1282
$C_{41}H_{38}Br_2CrN_9O_9$, Chromium(III) azo dye complex, 33B, 452
$C_{41}H_{39}N_5NiO_4Zn$, (μ-((5,5'-(1,2-Ethanediyldinitrilo)bis(1-phenyl-1,3-hexanedionato))(4-)))(nickel)pyridinezinc - dipyridine, 42B, 927
$C_{41}H_{39}N_5Ni_2O_4$, (μ-((5,5'-(1,2-Ethanediyldinitrilo)bis(1-phenyl-1,3-hexanedionato))(4-)))bispyridinedinickel - pyridine, 42B, 928
$C_{42}H_{26}Br_4N_4O_4W$, Tetrakis(5-bromo-8-quinolinolato)tungsten(IV) -

benzene, 37B, 612
$C_{42}H_{28}CoKN_6NiO_{14}$, Potassium tris(1,10-phenanthroline)nickel(II) tris(oxalato)cobaltate(III) dihydrate, 37B, 610
$C_{42}H_{38}Cl_4N_4ORe_2 \cdot 0.63\ CHCl_3$, Tetrahydrofuranbis(N,N'-diphenylbenzamidinato)bis(dichlororhenium(III)) chloroform solvate, 41B, 1094
$C_{42}H_{40}B_2F_8N_8O_4Ru_2$, μ-Oxalato-bis(tetrapyridineruthenium(II)) fluoroborate, 37B, 613
$C_{42}H_{46}N_2O_4Ti$, Bis(8-quinolato)bis(2,6-diisopropylphenoxo)titanium(IV), 39B, 669
$C_{42}H_{67}MoNO_4$, Tetrakis(1-adamantoxo)dimethylaminemolybdenum(IV), 46B, 1162
$C_{43}H_{35}N_3O_3P_2Pt$, [p-Nitrobenzoylhydrazido(2-)-N,O]bis(triphenylphosphine)platinum, 46B, 1163
$C_{44}H_{30}Cu_2N_2O_4$, β-Hydroxy-α-naphthaldehyde-(β-hydroxy-α-naphthylmethylimid)copper(II), 35B, 782
$C_{44}H_{40}Cu_2N_4O_6$, Bis(benzoylacetylacetonato)bispyridinedicopper(II) dipyridine solvate, 42B, 928
$C_{44}H_{44}Fe_2N_8O \cdot C_2H_3N$, μ-Oxo-bis[7,16-dihydro-6,8,16,17-tetramethyldibenzo[b,i][1,4,8,11]tetraazacyclotetradecinatoiron(III)] acetonitrile solvate, 45B, 1282
$C_{44}H_{50}N_2O_4Ti$, Bis(2-methyl-8-quinolato)bis(2,6-diisopropylphenoxo)-titanium(IV), 39B, 669
$C_{44}H_{56}Cr_2N_4O_5 \cdot C_7H_8$, Tetrakis(μ-N-2,6-xylyl-acetamidato)-tetrahydrofuran-dichromium toluene solvate, 46B, 1197
$C_{44}H_{60}Cl_4Cu_2N_4O_{12} \cdot 2\ H_2O$, Bis[(2,2'-bipyridyl-6,6'-diyl)(2,5,8,11,-14,17-hexaoxaoctadeca-1,18-diyl)(dichloro-copper(II)-N,N',O)] dihydrate, 46B, 1197
$C_{44}H_{60}Cr_2N_8O_5$, Tetrakis(μ-N-p-dimethylaniline-acetamidato)-tetrahydrofuran-dichromium, 46B, 1197
$C_{44}H_{88}Cu_4N_8O_4S_4$, Isothiocyanato(2-dibutylaminoethanolato)copper(II) tetramer, 43B, 1380
$C_{44}H_{88}Cu_4N_8O_8$, Isocyanato(2-dibutylaminoethanolato)copper(II) tetramer, 43B, 1380
$C_{48}H_{42}N_2O_2P_2Ru$, trans-Bis(6-methyl-2-pyridinolato)bis(triphenylphosphine)ruthenium(II), 46B, 1199
$C_{48}H_{46}Cl_4Cr_2N_8O_8$, Di-μ-hydroxobis(di(1,10-phenanthroline)-chromium(III)) chloride hexahydrate, 39B, 808
$C_{48}H_{64}Cr_2N_4O_6 \cdot C_4H_8O$, Tetrakis(μ-N-2,6-xylyl-acetamidato)-bis(tetrahydrofuran)-dichromium tetrahydrofuran solvate, 46B, 1197
$C_{50}H_{58}Cr_2N_6O_4 \cdot C_5H_5N$, Tetrakis(μ-N-2,6-xylyl-acetamido)-bis(pyridine) dichromium pyridine solvate, 46B, 1197
$C_{51}H_{42}Cl_{12}FeN_3NaO_{15}$, Ferroverdin (carbon tetrachloride - methanol solvate), 39B, 809
$C_{51}H_{42}FeN_3NaO_{14}$, Ferroverdin (acetone solvate), 39B, 809
$C_{52}H_{46}N_2O_3P_2Pt$, Bis(triphenylphosphine)(dibenzoylhydrazido)platinum ethanol solvate, 39B, 809
$C_{52}H_{52}Cu_2N_4O_8$, (N,N'-Bis(benzoylacetylacetonato)ethylenediamine)copper(II) dimer, 42B, 928
$C_{52}H_{76}Cu_3N_2O_{12}$, Bis(μ-(benzoato-O,O'))bis(benzoato)bis(μ-(2-dibutylaminoethanolato))bis(ethanol)tricopper(II), 46B, 1199
$C_{53}H_{40}As_2N_5NbOS_5$, Tetraphenylarsonium oxopentathiocyanatoniobate(V), 35B, 783
$C_{60}H_{87}Ag_3N_{12}Ni_3O_{12} \cdot 1.3\ CHCl_3$, Tris[bis(δ-camphorquinone dioximato)nickel-silver] chloroform, 46B, 1200
$C_{60}H_{87}Ag_3N_{12}O_{12}Pd_3 \cdot 1.1\ CHCl_3$, Tris[bis(δ-camphorquinone dioximato)palladium-silver] chloroform, 46B, 1200

$C_{62}H_{58}As_2N_2O_{16}U$, Bis(tetraphenylarsonium) bis(pyridine-2,6-dicarboxylato)uranyl hexahydrate, 40B, 995
$C_{66}H_{76}B_2Cl_2N_8Ni_2O_4$, μ-Chloranilato-bis(2,2',2"-triaminotriethylamine)dinickel bis(tetraphenylborate), 43B, 1160
$C_{72}H_{86}B_2Cl_2Cu_2N_6O_4$, μ-Chloranilato-bis(1,1,4,7,7-pentamethyldiethylenetriamine)dicopper(II) bis(tetraphenylborate), 43B, 1160
$C_{74}H_{64}N_8Ni_2O_6$, Bis(1,5-diphenyl-1,3,5-pentanetrionato)tetrapyridinedinickel(II)-tetrapyridine, 40B, 996
$C_{76}H_{60}N_4O_4P_4S_4U$, Tetrathiocyanatotetrakis(triphenylphosphine oxide)-uranium(IV), 45B, 1283
$C_{80}H_{66}Cu_4N_{20}O_4 \cdot 2.5\ C_6H_6 \cdot H_2O$, Tri-μ-hydroxy-(tri(1,3-bis(2-(4-methylpyridyl)-imino)-isoindolinato)-(1-(2-(4-methylpyridyl)-imino)-3-(2-(4-methylpyridyl-6-olato)-isoindolinato)))tetracopper-(@iI) benzene solvate monohydrate, 45B, 1281
$C_{100}H_{228}F_{10}Mo_{12}O_{24}$, Octakis(t-butoxy)tetrakis(μ-fluoro)tetramolybdenum octakis(t-butoxy)tris(μ-fluoro)-(μ-dimethylamido)tetramolybdenum (1:2), 45B, 1229

85. METAL COMPLEXES (SULFUR OR SELENIUM LIGAND)

CH_3BrHgS, Bromo-methylmercaptomercury(II), 45B, 1290
CH_3ClHgS, Methanethiolatomercury(II) chloride, 7, 217, 237; 45B, 1283
CH_5AgClN_3S, Monothiosemicarbazidesilver(I) chloride, 33B, 466
$CH_5Cl_2N_3SZn$, Thiosemicarbazidezinc chloride, 24, 290, 615
$C_2Cs_2MoO_4S_5$, Di-cesium bis(disulfido)-(monothio-oxalato)-oxomolybdate(VI), 45B, 1284
$C_2H_5CdCl_2NS$, Di-μ-chloro-bis(thioacetamide(chloro)cadmium(II)), 44B, 1014
$C_2H_6Cl_3KOPtS$, Potassium trichloro(dimethyl sulfoxide)platinate(II), 42B, 930
$C_2H_6HgS_2$, Mercury methylmercaptide, 29, 630
$C_2H_6N_4NiS_4$, Dimethyltetrathionitrosylnickel(II), 28, 661
$C_2H_8N_6NiS_2$, Bis(thiosemicarbazidato)nickel(II), 27, 837
$C_2H_9Cl_2NOPtS$, cis-Dichloro(ammine)(dimethyl sulfoxide)platinum(II)), 44B, 1014
$C_2H_{10}CuN_6O_4S_3$, Bis(thiosemicarbazide)copper(II) sulfate, 38B, 848
$C_2H_{10}CuN_8O_6S_2$, Bis(thiosemicarbazide)copper(II) nitrate, 38B, 849
$C_2H_{10}N_6NiO_4S_3$, Dithiosemicarbazidenickel(II) sulfate (β-form), 33B, 471
$C_2H_{10}N_8NiO_6S_2$, cis-Nickel(II) dithiosemicarbazide dinitrate, 38B, 847
$C_2H_{10}N_8NiO_6S_2$, trans-Nickel(II) dithiosemicarbazide dinitrate, 38B, 847
$C_2H_{12}CdCl_2N_8S_2$, Bis(thiocarbohydrazide-N,S)cadmium dichloride, 37B, 614
$C_2H_{14}N_8NiO_8S_2$, Diaquodithiosemicarbazidenickel(II) dinitrate, 33B, 472
$C_2H_{16}N_6NiO_7S_3$, Dithiosemicarbazidenickel sulfate trihydrate, 27, 663
$C_2H_{16}N_6O_7PtS_3$, Bis(thiosemicarbazide)platinum(II) sulfate trihydrate, 38B, 850
$C_2H_{21}F_{12}N_5OP_2RuS$, Dimethylsulfoxidepentammineruthenium(II) hexafluorophosphate(V), 37B, 614
$C_3H_6AgNO_3S_3$, 1,3,5-Trithian silver(I) nitrate, 33B, 473
$C_3H_6Cl_2HgS_3$, Dichloro(1,3,5-trithian)mercury(II), 31B, 448
$C_3H_6HgO_2S$, Acetato(methanethiolato)mercury(II), 44B, 1014

$C_3H_7N_5NiS_4$, N-Methyl-N,N-bis(methyleneamino-N'-(dithionitrite))amino-nickel(II), 35B, 786
$C_3H_8AgClO_5S_3$, 1,3,5-Trithian silver(I) perchlorate hydrate, 33B, 473
$C_3H_8AgNO_4S_3$, 1,3,5-Trithian silver(I) nitrate hydrate, 33B, 473
$C_3H_8Cl_2HgS$, (Ethyl methyl sulfide)dichloromercury(II), 46B, 1201
$C_4Cs_2MoO_5S_4 \cdot H_2O$, Dicesium bis(1,2-dithio-oxalato)-oxo-molybdate(IV) monohydrate, 45B, 1284
$C_4K_2NiO_4S_4$, Potassium bis(dithiooxalato)-nickelate(II) (black form), 45B, 1285
$C_4K_2NiO_4S_4$, Potassium bis(dithiooxalato)nickelate(II) (red form), 39B, 837
$C_4K_2O_4PtS_4$, Potassium bis(dithiooxalato)platinate(II), 3, 655, 694
$C_4Li_2NiO_4S_4 \cdot 2\ H_2O$, Lithium bis(dithioxalato)nickel(II) dihydrate, 46B, 1202
$C_4Na_2NiO_4S_4 \cdot 2\ H_2O$, Sodium bis(dithioxalato)nickel(II) dihydrate, 46B, 1202
$C_4H_5CuINOS_2$, 3-Methylrhodaninecopper(I) iodide, 38B, 1003
$C_4H_6S_4Pd \cdot CS_2$, Bis(dithioacetato)palladium(II) carbon disulfide solvate (monoclinic form I), 45B, 1285
$C_4H_6S_4Pd \cdot CS_2$, Bis(dithioacetato)palladium(II) carbon disulfide solvate (tetragonal form II), 45B, 1285
$C_4H_7CuN_2S$, N,N-Dimethylthioformamidecopper(I) cyanide, 43B, 951
$C_4H_8Cl_2HgS$, Tetrahydrothiophene mercury chloride, 29, 632
$C_4H_8CuN_2O_2S_2$, trans-Bis(N-methylformothiohydroxamato)copper(II), 44B, 1016
$C_4H_8Hg_2N_2O_6S_2$, 1,3-Dithiane-dimercury(I) dinitrate, 43B, 1382
$C_4H_8N_2NiO_2S_2$, Bis(thioglycinato)nickel(II), 38B, 1004
$C_4H_8N_2NiO_2S_2$, cis-Bis(thioacetohydroxamato)nickel(II), 33B, 480
$C_4H_9Cl_2N_3SZn$, Dichloro-(acetonethiosemicarbazone)zinc(II), 37B, 615
$C_4H_{10}Ag_3N_3O_3S$, Trisilver diethylsulphondiimine nitrate, 40B, 77
$C_4H_{10}ClCuS$, Copper(I) chloride diethylsulfide complex, 32B, 450
$C_4H_{10}ClCuS_2$, Chloro(2,5-dithiahexane)copper(I), 44B, 1016
$C_4H_{10}Cl_2CuO_2S$, Dichloro(β-thiodiglycol)copper(II), 45B, 1286
$C_4H_{10}Cl_2HgN_2S_2$, Bis(thioacetamide)mercury(II) chloride, 43B, 1382
$C_4H_{10}Cl_2N_2S_2Zn$, Dichlorobis(thioacetamide)zinc, 43B, 1383
$C_4H_{10}Cl_3S_3W$, Trichloro(2,5-dithiahexane)sulfidotungsten(V), 45B, 1286
$C_4H_{10}Cl_4Hg_2S$, Diethyl sulfide mercury chloride, 29, 632
$C_4H_{10}Fe_2N_4O_4S_2$, Roussin's red ethyl ester, 22, 579
$C_4H_{10}HgS_2$, Mercury ethylmercaptide, 5, 142, 151; 30B, 374
$C_4H_{10}N_4NiS_2$, Bis(thioacethydrazidato)nickel(II), 35B, 788
$C_4H_{10}NiO_2S_2$, Cyclohexakis(bis-(μ-1-hydroxyethane-2-thiolato)-nickel-(II)), 35B, 788
$C_4H_{12}Br_4Pd_2S_2$, Di-μ-bromo-dibromo-trans-bis(dimethyl sulfide)-dipalladium(II), 33B, 481
$C_4H_{12}Cl_2O_2PdS_2$, trans-Dichlorobis(dimethylsulfoxide)palladium(II), 32B, 450
$C_4H_{12}Cl_2O_2PtS_2$, cis-Dichlorobis(dimethylsulfoxide)platinum(II), 41B, 1144
$C_4H_{12}NiO_4P_2S_4$, Bis(dimethyldithiophosphato)nickel(II), 34B, 586
$C_4H_{12}NiO_8S$, Triaquo(3-thiapentanedioato)nickel(II) monohydrate, 44B, 1017
$C_4H_{12}NiP_2S_4$, Bis(dimethyldithiophosphinato)nickel(II), 34B, 587
$C_5H_5Cl_2CuN_4S$, catena-(Dichloro(8-mercaptopurine)-copper(I)), 42B, 930
$C_5H_6Cl_2F_6PtS_2$, cis-Dichloro(1,2-bis(trifluoromethylthio)propane)-platinum(II), 43B, 1384

$C_5H_7Cl_2CuN_4OS$, 6-Mercaptopurinecopper(I) chloride monohydrate dimer, 41B, 1144
$C_5H_7CoN_2O_2S_2$, Dinitrosyldithioacetylacetonatocobalt(-I), 42B, 931
$C_5H_{12}ClCuN_4OS_2$, Chloro(2,4-dithiobiuret)copper(I) N,N-dimethylformamide, 45B, 1287
$C_6H_8CaCoKO_{10}S_6$, Calcium potassium (+)-tris(dithiooxalato)cobaltate(III) tetrahydrate, 41B, 1145
$C_6H_8Cl_2CuN_4OS$, Dichloro-(6-thio-9-methylpurine)copper(II) monohydrate, 41B, 1146
$C_6H_9ClS_6Ti$, Chlorotris(dithioacetato)titanium(IV), 46B, 1203
$C_6H_{10}Br_2HgO_4S_2$, (3,6-Dithiolato-octanedioic acid-S,S')dibromomercury, 46B, 1203
$C_6H_{10}ClCuN_3OS_2$, Chloro(dimethylglyoxalmonooxime(methyldithiocarbonylhydrazone))copper(II), 44B, 1018
$C_6H_{10}N_4NiS_4$, Bis(thioacetamide)nickel(II) thiocyanate, 33B, 483
$C_6H_{11}AgS$, Silver(I) cyclohexanethiolate, 41B, 1147
$C_6H_{12}AgNO_3S_6$, 1,3,5-Trithian silver(I) nitrate, 33B, 476
$C_6H_{12}Cl_3N_2NbS_2$, Trichlorobis-(N-methylthioacetamido)niobium(V), 40B, 998
$C_6H_{12}FeN_3O_3S_3$, Tris(N-methylthioformohydroxamato)iron(III), 44B, 1018
$C_6H_{12}HgI_2N_2S_4$, Diiodo-N,N,N',N'-tetramethylthiuramdisulfidemercury-(II), 37B, 615
$C_6H_{14}Cu_2N_2O_8S_3$, catena-μ-Sulfato-(N,N'-bis(2-hydroxyethyl)dithiooxamido(2-))bis(aquocopper(II)), 44B, 1019
$C_6H_{14}N_4NiS_4$, Bis(ethyldithiocarbazate)nickel(II), 39B, 810
$C_6H_{14}N_6NiO_2S_4$, Bis(2,4-dithiobiureto)nickel(II) - glycol, 38B, 1005
$C_6H_{15}Cl_3Cu_3N_3S_3$, Cyclo-tri-μ-thioacetamide-tris(chlorocopper(I)), 43B, 1384
$C_6H_{16}ClCoNS_6 \cdot H_2O$, (Methyl dithiocarbazate-N^3,S)bis[methyl dithiocarbazato(1-)-N^3,S]cobalt(III) chloride monohydrate, 45B, 1287
$C_6H_{16}Cl_2CoN_4S_4 \cdot 0.5\ C_3H_8N_2S_2$, Dichlorobis(methyl 2-methyldithiocarbazato-N^3,S)cobalt(II) methyl 2-methyldithiocarbazate solvate, 46B, 1204
$C_6H_{16}Cl_2N_6NiO_9S_4$, Bis-(2,4-dithiobiureto)nickel(II) diperchlorate - ethanol, 39B, 811
$C_6H_{16}NiP_2S_4$, Bis(ethylmethyldithiophosphinato)nickel(II), 46B, 1205
$C_6H_{18}Cl_2CuN_2O_8S_2$, Bis(β-methylmercaptoethylamine)copper(II) diperchlorate, 42B, 931
$C_6H_{18}CoO_6P_3S_6$, Tris-(O,O'-dimethyldithiophosphato)cobalt(III), 38B, 1006
$C_6H_{18}N_2NiP_2S_4$, Bis[dimethylamido-(methyl)-dithiophosphonate]nickel-(II), 46B, 1205
$C_6H_{22}Cl_4CoN_5O_7SSn$, (Benzenesulfinato-S)pentaamminecobalt(III) trichloro(perchlorato)stannate(II) monohydrate, 44B, 1019
$C_7H_2Fe_2O_6S_2$, μ,μ'-Dithiolato-methanehexacarbonyldiiron(I), 45B, 1288
$C_7H_6F_8Fe_2N_2O_5P_4$, Di(methylaminobis(difluorophosphine))diiron pentacarbonyl, 43B, 1420
$C_7H_7N_5NiOS_4$, Nickel(II) S-phenylisocyanato aminodithionitride S-aminodithionitride, 31B, 451
$C_7H_{11}Cl_2NOPtS$, trans-Dichloro(dimethyl sulfoxide)(pyridine)platinum-(II), 46B, 1205
$C_7H_{18}Cl_2CoN_3O_2S_2$, (1,9-Diamino-3,7-dithianonane)chloronitrocobalt-(III) chloride, 39B, 812
$C_7H_{24}Cl_2CoN_5O_{11}S$, (p-Toluenesulfinato-S)pentaamminecobalt(III) perchlorate monohydrate, 44B, 1020
$C_8H_4K_2NiO_6S_4$, Potassium bis(dithiosquarato)nickelate dihydrate, 41B,

1147
$C_8H_5NO_5S_2W$, Thiazolidine-2-thionepentacarbonyltungsten, 40B, 998
$C_8H_6Br_2O_6Re_2S_2$, Di-μ-bromo-μ-(dimethyldisulfido)-bis(tricarbonylrhenium), 43B, 1385
$C_8H_6CrO_5S$, Pentacarbonyl(thioacetone)chromium, 43B, 1386
$C_8H_6F_8Fe_2N_2O_6P_4$, Methylaminobis(difluorophosphine)iron(0) tricarbonyl dimer, 43B, 1420
$C_8H_8FeO_4S_2$, 1,3-Dithianetetracarbonyliron, 41B, 1148
$C_8H_8Pd_2S_8$, Bis(ethylene-1,2-dithiolene)palladium dimer, 38B, 1007
$C_8H_8Pt_2S_8$, Bis(ethylene-1,2-dithiolene)platinum dimer, 38B, 1007
$C_8H_9CrO_5PS$, Pentacarbonyl(trimethylphosphinesulfide)chromium(0), 39B, 813
$C_8H_{10}HgN_2O_4S_4$, (3,6-Dithiolato-octanedioic acid-S,S')dithiocyanatomercury, 46B, 1203
$C_8H_{11}HgNO_2S$, Acetato(methanethiolato)pyridinemercury(II), 44B, 1014
$C_8H_{12}Cl_4Hg_2S_2$, Bis(mercury(II) chloride)-1,6-dithiacyclodeca-cis,cis-3,8-diene complex, 30B, 376
$C_8H_{12}Cl_6MoO_3S_2Sn$, 1,2-Bis(methylmercapto)ethane-tricarbonylmolybdenum-μ-chloro-trichlorotin methylene chloride solvate, 42B, 932
$C_8H_{12}FeO_2S_4$, Dicarbonyl(1,4,7,10-tetrathiadecane)iron, 46B, 1206
$C_8H_{12}MoS_8$, Tetrakis(dithioacetato)molybdenum(IV), 44B, 1020
$C_8H_{12}N_6NiS_4$, Bis(2-thioimidazolidine)nickel thiocyanate, 31B, 402
$C_8H_{12}Ni_3S_9$, Tris-μ-(dithioacetato)-$μ_3$-(trithio-o-acetato)-triangulo-trinickel(II) (form I), 41B, 1148
$C_8H_{12}Ni_3S_9$, Tris-μ-(dithioacetato)-$μ_3$-(trithio-o-acetato)-triangulo-trinickel(II) (form II), 41B, 1148
$C_8H_{12}Pt_2S_8$, Tetrakis(dithioacetato)diplatinum(II)(Pt-Pt), 46B, 1206
$C_8H_{12}S_8V$, Tetrakis(dithioacetato)vanadium(IV), 38B, 1008
$C_8H_{13}Cl_2NOPtS$, cis-Dichloro(dimethyl sulfoxide)(2-picoline)platinum-(II), 43B, 1386
$C_8H_{13}Cl_2NOPtS$, trans-Dichloro(dimethylsulfoxide)(2-picoline)platinum(II), 44B, 1021
$C_8H_{13}Cl_3O_3S_2SnW$, μ-Chloro-(dichloromethyltin)-2,5-dithiahexanetricarbonyltungsten, 34B, 585
$C_8H_{14}Cl_2N_2O_2PdS_2$, Dichlorobis(thiomorpholin-3-one)palladium(II), 44B, 1021
$C_8H_{14}CuN_6OS_2$, (2-Keto-3-ethoxybutyraldehyde bis(thiosemicarbazone))-copper(II), 31B, 454
$C_8H_{14}K_2NiO_{11}S_2$, Potassium bis-thio(diacetato)nickelate(II) trihydrate, 42B, 932
$C_8H_{14}N_2NiS_2$, Biacetylbis(mercaptoethylimine)nickel(II), 30B, 380
$C_8H_{14}PdS_4$, Bis(S-ethylethene-1,2-dithiolato)palladium(II), 45B, 1289
$C_8H_{14}PtS_4$, Bis(S-ethylethene-1,2-dithiolato)platinum(II), 45B, 1289
$C_8H_{16}Br_2O_2PtS_2$, trans-Dibromobis(1,4-oxathian)platinum(II), 43B, 1387
$C_8H_{16}Br_2O_2PtSe_2$, trans-Dibromobis(1,4-oxaselenan)platinum(II), 43B, 1387
$C_8H_{16}Cl_2CuN_2O_2S$, Copper(II) chloride - bis-(N,N-dimethylacetamido)-thioether, 35B, 795
$C_8H_{16}Cl_2HgO_2S_2$, Dichlorobis(1,4-thioxane)mercury(II), 32B, 452
$C_8H_{16}Cl_2O_2PdS_2$, Dichlorobis(1,4-thioxane)palladium(II), 44B, 1022
$C_8H_{16}Ni_2S_6$, 2,2'-Dimercaptodiethylsulfide-nickel(II), 34B, 591
$C_8H_{18}Cl_2PdSe_2$, 1,2-Bis(isopropylseleno)ethanepalladium(II) dichloride, 35B, 831
$C_8H_{18}FeN_3OS_2$, (N,N'-Dimethyl-N,N'-bis(2-thioethyl)ethylenediamine)-nitrosyliron, 43B, 1125
$C_8H_{18}N_4NiS_4$, N,N'-Dimethyl-S-methyldithiocarbazatonickel(II), 45B,

1289
$C_8H_{20}B_2CuF_8S_4$, Bis(2,5-dithiahexane)copper(II) bis(tetrafluoroborate), 43B, 1387
$C_8H_{20}Br_2CuN_2S_3 \cdot H_2O$, Bromo(1,11-diamino-3,6,9-trithiaundecane-(N,N',S,S',S")copper(II) bromide monohydrate, 46B, 1207
$C_8H_{20}Br_2N_2NiS_3$, Bromo(1,11-diamino-3,6,9-trithiaundecane-(N,N',S,S',S")nickel(II) bromide, 46B, 1207
$C_8H_{20}Br_2N_4NiS_4$, Tetrakis(thioacetamide)nickel(II) bromide, 37B, 617
$C_8H_{20}Br_4Pt_2S_2$, Tetrabromo-μ-bis(diethyl sulfide)-diplatinum(II), 33B, 481
$C_8H_{20}Cd_2O_4S_4$, Cadmium(II) bisthioglycolate, 40B, 1000
$C_8H_{20}ClCuN_4S_4$, Tetrakis(thioacetamide)copper(I) chloride, 4, 255, 282; 27, 841
$C_8H_{20}Cl_2CoO_4S_2$, Dichlorobis(2,2'-thiodiethanol)cobalt(II), 46B, 1208
$C_8H_{20}Cl_2N_4NiS_4$, trans-Dichlorotetrakis(thioacetamide)nickel(II), 38B, 1009
$C_8H_{20}Cl_2N_6NiOS_2$, Chlorobis(acetonethiosemicarbazone)nickel(II) chloride hydrate, 34B, 585
$C_8H_{20}Cl_2PdSe_2$, trans-Dichloro-bis(diethylselenide)palladium(II), 35B, 798
$C_8H_{20}Cl_5Re_2S_4$, Pentachloro-bis(1,5-dithiahexane)dirhenium, 33B, 483
$C_8H_{20}CoN_2S_2$, Bis(1-amino-2-mercapto-2-methylpropane)cobalt(II), 43B, 1388
$C_8H_{20}N_2NiS_2$, Bis(N,N-dimethyl-β-mercaptoethylamine)nickel(II), 32B, 453
$C_8H_{20}N_2S_2Zn$, Bis(2-amino-1,1-dimethylethanethiolato-N,S)zinc, 44B, 1022
$C_8H_{20}NiO_4P_2S_4$, Bis(diethyldithiophosphato)nickel(II), 32B, 454
$C_8H_{20}NiO_4P_2Se_4$, Bis(diethyldiselenophosphato)nickel(II), 41B, 1149
$C_8H_{20}NiP_2S_4$, Bis(diethyldithiophosphinato)nickel(II), 34B, 592
$C_8H_{20}O_4P_2S_4Zn$, Zinc diethyldithiophosphate, 34B, 592
$C_8H_{20}P_2S_4Zn$, Zinc(II) diethyldithiophosphinate, 35B, 796
$C_8H_{24}Cl_2N_4Ni_3S_4$, Tetrakis(2-aminoethanethiol)trinickel(II) chloride, 35B, 789
$C_8H_{24}Cl_3CoN_4S_2 \cdot 2.5\ H_2O$, (-)-μ-fac-Bis(di(2-aminoethyl)sulfide)cobalt(III) chloride dihydrate, 44B, 1023
$C_8H_{24}FeN_2P_4S_4$, Bis(imidotetramethyldithiodiphosphino-S,S)iron(II), 37B, 618
$C_8H_{24}N_2NiP_4S_4$, Bis(imidotetramethyldithiodiphosphino-S,S)nickel(II), 37B, 619
$C_8H_{26}Cl_3NO_3RuS_3$, Dimethylammonium trichlorotris(dimethylsulfoxide)-ruthenate(II), 41B, 1150
$C_8H_{26}N_2NiO_2P_2S_4$, Ethylammonium bis(ethyldithiophosphonato)nickel-(II), 44B, 1023
$C_9H_6CrO_6S$, Pentacarbonyl(2,5-dihydrothiophene 1-oxide)chromium, 43B, 1389
$C_9H_7NO_5S_2W$, Thiomorpholin-3-thione-pentacarbonyltungsten, 41B, 1150
$C_9H_{12}CuN_4S_2$, N-(S-Methyldithiocarbazato)-o-aminobenzaldiminatoamminecopper(II), 44B, 1024
$C_9H_{12}N_4NiOS_2$, [2,3,4-Pentanetrione-2,4-bis(thioacethydrazonato)]nickel(II), 45B, 1292
$C_9H_{13}HgNO_2S$, Acetato(ethanethiolato)-γ-pyridinemercury(II), 45B, 1290
$C_9H_{13}HgNO_2S$, Acetato(methanethiolato)-γ-picolinemercury(II), 45B, 1290
$C_9H_{14}N_4NiS_2 \cdot C_2H_3N$, [2,4-Pentanedionebis(thioacethydrazonato)]nickel(II) acetonitrile solvate, 45B, 1292

$C_9H_{15}CoS_6$, Tris(s-methylethene-1,2-dithiolato)cobalt, 45B, 1292
$C_9H_{15}RhS_6$, Tris-(S-methyl-ethylene-1,2-dithiolato)rhodium, 41B, 1151
$C_9H_{21}FeN_4O_2S_2$, [N,N'-Dimethyl-N,N'-bis(2-mercaptoethyl)-1,3-propanediaminato(1-)-(S,S')]dinitrosyliron, 46B, 1209
$C_9H_{24}Cl_2N_6NiS_6 \cdot 3\ H_2O \cdot 0.5\ C_2H_6O$, Tris(methyl 2-methyldithiocarbazate)nickel(II) chloride hemiethanol trihydrate, 46B, 1210
$C_9H_{27}Cl_3Cu_3P_3S_3$, Cyclo-tri-μ-(trimethylphosphine sulfide)-tris(chlorocopper(I)), 39B, 813
$C_{10}F_6Mn_2O_8Se_2$, Bis-μ-(trifluoromethyl)seleno-bis(manganese tetracarbonyl), 38B, 1010
$C_{10}H_8Cl_2HgN_8S_2$, Dichlorobis(6-mercaptopurine)mercury(II), 42B, 933
$C_{10}H_8F_6NiO_2S_2$, Bis(1,1,1-trifluoro-4-thiolopent-3-en-2-one)nickel-(II), 40B, 1001
$C_{10}H_8N_2O_2S_2Zn \cdot CHCl_3$, Bis(N-oxopyridine-2-thionato)zinc(II) chloroform, 43B, 1389
$C_{10}H_8N_2O_2S_2Zn$, Bis(N-oxopyridine-2-thionato)zinc(II), 43B, 1389
$C_{10}H_{10}Cl_2CoN_2S_2$, Dichlorobis(pyridinium-2-thiolate)cobalt(II), 45B, 1293
$C_{10}H_{10}Fe_2O_6S_2$, Ethylthiotricarbonyliron dimer, 28, 654
$C_{10}H_{11}ClCuN_4OS$, Chloro(thiosemicarbazonato-pyruvic acid anilide)copper(II), 45B, 1294
$C_{10}H_{12}CrO_4S_4$, Tetracarbonyl[tetrakis(methylthio)ethene-S,S"]-chromium, 45B, 1294
$C_{10}H_{12}NiP_2S_6$, Bis[(2-thienyl)methyldithiophosphinato]nickel(II) (form A), 45B, 1295
$C_{10}H_{12}NiP_2S_6$, Bis[(2-thienyl)methyldithiophosphinato]nickel(II) (form B), 45B, 1295
$C_{10}H_{14}Cl_4Cu_2N_8O_2S_2$, Bis(dichloro-6-mercaptopuriniumcopper(I)) dihydrate, 42B, 933
$C_{10}H_{14}CoS_4$, Bis(pentane-2,4-dithione)cobalt(II), 40B, 1001
$C_{10}H_{14}CrO_4S_2$, Tetracarbonyl(3,6-dithiaoctane)chromium(0), 42B, 934
$C_{10}H_{14}NiO_2S_2$, Bis(monothioacetylacetonato)nickel(II), 40B, 1002
$C_{10}H_{14}NiS_4$, Bis(pentane-2,4-dithione)nickel(II), 40B, 1001
$C_{10}H_{16}Br_2CoN_4S_2$, Dibromobis(1,3-dimethylimidazoline-2-thione)cobalt-(II), 42B, 934
$C_{10}H_{17}CoN_8O_2S_2 \cdot 2\ H_2O$, Bis(diacetylmonoxime-thiosemicarbazonato)cobalt(III) dihydrate, 45B, 1296
$C_{10}H_{18}CuO_4S_2 \cdot 2\ H_2O$, Bis[(isopropylthio)acetato]copper(II) dihydrate, 46B, 1210
$C_{10}H_{18}N_4NiS_4$, Bis(S-methyl-N-isopropylidenedithiocarbazato)nickel-(II), 44B, 1024
$C_{10}H_{18}NiS_4$, Bis(dithiopivalato)nickel(II), 39B, 815
$C_{10}H_{20}B_2F_8NiS_4$, 1,4,8,11-Tetrathiacyclotetradecanenickel(II) tetrafluoroborate, 41B, 1152
$C_{10}H_{20}Br_2HgN_2S_4$, Dibromotetraethylthiuramdisulfidemercury(II), 44B, 1025
$C_{10}H_{20}ClCuN_4S_2$, Chloro-bis(N,N'-dimethylimidazolidine-2-thione)copper(I), 46B, 1211
$C_{10}H_{20}ClCuO_4S_4$, 1,4,8,11-Tetrathiacyclotetradecanecopper(I) perchlorate, 43B, 1390
$C_{10}H_{20}Cl_2CuO_8S_4$, (1,4,8,11-Tetrathiacyclotetradecane)copper(II) perchlorate, 42B, 935
$C_{10}H_{20}Cl_4Hg_2S_4$, μ-(1,4,8,11-Tetrathiacyclotetradecane)-bis(dichloromercury(II)), 44B, 1025
$C_{10}H_{22}Cl_2CuN_4O_8S_2$, (1,8-Diamino-3,6-dithiaoctane)(1-methylimidazole)copper(II) perchlorate, 44B, 1026
$C_{10}H_{22}Cl_2HgO_9S_4$, Aqua(1,4,8,11-tetrathiacyclotetradecane)mercury(II)

perchlorate, 44B, 1025
$C_{10}H_{22}Mo_2N_2O_3S_4$, μ-Oxo-bis(oxo(2,2'-methyliminodiethanethiolato)molybdenum(V)), 44B, 1026
$C_{10}H_{22}N_4NiO_7S_2$, Nitrato-(1-oxa-7,10-diaza-4,13-dithiacyclopentadecane)nickel(II) nitrate, 45B, 1296
$C_{10}H_{22}N_4O_7PdS_2$, 1-Oxa-7,10-dithia-4,13-diazacyclopentadecanepalladium(II) nitrate, 40B, 1003
$C_{10}H_{24}MoN_2O_2S_2$, Bis(1,1-dimethyl-2-N-methylamino-ethane-thiolato)-dioxomolybdenum, 46B, 1211
$C_{10}H_{24}N_4NiO_8S_2$, 1-Oxa-7,10-dithia-4,13-diazacyclopentadecanenickel(II) nitrate monohydrate, 40B, 1004
$C_{10}H_{24}NiO_2P_2S_4$, Bis-(O-isopropylethyldithiophosphonato)nickel(II), 45B, 1297
$C_{10}H_{27}Cl_8S_5W_2$, Tris[μ-(ethanethiolato)]-bis[dichloro(dimethyl sulfide)tungsten(III,IV)], 46B, 1212
$C_{11}H_2O_{10}Os_3S_2$, (μ-Hydrido)-(μ-dithioformato)-decacarbonyltriosmium, 46B, 1212
$C_{11}H_5O_4ReS_2$, Tetracarbonyl-dithiobenzoato-rhenium, 37B, 623
$C_{11}H_{15}ClCuN_5O_3S_2$, Bis(2-thiouracil)chlorocopper(I) - dimethylformamide, 42B, 760
$C_{11}H_{21}Cl_2N_3O_2PtS$, trans-Dichloro(bis(isopropyl)sulfoxide-S)(1-methylcytosine-N)platinum(II), 42B, 820
$C_{11}H_{22}AgN_3OS_3$, 1-Oxa-7,10-diaza-4,13-dithiacyclopentadecane-silver thiocyanate, 43B, 1391
$C_{11}H_{22}AgN_3OS_3$, 1-Oxa-7,10-dithia-4,13-diazacyclopentadecanesilver thiocyanate, 42B, 935
$C_{11}H_{22}ClNiPS_2$, Chloro(pentane-2,4-dithionato)(triethylphosphine)nickel(II), 46B, 1213
$C_{11}H_{22}Hg_2I_4N_2S_4$, catena(μ-(2,6-Diethylamino-1,3,5,7-tetrathiahepta-1,6-diene)di-μ-iodo-di(iodomercury(II))), 43B, 1392
$C_{12}F_{18}MoSe_6$, Tris-(cis-1,2-di(trifluoromethyl)ethylene-1,2-diselenato)molybdenum, 37B, 624
$C_{12}Mn_2O_{12}S_2$, μ-1,2-Dithiooxalato(S,S')-bis(pentacarbonylmanganese(II)), 45B, 1301
$C_{12}H_4O_{12}Ru_3S$, Hydrido(thioacetic acid)decacarbonyltriruthenium, 44B, 1027
$C_{12}H_5Fe_2NO_6S$, o-Aminothiophenol-bis(tricarbonyliron), 39B, 817
$C_{12}H_6O_{10}Os_3S$, Hydrido(ethylsulfido)decacarbonyltriosmium, 43B, 1355
$C_{12}H_8Fe_3O_9S$, μ_3-Isopropylsulfido-μ-hydrido-nonacarbonyltriiron, 41B, 1152
$C_{12}H_{12}CrO_6S$, Pentacarbonyl(2,2,4,4-tetramethyl-3-thietanone)-chromium(0), 45B, 1297
$C_{12}H_{12}Mn_2N_2O_6S_4$, Di-μ-N-methylimino(methylthio)methanethiolato-bis-(tricarbonylmanganese), 44B, 1028
$C_{12}H_{14}CdN_4OS_2$, (N,N-Diethylnicotinamide)cadmium dithiocyanate, 38B, 1011
$C_{12}H_{16}Cl_3N_2ORhS$, Trichloro(dimethylsulfoxide)bispyridinerhodium(III), 42B, 936
$C_{12}H_{18}F_6MnO_2PS$, (η^5-Methylcyclopentadienyl)dicarbonyl(dimethylethylsulphonium)manganese hexafluorophosphate, 44B, 1028
$C_{12}H_{18}N_2NiS_2$, (N,N'-Ethylenebis(thioacetylacetoneiminato))nickel(II), 46B, 1214
$C_{12}H_{18}N_2S_2Zn$, (N,N'-Ethylenebis(thioacetylacetoneiminato))zinc(II), 46B, 1214
$C_{12}H_{18}O_2S_4Zn$, Bis(O-ethylthioacetatothioacetato)zinc(II), 41B, 1153
$C_{12}H_{18}O_4S_2Zn$, Bis(ethyl 3-mercaptobut-2-enoato)zinc(II), 43B, 1392
$C_{12}H_{19}CoN_6O_2S_4 \cdot CH_4O$, Bis[diacetylmonoxime-(N'-methyldithiocarboxy-

lato)hydrazonato]cobalt(III) methanol solvate, 45B, 1298
$C_{12}H_{19}CuO_8S_4$, Copper(I) hydrogen 2,5-dithiahexane-1,6-dicarboxylate, 39B, 818
$C_{12}H_{20}O_6Rh_2S_6$, Tetra-μ-monothioacetato-bis(monothioacetic acid rhodium), 43B, 1233
$C_{12}H_{22}ClCoN_2O_8S_2$, S,S-6,9-Diaza-2,13-dithiatetradecane-5,10-dicarboxylatocobalt(III) perchlorate, 43B, 1136
$C_{12}H_{22}Cl_2O_4SU$, cis-Dichloro-(meso-bis(trans-2-hydroxycyclohexyl)-sulfide-O,O,S)dioxouranium(VI), 41B, 1154
$C_{12}H_{24}Br_6Nb_2S_3$, Di-μ-bromo-μ-tetrahydrothiophene-bis(dibromotetrahydrothiopheneniobium(III))(Nb-Nb), 44B, 1029
$C_{12}H_{24}Br_6S_3Ta_2$, Di-μ-bromo-μ-tetrahydrothiophene-bis(dibromotetrahydrothiophenetantalum(III))(Ta-Ta), 44B, 1029
$C_{12}H_{24}Cl_2Cu_2N_8S_4$, Dichloro-μ-imidazolidine-2-thionato-tris(imidazolidine-2-thionato)dicopper(I), 42B, 936
$C_{12}H_{24}Cl_2NiS_4$, Bis(1,5-dithiacyclooctane)nickel(II) chloride, 40B, 1004
$C_{12}H_{24}Cl_2O_4PdS_2$, cis-Dichloro-(4,7,13,16-tetraoxa-1,10-dithiacyclooctadecane)palladium(II), 40B, 1005
$C_{12}H_{24}CuN_9O_3S_4$, Tetrakis(imidazoline-2-thionato)copper(I) nitrate, 42B, 936
$C_{12}H_{24}Fe_2O_4P_2S_2$, syn-Di-μ-methylsulfido-bis(dicarbonyltrimethylphosphineiron), 43B, 1393
$C_{12}H_{24}Fe_2O_6P_2S_3 \cdot 0.5\ C_4H_{10}O$, (μ-Sulfur dioxide)-bis(μ-methylthio)-bis[(trimethylphosphine)dicarbonyliron(I)] hemi(diethyl ether) solvate, 46B, 1215
$C_{12}H_{24}Pd_3S_9$, 2,2'-Dimercaptodiethylsulfide palladium(II) trimer, 34B, 594
$C_{12}H_{26}CdCl_4N_2O_8S_2 \cdot 2\ H_2O$, catena-Bis[μ-(N-methylpiperidinium-4-thiolato)]cadmium(II) perchlorate dihydrate, 45B, 1299
$C_{12}H_{26}Cl_3LaO_{17}S_2$, Aquadiperchlorato(1,4,10,13-tetraoxa-7,16-dithiacyclo-octadecane)lanthanum(III) perchlorate, 46B, 1215
$C_{12}H_{28}AgNO_3P_2PtS_2$, Bis(diethyl-(2-mercaptoethyl)-phosphine)platinum-(II) silver(I) nitrate complex, 34B, 596
$C_{12}H_{28}Au_2O_4P_2S_4$, catena-Bis(μ-(O,O'-diisopropyldithiophosphato)-digold(I)(Au-Au), 38B, 1020
$C_{12}H_{28}BCuF_4S_4$, Bis(3,6-dithiaoctane)copper(I) tetrafluoroborate, 43B, 1387
$C_{12}H_{28}Br_2Cu_2N_2O_2S_2$, Bromo[2-(2-dimethylamino)ethylthio)ethanolato)-copper(II) dimer, 46B, 1216
$C_{12}H_{28}CdO_4P_2S_4$, Cadmium O,O-diisopropylphosphorodithioate, 40B, 1127
$C_{12}H_{28}Cl_3Mo_2NO_4S_2$, Tetraethylammonium trichloro-μ-(2-oxoethanethiolato-S,μ-O)-μ-(2-oxoethanethiolato-μ-S-μ-O)bis(oxomolybdate(V)), 45B, 1299
$C_{12}H_{28}Cl_6Mo_2S_4$, Di-μ-chloro-bis(dichloro(3,6-dithiaoctane)molybdenum), 44B, 1036
$C_{12}H_{28}Cu_4N_4O_{16}S_6$, Di-μ-aqua-bis[μ-[N,N'-bis(2-hydroxyethyl)dithiooxamidato(2-)-N,O,S:N',O',S']-bis-[aquacopper(II) sulfatocopper-(II)] (B form), 45B, 1300
$C_{12}H_{28}HgO_4P_2S_4$, Mercury(II) O,O'-diisopropylphosphorodithionate, 37B, 625
$C_{12}H_{28}Mo_2O_6P_2S_6$, Di-μ-thio-bis(oxo-O,O-diisopropyldithiophosphato)-molybdenum, 44B, 1029
$C_{12}H_{28}NiO_4P_2S_4$, Bis(O,O'-diisopropyldithiophosphato)nickel(II), 43B, 1394
$C_{12}H_{28}O_4P_2S_4Zn$, Zinc O,O-diisopropylphosphorodithioate, 40B, 1127
$C_{12}H_{30}CrO_6P_3S_6$, Tris(diethyldithiophosphate)chromium(III), 38B, 1013

$C_{12}H_{30}Cu_4I_4S_3$, catena-μ-(Diethylsulfide)-bis(diethylsulfide)tetra-μ-iodo-tetracopper(I), 41B, 1154
$C_{12}H_{30}O_6P_3S_6V$, Vanadium(III) tris-(O,O-diethyl phosphorodithioate), 35B, 799
$C_{12}H_{36}B_2F_8O_6RuS_6$, Hexakis(dimethyl sulfoxide)ruthenium(II) tetrafluoroborate, 44B, 1030
$C_{13}H_9AuCl_6S_2$, Thianthrenegold(III) chloride chloroform solvate, 44B, 1031
$C_{13}H_9Fe_3NO_9S_2$, μ_3-Thio-μ_3-(N-t-butylthioimino)-tris(tricarbonyliron), 45B, 1168
$C_{13}H_{10}Cl_2HgN_4S$, Anhydro-5-mercapto-2,3-diphenyltetrazolium hydroxide mercury(II) chloride, 39B, 819
$C_{13}H_{12}Br_2N_2SZn$, Dibromo-N-2-methylthiophenyl-2'-pyridylmethyleneiminezinc, 38B, 1014
$C_{13}H_{12}Cl_2CuN_2S$, Dichloro-N-2-methylthiophenyl-2'-pyridylmethyleneiminecopper(II), 37B, 626
$C_{13}H_{12}Fe_2O_7S$, Hexacarbonyl-[μ-(2,4-dimethyl-4-sulfido-2-penten-3-olato)-μ-O,μ-S]-diiron(Fe-Fe), 42B, 882; 45B, 1301
$C_{13}H_{25}F_3MoO_4S_6$, (Hydrosulfido)(oxo)(1,5,9,13-tetrathiacyclohexadecane)molybdenum(IV) trifluoromethylsulphonate, 44B, 1031
$C_{14}H_6Fe_4O_{12}S_3$, Di(methylthio-di-iron-hexacarbonyl) sulfide, 32B, 455
$C_{14}H_{10}F_{12}NiS_4$, Nickel bis(1,2-bis(trifluoromethyl)-1,2-ethenedithiolate) - 2,3-dimethylbutadiene adduct, 39B, 820
$C_{14}H_{10}NiS_4$, Bis(dithiobenzoato)nickel(II) trimer, 41B, 1155
$C_{14}H_{10}NiS_4$, Bis(dithiotropolonato)nickel(II), 37B, 627
$C_{14}H_{10}NiS_6$, Bis(trithioperoxybenzoato)nickel(II), 37B, 627
$C_{14}H_{10}PdS_4$, Bis(dithiobenzoato)palladium(II) trimer, 41B, 1155
$C_{14}H_{10}S_4Zn$, Bis(dithiobenzoato)zinc(II), 38B, 1014
$C_{14}H_{10}S_6Zn$, Bis(trithioperoxybenzoato)zinc(II), 37B, 627
$C_{14}H_{12}CrO_5S$, Pentacarbonyl(ethylbenzylthio)chromium(0), 42B, 937
$C_{14}H_{14}Au_2Cl_2S_2$, μ-(1,2-Bis(phenylthio)ethane)-bis(chlorogold(I)), 39B, 821
$C_{14}H_{14}Cl_2O_2PtS_2$, cis-[meso-1,2-Bis(phenylsulfinyl)ethane]dichloroplatinum(II), 42B, 938; 45B, 55
$C_{14}H_{14}Cl_2O_2PtS_2$, cis-[rac-1,2-Bis(phenylsulfinyl)ethane]dichloroplatinum(II), 42B, 938; 45B, 55
$C_{14}H_{14}CoN_9O_3S_2$, Bis(pyridinaldehydethiosemicarbazonato)cobalt(III) nitrate, 39B, 822
$C_{14}H_{14}O_4S_2Zn$, Bisaquobis(thiobenzoato)zinc(II), 42B, 787
$C_{14}H_{16}ClCuN_2O_4S_2$, (Bis(2-(2-pyridyl)ethyl)disulfide)copper(I) perchlorate, 40B, 1006
$C_{14}H_{16}N_2NiO_2S_2$, Bis(pyridine)-bis(thioacetato)nickel(II), 42B, 939
$C_{14}H_{20}F_6FeN_2S_6$, Bis(N,N-diethyldithiocarbamato)(cis-1,2-bis(trifluoromethyl)ethylene-1,2-dithiolato)iron, 37B, 629
$C_{14}H_{20}F_6Fe_2S_4Sb$, μ-Disulfido-di-μ-ethylsulfido-bis(cyclopentadienyliron) hexafluoroantimonate, 42B, 939
$C_{14}H_{20}Fe_2S_4$, Bis(μ-ethylsulfido)-μ-disulfido-bis(pentahapto-cyclopentadienyliron), 39B, 822
$C_{14}H_{20}O_6P_2Re_2S_4$, Bis(tricarbonyl(diethyldithiophosphinato)-rhenium(I)), 40B, 1007
$C_{14}H_{22}N_2S_2Zn$, N,N'-Tetramethylenebis(thioacetylacetoniminato)zinc, 41B, 1156
$C_{14}H_{22}NiOS$, Bis(thioisobutyrylacetonato)nickel(II), 37B, 629
$C_{14}H_{22}O_4S_2W$, (μ-Ethanebis(t-butylsulfide))tetracarbonyltungsten(0), 44B, 1162
$C_{14}H_{25}Co_3O_4S_5$, Carbonylpentakis(ethylthio)tricarbonyltricobalt, 33B, 488

$C_{14}H_{32}ClMo_2NO_5S_3$, Tetraethylammonium chloro-μ-(2-oxoethanethiolato-S,μ-O)-μ-(2-oxoethanethiolato-μ-S,μ-O)-(2-oxoethanethiolato-S,O)-bis(oxomolybdate(V)), 45B, 1299
$C_{14}H_{32}N_4Ni_2PdS_4$ · 0.77 Br · 1.23 CNS, Bis[(3,7-diazanonane-1,9-dithiolato-N,N',S,S'-nickel)-S,S']palladium(II) bromide thiocyanate, 46B, 1217
$C_{14}H_{36}Cl_2P_2Pt_2S_4$, cis-μ-2,2-Bis(methylthio)ethene-1,1-dithiolato-S,S'':S',S'''-bis[chloro(triethylphosphine)platinum(II)], 45B, 1333
$C_{15}H_7F_{12}NiS_4$, Tropylium bis(cis-(1,2-perfluoromethylene-1,2-dithiolato))nickel, 35B, 804
$C_{15}H_8F_{12}NiS_4$, Norbornadiene-bis-cis-(1,2-perfluoromethylethene-1,2-dithiolato)nickel, 35B, 807
$C_{15}H_{24}O_4S_2W$, (μ-Propanebis(t-butylsulfide))tetracarbonyltungsten(0), 44B, 1162
$C_{16}H_5NO_9Ru_3S_2$, Hydrido(2-thiobenzthiazole)nonacarbonyltriruthenium, 44B, 1027
$C_{16}H_{10}Ir_2O_4S_2$, Bis(μ-benzenethiolato)-bis[dicarbonyliridium(I)], 45B, 1302
$C_{16}H_{12}O_{12}Re_4S_4$, Tetrakis(tricarbonyl-$\mu_3$-methanethiolato-rhenium), 38B, 1015
$C_{16}H_{14}Br_5S_5Pt$, Dibenzotetrathiofulvalene pentabromo(dimethylsulfido)platinate, 46B, 1217
$C_{16}H_{14}ClCuO_2S_4$, Chloro(4-tolylxanthogen)copper(I) polymer, 44B, 1032
$C_{16}H_{15}ClN_4NiO_4S_2$, (16,17-Dihydro-15H-dibenzo[f,m](1,11,4,5,7,8)dithiatetraazacyclotetradecinato-N^5,N^9,S^{14},S^{18})-nickel(II) perchlorate, 42B, 940
$C_{16}H_{16}CdO_3S_2$, Bis(monothiobenzoato)cadmium(II) ethanolate, 44B, 1032
$C_{16}H_{16}Fe_3O_8S_2$, Octacarbonyldi(tetrahydrothiophene)triiron, 40B, 1008
$C_{16}H_{16}HgN_2O_2S_2$, Mercury dimethoxide-phenylisothiocyanate complex, 32B, 456
$C_{16}H_{16}MoO_4S_2Ti$, Di-μ-methylthio-(bis-π-cyclopentadienyltitanium)tetracarbonylmolybdenum, 37B, 630
$C_{16}H_{20}Cl_2CuN_2O_8S_2$, Perchlorato[1,8-bis(2-pyridyl)-3,6-dithiaoctane]-copper(II) perchlorate, 45B, 1302
$C_{16}H_{20}CuF_6N_2PS_2$, [1,8-Bis(2-pyridyl)-3,6-dithiaoctane]copper(I) hexafluorophosphate, 45B, 1302
$C_{16}H_{20}N_2NiO_2S_2$, Bis(β-picoline)bis(thioacetato)nickel(II) (form a), 43B, 1241
$C_{16}H_{20}N_2NiO_2S_2$, Bis(β-picoline)bis(thioacetato)nickel(II) (form b), 43B, 1240
$C_{16}H_{20}N_2NiO_2S_2$, Bis(γ-picoline)bis(thioacetato)nickel(II) (form a), 44B, 1033
$C_{16}H_{20}N_2NiO_2S_2$, Bis(γ-picoline)bis(thioacetato)nickel(II) (form b), 43B, 1241
$C_{16}H_{20}N_2NiO_4P_2S_4$, Bis(O,O'-dimethyldithiophosphato)(1,10-phenanthroline)nickel, 35B, 808
$C_{16}H_{20}N_5NiS_4$, Tetraethylammonium bis(1,2-dicyanovinylene-1,2-dithiolato)nickelate(III), 43B, 1411
$C_{16}H_{22}BrClN_4NiO_4S_2$, Bromobis(2-((2-pyridylmethyl)amino)ethyl)-disulfidenickel(II) perchlorate, 41B, 1156
$C_{16}H_{24}Cl_2CoN_8O_8S_4$, Tetrakis[1-methylimidazoline-2(3H)-thione]cobalt-(II) diperchlorate, 45B, 1303
$C_{16}H_{24}CoN_{10}O_6S_4$ · H_2O, Tetrakis(1-methylimidazoline-2-(3H)-thione-S)cobalt(II) dinitrate hydrate, 46B, 1219
$C_{16}H_{24}Co_2S_8$, Bis-(cis-1,2-bis(trifluoromethyl)ethylene-1,2-dithiolate)cobalt, 30B, 372
$C_{16}H_{26}ClCoN_4O_{10}S_2$, Bis(2-((2-pyridylmethyl)amino)ethylsulfinato)co-

balt(III) perchlorate dihydrate, 44B, 1033
$C_{16}H_{26}Cl_{10}Nb_2S_4$, Decachloro-μ-1,4,8,11-tetrathiacyclotetradecanediniobium(V) benzene solvate, 41B, 1157; 42B, 941
$C_{16}H_{28}Mo_2O_2S_8$, Tetrakis-μ-dithioacetato-bis(tetrahydrofuran-molybdenum), 44B, 1034
$C_{16}H_{36}BCo_2F_4N_5OS_4$, μ-Nitrosyl-bis-μ-(N,N'-dimethyl-bis(2-thioethyl)-ethylenediamine)-dicobalt tetrafluoroborate, 43B, 1148
$C_{16}H_{36}Cl_4Cu_4O_8S_4$, Chloro(2-(2'-ethanolthio)ethanolato)copper(II) tetramer, 46B, 1219
$C_{16}H_{36}Fe_2N_4S_4$, Bis(μ-(N,N'-dimethyl-N,N'-bis(β-mercaptoethyl)ethylenediamine))diiron(II), 40B, 1012
$C_{16}H_{36}Ir_2O_8P_2S_2$, Di-μ-(t-butylthiolato)-bis(carbonyltrimethylphosphiteiridium), 45B, 1337
$C_{16}H_{36}Mo_2N_2O_6S_3$, Piperidinium μ-oxo-μ-(2-oxoethanethiolato-S,O)bis-[oxo(2-oxoethanethiolato)molybdate(V)], 45B, 1304
$C_{16}H_{36}O_2P_2S_2Zn$, Zinc(II) di-n-butylphosphinothionate, 35B, 810
$C_{16}H_{38}Ir_2O_8P_2S_2$, Di-μ-(t-butylthiolato)-bis(carbonylhydridotrimethylphosphiteiridium), 45B, 1337
$C_{16}H_{40}B_2Cu_2F_8N_4S_4$, cyclo-Di-μ-(bis(2-(N,N-dimethylamino)ethyl)-disulfide)-dicopper(I) tetrafluoroborate, 40B, 1009
$C_{16}H_{40}Cl_2FeMoN_2S_4$, Bis(tetraethylammonium) disulfidomolybdenum-bis(μ-sulfido)dichloroiron, 46B, 1229
$C_{16}H_{40}Cl_4Mo_2S_4$, Tetrachlorotetrakis(diethyl sulfide)dimolybdenum-(II), 46B, 1220
$C_{16}H_{40}Cl_6Ir_2S_4$, Di-μ-chloro-tetrachlorotetrakis(diethyl sulfide)diiridium(III), 46B, 1221
$C_{16}H_{40}Cl_6Ir_2S_4$, μ-Chloro-pentachloro-μ-(diethyl sulfide)-tris(diethyl sulfide)diiridium(III), 46B, 1220
$C_{16}H_{44}Cd_4Cl_4N_4O_2S_4$, Cadmium(II) aminomercaptide tetramer, 44B, 1035
$C_{17}H_{14}ClO_8Os_3PS_2$, (μ-Hydrido)-($μ_3$-sulfido)-($μ_3$-thioformaldehyde)-octacarbonyl-chloro-(dimethylphenylphosphine)triosmium, 46B, 1221
$C_{17}H_{17}MoNO_2S$, (-)-($η^5$-Cyclopentadienyl)dicarbonyl-N-(S)-α-phenylethylthioacetamidemolybdenum, 43B, 1074
$C_{17}H_{21}N_7NiS_6$, [2,6-Bis(1-methyl-4-methylthio-5-thia-2,3-diazahex-1,3-dienyl)pyridine]dithiocyanatonickel(II), 45B, 1304
$C_{17}H_{24}CuN_2S_4$, [N,N'-Trimethylenebis(methyl 2-amino-1-cyclopentenedithiocarboxylato)]copper(II), 45B, 1305
$C_{17}H_{25}ClCuN_3O_4S_2$, {2,15-Dimethyl-7,10-dithia-3,14,20-triazabicyclo-[14.3.1]eicosa-1(20),2,14,16,18-pentaene}copper(I) perchlorate, 46B, 1222
$C_{17}H_{27}NNiP_2S_4$, Bis(diethyldithiophosphinato)(quinoline)nickel, 35B, 808
$C_{17}H_{39}Mo_2NO_6S_4$, Tripropylammonium μ-(2-oxoethanethiolato-S,μ-O)-μ-(2-oxoethanethiolato-μ-S,μ-O)-(2-oxoethanethiolato-S,O)(2-hydroxylethanethiolato)bis(oxomolybdate(V)), 45B, 1299
$C_{18}O_{16}Re_4S_6$, μ-Bis(tetracarbonylrhenium(I))-bis(tetracarbonyl(trithiocarbonato)rhenium(I)), 40A, 124; 40B, 1013
$C_{18}H_{10}Br_2O_6Re_2S_2$, Di-μ-bromo-μ-diphenyldisulfido-bis(fac-tricarbonyl)dirhenium(I), 42B, 941
$C_{18}H_{10}Br_2O_6Re_2Se_2$, Di-μ-bromo-μ-diphenyldiselenide-bis[fac-tricarbonylrhenium(I)], 45B, 1305
$C_{18}H_{10}Fe_2O_6S_2$, Phenylthioiron tricarbonyl dimer, 38B, 1016
$C_{18}H_{10}N_2NiNa_2O_6S_4 \cdot 2\,H_2O$, Sodium 5-sulfo-8-mercaptoquinolinenickelate dihydrate, 46B, 1223
$C_{18}H_{12}Br_2NiO_2S_2$, Bis(cis-1-mercapto-2-p-bromobenzoylethylene)nickel-(II), 40B, 1010
$C_{18}H_{12}ClMoN_2OS_2$, cis-Oxochlorobis(8-mercaptoquinolinato)molybdenum-

(V), 45B, 1306
$C_{18}H_{12}HgN_2S_2$, Bis(8-mercaptoquinolinato)mercury(II), 39B, 823
$C_{18}H_{12}MoN_2O_2S_2$, cis-Dioxobis(8-mercaptoquinolinato)molybdenum(VI), 45B, 1306
$C_{18}H_{12}MoS_6$, Tris(benzene-1,2-dithiolato)molybdenum(VI), 42B, 942
$C_{18}H_{12}N_2PdS_2$, Bis(8-mercaptoquinolinato)palladium, 39B, 824
$C_{18}H_{12}N_2PtS_2$, Bis(8-mercaptoquinolinato)platinum, 39B, 824
$C_{18}H_{13}O_9Os_3PS_2$, (μ-Hydrido)-(μ-dithioformato)-bis(tricarbonyl-osmium)-tricarbonyl(dimethylphenylphosphine)osmium, 46B, 1212
$C_{18}H_{15}Cl_2HgPSe$, Triphenylphosphine selenide mercury(II) chloride, 40B, 1128
$C_{18}H_{15}Cl_3MoOPS$, Trichloro-oxo(triphenylphosphine sulfide)molybdenum-(V), 44B, 1035
$C_{18}H_{15}MoN_3S_3$, Tris(2-aminobenzenethiolato(2-)-N,S)molybdenum(VI), 44B, 1036
$C_{18}H_{20}MoN_2O_4S_2$, Dioxobis(N-methyl-p-tolylthiohydroxamato(O,S))molybdenum(VI), 46B, 1223
$C_{18}H_{21}CoN_6S_3 \cdot C_3H_6O$, Tris(4,6-dimethylpyrimidine-2-thionato(cobalt-(III) acetone solvate, 45B, 1307
$C_{18}H_{24}N_2NiO_4P_2S_4$, Bis(O,O'-dimethyldithiophosphato)(2,9-dimethyl-1,10-phenanthroline)nickel, 35B, 808
$C_{18}H_{26}Cl_2N_4Ni_2O_8S_2$, Bis-$\mu$-(2-((2-pyridylethyl)amino)ethylthiolato)-dinickel(II) perchlorate, 43B, 1395
$C_{18}H_{26}FeN_7O_2S_4 \cdot 4\ H_2O$, γ-Picolinium bis[S-methyl-dimethyldithio-carbazate-monoximato]ferrate(III) tetrahydrate, 45B, 1307
$C_{18}H_{30}N_2NiO_4P_2S_4$, Bis(O,O'-diethyldithiophosphato)bispyridinenickel-(II), 32B, 457
$C_{18}H_{36}AuBrN_2S_4$, Bis(N,N-di-n-butyldithiocarbamato)gold(III) bromide, 35B, 806
$C_{18}H_{36}AuBr_2CuN_2S_4$, Bis(N,N-di-n-butylthiocarbamato)gold(III) dibromocuprate(I), 35B, 810
$C_{18}H_{36}Cl_2N_2O_4PdS_2$, cis-Dichloro-(1,10-diaza-4,7,13,16-tetraoxa-21,24-dithiabicyclo[8.8.8]hexacosane)palladium(II), 40B, 1010
$C_{18}H_{38}Cl_4Cu_2N_2S_4$, μ-(1,7,13,19-Tetrathia-4,16-diazacyclopentacos-ane)-bis(dichlorocopper(II)), 45B, 1308
$C_{18}H_{38}Cu_2N_{14}S_4$, μ-(1,7,13,19-Tetrathia-4,16-diazacyclopentacos-ane)di-μ-azido-bis(azidocopper(II)), 45B, 1308
$C_{18}H_{39}BrCoF_6NPS_3$, Bromotris(2-t-butylthioethyl)aminecobalt(II) hexafluorophosphate, 41B, 1158
$C_{18}H_{40}Fe_2N_4S_4$, Bis(μ-(N,N'-dimethyl-N,N'-bis(β-mercaptoethyl)-1,3-propanediamine))diiron(II), 40B, 1012
$C_{18}H_{41}Fe_2N_7O_3S_4$, [N,N'-Dimethyl-N,N'-bis(2-mercaptoethyl)-1,3-propanediaminato(1-)-(S,S')]dinitrosyliron-[N,N'-dimethyl-N,N'-bis(2-mercaptoethyl)-1,3-propanediaminato(2-)-(S,S',N,N')]nitrosyliron (1:1), 46B, 1209
$C_{18}H_{57}O_{18}P_6Rh_3$, Tri-$\mu$-hydrido-tri(bis(trimethylphosphite)rhodium), 43B, 1430
$C_{19}H_{10}FeO_4S$, Tetracarbonyl(diphenylcyclopropenethione)iron, 44B, 1162
$C_{19}H_{20}Co_6O_{11}S_5$, Tetrakis(ethylthio)undecacarbonylsulfidohexacobalt, 33B, 490
$C_{19}H_{21}ClN_4NiO_5S_2$, 7-(3-Hydroxypropyl)-16,17-dihydro-15H-dibenzo(f,m)(1,11,4,5,7,8)dithiatetraazacyclotetradecinato-N^5,N^9,S^{14},S^{18}-nickel(II) perchlorate, 42B, 942
$C_{19}H_{25}Cl_2IrO_3S_2$, Dichloro(benzylacetophenone)bis(dimethylsulfoxide)-iridium(III), 35B, 811
$C_{19}H_{42}F_6Fe_2N_5O_2PS_4$, μ-Nitrosyl-di-μ-(N,N'-dimethyl-N,N'-bis(β-mer-

captoethyl)ethylenediamine)diiron(II) hexafluorophosphate - acetone, 40B, 1014
$C_{20}H_8Mn_2N_2O_6S_4$, Hexacarbonyldi-μ-mercaptobenzothiazolato(μ-S,N)dimanganese, 43B, 1396
$C_{20}H_8N_2O_6Re_2S_4$, Di-μ-mercaptobenzothiazolato(N,S)-hexacarbonyldirhenium(I), 42B, 943
$C_{20}H_{10}Fe_2O_6S_2$, μ,μ'-(cis-Stilbene-α,β-dithiolato)-bis(tricarbonyliron), 32B, 458
$C_{20}H_{16}CdN_2S_2$, Bis(2-methyl-8-mercaptoquinolinato)cadmium, 45B, 1309
$C_{20}H_{16}CuN_2S_4$, Bis(5-methylthio-8-mercaptoquinolinato)copper(II), 45B, 1309
$C_{20}H_{16}Mo_2N_2O_6S$, μ-(2-Mercaptoethanolato-S,O)-μ-oxo-bis(oxo-8-hydroxyquinolinatomolybdenum(V)), 41B, 1159
$C_{20}H_{16}Mo_2N_4O_3S_4$, μ-Oxo-bis(bis(pyridine-2-thiolato)molybdenum), 44B, 1036
$C_{20}H_{16}N_2NiS_2$, Bis(2-methyl-8-mercaptoquinolinato)nickel(II), 46B, 1224
$C_{20}H_{16}N_2PtS_2$, Bis(2-methyl-8-mercaptoquinolinato)platinum(II), 43B, 1396
$C_{20}H_{19}I_2NiPS_2$, Diiodo(bis(o-methylthiophenyl)phenylphosphine)nickel-(II), 34B, 588
$C_{20}H_{20}Co_4F_6PS_4$, Tetra(π-cyclopentadienyl)tetrathiotetracobalt hexafluorophosphosphate, 39B, 824
$C_{20}H_{20}Co_4S_4$, Tetra(π-cyclopentadienyl)tetrathiotetracobalt, 39B, 824
$C_{20}H_{20}N_4NiS_2$, (Butanedione di(phenylacetthiohydrazonato))nickel(II), 43B, 1397
$C_{20}H_{22}N_4NiS_2$, Bis(acetophenone-thioacethydrazonato)nickel(II), 40B, 1014
$C_{20}H_{22}NiS_5$, Dithiocumato(trithioperoxycumato)nickel(II), 38B, 1018
$C_{20}H_{22}O_2PdS_4$, Bis(O-2,4,6-trimethylphenyldithiocarbonato)palladium-(II), 44B, 1040
$C_{20}H_{22}S_6Zn$, Bis(trithioperoxycumato)zinc(II), 38B, 1019
$C_{20}H_{25}Co_5O_{10}S_5$, Pentakis(ethylthio)decacarbonylpentacobalt, 33B, 490
$C_{20}H_{28}N_2NiO_4P_2S_4$, Bis-(O,O'-diethyl-dithiophosphato)-1,10-phenanthrolinenickel(II), 37B, 633
$C_{20}H_{28}O_2P_2Rh_2S_2$, cis-Di(μ-phenylthiolato)-dicarbonylbis(trimethylphosphine)dirhodium(I), 43B, 1398
$C_{20}H_{30}Cl_2P_2PdS_2$, trans-Bis(diethylphenylthiophosphoryl)dichloropalladium(II), 41B, 1160
$C_{20}H_{30}CuKO_9S_4$, Potassium bis(1,1-carboethoxy-2,2-ethylenedithiolato)cuprate(III) etherate, 42B, 950
$C_{20}H_{34}AuO_9PS$, (2,3,4,6-Tetra-O-acetyl-1-thio-β-D-glycopyranosato-S)-(triethylphosphine)gold, 46B, 1224
$C_{20}H_{35}AgO_8S_4$, Trihydrogen bis(4,7-dithiadecane-1,10-dicarboxylato)-silver(I), 44B, 1037
$C_{20}H_{40}Co_4O_4S_8$, Tetracarbonyloctakis(ethylthio)tetracobalt, 39B, 826
$C_{20}H_{40}NO_3S_4Tc$, Tetra(n-butyl)ammonium bis(thiomercaptoacetato)oxotechnetate(V), 44B, 1038
$C_{20}H_{42}Cl_2N_2P_2Pt_2S_2$, α-Dichloro-di-μ-thiocyanato-bis(tri-n-propylphosphine)platinum(II), 24, 624; 35B, 814
$C_{20}H_{42}Cl_2N_2P_2Pt_2S_2$, β-Dichloro-di-μ-thiocyanato-bis(tri-n-propylphosphine)platinum(II), 35B, 814
$C_{20}H_{44}Cl_4Cu_2S_4$, Di-μ-chlorobis[chloro(5,8-dithiadodecane)copper-(II)], 45B, 1310
$C_{20}H_{48}Mo_2N_2S_8$, Tetraethylammonium anti-di-μ-sulfido-bis(sulfido-1,2-dimercaptoethanatomolybdate(V)), 43B, 1398; 44B, 1038
$C_{20}H_{48}Mo_2N_2S_8$, Tetraethylammonium syn-di-μ-sulfido-bis(sulfido-1,2-

dimercaptoethanatomolybdate(V)), 43B, 1398; 44B, 1038

$C_{21}H_4O_{20}Os_6S_2$, μ-Dithiolatomethane-bis(decacarbonyl-hydrido-tri-osmium), 45B, 1310

$C_{21}H_{13}ClCuN_3O_4S_7 \cdot 2\ CHCl_3$, Mercaptothiazole[bis(S-mercaptothiazole)sulfur]copper(I) perchlorate chloroform solvate, 45B, 1311

$C_{21}H_{17}N_3O_2S_2Zn$, (2-(2-Benzothiazolyl)-6-(2-(2-thiolophenyl)-2-azaethyl)pyridine)acetatozinc(II), 43B, 1243

$C_{21}H_{18}MoN_6S_3 \cdot C_3H_7NO$, Tris(thiobenzoyldiazinido)molybdenum(VI) dimethylformamide solvate, 45B, 1311

$C_{21}H_{21}ClCuN_3S_3$, Tris(thiobenzamide)copper(I) chloride, 43B, 1399

$C_{21}H_{21}Cl_2NiO_4PS_3$, Chlorotris(o-methylthiophenyl)phosphinenickel(II) perchlorate, 34B, 607

$C_{21}H_{22}BCoF_5N_6S$, Pentafluorophenylthiolato(hydrotris(3,5-dimethyl-1-pyrazolyl)borato)cobalt(II), 45B, 1312

$C_{21}H_{24}CoN_6OS_3$, Tris(4,6-dimethylpyrimidine-2-thionato)cobalt(III) acetone solvate, 43B, 1400

$C_{21}H_{24}F_3NO_6S_2W$, Tetraethylammonium pentacarbonyl(1,1,1-trifluoro-4-(2-thienyl)-4-thioxo-butan-2-onato)tungsten(0), 44B, 1039

$C_{21}H_{30}CoN_3OS_6$, Tris(2-aminocyclopentene-1-dithiocarboxylato)cobalt-(III) acetone solvate, 44B, 1039

$C_{22}H_{20}N_8NiOS_2$, Bis(isoquinoline-1-carboxaldehyde thiosemicarbazanato)nickel monohydrate, 40B, 1129

$C_{22}H_{24}N_6NiO_2S_4$, Bis(quinoxaline-2,3-dithiolato)nickel(II) - bis(N,N-dimethylformamide), 43B, 1401

$C_{22}H_{26}N_4NiS_4$, Bis(methyl 3-(1-(p-tolyl)ethylidene)dithiocarbazato)-nickel(II), 44B, 1043

$C_{22}H_{26}NiO_2S_4$, Bis(O-4-t-butylphenyldithiocarbonato)nickel(II), 44B, 1040

$C_{22}H_{29}F_6Fe_2O_4P_3S_2$, μ-Hydrido-bis(methylthiolato-dicarbonyldimethylphenylphosphineiron) hexafluorophosphate, 43B, 1436

$C_{22}H_{30}N_6NiS_2$, Bis(N,N-diethylphenylazo-thioformamide)nickel(II), 42B, 943

$C_{22}H_{31}BrHgO_2S$, 7α-Thiolato-(17R)-spiro[androst-4-ene-17,2(3H)-furan]bromomercury(II), 46B, 1225

$C_{22}H_{36}Cl_4Cu_2N_4O_8S_4$, μ-Chloro-chlorobis(3,4-bis(2-aminoethylthio)-toluene)dicopper(II) perchlorate, 41B, 1160

$C_{22}H_{38}NiO_2S_2$, Bis(2,2,6,6-tetramethylheptane-5-thiolo-3-onato)nickel(II), 37B, 634

$C_{22}H_{38}O_2PdS_2$, Bis-(2,2,6,6-tetramethylheptane-5-thiolo-3-onato)palladium(II), 42B, 944

$C_{23}H_{40}CuN_4O_2S \cdot H_2O$, [rac-5,7,7,12,14,14-Hexamethyl-1,4,8,11-tetraazacyclotetradecane]-o-mercaptobenzoato-copper(II) hydrate, 45B, 1313

$C_{24}H_{15}Cl_3MoN_6S_3$, (p-Chlorothiobenzoyl-hydrazone-diyl)-(p-chlorothiobenzoyl-hydrazone-triyl)-(N'-isopropylidene-p-chlorothiobenzoyl-hydrazone)molybdenum, 45B, 1313

$C_{24}H_{16}Cl_2HgO_2S_2$, Dichlorobis(phenoxathiin)mercury, 34B, 589

$C_{24}H_{16}Cl_6PtS_2$, cis-Dichlorobis(4,4'-dichlorodiphenyl sulfide)platinum(II), 37B, 635

$C_{24}H_{18}Cl_2N_4NiO_4S_2$, Chloro(bis(2-((2-pyridylmethyl)imino)phenyl)-disulfide)nickel(II) perchlorate, 40B, 1015

$C_{24}H_{18}CoN_4S_4$, Bis(benzothiazole-2-thiolato)bis(pyridine)cobalt(II), 43B, 1401

$C_{24}H_{18}N_4NiS_2$, Nickel α-thiopicolinanilidate, 43B, 1402

$C_{24}H_{19}Cl_3MoN_6S_3$, (4-Chlorobenzenecarbothioic acid-(1-methylethylidene)hydrazidato)((4-chlorophenyl)thioxomethyl)azo-N,S-((4-chlorophenyl)thioxomethyl)diazene-N^2,S^1-molybdenum, 46B, 1228

$C_{24}H_{20}CdN_2O_2S_2$, Bis(pyridine)bis(thiobenzoato)cadmium(II), 44B, 1041
$C_{24}H_{20}Cl_5NO_{10}SW_2$, Tetraethylammonium μ-pentachlorothiophenolato-bis-[pentacarbonyltungstate(0)], 46B, 1226
$C_{24}H_{20}CoO_2P_2S_2$, Cobalt(II) diphenylphosphinothionate, 35B, 810
$C_{24}H_{20}N_2NiO_2S_2$, Di(monothiobenzoato)dipyridinenickel(II), 43B, 1245
$C_{24}H_{20}N_2O_2S_2Zn$, Bis(monothiobenzoato)-bis(pyridine)zinc(II), 45B, 1314
$C_{24}H_{20}NiP_2S_4$, Bis(diphenyldithiophosphinato)nickel(II), 33B, 495
$C_{24}H_{20}NiP_2Se_4$, Bis(diphenyldiselenophosphinato)nickel(II), 43B, 1403
$C_{24}H_{21}FeS_7$, Thio-p-toluoyl-disulfido-bis(dithio-p-toluato)iron(II), 34B, 599
$C_{24}H_{22}CoN_4S_4$, Bis(2-mercapto-5-ethyl-3-thenylidene-α-aminopyridine)-cobalt(II), 43B, 1403
$C_{24}H_{24}BrCuO_6S_3 \cdot C_3H_6O \cdot H_2O$, Bromo(2,3,7,8,12,13-hexamethoxy-10,15-dihydro-5H-5,10,15-trithiatribenzo[a,d,g]-cyclononene)copper(I) acetone water (1:1:1), 45B, 1315
$C_{24}H_{24}CoN_3O_6$, Tris(N-methylthiobenzohydroxamato)cobalt(III), 45B, 1316
$C_{24}H_{24}CrN_3O_6$, Tris(N-methylthiobenzohydroxamato)chromium(III), 45B, 1316
$C_{24}H_{24}FeN_3O_6$, Tris(N-methylthiobenzohydroxamato)iron(III), 45B, 1316
$C_{24}H_{24}MnN_3O_6$, Tris(N-methylthiobenzohydroxamato)manganese(III), 45B, 1316
$C_{24}H_{24}N_2NiS_4$, Bis(5-S-propyl-8-mercaptoquinolinato)nickel(II), 43B, 1404; 44B, 1041
$C_{24}H_{24}N_3O_{15}RhS_3 \cdot 3\ C_4H_9NO$, (2,3,7,8,12,13-Hexamethoxy-10,15-dihydro-5H-5,10,15-trithia-tribenzo[a,d,g]-cyclononene)trinitratorhodium(III) N,N-dimethylacetamide (1:3), 45B, 1316
$C_{24}H_{28}Mo_2N_8S_4 \cdot CH_2Cl_2$, Tetrakis(4,6-dimethyl-2-mercaptopyrimidino)-dimolybdenum(II) dichloromethane solvate, 45B, 1317
$C_{24}H_{30}CuN_4S_2$, Bis(N-cyclohexylthiopicolinamidato)copper(II), 40B, 1016
$C_{24}H_{36}CuN_5S_4$, Tetra-n-butylammonium bis(maleonitriledithiolato)copper(III), 29, 560
$C_{24}H_{36}FeNS_4$, Tetraethylammonium bis(o-xylyl-α,α'-dithiolato)-ferrate(III), 43B, 1405
$C_{24}H_{36}FeN_5S_4$, Tetra-n-butylammonium bis(maleonitriledithiolato)iron-(III), 32B, 459
$C_{24}H_{36}N_3NiS_2$, Tetraethylammonium diacetyldihydrobis(2-mercaptoanil)-nickelate, 35B, 817
$C_{24}H_{40}Cu_2I_2N_4S_6$, Di-μ-(dipentamethylenethiurammonosulfide)-bis[iodo-copper(I)], 45B, 1318
$C_{24}H_{42}P_4RuS_4$, cis-Bis(diethylphosphinodithioato)bis(dimethylphenyl-phosphine)ruthenium(II), 40B, 1016
$C_{24}H_{56}Cu_4O_8P_4S_8$, Copper(I) O,O'-diisopropylphosphorodithioate tetramer, 38B, 1019
$C_{24}H_{56}Mo_2O_{11}P_4S_8$, μ-Oxo-bis(bis-(O,O'-diisopropyldithiophosphato)-oxomolybdenum(V)), 41B, 1128
$C_{24}H_{56}NbO_8P_4S_8$, Tetrakis(O,O'-diisopropyldithiophosphato)niobium-(IV), 43B, 1406
$C_{24}H_{56}O_8P_4S_8Zr$, Tetrakis(O,O'-diisopropyldithiophosphato)-zirconium(IV), 43B, 1406
$C_{24}H_{60}Cl_4Cu_3O_{16}S_{12}$, Hexakis(2,5-dithiahexane)dicopper(I)copper(II) perchlorate, 46B, 1226
$C_{24}H_{62}Cd_8I_4O_{13}S_{12}$, Cadmium(II) thioglycolate iodide hydrate, 42B, 944
$C_{25}H_9F_{24}Fe_4O_6S_{11}$, Tri-μ-methylmercapto-hexacarbonyldiiron(II) tetra-

kis(cis-1,2-di(perfluoromethyl)ethylene-1,2-dithiolato)diiron, 39B, 828
$C_{25}H_{32}BF_2N_4O_2RhSe_2$, trans-Bis(benzeneselenido)(difluoro-3,3'-(trimethylenedinitrilo)bis(2-pentanoneoximato)borate)rhodium(III), 42B, 944
$C_{25}H_{49}N_5NiS_2$, Tetrabutylammonium [2,4-pentanedionebis(thioacethydrazonato)]nickelate(II), 45B, 1292
$C_{26}H_{15}AsF_{12}FeS_4$, Bis(1,1,1,4,4,4-hexafluorobut-2-en-2,3-thiolene)-(triphenylarsine)iron, 43B, 1441
$C_{26}H_{15}AuClF_{12}PS_4$, Chlorotriphenylphosphonium bis(cis-1,2-bis(trifluoromethyl)ethene-1,2-dithiolato)gold, 33B, 496
$C_{26}H_{18}N_4O_2RuS_2$, Dicarbonyl(dimercaptobenzothiazolato-S)dipyridineruthenium(II), 42B, 945
$C_{26}H_{20}CrO_4S_2W$, Bis-π-cyclopentadienyltungsten(IV)(bis-μ-benzenethiolato)chromium(0) tetracarbonyl, 40B, 1019
$C_{26}H_{20}MoO_4S_2W$, Bis-π-cyclopentadienyltungsten(IV)(bis-μ-benzenethiolato)molybdenum(0) tetracarbonyl, 40B, 1019
$C_{26}H_{20}Mo_2O_4S_2$, Di-μ-(phenylthio)-bis((dicarbonyl)(η-cyclopentadienyl)molybdenum(II)), 44B, 1041
$C_{26}H_{20}N_2NiS_2$, Bis(benzylidene-2'-mercaptoanilato)nickel, 42B, 946
$C_{26}H_{20}O_4S_2W_2$, Bis-π-cyclopentadienyltungsten(IV)(bis-μ-benzenethiolato)tungsten(0) tetracarbonyl, 40B, 1019
$C_{26}H_{22}CuN_8S_2$, Copper dithizonate, 26, 663
$C_{26}H_{22}N_4NiO_2S_2 \cdot C_6H_6$, Bis(2-mercapto-2'-methoxyazobenzene)nickel-(II) benzene solvate, 46B, 1227
$C_{26}H_{22}N_8NiS_2$, Bis(phenylazo)thioformic acid 2-phenylhydrazidato nickel, 37B, 636
$C_{26}H_{22}N_8S_2Zn$, Zinc(II) dithizonate, 38B, 1021
$C_{26}H_{24}N_2NiS_4$, Bis(phenyldithioacetato)bis(pyridine)nickel(II), 42B, 946
$C_{26}H_{24}N_6NiO_4S_4$, Bis(1-ethyl-4-carbomethoxypyridinium) bis(maleonitriledithiolato)nickelate(II), 39B, 829
$C_{26}H_{26}CuN_2O_4S_2$, (3,3'-Ethylenedithiobis(o-phenyleneiminomethylidyne)bis(pentan-2,4-dionato)(2-))copper(II), 44B, 1042
$C_{26}H_{26}N_2NiO_4S_2$, (3,3'-Ethylenedithiobis(o-phenyleneiminomethylidyne)bis(pentan-2,4-dionato)(2-))nickel(II), 44B, 1042
$C_{26}H_{28}N_2NiS_4$, Bis(N-n-butyl-3-mercaptobenz[b]thiophene-2-aldiminato)nickel(II), 43B, 1406
$C_{26}H_{28}N_2S_4Zn$, Bis(N-n-butyl-3-mercaptobenz[b]thiophene-2-aldiminato)zinc(II), 43B, 1406
$C_{26}H_{32}N_2Ni$, Bis(3-thio-2-N-n-butylazomethinebenzothiophene)nickel-(II), 42B, 947
$C_{26}H_{36}N_2S_6Zr$, Bis(tetramethylammonium) tris(benzene-1,2-dithiolato)-zirconate(IV), 42B, 947
$C_{26}H_{46}Cl_4Hg_4N_2S_4$, Di-μ-chloro-bis(chloropyridinemercury-di-μ-(t-butylthiolato)-mercury), 44B, 1014
$C_{26}H_{54}F_6Mo_2O_8S_{12}$, sym-Dihydrosulfido-bis(μ-1,5,9,13-tetrathiacyclohexadecane)-dimolybdenum(II) trifluoromethanesulphonate dihydrate, 44B, 1043
$C_{26}H_{56}Cl_4Pd_2S_4 \cdot C_7H_8$, Bis(μ-1,5-di-t-butylthiopentane-S,S')-bis[dichloropalladium(II)] toluene solvate, 46B, 1227
$C_{27}H_{15}CsO_9Re_2S_3$, Cesium tri-μ-thiobenzoato-bis(tricarbonylrhenate), 45B, 1318
$C_{27}H_{18}N_4NiS_4P$, Triphenylmethylphosphonium bis(1,2-dicyanoethylene-1,2-dithiolato)nickelate(III), 31B, 452
$C_{27}H_{27}AsNiO_2P_2S_2 \cdot 0.5\ C_6H_6$, (1-(Diphenylarsino)-2-(diphenylphosphino)ethane)(O-methylphosphoro-dithioato)nickel(II) - hemibenzene,

40B, 1020
$C_{27}H_{28}MoN_6S_3$, (p-Tolylcarbothioic acid-(1-methylethylidene)-hydrazidato)((p-tolylthioxomethyl)azo-N,S)((p-tolylthioxomethyl)-diazene-N^2,S^1)molybdenum, 46B, 1228
$C_{27}H_{38}BCuKN_7O_4S$, Potassium p-nitrobenzenethiolato(hydrotris(3,5-dimethyl-1-pyrazolyl)borato)cuprate(I) diacetone, 45B, 1319
$C_{28}H_{20}MoS_8$, Tetrakis(dithiobenzoato)molybdenum(IV), 41B, 1161
$C_{28}H_{20}NiS_4$, Bis(dithiobenzil)nickel, 32B, 460
$C_{28}H_{20}S_8V$, Tetrakis(dithiobenzoato)vanadium(IV), 38B, 1022
$C_{28}H_{26}N_4NiS_4$, Bis(methyl 3-(1-(2-naphthyl)ethylidene)dithiocarbazato)nickel(II), 44B, 1043
$C_{28}H_{26}N_4NiS_4$, Bis(methyl-3-diphenyldithiocarbazato)nickel(II), 46B, 1229
$C_{28}H_{27}N_9OS_2Pd$, Bis(6-mercapto-9-benzylpurine)palladium(II) - dimethylacetamide, 40B, 1020
$C_{28}H_{28}AsOS_4Tc$, Tetraphenylarsonium bis(1,2-ethanedithiolato)oxotechnetate(V), 44B, 1044
$C_{28}H_{28}ClCuOP_2S_2$, Chlorobis(diphenylphosphinothioyl)methanecopper(I) acetone solvate, 42B, 948
$C_{28}H_{28}N_4NiP_2S_4$, Bis(1-diphenylphosphinothioyl-3-methylthioureato)-nickel(II), 45B, 1319
$C_{28}H_{33}CrN_4Na_2O_3S_3 \cdot 0.5\ NaH_3O_2 \cdot 18\ H_2O$, Disodium triethylmethylammonium tris(thiobenzohydroximato)chromate(III) hemikis-(sodium hydroxide hydrate) octadecahydrate, 45B, 1320
$C_{28}H_{48}Cl_4Mo_2O_{11}P_4S_8$, μ-Oxo-bis(bis(diethyldithiophosphato)oxomolybdenum(V)) di(1,2-dichlorobenzene) solvate, 34B, 598
$C_{28}H_{48}N_2NiS_4$, Bis(tetraethylammonium) bis(cyclopentadienedithiocarboxylato)nickelate(II), 42B, 948
$C_{28}H_{50}Cl_4Hg_4N_2S_4$, Tetrachlorotetra(2-methylpropane-2-thiolato)-di(γ-picoline)tetramercury(II), 45B, 1290
$C_{28}H_{50}FeMoN_2S_6$, Bis(tetraethylammonium) disulfidomolybdenum-bis(μ-sulfido)bis(phenylmercaptide)iron, 46B, 1229
$C_{29}H_{23}MnO_3P_2S_2$, Tricarbonylmanganese(I)-((diphenylphosphinomethyl)-diphenylphosphonium)-dithiolatoorthoformate, 38B, 1023
$C_{29}H_{53}F_9Mo_2O_{12}S_{11} \cdot H_2O$, (Ethoxy-(1,5,9,13-tetrathiacyclohexadecane)molybdenum(IV))-μ-oxo-(oxy-(1,5,9,13-tetrathiacyclohexadecane)-molybdenum(IV)) trifluoromethanesulphonate monohydrate, 45B, 1321
$C_{30}H_{18}CoF_9O_3S_3$, Tris(1,1,1-trifluoro-4-phenyl-4-thiolo-but-3-en-2-one)cobalt(III), 42B, 949
$C_{30}H_{18}F_9FeO_3S_3$, Tris(1,1,1-trifluoro-4-phenylbutan-2-one-4-thionato)iron(III), 41B, 1165
$C_{30}H_{22}O_2PdS_2$, Palladium monothiodibenzoylmethanate, 39B, 831
$C_{30}H_{26}Ni_2O_5S_4$, Ethanolatotetra-μ-thiobenzoato-dinickel(II), 35B, 800
$C_{30}H_{32}Cl_2FeN_4O_9S_2$, 15,18-Dithia-1,5,8,12-tetraaza-3,4:9,10:13,14:-19,20-tetrabenzocycloeicosane-1,11-dieneiron(II) perchlorate methanolate, 35B, 823
$C_{30}H_{33}CoO_3S_6$, Tris(O-2,4,6-trimethylphenyldithiocarbonato)cobalt-(III), 44B, 1040
$C_{31}H_{29}FeO_3PS_4$, Bis(1,2-diphenyl-1,2-ethylenedithiolato)trimethoxyphosphineiron(IV), 43B, 1407
$C_{32}H_{28}Cl_2CuN_2O_2S$, (N,N'-Bis((5-chloro-2-hydroxyphenyl)phenylmethylene)-4-thiaheptane-1,7-diamino)copper(II), 43B, 1408
$C_{32}H_{28}Ni_2S_8$, Bis(phenyldithioacetato)nickel(II) dimer, 43B, 1408
$C_{32}H_{28}Pd_2S_8$, Bis(phenyldithioacetato)palladium(II) dimer, 43B, 1409
$C_{32}H_{36}CoN_7S_4$, Tetra-n-propylammonium bis(maleonitriledithiolato)-1,10-phenanthrolinecobaltate, 35B, 821
$C_{32}H_{40}Cl_2Mo_2N_4P_2$, Bis($\mu$-7-azaindolyl)bis(chloro(triethylphosphine)-

molybdenum)(Mo-Mo), 44B, 1045
$C_{32}H_{44}Fe_4N_2S_4Se_4$, Bis(tetramethylammonium) tetrakis(μ_3-selenido)tetrakis(phenylthiolatoiron), 44B, 1045
$C_{32}H_{44}Fe_4N_2S_8$, Bis(tetramethylammonium) tetrakis(thiophenolatosulfidoiron, 40B, 1021
$C_{32}H_{44}Mo_2N_2O_4S_4$, Tetramethylammonium di-μ-oxo-di(di(benzenethiolato)oxomolybdate(V)), 44B, 1046
$C_{32}H_{44}P_5PrS_8$, Tetraphenylphosphonium tetrakis(dimethyldithiophosphinato)praseodymium(III), 42B, 949
$C_{32}H_{52}N_6NiO_2S_4$, Tetraethylammonium bis-(2,3-quinoxaline-dithiolato)-nickelate(II) dihydrate, 40B, 1022
$C_{32}H_{56}Fe_2N_2S_6$, Bis(tetraethylammonium) bis(o-xylyl-α,α'-dithiolato-μ-sulfido-ferrate(III)), 41B, 1162
$C_{33}H_{23}N_5O_4Ru_2S_4$, Di-$\mu$-mercaptobenzothiazolato-N,S-bis(dicarbonyl-(pyridine)ruthenium(I))(Ru-Ru) pyridine solvate, 42B, 949
$C_{33}H_{27}CoNP_2S_4$, Bis(diphenylphosphinodithioato)quinolinecobalt(II), 44B, 1046
$C_{33}H_{30}AsCoS_4 \cdot 0.5\ C_2H_5OH$, Triphenylmethylarsonium bis(toluene-3,4-dithiolato)cobaltate hemiethanolate, 33B, 514
$C_{33}H_{35}AsO_3PtS_6$, Tetraphenylarsonium tris(O-ethyldithiocarbonato)-platinate(II), 43B, 1409
$C_{34}H_{22}N_8NiS_4$, Bis(N-methylphenazinium) bis(maleodinitriledithiolato)nickelate(II), 45B, 1322
$C_{34}H_{28}PdS_4$, Bis(cis-1,2-diphenylethylene-1,2-dithiolato)palladium - cyclohexa-1,3-diene, 39B, 831
$C_{34}H_{30}N_2NiP_2S_4$, Bis(diphenyldithiophosphinato)nickel(II)-2-pyridine, 37B, 638
$C_{34}H_{48}N_2NiO_8S_4$, Bis(trimethylphenylammonium) bis(1,1-dicarboethoxy-2,2-ethylenedithiolato)nickelate(II), 42B, 950
$C_{36}H_{28}I_2P_2Pd_2S_2$, Diiodobis[o-(diphenylphosphino)benzenethiolato]dipalladium(II), 46B, 1230
$C_{36}H_{28}NiP_2Se_2$, trans-Bis(diphenyl-o-selenolatophenylphosphine)nickel(II), 35B, 832
$C_{36}H_{36}Mo_2O_2S_8$, Tetrakis-μ-dithiobenzoato-bis(tetrahydrofuran-molybdenum), 44B, 1034
$C_{36}H_{54}F_{10}P_2PtS_2$, Bis(pentafluorobenzenethiolato)bis(tributylphosphine)platinum(II), 35B, 824
$C_{36}H_{66}DyP_3S_6$, Tris(dicyclohexyldithiophosphinato)dysprosium(III), 46B, 1231
$C_{36}H_{66}LuP_3S_6$, Tris(dicyclohexyldithiophosphinato)lutecium(III), 46B, 1231
$C_{36}H_{66}P_3PrS_6$, Tris(dicyclohexyldithiophosphinato)praseodymium(III), 43B, 1410
$C_{36}H_{66}P_3S_6Sm$, Tris(dicyclohexyldithiophosphinato)samarium(III), 43B, 1410
$C_{36}H_{76}Cl_4N_8O_2S_8Zn_4$, Zinc(II) N,N'-dimethyl-N,N'-bis(β-mercaptoethyl)ethylenediamine chloride, 39B, 832
$C_{36}H_{84}S_{12}Pd_6$, Palladium n-propyl mercaptide, 33B, 497
$C_{36}H_{90}Fe_6Mo_2N_3S_{17}$, Tris(tetraethylammonium) tri-μ-sulfido-bis[tetra-μ_3-sulfido-tri(ethylthioiron)molybdate], 45B, 1323
$C_{37}H_{30}P_2PdS_2$, Bis(triphenylphosphine)(carbon disulfide)palladium, 33B, 500
$C_{37}H_{30}P_2PtS_2$, (Carbon disulfide)bis(triphenylphosphine)platinum(0), 35B, 825
$C_{37}H_{68}P_2PtS_2$, Hydrido(methane(dithioato)-S)bis(tricyclohexylphosphine)platinum, 42B, 951
$C_{38}H_{32}P_2RuS_4$, Bis(dithioformato)bis(triphenylphosphine)ruthenium-

(II), 39B, 833
$C_{38}H_{35}Cl_2CoN_3O_2S \cdot 0.5\ C_6H_7N$, N,N-bis((5-chloro-2-hydroxyphenyl)-phenylmethylene)-4-thiaheptane-1,7-diaminato-(3-methylpyridine)cobalt(II) hemi(3-methylpyridine), 43B, 1186
$C_{38}H_{35}Cl_2N_3NiO_2S \cdot 0.5\ C_6H_7N$, N,N-bis((5-chloro-2-hydroxyphenyl)-phenylmethylene)-4-thiaheptane-1,7-diaminato-(3-methylpyridine)nickel(II) hemi(3-methylpyridine), 43B, 1186
$C_{38}H_{72}N_2NiS_{10}$, Tetrabutylammonium bis(isotrithionedithiolato)nickelate(II), 45B, 1323
$C_{39}H_{31}O_2P_2ReS_2$, (Dithioformato)bis(triphenylphosphino)dicarbonylrhenium, 37B, 640
$C_{40}H_{24}NiO_4P_2S_4$, Bis[(S)-O,O'-(1,1'-binaphthyl-2,2'-diyl)dithiophosphato]nickel(II), 46B, 1232
$C_{40}H_{44}Pt_2S_8$, Bis(dithiocumato)platinum(II) dimer, 38B, 1023
$C_{40}H_{72}CoN_6S_4$, Di(tetra-n-butylammonium) bis(maleonitriledithiolato)-cobaltate(II), 29, 560
$C_{40}H_{72}CuN_6S_4$, Tetra-(n-butylammonium) bis(maleonitriledithiolato)-cuprate, 41B, 1163
$C_{40}H_{72}Mo_2N_2O_7S_5$, Tetrabutylammonium μ-oxo-μ-sulfido-bis[(1,2-dithiosquarato-S,S')oxomolybdate(V)], 46B, 1232
$C_{40}H_{72}Mo_2N_6O_2S_6$, Tetra-n-butylammonium di-μ-sulfido-bis(oxo-1,1-dicyanoethylene-2,2-dithiolatomolybdate(V)), 42B, 951
$C_{40}H_{72}N_6NiS_4$, Tetra-n-butylammonium bis(1,2-dicyanovinylene-1,2-dithiolato)nickelate(II), 43B, 1411
$C_{40}H_{88}Fe_2N_2S_8$, Bis(tetra-n-butylammonium) μ-S,S'-(tetrakis(ethane-1,2-dithiolate)diiron(III)), 39B, 834
$C_{40}H_{100}Fe_6Mo_2N_3S_{17}$, Tetraethylammonium (μ-sulfido)-di-μ-ethylthiolatobis(molybdenum-trisulfido-tris-(ethylthio)iron)), 45B, 1326
$C_{41}H_{30}F_6OP_2RuS_2$, Bis(triphenylphosphine)-bis(perfluoromethyl)-dithietineruthenium carbonyl, 40B, 1023
$C_{41}H_{30}F_6OP_2RuS_2$, Carbonylbis(triphenylphosphine)(hexafluorobut-2-en-2,3-dithiolo)ruthenium (violet isomer), 43B, 1411
$C_{41}H_{45}AgBN_3S_2$, {2,15-Dimethyl-7,10-dithia-3,14,20-triazabicyclo-[14.3.1]eicosa-1(20),2,14,16,18-pentaene}silver(I) tetraphenylborate, 46B, 1222
$C_{42}H_{30}ReS_6$, Tris(cis-1,2-diphenylethene-1,2-dithiolato)rhenium, 31B, 455
$C_{42}H_{30}S_6V$, Tris(cis-1,2-diphenylethene-1,2-dithiolato)vanadium, 31B, 456; 32B, 462
$C_{42}H_{32}AsNbS_6$, Tetraphenylarsonium tris(benzene-1,2-dithiolato)-niobate(V), 42B, 952
$C_{42}H_{32}AsS_6Ta$, Tetraphenylarsonium tris(benzenedithiolato)-tantalate(V), 41B, 1164
$C_{42}H_{40}N_2O_2P_2RuS_2$, cis,cis,cis-Bis(dimethylphenylphosphine)bis(monothiobenzoato)(1,10-phenanthroline)ruthenium(II), 46B, 1233
$C_{42}H_{40}N_2O_2P_2RuS_2$, cis,cis,trans-Bis(dimethylphenylphosphine)bis-(monothiobenzoato)(1,10-phenanthroline)ruthenium(II), 46B, 1233
$C_{42}H_{56}CoN_3S_4$, Tetra-n-butylammonium bis(toluene-3,4-dithiolato)-1,10-phenanthrolinecobaltate, 35B, 826
$C_{42}H_{105}Fe_6N_3S_{17}W_2$, Tris(tetraethylammonium) tris(μ_2-ethylthio)-bis-(tetrakis(μ_3-thio)-tris(μ_3-ethylthioiron))-di-tungstate, 45B, 1324
$C_{43}H_{35}CuP_2S_2$, Phenyldithiocarboxylatobis(triphenylphosphine)copper-(I), 46B, 1233
$C_{43}H_{35}CuP_2S_3 \cdot 0.5\ CS_2$, Bis(triphenylphosphine)(phenyl trithiocarbonato)copper(I) hemi(carbon disulfide), 43B, 1412
$C_{44}H_{37}NOP_2PtS$, 4-Methyl-N-thiobenzohydroximato-bis(triphenylphosphine)platinum(II), 46B, 1234

$C_{44}H_{42}AsFeN_2NaS_4$, Sodium tetraphenylarsonium bis(o-xylyl-α,α'-dithiolato)ferrate(II) bis(acetonitrile), 43B, 1405
$C_{44}H_{68}Fe_2N_2S_6$, Bis(tetraethylammonium) bis(bis(p-tolylthiolato)-μ-sulfido-ferrate(III)), 41B, 1162
$C_{44}H_{68}Fe_4N_2S_8$, Bis(tetraethylammonium) tetra(mercaptobenzyl-μ_3-sulfidoiron), 39B, 836
$C_{45}H_{33}FeO_3S_3$, Tris(1,3-diphenylpropan-1-one-3-thionato)iron(III), 41B, 1165
$C_{45}H_{43}Cl_2N_3NiO_2S$, N,N-bis((5-chloro-2-hydroxyphenyl)phenylmethylene)-4-thiaheptane-1,7-diaminato-(pyridine)nickel(II) m-xylene, 43B, 1186
$C_{45}H_{44}CuO_2P_3S_2$, Phenylthiotris(methyldiphenylphosphine)(sulfur dioxide)copper(I), 43B, 1413
$C_{45}H_{74}Fe_4N_3S_8$, Methyltriethylammonium tetrakis(phenylthiolato)tetrasulfurtetrairon, 44B, 1047
$C_{48}H_{40}AsMoOS_4$, Tetraphenylarsonium tetrakis(benzenethiolato)oxomolybdate(V), 44B, 1047
$C_{48}H_{40}Cl_2Hg_2O_8Se_4$, Tetrakis(diphenylseleno)dimercury(I) diperchlorate, 43B, 1413
$C_{48}H_{40}Cl_2Hg_2O_8Se_4$, Tetrakis(diphenylseleno)dimercury(I) diperchlorate (red form), 44B, 1048
$C_{48}H_{40}MnN_2P_4S_4$, Bis(tetraphenyldithioimidodiphosphinato)manganese-(II), 40B, 1024
$C_{48}H_{60}LaO_8P_5S_6$, Tris(O,O'-diethylphosphorodithioato)bis(triphenylphosphine oxide)lanthanum(III), 42B, 952
$C_{48}H_{72}ClCu_{14}O_{24}S_{12}Tl_5 \cdot 12\ H_2O$, Pentathallium(I) μ_8-chloro-dodecakis(α-mercaptoisobutyrato)octacuprate(I)hexacuprate(II) dodecahydrate, 45B, 1324
$C_{48}H_{120}ClCu_{14}N_{12}O_{14}S_{15.5} \cdot 20\ H_2O$, Chlorododeca(1,1-dimethyl-2-aminoethanethiolato)tetradecacopper sulfate eicosa-hydrate, 46B, 1235
$C_{48}H_{120}ClCu_{14}N_{12}S_{12} \cdot 3.5\ SO_4 \cdot 19\ H_2O$, Copper mercaptide cluster, 42B, 953
$C_{49}H_{44}OS_8Zn_4$, catena-μ-Benzenethiolato-(methanol-benzenethiolato-hexa-μ-benzenethiolato)-tetra-zinc(II), 46B, 1236
$C_{52}H_{48}Cl_2Cu_2P_4S_4$, Di-μ-chloro-bis[μ-1,2-bis(diphenylphosphinothioyl)ethane]copper(I), 45B, 1375
$C_{52}H_{88}Fe_4N_3S_8$, Tris(tetraethylammonium) tetra-μ_3-sulfido-tetra(benzylthiolatoferrate), 45B, 1325
$C_{53}H_{97}Fe_4N_6Na_5O_{13}S_8$, Pentasodium tetrabutylammonium tetrakis(β-mercaptopropionato)tetra-μ_3-thio-tetraferrate pentakis(N-methylpyrrolidone), 43B, 1414
$C_{54}H_{44}Cl_8NiO_4P_2S_4Sn_2$, Benzyltriphenylphosphonium bis(dithiooxalato)-nickel(II) - bis(tin(IV) chloride), 39B, 837
$C_{54}H_{46}Cl_4NiO_5P_2S_4Sn$, Benzyltriphenylphosphonium bis(dithiooxalato)-nickel(II) - tin(IV) chloride, 39B, 837
$C_{54}H_{52}Cu_2N_2P_4S_2$, μ-Dithiocyanatotetrakis(methyldiphenylphosphine)di-copper(I), 35B, 829
$C_{54}H_{108}Fe_7Mo_2N_3S_{20}$, Tris(trimethylbenzylammonium) octathiododeca-(thioethoxy)heptairondimolybdate, 46B, 1237
$C_{55}H_{45}OP_3Pt_2S$, μ-Sulfido-carbonyltris(triphenylphosphine)diplatinum, 40B, 1128
$C_{56}H_{40}FeO_4P_2S_4$, Tetraphenylphosphonium bis(dithiosquarate)ferrate(II), 42B, 954
$C_{56}H_{45}BF_4IrOP_3S_2$, Triphenylphosphoniodithiocarboxylato-S,S'-carbonylbis(triphenylphosphine)iridium(I) tetrafluoroborate, 46B, 1238
$C_{56}H_{72}Cl_{16}Co_2N_2S_8$, Tetra-n-butylammonium bis(1,2,3,4-tetrachlorobenzene-5,6-dithiolato)cobaltate, 33B, 499

$C_{56}H_{92}Fe_2N_4S_8$, Tetra-n-butylammonium μ-hydrazine-bis(di(benzene-1,2-dithiolato)iron(III)), 44B, 1048
$C_{57}H_{111}Fe_6Mo_2N_3S_{17}$, Triethylbenzylammonium tri-μ-ethylthiolatobis-(tetrakis-(μ_2-sulfido)-tris(ethylthiolatoiron)molybdenum), 45B, 1326
$C_{59}H_{48}Ir_2O_2P_2S_6$, Dicarbonylbis(triphenylphosphine)tris(toluene-3,4-dithiolato)diiridium(III), 38B, 1025
$C_{60}H_{30}F_{20}P_2Pd_2S_4$, Di-μ-pentafluorobenzenethiolatobis(pentafluorobenzenethiolato)-bis(triphenylphosphine)dipalladium(II), 35B, 830
$C_{60}H_{40}As_2FeN_6S_6$, Bis-tetraphenylarsonium tris-(cis-1,2-dicyano-1,2-ethylenedithiolato)ferrate(IV), 39B, 839
$C_{60}H_{40}As_2MoN_6S_6$, Bis(tetraphenylarsonium) tris(1,2-dicyanoethylene-dithiolato)molybdate(IV), 39B, 839
$C_{60}H_{40}As_2N_6NiSe_6$, Bis(tetraphenylarsonium) tris(2,2-diselenido-1,1-ethylenedicarbonitrile-Se,Se')nickelate(IV), 46B, 1239
$C_{60}H_{40}As_2N_6S_6W$, Bis(tetraphenylarsonium) tris(1,2-dicyanoethylenedi-thiolato)tungstate(IV), 39B, 839
$C_{62}H_{36}F_{20}OP_2Pd_2S_4$, Di-μ-pentafluorobenzenethiolato-trans-bis((pentafluorobenzenethiolato)(triphenylphosphine)palladium(II)) ethanol solvate, 38B, 1026
$C_{64}H_{60}Ag_2N_4NiP_2S_4$, Bis(bis(diphenylethylphosphine)silver(I)) bis-(1,2-dicyano-1,2-ethylenedithiolato)nickelate(II), 42B, 954
$C_{66}H_{55}CuP_2S_3$, Bis(tetraphenylphosphonium) tris(thiophenolato)copper-(I), 46B, 1239
$C_{66}H_{75}O_9P_6S_6Sm$, Bis(O,O'-diethylphosphorodithioato)tris(triphenyl-phosphine oxide)samarium(III) (O,O'-diethylphosphorodithioate), 42B, 952
$C_{68}H_{67}BCoNO_3P_3S \cdot 0.5\ CH_2Cl_2$, (Ethylsulfito)(tris(2-diphenylphos-phinoethyl)amine)cobalt(II) tetraphenylborate - 0.5 methylene chloride, 44B, 1049
$C_{68}H_{74}Co_4N_2S_{10}$, Bis(tetramethylammonium) hexa(μ-benzenethiolato)tetra(benzenethiolatocobaltate(II)), 45B, 1326
$C_{72}H_{60}CdP_2S_4$, Bis(tetraphenylphosphonium) tetrakis(thiophenolato)-cadmate(II), 44B, 1050
$C_{72}H_{60}CoP_2S_4$, Bis(tetraphenylphosphonium) tetrakis(thiophenolato)cobaltate(II), 44B, 1050
$C_{72}H_{60}FeP_2S_4$, Bis(tetraphenylphosphonium) tetrakis(thiophenolato)-ferrate(II), 42B, 954
$C_{72}H_{60}MnP_2S_4$, Bis(tetraphenylphosphonium) tetrakis(thiophenolato)-manganate(II), 44B, 1050
$C_{72}H_{60}NiP_2S_4$, Bis(tetraphenylphosphonium) tetrakis(thiophenolato)-nickelate(II), 44B, 1050
$C_{72}H_{60}P_2S_4Zn$, Bis(tetraphenylphosphonium) tetrakis(thiophenolato)-zincate(II), 44B, 1050
$C_{73}H_{60}Cl_6Cu_5N_3P_6S_6$, Tris(imidotetraphenyldithiodiphosphino-S,S)tetracopper(I) dichlorocuprate(I) carbon tetrachloride solvate, 43B, 1414; 44B, 1050
$C_{74}H_{74}FeO_{12}P_2S_6$, Bis(benzyltriphenylphosphonium) tris(1,1-dicarbo-ethoxyethylene-2,2-dithiolato)ferrate(IV), 43B, 1415
$C_{80}H_{60}Ag_2N_4NiP_4S_4$, Bis(bis(triphenylphosphine)silver(I)) bis(1,1-dicyano-2,2-ethylenedithiolato)nickelate(II), 40B, 1025
$C_{80}H_{60}Ag_2N_4NiP_4S_4$, Bis(bis(triphenylphosphine)silver(I)) bis(1,2-dicyano-1,2-ethylenedithiolato)nickelate(II), 40B, 1025
$C_{82}H_{64}P_4Pt_2S_4 \cdot 2.74\ C_3H_7NO$, Tetrakis(triphenylphosphine)diplatinum tetrathionaphthalene - 2.74 N,N-dimethylformamide, 44B, 1051
$C_{82}H_{78}Co_2P_6S_2 \cdot 1.4\ C_3H_7NO$, Bis[(μ-thio)(1,1,1-tris((diphenylphos-phino)methyl)ethane)cobalt] dimethylformamide solvate, 46B, 1241

$C_{84}H_{70}Cu_4P_2S_6$, Bis(tetraphenylphosphonium) hexakis(thiophenolato)-tetracopper(I), 46B, 1239
$C_{93}H_{90}Cl_2Ir_2O_7P_4S_2$, μ-Chloro-di-μ-(phenylthio)-di(hydridobis(triphenylphosphine)iridium(III)) perchlorate tris(acetone), 41B, 1165; 42B, 955
$C_{96}H_{150}Cu_{10}O_{32}S_{16}$, Copper di(t-butoxycarbonyl)dithiolatoethylene, 43B, 1416
$C_{100}H_{88}Cl_4Cu_4P_8S_8$, Bis{μ-[bis(diphenylthiophosphoryl)methane-S,μ,S']-chlorocopper(I)} [bis(diphenylthiophosphoryl)methane-S,S']-chlorocopper(I), 46B, 1241
$C_{106}H_{98}BCo_2P_6S_2 \cdot 0.5\ C_3H_7NO$, Bis[(μ-thio)(1,1,1-tris((diphenylphosphino)methyl)ethane)cobalt] tetraphenylborate dimethylformamide solvate, 46B, 1241
$C_{112}H_{204}Cu_8N_4O_{24}S_{12}$, Tetrakis(tetrabutylammonium) hexakis(1,1-dicarboethoxyethylene-2,2-dithiolato)octacuprate(I), 43B, 1483
$C_{122}H_{83}Cu_8NO_{12}P_4S_{12}$, Tetrakis(tetraphenylphosphonium) hexakis(dithiosquarato)octacuprate(I) acetonitrile, 43B, 1483
$C_{132}H_{124}B_2Co_2P_6S_2 \cdot 2\ C_3H_6O$, Bis[(μ-methylthio)(1,1,1-tris((diphenylphosphino)methyl)ethane)cobalt] bis(tetraphenylborate) acetone solvate, 46B, 1241
$C_{148}H_{228}Fe_7Mo_2N_4S_{20}$, Tetrakis(tetra-n-butylammonium) octathiododeca-(thiobenzyloxy)heptairondimolybdate, 46B, 1237
$C_{148}H_{228}Fe_7N_4S_{20}W_2$, Tetrakis(tetra-n-butylammonium) octathiododeca-(thiobenzyloxy)heptaironditungstate, 46B, 1237

86. METAL COMPLEXES (PHOSPHINE OR ARSINE LIGAND)

$C_3H_9AuBr_3P$, Tribromo(trimethylphosphine)gold, 10, 210
$C_3H_9Cl_{10}MoP_3 \cdot 0.33\ CS_2$, Tetrachlorotris(dichloro(methyl)phosphine)-molybdenum(IV) carbon disulfide solvate, 44B, 1051
$C_4F_{12}Fe_2N_4O_4P_2$, Bis-(μ-bis(trifluoromethyl)phosphido)-tetranitrosyldiiron, 42B, 956
$C_4H_{10}ClHgO_3P$, Chloro(diethyl phosphonato)mercury(II), 35B, 834
$C_4H_{12}Cl_2F_{16}Mo_2N_4P_8$, Tetrakis[bis(difluorophosphino)methylamine]dichlorodimolybdenum, 46B, 1242
$C_4H_{12}Cl_2HgP_2$, catena-Dichloro-μ-(tetramethyldiphosphine-P,P')mercury(II), 43B, 1417
$C_4H_{12}F_{16}FeN_4P_8$, Tetrakis(methylbis(difluorophosphino)amine)iron(0), 44B, 1052
$C_4H_{12}Fe_2N_4O_4P_2$, Di-μ-dimethylphosphino-bis(dinitrosyliron), 45B, 1327
$C_4H_{12}HgO_6P_2$, Bis(dimethylphosphonato)mercury(II), 39B, 841
$C_5F_3MoO_5P$, Pentacarbonyl(trifluorophosphine)molybdenum (gas-ed), 37B, 694
$C_5H_{12}F_{16}Fe_2N_4OP_8$, Tetrakis(methylaminobis(difluorophosphine))carbonyldiiron, 44B, 1052
$C_6H_9NO_4PReS$, Ammine(tetracarbonyl)-(dimethylthiophosphinito)rhenium, 45B, 1328
$C_6H_{15}AsCuI$, Triethylarsinocopper(I) iodide, 4, 258, 284
$C_6H_{15}Cl_2F_3P_2Pt$, cis-Dichloro(triethylphosphine)(trifluorophosphine)-platinum(II), 43B, 1417
$C_6H_{18}As_2Br_4Pd_2$, Di-μ-bromobis(bromo(trimethylarsine)palladium), 6, 201, 220
$C_6H_{18}As_2Cl_4Pt_2$, Di-μ-chloro-(dichlorobis(trimethylarsine))diplatinum(II), 35B, 865
$C_6H_{18}Cl_2P_2Pt$, cis-Dichloro-bis(trimethylphosphine)platinum(II), 32B,

464; 45B, 1328
$C_6H_{18}I_2P_2Pt$, cis-Diiodobis(trimethylphosphine)platinum(II), 43B, 1418
$C_6H_{18}I_2P_2Pt$, trans-Diiodobis(trimethylphosphine)platinum(II), 43B, 1418
$C_6H_{21}Cl_3NO_6P_2Ru$, Amminetrichlorobis(trimethyl phosphite)ruthenium(III), 46B, 1243
$C_7H_7CrO_5PS$, Pentacarbonyl(dimethylmercaptophosphane)chromium, 42B, 956
$C_7H_9O_7PRu$, Tetracarbonyl(trimethyl phosphite)ruthenium, 46B, 1243
$C_7H_{15}Cl_2OPPt$, cis-Dichlorocarbonyl(triethylphosphine)platinum(II), 43B, 1419
$C_7H_{18}I_2NiOP_2$, Diiodobis(trimethylphosphine)carbonylnickel(II), 46B, 1244
$C_7H_{26}B_5Br_2IrOP_2$, cis-Dibromo(pentaboranyl)bis(trimethylphosphine)-carbonyliridium(III), 41B, 1166
$C_8H_2F_6Fe_2O_6P_2$, trans-Di-μ-(hydridotrifluoromethylphosphido)hexacarbonyldiiron, 45B, 1329
$C_8H_{10}FFeN_2O_4P$, (2-Fluoro-1,3-dimethyl-1,3,2-diazaphospholidine)tetracarbonyliron(0), 44B, 1053
$C_8H_{11}CdCl_2P$, Dichloro(dimethylphenylphosphine)cadmium, 46B, 1244
$C_8H_{12}As_2Cl_4F_4Re$, Tetrachloro-(1,2-bis(dimethylarsino)-3,3,4,4-tetrafluorocyclobut-1-ene)rhenium(IV), 40B, 1026
$C_8H_{20}BrCuP_2$, Tetraethylbiphosphinecopper(I) bromide, 35B, 835; 42B, 957
$C_8H_{21}ClNiOP_2$, trans-Acetylchlorobis(trimethylphosphine)nickel, 42B, 958
$C_8H_{28}Cl_2P_4Ru$, trans-Dichlorotetrakis(dimethylphosphine)ruthenium-(II), 42B, 959
$C_9H_6AsCoFeO_7$, Tetracarbonyl-μ-(dimethylarsino)-(tricarbonylcobalt)-iron, 42B, 608
$C_9H_6CoFeO_7P$, μ-Dimethylphosphido-tetracarbonyliron-tricarbonylcobalt(Fe-Co), 43B, 1421
$C_9H_7CrN_2O_5P$, Pentacarbonyl (1,5-dimethyl-1,2,3-diazaphosphole)-chromium, 46B, 1245
$C_9H_{16}I_2O_3P_2W$, Tricarbonyldiiodo(bis(dimethylphosphine)ethane)tungsten (form I), 41B, 1167
$C_9H_{16}I_2O_3P_2W$, Tricarbonyldiiodo(bis(dimethylphosphine)ethane)tungsten (form II), 41B, 1167
$C_9H_{18}FeO_9P_2$, trans-Bis(trimethoxyphosphine)tricarbonyliron, 40B, 1027
$C_9H_{18}O_3P_2Ru$, Tricarbonylbis(trimethylphosphine)ruthenium(0), 46B, 815
$C_9H_{27}Br_2NiP_3$, Dibromotris(trimethylphosphine)nickel(II), 40B, 1028
$C_9H_{27}ClP_3Rh$, Chlorotris(trimethylphosphine)rhodium(I), 46B, 1250
$C_9H_{27}Cl_2P_3Pt$, Chloro-tris(trimethylphosphine)platinum(II) chloride, 46B, 1246
$C_9H_{27}I_2NiO_9P_3$, Diiodotris(trimethylphosphite)nickel(II), 41B, 1168
$C_{10}F_{12}Fe_2O_6P_2$, Bis(μ-bis(trifluoromethyl)phosphido)-hexacarbonyldiiron, 42B, 959
$C_{10}H_6AsFeMnO_8$, Tetracarbonyl-μ-(dimethylarsido)-(tetracarbonylferrio)manganese, 39B, 842
$C_{10}H_9As_3CrO_5$, (4-Methyl-1,2,6-triarsatricyclo[2.2.1.0$^{2-6}$]heptane)-pentacarbonylchromium(0), 44B, 1053
$C_{10}H_{12}As_2ClCrF_3O_4$, (1,2-Bis(dimethylarsino)-chlorotrifluoroethane)-tetracarbonylchromium, 38B, 1027
$C_{10}H_{12}As_2Fe_2O_6$, Di-μ-dimethylarsenido-bis(tricarbonyliron), 43B,

1421; 45B, 1330
$C_{10}H_{12}Fe_2I_2O_6P_2$, Di-μ-dimethylphosphido-bis(iodotricarbonyliron), 33B, 501
$C_{10}H_{13}As_2CrF_3O_4$, (1,2-Bis(dimethylarsino)trifluoroethane)tetracarbonylchromium, 38B, 1027
$C_{10}H_{13}As_2F_3MoO_4$, (1,2-Bis(dimethylarsino)trifluoroethane)tetracarbonylmolybdenum, 38B, 1027
$C_{10}H_{13}F_3MoO_4P_2$, (1,2-Bis(dimethylphosphino)trifluoroethane)tetracarbonylmolybdenum, 38B, 1027
$C_{10}H_{14}As_2CrF_2O_4$, (1,2-Bis(dimethylarsino)-1,1-difluoroethane)tetracarbonylchromium, 38B, 1027
$C_{10}H_{14}IMoO_5P$, π-Cyclopentadienyl-trans-iododicarbonyl(trimethylphosphite)molybdenum, 38B, 1028
$C_{10}H_{18}As_2ClMnO_3$, fac-Chloro-(1,3-bis(dimethylarsino)propane)tricarbonylmanganese, 39B, 842
$C_{10}H_{18}As_2Cl_3GeMnO_3$, fac-Trichlorogermano(bis(1,3-dimethylarsino)-propane)tricarbonylmanganese, 41B, 1168
$C_{10}H_{18}As_2MoO_4P_2$, cis-Tetracarbonyl-1,4-h(2)-(1,2-bis(dimethylphosphino)-1,2-dimethyldiarsine)molybdenum(0), 41B, 1169
$C_{10}H_{18}Fe_3N_3O_7P_3$, Dicarbonyl-(dicarbonylnitrosyliron)-tris-μ-(dimethylphosphido)(dinitrosyliron)iron, 45B, 1329
$C_{10}H_{18}MoO_4P_4$, cis-Tetracarbonyl(hexamethyltetraphosphine-P(1),P(4))-molybdenum(0), 41B, 1170
$C_{10}H_{18}O_4P_6W$, Hexamethylcyclohexaphosphinetungsten tetracarbonyl, 40B, 1029
$C_{10}H_{21}AgNPS$, Tri-n-propylphosphinesilver thiocyanate, 24, 613; 28, 655
$C_{11}F_{24}Ni_2O_3P_4S_2$, μ-Carbonyl-bis-μ-(bis(bis(trifluoromethyl)phosphino)sulfur-P,P')-bis(carbonylnickel), 40B, 1030
$C_{11}H_9N_2O_{13}Os_3P$, Octacarbonyl-1,1-dinitrosyl-1-(trimethyl phosphite)-triangulo-triosmium, 44B, 1054
$C_{11}H_{10}NO_{12}PW_2$, Nitrosyltricarbonyl(trimethoxyphosphine)tungsten-μ-hydrido-pentacarbonyltungsten, 42B, 959
$C_{11}H_{12}AsFe_2NO_7$, μ-Dimethylarsenido-μ-dimethylcarbamoyl-bis(tricarbonyliron)(Fe-Fe), 43B, 1421
$C_{11}H_{12}As_2CrFeO_7$, Di-μ-dimethylarsenido-(tetracarbonylchromium)(tricarbonyliron), 45B, 1330
$C_{11}H_{12}As_2F_6I_2O_3W$, (2,3-Bis(dimethylarsino)-1,1,1,4,4,4-hexafluorobut-2-ene-As,As)tricarbonyldiiodotungsten(II), 40B, 1031
$C_{11}H_{12}As_2F_6MoO_4$, (1,2-Bis(dimethylarsino)hexafluoropropane)tetracarbonylmolybdenum, 38B, 1027
$C_{11}H_{12}MoN_3O_5P$, Phosphatriazaadamantanemolybdenum pentacarbonyl, 42B, 960
$C_{11}H_{13}As_2ClCrF_4O_4$, (1,3-Bis(dimethylarsino)-2-chloro-1,1,3,3-tetrafluoropropane)tetracarbonylchromium, 38B, 1030
$C_{11}H_{15}FeNO_9P_4$, Bis(2,6,7-trioxa-1,4-diphosphabicyclo[2.2.2]octane)-tricarbonyliron - acetonitrile, 38B, 1031
$C_{11}H_{16}IMoO_5P$, π-(Methylcyclopentadienyl)-trans-iododicarbonyl(trimethylphosphite)molybdenum, 38B, 1028
$C_{11}H_{18}CoFe_2N_2O_7P_3$, Dicarbonyl-tris-μ-(dimethylphosphido)(dinitrosyliron)(tricarbonylcobalt)iron, 45B, 1329
$C_{11}H_{18}CrO_5P_4Si_3$, Hexamethyltrisilatetraphosphanortricyclenechromiumpentacarbonyl, 46B, 1246
$C_{12}H_6AsCrMnO_{10}$, Pentacarbonylmanganese(dimethylarsenide)pentacarbonylchromium, 38B, 1032
$C_{12}H_9N_2O_{14}Os_3P$, 1,1,1,2,2,2,2,3,3-Nonacarbonyl-1,3;1,3-di-μ-nitrosyl-3-(trimethyl phosphite)-triangulo-triosmium, 46B, 1247

$C_{12}H_{12}As_2Cr_2O_8$, Di-μ-dimethylarsenido-bis(tetracarbonylchromium), 45B, 1330
$C_{12}H_{12}Cr_2O_8P_2$, Di-μ-dimethylphosphido-bis(tetracarbonylchromium), 44B, 1055
$C_{12}H_{12}Fe_2O_8P_2$, μ-Tetramethyldiphosphine-bis(tetracarbonyliron), 33B, 503
$C_{12}H_{12}Mn_2O_8P_2$, Di-μ-dimethylphosphido-bis(tetracarbonylmanganese), 44B, 1055
$C_{12}H_{12}O_8P_2V_2$, Di-μ-dimethylphosphido-bis(tetracarbonylvanadium), 44B, 1055
$C_{12}H_{12}O_{10}P_2Re_2$, 1,4-Dioxa-2,5-tetramethyldiphospha-3,6-octacarbonyl-dirhenacyclohexane, 43B, 1422
$C_{12}H_{16}F_8Fe_2N_2P_4$, Methylaminobis(difluorophosphine)cyclopentadienyl-iron dimer, 44B, 1055
$C_{12}H_{18}Co_3O_6P_3$, Hexacarbonyl-tris-μ-(dimethylphosphido)-triangulo-tricobalt, 45B, 1329
$C_{12}H_{18}Co_5O_{11}P_3$, Undecacarbonyl-tris-μ-(dimethylphosphido)pentaco-balt, 45B, 1329
$C_{12}H_{19}CrN_2O_5PSi$, Pentacarbonyl(t-butylamino-trimethylsilyliminophos-phine)chromium(0), 43B, 1422
$C_{12}H_{21}ClPRh$, Chlorotri(3-butenyl)phosphinerhodium(I), 40B, 1032
$C_{12}H_{21}MoO_4P_3$, cis-Tetracarbonyl(1,1,3,3-tetramethyl-2-t-butyltri-phosphane-P(1),P(3))molybdenum(0), 42B, 961
$C_{12}H_{28}Au_2P_2S_2$, Di-μ-(2-diethylphosphinoethylthio)-digold(I), 44B, 1056
$C_{12}H_{28}Cl_2NPPt$, trans-Dichloro(triethylphosphine)(cis-2,3-dimethyl-pyrrolidine)platinum(II), 45B, 1331
$C_{12}H_{28}Cl_2NPPt$, trans-Dichloro(triethylphosphine)(cis-2,4-dimethyl-pyrrolidine)platinum(II), 45B, 1331
$C_{12}H_{30}Br_2NiP_2$, Bistriethylphosphinenickel(II) bromide, 22, 642
$C_{12}H_{30}Br_2P_2Pt$, trans-Dibromobis(triethylphosphine)platinum, 31B, 457
$C_{12}H_{30}Cl_2P_2Pt$, trans-Dichlorobis(triethylphosphine)platinum, 31B, 457
$C_{12}H_{30}Cl_3CoP_2$, Trichlorobis(triethylphosphine)cobalt, 43B, 1423
$C_{12}H_{30}Cl_4P_2Pt$, cis-Tetrachlorobis(triethylphosphine)platinum(IV), 43B, 1423
$C_{12}H_{30}Cl_6P_2Re_2$, Hexachlorobis(triethylphosphine)dirhenium(III), 33B, 504
$C_{12}H_{30}P_2PtS_4W$, Bis(triethylphosphine)platinumtetrathiotungsten, 46B, 1249
$C_{12}H_{31}BrP_2Pt$, Bis(triethylphosphine)bromoplatinum hydride, 24, 625
$C_{12}H_{31}ClP_2Pd$, trans-Chlorohydridobis(triethylphosphine)palladium, 39B, 843
$C_{12}H_{33}Cl_4Hg_2P_3$, Chlorobis(ethyldimethylphosphine)mercury(II) tri-chloro(ethyldimethylphosphine)mercurate(II), 46B, 1249
$C_{12}H_{36}Ag_2N_2O_{18}$, Di-μ-nitrato-di(bis(trimethylphosphite)silver(I)), 41B, 1171
$C_{12}H_{36}BBrF_4NiO_{12}P_4$, Bromotetrakis(trimethylphosphito)nickel(II) tet-rafluoroborate, 41B, 1171
$C_{12}H_{36}BBrF_4NiP_4$, Bromotetrakis(trimethylphosphine)nickel(II) tetra-fluoroborate, 44B, 1056
$C_{12}H_{36}ClP_4Rh$, Tetrakis(trimethylphosphine)rhodium(I) chloride, 46B, 1250
$C_{12}H_{36}Cl_6P_4Ta_2$, Di-μ-chloro-bis[dichlorobis(trimethylphosphine)-tantalum(III)](Ta-Ta), 46B, 1250
$C_{12}H_{38}Cl_6P_4Ta_2$, Di-μ-hydrido-di-μ-chloro-bis[dichlorobis(trimethyl-phosphine)tantalum(IV)](Ta-Ta), 46B, 1250

$C_{12}H_{41}BMoP_4$, Hydrido(tetrahydroborato)tetrakis(trimethylphosphine)-molybdenum(II), 46B, 1251
$C_{13}H_{11}AsMn_2O_6$, Dicarbonyl(cyclopentadienyl)-μ-(dimethylarsenido)-(tetracarbonylmanganio)manganese, 40B, 1033
$C_{13}H_{15}AsFeMnO_8P$, cis-Tetracarbonyliron-μ-dimethylarsino-tetracarbonyl(trimethylphosphine)manganese, 44B, 1057
$C_{13}H_{15}AsFeMnO_8P$, trans-Tetracarbonyliron-μ-dimethylarsino-tetracarbonyl(trimethylphosphine)manganese, 44B, 1057
$C_{13}H_{16}As_2FeO_3$, o-Phenylenebisdimethylarsinetricarbonyl iron(0), 32B, 464
$C_{13}H_{18}As_3ClMn_2O_7$, Tetracarbonyl-μ-(dimethylarsenido)-μ-(tetramethyldiarsenic)(tricarbonylchloromanganese)manganese, 46B, 1266
$C_{14}H_9O_{14}Os_3P$, Undecacarbonyl(trimethyl phosphite)-triangulo-triosmium, 44B, 1058
$C_{14}H_{12}As_2Co_2F_4O_6$, μ-(1,2-Bis(dimethylarsino)tetrafluorocyclobutene-As,As)-di-μ-carbonyl-bis-(dicarbonylcobalt)(Co-Co), 37B, 641
$C_{14}H_{12}As_2F_4Fe_2O_6$, 1,2-Bis(dimethylarsino)-3,3,4,4-tetrafluorocyclobutenebis(iron tricarbonyl), 31B, 466; 32B, 374
$C_{14}H_{16}As_2I_4O_4W$, Tetracarbonyliodo(o-phenylenebis(dimethylarsino))-tungsten(II) triiodide, 40B, 1034
$C_{14}H_{25}MoO_4P_5$, Pentaethylcyclopentaphosphinemolybdenum tetracarbonyl, 33B, 506
$C_{14}H_{27}ClIrO_2P$, Chloro-dicarbonyl-(tri-t-butylphosphine)iridium, 45B, 1332
$C_{14}H_{32}I_2O_2P_4W$, Iododicarbonylbis(bis(dimethylphosphino)ethane)tungsten(II) iodide, 43B, 1424
$C_{14}H_{33}O_2P_4Ta$, Dicarbonyl-bis(1,2-bis(dimethylphosphino)ethane)tantalum hydride, 40B, 1034
$C_{14}H_{35}Cl_3NPPt$, Tetraethylammonium trichloro(triethylphosphine)-platinate(II), 41B, 1246
$C_{14}H_{36}B_4P_2Pt$, 1,1-Bis(triethylphosphine)-2,4-dicarba-1-platina-closo-heptaborane, 46B, 1254
$C_{14}H_{38}B_4P_2Pt$, 4,5-μ-Hydridobis(triethylphosphine)platinum-2,3-dicarba-nido-hexaborane(8), 45B, 1334
$C_{14}H_{41}B_9P_2Pt$, 3,3-Bis(triethylphosphine)-1,2-dicarba-3-platina-dodecaborane(11), 44B, 1058
$C_{15}H_7Fe_3O_9P$, Bis(μ-hydrido)(μ$_3$-phenylphosphido)tris(tricarbonyl-iron), 46B, 1253
$C_{15}H_{20}As_2FeI_2NiO$, Diiodocarbonylferrocene-1,1'-bis(dimethylarsine)-nickel(II), 38B, 1032
$C_{15}H_{24}Cl_2NPPd$, [(S)-2-(Butylphenylphosphinomethyl)pyrrolidine]di-chloropalladium(II), 45B, 1334
$C_{15}H_{26}AsCl_2NPt$, trans-Dichloro(N-(p-xylenylidene)methylamine)triethylarsineplatinum(II), 41B, 1172
$C_{15}H_{27}CrN_2O_5PSi_2$, Pentacarbonyl(bis(trimethylsilyl)amino-t-butyliminophosphine)chromium(0), 43B, 1425
$C_{15}H_{27}NiO_3P$, Tricarbonyl(tri-t-butylphosphine)nickel, 42B, 961
$C_{15}H_{30}Cl_2MoO_3P_2$, Tricarbonyldichlorobis(triethylphosphine)molybdenum(II), 43B, 1426
$C_{16}H_3O_{16}Os_5P$, Pentadecacarbonyl(methoxyphosphinato)pentaosmium, 45B, 1335
$C_{16}H_{10}As_2FeO_4$, Tetracarbonyl(diphenylcycloferradiarsane), 44B, 1163
$C_{16}H_{11}AsCrO_7$, Pentacarbonyl(6-methyl-4-methylene-2-phenyl-4H-1,3-dioxa-2-arsorine)chromium, 46B, 1253
$C_{16}H_{11}CrO_7P$, Pentacarbonyl(6-methyl-4-methylene-2-phenyl-4H-1,3,2-dioxaphosphorin)chromium, 45B, 1336
$C_{16}H_{11}FeO_4P$, Diphenylphosphinetetracarbonyliron, 34B, 600

$C_{16}H_{12}As_2F_4Mn_2O_8$, Octacarbonyl-μ-(1,2-bis(dimethylarsino)-3,3,4,4-tetrafluorocyclobutene)-dimanganese(Mn-Mn), 39B, 844
$C_{16}H_{12}As_2F_4Mn_2O_8$, μ-(Dimethylarsino)-μ-(3,3,4,4-tetrafluoro-2-(dimethylarsino)cyclobut-1-enyl-As,C)-bis(tetracarbonylmanganese), 40B, 1035
$C_{16}H_{12}As_2F_8MoO_4$, 2,2'-Bis(dimethylarsino)octafluorobi-1-cyclobuten-1-ylmolybdenum tetracarbonyl, 43B, 1426
$C_{16}H_{12}F_8NiP_4S_4$, Tetrakis-(2-thienylfluorophosphine)nickel, 41B, 1173
$C_{16}H_{20}As_2PtS_2$, Bis(dimethyl-o-thiolophenylarsine)platinum(II), 37B, 642
$C_{16}H_{20}Fe_2O_{10}P_2$, Bis(μ-5,5-dimethyl-1,3,2-dioxaphosphorinano)-hexacarbonyldiiron(Fe-Fe), 44B, 1059
$C_{16}H_{20}Mo_2O_8P_2$, Di-μ-diethylphosphido-bis-(tetracarbonylmolybdenum), 39B, 845
$C_{16}H_{20}O_8P_2W_2$, Di-μ-diethylphosphido-bis-(tetracarbonyltungsten), 39B, 845
$C_{16}H_{22}AuCl_3P_2Sn$, Bis(dimethylphenylphosphine)trichlorostanniogold, 44B, 1059
$C_{16}H_{22}Br_3NiP_2 \cdot 0.5\ C_{16}H_{22}Br_2NiP_2 \cdot C_6H_6$, Tribromobis(phenyl(dimethyl)phosphine)nickel(III) dibromobis(phenyl(dimethyl)phosphine)nickel(II) benzene solvate, 34B, 601; 35B, 836
$C_{16}H_{22}ClMnN_2O_6P_2$, Chloro-dinitrosylbis(dimethyl phenylphosphonite)-manganese(I), 45B, 1336
$C_{16}H_{22}Cl_2P_2Pd$, cis-Dichlorobis(dimethylphenylphosphine)palladium-(II), 37B, 643
$C_{16}H_{22}Cl_4HgP_2Pt$, Di-μ-chloro-(dichloromercurio)bis(dimethylphenylphosphine)platinum(II), 38B, 1033
$C_{16}H_{22}I_2P_2Pd$, trans-Di-iodobis(dimethylphenylphosphine)palladium-(II), 33B, 507
$C_{16}H_{22}N_2P_2PdS_2$, cis-Bis(thiocyanato)bis(1,3,4-trimethylphosphole)-palladium(II), 46B, 1218
$C_{16}H_{26}As_2Cl_3O_2Rh$, Trichlorobis(o-methoxyphenyldimethylarsine)rhodium(III), 34B, 604
$C_{16}H_{26}As_2CrO_4$, (1,3-Dimethylarsino-2,2,4,4-tetramethylcyclobutane-As,As)tetracarbonylchromium, 39B, 846
$C_{16}H_{27}FeO_4P$, Tetracarbonyl(tri-t-butylphosphine)iron, 42B, 961
$C_{16}H_{28}BCl_3F_4NNiO_{10}P_3$, Nitrosonium tris(1,3,7-trioxa-2-phospha-5-methylbicyclo[2.2.2]octane)nickel(0) tetrafluoroborate chloroform, 41B, 1173
$C_{16}H_{28}NO_{17}P_3Ru_3$, 1,1,1,2,2,3,3-Heptacarbonyl-2,3-μ-hydrido-2,3-μ-nitrosyl-1,2,3-tris(trimethyl phosphite)-triangulo-triruthenium, 46B, 1254
$C_{16}H_{29}B_3P_2Pt$, Bis(dimethylphenylphosphine)platinum heptahydridotriborate(2-), 38B, 1033
$C_{16}H_{30}Fe_2HgN_2O_6P_2$, Bis(dicarbonyl(triethylphosphine)-nitrosylferrio)mercury, 38B, 1034
$C_{16}H_{36}As_3ClNNiO_4P$, Cyanotris(3-dimethylarsinopropyl)phosphinenickel-(II) perchlorate, 32B, 465
$C_{16}H_{36}Br_2F_2NiP_2$, trans-Dibromobis(di(t-butyl)fluorophosphine)nickel-(II), 39B, 847
$C_{16}H_{40}B_4P_2Pt$, 2,3-Dimethyl-1,1-bis(triethylphosphine)-2,3-dicarba-1-platina-closo-heptaborane, 46B, 1254
$C_{16}H_{42}Cl_4N_2O_{10}P_4Ru_2$, Di-μ-chlorobis[chloro[bis(diethyl phosphonito)]nitrosylruthenium(II)], 45B, 1338
$C_{17}H_5CrFe_2O_{11}P$, $μ_3$-Phenylphosphido-triangulo-(bis(tricarbonyliron)-pentacarbonylchromium), 44B, 1060
$C_{17}H_{12}As_2F_4Fe_3O_9$, 1,2-Bis(dimethylarsino)tetrafluorocyclobutenenona-

carbonyltriiron, 34B, 603; 37B, 644
$C_{17}H_{18}NiO_2P_2$, Dicarbonyl(1,3-bis(phenylphosphino)propane)nickel(0), 46B, 1256
$C_{17}H_{27}O_5PW$, Pentacarbonyl(tri-t-butylphosphine)tungsten, 42B, 962
$C_{17}H_{30}MoO_8P_2$, Bis(diisopropoxyphosphine)methylene-tetracarbonylmolybdenum, 46B, 1257
$C_{17}H_{35}B_2CoF_5O_9P_3$, (Cyclopentadienyl)(tris(diethylphosphito)fluoroborato-P,P',P'')cobalt(I) tetrafluoroborate, 43B, 1427
$C_{18}H_{10}ClFe_2O_6P$, μ-Chloro-μ-(diphenylphosphido)hexacarbonyldiiron(Fe-Fe), 46B, 1257
$C_{18}H_{12}As_2F_4Fe_3O_{10}$, Tetrafluorocyclobutenebis(dimethylarsine)triiron decacarbonyl, 35B, 866
$C_{18}H_{12}As_2F_4O_{10}Ru_3$, Decacarbonyl-μ-(1,2-bis(dimethylarsino)tetrafluorocyclobutene-As,As)-triangulo-triruthenium, 37B, 645
$C_{18}H_{12}As_2F_{12}MoO_4$, 2,2'-Bis(dimethylarsino)dodecafluoro(bi-1-cyclopenten-1-yl)molybdenum tetracarbonyl, 42B, 962
$C_{18}H_{15}AgNO_3P$, Triphenylphosphinesilver nitrate, 43B, 1428
$C_{18}H_{15}AsAuBr$, Bromo(triphenylarsine)gold(I), 41B, 1175
$C_{18}H_{15}AuClO_3P$, Chloro(triphenyl phosphite)gold(I), 43B, 1428
$C_{18}H_{15}AuClP$, Chloro(triphenylphosphine)gold(I), 42B, 963
$C_{18}H_{15}AuCl_3P$, Trichloro(triphenylphosphine)gold(III), 39B, 847
$C_{18}H_{15}Br_2CoF_6P_3$, Dibromotris(difluorophenylphosphine)cobalt(II), 43B, 1429
$C_{18}H_{15}ClCuP$, Triphenylphosphinecopper(I) chloride, 39B, 848
$C_{18}H_{15}ClFeN_2O_2P$, Chlorodinitrosyltriphenylphosphineiron(I), 41B, 1176
$C_{18}H_{15}HgN_2O_6P$, Dinitrato(triphenylphosphine)mercury(II), 40B, 940
$C_{18}H_{15}MnN_3O_3P$, Triphenylphosphine-trinitrosylmanganese, 46B, 1085
$C_{18}H_{22}Cl_2O_2P_2Rh_2$, cis-Di-μ-chloro-dicarbonylbis(dimethylphenylphosphine)dirhodium(I), 41B, 1177
$C_{18}H_{24}Br_2N_6NiP_2$, Dibromobis(tris(2-cyanoethyl)phosphine)nickel polymer, 43B, 1430
$C_{18}H_{24}Br_2N_6NiP_2$, Dibromobis(tris(2-cyanoethyl)phosphine)nickel, 43B, 1430
$C_{18}H_{24}Cl_2HgNP$, Dichloro(1-diphenylphosphino-2-diethylaminoethane)-mercury(II), 46B, 1258
$C_{18}H_{26}As_2I_2PdS_2$, Diiodobis(dimethyl-o-methylthiophenylarsine)palladium(II), 35B, 839
$C_{18}H_{28}Co_3FeO_{18}P_3$, μ_3-Hydrido-nonacarbonyl-tris(trimethylphosphite)-iron-tricobalt, 44B, 1061
$C_{18}H_{28}KO_5P_2RhS_4$, Potassium bis(dimethylphenylphosphine)bis(dithiocarbonato)rhodate(III) trihydrate, 42B, 963
$C_{18}H_{34}ClFN_2P_2Pt$, trans-Chlorobis(triethylphosphine)-(p-fluorophenyldiazo)platinum(II), 41B, 1178
$C_{18}H_{36}O_6P_4Re_2Se$, Bis(tricarbonylbis(trimethylphosphine)rhenium)selenide, 45B, 1338
$C_{18}H_{37}BClF_5N_2P_2Pt$, trans-Chlorobis(triethylphosphine)-(p-fluorophenylhydrazine)platinum(II) tetrafluoroborate, 41B, 1179
$C_{18}H_{40}Cl_2P_2Pt$, Dichloro-1,2-bis(di-t-butylphosphino)ethaneplatinum-(II), 42B, 964
$C_{18}H_{42}ClN_2P_2Rh$, trans-Chloro(dinitrogen)bis(tri-isopropylphosphine)-rhodium(I), 43B, 1043; 45B, 1339
$C_{18}H_{42}ClO_2P_2Rh$, trans-Chloro(dioxygen)bis(triisopropylphosphine)-rhodium(I), 43B, 1043
$C_{18}H_{42}Cl_4P_2Pt_2$, trans-Dichlorobis(tripropylphosphine)-μ,μ'-dichlorodiplatinum, 34B, 606
$C_{18}H_{42}Cl_7Mo_3O_{20}P_5$, Tri-μ-chloro-bis(dicarbonylbis(trimethylphos-

phite)molybdenum) tetrachloro(dimethylphosphito)oxomolybdate, 41B, 1179
$C_{18}H_{45}BClF_4P_3Pt$, Chloro-tris(triethylphosphine)platinum(II) tetrafluoroborate, 46B, 1258
$C_{18}H_{45}BF_5P_3Pt$, Fluoro-tris(triethylphosphine)platinum(II) tetrafluoroborate, 46B, 1258
$C_{18}H_{46}F_6P_4Pt$, Hydrido-tris(triethylphosphine)platinum(II) hexafluorophosphate, 46B, 1258
$C_{18}H_{57}BF_4O_3P_6Ru_2$, Tris($\mu$-hydroxo)-hexakis(trimethylphosphine)diruthenium(II) tetrafluoroborate, 46B, 815
$C_{18}H_{57}BF_4P_6Ru_2$, Tri-μ-hydrido-hexakis(trimethylphosphine)diruthenium(II) tetrafluoroborate, 46B, 1260
$C_{18}H_{57}O_{18}P_6Rh_3$, Tris($\mu$-hydrido)tris[bis(trimethyl phosphite)rhodium-(Rh-Rh), 45B, 1361; 46B, 1259
$C_{18}H_{58}P_6Ru_2$, Di-μ-hydrido-dihydridohexakis(trimethylphosphine)diruthenium(II), 46B, 1260
$C_{19}H_5CrFe_2O_{13}P$, μ-(Pentacarbonylchromium-phenylphosphido)bis(tetracarbonyliron), 44B, 1164
$C_{19}H_{10}Fe_2MnO_8P$, Di-μ-carbonyl-(η^5-2,4-cyclopentadien-1-yl)(hexacarbonyldiiron)(μ_3-(phenylphosphinidine))manganese, 42B, 965
$C_{19}H_{15}AuNP$, Cyanotriphenylphosphinegold, 34B, 604
$C_{19}H_{15}Cl_2OPPt$, cis-Carbonyldichlorotriphenylphosphineplatinum(II), 42B, 965
$C_{19}H_{15}FeN_2O_3P$, Dinitrosylcarbonyltriphenylphosphineiron, 40B, 1037
$C_{19}H_{19}BrN_4O_3PRe$, Bromotricarbonyl(phenyl-bis(3,5-dimethylpyrazolyl)-phosphine)rhenium(I), 45B, 1339
$C_{19}H_{19}N_4O_3PW$, Tricarbonyl(phenyl-bis(3,5-dimethylpyrazolyl)phosphine)tungsten(0), 45B, 1339
$C_{19}H_{22}BrMnO_7P_2$, fac-Bromotricarbonylbis(dimethoxyphenylphosphine)-manganese, 41B, 1180
$C_{19}H_{22}BrMnO_7P_2$, mer-trans-Bromotricarbonylbis(dimethoxyphenylphosphine)manganese, 41B, 1180
$C_{19}H_{22}ClIrO_7P_2$, Tricarbonylbis(dimethylphenylphosphine)iridium(I) perchlorate, 39B, 848
$C_{19}H_{22}N_3PPdS_2$, Isothiocyanatothiocyanato-(1-diphenylphosphino-3-dimethylaminopropane)palladium(II), 35B, 840
$C_{19}H_{28}ClCuIrN_3OP_2$, Bis(dimethylphenylphosphine)-carbonyl-iridium-copper-dimethyltriazene chloride, 42B, 966
$C_{19}H_{42}Cl_2P_2Pt \cdot 0.5\ C_6H_5Cl$, Dichloro[1,3-bis(di-t-butylphosphino)-propane]platinum(II) chlorobenzene solvate, 45B, 1340
$C_{20}H_{11}Mn_2O_8$, μ-Hydrido-μ-diphenylphosphido-bis(tetracarbonylmanganese), 32B, 466
$C_{20}H_{15}AsCoNO_3$, Dicarbonylnitrosyltriphenylarsinecobalt(0), 39B, 849
$C_{20}H_{15}AsHgN_2S_2$, Dithiocyanato(triphenylarsine)mercury(II), 41B, 1181
$C_{20}H_{15}CoNO_3P$, Triphenylphosphine-dicarbonyl-nitrosocobalt, 38B, 1035
$C_{20}H_{15}Fe_2O_7P$, μ-Hydroxo-μ-di(p-tolyl)phosphido-hexacarbonyldiiron, 39B, 849
$C_{20}H_{15}HgN_2PS_2$, Bis(thiocyanato)-(triphenylphosphine)mercury(II) (β-form), 45B, 1340, 1341
$C_{20}H_{16}As_2Fe_3O_{10}$, o-Phenylenebis(dimethylarsine)decacarbonyltriiron, 46B, 1261
$C_{20}H_{16}As_2Ir_4O_{10}$, [1,2-Bis(dimethylarsino)benzene]decacarbonyltetrairidium(6Ir-Ir), 45B, 1342; 46B, 1262
$C_{20}H_{17}ClMnN_2O_3P$, Chlorotricarbonyl(diphenyl-3,5-dimethylpyrazolylphosphine)manganese(I), 43B, 1431
$C_{20}H_{20}Co_4P_4$, Tetra(π-cyclopentadienyl)tetraphosphorustetracobalt, 39B, 850

$C_{20}H_{24}MoO_4P_2Si_2$, cis-Tetracarbonyl-1,4-η^2-1,2-bis(phenylphosphino-1,2-dimethyl)disilanemolybdenum(0), 43B, 1432
$C_{20}H_{26}B_9HgP \cdot 0.5\ C_4H_8O_2$, 3-Triphenylphosphine-3-mercura-1,2-dicarbadodecaborane(11) 1,4-dioxan solvate, 45B, 1342
$C_{20}H_{28}Cl_2NPPd$, Dichloro(N,N-dimethyl-α-methyl-o-(butylphenylphosphino)benzylamine)palladium(II), 43B, 1432
$C_{20}H_{28}N_8P_2Pd$, cis-Bis(dimethyl(phenyl)phosphine)bis(5-methyltetrazolato)palladium(II), 39B, 851
$C_{20}H_{30}ClF_2OP_3Pt$, trans-Bis(diethylphenylphosphine)(difluorophosphonato)chloroplatinum(II), 40B, 1038
$C_{20}H_{30}Cl_3OP_2Re$, trans-Oxotrichlorobis(diethylphenylphosphine)-rhenium(V), 28, 656
$C_{20}H_{30}Mn_2O_8P_2$, Bis(tetracarbonyltriethylphosphinemanganese(0)), 33B, 509
$C_{20}H_{32}As_2Cl_3N_2Rh$, mer-Trichlorobis-(o-dimethylaminophenyldimethylarsine)rhodium(III), 35B, 868
$C_{20}H_{32}As_2Cl_3O_4Tc$, Dichlorobis[o-phenylenebis(dimethylarsine)]-technetium(III) perchlorate, 46B, 1262
$C_{20}H_{32}As_2Cl_4F_6PTc$, Tetrachlorobis[o-phenylenebis(dimethylarsine)]-technetium(V) hexafluorophosphate, 46B, 1262
$C_{20}H_{32}As_2Cu_2I_2N_2$, Di-$\mu$-iodo-bis-((o-dimethylaminophenyl)dimethylarsine-As,N)-dicopper(I), 37B, 646
$C_{20}H_{32}As_4AuI_3$, Diiodobis(o-phenylenebis(dimethylarsine))gold(III) iodide, 34B, 606
$C_{20}H_{32}As_4Br_{10}Ta_2$, Tetrabromobis-(o-phenylenebis(dimethylarsine))-tantalum(V) hexabromotantalate(V), 41B, 1182
$C_{20}H_{32}As_4Cl_2CoNO_9$, Nitrosylbis(o-phenylenebis(dimethylarsine))cobalt(II) perchlorate, 41B, 1182
$C_{20}H_{32}As_4Cl_2CoO_8$, Bis(o-phenylenebis(dimethylarsine))cobalt(II) perchlorate, 32B, 468
$C_{20}H_{32}As_4Cl_2FeNO_9$, Nitrosylbis(o-phenylenebis(dimethylarsine))iron perchlorate, 43B, 1433
$C_{20}H_{32}As_4Cl_2Pt$, Dichlorodi-(o-phenylenebisdimethylarsine)platinum-(II), 29, 578
$C_{20}H_{32}As_4Cl_3Co$, Dichlorobis(o-phenylenebis(dimethylarsine))cobalt chloride, 38B, 1035
$C_{20}H_{32}As_4Cl_3CoO_4$, trans-Dichlorobis-(o-phenylene(bisdimethylarsine))cobalt(III) perchlorate, 35B, 869
$C_{20}H_{32}As_4Cl_3Ni$, Dichlorobis(o-phenylenebis(dimethylarsine))nickel chloride, 38B, 1035
$C_{20}H_{32}As_4Cl_3O_4Tc$, trans-Dichlorobis[o-phenylenebis(dimethylarsine)]-technetium(III) perchlorate, 46B, 1263
$C_{20}H_{32}As_4Cl_3Tc$, trans-Dichlorobis[o-phenylenebis(dimethylarsine)]-technetium(III) chloride, 46B, 1263
$C_{20}H_{32}As_4Cl_4I_3Mo$, Tetrachlorobis-(o-phenylenebis(dimethylarsino))molybdenum(V) triiodide, 40B, 1038
$C_{20}H_{32}As_4Cl_4Ti$, o-Phenylenebis(dimethylarsine)titanium tetrachloride, 26, 687
$C_{20}H_{32}As_4Cl_8Nb_2O$, Tetrachlorobis(o-phenylenebis(dimethylarsine))-niobium(V) tetrachlorooxoniobate(V), 41B, 1184
$C_{20}H_{32}As_4I_2Ni$, Diiododi-(o-phenylenebisdimethylarsine)nickel(II), 29, 576
$C_{20}H_{32}As_4I_2Pd$, Di-iododi(o-phenylenebisdimethylarsine)palladium(II), 27, 869
$C_{20}H_{32}As_4I_2Pt$, Di-iododi(o-phenylenebisdimethylarsine)platinum(II), 27, 869
$C_{20}H_{33}HgN_2PS_2$, Dithiocyanato(tricyclohexylphosphine)mercury(II),

43B, 1434
$C_{20}H_{36}Cr_2O_8P_8Si_6$, Bis(hexamethyltrisilatetraphosphanortricyclene-chromiumtetracarbonyl), 46B, 1246
$C_{20}H_{46}ClP_2Rh$, trans-Chloro(ethylene)bis(triisopropylphosphine)rhodium(I), 43B, 1043
$C_{21}H_{10}As_2Fe_3O_9$, Nonacarbonyl-bis(μ_3-phenylarsinidene)triiron(2Fe-Fe), 43B, 1435
$C_{21}H_{19}CrO_9P$, Pentacarbonyl[bis(1-methyl-3-oxo-1-butanyloxy)phenylphosphane]chromium, 45B, 1336
$C_{21}H_{23}AgAs_3CoO_4$, Tetracarbonyl((bis-(o-dimethylarsinophenyl)methylarsine)argentio)cobalt, 37B, 646
$C_{21}H_{25}As_2NPdS_2$, Bis(dimethyl-o-thiolophenylarsine)palladium(II) pyridine solvate, 38B, 1036
$C_{21}H_{32}As_4Br_2OOs$, Dibromocarbonylbis(o-phenylenebis(dimethylarsine))-osmium, 44B, 1062
$C_{21}H_{40}Fe_2O_{15}P_4$, Bis(tetraethyl pyrophosphite)pentacarbonyldiiron(Fe-Fe), 44B, 1063
$C_{21}H_{45}CrO_3P_3$, fac-Tricarbonyl-tris(trimethylphosphine)chromium, 46B, 1264
$C_{22}H_{10}Co_4O_{10}P_2$, Bis[($\mu$-carbonyl)-($\mu_4$-phenylphosphido)]tetrakis(dicarbonylcobalt) (monoclinic), 46B, 1264
$C_{22}H_{10}Co_4O_{10}P_2$, Bis[($\mu$-carbonyl)-($\mu_4$-phenylphosphido)]tetrakis(dicarbonylcobalt) (triclinic), 41B, 1183
$C_{22}H_{15}AuCoO_4P$, Tetracarbonyl(triphenylphosphineaurio)cobalt, 37B, 646
$C_{22}H_{15}ClMnO_4P$, Chlorotetracarbonyl(triphenylphosphine)manganese(I), 37B, 647
$C_{22}H_{15}FeO_4P$, (Triphenylphosphine)tetracarbonyliron, 46B, 1293
$C_{22}H_{15}O_4RuSb$, Tetracarbonyl(triphenylstibine)ruthenium, 45B, 1343
$C_{22}H_{16}AsF_4FeO_4P$, Tetracarbonyl-(1-(dimethylarsino)-2-(diphenylphosphino)-tetrafluorocyclobutene-P)-iron, 38B, 1036
$C_{22}H_{18}As_3ClFe_2Mn_2O_{16}$, Tetracarbonylbis($\mu$-(dimethylarsenido))(tetracarbonyliron)manganese tetracarbonylchloro-μ-(dimethylarsenido)-(tetracarbonyliron)manganese, 46B, 1265
$C_{22}H_{26}Cl_2NiP_2$, cis-Dichlorobis(1-benzyl-Δ^3-phospholen)nickel(II), 38B, 1037
$C_{22}H_{32}As_4CoN_3OS_2$, trans-Nitrosylisothiocyanatobis(o-phenylenebis(dimethylarsine))cobalt(II) thiocyanate, 41B, 1182
$C_{22}H_{34}ClNO_3P_2PtSi$, Nitrilotriethoxysilatranebis(phenyldimethylphosphine)chloroplatinum(II), 43B, 1436
$C_{22}H_{37}As_4Cl_9OTa_2$, Tetrachlorobis(o-phenylenebis(dimethylarsine))-tantalum(V) pentachloro(ethoxo)tantalate(V), 41B, 1184
$C_{22}H_{40}OsP_4$, Hydridobis-(1,2-bis(dimethylphosphino)ethane)naphthyl-osmium(II), 37B, 648
$C_{22}H_{40}P_4Ru$, Hydridobis-(1,2-bis(dimethylphosphino)ethane)naphthyl-ruthenium(II), 37B, 648
$C_{22}H_{41}CoF_3N_8O_4P$, trans-Bis(dimethylglyoximato)-2-(5-trifluoromethyl-tetrazolato)-(tri-n-butylphosphine)cobalt(III), 46B, 1266
$C_{22}H_{41}NiP$, Bis(ethylene)(tricyclohexylphosphine)nickel, 38B, 1038
$C_{22}H_{42}Mo_2O_6P_4$, Di-μ-dimethylphosphidobis(tricarbonyl)triethylphosphinemolybdenum, 32B, 469
$C_{22}H_{52}Cl_2P_2Pt_2S_2$, Dichloro-bis-$\mu$-ethanethiolato-bis(tripropylphosphine)diplatinum(II), 38B, 1039
$C_{23}H_{15}As_2Co_2O_5P$, (μ-Diarsenic)triphenylphosphine-pentacarbonyl-dicobalt, 45B, 1344
$C_{23}H_{15}Co_2O_5P_3$, (μ-Diphosphorus)(triphenylphosphine)pentacarbonyl-dicobalt, 45B, 1344

$C_{23}H_{15}CrO_5P$, Pentacarbonyltriphenylphosphinechromium(0), 39B, 852
$C_{23}H_{15}CrO_8P$, Pentacarbonyl(triphenyl phosphite)chromium(0), 39B, 852
$C_{23}H_{16}CoF_4O_3P$, Tricarbonyltriphenylphosphine-σ-tetrafluoroethyl cobalt(I), 32B, 470
$C_{23}H_{18}MnO_4P$, Methyltetracarbonyltriphenylphosphinemanganese(I), 38B, 1039
$C_{23}H_{20}As_2I_2MoO_3$, meso-Tricarbonyldiiodo(o-phenylenebis(methylphenylarsine))molybdenum(II), 41B, 1185
$C_{23}H_{47}BrP_2Pd_2$, μ-Cyclopentadienyl-μ-bromo-bis(triisopropylphosphine)dipalladium, 43B, 1086
$C_{24}H_{20}Fe_2N_4O_4P_2$, Di-μ-diphenylphosphino-bis(dinitrosyliron), 45B, 1327
$C_{24}H_{21}As_2Cl_3I_2MoO_3$, rac-Tricarbonyldiiodo(o-phenylenebis(methylphenylarsine))molybdenum(II) chloroform solvate, 41B, 1185
$C_{24}H_{23}Br_3NPZn$, 4-Methylpyridinium triphenylphosphinetribromozincate, 37B, 649
$C_{24}H_{24}As_4F_8O_8Ru_3$, Octacarbonylbis-μ-(bis(dimethylarsino)tetrafluorocyclobutene-AsAs)-triangulo-triruthenium, 33B, 845
$C_{24}H_{24}FeO_2P_2S_2$, (η^2-Carbon disulfide)dicarbonyl(trimethylphosphine)-(triphenylphosphine)iron(0), 44B, 1063
$C_{24}H_{26}ClCuP_2S$, Chloro[cis-2,10-diphenyl-2,10-diphospha-κ^2P-6-thia-κS-bicyclo[9.4.0]pentadeca-11(1),12,14-triene]copper(I), 46B, 1267
$C_{24}H_{26}Cl_2P_2Pd$, cis-Dichlorobis(1-phenyl-3,4-dimethylphosphole)palladium(II), 46B, 1267
$C_{24}H_{30}Br_2Cl_2NPPt$, 2-Chloroethyldiethylammonium trans-dibromochloro-(triphenylphosphine)platinate(II), 42B, 966
$C_{24}H_{30}Cl_2O_3P_2Pt$, cis-Dichloro(triethylphosphine)(triphenyl phosphite)platinum(II), 43B, 1437
$C_{24}H_{33}BF_4MnN_2O_8P_3$, Dinitrosyltris(dimethylphenylphosphonite)manganese(I) tetrafluoroborate, 45B, 1345
$C_{24}H_{33}Br_3P_3Rh$, Tribromotris(dimethylphenylphosphine)rhodium, 45B, 1345
$C_{24}H_{33}Cl_2MoOP_3$, Oxodichlorotris(dimethylphenylphosphine)molybdenum-(IV), 37B, 650
$C_{24}H_{33}Cl_3P_3Tc$, mer-Trichlorotris(dimethylphenylphosphine)-technetium(III), 42B, 967
$C_{24}H_{34}As_4Cl_7I_3Mo$, Dicarbonylchlorobis-(o-phenylenebis(dimethylarsino))molybdenum(II) tri-iodide - bischloroform, 39B, 853
$C_{24}H_{34}MoO_4P_2Si_4$, (1,4-Diphenyl-2,2',3,3',5,5',6,6'-octamethyl-cyclo-1,4-diphospha-2,3,5,6-tetrasilahexane)-tetracarbonyl-molybdenum, 45B, 1346
$C_{24}H_{36}Br_2ClN_2P_3W$, (Bromo-chloro-hydrazido-tri(dimethylphenylphosphine)tungsten) bromide, 45B, 1346
$C_{24}H_{37}OsP_3$, Tris(dimethylphenylphosphine)osmium tetrahydride, 43B, 1437
$C_{24}H_{38}As_4ClCoO_6$, (+)-Δ-cis-β-((2,13-Dimethyl-6,9-diphenyl-2,6,9,13-tetraarsatetradecane)(dioxygen)cobalt(III)) perchlorate, 43B, 1438
$C_{24}H_{39}ClN_2P_2Pd$, trans-Chloro-2-(phenylazo)phenylbis(triethylphosphine)palladium(II), 35B, 842
$C_{24}H_{45}Cl_3P_2Pt$, Dichloro-1,2-bis(di-t-butylphosphino)ethaneplatinum-(II) - chlorobenzene, 42B, 964
$C_{24}H_{46}Cl_2NiP_2$, cis-Bis(dicyclohexylphosphine)dichloronickel(II), 44B, 1064
$C_{24}H_{50}Cl_2NiP_2$, trans-Dichlorobis(dimethylneomenthylphosphine)nickel-(II), 43B, 1439
$C_{24}H_{50}Cl_2P_2Pd$, trans-Dichlorobis(dimethylmenthylphosphine)palladium-

(II), 44B, 1064
$C_{24}H_{50}Cl_2P_2Pd$, trans-Dichlorobis(dimethylneomenthylphosphine)palladium(II), 44B, 1064
$C_{24}H_{54}P_2Pt$, Bis(tri-t-butylphosphine)platinum(0), 45B, 1347
$C_{24}H_{55}Br_3NiP_2$, Tri-t-butylphosphonium tribromo(tri-t-butylphosphine)nickelate(II), 41B, 1186
$C_{24}H_{56}P_2Pt$, trans-Dihydrido-bis(tri-t-butylphosphine)platinum(II), 45B, 1347
$C_{24}H_{60}Ag_4Br_4P_4$, Triethylphosphinesilver(I) bromide, 42B, 968
$C_{24}H_{60}Ag_4Cl_4P_4$, Triethylphosphinesilver(I) chloride, 42B, 968
$C_{24}H_{60}Ag_4I_4P_4$, Tetrakis((triethylphosphine)silver(I) iodide), 41B, 1187
$C_{24}H_{60}As_4Cu_4I_4$, Triethylarsinecopper(I) iodide tetramer, 40B, 1040
$C_{24}H_{60}Br_4Cu_4P_4$, Tetrakis(triethylphosphinecopper(I) bromide), 41B, 1187
$C_{24}H_{60}Cl_4Cu_4P_4$, Tetrakis(triethylphosphinecopper(I) chloride), 41B, 1187
$C_{24}H_{60}Cl_4P_4Re_2$, Tetrachlorotetrakis(triethylphosphine)dirhenium(II), 42B, 968
$C_{24}H_{60}Cu_4I_4P_4$, Triethylphosphinecopper(I) iodide tetramer, 40B, 1040
$C_{24}H_{62}B_2F_8O_2P_4Pt_2$, Di-($\mu$-hydroxy)-tetrakis(triethylphosphine)diplatinum(II) tetrafluoroborate, 44B, 1065
$C_{25}H_{15}N_2OPRhS_2$, Carbonyl(triphenylphosphine)(1-(ethylthio)maleonitrile-2-thiolato)rhodium(I), 45B, 1348
$C_{25}H_{22}Au_2Cl_2P_2$, (Bis(diphenylphosphino)methane)dichlorodigold, 43B, 1440
$C_{25}H_{22}Br_2Cu_2P_2$, Bis(diphenylphosphino)methane-bis(copper(I) bromide), 41B, 1188
$C_{25}H_{22}Cl_2P_2Pd$, Dichloro(bis(diphenylphosphino)methane)palladium(II), 42B, 969
$C_{25}H_{22}Cu_2I_2P_2$, Bis(diphenylphosphino)methane-bis(copper(I) iodide), 41B, 1188
$C_{25}H_{22}NiO_{11}P_2$, Tricarbonyl-(trimethyl 1-methoxy-4-oxo-1,2-diphenyl-1,2-diphosphacyclopenten-3,3,5-tricarboxylate)nickel, 45B, 1348
$C_{25}H_{24}MnO_4PSi$, Tetracarbonyl(trimethylsilyl)(triphenylphosphine)manganese(I), 42B, 970
$C_{25}H_{26}As_4Co_4F_8O_9$, 2,2'-Dimethylarsino-octafluoro-1,1'-bicyclobut-1-enyl cobalt carbonyl complex, 38B, 1040
$C_{25}H_{35}Cl_4P_3Pd$, Dichlorotris(dimethylphenylphosphine)palladium dichloromethane solvate, 42B, 970
$C_{25}H_{49}ClNOP_2RhS$, Chlorobis(triisopropylphosphine)-4-methyl-N-sulfinyl-anilinerhodium, 45B, 1349
$C_{25}H_{54}ClOP_2Rh$, Carbonylchlorobis(tri-t-butylphosphine)rhodium(I), 43B, 1440
$C_{26}H_{20}Co_2N_2O_4P_2$, Di-$\mu$-diphenylphosphino-bis(carbonylnitrosylcobalt), 45B, 1327
$C_{26}H_{22}Au_2Cl_2P_2$, μ-cis-1,2-Bis(diphenylphosphino)ethylenebis[chlorogold(I)], 46B, 1268
$C_{26}H_{22}Br_2NiP_2$, trans-Dibromobis(5-methyl-5H-dibenzophosphole)nickel-(II), 42B, 971
$C_{26}H_{22}Mn_2O_4P_2$, Bis(cyclopentadienyldicarbonylmanganese)-1,2-diphenyldiphosphane, 42B, 971
$C_{26}H_{24}CoF_6N_2O_2P_3$, Dinitrosyl(1,2-bis(diphenylphosphino)ethane)cobalt hexafluorophosphate, 43B, 1442
$C_{26}H_{24}CoIN_2O_3P_2$, ((2-(Diphenylphosphino)ethyl)diphenylphosphine oxide)iodonitrosylcobalt(0), 40B, 916
$C_{26}H_{25}Cl_2NP_2Pd$, Dichlorobis(diphenylphosphino)ethylaminepalladium-

(II), 30B, 387; 39B, 854

$C_{26}H_{26}AuF_6P_3$, Bis(diphenylmethylphosphine)gold(I) hexafluorophosphate, 42B, 971

$C_{26}H_{26}Au_2P_2S_4W$, Bis(di-μ-sulfido-methyldiphenylphosphinegold)tungsten, 42B, 972

$C_{26}H_{26}Cl_2CoNOP_2$, Dichlorobis(methyldiphenylphosphine)nitrosylcobalt(I), 39B, 855

$C_{26}H_{26}Cl_2O_2P_2Pd$, cis-Dichlorobis(methyldiphenylphosphinite)palladium(II), 46B, 1268

$C_{26}H_{26}Cl_3NOP_2Ru$, Trichloronitrosylbis(methyldiphenylphosphine)ruthenium(II), 40B, 1041

$C_{26}H_{27}O_2P_3PdS_2$, Dimethylphosphinodithioato(diphenylphosphinito)(diphenylphosphinous acid)palladium(II), 43B, 1442

$C_{26}H_{29}ClCoN_4O_4P$, Chlorobis(dimethylglyoximato)triphenylphosphinecobalt(III), 41B, 1189

$C_{26}H_{29}ClN_4O_4PRh$, Chlorobis(dimethylglyoximato)triphenylphosphinerhodium(III), 37B, 650

$C_{26}H_{33}N_2NiP_3$, Dicyanotris(phenyldimethylphosphine)nickel(II), 34B, 609

$C_{26}H_{36}F_{12}N_2NiP_4 \cdot 0.5\ H_2O$, (14,16-Dimethyl-5,9-diphenyl-5,9-diphosphino-1,13-diazacyclohexadeca-13,16-diene)-nickel(II) bis(hexafluorophosphate) hemihydrate, 46B, 1269

$C_{26}H_{38}Cl_2NiO_4P_2$, trans-Dichlorobis-(4,4-dimethoxy-1-phenylphosphorinan)nickel(II), 38B, 1041

$C_{26}H_{39}Cl_4MoOP_3$, Tetrachlorotris(dimethylphenylphosphine)molybdenum(IV) - ethanol, 42B, 972

$C_{26}H_{42}Cl_2CoN_3O_4P$, Chloro(bis-(2-(diethylamino)ethyl)-2-(diphenylphosphino)ethylamine)cobalt(II) perchlorate, 38B, 1041

$C_{26}H_{46}ClIrOP_2$, Carbonylchloro(1,9-bis(di-t-butylphosphino)nona-1,3-diyne)iridium, 44B, 1066

$C_{26}H_{52}CoN_8O_9P$, Bis(dimethylglyoximato)(xanthinato)-(tri-n-butylphosphine)cobalt(III) dihydrate monomethanolate, 41B, 1190

$C_{27}H_{15}Fe_2O_9PPt$, μ-(Carbonyl(triphenylphosphine)platinio)-octacarbonyldiiron(2Pt-Fe,1Fe-Fe), 40B, 1041

$C_{27}H_{16}AsF_4Fe_3O_9P$, Nonacarbonyl-μ-dimethylarsino-μ-(2-(diphenylphosphino)tetrafluorocyclobut-1-enyl)-tri-iron, 38B, 1042

$C_{27}H_{17}O_9Os_3P$, Nonacarbonyl-di-μ-hydrido-(triphenylphosphine)-triangulo-triosmium, 45B, 1349

$C_{27}H_{22}As_2CrO_2$, Bis(diphenylarsino)methanechromium dicarbonyl, 40B, 1042

$C_{27}H_{22}N_2P_2PdS_2$, Dithiocyanato(bis(diphenylphosphino)methane)palladium(II), 41B, 1191

$C_{27}H_{26}Cl_2P_2Pd$, Dichloro(1,3-bis(diphenylphosphino)propane)palladium(II), 42B, 969

$C_{27}H_{26}Cl_4P_2Pd$, Dichloro(bis(diphenylphosphino)ethane)palladium(II) dichloromethane solvate, 42B, 969

$C_{27}H_{29}CoN_5O_4PS$, Isothiocyanatobis(dimethylglyoximato)(triphenylphosphine)cobalt(III), 44B, 1066

$C_{27}H_{32}Cl_4N_4O_4PRh$, Chlorodimethylglyoximatodimethylglyoximetriphenylphosphinerhodium(III) chloride methylene chloride solvate, 39B, 856

$C_{27}H_{33}HgN_2O_6P$, Dinitrato(trimesitylphosphine)mercury(II) dimer, 45B, 1350

$C_{27}H_{37}Cl_2MoO_3P_3$, Dichlorodicarbonyltris(dimethylphenylphosphine)molybdenum(II) methanol solvate, 38B, 1043

$C_{27}H_{37}Cl_3NOP_2Re$, Trichloro(p-acetylphenylimino)bis(diethylphenylphosphine)rhenium(V), 33B, 511

$C_{27}H_{39}Cl_3O_2P_3Tc$, Carbonyltrichlorotris(dimethylphenylphosphine)tech-

netium ethanol solvate, 44B, 1066
$C_{27}H_{52}ClOP_2Rh$, Carbonylchloro(dec-5-yne-1,10-diylbis(di-t-butylphosphine))rhodium(I), 44B, 1067
$C_{27}H_{64}P_3Rh$, Tris(triisopropylphosphine)hydridorhodium(I), 46B, 1270
$C_{28}H_{15}CrO_5P$, Pentacarbonyl(2,4,6-triphenylphosphorine)chromium(0), 39B, 857
$C_{28}H_{17}O_{10}Os_3P$, Decacarbonyl(triphenylphosphine)dihydridotriosmium, 43B, 1443
$C_{28}H_{18}Co_3O_8P$, Methinyltricobalt triphenylphosphine octacarbonyl, 35B, 846
$C_{28}H_{20}Ni_2P_2O_4$, Bis-μ-diphenylphosphido-bis(dicarbonylnickel), 35B, 848
$C_{28}H_{22}As_2I_2O_3W$, Diiodotricarbonylbis(diphenylarsino)methanetungsten(II), 43B, 1443
$C_{28}H_{22}FeO_3P_2$, (Bis(diphenylphosphino)methane)tricarbonyliron, 39B, 857
$C_{28}H_{22}I_2O_3P_2W$, [Bis(diphenylphosphino)methane-P,P']tricarbonyldiiodotungsten, 46B, 1273
$C_{28}H_{24}N_2P_2PdS_2$, Isothiocyanatothiocyanato-(1,2-bis(diphenylphosphino)ethane)palladium(II), 41B, 1191
$C_{28}H_{26}Br_2O_4P_2Pd$, Dibromobis(diphenylphosphinoacetic acid)palladium(II), 45B, 1350
$C_{28}H_{26}Cl_2N_2O_2PRe$, [Acetone benzoylhydrazonido(1-)-N',O]dichloro-oxo-(triphenylphosphine)rhenium(V), 45B, 1351
$C_{28}H_{26}CrINO_3P_2$, trans-Dicarbonyl-trans-bis(methyldiphenylphosphine)-iodo(nitrosyl)chromium, 42B, 973
$C_{28}H_{26}MoO_4P_2$, (η^2-7,10-Diphenyl-1,3,4,9-tetramethyl-7,10-diphosphatricyclo[4.3.0.1$^{2-5}$]deca-3,8-diene-P,P)tetracarbonylmolybdenum(0), 46B, 1270
$C_{28}H_{26}N_2NiP_2S_2$, Di-isothiocyanatobiš(methyldiphenylphosphine)nickel(II), 44B, 1068
$C_{28}H_{28}Cl_2NiOP_2$, Dichloro(oxydiethylenebis(diphenylphosphine))nickel(II), 35B, 849
$C_{28}H_{28}HgI_2P_2S$, Diiodo(bis(diphenylphosphinoethyl)sulfide)mercury(II), 39B, 858
$C_{28}H_{28}MoOP_2S_2$, Bis(2-diphenylphosphinoethane-1-thiolato)oxomolybdenum(IV), 45B, 1351
$C_{28}H_{29}Br_2NNiP_2$, Dibromobis-(2-diphenylphosphinoethyl)aminenickel(II), 35B, 850
$C_{28}H_{31}ClP_2Pt$, Hydridochlorobis(diphenylethylphosphine)platinum, 30B, 388
$C_{28}H_{37}Cl_3NOP_2Re$, Trichloro(p-methoxyphenylimino)bis(diethylphenylphosphine)rhenium(V), 33B, 511
$C_{28}H_{38}Cl_2NiP_2$, Bis(9-phenyl-9-phosphabicyclo[3.3.1]nonane)nickel chloride, 38B, 1044
$C_{28}H_{40}NiP_2$, trans-Bis(phenylethynyl)bis(triethylphosphine)nickel(II), 32B, 474
$C_{28}H_{41}B_{20}ClP_2Pt$, Chloro[1-diphenylphosphino-1,2-dicarba-closo-dodecaborane(12)-P][1-diphenylphosphino-1,2-dicarba-closo-dodecaboran-3-yl-B^3,P]platinum, 46B, 1271
$C_{28}H_{44}O_2P_2Pt$, trans-Bis(2-(di-t-butylphosphinato)phenoxy)platinum(II), 43B, 1444
$C_{28}H_{46}Cl_2P_2Pt$, cis-Bis(di-tert-butylphenylphosphine)dichloroplatinum(II), 46B, 1272
$C_{28}H_{46}O_2P_2Pd \cdot C_7H_8$, Dioxygen-bis[di(t-butyl)phenylphosphine]palladium(II) toluene solvate, 45B, 1352
$C_{28}H_{46}O_2P_2Pt \cdot C_7H_8$, Dioxygen-bis[di(t-butyl)phenylphosphine]platin-

um(II) toluene solvate, 45B, 1352
$C_{28}H_{46}P_2Pd$, Bis(phenyldi-t-butylphosphine)palladium(0), 42B, 973
$C_{28}H_{46}P_2Pt$, Bis(phenyldi-t-butylphosphine)platinum(0), 42B, 973
$C_{28}H_{47}N_2P_2Rh$, Hydrido(dinitrogen)bis(phenyl(di-t-butyl)phosphine)-rhodium(I), 42B, 974
$C_{28}H_{54}Br_2O_4P_2Ru_2$, Di-$\mu$-bromo-bis[dicarbonyl(tri-t-butylphosphine)-ruthenium(I)], 46B, 1272
$C_{29}H_{15}Fe_3O_{11}P$, Undecacarbonyl(triphenylphosphine)triiron, 33B, 512
$C_{29}H_{15}O_{11}PRu_3$, Undecacarbonyl(triphenylphosphine)-triangulo-triruth-enium, 45B, 1353
$C_{29}H_{20}F_6FeN_2O_2P_2$, (1,2-Bis(diphenylphosphino)hexafluorocyclopen-tene)dinitrosyliron, 37B, 651
$C_{29}H_{22}MoO_4P_2$, Bis(diphenylphosphino)methanetetracarbonylmolybdenum, 37B, 652
$C_{29}H_{24}I_2MoO_3P_2 \cdot CH_2Cl_2$, [1,2-Bis(diphenylphosphino)ethane-P,P']tri-carbonyldiiodomolybdenum methylene dichloride solvate, 46B, 1273
$C_{29}H_{26}N_2P_2PdS_2$, Diisothiocyanato-(1,3-bis(diphenylphosphino)-propane)palladium(II), 41B, 1191
$C_{29}H_{28}F_6O_2P_3Rh$, trans-(1,5-Bis(diphenylphosphino)-3-oxapentane-O,P,P')carbonylrhodium(I) hexafluorophosphate, 42B, 974
$C_{29}H_{30}Cl_2MoN_3OP$, (N-Benzoyl-N'-p-tolydiazene-N',O)dichloro(dimethyl-phenylphosphine)(p-tolylimido)-molybdenum, 45B, 1353
$C_{29}H_{30}HgI_2P_2$, Diiodo(1,5-bis(diphenylphosphino)pentane)mercury(II), 42B, 975
$C_{29}H_{33}Cl_3NP_2Re$, Trichloro(methylimino)bis(diphenylethylphosphine)-rhenium(V), 34B, 609
$C_{29}H_{39}Br_2MoO_3P_3$, Dibromodicarbonyltris(dimethylphenylphosphine)mo-lybdenum(II) - acetone, 43B, 1444
$C_{30}H_{20}Br_2O_6P_2Re_2$, Di-$\mu$-bromo-$\mu$-(tetraphenyldiphosphane)-bis(tricar-bonylrhenium(I)), 44B, 1069
$C_{30}H_{20}Ni_2O_6P_2$, μ-Tetraphenyldiphosphinebis(tricarbonylnickel), 32B, 475
$C_{30}H_{22}Cl_4F_6P_2Pd$, trans-Dichlorobis(2-chloro-3,3,3-trifluoropropenyl-diphenylphosphine)palladium(II), 41B, 1192
$C_{30}H_{24}CrO_4P_2$, (1,2-Bis-(diphenylphosphino)ethane)tetracarbonyl-chromium, 37B, 653
$C_{30}H_{25}Cl_5F_{10}Fe_2P_5$, Chloropentakis(difluorophenylphosphine)iron(II) tetrachloroferrate(III), 43B, 1446
$C_{30}H_{25}MoNO_4P_2$, Bis(diphenylphosphino)ethylaminetetracarbonylmolyb-denum, 30B, 387
$C_{30}H_{26}I_2MoO_3P_2$, [1,3-Bis(diphenylphosphino)propane-P,P']tricarbonyl-diiodomolybdenum, 46B, 1273
$C_{30}H_{28}Cl_3F_3O_2P_2Pd$, Chloro(1-(diphenylphosphino)-3,3,3-trifluoropro-pen-2-olato)(ethoxydiphenylphosphine)palladium(II) dichloromethane solvate, 42B, 976
$C_{30}H_{30}O_2P_2Pt$, Dicarbonylbis(diphenylethylphosphine)platinum(0), 38B, 1044
$C_{30}H_{32}I_2NiO_2P_2$, Di-iodo(ethylenebis(oxyethylene))bis(diphenylphos-phine)nickel(II), 37B, 653
$C_{30}H_{33}ClMoN_4O_2P_2$, (N-Benzoyl-diazene-N'O)[N-benzoyldiazeneido(1-)-N']chlorobis(dimethylphenylphosphine)molybdenum, 45B, 1354
$C_{30}H_{36}As_4Cl_2OPd$, o-Phenylenebis(o-dimethylarsino-phenylmethylar-sine)chloropalladium(II) perchlorate benzene solvate, 32B, 472
$C_{30}H_{45}Cl_2MoOP_3$, Dichlorotris(diethylphenylphosphine)oxomolybdenum-(IV), 38B, 1045
$C_{30}H_{45}Cl_2NP_3Re$, Nitridodichlorotris(diethylphenylphosphine)-rhenium(V), 32B, 477

$C_{30}H_{45}Cl_2NiO_{23}P_5$, Pentakis-2,8,9-trioxaphosphaadamantane-nickel(II) perchlorate, 35B, 852
$C_{30}H_{45}Cl_3NP_3Ru$, Trichlorobis(diethylphenylphosphine)(diethylphenylphosphineiminato)ruthenium(IV), 42B, 976
$C_{30}H_{45}Cl_3P_3Rh$, Trichlorotris(diethylphenylphosphine)rhodium(III), 39B, 858
$C_{30}H_{45}Cl_9P_3Re_3$, Tris(diethylphenylphosphine)nonachlorotrirhenium-(III), 29, 626
$C_{30}H_{54}Co_2O_6P_2$, Bis((tri-n-butylphosphine)tricarbonylcobalt), 33B, 513
$C_{31}H_{20}CoO_6P_2$, [2,3-Bis(diphenylphosphino)maleic anhydride]tricarbonylcobalt(0), 45B, 1355
$C_{31}H_{20}Co_3O_7P_2$, 1,2-μ-Carbonyl-1,1,2,2,3,3-hexacarbonyl-1,3:2,3-bis-μ-diphenylphosphido-triangulo-tricobalt, 44B, 1069
$C_{31}H_{20}FeO_6P_2$, [2,3-Bis(diphenylphosphino)maleic anhydride]tricarbonyliron(0), 45B, 1355
$C_{31}H_{22}As_2Cl_2O_6Re_2 \cdot 0.25\ C_6H_{14}$, μ-Bis(diphenylarsino)methane-hexacarbonyldi-μ-chloro-dirhenium(I) n-hexane, 41B, 1193
$C_{31}H_{24}ClN_6OP_2Rh$, trans-Carbonylchlorobis(tris(2-pyridylphosphine))-rhodium, 46B, 1275
$C_{31}H_{30}Cl_2N_2PRh \cdot 0.61\ CHCl_3$, Dichlorodipyridine-(o-(di-o-tolylphosphino)benzyl)rhodium(III)-0.61 chloroform, 35B, 852
$C_{31}H_{32}Cl_2NiO_2P_2$, (-)-2,2-Dimethyl-4,5-bis(diphenylphosphinomethyl)-1,3-dioxolanenickel chloride, 45B, 1356
$C_{31}H_{32}Cl_2O_2P_2Pd$, (-)-2,2-Dimethyl-4,5-bis(diphenylphosphinomethyl)-1,3-dioxolanepalladium chlorde, 45B, 1356
$C_{31}H_{32}Cl_2O_2P_2Pt$, (-)-2,2-Dimethyl-4,5-bis(diphenylphosphinomethyl)-1,3-dioxolaneplatinum chloride, 45B, 1356
$C_{31}H_{35}As_2I_2NNiO$, Diiodo-(N,N-bis(2-diphenylarsinoethyl)-2-methoxyethylamino)nickel(II), 39B, 859
$C_{31}H_{35}I_2NNiOP_2$, Iodo-(N,N-bis-(2-diphenylphosphinoethyl)-2-methoxyethylamino)-nickel(II) iodide, 35B, 853
$C_{31}H_{44}F_4FeO_3P_2$, (1,2-Bis(dicyclohexylphosphino)-3,3,4,4-tetrafluorocyclobut-1-ene-P,P')tricarbonyliron, 44B, 1070
$C_{31}H_{54}Co_2N_6O_{24}P_5$, Pentakis(4-methyl-1-phospha-2,6,7-trioxabicyclo-[2.2.2]octane)cobalt(I) (acetonitrile)trinitratocobaltate(II) bis(acetonitrile), 43B, 1447
$C_{32}H_{20}Mn_2O_{10}P_2$, 3,3,3,3,6,6,6,6-Octacarbonyl-2,2,5,5-tetraphenyl-1,4-dioxa-2,5-diphospha-3,6-dimanganacyclohexane, 43B, 880
$C_{32}H_{22}Fe_2O_7P_2$, (Tetraphenyldiphosphinomethane)heptacarbonyldiiron, 40B, 1043
$C_{32}H_{26}F_{12}O_3P_2Pt$, 2-Bis(methyldiphenylphosphine)-4,4,6,6-tetrakis-(trifluoromethyl)-1,3,5,2-trioxaplatinan, 41B, 1193
$C_{32}H_{30}Br_2MoO_4P_2$, Dibromotricarbonyl(1,2-bis(diphenylphosphino)ethane)molybdenum(II)-1-acetone, 38B, 1045
$C_{32}H_{35}F_6N_2NiOP_3S$, Isothiocyanato(N,N-bis-(2-diphenylphosphinoethyl)-2-methoxyethylamine-N,O,P,P)nickel(II) hexafluorophosphate, 37B, 654
$C_{32}H_{39}ClMoOP_2$, Chloro-1,2-bis(diphenylphosphino)ethanemonocarbonyl-π-cyclopentadienylmolybdenum(II), 35B, 854
$C_{32}H_{40}As_4Cl_2Ni$, Dichlorobis(2,2'-bis(dimethylarsine)biphenyl)nickel-(II), 46B, 1276
$C_{32}H_{40}Cl_2P_4Ru$, Dichlorobis(1,2-bis(methylphenylphosphino)ethane)-ruthenium(II), 43B, 1447
$C_{32}H_{44}As_4ClO_6Rh$, Dioxygentetra(dimethylphenylarsine)rhodium(I) perchlorate, 41B, 1194
$C_{32}H_{44}As_4Cl_2Cu_2$, Di-μ-chloro-di(bis(phenyldimethylarsine)copper(I)),

42B, 992

$C_{32}H_{44}ClN_2P_4Re$, Chlorodinitrogentetrakis(dimethylphenylphosphine)-rhenium(I), 37B, 655

$C_{32}H_{45}N_2NiO_6P_3$, Dicyanotris(phenyldiethoxyphosphine)nickel(II), 34B, 610

$C_{33}H_{23}IMnNO_6P_2$, [N-Methyl-2,3-bis(diphenylphosphino)maleimide]tetracarbonylmanganese iodide, 45B, 1355

$C_{33}H_{23}MnNO_6P_2 \cdot 0.5\ C_6H_6$, [N-Methyl-2,3-bis(diphenylphosphino)-maleimide]tetracarbonylmanganese benzene solvate, 45B, 1355

$C_{33}H_{24}Co_2NO_6P_2$, Di-μ-carbonyl-μ-bis(diphenylphosphino)amine-tetracarbonyldicobalt(0)(Co-Co) hemi(benzene), 43B, 1445

$C_{33}H_{33}CrNO_3P_2$, (N,N-Bis(2-diphenylphosphinoethyl)ethylamine)tricarbonylchromium, 40B, 1130

$C_{33}H_{33}Fe_3O_9P_3$, Di-μ-carbonyl-heptacarbonyltris((dimethyl)phenylphosphine)-triangulo-tri-iron, 37B, 656

$C_{33}H_{37}ClP_2PtSi$, (+)-trans-Chlorobis(dimethylphenylphosphine)(1-naphthylphenylmethylsilyl)platinum(II), 42B, 977

$C_{33}H_{38}F_6O_4P_3Rh$, trans-(1,8-Bis(diphenylphosphino)-3,6-dioxaoctane-P,P')-carbonyl(ethanol)rhodium(I) hexafluorophosphate, 43B, 1448

$C_{33}H_{38}F_6O_5P_3Rh$, trans-Aquo(1,11-bis(diphenylphosphino)-3,6,9-trioxaundecane-P,P')carbonylrhodium(I) hexafluorophosphate, 42B, 974

$C_{33}H_{45}F_6P_5RuS_2$, Dithioformato(tetrakis(dimethylphenylphosphine))-ruthenium(II), 44B, 1070

$C_{34}H_{32}As_4Co_4F_4O_8$, Di-μ-(1,2-bis(dimethylarsino)tetrafluorocyclobutene)-octa-carbonyl-tetrahedro-tetracobalt, 37B, 657

$C_{34}H_{33}Cl_2F_6NOP_2Pt$, cis-Dichloro(5,5,5-trifluoro-4-hydroxy-2-trifluoromethyl-1,3-diphenylphosphinopenta-1,3-diene)platinum(II) diethylamine, 42B, 978

$C_{34}H_{42}CoI_2N_2P_2$, N,N-Bis-(2-diphenylphosphinoethyl)-N-(2-diethylaminoethyl)amine-iodocobalt(II) iodide, 39B, 859

$C_{34}H_{42}I_2N_2NiP_2$, N,N-Bis-(2-diphenylphosphinoethyl)-N-(2-diethylaminoethyl)amine-iodonickel(II) iodide, 39B, 859

$C_{34}H_{44}B_2F_8N_2NiP_2S$, Thiolo(N,N-bis(2-diphenylphosphinoethyl)-2'-diethylammoniumethylamine)nickel(II) tetrafluoroborate, 44B, 1071

$C_{34}H_{44}F_6MnO_{10}P_5$, cis-Dicarbonyltetra(dimethoxyphenylphosphine)manganese(I) hexafluorophosphate, 42B, 978

$C_{34}H_{46}Cl_2OP_4Pd$, Bisethylene-1,2-bis(methylphenylphosphine)palladium-(II) chloride, 35B, 855

$C_{34}H_{52}Cl_6MoN_2O_2P_4Re$, Tetrachloro(chlorotetrakis(dimethyl(phenyl)-phosphine)rhenium(I))-μ-dinitrogen-methoxymolybdenum(V)-methanol-hydrochloric acid, 40B, 924

$C_{34}H_{62}N_{10}P_2Pd \cdot 4\ CH_4O$, trans-Bis(adeninato)bis(tri-n-butylphosphine)palladium(II) methanol solvate, 45B, 1357

$C_{34}H_{74}Cl_4N_2P_2Pt_2$, (μ-N,N'-Di-t-butylethylenediimine)bis(trans-dichlorotributylphosphineplatinum), 46B, 1277

$C_{35}H_{28}Cl_2N_2OP_2Rh_2$, (μ-Carbonyl)dichlorobis[2-(diphenylphosphino)pyridine-P,N]dirhodium(Rh-Rh), 46B, 1278

$C_{35}H_{28}Cl_3N_2OP_2PdRh$, Dichlorocarbonylrhodiumbis[μ-2-(diphenylphosphino)pyridine-P,N]chloropalladium(Rh-Pd), 46B, 1278

$C_{35}H_{36}Mo_2NO_9P$, Tetraethylammonium (pentacarbonylmolybdenum)-μ-hydrido-(tetracarbonyl-triphenylphosphinemolybdenum), 45B, 1357

$C_{36}H_{28}O_{10}P_2Ru_4$, Tetra-μ-hydrido-(bis-1,2-(diphenylphosphino)ethane)-decacarbonyltetraruthenium (form 1), 43B, 1449; 44B, 1072

$C_{36}H_{28}O_{10}P_2Ru_4$, Tetra-μ-hydrido-μ-[1,2-bis(diphenylphosphino)ethane]-decacarbonyl-tetraruthenium (form 2), 46B, 1278

$C_{36}H_{30}AuClP_2 \cdot 0.5\ C_6H_6$, Chlorobis(triphenylphosphine)gold(I) hemibenzenate, 40B, 1049

$C_{36}H_{30}Au_2Cl_2O_4P_2 \cdot CH_2Cl_2$, μ-Chloro-bis(triphenylphosphine)digold(I) perchlorate dichloromethane solvate, 46B, 1279
$C_{36}H_{30}Br_2NO_7P_2Rh$, cis-Dibromonitrosylbis(triphenylphosphite)rhodium, 42B, 979
$C_{36}H_{30}Br_2NiP_2$, Dibromobis(triphenylphosphine)nickel(II), 33B, 515
$C_{36}H_{30}CdCl_2P_2$, Bistriphenylphosphinecadmium(II) chloride, 37B, 658
$C_{36}H_{30}ClIr_2N_2O_6P_2$, Dinitrosylbis(triphenylphosphine)iridium perchlorate, 35B, 856
$C_{36}H_{30}ClNNiOP_2$, Chloronitrosylbis(triphenylphosphine)nickel, 44B, 1075
$C_{36}H_{30}ClNO_5P_2RuS$, Chlorosulfatonitrosylbis(triphenylphosphine)ruthenium(II), 40B, 1044
$C_{36}H_{30}ClN_2O_6P_2Rh$, Dinitrosylbis(triphenylphosphine)rhodium perchlorate, 41B, 1195
$C_{36}H_{30}Cl_2IrNOP_2$, Dichloronitrosylbis(triphenylphosphine)iridium, 37B, 658
$C_{36}H_{30}Cl_2Ir_2N_2O_3P_2$, μ-Oxido-bis(chlorotriphenylphosphinenitrosyliridium(I)), 41B, 1196
$C_{36}H_{30}Cl_2NOP_2Rh$, Dichloronitrosylbis(triphenylphosphine)rhodium, 41B, 1196
$C_{36}H_{30}Cl_2NP_2Re$, Nitridodichlorotris(triphenylphosphine)rhenium(V), 32B, 477
$C_{36}H_{30}Cl_2NiP_2$, Dichlorobis(triphenylphosphine)nickel(II), 28, 658
$C_{36}H_{30}Cl_3NOP_2Ru$, Trichloro(nitrosyl)bis(triphenylphosphine)ruthenium, 41B, 1197
$C_{36}H_{30}Cl_4Hg_2P_2$, Bis(μ-chloro)bis(triphenylphosphine)dichlorodimercury, 46B, 1290
$C_{36}H_{30}CoF_6N_2O_2P_3$, Dinitrosylbis(triphenylphosphine)cobalt hexafluorophosphate, 42B, 980
$C_{36}H_{30}CoNO_3P_2S$, Nitrosyl(sulfur dioxide)bis(triphenylphosphine)cobalt, 45B, 1358
$C_{36}H_{30}CuNO_3P_2$, Bis(triphenylphosphine)copper(I) nitrate, 40B, 1131
$C_{36}H_{30}FeN_2O_2P_2$, Dinitrosylbis(triphenylphosphine)iron, 40B, 1045
$C_{36}H_{30}NO_3P_2RhS$, Nitrosyl(sulfur dioxide)bis(triphenylphosphine)rhodium, 43B, 1450
$C_{36}H_{30}NO_5P_2RhS$, Nitrosylsulfatobis(triphenylphosphine)rhodium, 43B, 1451
$C_{36}H_{30}N_2O_2P_2Ru \cdot 0.5\ C_6H_6$, Dinitrosylbis(triphenylphosphine)ruthenium - hemibenzene, 40B, 1050
$C_{36}H_{30}N_2O_2P_2Ru$, Dinitrosylbis(triphenylphosphine)ruthenium, 41B, 1198
$C_{36}H_{30}N_4NiOP_2$, Azidonitrosylbis(triphenylphosphine)nickel, 37B, 659
$C_{36}H_{30}NiO_4P_2S_2$, Bis(sulfur dioxide)bis(triphenylphosphine)nickel(0), 45B, 1380
$C_{36}H_{31}F_6N_2O_3OsP_3$, Hydroxodinitrosylbis(triphenylphosphine)osmium(II) hexafluorophosphate, 41B, 1199
$C_{36}H_{31}IrN_2O_6P_2$, Hydridodinitratobis(triphenylphosphine)iridium(III), 41B, 1199
$C_{36}H_{32}Cl_6O_8P_2Pd_2 \cdot 2\ C_7H_8$, Bis[($\mu$-chloro)-bis(chloro(2-chloro-3-diphenylphosphinodimethyl maleate)palladium(II)) toluene solvate, 46B, 1255
$C_{36}H_{32}Cl_8MoO_5P_2Sn_2$, (1,2-Bis(diphenylphosphino)ethane)tetracarbonyl-(trichlorostannyl)molybdenum(I) aquopentachlorostannate benzene, 41B, 1200
$C_{36}H_{32}I_2N_2NiP_2$, Iodo-bis-(2-(diphenylphosphinomethyl)pyridine)nickel(II) iodide, 40B, 1045
$C_{36}H_{33}Br_2CoP_3$, Tris(diphenylphosphine)cobalt(II) bromide, 31B, 458

$C_{36}H_{33}Cl_3NO_3P_2$, Trichloroamminebis(triphenylphosphine)osmium(III), 34B, 611
$C_{36}H_{34}BCuP_2$, Borohydridobis(triphenylphosphine)copper(I), 32B, 478
$C_{36}H_{37}ClF_6NOP_4Rh$, Chlorobis(3-diphenylphosphinopropyl)phenylphosphine(nitrosyl)rhodium(I) hexafluorophosphate, 39B, 861
$C_{36}H_{37}ClO_2P_3RhS$, Chloro(sulfur dioxide)[bis(3-(diphenylphosphino)-propyl)phenylphosphine]rhodium(I), 46B, 1231
$C_{36}H_{37}ClP_3Rh$, Chlorobis(3-diphenylphosphinopropyl)phenylphosphine-rhodium(I), 39B, 861
$C_{36}H_{40}B_8P_2PtS$, 9,9-Bis(triphenylphosphine)-6-thia-9-platina-decaborane, 45B, 1096
$C_{36}H_{44}F_8MoO_4P_2$, Tetracarbonyl-(2,2'-bis(dicyclohexylphosphino)-octafluorobi(1-cyclobuten-1-yl))molybdenum, 44B, 1072
$C_{36}H_{46}N_2O_2P_2RuS_2$, (4-Amino-2-imino-4-methylpentane)bis(dimethylphenylphosphine)bis(monothiobenzoato)ruthenium(II), 44B, 1073
$C_{36}H_{50}F_6MnO_{11}P_5$, trans-Dicarbonyltetra(dimethoxyphenylphosphine)manganese(I) hexafluorophosphate ethanol, 42B, 978
$C_{36}H_{53}P_6Ta$, Bis((1,2-bis(dimethylphosphino)ethane)(diphenylphosphido))hydridotantantalum(III), 46B, 1279
$C_{36}H_{56}BNiP_4$, Tetrakis(trimethylphosphine)nickel(I) tetraphenylborate, 43B, 1452
$C_{36}H_{66}ClCuO_4P_2$, Perchloratobis(tricyclohexylphosphine)copper(I), 41B, 1201
$C_{36}H_{66}Cl_2Cu_2P$, (Tricyclohexylphosphine)copper(I) chloride dimer, 45B, 1359
$C_{36}H_{66}Cl_2HgO_8P_2$, Diperchloratobis(tricyclohexylphosphine)mercury-(II), 45B, 1359
$C_{36}H_{66}Cl_2NiP_2$, Dichlorobis(tri-cyclohexylphosphine)nickel(II), 28, 659
$C_{36}H_{66}Cl_2P_2Pt$, trans-Dichlorobis(tricyclohexylphosphine)platinum-(II), 46B, 1280
$C_{36}H_{66}Hg_2N_4O_{12}P_2$, Dinitrato(tricyclohexylphosphine)mercury(II) dimer, 43B, 1452
$C_{36}H_{66}I_2P_2Pt$, trans-Di-iodobis(tricyclohexylphosphine)platinum(II), 40B, 1046
$C_{36}H_{66}P_2Pt$, Bis(tricyclohexylphosphine)platinum(0), 41B, 1202
$C_{36}H_{70}P_2PtSi$, trans-Hydridosilyl(tricyclohexylphosphine)platinum-(II), 44B, 1073
$C_{36}H_{71}BCoP_2$, Hydridotetrahydroboratobis(tricyclohexylphosphine)cobalt, 43B, 1453
$C_{36}H_{71}BNiP_2$, Hydrido(tetrahydroborato)bis(tricyclohexylphosphine)-nickel(II), 44B, 1074
$C_{36}H_{81}Cl_6P_3Rh_2$, Trichloro-tri-μ-chloro-tris(tri-n-butylphosphine)dirhodium(III), 42B, 980
$C_{36}H_{86}N_2P_4Rh_2$, μ-Dinitrogen-bis(bis(tri-isopropylphosphine)-hydridorhodium(II)), 45B, 1360
$C_{36}H_{86}O_{12}P_4Rh_2$, Di-μ-hydrido-bis[bis(tri-iso-propylphosphite)rhodium], 45B, 1361
$C_{36}H_{88}O_{12}P_4W \cdot 0.66$ Ar, Tetrahydridotetrakis(triisopropyl phosphite)tungsten(IV) argon, 46B, 1280
$C_{37}H_{30}BCl_2F_4NO_2OsP_2$, Dichloro(carbonyl)(nitrosyl)bis(triphenylphosphine)osmium(II) tetrafluoroborate, 45B, 1361
$C_{37}H_{30}Br_2ClHgIrOP_2$, μ-Bromo-(carbonyl-bis(triphenylphosphine)chlorobromoiridium)mercury, 42B, 980
$C_{37}H_{30}ClIrO_3P_2$, Chlorocarbonyl bis(triphenylphosphine) iridium-oxygen complex, 30B, 386
$C_{37}H_{30}ClIrO_3P_2S$, Chlorocarbonyl(sulfur dioxide)bis(triphenylphos-

phine)iridium, 31B, 460
$C_{37}H_{30}ClO_3P_2RhS$, Chlorocarbonyl(sulfur dioxide)bis(triphenylphosphine)rhodium, 34B, 613
$C_{37}H_{30}Cl_3HgIrOP_2$, μ-Chloro-(carbonyl-bis(triphenylphosphine)dichloroiridium)mercury, 42B, 980
$C_{37}H_{30}CoNO_2P_2$, Bis(triphenylphosphine)-carbonyl-nitrosocobalt, 38B, 1035
$C_{37}H_{30}INO_2Ru$, Iodocarbonylbis(triphenylphosphine)nitrosylruthenium-(0), 40B, 1046
$C_{37}H_{30}IrNO_2P_2$, Nitrosylcarbonylbis(triphenylphosphine)iridium, 38B, 944
$C_{37}H_{30}N_2NiOP_2S$, (Isothiocyanato)nitrosylbis(triphenylphosphine)nickel, 44B, 1075
$C_{37}H_{30}N_3OP_2Rh$, Azidocarbonylbis(triphenylphosphine)rhodium(I), 43B, 1173
$C_{37}H_{30}OP_2PtS_2$, Dithiocarbonato(S,S)bis(triphenylphosphine)platinum-(II), 44B, 860
$C_{37}H_{31}Cl_2NO_2OsP_2 \cdot 0.5\ CH_2Cl_2$, Dichloro(carbonyl)(nitrosylhydride)-bis(triphenylphosphine)osmium(II) dichloromethane solvate, 45B, 1361
$C_{37}H_{32}Cl_3NO_3P_2RuS$, Chloronitrosyl($\eta^2$-sulfur dioxide)bis(triphenylphosphine)ruthenium - dichloromethane, 44B, 1075
$C_{37}H_{33}CrO_3P_3$, fac-[Bis(2-diphenylphosphinoethyl)phenylphosphine-P,P',P"]tricarbonyl-chromium(0), 46B, 1281
$C_{37}H_{33}IIrNOP_2$, Iodomethylnitrosylbis(triphenylphosphine)iridium, 37B, 660
$C_{37}H_{33}MoO_3P_3$, fac-[Bis(2-diphenylphosphinoethyl)phenylphosphine-P,P',P"]tricarbonyl-molybdenum(0), 46B, 1281
$C_{37}H_{34}CoO_3PSn$, (Methyl-phenyl-2-(phenylpropyl)tin)-tricarbonyl-triphenylphosphine-cobalt, 46B, 1282
$C_{37}H_{35}MoN_2O_3P_3$, Tricarbonyl-(2,4-diethyl-1,5-bis(diphenyl)-3-phenyl-1,3,5-triphospha-2,4-diazapentane-P,P,P)-molybdenum, 35B, 857
$C_{37}H_{37}AsF_6O_3P_3RhS$, Carbonyl(sulfur dioxide)[bis(3-(diphenylphosphino)propyl)phenylphosphine]rhodium(I) hexafluoroarsenate, 46B, 1231
$C_{37}H_{38}B_5IrOP_2$, 2-[Carbonylbis(triphenylphosphine)irida]-nido-hexaborane, 45B, 1362
$C_{37}H_{44}O_5P_4Pt_4$, Pentacarbonyltetra(dimethylphenylphosphine)tetraplatinum, 34B, 614
$C_{37}H_{66}AuNP_2S$, Bis(tricyclohexylphosphine)gold(I) thiocyanate, 46B, 1282
$C_{38}H_{29}ClP_2Pt$, Δ-(trans-1,2-Bis(o-(diphenylphosphino)phenyl)ethenyl)-chloroplatinum(II), 40B, 1047
$C_{38}H_{30}Cl_2OP_2RuSe$, Dichlorocarbonylselenocarbonylbis(triphenylphosphine)ruthenium(II), 43B, 1453
$C_{38}H_{30}FeO_{10}P_2S$, Dicarbonyl-bis(triphenyl phosphite)-(sulfur dioxide)iron, 46B, 1283
$C_{38}H_{30}HgN_2P_2S_2$, Dithiocyanatobis(triphenylphosphine)mercury(II), 39B, 863
$C_{38}H_{30}MnNO_3P_2$, Nitrosyldicarbonylbis(triphenylphosphine)manganese, 32B, 480
$C_{38}H_{30}N_2O_6P_2PdS_2$, trans-Di(thiocyanato)bis(triphenylphosphite)palladium(II), 42B, 981
$C_{38}H_{30}N_2P_2PdS_2$, trans-Di(isothiocyanato)bis(triphenylphosphine)palladium(II), 42B, 981
$C_{38}H_{30}NiO_2P_2$, Bis(triphenylphosphine)dicarbonylnickel, 40B, 1048
$C_{38}H_{30}O_2OsP_2Se_2$, Dicarbonylbis(triphenylphosphine)diselenium-

osmium(0), 45B, 1363
$C_{38}H_{30}O_6P_2RuS$, Bis(carbonyl)sulfatobis(triphenylphosphine)ruthenium, 42B, 982
$C_{38}H_{31}IrO_2P_2$, Hydridodicarbonylbis(triphenylphosphine)iridium(I), 42B, 982
$C_{38}H_{32}Cl_2IIrO_3P_2$, Dioxocarbonyliodobis(triphenylphosphine)iridium dichloromethane, 32B, 479
$C_{38}H_{32}Cl_6O_2P_2Pt$, Bis(triphenylphosphine)dioxygenplatinum - chloroform, 37B, 661
$C_{38}H_{33}Cl_3NOsP_2$, mer-Trichloro(acetonitrile)bis(triphenylphosphine)-osmium(III), 46B, 1283
$C_{38}H_{34}Br_2NiP_2$, Dibromobis-(benzyldiphenylphosphine)nickel(II), 35B, 858
$C_{38}H_{34}Cl_2O_2P_2Ru \cdot CH_2Cl_2$, Dichlorobis[2-(diphenylphosphino)anisole]-ruthenium(II) dichloromethane solvate, 45B, 1363
$C_{38}H_{34}Cl_4I_2P_2Pd$, Diiodobis(triphenylphosphine)palladium(II) (methylene chloride solvate), 39B, 863
$C_{38}H_{36}O_6P_2S_2Pt$, trans-Di(methyl sulfito)bis(triphenylphosphine)platinum(II), 45B, 1364
$C_{38}H_{38}N_2P_2PdS_2$, trans-Dithiocyanato-bis-((3,3-dimethylbutynyl)diphenylphosphine)palladium(II), 39B, 834
$C_{38}H_{38}N_2P_2PtS_2$, cis-Thiocyanatoisothiocyanatobis(3,3-dimethylbutynyldiphenylphosphine)platinum(II), 40B, 1048
$C_{38}H_{42}B_8OP_2PtS$, 8-Ethoxy-9,9-bis(triphenylphosphine)-6-thia-9-platina-decaborane, 45B, 1096
$C_{39}H_{30}ClIrN_2OP_2S$, Chlorocyano(thiocyanato-N)carbonylbis(triphenylphosphine)iridium(III), 39B, 864
$C_{39}H_{30}FeO_8P_2S$, Dicarbonylthiocarbonyl-bis(triphenyl phosphite)iron-(0), 46B, 1284
$C_{39}H_{30}O_3OsP_2$, Tricarbonylbis(triphenylphosphine)osmium(0), 34B, 617
$C_{39}H_{31}MnO_3P_2$, Hydrido(tricarbonyl)bis(triphenylphosphine)manganese-(I), 44B, 1076
$C_{39}H_{32}Cl_3NO_7OsP_2$, Dicarbonylnitrosylbis(triphenylphosphine)osmium(0) perchlorate dichloromethane, 41B, 1202
$C_{39}H_{33}BrCuP_2$, Bromobis(triphenylphosphine)copper(I) hemibenzenate, 39B, 861
$C_{39}H_{33}ClCuP_3$, Chlorotris(diphenylmethylphosphine)copper(I), 42B, 992
$C_{39}H_{33}ClO_6OsP_2S_2 \cdot 0.5\ C_6H_6$, Dicarbonyl-($\eta^2$-methyldisulfido)-bis-(triphenylphosphine)osmium perchlorate benzene solvate, 45B, 1364
$C_{39}H_{33}N_2O_2OsP_2$, Dinitrosylbis(triphenylphosphine)osmium hemibenzene, 41B, 1203
$C_{39}H_{35}CuP_2S_3$, Bis(triphenylphosphine)ethyltrithiocarbonatocopper(I), 41B, 1204
$C_{39}H_{36}ClCuN_3OP_2Rh$, Bis(triphenylphosphine)-carbonyl-rhodium-copper-dimethyltriazene chloride, 42B, 982
$C_{39}H_{38}B_5CuFeO_3P_2$, Tricarbonyliron-pentaborane-triphenylphosphine-copper, 45B, 1365
$C_{39}H_{39}N_3P_3ReS_3$, (1,2-Bis(diphenylphosphino)ethane)(diethylphenylphosphine)tri-isothiocyanatorhenium(III), 44B, 1076
$C_{39}H_{43}BCuP_3$, Tris(methyldiphenylphosphine)(tetrahydroborato)copper-(I), 44B, 1077
$C_{39}H_{69}Cl_7O_3OsP_2S$, Chlorohydridocarbonyl(sulfur dioxide)bis(tricyclohexylphosphine)osmium(II) bis(chloroform), 44B, 1077
$C_{39}H_{74}O_4P_2Pt$, trans-Hydridobis(tricyclohexylphosphine)monomethylcarbonatoplatinum(II) methanol solvate, 43B, 1250
$C_{40}H_{30}As_2Co_2O_4P$, (μ-Diarsenic)bis(triphenylphosphine)tetracarbonyl-dicobalt, 45B, 1344

$C_{40}H_{30}AuMnO_7P_2$, Triphenylphosphinegold-tetracarbonyltriphenoxyphosphinemanganese, 32B, 481
$C_{40}H_{30}CrO_{10}P_2$, trans-Bis(triphenyl phosphite)tetracarbonylchromium(0), 38B, 1046
$C_{40}H_{32}As_2F_8Fe_2O_4P_2$, 1-(Diphenylphosphino)-2-(dimethylarsino)tetrafluorocyclobut-1-ene-μ-(1-(diphenylphosphino)-2-(dimethylarsino)-tetrafluorocyclobut-1-ene)-tetracarbonyldiiron, 39B, 864
$C_{40}H_{32}N_2P_2Pt_2$, Fumaronitrilebis(triphenylphosphine)platinum, 35B, 859
$C_{40}H_{34}F_6Fe_2O_4P_3Rh$, Bis(μ-diphenylphosphido-μ'-carbonyl-π-methylcyclopentadienylcarbonyliron)rhodium(2Rh-Fe) hexafluorophosphate, 40B, 1051
$C_{40}H_{40}As_4Cl_2Pd \cdot 2\ C_2H_6O_2$, meso-Bis[RR,SS-o-phenylenebis(methylphenylarsine)]palladium(II) dichloride ethane-1,2-diol solvate, 46B, 1275
$C_{40}H_{40}As_4I_2Pd$, meso-Bis[RR,SS-o-phenylenebis(methylphenylarsine)]-palladium(II) diiodide, 46B, 1275
$C_{40}H_{51}BCoF_4O_6P_5$, Bis(2-diphenylphosphinoethyl)phenylphosphinebis-(trimethyl phosphite)cobalt tetrafluoroborate, 43B, 1460
$C_{40}H_{56}F_6P_6Ru$, Hydridopentakis(dimethylphenylphosphine)ruthenium(II) hexafluorophosphate, 43B, 1454
$C_{40}H_{60}Cl_2O_8P_4Tc$, Dichlorotetrakis(diethylphenylphosphonite)technetium(II), 40B, 1051; 43B, 1455
$C_{40}H_{62}FeO_8P_4$, cis-Dihydridotetrakis(diethyl phenylphosphonite)iron-(II), 38B, 1047
$C_{40}H_{62}O_8P_4Ru$, trans-Dihydridotetrakis(diethylphenylphosphonite)ruthenium(II), 39B, 865
$C_{40}H_{64}As_8Cl_{11}Nb_3O_2$, Tetrachlorobis(o-phenylenebis(dimethylarsine))-niobium(V) trichlorodioxoniobate(V), 41B, 1184
$C_{40}H_{68}P_4Re_2$, Octahydridotetrakis(diethylphenylphosphine)-dirhenium(IV), 43B, 1456
$C_{40}H_{80}P_2Pt_2Si_2$, Di-μ-hydridobis(dimethylsilyl)bis(tricyclohexylphosphine)diplatinum, 46B, 1285
$C_{41}H_{33}IrN_2OP_2$, Hydridocarbonyl(fumaronitrile)bis(triphenylphosphine)iridium, 34B, 615
$C_{41}H_{33}N_2NiP_3 \cdot 0.33\ H_2O$, Dicyanotris-(5-methyl-5H-dibenzophosphole)-nickel(II), 37B, 661
$C_{41}H_{39}CoO_4P_3S$, 1,1,1-Tris(diphenylphosphinomethyl)ethanecobalt(II) sulfate, 43B, 1456
$C_{41}H_{39}CoP_6$, (η^3-Cyclotriphosphorus)[1,1,1-tris((diphenylphosphino)-methyl)ethane]cobalt, 46B, 1285
$C_{41}H_{39}INiP_3$, Iodo(1,1,1-tris(diphenylphosphinomethyl)ethane)nickel-(I), 40B, 1052
$C_{41}H_{39}IrP_6$, (η^3-Triphosphorus)[1,1,1-tris(diphenylphosphinomethyl)-ethane]iridium, 45B, 1366
$C_{41}H_{39}NiO_2P_3S$, (Sulfur dioxide)(1,1,1-tris(diphenylphosphinomethyl)-ethane)nickel, 42B, 983
$C_{41}H_{39}NiO_4P_3Se$, 1,1,1-Tris(diphenylphosphinomethyl)ethanenickel selenate, 43B, 1456
$C_{41}H_{39}P_6Rh$, (η^3-Triphosphorus)[1,1,1-tris(diphenylphosphinomethyl)-ethane]rhodium, 45B, 1366
$C_{41}H_{43}BCoP_3$, Tetrahydroborato(1,1,1-tris(diphenylphosphinomethyl)-ethane)cobalt, 42B, 983
$C_{41}H_{43}IrO_2P_2Si_2 \cdot C_2H_6O$, Hydrido-1,3-(1,1,3,3-tetramethyldisiloxanediyl)carbonylbis(triphenylphosphine)iridium(III) ethanol solvate, 45B, 1366
$C_{41}H_{43}IrO_2P_2Si_2$, Hydrido-1,3-(1,1,3,3-tetramethyldisiloxanyl)carbon-

ylbis(triphenylphosphine)iridium(III), 43B, 1457
$C_{42}H_{30}Cl_4O_2P_2Pd$, 3,4,5,6-Tetrachlorocatecholatobis(triphenylphosphine)palladium(II), 41B, 1204
$C_{42}H_{30}F_{12}O_3P_2Pt$, 2-Bis(triphenylphosphine)-4,4,6,6-tetrakis(trifluoromethyl)-1,3,5,2-trioxaplatinan, 41B, 1193
$C_{42}H_{30}N_4P_2Pt$, Tetracyanoethylenebis(triphenylphosphine)platinum(0), 35B, 860
$C_{42}H_{32}As_2Co_2F_8O_6P_2$, Bis((2-dimethylarsino-3,3,4,4-tetrafluorocyclobut-1-enyl)diphenylphosphine)hexacarbonyldicobalt, 44B, 1078
$C_{42}H_{33}BrCrMnO_5P_3$, Bromotricarbonylmanganese(I)-(bis(2-diphenylphosphinoethyl)phenylphosphine)-pentacarbonylchromium(0) (α-isomer), 39B, 866
$C_{42}H_{35}AsBr_3NiP$, Tetraphenylarsonium tribromo(triphenylphosphine)nickelate(II), 46B, 1286
$C_{42}H_{35}AsI_3NiP$, Tetraphenylarsonium triiodo(triphenylphosphine)nickelate(II), 33B, 517
$C_{42}H_{36}ClF_6N_2O_2P_3Ru$, Chlorodinitrosylbis(triphenylphosphine)ruthenium hexafluorophosphate - benzene, 38B, 944
$C_{42}H_{36}F_6IrO_3P_3S$, Dicarbonyl(thiocarbonyl)bis(triphenylphosphine)iridium(I) hexafluorophosphate acetone, 38B, 1048
$C_{42}H_{37}N_2NiOP_3$, Dicyanotris-(5-methyl-5H-dibenzophosphole)nickel(II) methanolate, 37B, 661
$C_{42}H_{39}Br_2P_3Pd$, Dibromotris-(2-phenylisophosphindoline)palladium(II) (optically-resolved red form), 40B, 1054
$C_{42}H_{39}Cl_2NiP_3$, Dichlorotris(5-ethyl-5H-dibenzophosphole)nickel(II), 42B, 983
$C_{42}H_{42}As_3INNi$, (Tris-(2-diphenylarsinoethyl)amino)iodonickel(I), 46B, 1288
$C_{42}H_{42}BCl_2CoF_4NP_3$, Dichlorotris(2-diphenylphosphinoethyl)aminecobalt(III) tetrafluoroborate, 43B, 1458
$C_{42}H_{42}BCoF_4NP_3$, Tris(2-diphenylphosphinoethyl)aminocobalt(I) tetrafluoroborate, 40B, 1053
$C_{42}H_{42}BrCoF_6P_5$, Bromo(tris-(2-diphenylphosphinoethyl)phosphine)cobalt(II) hexafluorophosphate, 41B, 1205
$C_{42}H_{42}BrF_6FeNP_4 \cdot 0.5\ C_5H_{12}$, Bromo(tris(2-diphenylphosphinoethyl)amine)iron(II) hexafluorophosphate n-pentane solvate, 44B, 1079
$C_{42}H_{42}ClCoF_6NP_4$, Chloro(tris(2-diphenylphosphinoethyl)amine)cobalt-(II) hexafluorophosphate, 39B, 867
$C_{42}H_{42}ClF_6NNiP_4$, Chloro(tris-(2-diphenylphosphinoethyl)amine)nickel-(II) hexafluorophosphate, 41B, 1205
$C_{42}H_{42}ClNiO_4P_4$, Tris(2-diphenylphosphinoethyl)phosphinenickel(I) perchlorate, 46B, 1287
$C_{42}H_{42}Cl_4Hg_2O_8P_2$, Di-$\mu$-chloro-bis(perchlorato(tri-o-tolyl-phosphine)mercury(II)), 45B, 1367
$C_{42}H_{42}CoNP_6 \cdot 0.5\ C_4H_8O$, ($\eta^3$-Cyclo-triphosphorus)(tris(2-diphenylphosphinoethyl)amine)cobalt - tetrahydrofuran, 44B, 1079
$C_{42}H_{42}INNiP_3$, (Tris-(2-diphenylphosphinoethyl)amino)iodonickel(I), 46B, 1288
$C_{42}H_{42}I_2NNiP_3$, Iodo(tris-(2-diphenylphosphinoethyl)amino-NP)-nickel-(II) iodide, 35B, 862
$C_{42}H_{42}I_2P_2Pt \cdot CH_2Cl_2$, trans-Di-iodobis(tri-o-tolylphosphine)platinum(II) dichloromethane solvate, 45B, 1368
$C_{42}H_{42}NNiO_2P_3S$, (η^1-Sulfur dioxide-S)(tris(2-diphenylphosphino)ethyl-amine)nickel, 44B, 1080
$C_{42}H_{42}NNiP_3$, (Tris-(2-diphenylphosphinoethyl)amine)nickel(0), 41B, 1207
$C_{42}H_{42}N_2O_4P_2Pt$, trans-Dinitrobis(tri-p-tolylphosphine)platinum(II),

43B, 1458
$C_{42}H_{42}P_3NI_2Co$, Di-iodotris(2-diphenylphosphinoethyl)aminecobalt(II), 37B, 664
$C_{42}H_{42.5}BF_4NNiP_3$, Hydrido-tris(2-diphenylphosphinoethyl)aminonickel tetrafluoroborate, 40B, 1053
$C_{42}H_{43}BCoF_4P_4$, Hydrido(tris-(2-diphenylphosphinoethyl)phosphine)cobalt(II) tetrafluoroborate, 41B, 1206
$C_{42}H_{43}CoNP_3$, Hydridotris-(2-diphenylphosphinoethyl)aminecobalt(I), 40B, 1054; 41B, 1207
$C_{42}H_{43}CoNP_3 \cdot C_4H_8O$, Hydrido(tris(2-diphenylphosphinoethyl)amine)cobalt(I) tetrahydrofuran, 45B, 1368
$C_{42}H_{43}CoP_4 \cdot 0.5\ C_3H_6O$, Hydrido(tris-(2-diphenylphosphinoethyl)phosphine)cobalt(I) hemi(acetone), 41B, 1209
$C_{42}H_{43}CoP_4$, Hydrido(tris-(2-diphenylphosphinoethyl)phosphine)cobalt(I), 41B, 1208
$C_{42}H_{48}N_{12}Ni_4O_6P_4$, Tetranickel hexacarbonyl tetra(tripropionitrilephosphine), 32B, 482
$C_{42}H_{50}BBrClN_2P_3W \cdot CH_2Cl_2$, Bromochlorohydridotris(dimethylphenylphosphine)(triphenylboratodiazenido-N)tungsten dichloromethane, 46B, 1289
$C_{42}H_{60}ClO_{14}P_4Tc$, cis-Dicarbonyltetrakis(diethylphenylphosphonite)technetium(I) perchlorate, 41B, 1209
$C_{43}H_{35}BrCl_2CrMnO_5P_3$, Bromotricarbonylmanganese(I)-(bis(2-diphenylphosphinoethyl)phenylphosphine)-pentacarbonylchromium(0) (β-isomer methylene chloride solvate), 39B, 866
$C_{43}H_{35}CuN_2O_4P_2$, α-Nitro-α-aci-nitrotolueneatobis(triphenylphosphine)copper(I), 41B, 1210
$C_{43}H_{36}BF_4IIrNO_2P_2$, Iodocarbonylnitrosylbis(triphenylphosphine)iridium tetrafluoroborate - benzene, 34B, 614
$C_{43}H_{38}O_4P_2PtS_2$, Bis(sulfur dioxide)bis(triphenylphosphine)platinum toluene solvate, 42B, 984
$C_{43}H_{42}ClIrOP_2$, trans-Chlorocarbonylbis(tri-o-tolylphosphine)iridium(I), 41B, 1211
$C_{43}H_{42}Cl_4O_3P_2RuSn$, Acetone(carbonyl)chloro(trichlorostannio)bis(triphenylphosphine)ruthenium(II) acetone solvate, 44B, 1080
$C_{43}H_{42}NNiOP_3$, (Tris(2-diphenylphosphinoethyl)amine)nickel(0) carbonyl, 42B, 1000
$C_{43}H_{44}Cl_3F_6N_2P_4Rh$, Chloro(bis-(3-diphenylphosphinopropyl)phenylphosphine)phenyldiazorhodium hexafluorophosphate methylene chloride solvate, 41B, 1211
$C_{44}H_{30}Ni_8O_8P_6$, Hexa-μ_4-phenylphosphido-octacarbonyloctanickel, 42B, 985
$C_{44}H_{34}As_2Cl_2Pt$, trans-Dichloro-(2,11-bis(diphenylarsinomethyl)benzo[c]phenanthrene)platinum, 46B, 1289
$C_{44}H_{34}Cl_2P_2Pt \cdot CHCl_3$, cis-Dichloro(2,11-bis(diphenylphosphinomethyl)benzo[c]phenanthrene)platinum chloroform solvate, 46B, 1290
$C_{44}H_{34}Cl_4Hg_2P_2$, Bis(μ-chloro)bis(1,2,5-triphenylphosphole)dichlorodimercury, 46B, 1290
$C_{44}H_{35}BF_4FeN_2O_2P_2$, Phenyldiazobis(triphenylphosphine)dicarbonyliron tetrafluoroborate, 41B, 1212
$C_{44}H_{36}ClIrO_2P_2$, Chlorodicarbonylbis(triphenylphosphine)iridium(I) - benzene, 40B, 1131
$C_{44}H_{38}Cl_2N_2OOsP_2$, Hydrido(carbonyl)(phenyldiazo)bis(triphenylphosphine)osmium(0) dichloromethane solvate, 41B, 1213
$C_{44}H_{39}Cl_5N_2P_2Ru$, Trichloro(p-tolyldiazo)bis(triphenylphosphine)ruthenium dichloromethane solvate, 41B, 1197
$C_{44}H_{39}N_2NiP_3$, Dicyanotris-(5-ethyl-5H-dibenzophosphole)nickel(II),

37B, 661
$C_{44}H_{39}O_5P_3Pt_2Ru$, Di-μ-carbonyl-(μ-carbonylbis((methyldiphenylphosphine)platino))-dicarbonyl(methyldiphenylphosphine)ruthenium(2Ru-Pt)(Pt-Pt), 41B, 1213
$C_{44}H_{41}Br_2Cl_6NiP_3$, Dibromotris(5-ethyl-5H-dibenzophosphole)nickel(II) bis(chloroform), 43B, 1459
$C_{44}H_{42}FeO_{10}P_2S$, Dicarbonyl-bis(tri-o-tolyl phosphite)-(sulfur dioxide)iron, 46B, 1283
$C_{44}H_{47}BF_4NNiO_3P_3S \cdot 0.5\ C_2H_6O \cdot 0.5\ H_2O$, (Tris(2-diphenylphosphinoethyl)amine)(ethylsulfito)nickel tetrafluoroborate hemi(ethanol) hemihydrate, 43B, 1460
$C_{44}H_{48}BrCoF_6NOP_4$, Bromo(tris-(2-diphenylphosphinoethyl)amine)cobalt-(II) hexafluorophosphate ethanol, 41B, 1214
$C_{44}H_{49}BCoF_4O_2P_4$, Hydroxotris(2-diphenylphosphinoethyl)phosphinecobalt(II) tetrafluoroborate ethanol solvate, 42B, 985
$C_{44}H_{50}O_6P_4Pt_2S_2$, (Triphenylphosphito)(diethylphosphinesulfide)platinum(I) dimer, 41B, 1215
$C_{44}H_{54}BCoF_4O_5P_4$, Bis(3-diphenylphosphinopropyl)phenylphosphine(trimethyl phosphite)carbonylcobalt tetrafluoroborate tetrahydrofuran, 43B, 1460
$C_{44}H_{96}Cl_4P_4Pd_2$, Di-μ-[1,5-bis(di-t-butylphosphino)-3-methylpentane]-bis[dichloropalladium(II)], 46B, 1292
$C_{45}H_{34}ClNO_2P_2Ru \cdot CD_2Cl_2$, Carbonyl-chloro-nitrosyl-(2,11-bis(diphenylphosphinomethyl)-benzo[c]phenanthrene)ruthenium dideutero-dichloromethane solvate, 45B, 1369
$C_{45}H_{38}Br_3P_3Pt$, Dibromotris-(5-methyl-5H-dibenzophosphole)platinum-(II) - bromobenzene, 40B, 1057
$C_{45}H_{38}Cl_4N_2O_6P_2Ru$, Chlorodicarbonyl(cis-phenyldiazene)bis(triphenylphosphine)ruthenium(II) perchlorate methylene chloride solvate, 41B, 1216
$C_{45}H_{44}ClF_6IrN_2P_4$, Chloro(phenyldiazo)tris(methyldiphenylphosphine)-iridium hexafluorophosphate, 42B, 986
$C_{45}H_{45}Br_2OP_3Pd$, Dibromotris-(2-phenylisophosphindoline)palladium(II) - acetone (racemic orange form), 40B, 1054
$C_{45}H_{50}B_2CoF_8O_2P_4$, Aquotris(2-diphenylphosphinoethyl)phosphinecobalt-(II) bis(tetrafluoroborate) acetone solvate, 42B, 985
$C_{45}H_{90}Mn_6O_{36}P_9$, Tris(tricarbonylmanganese)-nonakis(μ-diethoxyphosphito)trimanganese, 46B, 1292
$C_{46}H_{34}O_{10}P_2Ru_4$, Tetra-μ-hydrido-decacarbonylbis(triphenylphosphine)-tetrahedro-tetraruthenium, 44B, 1081; 45B, 1369
$C_{46}H_{35}MnO_4P_2$, Tetraphenylphosphonium (triphenylphosphine)tetracarbonylmanganate, 46B, 1293
$C_{46}H_{37}Cl_2F_4IrO_3P_2$, Chloro(difluoromethyl)(O-chlorodifluoroacetato)-carbonylbis(triphenylphosphine)iridium(III) - benzene, 40B, 1056
$C_{46}H_{38}N_2P_2RuS_2$, Bis(pyridine-2-thiolato)bis(triphenylphosphine)ruthenium(II), 38B, 1049
$C_{46}H_{40}BClF_5IrN_2O_2P_2$, Carbonylchloro(4-fluorophenyl-diimide-2C,N')-bis(triphenylphosphine)iridium(III) tetrafluoroborate acetone solvate, 39B, 868
$C_{46}H_{43}Cl_3N_2OP_2Ru$, Trichloro(p-tolylazo)bis(triphenylphosphine)ruthenium(II) - acetone, 39B, 869
$C_{46}H_{96}P_4Pt_2$, 1,1'-Bis(2,2,6,6-tetra(t-butyl)-1-platina-2,6-diphosphacyclohexane) toluene solvate, 44B, 1081
$C_{46}H_{100}Cl_4P_4Pd_2$, Bis[μ-1,7-heptanediylbis(di-t-butylphosphine)]-bis-(dichloropalladium), 45B, 1370
$C_{47}H_{39}CoCrO_5P_3S_2$, [1,1,1-Tris(diphenylphosphinomethyl)ethane]cobalt-μ-carbon disulfide-pentacarbonyl-chromium, 45B, 1371

$C_{47}H_{58}Co_2N_4O_2P_5$, Dicyano-tris(dimethylphenylphosphine)cobalt-μ-cyano-dioxocyano-bis(dimethylphenylphosphine)cobalt benzene solvate, 41B, 1216
$C_{48}H_{30}F_{10}P_4Pt$, Bis(triphenylphosphine)decafluorophosphorobenzene-platinum, 42B, 987
$C_{48}H_{38}F_{12}IrN_3OP_4$, Nitrosyl(1,10-phenanthroline)bis(triphenylphosphine)iridium(I) dihexafluorophosphate, 46B, 1294
$C_{48}H_{41}MnOP_2$, Cyclopentadienylcarbonylbis(triphenylphosphine)manganese benzene solvate, 40B, 1057
$C_{48}H_{42}F_6N_2O_2P_5Rh \cdot 0.5\ C_3H_6O$, Bis-[bis(diphenylphosphino)amine-oxo]-rhodium(III) hexafluorophosphate acetone solvate, 46B, 1295
$C_{48}H_{44}Br_2ClP_3Pd$, Dibromotris-(5-ethyl-5H-dibenzophosphole)palladium-(II) - chlorobenzene, 40B, 1057
$C_{48}H_{44}Br_3P_3Pt$, Dibromotris-(5-ethyl-5H-dibenzophosphole)platinum(II) - bromobenzene, 40B, 1057
$C_{48}H_{48}F_{12}N_6P_4Pd_3$, Hexakis(methylisocyano)bis(triphenylphosphine)tri-palladium(II) di(hexafluorophosphate), 42B, 987
$C_{48}H_{58}As_4BFeN_2O_2S$, Nitrosylbis(o-phenylenebis(dimethylarsine))(thio-cyanato-N)iron tetraphenylborate acetone, 43B, 1433
$C_{48}H_{66}Cl_3F_6P_7Ru_2$, Tri-μ-chloro-hexakis(dimethylphenylphosphine)di-ruthenium(II) hexafluorophosphate, 42B, 988
$C_{48}H_{98}P_2Pt_2Si_2$, trans-Di-μ-hydrido-bis(tricyclohexylphosphine(tri-ethylsilyl)platinum), 44B, 1082
$C_{48}H_{108}Cl_5P_4Ru_2$, Tri-μ-chloro-dichlorotetrakis(tri-n-butylphosphine)diruthenium, 33B, 517
$C_{48}H_{108}Cl_6P_4Rh_2$, Tetrachloro-μ-dichloro-tetrakis(tri-n-butylphosphine)dirhodium(III), 40B, 1058
$C_{48}H_{108}Cl_8Hg_4P_4$, Bis(μ-chloro)bis(tributylphosphine)dichlorodimercu-ry dimer, 46B, 1290
$C_{49}H_{80}O_2P_2Pt$, trans-Hydridobis(tricyclohexylphosphine)formatoplatin-um(II) benzene solvate, 43B, 1250
$C_{50}H_{42}N_2O_4P_4Pd_2S_2$, Di-μ-thiocyanato-bis(hydrogen bis(diphenylphosphinato))dipalladium(II), 40B, 1060
$C_{50}H_{44}ClIrO_6P_4$, Bis(bis(diphenylphosphino)methane)dioxygeniridium(I) perchlorate, 42B, 893
$C_{50}H_{44}ClN_3P_2Pd$, Bis(triphenylphosphine)chloro-1,3-di-p-tolyltriazen-idopalladium(II), 42B, 989
$C_{50}H_{44}Cl_2O_2P_4Pd_2S \cdot CH_4O \cdot 0.5\ CH_2Cl_2$, μ-(Sulfur dioxide)-di[μ-(bis(diphenylphosphino)methane]bis(chloropalladium(II)) methanol dichloromethane solvate, 45B, 1371
$C_{50}H_{44}Cl_2P_4Pd_2S \cdot 3\ CH_2Cl_2$, μ-Thia-di[μ-(bis(diphenylphosphino)methane]bis(chloropalladium(II)) dichloromethane solvate, 45B, 1371
$C_{50}H_{44}Cl_2P_4Pt_2$, Bis-μ-[bis(diphenylphosphino)methane]-bis(chloro-platinum)(Pt-Pt), 43B, 1461; 45B, 1372
$C_{50}H_{44}Cl_4Cu_4P_4 \cdot x\ C_2H_4Cl_2$, Bis(diphenylphosphino)methane-bis(chlo-rocopper(I)) dimer dichloromethane solvate, 40B, 1059
$C_{50}H_{44}Cl_4P_4Pd_2Sn$, Di-μ-[bis(diphenylphosphinyl)methane]chloro(tri-chlorotin)dipalladium, 45B, 1373
$C_{50}H_{44}Cu_3I_3P_4 \cdot 0.5\ CH_2Cl_2$, Di-μ-(bis(diphenylphosphino)methane)-μ-iodo-di-μ_3-iodo-triangulo-tricopper(I) dichloromethane, 41B, 1217
$C_{50}H_{44}F_6IrO_2P_5$, Bis(bis(diphenylphosphino)methane)dioxygeniridium(I) hexafluorophosphate, 42B, 893
$C_{50}H_{45}CoNO_3PSn$, (Methyl-phenyl-(triphenylmethyl)tin)-tricarbonyl-(N-methyl-(1-phenylethylamino))-diphenylphosphine-cobalt, 46B, 1282
$C_{50}H_{46}Ag_4Br_4N_2P_4$, Di-μ-bromo-di-$\mu_4$-bromo-bis[μ-methylbis(diphenyl-phosphino)aminedisilver], 46B, 1311
$C_{50}H_{46}N_2O_2P_4Ru_2$, Dinitrosylbis(μ-diphenylphosphido)-bis(methyldi-

phenylphosphine)diruthenium, 39B, 870
$C_{50}H_{62}BBrN_3NiP$, Bromo(bis(2-diethylaminoethyl)(2-diphenylphosphinoethyl)amine)nickel(II) tetraphenylborate, 37B, 664
$C_{50}H_{75}Cl_4P_5Ru_2$, Tri-μ-chloro-chloropentakis(diethylphenylphosphine)-diruthenium(II), 33B, 518
$C_{50}H_{92}NiP_4$, Bis(bis(dicyclohexylphosphino)methane)nickel, 38B, 1049
$C_{51}H_{39}Cl_2NP_2Pd$, 2,11-Bis(diphenylphosphinomethyl)benzo[c]phenanthrene-dichloropalladium(II), 44B, 1083
$C_{51}H_{39}CoCr_2O_{10}P_6$, [1,1,1-Tris((diphenylphosphino)methyl)ethane]cobalt(μ-η^3-cyclotriphosphorus)bis(pentacarbonylchromium), 46B, 1285
$C_{51}H_{44}BF_4OP_4Rh$, Bis-[bis(diphenylphosphino)methane]carbonylrhodium-(I) tetrafluoroborate, 46B, 1295
$C_{51}H_{44}Br_2OP_4Rh_2$, (μ-Carbonyl)-bis[(μ-bis(diphenylphosphino)methane)-bromorhodium](Rh-Rh), 46B, 1296
$C_{51}H_{44}ClF_6OP_5Pt_2$, Carbonyl-platinum-di-$\mu$-(bis(diphenylphosphino)-methane)chloroplatinum hexafluorophosphate, 45B, 1373
$C_{51}H_{45}N_3OP_2Ru$, trans-Bis(triphenylphosphine)carbonyl(1,3-di-p-tolyl-triazenido)hydridoruthenium(II), 42B, 989
$C_{51}H_{53}BIO_3P_3W$, Tricarbonyliodotris(dimethylphenylphosphino)tungsten-(II) tetraphenylborate, 40B, 1060
$C_{51}H_{54}As_2Cl_2OPt_2$, Bis-$\mu$-[bis(diphenylarsino)methane]-μ-carbonyl-bis-[chloroplatinum(II)], 45B, 1374
$C_{52}H_{39}ClNOP_2Rh$, 2,11-Bis(diphenylphosphinomethyl)benzo[c]phenanthrene-carbonylchlororhodium(I), 44B, 1082
$C_{52}H_{44}As_4Br_2MoO_2$, (Bis(diphenylarsino)methane)dibromodicarbonylmolybdenum(II), 38B, 1050
$C_{52}H_{44}As_4Cl_2MoO_2$, Dicarbonyldichlorobis(bis(diphenylarsino)methane)-molybdenum(II), 43B, 1462
$C_{52}H_{44}As_4Cl_2O_2Rh_2$, Di-$\mu$-bis(diphenylarsino)methane-bis(trans-chlorocarbonylrhodium(I)), 34B, 616
$C_{52}H_{44}Cl_2FeP_4 \cdot 2\ C_3H_6O$, Dichlorobis[cis-1,2-bis(diphenylphosphino)-ethylene]iron(II) acetone, 46B, 1296
$C_{52}H_{44}Cl_2MoO_2P_4 \cdot 0.75\ C_6H_6$, Dicarbonyldichlorobis(bis(diphenylphosphino)methane)molybdenum(II) - benzene, 43B, 1462
$C_{52}H_{44}Cl_2O_2P_4Rh_2$, Bis[($\mu$-bis(diphenylphosphino)methane)chlorocarbonylrhodium], 46B, 1297
$C_{52}H_{44}O_2P_4Rh_2S$, (μ-Sulfido)bis[(μ-bis(diphenylphosphino)methane)carbonylrhodium(I)], 43B, 1462; 46B, 1297
$C_{52}H_{48}As_2Br_4Mo_2P_2$, Di-$\mu$-(1-diphenylphosphino-2-diphenylarsinoethane)bis(dibromomolybdenum), 45B, 1375
$C_{52}H_{48}ClP_4O_4Rh$, Bis-(1,2-bis(diphenylphosphino)ethane)rhodium(I) perchlorate, 35B, 864
$C_{52}H_{48}Cl_2MoP_4 \cdot CH_2Cl_2$, trans-Dichlorobis(1,2-bis(diphenylphosphino)ethano)molybdenum(II) dichloromethane solvate, 46B, 1298
$C_{52}H_{48}Cl_4CoP_4Sn$, Chlorobis(1,2-bis(diphenylphosphino)ethane)cobalt-(II) trichlorostannate(II), 38B, 1051; 39B, 871
$C_{52}H_{48}Cl_4P_4Re_2$, Tetrachlorobis(1,2-bis(diphenylphosphino)ethane)dirhenium, 44B, 1084
$C_{52}H_{48}CrFNOP_4$, Bis[1,2-bis(diphenylphosphino)ethane-P,P']fluoronitrosylchromium, 46B, 994
$C_{52}H_{48}F_6IrO_2P_5$, Bis(bis(diphenylphosphino)ethane)oxygeniridium(I) hexafluorophosphate, 40B, 1131; 41B, 1218
$C_{52}H_{48}F_6O_2P_5Rh$, Bis(bis(diphenylphosphino)ethane)oxygenrhodium(I) hexafluorophosphate, 40B, 1131
$C_{52}H_{48}IrO_2P_4S_2$, Bis(tetraphenyldiphosphinoethane)-disulfurdioxine-iridium cation, 42B, 990
$C_{52}H_{48}MoN_4P_4$, trans-Bis(dinitrogen)-bis(1,2-bis(diphenylphosphino)-

ethane)molybdenum(0), 41B, 1218
$C_{52}H_{49}BF_4MoO_2P_4$, Hydroxybis(1,2-bis(diphenylphosphino)ethane)oxomolybdenum(IV) tetrafluoroborate, 44B, 1084
$C_{52}H_{50}F_6IrP_5$, Bis(tetraphenyldiphosphinoethane)dihydridoiridium(III) hexafluorophosphate, 43B, 1463
$C_{52}H_{50}FeP_4$, Dihydrido-bis[1,2-bis(diphenylphosphino)ethane]iron(II), 46B, 1299
$C_{52}H_{50}P_4Ru \cdot C_6H_6$, cis-Dihydridobis[1,2-bis(diphenylphosphinyl)ethane]ruthenium benzene solvate, 45B, 1376
$C_{52}H_{51}P_4Re$, Trihydridobis(1,2-bisdiphenylphosphinoethane)-rhenium(III), 38B, 1052
$C_{52}H_{52}Ag_4P_4S_8W_2$, Tetrakis(methyldiphenylphosphinesilver)-bis(tetrathiotungsten), 45B, 1376
$C_{52}H_{52}Cu_2I_2O_2P_4S$, Tetrakis(methyldiphenylphosphine)di-μ-iodo-dicopper(I) - sulfur dioxide, 43B, 1464
$C_{52}H_{52}Cu_4I_4P_4$, (Methyldiphenylphosphine)copper(I) iodide tetramer, 43B, 1464
$C_{52}H_{52}I_8Ir_2P_4$, Tri-μ-iodo-diiodotetrakis(methyldiphenylphosphine)diiridium triiodide, 44B, 1085
$C_{52}H_{56}MoP_4$, Tetrahydridotetrakis(methyldiphenylphosphine)molybdenum-(IV), 39B, 872
$C_{52}H_{64}B_9IrP_4$, cis-Bis[1,2-bis(diphenylphosphino)ethane]dihydridoiridium(III) tetradecahydridononaborate(-1), 45B, 1377
$C_{53}H_{44}BClF_4O_2P_4Rh_2$, μ-Chloro-μ-bis[bis(diphenylphosphinyl)methane]-dicarbonyldirhodium tetrafluoroborate, 45B, 1378
$C_{53}H_{44}Ir_2O_3P_4S \cdot 2\ C_7H_8$, (μ-Carbonyl)(μ-thio)bis[(μ-bis(diphenylphosphino)methane)carbonyliridium](Ir-Ir) toluene, 46B, 1308
$C_{53}H_{48}MoOP_4$, Bis(1,2-bis(diphenylphosphino)ethane)carbonylmolybdenum(0), 44B, 1085
$C_{53}H_{50}Cl_2N_2O_9P_4Rh_2 \cdot CHCl_3$, Carbonyl-rhodium-di-μ-(ethyl-bis(diphenylphosphito)amine)chlororhodium chloroform solvate, 45B, 1378
$C_{53}H_{50}Cl_5IrP_4$, Bis(tetraphenyldiphosphinoethane)-trans-dichloro-iridium(III) chloride methylene chloride solvate, 42B, 991
$C_{53}H_{51}Br_2N_2P_4W$, Bis(1,2-bis(diphenylphosphino)ethane)-bromo-N-methylhydrazido(2-)-N'-tungsten(IV) bromide, 41B, 1001
$C_{53}H_{52}BCl_2F_5MoN_2P_4$, 1-(η-Hydrazido(2-))fluorobis(1,2-bis(diphenylphosphino)ethane)molybdenum tetrafluoroborate dichloromethane solvate, 42B, 991
$C_{54}H_{38}O_6P_2Pt$, (2,2'-Bis(indan-1,3-dione)-2,2'-diolato)bis(triphenylphosphine)platinum, 45B, 1379
$C_{54}H_{42}As_4Br_2Ru$, Tris(o-diphenylarsinophenyl)arsineruthenium dibromide, 30B, 391
$C_{54}H_{43}CoP_4$, Hydrido(tris(o-diphenylphosphinophenyl)phosphine)cobalt-(I), 41B, 1219; 42B, 991
$C_{54}H_{44}IrP_3$, Bis(ortho-metallated)tris(triphenylphosphine)iridium-(III) hydride, 40B, 1061
$C_{54}H_{45}Ag_2P_3S_4W$, Bis(triphenylphosphine)silver-bis(μ-thio)tungsten-bis(μ-thio)triphenylphosphinesilver, 45B, 1379
$C_{54}H_{45}Au_3BF_4OP_3$, Tris(triphenylphosphinegold)oxonium tetrafluoroborate, 46B, 1300
$C_{54}H_{45}Au_3F_6P_4S$, μ_3-Sulfido-tris[triphenylphosphinegold(I)] hexafluorophosphate, 46B, 1237
$C_{54}H_{45}BCuF_4P_3$, Tetrafluoroboratotris(triphenylphosphine)copper(I), 40B, 1062
$C_{54}H_{45}ClCuP_3$, Chlorotris(triphenylphosphine)copper(I), 42B, 992
$C_{54}H_{45}ClP_3Rh$, Chlorotris(triphenylphosphine)rhodium(I) (orange form), 43B, 1465

$C_{54}H_{45}ClP_3Rh$, Chlorotris(triphenylphosphine)rhodium(I) (red form), 43B, 1465
$C_{54}H_{45}Cl_2Cu_2P_3$, Di-μ-chloro-tris(triphenylphosphine)dicopper(I), 38B, 1053
$C_{54}H_{45}Cl_2P_3Ru$, Dichlorotris(triphenylphosphine)ruthenium(II), 30B, 390
$C_{54}H_{45}Cu_2I_2P_3$, Tris(triphenylphosphine)di-μ-iodo-dicopper(I), 43B, 1464
$C_{54}H_{45}Cu_2MoP_3S_4 \cdot 0.8\ CH_2Cl_2$, Bis(triphenylphosphine)copperbis(μ-thio)molybdenumbis(μ-thio)triphenylphosphinecopper methylene chloride solvate, 46B, 1300
$C_{54}H_{45}IrNOP_3$, Tris(triphenylphosphine)nitrosyliridium, 37B, 666
$C_{54}H_{45}NOP_3Rh$, Nitrosyltris(triphenylphosphine)rhodium, 43B, 1466
$C_{54}H_{45}NiO_2P_3S$, (Sulfur dioxide)tris(triphenylphosphine)nickel(0), 45B, 1380
$C_{54}H_{45}O_2P_3PtS \cdot 0.7\ SO_2$, Tris(triphenylphosphine)(sulfur dioxide)-platinum sulfur dioxide solvate, 42B, 994
$C_{54}H_{46}ClIrNO_5P_3$, Hydridonitrosyltris(triphenylphosphine)iridium(I) perchlorate (brown isomer), 40B, 1062
$C_{54}H_{46}ClIrNO_5P_3$, Hydridonitrosyltris(triphenylphosphine)iridium(I) perchlorate (black isomer), 37B, 667
$C_{54}H_{46}CoN_2P_3$, Hydridodinitrogentris(triphenylphosphine)cobalt(I), 34B, 618; 40B, 1132
$C_{54}H_{46}F_3P_4Rh \cdot 1.5\ C_6H_6$, Tris(triphenylphosphine)(trifluorophosphine)rhodium(I) hydride benzene solvate, 43B, 1472
$C_{54}H_{46}NORu$, Hydridonitrosyltris(triphenylphosphine)ruthenium, 38B, 947
$C_{54}H_{48}F_7MoO_2P_5$, Dicarbonyl-bis(diphenylphosphinoethane)fluoromolybdenum hexafluorophosphate, 46B, 1301
$C_{54}H_{51}ClIrNP_4S_2$, Disulfurbis(bis(diphenylphosphino)ethane)iridium(I) chloride acetonitrile solvate, 38B, 1024
$C_{54}H_{52}N_2NiP_2$, Bis(tri-p-tolylphosphine)(azobenzene)nickel(0), 39B, 873
$C_{54}H_{57}BF_4Ir_2P_4$, Tris(μ-hydrido)bis[hydrido(1,3-bis(diphenylphosphino)propane)iridium] tetrafluoroborate, 46B, 1301
$C_{54}H_{96}As_2Co_3N_8O_{10}P_2 \cdot 2\ H_2O$, Bis((tri-n-butylphosphine)-bis(dimethylglyoximato)-(methyl(phenyl)arsenyl)cobalt(III))-cobalt(II) dihydrate, 45B, 1380
$C_{54}H_{99}P_3Pt \cdot 1.5\ C_7H_{16}$, Tris(tricyclohexylphosphine)platinum heptane solvate, 43B, 1466
$C_{54}H_{102}Ni_2P_4$, Di-μ-hydrido-bis(1,3-propanebis(dicyclohexylphosphine))dinickel(Ni-Ni), 43B, 1467
$C_{54}H_{162}Co_6K_6N_{12}P_{18}$, Potassium dinitrogentris(trimethylphosphine)cobalt cluster, 44B, 1086
$C_{55}H_{44}As_4Br_2Cl_2Hg$, Tris(o-diphenylarsinophenyl)arsinemercury(II) bromide dichloromethane adduct, 31B, 463
$C_{55}H_{44}Cl_3O_5P_2Rh_3 \cdot CH_2Cl_2$, μ-Carbonyl-μ-chloro-di-μ-[bis(diphenylphosphino)methane]-bis[carbonyl-rhodium]dichloro-carbonyl-rhodium dichloromethane solvate, 45B, 1381
$C_{55}H_{45}OP_3Pt$, Tris(triphenylphosphine)carbonylplatinum, 40B, 1132
$C_{55}H_{46}CoOP_3$, trans-Carbonylhydridotris(triphenylphosphine)cobalt(I), 43B, 1468
$C_{55}H_{46}OP_3Rh$, Rhodium carbonyl hydride tris(triphenylphosphine), 30B, 389
$C_{55}H_{47}Cl_3O_4P_3Rh$, Tris(triphenylphosphine)rhodium(I) perchlorate methylene dichloride solvate, 43B, 1468
$C_{55}H_{54}BINiP_2S_2$, [1,9-Bis(diphenylphosphino)-3,7-dithianonane]mono-

iodonickel tetraphenylborate, 46B, 1302
$C_{55}H_{68}N_8O_{10}P_2Rh_2$, Bis(triphenylphosphine)tetrakis(dimethylglyoximato)dirhodium monohydrate mono-1-propanolate, 37B, 665
$C_{56}H_{40}Co_4O_8P_4 \cdot C_6H_6$, Bis[(μ-carbonyl)-(μ₄-phenylphosphido)triphenylphosphine]-hexacarbonyltetracobalt benzene solvate, 46B, 1264
$C_{56}H_{49}Cl_5O_2P_3Rh$, Chloro(dioxygen)tris(triphenylphosphine)rhodium(I) bis(dichloromethane), 43B, 1469
$C_{56}H_{51}MoN_2OP_4$, Bis(1,2-bis(diphenylphosphino)ethane)dinitrogencarbonylmolybdenum - hemibenzene, 44B, 1085
$C_{56}H_{54}Cl_6N_2P_4Re_2$, Di-μ-chloro-bis((bis(diphenylphosphino)ethane)dichlororhenium) bis(acetonitrile), 41B, 1220
$C_{56}H_{56}Ag_2Cl_2P_4S_2$, Chloro(bis(diphenylphosphinoethyl)sulfido)silver dimer, 40B, 1063
$C_{56}H_{56}Ag_2I_2P_4S_2$, Iodo(bis(diphenylphosphinoethyl)sulfido)silver dimer, 41B, 1221
$C_{56}H_{56}Au_2Cl_2O_2P_4$, (Bis(diphenylphosphino)methane)chlorogold(I) dimer bis(acetone), 43B, 1470
$C_{56}H_{56}Cl_4Cu_4O_2P_4$, Bis(diphenylphosphino)methane-bis(chlorocopper) dimer acetone solvate, 40B, 1064
$C_{56}H_{56}Cl_4Mo_2O_2P_4$, Tetrachlorobis(bis(diphenylphosphino)methane)dimolybdenum(II)(Mo-Mo) - bis(acetone), 44B, 1086
$C_{56}H_{64}BIrO_2P_4$, Dioxygentetra(dimethylphenylphosphine)iridium(I) tetraphenylborate, 42B, 893
$C_{56}H_{64}BO_2P_4Rh$, Dioxygentetra(dimethylphenylphosphine)rhodium(I) tetraphenylborate, 42B, 894
$C_{56}H_{64}P_4W$, Tetrahydrido-tetrakis(diphenylethylphosphino)tungsten, 45B, 1382
$C_{57}H_{48}O_2P_3PtS$, Tris(triphenylphosphine)(sulfur dioxide)platinum benzene solvate, 39B, 874
$C_{57}H_{50}NOP_3Re$, Dihydridonitrosyltris(triphenylphosphine)rhenium(I) benzene solvate, 42B, 994
$C_{57}H_{51}IrP_3$, mer-Trihydridotris(triphenylphosphine)iridium(III) hemi-(benzene), 41B, 1221
$C_{57}H_{52}O_5P_4Pd_4$, Tetra(methyldiphenylphosphine)pentacarbonyltetrapalladium, 46B, 1303
$C_{57}H_{99}O_3P_3Pt_3$, Tricarbonyltris(tricyclohexylphosphine)triplatinum, 43B, 1470
$C_{58}H_{44}O_8P_4Rh_4$, Bis(bis(diphenylphosphino)methane)tetrarhodium octacarbonyl, 42B, 995
$C_{58}H_{47}F_6O_4P_3Pt$, Hydrotris(triphenylphosphine)platinum(II) hydrogenbis(trifluoroacetate), 43B, 1471
$C_{58}H_{53}Cl_5CoP_4Sn$, Chlorobis(bis(1,2-diphenylphosphino)ethane)cobalt-(II) trichlorostannate chlorobenzene solvate, 38B, 1051
$C_{58}H_{56}Cl_2O_4P_4Rh_2 \cdot x\ CH_2Cl_2$, trans-(1,5-Bis(diphenylphosphino)-3-oxapentane-P,P')carbonylchlororhodium(I) dimer dichloromethane, 43B, 1448
$C_{58}H_{60}Ag_2Cl_2P_4$, Chloro-(1,5-bis(diphenylphosphino)pentane)silver dimer, 42B, 996
$C_{58}H_{64}BF_4IrP_4$, Tetrakis(methyldiphenylphosphine)iridium(I) tetrafluoroborate cyclohexane solvate, 41B, 1222
$C_{59}H_{45}FeO_{14}P_3Pt_2$, Tetracarbonylironcarbonyltris(triphenylphosphite)-diplatinum, 40B, 1065
$C_{59}H_{56}BNiO_2P_3S$, Bis(2-diphenylphosphinoethyl)phenylphosphinenickel-(O-methyl sulfinate) tetraphenylborate, 46B, 1303
$C_{60}H_{40}Cl_2F_{12}O_2P_4Rh_2$, Di(1,2-bis(diphenylphosphino)hexafluorocyclopentene)rhodium(I) cis-dicarbonyldichlororhodate(I), 37B, 668
$C_{60}H_{51}Cl_2Cu_2P_3$, Bis(triphenylphosphine)copper(I)-di-μ-chloro-tri-

phenylphosphinecopper(I) benzene solvate, 42B, 992
$C_{60}H_{52}Mo_2N_4O_4P_4S_4 \cdot 4\ CH_2Cl_2$, Bis(molybdenumdicarbonyl-P,P-diphenyl-N-methylphosphinothioformamido-μ-P,P-diphenyl-N-methylphosphinothioformamido) dichloromethane solvate, 46B, 1304
$C_{60}H_{56}Mo_2N_4O_2P_4S_4$, Tetrakis(isothiocyanato)bis(bis(diphenylphosphino)methane)dimolybdenum(II)(Mo-Mo) - bis(acetone), 44B, 1086
$C_{60}H_{57}As_3ClNNiO_4P$, (Triphenylphosphine)tris(2-diphenylarsinoethyl)aminenickel(I) perchlorate, 46B, 1305
$C_{60}H_{61}NOP_3Rh$, Hydridotris(triphenylphosphine)rhodium(I) - dimethylamine - tetrahydrofuran, 44B, 1087
$C_{61}H_{53}O_8P_3Pt_3S_4$, Tri-μ-(sulfur dioxide)-tris(triphenylphosphine)triplatinum toluene (sulfur dioxide), 43B, 1471
$C_{62}H_{51}CoMn_3O_6P_6$, 1,1,1-Tris(diphenylphosphinomethyl)ethanecobalt-μ-cyclotriphosphorus-tris(cyclopentadienyldicarbonylmanganese), 46B, 1305
$C_{62}H_{57}P_4Re$, Rhenium(III) hydride bis(triphenylphosphine) 1,2-bis(diphenylphosphino)ethane, 31B, 464
$C_{62}H_{60}Cl_2Mn_2O_5P_4$, Di-μ-bis(diphenylphosphino)methane-pentacarbonyldimanganese dichloromethane n-hexane solvate, 42B, 996
$C_{62}H_{107}BP_4Pt_2$, Trihydridobis[1,3-bis(di-t-butylphosphino)propane]diplatinum tetraphenylborate, 45B, 1382
$C_{64}H_{60}Cl_5P_4Re_2$, Bis(bis(diphenylphosphino)methane)pentachlorodirhenium(Re-Re) - toluene, 42B, 997
$C_{65}H_{37}Cl_5Mo_2N_2O_3P_4$, Bis[1,2-bis(diphenylphosphino)ethane]chloro-(oxo)molybdenum(IV) mer-trichloro(N'-p-chlorobenzoyl-N-phenyldiazene-N,O)-oxo-molybdate(IV), 45B, 1383
$C_{65}H_{50}LiO_5P_5Rh_4$, Lithium penta(μ-diphenylphosphido)pentacarbonyltetrarhodium, 46B, 1306
$C_{65}H_{90}Cl_8MoN_4P_8Re_2$, Di-μ-dinitrogen-tetrachlorobis(chlorotetrakis-(dimethylphenylphosphine)rhenio)molybdenum dichloromethane solvate, 42B, 998
$C_{66}H_{59}BBrCoOP_3$, Bromocarbonyl-1,1,1-tris(diphenylphosphinomethyl)-ethanecobalt(II) tetraphenylborate, 46B, 1307
$C_{66}H_{62}BBrFeP_4$, Bromo(tris(2-diphenylphosphinoethyl)phosphine)iron-(II) tetrafluoroborate, 44B, 1079
$C_{66}H_{62}BCoN_2OP_3$, Nitrosyl(tris(2-diphenylphosphinoethyl)amine)cobalt tetraphenylborate, 42B, 999
$C_{66}H_{62}BCoN_3O_3P_3$, Dinitrosyl(bis-(2-diphenylphosphinoethyl)-2-diphenylphosphineoxide-ethyl)aminecobalt tetraphenylborate, 41B, 1223
$C_{66}H_{62}BFeNOP_4$, Nitrosyltris-(2-diphenylphosphinoethyl)phosphineiron tetraphenylborate, 41B, 1224
$C_{66}H_{62}BFeN_2OP_3$, Nitrosyl(tris(2-diphenylphosphinoethyl)amine)iron tetraphenylborate, 42B, 999
$C_{66}H_{62}BN_2NiOP_3$, Nitrosyl(tris(2-diphenylphosphinoethyl)amine)nickel tetraphenylborate, 42B, 999
$C_{66}H_{63}BFeP_4S$, Thiolo(tris(2-diphenylphosphinoethyl)phosphine)iron-(II) tetraphenylborate, 43B, 1473
$C_{66}H_{63}BNiP_4S$, Thiolo(tris(2-diphenylphosphinoethyl)phosphine)nickel-(II) tetraphenylborate, 42B, 1000; 43B, 1473
$C_{66}H_{92}B_2N_4O_{12}P_4Ru$, trans-Bis(acetone hydrazone)tetrakis(trimethylphosphite)ruthenium(II) bis(tetraphenylborate), 40B, 1066
$C_{66}H_{92}Fe_2N_4Na_2O_{18}P_2$, Sodium di(μ-diphenylphosphido)-bis(tricarbonyliron) 1,10-diaza-4,7,13,16,21,24-hexaoxabicyclo[8.8.8]hexacosane, 45B, 1384
$C_{66}H_{94}B_2P_6Pd_3S_2$, Bis[($\mu_3$-sulfido)tris(trimethylphosphine)]tripalladium(II) bis(tetraphenylborate), 46B, 1307
$C_{67}H_{64}BBrCl_2FeP_4$, Bromo(hexaphenyl-1,4,7,10-tetraphosphadecane)iron-

(II) tetraphenylborate methylene chloride solvate, 40B, 1067
$C_{68}H_{66}Cl_2FeP_6 \cdot 2\ C_3H_6O$, Dichlorobis(2-diphenylphosphinoethyl)phenylphosphineiron(II) acetone, 46B, 1308
$C_{68}H_{69}ClFe_2O_4P_6S_3 \cdot C_3H_6O$, Tri-μ-mercaptobis{bis[2-(diphenylphosphino)ethyl]phenylphosphine}diiron(II) perchlorate acetone solvate, 45B, 1384
$C_{69}H_{86}As_4Cl_2OPd_2$, μ-Carbonyl-di-μ-bis(diphenylarsino)methyl-di(chloropalladium(I)) n-hexane solvate, 44B, 1088
$C_{70}H_{68}BCoNO_2P_3$, (Tris(2-diphenylphosphinoethyl)amine)cobalt(I) carbonyl tetraphenylborate - acetone, 42B, 1000
$C_{70}H_{68}BCoO_3P_3$, Acetato(1,1,1-tris(diphenylphosphinomethyl)ethane)cobalt(II) tetraphenylborate acetone solvate, 41B, 1229
$C_{70}H_{72}BCoNOP_3S$, Methylthio(tris(2-diphenylphosphinoethyl)amine)cobalt(II) tetraphenylborate acetone, 43B, 1473
$C_{70}H_{80}Cl_2Cu_2O_4P_4 \cdot 2\ CH_4O$, Bis(μ-chloro)-bis(((4R,5R)-trans-4,5-bis(bis((3-methylphenyl)phosphino)methyl)-2,2-dimethyl-1,3-dioxalane)copper(I)) methanol solvate, 46B, 1309
$C_{71}H_{55}O_5P_7Rh_4$, Tetrakis(μ-diphenylphosphido)(triphenylphosphine)pentacarbonyltetrarhodium, 46B, 1309
$C_{72}H_{30}F_{30}GeHgP_2PtSn \cdot C_6H_6$, [Tris(pentafluorophenyl)germanium-mercury]-bis(triphenylphosphine)-[tris(pentafluorophenyl)tin]platinum(0) benzene solvate, 46B, 1310
$C_{72}H_{60}Ag_2Cl_2P_4$, Chlorobis(triphenylphosphine)silver, 45B, 1385
$C_{72}H_{60}Ag_4Cl_4P_4$, Tetra((triphenylphosphine)silver chloride), 41B, 1225; 42B, 1001
$C_{72}H_{60}Ag_4I_4P_4$, Tetra((triphenylphosphine)silver iodide), 41B, 1225; 42B, 1002
$C_{72}H_{60}Ag_4I_4P_4 \cdot 1.5\ CH_2Cl_2$, Triphenylphosphinesilver iodide tetramer (form II), 42B, 1002
$C_{72}H_{60}Ag_4I_4P_4 \cdot n\ CH_2Cl_2$, Tetra((triphenylphosphine)silver iodide) methylene chloride solvate, 41B, 1225
$C_{72}H_{60}Ag_4Mo_2P_4S_8$, Tetrakis[(triphenylphosphine)silver(I)]bis[tetrathiomolybdate(VI)], 45B, 1385
$C_{72}H_{60}As_4Cu_4I_4 \cdot C_6H_6$, Tetrakis[(triphenylarsine)iodocopper(I)] benzene solvate, 45B, 1386
$C_{72}H_{60}Cl_4Cu_4P_4$, Triphenylphosphinecopper(I) chloride tetramer, 40B, 1067
$C_{72}H_{60}Cu_2N_6P_4$, μ-Diazido-tetrakis(triphenylphosphine)dicopper(I), 37B, 668
$C_{72}H_{60}Cu_4I_4P_4$, Triphenyphosphinecopper(I) iodide tetramer, 41B, 1226
$C_{72}H_{60}Cu_4O_2P_4S_6W_2$, Hexakis-$\mu_3$-sulfidobis(oxotungsten)tetra(triphenylphosphinecopper), 46B, 1240
$C_{72}H_{60}O_{12}P_4S_4Ru_2 \cdot C_7H_8$, Bis[(μ-sulfato-O,O',O")bis(triphenylphosphine)(sulfur dioxide)ruthenium(II)] toluene solvate, 45B, 1386
$C_{72}H_{60}P_4Pd \cdot 0.5\ C_6H_6$, Tetrakis(triphenylphosphine)palladium(0) benzene solvate, 42B, 1002
$C_{72}H_{61}ClO_7P_4Pt_2 \cdot 2\ C_6H_6$, μ-Hydroxo-μ-peroxo-bis[bis(triphenylphosphine)platinum(II)] perchlorate benzene solvate, 45B, 1387
$C_{72}H_{62}P_4Ru$, Dihydridotetrakis(triphenylphosphine)ruthenium(II), 38B, 1053
$C_{72}H_{132}N_2Ni_2P_4$, Bis(bis(tricyclohexylphosphine)nickel)dinitrogen, 37B, 669
$C_{73}H_{60}Cl_4P_4Ru_2S$, Thiocarbonylbis(triphenylphosphine)ruthenium-tri-μ-chloro-chlorobis(triphenylphosphine)ruthenium, 40B, 1068
$C_{73}H_{70}BCoINP_3$, Iodo(tris-(2-diphenylphosphinoethyl)amine)cobalt(II) tetraphenylborate toluene, 41B, 1214
$C_{74}H_{60}Ag_2N_2P_4S_2$, Thiocyanatobis(triphenylphosphine)silver(I), 39B,

874

$C_{74}H_{62}Br_4Cl_6Cu_4P_4$, Triphenylphosphinecopper(I) bromide tetramer chloroform adduct, 40B, 1069

$C_{74}H_{62}Cl_8O_4P_4Rh_2S_2$, Di-$\mu$-chloro-tetrakis(triphenylphosphine)bis(sulfur dioxide)dirhodium(I) bis(chloroform), 43B, 1473

$C_{74}H_{64}Cl_6O_4P_4Rh_2$, Bis(chloro(dioxygen)bis(triphenylphosphine)rhodium(I)) bis(dichloromethane), 43B, 1474

$C_{74}H_{65}BClP_4Rh$, Chlorohydridobis(bis(diphenylphosphino)methane)rhodium tetraphenylborate, 45B, 1388

$C_{74}H_{66}B_2Cu_2N_2P_4$, μ-Bis(cyanotrihydroborato)-tetrakis(triphenylphosphine)dicopper(I), 40B, 1070

$C_{74}H_{69}Cl_3O_6P_6Ru_2$, Tri-$\mu$-chloro-diphenylphosphinitotris(diphenylphosphinous acid)bis(methyldiphenylphosphinite)diruthenium(II), 43B, 1475

$C_{75}H_{60}O_3P_5Rh_3$, Tris(μ_2-diphenylphosphido)bis(triphenylphosphine)tricarbonyltrirhodium, 46B, 1311

$C_{75}H_{66}Ag_3Br_3P_6$, Bis(μ_3-bromo)-tris[(diphenylphosphino)methanesilver] bromide, 46B, 1311

$C_{76}H_{64}Cl_4O_2P_4Rh_2$, Tetrakis(triphenylphosphine)-di-μ-carbonyl-dirhodium(0) - bis(dichloromethane) solvate, 39B, 875

$C_{76}H_{68}Cl_2O_2P_4Rh_2$, Bis(triphenylphosphine)rhodium(I) chloride dimer - ethyl acetate, 44B, 1088

$C_{77}H_{64}BClO_3P_4Rh_2$, μ-Chloro-μ-carbonylbis-μ-[bis(diphenylphosphino)-methane]di(carbonylrhodium)(Rh-Rh) tetraphenylborate, 45B, 1388

$C_{78}H_{62}BClCoP_4$, Chloro(tris(o-diphenylphosphinophenyl)phosphine)cobalt(II) tetraphenylborate(III), 37B, 670

$C_{78}H_{65}AuBP_3$, Tris(triphenylphosphine)gold(I) tetraphenylborate, 46B, 1312

$C_{78}H_{72}BNOP_4Ru$, Nitrosylbis(1,3-bis(diphenylphosphino)propane)ruthenium tetraphenylborate, 43B, 1475

$C_{78}H_{72}CuN_6P_6$, Diazido-μ-1,2-bis(diphenylphosphino)ethane-bis(1,2-bis(diphenylphosphino)ethane)dicopper(I), 37B, 671

$C_{79}H_{74}BNO_2P_4Ru$, Nitrosylbis(1,2-bis(diphenylphosphino)ethane)ruthenium tetraphenylborate - acetone, 39B, 876

$C_{82}H_{72}Cu_2F_6N_8P_6$, Di-2-(5-perfluoromethyltetrazolato)-μ-1,2-bis(diphenylphosphino)ethane-bis(1,2-bis(diphenylphosphino)ethane)dicopper(I), 38B, 1054

$C_{82}H_{78}P_6Pt$, Bis(1,1,1-tris(diphenylphosphinomethyl)ethane)platinum-(0), 43B, 1476

$C_{82}H_{81}F_6Fe_2P_7 \cdot 1.5\ CH_2Cl_2$, Tri-$\mu$-hydrido-bis(1,1,1-tris(diphenylphosphinomethyl)ethaneiron(II)) hexafluorophosphate dichloromethane solvate, 41B, 1228

$C_{84}H_{60}N_6O_2P_4Rh_2$, μ-trans-1,1,2,3,4,4-Hexacyanobutenediido-bis((carbonyl)bis(triphenylphosphine)rhodium), 40B, 1071

$C_{84}H_{72}Co_2N_2P_6S$, μ-Thio-bis(tris(2-diphenylphosphinoethyl)aminecobalt(I)), 44B, 1090

$C_{84}H_{77}BNNiP_3Sn$, (Triphenylstannyl)(tris(2-diphenylphosphinoethyl)amine)nickel(II) tetraphenylborate, 43B, 1477

$C_{84}H_{84}Cl_2Cu_2O_2P_6$, Dichlorotris-(1,2-bis(diphenylphosphino)ethane)dicopper(I) bisacetone, 38B, 1054

$C_{84}H_{84}Cu_4O_2P_4S_6W_2$, Bis{bis(tri-p-tolylphosphine)copper(I))oxotrithiotungsten}, 45B, 1389

$C_{84}H_{124}B_2Cl_4N_2Ni_3P_6$, Dichlorobis(chlorotris(2-diethylphosphinoethyl)aminenickel(II))nickel(II) bis(tetraphenylborate), 46B, 1313

$C_{85}H_{93}B_2Cl_8F_8Ir_3P_6$, ($\mu_3$-Hydrido)tris($\mu$-hydrido)tris[hydrido(1,3-bis-(diphenylphosphino)propane)iridium] bis(tetrafluoroborate) tetrakis(chloroform) solvate, 46B, 1301

$C_{88}H_{80}Cu_2P_6$, Bis-μ-diphenylphosphido-bis(bis-1,2-diphenylphosphinoethanecopper(I)) benzene, 44B, 1089, 1166
$C_{88}H_{84}F_{12}O_4P_8Rh_2$, [μ-(1,4-Bis(diphenylphosphino)butane)]-[bis(1,4-bis(diphenylphosphino)butane)]tetracarbonyldirhodium hexafluorophosphate, 46B, 1313
$C_{88}H_{90}Cl_6Ni_3P_6 \cdot 0.7\ CHCl_3$, Bis(μ₃-(1,1,1-tris(diphenylphosphinoethyl)ethane))hexachlorotrinickel chloroform solvate, 46B, 1314
$C_{90}H_{135}Cl_6P_9Ru_3$, Tri-μ-chloro-hexakis(diethylphenylphosphine)diruthenium(II) trichloro-tris(diethylphenylphosphine)ruthenate(II), 34B, 619
$C_{96}H_{60}Cu_2I_2P_4$, Di-μ-iodo-tetrakis(tris(phenylethynyl)phosphine)dicopper, 43B, 1477
$C_{97}H_{81}O_3P_5Rh_2$, μ-Carbonato-pentakis(triphenylphosphine)dirhodium - benzene, 42B, 1003
$C_{99}H_{84}O_6P_4Pt_5$, Pentakis-μ-carbonyl-(carbonyltetra(triphenylphosphine))pentaplatinum toluene solvate, 44B, 1089
$C_{106}H_{101}As_6BCo_2$, Tri-μ-hydrido-bis(1,1,1-tris(diphenylarsinomethyl)-ethanecobalt(II)) tetraphenylborate, 41B, 1228
$C_{108}H_{80}BCo_4O_{12}P_4Sb \cdot CH_2Cl_2$, μ_4-Antimoniotetrakis[tricarbonyl(triphenylphosphine)cobalt] tetraphenylborate dichloromethane solvate, 45B, 1389
$C_{111}H_{103}Cu_6NOP_6$, Triphenylphosphinocopper(I) hydride hexamer dimethylformamide solvate, 38B, 1055
$C_{120}H_{128}As_6BIN_2Ni_2O_3$, μ-Iodo-bis(tris(2-diphenylarsinoethyl)amine-nickel(I)) tetraphenylborate tris(tetrahydrofuran), 42B, 1003; 43B, 1478
$C_{126}H_{84}Au_{11}F_{21}I_3P_7$, Triiodoheptakis(tri-p-fluorophenylphosphine)-undecagold, 38B, 1056
$C_{130}H_{118}As_3B_2Co_2P_6$, μ-Cyclo-triarsenic-bis[1,1,1-tris(diphenylphosphinomethyl)ethane-cobalt] tetraphenylborate, 45B, 1390
$C_{130}H_{118}B_2CoNiP_9 \cdot 2\ C_3H_6O$, (1,1,1-Tris(diphenylphosphinomethyl)ethane)cobalt-μ-(cyclo-triphosphorus)-(1,1,1-tris(diphenylphosphinomethyl)ethane)nickel tetraphenylborate acetone solvate, 45B, 1390
$C_{130}H_{118}B_2Ni_2P_9 \cdot 2.5\ C_3H_6O$, μ-Cyclo-triphosphorus-bis[(1,1,1-tris-(diphenylphosphinomethyl)ethane)nickel] tetraphenylborate acetone solvate, 45B, 1390
$C_{136}H_{150}B_2N_2Ni_2O_2S \cdot 1.6\ C_3H_7NO$, μ-Thio-bis(1,1,1-tris(diphenylphosphinomethyl)ethanenickel(II)) tetraphenylborate dimethylformamide solvate, 44B, 1090
$C_{142}H_{144}B_2Co_2O_6P_6$, Di-μ-hydroxo-bis(1,1,1-tris(diphenylphosphinomethyl)ethane)dicobalt(II) tetraphenylborate acetone solvate, 41B, 1229
$C_{174}H_{166}Au_6B_2P_6$, Hexakis(tris(p-tolyl)phosphine)-octahedro-hexagold bis(tetraphenylborate), 39B, 876

87. COMPLEX INORGANIC ANIONS

$CH_3F_{11}O_2SSb_2$, Methoxythionyl μ-fluoro-bis(pentafluoroantimonate(V)), 42B, 1004
$CH_{10}Cl_3Mn_2NO_2$, Manganese methylammonium trichloride dihydrate, 46B, 1314
$CH_{30}AlNO_{20}S_2$, Methylammonium alum, 1, 111, 455; 21, 522; 27, 813; 29, 537
$CH_{30}CrNO_{20}S_2$, Methylammonium chromium alum, 33A, 372; 33B, 519
$C_2H_{10}Cl_4CuN_2$, Ethylenediammonium copper tetrachloride, 38B, 36; 44B, 1090

$C_2H_{10}Cl_4MnN_2$, Ethylenediammonium manganese tetrachloride, 44B, 1090
$C_2H_{10}Cl_4N_2Pd$, Ethylenediammonium tetrachloropalladate(II), 42B, 1005
$C_2H_{10}Cl_6N_2Pb_2$, Ethylenediammonium lead(II) chloride, 42B, 1005
$C_2H_{10}F_5N_2OV$, Ethylenediammonium oxopentafluorovanadate, 42B, 1005
$C_2H_{12}Br_4CdN_2$, Methylammonium tetrabromo-cadmate(II), 45B, 1391
$C_2H_{12}Br_5FeN_2$, Bis(methylammonium) tetrabromoferrate(III) bromide, 38B, 37
$C_2H_{12}CdCl_4N_2$, Bis(methylammonium) cadmium tetrachloride (orthorhombic), 41B, 31
$C_2H_{12}CdCl_4N_2$, Bis(methylammonium) cadmium tetrachloride (tetragonal), 41B, 31
$C_2H_{12}Cd_2Cl_5NO_2$, Dicadmium dimethylammonium pentachloride dihydrate, 45B, 1391
$C_2H_{12}Cl_4MnN_2$, Bis(methylammonium) manganese tetrachloride (orthorhombic form 1), 41B, 32
$C_2H_{12}Cl_4MnN_2$, Bis(methylammonium) tetrachloromanganate(II) (orthorhombic form 2), 43B, 31
$C_2H_{12}Cl_4MnN_2$, Bis(methylammonium) manganese tetrachloride (tetragonal form 1), 41B, 32
$C_2H_{12}Cl_4MnN_2$, Bis(methylammonium) tetrachloromanganate(II) (tetragonal form 2), 43B, 31
$C_2H_{12}Cl_4MnN_2$, Methylammonium tetrachloromanganate(II), 44B, 1091
$C_2H_{12}Cl_4N_2Zn$, Monomethylammonium tetrachlorozincate(II), 42B, 1005
$C_2H_{12}Cl_6N_2Sn$, Methylammonium hexachlorostannate(IV), 45B, 1392
$C_2H_{12}Cr_3N_6O_{10}$, Guanidinium trichromate, 43B, 77
$C_2H_{14}F_8N_2O_4V_2$, Ethylenediammonium aquooxotetrafluorovanadate(V), 42B, 1006
$C_2H_{18}Cl_8N_8OReS_2 \cdot 2\ H_2O$, Bis(dithiobisformamidinium) aquapentachlororhenate(IV) trichloride dihydrate, 46B, 1315
$C_3H_9Cl_3HgS$, Trimethylsulphonium trichloromercurate(II), 43B, 94
$C_3H_9HgI_3S$, Trimethylsulphonium triiodomercurate(II), 31B, 450
$C_3H_{10}CdCl_3N$, Trimethylammonium cadmium trichloride, 46B, 1315
$C_3H_{12}CdCl_4N_2$, Propylenediammonium tetrachlorocadmate(II), 43B, 32
$C_3H_{12}Cl_4CuN_2$, Bis(propylene-1,3-diammonium) tetrachlorocuprate(II), 42B, 1006
$C_3H_{12}Cl_4N_2Zn$, 1,3-Propanediammonium tetrachlorozincate(II), 46B, 1316
$C_3H_{14}Cl_3MnNO_2$, Trimethylammonium manganese(II) chloride dihydrate, 42B, 1006
$C_4H_{12}Ag_2I_3N$, Tetramethylammonium triiododiargentate, 28, 392
$C_4H_{12}Br_3HgN$, Tetramethylammonium tribromomercurate(II), 28, 394
$C_4H_{12}Br_3MnN$, Tetramethylammonium manganese(II) tribromide, 44B, 1091
$C_4H_{12}Br_3NNi$, Tetramethylammonium tribromonickelate, 31B, 342
$C_4H_{12}CdCl_3N$, Tetramethylammonium cadmium chloride, 38B, 37
$C_4H_{12}CeCl_6N$, Tetramethylammonium hexachlorocerate, 18, 680
$C_4H_{12}Cl_2IN$, Tetramethylammonium dichloroiodate(I), 7, 215, 231
$C_4H_{12}Cl_3CuN$, Tetramethylammonium trichlorocuprate, 42B, 1007
$C_4H_{12}Cl_3GeN$, Tetramethylammonium trichlorogermanate(II), 46B, 1316
$C_4H_{12}Cl_3MnN$, Tetramethylammonium manganese(II) chloride, 32B, 30
$C_4H_{12}Cl_3NNi$, Tetramethylammonium trichloronickelate(II), 33B, 520
$C_4H_{12}I_9N$, Tetramethylammonium enneaiodide, 19, 507
$C_4H_{14}Cl_4MnN_2$, 1,4-Butanediyldiammonium manganese tetrachloride, 46B, 1317
$C_4H_{14}Cl_4MnN_2$, 1,4-Butanediyldiammonium manganese tetrachloride (high temperature form), 46B, 1317
$C_4H_{16}CdCl_4N_2$, Bis(ethylammonium) tetrachlorocadmate, 43B, 33
$C_4H_{16}CdCl_4N_2$, Bis(ethylammonium) tetrachlorocadmate (low-temp), 43B,

33
$C_4H_{16}Cl_4CuN_2$, Ethylammonium tetrachlorocuprate(II), 35B, 880
$C_4H_{16}Cl_4MnN_2$, Bis(ethylammonium) tetrachloromanganate(II) (β form), 42B, 1007; 44B, 1092
$C_4H_{16}Cl_4MnN_2$, Bis(ethylammonium) tetrachloromanganate(II) (γ form), 43B, 34
$C_4H_{16}Cl_4NO_2Rh$, Tetramethylammonium trans-diaquotetrachlororhodate-(III), 39B, 877
$C_4H_{16}Cl_5CuN_3$, Bis-(2-ammonioethyl)ammonium monochloride tetrachlorocuprate(II), 37B, 672; 43B, 34
$C_4H_{16}Cl_6N_2Sn$, Dimethylammonium hexachlorostannate, 3, 658, 694
$C_4H_{16}Cl_8N_2Re_2$, Bis(dimethylammonium) octachloro-dirhenate, 45B, 1399
$C_4H_{16}F_4MnNO_2$, Tetramethylammonium diaquatetrafluoromanganate(III), 45B, 1392
$C_4H_{20}Br_6CuN_4$, Bis(ethylenediammonium) dibromide tetrabromocuprate-(II), 37B, 673
$C_4H_{20}Cl_6CoN_4$, Bis(ethylenediammonium) tetrachlorocobaltate(II) chloride, 43B, 35
$C_4H_{24}Cl_8Mo_2N_4O_2$, Ethylenediammonium octachlorodimolybdate(II) dihydrate, 40B, 1133
$C_4H_{24}Mo_{12}N_{12}O_{40}Si \cdot H_2O$, Tetraguanidinium α-dodecamolybdosilicate monohydrate, 46B, 1317
$C_5H_6Ag_5I_6N$, Pyridinium hexaiodopentaargentate, 38B, 258
$C_5H_6Br_4NSb$, Pyridinium tetrabromoantimonate(III), 42B, 1008
$C_5H_6Cl_2IN$, Pyridinium dichloroiodide, 39B, 877
$C_5H_6Cl_4NSb$, Pyridinium tetrachloroantimonate(III), 35B, 875
$C_5H_6Cl_6NSb$, Pyridinium hexachloroantimonate(V), 38B, 681
$C_5H_7Br_4NRe$, Pyridine tetrabromorhenate(II) (α form), 30B, 293
$C_5H_7Br_4NRe$, Pyridine tetrabromorhenate(II) (β form), 30B, 293
$C_6H_8Cl_4NSb$, Anilinium tetrachloroantimonate, 46B, 1318
$C_6H_{10}Cl_4N_4O_2U$, Imidazolium tetrachlorodioxouranium(VI), 46B, 1319
$C_6H_{12}N_8O_{18}U_2$, Di-μ-aquo-bis(dioxobis(nitrato)uranium(VI))-diimidazole, 46B, 1141
$C_6H_{16}Cl_6N_4S_2Te$, N^1,N^2,N^3,N^4-Tetramethyl-α,α'-dithiobis(formamidinium) hexachlorotellurate(IV), 46B, 1319
$C_6H_{20}CdCl_4N_2$, Bis(propylammonium) tetrachlorocadmate, 44B, 1093
$C_6H_{20}Cl_4CuN_2$, Bis(isopropylammonium) tetrachlorocuprate(II), 40B, 28
$C_6H_{20}Cl_4CuN_2$, n-Propylammonium tetrachlorocuprate(II), 35B, 877
$C_6H_{20}Cl_4Mn$, Bis(propylammonium) manganese tetrachloride, 38B, 39
$C_6H_{20}Cl_4MnN_2$, Bis(n-propylammonium) tetrachloromanganate(II) (δ form), 44B, 1093
$C_6H_{20}Cl_4N_2Pd$, Bis(propylammonium) tetrachloropalladate(II), 43B, 37
$C_6H_{20}Cl_6N_2Sn$, Trimethylammonium hexachlorostannate(IV), 2, 805, 816
$C_6H_{20}Cl_{10}Cu_4N_2$, Bis(trimethylammonium) decachlorotetracuprate(II), 42B, 1008
$C_6H_{22}Cl_5CuN_3O_2$, (N-(2-Ammoniumethyl)piperazinium) pentachlorocuprate(II) dihydrate, 46B, 1320
$C_6H_{24}Cl_5CuN_3$, Tris(dimethylammonium) monochloride tetrachlorocuprate(II), 37B, 673
$C_6H_{28}Cl_6Mn_2N_2O_4$, Isopropylammonium di-μ-chlorobis(diaquo-dichloromanganate(II)), 45B, 1393
$C_6H_{32}Cl_8N_2O_{14}W_4$, Bis(trimethylammonium) tetra-μ-oxo-tetrakis(dichloroaquooxotungstate(V-VI)), 44B, 1094
$C_7H_{10}Cl_5Hg_2N$, Benzylammonium bis[dichloromercury(II)] chloride, 46B, 1320
$C_8D_{24}Cl_6N_2Pt$, Deuterated tetramethylammonium hexachloroplatinate, 46B, 1321

$C_8I_2O_8Re_2$, Bis[μ-iodo-tetracarbonylrhodium], 46B, 768
$C_8H_2O_7Os_3S_3$, Di-μ_3-thio-dihydroheptacarbonyl(thiocarbonyl)triosmium, 46B, 768
$C_8H_6Cu_2N_4S_3$, Pyridinium tris(thiocyanato)dicuprate(I), 45B, 1394
$C_8H_{19}AuCl_4N_2O_2$, Hydrogenbis(N,N-dimethylacetamide) tetrachloroaurate(III), 46B, 1321
$C_8H_{20}Br_6NSb$, Tetraethylammonium hexabromoantimonate(V), 37B, 673
$C_8H_{20}Br_6N_2O_4Te$, DL-α-Ammonio-n-butyric acid hexabromotellurate, 42B, 602
$C_8H_{20}Cl_3HgN$, Tetraethylammonium trichloromercurate(II), 44B, 1094
$C_8H_{20}Cl_4InN$, Tetraethylammonium tetrachloroindate, 34B, 619
$C_8H_{20}Cl_6Cu_3O_6S_2$, trans-Tetra-μ-chloro-bis(chloroaquocopper(II))copper(II)-bis(tetramethylene sulphone), 40B, 933
$C_8H_{20}Cl_6NW$, Tetraethylammonium hexachlorotungstate, 43B, 40
$C_8H_{22}Ag_2I_4N_2$, Disilver tetraiodide hexamethylethylenediamine, 41B, 34
$C_8H_{22}Ag_6I_8N_2$, Hexasilver octaiodide hexamethylethylenediamine, 41B, 35
$C_8H_{22}Br_4MoNO_2$, Tetraethylammonium aquooxotetrabromomolybdate(V), 45B, 1405
$C_8H_{22}Br_4NO_2Re$, Tetraethylammonium oxotetrabromoaquorhenate(V), 30B, 34
$C_8H_{22}I_4MoNO_2$, Tetraethylammonium aquooxotetraiodomolybdate(V), 45B, 1405
$C_8H_{23}Br_2NiO_8$, trans-Tetraaquodibromonickel(II)-1,4-dioxan (1:2), 42B, 564
$C_8H_{24}Ag_{13}I_{15}N_2$, Tetramethylammonium silver iodide, 35B, 881
$C_8H_{24}AlN$, Tetraethylammonium aluminum hydride, 39B, 878
$C_8H_{24}Br_4HgN_2$, Tetramethylammonium tetrabromomercurate(II), 42B, 1009
$C_8H_{24}Br_4N_2O_2U$, Di(tetramethylammonium) uranyl tetrabromide, 40B, 31
$C_8H_{24}Br_6N_2Te$, Tetramethylammonium hexabromotellurate(IV), 45B, 1394
$C_8H_{24}Cl_4CoN_2$, Tetramethylammonium tetrachlorocobaltate(II), 23, 569; 32B, 31
$C_8H_{24}Cl_4Co_{0.5}Cu_{0.5}N_2$, Tetramethylammonium copper-cobalt chloride, 42B, 1009
$C_8H_{24}Cl_4CuN_2$, Tetramethylammonium tetrachlorocuprate(II), 41B, 35
$C_8H_{24}Cl_4FeN_2$, Tetramethylammonium tetrachloroferrate(II), 41B, 36
$C_8H_{24}Cl_4MnN_2$, Bis(n-butylammonium) tetrachloromanganate(II), 45B, 1395
$C_8H_{24}Cl_4N_2Ni$, Tetramethylammonium tetrachloronickelate(II), 32B, 31
$C_8H_{24}Cl_4N_2O_2U$, Di(tetramethylammonium) uranyl tetrachloride, 40B, 31
$C_8H_{24}Cl_4N_2Zn$, Tetramethylammonium tetrachlorozincate(II), 23, 569; 32B, 32
$C_8H_{24}Cl_6Ga_2N_2$, Tetramethylammonium hexachlorodigallate(II), 39B, 878
$C_8H_{24}Cl_6N_2Pt$, Tetramethylammonium hexachloroplatinate(IV), 44B, 1095
$C_8H_{24}Cl_6N_2Sn$, Tetramethylammonium hexachlorostannate(IV), 2, 816; 40B, 31
$C_8H_{24}Cl_{18}N_2Nb_6$, Bis(tetramethylammonium) hexachloro(dodeca-μ-chlorohexaniobate), 38B, 40
$C_8H_{24}F_6N_2Si$, Tetramethylammonium hexafluorosilicate, 3, 656, 699
$C_8H_{24}Mo_2N_2O_2S_6$, Bis(tetramethylammonium) bis[(μ-thio)dithiooxomolybdate], 46B, 1322
$C_8H_{24}N_2O_4W$, t-Butylammonium monotungstate, 45B, 1403
$C_8H_{32}Cl_8N_6Pt$, Bis(diethylenetriammonium) tetrachloroplatinate(II) tetrachloride, 45B, 1396
$C_8H_{39}As_2MoN_2Na_3O_{39}$, Trisodium bis(tetramethylammonium) monohydrogenhexamolybdodiarsenate heptahydrate, 46B, 1322

$C_8H_{63}N_{24}O_{49}PV_{14}$, Hydrogen octakis(guanidinium) 1:14-vanadophosphate heptahydrate, 46B, 1323
$C_9H_{14}Cl_3CuN$, N-Methylphenethylammonium trichlorocuprate(II), 40B, 31
$C_9H_{14}Cl_3NNi$, N-Methylphenethylammonium trichloronickelate(II), 43B, 41
$C_9H_{30}Br_7Cd_2N_3$, Trimethylammonium heptabromodicadmate(II), 45B, 1396
$C_9H_{30}Cl_6InN_3$, Trimethylammonium hexachloroindate(III), 43B, 42
$C_9H_{30}Cl_7Cu_2N_3$, Tristrimethylammonium catena-tri-μ-chloro-cuprate(1-) tetrachlorocuprate(2-), 39B, 879
$C_9H_{30}Cl_7Mn_2N_3$, Tris(trimethylammonium) dimanganese heptachloride, 42B, 1010
$C_9H_{30}Cl_9N_3Sb_2$, Tris[trimethylammonium chloride] bis[antimony(III)-trichloride], 45B, 1397
$C_{10}H_{10}Cl_5N_2Sb$, 4,4'-Dipyridylium pentachloroantimonate, 46B, 1323
$C_{10}H_{10}Cl_6Cu_2N_2$, 4,4'-Bipyridylium di-μ-chlorotetrachlorodicuprate, 45B, 1397
$C_{10}H_{12}Br_4MnN_2$, Pyridinium tetrabromomanganate, 42B, 1010
$C_{10}H_{12}Cl_4MnN_2$, Pyridinium tetrachloromanganate, 42B, 1010
$C_{10}H_{12}Cl_6N_2Te$, Bis(pyridinium) hexachlorotellurate, 45B, 1398
$C_{10}H_{16}Br_6Mo_2N_2O_2$, Bis(pyridinium) hexabromodiaquodimolybdate(II), 45B, 1398
$C_{10}H_{16}Cl_4MoN_2O_4$, Dioxodichlorodiaquomolybdenum(VI)-bis(pyridinium chloride), 42B, 1011
$C_{10}H_{16}CuCl_4N_2$, N-Phenylpiperazinium tetrachlorocuprate(II), 45B, 1399
$C_{10}H_{16}I_6Mo_2N_2O_2$, Bis(pyridinium) hexaiododiaquodimolybdate(II), 44B, 1095
$C_{10}H_{18}Cl_4Mo_2N_2O_7$, Dipyridinium di-μ-oxo-bis(oxodichloroaquo-molybdate(V)) monohydrate, 42B, 1011; 43B, 1227
$C_{12}H_{2.6}F_{1.4}Mn_4O_{14.6}$ · 2 C_6H_6, Dodecacarbonyl-μ_3-fluoro-tri-μ_3-hydroxo-tetramanganese benzene solvate, 46B, 1324
$C_{12}H_{16}Cl_5N_2Sb$, Bis(anilinium) pentachloroantimonate, 46B, 1318
$C_{12}H_{20}Br_6Mo_2N_2O_2$, 4-Methylpyridinium bisaquohexabromo-dimolybdate(II), 42B, 1012
$C_{12}H_{20}I_6Mo_2N_2O_2$, Bis(4-methylpyridinium) hexaiododiaquodimolybdate-(II), 43B, 301
$C_{12}H_{22}CuO_{12}S_2$, Hexaaquocopper(II) benzenesulphonate, 44B, 967
$C_{12}H_{22}NO_4Re$, Tetraethylammonium dihydrotetracarbonylrhenate, 44B, 1167
$C_{12}H_{24}N_6Pd$, Diethylammonium tetracyanopalladate(II), 37B, 676
$C_{12}H_{28}MoN_2O_4$, Cyclohexylammonium monomolybdate, 45B, 1403
$C_{12}H_{30}Br_6N_4O_4Te$, Tetrakis(dimethylformamide) hexabromotellurate, 46B, 1324
$C_{12}H_{30}Cl_8N_4O_4Re_2$, Dihydrogen-tetrakis(dimethylformamide) octachloro-dirhenate, 45B, 1399
$C_{12}H_{32}Cl_4CuN_2$, Triethylammonium tetrachlorocuprate(II), 38B, 42; 39B, 880
$C_{12}H_{32}Cl_4N_2O_2U$, Bis(triethylammonium) tetrachlorodioxouranate(VI), 40B, 35; 42B, 1013
$C_{12}H_{36}Bi_2Br_9N_3$, Tetramethylammonium enneabromodibismuthate(III), 46B, 1325
$C_{12}H_{36}Bi_2I_9N_3$, Tetramethylammonium enneaiododibismuthate(III), 46B, 1325
$C_{12}H_{36}Br_{11}N_3Sb_2$, Tetramethylammonium nonabromodiantimonate(III)-dibromine, 38B, 43
$C_{12}H_{36}Cl_{15}Ge_5N_3Pt$, Tetramethylammonium pentakis(trichlorogermanyl)-platinate(II), 39B, 881

$C_{12}H_{36}Cl_{18}N_3Nb_6$, Tris(tetramethylammonium) hexachloro(dodeca-μ-chloro-hexaniobate), 40B, 35
$C_{12}H_{44}B_{12}N_2$, Triethylammonium closo-dodecaborane, 46B, 1326
$C_{14}H_{26}CuO_{12}S_2$, Hexaaquocopper(II) toluene-4-sulphonate, 44B, 967
$C_{14}H_{28}I_4NO_2Rh$, Tetra-n-propylammonium dicarbonyltetraiodorhodate-(III), 41B, 39
$C_{14}H_{52}Ce_2N_{24}O_{24}$, Guanidinium triscarbonatoperoxocerate dihydrate, 43B, 80
$C_{15}H_{22}Br_7Mo_2N_3O_2$, Pyridinium hexabromodiaquodimolybdate bromide, 44B, 1096
$C_{15}H_{22}ClMo_2N_3O_{16}P_4$, Pyridinium chlorotetrakis(hydrogen phosphato)-dimolybdate(III), 45B, 1401
$C_{16}H_{15}NO_{12}Os_4$, Tetramethylammonium tri-μ-hydrido-dodecacarbonyltetraosmate, 44B, 1096
$C_{16}H_{16}N_6Pt$, Dianilinium tetracyanoplatinate(II), 43B, 137
$C_{16}H_{36}Ag_3I_4N$, Tetrabutylammonium tetraiodotriargentate, 37B, 677
$C_{16}H_{40}Br_6N_2Pt_2$, Tetraethylammonium tetrabromo-μ,μ'-dibromodiplatinate(II), 29, 580; 41B, 40
$C_{16}H_{40}Cl_4Fe_2N_2S_2$, Tetraethylammonium tetrachlorodi-μ-sulfido-diferrate(III), 43B, 50
$C_{16}H_{40}Cl_4Fe_4N_2S_4$, Tetraethylammonium tetrakis(chloro-μ_3-sulfido-ferrate), 43B, 50
$C_{16}H_{40}Cl_4N_2Ni$, Tetraethylammonium tetrachloronickelate(II), 32B, 34
$C_{16}H_{40}Cl_4N_2O_2U$, Bis(tetraethylammonium) tetrachlorodioxouranate(VI), 42B, 1013; 45B, 1401
$C_{16}H_{40}Cl_5InN_2$, Tetraethylammonium pentachloroindate(III), 34B, 23
$C_{16}H_{40}Cl_5N_2OPa$, Bis(tetraethylammonium) oxopentachloroproactinate(V), 38B, 47
$C_{16}H_{40}Cl_6N_4O_2Pt_2$, Tetraethylammonium μ-chloro-abefh-pentachloro-μ-nitrosyl-nitrosyldiplatinate(II), 40B, 30
$C_{16}H_{40}F_{10}N_2OTa_2$, Bis(tetraethylammonium) μ-oxo-bis(pentafluoro-tantalate(V)), 43B, 51
$C_{16}H_{40}Mo_2N_2O_2S_6$, Bis(tetraethylammonium) di-μ-sulfidobis(oxo(disulfido)molybdate), 46B, 1326
$C_{16}H_{48}Cl_{10}Cu_4N_4O$, Tetramethylammonium μ_4-oxo-hexa-μ-chloro-tetra-(chlorocuprate(II)), 34B, 22; 45B, 1402
$C_{16}H_{48}Cl_{10}Cu_4N_4O$, Tetrakis(diethylammonium) μ_4-oxo-hexa-μ_2-chloro-tetra(chlorocuprate(II)), 43B, 51
$C_{16}H_{54}Cl_{12}N_2O_6Ru_3$, Bis(tetraethylammonium) bis(diaquooxonium) dodecachlorotriruthenate(4-), 46B, 1326
$C_{17}H_{36}Cl_3NOPt$, Tetrabutylammonium trichlorocarbonylplatinate(II), 42B, 621
$C_{18}H_{21}Cr_2NO_{10}$, Tetraethylammonium μ-hydrido-bis(pentacarbonyl-chromium), 43B, 53
$C_{18}H_{22}F_4N_2O_8W$, Bis(8-hydroxyquinolinium) oxoperoxotetrafluoro-tungstate(VI) trihydrate, 43B, 342
$C_{18}H_{22}F_5N_2NbO_7$, Bis(8-hydroxyquinolinium) peroxopentafluoroniobate-(V) trihydrate, 43B, 342
$C_{18}H_{28}Cl_4CuN_2$, Bis(N-methylphenethylammonium) tetrachlorocuprate(II) (green form), 40B, 39
$C_{18}H_{28}Cl_4CuN_2$, Bis(N-methylphenethylammonium) tetrachlorocuprate(II) (yellow form), 40B, 39
$C_{18}H_{36}FeN_9S_6$, Tetramethylammonium hexaisothiocyanatoferrate(III), 43B, 54
$C_{18}H_{66}Mo_8N_6O_{30}$, Hexakis(isopropylammonium) dihydrogenoctamolybdate(6-) dihydrate, 44B, 1097
$C_{19}H_{18}Cl_2O_2PV$, Triphenylmethylphosphonium dioxodichlorovanadate,

46B, 1327
$C_{19}H_{20}Co_4INO_{11}$, Tetraethylammonium tris(μ-carbonyl)-octacarbonyl-iodo-tetrahedro-tetracobalt, 46B, 795
$C_{20}H_{20}Br_6P_2U$, Bis(triphenylethylphosphonium) hexabromouranate(IV), 44B, 1097
$C_{20}H_{20}Cl_6P_2U$, Bis(triphenylethylphosphonium) hexachlorouranate(IV), 44B, 1097
$C_{20}H_{24}N_2Ni_6O_{12}$, Bis(tetramethylammonium) bis(hexacarbonyltrinickel), 40B, 43
$C_{20}H_{32}Cl_3N_2O_3U$, Benzyltrimethylammonium μ_2-peroxo-bis(trichlorodi-oxouranate(VI)), 43B, 55
$C_{20}H_{32}Cl_4CuN_2$, Bis((+)-N,α-dimethylphenethylammonium) tetrachloro-cuprate(II), 41B, 40
$C_{20}H_{32}Cl_4CuN_2$, Bis(trimethylbenzylammonium) tetrachlorocuprate(II) (α form), 32B, 35
$C_{20}H_{38}MoN_{10}O_2$, Bis(triethylammonium) dihydronium octacyanomolybdate-(IV), 46B, 1327
$C_{20}H_{40}FeN_7O$, Bis(tetraethylammonium) tetracyanonitrosylferrate(I), 43B, 987
$C_{20}H_{42}CuO_{14}S_2$, Hexaaquocopper(II) D-camphor-10-sulphonate, 44B, 967
$C_{20}H_{48}B_7FeNO_4$, Tetra-n-butylammonium heptaboranetetracarbonylfer-rate, 43B, 55
$C_{20}H_{48}Cl_4MnN_2$, Bis(monodecylammonium) tetrachloromanganate(II), 42B, 1014
$C_{20}H_{52}Cl_{10}Cu_4N_4O$, Bis(N,N,N',N'-tetraethylenediammonium) hexa-μ-chloro-μ_4-oxo-tetra(chlorocuprate(II)), 38B, 48
$C_{22}H_{12}Cd_4Fe_4O_{18}$, Tetrakis(tetracarbonylironcadmium) bis(acetone), 43B, 67
$C_{22}H_{24}Co_4N_2Ni_2O_{14}$, Bis(tetramethylammonium) octa-μ_3-carbonyl-hexa-carbonyl-octahedro-dinickeltetracobaltate(2-), 40B, 45
$C_{24}H_{20}AsCl_4CrO$, Tetraphenylarsonium tetrachlorooxochromate(V), 43B, 892
$C_{24}H_{20}AsCl_4MoN$, Tetraphenylarsonium nitridotetrachloromolybdate(VI), 43B, 892
$C_{24}H_{20}AsCl_4MoO$, Tetraphenylarsonium tetrachlorooxomolybdate(V), 43B, 892, 893
$C_{24}H_{20}AsCl_4ORe$, Tetraphenylarsonium tetrachlorooxorhenate(V), 43B, 893
$C_{24}H_{20}AsCl_4OV$, Tetraphenylarsonium tetrachlorooxovanadate(V), 46B, 1327
$C_{24}H_{20}AsF_4MoN$, Tetraphenylarsonium tetrafluoronitridomolybdate, 46B, 1327
$C_{24}H_{20}AsFe_4N_7O_7S_3$, Tetraphenylarsonium tri-μ_3-sulfido-heptanitrosyl-tetraferrate, 43B, 894
$C_{24}H_{20}AsMoN_{13}$, Tetraphenylarsonium tetraazidonitridomolybdate(VI), 46B, 1328
$C_{24}H_{20}Br_4MoNP$, Tetraphenylphosphonium tetrabromonitrilomolybdate-(VI), 46B, 1328
$C_{24}H_{20}Cl_5N_3NbP$, Tetraphenylphosphonium azidopentachloroniobate, 46B, 1329
$C_{24}H_{20}Cl_6NbP$, Tetraphenylphosphonium hexachloroniobate, 46B, 1329
$C_{24}H_{22}AsCl_4MoO_2$, Tetraphenylarsonium aquotetrachlorooxomolybdate(V), 43B, 894
$C_{24}H_{24}AsCl_4O_2Ru$, Tetraphenylarsonium cis-diaquotetrachlororuthenate monohydrate, 31B, 315
$C_{24}H_{36}Cl_4CoN_2$, Bis(4-benzylpiperidinium) tetrachlorocobaltate(II), 45B, 1402

$C_{24}H_{36}Cl_6Cu_2N_2$, Bis(4-benzylpiperidinium) hexachlorodicuprate(II), 46B, 1330
$C_{24}H_{42}N_2O_8W_2$, Bis(tetraethylammonium) di-μ-hydrido-octacarbonylditungstate, 40B, 46
$C_{24}H_{48}Co_2Mn_3N_{12}O_{12}$, Trimanganese(II) bis(hexacyanocobaltate(III)) - 12 methanol, 44B, 1097
$C_{24}H_{48}MoN_2O_4 \cdot H_2O$, Dicyclohexylammonium monomolybdate monohydrate, 45B, 1403
$C_{24}H_{54}Cl_{22}Cu_5N_{12}$, Tetrakis(N-(2-ammonioethyl)piperazinium) di-μ-chloro-dodecachlorotricuprate(II) bis(tetrachlorocuprate(II)), 46B, 1330
$C_{24}H_{56}Br_4N_2O_2U$, Di(tetrapropylammonium) uranyl tetrabromide, 40B, 47
$C_{24}H_{56}Br_6Ga_2N_2$, Tetra-n-propylammonium hexabromodigallate(II), 40B, 47
$C_{24}H_{56}Br_9N_2W_2$, Bis(tetrapropylammonium) tri-μ-bromo-hexabromoditungstate(2-), 43B, 56
$C_{24}H_{56}Cl_4N_2O_2U$, Di(tetrapropylammonium) uranyl tetrachloride, 40B, 47
$C_{24}H_{56}Hg_2I_6N_3$, Bis(tetrapropylammonium) bis(μ-iodo)tetraiododimercurate(II), 46B, 1331
$C_{24}H_{68}Cl_{12}Mo_3N_3O_4$, Tris(tetraethylammonium) hydrogen μ-hydrido-di-μ-chloro-bis(trichloro-molybdenum) aquo-tetrachloro-oxomolybdenum dihydrate, 45B, 1404
$C_{25}H_{24}N_2O_{16}Ru_6$, Tetramethylammonium carbidohexadecacarbonylhexaruthenate, 46B, 1331
$C_{25}H_{30}Br_{11}N_5Sb_2$, Pyridinium nonabromodiantimonate(III) dibromide, 35B, 876
$C_{25}H_{36}Br_{14}Mo_3N_5O_6$, Pentapyridinium tris(aquooxotetrabromomolybdate-(V)) dibromide, 45B, 1405
$C_{25}H_{43}N_2O_{10}Re_3$, Bis(tetraethylammonium) tri-μ-hydrido-μ_3-oxo-ennea-carbonyl-triangulo-trirhenate(2-), 43B, 58
$C_{26}H_{26}Mo_2N_{10}O_5S_6$, Pyridinium di-$\mu$-oxo-bis(oxotris(isothiocyanato)-molybdate(V)) monohydrate, 43B, 1227
$C_{27}H_{40}Fe_3N_2O_{11}$, Bis(tetraethylammonium) (μ_3-carbonyl)(μ-carbonyl)-nonacarbonyltriiron(3Fe-Fe), 46B, 1332
$C_{27}H_{42}Cl_9N_3Rh_2$, Trimethylphenylammonium nonachlorodirhodate(III), 38B, 50
$C_{28}H_{24}Mo_2N_8O_4S_6$, Tetrakis(pyridinium) di-μ-oxo-bis[oxotris(isothio-cyanato)molybdate(V)], 45B, 1405
$C_{28}H_{41}N_2O_{12}Re_3$, Bis(tetraethylammonium) μ-hydrido-dodecacarbonyl-triangulo-trirhenate(2-), 44B, 1167
$C_{29}H_{70}CoN_8O_7$, Tetraethylammonium dioxygenpentacyanocobaltate hydrate, 41B, 42
$C_{30}H_{36}Bi_4Cl_{18}N_6$, Pyridinium octadecachloro-tetrabismuth, 44N, 1097
$C_{30}H_{36}Br_{24}N_6Sb_4$, Pyridinium tetracosabromoantimon(III)triantimonate-(V), 37B, 679
$C_{31}H_{36}INO_{15}Rh_6$, Tetra-n-butylammonium iodopentadecacarbonylhexarhodate, 37B, 680
$C_{31}H_{44}N_2O_{15}Re_4$, Tetraethylammonium pentadecacarbonyltetrahydridotetrarhenate, 41B, 43
$C_{32}H_{24}MoN_{12}O_8$, Tetrakis(2-carboxypyridinium) octacyanomolybdate(IV), 46B, 1332
$C_{32}H_{38}N_2O_{12}Re_4$, Trimethylbenzylammonium hexa-μ_2-hydridodecacarbonyl-tetrahedro-tetrarhenate(2-), 43B, 59
$C_{32}H_{40}IN_2O_{16}Rh_7$, Tetraethylammonium μ-iodo-tetra-μ_3-carbonyl-di-μ-carbonyl-decacarbonyl-polyhedro-heptarhodate(2-), 41B, 44
$C_{32}H_{72}Br_4N_2O_2U$, Di(tetra-n-butylammonium) uranyl tetrabromide, 43B,

1448
$C_{32}H_{72}Cl_4N_2O_2U$, Di(tetrabutylammonium) uranyl tetrachloride, 40B, 48
$C_{32}H_{72}Cl_8N_2Re_2$, Tetra(n-butyl)ammonium octachlorodirhenate(III), 42B, 1014
$C_{32}H_{72}Cl_{13}Mo_5N_2$, Tetra-n-butylammonium chloromolybdate, 42B, 1015
$C_{32}H_{72}Mo_2N_2O_7$, Bis(tetra-n-butylammonium) dimolybdate, 43B, 60
$C_{32}H_{72}N_2O_{19}W_6$, Tetrabutylammonium hexatungstate, 40B, 1134; 44B, 1098
$C_{32}H_{96}N_8O_{20}Si_8 \cdot 64.8\ H_2O$, Tetramethylammonium silicate hydrate, 45B, 1406
$C_{34}H_{36}Cl_2FeMoNPS_4$, Benzyltrimethylammonium tetraphenylphosphonium bis(μ-sulfido)-dichloroirondisulfidomolybdenum, 46B, 1333
$C_{34}H_{36}Cl_2FeNPS_4W$, Benzyltrimethylammonium tetraphenylphosphonium bis(μ-sulfido)-dichloroirondisulfidotungsten, 46B, 1333
$C_{35}H_{20}BrIr_4O_{11}P$, Tetraphenylphosphonium bromo-tris-(μ-carbonyl)-octacarbonyl-tetrahedro-tetrairidate(-1), 46B, 1333
$C_{35}H_{27}N_{11}O_2S_8U$, 2-Pyridylthio-2-pyridinium pentakis(isothiocyanato)-dioxouranate(VI)(3-), 45B, 1406
$C_{35}H_{72}CoN_8$, Di(isopropyl)diethylammonium pentacyanocobaltate(II), 41B, 45
$C_{36}H_{30}Cl_4NOP_2Tc$, Bis(triphenylphosphin)iminium oxotetrachloro-technetate(V), 45B, 1407
$C_{36}H_{30}N_2P_2S_4$, Bis(triphenylphosphine)iminium tetrasulfur nitride, 46B, 1334
$C_{36}H_{32}As_2Cl_{18}Nb_6O_2$, Bis(hydroxo(triphenyl)arsonium(V)) dodeca-μ-chloro-hexachloro-octahedro-hexaniobate(2-), 39B, 882
$C_{38}H_{24}N_2O_{30}Rh_{12}$, Tetramethylammonium triacontacarbonyldodecarhodate, 34B, 436
$C_{38}H_{36}As_2Cl_4Ni$, Bis(triphenylmethylarsonium) tetrachloronickelate, 31B, 314
$C_{40}H_{34}BMoNO_4P_2$, Bis(triphenylphosphine)iminium tetrahydroboratotetracarbonylmolybdate, 43B, 884
$C_{40}H_{40}CeN_5O_{15}P_2$, Bistriphenyl(ethyl)phosphonium pentanitratocerate-(III), 39B, 882
$C_{40}H_{72}FeN_4Na_2O_{16}$, Bis(sodium crypt) tetracarbonylferrate(2-), 43B, 450
$C_{44}H_{31}Fe_2NO_8P_2$, Bis(triphenylphosphine)iminium hydridooctacarbonyl-diferrate, 44B, 1098
$C_{45}H_{31}NO_9Os_3S$, Bis(triphenylphosphine)iminium μ_2-hydrido-μ_3-sulfido-nonacarbonyltriosmate, 44B, 1099
$C_{46}H_{31}Cr_2NO_{10}$, Hexaphenyldiphosphazenium μ-deuterido-bis(pentacarbonylchromium), 45B, 1408
$C_{46}H_{31}Cr_2NO_{10}P_2$, Bis(triphenylphosphine)iminium μ-hydrido-bis(pentacarbonylchromate), 44B, 1099
$C_{47}H_{31}NO_{11}P_2Ru_3$, Bis(triphenylphosphine)iminium 1,2-μ-carbonyl-1,1,1,2,2,2,3,3,3-decacarbonyl-1,2-μ-hydrido-triangulo-triruthenate, 45B, 1408
$C_{47}H_{44}N_2O_{21}PRh_9$, Benzyltriethylammonium heneicosacarbonylphosphido-nonarhodium acetone solvate, 45B, 1408
$C_{48}H_{40}As_2Cl_4N_6Ti$, Bis(tetraphenylarsonium) tetrachlorodiazidotitanate(IV), 46B, 1335
$C_{48}H_{40}As_2Cl_5NbO \cdot 2\ CH_2Cl_2$, Bis(tetraphenylarsonium) pentachlorooxoniobate(V) methylene chloride solvate, 46B, 1335
$C_{48}H_{40}As_2Cl_8N_6Ti_2$, Bis(tetraphenylarsonium) μ-diazido-bis(tetrachlorotitanate(IV)), 46B, 1336
$C_{48}H_{40}As_2Cl_8Se_3W_2$, Bis(tetraphenylarsonium) μ-selenido-μ-diselenido-bis(tetrachloro)tungstate(V), 45B, 1409

$C_{48}H_{40}As_2CoN_4O_{12}$, Tetraphenylarsonium tetranitratocobaltate(II), 31B, 322
$C_{48}H_{40}As_2EuN_5O_{15}$, Bis(tetraphenylarsonium) pentanitratoeuropate-(III), 46B, 1336
$C_{48}H_{40}Br_{10}P_2Te_2$, Bis(tetraphenylphosphonium) decabromoditellurate-(IV), 46B, 1337
$C_{48}H_{40}Cl_6P_2PdSn$, Bis(tetraphenylphosphonium) trichloropalladiumtri-chlorotin, 46B, 1337
$C_{48}H_{40}CoP_3S_8W_2$, Bis(tetraphenylphosphonium) bis(tetrathiotungsten)-cobaltate, 44B, 1100
$C_{48}H_{40}Fe_2P_2S_{12}$, Bis(tetraphenylphosphonium) di-iron-dodecasulfide, 45B, 1410
$C_{48}H_{40}Mo_2NiP_2S_8$, Tetraphenylphosphonium bis(tetrathiomolybdato)nic-kelate(II), 42B, 1016
$C_{48}H_{40}P_2S_8W_2Zn$, Tetraphenylphosphonium bis(tetrathiotungstato)zinc-ate(II), 44B, 1100
$C_{48}H_{42}As_2Cl_{11}ORe_3$, Bis(tetraphenylarsonium) tri-μ-chloro-octachloro-trirhenate monohydrate, 31B, 319
$C_{48}H_{42}Cl_4CuOP_2$, Bis(tetraphenylphosphonium) tetrachlorocuprate(II) monohydrate, 42B, 1016
$C_{48}H_{72}N_2NiO_{16}Rh_6$, Bis(tetrabutylammonium) hexakis(μ-carbonyl)-tris(μ_3-carbonyl)-heptacarbonyl-hexarhodium-nickel, 46B, 1338
$C_{48}H_{112}N_4O_{32}W_{10}$, Tributylammonium decatungstate, 42B, 1016
$C_{49}H_{30}ClNO_{13}P_2Ru_4$, Bis(triphenylphosphine)iminium (μ-chloro)(μ-car-bonyl)tetrakis(tricarbonylruthenate), 46B, 1339
$C_{49}H_{30}CoNO_{13}P_2Ru_3$, Bis(triphenylphosphino)iminium tridecacarbonylco-balt-triruthenate, 46B, 1339
$C_{50}H_{44}CdCl_4P_2 \cdot 2\ C_2H_4Cl_2$, Bis(triphenylbenzylphosphonium) tetra-chlorocadmate dichloroethane, 46B, 1340
$C_{50}H_{44}Cd_2Cl_6P_2$, Bis(benzyltriphenylphosphonium) hexachlorodicadmate, 46B, 1340
$C_{51}H_{30}Co_6N_2O_{15}P_2$, Bis(triphenylphosphine)immonium nona-μ-carbonyl-nitrido-hexa(carbonylcobalt), 45B, 1410
$C_{51}H_{30}INO_{15}Os_5P_2$, Bis(triphenylphosphine)iminium pentadecacarbonyl-iodopentaosmate, 44B, 1101
$C_{51}H_{30}NO_{15}P_2Rh_5$, Bis(triphenylphosphine)iminium pentadecacarbonyl-pentarhodate, 46B, 1341
$C_{51}H_{31}NO_{15}Os_5P_2$, Bis(triphenylphosphine)iminium μ-hydrido-pentadeca-carbonylpentaosmate, 44B, 1102
$C_{52}H_{30}Mo_2N_2Ni_3O_{16}P_4$, Bis(triphenylphosphine)iminium tri-μ-carbonyl-tridecacarbonyltrinickelatedimolybdate(6Mo-Ni)(3Ni-Ni), 37B, 681
$C_{52}H_{30}N_2Ni_3O_{16}P_4W_2$, Bis(triphenylphosphine)iminium tri-μ-carbonyl-tridecacarbonyltrinickelateditungstate(6W-Ni)(3Ni-Ni), 37B, 681
$C_{53}H_{40}As_2N_6ReS_5$, Bis(tetraphenylarsonium) pentakis(isothiocyanato)-nitridorhenate(VI), 44B, 1102
$C_{53}H_{48}Cl_9OP_2Ru$, Benzyltriphenylphosphonium carbonylpentachloro-ruthenate(III) dichloromethane solvate, 44B, 792
$C_{53}H_{60}CoN_8O_2$, Dibenzyldimethylammonium dioxygenpentacyanocobaltate, 41B, 42
$C_{54}H_{31}NO_{18}P_2Ru_6$, Bis(triphenylphosphine)amine octadecacarbonylhydro-gen-octahedro-hexaruthenate, 46B, 1342
$C_{54}H_{108}N_9S_6Tc$, Tris(tetra-n-butylammonium) hexakis(isothiocyanato)-technetium(III), 46B, 1137
$C_{56}H_{108}MoN_{11}$, Tris(tetra-n-butylammonium) octacyanomolybdate(V), 35B, 879
$C_{59}H_{40}As_2Fe_3O_{11}$, Bis(tetraphenylarsonium) (μ_3-carbonyl)(μ-carbonyl)-nonacarbonyltriiron(3Fe-Fe), 46B, 1332

$C_{71}H_{66}N_3O_{32}Rh_{17}S_2$, Benzyltriethylammonium dithiodotriacontacarbonylheptadecarhodate, 44B, 1103
$C_{72}H_{60}Bi_2Br_9P_3$, Tetraphenylphosphonium enneabromodibismuthate(III), 43B, 885
$C_{78}H_{60}As_3InN_6S_6$, Tetraphenylarsonium hexakis(isothiocyanato)indate-(III), 42B, 586
$C_{84}H_{62}N_2O_{12}Os_4P_4$, Bis(triphenylphosphine)iminium dihydridododecacarbonyltetraosmate, 44B, 1109
$C_{86}H_{94}FeMo_2N_5O_2P_4S_8$, Tetraethylammonium bis[bis(triphenylphosphine)-iminium] bis(tetrathiomolybdenum)ferrate-N,N-dimethylformamide, 46B, 1342
$C_{99}H_{68}O_{24}P_3Rh_{13}$, Benzyltriphenylphosphonium dodeca-μ-carbonyl-dodecacarbonyldihydrido-polyhedro-tridecarhodate(3-), 45B, 1411
$C_{106}H_{83}Fe_2N_{11}P_4 \cdot 6\ H_2O$, Tetraphenylphosphonium pentacyanoferrate-(III)-μ-cyano-tetracyano-monoammine-ferrate(III) hexahydrate, 45B, 1412
$C_{126}H_{92}N_4O_{22}P_4Pt_{19}$, Tetrakis(tetraphenylphosphonium) platinumcarbonyl acetonitrile solvate, 45B, 1412

88. POLYMERS

CH_2, Polyethylene, 35B, 882
$C_2H_2F_2$, Poly(vinylidene fluoride), 31B, 37; 35B, 882; 38B, 1057
$C_2H_2O_2$, Polyglycolide, 33B, 521
C_2H_3Cl, Polyvinyl chloride, 24, 570; 26, 634
C_2H_3F, Polyvinyl fluoride, 26, 633
C_2H_4O, Polyacetaldehyde, 24, 576; 26, 635
C_2H_4O, Polyvinyl alcohol, 11, 588
C_3H_4O, Ethylene - carbon monoxide copolymer, 26, 632
C_3H_5ClO, Poly(chloromethyl ethylene oxide), 35B, 883
C_3H_6, Polypropylene, 20, 537
C_3H_6O, Poly(propylene oxide), 24, 572; 31B, 37
$C_3H_8O_2$, Polyoxacyclobutane hydrate, 35B, 884
C_4H_5Cl, Polychloroprene, 9, 364
C_4H_6, Isotactic 1,2-polybutadiene, 22, 667
C_4H_6, 1,2-Polybutadiene, 20, 540
C_4H_6, 1,4-cis-Polybutadiene, 20, 541
C_4H_7NO, Nylon 4, 31B, 31
C_4H_8, Poly-α-butene, 20, 542
C_4H_8O, Polybutyraldehyde, 24, 577
$C_4H_{10}Si$, Polyvinylethylsilane, 34B, 365
C_5H_8, Poly-cis-isoprene (rubber), 9, 354
C_5H_8, Poly-trans-isoprene (gutta percha), 9, 359; 10, 237
C_5H_8, trans-Polypentenamer, 32B, 44
C_5H_9Cl, Rubber hydrochloride, 9, 362
C_6H_4O, Poly(p-phenylene oxide), 34B, 177
$C_6H_{10}O_2$, Poly-ε-caprolactone, 35B, 885
$C_6H_{11}NO$, Polycaproamide, 19, 555
C_7H_{12}, trans-Polyheptenamer, 32B, 44
$C_8H_{12}O_4$, Polyethylene adipate, 27, 806
$C_8H_{13}NO_5$, Chitan (poly-N-acetyl-D-glucosamine), 33B, 522
$C_8H_{13}NO_5$, Chitin (poly-N-acetyl-D-glucosamine), 21, 567; 24, 728
C_8H_{14}, trans-Polyoctenamer, 32B, 45
$C_8H_{15}NO_6$, β-Chitin, 26, 821
$C_8H_{16}Cl_2HgO_4$, Poly(ethylene oxide) - mercuric chloride, 33B, 524
$C_{10}H_8O_4$, Polyethylene terephthalate, 41B, 1230

$C_{10}H_{16}O_4$, Polyethylene suberate, 27, 806
$C_{10}H_{18}$, trans-Polydecenamer, 32B, 45
$C_{10}H_{22}N_2O_4$, Tetramethylene diammonium adipate, 26, 644
$C_{12}H_{22}$, trans-Polydodecenamer, 32B, 46
$C_{12}H_{22}N_2O_2$, Polyhexamethylene adipamide (nylon 6.6), 11, 591
$C_{13}H_{14}NO_2$, Poly(1,2-bis(4-(phenylcarbamoyloxy)-n-butyl)-1-buten-3-ynylene), 44B, 1103
$C_{14}H_{24}NNaO_{17}$, Sodium hyaluronate tetrahydrate, 41B, 496
$C_{14}H_{26}N_2O_2$, Poly(heptamethylene pimelamide), 23, 629
$C_{16}H_{30}N_2O_2$, Polyhexamethylene sebacamide (nylon 6.10), 11, 594
$C_{18}H_{12}O$, Poly(2,6-diphenyl p-phenylene oxide), 35B, 886
$C_{20}H_{18}O_6S_2$, Poly(1,2-bis(p-tolylsulphonyloxymethylene)-1-buten-3-ynylene), 43B, 111
$C_{20}H_{42}O_4P_2Zn$, Zinc dialkylphosphinate copolymer, 33B, 525
$C_{30}H_{20}N_2$, Poly(1,6-di(N-carbazolyl)-2,4-hexadiyne), 44B, 1104
$C_{30}H_{24}N_2$, Poly[1,2-bis(diphenylaminomethyl)-1-buten-3-ynylene], 46B, 96

89. METAL COMPLEXES (OTHER LIGANDS)

$C_{13}H_8Br_4GeN_2O_3W$, Bromo(tribromogermyl)-2,2'-bipyridinetricarbonyltungsten, 35B, 870
$C_{20}H_{12}O_8Re_2Si$, Di-μ-hydrido-diphenylsiliconbis(tetracarbonylrhenium), 35B, 871
$C_{22}H_{15}FeO_4Sb$, Tetracarbonyl(triphenylstibine)iron, 40B, 1039
$C_{26}H_{29}ClN_4O_4RhSb$, Chlorobis(dimethylglyoximato)triphenylstibinerhodium(III), 39B, 856
$C_{31}H_{20}Fe_2Ge_2O_7$, μ-Carbonyl-bis-μ-diphenylgermanium-bis(tricarbonyliron), 40B, 1133
$C_{36}H_{31}Cl_4P_2RhSi \cdot x\ Cl_3HSi$, Hydridochloro(trichlorosilyl)bis(triphenylphosphine)rhodium, 35B, 873
$C_{39}H_{30}CoGeO_3P$, (Triphenylgermyl)tricarbonyl(triphenylphosphine)cobalt(I), 35B, 875